Natural Killer Cells

Basic Science and Clinical Application

ELSEVIER

science & technology books

Companion Web Site:

http://www.elsevierdirect.com/companions/9780123704542

Natural Killer Cells
Edited by Michael T. Lotze, Angus W. Thomson

Resources for Students and Researchers

- **All figures from the book available as .pdf files**

ACADEMIC PRESS

Natural Killer Cells

Basic Science and Clinical Application

Editors

Michael T. Lotze

Departments of Surgery and Bioengineering
University of Pittsburgh Cancer Institute
University of Pittsburgh Schools of the Health Sciences
Pittsburgh, Pennsylvania, United States

Angus W. Thomson

Departments of Surgery and Immunology
Thomas E. Starzl Transplantation Institute
University of Pittsburgh Schools of the Health Sciences
Pittsburgh, Pennsylvania, United States

ELSEVIER

AMSTERDAM • BOSTON • HEIDELBERG • LONDON
NEW YORK • OXFORD • PARIS • SAN DIEGO
SAN FRANCISCO • SINGAPORE • SYDNEY • TOKYO
Academic Press is an imprint of Elsevier

AP

Academic Press is an imprint of Elsevier
32 Jamestown Road, London NW1 7BY, UK
30 Corporate Drive, Suite 400, Burlington, MA 01803, USA
525 B Street, Suite 1900, San Diego, CA 92101-4495, USA

British Library Cataloguing-in-Publication Data
A catalogue record for this book is available from the British Library

Library of Congress Cataloging-in-Publication Data
A catalog record for this book is available from the Library of Congress

ISBN : 978-0-12-370454-2

For information on all Academic Press publications
visit our website at www.elsevierdirect.com

Typeset by Macmillan Publishing Solutions
www.macmillansolutions.com

Printed in the United States of America
Transferred to Digital Printing, 2012

Contents

PART V Special topics in NK biology

Contributors

Christopher E. Andoniou
Virology Program, Centre for Ophthalmology and Visual Science, The University of Western Australia, Western Australia, Australia and Centre for Experimental Immunology, Lions Eye Institute, Nedlands, Western Australia, Australia.

Anna Balato
Department of Dermatology, University of Naples Federico II, Naples, Italy.

Per H. Basse
Faculty of Immunology, Hillman Cancer Center, University of Pittsburgh, Pittsburgh, Pennsylvania, United States.

Vasileios Bekiaris
Division of Molecular Immunology, La Jolla Institute for Allergy & Immunology, La Jolla, California, United States.

Giovanni Bernardini
Department of Experimental Medicine and Pathology, Instituto Pasteur-Fondazione Cenci Bolognetti, University of Rome, Rome, Italy.

Edward L. Briercheck
Integrated Biomedical Graduate Program, Medical Scientist Program, Ohio State University College of Medicine, Columbus, Ohio, United States.

Maryanne A. Bryan
Department of Immunology, University of Pittsburgh School of Medicine, Pittsburgh, Pennsylvania, United States.

Lisa H. Butterfield
University of Pittsburgh Cancer Institute, Departments of Medicine and Immunology, University of Pittsburgh, Pittsburgh, Pennsylvania, United States.

Michael D. Cahalan
Department of Physiology and Biophysics, Center for Immunology, University of California, Irvine, California, United States.

Michael A. Caligiuri
Department of Microbiology, Virology, Immunology and Medical Genetics, James Cancer Hospital and Solove Research Institute, Ohio State University College of Medicine, Columbus, Ohio, United States.

William L. Camp
Department of Dermatology, University of Pittsburgh, Pittsburgh, Pennsylvania, United States.

Claudia Carlino
Department of Experimental Medicine and Pathology, Instituto Pasteur-Fondazione Cenci Bolognetti, University of Rome, Rome, Italy.

James R. Carlyle
Department of Immunology, Sunnybrook Research Institute, University of Toronto, Toronto, Canada.

Paolo Carrega
Istituto Scientifico Giannina Gaslini, Genoa, Italy.

Benedict J. Chambers
Department of Medicine Huddinge, Center for Infectious Medicine, Karolinska Institutet, Stockholm, Sweden.

Kenji Chamoto
Division of Immunoregulation, Hokkaido University, Hokkaido, Japan.

Sarah Cooley
Division of Hematology, Oncology, and Transplantation, University of Minnesota, Minneapolis, United States.

Régis Costello
Université de la Méditerranée, Institut de Cancérologie et d'Immunologie de Marseille, Service d'Hématologie, Hôpital Nord, Marseille, France.

Jerome D. Coudert
Virology Program, Centre for Ophthalmology and Visual Science, The University of Western Australia, Western Australia, Australia and Centre for Experimental Immunology, Lions Eye Institute, Nedlands, Western Australia, Australia.

Heike Daldrup-Link
Department of Radiology & Biomedical Imaging, University of California, San Francisco, San Francisco, California, United States.

Lesley R. de Armas
Department of Microbiology and Immunology, University of Miami, Miller School of Medicine, Miami, Florida, United States.

Mariapia Degli-Esposti
Virology Program, Centre for Ophthalmology and Visual Science, The University of Western Australia, Western Australia, Australia and Centre for Experimental Immunology, Lions Eye Institute, Nedlands, Western Australia, Australia.

Julie Y. Djeu
H. Lee Moffitt Cancer Center and Research Institute, Immunology Program, Tampa, Florida, United States.

P.K. Epling-Burnette
H. Lee Moffitt Cancer Center and Research Institute, Immunology Program, Tampa, Florida, United States.

Jia Fan
Department of Surgery, Liver Cancer Institute, Zhongshan Hospital, Fudan University Shanghai Medical School, Shanghai, China.

Contributors

Cyril Fauriat
Université de la Méditerranée, Institut de Cancérologie et d'Immunologie de Marseille, Marseille, France.

Guido Ferlazzo
Department of Immunology, Laboratory of Immunology and Biotherapy, University of Messina, Policlinico Universitario, Messina, Italy.

Aharon G. Freud
Department of Pathology, Stanford University of Medicine, Palo Alto, California, United States.

Kym R. Garrod
Department of Physiology and Biophysics, University of California, Irvine, California, United States.

Anthony A. Gaspari
Department of Dermatology, University of Maryland School of Medicine, Baltimore, Maryland, United States.

Godfrey S. Getz
Department of Pathology, University of Chicago, Chicago, Illinois, United States.

Angela Gismondi
Department of Experimental Medicine and Pathology, Instituto Pasteur-Fondazione Cenci Bolognetti, University of Rome, Rome, Italy.

Stephen R. Goding
Oncology Department, Johns Hopkins Medical Institution, Baltimore, Maryland, United States.

Segundo González
Department of Functional Biology, University of Oviedo, Oviedo, Spain.

Martin R. Goodier
Department of Immunology, Imperial College London, Chelsea and Westminster Hospital, London, United Kingdom.

Bartosz Grzywacz
University of Minnesota, Department of Pediatrics, Division of Blood and Marrow Transplantation, Minneapolis, Minnesota, United States.

Petter Höglund
Department of Microbiology and Cell Biology, Karolinska Institute, Stockholm, Sweden.

Anne Hosmalin
Département d'Immunologie, Institut Cochin, Université Paris-Descartes, Paris, France.

Franck Housseau
Oncology Department, Johns Hopkins University School of Medicine, Baltimore, Maryland, United States.

Takayuki Ikeda
Department of Food Science and Human Nutrition, Fuji Women's University, Hokkaido, Japan.

Taisuke Ito
Department of Dermatology, Hamamatsu University School of Medicine Handayama, Higashiku, Hamamatsu, Japan.

Priyanka Jha
Department of Radiology & Biomedical Imaging, University of California, San Francisco, San Francisco, California.

Veli-Matti Kähäri
Department of Dermatology, University of Turku, Turku, Finland.

Tatsuya Kanto
Department of Gastroenterology and Hepatology, Department of Dendritic Cell Biology and Clinical Applications, Osaka University Graduate School of Medicine, Osaka, Japan.

Alex Karlsson-Parra
Department of Microbiology and Immunology, Gothenburg University, Gothenburg, Sweden.

Rolf Kiessling
Experimental Oncology, IGT (Immune & Gene Therapy Laboratory), CCK (Cancer Center Karolinska), Stockholm, Sweden.

Hans Klingemann
Tufts Medical Center, Boston, Massachusetts, United States.

Christiane Knopp
Chemotherapeutisches Forschungsinstitut, Georg-Speyer-Haus, Frankfurt am Main, Germany.

Shinichi Koizumi
Division of Immunoregulation, Hokkaido University, Hokkaido, Japan.

Akira Komoriya
OncoImmunin. Inc., Gaithersburg, Maryland, United States.

Ming-Ling Kuo
Department of Microbiology and Immunology, Chang Gung University, Taoyuan, Taiwan.

Peter J.L. Lane
Medical Research Council for Immune Regulation, University of Birmingham Medical School, Birmingham, United Kingdom.

Yen-Chang Lee
Department of Pediatrics, Chang Gung Children's Hospital, Chang Gung University College of Medicine, Taoyuan, Taiwan.

Jussi Liippo
Department of Dermatology, University of Turku, Turku, Finland.

Syh-Jae Lin
Department of Pediatrics, Chang Gung Children's Hospital, Chang Gung University College of Medicine, Taoyuan, Taiwan.

Siyuan Liu
Department of Radiology & Biomedical Imaging, University of California, San Francisco, San Francisco, California, United States.

Hans-Gustaf Ljunggren
Center for Infectious Medicine, Department of Medicine, Karolinska University Hospital, Huddinge, Sweden.

Carlos López-Larrea
Department of Immunology, Hospital Universitario Central de Asturias, Oviedo, Spain, and Fundación Renal "Iñigo Álvarez de Toledo," Spain.

Alejandro López-Soto
Department of Functional Biology, University of Oviedo, Oviedo, Spain.

Michael T. Lotze
Departments of Surgery and Bioengineering, University of Pittsburgh Cancer Institute, University of Pittsburgh Schools of the Health Sciences, Pittsburgh, Pennsylvania, United States.

Tara J. Loux
Department of Surgery, University of Pittsburgh Medical Center, Pittsburgh, Pennsylvania, United States.

Lina Lu
Department of Immunology and General Surgery, Lerner Research Institute, Cleveland Clinic, Cleveland, Ohio, United States.

Michael Magee
University of Pittsburgh Cancer Institute, Department of Pathology, University of Pittsburgh, Pittsburgh, Pennsylvania, United States.

Robbie B. Mailliard
Celsense, Inc., Pittsburgh, Pennsylvania, United States.

Victoria H. Male
Department of Pathology, University of Cambridge, Cambridge, United Kingdom.

Ofer Mandelboim
Lautenberg Center for General and Tumor Immunology, Hebrew University, Biomedical Research Institute Israel Canada, Hadassah Medical School, Jerusalem, Israel.

Kazutaka Masuko
Division of Immunoregulation, Hokkaido University, Hokkaido, Japan.

Benjamin M. Matta
Department of Surgery and Immunology, University of Pittsburgh School of Medicine, Pittsburgh, Pennsylvania, United States.

Domenico Mavilio
Laboratory of Experimental Immunology, IRCCS Instituto Clinico Humanitas, Rozzano, Milan, Italy.

Borna Mehrad
Division of Pulmonary and Critical Care Medicine, Departments of Medicine and Microbiology, University of Virginia, Charlottesville, Virginia, United States.

Jeffery S. Miller
Division of Hematology, Oncology, and Transplantation, University of Minnesota, Minneapolis, Minnesota, United States.

Rieko Mitamura
Department of Food Science and Human Nutrition, Fuji Women's University, Hokkaido, Japan.

Ashley Moffett
Department of Pathology, University of Cambridge, Cambridge, United Kingdom.

Erika Montalto
Laboratory of Immunology and Biotherapy, University of Messina, Messina, Italy.

Alessandro Moretta
Department of Experimental Medicine, and Centre of Excellence for Biomedical Research, University of Genova, Genova, Italy.

Lorenzo Moretta
Istituto Giannina Gaslini, Genova, Department of Experimental Medicine, University of Genova, and Centre of Excellence for Biomedical Research, University of Genova, Genova, Italy.

William G. Morice
Division of Hematopathology, Mayo Clinic, Rochester, Minnesota, United States.

Christian Münz
Institute of Experimental Immunology, University Hospital of Zürich, Zürich, Switzerland.

Michael A. Nalesnik
Division of Transplantation and Hepatic Pathology, University of Pittsburgh Medical Center, Pittsburgh, Pennsylvania, United States.

Jerry Y. Niederkorn
Department of Ophthalmology, University of Texas Southwestern Medical Center, Dallas, Texas, United States.

Takashi Nishimura
Division of Immunoregulation, ROYCE Health Bioscience Institute for Genetic Medicine, Hokkaido University, Hokkaido, Japan.

Karen A. Norris
Department of Immunology, University of Pittsburgh School of Medicine, Pittsburgh, Pennsylvania, United States.

Daniel Olive
Université de la Méditerranée, Institut de Cancérologie et d'Immunologie de Marseille, Marseille, France.

John R. Ortaldo
Laboratory of Experimental Immunology, Cancer and Inflammation Program, National Cancer Institute, National Institutes of Health, Frederick, Maryland, United States.

Beverly Z. Packard
OncoImmunin, Inc., Gaithersburg, Maryland, United States.

Ralf Paus
Department of Dermatology, University Hospital Schleswig-Holstein, University of Lübeck, Lübeck, Germany.

Eckhard R. Podack
Department of Microbiology and Immunology, University of Miami, Miller School of Medicine, Miami, Florida, United States.

Shiguang Qian
Department of Immunology and General Surgery, Lerner Research Institute, Cleveland Clinic, Cleveland, Ohio, United States.

Shuang-Jian Qiu
Department of Surgery, Liver Cancer Institute, Zhongshan Hospital, Fudan University Shanghai Medical School, Shanghai, China.

Richard M. Ransohoff
Department of Neurosciences, Lerner Research Institute, Cleveland Clinic, Cleveland, Ohio, United States.

Catherine A. Reardon
Department of Pathology, University of Chicago, Chicago, Illinois, United States.

Verena Reinhart
Department of Radiology & Biomedical Imaging, University of California, San Francisco, San Francisco, California, United States.

Jérôme Rey
Université de la Méditerranée, Institut de Cancérologie et d'Immunologie de Marseille, Marseille, France.

Anna Rubartelli
Cell Biology Unit, National Cancer Research Institute, Genova, Italy.

Angela Santoni
Department of Experimental Medicine and Pathology, Instituto Pasteur-Fondazione Cenci Bolognetti, University of Rome, Rome, Italy.

Daniel Scott-Algara
Unité de Régulation des Infections Rétrovirales, Institut Pasteur, Paris, France.

Claudia Semino
Cell Biology Unit, National Cancer Research Institute, Genova, Italy.

Fu-Dong Shi
Center for Neurologic Diseases, Tranjin Medical University General Hospital, Tranjin 3000071, China and Department of Nuerology Institute, Barrow Neurological Institute, St. Joseph's Hospital and Medical Center, Phoenix, AZ 85013.

Helena Stabile
Department of Experimental Medicine and Pathology, Instituto Pasteur-Fondazione Cenci Bolognetti, University of Rome, Rome, Italy.

Noam Stern-Ginossar
Lautenberg Center for General and Tumor Immunology, Hebrew University, Biomedical Research Institute Israel Canada, Hadassah Medical School, Jerusalem, Israel.

Jeff Subleski
Laboratory of Experimental Immunology, Cancer and Inflammation Program, National Cancer Institute, National Institutes of Health, Frederick, Maryland, United States.

Angus W. Thomson
Departments of Surgery and Immunology, Thomas E. Starzl Transplantation Institute, University of Pittsburgh Schools of the Health Sciences, Pittsburgh, Pennsylvania, United States.

Mervi Toriseva
Department of Dermatology, University of Turku, Turku, Finland.

David S. Ucker
Department of Microbiology and Immunology, University of Illinois College of Medicine, Chicago, Illinois, United States.

Andrea Velardi
Division of Hematology and Clinical Immunology, University of Perugia, Perugia, Italy.

Michael R. Verneris
Pediatric Hematology/Oncology and Blood and Marrow Transplant, Minneapolis, Minnesota, United States.

Lazar N. Vujanovic
University of Pittsburgh Cancer Institute, Department of Medicine, University of Pittsburgh, Pittsburgh, Pennsylvania, United States.

Nikola L. Vujanovic
University of Pittsburgh Cancer Institute, Department of Pathology, University of Pittsburgh, Pittsburgh, Pennsylvania, United States.

Daiko Wakita
Division of Immunoregulation, Hokkaido University, Hokkaido, Japan.

Sheng Wei
H. Lee Moffitt Cancer Center and Research Institute, Immunology Program, Tampa, Florida, United States.

Jonathan M. Weiss
Laboratory of Experimental Immunology, Cancer and Inflammation Program, National Cancer Institute, National Institutes of Health, Frederick, Maryland, United States.

Winfried S. Wels
Chemotherapeutisches Forschungsinstitut, Georg-Speyer Haus, Frankfurt am Main, Germany.

Michael F. Wendland
Department of Radiology & Biomedical Imaging, University of California, San Francisco, San Francisco, California, United States.

Amy K. Wesa
Departments of Dermatology and Immunology, University of Pittsburgh School of Medicine, Pittsburgh, Pennsylvania, United States.

Robert H. Wiltrout
Laboratory of Experimental Immunology, Cancer and Inflammation Program, National Cancer Institute, National Institutes of Health, Frederick, Maryland, United States.

Yong-Sheng Xiao
Department of Surgery, Liver Cancer Institute, Zhongshan Hospital, Fudan University Shanghai Medical School, Shanghai, China.

Dah-Chin Yan
Department of Pediatrics, Chang Gung Children's Hospital, Chang Gung University College of Medicine, Taoyuan, Taiwan.

Makato Yawata
Institute for Stem Biology and Regenerative Medicine, Stanford University, Palo Alto, California, United States.

Nobuyo Yawata
Institute for Stem Biology and Regenerative Medicine, Stanford University, Palo Alto, California, United States.

Alexandra Y. Zhang
Department of Dermatology, University of Pittsburgh, Pittsburgh, Pennsylvania, United States.

Juan Carlos Zúñiga-Pflücker
Department of Immunology, Sunnybrook Research Institute, University of Toronto, Toronto, Canada.

Foreword

High background reactivity in a laboratory assay was the bane of my existence!

It was the early 1970s and I had joined the Department of Tumour Biology at the Karolinksa Institute with Eva Klein and Hans Wigzell as my Ph.D. supervisors. In the early 1960s, Hans did a postdoctoral stint at the Queen Victoria Hospital in Sussex and had established the ^{51}Cr release assay (Wigzell, 1965). It was subsequently used by Drs. Cerottini and Brunner (Brunner et al., 1968) to measure cellular lysis of tumour targets. My job as the new lab rat was to standardize this assay with YAC-1 cells to investigate the T cell reactivity against murine Moloney Virus induced leukemia. Much to my dismay, the background lysis of YAC-1 cells in the presence of mouse splenocytes remained high and persisted despite my best efforts, even with cells from non-immunized, control mice. By hindsight, it was through sheer serendipity that we chose YAC-1 cells but the choice was fortuitous indeed since even today, I cannot think of any cell that is more exquisitely sensitive to natural killer cell activity.

Gradually this "high background" activity evoked our interest and delineating the cell responsible for this phenomenon became the principal focus of my doctoral research. Our group and that of Ron Herberman at the National Institute of Health were the first to systematically characterize the effector cell responsible for the activity (Herberman et al., 1975a; Herberman et al., 1975b; Kiessling et al., 1975a; Kiessling et al., 1975b). The name "natural killer cells" came effortlessly as it was paraphrased on "natural antibodies", a field which was very active in the early 1970s. Nowadays, it is obvious that the cell we defined had several functions other than being a killer cell, such as secreting cytokines and recruiting other immune cells to the site of inflammation. Following its discovery and for several years later, the field of immunology continued to be dominated by research on T and B cells. NK cells were virtually relegated to the role of an interesting artifact or T cell type of unknown significance. Between 1975 and 1979 there were fewer than a hundred publications on NK cells. It was inconceivable that 30 years later NK cells would be investigated and found to have a role in all aspects of immunology including immunity to bacteria and viruses, transplantation, autoimmune disorders and hypersensitivity, as is obvious from the contributions to this volume.

The subject of target recognition by NK cells and their specificity arose very early on after their discovery and continues to be a subject of research even today. Three chapters in this book delve into the complex mechanisms and processes by which multiple ligands and signaling molecules regulate the activation and inhibition of NK target recognition and effector activity. At the onset, NK cell activity was defined by their ability to kill tumor targets and we looked upon NK cells as an alternative to T cell mediated immune surveillance (Kiessling and Haller, 1978). It became apparent soon, however, that the biological relevance of NK cells was much more complex and went beyond that of antitumor surveillance. As early as 1961, Snell and Stevens noticed that F1 hybrid mice derived from two inbred strains of mice often were relatively resistant to small tumor grafts of parental strain origin, compared to syngeneic recipients (Snell and Stevens, 1961). The late Gustavo Cudkowicz, then at University of Buffalo, had for many years pioneered studies on a similar type of "hybrid resistance" controlling rejection of hematopoietic grafts (Cudkowicz and Rossi, 1972). The resistance phenomenon had some rather distinctive characteristics that differed from the tenets of "conventional immunity" held at that time. In the summer of 1976, with America celebrating the bicentennial anniversary of its independence, I visited Gustavo's laboratory for a short and intense visit. As I worked with him and his colleagues it became apparent that there was a striking similarity between the mechanism of resistance to hematopoietic grafts and NK cell rejection (Kiessling et al., 1977).

Klas Kärre joined my laboratory as a doctoral student in the late 1970s. A soft-spoken, eloquent and slightly absent-minded guy, he made a succession of seminal discoveries which were summarized in 1981 in his doctoral thesis titled "On the immunobiology of Natural Killer Cells; studies of murine NK-cells and their interactions with T-cells and T-lymphomas". I delivered the galley proofs to him while he was at the hospital awaiting the birth of his first son. His doctoral defense was rather lively since a member of the advisory committee found his findings too speculative and not adequately substantiated with experimental evidence. However, the evidence for his "alternate immune defense" hypothesis accumulated rapidly in subsequent years. One of the key observations was that the RMA-S lymphoma selected to lack MHC class I expression (due to a mutation in the TAP2 gene (Yang et al., 1992)), was rejected in a T cell- independent, NK-cell dependent manner (Karre et al., 1986). This and other lines of verifications pointed to an inverse correlation between the expression of surface MHC class I molecules and susceptibility to NK-cell-mediated lysis of target cells. Later, Klas and his first doctoral student, Hans-Gustaf Ljunggren formulated the "missing self" hypothesis (Ljunggren and Karre, 1990). The hypothesis initially stirred quite a bit of controversy since it challenged the prevailing concept of how the immune system discriminated self and non self. To my mind, it was not until then and later following the molecular definition of NK receptors (Ciccone et al., 1992; Karlhofer et al., 1992), that NK biology truly became recognized and established as a bonafide domain of immunology.

The "missing self"-theory predicted the existence of inhibitory receptors that bind MHC class I. These receptors, now termed Ly49A, were identified on murine NK cells (Karlhofer et al., 1992). In parallel, antibodies against the human receptors were made (Ciccone et al., 1992). These discoveries provided a strong impetus to research on NK cells. During subsequent years it became clear that NK cells have a multitude of inhibitory and activating receptors that engage MHC class I as well as molecules similar to or entirely disparate from MHC class I. It is now common knowledge that the balance between these inhibitory and activating receptors ultimately regulates the cytotoxic function of NK cells, as will become apparent from several of the chapters in this book and continues to be the focus of research in several laboratories worldwide. The intricacy of this interaction is particularly perceptible in patients with MHC class I deficiencies. One would expect that these patients with "bare lymphocyte syndrome" would demonstrate high incidence of NK-mediated immunopathologic disease, but surprisingly they do not (Zimmer et al., 1998). One potential reason is that the NK receptors, never having encountered the MHC class I ligand, persists in an "uneducated" state, and is therefore unable to recognize MHC class I low target cells. The understanding of how NK cells are being "educated" is one of the more important aspects of NK research, closely associated with the question of how NK cells are maintained in a "tolerant" state to self, as will be discussed in several chapters of this volume.

Why have NK cells developed and why do we have them? Our early view was that NK cells were a vestigial remain of a primordial immune system which was a forerunner to the more refined adaptive immune system (Kiessling and Wigzell, 1981). However this theory is not compatible with several observations, including the fact that orthologs of most NK cell receptor families cannot be found earlier in evolution than mammals (Walzer et al., 2007). NK and T cells have complementary roles in host defense as well as have commonality in mechanisms of cytotoxicity, which rather suggests a common ancestral cell for NK cells and T cells. The most plausible explanation for why NK cells evolved is that they developed as a complementary system to adaptive T cell immunity for defense against viruses and transformed cells. The "virus activated killer cell" was studied by a handful of virologists in the seventies, which then merged with NK research when it proved that this killer cell was identical to the NK cell (Oldham et al., 1977; Welsh and Zinkernagel, 1977). The antiviral role of NK cells however is not universal and only extends to certain viruses like Herpes Viruses and influenza virus. Direct evidence for the protective effect comes from NK depletion or adoptive transfer experiments in mouse models of the herpesvirus MCMV (Bukowski et al., 1985). NK cell deficient mice infected with coxsackie B3 virus have higher titers of virus and more severe myocarditis compared to NK-replete control mice (Fairweather et al., 2001), demonstrating the importance of NK cells also in controlling immunopathology. There also exists a notable case report of severe Herpesvirus infection in a patient with selective NK cell defect (Biron et al., 1989).

The capacity of NK cells to react also with non-malignant activated or immature myelomonocytic cells was described in the early 1980s (Hansson et al., 1982). NK cells inhibited the development of granulocytic progenitor cells in colony forming assays performed in semisolid agar. This immunoregulatory function of NK cells and their interaction with cells of the myelomonocytic lineage has now been extensively verified, specifically by their ability to influence DC function. NK cells can kill both human and mouse DC, which may influence DC homeostasis and potentially also limit dendritic cell vaccination efficacy (Hayakawa et al., 2004). Paradoxically, however, NK cells can also facilitate antigen presentation by DC since antigens released by target cells following lysis by NK cells can be endocytosed and presented by DC.

Primary immunodeficiencies have frequently offered opportunities to study the input of distinct effector mechanisms towards resistance against microorganisms, but NK cell research has suffered from a paucity of animal models which selectively lack NK cells. Furthermore, patients with selective defects in NK functions are very rare. One may in fact dispute whether any truly selective NK immunodeficiency really exists in mouse or man (Fischer, 2007). Although several studies described deficiencies in NK cells primarily in conjunction with viral infections, a specific molecular defect leading to selective loss of NK function has never been identified. For example, the "beige" mouse, which is the murine equivalent of Chediak Higashi syndrome of man, also displays numerous defects in the monocyte and T cell compartment (Barak and Nir, 1987; Roder et al., 1979).

Regardless of the true biological role of NK cells, there is now much optimism in the NK field that the coming decades will see the development of NK cell based therapies in the clinical management of diverse diseases caused by infectious pathogens and cancer (Ljunggren and Malmberg, 2007). T cell based therapy of cancer and chronic viral infections has so far met with only limited success in the clinic. The efficacy of T cell therapies is restricted largely due to the strong tendency of tumors and viruses to develop "stealth" strategies based on loss of MHC class I expression. If we could harness the complementary role that NK cells have in eliminating MHC class I low tumor cells and utilize our rapidly increasing understanding of the NK receptors and their tumor ligands, it would have a significant impact on future immunotherapy. Pioneering studies were done with peripheral blood lymphocytes activated with IL-2 into "lymphokine-activated killer" (LAK) cells whose function can principally be attributed to activated NK cells. LAK cells combined with IL-2 can achieve very significant and long lasting responses in melanoma patients as well as other solid tumors (Rosenberg, 2000). Initially it was thought that the 15–20% response rate typically noted in patients could be improved by increasing the dose of IL-2 or LAK cells which was unfortunately not realized. It is presently known that several reasons including dose-limiting toxicity may curb the clinical response rate of IL-2 based therapies and it has become apparent that IL-2 also induces apoptosis in NK cells or expands the regulatory T cell subset which can directly inhibit NK function (Ghiringhelli et al., 2005; Rodella et al., 2001) . Additionally, tumor targets in most of the treated patients may lack the appropriate combination of activating and inhibitory receptors, and therefore cannot simply be eliminated by the IL-2 activated NK cells. While most NK assays are performed with long term in vitro cultured tumor lines, freshly explanted human tumors are relatively resistant to NK mediated cytotoxicity, although some non-cultured tumors such as ovarian carcinomas which frequently display various defects in MHC class I presentation (Norell et al., 2006) can be recognized and killed even with non-activated NK cells (Carlsten et al., 2007).

A high priority will be the discovery of methods to manipulate the activity of NK cells e.g. approaches tilting the balance in favor of activating versus inhibitory receptors or administration of the right combination of growth factors. Recently, there has been a major breakthrough in the treatment of leukemias which is directly related to the findings of NK cell mediated killing of lymphomas, the "missing self" hypothesis and the definition in molecular terms of HLA class I inhibitory receptors (Ruggeri et al., 2005). It is an attractive possibility to also utilize similar treatment modalities for solid tumors, as will be discussed in this volume in several chapters.

Clinical therapy with antibodies has been a real success story for modern biotherapy. Approximately 12 antibodies are currently approved for therapeutic use. ADCC by NK cells is known to be a major factor mediating the clinical effect of mAbs such as Rituximab reactive to CD20 and used for treatment of lymphoma and Herceptin specific for the oncogene Her2/neu expressed in a proportion of breast-cancers (Cartron et al., 2002). The ADCC is largely mediated by the CD56 dim NK cell subset, which has a high expression of the low-affinity Fc γ receptor IIIA, CD16. The conclusion that ADCC plays a major role in the efficacy of Rituximab and Herceptin is established from experiments with Fc receptor gamma deficient mice and more recently from the correlation between improved clinical efficacy and an Fc gamma IIIa gene polymorphism which results in a higher affinity for these antibodies (Cartron et al., 2002; Clynes et al., 2000). This knowledge has stimulated interest in combinatorial therapies with mAb administered together with therapies known to increase NK activity.

There is also an increasing awareness of the potential synergistic effects of combinatorial cancer therapies; not only focused on merging T cell and NK cell modalities but also those combining immunotherapies with conventional chemo-radiotherapy (Zitvogel et al., 2008). The "preconditioning" of patients with chemotherapy may have several effects which may increase NK- mediated tumor killing by boosting NK activity or increasing the target sensitivity to NK lysis. A non-myeloablative regimen with low dose cyclophosphamide and 5-fluorouracil was shown to preferentially eliminate regulatory T cells. Objective clinical responses were observed in 50% of advanced melanoma patients who received this regimen prior to adoptive transfer of tumor infiltrating lymphocytes (TIL). Since NK cells are also suppressed by regulatory T cells by a TGF-β-dependent mechanism (Ghiringhelli et al., 2005), the combination of the chemotherapy regimen with NK-based immunotherapy is particularly promising. Other combinatorial possibilities involve up-regulating ligands for activating NK receptors, such as the NKG2D, through low doses of chemotherapy or ionizing irradiation (Gasser et al., 2005). These regimens act through the DNA-damage response pathway, which may upregulate NKG2D, thereby "sensitizing" tumor cells to recognition by NK cells. Another example of combinatorial treatments based on NK cells and drugs is the recently approved combination of IL-2 and histamine in the treatment of patients with acute myelogenous leukemia (AML) (Brune et al., 2006). AML Patients in remission were shown to have a prolonged relapse free survival as a result of IL-2 and histamine treatment which activates both NK cells and CD8+ T cells. Histamine was shown to protect NK cells from oxidative stress-induced apoptosis, particularly the CD56 subpopulation which is of major importance for cytotoxicity against tumor targets.

In conclusion, there has been a tremendous increase in our knowledge of NK biology and function. We are beginning to develop approaches for utilizing NK cells for clinical therapy of malignancies, or in contrast neutralizing them to protect organ transplants or abrogate autoimmune disorders. These approaches are still in their infancy but will be greatly facilitated by the ever-expanding knowledge of the activation and inhibition pathways in NK cells. It has been a long and astonishing journey for a cell that started its life as a background noise in a laboratory assay.

Rolf Kiessling

References

Barak, Y. and Nir, E. (1987). Chediak-Higashi syndrome. The American journal of pediatric hematology/oncology **9**, 42–55.

Biron, C.A., Byron, K.S. and Sullivan, J.L. (1989). Severe herpesvirus infections in an adolescent without natural killer cells. The New England journal of medicine **320**, 1731–1735.

Brune, M., Castaigne, S., Catalano, J., Gehlsen, K., Ho, A.D., Hofmann, W. K., Hogge, D.E., Nilsson, B., Or, R., Romero, A.I., Rowe, J.M., Simonsson, B., Spearing, R., Stadtmauer, E.A., Szer, J., Wallhult, E. and Hellstrand, K. (2006). Improved leukemia-free survival after postconsolidation immunotherapy with histamine dihydrochloride and interleukin-2 in acute myeloid leukemia: results of a randomized phase 3 trial. Blood **108**, 88–96.

Brunner, K.T., Mauel, J., Cerottini, J.C. and Chapuis, B. (1968). Quantitative assay of the lytic action of immune lymphoid cells on 51-Cr-labelled allogeneic target cells in vitro; inhibition by isoantibody and by drugs. Immunology **14**, 181–196.

Bukowski, J.F., Warner, J.F., Dennert, G. and Welsh, R.M. (1985). Adoptive transfer studies demonstrating the antiviral effect of natural killer cells in vivo. The Journal of experimental medicine **161**, 40–52.

Carlsten, M., Bjorkstrom, N.K., Norell, H., Bryceson, Y., van Hall, T., Baumann, B. Ć., Hanson, M., Schedvins, K., Kiessling, R., Ljunggren, H.G. and Malmberg, K. J. (2007). DNAX accessory molecule-1 mediated recognition of freshly isolated ovarian carcinoma by resting natural killer cells. Cancer research **67**, 1317–1325.

Cartron, G., Dacheux, L., Salles, G., Solal-Celigny, P., Bardos, P., Colombat, P. and Watier, H. (2002). Therapeutic activity of humanized anti-CD20 monoclonal antibody and polymorphism in IgG Fc receptor FcgammaRIIIa gene. Blood **99**, 754–758.

Ciccone, E., Pende, D., Viale, O., Than, A., Donato, C.Di, Orengo, A.M., Biassoni, R., Verdiani, S., Amoroso, A., Moretta, A., et al. (1992). Involvement of HLA class I alleles in natural killer (NK) cell-specific functions: expression of HLA-Cw3 confers selective protection from lysis by alloreactive NK clones displaying a defined specificity (specificity 2). The Journal of experimental medicine **176**, 963–971.

Clynes, R.A., Towers, T.L., Presta, L.G. and Ravetch, J.V. (2000). Inhibitory Fc receptors modulate in vivo cytoxicity against tumor targets. Nature medicine **6**, 443–446.

Cudkowicz, G. and Rossi, G.B. (1972). Hybrid resistance to parental DBA-2 grafts: independence from the H-2 locus. I. Studies with normal hematopoietic cells. Journal of the National Cancer Institute **48**, 131–139.

Fairweather, D., Kaya, Z., Shellam, G.R., Lawson, C.M. and Rose, N.R. (2001). From infection to autoimmunity. Journal of autoimmunity **16**, 175–186.

Fischer, A. (2007). Human primary immunodeficiency diseases. Immunity **27**, 835–845.

Gasser, S., Orsulic, S., Brown, E.J. and Raulet, D.H. (2005). The DNA damage pathway regulates innate immune system ligands of the NKG2D receptor. Nature **436**, 1186–1190.

Ghiringhelli, F., Menard, C., Terme, M., Flament, C., Taieb, J., Chaput, N., Puig, P.E., Novault, S., Escudier, B., Vivier, E., Lecesne, A., Robert, C., Blay, J.Y., Bernard, J., Caillat-Zucman, S., Freitas, A., Tursz, T., Wagner-Ballon, O., Capron, C., Vainchencker, W., Martin, F. and Zitvogel, L. (2005). CD4 + CD25+ regulatory T cells inhibit natural killer cell functions in a transforming growth factor-beta-dependent manner. The Journal of experimental medicine **202**, 1075–1085.

Hansson, M., Beran, M., Andersson, B. and Kiessling, R. (1982). Inhibition of in vitro granulopoiesis by autologous allogeneic human NK cells. J Immunol **129**, 126–132.

Hayakawa, Y., Screpanti, V., Yagita, H., Grandien, A., Ljunggren, H.G., Smyth, M.J. and Chambers, B.J. (2004). NK cell TRAIL eliminates immature dendritic cells in vivo and limits dendritic cell vaccination efficacy. J Immunol **172**, 123–129.

Herberman, R.B., Nunn, M.E., Holden, H. T. and Lavrin, D.H. (1975a). Natural cytotoxic reactivity of mouse lymphoid cells against syngeneic and allogeneic tumors. II. Characterization of effector cells. International journal of cancer **16**, 230–239.

Herberman, R.B., Nunn, M.E. and Lavrin, D.H. (1975b). Natural cytotoxic reactivity of mouse lymphoid cells against syngeneic acid allogeneic tumors. I. Distribution of reactivity and specificity. International journal of cancer **16**, 216–229.

Karlhofer, F.M., Ribaudo, R.K. and Yokoyama, W.M. (1992). MHC class I alloantigen specificity of Ly-49+ IL-2-activated natural killer cells. Nature **358**, 66–70.

Karre, K., Ljunggren, H.G., Piontek, G. and Kiessling, R. (1986). Selective rejection of H-2-deficient lymphoma variants suggests alternative immune defence strategy. Nature **319**, 675–678.

Kiessling, R., Klein, E., Pross, H. and Wigzell, H. (1975a). "Natural" killer cells in the mouse. II. Cytotoxic cells with specificity for mouse Moloney leukemia cells. Characteristics of the killer cell. European journal of immunology **5**, 117–121.

Kiessling, R., Klein, E. and Wigzell, H. (1975b). "Natural" killer cells in the mouse. I. Cytotoxic cells with specificity for mouse Moloney leukemia cells. Specificity and distribution according to genotype. European journal of immunology **5**, 112–117.

Kiessling, R. and Haller, O. (1978). Natural killer cells in the mouse: an alternative immune surveillance mechanism?. Contemporary topics in immunobiology **8**, 171–201.

Kiessling, R., Hochman, P.S., Haller, O., Shearer, G.M., Wigzell, H. and Cudkowicz, G. (1977). Evidence for a similar or common mechanism for natural killer cell activity and resistance to hemopoietic grafts. European journal of immunology **7**, 655–663.

Kiessling, R. and Wigzell, H. (1981). Surveillance of primitive cells by natural killer cells. Current topics in microbiology and immunology **92**, 107–123.

Ljunggren, H.G. and Karre, K. (1990). In search of the 'missing self': MHC molecules and NK cell recognition. Immunology today **11**, 237–244.

Ljunggren, H.G. and Malmberg, K.J. (2007). Prospects for the use of NK cells in immunotherapy of human cancer. Nature reviews **7**, 329–339.

Norell, H., Carlsten, M., Ohlum, T., Malmberg, K.J., Masucci, G., Schedvins, K., Altermann, W., Handke, D., Atkins, D., Seliger, B. and Kiessling, R. (2006). Frequent loss of HLA-A2 expression in metastasizing ovarian carcinomas associated with genomic haplotype loss and HLA-A2-restricted HER-2/neu-specific immunity. Cancer research **66**, 6387–6394.

Oldham, R.K., Ortaldo, J.R. and Herberman, R.B. (1977). Natural cytotoxic reactivity of rat lymphocytes against gross virus-induced tumor cell lines as measured by [125I]iododeoxyuridine and tritiated proline microcytotoxicity assays. Cancer research **37**, 4467–4474.

Rodella, L., Zamai, L., Rezzani, R., Artico, M., Peri, G., Falconi, M., Facchini, A., Pelusi, G. and Vitale, M. (2001). Interleukin 2 and interleukin 15 differentially predispose natural killer cells to apoptosis mediated by endothelial and tumour cells. British journal of haematology **115**, 442–450.

Roder, J.C., Lohmann-Matthes, M.L., Domzig, W. and Wigzell, H. (1979). The beige mutation in the mouse. II. Selectivity of the natural killer (NK) cell defect. J Immunol **123**, 2174–2181.

Rosenberg, S.A. (2000). Interleukin-2 and the development of immunotherapy for the treatment of patients with cancer. The cancer journal from Scientific American **6**(Suppl 1), S2–S7.

Ruggeri, L., Mancusi, A., Capanni, M., Martelli, M.F. and Velardi, A. (2005). Exploitation of alloreactive NK cells in adoptive immunotherapy of cancer. Current opinion in immunology **17**, 211–217.

Snell, G.D. and Stevens, L.C. (1961). Histocompatibility genes of mice. III. H-1 and H-4, two histocompatibility loci in the first linkage group. Immunology **4**, 366–379.

Walzer, T., Jaeger, S., Chaix, J. and Vivier, E. (2007). Natural killer cells: from CD3(-)NKp46(+) to post-genomics meta-analyses. Current opinion in immunology **19**, 365–372.

Welsh , R.M. Jr. and Zinkernagel, R. M. (1977). Heterospecific cytotoxic cell activity induced during the first three days of acute lymphocytic choriomeningitis virus infection in mice. Nature **268**, 646–648.

Wigzell, H. (1965). Quantitative Titrations of Mouse H-2 Antibodies Using Cr-51-Labelled Target Cells.. Transplantation **3**, 423–431.

Yang, Y., Fruh, K., Chambers, J., Waters, J.B., Wu, L., Spies, T. and Peterson, P. A. (1992). Major histocompatibility complex (MHC)-encoded HAM2 is necessary for antigenic peptide loading onto class I MHC molecules. The Journal of biological chemistry **267**, 11669–11672.

Zimmer, J., Donato, L., Hanau, D., Cazenave, J.P., Tongio, M.M., Moretta, A. and de la Salle, H. (1998). Activity and phenotype of natural killer cells in peptide transporter (TAP)-deficient patients (type I bare lymphocyte syndrome). The Journal of experimental medicine **187**, 117–122.

Zitvogel, L., Apetoh, L., Ghiringhelli, F. and Kroemer, G. (2008). Immunological aspects of cancer chemotherapy. Nature reviews **8**, 59–73.

Preface

***Nature**, in the broadest sense, is equivalent to the **natural world**, **physical world** or **material world**. 'Nature' refers to the* phenomena *of the physical world, and also to* life *in general. Manufactured objects and human interaction generally are not considered part of nature, and are referred to as* artificial *or man-made. Nature is generally distinguished from the* supernatural*. It ranges in scale from the* subatomic *to the* galactic*. The word* nature *is derived from the Latin word* natura*, or 'essential qualities, innate disposition', but literally meaning 'birth'. Original sense is in 'human nature' (Harper, 2001a).* Natura *was a Latin translation of the Greek word* physis (φύσις)*, which originally related to the intrinsic characteristics that plants, animals, and other features of the world develop of their own accord.*[1] *This is shown in the first written use of the word* φύσις*, in connection with a plant.*[2] *The concept of nature as a whole, the physical* universe*, is one of several expansions of the original notion; it began with certain core applications of the word* φύσις *by pre-Socratic philosophers, and has steadily gained currency ever since. This usage was confirmed during the advent of modern* scientific method *in the last several centuries.*[3,4]

Indeed, there have been many changes in the source of information and quotes both ancient and more modern, decorate the beginnings of each of our chapters within this 'First Edition of Natural Killer Cells'. Given the changes that now occur in the means by which we derive information, it seemed particularly appropriate to use a rather democratic source of information available to all on the Internet, 'Wikipedia', to launch our preface. The scientific method, distilled as repeated observations attempting to nullify a central hypothesis, was applied in all of our efforts to launch this effort with many of our colleagues and mentors. They demeaned the process of writing books, now possibly considered dinosaurs of erudition, in favour of writing grants, giving talks, or crafting reviews for high-impact factor journals. Most of these are more remunerative directly or indirectly but we felt the democracy of a set of firm, experienced highly selected hands laying siege to individual topics, a steady editorial assistant in the person of Kristi Anderson, to whom we are indebted, and the iterative events of review and resubmission in a single volume was well-deserved labour for this most singular of cells, the so-called NK cell. Together, we have introduced other changes here: electronic availability through libraries for each chapter, the development of an abstract/construct to introduce each chapter comparable to what is available for journal articles and available through ScienceDirect for ready access and citation management systems. We thank our authors for their industry and willingness to commit to this volume and suffer our reminders and pursuit of them in their labours. For the value of this edition, we owe them everything; any defects remain with us.

Natural. It is now over 40 years since the first cytolytic assays were performed with dye exclusion and then ^{51}Cr release assays revealing in fine detail, the ability of lymphoid cells to mediate lytic activity against cultured tumour targets (Brunner et al., 1968). Flying in the face of conventional notions of immune specificity, it was subsequently found that some cells, so-called natural killer cells (Herberman, 1975a,b; Kiessling et al., 1975a,b) could kill tumour cells without prior sensitization and without MHC restriction. Since then the emphasis has rather been on the cytolytic capability of these cells, more than on their nature and their natural role. With this volume we explore the many other natural traits

[1]A useful though somewhat erratically presented account of the pre-Socratic use of the concept of φύσις may be found in Naddaf (2006). The word φύσις, while first used in connection with a plant in Homer, occurs very early in Greek philosophy, and in several senses. Generally, these senses match rather well the current senses in which the English word *nature* is used, as confirmed by Guthrie (1965).

[2]The first known use of *physis* was by Homer in reference to the intrinsic qualities of a plant:
ὣς ἄρα φωνήσας πόρε φάρμακον ἀργεϊφόντης ἐκ γαίης ἐρύσας, καί μοι φύσιν αὐτοῦ ἔδειξε . (So saying, Argeiphontes [=Hermes] gave me the herb, drawing it from the ground, and showed me its **nature**.) (The word is dealt with thoroughly in Liddell and Scott's *Greek Lexicon*.) For later but still very early Greek uses of the term, see earlier note.

[3]Isaac Newton's *Philosophiae Naturalis Principia Mathematica* (Newton, 1687), for example, is translated 'Mathematical Principles of Natural Philosophy', and reflects the then-current use of the words 'natural philosophy', akin to 'systematic study of nature'.

[4]The etymology of the word 'physical' shows its use as a synonym for 'natural' in about the mid-fifteenth century (Harper, 2001b). From Wikipedia, downloaded on Sunday, 29 March 2009.

of these cells, well expanding beyond simple killing of tumours, suggesting that in a less-innocent world we should consider these rather N cells, comparable to B and T cells, reflecting now on their ability to regulate TH1 responses, promote DC maturation, promote autophagy, and support the vascularization of the placenta, putting it in a critical role in perpetuation of our own species. Most recently the notions of NK cell progenitors being critical for the development of lymph nodes and all adaptive immune cells make them indeed the most natural of cells, critical for their emergent role throughout modern immunology. Although perhaps misplaced here, we should connote this important cell with a single letter like its brethren the T, the B, and the dendritic but perhaps we will have to wait for the next edition. Indeed, as abundantly revealed throughout this text, NK cells do far more than just kill and they have been revealed in a degree of complexity and importance rather unimagined in their debut almost 35 years ago. Thus 'N Cells' awaits your approbation and consideration.

Stress. Life is stressful. Nutrient loss, hypoxia, genomic stress, ER stress, infection, damage. Writing new textbooks in a difficult time with all scientists and clinicians oversubscribed with the promise and threat of instant information and manuscripts managed online and without the buffeting of the postal service is also stressful. Thus the tempo and temperament of the writing process has assumed the same hurried moment as the recruitment of inflammatory cells, including the NK cell to sites of stress. Perhaps following in the 'cell-steps' of the macrophage with which they interact early during inflammatory processes, on an evolutionary scale, NK cells are likely in our estimation the primordial adaptive cell. NK cells similarly have an ability to expand, contract and respond with, at the very least, short-term memory to microbial stress and, we suspect, tissue damage or injury (Sun et al., 2009). At a fundamental level, NK cells script and focus their myeloid brethren on a dangerous world full of pathogens and tissue damage. As such they define the quality and nature of the immune response in the setting of stress, serving as mobile paracrine agents releasing cytokines or inducing their production dependant on their integration of multiple signals, multi-tasking within the lymph node and peripheral tissues. The first section of our volume (Chapters 1–9) focuses on the development of NK cells from progenitors in rodents and humans, how they signal, how they identify stress in tissues and cancer, and how they interact coordinately with other cells within these tissues.

Recognition of stress. There are many ways for cells to communicate in multicellular organisms, through cytokines that indeed 'move' cells to change their shape or biology, through chemokines directing the to and fro of cells within tissues and secondary and tertiary lymphoid sites, and through direct cell–cell interactions. This goes beyond just their interaction with each other as inflammatory cells and is defined as the integrant of interactions with cells in the tissues including the endothelial cells, epithelial cells, stromal cells, and sessile inflammatory cells. NK cells are sophisticated communicators, sensing signals from all of these cells and providing necessary feedback, eliciting programmed cell death (apoptosis] when necessary or programmed cell survival under dire threats (autophagy). The rough and tumble of life makes NK cells exquisitely sensitive to signals emanating from the postcapillary venules to allow their rapid emigration into these sites and within lymph nodes across high endothelial venules, coordinately enhancing their response to stress. These important interactions are captured in Chapters 10–24.

Integration of signals within organs. As different are the NK cells serving as helper cells steering new blood vessels to the implanted embryo, are those pow-wowing with their dendritic cell intimates within the eye, the brain, the gut, and the liver. Thus, a nuanced and balanced role for their function is required. It is one thing to eliminate an hepatocyte which is stressed and readily regenerated as it is to over-react to the microbial flora within the gut. Indeed NK cells are necessary diplomats, cajoling when necessary, creating outright warfare when discussion is beyond reason, and defining the nature and shape of the negotiating table. This they do with remarkable insights into the various tissues within which they find themselves. Thus to understand NK cells in aggregate, more importantly, one should consider them within various tissues within which they share some central properties but not all. Focusing on their identification and varied roles at these sites are exemplified in Chapters 25–33.

Roles of NK cells in disease. With acute injury, there is a requisite need to mobilize resources to limit damage and repair, and in an informed way, prepare for similar insults with an enhanced response, what immunologists refer to as memory (Sun et al., 2009). NK cells are charged to (1) recognize damage; (2) limit further damage; (3) regenerate and vascularize damaged tissues; and (4) learn from the experience and in particular be prepared for further encounters with the same or similar pathogens (basically, the role of adaptive immunity). This characteristic originally relegated to a perceived more 'noble' T cell and B cell has now been recognized in NK cells. They too can commit sins of omission, with failure to recognize either cryptic microbial or neoplastic antigens or commission with the undesired aspects of autoimmunity and graft-versus-host or host-versus-graft disease. These can be readily defined as issues for both NK cells and T and B cells. It should at least be contemplated that NK cells' exuberant or deficient response to stress in the setting of disease may allow their promotion of a damaging TH1 response or a permissive and emergent role as NK regulatory cells, squelching

NE Book Fulfillment

THINGS TO KNOW ABOUT YOUR ORDER

Thanks! We really appreciate your order and aim to please.

ANY PROBLEMS WITH YOUR ORDER?
If you are less than satisfied with your order, please contact us right away and we will do our best to fix it. Complete customer satisfaction is very important to us, and we want to address and fix any issues you may have. Please keep in mind that damage may occur during shipping.

We're happy to accept returns:
After receiving your purchase you have 30 days to request and return the item. Please note that we charge a small restocking fee between 5% and 10% but will not exceed $10. **Don't forget to please pack carefully!**

If you ordered through Amazon and you're happy with your order, we'd really appreciate it if you'd give us feedback!
If you're happy with our service, your positive feedback helps us by letting potential customers know we can be trusted. Plus, we love to hear that you are happy with your purchase!

- Go to **www.amazon.com/feedback**. You'll be prompted to log into your account.
- After logging in, find the order and click the "Leave Seller Feedback" button on the right.

Please check out The Wisdom Shop for future purchases.
When purchasing future orders on Amazon please look to buy from us again at "The Wisdom Shop." Our Amazon storefront can be viewed at **www.amazon.com/shops/The_Wisdom_Shop**

immune reactivity where necessary and promoting programmed cell survival or autophagy. Their role in various disease states is considered in detail in Chapters 34–45.

Special issues in NK cells and 'Wicked Problems'. The role of new technologies in genetically modifying NK cells, assessing their cytolytic activities, or imaging them in vivo deserved special consideration, here assembled in Chapters 46–50. The 'dilemma' of understanding NK cells requires a deeper understanding of the problems of all 'social scientists' considering the emergent properties of complex organisms as presciently outlined in the journal *Policy Sciences*, published in 1973 by Elsevier, a Dutch company in Amsterdam and our publisher for this volume but, interesting to us, printed in Scotland (Rittel and Webber, 1973). Here Rittel and Webber wrestled with how to plan when neither the problems were well-understood nor suitable solutions with any well agreed upon and indisputable public good could be identified. This indeed, 'Dilemmas in a General Theory of Planning' deserves some comment for the social nature of NK cells in the organisms within which they evolved. Thus, ending with a quotation from their treatise seems only appropriate:

> *A great many barriers keep us from perfecting such a planning/governing system: theory is inadequate for decent forecasting; our intelligence is insufficient to our tasks; plurality of objectives held by pluralities of politics makes it impossible to pursue unitary aims; and so on. The difficulties attached to rationality are tenacious, and we have so far been unable to get untangled from their web. This is partly because the classical paradigm of science and engineering—the paradigm that has underlain modern professionalism—is not applicable to the problems of open societal systems. ...The error has been a serious one. The kinds of problems that planners deal with—societal problems—are inherently different from the problems that scientists and perhaps some classes of engineers deal with. Planning problems are inherently wicked. As distinguished from problems in the natural sciences, which are definable and separable and may have solutions that are findable, the problems of governmental planning—and especially those of social or policy planning—are ill-defined; and they rely upon elusive political judgment for resolution. (Not 'solution'. Social problems are never solved. At best they are only re-solved—over and over again.)*

And thus NK cells, thrust into their societal role within the organism, have this difficult problem of resolving the problems of complex biology over and over again. The relevance of the body politic and political bodies seem indeed congruent in this instance. And thus, undoubtedly, we will need to revisit this cell in a second edition, finding their solution impossible, but in their understanding, something wicked.

References

Brunner, K.T., Mauel, J., Cerottini, J.C. and Chapuis, B. (1968). Quantitative assay of the lytic action of immune lymphoid cells on ^{51}Cr-labelled allogeneic target cells *in vitro*; inhibition by isoantibody and by drugs. Immunology **14**, 181–196.

Guthrie, W.K.C. (1965). *Presocratic Tradition from Parmenides to Democritus* (volume 2 of his *History of Greek Philosophy*). Cambridge UP.

Harper, D. (2001a). Nature. *Online Etymology Dictionary*, http://www.etymonline.com/index.php?term=nature. Retrieved on 23 September 2006.

Harper, D. (2001b). Physical. *Online Etymology Dictionary*, http://www.etymonline.com/index.php?term=physical. Retrieved on 20 September 2006.

Herberman, R.B., Nunn, M.E., Holden, H.T. and Lavrin, D.H. (1975a). Natural cytotoxic reactivity of mouse lymphoid cells against syngeneic and allogeneic tumors. II. Characterization of effector cells. Int J Cancer **16**, 230–239.

Herberman, R.B., Nunn, M.E. and Lavrin, D.H. (1975b). Natural cytotoxic reactivity of mouse lymphoid cells against syngeneic acid allogeneic tumors. I. Distribution of reactivity and specificity. Int J Cancer **16**, 216–229.

Homer. In: Murray, A.T. (ed.), *Odyssey* 10.302–3.

Kiessling, R., Klein, E. and Wigzell, H. (1975a). "Natural" killer cells in the mouse. I. Cytotoxic cells with specificity for mouse Moloney leukemia cells. Specificity and distribution according to genotype. Eur J Immunol **5**, 112–117.

Kiessling, R., Klein, E., Pross, H. and Wigzell, H. (1975b). "Natural" killer cells in the mouse. II. Cytotoxic cells with specificity for mouse Moloney leukemia cells. Characteristics of the killer cell. Eur J Immunol **5**, 117–121.

Naddaf, G. (2006). *The Greek Concept of Nature*. SUNY Press.

Newton, I. (1687). *Philosophiae Naturalis Principia Mathematica*.

Rittel, H.W.J. and Webber, M.M. (1973). Dilemmas in a general theory of planning. Pol Sci **4**, 155–169.

Sun, J.C., Beilke, J.N. and Lanier, L.L. (2009). Adaptive immune features of natural killer cells. Nature **457**(7229), 557–561.

Dedication to our Institute Directors, Fadi Lakkis and Nancy Davidson. It is appropriate in this era of increasingly complex science, administration and clinical activities to dedicate this volume to our new directors. Both have dedicated themselves to the art and science of Transplantation Immunology and Cancer Biology, respectively. Our ability to construct this first volume has been dependant on their emergent support for the academic mission, creating and organizing knowledge during a period of extraordinary expansion of that knowledge, now during a period of unprecedented economic travail, and concerns about how this knowledge will be applied and expanded. Indeed, 'natural' progression of our innate understanding of immunity in the context of the problems of both acute and chronic inflammatory diseases mediated and modulated by NK cells will require their continued gentle ministrations.

PART I

NK biology

Chapter One

1

Developmental stages and pathways of NK cell maturation

Bartosz Grzywacz, Jeffery S. Miller, Michael R. Verneris

CHAPTER CONTENTS

Yes, there are two paths you can go by, but in the long run, there is still time to change the road you are on....

Robert Plant of Led Zeppelin (1970)

ABSTRACT

Hematopoietic stem cells (HSCs) by definition can differentiate into all types of blood cells. There are several factors and events that promote HSC differentiation into the natural killer (NK) cell lineage. These include soluble factors, with a prominent role for interleukin 15, as well as contact- or gradient-dependent signals, such as Gas6/Tyro family of ligands and factors activating Wnt pathway. A complete understanding of the factors that control NK cell differentiation may allow for manipulation of NK cell reconstitution following hematopoietic cell transplantation and efficient ex vivo generation of NK cells for adoptive immune therapy.

KEY WORDS

NK cells, Development, Hematopoiesis, Transcription factors, Differentiation, Lineage CHOICE, Cytokines and receptors, Morphogens, Innate immunity

The early events in hematopoiesis

Hematopoiesis sustains the production of blood cells throughout life and is the best understood model of stem cell differentiation. By way of asymmetric cell division, hematopoietic stem cells (HSCs) give rise to two daughter cells. One of these daughter cells maintains HSC characteristics, while the other goes on to generate progeny that ultimately develop into mature blood cells (Orkin and Zon, 2008). During stem cell differentiation two parallel processes are imposed upon the cell: (1) the gradual acquisition of features of a given cell lineage, and (2) the loss of the ability to give rise to other cell lineages. Differentiation entails the selective expression of proteins that determine the identity of a particular lineage. This is accompanied by the repression of genes that direct differentiation towards other lineages.

Hematopoiesis has been schematically depicted as a series of binary choices faced by a multipotent HSC. The most widely accepted model, according to Weissman and colleagues, describes the initial developmental choice between the myelo-erythroid vs. lymphoid fates. Accordingly, common lymphoid precursors (CLPs) have no myeloid potential (Kondo et al., 1997), and common myeloid progenitors (CMPs) lack lymphoid generating capacity (Akashi et al., 2000). This model is based on the prospective isolation of CLPs that give rise to lymphocytes but not myeloid cells upon transplantation. Similarly, CMPs can replenish the myeloid and erythrocyte compartments but do not generate lymphocytes after adoptive transfer. The simplicity of this model has been called into question by the demonstration of progenitors that lack erythroid potential but retain myeloid and lymphoid capacity (Adolfsson et al., 2005; Katsura, 2002). Thus, there is still some ambiguity regarding the order of choices during early hematopoiesis.

NK cells as a distinct cell type

The observations of natural killing (cytotoxicity without prior antigen priming) by a population of non-T lymphocytes and non-B lymphocytes have led to the identification of the natural killer (NK) cell lineage (Herberman et al., 1975; Kiessling et al., 1975). The advent of monoclonal antibodies and flow cytometry allowed for a phenotypic definition of NK cells (i.e. $CD56^+CD3^-$) (Lanier et al., 1986). Although a lymphoid vs. myeloid origin of NK cell development had been debated early in their discovery based on early application of monoclonal antibody typing (Li et al., 1994; Ortaldo and Herberman, 1984), the demonstration of a common T/NK precursor in human (Sanchez et al., 1994) and murine thymic tissue (Carlyle et al., 1997) placed NK cells developmentally close to T cells. The lymphoid origin of NK cells was more formally demonstrated by Akashi who isolated murine CLPs and demonstrated their ability to give rise to NK cells (in addition to T cells and B cells) upon transplantation into congenic animals (Kondo et al., 1997). Thus, NK cells can be derived from lymphoid progenitors.

Unlike the other progeny of CLPs (i.e. T cells and B cells), NK cells do not mediate conventional adaptive immunity. This is due to the lack of rearranged antigen-specific receptors, such as the T cell and B cell receptors that are generated following somatic recombination. Instead, NK cells express diverse sets of germline-encoded receptors responsible for antigen recognition (Lanier, 2005). This strategy (expression of multiple nonrearranged receptors) is commonly employed by the innate immune system. Recently however, NK cells have been shown to be closer to the adaptive immune system than previously appreciated. In this regard, NK cells can mediate recall or secondary immune responses, including contact dependent hypersensitivity to secondary challenge by chemical irritants (O'Leary et al., 2006). Recall responses by murine NK cells expressing Ly49H have also been observed after infection with mCMV (Sun et al., 2009). In human NK cells studies, expansion of NK cell clones expressing inhibitory receptors specific for ligands missing in the host have also been observed following HSC transplantation (Ruggeri et al., 1999, 2002). Despite this evidence for the expansion of reactive NK cell clones mediating secondary immune responses, NK cells lack the fine and exclusive specificity for the challenging antigen that is conferred by the B cell and T cell receptors. Likewise, questions still remain as to how long NK cell 'memory' persists.

While NK cells do not fully conform to the definition of adaptive immunity, they also differ from members of the innate immune system. For instance, NK cells do not mediate phagocytosis and lack bactericidal enzymatic systems. Rather, they express intracellular proteins associated with effector functions also used by cytotoxic T cells, including granzymes and perforin. As well, NK cells rapidly release a wide array of cytokines upon activation, including IFN-γ and TNF-α, which serve to shape adaptive immune responses. Consequently, owing to their CLP derivation, NK cells are developmentally close to the adaptive immune system, while functionally they retain features more in line with the innate immune system, perhaps suggesting a more ancient origin compared to T cells and B cells.

Lineage specific growth factors

A significant breakthrough in the understanding of hematopoiesis came with the demonstration of lineage-specific growth factors (reviewed by Kaushansky, 2006). Examples of such factors include G-CSF, which promotes the granulocytic lineage; M-CSF, which leads to monocytic development, or erythropoietin resulting in erythrocyte generation. Acquisition of a particular receptor renders precursors responsive to a particular lineage-specific growth factor, thereby marking an important event in lineage determination. Thus, precursors of distinct lineages can be identified by the presence of specific growth factor receptors. For example, the erythropoietin receptor marks erythroid precursors.

In the case of lymphoid progenitors, CD127 (IL-7R) has been used to define CLPs (Kondo et al., 1997), and IL-7 is necessary for murine T cell and B cell development (Akashi et al., 1999; Peschon et al., 1994). This supports the notion that IL-7 is a growth factor specific for CLPs. However, murine NK cells develop normally in the absence of IL-7 (He and Malek, 1996). Several cases of human severe combined immune deficiency (SCID), caused by a mutation in the IL-7R (CD127), have also been reported. These patients lack T cells, however, NK cells are present and functional (Giliani et al., 2005; Puel et al., 1998). Thus IL-7, a growth factor specific for the development of CLPs into T cells, is not required for NK cell development. In contrast, deficiency of the cytokine receptor common γ-chain (CD132) in both mice (Cao et al., 1995) and humans (Buckley et al., 1997) results in the lack of T cells, B cells, and NK cells. These observations led to the conclusion that some cytokines that signal through the common γ-chain (including IL-2, -4, -7, -9, -15 and -21) are required for NK cell development.

Early studies of human NK cell differentiation from hematopoietic precursors used IL-2 (Miller et al., 1992). Paradoxically, this cytokine is not abundant in the bone marrow (BM) microenvironment. This suggested that another growth factor present in the BM milieu is responsible for NK progenitor development and expansion. IL-15 is expressed by BM stroma and was a possible candidate (Mrozek et al., 1996; Puzanov et al., 1996). The receptor for IL-15 shares β and γ subunits with IL-2 receptor, explaining the redundancy between IL-2 and IL-15 in vitro. Murine studies identified IL-15 as an NK-specific growth factor since IL-15$^{-/-}$ mice show a near absence of NK cells (Kennedy et al., 2000) Similarly, the deficiency of the IL-15 receptor β-subunit (CD122, shared with IL-2R) also results in a profound decrease in NK cells (Gilmour et al., 2001; Lodolce et al., 1998) As well, mice lacking the α subunit of the IL-15 receptor (15R$\alpha^{-/-}$) have a reduction in NK cells due to the lack of IL-15 transpresentation (Kawamura et al., 2003; Lodolce et al., 1998), and IL-15 transpresentation (via IL-15Rα) supports human NK cell differentiation in a xenogeneic mouse model (Huntington et al., 2009). In summary, IL-15 has emerged as a requisite NK specific growth factor, although it is not entirely NK specific, as it also acts on CD8$^+$ T cells.

In line with this, CD122 (IL-15Rβ) has been used to isolate NK precursors (Rosmaraki et al., 2001). Primitive, nonlineage specific growth factors, including stem cell factor (SCF), FLT-3L and IL-3, also influence NK cell development (Williams et al., 1997). These growth factors act upon the early hematopoietic precursors, inducing IL-15Rβ (CD122) expression, thereby conferring IL-15 responsiveness (Yu et al., 1998). In line with this, IL-15$^{-/-}$ or 15R$\alpha^{-/-}$ mice are nearly devoid of mature NK cells (above), but they do have NK precursors (Vosshenrich et al., 2005). These findings suggest that at least one function of IL-15 is to provide survival signals to the developing NK cell. This has been confirmed by the demonstration that NK cell development in CD122 (IL-15Rβ)$^{-/-}$ mice can be restored by enforced, constitutive expression of the anti-apoptotic survival factor Bcl-2 (Minagawa et al., 2002). The downstream signalling through the IL-15/IL-2 receptor involves activation of JAK1/JAK3 and STAT3/STAT5b (Imada et al., 1998; Waldmann and Tagaya, 1999). Deficiencies of JAK3 or STAT5b also result in severe impairment in NK cell development (Imada et al., 1998; Park et al., 1995).

Sites of NK development: the importance of the developmental environment

BM HSCs give rise to lymphoid precursors and, at certain stages of development, these cells migrate to sites that facilitate terminal differentiation. Migration is an important factor in determining lineage fate. The environmental cues present at a particular site are required for the initiation of developmental programs. While it is well established that the thymus is the site for T cell development and B cells develop in the BM, we are just beginning to unravel the sites of NK cell development.

Initially, NK cell development was believed to occur exclusively in the BM (Kim et al., 2002). The critical role of BM for NK cell maturation in mice has been shown by using bone-seeking radioactive isotopes that injure the BM stroma, inducing a profound block in NK cell maturation (Mellen et al., 1982). However, recent studies show that NK cells also develop in human secondary lymphoid tissue (Freud et al., 2005). Freud and coworkers have identified consecutive stages of NK cell development starting from CD34$^+$ precursors, resulting in functional CD56$^+$ NK cells in lymph nodes (Freud et al., 2006). The relative importance of lymph nodes in NK cell development has not been completely established. Other tissues, such as intestinal epithelial layer (villous and crypt regions), also contain NK precursors (Chinen et al., 2007; Lynch et al., 2006) and are likely sites of NK cell development. The CD34$^+$ hematopoietic precursors isolated from gut tissue frequently co-express CD56 and differentiated into NK cells upon short-term culture with IL-15.

The thymus also appears to be a potential site of NK cell development. Thymocytes up to the double negative 2 stage retain the capacity to give rise to NK cells (Schmitt et al., 2004; Spits et al., 1998). Vosshenrich and colleagues (2006) have defined a thymus-dependent NK cell developmental pathway in mice. These thymic-derived

NK cells expressed high levels of CD127 (IL-7R), the transcription factor GATA-3 and differed functionally from the majority of murine (splenic) NK cells. This example (thymic NK cell development) underscores the importance of the particular environment (i.e. niche) in guiding the maturation of NK progenitors. Distinct niches likely provide unique combinations of developmental cues (discussed later) that shape NK cell function.

A peculiar population of NK cells is abundant in the decidua of the pregnant uterus, and the properties of these NK cells distinguish them from peripheral blood NK cells (Koopman et al., 2003; Yadi et al., 2008). These cells are characterized by a CD56^{bright} phenotype and a lack of cytotoxicity. It has not been clearly resolved whether uterine NK cells migrate from peripheral blood or are derived from precursors that differentiate locally within the uterine environment (Ashkar et al., 2003; Keskin et al., 2007; van den Heuvel et al., 2005). The latter possibility would explain their unique characteristics. Thus, one of the emerging views is that NK cells can complete differentiation in various organs outside of the BM, and depending upon the site, these cells differ functionally. Elucidation of the factors present in a particular site and mechanistic explanations for how they impact NK development requires further study. Furthermore, how various sites of differentiation contribute to the heterogeneity of NK cell subsets is not well established.

Fate determining interactions with stroma

Receptor-ligand systems that direct cellular development and differentiation have been collectively referred to as 'morphogens'. They are the means by which the environment shapes the fate of developing progenitors (Moore, 2004). Signalling systems, including Notch, Wnt and others, are highly conserved throughout evolution (Pires-daSilva and Sommer, 2003). These systems are important in many aspects of embryogenesis and, thus, in the differentiation of multiple organ systems. Importantly, the actions of morphogens are developmental stage- and context-dependent. For example, triggering of the Notch receptor represses B cell differentiation and skews lymphoid precursors to T-cell lineage (Schmitt et al., 2004). However at later stages of B-cell differentiation, Notch signalling is required for terminal B-cell maturation (Santos et al., 2007).

The difficulty in studying the importance of a particular morphogen in hematopoiesis is related to their vast redundancy, due to multiple homologues that mediate similar (or overlapping) functions. Thus, the elimination of a single factor may be compensated for by other homologues. Furthermore, morphogens have major roles in early embryogenesis and therefore genetic manipulations often result in lethal defects, prohibiting the study of their role in NK development. Methods to circumvent this include the creation of BM chimeras by transplantation of foetal liver hematopoietic precursors into irradiated wild type recipients. This method can be used if the mouse embryo with deleted genes survives beyond the initiation of hematopoiesis (approximately day 9 post conception). Alternative approaches include conditional knock-out strategies using tissue specific or drug-inducible promoters driving the CRE recombinase. In this way, a given gene can be eliminated from a selected tissue (or lineage) and/or at the desired time by administration of the promoter-inducing drug.

Notch signalling plays a critical role in directing hematopoiesis and lymphocyte development (MacDonald et al., 2001; Maillard et al., 2003). In particular, thymic stroma supports T cell differentiation by the expression of the Notch ligand, delta like ligand-1 (DLL1). Accordingly the murine BM stromal cell line OP9, engineered to express DLL1, efficiently promotes in vitro T-cell differentiation by providing continuous Notch engagement (Schmitt and Zuniga-Pflucker, 2002). Early Notch signalling induces the acquisition of CD7 (La Motte-Mohs et al., 2005) and CD161 by hematopoietic precursors. Both also mark NK precursors (Bennett et al., 1996; Miller et al., 1994) , perhaps suggesting a role for Notch in NK commitment. Additional evidence for Notch involvement in NK differentiation comes from studies on murine early progenitors with lymphoid and myeloid developmental potential (EPMLs), where Notch signalling favoured NK development (Rolink et al., 2006). While DLL1 is the most extensively investigated ligand, Notch engagement by a different ligand, Jagged 2, also promotes in vitro NK differentiation from hematopoietic precursors (DeHart et al., 2005). We interpret these (and our unpublished studies) to show that signalling through Notch induces the development of common T-NK precursors. Upon continuous Notch engagement progenitors advance towards the T-cell lineage (Schmitt et al., 2004), whereas early Notch signalling appears to be sufficient for NK cell development. Importantly, Notch signalling is not absolutely required for NK cell differentiation (Radtke et al., 2000). Hematopoietic precursors cultured with, but not without Notch triggering (De Smedt et al., 2007) expressed cytoplasmic CD3ε. Subsequently, NK cells resembling Notch-dependent in vitro derivatives could be found in human cord blood but not in adult blood. While Notch signalling is not absolutely required for NK cell differentiation, it is mandatory for T-cell development (Radtke et al., 2000). Collectively, the key factors required for T-cell differentiation, IL-7 and Notch signalling, are both dispensable for NK development, questioning whether the common T/NK precursor is the only pathway for NK cell development.

The Wnt signalling system represents another family of morphogens that influences NK cell development. Numerous Wnt proteins exist (18 identified members in mammals) and can interact with a receptor complex made up of the frizzled receptor and the LDL receptor-related protein (Staal et al., 2008). There are at least two intracellular Wnt signalling pathways, known as canonical and noncanonical. Canonical Wnt signalling leads to stabilization of β-catenin, which activates the LEF (lymphocyte enhancer factor) and TCF (T-cell factor, gene name tcf7) family of transcription factors. In the absence of LEF/TCF, neither T cells nor NK cells develop (Held et al., 2003). Gain-of-function and loss-of-function variants also demonstrate the importance of Wnt signalling in lymphopoiesis. In particular, a nondegradable, constitutively active β-catenin imposed lymphoid potential onto myeloid precursors (Baba et al., 2005). Despite the role of β-catenin in Wnt signalling, conditional deletion of β-catenin did not abolish lymphocyte development. Such cells showed sustained LEF/TCF expression, suggesting the contribution of an alternative, β-catenin independent pathway leading to LEF/TCF upregulation (Staal and Sen, 2008). Overall, these studies show that canonical WNT signalling via TCF/LEF is important in lymphopoiesis, including NK cell generation. In support of this, TCF expression was detected in NK cells, specifically in the $CD56^{bright}$ subset (Toor et al., 2001).

Another interaction recently shown to be involved in stroma-dependent NK cell differentiation is between GAS6 and protein S. These two highly homologous ligands, expressed on stroma, trigger the Tyro3/Axl/Mer protein tyrosine kinase receptors present on NK precursors. In animals lacking all three receptors, NK cells were phenotypically and functionally impaired (Caraux et al., 2006). Lack of only one receptor of this family (i.e. Axl) had a modest effect on NK cell development. Subsequently, fibroblasts expressing recombinant Gas6 could be shown to support in vitro NK cell differentiation (Caraux et al., 2006). Interestingly, the downstream signalling of Axl is reciprocally associated with IL-15 signalling (Hafizi and Dahlback, 2006) since Axl and IL-15Rα can heterodimerize. As a result, the IL-15 and GAS6-Axl pathways transactivate one another (Budagian et al., 2005). Indeed this association between IL-15 and Axl is operational in NK cell differentiation from human HPCs in vitro (Park et al., 2008). Perhaps the notion that GAS6–Axl interactions can transactivate the IL-15 pathway might provide the clue as to why IL-15-deficient mice have a residual population of NK cells (Vosshenrich et al., 2005).

Other receptor-ligand pairs with morphogenic functions in the differentiation of organ systems may also be involved in hematopoiesis. Members of the TNF superfamily of surface receptors are important in the development of lymphoid tissues (Mebius, 2003). Reciprocal interactions between hematopoietic progenitors and the nonhematopoietic stroma involve the expression of lymphotoxin-α on progenitors and lymphotoxin-β-receptor on stroma. This interaction (between LTα and LTβ-R) triggers IL-15 production by stroma, which in turn, supports NK cell development (Iizuka et al., 1999; Lian et al., 2004). Still other morphogenic receptors, including hedgehog, TGF—Smad, may also be involved in NK cell development; however, their roles have not been fully investigated.

Transcription factors involved in NK cell differentiation

Signalling through cell surface receptors results in a cascade of events that may ultimately lead to the activation of transcription factors. These DNA-binding proteins recognize consensus sequences in the promoter regions of target genes and influence gene transcription. The balance of multiple transcription factors, often with opposing functions, ultimately dictates whether initiation or repression of gene transcription occurs. Thus, depending upon the stimuli, certain genes are transcribed, while others are repressed. In this way, differentiation is directed towards a particular lineage. Consequently, several branching points in hematopoiesis are regulated by the balance of opposing transcription factors, such as Id proteins vs. E proteins or PU.1 vs. GATA-1 (see Figure 1.1). These opposing transcription factors regulate the expression of a number of genes, including receptors for lineage specific growth factors (discussed earlier). In this respect, PU.1 and C/EBPα drive expression of receptors for myeloid growth factors (G-, M- and GM-CSF), whereas GATA-1 induces erythropoietin receptor expression (Zhang et al., 1996). This opposition controls myeloid vs. erythroid lineage choice at a molecular level.

Transcription factors that play a dominant role in guiding differentiation towards a particular lineage have been referred to as master regulators. Expression of a master regulator is absolutely required for progression past a defined stage of differentiation. Precursors deficient in that factor are therefore unable to advance past a given checkpoint. PAX5 is an example of a master regulator of B-cell development. Cells deficient in PAX5 are unable to differentiate into mature B cells but retain myeloid, T cell and NK cell potential (Schaniel et al., 2002). Numerous transcription factors are involved in NK development and functional maturation. Mice lacking these factors have impaired NK cell generation. However, in most cases, the defect is not confined to

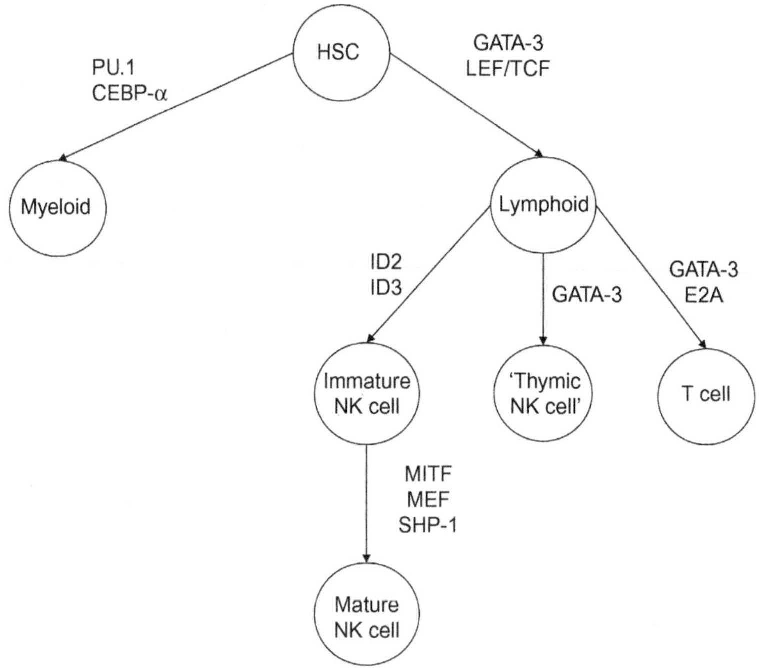

Figure 1.1 • Transcription factors important for lineage choice at pertinent branching points in differentiation of HSCs towards the NK lineage. Key transcription factors control the myeloid vs. lymphoid choice, such that PU.1 (dose dependent) and CEBPα favour myeloid lineage, whereas GATA-3 and LEF/TCF family members (downstream of Wnt) promote the lymphoid lineage. ID2 expression is required for NK cell development and antagonizes T cell fate promoted by E proteins (E2A, which is downstream of Notch). GATA-3 is necessary for both T cell and 'thymic NK cell' development. Other transcription factors, including MITF, MEF and SHP-1, are required for functional NK cell maturation.

the NK lineage. Along these lines, no master regulator (equivalent to PAX5 for B cells) has been identified or agreed upon for the NK lineage.

Consistent with the dominant role for IL-15 in NK cell development, transcription factors that are either down-stream of the IL-15R or are involved in IL-15 production by stroma are important for NK development. Accordingly, transcription factor defects influencing IL-15 signalling can be either intrinsic or extrinsic to the progenitor cell. Stat5a and Stat5b are transcription factors involved in signalling through the IL-2 and IL-15 receptor common β chain (i.e. CD122). Deficiency in Stat5b resulted in a ~50% reduction in splenic NK cells. These mice also showed a profound attenuation in NK cytotoxicity, which was not rescued by exogenous IL-2 or IL-15 (Imada et al., 1998). In parallel studies, Stat5a deficiency was associated with a marginal influence on either the number of NK cells or their function. As an example of hematopoietic progenitor cell-extrinsic factors, IRF-1 (interferon-regulatory factor-1) is required for IL-15 expression. Accordingly, irf-1$^{-/-}$ mice show an NK cell deficiency resembling IL-15$^{-/-}$ animals (Ogasawara et al., 1998). Hematopoietic progenitors from these irf-1$^{-/-}$ mice differentiate into NK cells with exogenous IL-15. Thus, the IRF-1 deficiency results in the impairment of the environmental support required for NK cell development. In contrast, irf-2$^{-/-}$ mice appear to have a defect that is intrinsic to NK precursors, resulting in a selective loss of mature NK cells in periphery, while BM NK cells are relatively unimpaired (Taki et al., 2005).

As described previously, Notch signalling has been implicated in NK cell development. Downstream effectors of Notch include the helix loop helix (HLH) proteins from the E protein family, including E2A and E47. These transcription factors are involved in the development towards the T-cell lineage (Ikawa et al., 2006). Similar to Notch deficiency (MacDonald et al., 2001), E47 deletion abrogates T-cell development, but NK cells and myeloid cells develop normally (Ikawa et al., 2006). A different set of transcription factors, which negatively regulates E proteins, is the set of dominant negative helix-loop-helix proteins, Id2 and Id3 (inhibitor of DNA binding 2 and 3). Both Id2 and Id3 are highly expressed by NK precursors. Upon further development, Id3 expression falls dramatically, while Id2 is sustained. These results support the involvement of Id2 in NK cell maturation. In fact, in the absence of Id2, NK cells fail to expand and mature, even though CD122^{+}DX5^{-} NK precursors are present in BM (Yokota et al., 1999). The activity of Id proteins in NK cell development involves the dominant negative regulation of T-cell differentiation, in favour of NK cell development (Heemskerk et al., 1997; Ikawa et al., 2001). The balance between Id2 and E2A dictates the ability of precursors to expand and terminally differentiate into NK cells. Moreover, the maturational defects seen in the absence of Id2 can be corrected by the deletion of E2A (Boos et al., 2007). Thus, Id transcription factors antagonize the Notch induced E proteins, diverting progenitors from the T-cell lineage towards the NK lineage (Fujimoto et al., 2007).

As described earlier, Wnt signalling results in activation of the transcription factors TCF-1 and LEF-1. In the absence of both, mice displayed a profound reduction in NK cells and T cells but not B cells (Held et al., 2003). The reduction in NK cells was seen both in the BM and spleen. Further studies showed that compared to LEF-1, TCF-1 played a more substantial role in NK cell differentiation and phenotypic maturation. Human studies confirmed that TCF-1 is acquired by HPCs as they differentiate into NK cells in vitro and that TCF-1 transcript and protein could be found in the CD56^{bright}, fraction of PB NK cells (Toor et al., 2001).

The transcription factor PU.1, a member of the Ets family, is strongly associated with myeloid differentiation (Dahl and Simon, 2003). PU.1 functions as a dose- and stage-dependent regulator of lineage fate in hematopoiesis. At an early branching point in hematopoiesis, PU.1 and GATA-1 antagonize one another, facilitating myeloid or erythroid differentiation, respectively (Stopka et al., 2005). At later developmental stages, PU.1 represses NK cell and T cell specific genes, favouring alternative lineages, including myeloid and B cells (Kamath et al., 2008). In PU.1$^{-/-}$ mice, NK cell differentiation is impaired, although not as severely as the T-cell and B-cell lineages (Colucci et al., 2001). Thus, the defect conferred by PU.1 deficiency is not NK specific. In contrast, a related transcription factor, Ets-1, is specifically required for NK development (Barton et al., 1998). In Ets-1$^{-/-}$ mice, the NK cell numbers were reduced (~threefold), and cytotoxicity was virtually absent, whereas T cells, B cells and other blood lineages were quantitatively normal. As a result, affected mice were more susceptible to lymphoid tumours upon challenge with the RMA-S cell line. MEF (myeloid Elf-1 like), another member of the Ets family, is also important for NK cell development. MEF appears to be involved at the latter stages when NK cells gain cytotoxicity. Deletion of MEF resulted in greatly impaired NK cell killing, as well as IFN-γ production. Mechanistically, the lack of cytotoxicity reveals a role for MEF in regulating perforin gene transcription (Lacorazza et al., 2002).

Similar effects on NK cell differentiation have been observed with C/EBPγ, a member of the leucine zipper family of transcription factors. These CCAAT/enhancer binding protein (C/EBP) transcription factors play a critical role in myeloid vs. lymphoid lineage determination. C/EBPα antagonizes the Tribbles homolog 2 (TRIB2) protein, one of the downstream effectors of Notch signalling (Wouters et al., 2007). Thus, C/EBPα antagonizes Notch signalling, favouring myeloid development. Another member of the C/EBP family, C/EBPγ plays a specific role in NK development. C/EBPγ$^{-/-}$ mice showed nearly normal numbers of NK cells, albeit with diminished cytotoxicity and cytokine production capacity, suggesting a maturational defect resembling MEF-deficient animals (described earlier). Likewise, perforin expression was greatly diminished in splenocytes from C/EBPγ$^{-/-}$ mice (Kaisho et al., 1999). Another transcription factor involved in NK cell functional maturation is MITF (microphthalmia transcription factor [MITF]), which similar to MEF, regulates perforin gene expression (Ito et al., 2001).

The family of distal-less (Dlx) homeobox proteins have been implicated in the development of multiple organ systems. This family contains multiple members, including Dlx 1 through Dlx 6. Dlx 3 is expressed preferentially at an immature stage of NK development (Sunwoo et al., 2008). The requirement for Dlx 3 in NK cell development has not been addressed since Dlx 3$^{-/-}$ mice die at an early embryonic stage, prior to the onset of hematopoiesis. However, over-expression of either Dlx 1 or Dlx 2 resulted in an arrest of NK development at an immature stage. Quite interestingly, over-expression of Dlx 1 also leads to a profound defect in the development of T lymphocytes and B lymphocytes, in addition to NK cells. These findings may suggest that this factor is important at the CLP stage.

Another TF involved in functional NK cell maturation is T-bet (Townsend et al., 2004), a factor known to control Th-1 lineage commitment and IFN-γ production by T cells (Szabo et al., 2000). T-bet$^{-/-}$ mice show reduced numbers of mature NK cells in the periphery and an increase in phenotypically and functionally immature NK cells (Townsend et al., 2004). This defect results from a higher rate of NK cell apoptosis in the periphery (i.e. spleen). T-bet is upregulated through STAT1 signalling, which, in turn, is triggered by IFN-γ. However, the requirement for STAT1 in NK cell maturation is less stringent than the requirement for T-bet, indicating that T-bet induced NK maturation can be STAT-1 independent (Townsend et al., 2004).

In contrast to T-bet, which drives Th1 polarization, the transcription factor GATA-3 is critically involved in Th-2 polarization of T cells. T-bet and GATA-3 are thus, expressed in a mutually exclusive fashion by Th1 and Th2 polarized CD4^{+} T cells, respectively (Ouyang et al., 1998). In the absence of GATA-3, T-cell differentiation is abrogated (Ting et al., 1996), whereas NK cells do develop. NK cells from GATA-3$^{-/-}$ animals produce less IFN-γ and appear phenotypically immature (Samson et al., 2003). Thus, it appears that GATA-3, a factor critical for Th2 polarized CD4^{+} T cells, is somewhat paradoxically required for IFN-γ production by NK cells. The impairment of NK cells from GATA-3$^{-/-}$ mice likely reflects the importance of GATA-3 in the thymic pathway of NK cell maturation, discussed later (Vosshenrich et al., 2006).

Second messenger signalling in NK cell development

NK cells are regulated by surface receptors with activating or inhibitory functions. The majority of activating receptors use adapter proteins (including DAP10, DAP12, FcεRIγ and CD3ζ) to link surface receptors with intracellular signalling pathways, resulting in effector functions. These signalling pathways involve second messengers, including Syk, ZAP70, Lck, PI3Kinases (phosphoinositide 3-kinases), SAP, Fyn, Vav, Grb-2, Phospholipase-Cγ-1 and -2 (reviewed in Lanier, 2008).

The importance of the individual elements of activating receptor signalling pathways on NK cell development has been partially investigated. Animals deficient in DAP10 have normal numbers of NK cells. Since DAP10 serves as adapter protein for NKG2D, these NK cells had reduced NKG2D expression and function. Otherwise, NK cell activity was not diminished. Unexpectedly, DAP10$^{-/-}$ mice showed enhanced resistance to skin carcinoma and did not have increased spontaneous tumour formation (Hyka-Nouspikel and Phillips, 2006). DAP10$^{-/-}$ mice were also more resistant to melanoma challenge, thorough studies indicated that cell subsets other than NK cells (including NK-T cells and Tregs) were also involved in this effect. Similar to DAP10, deficiency of ZAP70 did not result in the impairment of NK cell development (Negishi et al., 1995). To the contrary, NK1.1^{+}CD3^{-} cells were more numerous in ZAP70$^{-/-}$ animals (Iwabuchi et al., 2001). The constitutional lack of other kinases, including Syk, or SAP/Fyn in mice did not block NK cell development either (Colucci et al., 2002; Turner et al., 2000).

PI3Ks are a family of kinases composed of multiple isoforms that encode both regulatory and catalytic subunits. The p110δ isoform of the catalytic subunit of PI3K is required for NK cytokine secretion (IFN-γ, TNF-α and GM-CSF), while the other tested isoform p110γ was not. Neither of these individual isoforms were absolutely required for terminal NK maturation and/or acquisition of cytotoxicity (Kim et al., 2007; Tassi et al., 2007). However, in the absence of both isoforms, NK cell numbers and cytotoxicity were greatly reduced. Studying a related protein, BCAP (B-cell adapter for phosphatidylinositol 3-kinase), it was noted that in BCAP$^{-/-}$ mice the NK cells are overtly long-lived, mature and functionally active (MacFarlane et al., 2008). NK cells from these animals were more resistant to apoptosis, suggesting that BCAP mediated ITAM signalling (and activation of the Akt pathway) negatively impacts NK cell maturation and survival.

The reductionist approach of studying individual proteins in NK cell development is feasible in rodents; however, the nonfully overlapping functions of individual molecules may be a confounding factor when translating these results to humans. Several genetic defects in humans are informative as to the role of these individual proteins in NK development. For instance, DAP12 deficiency results in a rare syndrome known as Nasu–Hakola disease in which presents with impaired osteoclast activity (bone cysts) and dementia. NK cells are normal and functional in these patients (Paloneva et al., 2000). Deficiency of another adapter molecule, CD3ζ, was reported in a single human with T^{-}B^{+}NK^{+} severe combined immune deficiency (SCID) (Roberts et al., 2007). Curiously, the NK cell repertoire of this patient consisted of a peculiar population of CD56^{-}CD16^{+} NK cells. Expansion of CD56^{-}CD16^{+} cells has been previously observed in other situations (following allogeneic hematopoietic cell transplantation, HIV infection and in cord blood). Such cells are thought to be either dysfunctional or immature. In this patient, virtually all NK cells were, CD56^{-} and cytotoxic activity was diminished compared to healthy controls (Roberts et al., 2007). It is difficult to dissect whether this is due to a direct effect of CD3ζ deficiency on NK development or an indirect effect via severe impairment of T cells. T cells are the main source of IL-2, which next to IL-15, is the dominant survival factor for NK cells. The other clinical states characterized by an abundance of CD56^{-}CD16^{+} cells are also associated with functional T cell deficiency (i.e. after HCT, in HIV infections or in newborns).

While activating receptor triggering leads to kinase activation, NK inhibitory receptors signal through phosphatases. These phosphatases oppose kinase function. Phosphatases involved in NK cell signalling include SHP-1 (src homology region 2 domain-containing phosphatase 1; PTPN6), SHP-2 (src homology region 2 domain-containing phosphatase 1; PTPN11) and SHIP (SH2-containing inositol phosphatase; INPP5D) (Vely et al., 1997). SHP-1 deficient mice (motheaten-viable [*me-v*]) show impaired terminal NK cell maturation, characterized by reduced cytotoxicity and IFN-γ production (Clark et al., 1981; Lowin-Kropf et al., 2000). However, competitive transplantation into wild type hosts did not reveal intrinsic defects in NK cells from *me-v* mice (Kim et al., 2005). This was unexpected since inhibitory receptor signalling was important for NK licensing in this study. Perhaps in the absence of SHP-1, other phosphatases (i.e. SHP-2 and/or SHIP) may substitute, as the reverse occurs in SHIP deficient animals (Yusa and Campbell, 2003). The role of SHP-1 in the acquisition of NK function was also studied using a dominant negative variant of this molecule. NK cells expressing the dominant negative SHP-1 were defective at rejecting MHC deficient BM transplants but showed otherwise normal responsiveness

(Lowin-Kropf et al., 2000). SHP-2 deficient murine embryos did not survive to permit analysis of their NK cells. The role of SHIP in NK cell development and function was also tested in two independent models, which showed that in the absence of SHIP, NK cells were more numerous with improved survival (Wang et al., 2002) and higher IFN-γ production (Parihar et al., 2005). Interestingly, SHIP$^{-/-}$ NK cells also had limited expression of MHC-specific inhibitory receptors, resulting in an inability to reject allogeneic BM grafts. Inhibitory receptor skewing may be related to SHP-1 substituting for SHIP (Wahle et al., 2007). The increased IFN-γ production by the SHIP$^{-/-}$ NK cells provides insight into the functional dichotomy between human CD56^{bright} and CD56^{dim} NK cells, since the latter population expressed higher levels of SHIP when compared to CD56^{bright} cells (Trotta et al., 2005). To further prove the point, that differential SHIP expression underlies the observed functional differences between CD56^{bright} and CD56^{dim} subsets, SHIP was overexpressed in CD56^{bright} NK cells, resulting in a significant reduction in IFN-γ production.

The impact of activating and inhibitory surface receptor triggering on NK cell development has also been studied. Collectively, it appears that NK cells with competent inhibitory receptors have superior functionality. In contrast, a rare subset of NK cells that lack competent MHC-specific inhibitory receptors was found to be nonfunctional. This has led to the concept of NK cell licensing—a requirement for inhibitory receptor signalling to attain full functionality (Kim et al., 2005; Lowin-Kropf and Held, 2000; Raulet et al., 2001). In contrast, exposure to ubiquitously expressed ligands triggering activating receptors resulted in a reduced capacity of mature NK cells (Fauriat et al., 2007; Sun and Lanier, 2008; Tripathy et al., 2008). This has fundamental importance for our understanding of the process of NK cell development and functional maturation. While the exact roles of particular inhibitory and activating receptors have not been dissected, the emerging picture is that this mechanism is in place to assure tolerance. Inhibitory receptor signalling capacitates NK cells, while activating signals appears to incapacitate them (Lowin-Kropf and Held, 2000; Raulet et al., 2001). This contrasts with the developmental requirement of T cells and B cells, which principally require activating signalling to successfully progress in development and maturation.

The NK cell ontogeny—lessons from evolution

Evolutionarily primitive species, up to the jawless vertebrates, rely solely on the innate immune system and lack both MHC and RAG gene families. The adaptive immune system, based on the activity of the RAG genes, first appeared in jawed vertebrates. Thus, the hallmark of adaptive immunity is the expression of unique antigen recognition receptors generated by somatic recombination. These receptors, expressed by B cells (BCR) and T cells (TCR) confer specificity and memory. NK cells lack these highly variable receptors, and their activation is controlled by the integrated signalling from numerous germline encoded receptors. Such a strategy is characteristic for cells of the nonadaptive or innate immune system.

It is difficult to establish whether NK cells predated the development of an adaptive immune system. For obvious reasons, phenotypic markers used to define mammalian (human) NK cells cannot be applied to invertebrates. One method of NK cell identification in invertebrates would be to search for cells with functional characteristics of NK cells (i.e. perforin mediated cytotoxicity and IFN-γ production). It is interesting to note that immunocytes (i.e. macrophages) from a molluscan slug (*Incilaria fruhstorferi*) express perforin and can reject skin allografts. In line with this notion, the rejected tissue showed features of perforin-induced cell death (Furuta et al., 2006). Therefore, lack of lymphocytes in this invertebrate is potentially compensated for by immunocytes that mediate both perforin-dependent cytotoxicity and phagocytosis. In higher species (i.e. vertebrates), these functional characteristics are performed by separate types of cells. However, it has been recently documented that human dendritic cells (DCs) (which have phagocytic properties) can also acquire perforin and kill tumours (Stary et al., 2007).

NK cells use the 'missing self' strategy of immune recognition to identify targets for elimination. This approach is reminiscent of the rules governing mating by fusion of sea sponge colonies of the *Botryllus* species (De Tomaso et al., 2005). Individual colonies select fusion-partners on the basis of sharing of histocompatability antigens. Fusion with another colony that is missing a 'self' allele is prevented by a mechanism of rejection, likely immune in nature. Coincidentally, receptors with high homology to the mammalian NK receptors (CD94 and/or CD161) have been identified on hemocytes (i.e. blood cells) of *Botryllus* as well as a related sea sponge, *Ciona intestinalis* (Khalturin et al., 2003; Zucchetti et al., 2008). The CD94 homologues in *Ciona intestinalis* have a similar function to the mammalian receptors since they inhibit hemocyte activation, thereby reducing phagocytosis (Zucchetti et al., 2008). Hence, inhibitory receptors with homology to mammalian NK-related lectin-like receptors are expressed on phagocytic cells in jawless vertebrates. Therefore, the missing-self strategy could have evolved primarily as a selection criterion for mating. Obviously, the use of this same strategy by NK cells could be an independent phenomenon, but it is difficult to overlook the role of NK cells in the process of foetal implantation in mammals, which resembles the role of missing self recognition in implantation of marine invertebrate *Botryllus schlosseri* (Lightner et al., 2008). Moreover MHC, the target of NK recognition, has been implicated in partner selection in rodents and humans (Yamazaki and Beauchamp, 2007).

The evolution of NK cells as a cell type can also be considered from the perspective of NK-specific receptors. There are two main genomic clusters of receptors that encode proteins that regulate NK cell activation and inhibition. These are the leukocyte receptor cluster (LRC, chromosome 19q13.4) and NK gene complex (NKC, chromosome 12p13.1-2 in humans) (Trowsdale et al., 2001; Yokoyama and Plougastel, 2003). Encoded within the LRC are several gene families that share immunoglobulin-like structure and immune function. Included are the killer immunoglobulin-like receptors (KIR), as well as leukocyte Ig-like receptors or immunoglobin-like transcripts (LILR or ILT) (Cella et al., 2000). Both KIR and ILT clusters are believed to be related, perhaps derived from a common ancestor (Volz et al., 2001). Interestingly, KIR are expressed by NK cells and rare subsets of T cells (Uhrberg et al., 2001). In contrast, ILT receptors are found predominantly on NK cells, monocytes and DCs (Cella et al., 2000).

The NKC encodes a number of lectin-like receptors. Similar to the LRC, the NKC contains NK specific genes that are intermingled with receptors found on other cell types. Some of these lectin-like receptors are found mainly on dendritic and myeloid cells (Dectin1, CLEC4A/DCIR, Lox1), while several are distributed predominantly on NK cells and a subset of T cells (such as CD69, CD94, NKG2D), and still others are expressed by both myeloid/DC and NK cells (LLT1, MICL). The phylogenetic relationships between different lectin-like receptors within the NKC delineated 28 lineages of orthologous genes. The phylogenetic and physical clustering of NKC genes points to their origin by duplications, likely from a common precursor (Hao et al., 2006). As mentioned earlier, receptors encoded by LRC and NKC genes define functional activity of NK cells and are the means by which we classify these cells. The distribution of NK receptors argues that they can be placed evolutionarily between the myeloid and T cell lineages.

Lessons from embryogenesis

During foetal development, the first site of primitive hematopoiesis is the yolk sac. At a later time, it shifts to the AGM region (aorta–gonad–mesonephros) and continues in the foetal liver. Eventually, hematopoiesis is established in the BM. The contribution of the yolk sac to the intra-embryonic and definitive hematopoiesis is a matter of debate (Dzierzak and Speck, 2008; Tavian and Peault, 2005). More likely, these two regions (AGM and the yolk sac) are two independent sites of blood stem cell generation (Tavian and Peault, 2005; Yokota et al., 2006). Hematopoietic precursors from these two sites have been compared for their potential to generate distinct blood lineages (Tavian et al., 2001). This was done using in vitro culture assays for B cell (MS-5 stromal cell line) and T cell (foetal thymus organ culture [FTOC] assay) differentiation. Yolk sac precursors (i.e. extra-embryonic) could generate primitive nucleated erythrocytes, myeloid cells and NK cells, but lacked T cell and B cell generation potential. In contrast, progenitors from the AGM region (i.e. intra-embyronic hematopoietic cells) readily generated both B cells and T cells, as well as other lineages.

Thus NK cells, along with myeloid and erythroid cells, can be derived from yolk sac hematopoietic precursors, whereas T cells and B cells could not. Interestingly, the human embryonic stem cell line, H9, shows the same pattern of generating myeloid and NK cells but not T cells or B cells (Martin et al., 2008). In line with these findings, NK cells have been identified in human foetal liver as early as week 6 of gestation, whereas T cells are first observed in the foetal liver at 15–16 weeks (Phillips et al., 1992). These foetal NK cells had unconventional features, including the cytoplasmic expression of CD3ε and CD3δ subunits. However, they lacked membrane CD3 expression or TCR gene rearrangements, clearly distinguishing them from T cells. These findings demonstrate that NK development precedes that of T cells and B cells during embryogenesis. The presence of CD3ε and CD3δ subunits in the cytoplasm, may point to a common pathway of development for T cells and NK cells. Interestingly a proportion of cord blood NK cells also show cytoplasmic CD3ε, while this trait is not seen in adult PB NK cells. Studies of NK differentiation in vitro (De Smedt et al., 2007) show that cytoplasmic CD3ε expression is induced by DLL1-Notch triggering, perhaps implicating Notch signalling in the ontogeny of foetal liver NK cells but not adult NK cells.

Lessons from NK cell immune reconstitution after hematopoietic cell transplantation

Immune reconstitution is a process of rebuilding the immune system from transplanted HSCs. The reappearance of hematopoietic lineages follows a reproducible order, with monocytoid cells emerging first in the peripheral blood, followed by granulocytes and then NK cells. The recovery of NK cells significantly precedes T cells and B cells, with respect to both cell number and functional maturation (Storek et al., 2008). The timing of NK cell reconstitution coincides with myeloid recovery, similar to the pattern observed during embryonic development (discussed earlier). This supports the notion that posttransplant immune reconstitution recapitulates ontogeny.

The first wave of NK cells found in peripheral blood, after transplant have distinct features (Jacobs et al., 1992). They are $CD56^+$ but are predominantly $CD16^-$ and KIR^-. The vast majority of these, early recovering NK cells express the CD94/NKG2A inhibitory receptor heterodimer as their sole MHC-specific inhibitory receptor (Cooley et al., 2005; Nguyen et al., 2005; Shilling et al., 2003). In many respects, the early recovering NK cells closely resemble the $CD56^{bright}$ subset of peripheral blood NK cells, whereas the fraction of cells corresponding to the $CD56^{dim}CD16^+$ NK cell fraction increase and predominate at later times after transplant (Shilling et al., 2003). This orderly appearance of $CD56^{bright}$ and $CD56^{dim}CD16^+$ subsets supports the model that $CD56^{bright}$ cells are recently differentiated and upon further maturation they assume the characteristics of $CD56^{dim}CD16^+$ cells. However the confounding role of immunosuppressive agents (Cyclosporin A) cannot be ruled out, as the predominance of $CD56^{bright}$ NK cells could also reflect their relative resistance to Cyclosporin A (Wang et al., 2007). An important factor in the early posttransplant reconstitution of NK cells are high levels of IL-15 elicited by the pretransplant conditioning chemotherapy and irradiation (Miller et al., 2005).

Stages of NK cell development

The differentiation of every type of cell can be seen as a network of phenotypic and epigenetic changes that ultimately leads to the mature cell type. Stages of development represent semistable nodes in this network, at which developing precursors accumulate before traversing to the next juncture (Warren and Rothenberg, 2003). Since developmental stages are mostly defined by surface phenotype, the scheme of NK development elucidated in mice is not easily applicable to humans. This is related to the fact that key phenotypic markers of humans NK cells do not apply to mice (i.e. CD56, CD16, and KIR).

Systematic analysis of phenotypic and functional characteristics of NK cells has led the Yokoyama laboratory to propose five stages of murine NK development (Kim et al., 2002). The first stage is the NK precursor, marked by a $CD122^+NK1.1^-$ phenotype (Rosmaraki et al., 2001). These cells go on to acquire NK1.1 (NKRp1, CD161) and CD94/NKG2A at stage II, followed by Ly49 at stage III. Notably the acquisition of the two major types of MHC-specific inhibitory receptors (Veinotte et al., 2003) (CD94/NKG2A followed by Ly49) marks important steps in maturation, a pattern also observed in humans (Grzywacz et al., 2006) (discussed later). At the fourth stage, termed the 'expansion stage', NK cells undergo significant proliferation before reaching the final maturational stage, stage V, characterized by full cytotoxic ability and IFN-γ production. Stage V NK cells also acquire Mac-1 (CD11b) and CD43. More recently, cells at this final stage of maturation have been further subdivided according to the level of CD27 expression (Hayakawa and Smyth, 2006). Murine $CD27^{high}$ NK cells progress to a $CD27^{low}$ stage. In contrast to the $CD27^{low}$ subset, $CD27^{high}$ NK cells had lymph node migratory capacity and the ability to interact with DCs. These $CD27^{high}$ NK cells also had higher cytotoxicity and produced more IFN-γ in response to IL-12 and/or IL-18 stimulation compared to the $CD27^{low}$ counterparts. In support of their relative immaturity, the $CD27^{high}$ NK cells uniquely expressed c-kit receptor (CD117) and IL-7R (CD127) and showed a lower proportion of Ly49 receptor expressing cells. How these $CD27^{high}$ NK cells are related to the thymic pathway of NK maturation (Vosshenrich et al., 2006), also marked by CD127 expression, is not known.

The stages of human NK differentiation did not emerge as a direct correlate of murine studies. Beginning with the NK precursor subset, as defined in mice (Rosmaraki et al., 2001), the available reagents for staining of human CD122 (IL-2/IL-15Rβ) do not allow for a clear discrimination of a distinct $CD34^+CD122^+$ subset of human hematopoietic precursor cells (Grzywacz, unpublished). Instead, CD7 has been used to distinguish cells committed to the NK/T lineage (Miller et al., 1994). NK1.1 is an early marker of murine NK cells, which corresponds to human CD161. Human NK precursors with a $CD34^-CD161^+CD56^-$ phenotype have been characterized in vitro and are also found in peripheral blood and cord blood (Bennett et al., 1996). The loss of CD34 suggests that they have progressed in differentiation; however, these cells have not yet acquired cytotoxic ability. Culture with IL-2 promoted further NK differentiation, with the acquisition of CD56 and other NK receptors, cytotoxicity and IFN-γ production. Other investigators have also found that CD161 marks an early common NK/T precursor found in murine foetal thymus (Michie et al., 2000).

More recently the expression of CD45RA, along with integrin $\alpha4\beta7$ has been used to identify a subset of human peripheral blood $CD34^{+dim}$ hematopoietic precursors that preferentially differentiate into NK cells in vitro upon IL-15 stimulation (Freud et al., 2005). The presence of integrin $\alpha4\beta7$ has previously been used to distinguish progenitor cells with gut and lymph node homing capacity (Yoshida et al., 2001). The Caligiuri group proposed that $CD34^+CD45RA^+Integrin\beta7^+$ cells home to lymph nodes and differentiate into NK cells at this site. This was based on the identification of lymph node resident cells at four consecutive stages of NK development (Freud et al., 2006). These include the pro-NK cell ($CD34^+CD45RA^+$ Integrin$\beta7^+$, stage 1), pre-NK cells ($CD34^+CD117^+$, stage 2), iNK ($CD34^-CD117^+CD161^+CD94^-$, stage 3) and $CD117^{low}CD94^+$ $CD56^{bright}$ NK cells (stage 4). The authors go on to hypothesize that $CD56^{dim}$ NK cells constitute the final, fifth stage of NK differentiation, that is completed outside of the LNs (Freud and Caligiuri, 2006). Notably, CD56 acquisition is not an essential criterion in this model; however, consecutive NK developmental stages are marked by increasing CD56 expression. Our own studies on NK cell differentiation in vitro show that $CD34^+$ cells cultured on a monolayer of murine foetal liver cells and in the presence of cytokines (IL-3, IL-7, IL-15, FLT-3L and SCF) robustly develop into NK cells. Using CD56 as a marker of NK cells, we distinguished two $CD56^+$ subsets: immature NK cells ($CD56^+CD117^{high}CD94^-$) that could give rise to a more mature, functional NK cells ($CD56^+CD117^{low}CD94^+$) (Grzywacz et al., 2006). The transition from the $CD56^+CD117^{high}CD94^-$ to the $CD56^+CD117^{low}CD94^+$ stage was associated with the acquisition of activating receptors (NKp30, NKp46 and NKG2D), inhibitory receptors (CD94/NKG2A) and functionality (cytotoxicity and IFN-γ production). The immature NK cells did not express perforin or granzyme B but were $CD161^+$, which, as mentioned earlier, is an early marker of murine and human NK cells. Immature $CD56^+CD117^{high}CD94^-$ cells could be found in UCB and thus, represent physiological relevant intermediates of NK maturation. Moreover, the $CD56^+CD117^{high}CD94^- \rightarrow CD56^+CD117^{low}CD94^+$ transition is equivalent to the stage 3→stage 4 progression in lymph nodes (Freud et al., 2006).

The NK cells derived from in vitro culture rarely expressed KIR or CD16. Moreover, the acquisition of these two receptors was not coordinated at a single cell level, meaning that some developing NK cells acquired KIR but not CD16 (and vice versa) (Grzywacz, unpublished). This is somewhat unexpected considering that in vivo KIR expression is almost exclusive to the $CD16^+$ NK subset. Overall, in vitro derived NK cells as well as the NK cells developing in LN have features reminiscent of the $CD56^{bright}$ subset. The work by Ferlazzo et al. (2004) documented the abundance of $CD56^{bright}$ NK cells in secondary lymphoid tissue. Isolation of these cells and further in vitro culture with IL-2 lead to perforin, KIR and CD16 acquisition. Thus far, the conditions required to advance the in vitro derived NK cells to a $CD56^{dim}CD16^+$ stage are difficult to reproduce. Chan et al. (2007) had the interesting observation that $CD56^{bright}$ NK cells from the peripheral blood can develop into $CD56^{dim}$ cells upon interaction with synovial fibroblasts through a CD56:FGF-R1 interaction. It would be interesting to determine whether in vitro derived NK cells will follow a similar pattern. In another study transpresentation of IL-15 promoted acquisition of CD16 and KIR by $CD56^{bright}CD16^-$ NK cells in mice engrafted with human hematopoietic system (Huntington et al., 2009).

As with murine NK cells, human peripheral blood NK cells can be divided on the basis of CD27 staining intensity. $CD56^{bright}$ and $CD56^{dim}$ human NK cells show $CD27^{high}$ and $CD27^{low}$ expression, respectively. This has led the Smyth group to propose that murine $CD27^{high}$ NK cells correspond to $CD56^{bright}$ cells in humans (Silva et al., 2008). In fact, the LN homing, DC interaction, and IFN-γ production of the murine $CD27^{high}$ NK cells are also properties of $CD56^{bright}$ NK cells (Fehniger et al., 2003). Moreover, murine $CD27^{high}$ cells reconstitute early after BM transplantation, followed by the emergence of $CD27^{low}$ NK cells, reminiscent of the predominance of $CD56^{bright}$ cells early after transplant in humans. However, the $CD27^{high}$ subset in mice is more cytotoxic, as opposed to the relatively low cytotoxicity of human $CD56^{bright}$ ($CD27^{high}$) NK cells. Also, the expression of CD94/NKG2A, universally high on human $CD56^{bright}$ NK cells, does not distinguish murine $CD27^{high}$ and $CD27^{low}$ subsets. Despite these discrepancies, the dichotomy of NK cells subsets distinguished by CD27 expression can be observed in both mice and humans and appears to correspond reasonably well with a linear model of NK cell maturation.

Acquisition of inhibitory receptors during NK cell development

NK cells are restrained from auto-aggression by inhibitory receptors that are specific for MHC class I (HLA). Human MHC-specific inhibitory receptors belong to two structurally distinct families: (1) the lectin-like, CD94/NKG2A complex and (2) the immunoglobulin-like, KIR. Acquisition of CD94/NKG2A by NK progenitors marks an important step in NK development. This is because CD94/NKG2A expression is coordinated with attainment of functionality (activating receptor expression, cytotoxicity and IFN-γ production) (Grzywacz et al., 2006). Thus, the linking of inhibitory receptor expression with effector mechanisms appears to be a form of tolerance during NK development.

The order of inhibitory receptor acquisition during NK cell development appears not to be circumstantial. The ligand for CD94/NKG2A is HLA-E (Braud et al., 1998; Lee et al., 1998). This nonclassical HLA molecule has limited polymorphism. The conserved sequence of HLA-E and its ubiquitous expression, assures that CD94/NKG2A will find its ligand on all healthy cells and tissues (Kaiser et al., 2005). Similarly, CD94 and NKG2A both have strikingly conserved sequences in the human population, compared with other NK receptors (Shum et al., 2002). In contrast, different KIR receptors recognize only a selected set of HLA-C, B, or A molecules as their ligands (Parham, 2005). Thus, expression of a given KIR does not guarantee an effective inhibitory interaction with self-MHC. In this sense, it appears biologically justified that NK cells rely on the CD94/NKG2A as the first safety mechanism. In fact, all the CD56^{bright} NK cells express high levels of CD94/NKG2A, perhaps corresponding to their relatively recent developmental history. This is not true for the CD56^{dim}CD16^{+} NK cells, where the majority of cells express at least one self-specific inhibitory receptor, but frequently it is a KIR and *not* CD94/NKG2A. If CD56^{dim} NK cells are derived from CD56^{bright} NK cells, then this developmental transition would be associated with acquisition of KIR and the *loss* of CD94/NKG2A. Alternatively, the existence of such CD94/NKG2A^{-}KIR^{+} CD56^{dim} cells might support the existence of another developmental pathway (discussed later).

A peculiar subset of NK cells that lack self-specific inhibitory receptors has been identified in both mice (Fernandez et al., 2005) and humans (Anfossi et al., 2006; Cooley et al., 2007). Since licensing comes about through inhibitory interactions with self MHC (Kim et al., 2005), these 'not licensed' NK cells have weak, if any, cytotoxicity despite expression of perforin and granzyme B. Upon in vitro culture with IL-15 they could acquire CD94/NKG2A and/or KIR expression (Cooley et al., 2007). Moreover, phenotypically most of the cells are CD56^{dim}CD16^{+} NK cells, making it difficult to fit these finding in the linear model of NK cell development, according to which NK cells are derived from CD56^{bright} intermediates, highly positive for CD94/NKG2A (as described earlier). As well, the contribution of CD56^{+}CD117^{high}CD94^{-} developmental intermediates (Grzywacz et al., 2006) to the inhibitory receptor negative subset has not been verified nor excluded. Collectively, the outstanding question is whether CD56^{dim} NK cells, which do not express CD94/NKG2A or KIR are derived from CD56^{bright} NK cells.

Linear and branching models of human NK cell development

The most commonly accepted model of NK cell development depicts this process as a linear scheme (Figure 1.2). The dominant population of human peripheral blood NK cells—the CD56^{dim}CD16^{+} subset—corresponds to the

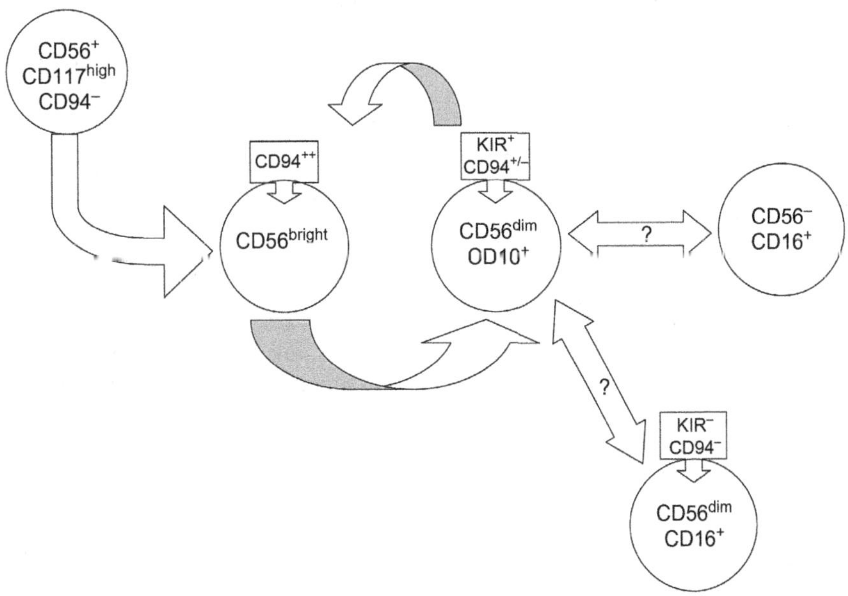

Figure 1.2 • The developmental relationship of NK cell subsets found in humans. The intermediate stage in NK cell development, characterized by CD56^{+}CD117^{high}CD94^{-} phenotype, can advance to the next developmental stage CD56^{bright}CD117^{low}CD94^{+} (Freud et al., 2006; Grzywacz et al., 2006). CD56^{bright} cells found in peripheral blood and secondary lymphoid organs can acquire CD16 and KIR expression after progressing to CD56^{dim}CD16^{+} stage (Chan et al., 2007; Ferlazzo et al., 2004; Huntington et al., 2009; Romagnani et al., 2007) (bottom, large arrow). However, the conversion in the opposite direction has also been demonstrated in certain conditions (Loza and Perussia, 2004; Mailliard et al., 2005) (top, smaller arrow). The CD56^{-}CD16^{+} subset (Gaddy and Broxmeyer, 1997), which includes CD56^{-}CD16^{+}CD122^{+} cells (Harada et al., 2004) could potentially represent intermediate stage of alternative pathway developing directly into CD56^{dim} NK cells, independent of a CD56^{bright} NK cell stage. The subset of predominantly CD56^{dim}CD16^{+} NK cells, that lack self-HLA specific KIR or CD94/NKG2A has been characterized as potentially immature (Anfossi et al., 2006; Cooley et al., 2007). Their relationship to the CD56^{bright}→CD56^{dim} progression of NK cell development is not resolved. They could be derived from the immature CD56^{bright}CD94/NKG2A^{+} subset or are possibly intermediates of an alternative NK differentiation pathway.

final stage of maturation, whereas CD56^{bright} NK cells are considered to be relatively immature and recently differentiated intermediates. Arguments in support of this notion have been presented previously, including: (1) the differentiation of CD34^{+} progenitors first into CD56^{bright}-like NK cells in vitro, as well as in lymph nodes, (2) the ability of CD56^{bright} NK cells to acquire phenotypic properties of CD56^{dim}CD16^{+} NK cells upon interaction with tissue-resident (synovial) fibroblasts (Chan et al., 2007) and (3) CD56^{+}CD16^{-} lymph node derived NK cells that acquire CD16 and/or KIR under the influence of IL-2 (Ferlazzo et al., 2004). Additional evidence supporting the previous assertion comes from studies on immune reconstitution after transplantation, where CD56^{bright} NK cells precede the emergence of the CD56^{dim}CD16^{+} counterparts. It has also been shown that CD56^{bright} NK cells have longer telomeres compared to CD56^{dim} cells (Romagnani et al., 2007). Lastly, transpresentation of IL-15 induced acquisition of CD16 and KIR by CD56^{bright}CD16^{-} NK cells in a xenogeneic mouse model engrafted with human hematopoietic system (Huntington et al., 2009).

However, an opposing view is that CD56^{bright} and CD56^{dim}CD16^{+} NK cells represent two distinct terminal differentiation states, in support of a branching model of NK cell development (Figure 1.2). Perhaps it would be premature to completely abandon this concept. The progression from a CD56^{bright} to CD56^{dim}CD16^{+} cell is not necessarily unidirectional, since the opposite direction (CD56^{dim} to CD56^{bright}) has also been documented. Others have proposed that CD56^{bright} NK cells resemble CD56^{dim}CD16^{+} NK cells that have been activated with IL-12 (Loza and Perussia, 2004). Likewise, under the influence of IL-18, CD56^{dim} NK cells acquire characteristics of CD56^{bright} NK cells, including downregulation of CD16, acquisition of LN migratory molecules (CCR7), production of large amounts of IFN-γ and the ability to interact with DCs (Mailliard et al., 2005). Moreover, activation of NK cells by susceptible target cells leads to a loss of CD16 by metalloproteinase-mediated shedding (Grzywacz et al., 2007; Harrison et al., 1991). Thus, the linear model of NK cell development describing CD56^{bright} NK cells as being an immature intermediate should be weighed against the alternative possibility that CD56^{bright} and CD56^{dim} NK cells represent derivatives of two different developmental pathways and are two different terminal differentiation states. The ability of one cell type to transition into another (Chan et al., 2007; Loza and Perussia, 2004) supports the notion that they correspond to distinct states of activation.

A number of NK cell subsets can be found in the peripheral blood that are difficult to understand how they arose from a linear model of NK cell development. For instance, a distinct population of CD56^{-}CD16^{+} NK cells has been identified in healthy adult peripheral blood. These cells are more prevalent in cord blood (Gaddy and Broxmeyer, 1997), in HIV infection (Mavilio et al., 2005), and after BM transplantation. This subset gives rise to CD56^{dim}CD16^{+} NK cells under the influence of IL-2 (Gaddy and Broxmeyer, 1997) and has been proposed to represent an immature population of recently differentiated NK cells. In particular, a subpopulation of CD56^{-}CD16^{+} cells coexpressing CD122 could very potently expand and differentiate into NK cells in culture with feeder cell line derived from Wilms tumour (Harada et al., 2004). The same cell line promoted engraftment of CD56$^{dim/-}$CD16^{+} NK population in xenogeneic mouse model of human NK development and strikingly CD56^{bright} NK cells were not observed in this model (Harada et al., 2005). CD56^{-}CD16^{+} NK cells could represent an intermediate stage in the CD56^{dim} NK cell development, and it is conceivable that they are developmentally independent of CD56^{bright} NK cells. An alternative possibility is that this NK subset is related to T cell deficiency or immaturity. Moreover the percentage of this fraction of NK cells correlates inversely with the percentage of CD3^{+} T cells in a group of post-transplant patients (Grzywacz, unpublished). The expansion of CD16^{+}CD56^{-} NK cells in T cell deficient individuals may be due to insufficient T cell help (perhaps inadequate IL-2 production). Moreover CD16^{+}CD56^{-} NK cells could represent improperly matured NK cells, but they could also correspond to an 'exhausted' and/or 'helpless' NK subset. Notably all the NK cells found in a patient with immunodeficiency due to a CD3ζ mutation were CD16^{+}CD56^{-}. This is a remarkable observation deserves further study to elucidate the ontongy of this subset. Reminiscent of the other instances where CD16^{+}CD56^{-} NK cell expansions were observed, T cells were absent in this patient. In summary, CD16^{+}CD56^{-} NK cells do not easily fit into the linear model of NK cell development. The existence of these cells supports the notion that distinct pathways of NK cell development exist and that they represent an intermediate stage of a pathway that gives rise to CD56^{dim} NK cells. Similarly, the rare subset of NK cells that lack self-specific inhibitory receptors (Anfossi et al., 2006; Cooley et al., 2007) do not conform to the linear model of NK cell development. Phenotypically, the majority of those cells are CD56^{dim}CD16^{+}, yet they do not express CD94/NKG2A or self-specific KIR. These cells are developmentally immature (Cooley et al., 2007) and it is hard to reconcile how these cells could be derived from CD56^{bright} NK cells, which are uniformly CD94/NKG2A^{+}.

Boundaries of NK cell lineage

Several recent reports have revealed the existence of cells that have sparked controversy regarding their assignment to the NK lineage. As described earlier, thymus-derived NK cells (Veinotte et al., 2006; Vosshenrich et al., 2006) are a rare subset of murine NK cells which express CD127, high levels of GATA-3 and differ functionally from majority of splenic NK cells. The NK cell character of these cells has been called into question by the demonstration of frequent TCR γ/δ gene rearrangements, cytoplasmic expression of CD3ε and/or TCR-β (Stewart et al., 2007; Veinotte et al., 2006). Thus, the thymic NK cells may emerge as cells that have aborted T cell differentiation. The ability of thymocytes (at least up to the DN2 stage) to become NK cells or DCs is well established (Masuda et al., 2007; Shen et al., 2003). This lineage promiscuity is lost later in the development (DN3 stage), when thymocytes express TCR β. The onset of TCR δ and γ rearrangement occurs at the DN2 stage, preceding TCR β rearrangement (Livak et al., 1999). NK cells in humans and in mice commonly express germline TCR δ (Biondi et al., 1989; Stewart et al., 2007). Thus, there is a theoretical possibility that after the initiation of TCR γ or δ rearrangement thymocytes can abort T-cell differentiation in favour of NK or DC lineage. This possibility appears to be realized in the form of thymic NK cells, or at least in a fraction of NK cells that show TCR γδ rearrangements. The argument that TCR rearrangement draws the line between T cells and NK cells (or other lineages) is hard to refute (Lanier, 2007). However, such definition is not compatible with practical designation of NK cell vs. T cell lineage, as it requires intracellular staining for detection of CD3 and/or TCR β, as well as molecular testing for TCR rearrangement. Purifying viable NK cells according to such definitions becomes simply impossible, as all these techniques involve killing the cells. Thus, the novel subset of thymic NK cells revived the discussion on the distinction between NK cells and T cells that appeared to be solved long ago (Lanier, 2007). Before we use the initiation of the TCR gene rearrangement to distinguish NK and T cells, it must be amenable to testing without destruction of the cells. Otherwise such definition, although biologically correct, will remain mute.

Other subsets of cells that reside on the boundaries of the NK lineage are interferon producing killer DCs (IKDC) (Chan et al., 2006; Taieb et al., 2006). Such cells share the properties of both NK cells and plasmacytoid DCs. (see Chapter 3) They are capable of both direct cytotoxicity and production of class I interferons. These functional features are considered to be unique to the two separate cell types (NK cells and plasmacytoid DCs, respectively). More recently two independent groups have questioned the existence of true IKDC, showing that distinct subsets produce IFN-α and perform cytotoxicity in this heterogeneous population consisting mostly of activated NK cells (Blasius et al., 2007; Vosshenrich et al., 2007). Other studies on the mechanism of action of TLR7 agonists in the treatment of skin cancer have revealed that DCs can acquire cellular cytotoxicity upon TLR triggering (Stary et al., 2007). Conversely, it has been also shown that NK cells can acquire efficient antigen presenting capacity upon stimulation (Hanna et al., 2004). Thus, DCs can acquire functional properties of NK cells and NK cells can acquire the function considered characteristic for DCs. Cells sharing properties of NK cells and DCs could reveal the plasticity and potential close developmental relationship of these cells (Spits and Lanier, 2007). Collectively, the controversies on the boundaries between NK cells and T cells as well as between NK cells and DCs underscore the complexity of developmental pathways and place NK cells between these two types of cells.

Summary

The development of multipotent hematopoietic cells into NK cells is a complex process. It is guided by environmental cues and intrinsic responsiveness of precursor cells to external signals. As hematopoietic progenitors progress in differentiation towards NK lineage two concomitant processes occur: (1) acquisition of NK specific gene expression pattern and (2) gradual loss of the ability to express genes characteristic for other lineages. Transcription factors play a critical role in guiding lineage determination, and even though much progress has been made, we are still far from disentanglement of the network of transcription factors that lead to the development of NK cells. Interactions of hematopoietic progenitors with the environment provides growth factors and morphogenic signals that affect lineage fate and guide functional maturation via the triggering of inhibitory and/or activating receptors. The progression from multipotent hematopoietic precursors to mature NK cells can be described on the basis of stages of NK cell development. While most commonly visualized as a linear process, we cannot rule out a branching model of NK cell development at the present time. The development of any hematopoietic lineage can follow heterogenous pathways that lead to a common final point—a mature cell. As precursors traverse an individual developmental pathway, they are exposed to distinct sets of transcription factors, differing in magnitude and time of exposure. These differences eventually affect the final phenotypic and functional features of terminally differentiated cell. Therefore the heterogeneity of NK cell subsets observed in both humans and mice may reflect distinct pathways of NK cell development

References

Adolfsson, J., Mansson, R., Buza-Vidas, N., Hultquist, A., Liuba, K., Jensen, C.T., Bryder, D., Yang, L., Borge, O.J., Thoren, L.A., et al. (2005). Identification of Flt3+ lympho-myeloid stem cells lacking erythro-megakaryocytic potential a revised road map for adult blood lineage commitment. Cell **121**, 295–306.

Akashi, K., Traver, D., Kondo, M. and Weissman, I.L. (1999). Lymphoid development from hematopoietic stem cells. Int J Hematol **69**, 217–226.

Akashi, K., Traver, D., Miyamoto, T. and Weissman, I.L. (2000). A clonogenic common myeloid progenitor that gives rise to all myeloid lineages. Nature **404**, 193–197.

Anfossi, N., Andre, P., Guia, S., Falk, C.S., Roetynck, S., Stewart, C.A., Breso, V., Frassati, C., Reviron, D., Middleton, D., et al. (2006). Human NK cell education by inhibitory receptors for MHC class I. Immunity **25**, 331–342.

Ashkar, A.A., Black, G.P., Wei, Q., He, H., Liang, L., Head, J.R. and Croy, B.A. (2003). Assessment of requirements for IL-15 and IFN regulatory factors in uterine NK cell differentiation and function during pregnancy. J Immunol **171**, 2937–2944.

Baba, Y., Garrett, K.P. and Kincade, P.W. (2005). Constitutively active beta-catenin confers multilineage differentiation potential on lymphoid and myeloid progenitors. Immunity **23**, 599–609.

Barton, K., Muthusamy, N., Fischer, C., Ting, C.N., Walunas, T.L., Lanier, L.L. and Leiden, J.M. (1998). The Ets-1 transcription factor is required for the development of natural killer cells in mice. Immunity **9**, 555–563.

Bennett, I.M., Zatsepina, O., Zamai, L., Azzoni, L., Mikheeva, T. and Perussia, B. (1996). Definition of a natural killer NKR-P1A+/CD56−/CD16− functionally immature human NK cell subset that differentiates in vitro in the presence of interleukin 12. J Exp Med **184**, 1845–1856.

Biondi, A., Allavena, P., Rossi, V., Bassan, R., Barbui, T., Champagne, E., Mak, T.W., Minden, M.D., Rambaldi, A. and Mantovani, A. (1989). T cell receptor delta gene organization and expression in normal and leukemic natural killer cells. J Immunol **143**, 1009–1014.

Blasius, A.L., Barchet, W., Cella, M. and Colonna, M. (2007). Development and function of murine B220+CD11c+NK1.1+ cells identify them as a subset of NK cells. J Exp Med **204**, 2561–2568.

Boos, M.D., Yokota, Y., Eberl, G. and Kee, B.L. (2007). Mature natural killer cell and lymphoid tissue-inducing cell development requires Id2-mediated suppression of E protein activity. J Exp Med **204**, 1119–1130.

Braud, V.M., Allan, D.S., O'Callaghan, C.A., Soderstrom, K., D'Andrea, A., Ogg, G.S., Lazetic, S., Young, N.T., Bell, J.I., Phillips, J.H., et al. (1998). HLA-E binds to natural killer cell receptors CD94/NKG2A, B and C. Nature **391**, 795–799.

Buckley, R.H., Schiff, R.I., Schiff, S.E., Markert, M.L., Williams, L.W., Harville, T.O., Roberts, J.L. and Puck, J.M. (1997). Human severe combined immunodeficiency: genetic, phenotypic, and functional diversity in one hundred eight infants. J Pediatr **130**, 378–387.

Budagian, V., Bulanova, E., Orinska, Z., Thon, L., Mamat, U., Bellosta, P., Basilico, C., Adam, D., Paus, R. and Bulfone-Paus, S. (2005). A promiscuous liaison between IL-15 receptor and Axl receptor tyrosine kinase in cell death control. EMBO J **24**, 4260–4270.

Cao, X., Shores, E.W., Hu-Li, J., Anver, M.R., Kelsall, B.L., Russell, S.M., Drago, J., Noguchi, M., Grinberg, A., Bloom, E.T., et al. (1995). Defective lymphoid development in mice lacking expression of the common cytokine receptor gamma chain. Immunity **2**, 223–238.

Caraux, A., Lu, Q., Fernandez, N., Riou, S., Di Santo, J.P., Raulet, D.H., Lemke, G. and Roth, C. (2006). Natural killer cell differentiation driven by Tyro3 receptor tyrosine kinases. Nat Immunol **7**, 747–754.

Carlyle, J.R., Michie, A.M., Furlonger, C., Nakano, T., Lenardo, M.J., Paige, C.J. and Zuniga-Pflucker, J.C. (1997). Identification of a novel developmental stage marking lineage commitment of progenitor thymocytes. J Exp Med **186**, 173–182.

Cella, M., Nakajima, H., Facchetti, F., Hoffmann, T. and Colonna, M. (2000). ILT receptors at the interface between lymphoid and myeloid cells. Curr Top Microbiol Immunol **251**, 161–166.

Chan, C.W., Crafton, E., Fan, H.N., Flook, J., Yoshimura, K., Skarica, M., Brockstedt, D., Dubensky, T.W., Stins, M.F., Lanier, L.L., et al. (2006). Interferon-producing killer dendritic cells provide a link between innate and adaptive immunity. Nat Med **12**, 207–213.

Chan, A., Hong, D.L., Atzberger, A., Kollnberger, S., Filer, A.D., Buckley, C.D., McMichael, A., Enver, T. and Bowness, P. (2007). CD56bright human NK cells differentiate into CD56dim cells: role of contact with peripheral fibroblasts. J Immunol **179**, 89–94.

Chinen, H., Matsuoka, K., Sato, T., Kamada, N., Okamoto, S., Hisamatsu, T., Kobayashi, T., Hasegawa, H., Sugita, A., Kinjo, F., et al. (2007). Lamina propria c-kit+ immune precursors reside in human adult intestine and differentiate into natural killer cells. Gastroenterology **133**, 559–573.

Clark, E.A., Shultz, L.D. and Pollack, S.B. (1981). Mutations in mice that influence natural killer (NK) cell activity. Immunogenetics **12**, 601–613.

Colucci, F., Samson, S.I., DeKoter, R.P., Lantz, O., Singh, H. and Di Santo, J.P. (2001). Differential requirement for the transcription factor PU.1 in the generation of natural killer cells versus B and T cells. Blood **97**, 2625–2632.

Colucci, F., Schweighoffer, E., Tomasello, E., Turner, M., Ortaldo, J.R., Vivier, E., Tybulewicz, V.L. and Di Santo, J.P. (2002). Natural cytotoxicity uncoupled from the Syk and ZAP-70 intracellular kinases. Nat Immunol **3**, 288–294.

Cooley, S., McCullar, V., Wangen, R., Bergemann, T.L., Spellman, S., Weisdorf, D.J. and Miller, J.S. (2005). KIR reconstitution is altered by T cells in the graft and correlates with clinical outcomes after unrelated donor transplantation. Blood **106**, 4370–4376.

Cooley, S., Xiao, F., Pitt, M., Gleason, M., McCullar, V., Bergemann, T.L., McQueen, K.L., Guethlein, L.A., Parham, P. and Miller, J.S. (2007). A subpopulation of human peripheral blood NK cells that lacks inhibitory receptors for self-MHC is developmentally immature. Blood **110**, 578–586.

Dahl, R. and Simon, M.C. (2003). The importance of PU.1 concentration in hematopoietic lineage commitment and maturation. Blood Cells Mol Dis **31**, 229–233.

De Smedt, M., Taghon, T., Van de Walle, I., De Smet, G., Leclercq, G. and Plum, J. (2007). Notch signaling induces cytoplasmic CD3 epsilon expression in human differentiating NK cells. Blood **110**, 2696–2703.

De Tomaso, A.W., Nyholm, S.V., Palmeri, K.J., Ishizuka, K.J., Ludington, W.B., Mitchel, K. and Weissman, I.L. (2005). Isolation and characterization of a protochordate histocompatibility locus. Nature **438**, 454–459.

DeHart, S.L., Heikens, M.J. and Tsai, S. (2005). Jagged2 promotes the development of natural killer cells and the establishment of functional natural killer cell lines. Blood **105**, 3521–3527.

Dzierzak, E. and Speck, N.A. (2008). Of lineage and legacy: the development of mammalian hematopoietic stem cells. Nat Immunol **9**, 129–136.

Fauriat, C., Just-Landi, S., Mallet, F., Arnoulet, C., Sainty, D., Olive, D. and Costello, R.T. (2007). Deficient expression of NCR in NK cells from acute myeloid leukemia: evolution during leukemia treatment and impact of leukemia cells in NCRdull phenotype induction. Blood **109**, 323–330.

Fehniger, T.A., Cooper, M.A., Nuovo, G.J., Cella, M., Facchetti, F., Colonna, M. and Caligiuri, M.A. (2003). CD56bright natural killer cells are present in human lymph nodes and are activated by T cell-derived IL-2: a potential new link between adaptive and innate immunity. Blood **101**, 3052–3057.

Ferlazzo, G., Thomas, D., Lin, S.L., Goodman, K., Morandi, B., Muller, W.A., Moretta, A. and Munz, C. (2004). The abundant NK cells in human secondary lymphoid tissues require activation to express killer cell Ig-like receptors and become cytolytic. J Immunol **172**, 1455–1462.

Fernandez, N.C., Treiner, E., Vance, R.E., Jamieson, A.M., Lemieux, S. and Raulet, D.H. (2005). A subset of natural killer cells achieves self-tolerance without expressing inhibitory receptors specific for self-MHC molecules. Blood **105**, 4416–4423.

Freud, A.G. and Caligiuri, M.A. (2006). Human natural killer cell development. Immunol Rev **214**, 56–72.

Freud, A.G., Becknell, B., Roychowdhury, S., Mao, H.C., Ferketich, A.K., Nuovo, G.J., Hughes, T.L., Marburger, T.B., Sung, J., Baiocchi, R.A., et al. (2005). A human CD34(+) subset resides in lymph nodes and differentiates into CD56bright natural killer cells. Immunity **22**, 295–304.

Freud, A.G., Yokohama, A., Becknell, B., Lee, M.T., Mao, H.C., Ferketich, A.K. and Caligiuri, M.A. (2006). Evidence for discrete stages of human natural killer cell differentiation in vivo. J Exp Med **203**, 1033–1043.

Fujimoto, S., Ikawa, T., Kina, T. and Yokota, Y. (2007). Forced expression of Id2 in fetal thymic T cell progenitors allows some of their progeny to adopt NK cell fate. Int Immunol **19**, 1175–1182.

Furuta, E., Seo, N. and Yamaguchi, K. (2006). Perforin-dependent cell death in skin allograft rejection of the terrestrial slug *Incilaria fruhstorferi*. Zoolog Sci **23**, 1093–1100.

Gaddy, J. and Broxmeyer, H.E. (1997). Cord blood CD16+ 56− cells with low lytic activity are possible precursors of mature natural killer cells. Cell Immunol **180**, 132–142.

Giliani, S., Mori, L., de Saint Basile, G., Le Deist, F., Rodriguez-Perez, C., Forino, C., Mazzolari, E., Dupuis, S., Elhasid, R., Kessel, A., et al. (2005). Interleukin-7 receptor alpha (IL-7Ralpha) deficiency: cellular and molecular bases. Analysis of clinical, immunological, and molecular features in 16 novel patients. Immunol Rev **203**, 110–126.

Gilmour, K.C., Fujii, H., Cranston, T., Davies, E.G., Kinnon, C. and Gaspar, H.B. (2001). Defective expression of the interleukin-2/interleukin-15 receptor beta subunit leads to a natural killer cell-deficient form of severe combined immunodeficiency. Blood **98**, 877–879.

Grzywacz, B., Kataria, N., Sikora, M., Oostendorp, R.A., Dzierzak, E.A., Blazar, B.R., Miller, J.S. and Verneris, M.R. (2006). Coordinated acquisition of inhibitory and activating receptors and functional properties by developing human natural killer cells. Blood **108**, 3824–3833.

Grzywacz, B., Kataria, N. and Verneris, M.R. (2007). CD56(dim)CD16(+) NK cells downregulate CD16 following target cell induced activation of matrix metalloproteinases. Leukemia **21**, 356–359 author reply 359.

Hafizi, S. and Dahlback, B. (2006). Gas6 and protein S. Vitamin K-dependent ligands for the Axl receptor tyrosine kinase subfamily. FEBS J **273**, 5231–5244.

Hanna, J., Gonen-Gross, T., Fitchett, J., Rowe, T., Daniels, M., Arnon, T.I., Gazit, R., Joseph, A., Schjetne, K.W., Steinle, A., et al. (2004). Novel APC-like properties of human NK cells directly regulate T cell activation. J Clin Invest **114**, 1612–1623.

Hao, L., Klein, J. and Nei, M. (2006). Heterogeneous but conserved natural killer receptor gene complexes in four major orders of mammals. Proc Natl Acad Sci U S A **103**, 3192–3197.

Harada, H., Watanabe, S., Saijo, K., Ishiwata, I. and Ohno, T. (2004). A Wilms tumor cell line, HFWT, can greatly stimulate proliferation of CD56+ human natural killer cells and their novel precursors in blood mononuclear cells. Exp Hematol **32**, 614–621.

Harada, H., Suzu, S., Ito, T. and Okada, S. (2005). Selective expansion and engraftment of human CD16+ NK cells in NOD/SCID mice. Eur J Immunol **35**, 3599–3609.

Harrison, D., Phillips, J.H. and Lanier, L.L. (1991). Involvement of a metalloprotease in spontaneous and phorbol ester-induced release of natural killer cell-associated Fc gamma RIII (CD16-II). J Immunol **147**, 3459–3465.

Hayakawa, Y. and Smyth, M.J. (2006). CD27 dissects mature NK cells into two subsets with distinct responsiveness and migratory capacity. J Immunol **176**, 1517–1524.

He, Y.W. and Malek, T.R. (1996). Interleukin-7 receptor alpha is essential for the development of gamma delta + T cells, but not natural killer cells. J Exp Med **184**, 289–293.

Heemskerk, M.H., Blom, B., Nolan, G., Stegmann, A.P., Bakker, A.Q., Weijer, K., Res, P.C. and Spits, H. (1997). Inhibition of T cell and promotion of natural killer cell development by the dominant negative helix loop helix factor Id3. J Exp Med **186**, 1597–1602.

Held, W., Clevers, H. and Grosschedl, R. (2003). Redundant functions of TCF-1 and LEF-1 during T and NK cell development, but unique role of TCF-1 for Ly49 NK cell receptor acquisition. Eur J Immunol **33**, 1393–1398.

Herberman, R.B., Nunn, M.E. and Lavrin, D.H. (1975). Natural cytotoxic reactivity of mouse lymphoid cells against syngeneic acid allogeneic tumors. I. Distribution of reactivity and specificity. Int J Cancer **16**, 216–229.

Huntington, N.D., Legrand, N., Alves, N.L., Jaron, B., Weijer, K., Plet, A., Corcuff, E., Mortier, E., Jacques, Y., Spits, H. and Di Santo, J.P. (2009). IL-15 trans-presentation promotes human NK cell development and differentiation in vivo. J Exp Med **206**, 25–34.

Hyka-Nouspikel, N. and Phillips, J.H. (2006). Physiological roles of murine DAP10 adapter protein in tumor immunity and autoimmunity. Immunol Rev **214**, 106–117.

Iizuka, K., Chaplin, D.D., Wang, Y., Wu, Q., Pegg, L.E., Yokoyama, W.M. and Fu, Y.X. (1999). Requirement for membrane lymphotoxin in natural killer cell development. Proc Natl Acad Sci U S A **96**, 6336–6340.

Ikawa, T., Fujimoto, S., Kawamoto, H., Katsura, Y. and Yokota, Y. (2001). Commitment to natural killer cells requires the helix-loop-helix inhibitor Id2. Proc Natl Acad Sci U S A **98**, 5164–5169.

Ikawa, T., Kawamoto, H., Goldrath, A.W. and Murre, C. (2006). E proteins and Notch signaling cooperate to promote T cell lineage specification and commitment. J Exp Med **203**, 1329–1342.

Imada, K., Bloom, E.T., Nakajima, H., Horvath-Arcidiacono, J.A., Udy, G.B., Davey, H.W. and Leonard, W.J. (1998). Stat5b is essential for natural killer cell-mediated proliferation and cytolytic activity. J Exp Med **188**, 2067–2074.

Ito, A., Kataoka, T.R., Kim, D.K., Koma, Y., Lee, Y.M. and Kitamura, Y. (2001). Inhibitory effect on natural killer activity of microphthalmia transcription factor encoded by the mutant mi allele of mice. Blood **97**, 2075–2083.

Iwabuchi, K., Iwabuchi, C., Tone, S., Itoh, D., Tosa, N., Negishi, I., Ogasawara, K., Uede, T. and Onoe, K. (2001). Defective development of NK1.1+ T-cell antigen receptor alphabeta+ cells in zeta-associated protein 70 null mice with an accumulation of NK1.1+ CD3− NK-like cells in the thymus. Blood **97**, 1765–1775.

Jacobs, R., Stoll, M., Stratmann, G., Leo, R., Link, H. and Schmidt, R.E. (1992). CD16− CD56+ natural killer cells after bone marrow transplantation. Blood **79**, 3239–3244.

Kaiser, B.K., Barahmand-Pour, F., Paulsene, W., Medley, S., Geraghty, D.E. and Strong, R.K. (2005). Interactions between NKG2x immunoreceptors and HLA-E ligands display overlapping affinities and thermodynamics. J Immunol **174**, 2878–2884.

Kaisho, T., Tsutsui, H., Tanaka, T., Tsujimura, T., Takeda, K., Kawai, T., Yoshida, N., Nakanishi, K. and Akira, S. (1999). Impairment of natural killer cytotoxic activity and interferon gamma production in CCAAT/enhancer binding protein gamma-deficient mice. J Exp Med **190**, 1573–1582.

Kamath, M.B., Houston, I.B., Janovski, A.J., Zhu, X., Gowrisankar, S., Jegga, A.G. and DeKoter, R.P. (2008). Dose-dependent repression of T-cell and natural killer cell genes by PU.1 enforces myeloid and B-cell identity. Leukemia **22**, 1214–1225.

Katsura, Y. (2002). Redefinition of lymphoid progenitors. Nat Rev **2**, 127–132.

Kaushansky, K. (2006). Lineage-specific hematopoietic growth factors. N Engl J Med **354**, 2034–2045.

Kawamura, T., Koka, R., Ma, A. and Kumar, V. (2003). Differential roles for IL-15R alpha-chain in NK cell development and Ly-49 induction. J Immunol **171**, 5085–5090.

Kennedy, M.K., Glaccum, M., Brown, S. N., Butz, E.A., Viney, J.L., Embers, M., Matsuki, N., Charrier, K., Sedger, L., Willis, C.R., et al. (2000). Reversible defects in natural killer and memory CD8T cell lineages in interleukin 15-deficient mice. J Exp Med **191**, 771–780.

Keskin, D.B., Allan, D.S., Rybalov, B., Andzelm, M.M., Stern, J.N., Kopcow, H. D., Koopman, L.A. and Strominger, J.L. (2007). TGFbeta promotes conversion of CD16+ peripheral blood NK cells into CD16− NK cells with similarities to decidual NK cells. Proc Natl Acad Sci U S A **104**, 3378–3383.

Khalturin, K., Becker, M., Rinkevich, B. and Bosch, T.C. (2003). Urochordates and the origin of natural killer cells: identification of a CD94/NKR-P1-related receptor in blood cells of *Botryllus*. Proc Natl Acad Sci U S A **100**, 622–627.

Kiessling, R., Klein, E. and Wigzell, H. (1975). "Natural" killer cells in the mouse. I. Cytotoxic cells with specificity for mouse Moloney leukemia cells. Specificity and distribution according to genotype. Eur J Immunol **5**, 112–117.

Kim, S., Iizuka, K., Kang, H.S., Dokun, A., French, A.R., Greco, S. and Yokoyama, W.M. (2002). In vivo developmental stages in murine natural killer cell maturation. Nat Immunol **3**, 523–528.

Kim, S., Poursine-Laurent, J., Truscott, S. M., Lybarger, L., Song, Y.J., Yang, L., French, A.R., Sunwoo, J.B., Lemieux, S., Hansen, T.H. and Yokoyama, W. M. (2005). Licensing of natural killer cells by host major histocompatibility complex class I molecules. Nature **436**, 709–713.

Kim, N., Saudemont, A., Webb, L., Camps, M., Ruckle, T., Hirsch, E., Turner, M. and Colucci, F. (2007). The p110delta catalytic isoform of PI3K is a key player in NK-cell development and cytokine secretion. Blood **110**, 3202–3208.

Kondo, M., Weissman, I.L. and Akashi, K. (1997). Identification of clonogenic common lymphoid progenitors in mouse bone marrow. Cell **91**, 661–672.

Koopman, L.A., Kopcow, H.D., Rybalov, B., Boyson, J.E., Orange, J.S., Schatz, F., Masch, R., Lockwood, C.J., Schachter, A.D., Park, P.J. and Strominger, J.L. (2003). Human decidual natural killer cells are a unique NK cell subset with immunomodulatory potential. J Exp Med **198**, 1201–1212.

La Motte-Mohs, R.N., Herer, E. and Zuniga-Pflucker, J.C. (2005). Induction of T-cell development from human cord blood hematopoietic stem cells by delta-like 1 in vitro. Blood **105**, 1431–1439.

Lacorazza, H.D., Miyazaki, Y., Di Cristofano, A., Deblasio, A., Hedvat, C., Zhang, J., Cordon-Cardo, C., Mao, S., Pandolfi, P.P. and Nimer, S.D. (2002). The ETS protein MEF plays a critical role in perforin gene expression and the development of natural killer and NK-T cells. Immunity **17**, 437–449.

Lanier, L.L. (2005). NK cell recognition. Annu Rev Immunol **23**, 225–274.

Lanier, L.L. (2007). Back to the future—defining NK cells and T cells. Eur J Immunol **37**, 1424–1426.

Lanier, L.L. (2008). Up on the tightrope: natural killer cell activation and inhibition. Nat Immunol **9**, 495–502.

Lanier , L.L., Phillips, J.H., Hackett, J. Jr., Tutt, M. and Kumar, V. (1986). Natural killer cells: definition of a cell type rather than a function. J Immunol **137**, 2735–2739.

Lee, N., Llano, M., Carretero, M., Ishitani, A., Navarro, F., Lopez-Botet, M. and Geraghty, D.E. (1998). HLA-E is a major ligand for the natural killer inhibitory receptor CD94/NKG2A. Proc Natl Acad Sci U S A **95**, 5199–5204.

Li, H., Pohler, U., Strehlow, I., Hertig, S., Baccarini, M., Emmendorffer, A., Tschopp, J. and Lohmann-Matthes, M. L. (1994). Macrophage precursor cells produce perforin and perform Yac-1 lytic activity in response to stimulation with interleukin-2. J Leukoc Biol **56**, 117–123.

Lian, R.H., Chin, R.K., Nemeth, H.E., Libby, S.L., Fu, Y.X. and Kumar, V. (2004). A role for lymphotoxin in the acquisition of Ly49 receptors during NK cell development. Eur J Immunol **34**, 2699–2707.

Lightner, A., Schust, D.J., Chen, Y.B. and Barrier, B.F. (2008). The fetal allograft revisited: does the study of an ancient invertebrate species shed light on the role of natural killer cells at the maternal–fetal interface?. Clin Dev Immunol **2008**, 631–920.

Livak, F., Tourigny, M., Schatz, D.G. and Petrie, H.T. (1999). Characterization of TCR gene rearrangements during adult murine T cell development. J Immunol **162**, 2575–2580.

Lodolce, J.P., Boone, D.L., Chai, S., Swain, R.E., Dassopoulos, T., Trettin, S. and Ma, A. (1998). IL-15 receptor maintains lymphoid homeostasis by supporting lymphocyte homing and proliferation. Immunity **9**, 669–676.

Lowin-Kropf, B. and Held, W. (2000). Positive impact of inhibitory Ly49 receptor–MHC class I interaction on NK cell development. J Immunol **165**, 91–95.

Lowin-Kropf, B., Kunz, B., Beermann, F. and Held, W. (2000). Impaired natural killing of MHC class I-deficient targets by NK cells expressing a catalytically inactive form of SHP-1. J Immunol **165**, 1314–1321.

Loza, M.J. and Perussia, B. (2004). The IL-12 signature: NK cell terminal CD56+high stage and effector functions. J Immunol **172**, 88–96.

Lynch, L., O'Donoghue, D., Dean, J., O'Sullivan, J., O'Farrelly, C. and Golden-Mason, L. (2006). Detection and characterization of hemopoietic stem cells in the adult human small intestine. J Immunol **176**, 5199–5204.

MacDonald, H.R., Wilson, A. and Radtke, F. (2001). Notch1 and T-cell development: insights from conditional knockout mice. Trends Immunol **22**, 155–160.

MacFarlane, A.W.t., Yamazaki, T., Fang, M., Sigal, L.J., Kurosaki, T. and Campbell, K.S. (2008). Enhanced NK-cell development and function in BCAP-deficient mice. Blood **112**, 131–140.

Maillard, I., Adler, S.H. and Pear, W.S. (2003). Notch and the immune system. Immunity **19**, 781–791.

Mailliard, R.B., Alber, S.M., Shen, H., Watkins, S.C., Kirkwood, J.M., Herberman, R.B. and Kalinski, P. (2005). IL-18-induced CD83+CCR7+ NK helper cells. J Exp Med **202**, 941–953.

Martin, C.H., Woll, P.S., Ni, Z., Zuniga-Pflucker, J.C. and Kaufman, D.S. (2008). Differences in lymphocyte developmental potential between human embryonic stem cell and umbilical cord blood-derived hematopoietic progenitor cells. Blood **112**, 2730–2737.

Masuda, K., Kakugawa, K., Nakayama, T., Minato, N., Katsura, Y. and Kawamoto, H. (2007). T cell lineage determination precedes the initiation of TCR beta gene rearrangement. J Immunol **179**, 3699–3706.

Mavilio, D., Lombardo, G., Benjamin, J., Kim, D., Follman, D., Marcenaro, E., O'Shea, M.A., Kinter, A., Kovacs, C., Moretta, A. and Fauci, A.S. (2005). Characterization of CD56−/CD16+ natural killer (NK) cells: a highly dysfunctional NK subset expanded in HIV-infected viremic individuals. Proc Natl Acad Sci U S A **102**, 2886–2891.

Mebius, R.E. (2003). Organogenesis of lymphoid tissues. Nat Rev **3**, 292–303.

Mellen, P.F., Lust, J.A., Bennett, M. and Kumar, V. (1982). Analysis of low natural killer cell activity in 89Sr-treated mice. Eur J Immunol **12**, 442–445.

Michie, A.M., Carlyle, J.R., Schmitt, T.M., Ljutic, B., Cho, S.K., Fong, Q. and Zuniga-Pflucker, J.C. (2000). Clonal characterization of a bipotent T cell and NK cell progenitor in the mouse fetal thymus. J Immunol **164**, 1730–1733.

Miller, J.S., Verfaillie, C. and McGlave, P. (1992). The generation of human natural killer cells from CD34+/DR− primitive progenitors in long-term bone marrow culture. Blood **80**, 2182–2187.

Miller, J.S., Alley, K.A. and McGlave, P. (1994). Differentiation of natural killer (NK) cells from human primitive marrow progenitors in a stroma-based long-term culture system: identification of a CD34+7+ NK progenitor. Blood **83**, 2594–2601.

Miller, J.S., Soignier, Y., Panoskaltsis-Mortari, A., McNearney, S.A., Yun, G. H., Fautsch, S.K., McKenna, D., Le, C., Defor, T.E., Burns, L.J., et al. (2005). Successful adoptive transfer and in vivo expansion of human haploidentical NK cells in patients with cancer. Blood **105**, 3051–3057.

Minagawa, M., Watanabe, H., Miyaji, C., Tomiyama, K., Shimura, H., Ito, A., Ito, M., Domen, J., Weissman, I.L. and Kawai, K. (2002). Enforced expression of Bcl-2 restores the number of NK cells, but does not rescue the impaired development of NKT cells or intraepithelial lymphocytes, in IL-2/IL-15 receptor beta-chain-deficient mice. J Immunol **169**, 4153–4160.

Moore, K.A. (2004). Recent advances in defining the hematopoietic stem cell niche. Curr Opin Hematol **11**, 107–111.

Mrozek, E., Anderson, P. and Caligiuri, M. A. (1996). Role of interleukin-15 in the development of human CD56+ natural killer cells from CD34^{+} hematopoietic progenitor cells. Blood **87**, 2632–2640.

Negishi, I., Motoyama, N., Nakayama, K., Nakayama, K., Senju, S., Hatakeyama, S., Zhang, Q., Chan, A.C. and Loh, D. Y. (1995). Essential role for ZAP-70 in both positive and negative selection of thymocytes. Nature **376**, 435–438.

Nguyen, S., Dhedin, N., Vernant, J.P., Kuentz, M., Al Jijakli, A., Rouas-Freiss, N., Carosella, E.D., Boudifa, A., Debre, P. and Vieillard, V. (2005). NK-cell reconstitution after haploidentical hematopoietic stem-cell transplantations: immaturity of NK cells and inhibitory effect of NKG2A override GvL effect. Blood **105**, 4135–4142.

Ogasawara, K., Hida, S., Azimi, N., Tagaya, Y., Sato, T., Yokochi-Fukuda, T., Waldmann, T.A., Taniguchi, T. and Taki, S. (1998). Requirement for IRF-1 in the microenvironment supporting development of natural killer cells. Nature **391**, 700–703.

O'Leary, J.G., Goodarzi, M., Drayton, D.L. and von Andrian, U.H. (2006). T cell- and B cell-independent adaptive immunity mediated by natural killer cells. Nat Immunol **7**, 507–516.

Orkin, S.H. and Zon, L.I. (2008). Hematopoiesis: an evolving paradigm for stem cell biology. Cell **132**, 631–644.

Ortaldo, J.R. and Herberman, R.B. (1984). Heterogeneity of natural killer cells. Annu Rev Immunol **2**, 359–394.

Ouyang, W., Ranganath, S.H., Weindel, K., Bhattacharya, D., Murphy, T.L., Sha, W. C. and Murphy, K.M. (1998). Inhibition of Th1 development mediated by GATA-3 through an IL-4-independent mechanism. Immunity **9**, 745–755.

Paloneva, J., Kestila, M., Wu, J., Salminen, A., Bohling, T., Ruotsalainen, V., Hakola, P., Bakker, A.B., Phillips, J.H., Pekkarinen, P., et al. (2000). Loss-of-function mutations in TYROBP (DAP12) result in a presenile dementia with bone cysts. Nat Genet **25**, 357–361.

Parham, P. (2005). MHC class I molecules and KIRs in human history, health and survival. Nat Rev **5**, 201–214.

Parihar , R., Trotta, R., Roda, J.M., Ferketich, A.K., Tridandapani, S., Caligiuri, M. A. and Carson, W.E. 3rd (2005). Src homology 2-containing inositol 5′-phosphatase 1 negatively regulates IFN-gamma production by natural killer cells stimulated with antibody-coated tumor cells and interleukin-12. Cancer Res **65**, 9099–9107.

Park, S.Y., Saijo, K., Takahashi, T., Osawa, M., Arase, H., Hirayama, N., Miyake, K., Nakauchi, H., Shirasawa, T. and Saito, T. (1995). Developmental defects of lymphoid cells in Jak3 kinase-deficient mice. Immunity **3**, 771–782.

Park, I.K., Giovenzana, C., Hughes, T. L., Yu, J., Trotta, R. and Caligiuri, M. A. (2008). The Axl/Gas6 pathway is required for optimal cytokine signaling during human natural killer cell development. Blood.

Peschon, J.J., Morrissey, P.J., Grabstein, K.H., Ramsdell, F.J., Maraskovsky, E., Gliniak, B.C., Park, L.S., Ziegler, S.F., Williams, D.E., Ware, C.B., et al. (1994). Early lymphocyte expansion is severely impaired in interleukin 7 receptor-deficient mice. J Exp Med **180**, 1955–1960.

Phillips, J.H., Hori, T., Nagler, A., Bhat, N., Spits, H. and Lanier, L.L. (1992). Ontogeny of human natural killer (NK) cells: fetal NK cells mediate cytolytic function and express cytoplasmic CD3 epsilon,delta proteins. J Exp Med **175**, 1055–1066.

Pires-daSilva, A. and Sommer, R.J. (2003). The evolution of signalling pathways in animal development. Nat Rev Genet **4**, 39–49.

Puel, A., Ziegler, S.F., Buckley, R.H. and Leonard, W.J. (1998). Defective IL7R expression in T(−)B(+)NK(+) severe combined immunodeficiency. Nat Genet **20**, 394–397.

Puzanov, I.J., Bennett, M. and Kumar, V. (1996). IL-15 can substitute for the marrow microenvironment in the differentiation of natural killer cells. J Immunol **157**, 4282–4285.

Radtke, F., Ferrero, I., Wilson, A., Lees, R., Aguet, M. and MacDonald, H.R. (2000). Notch1 deficiency dissociates the intrathymic development of dendritic cells and T cells. J Exp Med **191**, 1085–1094.

Raulet, D.H., Vance, R.E. and McMahon, C.W. (2001). Regulation of the natural killer cell receptor repertoire. Annu Rev Immunol **19**, 291–330.

Roberts, J.L., Lauritsen, J.P., Cooney, M., Parrott, R.E., Sajaroff, E.O., Win, C.M., Keller, M.D., Carpenter, J.H., Carabana, J., Krangel, M.S., et al. (2007). T−B+NK+ severe combined immunodeficiency caused by complete deficiency of the CD3zeta subunit of the T-cell antigen receptor complex. Blood **109**, 3198–3206.

Rolink, A.G., Balciunaite, G., Demoliere, C. and Ceredig, R. (2006). The potential involvement of Notch signaling in NK cell development. Immunol Lett **107**, 50–57.

Romagnani, C., Juelke, K., Falco, M., Morandi, B., D'Agostino, A., Costa, R., Ratto, G., Forte, G., Carrega, P., Lui, G., et al. (2007). CD56brightCD16− killer Ig-like receptor- NK cells display longer telomeres and acquire features of CD56dim NK cells upon activation. J Immunol **178**, 4947–4955.

Rosmaraki, E.E., Douagi, I., Roth, C., Colucci, F., Cumano, A. and Di Santo, J.P. (2001). Identification of committed NK cell progenitors in adult murine bone marrow. Eur J Immunol **31**, 1900–1909.

Ruggeri, L., Capanni, M., Casucci, M., Volpi, I., Tosti, A., Perruccio, K., Urbani, E., Negrin, R.S., Martelli, M.F. and Velardi, A. (1999). Role of natural killer cell alloreactivity in HLA-mismatched hematopoietic stem cell transplantation. Blood **94**, 333–339.

Ruggeri, L., Capanni, M., Urbani, E., Perruccio, K., Shlomchik, W.D., Tosti, A., Posati, S., Rogaia, D., Frassoni, F., Aversa, F., et al. (2002). Effectiveness of donor natural killer cell alloreactivity in mismatched hematopoietic transplants. Science **295**, 2097–2100.

Samson, S.I., Richard, O., Tavian, M., Ranson, T., Vosshenrich, C.A., Colucci, F., Buer, J., Grosveld, F., Godin, I. and Di Santo, J.P. (2003). GATA-3 promotes maturation, IFN-gamma production, and liver-specific homing of NK cells. Immunity **19**, 701–711.

Sanchez, M.J., Muench, M.O., Roncarolo, M.G., Lanier, L.L. and Phillips, J.H. (1994). Identification of a common T/natural killer cell progenitor in human fetal thymus. J Exp Med **180**, 569–576.

Santos, M.A., Sarmento, L.M., Rebelo, M., Doce, A.A., Maillard, I., Dumortier, A., Neves, H., Radtke, F., Pear, W.S., Parreira, L. and Demengeot, J. (2007). Notch1 engagement by delta-like-1 promotes differentiation of B lymphocytes to antibody-secreting cells. Proc Natl Acad Sci U S A **104**, 15454–15459.

Schaniel, C., Bruno, L., Melchers, F. and Rolink, A.G. (2002). Multiple hematopoietic cell lineages develop in vivo from transplanted Pax5-deficient pre-B I-cell clones. Blood **99**, 472–478.

Schmitt, T.M. and Zuniga-Pflucker, J.C. (2002). Induction of T cell development from hematopoietic progenitor cells by delta-like-1 in vitro. Immunity **17**, 749–756.

Schmitt, T.M., Ciofani, M., Petrie, H. T. and Zuniga-Pflucker, J.C. (2004). Maintenance of T cell specification and differentiation requires recurrent notch receptor–ligand interactions. J Exp Med **200**, 469–479.

Shen, H.Q., Lu, M., Ikawa, T., Masuda, K., Ohmura, K., Minato, N., Katsura, Y. and Kawamoto, H. (2003). T/NK bipotent progenitors in the thymus retain the potential to generate dendritic cells. J Immunol **171**, 3401–3406.

Shilling, H.G., McQueen, K.L., Cheng, N.W., Shizuru, J.A., Negrin, R.S. and Parham, P. (2003). Reconstitution of NK cell receptor repertoire following HLA-matched hematopoietic cell transplantation. Blood **101**, 3730–3740.

Shum, B.P., Flodin, L.R., Muir, D.G., Rajalingam, R., Khakoo, S.I., Cleland, S., Guethlein, L.A., Uhrberg, M. and Parham, P. (2002). Conservation and variation in human and common chimpanzee CD94 and NKG2 genes. J Immunol **168**, 240–252.

Silva, A., Andrews, D.M., Brooks, A.G., Smyth, M.J. and Hayakawa, Y. (2008). Application of CD27 as a marker for distinguishing human NK cell subsets. Int Immunol **20**, 625–630.

Spits, H. and Lanier, L.L. (2007). Natural killer or dendritic: what's in a name?. Immunity **26**, 11–16.

Spits, H., Blom, B., Jaleco, A.C., Weijer, K., Verschuren, M.C., van Dongen, J.J., Heemskerk, M.H. and Res, P.C. (1998). Early stages in the development of human T, natural killer and thymic dendritic cells. Immunol Rev **165**, 75–86.

Staal, F.J. and Sen, J.M. (2008). The canonical Wnt signaling pathway plays an important role in lymphopoiesis and hematopoiesis. Eur J Immunol **38**, 1788–1794.

Staal, F.J., Luis, T.C. and Tiemessen, M.M. (2008). WNT signalling in the immune system: WNT is spreading its wings. Nat Rev **8**, 581–593.

Stary, G., Bangert, C., Tauber, M., Strohal, R., Kopp, T. and Stingl, G. (2007). Tumoricidal activity of TLR7/8-activated inflammatory dendritic cells. J Exp Med **204**, 1441–1451.

Stewart, C.A., Walzer, T., Robbins, S.H., Malissen, B., Vivier, E. and Prinz, I. (2007). Germ-line and rearranged Tcrd transcription distinguish bona fide NK cells and NK-like gammadelta T cells. Eur J Immunol **37**, 1442–1452.

Stopka, T., Amanatullah, D.F., Papetti, M. and Skoultchi, A.I. (2005). PU.1 inhibits the erythroid program by binding to GATA-1 on DNA and creating a repressive chromatin structure. EMBO J **24**, 3712–3723.

Storek, J., Geddes, M., Khan, F., Huard, B., Helg, C., Chalandon, Y., Passweg, J. and Roosnek, E. (2008). Reconstitution of the immune system after hematopoietic stem cell transplantation in humans. Semin Immunopathol **30**, 425–437.

Sun, J.C. and Lanier, L.L. (2008). Tolerance of NK cells encountering their viral ligand during development. J Exp Med **205**, 1819–1828.

Sun, J.C., Beilke, J.N. and Lanier, L.L. (2009). Adaptive immune features of natural killer cells. Nature **457**, 557–561.

Sunwoo, J.B., Kim, S., Yang, L., Naik, T., Higuchi, D.A., Rubenstein, J.L. and Yokoyama, W.M. (2008). Distal-less homeobox transcription factors regulate development and maturation of natural killer cells. Proc Natl Acad Sci U S A **105**, 10877–10882.

Szabo, S.J., Kim, S.T., Costa, G.L., Zhang, X., Fathman, C.G. and Glimcher, L. H. (2000). A novel transcription factor, T-bet, directs Th1 lineage commitment. Cell **100**, 655–669.

Taieb, J., Chaput, N., Menard, C., Apetoh, L., Ullrich, E., Bonmort, M., Pequignot, M., Casares, N., Terme, M., Flament, C., et al. (2006). A novel dendritic cell subset involved in tumor immunosurveillance. Nat Med **12**, 214–219.

Taki, S., Nakajima, S., Ichikawa, E., Saito, T. and Hida, S. (2005). IFN regulatory factor-2 deficiency revealed a novel checkpoint critical for the generation of peripheral NK cells. J Immunol **174**, 6005–6012.

Tassi, I., Cella, M., Gilfillan, S., Turnbull, I., Diacovo, T.G., Penninger, J.M. and Colonna, M. (2007). p110gamma and p110delta phosphoinositide 3-kinase signaling pathways synergize to control development and functions of murine NK cells. Immunity **27**, 214–227.

Tavian, M. and Peault, B. (2005). The changing cellular environments of hematopoiesis in human development in utero. Exp Hematol **33**, 1062–1069.

Tavian, M., Robin, C., Coulombel, L. and Peault, B. (2001). The human embryo, but not its yolk sac, generates lympho-myeloid stem cells: mapping multipotent hematopoietic cell fate in intraembryonic mesoderm. Immunity **15**, 487–495.

Ting, C.N., Olson, M.C., Barton, K.P. and Leiden, J.M. (1996). Transcription factor GATA-3 is required for development of the T-cell lineage. Nature **384**, 474–478.

Toor, A.A., Lund, T.C. and Miller, J.S. (2001). T-cell factor-1 expression during human natural killer cell development and in circulating CD56(+) bright natural killer cells. Exp Hematol **29**, 499–506.

Townsend, M.J., Weinmann, A.S., Matsuda, J.L., Salomon, R., Farnham, P.J., Biron, C.A., Gapin, L. and Glimcher, L.H. (2004). T-bet regulates the terminal maturation and homeostasis of NK and Valpha14i NKT cells. Immunity **20**, 477–494.

Tripathy, S.K., Keyel, P.A., Yang, L., Pingel, J.T., Cheng, T.P., Schneeberger, A. and Yokoyama, W.M. (2008). Continuous engagement of a self-specific activation receptor induces NK cell tolerance. J Exp Med **205**, 1829–1841.

Trotta 2nd, R., Parihar, R., Yu, J., Becknell, B., Allard, J., Wen, J., Ding, W., Mao, H., Tridandapani, S., Carson, W.E. and Caligiuri, M.A. (2005). Differential expression of SHIP1 in CD56bright and CD56dim NK cells provides a molecular basis for distinct functional responses to monokine costimulation. Blood **105**, 3011–3018.

Trowsdale, J., Barten, R., Haude, A., Stewart, C.A., Beck, S. and Wilson, M.J. (2001). The genomic context of natural killer receptor extended gene families. Immunol Rev **181**, 20–38.

Turner, M., Schweighoffer, E., Colucci, F., Di Santo, J.P. and Tybulewicz, V.L. (2000). Tyrosine kinase SYK: essential functions for immunoreceptor signalling. Immunol Today **21**, 148–154.

Uhrberg, M., Valiante, N.M., Young, N.T., Lanier, L.L., Phillips, J.H. and Parham, P. (2001). The repertoire of killer cell Ig-like receptor and CD94:NKG2A receptors in T cells: clones sharing identical alpha beta TCR rearrangement express highly diverse killer cell Ig-like receptor patterns. J Immunol **166**, 3923–3932.

van den Heuvel, M.J., Chantakru, S., Xuemei, X., Evans, S.S., Tekpetey, F., Mote, P.A., Clarke, C.L. and Croy, B. A. (2005). Trafficking of circulating pro-NK cells to the decidualizing uterus: regulatory mechanisms in the mouse and human. Immunol Invest **34**, 273–293.

Veinotte, L.L., Wilhelm, B.T., Mager, D. L. and Takei, F. (2003). Acquisition of MHC-specific receptors on murine natural killer cells. Crit Rev Immunol **23**, 251–266.

Veinotte, L.L., Greenwood, C.P., Mohammadi, N., Parachoniak, C. A. and Takei, F. (2006). Expression of rearranged TCRgamma genes in natural killer cells suggests a minor thymus-dependent pathway of lineage commitment. Blood **107**, 2673–2679.

Vely, F., Olivero, S., Olcese, L., Moretta, A., Damen, J.E., Liu, L., Krystal, G., Cambier, J.C., Daeron, M. and Vivier, E. (1997). Differential association of phosphatases with hematopoietic co-receptors bearing immunoreceptor tyrosine-based inhibition motifs. Eur J Immunol **27**, 1994–2000.

Volz, A., Wende, H., Laun, K. and Ziegler, A. (2001). Genesis of the ILT/LIR/MIR clusters within the human leukocyte receptor complex. Immunol Rev **181**, 39–51.

Vosshenrich, C.A., Ranson, T., Samson, S. I., Corcuff, E., Colucci, F., Rosmaraki, E.E. and Di Santo, J.P. (2005). Roles for common cytokine receptor gamma-chain-dependent cytokines in the generation, differentiation, and maturation of NK cell precursors and peripheral NK cells in vivo. J Immunol **174**, 1213–1221.

Vosshenrich, C.A., Garcia-Ojeda, M.E., Samson-Villeger, S.I., Pasqualetto, V., Enault, L., Richard-Le Goff, O., Corcuff, E., Guy-Grand, D., Rocha, B., Cumano, A., et al. (2006). A thymic pathway of mouse natural killer cell development characterized by expression of GATA-3 and CD127. Nat Immunol **7**, 1217–1224.

Vosshenrich, C.A., Lesjean-Pottier, S., Hasan, M., Richard-Le Goff, O., Corcuff, E., Mandelboim, O. and Di Santo, J.P. (2007). CD11cloB220+ interferon-producing killer dendritic cells are activated natural killer cells. J Exp Med **204**, 2569–2578.

Wahle, J.A., Paraiso, K.H., Kendig, R. D., Lawrence, H.R., Chen, L., Wu, J. and Kerr, W.G. (2007). Inappropriate recruitment and activity by the Src homology region 2 domain-containing phosphatase 1 (SHP1) is responsible for receptor dominance in the SHIP-deficient NK cell. J Immunol **179**, 8009–8015.

Waldmann, T.A. and Tagaya, Y. (1999). The multifaceted regulation of interleukin-15 expression and the role of this cytokine in NK cell differentiation and host response to intracellular pathogens. Annu Rev Immunol **17**, 19–49.

Wang, J.W., Howson, J.M., Ghansah, T., Desponts, C., Ninos, J.M., May, S. L., Nguyen, K.H., Toyama-Sorimachi, N. and Kerr, W.G. (2002). Influence of SHIP on the NK repertoire and allogeneic bone marrow transplantation. Science **295**, 2094–2097.

Wang, H., Grzywacz, B., Sukovich, D., McCullar, V., Cao, Q., Lee, A.B., Blazar, B.R., Cornfield, D.N., Miller, J.S. and Verneris, M.R. (2007). The unexpected effect of cyclosporin A on CD56+CD16− and CD56+CD16+ natural killer cell subpopulations. Blood **110**, 1530–1539.

Warren, L.A. and Rothenberg, E.V. (2003). Regulatory coding of lymphoid lineage choice by hematopoietic transcription factors. Curr Opin Immunol **15**, 166–175.

Williams, N.S., Moore, T.A., Schatzle, J.D., Puzanov, I.J., Sivakumar, P.V., Zlotnik, A., Bennett, M. and Kumar, V. (1997). Generation of lytic natural killer 1.1+, Ly-49- cells from multipotential murine bone marrow progenitors in a stroma-free culture: definition of cytokine requirements and developmental intermediates. J Exp Med **186**, 1609–1614.

Wouters, B.J., Jorda, M.A., Keeshan, K., Louwers, I., Erpelinck-Verschueren, C.A., Tielemans, D., Langerak, A.W., He, Y., Yashiro-Ohtani, Y., Zhang, P., et al. (2007). Distinct gene expression profiles of acute myeloid/T-lymphoid leukemia with silenced CEBPA and mutations in NOTCH1. Blood **110**, 3706–3714.

Yadi, H., Burke, S., Madeja, Z., Hemberger, M., Moffett, A. and Colucci, F. (2008). Unique receptor repertoire in mouse uterine NK cells. J Immunol **181**, 6140–6147.

Yamazaki, K. and Beauchamp, G.K. (2007). Genetic basis for MHC-dependent mate choice. Adv Genet **59**, 129–145.

Yokota, Y., Mansouri, A., Mori, S., Sugawara, S., Adachi, S., Nishikawa, S. and Gruss, P. (1999). Development of peripheral lymphoid organs and natural killer cells depends on the helix-loop-helix inhibitor Id2. Nature **397**, 702–706.

Yokota, T., Huang, J., Tavian, M., Nagai, Y., Hirose, J., Zuniga-Pflucker, J.C., Peault, B. and Kincade, P.W. (2006). Tracing the first waves of lymphopoiesis in mice. Development **133**, 2041–2051.

Yokoyama, W.M. and Plougastel, B.F. (2003). Immune functions encoded by the natural killer gene complex. Nat Rev **3**, 304–316.

Yoshida, H., Kawamoto, H., Santee, S.M., Hashi, H., Honda, K., Nishikawa, S., Ware, C.F., Katsura, Y. and Nishikawa, S.I. (2001). Expression of alpha(4)beta(7) integrin defines a distinct pathway of lymphoid progenitors committed to T cells, fetal intestinal lymphotoxin producer, NK, and dendritic cells. J Immunol **167**, 2511–2521.

Yu, H., Fehniger, T.A., Fuchshuber, P., Thiel, K.S., Vivier, E., Carson, W.E. and Caligiuri, M.A. (1998). Flt3 ligand promotes the generation of a distinct CD34(+) human natural killer cell progenitor that responds to interleukin-15. Blood **92**, 3647–3657.

Yusa, S. and Campbell, K.S. (2003). Src homology region 2-containing protein tyrosine phosphatase-2 (SHP-2) can play a direct role in the inhibitory function of killer cell Ig-like receptors in human NK cells. J Immunol **170**, 4539–4547.

Zhang, D.E., Hohaus, S., Voso, M.T., Chen, H.M., Smith, L.T., Hetherington, C. J. and Tenen, D.G. (1996). Function of PU.1 (Spi-1), C/EBP, and AML1 in early myelopoiesis: regulation of multiple myeloid CSF receptor promoters. Curr Top Microbiol Immunol **211**, 137–147.

Zucchetti, I., Marino, R., Pinto, M.R., Lambris, J.D., Du Pasquier, L. and De Santis, R. (2008). ciCD94-1, an ascidian multipurpose C-type lectin-like receptor expressed in *Ciona intestinalis* hemocytes and larval neural structures. Differentiation **76**, 267–282.

Chapter Two

2

Developmental origins of thymus-derived natural killer cells

James R. Carlyle, Juan C. Zúñiga-Pflücker

CHAPTER CONTENTS

However beautiful the strategy, you should occasionally look at the results.

Winston Churchill, 1874–1965

ABSTRACT

Early progenitor populations within the thymus have long been known to contain cells with both T and natural killer (NK) lineage potential. Conditions that disfavour T-cell development frequently support the development, differentiation, expansion or outgrowth of NK cells. Nonetheless, the contribution of the thymus microenvironment towards the normal production of mature thymus-resident and/or emigrant NK cells has remained controversial or at least not well understood. Here, we revisit the ontogeny and developmental origins of thymus-derived NK cells, which are now known to comprise a significant and functionally distinct subset of the overall peripheral NK cell pool.

KEY WORDS

Ontogeny, Fetal thymus, NKR-P1 (NK1.1), Lymphocyte development, Lineage commitment

Introduction

The thymus is typically envisaged as the primary site of T-lymphopoiesis, in which the necessary molecular cues for the efficient differentiation of lymphocyte progenitors towards the T-cell lineage are provided. Although this is its principal function, the thymus is also capable of supporting the differentiation of other hematopoietic lineages, in particular natural killer (NK) and dendritic cells (DC) (Bhandoola et al., 2007; Petrie and Zúñiga-Pflücker, 2007). Of note, NK cells share several phenotypic and functional features with T cells, and both normally arise within the thymus, supporting the notion that these two lineages are closely related.

A number of studies investigating the developmental potential of precursor thymocytes discovered a close lineage relationship between early thymic progenitors and NK cells [reviewed in (Carlyle and Zúñiga-Pflücker, 1998a; Spits et al., 1998)]. Studies in both human and rodent models demonstrated that in vitro culture conditions suboptimally supportive of thymocyte development frequently resulted in a substantial outgrowth of lymphocytes with an NK cell phenotype (Lanier et al., 1992; Leclercq et al., 1996; Phillips et al., 1992; Rodewald et al., 1992; Sanchez et al., 1993).

Initial reports suggested this outgrowth was due to the default differentiation of NK cells from a bipotential T/NK progenitor capable of generating T-lineage cells in an intact thymic microenvironment yet retaining NK-lineage potential in vitro; such bipotent thymocytes were thus not yet fully T-lineage-committed (Leclercq et al., 1996; Rodewald et al., 1992; Sanchez et al., 1994). Other studies, in which NK cell potential (or DC potential) was not investigated, concluded that many intrathymic and pre-thymic T-cell progenitors were fully T-lineage-committed (characterized by their failure to give rise to B lymphocytes or other hematopoietic cells) (Hattori et al., 1996; Kawamoto et al., 1998; Rodewald et al., 1994). Nonetheless, later investigations revealed the important finding that significant populations of fully mature and functional NK cells normally exist within the thymic microenvironment during fetal (Carlyle et al., 1998; Ikawa et al., 1999; Phillips et al., 1992; Sanchez et al., 1993; Shen et al., 2003), neonatal (Veinotte et al., 2006), and adult thymopoiesis (Carlyle et al., 1998; Veinotte et al., 2006; Vosshenrich et al., 2006). These findings were significant for a number of reasons: (i) during embryonic ontogeny, within the context of a whole animal, fully functional NK cells first develop in situ within the fetal thymus (Carlyle et al., 1998), prior to their development elsewhere in the embryo (Carlyle and Zúñiga-Pflücker, 1998c; Phillips et al., 1992; Sanchez et al., 1993), and prior to the establishment of the bone marrow (BM) as a predominant site for NK-lymphopoiesis (Carlyle and Zúñiga-Pflücker, 1998a; Lanier et al., 1992; Spits et al., 1998); (ii) the thymic stromal microenvironment, like the BM stromal microenvironment, is fully capable of supporting functional NK-lineage differentiation (Carlyle and Zúñiga-Pflücker, 1998a,b; Carlyle et al., 1998; Michie et al., 2000; Phillips et al., 1992; Sanchez et al., 1993); (iii) the thymic subset of the mature NK cell pool possesses some unique characteristics that distinguish it from the BM-derived NK cell subset (Carlyle and Zúñiga-Pflücker, 1998a,c; Carlyle et al., 1998; Di Santo and Vosshenrich, 2006; Phillips et al., 1992; Sanchez et al., 1993; Veinotte et al., 2006; Vosshenrich et al., 2006). These findings also raised a number of questions, such as whether thymic NK cells simply represent a default or reserve cell fate, whether they fulfil a unique resident role during the regulation of thymopoiesis, or whether they represent an exportable subset of NK cells with unique functional and homing characteristics distinct from the BM-derived NK cell subset (Carlyle and Zúñiga-Pflücker, 1998a; Spits et al., 1998). Recent advances in genetic and molecular technologies have allowed a more precise phenotypic and functional delineation of the mature thymic and peripheral NK cell pools in adult animals and humans (Di Santo and Vosshenrich, 2006; Huntington et al., 2007). Here, we discuss new insights into thymic NK cell biology, from their developmental origins to unique characteristics.

Fetal thymic NK cell ontogeny

Early reports demonstrating that both human and rodent progenitor thymocytes exhibit considerable NK-lineage potential ex vivo could not rule out the possibility that some of this capacity originated from the outgrowth of pre-existing mature NK cell populations (Leclercq et al., 1996; Rodewald et al., 1992; Sanchez et al., 1994). This largely stemmed from a lack of available differentiation markers definitively outlining NK cell developmental stages, such as those characterizing the well defined T- and B-lineage differentiation pathways. However, the finding that cells with a mature NK cell phenotype are present in the early mouse fetal thymus by embryonic day 14–15 (e14–15) (Carlyle et al., 1998), at a time well before mature NK cells are found elsewhere in the periphery (i.e. fetal liver, blood, spleen, marrow) (Carlyle and Zúñiga-Pflücker, 1998c), strongly argues that NK cells develop in situ within the fetal thymus driven by developmental cues unique to the thymic microenvironment during fetal ontogeny (Carlyle and Zúñiga-Pflücker, 1998a,b). These mature fetal thymic NK cells differentiate via a precursor–progeny relationship from a defined population of fetal TNK progenitors (TNK) with at least some bipotential T/NK-lineage capacity and substantial NK-unipotency (Carlyle and Zúñiga-Pflücker, 1998b; Carlyle et al., 1997), via an intermediate that appears to be fully NK lineage-committed (Carlyle and Zúñiga-Pflücker, 1998a; Carlyle et al., 1998). All of these populations are characterized by a thymic double negative-1 (DN1; $CD44^+CD25^-$) precursor phenotype and expression of the NK1.1 marker (see below), yet these cells differ in expression of the developmental markers CD117 (c-kit), CD24 (HSA; heat-stable antigen), and CD49b (DX5), as follows (see Figure 2.1):

1. fetal TNK progenitors predominate at e13 with an $NK1.1^+CD117^+CD24^{lo}DX5^-$ phenotype, and they retain both T and NK lineage potential ex vivo (Carlyle and Zúñiga-Pflücker, 1998c; Carlyle et al., 1997; Michie et al., 2000);
2. thymic pre-NK cells develop by e14 displaying a $NK1.1^+CD117^-CD24^-DX5^-$ phenotype (conspicuously marked by the loss of CD117 and CD24, which are both incompatible with T-lineage differentiation in the absence of prior CD25 upregulation), and they exhibit only NK lineage potential ex vivo (Carlyle and Zúñiga-Pflücker, 1998a; Carlyle et al., 1997, 1998);

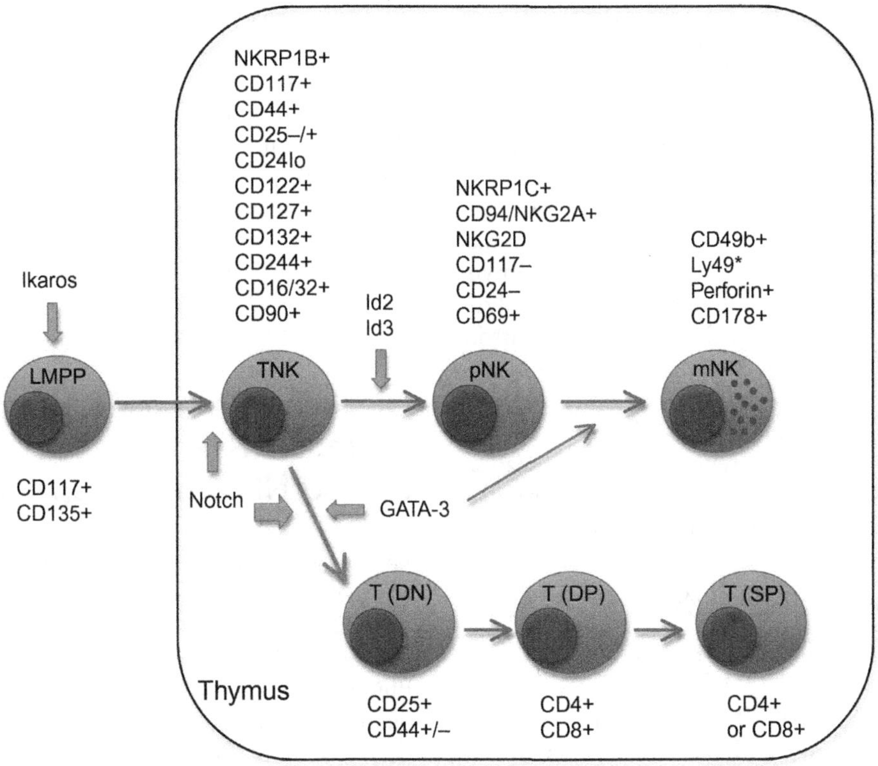

Figure 2.1 • Schematic overview of the developmental progression of thymic NK lineage cells. Lymphoid-primed multipotent progenitor (LMPP) cells expressing the Ikaros transcription factor along with high levels of CD117 (c-kit) and CD135 (Flt3/Flk2) (Yoshida et al., 2006) receive Notch signals either extra-thymically or upon entry into the fetal thymus and give rise to bipotent TNK progenitors (Carlyle and Zúñiga-Pflücker, 1998a). Since a nearly identical population of TNK cells is found enriched in the fetal blood and spleen, this represents a thymus-independent stage (Carlyle and Zúñiga-Pflücker, 1998c). Within the thymus, strong and continuous Notch signals direct commitment to the T lineage (Schmitt et al., 2004), leading to the sequential generation of CD4/CD8 DN, DP, and single-positive thymocytes, as depicted in the lower pathway. In contrast, transient or weak Notch signals permit specification to the NK lineage, while precluding B-lineage differentiation (Carotta et al., 2006; Rolink et al., 2006; Schmitt et al., 2004). High-level induction of Id2 and/or Id3 (Boos et al., 2008; Boos et al., 2007) promotes full commitment of clonogenic NK lineage precursors (pNK). GATA-3 is required for the development of thymic NK cells to a mature and functional NK cell stage (mNK) (Vosshenrich et al., 2006). Expression of the indicated markers reflects their ordered appearance at each stage of differentiation (Carlyle et al., 1998; Carlyle and Zúñiga-Pflücker, 1998a), and their respective maintenance at later stages unless otherwise denoted. Key developmental markers identifying each stage include NKR-P1B, NKR-P1C, and CD49b (DX5). The asterisk (*) indicates limited and selective expression of Ly49 family members on fetal thymic NK cells (Ly49E > G2 > others) (Fraser et al., 2002; Van Beneden et al., 2001).

3. mature thymic NK cells prevail by e15 with an NK1.1$^+$CD117$^-$CD24$^-$DX5$^+$ phenotype, notably highlighted by their ex vivo ability to direct cytotoxicity towards the canonical NK target, YAC-1, without a requirement for priming or activation using cytokines, antibodies, poly-inosinic/deoxycytidilic acid (pI:C), or extended lymphokine-activated killer (LAK) cell culture (Carlyle and Zúñiga-Pflücker, 1998a,c; Carlyle et al., 1998); this final competent stage is identifiable at e15 uniquely within the fetal thymus.

Thus, three distinct thymic stages were revealed, characterized functionally as follows (Figure 2.1):

1. TNK stage: NK-lineage specification of progenitors (loss of potential for B-lymphopoiesis with retention of at least some T-lineage potential and/or DC potential);
2. pNK (pre-NK) stage: full NK-lineage commitment of precursors (subsequent complete loss of T and DC-lineage potential);
3. mNK (mature NK) cells: NK-lineage maturation and differentiation (acquisition of cytotoxicity and/or other hallmark NK functions).

Importantly, due to the existence of a synchronous wave of NK cell development in the fetal thymus during ontogeny (Carlyle et al., 1997), this model system is free from many of the complications that arise when attempting to delineate sequential stages of NK cell developmental pathways in the adult thymus and BM, where all stages co-exist simultaneously under steady-state conditions and numerous distinct subsets exist at each stage (Kim et al., 2002; Vosshenrich et al., 2006). In contrast, each fetal thymic NK cell subset can be defined by the ordered appearance of hallmark phenotypic changes, functional precursor–progeny relationships, and unique NK-lineage developmental stages directly ex vivo during embryogenesis. In this regard, it should be emphasized that each of the above fetal thymic NK developmental stages are separated by only ~1 d during normal embryogenesis. Notwithstanding these findings, further analysis of 'earlier' extra-thymic fetal NK-lineage precursors (i.e. those found elsewhere in the embryo during ontogeny) provided an additional clue to NK-lineage specification events (Carlyle and Zúñiga-Pflücker, 1998c). Crucially, these findings were revealed only by analysis of mouse strains beyond the limitations of the canonical C57BL/6 strain (Carlyle and Zúñiga-Pflücker, 1998a).

Early thymus-independent stages of fetal NK cell development

Once mature and functional NK cells were identified in the mouse fetal thymus, a detailed staging and characterization of the requirements for NK cell development in vivo was facilitated. Since mature NK cells are absent elsewhere in the embryo at e15, NK cell differentiation could be fully characterized by the sequential ontogenic stages that arise within the fetal thymus in situ, in particular because thymic stromal cells appear to be uniquely capable of supporting efficient NK-lymphopoiesis during early fetal ontogeny (Carlyle and Zúñiga-Pflücker, 1998c; Carlyle et al., 1998). However, analysis of other fetal hematopoietic tissues, including fetal blood and spleen, revealed an earlier population of pre-thymic precursors that retain both NK-lineage and T-lineage potential ('prothymocytes': $CD117^+CD90^+$) (Carlyle and Zúñiga-Pflücker, 1998c; Rodewald et al., 1994). This bipotent developmental stage thus characterized a thymus-independent NK-lineage specification event, one that did not preclude T-lineage potential yet occurred immediately prior to full NK-lineage commitment (Raulet, 1999).

Strikingly, phenotypic analysis of this early embryonic (e13) population using available NK-lineage markers revealed that it differed phenotypically between two NK1.1-bearing mouse strains: Swiss/NIH (Sw) fetal blood TNK cells are $\mathbf{NK1.1}^+CD117^+CD90^+$, while the identical population in C57BL/6 (B6) mice is $\mathbf{NK1.1}^-CD117^+CD90^+$ (Carlyle and Zúñiga-Pflücker, 1998a,c; Rodewald et al., 1994). Because mature NK cells from both of these strains are known to express the NK1.1 marker, this seemed at first to represent a strain-related paradox (Carlyle and Zúñiga-Pflücker, 1998b; Ryan et al., 1992). However, subsequent detailed expression analysis of NKR-P1 gene products from these two strains ultimately led to the discovery that the NK1.1 marker actually identifies both the NKR-P1B inhibitory receptor (Carlyle et al., 1999) in addition to the better-known NKR-P1C stimulatory receptor (Ryan et al., 1992). These findings revealed a complex expression pattern of different NKR-P1 genes at early stages of NK cell development, in turn leading to additional questions regarding the significance of this expression pattern.

Dispelling the NK1.1 expression myth: NKR-P1B precedes NKR-P1C during NK lineage specification

Following an initial report using Sw-strain mice (Carlyle and Zúñiga-Pflücker, 1998c), further detailed analysis of NK1.1 expression on B6-strain fetal blood TNK cells revealed that they lack this marker (Carlyle and Zúñiga-Pflücker, 1998a). Moreover, it was subsequently demonstrated that the NK1.1 marker uniquely identifies the $NKR\text{-}P1B^{Sw}$ (Carlyle et al., 1999) and $NKR\text{-}P1C^{B6}$ (Ryan et al., 1992) alleles among these two mouse strains (reciprocally, $NKR\text{-}P1B^{B6}$ and $NKR\text{-}P1C^{Sw}$ are both $NK1.1^-$) (Carlyle et al., 1999; Kung et al., 1999; Ryan et al., 1992). Thus, during fetal ontogeny, NKR-P1B expression precedes NKR-P1C expression, such that circulating thymus-independent fetal TNK progenitors are $NK1.1^{hi/+}$ in Sw mice (Carlyle and Zúñiga-Pflücker, 1998c) but $NK1.1^{lo/-}$ in B6 mice (notably, a small subset of B6-derived fetal blood TNK progenitors do express NK1.1 at low but detectable levels) (Carlyle and Zúñiga-Pflücker, 1998a). Importantly, these findings suggest that the earliest NK lineage-committed precursors [also designated NKP (Di Santo and Vosshenrich, 2006)] likely express NKR-P1B at high levels, because the majority of circulating fetal TNK cells are $CD122^+$; conversely, the vast majority if not all circulating $CD122^+$ cells are $NKR\text{-}P1B^{hi}$ in fetal Sw mice (Carlyle and Zúñiga-Pflücker, 1998c). Thus, the NKR-P1B and CD122 markers likely identify an NK-lineage specification event characterized by the expression of at least some NK-lineage gene products as well as the potential for IL-15-responsiveness (Carlyle and Zúñiga-Pflücker, 1998c; Raulet, 1999; Reya et al., 1996; Rosmaraki et al., 2001).

Indeed, even though the majority of $NKR\text{-}P1B^+$ fetal blood TNK cells express CD122 at high levels, a small subset remains $CD122^{lo/-}$ (Carlyle and Zúñiga-Pflücker, 1998c). Consequently, these latter cells may represent a subset of TNK progenitors or prothymocytes with bona fide T-lineage potential (Bell and Zamoyska, 1991; Carlyle and Zúñiga-Pflücker, 1998c; Douagi et al., 2002a,b; Reya et al., 1996; Rodewald et al., 1997), whereas the

NKR-P1B$^+$CD122$^+$ phenotype may define true NK-lineage-committed precursors (pNK) at the clonal level. In any case, a significant CD122$^+$ subset lacking NKR-P1B expression could not be conclusively identified in fetal Sw mice (Carlyle and Zúñiga-Pflücker, 1998c), although such a miniscule subset is unlikely to represent the true majority population of clonogenic NKP (Rosmaraki et al., 2001).

These findings highlight a caveat in using serological markers to outline NK-developmental stages, as the use of a strict CD122$^+$NK1.1$^-$ definition for NKP (Kim et al., 2002; Rosmaraki et al., 2001) would suggest that fetal Sw mice are devoid of NKP, yet still fully competent in terms of mature NK cell function. Given the added NK1.1$^-$ phenotype of mature NK cells from numerous mouse strains, now known to be simply due to NKR-P1 allelic polymorphisms (Carlyle et al., 2006, 2008; Mesci et al., 2006), this strict definition is not broadly useful and likely misleading. In turn, combinations of specific gene product or molecular designations should be employed to uniquely identify functional stages demarcating NK-lineage specification, commitment, maturation, and differentiation events. In this regard, NKR-P1B may represent the first NK receptor delineating true NKP at the clonal level, as defined by an NKR-P1B$^+$CD122$^+$ phenotype (Carlyle and Zúñiga-Pflücker, 1998c; Raulet, 1999). Notably, another NK receptor, CD244 (2B4), was also found to be expressed on an early subset of BM cells in the adult mouse (Rosmaraki et al., 2001). Thus, NK inhibitory receptors for MHC-independent ligands may regulate the earliest stages of NK self-tolerance in vivo (Kumar and McNerney, 2005; Raulet, 1999).

NK-lineage specification versus commitment

As outlined above, the stringent use of cytokine receptor expression alone (among Lin$^-$ precursors), in the absence of positive expression of all known NK receptors, can be misleading in the identification of bona fide NK lineage-committed precursors (Kim et al., 2002; Rosmaraki et al., 2001). It is more likely that combinations of these functional markers actually characterize true NKP, defined as a homogeneous subset of precursor cells exhibiting exclusive NK-lineage potential at the population level, yet still retaining the highest NK progenitor frequency at the clonal level. Thus, the induction of certain early NK markers in isolation (such as NKR-P1B (Carlyle and Zúñiga-Pflücker, 1998a,c) or CD122 (Reya et al., 1996; Rosmaraki et al., 2001)) likely signifies NK-lineage specification events, without fully demarcating NK-lineage commitment. For example, CD122$^+$NK1.1$^-$ (NKR-P1C$^-$) B6-derived adult BM cells were found to give rise to NK cells with a frequency of ~1/12, whereas the corresponding CD122$^+$NK1.1$^+$ (NKR-P1C$^+$) subset retained an NK progenitor frequency of ~1/3 (Rosmaraki et al., 2001). This suggests that the CD122$^+$NKR-P1C$^-$ subset represents NK-specified progenitors, while the CD122$^+$NKR-P1C$^+$ subset likely represents true clonogenic NK-committed precursors.

Similarly, fetal TNK progenitors, defined as CD117$^+$NK1.1$^+$ (NKR-P1B$^+$) cells from Sw mice, were found to reproducibly give rise to T cells with a frequency of ~1/10–1/30 in limiting-dilution fetal thymic organ culture (FTOC) (Carlyle and Zúñiga-Pflücker, 1998c; Carlyle et al., 1997), or ~1/6 in single-cell FTOC (Michie et al., 2000), while these cells retained NK-unipotent progenitor frequencies in FTOC with equal or greater efficiencies. Moreover, in both investigations, these same precursors were found to give rise to NK cells upon OP9 stromal co-culture at higher frequencies than observed in FTOC (Carlyle and Zúñiga-Pflücker, 1998c; Carlyle et al., 1997, 1998; Michie et al., 2000). These results indicate that NK-lineage specification and commitment events likely represent real and sequential biological outcomes that reflect both stochastic cell fate determination events and cytokine/stroma-dependent inductive events, each revealed with different efficiencies under assay conditions capable of either supporting or skewing the commitment and full maturation of distinct lineages.

To illustrate this point further, recent re-investigations of T-lineage progenitor potential have obtained analogous results using the more well-controlled OP9/OP9-DL1 dual co-culture system (Schmitt et al., 2004). Here, CD117$^+$CD25$^+$ fetal thymic DN2 cells, once thought to represent the earliest T lineage-committed precursors, were found to give rise to NK cells on both OP9 and OP9-DL1 stromal cells, but with a greater frequency on OP9 cells (Schmitt et al., 2004). This likely indicates that CD25 expression actually signifies a T-lineage specification event in the absence of full T-lineage commitment. Thus, under conditions supporting T-lineage commitment (strong Notch activation by Delta-like-1), a T cell fate is induced at the expense of NK cell potential in a subset of precursors, while under conditions supporting NK-lineage commitment (BM stromal culture in the absence of strong Notch signals), NK cell potential can be realized as an alternative cell fate. Notably, all NK cell potential is subsequently lost in the DN3 (CD25$^+$CD44^{lo}) subset, indicative of a true T-lineage commitment event (Schmitt et al., 2004). By analogy then, CD122 and NKR-P1B expression may signify NK-lineage specification rather than commitment events, while a combination of these markers, or other markers (such as NKR-P1C), may better correlate with full NK-lineage commitment.

In keeping with this, two important findings stand out: (i) adult B6-strain BM-derived $CD122^{+}NK1.1^{+}$ ($NKR\text{-}P1C^{+}DX5^{-}Lin^{-}$) precursors display a greater clonogenic NK potential in vitro than their $CD122^{+}NK1.1^{-}$ ($NKR\text{-}P1C^{-}DX5^{-}Lin^{-}$) 'NKP' counterparts (Rosmaraki et al., 2001); (ii) adult B6-strain thymic precursors with a $CD122^{+}NK1.1^{+}$ ($NKR\text{-}P1C^{+}DX5^{-}Lin^{-}$) phenotype are greatly enriched in $RAG\text{-}2^{-/-}$ versus wild-type mice, whereas cells with a $CD122^{+}NK1.1^{-}$ ($NKR\text{-}P1C^{-}DX5^{-}Lin^{-}$) 'NKP' phenotype are actually depleted (Vosshenrich et al., 2006). This suggests that true clonogenic NK-committed progenitors are likely $NKR\text{-}P1C^{+}$. Furthermore, numerous NK receptors have been reported to be expressed on at least a subset of $CD122^{+}Lin^{-}$ 'NKP', including NKR-P1B (Carlyle and Zúñiga-Pflücker, 1998a–c; Carlyle et al., 1998), CD244 (2B4) (Rosmaraki et al., 2001), NKR-P1C (Carlyle and Zúñiga-Pflücker, 1998b; Carlyle et al., 1997; Michie et al., 1998), NKG2D (Di Santo and Vosshenrich, 2006; Jamieson et al., 2002), and even CD94 (Fraser et al., 2002), if the definition of NKP is adjusted to include functionally committed progenitors with the greatest clonogenic NK cell potential.

These are not just semantic differences, as they likely reflect real stochastic and inductive events in vivo. For example, while CD122 expression specifies NK-lineage development (Carlyle and Zúñiga-Pflücker, 1998c; Reya et al., 1996; Rosmaraki et al., 2001), $CD122^{+}$ precursors can be subdivided (yet only retrospectively) as an early subset with only the capacity to respond to IL-15 (IL-15-responsive) and a distinct subset that has already received an IL-15-dependent signal (IL-15-experienced). The latter subset is expected to be more NK-lineage committed if placed in a T-lineage-inductive microenvironment, such as OP9-DL1 co-culture or FTOC, whereas the former may possess a dual T/NK cell fate in vivo. Conversely, only a subset of Notch/DL1-experienced $CD25^{+}$ thymic DN2 cells appear to be fully T-lineage-committed when placed on OP9 cells ex vivo (Schmitt et al., 2004).

Consistent with this, a significant fraction of early fetal thymocytes with a $CD117^{hi}NK1.1^{-}$ multipotent phenotype dynamically upregulates NK1.1 expression (both NKR-P1B and NKR-P1C) ex vivo when simply removed from the thymic microenvironment (Carlyle and Zúñiga-Pflücker, 1998b). This 'spontaneous' $NK1.1^{+}$ transition likely marks those precursor cells that have already received an inductive Notch signal, because $CD117^{hi}NK1.1^{-}$ multipotent fetal liver cells do not exhibit this progression, unless exposed to thymic stromal cells in vitro (Carlyle and Zúñiga-Pflücker, 1998b). Moreover, this progression appears to be supported by IL-15-dependent signals, as exogenous addition of IL-15 greatly enhances NK1.1 upregulation on $CD117^{hi}NK1.1^{-}$ fetal thymocytes in vitro (J.R.C., unpublished observations). Thus, acquisition of NKR-P1C might delineate the subset of $CD122^{+}NKR\text{-}P1B^{+}$ NKP that have already received an inductive contact-dependent (Notch) and cytokine-dependent (IL-15) stromal signal in vivo. In support of this, recent findings indicate that transient exposure of adult BM-derived progenitors (or $Pax\text{-}5^{-/-}$ pro-B cells) to Notch signals on OP9-DL1 stromal cells promotes NK-lineage commitment (Carotta et al., 2006; Rolink et al., 2006). Another analogous example of such dynamic progression of a phenotypically homogenous population involves DN4 thymocytes that have already undergone β-selection and spontaneously differentiate to the double-positive (DP) stage (Wilson et al., 1989).

In any case, the finding that $CD122^{+}$ fetal NKP express NKR-P1B at high levels [e.g. Sw fetal blood TNK cells (Carlyle and Zúñiga-Pflücker, 1998c)], yet high-level NKR-P1C expression is only subsequently induced upon entry into the fetal thymus in vivo [e.g. B6 fetal thymic TNK cells (Carlyle and Zúñiga-Pflücker, 1998a; Carlyle et al., 1997)] or ex vivo [fetal thymocytes but not fetal liver cells (Carlyle and Zúñiga-Pflücker, 1998a; Carlyle and Zúñiga-Pflücker, 1998b)], delineates two sequential stages in NK lineage commitment. The first is thymus-independent ($NKR\text{-}P1B^{+}$), the second is thymus-induced ($NKR\text{-}P1C^{+}$) yet can be substituted by co-culture with thymic or BM-derived stromal cells ex vivo (enhanced by IL-15).

More importantly, these findings suggest that one of the first physiological NK recognition receptors to be expressed during NK cell development in vivo is an inhibitory NK receptor, NKR-P1B, specific for a broadly expressed, MHC-independent 'self' ligand, Clr-b (Carlyle et al., 2004; Iizuka et al., 2003). In turn, the finding that Clr-b appears to be monomorphic (i.e. well-conserved between strains, in contrast to its polymorphic NKR-P1B receptor (Carlyle et al., 2008)), and the fact that *Clr-b* (*Clec2d8*) is genetically linked to *Nkrp1b* (*Klrb1b*; *Klrb2*) within the NK gene complex (Carlyle et al., 2004, 2008; Iizuka et al., 2003; Mesci et al., 2006), thereby fulfils the requirements of the 'at-least-one' hypothesis governing the acquisition of self-specific inhibitory receptors during NK cell development (Carlyle et al., 2008; Raulet, 1999; Raulet et al., 1997). It remains to be determined whether NKR-P1B expression similarly marks the earliest NK developmental stages in the adult mouse (thymus and BM) and among other mouse strains. The reactivity of the PK136 mAb with several NKR-P1B alleles [including those of the Sw (Carlyle et al., 1999), SJL (Kung et al., 1999), FVB (Liu et al., 2000), and CD-1 (Carlyle et al., 2006) strains] and the availability of a specific mAb (2D12) to the $NKR\text{-}P1B^{B6}$ allele [a.k.a., NKR-P1D (Iizuka et al., 2003)], combined with the advantages of the OP9/OP9-DL1 dual co-culture system (Schmitt and Zúñiga-Pflücker, 2002), thus warrants further investigation.

Interestingly, the analogous subset of early human fetal liver and cord blood NKP ($CD34^{+/dim}CD38^{+}CD117^{+}CD122^{+/lo}CD56^{-}CD16^{-}$) appears to express human NKR-P1A (Bennett et al., 1996; Blom et al., 1997; Grzywacz et al., 2006; Jaleco et al., 1997; Lanier et al., 1994; Spits et al., 1998). Since the human NKR-P1A receptor (encoded by *CD161*, the likely ortholog of rodent *Nkrp1b*) is inhibitory in nature and functionally interacts with LLT-1 (encoded by *CLEC2D*, the likely ortholog of rodent *Clr-b*) (Aldemir et al., 2005; Carlyle et al., 2008; Lanier et al., 1994; Mesci et al., 2006; Rosen et al., 2005, 2008), this suggests that expression of NKR-P1 inhibitory receptors on early NKP is conserved between rodents and humans.

NK1.1^{+}CD117^{+}CD90^{+} precursors and their relationship to thymic NK cells

Historically, the unique phenotype and lineage potential of fetal TNK cells merited early investigations into their genetic and molecular expression patterns (Carlyle and Zúñiga-Pflücker, 1998c). While NK-like in terms of a largely $NK1.1^{+}CD122^{+}$ surface phenotype and T-like as suggested by their original characterization as $CD117^{+}CD90^{+}$ fetal 'prothymocytes', further analysis revealed a highly composite phenotype reminiscent of both T and NK lineage precursors (Carlyle and Zúñiga-Pflücker, 1998c; Raulet, 1999; Rodewald et al., 1994). These cells display surface markers characteristic of hematopoietic precursors ($CD117^{+}CD24^{lo}Lin^{-}$; Lin = $CD3^{-}CD11b^{-}CD19^{-}CD45R/B220^{-}Gr\text{-}1^{-}Ter119^{-}$) (Carlyle and Zúñiga-Pflücker, 1998c; Rodewald et al., 1994), they exhibit a characteristic DN1 thymocyte phenotype ($CD4^{-}CD8^{-}CD44^{+}CD25^{-}$) (Carlyle and Zúñiga-Pflücker, 1998c; Rodewald et al., 1994), they express lower levels of FcγRIII/II than other fetal blood cells ($CD16/32^{+/lo}$; as determined by 2.4G2 staining) (Carlyle and Zúñiga-Pflücker, 1998c; Moingeon et al., 1993; Rodewald et al., 1994), and they lack expression of the mature NK cell marker, CD49b (DX5) (Carlyle and Zúñiga-Pflücker, 1998c). In this regard, they differ by only a few markers from mature fetal thymic NK cells (which are $CD117^{-}CD24^{-}DX5^{+}$) (Carlyle and Zúñiga-Pflücker, 1998a,c; Carlyle et al., 1998).

Further analysis of their molecular signature by RT-PCR confirmed a composite T/NK expression pattern. They express high levels of the Ikaros transcription factor, possess higher levels of GATA-3 and TCF-1 (T cell factor-1; *Tcf7*) than total fetal thymocytes, express unrearranged germ-line TCR Cβ transcripts, low to negligible levels of pTα and RAG gene products, and high levels of Lck (Carlyle and Zúñiga-Pflücker, 1998c). In terms of cytokines, they are responsive to stem cell factor (SCF), c-Fms-like tyrosine kinase-ligand (Flt-3L), IL-2, IL-7 and IL-15 (J.R.C., unpublished observations), and they express high levels of IL-7Rα (CD127) but only low levels of IL-15Rα (Carlyle and Zúñiga-Pflücker, 1998a–c). They also express 2B4 (CD244), FcRγ (J.R.C., unpublished observations), low levels of transcripts for other NKR-P1 family members (NKR-P1A/C), and very low but detectable transcripts for perforin and Fas-L (CD95-L; CD178) (Carlyle and Zúñiga-Pflücker, 1998c). However, they lack expression of DAP-12 (DNAX-activation protein-12; *Tyrobp*), Ly-49, and other lymphocyte and myeloid lineage genes such as mb-1 (Igα; CD79a), and c-fms (CD115) (J.R.C., unpublished observations). They also lack functional cytotoxicity and IFN-γ production in vitro (J.R.C., unpublished observations).

Despite their expression of CD90 (Thy-1) and germline TCR Cβ transcripts, they do not possess genomic rearrangements at the TCRβ locus and lack CD3ε transcripts (Carlyle and Zúñiga-Pflücker, 1998c). In terms of lineage potential, they give rise to both T cells (αβ and γδ) and NK cells in FTOC, as well as functional cytotoxic $DX5^{+}$ NK cells (both $CD90^{+/-}$) and $NK1.1^{-}CD3^{+}CD90^{+}$ γδ T cells on OP9 stromal cell culture (in the presence of SCF, IL-2, IL-3, IL-6 and IL-7; J.R.C., unpublished observations) (Carlyle and Zúñiga-Pflücker, 1998c). Collectively, however, the question has always remained whether fetal TNK cells represent physiological T-lineage precursors with default NK-lineage potential, physiological NK precursors with residual 'prothymocyte' capacity, or another cell type fulfilling a unique role in lymphoid organogenesis (i.e. they closely resemble thymus-independent gut cryptopatch progenitors) (Carlyle and Zúñiga-Pflücker, 1998c; Kanamori et al., 1996; Raulet, 1999; Rodewald et al., 1994; Saito et al., 1998).

Consequently, while originally characterized as a population containing bipotential precursor cells with both T- and NK-lineage potential (Carlyle and Zúñiga-Pflücker, 1998c; Carlyle et al., 1997), subsequent clonal analyses revealed that a significant subset of TNK cells possess NK-unipotent activity in vitro (Carlyle and Zúñiga-Pflücker, 1998a,b; Carlyle et al., 1998; Michie et al., 2000). In particular, clonal analyses in single-cell FTOC reconstitution assays revealed an equal frequency of T/NK bipotent and NK-unipotent progenitors, with little T-unipotent precursor activity in vitro (Michie et al., 2000). Importantly, this revealed a greater NK-lineage versus T-lineage potential overall (~1/2.8 cloning frequency or 89% of growing wells containing NK cells, versus a frequency of ~1/4.4 cells or 56% growth of T cells), in fact more than might be expected from chance stochastic cell fate decisions (Michie et al., 2000).

Consistent with this, TNK cells co-cultured on OP9 stromal cells in vitro in the presence of exogenous cytokines (IL-3, IL-6, IL-7 and SCF) were found to routinely give rise to NK cells (Carlyle and Zúñiga-Pflücker, 1998a–c; Carlyle et al., 1997, 1998). While precise clonal NK progenitor frequencies have not been reported for fetal TNK cells co-cultured on stromal cells in vitro (Michie et al., 2000), similar clonal analyses to those reported for DN2 thymocytes using the OP9/OP9-DL1 dual co-culture system (Schmitt et al., 2004) would be expected to reveal their true NKP capacity. It remains likely that this NK lineage potential would be greater than that observed in FTOC, where reconstitution inefficiencies such as failed seeding and differential niches limit the practicability of the assay. Additionally, the differences between the T and NK potentials on OP9 versus OP9-DL1 likely approximate the true frequency of bipotent T/NK precursors in vivo (Schmitt et al., 2004). In any case, it is quite certain that subsets of circulating and thymic TNK precursors retain only unipotent NK-lineage potential and serve directly as precursors for fetal thymic NK cells in vivo (Carlyle and Zúñiga-Pflücker, 1998c; Carlyle et al., 1998; Michie et al., 2000; Reya et al., 1996; Rosmaraki et al., 2001).

NK cell development in vivo: developmental stages versus subsets of differentiation

If circulating fetal TNK cells represent the earliest extra-thymic NK-specified stage (containing a subset of committed NKP), paralleled or followed closely by fetal thymic TNK or DN2 cells, which have already been exposed to inductive thymic stromal signals, then the subsequent relationship of these precursors to mature fetal thymic NK cells defines the minimal requirements for a complete staging of NK cell development in vivo (Carlyle and Zúñiga-Pflücker, 1998c; Carlyle et al., 1997, 1998; Michie et al., 2000; Schmitt et al., 2004). Consequently, it follows that additional subsets visualized in the adult microenvironment (thymus or BM) may either represent true sequential stages of NK cell development from a distinct developmental precursor subset, or distinct subsets of NK cells developing from an analogous precursor population under different microenvironmental influences (adult thymic stroma or BM stroma, respectively) (Kim et al., 2002; Rosmaraki et al., 2001; Vosshenrich et al., 2006). In this regard, several models have been put forward as representing the sequential stages of NK cell 'development' in vivo that likely only reflect further diversification or differentiation of NK cell subsets in the adult system.

In particular, among the most commonly (but not exhaustively) ascribed differences between the phenotype of fetal thymic (Carlyle and Zúñiga-Pflücker, 1998c; Carlyle et al., 1997, 1998; Michie et al., 2000; Schmitt et al., 2004) versus adult BM-derived NK cells (Kim et al., 2002) pertain to expression the following: (i) the hematopoietic precursor marker, CD117 (c-kit); (ii) the often-used Lin marker, CD11b (Mac-1); and (iii) the functional receptors for classical MHC-Ia molecules in rodents, the Ly49 receptors.

In the fetal system, the developmental progression from CD117^{hi} (hematopoietic precursor) to CD117$^{+/lo}$ (TNK precursor) to CD117^{-} (pre-NK and mature NK) is quite clear and unambiguous (Carlyle and Zúñiga-Pflücker, 1998b,c; Carlyle et al., 1997, 1998). However, in the adult system, it has been suggested that the earliest NKP lack CD117 expression altogether, progress via a CD117^{+} intermediate, then return to a CD117^{-} mature NK stage (Kim et al., 2002). This is contradictory to results from the fetal system, and was largely based upon subsetting of BM-derived NK lineage cells using α_v integrin (CD51), a non-NK lineage marker. Specifically, while most of the BM data are consistent with a progression from CD117^{+} precursors to CD117^{-} mature NK cells, α_v integrin was found to be enriched on DX5^{-} cells (and vice versa), and NKG2A/C/E expression was enriched on $\alpha_v$$^{+}$CD117^{-} cells, while few Ly49 receptors were found on this subset (Kim et al., 2002). In turn, it was suggested that $\alpha_v$$^{+}$CD117^{-} cells must constitute the most immature NK subset (Kim et al., 2002). However, these data are also consistent with a pre-NK or mature NK cell subset that predominantly expresses CD94/NKG2 receptors for non-classical MHC-Ib, without yet expressing or ever functionally requiring Ly49 receptors for classical MHC-Ia molecules, such that the $\alpha_v$$^{+}$CD117^{-} subset may possess distinct developmental, functional, or homing properties. Indeed, in the fetal thymus, the majority of pre-NK cells (CD117^{-}) express CD94/NKG2A but lack expression of most Ly49 receptors (Carlyle and Zúñiga-Pflücker, 1998a,c; Carlyle et al., 1998; Fraser et al., 2002; Van Beneden et al., 1999, 2001; Van Den Broeck et al., 2008). In addition, upon co-culture with OP9 BM-derived stromal cells, sorted NK1.1^{+}CD117^{+} fetal precursors rapidly lose CD117 expression and gain CD94 expression, before they upregulate DX5 and Ly49 receptors (Carlyle and Zúñiga-Pflücker, 1998c; Carlyle et al., 1997, 1998) (J.R.C., unpublished observations). These insights suggest that $\alpha_v$$^{+}$CD117^{-} BM cells may represent a CD117^{-} transitional intermediate, a distinct DX5^{-} NK cell subset, or a parallel CD117^{-} lineage in the adult system. Alternatively, their CD94/NKG2^{+}Ly49^{-}DX5^{-} phenotype suggests they may possess skewed recognition of MHC-Ib versus MHC-Ia ligands, plus a potential for differential homing on α_v-dependent ligands. Nonetheless, the relevance of α_v integrin in NK function is unclear, and determination of its significance requires a re-evaluation of fetal NK cell development (thymus, spleen, marrow).

Similarly, all fetal thymic NK cells are CD11b$^-$ (Carlyle et al., 1998), while a significant subset of mature adult BM-derived NK cells have been defined as 'CD11b^{hi}' (Kim et al., 2002). First of all, CD11b^{hi} is a misnomer, as relative to bona fide Mac-1^{hi} (Lin$^+$) myeloid-lineage cells, all NK1.1$^+$CD3$^-$ NK cells are CD11b^{lo}, yet can be further subdivided into a CD11b$^-$ NK subset (less differentiated, enriched in the BM) and a CD11b$^{+/lo}$ subset (more differentiated, enriched in the periphery) (Boos et al., 2008; Di Santo and Vosshenrich, 2006). Importantly, both of these subsets represent 'mature' NK cells, as defined functionally by their ability to elicit cytotoxicity or produce IFN-γ, and phenotypically by their DX5$^+$ status and grossly normal NK receptor expression (Kim et al., 2002). Notably, even mature e15 fetal thymic NK cells with a DX5$^+$ phenotype (all of which are CD11b$^-$) exhibit cytotoxicity towards YAC-1 targets, albeit to a lesser extent than adult NK cells (Carlyle et al., 1998). Furthermore, DX5$^+$ adult thymic NK cells (which are similarly CD11b$^-$) produce more IFN-γ on a per-cell basis than adult BM-derived NK cells (Veinotte et al., 2008; Vosshenrich et al., 2006). We therefore propose that CD11b, another integrin marker ($\alpha_M\beta_2$), represents more of a differentiation or activation marker on mature NK cell subsets, rather than a true developmental marker delineating immature versus mature NK cells.

Finally, early fetal thymic NK cells have been reported to be deficient in Ly49 expression relative to adult BM-derived NK cells (Carlyle et al., 1998; Van Beneden et al., 1999). While this clearly has functional consequences, it does not affect the maturity or functional capacity of fetal thymic, or even adult thymic, NK cells (Carlyle et al., 1998; Veinotte et al., 2008; Vosshenrich et al., 2006). In fact, one major reason for this phenotype may be simply the utilization of distinct functional classes of receptors by fetal thymic NK cells, versus their adult BM-derived counterparts. Thymic NK cells in general express higher levels of CD94/NKG2A receptors than BM-derived NK cells (Fraser et al., 2002; Van Beneden et al., 2001; Vosshenrich et al., 2006). Furthermore, the percentage of CD94/NKG2A$^+$ thymic NK cells peaks during fetal NK ontogeny, decreasing proportionately during neonatal to adult development (Fraser et al., 2002; Van Beneden et al., 2001). In addition, while fetal thymic NK cells express fewer MHC-Ia-binding Ly49 receptors (Ly49G2 > Ly49C > Ly49D) (Van Beneden et al., 1999; Vosshenrich et al., 2006), they are highly enriched in expression of Ly49E (Fraser et al., 2002; Van Beneden et al., 1999, 2001), a unique Ly49 receptor recently reported to recognize a β_2m-independent non-MHC ligand, uPA (urokinase plasminogen activator) (Van Den Broeck et al., 2008). This phenotype, combined with early high-level expression of two other MHC-independent recognition receptors, NKR-P1B and CD244 (2B4), suggests that fetal thymic NK cells may be developmentally skewed towards recognition of MHC-independent 'self' determinants (Carlyle et al., 2004; Iizuka et al., 2003; Kumar and McNerney, 2005; Lee et al., 2004; Raulet, 1999). This may have important functional consequences regarding their enhanced potency to produce cytokines and diminished capacity to elicit cytotoxicity (Carlyle et al., 1998; Van Beneden et al., 1999; Veinotte et al., 2008; Vosshenrich et al., 2006). Collectively, these findings likely reflect unique phenotypic and functional differences that normally exist between mature NK cell subsets derived from the thymus versus BM of fetal versus adult mice (Carlyle et al., 1998; Kim et al., 2002; Rosmaraki et al., 2001; Veinotte et al., 2006, 2008; Vosshenrich et al., 2006).

Taken together, the expression of different NK-specific and non-NK lineage markers on developing NK cells under steady-state conditions, while useful for categorizing NK cell subsets, does not imply that developing NK cells follow such a linear program of step-wise transitions during normal development or ontogeny. In this light, although DX5 does represent another heterodimeric integrin (CD49b/CD29; VLA-2; $\alpha_2\beta_1$) (Arase et al., 2001), it is useful as a marker of NK cell developmental staging because it appears to signify the acquisition of functional maturity among both thymic and BM-derived NK cells, from both fetal and adult mice, and across many if not all mouse strains (Arase et al., 2001; Carlyle et al., 1998; Kim et al., 2002; Vosshenrich et al., 2006). In contrast, ontogenetic analysis of early NK cell development using markers such as CD11b, CD27, CD43, and many integrins suggests that these molecules more likely delineate distinct subsets of mature and functional NK cells in the adult BM (Boos et al., 2008; Di Santo and Vosshenrich, 2006). Thus, these markers more accurately reflect the functional diversification, parallel differentiation, or even extrinsic activation of already mature NK cell subsets, rather than serving to define obligate sequential stages during NK cell development. Similarly, attempts to define NK cell developmental stages by expression of subset-restricted, stochastically acquired NK receptors, or strain-dependent serological markers [e.g. NK1.1 (Carlyle et al., 2006)], can be misleading [NKR-P1B precedes NKR-P1C (Carlyle and Zúñiga-Pflücker, 1998a)]. As a result, a re-evaluation of NK cell development is warranted to delineate a composite model of the key sequential stages that govern the ordered appearance of NK markers en route to becoming mature functional NK cells (e.g. in the early fetal thymus, spleen, and BM rudiment).

Composite phenotype of fetal versus adult thymic NK cells

Recent advances have facilitated a detailed analysis of NK cell development in the adult thymus (Veinotte et al., 2006; Vosshenrich et al., 2006). Using a β-galactosidase reporter, knock-in mouse model of the *Gata3* locus, it was revealed that adult thymic NK cells express high levels of GATA-3 and CD127 (IL-7Rα) (Vosshenrich et al., 2006), much like fetal thymic NK cells and their TNK precursors (Carlyle and Zúñiga-Pflücker, 1998a,c). Furthermore, analysis of *Gata3*$^{-/-}$ mice and *Il7*$^{-/-}$ mice (on a *Rag2*$^{-/-}$ background to enrich NK cells) showed that adult thymic NK cells require GATA-3 and CD127 for their development (Vosshenrich et al., 2006), in contrast to adult BM-derived NK cells (Di Santo and Vosshenrich, 2006). This elegantly confirmed that the composite T/NK-like phenotype observed in the fetal system had direct functional consequences, including the finding that thymic NK cells express high levels of IL-7Rα but only low levels of IL-15Rα (Carlyle and Zúñiga-Pflücker, 1998a,c). Interestingly, adult thymic NK cells also closely resemble their fetal thymic NK counterparts in terms of most of the markers analyzed (Carlyle and Zúñiga-Pflücker, 1998c; Vosshenrich et al., 2006): they have a mature NK cell phenotype, characterized as CD122^{+}, NKR-P1C^{+} (NKR-P1B was not examined), DX5^{+}, NKG2D^{+}, and DN1-like (CD44^{hi}, CD25^{-}), yet they differ from adult BM-derived NK cells in that they are CD94/NKG2A^{hi}, Ly49^{lo} (G2 > A,C,D,I; Ly49E not tested), CD16^{lo}, CD69^{hi}, and more completely Lin^{-}, that is CD11b^{-}, CD45R/B220^{-}, and CD43^{-} (CD27 not tested). While CD90 (Thy-1) was not examined, the vast majority of fetal thymic NK cells express high levels of this marker (Carlyle et al., 1998), and this is likely also true for adult thymic NK cells. Interestingly, this correlation of higher levels of T-lineage markers such as CD90 on thymic NK cells, with lower levels of myeloid (Mac-1) and B-lineage markers (B220) than BM-derived NK cells, seems to be a product of their in situ development within the context of thymic stromal (Carlyle et al., 1998; Vosshenrich et al., 2006) versus BM stromal microenvironments (Kim et al., 2002), respectively.

Additionally, another molecular hallmark of T-lineage differentiation, discovered to be found exclusively among thymus-derived NK cells, is their curious expression of rearranged TCRγ chains, particularly Vγ2-Jγ1 rearrangements (Veinotte et al., 2006, 2008). Of note, limited TCRα and TCRδ rearrangements have also been demonstrated, but no TCRβ rearrangements have been detected; consistent with this, only germline TCR Cβ transcripts have been detected in fetal TNK and NK lineage cells (Carlyle and Zúñiga-Pflücker, 1998a,c). Interestingly, the thymic Vγ2-Jγ1^{+} NK phenotype was found in both neonatal and adult wild-type mice, yet lacking in athymic nude (*Foxn1*$^{nu/nu}$) mice (Veinotte et al., 2006, 2008), suggesting that the low level RAG gene expression detected in fetal TNK cells may be functional to some extent (Carlyle and Zúñiga-Pflücker, 1998c). Recent findings using a reporter transgene to assess RAG recombinase activity during lymphocyte development have established that a subset of mature NK cells do indeed display evidence of prior recombinase activity (Pilbeam et al., 2008). In addition, analysis of TCRδ-locus reporter mice revealed that a subset of thymic CD127^{+} NK cells may represent 'NK-like γδ T cells' possessing high level intracellular CD3 expression and low level surface TCRγδ expression (Stewart et al., 2007).

Collectively, one of the most striking findings that came as a result of the unique phenotype of thymus-derived NK cells was the demonstration that they were also found in the periphery, localized in an enriched manner within the lymph nodes (Veinotte et al., 2006, 2008; Vosshenrich et al., 2006). This export of a unique subset of thymic NK cells to the periphery had gone unnoticed in the fetal system, perhaps because fetal thymic NK cells may not be exported until the late fetal or neonatal stage, as revealed by the absence of NK1.1^{+}CD117^{-}DX5^{+} cells in the fetal blood or spleen, even as late as e16 (Carlyle and Zúñiga-Pflücker, 1998c; Carlyle et al., 1998); however, fetal lymphatic tissues were not examined in these studies. Thus, it was previously thought that fetal thymic NK cells fulfil a local resident function within the thymus, perhaps in regulating T cell development (Carlyle and Zúñiga-Pflücker, 1998a; Li et al., 2005; Schott et al., 2003). The export of thymic NK cells to the periphery, and particularly their homing to the lymph nodes, raised questions about whether they possess unique functional characteristics.

While fresh ex vivo fetal thymic NK cells were found to express perforin, CD178 (CD95-L; Fas-L), and could kill YAC-1 targets, they possess reduced cytotoxicity in comparison to adult RAG-2$^{-/-}$ thymic NK cells (Carlyle et al., 1998). Furthermore, fresh ex vivo adult thymic NK cells were found to possess lower YAC-1 cytotoxicity than adult BM-derived (splenic) NK cells (Vosshenrich et al., 2006), although IL-2-cultured adult thymic NK cells seem to possess equivalent cytotoxicity towards MHC-I-deficient RMA-S targets in comparison to adult splenic NK cells (Veinotte et al., 2008). However, it was also shown that mouse thymic NK cells, like the human CD56^{++}CD16^{-} NK cell subset, produce higher levels of IFN-γ and other cytokines (GM-CSF, TNF-α) on a per-cell basis than their BM-derived, splenic counterparts (Veinotte et al., 2008; Vosshenrich et al., 2006). These findings suggest that the primary function of thymus-derived NK cells may be cytokine production over

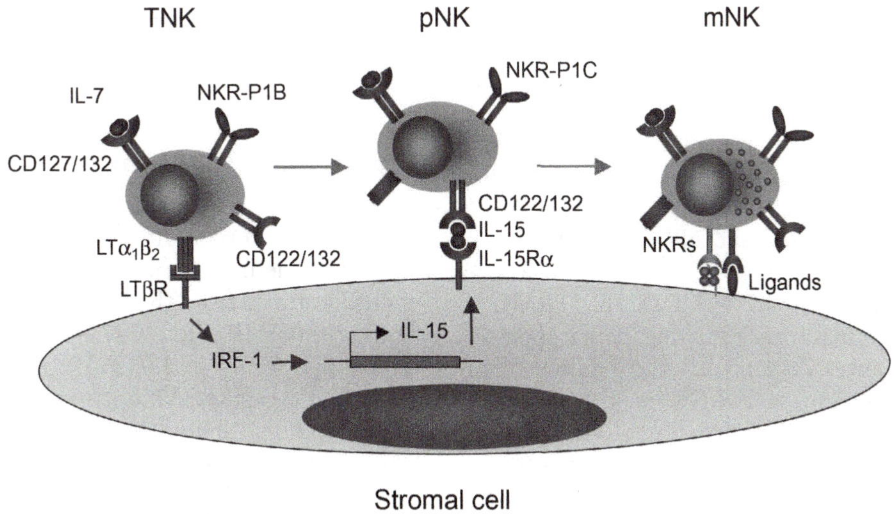

Figure 2.2 • Diagrammatic representation of NK precursor:stromal cell interactions supporting NK cell development. NK-lineage specified TNK progenitors (NKR-P1B$^+$) responding to stromal-derived IL-7 upregulate expression of surface $LT\alpha_1\beta_2$ (Yoshida et al., 2001), likely leading to contact-dependent induction of IL-15 via IRF-1-dependent transcriptional regulation (Ogasawara et al., 1998; Ohteki et al., 1998). Nascent IL-15 is presented in trans by IL-15Rα on stromal cells (Burkett et al., 2004; Mortier et al., 2008) to developing CD122$^+$ committed NK precursors (pNK). IL-15-experienced pNK cells (NKR-P1C$^+$) maintain stromal cell contact during subsequent stages of NK lineage maturation. Stromal cell interactions maintain LT-, IL-15-, and contact-dependent signals required for Ly49 receptor acquisition (Lian et al., 2004; Stevenaert et al., 2005). In turn, NK receptor (NKR) interactions with cognate ligands on stromal cells (Yokoyama and Plougastel, 2003) also promote NK cell education and self-tolerance (Raulet and Vance, 2006), culminating in the emergence of mature and functional NK cells (mNK; CD49b^{hi}).

cytotoxicity, akin to the uterine NK cell subset (Yadi et al., 2008).

TNK cells: physiological bipotent progenitor or lymphoid tissue inducer (LTi)-like cell?

The early ontogenic appearance of TNK cells in the fetal blood, their seeding of the fetal primary and secondary lymphoid tissues (thymus and spleen), and their close phenotypic resemblance to gut-colonizing progenitor cells suggests that they may have another physiological function in vivo (Carlyle and Zúñiga-Pflücker, 1998c; Eberl, 2005; Kanamori et al., 1996; Mebius, 2003; Saito et al., 1998). Several groups have characterized CD117$^+$CD127$^+$Lin$^-$ precursor cells that have variably been reported to express $LT\alpha_1\beta_2$ (Yoshida et al., 2001, 2002), both T and NK lineage markers (Carlyle and Zúñiga-Pflücker, 1998c), RORγt (retinoic acid receptor-related orphan receptor gamma; *Rorc*) (Cella et al., 2008; Cupedo et al., 2009; Luci et al., 2009; Sanos et al., 2009; Satoh-Takayama et al., 2008; Zenewicz et al., 2008), and can give rise to certain subsets of gut lymphocytes (Eberl, 2005), including mucosal IL-22-producing NKp46$^+$ cells. While there is likely more than one functional subset of gut-derived cells with the above characteristics, an intriguing possibility, due to their NK1.1^{lo-int} phenotype (in B6 strain mice; NKR-P1C), is that these cells likely express NKR-P1B at high levels. If so, it is possible that the physiological function of TNK cells in the fetal circulation is to establish the microenvironmental niches that initiate lymphoid organogenesis and subsequently lead to efficient in situ T and/or NK lineage development. In keeping with this, both NK lineage and LTi cells share a requirement for Id2-mediated suppression of E-protein activity for their development in vivo (Boos et al., 2007). While expression of $LT\alpha_1\beta_2$, RORγt, and NKp46 on TNK cells remains to be determined, since $LT\alpha_1\beta_2$ is upregulated by IL-7 via CD127 (Yoshida et al., 2001, 2002), LT-derived signals received (and delivered) upon interaction with stroma in the presence of IL-7 and IL-15 may promote in situ NK lineage development (Figure 2.2). Interestingly, human fetal LTi-like cells (CD117^{lo}CD127$^+$*RORC*$^+$) were recently reported to lack CD4 but express high levels of NKR-P1A (CD161) (Cupedo et al., 2009), the likely human equivalent of NKR-P1B (Carlyle et al., 2008; Mesci et al., 2006).

Concluding remarks

Significant progress has been made in the phenotypic and functional characterization of a unique population of thymus-derived NK cells in fetal (Carlyle et al., 1998; Ikawa et al., 1999; Phillips et al., 1992; Sanchez et al., 1993; Shen et al., 2003), neonatal (Veinotte et al., 2006), and adult mice (Carlyle et al., 1998; Veinotte et al., 2006; Vosshenrich et al., 2006). The early and in situ development of NK cells within the fetal thymus during embryonic ontogeny provides an unparalleled model system for outlining the minimal developmental stages required for NK lineage specification (NKR-P1B$^+$/CD122$^+$/CD244$^+$), commitment (NKR-P1C$^+$/CD94$^+$/NKG2D$^+$), and maturation (DX5$^+$/Ly49$^+$) of functional NK cells (Perforin$^+$/IFNγ^+/CD178$^+$/TRAIL$^+$) from their most immature circulating hematopoietic precursors (see Figures 2.1 and 2.2 for details). New insights into additional subsets of NK lineage cells with distinct functional characteristics found localized in other tissues (such as the lymph nodes, liver, uterus, and gut), supports the notion that NK cells represent an important and relatively unappreciated lymphocyte population. Further ontogenetic analysis will provide a more complete picture of the origins, developmental requirements, and functional differentiation and diversification of these novel NK subsets in vivo.

Acknowledgements

We thank Dr. David Allan for discussions and critical reading of the manuscript. This work was supported by operating grants from the Canadian Institutes of Health Research (CIHR FRN 74754 and 86630 to J.R.C.; CIHR FRN 12927 to J.C.Z.-P.). J.R.C. is supported by an Early Researcher Award from the Ontario Ministry of Research and Innovation, a New Investigator Award from the CIHR, and an Investigator in the Pathogenesis of Infectious Disease Award from the Burroughs Wellcome Fund. J.C.Z.-P. is a Canada Research Chair in Developmental Immunology.

References

Aldemir, H., Prod'homme, V., Dumaurier, M.J., Retiere, C., Poupon, G., Cazareth, J., Bihl, F. and Braud, V.M. (2005). Cutting edge: lectin-like transcript 1 is a ligand for the CD161 receptor. J Immunol **175**, 7791–7795.

Arase, H., Saito, T., Phillips, J.H. and Lanier, L.L. (2001). Cutting edge: the mouse NK cell-associated antigen recognized by DX5 monoclonal antibody is CD49b (alpha 2 integrin, very late antigen-2). J Immunol **167**, 1141–1144.

Bell, S.E. and Zamoyska, R. (1991). Identification of thymocyte precursors in murine fetal liver. Eur J Immunol **21**, 2931–2936.

Bennett, I.M., Zatsepina, O., Zamai, L., Azzoni, L., Mikheeva, T. and Perussia, B. (1996). Definition of a natural killer NKR-P1A+/CD56-/CD16- functionally immature human NK cell subset that differentiates in vitro in the presence of interleukin 12. J Exp Med **184**, 1845–1856.

Bhandoola, A., von Boehmer, H., Petrie, H.T. and Zúñiga-Pflücker, J.C. (2007). Commitment and developmental potential of extrathymic and intrathymic T cell precursors: plenty to choose from. Immunity **26**, 678–689.

Blom, B., Res, P., Noteboom, E., Weijer, K. and Spits, H. (1997). Prethymic CD34+ progenitors capable of developing into T cells are not committed to the T cell lineage. J Immunol **158**, 3571–3577.

Boos, M.D., Yokota, Y., Eberl, G. and Kee, B.L. (2007). Mature natural killer cell and lymphoid tissue-inducing cell development requires Id2-mediated suppression of E protein activity. J Exp Med **204**, 1119–1130.

Boos, M.D., Ramirez, K. and Kee, B.L. (2008). Extrinsic and intrinsic regulation of early natural killer cell development. Immunol Res **40**, 193–207.

Burkett, P.R., Koka, R., Chien, M., Chai, S., Boone, D.L. and Ma, A. (2004). Coordinate expression and trans presentation of interleukin (IL)-15Ralpha and IL-15 supports natural killer cell and memory CD8+ T cell homeostasis. J Exp Med **200**, 825–834.

Carlyle, J.R. and Zúñiga-Pflücker, J. C. (1998a). Lineage commitment and differentiation of T and natural killer lymphocytes in the fetal mouse. Immunol Rev **165**, 63–74.

Carlyle, J.R. and Zúñiga-Pflücker, J.C. (1998b). Regulation of NK1.1 expression during lineage commitment of progenitor thymocytes. J Immunol **161**, 6544–6551.

Carlyle, J.R. and Zúñiga-Pflücker, J.C. (1998c). Requirement for the thymus in alpha beta T lymphocyte lineage commitment. Immunity **9**, 187–197.

Carlyle, J.R., Michie, A.M., Furlonger, C., Nakano, T., Lenardo, M.J., Paige, C. J. and Zúñiga-Pflücker, J.C. (1997). Identification of a novel developmental stage marking lineage commitment of progenitor thymocytes. J Exp Med **186**, 173–182.

Carlyle, J.R., Michie, A.M., Cho, S.K. and Zúñiga-Pflücker, J.C. (1998). Natural killer cell development and function precede alpha beta T cell differentiation in mouse fetal thymic ontogeny. J Immunol **160**, 744–753.

Carlyle, J.R., Martin, A., Mehra, A., Attisano, L., Tsui, F.W. and Zúñiga-Pflücker, J.C. (1999). Mouse NKR-P1B, a novel NK1.1 antigen with inhibitory function. J Immunol **162**, 5917–5923.

Carlyle, J.R., Jamieson, A.M., Gasser, S., Clingan, C.S., Arase, H. and Raulet, D. H. (2004). Missing self-recognition of Ocil/Clr-b by inhibitory NKR-P1 natural killer cell receptors. Proc Natl Acad Sci U S A **101**, 3527–3532.

Carlyle, J.R., Mesci, A., Ljutic, B., Belanger, S., Tai, L.H., Rousselle, E., Troke, A. D., Proteau, M.F. and Makrigiannis, A.P. (2006). Molecular and genetic basis for strain-dependent NK1.1 alloreactivity of mouse NK cells. J Immunol **176**, 7511–7524.

Carlyle, J.R., Mesci, A., Fine, J.H., Chen, P., Belanger, S., Tai, L.H. and Makrigiannis, A.P. (2008). Evolution of the Ly49 and Nkrp1 recognition systems. Semin Immunol **20**, 321–330.

Carotta, S., Brady, J., Wu, L. and Nutt, S.L. (2006). Transient Notch signalling induces NK cell potential in Pax5-deficient pro-B cells. Eur J Immunol **36**, 3294–3304.

Cella, M., Fuchs, A., Vermi, W., Facchetti, F., Otero, K., Lennerz, J.K., Doherty, J.M., Mills, J.C. and Colonna, M. (2008). A human natural killer cell subset provides an innate source of IL-22 for mucosal immunity. Nature.

Cupedo, T., Crellin, N.K., Papazian, N., Rombouts, E.J., Weijer, K., Grogan, J. L., Fibbe, W.E., Cornelissen, J.J. and Spits, H. (2009). Human fetal lymphoid tissue-inducer cells are interleukin 17-producing precursors to RORC(+) CD127(+) natural killer-like cells. Nat Immunol **10**, 66–74.

Di Santo, J.P. and Vosshenrich, C.A. (2006). Bone marrow versus thymic pathways of natural killer cell development. Immunol Rev **214**, 35–46.

Douagi, I., Colucci, F., Di Santo, J.P. and Cumano, A. (2002a). Identification of the earliest prethymic bipotent T/NK progenitor in murine fetal liver. Blood **99**, 463–471.

Douagi, I., Vieira, P. and Cumano, A. (2002b). Lymphocyte commitment during embryonic development in the mouse. Semin Immunol **14**, 361–369.

Eberl, G. (2005). Inducible lymphoid tissues in the adult gut: recapitulation of a fetal developmental pathway? Nat Rev Immunol **5**, 413–420.

Fraser, K.P., Gays, F., Robinson, J.H., van Beneden, K., Leclercq, G., Vance, R. E., Raulet, D.H. and Brooks, C.G. (2002). NK cells developing in vitro from fetal mouse progenitors express at least one member of the Ly49 family that is acquired in a time-dependent and stochastic manner independently of CD94 and NKG2. Eur J Immunol **32**, 868–878.

Grzywacz, B., Kataria, N., Sikora, M., Oostendorp, R.A., Dzierzak, E.A., Blazar, B.R., Miller, J.S. and Verneris, M.R. (2006). Coordinated acquisition of inhibitory and activating receptors and functional properties by developing human natural killer cells. Blood **108**, 3824–3833.

Hattori, N., Kawamoto, H. and Katsura, Y. (1996). Isolation of the most immature population of murine fetal thymocytes that includes progenitors capable of generating T, B, and myeloid cells. J Exp Med **184**, 1901–1908.

Huntington, N.D., Vosshenrich, C.A. and Di Santo, J.P. (2007). Developmental pathways that generate natural-killer-cell diversity in mice and humans. Nat Rev Immunol **7**, 703–714.

Iizuka, K., Naidenko, O.V., Plougastel, B. F., Fremont, D.H. and Yokoyama, W. M. (2003). Genetically linked C-type lectin-related ligands for the NKRP1 family of natural killer cell receptors. Nat Immunol **4**, 801–807.

Ikawa, T., Kawamoto, H., Fujimoto, S. and Katsura, Y. (1999). Commitment of common T/natural killer (NK) progenitors to unipotent T and NK progenitors in the murine fetal thymus revealed by a single progenitor assay. J Exp Med **190**, 1617–1626.

Jaleco, A.C., Blom, B., Res, P., Weijer, K., Lanier, L.L., Phillips, J.H. and Spits, H. (1997). Fetal liver contains committed NK progenitors, but is not a site for development of CD34+ cells into T cells. J Immunol **159**, 694–702.

Jamieson, A.M., Diefenbach, A., McMahon, C.W., Xiong, N., Carlyle, J.R. and Raulet, D.H. (2002). The role of the NKG2D immunoreceptor in immune cell activation and natural killing. Immunity **17**, 19–29.

Kanamori, Y., Ishimaru, K., Nanno, M., Maki, K., Ikuta, K., Nariuchi, H. and Ishikawa, H. (1996). Identification of novel lymphoid tissues in murine intestinal mucosa where clusters of c-kit+ IL-7R+ Thy1+ lympho-hemopoietic progenitors develop. J Exp Med **184**, 1449–1459.

Kawamoto, H., Ohmura, K. and Katsura, Y. (1998). Presence of progenitors restricted to T, B, or myeloid lineage, but absence of multipotent stem cells in the murine fetal thymus. J Immunol **161**, 3799–3802.

Kim, S., Iizuka, K., Kang, H.S., Dokun, A., French, A.R., Greco, S. and Yokoyama, W.M. (2002). In vivo developmental stages in murine natural killer cell maturation. Nat Immunol **3**, 523–528.

Kumar, V. and McNerney, M.E. (2005). A new self: MHC-class-I-independent natural-killer-cell self-tolerance. Nat Rev Immunol **5**, 363–374.

Kung, S.K., Su, R.C., Shannon, J. and Miller, R.G. (1999). The NKR-P1B gene product is an inhibitory receptor on SJL/J NK cells. J Immunol **162**, 5876–5887.

Lanier, L.L., Spits, H. and Phillips, J.H. (1992). The developmental relationship between NK cells and T cells. Immunol Today **13**, 392–395.

Lanier, L.L., Chang, C. and Phillips, J.H. (1994). Human NKR-P1A. A disulfide-linked homodimer of the C-type lectin superfamily expressed by a subset of NK and T lymphocytes. J Immunol **153**, 2417–2428.

Leclercq, G., Debacker, V., de Smedt, M. and Plum, J. (1996). Differential effects of interleukin-15 and interleukin-2 on differentiation of bipotential T/natural killer progenitor cells. J Exp Med **184**, 325–336.

Lee, K.M., McNerney, M.E., Stepp, S.E., Mathew, P.A., Schatzle, J.D., Bennett, M. and Kumar, V. (2004). 2B4 acts as a non-major histocompatibility complex binding inhibitory receptor on mouse natural killer cells. J Exp Med **199**, 1245–1254.

Li, J., Rabinovich, B.A., Hurren, R., Cosman, D. and Miller, R.G. (2005). Survival versus neglect: redefining thymocyte subsets based on expression of NKG2D ligand(s) and I. Eur J Immunol **35**, 439–448.

Lian, R.H., Chin, R.K., Nemeth, H.E., Libby, S.L., Fu, Y.X. and Kumar, V. (2004). A role for lymphotoxin in the acquisition of Ly49 receptors during NK cell development. Eur J Immunol **34**, 2699–2707.

Liu, J., Morris, M.A., Nguyen, P., George, T.C., Koulich, E., Lai, W.C., Schatzle, J.D., Kumar, V. and Bennett, M. (2000). Ly49I NK cell receptor transgene inhibition of rejection of H2b mouse bone marrow transplants. J Immunol **164**, 1793–1799.

Luci, C., Reynders, A., Ivanov, II., Cognet, C., Chiche, L., Chasson, L., Hardwigsen, J., Anguiano, E., Banchereau, J., Chaussabel, D., Dalod, M., Littman, D.R., Vivier, E. and Tomasello, E. (2009). Influence of the transcription factor RORgammat on the development of NKp46(+) cell populations in gut and skin. Nat Immunol **10**, 75–82.

Mebius, R.E. (2003). Organogenesis of lymphoid tissues. Nat Rev Immunol **3**, 292–303.

Mesci, A., Ljutic, B., Makrigiannis, A.P. and Carlyle, J.R. (2006). NKR-P1 biology: from prototype to missing self. Immunol Res **35**, 13–26.

Michie, A.M., Carlyle, J.R., Schmitt, T.M., Ljutic, B., Cho, S.K., Fong, Q. and Zúñiga-Pflücker, J.C. (2000). Clonal characterization of a bipotent T cell and NK cell progenitor in the mouse fetal thymus. J Immunol **164**, 1730–1733.

Michie, A.M., Carlyle, J.R. and Zúñiga-Pflücker, J.C. (1998). Early intrathymic precursor cells acquire a CD4(low) phenotype. J Immunol **160**, 1735–1741.

Moingeon, P., Rodewald, H.R., McConkey, D., Mildonian, A., Awad, K. and Reinherz, E.L. (1993). Generation of natural killer cells from both Fc gamma RII/III+ and Fc gamma RII/III− murine fetal liver progenitors. Blood **82**, 1453–1462.

Mortier, E., Woo, T., Advincula, R., Gozalo, S. and Ma, A. (2008). IL-15R alpha chaperones IL-15 to stable dendritic cell membrane complexes that activate NK cells via trans presentation. J Exp Med **205**, 1213–1225.

Ogasawara, K., Hida, S., Azimi, N., Tagaya, Y., Sato, T., Yokochi-Fukuda, T., Waldmann, T.A., Taniguchi, T. and Taki, S. (1998). Requirement for IRF-1 in the microenvironment supporting development of natural killer cells. Nature **391**, 700–703.

Ohteki, T., Yoshida, H., Matsuyama, T., Duncan, G.S., Mak, T.W. and Ohashi, P.S. (1998). The transcription factor interferon regulatory factor 1 (IRF-1) is important during the maturation of natural killer 1.1+ T cell receptor-alpha/beta+ (NK1+ T) cells, natural killer cells, and intestinal intraepithelial T cells. J Exp Med **187**, 967–972.

Petrie, H.T. and Zúñiga-Pflücker, J.C. (2007). Zoned out: functional mapping of stromal signalling microenvironments in the thymus. Annu Rev Immunol **25**, 649–679.

Phillips, J.H., Hori, T., Nagler, A., Bhat, N., Spits, H. and Lanier, L.L. (1992). Ontogeny of human natural killer (NK) cells: fetal NK cells mediate cytolytic function and express cytoplasmic CD3 epsilon, delta proteins. J Exp Med **175**, 1055–1066.

Pilbeam, K., Basse, P., Brossay, L., Vujanovic, N., Gerstein, R., Vallejo, A.N. and Borghesi, L. (2008). The ontogeny and fate of NK cells marked by permanent DNA rearrangements. J Immunol **180**, 1432–1441.

Raulet, D.H. (1999). Development and tolerance of natural killer cells. Curr Opin Immunol **11**, 129–134.

Raulet, D.H. and Vance, R.E. (2006). Self-tolerance of natural killer cells. Nat Rev Immunol **6**, 520–531.

Raulet, D.H., Held, W., Correa, I., Dorfman, J.R., Wu, M.F. and Corral, L. (1997). Specificity, tolerance and developmental regulation of natural killer cells defined by expression of class I-specific Ly49 receptors. Immunol Rev **155**, 41–52.

Reya, T., Yang-Snyder, J.A., Rothenberg, E. V. and Carding, S.R. (1996). Regulated expression and function of CD122 (interleukin-2/interleukin-15R-beta) during lymphoid development. Blood **87**, 190–201.

Rodewald, H.R., Moingeon, P., Lucich, J.L., Dosiou, C., Lopez, P. and Reinherz, E.L. (1992). A population of early fetal thymocytes expressing Fc gamma RII/III contains precursors of T lymphocytes and natural killer cells. Cell **69**, 139–150.

Rodewald, H.R., Kretzschmar, K., Takeda, S., Hohl, C. and Dessing, M. (1994). Identification of pro-thymocytes in murine fetal blood: T lineage commitment can precede thymus colonization. Embo J **13**, 4229–4240.

Rodewald, H.R., Ogawa, M., Haller, C., Waskow, C. and DiSanto, J.P. (1997). Pro-thymocyte expansion by c-kit and the common cytokine receptor gamma chain is essential for repertoire formation. Immunity **6**, 265–272.

Rolink, A.G., Balciunaite, G., Demoliere, C. and Ceredig, R. (2006). The potential involvement of Notch signaling in NK cell development. Immunol Lett **107**, 50–57.

Rosen, D.B., Bettadapura, J., Alsharifi, M., Mathew, P.A., Warren, H.S. and Lanier, L.L. (2005). Cutting edge: lectin-like transcript-1 is a ligand for the inhibitory human NKR-P1A receptor. J Immunol **175**, 7796–7799.

Rosen, D.B., Cao, W., Avery, D.T., Tangye, S.G., Liu, Y.J., Houchins, J.P. and Lanier, L.L. (2008). Functional consequences of interactions between human NKR-P1A and its ligand LLT1 expressed on activated dendritic cells and B cells. J Immunol **180**, 6508–6517.

Rosmaraki, E.E., Douagi, I., Roth, C., Colucci, F., Cumano, A. and Di Santo, J.P. (2001). Identification of committed NK cell progenitors in adult murine bone marrow. Eur J Immunol **31**, 1900–1909.

Ryan, J.C., Turck, J., Niemi, E.C., Yokoyama, W.M. and Seaman, W.E. (1992). Molecular cloning of the NK1.1 antigen, a member of the NKR-P1 family of natural killer cell activation molecules. J Immunol **149**, 1631–1635.

Saito, H., Kanamori, Y., Takemori, T., Nariuchi, H., Kubota, E., Takahashi-Iwanaga, H., Iwanaga, T. and Ishikawa, H. (1998). Generation of intestinal T cells from progenitors residing in gut cryptopatches. Science **280**, 275–278.

Sanchez, M.J., Spits, H., Lanier, L.L. and Phillips, J.H. (1993). Human natural killer cell committed thymocytes and their relation to the T cell lineage. J Exp Med **178**, 1857–1866.

Sanchez, M.J., Muench, M.O., Roncarolo, M.G., Lanier, L.L. and Phillips, J.H. (1994). Identification of a common T/natural killer cell progenitor in human fetal thymus. J Exp Med **180**, 569–576.

Sanos, S.L., Bui, V.L., Mortha, A., Oberle, K., Heners, C., Johner, C. and Diefenbach, A. (2009). RORgammat and commensal microflora are required for the differentiation of mucosal interleukin 22-producing NKp46(+) cells. Nat Immunol **10**, 83–91.

Satoh-Takayama, N., Vosshenrich, C.A., Lesjean-Pottier, S., Sawa, S., Lochner, M., Rattis, F., Mention, J.J., Thiam, K., Cerf-Bensussan, N., Mandelboim, O., Eberl, G. and Di Santo, J.P. (2008). Microbial flora drives interleukin 22 production in intestinal NKp46(+) cells that provide innate mucosal immune defense. Immunity.

Schmitt, T.M. and Zúñiga-Pflücker, J.C. (2002). Induction of T cell development from hematopoietic progenitor cells by delta-like-1 in vitro. Immunity **17**, 749–756.

Schmitt, T.M., Ciofani, M., Petrie, H. T. and Zúñiga-Pflücker, J.C. (2004). Maintenance of T cell specification and differentiation requires recurrent notch receptor-ligand interactions. J Exp Med **200**, 469–479.

Schott, E., Bonasio, R. and Ploegh, H.L. (2003). Elimination in vivo of developing T cells by natural killer cells. J Exp Med **198**, 1213–1224.

Shen, H.Q., Lu, M., Ikawa, T., Masuda, K., Ohmura, K., Minato, N., Katsura, Y. and Kawamoto, H. (2003). T/NK bipotent progenitors in the thymus retain the potential to generate dendritic cells. J Immunol **171**, 3401–3406.

Spits, H., Blom, B., Jaleco, A.C., Weijer, K., Verschuren, M.C., van Dongen, J.J., Heemskerk, M.H. and Res, P.C. (1998). Early stages in the development of human T, natural killer and thymic dendritic cells. Immunol Rev **165**, 75–86.

Stevenaert, F., Van Beneden, K., De Colvenaer, V., Franki, A.S., Debacker, V., Boterberg, T., Deforce, D., Pfeffer, K., Plum, J., Elewaut, D. and Leclercq, G. (2005). Ly49 and CD94/NKG2 receptor acquisition by NK cells does not require lymphotoxin-beta receptor expression. Blood **106**, 956–962.

Stewart, C.A., Walzer, T., Robbins, S.H., Malissen, B., Vivier, E. and Prinz, I. (2007). Germ-line and rearranged Tcrd transcription distinguish bona fide NK cells and NK-like gammadelta T cells. Eur J Immunol **37**, 1442–1452.

Van Beneden, K., De Creus, A., Debacker, V., De Boever, J., Plum, J. and Leclercq, G. (1999). Murine fetal natural killer cells are functionally and structurally distinct from adult natural killer cells. J Leukoc Biol **66**, 625–633.

Van Beneden, K., Stevenaert, F., De Creus, A., Debacker, V., De Boever, J., Plum, J. and Leclercq, G. (2001). Expression of Ly49E and CD94/NKG2 on fetal and adult NK cells. J Immunol **166**, 4302–4311.

Van Den Broeck, T., Stevenaert, F., Taveirne, S., Debacker, V., Vangestel, C., Vandekerckhove, B., Taghon, T., Matthys, P., Plum, J., Held, W., Dewerchin, M., Yokoyama, W.M. and Leclercq, G. (2008). Ly49E-dependent inhibition of natural killer cells by urokinase plasminogen activator. Blood **112**, 5046–5051.

Veinotte, L.L., Greenwood, C.P., Mohammadi, N., Parachoniak, C.A. and Takei, F. (2006). Expression of rearranged TCRgamma genes in natural killer cells suggests a minor thymus-dependent pathway of lineage commitment. Blood **107**, 2673–2679.

Veinotte, L.L., Halim, T.Y. and Takei, F. (2008). Unique subset of natural killer cells develops from progenitors in lymph node. Blood **111**, 4201–4208.

Vosshenrich, C.A., Garcia-Ojeda, M.E., Samson-Villeger, S.I., Pasqualetto, V., Enault, L., Richard-Le Goff, O., Corcuff, E., Guy-Grand, D., Rocha, B., Cumano, A., Rogge, L., Ezine, S. and Di Santo, J.P. (2006). A thymic pathway of mouse natural killer cell development characterized by expression of GATA-3 and CD127. Nat Immunol **7**, 1217–1224.

Wilson, A., Petrie, H.T., Scollay, R. and Shortman, K. (1989). The acquisition of CD4 and CD8 during the differentiation of early thymocytes in short-term culture. Int Immunol **1**, 605–612.

Yadi, H., Burke, S., Madeja, Z., Hemberger, M., Moffett, A. and Colucci, F. (2008). Unique receptor repertoire in mouse uterine NK cells. J Immunol **181**, 6140–6147.

Yokoyama, W.M. and Plougastel, B.F. (2003). Immune functions encoded by the natural killer gene complex. Nat Rev Immunol **3**, 304–316.

Yoshida, H., Kawamoto, H., Santee, S.M., Hashi, H., Honda, K., Nishikawa, S., Ware, C.F., Katsura, Y. and Nishikawa, S.I. (2001). Expression of alpha(4)beta(7) integrin defines a distinct pathway of lymphoid progenitors committed to T cells, fetal intestinal lymphotoxin producer, NK, and dendritic cells. J Immunol **167**, 2511–2521.

Yoshida, H., Naito, A., Inoue, J., Satoh, M., Santee-Cooper, S.M., Ware, C.F., Togawa, A., Nishikawa, S. and Nishikawa, S. (2002). Different cytokines induce surface lymphotoxin-alphabeta on IL-7 receptor-alpha cells that differentially engender lymph nodes and Peyer's patches. Immunity **17**, 823–833.

Yoshida, T., Ng, S.Y., Zúñiga-Pflücker, J.C. and Georgopoulos, K. (2006). Early hematopoietic lineage restrictions directed by Ikaros. Nat Immunol **7**, 382–391.

Zenewicz, L.A., Yancopoulos, G.D., Valenzuela, D.M., Murphy, A.J., Stevens, S. and Flavell, R.A. (2008). Innate and adaptive interleukin-22 protects mice from inflammatory bowel disease. Immunity **29**, 947–957.

Chapter Three

3

Interferon-producing killer dendritic cells (IKDC)

Franck Housseau

CHAPTER CONTENTS

The IKDC has undergone more subtle and thorough inspection [than the Platypus], but immunologists will doubtless continue to wield their scissors to check whether all of its attributed functions remain firmly attached to a single cell type. Let's hope no scars will be left at the end of what we anticipate will be a prolonged endeavor.

Shortman and Villadangos, 2006

ABSTRACT

Interferon-producing killer dendritic cell (IKDC) is a recently described antigen-presenting cell (APC) whose affiliation with dendritic cells (DCs) fuelled a high level of controversy. IKDC is a multi-tasking cell that shares phenotypic and functional features of both natural killer (NK) cells and DC. Following activation by Toll-like receptor (TLR) ligands or tumour cells, IKDC developed cytotoxic properties and subsequently mature into a DC-type of cell able to present antigen to naïve T cells. This 'bi-typic' function is associated with innate and adaptive immunologic features, which mark these unique APCs as an attractive direct link between natural and acquired immunity. Since the original studies, new findings have established that so-called IKDCs should be defined as 'activated' NK cells. This conclusion was mainly based on phenotypic ($CD3^-NKp46^+$) and developmental considerations (IL-15- and Id2-dependency), in spite of other reports showing distinct morphology, distinct progenitors, distinct activation pathways, and distinct patterns of cytokine production between IKDC and NK cells. Beyond semantic and ontogenic considerations, this chapter focuses on the unique functions of IKDCs.

KEY WORDS

Antigen presentation, Cytotoxicity, DAMP, Dendritic cells, Granzyme, Imatinib mesylate, Interleukin-12, Interferon, Lymph nodes, Lipopolysacccharide, Murine cytomegalovirus, Natural killer cells, Oligodeoxynucleotide, Plasmacytoid dendritic cells, Perforin, Toll-like receptors, TNF-related apoptosis-inducing ligand

NK antigen-presenting cells

Dendritic cells (DCs) play a central role in the immune response by presenting antigens to naïve T cells and, therefore, triggering the adaptive arm of the immune response (Banchereau and Steinman, 1998). To do so, 'immature' DCs sample antigens at the periphery and re-circulate to the lymph nodes (LN) where they encounter T lymphocytes. Inflammatory stimuli provided either directly by recognition of damage-associated molecular patterns (DAMPs) or through cooperation with innate immune effectors, trigger the maturation of DC into immunogenic antigen-presenting cells (APCs), which express a high level of MHC class II (MHC-II) and co-stimulatory molecules (Banchereau and Steinman, 1998; Bianchi, 2007; Fernandez et al., 1999; Kadowaki et al., 2001). In the absence of inflammation, interactions between immature DCs and naïve T cells mediate peripheral tolerance to self antigens (Steinman et al., 2003). Cellular cooperation with natural killer (NK) cells has been established as a key process for the induction and selection of mature DCs (Moretta, 2002; Walzer et al., 2005). NKs represent a key element of the innate immune response and are responsible for the rapid recognition and killing of stressed cells (infected or tumour), which could threaten the integrity of tissues (Lanier, 2005). NK cells identify their targets by integrating a complex combination of signals provided by activating and inhibitory receptors, which interact with ligands expressed at the surface of stressed or normal cells, respectively (Cerwenka and Lanier, 2001). The decision to kill or not is thus the result of a subtle balance between positive signals (activating) and negative signals (inhibitory) (Lanier, 2005). Moreover, the activity of NK cells is also finely regulated through their cooperation with other immune cells, including DCs (Moretta, 2002; Walzer et al., 2005). The cross-talk evoked earlier between DCs and NKs is indeed bi-directional, and DCs contribute to arming NK cells.

Until recently, tasks were thought to be well distinguished between these two distinct populations, NKs being responsible for the killing of target cells and DCs being responsible for scavenging the resulting 'debris' and processing antigens into peptide epitopes which are recognized in the context of MHC-I or MHC-II by $CD8^+$ T cells or $CD4^+$ T cells, respectively (Walzer et al., 2005). However, there is accumulating evidence that innate cells, including NK cells as well as macrophages, plasmacytoid DCs (PDCs), and neutrophils, are able to fulfil the role of APCs in certain circumstances whose in vivo relevance is still poorly established (Beauvillain et al., 2007; Hanna et al., 2004; Hoeffel et al., 2007; Potter and Harding, 2001; Pozzi et al., 2005; Roncarolo et al., 1991; Zingoni et al., 2004). Human NK cells upregulate MHC-II and co-stimulatory molecules upon activation and are able to stimulate antigen specific $CD4^+$ T cells (Hanna et al., 2004; Roncarolo et al., 1991). Moreover, macrophages, neutrophils and PDCs have recently been shown to cross-present antigen to $CD8^+$ T cells, a function until then restricted to conventional DCs (CDCs) (Beauvillain et al., 2007; Hoeffel et al., 2007; Pozzi et al., 2005). Several groups have also reported the ability of DCs to kill targets, specifically tumour cells (for review see Chauvin et al., 2008 and Chan and Housseau, 2008). The killing properties are mostly described for ex vivo-generated immature DCs and rarely define a specific subset or mechanism. Nevertheless, perforin (*perf*)/granzyme $(grz)^+$ 'myeloid' DCs have been reported at the vicinity of human tumour and *grz*B^+CD2^+ PDCs endowed with cytotoxic properties have been described in vitro (Chan and Housseau, 2008; Stary et al., 2007). These DC populations did not display NK receptors and the mechanism by which they recognized their targets remained poorly understood. In addition to DCs with 'ectopic' killing function, Josien's group described a bona fide rat DC subset $CD103^+CD11b^+CD4^-MHC\text{-}II^+$ endowed with cytotoxic properties towards tumour cells (Chauvin et al., 2008). The killing mechanism remains largely unknown and does not use the classic *perf*/*grz* or TNF-related apoptosis inducing ligand (TRAIL) pathways. DCs phagocytosed cell fragments resulting from the killing of tumour cells. Interestingly, this DC subset was found to express NKR-P2, the rat ortholog of NKG2D, which is an activating NK receptor recognizing a stress-induced ligand (Alli et al., 2004; Srivastava et al., 2007). The cross-linking of NKR-P2 with an agonist monoclonal antibody (mAb) was shown to increase the tumouricidal activity of the $CD103^+$ DCs in vivo (Alli et al., 2004). These findings in the rat model have been extended to the mouse by Khar's group (Srivastava et al., 2007). Killer dendritic cells (KDCs) are clearly distinct from NK cells since they mostly lack the expression of NK receptors, especially NKp46 (encoded by Ncr-1), and they generally do not engage in exocytosis granule-mediated cytotoxicity. These results are in stark contrast to the recently discovered interferon-producing KDCs (IKDCs; Table 3.1), which unlike DCs and KDCs, express NK markers and secrete IFN-γ, and unlike NK cells express MHC-II and produce IL-12, suggesting that they may represent a separate subset that deserves a name of its own (Chan et al., 2006; Taieb et al., 2006).

Interferon-producing killer dendritic cells

Whereas DCs can exhibit killing properties ex vivo, we and others have identified IKDCs, a novel type of

Table 3.1 Mouse IKDC and rat KDC

	IKDC	KDC
Species	Mouse	Rat
Phenotype	$CD49b^{+}CD11c^{int}B220^{high}$**NKp46**$^{+}$ MHC-II$^{+/-}$ in spleen MHC-II^{+} in LN	$CD4^{-}CD11b^{+}CD103^{+}MHCII^{+}$ **NKp46**$^{-}$
Tissues	Spleen, LN	Spleen
Growth factors	IL-2, IL-15	Flt3 sensitive
NK receptor	NKG2D, KLRG1, Ly49, NCR (NKp46)	NKG2D
Targets	YAC-1, tumours, infected fibroblasts	YAC-1, tumour cells
Killing	TRAIL, Perforin	Ca^{2+}dependent, caspase-independent
Presentation	Naïve CD4+ and CD8+ T cells	?
Cytokines	IFN-I and II, IL-12p40	IL-12
Antigen uptake	Pinocytosis	Phagocytosis
Co-stimulation	CD80/CD86, CD40, PD-L1 when activated	CD80/CD86
In vivo function	Tumour regression	Tumour regression

innate cells endowed with natural killing activity and antigen-presenting properties (Chan et al., 2006; Taieb et al., 2006). Originally, attention was focused on $CD11c^{+}NK1.1^{+}$ bi-typic NK/DC cells, which were shown to exert CD40L-dependent tolerogenic activity in a model of LCMV-triggered autoimmune diabetes (Homann et al., 2002). DeMatteo's group further characterized $CD11c^{+}NK1.1^{+}$ NKDCs isolated from the spleen of C57BL/6 mice phenotypically and functionally and demonstrated the ability of this subset to kill and produce IFN-γ in the presence of tumour cells, as well as to present antigen to $CD4^{+}$ T cells (Chaudhry et al., 2006a,b; Pillarisetty et al., 2005). However, the marked heterogeneity of this population raised questions about functional duality at the single cell level, and the ability of NKDCs to cross-present antigen derived from the killed target.

Analysis of the morphology, phenotype, and function of IKDCs lead to the conclusion that this NK-like subset, included in the NKDCs (but largely dominated by NK cells), was probably responsible for the antigen-presenting function of $CD11c^{+}NK1.1^{+}$ cells, described by DeMatteo's group (Chan et al., 2006; Chaudhry et al., 2007). IKDCs fuelled a lot of interest, as the first example of a well-defined DC-like subset able to produce IFN-γ and to kill target cells. The lineage classification remains uncertain, with contradictory results, and deserves further attention (Blasius et al., 2007; Caminschi et al., 2007; Vosshenrich et al., 2007; Welner et al., 2007).

Isolation and molecular characterization

IKDCs were defined phenotypically as $CD3^{-}CD19^{-}$ $CD11c^{dim}B220^{+}NK1.1^{+}/CD49b^{+}$ when isolated from 'naïve' mice (Chan et al., 2006; Taieb et al., 2006). They were isolated originally from spleens of BALB/c and C57BL/6 mice, but were subsequently detected in every murine strain tested (Chan et al., 2006). We focused our attention on this subset while isolating PDCs as $CD11c^{dim}B220^{+}GR1^{+}$ cells and noticing that a considerable number of the $CD11c^{dim}B220^{+}$ cells from BALB/c spleen included $GR1^{-}$ cells (Figure 3.1). The mRNA microarray analysis highlighted the overexpression of a variety of NK markers by $GR1^{-}$ compared to $GR1^{+}$ cells. IKDCs were subsequently sorted as $CD11c^{dim}B220^{+}CD49b^{+}$ cells from the $CD11c^{+}$-enriched cell fraction of BALB/c spleen. They represent 1–5% of the $CD11c^{+}$ splenocytes depending on the strain and age of the mice. Importantly, these $CD11c^{dim}B220^{+}CD49b^{+}$ cells isolated from BALB/c mice express low but detectable level of MHC-II, which distinguishes them from NK cells (Figure 3.1). Lately, this point was challenged by several groups that did not find a correlation between the expression of CD11c/B220 and MHC-II. It was concluded that MHC-II expression, until then in mouse NK cells (Spits and Lanier, 2007), could be recognized as a hallmark of murine NK activation (Blasius et al., 2007; Caminschi et al., 2007; Vosshenrich et al., 2007). Although this point was plausible, transmission electron microscopy

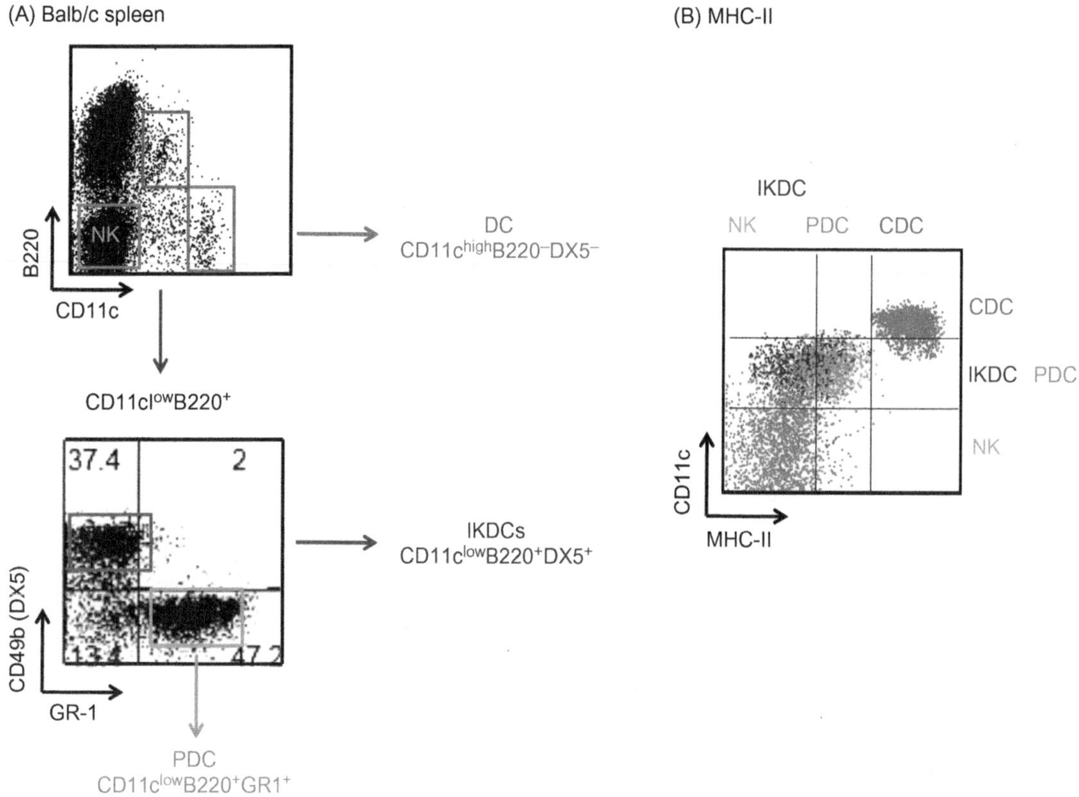

Figure 3.1 • Differential expression of CD11c, B220, CD49b and MHC-II molecules between IKDC, NK, CDC and PDC cell-sorted from BALB/c mouse spleens. (A) BALB/c splenocytes are labelled with CD11c, B220, DX5 (CD49b) and GR1 (Ly6C-Ly6G) mAb. IKDC ($CD11c^{low}B220^{+}DX5^{+}GR1^{-}$), PDC ($CD11c^{low}B220^{+}DX5^{-}GR1^{+}$), CDC ($CD11c^{low}B220^{+}DX5^{+}GR1^{-}$) and NK ($CD11c^{-}B220^{-}DX5^{+}GR1^{-}$) were cell-sorted and assessed for MHC-II molecules expression. (B) An overlay of the CD11c/MHC-II histograms for each sorted population. Very dark grey, IKDC; grey, NK; dark grey, CDC; light grey, PDC.

formally confirmed that freshly isolated IKDCs ($CD11c^{dim}B220^{+}CD49b^{+}$) have a strikingly distinct morphology from PDCs ($CD11c^{dim}B220^{+}CD49b^{-}$), CDCs ($CD11c^{hi}B220^{neg}CD49b^{-}$) and NKs ($CD11c^{-}B220^{-}CD49b^{+}$), even following activation (Figure 3.2) (Chan et al., 2006; Taieb et al., 2006). Gene expression profiling established that several genes involved in the individual steps of MHC-II processing machinery (enzymes, invariant chain, MHC alleles, co-stimulatory molecules, endocytosis) were found in IKDCs but not NK cells (Table 3.2), suggesting that instead of an ectopic expression of MHC-II, IKDCs, but not NKs, possess a fully competent MHC-II antigen presentation pathway.

FACS analysis of IKDCs show that, in addition to CD49b and NK1.1, IKDCs express a broad array of NK receptors, including Ly49 family members, NKG2D (CD314; encoded by killer cell lectin-like receptor subfamily K, member 1 or Klrk1), NKG2A/C/E, 2B4 (CD244), mast cell function-associated antigen (MAFA) (encoded by killer cell lectin-like receptor subfamily G, member 1 or Klrg1), and natural cytotoxicity triggering receptors (NCR) such as NKp46 (encoded by natural cytotoxicity triggering receptor 1 or Ncr1). It is also important to point out that IKDCs, like NKs but not DCs, express IL2-Rβ (CD122) and IL2-Rγ_c (CD132), two important components of the receptors for IL-2 and IL-15, which play a critical role in NK development and survival (Huntington et al., 2007).

IKDCs have been isolated from various lymphoid structures, including thymus, gut-associated lymphoid tissue (GALT), and a variety of other tissues, such as liver, skin, and lung (Chan et al., 2006). In LN, IKDCs are characterized by higher levels of MHC-II and co-stimulatory molecules associated with lower NKG2D expression than those observed in spleen. IKDCs express the chemokine receptor, CCR7 (Chan et al., 2006), which is a critical homing receptor, allowing DC to circulate from the blood to LN, where they encounter T cells (Randolph et al., 2005). It thus became tempting to postulate that LN IKDCs might represent a mature form of the IKDCs found at the periphery, that are able to present antigen to T cells and trigger the adaptive immune response when conditions allow it. Zitvogel's group showed that the trans-presentation of IL-15 to IKDCs induced the expression of CCR2, which

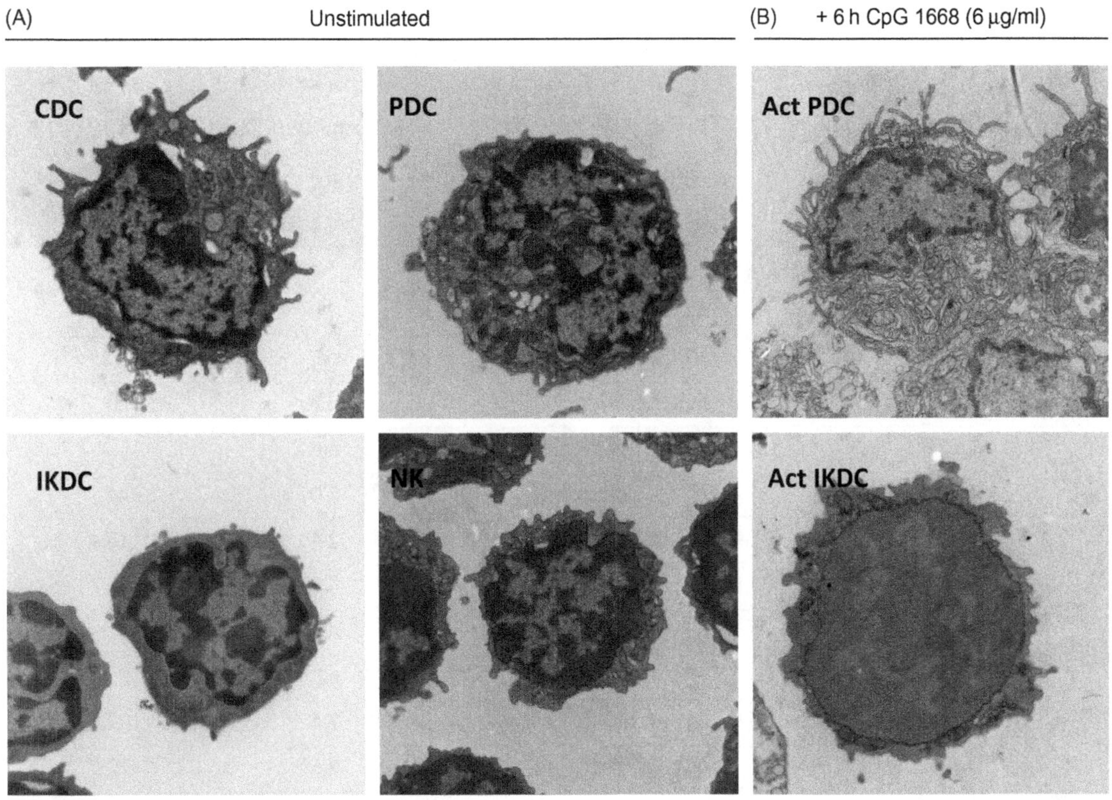

Figure 3.2 • Transmission electron microscopy of IKDC, NK, CDC and PDC. (A) Representative of ultrastructure details of freshly sorted splenic CDC (upper left), PDC (upper right), IKDC (lower left) and NK (lower right) using transmission electron microscopy. (B) Activated (Act) PDC (top panel) and Act IKDC (bottom panel) were obtained following 6-h culture in presence of CpG ODN 1668 prior to transmission electron microscopy.

is critical for their CCL2-dependent recruitment into the tumour bed (Mignot et al., 2008; Ullrich et al., 2008a).

IKDCs are 'multi-talented' cells endowed with NK and DC functions

Cytotoxicity: IKDC is a natural killing cell

Conforming to their phenotype described above, and highlighting the presence of activating and inhibitory NK receptors, IKDCs are endowed with potent cytotoxic properties towards typical NK targets, including tumour cells, infected cells and cell lines such as YAC-1, P815 or RMA-S (Chan et al., 2006 and unpublished data). In BALB/c mice, naïve IKDCs are poorly cytotoxic and require activation via CpG oligodeoxynucleotide (ODN) 1668, a Toll-like receptor (TLR)-9 ligand, or IL-2/IL-15 + IL-12 to kill targets (Chan et al., 2006). Activation of naïve BALB/c NK cells with CpG ODN 1668 did not induce their natural killing function. The killing mechanism is entirely Ca^{2+} and *perf/Gzm*-dependent (our unpublished results). The cytotoxicity of IKDC is dependant on the interaction of NK receptors with ligands expressed by stressed targets (tumour or infected) since the killing of YAC-1 is entirely blocked by anti-NKG2D and the killing of m157 expressing BaF3 (surrogate of murine cytomegalovirus [MCMV]-infected cells) is blocked by anti-Ly49H (Chan et al., 2006). Interestingly, Zitvogel's group has shown that C57BL/6 IKDCs kill B16F10 tumour cells in a TRAIL-dependent manner, in vivo and in vitro (Taieb et al., 2006). It is not clear yet whether the discrepancy between the findings of the two groups is a result of differences in the strains of mice used, or the nature of the targets. Indeed, YAC-1 used to test the cytotoxicity of BALB/c IKDCs is TRAIL-resistant (Kayagaki et al., 1999). Notably, we also found that in MCMV-infected mice, LN IKDCs upregulated TRAIL as compared to splenic IKDCs and NKs or LN NKs (unpublished results). LN IKDCs did not show cytotoxicity towards YAC-1 in a classic 4-h chromium release assay, however killing was effective after 8 h, which suggested that LN IKDC preferring TRAIL over *perf/gzm*. Pleiotropy of killing mechanisms could prevent targets escaping IKDC cytotoxicity by modifying their death pathway sensitivity (Chan and Housseau, 2008).

Table 3.2 Transcriptome analysis of antigen presentation—comparison of IKDC and NK cell mRNA microarrays performed on Balb/c mice

Gene symbol	Common name	Function	Fold change[1] IKDC vs NK in SPL	Fold change[1] IKDC vs NK in LN
H2-Ea	MHC-II	Ag presentation	3.3	28.5
H2-Eb1	MHC-II	Ag presentation	3.5	6.5
H2-Aa	MHC-II	Ag presentation	3.3	28.5
H2-Ab1	MHC-II	Ag presentation	2.6	8.2
C2ta	MHC-II transactivator	MHC II expression	nd	2.5
CD83	CD83	MHC II expression	2.7	nd
Ii	CD74	Ag and Ii processing	nd	6.7
Igm	Legumain	Ag and Ii processing	2.0	nd
Ctsl	Cathepsin L	Ag and Ii processing	2.4	13.4
Ctsh	Cathepsin H	Ag processing	4.1	19.7
Ctss	Cathepsin S	Ag processing	nd	2.2
CtsZ	Cathepsin Z	Ag processing .	nd	2.4
Ifi30	GILT	Ag processing	2.4	6.5
Cybb	Nox2	Ag processing (Lysosome)	2.5[2]	15.6
Cst3	Cystatin C	Ag processing (Protease inhibitor)	nd	2.5
H2-DMa	H2-DMa	Chaperone	2.5	3.6
H2-DMb1	H2-DMb1	Chaperone	nd	5.8
H2-Oa	H2-Oa	Chaperone	3.3	nd
H2-Ob	H2-Ob	Chaperone	nd	2.9
CD86	CD86	Co-stimulation	2.8	1.5
Mrc1	Mannose R (CD206)	Scavenger R. Endocytosis	6.1	nd
Marco	Marco	Scavenger R.	nd	2.4
Chl1		Clathrin-mediated endocytosis	5.8	12.7
EhD1		Clathrin-mediated endocytosis	2.5	nd
SnX9		Clathrin-mediated endocytosis	2.1	2.3
Scamp1		Clathrin-mediated endocytosis	2.1	5.7
Snap25		Clathrin-mediated endocytosis	nd	64

[1]Microarrays were performed from cell-sorted IKDC ($CD11c^{dim}B220^{+}CD49b^{+}$) and NK ($CD11c^{-}B220^{-}CD49b^{+}$) in spleen and LN of BALB/c mice. The table shows the ratio of the signal IKDC/NK.

[2]When the signal is stronger in NK compared to IKDC.

Cytokine secretion: IKDCs simultaneously produce IFNs and IL-12

IKDCs are unique in their ability to produce IFN-γ, IFN-α/β and IL-12 family members, including p40 (common subunit of IL-12p70 and IL-23) and EBi3 (common subunit of IL-27 and IL-35) (Chan et al., 2006 and unpublished observations). IKDCs produce high amounts of IFN-γ and upregulate MHC-II when stimulated with tumour cells, CpG ODN 1668, or MCMV-infected fibroblasts. As indicated above, we showed a clear dichotomy between IKDCs and NK cells regarding their response to TLR ligands. Indeed, upon stimulation with the TLR-9 ligand

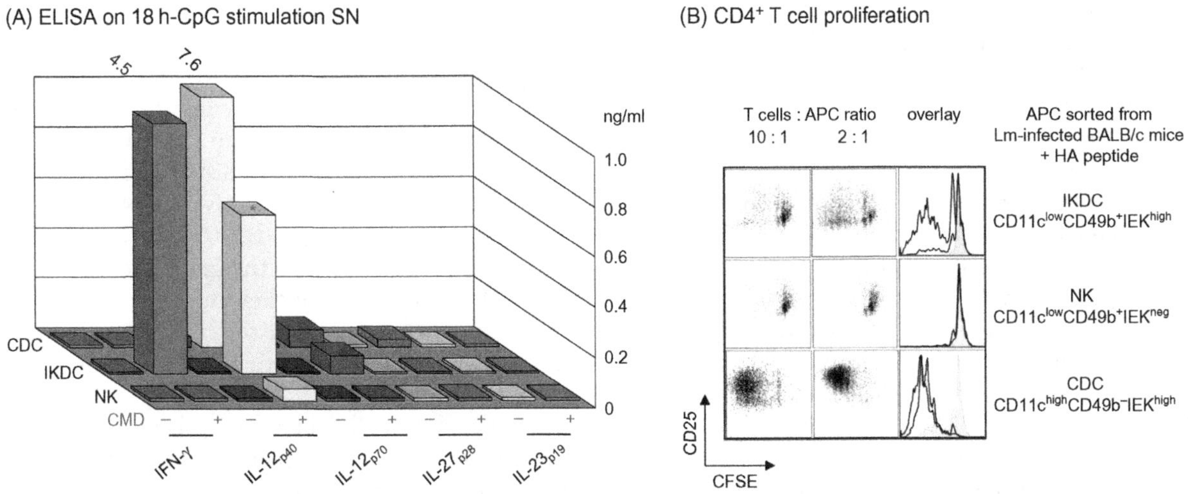

Figure 3.3 • Functional properties of activated IKDCs. (A) Cytokine secretion by cell-sorted IKDC, NK and CDC upon stimulation with CpG ODN 1668. Cytokine levels were measured in overnight culture supernatants using ELISA kits. (B) Cell-sorted activated IKDC, NK and CDC were pulsed in vitro with HA peptide and used as APC to stimulate CFSE-labelled HA-specific TCR transgenic 6.5 $CD4^+$ T cells. Proliferative response of activated T cells was measured by the dilution of CFSE using flow cytometry.

CpG ODN 1668, IFN-γ, IL-12p40, and IFN-α/β gene transcripts are detected only in IKDCs, whereas lipopolysaccharide (LPS), a TLR-4 ligand, induces the expression of IFN-γ and EBI3 in NK cells, but not IKDCs (unpublished results). Moreover, when measuring cytokines in the TLR-9 stimulation culture supernatant, we detected IFN-γ and IL-12p40/p70 simultaneously only for IKDCs, whereas DCs produced large amounts of IL-12p40 and some IL-12p70, but no IFN-γ, and NKs did not secrete any of these cytokines (Chan et al., 2006 and unpublished results) (Figure 3.3). Alternatively, upon stimulation with IL-2/IL-15 + IL-12, IKDCs and NK cells produced robust quantities of IFN-γ. However, under these conditions, IKDCs lose their MHC-II molecules to acquire a typical NK phenotype (Ullrich et al., 2008a). Altogether, these findings suggest that, although phenotypically close, IKDCs and NK cells represent two functionally distinct subsets. These results also illustrate the functional plasticity of IKDCs, which, according to the nature of the stimulation and microenvironment stimulation, are able to adapt their response and differentiate into functionally distinct effectors.

Antigen-presenting function of IKDC

Although recently challenged by other groups (Blasius et al., 2007; Caminschi et al., 2007), in our hands it is clear that activated IKDCs are fully competent APCs, able to engage cognate interactions with T lymphocytes and induce their activation (Chan et al., 2006). Indeed, when isolated from spleens of LM-infected BALB/c mice, activated IKDCs ($MHC\text{-}II^{hi}CD11c^+B220^+CD49b^+$) and CDCs ($MHC\text{-}II^{hi}CD11c^{hi}B220^+CD49b^-$), but not activated NK cells ($MHC\text{-}II^-CD11c^+B220^-CD49b^+$), were able to induce the proliferation of TCR-transgenic haemagglutinin (HA)-specific 6.5 $CD4^+$ T lymphocytes in the presence of the peptidic HA epitope (Figure 3.3). This result was recently reproduced in a model of C57BL/6 mice infected with MCMV (unpublished results). The discrepancy between our observations and others' (Blasius et al., 2007; Caminschi et al., 2007; Vosshenrich et al., 2007) was mainly explained by the sorting of the cells, since in our experiments, IKDCs were sorted as $MHC\text{-}II^{hi}$ cells, whereas the other groups used bulk $CD11c^+B220^+CD49b^+$ largely contaminated with NK cells when obtained from infected mice (NK cells upregulate CD11c and B220 following stimulation).

To demonstrate definitively the APC function of the LN IKDCs, we showed in a direct ex vivo antigen detection (DEAD) assay that LN $CD11c^{dim}B220^+CD49b^+MHC\text{-}II^{hi}$ but not splenic $CD11c^{dim}B220^+CD49b^+MHC\text{-}II^{lo}$ IKDCs sorted from ovalbumin (OVA)-expressing LM-infected mice induced the specific proliferation of OVA-specific $CD4^+$ T cells (Chan et al., 2006). When B16F10 melanoma-bearing mice were treated with STI571 (Imatinib mesylate, IM; inhibitor of c-kit tyrosine kinase) and IL-2, IKDCs represented three-fourths of the tumour-infiltrating $CD11c^+$ cells. Using blocking antibodies, Zitvogel's group showed that the tumouricidal activity of IM + IL-2 was TRAIL- but not TNF-dependent (Taieb et al., 2006). Adoptive cell transfer experiments performed in B16F10 melanoma-bearing $Rag^{-/-}IL\text{-}2R\gamma^{-/-}$ C57BL/6 mice showed that $B220^+NK1.1^+CD49b^+$ IKDCs but not $NK1.1^+CD49b^+$ NK cells induced regression of the tumour following IM + IL2 treatment (Taieb et al., 2006). Whether IKDCs induce immunogenic cancer cell death in vivo or directly exert an antigen-presenting function in this system is not established yet. The capacity of IKDC to kill tumour cells or infected cells and to cross-present antigens derived from the killed targets also remains to be further investigated. The expression of MHC-II and the ability to present antigens to $CD4^+$ T cells should not be considered as a curiosity since it is reminiscent of human NK (Hanna et al., 2004). In mouse, our results suggest that only IKDCs define a restricted subset of 'NK-like cells' with the ability to express MHC-II.

Importantly, expression of IFN-I and II were found to play a key role in this maturation process when Zitvogel's group showed that upregulation of TRAIL expression and MHC-II was dependant on the IFNγ-R signalling (personal communication). The migration and/or maturation of IKDCs is also dependant on the IFN-I-R signalling since we observed a profound defect in the recruitment of activated IKDC (MHC-$II^{hi}CD40^+$) in the LN when infecting IFN-I-R KO mice with MCMV (unpublished observation).

Lineage issues: IKDCs, NK cells with APC function, or DCs with NK function?

While IKDCs were originally assigned to the DC family because of their ability to (1) express MHC-II and co-stimulatory molecules, (2) produce IL-12 and (3) present antigen to $CD4^+$ T cells, several recent findings point to a closer relationship between IKDCs and NK cells.

Unlike DCs, IKDCs express IL-2Rβ (CD122) and IL-2Rγ_c (CD132), two critical components of the IL-2/IL-15R, which play an important role in the development of NK cells (Di Santo, 2006). In vitro, IL-2 and IL-15 improved IKDC survival and induced their proliferation and, in vivo, IKDCs were not detected in IL-2R$\beta^{-/-}$ mice and were found in IL-2R$\gamma_c^{-/-}$ mice in only minimal numbers (Chan et al., 2006). Similarly to NK cells, IKDC differentiation is dependant on inhibitor of DNA binding (Id) 2 and they express GATA binding protein 3 (GATA3) but not the transcription factor PU.1 and the Ets factor Spi-B (Caminschi et al., 2007; Spits and Lanier, 2007; Welner et al., 2007), both of the latter factors being indispensable for the development of DCs and PDCs, respectively (Nutt et al., 2005; Schotte et al., 2003) IKDCs express an extensive arsenal of NK receptors, inhibitory or activating (reviewed above). Since IKDCs comply with the recently established definition of NK cells as a $CD3^-NKp46^+$ subset, these cells were ultimately assigned to the NK lineage (Coulombel and Bensussan, 2008; Robbins et al., 2008; Walzer et al., 2007). Finally, genome-wide expression profiling performed on microarray available for a variety of DC subsets or NK cells in different species identified a transcriptional signature that is evolutionarily conserved and specific for each subset (Robbins et al., 2008). Studies of these genetic signatures in the transcriptional profile of IKDC allowed insights to be gained into their origin/lineage, and determined that IKDCs were more related to NKs than DCs. However, several facts can oppose the 'reductionist' paradigm that IKDCs represent a 'new suit' for activated NK cells because they express NK markers, B220, and IL-15R. After all, do NK and NKT belong to the same lineage? Can T cells function as NK because they upregulate NK receptors in certain circumstances? Are $CD4^+$ T cells and $CD8^+$ T cells identical because they are generated from the same progenitors?

Firstly, NK cells, like DCs, lack universal markers that are specific for their lineage and evolutionarily conserved, such as the T cell receptor (TCR) for the T cells/NKT or the B cell receptor (BCR) for B cells. NK-associated molecules have been reported on a variety of subsets, which are not classically affiliated to the NK lineage (Spits and Lanier, 2007). Examples abound: $CD8^+$ cytotoxic T cells (CTL) capable of upregulating NK markers such as NKG2D (Groh et al., 2001), tonsil PDCs expressing NKp44 (Fuchs et al., 2005), 'myeloid' DCs expressing NKG2D or CD49b (Caminschi et al., 2007; Srivastava et al., 2007) or $Lin^-ROR\gamma^+CD127^+NKP46^+$ cells recently identified as IL-22-producing cells in the gut (Sanos et al., 2009). There is, therefore, room for cells that phenotypically appear NK-like, but are generated from distinct progenitors. Kincade's group recently proposed that IKDCs may arise from a unique differentiation pathway distinct from those responsible for NKs, PDCs, or B/T lymphocytes (Welner et al., 2007). The authors identified c-$Kit^{hi}CD62L^+$ lymphoid progenitors of IKDCs. Secondly, microarrays used to identify NK or DC genetic signatures were generated from immature IKDCs, which display phenotypic and functional characteristics of NKs (Robbins et al., 2008). Therefore, it was not surprising that immature IKDCs are found to be closely related to the NK lineage. Activated IKDCs, which resemble DCs, should have been incorporated in this analysis. We established that, following activation, IKDCs switched on a genomic program similar to DCs, with expression of IL-12, MHC-II and co-stimulatory molecules, a program that is not usually described for NK cell bona fide.

IKDC, a unique link between innate and adaptive immunity

The uniqueness of IKDCs is exemplified by their functional duality. Following activation with TLR9 ligand or virus, IKDCs switch from an NK-type of cell to DC-related APCs. BALB/c IKDCs transiently acquire cytotoxic activity towards classical NK targets with upregulation of NKG2D and TRAIL, and development of granule Ca^{2+}-dependent exocytosis-mediating killing (Chan et al., 2006). They subsequently lose their cytotoxic potential and downregulate NKG2D, while upregulating MHC-II and co-stimulatory molecules, the hallmark of APCs (Figure 3.4). These activated IKDCs resemble those described in LN, suggesting that, like DCs, upon activation IKDCs mature and migrate to LN, where they can encounter $CD4^+$ T cells. We found that IKDCs expressed the chemokine receptor CCR7 (Chan et al., 2006), which is required for maturing DCs to migrate into LN (Randolph et al., 2005). Transfer of CFSE-labelled immature IKDCs ($CD11c^{dim}B220^+CD49b^+MHC\text{-}II^{dim}$,-sorted from the spleen) into the spleens of recipient mice, followed by LM infection, induced the recruitment into the LN of $CFSE^+CD11c^{dim}B220^+CD49b^+MHC\text{-}II^{hi}$ activated IKDCs (Chan et al., 2006). LN IKDCs are immunostimulatory APCs in vitro and are thus able to trigger adaptive immune responses (Chan et al., 2006).

The stimulation of tumour-specific T cells takes place in the LN, implying that recognition of tumour-associated antigens involves professional APC endowed with the capacity to shuttle tumour-derived material to the T cell area of lymphoid tissues (Huang et al., 1994). This phenomenon is dominated largely by DCs and engages a multi-step program with complex cellular cooperations: (1) lysis of tumour cells by cytotoxic effectors, (2) processing of the antigen from the dying tumour and (3) cross-presentation of the antigen to the T cells in the LN. The response results from the integration of a variety of signals delivered by the APCs to the T cells during the MHC/peptide/TCR interaction (Banchereau and Steinman, 1998). Only fully mature APCs have the ability to induce the activation of T cells. Cellular cooperation between DCs and NKs in the injured tissues, as well as in the LN, is thought to play a critical role in the maturation of DCs (Walzer et al., 2005) NK and DC start their interactions in the inflamed tissues where they established close contact at the early stage of the immune response (Moretta, 2002). Following stimulation via pattern recognition receptor (PRR), DC engage a maturation program, which influence NK cells function via, especially, the secretion of IL-12 and IFN-α/β, enhancing IFN-γ secretion and cytotoxicity, respectively (Walzer et al., 2005). Survival and proliferation of NK are also dependant on trans-presentation of membrane-bound IL-15 by DCs (Brilot et al., 2007; Huntington et al., 2009). Reciprocally, activated NK cells produced IFN-γ, TNF-α and GM-CSF, which are important mediators for the maturation of DCs (Moretta, 2002; Walzer et al., 2005). Moreover, cytotoxic activity of NK cells towards malignant or infected cells generates necrotic materials that trigger maturation of DC (Rock et al., 2005). NK cells are also unique in their ability to kill immature DC (iDCs) via the interaction of their

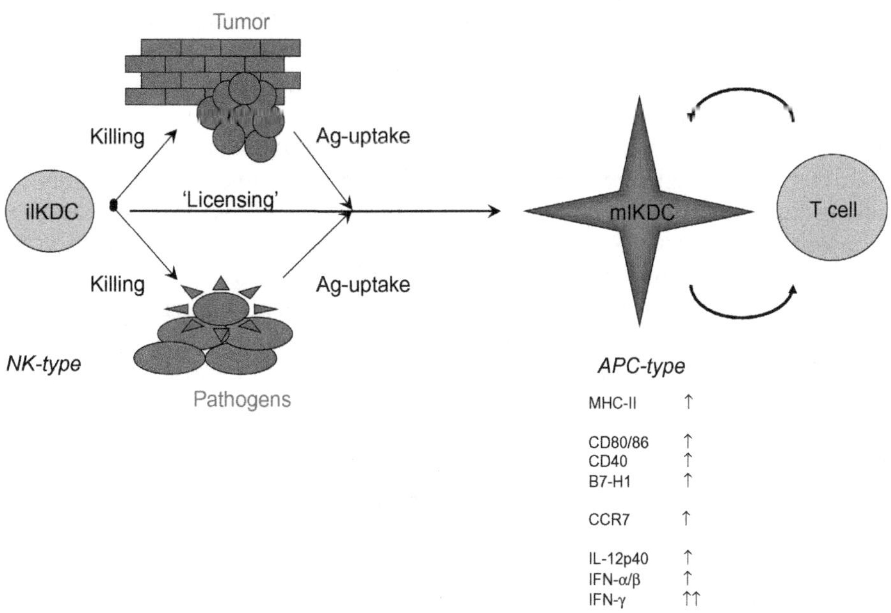

Figure 3.4 • Maturation of IKDC into APC-type of cells after target killing. Schematic representation of the maturation of IKDC into an APC endowed with immunogenic properties towards T cells. Immature IKDCs (iIKDCs) patrol at the periphery and, upon recognition of stressed cells (tumour or infected cells), become activated and kill the target cells. They uptake the antigen derived from the killed targets and migrate to the LN while differentiated into mature IKDCs (mIKDCs), which represent a DC-type of APC with upregulation of MHC-II, co-stimulatory molecules and production of inflammatory cytokines ('licensing' to present antigens).

NKp30 receptor with iDCs (Vitale et al., 2005). This process is supposed to lead to the selection of immunogenic mature DCs (mDCs) and the elimination of tolerogenic iDCs (Moretta, 2002). However, the role of this NKp30-mediated cytotoxicity towards DC is not fully elucidated. In human, only a subset of NK cells lacking the expression of MHC-I-specific NK receptor or KIR but expressing NKG2A/CD94 seems endowed with this DC selection property (Della Chiesa et al., 2003). Subsets of NK cells are also found in LN (Martin-Fontecha et al., 2004; Moretta, 2002; Sun et al., 2009). In lymphoid tissues, at later stages of the immune response, takes place a finely regulated cooperation between NK, DC and T cells, which is critical for the outcome of the adaptive immune response. In mice, NK were shown to be recruited into LN in a CCR7-independent and CXCR3-dependent manner to provide an early source of IFN-γ (Mailliard et al., 2005; Martin-Fontecha et al., 2004). After activation by DCs, T cells release IL-2, which is necessary for the activation and survival of NK cells. In turn, activated NK cells provide IFN-γ, which by acting on DC and T cells, may help to engage T cell differentiation towards a Th1 program (Martin-Fontecha et al., 2004). In human, non-cytotoxic IFN-γ-producing CD56brightCD16neg NK cells has been detected in LN. Interestingly, this subset expresses the LN homing markers CD62L and CCR7 (Fehniger et al., 2003). The perspective of a role for IKDCs in the presentation of antigen to T cells in LN unveiled an alternative, two-step process of immunosurveillance, where the 'killer APC' (1) uptakes antigens from its killed target and (2) presents the antigens directly to T cells (Figure 3.4). Indeed, Zitvogel's group described in vivo the tumouricidal effect of IKDCs and the infiltration of tumour by a large population of IKDCs (Taieb et al., 2006). However, in order to be immunogenic, these APCs have to be 'licensed' to undergo a program of maturation and acquire immunostimulatory properties (MHC-II and co-stimulatory molecules for instance). Interferon types I and II produced by IKDCs following the recognition of their target exerts critical feedback on the maturation and migration of IKDCs. However, this remains to be established. Their cooperation with DCs also deserves more attention since their early recruitment at inflammatory or tumour sites, their cytotoxic activity, and their ability to secrete IFNs may create an environment beneficial for the stimulation and maturation of local DCs. Alternatively, IKDCs are potentially endowed with killing and Ag capture properties in order to shuttle and transfer antigenic material to resident LN $CD8^+$ DCs, which are subsequently responsible for the $CD8^+$ T cell cross-priming (Allan et al., 2006). Whereas KDCs have the ability to use a broad array of mechanisms to take up antigens, including fluid-phase or receptor-mediated uptake of antigens freed from cellular debris or phagocytosis of apoptotic cells, it is still unclear how IKDCs capture antigen from their killed targets.

Translational implications and concluding remarks

DCs are promising vectors for the design of effective anti-tumour immunotherapies (Melief, 2008). They are potent adjuvants because of their immunostimulatory effects on T cells and their coordinated cellular cooperation with all the cellular elements of the immune response. However, tumours generate an immunosuppressive environment with over-expression of the signal transducer and activator of transcription (Stat)-3, leading to production of cytokines such as IL-10 or TGF-β (Kortylewski and Yu, 2008), recruitment of myeloid-derived suppressor cells (MDSC) and tumour-associated macrophages (TAM) (Gabrilovich, 2004), or expression of B7-H1, among other mechanisms (Azuma et al., 2008). This generates inadequately stimulated DCs, which suppress effector responses (and/or recruit regulatory T cells (Treg). Therefore, the new trend in cancer immunotherapy is the combination of DC vaccine or T cell therapy with a chemotherapy causing immunogenic tumour death, with DC activation, enhanced antigen cross-presentation or reduction of the immunosuppressive process (Melief, 2008; Zitvogel et al., 2008). In clinic, the use of DC-based therapy remains cumbersome, with technical, logistic and financial burdens regarding the ex vivo generation of patient's DCs as well as the choice of the antigens, which can vary from one patient to another, the number of HLA-binding peptides available, or ultimately the availability of the tumour tissue (Figdor et al., 2004; Melief, 2008; Steinman and Banchereau, 2007). Even though some encouraging results have been obtained, overall, clinical responses are rarely durable and/or complete (Itoh et al., 2009). Efforts have therefore been focused lately on the improvement of technologies to target antigenic material to 'endogenous' DC in vivo (Reddy et al., 2006; Swartz et al., 2008). The recent identification of subsets such as KDC and IKDC has brought an additional level of sophistication to the engineering of anti-tumour immunotherapies (Ullrich et al., 2008b). There are encouraging experimental findings suggesting that it is reasonable to conceive treatments targeting in vivo these 'tumour-killing DC-like cells' which can traffic into the tumour, induce 'immunogenic death' of the abnormal cells, and load their own antigenic material in vivo. Whereas both IL-15-expanded IKDCs (IL-15-IKDCs) and NKs (IL-15-NKs) trigger the translocation of calreticulin to the surface of B16 melanoma cells in vitro, only IL-15-IKDCs promote an immunogenic death of B16, which can mediate

T cell-mediated protective effect in vivo (Ullrich et al., 2008a). Moreover, in B16F10 melanoma-bearing mice IM (Gleevec®) in association with IL-2 induced the recruitment of activated IKDCs into the tumour and increased the survival of the mice in a TRAIL-dependent manner (Taieb et al., 2006). Of note, in humans, imantinib has shown efficacy for the treatment of gastrointestinal stromal tumours (GIST) in part, via the enhancement of the DC-mediated NK priming (Borg et al., 2004). It will thus be very interesting to translate the combination imantinib/IL-2 to human tumour treatment and to look among the immune effectors for a putative human IKDC. However, because most of the markers used to characterize DCs and NKs are not transferable from mice to human (i.e. CD11c, CD49b or NK1.1, B220), and because of the absence of specific markers for IKDCs, the identification of a human counterpart remains so far unsuccessful.

Recruitment of perforin-producing DCs at the vicinity of human skin carcinoma following TLR-7/8 stimulation (Stary et al., 2007), reports of antigen presentation by 'NK-like cells' (Chan et al., 2006), and identification of distinct progenitors between IKDCs and NKs (Welner et al., 2007), are evidence that subsets with phenotypes and functions overlapping those that are well-established for DCs and NKs, may not represent just an epiphenomenon related to ex vivo/in vitro experimentation, and potentially attractive bio-reagents for vaccination purpose.

References

Allan, R.S., Waithman, J., Bedoui, S., Jones, C.M., Villadangos, J.A., Zhan, Y., Lew, A.M., Shortman, K., Heath, W. R. and Carbone, F.R. (2006). Migratory dendritic cells transfer antigen to a lymph node-resident dendritic cell population for efficient CTL priming. Immunity **25**, 153–162.

Alli, R., Savithri, B., Das, S., Varalakshmi, C., Rangaraj, N. and Khar, A. (2004). Involvement of NKR-P2/NKG2D in DC-mediated killing of tumor targets: indicative of a common, innate, target-recognition paradigm? Eur J Immunol **34**, 1119–1126.

Azuma, T., Yao, S., Zhu, G., Flies, A.S., Flies, S.J. and Chen, L. (2008). B7-H1 is a ubiquitous antiapoptotic receptor on cancer cells. Blood **111**, 3635–3643.

Banchereau, J. and Steinman, R.M. (1998). Dendritic cells and the control of immunity. Nature **392**, 245–252.

Beauvillain, C., Delneste, Y., Scotet, M., Peres, A., Gascan, H., Guermonprez, P., Barnaba, V. and Jeannin, P. (2007). Neutrophils efficiently cross-prime naive T cells in vivo. Blood **110**, 2965–2973.

Bianchi, M.E. (2007). DAMPs, PAMPs and alarmins: all we need to know about danger. J Leukoc Biol **81**, 1–5.

Blasius, A.L., Barchet, W., Cella, M. and Colonna, M. (2007). Development and function of murine B220+CD11c+NK1.1+ cells identify them as a subset of NK cells. J Exp Med **204**, 2561–2568.

Borg, C., Terme, M., Taieb, J., Menard, C., Flament, C., Robert, C., Maruyama, K., Wakasugi, H., Angevin, E., Thielemans, K., et al. (2004). Novel mode of action of c-kit tyrosine kinase inhibitors leading to NK cell-dependent antitumor effects. J Clin Invest **114**, 379–388.

Brilot, F., Strowig, T., Roberts, S.M., Arrey, F. and Munz, C. (2007). NK cell survival mediated through the regulatory synapse with human DCs requires IL-15Ralpha. J Clin Invest **117**, 3316–3329.

Caminschi, I., Ahmet, F., Heger, K., Brady, J., Nutt, S.L., Vremec, D., Pietersz, S., Lahoud, M.H., Schofield, L., Hansen, D.S., et al. (2007). Putative IKDCs are functionally and developmentally similar to natural killer cells, but not to dendritic cells. J Exp Med **204**, 2579–2590.

Cerwenka, A. and Lanier, L.L. (2001). Natural killer cells, viruses and cancer. Nat Rev Immunol **1**, 41–49.

Chan, C.W., Crafton, E., Fan, H.N., Flook, J., Yoshimura, K., Skarica, M., Brockstedt, D., Dubensky, T.W., Stins, M.F., Lanier, L.L., et al. (2006). Interferon-producing killer dendritic cells provide a link between innate and adaptive immunity. Nat Med **12**, 207–213.

Chan, C.W. and Housseau, F. (2008). The 'kiss of death' by dendritic cells to cancer cells. Cell Death Differ **15**, 58–69.

Chaudhry, U.I., Katz, S.C., Kingham, T.P., Pillarisetty, V.G., Raab, J.R., Shah, A. B. and DeMatteo, R.P. (2006a). In vivo overexpression of Flt3 ligand expands and activates murine spleen natural killer dendritic cells. FASEB J **20**, 982–984.

Chaudhry, U.I., Kingham, T.P., Plitas, G., Katz, S.C., Raab, J.R. and DeMatteo, R. P. (2006b). Combined stimulation with interleukin-18 and CpG induces murine natural killer dendritic cells to produce IFN-gamma and inhibit tumor growth. Cancer Res **66**, 10497–10504.

Chaudhry, U.I., Plitas, G., Burt, B.M., Kingham, T.P., Raab, J.R. and DeMatteo, R.P. (2007). NK dendritic cells expanded in IL-15 exhibit antitumor responses in vivo. J Immunol **179**, 4654–4660.

Chauvin, C., Philippeau, J.M., Hemont, C., Hubert, F.X., Wittrant, Y., Lamoureux, F., Trinite, B., Heymann, D., Redini, F. and Josien, R. (2008). Killer dendritic cells link innate and adaptive immunity against established osteosarcoma in rats. Cancer Res **68**, 9433–9440.

Coulombel, L. and Bensussan, A. (2008). [IKDC: a new suit for active NK cells]. Med Sci (Paris) **24**, 521–524.

Della Chiesa, M., Vitale, M., Carlomagno, S., Ferlazzo, G., Moretta, L. and Moretta, A. (2003). The natural killer cell-mediated killing of autologous dendritic cells is confined to a cell subset expressing CD94/NKG2A, but lacking inhibitory killer Ig-like receptors. Eur J Immunol **33**, 1657–1666.

Di Santo, J.P. (2006). Natural killer cell developmental pathways: A Question of Balance. Annu Rev Immunol **24**, 257–286.

Fehniger, T.A., Cooper, M.A., Nuovo, G. J., Cella, M., Facchetti, F., Colonna, M. and Caligiuri, M.A. (2003). CD56bright natural killer cells are present in human lymph nodes and are activated by T cell-derived IL-2: a potential new link between adaptive and innate immunity. Blood **101**, 3052–3057.

Fernandez, N.C., Lozier, A., Flament, C., Ricciardi-Castagnoli, P., Bellet, D., Suter, M., Perricaudet, M., Tursz, T., Maraskovsky, E. and Zitvogel, L. (1999).

Dendritic cells directly trigger NK cell functions: cross-talk relevant in innate anti-tumor immune responses in vivo. Nat Med **5**, 405–411.

Figdor, C.G., de Vries, I.J., Lesterhuis, W.J. and Melief, C.J. (2004). Dendritic cell immunotherapy: mapping the way. Nat Med **10**, 475–480.

Fuchs, A., Cella, M., Kondo, T. and Colonna, M. (2005). Paradoxic inhibition of human natural interferon-producing cells by the activating receptor NKp44. Blood **106**, 2076–2082.

Gabrilovich, D. (2004). Mechanisms and functional significance of tumour-induced dendritic-cell defects. Nat Rev Immunol **4**, 941–952.

Groh, V., Rhinehart, R., Randolph-Habecker, J., Topp, M.S., Riddell, S.R. and Spies, T. (2001). Costimulation of CD8alphabeta T cells by NKG2D via engagement by MIC induced on virus-infected cells. Nat Immunol **2**, 255–260.

Hanna, J., Gonen-Gross, T., Fitchett, J., Rowe, T., Daniels, M., Arnon, T. I., Gazit, R., Joseph, A., Schjetne, K. W., Steinle, A., et al. (2004). Novel APC-like properties of human NK cells directly regulate T cell activation. J Clin Invest **114**, 1612–1623.

Hoeffel, G., Ripoche, A.C., Matheoud, D., Nascimbeni, M., Escriou, N., Lebon, P., Heshmati, F., Guillet, J.G., Gannage, M., Caillat-Zucman, S., et al. (2007). Antigen crosspresentation by human plasmacytoid dendritic cells. Immunity **27**, 481–492.

Homann, D., Jahreis, A., Wolfe, T., Hughes, A., Coon, B., van Stipdonk, M.J., Prilliman, K.R., Schoenberger, S.P. and von Herrath, M.G. (2002). CD40L blockade prevents autoimmune diabetes by induction of bitypic NK/DC regulatory cells. Immunity **16**, 403–415.

Huang, A.Y., Golumbek, P., Ahmadzadeh, M., Jaffee, E., Pardoll, D. and Levitsky, H. (1994). Role of bone marrow-derived cells in presenting MHC class I-restricted tumor antigens. Science **264**, 961–965.

Huntington, N.D., Legrand, N., Alves, N.L., Jaron, B., Weijer, K., Plet, A., Corcuff, E., Mortier, E., Jacques, Y., Spits, H. and Di Santo, J.P. (2009). IL-15 trans-presentation promotes human NK cell development and differentiation in vivo. J Exp Med **206**, 25–34.

Huntington, N.D., Vosshenrich, C.A. and Di Santo, J.P. (2007). Developmental pathways that generate natural-killer-cell diversity in mice and humans. Nat Rev Immunol **7**, 703–714.

Itoh, K., Yamada, A., Mine, T. and Noguchi, M. (2009). Recent advances in cancer vaccines: an overview. Jpn J Clin Oncol **39**, 73–80.

Kadowaki, N., Antonenko, S., Ho, S., Rissoan, M.C., Soumelis, V., Porcelli, S.A., Lanier, L.L. and Liu, Y.J. (2001). Distinct cytokine profiles of neonatal natural killer T cells after expansion with subsets of dendritic cells. J Exp Med **193**, 1221–1226.

Kayagaki, N., Yamaguchi, N., Nakayama, M., Takeda, K., Akiba, H., Tsutsui, H., Okamura, H., Nakanishi, K., Okumura, K. and Yagita, H. (1999). Expression and function of TNF-related apoptosis-inducing ligand on murine activated NK cells. J Immunol **163**, 1906–1913.

Kortylewski, M. and Yu, H. (2008). Role of Stat3 in suppressing anti-tumor immunity. Curr Opin Immunol **20**, 228–233.

Lanier, L.L. (2005). NK cell recognition. Annu Rev Immunol **23**, 225–274.

Mailliard, R.B., Alber, S.M., Shen, H., Watkins, S.C., Kirkwood, J.M., Herberman, R.B. and Kalinski, P. (2005). IL-18-induced CD83 + CCR7+ NK helper cells. J Exp Med **202**, 941–953.

Martin-Fontecha, A., Thomsen, L.L., Brett, S., Gerard, C., Lipp, M., Lanzavecchia, A. and Sallusto, F. (2004). Induced recruitment of NK cells to lymph nodes provides IFN-gamma for T(H)1 priming. Nat Immunol **5**, 1260–1265.

Melief, C.J. (2008). Cancer immunotherapy by dendritic cells. Immunity **29**, 372–383.

Mignot, G., Ullrich, E., Bonmort, M., Menard, C., Apetoh, L., Taieb, J., Bosisio, D., Sozzani, S., Ferrantini, M., Schmitz, J., et al. (2008). The critical role of IL-15 in the antitumor effects mediated by the combination therapy imatinib and IL-2. J Immunol **180**, 6477–6483.

Moretta, A. (2002). Natural killer cells and dendritic cells: rendezvous in abused tissues. Nat Rev Immunol **2**, 957–964.

Nutt, S.L., Metcalf, D., D'Amico, A., Polli, M. and Wu, L. (2005). Dynamic regulation of PU.1 expression in multipotent hematopoietic progenitors. J Exp Med **201**, 221–231.

Pillarisetty, V.G., Katz, S.C., Bleier, J.I., Shah, A.B. and Dematteo, R.P. (2005). Natural killer dendritic cells have both antigen presenting and lytic function and in response to CpG produce IFN-gamma via autocrine IL-12. J Immunol **174**, 2612–2618.

Potter, N.S. and Harding, C.V. (2001). Neutrophils process exogenous bacteria via an alternate class I MHC processing pathway for presentation of peptides to T lymphocytes. J Immunol **167**, 2538–2546.

Pozzi, L.A., Maciaszek, J.W. and Rock, K. L. (2005). Both dendritic cells and macrophages can stimulate naive CD8 T cells in vivo to proliferate, develop effector function, and differentiate into memory cells. J Immunol **175**, 2071–2081.

Randolph, G.J., Angeli, V. and Swartz, M. A. (2005). Dendritic-cell trafficking to lymph nodes through lymphatic vessels. Nat Rev Immunol **5**, 617–628.

Reddy, S.T., Swartz, M.A. and Hubbell, J.A. (2006). Targeting dendritic cells with biomaterials: developing the next generation of vaccines. Trends Immunol **27**, 573–579.

Robbins, S.H., Walzer, T., Dembele, D., Thibault, C., Defays, A., Bessou, G., Xu, H., Vivier, E., Sellars, M., Pierre, P., et al. (2008). Novel insights into the relationships between dendritic cell subsets in human and mouse revealed by genome-wide expression profiling. Genome Biol **9**, R17.

Rock, K.L., Hearn, A., Chen, C. J. and Shi, Y. (2005). Natural endogenous adjuvants. Springer Semin Immunopathol **26**, 231–246.

Roncarolo, M.G., Bigler, M., Haanen, J. B., Yssel, H., Bacchetta, R., de Vries, J. E. and Spits, H. (1991). Natural killer cell clones can efficiently process and present protein antigens. J Immunol **147**, 781–787.

Sanos, S.L., Bui, V.L., Mortha, A., Oberle, K., Heners, C., Johner, C. and Diefenbach, A. (2009). RORgammat and commensal microflora are required for the differentiation of mucosal interleukin 22-producing NKp46+ cells. Nat Immunol **10**, 83–91.

Schotte, R., Rissoan, M.C., Bendriss-Vermare, N., Bridon, J.M., Duhen, T., Weijer, K., Briere, F. and Spits, H. (2003). The transcription factor Spi-B is expressed in plasmacytoid DC precursors and inhibits T-, B-, and NK-cell development. Blood **101**, 1015–1023.

Shortman, K. and Villadangos, J.A. (2006). Is it a DC, is it an NK? No, it's an IKDC. Nat Med **12**, 167–168.

Spits, H. and Lanier, L.L. (2007). Natural killer or dendritic: what's in a name? Immunity **26**, 11–16.

Srivastava, R.M., Varalakshmi, C. and Khar, A. (2007). Cross-linking a mAb to NKR-P2/NKG2D on dendritic cells induces their activation and maturation leading to enhanced anti-tumor immune response. Int Immunol **19**, 591–607.

Stary, G., Bangert, C., Tauber, M., Strohal, R., Kopp, T. and Stingl, G. (2007). Tumoricidal activity of TLR7/8-activated inflammatory dendritic cells. J Exp Med **204**, 1441–1451.

Steinman, R.M. and Banchereau, J. (2007). Taking dendritic cells into medicine. Nature **449**, 419–426.

Steinman, R.M., Hawiger, D. and Nussenzweig, M.C. (2003). Tolerogenic dendritic cells. Annu Rev Immunol **21**, 685–711.

Sun, J.C., Beilke, J.N. and Lanier, L.L. (2009). Adaptive immune features of natural killer cells. Nature **457**, 557–561.

Swartz, M.A., Hubbell, J.A. and Reddy, S. T. (2008). Lymphatic drainage function and its immunological implications: from dendritic cell homing to vaccine design. Semin Immunol **20**, 147–156.

Taieb, J., Chaput, N., Menard, C., Apetoh, L., Ullrich, E., Bonmort, M., Pequignot, M., Casares, N., Terme, M., Flament, C., et al. (2006). A novel dendritic cell subset involved in tumor immunosurveillance. Nat Med **12**, 214–219.

Ullrich, E., Bonmort, M., Mignot, G., Jacobs, B., Bosisio, D., Sozzani, S., Jalil, A., Louache, F., Bulanova, E., Geissman, F., et al. (2008a). Trans-presentation of IL-15 dictates IFN-producing killer dendritic cells effector functions. J Immunol **180**, 7887–7897.

Ullrich, E., Chaput, N. and Zitvogel, L. (2008b). Killer dendritic cells and their potential role in immunotherapy. Horm Metab Res **40**, 75–81.

Vitale, M., Della Chiesa, M., Carlomagno, S., Pende, D., Arico, M., Moretta, L. and Moretta, A. (2005). NK-dependent DC maturation is mediated by TNFalpha and IFNgamma released upon engagement of the NKp30 triggering receptor. Blood **106**, 566–571.

Vosshenrich, C.A., Lesjean-Pottier, S., Hasan, M., Richard-Le Goff, O., Corcuff, E., Mandelboim, O. and Di Santo, J.P. (2007). CD11cloB220+ interferon-producing killer dendritic cells are activated natural killer cells. J Exp Med **204**, 2569–2578.

Walzer, T., Dalod, M., Robbins, S.H., Zitvogel, L. and Vivier, E. (2005). Natural killer cells and dendritic cells: "l'union fait la force". Blood **106**, 2252–2258.

Walzer, T., Jaeger, S., Chaix, J. and Vivier, E. (2007). Natural killer cells: from CD3(-)NKp46(+) to post-genomics meta-analyses. Curr Opin Immunol **19**, 365–372.

Welner, R.S., Pelayo, R., Garrett, K. P., Chen, X., Perry, S.S., Sun, X.H., Kee, B.L. and Kincade, P.W. (2007). Interferon-producing killer dendritic cells (IKDCs) arise via a unique differentiation pathway from primitive c-kitHiCD62L+ lymphoid progenitors. Blood **109**, 4825–4931.

Zingoni, A., Sornasse, T., Cocks, B.G., Tanaka, Y., Santoni, A. and Lanier, L.L. (2004). Cross-Talk between activated human NK cells and CD4+ T cells via OX40-OX40 ligand interactions. J Immunol **173**, 3716–3724.

Zitvogel, L., Apetoh, L., Ghiringhelli, F. and Kroemer, G. (2008). Immunological aspects of cancer chemotherapy. Nat Rev Immunol **8**, 59–73.

Chapter Four

4

Natural killer cells in cancer

Petter Höglund, Hans-Gustaf Ljunggren

CHAPTER CONTENTS

Things should be made as simple as possible, but not any simpler.

Albert Einstein

ABSTRACT

Data from experimental model systems have provided ample evidence for tumour cell recognition by natural killer (NK) cells. Here, following a historical denotation, we review recent insights into interactions between NK cells and human tumour cells, focusing on key receptor–ligand interactions. We present evidence for NK cell targeting of primary human cancer cells and provide a discussion on prospects for using NK cells to treat human malignancies.

KEY WORDS

Transgenic mice, Missing self, Inhibitory receptor, Cancer therapy, Solid tumours, Adoptive transfer, Stem cell transplantation

Introduction

Integrated with other immune cells, natural killer (NK) cells contribute to host anti-microbial and anti-tumour immunity (Moretta et al., 2002). The provision of early defence mechanisms against viral infections, particularly herpes viruses, is perhaps the most important clinical effect by NK cells, but it was their cytotoxic potency against tumour cells that brought about their discovery (Herberman et al., 1975a,b; Kiessling et al., 1975a,b). Their ability to lyse tumour cells in vitro without the requirement of prior immune sensitization of the host also gave them their name (Kiessling et al., 1975a). NK cells are now well characterized with respect to their origin, differentiation, receptor repertoire and effector functions; properties that are discussed elsewhere in this volume. Here, we focus on the role of NK cells in interaction with tumour cells, with an emphasis on human cancer.

NK cells in the host response against tumours

NK cells mediate killing of many different types of murine and human tumour cell lines in vitro. Several experimental studies in mice have also shown a role for NK cells in rejection responses against grafted murine tumour cell lines and against experimentally induced and spontaneously developing tumours in mice (Smyth et al., 2002; Wu and Lanier, 2003). Like cytotoxic T cells, NK cells possess different effector functions by which they mount anti-tumour responses (Wallace and Smyth, 2005). Two major mechanisms are used to induce target cell apoptosis, granule exocytosis and death receptor stimulation (Smyth et al., 2005). Granule exocytosis involves the release of perforin and granzymes (Trapani and Smyth, 2002), while the death receptor pathway is largely mediated by apoptosis-inducing members of the TNF superfamily such as FasL, TNF-α, LT and TRAIL (Screpanti et al., 2005; Smyth et al., 2005). NK cells can also produce many different cytokines (e.g. IFN-γ, TNF-α and GM-CSF) as well as chemokines, at least some of which have a direct effect on tumours. The best studied cytokine in this respect is IFN-γ, a cytokine which decreases proliferation, enhances autophagy, limits metabolic activity of tumour cells and inhibits angiogenesis (Hayakawa et al., 2002). IFN-γ produced by NK cells might also play a role in the regulation of killing by death receptors, either by downregulating anti-apoptotic proteins, or by upregulating caspases that are essential for death receptor-mediated apoptosis. By virtue of their IFN-γ production, NK cells also promote the development of T helper 1 (Th1) responses associated with the generation of cytotoxic T cells and activation of macrophages. Th1 responses are often thought to be beneficial to the host response to tumours. Th2 cytokines (e.g. IL-4, IL-10 and TGF-β) could antagonize Th1 responses. By promoting Th1 responses in this way, NK cells could counteract Th2-promoting tumour escape mechanisms mediated by cancer cells and regulatory T cells (Smyth et al., 2006).

Direct evidence for NK cell targeting of human cancer has come from studies of NK cell interactions with primary tumour cells tested for susceptibility to NK cells lysis ex vivo (Carlsten et al., 2009). It has also been shown that human NK cells adoptively transferred to mice participate in the rejection of grafted human tumours (Guimaraes et al., 2006). Evidence for NK cell targeting of human tumours has also come from clinical studies in settings of stem cell transplantation (SCT) and adoptive transfer of NK cells to cancer patients (Ljunggren and Malmberg, 2007; Miller et al., 2005; Ruggeri et al., 2002, 2006). Despite the large number of studies demonstrating the ability of NK cells to target tumour cells in vitro and in vivo, there is still only limited evidence for NK cell recognition of primary tumour cells ex vivo, and clinical studies involving transfer of activated NK cells to patients with cancer are still in their early days. Because of this, we devote some attention to these areas of research. Before this, however, we will put today's knowledge on molecular events involved in NK cell tumour-recognition into a historical context, followed by a section presenting recent insights into the molecular events involved in NK cell recognition of human cancer.

The early days—towards an understanding of NK cell tumour recognition

More than four decades ago, it was noted that F_1-hybrid mice (derived from a cross of two inbred strains) were often relatively resistant to bone marrow or tumour grafts of parental strain origin, compared to syngeneic recipients (Cudkowicz and Stimpfling, 1964). This phenomenon contrasted with the common laws of transplantation, which stated that graft rejection should only take place if the tumour carried transplantation antigens (major histocompatibility antigens, MHC) that were 'foreign' to the host. In F_1 hybrid resistance, no such 'foreignness' could be seen. Soon after the discovery of NK cells, the F_1 anti-parental reaction was correlated to NK cell activity (Kiessling et al., 1977).

The fact that NK cells could be demonstrated to be mediators of F_1 hybrid resistance was crucial for forthcoming discoveries, providing insights into the molecular specificity of NK cells. It was F_1 hybrid resistance that stimulated Klas Kärre to formulate the 'missing self' model for NK cell recognition (Karre, 2008; Karre et al., 1986; Ljunggren and Karre, 1990). The hypothesis postulated that absence, or reduced expression of, 'self' MHC class I products could be sufficient to allow a cell to be killed by NK cells. Possible reasons for such alterations were genetic differences, mutations, viral infection or tumour transformation.

To test the predictions of the missing-self hypothesis for tumour cell recognition, MHC class I-deficient variants (mutants) from MHC class I-sufficient NK cell-resistant tumour cell lines were generated by immunoselection, and their susceptibility to NK cell lysis was tested. Indeed, MHC class I-negative variants from RBL-5 and EL-4 lymphomas were more sensitive to NK cell lysis in vitro when compared with wild type cells (Karre et al., 1986; Ljunggren and Karre, 1985). When titrated numbers of mutants and wild type cells were inoculated in vivo in syngeneic C57BL/6 (B6) mice, there were striking differences in tumour takes. While the MHC class I-deficient cells only formed tumours when the highest numbers were administered, wild type cells formed tumours in almost all mice, even when very low doses of tumour cells (down to only 100 cells) were inoculated (Karre et al., 1986; Ljunggren and Karre, 1985). Subsequent studies demonstrated that the rejection of MHC class I-deficient cells was dependent on the presence of NK cells (Karre et al., 1986; Ljunggren and Karre, 1985; Ljunggren et al., 1988a,b).

A crucial step was to restore MHC class I expression to more directly link the NK cell-sensitive phenotype to the loss of MHC class I molecules. Cell lines that lacked β_2m expression were attractive because of the immediate possibility of restoring β_2m expression by transfection. Such critical experiments demonstrated directly that the NK cell-sensitive phenotype of the mutant cell lines was due to the loss of MHC class I expression (Glas et al., 1992; Ljunggren et al., 1989, 1990). In parallel, others described similar findings with human cell lines (Storkus et al., 1989). It is of historical interest to note that some of the more useful mutant cell lines that were derived in our laboratory for NK cell studies were shown to contain mutations also in genes other than β_2m, which were shown to be of critical importance for MHC class I expression. For example, in 1989, Alain Townsend used the RMA-S cell line to identify TAP genes, responsible for the transportation of peptides across the ER membrane before loading onto MHC class I molecules (Townsend et al., 1989).

The 'missing self' hypothesis predicted not only that one could render a target cell susceptible to NK cell lysis by deleting its MHC class I molecules but also that 'missing self' rejection would be induced if novel MHC class I alleles were introduced at the level of the host (e.g. by an MHC class I transgene). Through the generation of H2D^d transgenic mice on a B6 background, these predictions became possible to test. It was shown that NK cells in H2D^d transgenic B6 mice were able to reject B6-derived lymphomas (Hoglund et al., 1988) and bone marrow grafts (Ohlen et al., 1989). This experiment provided direct molecular evidence for the 'missing self' explanation for F_1-hybrid resistance, and definitively linked host MHC class I molecules to the development of functional specificity in the NK cell compartment.

A few years later, two lines of research opened up the field of MHC class I recognition by NK cells. The first was the generation of β_2m-deficient mice, which made it possible to study NK cell recognition of normal (i.e. untransformed) cells lacking expression of MHC class I molecules. Indeed, cells from these mice were susceptible to lysis by NK cells from corresponding wild type mice, and MHC class I-deficient bone marrow grafts from these mice were rejected by wild type mice (Bix et al., 1991; Hoglund et al., 1991; Liao et al., 1991). The second line of research was the identification of MHC class-specific inhibitory receptors (Karlhofer et al., 1992; Moretta and Moretta, 2004; Parham, 2005; Wagtmann et al., 1995). The generation of monoclonal antibodies against these MHC class I binding receptors on human and murine NK cells made it possible to test a critical prediction by one of the mechanistic models for 'missing self' recognition, the 'effector inhibition' model (Ljunggren and Karre, 1990). The model predicted that an inhibitory receptor blockade should lead to augmented killing of MHC class I expressing target cells, which was indeed observed.

While the identification of inhibitory receptors in part uncovered the molecular mechanisms used by NK cells to recognize tumour cells, sensing the absence of self MHC class I molecules is not sufficient to cause target cell killing. NK cells also need stimulation by target cell ligands to trigger activation via specific receptors. The identification of the latter remained elusive until a few years after the discovery of inhibitory receptors (Bauer et al., 1999; Pessino et al., 1998; Vitale et al., 1998). We now know that NK cell recognition of tumours is tightly regulated by processes involving the integration of signals delivered from multiple activating and inhibitory receptors (Lanier, 2005).

NK cell receptors involved in human tumour recognition

Natural cytotoxicity receptors (NCR) represent a group of human NK cell activation receptors that include NKp46, NKp30 and NKp44 (Pessino et al., 1998; Vitale et al., 1998). NKp46 and NKp30 are constitutively expressed on all peripheral blood NK cells, whereas NKp44 expression on NK cells requires activation by IL-2. Some T cells also express NKp44. Ligands for these receptors on tumour cells remain largely undefined. The NCRs have a major role in NK cell-mediated lysis of various human tumour cell lines, including melanomas, carcinomas, neuroblastomas, myeloid or lymphoblastic leukaemias and EBV-transformed B cells (Bottino et al., 2005). Other well-characterized activation receptors on NK cells are NKG2D and DNAM-1 (Bauer et al., 1999; Bottino et al., 2003). NKG2D recognizes the stress-induced molecules MICA and MICB as well as the UL16-binding proteins. NKG2D ligands are expressed on a number of human epithelial tumour and leukaemic cell lines and play significant roles in rendering these cells susceptible to NK cell-mediated lysis.

Recent studies in murine models have provided exciting new insights into the role of NKG2D and DNAM-1 receptors in tumour immunity. Early-arising aggressive prostate tumours are three times more frequent in mice lacking the NK cell receptor NKG2D compared with wild-type mice on a background susceptible to prostate cancer (Guerra et al., 2008). NKG2D deficiency is also associated with preB-B cell lymphoma development, as deduced from studies of Eμ-myc transgenic mice (Guerra et al., 2008). DNAM-1 recognizes PVR (CD155) and Nectin-2 (CD122), ligands highly expressed in human carcinomas, melanomas and neuroblastomas (Chang and Ferrone, 2006; Costello et al., 2004; Moretta et al., 2006). Experimentally, DNAM-1-deficiency promotes the development of fibrosarcomas and papillomas when mice are treated with the carcinogens MCA and 7, 12-dimethulbenz(a)anthracene (DMBA) (Iguchi-Manaka et al., 2008). These results are consistent with a role for NK cells in tumour surveillance, but they do not rule out a role for other lymphocytes expressing these receptors as well. CD16 on NK cells binds the Fc-portion of IgG on opsonized cells, thus mediating antibody-dependent cellular cytotoxicity (ADCC). In addition, several other receptors, including 2B4 (CD244), NTBA, NKp80, CD2, CD11a/CD18 and CD59, have important co-activating or co-stimulatory functions in NK cell activation and tumour cell recognition (Bryceson et al., 2006a).

MHC class I molecules are normally expressed on most healthy cells in the body but are often lost upon transformation or during tumour evolution (Mendez et al., 2008). In humans, KIR and CD94-NKG2A play major roles as HLA-class I-specific inhibitory NK cell receptors. KIR recognize groups of HLA-A, HLA-B, and HLA-C alleles (Moretta and Moretta, 2004; Parham, 2005; Wagtmann et al., 1995), whereas CD94-NKG2A/B receptors recognize HLA-E molecules presenting antigenic peptides derived from classical MHC class I alleles (Braud et al., 1998). Mismatches in any one of these receptor–ligand pairs may thus lead to NK cell-mediated killing of tumour cells by means of loss of NK cell tolerance to self. In humans, the KIR gene system is both polygenic and polymorphic. Specific KIR gene products are expressed on distinct subsets of NK cells, sometimes in overlapping combinations (Parham, 2005). NK cell tolerance in humans is controlled to a large extent by inhibitory KIR and NKG2A receptors, but if the stochastic mechanisms for KIR gene expression happens to generate an NK cell lacking an inhibitory receptor for self MHC class I, other mechanisms for self tolerance operate to avoid autoreactive responses (Anfossi et al., 2006). The diversity between individuals as well as the mechanisms for NK cell tolerance is potentially beneficial in settings of SCT and adoptive NK cell-based immunotherapy.

The importance of various NK cell activation receptors in the recognition of primary human tumours is only partially known. Interestingly, efficient natural cytotoxicity by ex vivo NK cells usually requires co-activation by several types of receptors (Bryceson et al., 2006a,b). Many tumour cells express ligands for several NK cell-activating receptors, which provides a likely explanation for the spontaneous tumour cell killing seen by primary, non-activated, NK cells. Tumour cells also frequently downregulate MHC class I, which contributes to NK cell sensitivity. Inhibition by MHC class I molecules is usually dominant, but in some situations, activation signals may override inhibition, as has been demonstrated for NKG2D-mediated triggering of some MHC class I-expressing tumour cell lines (Cerwenka et al., 2001; Diefenbach et al., 2001). It should be noted that the balance between positive and negative signals is not only affected by the presence or absence of ligands but also critically depends on the relative expression levels of ligands for activating and inhibitory receptors. This is important in genetically unstable tumour cells, in which immunoselection might rapidly generate escape mutants with such alterations.

NK cell targeting of primary human tumour cells

While NK cell-mediated killing of tumour cell lines can be readily demonstrated in vitro, fewer studies have more directly addressed the NK cell susceptibility of freshly isolated human cancer cells. Next, we briefly present a few studies that have aimed in the latter direction.

NK cell-mediated lysis of primary acute lymphatic leukaemia (ALL) blasts has been observed with autologous NK cells expanded in vitro (Torelli et al., 2005). The study used NK cells with single KIR specificities for HLA class I allotypes (Diermayr et al., 2008) and could demonstrate a beneficial effect of KIR ligand mismatched NK cells against freshly isolated acute myeloid leukemia (AML) blasts. In addition, tumour cells expressing NKG2D ligands were the most sensitive, while target cells lacking such ligands were resistant to lysis. Interestingly, induction of cell surface NKG2D ligands by valproic acid rendered the tumour cells more sensitive to NK cell-mediated lysis (Diermayr et al., 2008). Thus, alloreactive HLA class I-mismatched NK cells in combination with pharmacologic induction of NKG2D ligands suggests a possible approach to immunotherapy for AML.

In myeloma, several studies show promising results with respect to killing by allogeneic and autologous NK cells (Alici et al., 2008; Carbone et al., 2005; El-Sherbiny et al., 2007; Frohn et al., 2002). One study with allogeneic NK cells revealed a predominant role for the DNAM-1 receptor in tumor cell recognition (El-Sherbiny et al., 2007). In other studies, antibody blockade of autologous NK cells indicated the involvement of several activating receptors (Alici et al., 2008). The recognition of patient-derived multiple myeloma by autologous NK cells, as demonstrated with either IL-2 stimulated or long-term expanded autologous NK cells (Alici et al., 2008; Carbone et al., 2005), led to speculation that this tumour might be targeted by immunotherapeutic strategies involving autologous NK cells. It should be noted that patients with multiple myeloma frequently display reduced levels of HLA class I on the cell surface, which may explain the effectiveness of autologous NK cell preparation in this setting. Indeed, NK cell killing correlated inversely with the level of HLA-class I on the myeloma cells (Carbone et al., 2005).

Neuroblastoma represents a solid tumour that has been characterized with respect to NK cell susceptibility (Castriconi et al., 2004). Killing of freshly isolated neuroblastoma cells involves NKp30 and NKp46. A significant heterogeneity in susceptibility to lysis was found among neuroblastomas derived from different patients. Interestingly, susceptibility to lysis directly correlated with the surface expression of the DNAM-1 ligand PVR. PVR-expressing neuroblastoma cells were efficiently killed by NK cells, and monoclonal antibody (mAb) masking of either DNAM-1 on the NK cells or PVR on the tumour cells resulted in strong inhibition of tumour cell lysis. Thus, high cell surface PVR levels may represent a positive predictor for NK cell susceptibility of neuroblastoma.

Another solid tumour tested for NK cell-mediated recognition is ovarian carcinoma. This tumour type often gives rise to ascites containing tumour cells in suspension in the peritoneal cavity of affected individuals. Such cells represent a source of fresh tumour targets useful for studying NK cell susceptibility. Similar to fresh neuroblastomas, ovarian carcinoma cells are also sensitive to lysis by allogeneic NK cells (Carlsten et al., 2007). Also here, a dominant role for DNAM-1 and a complementary contribution of NKG2D-signalling as activating receptors was seen (Carlsten et al., 2007).

Treating patients with adoptive transfer of NK cells

Treating human cancer with NK cells could be envisaged by a number of different means (Table 4.1). As a few examples, one strategy could aim for blocking inhibitory KIR receptors using monoclonal antibodies, while another would focus on activation of endogenous NK cells in the cancer patient using cytokines. Another strategy could take advantage of the knowledge regarding missing self recognition and KIR-HLA mismatches in settings of allogeneic SCT, followed perhaps by the use of NK cell-based donor lymphocyte infusions post SCT. Table 4.1 lists these and a few other possible strategies that could be employed both in situations of SCT and adoptive immunotherapy using NK cells (Ljunggren and Malmberg, 2007; Malmberg et al., 2008).

Adoptive transfer of immune cells to cure cancer was pioneered by Rosenberg, Lotze and collaborators using autologous 'lymphokine activated killer' cells together with high-dose IL-2 (Rosenberg et al., 1985). Up to 20% of patients who suffered from metastatic renal cancer and melanoma responded to the infusions of these NK cell-containing populations, but subsequent studies showed that similar anti-tumour effects were achieved with high-dose IL-2 alone (Law et al., 1995). Because of the lack of significant clinical effects by autologous NK cells (Burns et al., 2003; Law et al., 1995; Rosenberg et al., 1985) and the promising effects observed in haploidentical T-cell-depleted SCT (Ruggeri et al., 2002), the focus has more recently shifted towards the potential of allogeneic NK cells in adoptive cell therapy for

Table 4.1 Possible ways NK cell efficiency against cancer cells could be enhanced

Strategy	How	Outcomes
Interfering with inhibitory receptor function	• Blocking mAb • siRNA	More efficient alloreactivity according to 'missing self' rejection due to blocking of inhibitory receptors at the cell surface, or by interfering with their expression using RNA interference.
Selection of appropriate donors for transplantation or adoptive transfer	• KIR and HLA typings	By selecting donors with a KIR–MHC mismatch, more efficient alloreactivity according to 'missing self' rejection would be seen.
Enhancing the expression of ligands for activating receptors	• Irradiation	When ligands against activating receptors are induced on tumors, the balance of positive and negative signals in NK cells is geared towards enhanced stimulation.
Enhancing NK cell activation and proliferation	• Stimulatory mAb • Cytokines • Drugs	Expansion of more activated NK cells via triggering of activating receptors and administration of cytokines or drugs that stimulate NK cells to become more active.
Enhancing ADCC	• Crosslinking mAb • Cytokines • Bispecific Ab	More efficient tumor cell killing by exogenously added normal or bispecific tumor-specific antibodies that bind crosslinks tumor cells and NK cells and thus mediates tumor cell killing. Cytokines may be supplied in parallel to enhance the response.
Breaking NK cell anergy	• Cytokines • Drugs	Induction of tumor cell killing by reversal of NK cell tolerance or anergy, which can be broken by cytokines such as IL-2.
Enhancing the immunoregulatory role of NK cells	• Cytokines	Activation of NK cells will lead to more efficient interactions with T cells and enhancement of T cell functions.

the treatment of cancer. In certain donor–recipient combinations, prospects for missing-self-reactivity prevail, providing better possibilities for anti-tumour reactivity.

In a pioneering study by Miller and collaborators, a total of 43 patients with advanced cancer were given haploidentical NK cell infusions together with IL-2 (Miller et al., 2005). Two low-intensity outpatient regimens and one high-dose inpatient immune suppressive regimen were tested. In patients given low-intensity regimens, donor NK cells were not detectable by day 7 or later, and there was no evidence of in vivo expansion. However, with the high dose regimen, long-term survival and in vivo expansion of donor-derived NK cells were seen in a majority of the patients. Notably, in one of the patients, there was preferential expansion of the alloreactive NK cell subset. Donor NK cell infusions were well tolerated without evidence for induction of Graft-*versus*-Host Disease (GVHD). Moreover, with this protocol, 5 out of 19 patients with AML achieved complete remission, a group that contained all 4 patients that received NK cells from a donor with a predictable alloreactive NK cell repertoire. Interestingly, 3 out of these 4 patients achieved complete remission (Miller et al., 2005), indicating that the choice of a KIR ligand-mismatched donor may be of critical importance to obtain successful results in future clinical trials (Miller et al., 2005; Ruggeri et al., 2002, 2005). The findings by Miller et al. further suggest that haploidentical NK cells can persist and expand in vivo and therefore may represent the most promising future role in the treatment of selected malignancies.

One important thing to keep in mind when discussing KIR ligand mismatches is that a particular donor–recipient mismatch on a genetic level poorly predicts the absolute numbers of truly alloreactive NK cells in the blood, lacking both KIR for self HLA class I molecules and CD94/NKG2A. In fact, the size of the alloreactive subset in different mismatched donors can vary from below 1% up to 60% of the NK cells (Fauriat et al., 2008). A prediction of effectiveness of therapy therefore cannot be based on genetics only but must also include an assessment of the NK cell repertoire at the cellular level. This could then lead to the selection of a donor with the largest alloreactive NK cell subset against the recipient in question and with the best predicted clinical outcome.

Several techniques have been developed for ex vivo expansion of NK cells. A few of these protocols have been developed to meet GMP conditions (Carlens et al., 2001; Klingemann, 2005). Expansion protocols provide greater numbers of activated NK cells to be used for adoptive therapy that might be desirable in some situations. However, some caution must be taken with respect to possible phenotypic changes, lineage deviation and selective expansion of specific subsets of NK cells. Another aspect to consider is to what extent in vitro manipulation may alter the ability of NK cells to mediate cell–cell interactions, trafficking and homing.

A potential problem with adoptive transfer of alloreactive NK cells is that they may be targeted by the host immune system. To minimize this risk, some type of pre-transfer conditioning will likely be required. A drawback of such conditioning, however, may be the destruction of normal host immune cells that may facilitate NK cell engraftment, such as regulatory T cells (Ghiringhelli et al., 2006). An advantage, on the other hand, may be reduced competition for growth factors as a result of lymphocyte depletion. For example, enhanced production of IL-15 may act in a beneficial way on infused NK cells by promoting their in vivo survival and expansion. NK cell expansion (Miller et al., 2005) was more efficient after a more intense preparative regimen, similar to what was used to induce long-term in vitro survival of adoptively transferred T cells (Dudley et al., 2002; Muranski et al., 2006). Future clinical studies will have to explore the ability of haematopoietic growth factors as well as cytokines to activate and expand NK cells in vivo to obtain the most efficient regimens (Farag et al., 2003). By developing new ways of activating endogenous NK cells, or modulating host tumour cells by drugs that increase the expression of ligands for activating NK cell receptors, NK cell tolerance against some tumours could be broken (Sheridan, 2006). Another strategy is to block inhibitory KIR with monoclonal antibodies, thereby augmenting tumour cell recognition by NK cells (Sheridan, 2006). Preclinical evidence in mouse models has shown that this strategy may enhance anti-tumour activity in autologous and allogeneic settings (Koh et al., 2001, 2003).

With respect to tumour-cell types, it is already evident from experimental and clinical studies that certain tumour types may be more responsive to NK cell therapy than others. One example of this is the inefficient killing by NK cells of lymphoid compared to myeloid leukaemia, which may in part be due to differences in the expression of ligands for NK cell adhesion and activation receptors (Pende et al., 2005; Ruggeri et al., 1999). Differences in MHC class I expression may also be a critical feature that will affect NK cell sensitivity of certain tumours and thus influence the potential of achieving good clinical responses. These aspects are particularly difficult to evaluate on primary human tumours, not the least because of difficulties in processing primary tumour samples for analysis by flow cytometry. In some cases of primary tumours, however, phenotypic characterizations are possible, as well as performing functional tests of the sensitivity of isolated tumour cells directly to NK cells lysis ex vivo (Carlsten et al., 2009). The latter may be important since tumours often acquire resistance to lysis by cellular effectors despite expression of an appropriate set of ligands for NK cell receptors (Malmberg and Ljunggren, 2006). The ability to study NK cell ligand expression and sensitivity to NK cell lysis in vitro may further predict the likelihood of achieving good clinical responses in adoptive NK cell therapy.

With respect to the quantitative effects of NK cell rejection of tumours, already the early work had demonstrated that NK cells were frequently overwhelmed by large tumour doses and fail to convey a clinical response in such situations (Ljunggren and Karre, 1985). The limited capacity of NK cells to proliferate in vivo after antigenic challenge (even if specific subsets clearly do proliferate following adoptive transfer to other hosts (Miller et al., 2005) or after virus infections (Sun et al., 2009), likely play a role. Thus, in a clinical setting, immunotherapy is most likely to be most effective in situations with small tumour burdens, for example, in patients with minimal residual disease, following surgery, chemotherapy or relapse after SCT. NK cell therapy against large solid tumours represents an extraordinary challenge, including the presumed necessity of NK cells to migrate to the tumour tissue, to infiltrate the tumour and to proliferate sufficiently (Albertsson et al., 2003). While knowledge is developing with respect to mechanisms that control trafficking of NK cells, we still have insufficient knowledge of the requirements for NK cell homing to, and infiltration into, tumours. New imaging technologies may provide insights into the trafficking of adoptively transferred NK cells (Morris and Ley, 2004). Likewise, host factors that downregulate NK cell function, such as cytokine-mediated downregulation of NK cell activation receptors (Castriconi et al., 2004; Chiesa et al., 2006), may exist in patients with large tumour burdens (Malmberg and Ljunggren, 2006). Interfering with these represents a particularly challenging task for therapeutic interventions.

In conclusion, data from haploidentical haematopoietic SCT and NK cell-based adoptive immunotherapy demonstrate clinical effects mediated by NK cells and show that NK cells can be used in therapeutic settings against cancer. Adverse side effects of adoptively transferred NK cells have not been reported, which represents a major clinical benefit of adoptive therapy with NK cells. With the present insights into the molecular specificities that regulate NK cell function and NK cell tolerance, novel NK cell-based immunotherapeutic strategies against human cancer are likely to emerge. Furthermore, in conditioned patients or through the help of cytokines and other growth factors, adoptively transferred NK cells may be expanded and activated in vivo. Combinatorial therapies, where NK cells represent one important mediator, may also become important against some forms of cancer.

Acknowledgements

We were both trained in Klas Kärre's laboratory at the Department of Tumor Biology, Karolinska Institutet, during the period when the first steps were taken to explore NK cell responses against tumour cells based on the missing self principle. We are grateful to Klas Kärre for his guidance during these years and to all colleagues, former and present, in our research groups for their contributions to stimulating and creative environments at the former Department of Tumor Biology, the present Department of Microbiology Tumor and Cell Biology (MTC) and the Center for Infectious Medicine (CIM), Karolinska Institutet.

References

Albertsson, P.A., Basse, P.H., Hokland, M., Goldfarb, R.H., Nagelkerke, J.F., Nannmark, U. and Kuppen, P.J. (2003). NK cells and the tumour microenvironment: implications for NK-cell function and anti-tumour activity. Trends Immunol **24**, 603–609.

Alici, E., Sutlu, T., Bjorkstrand, B., Gilljam, M., Stellan, B., Nahi, H., Quezada, H.C., Gahrton, G., Ljunggren, H.G. and Dilber, M.S. (2008). Autologous antitumor activity by NK cells expanded from myeloma patients using GMP-compliant components. Blood **111**, 3155–3162.

Anfossi, N., Andre, P., Guia, S., Falk, C.S., Roetynck, S., Stewart, C.A., Breso, V., Frassati, C., Reviron, D., Middleton, D., Romagne, F., Ugolini, S. and Vivier, E. (2006). Human NK cell education by inhibitory receptors for I. Immunity **25**, 331–342.

Bauer, S., Groh, V., Wu, J., Steinle, A., Phillips, J.H., Lanier, L.L. and Spies, T. (1999). Activation of NK cells and T cells by NKG2D, a receptor for stress-inducible MICA. Science **285**, 727–729.

Bix, M., Liao, N.S., Zijlstra, M., Loring, J., Jaenisch, R. and Raulet, D. (1991). Rejection of class I MHC-deficient haemopoietic cells by irradiated MHC-matched mice. Nature **349**, 329–331.

Bottino, C., Castriconi, R., Pende, D., Rivera, P., Nanni, M., Carnemolla, B., Cantoni, C., Grassi, J., Marcenaro, S., Reymond, N., Vitale, M., Moretta, L., Lopez, M. and Moretta, A. (2003). Identification of PVR (CD155) and Nectin-2 (CD112) as cell surface ligands for the human DNAM-1 (CD226) activating molecule. J Exp Med **198**, 557–567.

Bottino, C., Castriconi, R., Moretta, L. and Moretta, A. (2005). Cellular ligands of activating NK receptors. Trends Immunol **26**, 221–226.

Braud, V.M., Allan, D.S., O'Callaghan, C.A., Soderstrom, K., D'Andrea, A., Ogg, G.S., Lazetic, S., Young, N.T., Bell, J.I., Phillips, J.H., Lanier, L.L. and McMichael, A.J. (1998). HLA-E binds to natural killer cell receptors CD94/NKG2A, B and C. Nature **391**, 795–799.

Bryceson, Y.T., March, M.E., Ljunggren, H. G. and Long, E.O. (2006a). Activation, coactivation, and costimulation of resting human natural killer cells. Immunol Rev **214**, 73–91.

Bryceson, Y.T., March, M.E., Ljunggren, H.G. and Long, E.O. (2006b). Synergy among receptors on resting NK cells for the activation of natural cytotoxicity and cytokine secretion. Blood **107**, 159–166.

Burns, L.J., Weisdorf, D.J., DeFor, T.E., Vesole, D.H., Repka, T.L., Blazar, B.R., Burger, S.R., Panoskaltsis-Mortari, A., Keever-Taylor, C.A., Zhang, M.J. and Miller, J.S. (2003). IL-2-based immunotherapy after autologous transplantation for lymphoma and breast cancer induces immune activation and cytokine release: a phase I/II trial. Bone Marrow Transplant **32**, 177–186.

Carbone, E., Neri, P., Mesuraca, M., Fulciniti, M.T., Otsuki, T., Pende, D., Groh, V., Spies, T., Pollio, G., Cosman, D., Catalano, L., Tassone, P., Rotoli, B. and Venuta, S. (2005). HLA class I, NKG2D, and natural cytotoxicity receptors regulate multiple myeloma cell recognition by natural killer cells. Blood **105**, 251–258.

Carlens, S., Gilljam, M., Chambers, B.J., Aschan, J., Guven, H., Ljunggren, H.G., Christensson, B. and Dilber, M.S. (2001). A new method for in vitro expansion of cytotoxic human CD3-CD56+ natural killer cells. Hum Immunol **62**, 1092–1098.

Carlsten, M., Bjorkstrom, N.K., Norell, H., Bryceson, Y., van Hall, T., Baumann, B.C., Hanson, M., Schedvins, K., Kiessling, R., Ljunggren, H.G. and Malmberg, K.J. (2007). DNAX accessory molecule-1 mediated recognition of freshly isolated ovarian carcinoma by resting natural killer cells. Cancer Res **67**, 1317–1325.

Carlsten, M., Malmberg, K.J. and Ljunggren, H.G. (2009). Natural killer cell-mediated lysis of freshly isolated human tumor cells. Int J Cancer **124**, 757–762.

Castriconi, R., Dondero, A., Corrias, M.V., Lanino, E., Pende, D., Moretta, L., Bottino, C. and Moretta, A. (2004). Natural killer cell-mediated killing of freshly isolated neuroblastoma cells: critical role of DNAX accessory molecule-1-poliovirus receptor interaction. Cancer Res **64**, 9180–9184.

Cerwenka, A., Baron, J.L. and Lanier, L.L. (2001). Ectopic expression of retinoic acid early inducible-1 gene (RAE-1) permits natural killer cell-mediated rejection of a MHC class I-bearing tumor in vivo. Proc Natl Acad Sci U S A **98**, 11521–11526.

Chang, C.C. and Ferrone, S. (2006). NK cell activating ligands on human malignant cells: molecular and functional defects and potential clinical relevance. Semin Cancer Biol **16**, 383–392.

Chiesa, S., Mingueneau, M., Fuseri, N., Malissen, B., Raulet, D.H., Malissen, M., Vivier, E. and Tomasello, E. (2006). Multiplicity and plasticity of natural killer cell signaling pathways. Blood **107**, 2364–2372.

Costello, R.T., Fauriat, C., Sivori, S., Marcenaro, E. and Olive, D. (2004). NK cells: innate immunity against hematological malignancies? Trends Immunol **25**, 328–333.

Cudkowicz, G. and Stimpfling, J.H. (1964). Hybrid resistance to parental marrow grafts: association with the K region of H-2. Science **144**, 1339–1340.

Diefenbach, A., Jensen, E.R., Jamieson, A.M. and Raulet, D.H. (2001). Rae1 and H60 ligands of the NKG2D receptor stimulate tumour immunity. Nature **413**, 165–171.

Diermayr, S., Himmelreich, H., Durovic, B., Mathys-Schneeberger, A., Siegler, U., Langenkamp, U., Hofsteenge, J., Gratwohl, A., Tichelli, A., Paluszewska, M., Wiktor-Jedrzejczak, W., Kalberer, C.P. and Wodnar-Filipowicz, A. (2008). NKG2D ligand expression in AML increases in response to HDAC inhibitor valproic acid and contributes to allorecognition by NK-cell lines with single KIR-HLA class I specificities. Blood **111**, 1428–1436.

Dudley, M.E., Wunderlich, J.R., Robbins, P.F., Yang, J.C., Hwu, P., Schwartzentruber, D.J., Topalian, S.L., Sherry, R., Restifo, N.P., Hubicki, A.M., Robinson, M.R., Raffeld, M., Duray, P., Seipp, C.A., Rogers-Freezer, L., Morton, K.E., Mavroukakis, S.A., White, D.E. and Rosenberg, S.A. (2002). Cancer regression and autoimmunity in patients after clonal repopulation with antitumor lymphocytes. Science **298**, 850–854.

El-Sherbiny, Y.M., Meade, J.L., Holmes, T.D., McGonagle, D., Mackie, S.L., Morgan, A.W., Cook, G., Feyler, S., Richards, S.J., Davies, F.E., Morgan, G.J. and Cook, G.P. (2007). The requirement for DNAM-1, NKG2D, and NKp46 in the natural killer cell-mediated killing of myeloma cells. Cancer Res **67**, 8444–8449.

Farag, S.S., Fehniger, T.A., Becknell, B., Blaser, B.W. and Caligiuri, M.A. (2003). New directions in natural killer cell-based immunotherapy of human cancer. Expert Opin Biol Ther **3**, 237–250.

Fauriat, C., Andersson, S., Bjorklund, A.T., Carlsten, M., Schaffer, M., Bjorkstrom, N.K., Baumann, B.C., Michaelsson, J., Ljunggren, H.G. and Malmberg, K.J. (2008). Estimation of the size of the alloreactive NK cell repertoire: studies in individuals homozygous for the group A KIR haplotype. J Immunol **181**, 6010–6019.

Frohn, C., Hoppner, M., Schlenke, P., Kirchner, H., Koritke, P. and Luhm, J. (2002). Anti-myeloma activity of natural killer lymphocytes. Br J Haematol **119**, 660–664.

Ghiringhelli, F., Menard, C., Martin, F. and Zitvogel, L. (2006). The role of regulatory T cells in the control of natural killer cells: relevance during tumor progression. Immunol Rev **214**, 229–238.

Glas, R., Sturmhofel, K., Hammerling, G.J., Karre, K. and Ljunggren, H.G. (1992). Restoration of a tumorigenic phenotype by beta 2-microglobulin transfection to EL-4 mutant cells. J Exp Med **175**, 843–846.

Guerra, N., Tan, Y.X., Joncker, N.T., Choy, A., Gallardo, F., Xiong, N., Knoblaugh, S., Cado, D., Greenberg, N.M. and Raulet, D.H. (2008). NKG2D-deficient mice are defective in tumor surveillance in models of spontaneous malignancy. Immunity **28**, 571–580.

Guimaraes, F., Guven, H., Donati, D., Christensson, B., Ljunggren, H.G., Bejarano, M.T. and Dilber, M.S. (2006). Evaluation of ex vivo expanded human NK cells on antileukemia activity in SCID-beige mice. Leukemia **20**, 833–839.

Hayakawa, Y., Takeda, K., Yagita, H., Smyth, M.J., Van Kaer, L., Okumura, K. and Saiki, I. (2002). IFN-gamma-mediated inhibition of tumor angiogenesis by natural killer T-cell ligand, alpha-galactosylceramide. Blood **100**, 1728–1733.

Herberman, R.B., Nunn, M.E., Holden, H.T. and Lavrin, D.H. (1975a). Natural cytotoxic reactivity of mouse lymphoid cells against syngeneic and allogeneic tumors. II. Characterization of effector cells. Int J Cancer **16**, 230–239.

Herberman, R.B., Nunn, M.E. and Lavrin, D.H. (1975b). Natural cytotoxic reactivity of mouse lymphoid cells against syngeneic acid allogeneic tumors. I. Distribution of reactivity and specificity. Int J Cancer **16**, 216–229.

Hoglund, P., Ljunggren, H.G., Ohlen, C., Ahrlund-Richter, L., Scangos, G., Bieberich, C., Jay, G., Klein, G. and Karre, K. (1988). Natural resistance against lymphoma grafts conveyed by H-2Dd transgene to C57BL mice. J Exp Med **168**, 1469–1474.

Hoglund, P., Ohlen, C., Carbone, E., Franksson, L., Ljunggren, H.G., Latour, A., Koller, B. and Karre, K. (1991). Recognition of beta 2-microglobulin-negative (beta 2m-) T-cell blasts by natural killer cells from normal but not from beta 2m- mice: nonresponsiveness controlled by beta 2m- bone marrow in chimeric mice. Proc Natl Acad Sci U S A **88**, 10332–10336.

Iguchi-Manaka, A., Kai, H., Yamashita, Y., Shibata, K., Tahara-Hanaoka, S., Honda, S., Yasui, T., Kikutani, H., Shibuya, K. and Shibuya, A. (2008). Accelerated tumor growth in mice deficient in DNAM-1 receptor. J Exp Med **205**, 2959–2964.

Karlhofer, F.M., Ribaudo, R.K. and Yokoyama, W.M. (1992). MHC class I alloantigen specificity of Ly-49+ IL-2-activated natural killer cells. Nature **358**, 66–70.

Karre, K. (2008). Natural killer cell recognition of missing self. Nat Immunol **9**, 477–480.

Karre, K., Ljunggren, H.G., Piontek, G. and Kiessling, R. (1986). Selective rejection of H-2-deficient lymphoma variants suggests alternative immune defence strategy. Nature **319**, 675–678.

Kiessling, R., Klein, E., Pross, H. and Wigzell, H. (1975a). "Natural" killer cells in the mouse. II. Cytotoxic cells with specificity for mouse Moloney leukemia cells. Characteristics of the killer cell. Eur J Immunol **5**, 117–121.

Kiessling, R., Klein, E. and Wigzell, H. (1975b). "Natural" killer cells in the mouse. I. Cytotoxic cells with specificity for mouse Moloney leukemia cells. Specificity and distribution according to genotype. Eur J Immunol **5**, 112–117.

Kiessling, R., Hochman, P.S., Haller, O., Shearer, G.M., Wigzell, H. and Cudkowicz, G. (1977). Evidence for a similar or common mechanism for natural killer cell activity and resistance to hemopoietic grafts. Eur J Immunol **7**, 655–663.

Klingemann, H.G. (2005). Natural killer cell-based immunotherapeutic strategies. Cytotherapy **7**, 16–22.

Koh, C.Y., Blazar, B.R., George, T., Welniak, L.A., Capitini, C.M., Raziuddin, A., Murphy, W.J. and Bennett, M. (2001). Augmentation of antitumor effects by NK cell inhibitory receptor blockade in vitro and in vivo. Blood **97**, 3132–3137.

Koh, C.Y., Ortaldo, J.R., Blazar, B.R., Bennett, M. and Murphy, W.J. (2003). NK-cell purging of leukemia: superior antitumor effects of NK cells H2 allogeneic to the tumor and augmentation with inhibitory receptor blockade. Blood **102**, 4067–4075.

Lanier, L.L. (2005). NK cell recognition. Annu Rev Immunol **23**, 225–274.

Law, T.M., Motzer, R.J., Mazumdar, M., Sell, K.W., Walther, P.J., O'Connell, M., Khan, A., Vlamis, V., Vogelzang, N.J. and Bajorin, D.F. (1995). Phase III randomized trial of interleukin-2 with or without lymphokine-activated killer cells in the treatment of patients with advanced renal cell carcinoma. Cancer **76**, 824–832.

Liao, N.S., Bix, M., Zijlstra, M., Jaenisch, R. and Raulet, D. (1991). MHC class I deficiency: susceptibility to natural killer (NK) cells and impaired NK activity. Science **253**, 199–202.

Ljunggren, H.G. and Karre, K. (1985). Host resistance directed selectively against H-2-deficient lymphoma variants. Analysis of the mechanism. J Exp Med **162**, 1745–1759.

Ljunggren, H.G. and Karre, K. (1990). In search of the 'missing self': MHC molecules and NK cell recognition. Immunol Today **11**, 237–244.

Ljunggren, H.G. and Malmberg, K.J. (2007). Prospects for the use of NK-cells in immunotherapy against human cancer. Nat Rev Immunol **7**, 329–339.

Ljunggren, H.G., Ohlen, C., Hoglund, P., Yamasaki, T., Klein, G. and Karre, K. (1988a). Afferent and efferent cellular interactions in natural resistance directed against MHC class I deficient tumor grafts. J Immunol **140**, 671–678.

Ljunggren, H.G., Yamasaki, T., Collins, P., Klein, G. and Karre, K. (1988b). Selective acceptance of MHC class I-deficient tumor grafts in the brain. J Exp Med **167**, 730–735.

Ljunggren, H.G., Paabo, S., Cochet, M., Kling, G., Kourilsky, P. and Karre, K. (1989). Molecular analysis of H-2-deficient lymphoma lines. Distinct defects in biosynthesis and association of MHC class I heavy chains and beta 2-microglobulin observed in cells with increased sensitivity to NK cell lysis. J Immunol **142**, 2911–2917.

Ljunggren, H.G., Sturmhofel, K., Wolpert, E., Hammerling, G.J. and Karre, K. (1990). Transfection of beta 2-microglobulin restores IFN-mediated protection from natural killer cell lysis in YAC-1 lymphoma variants. J Immunol **145**, 380–386.

Malmberg, K.J. and Ljunggren, H.G. (2006). Escape from immune- and nonimmune-mediated tumor surveillance. Semin Cancer Biol **16**, 16–31.

Malmberg, K.J., Bryceson, Y.T., Carlsten, M., Andersson, S., Bjorklund, A., Bjorkstrom, N.K., Baumann, B. C., Fauriat, C., Alici, E., Dilber, M. S. and Ljunggren, H.G. (2008). NK cell-mediated targeting of human cancer and possibilities for new means of immunotherapy. Cancer Immunol Immunother **57**, 1541–1552.

Mendez, R., Rodriguez, T., Del Campo, A., Monge, E., Maleno, I., Aptsiauri, N., Jimenez, P., Pedrinaci, S., Pawelec, G., Ruiz-Cabello, F. and Garrido, F. (2008). Characterization of HLA class I altered phenotypes in a panel of human melanoma cell lines. Cancer Immunol Immunother **57**, 719–729.

Miller, J.S., Soignier, Y., Panoskaltsis-Mortari, A., McNearney, S.A., Yun, G.H., Fautsch, S.K., McKenna, D., Le, C., Defor, T.E., Burns, L.J., Orchard, P.J., Blazar, B.R., Wagner, J.E., Slungaard, A., Weisdorf, D.J., Okazaki, I.J. and McGlave, P.B. (2005). Successful adoptive transfer and in vivo expansion of human haploidentical NK cells in patients with cancer. Blood **105**, 3051–3057.

Moretta, L. and Moretta, A. (2004). Killer immunoglobulin-like receptors. Curr Opin Immunol **16**, 626–633.

Moretta, A., Bottino, C., Mingari, M.C., Biassoni, R. and Moretta, L. (2002). What is a natural killer cell? Nat Immunol **3**, 6–8.

Moretta, L., Bottino, C., Pende, D., Castriconi, R., Mingari, M.C. and Moretta, A. (2006). Surface NK receptors and their ligands on tumor cells. Semin Immunol **18**, 151–158.

Morris, M.A. and Ley, K. (2004). Trafficking of natural killer cells. Curr Mol Med **4**, 431–438.

Muranski, P., Boni, A., Wrzesinski, C., Citrin, D.E., Rosenberg, S.A., Childs, R. and Restifo, N.P. (2006). Increased intensity lymphodepletion and adoptive immunotherapy—how far can we go? Nat Clin Pract Oncol **3**, 668–681.

Ohlen, C., Kling, G., Hoglund, P., Hansson, M., Scangos, G., Bieberich, C., Jay, G. and Karre, K. (1989). Prevention of allogeneic bone marrow graft rejection by H-2 transgene in donor mice. Science **246**, 666–668.

Parham, P. (2005). MHC class I molecules and KIRs in human history, health and survival. Nat Rev Immunol **5**, 201–214.

Pende, D., Spaggiari, G.M., Marcenaro, S., Martini, S., Rivera, P., Capobianco, A., Falco, M., Lanino, E., Pierri, I., Zambello, R., Bacigalupo, A., Mingari, M.C., Moretta, A. and Moretta, L. (2005). Analysis of the receptor-ligand interactions in the natural killer-mediated lysis of freshly isolated myeloid or lymphoblastic leukemias: evidence for the involvement of the Poliovirus receptor (CD155) and Nectin-2 (CD112). Blood **105**, 2066–2073.

Pessino, A., Sivori, S., Bottino, C., Malaspina, A., Morelli, L., Moretta, L., Biassoni, R. and Moretta, A. (1998). Molecular cloning of NKp46: a novel member of the immunoglobulin superfamily involved in triggering of natural cytotoxicity. J Exp Med **188**, 953–960.

Rosenberg, S.A., Lotze, M.T., Muul, L.M., Leitman, S., Chang, A.E., Ettinghausen, S.E., Matory, Y.L., Skibber, J.M., Shiloni, E., Vetto, J.T., et al. (1985). Observations on the systemic administration of autologous lymphokine-activated killer cells and recombinant interleukin-2 to patients with metastatic cancer. N Engl J Med **313**, 1485–1492.

Ruggeri, L., Capanni, M., Casucci, M., Volpi, I., Tosti, A., Perruccio, K., Urbani, E., Negrin, R.S., Martelli, M.F. and Velardi, A. (1999). Role of natural killer cell alloreactivity in HLA-mismatched hematopoietic stem cell transplantation. Blood **94**, 333–339.

Ruggeri, L., Capanni, M., Urbani, E., Perruccio, K., Shlomchik, W.D., Tosti, A., Posati, S., Rogaia, D., Frassoni, F., Aversa, F., Martelli, M.F. and Velardi, A. (2002). Effectiveness of donor natural killer cell alloreactivity in mismatched hematopoietic transplants. Science **295**, 2097–2100.

Ruggeri, L., Mancusi, A., Capanni, M., Martelli, M.F. and Velardi, A. (2005). Exploitation of alloreactive NK cells in adoptive immunotherapy of cancer. Curr Opin Immunol **17**, 211–217.

Ruggeri, L., Aversa, F., Martelli, M.F. and Velardi, A. (2006). Allogeneic hematopoietic transplantation and natural killer cell recognition of missing self. Immunol Rev **214**, 202–218.

Screpanti, V., Wallin, R.P., Grandien, A. and Ljunggren, H.G. (2005). Impact of FASL-induced apoptosis in the elimination of tumor cells by NK cells. Mol Immunol **42**, 495–499.

Sheridan, C. (2006). First-in-class cancer therapeutic to stimulate natural killer cells. Nat Biotechnol **24**, 597.

Smyth, M.J., Hayakawa, Y., Takeda, K. and Yagita, H. (2002). New aspects of natural-killer-cell surveillance and therapy of cancer. Nat Rev Cancer **2**, 850–861.

Smyth, M.J., Cretney, E., Kelly, J.M., Westwood, J.A., Street, S.E., Yagita, H., Takeda, K., van Dommelen, S.L., Degli-Esposti, M.A. and Hayakawa, Y. (2005). Activation of NK cell cytotoxicity. Mol Immunol **42**, 501–510.

Smyth, M.J., Teng, M.W., Swann, J., Kyparissoudis, K., Godfrey, D.I. and Hayakawa, Y. (2006). CD4+ CD25+ T regulatory cells suppress NK cell-mediated immunotherapy of cancer. J Immunol **176**, 1582–1587.

Storkus, W.J., Alexander, J., Payne, J.A., Dawson, J.R. and Cresswell, P. (1989). Reversal of natural killing susceptibility in target cells expressing transfected class I HLA genes. Proc Natl Acad Sci U S A **86**, 2361–2364.

Sun, J.C., Beilke, J.N. and Lanier, L.L. (2009). Adaptive immune features of natural killer cells. Nature.

Torelli, G.F., Guarini, A., Maggio, R., Alfieri, C., Vitale, A. and Foa, R. (2005). Expansion of natural killer cells with lytic activity against autologous blasts from adult and pediatric acute lymphoid leukemia patients in complete hematologic remission. Haematologica **90**, 785–792.

Townsend, A., Ohlen, C., Bastin, J., Ljunggren, H.G., Foster, L. and Karre, K. (1989). Association of class I major histocompatibility heavy and light chains induced by viral peptides. Nature **340**, 443–448.

Trapani, J.A. and Smyth, M.J. (2002). Functional significance of the perforin/granzyme cell death pathway. Nat Rev Immunol **2**, 735–747.

Wagtmann, N., Rajagopalan, S., Winter, C.C., Peruzzi, M. and Long, E.O. (1995). Killer cell inhibitory receptors specific for HLA-C and HLA-B identified by direct binding and by functional transfer. Immunity **3**, 801–809.

Wallace, M.E. and Smyth, M.J. (2005). The role of natural killer cells in tumor control—effectors and regulators of adaptive immunity. Springer Semin Immunopathol **27**, 49–64.

Vitale, M., Bottino, C., Sivori, S., Sanseverino, L., Castriconi, R., Marcenaro, E., Augugliaro, R., Moretta, L. and Moretta, A. (1998). NKp44, a novel triggering surface molecule specifically expressed by activated natural killer cells, is involved in non-major histocompatibility complex-restricted tumor cell lysis. J Exp Med **187**, 2065–2072.

Wu, J. and Lanier, L.L. (2003). Natural killer cells and cancer. Adv Cancer Res **90**, 127–156.

Chapter **Five**

5

NK cell immune recognition: NKG2D ligands and stressed cells

Carlos López-Larrea, Alejandro López-Soto, Segundo González

CHAPTER CONTENTS

The major advances in science are made on the shoulders of giants.
Isaac Newton

ABSTRACT

Natural killer (NK) cells are important in the innate immune response against tumourigenic or virally infected cells. The mechanism involved in the recognition of these cells has been difficult to discern. During the past two decades, there has been a substantial increase in the understanding of how NK cells recognize diseased cells. One of these mechanisms is mediated by NKG2D, which is one of the best characterized NK cell activating receptors. NKG2D binds to a variety of ligands that are not expressed on normal cells, but up-regulated in response to cellular stress, which is frequently observed during microbial infection or cellular transformation. Engagement of NKG2D ligands by its receptor potently activates NK cells and co-stimulates effector T cells, favouring the elimination of the stressed cell by the immune system. This characteristic has sustained the development of the 'induced self' or 'stress self' theory, which is complementary to the 'missing self' theory. Some NKG2D ligands may also be constitutively expressed on some healthy cells, suggesting an additional role of this system that should be explored. NKG2D-signalling is also important in some other prevalent diseases including autoimmune ones. Inappropriate expression of NKG2D ligands may lead to an activation of the immune system against autologous cells, which might trigger or exacerbate a T cell-mediated autoimmune disease.

KEY WORDS

NKG2D, MICA, MICB, ULBPs, NKs

Introduction

Natural killer (NK) cells play an important role in the elimination of transformed and virus infected cells, but should not damage unstressed cells. For such a purpose, NK cells are regulated by both activating and inhibitory surface receptors. Inhibitory receptors bind to ubiquitously expressed major histocompatibility complex (MHC) class I molecules, thus avoiding elimination of autologous cells. However, target cells that have lost MHC class I expression by transformation or viral infection are lysed by NK cells. Thus, NK cells utilize the inhibitory receptors to differentiate 'self' from 'missing self' components (Cerwenka and Lanier, 2001; Diefenbach et al., 2001; González et al., 2006; Raulet 2003) (Figure 5.1). Additionally, NK cell functions are also regulated by activating receptors, including NKG2D, members of the leukocyte immunoglobulin-like receptor (LIR) family and the Nkp30, Nkp44 and Nkp46 proteins, which are also known as natural cytotoxicity receptors. Whereas most NK receptors bind to ubiquitous MHC (Diefenbach and Raulet, 2001) class I molecules, the activating receptor NKG2D recognizes distant relatives of the MHC class I that are inducibly expressed (Bacon et al., 2004; Bahram et al., 1994; Bauer et al., 1998; Chalupny et al., 2003; Cosman et al., 2001; Groh et al., 1998). In mice, NKG2D ligands are not expressed by normal cells, but are up-regulated on numerous tumour and virus infected cells (Cerwenka and Lanier, 2001; Diefenbach et al., 2001), which thus become susceptible to NK cell-killing in vivo (Bauer et al., 1999; Groh et al., 1998, 1999; Oppenheim et al., 2005; Smyth et al., 2005). This has originated the 'induced-' or 'stressed-self' hypothesis, which postulates that NKG2D ligands are not expressed in normal cells, but are up-regulated in response to the cellular damage (Figure 5.1). Thus, NKG2D ligands deliver an induced or damaged self signal that is coupled to cellular stress caused by microbes or malignant cell transformation. In humans, NKG2D was first identified as a receptor for MICA (MHC class I chain-related A) and MICB and, subsequently, for the UL16-binding proteins 1-5 (ULBP1-5). The distribution of MICA and MICB (MICA/B) is also highly restricted in benign tissues, but to date, little is known regarding the expression patterns of ULBPs (Groh et al., 1996). Human NKG2D ligands may be up-regulated by stressful situations, including some microbial infections, and they are frequently expressed in tumours of various origins. NKG2D engagement by its ligands potently activates NK cell functions and co-stimulates effector T cell responses (López-Larrea et al., 2008). A peculiarity of this system is the existence of many different ligands for the same receptor that could be explained, in the context of tumour transformation or infection, as a strategy to elude the immune system.

This system has also acquired an emerging relevance in disease. The immune system always has to maintain a delicate balance between rejecting the foreign and tolerating the self. Therefore, stimuli that induce the aberrant expression of NKG2D ligands in cells may trigger or exacerbate several T-cell–mediated autoimmune diseases (Caillat-Zucman, 2006). The importance of NKG2D in infection, cancer and autoimmunity suggests that there is great potential for manipulation of this system with therapeutic purposes.

NKG2D receptor

In humans, NKG2D is an activating receptor expressed on $CD8^+$ $TCR\alpha\beta$ and $\gamma\delta$ T cells and on all NK cells (López-Larrea et al., 2008). In mice, it is additionally expressed on activated NKT cells, macrophages and dendritic cells. This receptor is a member of the C-type lectin-activating receptor family that is evolutionarily conserved and encoded by a gene located on the human chromosome 12p12-p13 and on chromosome 6 in mice. It shares no direct relationship with other NKG2 receptors and it is not associated

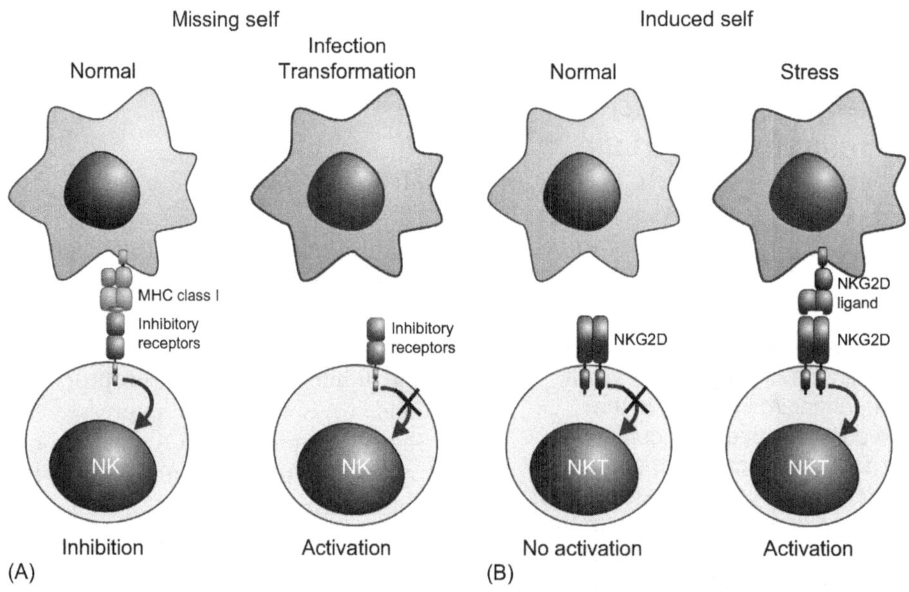

Figure 5.1 • Regulation of NK cell response. (A) NK inhibitory receptors bind to ubiquitous MHC class I molecules ('self') avoiding the elimination of autologous cells. The loss of MHC class I expression by transformation or viral infection ('missing self') leads to the elimination of these cells by NKs. (B) NKG2D recognizes a variety of ligands that are inducibly expressed ('induced-' or 'stressed-self'). NKG2D ligands are not present on most benign cells, but they are up-regulated on numerous tumour and virus infected cells.

with CD94. Instead, NKG2D signalling is mediated by specialized signalling adaptors that activate NK cells and co-stimulate T cells. Mouse NKG2D can associate with two distinct adaptors, DAP10 and DAP12 (KARAP), while human NKG2D exclusively binds to DAP10 (Diefenbach et al., 2002; Rosen et al., 2004; Wu et al., 1999). Alternative splicing of mouse NKG2D generates two different transcripts. The long isoform (NKG2D-L) is constitutively expressed on NK cells and only pairs with DAP10. The short form of NKG2D (NKG2D-S) is up-regulated following NK cell activation and may be associated with either DAP10 or DAP12. Association of NKG2D with these adaptors occurs non-covalently, via oppositely charged amino acids in the respective transmembrane domains. Stoichiometry analysis demonstrated that NKG2D homodimers assemble with two dimers of DAP10 proteins, thereby forming a hexameric structure similar to the T-cell receptor–CD3 complex (Garrity et al., 2005), which likely serves to reduce the threshold of activation. The DAP12 adaptor contains a traditional immunotyrosine-based activation motif (ITAM) in its cytoplasmic tail, similar to other receptors (e.g. TCR, BCR and FcR), whereas signalling through the non-ITAM DAP10 adaptor involves the phosphatidylinositol 3-kinase pathway (López-Larrea et al., 2008).

In general, NKG2D serves as the primary activating receptor in activated NK cells, where NKG2D engagement alone triggers cytotoxicity, even in the presence of NK inhibitory receptors and their respective MHC class I ligands (Bauer et al., 1999; Cerwenka et al., 2001). In contrast, NKG2D is a co-stimulatory receptor in $CD8^+$ T cells, requiring TCR-mediated stimulus for their activation. Conflicting results exist regarding whether NKG2D stimulation is capable of activating T cells in the absence of other stimuli. This discrepancy suggests that the functional outcome of NKG2D engagement may be determined by additional factors, such as the activating status of T cells, or may be modulated by several other factors, including cytokines. Some pro-inflammatory cytokines such as IL-12, IL-15 and IL-18 up-regulate expression of NKG2D and some of its ligands, accompanied by enhancement of NKG2D-mediated effector functions (González et al., 2006).

NKG2D ligand family members: structure and function

NKG2D ligands belong to two relatively distant families related to MHC class I molecules (5–10). One family is composed of MICA and MICB molecules and the other by ULBP1-5 proteins. *MICA/B* genes are encoded in the MHC region and they share structural and sequence similarity with MHC class I genes (28–35%). Like MHC class I proteins, MICA/B have α1-α2-α3 extracellular domains and transmembrane tails. However, they do not associate with β_2-microglobulin or peptides (Figure 5.2). Crystal structures of NKG2D in complex with its ligands show that MICA interacts with the NKG2D dimer through

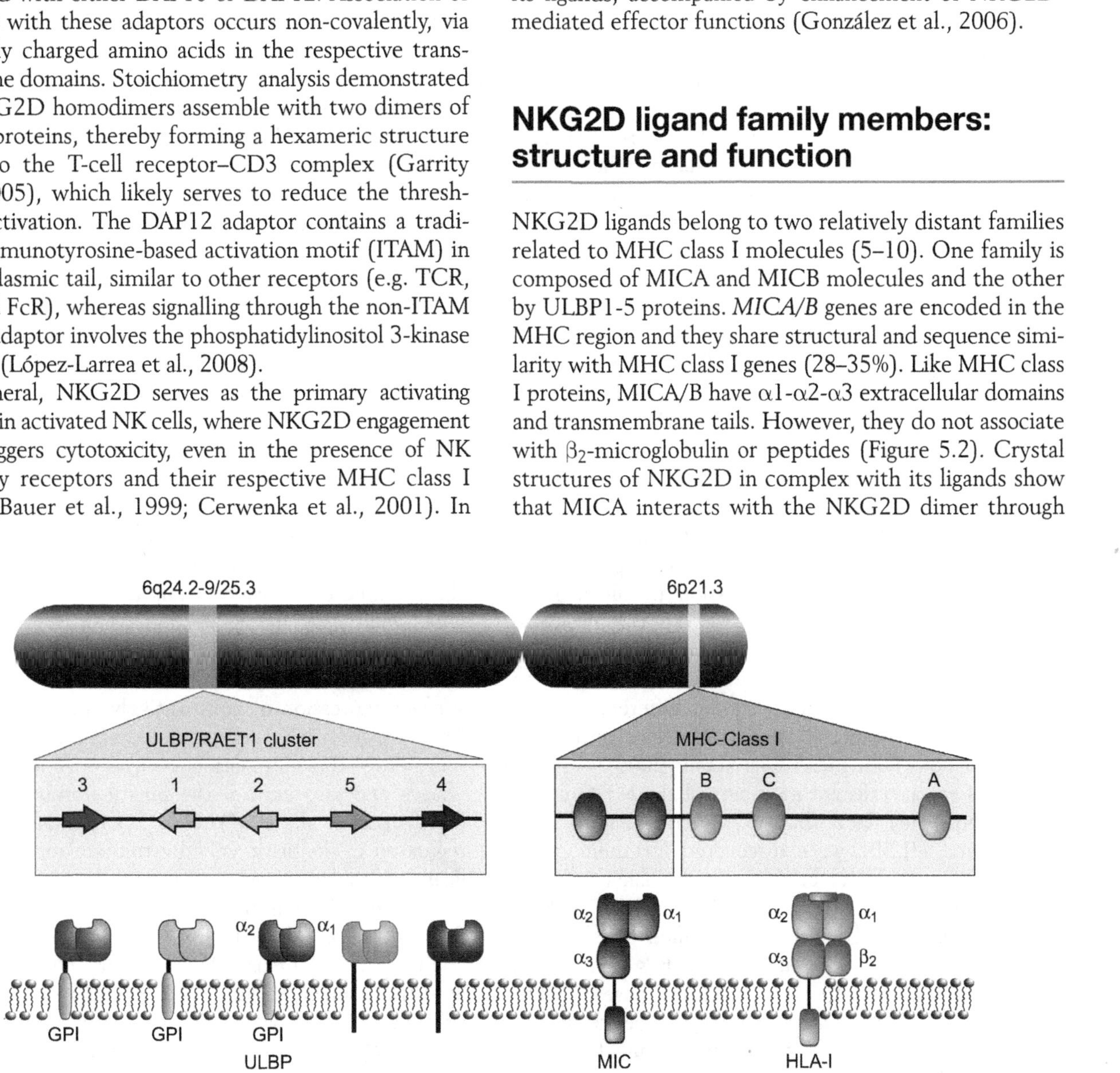

Figure 5.2 • Structure of NKG2D ligands. MICA and MICB are membrane glycoproteins with three extracellular domains (α1-α2-α3), similarly to MHC class I molecules. However, these glycoproteins do not associate with β_2-microglobulin or peptides. MICA and MICB genes are encoded on chromosome 6 in the MHC class I region. ULBP1-5 are membrane glycoproteins with only two extracellular domains (α1-α2), but they lack the α3 domain. ULBP1-3 proteins are bound to the membrane by a glycosylphosphatidylinositol (GPI) anchor, whereas ULPB4 and ULBP5 are transmembrane proteins. Their genes are located on chromosome 6q25, outside of the MHC region.

the $\alpha 1$-$\alpha 2$ domains (Bauer et al., 1998; Radaev et al., 2001). Sequences orthologous to MICA/B are encoded in genomes of all the mammalian species analysed with the exception of rodents (Bahram et al., 1994). The tissue distribution of MICA/B is restricted to variable areas of the intestinal epithelium. However, the functional activity of these molecules in this tissue is ill defined.

MICA/B are highly polymorphic, with close to 60 MICA and 25 MICB recognizable alleles. The significance of this variability is unknown, but MICA alleles may vary in their affinity for NKG2D binding, and these variations can affect the thresholds of recognition by NK cells and T lymphocytes (Li et al., 2000). Moreover, MICA also has a triplet repeat short-tandem polymorphism in its transmembrane region, and six distinct alleles have been reported. These are defined by a different number of GCT repeats that encode for 4–10 alanines in the transmembrane domain with another allele, A5.1, which contains an additional guanine insertion, resulting in a premature stop codon within the transmembrane region itself. The alleles, which carry the MICA A5.1 polymorphism, encode a glycoprotein lacking the cytoplasmic tail, which is relevant for targeting the protein at an apparently non-physiological basolateral localization in enterocytes (Suemizu et al., 2002). Interestingly, the particular MICA008/5.1 allele has been observed frequently in individual populations, with high gene frequencies of up to 50% in Caucasians. It has been described to be associated with a variety of autoimmune diseases such as Behcet's disease, celiac disease (CD), psoriasis vulgaris and Kawasaki disease (Bahram et al., 2005). Additionally, the promoter of MICB has several polymorphisms that modify the level of gene expression by altering the binding of the transcription factors to the promoter (Rodríguez-Rodero et al., 2007). This suggests an important variation in MICB expression among normal individuals, and this could imply relevant differences in the natural immune response against infections or tumour transformation (Rodríguez-Rodero et al., 2006).

ULBPs are also distant members of the MHC class I family, but they lack the $\alpha 3$ domain (Figure 5.2). ULBP1 and ULBP2 were discovered for their capacity to specifically interact with the human cytomegalovirus (HCMV) UL16 protein (Cosman et al., 2001). Three additional ULBPs were identified by sequence homology. The $\alpha 1$-$\alpha 2$ domains of ULBPs share about 50% homology and are equidistant from those of MHC class I and MICA/B, with about 25% sequence homology. ULBP1-3 are bound to the membrane by glycosylphosphatidylinositol (GPI) anchors whereas ULPB4 and ULBP5 are transmembrane proteins (Bacon et al., 2004; Chalupny et al., 2003; Cosman et al., 2001). Mice express the ULBP homologous proteins, namely, retinoic acid early inducible-1 (RAE-1), H60, and MULTI-1 (Cerwenka and Lanier, 2001; Diefenbach et al., 2001). These mouse ligands are structurally similar to MICA/B, but they lack the $\alpha 3$ extracellular domain. H60 is a transmembrane protein, whereas RAE-1s are attached to the membrane by GPI anchors. ULBPs have no direct relation with MICA/B and their genes are encoded on chromosome 6q25, outside of the MHC region. Significantly, it has been observed that MICA is S-acetylated at the transmembrane region (Eleme et al., 2004) and is expressed in lipid rafts. ULBP1-3 are also expressed in lipid rafts and accumulate at the activating NK immune-synapse cell surface region. A key question is whether or not the different ligands are equivalent in their capacity to trigger NKG2D signalling. Structural and binding studies have been carried out for some of the ligand interactions and suggest that such can compete among themselves for NKG2D binding.

Intriguingly, ULBP1-3 transcripts are quite ubiquitously expressed, whereas ULBP4 exhibits a more restricted tissue distribution, which is being expressed on the skin (Chalupny et al., 2003). However, there is not a direct correlation of mRNA expression of ULBP1-3 with surface protein expression, therefore suggesting an important role of post-transcriptional mechanisms in the regulation of these genes. Preliminary data indicate that ULBP proteins are expressed on the surface of some healthy cells, including epithelial, endothelial, antigen processing cells and hematopoietic cells (González et al., 2006). However, a detailed description of the pattern distribution and functional activity of ULBPs in healthy tissues remains to be established. Nevertheless, it suggests that they may also play a significant role in physiological conditions. In fact, NKG2D ligands participate in the immunoregulatory crosstalk between immune cells, which may regulate innate and adaptive immune responses. In macrophages, lipopolysaccharide (LPS) up-regulates the expression of ULBP1-3 and induces surface expression of constitutively transcribed MICA. This triggers NK cell-mediated cytotoxicity, which may allow the elimination of over-stimulated macrophages and also control the innate immune responses (Nedvetzki et al., 2007). NKG2D ligands are also up-regulated in dendritic cells by toll-like receptor (TLR) ligand stimulation, and they may participate in the activation of T cells and NK cells (Ebihara et al., 2007; Schrama et al., 2006). T cell activation in vitro up-regulates NKG2D ligand expression, which may provide the control of adaptive immune response by NK cells (Cerboni et al., 2007b; Diefenbach et al., 2000; Rabinovich et al., 2003). NK cells and dendritic cells also have the ability to reciprocally activate one another by cell–cell contact involving NKG2D receptor–ligand interaction and soluble mediators (Zwirner et al., 2007). In spite of such little knowledge existing regarding this particular matter, current data suggest that, similar to other apoptotic mechanisms, NKG2D may participate not only in the control of pathological or damaged cells, but may also regulate the physiology of normal cells.

NKG2D ligands are inducible: role in cancer

NKG2D ligands are frequently over-expressed on tumour cells from several origins and their expression renders tumour cells susceptible to NK cell-killing even if the transformed cells have normal MHC class I expression. NKG2D ligand expression in mice is induced by carcinogens and genotoxic stress, and tumour cells expressing these proteins are readily eliminated by NK and $CD8^+$ T cells in vivo (Cerwenka and Lanier, 2001; Diefenbach et al., 2001). Similarly, mice lacking $\gamma\delta$ T cells are highly susceptible to epithelial tumours, and NKG2D-expressing cells could eliminate skin carcinoma cells by an NKG2D-dependent mechanism in vitro (Girardi et al., 2001). In humans, MICA/B proteins are frequently expressed in epithelial tumours of multiple origins. However, they are less frequently found in hematopoietic malignancies. In contrast, ULBPs are not usually expressed in epithelial tumours, but are expressed in leukemias. The prominent role of NKG2D in the anti-tumour immune response is highlighted by the abundance of immune evasion mechanisms observed in cancer cells. To avoid the immune response, cancer cells repress the expression or function of NKG2D ligands or diminished NKG2D function on immune cells (Groh et al., 2002; Salih et al., 2002).

The mechanisms that up-regulate the expression of MICA/B and ULBPs on cancer cells are not well defined (Gasser et al., 2005; González et al., 2008). Some oncogenes may up-regulate the expression of NKG2D ligands. In chronic myeloid leukemia, the BCR/ABL fusion oncoprotein induces the expression of MICA on the surface of leukemic cells, whereas it is absent on healthy donor hematopoietic $CD34^+$ cells (Boissel et al., 2006; Cebo et al., 2006). The adenovirus serotype 5 E1A oncogene also up-regulates NKG2D ligand expression and sensitizes tumour cells to NK cell-mediated and tumour rejection in vivo (Routes et al., 2005). However, transformation alone is not sufficient to induce NKG2D ligand transcription since the expression of several oncogenes (*K-ras* and *c-myc* or *Akt* and *c-myc*), or the lack of tumour suppressor p53, are not sufficient to induce NKG2D ligand expression in primary ovarian epithelial cells (Gasser et al., 2005). Instead, there is increasing evidence that the main mechanism involved in the up-regulation of NKG2D ligands on cancer cells is cellular stress. Thus, the first insight into the regulation of NKG2D ligands was that heat shock induces MICA expression (Groh et al., 1998). NKG2D ligands are up-regulated in response to oxidative stress (Borchers et al., 2006; Yamamoto et al., 2001) and are also induced in non-tumour cell lines by genotoxic stress and stalled DNA replication (Gasser et al., 2005), which activates a major DNA damage checkpoint pathway initiated by ATM (ataxia telangiectasia, mutated) or ATR (ATM- and Rad3-related) protein kinases. The activation of ATM and ATR leads to cell cycle arrest and up-regulation of the DNA repair functions, and, if there is severe damage, this leads to apoptosis. Additionally, DNA damage also activates the expression of NKG2D ligands in, at least partially, a p53-independent fashion. It also favours the elimination of cancer cells by the immune system. DNA damage response is also constitutively activated in early carcinogenesis and in cell lines (Bartkova et al., 2005, 2006), which suggests that the chronic activity of the DNA damage response pathway may account for the constitutive expression of NKG2D ligands in primary tumours and cell lines. Nevertheless, the potential role of the NKG2D/NKG2D ligand system in tumour rejection and surveillance renders it as an interesting target for immunotherapy.

Tumour immune evasion

If the immune system is unable to eliminate the tumour, it sculpts or edits its phenotype, eliminating the most immunogenic cancer cells and, thereby, selecting the development of non-immunogenic tumours (Dunn et al., 2004). Consequently, advanced tumours accumulate several mechanisms to evade the immune system. Thus, although the expression of MICA/B may result in the elimination of the tumour, the shedding of MICA, probably by proteolysis, from the surface of the cells to the plasma is a common characteristic of many tumours expressing this protein (Groh et al., 2002; Salih et al., 2002). In accordance with the presence of soluble MICA on multiple primary tumours, diminished expression of NKG2D on circulating NK and T cells is observed (Groh et al., 1999). This soluble form reduces the amount of NKG2D ligand at the membrane of the tumour cells, causing endocytosis and degradation of the NKG2D receptor on $CD8^+$ T and NK cells and stimulates the expansion of $NKG2D^+$ $CD4^+$ T cells with immune suppressor-like functions (Groh et al., 2006). Significantly, an unexpected molecular mechanism that regulates MICA shedding has been recently described (Kaiser et al., 2007). MICA, but not ULBPs, interacts on the surface of tumour cells with the chaperone endoplasmic reticulum protein 5 (ERp5). Endoplasmic reticulum chaperones are also up-regulated in cancer cells and they may be translocated to the surface in response to cellular stress. The mechanisms and stimuli that transport ERp5 to the cell surface are unknown; however, it is probably independent of MICA, as intracellular interactions have not been observed. On the cell surface of tumour cells, ERp5 forms a transitory complex with MICA and reduces an inaccessible disulfide bond in the α3 domain, which must induce a conformational change that is essential for the proteolytic cleavage of MICA.

Localized high level expression of NKG2D ligands inhibits NK cell function (Oppenheim et al., 2005). The chronic exposure to tumour cells expressing NKG2D ligands alters NKG2D signalling and may facilitate the evasion of cancer cells from NK cell responses. The persistent expression of NKG2D ligands also results in a pronounced down-modulation of NKG2D on activated $CD8^+$ T cells, which may reduce co-stimulatory signals for proliferation and cell survival, mediated by the disruption of the DAP10/Akt pathway. Moreover, the existence of bilateral transfer of NKG2D and ligands (MICB) from NK cells to target cells in the context of a cytotoxic NK cell immune synapse has been observed (Roda-Navarro et al., 2006). Transfer of MICB from targets to NK cells is also observed in this synapse. These processes lead to a marked reduction in the capacity of the NK cells to mediate NKG2D-dependent cytotoxicity. Both, transfer of NK cell NKG2D to the target cell and the presence of soluble NKG2D ligands, may contribute to down-regulation of NKG2D expression and also to favouring the evasion of the tumour from the immune system.

The repression of the NKG2D ligand transcription by epigenetic mechanisms, in particular the structure of chromatin, is another relevant immune evasion mechanism. It has recently been reported that despite the fact that MICA/B and ULBP1 are both inducible genes, their transcription is mainly regulated by ubiquitous transcription factors including Sp1, Sp3 and NFY (CBF) (López-Soto et al., 2006; Roda-Navarro et al., 2006; Rodríguez-Rodero et al., 2007; Venkataraman et al., 2007). However, the binding of these transcription factors to their promoters may be regulated by the histone deacetylases (HDACs) and the structure of chromatin. Histones comprise the protein backbone of chromatin and HDACs remove the acetyl groups of histones, thus allowing the formation of compacted chromatin. This restrains the accessibility of transcription factors and the general transcription machinery to the promoter sequences, which repress the expression of the gene. HDACs are over-expressed in many cancer types and they participate in the repression of numerous proteins with tumour-suppression activities. In particular, histones associated with the promoter of MICA and MICB are hypoacetylated on leukemic cells and epithelial tumours and, consequently, treatment with HDAC inhibitors markedly induces the expression of MICA/B in these cells (Armeanu et al., 2005; Kato et al., 2007; Skov et al., 2005). Significantly, treatment with HDAC inhibitors did not increase the expression levels of MICA/B in mononuclear cells from healthy donors, suggesting that the epigenetic repression of MICA/B expression of MICA/B is specific to tumour cells, probably as a result of selection by the immune system or immunoediting (Dunn et al., 2004).

Importantly, we have recently gained insight into the epigenetic regulation of NKG2D ligands in cancer cells. In this sense, we demonstrated that the histone deacetylase 3 (HDAC3) is a key repressor of the expression of ULBPs in epithelial cancer cells, suggesting that the use of specific inhibitors of this HDAC may become a powerful strategy to enhance the immunogenicity of tumours through the activation of ULBPs expression in cancer cells (Lopez-Soto et al., 2009). Another epigenetic mechanism that may repress the expression of MICA/B has recently been described. This is mediated by microRNAs (miRs) that target the MICA/B 39-untranslated region, which is probably involved in the control of MICA/B protein expression in normal conditions, setting a threshold to limit killing of normal cells (Stern-Ginossar et al., 2007). However, during stress, the amount of miRs does not change significantly and the up-regulation of MICA and MICB mRNA overcomes the miR suppressive capacity. The existence of the viral hcmv-miR-UL112 that specifically targets MICB is also an immunoevasion mechanism used by CMV to evade the immune system (Stern-Ginossar et al., 2008). It is conceivable that a miR-based mechanism may also be used by some cancer cells to avoid immunity.

Transforming growth factor-β (TGF-β) may also account for the suppression of NKG2D ligand expression (Eisele et al., 2006; Friese et al., 2004). It is well established that TGF-β promotes cancer progression and immune evasion, in part, because it is a potent inhibitor of T-cell–mediated tumour clearance and NK activity. The release of TGF-β by tumour cells down-regulates NKG2D expression in $CD8^+$ T and NK cells, diminishing MICA and ULBP2 transcript expression and surface protein levels on malignant glioma cells. Additionally, TGF-β may play an important role in the conversion of normal T cells ($CD4^+$ $CD25^+$) into regulatory T cells (typically $CD4^+$ $CD25^+$ $FOXP3^+$ or T_{Reg}), a population of T cells that is found in high levels in the tumour microenvironment. T_{Reg} cells reduce NKG2D expression, suppress NKG2D-mediated NK cell cytotoxicity and accelerate the progression of tumours (Ghiringhelli et al., 2005; Smyth et al., 2006).

Role of NKG2D in infection

NK cells are important mediators of the innate immune response to infection of several pathogens, particularly the herpes virus, and NKG2D plays a crucial role in this anti-viral response. Several viral infections up-regulate

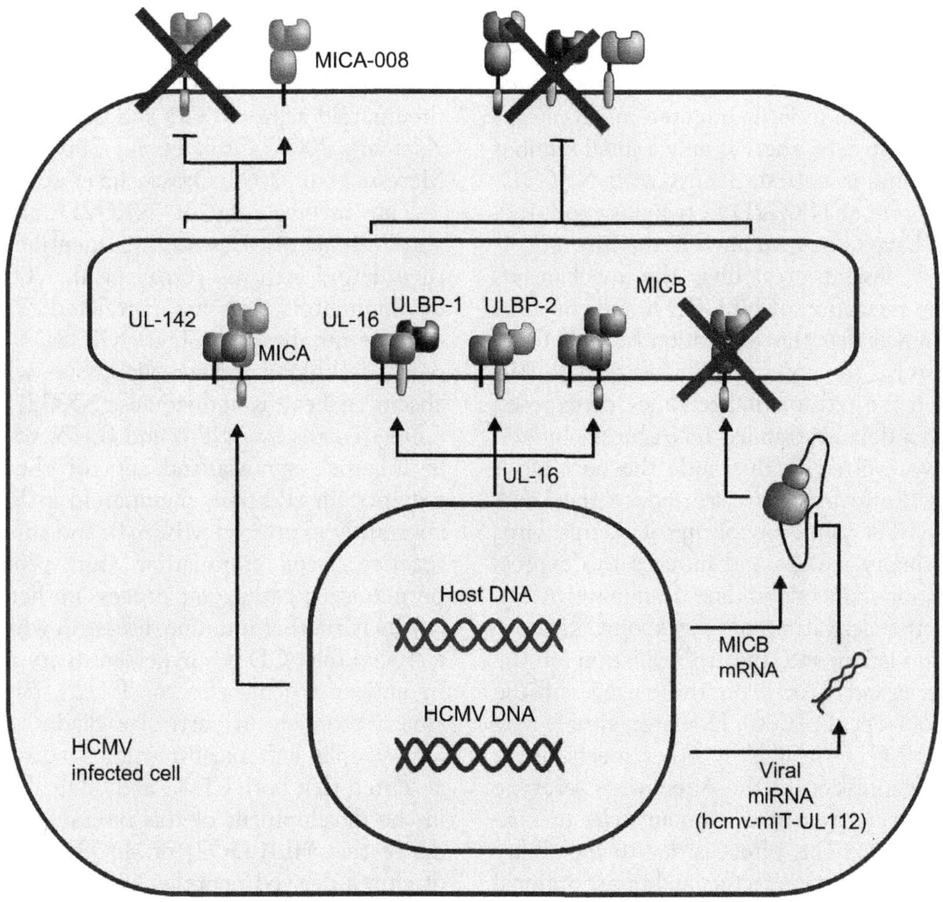

Figure 5.3 • Targeting NKG2D ligands in CMV infection. The down-regulation of expression of NKG2D ligands has been exploited by different viruses. Human and mouse cytomegalovirus (CMV) encode proteins that retain and block cell-surface expression of some NKG2D-ligands. HCMV-encoded glycoprotein UL16 sequesters three of these ligands (MICB, ULBP1 and ULBPs) in the endoplasmic reticulum but not the remaining ligands (ULBP3-4 and MICA). The viral product UL142 down-regulate full-length MICA but not the truncated allele MICA*008. In addition, miRs (hcm-miT-UL112) encoded by HCMV may also down-regulate specifically MICB expression during viral infection. Possessing multiple polymorphic ligands is clearly advantageous to the host during the development of viral immune evasion strategies.

the expression of NKG2D ligands in infected cells and stimulate a T-cell– and NK-cell–mediated response. In particular, the role of NKG2D in the immune response against HCMV has been widely studied (Dunn et al., 2003). HCMV evades virus-specific $CD8^+$ T cells responses via down-regulation of surface MHC class I molecules. Similarly, NKG2D ligands are up-regulated in response to infection by HCMV, and co-stimulate specific cytotoxic T cell response. However, the HCMV UL16 glycoprotein binds and retains MICB, ULBP1 and ULBP2 intracellularly, interfering with the NKG2D-mediated response (Dunn et al., 2003; Valés-Gómez et al., 2003; Welte et al., 2003). Likewise, the HCMV protein UL142 is also able to target (Figure 5.3) full-length MICA by down-regulating its cell surface expression and leading to protection from NK cytotoxicity (Chalupny et al., 2006). However, UL142 does not cause down-regulation of a truncated and prevalent allele MICA*008, which suggests that selective pressure on the host by CMV infection may have resulted in the evolutionary selection of alleles such as MICA*008. As mentioned above, HCMV also down-regulates MICB expression by a miR that targets the 3′-untranslated region of the *MICB* gene. NKG2D is also involved in the control of infection by other members of the herpes virus family, such as Epstein–Barr virus (EBV), up-regulating the expression of ULBP1, rendering the infected B cells susceptible to NK cell-mediated lysis (Pappworth et al., 2007). NKG2D-mediated response may also play an important role in host defence against human immunodeficiency virus 1 (HIV-1) infection (Ward et al., 2007). ULBPs are detected in HIV-infected $CD4^+$ T cells, which render infected cells susceptible to NKG2D-mediated cytotoxity. However, HIV-1 has evolved to escape from this cytotoxic response, since HIV-1 Nef protein down-modulates cell-surface expression of MICA, ULBP1 and ULBP2 (Cerboni et al., 2007a). MICA expression is also induced in T cells infected with another human retrovirus, the human T cell lymphotropic virus 1 (HTLV-1) (Azimi et al., 2006). Similarly, influenza A and Sendai viruses induce MICB mRNA expression in virus-infected macrophages (Sirén et al., 2004). To date, there is only a small number of studies connecting infectious agents with NKG2D. A more universal role of NKG2D in response to infectious disease will have to be examined in the future.

There are few insights regarding the mechanisms involved in the up-regulation of NKG2D ligands by virus infection. It is known that the replication of viral DNA in the nucleus also has the potential to activate the DNA damage response if the ends of viral genomes are exposed and recognized as a double stranded DNA break. In fact, many viruses have evolved to dismantle the host DNA damage response. Retrovirus may also induce the DNA damage response. The Abelson murine leukemia virus infects mouse primary B cells and induces the expression of the activation-induced cytidine deaminase (AID), which leads to the activation of checkpoint kinase-1 (Chk-1) phosphorylation and the up-regulation of the NKG2D murine ligand RAE-1 on the surface of the infected cells (Gourzi et al., 2006). However, viruses may also up-regulate NKG2D ligands by other mechanisms. For instance, the expression of the Adenovirus serotype 5 (Ad5) E1A oncogene sensitizes tumour cells to NK-mediated killing in vivo. This effect is due to the ability of E1A to bind to transcription factors and transcriptional co-adaptor molecules modulating the transcription of the host cells (Routes et al., 2005). In some cases, this is due, at least in part, to chromatin remodelling. Thus, E1A up-regulates NKG2D ligands by interacting with p300, which is a co-adaptor of several transcription factors. This molecule has histone acetyl transferase (HAT) activity and, as a consequence, it has the capacity of remodelling the chromatin structure and modifying the transcription profile of infected cells. The expression of MICA/B in HTLV-1 infected T cells is transactivated by viral Tax protein through the displacement of HDACs, which may also regulate the chromatin structure (Azimi et al., 2006). It is not yet clear whether the regulation of the transcription of the host genes by epigenetic mechanisms may be a common pathway used by other viral proteins.

NKG2D and disease

In predisposed individuals, either antibody or T cells can sustain an adaptive response against self antigens leading to autoimmunity. Since MICA/B can direct NK and cytotoxic T cells to recognize and lyse stressed cells, it has been predicted that the aberrant or inappropriate expression of NKG2D ligands in predisposed individuals may trigger or exacerbate T cell-mediated autoimmune diseases. In agreement with this notion, there is increasing evidence that NKG2D is involved in the pathogenesis of several T cell-mediated autoimmune diseases including rheumatoid arthritis, CD and diabetes mellitus (Caillat-Zucman, 2006; Groh et al., 2003; Hue et al., 2004; Meresse et al., 2004; Ogasawara et al., 2003, 2004).

The involvement of NKG2D and its ligands in autoimmune diseases was first identified in patients with rheumatoid arthritis (Groh et al., 2003). The severity of rheumatoid arthritis correlated with the presence of large numbers of $CD4^+$ $NKG2D^+$ cells in both the peripheral blood and synovial tissue, which are normally absent in healthy individuals. NKG2D was induced in $CD4^+$ T cells by TNF-α and IL-15, which are abundant in inflamed synovia and sera of rheumatoid arthritis patients. In addition, rheumatoid arthritis synoviocytes aberrantly expressed MICA/B, and this may cause autoreactive T-cell stimulation, thus promoting the self-perpetuating pathogenic process in rheumatoid arthritis.

CD is another immune disease in which NKG2D plays a crucial role. CD is a hypersensitivity reaction mediated by antigen-specific effector T cells, triggered by dietary gluten proteins, in particular gliadin, which causes villous atrophy and malabsorption syndrome. It is generally accepted that both $CD4^+$ and $CD8^+$ T cells are involved in the development of this disease. There is strong evidence that HLA-DQ2 (or -DQ8) mediated recognition of gluten-derived peptides by $CD4^+$ T cells plays a key role in the pathogenesis of CD (Sollid, 2002). These $CD4^+$ T cells do not express NKG2D, thus excluding a mechanism similar to that observed in patients with rheumatoid arthritis (Hue et al., 2004; Meresse et al., 2004). However, there is also a significant role of $CD8^+$ T cells in the development of this disease. There is a massive infiltration of intraepithelial lymphocytes in the mucosa of CD patients which may also be involved in antigen-independent epithelial destruction. These intraepithelial lymphocytes are $CD8^+$ T cells and, consequently, they constitutively express NKG2D. It has recently been shown that MICA is over-expressed on the surface of gut epithelial cells during active CD, whereas it remains intracellular in normal enterocytes. MICA disappears during gluten free diet and is re-expressed again after gliadin challenge, suggesting that gliadin itself is responsible for the up-regulation of MICA expression. Moreover, under conditions of deregulated IL-15 expression, frequently observed in CD patients, the intraepithelial lymphocytes change their phenotype to that of lymphokine-activated killer (LAK) cells capable of mediating epithelial cell damage by the recognition of MICA on epithelial cells in an antigen-independent pathway (Meresse et al., 2004).

NKG2D is also involved in the development of autoimmune diabetes in non-obese diabetic (NOD) mice, a model of human type I diabetes (Ogasawara et al., 2003, 2004). The murine ligands of NKG2D, that is, the RAE-1 proteins, are not expressed in the pancreas of non-diabetic mice, but they are expressed in the islet cells of the pre-diabetic and diabetic NOD mice. Concomitantly, the autoreactive $CD8^+$ T cells infiltrating the pancreas in NOD mice expressed NKG2D. In mice, NKG2D is not present in naïve $CD8^+$ T cells, which suggests that NKG2D is not involved in the initial antigen priming. Nevertheless, treatment with neutralizing anti-NKG2D antibody in vivo prevents, even when it is administrated relatively late during disease progression, the development of diabetes. This demonstrates that NKG2D is crucial for the progression of diabetes in NOD mice. In humans, the role of this receptor in the pathogenesis of this disease has not as yet been functionally addressed. Although polymorphisms of human MICA have been associated with type I diabetes, the functional relevance of such is far from clear (Caillat-Zucman, 2006).

MICA and, to a lesser extent, MICB are polymorphic in their extracellular domains. Numerous studies have investigated the relationship between MICA and MICB alleles and susceptibility to disease with closely linked HLA-B and -C alleles including Behcet's disease, ankylosing spondylitis and psoriasis (Stephes, 2001). However, positive associations are likely secondary, because of strong linkage disequilibrium between MICA with HLA-B and -C, and these have not been confirmed by analysis of haplotypes in different ethnic groups or by functional analysis. However, consistent with the role of NKG2D in the pathogenesis of the T-cell–mediated autoimmune diseases, MICA/B genes have also been associated with genetic susceptibility to class II-linked diseases such as rheumatoid arthritis, multiple sclerosis and CD (Fernández-Morera et al., 2008; González et al., 2004; López-Arbesú et al., 2007; López-Vázquez et al., 2002a,b). One convincing explanation for these associations has been recently published (Rodríguez-Rodero et al., 2007). The functional role of MICA and MICB polymorphisms has not been extensively addressed. However, in the MICB promoter, there is a deletion of two base pairs at −66 (AG/–) that diminishes MICB transcription. This polymorphism is quite frequent in the general population, which implies that there are important variations in MICB expression among individuals. Those who express higher levels of MICB are associated with a higher risk of suffering autoimmunity. It could be predicted that these individuals, who are more prone to develop autoimmunity, could be more protected from certain infections or even the development of cancer. However, this would need to be addressed experimentally.

NKG2D in graft and transplantation

In addition to the role described in autoimmunity, NKG2D has also been recently implicated in the outcome of organ transplantation (Collins, 2004). Similar to other related MHC molecules, MICA/B are highly polymorphic and may also be a target of allograft rejection. NKG2D ligands may act as stimulators or targets of an antibody response or they may induce an NKG2D-mediated cellular immune response. Presence of NKG2D ligands have been described in human transplant samples under circumstances of acute allograft rejection, chronic allograft nephropathy and renal acute tubular necrosis (Quiroga et al., 2006). Moreover, it has been suggested that expression of NKG2D in biopsies and urine during acute kidney allograft rejection implies that this molecule is an additionally informative biomarker of transplant rejection (Seiler et al., 2007). The presence of soluble MICA has also been correlated with a lower incidence of rejection (Suárez-Álvarez et al., 2006). This clearly suggests that better recipient–donor selection based on MICA/B compatibility will improve graft outcome. In fact, blockade of NKG2D with a neutralizing, non-depleting anti-NKG2D monoclonal antibody prevents NK cell-mediated bone marrow rejection in certain mouse strains (Ogasawara et al., 2005). Furthermore, blockade of NKG2D prolongs cardiac allograft survival in CD28-deficient mice and in wild-type mice if coupled with B7 blockade, representing the first demonstration of a functional role for NKG2D–NKG2D ligand interactions in the alloimmune response against solid organ grafts (Kim et al., 2007). These findings extend the functional reach of the NK system to include regulation of adaptive T cell responses and suggest that antibody-mediated blockade of NKG2D–NKG2D ligand interactions represents a new clinical approach for immunointervention on activated T cells in the context of transplantation.

Taken together, these findings suggest that, while the NKG2D pathway may be beneficial to protect us against infection and cancer, if this system is not properly regulated it may cause autoreactive T-cell stimulation, thus promoting T-cell–mediated autoimmune diseases or graft rejection. Furthermore, it also suggests that NKG2D and MICA could be promising therapeutic targets in these diseases.

Acknowledgement

This work was supported by the Spanish grants of Fondo de Investigaciones Sanitarias: PI-1080566, PI-6/0841, PI 080/566 and Red de Investigación Renal 'REDINREN' (RD06/0016); FICYT PC-06/010 and, 'Fundación Mútua Madrileña 2007–2009'. A.L.S. holds a predoctoral fellowship from FICYT of Asturias (BP06-99).

References

Armeanu, S., Bitzer, M., Lauer, U.M., Venturelli, S., Pathil, A., Krusch, M., Kaiser, S., Jobst, J., Smirnow, I., Wagner, A., Steinle, A. and Salih, H.R. (2005). Natural killer cell-mediated lysis of hepatoma cells via specific induction of NKG2D ligands by the histone deacetylase inhibitor sodium valproate. Cancer Res **65**, 6321–6329.

Azimi, N., Jacobson, S., Tanaka, Y., Corey, L., Groh, V. and Spies, T. (2006). Immunostimulation by induced expression of NKG2D and its MIC ligands in HTLV-1-associated neurologic disease. Immunogenetics **58**, 228–252.

Bacon, L., Eagle, R.A., Meyer, M., Easom, N., Young, N.T. and Trowsdale, J. (2004). Two human ULBP/RAET1 molecules with transmembrane regions are ligands for NKG2D. J Immunol **73**, 1078–1084.

Bahram, S., Bresnahan, M., Geraghty, D.E. and Spies, T. (1994). A second lineage of mammalian major histocompatibility complex class I genes. Proc Natl Acad Sci U S A **91**, 6259–6263.

Bahram, S., Inoko, H., Shiina, T. and Radosavljevic, M. (2005). MIC and other NKG2D ligands, from none to too many. Curr Opin Immunol **17**(5), 505–509.

Bartkova, J., Horejsi, Z., Koed, K., Kramer, A., Tort, F., Zieger, K., Guldberg, P., Sehested, M., Nesland, J.M., Lukas, C., Orntoft, T., Lukas, J. and Bartek, J. (2005). DNA damage response as a candidate anti-cancer barrier in early human tumourigenesis. Nature **434**, 864–870.

Bartkova, J., Rezaei, N., Liontos, M., Karakaidos, P., Kletsas, D., Issaeva, N., Vassiliou, L.V., Kolettas, E., Niforou, K., Zoumpourlis, V.C., Takaoka, M., Nakagawa, H., Tort, F., Fugger, K., Johansson, F., Sehested, M., Andersen, C.L., Dyrskjot, L., Ørntoft, T., Lukas, J., Kittas, C., Helleday, T., Halazonetis, T.D., Bartek, J. and Gorgoulis, V.G. (2006). Oncogene-induced senescence is part of the tumourigenesis barrier imposed by DNA damage checkpoints. Nature **444**, 633–637.

Bauer, S., Willie, S.T., Spies, T. and Strong, R.K. (1998). Expression, purification, crystallization and crystallographic characterization of the human MHC class I related protein MICA. Acta Crystallogr D Biol Crystallogr **54**, 451–453.

Bauer, S., Groh, V., Wu, J., Steinle, A., Phillips, J.H., Lanier, L.L. and Spies, T. (1999). Activation of NK cells and T cells by NKG2D, a receptor for stress-inducible MICA. Science **285**, 727–729.

Boissel, N., Rea, D., Tieng, V., Dulphy, N., Brun, M., Cayuela, J.M., Rousselot, P., Tamouza, R., Le Bouteiller, P., Mahon, F. X., Steinle, A., Charron, D., Dombret, H. and Toubert, A. (2006). BCR/ABL oncogene directly controls MHC class I chain-related molecule A expression in chronic myelogenous leukemia. J Immunol **176**, 5108–5116.

Borchers, M.T., Harris, N.L., Wesselkamper, S.C., Vitucci, M. and Cosman, D. (2006). NKG2D ligands are expressed on stressed human airway epithelial cells. Am J Physiol Lung Cell Mol Physiol **291**, L222–L231.

Caillat-Zucman, S. (2006). How NKG2D ligands trigger autoimmunity? Hum Immunol **67**(3), 204–207.

Cebo, C., Da Rocha, S., Wittnebel, S., Turhan, A.G., Abdelali, J., Caillat-Zucman, S., Bourhis, J.H., Chouaib, S. and Caignard, A. (2006). The decreased susceptibility of Bcr/Abl targets to NK cell-mediated lysis in response to imatinib mesylate involves modulation of NKG2D ligands, GM1 expression, and synapse formation. J Immunol **76**, 864–872.

Cerboni, C., Neri, F., Casartelli, N., Zingoni, A., Cosman, D., Rossi, P., Santoni, A. and Doria, M. (2007a). Human immunodeficiency virus 1 Nef protein downmodulates the ligands of the activating receptor NKG2D and inhibits natural killer cell-mediated cytotoxicity. J Gen Virol **88**, 242–250.

Cerboni, C., Zingoni, A., Cippitelli, M., Piccoli, M., Frati, L. and Santoni, A. (2007b). Antigen-activated human T lymphocytes express cell-surface NKG2D ligands via an ATM/ATR-dependent mechanism and become susceptible to autologous NK-cell lysis. Blood **110**, 606–615.

Cerwenka, A. and Lanier, L. (2001). Ligands for natural killer cell receptors: redundancy or specificity. Immunol Rev **181**, 158–169.

Cerwenka, A., Baron, J.L. and Lanier, L.L. (2001). Ectopic expression of retinoic acid early inducible-1 gene (RAE-1) permits natural killer cell-mediated rejection of a MHC class I-bearing tumour in vivo. Proc Natl Acad Sci U S A **98**, 11521–11526.

Chalupny, N.J., Sutherland, C.L., Lawrence, W.A., Rein-Weston, A. and Cosman, D. (2003). ULBP4 is a novel ligand for human NKG2D. Biochem Biophys Res Commun **305**, 129–135.

Chalupny, N.J., Rein-Weston, A., Dosch, S. and Cosman, D. (2006). Down-regulation of the NKG2D ligand MICA by the human cytomegalovirus glycoprotein UL142. Biochem Biophys Res Commun **346**, 175–181.

Collins, R.W. (2004). Human MHC class I chain related (MIC) genes: their biological function and relevance to disease and transplantation. Eur J Immunogenet **31**(3), 105–114.

Cosman, D., Mullberg, J., Sutherland, C. L., Chin, W., Armitage, R., Fanslow, W., Kubin, M. and Chalupny, N.J. (2001). ULBPs, novel MHC class I-related molecules, bind to CMV glycoprotein UL16 and stimulate NK cytotoxicity through the NKG2D receptor. Immunity **14**, 123–133.

Diefenbach, A., Jamieson, A.M., Liu, S.D., Shastri, N. and Raulet, D.H. (2000). Ligands for the murine NKG2D receptor: expression by tumour cells and activation of NK cells and macrophages. Nat Immunol **1**, 119–126.

Diefenbach, A., Jensen, E.R., Jamieson, A. M. and Raulet, D.H. (2001). Rae1 and H60 ligands of the NKG2D receptor stimulate tumour immunity. Nature **413**, 165–171.

Diefenbach, A., Tomasello, E., Lucas, M., Jamieson, A.M., Hsia, J.K., Vivier, E. and Raulet, D.H. (2002). Selective associations with signaling proteins determine stimulatory versus costimulatory activity of NKG2D. Nat Immunol **3**(12), 1142–1149.

Diefenbach, A. and Raulet, D.H. (2001). Strategies for target cell recognition by natural killer cells. Immunol Rev **181**, 170–184.

Dunn, C., Chalupny, N.J., Sutherland, C. L., Dosch, S., Sivakumar, P.V., Johnson, D.C. and Cosman, D. (2003). Human cytomegalovirus glycoprotein UL16 causes intracellular sequestration of NKG2D ligands, protecting against natural killer cell cytotoxicity. J Exp Med **197**, 1427–1439.

Dunn, G.P., Old, L.J. and Schreiber, R.D. (2004). The immunobiology of cancer immunosurveillance and immunoediting. Immunity **21**, 137–148.

Ebihara, T., Masuda, H., Akazawa, T., Shingai, M., Kikuta, H., Ariga, T., Matsumoto, M. and Seya, T. (2007). Induction of NKG2D ligands on human dendritic cells by TLR ligand stimulation and RNA virus infection. Int Immunol **10**, 1145–1155.

Eisele, G., Wischhusen, J., Mittelbronn, M., Meyermann, R., Waldhauer, I., Steinle,

A., Weller, M. and Friese, M.A. (2006). TGF-beta and metalloproteinases differentially suppress NKG2D ligand surface expression on malignant glioma cells. Brain **129**, 2416–2425.

Eleme, K., Taner, S.B., Onfelt, B., Collinson, L.M., McCann, F.E., Chalupny, N.J., Cosman, D., Hopkins, C., Magee, A.I. and Davis, DM. (2004). Cell surface organization of stress-inducible proteins ULBP and MICA that stimulate human NK cells and T cells via NKG2D. J Exp Med **199**(7), 1005–1010.

Fernández-Morera, J.L., Rodríguez-Rodero, S., Tuñón, A., Martínez-Borra, J., López-Vázquez, A., Rodrigo, L., Rodrigo, P., González, S., Lahoz, C.H. and López-Larrea, C. (2008). Genetic influence of the nonclassical major histocompatibility complex class I molecule MICB in multiple sclerosis susceptibility. Tissue Antigens **72**(1), 54–59.

Friese, M.A., Wischhusen, J., Wick, W., Weiler, M., Eisele, G., Steinle, A. and Weller, M. (2004). RNA interference targeting transforming growth factor-beta enhances NKG2D-mediated antiglioma immune response, inhibits glioma cell migration and invasiveness, and abrogates tumourigenicity in vivo. Cancer Res **64**, 7596–7603.

Garrity, D., Call, M.E., Feng, J. and Wucherpfennig, K.W. (2005). The activating NKG2D receptor assembles in the membrane with two signaling dimers into a hexameric structure. Proc Natl Acad Sci U S A **102**(21), 7641–7646.

Gasser, S., Orsulic, S., Brown, E.J. and Raulet, D.H. (2005). The DNA damage pathway regulates innate immune system ligands of the NKG2D receptor. Nature **436**, 1186–1190.

Ghiringhelli, F., Ménard, C., Terme, M., Flament, C., Taieb, J., Chaput, N., Puig, P.E., Novault, S., Escudier, B., Vivier, E., Lecesne, A., Robert, C., Blay, J.Y., Bernard, J., Caillat-Zucman, S., Freitas, A., Tursz, T., Wagner-Ballon, O., Capron, C., Vainchencker, W., Martin, F. and Zitvogel, L. (2005). CD4+CD25+ regulatory T cells inhibit natural killer cell functions in a transforming growth factor-beta-dependent manner. J Exp Med **202**, 1075–1085.

Girardi, M., Oppenheim, D.E., Steele, C. R., Lewis, J.M., Glusac, E., Filler, R., Hobby, P., Sutton, B., Tigelaar, R.E. and Hayday, A.C. (2001). Regulation of cutaneous malignancy by gammadelta T cells. Science **294**(5542), 605–609.

González, S., Rodrigo, L., López-Vázquez, A., Fuentes, D., Agudo-Ibáñez, L., Rodríguez-Rodero, S., Fdez-Morera, J. L., Martínez-Borra, J. and López-Larrea, C. (2004). Association of MHC class I related gene B (MICB) to celiac disease. Am J Gastroenterol **99**(4), 676–680.

González, S., Groh, V. and Spies, T. (2006). Immunobiology of human NKG2D and its ligands. Curr Top Microbiol Immunol **298**, 121–138.

González, S., López-Soto, A., Suárez-Álvarez, B., López-Vázquez, A. and López-Larrea, C. (2008). NKG2D ligands: key targets of the immune response. Trends Immunol **29**(8), 397–403.

Gourzi, P., Leonova, T. and Papavasiliou, F. N. (2006). A role for activation-induced cytidine deaminase in the host response against a transforming retrovirus. Immunity **24**(6), 779–786.

Groh, V., Bahram, S., Bauer, S., Herman, A., Beauchamp, M. and Spies, T. (1996). Cell stress-regulated human major histocompatibility complex class I gene expressed in gastrointestinal epithelium. Proc Natl Acad Sci U S A **93**, 12445–12450.

Groh, V., Steinle, A., Bauer, S. and Spies, T. (1998). Recognition of stress-induced MHC molecules by intestinal epithelial gammadelta T cells. Science **279**, 1737–1740.

Groh, V., Rhinehart, R., Secrist, H., Bauer, S., Grabstein, K.H. and Spies, T. (1999). Broad tumour-associated expression and recognition by tumour-derived gamma delta T cells of MICA and MICB. Proc Natl Acad Sci U S A **96**, 6879–6884.

Groh, V., Wu, J., Yee, C. and Spies, T. (2002). Tumour-derived soluble MIC ligands impair expression of NKG2D and T-cell activation. Nature **419**, 734–738.

Groh, V., Bruhl, A., El-Gabalawy, H., Nelson, J.L. and Spies, T. (2003). Stimulation of T cell autoreactivity by anomalous expression of NKG2D and its MIC ligands in rheumatoid arthritis. Proc Natl Acad Sci U S A **100**(16), 9452–9457.

Groh, V., Smythe, K., Dai, Z. and Spies, T. (2006). Fas-ligand-mediated paracrine T cell regulation by the receptor NKG2D in tumour immunity. Nat Immunol **7**, 755–762.

Hue, S., Mention, J.J., Monteiro, R.C., Zhang, S., Cellier, C., Schmitz, J., Verkarre, V., Fodil, N., Bahram, S., Cerf-Bensussan, N. and Caillat-Zucman, S. (2004). A direct role for NKG2D/MICA interaction in villous atrophy during celiac disease. Immunity **21**(3), 367–377.

Kaiser, B.K., Yim, D., Chow, I.T., Gonzalez, S., Dai, Z., Mann, H.H., Strong, R. K., Groh, V. and Spies, T. (2007). Disulphide-isomerase-enabled shedding of tumour-associated NKG2D ligands. Nature **447**, 482–486.

Kato, N., Tanaka, J., Sugita, J., Toubai, T., Miura, Y., Ibata, M., Syono, Y., Ota, S., Kondo, T., Asaka, M. and Imamura, M. (2007). Regulation of the expression of MHC class I-related chain A, B (MICA, MICB) via chromatin remodeling and its impact on the susceptibility of leukemic cells to the cytotoxicity of NKG2D-expressing cells. Leukemia **21**, 2103–2108.

Kim, J., Chang, C.K, Hayden, T., Liu, F., Benjamin, J., Hamerman, J.A., Lanier, L. L. and Kang, S.M. (2007). The activating immunoreceptor NKG2D and its ligands are involved in allograft transplant rejection. J Immunol **179**, 6416–6420.

Li, Z., Groh, V., Strong, R.K. and Spies, T. (2000). A single amino acid substitution causes loss of expression of a MICA allele. Immunogenetics **51**, 246–248.

López-Arbesú, R., Ballina-García, F.J., Alperi-López, M., López-Soto, A., Rodríguez-Rodero, S., Martínez-Borra, J., López-Vázquez, A., Fernández-Morera, J.L., Riestra-Noriega, J.L., Queiro-Silva, R., Quiñones-Lombraña, A., López-Larrea, C. and González, S. (2007). MHC class I chain-related gene B (MICB) is associated with rheumatoid arthritis susceptibility. Rheumatology (Oxford) **46**(3), 426–430.

López-Larrea, C., Suárez-Alvarez, B., López-Soto, A., López-Vázquez, A. and González, S. (2008). The NKG2D receptor: sensing stressed cells. Trends Mol Med **14**(4), 179–189.

López-Soto, A., Quiñones-Lombraña, A., López-Arbesú, R., López-Larrea, C. and González, S. (2006). Transcriptional regulation of ULBP1, a human ligand of the NKG2D receptor. J Biol Chem **281**, 30419–30430.

López-Soto, A., Folgueras, AR., Seto, E. and Gonzalez, S. (2009). HDAC3 represses the expression of NKG2D ligands ULBPs in epithelial tumour cells: potential implications for the immunosurveillance of cancer. Oncogene 21, 2370-2382.

López-Vázquez, A., Rodrigo, L., Fuentes, D., Riestra, S., Bousoño, C., García-Fernández, S., Martínez-Borra, J., González, S. and López-Larrea, C. (2002a). MHC class I chain related gene A (MICA) modulates the development of coeliac disease in patients with the high risk heterodimer DQA1*0501/ DQB1*0201. Gut **50**(3), 336–340.

López-Vázquez, A., Rodrigo, L., Fuentes, D., Riestra, S., Bousoño, C., García-Fernández, S., Martínez-Borra, J., González, S. and López-Larrea, C.

(2002b). MICA-A5.1 allele is associated with atypical forms of celiac disease in HLA-DQ2-negative patients. Immunogenetics **53**(10–11), 989–991.

Meresse, B., Chen, Z., Ciszewski, C., Tretiakova, M., Bhagat, G., Krausz, T. N., Raulet, D.H., Lanier, L.L., Groh, V., Spies, T., Ebert, E.C., Green, P.H. and Jabri, B. (2004). Coordinated induction by IL15 of a TCR-independent NKG2D signaling pathway converts CTL into lymphokine-activated killer cells in celiac disease. Immunity **21**(3), 357–366.

Nedvetzki, S., Sowinski, S., Eagle, R. A., Harris, J., Vély, F., Pende, D., Trowsdale, J., Vivier, E., Gordon, S. and Davis, D.M. (2007). Reciprocal regulation of human natural killer cells and macrophages associated with distinct immune synapses. Blood **109**, 3776–3785.

Ogasawara, K., Hamerman, J.A., Hsin, H., Chikuma, S., Bour-Jordan, H., Chen, T., Pertel, T., Carnaud, C., Bluestone, J.A. and Lanier, L.L. (2003). Impairment of NK cell function by NKG2D modulation in NOD mice. Immunity **18**(1), 41–51.

Ogasawara, K., Hamerman, J.A., Ehrlich, L.R., Bour-Jordan, H., Santamaria, P., Bluestone, J.A. and Lanier, L.L. (2004). NKG2D blockade prevents autoimmune diabetes in NOD mice. Immunity **20**(6), 757–767.

Ogasawara, K., Benjamin, J., Takaki, R., Phillips, J.H. and Lanier, L.L. (2005). Function of NKG2D in natural killer cell-mediated rejection of mouse bone marrow grafts. Nat Immunol **6**(9), 938–945.

Oppenheim, D.E., Roberts, S.J., Clarke, S. L., Filler, R., Lewis, J.M., Tigelaar, R.E., Girardi, M. and Hayday, A.C. (2005). Sustained localized expression of ligand for the activating NKG2D receptor impairs natural cytotoxicity in vivo and reduces tumour immunosurveillance. Nat Immunol **6**, 928–937.

Pappworth, I.Y., Wang, E.C. and Rowe, M. (2007). The switch from latent to productive infection in Epstein–Barr virus-infected B cells is associated with sensitization to NK cell killing. J Virol **81**(2), 474–482.

Quiroga, I., Salio, M., Koo, D.D., Cerundolo, L., Shepherd, D., Cerundolo, V. and Fuggle, S.V. (2006). Expression of MHC class I-related Chain B (MICB) molecules on renal transplant biopsies. Transplantation **81**(8), 1196–1203.

Rabinovich, B.A., Li, J., Shannon, J., Hurren, R., Chalupny, J., Cosman, D. and Miller, R.G. (2003). Activated, but not resting, T cells can be recognized and killed by syngeneic NK cells. J Immunol **170**, 3572–3576.

Radaev, S., Rostro, B., Brooks, A.G., Colonna, M. and Sun, P.D. (2001). Conformational plasticity revealed by the cocrystal structure of NKG2D and its class I MHC-like ligand ULBP3. Immunity **15**(6), 1039–1049.

Raulet, D.H. (2003). Roles of the NKG2D immunoreceptor and its ligands. Nat Rev Immunol **3**, 781–790.

Roda-Navarro, P., Vales-Gomez, M., Chisholm, S.E. and Reyburn, H.T. (2006). Transfer of NKG2D and MICB at the cytotoxic NK cell immune synapse correlates with a reduction in NK cell cytotoxic function. Proc Natl Acad Sci U S A **103**(30), 11258–11263.

Rodríguez-Rodero, S., Rodrigo, L., Fdez -Morera, J.L., Martínez-Borra, J., López-Vázquez, A., Fuentes, D., López-Arbesú, R., López-Soto, A., González, S. and López-Larrea, C. (2006). MHC class I chain-related gene B promoter polymorphisms and celiac disease. Hum Immunol **67**, 208–214.

Rodríguez-Rodero, S., González, S., Rodrigo, L., Fernández-Morera, J.L., Martínez-Borra, J., López-Vázquez, A. and López-Larrea, C. (2007). Transcriptional regulation of MICA and MICB, a novel polymorphism in MICB promoter alters transcriptional regulation by Sp1. Eur J Immunol **37**, 1938–1953.

Rosen, D.B., Araki, M., Hamerman, J.A., Chen, T., Yamamura, T. and Lanier, L. L. (2004). A structural basis for the association of DAP12 with mouse, but not human, NKG2D. J Immunol **173**(4), 2470–2478.

Routes, J.M., Ryan, S., Morris, K., Takaki, R., Cerwenka, A. and Lanier, L.L. (2005). Adenovirus serotype 5 E1A sensitizes tumour cells to NKG2D-dependent NK cell lysis and tumour rejection. J Exp Med **202**, 1477–1482.

Salih, H.R., Rammensee, H.G. and Steinle, A. (2002). Cutting edge: down-regulation of MICA on human tumours by proteolytic shedding. J Immunol **169**, 4098–4102.

Schrama, D., Terheyden, P., Otto, K., Kammerer, U., Brocker, E.B., Luhder, F., Cosman, D., Andersen, M.H. and Becker, J.C. (2006). Expression of the NKG2D ligand UL16 binding protein-1 (ULBP-1) on dendritic cells. Eur J Immunol **36**, 65–72.

Seiler, M., Brabcova, I., Viklicky, O., Hribova, P., Rosenberger, C., Pratschke, J., Lodererova, A., Matz, M., Schönemann, C., Reinke, P., Volk, H.D. and Kotsch, K. (2007). Heightened expression of the cytotoxicity receptor NKG2D correlates with acute and chronic nephropathy after kidney transplantation. Am J Transplant **7**(2), 423–433.

Sirén, J., Sareneva, T., Pirhonen, J., Strengell, M., Veckman, V., Julkunen, I. and Matikainen, S. (2004). Cytokine and contact-dependent activation of natural killer cells by influenza A or Sendai virus-infected macrophages. J Gen Virol **85**(8), 2357–2364.

Skov, S., Pedersen, M.T., Andresen, L., Straten, P.T., Woetmann, A. and Odum, N. (2005). Cancer cells become susceptible to natural killer cell killing after exposure to histone deacetylase inhibitors due to glycogen synthase kinase-3-dependent expression of MHC class I-related chain A and B. Cancer Res **65**, 11136–11145.

Smyth, M.J., Swann, J., Cretney, E., Zerafa, N., Yokoyama, W.M. and Hayakawa, Y. (2005). NKG2D function protects the host from tumour initiation. J Exp Med **202**, 583–588.

Smyth, M.J., Teng, M.W., Swann, J., Kyparissoudis, K., Godfrey, D.I. and Hayakawa, Y. (2006). CD4+CD25+ T regulatory cells suppress NK cell-mediated immunotherapy of cancer. J Immunol **176**, 1582–1587.

Sollid, L.M. (2002). Coeliac disease: dissecting a complex inflammatory disorder. Nat Rev Immunol **2**(9), 647–655.

Stephes, H.A. (2001). MICA and MICB genes: can the enigma of their polymorphism be resolved? Trends Immunol **22**(7), 378–385.

Stern-Ginossar, N., Elefant, N., Zimmermann, A., Wolf, D.G., Saleh, N., Biton, M., Horwitz, E., Prokocimer, Z., Prichard, M., Hahn, G., Goldman-Wohl, D., Greenfield, C., Yagel, S., Hengel, H., Altuvia, Y., Margalit, H. and Mandelboim, O. (2007). Host immune system gene targeting by a viral miRNA. Science **317**(5836), 376–381.

Stern-Ginossar, N., Gur, C., Biton, M., Horwitz, E., Elboim, M., Stanietsky, N., Mandelboim, M. and Mandelboim, O. (2008). Human microRNAs regulate stress-induced immune responses mediated by the receptor NKG2D. Nat Immunol **9**(9), 1065–1073.

Suárez-Álvarez, B., López-Vázquez, A., Díaz-Molina, B., Bernardo-Rodríguez, M.J., Álvarez-López, R., Pascual, D., Astudillo, A., Martínez-Borra, J., Lambert, J. L., González, S. and López-Larrea, C. (2006). The predictive value of soluble major histocompatibility complex class I chain-related molecule A (MICA) levels on heart allograft rejection. Transplantation **82**(3), 354–361.

Suemizu, H., Radosavljevic, M., Kimura, M., Sadahiro, S., Yoshimura, S., Bahram, S. and Inoko, H. (2002). A basolateral sorting motif in the MICA cytoplasmic tail. Proc Natl Acad Sci U S A **99**(5), 2971–2976.

Valés-Gómez, M., Browne, H. and Reyburn, H.T. (2003). Expression of the UL16 glycoprotein of Human Cytomegalovirus protects the virus-infected cell from attack by natural killer cells. BMC Immunol **4**, 4.

Venkataraman, G.M., Suciu, D., Groh, V., Boss, J.M. and Spies, T. (2007). Promoter region architecture and transcriptional regulation of the genes for the MHC class I-related chain A and B ligands of NKG2D. J Immunol **178**, 961–969.

Ward, J., Bonaparte, M., Sacks, J., Guterman, J., Fogli, M., Mavilio, D. and Barker, E. (2007). HIV modulates the expression of ligands important in triggering natural killer cell cytotoxic responses on infected primary T-cell blasts. Blood **110**(4), 1207–1214.

Welte, S.A., Sinzger, C., Lutz, S.Z., Singh-Jasuja, H., Sampaio, K.L., Eknigk, U., Rammensee, H.G. and Steinle, A. (2003). Selective intracellular retention of virally induced NKG2D ligands by the human cytomegalovirus UL16 glycoprotein. Eur J Immunol **33**, 194–203.

Wu, J., Song, Y., Bakker, A.B., Bauer, S., Spies, T., Lanier, L.L. and Phillips, J.H. (1999). An activating immunoreceptor complex formed by NKG2D and DAP10. Science **285**(5428), 730–732.

Yamamoto, K., Fujiyama, Y., Andoh, A., Bamba, T. and Okabe, H. (2001). Oxidative stress increases MICA and MICB gene expression in the human colon carcinoma cell line (CaCo-2). Biochim Biophys Acta **1526**, 10–12.

Zwirner, N.W., Fuertes, M.B., Girart, M. V., Domaica, C.I. and Rossi, L.E. (2007). Cytokine-driven regulation of NK cell functions in tumour immunity: role of the MICA-NKG2D system. Cytokine Growth Factor Rev **18**(1–2), 159–170.

Chapter Six

6

NK cell KIR heterogeneity and evolution

Makoto Yawata, Nobuyo Yawata

CHAPTER CONTENTS

Not only is the universe stranger than we imagine, it is stranger than we can imagine.

Sir Arthur Eddington (1882–1944), English Astronomer

ABSTRACT

Human natural killer (NK) cells in the peripheral blood are highly diverse in their expression of HLA class I-specific inhibitory receptors: Killer cell Immunoglobulin-like Receptors (KIR), NKG2A and LILRB1. Variegated expression of these receptors generates a NK-cell repertoire that is unique to each human individual. The mechanism that shapes human NK cell repertoires is distinct from the selection mechanisms operating on T-lymphocyte and B-lymphocyte repertoires. Polygenic and polymorphic KIR combine with diverse HLA class I to determine KIR expression frequencies in the NK cell population, levels of cell-surface expression and the strength of missing-self response for each NK cell subset. Human NK cell expression of KIR and NKG2A is balanced to calibrate the overall response of repertoires against missing-self stimulus. Functional heterogeneity of NK cells is a feature of innate immunity that has been actively maintained in mammalian species through genetic diversification of NK receptors and their ligands. Understanding NK cell heterogeneity will become crucial in clinical medicine as NK cells are increasingly used in immunotherapy.

KEY WORDS

Killer cell Immunoglobulin-like Receptors (KIR), HLA class I, NK cell repertoire, Missing-self, Polymorphism, Ly49, NKG2A

The roles of MHC class I inhibitory receptors in NK cell function

Natural killer (NK) cells in the peripheral blood are a diverse population (Anfossi et al., 2006; Fauriat et al., 2008; Moretta et al., 1995; Valiante et al., 1997; Voss et al., 1998; Yawata et al., 2008; Yu et al., 2007). NK cell heterogeneity is generated by variegated expression of polymorphic MHC class I specific-inhibitory receptors. In humans, these receptors comprise the polygenic and polymorphic Killer cell Immunoglobulin-like Receptors (KIR), the more conserved NKG2A/CD94 receptors and the LILRB1 receptor (Colonna and Samaridis, 1995; Colonna et al., 1997; D'Andrea et al., 1995; Houchins et al., 1991; Lanier, 2005; Moretta et al., 1995; Wagtmann et al., 1995). In mice, the Ly49 receptor family performs functions analogous to KIR (Karlhofer et al., 1992; Raulet et al., 2001; Takei et al., 2001).

'Missing-self' describes a capacity unique to NK cells: the ability to recognise and respond to lack or down-regulation of self-MHC class I on target cells (Karre et al., 1986). Subsequent research has shown that the inhibitory NK receptor families that recognise self-MHC class I in missing-self are those that define the effector response of an NK cell against lack of self-MHC class I, a concept termed 'education', 'licensing' or 'arming' by different laboratories. Both KIR and Ly49 are capable of defining NK cell response in this manner (Anfossi et al., 2006; Fernandez et al., 2005; Kim et al., 2005).

Human KIR variation

The KIR gene family

The 15 *KIR* genes encode receptors that have either three (KIR3D) or two (KIR2D) extracellular immunoglobulin-like domains (Figure 6.1) (Colonna and

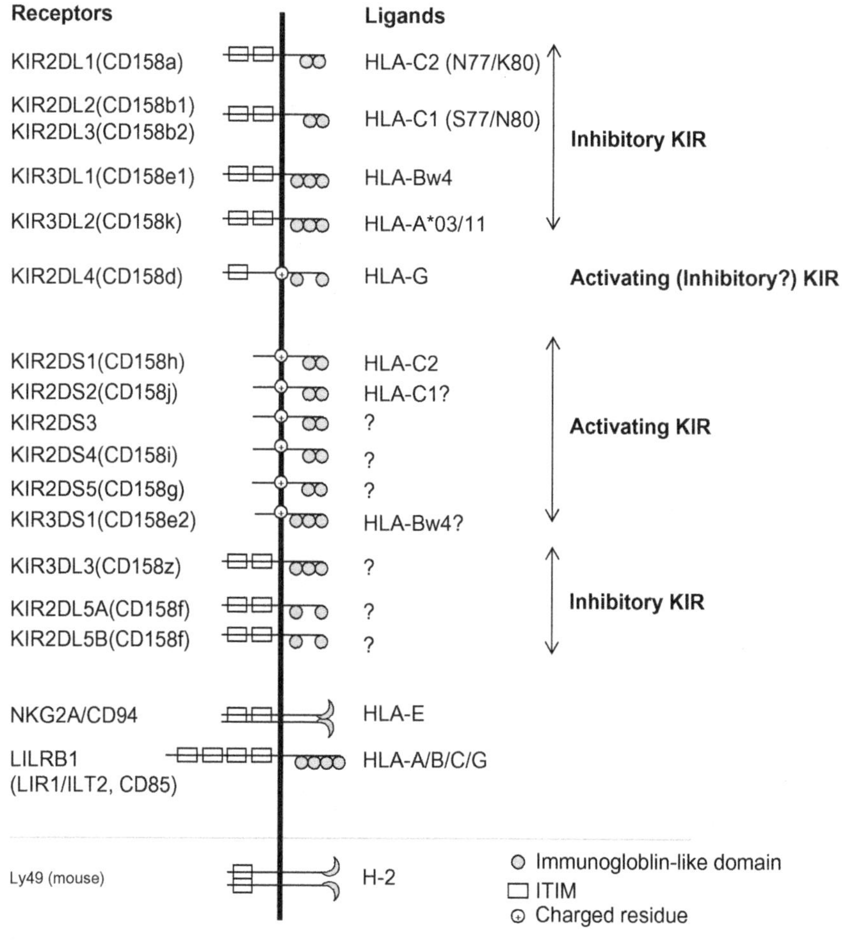

Figure 6.1 • MHC class-I specific inhibitory NK cell receptors and their ligands. KIR have either two or three extracellular immunogloblin-like domains. Inhibitory KIR have long cytoplasmic tails containing ITIM, while in activating KIR, the shorter cytoplasmic tails lack ITIMs but carry a charged residue in their transmembrane domain that enables association with an adapter molecule.

Samaridis, 1995; D'Andrea et al., 1995; Uhrberg et al., 1997; Valiante et al., 1997; Wagtmann et al., 1995). Receptors that carry two immunoreceptor tyrosine-based inhibitory motifs (ITIM) in their intracellular domains are considered inhibitory receptors (KIR2DL1/2/3/5, 3DL1/2/3) (Burshtyn et al., 1996; Wagtmann et al., 1995).

The receptors KIR3DS1 and KIR2DS1/2/3/4/5 lack ITIMs and instead carry a charged lysine residue in their transmembrane domain that enables the receptor to associate with DAP12, an adapter molecule that transduces activation signalling (Lanier et al., 1998). Knowledge of the biology of activating KIR is limited as compared with the inhibitory KIR. Clinical studies have suggested that activating KIR might contribute to reproductive success and viral response, and may affect susceptibility to autoimmune diseases (Alter et al., 2007; Hiby et al., 2004; Nelson et al., 2004).

KIR2DL4 is unique in that it possesses an ITIM in its cytoplasmic domain and an arginine residue in its transmembrane domain that enables association with an adapter molecule FceRI-g. Despite the ITIM, engagement of KIR2DL4 by a specific antibody induces NK cell activation (Kikuchi-Maki et al., 2003; Rajagopalan et al., 2006).

KIR expression on CD56^{dim} and CD56^{bright} NK cells

In peripheral blood, KIR are expressed primarily on CD56^{dim} NK cells in a variegated manner, while CD56^{bright} NK cells express NKG2A but little KIR (Voss et al., 1998). KIR expression is low in secondary lymphoid organs where CD56^{bright} NK cells dominate (Fehniger et al., 2003). Most uterine NK cells are CD56^{bright} and have different KIR phenotypes as compared to peripheral NK cells (Verma et al., 1997).

KIR2DL4 surface expression is limited to CD56^{bright} NK cells and uterine NK cells although KIR2DL4 transcripts are detected in the majority of peripheral NK cells (Kikuchi-Maki et al., 2003; Ponte et al., 1999; Valiante et al., 1997). KIR3DL3 transcripts are detected in CD56^{bright} NK cells in the peripheral circulation and in placental decidua, but no protein is found expressed on the cell surface (Trundley et al., 2006).

Ligand recognition by KIR

The ligand specificity of inhibitory KIR is described in Figure 6.1 (Lanier, 2005; Vilches and Parham, 2002). The conventional KIR ligands for KIR2DL2/2DL3 are group 1 HLA-C alleles (HLA-C1), the allotypes that have a serine at position 77 and asparagine at position 80. KIR2DL1 binds another subset of HLA-C alleles classified as group 2 ('HLA-C2'), which has asparagine at position 77 and lysine at position 80 (Colonna and Samaridis, 1995; Moretta et al., 1993). The HLA-C1 and HLA-C2 groups are mutually exclusive, and the frequencies of HLA-C ligands are variable amongst human populations (Single et al., 2007). KIR3DL1 binds a subset of HLA-B alleles that has the Bw4 motif in positions 77–83 (D'Andrea et al., 1995).

Exceptions to these conventional KIR ligand specificities have been described. HLA-B*46 and B*73, which are Bw6 alleles but carry an HLA-C1 motif, are both recognised by KIR2DL2/3 (Moesta et al., 2008; Winter et al., 1998). In addition, weak cross-reactivity has been demonstrated between HLA-C1 and -C2 for KIR2DL2/3 (Moesta et al., 2008; Pende et al., 2008; Winter et al., 1998). KIR3DL1 recognition of Bw4 motif-carrying HLA-A allotypes has been demonstrated in several reports (Foley et al., 2008b; Stern et al., 2008; Thananchai et al., 2007; Yawata et al., 2008). KIR3DL2 has been shown to bind dimeric HLA-B*27 (Kollnberger et al., 2007). Soluble HLA class I can also bind KIR. HLA-G molecules lacking the transmembrane domain are bound by KIR2DL4 and induce a cytokine response (Rajagopalan et al., 2006).

HLA-bound peptides modulate KIR binding to HLA class I. KIR3DL2 binds HLA-A*03 and -A*11 when EBNA3A peptides from EBV are loaded onto these HLA molecules (Hansasuta et al., 2004). Conversely, some virally derived peptides bound to HLA-A*24 molecules can abrogate binding of KIR3DL1 (Thananchai et al., 2007).

The physiological ligands for most activating KIR are not clear. Several studies have shown that KIR2DS1 recognises HLA-C2 with low affinity and that KIR2DS1$^+$NK cells from HLA-C1 homozygous individuals can lyse HLA-C2$^+$ targets, but the same NK cell subset from HLA-C2$^+$ individuals cannot (Chewning et al., 2007; Foley et al., 2008a; Moretta et al., 1995; Morvan et al., 2008; Pende et al., 2008; Stewart et al., 2005). KIR3DS1 recognises HIV-infected cells expressing HLA-B Bw4 alleles (Alter et al., 2007).

Gene content variation and the A- and B-haplotypes

KIR gene content varies substantially amongst human individuals and amongst world populations due to variations in the genomic region encoding the KIR genes, the so-called KIR complex (19q13.4) (Parham, 2005; Wilson et al., 2000; Uhrberg et al., 1997; Yawata et al., 2002a). Figure 6.2 depicts 128 distinct KIR genotypes that were identified in a compilation of KIR genotypes from 13 previous reports in 1701 individuals from 16 world populations (Gendzekhadze et al., 2006;

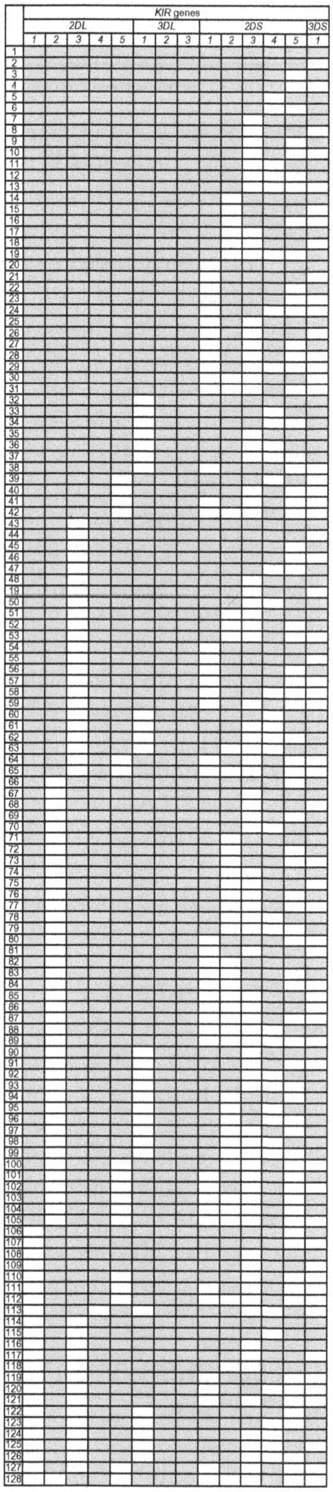

Figure 6.2 • KIR genotype variation in the world populations. Gene content variation in 16 human populations. Grey boxes indicate presence of a gene. Genotype #105 represents homozygotes for the group A haplotypes. The genotypes were compiled from previous reports on KIR distribution: (Gendzekhadze et al., 2006; Gutierrez-Rodriguez et al., 2006; Jiang et al., 2005; Norman et al., 2001, 2002; Rajalingam et al., 2002, 2008; Toneva et al., 2001; Uhrberg et al., 2002; Velickovic et al., 2006, 2008; Whang et al., 2005; Yawata et al., 2002a, 2006).

Gutierrez-Rodriguez et al., 2006; Jiang et al., 2005; Norman et al., 2001, 2002; Rajalingam et al., 2002, 2008; Uhrberg et al., 2002; Velickovic et al., 2006, 2008; Whang et al., 2005; Yawata et al., 2002a, 2006). KIR genotypes will likely become more complex as admixture proceeds amongst world populations.

Historical nomenclature classifies KIR gene content-haplotypes in two groups: the group A haplotypes, which are defined by a fixed set of 7 KIR genes (KIR3DL3, 2DL1, 2DL3, 2DL4, 3DL1, 2DS4 and 3DL2), and the group B haplotypes, which include all other KIR haplotypes (Figure 6.3a) (Shilling et al., 2002a; Uhrberg et al., 1997; Yawata et al., 2002a). Group B haplotypes were originally defined by the presence of a 24-Kb band on a Southern blot, a band which corresponds to the genome fragment carrying KIR2DL5 (Uhrberg et al., 1997). Although not all group B haplotypes carry KIR2DL5, the A/B KIR haplotype nomenclature remains in use in part due to the benefit in distinguishing the group A haplotype from other KIR haplotypes. The group A haplotype carries the fewest KIR genes and yet has the full array of inhibitory KIR for which MHC class I ligands are known and which contribute to NK cell education. The group B haplotypes often lack one or more inhibitory KIR that are present on the group A haplotypes and contain more activating KIR.

Frequencies of group A and B haplotypes vary substantially between human populations. Group A homozygotes are the most frequent KIR genotype in many world populations, especially in Northeast Asian populations. In other populations, such as the Australian Aborigine and some Indian subpopulations, group B haplotype frequencies prevail over the group A haplotypes (Jiang et al., 2005; Rajalingam et al., 2002, 2008; Toneva et al., 2001; Whang et al., 2005; Yawata et al., 2002a,b).

The basic structure of KIR haplotypes is configured by centromeric and telomeric genome blocks, each carrying different combinations of KIR genes (Figure 6.3b). A major recombination hotspot in the centre of the KIR complex has allowed centromeric and telomeric halves to shuffle and combine over human generations (Hsu et al., 2002; Shilling et al., 2002a; Uhrberg et al., 2002; Yawata et al., 2002a).

The wide variety of KIR haplotypes implies a history rich in genomic perturbations, including insertion or deletion of individual genes, block insertions, genome block recombination and nonhomologous crossover. The duplicated genes, KIR2DL5A and KIR2DL5B, and their unique insertions either in the telomeric or centromeric halves of the KIR complex, diversifies B haplotype structures (Gomez-Lozano et al., 2002). In some rare human KIR haplotypes, unequal recombination and segmental duplication has introduced two sets of KIR2DL4, 3DL1 and 3DP1 genes (Martin et al., 2003; Williams et al., 2003). Some KIR genes are nonvariable. KIR3DL3,

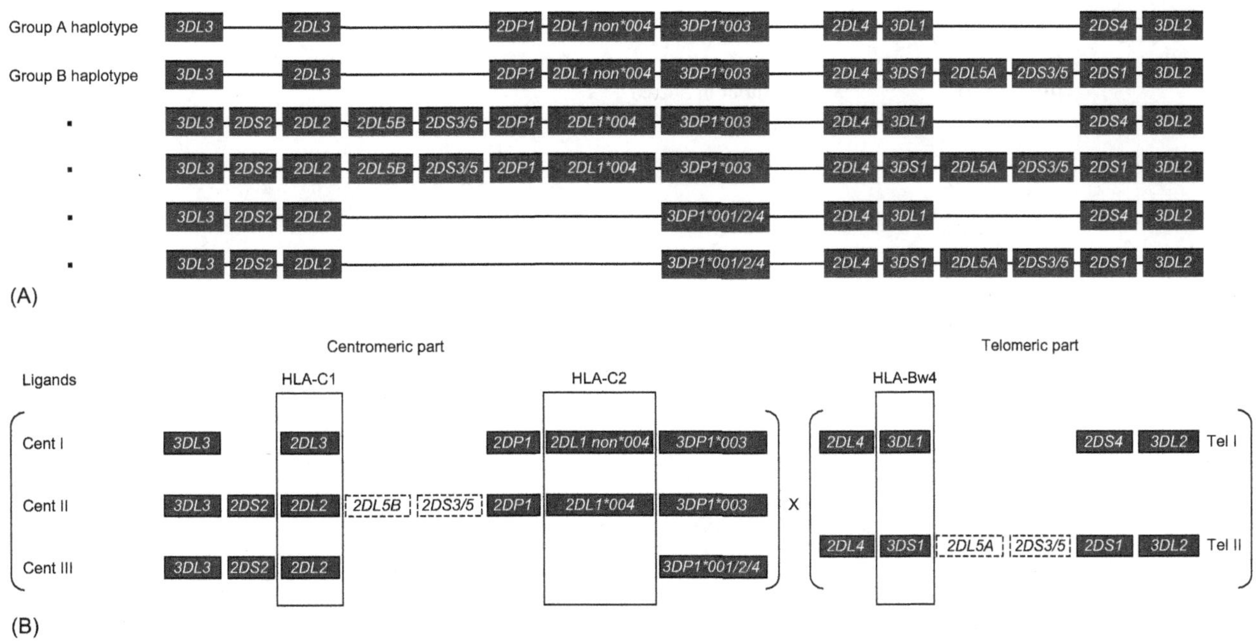

Figure 6.3 • Basic KIR haplotype structures. (A) Shown are representative human KIR haplotypes. The group A haplotype is shown at the top; some of the frequent B haplotypes are shown below. (B) KIR haplotypes are configured by a centromeric block (one of Cent I–III) and a telomeric block (one of Tel I–II). A major recombination hotspot at the centre of the complex has enabled individual centromeric and telomeric genome blocks to combine. KIR2DS3/5: presence of either KIR2DS3 or KIR2DS5. Frequently variable KIR genes are depicted by dotted lines.

2DL4 and 3DL2 are present on all human KIR haplotypes and are thus called framework genes.

In the context of HLA recognition, essentially all KIR haplotypes contain an inhibitory receptor gene for HLA-C1 (KIR2DL2 or 2DL3). In contrast, genes encoding inhibitory receptors for HLA-C2 (KIR2DL1) and HLA-Bw4 (KIR3DL1) are variable depending on the haplotype.

Three KIR pseudogenes exist in humans. KIR2DP1 and KIR3DP1 are present within the KIR complex; KIR3DX1 is a distantly related pseudogene that situates outside the KIR complex within the Leucocyte Receptor Complex (Sambrook et al., 2006; Wilson et al., 2000). A distinct form of KIR3DP1 is an important marker in determining KIR2DL1 zygosity because it is linked with the absence of the KIR2DL1 locus on the 'Cent III' segment in KIR haplotypes (Figure 6.3b) (Hsu et al., 2002; Yawata et al., 2002a).

KIR allotype functions and allele-level haplotypes

In addition to variation in gene content, KIR are highly polymorphic (Table 6.1A,B) (http://www.ebi.ac.uk/ipd/kir/alleles.html and http://www.allelefrequencies.net). KIR3DL1 and the framework genes, KIR3DL3, 2DL4 and 3DL2, are all highly polymorphic (Gedil et al., 2005, 2007; Hou et al., 2007; Meenagh et al., 2004; Norman et al., 2007; Yawata et al., 2006). In comparison, the activating KIR are oligomorphic except for KIR2DS4 (Hou et al., 2008; Maxwell et al., 2004; Yawata et al., 2006). KIR polymorphisms are dispersed throughout the entire length of each gene, and the majority encode nonsynonymous changes (Vilches and Parham, 2002; Yawata et al., 2002a).

Some KIR polymorphisms distinguish levels of cell-surface expression (Gardiner et al., 2001; Thomas et al., 2008; Yawata et al., 2006). For KIR3DL1, the hierarchy of mean fluorescence intensity (MFI) as assessed by antibody binding is in incremental order: 3DL1*028/053 < *005 < *007 < *008 < *001 < *002/015/020 (Gardiner et al., 2001; Thomas et al., 2008; Yawata et al., 2006). The bimodal histogram observed in individuals heterozygous for two KIR3DL1 variants of distinguishable expression levels infers also that expression of KIR3DL1 alleles on the two chromosomes are regulated independently, and most NK cells express a single allele (Chan et al., 2003; Gardiner et al., 2001).

KIR polymorphisms also define the strength of inhibitory signalling and the level of missing-self response, and determine the specificity of ligand recognition (Carr et al., 2005; Thananchai et al., 2007; Yawata et al., 2006, 2008). KIR3DL1*015, *005 and *002 are allotypes that display strong inhibitory capacity, whereas 3DL1*007 is a weak inhibitor (Carr et al., 2005; Yawata et al., 2006).

Table 6.1 Allelic variation of *KIR*

(A)

KIR	Number of alleles
2DL1	14
2DL2	5
2DL3	7
2DL4	23
2DL5A	2
2DL5B	5
3DL1	46
3DL2	17
3DL3	37
2DS1	4
2DS2	7
2DS3	4
2DS4	8
2DS5	5
3DS1	12
2DP1	3
3DP1	5

(B)

	Frequency in population group (%)		
KIR alleles	African[1]	Caucasian[2]	East Asian[3]
2DL3			
**001*	65.7	37.0	87.1
**002*	24.2	29.0	3.9
**005*	13.1	5.0	0.0
**007*	0.0	0.0	2.2
3DL1			
**001*	40.8	20.5	6.0
**002*	13.3	12.0	0.0
**004*	17.3	16.0	0.0
**005*	12.2	13.0	12.8
**007*	39.8	5.5	12.8
**008*	14.3	3.0	0.0
**009*	1.0	2.0	0.0
**015*	8.2	7.5	45.7

(B)

KIR alleles	Frequency in population group (%)		
	Afican[1]	**Caucasian[2]**	**East Asian[3]**
**019*	4.6	1.0	0.0
**020*	4.6	0.0	6.0
2DS4			
**001*	39.4	24.5	50.4
**003*	63.6	45.0	6.0
**004*	6.1	5.0	12.9
**006*	19.2	20.5	0.0
**007*	0.0	0.0	12.9

(C)

Non-expressed *KIR* alleles		**Frequencies in population group**
*2DL2*004*	Intracellular protein retention	3% (African[4])
*2DL4*008/11*	Truncated/secreted forms due to point deletion	*008:29%, *011:11% (Caucasian[5])
*2DL5*002/4*	Non-transcribed due to promoter mutation	*002:18% (Caucasian[6])
*3DL1*004*	Intracellular protein retention	16% (Caucasian[2])
*3DL1*024*	Premature stop codon due to point deletion	0.3% (African[7])
*2DS3*003*	Point mutation	0.8% (Caucasian[8])
*2DS4*003/4/6/7*	Truncated forms due to 22 bp deletion	70.5% (Caucasian[2]), 31.8% (East Asian[3])
*3DS1*049*	Premature stop codon due to point deletion	2% (Caucasian[8])

[1]Oman (Middleton et al., 2008), phenotype frequencies.
[2]Northern Irish (www.allelefrequencies.net).
[3]Japanese (Yawata et al., 2006).
[4]VandenBussche et al. (2006), phenotype frequencies.
[5]Shulse et al. (2006).
[6]Middleton et al. (2008), phenotype frequencies.
[7]Norman et al. (2007).
[8]Luo et al. (2007).

It is important to note that the inhibitory capacity of KIR is also affected by HLA class I polymorphisms (Carr et al., 2005; O'Connor et al., 2007; Sanjanwala et al., 2008).

Of equal importance are the KIR variants, which are not expressed on the cell-surface (Table 6.1C). Disruption of function is caused by frame-shift insertion/deletions, premature termination and protein misfolding (Goodridge et al., 2003; Luo et al., 2007; Maxwell et al., 2002; Pando et al., 2003; VandenBussche et al., 2006; Vilches et al., 2000). Some of these KIR variants are relatively frequent; allele frequency of

KIR3DL1*004 in a Caucasian population is 17%, and KIR2DS4*003 frequency is 45% (Shilling et al., 2002a) (http://www.allelefrequencies.net).

Linkage disequilibrium between KIR alleles forms distinct allele-level haplotypes that encode a functional set of receptors with distinct character (Middleton et al., 2007; Norman et al., 2004; Shilling et al., 2002a; Yawata et al., 2006). Of note, the group A haplotypes display the most variation at the allele level.

Human NK cell repertoires

Mechanisms generating the variegated expression of KIR: epigenetics and bidirectional promoters

NK cell KIR expression is variegated and maintains stable phenotypes (Shilling et al., 2002b; Valiante et al., 1997). KIR3DL1 alleles on the two chromosomes are expressed independently but not always in a mutually exclusive manner. The same was observed for KIR3DL2 but not for KIR2DL4 where the two alleles in an individual are both transcribed in all NK cells (Chan et al., 2005).

DNA methylation is important in variegated KIR expression (Chan et al., 2005; Santourlidis et al., 2002). Several laboratories have reported that KIR genes in hematopoietic progenitor cells are hypermethylated. KIR expression on mature NK cells correlates with the demethylation status in the promoter of each KIR gene, and treatment by demethylase induces KIR expression (Chan et al., 2005; Santourlidis et al., 2002, 2008). Further, monoallelic KIR3DL1 expression correlates with the methylation status of each KIR3DL1 gene (Chan et al., 2005). This infers that the KIR genes on the two chromosomes are controlled independently. This is why homozygotes for a KIR gene locus have double the expression frequency in the NK cell population as compared to individuals with one copy of the gene (Li et al., 2008; Yawata et al., 2006). Histone modification, another mode of epigenetic control, is less likely to contribute to variegated KIR expression since active signatures of modification are found in all mature NK cells regardless of their KIR expression. This is not the case for the Ly49 system (Chan et al., 2005; Rouhi et al., 2006; Santourlidis et al., 2008).

KIR genes are controlled by a set of distal and bidirectional proximal promoters upstream of each locus (Davies et al., 2007; Stewart et al., 2003; Trompeter et al., 2005). Reverse transcription from the proximal promoter inhibits KIR expression in immature NK cells, and the relative strength of the competing promoters determines the degree of KIR expression as a probabilistic switch. Contrasting to the KIR system, the bidirectional promoter for Ly49 is the distal promoter (Pascal et al., 2006; Saleh et al., 2004). Promoter sequences are distinct amongst KIR genes. Some promoters are unique to KIR allotypes and correlate with distinct frequencies of KIR expression within the NK cell population (Davies et al., 2007; Li et al., 2008; Stewart et al., 2003; Trompeter et al., 2005).

KIR and NKG2A co-expression in the NK cell repertoire

Systematic characterisation of KIR, NKG2A and LILRB1 expression in diverse human NK cell repertoires has shown that the majority of CD56^{dim} NK cells in peripheral blood co-express multiple HLA-specific receptors (Figure 6.4) (Fauriat et al., 2008; Yawata et al., 2008).

Is KIR co-expression a random process? Under the 'product rule', the frequency of randomly co-expressed receptors is calculated as the product of the expression frequencies of each receptor (Raulet et al., 2001; Valiante et al., 1997). Comparison of KIR repertoires in a panel of human individuals with diverse combinations of KIR and HLA showed that the degree of skewing from stochastic KIR co-expression varies substantially by the donor. The resulting continuum of phenotype variation and patterns of receptor co-expression indicated that this is a complex process where the product rule predicts the repertoire structures of some donors but not of others (Yawata et al., 2008). Of note, co-expression of both NKG2A and KIR is consistently lower in all donors than the predictions under the product rule in CD56^{dim} NK cells (Valiante et al., 1997; Yawata et al., 2008).

Genetic determinants of diverse NK subset responses against missing-self

HLA class I is the most polymorphic gene family in humans (Horton et al., 2004) (http://www.anthonynolan.org.uk/rsearch/hlainformaticsgroup/hla-informatics-group.html). With new understanding of the roles of allelic polymorphisms in NK cell education, the importance of class I variation now extends beyond its well-known function in antigen presentation to CD8 T lymphocytes.

The strength of NK cell education conferred by KIR or NKG2A can be quantified and compared between NK cell subsets that express a single receptor. Table 6.2 depicts the capacity of KIR and HLA class I allotypes to regulate NK cell education. Enhancement of missing-self responses (NK cell education) in each NK cell subset is

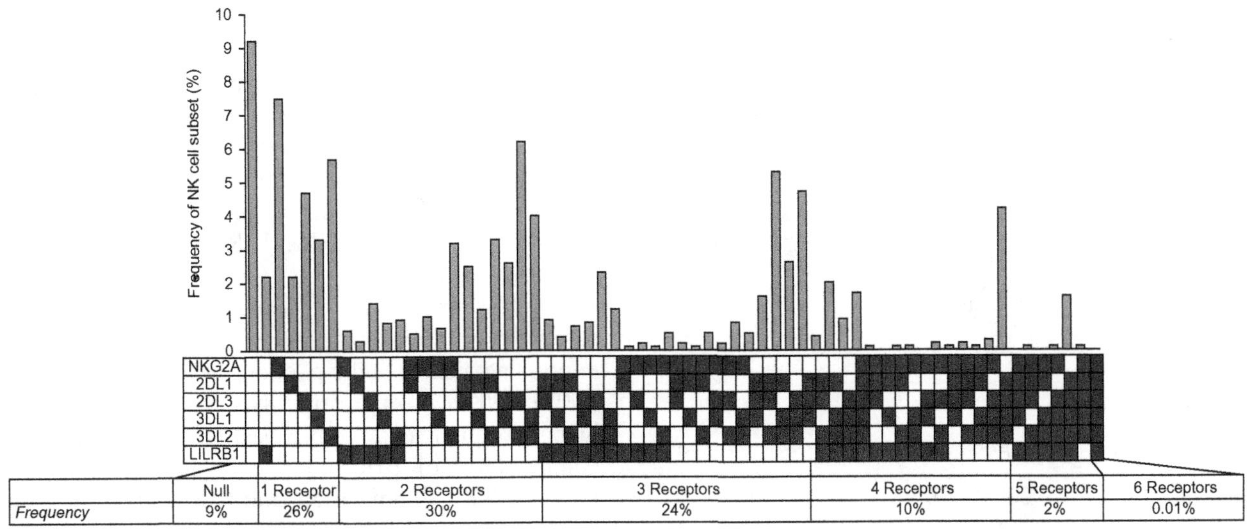

Figure 6.4 • Variegated expression of inhibitory KIR, NKG2A and LILRB1 in CD56dimNK cells. Bars indicate frequencies of the 64 NK subsets in one individual. Black boxes show the receptors expressed on each subset. Frequencies of receptor co-expression in this individual are indicated below (Yawata et al., 2008. This research was originally published in *Blood*. © by the American Society of Hematology).

precisely defined by the KIR and HLA class I polymorphisms present in the genome of each human individual.

Even within a particular KIR ligand group is large variation in the capacity for NK cell education. Some HLA allotypes confer strong NK cell education; other allotypes that have been considered KIR ligands lack educational capacity altogether. Examples are the allotypes of the HLA-C1 group, where the presence of HLA-Cw*07 confers strong education, whereas HLA-Cw*1402 confers minimal education (Yawata et al., 2008). HLA-Bw4 allotypes also contribute to differences in KIR3DL1 recognition (Foley et al., 2008b; Martin et al., 2002; Sanjanwala et al., 2008; Yawata et al., 2008).

KIR polymorphisms are equally important. Examples are the KIR3DL1 polymorphisms, where the KIR3DL1*-1502 allotype has a stronger capacity both in inhibition and acquisition of missing-self as compared with KIR3DL1*007 (Yawata et al., 2006, 2008). Also of note is the intermediate level of education conferred by the monomorphic NKG2A receptor (Yawata et al., 2008).

Co-expression of multiple KIR alone or with NKG2A comes with functional consequences. The level of missing-self response is enhanced additively when an NK cell expresses multiple self-class I specific KIR. In these subsets that display stronger levels of education, inhibition through only one of the multiple receptors is not sufficient to fully inhibit the subset (Fauriat et al., 2008; Yawata et al., 2008). This enhancement of education is attenuated when more than two strong receptors confer education (Yawata et al., 2008).

Not all MHC-class I specific NK receptors confer education. As expected from previous reports describing the lack of binding between KIR3DL2 and HLA-A*03/11, except when the HLA molecule has bound EBV peptide (Hansasuta et al., 2004; Valiante et al., 1997), these ligands do not confer education through KIR3DL2 (Fauriat et al., 2008; Yawata et al., 2008).

NK cell education seems to occur only above a certain threshold of signalling through an HLA–KIR interaction. LILRB1, which has relatively low affinity for HLA-A, HLA-B and HLA-C (Chapman et al., 1999; Colonna et al., 1997), does not confer measurable levels of NK cell education. The weak cross-reactivity of HLA-C2 to KIR2DL3 is likewise incapable of education.

Effect of HLA class I ligands on KIR expression

The observation that KIR identity alone is insufficient to produce similar KIR phenotypes and that KIR phenotypes are more similar between KIR- and HLA-identical siblings suggested an effect of HLA class I on KIR expression (Shilling et al., 2002b). When the effect of HLA class I was investigated in more detail in a panel of unrelated individuals, KIR2DL1 and KIR3DL1 were expressed more frequently in the NK population in individuals that carried the cognate ligand as compared to individuals that did not (Yawata et al., 2006). Interestingly, in the presence of ligands for other KIR,

Table 6.2 Hierarchy in the strength of NK cell education

KIR	HLA	Level of response
2DL3	Cw*07	++++
3DL1^high	Bw4^strong	++++
2DL3	Cw*12	+++
2DL3	C1 + B*46	+++
2DL1 non*004	Cw*02,4,5,6,15	++
3DL1*007	Bw4^strong	++
NKG2A	HLA-E	++
2DL1*004	Cw*02,4,5,6,15	+
2DL3	Cw*01,3,8,1403	+
3DL1^high	B*27	+
3DL1^high	A*24	+
2DL1	Cw*01,3,7,8,12,1403	–
2DL3	Cw*1402	–
2DL3	Cw*02,4,5,6,15	–
3DL1^high	B*13	–
3DL1^high	B*37	–
3DS1	Bw4^strong	–
3DL2	A*3,11	–

3DL1^high; 3DL1 *001/002/015/020
Bw4^strong; HLA-B*38/44/51/52/57/58/59

expression frequency was in turn reduced. This suggested that KIR expression is an integrated system whereby KIR ligands interactively affect receptor phenotypes.

'Repertoire calibration': a mechanism unique to NK cells

MHC-specific NK receptors are acquired in a stochastic and stepwise manner during a window of time in NK cell development (Raulet et al., 2001; Takei et al., 2001). Systematic characterisation of the human NK cell repertoire has demonstrated that all NK cell subsets are retained in the peripheral NK population, from the subset lacking inhibitory receptors (the 'receptor-null' cells), to the subset expressing all possible combinations of inhibitory receptors. From the perspective of NK cell education, NK cells lacking all HLA-specific inhibitory receptors are allowed because they are rendered hyporesponsive and thus self-tolerant (Anfossi et al., 2006).

The retention of all possible forms of receptor expression (NK subsets) and the tight, genetic control of the NK response against missing-self necessitates a mechanism to prevent the formation of NK cell repertoires with excessively strong or weak overall response defined by the specific HLA–KIR genotypes inherited in the individual.

NK cell repertoires of human individuals differ substantially in their frequencies of NKG2A- and KIR-expressing cells (Shilling et al., 2002b; Yawata et al., 2008). NKG2A-expressing NK cells are present at higher frequencies in individuals who have only genetic combinations of KIR and HLA that encode for weak levels of NK cell education, and KIR expression is dominant in individuals carrying a single strong KIR–HLA combination. Of note, in individuals that carry multiple strong combinations of KIR and HLA, NKG2A expression is dominant.

This mechanism, 'repertoire calibration', is distinctively unique from the selection processes that shape T-cell and B-cell repertoires. In the NK cell system, the strength of missing-self response is the measure. KIR and NKG2A expression is balanced to calibrate the overall response of repertoires against missing-self stimulus. NKG2A is a receptor particularly suited for this role because the level of education it confers is moderate regardless of the individual's HLA type (Yawata et al., 2008).

Classifying diverse human NK cell repertoires into five groups

Five types of NK cell repertoires are classified in human individuals (Figure 6.5). Human NK cell repertoires differ in their NKG2A expression, degree of receptor co-expression and frequency of receptor-null cells. Differences in receptor co-expression are particularly important because they enhances the strength of missing-self response while broadening the spectrum of binding to HLA class I. Inhibition by multiple HLA class I ligands is a particularly important aspect in understanding the NK cell response in cancer immunology, allogeneic stem cell transplantation and NK cell-based immunotherapy.

NK cell repertoire structures and responses are largely governed by the genetic polymorphisms of KIR and HLA class I of an individual, but other factors may contribute to altering NK cell repertoire structure and function (Epling-Burnette et al., 2004). Pathogen history and environmental factors can alter NK cell repertoires, such as in HCMV infection when LILRB1 expression increases on NK cells and T cells (Guma et al., 2004).

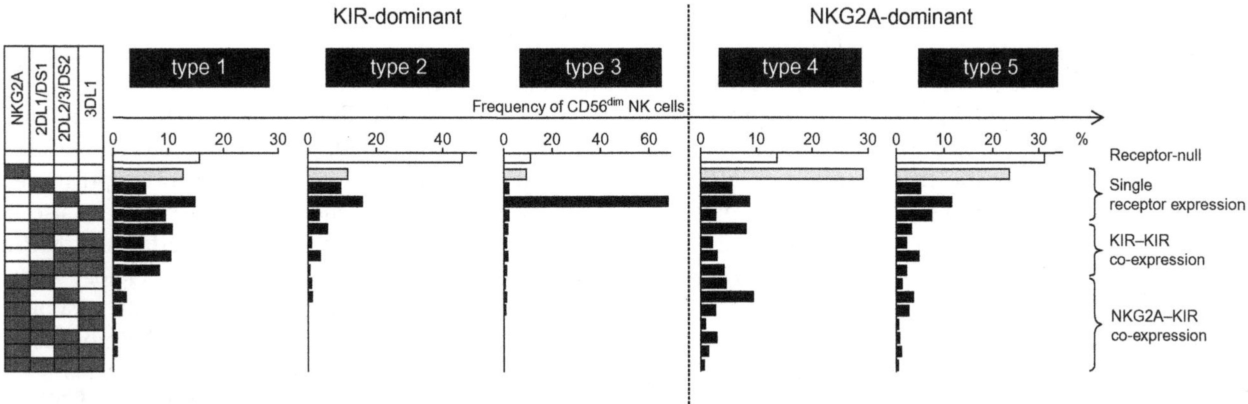

	NKG2A⁻/NKG2A⁺	Receptor-null (%)	KIR co-expression/single expression
type 1	≥2.4	<22	≥0.7
type 2	≥2.4	≥22	
type 3	≥2.4	<22	<0.4
type 4	<2.4	<22	
type 5	<2.4	≥22	

Figure 6.5 • Diverse human NK cell repertoires grouped into 5 types. Representative profiles of NK cell repertoires. Bars indicate frequencies of subsets in the CD56^{dim} NK population; black boxes indicate the receptors expressed on NK subsets. The classification is based on NKG2A expression frequency, frequency of the receptor-null cells and the degree of KIR co-expression in the repertoire (Yawata et al., 2008. This research was originally published in *Blood*. © by the American Society of Hematology).

NK cell heterogeneity as a common feature in mammalian NK cell systems

In many mammalian species, one family of polygenic and polymorphic MHC class I NK receptors has evolved: *KIR* in hominoids, Old World monkeys, cattle, pigs, domestic cats and dogs; *Ly49* in mouse, rat, and horse; and *NKG2A* in New World monkeys (Averdam et al., 2007; Guethlein et al., 2002; Khakoo et al., 2000; LaBonte et al., 2001; McQueen et al., 2002; Rajalingam et al., 2001; Takahashi et al., 2004; Yokoyama et al., 1989). This implies that maintaining diversity in self-, nonself recognition is an essential and intrinsic feature of mammalian NK cell systems, and this has driven the evolution of MHC-specific NK receptor genes (De Tomaso et al., 2005; Hao and Nei, 2005; Hughes, 2002; Parham, 2005).

Although human *KIR* and mouse *Ly49* are nonhomologous genes, they do share many common features: their variegated expression, allele-specific expression, epigenetic control of expression, regulation by bidirectional promoter systems and variation in gene content (Held and Raulet, 1997; Kubota et al., 1999; Proteau et al., 2004; Rouhi et al., 2006; Saleh et al., 2004). Amongst the three *Ly49* haplotypes sequenced to date, gene content varies substantially. Allelic polymorphisms are also a common feature (Mehta et al., 2001). In the context of ligand specificity, Ly49 receptors tend to have more overlap in their recognition of H-2 (Hanke et al., 1999). The effect of MHC-bound peptides on receptor binding will also differ between KIR and Ly49 because of their differences in the site of binding to the MHC molecule (Boyington et al., 2000; Dam et al., 2003; Maenaka et al., 1999; Matsumoto et al., 2001; Natarajan et al., 2002).

In present day human populations, *KIR* and *HLA class I* are distributed unevenly. Analyses employing population genetics on *KIR* distributions have inferred that diversity in some *KIR*, such as *KIR3DL1*, is actively maintained by balancing selection (Norman et al., 2007; Single et al., 2007; Yawata et al., 2006). Within a population, the overall function encoded by a set of *KIRs* on allele-level haplotypes is preferred over others and is driven to high frequencies by positive selection. Analyses have shown also that many other allele-level *KIR* haplotypes are not eliminated but are maintained in the population at lower frequencies. The resulting large gene pool of allele-level *KIR* haplotypes enables natural selection to choose from a wide range of functionality should the environment for a human population change (Yawata et al., 2006). Pathogens, other factors present in the environment and reproductive success are all factors that might have shaped *KIR* distribution during the evolution of human populations (Hiby et al., 2004; Parham, 2005).

Impact of KIR diversity on human health and disease

Numerous studies have investigated genetic associations of *KIR* with specific disease and have correlated *KIR–HLA* combinations with resistance or susceptibility to viral infection, cancer and autoimmune disease. HCV infection is cleared more efficiently in individuals homozygous for *KIR2DL3* and *HLA-C1* (Khakoo et al., 2004). *KIR3DS1*, when matched with certain *Bw4* motifs, delays progression of AIDS (Martin et al., 2002). *KIR2DS1* and *2DS2* confer susceptibility to psoriatic arthropathy with *HLA-C1* or *-C2* homozygosity (Nelson et al., 2004). The presence of activating *KIR* in the maternal genotype combined with *HLA-C2* in the foetal genotype correlates with reproductive success (Hiby et al., 2004).

A previous report described a benefit of NK cell alloreactivity in T cell-depleted HLA haploidentical HSCT, when KIR ligands are mismatched in the GVL direction in AML patients (Ruggeri et al., 2002). Subsequent reports in the field have either supported or contradicted this finding, and many questions remain (Dupont and Hsu, 2004; Velardi, 2008).

Because *KIR* distribution is strongly dependent on population background, it is important to match the demographics of patient and control populations in association studies. Gender differences might also be a concern. Investigating the combined effects of KIR and HLA is more likely to reveal the true biology because they act in concert to determine the cellular response of NK cells. Understanding the functional heterogeneity of human NK cells encoded by KIR and HLA class I will become crucial in effective implementation of NK cell-based immunotherapy.

References

Alter, G., Martin, M.P., Teigen, N., Carr, W.H., Suscovich, T.J., Schneidewind, A., Streeck, H., Waring, M., Meier, A., Brander, C., Lifson, J.D., Allen, T.M., Carrington, M. and Altfeld, M. (2007). Differential natural killer cell-mediated inhibition of HIV-1 replication based on distinct KIR/HLA subtypes. J Exp Med **204**, 3027–3036.

Anfossi, N., Andre, P., Guia, S., Falk, C.S., Roetynck, S., Stewart, C.A., Breso, V., Frassati, C., Reviron, D., Middleton, D., Romagne, F., Ugolini, S. and Vivier, E. (2006). Human NK cell education by inhibitory receptors for MHC class I. Immunity **25**, 331–342.

Averdam, A., Kuhl, H., Sontag, M., Becker, T., Hughes, A.L., Reinhardt, R. and Walter, L. (2007). Genomics and diversity of the common marmoset monkey NK complex. J Immunol **178**, 7151–7161.

Boyington, J.C., Motyka, S.A., Schuck, P., Brooks, A.G. and Sun, P.D. (2000). Crystal structure of an NK cell immunoglobulin-like receptor in complex with its class I MHC ligand. Nature **405**, 537–543.

Burshtyn, D.N., Scharenberg, A.M., Wagtmann, N., Rajagopalan, S., Berrada, K., Yi, T., Kinet, J.P. and Long, E.O. (1996). Recruitment of tyrosine phosphatase HCP by the killer cell inhibitor receptor. Immunity **4**, 77–85.

Carr, W.H., Pando, M.J. and Parham, P. (2005). KIR3DL1 polymorphisms that affect NK cell inhibition by HLA-Bw4 ligand. J Immunol **175**, 5222–5229.

Chan, H.W., Kurago, Z.B., Stewart, C.A., Wilson, M.J., Martin, M.P., Mace, B. E., Carrington, M., Trowsdale, J. and Lutz, C.T. (2003). DNA methylation maintains allele-specific KIR gene expression in human natural killer cells. J Exp Med **197**, 245–255.

Chan, H.W., Miller, J.S., Moore, M.B. and Lutz, C.T. (2005). Epigenetic control of highly homologous killer Ig-like receptor gene alleles. J Immunol **175**, 5966–5974.

Chapman, T.L., Heikeman, A.P. and Bjorkman, P.J. (1999). The inhibitory receptor LIR-1 uses a common binding interaction to recognize class I MHC molecules and the viral homolog UL18. Immunity **11**, 603–613.

Chewning, J.H., Gudme, C.N., Hsu, K.C., Selvakumar, A. and Dupont, B. (2007). KIR2DS1-positive NK cells mediate alloresponse against the C2 HLA-KIR ligand group in vitro. J Immunol **179**, 854–868.

Colonna, M. and Samaridis, J. (1995). Cloning of immunoglobulin-superfamily members associated with HLA-C and HLA-B recognition by human natural killer cells. Science **268**, 405–408.

Colonna, M., Navarro, F., Bellon, T., Llano, M., Garcia, P., Samaridis, J., Angman, L., Cella, M. and Lopez-Botet, M. (1997). A common inhibitory receptor for major histocompatibility complex class I molecules on human lymphoid and myelomonocytic cells. J Exp Med **186**, 1809–1818.

Dam, J., Guan, R., Natarajan, K., Dimasi, N., Chlewicki, L.K., Kranz, D.M., Schuck, P., Margulies, D.H. and Mariuzza, R.A. (2003). Variable MHC class I engagement by Ly49 natural killer cell receptors demonstrated by the crystal structure of Ly49C bound to H-2K(b). Nat Immunol **4**, 1213–1222.

D'Andrea, A., Chang, C., Franz-Bacon, K., McClanahan, T., Phillips, J.H. and Lanier, L.L. (1995). Molecular cloning of NKB1. A natural killer cell receptor for HLA-B allotypes. J Immunol **155**, 2306–2310.

Davies, G.E., Locke, S.M., Wright, P.W., Li, H., Hanson, R.J., Miller, J.S. and Anderson, S.K. (2007). Identification of bidirectional promoters in the human KIR genes. Genes Immun **8**, 245–253.

De Tomaso, A.W., Nyholm, S.V., Palmeri, K.J., Ishizuka, K.J., Ludington, W.B., Mitchel, K. and Weissman, I.L. (2005). Isolation and characterization of a protochordate histocompatibility locus. Nature **438**, 454–459.

Dupont, B. and Hsu, K.C. (2004). Inhibitory killer Ig-like receptor genes and human leukocyte antigen class I ligands in haematopoietic stem cell transplantation. Curr Opin Immunol **16**, 634–643.

Epling-Burnette, P.K., Painter, J.S., Chaurasia, P., Bai, F., Wei, S., Djeu, J.Y. and Loughran, T.P., Jr. (2004). Dysregulated NK receptor expression in patients with lymphoproliferative disease of granular lymphocytes. Blood **103**, 3431–3439.

Fauriat, C., Andersson, S., Bjorklund, A.T., Carlsten, M., Schaffer, M., Bjorkstrom, N.K., Baumann, B.C., Michaelsson, J., Ljunggren, H.G. and Malmberg, K.J. (2008). Estimation of the size of the alloreactive NK cell repertoire: studies in individuals homozygous for the group A KIR haplotype. J Immunol **181**, 6010–6019.

Fehniger, T.A., Cooper, M.A., Nuovo, G. J., Cella, M., Facchetti, F., Colonna, M. and Caligiuri, M.A. (2003). CD56bright natural killer cells are present in human lymph nodes and are activated by T cell-derived IL-2: a potential new link between adaptive and innate immunity. Blood **101**, 3052–3057.

Fernandez, N.C., Treiner, E., Vance, R. E., Jamieson, A.M., Lemieux, S. and Raulet, D.H. (2005). A subset of natural killer cells achieves self-tolerance without expressing inhibitory receptors specific for self-MHC molecules. Blood **105**, 4416–4423.

Foley, B., De Santis, D., Lathbury, L., Christiansen, F. and Witt, C. (2008a). KIR2DS1-mediated activation overrides NKG2A-mediated inhibition in HLA-C C2-negative individuals. Int Immunol **20**, 555–563.

Foley, B.A., De Santis, D., Van Beelen, E., Lathbury, L.J., Christiansen, F.T. and Witt, C.S. (2008b). The reactivity of Bw4-positive HLA-B and HLA-A alleles with KIR3DL1: implications for patient and donor suitability for haploidentical stem cell transplants. Blood.

Gardiner, C.M., Guethlein, L.A., Shilling, H.G., Pando, M., Carr, W. H., Rajalingam, R., Vilches, C. and Parham, P. (2001). Different NK cell surface phenotypes defined by the DX9 antibody are due to KIR3DL1 gene polymorphism. J Immunol **166**, 2992–3001.

Gedil, M.A., Steiner, N.K. and Hurley, C. K. (2005). Genomic characterization of KIR2DL4 in families and unrelated individuals reveals extensive diversity in exon and intron sequences including a common frameshift variation occurring in several alleles. Tissue Antigens **65**, 402–418.

Gedil, M.A., Steiner, N.K. and Hurley, C.K. (2007). KIR3DL2: diversity in a hematopoietic stem cell transplant population. Tissue Antigens **70**, 228–232.

Gendzekhadze, K., Norman, P.J., Abi-Rached, L., Layrisse, Z. and Parham, P. (2006). High KIR diversity in Amerindians is maintained using few gene-content haplotypes. Immunogenetics **58**, 474–480.

Gomez-Lozano, N., Gardiner, C.M., Parham, P. and Vilches, C. (2002). Some human KIR haplotypes contain two KIR2DL5 genes: KIR2DL5A and KIR2DL5B. Immunogenetics **54**, 314–319.

Goodridge, J.P., Witt, C.S., Christiansen, F. T. and Warren, H.S. (2003). KIR2DL4 (CD158d) genotype influences expression and function in NK cells. J Immunol **171**, 1768–1774.

Guethlein, L.A., Flodin, L.R., Adams, E.J. and Parham, P. (2002). NK cell receptors of the orangutan (*Pongo pygmaeus*): a pivotal species for tracking the coevolution of killer cell Ig-like receptors with MHC-C. J Immunol **169**, 220–229.

Guma, M., Angulo, A., Vilches, C., Gomez-Lozano, N., Malats, N. and Lopez-Botet, M. (2004). Imprint of human cytomegalovirus infection on the NK cell receptor repertoire. Blood **104**, 3664–3671.

Gutierrez-Rodriguez, M.E., Sandoval-Ramirez, L., Diaz-Flores, M., Marsh, S. G., Valladares-Salgado, A., Madrigal, J. A., Mejia-Arangure, J.M., Garcia, C. A., Huerta-Zepeda, A., Ibarra-Cortes, B., Ortega-Camarillo, C. and Cruz, M. (2006). KIR gene in ethnic and Mestizo populations from Mexico. Hum Immunol **67**, 85–93.

Hanke, T., Takizawa, H., McMahon, C.W., Busch, D.H., Pamer, E.G., Miller, J.D., Altman, J.D., Liu, Y., Cado, D., Lemonnier, F.A., Bjorkman, P.J. and Raulet, D.H. (1999). Direct assessment of MHC class I binding by seven Ly49 inhibitory NK cell receptors. Immunity **11**, 67–77.

Hansasuta, P., Dong, T., Thananchai, H., Weekes, M., Willberg, C., Aldemir, H., Rowland-Jones, S. and Braud, V. M. (2004). Recognition of HLA-A3 and HLA-A11 by KIR3DL2 is peptide-specific. Eur J Immunol **34**, 1673–1679.

Hao, L. and Nei, M. (2005). Rapid expansion of killer cell immunoglobulin-like receptor genes in primates and their coevolution with MHC class I genes. Gene **347**, 149–159.

Held, W. and Raulet, D.H. (1997). Expression of the Ly49A gene in murine natural killer cell clones is predominantly but not exclusively mono-allelic. Eur J Immunol **27**, 2876–2884.

Hiby, S.E., Walker, J.J., O'Shaughnessy K, M., Redman, C.W., Carrington, M., Trowsdale, J. and Moffett, A. (2004). Combinations of maternal KIR and fetal HLA-C genes influence the risk of preeclampsia and reproductive success. J Exp Med **200**, 957–965.

Horton, R., Wilming, L., Rand, V., Lovering, R.C., Bruford, E.A., Khodiyar, V.K., Lush, M.J., Povey, S., Talbot, C.C., Jr., Wright, M.W., Wain, H.M., Trowsdale, J., Ziegler, A. and Beck, S. (2004). Gene map of the extended human MHC. Nat Rev Genet **5**, 889–899.

Hou, L., Chen, M., Steiner, N.K., Belle, I., Turino, C., Ng, J. and Hurley, C. K. (2007). Seventeen novel alleles add to the already extensive KIR3DL3 diversity. Tissue Antigens **70**, 449–454.

Hou, L., Steiner, N.K., Chen, M., Belle, I., Kubit, A.L., Ng, J. and Hurley, C. K. (2008). Limited allelic diversity of stimulatory two-domain killer cell immunoglobulin-like receptors. Hum Immunol **69**, 174–178.

Houchins, J.P., Yabe, T., McSherry, C. and Bach, F.H. (1991). DNA sequence analysis of NKG2, a family of related cDNA clones encoding type II integral membrane proteins on human natural killer cells. J Exp Med **173**, 1017–1020.

Hsu, K.C., Chida, S., Geraghty, D.E. and Dupont, B. (2002). The killer cell immunoglobulin-like receptor (KIR) genomic region: gene-order, haplotypes and allelic polymorphism. Immunol Rev **190**, 40–52.

Hughes, A.L. (2002). Evolution of the human killer cell inhibitory receptor family. Mol Phylogenet Evol **25**, 330–340.

Jiang, K., Zhu, F.M., Lv, Q.F. and Yan, L. X. (2005). Distribution of killer cell immunoglobulin-like receptor genes in the Chinese Han population. Tissue Antigens **65**, 556–563.

Karlhofer, F.M., Ribaudo, R.K. and Yokoyama, W.M. (1992). MHC class I alloantigen specificity of Ly-49+ IL-2-activated natural killer cells. Nature **358**, 66–70.

Karre, K., Ljunggren, H.G., Piontek, G. and Kiessling, R. (1986). Selective rejection of H-2-deficient lymphoma variants suggests alternative immune defence strategy. Nature **319**, 675–678.

Khakoo, S.I., Rajalingam, R., Shum, B.P., Weidenbach, K., Flodin, L., Muir, D.G., Canavez, F., Cooper, S.L., Valiante, N. M., Lanier, L.L. and Parham, P. (2000). Rapid evolution of NK cell receptor systems demonstrated by comparison of chimpanzees and humans. Immunity **12**, 687–698.

Khakoo, S.I., Thio, C.L., Martin, M.P., Brooks, C.R., Gao, X., Astemborski, J., Cheng, J., Goedert, J.J., Vlahov, D., Hilgartner, M., Cox, S., Little, A.M., Alexander, G.J., Cramp, M.E., O'Brien, S.J., Rosenberg, W.M., Thomas, D.L. and Carrington, M. (2004). HLA and NK cell inhibitory receptor genes in resolving hepatitis C virus infection. Science **305**, 872–874.

Kikuchi-Maki, A., Yusa, S., Catina, T.L. and Campbell, K.S. (2003). KIR2DL4 is an IL-2-regulated NK cell receptor that exhibits limited expression in humans but triggers strong IFN-gamma production. J Immunol **171**, 3415–3425.

Kim, S., Poursine-Laurent, J., Truscott, S. M., Lybarger, L., Song, Y.J., Yang, L.,

French, A.R., Sunwoo, J.B., Lemieux, S., Hansen, T.H. and Yokoyama, W. M. (2005). Licensing of natural killer cells by host major histocompatibility complex class I molecules. Nature **436**, 709–713.

Kollnberger, S., Chan, A., Sun, M.Y., Chen, L.Y., Wright, C., di Gleria, K., McMichael, A. and Bowness, P. (2007). Interaction of HLA-B27 homodimers with KIR3DL1 and KIR3DL2, unlike HLA-B27 heterotrimers, is independent of the sequence of bound peptide. Eur J Immunol **37**, 1313–1322.

Kubota, A., Kubota, S., Lohwasser, S., Mager, D.L. and Takei, F. (1999). Diversity of NK cell receptor repertoire in adult and neonatal mice. J Immunol **163**, 212–216.

LaBonte, M.L., Hershberger, K.L., Korber, B. and Letvin, N.L. (2001). The KIR and CD94/NKG2 families of molecules in the rhesus monkey. Immunol Rev **183**, 25–40.

Lanier, L.L. (2005). NK cell recognition. Annu Rev Immunol **23**, 225–274.

Lanier, L.L., Corliss, B.C., Wu, J., Leong, C. and Phillips, J.H. (1998). Immunoreceptor DAP12 bearing a tyrosine-based activation motif is involved in activating NK cells. Nature **391**, 703–707.

Li, H., Pascal, V., Martin, M.P., Carrington, M. and Anderson, S.K. (2008). Genetic control of variegated KIR gene expression: polymorphisms of the bi-directional KIR3DL1 promoter are associated with distinct frequencies of gene expression. PLoS Genet **4**, e1000254.

Luo, L., Du, Z., Sharma, S.K., Cullen, R., Spellman, S., Reed, E.F. and Rajalingam, R. (2007). Chain-terminating natural mutations affect the function of activating KIR receptors 3DS1 and 2DS3. Immunogenetics **59**, 779–792.

Maenaka, K., Juji, T., Stuart, D.I. and Jones, E.Y. (1999). Crystal structure of the human p58 killer cell inhibitory receptor (KIR2DL3) specific for HLA-Cw3-related MHC class I. Structure **7**, 391–398.

Martin, M.P., Gao, X., Lee, J.H., Nelson, G.W., Detels, R., Goedert, J.J., Buchbinder, S., Hoots, K., Vlahov, D., Trowsdale, J., Wilson, M., O'Brien, S. J. and Carrington, M. (2002). Epistatic interaction between KIR3DS1 and HLA-B delays the progression to AIDS. Nat Genet **31**, 429–434.

Martin, M.P., Bashirova, A., Traherne, J., Trowsdale, J. and Carrington, M. (2003). Cutting edge: expansion of the KIR locus by unequal crossing over. J Immunol **171**, 2192–2195.

Matsumoto, N., Mitsuki, M., Tajima, K., Yokoyama, W.M. and Yamamoto, K. (2001). The functional binding site for the C type lectin like natural killer cell receptor Ly49A spans three domains of its major histocompatibility complex class I ligand. J Exp Med **193**, 147–158.

Maxwell, L.D., Wallace, A., Middleton, D. and Curran, M.D. (2002). A common KIR2DS4 deletion variant in the human that predicts a soluble KIR molecule analogous to the KIR1D molecule observed in the rhesus monkey. Tissue Antigens **60**, 254–258.

Maxwell, L.D., Williams, F., Gilmore, P., Meenagh, A. and Middleton, D. (2004). Investigation of killer cell immunoglobulin-like receptor gene diversity: II. KIR2DS4. Hum Immunol **65**, 613–621.

McQueen, K.L., Wilhelm, B.T., Harden, K. D. and Mager, D.L. (2002). Evolution of NK receptors: a single Ly49 and multiple KIR genes in the cow. Eur J Immunol **32**, 810–817.

Meenagh, A., Williams, F., Sleator, C., Halfpenny, I.A. and Middleton, D. (2004). Investigation of killer cell immunoglobulin-like receptor gene diversity V. KIR3DL2. Tissue Antigens **64**, 226–234.

Mehta, I.K., Wang, J., Roland, J., Margulies, D.H. and Yokoyama, W. M. (2001). Ly49A allelic variation and MHC class I specificity. Immunogenetics **53**, 572–583.

Middleton, D., Meenagh, A. and Gourraud, P.A. (2007). KIR haplotype content at the allele level in 77 Northern Irish families. Immunogenetics **59**, 145–158.

Moesta, A.K., Norman, P.J., Yawata, M., Yawata, N., Gleimer, M. and Parham, P. (2008). Synergistic polymorphism at two positions distal to the ligand-binding site makes KIR2DL2 a stronger receptor for HLA-C than KIR2DL3. J Immunol **180**, 3969–3979.

Moretta, A., Vitale, M., Bottino, C., Orengo, A.M., Morelli, L., Augugliaro, R., Barbaresi, M., Ciccone, E. and Moretta, L. (1993). P58 molecules as putative receptors for major histocompatibility complex (MHC) class I molecules in human natural killer (NK) cells. Anti-p58 antibodies reconstitute lysis of MHC class I-protected cells in NK clones displaying different specificities. J Exp Med **178**, 597–604.

Moretta, A., Sivori, S., Vitale, M., Pende, D., Morelli, L., Augugliaro, R., Bottino, C. and Moretta, L. (1995). Existence of both inhibitory (p58) and activatory (p50) receptors for HLA-C molecules in human natural killer cells. J Exp Med **182**, 875–884.

Morvan, M., David, G., Sebille, V., Perrin, A., Gagne, K., Willem, C., Kerdudou, N., Denis, L., Clemenceau, B., Follea, G., Bignon, J.D. and Retiere, C. (2008). Autologous and allogeneic HLA KIR ligand environments and activating KIR control KIR NK-cell functions. Eur J Immunol

Natarajan, K., Dimasi, N., Wang, J., Mariuzza, R.A. and Margulies, D.H. (2002). Structure and function of natural killer cell receptors: multiple molecular solutions to self, nonself discrimination. Annu Rev Immunol **20**, 853–885.

Nelson, G.W., Martin, M.P., Gladman, D., Wade, J., Trowsdale, J. and Carrington, M. (2004). Cutting edge: heterozygote advantage in autoimmune disease: hierarchy of protection/susceptibility conferred by HLA and killer Ig-like receptor combinations in psoriatic arthritis. J Immunol **173**, 4273–4276.

Norman, P.J., Stephens, H.A., Verity, D.H., Chandanayingyong, D. and Vaughan, R.W. (2001). Distribution of natural killer cell immunoglobulin-like receptor sequences in three ethnic groups. Immunogenetics **52**, 195–205.

Norman, P.J., Carrington, C.V., Byng, M., Maxwell, L.D., Curran, M.D., Stephens, H.A., Chandanayingyong, D., Verity, D.H., Hameed, K., Ramdath, D. D. and Vaughan, R.W. (2002). Natural killer cell immunoglobulin-like receptor (KIR) locus profiles in African and South Asian populations. Genes Immun **3**, 86–95.

Norman, P.J., Cook, M.A., Carey, B.S., Carrington, C.V., Verity, D.H., Hameed, K., Ramdath, D.D., Chandanayingyong, D., Leppert, M., Stephens, H.A. and Vaughan, R.W. (2004). SNP haplotypes and allele frequencies show evidence for disruptive and balancing selection in the human leukocyte receptor complex. Immunogenetics **56**, 225–237.

Norman, P.J., Abi-Rached, L., Gendzekhadze, K., Korbel, D., Gleimer, M., Rowley, D., Bruno, D., Carrington, C.V., Chandanayingyong, D., Chang, Y. H., Crespi, C., Saruhan-Direskeneli, G., Fraser, P.A., Hameed, K., Kamkamidze, G., Koram, K.A., Layrisse, Z., Matamoros, N., Mila, J., Park, M.H., Pitchappan, R.M., Ramdath, D.D., Shiau, M.Y., Stephens, H.A., Struik, S., Verity, D.H., Vaughan, R.W., Tyan, D., Davis, R.W., Riley, E.M., Ronaghi, M. and Parham, P. (2007). Unusual selection on the KIR3DL1/S1 natural killer cell receptor in Africans. Nat Genet **39**, 1092–1099.

O'Connor, G.M., Guinan, K.J., Cunningham, R.T., Middleton, D., Parham, P. and Gardiner, C.M. (2007). Functional polymorphism of the KIR3DL1/S1 receptor on human NK cells. J Immunol **178**, 235–241.

Pando, M.J., Gardiner, C.M., Gleimer, M., McQueen, K.L. and Parham, P. (2003). The protein made from a common allele of KIR3DL1 (3DL1*004) is poorly expressed at cell surfaces due to substitution at positions 86 in Ig domain 0 and 182 in Ig domain 1. J Immunol **171**, 6640–6649.

Parham, P. (2005). MHC class I molecules and KIRs in human history, health and survival. Nat Rev Immunol **5**, 201–214.

Pascal, V., Stulberg, M.J. and Anderson, S. K. (2006). Regulation of class I major histocompatibility complex receptor expression in natural killer cells: one promoter is not enough!. Immunol Rev **214**, 9–21.

Pende, D., Marcenaro, S., Falco, M., Martini, S., Bernardo, M.E., Montagna, D., Romeo, E., Cognet, C., Martinetti, M., Maccario, R., Mingari, M.C., Vivier, E., Moretta, L., Locatelli, F. and Moretta, A. (2008). Anti-leukemia activity of alloreactive NK cells in KIR ligand-mismatched haploidentical HSCT for pediatric patients: evaluation of the functional role of activating KIR and re-definition of inhibitory KIR specificity. Blood.

Ponte, M., Cantoni, C., Biassoni, R., Tradori-Cappai, A., Bentivoglio, G., Vitale, C., Bertone, S., Moretta, A., Moretta, L. and Mingari, M.C. (1999). Inhibitory receptors sensing HLA-G1 molecules in pregnancy: decidua-associated natural killer cells express LIR-1 and CD94/NKG2A and acquire p49, an HLA-G1-specific receptor. Proc Natl Acad Sci U S A **96**, 5674–5679.

Proteau, M.F., Rousselle, E. and Makrigiannis, A.P. (2004). Mapping of the BALB/c Ly49 cluster defines a minimal natural killer cell receptor gene repertoire. Genomics **84**, 669–677.

Rajagopalan, S., Bryceson, Y.T., Kuppusamy, S.P., Geraghty, D.E., van der Meer, A., Joosten, I. and Long, E. O. (2006). Activation of NK cells by an endocytosed receptor for soluble HLA-G. PLoS Biol **4**, e9.

Rajalingam, R., Hong, M., Adams, E.J., Shum, B.P., Guethlein, L.A. and Parham, P. (2001). Short KIR haplotypes in pygmy chimpanzee (Bonobo) resemble the conserved framework of diverse human KIR haplotypes. J Exp Med **193**, 135–146.

Rajalingam, R., Krausa, P., Shilling, H.G., Stein, J.B., Balamurugan, A., McGinnis, M.D., Cheng, N.W., Mehra, N.K. and Parham, P. (2002). Distinctive KIR and HLA diversity in a panel of north Indian Hindus. Immunogenetics **53**, 1009–1019.

Rajalingam, R., Du, Z., Meenagh, A., Luo, L., Kavitha, V.J., Pavithra-Arulvani, R., Vidhyalakshmi, A., Sharma, S.K., Balazs, I., Reed, E.F., Pitchappan, R. M. and Middleton, D. (2008). Distinct diversity of KIR genes in three southern Indian populations: comparison with world populations revealed a link between KIR gene content and pre-historic human migrations. Immunogenetics **60**, 207–217.

Raulet, D.H., Vance, R.E. and McMahon, C.W. (2001). Regulation of the natural killer cell receptor repertoire. Annu Rev Immunol **19**, 291–330.

Rouhi, A., Gagnier, L., Takei, F. and Mager, D.L. (2006). Evidence for epigenetic maintenance of Ly49a monoallelic gene expression. J Immunol **176**, 2991–2999.

Ruggeri, L., Capanni, M., Urbani, E., Perruccio, K., Shlomchik, W.D., Tosti, A., Posati, S., Rogaia, D., Frassoni, F., Aversa, F., Martelli, M.F. and Velardi, A. (2002). Effectiveness of donor natural killer cell alloreactivity in mismatched hematopoietic transplants. Science **295**, 2097–2100.

Saleh, A., Davies, G.E., Pascal, V., Wright, P.W., Hodge, D.L., Cho, E.H., Lockett, S.J., Abshari, M. and Anderson, S.K. (2004). Identification of probabilistic transcriptional switches in the Ly49 gene cluster: a eukaryotic mechanism for selective gene activation. Immunity **21**, 55–66.

Sambrook, J.G., Bashirova, A., Andersen, H., Piatak, M., Vernikos, G.S., Coggill, P., Lifson, J.D., Carrington, M. and Beck, S. (2006). Identification of the ancestral killer immunoglobulin-like receptor gene in primates. BMC Genomics **7**, 209.

Sanjanwala, B., Draghi, M., Norman, P.J., Guethlein, L.A. and Parham, P. (2008). Polymorphic sites away from the Bw4 epitope that affect interaction of Bw4+ HLA-B with KIR3DL1. J Immunol **181**, 6293–6300.

Santourlidis, S., Trompeter, H.I., Weinhold, S., Eisermann, B., Meyer, K. L., Wernet, P. and Uhrberg, M. (2002). Crucial role of DNA methylation in determination of clonally distributed killer cell Ig-like receptor expression patterns in NK cells. J Immunol **169**, 4253–4261.

Santourlidis, S., Graffmann, N., Christ, J. and Uhrberg, M. (2008). Lineage-specific transition of histone signatures in the killer cell Ig-like receptor locus from hematopoietic progenitor to NK cells. J Immunol **180**, 418–425.

Shilling, H.G., Guethlein, L.A., Cheng, N.W., Gardiner, C.M., Rodriguez, R., Tyan, D. and Parham, P. (2002a). Allelic polymorphism synergizes with variable gene content to individualize human KIR genotype. J Immunol **168**, 2307–2315.

Shilling, H.G., Young, N., Guethlein, L.A., Cheng, N.W., Gardiner, C.M., Tyan, D. and Parham, P. (2002b). Genetic control of human NK cell repertoire. J Immunol **169**, 239–247.

Single, R.M., Martin, M.P., Gao, X., Meyer, D., Yeager, M., Kidd, J.R., Kidd, K. K. and Carrington, M. (2007). Global diversity and evidence for coevolution of KIR and HLA. Nat Genet **39**, 1114–1119.

Stern, M., Ruggeri, L., Capanni, M., Mancusi, A. and Velardi, A. (2008). Human leukocyte antigens A23, A24, and A32 but not A25 are ligands for KIR3DL1. Blood **112**, 708–710.

Stewart, C.A., Van Bergen, J. and Trowsdale, J. (2003). Different and divergent regulation of the KIR2DL4 and KIR3DL1 promoters. J Immunol **170**, 6073–6081.

Stewart, C.A., Laugier-Anfossi, F., Vely, F., Saulquin, X., Riedmuller, J., Tisserant, A., Gauthier, L., Romagne, F., Ferracci, G., Arosa, F.A., Moretta, A., Sun, P.D., Ugolini, S. and Vivier, E. (2005). Recognition of peptide-MHC class I complexes by activating killer immunoglobulin-like receptors. Proc Natl Acad Sci U S A **102**, 13224–13229.

Takahashi, T., Yawata, M., Raudsepp, T., Lear, T.L., Chowdhary, B.P., Antczak, D.F. and Kasahara, M. (2004). Natural killer cell receptors in the horse: evidence for the existence of multiple transcribed LY49 genes. Eur J Immunol **34**, 773–784.

Takei, F., McQueen, K.L., Maeda, M., Wilhelm, B.T., Lohwasser, S., Lian, R.H. and Mager, D.L. (2001). Ly49 and CD94/NKG2: developmentally regulated expression and evolution. Immunol Rev **181**, 90–103.

Thananchai, H., Gillespie, G., Martin, M. P., Bashirova, A., Yawata, N., Yawata, M., Easterbrook, P., McVicar, D.W., Maenaka, K., Parham, P., Carrington, M., Dong, T. and Rowland-Jones, S. (2007). Cutting edge: allele-specific and peptide-dependent interactions between KIR3DL1 and HLA-A and HLA-B. J Immunol **178**, 33–37.

Thomas, R., Yamada, E., Alter, G., Martin, M.P., Bashirova, A.A., Norman, P.J.,

Altfeld, M., Parham, P., Anderson, S. K., McVicar, D.W. and Carrington, M. (2008). Novel KIR3DL1 alleles and their expression levels on NK cells: convergent evolution of KIR3DL1 phenotype variation?. J Immunol **180**, 6743–6750.

Toneva, M., Lepage, V., Lafay, G., Dulphy, N., Busson, M., Lester, S., Vu-Trien, A., Michaylova, A., Naumova, E., McCluskey, J. and Charron, D. (2001). Genomic diversity of natural killer cell receptor genes in three populations. Tissue Antigens **57**, 358–362.

Trompeter, H.I., Gomez-Lozano, N., Santourlidis, S., Eisermann, B., Wernet, P., Vilches, C. and Uhrberg, M. (2005). Three structurally and functionally divergent kinds of promoters regulate expression of clonally distributed killer cell Ig-like receptors (KIR), of KIR2DL4, and of KIR3DL3. J Immunol **174**, 4135–4143.

Trundley, A.E., Hiby, S.E., Chang, C., Sharkey, A.M., Santourlidis, S., Uhrberg, M., Trowsdale, J. and Moffett, A. (2006). Molecular characterization of KIR3DL3. Immunogenetics **57**, 904–916.

Uhrberg, M., Valiante, N.M., Shum, B.P., Shilling, H.G., Lienert-Weidenbach, K., Corliss, B., Tyan, D., Lanier, L.L. and Parham, P. (1997). Human diversity in killer cell inhibitory receptor genes. Immunity **7**, 753–763.

Uhrberg, M., Parham, P. and Wernet, P. (2002). Definition of gene content for nine common group B haplotypes of the Caucasoid population: KIR haplotypes contain between seven and eleven KIR genes. Immunogenetics **54**, 221–229.

Valiante, N.M., Uhrberg, M., Shilling, H. G., Lienert-Weidenbach, K., Arnett, K. L., D'Andrea, A., Phillips, J.H., Lanier, L.L. and Parham, P. (1997). Functionally and structurally distinct NK cell receptor repertoires in the peripheral blood of two human donors. Immunity **7**, 739–751.

VandenBussche, C.J., Dakshanamurthy, S., Posch, P.E. and Hurley, C.K. (2006). A single polymorphism disrupts the killer Ig-like receptor 2DL2/2DL3 D1 domain. J Immunol **177**, 5347–5357.

Velardi, A. (2008). Role of KIRs and KIR ligands in hematopoietic transplantation. Curr Opin Immunol **20**, 581–587.

Velickovic, M., Velickovic, Z. and Dunckley, H. (2006). Diversity of killer cell immunoglobulin-like receptor genes in Pacific Islands populations. Immunogenetics **58**, 523–532.

Velickovic, M., Velickovic, Z., Panigoro, R. and Dunckley, H. (2008). Diversity of killer cell immunoglobulin-like receptor genes in Indonesian populations of Java, Kalimantan, Timor and Irian Jaya. Tissue Antigens.

Verma, S., King, A. and Loke, Y.W. (1997). Expression of killer cell inhibitory receptors on human uterine natural killer cells. Eur J Immunol **27**, 979–983.

Vilches, C. and Parham, P. (2002). KIR: diverse, rapidly evolving receptors of innate and adaptive immunity. Annu Rev Immunol **20**, 217–251.

Vilches, C., Gardiner, C.M. and Parham, P. (2000). Gene structure and promoter variation of expressed and nonexpressed variants of the KIR2DL5 gene. J Immunol **165**, 6416–6421.

Voss, S.D., Daley, J., Ritz, J. and Robertson, M.J. (1998). Participation of the CD94 receptor complex in costimulation of human natural killer cells. J Immunol **160**, 1618–1626.

Wagtmann, N., Biassoni, R., Cantoni, C., Verdiani, S., Malnati, M.S., Vitale, M., Bottino, C., Moretta, L., Moretta, A. and Long, E.O. (1995). Molecular clones of the p58 NK cell receptor reveal immunoglobulin-related molecules with diversity in both the extra- and intracellular domains. Immunity **2**, 439–449.

Whang, D.H., Park, H., Yoon, J.A. and Park, M.H. (2005). Haplotype analysis of killer cell immunoglobulin-like receptor genes in 77 Korean families. Hum Immunol **66**, 146–154.

Williams, F., Maxwell, L.D., Halfpenny, I. A., Meenagh, A., Sleator, C., Curran, M.D. and Middleton, D. (2003). Multiple copies of KIR 3DL/S1 and KIR 2DL4 genes identified in a number of individuals. Hum Immunol **64**, 729–732.

Wilson, M.J., Torkar, M., Haude, A., Milne, S., Jones, T., Sheer, D., Beck, S. and Trowsdale, J. (2000). Plasticity in the organization and sequences of human KIR/ILT gene families. Proc Natl Acad Sci U S A **97**, 4778–4783.

Winter, C.C., Gumperz, J.E., Parham, P., Long, E.O. and Wagtmann, N. (1998). Direct binding and functional transfer of NK cell inhibitory receptors reveal novel patterns of HLA-C allotype recognition. J Immunol **161**, 571–577.

Yawata, M., Yawata, N., Abi-Rached, L. and Parham, P. (2002a). Variation within the human killer cell immunoglobulin-like receptor (KIR) gene family. Crit Rev Immunol **22**, 463–482.

Yawata, M., Yawata, N., McQueen, K. L., Cheng, N.W., Guethlein, L.A., Rajalingam, R., Shilling, H.G. and Parham, P. (2002b). Predominance of group A KIR haplotypes in Japanese associated with diverse NK cell repertoires of KIR expression. Immunogenetics **54**, 543–550.

Yawata, M., Yawata, N., Draghi, M., Little, A.M., Partheniou, F. and Parham, P. (2006). Roles for HLA and KIR polymorphisms in natural killer cell repertoire selection and modulation of effector function. J Exp Med **203**, 633–645.

Yawata, M., Yawata, N., Draghi, M., Partheniou, F., Little, A.M. and Parham, P. (2008). MHC class I-specific inhibitory receptors and their ligands structure diverse human NK-cell repertoires toward a balance of missing self-response. Blood **112**, 2369–2380.

Yokoyama, W.M., Jacobs, L.B., Kanagawa, O., Shevach, E.M. and Cohen, D.I. (1989). A murine T lymphocyte antigen belongs to a supergene family of type II integral membrane proteins. J Immunol **143**, 1379–1386.

Yu, J., Heller, G., Chewning, J., Kim, S., Yokoyama, W.M. and Hsu, K.C. (2007). Hierarchy of the human natural killer cell response is determined by class and quantity of inhibitory receptors for self-HLA-B and HLA-C ligands. J Immunol **179**, 5977–5989.

Chapter Seven

7

Signalling events in natural killer cells

P.K. Epling-Burnette, Sheng Wei, Julie Y. Djeu

CHAPTER CONTENTS

A thousand mile trip begins with the first step.

Ancient Chinese Proverb

ABSTRACT

Our understanding of natural killer (NK) cells over the last decade dramatically advanced following the seminal discovery of inhibitory and activating receptors that control NK cell effector function. Subsequent elucidation of intracellular signalling events that couple external stimulation to a sophisticated repertoire of functions demonstrated the intricate mechanisms in which innate immunity provides protection against malignant and virally infected cells. Positive and negative signals must be balanced during these interactions to provide an adequate innate immune response that also protects against development of deleterious autoimmune reactivity.

KEY WORDS

Signalling, Natural killer cells, Cytotoxicity, Immunosurveillance, Killer cell immunoglobulin-like receptors, Natural cytotoxicity receptors, C-lectin-like receptors, Adaptor molecules, Antigen-presenting cells

Introduction

Natural killer (NK) cells provide immune protection by spontaneously lysing abnormal cells and secreting cytokines to provide immune surveillance against MHC class I (MHC-I)-deficient tumour cells and virus-infected cells that have escaped detection by cytotoxic T-cells. NK cell activation is controlled by balanced signals from activating and inhibitory receptors. Activating receptors that induce effector functions include the C-lectin-like family (NKG2), natural cytotoxicity family receptors (NKp46, NKp44, NKp30), CD244 (2B4) and killer cell immunoglobulin-like receptors (KIR). In contrast, engagement of inhibitory receptors blocks direct cytotoxicity and antibody-dependent cellular cytotoxicity (ADCC). Maintenance of normal NK function relies on the balance between these activating and inhibitory signals (Figure 7.1).

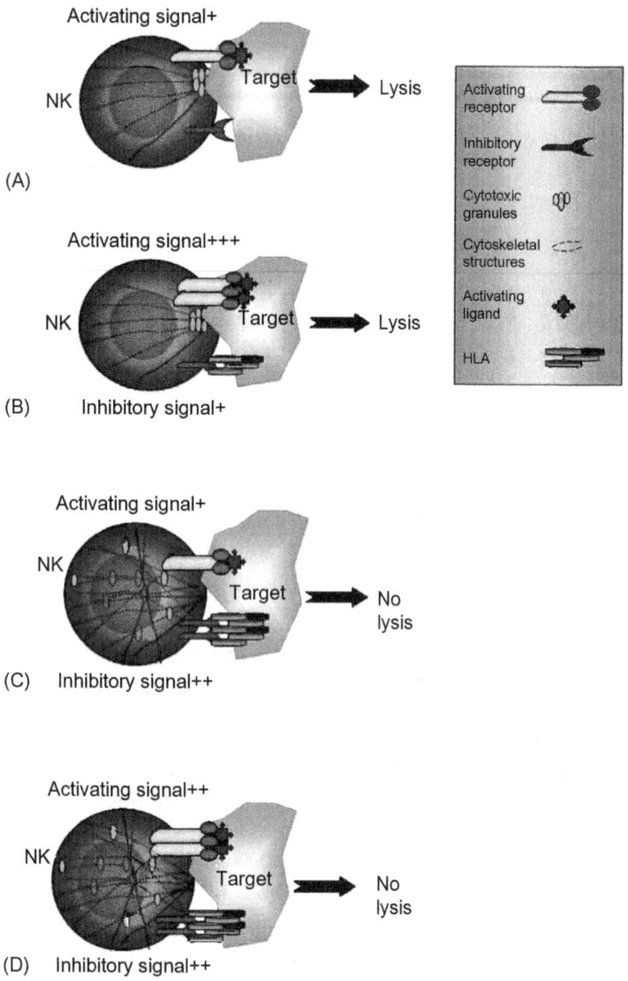

Figure 7.1 • Balance of NK receptor signallings. There is a critical relationship between the interaction and signalling events that lead to lysis or protection of target cells. NK cells with polarized cytotoxic granules move along cytoskeletal structures toward the contact site of a target cell and lysis ensues when the activating receptor signals dominate as shown in (A) and (B). This occurs when there is downregulation or absence of HLA recognition (A) or less potent interaction by the inhibitory signal (B) compared to the activating signal. Alternatively, lysis is blocked and polarized granule mobilization is inhibited by engagement and signalling of inhibitory receptors as shown in (C) and (D).

Table 7.1 Characteristics of KIRs and ligands

KIR	Structure	Function	Ligand
2DL1	2 Ig domains, ITIM	Inhibitory	*C2-epitope$^{lys\ 80}$
2DS1	2 Ig domains, ITAM	Activating	*C2-epitope$^{lys\ 80}$
2DL2	2 Ig domains, ITIM	Inhibitory	**C1-epitope ^{Asn80}
2DL3	2 Ig domains, ITIM	Inhibitory	**C1-epitope^{Asn80}
2DS2	2 Ig domains, ITAM	Activating	**C1-epitope^{Asn80}
2DL4	2 Ig domains, single ITIM	Inhibitory	HLA-G
2DL5	2 Ig domains, ITIM	Inhibitory	?
2DS3	2 Ig domains, ITAM	Activating	**C1-epitope^{Asn80}
2DS4	2 Ig domains, ITAM	Activating	?
2DS5	2 Ig domains, ITAM	Activating	?
3DL1	3 Ig domains, ITIM	Inhibitory	***HLA-Bw4 epitope
3DS1	3 Ig domains, activating ITAM	Activating	***HLA-Bw4 epitope
3DL2	3 Ig domains, ITIM	Inhibitory	HLA-Aw3 and HLA-Aw11
3DL3		Inhibitory	?

*C2-epitope lys 80:HLA-Cw02, Cw04, Cw05, Cw06, Cw15, CW1602, Cw17, Cw18.
**C1-epitopeAsn80: HLA-Cw01, Cw03, Cw07, Cw08, Cw12, Cw14, Cw16 (not Cw1602).
***HLA-Bw4 epitope: B4, B5102, B5103, B13, B17, B27, B37, B38(16),B44(12), B47, B49(21), B51(5), B52(5), B53, B57(17), B58(17), B59, B63(15), B77(15), A9, A23(9), A24(9), A2403, A25(10), A32(19).

Inhibitory NK receptors

Signals by inhibitory KIR

Inhibitory KIR molecules, located on human chromosome 19q13.4, play the primary role of self discrimination. Classical MHC class Ia (HLA-A, HLA-B and HLA-C) molecules serve as ligands for KIR with specificities shown in Table 7.1. Most KIRs with two immunoglobulin domains (KIR2DL) recognize subsets of HLA-C with specific epitope recognition patterns containing either an Asn at position 80 (C1 epitope) or Lys at position 80 (C2 epitope); whereas, those expressing three immunoglobulin domains (KIR3DL) recognize HLA-A and HLA-B (Figure 7.2) (Lanier et al., 1997). Nonclassical HLA-G, which is normally expressed only on fetal-derived trophoblast cells, is uniquely recognized by KIR2DL4 (Rajagopalan and Long, 1999; Uhrberg et al., 2002). The universal expression of KIR2DL4 on all human NK cells suggests a critical and nonredundant purpose for this receptor in human development. The mechanism of KIR-mediated inhibition involves recruitment of protein tyrosine phosphatases (PTP) to intracellular motifs within activating receptor/adaptor complexes (Bakker et al., 2000). PTPs with proven inhibitory NK function include Src homology region 2-containing protein tyrosine phosphatase (SHP)-1 and SHP-2. Recruitment of these PTPs occurs following phosphorylation of a conserved immunoreceptor tyrosine-based inhibitory motif (ITIM) located within the intracellular membrane. The canonical ITIM motif [(I/V)xYxx(L/V)] contains a tyrosine residue that is critical for recruitment and docking by specific phosphatases that functionally block the progression of activating signals through dephosphorylation (Burshtyn et al., 1997). Multiple ITIMs are generally present within the receptor intracellular domain (Bruhns et al., 1999). Studies using cells that lack one or more phosphatases demonstrated that SHP-1 is preferentially recruited to the proximal KIR ITIM (Ono et al., 1997; Vely et al., 1997). In contrast, specific site-directed mutagenesis in the distal ITIM of the KIR molecule proved to be less effective at regulating the activation signals (Bruhns et al., 1999). Resolution of the crystal structure of SHP-2-complexed with a KIR ITIM showed an antiparallel configuration suggesting that binding of this phosphatase probably occurs in pairs of two molecules possibly favouring a trans orientation (Eck et al., 1996). In general, the inhibitory function of a receptor with only a single ITIM has been questioned. The efficacy of a single-ITIM-containing receptor was unequivocally confirmed (Yusa et al., 2002). Using NK-92 cells, which naturally express NKp46, NKp44 and CD244, the inhibitory function of 3DL1 fused to 2DL4 (3DL1/L4 chimera) was examined. Results of these studies showed that two ITIMs of 3DL1 and one ITIM of 2DL4 are comparable in their ability to block activation signals generated by the FcR. While clearly both ITIMs of KIR3DL1 are necessary for function, the degree of contribution and the role of the distal ITIM in KIR regulation remain unclear. In contrast to NK cells, SHP-2 has shown preferential association with chimeric KIR receptors expressed in both B cells and mast cells, suggesting that the cellular context may also affect the signalling pathway (Bruhns et al., 1999; Vely et al., 1997; Yamashita et al., 1998). In experiments using the 3DL1/L4 chimera expressed in NK cells, SHP-2 but not SHP-1 displayed selective association with the 2DL4 ITIM, suggesting that the ITIM sequence may also contribute to binding partner selection (Yusa et al., 2002).

During NK–target cell interactions, co-engagement of inhibitory KIR and activating receptors occurs at the contact zone within the cellular immune synapse. Similar supramolecular activation complexes are important for T-cell activation and adhesion molecules are also important for complex formation and KIR recruitment (Monks et al., 1998; Qi et al., 2001). Mutation of the Zn^{2+} motif in the N-terminus of KIR demonstrated the mechanism responsible for KIR enrichment in the contract zone. Zn^{2+} chelation prevented clustering of KIR at the cellular interface and blocked dimerization of soluble KIR (Fan et al., 2000; Rajagopalan and Long, 1998; Rajagopalan et al., 1995). Clustering of KIR through the Zn^{2+} N-terminal domain is therefore an important step in the process induced by NK receptors.

Ly49 molecules

In the murine genome, MHC-class I-binding receptors are C-lectin-like molecules located on chromosome 6 in the NK gene complex (Yokoyama and Seaman, 1993) and consist of Ly49*a, c, d, f, j, i* and *j* genes. Like KIR, Ly49 molecules exist in both inhibitory and activating variants. There are at least 14 functional inhibitory Ly49 genes reported with different types of genes found in each mouse strain (Pascal, Stulberg, and Anderson). Ly49A-function is severely compromised in SHP-1-deficient mice (Nakamura et al., 1997) and ITIMs of the inhibitory Ly49 receptors bind SHP-1, suggesting that Ly49-mediated inhibition parallels that of KIR (Olcese et al., 1996).

Downstream signalling targets of PTPs following KIR and Ly49-recruitment

After an activating receptor complex phosphorylates the KIR or Ly49 ITIMs and SHP-1 and SHP-2 is recruited, these phosphatases dephosphorylate the early tyrosine kinase substrates to blunt the activating signal. Substrates of SHP-1 and/or SHP-2 include many tyrosine-containing signalling molecules such as T-cell receptor (TCR), Syk, Zap-70, phospholipase Cγ (PLCγ), LAT and SLP-76 (Figure 7.2) (Binstadt et al., 1998; Jevremovic et al., 1999; McVicar and Burshtyn, 2001). The primary goal of substrate dephosphorylation and KIR engagement is to block NK cytotoxicity and cytokine production. SHP-1 dephosphorylation of the downstream signalling intermediate Vav-1 demonstrates the promiscuous power of inhibitory KIR to regulate signalling from multiple activating NK receptors (Long et al., 2001; Stebbins et al., 2003; Watzl et al., 2000). Dephosphorylation events also have profound effects on lipid rafts. Ligation of self-MHC class I by KIR blocks lipid raft formation along with polarization of granules at the target contact zone, suggesting that SHP-mediated dephosphorylation of cellular substrates prevents the migration of molecules that align after engagement of activating NK receptors (Lou et al., 2000).

Activating NK receptors

Activating KIR

Activating forms of KIR share 95–98% sequence identity with the inhibitory forms but possess a truncated cytoplasmic domain and lack functional ITIMs (Uhrberg et al., 1997). The charged acidic amino acid within the transmembrane regions of the activating KIR recruits adaptors with immunoreceptor tyrosine-based activation motifs (ITAMs). ITAMs comprise two copies of the motif Yxx(I/L) precisely spaced six or seven residues apart within the cytoplasmic domain of the activating receptors (Lanier et al., 1998; Moretta et al., 2000). In addition to KIR, an ITAM activation motif is found in the FcRγ,

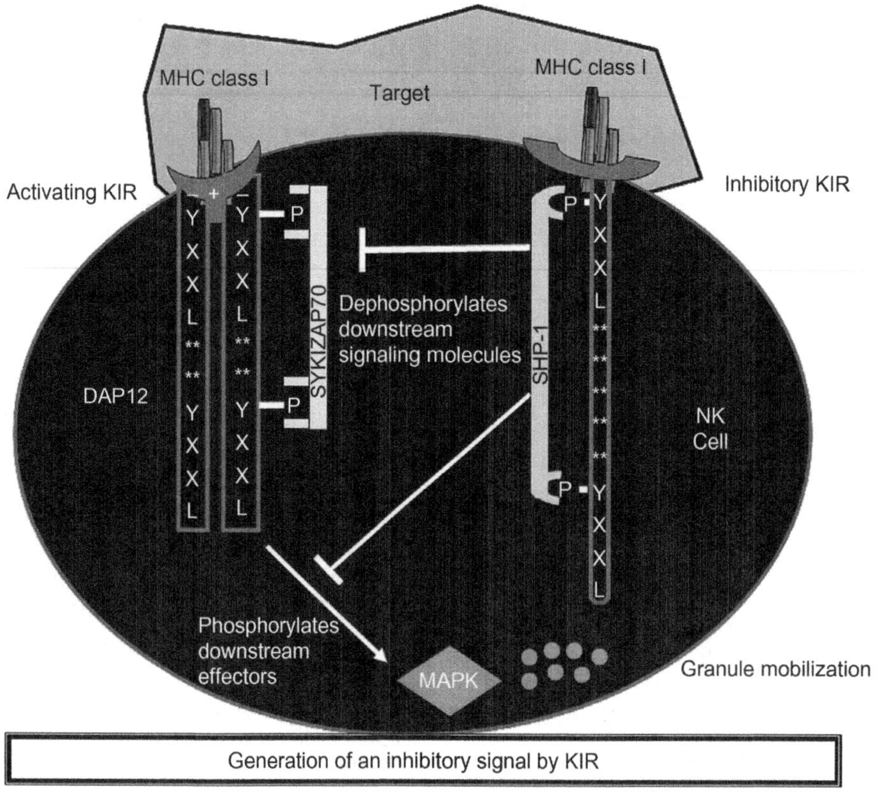

Figure 7.2 • Model of inhibitory KIR signalling. In the case of inhibitory KIRs, binding to an HLA-class I ligand leads to phosphorylation of the tyrosines in the cytoplasmic ITIM (YxxL x $_{28-32}$ YxxL) motif. These phosphotyrosines specifically bind the SH2 domains of the tyrosine phosphatase, SHP-1 or SHP-2. Recruitment of SHP-1 or SHP-2 then leads to activation and inhibition of NK-mediated cytotoxicity by dephosphorylating of essential signal molecules such as Syk/ZAP70. The interaction between inhibitory KIR and activating receptor requires close proximity between the two classes of receptors, which occurs through KIR clustering.

the TCR–CD3 complex and several other NK receptors with activating function. Activating KIRs and other NK activating receptors associate with a unique 12 kDa ITAM-containing adaptor originally designated killer-cell activating receptor-associated protein (KARAP) or DAP12, as shown in Figure 7.3. The gene encoding DAP12 is located roughly 1 Mb from the KIR locus on human chromosome 19q13.1 (Lanier et al., 1998). Analogous to human KIR, the murine *dap12* (*tyropb*) gene product is located on chromosome 7 and binds activating Ly49D and Ly49H which recognize H-2 class I ligands (Lanier et al., 1998; Smith et al., 1998). DAP12 is a type I transmembrane protein expressed in NK cells, myeloid cells and a subpopulation of T-cells. DAP12 phosphorylation leads to an association between DAP12 and ZAP-70 and Syk protein tyrosine kinases in humans and preferentially to Syk in the mouse (McVicar et al., 1998). Unlike other adaptor molecules that signal through heterodimers, DAP12 signals are mediated by homodimerization.

NKG2-family of C-lectin-like receptors

The NKG2-family is encoded by genes within the NK gene complex located on human chromosome 12 (Houchins et al., 1990; Yabe et al., 1993, 1990). This family consists of NKG2A-F with a high degree of sequence homogeneity. NKG2A, B, E and F are inhibitory receptors that bind as a heterodimer with CD94 to the nonclassical MHC class Ib molecule HLA-E (Braud et al., 1997). The mouse Qa-1 is syntenic with the human HLA-E and both are notable for their strong expression in most tissues. Nonclassical HLA-E, like the classical human MHC class 1a molecules, possesses the ability to present antigen in conjunction with the heavy chain and β_2-microglobulin invariant chain. Unlike the highly polymorphic structure and diverse peptide presentation, HLA-E and murine Qa-1 shows limited polymorphisms and binds nonameric peptides from the leader sequence of classical MHC class I molecules, HLA-A, B and C with methionine at position 2, valine at position 7, and leucine at position 9 (O'Callaghan et al., 1998). The human mutant lymphoblastoid cell line 721.221, which serves as a target for direct cytolysis in many NK functional studies due to its lack of expression of classical MHC class I molecules, expresses only HLA-E, suggesting that engagement of HLA-E in the absence of the peptide leader sequence fails to provide protection from direct cytotoxicity (Braud et al., 1997). NKG2-C,

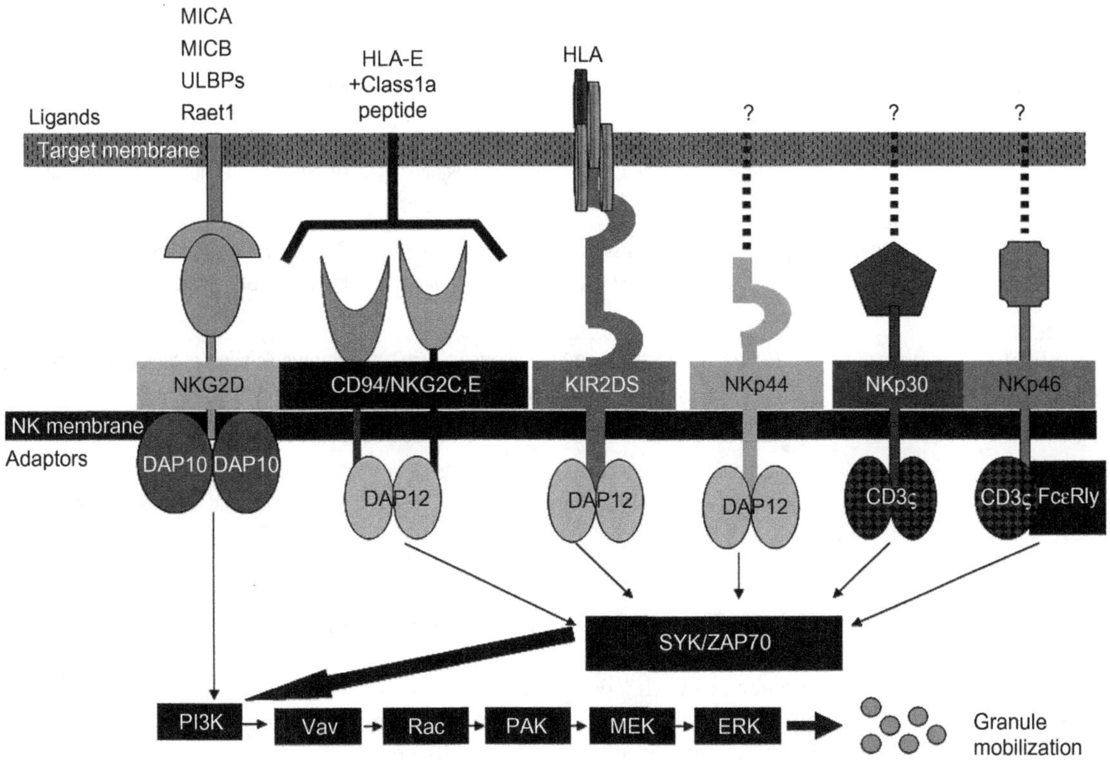

Figure 7.3 • Model of NK activating receptors. Activating NK receptors possess short cytoplasmic tails and an ITAM for adaptor protein binding. DAP12 homodimers associate with KIR activating receptors. Upon ligand engagement, the tyrosines in the ITAM (YxxL x $_{6-8}$ YxxL) of DAP12 becomes phosphorylated, and recruit SH2 domains of SYK/ZAP70 kinase. Activation of SYK/ZAP70 then ultimately leads to ERK activation and granule mobilization. NKG2D, NKG2C, and the NCR family (NKp30, NKp44, NKp46) are important activating receptors for tumour cell recognition. Upon NK receptor–ligand binding, adaptor proteins then associate with the intracellular domain of each receptor through charged interactions to activate PI3K or Syk/Zap70. Activation of these signalling events then leads to a cascade that culminates in ERK phosphorylation, granule exocytosis and target cell lysis.

in contrast to the other NKG2-family, heterodimerizes with CD94 and recognizes HLA-E, but activates NK cytotoxic function by complex formation with DAP12, as shown in Figure 7.3 (Braud et al., 1998).

NKG2D has the ability to induce killing of tumours and virus-infected cells and to produce cytokines (see Chapter 5) and is uniquely independent of CD94 and dependent on the adaptor molecule DAP10 that has ~20% amino acid sequence homology with DAP12 (Wu et al., 1999). DAP10 is a 93 amino acid type I transmembrane protein with a Tyr-X-X-Met (YXXM) motif, as shown in Figure 7.3. The human DAP10 gene, like DAP12, is located on human chromosome 19q13.1 (Wu et al., 1999). DAP10 mRNA expression is found by reverse transcription polymerase chain reaction (RT-PCR) in NK cells, CD4$^+$ and CD8$^+$ T-cells, monocytes, granulocytes, and dendritic cells (DCs) but not in fibroblasts or other tissues. The YXXM motif is a conical docking site for the SH2 domain of the PI3-kinase p85 subunit (Ogasawara et al., 2003). The presence of this site prompted investigators to quickly link NKG2D/DAP10 activation to PI3-kinase signalling, where a synthetic tyrosine-phosphorylated peptide derived from the DAP10 cytoplasmic domain containing this YXXM motif was found to bind the p85 subunit (Wu et al., 1999).

The best characterized of all the NK receptor ligands is the stress-inducible MHC class I-related chain A (MICA), MICB, and UL16-binding proteins (ULBPs) in humans and Rae-1 in mouse, which constitute the major cellular ligands for human NKG2D (Arase et al., 2002; Cosman et al., 2001; Leong et al., 1998; Sutherland et al., 2001, 2002). Specific features of NKG2D ligands are described in Table 7.2 and in Chapter 5. In addition to viral-induced expression, NKG2D ligands are often expressed by tumour cells (Bauer et al., 1999; Wu et al., 1999). Cytotoxicity by NKG2D is coupled to the transmembrane adaptor DAP10 for intracellular signalling, which can deliver a full cytotoxic response. Instead of activating the traditional ITAM-associated signalling events such as Syk and Zap70, DAP10-associated recruitment and activation of PI3-kinase bypasses this step and directly activates downstream effector proteins such as SLP-76, PLC-γ2 and Vav-1 during the

Table 7.2 Specific features of NKG2D ligands

Ligand	Species	Structure	Expression
MICA MICB	Human	Transmembrane proteins with MHC class I related a1, a2 and a3 domains with highly polymorphic characteristics	Gut epithelium, epithelial tumors, normal bone marrow, hematologic malignancies
ULBP1,2,3	Human	GPI-anchored transmembrane proteins with MHC class I related a1, a2 and a3 domains	Some normal and malignant tissues, ULBP3 in kidney, ULBP1 and 3 in thyroid, RAET1 absent in hematopoietic tissues
Rae1α, Rae1β, Rae1γ, Rae1δ, Rae1βε, RAE-1B6	Mouse	GPI-anchored transmambrane proteins with MHC class I related a1, a2, and a3 domains Inducible by retinoic acid and carcinogens	Embryonic tissues, tumor cell lines not expressed in most adult tissues
H60	Mouse	Transmembrane proteins with MHC class I related a1, a2 and a3 domains	Strain specific in activated PBMCs and splenocytes and after carcinogens

cytotoxic response (Upshaw et al., 2006). The NKG2D receptor complex is hexameric in structure, containing two NKG2D and four DAP10 molecules with four potential YXXM binding sites. This receptor complex may therefore initiate multiple signalling events simultaneously. Although the PI3-kinase inhibitor wortmannin abrogates NKG2D-mediated cytotoxicity, activation of recombinant p85 alone fails to initiate the phosphorylation of downstream effectors or trigger calcium release (Billadeau et al., 2003). Binding of DAP10 to growth factor receptor–bound protein 2 (Grb2) fully restored the signalling events that failed to be initiated by p85 including phosphorylation of SLP-76, PLC-γ2 and the guanine nucleotide exchange factor Vav1 which are necessary for calcium release and cytotoxicity (Upshaw et al., 2006, 2005). These results suggested that PI3-kinase phosphorylation and recruitment of Grb2 are essential events in NKG2D signalling. The binding of Grb2-Vav1 during this interaction has also been confirmed by resolution of the crystal structure.

NKG2D function is recognized as an important factor in bone marrow transplantation since NKG2D blocking antibodies prevents graft rejection. NKG2D is also important in the pathogenesis of autoimmune diseases (Ogasawara et al., 2004). For example, NOD mice with forced overexpression of MICA or endogenous expression of Rae-1 mediate diabetes progression (Ogasawara et al., 2003).

Natural cytotoxicity receptors (NCR)

NCRs, a name first coined by Pende and colleagues (Pende et al., 1999), consist of two constitutively expressed receptors, NKp46 and NKp30, and an inducible receptor NKp44 (Pende et al., 1999; Sivori et al., 1997; Vitale et al., 1998). NKp46 was the first of the NCR to be discovered in 1997 by Sivori and associates using a functional cloning approach. Unlike all other NK cell cytotoxicity-associated molecules, NKp46 and the other NCRs are exclusively expressed by NK cells (Sivori et al., 1997; Vitale et al., 1998). While NKp30 and NKp46 share only 13% sequence identity and are encoded by genes located on different chromosomes, they clearly play complementary roles in innate immunity as the primary natural cytotoxicity receptors expressed on NK cells. These receptors are expressed in low and higher density and trigger tumour cell lysis of a variety of targets by associating with more diverse ITAM-containing adaptors than KIR or NKG2-family receptors. NKp46 binds and signals through CD3ζ/FcγR heterodimers, NKp30 to CD3ζ homodimers, and NKp44 directly to DAP12, as shown in Figure 7.3. These receptors also show strong cooperation to widen the target range of cytotoxic killing. Moreover, it is the combination of NKp30 and NKp46 that mediates most of the cytotoxic function so characteristic of freshly isolated NK cells (Moretta et al., 2001), whereas, NKp44 is expressed on NK cells following interleukin 2 activation (Cantoni et al., 1999; Vitale et al., 1998). While these receptors play critical roles in natural cytotoxicity, the ligands have remained elusive. In one report, hemagglutinin and hemagglutinin-neuraminidase viral-encoded protein products of influenza and parainfluenza, respectively, were shown to be recognized by NKp46 (Mandelboim et al., 2001). These two virally encoded products, however, fail to link the NCRs to tumour recognition and suggest that the receptor recognizes peptide structures and the elucidation of ligands for the NCRs should be an important focus of future studies.

Antibody-dependent cell cytotoxicity (ADCC) by CD16 couples to CD3ζ and FCεRIγ

In addition to direct lysis of target cells, NK cells also possess the ability to mediate activation-dependent cell cytotoxicity via expression of FcRγIII (CD16) (Leibson, 1997). Many of the same signalling molecules participate in both direct NK lysis and ADCC due most likely to the conserved CD3ζ and FcεRIγ usage.

Two faces of CD244

Other triggering surface molecules CD244 (2B4) and NKp80 appear to function as co-receptors with the capacity to induce either inhibition or cytotoxicity. Human and murine CD244 genes are located on chromosome 1 in the CD2 subfamily cluster also including CD48, CD58, CD84, signalling lymphocytic activation molecule (SLAM, CD50) and Ly-9. CD244 is a 70 kDa surface receptor on NK cells, which is also expressed by a subset of CD8+ T-cells, monocytes, and granulocytes. Redirected killing by ligation of CD244 against P815 target cells requires the expression NKp46. As a member of the SLAM family, human CD244 in humans and mouse share 70% identity and is characterized by one membrane-distal immunoglobulin V-type domain in the extracellular region (characterized by the absence of cysteines involved in the formation of disulfide bridges) and one membrane-proximal immunoglobulin C2 type domain. The transmembrane portion does not contain any charged amino acid, while the long cytoplasmic tail displays four tyrosine-based motifs (TxYxxI/V) that undergo phosphorylation. The ligand for CD244, CD48, in both mouse and humans is a glycophosphatidylinositol (GPI)-linked molecule expressed on all nucleated hematopoetic cell (Moretta et al., 2001). Engagement of CD244 with antibodies or CD48 promotes cytotoxicity and IFN-γ secretion and also augments TCR-dependent stimulation (Kambayashi et al., 2001). The key to CD244 signalling is the interaction with a small intracellular Src homology 2 (SH2)-domain containing adaptor SAP and the Src tyrosine-family member Fyn that are expressed in NK cells, T-cells, and B-cells (Engel et al., 2003).

The critical function of CD244 is evident since a molecular loss-of-function mutation is associated with a severe form of immunodeficiency known as X-linked lymphoproliferative (XLP) disease associated with fetal infectious mononucleosis after infection with Epstein–Barr virus (Engel et al., 2003). Mutations in the gene encoding SAP and also loss of CD244 have been linked to this and other similar immunodeficiency, which have impaired ability to class switch to IgG antibodies and therefore leads to IgA and IgM isotype accumulation (Nichols et al., 2005). CD244 also induces a strong cytolytic response in NK cells displaying high surface expression of NKp46 (Moretta et al., 2001; Sivori et al., 2000). Evidence suggests that NKp46, CD16 and NKp44 serve as the primary receptor and CD244 the co-receptor to induce optimal cytotoxic responses.

SLAM–SLAM homotypic interactions, however, inhibit IFNγ secretion during T-cell activation (Engel et al., 2003). In addition to SAP and Fyn, the intracellular substrates of CD244 signalling include Vav-1 and phospholipase C-γ1 (Watzl et al., 2000). Interestingly, CD244 is one of the few SLAM members with the capacity to induce both IFNγ production and cytotoxicity. This has been attributed to an extra tyrosine motif in the intracytoplasmic tail of CD244 that alters the signalling intermediates that are recruited (Watzl et al., 2000). Differential recruitment of SHIP-1, Doc1/2, and Ras-GAP activates cytokine secretion; whereas, recruitment of Vav-1, SHIP-1, and c-Cbl trigger cytotoxicity (Chen et al., 2004). Therefore, the nature of tyrosine phosphorylation signal intermediates recruited by CD244 may act to elicit specific functional events.

In addition to its co-stimulatory activity, CD244 mediates inhibitory function. To determine if CD244-mediated activation or inhibition is required for optimal NK functions, the cytotoxicity of wild-type and perforin knock-out NK cells were compared in the absence and presence of CD244–CD48 interactions. Results from these studies revealed that CD244 inhibits NK–NK fratricide, possibly explaining the apparent dual functions of CD244 (Taniguchi et al., 2007).

Transcriptional NK receptor control

Both the KIR and Ly49 proteins are expressed in a selective, stochastic pattern in human and mouse NK cells, respectively. The transcriptional control of these receptors is therefore a matter of intense interest since modulation of the expression pattern has potential application in the setting of transplantation, autoimmunity, control of infections, and immunotherapy. One of the most interesting scientific discoveries in the last decade is that of the complex promoter regulation of the Ly49 family. The promoter regions of these C-lectin-like proteins contain up to three separate promoters that are not only expressed in a random pattern but also differentially regulated during NK cell development. Moreover, these genes are not expressed in an 'all or none' fashion, suggesting that the routine promoter and locus control regions may be too simplistic for such complex expression patterns.

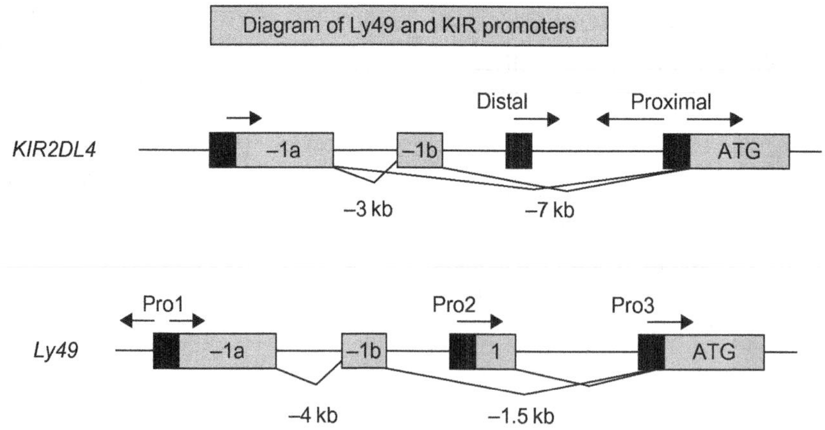

Figure 7.4 • Model of KIR gene promoter assembly and activation. Several RNA transcriptional start sites were identified within the *Ly49g* and *KIR2DL4* genes. Promoter regions, including bi-directional promoters, are indicated in black and the numbered regions indicate exons. The orientation of the transcripts and genomic localization are indicated by arrows.

The Ly49 molecule represents the most thoroughly studied of the NK receptor superfamilies with the Ly49a promoter region first identified in 1993 (Kubo et al., 1993). Ly49a is an inhibitory receptor that is expressed later in NK cell development (Dam et al., 2003; Saurer et al., 2003; Williams et al., 2000). The first functional studies were not published until 1999 when the ATF-2 and TCF-1 transcription factors were shown to play a central role in Ly49a gene transcription in EL-4 cells (Held et al., 1999; Kubo et al., 1999). Soon to follow also in EL-4 cells, studies were performed to determine the core promoter elements in the Ly49c, i and j promoters (Gosselin et al., 2000; McVicar and Burshtyn, 2001). These studies revealed that there were both strong activating and inhibitory regions within the core promoter regions, as shown in Figure 7.4. Comparison of the individual Ly49 family exposed three distinct classes of *Ly49* promoters: the *Ly49a/g*-related family, the *Ly49e/c*-related family, and the activating *Ly49* promoter family. The role of TCF-1 was determined in mice with homozygous and heterozygous deletions of the *tcf-1* gene (Held et al., 2003) In this study, homozygous deletion altered the NK receptor repertoire in these mice; reducing the expression of Ly49a from 20% of NK cells to 1% of NK cells. The control of the repertoire distribution was dependent on gene dosage since the *tcf-1* heterozygotes displayed only 10% of NK cells positive for Ly49a in the peripheral blood. Therefore, the expression of TCF-1 was thought to control the probability of the Ly49a gene expression. Ly49d expression in these TCF-1-null mice was also altered but TCF-1 sites were absent in the promoter. After many years of study, each of the inhibitory Ly49 genes was found to contain a pair of overlapping, opposing promoters that are upstream of the Ly49 gene transcriptional start site, as shown in Figure 7.4. These overlapping promoter sites act as a probabilistic 'switch' to determine the stable expression pattern in mature NK cells, as shown using promoters conjugated to the yellow fluorescent protein (YFP) and the reverse transcription from a cyan fluorescent protein (CFP) (Saleh et al., 2004). A single-cell clone with a two-colour vector under the control of the *Ly49g* Pro1 element, which is normally present in 45% of adult splenic NK cells, produced a nonsegregated population that were 50% blue (CFP) or 50% yellow (YFP). The stability of this expression pattern was maintained into the adult NK population and switching was dependent on cell proliferation when the parent cell gave rise to an equal distribution of yellow and blue-expressing daughter cells. Signalling from the forward promoter was found to control repertoire diversity in immature NK cells and then the unidirectional promoter assumes control of Ly49 expression in mature cells. It is in the immature state that the forward promoter prevents chromatin changes that would close the transcription by the relative concentration of the transcription factors present in the nucleus. Unlike the *Ly49g* gene, the *Ly49a* gene is induced only in mature NK cells that are proliferating. Dysregulation in the balance of Ly49a was uniquely associated with SHIP-deficiency (Wang et al., 2002). SHIP-null mice had unusually high percentages of splenic NK cells that expressed the Ly49a receptor (80%) when compared with wild-type mice. These results suggest that NK cell activation and the balance of negative and positive signals mediated by SHIP may control the decision to express Ly49A during the activation process.

Less is known about the transcriptional control of human KIR genes in contrast to Ly49 but the KIR2DL4 basic promoter elements are similar to that of Ly49g as shown in Figure 7.4. Multiple transcription factors have been shown to participate in KIR gene expression based on DNAse I footprinting (Santourlidis et al.,

2008; Stewart et al., 2003). A core promoter region in this study, located 262 bp from the transcriptional start site of the KIR2DL4 promoter, was found to contain binding sites for GATA-3, TCF-2, MYC/MAX, AP-1, RUNX/AML and c-Ets1 transcription factors. KIR2DL4 is expressed universally, whereas KIR3DL1 displays variegated expression and has involvement of STAT, YY-1 and SP-1 in addition to the core transcription factors (Stewart et al., 2003). After analysis of all KIR gene families, it was recognized that only the RUNX/AML site was conserved in all KIR genes (Trowsdale, 2001; Trowsdale et al., 2001). Expression from these promoters is controlled by a small CpG island surrounding the transcriptional start site in which consistently methylated regions silence the expression of the KIR genes (Santourlidis et al., 2002). Treatment of polyclonal NK cells with demethylating agents such as 5-aza-2′-deoxycytidine results in the expression of KIRs (Santourlidis et al., 2002). Consistent evidence points to methylation and demethylation in the control of KIR gene expression but the question of whether methylation controls the probabilistic expression in the KIR repertoire remains unanswered along with the question of additional upstream bi-directional promoter elements that are observed in the Ly49 family.

Unlike NK cells, KIR expression in T-cells is an acquired phenotype that is associated with TCR activation in a subset of both CD4+ and CD8+ T-cells (Young and Uhrberg, 2002). KIR expression on memory cells increases with age but this may be due to expansion of the specific KIR+ population of memory cells with age rather than increased transcriptional activation of the KIR genes (unpublished data). The transcriptional machinery leading to KIR expression in T-cells occurs by activators that bind to position −52 to −61 of the KIR promoter, whereas the RUNX/AML site that is critical for KIR expression in NK cells is located around position −98 (Xu et al., 2005). Transfection of the minimal KIR promoter in primary human naïve and memory T-cells showed that the promoter was active in both cell types; supporting the hypothesis that the appropriate transcription factors are present in T-cells prior to activation. Further evidence supported that KIR gene expression in T-cells may be controlled by epigenetic regulation such as changes in DNA methylation in naïve cells. The control of KIR expression and diversity are far from completely understood in human NK and T-cells.

Signalling for NK-cell recruitment

Recruitment in response to IFNs (for viruses) or chemokines represents the first step to the innate immune response initiated by NK cells. Important for this process are the chemokines such as CCL22 (Gong and Clark-Lewis, 1995), CX_3CL1 (Loetscher et al., 1996) and CXCL8 (Bochner et al., 1999; Chantry et al., 1998; Imai et al., 1998), which are released by tissue-resident macrophages, endothelial cells and neutrophils during an inflammatory response. Most NK cells constitutively express the necessary receptors, such as CXCR1 and CX_3CR1, for this initial response. CXCR1 might be of particular importance since this receptor has the ability to arrest rolling lymphocytes under flow conditions (Chantry et al., 1998). The chemokine-receptor repertoire of NK cells is similar to that of neutrophils, and neutrophils may regulate the NK recruitment by releasing CXCL8. Extravasation of NK cells depends on selectins in contact with the vascular endothelium, interaction with chemokines, integrin adhesion, and transmigration. Together, these signalling mechanisms initiate recruitment of NK cells to sites of infection within minutes of pathogen exposure.

Signals that mediate lytic granule mobilization

The most intense exploration into NK signalling has focused on deciphering the molecular mechanisms associated with direct cytolytic function. Key to these signalling events is the movement of perforin together with granzymes localized in lytic granules that migrate at the target: NK interface within supramolecular structures. Several inherited immune disorders are associated with defective perforin expression or transport and these disorders highlight the critical role of the secretory cytotoxic pathway in NK and T-cell immune response and in lymphocyte homeostasis (Arico et al., 2002). Perforin deficiency, resulting from gene mutations, results in deficiency in cytotoxic granules and impairment in both T- and NK-cell cytotoxic activities. Perforin defects account for one-third of the cases of familial hemophagocytic lymphohistiocytosis (FHL) (de Saint Basile and Fischer, 2001). Other secretory diseases occur in patients with Chediak–Higashi syndrome (CHS), Griscelli syndrome (GS) and in X-linked lymphoproliferative syndrome (XLP), which are characterized by lymphoid organ and extranodal infiltration by polyclonal T-cells, mostly of the CD8+ subset, suggesting that granule-dependent function is important for immune regulation (de Saint Basile and Fischer, 2001).

Our laboratory has performed many critical experiments to characterize the signals that control perforin and granzyme-mediated mobilization during direct cytotoxicity against tumour targets. The first signalling intermediate linked to granule mobilization was the mitogen-activated protein kinases (MAPK) pathway, which integrates multiple intracellular signals including c-Jun, N-terminal kinase/stress-activated protein kinase, and p38/MAPK.

ERKs display a conserved TEY motif that is dually phosphorylated at the Tyr and Thr sites upon activation (Nishimoto and Nishida, 2006). They play critical roles in gene expression related not only to cytotoxic function, but also to cell growth and differentiation and possess the ability to activate nuclear transcription factors, including Elk-1, c-Jun, c-Myc, NF-IL-6 and TAL1 (Cha and Shapiro, 2001; Ligeza et al., 2008). Activation of ERK follows a highly conserved flow of molecular events, connected to upstream signals associated with the receptor activation in which the GRB-2 complex recruits Ras to the plasma membrane, Raf-1 binding, and then activation of Ras GDP:GTP exchange. Phosphorylation of MAPK/ERK kinase 1 (MEK-1) then activates a dual kinase (MEKK) with the ability to phosphorylate the TEY motif in ERK1/2. We were the first to demonstrate that intimate NK interaction with tumour results in the rapid activation of ERK phosphorylation and kinase function in human NK cells (Wei et al., 1998). To link ERK phosphorylation to NK function, we treated NK cells with PD098059, a MEK1 inhibitor, which suppresses NK cell lysis of Raji tumour cells in 5-h ^{51}Cr release assays. Immunofluorescence staining with anti-perforin and anti-granzyme B antibodies showed that PD098059 blocked polarization and mobilization of both perforin- and granzyme B-containing granules. Transient transfection of human NK cells with dominant negative ERK2, but not wild-type ERK2, interfered with lytic activity and perforin and granzyme B mobilization, confirming the critical role of ERK signalling in this functional process. In addition to the role of ERK in direct tumour cell lysis, other laboratories have conclusively documented the importance of ERK signalling in direct lysis and ADCC (Djeu et al., 2002).

While most canonical signalling events regulated by ERK involve the activation of Ras, we conclusively showed that granule mobilization is independent of Ras function (Wei et al., 1998, 2000). Employing a GST fusion protein linked to the critical segment of the Raf binding domain, GST-Raf-RBD, we demonstrated that tumour target:NK ligation results in activation of Ras. Under these experimental conditions, activated GTP-bound Ras was captured by binding to GST-Raf-RBD. Pharmacological inhibition of Ras, with a farnesyltransferase inhibitor that prevents the accumulation of Ras at the cell membrane, failed to prevent granule mobilization and direct NK cytotoxicity. To identify the role of Ras in MAPK activation and lytic function, a dominant negative form of Ras (N17-Ras) was introduced into NK-92 cells via vaccinia viral gene delivery. While N17-Ras abolished Ras activation triggered by either IL-2 or Raji tumour cell co-incubation, granule mobilization and cytotoxicity were intact. These results suggested that the lytic process triggered by direct binding of NK cells and tumour cells is controlled by a Ras-independent MAPK-dependent signalling event that occurs nearly instantaneously after triggering the NK receptor.

Downstream of activating NK receptors, PI3-kinase and Syk are connected to DAP12 signalling. Syk-mediated signalling events were also examined in the NK-92 cell line (Jiang et al., 2000, 2002). We confirmed the role of Syk in DAP12-mediated granule mobilization following Raji tumour cell interaction. Syk was immunoprecipitated with DAP12 after 5 minutes and pretreatment with piceatannol, a Syk-specific inhibitor and expression of a truncated Syk mutant gene lacking the kinase domain (SykT), effectively suppressed the cytotoxicity of NK-92 cells against Raji tumour cells. CD56, which is an irrelevant gene product in these assays, had no deleterious effect on NK cell function. An in vitro kinase assay with Syk immunoprecipitates using a nonspecific substrate such as myelin basic protein (MBP) ensured that piceatannol treatment and SykT expression effectively inhibited Syk activation triggered by target engagement in the NK-92 cells. Under these experimental conditions, perforin granule redistribution and polarized movement toward the contact site of the conjugated target cells were blocked. Rac 1, which often operates with Syk via the PI3K pathway, was also stimulated by the engagement of NK and target cells. In addition, Vav, the upstream activator of Rac1, also participated in granule mobilization and direct NK cytotoxicity (Salojin et al., 1999). Using SykT and piceatannol, we found reduced target-induced Rac1 activation, as demonstrated by less binding between Rac1 and GTP. Thus, these results showed that Syk functions as an upstream modulator of Rac1 during the NK lytic process. We then used successive functional analyses to identify the PI3K→Rac1→PAK1→MEK→ERK pathway as the critical signalling cascade to mediate granule mobilization and direct NK cytotoxicity, as shown in Figure 7.3 (Djeu et al., 2002).

Further evidence substantiated the participation of these signalling molecules in NK cells. Vaccinia-virus-overexpressed mutant Syk, Rac and mitogen-activated protein kinase (MAPK)/ERK were all independently found to block lytic function in separate studies (Gismondi et al., 2003; Jabril-Cuenod et al., 1996; Scharenberg et al., 1995; Ting et al., 1995). Also confirmed in this process was LAT (Inoue et al., 2002; Umehara et al., 2002) and SLP76 (Nishida et al., 2005), where expression of the wild-type but not the dominant-negative form increased the lytic ability of NK cells. Using biochemical analysis of kinase enzyme activation, gene transfer of dominant-negative and constitutively active signal molecules, accompanied by parallel examination of lytic function against ^{51}Cr-labelled tumour cells, it became clear that a similar signalling cascade was triggered in normal human activated NK cells following tumour cell interaction. It is the involvement of ERK phosphorylation that promotes the movement of actin and tubulin in the polarized direction of the immune synapse culminating in the critical killing process (Andzelm et al., 2007; Bovenschen et al., 2008; Kopcow et al., 2005; Orange et al., 2003; Roda-Navarro et al., 2006).

Target destruction is mediated by the release of preformed granular proteins including perforin and granzymes. These granules reside in the cytoplasm of the cells, where they await contact and activation-induced release. Upon target cell contact, generation of a specific activation signal induces the polarized movement of granules toward the plasma cell membrane where fusion between the effector and target cell releases granular content into the tight intracellular junction formed between the two cells. In the presence of calcium, perforin then polymerizes and attaches to the target cells. Granzymes, including granzyme A and B, are death inducing granule proteins that activate the apoptotic machinery by cleaving cellular substrates after an aspartic acid residue with similar specificity to the caspases. Granzyme B can directly cleave caspase 3, an effector caspase leading directly to apoptosis. Several groups have created perforin-deficient mice. CD8+ T-cells and NK cells from these mice are dramatically reduced in their ability to induce apoptosis in target cells and marked susceptibility to a variety of viral infections. Interestingly, perforin deficient mice display improved responses following bone marrow transplantation due to reduced frequency of graft-versus-host disease (GVHD). Humans with a homozygous loss of the perforin gene have been reported, presenting with the disease, FHL. In this setting, uncontrolled activation of T-cells and macrophages and overproduction of inflammatory cytokines are found. The exact role of perforin deficiency in these patients is incompletely understood. Granzyme A and granzyme B knockouts have been generated with marked deficiency in inducing DNA fragmentation and apoptosis in target cells even in the presence of perforin suggesting that granzymes play a critical role in target elimination and perforin may be solely involved in the regulation of this response. Using active recombinant granzymes, the death pathway initiated by these proteins was recently established (Martinvalet et al., 2008). Granzymes are structurally related to trypsin and other serine proteases that target key molecular molecules in the cytosol, nucleus and mitochondrion for proteolysis. Granzyme A was recently shown to migrate to the mitochondrial matrix where it induces caspase-independent mictochondrial damage and apoptosis (Martinvalet et al., 2008). Entry of granzymes are mediated by a perforin-dependent process that is incompletely understood (Chowdhury and Lieberman, 2008). Although the original hypothesis that perforin creates pro-forming units in the membrane was disproved, entry of granzymes is definitively perforin-dependent. A revised model was proposed suggesting that perforin creates microscopic holes in the plasma membrane that then allow Ca+ influx and granzyme endocytosis (Keefe et al., 2005).

NK receptor signalling in T-cells

KIR expression has also been reported on peripheral blood T-cell subsets; including the constitutive expression of KIR in $\gamma\delta$ T-cells and inducible expression in effector memory CD8$^+$ $\alpha\beta$ T-cells. The mechanism governing the induction of these receptors on T-cells is unknown. KIRs have been postulated to be induced in response to self-antigen and increase with aging (Goronzy et al., 2005). The fact that tumour-specific CD8+ T-cells express KIR suggests that this receptor may be induced when MHC-class I is diminished within the tumour microenvironment (Gati et al., 2001). Moreover, inhibitory KIR expression in the tumour environment is strongly associated with functional anergy.

While cytolytic NK and T-cells use distinct antigen receptors, these cells employ identical lytic processes to mediate tumour cell death. Release of perforin and granzyme B is critical not only in NK cell cytotoxicity but also for cytotoxicity mediated by T-cells (Fischer et al., 2007). Upon contact with tumour cells, the lytic granules in T-cells are mobilized toward the tumour cell, and perforin forms pores in the target cell membrane that then allows granzyme B and activated pro-apoptotic cysteine proteases, caspases, to induce apoptosis in the target cell. Granule exocytosis potently activates cell death in the target cells through the activation of these caspases, but can also cause target cell death in the absence of activated caspases, as shown in caspase-deficient mice (Smyth et al., 2005). The second pathway involves engagement of the Fas/CD95 death receptor on target cells by their cognate ligands FasL, expressed on NK cells, resulting in classical caspase-dependent apoptosis in the target cell (reviewed, (Smyth et al., 2005).

In addition to the role of NKR in cytotoxicity, KIR expression on memory T-cells may be important for T-cell survival (Young and Uhrberg, 2002). Engagement of KIRs with antibodies showed that KIR+ T-cells were more resistant to activation-induced cell death (AICD) in comparison to KIR- T-cells (Gati et al., 2003). Overexpression of recombinant KIR3DL in Jurkat T-cells showed that AICD was blocked by a process involving protein kinase C activation (Kwon et al., 2000). AICD is mediated by a pathway of receptor-mediated apoptosis involving the activation of a molecular death-inducing signalling complex (DISC), which involves ligation of Fas, recruitment of FADD and activation and autolytic cleavage of caspase 8 (Medema et al., 1997; Scaffidi et al., 1998). KIR+ cells activate an inhibitory molecule, c-FLIP, that inhibits the complex by binding to FADD and preventing the recruitment of caspase 8, which then blunts the apoptotic response (Gati et al., 2003). The role of c-FLIP has been documented in vitro (Miller and McCullar, 2001) and in vivo (Jansen et al., 2007) and it has been well established that this recruitment depends on phosphorylation of the inhibitory ITIM and recruitment of SHP-1 and SHP-2 (Verbrugge et al., 2006). Homozygous deletion of SHP-1 results in diminished AICD in activated T-cells and lymphoproliferation (Zhou et al., 1996).

Since KIR expression may play a complex biological role in T-cell function, studies were performed on KIR-HLA transgenic (Tg) mice (Cambiaggi et al., 1999). The selective expression of NK receptors were analyzed in wild-type mice, HLA Tg, KIR Tg, and KIR-HLA Tg mice. Splenocytes were analyzed for the cell surface expression of a marker of mouse memory CD44 in CD8+ T-cells. Absolute numbers of CD4+ and CD8+ T-cells were equivalent in control and KIR-HLA Tg mice but the size of the CD8+ memory compartment was increased in KIR-HLA mice. The memory cells were characterized by the absence of CD25 but maintained the ability to respond quickly to TCR ligation and possessed a higher number of cells capable of IFN-γ secretion in the peripheral blood. Interestingly, a two- to threefold increase in the number of cycling CD8$^+$ T-cells was observed in KIR-HLA Tg mice compared to control mice. Taken together, these results showed that the in vivo engagement of inhibitory NKR with cognate MHC class I ligand leads to the selective accumulation of terminal memory cells. In humans, a syndrome of lymphoproliferation of T-cells or NK cells called lymphoproliferative disease of large granular lymphocytes (LDGL) or large granular lymphocyte (LGL) leukemia in association with myeloid bone marrow failure that is linked to functional KIR and dysregulated NCR expression. Dysregulation in the NKR signals may contribute to T-cell and NK-cell accumulation and disease pathogenesis (Epling-Burnette et al., 2004; Zambello et al., 2003).

Less is known about role of KIR expression and function in CD4$^+$ T-cells (van Bergen et al., 2004). A few studies have shown memory KIR$^+$ CD4$^+$ T-cells to be present in healthy individuals, but the acquisition and expansion of these cells is more frequently detected in patients with acute coronary syndrome or rheumatoid arthritis. In these settings, KIR2DL2, KIR2DL3 and KIR2DS2 show preferential expression. To test the function of KIR in CD4$^+$ T-cells, KIR2DL1 was transfected into Jurkat T-cells. Binding of the HLA-ligands for KIR2DL1 (HLA-C2-containing epitopes; see Table 7.1) in these transfected cells, led to co-stimulator function that resulted in IL-2 production but also to SHP-2 recruitment. KIR expression on CD4+ T-cell therefore has a highly pleotrophic effect on function that involves activation of lytic activity, secretion of cytokines, and enhancement of survival to improve the potential long-term accumulation of memory cells. Tumour-associated ligands of the activating NKG2D receptor can effectively stimulate T-cell responses at early but not late stages of tumour growth.

A critical determinant of NK and T-cell intracellular signalling is controlled by the Vav-1 molecule. The Vav molecule contains several complex structural domains including calponin homology (CH) domain, an acidic domain (AD), a pleckstrin homology (PH) domain, SH2, SH3, and cysteine-rich(CR) and praline-rich (PR) regions. Specific protein sub-domains of Vav-1 mediate distinct cellular processes (Billadeau et al., 2000). Examination of a CH-deficient mutant revealed that this domain is essential for NF-AT/AP-1-mediated transcription secondary to a defect in intracellular calcium mobilization. Expression of this mutant dramatically altered T-cell-mediated cytotoxicity. The PH domain mutant altered not only T-cell-dependent function but also impaired FcR-mediated killing by NK cells. In contrast, the PH domain mutant failed to regulate direct cytotoxicity, suggesting that the PH domain may differentially regulate distinct forms of killing. Mutation of three tyrosine residues within the AD revealed a negative regulatory site and hyperactive NF-AT/AP-1-mediated gene transcription and enhanced cell-mediated cytotoxicity. Together, these data demonstrate the importance of structural subdomains within the Vav-1 molecule that mediated domain-specific lymphocyte functions.

Signals involved in NK-mediated immunoediting

Tissue damage and/or release of pro-inflammatory cytokines after virus infection, tumour development or transplanted tissues leads to activation of macrophages and antigen-presenting cells such as DCs that bind to NK cells. DCs release cytokines such as interleukin-12 (IL-12) that leads to the induction of IFNγ production by NK cells. Recent data about the interaction of NK cells and DCs have important implications for a role of immunoediting conducted by the interaction of these two cells during the initial phases of an innate immune response. This was first demonstrated in a bone marrow transplant setting when a group of patients treated with a haploidentical bone marrow transplant displayed 100% engraftment and survival when HLA-C was fully mismatched (Ruggeri et al., 1999, 2002). In this allogeneic setting, a fraction of the NK cells of the donor expressed KIRs that fail to recognize one or more MHC class I alleles of the host. These so-called 'KIR epitope-mismatched NK cells' were able to induce a more effect alloreactive graft-versus-leukemia (GVL) effect compared to patients that received HLA-class 1a matched alleles at the C-locus. Interestingly, the incidence of GVHD, early after transplant, was strikingly reduced in these mismatched transplants. This unexpected result revolutionized our thinking about the possible role of DCs and NK cells in an allogeneic and in an autologous setting. Taken together, these results suggest that the interaction between DCs and NK cells function to edit the final immune response and that the signals mediated by KIR and possibly other NK receptors may regulate the magnitude and the type of DC priming that occurs.

After pathogen invasion, DC precursors accumulate at the site of the inflammatory response due to the release of chemokines. DC precursors, such as monocytes, then transform into immature DCs (iDCs), which are particularly efficient at phagocytosis to capture the antigen that will then be processed and presented after full maturation of the DCs (mDCs). Triggering by members of the Toll-like receptor family widely expressed of pathogens, including for example lipopolysaccharide (LPS), induced this final developmental program of maturation. It is the fully mature DCs that mediate T-cell priming, but at the expense of this maturation, DCs lose the endocytic capability and downregulate the surface receptors that are involved in antigen capture.

Recruitment occurs in response to stimuli such as type I IFNs and chemokines. A surge of chemokines is released by iDCs after antigen uptake that acts on NK cells including CCL3 and CXCL8 (IL-8). Mature DCs (mDCs) express high levels of HLA class I molecules; whereas, iDCs are characterized by low amounts of surface HLA class I molecules. Low expression of HLA class I renders iDCs susceptible to NK-mediated lysis (Della Chiesa et al., 2003). The ability to kill iDCs leads then to final immunoediting and removal of DCs that fail to acquire the optimal capability of antigen presentation and T-cell priming. It is the exploitation of this immunoediting function that provides improved engraftment for haploidentical bone marrow transplants when the HLA-C locus is fully mismatched (Ruggeri et al., 2001). Alloreactive NK cells are able to eliminate residual leukemic cells, providing better therapeutic responses, but also depletion of donor T-cell priming that protects against the development of GVHD.

With this powerful application to clinical medicine, the hunt for the NK receptors that mediate lysis of iDCs lead to the discovery of an NKp30-dependent process (Castriconi et al., 2004; Ferlazzo et al., 2002). Interestingly, IL-2 activated NK cells kill iDCs efficiently, even though they express substantial amounts of HLA class I, but fail to kill mDCs. Triggering receptors and co-receptors were analyzed for participation in iDC cytotoxicity by masking each of the receptors with monoclonal antibodies and then determining the effect on NK-mediated cytolysis. Anti-NKp44 antibody had no effect on lysis, while inhibition was observed with anti-NKp46 and anti-NKG2D antibodies. Marked inhibition was detected after addition of the anti-NKp30 antibody. In these studies, the protective role of DC maturation was eliminated by blocking antibodies to HLA A, B or C, suggesting that the differential susceptibility of iDCs is at least partially related to the dosage of HLA-class I. This however failed to completely explain the susceptibility of DCs to NK-mediated lysis. DNAM-1 is an activating receptor that is expressed by virtually all human NK cells and T-cells (Chen et al., 2007). Activation of DNAM-1 signalling was found to participate with NKp30 and contribute to NK-mediated lysis of both iDCs and mDCs (Pende et al., 2005).

Bi-directional cross-talk

On the basis of ground-breaking studies in mice (Fernandez et al., 1999), several groups have highlighted the importance of the interaction between DCs and NK cells in humans (Ferlazzo et al., 2002; Piccioli et al., 2002; Wilson et al., 1999). Early studies on NK–DC interactions were focused on the ability of NK cells to lyse and edit the survival of iDCs (Spaggiari et al., 2001; Wilson et al., 1999), but more recent studies have provided evidence that the interaction between NK cells and DCs functions as an important regulator of the intensity of an innate immune responses (Ferlazzo et al., 2000). These two types of cell that both act during the initial phases of the innate response influence the maturation of each other; DCs act during the priming phase of NK cell activation, and NK cells promote DC maturation through cytokine production (Zitvogel, 2002). The cross-talk between NK cells and DCs might take place at different stages of the innate and adaptive immune responses, which indicates that the interaction of these cells has a role in controlling the links between innate and adaptive immunity. It is possible that IL-2 promotes DC–NK cell activation and that IL-15 produced by iDCs controls NK cell expansion, activation and CD8+ T-cell distribution that results in the potentiation of cytotoxicity (Fernandez et al., 1999; Piccioli et al., 2002; Schonland et al., 2003; Ullrich et al., 2008; Zitvogel, 2002). It is possible to speculate that that the lack of NK cell interaction with iDCs may influence the expansion of annergy-inducing immature populations of DCs that are so critical for the immunosuppressive tumour microenvironment (Nagaraj and Gabrilovich, 2008; Nefedova et al., 2007). The cognate interaction between the two cell types was found to be crucial for inducing the release of TNFα and GM-CSF by iDCs (Moretta, 2002). A careful analysis of the immunological synapse at the interface between NK cells and iDCs will help to clarify which receptor and ligand interactions that are essential to mediate cross-talk between NK cells and DCs. However, prevalent mismatches between activating KIR gene, expansion of iDCs and autoimmunity suggests that the immunoediting function provided by KIR may contribute to this process.

References

Andzelm, M.M., Chen, X., Krzewski, K., Orange, J.S. and Strominger, J.L. (2007). Myosin IIA is required for cytolytic granule exocytosis in human NK cells. J Exp Med **204**, 2285–2291.

Arase, H., Mocarski, E.S., Campbell, A.E., Hill, A.B. and Lanier, L.L. (2002). Direct recognition of cytomegalovirus by activating and inhibitory NK cell receptors. Science **296**, 1323–1326.

Arico, M., Allen, M., Brusa, S., Clementi, R., Pende, D., Maccario, R., Moretta, L. and Danesino, C. (2002). Haemophagocytic lymphohistiocytosis: proposal of a diagnostic algorithm based on perforin expression. Br J Haematol **119**, 180–188.

Bakker, A.B., Wu, J., Phillips, J.H. and Lanier, L.L. (2000). NK cell activation: distinct stimulatory pathways counterbalancing inhibitory signals. Hum Immunol **61**, 18–27.

Bauer, S., Groh, V., Wu, J., Steinle, A., Phillips, J.H., Lanier, L.L. and Spies, T. (1999). Activation of NK cells and T cells by NKG2D, a receptor for stress-inducible MICA. Science **285**, 727–729.

Billadeau, D.D., Mackie, S.M., Schoon, R. A. and Leibson, P.J. (2000). Specific subdomains of Vav differentially affect T cell and NK cell activation. J Immunol **164**, 3971–3981.

Billadeau, D.D., Upshaw, J.L., Schoon, R. A., Dick, C.J. and Leibson, P.J. (2003). NKG2D-DAP10 triggers human NK cell-mediated killing via a Syk-independent regulatory pathway. Nat Immunol **4**, 557–564.

Binstadt, B.A., Billadeau, D.D., Jevremovic, D., Williams, B.L., Fang, N., Yi, T., Koretzky, G.A., Abraham, R.T. and Leibson, P.J. (1998). SLP-76 is a direct substrate of SHP-1 recruited to killer cell inhibitory receptors. J Biol Chem **273**, 27518–27523.

Bochner , B.S., Bickel, C.A., Taylor, M. L., MacGlashan, D.W., Gray, P.W., Raport, C.J. and Godiska, R. Jr. (1999). Macrophage-derived chemokine induces human eosinophil chemotaxis in a CC chemokine receptor 3- and CC chemokine receptor 4-independent manner. J Allergy Clin Immunol **103**, 527–532.

Bovenschen, N., de Koning, P.J., Quadir, R., Broekhuizen, R., Damen, J.M., Froelich, C.J., Slijper, M. and Kummer, J.A. (2008). NK cell protease granzyme M targets alpha-tubulin and disorganizes the microtubule network. J Immunol **180**, 8184–8191.

Braud, V., Jones, E.Y. and McMichael, A. (1997). The human major histocompatibility complex class Ib molecule HLA-E binds signal sequence-derived peptides with primary anchor residues at positions 2 and 9. Eur J Immunol **27**, 1164–1169.

Braud, V.M., Allan, D.S., O'Callaghan, C.A., Soderstrom, K., D'Andrea, A., Ogg, G.S., Lazetic, S., Young, N.T., Bell, J.I., Phillips, J.H., Lanier, L.L. and McMichael, A.J. (1998). HLA-E binds to natural killer cell receptors CD94/NKG2A, B and C. Nature **391**, 795–799.

Bruhns, P., Marchetti, P., Fridman, W. H., Vivier, E. and Daeron, M. (1999). Differential roles of N- and C-terminal immunoreceptor tyrosine-based inhibition motifs during inhibition of cell activation by killer cell inhibitory receptors. J Immunol **162**, 3168–3175.

Burshtyn, D.N., Yang, W., Yi, T. and Long, E.O. (1997). A novel phosphotyrosine motif with a critical amino acid at position −2 for the SH2 domain-mediated activation of the tyrosine phosphatase SHP-1. J Biol Chem **272**, 13066–13072.

Cambiaggi, A., Darche, S., Guia, S., Kourilsky, P., Abastado, J.P. and Vivier, E. (1999). Modulation of T-cell functions in KIR2DL3 (CD158b) transgenic mice. Blood **94**, 2396–2402.

Cantoni, C., Bottino, C., Vitale, M., Pessino, A., Augugliaro, R., Malaspina, A., Parolini, S., Moretta, L., Moretta, A. and Biassoni, R. (1999). NKp44, a triggering receptor involved in tumor cell lysis by activated human natural killer cells, is a novel member of the immunoglobulin superfamily. J Exp Med **189**, 787–796.

Castriconi, R., Della Chiesa, M. and Moretta, A. (2004). Shaping of adaptive immunity by innate interactions. C R Biol **327**, 533–537.

Cha, H. and Shapiro, P. (2001). Tyrosine-phosphorylated extracellular signal–regulated kinase associates with the Golgi complex during G2/M phase of the cell cycle: evidence for regulation of Golgi structure. J Cell Biol **153**, 1355–1367.

Chantry, D., DeMaggio, A.J., Brammer, H., Raport, C.J., Wood, C.L., Schweickart, V. L., Epp, A., Smith, A., Stine, J.T., Walton, K., Tjoelker, L., Godiska, R. and Gray, P.W. (1998). Profile of human macrophage transcripts: insights into macrophage biology and identification of novel chemokines. J Leukoc Biol **64**, 49–54.

Chen, R., Relouzat, F., Roncagalli, R., Aoukaty, A., Tan, R., Latour, S. and Veillette, A. (2004). Molecular dissection of 2B4 signaling: implications for signal transduction by SLAM-related receptors. Mol Cell Biol **24**, 5144–5156.

Chen, X., Trivedi, P.P., Ge, B., Krzewski, K. and Strominger, J.L. (2007). Many NK cell receptors activate ERK2 and JNK1 to trigger microtubule organizing center and granule polarization and cytotoxicity. Proc Natl Acad Sci U S A **104**, 6329–6334.

Chowdhury, D. and Lieberman, J. (2008). Death by a thousand cuts: granzyme pathways of programmed cell death. Annu Rev Immunol **26**, 389–420.

Cosman, D., Mullberg, J., Sutherland, C.L., Chin, W., Armitage, R., Fanslow, W., Kubin, M. and Chalupny, N.J. (2001). ULBPs, novel MHC class I-related molecules, bind to CMV glycoprotein UL16 and stimulate NK cytotoxicity through the NKG2D receptor. Immunity **14**, 123–133.

Dam, J., Guan, R., Natarajan, K., Dimasi, N., Chlewicki, L.K., Kranz, D.M., Schuck, P., Margulies, D.H. and Mariuzza, R.A. (2003). Variable MHC class I engagement by Ly49 natural killer cell receptors demonstrated by the crystal structure of Ly49C bound to H-2K(b). Nat Immunol **4**, 1213–1222.

de Saint Basile, G. and Fischer, A. (2001). The role of cytotoxicity in lymphocyte homeostasis. Curr Opin Immunol **13**, 549–554.

Della Chiesa, M., Vitale, M., Carlomagno, S., Ferlazzo, G., Moretta, L. and Moretta, A. (2003). The natural killer cell-mediated killing of autologous dendritic cells is confined to a cell subset expressing CD94/NKG2A, but lacking inhibitory killer Ig-like receptors. Eur J Immunol **33**, 1657–1666.

Djeu, J.Y., Jiang, K. and Wei, S. (2002). A view to a kill: signals triggering cytotoxicity. Clin Cancer Res **8**, 636–640.

Eck, M.J., Pluskey, S., Trub, T., Harrison, S.C. and Shoelson, S.E. (1996). Spatial constraints on the recognition of phosphoproteins by the tandem SH2 domains of the phosphatase SH-PTP2. Nature **379**, 277–280.

Engel, P., Eck, M.J. and Terhorst, C. (2003). The SAP and SLAM families in immune responses and X-linked lymphoproliferative disease. Nat Rev Immunol **3**, 813–821.

Epling-Burnette , P.K., Painter, J.S., Chaurasia, P., Bai, F., Wei, S., Djeu, J.Y. and Loughran, T.P. Jr. (2004). Dysregulated NK receptor expression in patients with lymphoproliferative disease of granular lymphocytes. Blood **103**, 3431–3439.

Fan, Q.R., Long, E.O. and Wiley, D.C. (2000). Cobalt-mediated dimerization of the human natural killer cell inhibitory receptor. J Biol Chem **275**, 23700–23706.

Ferlazzo, G., Spaggiari, G.M., Semino, C., Melioli, G. and Moretta, L. (2000). Engagement of CD33 surface molecules prevents the generation of dendritic cells from both monocytes and CD34+ myeloid precursors. Eur J Immunol **30**, 827–833.

Ferlazzo, G., Tsang, M.L., Moretta, L., Melioli, G., Steinman, R.M. and Munz, C. (2002). Human dendritic cells activate resting natural killer (NK) cells and are recognized via the NKp30 receptor by activated NK cells. J Exp Med **195**, 343–351.

Fernandez, N.C., Lozier, A., Flament, C., Ricciardi-Castagnoli, P., Bellet, D., Suter, M., Perricaudet, M., Tursz, T., Maraskovsky, E. and Zitvogel, L. (1999). Dendritic cells directly trigger NK cell functions: cross-talk relevant in innate anti-tumor immune responses in vivo. Nat Med **5**, 405–411.

Fischer, A., Latour, S. and de Saint Basile, G. (2007). Genetic defects affecting lymphocyte cytotoxicity. Curr Opin Immunol **19**, 348–353.

Gati, A., Guerra, N., Giron-Michel, J., Azzarone, B., Angevin, E., Moretta, A., Chouaib, S. and Caignard, A. (2001). Tumor cells regulate the lytic activity of tumor-specific cytotoxic t lymphocytes by modulating the inhibitory natural killer receptor function. Cancer Res **61**, 3240–3244.

Gati, A., Guerra, N., Gaudin, C., Da Rocha, S., Escudier, B., Lecluse, Y., Bettaieb, A., Chouaib, S. and Caignard, A. (2003). CD158 receptor controls cytotoxic T-lymphocyte susceptibility to tumor-mediated activation-induced cell death by interfering with Fas signaling. Cancer Res **63**, 7475–7482.

Gismondi, A., Jacobelli, J., Strippoli, R., Mainiero, F., Soriani, A., Cifaldi, L., Piccoli, M., Frati, L. and Santoni, A. (2003). Proline-rich tyrosine kinase 2 and Rac activation by chemokine and integrin receptors controls NK cell transendothelial migration. J Immunol **170**, 3065–3073.

Gong, J.H. and Clark-Lewis, I. (1995). Antagonists of monocyte chemoattractant protein 1 identified by modification of functionally critical NH2-terminal residues. J Exp Med **181**, 631–640.

Goronzy, J.J., Henel, G., Sawai, H., Singh, K., Lee, E.B., Pryshchep, S. and Weyand, C.M. (2005). Costimulatory pathways in rheumatoid synovitis and T-cell senescence. Ann N Y Acad Sci **1062**, 182–194.

Gosselin, P., Makrigiannis, A.P., Nalewaik, R. and Anderson, S.K. (2000). Characterization of the Ly49I promoter. Immunogenetics **51**, 326–331.

Held, W., Kunz, B., Lowin-Kropf, B., van de Wetering, M. and Clevers, H. (1999). Clonal acquisition of the Ly49A NK cell receptor is dependent on the trans-acting factor TCF-1. Immunity **11**, 433–442.

Held, W., Clevers, H. and Grosschedl, R. (2003). Redundant functions of TCF-1 and LEF-1 during T and NK cell development, but unique role of TCF-1 for Ly49 NK cell receptor acquisition. Eur J Immunol **33**, 1393–1398.

Houchins, J.P., Yabe, T., McSherry, C., Miyokawa, N. and Bach, F.H. (1990). Isolation and characterization of NK cell or NK/T cell-specific cDNA clones. J Mol Cell Immunol **4**, 295–304 discussion 305–306.

Imai, T., Chantry, D., Raport, C.J., Wood, C. L., Nishimura, M., Godiska, R., Yoshie, O. and Gray, P.W. (1998). Macrophage-derived chemokine is a functional ligand for the CC chemokine receptor 4. J Biol Chem **273**, 1764–1768.

Inoue, H., Miyaji, M., Kosugi, A., Nagafuku, M., Okazaki, T., Mimori, T., Amakawa, R., Fukuhara, S., Domae, N., Bloom, E.T. and Umehara, H. (2002). Lipid rafts as the signaling scaffold for NK cell activation: tyrosine phosphorylation and association of LAT with phosphatidylinositol 3-kinase and phospholipase C-gamma following CD2 stimulation. Eur J Immunol **32**, 2188–2198.

Jabril-Cuenod, B., Zhang, C., Scharenberg, A.M., Paolini, R., Numerof, R., Beaven, M.A. and Kinet, J.P. (1996). Syk-dependent phosphorylation of Shc. A potential link between FcepsilonRI and the Ras/mitogen-activated protein kinase signaling pathway through SOS and Grb2. J Biol Chem **271**, 16268–16272.

Jansen, C.A., Cruijsen, C.W., de Ruiter, T., Nanlohy, N., Willems, N., Janssens-Korpela, P.L. and Meyaard, L. (2007). Regulated expression of the inhibitory receptor LAIR-1 on human peripheral T cells during T cell activation and differentiation. Eur J Immunol **37**, 914–924.

Jevremovic, D., Billadeau, D.D., Schoon, R.A., Dick, C.J., Irvin, B.J., Zhang, W., Samelson, L.E., Abraham, R.T. and Leibson, P.J. (1999). Cutting edge: a role for the adaptor protein LAT in human NK cell-mediated cytotoxicity. J Immunol **162**, 2453–2456.

Jiang, K., Zhong, B., Gilvary, D.L., Corliss, B.C., Hong-Geller, E., Wei, S. and Djeu, J.Y. (2000). Pivotal role of phosphoinositide-3 kinase in regulation of cytotoxicity in natural killer cells. Nat Immunol **1**, 419–425.

Jiang, K., Zhong, B., Gilvary, D.L., Corliss, B.C., Vivier, E., Hong-Geller, E., Wei, S. and Djeu, J.Y. (2002). Syk regulation of phosphoinositide 3-kinase-dependent NK cell function. J Immunol **168**, 3155–3164.

Kambayashi, T., Assarsson, E., Chambers, B. J. and Ljunggren, H.G. (2001). Cutting edge: regulation of CD8(+) T cell proliferation by 2B4/CD48 interactions. J Immunol **167**, 6706–6710.

Keefe, D., Shi, L., Feske, S., Massol, R., Navarro, F., Kirchhausen, T. and Lieberman, J. (2005). Perforin triggers a plasma membrane-repair response that facilitates CTL induction of apoptosis. Immunity **23**(3), 249–262.

Kopcow, H.D., Allan, D.S., Chen, X., Rybalov, B., Andzelm, M.M., Ge, B. and Strominger, J.L. (2005). Human decidual NK cells form immature activating synapses and are not cytotoxic. Proc Natl Acad Sci U S A **102**, 15563–15568.

Kubo, S., Itoh, Y., Ishikawa, N., Nagasawa, R., Mitarai, T. and Maruyama, N. (1993). The gene encoding mouse lymphocyte antigen Ly-49: structural analysis and the 5′-flanking sequence. Gene **136**, 329–331.

Kubo, S., Nagasawa, R., Nishimura, H., Shigemoto, K. and Maruyama, N. (1999). ATF-2-binding regulatory element is responsible for the Ly49A expression in murine T lymphoid line, EL-4. Biochim Biophys Acta **1444**, 191–200.

Kwon, D., Chwae, Y.J., Choi, I.H., Park, J.H., Kim, S.J. and Kim, J. (2000). Diversity of the p70 killer cell inhibitory receptor (KIR3DL) family members in a single individual. Mol Cells **10**, 54–60.

Lanier, L.L., Corliss, B. and Phillips, J.H. (1997). Arousal and inhibition of human NK cells. Immunol Rev **155**, 145–154.

Lanier, L.L., Corliss, B.C., Wu, J., Leong, C. and Phillips, J.H. (1998). Immunoreceptor DAP12 bearing a tyrosine-based activation motif is involved in activating NK cells. Nature **391**, 703–707.

Leibson, P.J. (1997). Signal transduction during natural killer cell activation: inside the mind of a killer. Immunity **6**, 655–661.

Leong, C.C., Chapman, T.L., Bjorkman, P.J., Formankova, D., Mocarski, E.S., Phillips, J.H. and Lanier, L.L. (1998). Modulation of natural killer cell cytotoxicity in human cytomegalovirus infection: the role of endogenous class I major histocompatibility complex and a viral class I homolog. J Exp Med **187**, 1681–1687.

Ligeza, A., Wawrzczak-Bargiela, A., Kaminska, D., Korostynski, M. and Przewlocki, R. (2008). Regulation of ERK1/2 phosphorylation by acute and chronic morphine-implications for the role of cAMP-responsive element binding factor (CREB)-dependent and Ets-like protein-1 (Elk-1)-dependent transcription; small interfering RNA-based strategy. Febs J **275**, 3836–3849.

Loetscher, P., Seitz, M., Clark-Lewis, I., Baggiolini, M. and Moser, B. (1996). Activation of NK cells by CC chemokines. Chemotaxis, $Ca^{2}+$ mobilization, and enzyme release. J Immunol **156**, 322–327.

Long, E.O., Barber, D.F., Burshtyn, D.N., Faure, M., Peterson, M., Rajagopalan, S., Renard, V., Sandusky, M., Stebbins, C.C., Wagtmann, N. and Watzl, C. (2001). Inhibition of natural killer cell activation signals by killer cell immunoglobulin-like receptors (CD158). Immunol Rev **181**, 223–233.

Lou, Z., Jevremovic, D., Billadeau, D.D. and Leibson, P.J. (2000). A balance between positive and negative signals in cytotoxic lymphocytes regulates the polarization of lipid rafts during the development of cell-mediated killing. J Exp Med **191**, 347–354.

Mandelboim, O., Lieberman, N., Lev, M., Paul, L., Arnon, T.I., Bushkin, Y., Davis, D.M., Strominger, J.L., Yewdell, J.W. and Porgador, A. (2001). Recognition of haemagglutinins on virus-infected cells by NKp46 activates lysis by human NK cells. Nature **409**, 1055–1060.

Martinvalet, D., Dykxhoorn, D.M., Ferrini, R. and Lieberman, J. (2008). Granzyme A cleaves a mitochondrial complex I protein to initiate caspase-independent cell death. Cell **133**(4), 681–689.

McVicar, D.W. and Burshtyn, D.N. (2001). Intracellular signaling by the killer immunoglobulin-like receptors and Ly49. Sci STKE **2001**, RE1.

McVicar, D.W., Taylor, L.S., Gosselin, P., Willette-Brown, J., Mikhael, A.I., Geahlen, R.L., Nakamura, M.C., Linnemeyer, P., Seaman, W.E., Anderson, S.K., Ortaldo, J.R. and Mason, L.H. (1998). DAP12-mediated signal transduction in natural killer cells. A dominant role for the Syk protein-tyrosine kinase. J Biol Chem **273**, 32934–32942.

Medema, J.P., Scaffidi, C., Kischkel, F.C., Shevchenko, A., Mann, M., Krammer, P.H. and Peter, M.E. (1997). FLICE is activated by association with the CD95 death-inducing signaling complex (DISC). Embo J **16**, 2794–2804.

Miller, J.S. and McCullar, V. (2001). Human natural killer cells with polyclonal lectin and immunoglobulinlike receptors develop from single hematopoietic stem cells with preferential expression of NKG2A and KIR2DL2/L3/S2. Blood **98**, 705–713.

Monks, C.R., Freiberg, B.A., Kupfer, H., Sciaky, N. and Kupfer, A. (1998). Three-dimensional segregation of supramolecular activation clusters in T cells. Nature **395**, 82–86.

Moretta, A. (2002). Natural killer cells and dendritic cells: rendezvous in abused tissues. Nat Rev Immunol **2**, 957–964.

Moretta, A., Biassoni, R., Bottino, C. and Moretta, L. (2000). Surface receptors delivering opposite signals regulate the function of human NK cells. Semin Immunol **12**, 129–138.

Moretta, A., Bottino, C., Vitale, M., Pende, D., Cantoni, C., Mingari, M.C., Biassoni, R. and Moretta, L. (2001). Activating receptors and coreceptors involved in human natural killer cell-mediated cytolysis. Annu Rev Immunol **19**, 197–223.

Nagaraj, S. and Gabrilovich, D.I. (2008). Tumor escape mechanism governed by myeloid-derived suppressor cells. Cancer Res **68**, 2561–2563.

Nakamura, M.C., Niemi, E.C., Fisher, M.J., Shultz, L.D., Seaman, W.E. and Ryan, J.C. (1997). Mouse Ly-49A interrupts early signaling events in natural killer cell cytotoxicity and functionally associates with the SHP-1 tyrosine phosphatase. J Exp Med **185**, 673–684.

Nefedova, Y., Fishman, M., Sherman, S., Wang, X., Beg, A.A. and Gabrilovich, D.I. (2007). Mechanism of all-trans retinoic acid effect on tumor-associated myeloid-derived suppressor cells. Cancer Res **67**, 11021–11028.

Nichols, K.E., Ma, C.S., Cannons, J.L., Schwartzberg, P.L. and Tangye, S.G. (2005). Molecular and cellular pathogenesis of X-linked lymphoproliferative disease. Immunol Rev **203**, 180–199.

Nishida, K., Yamasaki, S., Ito, Y., Kabu, K., Hattori, K., Tezuka, T., Nishizumi, H., Kitamura, D., Goitsuka, R., Geha, R.S., Yamamoto, T., Yagi, T. and Hirano, T. (2005). FcεRI-mediated mast cell degranulation requires calcium-independent microtubule-dependent translocation of granules to the plasma membrane. J Cell Biol **170**, 115–126.

Nishimoto, S. and Nishida, E. (2006). MAPK signalling: ERK5 versus ERK1/2. Embo Rep **7**, 782–786.

O'Callaghan, C.A., Tormo, J., Willcox, B.E., Braud, V.M., Jakobsen, B.K., Stuart, D.I., McMichael, A.J., Bell, J.I. and Jones, E. Y. (1998). Structural features impose tight peptide binding specificity in the nonclassical MHC molecule HLA-E. Mol Cell **1**, 531–541.

Ogasawara, K., Hamerman, J.A., Hsin, H., Chikuma, S., Bour-Jordan, H., Chen, T., Pertel, T., Carnaud, C., Bluestone, J.A. and Lanier, L.L. (2003). Impairment of NK cell function by NKG2D modulation in NOD mice. Immunity **18**, 41–51.

Ogasawara, K., Hamerman, J.A., Ehrlich, L.R., Bour-Jordan, H., Santamaria, P., Bluestone, J.A. and Lanier, L.L. (2004). NKG2D blockade prevents autoimmune diabetes in NOD mice. Immunity **20**, 757–767.

Olcese, L., Lang, P., Vely, F., Cambiaggi, A., Marguet, D., Blery, M., Hippen, K.L., Biassoni, R., Moretta, A., Moretta, L., Cambier, J.C. and Vivier, E. (1996). Human and mouse killer-cell inhibitory receptors recruit PTP1C and PTP1D protein tyrosine phosphatases. J Immunol **156**, 4531–4534.

Ono, M., Okada, H., Bolland, S., Yanagi, S., Kurosaki, T. and Ravetch, J.V. (1997). Deletion of SHIP or SHP-1 reveals two distinct pathways for inhibitory signaling. Cell **90**, 293–301.

Orange, J.S., Harris, K.E., Andzelm, M.M., Valter, M.M., Geha, R.S. and Strominger, J.L. (2003). The mature activating natural killer cell immunologic synapse is formed in distinct stages. Proc Natl Acad Sci U S A **100**, 14151–14156.

Pende, D., Parolini, S., Pessino, A., Sivori, S., Augugliaro, R., Morelli, L., Marcenaro, E., Accame, L., Malaspina, A., Biassoni, R., Bottino, C., Moretta, L. and Moretta, A. (1999). Identification and molecular characterization of NKp30, a novel triggering receptor involved in natural cytotoxicity mediated by human natural killer cells. J Exp Med **190**, 1505–1516.

Pende, D., Spaggiari, G.M., Marcenaro, S., Martini, S., Rivera, P., Capobianco, A., Falco, M., Lanino, E., Pierri, I., Zambello, R., Bacigalupo, A., Mingari, M.C., Moretta, A. and Moretta, L. (2005). Analysis of the receptor-ligand interactions in the natural killer-mediated lysis of freshly isolated myeloid or lymphoblastic leukemias: evidence for the involvement of the Poliovirus receptor (CD155) and Nectin-2 (CD112). Blood **105**, 2066–2073.

Piccioli, D., Sbrana, S., Melandri, E. and Valiante, N.M. (2002). Contact-dependent stimulation and inhibition of dendritic cells by natural killer cells. J Exp Med **195**, 335–341.

Qi, S.Y., Groves, J.T. and Chakraborty, A.K. (2001). Synaptic pattern formation during cellular recognition. Proc Natl Acad Sci U S A **98**, 6548–6553.

Rajagopalan, S. and Long, E.O. (1998). Zinc bound to the killer cell-inhibitory receptor modulates the negative signal in human NK cells. J Immunol **161**, 1299–1305.

Rajagopalan, S. and Long, E.O. (1999). A human histocompatibility leukocyte antigen (HLA)-G-specific receptor expressed on all natural killer cells. J Exp Med **189**, 1093–1100.

Rajagopalan, S., Winter, C.C., Wagtmann, N. and Long, E.O. (1995). The Ig-related killer cell inhibitory receptor binds zinc and requires zinc for recognition of HLA-C on target cells. J Immunol **155**, 4143–4146.

Roda-Navarro, P., Vales-Gomez, M., Chisholm, S.E. and Reyburn, H.T. (2006). Transfer of NKG2D and MICB at the cytotoxic NK cell immune synapse correlates with a reduction in NK cell cytotoxic function. Proc Natl Acad Sci U S A **103**, 11258–11263.

Ruggeri, L., Capanni, M., Casucci, M., Volpi, I., Tosti, A., Perruccio, K., Urbani, E., Negrin, R.S., Martelli, M.F. and Velardi, A. (1999). Role of natural killer cell alloreactivity in HLA-mismatched hematopoietic stem cell transplantation. Blood **94**, 333–339.

Ruggeri, L., Capanni, M., Martelli, M.F. and Velardi, A. (2001). Cellular therapy: exploiting NK cell alloreactivity in transplantation. Curr Opin Hematol **8**, 355–359.

Ruggeri, L., Capanni, M., Urbani, E., Perruccio, K., Shlomchik, W.D., Tosti, A., Posati, S., Rogaia, D., Frassoni, F., Aversa, F., Martelli, M.F. and Velardi, A. (2002). Effectiveness of donor natural killer cell alloreactivity in mismatched hematopoietic transplants. Science **295**, 2097–2100.

Saleh, A., Davies, G.E., Pascal, V., Wright, P.W., Hodge, D.L., Cho, E.H., Lockett, S.J., Abshari, M. and Anderson, S.K. (2004). Identification of probabilistic transcriptional switches in the Ly49 gene cluster: a eukaryotic mechanism for selective gene activation. Immunity **21**, 55–66.

Salojin, K.V., Zhang, J. and Delovitch, T.L. (1999). TCR and CD28 are coupled via ZAP-70 to the activation of the Vav/Rac-1-/PAK-1/p38 MAPK signaling pathway. J Immunol **163**, 844–853.

Santourlidis, S., Trompeter, H.I., Weinhold, S., Eisermann, B., Meyer, K.L., Wernet, P. and Uhrberg, M. (2002). Crucial role of DNA methylation in determination of clonally distributed killer cell Ig-like receptor expression patterns in NK cells. J Immunol **169**, 4253–4261.

Santourlidis, S., Graffmann, N., Christ, J. and Uhrberg, M. (2008). Lineage-specific transition of histone signatures in the killer cell Ig-like receptor locus from hematopoietic progenitor to NK cells. J Immunol **180**, 418–425.

Saurer, L., Seibold, I., Vallan, C., Held, W. and Mueller, C. (2003). Cutting edge: stimulation with the cognate self-antigen induces expression of the Ly49A receptor on self-reactive T cells which modulates their responsiveness. J Immunol **171**, 6334–6338.

Scaffidi, C., Fulda, S., Srinivasan, A., Friesen, C., Li, F., Tomaselli, K.J., Debatin, K.M., Krammer, P.H. and Peter, M.E. (1998). Two CD95 (APO-1/Fas) signaling pathways. Embo J **17**, 1675–1687.

Scharenberg, A.M., Lin, S., Cuenod, B., Yamamura, H. and Kinet, J.P. (1995). Reconstitution of interactions between tyrosine kinases and the high affinity IgE receptor which are controlled by receptor clustering. Embo J **14**, 3385–3394.

Schonland, S.O., Zimmer, J.K., Lopez-Benitez, C.M., Widmann, T., Ramin, K.D., Goronzy, J.J. and Weyand, C. M. (2003). Homeostatic control of T-cell generation in neonates. Blood **102**, 1428–1434.

Sivori, S., Vitale, M., Morelli, L., Sanseverino, L., Augugliaro, R., Bottino, C., Moretta, L. and Moretta, A. (1997). p46, a novel natural killer cell-specific surface molecule that mediates cell activation. J Exp Med **186**, 1129–1136.

Sivori, S., Parolini, S., Falco, M., Marcenaro, E., Biassoni, R., Bottino, C., Moretta, L. and Moretta, A. (2000). 2B4 functions as a co-receptor in human NK cell activation. Eur J Immunol **30**, 787–793.

Smith, K.M., Wu, J., Bakker, A.B., Phillips, J.H. and Lanier, L.L. (1998). Ly-49D and Ly-49H associate with mouse DAP12 and form activating receptors. J Immunol **161**, 7–10.

Smyth, M.J., Cretney, E., Kelly, J.M., Westwood, J.A., Street, S.E., Yagita, H., Takeda, K., van Dommelen, S.L., Degli-Esposti, M.A. and Hayakawa, Y. (2005). Activation of NK cell cytotoxicity. Mol Immunol **42**, 501–510.

Spaggiari, G.M., Carosio, R., Pende, D., Marcenaro, S., Rivera, P., Zocchi, M.R., Moretta, L. and Poggi, A. (2001). NK cell-mediated lysis of autologous antigen-presenting cells is triggered by the engagement of the phosphatidylinositol 3-kinase upon ligation of the natural cytotoxicity receptors NKp30 and NKp46. Eur J Immunol **31**, 1656–1665.

Stebbins, C.C., Watzl, C., Billadeau, D.D., Leibson, P.J., Burshtyn, D.N. and Long, E.O. (2003). Vav1 dephosphorylation by the tyrosine phosphatase SHP-1 as a mechanism for inhibition of cellular cytotoxicity. Mol Cell Biol **23**, 6291–6299.

Stewart, C.A., Van Bergen, J. and Trowsdale, J. (2003). Different and divergent regulation of the KIR2DL4 and KIR3DL1 promoters. J Immunol **170**, 6073–6081.

Sutherland, C.L., Chalupny, N.J. and Cosman, D. (2001). The UL16-binding proteins, a novel family of MHC class I-related ligands for NKG2D, activate natural killer cell functions. Immunol Rev **181**, 185–192.

Sutherland, C.L., Chalupny, N.J., Schooley, K., VandenBos, T., Kubin, M. and Cosman, D. (2002). UL16-binding proteins, novel MHC class I-related proteins, bind to NKG2D and activate multiple signaling pathways in primary NK cells. J Immunol **168**, 671–679.

Taniguchi, R.T., Guzior, D. and Kumar, V. (2007). 2B4 inhibits NK-cell fratricide. Blood **110**, 2020–2023.

Ting, A.T., Dick, C.J., Schoon, R.A., Karnitz, L.M., Abraham, R.T. and Leibson, P.J. (1995). Interaction between lck and syk family tyrosine kinases in Fc gamma receptor-initiated activation of natural killer cells. J Biol Chem **270**, 16415–16421.

Trowsdale, J. (2001). Genetic and functional relationships between MHC and NK receptor genes. Immunity **15**, 363–374.

Trowsdale, J., Barten, R., Haude, A., Stewart, C.A., Beck, S. and Wilson, M.J. (2001). The genomic context of natural killer receptor extended gene families. Immunol Rev **181**, 20–38.

Uhrberg, M., Valiante, N.M., Shum, B.P., Shilling, H.G., Lienert-Weidenbach, K., Corliss, B., Tyan, D., Lanier, L.L. and Parham, P. (1997). Human diversity in killer cell inhibitory receptor genes. Immunity **7**, 753–763.

Uhrberg, M., Parham, P. and Wernet, P. (2002). Definition of gene content for nine common group B haplotypes of the Caucasoid population: KIR haplotypes contain between seven and eleven KIR genes. Immunogenetics **54**, 221–229.

Ullrich, E., Bonmort, M., Mignot, G., Jacobs, B., Bosisio, D., Sozzani, S., Jalil, A., Louache, F., Bulanova, E., Geissman, F., Ryffel, B., Chaput, N., Bulfone-Paus, S. and Zitvogel, L. (2008). Trans-presentation of IL-15 dictates IFN-producing killer dendritic cells effector functions. J Immunol **180**, 7887–7897.

Umehara, H., Inoue, H., Huang, J., Kono, T., Minami, Y., Tanaka, Y., Okazaki, T., Mimori, T., Bloom, E.T. and Domae, N. (2002). Role for adapter proteins in costimulatory signals of CD2 and IL-2 on NK cell activation. Mol Immunol **38**, 587–596.

Upshaw, J.L., Schoon, R.A., Dick, C.J., Billadeau, D.D. and Leibson, P.J. (2005). The isoforms of phospholipase C-gamma are differentially used by distinct human NK activating receptors. J Immunol **175**, 213–218.

Upshaw, J.L., Arneson, L.N., Schoon, R.A., Dick, C.J., Billadeau, D.D. and Leibson, P.J. (2006). NKG2D-mediated signaling requires a DAP10-bound Grb2-Vav1 intermediate and phosphatidylinositol-3-kinase in human natural killer cells. Nat Immunol **7**, 524–532.

van Bergen, J., Thompson, A., van der Slik, A., Ottenhoff, T.H., Gussekloo, J. and Koning, F. (2004). Phenotypic and functional characterization of CD4 T cells expressing killer Ig-like receptors. J Immunol **173**, 6719–6726.

Vely, F., Olivero, S., Olcese, L., Moretta, A., Damen, J.E., Liu, L., Krystal, G., Cambier, J.C., Daeron, M. and Vivier, E. (1997). Differential association of phosphatases with hematopoietic co-receptors bearing immunoreceptor tyrosine-based inhibition motifs. Eur J Immunol **27**, 1994–2000.

Verbrugge, A., Rijkers, E.S., de Ruiter, T. and Meyaard, L. (2006). Leukocyte-associated Ig-like receptor-1 has SH2 domain-containing phosphatase-independent function and recruits C-terminal Src kinase. Eur J Immunol **36**, 190–198.

Vitale, M., Bottino, C., Sivori, S., Sanseverino, L., Castriconi, R., Marcenaro, E., Augugliaro, R., Moretta, L. and Moretta, A. (1998). NKp44, a novel triggering surface molecule specifically expressed by activated natural killer cells, is involved in non-major histocompatibility complex-restricted tumor cell lysis. J Exp Med **187**, 2065–2072.

Wang, J.W., Howson, J.M., Ghansah, T., Desponts, C., Ninos, J.M., May, S. L., Nguyen, K.H., Toyama-Sorimachi, N. and Kerr, W.G. (2002). Influence of SHIP on the NK repertoire and allogeneic bone marrow transplantation. Science **295**, 2094–2097.

Watzl, C., Stebbins, C.C. and Long, E.O. (2000). NK cell inhibitory receptors prevent tyrosine phosphorylation of the activation receptor 2B4 (CD244). J Immunol **165**, 3545–3548.

Wei, S., Gamero, A.M., Liu, J.H., Daulton, A.A., Valkov, N.I., Trapani, J.A., Larner, A.C., Weber, M.J. and Djeu, J.Y. (1998). Control of lytic function by mitogen-activated protein kinase/extracellular regulatory kinase 2 (ERK2) in a human natural killer cell line: identification of perforin and granzyme B mobilization by functional ERK2. J Exp Med **187**, 1753–1765.

Wei, S., Gilvary, D.L., Corliss, B.C., Sebti, S., Sun, J., Straus, D.B., Leibson, P.J., Trapani, J.A., Hamilton, A.D., Weber, M.J. and Djeu, J.Y. (2000). Direct tumor lysis by NK cells uses a Ras-independent mitogen-activated protein kinase signal pathway. J Immunol **165**, 3811–3819.

Williams, N.S., Kubota, A., Bennett, M., Kumar, V. and Takei, F. (2000). Clonal analysis of NK cell development from bone marrow progenitors in vitro: orderly acquisition of receptor gene expression. Eur J Immunol **30**, 2074–2082.

Wilson, J.L., Heffler, L.C., Charo, J., Scheynius, A., Bejarano, M.T. and Ljunggren, H.G. (1999). Targeting of human dendritic cells by autologous NK cells. J Immunol **163**, 6365–6370.

Wu, J., Song, Y., Bakker, A.B., Bauer, S., Spies, T., Lanier, L.L. and Phillips, J.H. (1999). An activating immunoreceptor complex formed by NKG2D and DAP10. Science **285**, 730–732.

Xu, J., Vallejo, A.N., Jiang, Y., Weyand, C. M. and Goronzy, J.J. (2005). Distinct transcriptional control mechanisms of killer immunoglobulin-like receptors in natural killer (NK) and in T cells. J Biol Chem **280**, 24277–24285.

Yabe, T., McSherry, C., Bach, F.H. and Houchins, J.P. (1990). A cDNA clone expressed in natural killer and T cells that likely encodes a secreted protein. J Exp Med **172**, 1159–1163.

Yabe, T., McSherry, C., Bach, F.H., Fisch, P., Schall, R.P., Sondel, P.M. and Houchins, J. P. (1993). A multigene family on human chromosome 12 encodes natural killer-cell lectins. Immunogenetics **37**, 455–460.

Yamashita, Y., Fukuta, D., Tsuji, A., Nagabukuro, A., Matsuda, Y., Nishikawa, Y., Ohyama, Y., Ohmori, H., Ono, M. and Takai, T. (1998). Genomic structures and chromosomal location of p91, a novel murine regulatory receptor family. J Biochem **123**, 358–368.

Yokoyama, W.M. and Seaman, W.E. (1993). The Ly-49 and NKR-P1 gene families encoding lectin-like receptors on natural killer cells: the NK gene complex. Ann Rev Immunol **II**, 613–635.

Young, N.T. and Uhrberg, M. (2002). KIR expression shapes cytotoxic repertoires: a developmental program of survival. Trends Immunol **23**, 71–75.

Yusa, S., Catina, T.L. and Campbell, K.S. (2002). SHP-1- and phosphotyrosine-independent inhibitory signaling by a killer cell Ig-like receptor cytoplasmic domain in human NK cells. J Immunol **168**, 5047–5057.

Zambello, R., Falco, M., Della Chiesa, M., Trentin, L., Carollo, D., Castriconi, R., Cannas, G., Carlomagno, S., Cabrelle, A., Lamy, T., Agostini, C., Moretta, A., Semenzato, G. and Vitale, M. (2003). Expression and function of KIR and natural cytotoxicity receptors in NK-type lymphoproliferative diseases of granular lymphocytes. Blood **102**, 1797–1805.

Zhou , T., Edwards, C.K., Yang, P., Wang, Z., Bluethmann, H. and Mountz, J. D. 3rd (1996). Greatly accelerated lymphadenopathy and autoimmune disease in lpr mice lacking tumor necrosis factor receptor I. J Immunol **156**, 2661–2665.

Zitvogel, L. (2002). Dendritic and natural killer cells cooperate in the control/switch of innate immunity. J Exp Med **195**, F9–F14.

Chapter **Eight**

8

Human natural killer cell development

Edward L. Briercheck, Aharon G. Freud, Michael A. Caligiuri

CHAPTER CONTENTS

If I can see an ending I can work backward.

Arthur Miller, American playwright
(New York Times, 9 February 1986)

ABSTRACT

The development of natural killer (NK) cells in humans can be characterized into five distinct stages according to cell surface expression as pro-NK, pre-NK, immature iNK, CD56^{bright} NK and CD56^{dim} NK. The location of human NK cell development appears to be multi-centred as NK cells are derived from hematopoietic stem cells in the bone marrow, which travel in the blood to secondary lymphoid tissue where maturation likely occurs and functional responsiveness to the environment begin. The progression in NK cell development involves distinct changes in cell surface phenotype as well as cell function. Genetic disruption studies in the mouse demonstrate that cytokines such as interleukin-15, c-kit ligand (KL) and flt3 ligand (FL) are essential for normal NK cell development and are likely important for changes in stage-specific requisite transcription factor expression.

KEY WORDS

Lymphocyte development, Innate immunity, Hematopoiesis, CD56^{bright} NK, CD56^{dim} NK, Interleukin-15, c-kit Ligand, flt3 Ligand

Introduction

Congenital defects in the development of T cells and B cells lead in some instances to profound immune deficiencies (Edgar, 2008; LeBien and Tedder, 2008). Understanding the developmental defects in adaptive immunity has led to the subsequent development of innovative, life-saving therapies for those afflicted (Hacein-Bey-Abina et al., 2002). While the adaptive immune response has both amazing specificity and memory, the time required following exposure to effectively respond to primary antigen is several days to weeks. Host defence against invading pathogens and malignant transformation requires a 'first line' that can quickly place the foreign invasion in check while the primary adaptive immune response is being developed. This is where the effector arm of the innate immune system comes into play. natural killer (NK) cells form part of the hosts' first line of defence against such pathogens with the ability to kill targets or prime the immune system with cytokines such as IFN-γ within minutes of directly encountering a cellular target, such as a virus-infected cell, or receiving an activation signal via the provision of cytokines from dendritic cells within secondary lymphoid tissues (SLT). Indeed, defects in immune mediators such as IFN-γ or in NK cells also lead to profound and often fatal susceptibility to foreign pathogens (Dorman et al., 2004; Orange, 2006). Here, we explore the development of human NK cells, trying to discern where they develop, how they develop, and whether certain cellular stages of NK development already acquire important defence mechanisms even prior to reaching what is generally considered full functional maturation.

Stages of human NK development

The progression of NK cells through the five stages of differentiation described later in this section does not infer that all cells categorized as, for example, pro-NK cells are restricted to development along the NK lineage. Rather, it implies that if a NK cell precursor is to eventually become a mature NK cell, it will likely have at one time demonstrated the characteristics of these five stages. Individual cell types may share the pathway, in the form of antigen surface expression or function, but eventually each cell type differentiates along paths that other cell types do not. Terminal differentiation occurs through a series of stochastic events, mediated by both inductive and inhibitory influences such as cytokines and cognate signals for cell surface receptors dictated by changing conditions.

For hematopoietic stem cells, the first step towards lineage differentiation is the lymphoid or myeloid pathway (Stehling-Sun et al., 2009) (Figure 8.1). This myeloid versus lymphoid model is not without question, and studies have shown deviation from such simple models (Blom and Spits, 2006). The common lymphoid progenitor cell in humans is defined as CD34(+)Lin(−)CD10(+) (Di Santo, 2006; Galy et al., 1995). The common lymphoid progenitor cell is where NK cells begin their differentiation pathway (Figure 8.1).

Through analysis of cell surface antigen expression and ex vivo culture studies, five stages of developing human NK cells can be identified within SLT: pro-NK, pre-NK, immature iNK, CD56^{bright} NK and CD56^{dim} NK (Freud et al., 2006) (Figure 8.1). The transition from pro-NK to pre-NK is phenotypically highlighted by the gain in expression of CD117, also known as c-kit, and functionally by the ability to respond to IL-15 (Freud et al., 2006) (Figure 8.1).

Culture studies have demonstrated that pro-NK and pre-NK cells have ex vivo potential for non-NK lineage differentiation, whereas iNK cells are committed to the NK lineage. The latter are CD34(−)CD117(+)CD94(−), and this immunophenotype was determined by noting that CD34 (expressed on pro-NK and pre-NK cells) and CD94 (expressed on CD56^{bright} NK cells) are mutually exclusive antigens (Figure 8.2), implying that an intermediate cell type, which no longer expresses CD34 but does not yet express CD94, must exist (Freud et al., 2006). In addition to NK lineage restriction that occurs at the iNK stage of development, there is also an increase in expression of some NK cell-associated antigens such as CD11b, which is an NK maturity marker in mice (Freud et al., 2006; Kim et al., 2002).

CD56 expression gradually accumulates at the population level as cells progress from the pre-NK stage to the iNK stage of maturation. Moreover CD56 expression is uniformly high within the CD56^{bright} NK cell population (Freud et al., 2006). Notably, although CD56 is typically considered a marker of mature NK cells, the final stage of human NK cell maturation is marked by a decrease in CD56 and CD94 expression and an increase in CD16 (FcRγIII receptor) and killer immunoglobulin-like receptors (KIR) (Figure 8.1) (Caligiuri, 2008; Freud and Caligiuri, 2006; Freud et al., 2006).

The phenotypic distinction between CD56^{bright} and CD56^{dim} NK subsets in blood is relevant to their seemingly divergent roles in immunity. Circulating CD56^{bright} NK cells can migrate into SLT due to relatively high expression of CD62L and CCR7. In SLT, these cells can interact with macrophages and dendritic cells whose monokines induce CD56^{bright} NK cells to secrete cytokines that can activate antigen-presenting cells and epithelial cells. In contrast, CD56^{dim} NK cells likely rarely enter SLT and more likely function to survey for evidence of abnormal MHC class I expression secondary to viral infection or malignant transformation (Caligiuri, 2008; Cooper et al., 2001a).

The location of human NK development

While the location of hematopoiesis in embryonic and fetal development occurs in the yolk sac, aorta–gonad–mesonephros region and liver, the primary location in adults is the bone marrow (Godin and Cumano, 2002). Consequently, one might assume that this is the primary location of NK cell development, especially since this seems to be the case in mice (Kim et al., 2002). Indeed, the common lymphoid progenitor defined as CD34(+)Lin(−)CD10(+) exists in the bone marrow (Galy et al., 1995). The pro-NK cell described previously is CD34^{dim}CD10(+)CD45RA(+) (Freud et al., 2006; Galy et al., 1995). However, the majority of pro-NK cells co-express CD62L and β_7 integrin, which allow them to home to SLT where they are highly enriched among total CD34(+) HPCs compared to the blood (Freud et al., 2005). SLT includes, among others, lymph nodes, tonsils, spleen and Peyer's patches. Thus, the

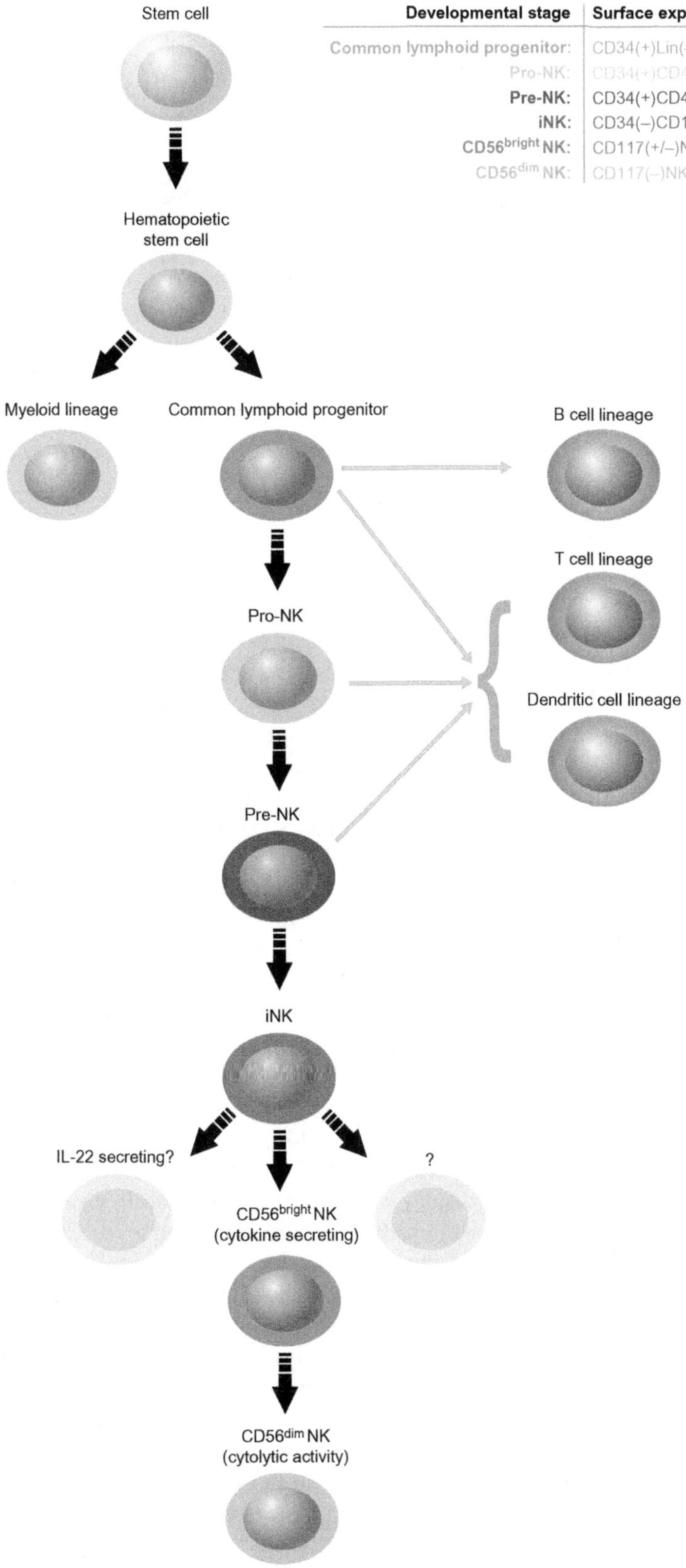

Developmental stage	Surface expression
Common lymphoid progenitor:	CD34(+)Lin(–)CD10(+)
Pro-NK:	CD34(+)CD45RA(+)CD10(+)CD117(–)CD161(–)
Pre-NK:	CD34(+)CD45RA(+)CD10(–)CD117(+)CD161(+/–)
iNK:	CD34(–)CD117(+)CD161(+)NKp46(–)CD94/NKG2A(–)
CD56^{bright} NK:	CD117(+/–)NKp46(+)CD94/NKG2A(+)CD16(–)KIR(+/–)
CD56^{dim} NK:	CD117(–)NKp46(+)CD94/NKG2A(+/–)CD16(+)KIR(+)

Figure 8.1 • NK development progresses through a series of stages characterized by distinct expression of cell surface antigens. NK cells are derived from the common lymphoid progenitor, which gives rise to the pro-NK, pre-NK, iNK, CD56^{bright} NK and CD56^{dim} NK cell types. Developmental stage and corresponding cell surface expression are shown to the right. Specific stages shown by grey arrows may divert away from the NK cell lineage to become B, T or dendritic cells. The immature iNK cells show potential for NK cell-specific branch points in development that are currently unresolved.

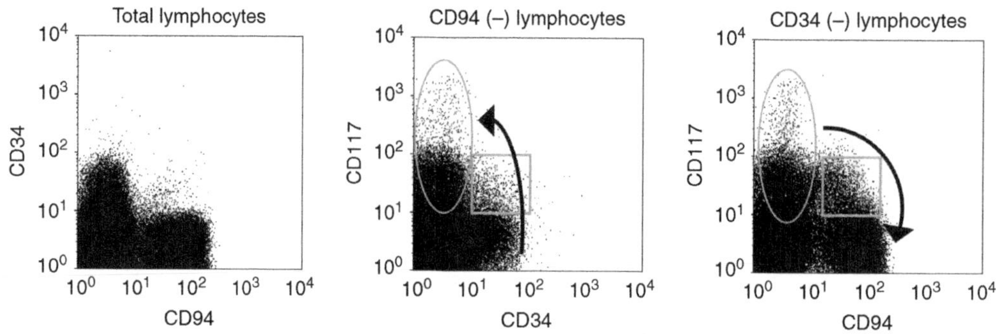

Figure 8.2 • Flow cytometric analyses shows that CD34 expression and CD94 expression are largely mutually exclusive (left). However, CD34(+) cells may express CD117 (centre, grey square), and CD94(+) may express CD117 (right, grey square) suggesting that there is a stage of transition between these two populations. The CD34(−)CD117(+)CD94(−) (oval) population identifies the iNK cell shown in the centre and right panels. The arrow shown in the two panels indicates the direction of human NK cell development going from pro/pre to iNK (centre) and from iNK to NK (right).

earliest NK cell progenitors appear to migrate from human bone marrow to SLT.

The pro-NK cell, pre-NK cell, iNK cell and CD56^{bright} NK cell populations have each been found to be enriched within SLT, whereas CD56^{dim} NK cells are relatively enriched in bone marrow, blood and spleen (Fehniger et al., 2003; Ferlazzo et al., 2004; Freud et al., 2006). Thus, while as of yet there is no formal proof that NK cell developmental intermediates mature within SLT, the fact that each is enriched within SLT where there is an abundance of DC with membrane bound IL-15 strongly suggests that SLT is the location for human NK cell development. The expression of L-selectin and CCR7, both critical to migration into SLT (Campbell et al., 2001; Frey et al., 1998), diminishes on CD56^{dim} NK cells and likely contributes to their egression from SLT. The fact that CD56^{bright} NK cells do exist in the blood along with the more abundant CD56^{dim} NK cells suggests that the final step of maturation from CD56^{bright} to CD56^{dim} does not necessarily have to occur in SLT and may occur within the circulation.

Diversity in development and function

As noted earlier, both pro-NK cells and pre-NK cells can vary in their final lineage or destination, yet they can be differentiated into NK cells in vitro (Freud et al., 2006). Further, during this progression in NK cell maturation, functional characteristics may be absent, gained and lost once again within cells of a discrete stage or throughout the process of differentiation. Indeed, flow cytometric analyses of different stages show a continuum of antigen acquisition and loss, such as CD117 expression appearing in pre-NK cells, remaining expressed on iNK cells and diminished to absent on CD56^{bright} NK cells (Freud et al., 2006). This suggests that the discrete phenotypic and functional stages that we commonly diagram are more likely to actually be a continuum in vivo and more useful for the purposes of modelling human NK cell development.

Pro-NK cells appear unique compared to other stages of NK cell development by their lack of responsiveness to IL-15 (Freud et al., 2004). While IL-15 responsiveness is a critical step on the road to NK cell maturity, and the CD117 antigen co-expressed with CD34 can be used to identify IL-15-responsive pre-NK cells within this population, the pre-NK cells have yet to fully commit to the NK lineage. Pre-NK cells, like pro-NK cells, can still become T cells or dendritic cells (Figure 8.1). However, when cultures of pro-NK cells are grown in flt3 ligand (FL) and IL-7, and co-cultured with the OP9-DL1 cell line, the pro-NK cells generate a 24-fold greater number of T cells compared to pre-NK cells (Freud et al., 2006).

Unlike pre-NK and pro-NK cells, iNK cells do not display multi-lineage potential. In vivo-derived iNK cells can produce GM-CSF (Freud et al., 2006), and in vitro-derived iNK produce type 2 cytokines IL-5 and IL-13, which is lost following maturation to CD56^{bright} NK cells. Furthermore, these in vitro derived IL-2-stimulated iNK cells lyse the T-cell leukemia Jurkat cell line via a TRAIL-dependent mechanism (tumour-necrosis factor-related apoptosis inducing ligand) (Loza and Perussia, 2001; Loza et al., 2002; Zamai et al., 1998). These studies on in vitro-derived iNK cells suggest that the latter may be the first stage at which NK cells have the capacity to produce cytokines and the ability to acquire cytolytic activity.

IL-22 is an inflammatory cytokine that is produced within the T_H17-cell lineage. IL-22 can activate epithelial cells such as those that line the mucosa to secrete anti-microbial peptides (Liang et al., 2006; Wolk et al.,

2004). Recently, a small subset of NKp44(+) NK cells has been identified in the mucosa-associated lymphoid tissue (MALT) to produce IL-22 following activation (Cella et al., 2008). Two other groups have since shown that in addition to this small activated NK cell population, the freshly isolated iNK cell within human tonsil constitutively produces abundant IL-22, suggesting this cell type might have a functional role in mucosal immunity, in addition to being an intermediary cell in human NK cell development (Cupedo et al., 2009; Hughes et al., 2009). It is also conceivable that subsets within the iNK cell population have divergent pathways with at least some of these iNK cells terminally differentiating as IL-22-secreting NK cells that never mature to CD56^{bright} NK cells.

The phenotypic and genotypic analyses performed on iNK to date also suggest that the iNK cell has similarities to the lymphoid tissue-inducing (LTi) cell found in mice (Cupedo et al., 2009; Hughes et al., 2009). Thus, there may indeed be a third function to this iNK cell type that is the most abundant of the five stages of NK cells found in human SLT. All three pathways of iNK cells, shown in Figure 8.1, require continued investigation.

The acquisition of CD94 expression among the CD56^{bright} NK cell in SLT carries functional significance. CD94 complexes with NKG2A and gives these CD56^{bright} NK cells the ability to recognize MHC class I molecules, specifically HLA-E, and to activate an inhibitory program (Brooks et al., 1999; Lazetic et al., 1996; Phillips et al., 1996). The acquisition of these inhibitory properties is important because at this stage, developing NK cells begin to acquire their cytolytic weaponry, and therefore to prevent indiscriminate activation of this arsenal, an inhibitory system must be in place. This arsenal includes intracellular perforin with concomitant ability to undergo perforin-mediated killing, the acquisition of Fas ligand and the ability to undergo Fas-ligand-mediated killing and the primary function of these cells, cytokine production, specifically IFNγ as well as GM-CSF, TNF-β, IL-10 and IL-13 (Cooper et al., 2001b). Researchers have noted that while acquisition of this full arsenal is indicative of mature NK cells, the expression of such weaponry occurs during, not prior to, reaching the CD56^{bright} NK stage. Therefore, CD56^{bright} NK cells may initially show minimal expression of these markers but still meet the characteristic criteria of a mature NK cell (Freud et al., 2006). Similar to classical T cells, there is some evidence in murine models that mature NK cells may exhibit memory. Murine NK cells primed with NK-activating cytokines IL-12 and IL-18 show a heritable increase in IFN-γ production upon restimulation (Cooper et al., 2009). A similar pattern of response has also been demonstrated against NK-specific targets (Sun et al., 2009). As of yet, neither of these recent findings has been investigated in human NK cells.

Cytokines driving development

Genetic disruption experiments in the mouse have confirmed the critical role that certain cytokines and growth factors play in human NK development (Caligiuri, 2008). Among these, IL-15 appears at this time to be the most important cytokine. As noted previously, the earliest pro-NK cell is unresponsive to IL-15 when directly isolated from human SLT, yet earlier studies in humans showed that exposure of this cell to either one of two tyrosine receptor kinase ligands, c-kit ligand (KL) or FL, will result in the induction of the signal transducing α and β components of the IL-15 receptor (IL-15R) and subsequent responsiveness to IL-15 (Yu et al., 2006). Hence, it is likely that stromal cell factors such as KL and FL are the first to direct the more pluripotent pro-NK cell towards NK cell differentiation.

Once these CD34^{dim}CD10(+)CD45RA(+) pro-NK cells acquire IL-15R$\beta\gamma$, they transition to an IL-15-responsive pre-NK cell. IL-15 is then presented in trans from other cells that express both IL-15 and the high-affinity IL-15Rα chain and are in close proximity to the pre-NK cells. While not yet definitively proven, the dendritic cell, which also resides in the parafollicular T cell rich region of SLT, is the most likely source of membrane-bound IL-15 being presented in trans via the IL-15Rα, along with other antigen-presenting cells (Caligiuri, 2008). Interestingly, some CD34(+) pre-NK cells also express the heterotrimeric high-affinity IL-2R (Freud et al., 2005). As IL-2 is only expressed by activated T cells, it is tempting to speculate that at times of T cell activation within SLT, these CD34(+)IL-2R$\alpha\beta\gamma$(+) pre-NK cells can successfully compete for IL-2 derived from T cells, which in turn can drive NK cell development. (Caligiuri et al., 1990; Fehniger et al., 2003; Ferlazzo et al., 2004; Freud et al., 2005). The cytokines that continue to drive the pre-NK cell towards the iNK cell and the CD56^{bright} NK cell include IL-15 likely as both a survival factor and a maturation factor, but probably include other cytokines that have not yet been identified. It will be interesting to determine if any of the cytokines or other co-factors involved in the genesis of IL-22-producing T_H17 cells are involved in the differentiation of IL-22-producing iNK cells.

Gene regulation of NK cell development

The progression of NK cell development in mice and humans has many similarities. The majority of work on gene regulation in NK development has been done in mice via genetic disruption studies. The transcription factors (TF) of NK development can be placed into three groups: (1) TF with a role in generation of normal NK cell expansion, (2) TF with a role in NK cell differentiation and (3) TF with a role in NK cell effector functions (Vosshenrich et al., 2005) (Table 8.1). The first group includes Ikaros, Ets-1, VDUP-1, PU.1 and the Id proteins Id2 and Id3 (Lee et al., 2005; Vosshenrich et al., 2005). Ikaros knockout mice have drastically reduced numbers of NK cells (Boggs et al., 1998). In contrast to Ikaros, which has also been shown to cause deficiencies in B and T cell lineages, knockout of Ets-1 has been shown to specifically deplete NK cells (Barton et al., 1998; Georgopoulos et al., 1994). The mechanism of Ets-1 may be related to its critical role in the induction of the IL-15Rβ chain, which is required for signalling via the IL-15R (Ye et al., 2005). Further L-15 promotes the induction of Ets-1, inducing a positive feedback loop that would favour the NK cell lineage (Grund et al., 2005). Other TF such as Vitamin D3 upregulated protein-1 (VDUP-1) regulate NK cell development via the control of the IL-15Rβ chain (Lee et al., 2005). PU.1 is essential for the B and T cell lineage, and while it has a less dramatic effect on NK cell development, its early role in multi-lymphoid progenitor cell differentiation leads to lower mature NK cell numbers with decreased expression of IL-7Rα, Ly49 and decreased response to cytokines IL-2 and IL-12 (Colucci et al., 2001). The over-expression of several TF, most notably some of the Id proteins (i.e. Id2 and Id3), may actually delay B cell or T cell development by disproportionately driving NK cell development (Di Santo, 2006). In contrast, over-expression of E2A and HEB TF can disproportionately drive B and T cell development over NK cell development (Engel and Murre, 2001). The Id proteins cannot directly bind to DNA. However, Id proteins exhibit high-affinity binding to the NK lineage-inhibiting E proteins. It is likely that Id proteins bind to and inhibit the E proteins' NK cell-suppressing activity. A lack of release from E protein inhibition accounts for a disruption in Id protein expression leading to a decrease in NK cell differentiation (Boos et al., 2007).

The second group of TF include Gata-3, T-bet and IRF-2 (Vosshenrich et al., 2005) (Table 8.1). Gata-3 deficiency results in decreased maturation, decreased IFNγ production and decreased liver-specific NK cell homing (Colucci et al., 2001). T-bet knockout mice also show decreased maturity and lower IFNγ production but a more global deficiency in peripheral NK cells (Townsend et al., 2004). Mice deficient in IRF-2 show

Table 8.1 Important transcription factors for NK cell development: generation of normal NK numbers (*), promote NK maturity (**) and promote effector functions (***)

Transcription factor	Function	References
Ikaros*	Early multi-lymphoid progenitor factor, early IL-15Rβ promoter	Boggs et al., 1998
Ets-1*	Promotes early NK development, IL-15Rβ expression	Barton et al., 1998, Georgopoulos et al., 1994
VDUP-1*	Promotes early NK development, IL-15Rβ expression	Lee et al., 2005
E2A/EBF-1*	Promotes B cell lineage	Engel and Murre, 2001
HEB/HEBAlt*	Promotes T cell lineage	Engel and Murre, 2001
PU.1*	Early multi-lymphoid progenitor TF, essential for B cell development, NK expression of IL-7R, Ly49 and response to IL-2/15 and IL-12	Colucci et al., 2001
Id2/Id3*	Binds E proteins to drive NK lineage	Di Santo, 2006, Boos et al., 2007
Gata-3**	NK homing to liver, maturation, antagonize T-bet driven IFNγ	Colucci et al., 2001, Vosshenrich et al., 2005, Freud et al., 2006
T-Bet**	Promotes NK maturity, promotes IFNγ production to promotes IL-15Rβ	Townsend et al., 2004, Vosshenrich et al., 2005, Freud et al., 2006
IRF-2**	Promotes NK maturity, survival and generation of peripheral NK	Taki et al., 2005
MEF***	Promotes cytotoxicity and IFNγ production	Lacorazza et al., 2002
MITF***	Promotes IL-12R and IL-18R	Kataoka et al., 2005
CEBPγ***	Promotes cytotoxicity and IFNγ production downstream of IL-12 and IL-18 signalling	Kaisho et al., 1999

lower levels of IFNγ, deficits in peripheral maturity, while bone marrow NK cells show a relatively higher maturity. IRF-2 may play an anti-apoptotic role in late NK maturity (Taki et al., 2005).

The final group of TF includes Myeloid Elf Factor (MEF), microphthalmia-associated transcription factor (MITF), and CCAAT/enhancer binding protein-γ (CEBP-γ) (Vosshenrich et al., 2005) (Table 8.1). MEF has been shown to have an NK cell-specific role in perforin expression explaining killing defects for these cells (Lacorazza et al., 2002). CEBP-γ promotes IFNγ secretion downstream of IL-12 and IL-18 signalling (Kaisho et al., 1999). In contrast, MITF's role as a promoter of the IL-12R and IL-18R may explain the loss of NK cytolytic activity and IFNγ secretion in MITF null mice (Kataoka et al., 2005).

Limited work on TF in human NK cell development has focused on regulation of IFNγ production. Acquisition of the ability to secrete IFNγ appears in the CD56^{bright} NK cell population. The appearance of this property is under the control of the TF T-bet, which in turn appears to be controlled at least in part by GATA-3 (Vosshenrich et al., 2005). NK cells transitioning from the iNK cell stage to CD56^{bright} NK cell stage show a large drop in GATA-3, an increase in T-bet and a concomitant increase in IFN-γ production (Freud et al., 2006). T-bet expression is negatively regulated by TGF-β via SMAD2, SMAD3 and SMAD4 signalling (Yu et al., 2006).

NK education and regulation beyond final maturation

Killer immunoglobulin-like receptors (KIR) help NK cells discriminate between normal sell and target cells by recognition of MHC class I molecules (Colonna and Samaridis, 1995; Karlhofer et al., 2006; Wagtmann et al., 1995). Binding of MHC class I molecules to KIR inhibits NK cell activation by signalling through associated immuno-receptor tyrosine-based inhibitory motifs (ITIM). In the presence of a human MHC class I deficiency, NK cells are inactive (Furukawa et al., 1999; Vitale et al., 2002; Zimmer et al., 1998). Therefore, not only is MHC class I recognition important for inactivation of mature NK cells, but it is also important in the progression towards functional maturity. Two theories have been proposed for the role of KIR in the education or 'licensing' of the NK cell. The 'arming' or 'stimulatory' model attributes licensing directly to engagement of KIR (Anfossi et al., 2006). However, arguments against this theory note that licensing a NK cell requires an activation signal, whereas most KIR are inhibitory so these receptors would have to switch from activator to inhibitor during NK development. Considering the complexity of signalling pathways and the likely alternative intracellular environment of a developing versus a mature NK cell, this activator–inhibitor cell-signalling switch is plausible. The opposing 'disarming' or 'inhibitory' model describes an additional unknown activating signal that if left unopposed would over stimulate the unlicensed NK cell leading to anergy, thus necessitating a balance of a second, inhibitory signal from KIR-MHC class I engagement (Fernandez et al., 2005; Yokoyama and Kim, 2006).

The Ly49 receptors are analogous to KIR in humans. While structurally different, their functional signalling is remarkably similar (Long, 1999). Mice deficient in MHC class I have increased Ly49 and are not autoreactive but are hyporesponsive to MHC class I-deficient targets (Held et al., 1996; Liao et al., 1991; Salcedo et al., 1997). Indeed further work has shown mouse NK cells become licensed only after Ly49 receptor engagement of MHC class I ligands and signalling through the ITIM (Kim et al., 2005).

KIR expression and therefore NK cell education does not appear to occur before the CD56^{bright} NK stage (Freud et al., 2006), but this does not rule out KIR regulation beginning at an earlier stage of development. In addition, KIR expression patterns are not identical on all NK cells but highly variable both within and among individuals. In contrast, clonal human NK cell lines maintain identical patterns of KIR expression (Colonna and Samaridis, 1995). At the transcriptional level, DNA methylation is primarily responsible for determining which KIR are expressed. Active KIR transcription occurs at the hypomethylated CpG regions of the KIR cluster. Further, this activation can be monoallelic or biallelic, and other epigenetic mechanisms such as histone acetylation may play a role (Chan et al., 2003, 2005). The interplay that determines which KIR in this densely packed gene locus are hypomethylated and expressed has yet to be elucidated.

There is reasonable evidence that NK cell development proceeds from the CD56^{bright} NK cell to the CD56^{dim} NK cell (Chan et al., 2007; Ouyang et al., 2007; Romagnani et al., 2007; Takahashi et al., 2007); however, it is not entirely clear as to what factors drive this process. Multiple studies have shown IL-15 and IL-2 to drive NK cell cytotoxicity, a characteristic of the assumed more mature CD56^{dim} NK cell population (Freud and Caligiuri, 2006). However, little is known about the exact mechanism through which this transition takes place. Multiple factors are likely involved, as there is a wide variation in gene expression between these two populations (Wendt et al., 2006). It is important to note that while a dichotomy has been shown in the function of these two populations, neither of these functions has been attributed to the CD56 molecule itself. Therefore, it remains for now only a marker for these two subsets.

Conclusion

Substantial advances have been made in unlocking the development of human NK cells from the hematopoietic stem cell to the mature CD56^{dim} NK cell. Considering the important roles of NK cells in the innate immune response to infection, tumour surveillance and chronic inflammatory disease, further elucidation of the discrete steps of human NK cell maturation will allow a greater understanding of how such cells may be manipulated for the prevention and treatment of such conditions. Further, NK cells are not an entity of their own. They appear to be closely related to T cells and also share functions with other effector cells of the immune system.

Acknowledgements

The authors would like to thank Ani A. Avanian of http://www.aniavanian.com for design assistance and final illustrations.

References

Anfossi, N., Andre, P., Guia, S., Falk, C.S., Roetynck, S., Stewart, C.A., Breso, V., Frassati, C., Reviron, D., Middleton, D., Romagne, F., Ugolini, S. and Vivier, E. (2006). Human NK cell education by inhibitory receptors for I. Immunity **25**, 331–342.

Barton, K., Muthusamy, N., Fischer, C., Ting, C.N., Walunas, T.L., Lanier, L. L. and Leiden, J.M. (1998). The Ets-1 transcription factor is required for the development of natural killer cells in mice. Immunity **9**, 555–563.

Blom, B. and Spits, H. (2006). Development of human lymphoid cells. Annu Rev Immunol. **24**, 287–320.

Boggs, S.S., Trevisan, M., Patrene, K. and Geogopoulos, K. (1998). Lack of natural killer cell precursors in fetal liver of Ikaros knockout mutant mice. Nat Immun **16**, 137–145.

Boos, M.D., Yokota, Y., Eberl, G. and Kee, B.L. (2007). Mature natural killer cell and lymphoid tissue-inducing cell development requires Id2-mediated suppression of E protein activity. J Exp Med **204**, 1119–1130.

Brooks, A.G., Borrego, F., Posch, P. E., Patamawenu, A., Scorzelli, C.J., Ulbrecht, M., Weiss, E.H. and Coligan, J.E. (1999). Specific recognition of HLA-E, but not classical, HLA class I molecules by soluble CD94/NKG2A and NK cells. J Immunol **162**, 305–313.

Caligiuri, M.A. (2008). Human natural killer cells. Blood **112**, 461–469.

Caligiuri, M.A., Zmuidzinas, A., Manley, T.J., Levine, H., Smith, K.A. and Ritz, J. (1990). Functional consequences of interleukin 2 receptor expression on resting human lymphocytes. Identification of a novel natural killer cell subset with high affinity receptors. J Exp Med **171**, 1509–1526.

Campbell, J.J., Qin, S., Unutmaz, D., Soler, D., Murphy, K.E., Hodge, M.R., Wu, L. and Butcher, E.C. (2001). Unique subpopulations of CD56+ NK and NK-T peripheral blood lymphocytes identified by chemokine receptor expression repertoire. J Immunol **166**, 6477–6482.

Cella, M., Fuchs, A., Vermi, W., Facchetti, F., Otero, K., Lennerz, J.K., Doherty, J. M., Mills, J.C. and Colonna, M. (2008). A human natural killer cell subset provides an innate source of IL-22 for mucosal immunity. Nature.

Chan, H.W., Kurago, Z.B., Stewart, C.A., Wilson, M.J., Martin, M.P., Mace, B. E., Carrington, M., Trowsdale, J. and Lutz, C.T. (2003). DNA methylation maintains allele-specific KIR gene expression in human natural killer cells. J Exp Med **197**, 245–255.

Chan, H.W., Miller, J.S., Moore, M.B. and Lutz, C.T. (2005). Epigenetic control of highly homologous killer Ig-like receptor gene alleles. J Immunol **175**, 5966–5974.

Chan, A., Hong, D.L., Atzberger, A., Kollnberger, S., Filer, A.D., Buckley, C.D., McMichael, A., Enver, T. and Bowness, P. (2007). CD56bright human NK cells differentiate into CD56dim cells: role of contact with peripheral fibroblasts. J Immunol **179**, 89–94.

Colonna, M. and Samaridis, J. (1995). Cloning of immunoglobulin-superfamily members associated with HLA-C and HLA-B recognition by human natural killer cells. Science **268**, 405–408.

Colucci, F., Samson, S.I., DeKoter, R.P., Lantz, O., Singh, H. and Di Santo, J. P. (2001). Differential requirement for the transcription factor PU.1 in the generation of natural killer cells versus B and T cells. Blood **97**, 2625–2632.

Cooper, M.A., Fehniger, T.A. and Caligiuri, M.A. (2001a). The biology of human natural killer-cell subsets. Trends Immunol **22**, 633–640.

Cooper, M.A., Fehniger, T.A., Turner, S.C., Chen, K.S., Ghaheri, B.A., Ghayur, T., Carson, W.E. and Caligiuri, M.A. (2001b). Human natural killer cells: a unique innate immunoregulatory role for the CD56(bright) subset. Blood **97**, 3146–3151.

Cooper, M.A., Elliott, J.M., Keyel, P.A., Yang, L., Carrero, J.A. and Yokoyama, W. M. (2009). Cytokine-induced memory-like natural killer cells. Proc Natl Acad Sci USA **106**, 1915–1919.

Cupedo, T., Crellin, N.K., Papazian, N., Rombouts, E.J., Weijer, K., Grogan, J.L., Fibbe, W.E., Cornelissen, J.J. and Spits, H. (2009). Human fetal lymphoid tissue-inducer cells are interleukin 17-producing precursors to RORC+ CD127+ natural killer-like cells. Nat Immunol **10**, 66–74.

Di Santo, J.P. (2006). Natural killer cell developmental pathways: a question of balance. Annu Rev Immunol **24**, 257–286.

Dorman, S.E., Picard, C., Lammas, D., Heyne, K., van Dissel, J.T., Baretto, R., Rosenzweig, S.D., Newport, M., Levin, M., Roesler, J., Kumararatne, D., Casanova, J.L. and Holland, S.M. (2004). Clinical features of dominant and recessive interferon gamma receptor 1 deficiencies. Lancet **364**, 2113–2121.

Edgar, J.D. (2008). T cell immunodeficiency. J Clin Pathol **61**, 988–993.

Engel, I. and Murre, C. (2001). The function of E- and Id proteins in lymphocyte development. Nat Rev Immunol **1**, 193–199.

Fehniger, T.A., Cooper, M.A., Nuovo, G. J., Cella, M., Facchetti, F., Colonna, M. and Caligiuri, M.A. (2003). CD56^{bright} natural killer cells are present in human lymph nodes and are activated by T cell-derived IL-2: a potential new link between adaptive and innate immunity. Blood **101**, 3052–3057.

Ferlazzo, G., Thomas, D., Lin, S.L., Goodman, K., Morandi, B., Muller, W.A., Moretta, A. and Munz, C. (2004). The abundant NK cells in human secondary lymphoid tissues require activation to express killer cell Ig-like receptors and become cytolytic. J Immunol **172**, 1455–1462.

Fernandez, N.C., Treiner, E., Vance, R.E., Jamieson, A.M., Lemieux, S. and Raulet, D.H. (2005). A subset of natural killer cells achieves self-tolerance without expressing inhibitory receptors specific for self-MHC molecules. Blood **105**, 4416–4423.

Freud, A.G. and Caligiuri, M.A. (2006). Human natural killer cell development. Immunol Rev **214**, 56–72.

Freud, A.G., Becknell, B., Roychowdhury, S., Mao, H.C., Ferketich, A.K., Nuovo, G.J., Hughes, T.L., Marburger, T.B., Sung, J., Baiocchi, R.A., Guimond, M. and Caligiuri, M.A. (2005). A human CD34(+) subset resides in lymph nodes and differentiates into CD56bright natural killer cells. Immunity **22**, 295–304.

Freud, A.G., Yokohama, A., Becknell, B., Lee, M.T., Mao, H.C., Ferketich, A.K. and Caligiuri, M.A. (2006). Evidence for discrete stages of human natural killer cell differentiation in vivo. J Exp Med **203**, 1033–1043.

Frey, M., Packianathan, N.B., Fehniger, T.A., Ross, M.E., Wang, W.C., Stewart, C.C., Caligiuri, M.A. and Evans, S.S. (1998). Differential expression and function of L-selectin on $CD56^{bright}$ and $CD56^{dim}$ natural killer cell subsets. J Immunol **161**, 400–408.

Furukawa, H., Yabe, T., Watanabe, K., Miyamoto, R., Miki, A., Akaza, T., Tadokoro, K., Tohma, S., Inoue, T., Yamamoto, K. and Juji, T. (1999). Tolerance of NK and LAK activity for HLA class I-deficient targets in a TAP1-deficient patient (bare lymphocyte syndrome type I). Hum Immunol **60**, 32–40.

Galy, A., Travis, M., Cen, D. and Chen, B. (1995). Human T, B, natural killer, and dendritic cells arise from a common bone marrow progenitor cell subset. Immunity **3**, 459–473.

Georgopoulos, K., Bigby, M., Wang, J.H., Molnar, A., Wu, P., Winandy, S. and Sharpe, A. (1994). The Ikaros gene is required for the development of all lymphoid lineages. Cell **79**, 143–156.

Godin, I. and Cumano, A. (2002). The hare and the tortoise: an embryonic haematopoietic race. Nat Rev Immunol **2**, 593–604.

Grund, E.M., Spyropoulos, D.D., Watson, D.K. and Muise-Helmericks, R.C. (2005). Interleukins 2 and 15 regulate Ets1 expression via ERK1/2 and MNK1 in human natural killer cells. J Biol Chem **280**, 4772–4778.

Hacein-Bey-Abina, S., Le Deist, F., Carlier, F., Bouneaud, C., Hue, C., De Villartay, J.P., Thrasher, A.J., Wulffraat, N., Sorensen, R., Dupuis-Girod, S., Fischer, A., Davies, E.G., Kuis, W., Leiva, L. and Cavazzana-Calvo, M. (2002). Sustained correction of X-linked severe combined immunodeficiency by ex vivo gene therapy. N Engl J Med **346**, 1185–1193.

Held, W., Dorfman, J.R., Wu, M.F. and Raulet, D.H. (1996). Major histocompatibility complex class I-dependent skewing of the natural killer cell Ly49 receptor repertoire. Eur J Immunol **26**, 2286–2292.

Hughes, T., Becknell, B., McClory, S., Briercheck, E., Freud, A.G., Zhang, X., Mao, H., Nuovo, G., Yu, J. and Caligiuri, M.A. (2009). Stage three immature human natural killer cells found in secondary lymphoid tissue constitutively and selectively express the TH17 cytokine interleukin-22. Blood.

Kaisho, T., Tsutsui, H., Tanaka, T., Tsujimura, T., Takeda, K., Kawai, T., Yoshida, N., Nakanishi, K. and Akira, S. (1999). Impairment of natural killer cytotoxic activity and interferon gamma production in CCAAT/enhancer binding protein gamma-deficient mice. J Exp Med **190**, 1573–1582.

Karlhofer, F.M., Ribaudo, R.K. and Yokoyama, W.M. (2006). MHC Class I alloantigen specificity of Ly-49+ IL-2-activated natural killer cells. J Immunol **177**, 5761–5765 Nature **358**, 66–70, 1992.

Kataoka, T.R., Komazawa, N., Oboki, K., Morii, E. and Nakano, T. (2005). Reduced expression of IL-12 receptor beta2 and IL-18 receptor alpha genes in natural killer cells and macrophages derived from B6-mi/mi mice. Lab Invest **85**, 146–153.

Kim, S., Iizuka, K., Kang, H.S., Dokun, A., French, A.R., Greco, S. and Yokoyama, W.M. (2002). In vivo developmental stages in murine natural killer cell maturation. Nat Immunol **3**, 523–528.

Kim, S., Poursine-Laurent, J., Truscott, S.M., Lybarger, L., Song, Y.J., Yang, L., French, A.R., Sunwoo, J.B., Lemieux, S., Hansen, T.H. and Yokoyama, W.M. (2005). Licensing of natural killer cells by host major histocompatibility complex class I molecules. Nature **436**, 709–713.

Lacorazza, H.D., Miyazaki, Y., Di Cristofano, A., Deblasio, A., Hedvat, C., Zhang, J., Cordon-Cardo, C., Mao, S., Pandolfi, P.P. and Nimer, S.D. (2002). The ETS protein MEF plays a critical role in perforin gene expression and the development of natural killer and NK-T cells. Immunity **17**, 437–449.

Lazetic, S., Chang, C., Houchins, J.P., Lanier, L.L. and Phillips, J.H. (1996). Human natural killer cell receptors involved in MHC class I recognition are disulfide-linked heterodimers of CD94 and NKG2 subunits. J Immunol **157**, 4741–4745.

LeBien, T.W. and Tedder, T.F. (2008). B lymphocytes: how they develop and function. Blood **112**, 1570–1580.

Lee, K.N., Kang, H.S., Jeon, J.H., Kim, E.M., Yoon, S.R., Song, H., Lyu, C.Y., Piao, Z.H., Kim, S.U., Han, Y.H., Song, S.S., Lee, Y.H., Song, K.S., Kim, Y.M., Yu, D.Y. and Choi, I. (2005). VDUP1 is required for the development of natural killer cells. Immunity **22**, 195–208.

Liang, S.C., Tan, X.Y., Luxenberg, D.P., Karim, R., Dunussi-Joannopoulos, K., Collins, M. and Fouser, L.A. (2006). Interleukin (IL)-22 and IL-17 are coexpressed by Th17 cells and cooperatively enhance expression of antimicrobial peptides. J Exp Med **203**, 2271–2279.

Liao, N.S., Bix, M., Zijlstra, M., Jaenisch, R. and Raulet, D. (1991). MHC class I deficiency: susceptibility to natural killer (NK) cells and impaired NK activity. Science **253**, 199–202.

Long, E.O. (1999). Regulation of immune responses through inhibitory receptors. Annu Rev Immunol **17**, 875–904.

Loza, M.J. and Perussia, B. (2001). Final steps of natural killer cell maturation: a model for type 1-type 2 differentiation? Nat Immunol **2**, 917–924.

Loza, M.J., Zamai, L., Azzoni, L., Rosati, E. and Perussia, B. (2002). Expression of type 1 (interferon gamma) and type 2 (interleukin-13, interleukin-5) cytokines at distinct stages of natural killer cell differentiation from progenitor cells. Blood **99**, 1273–1281.

Orange, J.S. (2006). Human natural killer cell deficiencies. Curr Opin Allergy Clin Immunol **6**, 399–409.

Ouyang, Q., Baerlocher, G., Vulto, I. and Lansdorp, P.M. (2007). Telomere length in human natural killer cell subsets. Ann NY Acad Sci **1106**, 240–252.

Phillips, J.H., Chang, C., Mattson, J., Gumperz, J.E., Parham, P. and Lanier, L. L. (1996). CD94 and a novel associated protein (94AP) form a NK cell receptor involved in the recognition of HLA-A,

HLA-B, and HLA-C allotypes. Immunity **5**, 163–172.

Romagnani, C., Juelke, K., Falco, M., Morandi, B., D'Agostino, A., Costa, R., Ratto, G., Forte, G., Carrega, P., Lui, G., Conte, R., Strowig, T., Moretta, A., Munz, C., Thiel, A., Moretta, L. and Ferlazzo, G. (2007). CD56brightCD16- killer Ig-like receptor- NK cells display longer telomeres and acquire features of CD56dim NK cells upon activation. J Immunol **178**, 4947–4955.

Salcedo, M., Diehl, A.D., Olsson-Alheim, M.Y., Sundback, J., Van Kaer, L., Karre, K. and Ljunggren, H.G. (1997). Altered expression of Ly49 inhibitory receptors on natural killer cells from MHC class I-deficient mice. J Immunol **158**, 3174–3180.

Stehling-Sun, S., Dade, J., Nutt, S.L., DeKoter, R.P. and Camargo, F.D. (2009). Regulation of lymphoid versus myeloid fate 'choice' by the transcription factor Mef2c. Nat Immunol **10**, 289–296.

Sun, J.C., Beilke, J.N. and Lanier, L.L. (2009). Adaptive immune features of natural killer cells. Nature **457**, 557–561.

Takahashi, E., Kuranaga, N., Satoh, K., Habu, Y., Shinomiya, N., Asano, T., Seki, S. and Hayakawa, M. (2007). Induction of CD16+ CD56bright NK cells with antitumour cytotoxicity not only from CD16- CD56bright NK Cells but also from CD16- CD56dim NK cells. Scand J Immunol **65**, 126–138.

Taki, S., Nakajima, S., Ichikawa, E., Saito, T. and Hida, S. (2005). IFN regulatory factor-2 deficiency revealed a novel checkpoint critical for the generation of peripheral NK cells. J Immunol **174**, 6005–6012.

Townsend, M.J., Weinmann, A.S., Matsuda, J.L., Salomon, R., Farnham, P.J., Biron, C.A., Gapin, L. and Glimcher, L.H. (2004). T-bet regulates the terminal maturation and homeostasis of NK and Valpha14i NKT cells. Immunity **20**, 477–494.

Vitale, M., Zimmer, J., Castriconi, R., Hanau, D., Donato, L., Bottino, C., Moretta, L., de la Salle, H. and Moretta, A. (2002). Analysis of natural killer cells in TAP2-deficient patients: expression of functional triggering receptors and evidence for the existence of inhibitory receptor(s) that prevent lysis of normal autologous cells. Blood **99**, 1723–1729.

Vosshenrich, C.A., Samson-Villeger, S.I. and Di Santo, J.P. (2005). Distinguishing features of developing natural killer cells. Curr Opin Immunol **17**, 151–158.

Wagtmann, N., Biassoni, R., Cantoni, C., Verdiani, S., Malnati, M.S., Vitale, M., Bottino, C., Moretta, L., Moretta, A. and Long, E.O. (1995). Molecular clones of the p58 NK cell receptor reveal immunoglobulin-related molecules with diversity in both the extra- and intracellular domains. Immunity **2**, 439–449.

Wendt, K., Wilk, E., Buyny, S., Buer, J., Schmidt, R.E. and Jacobs, R. (2006). Gene and protein characteristics reflect functional diversity of CD56dim and CD56bright NK cells. J Leukoc Biol **80**, 1529–1541.

Wolk, K., Kunz, S., Witte, E., Friedrich, M., Asadullah, K. and Sabat, R. (2004). IL-22 increases the innate immunity of tissues. Immunity **21**, 241–254.

Ye, S.K., Kim, T.J., Won, S.S., Yoon, T.J., Park, T.K., Yoo, Y.C., Kim, Y.N., Lee, H. C., Ikuta, K., Chung, M.H. and Lee, K. H. (2005). Transcriptional regulation of the mouse interleukin-2 receptor beta chain gene by Ets and Egr-1. Biochem Biophys Res Commun **329**, 1094–1101.

Yokoyama, W.M. and Kim, S. (2006). How do natural killer cells find self to achieve tolerance?. Immunity **24**, 249–257.

Yu , J., Wei, M., Becknell, B., Trotta, R., Liu, S., Boyd, Z., Jaung, M.S., Blaser, B.W., Sun, J., Benson, D.M., Mao, H., Yokohama, A., Bhatt, D., Shen, L., Davuluri, R., Weinstein, M., Marcucci, G. and Caligiuri, M.A. Jr. (2006). Pro- and antiinflammatory cytokine signaling: reciprocal antagonism regulates interferon-gamma production by human natural killer cells. Immunity **24**, 575–590.

Zamai, L., Ahmad, M., Bennett, I.M., Azzoni, L., Alnemri, E.S. and Perussia, B. (1998). Natural killer (NK) cell-mediated cytotoxicity: differential use of TRAIL and Fas ligand by immature and mature primary human NK cells. J Exp Med **188**, 2375–2380.

Zimmer, J., Donato, L., Hanau, D., Cazenave, J.P., Tongio, M.M., Moretta, A. and de la Salle, H. (1998). Activity and phenotype of natural killer cells in peptide transporter (TAP)-deficient patients (type I bare lymphocyte syndrome). J Exp Med **187**, 117–122.

PART II

NK interactions with other cells

Chapter Nine

9

Isolation, culture and propagation of natural killer cells

Michael Magee, Lazar N. Vujanovic, Lisa H. Butterfield, Nikola L. Vujanovic

CHAPTER CONTENTS

The beginning of knowledge is the discovery of something we do not understand.

Frank Herbert

ABSTRACT

Natural killer (NK) cells are essential mediators of the central innate immune functions such as spontaneous killing of infected and transformed cells, inflammation, engraftment of hematopoetic stem cells and regulation of immune function. NK cells have been used to treat viral infections and malignancies as well as to increase engraftment of bone marrow stem cell transplants in experimental animal models and humans. NK cells represent a relatively small population of cells that is mixed with other populations of immune cells within the immune system. In order to study or utilize NK cells in therapy, they are often isolated from other immune cells, activated, and expanded. Based on their specific buoyant, phenotypic and functional characteristics, several effective ex vivo techniques for isolation and expansion of NK cells have been developed and successfully utilized. Here we review the most frequently used techniques for isolation and expansion of NK cells, and provide detailed protocols for purification of NK cells using magnet activated cell sorting (MACS) and separation and expansion of IL-2-activated adherent NK (A-NK) cells.

KEY WORDS

Natural killer (NK) cells, Isolation, Purification, Activation, Culture, Expansion

Introduction

Natural killer (NK) cells are essential effector cells of the innate immune system, which rapidly recognize and directly destroy virally infected and transformed cells, mediate inflammation, and regulate innate and adaptive immune function via interaction with other immune cells (Andrews et al., 2003; Banchereau and Steinman, 1998; Farag et al., 2002; Ferlazzo et al., 2002; Fernandez et al., 1999; Gerosa et al., 2002; Goodier and Londei, 2000; Jinushi et al., 2003; Kapsenberg, 2003; Lanier and Phillips, 1992; Maillard et al., 2003; Moretta, 2002; Moretta et al., 2003; Osada et al., 2001; Piccioli et al., 2002; Poggi et al., 2002; Robertson and Ritz, 1990; Ruggeri et al., 2002; Trinchieri, 1989; Yu et al., 2001). NK cells represent one of the first lines of host immune defense against viruses and cancer as well as central regulators of immune function (Moretta 2002; Moretta et al., 2003; Ruggeri et al., 2002). NK cells have a remarkable ability to increase the engraftment of hematopoietic stem cells and to respond to cytokines such as IL-2 by activation, expansion and increases in antiviral and antitumor activity (Moretta 2002; Moretta et al., 2003; Ruggeri et al., 2002). In vivo manipulation or in vitro purification and/or activation followed by adoptive transfer of NK cells is used to increase engraftment of bone marrow stem cell transplants, and to treat viral infections and malignancies in experimental animal models and humans (Farag et al., 2003; McKenna et al., 2007; Moretta 2002; Moretta et al., 2003; Passweg et al., 2006; Ruggeri et al., 2002; Vujanovic and Whiteside, 1995; Vujanovic et al., 1996a,b). Within the immune system, NK cells represent a relatively small population of cells admixed with other populations of immune cells, and comprise 5–15% of human peripheral blood mononuclear leucocytes (PBMNL) and 1–3% of murine splenocytes. NK cells may be conveniently isolated to high purity and sufficient numbers and/or be activated and expanded in vitro. Because NK cells are present in the blood and tissues as a small population of cells admixed with other immune cells, these tasks are not easy. Here we review the most commonly used methods for isolation, activation and expansion of NK cells.

Human NK cells are phenotypically defined by their cell surface expression of CD56 (neural-cell adhesion molecule, N-CAM), CD161 (NKR-P1A), CD122 (β chain of IL-2/15R complex), CD132 (common γ chain of IL-2R complex, γc), CD16 (FcγRIII, low affinity receptor for IgG), killer cell Ig-like or inhibitory receptors (KIR), killer cell activating receptors (KAR), and by the absence of antigen-specific receptors and lineage markers of T cells, B cells or monocytes/macrophages. In addition, NK cells contain lysosomal granules in their cytoplasm and have large granular lymphocyte (LGL) morphology (Moretta, 2002; Moretta et al., 2003; Pilbeam et al., 2007; Ruggeri et al., 2002). Human NK cells are heterogeneous and consist of two major subsets. As for human peripheral blood NK cells, 90–95% are **CD56^{low}CD16^{+}CD161^{+}**IL-2R$\beta\gamma^{+}$ cytotoxic cells that poorly proliferate and produce immunoregulatory cytokines in the presence of high concentrations (6000 international units [IU] or 22 nM) of IL-2. The remaining human peripheral blood NK cells are **CD56^{high}CD16^{-}**IL-2R$\beta\gamma\alpha^{+}$, lack cytotoxic activity, are highly proliferative in the presence of low concentration (6–60 IU, 0.2–2 nM) of IL-2, and produce high levels of immunoregulatory cytokines such as IFN-γ, TNF and GM-CSF (Pilbeam et al., 2007). An additional subset of human peripheral blood NK cells, named pre-A-NK cells, has been defined in our laboratory on the basis of selective expression of the N-CAM isoform **ANK-1**, high levels of CAMs, adhesiveness and proliferation in response to high concentration (22 nM) of IL-2, with the ability to penetrate and kill cancer cells in solid tumour tissues (Pilbeam et al., 2007). In contrast to human NK cells, mouse NK cells do not express CD56 and ANK-1 and thus appear less heterogeneous. However, murine NK cells have other similar phenotypic and functional characteristics to human NK cells. Murine NK cells are CD3^{-}TCR^{-}CD14^{-}CD19^{-}CD16^{+} **CD161(NK1)$^{+}$CD49b(DX5)$^{+}$**IL-2R$\beta\gamma^{+}$ LGL, have cytotoxic activity, respond to high concentration (22 nM) of IL-2 with activation, proliferation and development of adhesive properties, and produce immunoregulatory cytokines (Pilbeam et al., 2007). CD161 is expressed on both immature and mature murine NK cells, and is considered a good marker of NK cells. However, it appears in two isoforms, NK1.1 and NK1.2, which are selectively expressed in various mouse strains. As only anti-NK1.1, but not anti-NK1.2, antibodies are commercially available, CD161 can serve as a specific marker of NK cells only in mouse strains whose NK cells selectively express NK1.1 (i.e. C57BL/6), but not in mouse strains in which their NK cells selectively express NK1.2 (i.e. BALB/c). In contrast to CD161, CD49b is expressed on mature NK cells of all mouse strains and therefore is a better marker of mouse NK cells than NK1.1 (Pilbeam et al., 2007).

Isolation of NK cells

Based on their specific buoyant, phenotypic and functional characteristics, several techniques have been developed and successfully used for isolation of NK cells. The most frequently used techniques include Percoll density gradient centrifugation (Reynolds et al., 1981; Timonen et al., 1981), panning (Cosentino and Cathcart, 1987; Ritson and Bulmer, 1989), sheep red blood cell rosetting (Cooper and Caligiuri, 2004), fluorescence activated cell sorting (FACS) (Cooper and Caligiuri, 2004), magnet activated cell sorting (MACS) (Farace et al., 1992; Miltenyi et al., 1990; Vujanovic and Whiteside, 1995), and IL-2-induced cell adherence selection (Gunji et al., 1989; Melder et al., 1988; Rabinowich et al., 1991; Vujanovic et al., 1988, 1993).

Percoll density gradient centrifugation

When NK cells were initially defined as 'LGL' and before the ready availability of specific NK-cell surface markers, the Percoll density gradient centrifugation technique was developed and successfully applied to purification of NK cells in both humans (Timonen et al., 1981) and rats (Reynolds et al., 1981). This technique is based on the low buoyant density of LGL that enables separation of NK cells from the higher buoyant density other immune cells by centrifugation on Percoll multilayer density gradients. It can be used to isolate human or rat peripheral blood LGL up to 90% purity. However, it is poorly efficient in purification of NK cells from murine splenic cell populations (unpublished data). Because monocytes and macrophages have a similar density to LGL, they can be the primary contaminating cell population in Percoll isolated populations of NK cells, and it is recommended as a first step in Percoll separation of NK cells to eliminate monocytes and macrophages by plastic-surface or nylon-wool adherence. The Percoll technique is inexpensive and relatively simple. However, it is technically very demanding, only yielding consistently good results in specialized and experienced laboratories, and is only efficient in isolation of peripheral blood NK cells.

Panning

When cell surface differentiation markers were discovered and specific antibodies for different populations of immune cells became available, panning techniques were introduced for isolation of NK cells. Both positive selection techniques, which are based on labelling of NK cells with specific antibodies, and negative selection techniques, which are based on labelling of other immune cells, but not NK cells, have been widely tested (Cosentino and Cathcart, 1987; Ritson and Bulmer, 1989). However, there are only a few reports that suggest significant efficiency of the method. Among the more successful methods is a multi-step negative selection (Cosentino and Cathcart, 1987), which has been described for isolation of human peripheral blood NK cells. This method consists of three consecutive steps: (1) depletion of B cells and monocytes by a nylon wool column, (2) depletion of T cells that form rosettes with sheep red blood cells at 29°C by centrifugation on Ficoll-Hypaque density gradient and (3) depletion of the remaining CD3$^+$ T cells by panning. The panning step is performed by labelling T cells with a mouse anti-human CD3 monoclonal antibody and incubation of cells in a plastic dish which surface is coated with polyclonal anti-mouse Ig antibodies. Using this method, the authors reportedly obtained 70–80% CD16$^+$ NK cells that were contaminated with only 1–3% CD3$^+$ T cells and 5–7% Ig$^+$ B cells. In their hands, the method was significantly better than Percoll density gradient centrifugation, yielding higher recovery (30% vs. 7%), purity and cytotoxic activity of NK cells. Unfortunately, similar to the Percoll density gradient technique, panning techniques have not been consistently successful in obtaining high purity NK cells.

RosetteSep

Recent studies have led to the development of high affinity monoclonal antibodies specific for cell differentiation antigens of NK cells and other populations of immune cells. Based on this availability, more simple and straightforward techniques that effectively enrich or purify NK cells have been developed. One such technique is 'RosetteSep' from Stem Cell Technologies (Cooper and Caligiuri, 2004). RosetteSep is a rapid technique for negative selection of NK cells that can enrich NK cells directly from whole human peripheral blood or mouse spleen cell suspensions using a simple one step procedure. It is based on the cocktail of bispecific tetrameric complexes of monoclonal antibodies specific to CD3 (T cells), CD4 (T cells and some dendritic cells [DCs]), CD19 (B cells), CD36 (monocytes, macrophages), CD66b (granulocytes), and glycophorin A (erythrocytes). The tetrameric complex cocktail is added to cell suspensions containing mixtures of leucocytes and erythrocytes. The antibody complexes bind to both erythrocytes and leucocytes making high density rosettes of unwanted mononuclear cells and erythrocytes. NK cells are separated from the rosettes of other cells by Ficoll-Hypaque gradient centrifugation. Reportedly, the method provides ~44% recovery and ≥70% purity of CD56$^+$CD3$^-$ cells from human peripheral blood and can be used in experiments when high numbers, but not highest purity, of NK cells is required, or as an initial step in obtaining high purity NK cells by cell sorting. Although the similar reagents are available for purification of murine splenic NK cells, the effectiveness of the method has not been fully validated yet.

Flow cytometric separation

Flow cytometric separation is the method of choice for obtaining highly pure populations of NK cells and their major subsets (Cooper and Caligiuri, 2004). FACS is usually used to perform positive selection of NK cells. Most mature human NK cells are phenotypically CD56$^+$CD3$^-$. To purify NK cells, PBMNL are isolated by Ficoll-Hypaque gradient centrifugation of whole blood, stained with anti-CD56 and anti-CD3 antibodies that are conjugated with various fluorophores, and CD56$^+$CD3$^-$ cells are sorted using FACS. Using a similar method, CD56^{high} and CD56^{low} NK-cell subsets can be also sorted. Similarly, NK-1.1 or CD49b and CD3 specific antibodies, conjugated with different fluorophores, can be used to purify mouse NK cells by flow cytometry. With this method, NK cells and their subsets are typically ≥96% pure. This purification of NK cells is an expensive and time consuming method that requires expensive equipment and specialized expertise. It can also lead to significant loss and low yield of NK cells. An additional serious concern regarding positive FACS purification of NK cells is the potential functional alteration of NK cells due to the presence of fluorochrome-conjugated antibodies on the surface of NK cells and both their exposure to laser beam and interaction with CD16 (low affinity FcγRIII) receptor. To increase the yield and decrease the duration of the cell sorting procedure, it is recommended that an initial enrichment step be performed with the NK cells before sorting. This can be achieved using other available simpler, faster and less expensive methods for isolation of NK cells, such as negative selection by 'RosetteSep' or 'MACS' kits. In addition, higher speed flow cytometric instruments, such as MoFlo Cytomation (Fort Collins, CO), that are now available can be used to obtain high purity NK cells and their subsets in a relatively short period of time.

Magnet activated cell sorting

MACS is a highly efficient method for isolation of NK cells to high purity, numbers and functionality (Vujanovic and Whiteside, 1995; Farace et al., 1992; Miltenyi et al., 1990). It is based on the labelling of cells of interest using antibody-paramagnetic bead complexes and direct isolation or elimination of the labelled cells by their exposure to a strong magnetic field. Both positive and negative selection of NK cells from human peripheral blood or rodent spleens can be performed using MACS. It has been in use for more than 20 years and has evolved into a simple, well characterized, precise and highly reproducible method, due to the development of high affinity antibodies, improvements of paramagnetic particles, improvements of magnets, development of magnetizing columns, and definition and standardization of buffers and conditions for optimal cell isolation. Our laboratories have continuously used, for 18 years, MACS for purification of human peripheral blood and/or mouse splenic NK cells. We have the most frequently applied the negative selection methods, to study phenotype and function of NK cells, without the concern of possible effects of antibody labelling (Kashii et al., 1999; Li et al., 2004; Vitolo et al., 1993; Vujanovic et al., 1993, 1995, 1996a; Wahlberg et al., 2001; Xu et al., 2007).

Purification of fresh mouse splenic NK cells by negative selection

Materials and equipment

There are several strategies to obtain purified murine NK cells, but the following procedure has been used successfully. In a suspension of lymphoid cells such as splenocytes, non-NK cells (T cells, B cells, macrophages, DCs, myeloid and erythroid cells) are labelled with biotin-conjugated monoclonal antibodies and then with paramagnetic microbeads conjugated with anti-biotin monoclonal antibody. The cells are passed through LS magnetizing column while exposed to a strong magnetic field. The magnetizing column retains non-NK cells labelled with paramagnetic microbeads, and unlabelled purified NK cells pass through the column.

It is important that all materials and equipments are cooled at 0 or 4°C and/or sterile.

- Freshly obtained spleen from a C57BL/6 mouse
- Cocktail of biotin-conjugated monoclonal antibodies against CD4, CD5, CD8α, CD19, Gr-1 and Ter-119 (Miltenyi Biotec Co. kit catalogue #120-001-501) for depletion of T cells, DCs, B cells, granulocytes, macrophages and erythroid cells
- Micro-beads conjugated to monoclonal anti-biotin antibody (Miltenyi Biotec Co. kit catalogue #120-000-900)
- Phosphate buffered saline (PBS) pH 7.2, supplemented with 10% foetal calf serum (FCS) and 2 mM EDTA (PBS-10)
- Degassed PBS pH 7.2, supplemented with 1% FCS and 2 mM EDTA (PBS-1)
- RPMI-1640 medium supplemented with 10% FCS (RPMI-10)
- Cell strainer with 70 μm pore size (BD Falcon, catalogue #352350)
- MACS pre-separation filter with 30 μm pore size (Milteneyi Biotec Co., catalogue #130-041-407)
- Magnetizing LS columns (Miltenyi Biotec Co. catalogue #120-000-475)
- MidiMACS or QuadroMACS magnet (Miltenyi Biotec Co.)

(1) All procedures must be performed on ice at (0°C) or in refrigerator (at 4°C) and under sterile conditions. (2) Splenocytes must be highly viable (≥95%), as dead cells can nonspecifically bind to paramagnetic beads and interfere with specific binding of antibody/paramagnetic bead-labelled cells preventing their elimination and resulting in low purity of NK cells. (3) It is also critically important to work with well degassed MACS PBS-1 buffer, as gas bubbles can block the column and drastically decrease purification efficiency. Efficient degassing of PBS-1 is performed using 50 ml Millipore Steriflip filter tubes (cat # SCGP00525), by first filtering the buffer and then exposing it to negative pressure for an extended period of time (≥30min).

Preparation of single cell suspension of viable splenocytes

Place a spleen immersed in 0.5 ml of PBS-10 on an ice-cooled cell strainer. Mince the spleen with scissors. Gently and repeatedly press the spleen tissue in circular motions with a 5 ml syringe plug. Continuously rinse the cell strainer with PBS-10 and collect the produced cell suspension in a 50 ml tube placed on ice. When only the white colour-associated splenic connective tissue, and no red colour-associated lymphoid tissue is left on the strainer, the preparation of cell suspension is completed. Filter the cell suspension through a pre-separation filter to remove cell clumps, and collect cells in a 14 ml polypropylene round bottom tube. Centrifuge the cells for 10 min on 300*g*. Carefully remove the supernatant and completely re-suspend the cell pellet by vortexing. Lyse the erythrocytes by adding to the cell suspension of 900 μl of ice-cooled distilled water and gently vortexing for exactly 10 s. Immediately reconstitute in the cell suspension the physiological salt concentration by adding 100 μl of ice-cooled 10× PBS. Add 4 ml of RPMI-10, gently vortex and filter the cell suspension through a pre-separation filter into a 14 ml round-bottom tube, to remove cell clumps and debris. Count the cells resuspended in trypan blue solution on a light microscope, to determine their number and viability. Centrifuge cells for 5 min at 300*g*. Wash cells two times in 5 ml PBS-1 using 5-minute centrifugations at 300*g*. Count cells carefully before the last centrifugation, to determine the number of cells to be used for purification of NK cells.

Although the original Miltenyi protocol for purification of NK cells suggests not to remove erythrocytes as the biotin-antibody cocktail contains an antibody specific for erythrocytes, we favour the removal of erythrocytes because they are a large population of unwanted cells in suspension of splenocytes that can compete with other unwanted cells for space in magnetizing column, and can add to the saturation of column and lead to inefficient NK cell purification. In addition, the concentration of wanted NK cells is increased, while total number of cells is decreased by removal of erythrocytes. Therefore, the removal of erythrocytes from a suspension of splenocytes enables the use of smaller amounts of reagents and fewer columns, and results in a more efficient purification and better purity of NK cells. However, the removal of erythrocytes must be performed very carefully so that the cell viability remains high.

Labelling of unwanted cells with antibodies and paramagnetic beads

After washing splenocytes in PBS-1, remove completely the supernatant using a transfer pipette. Completely resuspend the cell pellet by vortexing. Add 400 μl of PBS-1 and 100 μl of biotin-antibody cocktail per 10^8 splenocytes. Mix by gently vortexing, and incubate the suspension of cells and antibodies at 4°C for exactly 10 min. Add 300 μl of PBS-1 and 200 μl of anti-biotin antibody beads. Gently vortex and incubate the cell-antibody-paramagnetic bead suspension at 4°C for exactly 15 min. During this incubation, mix gently by hand the suspension every 5 min. Wash the cells with 14 ml of PBS-1 using centrifugation at 300*g* for 10 min. Remove completely the supernatant using transfer pipette. From now on, make sure that cell suspension does not contain any air bubbles or cell clumps as they can jam the magnetizing column. Resuspend completely the cell pellet by continuous gentle vortexing. Add carefully to the cell suspension 1 ml of degassed PBS-1. Resuspend the cells by carefully pipetting up and down using 1 ml pipettor.

Separation of unwanted cells from NK cells using magnet

During the last washing of cells, wash LS magnetizing column. Place the column on a MidiMACS or QuadroMACS magnet and wash it 3 times each with 3 ml of degassed PBS-1. It is important to wash the column well, as the column may contain toxic substances, which can decrease both viability and functionality of NK cells. After washing, add the suspension of labelled splenocytes carefully to the column. Collect the flow through suspension of purified NK cells in a 14 ml round bottom tube. Wash the column 3 times each with 3 ml PBS-1 and collect the remaining NK cells. The paramagnetic bead-labelled cells retained in the column can be collected by washing the column outside of the magnetic field using PBS-1.

Testing purity of NK cells

After their collection from the magnetized column, centrifuge NK cells at 300*g* for 10 min. Resuspend the cells in RPMI-10 and test their number and viability using trypan blue solution and microscopic counting. If the purification is successful, from 10^8 C57BL/6 mouse splenocytes, ~2–3 × 10^6 cells which are >95% viable should be obtained. To determine the purity of NK cells, stain the cells with different fluorochrome-conjugated anti-NK-1.1 or anti-DX5 (CD49b)/anti-CD3, anti-CD19/anti-CD14 and anti-Gr-1 monoclonal antibodies, and perform flow cytometry analysis. A well purified population of NK cells should contain ≥70% $NK1.1^+$ or $DX5^+$ and <3% of $CD3^+$, $CD14^+$, $CD19^+$ or <10% of $Gr\text{-}1^+$ cells. If the purified NK cells should be cultured and/or used for functional studies, they must be washed two times in RPMI-10.

NOTE: (1) As the number of purified NK cells is usually small, their washing has to be performed with particular caution, such as cell centrifugation in 5-ml tubes and using 3 ml of RPMI-10, and careful removal of the supernatants after centrifugation using a transfer pipette. (2) If either the reagents is deficient or the purification is not optimally performed, 5–15 × 10^6 'purified' cells could be obtained from 10^8 splenocytes. Thus purified cells usually contain only 10–20% of $NK1.1^+CD3^-$. The purification can be improved by passing the cells through a MS or LD magnetizing column.

Purification of fresh human peripheral blood NK cells using negative selection

Similar procedure to the one described above for purification of fresh mouse NK cells by negative selection is available for purification of fresh human peripheral blood NK cells. NK cells are isolated by MACS from PBMNC that are separated from peripheral blood by Ficoll-Hypaque density gradient centrifugation. Using the LS magnetizing column, the Miltenyi negative selection kit for isolation of human NK cells, and MidiMACS or QuadroMACS magnet, usually 70–90% pure populations of $CD56^+CD3^-$ cells are obtained. The purification of human NK cells can be dramatically improved by using LD column instead of LS column. LD columns have greater volume, higher density of iron resin and thinner nozzle than LS columns. These differences enable the increase in magnetic power and decrease in cell flow in column, which enable more efficient retention of weakly labelled cells with paramagnetic beads and more efficient selection of unlabelled NK cells. In our hands, this simple and small change in the human NK cell purification procedure consistently yeilds ≥95% pure populations of $CD56^+CD3^-$ cells, which matches the level of NK cell purity obtained by FACS sorting. Recently, the Miltenyi Company has developed the automated CliniMACS Plus instrument, a closed sterile system for clinical use, which integrates a micro-computer, a magnetic separation unit and a peristaltic pump, and can be applied for purification of human peripheral blood NK cells on a clinical scale for clinical application (Passweg et al., 2004).

Other paramagnetic bead-based methods for purification of NK cells

We have successfully used a cocktail of commercially available mouse monoclonal antibodies specific to human T cells, B cells, monocytes, B cells and erythrocytes (pretested by us) in conjunction with Advanced Magnetics and/or Dynal paramagnetic beads coated with goat anti-mouse IgG (Kashii et al., 1999; Li et al., 2004; Vujanovic et al., 1995, 1996a; Wahlberg et al., 2001). In this earlier work, we consistently obtained highly enriched (80–95% $CD56^+CD3^-$) populations of human peripheral blood NK cells. Several laboratories have used similar methods at that time with variable success. Many different kinds of paramagnetic beads have been made. However, Dynal beads have shown superiority in purification of NK cells (Farace et al., 1992). Recently, StemCell Technologies have generated new reagents and now offer kits for purification of both murine and human NK cells. The initial testing of these reagents has provided promising results and indicated usefulness of these new kits. In our hands, one of these kits, EasySep, consistently provides solid purity of mouse splenic NK cells (≥65% of $NK.1.1^+CD3^-$). The method is simple and reliable for obtaining fully functional NK cells.

Activation and expansion of NK cells

Purification and propagation of human IL-2-activated adherent NK (A-NK) cells

A-NK cells represent a phenotypically and functionally distinct subset of activated NK cells, which are selected and purified by IL-2-induced rapid adherence, and are activated and expanded in the presence of IL-2 (Li et al., 2004; Melder et al., 1988; Pilbeam et al., 2007; Vitolo et al., 1993; Vujanovic et al., 1993, 1995, 1996a). A-NK cells are phenotypically $CD3^-CD56^{low}CD16^{low}$ $CAMs^{high}$ LGLs. They represent one of the most powerful anti-cancer immune effector cells, and display high NK and lymphokine-activated killer (LAK) activities against a large variety of cultured and freshly isolated tumour cells. They also exhibit the specific ability to migrate, infiltrate and efficiently function in solid tumour tissues, which leads to rapid destruction of tumour tissues and elimination of metastases (Li et al., 2004; Pilbeam et al., 2007; Vujanovic et al., 1995, 1996a). Because of these characteristics, A-NK cells have been used at our institution in adoptive immunotherapy (AIT) of cancer and showed promising beneficial clinical effects in patients with metastatic renal cell carcinoma and melanoma (Vujanovic et al., 1996a). Selection and expansion of A-NK cells is based on a selective ability of a small subset (~27%) of NK cells to rapidly respond to IL-2 by developing both adherence to plastic surfaces and high proliferation (Li et al., 2004; Pilbeam et al., 2007; Vujanovic et al., 1993, 1996a,b). Thus, following stimulation of monocyte-depleted PBMNL with 22 nM IL-2, a rapid (5 h) adherence of a small proportion of NK cells occurs. In addition, removal of non-adherent lymphocytes and further culture of separated adherent NK cells in the presence of IL-2 stimulates their vigorous proliferation and results in generation of large numbers of highly pure activated NK cells (Melder et al., 1988; Vujanovic et al., 1996a,b). The method of purification and expansion of A-NK cells has been modified and improved in comparison to the originally described method. The modifications and improvements are included below.

All works with human blood or cells must be performed with full implementation of biosafety level 2 (BSL-2) standards. All solutions, reagents and equipments coming in contact with cells must be sterile, and proper sterile techniques must be used accordingly and all incubations are performed in a humidified 37°C, 5% CO_2 incubator.

Essential materials

- Heparinized whole venous blood, leukopak or buffy coat (Central Blood Bank)
- RPMI 1640 medium (e.g. Life Technologies)
- Heat-inactivated pooled human AB serum (Cambrex)
- Phenylalanine-methyl ester (PME, duPont de Nemours, Glenolden, PA)
- Human recombinant interleukin-2 (IL-2, Chiron Corp., Emeryville, CA) (stock solution contains 6×10^5 IU or 10^5 Cetus U/ml of RPMI 1640)
- Heparin (Sigma) (stock solution contains 10^4 U/ml of PBS)
- $CaCl_2 \times 2H_2O$ (e.g. Sigma) (stock solution contains 1 M in H_2O)
- $MgCl_2 \times 6H_2O$ (e.g. Sigma) (stock solution contains 1 M in H_2O)
- Concavalin A (Con A) (e.g. CALBIOCHEM) (stock solution contains 1 mg/ml of RPMI 1640)
- Fluorochrome-conjugated isotype control and marker-specific monoclonal antibodies for testing phenotype of A-NK cells: FITC-IgG1, PE-IgG2a, FITC-anti-CD3, PE-anti-CD56, FITC-anti-CD16, FITC-anti-CD14 and PE-anti-CD19 (BD-Pharmingen)
- Cultured K562 and Daudi target cells
- [^{51}Cr] sodium chromate (NEN, Boston, MA)

Cell culture media

- **10% AB medium**: RPMI 1640 supplemented with 10% heat inactivated human AB serum, 2 mM L-glutamine, 1 U/ml penicillin, 1 μg/ml streptomycin, 2.5 μg/ml Fungizone and 25 mM Hepes buffer
- **NK cell adherence inducing medium**: 10% AB medium supplemented with 1 mM of $CaCl_2$, 1 mM of $MgCl_2$, 22 nM (6000 IU/ml) IL-2 and 100 U/ml Heparin
- **A-NK cell culture medium**: 10% AB medium supplemented with 22 nM (6000 IU/ml) of IL-2
- **1% AB medium**: RPMI 1640 medium supplemented with 1% heat inactivated human AB serum, 1 U/ml penicillin, 1 μg/ml streptomycin

Separation of PBMNL

All media, reagents, equipments and procedures are at room temperature. Dilute 10 ml of heparinized blood with 20 ml of RPMI 1640 or a leukopak or a buffy coat (containing $1–3 \times 10^9$ cells) with 180 ml of RPMI 1640. Place 30 ml of diluted blood, leukopak or buffy coat into 50 ml tubes. Using 10 ml serological pipettes underlay the diluted blood or leukopak with 12.5 ml of Ficoll-Hypaque. Separate PBMNL by centrifugation of the tubes at 600*g* for 30 min. Collect PBMNL from the gradient interface. Wash PBMNL three times in RPMI 1640 medium using centrifugations at 300*g* for 5 min. Resuspend washed PBMNL in RPMI 1640. Count cells and check their viability by trypan blue exclusion. Adjust the cell concentration to be 10^7 cells/ml.

Depletion of monocytes

Incubate the suspension of PBMNL (10^7/ml) in RPMI 1640 medium supplemented with 1 μg/ml PME (a monocyte/macrophage-toxic chemical) for 40 min at room temperature. Remove large clumps of dead cells by 1*g* sedimentation at room temperature for 5 min. Wash monocyte-depleted PBMNL 3 times in RPMI 1640 using centrifugations at 300*g* for 5 min. Alternatively, incubate suspension of PBMNL in RPMI 1640 medium containing 10% FCS in nylon wool columns or in horizontally positioned tissue culture flasks at 37°C for 1 h. The later procedures deplete not only monocytes but also B cells.

Generation of feeder cells

Resuspend autologous or allogeneic PBMNL (1.5×10^6/ml) in 10% AB medium supplemented with 10 μg/ml Con A and 1.1 nM IL-2, and culture for 3–5 days in horizontally-positioned tissue culture flasks. Feeder cells can be cultured up to 10–14 days without loosing their feeder property. However, during this long-term culture, they need to be maintained in density of $1.5–2.0 \times 10^6$ cells/ml by adding warm (37°C) 10% AB medium supplemented with 1.1 nM (300 IU) of IL-2. They can be also generated from frozen PBMNC. In addition, feeder cells can be frozen after several days of their culture and utilized either immediately after thawing or following 2–3 days of additional culture. Wash Con A-activated feeder cells 3 times in RPMI 1640 medium, to remove Con A. Resuspend feeder cell in A-NK cell culture medium. Count feeder cells and determine their viability using trypan blue exclusion test. Adjust the number of feeder cells to be 10×10^7 cells/ml of A-NK cell culture medium. Irradiate feeder cells with 3000 rads.

Selection of A-NK cells

Resuspend 10×10^6/ml monocyte-depleted PBMNL in NK cell adherence–inducing medium. Purified fresh NK cells can be also used, at 1×10^6/ml NK cells. Place the cell suspension into tissue culture flasks (3 ml/T25, 10 ml/T75, 20 ml/T150, and 30 ml/T220 flasks). Incubate horizontally positioned flasks containing the suspension of monocyte-depleted PBMNL without disturbance at 37°C for 5 h. During this incubation, a small proportion of lymphocytes adheres to plastic. They are A-NK cells. Resuspend non-adherent cells by gentle horizontal back and forth rocking of flasks for 1 min. Decant medium and non-adherent cells as much as possible. Wash A-NK cells in the flasks 5 times by adding warm (37°C) 1% AB medium (8 ml/T25, 25 ml/T75,

50 ml/T150 and 75 ml/T220 flasks), rocking flasks and decanting the medium containing remaining non-adherent cells, as described above. After adding for the fifth time washing medium, view the flasks using an inverted microscope and phase contrast. If the procedure is well performed, almost all adherent cells would have a dark crescent-like shape characteristic of A-NK cells. Count A-NK cells on 6 to 10 microscopic fields using an ocular grid and determine the total number of A-NK cells per flask using the following formula: *Mean A-NK cell number per grid divided by 0.0025 and multiplied by flask surface in cm*^{2}.

Harvesting A-NK cells

Fresh A-NK cells can be detached and collected for studies. Following elimination of nonadherent cells, additionally wash A-NK cells two times with cold PBS, and incubate in the presence of cold PBS at 4°C for 15 min. This incubation results in detachment of A-NK cells. Collect A-NK cells by pipetting.

Expansion of A-NK cells

Add the A-NK cell culture medium containing 1×10^6/ml of irradiated feeder cells to the flasks with selected A-NK cells. The final volume of the medium including feeder cells should be the same as that of A-NK cell inducing medium (e.g. 10 ml/T75 flask). Culture horizontally positioned flasks containing A-NK cells and feeder cells without disturbance for 5–7 days. During this period of time, A-NK cells detach from the plastic surface, form floating clumps with feeder cells and begin to proliferate vigorously. Usually, between day 5 and day 7 of culture, the culture medium becomes yellow and A-NK cells need to be fed for the first time. Add fresh warm (37°C) A-NK cell culture medium in a volume that doubles the original A-NK cell culture volume. From this time point of A-NK cell culture, observe daily and feed regularly A-NK cells, to keep their number at $1.5–2.0 \times 10^6$ cells/ml at all time. The maximal volume of A-NK cell cultures should not exceed 10 ml, 30 ml, 60 ml or 90 ml for T25, T75, T150 or T220 flasks, respectively. Activation and expansion of A-NK cells are completed by day 14 of culture. At that point of A-NK cell culture, feeder cells are completely destroyed and used up, without any trace remaining. In addition, at that time, most of human A-NK cells are floating and they can be easily collected from flasks by pipetting.

Testing of A-NK cells

Collect A-NK cells and determine their number and viability by trypan blue exclusion test. Determine expansion of A-NK cells by dividing the total number of A-NK cells at day 14 of culture with the total number of A-NK cells after 5 h of IL-2-induced adherence selection. It is expected that A-NK cells expand between 100- and 1000-fold during 14 days of their culture in the presence of IL-2. Examine the phenotype of A-NK cells by flow cytometry and determine the percentage of $CD3^-CD56^+CD16^+$ cells. It is expected that, after 14 days of culture, A-NK cells are ≥90% $CD3^-CD56^+CD16^+$ cells. Assess antitumor cytotoxic activity of A-NK cells by testing their killing ability against K562 (NK susceptible) and Daudi (LAK susceptible) cell targets, using 4-h ^{51}Cr release cytotoxicity assays. It is expected that, at day 14 of culture, A-NK cells have high cytotoxic activity against both targets (4000–20000 $LU_{20}/10^7$ against K562, 3000–8000 $LU_{20}/10^7$ against Daudi).

Generation and expansion of rat and mouse A-NK cells

A-NK cells can also be generated from rat (Vujanovic et al., 1988) and mouse (Gunji et al., 1989) splenocytes using a similar method to that of human A-NK cells. However, rat and mouse splenic A-NK cells differ from human peripheral blood A-NK cells in their kinetics of adherence and growth. Therefore, the optimal times for rat and mouse splenic A-NK cell selection are 48 and 72 h, while for expansion are 5 and 9 days of culture, respectively. To generate murine A-NK cells, the complete murine A-NK cell culture medium containing RPMI-1640 medium, 10% FCS (selected for high ability to support growth of NK-cells), 2-mercaptoethanol (5×10^{-5} M), L-glutamin (1 mM), sodium pyruvate (2 mM), non-essential amino acids (0.1 mM), penicillin (1 U/ml), streptomycin (1 μg/ml), Hepes buffer (25 mM) and 22 nM IL-2 should be utilized. At the end of culture, it is expected that purity of rat and mouse A-NK cells is ≥95 ($CD3^-NKRP1^+$) and ≥70% ($CD3^-NK1.1^+$), respectively, while their expansion 100- to 1000-fold. In mice, the number of selected A-NK cells, their growth and purity can be significantly increased by adding 100 U/ml of Heparin, 1 mM of $CaCl_2$ and 1 mM of $MgCl_2$ to the culture of splenocytes, for the last 5 h of A-NK cell adherence selection. After this additional step, non-adherent cells, Heparin, and excess of $CaCl_2$ and $MgCl_2$ should be removed by washing, and thus selected A-NK cells should be supplemented with and grown in the complete murine A-NK cell culture medium.

Activation and culture of purified fresh whole population of NK cells

Both human (Vujanovic et al., 1993) and mouse (Xu et al., 2007) purified fresh whole population of NK cells can be also efficiently activated and expanded using the cell culture media for human or mouse A-NK cells, respectively. For an efficient expansion of human NK cells, it is also recommended to utilize the feeder cells described in the protocol for generation of human A-NK cells. These cultures are performed in horizontally positioned flasks and contain an initial density of purified NK cells of 0.1×10^6/ml of culture medium. The cultures lead to highly pure populations of activated NK cells.

Conclusions

Several effective methods for isolation and expansion of whole population of human and murine NK cells have been developed. The separation of major $CD56^{high}$, $CD56^{low}$ and ANK-1^+ human NK cell subsets have also become possible using fluorescence activated sell sorting. Methods have been established for isolation and expansion of A-NK cell subpopulation. Among the methods for isolation of NK cells, Miltenyi MACS is the most advanced system for consistent and rapid separation of highly pure, viable, functional and in high numbers fresh NK cells for both research and clinical use. The current MACS methods have the capability to purify whole populations of mature NK cells from either human peripheral blood or murine spleen, but have no ability to isolate subpopulations of NK cells. Therefore, one of the important future directions in the development of MACS for NK cell isolation is to define the cell surface antigenic markers of NK cell subsets, produce specific antibodies against them, and utilize these antibodies to make standardized kits for isolation of NK cell subsets.

During several years of continuous utilization of MACS kits for purification of both human and mouse NK cells, we have observed that different lots of MACS kits differ in their efficiency to isolate NK cells, indicating that the preparations of reagents and kits are not yet fully standardized. Therefore, there is a need for further standardization and improvements of MACS kits.

Using the existing MACS kit for negative selection of mouse NK cells, we have also noticed that, 80% pure populations of $CD3^-NK1.1^+$ cells can be obtained, at best. Thus purified populations of NK cells are depleted of T cells, B cells and macrophages, but contaminated with 5–10% of Gr-1^+ (myeloid) cells and 5–10% of 'null' cells that cannot be classified using commercially available antibodies to cell differentiation antigens. The 'null' cells may be hematopoietic progenitor cells and/or stem cells. Therefore, to improve the purification of NK cells, a higher affinity anti-Gr-1 antibody should be included in the kit. In addition, cell differentiation markers for 'null' cells should be identified and antibodies specific for these markers should be developed and added to the existing kit.

Low labelling of unwanted cells by antibodies due to their insufficient affinity, and nonspecific binding of antibodies to Fc receptors on NK cells due to their Fc fragment are among the main causes of low yield and low purity of NK cells that can occur during MACS purification of NK cells. To improve the yield and purity of NK cells using MACS system, the antigen-binding ability and specificity of antibodies in MACS NK-cell isolation kits should be ameliorated by generating and including into the isolation kits the antibodies both having higher affinity for cell differentiation antigens of interest and lacking Fc fragment. In addition, a standardized device for degassing of buffer, higher quality paramagnetic beads that can bind more specifically and more efficiently to the cells of interest and can be easier magnetized; and higher capacity and efficiency magnetizing columns should be developed and provided in MACS kits.

The current effective methods for ex vivo activation and propagation of NK cells are based on bulky cell cultures performed in specifically formulated complete cell culture media supplemented with sera, IL-2 and/or feeder cells, and carried on in standard tissue culture flasks. It is expected that the future improvements will be primarily made for activation and expansion of NK cells for clinical use. That may include the development of serum-free synthetic culture media that well support NK cell activation and growth; development of particular compact cell culture devices for closed automated selection, activation, culturing and feeding of large numbers of NK cells, in order to reduce bulkiness of cultures; and use of additional cytokines and ligands for NK cell stimulation, to make NK cell growth independent of feeder cells.

Production of synthetic, serum-free culture media that are able to fully support NK cell activation and growth might be an important direction of the development. Several serum-free cell culture media are already commercially available. A synthetic medium produced by CellGro Company (catalogue number 2001) is even designated by the company for growth of human NK cells. The ability of available synthetic cell culture media to support activation and growth of NK cells in culture awaits rigorous laboratory testing. In addition, new synthetic cell culture media that are specifically formulated to optimally support NK cell cultures should be made.

An improvement of the technology that will enable automatic selection, activation, growth and feeding of NK cells in relatively small closed devices is under development. There are several commercially available automated and compact, but relatively complicated for use, cell culture devices, including Hollow Fiber Cartridge from FiberCell Systems Inc. and TotiCell Bioreactor from Biotron. Unfortunately, their testing has not convincingly shown that they can properly support activation and growth of NK cells. Therefore, the available devices should be specifically improved or new devices having the ability to support activation and growth of NK cells should be developed.

Growth of human NK cells is relatively slow and highly dependant on feeder cells, which are believed to provide additional growth factors to NK cells. Preparation of feeder cells for human NK cell cultures significantly add to the complexity and labour intensity of the technology. In addition, feeder cells can vary in their ability to support growth of NK cells and, if they are allogeneic, can potentially be vectors of blood-borne pathogens. Therefore, an important future direction for improvement of the technology for activation and growth of NK cells for clinical use should include the replacement of feeder cells with recombinant growth factors that may include some of the known NK cell stimulating cytokines and ligands such as IL-15, IL-12, IL-18, IL-1, interferons and tumour necrosis factor superfamily ligands.

Financial support

This study was supported by research funding from the National Institute of Health grants PO1 DE13059, RO1 DE14775 and RO1 DE17150, and from the Pennsylvania Department of Health Funding to N.L.V.

References

Andrews, D.M., Scalzo, A.A., Yokoyama, W. M., Smyth, M.J. and Degli-Esposti, M.A. (2003). Functional interactions between dendritic cells and NK cells during viral infection. Nat Immunol **4**, 175–181.

Banchereau, J. and Steinman, R.M. (1998). Dendritic cells and the control of immunity. Nature **392**, 245–252.

Cooper, M.A. and Caligiuri, M.A. (2004). Isolation and characterization of human natural killer cell subsets. Current Protoc Immunol **60**, 7.34.1–7.34.12.

Cosentino, L.M. and Cathcart, M.K. (1987). A multi-step isolation scheme for obtaining CD16+ human natural killer cells. J Immunol Methods **103**, 195–204.

Farace, F., Le Ridant, A.M., Escudier, B., Hercend, T. and Tribel, F. (1992). Studies on NK cell purification by negative selection in human peripheral blood. Biotherapy **5**, 11–20.

Farag, S.S., Fehniger, T.A., Ruggeri, L., Velardi, A. and Caligiuri, M.A. (2002). Natural killer cell receptors: new biology and insights into the graft-versus-leukemia effect. Blood **100**, 1935–1947.

Farag, S.S., VanDeusen, J.B., Fehniger, T.A. and Caligiuri, M.A. (2003). Biology and clinical impact of human natural killer cells. Int J Hematol **78**, 7–17.

Ferlazzo, G., Tsang, M.L., Morretta, L., et al. (2002). Human dendritic cells activate resting natural killer (NK) cells and are recognized via NKp30 receptor by activated NK cells. J Exp Med **195**, 343–351.

Fernandez, N.C., Lozier, A., Flament, C., et al. (1999). Dendritic cells directly trigger NK cell functions: cross-talk relevant in innate anti-tumor immune responses in vivo. Nat Med **5**, 405–411.

Gerosa, F., Baldani-Guerra, B., Nisii, C., et al. (2002). Reciprocal activating interaction between natural killer cells and dendritic cells. J Exp Med **195**, 327–333.

Goodier, M. and Londei, M. (2000). Lipopolysaccharide stimulates the proliferation of human CD56+CD3− NK cells: a regulatory role of monocytes and IL-10. J Immunol **165**, 139–147.

Gunji, Y., Vujanovic, N.L., Hiserodt, J. C., Herberman, R.B. and Gorelik, E. (1989). Generation and characterization of purified adherent lymphokine-activated killer cells in mice. J Immunol **142**, 1748–1754.

Jinushi, M., Takehara, T., Kanto, T., et al. (2003). Critical role of MHC class I-related chain A and B expression on IFN-alpha-stimulated dendritic cells in NK cell activation: impairment in chronic hepatitis C virus infection. J Immunol **170**, 1249–1256.

Kapsenberg, M.L. (2003). Dendritic-cell control of pathogen-driven T-cell polarization. Nat Rev Immunol **3**, 984–993.

Kashii, Y., Giorda, R., Herberman, R. B., Whiteside, T.L. and Vujanovic, N. L. (1999). Constitutive expression and role of the tumor necrosis factor family ligands in apoptotic killing by human natural killer cells. J Immunol **163**, 5358–5366.

Lanier, L.L. and Phillips, J.H. (1992). Natural killer cells. Curr Opin Immunol **4**, 38–42.

Li, S., Xu, J., Makarenkova, V., et al. (2004). A novel N-CAM epitope defines precursors of human adherent NK cells. J Leukoc Biol **76**, 1187–1199.

Maillard, R.B., Son, Y.-I., Redlinger, R., et al. (2003). Dendritic cells mediate NK cell help for Th1 and CTL responses: two-signal requirement for the induction of NK cell helper function. J Immunol **171**, 2366–2373.

McKenna, D.H., Sumstad, D., Bostrom, N., Kadidlo, D.M., et al. (2007). Good manufacturing practices production of natural killer cells for immunotherapy: a six-year single-institution experience. Transfusion **47**, 520–528.

Melder, R.J., Whiteside, T.L., Vujanovic, N. L., Hiserodt, J.C. and Herberman, R.B. (1988). A new approach to generating antitumor effectors for adoptive immunotherapy using human adherent lymphokine-activated killer cells. Cancer Res **48**, 3461–3469.

Miltenyi, S., Muller, W., Weichel, W. and Radbruch, A. (1990). High gradient magnetic cell separation with MACS. Cytometry **11**, 231–238.

Moretta, A. (2002). Natural killer cells and dendritic cells: rendezvous in abused tissues. Nat Rev Immunol **2**, 957–964.

Moretta, L., Bottino, C., Ferlazzo., et al. (2003). Surface receptors and functional interactions of human natural killer cells: from bench to the clinic. Cell Mol Life Sci **60**, 2139–2146.

Osada, T., Nagawa, H., Kitayama, J., et al. (2001). Peripheral blood dendritic cells, but not monocyte-derived dendritic cells, can augment human NK cell function. Cell Immunol **213**, 14–23.

Passweg, J.R., Tichelli, A., Meyer-Monard, S., et al. (2004). Purified donor NK-lymphocyte infusion to consolidate engraftment after haploidentical stem cell transplantation. Leukemia **18**, 1835–1838.

Passweg, J.R., Koehl, U., Uharek, L., Meyer-Monard, S. and Tichelli, A. (2006). Natural-killer-cell-based treatment in haematopoietic stem-cell transplantation. Best Pract Clin Haematol **19**, 811–824.

Piccioli, D., Sbrana, S., Melandri, E. and Valiante, N.M. (2002). Contact-dependent stimulation and inhibition of dendritic cells by natural killer cells. J Exp Med **195**, 335–341.

Pilbeam, K., Vujanovic, N.L. and Borghesi, L. (2007). Ontogeny of natural killer cells in man and mouse.

In: Brossy, L. (ed), Everything You Always Wanted to Know About NK Cells but Were Afraid to Ask: Transworld Research Network, pp. 1–26.

Poggi, A., Carosio, R., Spaggiari, G.M., et al. (2002). NK cell activation of dendritic cells is dependent on LFA-1-mediated induction of calcium-calmodulin kinase II: inhibition by HIV-1 Tar C-terminal domain. J Immunol **168**, 95–101.

Rabinowich, H., Sedlmayr, P., Herberman, R.B. and Whiteside, T.L. (1991). Increased proliferation, lytic activity, and purity of human natural killer cells cocultured with mitogen-activated feeder cells. Cell Immunol **135**, 454–470.

Reynolds, C.W., Timonen, T. and Herberman, R.B. (1981). Natural killer (NK) cell activity in the rat. I. Isolation and characterization of the effector cells. J Immunol **127**, 282–287.

Ritson, A. and Bulmer, J.N. (1989). Isolation and functional studies of granulated lymphocytes in first trimester human decidua. Clin Exp Immunol **77**, 263–268.

Robertson, M.J. and Ritz, J. (1990). Biology and clinical relevance of human NK cells. Blood **76**, 1990–2421.

Ruggeri, L., Capanni, M., Urbani, E., Perruccio, K., et al. (2002). Effectiveness of donor natural killer cell alloreactivity in mismatched hematopietic transplants. Science **295**, 2097–2100.

Timonen, T., Ortaldo, J.R. and Herberman, R.B. (1981). Characteristics of human large granular lymphocytes and relationship to natural killer and K cells. J Exp Med **153**, 569–582.

Trinchieri, G. (1989). Biology of natural killer cells. Adv Immunol **47**, 187–376.

Vitolo, D., Vujanovic, N.L., Rabinowich, H., Schlesinger, M., Herberman, R. B. and Whiteside, T.L. (1993). Rapid interleukin-2-induced adherence of human natural killer (NK) cells. Expression of mRNA for cytokines and IL2 receptors in adherent NK cells. J Immunol **151**, 1926–1937.

Vujanovic, N.L. and Whiteside, T.L. (1995). Selection of lymphocytes for immunotherapy. In: Ricordi, C. (ed.) *Methods in Cell Transplantation*. Austin, TX: R.G. Landes Co., 271–276.

Vujanovic, N.L., Herberman, R.B., Maghazachi, A.A. and Hiserodt, J.C. (1988). Lymphokine-activated killer cells in rats: III. A simple method for the purification and rapid expansion of lymphokine-activated killer cells. J Exp Med **167**, 15–29.

Vujanovic, N.L., Rabinowich, H., Lee, Y.J., Jost, L., Herberman, R.B. and Whiteside, T.L. (1993). Distinct phenotypic and functional characteristics of human natural killer cells obtained by rapid interleukin-2-induced adherence to plastic. Cell Immunol **151**, 133–157.

Vujanovic, N.L., Yasumura, S., Hirabayashi, H., Lin, W.-C., Watkins, S., Herberman, R.B. and Whiteside, T.L. (1995). Antitumor activity of subsets of human IL2-activated natural killer cells in solid tissues. J Immunol **154**, 281–289.

Vujanovic, N.L., Basse, P., Herberman, R.B. and Whiteside, T.L. (1996). Antitumor functions of natural killer cells and control of metastases. Methods **9**, 394–408.

Vujanovic, N.L., Nagashima, S., Herberman, R.B. and Whiteside, T.L. (1996). Non-secretory apoptotic killing by human natural killer cells. J Immunol **157**, 1117–1126.

Wahlberg, B.J., Burholt, D.R., Kornblith, P., Richards, T., Bruffsky, A., Herberman, R.B. and Vujanovic, N.L. (2001). Measurement of NK activity by the microcytotoxicity assay (MCA): a new application for an old assay. J Immunol Methods **253**, 69–81.

Xu, J., Chakrabarti, A.K., Tan, J.A., Ge, L., Gambotto, A. and Vujanovic, N. L. (2007). Essential role of the TNF–TNFR2 cognate interaction in mouse dendritic cell–natural killer cell cross-talk. Blood **109**, 3333–3341.

Yu, Y., Hagihara, M., Ando, K., et al. (2001). Enhancement of human cord blood CD34+cell-derived NK cell cytotoxicity by dendritic cells. J Immunol **166**, 1590–1600.

Chapter Ten

10

The localization and migration of natural killer cells in health and disease

Vasileios Bekiaris, Peter J.L. Lane

CHAPTER CONTENTS

As you set out for Ithaca hope your road is a long one, full of adventure, full of discovery.

Ithaca, by Konstantinos P. Kavafis

ABSTRACT

Natural killer (NK) cells comprise a finite lymphocyte lineage with distinctive gene expression patterns. The unique transcriptome of NK cells renders them capable of protecting the host from a vast array of disease states. Their undisputed importance in host protection is conferred by their ability to eliminate unhealthy cells. However, in order for NK cells to exert their effects, they need to be strategically located at the right places. Furthermore, within their anatomical microenvironment, NK cells must be able to freely move and receive stimuli that will direct their migration between different organs. This chapter will provide an overview of our current understanding of the localization of NK cell populations and their ability to migrate in response to homeostatic and pathological conditions.

KEY WORDS

Natural killer cells, Migration, Chemokines, Lymphoid tissue, Localization

Introduction

Cell migration occurs mainly due to the action of cellular chemo-attractants that are usually in the form of small proteins, collectively known as chemokines. Chemokines bind to specific receptors that are expressed by immune cells and can only exert their effects if the cell expresses the specific receptor. However, immune cells express an array of chemokine receptors at differing levels, depending on their activation status, which makes them multi-chemokine responders. Thus, the ability to react to a number of chemokines provides cells of immunological significance with the capacity to be recruited to wherever they are needed.

Controlled migration of immune cells is a fundamental process for the development, homeostasis and function of the immune system. B and T lymphocytes undergo continuous re-circulation via blood and lymph to secondary lymphoid tissues (spleen, lymph nodes and mucosal-associated lymphoid tissues) in response to chemokines expressed in fixed stromal cell populations that both attract and guide their specific interactions. For example, during lymph node (LN) organogenesis, lymphoid tissue inducer (LTi) cells populate the LN (Mebius, 2003) or spleen (Withers et al., 2007) anlagen where they interact with stromal cells and activate them to produce B and T cell-specific chemokines. This attracts B and T lymphocytes to defined B and T zones and initiates the formation of secondary lymphoid organs. This example clearly demonstrates the need for controlled cell migration.

Natural killer (NK) cells develop in the bone marrow (BM) (Haller et al., 1977) and are not static but populate secondary and primary lymphoid organs. A unique feature of NK cells is their expression of activating and inhibitory receptors, which allow them to respond either when ligands for activating receptors are upregulated or when ligands for inhibitory receptors are downregulated (Yokoyama and Plougastel, 2003).

There have been significant advances in our understanding of NK cell migration. This was primarily achieved by the progress of in situ cell-tracking systems, real-time live microscopy and flow methods that allow identification of NK cells in lymphoid- and non-lymphoid organs. In this chapter, we will describe the anatomical location of different NK cell subsets and how they are recruited, with emphasis on chemokines and their receptors that are involved in mobilizing NK cells during health and disease.

NK cells in secondary lymphoid organs

While trafficking around the body, B and T cells perform necessary 'pit-stops' within secondary lymphoid organs, such as the LNs, the spleen, and the mucosal associated lymphoid tissue (MALT), where they scan for trapped antigen and engage in cognate interactions with accessory cells, such as dendritic cells (DCs). Within the lymphocyte rich areas, B and T cells respond to chemokine signals and segregate to distinct B follicles and T cell zones, respectively. This B/T segregation allows lymphocytes to move in an organized fashion (Bajenoff et al., 2006b) and to help find each other for the development of effective adaptive immune responses. NK cells are classically characterized as innate lymphocytes, implicating that they can be readily activated without the need for a high degree of organization in the surrounding lymphoid environment. However, there is compelling evidence that secondary lymphoid organs contain unique NK cell subsets with distinct phenotype and function. Here we describe NK cell subpopulations in the LN and the spleen. MALT will not be discussed in this section because there is little evidence that NK cells are directly associated with the lymphoid follicles found in such tissues, despite their undisputed presence and importance in the surrounding microenvironments. In later sections, however, we will discuss gut and lung NK cell populations.

Lymph node NK cells

Peripheral blood human NK cells can be largely subdivided into $CD56^{dim}CD16^{+}$ and $CD56^{bright}CD16^{-/dim}$ (Benedict et al., 2001; Campbell et al., 2001). The former are highly cytotoxic, whereas the latter have lower natural cytotoxicity but produce an abundance of cytokines (Benedict et al., 2001). Early attempts to further characterize differences between the two subsets showed that only $CD56^{bright}$ cells expressed constitutively and at high levels the adhesion molecule L-selectin (Frey et al., 1998). Moreover, L-selectin expressing $CD56^{bright}$ NK cells bind to high endothelial venules (HEVs), which constitute the point of entry of lymphocytes into the LN (Frey et al., 1998). L-selectin, together with the chemokine ligand for CCR7, CCL21 facilitate cell transmigration within the LN through HEVs (Ansel et al., 1999; Warnock et al., 1998, 2000). Expression of L-selectin by peripheral $CD56^{bright}$ NK cells was later confirmed independently, and it was additionally shown that this population, and not $CD56^{dim}$ NK cells, expressed CCR7 and were capable of migrating towards CCL21 gradients (Campbell et al., 2001). It has also been shown that stimulation of human peripheral blood NK cells with interleukin-18 (IL-18) upregulates their expression of CCR7 (Mailliard et al., 2005). This provided evidence that NK cells might enter the LN, a specialized secondary lymphoid organ. Immunohistological and in situ hybridization screening of normal human LN tissue sections revealed that the T cell areas contained $CD56^{+}$ cells that lacked expression of CD3 (Fehniger et al., 2003; Ferlazzo and Munz, 2004). Further analysis showed that NK cells in LN tissue were $CD56^{bright}CD16^{-}$, confirming the hypothesis that these cells, through their expression of CCR7 and L-selectin, could home to LNs. Importantly, $CD56^{bright}$ NK cells in the LN expressed different sets of activating and inhibitory receptors, and their low cytotoxicity could be restored upon stimulation with IL-2 (Ferlazzo and Munz, 2004), providing further evidence that NK cells in secondary lymphoid organs were distinct from those populations found in other locations. Endogenous IL-2 is believed to be provided by resident T cells (Fehniger et al., 2003; Ferlazzo and Munz, 2004), whereas DCs residing in the T cell areas play a role in promoting interferon-γ (IFNγ) expression and proliferation of $CD56^{bright}$ NK cells through IL-12 and IL-15, respectively (Ferlazzo and Munz, 2004).

Following their discovery, two hypotheses were proposed for the derivation and localization of CD56^{bright} NK cells within the LN. Firstly, that CD56^{bright} cells developed in the BM and by virtue of their expression of L-selectin and CCR7 migrated to LN. The alternative hypothesis suggested that a precursor residing in the LN was giving rise to CD56^{bright} NK cells in response to stimuli. Whereas there is no evidence rejecting the first hypothesis, Caligiuri and co-workers have identified a CD34^{dim}CD45RA(+)$\alpha 4\beta 7^{bright}$ hematopoietic progenitor that develops in the BM and migrates to the LN, where it gives rise exclusively to CD56^{bright} NK cells in response to T cell produced IL-2 or IL-15 (Freud et al., 2005).

Mouse LNs contain less NK cells compared to man. However, in vivo activation studies have shown that they can rapidly migrate into antigen-draining LNs (Martin-Fontecha et al., 2004). Unlike in humans, the migration of mouse NK cells to the LNs depends on the chemokine receptor CXCR3 and not CCR7, although L-selectin is required for NK cell entry into the LNs of both species (Adam et al., 2005; Martin-Fontecha et al., 2004). Thus, CXCR3-deficient NK cells poorly migrated into LNs, whereas CCR7-deficient did not show any migratory defect (Martin-Fontecha et al., 2004).

As described previously, CD56 expression distinguishes two human NK cell subsets. In mice, expression of CD27 gives a similar picture, though distinctly different functionally (Hayakawa and Smyth, 2006). Thus, CD27^{high} NK cells are very effective killers and produce high amounts of cytokines in response to activation. In contrast, CD27^{low} NK cells present with low cytotoxicity and cytokine secretion, which could be attributed to their increased expression of inhibitory receptors (Hayakawa and Smyth, 2006). Markedly, CD27^{high} and not CD27^{low} mouse NK cells express CXCR3, respond to its ligands, CXCL10 and CXCL11, and can be found residing in or recruited to the LN (Hayakawa and Smyth, 2006; Watt et al., 2008). Detailed fluorescence and intra-vital 2-photon microscopy showed that LN NK cells are motile, they are located both in the T zone and the medulla and actively engage in interactions with DCs and MHC-mismatched cells (Bajenoff et al., 2006a; Garrod et al., 2007). Furthermore, in response to infection, LN NK cells were redistributed, locally produced cytokines and productively interacted with infected cells (Bajenoff et al., 2006a). Collectively, the significance of these studies lies in the fact that, like the T cell, which continuously scans for antigen, the life of the NK cell within the LN is highly dynamic.

Splenic NK cells

The spleen is the prototype secondary lymphoid tissue found in all vertebrates with an adaptive immune system in contrast to LNs that are almost exclusive to mammals. In addition to adaptive immunity that develops in the lymphocyte-rich white pulp areas, the red pulp of the spleen is a meshwork of sinusoids lined by macrophages that filter the blood of dead or dying erythrocytes and phagocytose pathogens (Mebius and Kraal, 2005). The lymphoid compartment of the spleen, the white pulp (named after its white colour in appearance due to the presence of white blood cells) is embedded within the red pulp. Lymphocytes enter through the central arteriole and marginal sinus and exit via blood veins in the red pulp. Whereas CCR7 and CCL21 facilitate cell migration in the splenic red pulp, L-selectin (which is important for mouse and human NK cell recruitment to the LN) is not necessary.

Consequently, while characterizing differences in NK cells between blood and secondary lymphoid organs, it became apparent that the spleen contained a mix of populations that could be present in both blood and LN. Researchers studying human NK cells found that the spleen contained both CD56^{bright} and CD56^{dim} populations, although on average CD56^{dim} cells were more abundant. Furthermore, CD56^{bright} cells were more prevalent in the spleen than in blood, suggesting preferential recruitment to this organ (Ferlazzo and Munz, 2004). In accordance, mouse spleen contains both CD27^{bright} and CD27^{low} NK cells, whereas as mentioned before, the LN contains almost exclusively CD27^{high} NK cells.

Early work with human splenic tissue showed that NK cells localize mainly in the red pulp and occasionally in the marginal zone of the spleen (Garni-Wagner et al., 1990; Kummer et al., 1995). An analogous picture is observed in the spleen of rodents (Andrews et al., 2001; Bekiaris et al., 2008; Brown et al., 2001; Gregoire et al., 2007, 2008; Salazar-Mather et al., 1996; van den Brink et al., 1991). In situ identification of mouse NK cells shows that although under naïve conditions their preferred location is the red pulp, they can migrate to the marginal zone upon type I IFN signals (Salazar-Mather et al., 1996). In murine cytomegalovirus (MCMV) infection where NK cells play a crucial role in early defence, infection results in the specific migration of NK cells from the red pulp to the marginal zone and the white pulp (Andrews et al., 2001; Bekiaris et al., 2008; Brown et al., 2001; Gregoire et al., 2008) (Figure 10.1). Whereas recruitment of NK cells to splenic red pulp is independent of chemokine receptor signalling (Gregoire et al., 2008), migration from the red pulp to the white pulp depends on CXCR3, both during MCMV infection (Bekiaris et al., 2008) and during activation of the IFNα/β pathway (Gregoire et al., 2008). Due to the heterogeneity of splenic NK cell populations, it remains to be elucidated whether a specific subset is more potent at migrating towards the lymphoid white pulp, where it can interact with T zone resident DCs and primed B and T cells.

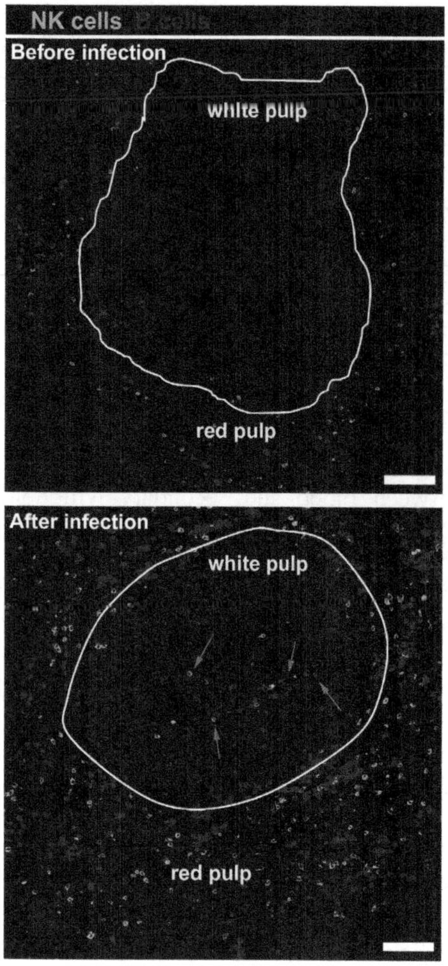

Figure 10.1 • NK cells migrate from the red to the white pulp of the spleen following infection with murine CMV. B cell (blue) staining reveals the splenic white pulp (structure enclosed by the white line). Before infection (upper micrograph), NK cells (green) are located in the red pulp. After infection (lower micrograph), NK cells migrate to the white pulp (red arrows). The scale bar is 100 μm.

NK cells in other organs

Gut NK cells

The gut contains organized lymphoid structures called Peyer's Patches (PPs) immediately under the epithelial surface of the lamina propria. Like the spleen and LN, PPs have distinct T and B cell zones and promote strong immune responses. An important feature of the gut is that lymphocytes, including NK cells, reside within the intraepithelial space and are thus termed intraepithelial lymphocytes (IELs).

Initial studies by Tagliabue et al. showed that human IELs as well as lamina propria lymphocytes were able to eliminate NK-sensitive targets but were non-cytotoxic against NK-insensitive targets (Tagliabue et al., 1981, 1982), providing early evidence that gut contained NK cells. Besides, the same investigators could detect cytotoxicity but not NK cell activity in PPs (Tagliabue et al., 1983). Independent reports also supported the presence of intraepithelial NK cells and suggested possible differences from peripheral NK cells, such as a higher sensitivity to IL-2 stimulation and the requirement for prolonged contact with the target to exert cytotoxicity (Gibson and Jewell, 1985; Mowat et al., 1983).

Later analyses also identified NK cell activity in IELs (Kato et al., 1995), whereas IL-2 could induce the expression of the gut-homing integrin $\alpha 4\beta 7$ (Baker and Wood, 1992; Berlin et al., 1993; Schweighoffer et al., 1993) on the surface of human blood NK cells (Perez-Villar et al., 1996) indicating that upon activation in the periphery, NK cells can gain the capacity to migrate to the gut. Furthermore, CCL21 was shown to be partly responsible for recruiting lymphocyte subsets, including NK cells, to PPs via a mechanism that involves binding of $\alpha 4\beta 7$ to MAdCAM-1 (mucosal address in cell adhesion molecule-1) (Pachynski et al., 1998).

Intestinal epithelial cells are able to respond to infection and produce a number of cytokines. Among them, IL-15 has been shown to be a major activator of intraepithelial NK cells inducing potent perforin-mediated killing (Fukuyama et al., 2002). Thus, an attractive regulatory mechanism was proposed according to which infected epithelium could immediately and locally activate neighbouring NK cells through the production of IL-15 and quite possibly other inflammatory cytokines. Furthermore, the normal function and maintenance of intraepithelial NK cells also depended on normal expression of the surface phosphatase CD45 (Martin et al., 2001).

Recent evidence has identified potential roles of intraepithelial NK cells in both gut infections and autoimmune disease. Data from the mouse model of the helminth *Trichinella spiralis* show that infection is associated with IL-13-mediated gut pathology, and that the main IL-13-producing cell was the intraepithelial NK cell (Bienenstock and McDermott, 2005), linking thus infection-induced intestinal immunopathology with gut-resident NK cells. Furthermore, in the absence of protective Th2 immunity against infection with *Trichuris muris*, NK cells (and CD4 T cells) in the gut produce IL-13, which contributes to protection, in an IL-18-depenent mechanism (Fan et al., 2006).

In addition, intraepithelial NK cell have been associated with intestinal autoimmune disease, such as inflammatory bowel disease (IBD) and colitis. Mouse models have indicated that dysregulated immune responses to gut commensal bacteria can cause IBD-like disease (Keilbaugh et al., 2005). To this end, it was shown that colonization of the gut by commensal bacteria induces upregulation of IFNγ by NK cells (Keilbaugh et al., 2005). Interestingly, this only occurred in severe combined immunodeficient and not in immunosufficient animals

(Keilbaugh et al., 2005), suggesting that adaptive immune cells could negatively regulate NK cell activation. The opposite is also true. Thus, in a model of colitis, NK cells produce (1) the early IFNγ that polarizes CD4 T cells towards the pathological Th1 phenotype and (2) the chemokine CXCL10, which attracts CXCR3$^+$ cells, including more Th1 cells (Singh et al., 2008).

Exciting new research has identified unique gut NK cell subsets in the lamina propria as well as in cryptopatches (Luci et al., 2008; Sanos et al., 2008). These NK cells expressed low levels of CD122 and CD11b, but they were bright for CD27 and could be either NK1.1^{low} or NK1.1$^{+/high}$ (Luci et al., 2008; Sanos et al., 2008). Functionally, they presented with almost undetectable cytotoxicity and very low capacity to produce IFNγ, but they could secrete IL-22, which is involved in tissue repair. Interestingly, these NK cells expressed molecules related to LTi cells (Kim et al., 2003), including IL-7Rα, c-Kit, a number of tumor necrosis factor (TNF) ligands, and the transcription factor RORγt. Further research will characterize the importance of these NK cell subsets in health and disease.

Lung NK cells

Cytotoxicity assays from crude lymphocyte preparations of murine lung indicated the presence of NK cells (Puccetti et al., 1980). It was then further shown that the use of anti-NK cell depleting antibodies decreased lung cell cytotoxicity, whereas influenza infection increased cytotoxicity, providing evidence for resident NK cells in the mouse lung (Stein-Streilein et al., 1983). Further experiments with mice showed that lung NK cells can present with higher cytotoxic activity than blood and spleen populations (Wiltrout et al., 1985), while depletion of NK cells from the lung led to the increased susceptibility and mortality of mice infected with the influenza virus (Stein-Streilein and Guffee, 1986). In addition, human lung lymphocytes displayed cytotoxic activity (Bordignon et al., 1982), although resting lung NK cells were only capable of killing the target following stimulation (Robinson et al., 1984). It was also shown that alveolar macrophages could potently inhibit lung NK cell activity but had little effect on peripheral blood NK cells (Roth and Golub, 1989), suggesting that in the lung, NK cells might represent a different subtype that responds to different regulatory mechanisms.

In the mouse, NK cells can also be actively recruited to the lung following immunization (Fogler et al., 1996) and have been implicated in a vast array of bronchial disease models. Thus, in allergen-induced pulmonary airway inflammation, NK cells were found to induce the infiltration of both eosinophils and T cells (Korsgren et al., 1999). NK cells could also induce pulmonary inflammation during infection with *Streptococcus pneumoniae* (Kerr et al., 2005). Moreover, the cytokine IL-18 in combination with IL-2 is associated with lethal lung injury due to uncontrolled NK cell infiltration (Okamoto et al., 2002). IL-18 has also been shown to promote lung NK cell activity during influenza virus infection (Fan et al., 2006) as well as after intranasal challenge of herpes simplex virus (HSV)-1 (Reading et al., 2007). Infection with the intracellular pathogen *Mycobacterium tuberculosis* also results in local lung NK cell activation, however, their importance in clearing the infection is not clear yet (Feng et al., 2006; Junqueira-Kipnis et al., 2003). Nevertheless, lung NK cells appear to be very important for resistance against *Staphylococcus aureus* infection (Desanti et al., 2008).

In addition to controlling infection, NK cells in the lung present a very important line of defense against cancer. Thus, there is a direct correlation between the number of lung NK cells in patients with primary squamous cell adenocarcinoma: The higher the NK cell numbers, the better the survival (Villegas et al., 2002). Interestingly, when lung adenocarcinoma was deficient for MHC-I expression, the result was inhibition of killing this tumour but not other tumour cell lines or targets (Le Maux Chansac et al., 2005), suggesting that the tumour itself or its surrounding microenvironment can influence the functional capacity of NK cells.

NK cells in the lung therefore comprise a very important line of defense both against infection and malignancy.

Liver NK cells

One of the main NK cell reservoirs in humans and rodents is the liver. In fact, NK cells were first described in the rat liver, as large granular cells residing in the sinusoids and were named Pit cells (Doherty and O'Farrelly, 2000; Wisse et al., 1976). Their presence has been confirmed in both mouse (Doherty and O'Farrelly, 2000; Wiltrout et al., 1984) and human (Doherty and O'Farrelly, 2000; Doherty et al., 1999). NK cells comprise almost 30% and 20% of total hepatic lymphocytes in human (Doherty and O'Farrelly, 2000) or mouse, respectively (Smyth et al., 2001). Normal human liver contains approximately equal numbers of NK and T cells (Doherty and O'Farrelly, 2000; Doherty et al., 1999; Hata et al., 1991, 1992); however, during malignancy, NK cells can constitute more than 50% of total liver lymphocytes (Takii et al., 1994), suggesting an important role of NK cells in liver cancer. Hence, early experiments revealed that lack of NK cells is associated with significantly increased liver metastasis (Wiltrout et al., 1985). It has been shown that besides utilizing chemokines (discussed later), NK cells migrate and enter the liver through interactions with vascular cell adhesion molecule-1 (VCAM-1) (Fogler et al., 1996, 1998), while NK cells that are deficient in their expression of the transcription factor GATA-3 have impaired capacity to migrate to the liver (Samson et al., 2003).

Besides the fact that the liver is enriched in NK cells, there is significant evidence showing that hepatic NK cells are functionally distinct from peripheral blood or splenic NK cells. Thus, murine hepatic but not splenic NK cells could lyse otherwise NK-insensitive target cell lines (Cohen et al., 1985) and be presented with a generalized increased activity (Magilavy et al., 1987; Wiltrout et al., 1985). Experiments using primary freshly isolated human NK cells from liver samples have also concluded that hepatic NK cells have a substantial increase in activation when compared to peripheral blood populations (Hata et al., 1991, 1992). It has been proposed that the differences in liver NK cell activity is due to the different microenvironment by comparison to blood and spleen, and especially due to the presence of liver-resident macrophages, the Kupffer cells (Alba et al., 2008; Vanderkerken et al., 1995). In agreement with this hypothesis, depletion of Kupffer cells results in reduced NK cell numbers, while culture of NK cells with media that were pre-conditioned with Kupffer cells showed increased cytolytic capacity (Vanderkerken et al., 1995). Furthermore, Kupffer cells isolated from human liver responded to Toll-like receptors (TLRs) and could induce the production of IFNγ by NK cells in a contact-dependent manner (Alba et al., 2008). Interestingly, certain TLR ligands could stimulate Kupffer cells to secrete IL-10, which could in turn inhibit NK cell IFNγ production (Alba et al., 2008). Therefore, Kupffer cells have an active and very important role in regulating NK cell activation in the liver.

The presence of NK cells in the liver indicates a role for anti-viral protection. Subsequently, it has been shown that NK cell depletion is associated with increased replication of MCMV, vaccinia virus and mouse hepatitis virus (Bukowski et al., 1983). Extensive in vivo mouse work has linked MCMV replication in the liver with NK cell activation and migration by type I IFNs and a family of chemokines (see Section CC chemokines). Strong associations between NK cells and hepatitis C virus (HCV) infection have also been reported. Accordingly, it has been shown that inhibitory NK receptors can confer relative resistance to HCV if their ability to inhibit is diminished (Khakoo et al., 2004). In addition, by comparison to healthy individuals, the numbers of NK cell populations in cirrhotic livers of chronic HCV-infected patients were decreased, whereas blood NK cell numbers were not affected (Kawarabayashi et al., 2000). On the other hand, peripheral NK cells from HCV-infected people were incapable of inducing the maturation of DCs because they produced higher levels than normal IL-10 and TGFβ, two important anti-inflammatory cytokines (Jinushi et al., 2004). These NK cells expressed increased levels of the inhibitory receptor complex CD94/NKG2A, which when blocked, resulted in reduction of both IL-10 and TGFβ and restoration of NK cell mediated DC maturation (Jinushi et al., 2004). It is apparent therefore that a liver-specific virus, HCV, is firstly actively controlled by NK cells and secondly can cause disturbances in NK cell subsets both locally and in the periphery.

In contrast to HCV causing liver NK cell numbers to drop (Kawarabayashi et al., 2000), infection with *Plasmodium yoelii* is followed by higher numbers of liver NK cells but reduced splenic NK numbers, which also lose their ability to protect (Roland et al., 2006) and further demonstrate the significant differences between NK cells in the liver and other organs.

NK cells have also been associated with protection against hepatitis B virus (HBV) (Webster et al., 2000), though a direct correlation is lacking. Nevertheless, there is evidence implicating hepatic NK cells in immune-mediated liver pathology that is common in HBV infection (Dunn et al., 2007). Liver NK cells uniquely express surface TRAIL (TNF-related apoptosis inducing ligand) (Smyth et al., 2001), which binds the TRAIL death-inducing receptors and induces apoptosis of target cells (Dunn et al., 2007). Thus, chronic HBV patients have elevated TRAIL expression on NK cells, whereas hepatocytes show induction of TRAIL receptors that renders them sensitive to NK killing, which in turn will cause liver damage (Dunn et al., 2007). Importantly, mouse studies have shown that a combination of TRAIL expression and lack of inhibitory receptors from liver NK cells makes them capable of syngeneic hepatocyte killing (Ochi et al., 2004), a property that is likely to contribute to liver injury. On the other hand, TRAIL-expressing hepatic NK cells can eliminate TRAIL-sensitive tumours, and as a result, they can protect against cancer metastasis into the liver, which is significantly increased when TRAIL is blocked (Smyth et al., 2001). In human liver NK cells, TRAIL has been reported to be expressed after IL-2 stimulation (Ishiyama et al., 2006).

The liver therefore contains phenotypically [TRAIL expression and lack of CD16 in humans (Hata et al., 1991)] and functionally distinct NK cell populations. The significance of this finding is that it suggests either specific recruitment of subpopulations of NK cells, or alternatively, modification of their phenotype and function by the local microenvironment in response to infection and malignancy.

Thymic NK cells

Early studies suggested that in vitro culture of human thymocytes resulted in the generation of NK-like cells that could eliminate NK-sensitive targets (Blue et al., 1987), while culture with recombinant IL-2 generated CD56$^+$ cells that lacked CD16 or CD3 and possessed NK cell activity (Ramsdell and Golub, 1987). Cells with NK cell activity and phenotype were also later described in mouse thymus, using animals that lacked all B and T cells (Garni-Wagner et al., 1990). In situ identification of NK cells in the adult murine thymus locates them both in the cortex and the medulla (Brown et al., 2001). Recent work in the mouse has provided substantial evidence that thymic NK cells constitute a unique population (Di Santo and Vosshenrich, 2006). Thus, NK cells in the thymus express high levels of IL-7Rα and the transcription factor GATA-3, and, consequently, a deficiency in IL-7 or GATA-3 diminishes their production and homeostasis. Furthermore, they are negative for CD16, and low for CD11b and Ly49 receptors, but high for CD69, whereas functionally they resemble human CD56^{bright}CD16$^-$ NK cells—that is decreased natural cytotoxicity but increased cytokine production. The authors additionally discovered that human CD56^{bright}CD16$^-$ NK cells expressed more IL-7Rα and GATA-3, while the development of murine thymic NK cells completely depended on GATA-3 and IL-7 expression.

Moreover, human fetal thymus contained committed NK cell progenitors (Sanchez et al., 1993) as well as T/NK bi-potent progenitors (Ikawa et al., 1999; Sanchez et al., 1994) that can be identified by their expression of CD34 (Sanchez et al., 1994). Bi-potent T/NK progenitors in the thymus differentiate preferentially to the T lineage. NK cell differentiation depends on the availability of the cytokines IL-2 and IL-15 (Leclercq et al., 1996) and the expression of the transcription factor Id3 (Heemskerk et al., 1997). Thus, IL-15 (Leclercq et al., 1996) and Id3 (Heemskerk et al., 1997) are positive regulators of NK cell differentiation from thymic T/NK bi-potent progenitors. Surprisingly, the fetal murine thymus contains NK cells that express activating and inhibitory receptors and are capable of direct target lysis ex vivo (Carlyle et al., 1998).

Despite the fact that considerable work has elucidated the developmental requirements for thymic NK cells, their precise function during health and disease remains to be determined.

Table 10.1 Anatomical location of mature NK cells

Organ	Specific location	Special features
Lymph node	Medulla & T zone	CD56^{bright} in human CD27^{high} in mouse
Spleen	Red pulp (marginal zone)	Mixed populations Migrate to white pulp after activation
Gut	Intraepithelial, lamina propria & cryptopatches	Low cytotoxicity & IFNγ Resemble LTi cells Some make IL-22, and some can participate in Th2 responses
Lung	Bronchoalveolar space, lung interstitium	Locally regulated by alveolar macrophages
Liver	Sinusoids	Express TRAIL CD16$^-$ in human Lower activation threshold Regulated by Kupffer cells
Thymus	Cortex & medulla	IL7Rα, GATA-3$^+$, CD16$^-$, CD11b/Ly49^{low} Low cytotoxicity, high IFNγ

LTi, lymphoid tissue inducer cell

NK cell migration

All the preceding information suggests that the typical NK cell stems from the BM and then migrates to various organs where it rests until it encounters a danger signal (Table 10.1). This, however, is oversimplified. First, it is now well appreciated that NK cells undergo homeostatic proliferation. Thus, there is constant NK cell turnover (Jamieson et al., 2004), which is increased substantially in lymphopenic environments (Jamieson et al., 2004; Prlic et al., 2003). Importantly, NK cells express a vast array of chemokine receptors (Berahovich et al., 2006; Gregoire et al., 2007) and respond to chemokines (Inngjerdingen et al., 2001). An extensive amount of work has characterized important chemokine signatures that help NK cells to move in response to infectious, tumourigenic or inflammatory challenges.

CC chemokines

Chemokines contain one or two conserved NH_2-terminal cysteines (C), which can be adjacent (CC chemokines) or separated by amino acids (CXC or CX_3C chemokines; where X = any amino acid) (Cyster, 1999). Specialized cell membrane-expressed receptors can recognize and bind chemokines; however, a single chemokine can interact with more than one receptor. A receptor can only bind either a CC chemokine [CC-receptor(R)] or a CXC chemokine (CXCR), but never both. Thus, CC chemokines can bind one or more CCRs but not CXCRs, whereas the opposite stands for CXC chemokines.

Accordingly, the monocyte chemotactic proteins MCP-1 or CCL2, MCP-2 or CCL8, and MCP-3 or

CCL7 share a common receptor, CCR2, which is expressed at a high level on IL-2 activated NK cells (Godiska et al., 1997; Polentarutti et al., 1997). Hence, all three chemokines are capable of inducing in vitro NK cell chemotaxis (Allavena et al., 1994). However, CCL2 was found to be the strongest NK cell chemo-attractant (Allavena et al., 1994; Maghazachi et al., 1994), and its contribution to NK cell migration has been more carefully characterized, in comparison to CCL7 or CCL8.

To experimentally test the in vivo importance of CCL2 in NK cell migration, tumour cells were engineered to express CCL2 and were then transplanted into mice (Nokihara et al., 2000). $CCL2^+$ tumour cells were very inefficient in metastasizing to the lung compared to $CCL2^-$ tumours, and the recipients survived for longer. This however, was the case only in NK-sufficient but not in NK-depleted hosts, suggesting that CCL2 can induce NK cell-mediated eradication of metastatic tumours (Nokihara et al., 2000).

Aspergillosis can be a devastating fungal-induced disease in immunocompromised, especially neutropenic, patients (Mehrad et al., 1999; Morrison et al., 2003). Murine models have showed that infection with *Aspergillus fumigatus* results in increased levels of CCL2 in lung and serum and that depletion of NK cells correlates with increased susceptibility (Mehrad et al., 1999; Morrison et al., 2003). It was then demonstrated that neutralization of CCL2 firstly diminishes NK cell accumulation in the lungs of infected mice and secondly reduces the ability of fungal clearance, a phenotype that was replicated when CCR2 was absent from NK cells (Morrison et al., 2003). Therefore, there is evidence implicating the requirement for CCL2/CCR2-mediated recruitment of NK cells in the lung, both during tumour metastasis (Nokihara et al., 2000) and during fungal infection (Morrison et al., 2003).

In addition to lung infections, CCL2 was found to be strongly expressed in the liver of mice that were infected with MCMV (Hokeness et al., 2005). Further experiments provided clear evidence that this CCL2 production was primarily coming from liver-resident leukocytes, and it was dependent on IFNα/β (Hokeness et al., 2005). Significantly, mice lacking either CCL2 or CCR2 displayed deficient NK cell recruitment to the liver during MCMV infection (but not in uninfected mice), and this was associated with increased hepatic viral titres and exacerbated liver damage (Hokeness et al., 2005). Moreover, a model of adenoviral-based gene therapy against hepatocellular carcinoma showed that usage of recombinant adenovirus carrying CCL2 is more effective at tumour eradication via a mechanism involving the preferential recruitment of NK cells to the liver, as shown by increased NK cell infiltration and abrogation of protection when NK cells were depleted (Tsuchiyama et al., 2007).

Besides preventing infection or tumour formation, NK cells may also be the cause of disease in certain clinical settings. Thus, patients deficient in TAP2 (transporter associated with antigen presentation) can develop lethal lung granulomatous lesions due to increased and chronic NK cell accumulation (Moins-Teisserenc et al., 1999). To this end, it was recently shown that TAP2-deficient patients presented with increased levels of CCL2 in bronchioalveolar lavage samples, and their NK cells expressed unnaturally high levels of CCR2—and migrated towards gradients of CCL2—and the other MCP chemokines and CCR2 ligands, CCL7 and CCL8 (Hanna et al., 2005).

A crucial difference between CCL2 and CCL7 and CCL8 is that the latter two chemokines also interact with CCR5, which has the alternative ligand, CCL5. CCL5 or RANTES (regulated upon activation, normal T cell expressed and secreted) in turn can also interact with CCR1 and CCR3, and can be a potent NK cell chemo-attractant (Loetscher et al., 1996). Accordingly, injection of a CCL5-expressing thymoma cell line into syngeneic mice caused significant reduction in tumour growth and resulted in NK cell influx (Lavergne et al., 2004). However, the role of NK cells in this model remains obscure because mice lacking only B and T cells could not clear CCL5-expressing tumours, suggesting a more dominant role for lymphocytes other than NK cells (Lavergne et al., 2004). Furthermore, infection of $CCR5^{-/-}$ mice with HSV-1 results in decreased ability to control viral replication, which is associated with diminished mobilization of NK cells towards the spleen, brain stem and spinal chord (Thapa et al., 2007). Surprisingly, concanavalin A- mediated hepatitis in CCR5-deficient mice resulted in severe disease due to uncontrolled migration of NK cells to the liver (Ajuebor et al., 2007). It was then shown that in the liver of these animals, there were markedly elevated CCL5 levels, which actively attracted NK cells by a mechanism involving CCR1, the other receptor that recognizes CCL5 and is expressed on NK cells (Ajuebor et al., 2007). Besides the implications of the latter study in possible treatments against liver disease, a secondary outcome is that it presents a good example of the complex interactions between different chemokines with different receptors.

Therefore, the chemokines CCL3 (macrophage inflammatory protein-1 or MIP-1α) and CCL4 (MIP-1β) also bind CCR5, while CCL3 can interact with CCR1, and early on it was established that CCL3 is a more potent chemotactic molecule for NK cells (Maghazachi et al., 1994; Taub et al., 1995). Similar to CCL2, expression levels of CCL3 are markedly upregulated in the liver during the early stages of MCMV infection (Salazar-Mather et al., 1998). Hence, mice deficient in CCL3 displayed impaired NK cell recruitment to the liver, which was followed by decreased inflammatory foci but elevated viral replication (Salazar-Mather et al., 1998) and reduced host survival (Salazar-Mather et al., 2000). Interestingly, in the absence of CCL3, NK cell numbers decreased in the livers of MCMV infected mice, and blood NK cell numbers increased significantly compared to uninfected animals (Salazar-Mather et al., 2000). Work from the same research group also demonstrated that CCL3 induction in the liver and subsequent NK cell accumulation depends on type I IFNs (Dalod et al., 2002), suggesting that CCL2 as well as CCL3 play an interconnecting role in NK cell recruitment towards the liver during CMV infection. In this infection setting, CCL2 is expressed earlier than CCL3 (Hokeness et al., 2005). This and the fact that CCL3$^{-/-}$ mice are able to produce CCL2 (Hokeness et al., 2005) depicts that the effects of these chemokines are synergistic but not redundant and would suggest that they act in sequence—that is CCL2 mediates the very early NK cell influx that needs sustained CCL3 production for efficient migration to the liver. It would be therefore of great interest to study NK cell migration in the absence of both CCL2 and CCL3 either by creating double-deficient animals or by antibody blockade of either chemokine in single-deficient mice.

The use of *Klebsiella pneumoniae* as an infection model showed that induced expression of CCL3 in the lung correlated very well with reduced bacteria replication and a significant increase of NK cell numbers in the lungs of infected mice (Zeng et al., 2003). Of interest is a study that showed that injection of rituximab (anti-CD20 antibody) induced the expression of CCL3 and this was required for efficient tumour cell eradication (Cittera et al., 2007). Subsequent experiments revealed the action of Rituximab and CCL3 could be reversed by depletion of leukocyte populations, including NK cells (Cittera et al., 2007). The diverse role of CCL3 is also evident by the observations that during pregnancy specialized placental cells can secrete CCL3 and attract maternal NK cells (Drake et al., 2001).

Phenotypic characterization both at the protein and transcript level provided evidence that NK cells can express the chemokine receptors CCR4 and CCR8 following activation with IL-2 (Inngjerdingen et al., 2000). Both of these receptors can induce the chemotaxis of NK cells through the action of CCL17, which binds both receptors; CCL1, which binds CCR8; and CCL22, which interacts with CCR4 (Godiska et al., 1997; Inngjerdingen et al., 2000, 2001).

It is apparent that CC chemokines and their receptors have a very important role to play in the biology of NK cells by inducing their migration and mobilization at situations of host evasion by means of inflammation, infection or malignancy.

CXC chemokines

Besides members of the CC chemokine family, the microenvironment that is created by inflammation, infection and malignancy induces the local production of a number of CXC chemokines, some of which have been shown to promote NK cell migration towards the affected tissue. Of these chemokines, CXCL9, CXCL10, and CXCL11 play a very prominent and often interchangeable role in NK cell mobilization by sharing CXCR3 as their binding counterpart (Weng et al., 1998) and by inducing strong NK cell chemotaxis (Maghazachi et al., 1997).

Previously we have discussed that CXCR3 binding chemokines can promote the active recruitment of NK cells from the red to the white pulp of the spleen during infectious stimuli (Bekiaris et al., 2008; Gregoire et al., 2008), as well as their recruitment to LNs (Martin-Fontecha et al., 2004). The study of *Leishmania major* in mice showed that resistant but not susceptible strains expressed high levels of CXCL10 in infected LNs while injection of recombinant CXCL10 in infected mice increased LN NK cell activity (Vester et al., 1999). In addition to *L. major*, infection of DCs with *M. tuberculosis* renders them chemo-attractant for NK cells (Lande et al., 2003). This chemotactic property of DCs towards NK cells was attributed to their production of CXCL9 and CXCL10, and it was further shown that induction of CXCL10 depended on type I IFN (Lande et al., 2003).

The classical type I IFN-producing cells, the plasmacytoid DCs (pDCs), can also attract NK cells following their infection with HSV utilizing a mechanism involving the production of CXCL10 (Megjugorac et al., 2004). Interestingly, there was some evidence suggesting that IFNα could act on pDCs in an autocrine fashion to promote CXCL10 expression (Megjugorac et al., 2004). Ocular infection with HSV-1 in the absence of CXCR3 is associated with a significant reduction in NK cell recruitment and an elevation in the viral titre as marked in NK-depleted mice (Alba et al., 2008), providing evidence for the active participation of NK cells in ocular protection. Moreover, when either CXCL9- or CXCL10-deficient mice were infected with genital HSV-2, they presented with acute reduction in NK cell numbers from the brain and spinal chord, which was associated with elevated viral replication and impaired survival (Thapa et al., 2008).

In addition, infection of mice with mouse coronavirus can result in 100% mortality and very high titres in the brain unless the virus is engineered to express CXCL10 (Trifilo et al., 2004). Hence, CXCL10-expressing coronavirus induced substantial NK cell infiltration in the brain with subsequent resolution of infection and host survival (Trifilo et al., 2004).

Therefore, there is a very tight association between recruitment of NK cells in pathogen-infected tissues and the expression of CXCR3 ligands. A similar association has also been described for tumour- and inflammation-induced cell infiltration. Thus, tumours from

$CXCR3^{-/-}$ mice have reduced NK cell numbers, while $CXCR3^{-/-}$ NK cells present with impaired capacity to infiltrate tumours when adoptively transferred into wild type recipients (Wendel et al., 2008). Besides recruitment, CXCL10 can directly stimulate NK cell tumour killing as well as induce B7-H1 on the surface, which acts to activate anti-tumour T cells and thus confer long-term protection (Saudemont et al., 2005).

Furthermore, a mouse model of non-infectious pulmonary fibrosis showed that in CXCR3 deficiency, there is significant acceleration of disease on one hand and significant decrease in NK and CD8 T cell accumulation in the lungs on the other (Jiang et al., 2004). Importantly, the investigators found that $CXCR3^{-/-}$ mice had very reduced NK cell numbers in lung, liver and blood, and went on to demonstrate that early IFNγ production by lung cells is responsible for limiting lung fibrosis (Jiang et al., 2004). Besides, $CD56^{bright}CD16^{-}$ NK cells that can be identified in and isolated from psoriatic human skin express CXCR3 and are able to migrate towards CXCL10 gradients (Ottaviani et al., 2006), suggesting a role of NK cells in psoriasis. Of importance is the observation that human uterine NK cells express high levels of CXCR3, and the endometrium upregulates both CXCL10 and CXCL11 in response to sex hormones, such as progesterone and estradiol (Sentman et al., 2004), suggesting that recruitment of NK cells to the endometrium during pregnancy might occur via CXCR3.

Thus, CXCR3 through interacting with its three ligands, CXCL9, CXCL10, CXCL11, regulates NK cell migration during a plethora of pathological conditions, including cancer, infection and inflammation, which renders it a potential therapeutic target. However, due to the diversity of CXCR3 functions on NK cells, we need to expand our knowledge in mechanisms that regulate its expression and the expression of its ligands. It is worth noting, for example, a study that provided evidence that the potent NK-activating cytokines IL-2 and IL-12 reduced membrane CXCR3 levels and hence chemotaxis towards CXCL10 (Hodge et al., 2002). It follows that generalized NK cell activation might not necessarily lead to CXCR3 responsiveness. Moreover, our group has also observed downregulation of CXCR3 from NK cells during MCMV infection, and we attributed this to receptor/ligand downmodulation (Bekiaris et al., 2008). However, it is very likely that expression of CXCR3 is regulated actively through the action of cytokines or other immunological mediators in order to maintain a balance between excessive NK-mediated damage and resolution of infection or malignancy.

In addition to CXCR3 ligands, other members of the CXC family of chemokines are involved in the mobilization of NK cells. It has been recently shown that mast cells, which have been previously activated with reovirus express high levels of CXCL8 (Burke et al., 2008). CXCL8 can bind either to CXCR1 or CXCR2, and its production from mast cells is sufficient to attract NK cells mainly via interactions with CXCR1 (Burke et al., 2008). Similarly, the chemokine CXCL14 has also been found to induce the chemotaxis of NK cells (Starnes et al., 2006). It is of interest that while most tissues can express CXCL14, a number of tumour cells fail to do so, indicating that downregulation of CXCL14 might be a mechanism to prevent NK cell recruitment (Starnes et al., 2006).

CXCL12, also known as SDF-1α (stromal cell derived factor-1α) interacts with CXCR4, which is expressed on mouse (Bernardini et al., 2008) as well as human (Beider et al., 2003) NK cells. It has been shown that adoptive transfer of human NK cells into immunodeficient mice results in their migration towards both the spleen and the BM. However, only migration to the BM depended on CXCR4 (Beider et al., 2003). Thus, CXCR4 and CXCL12 appear to regulate homing of NK cells to the BM. CXCL12 is expressed by BM endothelium, which also expresses the adhesion molecule VCAM-1, the ligand for which, the integrin $\alpha4\beta1$ or VLA-4 (very late antigen-4) is expressed on NK cells (Franitza et al., 2004). There is evidence demonstrating that CXCL12 induces the firm adhesion of NK cells on $VCAM\text{-}1^{+}$ cells through interactions with VLA-4, while in vivo blocking of VLA-4 prevents NK cell recruitment to the BM (Franitza et al., 2004).

In the mouse, CXCR4 is expressed by blood and splenic NK cells as well as by BM NK populations irrespective of whether they are at the progenitor, immature or mature stage of their development (Bernardini et al., 2008). However, BM mature NK cells show the highest responsiveness to CXCL12, and when CXCR4 signalling is blocked, BM NK cell numbers decrease, whereas blood and splenic NK numbers increase accordingly, suggesting therefore that the CXCR4/CXCL12 interactions can promote retention of NK cells in the BM (Bernardini et al., 2008).

An important question however, is what regulates the exit of NK cells from the BM. Recent evidence suggests that BM emigration is not regulated by a chemokine/chemokine receptor pair but by the sphingosine 1-phosphate receptor 5 (S_1P_5) (Gregoire et al., 2007). Thus, mice deficient in S_1P_5 present with decreased NK numbers in blood, spleen and lung but increased numbers in the BM (Gregoire et al., 2007), i.e. the opposite of CXCR4 blockade.

In addition, CXCL12 is expressed by endothelial cells of adenoids (a MALT site), while adenoid NK cells express high levels of CXCR4, and in vitro, they can migrate towards CXCL12 gradients, suggesting that homing of NK cells to adenoids may be partly controlled by CXCL12/CXCR4 dependent migration (Mizrahi et al., 2007). Moreover, there is evidence to suggest that $CD56^{bright}CD16^{-}$ human decidual NK cells express

CXCR4, and respond and migrate towards CXCL12-expressing trophoblasts (Beider et al., 2003; Infante-Duarte et al., 2005).

Taken together, there is strong experimental evidence that links directly the mobilization of NK cells during disease and steady-state conditions with chemokines of the CXC family, particularly with CXCR3 and CXCR4 ligands. Further research will elucidate the exact mechanisms by which these and other chemokine systems organize how NK cells move throughout the body, a process that is essential for immune surveillance and protection against infection.

CX_3CL1, fractalkine

Fractalkine was identified as a unique chemokine containing a three amino acid motif between the two cysteines (C-x-x-xC, CX_3CL1) and a mucin-like domain (Bazan et al., 1997). In addition, CX_3CL1 can exist as a soluble form inducing chemo-attraction and as a membrane-bound form promoting the firm adhesion of cells on endothelium (Bazan et al., 1997). Both of these functions require the presence of its specific receptor, CX_3CR1, which is expressed at high levels on NK cells (Imai et al., 1997). Thus, CX_3CL1 expression by endothelial cells can promote the adhesion of and subsequent lysis by NK cells (Yoneda et al., 2000).

Experiments to assess its role on NK cell biology showed a strong correlation between CX_3CL1 and anti-tumour responses. Hence, when CX_3CL1-expressing lymphoma cell lines were injected into mice, tumour growth was suppressed due to the recruitment of NK cells that controlled the tumour via production of IFNγ and perforin (Lavergne et al., 2003). Similar experiments utilizing a lung carcinoma cell line also concluded that CX_3CL1 is required for NK cell recruitment at the site of tumour growth, leading thus to its eradication (Guo et al., 2003). Infiltration of NK cells associated with tumour clearance was also observed when an adenoviral vector expressing CX_3CL1 was used to deliver the chemokine at the tumour site (Banks et al., 2005). Furthermore, mice deficient for CX_3CR1 were substantially repressed in their ability to recruit NK cells in the lung both in the presence of cancer and under homeostatic conditions (Giroux et al., 2007).

In addition to promoting their migration, CX_3CR1 is also involved in IFNγ production by NK cells. Thus, when CX_3CR1 expression is lost, NK cells present with reduced capacity to secrete IFNγ (Giroux et al., 2007), while IFNγ production is induced in vitro by human peripheral blood NK cells upon incubation with CX_3CL1 (Yoneda et al., 2003). Moreover, CX_3CL1 expressed by mature DCs is involved in DC-induced NK cell activation (Pallandre et al., 2008). The effects of CX_3CL1 can be modulated by IL-15 since it has been shown to downregulate the expression of CX_3CR1 by NK cells (Sechler et al., 2004).

Therefore, CX_3CL1 and its receptor CX_3CR1 regulate the migration of NK cells to tumour sites as well as their effector function.

Do NK cells traffic?

From the preceding information, it becomes apparent that there is an avalanche of information regarding the migration of NK cells during disease states. We saw, for example, how CC and CXC chemokines regulate the mobilization of NK cells towards various organs that are affected by infection or cancer (Table 10.2). Similarly, chemokine/chemokine receptor codes localize NK cells in the organs around the body, while in the absence of infection, CXCR4 appears to regulate recruitment to and retention in the BM.

The fact that NK cells are seen in many organs would suggest two alternatives:

1. They leave the BM and migrate to various organs where they take on certain activities and phenotype

Table 10.2 Examples of chemokines required for NK cell migration

Chemokine	Receptor	NK migratory capacity
CCL2	CCR2	Lung (malignancy, fungal infection, chronic inflammation)
CCL3	CCR1, CCR5	Lung (bacterial infection, inflammation) Liver (viral infection) Placenta (pregnancy)
CCL5	CCR1, CCR3, CCR5	Spleen (viral infection) Liver (viral infection)
CCL21	CCR7	Lymph node (homeostasis)
CXCL9 CXCL10 CXCL11	CXCR3	Splenic white pulp (viral infection) Lymph node (homeostasis) Eye & brain (viral infection) Various tumour sites Lung (inflammation) Skin (psoriasis) Endometrium (pregnancy)
CXCL12	CXCR4	Bone marrow (homeostasis) Adenoids (homeostasis) Trophoblasts (pregnancy)
CX_3CL1	CX_3CR1	Various tumour sites

imprinted by the local environment. They then can get mobilized again during disease that causes alterations in chemokine gradients.

2. NK cells are constantly 'on the go' trafficking around tissues.

Unfortunately, it is still uncertain which of these alternatives is true, or whether there is a mixture of both scenarios. Work from Vivier and colleagues suggest the latter. Thus, when transferred into a naïve syngeneic host, spleen-derived murine NK cells were found in all the organs where NK cells localize and at the same proportions as host populations (Gregoire et al., 2007). This suggests that NK cells from one anatomical location are not restricted to that environment and can re-circulate between organs. It is possible, however, that due to the direct connection between spleen and blood, splenic NK cells contain recent BM emigrants that have the capacity to migrate everywhere. It would be of interest to repeat the preceding transfer experiments with NK cells from different organs, as there is still the possibility that a certain microenvironment can imprint changes on the NK cell that affect its ability to traffic. It is also plausible that while certain NK cell subsets traffic continuously, some other subsets do not or do so during host evasion by pathogens, cancer or inflammation. The determination and characterization of these possible functions will be seminal in furthering our understanding of NK cell biology.

Concluding remarks

NK cells develop in the BM, which they exit using specific molecular interactions. Exit from the BM is followed by localization to a number of tissues, including secondary lymphoid organs. Within each tissue, NK cells often acquire unique function and phenotype that is regulated by the local microenvironment. Their localization is primarily directed by the action of chemokines and therefore is in tight association with the activation status of the organism. Changes in chemokine expression during disease results in further NK cell mobilization and allows them to protect the host from infection and malignancy. Thus, from their time of production until their end, NK cells travel exhaustively over long distances and visit places that influence their already dynamic life. The future promises to uncover a lot more truths about their nature.

References

Adam, C., King, S., Allgeier, T., Braumuller, H., Luking, C., Mysliwietz, J., Kriegeskorte, A., Busch, D.H., Rocken, M. and Mocikat, R. (2005). DC-NK cell cross-talk as a novel CD4+ T cell-independent pathway for antitumor CTL induction. Blood **106**, 338–344.

Ajuebor, M.N., Wondimu, Z., Hogaboam, C.M., Le, T., Proudfoot, A.E. and Swain, M.G. (2007). CCR5 deficiency drives enhanced natural killer cell trafficking to and activation within the liver in murine T cell-mediated hepatitis. Am J Pathol **170**, 1975–1988.

Alba, A., Planas, R., Clemente, X., Carrillo, J., Ampudia, R., Puertas, M.C., Pastor, X., Tolosa, E., Pujol-Borrell, R., Verdaguer, J. and Vives-Pi, M. (2008). Natural killer cells are required for accelerated type 1 diabetes driven by interferon-beta. Clin Exp Immunol **151**, 467–475.

Allavena, P., Bianchi, G., Zhou, D., van Damme, J., Jilek, P., Sozzani, S. and Mantovani, A. (1994). Induction of natural killer cell migration by monocyte chemotactic protein-1, -2 and -3. Eur J Immunol **24**, 3233–3236.

Andrews, D.M., Andoniou, C.E., Granucci, F., Ricciardi-Castagnoli, P. and Degli-Esposti, M.A. (2001). Infection of dendritic cells by murine cytomegalovirus induces functional paralysis. Nat Immunol **2**, 1077–1084.

Ansel, K.M., McHeyzer-Williams, L.J., Ngo, V.N., McHeyzer-Williams, M.G. and Cyster, J.G. (1999). In vivo-activated CD4 T cells upregulate CXC chemokine receptor 5 and reprogram their response to lymphoid chemokines. J Exp Med **190**, 1123–1134.

Bajenoff, M., Breart, B., Huang, A.Y., Qi, H., Cazareth, J., Braud, V.M., Germain, R.N. and Glaichenhaus, N. (2006a). Natural killer cell behavior in lymph nodes revealed by static and real-time imaging. J Exp Med **203**, 619–631.

Bajenoff, M., Egen, J.G., Koo, L.Y., Laugier, J.P., Brau, F., Glaichenhaus, N. and Germain, R.N. (2006b). Stromal cell networks regulate lymphocyte entry, migration and territoriality in lymph nodes. Immunity **25**, 989–1001.

Baker, D.H. and Wood, R.J. (1992). Cellular antioxidant status and human immunodeficiency virus replication. Nutr Rev **50**, 15–18.

Banks, T.A., Rickert, S., Benedict, C.A., Ma, L., Ko, M., Meier, J., Ha, W., Schneider, K., Granger, S.W., Turovskaya, O., Elewaut, D., Otero, D., French, A.R., Henry, S.C., Hamilton, J.D., Scheu, S., Pfeffer, K. and Ware, C.F. (2005). A lymphotoxin-IFN-beta axis essential for lymphocyte survival revealed during cytomegalovirus infection. J Immunol **174**, 7217–7225.

Bazan, J.F., Bacon, K.B., Hardiman, G., Wang, W., Soo, K., Rossi, D., Greaves, D.R., Zlotnik, A. and Schall, T.J. (1997). A new class of membrane-bound chemokine with a CX3C motif. Nature **385**, 640–644.

Beider, K., Nagler, A., Wald, O., Franitza, S., Dagan-Berger, M., Wald, H., Giladi, H., Brocke, S., Hanna, J., Mandelboim, O., Darash-Yahana, M., Galun, E. and Peled, A. (2003). Involvement of CXCR4 and IL-2 in the homing and retention of human NK and NK T cells to the bone marrow and spleen of NOD/SCID mice. Blood **102**, 1951–1958.

Bekiaris, V., Timoshenko, O., Hou, T. Z., Toellner, K., Shakib, S., Gaspal, F., McConnell, F.M., Parnell, S.M., Withers, D., Buckley, C.D., Sweet, C., Yokoyama, W.M., Anderson, G. and Lane, P.J. (2008). Ly49H+ NK cells migrate to and protect splenic white pulp stroma from murine cytomegalovirus infection. J Immunol **180**, 6768–6776.

Benedict, C.A., Banks, T.A., Senderowicz, L., Ko, M., Britt, W.J., Angulo, A.,

Ghazal, P. and Ware, C.F. (2001). Lymphotoxins and cytomegalovirus cooperatively induce interferon-beta, establishing host-virus detente. Immunity **15**, 617–626.

Berahovich, R.D., Lai, N.L., Wei, Z., Lanier, L.L. and Schall, T.J. (2006). Evidence for NK cell subsets based on chemokine receptor expression. J Immunol **177**, 7833–7840.

Berlin, C., Berg, E.L., Briskin, M.J., Andrew, D.P., Kilshaw, P.J., Holzmann, B., Weissman, I.L., Hamann, A. and Butcher, E.C. (1993). Alpha 4 beta 7 integrin mediates lymphocyte binding to the mucosal vascular addressin MAdCAM-1. Cell **74**, 185–195.

Bernardini, G., Sciume, G., Bosisio, D., Morrone, S., Sozzani, S. and Santoni, A. (2008). CCL3 and CXCL12 regulate trafficking of mouse bone marrow NK cell subsets. Blood **111**, 3626–3634.

Bienenstock, J. and McDermott, M.R. (2005). Bronchus- and nasal-associated lymphoid tissues. Immunol Rev **206**, 22–31.

Blue, M.L., Levine, H., Daley, J.F., Craig, K.A. and Schlossman, S.F. (1987). Development of natural killer cells in human thymocyte culture: regulation by accessory cells. Eur J Immunol **17**, 669–673.

Bordignon, C., Villa, F., Vecchi, A., Giavazzi, R., Introna, M., Avallone, R. and Mantovani, A. (1982). Natural cytotoxic activity in human lungs. Clin Exp Immunol **47**, 437–444.

Brown, M.G., Dokun, A.O., Heusel, J. W., Smith, H.R., Beckman, D.L., Blattenberger, F.A., Dubbelde, C.E., Stone, L.R., Scalzo, A.A. and Yokoyama, W.M. (2001). Vital involvement of a natural killer cell activation receptor in resistance to viral infection. Science **292**, 934–937.

Bukowski, J.F., Woda, B.A., Habu, S., Okumura, K. and Welsh, R.M. (1983). Natural killer cell depletion enhances virus synthesis and virus-induced hepatitis in vivo. J Immunol **131**, 1531–1538.

Burke, S.M., Issekutz, T.B., Mohan, K., Lee, P.W., Shmulevitz, M. and Marshall, J. S. (2008). Human mast cell activation with virus-associated stimuli leads to the selective chemotaxis of natural killer cells by a CXCL8-dependent mechanism. Blood **111**, 5467–5476.

Campbell, J.J., Qin, S., Unutmaz, D., Soler, D., Murphy, K.E., Hodge, M.R., Wu, L. and Butcher, E.C. (2001). Unique subpopulations of CD56+ NK and NK-T peripheral blood lymphocytes identified by chemokine receptor expression repertoire. J Immunol **166**, 6477–6482.

Carlyle, J.R., Michie, A.M., Cho, S.K. and Zuniga-Pflucker, J.C. (1998). Natural killer cell development and function precede alpha beta T cell differentiation in mouse foetal thymic ontogeny. J Immunol **160**, 744–753.

Cittera, E., Leidi, M., Buracchi, C., Pasqualini, F., Sozzani, S., Vecchi, A., Waterfield, J.D., Introna, M. and Golay, J. (2007). The CCL3 family of chemokines and innate immunity cooperate in vivo in the eradication of an established lymphoma xenograft by rituximab. J Immunol **178**, 6616–6623.

Cohen, S.A., Salazar, D., von Muenchhausen, W., Werner-Wasik, M. and Nolan, J.P. (1985). Natural antitumor defense system of the murine liver. J Leukoc Biol **37**, 559–569.

Cyster, J.G. (1999). Chemokines and cell migration in secondary lymphoid organs. Science **286**, 2098–2102.

Dalod, M., Salazar-Mather, T.P., Malmgaard, L., Lewis, C., Asselin-Paturel, C., Briere, F., Trinchieri, G. and Biron, C.A. (2002). Interferon alpha/beta and interleukin 12 responses to viral infections: pathways regulating dendritic cell cytokine expression in vivo. J Exp Med **195**, 517–528.

Desanti, G.E., Cumano, A. and Golub, R. (2008). Identification of CD4int progenitors in mouse foetal spleen, a source of resident lymphoid cells. J Leukoc Biol **83**, 1145–1154.

Di Santo, J.P. and Vosshenrich, C.A. (2006). Bone marrow versus thymic pathways of natural killer cell development. Immunol Rev **214**, 35–46.

Doherty, D.G. and O'Farrelly, C. (2000). Innate and adaptive lymphoid cells in the human liver. Immunol Rev **174**, 5–20.

Doherty, D.G., Norris, S., Madrigal-Estebas, L., McEntee, G., Traynor, O., Hegarty, J.E. and O'Farrelly, C. (1999). The human liver contains multiple populations of NK cells, T cells, and CD3 + CD56+ natural T cells with distinct cytotoxic activities and Th1, Th2, and Th0 cytokine secretion patterns. J Immunol **163**, 2314–2321.

Drake, P.M., Gunn, M.D., Charo, I.F., Tsou, C.L., Zhou, Y., Huang, L. and Fisher, S.J. (2001). Human placental cytotrophoblasts attract monocytes and CD56(bright) natural killer cells via the actions of monocyte inflammatory protein 1alpha. J Exp Med **193**, 1199–1212.

Dunn, C., Brunetto, M., Reynolds, G., Christophides, T., Kennedy, P.T., Lampertico, P., Das, A., Lopes, A.R., Borrow, P., Williams, K., Humphreys, E., Afford, S., Adams, D.H., Bertoletti, A. and Maini, M.K. (2007). Cytokines induced during chronic hepatitis B virus infection promote a pathway for NK cell-mediated liver damage. J Exp Med **204**, 667–680.

Fan, Z., Yu, P., Wang, Y., Fu, M.L., Liu, W., Sun, Y. and Fu, Y.X. (2006). NK-cell activation by LIGHT triggers specific-specific CD8+ T-cell immunity to reject established tumours. Blood **107**, 1342–1351.

Fehniger, T.A., Cooper, M.A., Nuovo, G. J., Cella, M., Facchetti, F., Colonna, M. and Caligiuri, M.A. (2003). CD56bright natural killer cells are present in human lymph nodes and are activated by T cell-derived IL-2: a potential new link between adaptive and innate immunity. Blood **101**, 3052–3057.

Feng, C.G., Kaviratne, M., Rothfuchs, A. G., Cheever, A., Hieny, S., Young, H. A., Wynn, T.A. and Sher, A. (2006). NK cell-derived IFN-gamma differentially regulates innate resistance and neutrophil response in T cell-deficient hosts infected with Mycobacterium tuberculosis. J Immunol **177**, 7086–7093.

Ferlazzo, G. and Munz, C. (2004). NK cell compartments and their activation by dendritic cells. J Immunol **172**, 1333–1339.

Fogler, W.E., Volker, K., McCormick, K.L., Watanabe, M., Ortaldo, J. R. and Wiltrout, R.H. (1996). NK cell infiltration into lung, liver, and subcutaneous B16 melanoma is mediated by VCAM-1/VLA-4 interaction. J Immunol **156**, 4707–4714.

Fogler, W.E., Volker, K., Watanabe, M., Wigginton, J.M., Roessler, P., Brunda, M.J., Ortaldo, J.R. and Wiltrout, R.H. (1998). Recruitment of hepatic NK cells by IL-12 is dependent on IFN-gamma and VCAM-1 and is rapidly down-regulated by a mechanism involving T cells and expression of Fas. J Immunol **161**, 6014–6021.

Franitza, S., Grabovsky, V., Wald, O., Weiss, I., Beider, K., Dagan, M., Darash-Yahana, M., Nagler, A., Brocke, S., Galun, E., Alon, R. and Peled, A. (2004). Differential usage of VLA-4 and CXCR4 by CD3 + CD56+ NKT cells and CD56 + CD16+ NK cells regulates their interaction with endothelial cells. Eur J Immunol **34**, 1333–1341.

Freud, A.G., Becknell, B., Roychowdhury, S., Mao, H.C., Ferketich, A.K., Nuovo, G.J., Hughes, T.L., Marburger, T.B., Sung, J., Baiocchi, R.A., Guimond, M. and Caligiuri, M.A. (2005). A human CD34(+) subset resides in lymph nodes and differentiates into CD56bright natural killer cells. Immunity **22**, 295–304.

Frey, M., Packianathan, N.B., Fehniger, T.A., Ross, M.E., Wang, W.C., Stewart, C.C., Caligiuri, M.A. and Evans, S.S. (1998). Differential expression and function of L-selectin on CD56bright and CD56dim natural killer cell subsets. J Immunol **161**, 400–408.

Fukuyama, S., Hiroi, T., Yokota, Y., Rennert, P.D., Yanagita, M., Kinoshita, N., Terawaki, S., Shikina, T., Yamamoto, M., Kurono, Y. and Kiyono, H. (2002). Initiation of NALT organogenesis is independent of the IL-7R, LTbetaR, and NIK signaling pathways but requires the Id2 gene and CD3(−)CD4(+)CD45(+) cells. Immunity **17**, 31–40.

Garni-Wagner, B.A., Witte, P.L., Tutt, M. M., Kuziel, W.A., Tucker, P.W., Bennett, M. and Kumar, V. (1990). Natural killer cells in the thymus. Studies in mice with severe combined immune deficiency. J Immunol **144**, 796–803.

Garrod, K.R., Wei, S.H., Parker, I. and Cahalan, M.D. (2007). Natural killer cells actively patrol peripheral lymph nodes forming stable conjugates to eliminate MHC-mismatched targets. Proc Natl Acad Sci U S A **104**, 12081–12086.

Gibson, P.R. and Jewell, D.P. (1985). The nature of the natural killer (NK) cell of human intestinal mucosa and mesenteric lymph node. Clin Exp Immunol **61**, 160–168.

Giroux, M., Yurchenko, E., St-Pierre, J., Piccirillo, C.A. and Perreault, C. (2007). T regulatory cells control numbers of NK cells and CD8alpha+ immature dendritic cells in the lymph node paracortex. J Immunol **179**, 4492–4502.

Godiska, R., Chantry, D., Raport, C.J., Sozzani, S., Allavena, P., Leviten, D., Mantovani, A. and Gray, P.W. (1997). Human macrophage-derived chemokine (MDC), a novel chemoattractant for monocytes, monocyte-derived dendritic cells, and natural killer cells. J Exp Med **185**, 1595–1604.

Gregoire, C., Chasson, L., Luci, C., Tomasello, E., Geissmann, F., Vivier, E. and Walzer, T. (2007). The trafficking of natural killer cells. Immunol Rev **220**, 169–182.

Gregoire, C., Cognet, C., Chasson, L., Coupet, C.A., Dalod, M., Reboldi, A., Marvel, J., Sallusto, F., Vivier, E. and Walzer, T. (2008). Intrasplenic trafficking of natural killer cells is redirected by chemokines upon inflammation. Eur J Immunol **38**, 2076–2084.

Guo, J., Chen, T., Wang, B., Zhang, M., An, H., Guo, Z., Yu, Y., Qin, Z. and Cao, X. (2003). Chemoattraction, adhesion and activation of natural killer cells are involved in the antitumor immune response induced by fractalkine/CX3CL1. Immunol Lett **89**, 1–7.

Haller, O., Kiessling, R., Orn, A. and Wigzell, H. (1977). Generation of natural killer cells: an autonomous function of the bone marrow. J Exp Med **145**, 1411–1416.

Hanna, J., Mussaffi, H., Steuer, G., Hanna, S., Deeb, M., Blau, H., Arnon, T. I., Weizman, N. and Mandelboim, O. (2005). Functional aberrant expression of CCR2 receptor on chronically activated NK cells in patients with TAP-2 deficiency. Blood **106**, 3465–3473.

Hata, K., Van Thiel, D.H., Herberman, R. B. and Whiteside, T.L. (1991). Natural killer activity of human liver-derived lymphocytes in various liver diseases. Hepatology **14**, 495–503.

Hata, K., Van Thiel, D.H., Herberman, R.B. and Whiteside, T.L. (1992). Phenotypic and functional characteristics of lymphocytes isolated from liver biopsy specimens from patients with active liver disease. Hepatology **15**, 816–823.

Hayakawa, Y. and Smyth, M.J. (2006). CD27 dissects mature NK cells into two subsets with distinct responsiveness and migratory capacity. J Immunol **176**, 1517–1524.

Heemskerk, M.H., Blom, B., Nolan, G., Stegmann, A.P., Bakker, A.Q., Weijer, K., Res, P.C. and Spits, H. (1997). Inhibition of T cell and promotion of natural killer cell development by the dominant negative helix loop helix factor Id3. J Exp Med **186**, 1597–1602.

Hodge, D.L., Schill, W.B., Wang, J.M., Blanca, I., Reynolds, D.A., Ortaldo, J. R. and Young, H.A. (2002). IL-2 and IL-12 alter NK cell responsiveness to IFN-gamma-inducible protein 10 by down-regulating CXCR3 expression. J Immunol **168**, 6090–6098.

Hokeness, K.L., Kuziel, W.A., Biron, C. A. and Salazar-Mather, T.P. (2005). Monocyte chemoattractant protein-1 and CCR2 interactions are required for IFN-alpha/beta-induced inflammatory responses and antiviral defense in liver. J Immunol **174**, 1549–1556.

Ikawa, T., Kawamoto, H., Fujimoto, S. and Katsura, Y. (1999). Commitment of common T/Natural killer (NK) progenitors to unipotent T and NK progenitors in the murine foetal thymus revealed by a single progenitor assay. J Exp Med **190**, 1617–1626.

Imai, T., Hieshima, K., Haskell, C., Baba, M., Nagira, M., Nishimura, M., Kakizaki, M., Takagi, S., Nomiyama, H., Schall, T. J. and Yoshie, O. (1997). Identification and molecular characterization of fractalkine receptor CX3CR1, which mediates both leukocyte migration and adhesion. Cell **91**, 521–530.

Infante-Duarte, C., Weber, A., Kratzschmar, J., Prozorovski, T., Pikol, S., Hamann, I., Bellmann-Strobl, J., Aktas, O., Dorr, J., Wuerfel, J., Sturzebecher, C. S. and Zipp, F. (2005). Frequency of blood CX3CR1-positive natural killer cells correlates with disease activity in multiple sclerosis patients. Faseb J **19**, 1902–1904.

Inngjerdingen, M., Damaj, B. and Maghazachi, A.A. (2000). Human NK cells express CC chemokine receptors 4 and 8 and respond to thymus and activation-regulated chemokine, macrophage-derived chemokine, and I-309. J Immunol **164**, 4048–4054.

Inngjerdingen, M., Damaj, B. and Maghazachi, A.A. (2001). Expression and regulation of chemokine receptors in human natural killer cells. Blood **97**, 367–375.

Ishiyama, K., Ohdan, H., Ohira, M., Mitsuta, H., Arihiro, K. and Asahara, T. (2006). Difference in cytotoxicity against hepatocellular carcinoma between liver and periphery natural killer cells in humans. Hepatology **43**, 362–372.

Jamieson, A.M., Isnard, P., Dorfman, J.R., Coles, M.C. and Raulet, D.H. (2004). Turnover and proliferation of NK cells in steady state and lymphopenic conditions. J Immunol **172**, 864–870.

Jiang, D., Liang, J., Hodge, J., Lu, B., Zhu, Z., Yu, S., Fan, J., Gao, Y., Yin, Z., Homer, R., Gerard, C. and Noble, P.W. (2004). Regulation of pulmonary fibrosis by chemokine receptor CXCR3. J Clin Invest **114**, 291–299.

Jinushi, M., Takehara, T., Tatsumi, T., Kanto, T., Miyagi, T., Suzuki, T., Kanazawa, Y., Hiramatsu, N. and Hayashi, N. (2004). Negative regulation of NK cell activities by inhibitory receptor CD94/NKG2A leads to altered NK cell-induced modulation of dendritic cell functions in chronic hepatitis C virus infection. J Immunol **173**, 6072–6081.

Junqueira-Kipnis, A.P., Kipnis, A., Jamieson, A., Juarrero, M.G., Diefenbach, A., Raulet, D.H., Turner, J. and Orme, I.M. (2003). NK cells respond to pulmonary infection with *Mycobacterium tuberculosis*, but play a minimal role in protection. J Immunol **171**, 6039–6045.

Kato, Y., Yokochi, T., Maeda, K., Sasaki, K., Kawamoto, Y., Tsuji, T. and Miyama, A. (1995). Natural killer (NK)-like cytotoxicity of murine intraepithelial lymphocytes in the small intestine (iIEL) and the effect of the serine proteases. Microbiol Immunol **39**, 291–294.

Kawarabayashi, N., Seki, S., Hatsuse, K., Ohkawa, T., Koike, Y., Aihara, T., Habu, Y., Nakagawa, R., Ami, K., Hiraide, H. and Mochizuki, H. (2000). Decrease of CD56(+)T cells and natural killer cells in cirrhotic livers with hepatitis C may be involved in their susceptibility to hepatocellular carcinoma. Hepatology **32**, 962–969.

Keilbaugh, S.A., Shin, M.E., Banchereau, R.F., McVay, L.D., Boyko, N., Artis, D., Cebra, J.J. and Wu, G.D. (2005). Activation of RegIIIbeta/gamma and interferon gamma expression in the intestinal tract of SCID mice: an innate response to bacterial colonisation of the gut. Gut **54**, 623–629.

Kerr, A.R., Kirkham, L.A., Kadioglu, A., Andrew, P.W., Garside, P., Thompson, H. and Mitchell, T.J. (2005). Identification of a detrimental role for NK cells in pneumococcal pneumonia and sepsis in immunocompromised hosts. Microbes Infect **7**, 845–852.

Khakoo, S.I., Thio, C.L., Martin, M.P., Brooks, C.R., Gao, X., Astemborski, J., Cheng, J., Goedert, J.J., Vlahov, D., Hilgartner, M., Cox, S., Little, A.M., Alexander, G.J., Cramp, M.E., O'Brien, S.J., Rosenberg, W.M., Thomas, D.L. and Carrington, M. (2004). HLA and NK cell inhibitory receptor genes in resolving hepatitis C virus infection. Science **305**, 872–874.

Kim, M.Y., Gaspal, F.M., Wiggett, H.E., McConnell, F.M., Gulbranson-Judge, A., Raykundalia, C., Walker, L.S., Goodall, M.D. and Lane, P.J. (2003). CD4(+)CD3(−) accessory cells costimulate primed CD4 T cells through OX40 and CD30 at sites where T cells collaborate with B cells. Immunity **18**, 643–654.

Korsgren, M., Persson, C.G., Sundler, F., Bjerke, T., Hansson, T., Chambers, B. J., Hong, S., Van Kaer, L., Ljunggren, H.G. and Korsgren, O. (1999). Natural killer cells determine development of allergen-induced eosinophilic airway inflammation in mice. J Exp Med **189**, 553–562.

Kummer, J.A., Kamp, A.M., Tadema, T. M., Vos, W., Meijer, C.J. and Hack, C. E. (1995). Localization and identification of granzymes A and B-expressing cells in normal human lymphoid tissue and peripheral blood. Clin Exp Immunol **100**, 164–172.

Lande, R., Giacomini, E., Grassi, T., Remoli, M.E., Iona, E., Miettinen, M., Julkunen, I. and Coccia, E.M. (2003). IFN-alpha beta released by *Mycobacterium tuberculosis*-infected human dendritic cells induces the expression of CXCL10: selective recruitment of NK and activated T cells. J Immunol **170**, 1174–1182.

Lavergne, E., Combadiere, B., Bonduelle, O., Iga, M., Gao, J.L., Maho, M., Boissonnas, A., Murphy, P.M., Debre, P. and Combadiere, C. (2003). Fractalkine mediates natural killer-dependent antitumor responses in vivo. Cancer Res **63**, 7468–7474.

Lavergne, E., Combadiere, C., Iga, M., Boissonnas, A., Bonduelle, O., Maho, M., Debre, P. and Combadiere, B. (2004). Intratumoral CC chemokine ligand 5 overexpression delays tumour growth and increases tumour cell infiltration. J Immunol **173**, 3755–3762.

Le Maux Chansac, B., Moretta, A., Vergnon, I., Opolon, P., Lecluse, Y., Grunenwald, D., Kubin, M., Soria, J.C., Chouaib, S. and Mami-Chouaib, F. (2005). NK cells infiltrating a MHC class I-deficient lung adenocarcinoma display impaired cytotoxic activity toward autologous tumour cells associated with altered NK cell-triggering receptors. J Immunol **175**, 5790–5798.

Leclercq, G., Debacker, V., de Smedt, M. and Plum, J. (1996). Differential effects of interleukin-15 and interleukin-2 on differentiation of bipotential T/natural killer progenitor cells. J Exp Med **184**, 325–336.

Loetscher, P., Seitz, M., Clark-Lewis, I., Baggiolini, M. and Moser, B. (1996). Activation of NK cells by CC chemokines. Chemotaxis, Ca2+ mobilization, and enzyme release. J Immunol **156**, 322–327.

Luci, C., Reynders, A., Ivanov, II., Cognet, C., Chiche, L., Chasson, L., Hardwigsen, J., Anguiano, E., Banchereau, J., Chaussabel, D., Dalod, M., Littman, D. R., Vivier, E. and Tomasello, E. (2008). Influence of the transcription factor RORgammat on the development of NKp46(+) cell populations in gut and skin. Nat Immunol **10**, 75–82.

Maghazachi, A.A., al-Aoukaty, A. and Schall, T.J. (1994). C-C chemokines induce the chemotaxis of NK and IL-2-activated NK cells. Role for G proteins. J Immunol **153**, 4969–4977.

Maghazachi, A.A., Skalhegg, B.S., Rolstad, B. and Al-Aoukaty, A. (1997). Interferon-inducible protein-10 and lymphotactin induce the chemotaxis and mobilization of intracellular calcium in natural killer cells through pertussis toxin-sensitive and -insensitive heterotrimeric G-proteins. Faseb J **11**, 765–774.

Magilavy, D.B., Steinberg, A.D. and Latta, S.L. (1987). High hepatic natural killer cell activity in murine lupus. J Exp Med **166**, 271–276.

Mailliard, R.B., Alber, S.M., Shen, H., Watkins, S.C., Kirkwood, J.M., Herberman, R.B. and Kalinski, P. (2005). IL-18-induced CD83 + CCR7+ NK helper cells. J Exp Med **202**, 941–953.

Martin, S.M., Mehta, I.K., Yokoyama, W.M., Thomas, M.L. and Lorenz, R. G. (2001). Development of intestinal intraepithelial lymphocytes, NK cells, and NK 1.1+ T cells in CD45-deficient mice. J Immunol **166**, 6066–6073.

Martin-Fontecha, A., Thomsen, L.L., Brett, S., Gerard, C., Lipp, M., Lanzavecchia, A. and Sallusto, F. (2004). Induced recruitment of NK cells to lymph nodes provides IFN-gamma for T(H)1 priming. Nat Immunol **5**, 1260–1265.

Mebius, R.E. (2003). Organogenesis of lymphoid tissues. Nat Rev Immunol **3**, 292–303.

Mebius, R.E. and Kraal, G. (2005). Structure and function of the spleen. Nat Rev Immunol **5**, 606–616.

Megjugorac, N.J., Young, H.A., Amrute, S.B., Olshalsky, S.L. and Fitzgerald-Bocarsly, P. (2004). Virally stimulated plasmacytoid dendritic cells produce chemokines and induce migration of T and NK cells. J Leukoc Biol **75**, 504–514.

Mehrad, B., Strieter, R.M. and Standiford, T.J. (1999). Role of TNF-alpha in pulmonary host defense in murine invasive aspergillosis. J Immunol **162**, 1633–1640.

Mizrahi, S., Yefenof, E., Gross, M., Attal, B., Ben Yaakov, A., Goldman Wohl, D., Maly, B., Stern, N., Katz, G., Gazit, R., Sionov, R.V., Mandelboim, O. and Chaushu, S. (2007). A phenotypic and functional characterization of NK cells in adenoids. J Leukoc Biol **82**, 1095–1105.

Moins-Teisserenc, H.T., Gadola, S.D., Cella, M., Dunbar, P.R., Exley, A., Blake, N., Baykal, C., Lambert, J., Bigliardi, P., Willemsen, M., Jones, M., Buechner, S., Colonna, M., Gross, W.L. and Cerundolo, V. (1999). Association of a syndrome resembling Wegener's granulomatosis with low surface expression of HLA class-I molecules. Lancet **354**, 1598–1603.

Morrison, B.E., Park, S.J., Mooney, J.M. and Mehrad, B. (2003). Chemokine-mediated recruitment of NK cells is a critical host defense mechanism in invasive aspergillosis. J Clin Invest **112**, 1862–1870.

Mowat, A.M., Tait, R.C., MacKenzie, S., Davies, M.D. and Parrott, D.M. (1983).

Analysis of natural killer effector and suppressor activity by intraepithelial lymphocytes from mouse small intestine. Clin Exp Immunol **52**, 191–198.

Nokihara, H., Yanagawa, H., Nishioka, Y., Yano, S., Mukaida, N., Matsushima, K. and Sone, S. (2000). Natural killer cell-dependent suppression of systemic spread of human lung adenocarcinoma cells by monocyte chemoattractant protein-1 gene transfection in severe combined immunodeficient mice. Cancer Res **60**, 7002–7007.

Ochi, M., Ohdan, H., Mitsuta, H., Onoe, T., Tokita, D., Hara, H., Ishiyama, K., Zhou, W., Tanaka, Y. and Asahara, T. (2004). Liver NK cells expressing TRAIL are toxic against self hepatocytes in mice. Hepatology **39**, 1321–1331.

Okamoto, M., Kato, S., Oizumi, K., Kinoshita, M., Inoue, Y., Hoshino, K., Akira, S., McKenzie, A.N., Young, H.A. and Hoshino, T. (2002). Interleukin 18 (IL-18) in synergy with IL-2 induces lethal lung injury in mice: a potential role for cytokines, chemokines, and natural killer cells in the pathogenesis of interstitial pneumonia. Blood **99**, 1289–1298.

Ottaviani, C., Nasorri, F., Bedini, C., de Pita, O., Girolomoni, G. and Cavani, A. (2006). CD56brightCD16(−) NK cells accumulate in psoriatic skin in response to CXCL10 and CCL5 and exacerbate skin inflammation. Eur J Immunol **36**, 118–128.

Pachynski, R.K., Wu, S.W., Gunn, M.D. and Erle, D.J. (1998). Secondary lymphoid-tissue chemokine (SLC) stimulates integrin alpha 4 beta 7-mediated adhesion of lymphocytes to mucosal addressin cell adhesion molecule-1 (MAdCAM-1) under flow. J Immunol **161**, 952–956.

Pallandre, J.R., Krzewski, K., Bedel, R., Ryffel, B., Caignard, A., Rohrlich, P.S., Pivot, X., Tiberghien, P., Zitvogel, L., Strominger, J.L. and Borg, C. (2008). Dendritic cell and natural killer cell cross-talk: a pivotal role of CX3CL1 in NK cytoskeleton organization and activation. Blood **112**, 4420–4424.

Perez-Villar, J.J., Zapata, J.M., Melero, I., Postigo, A., Sanchez-Madrid, E. and Lopez-Botet, M. (1996). Expression and function of alpha 4/beta 7 integrin on human natural killer cells. Immunology **89**, 96–104.

Polentarutti, N., Allavena, P., Bianchi, G., Giardina, G., Basile, A., Sozzani, S., Mantovani, A. and Introna, M. (1997). IL-2-regulated expression of the monocyte chemotactic protein-1 receptor (CCR2) in human NK cells: characterization of a predominant 3.4-kilobase transcript containing CCR2B and CCR2A sequences. J Immunol **158**, 2689–2694.

Prlic, M., Blazar, B.R., Farrar, M.A. and Jameson, S.C. (2003). In vivo survival and homeostatic proliferation of natural killer cells. J Exp Med **197**, 967–976.

Puccetti, P., Santoni, A., Riccardi, C. and Herberman, R.B. (1980). Cytotoxic effector cells with the characteristics of natural killer cells in the lungs of mice. Int J Cancer **25**, 153–158.

Ramsdell, F.J. and Golub, S.H. (1987). Generation of lymphokine-activated killer cell activity from human thymocytes. J Immunol **139**, 1446–1453.

Reading, P.C., Whitney, P.G., Barr, D.P., Wojtasiak, M., Mintern, J.D., Waithman, J. and Brooks, A.G. (2007). IL-18, but not IL-12, regulates NK cell activity following intranasal herpes simplex virus type 1 infection. J Immunol **179**, 3214–3221.

Robinson, B.W., Pinkston, P. and Crystal, R.G. (1984). Natural killer cells are present in the normal human lung but are functionally impotent. J Clin Invest **74**, 942–950.

Roland, J., Soulard, V., Sellier, C., Drapier, A.M., Di Santo, J.P., Cazenave, P.A. and Pied, S. (2006). NK cell responses to *Plasmodium* infection and control of intrahepatic parasite development. J Immunol **177**, 1229–1239.

Roth, M.D. and Golub, S.H. (1989). Inhibition of lymphokine-activated killer cell function by human alveolar macrophages. Cancer Res **49**, 4690–4695.

Salazar-Mather, T.P., Ishikawa, R. and Biron, C.A. (1996). NK cell trafficking and cytokine expression in splenic compartments after IFN induction and viral infection. J Immunol **157**, 3054–3064.

Salazar-Mather, T.P., Orange, J.S. and Biron, C.A. (1998). Early murine cytomegalovirus (MCMV) infection induces liver natural killer (NK) cell inflammation and protection through macrophage inflammatory protein 1alpha (MIP-1alpha)-dependent pathways. J Exp Med **187**, 1–14.

Salazar-Mather, T.P., Hamilton, T.A. and Biron, C.A. (2000). A chemokine-to-cytokine-to-chemokine cascade critical in antiviral defense. J Clin Invest **105**, 985–993.

Samson, S.I., Richard, O., Tavian, M., Ranson, T., Vosshenrich, C.A., Colucci, F., Buer, J., Grosveld, F., Godin, I. and Di Santo, J.P. (2003). GATA-3 promotes maturation, IFN-gamma production, and liver-specific homing of NK cells. Immunity **19**, 701–711.

Sanchez, M.J., Spits, H., Lanier, L.L. and Phillips, J.H. (1993). Human natural killer cell committed thymocytes and their relation to the T cell lineage. J Exp Med **178**, 1857–1866.

Sanchez, M.J., Muench, M.O., Roncarolo, M.G., Lanier, L.L. and Phillips, J.H. (1994). Identification of a common T/natural killer cell progenitor in human foetal thymus. J Exp Med **180**, 569–576.

Sanos, S.L., Bui, V.L., Mortha, A., Oberle, K., Heners, C., Johner, C. and Diefenbach, A. (2008). RORgammat and commensal microflora are required for the differentiation of mucosal interleukin 22-producing NKp46(+) cells. Nat Immunol **10**, 83–91.

Saudemont, A., Jouy, N., Hetuin, D. and Quesnel, B. (2005). NK cells that are activated by CXCL10 can kill dormant tumour cells that resist CTL-mediated lysis and can express B7-H1 that stimulates T cells. Blood **105**, 2428–2435.

Schweighoffer, T., Tanaka, Y., Tidswell, M., Erle, D.J., Horgan, K.J., Luce, G.E., Lazarovits, A.I., Buck, D. and Shaw, S. (1993). Selective expression of integrin alpha 4 beta 7 on a subset of human CD4+ memory T cells with Hallmarks of gut-trophism. J Immunol **151**, 717–729.

Sechler, J.M., Barlic, J., Grivel, J.C. and Murphy, P.M. (2004). IL-15 alters expression and function of the chemokine receptor CX3CR1 in human NK cells. Cell Immunol **230**, 99–108.

Sentman, C.L., Meadows, S.K., Wira, C.R. and Eriksson, M. (2004). Recruitment of uterine NK cells: induction of CXC chemokine ligands 10 and 11 in human endometrium by estradiol and progesterone. J Immunol **173**, 6760–6766.

Singh , U.P., Singh, S., Singh, R., Cong, Y., Taub, D.D. and Lillard, J.W. Jr. (2008). CXCL10-producing mucosal CD4+ T cells, NK cells, and NKT cells are associated with chronic colitis in IL-10(−/−) mice, which can be abrogated by anti-CXCL10 antibody inhibition. J Interferon Cytokine Res **28**, 31–43.

Smyth, M.J., Cretney, E., Takeda, K., Wiltrout, R.H., Sedger, L.M., Kayagaki, N., Yagita, H. and Okumura, K. (2001). Tumour necrosis factor-related apoptosis-inducing ligand (TRAIL) contributes to interferon gamma-dependent natural killer cell protection from tumour metastasis. J Exp Med **193**, 661–670.

Starnes, T., Rasila, K.K., Robertson, M.J., Brahmi, Z., Dahl, R., Christopherson, K. and Hromas, R. (2006). The chemokine CXCL14 (BRAK) stimulates activated

NK cell migration: implications for the downregulation of CXCL14 in malignancy. Exp Hematol **34**, 1101–1105.

Stein-Streilein, J. and Guffee, J. (1986). In vivo treatment of mice and hamsters with antibodies to asialo GM1 increases morbidity and mortality to pulmonary influenza infection. J Immunol **136**, 1435–1441.

Stein-Streilein, J., Bennett, M., Mann, D. and Kumar, V. (1983). Natural killer cells in mouse lung: surface phenotype, target preference, and response to local influenza virus infection. J Immunol **131**, 2699–2704.

Tagliabue, A., Luini, W., Soldateschi, D. and Boraschi, D. (1981). Natural killer activity of gut mucosal lymphoid cells in mice. Eur J Immunol **11**, 919–922.

Tagliabue, A., Befus, A.D., Clark, D.A. and Bienenstock, J. (1982). Characteristics of natural killer cells in the murine intestinal epithelium and lamina propria. J Exp Med **155**, 1785–1796.

Tagliabue, A., Villa, L., Scapigliati, G. and Boraschi, D. (1983). Peyer's patch lymphocytes express natural cytotoxicity but not natural killer activity. Nat Immun Cell Growth Regul **3**, 95–101.

Takii, Y., Hashimoto, S., Iiai, T., Watanabe, H., Hatakeyama, K. and Abo, T. (1994). Increase in the proportion of granulated CD56+ T cells in patients with malignancy. Clin Exp Immunol **97**, 522–527.

Taub, D.D., Proost, P., Murphy, W.J., Anver, M., Longo, D.L., van Damme, J. and Oppenheim, J.J. (1995). Monocyte chemotactic protein-1 (MCP-1), -2, and -3 are chemotactic for human T lymphocytes. J Clin Invest **95**, 1370–1376.

Thapa, M., Kuziel, W.A. and Carr, D.J. (2007). Susceptibility of CCR5-deficient mice to genital herpes simplex virus type 2 is linked to NK cell mobilization. J Virol **81**, 3704–3713.

Thapa, M., Welner, R.S., Pelayo, R. and Carr, D.J. (2008). CXCL9 and CXCL10 expression are critical for control of genital herpes simplex virus type 2 infection through mobilization of HSV-specific CTL and NK cells to the nervous system. J Immunol **180**, 1098–1106.

Trifilo, M.J., Montalto-Morrison, C., Stiles, L.N., Hurst, K.R., Hardison, J. L., Manning, J.E., Masters, P.S. and Lane, T.E. (2004). CXC chemokine ligand 10 controls viral infection in the central nervous system: evidence for a role in innate immune response through recruitment and activation of natural killer cells. J Virol **78**, 585–594.

Tsuchiyama, T., Nakamoto, Y., Sakai, Y., Marukawa, Y., Kitahara, M., Mukaida, N. and Kaneko, S. (2007). Prolonged, NK cell-mediated antitumor effects of suicide gene therapy combined with monocyte chemoattractant protein-1 against hepatocellular carcinoma. J Immunol **178**, 574–583.

van den Brink, M.R., Palomba, M.L., Basse, P.H. and Hiserodt, J.C. (1991). In situ localization of 3.2.3+ natural killer cells in tissues from normal and tumor-bearing rats. Cancer Res **51**, 4931–4936.

Vanderkerken, K., Bouwens, L., Van Rooijen, N., Van den Berg, K., Baekeland, M. and Wisse, E. (1995). The role of Kupffer cells in the differentiation process of hepatic natural killer cells. Hepatology **22**, 283–290.

Vester, B., Muller, K., Solbach, W. and Laskay, T. (1999). Early gene expression of NK cell-activating chemokines in mice resistant to *Leishmania major*. Infect Immun **67**, 3155–3159.

Villegas, F.R., Coca, S., Villarrubia, V. G., Jimenez, R., Chillon, M.J., Jareno, J., Zuil, M. and Callol, L. (2002). Prognostic significance of tumour infiltrating natural killer cells subset CD57 in patients with squamous cell lung cancer. Lung Cancer **35**, 23–28.

Warnock, R.A., Askari, S., Butcher, E. C. and von Andrian, U.H. (1998). Molecular mechanisms of lymphocyte homing to peripheral lymph nodes. J Exp Med **187**, 205–216.

Warnock, R.A., Campbell, J.J., Dorf, M. E., Matsuzawa, A., McEvoy, L.M. and Butcher, E.C. (2000). The role of chemokines in the microenvironmental control of T versus B cell arrest in Peyer's patch high endothelial venules. J Exp Med **191**, 77–88.

Watt, S.V., Andrews, D.M., Takeda, K., Smyth, M.J. and Hayakawa, Y. (2008). IFN-gamma-dependent recruitment of mature CD27(high) NK cells to lymph nodes primed by dendritic cells. J Immunol **181**, 5323–5330.

Webster, G.J., Reignat, S., Maini, M. K., Whalley, S.A., Ogg, G.S., King, A., Brown, D., Amlot, P.L., Williams, R., Vergani, D., Dusheiko, G.M. and Bertoletti, A. (2000). Incubation phase of acute hepatitis B in man: dynamic of cellular immune mechanisms. Hepatology **32**, 1117–1124.

Wendel, M., Galani, I.E., Suri-Payer, E. and Cerwenka, A. (2008). Natural killer cell accumulation in tumours is dependent on IFN-gamma and CXCR3 ligands. Cancer Res **68**, 8437–8445.

Weng, Y., Siciliano, S.J., Waldburger, K. E., Sirotina-Meisher, A., Staruch, M.J., Daugherty, B.L., Gould, S.L., Springer, M.S. and DeMartino, J.A. (1998). Binding and functional properties of recombinant and endogenous CXCR3 chemokine receptors. J Biol Chem **273**, 18288–18291.

Wiltrout, R.H., Mathieson, B.J., Talmadge, J.E., Reynolds, C.W., Zhang, S.R., Herberman, R.B. and Ortaldo, J.R. (1984). Augmentation of organ-associated natural killer activity by biological response modifiers. Isolation and characterization of large granular lymphocytes from the liver. J Exp Med **160**, 1431–1449.

Wiltrout , R.H., Herberman, R.B., Zhang, S.R., Chirigos, M.A., Ortaldo, J. R., Green, K.M. and Talmadge, J.E. Jr. (1985). Role of organ-associated NK cells in decreased formation of experimental metastases in lung and liver. J Immunol **134**, 4267–4275.

Wisse, E., van't Noordende, J.M., van der Meulen, J. and Daems, W.T. (1976). The pit cell: description of a new type of cell occurring in rat liver sinusoids and peripheral blood. Cell Tissue Res **173**, 423–435.

Withers, D.R., Kim, M.Y., Bekiaris, V., Rossi, S.W., Jenkinson, W.E., Gaspal, F., McConnell, F., Caamano, J.H., Anderson, G. and Lane, P.J. (2007). The role of lymphoid tissue inducer cells in splenic white pulp development. Eur J Immunol **37**, 3240–3245.

Yokoyama, W.M. and Plougastel, B.F. (2003). Immune functions encoded by the natural killer gene complex. Nat Rev Immunol **3**, 304–316.

Yoneda, O., Imai, T., Goda, S., Inoue, H., Yamauchi, A., Okazaki, T., Imai, H., Yoshie, O., Bloom, E.T., Domae, N. and Umehara, H. (2000). Fractalkine-mediated endothelial cell injury by NK cells. J Immunol **164**, 4055–4062.

Yoneda, O., Imai, T., Nishimura, M., Miyaji, M., Mimori, T., Okazaki, T., Domae, N., Fujimoto, H., Minami, Y., Kono, T., Bloom, E.T. and Umehara, H. (2003). Membrane-bound form of fractalkine induces IFN-gamma production by NK cells. Eur J Immunol **33**, 53–58.

Zeng, X., Moore, T.A., Newstead, M.W., Hernandez-Alcoceba, R., Tsai, W.C. and Standiford, T.J. (2003). Intrapulmonary expression of macrophage inflammatory protein 1alpha (CCL3) induces neutrophil and NK cell accumulation and stimulates innate immunity in murine bacterial pneumonia. Infect Immun **71**, 1306–1315.

Chapter Eleven

11

Receptors on NK cells

Noam Stern-Ginossar, Ofer Mandelboim

CHAPTER CONTENTS

True power is when we have every justification to kill, and don't.

Oscar Schindler

ABSTRACT

Natural killer (NK) cells are bone marrow derived lymphocytes that are well-equipped for the destruction of tumour and virally infected cells without the need for prior antigen stimulation. Their killing machinery is quite complex and is determined by integrated signals obtained from activating and inhibitory receptors. Inhibitory receptors recognize molecules that are expressed on normal cells as a means of protecting healthy cells from attack by NK. The prototypic inhibitory receptors recognize MHC class I molecules, but recent work has identified inhibitory receptors that have non-MHC-molecule ligands. The NK-activating receptors recognize pathogen-derived, stress-induced molecules and a series of still uncharacterized cellular ligands. Here, we review the known NK receptors, their function, ligand specificity and their overall contribution to NK cell recognition.

KEY WORDS

Natural killer, Natural killer activating and inhibitory receptors, Natural cytotoxicity receptors, MicroRNA, Killer ligands

When initially discovered, natural killer (NK) cells were considered as non-specific in their interaction with target cells. A significant breakthrough in the understanding of NK regulation was achieved when Klas Karre and colleagues first postulated their 'missing self' hypothesis in 1986 (Karre et al., 1986). According to this theory, which was quite radical at that time, NK cells recognize specifically the self-MHC class I molecules and this recognition leads to the inhibition of their activity. Thus, NK cells and cytotoxic T lymphocytes (CTL) both recognize MHC class I; however, while CTL use MHC class I to activate their killing, NK cells use it to inhibit their activity. The NK activity in this regard is complementary to that of CTL, as tumours and several viruses (like herpes viruses) reduce the expression of MHC class I protein in order to evade CTL response (Cordon-Cardo et al., 1991; Hewitt, 2003). However, since the MHC class I molecules are also used to inhibit NK cytotoxicity, this downregulation of MHC class I renders these hazardous cells sensitive for NK cell destruction.

The 'missing self' hypothesis predicted that NK cells must express inhibitory receptors that interact with self MHC class I molecules and deliver an inhibitory signal upon interaction. The molecular basis of this phenomenon was revealed several years later by the identification and cloning of novel MHC class I-binding NK receptors (Colonna and Samaridis, 1995; Moretta et al., 1995; Yokoyama et al., 1995).

The concept of the 'missing self' hypothesis and the pioneering work of Karre and colleagues were proven correct, and survived the challenges presented over the years. Nevertheless, we know today that an additional mechanism also exists and that non-MHC ligands inhibit NK-mediated killing. Why another non-MHC–related mechanism was developed is still an open question, but one possible explanation is that these 'alternative self' mechanisms were developed as a back-up system to inhibit NK activity under certain unique conditions, for example, the inhibitory receptor carcinoembryonic antigen-related cell adhesion molecule 1 (CEACAM1) plays a major role during pregnancy in decidual NK cells when most of the MHC class I proteins are absent (Markel et al., 2002b).

We know today that NK cell cytotoxicity is not evident only in the absence of self inhibition. Indeed, certain NK-susceptible targets express normal MHC class I repertoire, while some MHC-negative cells are resistant to NK attack. Thus, the activation of NK cells requires more than just the absence of inhibitory signals and in order to tip the balance in favour of NK cell activation, targets must express specific ligands for NK activating receptors.

The following sections will describe the central mechanisms that participate in the delivery of both inhibitory and activating signals and determine the overall regulation of NK activity, as known today.

Inhibitory NK receptors

All of the well-defined NK inhibitory receptors contain in their cytoplasmic tails one or more immunoreceptor tyrosine-based inhibitory motif (ITIM) that includes the consensus sequence: Ile/Val/Leu/Ser–X–Tyr–X–X–Leu/Val (where 'X' represents any amino acid). Upon ligation, the Tyr residues in the ITIM are phosphorylated (probably by Src family kinase) and recruit beta-arrestin 2 (Yu et al., 2008), which mediate the binding of tyrosine phosphatases. The recruitment of the phosphatases to the membrane proximal location hampers or prevents activating signals. The primary NK inhibitory receptors in humans are summarized in Table 11.1.

MHC class I-mediated inhibition

NK receptors can recognize 'classical' and 'non-classical' MHC class I proteins. Receptors for MHC class II proteins have not been identified. However, since NK cells

Table 11.1 Inhibitory NK receptors

Gene	Other names	Structure	Signalling	Known ligands
KLRG1	Mafa	C-type lectin homodimer	1 ITIM	E-cadherin, N-cadherin, R-cadherin
KLRD1-KLRC1	CD94-NKG2A (CD159a)	C-type lectin heterodimer	1 ITIM	HLAE
KLRB1	NKR-P1A, CD161	C-type lectin homodimer	1 ITIM	LLT1 (OCIL, CLEC2D)
LAIR1	CD305	Ig monomer	2 ITIM	Collagen
LILRB1	ILT2, LIR1, CD85j	Ig monomer	4 ITIM	HLA class I
KIR2DL1-3	CD158b	Ig monomer	2 ITIM	HLA-C
KIR3DL1	CD158e1	Ig monomer	2 ITIM	HLA- Bw4
KIR3DL2	CD158k	Ig homodimer	2 ITIM	HLA-A3, HLA-A11
CEACAM1	CD66a	Ig monomer	2 ITIM	CEACAM1, CEA
SIGLEC7	CDw328, p75	Ig monomer	1 ITIM	α2,8 disialic acid
SIGLEC9	CD329	Ig monomer	1 ITIM + 1 ITSM	Terminal α(2,3)- and α(2,6) disialic acids
IRp60	CD300a	Ig monomer	4 ITIM	?

can interact with dendritic cell (DC), macrophages and B cells, it is possible that MHC class II–recognizing NK receptors also exist. The two major families of MHC-specific inhibitory receptors identified in humans include: the Ig superfamily (which includes the killer and leukocyte immunoglobulin-like receptors designated as killer Ig-like receptors [KIR] and leukocyte Ig-like inhibitory receptors [LIR], respectively) and the C-type lectin receptor superfamily (Carretero et al., 1998; Colonna and Samaridis, 1995; Raulet et al., 2001). The individual inhibitory receptors display a variety of specificities towards MHC class I proteins. Some of the NK inhibitory receptors can detect shared allelic determinants of MHC class I proteins, while others are able to recognize a broad spectrum of MHC class I proteins.

An important feature of all inhibitory NK receptors is that they are expressed in a variegated fashion, so that each NK cell expresses multiple receptors in a complex combinatorial repertoire that results in various subpopulations of NK cells able to sense the loss of even a single MHC class I protein (Gazit et al., 2004; Moretta et al., 1996). Understanding the regulatory events that lead to the acquisition of the combinatorial pattern of the inhibitory receptor repertoire is a subject of intense investigation.

Surprisingly, NK cell-mediated autoimmune diseases have not been observed in mouse models or in TAP2-deficient human patients when MHC class I proteins are either absent or low, and the overall killing of NK cells in these cases is poor (Liao et al., 1991). These observations suggest that, in the course of their development, NK cells are educated to become tolerant towards self, based on their interactions with MHC class I molecules (Fernandez et al., 2005; Kim et al., 2005). Two theories were postulated to explain the requirement of MHC class I for the generation of functional NK cells. In the 'licensing' model, Yokoyama and colleagues suggested an instructive mode of action for MHC class I proteins in determining NK cell cytotoxicity. In the 'disarming' model postulated by Raulet and colleagues, it was suggested that in the absence of MHC class I, NK cells may be chronically stimulated, resulting in NK cells that are hyporesponsive (Fernandez et al., 2005). Though further understanding of its mechanism is required, recent studies demonstrate that constant engagement of a specific activation receptor in vivo induces NK cell tolerance that is not affected by self-MHC-specific inhibition. This suggests that the disarming model is probably more accurate (Sun and Lanier, 2008; Tripathy et al., 2008). Still, it is agreed that the engagement of inhibitory receptors by self-MHC class I plays an important role in the acquisition of NK cell function.

Killer Ig-like receptors

This family of receptors belongs to the Ig superfamily and is hence termed KIRs. The receptors in this family are encoded by 15 genes and 2 pseudogenes located within the leukocyte Ig-like receptor complex (LRC) on chromosome 19q13.4, that probably has evolved by gene duplication (Lanier, 2005). Among the NK cell receptors, the KIRs comprise the most versatile and polymorphic family of receptors.

The KIRs are type I transmembrane glycoproteins that can be divided into two subfamilies based on the number of Ig-like domains in the extracellular portion of the protein. The KIR3D subfamily contains three Ig-like domains, whereas the KIR2D proteins contain only two Ig-like domains (Moretta et al., 1996). The KIR family contains both inhibitory and activating receptors. The activating KIR receptors will be discussed below. As mentioned above, the inhibitory KIRs contain ITIM motives in their cytoplasmic domains, which are responsible for the delivery of the inhibitory signal upon receptor ligation. While all inhibitory KIRs interact with MHC class I molecules, they manifest various allelic specificity. The KIR3D receptors bind HLA-A3 and -A11 and some HLA-B proteins bearing the Bw4 motif, whereas the KIR2D receptors recognize predominantly the HLA-C alleles (Lanier, 2005). Remarkably, it was demonstrated that a single amino acid at position 80 of the HLA-C proteins determines the recognition of the HLA-C proteins by the appropriate KIR2D receptor (Mandelboim et al., 1996). Thus, the entire KIR2D subfamily is subdivided into receptors recognizing HLA-C proteins containing either K or N residue at position 80 (Mandelboim et al., 1996). The fact that many of the HLA-A and HLA-B allotypes could not be recognized by any KIR receptor along with the recognition of all HLA-C alleles by the KIR2D receptors implies that HLA-C was probably evolved to provide protection from NK cells.

Leukocyte Ig-like inhibitory receptors

The second group of NK inhibitory receptors belonging to the Ig superfamily is the LIRs or Ig-like transcript (ILTs). This family contains 13 genes (which are also located within the LRC on chromosome 19q13.4) from which only two encode for inhibitory receptors (LIR1 and LIR2), and only one (LIR1) is expressed on NK cells. LIR1 is expressed also on certain subsets of T cells, as well as on B cells and monocytes (Colonna et al., 1997).

The LIR1 receptor contains 4 ITIM motives in its cytoplasmic tail that delivers inhibitory signals upon engagement by its MHC class I ligands. However, unlike the inhibitory KIRs, LIR1 binds with low affinity to a conserved region in the $\alpha 3$ domain of MHC class I proteins (Chapman et al., 2000). This enables the recognition of a very broad spectrum of MHC class I proteins including most HLA-A, HLA-B and HLA-C alleles, as well as the non-classical HLA-G (Colonna et al., 1999).

The most efficient binding of LIR1 is observed to HLA-G. Complexes of HLA-G proteins are found on the cell surface, formed by two cysteines unique to HLA-G (Gonen-Gross et al., 2003). Despite the fact that the α3 domain is conserved among various HLA proteins, differences can still be observed in the binding of LIR1 to various HLA proteins (Stern-Ginossar et al.,). The molecular explanation for these differences is currently unknown. In addition, it was reported that the LIR1 recognizes the human cytomegalovirus (HCMV)-encoded protein UL18, that has structural homology to MHC class I (Cosman et al., 1997). Thus, UL18 may engage inhibitory signals through the LIR1 receptor and this may suppress NK cell-mediated antiviral responses.

C-type lectin receptors (CD94/NKG2A)

CD94 and NKG2 are type II integral membrane glycoproteins that contain an extracellular C-type carbohydrate recognition domain and are encoded by genes clustered together in human chromosome 12p12.3-p13.2 (Carretero et al., 1998). The CD94 protein covalently assembles with distinct members of the NKG2 family to form heterodimers, which are selectively expressed by NK cells and a subset of CTLs (Carretero et al., 1998). Association of CD94 with the NKG2A (which contains an ITIM in its cytoplasmic tail) constitutes another MHC class I-specific inhibitory receptor (Carretero et al., 1998; Le Drean et al., 1998). The CD94/NKG2A heterodimer has very limited polymorphism and unlike the KIR and LIR receptors, which recognize various HLA proteins, it recognizes a single non-classical MHC class I molecule, HLA-E (Braud et al., 1998; Lee et al., 1998b). Interestingly, the HLA-E molecule assembles at the endoplasmic reticulum (ER) with peptides derived from the leader peptides of HLA-A, B, C and G (Braud et al., 1997; Lee et al., 1998a). Consequently, the expression of HLA-E depends on the expression of the classical MHC molecules and therefore reflects the overall biogenesis of MHC class I proteins in the cells. It is tempting to speculate that this unique expression of HLA-E may enable NK cells to sense general changes in MHC class I expression, which may be particularly important in situations where selective downregulation of specific MHC class I alleles has occurred. Interestingly, it was demonstrated that HCMV, which is known to actively downregulate MHC class I expression, also encodes a protein, UL40, which contains a segment that assembles with HLA-E, resulting in upregulation of surface HLA-E and consequently a reduction in NK cell activity (Tomasec et al., 2000). Another example of the importance of HLA-E in NK cell inhibition is provided by HIV, which developed a nef-dependent mechanism to downregulate HLA-A and HLA-B but not HLA-C and HLA-E probably to avoid CTL attack and at the same time to inhibit NK activity (Cohen et al., 1999).

MHC class I-independent inhibition

As was suggested by the missing self hypothesis, MHC class I molecules play a major role in NK cell inhibition. However, as stated above, in some cases, even the loss of MHC class I proteins is not enough to render the cells susceptible for NK cell attack. The most remarkable example is the surprising finding that humans and mice with abnormalities in MHC class I expression develop normal numbers of NK cells. These NK cells can kill tumour cells, but remarkably, they are still tolerant to autologous cells that express low levels of MHC class I molecules (Sivori et al., 1997). In addition, certain NK-susceptible target cells express a normal MHC class I repertoire, while some MHC-negative cells are resistant to NK attack. Therefore, it has been speculated that other class I-independent inhibitory mechanisms must exist.

Carcinoembryonic antigen-related cell adhesion molecule 1

The carcinoembryonic antigen (CEA) family is a large and multifunctional family that belongs to the Ig superfamily. It is composed of 29 genes tandemly arranged on chromosome 19q13.2. Its founding member, CEA-related cell-adhesion molecule 5 (CEACAM5; also known as CEA), is expressed on a wide range of carcinomas (Hammarstrom, 1999). Based on nucleotide homologies, these genes are classified into two major subfamilies, the CEACAM and the pregnancy-specific glycoprotein (PSG) subgroups. The CEACAM subgroup contains seven members and CEACAM1 (also known as CD66a or BGP) is the only CEACAM member identified so far that is expressed by NK cells. The CEACAM1 protein is also expressed on a variety of immune cells, including monocytes, granulocytes, activated T cells and B cells (Beauchemin et al., 1999; Boulton and Gray-Owen, 2002). CEACAM1 has two ITIM motifs in its cytoplasmic tail and our laboratory has demonstrated that the CEACAM1 homophilic or heterophilic interactions with CEACAM1 or CEA proteins inhibit NK killing, independently of MHC class I recognition (Markel et al., 2002a; Stern et al., 2005). Remarkably, NK cells from TAP2-deficient patients that express low levels of MHC class I molecules show elevated levels of CEACAM1 expression following activation, suggesting that this alternative inhibitory mechanism contributes to prevention of autoimmunity and maintains self tolerance in the absence of MHC class I inhibition (Markel et al., 2002a). Similar observations were made during pregnancy (Markel et al., 2002b). Thus, we suggest that CEACAM1 interactions may represent an 'alternative self', which is important in special circumstances in the absence of MHC class I (such as in TAP2-deficient patients or during pregnancy).

KLRG1

The killer cell lectin-like receptor G1 (KLRG1) is a C-type lectin-like receptor containing one ITIM motif. It is expressed on subsets of NK cells and T cells in both mouse and human (Butcher et al., 1998; Hanke et al., 1998). KLRG1 has been shown to interact with three members of the cadherin family: epithelial (E)-, neuronal (N)- and retinal (R)-cadherin (Grundemann et al., 2006; Ito et al., 2006), and these interactions were shown to inhibit NK cytotoxicity. Recognition of CEACAMs and cadherin molecules by NK inhibitory receptors suggests that the immune system has the capacity to detect molecules involved in tissue organization, and thus it might suggest that NK cell activity is also controlled by this process.

NK-cell-receptor protein 1

Human NK-cell-receptor protein 1A (NKR-P1A) (CD161), encoded by the *KLRB1* gene, represents the only human homolog of the rodent NKR-P1 family, which includes NKR-P1C, the prototypic NK1.1 molecule. This receptor is a type II disulfide-linked homodimer, which belongs to the C-type lectin superfamily (Lanier et al., 1994). Human NKR-P1A is expressed on immature NK cells, and the expression of NKR-P1A can be upregulated on IL-12-activated NK cells (Poggi et al., 1998). NKR-P1A is also expressed on different subsets of T cells, B cells and TLR-activated DCs (Rosen et al., 2008). Crosslinking of NKR-P1A inhibits human NK cell-mediated cytotoxicity (Lanier et al., 1994). The lectin-like transcript-1 (LLT1), is a ligand for NKR-P1A (Aldemir et al., 2005; Rosen et al., 2005). Interestingly, the gene encoding LLT1, *CLEC2D*, is located directly adjacent to the *KLRB1* gene in the NK genomic complex, on human chromosome 12. Functional studies indicate that the interaction of LLT1 on target cells with NKR-P1A on NK cells inhibit NK cell-mediated cytotoxicity. These effects could not be correlated with any known signalling motifs in the NKR-P1A sequence and thus its inhibitory signalling pathway is still elusive.

Inhibitory receptor protein 60

This inhibitory receptor belongs to the Ig superfamily and is encoded on human chromosome 17. It is a type I transmembrane glycoprotein of 60 kDa, composed of a single extracellular V-type Ig-like domain and a cytoplasmic tail that contains four ITIMs. In accordance with this initial finding, Inhibitory receptor protein 60 (IRp60) was shown to be an inhibitory receptor as its crosslinking resulted in NK cell inhibition (Cantoni et al., 1999a). Unlike other inhibitory receptors, IRP60 is expressed on all activated and resting human NK cells. The ligand(s) recognized by IRp60 are still unknown (Cantoni et al., 1999a).

Leukocyte-associated Ig-like receptor

Leukocyte-associated Ig-like receptor (LAIR-1) is type I transmembrane glycoprotein containing a single extracellular C2-type Ig-like domain and two ITIMs. It is structurally related to other Ig inhibitory receptors encoded within the LRC on chromosome 19q13.4, suggesting that these molecules have evolved from a common ancestor (Martin et al., 2002). LAIR-1 is a unique member of the inhibitory receptors as it is expressed on almost all cells in the immune system, including all NK cells. LAIR-1 was shown to be an inhibitory receptor as its cross-linking inhibits target cell lysis by both resting and activated NK cells. It was recently demonstrated that collagens serve as high affinity functional ligands of LAIR-1 (Meyaard, 2008), however the physiological role of these interactions still needs to be studied.

Sialic-acid-binding immunoglobulin-like lectins

The sialic-acid-binding immunoglobulin-like lectins (SIGLECs) are type I lectins displaying an amino-terminal V-like Ig domain that binds sialic acid and variable numbers of C2-like Ig domains. In humans, 11 SIGLECs have been identified and human NK cells express SIGLEC7 (also known as p75 or AIRM1) and SIGLEC9 (Crocker and Varki, 2001). SIGLEC7 has two ITIM motifs and SIGLEC9 has an immunoreceptor tyrosine-based switch motif (ITSM) (similar to the 2B4 receptor, see below) and an ITIM motif. SIGLEC7 was originally identified as an inhibitory receptor expressed by all NK cells, monocytes and some $CD8^+$ T cells (Nicoll et al., 1999). SIGLEC7 inhibits NK-cell killing when its extracellular domain is crosslinked by antibody. One ligand that has been identified for SIGLEC7 is the ganglioside GD3, a glycosphingolipid that contains an α2,8-linked disialic acid (Nicoll et al., 2003). SIGLEC9 is an inhibitory receptor expressed by only 50% of NK cells and is also expressed on T cells and on basophils (Avril et al., 2004). Little is known about the physiological function of these receptors and further work is needed to address this point.

Activating NK receptors

For many years NK cells were thought to be controlled only by inhibitory mechanisms. Killing of target cells was considered a default, resulting from the absence of inhibitory signals. However, it is now well established that NK cells require specific activating signals in order to exhibit killing activity. Direct evidence for the existence of such receptors was first provided by the development of specific monoclonal antibodies that blocked the NK-mediated killing of many tumour lines (Pende et al., 1999; Pessino et al., 1998; Sivori et al., 1997; Vitale et al., 1998). In humans, the major NK triggering receptors identified so far include the NKG2D (Pende et al., 2001), CD16 (Mandelboim et al., 1999) and the natural cytotoxic receptors (collectively named NCRs), which include NKp46 (Sivori et al., 1997), NKp44 (Vitale et al., 1998) and NKp30 (Pende et al., 1999). The molecular cloning of the NCRs revealed that they share no homology with each other and only a low degree of identity with other human proteins (Moretta et al., 2002). The NCRs represent the most specific surface marker known today for human and mouse NK cells (Moretta and Moretta, 2004).

The activating receptors are defined by their ability to directly mediate killing of targets. Nevertheless, recent findings demonstrate that the activation of some of the NK triggering receptors, in resting NK cells, requires the synergistic stimulation of more than one receptor (Bryceson et al., 2006). In addition, some NK receptors are thought to serve as costimulatory receptors that reduce the threshold for activation but cannot mediate direct lysis on their own (Moretta et al., 2002).

Unlike the inhibitory receptors that contain an inhibitory motif in their cytoplasmic tail, the transduction of the activating signals is mediated by the association of the activating receptors with an Immunodominant tyrosine based activation motif (ITAM)-containing adaptor protein, such as CD3ζ, FcεRIγ, DAP10 and DAP12 (Lanier, 2003). The fact that these receptors do not signal through their own tail, but are rather coupled to different adaptor proteins, can provide a more flexible machinery to control the killing activity in diverse physiological conditions (Horng et al., 2007).

Thus, we know that the killing of target cells by NK cells requires not only a reduction in the inhibitory ligands expression but also increased expression of multiple ligands for the NK activating receptors. The major NK activating receptors in humans and their known ligands are described in the following section and summarized in Table 11.2.

NKG2D

NKG2D is a type II transmembrane glycoprotein encoded by a single gene and is expressed on the surface of all human (and mouse) NK cells, but in contrast to the NCRs, its expression is not unique to NK cells. In humans, NKG2D is found on all CD8-positive T cells as well as on all $\gamma\delta$ T cells (Eagle and Trowsdale, 2007). It is a genuine activating receptor on NK cells, as specific

Table 11.2 Activating NK receptors

Gene	Other names	Structure	Signalling	Known ligands
KLRK1	NKG2D, CD314	C-lectin homodimer	DAP10	MICA, MICB, ULBP1-5
NCR1	NKp46, CD335	Ig monomer	FcεRIγ, CD3ζ	Influenza Hemagglutinin
NCR2	NKp44, CD336	Ig monomer	DAP12	Influenza Hemagglutinin
NCR3	NKp30, CD337	Ig monomer	FcεRIγ, CD3ζ	Viral pp65, BAT3
FCGR3	CD16	Ig monomer	FcεRIγ, CD3ζ	IgG
CD226	DNAM-1	Ig monomer	Protein kinase C	CD112, CD155
KLRF1	NKp80, CLEC5C	C-lectin homodimer	?	AICL
CD244	2B4	Ig monomer	ITSM, SAP	CD48
SLAMF7	CRACC, CD319	Ig monomer	ITSM, EAT2, SAP	CRACC
SLAMF6	NTB-A	Ig monomer	ITSM, SAP, EAT2	NTB-A
KIR2DS1	CD158h	Ig monomer	DAP12	HLA-C
KIR2DS4	CD158i	Ig monomer	DAP12	HLA-Cw4
KIR3DS1 2	CD158e	Ig monomer	DAP12	?

engagement of NKG2D activates NK cells, independent of other activating stimuli (Jamieson et al., 2002; Pende et al., 2001). The expression of NKG2D on both NK and T cells can be increased by culturing with IL-15 or TNF-α and is significantly reduced in the presence of TGF-β (Castriconi et al., 2003). Despite its name, NKG2D shows very little homology to the NKG2 gene family. It does not form heterodimers with CD94, and it is expressed as disulfide-linked homodimers. Stimulation of NK cells via NKG2D leads to a potent cytotoxic response that is mediated via association of NKG2D with the adaptor molecule DAP10 (Bauer et al., 1999).

Several ligands have been identified for human NKG2D, including MICA, MICB and ULBP1–5 (Eagle and Trowsdale, 2007). These proteins constitute a family of stress-inducible cellular ligands with structural homology to MHC class I proteins. Interestingly, a comparison between the different ULBPs and MIC proteins shows no significant homology (<20%). However, the crystal structure of NKG2D complexes with ULBP3 and MICA has revealed that the receptor–ligand interface is remarkably similar, despite the use of completely different residues for binding (Li et al., 2001; Radaev et al., 2001).

NKG2D ligands are expressed on diseased cells and numerous stress pathways lead to the upregulation of these ligands. The first insight into the regulation of NKG2D ligands, presenting the concept that the NKG2D receptor is able to detect cellular stress, came from the finding that the MICA/B promotors contain heat shock elements similar to those found in HSP70 and that heat shock induces MICA expression (Groh et al., 1996). Similarly NKG2D ligand expression is also upregulated in response to oxidative stress (Groh et al., 1998; Venkataraman et al., 2007), genotoxic stress (Gasser et al., 2005) and viral infection (Groh et al., 2001).

The best established role for NKG2D as an anti-viral defence mechanism is in HCMV infection. Infection with HCMV leads to upregulation of NKG2D ligands transcripts, including MICA, MICB and the ULBPs (Groh et al., 2001). On the other hand, in order to cope with NKG2D mediated killing, HCMV developed several counter attack mechanisms. The UL16 protein of HCMV binds to MICB, ULBP1 and 2 resulting in the retention and sequestration of these ligands and consequently preventing their surface expression (Dunn et al., 2003; Wu et al., 2003). Subsequent studies demonstrated that the viral UL142 protein downregulates the expression of MICA protein (Wills et al., 2005). We have recently demonstrated that a microRNA encoded by the HCMV, hcmv-miR-UL112, downregulates MICB expression by targeting a specific site in the MICB-3′UTR and that this downregulation protects infected cells from NKG2D-dependent killing (Stern-Ginossar et al., 2007). The fact that MICB protein is also targeted by HCMV UL16 and the observation that other viral microRNAs also target MICB (Nachmani, Stern-Ginossar and Mandelboim, unpublished observations) suggest a strong selective pressure for the development of several viral mechanisms to reduce NKG2D ligand expression.

The *MIC* genes are highly polymorphic; approximately 60 alleles of MICA and 25 alleles of MICB have so far been identified (Eagle and Trowsdale, 2007). The reasons for the existence of such polymorphisms are unknown, but it has been postulated that such polymorphisms provide an advantage for the host against viruses (Eagle and Trowsdale, 2007). Indeed, two studies (Chalupny et al., 2006; Zou et al., 2005) have shown that a common MICA allele that lacks the cytoplasmic tail is resistant to downregulation during HCMV infection and it has been suggested that this human variant escaped from the binding of HCMV UL142 protein.

NKG2D ligands are also induced by genotoxic stress and stalled DNA replication (Gasser et al., 2005). In tumours, NKG2D ligands have been often shown to be upregulated (Diefenbach et al., 2000; Groh et al., 1999) and it was therefore suggested that NKG2D ligands expression is induced by the DNA-damage response, which is known to be activated in early stages of tumorigenesis. Upregulation of NKG2D ligands during tumorigenesis may sensitize the emerging tumour cells to NKG2D-dependent elimination. Indeed, expression of NKG2D ligands on NK-resistance tumour lines was shown to mediate the in vivo elimination of these tumours (Cerwenka et al., 2000) and NKG2D plays a critical role in tumour immunosurveillancer in vivo (Guerra et al., 2008).

Similar to viruses, tumours have developed multiple ways to evade NKG2D-mediated immune response. At advanced tumour stages, sustained surface expression of NKG2D ligands and shedding of soluble MICA induces the internalization and degradation of NKG2D, thus promoting tumour immune evasion (Doubrovina et al., 2003; Groh et al., 2002). In addition, the frequent production of TGF-β by tumours may also contribute to downregulation of NKG2D (Castriconi et al., 2003).

NKG2D ligands also participate in the crosstalk between immune cells. These ligands were shown to be upregulated on TLR-stimulated DCs and macrophages and to regulate innate and adaptive immune responses (Ebihara et al., 2007; Nedvetzki et al., 2007).

Several reports demonstrate the existence of MICA and MICB transcripts in several normal cells and tissues, but surprisingly, these cells lack the expression of the corresponding proteins. These observations, together with the surprising finding that the site which is targeted by HCMV-miR-UL112 is conserved in the 3′ UTR sequences of various MICB alleles and that a similar site also exists in MICA 3′ UTR, lead us to hypothesize that MICA and MICB expression is controlled by cellular microRNAs. Indeed, we recently identified a group of endogenous cellular microRNAs that control MICA and MICB expression by binding to *MICA* and *MICB* 3′ UTR sites that overlap with those bound by HCMV-miR-UL112. We demonstrated that under normal conditions, these microRNAs maintain MICA and MICB protein expression under a certain threshold, yet facilitate acute MICA and MICB upregulation during cellular stress (Stern-Ginossar et al., 2008). This discovery suggests a novel post-transcriptional regulation on MICA and MICB expression and can explain some of the reported discrepancies between the presence mRNA of NKG2D ligands and the absence of protein expression in different cells and tissues (Boissel et al., 2006; Groh et al., 1996; Nedvetzki et al., 2007). Importantly, a large fraction of the MICA and MICB targeting microRNAs are overexpressed in various tumours. We have also demonstrated that they help tumours avoid immune recognition.

The concept arising from these findings is that viral microRNAs do not target random sequences in the 3′ UTR of human genes, but rather use sequences that are also used by cellular microRNAs and are therefore indispensable for the host. The fact that both tumours and viruses developed so many different mechanisms to interfere with stress-induced ligand expression indicates the importance of these ligands in immune cell attack.

Natural cytotoxic receptors

The natural cytotoxic receptors (NCRs) include NKp46, NKp44 and NKp30. Orthologues for NCR genes were identified in mice, rats and primates (Biassoni et al., 1999; De Maria et al., 2001; Falco et al., 1999). However, NKp46 is the only NCR with functional orthologues in mouse (Biassoni et al., 1999) and rat (Falco et al., 1999). Antibody-blocking experiments suggested that each NCR probably recognizes a different set of ligands, (Bottino et al., 2000). Although the cellular ligands of the NCRs are unknown, heparan sulfate proteoglycans (HSP) were shown to participate in the NCR recognition of their ligands (Bloushtain et al., 2004).

The NCRs belong to the Ig superfamily and contain a charged amino acid in their transmembrane domain which associates with ITAM-bearing adaptor molecules. Importantly, both NKp46 and NKp30 are expressed exclusively on activated and resting NK cells, while NKp44 is upregulated after activation and is also reported to be expressed on plasmacytoid DCs (Fuchs et al., 2005). This makes NKp46 and NKp30 the only NK-specific markers known today. Although recent evidence suggests that a very small subset of T cells express NKp46 (Caligiuri, 2008), the importance of these T cells and whether these cells are bona fide T cells still needs to be addressed.

It was suggested that functional crosstalk exists between the three NCRs since engagement of a single NCR also leads to activation of the signalling cascade associated with the others. Thus, it seems that the NCRs operate as one unit displaying functional crosstalk between themselves (Bryceson et al., 2006).

The NCRs are involved in recognition and killing of tumour cells. Anti-NCR monoclonal antibodies block NK-mediated killing of most tumour lines (Pende et al., 1999; Pessino et al., 1998; Sivori et al., 1997; Vitale et al., 1998). Despite accumulating evidence for the important role played by the NCRs in anti-tumour activity, little is known about the nature and distribution of their ligands. Further work should be performed to elucidate the contribution of HSP in this regard.

NKp46

NKp46 was the first NCR to be identified (Pessino et al., 1998; Sivori et al., 1997). Analysis of the tissue distribution of NKp46 revealed that it is expressed on both activated as well as resting NK cells, but is not found on other cell types tested. Crosslinking of NKp46 led to calcium mobilization, cytotoxicity and cytokine release (Sivori et al., 1997). Indeed, NKp46 is considered a major NK lysis receptor and plays a dominant role in the activation of NK cells against various targets (Sivori et al., 1999). NKp46 is a 46 kDa glycoprotein that contains two C2 Ig-like extracellular domains. Activation of NKp46 is mediated via the association of NKp46 with the adaptor molecules CD3z and FcεRγ that contain the activating ITAM motif (Lanier, 2003).

While the identity of the cellular ligand(s) for NKp46 is still obscure, it was suggested that HSP are involved in the recognition of tumour cells by NKp46 and NKp30 (Bloushtain et al., 2004). Tumour cells expressing cell surface heparinase are less susceptible to NK-mediated killing, compared to parental cells that do not express heparanase. Similarly, the absence of heparan sulfate expression reduced NK-mediated killing (Bloushtain et al., 2004).

Hemagglutinin molecules of different influenza strains were identified as the first specific NKp46 and NKp44 ligands (Arnon et al., 2001, 2004; Mandelboim et al., 2001). In a substantial subset of NK cells, hemagglutinin recognition by NKp46 was needed to kill the infected cells (Achdout et al., 2008; Mandelboim et al., 2001). The interaction was demonstrated to be direct and was mediated mainly via α2,6-linked sialic acid carried by NKp46. In addition, it was demonstrated that the ability of NKp46 to recognize infected target cells is confined to the membrane proximal domain, and largely relies on the highly conserved sugar-carrying residue, Thr 225, which is located in the stalk region (Arnon et al., 2004).

The crystal structure of NKp46 has been resolved (Foster et al., 2003). The overall structure showed resemblance to the LIR1 and KIR2D receptors that bind MHC class I proteins. The known ligand-binding sites of KIR and LIR were compared with the corresponding structural regions of NKp46 and a potential binding site for the unknown cellular ligand was predicted to be near the interdomain hinge, a region that mediates ligand binding in the KIR receptor (Foster et al., 2003).

NKp44

NKp44 is the second NCR identified on human NK cells. It encodes a 44Dka surface glycoprotein and blocking of NKp44 with specific mAbs partially inhibited cytotoxicity against MHC class I-deficient targets (Cantoni et al., 1999b; Vitale et al., 1998). The simultaneous blocking of both NKp44 and NKp46 led to a significantly increased inhibition (Vitale et al., 1998), supporting a functional crosstalk among these receptors (Augugliaro et al., 2003).

The activating signal of NKp44 is delivered via the association and phosphorylation of the DAP12 adaptor molecule (Cantoni et al., 1999b; Vitale et al., 1998). NKp44 is not expressed on resting NK cells, but rather requires (at least in vitro) activation for its expression (Vitale et al., 1998). In vivo, NKp44 was also found on a subset of NK cells in human tonsils (Ferlazzo et al., 2004), in adenoids (Mizrahi et al., 2007) and in decidual NK cells during pregnancy (Hanna et al., 2006), possibly representing the in vivo activated status of NK cells in these organs.

The crystal structure of the extracellular domain of NKp44 was also determined. The overall structure of NKp44 adopts a standard IgV-like fold. However, the crystal structure revealed a putative ligand binding site for anionic ligands (Cantoni et al., 2003). Similar to NKp46, very little is known regarding the cellular ligands of NKp44 and the only ligand identified so far for this receptor is the influenza hemagglutinin protein which directly binds in a sialic acid dependent manner and activates NK cell cytotoxicity (Arnon et al., 2001, 2004).

NKp30

NKp30 was the third NCR to be identified, functioning in the killing of targets that are relatively resistant to NKp46/44-mediated killing, demonstrating that it probably recognizes ligands other than those recognized by NKp46/44 (Moretta et al., 2000; Pende et al., 1999).

NKp30 is a 30kDa glycoprotein that contains one V-type Ig-like extracellular domain. With the exception of NK cells present in the lymph nodes and in the endometrium during the menstrual cycle (Manaster et al., 2008), NKp30 is selectively expressed on all human NK cells (Pende et al., 1999). The transmembrane portion of NKp30 contains an arginine residue, which is probably involved in the association with CD3ζ chains for the transduction of the downstream activating signals (Pende et al., 1999).

Accumulating evidence indicates that NKp30 is involved in regulation of the immune response via interaction with DCs. NKp30 was shown to recognize an unknown ligand expressed on immature (imDCs), but not on mature (mDCs), DCs (Cooper et al., 2004; Ferlazzo et al., 2002). The engagement of NKp30 was shown to result in either killing or maturation of autologous imDCs (Degli-Esposti and Smyth, 2005). The DC–NKp30 crosstalk leads to the priming and maturation of NK cells that in turn start to secrete cytokines and other factors that activate and promote the maturation of DCs (Ferlazzo, 2005). Further research should elucidate the in vivo significance of these observations.

NKp30 is also suggested to interact with HSP (Bloushtain et al., 2004). However, based on the differential susceptibility of targets to the killing mediated by NKp30 and NKp46, it is likely that these receptors recognize different cellular ligands, (Moretta et al., 2000). Thus, HSP are probably co-ligands for these receptors.

The HCMV tegument protein, pp65, interacts with NKp30 (Arnon et al., 2005). Surprisingly, pp65 was found to antagonize NKp30 activation by causing a dissociation of the adaptor molecule CD3ζ from NKp30. The viral pp65 is an intercellular protein and it was therefore suggested that soluble pp65 derived from the direct lysis of virus-infected cells or from apoptotic cells may bind to NKp30 and hamper its activity (Arnon et al., 2005).

The possibility that NKp30 can recognize intracellular ligands was supported by the finding that the nuclear factor HLA-B-associated transcript 3 (BAT3) is released from tumour cells, bound directly to NKp30, and engaged NKp30-mediated cytotoxicity. BAT3 is thus the first cellular ligand identified for NKp30 (Pogge von Strandmann et al., 2007).

NKp80

NKp80 was identified during a search for NK cell–specific surface markers and was shown to stimulate NK cell cytotoxicity and induces calcium influx in human NK cells after triggering by appropriate antibodies (Vitale et al., 2001). It is a type II transmembrane protein belonging to the C-type lectin superfamily, expressed on the cell surface as a dimer of 80kDa. The gene coding for NKp80 maps on chromosome 12 in the NK-gene complex region (Biassoni et al., 2001). The transmembrane region is characterized by non-polar amino acids and no classical ITAM-containing adaptors were found to associate with NKp80. Although sequence analysis showed that the cytoplasmic tail contains two tyrosine-based motifs, the signal transduction pathway of NKp80 is still unknown.

The activation-induced C-type lectin (AICL) (which is also encoded in the NK-gene complex) was identified as a ligand of NKp80. AICL is a myeloid-specific receptor expressed by monocytes, macrophages and granulocytes. Interaction between NK cells and monocytes resulted in a cytokine release that was partially dependant on NKp80 engagement. Hence, it seems that NKp80–AICL interaction is involved in the crosstalk between NK cells and myeloid cells and thus may influence the initiation and maintenance of immune responses in humans (Welte et al., 2006).

CD16

CD16 (FcγRIIIa) is type I transmembrane receptor containing two extracellular Ig-like domains. CD16 function is a low affinity IgG receptor mediating antibody-dependent cellular cytotoxicity (ADCC) by NK cells and was also demonstrated to directly recognize unknown tumour ligand (Mandelboim et al., 1999). CD16 signals via association with the adaptor molecules CD3z and FcεRγ that contain the activation ITAM motif (Vivier et al., 1991). Recently CD16 was reported to be the most potent activating receptor on freshly isolated human NK cells, able to elicit strong cytotoxicity and cytokine production (Bryceson et al., 2006).

Activating KIRs

Apart from KIR receptors that inhibit NK cytotoxicity, other KIRs were found to enhance NK cytotoxicity (Moretta et al., 1995). The activating KIRs, like the inhibitory KIRs, are type I transmembrane glycoproteins that consist of either two (KIR2D) or three (KIR3D) extracellular C2-type Ig-like domains. However, they posses a charged amino acid in their transmembrane and contain a short cytoplasmic tail without any known signalling motif (Biassoni et al., 1996). Instead, they are associated (via the charged amino acid) with the ITAM-containing signalling protein, DAP12. There are six genes encoding for the activating KIRs clustered on chromosome 19 within the LRC (Lanier, 2005). The ligands that are recognized by the activating KIRs are only poorly defined and the interaction of the activating KIRs with MHC class I molecules are generally of low affinity (Katz et al., 2004; Vales-Gomez et al., 1998). KIR2DS4 was shown to also interact with both MHC class I proteins and a non-MHC class I protein expressed on melanomas, resulting in enhanced NK killing (Katz et al., 2004). In addition, by using tetramers of class I, it was shown that activating KIR binding to MHC class I was dependant on peptides presented by MHC class I (Stewart et al., 2005). As only a small fraction of MHC class I proteins are occupied by a particular peptide, the importance of these findings is still unclear.

2B4

2B4 (also called CD244) and its ligand CD48 (Brown et al., 1998) are members of the CD2 family of the Ig superfamily. In humans 2B4 is expressed on all NK cells, $\gamma\delta$ T cells, CD8+ T cells, monocytes and basophils (Lanier, 2005). 2B4 contains two Ig-like domains and its cytoplasmic tail contains four Thr–x–Tyr–x–x–Leu/Ile motifs called ITSM motif. This motif defines a family of receptors sharing a common signalling pathway and includes the NTB-A and CRACC receptors that are also expressed by NK cells and are also able to activate their killing (Veillette, 2006). Upon phosphorylation of the tyrosine in the ITSM motif, probably by the Src family kinase Fyn, 2B4 binds to the SAP intracellular adaptor protein, or under special circumstances to the Src homology 2 domain-containing protein tyrosine phosphatase (SHP)-1 and SHP-2 tyrosine phosphatases. The function of 2B4 may be different in human and mouse NK cells. In mice, 2B4 apparently functions as an inhibitory NK cell receptor. By contrast, in mature human NK cells, 2B4 behaves as an activating receptor. Whether 2B4 is capable of independently triggering effector cell functions or rather serves as a costimulatory receptor is still unresolved (Nakajima et al., 1999; Sivori et al., 2000).

The strongest evidence supporting an activating role for 2B4 in human NK cells comes from the demonstration that transfection of CD48 into certain NK-resistant target cells renders them susceptible to NK cell–mediated cytotoxicity and triggers the production of IFN-γ (Tangye et al., 2000). The functional requirement for SAP in 2B4 mediated activation of human NK cells is revealed in patients with a loss-of-function mutation in the X-linked SAP gene; NK cells from these patients can no longer able be activated through the 2B4 receptor (Nakajima et al., 2000; Parolini et al., 2000).

DNAM-1

DNAM-1 receptor (also called CD226) is a member of the Ig superfamily, encoded by a gene on human chromosome 18q22.3, is expressed by human NK cells, T cells, a subset of B cells, monocytes and platelets (Lanier, 2005). CD155 (also known as polio virus receptor [PVR]) and CD112 (also called nectin-2) have been identified as ligands for DNAM-1 (Bottino et al., 2003). Interactions between DNAM-1 on NK cells and its ligands on tumour cells augments the NK cell–mediated cytotoxicity and cytokine production (Bottino et al., 2003; Tahara-Hanaoka et al., 2004). Moreover, it was shown that the interaction of DNAM-1 with CD112 and CD155 contributes to the NK-mediated lysis of both imDCs and mDCs (Pende et al., 2006). NK cells also express CD96, which shows only ~20% homology to DNAM-1 but was also shown to recognize CD155 and promote NK cell adhesion and activation (Fuchs et al., 2004). An indication that CD155 plays a key role in regulating NK cell function was evident by the demonstration that the HCMV protein UL141 blocks the surface expression of CD155 and help the virus to evade NK cell killing (Tomasec et al., 2005).

Summary

The killing mediated by NK cells is controlled by many different inhibitory and activating NK receptors. This complex repertoire of NK cell receptors results in various NK cell subpopulations that can respond to bacteria, viruses, parasites and tumours. Thus, although NK cells (unlike T and B cells) express only a limited number of receptors, the various combinations of receptors enable a remarkable diversity in NK cell activity. Some of the future challenges in the NK field might include population dynamics, elucidating the precise threshold needed for NK cell activation and inhibition, and identifying the entire spectrum of NK cell inhibitory and activating receptors following by the identification of their various physiological ligands.

References

Achdout, H., et al. (2008). Influenza virus infection augments NK cell inhibition through reorganization of major histocompatibility complex class I proteins. J Virol **82**, 8030–8037.

Aldemir, H., et al. (2005). Cutting edge: lectin-like transcript 1 is a ligand for the CD161 receptor. J Immunol **175**, 7791–7795.

Arnon, T.I., et al. (2005). Inhibition of the NKp30 activating receptor by pp65 of human cytomegalovirus. Nat Immunol **6**, 515–523.

Arnon, T.I., et al. (2004). The mechanisms controlling the recognition of tumor- and virus-infected cells by NKp46. Blood **103**, 664–672.

Arnon, T.I., et al. (2001). Recognition of viral hemagglutinins by NKp44 but not by NKp30. Eur J Immunol **31**, 2680–2689.

Augugliaro, R., et al. (2003). Selective cross-talk among natural cytotoxicity receptors in human natural killer cells. Eur J Immunol **33**, 1235–1241.

Avril, T., et al. (2004). The membrane-proximal immunoreceptor tyrosine-based inhibitory motif is critical for the inhibitory signaling mediated by siglecs-7 and -9, CD33-related siglecs expressed on human monocytes and NK cells. J Immunol **173**, 6841–6849.

Bauer, S., et al. (1999). Activation of NK cells and T cells by NKG2D, a receptor for stress-inducible MICA. Science **285**, 727–729.

Beauchemin, N., et al. (1999). Redefined nomenclature for members of the carcinoembryonic antigen family. Exp Cell Res **252**, 243–249.

Biassoni, R., et al. (1996). The human leukocyte antigen (HLA)-C-specific "activatory" or "inhibitory" natural killer cell receptors display highly homologous extracellular domains but differ in their transmembrane and intracytoplasmic portions. J Exp Med **183**, 645–650.

Biassoni, R., et al. (2001). Human natural killer cell receptors and co-receptors. Immunol Rev **181**, 203–214.

Biassoni, R., et al. (1999). The murine homologue of the human NKp46, a triggering receptor involved in the induction of natural cytotoxicity. Eur J Immunol **29**, 1014–1020.

Bloushtain, N., et al. (2004). Membrane-associated heparan sulfate proteoglycans are involved in the recognition of cellular targets by NKp30 and NKp46. J Immunol **173**, 2392–2401.

Boissel, N., et al. (2006). BCR/ABL oncogene directly controls MHC class I chain-related molecule A expression in chronic myelogenous leukemia. J Immunol **176**, 5108–5116.

Bottino, C., et al. (2000). The human natural cytotoxicity receptors (NCR) that induce HLA class I-independent NK cell triggering. Hum Immunol **61**, 1–6.

Bottino, C., et al. (2003). Identification of PVR (CD155) and nectin-2 (CD112) as cell surface ligands for the human DNAM-1 (CD226) activating molecule. J Exp Med **198**, 557–567.

Boulton, I.C. and Gray-Owen, S.D. (2002). Neisserial binding to CEACAM1 arrests the activation and proliferation of CD4+ T lymphocytes. Nat Immunol **3**, 229–236.

Braud, V., et al. (1997). The human major histocompatibility complex class Ib molecule HLA-E binds signal sequence-derived peptides with primary anchor residues at positions 2 and 9. Eur J Immunol **27**, 1164–1169.

Braud, V.M., et al. (1998). HLA-E binds to natural killer cell receptors CD94/NKG2A, B and C. Nature **391**, 795–799.

Brown, M.H., et al. (1998). 2B4, the natural killer and T cell immunoglobulin superfamily surface protein, is a ligand for CD48. J Exp Med **188**, 2083–2090.

Bryceson, Y.T., et al. (2006). Synergy among receptors on resting NK cells for the activation of natural cytotoxicity and cytokine secretion. Blood **107**, 159–166.

Butcher, S., et al. (1998). MAFA-L, an ITIM-containing receptor encoded by the human NK cell gene complex and expressed by basophils and NK cells. Eur J Immunol **28**, 3755–3762.

Caligiuri, M.A. (2008). Human natural killer cells. Blood **112**, 461–469.

Cantoni, C., et al. (1999a). Molecular and functional characterization of IRp60, a member of the immunoglobulin superfamily that functions as an inhibitory receptor in human NK cells. Eur J Immunol **29**, 3148–3159.

Cantoni, C., et al. (1999b). NKp44, a triggering receptor involved in tumor cell lysis by activated human natural killer cells, is a novel member of the immunoglobulin superfamily. J Exp Med **189**, 787–796.

Cantoni, C., et al. (2003). The three-dimensional structure of the human NK cell receptor NKp44, a triggering partner in natural cytotoxicity. Structure (Camb) **11**, 725–734.

Carretero, M., et al. (1998). Specific engagement of the CD94/NKG2-A killer inhibitory receptor by the HLA-E class Ib molecule induces SHP-1 phosphatase recruitment to tyrosine-phosphorylated NKG2-A: evidence for receptor function in heterologous transfectants. Eur J Immunol **28**, 1280–1291.

Castriconi, R., et al. (2003). Transforming growth factor beta 1 inhibits expression of NKp30 and NKG2D receptors: consequences for the NK-mediated killing of dendritic cells. Proc Natl Acad Sci U S A **100**, 4120–4125.

Cerwenka, A., et al. (2000). Retinoic acid early inducible genes define a ligand family for the activating NKG2D receptor in mice. Immunity **12**, 721–727.

Chalupny, N.J., et al. (2006). Down-regulation of the NKG2D ligand MICA by the human cytomegalovirus glycoprotein UL142. Biochem Biophys Res Commun **346**, 175–181.

Chapman, T.L., et al. (2000). Crystal structure and ligand binding properties of the D1D2 region of the inhibitory receptor LIR-1 (ILT2). Immunity **13**, 727–736.

Cohen, G.B., et al. (1999). The selective downregulation of class I major histocompatibility complex proteins by HIV-1 protects HIV-infected cells from NK cells. Immunity **10**, 661–671.

Colonna, M., et al. (1999). Inhibitory and activating receptors involved in immune surveillance by human NK and myeloid cells. J Leukoc Biol **66**, 718–722.

Colonna, M., et al. (1997). A common inhibitory receptor for major histocompatibility complex class I molecules on human lymphoid and myelomonocytic cells. J Exp Med **186**, 1809–1818.

Colonna, M. and Samaridis, J. (1995). Cloning of immunoglobulin-superfamily members associated with HLA-C and HLA-B recognition by human natural killer cells. Science **268**, 405–408.

Cooper, M.A., et al. (2004). NK cell and DC interactions. Trends Immunol **25**, 47–52.

Cordon-Cardo, C., et al. (1991). Expression of HLA-A,B,C antigens on primary and metastatic tumor cell populations of human carcinomas. Cancer Res **51**, 6372–6380.

Cosman, D., et al. (1997). A novel immunoglobulin superfamily receptor for cellular and viral MHC class I molecules. Immunity **7**, 273–282.

Crocker, P.R. and Varki, A. (2001). Siglecs, sialic acids and innate immunity. Trends Immunol **22**, 337–342.

De Maria, A., et al. (2001). Identification, molecular cloning and functional characterization of NKp46 and NKp30 natural cytotoxicity receptors in *Macaca fascicularis* NK cells. Eur J Immunol **31**, 3546–3556.

Degli-Esposti, M.A. and Smyth, M.J. (2005). Close encounters of different kinds: dendritic cells and NK cells take centre stage. Nat Rev Immunol **5**, 112–124.

Diefenbach, A., et al. (2000). Ligands for the murine NKG2D receptor: expression by tumor cells and activation of NK cells and macrophages. Nat Immunol **1**, 119 126.

Doubrovina, E.S., et al. (2003). Evasion from NK cell immunity by MHC class I chain-related molecules expressing colon adenocarcinoma. J Immunol **171**, 6891–6899.

Dunn, C., et al. (2003). Human cytomegalovirus glycoprotein UL16 causes intracellular sequestration of NKG2D ligands, protecting against natural killer cell cytotoxicity. J Exp Med **197**, 1427–1439.

Eagle, R.A. and Trowsdale, J. (2007). Promiscuity and the single receptor: NKG2D. Nat Rev Immunol **7**, 737–744.

Ebihara, T., et al. (2007). Induction of NKG2D ligands on human dendritic cells by TLR ligand stimulation and RNA virus infection. Int Immunol **19**, 1145–1155.

Falco, M., et al. (1999). Identification of the rat homologue of the human NKp46 triggering receptor. Immunol Lett **68**, 411–414.

Ferlazzo, G. (2005). Natural killer and dendritic cell liaison: recent insights and open questions. Immunol Lett **101**, 12–17.

Ferlazzo, G., et al. (2004). The abundant NK cells in human secondary lymphoid tissues require activation to express killer cell Ig-like receptors and become cytolytic. J Immunol **172**, 1455–1462.

Ferlazzo, G., et al. (2002). Human dendritic cells activate resting natural killer (NK) cells and are recognized via the NKp30 receptor by activated NK cells. J Exp Med **195**, 343–351.

Fernandez, N.C., et al. (2005). A subset of natural killer cells achieves self-tolerance without expressing inhibitory receptors specific for self-MHC molecules. Blood **105**, 4416–4423.

Foster, C.E., et al. (2003). Crystal structure of the human natural killer (NK) cell activating receptor NKp46 reveals structural relationship to other leukocyte receptor complex immunoreceptors. J Biol Chem **278**, 46081–46086.

Fuchs, A., et al. (2004). Cutting edge: CD96 (tactile) promotes NK cell–target cell adhesion by interacting with the poliovirus receptor (CD155). J Immunol **172**, 3994–3998.

Fuchs, A., et al. (2005). Paradoxic inhibition of human natural interferon-producing cells by the activating receptor NKp44. Blood **106**, 2076–2082.

Gasser, S., et al. (2005). The DNA damage pathway regulates innate immune system ligands of the NKG2D receptor. Nature **436**, 1186–1190.

Gazit, R., et al. (2004). Expression of KIR2DL1 on the entire NK cell population: a possible novel immunodeficiency syndrome. Blood **103**, 1965–1966.

Gonen-Gross, T., et al. (2003). Complexes of HLA-G protein on the cell surface are important for leukocyte Ig-like receptor-1 function. J Immunol **171**, 1343–1351.

Groh, V., et al. (1996). Cell stress-regulated human major histocompatibility complex class I gene expressed in gastrointestinal epithelium. Proc Natl Acad Sci U S A **93**, 12445–12450.

Groh, V., et al. (2001). Costimulation of CD8alphabeta T cells by NKG2D via engagement by MIC induced on virus-infected cells. Nat Immunol **2**, 255–260.

Groh, V., et al. (1999). Broad tumor-associated expression and recognition by tumor-derived gamma delta T cells of MICA and MICB. Proc Natl Acad Sci U S A **96**, 6879–6884.

Groh, V., et al. (1998). Recognition of stress-induced MHC molecules by intestinal epithelial gammadelta T cells. Science **279**, 1737–1740.

Groh, V., et al. (2002). Tumour-derived soluble MIC ligands impair expression of NKG2D and T-cell activation. Nature **419**, 734–738.

Grundemann, C., et al. (2006). Cutting edge: identification of E-cadherin as a ligand for the murine killer cell lectin-like receptor G1. J Immunol **176**, 1311–1315.

Guerra, N., et al. (2008). NKG2D-deficient mice are defective in tumor surveillance in models of spontaneous malignancy. Immunity **28**, 571–580.

Hammarstrom, S. (1999). The carcinoembryonic antigen (CEA) family: structures, suggested functions and expression in normal and malignant tissues. Semin Cancer Biol **9**, 67–81.

Hanke, T., et al. (1998). 2F1 antigen, the mouse homolog of the rat "mast cell function-associated antigen", is a lectin-like type II transmembrane receptor expressed by natural killer cells. Eur J Immunol **28**, 4409–4417.

Hanna, J., et al. (2006). Decidual NK cells regulate key developmental processes at the human fetal–maternal interface. Nat Med **12**, 1065–1074.

Hewitt, E.W. (2003). The MHC class I antigen presentation pathway: strategies for viral immune evasion. Immunology **110**, 163–169.

Horng, T., et al. (2007). NKG2D signaling is coupled to the interleukin 15 receptor signaling pathway. Nat Immunol **8**, 1345–1352.

Ito, M., et al. (2006). Killer cell lectin-like receptor G1 binds three members of the classical cadherin family to inhibit NK cell cytotoxicity. J Exp Med **203**, 289–295.

Jamieson, A.M., et al. (2002). The role of the NKG2D immunoreceptor in immune cell activation and natural killing. Immunity **17**, 19–29.

Karre, K., et al. (1986). Selective rejection of H-2-deficient lymphoma variants suggests alternative immune defence strategy. Nature **319**, 675–678.

Katz, G., et al. (2004). MHC class I-independent recognition of NK-activating receptor KIR2DS4. J Immunol **173**, 1819–1825.

Kim, S., et al. (2005). Licensing of natural killer cells by host major histocompatibility complex class I molecules. Nature **436**, 709–713.

Lanier, L.L. (2003). Natural killer cell receptor signaling. Curr Opin Immunol **15**, 308–314.

Lanier, L.L. (2005). NK cell recognition. Annu Rev Immunol **23**, 225–274.

Lanier, L.L., et al. (1994). Human NKR-P1A. A disulfide-linked homodimer of the C-type lectin superfamily expressed by a subset of NK and T lymphocytes. J Immunol **153**, 2417–2428.

Le Drean, E., et al. (1998). Inhibition of antigen-induced T cell response and antibody-induced NK cell cytotoxicity by NKG2A: association of NKG2A with SHP-1 and SHP-2 protein-tyrosine phosphatases. Eur J Immunol **28**, 264–276.

Lee, N., et al. (1998a). HLA-E surface expression depends on binding of TAP-dependent peptides derived from certain HLA class I signal sequences. J Immunol **160**, 4951–4960.

Lee, N., et al. (1998b). HLA-E is a major ligand for the natural killer inhibitory receptor CD94/NKG2A. Proc Natl Acad Sci U S A **95**, 5199–5204.

Li, P., et al. (2001). Complex structure of the activating immunoreceptor NKG2D and its MHC class I-like ligand MICA. Nat Immunol **2**, 443–451.

Liao, N.S., et al. (1991). MHC class I deficiency: susceptibility to natural killer (NK) cells and impaired NK activity. Science **253**, 199–202.

Manaster, I., et al. (2008). Endometrial NK cells are special immature cells that await pregnancy. J Immunol **181**, 1869–1876.

Mandelboim, O., et al. (2001). Recognition of haemagglutinins on virus-infected cells by NKp46 activates lysis by human NK cells. Nature **409**, 1055–1060.

Mandelboim, O., et al. (1999). Human CD16 as a lysis receptor mediating direct natural killer cell cytotoxicity. Proc Natl Acad Sci U S A **96**, 5640–5644.

Mandelboim, O., et al. (1996). Protection from lysis by natural killer cells of group 1 and 2 specificity is mediated by residue 80 in human histocompatibility leukocyte antigen C alleles and also occurs with empty major histocompatibility complex molecules. J Exp Med **184**, 913–922.

Markel, G., et al. (2002a). CD66a interactions between human melanoma and NK cells: a novel class I MHC-independent inhibitory mechanism of cytotoxicity. J Immunol **168**, 2803–2810.

Markel, G., et al. (2002b). Pivotal role of CEACAM1 protein in the inhibition of activated decidual lymphocyte functions. J Clin Invest **110**, 943–953.

Martin, M.P., et al. (2002). Epistatic interaction between KIR3DS1 and HLA-B delays the progression to AIDS. Nat Genet **31**, 429–434.

Meyaard, L. (2008). The inhibitory collagen receptor LAIR-1 (CD305). J Leukoc Biol **83**, 799–803.

Mizrahi, S., et al. (2007). A phenotypic and functional characterization of NK cells in adenoids. J Leukoc Biol **82**, 1095–1105.

Moretta, A., et al. (2000). Natural cytotoxicity receptors that trigger human NK-cell-mediated cytolysis. Immunol Today **21**, 228–234.

Moretta, A., et al. (1996). Receptors for HLA class-I molecules in human natural killer cells. Annu Rev Immunol **14**, 619–648.

Moretta, A., et al. (1995). Existence of both inhibitory (p58) and activatory (p50) receptors for HLA-C molecules in human natural killer cells. J Exp Med **182**, 875–884.

Moretta, L., et al. (2002). Human natural killer cells: their origin, receptors and function. Eur J Immunol **32**, 1205–1211.

Moretta, L. and Moretta, A. (2004). Unravelling natural killer cell function: triggering and inhibitory human NK receptors. EMBO J **23**, 255–259.

Nakajima, H., et al. (2000). Patients with X-linked lymphoproliferative disease have a defect in 2B4 receptor-mediated NK cell cytotoxicity. Eur J Immunol **30**, 3309–3318.

Nakajima, H., et al. (1999). Activating interactions in human NK cell recognition: the role of 2B4-CD48. Eur J Immunol **29**, 1676–1683.

Nedvetzki, S., et al. (2007). Reciprocal regulation of human natural killer cells and macrophages associated with distinct immune synapses. Blood **109**, 3776–3785.

Nicoll, G., et al. (2003). Ganglioside GD3 expression on target cells can modulate NK cell cytotoxicity via siglec-7-dependent and -independent mechanisms. Eur J Immunol **33**, 1642–1648.

Nicoll, G., et al. (1999). Identification and characterization of a novel siglec, siglec-7, expressed by human natural killer cells and monocytes. J Biol Chem **274**, 34089–34095.

Parolini, S., et al. (2000). X-linked lymphoproliferative disease. 2B4 molecules displaying inhibitory rather than activating function are responsible for the inability of natural killer cells to kill Epstein–Barr virus-infected cells. J Exp Med **192**, 337–346.

Pende, D., et al. (2001). Role of NKG2D in tumor cell lysis mediated by human NK cells: cooperation with natural cytotoxicity receptors and capability of recognizing tumors of nonepithelial origin. Eur J Immunol **31**, 1076–1086.

Pende, D., et al. (2006). Expression of the DNAM-1 ligands, nectin-2 (CD112) and poliovirus receptor (CD155), on dendritic cells: relevance for natural killer–dendritic cell interaction. Blood **107**, 2030–2036.

Pende, D., et al. (1999). Identification and molecular characterization of NKp30, a novel triggering receptor involved in natural cytotoxicity mediated by human natural killer cells. J Exp Med **190**, 1505–1516.

Pessino, A., et al. (1998). Molecular cloning of NKp46: a novel member of the immunoglobulin superfamily involved in triggering of natural cytotoxicity. J Exp Med **188**, 953–960.

Pogge von Strandmann, E., et al. (2007). Human leukocyte antigen-B-associated transcript 3 is released from tumor cells and engages the NKp30 receptor on natural killer cells. Immunity **27**, 965–974.

Poggi, A., et al. (1998). IL-12-induced up-regulation of NKRP1A expression in human NK cells and consequent NKRP1A-mediated down-regulation of NK cell activation. Eur J Immunol **28**, 1611–1616.

Radaev, S., et al. (2001). Conformational plasticity revealed by the cocrystal structure of NKG2D and its class I MHC-like ligand ULBP3. Immunity **15**, 1039–1049.

Raulet, D.H., et al. (2001). Regulation of the natural killer cell receptor repertoire. Annu Rev Immunol **19**, 291–330.

Rosen, D.B., et al. (2005). Cutting edge: lectin-like transcript-1 is a ligand for the inhibitory human NKR-P1A receptor. J Immunol **175**, 7796–7799.

Rosen, D.B., et al. (2008). Functional consequences of interactions between human NKR-P1A and its ligand LLT1 expressed on activated dendritic cells and B cells. J Immunol **180**, 6508–6517.

Sivori, S., et al. (2000). 2B4 functions as a co-receptor in human NK cell activation. Eur J Immunol **30**, 787–793.

Sivori, S., et al. (1999). NKp46 is the major triggering receptor involved in the natural cytotoxicity of fresh or cultured human NK cells. Correlation between surface density of NKp46 and natural cytotoxicity against autologous, allogeneic or xenogeneic target cells. Eur J Immunol **29**, 1656–1666.

Sivori, S., et al. (1997). p46, a novel natural killer cell-specific surface molecule that mediates cell activation. J Exp Med **186**, 1129–1136.

Stern-Ginossar, N., et al. (2007). Host immune system gene targeting by a viral miRNA. Science **317**, 376–381.

Stern-Ginossar, N., et al. (2008). Human microRNAs regulate stress-induced immune responses mediated by the receptor NKG2D. Nat Immunol.

Stern, N., et al. (2005). Carcinoembryonic antigen (CEA) inhibits NK killing via interaction with CEA-related cell adhesion molecule 1. J Immunol **174**, 6692–6701.

Stewart, C.A., et al. (2005). Recognition of peptide-MHC class I complexes by activating killer immunoglobulin-like receptors. Proc Natl Acad Sci U S A **102**, 13224–13229.

Sun, J.C. and Lanier, L.L. (2008). Tolerance of NK cells encountering their viral ligand during development. J Exp Med **205**, 1819–1828.

Tahara-Hanaoka, S., et al. (2004). Functional characterization of DNAM-1 (CD226) interaction with its ligands PVR (CD155) and nectin-2 (PRR-2/CD112). Int Immunol **16**, 533–538.

Tangye, S.G., et al. (2000). The CD2-subset of the Ig superfamily of cell surface molecules: receptor-ligand pairs expressed by NK cells and other immune cells. Semin Immunol **12**, 149–157.

Tomasec, P., et al. (2000). Surface expression of HLA-E, an inhibitor of natural killer cells, enhanced by human cytomegalovirus gpUL40. Science **287**, 1031.

Tomasec, P., et al. (2005). Downregulation of natural killer cell-activating ligand CD155 by human cytomegalovirus UL141. Nat Immunol **6**, 181–188.

Tripathy, S.K., et al. (2008). Continuous engagement of a self-specific activation receptor induces NK cell tolerance. J Exp Med **205**, 1829–1841.

Vales-Gomez, M., et al. (1998). Differential binding to HLA-C of p50-activating and p58-inhibitory natural killer cell receptors. Proc Natl Acad Sci U S A **95**, 14326–14331.

Veillette, A. (2006). NK cell regulation by SLAM family receptors and SAP-related adapters. Immunol Rev **214**, 22–34.

Venkataraman, G.M., et al. (2007). Promoter region architecture and transcriptional regulation of the genes for the MHC class I-related chain A and B ligands of NKG2D. J Immunol **178**, 961–969.

Vitale, M., et al. (1998). NKp44, a novel triggering surface molecule specifically expressed by activated natural killer cells, is involved in non-major histocompatibility complex-restricted tumor cell lysis. J Exp Med **187**, 2065–2072.

Vitale, M., et al. (2001). Identification of NKp80, a novel triggering molecule expressed by human NK cells. Eur J Immunol **31**, 233–242.

Vivier, E., et al. (1991). Tyrosine phosphorylation of the Fc gamma RIII(CD16): zeta complex in human natural killer cells. Induction by antibody-dependent cytotoxicity but not by natural killing. J Immunol **146**, 206–210.

Welte, S., et al. (2006). Mutual activation of natural killer cells and monocytes mediated by NKp80–AICL interaction. Nat Immunol **7**, 1334–1342.

Wills, M.R., et al. (2005). Human cytomegalovirus encodes an MHC class I-like molecule (UL142) that functions to inhibit NK cell lysis. J Immunol **175**, 7457–7465.

Wu, J., et al. (2003). Intracellular retention of the MHC class I-related chain B ligand of NKG2D by the human cytomegalovirus UL16 glycoprotein. J Immunol **170**, 4196–4200.

Yokoyama, W.M., et al. (1995). A family of murine NK cell receptors specific for target cell MHC class I molecules. Semin Immunol **7**, 89–101.

Yu, M.C., et al. (2008). An essential function for beta-arrestin 2 in the inhibitory signaling of natural killer cells. Nat Immunol **9**, 898–907.

Zou, Y., et al. (2005). Effect of human cytomegalovirus on expression of MHC class I-related chains A. J Immunol **174**, 3098–3104.

Chapter Twelve

12

Development and testing of NK cell lines

Hans Klingemann*

CHAPTER CONTENTS

Never get discouraged or have doubts when you want to protect and commercialize your invention.

Unknown

ABSTRACT

Infusion of enriched natural killer (NK) cell preparations have now become a therapeutic possibility with the advent of magnetic columns that can deplete unwanted cells and enrich for NK cells. However, autologous NK cells are generally inactive as they express KIR receptors that block their activation. Allogeneic NK cells are preferred but it requires removal of T-lymphocytes resulting in a limited cell product. Clonal allogeneic cell lines could be of more benefit but only very few true NK cell lines have been cloned mostly form patients with large granular lymphocytes (LGL) leukemia. The NK-92 cell line has gone through preclinical development and has finished two phase 1 studies confirming safety of this cell-based treatment. Cell lines can be genetically manipulated which has been shown for NK-92 by inserting a gene to express chimeric antigen receptors recognizing tumour cells and high affinity Fc receptors allowing to combine antibody treatment and NK cell infusions for augmented ADCC.

KEY WORDS

Allogeneic NK cells, Targeted NK cells, Chimeric antigen receptors, ADCC

Clinicians have taken different approaches when developing natural killer (NK) cells into a therapeutic cell product. Initial studies used autologous NK cells, obtained by leukapheresis, that were expanded ex vivo and activated with IL-2 (Rosenberg et al., 1985). These cells—referred to under the term lymphokine-activated killer (LAK) cells—consisted of a 'gemisch' of cytotoxic lymphocytes that in addition to NK cells, contained essentially any cell that would be responsive to IL-2, such as NK-T cells, T cells and even monocytes. These initial LAK cell studies using autologous lymphocytes showed responses in the occasional patient with renal cell cancer or melanoma. Since patients were also given high doses of IL-2 post-infusion, it was never resolved whether the LAK cells contributed in a significant way to the therapeutic benefit in those patients. Randomized studies were indeed ambiguous (Rosenberg et al., 1993). LAK cells were subsequently tested widely, but because of the technical challenges to expand them and the borderline efficacy data, their further development stalled (Law et al., 1995). Now some 20 years later, we are beginning to know why the infusion of autologous NK cells had limited therapeutic effect. The 'self' major histocompatibility complex (MHC) antigens expressed

*Conflict of interest: The author is the co-founder of ZelleRx Corp, a company that pursues the worldwide development of NK-92-based therapies.

by the cancer cells prevent their recognition by autologous NK cells and in fact inactivate them through the KIR receptor mechanism (Karre, 1995; Rosen et al., 2008), in spite of the presence of activating ligands for NKG2D. Unless a tumour expresses mutated or no MHC antigens, it is unlikely that autologous NK cells can be of substantial benefit. One possibility to make autologous NK cells lyse the patient's own cancer cell is by engineering them to express ligands that recognize specific antigens on the tumour, thereby overriding any inhibitory KIR activation by the 'self' MHC. This strategy has been employed with NK cells that express chimeric antigen receptors (CARs) against Her2/neu (Uherek et al., 2002) or the lymphoid antigens CD19 (Boissel et al., 2009) and CD20 (Mueller, 2008).

In contrast to autologous NK cells, *allogeneic* NK cells can recognize and kill MHC-disparate tumour targets. However, it would be too simple to assume that the 'non-self' situation would be sufficient to kill all malignant cells. In addition to KIR receptor-mediated 'inactivation', *activating* receptors on NK cells need to bind to corresponding ligands on malignant cells, which can be variably missing and this can account for lack of killing by autologous or allogeneic NK cells (Sivori et al., 1999). Some tumour cells are resistant to perforin and granzymes (Van den Broek et al., 1995). Further, tumour cells can also be difficult for NK cells to reach due to tissue anatomy and aberrant vasculature. However, clinical data support the observation that allogeneic and KIR mismatched NK cells can prevent relapse following stem cell transplant for acute myeloid leukemia (AML) (Ruggeri, 1999), and that even patients with advanced leukemia may show response and enter remission (Miller et al., 2005). It remains to be seen whether allogeneic NK cells can be equally efficacious in patients with solid tumours. The recognition of the potential of allogeneic NK cells and recent positive reports on infusions of allogeneic NK cells, have sparked interest in allogeneic NK cell-based therapies. Since a leukapheresis collection to obtain NK cells also enriches for alloreactive T-lymphocytes, these need to be removed before infusion to prevent graft-versus-host disease (GvHD) of the recipient. This can be accomplished by immunomagnetic absorption with subsequent removal of the T-cells via a magnet (Boissel et al., 2007; Miller et al., 2005). Some investigators have added a positive selection step for CD56+ cells to this process (Yangar, 2003). However, the loss of NK cells with this dual enrichment step is substantial, and the process is time consuming and expensive. In addition, the yield of allogeneic CD56+ NK cells after CD3 depletion can be highly variable, ranging from 10–30%. Unfortunately, expanding NK cells does not occur in a linear fashion and it can take at least 2 weeks to obtain meaningful cell numbers (Klingemann et al., 2004). It also requires expansion protocols that have to be in compliance with stringent conditions of good manufacturing practices (GMP). Considering all those obstacles, investigators have established clonal NK cell lines derived from normal donors' NK cells. This has been rather unsuccessful due to the limited life span of these cells (Hercend et al., 1983; Van de Griend et al., 1984).

Only a few 'true' NK cell lines so far have been established that can be maintained in culture (Chen et al., 2004; Gong et al., 1984; Kagami et al., 1998; Nagata et al., 2001; Robertson et al., 1996; Tsuchiyama et al., 1998; Yagita et al., 2000; Yodoi et al., 1985; and review Suck, 2006 and Table 12.1). Legitimate NK cell lines are defined by the expression of CD56 and being negative for CD3 expression and T-cell receptor rearrangement with the exception of a truncated TCRβ chain observed in some cells. Most of the NK cell lines were established from patients with NK lymphoma or large granular lymphocytes (LGL) leukemia and the majority are Epstein–Barr Virus (EBV) positive, which probably contributed to their clonal outgrowth. The NK-92 cell line was initially EBV negative, but over time acquired the virus although no active virions are produced. These cell lines have become useful to investigators to study the biology of NK cells and NK lymphomas/leukemias. Few of the NK cell lines have been shown to have cytolytic activity against tumour targets; only the NK-92 cell line has shown consistently high cytotoxicity, even at low effector-to-target ratios. A number of cytotoxic cell lines have also been described (summarized in Matsuo et al., 2003) that express CD3 (and variably CD56) that either are T-cell derived or belong to the 'intermediate' group of NK/T cells (Table 12.2).

The only NK cell line that has undergone extensive preclinical development and has completed phase I clinical trials is **NK-92** (Gong, 1984). The cell line was established in 1992 from a patient with NK cell lymphoma, has been well characterized and has been taken through phase I trials in the United States and Europe (Arai et al., 2008; Tonn et al., 2001). The consistently high cytotoxic activity of NK-92 cells is due to the fact that they do not express any currently known inhibitory KIR. When NK-92 cells are transfected with KIR, they lose their cytolytic activity, further supporting the importance of KIR in NK–tumour target cell interaction (Lutz et al., 1999).

Establishing an NK cell line as a clinical-grade product

In order for an NK cell line to be used for treatment of patients, it has to be demonstrated that the cells do not have substantial toxicity in pre-clinical models, but at

Table 12.1 Currently known NK cell lines

Designation	Origin	EBV status	CD56	CD3	CD16	Cytokine dependence	Cytotoxic activity[2]	ADCC	Reference
NK-92	LGL-NHL	Neg[1]	Pos	Neg	Neg	IL-2	Consistently high: 75%	No	Gong
YT	ALL, thymoma	Pos	Neg	Neg	Pos	None	25%	No	Yodoi
NKL	LGL-leukemia	Neg	Neg	Neg	Pos	IL-2	Variable, 20% at 40:1	Yes	Robertson
HANK-1	NK-T NHL (nasal)	Pos	Pos	CD3e	Neg	IL-2	Not reported	Not reported	Kagami
NK-YS	NK-NHL (nasal, angiocentric)	Pos	Pos	Neg	Neg	IL-2, initially grown on mouse stromal cell line	<10%	No	Tsuchiyama
KHYG-1	LGL leukemia	Neg	Pos	Neg	Neg	IL-2	75%		Yagita
SNK-6	NK-T NHL	Pos	Pos	Neg	Neg	IL-2	Not reported	Not reported	Nagata
IMC-1	LGL leukemia	Neg	Pos	Neg	Low levels	IL-2	50%	No	Chen

[1]NK-92 cells were initially EBV negative but acquired EBV.
[2]Cytotoxicity determined by lysis of K562 at effector:target ratios (E:T) of 3–5:1.

the same time display significant, maintained and broad cytotoxic activity to warrant their development into cellular therapy. Issues related to the value of cytolysis when compared to other NK functions such as interferon γ production are also important. The safety issue is of particular importance as NK-92 cells were obtained from the peripheral blood of a patient who suffered from an aggressive NK cell lymphoma (Gong, 1984). The ficoll-separated mononuclear cells were initially placed in Myelocult® medium, which contains 12.5% fetal bovine serum and 12.5% horse serum supplemented with hydrocortisone. The medium was not chosen by design, but at that time (1992), was widely used at the Terry Fox Laboratory (Vancouver) for long-term culture of hematopoietic stem cells. The NK-92 cells require IL-2 to be present (minimum of 100 units/ml) and medium change every 3–4 days. They grow easily in flasks in suspension culture and over the years have maintained their phenotype: CD56 bright+, CD3−, ICAM-1+ and LFA-1+ (Gong, 1984; Maki et al., 2001). As the next step in the development of NK-92 cells into a potential cellular treatment, the spectrum of cytotoxicity against cancer cell lines, primary cancer cells and their anti-tumour effect in small animals needed to be mapped out. Cytotoxicity of NK-92 cells was tested against a host of tumour target cell lines, all of which, with the exception of a few, were lysed very effectively (Klingemann et al., 1996). NK-92 cells were also shown to effectively purge ex vivo bone marrow preparations that were spiked with chronic myelogenous leukemia (CML) cells (Maki et al., 2003). They were further tested by various investigators against primary tumour cells and in SCID mice inoculated with human tumours and their significant in vivo cytotoxic activity confirmed. Yan and colleagues (Yan et al., 1998) injected NK-92 cells into SCID mice that had been inoculated with human primary AML. All animals in the control group died within 2 weeks, whereas about 50% of the mice injected with NK-92 survived long term. Similar observations were made in an SCID mouse model of human melanoma, confirming that injection of NK-92 could control the metastatic outgrowth of human melanoma (Tam et al., 1999). Having shown significant and consistent cytotoxicity against malignant cells, it was essential to demonstrate that NK-92 cells would not affect normal, non-transformed human cells. As target cells, human foreskin fibroblasts and hematopoietic progenitor cells were tested. It was shown that neither the differentiation nor the growth in culture of these two cell types was affected by NK-92 (Klingemann et al., 1996). Importantly, immunocompromised SCID mice did not form tumours following injection (both

Table 12.2 Allogeneic NK cell line development strategy for therapeutic application

- Establish cell line (appropriate source, appropriate culture conditions, serendipity)
- Characterize the cells:
  - surface marker expression (CD56, CD3, CD16, adhesion molecules, KIR, activating NK receptors)
  - cytokine production
  - cytokine dependence of cell growth
- Efficacy testing:
  - in vitro killing of malignant cell lines and primary malignant cells from patients
  - anti-tumour effect against human cancers in SCID mice
- Safety testing:
  - using normal human cells as targets
  - injection into immunocompromised mice to rule out tumour formation
- Translational development into a clinical grade product:
  - Develop animal serum free culture conditions and use FDA-sanctioned materials
  - Confirm that radiation of the cells maintains cytotoxicity but prevents proliferation
  - Scale up production conforming to GMP conditions
- Establish master cell bank with microbial and viral testing
- Create investigator brochure
- Design clinical trial—write protocol and submit IND to FDA
- Conduct clinical phase I studies

intravenously or subcutaneously), even when non-irradiated NK-92 cells were injected (Tam et al., 1999; Yan et al., 1998).

In order for NK-92 cells to be developed into a clinical grade product, they had to be 'switched' to a culture medium free of animal components and all culture steps had to be compliant with US FDA guidelines. At the same time, the cell line had to maintain its cytotoxic activity. The NK-92 cells have been shown to grow in serum free X-Vivo 10 supplemented with 2.5% human serum and IL-2 (500 units/ml) and to maintain their cytotoxic activity. Different culture containers were tested, including the Wave bioreactor, various stirred bioreactors, flasks and bags from different providers. The AFC bags (American Fluoroseal Company, Gaithersburg, MD, USA) were found to be the most suitable for a 2-week NK expansion protocol (Tam et al., 2003). FDA requirements called for irradiating NK-92 cells prior to any infusion into patients to prevent in vivo proliferation, since the line had been generated from an individual who suffered from an aggressive non-Hodgkin's lymphoma (NHL) and also carried multiple cytogenetic abnormalities (MacLeod et al., 2002). It was therefore important to determine a radiation dose that prevented in vitro proliferation (as assessed by thymidine incorporation) but at the same time still maintained cytotoxicity. This radiation dose was determined to be between 500 and 1000 cGy and clinical trials have used the dose of 1000 cGy prior to infusion into the patient.

In order to use NK-92 cells in clinical trials, a master cell bank had to be established and, was contracted to BioReliance (Rockville, MD, USA). In addition to the usual sterility and efficacy testing, comprehensive viral testing was performed. A single vial from the master cell bank serves as a starting material for creating a working cell bank from which cells are frozen after expansion to provide a consistent lot-specific population of cells for patient treatment.

In 2002 the FDA granted an IND for a phase I study with NK-92 for the treatment of patients with therapy-resistant renal cell cancer and melanoma. Twelve patients were treated in the United States at different dose levels (Arai, 2008). The infusions were well tolerated with fever as the only grade 3 side effect at the highest dose level ($3 \times 10^9/m^2$). Although efficacy assessment was not the objective of this phase I trial, some patients experienced partial responses and prolonged stable disease. At about the same time, a study in Frankfurt (Germany) infused NK-92 cells to patients with various advanced cancers (Tonn et al., 2001). Again, the infusions were well tolerated and two patients with advanced lung cancer showed partial regression of metastatic lesions. A third clinical trial is currently ongoing at the University of Toronto (Armand Keating) that so far has confirmed the favourable safety profile of NK-92. After having confirmed its safety profile, NK-92 cells are now to be tested in phase II studies, to determine if they can control primary tumour growth or metastatic spread.

Targeting of NK cell lines to tumours

NK-92 cells have been tested against an array of malignant cell lines that generally were killed very effectively when utilized even at low effector:target ratios, including cell lines from hematopoietic and solid organ tumours (Klingemann et al., 1996; Yan et al., 1998). Although malignant cell lines are killed readily, NK-92 cells are variably efficacious killers for some primary malignant cells (Yan et al., 1998). Clonal cells of myeloid origin are killed quite consistently, but many primary *lymphoid* malignancies are resistant, or show only moderate sensitivity, as reported by Mueller and associates (Mueller, 2008), Reid (Reid et al., 2002) and our group that tested a panel of primary CLL cells and found no cytolysis by NK-92 (Boissel et al., 2009).

The lack of killing of malignancies of lymphoid lineage by allogeneic NK cells was also observed in patients receiving an MHC haplo-mismatched transplant in which the NK cells were administered as part of the transplant product (Ruggeri, 1999). Patients transplanted for AML had a low relapse rate, but there was no such benefit for patients with acute lymphocytic leukemia (ALL). The lack of killing by allogeneic NK cells (and cell lines) of certain tumour targets has triggered the development of engineered, tumour-targeted NK-92 cells that recognize targets on malignant cells and upon engagement with the tumour cell, release their cytotoxic granule content. Such target specificity has been accomplished by introducing CARs for CD19 (Boissel et al., 2009) and CD20 (Mueller, 2008), but also for Her-2/neu (Uherek et al., 2002), CD38 (Yang et al., 2005) and EPCAM (Uherek et al., 2004). Those CARs have been transfected into NK-92 using retrovirus-based gene vectors. Although those studies have shown that specific CARs can restore killing of initially resistant targets, retroviral gene transfer poses the risk of insertional mutagenesis. Our group therefore has developed non-viral–based plasmids that can be transfected by physical methods, such as electroporation or lipofection (Boissel et al., 2009; Tam et al., 1999a,b). The transfected mRNA has recently been shown to express the CAR for CD19 on NK-92 and to restore their ability to effectively lyse primary CLL cells (Boissel et al., 2009). In contrast, cDNA by electroporation resulted in much lower transfection efficiency.

Transfecting IL-2 and FcR into NK cell lines

To enable NK-92 proliferation and expansion independent of exogenous IL-2, variants were created by transfection of the IL-2 gene, that rendered them able to produce low (NK-92ci) or high (NK-92mi) concentrations of IL-2 (Konstantinidis et al., 2005; Tam et al., 1999). Both variants do not require IL-2 in the culture medium for expansion. Pre-clinical data confirmed that the transfection did not impact cytotoxic ability and did not render the cells more radiation sensitive as cytotoxicity was maintained after 1000 cGy gamma radiation (Tam et al., 1999). The cells were effective in the SCID mouse model of metastatic melanoma (Tam et al., 1999), but have not been put into clinical trials as the parent NK-92 has priority for clinical development. Another cell line that has been transfected with a cytokine gene is NKL, a cell line that has only moderate and variable cytotoxicity. After transfection with an IL-15 gene, cytotoxicity against some targets improved (hepatoma cell lines were tested) and cell proliferation became less dependant on exogenous IL-2 (Jiang et al., 2008).

The parent NK-92 cell line does not express the Fc receptor and consequently mediates no ADCC. Binyamin and colleagues (Binyamin et al., 2008) transfected a high affinity Fc receptor into the parent cell line and could show that there was significant ADCC when anti-CD20 [Rituximab] or anti-Her2/neu [Herceptin] were used in the assay. Our group has confirmed that the NK-92 cells (expressing no FcR) and the NK-92Fc expressing variant (original name: NK-92.26.5) represent an ideal test system to determine the contribution to ADCC of various monoclonal antibodies. Using these two cell lines, we could show that Rituximab and Veltuzumab (both recognizing the CD20 antigen) (ImmunoMedics) display significant ADCC against primary CLL cells, whereas Epratuzumab (recognizing CD22) and Lumiliximab (recognizing CD23) display only minimal ADCC and largely are cytoreductive through other mechanisms (Weitzman et al., 2009). The Fc receptor expressing a variant of NK-92 will likely be developed into a clinical product as the administration of NK-92Fc, together with monoclonal antibodies that require ADCC for anti-tumour activity, would be an obvious strategy to improve efficacy of monoclonal antibody therapy.

At the present time, NK-92 cells have to be utilized directly from culture for infusion into the patient. Research is ongoing to cryopreserve the cells. Having NK-92 cells customized for different tumour types (via expression of specific CARs) or used in combination with a monoclonal antibody may serve as a model for targeted cell therapy.

References

Arai, S., Meagher, R., Swearingen, M., Myint, H., Rich, E., Martinson, J., et al. (2008). Infusion of the allogeneic cell line NK-92 in patients with advanced renal cell cancer or melanoma – A phase I trial. Cytotherapy **10**, 625–632.

Binyamin, L., Alpaugh, R.K., Hughes, T.L., Lutz, C.T., Campbell, K.S. and Weiner, LM. (2008). Blocking NK cell inhibitory self-recognition promotes antibody-dependent cellular cytotoxicity in a model of anti-lymphoma therapy. J Immunol **180**, 6392.

Boissel, L., Tuncer, H.H., Betancur, M., Wolfberg, A. and Klingemann, H. (2007). Umbilical cord mesenchymal stem cells increase expansion of cord blood natural killer cells. Biol Blood Marrow Transplant **13**, 1477–1486.

Boissel, L., Betancur, M., Wels, W.S., Tuncer, H. and Klingemann, H. (2009). Transfection with mRNA for CD19 specific chimeric antigen receptor restores NK cell mediated killing of CLL cells. Leuk Res **33**, 1255–1259.

Chen, I.-M., Whalen, M., Bankhurst, A., Sever, C.E., Doshi, R., Hardekopf, D., Montgomery, K. and Willman, CL. (2004). A new human natural killer leukemia cell line, IMC-1. A complex chromosomal arrangement defined by spectral karyotyping: functional and cytogenetic characterization. Leuk Res **28**, 275–284.

Gong, J.H., Maki, G. and Klingemann, HG. (1984). Characterization of a human cell line (NK-92) with phenotypical and functional characteristics of activated natural killer cells. Leukemia **8**, 652–658.

Hercend, T., Reinherz, E.L., Meuer, S., Schlossman, S.F. and Ritz, J. (1983). Phenotypic and functional heterogeneity of human cloned natural killer cell lines. Nature **301**, 158.

Jiang, W., Zhang, J. and Tian, Z. (2008). Functional characterization of interleukin-15 gene transduction into the human natural killer cell line NKL. Cytotherapy **10**, 265–274.

Kagami, Y., Nakamura, S., Suzuki, R., et al. (1998). Establishment of n IL-2 dependent cell line derived from "nasal-type" NK/T lymphoma of CD2+, sCD3−, CD3e+, CD56+ phenotype and associated with the Epstein–Barr virus. Br J Haematol **103**, 669–677.

Karre, K. (1995). Express yourself or die: peptides, MHC molecules, and NK cells. Science **267**, 978.

Klingemann, H.-G. and Martinson, J. (2004). Ex vivo expansion of natural killer cells for clinical application. Cytotherapy **6**(1), 15–22.

Klingemann, H.-G., Wong, E. and Maki, G. (1996). A cytotoxic NK-cell line (NK-92) for ex vivo purging of leukemia from blood. Biol Blood Marrow Transplant **2**, 68–75.

Konstantinidis, K.V., Alici, E., Aints, A., Christensson, B., Ljunggren, H.-G. and Dilber, MS. (2005). Targeting IL-2 to the endoplasmatic reticulum confines autocrine growth stimulation to NK-92 cells. Exp Hematol **33**, 159–164.

Law, T.M., Motzer, R.J., Mazumdar, M., et al. (1995). Phase III randomized trial of interleukin-2 with or without lymphokine-activated killer cells in the treatment of patients with advanced renal cell carcinoma. Cancer **76**, 824–832.

Lutz, C.T. and Kurago, ZB. (1999). Human leukocyte antigen class I expression on squamous cell carcinoma cells regulates natural killer cell activity. Cancer Res **59**, 5793–5799.

MacLeod, R.A.F., Nagel, F., Kaufman, M., Greulich-Bode, K. and Drexler, HG. (2002). Multicolor-FISH analysis of a natural killer cell line (NK-92). Leuk Res **26**, 1027–1033.

Maki, G., Klingemann, H-G., Martinson, J.A. and Tam, YK. (2001). Factors regulating the cytotoxic activity of the human natural killer cell line, NK-92. J Hematother Stem Cell Res **10**, 369–383.

Maki, G., Tam, Y., Berkahn, L. and Klingemann, H-G. (2003). Ex-vivo purging with NK-92 cells prior to autografting for chronic myelogenous leukemia. Bone Marrow Transplant **31**, 1119–1125.

Matsuo, Y. and Drexler, H.G. (2003). Immunoprofiling of cell lines derived from natural killer-cell and natural killer-like T-cell leukemia-lymphoma. Leuk Res **27**, 935–945.

Miller, J.S., Soignier, Y., Panoskaltsis-Mortari, A., et al. (2005). Successful adoptive transfer and in vivo expansion of human haploidentical NK cells in patients with cancer. Blood **105**, 3051.

Muller, T., Uherek, C., Maki, G., Chow, K.U., Schimpf, A., Klingemann, H.G., et al. (2008). Expression of a CD20-specific chimeric antigen receptor enhances cytotoxic activity of NK cells and overcomes NK-resistance of lymphoma and leukemia cells. Cancer Immunol Immunother **57**, 411–412.

Nagata, H., Konno, A., Kimura, N., et al. (2001). Characterization of novel natural killer (NK)-cell and γδ T-cell lines established from primary lesions of nasal T/NK – cell lymphomas associated with the Epstein–Barr virus. Blood **97**, 708–713.

Reid, G.S., Bharya, S., Klingemann, H.G. and Schultz, KR. (2002). Differential killing of pre-B acute lymphoblastic leukaemia cells by activated NK cells and the NK-92ci cell line. Clin Exp Immunol **129**, 265–271.

Robertson, M.J., Cochran, K.J., Cameron, C., Le, J.-M., Tantravahi, R. and Ritz, J. (1996). Characterization of a cell line, NKL, derived from an aggressive human natural killer cell leukemia. Exp Hematol **24**, 406–415.

Rosen, D.B., Cao, W., Avery, D.T., et al. (2008). Up on the tightrope: natural killer cell activation and inhibition. Nat Immunol **9**, 495–502.

Rosenberg, S.A., Lotze, M.T., Muul, L.M., et al. (1985). Observations on the systemic administration of autologous lymphokine – activated killer cells and recombinant IL-2 to patients with metastatic cancer. N Engl J Med **313**, 1485–1492.

Rosenberg, S.A., Lotze, M.T., Yang, J. C., Topalian, S.L., Chang, A.E., Schwartzentruber, D.J., Aebersold, P., Leitman, S., Linehan, W.M., Seipp, C.A., et al. (1993). Prospective randomized trial of high-dose interleukin-2 alone or in conjunction with lymphokine-activated killer cells for the treatment of patients with advanced cancer. J Natl Cancer Inst **85**(8), 622–632.

Ruggerei, L., Campanni, M., Casucci, M., et al. (1999). Role of natural killer cell alloreactivity in HLA mismatched hematopoietic stem cell transplantation. Blood **94**, 333.

Sivori, S., Pende, D., Bottino, C., Marcenaro, E., Pessino, A., Biassoni, R., Moretta, L. and Moretta, A. (1999). NKp46 is the major triggering receptor involved in the natural cytotoxicity of fresh or cultured human NK cells. Correlation between surface density of NKp46 and natural cytotoxicity against autologous, allogeneic or xenogeneic target cells. Eur J Immunol **29**, 1656.

Suck, G. (2006). Novel approaches using natural killer cells in cancer therapy. Semin Cancer Biol **16**, 412–418.

Tam, Y.K., Maki, G., Miyagawa, B., Hennemann, B., Tonn, T. and Klingemann, H-G. (1999a). Characterization of genetically altered, interleukin 2 independent natural killer cell lines suitable for adoptive cellular immunotherapy. Hum Gene Ther **10**, 1359–1373.

Tam, Y.K., Miyagawa, B., Ho, V.C. and Klingemann, H-G. (1999b). Immunotherapy of malignant melanoma in a SCID mouse model using the highly cytotoxic natural killer cell line NK-92. J Hematother **8**, 281–290.

Tam, Y., Martinson, J.A., Doligosa, K. and Klingemann, H-G. (2003). Ex vivo expansion of the highly cytotoxic human NK-92 cell line under cGMP conditions for clinical adoptive cellular immunotherapy. Cytotherapy **5**, 259–272.

Tonn, T., Becker, S., Esser, R., Schwabe, D. and Seifried, E. (2001). Cellular immunotherapy of malignancies using the clonal natural killer cell line NK-92. J Hematother Stem Cell Res **10**, 535–544.

Tsuchiyama, J., Yoshino, T., Mori, A., et al. (1998). Characterization of a novel human natural killer cell line (NK-YS) established from natural killer cell lymphoma/leukemia associated with Epstein–Barr virus infection. Blood **92**, 1374–1383.

Uherek, C., Tonn, T., Uherek, B., Becker, S., Schnierle, B., Klingemann, H., et al. (2002). Retargeting of natural killer-cell cytolytic activity to ErbB2-expressing cancer cells results in efficient and selective tumor cell destruction. Blood **100**, 1265–1273.

Uherek, C., Mueller, T., Tonn, T., Uherek, B., Klingemann, H.-G. and Wels, WS. (2004). Genetically modified natural killer cells specifically recognizing the tumor-associated antigens ErbB2/HER2 and EpCAM. Cancer Cell Int **4**(Suppl. 1), S7.

Van de Griend, R.J., van Krimpen, B.A., Ronteltap, C.P.M. and Bolhuis, R.L.H. (1984). Rapidly expanded activated human killer cell clones have strong antitumor activity and have the surface phenotype of either Tg, T-non-g, or null cells. J Immunol **132**, 3185.

Van den Broek, M.F., Kagi, D., Zinkernagel, R.M and Hengartner, H. (1995). Perforin dependence of natural killer cell mediated tumor control *in vivo*. Eur J Immunol **25**, 3514–3516.

Weitzman, J., Betancur, M., Boissel, L., Rabinowitz, A.P., Klein, A. and Klingemann, H. (2009). Variable contribution of monoclonal antibodies to ADCC in CLL patients. Leuk Lymphoma **26**, 1–8 [Epub ahead of print].

Yagita, M., Huang, C.L., Umehara, H., et al. (2000). A novel natural killer-cell line (KHYG-1) from a patient with aggressive natural killer-cell leukemia carrying a p53 point mutation. Leukemia **14**, 922–930.

Yan, Y., Steinherz, P., Klingemann, H.-G., Denning, D., Childs, B.H., McGuirk, J. and O'Reilly, RJ. (1998). Antileukemia activity of a natural killer cell line against human leukemia. Clin Cancer Res **4**, 2859–2868.

Yang, S., An, X., Brown, R.D., Ho, J., et al. (2005). Development of retargeted CD38 specific NK-92 cell line for potential anti-myeloma immunotherapy. Blood **106**, 5104.

Yengar, R., Handgretinger, R., Babarin-Dorner, A., et al. (2003). Purification of human natural killer cells using a clinical-scale immunomagnetic method. Cytotherapy **5**, 479.

Yodoi, J., Teshigawara, K., Nikaido, T., et al. (1985). TCGF (IL-2) receptor inducing factor(s). Regulation of IL-2 receptor on a natural killer-like cell line (YT cells). J Immunol **134**, 1623–1630.

Chapter Thirteen

13

NK cell-derived cytokines and delivery: NK cell synapses

Claudia Semino, Anna Rubartelli

CHAPTER CONTENTS

It was thy kiss, Love, that made me immortal.

Margaret Fuller

ABSTRACT

Natural killer (NK) cells have long been recognized as effector lymphocytes of the innate immune system that mediate numerous anti-tumour and anti-viral effector functions. More recently, many studies have shown that NK cells also play important regulatory roles. Indeed, NK cells elaborate cytokines and chemokines that participate in pathogen clearance. Furthermore, NK cells respond to a plethora of cytokines generated by numerous cell types by activating different immunomodulatory programs (See Chapter 14). Various soluble mediators act on NK cells, delivered to control NK cell activation and limit crosstalk to other cells in the lymph node or presumably the tissue. The interactions between NK and other immune cells as bi-directional crosstalk is controlled by soluble factors released at the immune synapse, resulting in modulation of the adaptive immune response.

KEY WORDS

IL-1, HMGB1, IL-18, Crosstalk, Cytokine, Delivery, Immune synapse, Immune regulation, Monokine, Polarization

Cytokine production by NK cell subsets

Natural killer (NK) cells represent an important source of immunoregulatory cytokines and chemokines. They bind other immune cells during the early phases of inflammatory responses and the resulting functional interactions shape both the innate immune response within inflamed peripheral tissues and the adaptive immune response found in secondary lymphoid organs (Biron et al., 1999; Vivier et al., 2008).

The effector or immunoregulatory functions of NK cells correlate with the levels of expression of the surface marker CD56. CD56^{dim} cells compose the majority of NK cells, express high levels of CD16 and function as effectors of natural cytotoxicity and antibody-dependent cellular cytotoxicity. Conversely, CD56^{bright} cells, that represent the minor NK subset in blood (~10%), are CD16$^{dim/neg}$ and display immunomodulatory function through secretion of cytokines (Cooper et al., 2001;

Jacobs et al., 2001). The regulation of the development of human NK cells in CD56^{bright} or CD56^{dim} subsets has been partially clarified with the functional characterization of a novel cytokine, IL-21 (Parrish-Novak et al., 2000). In synergy with Flt3L and IL-15, IL-21 promotes expansion and differentiation of NK cells from bone marrow progenitors in vitro. While the combination of the three cytokines supports the development of CD56^{dim}CD16^{bright}, highly lytic NK cells, in the absence of IL-21 the generation of CD56^{bright} NK cells lacking CD16 and KIRs is favoured (Sivori et al., 2003). Thus, the differentiation of NK cells to CD56^{bright} or CD56^{dim} NK cells is likely to depend on the relative presence and abundance of growth factors such as IL-15 and IL-21 within the bone marrow microenvironment.

Human NK cells express chemokine receptors and adhesion molecules involved in cell–cell interactions, trafficking, and homing. The expression pattern of these molecules differs in CD56^{bright} and CD56^{dim} NK cells, consistent with their functional differences. CD56^{bright} NK cells express high levels of CC-chemokine receptor 7 (CCR7) and L-selectin (CD62L) (Campbell et al., 2001; Frey et al., 1998), both molecules implicated in the homing of immune cells to secondary lymphoid organs. In keeping, CD56^{bright} NK cells are ten times more abundant in parafollicular (T-cell) regions of normal lymph nodes than in blood (Fehniger et al., 2003). Therefore, dendritic cells (DCs) migrating to secondary lymphoid tissues upon activation will encounter, before T lymphocytes, NK cells of the CD56^{bright}CD16^{-} subset. The interaction with DCs will then lead NK cells to locally release cytokines which influence APC functions as well as the emergent adaptive immune response. Conversely, the chemokine receptor repertoire and the chemokine responsiveness of CD56^{dim}CD16^{+} NK cells suggest that this subset may efficiently home to sites of inflammation (Campbell et al., 2001). CD56^{dim}CD16^{+} NK cells express CXCR1 and CX3CR1, the receptors for IL-8 and soluble fractalkine. Both chemokines are induced by proinflammatory cytokines, like IL-1 and TNF-α. Fractalkine mediates adhesion to endothelia and emigration of NK cells from the blood stream, whereas IL-8 mediates further migration to inflamed tissues (Campbell et al., 2001).

Interferon γ. The most abundant cytokine produced by all NK cells is IFN-γ, a pleiotropic cytokine that promotes macrophage activation, mediates antiviral and antibacterial immunity, promotes autophagy, enhances antigen presentation, orchestrates activation of the innate immune system, coordinates lymphocyte–endothelium interaction, regulates Th1/Th2 balance, and controls cellular proliferation and apoptosis (Schoenborn and Wilson, 2007). Although the other NK cell-derived cytokines are produced at lower levels than IFN-γ, they play important roles in different regulatory functions of NK cells, including the modulation of inflammatory and allergic responses, APC function, extramedullary hematopoiesis and B-cell differentiation.

The CD56^{bright} NK cell subset is the primary source of NK-derived immunoregulatory cytokines. In addition to IFN-γ, CD56^{bright} NK cells also secrete TNF-α, TNF-β, GM-CSF, IL-10, and IL-13, at levels higher than the CD56^{dim} NK cell subset (Cooper et al., 2001).

The difference in cytokine production between the two subsets may be related to differential expression and/or density of receptors for monokines which, in turn, induce cytokine secretion by NK cells. However, the stimulation of NK cell subsets with phorbol esters plus ionomycin, which is independent of monokine receptor activation, results in significantly greater production of IFN-γ, TNF-β, and GM-CSF by CD56^{bright} NK cells than by CD56^{dim} (Cooper et al., 2001). Thus the CD56^{bright} NK subset has a significantly higher capacity for cytokine production than the CD56^{dim} subset.

Cytokine secretion pattern of decidual NK cells

In human pregnancy, placental NK cells are massively recruited at the site of embryonic implantation (Carlino et al., 2008; Moffett-King, 2002). These decidual NK cells diverge in many ways from their peripheral blood NK cell counterparts in terms of gene expression, phenotype and function (Higuma-Myojo et al., 2005). Differently from the peripheral blood, the major subpopulation of decidual NK cells is CD56^{bright}. Decidual NK cell cytolytic function is much reduced despite the presence of activating receptors and of the lytic machinery (Kopcow et al., 2005) but they produce soluble mediators, such as IFN-γ, and TNF-α, macrophage inflammatory protein (MIP)-1α, GM-CSF and CFS1, at higher levels than peripheral blood CD56^{bright} NK cells (Higuma-Myojo et al., 2005). Moreover, decidual NK cells produce cytokines that are not normally secreted by peripheral blood NK cells, including leukemia inhibitory factor (LIF), and angiogenic growth factors such as angiopoietin-2, vascular endothelial growth factor-C (VEGF-C), or placental growth factor (PIGF) (Moffett-King, 2002). Interestingly, freshly isolated decidual NK cells secrete cytokines even if not stimulated in vitro, suggesting that they be activated in situ, but cytokine production may be further increased in vitro by addition of IL-15 or IL-18, both cytokines normally present in gestational endometrium (Laskarin et al., 2005). Several functions for these decidual NK cell-produced cytokines have been proposed. Cytokines could control extravillous trophoblast invasion by a non-cytotoxic mechanism, as suggested by the presence of receptors for GM-CSF, CSF-1, IFN-γ and TNF-α on human trophoblast cells (Jokhi et al., 1994). In keeping, promotion of angiogenesis, recognizing paternal alloantigen has been postulated as deficient in the setting of eclampsia (Eastabrook et al., 2008). Another important function could be modulation of the activity of local DCs and T cells (Hanna et al., 2005): Indeed, activated NK cells can prime DCs to induce a protective $CD8^+$ T cell response. A third non-exclusive function of decidual NK-produced cytokines could be the prevention of uterine viral spreading to the foetus, that can be achieved by the secretion of large amounts of the anti-viral IFN-γ cytokine (Cooper et al., 2001).

Stimuli inducing cytokine secretion by NK cells

Inflammatory mediators

NK cells are strongly induced to secrete cytokines and chemokines by soluble factors, including cytokines and monokines that trigger their relevant receptors constitutively expressed by NK cells (Cooper et al., 2001; Fehniger et al., 1999; Wang et al., 2000). The repertoire as well as the amount of inflammatory mediators produced following host infection influence $CD56^{bright}$ NK cells toward production of type 1 or type 2 cytokines (Morel and Oriss, 1998; Romagnani, 1992). For instance, although either IL-12 or IL-15 has the capacity to stimulate both type 1 and type 2 responses, the relative quantity of each and the presence of other monokines (e.g. IL-1 or IL-18) can shape the predominant $CD56^{bright}$ cytokine response (Cooper et al., 2001). Similarly, the presence and the relative abundance of other cytokines, including IL-4, IL-10, and IL-12 may modulate the production of cytokines by NK cells. For example, production of IFN-γ in response to IL-2 is significantly inhibited by IL-4 (Hsu et al., 1992) while it is induced by IL-12 (Wang et al., 2000). Thus, the cytokine repertoire of the inflammatory milieu may dictate the production of type 1 or type 2 cytokines by $CD56^{bright}$ NK cells, and modulate the amplitude of cytokine secretion. Obviously, responsiveness to the various inflammatory mediators also depends upon the NK cell subsets present and the surface expression of the relevant receptors.

Although the optimal production of some NK cell cytokines requires stimulation of monokine receptors in the presence of growth factors such as IL-2, cytokine secretion is independent of growth and cytotoxicity, with the exception of IL-5 production, which correlates with NK cells proliferation (Perussia, 1996).

TLR ligands

An alternative mode of NK-cell activation has recently been identified thanks to the discovery that human NK cells can express toll-like receptors (TLRs) (Sivori et al., 2004). TLRs are pattern recognition receptors (PRRs), which trigger innate immune responses, providing both immediate protection against various pathogens and instructing the adaptive immune system through the induction of DC recruitment and maturation. Ten different TLRs have been described in humans, and most of their specific ligands have been identified (Takeda and Akira, 2005). The best known ligands of TLRs are the pathogen-associated molecular patterns (PAMP) molecules. These include lipopolysaccharide (LPS), recognized by TLR4; bacterial lipoproteins and lipoteichoic acids, recognized by TLR2; flagellin, recognized by TLR5; unmethylated CpG typical of bacterial and viral DNA, recognized by TLR9; double-stranded RNA (dsRNA) recognized by TLR3 and single-stranded RNA recognized by TLR7. Recently, some damage-associated molecular pattern (DAMP) molecules have also been found to bind and activate TLRs (Rubartelli and Lotze, 2007).

Human NK cells, independent of their status of activation, express functional TLR2 (Becker et al., 2003), TLR3 (Pisegna et al., 2004; Schmidt et al., 2004; Sivori et al., 2004) and TLR9 (Sivori et al., 2004) that enable their response to both viral and bacterial products. For instance, recognition of *Mycobacterium bovis* via TLR2 leads NK cells to release IFN-γ and TNF-α and to kill different targets more efficiently (Marcenaro et al., 2008).

Triggering of TLR3 by viral dsRNA increases NK cytotoxicity and cytokine production (Sivori et al., 2004). In particular, dsRNA or CpG can induce NK-cell priming, which, in the presence of IL-12 secreted by myeloid DCs, results in the release of abundant IFN-γ and TNF-α. Moreover, under these conditions, NK cells upregulate their cytolytic activity against tumour cells and acquire the ability to kill immature myeloid DCs (iDCs). Thus, the simultaneous engagement of TLR3 expressed by both NK cells and myeloid DC might be sufficient to initiate the series of events characterizing the early phases of innate immune responses.

Recently, the DAMP protein HMGB1 has been proposed to enhance IFN-γ release from macrophage-stimulated NK cells. Again, this is effective only when coupled with other pro-inflammatory cytokines, particularly with IL-2 in combination with IL-1 or IL-12 (DeMarco et al., 2005).

Interaction with target cells

Many of the known NK cell-activating receptors, such as CD16 and the activating isoform of NKG2D, associate with a number of adaptor proteins, which contain immunoreceptor tyrosine-based activation motifs (ITAMs) (Lanier, 2005; Moretta and Moretta, 2004; Vivier et al., 2004). Ligation of an ITAM-associated NK cell receptor trigger a cascade of intracellular events common to both cytokine production and killing of the target cell. Then, in principle, cytokine expression upon activating receptor triggering and killing should be coupled. In some cases stimulation of NK cell receptors leads to cytokine release independent of degranulation. For instance, Rajagopalan and associates (2006) reported that, in decidual NK cells, the rather unique receptor KIR2DL4 responds to soluble HLA-G and signals from endosomes by inducing a proinflammatory and proangiogenic profile of cytokines uncoupled from degranulation.

Similarly, the tyrosine phosphatase CD45 is essential for cytokine and chemokine production but is not absolutely required for cytotoxicity: NK cells from CD45-deficient mice are relatively competent for ITAM receptor-induced cell-mediated cytotoxicity, yet completely deficient for cytokine secretion after receptor triggering (Hesslein et al., 2006). It is likely that strength and/or duration of the signal may determine the requirement for CD45. Release of cytotoxic granules occurs near the cell surface, in close proximity to many cell signalling components that are activated upon receptor/ITAM activation. This process requires little time (minutes) between receptor crosslinking and granule release. Conversely, studies on CTL showed that the induction of cytokines by receptor engagement is a slow process (lasting hours), involving signal transduction, gene transcription, RNA processing, translation, and secretion (Valitutti et al., 1995). Sustained signalling may be needed for induction of cytokines also in NK cells, whereas brief stimulation may be sufficient for a robust cytotoxicity response by NK cells.

The NK cell immune synapse

NK cell activity results from a balance between inhibitory and activating signals originated from the interactions of NK receptors with ligands expressed by target cells that give rise to the immunological synapse (NK-IS) (Davis et al., 1999). The balance between activating and inhibitory signals also controls the supramolecular organization of proteins at the NK cell–target cell contact site, defining the formation of activating or inhibitory NK-IS (Vyas et al., 2002a).

During cell-to-cell contact, triggering of NK receptors including LFA-1 or CD2 leads to formation of supramolecular activating clusters (SMACs) or supramolecular inhibitory clusters (SMICs) that accumulate at the contacting plasma membrane between NK cells and target cells within the NK-IS (Vyas et al., 2002a). Some components (i.e. talin, a cytoskeletal protein that links the actin-cytoskeletal network to the extracellular matrix and LFA-1) are shared by SMAC or SMIC (Vyas et al., 2002a,b). However, other molecules or complexes, such as perforin and F-actin segregate differently in cytolytic and noncytolytic interactions (Krzewski et al., 2006; Sanni et al., 2004). Therefore, the outcome of the interactions between NK and target cells is determined by the NK-IS composition.

Activating NK-IS

When NK cells bind to target cells lacking self MHC, a cytolytic NK-IS is formed at the contact site and involves on the NK cell side, β-integrins and/or activating receptors. The NK cell–target cell contact area assembles into topologically and spatially distinct regions, the peripheral SMAC (pSMAC) and the central SMAC (cSMAC). In particular, the activating NK-IS shows, on the NK side, clusters of filamentous actin (F-actin) as well as of other cytoskeleton-related proteins such as talin and of surface molecules, including CD2, LFA-1 and CD11b, in a peripheral ring (pSMAC) (Orange et al., 2003; Vyas et al., 2002a). The accumulation of F-actin and surface receptors is rapid and depends on Wiskott–Aldrich syndrome protein (WASp)-driven actin polymerization. Both actin polymerization and WASp function are required not only for effective localization of CD2 and LFA-1 in the pSMAC, but also for targeting perforin-containing granules in the cSMAC (Orange et al., 2003). Direct imaging of the signalling molecules included in the cytolytic cSMAC has provided great insight into the spatial organization of enzymes and adaptor molecules needed for initiation of cytolytic effector function (Vyas et al., 2001). A multimolecular signalling complex including the phosphatase SHP-1 (src homology 2 domain-bearing protein tyrosine phosphatase-1) is assembled in the cSMAC. The kinetics of SHP-1 movement from pSMAC to cSMAC determines its activatory function. Translocation of signalling molecules from the cytosol to the NK cell plasma membrane occurs within a limited region of contact with the target cell and is mediated by glycolipid-enriched microdomains (GEMs), enriched in signalling molecules. This redistribution requires activation of Src and Syk kinases (Lou et al., 2000).

Interestingly, formation of the pSMAC is highly dependant on the actin cytoskeleton but not on microtubules. Surface receptors move to the pSMAC of the

NK-IS laterally through the cell membrane, and are brought to the surface in a microtubule-independent manner. In contrast, perforin polarization to the cSMAC requires microtubule polymerization which follows actin reorganization. Thus, the activating NK-IS is formed in a distinct sequence involving the actin cytoskeleton first and the microtubules second (Orange et al., 2003).

A well-studied stimulatory pathway for cytotoxicity of NK cells is represented by interaction of the activating receptor NKG2D with its ligands. This interaction leads to formation of an activating NK-IS where localization of NKG2D to cSMAC correlates with activation of NK cytotoxicity, suggesting that the segregation of activating receptors to the cSMAC is an important step for the cytotoxic process of NK cells (Chen et al., 2007). It has been shown recently that NKG2D ligation induces a strong activation of MAP kinases (Li et al., 2008). This activation is necessary for NK cytotoxicity activation: selective inhibition of JNK MAP kinase blocks the movement of the microtubule organizing centre (MTOC), granzyme B, and the scaffold protein paxillin to the NK-IS and greatly reduces NKG2D-mediated cytotoxicity toward target cells (Li et al., 2008).

The ligation of different NK receptors, such as β2-integrins or CD2, does not always lead to NK-IS formation. Triggering of these molecules in small numbers before their clustering may rather generate signals inducing further NK cell maturation. This ensures correct activation of the NK cell before performing its effector function (Orange et al., 2003).

While the majority of studies of the NK-IS have been carried out using NK cells that have been cultured in vitro with IL-2, resting NK cells can also form cytotoxic synapses with MHC class I negative targets (Orange et al., 2003; Roda-Navarro et al., 2004). Dynamic studies have shown that accumulation of the NK surface receptor 2B4 at these synapses occurs rapidly, suggesting an important role for this receptor in the initial adhesion of resting NK cells to targets before initiation of the killing process. Moreover, 2B4 forms stable complexes with its intracellular partner SAP clustered at the synapse, consistent with the idea that sustained signalling triggered by this receptor complex is required for activation of lysis by resting NK cells (Roda-Navarro et al., 2004). Fresh peripheral blood NK cells use LFA-1 to polarize granules but require activating receptors for exocytosis of the granules and hence for completing the cytotoxic process (Bryceson et al., 2005).

In the case of decidual NK cells, the process of synapse formation is initiated with a laying down of actin but NK cells are then unable to polarize their granules towards ICAM-bearing targets, even though they express LFA-1. This suggests that the ability of decidual NK cells, and perhaps of other NK subsets that have a similar phenotype, to secrete cytokines without degranulation might be achieved by specific uncoupling of molecules such as LFA-1 from granule polarization (Kopcow et al., 2005).

Inhibitory NK-IS

When target cells express high levels of MHC class I, an inhibitory NK-IS takes place and NK cell cytotoxicity is prevented. As in activating NK-IS, interactions between target cells and NK cells lead to polarization of talin and LFA-1 at the synapse within 1 min of interaction (Vyas et al., 2001). However, similarly to LFA-1, inhibitory NK receptors (KIRs) organize into a ring-shaped structure opposite the MHC proteins. Thus MHC and ICAM-1 proteins on the target cell correspond exactly to their receptors, KIR1 and LFA-1, on the NK cell side of the synapse (Davis et al., 1999). KIR-MHC binding blocks raft polarization (Fassett et al., 2001), and down-regulates integrin function, leading to temporal shortening of contact between effector and target cells.

SHP-1 is the only component of cSMIC in iNK-IS: the activating signalling molecule Lck is seen dispersed in multiple small clusters at the inhibitory NK-IS (Vyas et al., 2002b). Moreover, while in the cytolytic synapse SHP-1 first clusters at the periphery, in noncytolytic synapse it clusters in the centre, recruited by MHC–KIR interaction. The early translocation of SHP-1 to the cSMAC can be required to prevent cytotoxicity. Indeed, the perimembraneous components of Lck and SHP-1 would favour rapid initiation of the inhibitory signal transduction pathway. This could then facilitate interruption of the cytolytic signalling cascade, which also depends on Src kinase activation. As a consequence, after 5 min, rapid dissociation of talin clusters is observed with no further downstream signalling, resulting in unstable, short-lived, noncytolytic conjugates (Vyas et al., 2001).

The inhibitory signals within a single NK cell are spatially and temporally restricted and are limited to interactions with insusceptible, resistant target cells. This localized inhibition does not lead to a general inactivation of the cytolytic effector function of the cell (Vyas et al., 2001).

NK–DC synapse

In the last 10 years many observations converged to indicate a crucial role of NK interaction with autologous DCs in the first phases of the immune response (Ferlazzo et al., 2002; Fernandez et al., 1999; Poggi et al., 2002). NK–DC interaction may lead to NK cell activation, DC activation, or DC killing depending on the activation status of both cell types (Gerosa et al., 2002; Piccioli et al., 2002). Thus, the outcome of NK–DC crosstalk is likely to influence both innate and adaptive immune responses.

Among NK cell subsets, both the $CD56^{dim}$ and the minor blood NK subset of $CD56^{bright}CD16^{-}$ cells have been shown to interact with DCs. Interestingly immunoregulatory $CD56^{bright}CD16^{-}$ NK cells form conjugates with DCs more readily than cytotoxic $CD56^{dim}CD16^{+}$ NK cells (Brilot et al., 2007; Vitale et al., 2004).

NK-IS leading to DC killing shows features similar to activating NK-IS. At variance, a different regulatory NK-IS has been described to take place in NK–DC conjugates, leading to cell activation or maturation.

This regulatory synapse contains features of inhibitory NK-IS, but also displays classic hallmarks of activation (i.e. intracellular calcium mobilization, sustained physical interactions, upregulation of CD69 (Brilot et al., 2007). Recruitment of CD94 and KIR on NK cells and MHC class I on DCs is induced at the site of contact between cells, resulting in the formation at the centre of the synapse of inhibitory interactions which protect DCs from NK cell lysis (Brilot et al., 2007). In contrast, talin and LFA-1 are found in the periphery of the DC–NK cell synapse. This suggests that activating and inhibitory signals are transmitted from distinct areas of the DC–NK cell synapse, allowing initiation of activating and inhibitory interactions in parallel. IL-15Ra accumulates at the cSMAC of the NK cell synapse and interacts with IL-15 on the DC surface, thereby assuring NK cell survival and apoptosis escape after interactions with DCs (Brilot et al., 2007).

Regulatory synapses between NK cells and DCs take place preferably in areas of close DC–NK cell encounter like the parafollicular T cell zone of secondary lymphoid organs and sites of inflammation and ensure optimal NK cell activation (Brilot et al., 2007).

DC-mediated NK cell activation

A role for DC in the process of NK cell-activation has been clearly demonstrated in many different experimental systems (Ferlazzo et al., 2002; Fernandez et al., 1999; Gerosa et al., 2002; Zanoni et al., 2005). Starting with the pioneering work of Fernandez and colleagues (1999), many other authors described the key role of DC-derived cytokines and membrane-bound molecules in the activation process of NK cells (Figure 13.1).

These studies generally agree on a major role for DC-derived type I IFN in driving NK cell cytotoxic activity (Dalod et al., 2003; Gerosa et al., 2005; Granucci et al., 2004). However DC-mediated NK cell activation both in humans and mice is also triggered by other DC-derived cytokines, including IL-12, IL-15, IL-18, and IL-2 (Alli and Khar, 2004; Borg et al., 2004; Ferlazzo et al., 2004; Lucas et al., 2007; Mailliard et al., 2005; Munz et al., 2005; Semino et al., 2005). Moreover, it has been proposed that DC–NK cell crosstalk in the mouse enhances NK-cell functions via triggering NK-cell trans-membrane TNF receptor 2 by DC trans-membrane TNF, but does not require secreted products (Xu et al., 2007).

It must be borne in mind that different human DC subsets show distinct patterns of cytokine secretion that correlate with their ability to activate NK cells in vitro (Munz et al., 2005). While monocyte-derived DC produce IL-12p70 and show the capacity to induce NK cell activation in terms of NK cell proliferation, cytotoxicity and up-regulation of CD56, Langerhans cells derived from $CD34^{+}$ hematopoietic progenitor cell (HPC) lack sufficient IL-12p70 secretion and IL-15 receptor expression to induce NK cell activation. Still, once activated by recombinant IL-12, Langerhans cells provide additional factors (possibly IL-15 and IL-18 that are produced at higher levels by these cells) that promote NK cell proliferation and survival, sustaining their IL-12-induced activation. $CD34^{+}$ HPC-derived dermal-interstitial DCs have intermediate capacity to produce these cytokines and to activate NK cells.

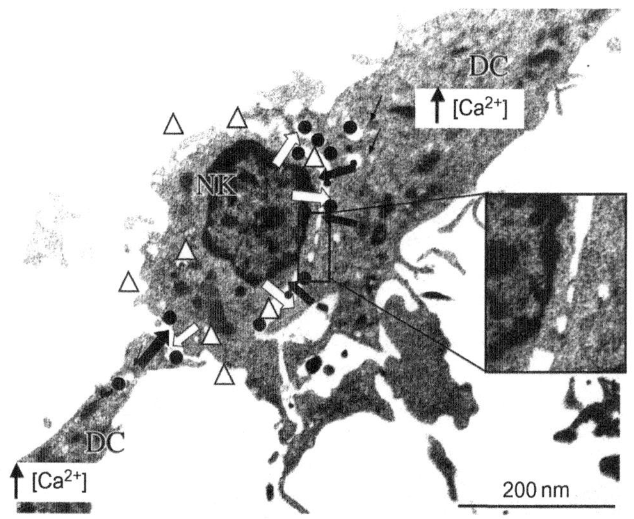

Figure 13.1 • Ultrastructural analysis of the interaction between DC and NK cells. Cells were co-cultured for 3h and processed for conventional thin section electron microscopy (Semino et al., 2007). Interaction between cells occurs primarily in correspondence with DC areas enriched by mitochondria and granules (black arrows). The dark grey square includes an area of tight contact between the plasma membranes of the interacting cells. A magnification is shown in the inset. Interaction between DCs and NK cells results in reciprocal activation (see text). The cartoon shows DC-derived cytokines (very dark grey, including IL-12, IL-18, IL-2, IL-15) and NK derived cytokines (red, including TNF-α, IFN-γ, or HMGB1). In DCs, cytokines such as IL-12 (Borg et al., 2004) and IL-18 (Semino et al., 2005) have been shown to polarize toward the interacting NK cells and undergo secretion restricted to the synaptic cleft (very dark grey arrows). With permission of Dr M. Torrisi and Dr L. Lotti, University Roma La Sapienza, Italy.

NK-cell activating cytokines produced by DCs

IL-12

Functional IL-12 is delivered to NK cells through a tight paracrine mechanism with enrichment of IL-12-containing secretory granules toward the NK–DC interface area. IL-12 is then released at the synaptic cleft such that low doses of IL-12 are efficiently presented to NK cells (Borg et al., 2004). Secreted IL-12 activates NK cells to produce cytokine. In some experimental settings, IL-12 is found to be a key regulator also of NK cell cytotoxicity (Alli and Khar, 2004) while in others, in which peripheral blood DCs were used, IL-12 and cell–cell contacts seem to have only a marginal role in induction of cytotoxicity, indicating that other soluble factors are involved (Osada et al., 2001). As IL-12 is not expressed by immature DCs, but is induced upon maturation, this cytokine is likely to play a major role when NK cells interact with mature DCs, while other constitutively expressed cytokines such as IL-18 (Gardella et al., 1999) may have a greater importance in NK cell activation following interaction with immature DCs.

IL-18

IL-18 is not a classical secretory protein in that it lacks a secretory signal sequence, a feature shared with other cytokines, such as IL-1β (Dinarello, 2007). Like IL-1β, IL-18 is synthesized as a precursor protein, which is processed proteolytically by the IL-1β-converting enzyme (ICE) in turn activated by the inflammasome. DC accumulate the precursor form of IL-18 (pro-IL-18) in the cell cytosol and in organelles co-fractionating with endolysosomes, called secretory lysosomes (Gardella et al., 1999, 2000). Secretion of IL-18 is induced by Ca2+ influx into cells and is accompanied by secretion of lysosomal enzymes, such as cathepsin D, suggesting that extracellular IL-18 derives from exocytosis of pro-IL-18-containing organelles (Gardella et al., 1999, 2000). We have recently observed that following conjugate formation with autologous NK cells, immature DCs undergo a functional polarization, with increases in

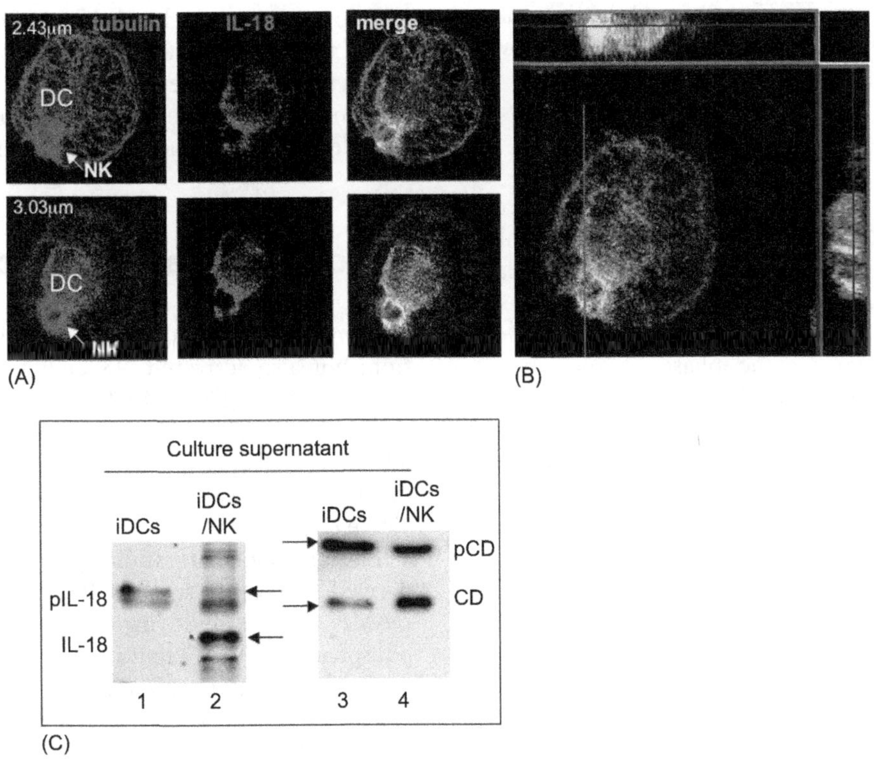

Figure 13.2 • (A) Confocal microscopy analysis of tubulin (grey) and IL-18 in an immature DC–NK conjugate after 3h of interaction. Two focus layers of the same conjugate are shown to make clear the rearrangement of the microtubular cytoskeleton at the synapse. Notice in the 'merge' panel, the strong co-staining of IL-18 and tubulin, supporting the role of tubulin in mediating the transport of IL-18-containing vesicles to the NK–DC synapse (Semino et al., 2005). With permission of Dr M. Torrisi and Dr F. Belleudi, University Roma La Sapienza, Italy. (B) Orthogonal and sagittal layers of the NK–DC synapse showing details of the tubulin-mediated IL-18 release process at the synapse. (C) Western blot analyses with anti-IL-18 (lanes 1 and 2) or anti-cathepsin D (CD; lanes 3 and 4) of 3-h supernatants of immature DCs cultured alone (iDCs; lane 1 and 3) or with NK cells (NK/iDCs; lanes 2 and 4). Arrows point to the precursors (pIL-18 and pCD) and mature (IL-18 and CD) forms of the two proteins.

intracellular free Ca^{2+} concentration ($[Ca^{2+}]_i$), cytoskeleton rearrangement, accumulation of IL-18-containing secretory lysosomes at the NK–DC synapse, and regulated secretion of IL-18 toward the interacting NK cell (Figure 13.2) (Semino et al., 2005, 2007).

IL-18-exposed human NK cells show rapid induction of CCR7 which induces migration to secondary lymphoid organs (Mailliard et al., 2005), but not augmented lytic activity. In a different experimental setting however, IL-18 also enhances NK cell cytotoxicity (Yu et al., 2001). IL-12 and IL-18 exert striking synergistic activities for NK cell proliferation and activation. NK cells stimulated with IL-12 and IL-18 in combination (but not with other cytokines such as IL-2 or IL-15), produce the highest levels of IFN-γ, IL-3, IL-6 and TNF-α.

IL-15

A membrane-bound form of DC-derived IL-15 appears to be necessary to induce activation or at least proliferation of NK cells (Ferlazzo et al., 2004). The effect of IL-15 requires the presence of the IL-15 receptor on the surface of DC, suggesting that DC can present in *trans*-IL-15 to NK cells. Enrichment of NK IL-15Ra and DC-membrane-bound IL-15 at the DC–NK synapse assures NK cell survival and apoptosis escape following interaction with mDCs (Brilot et al., 2007).

IL-2

IL-2 is classically used in NK cell cultures in vitro to obtain hyper-responsive cells. However, this cytokine was not considered important for NK cell-mediated anti-tumour or anti-microbial responses in vivo as its production was thought to be restricted to T cells during the late, antigen-specific phase of the immune response, when the peak of NK cell activation is already exhausted (Biron et al., 1999). However, in 2004, Ricciardi-Castagnoli and co-workers observed that activated murine DCs produced IL-2 during the first hours following their stimulation (Granucci et al., 2004) and subsequently demonstrated that mature DC-derived IL-2 was required for NK cell activation both in vitro and in vivo, together with other soluble mediators (Granucci et al., 2004). Interestingly, only TLR-dependent but not TLR-independent DC maturation can elicit IL-2-mediated IFN-γ production by NK cells (Zanoni et al., 2005).

Recently, a role for monocyte/DC derived IL-2 in the activation of NK cells has also been described in humans. Newman and co-workers (2006) demonstrated that the capacity of human NK cells to produce IFN-γ in response to stimulation by *Plasmodium falciparum*-infected red blood cells was strictly dependant upon cell contact and IL-2/type I IFN-mediated signal derived from monocytes and myeloid DC. The authors excluded a major role for IL-12 and IL-15 in activating NK cells in their experimental setting.

Cell–cell contact

In addition to soluble factors, many studies indicate a role for cell–cell contact during DC-mediated NK cell activation (Borg et al., 2004; Gerosa et al., 2002; Granucci et al., 2004; Xu et al., 2007). On the one hand, cell–cell contact is likely to reflect the necessity for the formation of a synapse between DC and NK cells that could lead to local delivery of known or yet unknown cytokines at high concentration. Indeed, the formation of a synapse with calcium influx, remodelling of the DC cytoskeleton, raft mobilization, and redistribution of adhesion molecules is required for IL-12 and IL-18 granule polarization (Borg et al., 2004; Semino et al., 2005). On the other hand, it cannot be excluded that surface receptor–ligand interactions are directly necessary for optimal NK cell activation. Support for the latter hypothesis came from studies by Ortaldo and co-workers (2006), who showed that greatly enhanced IFN-γ production is obtained when NK cells are simultaneously stimulated by IL-12 and IL-18 and by triggering of their surface receptors. Taken together, all these observations indicate that multiple convergent signals maximize the innate immune response by triggering complementary biochemical signalling pathways.

NK cell-mediated DC activation

Not only NK–DC interaction results in NK cell activation, but also activated NK cells provide maturational stimuli to DCs (Figure 13.1). This process requires both cell–cell contact, mediated via the receptors NKp30, KIR and NKG2A (Vitale et al., 2005) and production of cytokines such as TNF-α or IFN-γ (Piccioli et al., 2002; Vitale et al., 2005).

DC undergo maturation upon exposure to pathogen derived molecules such as PAMPs. Also a number of DAMPs released by dying cells, including adenosine triphosphate (ATP) (Schnurr et al., 2000), uric acid (Shi et al., 2003), and high mobility group B1 (HMGB1) (Messmer et al., 2004, Rovere-Querini et al., 2004), promote DC maturation. Although most DAMP molecules are passively released by necrotic cells (Gallucci and Matzinger, 2001; Rubartelli and Lotze, 2007), the nuclear protein HMGB1 can also be actively secreted, in the absence of cell death, by inflammatory cells (Gardella et al., 2002) and behaves as a powerful proinflammatory cytokine (Wang et al., 1999). Interestingly, upon co-culture with DCs, NK cells secrete large amounts of

HMGB1, which induces DC maturation, protecting DCs from NK cell cytotoxicity (Semino et al., 2005).

As discussed above, interaction between iDCs and NK cells leads to either DC maturation or death, raising the question of whether effector and maturation-inducing properties may coexist or segregate in individual NK subsets. Clonal analyses of human NK cells showed that the ability of individual NK cell clones to induce iDC maturation is not linked to their phenotypic or cytolytic features, but rather correlates with the relocation of HMGB1 from the nucleus to the cytoplasm (Semino et al., 2007). Maturation-inducing NK cell clones secrete HMGB1 spontaneously, strongly enhanced by engagement of the surface molecule NKp30, but only slightly induced by triggering of the activating NK receptor CD16. However, culturing freshly-isolated NK cells for 1 week with low doses of anti-CD16 triggers the relocation of HMGB1 from nucleus to cytoplasm and its spontaneous secretion, resulting in a stronger maturation potential of the NK cells (Semino et al., 2007). Together, these data indicate that NK cells comprise functionally distinct subsets, endowed with varying capacities to secrete HMGB1 and to induce maturation of autologous iDC. Nonetheless, maturation properties can be modulated by different stimuli. This suggests that, depending on the environmental stimuli, NK–iDC interaction can lead to different outcomes, thus influencing immune response.

NK cells have been found recently able to induce an early stage of DC differentiation from CD14+ monocyte precursors (Zhang et al., 2007). Co-culturing these cells with autologous NK cells in the presence of IL-15 induced morphological and phenotypic changes associated with DC. The process requires cell–cell contact between NK and monocytes, as well as soluble factors, such as GM-CSF and CD40L (CD154), produced by NK cells (Zhang et al., 2007).

References[1]

Alli, R.S. and Khar, A. (2004). Interleukin-12 secreted by mature dendritic cells mediates activation of NK cell function. FEBS Lett **559**, 71–76.

Becker, I., Salaiza, N., Aguirre, M., Delgado, J., Carrillo-Carrasco, N., Kobeh, L.G., Ruiz, A., Cervantes, R., Torres, A.P., Cabrera, N., González, A., Maldonado, C. and Isibasi, A. (2003). *Leishmania* lipophosphoglycan (LPG) activates NK cells through toll-like receptor-2. Mol Biochem Parasitol **130**, 65–74.

Biron, C.A., Nguyen, K.B., Pien, G.C., Cousens, L.P. and Salazar-Mather, T.P. (1999). Natural killer cells in antiviral defense: function and regulation by innate cytokines. Annu Rev Immunol **17**, 189–220.

Borg, C., Jalil, A., Laderach, D., Maruyama, K., Wakasugi, H., Charrier, S., Ryffel, B., Cambi, A., Figdor, C., Vainchenker, W., Galy, A., Caignard, A. and Zitvogel, L. (2004). NK cell activation by dendritic cells (DCs) requires the formation of a synapse leading to IL-12 polarization in DCs. Blood **104**, 3267–3275.

Brilot, F., Strowig, T., Roberts, S.M., Arrey, F. and Münz, C. (2007). NK cell survival mediated through the regulatory synapse with human DCs requires IL-15Rα. J Clin Invest **117**, 3316–3329.

Bryceson, Y.T., March, M.E., Barber, D.F., Ljunggren, H.G. and Long, E.O. (2005). Cytolytic granule polarization and degranulation controlled by different receptors in resting NK cells. J Exp Med **202**, 1001–1012.

Campbell, J.J., Qin, S., Unutmaz, D., Soler, D., Murphy, K.E., Hodge, M.R., Wu, L. and Butcher, E.C. (2001). Unique subpopulations of CD56+ NK and NK-T peripheral blood lymphocytes identified by chemokine receptor expression repertoire. J Immunol **166**, 6477–6482.

Carlino, C., Stabile, H., Morrone, S., Bulla, R., Soriani, A., Agostinis, C., Bossi, F., Mocci, C., Sarazani, F., Tedesco, F., Santoni, A. and Gismondi, A. (2008). Recruitment of circulating NK cells through decidual tissue: a possible mechanism controlling NK cell accumulation in the uterus during early pregnancy. Blood **111**, 3108–3115.

Chen, X., Trivedi, P.P., Ge, B., Krzewski, K. and Strominger, J.L. (2007). Many NK cell receptors activate ERK2 and JNK1 to trigger microtubule organizing center and granule polarization and cytotoxicity. Proc Natl Acad Sci U S A **104**, 6329–6334.

Cooper, M.A., Fehniger, T.A., Turner, S. C., Chen, K.S., Ghaheri, B.A., Ghayur, T., Carson, W.E. and Caligiuri, M.A. (2001). Human natural killer cells: a unique innate immunoregulatory role for the CD56(bright) subset. Blood **97**, 3146–3151.

Dalod, M., Hamilton, T., Salomon, R., Salazar-Mather, T.P., Henry, S.C., Hamilton, J.D. and Biron, C.A. (2003). Dendritic cell responses to early murine cytomegalovirus infection: subset functional specialization and differential regulation by interferon/β. J Exp Med **197**, 885–898.

Davis, D.M., Chiu, I., Fassett, M., Cohen, G.B., Mandelboim, O. and Strominger, J. L. (1999). The human natural killer cell immune synapse. Proc Natl Acad Sci USA **96**, 15062–15067.

DeMarco, R.A., Fink, M.P. and Lotze, M. T. (2005). Monocytes promote natural killer cell interferon gamma production in response to the endogenous danger signal HMGB1. Mol Immunol **42**, 433–444.

Dinarello, CA. (2007). Historical insights into cytokines. Eur J Immunol **37**, S34–S45.

Eastabrook, G., Hu, Y. and von Dadelszen, P. (2008). The role of decidual natural killer cells in normal placentation and in the pathogenesis of preeclampsia. J Obstet Gynaecol Can J **30**(6), 467–476.

Fassett, M.S., Davis, D.M., Valter, M.M., Cohen, G.B. and Strominger, J.L. (2001). Signaling at the inhibitory natural killer cell immune synapse regulates lipid raft polarization but not class I MHC clustering. Proc Natl Acad Sci U S A **98**, 14547–14552.

Fehniger, T.A., Shah, M.H., Turner, M.J., VanDeusen, J.B., Whitman, S.P., Cooper, M.A., Suzuki, K., Wechser, M., Goodsaid, F. and Caligiuri, M.A. (1999). Differential cytokine and

[1]We apologize with the many colleagues whose seminal work we could not cite solely for space limitations.

chemokine gene expression by human NK cells following activation with IL-18 or IL-15 in combination with IL-12: implications for the innate immune response. J Immunol **162**, 4511–4520.

Fehniger, T.A., Cooper, M.A., Nuovo, G. J., Cella, M., Facchetti, F., Colonna, M. and Caligiuri, M.A. (2003). CD56^{bright} natural killer cells are present in human lymph nodes and are activated by T cell derived IL-2: a potential new link between adaptive and innate immunity. Blood **101**, 3052–3057.

Ferlazzo, G., Tsang, M.L., Moretta, L., Melioli, G., Steinman, R.M. and Münz, C. (2002). Human dendritic cells activate resting natural killer (NK) cells and are recognized via the NKp30 receptor by activated NK cells. J Exp Med **195**, 341–351.

Ferlazzo, G., Pack, M., Thomas, D., Paludan, C., Schmid, D., Strowig, T., Bougras, G., Muller, W.A., Moretta, L. and Münz, C. (2004). Distinct roles of IL-12 and IL-15 in human natural killer cell activation by dendritic cells from secondary lymphoid organs. Proc Natl Acad Sci U S A **101**, 16606–16611.

Fernandez, N.C., Lozier, A., Flament, C., Ricciardi-Castagnoli, P., Bellet, D., Suter, M., Perricaudet, M., Tursz, T., Maraskovsky, E. and Zitvogel, L. (1999). Dendritic cells directly trigger NK cell functions: crosstalk relevant in innate anti-tumor immune responses in vivo. Nat Med **5**, 405–411.

Frey, M., Packianathan, N.B., Fehniger, T.A., Ross, M.E., Wang, W.C., Stewart, C.C, Caligiuri, M.A. and Evans, S.S. (1998). Differential expression and function of L-selectin on CD56^{bright} and CD56^{dim} natural killer cell subsets. J Immunol **161**, 400–408.

Gallucci, S. and Matzinger, P. (2001). Danger signals: SOS to the immune system. Curr Opin Immunol **13**, 114–119.

Gardella, S., Andrei, C., Costigliolo, S., Poggi, A., Zocchi, M.R. and Rubartelli, A. (1999). Interleukin-18 synthesis and secretion by dendritic cells are modulated by interaction with antigen-specific T cells. J Leukoc Biol **66**, 237–241.

Gardella, S., Andrei, C., Poggi, A., Zocchi, M.R. and Rubartelli, A. (2000). Control of interleukin-18 secretion by dendritic cells: role of calcium influxes. FEBS Lett **481**, 245–248.

Gardella, S., Andrei, C., Ferrera, D., Lotti, L.V., Torrisi, M.R., Bianchi, M.E. and Rubartelli, A. (2002). The nuclear protein HMGB1 is secreted by monocytes via a non-classical, vesicle-mediated secretory pathway. EMBO Rep **3**, 995–1001.

Gerosa, F., Baldani-Guerra, B., Nisii, C., Marchesini, V., Carra, G. and Trinchieri, G. (2002). Reciprocal activating interaction between natural killer cells and dendritic cells. J Exp Med **195**, 327–333.

Gerosa, F., Gobbi, A., Zorzi, P., Burg, S., Briere, F., Carra, G. and Trinchieri, G. (2005). The reciprocal interaction of NK cells with plasmacytoid or myeloid dendritic cells profoundly affects innate resistance functions. J Immunol **174**, 727–734.

Granucci, F., Zanoni, I., Pavelka, N., Van Dommelen, S.L., Andoniou, C.E., Belardelli, F., Degli Esposti, M.A. and Ricciardi-Castagnoli, P. (2004). A contribution of mouse dendritic cell-derived IL-2 for NK cell activation. J Exp Med **200**, 287–295.

Hanna, J., Fitchett, J., Rowe, T., Daniels, M., Heller, M., Gonen-Gross, T., Manaster, E., Cho, S.Y., LaBarre, M.J. and Mandelboim, O. (2005). Proteomic analysis of human natural killer cells: insights on new potential NK immune functions. Mol Immunol **42**, 425–431.

Hesslein, D.G., Takaki, R., Hermiston, M. L., Weiss, A. and Lanier, L.L. (2006). Dysregulation of signaling pathways in CD45-deficient NK cells leads to differentially regulated cytotoxicity and cytokine production. Proc Natl Acad Sci U S A **103**, 7012–7017.

Higuma-Myojo, S., Sasaki, Y., Miyazaki, S., Sakai, M., Siozaki, A., Miwa, N. and Saito, S. (2005). Cytokine profile of natural killer cells in early human pregnancy. Am J Reprod Immunol **54**, 21–29.

Hsu, D.H., Moore, K.W. and Spits, H. (1992). Differential effects of IL-4 and IL-10 on IL-2-induced IFN-gamma synthesis and lymphokine-activated killer activity. Int Immunol **4**, 556–563.

Jacobs, R., Hintzen, G., Kemper, A., Beul, K., Kempf, S., Behrens, G., Sykora, K. W. and Schmidt, R.E. (2001). CD56^{bright} cells differ in their KIR repertoire and cytotoxic features from CD56^{dim} NK cells. Eur J Immunol **31**, 3121–3127.

Jokhi, P.P., King, A., Jubinsky, P.T. and Loke, Y.W. (1994). Demonstration of the low affinity alpha subunit of the granulocyte-macrophage colony-stimulating factor receptor (GM-CSF-R alpha) on human trophoblast and uterine cells. J Reprod Immunol **26**, 147–164.

Kopcow, H.D., Allan, D.S., Chen, X., Rybalov, B., Andzelm, M.M., Ge, B. and Strominger, J.L. (2005). Human decidual NK cells form immature activating synapses and are not cytotoxic. Proc Natl Acad Sci U S A **102**, 15563–15568.

Krzewski, K., Chen, X., Orange, J.S. and Strominger, J.L. (2006). Formation of a WIP-, WASp-, actin-, and myosin IIA-containing multiprotein complex in activated NK cells and its alteration by KIR inhibitory signaling. J Cell Biol **173**, 121–132.

Lanier, L.L. (2005). NK cell recognition. Annu Rev Immunol **23**, 225–274.

Laskarin, G., Strbo, N., Bogovic-Crncic, T., Juretic, K., Ledee-Bataille, N., Chaouat, G. and Rukavina, D. (2005). Physiological role of IL-15 and IL-18 at the maternal–fetal interface. In: Markert, U. (ed), Immunology of PregnancyVol. 89. Basel: Karger, pp. 10–25.

Li, C., Ge, B., Nicotra, M., Stern, J. N., Kopcow, H.D., Chen, X. and Strominger, J.L. (2008). JNK MAP kinase activation is required for MTOC and granule polarization in NKG2D-mediated NK cell cytotoxicity. Proc Natl Acad Sci U S A **105**, 3017–3022.

Lou, Z., Jevremovic, D., Billadeau, D. D. and Leibson, P.J. (2000). A balance between positive and negative signals in cytotoxic lymphocytes regulates the polarization of lipid rafts during the development of cell-mediated killing. J Exp Med **191**, 347–354.

Lucas, M., Schachterle, W., Oberle, K., Aichele, P. and Diefenbach, A. (2007). Dendritic cells prime natural killer cells by trans-presenting interleukin 15. Immunity **26**, 503–517.

Mailliard, R.B., Alber, S.M., Shen, H., Watkins, S.C., Kirkwood, J.M., Herberman, R.B. and Kalinski, P. (2005). IL-18-induced CD83+CCR7+ NK helper cells. J Exp Med **202**, 941–953.

Marcenaro, E., Ferranti, B., Falco, M., Moretta, L. and Moretta, A. (2008). Human NK cells directly recognize *Mycobacterium bovis* via TLR2 and acquire the ability to kill monocyte-derived DC. Int Immunol **20**, 1155–1167.

Messmer, D., Yang, H., Telusma, G., Knoll, F., Li, J., Messmer, B., Tracey, K.J. and Chiorazzi, N. (2004). High mobility group box protein 1: an endogenous signal for dendritic cell maturation and Th1 polarization. J Immunol **173**, 307–313.

Moffett-King, A. (2002). Natural killer cells and pregnancy. Nat Rev Immunol **9**, 656–663.

Morel, P.A. and Oriss, T.B. (1998). Cross-regulation between Th1 and Th2 cells. Crit Rev Immunol **18**, 275–303.

Moretta, L. and Moretta, A. (2004). Unravelling natural killer cell function: triggering and inhibitory human NK receptors. EMBO J **23**, 255–259.

Munz, C., Dao, T., Ferlazzo, G., de Cos, M.A., Goodman, K. and Young, J.W. (2005). Mature myeloid dendritic cell subsets have distinct roles for activation and viability of circulating human natural killer cells. Blood **105**, 266–273.

Newman, K.C., Korbel, D.S., Hafalla, J.C. and Riley, E.M. (2006). Crosstalk with myeloid accessory cells regulates human natural killer cell interferon-α responses to malaria. PLoS Pathog **2**, e118.

Orange, J.S., Harris, K.E., Andzelm, M. M., Valter, M.M., Geha, R.S. and Strominger, J.L. (2003). The mature activating natural killer cell immunologic synapse is formed in distinct stages. Proc Natl Acad Sci U S A **100**, 14151–14156.

Ortaldo, J.R., Winkler-Pickett, R., Wigginton, J., Horner, M., Bere, E. W., Mason, A.T., Bhat, N., Cherry, J., Sanford, M., Hodge, D.L. and Young, H. A. (2006). Regulation of ITAM-positive receptors: role of IL-12 and IL-18. Blood **107**, 1468–1475.

Osada, T., Nagawa, H., Kitayama, J., Tsuno, N.H., Ishihara, S., Takamizawa, M. and Shibata, Y. (2001). Peripheral blood dendritic cells, but not monocyte-derived dendritic cells, can augment human NK cell function. Cell Immunol **213**, 14–23.

Parrish-Novak, J., Dillon, S.R., Nelson, A., Hammond, A., Sprecher, C., Gross, J. A., Johnston, J., Madden, K., Xu, W., West, J., Schrader, S., Burkhead, S., Heipel, M., Brandt, C., Kuijper, J.L., Kramer, J., Conklin, D., Presnell, S.R., Berry, J., Shiota, F., Bort, S., Hambly, K., Mudri, S., Clegg, C., Moore, M., Grant, F.J., Lofton-Day, C., Gilbert, T., Rayond, F., Ching, A., Yao, L., Smith, D., Webster, P., Whitmore, T., Maurer, M., Kaushansky, K., Holly, R.D. and Foster, D. (2000). Interleukin 21 and its receptor are involved in NK cell expansion and regulation of lymphocyte function. Nature **408**, 57–63.

Perussia, B. (1996). The cytokine profile of resting and activated NK cells. Methods **9**, 370–378.

Piccioli, D., Sbrana, S., Melandri, E. and Valiante, N.M. (2002). Contact-dependent stimulation and inhibition of dendritic cells by natural killer cells. J Exp Med **195**, 335–341.

Pisegna, S., Pirozzi, G., Piccoli, M., Frati, L., Santoni, A. and Palmieri, G. (2004). p38MAPK activation controls the TLR3-mediated upregulation of cytotoxicity and cytokine production in human NK cells. Blood **104**, 4157–4164.

Poggi, A., Carosio, R., Spaggiari, G.M., Fortis, C., Tambussi, G., Dell'Antonio, G., Dal Cin, E., Rubartelli, A. and Zocchi, M.R. (2002). NK cell activation by dendritic cells is dependent on LFA-1-mediated induction of calcium-calmodulin kinase II: inhibition by HIV-1 Tat C-terminal domain. J Immunol **168**, 95–101.

Rajagopalan, S., Bryceson, Y.T., Kuppusamy, S.P., Geraghty, D.E., van der Meer, A., Joosten, I. and Long, E.O. (2006). Activation of NK cells by an endocytosed receptor for soluble HLA-G. PLoS Biol **4**, e9.

Roda-Navarro, P., Mittelbrunn, M., Ortega, M., Howie, D., Terhorst, C., Sanchez-Madrid, F. and Fernandez-Ruiz, E. (2004). Dynamic redistribution of the activating 2B4/SAP complex at the cytotoxic NK cell immune synapse. J Immunol **173**, 3640–3646.

Romagnani, S. (1992). Induction of TH1 and TH2 responses: a key role for the "natural" immune response? Immunol Today **13**, 379–381.

Rovere-Querini, P., Capobianco, A., Scaffidi, P., Valentinis, B., Catalanotti, F., Giazzon, M., Dumitriu, I.E., Müller, S., Iannacone, M., Traversari, C., Bianchi, M.E. and Manfredi, A.A. (2004). HMGB1 is an endogenous immune adjuvant released by necrotic cells. EMBO Rep **5**, 825–830.

Rubartelli, A. and Lotze, M.T. (2007). Inside, outside, upside down: damage-associated molecular-pattern molecules (DAMPs) and redox. Trends Immunol **28**, 429–436.

Sanni, T.B., Masilamani, M., Kabat, J., Coligan, J.E. and Borrego, F. (2004). Exclusion of lipid rafts and decreased mobility of CD94/NKG2A receptors at the inhibitory NK cell synapse. Mol Biol Cell **15**, 3210–3223.

Schmidt, K.N, Leung, B., Kwong, M., Zarember, K.A, Satyal, S., Navas, T. A, Wang, F. and Godowski, P.J. (2004). APC-independent activation of NK cells by the Toll-like receptor 3 agonist double-stranded RNA. J Immunol **172**, 138–143.

Schnurr, M., Then, F., Galambos, P., Scholz, C., Siegmund, B., Endres, S. and Eigler, A. (2000). Extracellular ATP and TNF-alpha synergize in the activation and maturation of human dendritic cells. J Immunol **165**, 4704–4709.

Schoenborn, J.R. and Wilson, C.B. (2007). Regulation of interferon-gamma during innate and adaptive immune responses. Adv Immunol **96**, 41–101.

Semino, C., Angelini, G., Poggi, A. and Rubartelli, A. (2005). NK/iDC interaction results in IL-18 secretion by DCs at the synaptic cleft followed by NK cell activation and release of the DC maturation factor HMGB1. Blood **15**, 609–616.

Semino, C., Ceccarelli, J., Lotti, L.V., Torrisi, M.R., Angelici, G. and Rubartelli, A. (2007). The maturation potential of NK cell clones toward autologous dendritic cells correlates with HMGB1 secretion. J Leukoc Biol **81**, 92–99.

Shi, Y., Evans, J.E. and Rock, K.L. (2003). Molecular identification of a danger signal that alerts the immune system to dying cells. Nature **425**, 516–521.

Sivori, S., Cantoni, C., Parolini, S., Marcenaro, E., Conte, R., Moretta, L. and Moretta, A. (2003). IL-21 induces both rapid maturation of human CD34+ cell precursors towards NK cells and acquisition of surface killer Ig-like receptors. Eur J Immunol **33**, 3439–3447.

Sivori, S., Falco, M., Della Chiesa, M., Carlomagno, S., Vitale, M., Moretta, L. and Moretta, A. (2004). CpG and double-stranded RNA trigger human NK cells by Toll-like receptors: induction of cytokine release and cytotoxicity against tumors and dendritic cells. Proc Natl Acad Sci U S A **101**, 10116–10121.

Takeda, K. and Akira, S. (2005). Toll-like receptors in innate immunity. Int Immunol **17**, 1–14.

Valitutti, S., Dessing, M., Aktories, K., Gallati, H. and Lanzavecchia, A. (1995). Sustained signaling leading to T cell activation results from prolonged T cell receptor occupancy. Role of T cell actin cytoskeleton. J Exp Med **181**, 577–584.

Vitale, M., Della Chiesa, M., Carlomagno, S., Romagnani, C., Thiel, A., Moretta, L. and Moretta, A. (2004). The small subset of CD56brightCD16− natural killer cells is selectively responsible for both cell proliferation and interferon-γ production upon interaction with dendritic cells. Eur J Immunol **34**, 1715–1722.

Vitale, M., Della Chiesa, M., Carlomagno, S., Pende, D., Aricò, M., Moretta, L. and Moretta, A. (2005). NK-dependent DC maturation is mediated by TNFalpha and IFNgamma released upon engagement of the NKp30 triggering receptor. Blood **106**, 566–571.

Vivier, E., Nunes, J.A. and Vely, F. (2004). Natural killer cell signaling pathways. Science **306**, 1517–1519.

Vivier, E., Tomasello, E., Baratin, M., Walzer, T. and Ugolini, S. (2008).

Functions of natural killer cells. Nat Immunol **9**, 503–510.

Vyas, Y.M., Mehta, K.M., Morgan, M., Maniar, H., Butros, L., Jung, S., Burkhardt, J.K. and Dupont, B. (2001). Spatial organization of signal transduction molecules in the NK cell immune synapses during MHC class I-regulated noncytolytic and cytolytic interactions. J Immunol **167**, 4358–4367.

Vyas, Y.M., Maniar, H. and Dupont, B. (2002a). Visualization of signaling pathways and cortical cytoskeleton in cytolytic and noncytolytic natural killer cell immune synapses. Immunol Rev **189**, 161–178.

Vyas, Y.M., Maniar, H. and Dupont, B. (2002b). Cutting edge: differential segregation of the SRC homology 2-containing protein tyrosine phosphatase-1 within the early NK cell immune synapse distinguishes noncytolytic from cytolytic interactions. J Immunol **168**, 3150–3154.

Wang, H., Bloom, O., Zhang, M., Vishnubhakat, J.M., Ombrellino, M., Che, J., Frazier, A., Yang, H., Ivanova, S., Borovikova, L., Manogue, K.R., Faist, E., Abraham, E., Andersson, J., Andersson, U., Molina, P.E., Abumrad, N.N., Sama, A. and Tracey, K.J. (1999). HMG-1 as a late mediator of endotoxin lethality in mice. Science **285**, 248–251.

Wang, K.S., Frank, D.A. and Ritz, J. (2000). Interleukin-2 enhances the response of natural killer cells to interleukin-12 through up-regulation of the interleukin-12 receptor and STAT4. Blood **95**, 3183–3190.

Xu, J., Chakrabarti, A.K., Tan, J.L., Ge, L., Gambotto, A. and Vujanovic, N. L. (2007). Essential role of the TNF–TNFR2 cognate interaction in mouse dendritic cell–natural killer cell crosstalk. Blood **109**, 3333–3341.

Yu, Y., Hagihara, M., Ando, K., Gansuvd, B., Matsuzawa, H., Tsuchiya, T., Ueda, Y., Inoue, H., Hotta, T. and Kato, S. (2001). Enhancement of human cord blood CD34+ cell-derived NK cell cytotoxicity by dendritic cells. J Immunol **166**, 1590–1600.

Zanoni, I., Foti, M., Ricciardi-Castagnoli, P. and Granucci, F. (2005). TLR-dependent activation stimuli associated with Th1 responses confer NK cell stimulatory capacity to mouse dendritic cells. J Immunol **175**, 286–292.

Zhang, A.L., Colmenero, P., Purath, U., Teixeira de Matos, C., Hueber, W., Klareskog, L., Tarner, I.H., Engleman, E. G. and Söderström, K. (2007). Natural killer cells trigger differentiation of monocytes into dendritic cells. Blood **110**, 2484–2493.

Chapter **Fourteen** 14

NK cells as recipients of cytokine signals

Tara J. Loux, Michael T. Lotze, Herbert J. Zeh III

CHAPTER CONTENTS

Method is more important than strength, when you wish to control your enemies.

Henry Wadsworth Longfellow

ABSTRACT

The response of natural killer (NK) cells to the cytokine signals found in their microenvironment appears to be dependent upon many factors: the specific combination and concentration of secreted cytokines, the expression of receptors on the NK cell itself, the complement of neighbouring immune and stromal cells, the availability of cell–cell contact and the pathologic process in question, whether inflammatory, infectious, malignant or even autoimmune. Interleukin (IL)-2, IL-12, IL-15, IL-18, IL-21 and IL-27, as well as the type I interferons, all appear to have roles in stimulation of proliferation, activation of effector function and homeostasis of peripheral NK populations. Many different chemokines have indispensable roles in recruitment of NK cells to various locations of disease. Other cytokines such as TNF-α, TGF-β, IL-4, IL-7 and IL-10 may have complementary or even opposing functions in regulation of NK cell development and function.

KEY WORDS

Cytokine, Receptor, Chemokine, Interferon, Interleukin

Introduction

Cytokines are polypeptide mediators that play pivotal roles in communication between cells and can be broadly distinguished as leadered cytokines, including hematopoietins, colony stimulating factors, chemokines, tumour necrosis factor (TNF) family members, and leaderless interleukin (IL)-1 extended family, fibroblast growth factor family, high mobility group box 1 protein (HMGB1) and others. They are pleiotropic, synergistic and redundant, conferring substantial evolutionary advantages. Cytokine expression is disturbed in many infectious, inflammatory and autoimmune disease states. Autocrine activities mediated by cytokines in natural killer (NK) cells include the roles of self-secreted interferon (IFN)-γ and IL-10; paracrine mediators include the cross-talk of HMGB1, IL-12 and IL-15 between NK cells and dendritic cells (DCs) at the immunologic synapse (Ferlazzo et al., 2004; Semino et al., 2005), while endocrine activity received by NKs from release of chemokines into the vascular circulation by distant cells is vital for migration of the innate immune response from the bone marrow into the peripheral blood to an area of disease.

NK cells recognize stressed cells, entertaining a constitutive ability to mediate cytotoxicity in target cells and secrete cytokines rather rapidly in response. They participate in the innate resistance to intracellular pathogens and malignancies. Their role is critical to expansion of the Th1 biased adaptive immune response and development of secondary lymphatic sites. They also influence hematopoiesis, increasing myelopoiesis and decreasing megacytopoiesis and erythropoiesis.

They respond to multiple cytokines as part of the innate immune response, including IL-2, IL-12, IL-15, IL-18, TNF-α, IFNs-α and β and various chemokines. In response to cytokine stimulation, they then secrete additional cytokines as part of their immuno-modulatory function. These include IFN-γ and GM-CSF, and in lesser quantities IL-3, IL-5, IL-10, IL-13, IL-18, M-CSF and TNF-α.

Cells influencing NK behaviour

NK cells engage in extensive communication with other heterogeneous cell lines via both cytokine signalling and direct cell–cell contact. Of particular importance is the NK/DC relationship, as these two types of cells engage in a significant amount of 'cross-talk' necessary for mutual tailoring of immune function. DCs secrete large amounts of IFNs α and β, IL-2, IL-12 and IL-18, all of which stimulate NK cells (Biron, 2001; Ferlazzo and Munz, 2004; Nguyen et al., 2002). NK cells co-localize with DCs in the T-cell areas of peripheral lymph nodes (Ferlazzo et al., 2004). Not only peripheral blood monocyte-derived DCs but also dermal-interstitial derived and other subsets of DCs and their culture supernatants alone increase NK cell proliferation, cytotoxicity, IFN-γ secretion and CD56 expression, although individual DC subsets appear to have differing efficacies (Munz et al., 2005; Nishioka et al., 2001).

DC release of IL-12 within the immunologic synapse appears to be essential for NK cell secretion of IFN-γ, while DC secretion of IL-15 appears necessary for NK proliferation and survival (Ferlazzo et al., 2004). In the human, IL-12 secretion by DCs (CD14+ monocytes) is accompanied by an increase in CD40 expression, along with increased CD40L expression on CD56+ cells, implying that NK/DC cross-talk results in cross-stimulation between the two cell lines (Bose and Baral, 2007). Plasmacytoid DCs secreting IL-18 are critical in stimulation of NK cells to elicit the effector function against herpes simplex virus type 1-infected cells (Barr et al., 2007).

Activated monocytes and macrophages produce IL-12, IL-15 and TNF-α, all associated with NK cell activation (Carson et al., 1994; Tripp et al., 1993). Neutrophils produce IL-18 in response to *Legionella* infection, and this is necessary for bacterial clearance by NK cells via secretion of IFN-γ (Sporri et al., 2008). T cells in human lymph nodes secrete IL-2, which further stimulates CD56+ (bright) NK cells localized there to secrete IFN-γ (Fehniger et al., 2003). This sort of immune cell cross-talk (see Figure 14.1) is essential for communication during recognition of infection or other cellular abnormalities, including malignancy or microenvironmental stress, linking the innate and adaptive immune response.

Cytokines that affect NK cells

IL-2

In supraphysiologic doses, IL-2 causes NK cell activation and proliferation and increases cytotoxicity and IFN-γ production (London et al., 1986; Trinchieri et al., 1984; Young et al., 1987). The IL-2 receptor is composed of IL2-specific α and β subunits in addition to the common cytokine receptor gamma subunit. CD56+ (bright) NK cells express a high affinity IL-2 receptor (Fehniger et al., 2003), while CD56+ (dim) NK cells express a more intermediate-affinity IL-2 receptor (Carson et al., 1994). IL-2 receptor α chain deficiency does not affect T cell or B cell development; however, adult deficient mice develop overexpansion of peripheral lymphoid populations and autoimmune diseases such as haemolytic anaemia and inflammatory bowel disease (Willerford et al., 1995). The IL-2 β chain is shared with the IL-15 receptor, and IL-2 receptor β chain deficiency in the mouse results in reduced numbers of circulating NK cells and abolition of cytotoxic capabilities (Carson et al., 1994; Suzuki et al., 1997). Enforced expression of bcl-2 in β chain deficient animals rescues proliferation, likely via the effect of IL-15, but does not improve cytotoxicity (Minagawa et al., 2002).

The common cytokine γ chain is used in IL-2, IL-4, IL-7, IL-9 and IL-15 receptors. Deficiency of this subunit in the mouse results in absence of NK cells, peripheral and gut-associated lymph nodes, intraepithelial DCs and γ-δ T lymphocytes, as well as deficiencies in T cell and B cell populations (Cao et al., 1995; DiSanto et al., 1995). In humans, a homologous mutation of the common chain is present in 75% of children with severe combined immunodeficiency (SCID).

IL-2 causes increases in transcription and post-transcriptional stabilization of granzyme mRNA and to a lesser extent perforin mRNA (Salcedo et al., 1993). It also mediates up-regulation of the SAP (SLAM-associated

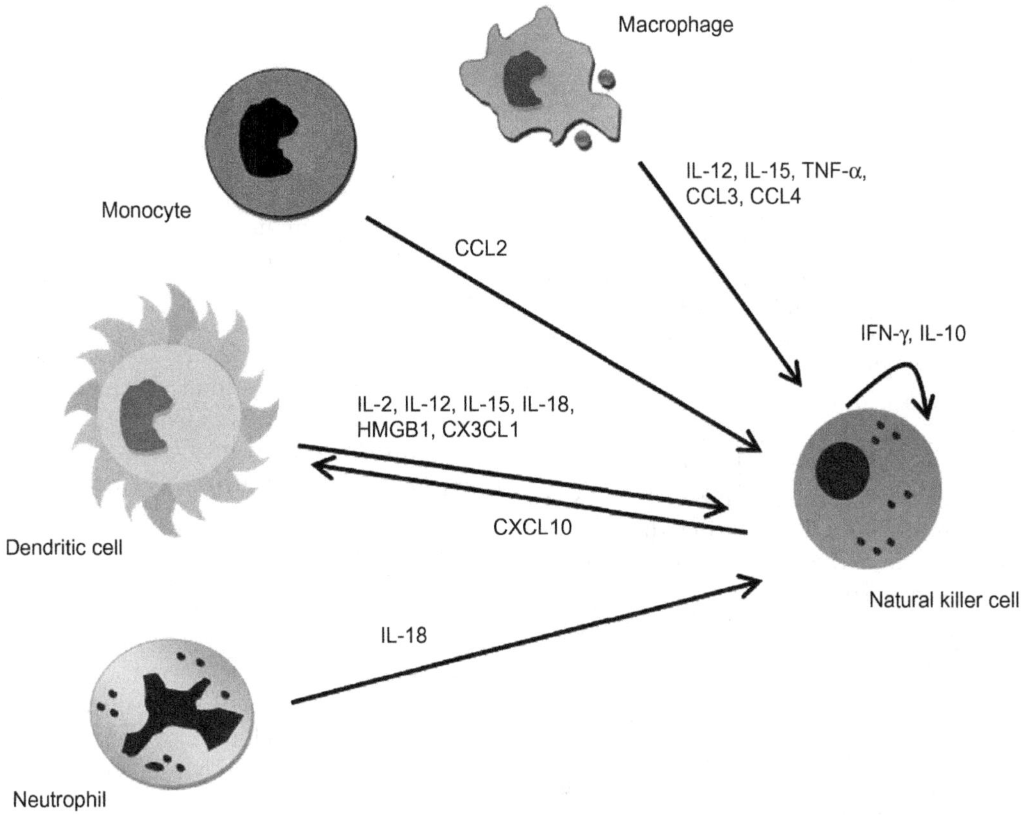

Figure 14.1 • Cytokine signals elaborated by monocytes, macrophages, DCs and neutrophils produce effects on NK cells. In addition, certain signals elaborated by NK cells have autocrine effects. Not pictured in this figure are the multiple stromal or epithelial cell types that express chemokines or their receptors and as such additionally play roles in recruitment of NKs to the inflammatory response.

protein) gene product, which in conjunction with 2B4 and SLAM (signalling lymphocyte activation molecule) on the NK cell surface, are important in NK cell response to EBV infection. Abnormalities in this gene product are found in X-linked lymphoproliferative syndrome (Sayos et al., 2000).

IL-12

IL-12 was initially identified and purified as NK cell stimulatory factor (NKSF) (Kobayashi et al., 1989). It functions synergistically with IL-2 to augment NK cell cytotoxicity and induce IFN-γ production (Chan et al., 1991). It appears to be required for IFN-γ secretion but not blastogenesis or proliferation (Orange and Biron, 1996). It induces up-regulation of the IL-18 receptor on lymphocytes to create a synergistic increase in IFN-γ production with IL-18 (Yoshimoto et al., 1998).

With regard to cytotoxicity, IL-12 up-regulates transcription of granzyme and perforin mRNAs (Salcedo et al., 1993). Augmentation of NK cytotoxicity is dependent upon this increase in perforin, as it is not found in perforin-deficient mice or upon using a perforin inhibitor (Hyodo et al., 1999). IL-12 also up-regulates TRAIL and NKG2D cell surface expression. Interaction of MIC-A with NKG2D increases the cytotoxic capacity of the NK cell, and blockade of either NKG2D or MIC-A attenuates the effect of IL-12 on NK cytolysis (Zhang et al., 2008). Humans with congenital deficiency of the IL-12 receptor β- chain show diminished innate cytotoxicity in NK and NKT cell populations (Guia et al., 2008).

In humans, IL-12 preferentially induces the NK1 subset of cells, which secrete IL-10 and IFN-γ and express higher levels of surface CD95 (Fas), rendering them more sensitive to FasL-induced apoptosis (Peritt et al., 1998). IL-12 signals via STAT4 (signal transduction and activator of transcription-4), and in conjunction with IL-2, this increases NK cell IL-10 expression (Grant et al., 2008; Nguyen et al., 2002). In vivo data from operations performed in rats suggest that stress associated with surgery may cause suppression of the innate immune axis. Prophylactic IL-12 administration prior to operation in rats led to decreased post-operative tumour metastasis, concomitant with an increase in NK cell numbers in the circulation and marginating pulmonary pools (Schwartz et al., 2008). Additionally, intraocular injection of recombinant IL-12 in mice with endotoxin-induced uveitis results in recruitment of NK cells and secretion of IFN-γ with protection from disease, implying that NK cells stimulated by IL-12 may have a protective effect against unruly autoimmunity (Figueiredo et al., 2007).

IL-15

IL-15 is an essential mediator of peripheral NK cell homeostasis, without an effect on proliferation per se (Kennedy et al., 2000). Investigation into the mechanism of this cytokine shows that it supports and maintains production of the anti-apoptotic factor Bcl-2 in NK cells (Jamieson et al., 2004; Ranson et al., 2003). Wild type (WT) NK cells injected into IL-15 knockout (KO) mice survive for up to 36 hours following injection but are completely absent by 5 days following injection (Cooper et al., 2002).

IL-15 also augments cytotoxicity of NK cells and synthesis and secretion of IFN-γ and increases their ability to kill cells infected with herpes viruses (Gosselin et al., 1999). In fact, virus-induced activation by several human herpes viruses is entirely dependent on IL-15 production and was totally neutralized by antibodies against IL-15 but not antibodies to IFN-γ, IL-2, TNF-α or IL-12 (Ahmad et al., 2000). In humans with HIV infection, CD56+ cells are diminished in total number; however, they respond equally to IL-15 as cells from non-infected humans, with increased cytotoxicity, increased CD107a and IFN-γ expression and up-regulation of perforin expression in the CD56+ (bright) subset. These IL-15 effects are mediated via STAT5 (Strbo et al., 2008).

The IL-15 receptor on NK cells is composed of IL-2/IL-15R β-chain and the common cytokine γ-chain, paired with an IL-15 receptor-specific α subunit. This receptor signals via the Jak1 (Janus kinase-1)/Jak3 and STAT5/STAT3 systems (Waldmann and Tagaya, 1999). In order to signal, IL-15 must interact with the IL-15R-α subunit, which is ubiquitously expressed, then presented in trans via cell–cell contact to an opposing cell's IL-15R-β and common γ-chain subunits (Burkett et al., 2004). IL-15R-α chain deficiency leads to a significant decrease in NK cell number, although it is not entirely clear whether this is a primary effect of lack of the receptor on the NK cell surface or more likely a secondary effect due to lack of presentation of IL-15 by other supporting hematopoietic cells to the NK cell (Burkett et al., 2004; Lodolce et al., 1998; Prlic et al., 2003).

Reciprocal trans-presentation of membrane bound IL-15 between human CD34+ peripheral blood hematopoietic progenitor cells leads to maturation of a non-cytolytic NK cell line which expresses HLA-G and secretes mainly IL-10 and IL-21. The downstream effects of this cell line are down-regulation of the immune response and reversal of DC maturation, blocking cytolytic functions of conventional NK cells and inducing HLA-G expression on peripheral blood mononuclear cells (PBMC). In vivo, a small subset of uterine/placental decidual NK cells also expresses this immuno-regulatory phenotype (Giuliani et al., 2008).

IL-18

IL-18 enhances NK cell cytotoxicity and IFN-γ production in the serum and spleen; however, there are conflicting reports as to its activity in the liver (Akira, 2000; Pien et al., 2000; Tsutsui et al., 1997). Its action is synergistic with that of IL-12, as shown in double deficient IL-12/IL-18 KO mice (Takeda et al., 1998). IL-18 secretion may be a priming event required for subsequent IL-12 stimulation (Chaix et al., 2008). Some of its cytotoxic activity may be due to up-regulation of FasL on the NK cell surface (Tsutsui et al., 1996); however, it is unable to effect cytotoxicity in the perforin-deficient state (perforin inhibition or perforin KO mice), thus granzyme-perforin mediated cytotoxicity is likely the most important mechanism that it regulates (Hyodo et al., 1999). It may also suppress NK cell secretion of TNF-α (Tsutsui et al., 1997).

IL-18 appears to induce the maturation of a 'helper' phenotype in NK cells, which displays increased migratory responsiveness to lymph node-associated chemokines and improved ability to produce IFN-gamma in response to DC or Th1 signalling. These NK cells express CD83, CCR7 and CD25. Prostaglandin E2 blocks this pathway of maturation (Mailliard et al., 2005).

IL-18 also causes the NK cell to secrete HMGB1, which leads to DC maturation and promotion of inflammation. DC secretion of IL-18 may occur within an interactive synapse, so as not to release the cytokine to the surrounding milieu, allowing rapid oxidative denaturation of the protein (Semino et al., 2005, 2007).

The IL-18 receptor is made up of extracellular ligand-binding and signal-transducing subunits, which are complexed with MyD88 intracytoplasmically. Downstream signalling occurs via IRAK, TRAF-6, NIK, IKK and the NF-κB transcription factor. IL-18 shares this signalling path-way with IL-1 (Adachi et al., 1998; Kojima et al., 1998).

IL-21

IL-21, another IL-2 family member, stimulates proliferation of CD56+ (bright) NK cells and augments cytotoxicity of CD56+ (dim) NK cells, where each of the individual subsets express identical numbers of IL-21 receptors. IL-21 alone does not induce cytokine production (Wendt et al., 2007). IL-21 up-regulates perforin, granzyme, CD107a, and IFN-γ expression via STAT3 and STAT5 (Strbo et al., 2008). IL-21 is synergistic with $F_C\gamma$ receptor stimulation in augmenting NK cell anti-tumour lytic activity, and when IL-21 stimulation is combined with $F_c\gamma$ receptor stimulation, NK cells produce copious quantities of IFN-γ. These effects are absent in STAT1-deficient NK cells (Roda et al., 2006). For human, but not murine, NK cells, IL-21 alone is sufficient to sustain survival at the same level as IL-2. It synergizes with IL-2 to up-regulate NKG2A, CD25, CD86 and CD69 surface expression (Skak et al., 2008).

IL-21 is synergistic with IFN-α in increasing NK cell cytotoxicity. Treatment of an in vivo murine renal carcinoma model with the combination of IL-21 and IFN-α induced STAT3 activation, significantly decreased tumour progression and improved host survival (Eriksen et al., 2008). In a murine tumour model with HER-1 positive malignancy, the effect of cetuximab (an anti-HER-1 monoclonal Ab) was significantly enhanced by combination with IL-21 treatment, leading to enhanced cytokine secretion and antibody-dependent cytotoxicity (Roda et al., 2007). IL-21 given intravenously to patients with metastatic melanoma or renal cell carcinoma caused increases of perforin and granzyme B mRNA in both NK and CD8+ T cells, in addition to decreases in circulating NK cell and T cell levels during treatment, and enhanced ex vivo cytotoxicity of NK cells (Frederiksen et al., 2008). Despite these changes in the patients' immune systems, the tumour burden was largely unaffected by IL-21 treatment (Davis et al., 2007).

IL-27

IL-27 is a member of the IL-12 extended family. Induced expression of IL-27 in human aesophageal cancer cells inoculated into nude mice retarded growth of the cancer cells. This effect was secondary to increased IFN-γ expression and cytotoxicity of splenocytes in the tumour-bearing mouse (Liu et al., 2008). Murine melanoma B16 cells overexpressing IL-27 showed significant early growth retardation when injected into a mouse model, mainly due to NK cell mediated effects in an IFN-γ independent manner (Oniki et al., 2006). Mice injected with colon 26 murine colon carcinoma cells, which transgenically overexpressed IL-27, rejected their tumour and developed tumour-specific protective immunity dependent upon the presence of NK cells (Chiyo et al., 2005); however, other studies in mice with IL-27 overexpressing neuroblastomas found decreased tumour growth with tumour-specific protective immunity to be dependent on CD8+ T cells and not on NK cells (Salcedo et al., 2004).

The IL-27 receptor contains a ligand-specific component called WSX-1, which is expressed on many immune cells of both innate and adaptive branches. NK cell activation by *Toxoplasma* infection or IL-2 stimulation induces a reduction in WSX-1 cell surface expression, potentially decreasing NK cell responsiveness to IL-27 in a negative feedback manner (Villarino et al., 2005; Table 14.1).

Table 14.1 Families of interleukins and their effects on natural killer cells

Family	Members	Major Effects
IL-1	IL-1, IL-18, IL-33	Potentiation of IL-2 and IL-12 effects
IL-2	IL-2, IL-4, IL-15, IL-21	Proliferation, homeostasis, cytotoxicity, cytokine secretion
IL-12	IL-12, IL-23, IL-27	Proliferation, cytotoxicity, cytokine secretion

Table 14.1 delineates the major interleukin families and their broad effects on NK behavior.

Interferon α/β

The type I IFNs negatively regulate IL-12 secretion and block the NK cell response to IL-12; however, they also up-regulate expression of a high-affinity IL-12 receptor on the NK cell surface (Nguyen et al., 2000). They suppress IFN-γ secretion but induce and activate NK cell cytotoxicity, blastogenesis and proliferation (Orange and Biron 1996; Trinchieri and Santoli, 1978). Suppression of IFN-γ is dependent on STAT1 signalling, and when STAT1 is absent, IFN-α/β actually induces IFN-γ secretion (Nguyen et al., 2000).

The type I IFNs increase IL-15 expression, which, as noted earlier, is critical for long-term survival of NK cells (Biron, 2001; Nguyen et al., 2002) and indirectly increase levels of MIP-1α in the liver, which is necessary for clearance of certain hepatic viral infections (Salazar-Mather and Hokeness, 2006). In human patients with hepatitis B virus infection, flares of liver inflammation correlate with increases in IFN-α and IL-8 levels. NK cells from these patients express higher levels of TRAIL when cultured ex vivo (Dunn et al., 2007), such that hepatocyte death may be a direct or indirect effect of NK cells responding to the infection.

The IFN-α receptors 1 and 2 (IFNAR-1 and IFNAR-2) signal via Jak1/Tyk1 tyrosine kinases and STAT1/STAT2 heterodimers (Biron, 2001). IFNAR-1 and IFNAR-2 deficient mice exhibit poor NK cell mediated anti-tumour responses (Swann et al., 2007).

Chemokines

NK cells have been shown to be responsive to a number of chemokines and express multiple different chemokine receptors. Freshly isolated NK cells all express CXCR1, CXCR3 and CXCR4, while subsets of NK cells can express CCR1, CCR4, CCR5, CCR6, CCR7, CCR9, CXCR5 and CXCR6. Most of these receptors are expressed at higher levels on the CD56+ (bright) subset of NK cells, especially CCR7, which is found exclusively on this subset. CCR4, on the other hand, is expressed more highly on the CD56+ (dim) subset (Berahovich et al., 2006). Chemokines are vital signals for migration of NK cells to sites of inflammation or infection, and disruption of these signals leads to impaired response to infection and malignancy, or unruly autoimmunity.

CXCL9/CXCL10/CXCR3

IFN-γ inducible protein-10 (IP-10, also known as CXCL10) and monokine induced by IFN-γ, (Mig, also known as CXCL9) are important ligands of CXCR3. CXCL10 is important for NK cell recruitment to the myocardium and limitation of viral replication during murine coxsackievirus infection. Increased levels of CXCL10 in the early phase of this infection correlate with decreased viral titre and improved cardiac function (Yuan et al., 2009). CXCL10 or CXCR3 KO mice are deficient in the NK cell response to murine HSV-1 infection in the nervous system (Wuest et al., 2008), and CXCL10 or CXCL9 KO mice have significantly decreased ability to clear genital HSV type 2 infection due to decreased recruitment of both NK and cytotoxic virus-specific T cells to the site of infection. CXCL10 appears to be more important in acute infection, while CXCL9 importance emerges in subacute infection (Thapa et al., 2008). CXCL9 is protective from murine hepatitis virus infection, with clearance of virus from the liver mediated by NK cells (Wuest et al., 2008).

CXCR3 KO mice with corneal HSV-1 infection had deficient NK cell recruitment, with normal recruitment of all other immune populations, to the anterior chamber of the eye, despite elevated levels of various chemokines ligands (Carr et al., 2008).

CXCR3 KO mice with subcutaneous implanted tumours had less intratumoural NK cell infiltration and poorer host survival than WT mice. Intratumoural injection or ectopic expression of CXCL10 or IFN-γ improved NK cell infiltration and host survival (Wendel et al., 2008). Mobilization and migration of splenic NK cells to the peritoneum, liver and tumour-bearing lung tissue in response to intraperitoneal injection of IFN-γ is dependent upon the expression of CXCR3 (Wald et al., 2006). CXCR3 antagonism with a small molecule inhibitor in a malignant murine mammary tumour model inhibited the tumours' ability to metastasize to the lung after orthotopic injection, while local tumour growth was not affected. This anti-metastatic activity was reversed in mice depleted of NK cells (Walser et al., 2006).

CXCL9 over-expressing murine mammary tumours injected orthotopically grew more slowly, caused fewer lung metastases and were associated with prolonged host survival. NK cells were critical in the model to the mechanism by which CXCL9 limits metastasis, while local tumour growth appeared more dependent on the T cell response (Walser et al., 2007).

In an IL-10 KO mouse model of chronic colitis, NK cells express significant amounts of CXCL10, the blockade of which is associated with resolution of colitis by decreasing intestinal recruitment of CXCR3+ DCs (Singh et al., 2008). Blockade of CXCL10 results in decreased NK cell proliferation both systemically and locally after murine cyclophosphamide-induced interstitial cystitis and protects the host from disease (Sakthivel et al., 2008).

CX3CL1/CX3CR1

CX3CL1, also known as fractalkine, significantly augments cytotoxicity and IFN-γ expression and secretion of NK cells in culture (Zhang et al., 2007). Mature DCs that express fractalkine promote NK cell activation (Pallandre et al., 2008). Anti-CX3CR1 antibody strongly neutralizes the cytotoxicity of NK cells against NK-sensitive tumour targets, and pretreatment of NK cells with recombinant CX3CL1 improves NK cytolytic function (Zhang et al., 2006). CX3CL1 is expressed to a high degree in the uterine decidua and is responsible for recruiting NK cells during the first trimester of pregnancy, along with CXCL12 and CCL3 (Santoni et al., 2008).

CX3CR1 KO mice have increased lung tumour burden and cachexia when treated with the B16 melanoma tumour line. There were reduced numbers of NK cells and monocytes in the lung of these animals, whether control or tumour-bearing. CX3CR1 KO NK cells exhibited equal cytotoxicity against B16 cells in vitro; however, they had a tumourigenic cytokine expression profile, with defective IFN-γ expression and enhanced IL-6 production in response to TLR3 activation with poly I:C (Yu et al., 2007). Fractalkine is highly expressed in a great majority of neuroblastomas; however, this does not correlate with anti-tumour efficacy of migrating immune cells unless IL-2 is added to the tumour microenvironment, in which case, a highly effective anti-tumour response is elicited, reversible upon depletion of both NK and T cells (Zeng et al., 2007).

CX3CR1 expression is essential for recruitment of NK cells to the CNS in a murine model of autoimmune encephalitis but not essential for NK cell recruitment to the liver in murine CMV. T cell, NKT cell, monocyte and macrophage recruitment to the CNS were not affected by CX3CR1 deficiency in autoimmune encephalitis; however, CX3CR1 KO animals experience significant increased morbidity and mortality from this condition (Huang et al., 2006).

In resection specimens of human gastric adenocarcinoma, fractalkine expression by the tumour correlated highly with NK and CD8+ T cell infiltration, in addition to more favourable prognosis, including improved disease-free survival (Hyakudomi et al., 2008).

CCL5/CCR5/CCR1

CCL5 (also known as RANTES) may promote infiltration of NK cells into the liver during acute and chronic liver injury via interaction with CCR1 expressed on NK cells (Karlmark et al., 2008). CCR5 expression on NK cells controls trafficking into areas of parasitic infection in the liver and spleen, without which hosts are unable to clear infection (Khan et al., 2006). NK cell homing to the red pulp of the spleen during inflammation or CMV infection is dependent on the synergistic action of CXCR3 and CCR5 (Gregoire et al., 2008). CCL5 (and CXCL10) expressed by keratinocytes attract a subset of CD56+ (bright), CD16-, CD158b- NK cells to psoriatic skin lesions, based on their expression of CCR5 and CXCR3, respectively. NK cell recruitment exacerbates inflammation of psoriatic plaques (Ottoviani et al., 2006).

HIV+ humans with NK cells exhibiting a higher frequency of CCR5 expression also had higher viral loads and lower CD4+ T cell counts. Frequency of CCR5 expression in NKs was lower in patients treated with HAART and in those patients classified as long-term slow progressors, implying a more aggressive phenotype of HIV progression associated with high CCR5 expression (Jiang et al., 2008).

CCR5 KO mice exhibit increased NK cell recruitment into the liver affected with chemically induced T-cell mediated hepatitis, in a CCL5/CCR1 dependent manner, and severe hepatitis in these mice was ameliorated with NK cell depletion. It is thought that the chemokine receptor imbalance in the liver in this model may induce NK cells to take on an effector role (Ajuebor et al., 2007). Interestingly, CCR5 KO mice have decreased recruitment of NK cells into the spleen and brainstem during genital HSV type 2 infection, leading to significantly increased mortality in infected KO mice as compared to WT controls (Thapa et al., 2007).

CCL2/CCR2

Monocyte chemo-attractant protein-1 (MCP-1), also known as CCL2, precedes MIP-1α secretion by macrophages in murine CMV infection, via stimulation of CCR2, and deficiency of either MCP-1 or CCR2 leads to markedly diminished NK response in the liver (Hokeness et al., 2005). NK cells from pleural fluid in patients with *Mycobacterium tuberculosis* infection had significantly elevated expression of chemokine receptors CCR1, CCR2 and CCR7, which mediated differential transmigration of a predominantly CD56+ (bright) subset of NK cells to the site of infection (Pokkali et al., 2009).

CCR2 is aberrantly overexpressed on chronically activated NK cells in patients with TAP-2 (transporter associated with antigen processing-2) deficiency, a disease characterized by the formation of chronic granulomas in the respiratory tract and skin. CCL2 was also elevated in serum and bronchoalveolar lavage samples from these patients, indicating that inappropriate chemo-attraction may be taking place (Hanna et al., 2005).

CCL21

CCL21 overexpression has been tested in many murine models of orthotopic tumour. In murine mammary tumours, intratumoural injection of CCL21 resulted in decreased tumour growth, increased tumour infiltration by NK cells, T cells and DCs, and prolonged survival of both control tumour-bearing mice and those undergoing surgical resection of their tumour (Ashour et al., 2007). Intratumoural injection of CCL21 into murine subcutaneously injected pancreatic tumours recruited significant numbers of immune cells, including NK cells and inhibited local tumour growth and growth of subsequently placed distant tumours, likely via an adaptive immune mechanism (Turnquist et al., 2007). Intrapulmonary injection of DCs overexpressing CCL21 into transgenic mice with a propensity to develop multiple bilateral bronchoalveolar carcinomas was associated with a marked increase in survival, reduction in tumour burden and increased NK and CD8+ T cell anti-tumour responses (Yang et al., 2006).

Other chemokines

CCL3, also known as macrophage inflammatory protein 1-alpha (MIP-1α), is critical in the murine response to cytomegalovirus infection. MIP-1α KO mice had profoundly decreased resistance to murine CMV, associated with dramatically reduced NK cell accumulation and IFN-γ secretion in the liver but not in the blood or spleen (Salazar Mather et al., 2000).

Intratumoural injection of CCL4 (also known as MIP1-β or macrophage inflammatory protein 1-β) into subcutaneously placed murine colorectal tumours significantly inhibited tumour growth and prolonged survival of tumour-bearing mice via both NK cell and T cell effects (Luo et al., 2004)

CCL6 overexpression in a mouse model of lethal septic peritonitis abolished mortality via recruitment of IFN-γ producing NK cells and DCs into the peritoneum. This protection was reversed with administration of IFN-γ neutralizing antibodies (Coelho et al., 2007).CCL27 attracts NK cells when intratumourally injected. Its receptor, CCR10, is strongly expressed on NK cells. CCL27's chemo-attractant ability is synergistically enhanced when combined with IL-12 (Gao et al., 2009).

In the placental decidua, a subset of CD16(-) NK cells expresses CXCR4, mostly resulting from stimulation by IL-15. The CXCR4 ligand CXCL12 is secreted by fetal extravillous trophoblasts attempting endovascular invasion and may play a role in recruiting NK cells to regulate the precise amount of fetal invasion into maternal tissues (Hanna et al., 2003). CXCL14, also known as BRAK, stimulates in vitro migration of activated human NK cells, without affecting their proliferation or cytotoxic activity. Almost all normal human tissues express CXCL14; however, it is absent in almost all malignant tissues, indicating that down-regulation of CXCL14 in malignant cells may confer protection from NK cell infiltration (Starnes et al., 2006).CXCL16 expression on colorectal tumours correlates with increased tumour infiltration by lymphocytes and improved cancer prognosis. The CXCL16 receptor, CXCR6, is present on NK cells (Hojo et al., 2007).

Other cytokine effects

IL-4 appears to inhibit IL-2 induced activation of NK cells at low concentrations but to enhance it at higher concentrations and synergizes with IL-2 to expand tumour-infiltrating lymphocytes (Kawakami et al., 1988, 1989). Exposure of NK cells to IL-4 preferentially induces the NK2 subset, which secretes IL-5 and IL-13 and is less susceptible to FasL-induced apoptosis (Peritt et al., 1998). IL-4 stimulation results in an increase in murine NK cell IFN-γ production via interaction with the Type 1 IL-4 receptor. IL-13, which acts via the Type 2 IL-4 receptor, causes a decrease in basal IFN-γ production (Morris et al., 2006). Stimulation of DCs with IL-4 confers upon them the ability to increase subsequent activation of NK cells but not T cells (Terme et al., 2004).

IL-7 enhances IL-15 responsiveness of bone marrow NK progenitor cells, along with stem cell factor (SCF) and flt3 ligand. SCF augments NK cell expansion when in combination with IL-2 or IL-15 but has no stand-alone effect (Williams et al., 1999). Flt3 ligand shares strong homology with SCF and also appears to be necessary for expansion and maturation of NK cell populations, as flt3L KO mice exhibit absent or very few mature NK cells (McKenna et al., 1996), but on its own is insufficient to induce Class I MHC receptors (Williams et al., 1999).

IL-10 potently inhibits NK cell proliferation and IFN-γ production in a manner not reversed by supraphysiologic exogenous IL-2 (Bejarano et al., 1992; Tripp et al., 1993). However, IL-10 produced by differentiated regulatory DCs in response to bacterial lipopolysaccharide (LPS) stimulation causes an increase in NK cytotoxicity, leading to killing of those surrounding DCs (Qian et al., 2006).

IL-28, a member of the extended IFN family and also known as IFN-λ, increases the number of splenic NK cells in SCID mice and enhances IL-12 induced IFN-γ production in wild-type mice (Numasaki et al., 2007).

IL-33, an IL-1 family member, induces IFN-γ secretion from human NK cells in cooperation with IL-12 (Smithgall et al., 2008).

TNF-α increases NK cell IFN-γ production but decreases cytotoxicity (Orange and Biron, 1996). This effect requires costimulation with IL-12 (Tripp et al., 1993).

Transforming growth factor β (TGF-β) causes an immediate decrease in production of IFN-γ and a delayed decrease in granzyme A and B expression in the NK cell, resulting in decreased antibody-dependent cellular cytotoxicity via SMAD 3 phosphorylation and T-BET suppression (Trotta et al., 2008). Blockade of TGF-β results in significant IFN-γ production even under suboptimal stimulation conditions (Laouar et al., 2005; Meadows et al., 2006).

In the mouse, the Gas6 and protein S ligands of the Tyro3 receptor tyrosine kinase family (Tyro3, Axl and Mer) additionally stimulate NK cell growth and differentiation (Caraux et al., 2006a).

Effector signalling pathways

A number of individual downstream signalling pathways have been shown to be important mediators of cytokine signals on the NK cell. In particular, several of the JAK/STAT pathways, phospholipase C-γ, p38 MAP kinase, ERK-dependent pathways and protein kinase C-θ mediate cytokine signals that are important for NK cell activation and function.

JAK/STAT

Many interactions of the cytokine profile with the JAK/STAT pathways have been previously mentioned. In addition, at low doses, IL-2 stimulates phosphorylation of STAT5 in the CD56+ (bright) subset of human NK cells (Wendt et al., 2007). IL-12 increases phosphorylation of STAT1 and STAT4 (Zhang et al., 2008). IL-21 preferentially increases phosphorylation of STAT3 (Wendt et al., 2007). Upon stimulation of the IFNAR, STAT1/STAT2 heterodimers complex with a p48 protein known as IFN response factor 9 (IRF9), to form the IFN-stimulated gene factor 3 (ISGF-3). ISGF-3 induces transcription as a result of recognizing IFN stimulated response elements (ISREs) in promoter regions of IFN-responsive genes (Biron, 2001).

Other important signalling components

NK cells deficient in phospholipase C-γ2 (PLC-γ2) are unable to secrete cytotoxic granules when stimulated, due to defective calcium mobilization. Despite this, secretion of IFN-γ upon stimulation with IL-12 is unaffected (Caraux et al., 2006b). PLC-γ2 deficient NK cell precursors are significantly impaired in Ly49 acquisition and terminal maturation. Overexpression of PLC-γ1 in PLC-γ2 deficient cells restored Ly49 acquisition but could only partially rescue NKG2D-mediated cytotoxicity and resulted in no cytokine production whatsoever (Regunathan et al., 2006).

The p38 MAP kinase pathway transduces signals involved in release and reception of IL-12 (Bose and Baral, 2007). IL-12 also increases phosphorylation of ERK1/2 (Zhang et al., 2008).

NK cells from mice deficient in protein kinase C-θ have a diminished ability to secrete IFN-γ in response to IL-12. However, IL-18 can still elicit a normal IFN-γ response from these cells, and this differential seems to be regulated at a post-transcriptional level (Page et al., 2008).

Conclusion

The response of NK cells to the cytokine signals found in their microenvironment appears to be dependent upon many factors: the specific combination and concentration of secreted cytokines, the expression of receptors on the NK cell itself, the complement of neighbouring immune and stromal cells, the availability of cell–cell contact and the pathologic process in question, whether inflammatory, infectious, malignant or even autoimmune. Although many of the cytokine signalling pathways and responses have yet to be fully elicited, it is clear that NK cell function is dependent on a plethora of signals in its surrounding milieu, interacting jointly to produce the desired response, which itself results in addition to the cytokine medley in the surrounding microenvironment.

References

Adachi, O., Kawai, T., Takeda, K., Matsumoto, M., Tsutsui, H., Sakagami, M., Nakanishi, K. and Akira, S. (1998). Targeted disruption of the MyD88 gene results in loss of IL-1- and IL-18-mediated function. Immunity **9**, 143–150.

Ahmad, A., Sharif-Askari, E., Fawaz, L. and Menezes, J. (2000). Innate immune response of the human host to exposure with herpes simplex virus type 1: in vitro control of the virus infection by enhanced natural killer activity via interleukin-15 induction. J Virol **74**, 7196–7203.

Ajuebor, M.N., Wondimu, Z., Hogaboam, C.M., Le, T., Proudfoot, A.E. and Swain, M.G. (2007). CCR5 deficiency drives enhanced natural killer cell trafficking to and activation within the liver in murine T cell-mediated hepatitis. Am J Pathol **170**, 1975–1988.

Akira, S. (2000). The role of IL-18 in innate immunity. Curr Opin Immunol **12**, 59–63.

Ashour, A.E., Lin, X., Wang, X., Turnquist, H.R., Burns, N.M., Tuli, A., Sadanandam, A., Suleiman, K., Singh, R.K., Talmadge, J.E. and Solheim, J.C. (2007). CCL21 is an effective surgical neoadjuvant for treatment of mammary tumors. Cancer Biol Ther **6**, 1206–1210.

Barr, D., Belz, G., Reading, P., Wojtasiak, M., Whitney, P., Heath, W., Carbone, F. and Brooks, A. (2007). A role for plasmacytoid dendritic cells in the rapid IL-18-dependent activation of NK cells following HSV-1 infection. Eur J Immunol **37**, 1334–1342.

Biron, C. (2001). Interferons alpha and beta as immune regulators – a new look. Immunity **14**, 661–664.

Bejarano, M., de Waal Malefyt, R., Abrams, J., Bigler, M., Bacchetta, R., de Vries, J. and Roncarolo, M. (1992). Interleukin 10 inhibits allogeneic proliferative and cytotoxic T cell responses generated in primary mixed lymphocyte cultures. Int Immunol **4**, 1389–1397.

Berahovich, R.D., Lai, N.L., Wei, Z., Lanier, L.L. and Schall, T.J. (2006). Evidence for NK cell subsets based on chemokine receptor expression. J Immunol **177**, 7833–7840.

Bose, A. and Baral, R. (2007). Natural killer cell mediated cytotoxicity of tumor cells initiated by neem leaf preparation is associated with CD40-CD40L-mediated endogenous production of interleukin-12. Hum Immunol **68**, 823–831.

Burkett, P., Koka, R., Chien, M., Chai, S., Boone, D. and Ma, A. (2004). Coordinate expression and trans presentation of interleukin (IL)-15Ralpha and IL-15 supports natural killer cell and memory CD8+ T cell homeostasis. J Exp Med **200**, 825–834.

Cao, X., Shores, E., Hu-Li, J., Anver, M., Kelsall, B., Russell, S., Drago, J., Noguchi, M., Grinberg, A., Bloom, E., et al. (1995). Defective lymphoid development in mice lacking expression of the common cytokine receptor gamma chain. Immunity **2**, 223–238.

Caraux, A., Lu, Q., Fernandez, N., Riou, S., DiSanto, J., Raulet, D., Lemke, G. and Roth, C. (2006a). Natural killer cell differentiation driven by Tyro3 receptor tyrosine kinases. Nat Immunol **7**, 747–754.

Caraux, A., Kim, N., Bell, S., Zompi, S., Ranson, T., Lesjean-Pottier, S., Garcia-Ojeda, M., Turner, M. and Colucci, F. (2006b). Phospholipase C-gamma2 is essential for NK cell cytotoxicity and innate immunity to malignant and virally infected cells. Blood **107**, 994–1002.

Carr, D.J., Wuest, T. and Ash, J. (2008). An increase in herpes simplex virus type 1 in the anterior segment of the eye is linked to a deficiency of NK cell infiltration in mice deficient in CXCR3. J Interferon Cytokine Res **28**, 245–251.

Carson, W., Giri, J., Lindemann, M., Linett, M., Ahdieh, M., Paxton, R., Anderson, D., Eisemann, J., Grabstein, K. and Caligiuri, M. (1994). Interleukin (IL) 15 is a novel cytokine that activates human natural killer cells via components of the IL-2 receptor. J Exp Med **180**, 1395–1403.

Chaix, J., Tessmer, M., Hoebe, K., Fuseri, N., Ryffel, B., Dalod, M., Alexopoulou, L., Beutler, B., Brossay, L., Vivier, E. and Walzer, T. (2008). Cutting edge: priming of NK cell by IL-18. J Immunol **181**, 1627–1631.

Chan, S., Perussia, B., Gupta, J., Kobayashi, M., Pospisil, M., Young, H., Wolf, S., Clark, S. and Trinchieri, G. (1991). Induction of interferon gamma production by natural killer cell stimulatory factor: characteristics of the responder cells and synergy with other inducers. J Exp Med **173**, 869–879.

Chiyo, M., Shimozato, O., Yu, L., Kawamura, K., Iizasa, T., Fujisawa, T. and Tagawa, M. (2005). Expression of IL-27 in murine carcinoma cells produces antitumor effects and induces protective immunity in inoculated host animals. Int J Cancer **115**, 437–442.

Coelho, A.L., Schaller, M.A., Benjamin, C. F., Orlofsky, A.Z., Hogaboam, C.M. and Kunkel, S.L. (2007). The chemokines CCL6 promotes innate immunity via immune cell activation and recruitment. J Immunol **179**, 5474–5482.

Cooper, M., Bush, J., Fehniger, T., VanDeusen, J., Waite, R., Liu, Y., Aguila, H. and Caligiuri, M. (2002). In vivo evidence for a dependence on interleukin 15 for survival of natural killer cells. Blood **100**, 3633–3638.

Davis, I.D., Skrumsager, B.K., Cebon, J., Nicholaou, T., Barlow, J.W., Moller, N.P., Skak, K., Lundsgaard, D., Frederiksen, K.S., Thygesen, O. and McArthur, G.A. (2007). An open-label, two-arm, phase I trial of recombinant human interleukin-21 in patients with metastatic melanoma. Clin Cancer Res **13**, 3630–3636.

DiSanto, J., Muller, W., Guy-Grand, D., Fischer, A. and Rajewsky, K. (1995). Lymphoid development in mice with a targeted deletion of the interleukin 2 receptor gamma chain. Proc Natl Acad Sci U S A **92**, 377–381.

Dunn, C., Brunetto, M., Reynolds, G., Christophides, T., Kennedy, P.T., Lampertico, P., Das, A., Lopes, A.R., Borrow, P., Williams, K., Humphreys, E., Afford, S., Adams, D.H., Bertoletti, A. and Maini, M.K. (2007). Cytokines induced during chronic hepatitis B virus infection promote a pathway for NK cell-mediated liver damage. J Exp Med **204**, 667–680.

Eriksen, K.W., Sondergaard, H., Woetmann, A., Krejsgaard, T., Skak, K., Geisler, C., Wasik, M.A. and Odum, N. (2008). The combination of IL-21 and IFN-alpha boosts STAT3 activation, cytotoxicity and experimental tumor therapy. Mol Immunol **46**, 812–820.

Fehniger, T., Cooper, M., Nuovo, G., Cella, M., Facchetti, F., Colonna, M. and Caligiuri, M. (2003). CD56bright natural killer cells are present in human lymph nodes and are activated by T cell-derived IL-2: a potential new link between adaptive and innate immunity. Blood **101**, 3052–3057.

Ferlazzo, G. and Munz, C. (2004). NK cell compartments and their activation by dendritic cells. J Immunol **172**, 1333–1339.

Ferlazzo, G., Pack, M., Thomas, D., Paludan, C., Schmid, D., Strowig, T., Bougras, G., Muller, W., Moretta, L. and Munz, C. (2004). Distinct roles of IL-12 and IL-15 in human natural killer cell activation by dendritic cells from secondary lymphoid organs. Proc Natl Acad Sci U S A **101**, 16606–16611.

Figueiredo , F., Commodaro, A.G., de Camargo, M.M., Rizzo, L.V. and Belfort, R. Jr. (2007). NK1.1 cells downregulate murine endotoxin-induced uveitis following intraocular administration of interleukin-12. Scand J Immunol **66**, 329–334.

Frederiksen, K.S., Lundsgaard, D., Freeman, J.A., Hughes, S.D., Holm, T. L., Skrumsager, B.K., Petri, A., Hansen, L.T., McArthur, G.A., Davis, I.D. and Skak, K. (2008). IL-21 induces in vivo immune activation of NK cells and CD8(+) T cells in patients with metastatic melanoma and renal cell carcinoma. Cancer Immunol Immunother **57**, 1439–1449.

Gao, J., Tsuda, Y., Han, M., Xu, S., Kanagawa, N., Hatanaka, Y., tani, Y., Mizuguchi, H., Tsutsumi, Y., Mayumi, T., Okada, N. and Nakagawa, S. (2009). NK cells are migrated and indispensable in the anti-tumor activity induced by CCL27 gene therapy. Cancer Immunol Immunother **58**, 291–299.

Giuliani, M., Giron-Michel, J., Negrini, S., Vacca, P., Durali, D., Caignard, A., Le Bousse-Kerdiles, C., Chouaib, S., Devocelle, A., Bahri, R., Durrbach, A., Taoufik, Y., Ferrini, S., Croce, M., Mingari, M.C., Moretta, L. and Azzarone, B. (2008). Generation of a novel regulatory Nk cell subset from peripheral blood CD34+ progenitors promoted by membrane-bound IL-15. PLoS One **3**, e2441.

Gosselin, J., Tomolu, A., Gallo, R. and Flamand, L. (1999). Interleukin-15 as an activator of natural killer cell-mediated antiviral response. Blood **94**, 4210–4219.

Grant, L., Yao, Z., Hedrich, C., Wang, F., Moorthy, A., Wilson, K., Ranatunga, D. and Bream, J. (2008). Stat4-dependent, T-bet-independent regulation of IL-10 in NK cells. Genes Immun **9**, 316–327.

Gregoire, C., Cognet, C., Chasson, L., Coupet, C.A., Dalod, M., Reboldi, A., Marvel, J., Sallusto, F., Vivier, E. and Walzer, T. (2008). Intrasplenic trafficking of natural killer cells is redirected by chemokines upon inflammation. Eur J Immunol **38**, 2076–2084.

Guia, S., Cognet, C., de Beaucoudrey, L., Tessmer, M., Jouanguy, E., Berger, C., Filipe-Santos, O., Feinberg, J., Camcioglu, Y., Al Jumaar, S., Al-Hajjar, S., Stephan, J., Fieschi, C., Abel, L., Brossay, L., Casanova, J. and Vivier, E. (2008). A role for interleukin-12/23 in the maturation of human natural killer and Cd56+ T cells in vivo. Blood **111**, 5008–5016.

Hanna, J., Wald, O., Goldman-Wohl, D., Prus, D., Markel, G., Gazit, R., Katz, G., Haimov-Kochman, R., Fujii, N., Yagel, S., Peled, A. and Mandelboim, O. (2003). CXCL12 expression by invasive trophoblasts induces the specific migration of CD16- human natural killer cells. Blood **102**, 1569–1577.

Hanna, J., Mussaffi, H., Steuer, G., Hanna, S., Deeb, M., Blau, H., Arnon, T. I., Weizman, N. and Mandelboim, O. (2005). Functional aberrant expression of CCR2 receptor on chronically activated NK cells in patients with TAP-2 deficiency. Blood **106**, 3465–3473.

Hojo, S., Koizumi, K., Tsuneyama, K., Arita, Y., Cui, Z., Shinohara, K., Minami, T., Hasimoto, I., Nakayama, T., Sakurai, H., Takano, Y., Yoshie, O., Tsukada, K. and Saiki, I. (2007). High-level expression of chemokines CXCL16 by tumor cells correlates with a good prognosis and increased tumor-infiltrating lymphocytes in colorectal cancer. Cancer Res **67**, 4725–4731.

Hokeness, K., Kuziel, W., Biron, C. and Salazar-Mather, T. (2005). Monocyte chemoattractant protein-1 and CCR2 interactions are required for IFN-alpha/beta-induced inflammatory responses and antiviral defense in liver. J Immunol **174**, 1549–1556.

Huang, D., Shi, F.D., Jung, S., Pien, G.C., Wang, J., Salazar-Mather, T.P., He, T. T., Weaver, J.T., Ljunggren, H.G., Biron, C.A., Littman, D.R. and Ransohoff, R. M. (2006). The neuronal chemokines CX3CL1/fractalkine selectively recruits NK cells that modify experimental autoimmune encephalitis within the central nervous system. FASEB J **20**, 896–905.

Hyakudomi, M., Matsubara, T., Hyakudomi, R., Yamamoto, T., Kinugasa, S., Yamanoi, A., Maruyama, R. and Tanaka, T. (2008). Increased expression of fractalkine is correlated with a better prognosis and an increased number of both CD8+ T cells and natural killer cells in gastric adenocarcinoma. Ann Surg Oncol **15**, 1775–1782.

Hyodo, Y., Matsui, K., Hayashi, N., Tsutsui, H., Kasiwamura, S., Yamauchi, H., Hiroishi, K., Takeda, K., Tagawa, Y., Iwakura, Y., Kayagaki, N., Kurimoto, M., Okamura, H., Hada, T., Yagita, H., Akira, S., Nakanishi, K. and Higashino, K. (1999). IL-18 upregulates perforin-mediated NK activity without increasing perforin messenger RNA expression by binding to constitutively expressed IL-18 receptor. J Immunol **162**, 1662–1668.

Jamieson, A., Isnard, P., Dorfman, J., Coles, M. and Raulet, D. (2004). Turnover and proliferation of NK cells in steady state and lymphopenic conditions. J Immunol **172**, 864–870.

Jiang, Y., Zhang, Z., Diao, Y., Jin, X., Shi, W., Geng, W., Dai, D., Zhang, M., Han, X., Liu, J., Wang, y. and Shang, H. (2008). Expression of chemokines receptors on natural killer cells in HIV-infected individuals. Cell Immunol **251**, 19–24.

Karlmark, K.R., Wasmuth, H.E., Trautwien, C. and Tacke, F. (2008). Chemokine-directed immune cell infiltration in acute and chronic liver disease. Expert Rev Gastroenterol Hepatol **2**, 233–242.

Kawakami, Y., Rosenberg, S.A. and Lotze, M.T. (1988). Interleukin 4 promotes the growth of tumor-infiltrating lymphocytes cytotoxic for human autologous melanoma. J Exp Med **168**, 2183–2191.

Kawakami, Y., Custer, M.C., Rosenberg, S. A. and Lotze, M.T. (1989). IL-4 regulates IL-2 induction of lymphokine-activated killer activity from human lymphocytes. J Immunol **142**, 3452–3461.

Kennedy, M., Glaccum, M., Brown, S., Butz, E., Viney, J., Embers, M., Matsuki, N., Charrier, K., Sedger, L., Willis, C., Brasel, K., Morrissey, P., Stocking, K., Schuh, J., Joyce, S. and Peschon, J. (2000). Reversible defects in natural killer and memory CD8 T cell lineages in interleukin 15-deficient mice. J Exp Med **191**, 753–756.

Khan, I.A., Thomas, S.Y., Moretto, M. M., Lee, F.S., Islam, S.A., Combe, C., Schwartzman, J.D. and Luster, A.D. (2006). CCR5 is essential for NK cell trafficking and host survival following *Toxoplasma gondii* infection. PLoS Pathog **2**, e49.

Kobayashi, M., Fitz, L., Ryan, M., Hewick, R., Clark, S., Chan, S., Loudon, R., Sherman, F., Perussia, B. and Trinchieri, G. (1989). Identification and purification of natural killer cell stimulatory factor (NKSF), a cytokine with multiple biologic effects on human lymphocytes. J Exp Med **170**, 827–845.

Kojima, H., Takeuchi, M., Ohta, T., Nishida, Y., Arai, N., Ikeda, M., Ikegami, H. and Kurimoto, M. (1998). Interleukin-18 activates the IRAK-TRAF6 pathway in mouse EL-4 cells. Biochem Biophys Res Commun **244**, 183–186.

Laouar, Y., Sutterwala, F.S., Gorelik, L. and Flavell, R.A. (2005). Transforming growth factor-beta controls T helper type 1 cell development through regulation of natural killer cell interferon-gamma. Nat Immunol **6**, 600–607.

Liu, L., Wang, S., Shan, B., Shao, L., Sato, A., Kawamura, K., Li, Q., Ma, G. and Tagawa, M. (2008). IL-27-mediated activation of natural killer cells and inflammation produced antitumor effects for human oesophageal carcinoma cells. Scand J Immunol **68**, 22–29.

Lodolce, J., Boone, D., Chai, S., Swain, R., Dassopoulos, T., Trettin, S. and Ma, A. (1998). IL-15 receptor maintains lymphoid homeostasis by supporting lymphocyte homing and proliferation. Immunity **9**, 669–676.

London, L., Perussia, B. and Trinchieri, G. (1986). Induction of proliferation in vitro of resting human natural killer cells: IL 2 induces into cell cycle most peripheral blood NK cells, but only a minor subset of low density T cells. J Immunol **137**, 3845–3854.

Luo, X., Yu, Y., Liang, A., Xie, Y., Liu, S., Guo, J., Wang, W., Qi, R., An, H., Zhang, M., Xu, H., Guo, Z. and Cao, X. (2004). Intratumoral expression of MIP-1beta induces antitumor responses in a pre-established tumor model through chemoattracting T cells and NK cells. Cell Mol Immunol **1**, 199–204.

Mailliard, R.B., Alber, S.M., Shen, H., Watkins, S.C., Kirkwood, J.M., Herberman, R.B. and Kalinski, P. (2005). IL-18-induced CD83+ CCR7+ NK helper cells. J Exp Med **202**, 941–953.

McKenna, H., Miller, R., Brasel, K., Maraskovsky, E., Maliszewski, C., Pulendran, B., Lynch, D., Teepe, M., Roux, E., Smith, J., et al. (1996). Targeted disruption of the flt3 ligand gene in mice affects multiple hematopoietic lineages, including natural killer cells, B lymphocytes, and dendritic cells. Blood **88**, 474a.

Meadows, S.K., Eriksson, M., Barber, A. and Sentman, C.L. (2006). Human NK cell IFN-gamma production is regulated by endogenous TGF-beta. Int Immunopharmacol **6**, 1020–1028.

Minagawa, M., Watanabe, H., Miyaji, C., Tomiyama, K., Shimura, H., Ito, A., Ito, M., Domen, J., Weissman, I. and Kawai, K. (2002). Enforced expression of Bcl-2 restores the number of NK cells, but does not rescue the impaired development of NKT cells or intraepithelial lymphocytes, in IL-2/IL-15 receptor beta-chain-deficient mice. J Immunol **169**, 4153–4160.

Morris, S.C., Orekhova, T., Meadows, M.J., Heidorn, S.M., Yang, J. and Finkelman, F.D. (2006). IL-4 induces in vivo production of IFN-gamma by NK and NKT cells. J Immunol **176**, 5299–5305.

Munz, C., Dao, T., Ferlazzo, G., de Cos, M., Goodman, K. and Young, J. (2005). Mature myeloid dendritic cell subsets have distinct roles for activation and viability of circulating human natural killer cells. Blood **105**, 266–273.

Nguyen, K., Cousens, L., Doughty, L., Pien, G., Durbin, J. and Biron, C. (2000). Interferon alpha/beta-mediated inhibition and promotion of interferon gamma: STAT1 resolves a paradox. Nat Immunol **1**, 70–76.

Nguyen, K., Salazar-Mather, T., Dalod, M., Van Deusen, J., Wei, X., Liew, F., Caligiuri, M., Durbin, J. and Biron, C. (2002). Coordinated and distinct roles for IFN-alpha beta, IL-12 and IL-15 regulation of NK cell responses to viral infection. J Immunol **169**, 4279–4287.

Nishioka, Y., Nishimura, N., Suzuki, Y. and Sone, S. (2001). Human monocyte-derived and CD83(+) blood dendritic cells enhance NK cell-mediated cytotoxicity. Eur J Immunol **31**, 2633–2641.

Numasaki, M., Tagawa, M., Iwata, F., Suzuki, T., Nakamura, A., Okada, M., Iwakura, Y., Aiba, S. and Yamaya, M. (2007). IL-28 elicits antitumor responses against murine fibrosarcoma. J Immunol **178**, 5086–5098.

Oniki, S., Nagai, H., Horikawa, T., Furukawa, J., Belladonna, M.L., Yoshimoto, T., hara, I. and Nishigori, C. (2006). Interleukin-23 and interleukin-27 exert quite different antitumor and vaccine effects on poorly immunogenic melanoma. Cancer Res **66**, 6395–6404.

Orange, J. and Biron, C. (1996). Characterization of early IL-12, IFN-alphabeta, and TNF effects on antiviral state and NK cell responses during murine cytomegalovirus infection. J Immunol **156**, 4746–4756.

Ottoviani, C., Nasorri, F., Bedini, C., de Pita, O., Girolomoni, G. and Cavani, A. (2006). CD56brightCD16(−) NK cells accumulate in psoriatic skin in response to CXCL10 and CCL5 and exacerbate skin inflammation. Eur J Immunol **36**, 118–128.

Page, K., Chaudhary, D., Goldman, S. and Kasaian, M. (2008). Natural killer cells from protein kinase C theta −/− mice stimulated with interleukin-12 are deficient in production of interferon-gamma. J Leukoc Biol **83**, 1267–1276.

Pallandre, J.R., Krzewski, K., Bedel, R., Ryffel, B., Caignard, A., Rohrlich, P.S., Pivot, X., Tiberghien, P., Zitvogel, L., Strominger, J.L. and Borg, C. (2008). Dendritic cell and natural killer cell cross-talk: a pivotal role of CX3CL1 in NK cytoskeleton organization and activation. Blood **112**, 4420–4424.

Peritt, D., Robertson, S., Gri, G., Showe, L., Aste-Amezaga, M. and Trinchieri, G. (1998). Differentiation of human NK cells into NK1 and NK2 subsets. J Immunol **161**, 5821–5824.

Pien, G., Satoskar, A., Takeda, K., Akira, S. and Biron, C. (2000). Cutting edge: selective IL-18 requirements for induction of compartmental IFN-gamma responses during viral infection. J Immunol **165**, 4787–4791.

Pokkali, S., Das, S.D. and Selvaraj, A. (2009). Differential upregulation of chemokines receptors on CD56(+) NK cells and their transmigration to the site of infection in tuberculous pleurisy. FEMS Immunol Med Microbiol **55**, 352–360.

Prlic, M., Blazar, B., Farrar, M. and Jameson, S. (2003). In vivo survival and homeostatic proliferation of natural killer cells. J Exp Med **197**, 967–976.

Qian, C., Jiang, X., An, H., Yu, Y., Guo, Z., Liu, S., Xu, H. and Cao, X. (2006). TLR agonists promote ERK-mediated preferential IL-10 production of regulatory dendritic cells (diffDCs), leading to NK cell activiation. Blood **108**, 2307–2315.

Ranson, T., Vosshenrich, C., Corcuff, E., Richard, O., Muller, W. and di Santo, J. (2003). IL-15 is an essential mediator of peripheral NK-cell homeostasis. Blood **101**, 4887–4893.

Regunathan, J., Chen, y., Jutlesa, S., Dai, X., Bai, L., Wen, R., Wang, D. and Malarkannan, S. (2006). Differential and nonredundant roles of phospholipase Cgamma2 and phospholipase Cgamma1 in the terminal maturation of NK cells. J Immunol **177**, 5365–5376.

Roda, J.M., Parihar, R., Lehman, A., Mani, A., Tridandapani, S. and Carson, W.E. 3rd. (2006). Interleukin-21 enhances NK cell activation in response to antibody-coated targets. J Immunol **177**, 120–129.

Roda, J.M., Joshi, T., Butchar, J.P., McAlees, J.W., Lehman, A., Tridandapani, S. and Carson, W.E. 3rd. (2007). The activation of natural killer cell effector functions by cetuximab-coated, epidermal growth factor receptor positive tumor cells is enhanced by cytokines. Clin Cancer Res **13**, 6419–6428.

Sakthivel, S.K., Singh, U.P., Singh, S., Taub, D.D., Novakovic, K.R. and Lillard, J.W. Jr. (2008). CXCL10 blockade protects mice from cyclophosphamide-induced cystitis. J Immune Based Ther Vaccines **6**, 6.

Salazar-Mather, T. and Hokeness, K. (2006). Calling in the troops: regulation of inflammatory cell trafficking through innate cytokine/chemokines networks. Viral Immunol **16**, 291–306.

Salazar-Mather, T., Hamilton, T. and Biron, C. (2000). A chemokine-to-cytokine-to-chemokine cascade critical in antiviral defense. J Clin Invest **105**, 985–993.

Salcedo, T., Azzoni, L., Wolf, S. and Perussia, B. (1993). Modulation of perforin and granzyme messenger RNA expression in human natural killer cells. J Immunol **151**, 2511–2520.

Salcedo, R., Stauffer, J.K., Lincoln, E., Back, T.C., Hixon, J.A., Hahn, C., Shafer-Weaver, K., Malyguine, A., Kastelein, R. and Wigginton, J.M. (2004). IL-27 mediates complete regression of orthotopic primary and metastatic murine neuroblastoma tumors: role for CD8+ T cells. J Immunol **173**, 7170–7182.

Santoni, A., Carlino, C., Stabile, H. and Gismondi, A. (2008). Mechanisms underlying recruitment and accumulation of decidual NK cells in uterus during pregnancy. Am J Reprod Immunol **59**, 417–424.

Sayos, J., Nguyen, K., Wu, C., Stepp, S., Howie, D., Schatzle, J., Kumar, V., Biron, C. and Terhorst, C. (2000). Potential pathways for regulation of NK and T cell responses: differential X-linked lymphoproliferative syndrome gene product SAP interactions with SLAM and 2B4. Int Immunol **12**, 1749–1757.

Schwartz, Y., Avraham, R., Benish, M., Rosenne, E. and Ben-Eliyahu, S. (2008). Prophylactic IL-12 treatment reduces postoperative metastasis: mediation by increased numbers but not cytotoxicity of NK cells. Breast Cancer Res Treat **107**, 211–223.

Semino, C., Angelini, G., Poggi, A. and Rubartelli, A. (2005). NK/iDC interaction results in IL-18 secretion by DCs at the synaptic cleft followed by NK cell activation and release of the DC maturation factor HMGB1. Blood **106**, 609–616.

Semino, C., Ceccarelli, J., Lotti, L., Torrisi, M., Angelini, G. and Rubartelli, A. (2007). The maturation potential of NK cell clones toward autologous dendritic cells correlates with HMGB1 secretion. J Leukoc Biol **81**, 92–99.

Singh, U.P., Singh, S., Singh, R., Cong, Y., Taub, D.D. and Lillard, J.W. Jr. (2008). CXCL10-producing mucosal CD4+ T cells, NK cells, and NKT cells are associated with chronic colitis in IL-10 (−/−) mice, which can be abrogated by anti-CXCL10 antibody inhibition. J Interferon Cytokine Res **28**, 31–43.

Skak, K., Frederiksen, K.S. and Lundsgaard, D. (2008). Interleukin-21 activates human natural killer cells and modulates their surface receptor expression. Immunology **123**, 575–583.

Smithgall, M., Comeau, M., Yoon, B., Kaufman, D., Armitage, R. and Smith, D. (2008). IL-33 amplifies both Th1- and Th2-type responses through its activity on human basophils, allergen-reactive Th2 cell, iNKT and NK cells. Int Immunol **20**, 1019–1030.

Sporri, R., Joller, N., Hilbi, H. and Oxenius, A. (2008). A novel role for neutrophils as critical activators of NK cells. J Immunol **181**, 7121–7130.

Starnes, T., Rasila, K.K., Robertson, M.J., Brahmi, Z., Dahl, R., Christopherson, K. and Hromas, R. (2006). The chemokines CXCL14 (BRAK) stimulates activated NK cell migration: implications for the downregulation of CXCL14 in malignancy. Exp Hematol **34**, 1101–1105.

Strbo, N., de Armas, L., Liu, H., Kolber, M.A., Lechtenheld, M. and Pahwa, S. (2008). IL-21 augments natural killer effector functions in chronical HIV-infected individuals. AIDS **22**, 1551–1560.

Suzuki, H., Duncan, G., Takimoto, H. and Mak, T. (1997). Abnormal development of intestinal intraepithelial lymphocytes and peripheral natural killer cells in mice lacking the IL-2 receptor beta chain. J Exp Med **185**, 499–505.

Swann, J., Hayakawa, Y., Zerafa, N., Sheehan, K., Scott, B., Schreiber, R., Hertzog, P. and Smyth, M. (2007). Type I IFN contributes to NK cell homeostasis, activation, and antitumor function. J Immunol **178**, 7540–7549.

Takeda, K., Tsutsui, H., Yoshimoto, T., Adachi, O., Yoshida, N., Kishimoto, T., Okamura, H., Nakanishi, K. and Akira, S. (1998). Defective NK cell activity and Th1 response in IL-18-deficient mice. Immunity **8**, 383–390.

Terme, M., Tomasello, E., Maruyama, K., Crepineau, F., Chaput, N., Flament, C., Marolleau, J., Angevin, E., Wagner, E., Salomon, B., Lemmonier, F., Wakasugi, H., Colonna, M., Vivier, E. and Zitvogel, L. (2004). IL-4 confers NK stimulatory capacity to murine dendritic cells: a signaling pathway involving KARAP/DAP12-triggering receptor expressed on myeloid cell 2 molecules. J Immunol **172**, 5957–5966.

Thapa, M., Kuziel, W.A. and Carr, D.J. (2007). Susceptibility of CCR5-deficient mice to genital herpes simplex virus type 2 is linked to NK cell mobilization. J Virol **81**, 3704–3713.

Thapa, M., Welner, R.S., Pelayo, R. and Carr, D.J. (2008). CXCL9 and CXCL10 expression are critical for control of genital herpes simplex virus type 2 infection through mobilization of HSV-specific CTL and NK cells to the nervous system. J Immunol **180**, 1098–1106.

Trinchieri, G. and Santoli, D. (1978). Anti-viral activity induced by culturing lymphocytes with tumor-derived or virus-transformed cell. Enhancement of human natural killer cell activity by interferon and antagonistic inhibition of susceptibility of target cells to lysis. J Exp Med **147**, 1314–1333.

Trinchieri, G., Matsumoto-Kobayashi, M., Clark, S., Seehra, J., London, L. and Perussia, B. (1984). Response of resting human peripheral blood natural killer cells to interleukin 2. J Exp Med **160**, 1147–1169.

Tripp, C., Wolf, S. and Unanue, E. (1993). Interleukin 12 and tumor necrosis factor alpha are costimulators of interferon gamma production by natural killer cells in severe combined immunodeficiency mice with listeriosis, and interleukin 10 is a physiologic antagonist. Proc Natl Acad Sci U S A **90**, 3725–3729.

Trotta 2nd, R., Col, J.D., Yu, J., Ciarlariello, D., Thomas, B., Zhang, X., Allard, J., Wei, M., Mao, H., Byrd, J.C., Perrotti, D. and Caligiuri, M.A. (2008). TGF-beta utilizes SMAD3 to inhibit CD16-mediated IFN-gamma production and antibody-dependent cellular cytoxicity in human NK cells. J Immunol **181**, 3784–3792.

Tsutsui, H., Nakanishi, K., Matsui, K., Higashino, K., Okamura, H., Miyazawa, Y. and Kaneda, K. (1996). IFN-g-inducing factor up-regulates Fas ligand-mediated cytotoxic activity of murine natural killer cell clones. J Immunol **157**, 3967–3973.

Tsutsui, H., Matsui, K., Kawada, N., Hyodo, Y., Hayashi, N., Okamura, H., Higashino, K. and Nakanishi, K. (1997). IL-18 accounts for both TNF-a- and FasL-mediated hepatotoxic pathways in endotoxin-induced liver injury. J Immunol **159**, 3961.

Turnquist, H.R., Lin, X., Ashour, A.E., Hollingsworth, M.A., Singh, R.K., Talmadge, J.E. and Solheim, J.C. (2007). CCL21 induces extensive intratumoral immune cell infiltration and specific anti-tumor cellular immunity. Int J Oncol **30**, 631–639.

Villarino, A.V., Larkin, J., 3rd., Saris, C. J., Caton, A.J., Lucas, S., Wong, T., de Sauvage, F.J. and Hunter, C.A. (2005). Positive and negative regulation of the IL-27 receptor during lymphoid cell activation. J Immunol **174**, 7684–7691.

Wald, O., Weiss, I.D., Wald, H., Shoham, H., Bar-Shavit, Y., Beider, K., Galun, E., Weiss, L., Flaishon, L., Shachar, I., Nagler, A., Lu, B., Gerard, C., Gao, J. L., Mishani, E., Farber, J. and Peled, A. (2006). IFN-gamma acts on T cells to induce NK cell mobilization and accumulation in target organs. J Immunol **176**, 4716–4729.

Waldmann, T. and Tagaya, Y. (1999). The multifaceted regulation of interleukin-15 expression and the role of this cytokine in NK cell differentiation and host response to intracellular pathogens. Annu Rev Immunol **17**, 19–49.

Walser, T.C., Rifat, S., Ma, X., Kundu, N., Ward, C., Goloubeva, O., Johnson, M.G., Medina, J.C., Collins, T.L. and Fulton, A.M. (2006). Antagonism of CXCR3 inhibits lung metastasis in a murine model of metastatic breast cancer. Cancer Res **66**, 7701–7707.

Walser, T.C., Ma, X., Kundu, N., Dorsey, R., Goloubeva, O. and Fulton, A.M. (2007). Immune-mediated modulation of breast cancer growth and metastasis by the chemokines Mig (CXCL9) in a murine model. J Immunother **30**, 490–498.

Wendel, M., Galani, I.E., Suri-Payer, E. and Cerwenka, A. (2008). Natural killer cell accumulation in tumors is dependent of IFN-gamma and CXCR3 ligands. Cancer Res **68**, 8437–8445.

Wendt, K., Wilk, E., Buyny, S., Schmidt, R. and Jacobs, R. (2007). Interleukin-21 differentially affects human natural killer cell subsets. Immunology **122**, 486–495.

Willerford, D., Chen, J., Ferry, J., Davidson, L., Ma, A. and Alt, F. (1995). Interleukin-2 receptor alpha chain regulates the size and content of the peripheral lymphoid compartment. Immunity **3**, 521–530.

Williams, N., Klem, J., Puzanov, I., Sivakumar, O., Bennett, M. and Kumar, V. (1999). Differentiation of NK1.1+, Ly49+ NK cells from flt3+ multipotent marrow progenitor cells. J Immunol **163**, 2648–2656.

Wuest, T.R. and Carr, D.J. (2008). Dysregulation of CXCR3 signaling due to CXCL10 deficiency impairs the antiviral response to herpes simplex virus 1 infection. J Immunol **181**, 7985–7993.

Yang, S.C., Batra, R.K., Hillinger, S., Reckamp, K.L., Strieter, R.M., Dubinett, S.M. and Sharma, S. (2006). Intrapulmonary administration of CCL21 gene-modified dendritic cells reduces tumor burden in spontaneous murine bronchoalveolar cell carcinoma. Cancer Res **66**, 3205–3213.

Yoshimoto, T., Takeda, K., Tanak, T., Ohkusu, K., Kasiwamura, S., Okamura, H., Akira, S. and Nakanishi, K. (1998). IL-12 upregulates IL-18 receptor expression on T cells, Th1 cells, and B cells: synergism with IL-18 for IFN-gamma production. J Immunol **161**, 3400–3407.

Young, H. and Ortaldo, J. (1987). One-signal requirement for interferon-gamma production by human large granular lymphocytes. J Immunol **139**, 724–727.

Yu, Y.R., Fong, A.M., Combadiere, C., Gao, J.L., Murphy, P.M. and Patel, D.D. (2007). Defective antitumor responses in CX3CR1-deficient mice. Int J Cancer **121**, 316–322.

Yuan, J., Liu, Z., Lim, T., Zhang, H., He, J., Walker, E., Shier, C., Wang, Y., Su, Y., Sall, A., McManus, B. and Yang, D. (2009). CXCL10 inhibits viral replication through recruitment of natural killer cells in coxsackievirus B3-induced myocarditis. Circ Res **104**, 628–638.

Zeng, Y., Huebener, N., Fest, S., Weixler, S., Schroeder, U., Gaedicke, G., Ziang, R., Schramm, A., Eggert, A., Reisfeld, R. A. and Lode, H.N. (2007). Fractalkine (CX3CL1)- and interleukin-2-enriched neuroblastoma microenvironment induces eradication of metastases mediated by T cells and natural killer cells. Cancer Res **67**, 2331–2338.

Zhang, X., Wei, H., Wang, H. and Tian, Z. (2006). Involvement of interaction between fractalkine and CX3CR1 in cytotoxicity of natural killer cells against tumor cells. Oncol Rep **15**, 485–488.

Zhang, X., Wei, H., Chan, Q. and Tian, Z. (2007). Activation of human natural killer cells by recombinant membrane-expressed fractalkine on the surface of tumor cells. Oncol Rep **17**, 1371–1375.

Zhang, C., Zhang, J., Niu, J., Zhou, Z., Zhang, J. and Tian, Z. (2008). Interleukin-12 improves cytotoxicity of natural killer cells via upregulated expression of NKG2D. Hum Immunol **69**, 490–500.

Chapter Fifteen

15

NK cells and chemokines

Angela Gismondi*, Giovanni Bernardini*, Angela Santoni

CHAPTER CONTENTS

Considerate la vostra semenza: fatti non foste a viver come bruti ma per seguir virtute e canoscenza.

Dante Alighieri

Consider ye your origin; ye were not made to live as brutes, but for pursuit of virtue and of knowledge.

Translation by Charles Eliot Norton

ABSTRACT

Natural killer (NK) cells represent a distinct population of circulating and tissue-resident lymphocytes that play an important role in the early phases of immune responses against microbial pathogens by exhibiting cytotoxic functions and secreting a number of cytokines and chemokines.

Studies in man and mouse have shown that NK cells are distributed in several organs under normal conditions. Their frequency is comparatively high in nonlymphoid organs such as the lung, the liver and the mucosal tissue of maternal uterus, and rare in thymus and lymph nodes. Yet, these cells are rapidly recruited to the parenchyma of injured organs during inflammation, viral infections and tumour growth.

Chemotactic factors, including chemokines, play critical roles in the regulation of NK cell migration across endothelium and into the tissues. The differences in chemokine receptor expression together with distinct adhesive properties of different NK cell subsets as well as activated NK cells, imply that they have multiple routes of circulation and trafficking patterns.

Besides their role in the regulation of NK cell trafficking, chemotactic molecules can also affect NK cell effector functions by regulating their priming and their ability to kill and secrete cytokines.

KEY WORDS

Chemokine receptors, Chemokines, Chemotaxis, Lymphocyte homing, Signal transduction

*These authors equally contributed.

Introduction

Natural killer (NK) cells belong to a distinct lineage of lymphocytes that play an important role in the early phase of immune responses against certain microbial pathogens by exhibiting cytotoxic functions and secreting a number of cytokines and chemokines (Trinchieri, 1989).

NK cells develop from a common lymphoid precursor resident in the bone marrow (BM) that is considered the main site of their generation. The BM microenvironment provides a rich source of cytokines and growth factors and allows intimate contact between developing NK cells and stromal cells, which is required for their full maturation (Roth et al., 2007). In the mouse, a lymphoid precursor (NKP) committed to the NK cell lineage has been identified (Di Santo, 2006; Rosmaraki et al., 2001). NKP cells express the common IL-2R/IL-15R beta chain (CD122) and the activating receptor NKG2D, and lack lineage markers. At the next stage of maturation, immature NK cells (iNK) express NK1.1 and CD94, along with the alpha$_v$ integrin chain (CD51) and low levels of the integrin chain CD11b, and have poor cytotoxic and cytokine secretion capacity (Kim et al., 2002; Vosshenrich et al., 2005). The subsequent acquisition of Ly49 receptors is followed by an expansion stage that is characterized by upregulation of the integrin CD49b (defined by the DX5 mAb clone) and by the loss of CD51 expression; among this population, cells with increased expression of CD11b and of the mucin-like molecule CD43 are considered fully functional/mature (mNK). CD11b^{low} NK cells predominate in BM and lymph nodes (LN), while CD11b^{high} NK cells prevail in the blood, liver, spleen and lungs. The observation that in addition to mNK cells, NKPs and iNK are found in the spleen, LN and liver, suggests that also the latter populations can exit the BM under normal conditions and that final maturation of BM-derived NK cell precursors can occur also in the periphery (Freud et al., 2005). In this regard, we recently demonstrated that iNK as well as mNK cells can exit the BM in response to mobilizing stimuli, such as the CXCR4 pharmacological antagonist AMD3100 (Bernardini et al., 2008).

Recent evidence indicates that a population of mouse NK cells that express the IL-7 receptor alpha chain (CD127) differentiates in the thymus and expresses low levels of CD11b and CD43 and a reduced repertoire of NK cell receptors than splenic NK cells. Differently from BM-derived NK cells, thymic-derived NK cells preferentially home to LN, suggesting that they selectively respond to LN homing molecules (Vosshenrich et al., 2006).

Mature NK cells mainly circulate in the peripheral blood but are also resident in several lymphoid and non-lymphoid organs, such as spleen, tonsils, liver, lungs, intestine and uterine decidua. In the mouse spleen, the majority of NK cells are found in the red pulp, while NK cells are preferentially present inside vessels or sinuses in the LN (Dokun et al., 2001; Walzer et al., 2007b). The same tissue distribution is also evident in the liver, where NK cells can be found in high frequency within sinusoid mononuclear cells (Bouwens et al., 1987).

During viral infections, inflammation, tumour growth and invasion, NK cells are rapidly recruited from the blood and accumulate in the parenchymas of injured organs (Biron, 1997; Fogler et al., 1996; Trinchieri, 1989); tissue-recruited NK cells can kill target cells and release inflammatory cytokines and chemokines, thus participating in the recruitment and activation of other leukocytes and in the modulation of dendritic cell (DC) function. In addition, homing to LN of a particular subset of activated NK cells has been described both in human and mouse (Fehniger et al., 2003; Martin-Fontecha et al., 2004). NK cells enter into LN stimulated with LPS-matured DCs or with selected adjuvants, become activated and provide an early source of IFN-γ that is necessary for T helper 1 polarization (Lucas et al., 2007; Martin-Fontecha et al., 2004). This evidence suggests that NK cells enter into LN to acquire effector capabilities through interacting with other cells of the innate immune system, such as DCs, and to modulate adaptive T-cell responses.

Unlike B cells and T cells that express a single antigen-specific receptor, NK cells are endowed with a multiple cell surface receptor system encoded by genes that do not undergo recombination or sequence diversification. This complex receptor system is acquired during NK cell development and consists of both activating and inhibitory receptors (Lanier, 2005; Raulet et al., 2001). Therefore, activation of NK cell functions is the result of concomitant engagement of various activating and inhibitory receptors by the particular set of ligands on target cells (McQueen and Parham, 2002).

All the receptors expressed by NK cells are not unique to this cell type but are also present on cells of other lineages, such as T cells or myeloid cells. Their expression on NK cells is highly regulated, and some receptors are oligoclonally distributed, or expressed on subsets of NK cells. Unlike peripheral blood human NK cells, some tissue-resident NK cells do not express CD16 but show high levels of the Neural Cell Adhesion Molecule (NCAM), CD56.

Based on the receptor repertoire and surface receptor levels, phenotypically distinct NK cell populations have been identified and suggested to represent specialized subsets capable of performing different functions and endowed with distinct migratory properties. Two major subsets of human peripheral blood NK cells have been described: About 90% are CD56^{low}CD16^{high}, whereas about 10% are CD56^{high}CD16^{low}. It has been proposed that CD56^{high} NK cells have a unique functional role in the innate immune response as a primary source of NK cell-derived immunoregulatory cytokines, whereas the CD56^{low}CD16^{high} subset represents the principal cytotoxic population (Cooper et al., 2001).

Populations of NK cells similar to the two main human NK cell subsets have been described in mice. Mouse CD11b^{high} NK cells can be dissected into CD27^{high} and CD27^{low} fractions that differ in terms of expression of NK cell inhibitory receptors and of chemokine receptors and are functionally different. CD27^{low} cells are mostly found in nonlymphoid organs (blood, liver and lungs). As compared to CD27^{high} cells, they express a higher percentage of inhibitory receptors Ly49 I and C isoform and almost exclusively the inhibitory killer cell lectin-like receptor G1 (KLRG1), and their cytotoxic and cytokine production activities are more tightly regulated (Hayakawa and Smyth, 2006; Hayakawa et al., 2006). The CD27^{high} cells have many features similar to that of CD56^{high} cells. They predominate in the LN and are extremely responsive to IL-12 and IL-18; but unlike CD56^{high} cells, CD27^{high} cells are also highly cytotoxic. Considering that CD56^{high} and CD56^{low} NK cells were also recently shown to express high and low levels of CD27, respectively, it has been suggested that this marker can also be used to better identify the two major subsets of human NK cells (Silva et al., 2008).

The ability of leukocytes to traffic coordinately throughout the body is an essential requirement for the maintenance of immunosurveillance. NK cell migration across endothelium, as for other leukocytes, is a spatially and temporally integrated multi-step process regulated by a plethora of chemoattractants and adhesive molecules belonging to the selectin, integrin and immunoglobulin families, as well as chemokines (Kunkel and Butcher, 2002; Springer, 1994).

Among adhesion molecules, both selectins and integrins contribute to the initial leukocyte tethering and rolling along vessel endothelium, while firm adhesion of the leukocyte to vascular endothelium and subsequent diapedesis into the underlying extravascular tissue is mainly mediate by integrins. The various steps of migration are tightly regulated; in fact, for migration to be effective, adhesion receptors must undergo cycles of attachment and detachment from their endothelial ligands.

Chemokines are a superfamily of inflammatory mediators that properly guide leukocyte recruitment and positioning into healthy or diseased tissues. Their action is mediated by interaction with G-protein-coupled seven-transmembrane-domain receptors that initiate complex signalling events that govern leukocyte migration not only by eliciting a chemotactic response but also through a dynamic regulation of integrin adhesiveness for endothelial and extracellular matrix ligands (Baggiolini et al., 1997; Clark and Brugge, 1995; Mantovani, 1999; Rossi and Zlotnik, 2000). In addition, by exerting a pro-adhesive function, a few chemokines can also modulate other steps of leukocyte migration into tissues, including firm adhesion to the endothelial layer (Fong et al., 1998).

Depending on the number and spacing of conserved cysteine residues in their amino acid sequence, chemokines have been classified into four major groups: the CXC (or alpha), CC (or beta), CX3C and C subfamilies (Rossi and Zlotnik, 2000).

The differential expression of chemokine receptors on distinct NK cell populations, together with their ability to regulate integrin expression and function, can be responsible for the recruitment of specialized NK cell subsets during inflammation.

Chemokine receptor expression by NK cells and chemokine-regulated NK cell functions in vitro

A large body of evidence indicates that NK cells can express several receptors for CXC, CC, C and CX3C chemokines, with great heterogeneity in the chemokine receptor repertoire among different NK cell populations and between resting versus activated NK cells (Gregoire et al., 2007; Hayakawa et al., 2006; Robertson, 2002).

With respect to the CXCR and CX3CR families, it has been reported previously that human peripheral blood NK cells express both CXCR1 and CXCR2 as CXCL8 (IL-8) receptor (Casilli et al., 2005; Chuntharapai et al., 1994; Morohashi et al., 1995) and CX3CR1 as CX3CL1 (fractalkine) receptor (Imai et al., 1997; Yoneda et al., 2000). These observations have been further extended by Campbell and colleagues (2001) who provided the first evidence that distinct (CD56^{pos}CD16^{neg} and CD56^{pos}CD16^{pos}) peripheral blood NK cell subsets have a unique repertoire of chemokine receptors. CD16^{pos} NK cells uniformly express high levels of CXCR1 and CX3CR1, low levels of CXCR2 and CXCR3 and no detectable levels of CXCR5. In contrast, CD16^{neg} NK cells express high levels of CXCR3, express low levels of CX3CR1 and are negative for CXCR1, CXCR2 and CXCR5, moreover, both NK cell subsets express high levels of CXCR4, the receptor for CXCL12 (SDF-1alpha/beta). With respect to the CC chemokine receptor family, the majority of NK cells lack expression of CCR1-7 and CCR9, and only the CD16^{neg} NK cell subset expresses high levels of CCR5 and CCR7, with the latter molecule mainly involved in the homing of lymphocytes to secondary lymphoid organs (Campbell et al., 2001).

Consistent with this expression profile, CXCL8 and soluble CX3CL1 preferentially attract the CD16^{pos} NK cell subset, which can also respond moderately to the CXCR3 ligands, CXCL11 (I-TAC) and CXCL10 (IP-10); by contrast, CD16^{neg} NK cells respond more dramatically to the CCR7 ligands, CCL19 (ELC/MIP-3beta) and CCL21 (SLC), as well as to the CXCR3 ligands, CXCL11 and CXCL10, and poorly to a CCR2 ligand, CCL2 (MCP-1), or CCR5 ligands, CCL4 (MIP-1beta) and CCL5 (RANTES). Both NK cell subsets strongly migrate in response to the ligand for CXCR4, CXCL12 (Campbell et al., 2001; Taub et al., 1995).

Moreover, Kim et al. (1999) have found that CD56^{high} CD16^{neg} cells respond more than CD56^{low} CD16^{pos} cells to CCL21 and CCL19 when used at high concentrations, although they observed that the two NK cell subsets express equal levels of CCR7 mRNA.

Besides chemokine receptors, human NK cells also express receptors for chemotactic molecules that do not belong to the chemokine superfamily and regulate NK cell trafficking under steady state and inflammatory conditions. In this regard, NK cells migrate in response to the proinflammatory plasma protein chemerin, and that CD56^{low}CD16^{pos}, but not CD56^{high}CD16^{neg}, expresses its receptor, ChemR23. Recently, a silent receptor for chemerin, CCRL2, has been also identified. CCRL2 is able to bind chemerin and to increase its local concentration and availability. Its expression and function on NK cells has not been clarified yet (Zabel et al., 2008). Sphingosine 1-phosphate (S1P) is a sphingophospholipid that influences lymphocyte trafficking as well as proliferation, adherence, and morphogenesis and is generated by the conversion of sphingomyelin into ceramide by sphingomyelinase (Matloubian et al., 2004). It was recently reported that human NK cells express the mRNA for S1P1, one of the five S1P G protein-coupled receptors, and that activation with IL-2 increases S1P1 and promotes S1P4 and S1P5 but not S1P2 expression; in addition, S1P but not sphingosine induces the chemotaxis of these cells (Kveberg et al., 2002; Maghazachi, 2003).

Interestingly, the selective expression of CXCR3 on CD56^{high} and of CX3CR1 on CD56^{low} NK cells is very similar to their expression pattern on the two main subsets of mouse mature NK cells defined as CD27^{high} and CD27^{low}. Indeed, as compared to CD27^{low} NK cells, CD27^{high} (both CD11b^{low} and CD11b^{high}) cells have higher expression of the chemokine receptor CXCR4, selectively express CXCR3 and has much lower expression of the chemoattractant receptor S1P5, a mouse NK cell-specific receptor for S1P (Hayakawa and Smyth, 2006; Walzer et al., 2007a). We recently extended this observation, showing that CXCR4 expression is very high on NKP cells and progressively decreases during NK cell differentiation, and that NKP cells functionally express also CXCR3 and CCR1 that may participate to their recruitment from BM to specific organs (Bernardini et al., 2008). In addition, using C57BL/6 mice in which a green fluorescent protein cDNA was knocked in to genes encoding CX3CR1, we and others observed selective expression of CX3CR1 on the KLRG1^{pos}/CD27^{low} NK cell subset (Gregoire et al., 2007; Bernardini, unpublished observations) (Table 15.1).

The expression of chemokine receptors on NK cells can be modulated upon cytokine stimulation. A significant decrease of CXCR3 expression on human NK cells treated for 6 h or 24 h with IL-2 and IL-12 alone or in combination has been reported, and the decreased expression was associated with reduced chemotaxis to CXCL10. Similarly, CCR7 receptor is down-regulated on NK cells after IL-2 activation, whereas the CCR4 and CX3CR1 molecule is induced on IL-2 activated NK cells. The same treatment did not affect the expression of other chemokine receptors such as CCR1, CCR2 or CXCR4 (Hodge et al., 2002). However, previous reports have shown that short-term exposure of freshly isolated NK cells to IL-2 can positively modulate CCR2 mRNA expression (Polentarutti et al., 1997), and long-term (8–10 days) IL-2 stimulation results in increased expression of CCR1, CCR2, CCR4, CCR5 and CCR8 (Inngjerdingen et al., 2001). CXCR4 and CX3CR1 have been also shown to be down-regulated through stimulation with different cytokines, including IL-15, whereas TGF-beta has been proven to be a potent inducer of CXCR4 on NK cell subsets (Barlic et al., 2003; Inngjerdingen et al., 2001).

In agreement with these observations, IL-2 activated NK cells can migrate in response to many CC chemokines, such as CCL2, CCL8 (MCP-2), CCL7 (MCP-3), CCL3 (MIP-1 alpha), CCL4, CCL5 and CCL22 (MDC) (Allavena et al., 1994; Godiska et al., 1997; Inngjerdingen et al., 2001; Loetscher et al., 1996).

NK cell treatment with IL-18, differently from IL-2, results in selective induction of CCR7 expression on the CD56^{low} NK cell subset but does not affect CCR7 expression on the CD56^{high} subset; increased expression of CCR7 on CD56^{low} NK cells is associated with reduced levels of CD16 and enhanced capability to migrate in response to the LN associated chemokine CCL21 (Mailliard et al., 2005).

Besides migration, other human and mouse NK cell functions can be affected by chemokines. Indeed, chemokines can activate an NK cell defence machinery that may directly counteract an infectious agent by performing cytotoxicity or by secreting pro-inflammatory cytokines that recruit and activate other effector cells (Taub et al., 1995). Those effects are well exemplified by the multifunctional role of the chemokine CX3CL1 in NK cell activation. Soluble and membrane-bound fractalkine induce IFN-γ production by NK cells and affect NK cell ability to kill tumour cells both in vitro and in vivo (Guo et al., 2003; Yoneda et al., 2003; Zeng et al., 2007). The CX3CL1-mediated pro-adhesive function is relevant in CX3CL1-promoted granule exocytosis and IFN-γ production by NK cells during endothelial cell (EC) contact (Yoneda et al., 2000). These findings suggest that the expression of fractalkine at the site of inflammation can attract and activate NK cells through CX3CR1 and that NK cells, once activated, can lyse neighbouring ECs, promoting vascular injury (Umehara et al., 2004). This effect may have important implications in several pathological conditions such as graft rejection and tissue damage promoted by chronic infections

Table 15.1 Chemoattractant receptor expression on the main subsets of human and mouse NK cells subsets

	Human		Mouse	
	CD56[low] CD16[high]	CD56[high] CD16[low]	CD27[high]	CD27[low]
CXCR1	++[1]	–[3]	ND[4]	ND
CXCR2	+[2]	–	ND	ND
CXCR3	+	++	++	+/–[5]
CXCR4	++	++	++	+
CCR1	–	–	ND	++
CCR2	–	–	++	+
CCR3	–	–	ND	ND
CCR4	–	–	ND	ND
CCR5	–	++	+	++
CCR6	–	–	ND	ND
CCR7	–	++	ND	ND
CCR9	–	–	ND	ND
CX3CR1	++	+	–	++
ChemR23	++	+/–	?	?
S1P5	?	?	+/–	++

[1]Indicates high levels of expression.
[2]Indicates intermediate levels of expression.
[3]Indicates undetectable levels of expression.
[4]Not done.
[5]Indicates low levels of expression.

(Bolovan-Fritts and Spector, 2008; Robinson et al., 2000). Enhancement of NK cell degranulation has also been demonstrated in response to other chemokines belonging to the CC family and to CXCL10 (Maghazachi et al., 1996; Taub et al., 1996). In addition, CC chemokines-mediated redistribution of adhesion molecules on NK cell surfaces may also increase NK cell cytotoxicity by promoting effector-target cell interaction (Nieto et al., 1998).

CX3CL1 has been found to be a pivotal molecule driving immune synapse formation during DC/NK cell interaction leading to efficient NK cell activation (Pallandre et al., 2008). This is similar to what is described for other chemokines that can act as T cell costimulators by prolonging the duration of T cell-antigen presenting cell interaction and by avoiding premature cell splitting. Importantly, the absence of CX3CR1, or CX3CL1-blockade, abrogated the ability of NK cells to produce IFN-γ when exposed to DC, underlying a key role of this chemokine during DC-mediated NK cell priming.

Thus, the CX3CR1/CX3CL1 axis regulates NK cell functions at different levels promoting their migration through endothelial vessels, their activation and their ability to kill or secrete cytokines.

Signalling events controlling chemokine-regulated NK cell functions

Despite increasing evidence of a prominent role of chemokines and integrins in the dynamic regulation of leukocyte adhesion and migration, the signalling pathways responsible for the integrin-supported leukocyte migration elicited by chemokines are not yet completely defined. The propagation of the migratory signals depends on a complex interplay among molecules that regulate actin, myosin, and other cytoskeleton components, and results in the formation of protrusive structures at the front of the migrating cell and retraction at the cell rear (Ridley et al., 2003; Vicente-Manzanares and Sanchez-Madrid, 2004).

Thus, NK cell migration, as for all leukocytes, depends on a highly integrated signalling network culminating in coordinate activation and functional cooperation between different pathways triggered by integrin and chemokine receptors. Chemokine receptors are mainly coupled to Galpha$_i$-dependent heterotrimeric G proteins, as in most cases Pertussis Toxin X (PTX) inhibits the biological activities induced by chemokines, including cell migration. A PTX-sensitive signalling pathway has been shown to be involved in CX3CL1-activated NK cell polarization leading to NK cell priming by DC- or NK cell-mediated target cell lysis (Pallandre et al., 2008). IFN-γ production by NK cells exposed to mature DCs was shown to be dependant on the CX3CL1 ability to promote redistribution of lipid rafts on NK cell membrane that excluded KIR2DL1 inhibitory receptor from the immune synapse leading to inhibition of ligand-induced KIR phosphorylation and recruitment of SHP1.

However, chemokine-induced NK cell chemotaxis is coupled also to PTX-insensitive G-protein such as Galpha$_q$ (Maghazachi, 1997; Soede et al., 2001).

PI3K and its products are signalling intermediates that play a crucial role in cell migration. In this regard, evidence is available on the involvement of PI3K in chemokine-mediated NK cell chemotaxis. In particular, it has been reported that wortmannin, as well as Ab directed against PI3K-γ but not PI3K-α can inhibit C, CC, and CXC chemokine-induced NK cell chemotaxis, suggesting that PI3K IB plays a crucial role in chemokine-induced activation of NK cells. In agreement with these results, recruitment of PI3K-γ into NK cell membranes in response to CCL5 stimulation has been reported (al-Aoukaty et al., 1999).

Activation of protein tyrosine kinases (PTKs) is a prerequisite event for leukocyte migration controlling both integrin adhesiveness and chemotactic responses. The involvement of PTKs belonging to the Src and Syk/Zap families in cell migration has been largely documented for T lymphocytes and cells of myeloid lineage.

Using PTK inhibitors, such as the general tyrosine kinase inhibitor herbimycin A, the specific Lck inhibitor damnacanthal, and the Syk inhibitor piceatannol, a role for the Src kinase Lck but not for Syk in CXCL12-induced NK cell chemotaxis has been described. In accordance with these results, NK cell stimulation with CXCL12 leads to tyrosine phosphorylation and activation of Lck (Inngjerdingen et al., 2002). Activation of Lck can be PI3Kgamma-dependent or -independent, as Galpha$_s$ can directly couple to Src family PTKs, and lead to Vav phosphorylation, leading to activation of Cdc42 and Rac small GTPases and lamellipodia formation.

More recently, a role for the focal adhesion kinases as cytoplasmic mediators of motility events in multiple cell types has been reported. The focal adhesion kinase family comprises two members that share an amino acid identity of almost 50%, the p125 focal adhesion kinase (p125Fak) and the proline-rich tyrosine kinase 2 (Pyk-2) also known as cell adhesion kinase-β (CAK-β), or related adhesion focal tyrosine kinase (RAFTK). They are non-receptor PTKs capable of coupling several receptors, including integrins and chemokine receptors, with a variety of downstream effectors, such as small GTP binding proteins belonging to the Ras and Rho families, MAP kinases, PKC and inositol phosphate metabolism (Lev et al., 1995).

The expression of Fak family members on NK is controversial. p125Fak is expressed on NK cells with β1 integrin engagement results in activation of this kinase and its association with Fyn and Zap-70 PTKs (Rabinowich et al., 1996). In contrast, we demonstrated that human peripheral blood NK cells express Pyk-2 that is constitutively associated with the cytoskeletal protein paxillin but not p125 FAK. More recently, we have reported that NK cell binding to endothelium activates Pyk-2 and the small GTP binding protein Rac, a key regulator of actin cytoskeleton dynamics. Both Pyk-2 and Rac activation are coupled to integrins and chemokine receptors. By using recombinant vaccinia viruses encoding dominant negative mutants of Pyk-2 and Rac, we demonstrated that both Pyk-2 and Rac are functionally involved in chemokine-induced NK cell migration through endothelium or ICAM-1 or VCAM-1 adhesive proteins. We also found that Pyk-2 is associated with the Rac guanine nucleotide exchange factor Vav that undergoes tyrosine phosphorylation upon integrin triggering but not with PIX, another exchange factor for Rac that is associated with paxillin through p95 PKL. Collectively, these results indicate that Pyk-2 acts as a receptor-proximal link between integrin and chemokine receptor signalling, and Pyk-2/Rac pathway plays a pivotal role in the control of NK cell transendothelial migration (Gismondi et al., 2003).

These results are consistent with findings indicating that Pyk-2 can colocalize with the microtubule-organizing centre at the trailing edge of migrating NK cells and in the area of the NK cell membrane that faces target cells (Sancho et al., 2000).

In vivo regulation of NK cell functions by chemokines

NK cells express several chemotactic receptors that are involved in the control of NK cell migration across endothelium and in the correct tissue positioning of lymphocytes, but only recently, we are beginning to appreciate the contribution of chemotactic factors on NK cell trafficking and tissue distribution under steady-state and pathological conditions (Figure 15.1).

Although the differential expression of chemokine receptors and chemotactic responsiveness of the CD27^{low} and CD27^{high} subsets strongly suggests a correlation with their tissue distribution under normal conditions, few chemoattractant receptors have been demonstrated to play a role in NK cell subset distribution in vivo at present. A drastic decrease in NK cell numbers has been observed in the blood, spleen, and lungs of S1P5-deficient mice, associated with an increased number of NK cells in the BM and LNs. This altered distribution has suggested that S1P5 provides an egress signal to NK cells, promoting their exit from BM and LN (Walzer et al., 2007a). In addition, the observation that CX3CR1-deficient mice display a selective reduction of NK cell number in the lung strongly suggests that this receptor is important for the accumulation of at least one subset of NK cells in this organ under steady-state conditions (Yu et al., 2007). Also the expression of CXCR4 profoundly affects NK cell subset distribution likely by contributing to the maintenance of an NK cell pool within BM, as shown by the recruitment of BM iNK and mNK cells into circulation and spleen following in vivo delivery of the CXCR4 pharmacological antagonist AMD-3100 (Bernardini et al., 2008).

Interestingly, altered distribution of NK cell subsets has been revealed also in mice defective for transcription factors (Boos et al., 2007). Indeed, although normal number and function of BM NK cells were found in Id2$^{-/-}$E2a$^{-/-}$ mice, very few mature NK cells were found in the spleen, whereas liver-specific homing of BM NK cells was reduced in the absence of Gata-3 (Samson et al., 2003). Overall, these findings suggest

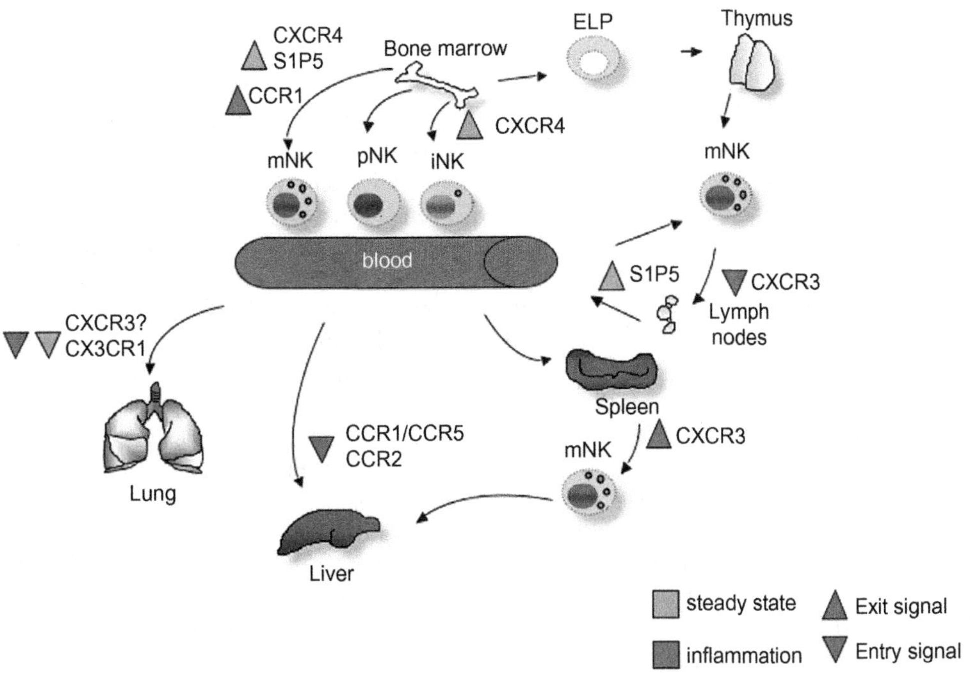

Figure 15.1 • Chemoattractant receptor involvement in the trafficking of mouse NK cells. The majority of NK cells develop in the BM, and a subset that is characterized by CD127 expression develops into the thymus. Recruitment into circulation from different locations of NK cells at different stages of maturation is directed by chemoattractants both under normal (light grey triangles) or inflammatory (dark grey triangles) conditions. Once into circulation, the differential expression of chemoattractant receptors on different NK cell subsets or differently activated NK cells regulate their homing into normal or inflamed tissues.

that these transcription factors are strongly involved in the expression of tissue homing receptors on NK cells either by directly regulating their transcription or by affecting the differentiation of selected NK cell subsets expressing such receptors.

While the contribution of chemokines in the regulation of NK cell tissue distribution during homeostasis has been poorly addressed so far, a number of studies indicate that selected chemokines play a critical role in the orchestration of NK cell trafficking during inflammation. During murine cytomegalovirus infection, NK cells migrate through a CCL3-dependent mechanism to sites of liver infection where they contribute to antiviral defence (Salazar-Mather et al., 1998). This evidence was further supported by the demonstration that CCL3-deficient mice show decreased resistance to cytomegalovirus infection that is associated with a dramatic reduction of NK cell accumulation and IFN-γ production in the liver. The recruitment of NK cells to the liver during infection required the receptor for CCL3, CCR5, and the receptor for CCL2, CCR2 (Hokeness et al., 2005; Salazar-Mather et al., 1998). IFN-γ-dependent recruitment of NK cells to liver has also been described in a model of Concanavalin A-induced hepatitis. In this case, NK cell entry in the liver was reduced in the absence of CCR1 and involved NK cells that used CXCR3 to exit the spleen (Wald et al., 2006).

A role for CCL3 in recruitment of NK cells has also been demonstrated by intrapulmonary transient transgenic expression of this chemokine that resulted in increased *Klebsiella pneumonia* lung clearance associated with NK cell activation and accumulation in this organ (Zeng et al., 2003). Accumulation of NK cells in the lungs has been also observed in mice with invasive aspergillosis. In this model, however, NK cell recruitment was mediated by CCL2, as neutralization of this chemokine resulted in a reduced NK cell number in the lungs and impaired clearance of the pathogen from this organ (Morrison et al., 2003). In addition, using an in vivo model of NK cell-mediated lung tumour cell clearance in CX3CR1-deficient mice or after treatment with blocking antibodies against CX3CL1 or CX3CR1, it has been demonstrated that decreased clearance of tumour cells following perturbation of CX3CL1/CX3CR1 interaction is attributable to defective NK cell recruitment to the lung (Robinson et al., 2003; Yu et al., 2007). Using CXCR3-knockout mice, recruitment of NK cells in the lungs also participates in the pulmonary host defence against *Bordetella bronchiseptica* (Widney et al., 2005). In addition, a specific defect of NK cell recruitment to pulmonary granulomas was observed in CCR1-deficient mice (Shang et al., 2000).

In the central nervous system (CNS), CXCL10 promotes innate defence mechanisms following coronavirus infection by recruiting and activating NK cells, while

CX3CR1GFP/GFP mice showed a selective deficiency of NK cell recruitment in the CNS during experimental autoimmune encephalomyelitis that was accompanied by very severe disease (Trifilo et al., 2004). The results of this study suggested that loss of the regulatory influence of NK cells accounts for the severe EAE phenotype in CX3CR1GFP/GFP mice. Coincidently, alterations of CX3CR1^{pos} NK cell numbers are observed in patients with multiple sclerosis. Whether or not CX3CR1 can be used in multiple sclerosis therapy remains to be seen, but the beneficial effects observed during anti-CD25-based therapy of multiple sclerosis have been attributed to activated CD56^{high} cells (Infante-Duarte et al., 2005).

The involvement of CXCL10 and CX3CL1 in supporting NK cell function in vivo has also been provided in antitumour immunity. CXCL10 promotes a strong antileukemic NK cell-mediated response via enhanced induction of cytotoxicity and expression of the T cell costimulatory molecule B7-H1 (Saudemont et al., 2005). In addition, a link between IFN-γ induction of CXCR3 ligand expression during the antitumour immune response and CD27^{high} NK cell recruitment within the tumour mass was found and shown to be critical for host survival (Wendel et al., 2008). CX3CL1-transfected tumour cells exhibit a reduced growth capability that is mediated by an increased recruitment and activation of NK cells (Robinson et al., 2003).

The selective role of chemokine/chemokine receptor interaction in NK cell migration in vivo suggests that different NK cell subsets may be independently recruited in distinct inflammatory settings, as the two major subsets of mature NK cells, CD27^{high} and CD27^{low}, can be characterized by the mutually exclusive expression of CXCR3 and CX3CR1.

Recently, by using selective depletion and adoptive transfer experiments, Martin-Fontecha et al. have reported that DC-induced recruitment of NK cells into LN occurs in a CXCR3- but not CCR7-dependent manner (2004). The role of CXCR3 has also been emphasized within the spleen in the recruitment of red pulp-localized NK cells to the white pulp during TLR ligand stimulation in vivo (Gregoire et al., 2008). Interestingly, the entry of NK cells into the splenic red pulp is PTX-insensitive, thus suggesting that NK cell localization into this organ is regulated by a chemokine-independent or a PTX-insensitive chemokine receptor. The entry of NK cells within the T cell area of secondary lymphoid organs has several implications allowing their correct priming by DCs as well as an efficient T-cell polarization. Using a genetic approach to in vivo deplete CD11c^{high} DC, Lucas and coworkers showed that NK cell responses to viral and bacterial pathogens in vivo depend on their interaction with CD11c^{high} DC activated by IFN type I signals within the secondary lymphoid organs (2007).

These data collected in mice strongly support the in vivo relevance of a number of chemokine receptor-ligand interactions, including CX3CR1-CX3CL1, CXCR3-CXCL10/CXCL11, CCR5-CCL3/CCL4 and CCR5-CCL5, which have been shown to mediate human NK cell chemotactic response in vitro.

Inflammatory conditions can induce drastic changes in the pattern of chemokine receptor expression on human NK cells and the regulation of the expression of their ligands in inflamed or injured tissues, thus altering NK cell subset recruitment and redistribution. Indeed, granulomatous lesions in the skin and the respiratory tract of patients with transporter associated with antigen processing 2 (TAP-2) (transporter of activated peptide 2) deficiency display an accumulation of activated NK cells hyper responsive to CCR2 ligands (Hanna et al., 2005). In addition, a strong correlation between ChemR23 expression and colocalization of NK cells with DCs was observed in human biopsies of lichen planus in areas where chemerin is present (Parolini et al., 2007).

Conclusions

Individual NK cell subsets displaying unique functional features, and tissue locations have been identified both in mouse and humans. Although this suggests that the expression of specific homing receptors is involved, the role of chemokines in the trafficking of NK cells in normal and disease conditions is only now starting to be elucidated.

In the steady state, NK cells are present at a high frequency in the circulation, ready to extravasate to tissues under inflammatory conditions. The recent findings reviewed herein highlight that NK cells respond to several chemoattractants regulating the maintenance of resting NK cells in the circulation (i.e. S1P) or the recruitment of activated NK cells into the sites of diseases and inflammation. In these locations, NK cells can play an important role as active participants in directing DC maturation and T-cell response polarization and/or as cytotoxic effector cells.

Involvement of chemokines in the regulation of DC-mediated NK cell priming and effector functions has also been documented and should be taken into account when analyzing the role of chemokines in NK cell-dependent immune responses.

Acknowledgements

This work was supported by grants from Istituto Pasteur Fondazione Cenci Bolognetti, MIUR-PRIN, Centro di Eccellenza BEMM, EC NoE EMBIC contract # 512040.

References

al-Aoukaty, A., Rolstad, B. and Maghazachi, A.A. (1999). Recruitment of pleckstrin and phosphoinositide 3-kinase gamma into the cell membranes, and their association with G beta gamma after activation of NK cells with chemokines. J Immunol **162**, 3249–3255.

Allavena, P., Bianchi, G., Zhou, D., van Damme, J., Jilek, P., Sozzani, S. and Mantovani, A. (1994). Induction of natural killer cell migration by monocyte chemotactic protein-1, -2 and -3. Eur J Immunol **24**, 3233–3236.

Baggiolini, M., Dewald, B. and Moser, B. (1997). Human chemokines: an update. Annu Rev Immunol **15**, 675–705.

Barlic, J., Sechler, J.M. and Murphy, P.M. (2003). IL-15 and IL-2 oppositely regulate expression of the chemokine receptor CX3CR1. Blood **102**, 3494–3503.

Bernardini, G., Sciume, G., Bosisio, D., Morrone, S., Sozzani, S. and Santoni, A. (2008). CCL3 and CXCL12 regulate trafficking of mouse bone marrow NK cell subsets. Blood **111**, 3626–3634.

Biron, C.A. (1997). Natural killer cell regulation during viral infection. Biochem Soc Trans **25**, 687–690.

Bolovan-Fritts, C.A. and Spector, S.A. (2008). Endothelial damage from cytomegalovirus-specific host immune response can be prevented by targeted disruption of fractalkine–CX3CR1 interaction. Blood **111**, 175–182.

Boos, M.D., Yokota, Y., Eberl, G. and Kee, B.L. (2007). Mature natural killer cell and lymphoid tissue-inducing cell development requires Id2-mediated suppression of E protein activity. J Exp Med **204**, 1119–1130.

Bouwens, L., Remels, L., Baekeland, M., Van Bossuyt, H. and Wisse, E. (1987). Large granular lymphocytes or "pit cells" from rat liver: isolation, ultrastructural characterization and natural killer activity. Eur J Immunol **17**, 37–42.

Campbell, J.J., Qin, S., Unutmaz, D., Soler, D., Murphy, K.E., Hodge, M.R., Wu, L. and Butcher, E.C. (2001). Unique subpopulations of CD56+ NK and NK-T peripheral blood lymphocytes identified by chemokine receptor expression repertoire. J Immunol **166**, 6477–6482.

Casilli, F., Bianchini, A., Gloaguen, I., Biordi, L., Alesse, E., Festuccia, C., Cavalieri, B., Strippoli, R., Cervellera, M.N., Di Bitondo, R., Ferretti, E., Mainiero, F., Bizzarri, C., Colotta, F. and Bertini, R. (2005). Inhibition of interleukin-8 (CXCL8/IL-8) responses by repertaxin, a new inhibitor of the chemokine receptors CXCR1 and CXCR2. Biochem Pharmacol **69**, 385–394.

Chuntharapai, A., Lee, J., Hebert, C.A. and Kim, K.J. (1994). Monoclonal antibodies detect different distribution patterns of IL-8 receptor A and IL-8 receptor B on human peripheral blood leukocytes. J Immunol **153**, 5682–5688.

Clark, E.A. and Brugge, J.S. (1995). Integrins and signal transduction pathways: the road taken. Science **268**, 233–239.

Cooper, M.A., Fehniger, T.A., Turner, S.C., Chen, K.S., Ghaheri, B.A., Ghayur, T., Carson, W.E. and Caligiuri, M.A. (2001). Human natural killer cells: a unique innate immunoregulatory role for the CD56(bright) subset. Blood **97**, 3146–3151.

Di Santo, J.P. (2006). Natural killer cell developmental pathways: a question of balance. Annu Rev Immunol **24**, 257–286.

Dokun, A.O., Chu, D.T., Yang, L., Bendelac, A.S. and Yokoyama, W. M. (2001). Analysis of in situ NK cell responses during viral infection. J Immunol **167**, 5286–5293.

Fehniger, T.A., Cooper, M.A., Nuovo, G.J., Cella, M., Facchetti, F., Colonna, M. and Caligiuri, M.A. (2003). CD56bright natural killer cells are present in human lymph nodes and are activated by T cell-derived IL-2: a potential new link between adaptive and innate immunity. Blood **101**, 3052–3057.

Fogler, W.E., Volker, K., McCormick, K.L., Watanabe, M., Ortaldo, J.R. and Wiltrout, R.H. (1996). NK cell infiltration into lung, liver, and subcutaneous B16 melanoma is mediated by VCAM-1/VLA-4 interaction. J Immunol **156**, 4707–4714.

Fong, A.M., Robinson, L.A., Steeber, D. A., Tedder, T.F., Yoshie, O., Imai, T. and Patel. (1998). Fractalkine and CX3CR1 mediate a novel mechanism of leukocyte capture, firm adhesion, and activation under physiologic flow. J Exp Med **188**, 1413–1419.

Freud, A.G., Becknell, B., Roychowdhury, S., Mao, H.C., Ferketich, A.K., Nuovo, G.J., Hughes, T.L., Marburger, T.B., Sung, J., Baiocchi, R.A., Guimond, M. and Caligiuri, M.A. (2005). A human CD34(+) subset resides in lymph nodes and differentiates into CD56bright natural killer cells. Immunity **22**, 295–304.

Gismondi, A., Jacobelli, J., Strippoli, R., Mainiero, F., Soriani, A., Cifaldi, L., Piccoli, M., Frati, L. and Santoni, A. (2003). Proline-rich tyrosine kinase 2 and Rac activation by chemokine and integrin receptors controls NK cell transendothelial migration. J Immunol **170**, 3065–3073.

Godiska, R., Chantry, D., Raport, C.J., Sozzani, S., Allavena, P., Leviten, D., Mantovani, A. and Gray, P.W. (1997). Human macrophage-derived chemokine (MDC), a novel chemoattractant for monocytes, monocyte-derived dendritic cells, and natural killer cells. J Exp Med **185**, 1595–1604.

Gregoire, C., Chasson, L., Luci, C., Tomasello, E., Geissmann, F., Vivier, E. and Walzer, T. (2007). The trafficking of natural killer cells. Immunol Rev **220**, 169–182.

Gregoire, C., Cognet, C., Chasson, L., Coupet, C.A., Dalod, M., Reboldi, A., Marvel, J., Sallusto, F., Vivier, E. and Walzer, T. (2008). Intrasplenic trafficking of natural killer cells is redirected by chemokines upon inflammation. Eur J Immunol **38**, 2076–2084.

Guo, J., Chen, T., Wang, B., Zhang, M., An, H., Guo, Z., Yu, Y., Qin, Z. and Cao, X. (2003). Chemoattraction, adhesion and activation of natural killer cells are involved in the antitumor immune response induced by fractalkine/CX3CL1. Immunol Lett **89**, 1–7.

Hanna, J., Mussaffi, H., Steuer, G., Hanna, S., Deeb, M., Blau, H., Arnon, T.I., Weizman, N. and Mandelboim, O. (2005). Functional aberrant expression of CCR2 receptor on chronically activated NK cells in patients with TAP-2 deficiency. Blood **106**, 3465–3473.

Hayakawa, Y. and Smyth, M.J. (2006). CD27 dissects mature NK cells into two subsets with distinct responsiveness and migratory capacity. J Immunol **176**, 1517–1524.

Hayakawa, Y., Huntington, N.D., Nutt, S.L. and Smyth, M.J. (2006). Functional subsets of mouse natural killer cells. Immunol Rev **214**, 47–55.

Hodge, D.L., Schill, W.B., Wang, J.M., Blanca, I., Reynolds, D.A., Ortaldo, J.R. and Young, H.A. (2002). IL-2 and IL-12 alter NK cell responsiveness to IFN-gamma-inducible protein 10 by down-regulating CXCR3 expression. J Immunol **168**, 6090–6098.

Hokeness, K.L., Kuziel, W.A., Biron, C.A. and Salazar-Mather, T.P. (2005). Monocyte chemoattractant protein-1 and CCR2 interactions are required for IFN-alpha/beta-induced inflammatory responses and antiviral defense in liver. J Immunol **174**, 1549–1556.

Imai, T., Hieshima, K., Haskell, C., Baba, M., Nagira, M., Nishimura, M., Kakizaki, M., Takagi, S., Nomiyama, H., Schall, T.J. and Yoshie, O. (1997). Identification and molecular characterization of fractalkine receptor CX3CR1, which mediates both leukocyte migration and adhesion. Cell **91**, 521–530.

Infante-Duarte, C., Weber, A., Kratzschmar, J., Prozorovski, T., Pikol, S., Hamann, I., Bellmann-Strobl, J., Aktas, O., Dorr, J., Wuerfel, J., Sturzebecher, C.S. and Zipp, F. (2005). Frequency of blood CX3CR1-positive natural killer cells correlates with disease activity in multiple sclerosis patients. FASEB J **19**, 1902–1904.

Inngjerdingen, M., Damaj, B. and Maghazachi, A.A. (2001). Expression and regulation of chemokine receptors in human natural killer cells. Blood **97**, 367–375.

Inngjerdingen, M., Torgersen, K.M. and Maghazachi, A.A. (2002). Lck is required for stromal cell-derived factor 1 alpha (CXCL12)-induced lymphoid cell chemotaxis. Blood **99**, 4318–4325.

Kim, C.H., Pelus, L.M., Appelbaum, E., Johanson, K., Anzai, N. and Broxmeyer, H.E. (1999). CCR7 ligands, SLC/6Ckine/Exodus2/TCA4 and CKbeta-11/MIP-3beta/ELC, are chemoattractants for CD56(+)CD16(−) NK cells and late stage lymphoid progenitors. Cell Immunol **193**, 226–235.

Kim, S., Iizuka, K., Kang, H.S., Dokun, A., French, A.R., Greco, S. and Yokoyama, W.M. (2002). In vivo developmental stages in murine natural killer cell maturation. Nat Immunol **3**, 523–528.

Kunkel, E.J. and Butcher, E.C. (2002). Chemokines and the tissue-specific migration of lymphocytes. Immunity **16**, 1–4.

Kveberg, L., Bryceson, Y., Inngjerdingen, M., Rolstad, B. and Maghazachi, A.A. (2002). Sphingosine 1 phosphate induces the chemotaxis of human natural killer cells. Role for heterotrimeric G proteins and phosphoinositide 3 kinases. Eur J Immunol **32**, 1856–1864.

Lanier, L.L. (2005). NK cell recognition. Annu Rev Immunol **23**, 225–274.

Lev, S., Moreno, H., Martinez, R., Canoll, P., Peles, E., Musacchio, J.M., Plowman, G.D., Rudy, B. and Schlessinger, J. (1995). Protein tyrosine kinase PYK2 involved in Ca(2+)-induced regulation of ion channel and MAP kinase functions. Nature **376**, 737–745.

Loetscher, P., Seitz, M., Clark-Lewis, I., Baggiolini, M. and Moser, B. (1996). Activation of NK cells by CC chemokines. Chemotaxis, Ca2+ mobilization, and enzyme release. J Immunol **156**, 322–327.

Lucas, M., Schachterle, W., Oberle, K., Aichele, P. and Diefenbach, A. (2007). Dendritic cells prime natural killer cells by trans-presenting interleukin 15. Immunity **26**, 503–517.

Maghazachi, A.A. (1997). Role of the heterotrimeric G proteins in stromal-derived factor-1alpha-induced natural killer cell chemotaxis and calcium mobilization. Biochem Biophys Res Commun **236**, 270–274.

Maghazachi, A.A. (2003). G protein-coupled receptors in natural killer cells. J Leukoc Biol **74**, 16–24.

Maghazachi, A.A., Al-Aoukaty, A. and Schall, T.J. (1996). CC chemokines induce the generation of killer cells from CD56+ cells. Eur J Immunol **26**, 315–319.

Mailliard, R.B., Alber, S.M., Shen, H., Watkins, S.C., Kirkwood, J.M., Herberman, R.B. and Kalinski, P. (2005). IL-18-induced CD83+CCR7+ NK helper cells. J Exp Med **202**, 941–943.

Mantovani, A. (1999). The chemokine system: redundancy for robust outputs. Immunol Today **20**, 254–257.

Martin-Fontecha, A., Thomsen, L.L., Brett, S., Gerard, C., Lipp, M., Lanzavecchia, A. and Sallusto, F. (2004). Induced recruitment of NK cells to lymph nodes provides IFN-gamma for T(H)1 priming. Nat Immunol **5**, 1260–1265.

Matloubian, M., Lo, C.G., Cinamon, G., Lesneski, M.J., Xu, Y., Brinkmann, V., Allende, M.L., Proia, R.L. and Cyster, J.G. (2004). Lymphocyte egress from thymus and peripheral lymphoid organs is dependent on S1P receptor 1. Nature **427**, 355–360.

McQueen, K.L. and Parham, P. (2002). Variable receptors controlling activation and inhibition of NK cells. Curr Opin Immunol **14**, 615–621.

Morohashi, H., Miyawaki, T., Nomura, H., Kuno, K., Murakami, S., Matsushima, K. and Mukaida, N. (1995). Expression of both types of human interleukin-8 receptors on mature neutrophils, monocytes, and natural killer cells. J Leukoc Biol **57**, 180–187.

Morrison, B.E., Park, S.J., Mooney, J.M. and Mehrad, B. (2003). Chemokine-mediated recruitment of NK cells is a critical host defense mechanism in invasive aspergillosis. J Clin Invest **112**, 1862–1870.

Nieto, M., Navarro, F., Perez-Villar, J.J., del Pozo, M.A., Gonzalez-Amaro, R., Mellado, M., Frade, J.M., Martinez, A.C., Lopez-Botet, M. and Sanchez-Madrid, F. (1998). Roles of chemokines and receptor polarization in NK – target cell interactions. J Immunol **161**, 3330–3339.

Pallandre, J.R., Krzewski, K., Bedel, R., Ryffel, B., Caignard, A., Rohrlich, P.S., Pivot, X., Tiberghien, P., Zitvogel, L., Strominger, J.L. and Borg, C. (2008). Dendritic cell and natural killer cell cross-talk: a pivotal role of CX3CL1 in NK cytoskeleton organization and activation. Blood.

Parolini, S., Santoro, A., Marcenaro, E., Luini, W., Massardi, L., Facchetti, F., Communi, D., Parmentier, M., Majorana, A., Sironi, M., Tabellini, G., Moretta, A. and Sozzani, S. (2007). The role of chemerin in the colocalization of NK and dendritic cell subsets into inflamed tissues. Blood **109**, 3625–3632.

Polentarutti, N., Allavena, P., Bianchi, G., Giardina, G., Basile, A., Sozzani, S., Mantovani, A. and Introna, M. (1997). IL-2-regulated expression of the monocyte chemotactic protein-1 receptor (CCR2) in human NK cells: characterization of a predominant 3.4-kilobase transcript containing CCR2B and CCR2A sequences. J Immunol **158**, 2689–2694.

Rabinowich, H., Manciulea, M., Herberman, R.B. and Whiteside, T.L. (1996). Beta1 integrin-mediated activation of focal adhesion kinase and its association with and Zap-70 in human NK cells. J Immunol **157**, 3860–3868.

Raulet, D.H., Vance, R.E. and McMahon, C.W. (2001). Regulation of the natural killer cell receptor repertoire. Annu Rev Immunol **19**, 291–330.

Ridley, A.J., Schwartz, M.A., Burridge, K., Firtel, R.A., Ginsberg, M.H., Borisy, G., Parsons, J.T. and Horwitz, A.R. (2003). Cell migration: integrating signals from front to back. Science **302**, 1704–1709.

Robertson, M.J. (2002). Role of chemokines in the biology of natural killer cells. J Leukoc Biol **71**, 173–183.

Robinson, L.A., Nataraj, C., Thomas, D.W., Howell, D.N., Griffiths, R., Bautch, V., Patel., Feng, L. and Coffman, T.M. (2000). A role for fractalkine and its receptor (CX3CR1) in cardiac allograft rejection. J Immunol **165**, 6067–6072.

Robinson, L.A., Nataraj, C., Thomas, D.W., Cosby, J.M., Griffiths, R., Bautch, V.L., Patel. and Coffman, T.M. (2003). The chemokine CX3CL1 regulates NK cell activity in vivo. Cell Immunol **225**, 122–130.

Rosmaraki, E.E., Douagi, I., Roth, C., Colucci, F.,Cumano, A. and Di Santo, J.P. (2001). Identification of committed NK cell progenitors in adult murine bone marrow. Eur J Immunol **31**, 1900–1909.

Rossi, D. and Zlotnik, A. (2000). The biology of chemokines and their receptors. Annu Rev Immunol **18**, 217–242.

Roth, C., Rothlin, C., Riou, S., Raulet, D.H. and Lemke, G. (2007). Stromal-cell regulation of natural killer cell differentiation. J Mol Med **85**, 1047–1056.

Salazar-Mather, T.P., Orange, J.S. and Biron, C.A. (1998). Early murine cytomegalovirus (MCMV) infection induces liver natural killer (NK) cell inflammation and protection through macrophage inflammatory protein 1alpha (MIP-1alpha)-dependent pathways. J Exp Med **187**, 1–14.

Samson, S.I., Richard, O., Tavian, M., Ranson, T., Vosshenrich, C.A., Colucci, F., Buer, J., Grosveld, F., Godin, I. and Di Santo, J.P. (2003). GATA-3 promotes maturation, IFN-gamma production, and liver-specific homing of NK cells. Immunity **19**, 701–711.

Sancho, D., Nieto, M., Llano, M., Rodriguez-Fernandez, J.L., Tejedor, R., Avraham, S., Cabanas, C., Lopez-Botet, M. and Sanchez-Madrid, F. (2000). The tyrosine kinase PYK-2/RAFTK regulates natural killer (NK) cell cytotoxic response, and is translocated and activated upon specific target cell recognition and killing. J Cell Biol **149**, 1249–1262.

Saudemont, A., Jouy, N., Hetuin, D. and Quesnel, B. (2005). NK cells that are activated by CXCL10 can kill dormant tumor cells that resist CTL-mediated lysis and can express B7-H1 that stimulates T cells. Blood **105**, 2428–2435.

Shang, X., Qiu, B., Frait, K.A., Hu, J.S., Sonstein, J., Curtis, J.L., Lu, B., Gerard, C. and Chensue, S.W. (2000) Chemokine receptor 1 knockout abrogates natural killer cell recruitment and impairs type-1 cytokines in lymphoid tissue during pulmonary granuloma formation. Am J Pathol **157**, 2055–2063.

Silva, A., Andrews, D.M., Brooks, A.G., Smyth, M.J. and Hayakawa, Y. (2008). Application of CD27 as a marker for distinguishing human NK cell subsets. Int Immunol **20**, 625–630.

Soede, R.D., Zeelenberg, S., Wijnands, Y.M., Kamp, M. and Roos, E. (2001). Stromal cell-derived factor-1-induced LFA-1 activation during in vivo migration of T cell hybridoma cells requires Gq/11, RhoA, and myosin, as well as Gi and Cdc42. J Immunol **166**, 4293–4301.

Springer, T.A. (1994). Traffic signals for lymphocyte recirculation and leukocyte emigration: the multistep paradigm. Cell **76**, 301–314.

Taub., Sayers, T.J., Carter, C.R. and Ortaldo, J.R. (1995). Alpha and beta chemokines induce NK cell migration and enhance NK-mediated cytolysis. J Immunol **155**, 3877–3888.

Taub., Ortaldo, J.R., Turcovski-Corrales, S.M., Key, M.L., Longo, D.L. and Murphy, W. J. (1996). Beta chemokines costimulate lymphocyte cytolysis, proliferation, and lymphokine production. J Leukoc Biol **59**, 81–89.

Trifilo, M.J., Montalto-Morrison, C., Stiles, L.N., Hurst, K.R., Hardison, J.L., Manning, J.E., Masters, P.S. and Lane, T.E. (2004). CXC chemokine ligand 10 controls viral infection in the central nervous system: evidence for a role in innate immune response through recruitment and activation of natural killer cells. J Virol **78**, 585–594.

Trinchieri, G. (1989). Biology of natural killer cells. Adv Immunol **47**, 187–376.

Umehara, H., Bloom, E.T., Okazaki, T., Nagano, Y., Yoshie, O. and Imai, T. (2004). Fractalkine in vascular biology: from basic research to clinical disease. Arterioscler Thromb Vasc Biol **24**, 34–40.

Vicente-Manzanares, M. and Sanchez-Madrid, F. (2004). Role of the cytoskeleton during leukocyte responses. Nat Rev Immunol **4**, 110–122.

Vosshenrich, C.A., Samson-Villeger, S.I. and Di Santo, J.P. (2005). Distinguishing features of developing natural killer cells. Curr Opin Immunol **17**, 151–158.

Vosshenrich, C.A., Garcia-Ojeda, M.E., Samson-Villeger, S.I., Pasqualetto, V., Enault, L., Richard-Le Goff, O., Corcuff, E., Guy-Grand, D., Rocha, B., Cumano, A., Rogge, L., Ezine, S. and Di Santo, J.P. (2006). A thymic pathway of mouse natural killer cell development characterized by expression of GATA-3 and CD127. Nat Immunol **7**, 1217–1224.

Wald, O., Weiss., Wald, H., Shoham, H., Bar-Shavit, Y., Beider, K., Galun, E., Weiss, L., Flaishon, L., Shachar, I., Nagler, A., Lu, B., Gerard, C., Gao, J.L., Mishani, E., Farber, J. and Peled, A. (2006). IFN-gamma acts on T cells to induce NK cell mobilization and accumulation in target organs. J Immunol **176**, 4716–4729.

Walzer, T., Chiossone, L., Chaix, J., Calver, A., Carozzo, C., Garrigue-Antar, L., Jacques, Y., Baratin, M., Tomasello, E. and Vivier, E. (2007a). Natural killer cell trafficking in vivo requires a dedicated sphingosine 1-phosphate receptor. Nat Immunol **8**, 1337–1344.

Walzer, T., Jaeger, S., Chaix, J. and Vivier, E. (2007b). Natural killer cells: from CD3(−)NKp46(+) to post-genomics meta-analyses. Curr Opin Immunol **19**, 365–372.

Wendel, M., Galani, I.E., Suri-Payer, E. and Cerwenka, A. (2008). Natural killer cell accumulation in tumors is dependent on IFN-gamma and CXCR3 ligands. Cancer Res **68**, 8437–8445.

Widney, D.P., Hu, Y., Foreman-Wykert, A.K., Bui, K.C., Nguyen, T.T., Lu, B., Gerard, C., Miller, J.F. and Smith, J.B. (2005). CXCR3 and its ligands participate in the host response to *Bordetella bronchiseptica* infection of the mouse respiratory tract but are not required for clearance of bacteria from the lung. Infect Immun **73**, 485–493.

Yoneda, O., Imai, T., Goda, S., Inoue, H., Yamauchi, A., Okazaki, T., Imai, H., Yoshie, O., Bloom, E.T., Domae, N. and Umehara, H. (2000). Fractalkine-mediated endothelial cell injury by NK cells. J Immunol **164**, 4055–4062.

Yoneda, O., Imai, T., Nishimura, M., Miyaji, M., Mimori, T., Okazaki, T., Domae, N., Fujimoto, H., Minami, Y., Kono, T., Bloom, E.T. and Umehara, H. (2003). Membrane-bound form of fractalkine induces IFN-gamma production by NK cells. Eur J Immunol **33**, 53–58.

Yu, Y.R., Fong, A.M., Combadiere, C., Gao, J.L., Murphy, P.M. and Patel. (2007). Defective antitumor responses in CX3CR1-deficient mice. Int J Cancer **121**, 316–322.

Zabel, B.A., Nakae, S., Zuniga, L., Kim, J.Y., Ohyama, T., Alt, C., Pan, J., Suto, H., Soler, D., Allen, S.J., Handel, T.M., Song, C.H., Galli, S.J. and Butcher, E.C. (2008). Mast cell-expressed orphan receptor CCRL2 binds chemerin and is required for optimal induction of IgE-mediated passive cutaneous anaphylaxis. J Exp Med **205**, 2207–2220.

Zeng, X., Moore, T.A., Newstead, M.W., Hernandez-Alcoceba, R., Tsai, W.C. and Standiford, T.J. (2003). Intrapulmonary expression of macrophage inflammatory protein 1alpha (CCL3) induces neutrophil and NK cell accumulation and stimulates innate immunity in murine bacterial pneumonia. Infect Immun **71**, 1306–1315.

Zeng, Y., Huebener, N., Fest, S., Weixler, S., Schroeder, U., Gaedicke, G., Xiang, R., Schramm, A., Eggert, A., Reisfeld, R. A. and Lode, H.N. (2007). Fractalkine (CX3CL1)- and interleukin-2-enriched neuroblastoma microenvironment induces eradication of metastases mediated by T cells and natural killer cells. Cancer Res **67**, 2331–2338.

Chapter **Sixteen**

16

Natural killer cytolytic activity

Lesley R. de Armas, Eckhard R. Podack

CHAPTER CONTENTS

And come he slow, or come he fast, it is but Death who comes at last.

Sir Walter Scott

ABSTRACT

Natural killer (NK)–mediated cellular killing has been extensively studied over the past 35 years and the current 'granule exocytosis' model integrates work from many investigators to outline the primary means of NK cytotoxicity. A secondary pathway is mediated through so-called death-receptors, and exhibits less relevance for host defense against pathogens in vivo but is important for the elimination of auto-reactive lymphoid cells and also homeostasis. NK cells directly execute the destruction of both autologous and allogeneic cells. NK cytotoxicity has a direct role in host defense and immunopathology in response to viral infection and malignant transformation.

KEY WORDS

Perforin, Granzyme, Cytotoxicity, Lytic granule, Pore, Death receptor

Introduction

Natural killer (NK) cells were named plainly in 1975 for their 'innate' and 'cytotoxic' character. They were discovered as a population of splenic-derived lymphocytes that were neither B nor T cell subsets and exhibited cytolytic activity equal to that of the entire spleen (Herberman et al., 1975; Kiessling et al., 1975). Now, NK cells are heralded as the first wave of defense against a variety of pathogens in which they employ potent weaponry to kill infected and malignant cells. In addition to killing these 'stressed' cells, NK are implicated in the elimination of autologous activated immune cells (T cells and macrophages) following an inflammatory response (Lu et al., 2007; Rabinovich et al., 2003; Spielman et al., 1998; Van Dommelen et al., 2006). Furthermore, NK are critical producers of cytokines, most notably the pro-inflammatory interferon-γ (IFN-γ), which has pleiotropic effects on cells that modulate the adaptive immune response (Schoenborn and Wilson, 2007). NK cells are the innate constituents of cytotoxic lymphocytes and they utilize identical cytolytic pathways as the CD8+ cytotoxic T lymphocytes (CTL), namely via granule exocytosis and death-receptor mediated killing. Interestingly, recent evidence suggests that NK have an adaptive component as well, marked by persistence of previously expanded and contracted virus-specific NK following murine cytomegalovirus (MCMV) infection (Sun et al., 2009). These 'memory' NK are capable of mediating protective immunity in naïve mice. Regardless, the distinction between CTL and NK lies in the fact that the latter are equipped with pre-formed granules containing lytic proteins, while CTL turn on granule formation upon antigen-recognition, subsequent activation and differentiation. NK cells are considered part of the innate immune system due to their germ-line encoded, early and rapid cytolytic response to a diverse array of pathogens, although they do not kill 'non-specifically'. The simple fact that NK cytotoxicity can cause cell death *demands* strict regulation upon the activation of NK by potential target cells. This regulation is achieved by a multitude of surface receptors expressed by NK cells that integrate activating and inhibitory signals to translate the 'threat' of the target cell into an appropriate response. The receptors that dictate NK cytolytic function are described in a separate chapter, thus focus here will be placed on the effector mechanisms leading to cell destruction.

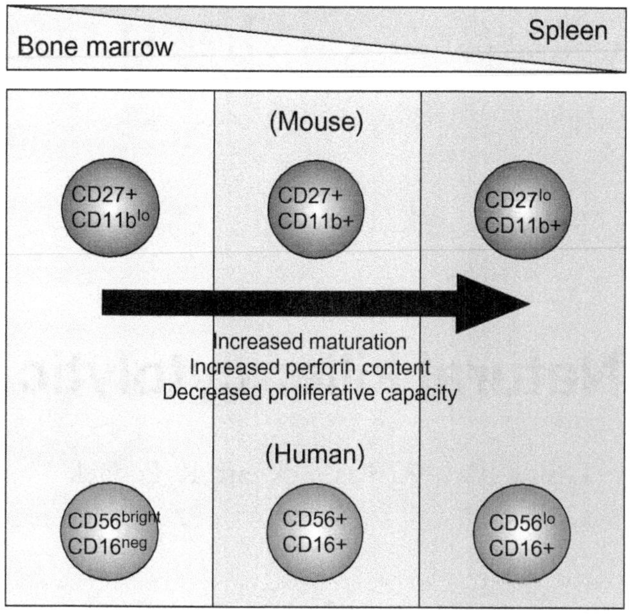

Figure 16.1 • NK cell phenotyping and trafficking. A model of mouse and human NK cell trafficking is proposed in which NK primarily develop in the bone marrow (but also lymph nodes) and migrate to the spleen (or periphery) by undefined signals. NK undergo maturation through discrete stages defined by cell surface expression of CD11b and CD27 (mouse) and CD56 and CD16 (human) and characterized by increased perforin content and decreased proliferative potential.

NK cell subsets and localization

Similar to all immune cell types, both mouse and human NK cells can be divided into subsets defined phenotypically, functionally and on the basis of anatomical localization (Figure 16.1). In the mouse, 'immature' NK cells express the surface receptor, CD11b at low levels and are CD27 positive (Hayakawa and Smyth, 2006). These immature cells are most abundant in the bone marrow and lymph nodes and have low intracytoplasmic perforin and a high proliferative potential. As NK cells move from the bone marrow to the periphery they become CD11b, CD27 double positive (i.e. upregulate CD11b) and when fully mature, they downregulate CD27 expression whilst maintaining CD11b expression. CD11b+CD27lo NK cells are mainly distributed throughout the spleen, lung, liver and blood and they express high levels of perforin but are essentially end stage cells. In humans, similar phenotypes are observed, although with different defining surface markers. CD16+ and CD56dim expression is evident on the cytotoxic (i.e. high perforin-expressing) NK cells that circulate in the blood and spleen, and these cells exhibit tumouricidal properties (Penack et al., 2005). CD16neg CD56bright NK cells lack perforin and are abundant in lymph nodes and tonsils, both of which are areas of immune cell production and maturation (Ferlazzo and Munz, 2004).

Granule exocytosis model overview

The granule exocytosis model describes the major pathway utilized by NK cells to carry out cytotoxic function (Henkart, 1985; Podack and Konigsberg, 1984). Upon recognition of the NK with its target via specific receptor–ligand interactions, subsequent adhesion molecule interactions (i.e. LFA-1 and ICAM-1) initiate immunological synapse (IS) formation. The NK cell IS provides a platform for intense cell to cell signalling as well as a physically defined area for accurate targeting of cytotoxic machinery and cytokines to the target cell (McCann et al., 2003). If the target either sufficiently activates NK cell activating receptors OR fails to display the appropriate major histocompatibility complex (MHC) 'self' molecules to inhibit NK activation via inhibitory receptors, then the NK cell will be sufficiently activated to initiate an irreversible, unidirectional cytolytic response. At this point within the activated NK cell, lytic granules are mobilized toward the IS via polarization of the microtubule organizing center (MTOC) and drastic cytoskeletal rearrangements. Upon fusion of the granular vesicles with the plasma membrane, the lytic contents, namely the pore-forming protein perforin (Dennert and Podack, 1983; Podack and Dennert, 1983) and serine proteases called granzymes (Jenne and Tschopp, 1988), are released into the intercellular space. Perforin monomers are exposed to extracellular levels of free calcium ions ($Ca^{+}2$) in the synaptic space, which elicits simultaneous unfolding, polymerization and insertion of the proteins into apposing target cell plasma membranes (Young et al., 1986). The resulting poly-perforin pores lead to a transient $Ca^{+}2$ influx, which triggers a membrane repair response in the target cell. In the effort to repair the cellular wound, the target cell upregulates vesicular trafficking in which lysosomes and endosomes are rapidly exocytosed to provide a source of healthy membrane donors, while simultaneous pinocytosis occurs in the regions containing damaged membrane (Keefe et al., 2005; Podack, 1999). The repair process inadvertently leads to endocytosis of granzymes and other lytic granule components present within the synaptic space. The final step is delivery of the granzymes into the cytoplasm where they target cellular proteins to initiate programmed cell death by both caspase-dependent and caspase-independent pathways. Thus, the target cell faces a difficult 'catch-22' situation in which repairing damaged membranes leads to the introduction of harmful proteases and toxins that mediate cell death, while the failure to repair damaged membranes leads to colloid osmotic lysis due to the presence of pores in the membrane. In plain terms, the target cell is 'damned if it repairs the membrane, and damned if it does not'.

Lytic granule components

Perforin

Perforin was first characterized as a component present in dense cytoplasmic granules of both NK and CTL (Dennert and Podack, 1983; Podack and Dennert, 1983). Isolated granules from cytolytic lymphocyte cell lines demonstrated rapid and unspecific lysis of any tumour cell, while the NK cell lines themselves showed restriction as to which tumour cells would be lysed. Rapid cytolysis of tumour cells and sheep erythrocytes by the isolated granules was found to be both $Ca^{+}2$-dependent and temperature sensitive (optimal at 37°C and neutral pH). Ultimately, perforin protein was isolated from the cytoplasmic granules and exhibited all the characteristics of the granules themselves (i.e. haemolytic activity against sheep erythrocytes in the presence of $Ca^{+}2$ and at the optimal temperature). Poly-perforin pores are visible on the lysed membranes under an electron microscope (Figure 16.2). The pores are 16nm in diameter and are comprised of up to 16 protomers (Lichtenheld et al., 1988; Podack and Dennert, 1983). Early observations of a second tubular molecule believed to be perforin-2 probably were due to the presence of proteasomes that have similar structures in the electron microscope. The potent effects of perforin-1 and granules in these seminal studies demonstrated an important feature of NK cytotoxic function; likened to the bullets in a gun, perforin itself is not specific—once the 'trigger' is pulled (i.e. granules are aimed and exocytosed) the cell fate is the same regardless of the target cell type.

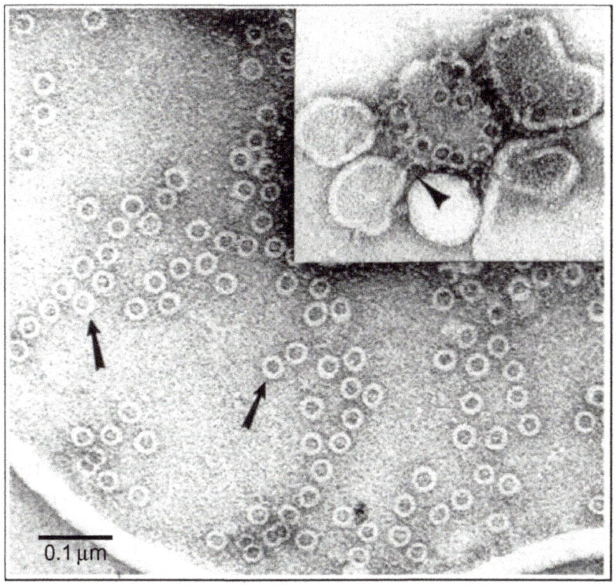

Figure 16.2 • Electron micrograph of membrane-associated perforin pores. Main panel shows top views of pores formed following incubation of NK cells with rabbit erythrocytes and inset shows side views.

Perforin is a 70kDa protein that contains multiple domains, most with important biological properties that have been perfected and conserved throughout evolution from fish to humans (Hwang et al., 2004). The first domain consists of a leader peptide assigning the protein to the secretory pathway. The domain spanning amino acids 163–368 makes up the membrane attack complex of Complement/Perforin (MACPF) domain. The MACPF protein family is made up of pore-forming proteins, such as C6, C7, C8α,C8β and C9 from the complement system that make up the membrane attack complex (DiScipio et al., 1984; Tschopp et al., 1985) and sea anemone toxins, such as AvTX-60A (Oshiro et al., 2004) that contain this conserved domain of approximately 210 amino acids. The MACPF domain is believed to be required for the synchronized polymerization and unfolding of complement (C8α and C9) and perforin proteins into the lipid bilayer of membranes. C9 by itself forms lytic pores similar in structure to poly-perforin, although physiologically the earlier complement components accelerate, in a catalytic fashion, C9 polymerization and direct it at the appropriate targets (Young et al., 1986). Targeting for perforin is achieved by the killer cell specificity prior to granule exocytosis. Recent crystallographic studies of the MACPF-domain containing proteins, Plu-MACPF from *Photorhabdus luminescens* (Rosado et al., 2007) and the human complement protein C8α (Hadders et al., 2007; Slade et al., 2008) demonstrate structural similarity with the well-described pore-forming cholesterol-dependent cytolysins (CDC) from gram-positive bacteria, thereby suggesting similar molecular mechanisms of action. The majority of CDCs are secreted toxins that permeabilize membranes of host cells and organelles during infection by assembling a ring-shaped pre-pore that is inserted into the membrane forming an amphipathic β-barrel pore. It was long hypothesized that conserved α-helices within the MACPF domain were required for membrane insertion (Peitsch et al., 1990), however it has recently been shown to be a conserved fold similar to that in the CDC proteins that dictates membrane insertion and β-barrel pore formation (Rosado et al., 2008). Perforin, itself has not been crystallized due to its instability and inherent cytotoxicity.

The C-terminal domain of perforin contains a proteolytic cleavage site, which upon enzymatic cleavage renders the protein active by revealing a phospholipid-binding C2 domain (Uellner et al., 1997). Cleavage of perforin occurs during the biosynthesis of the lytic granules, so that once the secretory granules are formed the proteins contained within them will not necessitate further modification—allowing rapid killing upon granule exocytosis. However, the protease responsible for cleaving perforin into its 60kD active form (from a 70kD pre-protein) is still unknown. Interestingly, the cleavage product consists of only the last 12–20 amino acids of the C-terminus, constituting ~1–2 kDa. So how does the protein molecular weight decrease by 10 kDa? Asp (548) is heavily glycosylated during the protein's biogenesis in the Golgi, serving to obscure the cleavage site and prevent premature perforin activation (Voskoboinik and Trapani, 2006). Another mechanism shown to prevent inadvertent pore formation via perforin activation in the killer cell is maintaining an acidic pH in the granules (pH 4.9–5.2) (Young et al., 1987). The lower pH ensures protonation of aspartate residues of the C2 domain that would otherwise bind free calcium (Voskoboinik et al., 2005). Upon release of the granule contents into the synaptic space, active perforin is exposed to neutral pH and extracellular concentrations of free $Ca^{+}2$ (~6500 times higher than inside the cell) eliciting the conformational change and polymerization required for insertion into the target cell plasma membrane. Balaji and associates (2002) proposed a third protective mechanism utilized by killer cells to prevent 'self'-targeted pore formation—a membrane-bound, lysosomal protease, cathepsin B that is stored within lytic granules. Upon granule exocytosis, the protease would effectively shield the NK cell membrane by inactivating perforin monomers that have aimlessly diffused back toward the killer cell.

Granzymes

A family of lymphocytic serine proteases, known as granzymes, make up the bulk protein mass (90%) of the cytolytic granule contents. Five granzymes have been identified in humans: A, B, H, M and tryptase-2/granzyme 3 (Hameed et al., 1988), and eleven have been identified in mice (Grossman et al., 2003; Jenne and Tschopp, 1988; Kaiserman et al., 2006). Granzyme B is by far the most well characterized protein in the granzyme family. Substrate specificity of granzyme B is slightly varied by species (human, mouse and rat), but was first classified as an asp-ase, with similarities to the apoptotic cysteine protease family (caspases) because it prefers to cleave following aspartic acid residues in the substrate P1 position. In all species, cleavage of target peptides by granzyme B is highly restricted by peptide length and sequence (Harris et al., 1998). Additional studies of human and mouse granzyme B specificity revealed the recognition sequence in humans to be (I/V)-(G/E)-A-D-X-L-V, while in mice it differs as such, (I/L)-X-(F/Y)-D-X-G-V (Kaiserman et al., 2006). It turns out that the recognition sequences are divergent enough to impart functional differences to granzyme B amongst the two species, which was further supported following identification of their crystal structures (Rotonda et al., 2001). Interestingly, human granzyme B is 30 times more cytotoxic than the mouse protein, as determined by in vitro co-incubation of various tumour cells with purified perforin and granzymes.

The second most well studied granzyme, A, behaves as a tryptase owing to its preference for cleaving substrates with basic residues in the P1 position (arginine and lysine). The crystal structure data for granzyme A indicates that mouse and human proteins have distinct extended substrate specificities, suggesting functional divergence as well. Indeed, Kaiserman and colleagues showed that human granzyme A was not cytotoxic in their system when introduced to human or mouse tumour cells, while the murine protein displayed potent cytolytic activity. The differences in granzyme A and B substrate specificity and function between species highlight their intrinsic redundancy as a protein family. Granzymes seem to have evolved in a species-specific manner to overcome immune assaults through both gene duplication and substrate specificity variation. Moreover, this substantiates the use of caution when making generalizations about a specific granzyme function based on results from only one species.

The functions of the remaining granzymes (not A or B) are essentially unknown, although their general biochemical specificities have been determined. In humans, the three remaining granzymes have the following characteristics: H is a chymase, K is a tryptase (like A) and M is a metase (Kam et al., 2000). The granule protease genes are distributed into clusters within mammalian genomes. Granzymes A and K are linked on chromosomes 13 and 5 in the mouse and human, respectively. The granzyme B cluster in mice and humans are both located on chromosome 14 (Grossman et al., 2003).

Once taken into the cell and released into the cytoplasm granzymes rapidly target a number of substrates, some of which are directly involved in apoptotic pathways, including procaspases (Russell and Ley, 2002). There are two published and highly debated models concerning granzyme delivery into the cytoplasm. The vesicle burst model argues that granzymes released into the synaptic space can bind mannose 6-phosphate (M6P) receptors directly on the surface of target cells, and are endocytosed and released into the cytoplasm to induce apoptosis (Motyka et al., 2000). Further studies supporting this model showed that heparin sulfate (among other proteoglycans) bound to the M6P receptors both stabilized and enhanced granzyme B uptake in vitro in the absence of perforin (Bird et al., 2005; Kurschus et al., 2005; Raja et al., 2005). However, in vivo, this observation holds questionable relevance, as perforin is absolutely required for cytotoxicity via the granule exocytosis pathway. The heparin sulfate-M6P receptor uptake of granzyme B appears to be non-specific as demonstrated by mutating the binding capabilities of granzyme B and diminished surface expression of heparin sulfate, both of which had no effect on granzyme B delivery and subsequent killing of target cells by NK cell lines (Kurschus et al., 2008). On the other hand, the second model called the pore delivery model is consistent with a requirement for perforin in efficient granule-mediated killing. It is characterized by the diffusion of granzymes through poly-perforin pores into the cytoplasm from within de novo vesicles created by the repair endocytosis process.

As mentioned before, granzymes as a family have a wide range of substrate specificities. Granzyme B induces DNA damage through direct and indirect procaspase-activation thereby initiating apoptotic death in a caspase-like fashion (Trapani and Sutton, 2003). Granzyme A does not have the ability to activate caspases, rather it targets nuclear proteins directly to induce DNA fragmentation via single-stranded DNA breaks (Beresford et al., 1999). Granzyme A can also target mitochondria in target cells and cause non-apoptotic death, which is dependant on reactive oxygen species (ROS) generation (Martinvalet et al., 2008). Granzymes are capable of cleaving multiple apoptotic and non-apoptotic substrates leading to uncontrolled, rapid death. This type of death may also include the failure to upregulate receptors for orderly removal characteristic for apoptosis. Interestingly, in vitro tumouricidal assays with NK and lymphomas in which the transformed cells have upregulated anti-apoptotic proteins like B-cell leukaemia/lymphoma 2 (bcl-2), granule exocytosis mediated killing still results in apoptosis of the target cell (Godal et al., 2006). Therefore, killing by NK-mediated granule exocytosis can circumvent cellular apoptotic pathways to induce target cell death that is distinct from programmed cell death.

Granulysin

Granulysin is a member of the saposin-like protein (SAPLIP) family that is also a component of human CTL and NK lytic granules. Consistent with other proteins in the SAPLIP family, including NK-lysin, an antibacterial porcine protein, and amoebapores, the granulysin structure suggests it has lytic properties (Clayberger and Krensky, 2003). A seminal study on the role of granulysin in anti-bacterial immunity (Stenger et al., 1998) demonstrated that granulysin was potently adept in lysing various bacterial species extracellularly (i.e. eukaryotic cell-free systems), such as *S. aureus*, *E. coli*, *L. monocytogenes* and *S. typhimurium*. However, granulysins anti-bacterial effect was abrogated in the presence of *M. tuberculosis*-infected macrophages. The co-addition of perforin and granulysin, however, led to efficient killing of intracellular *M. tuberculosis*, defining a clear role for perforin in providing access of anti-microbial granule proteins to their targets. Further, the perforin-dependency of granulysin was confirmed in a model of *Listeria innocua* infection of dendritic cells (Walch et al., 2007). Most reports on granulysin focus on its mechanism in CTL cytotoxicity, not NK. This can be attributed to the observation that perforin-deficient mice do not exhibit defects in early control of *M. tuberculosis* infection, while long-term protection requires perforin (Stenger et al., 1998). Thus, the role of granulysin in NK-mediated function has yet to be determined. Indeed, a granulysin knockout mouse would elucidate some of these questions; however, a murine homologous protein of granulysin does not exist, and as such granulysin-deficiency in humans has yet to be characterized.

Serglycin

Serglycin is an 'intracellular' proteoglycan important for the retention of key effector or inflammatory molecules within storage and secretory granules of endothelial cells, mast cells, neutrophils, macrophages and cytotoxic lymphocytes (NK and CTL) (Kolset and Tveit, 2008). Serglycin was identified as a scaffolding protein for granzyme B and perforin monomers within lytic granules (Metkar et al., 2002), and it appeared that the complex was taken into cells as an intact trimolecular complex. Whether this mechanism of granzyme delivery naturally occurs in vivo is still controversial (Kurschus et al., 2008); however, the generation of serglycin-deficient mice displayed specific defects in granzyme B storage within lytic granules (Grujic et al., 2005), confirming the importance of serglycin–granzyme B interactions.

Pore delivery model

The original model for NK and CTL cytotoxicity described perforin as the principle effector molecule able to kill cells via pore formation, although, it was known at the time that nucleated cells killed by CTL also exhibited a nuclear degradation not seen after death by complement. Upon discovery and isolation of granzymes and other granule toxins, perforins role was speculated to create transmembrane pores for the 'passive diffusion' of the proteolytic, DNA-damage inducing effector molecules. Under the current granule-exocytosis model, we postulate that, while perforin pores can cause osmotic lysis of target cells, the system has evolved with the acquisition of granzymes and other granule-associated toxins to induce cell death of stressed cells. Granzymes target both caspase-dependent apoptotic signalling pathways as well as other pathways directly. Therefore, the cell is not voluntarily committing suicide or following programmed cell death protocols, as occurs in death-receptor cytotoxic pathways. Nor are the death signals derived from target cell proteins. Rather, granule exocytosis via perforin induces target cells to undergo 'involuntary' suicide by physically and forcibly introducing several different potent death mediators into the cytoplasm. This is a useful system especially in cases of stressed or transformed and viral-infected epithelial cells in which apoptotic pathways are inhibited in order for the virus to survive, and is probably why there are so many isoforms of granzyme proteins (summarized in Figure 16.3).

Consequences of perforin deficiency

As discussed previously, perforin is the rate-limiting reagent in the granule exocytosis pathway of NK-mediated cellular killing. Therefore, the generation of perforin knockout mice has greatly helped the elucidation of the contribution of granule exocytosis to the immune system as well as the mechanisms by which it acts. However, these mice are deficient in NK and CTL-mediated cytotoxicity (not to mention NKT and γδ T cell types, which both also express perforin) so the direct effect of NK cell perforin function is often obscured. The perforin knockout mouse was introduced in 1994 (Kagi et al., 1994) and since then numerous challenge studies have been performed. Homozygous perforin-deficient mice exhibit a null mutation in the

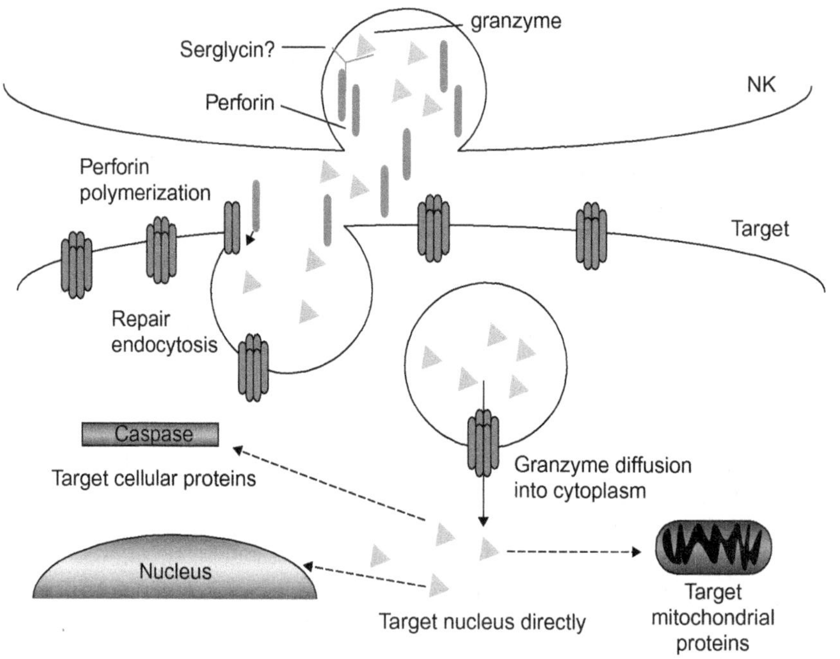

Figure 16.3 • Pore-delivery model of granzyme release. Perforin and granzymes are stored in granules within an NK cell (possibly stabilized by a tethering molecule such as serglycin) until granule exocytosis is initiated and granule contents are expelled toward the target cell. Extracellular calcium levels trigger perforin pore formation into the lipid bilayers of apposing cells. Membrane pores in the target cell elicit rapid endocytosis to repair the membrane, in which intact pores and other soluble granule components are endocytosed. Granule toxins (mainly granzymes) can then diffuse out of the endosome through the poly-perforin pore to access the cytosol and target any number of substrates to induce cell death.

PRF1 gene, caused by disruption of the MACPF domain (residues 167–221) (Lowin et al., 1994).

Perforin and tumour protection and surveillance

NK cells have been postulated as necessary for the clearance of tumour cells since their initial discovery due to their potent in vitro tumour-killing abilities (Trinchieri, 1989). Van den Broek et al. (1996, 2000) demonstrated systematically that perforin knockout mice displayed examples of impaired tumour control in all of three different tumourigenic approaches; syngeneic tumour cell line injection, carcinogenic substance exposure, or injection of oncogenic viruses. Interestingly, perforin-deficient mice exhibited increased dose sensitivity to methylcolanthrene (MCA)-induced fibrosarcomas compared to wild-type and CD8-depleted mice, suggesting that NK cell-derived perforin was especially important for surveillance of tumours induced via carcinogen exposure. In a separate model, 12-O-tetradecanoyl phorbol-13-acetate (TPA) + 7,12-dimethylbenzanthracene (DMBA)-induced papillomas, perforin expression did not impart reduced incidence or slower kinetics of papilloma formation to the mice suggesting that NK cytotoxicity may be important only for tumour surveillance relative to particular carcinogenic substances.

Depletion of NK cells in mouse studies via antibody-targeted depletion to various NK-specific receptors such as the glycolipid asialo-GM1 or NK1.1 have supported the results from perforin knockout mice showing the importance of NK cells for anti-tumour immune responses in vivo. The depletion in these studies lack absolute specificity, since other cell types can also express asialo-GM1 and NK1.1. However, NK cell-deficient mice are more susceptible to MCA-induced sarcomas than wild-type mice (Smyth et al., 2005a,b), similar to perforin-deficient mice.

Perforin and viral infection

NK cells play a critical role in containing viral infection especially at early time points as the adaptive immune system needs a considerable amount of time (up to a week) to mature and migrate to the site of infection. Cytopathic and non-cytopathic viral infections are controlled by NK cytotoxicity, including herpesviruses such as cytomegamolvirus (CMV), herpes simplex virus (HSV), Epstein–Barr virus (EBV) and HIV. NK cell deficiencies in humans, though rare, point to a profound involvement in control of herpesvirus infections (Orange, 2006). Type I IFNs (α/β) have been implicated in restoring NK cell perforin expression in vivo from HIV patients, whose NK activity is compromised (Fogli et al., 2008; Portales et al., 2003). In the murine model of CMV infection (MCMV), perforin−/− mice display increased viraemia and decreased survival compared to wild type mice (Loh et al., 2005; Tay and Welsh, 1997). Furthermore, in the MCMV infection model, granzyme (A and B cluster) expression is not required for survival but seems to play a role in controlling viral replication at early time points (Van Dommelen et al., 2006). The same study demonstrated that NK cell-derived perforin is crucial for survival from MCMV infection not only because of its early effects on controlling viral replication but also due to its containment of the resulting immunopathology. Perforin−/− mice display increased organ pathology (spleen, liver and lung) based on haemorrhaging and architectural remodelling compared to granzyme A/B−/− mice, which is caused by tumour necrosis factor (TNF)-α-producing macrophages in a perforin-dependent manner. In studies with granzyme–deficient (including A X B cluster deficient mice), it is difficult to conclude their contribution to the cytotoxic response because of the possibility that other granule proteases are present to carry out their apoptotic function. However, the results of the Van Dommelen study beg the question of whether specific granzymes (or other granule proteases) might mediate NK immunoregulatory cytotoxic functions versus NK anti-viral cytotoxic functions. Granzymes were only important for control of MCMV infection during the acute phase. It is possible that granzyme A and B target specific genes or pathways crucial to MCMV survival, while another NK-derived perforin-dependent toxin regulates elimination of macrophages.

In response to lymphocytic choriomeningitis virus (LCMV) infection, perforin-deficient mice develop the fatal haemophagocytic lymphohistiocytosis (HLH)-like disease characterized by an excessive production of IFN-γ leading to haemophagocytosis by over stimulated macrophages, consequently resulting in cytopenia (Jordan et al., 2004). This is profoundly similar to the disease presented by perforin-deficient humans, familial HLH (FHL) type 2. FHL is a genetic disorder (Stepp et al., 1999) with 60% of reported cases caused by mutations in the gene encoding perforin, *PRF1*, which can be categorized as follows: (1) both alleles with nonsense or frame-shift mutations, resulting in premature termination of the protein, (2) compound heterozygotes with one allele carrying a nonsense or frame-shift mutation and the other having a missense mutation, (3) homozygotes with both alleles having the same missense mutation, and (4) compound heterozygotes with both alleles bearing missense mutations (categories taken directly from Voskoboinik and Trapani, 2006). Interestingly, the remaining FHL patients do not have mutations in perforin. This observation along with the fact that perforin is highly conserved and has no known isoforms suggests that the remaining mutations in FHL patients must map to genes whose proteins directly regulate perforin function and release. Indeed, a mutation in Munc13-4, a cytoplasmic granule trafficking protein required for the fusion of vesicles with the plasma membrane upon degranulation, was discovered to characterize at least 20% of FHL patients (Feldmann et al., 2003).

At first there was confusion over how the disease manifestation of FHL could be related to the LCMV-induced HLH-like disease in perforin-deficient mice, since the human disease is genetic and was thought to be a spontaneous lymphoproliferative disorder based on the dysregulation of homeostatic mechanisms conferred by NK cytotoxicity (Russell and Ley, 2002). Interestingly, perforin knockout mice are healthy when raised in controlled, pathogen-free conditions, except for a small percentage of otherwise unchallenged mice that succumbed to adenocarcinoma in their old age (Street et al., 2002). However, perforin/Fas ligand (FasL) double deficient mice succumb to an auto-aggressive disorder at 10 weeks under pathogen free conditions. This finding has been interpreted to indicate that perforin is required to kill antigen presenting cells (APCs) thereby preventing excessive CTL activation (Spielman et al., 1998). The role of NK-derived perforin in protection against immunopathology during MCMV infection demonstrated by Van Dommelen et al., at last, experimentally alleviates the confusion. It would be almost impossible for anyone, including someone with a mutation within the *PRF1* gene, to avoid a viral infection at some point in his or her life. Therefore, the manifestation of HLH disorders in humans is likely triggered by a viral or some other infection.

Perforin and autoimmunity

Given the intricate signalling requirements for activation and potent killer function that characterize NK cytolytic activity, it is not surprising that NK cells could also provide a regulatory, immuno-suppressive function in vivo. Indeed, NK cells are responsible for eliminating activated CD4+ T cells in a perforin-dependent manner (Lu et al., 2007). The mechanism involves Qa1, a ligand for the inhibitory NK receptor, NKG2A. The absence of Qa1 transmits a death signal for T cells, similar to the absence of MHC class I molecules. In the experimental autoimmune encephalomyelitis (EAE) model, NK cells were responsible for a higher clinical score compared to NK-depleted mice and blocking Qa1 using antibody ameliorated the disease symptoms.

Granzyme-deficient mice

As demonstrated by gene knockout technology in mice, without perforin the function of granzymes alone does not lead to target cell death in vivo, despite several reports that demonstrate perforin-independent in vitro endocytosis of granzymes (Froelich et al., 1996; Motyka et al., 2000; Pinkoski et al., 1998; Shi et al., 1997). The fate of a mouse deficient in a single granzyme protein is that of a wild-type mouse due to the redundancy of function within granzyme isoforms present in the genome. Nevertheless, granzyme A-deficient mice were found to display normal NK cell cytotoxicity (as well as CTL-mediated cytotoxicity), and granzyme B cluster-deficient mice exhibit delayed, but otherwise normal cytotoxicity as determined by examining DNA fragmentation (Ebnet et al., 1995; Heusel et al., 1994). The use of a double deficient mouse (granzymes A and B cluster) revealed comparable susceptibility to ectromelia virus, the cytopathic agent of mousepox, as in the perforin-deficient mice (Mullbacher et al., 1999), and the susceptibility was in part attributed to a decreased ability of CTL to specifically lyse ectromelia virus-infected cells ex vivo. The CTL from these mice were not deficient, however, in their ability to lyse other virus-infected targets or tumour cells (NK cytotoxicity was not directly examined in this study). Granzyme A/B double deficient mice displayed few other defects in tumour or virus infection models, further corroborating the idea that individual granzymes evolved in a species-specific manner to combat specialized pathogens.

In an attempt to determine the effect of a pan-granzyme deficiency while maintaining functional perforin, Pham and Ley (1999) generated a dipeptidyl peptidase I (DPPI) knockout mouse. DPPI (also known as cathepsin C) is necessary for the enzymatic cleavage of the N-terminal dipeptide of all granzyme proteins and is stored in the lysosomal granules. DPPI-deficient mice are lacking functional granzymes and all bone marrow derived serine proteases. NK cells were not directly examined in the study, however, CTL from DPPI−/− mice showed severe defects in death receptor-independent cytotoxicity due to an inability to induce DNA degradation, finally supporting an essential role granule serine proteases in granule exocytosis cytotoxicity.

Death-receptor mediated cytotoxicity

While perforin-deficient mice and humans are absolutely impaired in their cytotoxic functions, alternative cytolytic pathways can be utilized for the destruction of tumour cells. Here, we introduce the function of apoptosis mediated by death ligands in NK cytotoxicity. Many so-called death ligands and receptors belong to the TNF family of cytokines and TNF receptor (TNFR) family, respectively. FasL is a type II membrane protein and is expressed almost exclusively by NK and CTL. FasL can exist as a membrane-bound protein or it can be cleaved by a metalloproteinase to generate a soluble ligand (Tanaka et al., 1996). FasL is expressed constitutively by NK cells and is packaged in the cytoplasmic granules. The receptor for FasL, Fas (CD95) is widely distributed in various tissues and was found to cross-link CTL in the absence of perforin, granule exocytosis and $Ca^{+}2$ (Rouvier et al., 1993). Fas trimerizes upon ligand recognition and causes aggregation of death domains in the cytoplasmic region of the receptor, which leads to endocytosis of the receptor: ligand complex. Following internalization, the adaptor molecule FADD (Fas associated death domain) binds the death domains of the receptor complex and the classical apoptotic signalling cascade is initiated. Tumour cells have been shown to upregulate Fas expression in response to NK-produced IFN-γ, thus rendering them susceptible to NK and CTL-mediated Fas-dependent killing (Screpanti et al., 2001). Although, in an apparent mechanism to thwart the anti-tumour response, a variety of tumour cells also express FasL which has been shown to result in NK cell depletion (Khar et al., 1998), decreased numbers of tumour-infiltrating lymphocytes (TIL) (Bennett et al., 1998; Okada et al., 2000) and subsequent immune evasion in vivo. However, plasmid-directed overexpression of FasL in tumour cell lines and allografts leads to their swift rejection in animal models, which is mediated mostly by neutrophils (Chen et al., 2004; Kang et al., 1997; Takeuchi et al., 1999). Although it is still controversial, current evidence points toward a role for FasL in immune evasion rather than immune protection especially since the overexpression experiments rely primarily on inflammatory means of rejection, which are known to be greatly suppressed in the tumour microenvironment (Dumont and Arteaga, 2002). Furthermore, evidence by Ryan et al., (2005) showed that silencing of the FasL gene in a colon cancer cell line led to an increased anti-tumour response in vivo.

TNF-related apoptosis-inducing ligand (TRAIL) is another type II membrane protein expressed by NK and CTL. TRAIL receptors are evident in both mice and humans, and signalling through functional receptors on target cells also integrate on the FADD-dependent signalling cascade. In addition to the apoptosis-signalling TRAIL receptors, DR4 and DR5, there are 'decoy' receptors that lack a functional death domain but serve to regulate TRAIL-mediated programmed cell death (i.e. to effectively absorb excess ligand). TRAIL is upregulated on NK cells following stimulation with pro-inflammatory cytokines, IL-2, IL-15 and IFNs (Smyth et al., 2005a,b).

Deficiency of death receptors and ligands

The physiological function of apoptosis by death ligands and receptors is largely confined to homeostatic regulation of lymphocytes and other hematopoietic cells. The generation of FasL−/− (*gld*) and Fas−/− (*lpr*) mice has been instrumental in determining the role of death-receptor mediated pathways of cytotoxicity (Takahashi et al., 1994; Watanabe et al., 1992). Mice with either of these mutations develop autoimmune nephritis and other symptoms consistent with the disease manifestation of human systemic lupus erythematosus (SLE) (Matiba et al., 1997). The lymphoproliferation and autoantibody production displayed in the *lpr* and *gld* mice confirms their role in the control and depletion of 'self' reactive adaptive immune lymphocytes. Fas receptor mutants occur in humans as well, and they exhibit autosomal lymphoproliferative disorder (ref).

TRAIL deficient mice have been generated (Cretney et al., 2002), and their defects are more consistent with a loss of NK cytotoxic function. For example, TRAIL deficiency yields a more rapid onset of fibrosarcoma formation in response to the carcinogen, MCA. Further, NK cells in the liver constitutively express TRAIL and the loss of expression via addition of blocking antibodies or gene disruption revealed an anti-metastatic role for this receptor in the Renca tumour metastases model (Cretney et al., 2002; Takeda et al., 2001). This group also reported that comparable to perforin−/− mice, TRAIL deficient cells were unable to kill tumour cell lines in vitro. The coordinated mechanisms of TRAIL function with granule exocytosis and perforin function have not yet been elucidated.

Immunological consequences of NK cytotoxicity

All potential targets of NK cytotoxicity, determine their own fate so to speak, by the expression of ligands to NK receptors on their surface. The target cell can modulate NK function in two respects: (1) expression of ligands with specificity to NK activating receptors will lead to cell death, and (2) expression of adequate 'self' MHC molecules will lead to survival through interaction with NK inhibitory receptors. It is well understood through a vast amount of research (Lanier, 2008; Moretta et al., 2008), that the 'overall' message delivered by the target cell to the NK cell determines the fate (i.e. activation > inhibition yields activation; and activation < inhibition yields inhibition). Thus, it is in the best interest for intracellular pathogens, in the evolutionary sense, to alter ligand expression accordingly to promote growth and survival, and vice versa for the killer cell.

An early misconception concerning NK cell function is that they do not require priming like CTL, owing to the pre-formed lytic granules and rapid cytolysis, although compelling evidence indicates that the cytokine milieu along with ligation of activating receptors (and absence of ligation of inhibitory receptors) can activate NK subsets to proliferate, secrete IFN-γ and/or increase their cytotoxic potential (Biron et al., 1999). Cytotoxic potential refers to the amount of lytic proteins expressed by the NK cell, that is, higher expression of perforin and granzymes means greater cytotoxic potential. The cytokines affecting NK function are primarily pro-inflammatory cytokines such as IL-12, IL-15, IL-18, IL-21 and IFN-α/β(Lee et al., 2007).

How does NK-mediated cell killing affect the overall immune response? Nucleated target cells that fall prey to NK cell killing undergo cell death that can frequently be characterized by the utilization of well-known apoptotic mechanisms (such as caspase-dependent proteolytic cascades). However, alternative mechanisms have been described in which granule effector components directly target the DNA and mitochondria to yield target cell death. Furthermore, the presence of pro-apoptotic proteins and specific granzyme inhibitors does not block death by granule exocytosis. Either way you slice it, NK-mediated granule exocytosis is an extremely efficient means of cell destruction. The accompanied cytokine storm (especially IFN-γ) from the cytolytic NK cell is responsible for activating local and recruited APCs to control the spread of microbial infections and also induce an adaptive response to mediate memory responses from CD4+ and CD8+ T cells and B cells. NK also kill uninfected, activated monocyte/macrophages and dendritic cells to curtail the inflammatory response, as they are important producers of inflammatory cytokines like TNF and IL-12 (Moretta et al., 2006). This function of NK cytotoxicity is essential in preventing overt inflammation and induction of systemic inflammatory syndrome. These mechanisms of NK function are manifested most clearly in the HLH-like syndrome caused by LCMV infection in mice and FHL disease exhibited in patients with mutations in the *PRF1* gene, in which the loss of perforin and consequently granule exocytosis leads to fatal immunopathology.

References

Balaji, K.N., Schaschke, N., Machleidt, W., Catalfamo, M. and Henkart, P.A. (2002). Surface cathepsin B protects cytotoxic lymphocytes from self-destruction after degranulation. J Exp Med **196**, 493–503.

Bennett, M.W., O'Connell, J., O'Sullivan, G.C., Brady, C., Roche, D., Collins, J.K. and Shanahan, F. (1998). The Fas counterattack in vivo: apoptotic depletion of tumor-infiltrating lymphocytes associated with Fas ligand expression by human esophageal carcinoma. J Immunol **160**, 5669–5675.

Beresford, P.J., Xia, Z., Greenberg, A.H. and Lieberman, J. (1999). Granzyme A loading induces rapid cytolysis and a novel form of DNA damage independently of caspase activation. Immunity **10**, 585–594.

Bird, C.H., Sun, J., Ung, K., Karambalis, D., Whisstock, J.C., Trapani, J.A. and Bird, P.I. (2005). Cationic sites on granzyme B contribute to cytotoxicity by promoting its uptake into target cells. Mol Cell Biol **25**, 7854–7867.

Biron, C.A., Nguyen, K.B., Pien, G.C, Cousens, L.P. and Salazar-Mather, T.P. (1999). Natural killer cells in antiviral defense: function and regulation by innate cytokines. Annu Rev Immunol **17**, 189–220.

Chen, C.M., Song, W., Kao, J.Y., Zheng, Q. D. and Chen, J.J. (2004). Expression of Fas ligand is not a main mechanism used by tumors to counteract antitumor immunity. Front Biosci **9**, 448–456.

Clayberger, C. and Krensky, A.M. (2003). Granulysin. Curr Opin Immunol **5**, 560–565.

Cretney, E., Takeda, K., Yagita, H., Glaccum, M., Peschon, J.J. and Smyth, M.J. (2002). Increased susceptibility to tumor initiation and metastasis in TNF-related apoptosis-inducing ligand-deficient mice. J Immunol **168**, 1356–1361.

Dennert, G. and Podack, E.R. (1983). Cytolysis by H-2-specific T killer cells. Assemby of tubular complexes on target membranes. J Exp Med **157**, 1483–1495.

DiScipio, R.G., Gehring, M.R., Podack, E. R., Kan, C.C., Hugli, T.E. and Fey, G.H. (1984). Nucleotide sequence of cDNA and derived amino acid sequence of human complement component C9. Proc Natl Acad Sci U S A **81**, 7298–7302.

Dumont, N. and Arteaga, C.L. (2002). The tumor microenvironment: a potential arbitrator of the tumor suppressive and promoting actions of TGFbeta. Differentiation **70**, 574–582.

Ebnet, K., Hausmann, M., Lehmann-Grube, F., Mullbacer, A., kopf, M., Lamers, M. and Simon, M.M. (1995). Granzyme A-deficient mice retain potent cell-mediated cytotoxicity. EMBO J **14**, 4230–4239.

Feldmann, J., Callebaut, I., Raposo, G., Certain, S., Bacq, D., Dumont, C., Lambert, N., Ouachee-Chardin, M., Chedeville, G., Tamary, H., Minard-Colin, V., Vilmer, E., Blanche, S., Le Deist, F., Fischer, A. and de Saint Basile, G. (2003). Munc13-4 is essetial for cytolytic granules fusion and is mutated in a form of familial hemophagocytic lymphohistiocytosis (FHL3). Cell **115**, 461–473.

Ferlazzo, G. and Munz, C. (2004). NK cell compartments and their activation by dendritic cells. J Immunol **172**, 1333–1339.

Fogli, M., Mavilio, D., Brunetta, E., Varchetta, S., Ata, K., Roby, G., Kovacs, C., Follmann, D., Pende, D., Ward, J., Barker, E., Marcenaro, E., Moretta, A. and Fauci, A.S. (2008). Lysis of endogenously infected CD4+ T cell blasts by rIL-2 activated autologous natural killer cells from HIV-infected viremic individuals. PLoS Pathog **4**(7), e1000101.

Froelich, C.J., Orth, K., Turbov, J., Seth, P., Gottlieb, R., Babior, B., Shah, G.M., Bleackley, R.C., Dixit, V.M. and Hanna, W. (1996). New paradigm for lmphocyte granule-mediated cytotoxicity. Target cells bind and internalize granzyme B, but an endosomolytic agent is necessary for cytosolic delivery and subsequent apoptosis. J Biol Chem **271**, 29073–29079.

Godal, R., Keilholz, U., Uharek, L., Letsch, A., Asemissen, A.M., Busse, A., Na, I.K., Thiel, E. and Scheibenbogen, C. (2006). Lymphomas are sensitive to perforin-dependent cytotoxic pathways despite expression of PI-9 and overexpression of bcl-2. Blood **107**, 3205–3211.

Grossman, W.J., Revell, P.A., Lu, Z.H., Johnson, H., Bredemeyer, A.J. and Ley, T.J. (2003). The orphan granzymes of humans and mice. Curr Opin Immunol **15**, 544–552.

Grujic, M., Braga, T., Lukinius, A., Eloranta, M.L., Knight, S.D., Pejler, G. and Abrink, M. (2005). Serglycin-deficient cytotoxic T lymphocytes display defective secretory granule maturation and granzyme B storage. J Biol Chem **280**(39), 33411–33418.

Hadders, M.A., Beringer, D.X. and Gros, P. (2007). Structure of C8alpha-MACPF reveals mechanism of membrane attack in complement immune defense. Science **317**, 1552–1554.

Hameed, A., Lowrey, D.M., Lichtenheld, M. and Podack, E.R. (1988). Characterization of three serine esterases isolated from human IL-2 activated killer cells. J Immunol **141**, 3142–3147.

Harris, J.L., Peterson, E.P., Hudig, D., Thornberry, N.A. and Craik, C.S. (1998). Definition and redesign of the extended substrate specificity of granzyme B. J Biol Chem **273**(27), 364–373.

Hayakawa, Y. and Smyth, M.J. (2006). CD27 dissects mature NK cells into two subsets with distinct responsiveness and migratory capacity. J Immunol **176**, 1517–1524.

Henkart, P.A. (1985). Mechanism of lymphocyte-mediated cytotoxicity. Annu Rev Immunol **3**, 31–58.

Herberman, R.B., Nunn, M.E., Holden, H.T. and Lavrin, DH. (1975). Natural cytotoxic reactivity of mouse lymphoid cells against syngeneic and allogeneic tumors. II. Characterization of effector cells. Int J Cancer **16**, 230–239.

Heusel, J.W., Wesselschmidt, R.L., Shresta, S., Russell, J.H. and Ley, T.J. (1994). Cytotoxic lymphocytes require granzyme B for the rapid induction of DNA fragmentation and apoptosis in allogeneic target cells. Cell **76**, 977–987.

Hwang, J.Y., Ohira, T., Hirono, I. and Aoki, T. (2004). A pore-forming protein, perforin, from a non-mammalian organism, Japanese flounder, Paralichthys olivaceus. Immunogenetics **56**(5), 360–367.

Jenne, D.E. and Tschopp, J. (1988). Granzymes, a family of serine proteases released from granules of cytolytic T lymphocytes upon T cell receptor stimulation. Immunol Rev **103**, 53–71.

Jordan, M.B., Hildeman, D., Kappler, J. and Marrack, P. (2004). An animal model of hemophagocytic lymphohistiocytosis (HLH): CD8+ T cells and interferon gamma are essential for the disorder. Blood **104**, 735–743.

Kagi, D., Ledermann, B., Burki, K., Seiler, P., Odermatt, B., Olsen, K.J., Podack, E. R., Zinkernagel, R.M. and Hengartner, H. (1994). Cytotoxicity mediated by T cells and natural killer cells is greatly impaired in perforin-deficient mice. Nature **369**, 31–37.

Kaiserman, D., Bird, C.H., Sun, J., Matthews, A., Ung, K., Whisstock, J.C., Thompson, P.E., Trapani, J.A. and Bird, P.I. (2006). The major human and mouse granzymes are structurally and functionally divergent. J Cell Biol **175**, 619–630.

Kam, C.M., Hudig, D. and Powers, J.C. (2000). Granzymes (lymphocyte serine proteases): characterization with natural and synthetic substrates and inhibitors. Biochem Biophys Acta **1477**, 301–323.

Kang, S.M., Schneider, D.B., Lin, Z., Hanahan, D., Dichek, D.A., Stock, P.G. and Baekkeskov, S. (1997). Fas ligand expression in islets of Langerhans does not confer immune privilege and instead targets them for rapid destruction. Nat Med **3**, 738–743.

Keefe, D., Shi, L., Feske, S., Massol, R., Navarro, F., Kirchhausen, T. and Lieberman, J. (2005). Perforin triggers a plasma membrane-repair response that facilitates CTL induction of apoptosis. Immunity **23**, 249–262.

Khar, A., Varalakshmi, C., Pardhasaradhi, B.V., Mubarak Ali, A. and Kumari, A.L. (1998). Depletion of the natural killer cell population in the peritoneum by AK-5 tumor cells overexpressing fas-ligand: a mechanism of immune evasion. Cell Immunol **189**, 85–91.

Kiessling, R., Klein, E., Pross, H. and Wigzell, H. (1975). Natural killer cells in the mouse. II. Cytotoxic cells with specificity for mouse Moloney leukemia cells. Characteristics of the killer cell. Eur J Immunol **5**, 117–121.

Kolset, S.O. and Tveit, H. (2008). Serglycin—structure and biology. Cell Mol Life Sci **65**, 1073–1085.

Kurschus, F.C., Bruno, R., Fellows, E., Falk, C.S. and Jenne, D.E. (2005). Membrane receptors are not required to deliver granzyme B during killer cell attack. Blood **105**, 2049–2058.

Kurschus, F.C., Fellows, E., Stegmann, E. and Jenne, D.E. (2008). Granzyme B delivery via perforin is restricted by size but not by heparan sulfate-dependent endocytosis. Proc Natl Acad Sci U S A **105**, 13799–13804.

Lanier, L.L. (2008). Up on the tightrope: natural killer cell activation and inhibition. Nat Immunol **9**, 495–502.

Lee, S.H., Miyagi, T. and Biron, C.A. (2007). Keeping NK cells in highly regulated antiviral warfare. Trends Immunol **28**, 252–259.

Lichtenheld, M.G., Olsen, K.J., Lu, P., Lowrey, D.M., Hameed, A., Hengartner, H. and Podack, E.R. (1988). Structure and function of human perforin. Nature **335**, 448–451.

Lowin, B., Beermann, F., Schmidt, A. and Tschopp, J. (1994). A null mutation in the perforin gene impairs cytolytic T lymphocyte- and natural killer cell-mediated cytotoxicity. Proc Natl Acad Sci U S A **91**, 11571–11575.

Loh , J., Chu, D.T., O'Guin, A.K., Yokoyama, W.M and Virgin, H.W. IV (2005). Natural killer cells utilize both perforin and gamma interferon to regulate murine cytomegalovirus infection in the spleen and liver. J Virol **79**, 661–667.

Lu, L., Ikizawa, K., Hu, D., Werneck, M. B., Wucherpfennig, K.W. and Cantor, H. (2007). Regulation of activated CD4+ T cells by NK cells via the Qa-1-NKG2A inhibitory pathway. Immunity **26**, 593–604.

Martinvalet, D., Dykxhoorn, D.M., Ferrini, R. and Lieberman, J. (2008). Granzyme A cleaves a mitochondrial complex I protein to initiate caspase-independent cell death. Cell **133**, 681–692.

Matiba, B., Mariani, S.M. and Krammer, P. H. (1997). The CD95 system and the death of a lymphocyte. Semin Immunol **9**, 59–68.

McCann, F.E., Vanherberghen, B., Eleme, K., Carlin, L.M., Newsam, R.J., Goulding, R.J. and Davis, D.M. (2003). The size of the synaptic cleft and distinct distributions of filamentous actin, ezrin, CD43, and CD45 at activating and inhibitory human NK cell immune synapses. J Immunol **170**, 2862–2870.

Metkar, S.S, Wang, B., Aguilar-Sentelises, M., Raja, S.M., Uhlin-Hansen, L., Podack, E., Trapani, J.A. and Froelich, C.J. (2002). Cytotoxic cell granule-mediated apoptosis: perforin delivers granzyme B–serglycin complexes into target cells without plasma membrane pore formation. Immunity **16**, 417–429.

Moretta, L. (2006). Effector and regulatory events during natural killer–dendritic cell interactions. Immunol Rev **214**, 219–228.

Moretta, A., Marcenaro, E., Parolini, S., Ferlazzo, G. and Moretta, L. (2008). NK cells at the interface between innate and adaptive immunity. Cell Death Differ **15**, 226–233.

Motyka, B., Korbutt, G., Pinkoski, M.J., Heibein, J.A., Caputo, A., Hobman, M., Barry, M., Shostak, I., Sawchuk, T., Holmes, C.F.B., Gauldie, J. and

Bleackley, R.C. (2000). Mannose 6-phosphate/insulin-like growth factor II receptor is a death receptor for granzyme B during cytotoxic T cell-induced apoptosis. Cell **103**, 491–500.

Mullbacher, A., Waring, P., Tha Hla, R., Tran, T., Chin, S., Stehle, T., Museteanu, C. and Simon, M.M. (1999). Granzymes are the essential downstream effector molecules for the control of primary virus infections by cytolytic leukocytes. Proc Natl Acad Sci U S A **96**, 13950–13955.

Okada, K., Komuta, K., Hashimoto, S., Matsuzaki, S., Kanematsu, T. and Koji, T. (2000). Frequency of apoptosis of tumor-infiltrating lymphocytes induced by fas counterattack in human colorectal carcinoma and its correlation with prognosis. Clin Cancer Res **6**, 3560–3564.

Orange, J.S. (2006). Human natural killer cell deficiencies. Curr Opin Allergy Clin Immunol **6**(6), 399–409.

Oshiro, N., Kobayashi, C., Iwanaga, S., Nozaki, M., Namikoshi, M., Spring, J. and Nagai, H. (2004). A new membrane-attack complex/perforin (MACPF) domain lethal toxin from the nematocyst venom of the Okinawan sea anemone *Actineria villosa*. Toxicon **43**, 225–228.

Peitsch , M.C., Amiguet, P., Guy, R., Brunner, J., Maizel, J.V. and Tschopp, J. Jr (1990). Localization and molecular modelling of the membrane-inserted domain of the ninth component of human complement and perforin. Mol Immunol **27**, 589–602.

Penack, O., Gentilini, C., Fischer, L., Asemissen, A.M., Scheibenbogen, C., Thiel, E. and Uharek, L. (2005). CD56dimCD16neg cells are responsible for natural cytotoxicity against tumor targets. Leukemia **19**, 835–840.

Pham, C.T.N. and Ley, T.J. (1999). Dipeptidyl peptidase I is required for the processing and activation of granzymes A and B in vivo. Proc Natl Acad Sci U S A **96**, 8627–8632.

Pinkoski, M.J., Hobman, M., Heibein, J.A., Tomaselli, K., Li, F., Seth, P., Froelich, C. J. and Bleackley, R.C. (1998). Entry and trafficking of granzyme B in target cells during granzyme B–perforin-mediated apoptosis. Blood **92**, 1044–1054.

Podack, E. (1999). How to induce involuntary suicide: the need for dipeptidyl peptidase I. Proc Natl Acad Sci U S A **96**, 8312–8314.

Podack, E.R. and Dennert, G. (1983). Assembly of two types of tubules with putative cytolytic function by cloned natural killer cells. Nature **302**, 442–452.

Podack, E.R. and Konigsberg, P.J. (1984). Cytolytic T cell granules. Isolation, structural, biochemical, and functional characterization. J Exp Med **160**, 695–710.

Portales, P., Reynes, J., Pinet, V., Rouzier-Panis, R., Baillat, V., Clot, J. and Corbeau, P. (2003). Interferon-alpha restores HIV-induced alteration of natural killer cell perforin expression in vivo. AIDS **17**(4), 495–504.

Rabinovich, B.A., Li, J., Shannon, J., Hurren, R., Chalupny, J., Cosman, D. and Miller, R.G. (2003). Activated, but not resting, T cells can be recongnized and killed by syngeneic NK cells. J Immunol **170**, 3572–3576.

Raja, S.M., Metkar, S.S., Honing, S., Wang, B., Russin, W.A., Pipalia, N.H., Menaa, C., Belting, M., Cao, X., Dressel, R. and Froelich, C.J. (2005). A novel mechanism for protein delivery: granzyme B undergoes electrostatic exchange from serglycin to target cells. J Biol Chem **280**, 20752–20761.

Rosado, C.J., Buckle, A.M., Law, R.H., Butcher, R.E., Kan, W.T., Bird, C.H., Ung, K., Browne, K.A., Baran, K., Bashtannyk-Puhalovich, T.A., Faux, N.G., Wong, W., Porter, C.J., Pike, R.N., Ellisdon, A.M., Pearce, M.C., Bottomley, S.P., Emsley, J., Smith, A.I., Rossjohn, J., Hartland, E. L., Voskoboinik, I., Trapani, J.A., Bird, P. I., Dunstone, M.A. and Whisstock, J. C. (2007). A common fold mediates vertebrate defense and bacterial attack. Science **317**, 1548–1551.

Rosado, C.J., Kondos, S., Bull, T.E., Kuiper, M.J., Law, R.H.P, Buckle, A. M., Voskoboinik, I., Bird, P.I., Trapani, J. A., Whisstock, J.C. and Dunstone, M. A. (2008). The MACPF/CDC family of pore-forming toxins. Cell Microbiol **10**, 1765–1774.

Rotonda, J., Garcia-Calvo, M., Bull, H. G., Geissler, W.M., McKeever, B. M., Willoughby, C.A., Thornberry, N.A. and Becker, J.W. (2001). The three-dimensional structure of human granzyme B compared to caspase-3, key mediators of cell death with cleavage specificity for aspartic acid in P1. Chem Biol **8**, 357–368.

Rouvier, E., Luciani, M.F. and Goldstein, P. (1993). Fas involvement in Ca+2-independent T cell-mediated cytotoxicity. J Exp Med **177**, 195–200.

Russell, J.H. and Ley, T.J. (2002). Lymphocyte-mediated cytotoxicity. Annu Rev Immunol **20**, 323–370.

Ryan, A.E., Shanahan, F., O'Connell, J. and Houston, A.M. (2005). Addressing the "fas counterattack" controversy: blocking fas ligand expression suppresses tumor immune evasion of colon cancer in vivo. Cancer Res **65**, 9817–9823.

Schoenborn, J.R. and Wilson, C.B. (2007). Regulation of interferon-gamma during innate and adaptive immune responses. Adv Immunol **96**, 41–101.

Screpanti, V., Wallin, R.P., Ljunggren, H.G. and Grandien, A. (2001). A central role for death receptor-mediated apoptosis in the rejection of tumors by NK cells. J Immunol **167**, 2068–2073.

Shi, L., Mai, S., Israels, S., Browne, K., Trapani, J.A. and Greenberg, A. H. (1997). Granzyme B (GraB) autonomously crosses the cell membrane and perforin initiates apoptosis and GraB nuclear localization. J Exp Med **186**, 855–866.

Slade, D.J., Lovelace, L.L, Chruszcz, M., Minor, W., Lebioda, L. and Sodetz, J. M. (2008). Crystal structure of the MACPF domain of human complement protein C8 alpha in complex with the C8 gamma subunit. J Mol Biol **379**, 331–342.

Smyth, M.J., Swann, J., Cretney, E., Zerafa, N., Yokoyama, W.M. and Hayakawa, Y. (2005). NKG2D function protects the host from tumor initiation. J Exp Med **202**, 583–588.

Smyth, M.J., Cretney, E., Kelly, J.M, Westwood, J.A., Street, S.E.A., Yagita, H., Takeda, K., Van Dommelen, S.L.H., Degli-Esposti, M.A. and Hayakawa, Y. (2005). Activation of NK cell cytotoxicity. Mol Immunol **42**, 501–510.

Spielman, J., Lee, R.K. and Podack, E.R. (1998). Perforin/Fas-ligand double deficiency is associated with macrophage expansion and severe pancreatitis. J Immunol **161**, 7063–7070.

Stenger, S., Hanson, D.A., Teitelbaum, R., Dewan, P., Niazi, K.R., Froelich, C.J., Ganz, T., Thoma-Uszynski, S., Melian, A., Bogdan, C., Porcelli, S.A., Bloom, B.R., Krensky, A.M. and Modlin, R.L. (1998). An antimicrobial activity of cytolytic T cells mediated through granulysin. Science **282**, 121–125.

Stepp, S.E., Dufourcq-Lagelouse, R., Le Deist, F., Bhawan, S., Certain, S., Mathew, P.A., Henter, J.L., Bennett, M., Fischer, A., de Saint Basile, G. and Kumar, V. (1999). Perforin gene defects in familial hemophagocytic lymphohistiocytosis. Science **286**, 1957–1959.

Street, S.E., Trapani, J.A., MacGregor, D. and Smyth, M.J. (2002). Suppression of lymphoma and epithelial malignancies effected by interferon gamma. J Exp Med **196**, 129–134.

Sun, J.C., Beilke, J.N. and Lanier, L.L. (2009). Adaptive immune features of natural killer cells. Nature **457**, 557–561.

Takahashi, T., Tanaka, M., Brannan, C.I., Jenkins, N.A., Copeland, N.G., Suda,

T. and Nagata, S. (1994). Generalized lymphoproliferative disease in mice, caused by a point mutation in the Fas ligand. Cell **76**, 969–976.

Takeda, K., Smyth, M.J., Cretney, E., Hayakawa, Y., Yamaguchi, N., Yagita, H. and Okumura, K. (2001). Involvement of tumor necrosis factor-related apoptosis-inducing ligand in NK cell-mediated and IFN-gamma-dependent suppression of subcutaneous tumor growth. Cell Immunol **214**, 194–200.

Takeuchi, T., Ueki, T., Nishimatsu, H., Kajiwara, T., Ishida, T., Jishage, K., Ueda, O., Suzuki, H., Li, B., Moriyama, N. and Kitamura, T. (1999). Accelerated rejection of Fas ligand-expressing heart grafts. J Immunol **162**, 518–522.

Tanaka, M., Suda, T., Haze, K., Nakamura, N., Sato, K., Kimura, F., Motoyoshi, K., Mizuki, M., Tagawa, S., Ohga, S., Hatake, K., Drummond, A.H. and Nagata, S. (1996). Fas ligand in human serum. Nat Med **2**, 317–322.

Tay, C.H. and Welsh, R.M. (1997). Distinct organ-dependent mechanisms for the control of murine cytomegalovirus infection by natural killer cells. J Virol **71**, 265–275.

Trapani, J.A. and Sutton, V.R. (2003). Granzyme B: pro-apoptotic, anti-viral and anti-tumor functions. Curr Opin Immunol **15**, 533–543.

Trinchieri, G. (1989). Biology of natural killer cells. Adv Immunol **47**, 187–376.

Tschopp, J., Podack, E.R. and Muller-Eberhard, H.J. (1985). The membrane attack complex of complement:C5b-8 complex as accelerator of C9 polymerization. J Immunol **134**, 495–499.

Uellner, R., Zvelebil, M.J., Hopkins, J., Jones, J., MacDougall, L.K., Morgan, B.P., Podack, E., Waterfield, M.D. and Griffiths, G.M. (1997). Perforin is activated by a proteolytic cleavage during biosynthesis which reveals a phospholipids-binding C2 domain. EMBO J **16**, 7287–7296.

Van den Broek, M., Kagi, D. and Hengartner, H. (2000). Tumor surveillance in vivo depends on perforin-mediated cytolysis. In: Sitkovsky, M. and Henkart, P. (eds), Cytotoxic T Cells: Basic Mechanisms and Medical Applications. Philadelphia: Lippincott, Williams & Wilkins.

Van den Broek, M., Kagi, D., Ossendorp, F., Toes, R., Vamvakas, S., Lutz, W., Melief, C., Zinkernagel, R. and Hengartner, H. (1996). Decreased tumor surveillance in perforin-deficient mice. J Exp Med **184**, 1781–1790.

Van Dommelen, S.L.H., Humaria, N., Schreiber, R.D., Scalzo, A.A., Smyth, M.J. and Degli-Esposti, M.A. (2006). Perforin and granzymes have distinct roles in defensive immunity and immunopathology. Immunity **25**, 835–848.

Voskoboinik, I. and Trapani, J.A. (2006). Addressing the mysteries of perforin function. Immunol Cell Biol **84**, 66–71.

Voskoboinik, I., Thia, M.C., Fletcher, J., Ciccone, A., Browne, K., Smyth, M. J. and Trapani, J.A. (2005). Calcium-dependent plasma membrane binding and cell lysis by perforin are mediated through its C2 domain: a critical role for aspartate residues 429, 435, 483, and 485 but not 491. J Biol Chem **280**, 8426–8434.

Walch, M., Latinovic-Golic, S., Velic, A., Sundstrom, H., Dumrese, C., Wagner, C.A., Groscurth, P. and Ziegler, U. (2007). Perforin enhances the granulysin-induced lysis of *Listeria innocua* in human dendritic cells. BMC Immunol **16**, 8–14.

Watanabe, F.R., Brannan, C.I., Copeland, N.G., Jenkins, N.A. and Nagata, S. (1992). Lymphoproliferation disorder in mice explained by defects in fas antigen that mediates apoptosis. Nature **356**, 314–317.

Young, J.D., Cohn, Z.A. and Podack, E. R. (1986). The ninth component of complement and the pore-forming protein (perforin 1) from cytotoxic T cells: structural, immunological, and functional similarities. Science **4760**, 184–190.

Young, J.D., Damiano, A., DiNome, M. A., Leong, L.G. and Cohn, Z.A. (1987). Dissociation of membrane binding and lytic activities of the lymphocyte pore-forming protein (perforin). J Exp Med **165**, 1371–1382.

Chapter Seventeen

17

Illuminating intranodal natural killer cell behaviour using two-photon microscopy

Kym R. Garrod, Michael D. Cahalan

CHAPTER CONTENTS

It is not the form that dictates the color, but the color that brings out the form.
Hans Hoffman

ABSTRACT

Two-photon imaging to investigate fundamental immune processes as they occur in real time has revealed the choreography of cell movement and interactions that take place hidden from view inside lymphoid organs. This technique is now providing immunologists with the ideal tool for characterizing, at single-cell resolution in real time, leukocyte motility, migration patterns, and cell interaction dynamics within the native lymphoid-tissue environment. In this chapter we review advances in the field of natural killer (NK) cell imaging, with particular attention to a comparison of steady-state and activating conditions. In addition, we address key questions regarding the physiological significance of NK cells within the lymph node compartment.

KEY WORDS

Natural killer (NK) cell, Lymph node, Two-photon microscopy, Imaging, Cellular dynamics

NK cell development and tissue distribution

NK cell development in the bone marrow proceeds from the initial commitment of a hematopoietic stem cell to the lymphoid cell lineage (Di Santo, 2006; Yokoyama et al., 2004). Further NK cell-specific differentiation is defined by the acquisition of specific markers that qualitatively distinguish each stage of NK cell development. CD122 expression marks the transition from the bipotent T/NK progenitor to the NK cell progenitor stage. CD122 (IL-2R/15Rβ) trimerizes with IL-15Rα and common gamma chain (γc or CD132) to form the functional IL-15 receptor. A critical role of IL-15 signalling in promoting NK cell differentiation and survival is underscored by the dramatic defect in mature NK cells in IL-15-deficient animals (Kennedy et al., 2000). Although NK cell progenitors lack cytolytic activity at this stage in development, ~10% of the cells express the activating receptor NKG2D (Di Santo, 2006; Vosshenrich et al., 2005).

NK cell progenitor progression to the mature NK cell stage is characterized by further phenotypic and functional changes. Specifically, phenotypic features include the acquisition of CD49b (DX5, a pan-NK cell marker) and NK1.1 (also known as Nkrp1c; expressed on the C57BL/6 background), as well as the major histocompatibility complex (MHC) class I-specific inhibitory Ly49 molecules and CD94/NKG2 receptors that are required for target cell recognition (Lanier, 2005). Concurrently with expression of the late markers of NK cell differentiation, maturing cells gain cytolytic effector function as well as the ability to secrete pro-inflammatory cytokines (Kim et al., 2002).

Although bone marrow has traditionally been considered the main site for NK cell development, recent evidence suggests NK cells can originate from the thymus, spleen and lymph node (LN) compartments. For example, a unique population of $CD127^{+}$ NK cells was demonstrated to specifically arise in the thymus of C57BL/6 mice (Vosshenrich et al., 2006). These $CD127^{+}$ thymic NK cells were shown to be a phenotypically and functionally distinct population resembling the subset of immunoregulatory NK cells ($CD56^{hi}CD16^{-}$) identified in humans.

Once functional maturation is achieved, NK cells enter the circulation to reside within the periphery. Although widely disseminated throughout lymphoid and nonlymphoid tissue, NK cells constitute only a minor subset of lymphoid-resident lymphocytes (mouse: 1–2% bone marrow, 0.1% thymus, 2–4% spleen, 0.3–0.5% LN, 4–8% liver, 8.5% blood and 5–12% lung) (Chen et al., 2005; Gregoire et al., 2007; Vosshenrich et al., 2006). The spleen contains the largest pool of NK cells (mouse: $2–3 \times 10^{6}$), whereas in non-lymphoid tissue at steady-state NK cells represent the highest frequency of lymphocytes (Gregoire et al., 2007). NK cell distribution is governed by the expression of various G protein-coupled receptors and adhesion molecules (Gregoire et al., 2007). Under homeostatic conditions, NK cells recirculate through blood, liver, lung, and secondary lymphoid organs. Although the mechanisms regulating steady-state homing still require further investigation, efficient trafficking and recirculation have been shown to depend upon the expression of the lysophospholipid sphingosine 1-phosphate receptor $S1P_5$ (Walzer et al., 2007b). Upon encounter with inflammatory signals, NK cell recruitment is mediated by surface expression of the chemokine receptors CCR2, CCR5, CXCR3 and CX_3CR1 (Gregoire et al., 2007). Thus, the mechanism of NK cell trafficking is a highly regulated process, with unique signalling events and environmental cues that coordinate tissue distribution.

NK cell recruitment to LN

Although traditionally considered relatively excluded from LN, a recent study has investigated the 'induced' recruitment and functional significance of nodal NK cells (Martin-Fontecha et al., 2004). The authors demonstrated enhanced NK cell homing from the blood to LN draining dendritic cells (DC) via CD62L- and CXCR3-dependent, but CCR7-independent, extravasation. Of note, similar NK cell migration was achieved when animals were immunized with adjuvants, such as R848 and Ribi. Enhanced recruitment of NK cells to activated LN was correlated with robust antigen-specific T-cell activation. Importantly, further investigation revealed that NK cells provide an early source of IFN-γ that promoted efficient induction of T_H1 polarized responses.

More recently, Chen and colleagues (Chen et al., 2005) characterized NK cells resident in peripheral LN and identified a unique subset functionally and phenotypically distinct from NK cells isolated from spleen or blood. Specifically, LN-derived NK cells exhibited a diminished IFN-γ response to LPS stimulation and expressed fewer Ly49C/I, Ly49D, and Ly49H receptors than those in the spleen or blood. Notably, however, each subset displayed similar lytic activity toward YAC-1 and RMA/S tumour cells. Further examination demonstrated that L-selectin-dependent recruitment of NK cells to tumour bearing LN inhibited B16 melanoma tumour progression. Taken together, these reports uncover a previously unappreciated role for intranodal NK cells in promoting acquired T-cell responses and antitumour immunity, respectively.

NK cell distribution within LN

Several reports have examined the localization of NK cells within the LN of mice (Bajenoff et al., 2006a; Walzer et al., 2007a) and humans (Ferlazzo et al., 2004). Histologic examination of LN at steady-state has shown that NK cells localize to the paracortex (just underneath and between B cell follicles) and the medulla (Figure 17.1A, B). A closer examination found NK cells to be in close proximity to DC (Figure 17.1C) and macrophages (Bajenoff et al., 2006a; Walzer et al., 2007a). Quantification of NK cell density in each of the two regions showed higher densities in the medulla (62.3 cells/mm^2) compared to the T cell area (22.8 cells/mm^2) (Bajenoff et al., 2006a). To determine if the pattern of homeostatic NK cell localization was altered with the enhanced recruitment of NK cells under activating conditions, NK cells were enumerated following *Leishmania major* infection. Interestingly, NK cells preferentially accumulated to threefold higher density in the T cell areas of inflamed LN with no apparent increase in NK cell frequency in the medullary region (Bajenoff et al., 2006a).

The strategic localization of NK cells within the cortical ridge and close juxtaposition with antigen-presenting cells highlights a potential central role for intranodal NK cells in bridge innate and adaptive immunity. Applying real-time imaging techniques, such as two-photon microscopy, to visualize LN-resident NK cells is starting to provide further insight into the physiological significance of these cells.

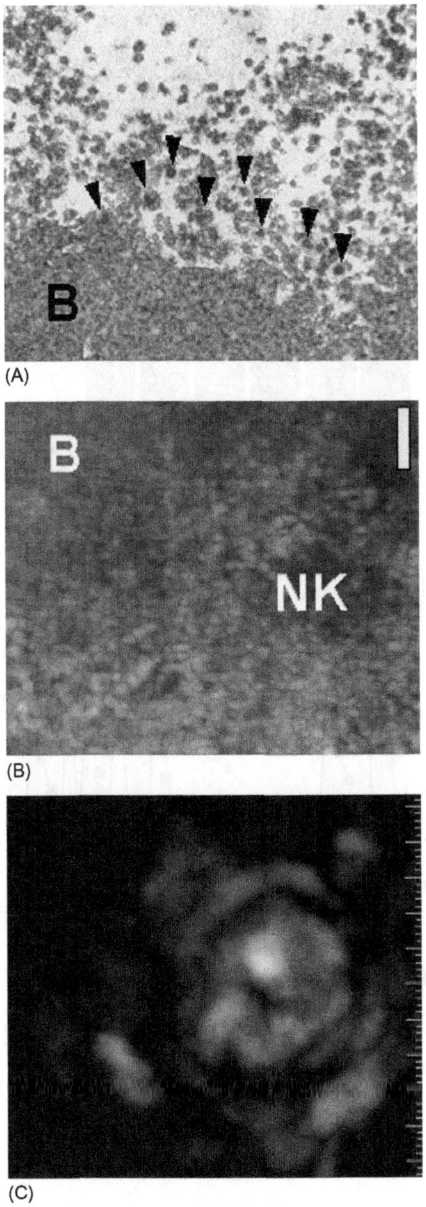

Figure 17.1 • Intranodal NK cell localization. (A) Photomicrograph depicting a lymph node with NK cells (dark grey) localized to the edge of a B cell follicle. Black arrowheads indicate individual NK cells (magnification, 20×). (B) Two-photon images depicting NK cell (grey) migration along the periphery of a B cell follicle (dark grey). Cumulative fluorescence was formed by the summation of successive images over time. Scale bar, 20 μm. (C) Snapshot visualization of an NK cell (dark grey) interacting with several DC. Individual DC are labelled grey and very dark grey. Large tick marks, 5 μm.

Principles of two-photon immunoimaging

Two-photon excitation microscopy is an attractive method for three-dimensional (3D) imaging in lymphoid and other organs for several reasons. The scanning laser two-photon microscope has the ability to image deep (several hundred microns) within living tissue for extended periods of time with negligible phototoxicity and minimal photobleaching. This advantage over conventional confocal microscopy is due to the basic physical properties of two-photon excitation (Cahalan et al., 2002). Two-photon excitation relies on concentrated femtosecond pulsed laser illumination from a light source (i.e. Ti-Sapphire laser) to produce sufficient photon density for the near-simultaneous absorption of two infrared photons by a fluorescent molecule. Pulsed laser excitation reduces the total energy that tissue is exposed to without reducing the probability of fluorophore excitation. Furthermore, focusing of the light by an objective lens restricts two-photon excitation to the focal point; therefore, two-photon excitation occurs only in the focal plane. The benefit of localized excitation is that emission is restricted to the narrow focal region, providing optical sectioning without the use of a pinhole. In addition, two-photon microscopy increases sample penetration because of the reduced absorption and scattering of near-infrared radiation (Piston, 1999). Moreover, the limited excitation region reduces phototoxicity because photodamage is largely confined to the focal volume. Eliminating the confocal pinhole increases sensitivity of capturing the emitted light since the entire signal reaches the photomultiplier detector even if it is scattered by tissue. The combination of advantages afforded by two-photon imaging allows intact secondary lymphoid organs to be imaged at reasonable spatial resolution with little damage to the tissue environment. Scanning the laser rapidly in the *x–y* plane while moving the objective stepwise in the *z* axis enables 3D imaging with acquisition times of several seconds that are fast enough to accommodate tracking of cell migration. Results from LN explant and intravital preparations of inguinal or popliteal LN agree quite well (Miller et al., 2003). Thus, two-photon microscopy is providing immunologists with an ideal tool for investigating fundamental cellular processes as they occur in real time.

The application of two-photon microscopy to image leukocytes within lymphoid tissue has had a decisive role in providing immunologists with a clear understanding of the cellular processes involved in eliciting productive immunity as well as maintaining tolerance to self-antigens. Over the past several years, pioneering work using two-photon analysis to describe the dynamics of immune cells within secondary lymphoid organs has revealed an elaborate network of interacting partners (Bousso et al., 2002; Cahalan and Parker, 2005; Cahalan et al., 2003; Sumen et al., 2004). In general, one of the more striking results from two-photon immunoimaging is the revelation that lymphoid organs are a beehive of cellular activity, with cells actively migrating by amoeboid motility. T and B cells migrate within their compartments—the diffuse cortex and the follicle, respectively—at 3D speeds of 10–15 and 7–9 μm/min, respectively (Miller et al., 2002). Rather than migrating *en masse*, cell trajectories are well described by a random walk. The seemingly chaotic pattern is well suited for detecting antigen by a stochastic process of random collisions (Miller et al., 2002, 2003). DC with agile dendrites are sampled efficiently in the diffuse cortex by T cells (Miller et al., 2004a). Meanwhile, in the follicle, B cells capture antigen initially from macrophages that line the subcapsular sinus and exchange it with follicular dendritic cells that provide a reservoir of antigen (Phan et al., 2007). Once an antigen is detected, T cells undergo a fairly stereotyped pattern of behaviour (Mempel et al., 2004; Miller et al., 2004b). During phase I, T cells continue to make transient contacts with antigen-presenting DC, but then after a few hours become immobilized while in stable contact with DC (Phase II). Later, in Phase III, after several hours and depending on antigen quality and quantity, T cells begin to separate from DC and resume their motility and intermittent contacts with DC, and then undergo several rounds of proliferation. While the T cell dance with DC is taking place in the diffuse cortex, B cells that have captured specific antigen undergo a non-random directed walk by chemotaxis (via CCR7) to the follicle edge (Okada et al., 2005). The migration of antigen-engaged B cells poises them in position to contact activated $CD4^+$ helper T cells at the edge of the follicle, where they pair off and migrate as conjugate partners, with the B cell leading the way.

Imaging intranodal NK cell dynamics

Recently, real-time imaging has been applied to characterize intranodal NK cell behaviour (Bajenoff et al., 2006a; Garrod et al., 2007). Analysis of individual NK cell mobility has uncovered prototypic lymphocyte behaviour (Garrod et al., 2007). NK cells appeared remarkably motile (Figure 17.2A, B), exhibiting average speeds of 6–7 μm/min [2D analysis: 6.7 ± 0.2 μm/min (Garrod et al., 2007)]; and as expected, 3D analysis has revealed even greater velocities (9–10 μm/min; Garrod et al., 2007; Beuneu et al., in press). NK cells exhibited a pattern of 'stop-and-go' movement similar to other lymphocyte subsets (Miller et al., 2002). With the qualification that NK cells appeared to be preferentially retained near the cortical ridge underlying B cell follicles, it was determined that within these areas individual NK cells followed apparently random migration (Figure 17.2C). Further experiments are required

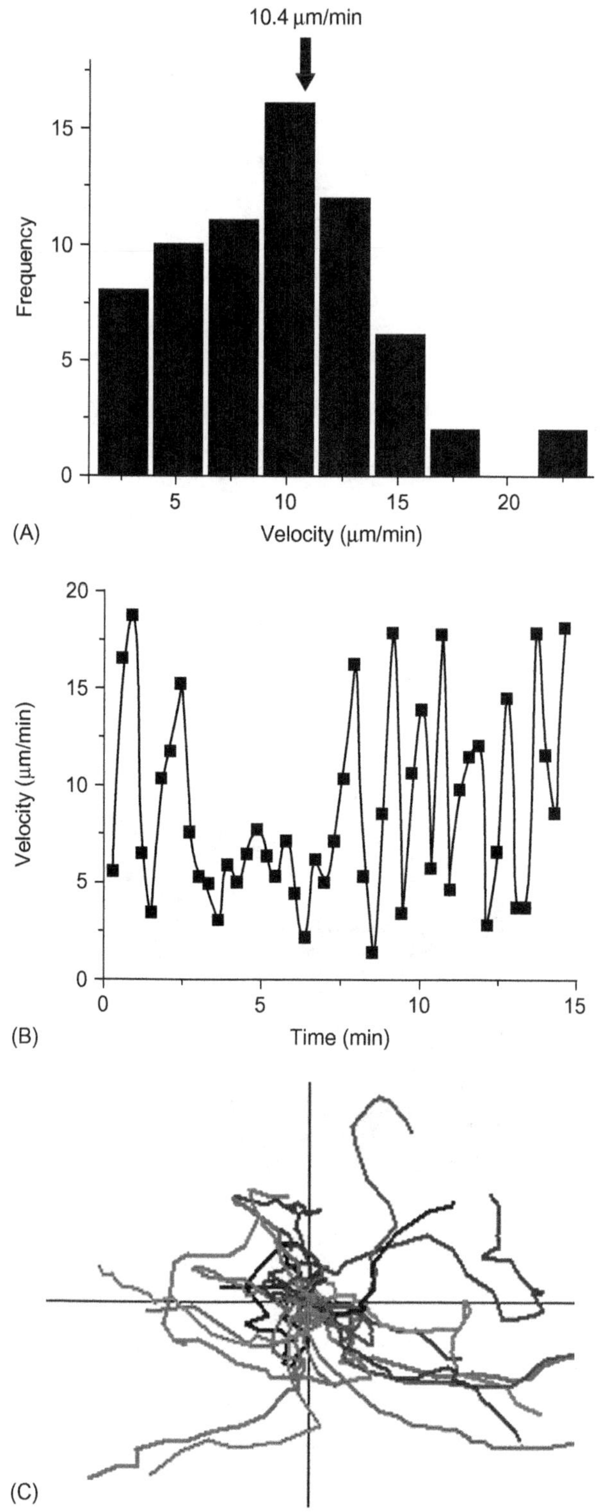

Figure 17.2 • NK cell motility. (A) Frequency distribution of mean velocity for 67 individual NK cells. Black arrow indicates average velocity of 10.4 μm/min. (B) Time course of velocity changes of a representative NK cell, demonstrating the characteristic 'stop and-go' manner of lymphocyte locomotion. (C) Tracks representing 2D trajectories of 24 individual NK cells after normalization to starting coordinates.

to investigate the possibility that different subregions of the LN (i.e. medulla) may elicit distinct NK cell behaviour. Taken together, these results underscore the dynamic nature of NK cell motility in peripheral LN tissue.

Although the precise contribution of specific integrin-ligand pairs in regulating lymphocyte trafficking within secondary lymphoid organs awaits further investigation, it is well established that integrins are of prime importance in cell adhesion, migration, and target cell recognition (Schwartz et al., 1995). Lymphocyte migration within secondary lymphoid tissue has been shown to be regulated by confined movement along nonhematopoietic structural cells, such as fibroblastic reticular cells (FRC) and follicular dendritic cells (FDC) for T and B cells, respectively (Bajenoff et al., 2006b). It has been proposed that the FRC and FDC stromal cell networks independently act to define the respective spatial organization of paracortical T cells and follicular B cells through the expression of a unique profile of adhesion molecules and chemokines specifically recognized by each lymphocyte subset (Bajenoff et al., 2006b). A key integrin expressed during NK cell maturation is CD49b (recognized by clone DX5 monoclonal antibody). CD49b heterodimerizes with CD29 to form the functional integrin receptor VLA-2 (Shimaoka et al., 2002). One of the principal ligands for VLA-2 is the extracellular matrix protein collagen. Direct in vivo analysis of NK cell migration following CD49b activation using either antibody-mediated stimulation (Garrod et al., 2007) or positive isolation techniques (Bajenoff et al., 2006b) demonstrated impaired NK cell motility. Further analysis using in vitro assays showed enhanced NK cell adhesion to collagen-coated plates following CD49b-crosslinking. In addition, in vivo imaging of CD49b-stimulated NK cells covisualized in contact with collagen fibres within LN showed irreversible adhesion (Garrod et al., 2007). Thus, the observed decrease in NK cell motility following CD49b-crosslinking could provide important insight into the mechanical mechanisms driving lymphocyte migration and identify a functional role of CD49b expression on NK cells.

Experimental evidence indicates that integrin engagement can result in the activation of signalling events such as stimulation of phosphoinostitide metabolism, elevation of Ca^{2+} transients, and activation of a number of tyrosine kinases (Juliano and Haskill, 1993). Moreover, VLA-2 activation in Jurkat cells stimulated tyrosine phosphorylation of multiple proteins (i.e. paxillin, focal adhesion kinase (FAK)) as well as resulted in the accumulation of active $p21^{ras}$-GTP (Kapron-Bras et al., 1993). Therefore, there is a potential role for integrin stimulation in modulating NK cell effector function. Analysis of CD49b-crosslinked NK cells revealed impaired cytolytic activity as well as augmented IFN-γ production under IL-2 activating conditions (Garrod et al., 2007). Taken together, CD49b activation skews NK cell effector function to a phenotype that exhibits elevated IFN-γ production with a concomitant inhibition of motility and cytolytic function.

Cognate NK–target cell interactions

The distinguishing attribute of NK cells is their unique ability to lyse target cells without the need for prior sensitization. Ample in vitro experimental evidence demonstrates the requisite of cell–cell contact for efficient killing, but little has been known regarding the duration of interaction or the pattern of engagement with target cells in vivo. Specifically, it has been unclear whether NK cells preferentially form monogamous pairs or cooperative swarms in vivo prior to eliciting lytic effector function. Furthermore, although in vitro analysis of cognate NK–target cell interactions has revealed the formation of an activating immunological synapse (Orange, 2008; Vyas et al., 2001, 2002), the stability of in vivo interaction has not yet been documented.

Previous work suggests that a lack of inhibitory signals (MHC class I-deficiency) is insufficient by itself to induce efficient killing of target cells. Rather, NK cell-mediated cytolysis requires both the coordinated release of inhibition and engagement of activating signals (Lanier, 2005). While the stress-induced activating NKG2D ligands have been shown to be up-regulated upon transplantation (Hankey et al., 2002; Ogasawara et al., 2005), the cellular dynamics of NK cell-mediated allorecognition had remained ill-defined until only recently. With the application of two-photon microscopy, imaging NK cell interactions with MHC-mismatched targets in vivo has provided important information regarding the nature and duration of conjugate formation (Garrod et al., 2007). Adoptively transferred NK cells and allogeneic B cells were imaged in LN explants. In most cases, NK cells formed monogamous pairs with allogeneic B cells (Figure 17.3A), but on occasion NK cells were seen in brief contact with two B cells during transition from one to the other. In some instances, however, multiple NK cells 'swarmed' a single allogeneic B cell. These swarming events appeared to result in the elimination of the target cell, as evidenced by diminished velocity and membrane blebbing of the target. The majority of NK–target cell interactions (~77%) were transient, lasting less than 5 min. These transient interactions did not appear to result in a productive lytic event. In contrast, stable interactions, lasting greater than 5 min, preferentially resulted in NK cell-mediated killing.

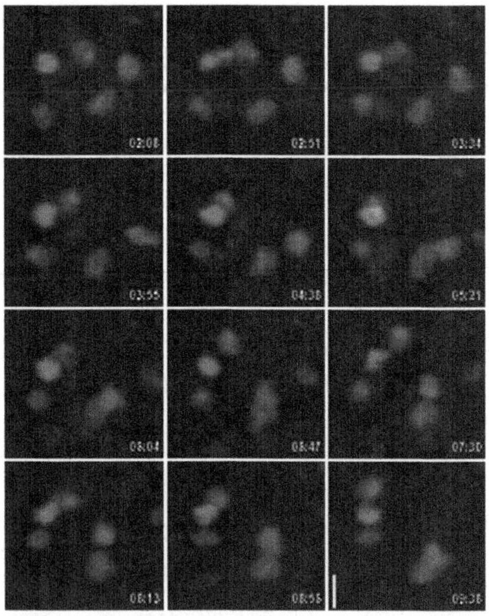

(A)

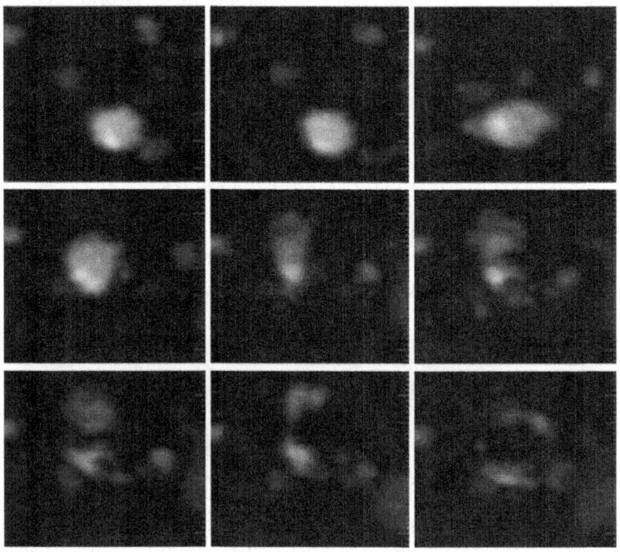

(B)

Figure 17.3 • NK–target cell interaction dynamics. (A) Consecutive images of an NK cell (grey) making serial contacts with the same B cell (dark grey). Scale bar, 10 μm. (B) Time-lapse images of an NK cell engaging (dark grey) in a productive interaction with an individual allogeneic DC (grey) which results in lysis of the DC target as evidenced by membrane blebbing. Large tick marks, 5 μm.

A role for intranodal NK cells in elimination of MHC-mismatched cells is further supported from studies using allogeneic DC targets (Garrod et al., 2007). Using two-photon microscopy to visualize interaction dynamics, NK cells appeared to engage in transient contacts with both syngeneic and allogeneic DC (4 min 52 sec ± 36 sec and 5 min 54 sec ± 42 sec, respectively). Occasionally, however, NK cells formed longer-lasting conjugates with allogeneic DC (20 min 20 sec ± 4 min 17 sec) that resulted in a productive, lytic event (Figure 17.3B). Thus, NK cells form both transient and stable conjugates with allogeneic cells, with prolonged interactions resulting in elimination of the target cell.

An important caveat to these observations is that due to the limited time window for observation (1–2 h), the history of all interactions preceding a lytic event cannot be assessed by two-photon microscopy. Therefore, a potential functional role for transient interactions in priming target cells prior to prolonged, lytic interactions cannot be adequately addressed.

Noteworthy are the striking qualitative differences in NK cell interactions compared to other lymphocytes, such as cytotoxic T lymphocytes (CTL). For example, in comparison to the ability of NK cells to swarm a single allogeneic target, CTLs form strictly monogamous pairs with cognate B cell targets (Mempel et al., 2006). In addition, CTL detachment follows target cell arrest and apoptosis (Mempel et al., 2006). In contrast, NK cell dissociation precedes target cell arrest (Garrod et al., 2007). Furthermore, a recent report noted intratumoural CTLs formed stable conjugates with tumour cells, which appeared to persist for hours (Breart et al., 2008) as opposed to the relatively brief NK–target cell interactions that last on the order of minutes (Garrod et al., 2007). These differences perhaps connote unique molecular mechanisms for target cell recognition and/or killing strategies. Therefore, although NK cells share similarities with T and B cell (i.e. stop-and-go motility, amoeboid locomotion, nondirectional migration), distinct behavioural differences set NK cells apart from other lymphocyte subsets.

Imaging NK–DC interactions

As an essential component of innate immunity, NK cells are hard-wired to recognize and respond to invading pathogens. The expression of activating and inhibitory receptors equips NK cells with the ability to discern microbially infected cells from healthy tissue. However, it appears that receptor recognition alone is insufficient to trigger NK cells efficiently. Accumulating evidence suggests a requirement for DC-dependent help to prime NK cell stimulation. The significance of DC in augmenting NK cell effector function was originally proposed based upon a prominent role for DC in promoting effective NK cell tumouricidal activity (Fernandez et al., 1999). More recently, Lucas and associates have shown that DC directly regulate NK cell function via DC-derived *trans*-presentation of IL-15 (Lucas et al., 2007). Reciprocally, activated NK cells stimulate DC maturation (Gerosa et al., 2002). Thus, experimental evidence proposes a role for NK–DC 'cross-talk' in the induction of innate resistance to tumours and infection

(Fernandez et al., 1999; Moretta, 2002). Complicating this interpretation, however, has been the paucity of in vivo analysis of NK–DC interactions.

The application of two-photon microscopy to directly assess the pattern of NK–DC engagement during the course of an immune response could provide insight into this complex relationship. Hélène Beuneu, Philippe Bousso and colleagues (Beuneu et al., in press) have developed a system to simultaneously visualize endogenous NK cells and DC using the respective Ncr1$^{GFP/+}$ knock-in mice (Gazit et al., 2006) backcrossed to the CD11c-EYFP reporter mice (Lindquist et al., 2004). During imaging of the paracortical region of LN at steady-state, NK cells appeared to form brief transient contacts with resident DC (lasting less than 5 min in duration). Similarly, NK cells adoptively transferred into CD11c-EYFP hosts preferentially engaged in brief interactions with endogenous DC (Figure 17.4). Surprisingly, however, using poly I:C activation to mimic viral infection does not induce stable NK–DC conjugate formation (Beuneu et al., in press). In fact, NK cell velocity and a straightness index increased following poly I:C stimulation. Thus, it appears that under DC-dependent activating conditions (poly I:C treatment), NK cells do not form stable pairs with cognate DC. Instead, NK cell activation is likely achieved by the integration of signals generated through a series of transient interactions within the network of resident DC (Beuneu et al., in press).

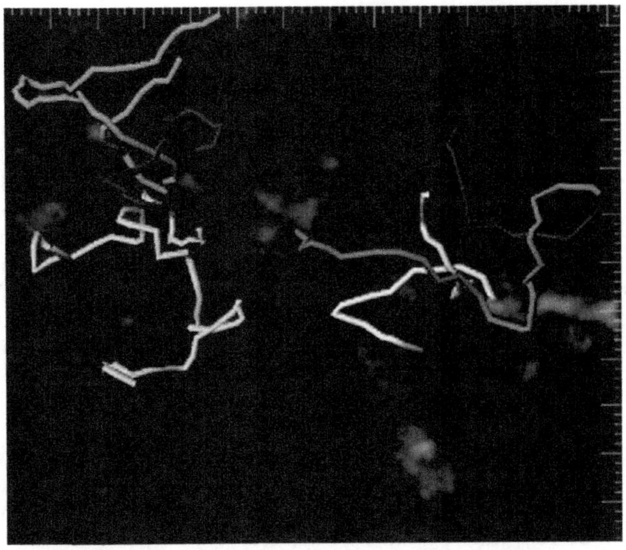

Figure 17.4 • NK–DC interactions. Two-photon image of individual NK cells (dark grey) interacting with CD11c-yfp$^+$ endogenous DC (grey). Tracks of 7 representative NK cells are superimposed. Large tick marks, 20 μm.

Concluding remarks

The basic characterization of NK cell behaviour using two-photon microscopy has contributed fundamental information about their migration, distribution pattern, motility, cellular interaction dynamics, and cytolytic effector function. These insights now provide a framework for understanding the role of intranodal NK cells in eliminating MHC-mismatched targets and establish the feasibility of functional immunoimaging approaches to investigate the molecular mechanism(s) that underlie NK cell-mediated immunosurvellience.

Acknowledgements

Authors wish to thank Hélène Beuneu and Philippe Bousso (Institut Pasteur) for kindly sharing unpublished material. K.R.G. acknowledges generous support from the Arnold and Mabel Beckman Foundation and Achievement Rewards for College Scientists Foundation. This work is supported by grants from the National Institute of Health (T32 AI-060573 to K.R.G. and GM-41514 to M.D.C.).

References

Bajenoff, M., Breart, B., Huang, A.Y., Qi, H., Cazareth, J., Braud, V.M., Germain, R.N. and Glaichenhaus, N. (2006a). Natural killer cell behavior in lymph nodes revealed by static and real-time imaging. J Exp Med **203**, 619–631.

Bajenoff, M., Egen, J.G., Koo, L.Y., Laugier, J.P., Brau, F., Glaichenhaus, N. and Germain, R.N. (2006b). Stromal cell networks regulate lymphocyte entry, migration, and territoriality in lymph nodes. Immunity **25**, 989–1001.

Beuneu, H., Deguine, J., Breart, B., Mandelboim, O., Di Santo, J.P. and Bousso, P. Dynamic behavior of NK cells during activation in lymph nodes. Blood (in press).

Bousso, P., Bhakta, N.R., Lewis, R.S. and Robey, E. (2002). Dynamics of thymocyte–stromal cell interactions visualized by two-photon microscopy. Science **296**, 1876–1880.

Breart, B., Lemaitre, F., Celli, S. and Bousso, P. (2008). Two-photon imaging of intratumoral CD8+ T cell cytotoxic activity during adoptive T cell therapy in mice. J Clin Invest **118**, 1390–1397.

Cahalan, M.D. and Parker, I. (2005). Close encounters of the first and second kind: T-DC and T-B interactions in the lymph node. Semin Immunol **17**, 442–451.

Cahalan, M.D., Parker, I., Wei, S.H. and Miller, M.J. (2002). Two-photon tissue imaging: seeing the immune system in a fresh light. Nat Rev Immunol **2**, 872–880.

Cahalan, M.D., Parker, I., Wei, S.H. and Miller, M.J. (2003). Real-time imaging of lymphocytes in vivo. Curr Opin Immunol **15**, 372–377.

Chen, S., Kawashima, H., Lowe, J.B., Lanier, L.L. and Fukuda, M. (2005). Suppression of tumor formation in lymph nodes by L-selectin-mediated natural killer cell recruitment. J Exp Med **202**, 1679–1689.

Di Santo, J.P. (2006). Natural killer cell developmental pathways: a question of balance. Annu Rev Immunol **24**, 257–286.

Ferlazzo, G., Pack, M., Thomas, D., Paludan, C., Schmid, D., Strowig, T., Bougras, G., Muller, W.A., Moretta, L. and Munz, C. (2004). Distinct roles of IL-12 and IL-15 in human natural killer cell activation by dendritic cells from secondary lymphoid organs. Proc Natl Acad Sci U S A **101**, 16606–16611.

Fernandez, N.C., Lozier, A., Flament, C., Ricciardi-Castagnoli, P., Bellet, D., Suter, M., Perricaudet, M., Tursz, T., Maraskovsky, E. and Zitvogel, L. (1999). Dendritic cells directly trigger NK cell functions: cross-talk relevant in innate anti-tumor immune responses in vivo. Nat Med **5**, 405–411.

Garrod, K.R., Wei, S.H., Parker, I. and Cahalan, M.D. (2007). Natural killer cells actively patrol peripheral lymph nodes forming stable conjugates to eliminate MHC-mismatched targets. Proc Natl Acad Sci U S A **104**, 12081–12086.

Gazit, R., Gruda, R., Elboim, M., Arnon, T.I., Katz, G., Achdout, H., Hanna, J., Qimron, U., Landau, G., Greenbaum, E., Zakay-Rones, Z., Porgador, A. and Mandelboim, O. (2006). Lethal influenza infection in the absence of the natural killer cell receptor gene Ncr1. Nat Immunol **7**, 517–523.

Gerosa, F., Baldani-Guerra, B., Nisii, C., Marchesini, V., Carra, G. and Trinchieri, G. (2002). Reciprocal activating interaction between natural killer cells and dendritic cells. J Exp Med **195**, 327–333.

Gregoire, C., Chasson, L., Luci, C., Tomasello, F., Geissmann, F., Vivier, E. and Walzer, T. (2007). The trafficking of natural killer cells. Immunol Rev **220**, 169–182.

Hankey, K.G., Drachenberg, C.B., Papadimitriou, J.C., Klassen, D.K., Philosophe, B., Bartlett, S.T., Groh, V., Spies, T. and Mann, D.L. (2002). MIC expression in renal and pancreatic allografts. Transplantation **73**, 304–306.

Juliano, R.L. and Haskill, S. (1993). Signal transduction from the extracellular matrix. J Cell Biol **120**, 577–585.

Kapron-Bras, C., Fitz-Gibbon, L., Jeevaratnam, P., Wilkins, J. and Dedhar, S. (1993). Stimulation of tyrosine phosphorylation and accumulation of GTP-bound p21ras upon antibody-mediated alpha 2 beta 1 integrin activation in T-lymphoblastic cells. J Biol Chem **268**, 20701–20704.

Kennedy, M.K., Glaccum, M., Brown, S. N., Butz, E.A., Viney, J.L., Embers, M., Matsuki, N., Charrier, K., Sedger, L., Willis, C.R., Brasel, K., Morrissey, P.J., Stocking, K., Schuh, J.C., Joyce, S. and Peschon, J.J. (2000). Reversible defects in natural killer and memory CD8 T cell lineages in interleukin 15-deficient mice. J Exp Med **191**, 771–780.

Kim, S., Iizuka, K., Kang, H.S., Dokun, A., French, A.R., Greco, S. and Yokoyama, W.M. (2002). In vivo developmental stages in murine natural killer cell maturation. Nat Immunol **3**, 523–528.

Lanier, L.L. (2005). NK cell recognition. Annu Rev Immunol **23**, 225–274.

Lindquist, R.L., Shakhar, G., Dudziak, D., Wardemann, H., Eisenreich, T., Dustin, M.L. and Nussenzweig, M.C. (2004). Visualizing dendritic cell networks in vivo. Nat Immunol **5**, 1243–1250.

Lucas, M., Schachterle, W., Oberle, K., Aichele, P. and Diefenbach, A. (2007). Dendritic cells prime natural killer cells by trans-presenting interleukin 15. Immunity **26**, 503–517.

Martin-Fontecha, A., Thomsen, L.L., Brett, S., Gerard, C., Lipp, M., Lanzavecchia, A. and Sallusto, F. (2004). Induced recruitment of NK cells to lymph nodes provides IFN-gamma for T(H)1 priming. Nat Immunol **5**, 1260–1265.

Mempel, T.R., Henrickson, S.E. and Von Andrian, U.H. (2004). T-cell priming by dendritic cells in lymph nodes occurs in three distinct phases. Nature **427**, 154–159.

Mempel, T.R., Pittet, M.J., Khazaie, K., Weninger, W., Weissleder, R., von Boehmer, H. and von Andrian, U.H. (2006). Regulatory T cells reversibly suppress cytotoxic T cell function independent of effector differentiation. Immunity **25**, 129–141.

Miller, M.J., Wei, S.H., Parker, I. and Cahalan, M.D. (2002). Two-photon imaging of lymphocyte motility and antigen response in intact lymph node. Science **296**, 1869–1873.

Miller, M.J., Wei, S.H., Cahalan, M.D. and Parker, I. (2003). Autonomous T cell trafficking examined in vivo with intravital two-photon microscopy. Proc Natl Acad Sci U S A **100**, 2604–2609.

Miller, M.J., Hejazi, A.S., Wei, S.H., Cahalan, M.D. and Parker, I. (2004a). T cell repertoire scanning is promoted by dynamic dendritic cell behavior and random T cell motility in the lymph node. Proc Natl Acad Sci U S A **101**, 998–1003.

Miller, M.J., Safrina, O., Parker, I. and Cahalan, M.D. (2004b). Imaging the single cell dynamics of CD4+ T cell activation by dendritic cells in lymph nodes. J Exp Med **200**, 847–856.

Moretta, A. (2002). Natural killer cells and dendritic cells: rendezvous in abused tissues. Nat Rev Immunol **2**, 957–964.

Ogasawara, K., Benjamin, J., Takaki, R., Phillips, J.H. and Lanier, L.L. (2005). Function of NKG2D in natural killer cell-mediated rejection of mouse bone marrow grafts. Nat Immunol **6**, 938–945.

Okada, T., Miller, M.J., Parker, I., Krummel, M.F., Neighbors, M., Hartley, S.B., O'Garra, A., Cahalan, M.D. and Cyster, J.G. (2005). Antigen-engaged B cells undergo chemotaxis toward the T zone and form motile conjugates with helper T cells. PLoS Biol **3**, 1047–1061.

Orange, J.S. (2008). Formation and function of the lytic NK-cell immunological synapse. Nat Rev Immunol.

Phan, T.G., Grigorova, I., Okada, T. and Cyster, J.G. (2007). Subcapsular encounter and complement-dependent transport of immune complexes by lymph node B cells. Nat Immunol **8**, 992–1000.

Piston, D.W. (1999). Imaging living cells and tissues by two-photon excitation microscopy. Trends Cell Biol **9**, 66–69.

Schwartz, M.A., Schaller, M.D. and Ginsberg, M.H. (1995). Integrins: emerging paradigms of signal transduction. Annu Rev Cell Dev Biol **11**, 549–599.

Shimaoka, M., Takagi, J. and Springer, T.A. (2002). Conformational regulation of integrin structure and function. Annu Rev Biophys Biomol Struct **31**, 485–516.

Sumen, C., Mempel, T.R., Mazo, I.B. and von Andrian, U.H. (2004). Intravital microscopy: visualizing immunity in context. Immunity **21**, 315–329.

Vosshenrich, C.A., Ranson, T., Samson, S.I., Corcuff, E., Colucci, F., Rosmaraki, E.E.

and Di Santo, J.P. (2005). Roles for common cytokine receptor gamma-chain-dependent cytokines in the generation, differentiation, and maturation of NK cell precursors and peripheral NK cells in vivo. J Immunol **174**, 1213–1221.

Vosshenrich, C.A., Garcia-Ojeda, M.E., Samson-Villeger, S.I., Pasqualetto, V., Enault, L., Richard-Le Goff, O., Corcuff, E., Guy-Grand, D., Rocha, B., Cumano, A., Rogge, L., Ezine, S. and Di Santo, J.P. (2006). A thymic pathway of mouse natural killer cell development characterized by expression of GATA-3 and CD127. Nat Immunol **7**, 1217–1224.

Vyas, Y.M., Mehta, K.M., Morgan, M., Maniar, H., Butros, L., Jung, S., Burkhardt, J.K. and Dupont, B. (2001). Spatial organization of signal transduction molecules in the NK cell immune synapses during MHC class I-regulated noncytolytic and cytolytic interactions. J Immunol **167**, 4358–4367.

Vyas, Y.M., Maniar, H. and Dupont, B. (2002). Cutting edge: differential segregation of the SRC homology 2-containing protein tyrosine phosphatase-1 within the early NK cell immune synapse distinguishes noncytolytic from cytolytic interactions. J Immunol **168**, 3150–3154.

Walzer, T., Blery, M., Chaix, J., Fuseri, N., Chasson, L., Robbins, S.H., Jaeger, S., Andre, P., Gauthier, L., Daniel, L., Chemin, K., Morel, Y., Dalod, M., Imbert, J., Pierres, M., Moretta, A., Romagne, F. and Vivier, E. (2007a). Identification, activation, and selective in vivo ablation of mouse NK cells via NKp46. Proc Natl Acad Sci U S A **104**, 3384–3389.

Walzer, T., Chiossone, L., Chaix, J., Calver, A., Carozzo, C., Garrigue-Antar, L., Jacques, Y., Baratin, M., Tomasello, E. and Vivier, E. (2007b). Natural killer cell trafficking in vivo requires a dedicated sphingosine 1-phosphate receptor. Nat Immunol **8**, 1337–1344.

Yokoyama, W.M., Kim, S. and French, A.R. (2004). The dynamic life of natural killer cells. Annu Rev Immunol **22**, 405–429.

Chapter Eighteen 18

Dendritic cells and NK cells

Amy K. Wesa, Robbie B. Mailliard

CHAPTER CONTENTS

What is the answer? In that case, what is the question?

Gertrude Stein

ABSTRACT

Dendritic cells (DCs) play a central role in the initiation and regulation of the immune response. Acting as the bridge between the innate and adaptive branches of immunity, DCs convert environmental cues they gather at sites of inflammation into the development of polarized adaptive immune responses once they reach draining lymph nodes. An important aspect in this system lies with the ability of DCs and natural killer (NK) cells to exchange bi-directional signals capable of influencing the functional status of each cell type as well as the overall nature of the immune response. DC activation of NK cells through their production of such factors as IL-12, IL-18, IL-15 and type-1 interferons (IFNs), and the NK cell modulation of DC function through their production of TNF-α and IFN-γ can result from direct cell-to-cell contact involving NK cell receptors such as NKp30, NKp46 and NKG2D or through indirect means of bystander activation. Moreover, while NK cells can act as 'helper' cells to promote DC function, they can also serve to police DC activity through a selective elimination process resulting in DC death. Importantly, the net immunological effects of the interactions of these cells are largely dependant on the activation status of both cell types as well as the context in which they meet. Here we discuss the functional properties of the DC, with particular attention focusing on the immunoregulatory impact of their interactions with NK cells.

KEY WORDS

Dendritic cells, TH1, TH2, Helper activity, Cancer immunotherapy, Cytokines; IL-12, IFN-γ, HMGB1, IFN-α

Dendritic cells: the professional antigen-presenting cell

Dendritic cells (DCs) are antigen-presenting cells (APCs) controlling two reciprocal functions: maintenance of self-tolerance and initiation of immune responses to foreign antigens. Cellular immunity is initiated by the presentation of antigen to T lymphocytes (signal 1; identity and restriction) in the presence of appropriate costimulatory molecules (signal 2; strength of signal), and cytokines (signal 3; polarization) (Kalinski et al., 1999), which promote T cell differentiation into effectors capable of mediating the desired response. DCs are uniquely endowed with the capacity to orchestrate the priming of cellular immunity as they effectively and efficiently perform each of these tasks: antigen capture, processing into peptide epitopes for presentation to CD4 and CD8 T cells onto MHC class II and class I molecules, respectively, provision of costimulatory signals and secretion of T cell modulating cytokines (Banchereau and Steinman, 1998).

A suitable paradigm for the role of DCs in the induction of cellular immunity is represented by the function of skin Langerhans cells (LC) (Steinman, 1991). These sentinels of the immune system, located in the epidermis, persistently sample antigens from their environment using pinocytosis, receptor-mediated endocytosis, direct infection or phagocytosis, localizing these in Birbeck granules. In their resting immature form, they acquire antigens, but following activation signals, including local inflammatory cytokines and pathogen products, they process and present them. In addition, following encounters with such activating signals, they are receptive for chemokine-mediated migration into the draining lymph nodes as well as the acquisition of distinct immunostimulatory properties. As they mature, DCs undergo a dramatic change in their morphological appearance where they begin to develop long processes called dendrites (see Figure 18.1). This process of maturation diminishes antigen uptake capacity while simultaneously upregulating their ability to interact with naïve T lymphocytes through enhanced surface expression of MHC class I and class II molecules as well as the costimulatory molecules CD86 and CD80. In the lymph node, reciprocal interactions with T cells or natural killer (NK) cells, recruited across high endothelial venules, via cognate surface receptors such as CD40-CD40L (Schoenberger et al., 1998) and cytokines (see Figure 18.2) perpetuate the process with terminal maturation of DCs (Shreedhar et al., 1999), and during both the initial and final maturation process, DCs secrete T cell and NK cell activating cytokines and chemokines.

The effective elimination of a particular pathogen type requires the activation of specific types of immune response that most suitably match the particular character of the target as well as the affected tissue type. Cellular (type-1) mechanisms of immunity provide the best defence for eliminating intracellular pathogens as well as transformed cells, while humoural (type-2) immune responses are most effective for controlling pathogens of an extracellular nature. A fine balance must be maintained to enable the host to selectively mount these protective responses when necessary while minimizing the risk of damage to healthy tissue. The regulation of the adaptive immune response is tightly governed by the actions of T helper (T_H) cell subsets, with many diseases being associated with ineffective or inappropriate T_H cell responses (Mosmann and Sad, 1996).

DCs can direct the overall nature of the immune response, playing a pivotal role in the differentiation of naïve $CD4^+$ T cells into their polarized effector T_H subsets (Kalinski et al., 1999; Kapsenberg, 2003; Mosmann and Sad, 1996). DCs act as carriers of information obtained in the peripheral environment from which they were themselves activated, to in turn provide lymph node residing naïve T_H cells with additional signals

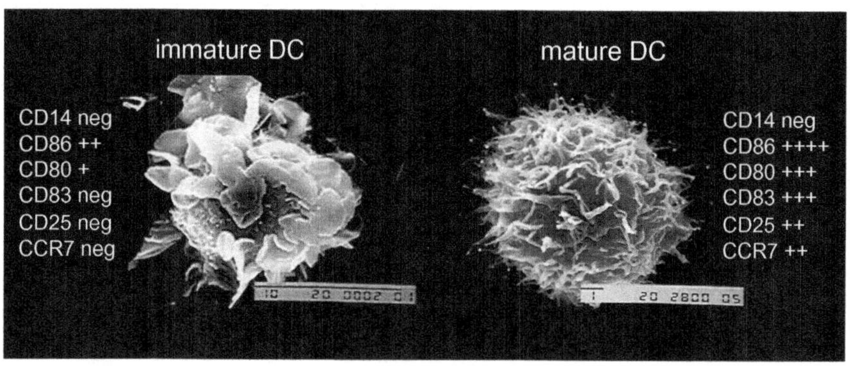

Figure 18.1 • Stages of DC maturation. Images generated using scanning electron microscopy (SEM) showing morphological differences between immature (left) and mature (right) human DCs. The major difference is the longer, thinner dendrites appearing with maturity. Careful analysis of dendrites on DCs has not been carried out on the various DC subsets that have been identified (Images courtesy of Simon Watkins and Robbie Mailliard, University of Pittsburgh, Pittsburgh, PA).

(Signal 3) that are subsequently translated into the development of polarized T_H (T_H1, T_H2, T_H3, T_H9 or T_H17) responses (Dhodapkar et al., 2008; Ghiringhelli et al., 2005; Kalinski et al., 1999; Veldhoen et al., 2008; Kapsenberg, 2003). Indeed, the instructions DCs gather can come directly from pathogen associated molecular pattern molecules (PAMPs) as well as from endogenous signals such as damage associated molecular pattern molecules (DAMPs) released by the surrounding tissue, including those provided by other immune cells. They may also actively suppress the induction of such immune responses by eliciting regulatory T-cell responses capable of further dampening an evolving immune response, in part mediated through cytokines such as TGF-β and IL-10. Many DC-mediated factors have been suggested to contribute to the development and overall quality of these DC-induced polarized T_H responses. To date, the actions of DC-produced soluble factors in T_H cell differentiation seem to be more clearly defined, with members of the IL-12 family (IL-12, IL-23, IL-27 and IL-35) being the most well described of the T_H1 factors (Del Vecchio et al., 2007).

Decision making by DCs

The question of how DCs actually discern which type of immune response is most appropriate to effectively deal with a particular agent and tissue type has perplexed researchers for many years. One early concept introduced was based on evolutionary selection by which different subsets of DCs, originating from either the myeloid or the lymphoid lineage had a selective ability to interact with distinct classes of pathogens, which coincided with observations that there were intrinsic differences in the ability of individual DC subsets to mount either T_H1 or T_H2 responses (Maldonado-Lopez et al., 2001; Pulendran et al., 1999).

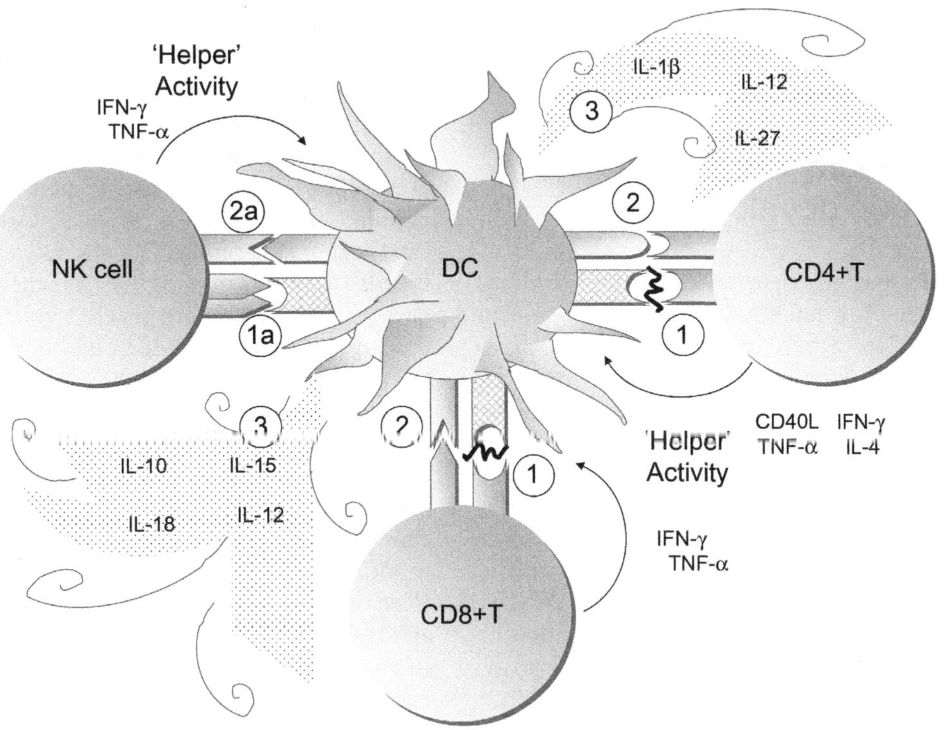

Figure 18.2 • Reciprocal activation of DCs and complementary priming/activation of T cells and NK cells. Lymphocyte activation is dependant upon a coordinated assemblage of signals, as well as the proper activation state of the DCs (Lapointe et al., 2000; Shreedhar et al., 1999). DC–T cell interactions involved in T cell activation includes (1) Signal 1, (2) Signal 2 and (3) Signal 3, the secretion of various immunomodulatory cytokines by DCs (Tominaga et al., 2000; Wesa et al., 2007). Similarly Signal 4, the presence of integrins relating to the tissue of origin, and an inchoate group of Signal 5s can be found important for integration of the effector immune response within and dictated by the tissue (Lotze et al., 2007). T cells can reciprocally activate DCs for enhanced cytokine production, DC maturation and type-1 T cell polarization (Mailliard et al., 2002; Schoenberger et al., 1998; Wesa and Galy, 2002), which amplifies the cycle of T cell activation and priming. A similar paradigm can be observed in the activation of NK cells. While NK cells are responsive to the lack of MHC Class I on NK-sensitive targets, the presence of MHC Class I on DCs can keep NK cells in check (1a). In this setting, NK cells, however, can still be triggered by DC in part through the interaction of (2a) costimulatory ligands with NKG2D and other activating receptors expressed on NK cells (Draghi et al., 2007). Like T cells, cytokines derived from DC (3) may provide additional stimuli to enhance NK activation and effector functions (Ferlazzo et al., 2004a). Properly activated NK cells may consequently provide 'help' to further activate DCs, including release of HMGB1 (Semino et al., 2007) and cell to cell or soluble factors (Mailliard et al., 2003; Vitale et al., 2005).

A complementary notion to the DC lineage model emerged suggesting that through environmental instruction, each DC subset can display a certain degree of plasticity in its ability to drive polarized T_H responses depending on its activation by either exogenous signals derived from the pathogen itself, or from endogenous signals provided by the cells in the surrounding tissue microenvironment in response to the pathogen (Kalinski et al., 1999). While DC subsets differ in their ability to recognize individual pathogen motifs and subsequent responses through expression of Toll-like receptors (TLRs), NOD-like receptors (NLRs) and RIG-I-like receptors (RLRs), it is now clear that they are for the most part flexible, each contributing to the overall quality of the adaptive immune response. Therefore, it is likely that pathogen-associated environmental instruction provided either directly through the activation of their germ-line encoded pattern recognition receptors (PRRs) (Iwasaki and Medzhitov, 2004) or indirectly through the induction of endogenous inflammatory mediators in neighbouring tissues (Kalinski et al., 1999), as well as factors associated with specific DC lineage (Maldonado-Lopez et al., 2001; Pulendran et al., 1999), each contribute in a complementary fashion to the T_H-polarizing function of the DCs.

DC heterogeneity and hematopoietic development

While DCs represent less than 1% of the cells in lymphoid organs, they are nonetheless highly variable in their phenotype. Up to five discrete DC subsets have been identified within lymph nodes and three within the spleen in mice on the basis of surface markers, including CD4, CD8α, DEC205, CD11b, CD11c, Gr1 and B220 (Henri et al., 2001), while five subtypes of DCs have been found in human tonsils (Summers et al., 2001). The diverse subsets of DCs found throughout the adult body in both lymphoid and non-lymphoid compartments are descendents of hematopoietic stem cells (HSCs) originating in the bone marrow, with the exception of follicular DCs, found in lymphoid germinal centres, which are most closely related to stromal cells. Evidence of the hematopoietic origin of DCs comes from in vivo studies of hematopoietic reconstitution of DCs in mice and in vitro studies of human DC differentiation from multipotent HSCs following instruction provided by cocktails of key cytokines (including Flt-3 ligand, c-kit ligand and/or GM-CSF, among others) (Caux et al., 1996; Wesa and Galy, 2001). HSCs can repopulate all cells of the immune system, including erythroid, myeloid (macrophages and granulocytes) and lymphoid (NK cells, B cells and T cells). The lineage commitments to these diverse fates are tightly governed by distinct transcription factors and environmental signals, so an early question emerged: To which lineage do these distinct DC subsets belong?

DCs: lymphoid and/or myeloid origin?

Development of hematopoietic cells from HSCs in the bone marrow appears to occur along two distinct pathways involving either an intermediate common lymphoid progenitor (CLP) or the common myeloid progenitor (CMP) (Katsura and Kawamoto, 2001). DCs were found to develop along pathways using either the CLP (Galy et al., 1995) or the CMP (Wu et al., 2001). DCs have also been reported to develop from pro-B cells as well as monocytes (Björck and Kincade, 1998; Leon and Ardavin, 2008). Distinct cytokine requirements also regulate the development of separate DC subsets, with plasmacytoid DCs (pDCs) dependant upon Flt-3 ligand but not GM-CSF, which is required for myeloid conventional DCs (cDCs) (Gilliet et al., 2002; Naik et al., 2007), suggesting distinct pathways of development. However, adoptive transfer of highly purified CLP or CMP cause the development of both CD8α^+ 'lymphoid', 'CD8α^- myeloid DCs' as well as pDCs (Manz et al., 2001; Shigematsu et al., 2004; Traver et al., 2000). Studies of early thymic progenitors, which populate T cells in the thymus, were found to have capacity for generating NK cells, as well as DCs, but not B cells (Marquez et al., 1998; Sanchez et al., 1994; Wu et al., 1996), and more recent evidence indicates that these progenitor cells retain additional myeloid potential (Bell and Bhandoola, 2008; Wada et al., 2008). As a consequence, the concepts of CLP and CMP are being amended as evidence indicates that B cell development branches off separately from T cell and NK cell and myeloid development, at least for thymic progenitor cells. The evidence for a common CLP that can give rise to DCs but not myeloid has not been entirely refuted (Karsunky et al., 2008; Welner et al., 2008), however, and contributes to the ongoing controversy in the field. In addition, the development of DC subsets from precursors can be altered during infection and inflammation, a process referred to as 'transdifferentation', adding an additional layer of complexity to this highly intricate system (Shortman and Naik, 2007; Welner et al., 2008; Xu et al., 2007). Thus the entire schema of hematopoiesis (and dendropoiesis) is under revision (Buza-Vidas et al., 2007; Rathinam and Flavell, 2008), as evidence suggests that hematopoietic development occurs not along a series of binary decision points but rather reflects a paradigm where commitment occurs in a graded fashion, with loss of potential for specific cell types coinciding with the classification of progenitor phenotypes that are somewhat more flexible in regards to their progeny than previously considered. Furthermore, DC development under steady-state conditions may be distinct from that occurring in the presence of an inflammatory/infectious environment.

More recent evidence indicates that the murine pDCs, CD8α^+ DCs and cDCs diverge from a common precursor, isolated from bone marrow (Naik et al., 2007; Onai et al., 2007). This pro-DC subset can give rise to all three DC subtypes in vitro and in vivo, retains some myeloid colony-forming potential and does not generate B cells, T cells, or NK cells, while a more differentiated pre-DC subset exhibits pure commitment to DCs (Naik et al., 2007). Thus, these distinct DC subsets share a common developmental pathway. This new vision of hematopoiesis and dendropoiesis has not yet been confirmed in humans, although it is likely that an equivalent of this common DC precursor cell will be identified in the near future, and that the human development of DCs from progenitors may be similarly responsive to the presence of inflammatory or infectious stimuli.

Functional niches for DC subsets

Given the highly heterogeneous phenotypes of DCs, it is not surprising that they exhibit distinct properties. In contrast to human cDCs, and epidermal LC, which secrete IL-12 upon activation, pDCs secrete high levels of interferon (IFN)-α upon viral infection or exposure to certain bacterial products (Cella et al., 2000; Krug et al., 2001). Direct comparisons of their functional properties following ex-vivo isolation indicate that various DC subsets from both mice and humans can induce distinct types of immune responses (Maldonado-Lopez et al., 1999; Rissoan et al., 1999). However, even these DCs can be modulated by their cytokine milieu (Liu et al., 2001; Maldonado-Lopez et al., 2001). Additionally, the locality further influences DC function: DCs isolated from different organs have distinct activities. DCs that originate from the lung and Peyer's patch have a T_H2 biasing function (Iwasaki and Kelsall, 1999; Stumbles et al., 1998), while liver-derived DCs bias T cell development towards T_H1 (O'Connell et al., 2003).

Direct instructions from pathogens: DC activation through PRRs

Immature DCs (iDCs) sense invading pathogens through a set of evolutionary conserved PRRs that can recognize distinct sets of conserved microbial molecules (Janeway, 1989). TLRs, one set of PRRs, are widely distributed throughout the body and are differentially expressed among cells of the innate immune system. TLR signalling of iDCs promotes their maturation, causing migration to draining lymph nodes where they can initiate primary T_H responses. The ability of DCs to distinguish between different classes of pathogens based on differential activation of these specific PRRs also contributes to their ability to selectively drive polarized T_H1 and T_H2 responses (Iwasaki and Medzhitov, 2004).

Interestingly, DC subsets differentially express certain TLRs, supporting the notion that individual DC subsets may have an intrinsic ability to drive polarized T_H responses upon recognition of certain classes of pathogens. For example, while cDCs in humans can express TLRs 1–8, pDCs to date have been described as only expressing TLR1, TLR6, TLR7 and TLR9. However, the expression of these molecules is highly regulated by the maturation and activation status of DCs (Visintin et al., 2001). There is also some interspecies variation, as only pDCs appear to be sensitive to ligands for TLR9 in humans (Krug et al., 2001), while in mice, multiple DC subtypes respond to TLR9 signalling (Shah et al., 2003). In addition, while separate DC subsets may respond to the same TLR ligand, they may have distinct cytokine profiles (Ito et al., 2002; Shah et al., 2003).

Indirect instructions: DC activation through endogenous 'danger' signals

In addition to directly modifying DC function, pathogens can indirectly influence DC function by inducing the release of a variety of either pro-inflammatory or anti-inflammatory mediators as a result of their insult on affected tissue. Such factors include cytokines, growth factors, chemokines, histamines, prostaglandins and products released by dying cells, such as heat-shock proteins and HMGB1 (Bianchi, 2007). Many of these endogenous factors can contribute greatly to the activation status of the DCs as well as the T_H cell polarizing abilities, which is consistent with the notion that DCs have the innate ability to recognize 'danger' signals resulting from damage or assault on surrounding tissue (Gallucci and Matzinger, 2001). The same signals likely to be perceived by iDCs as 'danger' are likely to affect their subsequent ability to induce appropriate immune responses.

Although microbial stimuli can promote DC maturation, tissue-derived factors such as IFN-γ and PGE2 can define their T_H cell polarizing function (Kalinski et al., 1999). These polarizing factors act by modulating the IL-12 family producing capacity of DCs. DC maturation in the presence of IFN-γ, for example, results in the induction of an effector memory type-1 polarized DC (DC1) characterized by an enhanced IL-12p70 producing capacity and a strong bias towards promoting the development of T_H1 responses (Kalinski et al., 1999; Vieira et al., 2000). Conversely, exposure of maturing DCs to PGE2 selectively promotes their T_H2 driving function. PGE2 does this by not only diminishing their IL-12p70 producing capacity but also by enhancing their production of IL-12p40, which acts as a competitive inhibitor of the biologically active IL-12p70 (Kalinski et al., 1999).

Killer DCs: another function for DCs

While DCs are best known by serving as professional APCs, there is also now evidence that they may mediate a more primordial role as innate tumouricidal effectors. Such 'killer' DCs (KDCs) may represent a 'multi-tasking' 'hunter/gatherer' cell type that sequentially may gather antigen through killing targets, and then instruct and enforce T cell responses in vivo (Wesa and Storkus, 2008).

Killing, like many other DC functions, is conditional upon the DC subtype and is tightly regulated by distinct signals. Thus, human cDC but not pDC express tumour necrosis factor-related apoptosis-inducing ligand (TRAIL) after treatment with IFNs or TLR7/8 ligands, enabling apoptosis/autophagy of tumour cells (Fanger et al., 1999; Stary et al., 2007). Another subtype of cDC, M-DC8$^+$ cells from the peripheral blood, promotes target cell apoptosis or antibody-dependent cellular cytotoxicity (Schmitz et al., 2002, 2005). KDC may also impose necrotic cell death on tumour cells: A subset of cDCs exposed to TLR7/8 agonists express perforin and granzyme B, licensing them to kill tumour cell lines (Stary et al., 2007). Although freshly isolated human DC subsets are not cytotoxic, they may become killers with an activation requirement, which may serve as a safety measure to prevent unintended deletion of normal or effector/memory T cells.

In mice, a subset of NK cells resembling both NK and DCs (termed 'IKDC') has been identified (as described previously in Chapter 3). Identified in rats, KDCs can both kill tumour targets and subsequently uptake antigen for T cell presentation (Trinite et al., 2005), suggesting that this subset belongs to the DC lineage, although this still requires more definitive proof.

One of the important implications of the ability of DCs to kill their targets relates to acquisition of tumour antigens. Indeed, evidence suggests that subsequent to DC-killing of target cells, these APCs may engulf antigens from tumour apoptotic bodies or autophagic cells (Uhl et al., 2009) and cross-prime T cell responses (Huang et al., 2005; Trinite et al., 2005). The convergence of these operations in KDCs is an efficient anti-tumour paradigm, which generates strong appeal for clinical translation in the cancer setting, using either intratumoural injection of ex vivo generated KDCs or treatment of cancer patients with modalities that augment KDC function, generation and recruitment in vivo within the tumour microenvironment (Stary et al., 2007).

DC activation of NK cells

There have been a number of studies demonstrating DC-mediated enhanced NK cell activation and function resulting from dialogue between the two cell types. The first in vivo evidence of such crosstalk came from a report by Fernandez et al. (1999). This study revealed NK cell-dependent anti-tumour responses in mice bearing MHC class-I negative tumours following either adoptive transfer of DCs or induced in vivo expansion of DCs. More recent studies have shown that these NK cell to DC interactions are also important for the activation, expansion and maintenance of NK cells during microbial infections (Andrews et al., 2003; Hochweller et al., 2008). This DC-mediated NK cell activation can occur through mechanisms involving direct contact or by way of soluble factors released by DCs.

DC activation of NK cells through cell-to-cell contact

The study by Fernandez et al. demonstrated that DCs were capable of inducing NK cell activation in a contact-dependent fashion (1999). While this study did not fully characterize the mechanisms of this activation, subsequent reports showed that DCs can activate NK cells through the activating receptors NKp30, NKp46 and NKG2D (Draghi et al., 2007; Ferlazzo et al., 2002). Other reports describe DCs activating CD28 expressing NK cells by providing the costimulatory ligand CD86 (Wilson et al., 1999; Galea-Lauri et al., 1999; Gao et al., 2003; Martin-Fontecha et al., 1999). DC surface expression of MHC class I-related (MIC) A and B could be induced with IFN-α treatment and subsequently lead to the activation NK cells via NKG2D signalling (Jinushi et al., 2003). This dialogue between DCs and NK cells is disrupted in the setting of chronic hepatitis infections, where NK cells often demonstrate an inability to provide the reciprocal DC activation signals necessary to effectively combat this condition due to an inhibition of the induced surface expression of MICA/B on patient DCs (Jinushi et al., 2003).

Instead of acting merely as APCs, DCs may also play a role as cytokine-presenting cells, being capable of activating effector cells, such as CTLs and NK cells, by directly delivering to them surface-bound cytokines upon contact. Through their surface expression of IL-15Rα, murine DCs can present IL-15 in *trans* to enhance NK cell function and proliferation (Ferlazzo et al., 2004a; Koka et al., 2004). A similar phenomenon may occur with DC presentation of IL-2 via the expression of the IL-2Rα (CD25) on mature DCs. Although it is not clear what functional role CD25 plays on mature DCs, their surface expression of CD25 in the absence of the signalling IL-2Rβ suggests that instead of having a direct use for IL-2 themselves, they may use CD25 to capture and deliver IL-2 to other cells.

Direct interactions between NK cells and DCs can also result in the formation of stimulatory synapses between the two cells types. This can allow for the delivery of preassembled stores of cytokines such as IL-12 from DCs to NK cells, which has been previously reported (Borg et al., 2004). A similar synaptic junction between the two cell types has also been shown to allow immature DCs to provide IL-18 to NK cells (Semino et al., 2005). In return, these IL-18 activated NK cells respond by secreting HMGB1 to promote DC maturation (Semino et al., 2005).

DC activation of NK cells by soluble factors

Various types of microbial stimulation and TLR activation of DCs can induce the production of a number of cytokines that activate NK cells. One example is IL-12, a key factor responsible for driving T_H1-mediated cellular immunity, which directly promotes enhanced NK cell survival, cytotoxic function and IFN-γ production (Trinchieri, 2003). DCs produce IL-12 in response to a wide variety of pathogen-related agents, including LPS and nucleic acids: CpG motifs double-stranded RNA or poly-I:C. In vitro, LPS-, actinobacillus- and poly I:C-stimulated DCs activate freshly isolated human NK cells in an IL-12 dependent manner, promoting their cytolytic activity and production of IFN-γ (Gerosa et al., 2002; Kikuchi et al., 2004). In mice, MCMV-activated pDC produces IL-12, contributing to NK cell activation (Dalod et al., 2003).

DCs can also produce IL-12 in response to endogenous signals from other cell types. A good example of such a factor would be the T helper cell derived signal CD40L (CD154) (Josien et al., 1999; Koch et al., 1996). While they are unable to induce IL-12 by themselves, the T helper produced cytokines IFN-γ and IL-4 can each act as powerful costimulators to enhance IL-12 production; (Hilkens et al., 1997; Snijders et al., 1998) (refer to Figure 18.2). During T cell priming in the lymph node, IL-12 produced by DCs during their interaction with CD40L expressing T helper cells would likely lead to the activation of the high IFN-γ producing lymph node residing NK cells (Cooper et al., 2001c; Ferlazzo et al., 2004b; Mailliard et al., 2005; Vitale et al., 2005).

IL-18, a member of the IL-1 family, is yet another cytokine produced by DCs in response to a number of microbial factors (Akira, 2000). It is most well known for its ability to act as a costimulator with IL-12 for the induction of IFN-γ (Trinchieri, 1998). While IL-1β synergizes with IL-12 for the activation of NK cells (Cooper et al., 2001b), the costimulatory abilities of IL-18 seem to be more potent. Crosstalk between NK cells and IL-18 and IL-12 expressing DCs enhances the NK cell cytotoxic function (Yu et al., 2001). While IL-18 is not by itself a particularly strong inducer of IFN-γ production, it acts in a synergistic manner with a number of other cytokines, including IL-2, IL-15, IL-10 and type-1 IFNs as costimulator by upregulating NK cell expression of the respective receptors for these cytokines (Kawakami, 2002; Mailliard et al., 2005; Son et al., 2001). IL-18 can also prime NK cells for high IFN-γ production in response to secondary signals they may receive at later times (Chaix et al., 2008; Mailliard et al., 2005). Moreover, exposure to IL-18 can induce the lymph node homing CCR7$^+$, CD83$^+$ and CD25$^+$ DC-like phenotype on the more abundant CD56^{dim} population of NK cells (Mailliard et al., 2005) (see Figure 18.3).

A number of in vitro studies have also demonstrated an important role for DC-produced type-1 IFNs in the induction of NK cell activation following DC activation by viruses or TLR4/7/8/9 ligands (Dalod et al., 2003; Gerosa et al., 2002). While IFN-α/β have been reported to be products of most DC types or DC subsets, pDCs appear to be more specialized as being the highest producers of these cytokines.

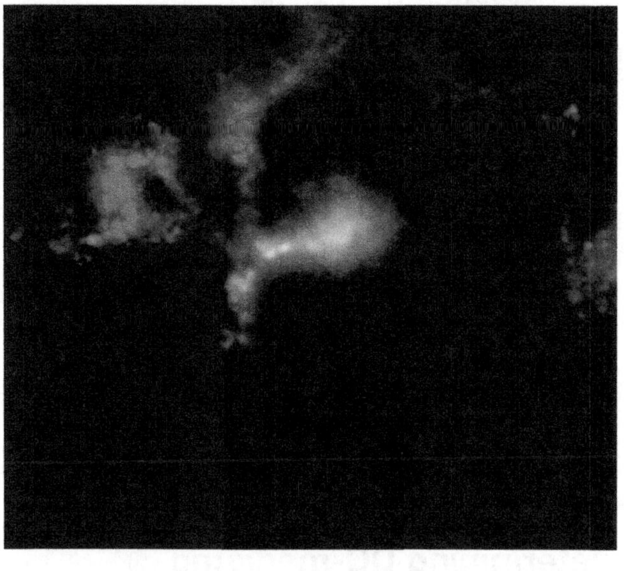

Figure 18.3 • CD83+ lymph node residing human NK cells. Three-colour fluorescent microscopy demonstrating the presence of CD3− cells (CD3+ cells in very dark grey), co-expressing CD56(dark grey) and CD83(grey) cells within the T cell areas of a lymph node (very light grey colour representing co-expression of CD56 and CD83), juxtaposed to single-positive CD3+ T cells and CD83+ DC (Mailliard et al., 2005. Originally published in *The Journal of Experimental Medicine*. doi:10.1084/jem.20050128).

DC selection by NK cells

In some studies, interactions between NK cells and autologous DCs result in the death of the DCs (Carbone et al., 1999; Piccioli et al., 2002). In these studies, it appears that the fate of the DCs interacting with NK cells relates in part to the relative cell numbers involved; low NK cell to DC ratios result in DC activation, and high NK cell to DC ratios lead to DC killing. In addition, the activation status of both cell types seems to be critical. Immature DCs have been shown to be more susceptible to cytotoxicity than mature DCs (Della Chiesa et al., 2003). On the other side, in vitro studies have shown that resting NK cells are less likely to kill DCs and more likely to activate them, while IL-2 activated NK cells are more likely to induce DCs lysis (Mailliard et al., 2003, 2005; Piccioli et al., 2002). The NK cell activating receptors NKp30 and NKp46 (Moretta et al., 2001; Ferlazzo et al., 2002; Poggi et al., 2005) have been described as critical components of NK-mediated lysis of DCs. Protection from NK cell-mediated lysis of DCs depends on the upregulated surface expression of MHC-class I molecules (Ferlazzo et al., 2001). HLA-E expression appears to be particularly critical for DC survival (Della Chiesa et al., 2003). Immature DCs are especially vulnerable to NK cell killing due to the fact that, unlike macrophages and mature DCs, they express levels of HLA-E that are too low to engage the inhibitory receptor CD94/NKG2A expressed on a subset of NK cells that also lack the inhibitory killer Ig-like receptors (KIRs). Therefore, this lack of the inhibitory signal results in their 'missing-self' cytolytic response.

In vivo, immature DCs are rapidly eliminated by NK cells in a TRAIL-dependent manner, thus limiting their particular effectiveness as a DC vaccine (Hayakawa et al., 2004). In the setting of transplantation, Ruggeri et al. (2002) demonstrated that intentional NK-cell receptor mismatches could suppress T cell mediated graft-versus-host disease (GvHD) through NK cell recognition and lysis of host DCs (2002). While it remains unclear whether DC killing by NK cells occurs under normal physiological conditions, it is conceivable that NK cell elimination of DCs may in fact be a DC selection mechanism built in to help regulate the immune response.

Role of killer effector cells in determining DC-mediated polarized immune responses

The ability of DCs to act as flexible mediators of environmental signals offers an additional means of protection, reducing a pathogen's ability to evade the immune system by allowing other cell types such as NK cells and $CD8^+$ T cells to participate in the immunoregulatory decision-making process (Kalinski and Moser, 2005). These cell types have the natural ability to detect minor cellular changes associated with intracellular infection and/or cellular transformation, allowing them not only to provide local DCs with a source of antigen through the lysis of target cells but also to provide a variety of DCs with activating and polarizing factors (Mailliard et al., 2002, 2003). Effector cell-based help provides DCs with a simple yet adaptable nonpathogen-directed mechanism for determining the appropriate response needed to effectively match the character of the pathogen based simply on whether or not the problem is originating from an intracellular source, making it extremely difficult for a pathogen to evade the immune system (Kalinski and Moser, 2005).

Reverse 'help' from NK cells

NK cells rapidly localize to sites of pathogen entry in peripheral tissues and in the marginal zones of the spleen and lymph nodes, acting as sentinels against viral challenge (Biron, 1997). NK cells are not only important as an early line of defence for the elimination of virally infected cells, they also clearly play an immunoregulatory role, with NK cell deficiencies being associated with recurrent viral and mycobacterial infections indicative of the host's inability to mount effective T cell memory responses (Biron et al., 1989; Joncas et al., 1989). Having long been known to support the induction of T_H1 and CTL responses in animal models (Doherty and Allan, 1987; Scharton and Scott, 1993), NK cells have been more recently shown to play a critical role as 'helpers' in the induction of anti-tumour-specific T cell responses (Kelly et al., 2002) as well as allo-specific T_H1 responses to transplant antigens (Chargui et al., 2000; Murphy et al., 2001; Tiberghien et al., 1990). Other studies also show that NK cells can promote the induction of T_H1-associated antibody responses against polyomavirus (Szomolanyi-Tsuda et al., 2001) and can reduce the T_H2-associated development of eosinophilia during respiratory syncytial virus infection (Hussell and Openshaw, 2000). It is now clear that at least part of NK cell 'helper' activity is associated with their modulatory impact on DC function resulting from interactions between the two cell types that can occur at inflammatory sites as well as in draining lymph nodes (Ferlazzo et al., 2002; Gerosa et al., 2002; Mailliard et al., 2003; Piccioli et al., 2002).

DC: NK crosstalk in peripheral tissues

While DCs may support the function of NK cells (Fernandez et al., 1999), more recent in vitro studies demonstrated that NK cells can provide DCs with reciprocal activating signals to enhance their activation status as well (Ferlazzo et al., 2002; Gerosa et al., 2002; Piccioli et al., 2002; Mailliard et al., 2003). In addition to their sensitivity to microbial-induced cytokines, NK cells themselves produce a variety of pro-inflammatory factors such as IFN-γ, TNF-α, GM-CSF, CCL3 (MIP1α), CCL4 (MIP1β) and CCL5 (RANTES) (Moretta, 2002), allowing them to recruit and activate other inflammatory cells such as infiltrating macrophages and immature DCs. NK cells themselves are capable of inducing the differentiation of macrophages into DCs (Zhang et al., 2007) in a GM-CSF and CD154 (CD40L)-dependent manner. NK cell–DC interaction in peripheral sites of inflammation can result in the NK-mediated IFN-γ and TNF-α dependent induction of type-1 polarized DCs (DC1s) (Mailliard et al., 2003). While NK cells can kill without prior activation, analogous to the two-signal paradigm found with the full activation of T cells and B cells, NK cells require at least one additional co-activating signals to carry out their more stringent DC-polarizing immunoregulatory functions (Mailliard et al., 2003) (refer to Figure 18.2). Type-1 IFNs produced by pDCs and virally infected cells, or other cytokines such as IL-2, IL-15 and IL-18 can provide the necessary costimulation required by the NK cells to produce cytokines such as TNF-α and IFN-γ, which may reciprocally activate DCs (Mailliard et al., 2003). While these soluble factors appear to be the key aspects involved in the DC1 polarizing activity of NK cells, additional cytokines and contact-related factors are likely to also play a role.

DC–NK crosstalk in lymph nodes

Human NK cells can be divided into functional subsets based on their surface expression of CD56 and CD16. In peripheral blood, most of the NK cells are $CD56^{dim}/CD16^{+}$, with approximately 5–10% of the NK cells having a $CD56^{bright}/CD16^{neg}$ phenotype (Cooper et al., 2001a). The latter are thought to play more of an immunoregulatory role due to the fact that they can produce high levels of IFN-γ and are less cytolytic than their $CD56^{dim}/CD16^{+}$ counterpart, and during their resting state, they express low levels of the lymph node homing chemokine receptor CCR7, suggesting that they may be capable of influencing the T cell priming function of DCs within the lymph nodes. The similar $CD56^{bright}/CD16^{neg}$ NK cell type has been described as residing within lymph nodes lending support to the idea that these two cell types may indeed be related and capable of supplying DCs with important type-1 polarizing signals (Campbell et al., 2001; Fehniger et al., 2003; Ferlazzo et al., 2004b).

NK cell recruitment into secondary lymphoid tissue supports DC priming of T_H1 cells in mice. Unlike in humans where CCR7 appears to be the critical lymph node homing chemokine receptor for NK cells (Fehniger et al., 2003; Ferlazzo et al., 2004b), CXCR3 seems to also play an important role in mice (Martin-Fontecha et al., 2004). In humans, IL-18 appears to promote the lymph node homing 'helper' rather than 'killer' pathway of NK cell differentiation (Mailliard et al., 2005). In vitro, induces a lymph node homing, DC-like, $CCR7^{+}$ $CD83^{+}$ NK cell phenotype with demonstrated migratory responsiveness to the lymph node associated chemokine CCL21 (Mailliard et al., 2005) (refer to Figure 18.3). Upon migration in response to CCL21, these NK cells can produce high levels of IFN-γ upon secondary exposure to T cell-, mDC- and pDC-related factors they would likely encounter in the lymph node such as IL-2, IL-12 and IFN-α, promoting DCs mediated polarized T_H1 responses (Mailliard et al., 2005).

While there are no clear subsets of mouse NK cells to directly compare with the lymph node DCs interacting $CD56^{bright}$, or $CCR7^{+}$ NK cell populations described in human studies, the murine $CD27^{bright}$ population seems to be a close fit. However, recruitment of the $CD27^{bright}$ population to lymph nodes is mediated by an IFN-γ-dependent mechanism (Watt et al., 2008), suggesting the involvement of CXCR3, the receptor for IFN-γ inducible chemokines (CCL9, 10 and 11), rather than CCR7. Most recently, the ability and requirement of NK precursors to promote the development of lymph nodes has been reported (Cupedo et al., 2009; Hughes et al., 2009).

Exploiting NK 'helper'/DC1 driving function in cancer

Use of NK cell 'help' to induce DC-mediated type-1 polarized responses may be key to improving the efficacy of future DC-based strategies to treat diseases such as cancer. One way to possibly achieve the desired interactions between NK cells and DCs in the setting of cancer is through direct co-administration of both cell types into tumour sites. Their co-delivery would likely result in the enhancement of the tumouricidal effects of the NK cells since DCs have been shown to support their killing activity (Fernandez et al., 1999). In addition, with DCs being in close proximity, the NK cell-mediated lysis of the tumour cells may contribute to the development of tumour-specific immunity by providing DCs with a source of antigenic material, including apoptotic bodies (see Figure 18.4), to be cross-presented to CTL precursors (Albert et al., 1998).

Another option is to facilitate their interactions in vivo by administrating either NK and/or DC activating factors systemically, or locally at preferred sites. While activated NK cells, either through ex vivo manipulation or by systemic cytokine therapy, may help to control

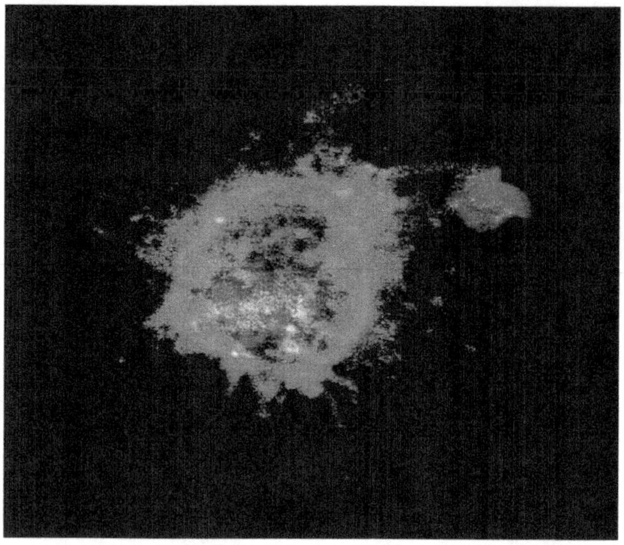

Figure 18.4 • DC captures and engulfs tumour cell antigen. Confocal microscopy showing a CD86 positive (grey) mouse DC having intracellular contents derived from UV-treated apoptotic MCA205 mouse fibrosarcoma cells labelled with the lipophilic carbocyanine dye DiI (dark grey) following their short-term co-incubation (Picture courtesy of Borys Mascarenhas, Joseph Baar and Simon Watkins, University of Pittsburgh, Pittsburgh, PA).

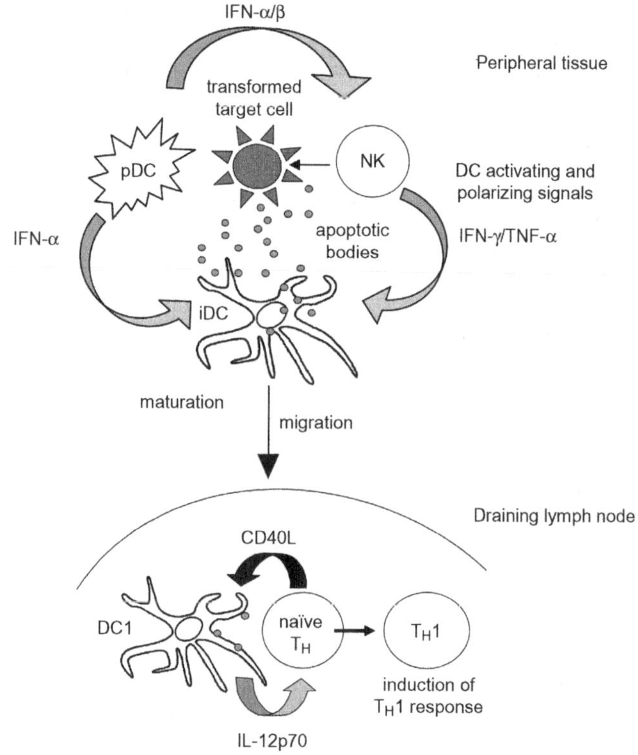

Figure 18.5 • NK cells provide immune 'help' for DC induction of primary T_H1 responses. Production of IFN-α by pDCs, caused by viral infection or stimulation with CpGs (Liu et al., 2008), can promote crosstalk between NK cells and immature DCs resulting in the TNF-α and IFN-γ dependant NK cell-induced maturation and type-1 polarization of DCs (Mailliard et al., 2003). Transformed target cells signal activating receptors on NK to induce their lytic activity. Neighbouring DCs can then act as carriers of NK cell-derived instructions, process captured apoptotic target cells and migrate to draining lymph nodes where they and present processed antigen to T cells (Albert et al., 1998). Subsequent encounters with antigen-specific CD40 ligand (CD40L) expressing naïve $CD4^+$ T cells results in DC production of high levels of IL-12p70, thus promoting T_H1 responses. In this model, NK cells can promote type-1 immune responses without themselves being required migrate to draining lymph nodes.

the initial growth of tumours, the lack of additional activating signals likely present during viral infections (see Figure 18.5), such as IFN-α/β and or IL-18, would probably limit their ability to support the development of type-1 immunity against transformed cells. Because NK cells have a two-signal requirement to activate their helper activity (Mailliard et al., 2003), there is rationale for applying systemic therapies using factors such as IL-18 and IFN-α in combination or in addition to adoptively transferred NK cells to promote NK cell 'help' in vivo.

Another strategy is to target the activation of the type-1 IFN producing pDCs by including CpGs to stimulate their release of IFN-α. A recent study demonstrated robust therapeutic responses in a mouse melanoma B16 model by direct injection of TLR9 activated pDCs into subcutaneous tumours (Liu et al., 2008). This treatment induced CTL responses to multiple B16 tumour-associated antigens, causing regression of both treated tumours as well those distal from the injection site. In this model, T cell cross-priming is mediated by cDCs and was totally dependant on the early local recruitment of NK cells at the tumour injection site by the pDCs (Liu et al., 2008). Thus, activation of pDCs can induce an effective systemic anti-tumour response against established metastatic tumours by initiating a cascade of events involving the activation of and crosstalk between pDCs, NK cells, cDCs and T cells (Liu et al., 2008) (refer to Figure 18.5).

Similarly, combination therapies using antibodies might also be considered. Since certain therapeutic antibodies, such as Herceptin (trastuzumab), can also act in a costimulatory fashion to induce NK cell production of DC-polarizing cytokines (Carson et al., 2001; Parihar et al., 2002), the effectiveness of certain anti-cancer antibody therapies may also be improved if used in combination with other NK–DC polarizing co-activating factors such as type-1 IFNs, IL-2 or IL-18 (Mailliard et al., 2003). Besides providing the DC1 polarizing signals, the NK cell lysis of antibody tumour cells would also likely facilitate DC uptake of tumour material and promote DC activation and cross-presentation as mentioned earlier (Albert et al., 1998).

Generating mature type-1 polarized DCs ex vivo for vaccine development, whether using two-signal activated NK cells to generate them (Mailliard et al., 2003) or factors that would mimic a natural scenario involving NK:DC crosstalk during viral infection—such as the method described for generating the high IL-12 producing mature alpha type-1 polarized DCs (αDC1) (Mailliard et al., 2004)—could be considered. This protocol uses a defined five-component cocktail consisting of TNF-α, IL-1β, IFN-α, IFN-γ and poly-I:C. Data from in vitro experiments clearly demonstrated significant advantages to using NK cell-induced DC1 or αDC1 for inducing T_H1 and CTL responses as compared to using the conventional low IL-12 producing DCs typically used and generated with the standard cytokine cocktail consisting of TNF-α/IL-1β/IL-6/and PGE_2 (Jonuleit et al., 1997). While it is unclear if the in vitro differences noted between these DC types would translate in a similar manner in vivo, these findings nevertheless provide rationale for designing anti-cancer immunotherapies to implement the use of type-1 polarized DCs as a vehicle to carry not only the antigen-specific signal 1 and costimulatory signal 2 but also the T_H1-polarizing signal 3 as an additional novel component.

References

Albert, M.L., Sauter, B. and Bhardwaj, N. (1998). Dendritic cells acquire antigen from apoptotic cells and induce class I-restricted CTLs. Nature **392**, 86–89.

Andrews, D.M., Scalzo, A.A., Yokoyama, W.M., Smyth, M.J. and Degli-Esposti, M.A. (2003). Functional interactions between dendritic cells and NK cells during viral infection. Nat Immunol **4**, 175–181.

Banchereau, J. and Steinman, R.M. (1998). Dendritic cells and the control of immunity. Nature **392**, 245–252.

Bell, J.J. and Bhandoola, A. (2008). The earliest thymic progenitors for T cells possess myeloid lineage potential. Nature **452**, 764.

Bianchi, M.E. (2007). DAMPs, PAMPs and alarmins: all we need to know about danger. J Leukoc Biol **81**, 1–5.

Biron, C.A. (1997). Activation and function of natural killer cell responses during viral infections. Curr Opin Immunol **9**, 24–34.

Biron, C.A., Byron, K.S. and Sullivan, J.L. (1989). Severe herpesvirus infections in an adolescent without natural killer cells. N Engl J Med **320**, 1731–1735.

Björck, P. and Kincade, P.W. (1998). CD19+ pro-B cells can give rise to dendritic cells in vitro. J Immunol **161**, 5795–5799.

Borg, C., Jalil, A., Laderach, D., Maruyama, K., Wakasugi, H., Charrier, S., Ryffel, B., Cambi, A., Figdor, C., Vainchenker, W., Galy, A., Caignard, A. and Zitvogel, L. (2004). NK cell activation by dendritic cells (DCs) requires the formation of a synapse leading to IL-12 polarization in DCs. Blood **104**, 3267–3275.

Buza-Vidas, N., Luc, S. and Jacobsen, S.E. (2007). Delineation of the earliest lineage commitment steps of haematopoietic stem cells: new developments, controversies and major challenges. Curr Opin Hematol **14**, 315–321.

Campbell, J.J., Qin, S., Unutmaz, D., Soler, D., Murphy, K.E., Hodge, M.R., Wu, L. and Butcher, E.C. (2001). Unique subpopulations of CD56+ NK and NK-T peripheral blood lymphocytes identified by chemokine receptor expression repertoire. J Immunol **166**, 6477–6482.

Carbone, E., Terrazzano, G., Ruggiero, G., Zanzi, D., Ottaiano, A., Manzo, C., Karre, K. and Zappacosta, S. (1999). Recognition of autologous dendritic cells by human NK cells. Eur J Immunol **29**, 4022–4029.

Carson, W.E., Parihar, R., Lindemann, M.J., Personeni, N., Dierksheide, J., Meropol, N.J., Baselga, J. and Caligiuri, M.A. (2001). Interleukin-2 enhances the natural killer cell response to Herceptin-coated Her2/neu-positive breast cancer cells. Eur J Immunol **31**, 3016–3025.

Caux, C., Vanbervliet, B., Massacrier, C., Dezutter-Dambuyant, C., de Saint-Vis, B., Jacquet, C., Yoneda, K., Imamura, S., Schmitt, D. and Banchereau, J. (1996). CD34+ hematopoietic progenitors from human cord blood differentiate along two independent dendritic cell pathways in response to GM-CSF + TNF alpha. J Exp Med **184**, 695–706.

Cella, M., Facchetti, F., Lanzavecchia, A. and Colonna, M. (2000). Plasmacytoid dendritic cells activated by influenza virus and CD40L drive a potent TH1 polarization. Nat Immunol **1**, 305–310.

Chaix, J., Tessmer, M.S., Hoebe, K., Fuseri, N., Ryffel, B., Dalod, M., Alexopoulou, L., Beutler, B., Brossay, L., Vivier, E. and Walzer, T. (2008). Cutting edge: priming of NK cells by IL-18. J Immunol **181**, 1627–1631.

Chargui, J., Oyama, A., Yoshimura, R., Wada, S., Hase, T. and Kishimoto, T. (2000). Inhibition of NK cell activity induces improvement and stable chimerism after allogeneic transplantation. Transplant Proc **32**, 2462–2463.

Cooper, M.A., Fehniger, T.A. and Caligiuri, M.A. (2001a). The biology of human natural killer-cell subsets. Trends Immunol **22**, 633–640.

Cooper, M.A., Fehniger, T.A., Ponnappan, A., Mehta, V., Wewers, M.D. and Caligiuri, M.A. (2001b). Interleukin-1beta costimulates interferon-gamma production by human natural killer cells. Eur J Immunol **31**, 792–801.

Cooper, M.A., Fehniger, T.A., Turner, S.C., Chen, K.S., Ghaheri, B.A., Ghayur, T., Carson, W.E. and Caligiuri, M.A. (2001c). Human natural killer cells: a unique innate immunoregulatory role for the CD56(bright) subset. Blood **97**, 3146–3151.

Cupedo, T., Crellin, N.K., Papazian, N., Rombouts, E.J., Weijer, K., Grogan, J.L., Fibbe, W.E., Cornelissen, J.J. and Spits, H. (2009). Human fetal lymphoid tissue-inducer cells are interleukin 17-producing precursors to RORC+ CD127+ natural killer-like cells. Nat Immunol **10**, 66–74.

Dalod, M., Hamilton, T., Salomon, R., Salazar-Mather, T.P., Henry, S.C., Hamilton, J.D. and Biron, C.A. (2003). Dendritic cell responses to early murine cytomegalovirus infection: subset functional specialization and differential regulation by interferon alpha/beta. J Exp Med **197**, 885–898.

Del Vecchio, M., Bajetta, E., Canova, S., Lotze, M.T., Wesa, A., Parmiani, G. and Anichini, A. (2007). Interleukin-12: biological properties and clinical application. Clin Cancer Res **13**, 4677–4685.

Della Chiesa, M., Vitale, M., Carlomagno, S., Ferlazzo, G., Moretta, L. and Moretta, A. (2003). The natural killer cell-mediated killing of autologous dendritic cells is confined to a cell subset expressing CD94/NKG2A, but lacking inhibitory killer Ig-like receptors. Eur J Immunol **33**, 1657–1666.

Dhodapkar, K.M., Barbuto, S., Matthews, P., Kukreja, A., Mazumder, A., Vesole, D., Jagannath, S. and Dhodapkar, M.V. (2008). Dendritic cells mediate the induction of polyfunctional human IL17-producing cells (Th17-1 cells) enriched in the bone marrow of patients with myeloma. Blood **112**, 2878–2885.

Doherty, P.C. and Allan, J.E. (1987). Anti-asialo GM1 eliminates both inflammatory process and cytotoxic T-cell function in the lymphocytic choriomeningitis adoptive transfer model. Cell Immunol **107**, 1–7.

Draghi, M., Pashine, A., Sanjanwala, B., Gendzekhadze, K., Cantoni, C., Cosman, D., Moretta, A., Valiante, N.M. and Parham, P. (2007). NKp46 and NKG2D recognition of infected dendritic cells is necessary for NK cell activation in the human response to influenza infection. J Immunol **178**, 2688–2698.

Fanger, N.A., Maliszewski, C.R., Schooley, K. and Griffith, T.S. (1999). Human dendritic cells mediate cellular apoptosis via tumor necrosis factor-related apoptosis-inducing ligand (TRAIL). J Exp Med **190**, 1155–1164.

Fehniger, T.A., Cooper, M.A., Nuovo, G.J., Cella, M., Facchetti, F., Colonna, M. and Caligiuri, M.A. (2003). CD56bright natural killer cells are present in human lymph nodes and are activated by T cell-derived IL-2: a potential new link between adaptive and innate immunity. Blood **101**, 3052–3057.

Ferlazzo, G., Semino, C. and Melioli, G. (2001). HLA class I molecule expression is up-regulated during maturation of dendritic cells, protecting them from natural killer cell-mediated lysis. Immunol Lett **76**, 37–41.

Ferlazzo, G., Tsang, M.L., Moretta, L., Melioli, G., Steinman, R.M. and Munz, C. (2002). Human dendritic cells activate resting natural killer (NK) cells and are recognized via the NKp30 receptor by activated NK cells. J Exp Med **195**, 343–351.

Ferlazzo, G., Pack, M., Thomas, D., Paludan, C., Schmid, D., Strowig, T., Bougras, G., Muller, W.A., Moretta, L. and Munz, C. (2004a). Distinct roles of IL-12 and IL-15 in human natural killer cell activation by dendritic cells from secondary lymphoid organs. Proc Natl Acad Sci U S A **101**, 16606–16611.

Ferlazzo, G., Thomas, D., Lin, S.L., Goodman, K., Morandi, B., Muller, W.A., Moretta, A. and Munz, C. (2004b). The abundant NK cells in human secondary lymphoid tissues require activation to express killer cell Ig-like receptors and become cytolytic. J Immunol **172**, 1455–1462.

Fernandez, N.C., Lozier, A., Flament, C., Ricciardi-Castagnoli, P., Bellet, D., Suter, M., Perricaudet, M., Tursz, T., Maraskovsky, E. and Zitvogel, L. (1999). Dendritic cells directly trigger NK cell functions: cross-talk relevant in innate anti-tumor immune responses in vivo. Nat Med **5**, 405–411.

Galea-Lauri, J., Darling, D., Gan, S.U., Krivochtchapov, L., Kuiper, M., Gaken, J., Souberbielle, B. and Farzaneh, F. (1999). Expression of a variant of CD28 on a subpopulation of human NK cells: implications for B7-mediated stimulation of NK cells. J Immunol **163**, 62–70.

Gallucci, S. and Matzinger, P. (2001). Danger signals: SOS to the immune system. Curr Opin Immunol **13**, 114–119.

Galy, A., Travis, M., Cen, D. and Chen, B. (1995). Human T, B, natural killer, and dendritic cells arise from a common bone marrow progenitor cell subset. Immunity **3**, 459–473.

Gao, J.X., Liu, X., Wen, J., Caligiuri, M.A., Stroynowski, I., Zheng, P. and Liu, Y. (2003). Two-signal requirement for activation and effector function of natural killer cell response to allogeneic tumor cells. Blood **102**, 4456–4463.

Gerosa, F., Baldani-Guerra, B., Nisii, C., Marchesini, V., Carra, G. and Trinchieri, G. (2002). Reciprocal activating interaction between natural killer cells and dendritic cells. J Exp Med **195**, 327–333.

Ghiringhelli, F., Puig, P.E., Roux, S., Parcellier, A., Schmitt, E., Solary, E., Kroemer, G., Martin, F., Chauffert, B. and Zitvogel, L. (2005). Tumor cells convert immature myeloid dendritic cells into TGF-beta-secreting cells inducing CD4+ CD25+ regulatory T cell proliferation. J Exp Med **202**, 919–929.

Gilliet, M., Boonstra, A., Paturel, C., Antonenko, S., Xu, X.L., Trinchieri, G., O'Garra, A. and Liu, Y.J. (2002). The development of murine plasmacytoid dendritic cell precursors is differentially regulated by FLT3-ligand and granulocyte/macrophage colony-stimulating factor. J Exp Med **195**, 953–958.

Hayakawa, Y., Screpanti, V., Yagita, H., Grandien, A., Ljunggren, H.G., Smyth, M.J. and Chambers, B.J. (2004). NK cell TRAIL eliminates immature dendritic cells in vivo and limits dendritic cell vaccination efficacy. J Immunol **172**, 123–129.

Henri, S., Vremec, D., Kamath, A., Waithman, J., Williams, S., Benoist, C., Burnham, K., Saeland, S., Handman, E. and Shortman, K. (2001). The dendritic cell populations of mouse lymph nodes. J Immunol **167**, 741–748.

Hilkens, C.M., Kalinski, P., de Boer, M. and Kapsenberg, M.L. (1997). Human dendritic cells require exogenous interleukin-12-inducing factors to direct the development of naive T-helper cells toward the Th1 phenotype. Blood **90**, 1920–1926.

Hochweller, K., Striegler, J., Hammerling, G.J. and Garbi, N. (2008). A novel CD11c.DTR transgenic mouse for depletion of dendritic cells reveals their requirement for homeostatic proliferation of natural killer cells. Eur J Immunol **38**, 2776–2783.

Huang, J., Tatsumi, T., Pizzoferrato, E., Vujanovic, N. and Storkus, W.J. (2005). Nitric oxide sensitizes tumor cells to dendritic cell-mediated apoptosis, uptake, and cross-presentation. Cancer Res **65**, 8461–8470.

Hughes, T., Becknell, B., McClory, S., Briercheck, E., Freud, A.G., Zhang, X., Mao, H., Nuovo, G., Yu, J. and Caligiuri, M.A. (2009). Stage three immature human natural killer cells found in secondary lymphoid tissue constitutively and selectively express the TH17 cytokine interleukin-22. Blood **113**, 4008–4010.

Hussell, T. and Openshaw, P.J. (2000). IL-12-activated NK cells reduce lung eosinophilia to the attachment protein of respiratory syncytial virus but do not enhance the severity of illness in CD8 T cell-immunodeficient conditions. J Immunol **165**, 7109–7115.

Ito, T., Amakawa, R., Kaisho, T., Hemmi, H., Tajima, K., Uehira, K., Ozaki, Y., Tomizawa, H., Akira, S. and Fukuhara, S. (2002). Interferon-alpha and interleukin-12 are induced differentially by Toll-like receptor 7 ligands in human blood dendritic cell subsets. J Exp Med **195**, 1507–1512.

Iwasaki, A. and Kelsall, B.L. (1999). Freshly isolated Peyer's patch, but not spleen, dendritic cells produce interleukin 10 and induce the differentiation of T helper type 2 cells. J Exp Med **190**, 229–239.

Iwasaki, A. and Medzhitov, R. (2004). Toll-like receptor control of the adaptive immune responses. Nat Immunol **5**, 987–995.

Janeway , C.A. Jr. (1989). Approaching the asymptote? Evolution and revolution in immunology. Cold Spring Harb Symp Quant Biol **54**(Pt 1), 1–13.

Jinushi, M., Takehara, T., Kanto, T., Tatsumi, T., Groh, V., Spies, T., Miyagi, T., Suzuki, T., Sasaki, Y. and

Hayashi, N. (2003). Critical role of MHC class I-related chain A and B expression on IFN-alpha-stimulated dendritic cells in NK cell activation: impairment in chronic hepatitis C virus infection. J Immunol **170**, 1249–1256.

Joncas, J., Monczak, Y., Ghibu, F., Alfieri, C., Bonin, A., Ahronheim, G. and Rivard, G. (1989). Brief report: killer cell defect and persistent immunological abnormalities in two patients with chronic active Epstein-Barr virus infection. J Med Virol **28**, 110–117.

Jonuleit, H., Kuhn, U., Muller, G., Steinbrink, K., Paragnik, L., Schmitt, E., Knop, J. and Enk, A.H. (1997). Pro-inflammatory cytokines and prostaglandins induce maturation of potent immunostimulatory dendritic cells under fetal calf serum-free conditions. Eur J Immunol **27**, 3135–3142.

Josien, R., Wong, B.R., Li, H.L., Steinman, R.M. and Choi, Y. (1999). TRANCE, a TNF family member, is differentially expressed on T cell subsets and induces cytokine production in dendritic cells. J Immunol **162**, 2562–2568.

Kalinski, P. and Moser, M. (2005). Consensual immunity: success-driven development of T-helper-1 and T-helper-2 responses. Nat Rev Immunol **5**, 251–260.

Kalinski, P., Hilkens, C.M., Wierenga, E.A. and Kapsenberg, M.L. (1999). T-cell priming by type-1 and type-2 polarized dendritic cells: the concept of a third signal. Immunol Today **20**, 561–567.

Kapsenberg, M.L. (2003). Dendritic-cell control of pathogen-driven T-cell polarization. Nat Rev Immunol **3**, 984–993.

Karsunky, H., Inlay, M.A., Serwold, T., Bhattacharya, D. and Weissman, I.L. (2008). Flk2+ common lymphoid progenitors possess equivalent differentiation potential for the B and T lineages. Blood **111**, 5562–5570.

Katsura, Y. and Kawamoto, H. (2001). Stepwise lineage restriction of progenitors in lympho-myelopoiesis. Int Rev Immunol **20**, 1–20.

Kawakami, K. (2002). Interleukin-18 and host defense against infectious pathogens. J Immunother **25**(Suppl. 1), S12–S19.

Kelly, J.M., Darcy, P.K., Markby, J.L., Godfrey, D.I., Takeda, K., Yagita, H. and Smyth, M.J. (2002). Induction of tumor-specific T cell memory by NK cell-mediated tumor rejection. Nat Immunol **3**, 83–90.

Kikuchi, T., Hahn, C.L., Tanaka, S., Barbour, S.E., Schenkein, H.A. and Tew, J.G. (2004). Dendritic cells stimulated with Actinobacillus actinomycetemcomitans elicit rapid gamma interferon responses by natural killer cells. Infect Immun **72**, 5089–5096.

Koch, F., Stanzl, U., Jennewein, P., Janke, K., Heufler, C., Kampgen, E., Romani, N. and Schuler, G. (1996). High level IL-12 production by murine dendritic cells: upregulation via MHC class II and CD40 molecules and downregulation by IL-4 and IL-10. J Exp Med **184**, 741–746.

Koka, R., Burkett, P., Chien, M., Chai, S., Boone, D.L. and Ma, A. (2004). Cutting edge: murine dendritic cells require IL-15R alpha to prime NK cells. J Immunol **173**, 3594–3598.

Krug, A., Towarowski, A., Britsch, S., Rothenfusser, S., Hornung, V., Bals, R., Giese, T., Engelmann, H., Endres, S., Krieg, A.M. and Hartmann, G. (2001). Toll-like receptor expression reveals CpG DNA as a unique microbial stimulus for plasmacytoid dendritic cells which synergizes with CD40 ligand to induce high amounts of IL-12. Eur J Immunol **31**, 3026–3037.

Lapointe, R., Toso, J.F., Butts, C., Young, H.A. and Hwu, P. (2000). Human dendritic cells require multiple activation signals for the efficient generation of tumor antigen-specific T lymphocytes. Eur J Immunol **30**, 3291–3298.

Leon, B. and Ardavin, C. (2008). Monocyte-derived dendritic cells in innate and adaptive immunity. Immunol Cell Biol **86**, 320.

Liu, Y.J., Kanzler, H., Soumelis, V. and Gilliet, M. (2001). Dendritic cell lineage, plasticity and cross-regulation. Nat Immunol **2**, 585–589.

Liu, C., Lou, Y., Lizee, G., Qin, H., Liu, S., Rabinovich, B., Kim, G.I., Wang, Y.H., Ye, Y., Sikora, A.G., Overwijk, W.W., Liu, Y.J., Wang, G. and Hwu, P. (2008). Plasmacytoid dendritic cells induce NK cell-dependent, tumor antigen-specific T cell cross-priming and tumor regression in mice. J Clin Invest **118**, 1165–1175.

Lotze, M.T., Zeh, H.J., Rubartelli, A., Sparvero, L.J., Amoscato, A.A., Washburn, N.R., Devera, M.E., Liang, X., Tor, M. and Billiar, T. (2007). The grateful dead: damage-associated molecular pattern molecules and reduction/oxidation regulate immunity. Immunol Rev **220**, 60–81.

Mailliard, R.B., Egawa, S., Cai, Q., Kalinska, A., Bykovskaya, S.N., Lotze, M.T., Kapsenberg, M.L., Storkus, W.J. and Kalinski, P. (2002). Complementary dendritic cell-activating function of CD8+ and CD4+ T cells: helper role of CD8+ T cells in the development of T helper type 1 responses. J Exp Med **195**, 473–483.

Mailliard, R.B., Son, Y.I., Redlinger, R., Coates, P.T., Giermasz, A., Morel, P.A., Storkus, W.J. and Kalinski, P. (2003). Dendritic cells mediate NK cell help for Th1 and CTL responses: two-signal requirement for the induction of NK cell helper function. J Immunol **171**, 2366–2373.

Mailliard, R.B., Wankowicz-Kalinska, A., Cai, Q., Wesa, A., Hilkens, C.M., Kapsenberg, M.L., Kirkwood, J.M., Storkus, W.J. and Kalinski, P. (2004). Alpha-type-1 polarized dendritic cells: a novel immunization tool with optimized CTL-inducing activity. Cancer Res **64**, 5934–5937.

Mailliard, R.B., Alber, S.M., Shen, H., Watkins, S.C., Kirkwood, J.M., Herberman, R.B. and Kalinski, P. (2005). IL-18-induced CD83+ CCR7+ NK helper cells. J Exp Med **202**, 941–953.

Maldonado-Lopez, R., De Smedt, T., Pajak, B., Heirman, C., Thielemans, K., Leo, O., Urbain, J., Maliszewski, C.R. and Moser, M. (1999). Role of CD8alpha+ and CD8alpha− dendritic cells in the induction of primary immune responses in vivo. J Leukoc Biol **66**, 242–246.

Maldonado-Lopez, R., Maliszewski, C., Urbain, J. and Moser, M. (2001). Cytokines regulate the capacity of CD8alpha(+) and CD8alpha(−) dendritic cells to prime Th1/Th2 cells in vivo. J Immunol **167**, 4345–4350.

Manz, M.G., Traver, D., Miyamoto, T., Weissman, I.L. and Akashi, K. (2001). Dendritic cell potentials of early lymphoid and myeloid progenitors. Blood **97**, 3333–3341.

Marquez, C., Trigueros, C., Franco, J.M., Ramiro, A.R., Carrasco, Y.R., Lopez-Botet, M. and Toribio, M.L. (1998). Identification of a common developmental pathway for thymic natural killer cells and dendritic cells. Blood **91**, 2760–2771.

Martin-Fontecha, A., Assarsson, E., Carbone, E., Karre, K. and Ljunggren, H.G. (1999). Triggering of murine NK cells by CD40 and CD86 (B7-2). J Immunol **162**, 5910–5916.

Martin-Fontecha, A., Thomsen, L.L., Brett, S., Gerard, C., Lipp, M., Lanzavecchia, A. and Sallusto, F. (2004). Induced recruitment of NK cells to lymph nodes provides IFN-gamma for T(H)1 priming. Nat Immunol **5**, 1260–1265.

Moretta, A. (2002). Natural killer cells and dendritic cells: rendezvous in abused tissues. Nat Rev Immunol **2**, 957–964.

Moretta, A., Bottino, C., Vitale, M., Pende, D., Cantoni, C., Mingari, M.C., Biassoni, R. and Moretta, L. (2001). Activating receptors and coreceptors involved in human natural killer cell-mediated cytolysis. Annu Rev Immunol **19**, 197–223.

Mosmann, T.R. and Sad, S. (1996). The expanding universe of T-cell subsets: Th1, Th2 and more. Immunol Today **17**, 138–146.

Murphy, W.J., Koh, C.Y., Raziuddin, A., Bennett, M. and Longo, D.L. (2001). Immunobiology of natural killer cells and bone marrow transplantation: merging of basic and preclinical studies. Immunol Rev **181**, 279–289.

Naik, S.H., Sathe, P., Park, H.Y., Metcalf, D., Proietto, A.I., Dakic, A., Carotta, S., O'Keeffe, M., Bahlo, M., Papenfuss, A., Kwak, J.Y., Wu, L. and Shortman, K. (2007). Development of plasmacytoid and conventional dendritic cell subtypes from single precursor cells derived in vitro and in vivo. Nat Immunol **8**, 1217–1226.

O'Connell, P.J., Son, Y.I., Giermasz, A., Wang, Z., Logar, A.J., Thomson, A.W. and Kalinski, P. (2003). Type-1 polarized nature of mouse liver CD8alpha− and CD8alpha+dendritic cells: tissue-dependent differences offset CD8alpha-related dendritic cell heterogeneity. Eur J Immunol **33**, 2007–2013.

Onai, N., Obata-Onai, A., Schmid, M.A., Ohteki, T., Jarrossay, D. and Manz, M.G. (2007). Identification of clonogenic common Flt3+ M-CSFR+ plasmacytoid and conventional dendritic cell progenitors in mouse bone marrow. Nat Immunol **8**, 1207–1216.

Parihar, R., Dierksheide, J., Hu, Y. and Carson, W.E. (2002). IL-12 enhances the natural killer cell cytokine response to Ab-coated tumor cells. J Clin Invest **110**, 983–992.

Piccioli, D., Sbrana, S., Melandri, E. and Valiante, N.M. (2002). Contact-dependent stimulation and inhibition of dendritic cells by natural killer cells. J Exp Med **195**, 335–341.

Poggi, A., Massaro, A.M., Negrini, S., Contini, P. and Zocchi, M.R. (2005). Tumor-induced apoptosis of human IL-2-activated NK cells: role of natural cytotoxicity receptors. J Immunol **174**, 2653–2660.

Pulendran, B., Smith, J.L., Caspary, G., Brasel, K., Pettit, D., Maraskovsky, E. and Maliszewski, C.R. (1999). Distinct dendritic cell subsets differentially regulate the class of immune response in vivo. Proc Natl Acad Sci U S A **96**, 1036–1041.

Rathinam, C. and Flavell, R.A. (2008). The hematopoiesis paradigm: clarity or ambiguity? Blood **112**, 3534–3535.

Rissoan, M.C., Soumelis, V., Kadowaki, N., Grouard, G., Briere, F., de Waal Malefyt, R. and Liu, Y.J. (1999). Reciprocal control of T helper cell and dendritic cell differentiation. Science **283**, 1183–1186.

Ruggeri, L., Capanni, M., Urbani, E., Perruccio, K., Shlomchik, W.D., Tosti, A., Posati, S., Rogaia, D., Frassoni, F., Aversa, F., Martelli, M.F. and Velardi, A. (2002). Effectiveness of donor natural killer cell alloreactivity in mismatched hematopoietic transplants. Science **295**, 2097–2100.

Sanchez, M.J., Muench, M.O., Roncarolo, M.G., Lanier, L.L. and Phillips, J.H. (1994). Identification of a common T/natural killer cell progenitor in human fetal thymus. J Exp Med **180**, 569–576.

Scharton, T.M. and Scott, P. (1993). Natural killer cells are a source of interferon gamma that drives differentiation of CD4+ T cell subsets and induces early resistance to *Leishmania major* in mice. J Exp Med **178**, 567–577.

Schmitz, M., Zhao, S., Schakel, K., Bornhauser, M., Ockert, D. and Rieber, E.P. (2002). Native human blood dendritic cells as potent effectors in antibody-dependent cellular cytotoxicity. Blood **100**, 1502–1504.

Schmitz, M., Zhao, S., Deuse, Y., Schakel, K., Wehner, R., Wohner, H., Holig, K., Wienforth, F., Kiessling, A., Bornhauser, M., Temme, A., Rieger, M.A., Weigle, B., Bachmann, M. and Rieber, E.P. (2005). Tumoricidal potential of native blood dendritic cells: direct tumor cell killing and activation of NK cell-mediated cytotoxicity. J Immunol **174**, 4127–4134.

Schoenberger, S.P., Toes, R.E., van der Voort, E.I., Offringa, R. and Melief, C.J. (1998). T-cell help for cytotoxic T lymphocytes is mediated by CD40-CD40L interactions. Nature **393**, 480–483.

Semino, C., Angelini, G., Poggi, A. and Rubartelli, A. (2005). NK/iDC interaction results in IL-18 secretion by DCs at the synaptic cleft followed by NK cell activation and release of the DC maturation factor HMGB1. Blood **106**, 609–616.

Semino, C., Ceccarelli, J., Lotti, L.V., Torrisi, M.R., Angelini, G. and Rubartelli, A. (2007). The maturation potential of NK cell clones toward autologous dendritic cells correlates with HMGB1 secretion. J Leukoc Biol **81**, 92–99.

Shah, J.A., Darrah, P.A., Ambrozak, D.R., Turon, T.N., Mendez, S., Kirman, J., Wu, C.-Y., Glaichenhaus, N. and Seder, R.A. (2003). Dendritic cells are responsible for the capacity of CpG oligodeoxynucleotides to act as an adjuvant for protective vaccine immunity against *Leishmania major* in mice. J Exp Med **198**, 281–291.

Shigematsu, H., Reizis, B., Iwasaki, H., Mizuno, S., Hu, D., Traver, D., Leder, P., Sakaguchi, N. and Akashi, K. (2004). Plasmacytoid dendritic cells activate lymphoid-specific genetic programs irrespective of their cellular origin. Immunity **21**, 43–53.

Shortman, K. and Naik, S.H. (2007). Steady-state and inflammatory dendritic-cell development. Nat Rev Immunol **7**, 19–30.

Shreedhar, V., Moodycliffe, A.M., Ullrich, S.E., Bucana, C., Kripke, M.L. and Flores-Romo, L. (1999). Dendritic cells require T cells for functional maturation in vivo. Immunity **11**, 625–636.

Snijders, A., Kalinski, P., Hilkens, C.M. and Kapsenberg, M.L. (1998). High-level IL-12 production by human dendritic cells requires two signals. Int Immunol **10**, 1593–1598.

Son, Y.I., Dallal, R.M., Mailliard, R.B., Egawa, S., Jonak, Z.L. and Lotze, M.T. (2001). Interleukin-18 (IL-18) synergizes with IL-2 to enhance cytotoxicity, interferon-gamma production, and expansion of natural killer cells. Cancer Res **61**, 884–888.

Stary, G., Bangert, C., Tauber, M., Strohal, R., Kopp, T. and Stingl, G. (2007). Tumoricidal activity of TLR7/8-activated inflammatory dendritic cells. J Exp Med **204**, 1441–1451.

Steinman, R.M. (1991). The dendritic cell system and its role in immunogenicity. Annu Rev Immunol **9**, 271–296.

Stumbles, P.A., Thomas, J.A., Pimm, C.L., Lee, P.T., Venaille, T.J., Proksch, S. and Holt, P.G. (1998). Resting respiratory tract dendritic cells preferentially stimulate T helper cell type 2 (Th2) responses and require obligatory cytokine signals for induction of Th1 immunity. J Exp Med **188**, 2019–2031.

Summers, K.L., Hock, B.D., McKenzie, J.L. and Hart, D.N.J. (2001). Phenotypic characterization of five dendritic cell subsets in human tonsils. Am J Pathol **159**, 285–295.

Szomolanyi-Tsuda, E., Brien, J.D., Dorgan, J.E., Garcea, R.L., Woodland, R.T. and Welsh, R.M. (2001). Antiviral T-cell-independent type 2 antibody responses induced in vivo in the absence of T and NK cells. Virology **280**, 160–168.

Tiberghien, P., Longo, D.L., Wine, J. W., Alvord, W.G. and Reynolds, C.W. (1990). Anti-asialo GM1 antiserum treatment of lethally irradiated recipients before bone marrow transplantation: evidence that recipient natural killer depletion enhances survival, engraftment, and hematopoietic recovery. Blood **76**, 1419–1430.

Tominaga, K., Yoshimoto, T., Torigoe, K., Kurimoto, M., Matsui, K., Hada, T., Okamura, H. and Nakanishi, K. (2000). IL-12 synergizes with IL-18 or IL-1beta for IFN-gamma production from human T cells. Int Immunol **12**, 151–160.

Traver, D., Akashi, K., Manz, M., Merad, M., Miyamoto, T., Engleman, E.G. and Weissman, I.L. (2000). Development of CD8alpha-positive dendritic cells from a common myeloid progenitor. Science **290**, 2152–2154.

Trinchieri, G. (1998). Interleukin-12: a cytokine at the interface of inflammation and immunity. Adv Immunol **70**, 83–243.

Trinchieri, G. (2003). Interleukin-12 and the regulation of innate resistance and adaptive immunity. Nat Rev Immunol **3**, 133–146.

Trinite, B., Chauvin, C., Peche, H., Voisine, C., Heslan, M. and Josien, R. (2005). Immature CD4− CD103+ rat dendritic cells induce rapid caspase-independent apoptosis-like cell death in various tumor and nontumor cells and phagocytose their victims. J Immunol **175**, 2408–2417.

Uhl, M., Kepp, O., Jusforgues-Saklani, H., Vicencio, J.M., Kroemer, G. and Albert, M.L. (2009). Autophagy within the antigen donor cell facilitates efficient antigen cross-priming of virus-specific CD8(+) T cells. Cell Death Differ **16**, 991–1005.

Veldhoen, M., Uyttenhove, C., van Snick, J., Helmby, H., Westendorf, A., Buer, J., Martin, B., Wilhelm, C., Stockinger, B. (2008). Transforming growth factor-beta 'reprograms' the differentiation of T helper 2 cells and promotes an interleukin 9-producing subset. Nat Immunol **9**, 1341-1346.

Vieira, P.L., de Jong, E.C., Wierenga, E.A., Kapsenberg, M.L. and Kalinski, P. (2000). Development of Th1-inducing capacity in myeloid dendritic cells requires environmental instruction. J Immunol **164**, 4507–4512.

Visintin, A., Mazzoni, A., Spitzer, J.H., Wyllie, D.H., Dower, S.K. and Segal, D.M. (2001). Regulation of Toll-like receptors in human monocytes and dendritic cells. J Immunol **166**, 249–255.

Vitale, M., Della Chiesa, M., Carlomagno, S., Pende, D., Arico, M., Moretta, L. and Moretta, A. (2005). NK-dependent DC maturation is mediated by TNFalpha and IFNgamma released upon engagement of the NKp30 triggering receptor. Blood **106**, 566–571.

Wada, H., Masuda, K., Satoh, R., Kakugawa, K., Ikawa, T., Katsura, Y. and Kawamoto, H. (2008). Adult T-cell progenitors retain myeloid potential. Nature **452**, 768–772.

Watt, S.V., Andrews, D.M., Takeda, K., Smyth, M.J. and Hayakawa, Y. (2008). IFN-gamma-dependent recruitment of mature CD27(high) NK cells to lymph nodes primed by dendritic cells. J Immunol **181**, 5323–5330.

Welner, R.S., Pelayo, R., Nagai, Y., Garrett, K.P., Wuest, T.R., Carr, D.J., Borghesi, L.A., Farrar, M.A. and Kincade, P.W. (2008). Lymphoid precursors are directed to produce dendritic cells as a result of TLR9 ligation during herpes infection. Blood **112**, 3753–3761.

Wesa, A.K. and Galy, A. (2001). Regulation of T cell cytokine production by dendritic cells generated in vitro from hematopoietic progenitor cells. Cell Immunol **208**, 115–124.

Wesa, A. and Galy, A. (2002). Increased production of pro-inflammatory cytokines and enhanced T cell responses after activation of human dendritic cells with IL-1 and CD40 ligand. BMC Immunol **3**, 14.

Wesa, A.K. and Storkus, W.J. (2008). Killer dendritic cells: mechanisms of action and therapeutic implications for cancer. Cell Death Differ **15**, 51–57.

Wesa, A., Kalinski, P., Kirkwood, J.M., Tatsumi, T. and Storkus, W.J. (2007). Polarized type-1 dendritic cells (DC1) producing high levels of IL-12 family members rescue patient TH1-type antimelanoma CD4+ T cell responses in vitro. J Immunother **30**, 75–82.

Wilson, J.L., Charo, J., Martin-Fontecha, A., Dellabona, P., Casorati, G., Chambers, B.J., Kiessling, R., Bejarano, M.T. and Ljunggren, H.G. (1999a). NK cell triggering by the human costimulatory molecules CD80 and CD86. J Immunol **163**, 4207–4212.

Wu, L., Li, C.L. and Shortman, K. (1996). Thymic dendritic cell precursors: relationship to the T lymphocyte lineage and phenotype of the dendritic cell progeny. J Exp Med **184**, 903–911.

Wu, L., Vandenabeele, S. and Georgopoulos, K. (2001). Derivation of dendritic cells from myeloid and lymphoid precursors. Int Rev Immunol **20**, 117–135.

Xu, Y., Zhan, Y., Lew, A.M., Naik, S.H. and Kershaw, M.H. (2007). Differential development of murine dendritic cells by GM-CSF versus Flt3 ligand has implications for inflammation and trafficking. J Immunol **179**, 7577–7584.

Yu, Y., Hagihara, M., Ando, K., Gansuvd, B., Matsuzawa, H., Tsuchiya, T., Ueda, Y., Inoue, H., Hotta, T. and Kato, S. (2001). Enhancement of human cord blood CD34+ cell-derived NK cell cytotoxicity by dendritic cells. J Immunol **166**, 1590–1600.

Zhang, A.L., Colmenero, P., Purath, U., Teixeira de Matos, C., Hueber, W., Klareskog, L., Tarner, I.H., Engleman, E.G. and Soderstrom, K. (2007). Natural killer cells trigger differentiation of monocytes into dendritic cells. Blood **110**, 2484–2493.

Chapter Nineteen

19

NK and NKT cells: the innate–adaptive interface including humoral responses

Jeff Subleski, Jonathan M. Weiss, Robert H. Wiltrout, John R. Ortaldo

CHAPTER CONTENTS

Many a beautiful hypothesis has been ruined by an ugly fact.

T.H. Huxley

ABSTRACT

Natural killer (NK) and natural killer T (NKT) cells represent unique lymphoid subsets of the innate immune system that are both critical for some aspects of initiating and directing host adaptive immune responses. These cells share numerous phenotypic markers and functional features, yet are lymphocytes of distinct developmental lineages. As first responders, these innate cells rapidly produce cytokines that modulate various activities of other leukocytes which subsequently influence the ensuing inflammatory response. Moreover, through the expression of effector molecules such as Fas and tumour necrosis factor (TNF)-related apoptosis-inducing ligand (TRAIL) these cells are capable of directly targeting infected or transformed cells for elimination through cytotoxic pathways. Furthermore, both NK and NKT cells polarize T (helper) cell responses and enhance antigen presentation to cytotoxic T-lymphocytes.

KEY WORDS

NK, NKT, receptors, innate, adaptive, immunity

Introduction

Tissue inflammation induced by antigens and pathogens varies considerably, but many diseases have a common pathology including stress and activation of epithelial cells that promotes innate immune responses, providing the host with its first line of defense against infections. Signals generated by subsets of innate lymphocytes, including natural killer (NK) cells, natural killer T (NKT) cells, and antigen presenting cells (APCs) during this early host response determine the nature of downstream adaptive immune responses. NK cells represent a population of specialized lymphocytes capable of recognizing and eliminating a wide range of cancer and virus-infected cells but not normal cells (Ortaldo and Herberman, 1984; Trinchieri, 1989). The function of NK cells is regulated by a fine balance of inhibitory and activating signals, which are mediated by a diverse array of cell-surface receptors. A variety of in vivo and in vitro studies have described these receptors (Daniels et al., 2001; Dokun et al., 2001; Ortaldo and Herberman, 1984; Ortaldo et al., 2001, 2004; Scalzo et al., 1992; Smith et al., 1998, 2000, 2002; Trinchieri, 1989). NK cells develop in the bone marrow using a process that involves sequential modulation of signalling receptors. Through in vitro culture systems and gene knock-out (KO) mice it was found that committed NK progenitors express IL-2Rβ, cKit, FLT3 and IL-7 receptors and develop into NKR-P1 expressing cells that lack mature Ly49 (killer cell immunoglobulin-like receptor [KIR]) and CD94/NKG family members (Di Santo, 2006; Lian and Kumar, 2002; Yu et al., 2006). In the next stage of development, NK cells acquire a combination of inhibitory and activating Ly49 receptors in a controlled manner such that inhibitory Ly49 receptors counterbalance signals from activating Ly49 receptors to maintain self-tolerance in a process termed NK cell education (Lian and Kumar, 2002). Using these receptors, NK cells survey host tissue for MHC class I molecules. Tissues that have missing or down-modulated MHC class I molecules (missing self) release NK cells from inhibitory Ly49 signals allowing them to kill target tissues. Furthermore, these newly activated NK cells can produce immunomodulatory cytokines that can affect the type and magnitude of the subsequent immune response. The purpose of this review section is to discuss studies of molecular and cellular mechanisms by which NK cells help to shape the adaptive and innate immune responses to disease. Evidence will be discussed that suggests that both cellular responses and secreted products of NK cells can profoundly influence both immunity and inflammation.

Functions of NK cells

Originally, NK cells were described as large granular lymphocytes with natural or non-specific cytotoxicity against tumour cells. NK cells were postulated as a separate lymphocyte lineage, and although initially studied for their cytotoxic function, they were later found to be potent cytokine-producing effector cells (summarized in Table 19.1) (Herberman and Ortaldo, 1981; Herberman et al., 1979; Ortaldo and Herberman, 1984; Trinchieri, 1989). Through the coordinated interaction of receptor signalling, cytokine production and their involvement in regulating antigen presentation, NK cells are key in the development of adaptive immunity or as effector cells in mediating immunity against pathogens and transformed cells.

NK receptors

Similar to T cells, where two signals are required for optimal cell activation, it has been recently shown (analyzing Ly49D receptor expression) that NK cells require two positive signals to override the ever vigilant inhibitory receptor blockade. Murine NK cells express multiple Ly49 receptors that are type II transmembrane receptors (Leibson, 1995; Long et al., 1996, 1997; Moretta et al., 1997; Ortaldo and McVicar, 1999; Raulet et al., 1995; Ryan and Seaman, 1997; Yokoyama, 1995). These receptors either inhibit or activate NK cell functions, which include cytolysis and cytokine secretion. A functionally similar family of molecules exists on human NK cells, the KIRs. However, the human KIRs are structurally different from murine Ly49 receptors, as they belong to the immunoglobulin receptor gene superfamily.

The inhibitory Ly49 receptors (Ly49A, C, G and I), inhibit NK cell function upon binding of class I ligands present on target cells (Mason et al., 1997; Murphy et al., 1995; Ryan and Seaman, 1997). These Ly49 inhibitory receptors (as well as inhibitory KIRs) contain cytoplasmic immune receptor tyrosine-based inhibitory motifs (ITIMs) that are phosphorylated upon stimulation, leading to the recruitment of SHP-1 phosphatase and attenuation of intracellular signals (Burshtyn et al., 1996; Mason et al., 1997; Nakamura et al., 1997). In contrast, the amino acid sequences for the activating receptors (e.g. Ly49D and Ly49H) do not contain any ITIMs in their cytoplasmic domains, confirming that these are not inhibitory receptors (Burshtyn et al., 1996; Gosselin et al., 1999; Lanier, 1998; Lanier et al., 1998a; Mason et al., 1996, 1998; Nakamura et al., 1997; Olcese et al., 1997). Activating Ly49 molecules associate with a 12 kD homodimeric protein, DAP12, that contains an immunoreceptor tyrosine-based activation motif (ITAM) critical for positive signalling by these receptors (Mason et al., 1998; Smith et al., 1998). Ly49D cross-linking results in the mobilization of intracellular Ca^{2+} and mediates reverse antibody-dependent cellular cytotoxicity (ADCC) (Mason et al., 1998; Smith et al., 1998). Ly49D weakly recognizes and is activated by H-2d; however, other ligands for this receptor are expected to be present and more physiological. Another activating Ly49, Ly49H evolved to counter mouse cytomegalovirus (MCMV) infection in C57BL/6 mice. MCMV evades elimination in susceptible mice by encoding a MHC-I-like protein (m157) that binds to at least one NK cell inhibitory receptor Ly49I (Adams et al., 2007). However, in C57BL/6 mice m157 binds to an activating Ly49H receptor that strongly

Table 19.1 Summary of NK receptors and functions				
Function	**Specific**	**Induction**		**Reference**
NK lysis				
	NKG2D	Rae-1, H60, Multi1	Activating	(Diefenbach et al., 2002; Kriegeskorte et al., 2005; Lanier, 2005; Raulet, 2003)
	Ly49D	H-2^d	Activating	(George et al., 1999; Mason et al., 2000; Smith et al., 1998)
	Ly49H	CMV-M157	Activating	(Daniels et al., 2001; Smith et al., 2002)
	NKRp1	Clr	Activating	(Iizuka et al., 2003; Plougastel and Yokoyama, 2006; Plougastel et al., 2001)
	Ly49G2	MHC-H-2d	Inhibitory	(Mason et al., 1995)
	Ly49C/I	MHC-H-2b	Inhibitory	(Stoneman et al., 1995; Yu et al., 1996)
	Ly49A	MHC-H-2d	Inhibitory	(Chung et al., 2000; Kim and Yokoyama, 1998; Natarajan et al., 1999)
ADCC lysis	**$FcR_\gamma III$**	**IgG-Fc**	**Activating**	
Cytokine production			**Targets**	
	IFNγ	NKRs and cytokines	Immune cells	(Allavena et al., 1985; Djeu et al., 1982; Trinchieri, 1989)
	IFNα	NKRs and cytokines	All cells	
	IL-1	NKRs and cytokines	Immune cells	(Allavena et al., 1985; Scala et al., 1984)
	IL-5	NKRs and cytokines	Immune cells	(Allavena et al., 1985)
	IL-13	NKRs and cytokines	Immune cells	(Hoshino et al., 1999)
	TNFα/β	NKRs and cytokines	All cells	(Ortaldo et al., 2001)
	IL-8	NKRs and cytokines	Immune cells	(Mariani et al., 2001)
	IL-10	Pathogen, ??	Immune cells	(Maroof et al., 2008)
Chemokine production				
	MIP1α	NKRs and cytokines		(Ortaldo et al., 2001)
	MIP1β	NKRs and cytokines		(Ortaldo et al., 2001)
	IP10	NKRs and cytokines		(Ortaldo et al., 2001)
	TCA3	NKRs and cytokines		
	Lymphotactin	NKRs and cytokines		(Ortaldo et al., 2001)
	Rantes	NKRs and cytokines		(Ortaldo et al., 2001)
Antigen presentation				
	Staph. Pr.A (SpA) Streptolysin O (SLO)	Positive regulation		(Scala et al., 1985)
	MHC	Negative regulation		(Barber et al., 2007)

activates NK cells to kill cytomegalovirus (CMV)-infected targets (Arase et al., 2002).

In addition to KIR and Ly49, both human and mouse share the NKG family of activating and inhibitory receptors (Cantoni et al., 1998; Houchins et al., 1991; Lanier et al., 1998b; Lian et al., 2002). The NKG2D receptors recognize antigens not expressed on normal cells but often present on immature, activated or infected cells. These ligands have been well characterized in studies by Raulet and Lanier (Lanier, 2005; Lanier et al., 1998b; Raulet, 2003). Another activating receptor found in both human and mouse is the FcRγIII that recognizes antibody coated targets or pathogens via the Fc receptor of IgG. Table 19.1 summarizes the numerous functions of mouse NK cell receptors (NKRs) and functions.

Activated NK cells are able to directly perform effector functions using cell surface and secreted factors (Smyth et al., 2002b). A key mechanism for the removal of target cells by activated NK cells is through interaction of TNF super-family members TNF-related apoptosis-inducing ligand (TRAIL) and FAS with their respective ligands. Another way NK cells can induce target cell death is by the release of pore forming perforin and apoptosis-inducing proteases, granzymes A/B (Smyth et al., 2002).

NK cytokines and chemokines

NK cells have potent ability to secrete factors, within minutes of activation, which are important in initial activation or suppression of cells from the innate and adaptive immune system. Perhaps one of the earliest and most studied cytokines produced by NK cells is IFNγ NK cells rapidly produce IFNγ following IL-2, IL-12, IL-15 or IL-18 cytokine activation or following NKR triggering. This production of IFNγ is well established to be important in immunity against viruses (Biron et al., 2002; Tay et al., 1998) and tumours by enhancing NK cell lytic activity (Dalton et al., 1993), T helper type 1 (Th1) development (Martin-Fontecha et al., 2004), and cytotoxic T-lymphocyte (CTL) activity (Robbins et al., 2007). In addition to IFNγ, NK cells also produce a variety of other cytokines in response to NKR triggering or cytokine stimulation; such as IL-1, IL-5, IL-8, IL-13 and TNFα (Table 19.1). It is evident that NK cells, capable of early immune responses to various pathogens, function to amplify and regulate the ongoing inflammatory response through the coordinated expression of numerous key cytokines. The production of IFNγ by NK cells, for example, upregulates expression of the chemokine receptor CXCR3 that was important for the recruitment of CXCR3$^+$ T cells to infected tissues (Sallusto and Baggiolini, 2008). In turn, NK cells themselves are potentially subject to autocrine activation by these cytokines for enhanced cytotoxic and effector functions (Dalton et al., 1993). Similarly, NK cells can also be stimulated via NKR or cytokines to produce chemokines that are important for the recruitment of T cells, B cells, neutrophils and other activated NK cells, thereby amplifying the inflammatory response. IL-2 or IL-12-activated NK cells secrete several T cell–recruiting chemokines, including MIP-1α, MIP-1β, IL-8, macrophage-derived chemokine (MDC), and regulated on activation, normal T-cell expressed and secreted (RANTES) (Bluman et al., 1996).

A further enhancement of chemokine production was observed when IL-2 or IL-12 was combined with TLR agonists (Sawaki et al., 2007). NK cells costimulated by antibody-coated tumours and activated with IL-2 or IL-12 expressed MIG and IP-10, which may also facilitate T cell recruitment (Roda et al., 2006). Additionally, NKR ligation induced the coexpression of RANTES, MIP-1α and MIP-1β along with IFNγ that was important for the recruitment of CD8$^+$ T cells (Dorner et al., 2004). Taken together, these studies support a role for NK cells in the initiation of specific immune responses by facilitating T-cell recruitment and subsequent activation. Consistently, Dorner and colleagues suggested that this may in turn facilitate the transition to an antigen-specific, adaptive phase of immune responses (Dorner et al., 2004).

In some cases, NK cells suppress an immune response. In one regard, this is due to their production of immunoregulatory cytokines, such as TGF-β, IL-4 and IL-10. NK cells produce both active and latent forms of TGF-β, to levels comparable to monocytes (Gray and Horwitz, 1995). Moreover, anti-CD2 stimulation strongly enhanced the capacity of purified NK cells to produce active TGF-β (Gray and Horwitz, 1995) that is likely to regulate T cell responses and may also facilitate the generation of regulatory T cells. Recently, NK-derived IL-10 production during *Leishmania* infection was shown to be responsible for the inhibition of host protective immune responses in the spleen and liver (Maroof et al., 2008).

The role of NK cells at the innate-adaptive interface

The potential role for NK cells in the regulation of adaptive immune responses has long been postulated (Herberman and Ortaldo, 1981; Ortaldo and Herberman, 1984) due to the wide array of non-cytotoxic functions of NK cells. The ability to clearly define the role of NK cells in adaptive responses to tumours may have been more difficult due to poorly defined tumour antigens in mouse model systems. NK cell depletion can alter the size of tumours, as well as the number of metastases, but the precise role of NK cells in altering T cell responses has been difficult to define. The best and most definitive studies regarding NK cells and their involvement at the adaptive interface arose from their ability to kill virus infected targets. Early studies in the CMV infection models clearly demonstrated the important regulatory role of NK cells in the CD8-mediated T cell responses (Biron et al., 2002; Tay et al., 1998). Activated NK cells were identified as critical leukocyte populations and sources of significant levels of IFNγ between days 2 and 8 post-infection, whereas virus-specific cytotoxic T-cell activity was more readily detected between days 6 and 14. Other studies also defined the role of NK cells in the natural resistance of mice to infections. Upon infection, mice depleted of NK cells had significantly higher (up to 500-fold) titres of MCMV, MHV, or vaccinia virus in their livers and spleens as compared to control mice (Bukowski et al., 1983). NK cell-depleted, virus-infected mice had more extensive hepatitis and inflammatory foci in their livers and diminished T cell immunity. In addition to a role in controlling viral infections, NK cells producing IFNγ have recently been shown to also regulate generation of T cell immunity (Goldszmid et al., 2007) in a parasitic infection (*Toxoplasma gondii*). This demonstrates an important role for NK cells in developing T cell immunity against infection.

A role for NK cells in regulating the maturation of dendritic cells (DC) has been demonstrated in both human (Moretta et al., 2007; Walzer et al., 2005) and mouse (Makarenkova et al., 2005; Walzer et al., 2005) studies. NK cells are classically capable of promoting DC maturation. Indeed, bi-directional interactions between NK cells and DC usually results in more optimal effector functions by both cell types. DC-NK interactions cause the proliferation and production of IFNγ by NK cells. The production of IFNγ by NK cells, in turn, promotes DC maturation and facilitates their priming of Th1 cells (Kalinski et al., 2005). In a malignant melanoma tumour vaccine model, the depletion of NK cells prior to DC vaccination significantly reduced IFNγ production and resulted in impaired tumour-specific CTL activity and vaccine potency (Kim et al., 2000a). In turn, DC induce NK cells to proliferate and mediate critical effector functions (Lucas et al., 2007). The trans-presentation of IL-15, in association with IL-15Rα, by DC further supports NK cell expansion and activation in vivo (Mortier et al., 2008). On the other hand, some recent studies have also identified the potential for negative regulation of antigen presentation by NK cells (Barber et al., 2007). In that study, NK depletion resulted in the development of increased numbers of tumour-specific $CD8^+$ T cells, IFNγ production and greater antigen presentation in draining lymph nodes as compared to NK-bearing mice. Another example of the potential for immunoregulation involves the NK-mediated cell lysis of APCs, including DC, through a process that appears to involve TRAIL (Hayakawa et al., 2004). Taken together, these observations indicate that NK cells are capable of both stimulating and limiting the development of adaptive immune responses, the latter aspect of which may indicate a role for NK cells in dampening immune responses as an infection becomes controlled and viral antigens cleared.

Several studies have also described the ability of NK cells to exhibit antigen presentation function themselves, albeit of more limited efficiency (Scala et al., 1985). Additionally, a rare population of cells, termed interferon-producing killer dendritic cells (IKDC) or natural killer dendritic cells (NKDC) co-express NK1.1 and CD11c (Chan et al., 2006); (Pillarisetty et al., 2005). These cells can kill typical NK targets using NK-activating receptors and also express class II MHC and possess antigen-presenting capability. More recently, Vosshenrich and associates showed that the development of IKDC parallels that of NK cells and that these cells most likely represent activated NK cells (Vosshenrich et al., 2007). Taken together, these results highlight an unappreciated role of NK cells in the direct and indirect regulation of antigen presentation.

NK cells in cancer

NK cells are involved in the surveillance of malignant cells and the elimination of established metastases. Unlike T cells, NK cells are activated in the absence of self-MHC class I molecules and are therefore uniquely capable of rapid cellular and cytotoxic responses upon encounter with transformed cells that may express no or altered MHC class I molecules. Through the coordinated regulation of NK activation by both activating and inhibitory receptors, NK cells are usually capable of distinguishing malignant from healthy cells. The anti-tumour activities of NK cells may directly lead to tumour eradication by means of cytolysis or IFNγ secretion, but may also indirectly contribute to tumour control by inducing an efficient T-cell-mediated anti-tumour response (Kelly et al., 2002). Additionally, NK-derived IFNγ expression is critical for the activation of DC and generation of adaptive, T cell, mediated anti-tumour responses (Adam et al., 2005; Kelly et al., 2002). Consistently, numerous studies utilizing mouse tumour models have demonstrated that NK cells play an important role in host anti-tumour responses (Hayakawa and Smyth, 2006; Kim et al., 2000; Wu and Lanier, 2003). In these models, the depletion or functional deficiency of NK cells results in more aggressive tumour growth and metastasis. In humans, the infiltration of tumours by NK cells represents a positive prognostic marker in several carcinomas (Coca et al., 1997; Ishigami et al., 2000) and this migratory process can be further enhanced by immunotherapeutic regimens that enhance NK recruitment to the tumour microenvironment (Choi et al., 2007; Pan et al., 2004). However, some tumours may evade NK-mediated surveillance by downregulating their expression of NKG2D ligands (Eisele et al., 2006). The role for NK cells in regulating an adaptive anti-tumour response has been difficult to demonstrate. Recently, using a highly tumourigenic murine cell line, MC38, and a replication-deficient recombinant avian (fowlpox) virus that expresses murine IFN-γ (rF-IFN-γ), primary tumour growth in a vaccine model was evaluated (Zeytin et al., 2008). Mice depleted of NK cells prior to the injection with rF-IFN-γ based tumour cell vaccine were not protected from primary MC38 tumour growth. The findings provide direct evidence that NK cells can provide immunoregulatory signals required for the development of an adaptive immune response and for vaccine-induced protection of tumour growth. As we continue to expand our understanding of the regulation of NK cytotoxic functions, we will hopefully be able to develop more effective immunotherapeutic approaches that capitalize on the effector functions of NK cells in ways that may, in some cases, help overcome these tumour-derived immunosuppressive countermeasures.

NK cells in autoimmunity

In many autoimmune diseases, such as uveitis, haemolytic anaemia, colitis, myasthenia gravis, lupus and rheumatoid arthritis, T or B lymphocytes have been shown to be the principal effector cells responsible for damage to the tissue target (Huang et al., 1977; Mor et al., 2003; Mountz et al., 1991; Playfair and Marshall-Clarke, 1973; Radford-Smith, 1997; Samuels et al., 2005). NK cells have also been implicated in such autoimmune diseases as diabetes and insulitis, as well as rheumatoid arthritis (Dalbeth et al., 2004; Flodstrom et al., 2002; Poirot et al., 2004). Previous studies have been somewhat limited in examining the interactive role of these different lymphocyte subsets in shifting the balance of immunity toward an attack on normal host tissues.

Evidence of a regulatory role for NK cells in some T cell-mediated experimental autoimmune diseases, including murine models of encephalomyelitis (Matsumoto et al., 1998; Zhang et al., 1997), colitis (Fort et al., 1998) and diabetes (Lehuen et al., 1998a) has been shown. Studies by Ljunggren (Shi et al., 2000b, 2001) have detailed the contribution of NK cells in regulating the production of autoantibodies by B cells that are the primary cause of disease in a variety of autoimmune conditions, including haemolytic anaemia, thyroiditis, stiff man syndrome, pemphigus vulgaris and systemic lupus erythematosis. Zhang and colleagues demonstrated that the in vivo depletion of NK cells improved clinical symptoms of experimental autoimmune encephalomyelitis (EAE) in wild-type (WT) C57BL/6 mice (Zhang et al., 1997). Other studies by Shi and associates (Shi et al., 2000a) and Huang and colleagues (Huang et al., 2006) demonstrated that the development of EAE was dependant upon IL-18 expression and IFNγ production by NK cells. These findings established an important, but previously unrecognized link between NK cells and autoimmunity in a primary in vivo model system. More recently, mice lacking CX3CR1 that developed myelin oligodendrocyte glycoprotein (MOG)-induced EAE demonstrated a reduced recruitment of NK cells to the central nervous system (CNS) (Huang et al., 2006).

Our recent studies (Winkler-Pickett et al., 2008) have focused on the EAE model, which is a prototypic autoimmune disease model for multiple sclerosis in many clinical and histopathological aspects (Alvord, 1984; Tabira and Kira, 1992). EAE is known to be mediated by $CD4^+$ T cells that recognize peptides derived from encephalitogenic proteins of the CNS. Cytokines, particularly TNFα, are considered to be the mediators of the pathology observed in the CNS with conflicting analyses of their effects reported (Merrill and Benveniste, 1996). In the mouse, the disease is characterized by a paralysis proceeding from the hind limbs to the forelimbs. Paralysis initiates within 2 weeks of injection of MOG peptide (Chen et al., 1994; Kumar and Sercarz, 1993). The results of many of these studies led to the hypothesis that NK cells appear to be involved in changing the balance of immunity by initiating or regulating the intensity of autoimmune reactions and/or modifying the effector cells that can accumulate in the target organ(s). These mechanisms might include: (1) production of cytokines that alter DC or T cell activation and/or proliferation; (2) direct interactions with APC that could alter antigen presentation; and (3) alteration of regulatory cells by direct or indirect mechanisms. The primary EAE mouse model provides an in vivo system to examine the regulatory role of NK cells as they may provide either protection from or exacerbation of the clinical course of autoimmune disease(s). In addition, the ability to selectively deplete NK cell subsets in a primary model allows for a more specific interpretation of the role of NK cells in the initiation of autoimmune disease.

Using a model of primary EAE induced in B6 mice with MOG 35–55 peptide (Mendel et al., 1995), our recent studies (summarized in Table 19.2) evaluated the effects of depleting populations of NK cells using anti-NK1.1, anti-asialo-GM1 and select anti-Ly49 antibodies upon the clinical course of EAE in vivo. Our studies clearly demonstrated that NK depletion prior to immunization diminished the onset of EAE and extent of paralysis. The use of anti-asialo-GM1 and select anti-Ly49s ruled out a role for NKT cells since these cells remain unaffected by these treatments. Our findings that elimination of NK cells expressing Ly49D and Ly49H, a highly overlapping but minor subset of NK cells, had an effect similar to that of total NK depletion, implicated this subset of the NK cells in the clinical response. Furthermore, when our

Table 19.2 Summary of NK-mediated modulation of EAE

Parameter*	NK1.1 depleted	Ly-49H depleted	Ly49D depleted
Mean clinical score	Decreased	Decreased	Decreased
Percent survival	Unchanged	Unchanged	Unchanged
Percent with EAE	Unchanged	Unchanged	Unchanged
TNFα	Decreased	??	??
IFNγ	Decreased	??	??
IL-2	Decreased	??	??
IL-6	Decreased	??	??
MOG proliferation	Decreased	??	??
DC #s in brain	Decreased	??	??
DC #s in cervical LN	Decreased	??	??
DC #s in draining LN	Unchanged	??	??
CD80/86 levels in brain	Unchanged	??	??
TcR-Vβ 8.1. 14 & 15	Decreased	??	??

*Data from Winkler-Pickett et. al. (2008). "??" indicates unknown.

studies were translated into an in vitro evaluation of T cell responses to cognate MOG antigen, NK depleted lymphoid tissues demonstrated a diminished T cell response as measured by proliferation and cytokine production (IFNγ and TNFα). Although, recent studies evaluating IL-17 production by T cells have implicated this cytokine as a major autoimmune regulator, the removal of NK cells did not consistently alter the IL-17 or IL-23 levels induced by MOG stimulation, although a dramatic decrease in the ratio of IL-17/IFNγ production was observed. These findings suggested that the target of NK effects is not within the pathway of IL-17/23 regulation of effector T cells.

Evaluation of the CNS leukocytes, unlike the draining lymph nodes, indicated both a qualitative alteration (TcR Vβ) and a quantitative change, as a reduced frequency of MOG reactive T cells that emigrate to the CNS and mediate autoimmune reaction was observed. Analysis of TcR vβ usage after NK cell depletion revealed a qualitative shift in the T cell population as T cells containing vβ 8, 14 and 15 were decreased. As these T cells have been shown to be the major effector repertoire in C57BL/6 mice (Mendel et al., 1995), loss of this population as a result of NK cell depletion would explain the altered clinical outcome observed in our study. The evaluation of lymph node DC demonstrated an alteration in mature DC levels (based on CD80 and CD86 expression) indicating that NK cells may prevent optimal secondary T cell responses and antigen presentation. Thus the role of NK cells may be non-lytic as studies with perforin KO mice demonstrated no regulatory role in primary EAE induction and disease. Collectively, these data support the hypothesis (Farag et al., 2003) that in addition to being anti-tumour and anti-viral effector cells, NK cells mediate a critical link between NK cells and adaptive T and B cell responses as indicated in previous reports in the CMV infection model system (Biron et al., 2002; Tay et al., 1999) and in the MS model (Huang et al., 2006; Shi and Van, 2006; Shi et al., 2001).

Role of NK cells in primary B cell responses

In addition, the role of NK cells in primary B cell responses was demonstrated by Michael and associates (Michael et al., 1989) in a murine in vitro co-culture system. In this system NK cells were shown to be stimulated by B cells to produce IFNγ that inhibited polyclonally B cell proliferation in a paracrine loop. These results indicated an important regulatory link between NK cells and B cells in humoral responses (Figure 19.1). Evaluation of the in vivo relevance of these interactions revealed that activated NK cells can increase the IgG2a response (Yuan et al., 1994). Our studies (Blanca et al., 2001) demonstrated that human NK cells can induce autologous resting peripheral B cells to synthesize Ig, including switching to IgG and IgA, reminiscent of a secondary Ab response. This Ig switch in humans was dependant on CD40–CD40L cell contact. Gao and colleagues demonstrated an IFNγ and CD40–40L independent switch of germ line B cells to produce immunoglobulin (IgG2a) (Gao et al., 2001). However, this activation was CD40–CD40L independent. Furthermore, studies indicated that NK cells require additional signals to preferentially switch germ line B cells. Recently, Gao and colleagues employed nitrophenol-specific transgenic B cells to demonstrate a NK cell dependent mechanism for the induction of IgG2a and IgG1 (Gao et al., 2008). The IgG2a and IgG1 induction required the NK activation of NK cells that was dependant on CD48 and direct cell contact. These authors concluded that NK cells can provide necessary signals that substitute for cytokines in the induction of IgG2a as well as IgG1 expression. These in vitro analyses provide a mechanistic basis for understanding how NK cells regulate B cell responses that are independent of T-cells and the documented NK cell effects on T-independent B cell responses in vivo.

These findings were extended to a mouse model using primary immunization with TNP–KLH. Our results from this model (Ortaldo et al., unpublished observations) indicated that in vivo depletion of NK cells can significantly alter the quality and quantity (specific for isotypes IgG1, IgG2a and IgG3) of anti-TNP antibody produced in vivo. The removal of NK cells prior to immunization resulted in significant increases in antigen specific T and

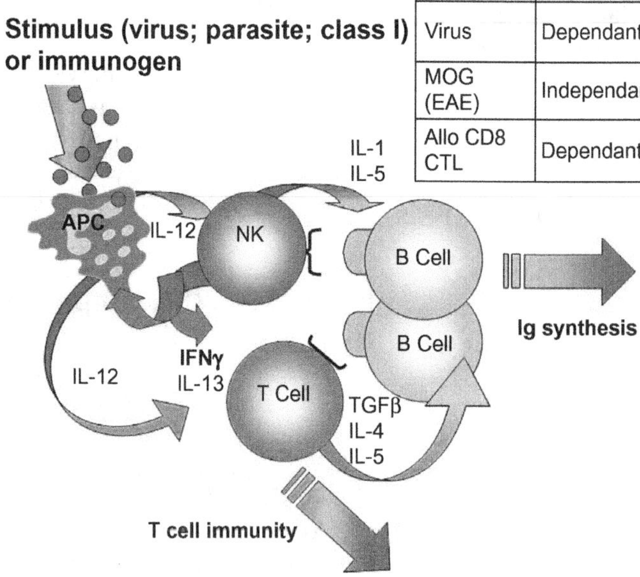

Antigen	Perforin pathway	Factor requirement
Virus	Dependant	IFNγ,α
MOG (EAE)	Independant	IL-12
Allo CD8 CTL	Dependant	????

Figure 19.1 • Model for diverse functional role of NK cells in innate to adaptive interface.

B cells in the host. Much of this regulation of Ig production appeared to be through the regulated production of soluble factors in the host's immune organs. The existence of this novel mechanism of B cell regulation suggests important implications in innate immunity, B cell-mediated autoimmunity, and B cell neoplasia.

The mechanisms by which NK cells regulate these adaptive responses are not yet fully elucidated but their site of regulation has been clearly shown to involve the draining lymph nodes after MOG peptide immunization, an immunoregulatory site not generally thought to involve NK cells until recently. The cytokines that are critical for the maturation of EAE responses and that determine the differentiation of T cells into MOG specific Th1 (IL-2, IFNγ, TNFα) T cells would be expected to have regulatory impact on numerous other adaptive responses. Our findings are consistent with the proposed hypothesis that through cytokine production, NK cells can regulate the ability of intracellular bacteria and viruses to induce and regulate antigen-specific T cells (Fearon and Locksley, 1996; Romagnani, 1992). Our data suggest an important lymph node interaction between NK cells, DC and the development of antigen-specific T and perhaps, B cells. However, further experimentation is required to more fully define all the signals that regulate NK–DC interactions.

NK cells in allogeneic transplantation responses

Another aspect of NK cell biology that has been studied since their discovery is their ability to alter allo-transplantation or semi-allo-transplantation (hybrid resistance). NK cell alloreactivity is demonstrated in mice and can be exploited in bone marrow transplantation to improve clinical outcome (Beilke et al., 2005; Yu et al., 2006). Likewise, in organ transplantation, (McNerney et al., 2006) recipient NK cells may limit alloreactive T-cell responses through their capacity to regulate the persistence of graft-derived allogeneic DCs.

In recent studies (Laffont et al., 2008), the mechanisms governing the killing of allogeneic DCs by host NK cells was evaluated. These results, in NK depleted mice, demonstrated impaired development of alloreactive T-cell responses. Allogeneic (H-2d) DC elimination was perforin-dependent, occurred in draining lymph nodes, and required the recruitment of CD127 Ly49D+ blood-borne NK cells. This study, together with recent work (Yu et al., 2006) suggests that alloreactive NK cells mediate their rejection through their capacity to rapidly kill donor-derived allogeneic DC.

Conclusions regarding the role of NK cells in regulating adaptive immune responses

NK cells play critical, nonredundant roles during the generation of adaptive immune responses. First, they produce a variety of cytokines and chemokines that can recruit and activate specific leukocyte subsets, thereby regulating and potentially polarizing adaptive immunity. Second, NK cells mediate lytic functions that create availability of antigen for presentation by DC and other APCs. Third, NK cells have direct interactions with APCs (e.g. monocytes, macrophages or DC) that alter function in inflammation and antigen presentation. Initial studies regarding NK cells did not further appreciate the ability to migrate into secondary lymphoid tissues, such as lymph nodes, during active infection and/or immunization. This event is often accompanied by alterations in DC presence or activation that further defines a role for NK cells in the maintenance of Th1-mediated inflammatory responses.

NKT cells

An effective immune response against disease requires activation of innate and adaptive arms of the immune system (Biron et al., 1998). NKT cells are part of a small group of cells that bridge innate and adaptive responses (Getz, 2005; Taniguchi et al., 2003b). They share functional properties with both NK cells of the innate and T cells of the adaptive components of immunity. NKT cells express markers of both NK (DX5 and NK1.1 in mouse and CD56 in human) and T cell (CD3, CD5 and CD4) lineages; however, they lack NK-specific receptors such as NKp46 and activating Ly49s. Consistent with an innate response, NKT cells rapidly secrete cytokines following activation. In fact NKT cells along with NK cells, constitutively express IFN-γ mRNA (but not protein), allowing for an immediate response to activation (Stetson et al., 2003). Activated NKT cells are able to kill tumour targets using the perforin-mediated pathway (Kawano et al., 1998) a hallmark of NK cell function. Similar to conventional T cells of the adaptive response, NKT cells can also respond to antigens via their TCR. In stark contrast to conventional T cells that are MHC I and II restricted and respond to peptide antigens (Robey and Fowlkes, 1994), NKT cells respond to glycolipid antigens and are CD1d (a MHC I like molecule) restricted (Smiley et al., 1997). NKT cells therefore not only share functions with other innate and adaptive cells, they themselves bridge innate and adaptive immune responses by influencing effector functions of macrophages, DCs and NK cells as well as T (Bendelac et al., 2007) and B cells (Kitamura et al., 2000).

NKT cells have diverse regulatory abilities on the development of adaptive immune responses. NKT cells suppress autoimmunity (Godfrey and Kronenberg, 2004) but can also promote asthma (Akbari et al., 2003), atherosclerosis (Nakai et al., 2004), arthritis (Kim et al., 2005) and resistance to viral (van Dommelen and Degli-Esposti, 2004) and bacterial (Brigl et al., 2003) infections. These opposing functions are also apparent in studies of tumour immunity (Berzofsky and Terabe, 2008).

NKT cell characterization, tissue and species distribution

Early studies defined cells that expressed both αβTCR and NK receptors as NKT cells. However, we now know that chronically activated or aged T-cells can co-express these markers (Tarazona et al., 2000) and that NKT cells do not always express NK markers (McNab et al., 2007). Subsequent studies found that NKT cells, but not conventional T-cells, were dependant on CD1d for development (Smiley et al., 1997), identifying them as a separate lymphocyte lineage. Thus, early studies may have attributed T cell functions to NKT cells, prior to more definitive work that more clearly delineated their unique nature. NKT cells are now categorized into two groups based on their T cell receptor gene arrangement (Table 19.3). Type I-NKT cells, also termed invariant NKT cells (*i*NKT cells) constitutes the larger of the two groups and the most studied. They express an invariant Vα18Jα18 in humans or a Vα14Jα18 αTCR chain in mice. A synthetic glycolipid, α-galactosylceramide (α-GalCer), originally extracted from a marine sponge, was shown to specifically activate (Kawano et al., 1997, 1999) and detect *i*NKT cells (Gumperz et al., 2002; Liu et al., 2006). Based on the chemical structure of α-GalCer, other *i*NKT cell ligands have been characterized (Brutkiewicz, 2006). Within the *i*NKT cell category there are distinct phenotypes. In humans and

Table 19.3 Comparison of type I and type II NKT cells

	Type I or invariant NKT-cells (*i*NKT cell)	Type II or variant NKT cells
CD1d-dependent	Yes	Yes
α-TCR chain	Vα24-Jα18 (humans) Vα14-Jα18 (mice)	Diverse
Deficient in CD1d$^{(-/-)}$ mice?	Yes	Yes
Deficient in Jα18$^{(-/-)}$ mice?	Yes	No
Deficient in RAG$^{(-/-)}$ mice?	Yes	Yes
Glycolipid known to detect and activate	αGalCer and its derivatives	Sulfatide (select subset?)
Known immune function	Effector, immunomodulatory, suppressor	Suppressor

Adapted from Berzins et al. (2005) and Godfrey et al. (2004).

mice, *i*NKT cells can be $CD4^+CD8^-$ or $CD4^-CD8^-$ and in humans a $CD4^-CD8^+$ subset exists (Godfrey et al., 2004). In mice *i*NKT cells are further divided by expression of the NK receptor NK1.1 (Pellicci et al., 2002). Taken together, this defines four distinct phenotypic *i*NKT cell subsets in mice: $CD4^+NK1.1^+$, $CD4^+NK1.1^-$, $CD4^-NK1.1^+$ and $CD4^-NK1.1^-$ and three subsets in humans: $CD4^+$, $CD8^+$ and $CD4^-CD8^-$. Type II NKT cells (variant NKT cells) encompass the non-*i*NKT cells and express a more diverse TCR repertoire. Sulfatide, an abundant glycolipid of the myelin compartment of the CNS, activates and marks at least a subset of mouse Type II NKT cells (Jahng et al., 2004).

Mouse *i*NKT cells are detected in blood, lung, thymus, bone marrow, lymph nodes, liver (Matsuda et al., 2000), lamina propria (Ronet et al., 2005) and decidua of pregnant mice (Ito et al., 2000). Invariant NKT cells are enriched in the liver and bone marrow in mice. Distribution of human *i*NKT cells is less studied but they are found at lower frequencies in the blood (Lee et al., 2002a,b) and liver (Kita et al., 2002) and decidua of pregnant women (Boyson et al., 2002). Less is known about variant NKT cells because of the limiting reagents to detect these cells. Variant NKT cells are found in the thymus, spleen and the liver of mice (Jahng et al., 2004). Although non-*i*NKT cells are known to exist in humans, their tissue distribution has not been carefully studied.

NKT cell development

Insight into NKT cell function comes from their development into a specialized lineage of self-reactive cells. To prevent perilous autoimmunity, mechanisms have evolved during development to limit T cells that recognize self-components (Goodnow et al., 2005). Despite these safeguards, self-reactive cells still escape to the periphery. One mechanism by which self-reactive T cells are restrained from causing autoimmunity is through active suppression of these cells by regulatory T cells termed T-regs. Interestingly, T-regs also develop as self-reactive cells, but unlike their potentially harmful effector counterparts, T-regs are instructed to develop into suppressive cells (Sakaguchi, 2004).

In addition to T-regs, NKT cells also develop as a lineage of self-reactive T cell. However, NKT cells develop differently than either T-reg or conventional T cells, and eventually become specialized cells that recognize self-glycolipids (Zhou et al., 2004). Like conventional T cells, NKT cells develop in the thymus (Godfrey and Berzins, 2007). NKT cells segregate from conventional T cells at the common $CD4^+CD8^+$ double positive (DP) precursor stage (Egawa et al., 2005; Gapin et al., 2001). Whereas conventional T cells are selected by cortical epithelial cells expressing peptide antigens in the context of MHC molecules, NKT cells are selected by DP thymocytes (Coles and Raulet, 2000) expressing self-glycolipids in the context of CD1d molecules (Gapin et al., 2001). Thus, NKT cells are able to interact with T cells expressing self-antigens. Controversy exists regarding the nature of the endogenous self-glycolipid(s) responsible for selecting NKT cells (Godfrey and Berzins, 2007). Biochemical signalling during NKT cell development is distinct from that seen in conventional T cells. Conventional T cells do not develop in mice with interrupted Ras and Mek1 signalling, while NKT cells differentiate normally in this setting (Alberola-Ila et al., 1996). Conversely, NKT cells do not develop in Fyn deficient mice (Eberl et al., 1999) or mice defective in NF-κB signalling (Sivakumar et al., 2003), whereas conventional T cells develop normally in these mice. T-regs and to a lesser extent activated/memory T cells are also absent in mice with defective NF-κB signalling (Schmidt-Supprian et al., 2004), demonstrating that NF-κB signalling is necessary for generating antigen-educated T cells. The failure of NKT cells and T-regs to develop in mice defective in NF-κB signalling demonstrates that these cells develop as self-reactive cells.

NKT cells constitutively express activation/memory markers ($CD69^+$ $CD62L^-$ $CD44^{high}$ $IL\text{-}2R\beta^{high}$) providing further evidence that these cells are self-reactive (Kronenberg and Gapin, 2002). Clear evidence of the role of T-regs as suppressor cells in preventing autoimmunity comes from the observation that autoimmunity spontaneously develops in humans (Bennett et al., 2001) and mice (Brunkow et al., 2001) lacking T-regs. The role for self-reactive NKT cells in autoimmunity is not as clear. CD1d KO mice that lack NKT cells display reduced IL-4 production, but they do not spontaneously develop autoimmune diseases. While numerous studies in mice showed the reduction of activated *i*NKT cells in mouse models of collagen-induced arthritis, diabetes and EAE, other studies implicated *i*NKT cells in exacerbating airway hypersensitivity (mouse asthma model) (Akbari et al., 2003), atherosclerosis (Tupin et al., 2004) and inflammatory bowel response (Ronet et al., 2005). Thus, the role of NKT cells in autoimmunity may depend on the nature of the disease and by the mechanisms by which it occurs. For example, restoring the ability of *i*NKT cells to produce IL-4 was beneficial in reducing clinical manifestations of diabetes in non-obese diabetic (NOD) mice (Falcone et al., 2004). In contrast, IL-4 production by *i*NKT cells was required for inducing airway hypersensitivity in mice (Akbari et al., 2003). A better understanding of the complex functions of NKT cells in different settings of autoimmunity is needed before NKT-mediated effects can be influenced to alter the course of the disease.

NKT cell mechanisms of action

NKT cell mobilization and homeostasis

Effective responses to disease require cell activation and mobilization to the site of insult. The coordinated expression of adhesion molecules and chemokines regulates the trafficking and arrest of NKT cells in various tissues where they regulate homeostasis of the microenvironment. In allograft tolerance, it was found that the blockade of lymphocyte function-associated antigen-1 (LFA-1)/intercellular adhesion molecule-1 (ICAM-1) interactions induced graft tolerance that was dependant on NKT cells (Seino et al., 2001). Interestingly, activating NKT cells from LFA-1$^{(-/-)}$ mice with α-GalCer increased tolerizing cytokines IL-4 and IL-13 (Matsumoto et al., 2004), whereas IL-12 activated NKT cells from LFA-1$^{(-/-)}$ mice were less cytolytic against YAC and EL-4 mouse tumour lines (Matsumoto et al., 2000). Chemokines expressed locally further regulate migration of leukocytes expressing their respective chemokine receptors. Differential tissue distribution and enrichment of NKT cells could be due to the specific chemokine receptor profile of NKT cells and the chemokine(s) expressed in the microenvironment. The chemokine receptor profile or NKT cells is similar to that of Th1 homing cells (Thomas et al., 2003) and consistent with cells that carry out functions in peripheral tissues rather than lymphoid tissue. Most NKT cells lack the lymph node homing receptor CCR7, accounting for their low concentration in the lymph nodes. NKT cells express CXCR6 and liver and lung tissue highly express CXCL16, which is the ligand for CXCR6 (Germanov et al., 2008; Wilbanks et al., 2001). Mice lacking CXCR6 have reduced levels of NKT cells in the liver and lung but normal levels in the thymus, showing CXCR6 is not needed for development. Interestingly, the livers of CXCR6$^{(-/-)}$ mice were not deficient in NK1.1$^{(-)}$ NKT cells but were deficient in NK1.1$^{(+)}$ NKT cells, suggesting CXCR6 may be important for NKT cell maturation.

Functional relevance for CXCR6 expression by NKT cells was shown in a mouse model of atherosclerosis where mice that were deficient in CXCR6 had reduced atherosclerotic pathogenesis (Galkina et al., 2007). Conversely, blocking CXCL16–CCR6 interactions between graft and NKT cells, respectively, leads to graft rejection and further demonstrates a role for this receptor in NKT cell induction of tolerance (Jiang et al., 2005). Taken together, these experiments indicate CXCR6 targets NKT cells to tissue sites, while the LFA-1 interactions control the type of response that ensues.

NKT cell activation

Cytokine release profiles may determine what function NKT cells play in immune promotion or tolerance (Table 19.4). NKT cells are activated by cytokines or by antigen signals delivered through their TCR. NKT cells are a major target for IL-12 activation because NKT cells express the mature form of the IL-12 receptor while resting CD8 and NK cells do not (Taniguchi et al., 2003a). IL-12 induced IFN-γ production is critical in mediating immune responses to infection (Watford et al., 2003) and against tumours (Weiss et al., 2007). Mice lacking *i*NKT cells (Jα18$^{(-/-)}$ mice) initially produce little IFN-γ in response to IL-12 (Taniguchi et al., 2003a) and lack IL-12 initiated anti-tumour responses (Cui et al., 1997).

NKT cells can also be activated with antigens. NKT cells acquire inhibitory Ly49 receptors late in development (Voyle et al., 2003). Splenic NKT cells from Ly49$^{(-)}$ but not Ly49$^{(+)}$ mice were activated in vitro by immature DCs loaded with *i*NKT cell antigen α-GalCer, suggesting Ly49 ligation prevents NKT cell activation and a second single is needed (Maeda et al., 2001). Furthermore, Ly49 engagement of NKT cells was shown to induce tolerance through IL-10 induction in a mouse model of anterior chamber-associated immune deviation (ACAID) (Watte et al., 2008). Engaging costimulatory molecule CD28 on *i*NKT cells overcomes the inability of α-GalCer loaded immature DCs to induce *i*NKT cells to secrete IFN-γ and IL-4 (Ikarashi et al., 2001). Additionally, treating CD28$^{(-/-)}$ or ICOS$^{(-/-)}$ deficient mice with α-GalCer failed to induce *i*NKT cells to produce IFN-γ, IL-13 and IL-4, whereas blocking CD40–CD154 interaction in this modelled only failed to induce *i*NKT cell derived IFN-γ (Hayakawa et al., 2001). These studies reveal cross-talk between NKT cells and APCs, where interaction of NKT cells with antigens on APCs induces CD40L surface expression on NKT cells. Subsequent upregulation of CD40L on NKT cells drives maturation, IL-12 production and the acquisition of costimulatory molecules by APCs, which then trigger NKT cells to release cytokines. The profile of cytokines released from this carefully orchestrated cascade signals bystander cells to mobilize suppression or induction of inflammation. This complex series of events is depicted in Figure 19.2.

NKT cell effector functions

Once activated, NKT cells mediate their regulatory or effector functions directly or through secretion of cytokines and chemokines (see Table 19.4). NKT cells can directly induce cytotoxicity of tumours and infected cells using cell death effector molecules FasL (Nakagawa et al., 2001), TRAIL (Nieda et al., 2001) and the

Table 19.4 Potential cellular targets of NKT-cell derived immunomodulatory molecules

Target cell	Cytokine	Chemokine	Cytotoxic molecules
NK cells	IL-2, IL-4, IL-21, IFNγ, TGFβ(i)	RANTES, LT, MIP-1α/β	FASL, perforin/granzyme
T-reg	IL-2, IL-6(i), IL-10, TGFβ	RANTES, MIP-1α/β	FASL, perforin/granzyme
Th1	IL-2, IL-10(i), IL-12, IL-17(i), IFNγ, TGFβ(i)	RANTES, LT, MIP-1α/β	FASL, perforin/granzyme
Th2	IL-2, IL-4, IL-9, IL-10(i), IL-21, IFNγ(i),TGFβ(i)	RANTES, LT, MIP-1α/β, Eotaxin	FASL, perforin/granzyme
Th17	IL-2(i), IL-4(i), IL-6, IL-10(i), IFNγ(i), TGFβ	RANTES?, MIP-1α/β?	FASL, perforin/granzyme
CD8 T-cells	IL-2, IL-6, IL-21, IFNγ, TGFβ(i)	RANTES, LT, MIP-1α/β	TRAIL, perforin/granzyme
B-cells	IL-2,IL-4,IL-5,IL-6,IL-13,IL-21, IFNγ, TGFβ, CD40L	RANTES, MIP-1α/β	FASL, perforin/granzyme
Dendritic cells	IL-4, IL-10(i), IL-21, IFNγ, GM-CSF, TGFβ(i), TNFα, CD40L	RANTES, MIP-1α/β, Eotaxin	TRAIL, perforin/granzyme
Monocyte/ Macrophage	IL-3, IL-4, IL-6, IL-10, IL-13, IFNγ,GM-CSF,TGFβ,TNFα, CD40L	RANTES, MIP-1α/β	FASL, perforin/granzyme
Neutrophils	IL-9, IFNγ,GM-CSF, TGFβ, TNFα	?	FASL, perforin/granzyme
Eosinophils	IL-3, IL-5, IFNγ,GM-CSF, TGFβ	RANTES, MIP-1α, Eotaxin	FASL, perforin/granzyme
Mast cells	IL-3, IL-4, IL-9, IFNγ,TGFβ	?	FASL, perforin/granzyme
Epithelial/ Endothelial	IL-9, IL-17, IFNγ, TGFβ, TNFα	RANTES, MIP-1α/β, Eotaxin	TRAIL, perforin/granzyme

(i) Inhibits target cell activation. Adapted from Coquet et al. (2008) and Matsuda et al. (2008).

perforin/granzyme pathway (Kawano et al., 1998). NKT cells also have the distinct ability to direct a response by producing Th1, Th2 and Th17 cytokines without being polarized. Secretion of IFN-γ by NKT cells directs NK cells to activate $CD8^+$ CTLs that are critical for anti-tumour responses and fighting infections (Kitamura et al., 1999; Tomura et al., 1999). The secretion of Th2 cytokines IL-4 and IL-13 by NKT cells was shown to induce T cell tolerance (Taniguchi et al., 2003a). On the other hand, secretion of IL-17 by NKT cells induced mouse airway hyper-reactivity (Pichavant et al., 2008; Michel et al., 2007), and mediated EAE, collagen induced arthritis (Furuzawa-Carballeda et al., 2007) and the elimination of pathogens. NKT cells can also affect an immune response by their ability to mobilize other effector cells through the secretion of chemokines. Future studies examining NKT cell derived chemokines should elucidate the contributions these chemokines play in disease pathogenesis.

CD1d regulates NKT cell tolerance and pathogenesis

In mice, *i*NKT cells account for 20–40% of the total lymphocyte population in the liver, while in humans they are present at much lower frequency and numbers (Kita et al., 2002). This inter-species difference in *i*NKT cell frequency may be due to differing levels of CD1d expression in the liver. In mice, CD1d is constitutively expressed on hepatocytes (Trobonjaca et al., 2001), whereas only very low levels of CD1d were found on healthy human hepatocytes (Bleicher et al., 1990). However, when CD1d expression is increased in the liver of patients with inflammatory diseases such as HCV (Durante-Mangoni et al., 2004) and primary biliary cirrhosis (Kita et al., 2002) an increase in *i*NKT cell numbers is observed. While it is currently unclear what role modulation of CD1d and *i*NKT cells may play in these diseases it is possible liver inflammation upregulates CD1d to induce suppression of this pathology by *i*NKT cells or that increased expression of CD1d could contribute to chronic *i*NKT cells induced inflammation. A predisposition to liver damage from pro-inflammatory signals is seen in ob/ob mice that have fatty livers and reduced CD1d expression on hepatocytes and have reduced levels of liver *i*NKT cells (Yang et al., 2007). This study suggests local CD1d expression regulates local NKT cell tolerance. Similarly, CD1d is expressed on trophoblast cells at the foctal–maternal interface in both mice and humans (Boyson et al., 2002), coincided with enrichment of *i*NKT cells that are suggested

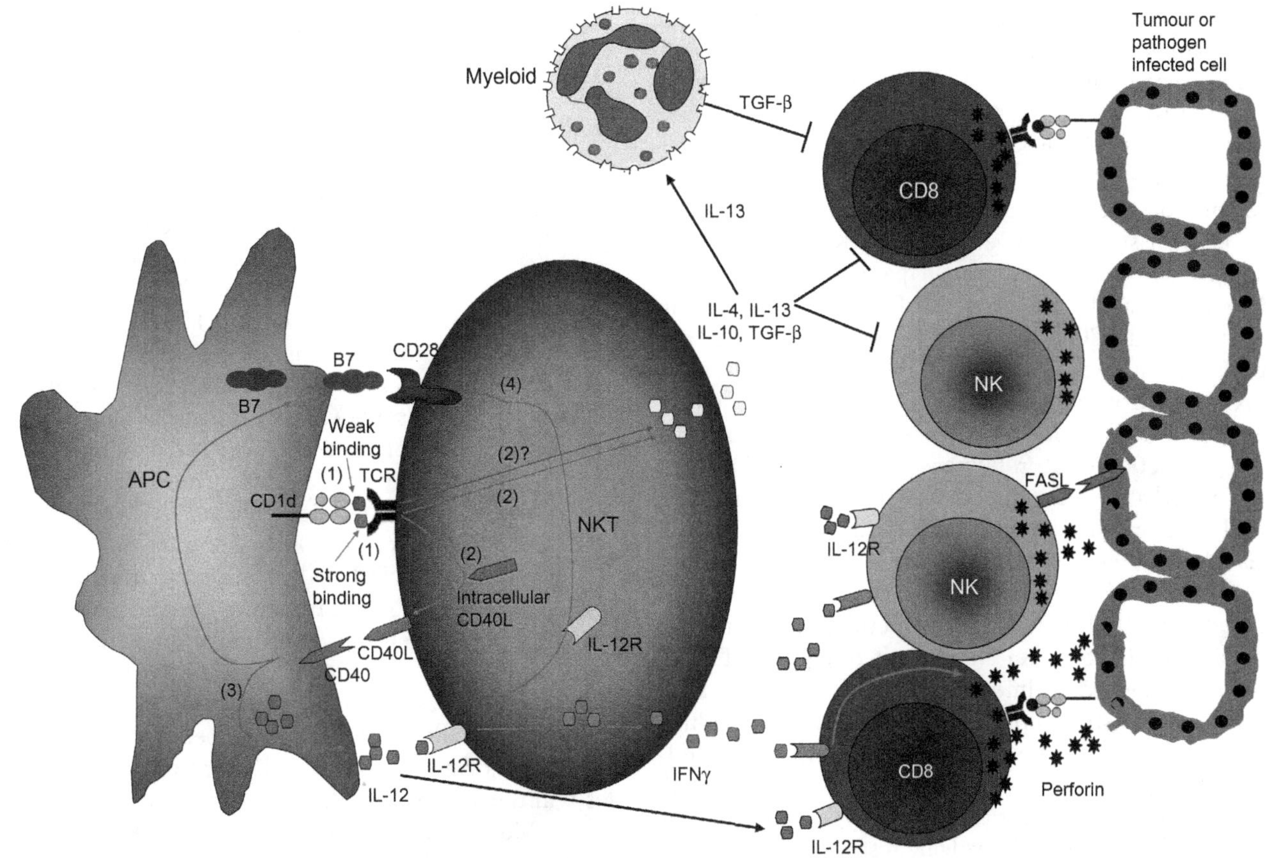

Figure 19.2 • Schematic depicting NKT-cell activation and immunoregulation of effector cells following interaction with APC. (1) Strong antigen triggering (dark grey arrows) of NKT-cells induces production of Th2 cytokines and (2) upregulates cell surface CD40L. (3) CD40L next induces the APC to express B7 costimulatory molecule and to release IL-12. (4) Costimulation induces further expression of IL-12 receptors on NKT cell allowing for increased IL-12 signalling of NKT-cells and (5) production of IFN-γ that can activate NK and CD8 T cells. In contrast, (1) weak antigens (very dark grey) are unable to upregulate CD40L on NKT-cells. Subsequently (2) NKT cells have limited cytokine production or are skewed toward producing Th2 or suppressive cytokines that inhibit NK or CD8 effector T cells.

to induce tolerance by producing immunosuppressive cytokines (Boyson et al., 2008). In addition to inducing local NKT cell tolerance, CD1d expression has also been reported to contribute to induction of systemic tolerance. Peripheral tolerance can be induced by delivering antigen to the anterior chamber of the eye in a model termed anterior chamber-associated immune deviation (ACAID). In this model, $CD8^{(+)}$ regulatory T cells are generated in the spleen and NKT cells were shown to be necessary for inducing tolerance (Sonoda et al., 1999). This carefully orchestrated process involved production of IL-10, following NKT cell interactions with eye-derived APCs and splenic B-cells expressing CD1d (Sonoda and Stein-Streilein, 2002; Sonoda et al., 2001). Alternatively, if CD1d expression induces tolerance, then blocking CD1d signalling should diminish tolerance. Indeed, this was found to be the case in a murine tumour model, where blocking CD1d antigen presentation to NKT cells using a non-depleting antibody, reduced tumour burden in mice (Terabe et al., 2005). These results suggest CD1d was necessary for NKT-mediated suppression of tumour immune surveillance.

Type-I diabetes is the most studied disease associated with the loss of *i*NKT cells. NOD mice, a murine model of type-I diabetes, have a reduced number of *i*NKT cells (Hammond and Kronenberg, 2003). Adoptive transfer of *i*NKT cells (Baxter et al., 1997) or over-expression of the *i*NKT cell restricted TCR chain Vα14-Jα18 (Lehuen et al., 1998a,b) ameliorated autoimmune diabetes in these mice. Another study found *i*NKT cells in the pancreas of NOD mice had altered IL-4 production (Falcone et al., 2004). In this study, a critical role for CD1d-mediated regulation of NKT cell induced tolerance was found whereby over-expression of CD1d in pancreatic islets prevented diabetes in NOD mice. This induction of tolerance did not depend on increasing *i*NKT cells but rather by increasing IL-4 production from *i*NKT cells in pancreatic lymph nodes. Examination of *i*NKT cells from the blood of NOD and normal mice found no differences in *i*NKT cell numbers or ability to produce IL-4 (Berzins et al., 2004). Taken together, these studies suggest defects in *i*NKT cell numbers and altered ability to produce IL-4 local to the site of autoimmunity may contribute to diabetes. Human studies that examined *i*NKT cell numbers and IL-4 production in human patients with insulin-dependent diabetes mellitus (IDMM) were inconsistent (Kukreja et al., 2002; Lee et al., 2002b; Oikawa et al., 2002; Wilson et al., 1998). A more detailed analysis of *i*NKT cells from pancreatic lymph nodes is needed to determine if the same alterations in cell numbers and IL-4 production exist in humans as seen in NOD mice. CD1d-mediated modulation of IL-4 production by *i*NKT cells was also observed in humans with viral hepatitis (de Lalla et al., 2004). In this study, strong CD1d expression on liver APCs from infected individuals correlated with IL-4 and IL-13 production, and increased fibrosis and cirrhosis. An enrichment of liver NKT cells was also observed in infected livers. The authors suggested viral or stress lipid antigens may trigger production of profibrotic cytokines IL-4 and IL-13 from NKT cells leading to fibrosis. Whereas over-expression of CD1d in pancreatic islets protects mice from disease in a murine diabetic model involving restoration of protective IL-4 production by *i*NKT cells, over-expression of CD1d in the liver of patients with viral hepatitis is associated with pathogenesis also involving IL-4 in addition to IL-13 expression. Thus *i*NKT cell derived IL-4, regulated by CD1d expression, can prevent or cause pathogenesis depending on disease.

The relationship between CD1d expression in the liver and *i*NKT cell survival has also been studied by transfer of immature *i*NKT cells from WT mice into the thymus of CD1d–KO mice. The authors found cell migration and proliferation of NKT cells in the liver were normal (McNab et al., 2005), which suggests that CD1d expression was not needed for survival or recruitment of *i*NKT cells to the liver. Another study confirmed this conclusion by using mice expressing CD1d exclusively on cortical thymocytes, where *i*NKT cells were shown to be normal in cell differentiation and tissue distribution (Wei et al., 2005). However, in both studies a severe reduction was observed in *i*NKT cells expressing NK1.1^{+}, a marker often associated with NKT cell maturation (Pellicci et al., 2002). Since most NKT cells leave the thymus as NK1.1^{-} this suggests NKT cells need CD1d signalling in the periphery for maturation. Taken together the studies reviewed in this section indicate that while CD1d signalling is not needed for NKT cell survival and tissue distribution, it is important for NKT cell tolerance, which in some models is through the induction of suppressive cytokines.

Functional subsets of NKT cells

NKT cells are a heterogeneous population and numerous studies have examined subsets for functional differences. A detailed study compared human *i*NKT cell CD4$^{(+)}$ and CD4$^{(-)}$ subsets for the expression of cytokines (Lee et al., 2002a). These authors found the CD4$^{(+)}$ *i*NKT cell subset was the sole producer of regulatory cytokines IL-4 and IL-13 following activation with PMA/ionomycin, while both subsets produced TNF-α and IFN-γ. However these distinctions in cytokine production by CD4$^{(+)}$ and CD4$^{(-)}$ subsets were not discernible in mice. This is a somewhat surprising finding because adoptive transfer of liver CD4$^{(-)}$ *i*NKT cells rejected methylcholanthrene (MCA)-1 sarcoma and B16F10 melanoma better then did transfer of liver CD4$^{(+)}$ *i*NKT cells (Crowe et al., 2005). Adoptively transferring splenic or thymic *i*NKT cells revealed an impaired ability of these cells to reject tumours. Diverse anti-tumour activity of *i*NKT cell subsets from different tissues suggests the microenvironment is important in shaping *i*NKT cell function. Careful analysis and comparisons of tissue-specific NKT cells are needed before broad conclusions can be drawn, and adoptive transfer studies will need to be modified in context of these differences. Mouse NKT cells can also be further subdivided by their expression of the NK cell marker, NK1.1. In light of differences in anti-tumour activity within *i*NKT cell subsets, a comprehensive study examined cytokines released by CD4$^{(+)}$NK1.1$^{(+)}$, CD4$^{(-)}$NK1.1$^{(+)}$ and NK1.1$^{(-)}$ subsets from mouse spleen, liver and thymus (Coquet et al., 2008). In general thymic *i*NKT cells released higher levels of cytokines following activation with plate bound anti-CD3 and anti-CD28. Analysis of these subsets found activated NK1.1$^{(+)}$ *i*NKT cells released higher levels of IFN-γ and lower levels of TH2 cytokines IL-4, IL-13 and IL-10, compared to NK1.1$^{(-)}$ *i*NKT cells. An explosion of recent papers revealed NK1.1$^{(-)}$ *i*NKT cells are exclusive producers of IL-17 among the total NKT cell population (Pichavant et al., 2008; Michel et al., 2007; Rachitskaya et al., 2008). One group further identified CD4$^{(-)}$NK1.1$^{(-)}$ *i*NKT cells as the main producer of IL-17 (Coquet et al., 2008). Because NK1.1$^{(-)}$ NKT cells are known to contain immature cells the authors used a foetal thymus culture system and found NK1.1$^{(-)}$ *i*NKT cells matured and acquired NK1.1$^{(+)}$. This raises the question of whether IL-17 producing *i*NKT cells lose the ability to produce IL-17 once they acquire NK1.1. Also is it possible that the immature *i*NKT cell subset is the only NKT subset to produce IL-17? A report does suggest the existence of a mature population of *i*NKT cells that are NK1.1$^{(-)}$ (McNab et al., 2007). It remains to be determined if this mature population of NK1.1$^{(-)}$ *i*NKT cell are able to produce IL-17. Recently it was found that human NKT cells produce IL-17; however, a more detailed analysis is needed to determine if these cells are part of an immature or other subset (Rachitskaya et al., 2008).

Perhaps the best example of conflicting functional attributes of NKT cells is in tumour immunology. Numerous functional studies that examined NKT cells activated with factors such as α-GalCer and IL-12 showed these cells promoted tumour immunity in mouse models and against human tumours (Godfrey and Kronenberg, 2004; Smyth et al., 2000). However, the physiological role of NKT cells in natural tumour surveillance and immunity is less clear. Our lab and others sought to examine this issue using mice deficient in all NKT cells or just *i*NKT cells. Transplantation of the murine renal cell carcinoma Renca and fibrosarcoma line 15-12RM tumours transplanted into $CD1d^{(-/-)}$ mice, deficient in all NKT cells, had reduced tumour growth (Subleski et al., 2006) or rejected tumours (Terabe et al., 2000) respectively as compared to WT mice. IL-13 was shown to be critical for the suppression of tumour rejection (Terabe et al., 2000). In contrast, other investigators that induced tumour formation with the chemical carcinogen MCA found increased tumour formation in *i*NKT cell deficient $J\alpha18^{(-/-)}$ mice as compared to WT mice (Smyth et al., 2000). Differences in tumour models may at least partially explain these differences. Regulatory NKT cells might suppress immunity toward the transplanted tumours, allowing for tumour growth, while in the chemical carcinogen model NKT cells could be suppressing cancer-causing inflammation that would decrease tumour formation. It is also possible that disparate effects of functionally distinct NKT cell subsets may contribute to these apparent discrepancies since different KO mice with defects in different NKT subsets were used in these studies. This possibility was examined by comparing the growth of several tumour lines in $CD1d^{(-/-)}$ and $J\alpha18^{(-/-)}$ mice. Terabe and associates found that tumour burden was reduced in $CD1d^{(-/-)}$ mice as compared to WT and $J\alpha18^{(-/-)}$ mice (Terabe et al., 2005). From these results the authors concluded that non-*i*NKT cells were sufficient for suppressing tumour immunity and that *i*NKT cells were not necessary.

With no reagents currently available to study *i*NKT cells in the absence of non-*i*NKT cells, it is impossible to determine if *i*NKT cells also contribute to suppressing tumour immunity. A subsequent study by these authors found that specific activation of non-*i*NKT cells with weak non-*i*NKT cells agonist sulfatide increased tumour burden, whereas activating *i*NKT cells with the strong agonists α-GalCer or OCH reduced tumour burden (Ambrosino et al., 2007). Mice treated simultaneously with the *i*NKT cell agonist α-GalCer and non-*i*NKT cell agonist sulfatide slightly reduced IFN-γ and slightly increased IL-13 when compared to mice treated with α-GalCer alone. From this data the authors concluded that in tumour immunity *i*NKT cells are protective and non-*i*NKT cells are suppressive. However, this conclusion may not take into consideration the different potencies of the agonists that would be expected to activate each NKT subset disproportionately. For example the authors showed that α-GalCer and OCH each induced a cytokine storm whereas sulfatide had minimal effect on cytokine expression. It is well established that *i*NKT cell glyco-antigens vary in intensity and type of immune modulating cytokines they induce (Wu et al., 2005). The immunological differences in NKT cell subsets could be more a function of the agonist used in the study then their inherent regulatory abilities. Our lab found that a weak *i*NKT cell ligand, β-GalCer, reduced *i*NKT cells in mice without producing a cytokine storm (Ortaldo et al., 2004). Treating mice with β-GalCer prior to tumour implantation reduced *i*NKT cells but increased anti-tumour activity in the liver, suggesting that *i*NKT cells suppress immune surveillance in this model (Subleski et al., 2006). A concern with using these agonists to ascribe biological functions to NKT cells is that they are not physiological ligands and may not be informative as to the natural function of NKT cells in tumour immunology in vivo.

Conclusion

Although *i*NKT cells are a minor population of lymphocytes in humans, several factors support their broad importance as immune regulating cells. First, they are decreased in number and function in cancer patients (Muhammad Ali Tahir et al., 2001). Conversely, NKT cells are increased in patients with chronic hepatitis C viral (HCV) infection (Durante-Mangoni et al., 2004) and primary biliary cirrhosis (Kita et al., 2002). Taken together, these data suggests the potential importance of NKT cells in disease. Second, mouse and human *i*NKT cells recognize the same antigen α-GalCer demonstrating evolutionary conservation. This conservation allows close comparison of data generated in mouse disease models with patient samples. Third, the adoptive transfer of manipulated immune cells for treating various patient diseases has seen an upward trend. NKT cells are a logical target for adoptive transfer therapies because they can be induced to produce large amounts of immune-modulating cytokines and because they are implicated in regulating progression or resolution of a variety of disease states. Further study of NKT cells is warranted to better understand their role in human diseases and to determine whether manipulation of the numbers and/or functions of these cells may be beneficial for the prevention or treatment of cancer, autoimmunity or infectious diseases.

References

Adam, C., King, S., Allgeier, T., Braumuller, H., Luking, C., Mysliwietz, J., Kriegeskorte, A., Busch, D.H., Rocken, M. and Mocikat, R. (2005). DC–NK cell cross talk as a novel CD4+ T-cell-independent pathway for antitumor CTL induction. Blood 106, 338–344.

Adams, E.J., Juo, Z.S., Venook, R.T., Boulanger, M.J., Arase, H., Lanier, L. L. and Garcia, K.C. (2007). Structural elucidation of the m157 mouse cytomegalovirus ligand for Ly49 natural killer cell receptors. Proc Natl Acad Sci U S A **104**, 10128–10133.

Akbari, O., Stock, P., Meyer, E., Kronenberg, M., Sidobre, S., Nakayama, T., Taniguchi, M., Grusby, M.J., DeKruyff, R.H. and Umetsu, D.T. (2003). Essential role of NKT cells producing IL-4 and IL-13 in the development of allergen-induced airway hyperreactivity. Nat Med **9**, 582–588.

Alberola-Ila, J., Hogquist, K.A., Swan, K.A., Bevan, M.J. and Perlmutter, R.M. (1996). Positive and negative selection invoke distinct signaling pathways. J Exp Med **184**, 9–18.

Allavena, P., Scala, G., Djeu, J.Y., Procopio, A.D., Oppenheim, J.J., Herberman, R.B. and Ortaldo, J.R. (1985). Production of multiple cytokines by clones of human large granular lymphocytes. Cancer Immunol Immunother **19**, 121–126.

Alvord Jr., E.C. (1984). Experimental Allergic Encephalomyelitis: A Useful Model for Multiple Sclerosis. New York: Alan R. Liss.

Ambrosino, E., Terabe, M., Halder, R.C., Peng, J., Takaku, S., Miyake, S., Yamamura, T., Kumar, V. and Berzofsky, J.A. (2007). Cross-regulation between type I and type II NKT cells in regulating tumor immunity: a new immunoregulatory axis. J Immunol **179**, 5126–5136.

Arase, H., Mocarski, E.S., Campbell, A.E., Hill, A.B. and Lanier, L.L. (2002). Direct recognition of cytomegalovirus by activating and inhibitory NK cell receptors. Science **296**, 1323–1326.

Barber, M.A., Zhang, T., Gagne, B.A. and Sentman, C.L. (2007). NK cells negatively regulate antigen presentation and tumor-specific CTLs in a syngeneic lymphoma model. J Immunol **178**, 6140–6147.

Baxter, A.G., Kinder, S.J., Hammond, K.J., Scollay, R. and Godfrey, D.I. (1997). Association between alphabetaTCR+CD4−CD8− T-cell deficiency and IDDM in NOD/Lt mice. Diabetes **46**, 572–582.

Beilke, J.N., Kuhl, N.R., Van, K.L. and Gill, R.G. (2005). NK cells promote islet allograft tolerance via a perforin-dependent mechanism. Nat Med **11**, 1059–1065.

Bendelac, A., Savage, P.B. and Teyton, L. (2007). The biology of NKT cells. Annu Rev Immunol **25**, 297–336.

Bennett, C.L., Christie, J., Ramsdell, F., Brunkow, M.E., Ferguson, P.J., Whitesell, L., Kelly, T.E., Saulsbury, F.T., Chance, P.F. and Ochs, H.D. (2001). The immune dysregulation, polyendocrinopathy, enteropathy, X-linked syndrome (IPEX) is caused by mutations of FOXP3. Nat Genet **27**, 20–21.

Berzins, S.P., Kyparissoudis, K., Pellicci, D.G., Hammond, K.J., Sidobre, S., Baxter, A., Smyth, M.J., Kronenberg, M. and Godfrey, D.I. (2004). Systemic NKT cell deficiency in NOD mice is not detected in peripheral blood: implications for human studies. Immunol Cell Biol **82**, 247–252.

Berzins, S.P., Smyth, M.J. and Godfrey, D.I. (2005). Working with NKT cells—pitfalls and practicalities. Curr Opin Immunol **17**(4), 448–454 Review. Erratum in: Curr Opin Immunol 2005 Oct. 17(5), 565–566.

Berzofsky, J.A. and Terabe, M. (2008). NKT cells in tumor immunity: opposing subsets define a new immunoregulatory axis. J Immunol **180**, 3627–3635.

Biron, C.A., Cousens, L.P., Ruzek, M.C., Su, H.C. and Salazar-Mather, T.P. (1998). Early cytokine responses to viral infections and their roles in shaping endogenous cellular immunity. Adv Exp Med Biol **452**, 143–149.

Biron, C.A., Nguyen, K.B. and Pien, G.C. (2002). Innate immune responses to LCMV infections: natural killer cells and cytokines. Curr Top Microbiol Immunol **263**, 7–27.

Blanca, I.R., Bere, E.W., Young, H.A. and Ortaldo, J.R. (2001). Human B cell activation by autologous NK cells is regulated by CD40-CD40 ligand interaction: role of memory B cells and CD5+ B cells. J Immunol **167**, 6132–6139.

Bleicher, P.A., Balk, S.P., Hagen, S.J., Blumberg, R.S., Flotte, T.J. and Terhorst, C. (1990). Expression of murine CD1 on gastrointestinal epithelium. Science **250**, 679–682.

Bluman, E.M., Bartynski, K.J., Avalos, B.R. and Caligiuri, M.A. (1996). Human natural killer cells produce abundant macrophage inflammatory protein-1 alpha in response to monocyte-derived cytokines. J Clin Invest **97**, 2722–2727.

Boyson, J.E., Rybalov, B., Koopman, L.A., Exley, M., Balk, S.P., Racke, F.K., Schatz, F., Masch, R., Wilson, S.B. and Strominger, J.L. (2002). CD1d and invariant NKT cells at the human maternal–fetal interface. Proc Natl Acad Sci U S A **99**, 13741–13746.

Boyson, J.E., Aktan, I., Barkhuff, D.A. and Chant, A. (2008). NKT cells at the maternal–fetal interface. Immunol Invest **37**, 565–582.

Brigl, M., Bry, L., Kent, S.C., Gumperz, J.E. and Brenner, M.B. (2003). Mechanism of CD1d-restricted natural killer T cell activation during microbial infection. Nat Immunol **4**, 1230–1237.

Brunkow, M.E., Jeffery, E.W., Hjerrild, K.A., Paeper, B., Clark, L.B., Yasayko, S.A., Wilkinson, J.E., Galas, D., Ziegler, S.F. and Ramsdell, F. (2001). Disruption of a new forkhead/winged-helix protein, scurfin, results in the fatal lymphoproliferative disorder of the scurfy mouse. Nat Genet **27**, 68–73.

Brutkiewicz, R.R. (2006). CD1d ligands: the good, the bad, and the ugly. J Immunol **177**, 769–775.

Bukowski, J.F., Woda, B.A., Habu, S., Okumura, K. and Welsh, R.M. (1983). Natural killer cell depletion enhances virus synthesis and virus-induced hepatitis in vivo. J Immunol **131**, 1531–1538.

Burshtyn, D.N., Scharenberg, A.M., Wagtmann, N., Rajagopalan, S., Berrada, K., Yi, T., Kinet, J.P. and Long, E.O. (1996). Recruitment of tyrosine phosphatase HCP by the killer cell inhibitor receptor. Immunity **4**, 77–85.

Cantoni, C., Biassoni, R., Pende, D., Sivori, S., Accame, L., Pareti, L., Semenzato, G., Moretta, L., Moretta, A. and Bottino, C. (1998). The activating form of CD94 receptor complex: CD94 covalently associates with the Kp39 protein that represents the product of the NKG2-C gene. Eur J Immunol **28**, 327–338.

Chan, C.W., Crafton, E., Fan, H.N., Flook, J., Yoshimura, K., Skarica, M., Brockstedt, D., Dubensky, T.W., Stins, M.F., Lanier, L.L., Pardoll, D.M. and Housseau, F. (2006). Interferon-producing killer dendritic cells provide a link between innate and adaptive immunity. Nat Med **12**, 207–213.

Chen, Y., Kuchroo, V.K., Inobe, J., Hafler, D.A. and Weiner, H.L. (1994). Regulatory T cell clones induced by oral tolerance: suppression of autoimmune encephalomyelitis. Science **265**, 1237–1240.

Choi, B.K., Kim, Y.H., Kang, W.J., Lee, S.K., Kim, K.H., Shin, S.M., Yokoyama, W.M., Kim, T.Y. and Kwon, B.S. (2007). Mechanisms involved in synergistic anticancer immunity of anti-4-1BB and anti-CD4 therapy. Cancer Res **67**, 8891–8899.

Chung, D.H., Natarajan, K., Boyd, L.F., Tormo, J., Mariuzza, R.A., Yokoyama, W.M. and Margulies, D.H. (2000). Mapping the ligand of the NK inhibitory receptor Ly49A on living cells. J Immunol **165**, 6922–6932.

Coca, S., Perez-Piqueras, J., Martinez, D., Colmenarejo, A., Saez, M.A., Vallejo, C., Martos, J.A. and Moreno, M. (1997). The prognostic significance of intratumoral natural killer cells in patients with colorectal carcinoma. Cancer **79**, 2320–2328.

Coles, M.C. and Raulet, D.H. (2000). NK1.1+ T cells in the liver arise in the thymus and are selected by interactions with class I molecules on CD4 + CD8+ cells. J Immunol **164**, 2412–2418.

Coquet, J.M., Chakravarti, S., Kyparissoudis, K., McNab, F.W., Pitt, L.A., McKenzie, B.S., Berzins, S.P., Smyth, M.J. and Godfrey, D.I. (2008). Diverse cytokine production by NKT cell subsets and identification of an IL-17-producing $CD4^-NK1.1^-$ NKT cell population. Proc Natl Acad Sci U S A **105**, 11287–11292.

Crowe, N.Y., Coquet, J.M., Berzins, S.P., Kyparissoudis, K., Keating, R., Pellicci, D.G., Hayakawa, Y., Godfrey, D.I. and Smyth, M.J. (2005). Differential antitumor immunity mediated by NKT cell subsets in vivo. J Exp Med **202**, 1279–1288.

Cui, J., Shin, T. and Kawano, T. (1997). Requirement for V[alpha]14 NKT cells in IL-12-mediated rejection of tumors. Science **278**, 1623–1626.

Dalbeth, N., Gundle, R., Davies, R.J., Lee, Y.C., McMichael, A.J. and Callan, M.F. (2004). CD56right NK cells are enriched at inflammatory sites and can engage with monocytes in a reciprocal program of activation. J Immunol **173**, 6418–6426.

Dalton, D.K., Pitts-Meek, S., Keshav, S., Figari, I.S., Bradley, A. and Stewart, T.A. (1993). Multiple defects of immune cell function in mice with disrupted interferon-gamma genes. Science **259**, 1739–1742.

Daniels, K.A., Devora, G., Lai, W.C., O'Donnell, C.L., Bennett, M. and Welsh, R.M. (2001). Murine cytomegalovirus is regulated by a discrete subset of natural killer cells reactive with monoclonal antibody to Ly49H. J Exp Med **194**, 29–44.

de Lalla, C., Galli, G., Aldrighetti, L., Romeo, R., Mariani, M., Monno, A., Nuti, S., Colombo, M., Callea, F., Porcelli, S.A., Panina-Bordignon, P., Abrignani, S., Casorati, G. and Dellabona, P. (2004). Production of profibrotic cytokines by invariant NKT cells characterizes cirrhosis progression in chronic viral hepatitis. J Immunol **173**, 1417–1425.

Di Santo, J.P. (2006). Natural killer cell developmental pathways: a question of balance. Annu Rev Immunol **24**, 257–286.

Diefenbach, A., Tomasello, E., Lucas, M., Jamieson, A.M., Hsia, J.K., Vivier, E. and Raulet, D.H. (2002). Selective associations with signaling proteins determine stimulatory versus costimulatory activity of NKG2D. Nat Immunol **3**, 1142–1149.

Djeu, J.Y., Stocks, N., Zoon, K., Stanton, G.J., Timonen, T. and Herberman, R.B. (1982). Positive self regulation of cytotoxicity in human natural killer cells by production of interferon upon exposure to influenza and herpes viruses. J Exp Med **156**, 1222–1234.

Dokun, A.O., Kim, S., Smith, H.R., Kang, H.S., Chu, D.T. and Yokoyama, W.M. (2001). Specific and nonspecific NK cell activation during virus infection. Nat Immunol **2**, 951–956.

Dorner, B.G., Smith, H.R., French, A.R., Kim, S., Poursine-Laurent, J., Beckman, D.L., Pingel, J.T., Kroczek, R.A. and Yokoyama, W.M. (2004). Coordinate expression of cytokines and chemokines by NK cells during murine cytomegalovirus infection. J Immunol **172**, 3119–3131.

Durante-Mangoni, E., Wang, R., Shaulov, A., He, Q., Nasser, I., Afdhal, N., Koziel, M.J. and Exley, M.A. (2004). Hepatic CD1d expression in hepatitis C virus infection and recognition by resident proinflammatory CD1d-reactive T cells. J Immunol **173**, 2159–2166.

Eberl, G., Lowin-Kropf, B. and MacDonald, H.R. (1999). Cutting edge: NKT cell development is selectively impaired in Fyn-deficient mice. J Immunol **163**, 4091–4094.

Egawa, T., Eberl, G., Taniuchi, I., Benlagha, K., Geissmann, F., Hennighausen, L., Bendelac, A. and Littman, D.R. (2005). Genetic evidence supporting selection of the Valpha14i NKT cell lineage from double-positive thymocyte precursors. Immunity **22**, 705–716.

Eisele, G., Wischhusen, J., Mittelbronn, M., Meyermann, R., Waldhauer, I., Steinle, A., Weller, M. and Friese, M.A. (2006). TGF-beta and metalloproteinases differentially suppress NKG2D ligand surface expression on malignant glioma cells. Brain **129**, 2416–2425.

Falcone, M., Facciotti, F., Ghidoli, N., Monti, P., Olivieri, S., Zaccagnino, L., Bonifacio, E., Casorati, G., Sanvito, F. and Sarvetnick, N. (2004). Up-regulation of CD1d expression restores the immunoregulatory function of NKT cells and prevents autoimmune diabetes in nonobese diabetic mice. J Immunol **172**, 5908–5916.

Farag, S.S., VanDeusen, J.B., Fehniger, T.A. and Caligiuri, M.A. (2003). Biology and clinical impact of human natural killer cells. Int J Hematol **78**, 7–17.

Fearon, D.T. and Locksley, R.M. (1996). The instructive role of innate immunity in the acquired immune response. Science **272**, 50–53.

Flodstrom, M., Maday, A., Balakrishna, D., Cleary, M.M., Yoshimura, A. and Sarvetnick, N. (2002). Target cell defense prevents the development of diabetes after viral infection. Nat Immunol **3**, 373–382.

Fort, M.M., Leach, M.W. and Rennick, D.M. (1998). A role for NK cells as regulators of CD4+ T cells in a transfer model of colitis. J Immunol **161**, 3256–3261.

Furuzawa-Carballeda, J., Vargas-Rojas, M.I. and Cabral, A.R. (2007). Autoimmune inflammation from the Th17 perspective. Autoimmun Rev **6**, 169–175.

Galkina, E., Harry, B.L., Ludwig, A., Liehn, E.A., Sanders, J.M., Bruce, A., Weber, C. and Ley, K. (2007). CXCR6 promotes atherosclerosis by supporting T-cell homing, interferon-gamma production, and macrophage accumulation in the aortic wall. Circulation **116**, 1801–1811.

Gao, N., Dang, T. and Yuan, D. (2001). IFN-gamma-dependent and -independent initiation of switch recombination by NK cells. J Immunol **167**(4), 2011–2018.

Gao, N., Jennings, P. and Yuan, D. (2008). Requirements for the natural killer cell-mediated induction of IgG1 and IgG2a expression in B lymphocytes. Int Immunol **20**(5), 645–657.

Gapin, L., Matsuda, J.L., Surh, C.D. and Kronenberg, M. (2001). NKT cells derive from double-positive thymocytes that are positively selected by CD1d. Nat Immunol **2**, 971–978.

George, T.C., Mason, L.H., Ortaldo, J.R., Kumar, V. and Bennett, M. (1999). Positive recognition of MHC class I molecules by the Ly49D receptor of murine NK cells. J Immunol **162**, 2035–2043.

Germanov, E., Veinotte, L., Cullen, R., Chamberlain, E., Butcher, E.C. and Johnston, B. (2008). Critical role for the chemokine receptor CXCR6 in homeostasis and activation of CD1d-restricted NKT cells. J Immunol **181**, 81–91.

Getz, G.S. (2005). Thematic review series: the immune system and atherogenesis. Bridging the innate and adaptive immune systems. J Lipid Res **46**, 619–622.

Godfrey, D.I. and Berzins, S.P. (2007). Control points in NKT-cell development. Nat Rev Immunol **7**, 505–518.

Godfrey, D.I. and Kronenberg, M. (2004). Going both ways: immune regulation via CD1d-dependent NKT cells. J Clin Invest **114**, 1379–1388.

Godfrey, D.I., MacDonald, H.R., Kronenberg, M., Smyth, M.J. and Van, K.L. (2004). NKT cells: what's in a name. Nat Rev Immunol **4**, 231–237.

Goldszmid, R.S., Bafica, A., Jankovic, D., Feng, C.G., Caspar, P., Winkler-Pickett, R., Trinchieri, G. and Sher, A. (2007). TAP-1 indirectly regulates CD4+ T cell priming in *Toxoplasma gondii* infection by controlling NK cell IFN-gamma production. J Exp Med **204**, 2591–2602.

Goodnow, C.C., Sprent, J., Fazekas de St, G.B. and Vinuesa, C.G. (2005). Cellular and genetic mechanisms of self tolerance and autoimmunity. Nature **435**, 590–597.

Gosselin, P., Mason, L.H., Willette-Brown, J., Ortaldo, J.R., McVicar, D.W. and Anderson, S.K. (1999). Induction of DAP12 phosphorylation, calcium mobilization, and cytokine secretion by Ly49H. J Leukoc Biol **66**, 165–171.

Gray, J.D. and Horwitz, D.A. (1995). Activated human NK cells can stimulate resting B cells to secrete immunoglobulin. J Immunol **154**, 5656–5664.

Gumperz, J.E., Miyake, S., Yamamura, T. and Brenner, M.B. (2002). Functionally distinct subsets of CD1d-restricted natural killer T cells revealed by CD1d tetramer staining. J Exp Med **195**, 625–636.

Hammond, K.J. and Kronenberg, M. (2003). Natural killer T cells: natural or unnatural regulators of autoimmunity. Curr Opin Immunol **15**, 683–689.

Hayakawa, Y. and Smyth, M.J. (2006). Innate immune recognition and suppression of tumors. Adv Cancer Res **95**, 293–322.

Hayakawa, Y., Takeda, K., Yagita, H., Van, K.L., Saiki, I. and Okumura, K. (2001). Differential regulation of Th1 and Th2 functions of NKT cells by CD28 and CD40 costimulatory pathways. J Immunol **166**, 6012–6018.

Hayakawa, Y., Screpanti, V., Yagita, H., Grandien, A., Ljunggren, H.G., Smyth, M.J. and Chambers, B.J. (2004). NK cell TRAIL eliminates immature dendritic cells in vivo and limits dendritic cell vaccination efficacy. J Immunol **172**, 123–129.

Herberman, R.B. and Ortaldo, J.R. (1981). Natural killer cells: their roles in defenses against disease. Science **214**, 24–30.

Herberman, R.B., Djeu, J., Kay, H.D., Ortaldo, J.R., Riccardi, C., Bonnard, G.D., Holden, H.T., Fagnani, R., Santoni, A. and Puccetti, P. (1979). Natural killer cells: characteristics and regulation of activity. Immunol Rev **44**, 43–70.

Hoshino, T., Winkler-Pickett, R.T., Mason, A.T., Ortaldo, J.R. and Young, H.A. (1999). IL-13 production by NK cells: IL-13-producing NK and T cells are present in vivo in the absence of IFN-gamma. J Immunol **162**, 51–59.

Houchins, J.P., Yabe, T., McSherry, C. and Bach, F.H. (1991). DNA sequence analysis of NKG2, a family of related cDNA clones encoding type II integral membrane proteins on human natural killer cells. J Exp Med **173**, 1017–1020.

Huang, S.W., Rose, J.W. and Mayer, R.F. (1977). Assessment of cellular and humoral immunity of myasthenics. J Neurol Neurosurg Psychiatry **40**, 1053–1059.

Huang, D., Shi, F.D., Jung, S., Pien, G.C., Wang, J., Salazar-Mather, T.P., He, T.T., Weaver, J.T., Ljunggren, H.G., Biron, C.A., Littman, D.R. and Ransohoff, R.M. (2006). The neuronal chemokine CX3CL1/fractalkine selectively recruits NK cells that modify experimental autoimmune encephalomyelitis within the central nervous system. FASEB J **20**, 896–905.

Iizuka, K., Naidenko, O.V., Plougastel, B.F., Fremont, D.H. and Yokoyama, W.M. (2003). Genetically linked C-type lectin-related ligands for the NKRP1 family of natural killer cell receptors. Nat Immunol **4**, 801–807.

Ikarashi, Y., Mikami, R., Bendelac, A., Terme, M., Chaput, N., Terada, M., Tursz, T., Angevin, E., Lemonnier, F.A., Wakasugi, H. and Zitvogel, L. (2001). Dendritic cell maturation overrules H-2D-mediated natural killer T (NKT) cell inhibition: critical role for B7 in CD1d-dependent NKT cell interferon gamma production. J Exp Med **194**, 1179–1186.

Ishigami, S., Natsugoe, S., Tokuda, K., Nakajo, A., Che, X., Iwashige, H., Aridome, K., Hokita, S. and Aikou, T. (2000). Prognostic value of intratumoral natural killer cells in gastric carcinoma. Cancer **88**, 577–583.

Ito, K., Karasawa, M., Kawano, T., Akasaka, T., Koseki, H., Akutsu, Y., Kondo, E., Sekiya, S., Sekikawa, K., Harada, M., Yamashita, M., Nakayama, T. and Taniguchi, M. (2000). Involvement of decidual Valpha14 NKT cells in abortion. Proc Natl Acad Sci U S A **97**, 740–744.

Jahng, A., Maricic, I., Aguilera, C., Cardell, S., Halder, R.C. and Kumar, V. (2004). Prevention of autoimmunity by targeting a distinct, noninvariant CD1d-reactive T cell population reactive to sulfatide. J Exp Med **199**, 947–957.

Jiang, X., Shimaoka, T., Kojo, S., Harada, M., Watarai, H., Wakao, H., Ohkohchi, N., Yonehara, S., Taniguchi, M. and Seino, K.I. (2005). Cutting edge: critical role of CXCL16/CXCR6 in NKT cell trafficking in allograft tolerance. J Immunol **175**, 2051–2055.

Kalinski, P., Mailliard, R.B., Giermasz, A., Zeh, H.J., Basse, P., Bartlett, D.L., Kirkwood, J.M., Lotze, M.T. and Herberman, R.B. (2005). Natural killer–dendritic cell cross-talk in cancer immunotherapy. Expert Opin Biol Ther **5**, 1303–1315.

Kawano, T., Cui, J., Koezuka, Y., Toura, I., Kaneko, Y., Motoki, K., Ueno, H., Nakagawa, R., Sato, H., Kondo, E., Koseki, H. and Taniguchi, M. (1997). CD1d-restricted and TCR-mediated activation of V{alpha}14 NKT cells by glycosylceramides. Science **278**, 1626–1629.

Kawano, T., Cui, J., Koezuka, Y., Toura, I., Kaneko, Y., Sato, H., Kondo, E., Harada, M., Koseki, H., Nakayama, T., Tanaka, Y. and Taniguchi, M. (1998). Natural killer-like nonspecific tumor cell lysis mediated by specific ligand-activated Valpha14 NKT cells. Proc Natl Acad Sci U S A **95**, 5690–5693.

Kawano, T., Tanaka, Y., Shimizu, E., Kaneko, Y., Kamata, N., Sato, H., Osada, H., Sekiya, S., Nakayama, T. and Taniguchi, M. (1999). A novel recognition motif of human NKT antigen receptor for a glycolipid ligand. Int Immunol **11**, 881–887.

Kelly, J.M., Darcy, P.K., Markby, J.L., Godfrey, D.I., Takeda, K., Yagita, H. and Smyth, M.J. (2002). Induction of tumor-specific T cell memory by NK cell-mediated tumor rejection. Nat Immunol **3**, 83–90.

Kim, S. and Yokoyama, W.M. (1998). NK cell granule exocytosis and cytokine production inhibited by Ly-49A engagement. Cell Immunol **183**, 106–112.

Kim, S., Iizuka, K., Aguila, H.L., Weissman, I.L. and Yokoyama, W.M. (2000). In vivo natural killer cell activities revealed by natural killer cell-deficient mice. Proc Natl Acad Sci U S A **97**, 2731–2736.

Kim, H.Y., Kim, H.J., Min, H.S., Kim, S., Park, W.S., Park, S.H. and Chung, D.H. (2005). NKT cells promote antibody-induced joint inflammation by suppressing transforming growth factor {beta} 1 production. J Exp Med **201**, 41–47.

Kita, H., Naidenko, O.V., Kronenberg, M., Ansari, A.A., Rogers, P., He, X.S., Koning, F., Mikayama, T., Van De, W.J., Coppel, R.L., Kaplan, M. and Gershwin, M.E. (2002). Quantitation and phenotypic analysis of natural killer T cells in primary biliary cirrhosis using a human CD1d tetramer. Gastroenterology **123**, 1031–1043.

Kitamura, H., Iwakabe, K. and Yahata, T. (1999). The natural killer T (NKT) cell ligand [alpha]-galactosylceramide demonstrates its immunopotentiating effect by inducing interleukin (IL)-12 production by dendritic cells and IL-12 receptor expression on NKT cells. J Exp Med **189**, 1121–1128.

Kitamura, H., Ohta, A. and Sekimoto, M. (2000). [alpha]-galactosylceramide induces early B-cell activation through IL-4 production by NKT cells. Cell Immunol **199**, 37–42.

Kriegeskorte, A.K., Gebhardt, F.E., Porcellini, S., Schiemann, M., Stemberger, C., Franz, T.J., Huster, K.M., Carayannopoulos, L.N., Yokoyama, W.M., Colonna, M., Siccardi, A.G., Bauer, S. and Busch, D.H. (2005). NKG2D-independent suppression of T cell proliferation by H60 and MICA. Proc Natl Acad Sci U S A **102**, 11805–11810.

Kronenberg, M. and Gapin, L. (2002). The unconventional lifestyle of NKT cells. Nat Rev Immunol **2**, 557–568.

Kukreja, A., Cost, G., Marker, J., Zhang, C., Sun, Z., Lin-Su, K., Ten, S., Sanz, M., Exley, M., Wilson, B., Porcelli, S. and Maclaren, N. (2002). Multiple immuno-regulatory defects in type-1 diabetes. J Clin Invest **109**, 131–140.

Kumar, V. and Sercarz, E.E. (1993). The involvement of T cell receptor peptide-specific regulatory CD4+ T cells in recovery from antigen-induced autoimmune disease. J Exp Med **178**, 909–916.

Laffont, S., Seillet, C., Ortaldo, J., Coudert, J.D. and Guery, J.C. (2008). Natural killer cells recruited into lymph nodes inhibit alloreactive T-cell activation through perforin-mediated killing of donor allogeneic dendritic cells. Blood **112**, 661–671.

Lanier, L.L. (1998). NK cell receptors. Annu Rev Immunol **16**, 359–393.

Lanier, L.L. (2005). NKG2D in innate and adaptive immunity. Adv Exp Med Biol **560**, 51–56.

Lanier, L.L., Corliss, B.C., Wu, J., Leong, C. and Phillips, J.H. (1998a). Immunoreceptor DAP12 bearing a tyrosine-based activation motif is involved in activating NK cells. Nature **391**, 703–707.

Lanier, L.L., Corliss, B.C., Wu, J. and Phillips, J.H. (1998b). Association of DAP12 with activating CD94/NKG2C NK cell receptors. Immunity **8**, 693–701.

Lee, P.T., Benlagha, K., Teyton, L. and Bendelac, A. (2002a). Distinct functional lineages of human V[alpha]24 natural killer T cells. J Exp Med **195**, 637–641.

Lee, P.T., Putnam, A., Benlagha, K., Teyton, L., Gottlieb, P.A. and Bendelac, A. (2002b). Testing the NKT cell hypo-thesis of human IDDM pathogenesis. J Clin Invest **110**, 793–800.

Lehuen, A., Lantz, O., Beaudoin, L., Laloux, V., Carnaud, C., Bendelac, A., Bach, J.F. and Monteiro, R.C. (1998a). Overexpression of natural killer T cells protects Valpha14-Jalpha281 transgenic nonobese diabetic mice against diabetes. J Exp Med **188**, 1831–1839.

Lehuen, A., Lantz, O., Beaudoin, L., Laloux, V., Carnaud, C., Bendelac, A., Bach, J.F. and Monteiro, R.C. (1998b). Overexpression of natural killer T cells protects Valpha14-Jalpha281 transgenic nonobese diabetic mice against diabetes. J Exp Med **188**, 1831–1839.

Leibson, P.J. (1995). MHC-recognizing receptors: they're not just for T cells anymore. Immunity **3**, 5–8.

Lian, R.H. and Kumar, V. (2002). Murine natural killer cell progenitors and their requirements for development. Semin Immunol **14**, 453–460.

Lian, R.H., Maeda, M., Lohwasser, S., Delcommenne, M., Nakano, T., Vance, R.E., Raulet, D.H. and Takei, F. (2002). Orderly and nonstochastic acquisition of CD94/NKG2 receptors by developing NK cells derived from embryonic stem cells in vitro. J Immunol **168**, 4980–4987.

Liu, Y., Goff, R.D., Zhou, D., Mattner, J., Sullivan, B.A., Khurana, A., Cantu, C., Ravkov, E.V., Ibegbu, C.C., Altman, J.D., Teyton, L., Bendelac, A. and Savage, P.B. (2006). A modified [alpha]-galactosyl ceramide for staining and stimulating natural killer T cells. J Immunol Methods **312**, 34–39.

Long, E.O., Burshtyn, D.N., Clark, W.P., Peruzzi, M., Rajagopalan, S., Rojo, S., Wagtmann, N. and Winter, C.C. (1997). Killer cell inhibitory receptors: diversity, specificity, and function. Immunol Rev **155**, 135–144.

Long, E.O., Colonna, M. and Lanier, L.L. (1996). Inhibitory MHC class I receptors on NK and T cells: a standard nomenclature. Immunol Today **17**, 100.

Lucas, M., Schachterle, W., Oberle, K., Aichele, P. and Diefenbach, A. (2007). Dendritic cells prime natural killer cells by trans-presenting interleukin 15. Immunity **26**, 503–517.

Maeda, M., Lohwasser, S., Yamamura, T. and Takei, F. (2001). Regulation of NKT cells by Ly49: analysis of primary NKT cells and generation of NKT cell line. J Immunol **167**, 4180–4186.

Makarenkova, V., Chakrabarti, A.K., Liberatore, J.A., Popovic, P., Lu, G., Watkins, S. and Vujanovic, N.L. (2005). Dendritic cells and natural killer cells interact via multiple TNF family molecules. J Leukoc Biol **77**, 408–413.

Mariani, E., Pulsatelli, L., Meneghetti, A., Dolzani, P., Mazzetti, I., Neri, S., Ravaglia, G., Forti, P. and Facchini, A. (2001). Different IL-8 production by T and NK lymphocytes in elderly subjects. Mech Ageing Dev **122**, 1383–1395.

Maroof, A., Beattie, L., Zubairi, S., Svensson, M., Stager, S. and Kaye, P.M. (2008). Posttranscriptional regulation of il10 gene expression allows natural killer cells to express immunoregulatory function. Immunity **29**, 295–305.

Martin-Fontecha, A., Thomsen, L.L., Brett, S., Gerard, C., Lipp, M., Lanzavecchia, A. and Sallusto, F. (2004). Induced recruitment of NK cells to lymph nodes provides IFN-gamma for T(H)1 priming. Nat Immunol **5**, 1260–1265.

Mason, L.H., Ortaldo, J.R., Young, H.A., Kumar, V., Bennett, M. and Anderson, S.K. (1995). Cloning and functional characteristics of murine large granular lymphocyte-1: a member of the Ly-49 gene family (Ly-49G2). J Exp Med **182**, 293–303.

Mason, L.H., Anderson, S.K., Yokoyama, W.M., Smith, H.R., Winkler-Pickett, R. and Ortaldo, J.R. (1996). The Ly-49D receptor activates murine natural killer cells. J Exp Med **184**, 2119–2128.

Mason, L.H., Gosselin, P., Anderson, S.K., Fogler, W.E., Ortaldo, J.R. and McVicar, D.W. (1997). Differential tyrosine phosphorylation of inhibitory versus activating Ly-49 receptor proteins and their recruitment of SHP-1 phosphatase. J Immunol **159**, 4187–4196.

Mason, L.H., Willette-Brown, J., Anderson, S.K., Gosselin, P., Shores, E.W., Love, P.E., Ortaldo, J.R. and McVicar, D.W. (1998). Cutting edge: characterization of an associated 16-kDa tyrosine phosphoprotein required for Ly-49D signal transduction. J Immunol **160**, 4148–4152.

Mason, L.H., Willette-Brown, J., Mason, A.T., McVicar, D. and Ortaldo, J.R. (2000). Interaction of Ly-49D+ NK cells with H-2Dd target cells leads to Dap-12 phosphorylation and IFN-gamma secretion. J Immunol **164**, 603–611.

Matsuda, J.L., Naidenko, O.V., Gapin, L., Nakayama, T., Taniguchi, M., Wang, C.R., Koezuka, Y. and Kronenberg, M. (2000). Tracking the response of natural killer T cells to a glycolipid antigen using CD1d tetramers. J Exp Med **192**, 741–754.

Matsuda, J.L., Mallevaey, T., Scott-Browne, J. and Gapin, L. (2008). CD1d-restricted iNKT cells, the 'Swiss-Army knife' of the immune system. Curr Opin Immunol **20**(3), 358–368.

Matsumoto, Y., Kohyama, K., Aikawa, Y., Shin, T., Kawazoe, Y., Suzuki, Y. and Tanuma, N. (1998). Role of natural killer cells and TCR gamma delta T cells in acute autoimmune encephalomyelitis. Eur J Immunol **28**, 1681–1688.

Matsumoto, G., Omi, Y., Lee, U., Nishimura, T., Shindo, J. and Penninger, J.M. (2000). Adhesion mediated by LFA-1 is required for efficient IL-12-induced NK and NKT cell cytotoxicity. Eur J Immunol **30**, 3723–3731.

Matsumoto, G., Kubota, E., Omi, Y., Lee, U. and Penninger, J.M. (2004). Essential role of LFA-1 in activating Th2-like responses by alpha-galactosylceramide-activated NKT cells. J Immunol **173**, 4976–4984.

McNab, F.W., Berzins, S.P., Pellicci, D.G., Kyparissoudis, K., Field, K., Smyth, M.J. and Godfrey, D.I. (2005). The influence of CD1d in postselection NKT cell maturation and homeostasis. J Immunol **175**, 3762–3768.

McNab, F.W., Pellicci, D.G., Field, K., Besra, G., Smyth, M.J., Godfrey, D.I. and Berzins, S.P. (2007). Peripheral NK1.1 NKT cells are mature and functionally distinct from their thymic counterparts. J Immunol **179**, 6630–6637.

McNerney, M.E., Lee, K.M., Zhou, P., Molinero, L., Mashayekhi, M., Guzior, D., Sattar, H., Kuppireddi, S., Wang, C.R., Kumar, V. and Alegre, M.L. (2006). Role of natural killer cell subsets in cardiac allograft rejection. Am J Transplant **6**, 505–513.

Mendel, I., Kerlero, D.R. and Ben Nun, A. (1995). A myelin oligodendrocyte glycoprotein peptide induces typical chronic experimental autoimmune encephalomyelitis in H-2b mice: fine specificity and T cell receptor V beta expression of encephalitogenic T cells. Eur J Immunol **25**, 1951–1959.

Merrill, J.E. and Benveniste, E.N. (1996). Cytokines in inflammatory brain lesions: helpful and harmful. Trends Neurosci **19**, 331–338.

Michael, A., Hackett, J.J., Bennett, M., Kumar, V. and Yuan, D. (1989). Regulation of B lymphocytes by natural killer cells. Role of IFN-gamma. J Immunol **42**(4), 1095–1101.

Michel, M.L., Keller, A.C., Paget, C., Fujio, M., Trottein, F., Savage, P.B., Wong, C.H., Schneider, E., Dy, M. and Leite-de-Moraes, M.C. (2007). Identification of an IL-17-producing NK1.1(neg) iNKT cell population involved in airway neutrophilia. J Exp Med **204**, 995–1001.

Mor, F., Quintana, F., Mimran, A. and Cohen, I.R. (2003). Autoimmune encephalomyelitis and uveitis induced by T cell immunity to self beta-synuclein. J Immunol **170**, 628–634.

Moretta, A., Biassoni, R., Bottino, C., Pende, D., Vitale, M., Poggi, A., Mingari, M.C. and Moretta, L. (1997). Major histocompatibility complex class I specific receptors on human natural killer and T lymphocytes. Immunol Rev **155**, 105–117.

Moretta, A., Marcenaro, E., Parolini, S., Ferlazzo, G. and Moretta, L. (2007). NK cells at the interface between innate and adaptive immunity. Cell Death Differ **15**(2), 226–233.

Mortier, E., Woo, T., Advincula, R., Gozalo, S. and Ma, A. (2008). IL-15Ralpha chaperones IL-15 to stable dendritic cell membrane complexes that activate NK cells via trans presentation. J Exp Med **205**, 1213–1225.

Mountz, J.D., Gause, W.C. and Jonsson, R. (1991). Murine models for systemic lupus erythematosus and Sjogren's syndrome. Curr Opin Rheumatol **3**, 738–756.

Muhammad Ali Tahir, S., Cheng, O., Shaulov, A., Koezuka, Y., Bubley, G.J., Wilson, S.B., Balk, S.P. and Exley, M.A. (2001). Loss of IFN-{gamma} production by invariant NK T cells in advanced cancer. J Immunol **167**, 4046–4050.

Murphy, W.J., Raziuddin, A., Mason, L., Kumar, V., Bennett, M. and Longo, D.L. (1995). NK cell subsets in the regulation of murine hematopoiesis. I. 5E6+ NK cells promote hematopoietic growth in H-2d strain mice. J Immunol **155**, 2911–2917.

Nakagawa, R., Nagafune, I., Tazunoki, Y., Ehara, H., Tomura, H., Iijima, R., Motoki, K., Kamishohara, M. and Seki, S. (2001). Mechanisms of the antimetastatic effect in the liver and of the hepatocyte injury induced by alpha-galactosylceramide in mice. J Immunol **166**, 6578–6584.

Nakai, Y., Iwabuchi, K., Fujii, S., Ishimori, N., Dashtsoodol, N., Watano, K., Mishima, T., Iwabuchi, C., Tanaka, S., Bezbradica, J.S., Nakayama, T., Taniguchi, M., Miyake, S., Yamamura, T., Kitabatake, A., Joyce, S., Van Kaer, L. and Onoe, K. (2004). Natural killer T cells accelerate atherogenesis in mice. Blood **104**, 2051–2059.

Nakamura, M.C., Niemi, E.C., Fisher, M.J., Shultz, L.D., Seaman, W.E. and Ryan, J.C. (1997). Mouse Ly-49A interrupts early signaling events in natural killer cell cytotoxicity and functionally associates with the SHP-1 tyrosine phosphatase. J Exp Med **185**, 673–684.

Natarajan, K., Boyd, L.F., Schuck, P., Yokoyama, W.M., Eliat, D. and Margulies, D.H. (1999). Interaction of the NK cell inhibitory receptor Ly49A with H-2Dd: identification of a site distinct from the TCR site. Immunity **11**, 591–601.

Nieda, M., Nicol, A., Koezuka, Y., Kikuchi, A., Lapteva, N., Tanaka, Y., Tokunaga, K., Suzuki, K., Kayagaki, N., Yagita, H., Hirai, H. and Juji, T. (2001). TRAIL expression by activated human CD4(+)V alpha 24NKT cells induces in vitro and in vivo apoptosis of human acute myeloid leukemia cells. Blood **97**, 2067–2074.

Oikawa, Y., Shimada, A., Yamada, S., Motohashi, Y., Nakagawa, Y., Irie, J., Maruyama, T. and Saruta, T. (2002). High frequency of valpha24(+) vbeta11(+) T-cells observed in type 1 diabetes. Diabetes Care **25**, 1818–1823.

Olcese, L., Cambiaggi, A., Semenzato, G., Bottino, C., Moretta, A. and Vivier, E. (1997). Human killer cell activatory receptors for MHC class I molecules are included in a multimeric complex expressed by natural killer cells. J Immunol **158**, 5083–5086.

Ortaldo, J. and McVicar, D.W. (1999). Murine NK receptors: Ly-49 expression, function and intracellular signaling. In: Sitkovsky, M.V. and Henkart, P.A. (eds) Cytotoxic Cells: Basic Mechanisms and Medical Applications: Lippincott, Williams and Wilkins, pp. 45–63.

Ortaldo, J.R. and Herberman, R.B. (1984). Heterogeneity of natural killer cells. Annu Rev Immunol **2**, 359–394.

Ortaldo, J.R., Bere, E.W., Hodge, D. and Young, H.A. (2001). Activating Ly-49 NK receptors: central role in cytokine and chemokine production. J Immunol **166**, 4994–4999.

Ortaldo J.R., Young, H.A., Winkler-Pickett, R.T., Bere, E.W., Jr., Murphy, W.J. and Wiltrout, R.H. (2004). Dissociation of NKT stimulation, cytokine induction, and NK activation in vivo by the use of distinct TCR-binding ceramides. J Immunol **172**, 943–953.

Pan, P.Y., Gu, P., Li, Q., Xu, D., Weber, K. and Chen, S.H. (2004). Regulation of dendritic cell function by NK cells: mechanisms underlying the synergism in the combination therapy of IL-12 and 4-1BB activation. J Immunol **172**, 4779–4789.

Pellicci, D.G., Hammond, K.J., Uldrich, A.P., Baxter, A.G., Smyth, M.J. and Godfrey, D.I. (2002). A natural killer T (NKT) cell developmental pathway involving a thymus-dependent NK1.1(−)CD4(+) CD1d-dependent precursor stage. J Exp Med **195**, 835–844.

Pichavant, M., Goya, S., Meyer, E.H., Johnston, R.A., Kim, H.Y., Matangkasombut, P., Zhu, M., Iwakura, Y., Savage, P.B., DeKruyff, R.H., Shore, S.A. and Umetsu, D.T. (2008). Ozone exposure in a mouse model induces airway hyperreactivity that requires the presence of natural killer T cells and IL-17. J Exp Med **205**, 385–393.

Pillarisetty, V.G., Katz, S.C., Bleier, J.I., Shah, A.B. and DeMatteo, R.P. (2005). Natural killer dendritic cells have both antigen presenting and lytic function and in response to CpG produce IFN-gamma via autocrine IL-12. J Immunol **174**, 2612–2618.

Playfair, J.H. and Marshall-Clarke, S. (1973). Induction of red cell autoantibodies in normal mice. Nat New Biol **243**, 213–214.

Plougastel, B.F. and Yokoyama, W.M. (2006). Extending missing-self? Functional interactions between lectin-like NKrp1 receptors on NK cells with lectin-like ligands. Curr Top Microbiol Immunol **298**, 77–89.

Plougastel, B., Matsumoto, K., Dubbelde, C. and Yokoyama, W.M. (2001). Analysis of a 1-Mb BAC contig overlapping the mouse Nkrp1 cluster of genes: cloning of three new Nkrp1 members, Nkrp1d, Nkrp1e, and Nkrp1f. Immunogenetics **53**, 592–598.

Poirot, L., Benoist, C. and Mathis, D. (2004). Natural killer cells distinguish innocuous and destructive forms of pancreatic islet autoimmunity. Proc Natl Acad Sci U S A **101**, 8102–8107.

Rachitskaya, A.V., Hansen, A.M., Horai, R., Li, Z., Villasmil, R., Luger, D., Nussenblatt, R.B. and Caspi, R.R. (2008). Cutting edge: NKT cells constitutively express IL-23 receptor and RORgammat and rapidly produce IL-17 upon receptor ligation in an IL-6-independent fashion. J Immunol **180**, 5167–5171.

Radford-Smith, G. (1997). Ulcerative colitis: an immunological disease. Baillieres Clin Gastroenterol **11**, 35–52.

Raulet, D.H. (2003). Roles of the NKG2D immunoreceptor and its ligands. Nat Rev Immunol **3**, 781–790.

Raulet, D.H., Correa, I., Corral, L., Dorfman, J. and Wu, M.F. (1995). Inhibitory effects of class I molecules on murine NK cells: speculations on function, specificity and self-tolerance. Semin Immunol **7**, 103–107.

Robbins, S.H., Bessou, G., Cornillon, A., Zucchini, N., Rupp, B., Ruzsics, Z., Sacher, T., Tomasello, E., Vivier, E., Koszinowski, U.H. and Dalod, M. (2007). Natural killer cells promote early CD8 T cell responses against cytomegalovirus. PLoS Pathog **3**, e123.

Robey, E. and Fowlkes, B.J. (1994). Selective events in T cell development. Annu Rev Immunol **12**, 675–705.

Roda, J.M., Parihar, R., Magro, C., Nuovo, G.J., Tridandapani, S. and Carson, W.E. (2006). Natural killer cells produce T cell-recruiting chemokines in response to antibody-coated tumor cells. Cancer Res **66**, 517–526.

Romagnani, S. (1992). Induction of TH1 and TH2 responses: a key role for the 'natural' immune response. Immunol Today **13**, 379–381.

Ronet, C., Darche, S., de Moraes, M.L., Miyake, S., Yamamura, T., Louis, J.A., Kasper, L.H. and Buzoni-Gatel, D. (2005). NKT cells are critical for the initiation of an inflammatory bowel response against Toxoplasma gondii. J Immunol **175**, 899–908.

Ryan, J.C. and Seaman, W.E. (1997). Divergent functions of lectin-like receptors on NK cells. Immunol Rev **155**, 79–89.

Sakaguchi, S. (2004). Naturally arising CD4+ regulatory t cells for immunologic self-tolerance and negative control of immune responses. Annu Rev Immunol **22**, 531–562.

Sallusto, F. and Baggiolini, M. (2008). Chemokines and leukocyte traffic. Nat Immunol **9**, 949–952.

Samuels, J., Ng, Y.S., Coupillaud, C., Paget, D. and Meffre, E. (2005). Impaired early B cell tolerance in patients with rheumatoid arthritis. J Exp Med **201**, 1659–1667.

Sawaki, J., Tsutsui, H., Hayashi, N., Yasuda, K., Akira, S., Tanizawa, T. and Nakanishi, K. (2007). Type 1 cytokine/chemokine production by mouse NK cells following activation of their TLR/MyD88-mediated pathways. Int Immunol **19**, 311–320.

Scala, G., Allavena, P., Djeu, J.Y., Kasahara, T., Ortaldo, J.R., Herberman, R.B. and Oppenheim, J.J. (1984). Human large granular lymphocytes are potent producers of interleukin-1. Nature **309**, 56–59.

Scala, G., Allavena, P., Ortaldo, J.R., Herberman, R.B. and Oppenheim, J.J. (1985). Subsets of human large granular lymphocytes (LGL) exhibit accessory cell functions. J Immunol **134**, 3049–3055.

Scalzo, A.A., Fitzgerald, N.A., Wallace, C.R., Gibbons, A.E., Smart, Y.C., Burton, R.C. and Shellam, G.R. (1992). The effect of the Cmv-1 resistance gene, which is linked to the natural killer cell gene complex, is mediated by natural killer cells. J Immunol **149**, 581–589.

Schmidt-Supprian, M., Tian, J., Grant, E.P., Pasparakis, M., Maehr, R., Ovaa, H., Ploegh, H.L., Coyle, A.J. and Rajewsky, K. (2004). Differential dependence of CD4 + CD25+ regulatory and natural killer-like T cells on signals leading to NF-kappaB activation. Proc Natl Acad Sci U S A **101**, 4566–4571.

Seino, K.I., Fukao, K., Muramoto, K., Yanagisawa, K., Takada, Y., Kakuta, S., Iwakura, Y., Van, K.L., Takeda, K., Nakayama, T., Taniguchi, M., Bashuda, H., Yagita, H. and Okumura, K. (2001). Requirement for natural killer T (NKT) cells in the induction of allograft tolerance. Proc Natl Acad Sci U S A **98**, 2577–2581.

Shi, F.D. and Van, K.L. (2006). Reciprocal regulation between natural killer cells and autoreactive T cells. Nat Rev Immunol **6**, 751–760.

Shi, F.D., Takeda, K., Akira, S., Sarvetnick, N. and Ljunggren, H.G. (2000a). IL-18 directs autoreactive T cells and promotes autodestruction in the central nervous system via induction of IFN-gamma by NK cells. J Immunol **165**, 3099–3104.

Shi, F.D., Wang, H.B., Li, H., Hong, S., Taniguchi, M., Link, H., Van Kaer, L. and Ljunggren, H.G. (2000b). Natural killer cells determine the outcome of B cell-mediated autoimmunity. Nat Immunol **1**, 245–251.

Shi, F., Ljunggren, H.G. and Sarvetnick, N. (2001). Innate immunity and autoimmunity: from self-protection to self-destruction. Trends Immunol **22**, 97–101.

Sivakumar, V., Hammond, K.J., Howells, N., Pfeffer, K. and Weih, F. (2003). Differential requirement for Rel/nuclear factor kappa B family members in natural killer T cell development. J Exp Med **197**, 1613–1621.

Smiley, S.T., Kaplan, M.H. and Grusby, M.J. (1997). Immunoglobulin E production in the absence of interleukin-4-secreting CD1-dependent cells. Science **275**, 977–979.

Smith, K.A., Wu, J., Bakker, A.B.H., Phillips, J.H. and Lanier, L.L. (1998). Ly49D and Ly49H associate with mouse DAP12 and form activating receptors. J Immunol **161**, 7–10.

Smith, H.R., Chuang, H.H., Wang, L.L., Salcedo, M., Heusel, J.W. and Yokoyama, W.M. (2000). Nonstochastic coexpression of activation receptors on murine natural killer cells. J Exp Med **191**, 1341–1354.

Smith, H.R., Heusel, J.W., Mehta, I.K., Kim, S., Dorner, B.G., Naidenko, O.V., Iizuka, K., Furukawa, H., Beckman, D.L., Pingel, J.T., Scalzo, A.A., Fremont, D.H. and Yokoyama, W.M. (2002). Recognition of a virus-encoded ligand by a natural killer cell activation receptor. Proc Natl Acad Sci U S A **99**, 8826–8831.

Smyth, M.J., Thia, K.Y., Street, S.E., Cretney, E., Trapani, J.A., Taniguchi, M., Kawano, T., Pelikan, S.B., Crowe, N.Y. and Godfrey, D.I. (2000). Differential tumor surveillance by natural killer (NK) and NKT cells. J Exp Med **191**, 661–668.

Smyth, M.J., Hayakawa, Y., Takeda, K. and Yagita, H. (2002). New aspects of natural-killer-cell surveillance and therapy of cancer. Nat Rev Cancer **2**, 850–861.

Sonoda, K.H. and Stein-Streilein, J. (2002). CD1d on antigen-transporting APC and splenic marginal zone B cells promotes NKT cell-dependent tolerance. Eur J Immunol **32**, 848–857.

Sonoda, K.H., Exley, M., Snapper, S., Balk, S.P. and Stein-Streilein, J. (1999). CD1-reactive natural killer T cells are required for development of systemic tolerance through an immune-privileged site. J Exp Med **190**, 1215–1226.

Sonoda, K.H., Faunce, D.E., Taniguchi, M., Exley, M., Balk, S. and Stein-Streilein, J. (2001). NK T cell-derived IL-10 is essential for the differentiation of antigen-specific T regulatory cells in systemic tolerance. J Immunol **166**, 42–50.

Stetson, D.B., Mohrs, M., Reinhardt, R.L., Baron, J.L., Wang, Z.E., Gapin, L., Kronenberg, M. and Locksley, R.M. (2003). Constitutive cytokine mRNAs mark natural killer (NK) and NK T cells poised for rapid effector function. J Exp Med **198**, 1069–1076.

Stoneman, E.R., Bennett, M., An, J., Chesnut, K.A., Wakeland, E.K., Scheerer, J.B., Siciliano, M.J., Kumar, V. and Mathew, P.A. (1995). Cloning and characterization of 5E6(Ly-49C), a receptor molecule expressed on a subset of murine natural killer cells. J Exp Med **182**, 305–313.

Subleski, J.J., Hall, V.L., Back, T.C., Ortaldo, J.R. and Wiltrout, R.H. (2006). Enhanced antitumor response by divergent modulation of natural killer and natural killer T cells in the liver. Cancer Res **66**, 11005–11012.

Tabira, T. and Kira, J. (1992). Strain and species differences of encephalitogenic determinants of myelin basic protein and proteolipid apoprotein. In: Martenson, R.E. (ed), Myelin: Biology and Chemistry. Boca Raton, FL: CRC Press Inc, pp. 783–799.

Taniguchi, M., Harada, M., Kojo, S., Nakayama, T. and Wakao, H. (2003a). The regulatory role of Valpha14 NKT cells in innate and acquired immune response. Annu Rev Immunol **21**, 483–513.

Taniguchi, M., Seino, K. and Nakayama, T. (2003b). The NKT cell system: bridging innate and acquired immunity. Nat Immunol **4**, 1164–1165.

Tarazona, R., DelaRosa, O., Alonso, C., Ostos, B., Espejo, J., Pena, J. and Solana, R. (2000). Increased expression of NK cell markers on T lymphocytes in aging and chronic activation of the immune system reflects the accumulation of effector/senescent T cells. Mech Ageing Dev **121**, 77–88.

Tay, C.H., Szomolanyi-Tsuda, E. and Welsh, R.M. (1998). Control of infections by NK cells. Curr Top Microbiol Immunol **230**, 193–220.

Tay, C.H., Yu, L.Y., Kumar, V., Mason, L., Ortaldo, J.R. and Welsh, R.M. (1999). The role of LY49 NK cell subsets in the regulation of murine cytomegalovirus infections. J Immunol **162**, 718–726.

Terabe, M., Matsui, S., Noben-Trauth, N., Chen, H., Watson, C., Donaldson, D.D., Carbone, D.P., Paul, W.E. and Berzofsky, J.A. (2000). NKT cell-mediated repression of tumor immunosurveillance by IL-13 and the IL-4R-STAT6 pathway. Nat Immunol **1**, 515–520.

Terabe, M., Swann, J., Ambrosino, E., Sinha, P., Takaku, S., Hayakawa, Y., Godfrey, D.I., Ostrand-Rosenberg, S., Smyth, M.J. and Berzofsky, J.A. (2005). A nonclassical non-Valpha14Jalpha18 CD1d-restricted (type II) NKT cell is sufficient for down-regulation of tumor immunosurveillance. J Exp Med **202**, 1627–1633.

Thomas, S.Y., Hou, R., Boyson, J.E., Means, T.K., Hess, C., Olson, D.P., Strominger, J.L., Brenner, M.B., Gumperz, J.E., Wilson, S.B. and Luster, A.D. (2003). CD1d-restricted NKT cells express a chemokine receptor profile indicative of Th1-type inflammatory homing cells. J Immunol **171**, 2571–2580.

Tomura, M., Yu, W.G. and Ahn, H.J. (1999). A novel function of V[alpha]14 + CD4+ NKT cells: stimulation of IL-12 production by antigen-presenting cells in the innate immune system. J Immunol **163**, 93–101.

Trinchieri, G. (1989). Biology of natural killer cells. Adv Immunol **47**, 187–376.

Trobonjaca, Z., Leithauser, F., Moller, P., Schirmbeck, R. and Reimann, J. (2001). Activating immunity in the liver. I. Liver dendritic cells (but not hepatocytes) are potent activators of IFN-{gamma} release by liver NKT cells. J Immunol **167**, 1413–1422.

Tupin, E., Nicoletti, A., Elhage, R., Rudling, M., Ljunggren, H.G., Hansson, G.K. and Berne, G.P. (2004). CD1d-dependent activation of NKT cells aggravates atherosclerosis. J Exp Med **199**, 417–422.

Van Dommelen, S.L. and gli-Esposti, M.A. (2004). NKT cells and viral immunity. Immunol Cell Biol **82**, 332–341.

Vosshenrich, C.A., Lesjean-Pottier, S., Hasan, M., Richard-Le, G.O., Corcuff, E., Mandelboim, O. and Di Santo, J.P. (2007). CD11cloB220+ interferon-producing killer dendritic cells are activated natural killer cells. J Exp Med **204**, 2569–2578.

Voyle, R.B., Beermann, F., Lees, R.K., Schumann, J., Zimmer, J., Held, W. and MacDonald, H.R. (2003). Ligand-dependent inhibition of CD1d-restricted

NKT cell development in mice transgenic for the activating receptor Ly49D. J Exp Med **197**, 919–925.

Walzer, T., Dalod, M., Vivier, E. and Zitvogel, L. (2005). Natural killer cell–dendritic cell crosstalk in the initiation of immune responses. Expert Opin Biol Ther **5**(Suppl. 1), S49–S59.

Watford, W.T., Moriguchi, M., Morinobu, A. and O'Shea, J.J. (2003). The biology of IL-12: coordinating innate and adaptive immune responses. Cytokine Growth Factor Rev **14**, 361–368.

Watte, C.M., Nakamura, T., Lau, C.H., Ortaldo, J.R. and Stein-Streilein, J. (2008). Ly49 C/I-dependent NKT cell-derived IL-10 is required for corneal graft survival and peripheral tolerance. J Leukoc Biol **83**, 928–935.

Wei, D.G., Lee, H., Park, S.H., Beaudoin, L., Teyton, L., Lehuen, A. and Bendelac, A. (2005). Expansion and long-range differentiation of the NKT cell lineage in mice expressing CD1d exclusively on cortical thymocytes. J Exp Med **202**, 239–248.

Weiss, J.M., Subleski, J.J., Wigginton, J.M. and Wiltrout, R.H. (2007). Immunotherapy of cancer by IL-12-based cytokine combinations. Expert Opin Biol Ther **7**, 1705–1721.

Wilbanks, A., Zondlo, S.C., Murphy, K., Mak, S., Soler, D., Langdon, P., Andrew, D.P., Wu, L. and Briskin, M. (2001). Expression cloning of the STRL33/BONZO/TYMSTR ligand reveals elements of CC, CXC, and CX3C chemokines. J Immunol **166**, 5145–5154.

Wilson, S.B., Kent, S.C., Patton, K.T., Orban, T., Jackson, R.A., Exley, M., Porcelli, S., Schatz, D.A., Atkinson, M.A., Balk, S.P., Strominger, J.L. and Hafler, D.A. (1998). Extreme Th1 bias of invariant Valpha24JalphaQ T cells in type 1 diabetes. Nature **391**, 177–181.

Winkler-Pickett, R., Young, H.A., Cherry, J.M., Diehl, J., Wine, J., Back, T., Bere, W.E., Mason, A. T. and Ortaldo, J.R. (2008). In vivo regulation of experimental autoimmune encephalomyelitis by NK cells: alteration of primary adaptive responses. J Immunol **180**, 4495–4506.

Wu, J. and Lanier, L.L. (2003). Natural killer cells and cancer. Adv Cancer Res **90**, 127–156.

Wu, D., Xing, G.W., Poles, M.A., Horowitz, A., Kinjo, Y., Sullivan, B., Bodmer-Narkevitch, V., Plettenburg, O., Kronenberg, M., Tsuji, M., Ho, D.D. and Wong, C.H. (2005). Bacterial glycolipids and analogs as antigens for CD1d-restricted NKT cells. Proc Natl Acad Sci U S A **102**, 1351–1356.

Yang, L., Jhaveri, R., Huang, J., Qi, Y. and Diehl, A.M. (2007). Endoplasmic reticulum stress, hepatocyte CD1d and NKT cell abnormalities in murine fatty livers. Lab Invest **87**, 927–937.

Yokoyama, W.M. (1995). Natural killer cell receptors. Curr Opin Immunol **7**, 110–120.

Yu, Y.Y., George, T., Dorfman, J.R., Roland, J., Kumar, V. and Bennett, M. (1996). The role of Ly49A and 5E6(Ly49C) molecules in hybrid resistance mediated by murine natural killer cells against normal T cell blasts. Immunity **4**, 67–76.

Yu, G., Xu, X., Vu, M.D., Kilpatrick, E.D. and Li, X.C. (2006). NK cells promote transplant tolerance by killing donor antigen-presenting cells. J Exp Med **203**, 1851–1858.

Yuan, D., Koh, C.Y. and Wilder, J.A. (1994). Interactions between B lymphocytes and NK cells. FASEB J **8**, 1012–1018.

Zeytin, H., Reali, E., Zaharoff, D.A., Rogers, C.J., Schlom, J. and Greiner, J. W. (2008). Targeted delivery of murine IFN-gamma using a recombinant fowlpox virus: NK cell recruitment to regional lymph nodes and priming of tumor-specific host immunity. J Interferon Cytokine Res **28**, 73–87.

Zhang, B., Yamamura, T., Kondo, T., Fujiwara, M. and Tabira, T. (1997). Regulation of experimental autoimmune encephalomyelitis by natural killer (NK) cells. J Exp Med **186**, 1677–1687.

Zhou III, D., Mattner, J., Cantu, C., Schrantz, N., Yin, N., Gao, Y., Sagiv, Y., Hudspeth, K., Wu, Y.P., Yamashita, T., Teneberg, S., Wang, D., Proia, R.L., Levery, S.B., Savage, P.B., Teyton, L. and Bendelac, A. (2004). Lysosomal glycosphingolipid recognition by NKT cells. Science **306**, 1786–1789.

Chapter Twenty

20

NK cell and endothelial cell interactions

Angela Gismondi, Claudia Carlino, Helena Stabile, Angela Santoni

CHAPTER CONTENTS

Preme ergo quod coepisti, et fortasse perduceris aut ad summum aut eo quod summum nondum esse solus intellegas.
Press on, therefore, as you have begun; perhaps you will be led to perfection, or to a point which you alone understand is still short of perfection.

Lucio Anneo Seneca EpII, 20.6

ABSTRACT

Endothelia represent a barrier that natural killer (NK) cells have to cross to localize within the parenchyma of individual tissues during immune surveillance, inflammatory responses, tumour growth or wound healing. This event is regulated by many adhesive proteins and chemotactic factors present on the endothelium as well as by their corresponding counter-receptors present on the NK cell. The differential expression of these molecular components and their regulation by individual stimuli contribute to the recruitment of specialized NK cell effectors in both physiological and pathological conditions. As a consequence of NK cell–endothelial cell interactions, modulation of endothelial cell functions can also occur, and a role for NK cells in promoting angiogenesis or vascular remodelling as well as in causing endothelial injury or death has been documented in many physiopathological conditions.

KEY WORDS

Endothelial cells, Adhesion molecules (VCAM, ICAM, MadCAM, L-selectin, E-selectin, P-selectin, PSGL-1, sLex), Integrins, NK cell–endothelial cell interaction, Pro-angiogenesis, Anti-angiogenesis, Endothelial cell damage

NK cell tissue distribution

Natural killer (NK) cells are a small subset of lymphocytes that represent critical effector cells of innate immunity as they are endowed with the ability to mediate cytotoxic activity and to secrete a variety of cytokines and chemokines. NK cells' prompt response, without the need for prior activation, represents a first line of defence against microbial infections, early cellular transformation and tumour growth as well as bone marrow transplantation. Moreover, accumulating evidences indicate that NK cells, by interacting with many cellular components of the immune system, play a crucial role also in the instruction and modulation of adaptive immune responses (Christopher et al., 2008; Lanier, 2005; Trinchieri, 1989).

NK cells develop from a lymphoid precursor resident in the bone marrow that is considered the main site of their generation, even if final maturation of NK cell precursors can occur also in the periphery (Di Santo and Vosshenrich, 2006). A thymic subset of NK cells bearing molecular markers and functional capabilities distinct from most peripheral NK cells and exported from the thymus into the secondary lymphoid organs has been recently described, thus suggesting the existence of a pathway of NK cell development within the thymus (Di Santo, 2006). Mature NK cells predominantly circulate in the peripheral blood but are also resident in several lymphoid and non-lymphoid organs, such as the spleen, tonsils, lymph nodes, liver, lungs and intestine (Gismondi and Santoni, 2006). They are the most abundant class of lymphocytes found within the mucosal tissues of the uterus (King et al., 1991, 1996–1997; Whitelaw and Croy, 1996). In these organs, NK cells have been found in close proximity of vasculature, and in most instances, in areas distinct from those of T cells or B cells. In the spleen, NK cells are primarily found in the red pulp and only occasionally in the white pulp. The NK cell co-localization with endothelial cells in the red pulp has suggested that splenic NK cells are mainly distributed within blood sinuses or vessels (Andrews et al., 2001; Dokun et al., 2001; Grégoire et al., 2007; Salazar-Mather et al., 1996). Using immunohistochemical analysis and dynamic intravital imaging, it has been shown that inside the lymph nodes, NK cells reside in the medulla, in the perifollicular regions as well as in the paracortex where they co-localize with dendritic cells (DC). In a steady state condition, the density of lymph node NK cells is higher in the medulla than in T-cell or B-cell areas, and the NK cells are mainly located within lymphatic sinuses (Bajénoff et al., 2006; Walzer et al., 2007).

Recently, Cella and co-workers have described in the mucosa-associated lymphoid tissues (MALT) such as the tonsils and the Peyer's Patches of the ileum and appendix, a human NK cell subset that is predominantly located in the mucosa surrounding the lymphoid follicles and only barely detected in the interfollicular areas. This NK cell subset expresses the NKp44 receptor and is referred to as NK-22 cells for its ability to release IL-22, IL-26 and leukaemia inhibitory factor (LIF) particularly following acute activation with IL-23. In the mouse, NK-22 cells can be found in the lamina propria of the small intestine during bacterial infection thus suggesting that NK cells can play a role in the protection of mucosa by producing IL-22 (Cella et al., 2008; Ferlazzo and Münz, 2004; Münz, 2008).

Among the non-lymphoid organs, the liver and uterus are selectively enriched in NK cells. Liver NK cells have been originally defined as PIT cells based on their spherical dense granules in the cytoplasm and rod-cored vesicles. Mouse hepatic lymphocytes contain about 10% NK cells, whereas rat and human hepatic NK cells represent about 30–50% of the lymphocyte population. Liver NK cells are mainly situated in the lumen of the intrahepatic sinusoids in close contact with Kupffer and endothelial cells (Doherty and O'Farrelly, 2000; Kaneda and Wake, 1983).

In the uterus, the NK cell number increases drastically in the late secretory phase of the menstrual cycle and early pregnancy when their frequency reaches 60–70% (King et al., 1996–1997; Whitelaw and Croy, 1996). Inside the uterine compartment, NK cells accumulate as single cells or aggregates around endometrial glands and vessels playing a crucial role for the normal development of placenta and its vasculature (King, 2000).

Even if specialized NK cell subsets have been found in individual non-lymphoid organs, their relationship with the peripheral blood counterpart and their origin are still poorly defined. In addition, it is still unclear if they are recruited from the blood or from cells resident within the tissue, or if they develop from a precursor recruited from other tissues or the blood. Tissue-resident NK cells have many functions, including cytotoxicity or cytokine release, influencing the recruitment and activation of other leukocytes and also affecting vessel behaviour (Biron, 1997; Fogler et al., 1996; Trinchieri, 1989).

Adhesion molecules involved in the control of NK cell–endothelial cell interactions

Interaction between NK cells and endothelial cells is a crucial step leading to NK cell recruitment from blood vessels into extravascular sites during viral infections, inflammation or other pathophysiological conditions. Moreover, endothelium has been shown to be a special target for NK cell-mediated functions in many physiological or pathological circumstances. Similarly to other leukocytes, NK cell migration across endothelial cells is a spatially and temporally integrated multi-step process regulated by several chemo-attractants and adhesive molecules. Among the adhesion molecules involved in this process, those belonging to the selectin, integrin and immunoglobulin families are crucial (Kunkel and Butcher, 2002; Springer, 1994).

Both selectins and integrins contribute to the initial leukocyte tethering and rolling along vessel endothelium, while firm adhesion to vascular endothelium and subsequent diapedesis into the underlying extravascular tissue is mainly mediated by integrins.

The dynamic regulation of integrin adhesiveness for endothelial cells and extracellular matrix ligands is governed by complex signalling events elicited by chemokines or chemotactic factors (Baggiolini et al., 1997; Mantovani, 1999; Rossi and Zlotnik, 2000). Integrin engagement by ligands can also contribute to trigger intracellular signalling events that cooperate and integrate chemokine-induced signalling pathways leading to cell migration (Clark and Brugge, 1995; De Filippi et al., 1997).

Selectin family

Human NK cells express L-selectin (CD62L), a molecule involved in the initial adhesion of leukocytes to peripheral lymph node (pLN) high endothelial venules (HEVs) (Maenpaa et al., 1993; Uksila et al., 1997). L-selectin is uniquely expressed on the CD56^{high} subset of peripheral blood human NK cells at a density higher than that of all other peripheral blood leukocytes, including CD56^{low} NK cells. NK cell activation results in modulation of L-selectin expression depending on the stimulus: phorbol esters, IL-2, IL-15 and TGF-β down-regulate L-selectin on CD56^{high} NK cell subset, whereas increased levels can be observed on both NK cell subsets in response to IL-12, IL-10 and IFN-α. In accordance with these observations, CD56^{high} NK cells bind to the physiologic L-selectin ligands on peripheral lymph node HEVs with higher efficiency as compared to the CD56^{low} subpopulation, thus resulting in a selective advantage of this population in the extravasation across HEVs (Frey et al., 1998). There is also evidence by Uksila et al. showing that a proportion of CD16$^+$ NK cells express L-selectin, and that IL-2 treatment diminishes the expression of this molecule and concomitantly increases the levels of α4 integrin and CD44, two major receptors involved in lymphocyte binding to mucosal HEVs. Thus, IL-2 activation of NK cells results in decreased adherence to pLN HEVs while adherence to mucosal HEVs is increased (Uksila et al., 1997). In the mouse, L-selectin and L-selectin ligands generated from fucosyltransferase IV and VII are crucial for the recruitment of NK cells to lymph nodes in both resting and activated conditions. L-selectin-mediated interactions can facilitate NK cell recruitment into lymph nodes undergoing inflammation following complete Freund's adjuvant stimulation or in the presence of metastatic tumour cells (Chen et al., 2005; Martin-Fontecha et al., 2004). In contrast, L-selectin is not required for the recruitment of NK cells to the spleen, liver and lung (Chen et al., 2005; Fogler et al., 1996).

NK cells can also express selectin ligands such as the stage-specific embryonic antigen 1, sialyl-Lewis^x (sLe^x) ligand and the PSGL-1 (P-selectin glycoprotein ligand-1) that bind to E-selectin and P-selectin under static and flow conditions. Interestingly, this binding is up-regulated on NK cells by IL-12 (Moore and Thompson, 1992; Pinola et al., 1994a; Snapp et al., 1998; Yago et al., 1998). These results suggest that PSGL-1-P-selectin or PSGL-1-E-selectin pair may promote cell–cell interactions and amplify the accumulation of NK cells at sites of inflammation other than lymph nodes (Table 20.1 and Figure 20.1). Other carbohydrate modifications of PSGL-1, such as PEN-5 that confers to PSGL-1 the ability to bind to L-selectin, and CLA, which is a marker for skin infiltrating leukocytes, have also been found on the NK cell surface. Notably, expression of PEN5 and CLA on NK cells is mutually exclusive suggesting that individual NK cell subsets exhibit distinct trafficking properties (Andre et al., 2000).

Integrins

Human NK cells express various members of the β1, β2 and β7 integrin families. Among the β1 integrins, freshly isolated peripheral blood NK cells express α5β1 and α4β1 as fibronectin and VCAM-1 receptors, and α6β1 as laminin receptor (Gismondi et al., 1991). The pattern of β1 integrin expression changes upon NK cell activation, in that activated NK cells acquire α1β1 and α2β1 integrins and down-regulate the expression of α6β1 (Gismondi et al., 1992; Maenpaa et al., 1993; Mainiero et al., 1994; Perez-Villar et al., 1996a). The α4β1-VCAM-1 adhesive pathway is involved in the adhesion and migration of resting or IL-2 activated NK cells across IL-1β-, IFN-γ-, TNF-α-activated but not resting endothelial cells. The ability of NK cells to bind to resting or IL-1 activated endothelial cells was inhibited by

Table 20.1 Adhesion molecules expressed on peripheral blood NK cell subsets

	CD56^{low} CD16^{high}	CD56^{high} CD16^{low}
CD62L (L-selectin)	+[1]	++[2]
PSGL-1/PEN5	+	−[3]
PSGL-1/CLA	−	+
CD49dCD29 (α4β1)	+	+
CD49eCD29 (α5β1)	+	+
CD49fCD29 (α6β1)	+	+
CD49dβ7 (α4β7)	+/−[4]	+
CD11aCD18 (αLβ2)	++	+
CD11bCD18 (αMβ2)	+	+
CD11cCD18 (αXβ2)	+/−	+/−

[1]Indicates intermediate levels of expression.
[2]Indicates high level of expression.
[3]Indicates undetectable levels of expression.
[4]Indicates low levels of expression.

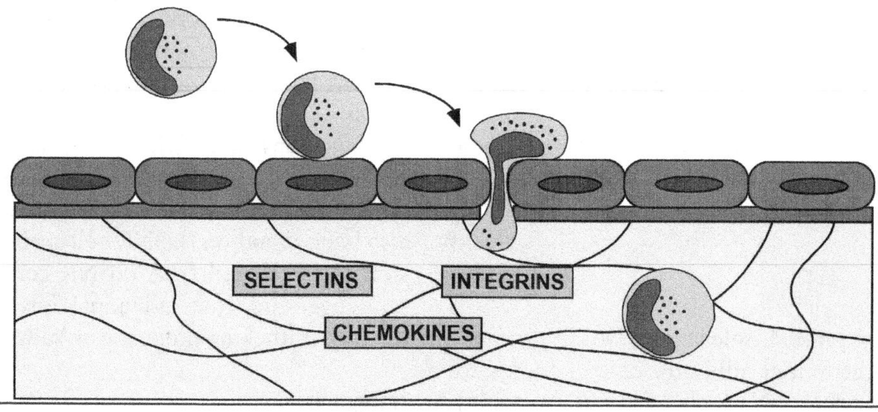

Phase	NK cell receptor	Endothelial cell ligand	Function	Reference
Rolling	L-selectin	PSGL-1	LN-HEV, uterus	Maenpaa et al. (1993); Uksila et al. (1997); Frey et al. (1998); Chen et al. (2005); Martin-Fontecha et al. (2004); Chantakru et al. (2003); Yamaguchi et al. (2006); van den Heuvel et al. (2005)
	sLex, PSGL-1	E-and P-selectin	Inflamed endothelium	Moore and Thompson (1992); Pinola et al. (1994a); Snapp et al. (1998); Yago et al. (1998)
Adhesion/ Transmigration	α4β7	VCAM-1, MadCAM-1	Mucosal HEV, uterus	Uksila et al. (1997); Pinola et al. (1994b); Perez-Villar (1996b); Fernekorn et al. (2004)
	β2 (LFA-1)	ICAM-1,-2, -3	Inflamed endothelium (liver)	Allavena et al. (1991); Luo et al. (1996); Timonen (1997)
	α4β1 (VLA-4)	VCAM-1	Inflamed endothelium (lung, liver, tumor lesions)	Allavena et al. (1991); Fogler et al. (1996) and (1998); Schneider et al. (2002)

Figure 20.1 • Adhesion molecules involved in the control of NK cell transendothelial migration. LN, lymph node; HEV, high endothelial venule.

IL-4 thus indicating that this cytokine, unlike IL-2, can prevent NK cell recruitment (Paganini et al., 1994).

Integrin-mediated NK cell interaction with endothelial cells is characterized by a peculiar structural feature: the formation of podosomes that represent highly dynamic conical protrusions of the cellular ventral membrane provided with adhesive properties and formed by particular cytoskeletal architecture (Allavena et al., 1991). Interaction of α4β1 integrin with VCAM-1 on porcine endothelial cells is also required for both rolling and firm adhesion of human NK cells on porcine endothelial cells (Schneider et al., 2002).

The α4 integrin subunit also associates with another β chain, the β7, to give a functionally distinct integrin receptor capable of binding the mucosal vascular addressin MAdCAM-1. α4β7 is expressed on NK cells and mediates NK cell binding to mucosal HEVs (Perez-Villar et al., 1996b; Pinola et al., 1994b; Uksila et al., 1997). Functional evidences indicate, however, that NK cells expressing both α4β7 and α4β1 bind well to VCAM-1 but poorly to MAdCAM-1 suggesting that regulation of MAdCAM-1 versus VCAM-1 expression might critically control the recruitment of NK cell subsets to distinct tissues (Rott et al., 1996).

NK cells express all members of the β2 integrin family (CD11a-d/CD18), which are leukocyte-associated adhesion molecules mainly involved in the regulation of cell–cell interactions (Timonen, 1997). The leukocyte function-associated antigen 1 (CD11a/CD18 also known as LFA-1) is the receptor for the intercellular adhesion molecules (ICAM-1, 2, 3) and plays a crucial role in mediating NK cell adhesion to target cells as well as NK cell binding and migration across endothelial cells (Allavena et al., 1991). The expression and function of β2 integrins on NK cells is highly regulated. The levels of LFA-1 are higher on the CD56^{low} subset when compared with the CD56^{high}, while αMβ2 (CD11b/CD18 also known as Mac-1) and αXβ2 (CD11c/CD18) integrins are expressed on all and one-half of the NK cell population, respectively (Frey et al., 1998; Uksila et al., 1997). NK cell activation by cytokines, such as IL-2 or IL-12, results in up-regulation of LFA-1 expression and function, while CD11b and CD11c are down-regulated (Allavena et al., 1994; Maenpaa et al., 1993).

By using concomitant treatment of mice with mAb to either α4β1 or its ligand VCAM-1, or mAb directed against LFA-1 or ICAM-1, poly-L-lysine stabilized in carboxylmethyl cellulose (poly-ICLC) and IL-2-induced NK cell recruitment to pulmonary or hepatic parenchyma as well as to tumour lesions was abrogated by α4β1/VCAM-1 blockade. Similar results were observed upon IL-12-induced NK cell recruitment to the hepatic compartment. In contrast, α4β1/VCAM-1 interaction does not affect NK cell recruitment to the spleen. These results indicate that α4β1/VCAM-1 interaction at the interface between peripheral blood and endothelium is critical for the recruitment of NK cells into the lung, liver or tumour lesions (Fogler et al., 1996, 1998). Disruption of LFA-1/ICAM-1 interaction, however, through intravenous injections of anti-CD11a, anti-CD18 or anti-CD54 antibodies into rats, results in decreased numbers of PIT cells in the liver. This indicates that the LFA-1/CAM-1 adhesive pathway plays a role in rat NK cell recruitment in this organ (Luo et al., 1996) (refer to Table 20.1 and Figure 20.1).

Uterine NK cells

NK cells represent the predominant leukocyte population found in the uterus during the first trimester of human pregnancy or in the late secretory phase of the menstrual cycle (King, 2000; Whitelaw and Croy, 1996). A number of studies aimed at understanding the mechanisms underlying the accumulation of NK cells during first trimester pregnancy or their numerical fluctuation during the menstrual cycle, suggest that NK cell recruitment from blood could contribute to this event (Chantakru et al., 2002). A crucial role for L-selectin-CSPG-2 interaction in the control of tethering and rolling of peripheral blood CD56^{high} NK cells on endometrial endothelial cells has been suggested (Yamaguchi et al., 2006). Interestingly, studies in the mouse have demonstrated that MAdCAM-1 and P-selectin are co-expressed on the uterine vessels of the vascular zone, while VCAM-1 is mainly present in the vessels of decidua basalis. The unusual co-expression of P-selectin and MAdCAM-1 at this site has been suggested to provide a mechanism for selecting specialized subsets of leukocytes displaying a combination of P-selectin and MAdCAM-1 binding activities (Fernekorn et al., 2004). In addition, Fernekorn and co-workers also showed that at day 11 of pregnancy, implantation sites of mice lacking P-selectin or β7 integrin, or treated with blocking antibodies against MAdCAM-1 or α4β7 integrin, lack α4β7^{positive} leukocytes and are characterized by altered size and frequency of decidual NK cells (refer to Figure 20.1).

Based on the expression of VCAM-1 between the mid- and late secretory phase of the menstrual cycle (Rees et al., 1993) and on the constitutively and constant expression of ICAM-1 by all decidual vascular endothelium (Burrows et al., 1994; Tawia et al., 1993), a role has been attributed for β2 and α4β1 integrin-mediated NK cell interaction with uterine endothelial cells. Interestingly, exposure of peripheral blood leukocytes to pregnancy-associated hormones results in enhanced adhesiveness of peripheral blood NK cells to decidual vascular endothelial cells through L-selectin-dependent and α4 integrin-dependent mechanisms (Chantakru et al., 2003; van den Heuvel et al., 2005). In agreement with this observation, we found that peripheral blood NK cells from first trimester pregnant women migrate through primary cultures of decidual endothelial cells more efficiently than peripheral blood NK cells from non-pregnant women or male donors (Carlino et al., 2008). Thus, it is likely that pregnancy-associated factors such as sex hormones or cytokines acting at the systemic level, can enhance the adhesive and migratory capacity of peripheral blood NK cells through decidual endothelial cells.

Based on all these evidences, it can be postulated that differential expression of adhesion molecules on endothelium of different anatomical regions in normal or pathological conditions together with quantitative and qualitative regulation of integrin expression and function occurring following NK cell activation, are responsible for the recruitment of specialized NK cell subsets in lymphoid and non-lymphoid organs during steady state or inflammatory conditions, or during pregnancy.

Functional consequences of NK cell–endothelial cell interaction

NK cell interactions with endothelial cells are crucial early events during the immune surveillance of tissues, inflammatory responses as well as wound healing. Accumulating evidence indicates that NK cell interaction with endothelial cells is not only important for their localization in extravascular areas, but modulation of endothelial cell functions can also occur as a consequence of this interaction. In this regard, a role for NK cells in promoting angiogenesis or vascular remodelling as well as in causing endothelial injury or death primarily associated with endothelial dysfunction, has been documented in many physiopathological conditions. Thus, endothelial cells can be considered one of the primary targets of NK cells, and this is especially true for some tissue-resident NK cells such as the decidual NK cells that are localized in close contact with the vasculature.

NK cell-mediated promotion of vascular remodelling

A clear example of the ability of NK cells to promote angiogenesis and vascular remodelling is represented by decidual NK cells. Decidual NK cells significantly contribute to the maintenance of pregnancy (Croy et al., 2006; Moffett-King, 2002). Altered numbers and distribution of NK cells have been reported in spontaneous abortion, in-vitro fertilization failure and serious foetal growth restriction (Dosiou and Giudice, 2005; Wold and Arici, 2005). In contrast to their peripheral blood counterpart, decidual NK cells are poorly cytotoxic, and although in humans they express a number of activating receptors, including NKG2D, CD244, NKp30, NKp44 and NKp46, and are endowed with an intact cytolytic machinery (Tabiasco et al., 2006), they fail to polarize the microtubule-organizing centre and the cytolytic granules towards the synapse, thus exhibiting a defective secretory pathway (Kopcow et al., 2005). Thus, decidual NK cells are poor killers, but they are capable of secreting a wide array of cytokines and chemokines without stimulation, suggesting that they have undergone activation in the decidua. It is likely that through the release of cytokines and chemokines, NK cells control extravillous trophoblast invasion, as well as the recruitment and functions of other immune cells such as DC and T lymphocytes (Hanna et al., 2006; Moffett-King, 2002).

The close encirclement of spiral arteries by decidual NK cells together with their ability to produce angiogenic factors such as VEGF, PlGF, IL-8 and angiopoietin 2 in both mice and humans, support the notion that the major role of decidual NK cells is the control of mucosal vascularization and placental development (Croy et al., 2006; Hanna et al., 2006; Leonard et al., 2006). Based on the ability of NK cells to produce VEGF, it has been suggested that they act as guidance of the endothelial tip cells towards implantation sites or trophoblast. In addition, decidual NK cells appear to be directly involved in initiating spiral artery structural changes that are characterized by loss of vascular smooth muscle and lumen dilatation. This function is supported by a number of findings showing that decidual NK cell-deficient mice that however retain fertility, have profound vascular alterations in spiral arteries that are more constricted, thicker walled and shorter than in control animals. Moreover, decidual NK cells can promote choriocarcinoma tumour growth and vascularization when co-injected with tumour cells into nude mice (Hanna et al., 2006). This decidual NK cell-mediated activity on vascular remodelling is likely to involve production of cytokines rather than endothelial cell killing. In addition, a role of decidual NK cells in transient lymphatic drainage has also been hypothesized based on their ability to synthesize the lympho-angiogenic molecule VEGF-C and based on the finding of excessive decidual edema in uterine NK cell-deficient mice.

Although purified decidual NK cells constitutively express transcripts encoding for angiogenic and endothelial cell mitogenic factors, the triggering signals leading to the release of these factors in vivo are still under investigation. In this regard a role for IL-15 has been suggested, and evidence indicates that conditioned medium from IL-15-stimulated decidual NK cells can promote human umbilical vascular endothelial cell migration and formation in vitro (Hanna et al., 2006). Moreover, the interaction of soluble or membrane-bound HLA-G with its KIR2DL4 receptor on NK cells can induce the release of many cytokines and chemokines as well as angiogenic factors thus suggesting that NK cells may promote vasculature remodelling at the site of HLA-G expression such as in maternal decidua during pregnancy (Rajagopalan et al., 2006).

Aberrant NK cell activation locally in the decidua as well as at systemic levels together with a shift in the decidual NK cell cytokine production has been associated with pre-eclampsia, a pregnancy-specific pathological condition that is characterized by hypertension, proteinuria and edema, and represents a cause of maternal and foetal morbidity and mortality (Borzychowski et al., 2005; Sargent et al., 2007). Reduced expression of HLA-G on invasive extravillous cytotrophoblast correlates with pre-eclampsia. This can lead to an impaired activation of decidual NK cells and thus to altered invasion and remodelling of spiral artery (Goldman-Wohl et al., 2000). Particular combinations of maternal KIR (KIRAA) and foetal MHC class I (HLA-C2 group of alleles) are associated with an increased risk of pre-eclampsia. Based on this finding, it has been suggested that interaction between maternal KIR present on decidual NK cells and paternal derived HLA-C alleles expressed by extravillous cytotrophoblast, which are poorly stimulatory for decidual NK cells, has important functional consequences in term of regulation of placental development and vascular remodelling (Hiby et al., 2004; Parham, 2004).

NK cell-mediated endothelial cell damage and anti-angiogenic effects

In contrast to the pro-angiogenic effect exerted by decidual NK cells in early pregnancy, NK cell interaction with endothelial cells in situations other than pregnancy can result in endothelial cell injury and death leading to vascular dysfunction. During an immune response, an endothelial damage is not generally observed, but in certain conditions such as chronic inflammation related to a variety of infections, autoimmune diseases, graft-versus-host disease and vascular leak syndrome (observed during IL-2 immunotherapy), injured endothelial cells have been found.

CX3CL1/Fractalkine, one of the major chemokines controlling NK cell transendothelial migration, not only regulates NK cell binding to endothelial cells but also the induction of NK cell-mediated damage of endothelium. Transfection of fractalkine cDNA into ECV304 cells or human umbilical vein endothelial cells (HUVEC) was able to increase adhesion of NK cells to transfected endothelial cells as well as their susceptibility to NK cell-mediated cytotoxicity; moreover, stimulation of NK cells with soluble fractalkine was shown to enhance NK cell granule exocytosis (Imai et al., 1997; Yoneda et al., 2000). Fractalkine expressed on endothelial cell possibly functions as a triggering molecule for NK cells. Once NK cells are activated, they are able to lyse the endothelial cells that they have contacted through the release of lytic granules.

Fractalkine-induced NK cell interaction with endothelial cells as well as NK cell activation has also been shown to play a critical role in mediating the endothelial damage associated with inflammatory diseases linked to human cytomegalovirus (CMV) infection such as vascular disease and chronic transplant rejection (Bolovan-Fritts and Spector, 2008).

NK cell–endothelial cell interactions can play a primary role in the rejection of non-compatible tissues as emerged by the observation that NK cells infiltrate pig organs perfused with human blood (Khalfoun et al., 2000; Kirk et al., 1993) and by the in vitro data indicating that NK cells can adhere and lyse porcine endothelial cells (Kwiatkowski et al., 1997; Schneider et al., 2001; Seebach et al., 1996). Moreover, antibody-dependent hyperacute vascular xenograft rejection is associated with IgG-mediated vascular damage. The mechanisms involved in IgG-induced vascular injury are dependent not only on the ability of IgG to activate the complement cascade but also on NK cell activity through engagement of FcγR (CD16) (Yin et al., 2004).

NK cell–endothelial cell interactions in cancer

The vascular leak syndrome, observed within a few hours of IL-2 treatment, is characterized by increased vascular permeability and impaired microcirculatory perfusion that results in generalized interstitial edema and multiple organ dysfunctions. NK cells play a crucial role both in the early and late events in the pulmonary vascular leak syndrome induced by IL-2 (Assier et al., 2004). In addition, IL-2 can enhance the ability of NK cells to kill endothelial cells. Nonetheless, the NK cell-mediated mechanisms responsible for pulmonary vascular leak induced by IL-2 in vivo are still not fully clear.

Accumulating evidences also indicate that NK cell-mediated functions, by acting at vascular levels, could play a role in the control of tumour growth. In a model in which human Burkitt tumours are reproducibly established subcutaneously in athymic mice, treatment with IL-12 induces tumour tissue necrosis, vascular damage and accumulation of activated NK cells in surrounding small vessels at site where tumour tissue necrosis progresses (Yao et al., 1999). The ability of NK cells to mediate cytotoxicity against endothelial cells together with their ability to release IFN-γ, which in turn enhances the release of angiostatic chemokines such as IP10 and MIG, has been proposed as potential mechanisms by which IL-12 suppresses neovascularization in human and murine tumours (Strasly et al., 2001; Yao et al., 1999). Moreover, using a naked IL-12 DNA expression vector to achieve constant low levels of systemic cytokine expression to inhibit breast tumour growth in MMTVneu transgenic and transplanted models, it has been demonstrated that IL-12 tumour-suppressive effects are completely lost in SCID/common gamma chain knock-out mice lacking NK cells, thus supporting the notion that NK cells are critical mediators of the anti-angiogenic effects exerted by IL-12 (Faggioli et al., 2008).

Conclusion

NK cells not only represent crucial effector mechanisms of innate immunity but are also implicated in the regulation of adaptive immune responses. They can also act on endothelial cells by regulating their differentiation and proliferation, or inducing their lysis. These opposing effects on the endothelium may explain the ability of NK cells to contribute to both the maintenance of tissue homeostasis and the exacerbation of tissue damage. The molecular events underlying various effects exerted by NK cells on the endothelium are complex and still under investigation. The endothelium represents a barrier that peripheral blood NK cells need to cross to extravasate and localize in the parenchyma of many lymphoid and non-lymphoid organs. The expression of specific adhesion molecules and chemotactic factors on distinct endothelial vessels under pathophysiological conditions may contribute to the trafficking of highly specialized NK cell subsets to the proper tissue sites. The presence of distinct NK cell effector functions, however, in various organs in normal or pathological conditions is not only attributable to the specific properties of organ-associated endothelium but can also be dependent on the tissue microenvironment that provides specific signals leading to the generation of specialized NK cell effector functions. In this regard, poorly cytotoxic decidual NK cells play a crucial role in the control of spiral artery formation and vascular remodelling by releasing angiogenic factors during normal pregnancy, while the aberrant activation of decidual NK cells during pre-eclampsia leads to a shift in the pattern of cytokines released that is associated with a loss of NK cell beneficial function. In many pathological conditions, NK cell activation can lead to endothelial cell damage and vascular dysfunctions, which contributes to disease severity. At present, the role played by NK cells on tumour angiogenesis is still poorly investigated, and the ability of NK cells to affect vascular remodelling is well established only during pregnancy. Further insights on the molecular mechanisms that govern NK cell activation and differentiation in the tissues are essential to better understand NK cell effector functions on endothelial cells and will be important to design therapeutic strategies to redirect abnormal NK cell functional activities within individual pathological conditions.

Acknowledgements

This work was supported by grants from Istituto Pasteur Fondazione Cenci Bolognetti, MIUR-PRIN, EC NoE EMBIC contract # 512040, Centro di Eccellenza BEMM.

References

Allavena, P., Paganin, C., Martin-Padura, I., Peri, G., Gaboli, M., Dejana, E., Marchisio, P.C. and Mantovani, A. (1991). Molecules and structures involved in the adhesion of natural killer cells to vascular endothelium. J Exp Med **173**, 439–448.

Allavena, P., Paganin, C., Zhou, D., Bianchi, G., Sozzani, S. and Mantovani, A. (1994). Interleukin-12 is chemotactic for natural killer cells and stimulates their interaction with vascular endothelium. Blood **84**, 2261–2268.

Andre, P., Spertini, O., Guia, S., Rihet, P., Dignat-George, F., Brailly, H., Sampol, J., Anderson, P.J. and Vivier, E. (2000). Modification of P-selectin glycoprotein ligand-1 with a natural killer cell-restricted sulfated lactosamine creates an alternate ligand for L-selectin. Proc Natl Acad Sci U S A **97**, 3400–3405.

Andrews, D.M., Farrell, H.E., Densley, E.H., Scalzo, A.A., Shellam, G.R. and Degli-Esposti, M.A. (2001). NK1.1+ cells and murine cytomegalovirus infection: what happens in situ?. J Immunol **166**, 1796–1802.

Assier, E., Jullien, V., Lefort, J., Moreau, J.L., Di Santo, J.P., Vargaftig, B.B., Lapa e Silva, J.R. and Thèze, J. (2004). NK cells and polymorphonuclear neutrophils are both critical for IL-2-induced pulmonary vascular leak syndrome. J Immunol **172**, 7661–7668.

Baggiolini, M., Dewald, B. and Moser, B. (1997). Human chemokines: an update. Annu Rev Immunol **15**, 675–705.

Bajénoff, M., Breart, B., Huang, A.Y., Qi, H., Cazareth, J., Braud, V.M., Germain, R.N. and Glaichenhaus, N. (2006). Natural killer cell behavior in lymph nodes revealed by static and real-time imaging. J Exp Med **203**, 619–631.

Biron, C.A. (1997). Activation and function of natural killer cell responses during viral infection. Curr Opin Immunol **9**, 24–34.

Bolovan-Fritts, C.A. and Spector, S.A. (2008). Endothelial damage from cytomegalovirus-specific host immune response can be prevented by targeted disruption of fractalkine-CX3CR1 interaction. Blood **111**, 175–182.

Borzychowski, A.M., Croy, B.A., Chan, W.L., Redman, C.W. and Sargent, I.L. (2005). Changes in systemic type 1 and type 2 immunity in normal pregnancy and pre-eclampsia may be mediated by natural killer cells. Eur J Immunol **35**, 3054–3063.

Burrows, T.D., King, A. and Loke, Y.W. (1994). Expression of adhesion molecules by endovascular trophoblast and decidual endothelial cells: implications for vascular invasion during implantation. Placenta **15**, 21–33.

Carlino, C., Stabile, H., Morrone, S., Bulla, R., Soriani, A., Agostinis, C., Bossi, F., Mocci, C., Sarazani, F., Tedesco, F., Santoni, A. and Gismondi, A. (2008). Recruitment of circulating NK cells through decidual tissues: a possible mechanism controlling NK cell accumulation in the uterus during early pregnancy. Blood **111**, 3108–3115.

Cella, M., Fuchs, A., Vermi, W., Bacchetti, F., Otero, K., Lennerz, J.K., Doherty, J.M., Mills, J.C. and Colonna, M. (2008). A human natural killer cell subset provides an innate source of IL-22 for mucosal immunity. Nature **Nov 2** [Epub ahead of print].

Chantakru, S., Miller, C., Roach, L.E., Kuziel, W.A., Maeda, N., Wang, W.C., Evans, S.S. and Croy, B.A. (2002). Contributions from self-renewal and trafficking to the uterine NK cell population of early pregnancy. J Immunol **168**, 22–28.

Chantakru, S., Wang, W.C., van den Heuvel, M., Bashar, S., Simpson, A., Chen, Q., Croy, B.A. and Evans, S.S. (2003). Coordinate regulation of lymphocyte-endothelial interactions by pregnancy associated hormones. J Immunol **171**, 4011–4019.

Chen, S., Kawashima, H., Lowe, J.B., Lanier, L.L. and Fukuda, M. (2005). Suppression of tumor formation in lymph nodes by L-selectin-mediated natural killer cell recruitment. J Exp Med **202**, 1679–1689.

Christopher, E.A., Coudert, J.D. and Degli-Esposti, M.A. (2008). Killers and beyond: NK-cell-mediated control of immune responses. Eur J Immunol **38**, 2938–2942.

Clark, E.A. and Brugge, J.S. (1995). Integrins and signal transduction pathways: the road taken. Science **268**, 233–239.

Croy, B.A., van den Heuvel, M.J., Borzychowski, A.M. and Tayade, C. (2006). Uterine natural killer cells: a specialized differentiation regulated by ovarian hormones. Immunol Rev **214**, 161–185.

De Filippi, P., Gismondi, A., Santoni, A. and Tarone, G. (1997). Signal transduction by integrins. Austin, TX: Springer Publishing Company.

Di Santo, J.P. (2006). Natural killer cell developmental pathways: a question of balance. Annu Rev Immunol **24**, 257–286.

Di Santo, J.P. and Vosshenrich, C. A.J. (2006). Bone marrow versus thymic pathways of natural killer cell development. Immunol Rev **214**, 35–46.

Doherty, D.G. and O'Farrelly, C. (2000). Innate and adaptive lymphoid cells in the human liver. Immunol Rev **174**, 5–20.

Dokun, A.O., Chu, D.T., Yang, L., Bendelac, A.S. and Yokoyama, W.M. (2001). Analysis of in situ NK cell responses during viral infection. J Immunol **167**, 5286–5293.

Dosiou, C. and Giudice, L.C. (2005). Natural killer cells in pregnancy and recurrent pregnancy loss: endocrine and immunologic perspectives. Endocrinol Rev **26**, 44–62.

Faggioli, F., Soldati, S., Scanziani, E., Catò, E. M., Adorni, F., Vezzoni, P., Noonan, D.M. and Sacco, M.G. (2008). Effects of IL-12 gene therapy on spontaneous transgenic and transplanted breast tumors. Breast Cancer Res Treat **110**, 223–226.

Ferlazzo, G. and Münz, C. (2004). NK cell compartments and their activation by dendritic cells. J Immunol **172**, 1333–1339.

Fernekorn, U., Butcher, E.C., Behrends, J., Hasrtz, S. and Kruse, A. (2004). Functional involvement of P-selectin and MAdCAM-1 in the recruitment of alpha4beta7-integrin-expressing monocyte-like cells to the pregnant mouse uterus. Eur J Immunol **34**, 3423–3433.

Fogler, W.E., Volker, K., McCormick, K.L., Watanabe, M., Ortaldo, J.R. and Wiltrout, R.H. (1996). NK cell infiltration into lung, liver, and subcutaneous B16 melanoma is mediated by VCAM-1/VLA-4 interaction. J Immunol **156**, 4707–4714.

Fogler, W.E., Volker, K., Watanabe, M., Wigginton, J.M., Roessler, P., Brunda, M.J., Ortaldo, J.R. and Wiltrout, R.H. (1998). Recruitment of hepatic NK cells by IL-12 is dependent on IFN-gamma and VCAM-1 and is rapidly down-regulated by a mechanism involving T cells and expression of Fas. J Immunol **161**, 6014–6021.

Frey, M., Packianathan, N.B., Fehniger, T.A., Ross, M.E., Wang, W.C., Stewart, C.C., Caligiuri, M.A. and Evans, S.S. (1998). Differential expression and function of L-selectin on CD56bright and CD56dim natural killer cell subsets. J Immunol **161**, 400–408.

Gismondi, A., Mainiero, F., Morrone, S., Palmieri, G., Piccoli, M., Frati, L. and Santoni, A. (1992). Triggering through CD16 or phorbol esters enhances adhesion of NK cells to laminin via very late antigen 6. J Exp Med **176**, 1251–1257.

Gismondi, A., Morrone, S., Humphries, M.J., Piccoli, M., Frati, L. and Santoni, A. (1991). Human natural killer cells express VLA-4 and VLA-5, which mediate their adhesion to fibronectin. J Immunol **146**, 384–392.

Gismondi, A. and Santoni, A. (2006). Migration of NK cells. In: Badolato, R. and Sozzani, S. (eds) Lymphocyte trafficking in Health and Disease. Basel: Birkhauser Verlag, p. 95.

Goldman-Wohl, D.S., Ariel, I., Greenfield, C., Hochner-Celnikier, D., Cross, J., Fisher, S. and Yagel, S. (2000). Lack of human leukocyte antigen-G expression in extravillous trophoblasts is associated with pre-eclampsia. Mol Hum Reprod **6**, 88–95.

Grégoire, C., Chasson, L., Luci, C., Tomasello, E., Geissmann, F., Vivier, E. and Walzer, T. (2007). The trafficking of natural killer cells. Immunol Rev **220**, 169–182.

Hanna, J., Goldman-Wohl, D., Hamani, Y., Avraham, I., Greenfield, C., Natanson-Yaron, S., Prus, D., Cohen-Daniel, L., Arnon, T.I., Manaster, I., Gazit, R., Yutkin, V., Benharroch, D., Porgador, A., Keshet, E., Yagel, S. and Mandelboim, O. (2006). Decidual NK cells regulate key developmental processes at the human fetal-maternal interface. Nat Med **12**, 1065–1074.

Hiby, S.E., Walzer, J.J., O'shaughnessy, K.M., Redman, C.W., Carrington, M., Trowsdale, J. and Moffett, A. (2004). Combinations of maternal KIR and fetal HLA-C genes influence the risk of preeclampsia and reproductive success. J Exp Med **200**, 957–965.

Imai, T., Hieshima, K., Haskell, C., Baba, M., Nagira, M., Nishimura, M., Kakizaki, M., Takagi, S., Nomiyama, H., Schall, T.J. and Yoshie, O. (1997). Identification and molecular characterization of fractalkine receptor CX3CR1, which mediates both leukocyte migration and adhesion. Cell **91**, 521–530.

Kaneda, K. and Wake, K. (1983). Distribution and morphological characteristics of the pit cells in the liver of the rat. Cell Tissue Res **233**, 485–505.

Khalfoun, B., Barrat, D., Watier, H., Machet, M.C., Arbeille-Brassart, B., Riess, J.G., Salmon, H., Gruel, Y., Bardos, P. and Lebranchu, Y. (2000). Development of an ex vivo model of pig kidney perfused with human lymphocytes. Analysis of xenogeneic cellular reactions. Surgery **128**, 447–457.

King, A. (2000). Uterine leukocytes and decidualization. Hum Reprod Update **6**, 28–36.

King, A., Balendran, N., Wooding, P., Carter, N.P. and Loke, Y.W. (1991). CD3- leukocytes present in the human uterus during early placentation: phenotypic and morphologic characterization of the CD56 + + population. Dev Immunol **1**, 169–190.

King, A., Burrows, T. and Loke, Y.W. (1996–1997). Human uterine natural killer cells. Nat Immunol **15**, 41–52.

Kirk, A.D., Heinle, J.S., Mault, J.R. and Sanfilippo, F. (1993). Ex vivo characterization of human anti-porcine hyperacute cardiac rejection. Transplantation **56**, 785–793.

Kopcow, H.D., Allan, D.S., Chen, X., Rybalov, B., Andzelm, M.M., Ge, B. and Strominger, J.L. (2005). Human decidual NK cells form immature activating synapses and are not cytotoxic. Proc Natl Acad Sci U S A **102**, 15563–15568.

Kunkel, E.J. and Butcher, E.C. (2002). Chemokines and the tissue-specific migration of lymphocytes. Immunity **16**, 1–4.

Kwiatkowski, B., Artrip, J.H., Michler, R.E. and Itescu, S. (1997). Human NK cells demonstrate increased adherence and lysis of porcine endothelium following activation with IL-2. Ann Transplant **2**, 21–25.

Lanier, L.L. (2005). NK cell recognition. Annu Rev Immunol **23**, 225–274.

Leonard, S., Murrant, C., Tayade, C., van den Heuvel, M., Watering, R. and Croy, B.A. (2006). Mechanisms regulating immune cell contributions to spiral artery modification—facts and hypotheses—a review. Placenta **27**(Suppl A), S40–S46.

Luo, D., Vanderkerken, K., Bouwens, L., Kuppen, P.J., Baekeland, M., Seynaeve, C. and Wisse, E. (1996). The role of adhesion molecules in the recruitment of hepatic natural killer cells (pit cells) in rat liver. Hepatology **24**, 1475–1480.

Maenpaa, A., Jaaskelainen, J., Carpen, O., Patarroyo, M. and Timonen, T. (1993). Expression of integrins and other adhesion molecules on NK cells; impact of IL-2 on short- and long-term cultures. Int J Cancer **53**, 850–855.

Mainiero, F., Gismondi, A., Milella, M., Morrone, S., Palmieri, G., Piccoli, M., Frati, L. and Santoni, A. (1994). Long-term activation of natural killer cells results in modulation of beta1-integrin expression and function. J Immunol **152**, 446–454.

Mantovani, A. (1999). The chemokine system: redundancy for robust outputs. Immunol Today **20**, 254–257.

Martín-Fontecha, A., Thomsen, L.L., Brett, S., Gerard, C., Lipp, M., Lanzavecchia, A. and Sallusto, F. (2004). Induced recruitment of NK cells to lymph nodes provides IFN-gamma for T(H)1 priming. Nat Immunol **5**, 1260–1265.

Moffett-King, A. (2002). Natural killer cells and pregnancy. Nat Rev Immunol **2**, 656–663.

Moore, K.L. and Thompson, L.F. (1992). P-selectin (CD62) binds to subpopulations of human memory T lymphocytes and natural killer cells. Biochem Biophys Res Commun **186**, 173–181.

Münz, C. (2008). Non-cytotoxic protection by human NK cells in mucosal secondary lymphoid tissues. Eur J Immunol **38**, 2946–2948.

Paganini, C., Matteucci, C., Cenzuales, S., Mantovani, A. and Allavena, P. (1994). IL-4 inhibits binding and cytotoxicity of NK cells to vascular endothelium. Cytokine **6**, 135–140.

Parham, P. (2004). NK cells and trophoblasts: partners in pregnancy. J Exp Med **200**, 951–955.

Perez-Villar, J.J., Melero, I., Gismondi, A., Santoni, A. and Lopez-Botet, M. (1996a). Functional analysis of alpha1beta1 integrin in human natural killer cells. Eur J Immunol **26**, 2023–2029.

Perez-Villar, J.J., Zapata, J.M., Melero, I., Postigo, A., Sanchez-Madrid, E. and Lopez-Botet, M. (1996b). Expression and function of alpha 4/beta 7 integrin on human natural killer cells. Immunology **89**, 96–104.

Pinola, M., Renkonen, R., Majuri, M.L., Tiisala, S. and Saksela, E. (1994a). Characterization of the E-selectin ligand on NK cells. J Immunol **152**, 3586–3594.

Pinola, M., Saksela, E., Tiisala, S. and Renkonen, R. (1994b). Human NK cells expressing alpha 4 beta 1/beta 7 adhere to VCAM-1 without preactivation. Scand J Immunol **39**, 131–136.

Rajagopalan, S., Bryceson, Y.T., Kuppusamy, S.P., Geraghty, D.E., van der Meer, A., Joosten, I. and Long, E.O. (2006). Activation of NK cells by an endocytosed receptor for soluble HLA-G. PLoS Biol **4**, e9.

Rees, M.C., Heryet, A.R. and Bicknell, R. (1993). Immunohistochemical properties of the endothelial cells in the human uterus during the menstrual cycle. Hum Reprod **8**, 1173–1178.

Rossi, D. and Zlotnik, A. (2000). The biology of chemokines and their receptors. Annu Rev Immunol **18**, 217–242.

Rott, L.S., Briskin, M.J., Andrew, D.P., Berg, E.L. and Butcher, E.C. (1996). A fundamental subdivision of circulating lymphocytes defined by adhesion to mucosal addressin cell adhesion molecule-1. Comparison with vascular cell adhesion molecule-1 and correlation with beta 7 integrins and memory differentiation. J Immunol **156**, 3727–3736.

Salazar-Mather, T.P., Ishikawa, R. and Biron, C.A. (1996). NK cell trafficking and cytokine expression in splenic compartments after IFN induction and viral infection. J Immunol **157**, 3054–3064.

Sargent, I.L., Borzychowski, A.M. and Redman, C.W. (2007). NK cells and pre-eclampsia. J Reprod Immunol **76**, 40–44.

Schneider, M.K., Forte, P. and Seebach, J.D. (2001). Adhesive interactions between human NK cells and porcine endothelial cells. Scand J Immunol **54**, 70–75.

Schneider, M.K., Strasser, M., Gilli, U.O., Kocher, M., Moser, R. and Seebach, J.D. (2002). Rolling adhesion of human NK cells to porcine endothelial cells mainly relies on CD49d-CD106 interactions. Transplantation **73**, 789–796.

Seebach, J.D., Yamada, K., McMorrow, I. M., Sachs, D.H. and DerSimonian, H. (1996). Xenogeneic human anti-pig cytotoxicity mediated by activated natural killer cells. Xenotransplantation **3**, 188–197.

Snapp, K.R., Ding, H., Atkins, K., Warnke, R., Luscinskas, F.W. and Kansas, G.S. (1998). A novel P-selectin glycoprotein ligand-1 monoclonal antibody recognizes an epitope within the tyrosine sulfate motif of human PSGL-1 and blocks recognition of both P- and L-selectin. Blood **91**, 154–164.

Springer, T.A. (1994). Traffic signals for lymphocyte recirculation and leukocyte emigration: the multistep paradigm. Cell **76**, 301–314.

Strasly, M., Cavallo, F., Geuna, M., Mitola, S., Colombo, M.P., Forni, G. and Bussolino, F. (2001). IL-12 inhibition of endothelial cell functions and angiogenesis depends on lymphocyte-endothelial cell cross-talk. J Immunol **166**, 3890–3899.

Tabiasco, J., Rabot, M., Aguerre-Girr, M., El Costa, H., Berrebi, A., Parant, O., Laskarin, G., Juretic, K., Bensussan, A., Rukavina, D. and Le Bouteiller, P. (2006). Human decidual NK cells: unique phenotype and functional properties—a review. Placenta **Apr 27**(Suppl A), S34–S39.

Tawia, S.A., Beaton, L.A. and Rogers, P.A. (1993). Immunolocalization of the cellular adhesion molecules, intercellular adhesion molecule-1 (ICAM-1) and platelet endothelial cell adhesion molecule (PECAM), in human endometrium throughout the menstrual cycle. Hum Reprod **8**, 175–181.

Timonen, T. (1997). Natural killer cells: endothelial interactions, migration and target cell recognition. J Leukoc Biol **62**, 693–701.

Trinchieri, G. (1989). Biology of natural killer cells. Adv Immunol **47**, 187–376.

Uksila, J., Salmi, M., Butcher, E.C., Tarkkanen, J. and Jalkanen, S. (1997). Function of lymphocyte homing-associated adhesion molecules on human natural killer and lymphokine-activated killer cells. J Immunol **158**, 1610–1617.

van den Heuvel, M.J., Horrocks, J., Bashar, S., Taylor, S., Burke, S., Hatta, K., Lewis, J.E. and Croy, B.A. (2005). Menstrual cycle hormones induce changes in functional interactions between lymphocytes and decidual vascular endothelial cells. J Clin Endocrinol Metab **90**, 2835–2842.

Walzer, T., Bléry, M., Chaix, J., Fuseri, N., Chasson, L., Robbins, S.H., Jaeger, S., André, P., Gauthier, L., Daniel, L., Chemin, K., Morel, Y., Dalod, M., Imbert, J., Pierres, M., Moretta, A., Romagné, F. and Vivier, E. (2007). Identification, activation, and selective in vivo ablation of mouse NK cells via NKp46. Proc Natl Acad Sci U S A **104**, 3384–3389.

Whitelaw, J.F. and Croy, B.A. (1996). Granulated lymphocytes of pregnancy. Placenta **17**, 533–543.

Wold, A.S. and Arici, A. (2005). Natural killer cells and reproductive failure. Curr Opin Obstet Gynecol **17**, 237–241.

Yago, T., Tsukuda, M., Fukushima, H., Yamaoka, H., Kurata-Miura, K., Nishi, T. and Minami, M. (1998). IL-12 promotes the adhesion of NK cells to endothelial selectins under flow conditions. J Immunol **161**, 1140–1145.

Yamaguchi, T., Kitaya, K., Daikoku, N., Yasuo, T., Fushiki, S. and Honjo, H. (2006). Potential selectin L ligands involved in selective recruitment of peripheral blood CD16(−) natural killer cells into human endometrium. Biol Reprod **74**, 35–40.

Yao, L., Sgadari, C., Furuke, K., Bloom, E.T., Teruya-Feldstein, J. and Tosato, G. (1999). Contribution of natural killer cells to inhibition of angiogenesis by interleukin-12. Blood **93**, 1612–1621.

Yin, D., Zeng, H., Ma, L., Shen, J., Xu, H., Byrne, G.W. and Chong, A.S. (2004). NK cells mediate IgG1-dependent hyperacute rejection of xenografts. J Immunol **172**, 7235–7238.

Yoneda, O., Imai, T., Goda, S., Inoue, H., Yamauchi, A., Okazaki, T., Imai, H., Yoshie, O., Bloom, E.T., Domae, N. and Umehara, H. (2000). Fractalkine-mediated endothelial cell injury by NK cells. J Immunol **164**, 4055–4062.

Chapter Twenty-One

21

Natural killer cell/epithelial interactions

Taisuke Ito, Masao Takei, Ralf Paus

CHAPTER CONTENTS

ABSTRACT

Natural killer (NK) cells and epithelial cells can interact closely at sites of inflammation, contributting to the pathogenesis of inflammatory diseases such as contact hypersensitivity. For example, NK cells are rarely found in the dermis and are not found in normal epidermis. However, NK cells can be found to infiltrate lesions of several skin diseases and may play an important role in the innate immune response. The interaction between NK cells and the skin (the major epithelium of the human and murine soma) are particularly interesting. Epithelial tissues, including skin and airway, are continuously exposed to pathogens, allergens and environmental contaminants, requiring profound regulation of the innate and adaptive immune systems. Skin, the largest immune tissue in the body, is composed of epidermis, dermis and appendices, including hair follicles, sebaceous glands and sweat glands. To accumulate and infiltrate within epithelial lesions, chemokine and adhesion molecules are induced by NK cells releasing several cytokines and cytotoxic granules such as interferon (IFN)-γ, tumour necrosis factor (TNF)-α, granulocyte macrophage-colony stimulating factor (GM-CSF), IL-5 and IL-8.

KEY WORDS

Epithelium, Immune privilege, NKG2D, CD16, CD56

Phenotype of NK cells

Human natural killer (NK) cells constitute about 10–15% of the total blood lymphoid cells. In the peripheral blood, the phenotype of NK cells is charactrized by CD3^{-}CD56^{+}, and can be classified into two major subsets: those that express CD56 but not CD16 (known as CD56^{bright}CD56^{-}), and those that express CD16 and low CD56 (known as CD56^{dim}CD16^{+}). Approximately 90% of human NK cells are phenotypically CD56^{dim}CD16^{+}, and the rest are CD56^{bright}CD16^{-} or CD56^{-}CD16^{-} that represent immature precursor forms (Nagler et al., 1989; Bennett et al., 1996). These two subsets of NK cells show differences of function. For example, CD56^{bright}CD16^{-} NK cells are highly cytolytic when compared with CD56^{dim}CD16^{+} but secrete cytokines (Cooper et al., 2001). CD56^{bright}CD16^{-} is also defined as immunoregulatory NK cells characterized by prominent recirculation in secondary lymphoid organs, limited cytotoxic function and high interferon

(IFN)-γ production depend on activation (Ottaviani et al., 2006). The CD56^{dim}CD16^{+} cells are regarded as effector NK cells, which express CD94/NKG2, KIR, natural cytotoxicity (NC) and perforin. On the other hand, CD56^{bright}CD16^{-} cells are KIR$^{low\ or\ -}$, NC^{low} and perforin^{-} (Chiesa et al., 2003). Furthermore, the expression of adhesion molecules is different between CD56^{bright}CD16^{-} cells and CD56^{dim}CD16^{+} cells. CD56^{bright}CD16^{-} cells express L-selection and integrin but no PSGL-1/PEN5 (Andre et al., 2000).

Chemokine receptors of NK Cell

For the migration and accumulation of immune cells such as NK cells in inflamed tissues, chemoattractants play an important role (Maghazachi, 2003). Chemokine and its receptors are categorized into inflammatory, constitutive and dual groups. The first group is inflammatory chemokines and its receptors, such as CXCR3/immune privilege (IP)-10 and CCR5/RANTES. The second group includes constitutive chemokines and its receptors such as CCR7 and ELC (MIP-3β). The third group is inflammatory and constitutive chemokines and its receptors (i.e. CCR4/TARC and CX3CR1/Fractalkine). Resting NK cells express the constitutive chemokine receptors (CCR7, CXCR4) and dual chemokine receptors (CCR4 and CX3CR1). However, inflammatory chemokine receptors CCR1 and CXCR3 are low expression in resting NK cells (Inngjerdingen et al., 2001). CCR7 is important to migrate into the peripheral lymph nodes (Baekkevold et al., 2001). Recent work reported that CCR7 is equally expressed on CD16+ and CD16− NK cells (Maghazachi, 2005).

Adhesion molecules on NK cells

NK cells are only rarely detected in the dermis under physiological conditions, but are often found to accumulate here in inflammatory skin diseases, such as psoriasis vulgaris, atopic dermatitis and alopecia areata. Therefore, NK cells are capable of extravasation and migrate into the inflamed skin lesions. Then, NK cells may contact with keratinocytes followed by immune reactions. Actually, there are several reports that studied the expression of adhesion molecules on NK cells. Integrin and selection families are known as adhesion molecules. Integrin molecules are constructed by α and β chains, and the combination of α and β chain contributes the specificity against extracellular matrix, serum proteins and surface antigens of the cells. For example, the β1 subfamily is called VLA (very late antigen) that can combine to α1–α9 and αv subfamilies. β1 subfamilies can adhere extracellular matrix such as collagen, laminin and fibronectin. In the β1 subfamily, VLA-4 (α4β1) only combines extracellular antigens (ligands) such as VCAM1 and MAdCAM-1. β2 (CD18) subfamilies include LFA-1 (αLβ2), Mac-1 (αMβ2), CR3 (αMβ2) and αDβ2. LFA-1 can combines ligands called ICAM-1, ICAM-2 and ICAM-3. The selectin family that is expressed in the inflamed site is subdivided into L, E and P, which are mainly expressed on leukocytes, vascular endothelial cells and platelets, respectively. Sialyl LewesX is the ligand for E-selectins and P-selectins. P-selectin can also combine PSGL-1.

Recently, the expression of these families and ligands has been studied. For example, NK cells express the carbohydrate sLex (Sialyl LewesX) as well as L-selectin detection carbohydrate ligands on endothelium (Pinola et al., 1994; Mäenpää et al., 1993). Furthermore, NK cells express abundantly the α4β1 (VLA-4) integrin and their ligand VCAM-1 that may be a target in the adhesion of NK cells to endothelium. This was supported by the inhibition of NK cell homing into experimental melanoma metastasis in vivo by anti-VCAM-1 antibodies (Fogler et al., 1996). In addition, antibody against β2-integrin and ICAM-1 also inhibit NK cell adhesion to cultured epithelial cells. Extracellular matrix is also involved in NK cell migration. Fibronectin facilitates the migration of NK cells, and NK cells recognize fibronectin through α4β1 (VLA-4) and α5β1-integrins (VLA-5) (Somersalo and Saksela, 1991). α1β1 integrin (VLA-1) are expressed by long-term activated NK cells. In addition, laminin and collagen I are recognized by activated NK cells, probably through α1β1 integrin (VLA-1) (Mäenpää et al., 1993; Pérez-Villar et al., 1996a,b). Upon IL-2-activation, expression of functional α4β7-integrin (LPAM-1) is induced on NK cells potentially participating in their interaction with both extracellular matrix and endothelial cells (Pérez-Villar et al., 1996a,b). Additional adhesion molecules are also suggested to contribute on NK cell extravasation such as CD44 (Galandrini et al., 1994), CD2 (Anasetti et al., 1987) and CD31 (PECAM-1) (Berman et al., 1996).

NK cell distribution and characteristics in normal skin in mouse and human

Few CD56+ NK cells scatter in dermis close to epidermis (Buentke et al., 2002), including around anagen hair follicles (HFs) in normal human skin (Ito et al., 2008). CD56+ cells were found extremely rarely in the outer root sheath (ORS) (approximately 2 cells per 100 HFs in the distal ORS only) and in the connective tissue sheath (CTS) (approximately 4 cells per 100 HFs in the distal CTS vs.

2 cells per 100 HFs in the proximal CTS) (Christoph et al., 2000). Normal human nail apparatus also shows scarce distribution of NK cells (Ito et al., 2005). NK cells normally recognize major histocompatibility complex (MHC) class I negative tissues. However, NK cells around do not attack the proximal ORS of HFs that have low or absent MHC class I expression. Uninvolved psoriatic skin and normal skin also shows no CD56+ cells distribution. In addition, CD16+ cells were rarely detected in uninvolved skin in psoriasis patients (Cameron et al., 2002). The exact distribution and composition of NK cell subpopulations in normal murine skin remains to be systematically characterized.

Abnormalities in NK cell distribution, activity and number in inflamed skin

Skin is the first defence for the host against a danger signal by physical and immunological barrier. In the immunological barrier system, the NK cell is categorized in the innate immune system that does not depend on immune recognition by lymphocytes (David, 2006). The function of NK cells is to recognize and kill the target cells such as virus-infected cells and certain tumour cells through apoptosis. Therefore, NK cells may play an important role to protect the host's body against an invasion of microorganisms thorough the endothelium, especially skin. While NK cells are rarely found in the dermis of normal state epidermis and dermis, including appendix such as HFs, and the role of NK cells in the normal state may be one of the sentry in skin immune systems, once skin suffers from some diseases, such as psoriasis vulgaris, virus infection and alopecia areata, NK cells infiltrate the lesions and take part in the pathogenesis of the skin diseases. The following section will interpret interactions in the skin and its diseases, before briefly covering NK cell interactions with lung, kidney and mucosal epithelium.

NK cell–keratinocyte interactions

In the normal state, epidermal keratinocytes strongly express MHC class I molecules so that NK cells do not recognize MHC class I+ keratinocytes. Actually, NK cells are not found in/around MHC class I+ epidermis, and there may be no remarkable interaction between NK cells and epidermal keratinocytes in the normal state. However, epidermal keratinocytes interacts with NK cells in inflamed sites such as psoriasis vulgaris, drug eruption, contact dermatitis via chemokines, cytokines and adhesion molecules. The following sections explain the detailed interaction between NK cells and epidermal keratinocytes.

The role of NK cells in psoriasis

Psoriasis is characterized by abnormal keratinocyte proliferation and differentiation and by an inflammatory infiltrate of mononuclear cells in the dermis. This infiltrate consists mainly of Th1 and Th17 cells that secrete IFN-γ, IL-2, tumour necrosis factor (TNF)-α, IL-17 and IL-22. However, the contribution of NK cells to psoriasis is still little known. The number of infiltrated CD16+ NK cells is significantly lower in the psoriatic lesions compared to normal skin. In addition, the percentage of perforin and CD16+ cells was significantly lower in lesional dermis compared to healthy dermis (Kastelan et al., 2004). Furthermore, there were significantly fewer cells expressing the NK cell markers CD16 and CD56 in patients with psoriasis compared with normal controls in PBMCs. These results are similar to the study of psoriatic lesions. Therefore, it has been proposed that NK cells play a regulatory role in autoimmunity in these conditions (Lanier, 2000; Seaman, 2000).

On the other hand, Cameron et al. reported that the number of infiltrating CD16+, CD94+ and CD158a+ cells were significantly higher in the papillary dermis and reticular dermis of psoriatic lesions compared to control (Cameron et al., 2002). CD16 expression demarcates activated NK cells, while CD94 and CD158a expression reflects inhibitory NK cell receptors. Furthermore, Cameron et al. studied the phenotype of NK cells in PBMC, and they found significant decrease of CD16, CD56, CD95 and CD158a in PBMC of psoriasis compared to the healthy control (Cameron et al., 2003). NK cell are activated by IL-2, IL-12 and IL-15, which are highly expressed in psoriatic plaques (Waldmann and Tagaya, 1999; Valdimarsson et al., 1995; Ortonne, 1999; Yawalkar et al., 1998; D'Auria et al., 1999). Recent study of genetic alleles reveals that psoriasis susceptibility has been described in the MHC class I chain-related gene A (MICA) (Cheng et al., 2000). The NKG2D protein, expressed on NK cells, is a receptor for MICA (Bauer et al., 1999). That is highly expressed in human epithelial tumor (Groh et al., 2001). Therefore, some NK cell interactions may be genetically influenced in a way that may predispose to psoriasis. Activated NK cells produce Th1 cytokine such as IFN-γ that can initiate psoriatic lesions with consequent T-cell activation and infiltration (Fehniger et al., 1999).

Natural killer cells isolated from psoriatic lesions co-expressed CD161 and NKG2A but were negative for CD16 and CD158b, suggesting that they belong to the CD56^{bright}CD16^{-} NK cell subset. These skin-infiltrating NK cells release a remarkable amount of IFN-γ that activate keratinocytes from psoriatic lesions and increase MHC class I, MHC class II and ICAM-1 expression (Ottaviani et al., 2006). Chemokine and chemokin receptors on NK cells are also related with the pathogenesis of psoriasis vulgaris. Psoriatic NK cells highly express chemokine receptors CXCR3 and CCR5. In addition, they showed intermediate levels of CXCR1, CCR6 and CCR8. Psoriatic NK cells show strong migration to CCL5 and CXCL10 produced by psoriatic keratinocytes. IP-10 (CXCL10) is produced by inflamed keratinocytes with IFN-γ stimulation (Kanda et al., 2007). In conclusion, psoriatic keratinocytes appear to interact with NK cells primarily through chemokines and their receptors.

The perforin+ cells have also been studied in psoriatic lesions. Perforin is a pore-forming protein found in NK cells and CD8+ T cells, which induces osmotic lysis and subsequently apoptosis of target cells. Although there is no difference in the distribution of CD56+ NK cells between lesional and non-lesional psoriatic skin, perforin+ cells are significantly increased in the spongiotic lesions of psoriatic epidermis (Kastelan et al., 2004), suggesting a contribution of NK cells to the pathogenesis of psoriasis vulgaris.

NK cells and allergic contact hypersensitivity

While it is widely recognized that T cells and Langerhans cells play a key role in allergic contact hypersensitivity (CHS)-associated immune responses, the contributions of NK cells to CHS are much less well-studied (O'Leary et al., 2006; Boehncke et al., 2005). Even in the absence of T cells and B cells, CHS can be elaborated by skin-homing NK cells through a recruitment mechanism via dermal E-selectin, P-selectin and β2 integrins (O'Leary et al., 2006). E-selectin and P-selectin are expressed on dermal endothelium and are necessary for T cell and NK cell recruitment into inflamed lesions of the skin. T cell and NK cell traffic into cutaneous sites through adhesive and chemotactic factors such as P-selectin/E-selectin and chemokines (Luster et al., 2005). These factors tether leukocytes that roll and adhere dermal capillary venules and infiltrate into dermis via endothelial surfaces (Gainers et al., 2007). Of course, E-selectin and P-selectin ligands, which are called skin-homing receptors, are also important for the T cell and NK cell recruitment. The ubiquitous P-selectin ligand on T cells and NK cells is P-selectin glycoprotein ligand-1 (PSGL-1), which is also a major E-selectin ligand on T cells. The predominant E-selectin ligands on NK cells are PSGL-1 and protease-resistant glycolipids (Gainers et al., 2007). Therefore, E-selectin and P-selectin are the targets of the new strategy for the treatment of inflammatory skin diseases.

NK cells and the hair follicle epithelium in health and disease

Immunoprivileged sites are observed in a few, well-defined tissue compartments in the mammalian body. These sites include the anterior chamber of the eye, testis, the central nervous system behind blood–brain–barrier and the hamster cheek pouch (Head and Billingham, 1985; Niederkorn, 2002; Paus et al., 2003; Mellor and Munn, 2006; Simpson, 2006). Immunoprivileged tissues suppress a cytotoxic immune attack on cells and antigens harboured inside these sites by a whole range of different mechanisms. The potential mechanisms of immune privilege (IP) are (1) down-regulation or turn off of classical MHC class I expression, thereby sequestering (auto-)antigens in these sites and hindering their presentation to CD8+ T cells; (2) local production of potent immunosuppressants such as TGF-β1, IL-10 and α-MSH, thereby creating an immune-friendly milieu; (3) functional impairment of antigen-presenting cells, thereby reducing adaptive immune response; (4) absence of lymphatics, thereby reducing migrating immune cells; (5) establishment of extracellular matrix barriers, thereby hindering immune cell trafficking; (6) expression of non-classical MHC class I molecules (such as the MHC class Ib molecules HLA-G in humans, and Qa-2 in mice), thereby impairing cytotoxic T lymphocyte (CTL) function and inhibiting NK cell lysis; and (7) expression of Fas ligand (FasL), thereby enabling the killing of autoreactive, Fas-expressing T-cells and B-cells.

HFs also maintain immunoprivileged sites in the proximal ORS and hair matrix, since both epithelial compartments are characterized by very low or absent MHC class I expression. Reporting the distribution of MHC antigens in normal human skin, including human terminal HFs, (Harrist et al., 1983) showed negative MHC class I expression in the dermal papilla, proximal ORS and inner root shealth (IRS) of normal human HFs. Also, while the distal ORS shows strong expression of MHC class I and MHC class II-positive dendritic cells, the later are almost absent within the proximal ORS epithelium. The concept of anagen HF IP has been further supported by the following additional features such as expression of potent immunosuppressants [TGF-β1 (Welker et al., 1997; Foitzik et al., 2000), ACTH (Slominski et al., 1993; Slominski et al., 2000) and α-MSH (Slominski et al., 2000; Paus, et al., 1999; Botchkarev et al., 1999)], reduced number of apparently non-functional MHC class II antigen-negative, Langerhans cells in proximal parts of HFs (Christoph et al., 2000), absence of γδTCR+ lymphocytes below the bulge region (Paus et al., 1994), rare distribution of CD4+ T cells and absence of CD8+ T cells around anagen VI HFs (Christoph et al., 2000).

In this HF-IP milieu, CD8+ T cells NK cells can attack the MHC class I negative HF tissues. However, only very few perifollicular NK cells are found around healthy human anagen HFs (Christoph et al., 2000), which suggests that HF indeed inhibit or contain NK functions within tightly controlled limits of activity. For example, the NK cell inhibitor, MIF, is strongly expressed by the HF epithelium, and only very few CD56+/NKG2D+ NK cells are observed in and around normal anagen HFs compared to AA with prominent aggregations of CD56+/NKG2D+ NK around AA-HFs. Much fewer NK function-activating receptors (NKG2D, NKG2C) and more KIR-2D2/2D3 were significantly found to be expressed on peripheral blood CD56+ NK of healthy controls than of AA patients. In addition, only weak immunoreactivity for MICA was observed in normal anagen HFs compared to AA (Ito et al., 2008). Therefore, alopecia areata, an organ-specific autoimmune disease thought to result from a collapse of HF-IP, shows striking defects in NK cell inhibition/containment.

The interactions of NK cells with defined adhesion molecules in human skin epithelium remain to be systematically dissected. In this repect, β1 integrin, which is prominently expression by basal layer ORS keratinocytes, including by epithelial hair follicle stem cells, may be particulary interesting (Kloepper et al., 2008). Integrin-linked kinase plays important roles for epidermis and HF morphogenesis by modulating integrin-mediated adhesion, actin reorganization and plasma membrane dynamics in keratinocytes (Lorenz et al., 2007). Further study of the integrin expression, hair diseases and NK cells is required.

Epithelium of the kidney and NK cells

Cultured renal epithelial cells highly express MHC class I antigens. This high-level expression of MHC class I antigen limits lysis mediated by NK cells. Cytokines present within the renal microenvironment during rejection protect graft cells from lysis by NK cells by causing local up-regulation of the expression of class I MHC molecules. Up-regulation of MHC class I antigens has no effect on CTL-mediated lysis or the ADCC effector mechanism (Lin et al., 1993).

Waldeyer's ring and NK cells

Waldeyer's ring is one of the adenoid tissues that belong to the lymphoepithelial secretory immune system. This is located at the gateway of the respiratory and alimentary tract and belongs to the MALT. Innate immune responses within the MALT appear to be dominated by NK cell activities (Ashkar and Rosenthal, 2003; Matsuo et al., 2000; Yuan et al., 2004). The Adenoid-NK cells (A-NK cells) are mostly $CD56^{bright}CD16^{-}$ and display an activated phenotype. On the other hand, blood-NK cells are $CD16^{-}$. A-NK cells express several activation-induced receptors, such as NKp44, CD25, HLA-DR, carcinoembryonic antigen-related cell adhesion molecule 1 (CEACAM1) and CXCR4, which is induced by chronic interaction with antigens penetrating the body through the nasal route (Mizrahi et al., 2007). A-NK cells migrated in vitro toward a gradient of the CXCR4 ligand, CXCL12, which is found in endothelial cells of adenoid blood vessels and in the epithelial lining of the adenoids. Respiratory epithelial cells might attract A-NK cells towards the surface of the adenoids. Then, $CD56^{bright}$ A-NK cells encounter with the invading pathogens and secrete immunoregulatory cytokines, such as IFN-γ, TNF-α and IL-5, which regulate the adaptive immune system.

Airway epithelial cells and NK cells

The airways are always exposed to pathogens, allergens and environmental contaminants. The respiratory tract may be damaged by excessive immune reactions if the airway does not efficiently and effectively neutralize the potentially deleterious effects of these exposures on the delicate tissue of the air spaces. Epithelial cells undergoing physical or chemical stress and compromised function must be efficiently removed to control inflammation and promote cellular repair. Multiple mechanisms for the elimination of damaged cells have been identified, including immune cell activation and NKG2D receptor activation (Gleimer and Parham, 2003). NKG2D receptor is constitutively expressed on the cell surface of human NK cells (Diefenbach et al., 2000), γδT cells and $CD8^{+}$ T cells (Bauer et al., 1999). Ligand for NKG2D is the MHC class I chain-related (MIC) molecules, MICA and MICB (Bauer et al., 1999) and the UL-16 binding proteins 1, 2, 3 and 4 (ULBP1, 2, 3, 4) (Chalupny et al., 2003; Cosman et al., 2001). MICA and MICB expression is low or absent on normal tissues but is induced by various stresses or in some pathological conditions (Groh et al., 1999, Jinushi et al., 2003). MICA/B protein expression is absent on the primary airway epithelial cells. NKG2D ligand transcripts are also expressed in primary airway epithelial cells. ULBP2 is consistently expressed at the highest levels in these cells, whereas ULBP4 is consistently present at a level two orders of magnitude below the other ligands. Exposure to 0.3 mM H_2O_2 for 48 h induced the expression of MICA/B and

ULBP2 in immersed and differentiated normal human bronchial epithelial cells. Then, NK cells may react with NKG2D ligand^{+} epithelial cells for repairing and preserving the pulmonary tissues.

Conclusions

The complex, bi-directional interactions between epithelium and NK cells are as yet far from understood, but deserve much more careful future scrutiny. Under physiological conditions, intra- and peri-epithelial NK cells are very scarce (e.g. in human epidermis and papillary dermis), while their number often sharply increases when epithelia undergo inflammatory changes: Apparently, NK cells and epithelial cells start their interactions after inflammation is induced. However, NK cells and epithelial cells start interactions after inflammation is induced. The interaction is initiated by adhesion molecules, chemokine and its receptors and cytolytic proteins. NK cells accumulate into inflamed lesions without T cell help using chemokine, chemokine receptors, adhesion molecules and cytokines. Then, NK cells release cytolytic proteins that induce apoptosis of epithelium. Thus, the conventional strategy of treating inflammatory diseases of the epithelium (e.g. in inflammatory skin disorders) by inhibiting T cell responses should be complemented by modulating NK cell-epithelial cell interactions.

References

Anasetti, C., Martin, P.J., June, C.H., Hellstrom, K.E., Ledbetter, J.A., Rabinovitch, P.S., Morishita, Y., Hellstrom, I. and Hansen, J.A. (1987). Induction of calcium flux and enhancement of cytolytic activity in natural killer cells by cross-linking of the sheep erythrocyte binding protein (CD2) and the Fc-receptor (CD16). J Immunol **139**(6), 1772–1779.

Andre, P., Spertini, O., Guia, S., Rihet, P., Dignat-George, F., Brailly, H., Sampol, J., Anderson, P.J. and Vivier, E. (2000). Modification of P-selectin glycoprotein ligand-1 with a natural killer cell-restricted sulfated lactosamine creates an alternate ligand for L-selectin. Proc Natl Acad Sci U S A **97**, 3400–3405.

Ashkar, A.A. and Rosenthal, K.L. (2003). Interleukin-15 and natural killer and NKT cells play a critical role in innate protection against genital herpes simplex virus type 2 infection. J Virol **77**, 10168–10171.

Baekkevold, E.S., Yamanaka, T., Palframan, R.T., Carlsen, H.S., Reinholt, F.P., von Andrian, U.H., Brandtzaeg, P. and Haraldsen, G. (2001). The CCR7 ligand ELC (CCL19) is transcytosed in high endothelial venules and mediates T cell recruitment. J Exp Med **193**, 1105–1111.

Bauer, S., Groh, V., Wu, J., Steinle, A., Phillips, J.H., Lanier, L.L. and Spies, T. (1999). Activation of NK cells and T cells by NKG2D, a receptor for stress-inducible MICA. Science **285**, 727–729.

Bennett, I.M., Zatsepina, O., Zamai, L., Azzoni, L., Mikheeva, T. and Perussia, B. (1996). Definition of a natural killer NKR-P1A+/CD56-/CD16- functionally immature human NK cell subset that differentiates in vitro in the presence of interleukin 12. J Exp Med **184**(5), 1845–1856.

Berman, M.E., Xie, Y. and Muller, W.A. (1996). Roles of platelet/endothelial cell adhesion molecule-1 (PECAM-1, CD31) in natural killer cell transendothelial migration and beta 2 integrin activation. J Immunol **156**(4), 1515–1524.

Boehncke, W.H., Schon, M.P., Giromolomi, G., Griffiths, C., Bos, J.D., Thestrup-Pedersen, K., Cavani, A., Nestle, F., Bonish, B.K., Campbell, J.J., Brakebusch, C. and Nickoloff, B. (2005). Leukocyte extravasation as a target for anti-inflammatory therapy: which molecule to choose? Exp Dermatol **14**, 70–80.

Botchkarev, V.A., et al. (1999). Developmentally regulated expression of alpha-MSH and MC-1 receptor in C57BL/6 mouse skin suggests functions beyond pigmentation. Ann N Y Acad Sci **885**, 433.

Buentke, E., Heffler, L.C., Wilson, J.L., Wallin, R.P., Löfman, C., Chambers, B.J., Ljunggren, H.G. and Scheynius, A. (2002). Natural killer and dendritic cell contact in lesional atopic dermatitis skin—Malassezia-influenced cell interaction. J Invest Dermatol **119**(4), 850–857.

Cameron, A.L., Kirby, B., Fei, W. and Griffiths, C.E. (2002). Natural killer and natural killer-T cells in psoriasis. Arch Dermatol Res **294**(8), 363–369.

Cameron, A.L., Kirby, B. and Griffiths, C.E. (2003). Circulating natural killer cells in psoriasis. Br J Dermatol **49**(1), 160–164.

Cheng, L., Zhang, S.Z., Xiao, C.Y., et al. (2000). The A5.1 allele of the major histocompatibility complex class I chain-related gene A is associated with psoriasis vulgaris in Chinese. Br J Dermatol **143**, 324–329.

Chiesa, M.D., Vitale, M., Carlomagno, S., Ferlazzo, G., Moretta, L. and Moretta, A. (2003). The natural killer cell-mediated killing of autologous dendritic cells is confined to a cell subset expressing CD94/NKG2A, but lacking inhibitory killer Ig-like receptors. Eur J Immunol **33**, 1657–1666.

Christoph, T., Müller-Röver, S., Audring, H., Tobin, D.J., Hermes, B., Cotsarelis, G., Rückert, R. and Paus, R. (2000). The human hair follicle immune system: cellular composition and immune privilege. Br J Dermatol **142**(5), 862–873.

Cooper, M.A., Fehniger, T.A., Turner, S.C., Chen, K.S., Ghaheri, B.A., Ghayur, T., Carson, W.E. and Caligiuri, M.A. (2001). Human natural killer cells: a unique innate immunoregulatory role for the CD56bright subset. Blood **97**, 3146–3151.

Cosman, D., Mullberg, J., Sutherland, C.L., Chin, W., Armitage, R., Fanslow, W., Kubin, M. and Chalupny, N.J. (2001). ULBPs, novel MHC class I-related molecules, bind to CMV glycoprotein UL16 and stimulate NK cytotoxicity through the NKG2D receptor. Immunity **14**, 123–133.

D'Auria, L., Bonifati, C., Cordiali-Fei, P., Leone, G., Picardo, M., Pietravalle, M., Giacalone, B. and Ameglio, F. (1999). Increased serum interleukin-15 levels in bullous skin diseases: correlation with disease intensity. Arch Dermatol Res **291**(6), 354–356.

David, M. (2006). Mechanisms of innate immunityChapter 6. In: Male, D., Brostoff, J., Roth, D.B. and Roitt, I.

(eds) Immunology 7th ed. London: Mosby, pp. 127–144.

Diefenbach, A., Jamieson, A.M., Liu, S.D., Shastri, N. and Raulet, D.H. (2000). Ligands for the murine NKG2D receptor: expression by tumor cells and activation of NK cells and macrophages. Nat Immun **1**, 119–126.

Fehniger, T.A., Carson, W.E. and Caligiuri, M.A. (1999). Costimulation of human natural killer cells is required for interferon gamma production. Transplant Proc **31**(3), 1476–1478.

Foitzik, K., et al. (2000). Control of murine hair follicle regression (catagen) by TGF-β1 in vivo. FASEB J **14**, 752.

Fogler, W.E., Volker, K., McCormick, K.L., Watanabe, M., Ortaldo, J.R. and Wiltrout, R.H. (1996). NK cell infiltration into lung, liver, and subcutaneous B16 melanoma is mediated by VCAM-1/VLA-4 interaction. J Immunol **156**(12), 4707–4714.

Gainers, M.E., Descheny, L., Barthel, S.R., Liu, L., Wurbel, M.A. and Dimitroff, C.J. (2007). Skin-homing receptors on effector leukocytes are differentially sensitive to glyco-metabolic antagonism in allergic contact dermatitis. J Immunol **179**(12), 8509–8518.

Galandrini, R., De Maria, R., Piccoli, M., Frati, L. and Santoni, A. (1994). CD44 triggering enhances human NK cell cytotoxic functions. J Immunol **153**(10), 4399–4407.

Gleimer, M. and Parham, P. (2003). Stress management: MHC class I and class I-like molecules as reporters of cellular stress. Immunity **19**, 469–477.

Groh, V., Rhinehart, R., Secrist, H., Bauer, S., Grabstein, K.H. and Spies, T. (1999). Broad tumor-associated expression and recognition by tumor-derived gamma delta T cells of MICA and MICB. Proc Natl Acad Sci U S A **96**, 6879–6884.

Groh, V., et al. (2001). Costimulation of CD8αβ T cells by NKG2D via engagement by MIC induced on virus-infected cells. Nat Immunol **2**, 255–260.

Harrist, T.J., et al. (1983). Distribution of major histocompatibility antigens in normal skin. Br J Dermatol **109**, 623.

Head, J.R. and Billingham, R.E. (1985). Immunologically privileged sites in transplantation immunology and oncology. Perspect Biol Med **29**, 115–131.

Inngjerdingen, M., Damaj, B. and Maghazachi, A.A. (2001). Expression and regulation of chemokine receptors in human natural killer cells. Blood **97**, 367–375.

Ito, T., Ito, N., Saathoff, M., Stampachiacchiere, B., Bettermann, A., Bulfone-Paus, S., Takigawa, M., Nickoloff, B.J. and Paus, R. (2005). Immunology of the human nail apparatus: the nail matrix is a site of relative immune privilege. J Invest Dermatol **125**(6), 1139–1148.

Ito, T., Ito, N., Saatoff, M., Hashizume, H., Fukamizu, H., Nickoloff, B.J., Takigawa, M. and Paus, R. (2008). Maintenance of hair follicle immune privilege is linked to prevention of NK cell attack. J Invest Dermatol **128**(5), 1196–1206.

Chalupny, N.J., Sutherland, C.L., Lawrence, W.A., Rein-Weston, A. and Cosman, D. (2003). ULBP4 is a novel ligand for human NKG2D. Biochem Biophys Res Commun **305**, 129–135.

Jinushi, M., Takehara, T., Tatsumi, T., Kanto, T., Groh, V., Spies, T., Kimura, R., Miyagi, T., Mochizuki, K., Sasaki, Y. and Hayashi, N. (2003). Expression and role of MICA and MICB in human hepatocellular carcinomas and their regulation by retinoic acid. Int J Cancer **104**, 354–361.

Kanda, N., Shimizu, T., Tada, Y. and Watanabe, S. (2007). IL-18 enhances IFN-gamma-induced production of CXCL9, CXCL10, and CXCL11 in human keratinocytes. Eur J Immunol **37**(2), 338–350.

Kastelan, M., Prpi Massari, L., Gruber, F., Zamolo, G., Zauhar, G., Coklo, M. and Rukavina, D. (2004). Perforin expression is upregulated in the epidermis of psoriatic lesions. Br J Dermatol **151**(4), 831–836.

Kloepper, J.E., Tiede, S., Brinckmann, J., Reinhardt, D.P., Meyer, W., Faessler, R. and Paus, R. (2008). Immunophenotyping of the human bulge region: the quest to define useful in situ markers for human epithelial hair follicle stem cells and their niche. Exp Dermatol **17**(7), 592–609.

Lanier, L.L. (2000). The origin and functions of natural killer cells. Clin Immunol **95**, S14–S18.

Lorenz, K., Grashoff, C., Torka, R., Sakai, T., Langbein, L., Bloch, W., Aumailley, M. and Fässler, R. (2007). Integrin-linked kinase is required for epidermal and hair follicle morphogenesis. J Cell Biol **177**(3), 501–513.

Lin, Y., Proud, G., Taylor, R.M. and Kirby, J.A. (1993). Renal allograft rejection: protection of renal epithelium from natural killer cells by cytokine-induced up-regulation of class I major histocompatibility antigens. Immunology **79**, 290–297.

Luster, A.D., Alon, R. and von Andrian, U.H. (2005). Immune cell migration in inflammation: present and future therapeutic targets. Nat Immunol **6**, 1182–1190.

Mäenpää, A., Jääskeläinen, J., Carpén, O., Patarroyo, M. and Timonen, T. (1993). Expression of integrins and other adhesion molecules on NK cells; impact of IL-2 on short- and long-term cultures. Int J Cancer **53**(5), 850–855.

Maghazachi, A.A. (2003). G protein-coupled receptors in natural killer cells. J Leukoc Biol **74**, 16–24.

Maghazachi, A.A. (2005). Compartmentalization of human natural killer cells. Mol Immunol **42**(4), 523–529.

Matsuo, K., Yoshikawa, T., Asanuma, H., Iwasaki, T., Hagiwara, Y., Chen, Z., Kadowaki, S.E., Tsujimoto, H., Kurata, T. and Tamura, S.I. (2000). Induction of innate immunity by nasal influenza vaccine administered in combination with an adjuvant (cholera toxin). Vaccine **18**, 2713–2722.

Mellor, A.L. and Munn, D.H. (2000). Immunology at the maternal-fetal interface: lessons for T cell tolerance and suppression. Ann Rev Immunol **18**, 367–391.

Mizrahi, S., Yefenof, E., Gross, M., Attal, P., Ben Yaakov, A., Goldman-Wohl, D., Maly, B., Stern, N., Katz, G., Gazit, R., Sionov, R.V., Mandelboim, O. and Chaushu, S. (2007). Aphenotypic and functional characterization of NK cells in adenoids. J Leukoc Biol **82**, 1095–1105.

Nagler, A., Lanier, L.L., Cwirla, S. and Phillips, J.H. (1989). Comparative studies of human FcRIII-positive and negative natural killer cells. J Immunol **143**(10), 3183–3191.

Niederkorn, J.Y. (2002). Immune privilege in the anterior chamber of the eye. Crit Rev Immunol **22**, 13–46.

O'Leary, J.G., Goodarzi, M., Drayton, D.L. and von Andrian, U.H. (2006). T cell- and B cell-independent adaptive immunity mediated by natural killer cells. Nat Immunol **7**, 507–516.

Ortonne, J.P. (1999). Recent developments in the understanding of the pathogenesis of psoriasis. Br J Dermatol **140**(Suppl. 54), 1–7.

Ottaviani, C., Nasorri, F., Bedini, C., de Pità, O., Girolomoni, G. and Cavani, A. (2006). CD56brightCD16(−) NK cells accumulate in psoriatic skin in response to CXCL10 and CCL5 and exacerbate skin inflammation. Eur J Immunol **36**(1), 118–128.

Paus, R., et al. (1994). Distribution and changing density of gamma-delta T cells in murine skin during the induced hair cycle. Br J Dermatol **130**, 281.

Paus, R., et al. (1999). The skin POMC system (SPS): leads and lessons from the hair follicle. Ann N Y Acad Sci **885**, 350.

Paus, R., Ito, N., Takigawa, M. and Ito, T. (2003). The hair follicle and immune privilege. J Investig Dermatol Symp Proc **8**, 188–194.

Pérez-Villar, J.J., Melero, I., Gismondi, A., Santoni, A. and López-Botet, M. (1996a). Functional analysis of alpha 1 beta 1 integrin in human natural killer cells. Eur J Immunol **26**(9), 2023–2029.

Pérez-Villar, J.J., Zapata, J.M., Melero, I., Postigo, A., Sánchez-Madrid, E. and López-Botet, M. (1996b). Expression and function of alpha 4/beta 7 integrin on human natural killer cells. Immunology **89**(1), 96–104.

Pinola, M., Renkonen, R., Majuri, M.L., Tiisala, S. and Saksela, E. (1994). Characterization of the E-selectin ligand on NK cells. J Immunol **152**(7), 3586–3594.

Seaman, W.E. (2000). Natural killer and natural killer T cells. Arthritis Rheum **43**, 1204–1217.

Simpson, E.A. (2006). Historical perspective on immunological privilege. Immunol Rev **213**, 12–22.

Slominski, A., et al. (1993). Detection of proopiomelanocortin-derived antigens in normal and pathologic human skin. J Lab Clin Med **122**, 658.

Slominski, A., et al. (2000). Corticotropin releasing hormone and proopiomelanocortin involvement in the cutaneous response to stress. Physiol Rev **80**, 979.

Somersalo, K. and Saksela, E. (1991). Fibronectin facilitates the migration of human natural killer cells. Eur J Immunol **21**(1), 35–42.

Valdimarsson, H., Baker, B.S., Jónsdóttir, I., Powles, A. and Fry, L. (1995). Psoriasis: a T-cell-mediated autoimmune disease induced by streptococcal superantigens? Immunol Today **16**(3), 145–149.

Waldmann, T.A. and Tagaya, Y. (1999). The multifaceted regulation of interleukin-15 expression and the role of this cytokine in NK cell differentiation and host response to intracellular pathogens. Annu Rev Immunol **17**, 19–49.

Welker, P., et al. (1997). Hair cycle-dependent changes in the gene expression and protein content of transforming factor-β1 and β3 in murine skin. Arch Dermatol Res **289**, 554.

Yawalkar, N., Karlen, S., Hunger, R., Brand, C.U. and Braathen, L.R. (1998). Expression of interleukin-12 is increased in psoriatic skin. J Invest Dermatol **111**(6), 1053–1057.

Yuan, Q. and Walker, W.A. (2004). Innate immunity of the gut: mucosal defense in health and disease. J Pediatr Gastroenterol Nutr **38**, 463–473.

Chapter Twenty-Two

22

NK cell–T cell interactions

Benedict J. Chambers

CHAPTER CONTENTS

Don't walk in front of me, I may not follow. Don't walk behind me, I may not lead. Walk beside me and be my friend.

Albert Camus

ABSTRACT

As lymphocytes of the innate immune system, natural killer (NK) cells play an important role in the immunosurveillance against infection and cancers. How the NK cells respond to a specific infection can have knock-on effects on the development of the adaptive immune responses. However, T cells in the later stages of the immune response may also play a role in regulating NK cells. NK cells can stimulate T cells indirectly by activating dendritic cells (DCs). IL-12 release by DCs leads to IFNγ production by NK cells, which increases MHC I and MHC II expression and IL-12 production. These events enhance T-cell activation and T-cell polarization. Activated NK cells may also stimulate T cells directly by expressing co-stimulatory molecules and MHC molecules. IFNγ from NK cells may directly polarize Th1 cell response by inducing T-bet. NK cell-mediated killing of DCs limits stimulation of T cells. NK cells can limit immunity by directly killing activated T cells.

KEY WORDS

TH1, TH2, TH17, Tregs, TGFβ, IL21, IFNγ

Introduction

Natural killer (NK) cells are lymphocytes of the innate immune system that play a vital role in host defence. The primary function of NK cells is to kill infected cells and produce cytokines and chemokines during the early phase of pathogen infection (Vivier et al., 2008). These NK cell functions are regulated by the expression of receptors that can lead to activating or inhibiting signals (Bryceson and Long, 2008). Activating receptors on NK cells recognize a variety of molecular structures and use a variety of signalling pathways. These receptors include DNAX-activating protein (DAP)12-associated receptors such as CD158c/e2/f/g/h/i/j (human NK cells), CD94/NKG2C (mouse and human), Ly49D/H (mouse), CD16, (mouse and human), FcεRIγ and CD3-associated receptors NKp30, 44 and 46 (human) receptors and DAP-10-associated receptor NKG2D (mouse and human). Inhibitory receptors such as CD158a/b1/b2/e1/f/k (human), CD94/NKG2A (mouse and human) and Ly49A/C/G2/I (mouse) on NK cells contain immunoreceptor tyrosine-based inhibition motifs (ITIMs) in their tail which recruit tyrosine phosphatases upon ligation of either classical or non-classical MHC class I molecules. Taken together, NK cell responses to healthy or infected cells are based upon recognition of a broad range of ligands, which tightly control both activating and inhibitory signals.

As part of the adaptive immune system, T cells play a role in mounting cytotoxic or cytokine responses to both intracellular and extracellular infections. T cells can recognize self from non-self by the T cell receptor (TCR) discerning the differences in peptides presented on MHC class I ($CD8^+$ T cells) or MHC class II ($CD4^+$ T cells) (Viret and Janeway, 1999). T cell diversity is generated by rearrangement of the genes that make up the TCR. In order to prevent auto-reactivity to self-peptides, most T cells require education within the thymus where the relative binding to host MHC molecules leads to death or survival. Naïve T cells that leave the thymus still require further signals to become fully activated. Thus, activation of naïve T cells requires two signals, one via MHC molecules, and the other by co-stimulatory molecules such as CD80 and CD86. Furthermore, cytokines can induce phenotypic changes in the T cells. After all these signals, antigen-specific T cells can become effector cells, which can lead to the development of specific antigen memory.

T lymphocyte populations

$\alpha\beta^+$ T cells can be divided into at least four major groups: $CD8^+$ cytotoxic T cells (Tc cells), $CD4^+$ helper T cells (Th cells), $CD4^+$ regulatory T cells (Treg) and $CD4^+$ IL-17 producing T cells (Th17 cells) (Zhu and Paul, 2008). A fifth subset could also be added that expresses an invariant α chain of the TCR, namely the NKT cells (Godfrey et al., 2004). All these T cells require thymus education before going out into the periphery.

Th, Treg and Th17 cells are the primary cytokine producers that can affect further development of adaptive immune responses. Originally, Mosmann and Coffman described that T cells could be divided into two populations based on their cytokine production (Mosmann et al., 1986). IFNγ-producing cells became known as Th1 cells, and IL-4-producing cells were called Th2. Th1 cell differentiation is dependant on the transcription factor T-bet (Lighvani et al., 2001). Thus T-bet deficient mice have defects in IFNγ production in both Th cells and Tc cells (Lugo-Villarino et al., 2003; Sullivan et al., 2003). Th2 cell differentiation is dependant on the transcription factors STAT6 and GATA-3 (Kaplan et al., 1996; Zhang et al., 1997; Zheng and Flavell, 1997). In general, Th1 responses are then associated with immune responses against intracellular infections, while Th2 responses aid in the generation of antibodies specific for extracellular infections (Zhu and Paul, 2008).

Treg and Th17 cells were discovered more recently. Some Treg cells (nTreg) express constitutively the transcriptional factor Foxp3 and CD25 (Vignali et al., 2008). However, Foxp3 expression can be induced by incubating $CD4^+$ T cells with TGFβ (iTreg) (Vignali et al., 2008). Treg cells, similar to Th1 and Th2 cells, can make IL-10, but they are also major producers of TGFβ. Both of these cytokines can suppress immune responses (Li et al., 2006; Moore et al., 2001). Th17 cells produce IL-17, a proinflammatory cytokine that is involved in the host defence against extracellular infections, such as bacteria and fungi, and in autoimmune diseases (Weaver et al., 2007). Th17 cell development is dependant on IL-6, IL-23 and TGFβ. The orphan nuclear receptor RORγt is a master regulator for the development of Th17 cells (Ivanov et al., 2006). Th17 cells are also a source for IL-21, immunomodulatory cytokine (Mehta et al., 2004), and IL-22, which promotes anti-microbial defence, protecting against damage and re-organizing non-immune tissues (Wolk and Sabat, 2006).

The function of the Tc cells is primarily to kill cells expressing MHC class I complexes containing foreign derived peptides such as that from viruses. Tc cells can also produce IFNγ to enhance perforin production in T cells as well as induce intracellular cytotoxic functions in infected cells. However, like Th cells, some Tc cells can also express IL-4 (Tc2 cells), but these cells are not as cytotoxic as IFNγ-producing Tc cells (Tc1 cells) (Mosmann et al., 1997).

NKT cells are specialized T cells that express a restricted invariant TCR Vα chain that recognizes lipids bound to the non-classical MHC class I molecule CD1d (Godfrey et al., 2004). Typically, NKT cells have a memory-like phenotype as well as expressing molecules normally associated with NK cells. Interestingly, NKT cells can express either Th1 or Th2 cytokines and may be an early source for both of these cytokines during an immune response (Godfrey et al., 2004).

NK cell populations

NK cells are lymphocytes of the innate immune system and have an important function in the early defence against certain viruses, microbial infections and cancer (Dunn et al., 2004; Korbel et al., 2004; Lee et al., 2007; Lodoen and Lanier, 2006; Scalzo, 2002). NK cells were originally identified for their ability to kill tumour cells in vitro without the need for prior sensitization (Kiessling et al., 1975). In addition, NK cells are major producers of cytokines and chemokines. In humans, increased cytokine production is found in the $CD56^{bright}$ population of NK cells, while $CD56^{dim}$ NK cells express more perforin and so are more cytotoxic. In mice, the $CD27^{bright}$ population can express high levels

Table 22.1 NK cell effects on T cells

NK cell function	Effect on T cells	References
Th1 cytokine production	Induce Th1, protection against infection and tumours	Agaugue et al. (2008), Biron et al. (1999), Combe et al. (2005), Ferlazzo et al. (2004), Kos and Engleman (1996), Mailliard et al. (2003), Martin-Fontecha et al. (2004), Robbins et al. (2007), Vankayalapati et al. (2004)
	Inhibiting effects of Th17	Ashkar et al. (2009), Lo et al. (2008)
Th2 cytokine production	Induction of Th2 responses	Agaugue et al. (2008)
IL-10 and TGFβ production	Inhibit T cell responses	Li et al. (2006), Moore et al. (2001)
Antigen presentation	Presentation of antigen on MHC class II	Scala et al. (1985), Brooks and Moore (1986), D'Orazio and Stein-Streilein (1996), Roncarolo et al. (1991), Hanna et al. (2004)
Co-stimulation	B7 family members, OX40L, CD244	Azuma et al. (1993), Lee et al. (2003), Hanna et al. (2004), Zingoni et al. (2004), Assarsson et al. (2004), Saudemont et al. (2005)
Cytotoxicity	Elimination of T cells	Cerboni et al. (2007), Rabinovich et al. (2003), Roy et al. (2008)
	Elimination of DC inhibits T cell activation	Gilbertson et al. (1986), Hayakawa et al. (2004), Laffont et al. (2008), Ruggeri et al. (2002), Yu et al. (2006)

of cytokines (Hayakawa and Smyth, 2006). While IFNγ is the chief cytokine produced by NK cells, other cytokines such as IL-2, IL-3, IL-5, IL-10, IL-13, granulocyte/macrophage colony-stimulating factor (GM-CSF), tumour necrosis factor (TNF) and TGF-β are made by NK cells. NK cells can also produce the chemokines MIP1α (CCL3) and IL-8 (CXCL8) (Biron et al., 1999; Cooper et al., 2001; Loza and Perussia, 2001; Strowig et al., 2008). Similar to T cells, NK cells may also be defined as either NK1 or NK2 cells (Loza and Perussia, 2001; Peritt et al., 1998). TGF-β and IL-10 production by a subset of NK cells suggests that regulatory NK cells (NKreg) may also exist. However, unlike T cells in which transcription factors can clearly define Th subsets, this does not seem to be case for NK cells (Katsumoto et al., 2004; Samson et al., 2003; Townsend et al., 2004).

NK cell cytokine production and T cell activation

The IFNγ/IL-12 axis plays an important role in host defence against intracellular infections. Upon infection with pathogens, IL-12 and L-18 derived from dendritic cells (DCs)/macrophages can induce IFNγ production by NK cells (Degli-Esposti and Smyth, 2005). In DCs and macrophages, IFNγ induces the production of free radicals, which are important in intracellular killing of pathogens, and IFNγ increases the surface expression of MHC class I and MHC class II, which enhances their recognition by T cells. IFNγ can also induce T-bet expression in lymphoid and myeloid cells (Lighvani et al., 2001), suggesting that NK cells play an early role in Th1 cell polarization. IL-12 production is also enhanced by IFNγ feedback on DCs and macrophages. IL-12 activation of T cells leads to the polarization of Th1 cells and the activation of CD8$^+$ Tc cells leading to the development of cytotoxic and memory T cells (Agaugue et al., 2008; Biron et al., 1999; Combe et al., 2005; Fan et al., 2006; Ferlazzo et al., 2004; Kos and Engleman, 1996; Mailliard et al., 2003; Martin-Fontecha et al., 2004; Robbins et al., 2007; Tosi et al., 2004; Vankayalapati et al., 2004). Indeed, depletion of NK cells, and so loss of IFNγ-producing cells, during infection can skew the immune responses from a Th1 response to a Th2 response (Byrne et al., 2004; Hoshino et al., 1999). Furthermore, IFNγ production by NK cells suppresses IL-17 protecting the host from inflammation and infection (Ashkar et al., 2009; Lo et al., 2008) (Table 22.1).

NK2 cells have been shown to exist as a stage in NK cell maturation and upon Th2 cytokine stimulation of NK cells (Loza and Perussia, 2001; Peritt et al., 1998). NK2 induction is in part dependant on STAT6 (Katsumoto et al., 2004; Yu et al., 1998). However, unlike Th2 cells, NK2 cells do not appear to make IL-4, but NK2 cells do produce IL-5 and IL-13 (Deniz et al., 2002; Peritt et al., 1998). IL-5 and IL-13 produced by NK2 cells may then contribute to the generation of antibodies to extracellular pathogens by inducing Th2 cells. However, IL-5 production may also have consequences with regards to the induction of allergic inflammation (Korsgren et al., 1999; Walker et al., 1998) and in patients with hyper-IgE (Aktas et al., 2005). IL-5 and IL-13 can be induced in NK cells by IL-4. Since mast cells and NKT cells are early sources of IL-4 and can respond to extracellular infections, these two cell types may be able to induce NK2 responses in vivo. IL-4-treated NK cells may also enhance Th2 responses by preventing polarization of T cells to Th1 cells (Agaugue et al., 2008) (Table 22.1).

NK cells also can play a regulatory role by secreting TGFβ (Gray et al., 1998; Horwitz et al., 1999; Laouar et al., 2005; Meadows et al., 2006; Su et al., 1993) and IL-10 (Cai et al., 1999; Cooper et al., 2001; Deniz et al., 2008; Mehrotra et al., 1998). Both TGFβ and IL-10 inhibit Th1 responses by inhibiting DC maturation or antagonizing proinflammatory cytokines (Li et al., 2006; Moore et al., 2001). However, because of their ability to make TGFβ, NK cells may be involved in the development of Treg and Th17 cells since TGFβ can induce both Treg and Th17 cells (Zhu and Paul, 2008). Thus theoretically, such regulatory NK cells, NKreg, could essentially 'hand the torch' to Treg and Th17 cells during the immune response (Table 22.1).

NK cells as antigen-presenting cells

Antigen-presenting cells (APC), such as DCs, macrophages and B cells, have the ability to take up, process and then present antigen on MHC class I and MHC class II, and with the help of co-stimulatory molecules, to stimulate naïve T cells. However, NK cells may also act as APC. MHC class II$^+$ NK cells were originally described as a population that arose in human mixed lymphocyte cultures (Phillips et al., 1984). Scala et al. demonstrated that a subset of fresh human NK cells could express MHC class II (Scala et al., 1985). In that study, NK cells were found to be equally efficient as monocytes in presenting soluble antigens on MHC class II and stimulating T cells in mixed lymphocyte cultures. However, a later study did not find that NK cells could present soluble antigen but could confirm indeed that these NK cells were able to stimulate T cells (Brooks and Moore, 1986). Using MHC class II expressing NK cell clones, Roncarolo et al. found the ability of NK cells to present soluble antigens was dependant on the type of antigen used, suggesting that NK cells were not as efficient in processing and presenting antigens as monocyte-derived cells (Roncarolo et al., 1991). Presentation of superantigens on MHC class II expressing NK cells could also induce non-specific T cell expansion, which was suggested to play a role in the elimination of superantigen elaborating pathogens (D'Orazio and Stein-Streilein, 1996). More recently, MHC class II expression was also observed on NK cells from inflamed tissues (Hanna et al., 2004). These authors also demonstrated that MHC class II$^+$ NK cells could present antigen derived from target cells that they had lysed previously. Furthermore, CD16$^+$ MHC class II$^+$ NK cells could efficiently process and present immune complexes (Hanna et al., 2004).

In order to function as APC, NK cells also must express co-stimulatory molecules. Originally, NK cell clones were found to express CD80 (Azuma et al., 1993). NK cells activated in vitro express a number of co-stimulatory molecules, including CD70, CD80, CD86 and CD134L (OX40L) (Hanna et al., 2004; Zingoni et al., 2004). NK cell-induced T cell proliferation can be inhibited by blocking these co-stimulatory molecules, substantiating the APC-like function of activated NK cells (Hanna et al., 2004; Zingoni et al., 2004). Expression of co-stimulatory molecules on NK cells is dependant on activating cytokines such as IL-2, IL-12 and IL-15, as well as cross-linking activating receptors such as NKG2D or CD16 (Hanna et al., 2004). In the case of CD16 cross-linking, it is interesting to note that immune complexes cannot only deliver antigen to MHC class II but also induce co-stimulatory molecules on the cell surface of NK cells (Hanna et al., 2004; Zingoni et al., 2004). This would allow NK cells to potentially enhance T cell responses to immune complexes in a manner similar to macrophages and DCs. In a similar vein, NKG2D mediated lysis of tumour cells or pathogen-infected cells expressing NKG2D ligands could induce the APC-like activity in NK cells allowing them to stimulate or enhance antigen-specific T cell responses at the site of the killing. Recently, MHC class II$^+$ NK cells were found in mice, however, their ability to present antigen to T cells still needs to be unravelled (Blasius et al., 2007; Caminschi et al., 2007; Chan et al., 2006; Vosshenrich et al., 2007).

NK cells express other molecules that can provide co-stimulation to T cells. These include CD244 and CD274 (Assarsson et al., 2004; Lee et al., 2003; Saudemont et al., 2005). CD244 expression on mouse NK cells acts as a co-stimulator by binding its ligand CD48 on T cells. CD244 could enhance the proliferation of both CD4$^+$ and CD8$^+$ T cells upon cytokine stimulation or CD3 cross-linking (Assarsson et al., 2004). We proposed a model by which NK cells in the proximity of DC–T cell stimulation would enhance the T cell responses by not only secreting cytokines but also providing co-stimulation via CD48–CD244 interactions (Assarsson et al., 2005). CD274 (PD-L1, B7-H1) is expressed on mouse NK cells during *Listeria monocytogenes* infection and can be induced by the chemokine CXCL10 (Saudemont et al., 2005; Seo et al., 2008). CD274 has been associated with having inhibitory effects by binding PD-1 on T cells (Greenwald et al., 2005). However, the presence of CD274$^+$ NK cells was found to induce T-cell proliferation and cytokine production. Interestingly, this activation appeared to be independent of PD-1 and may be associated with the ability of PD-L1 to bind other members of the B7 family (Butte et al., 2007, 2008) (Table 22.1).

While NK cells can play a role as APCs and co-stimulators of T cell responses, the question remains as to where these events would take place in vivo. NK cells can be recruited to mouse lymph nodes during inflammation via CXCR3 (Martin-Fontecha et al., 2004). NK cells are found in the T-cell areas of the human lymph nodes, suggesting that they are involved in aiding DCs in stimulating T cells (Fehniger et al., 2003). Intravital staining of murine lymph nodes has shown that under normal conditions, NK cells are found in both the paracortex and medulla (Bajenoff et al., 2006). During infection with *Leishmania major*, NK cells were found to be more associated with the paracortical region. Interestingly, NK cells appear to be less motile than T cells and had prolonged contacts with MHC class II$^+$ cells within the lymph nodes suggesting that these interactions may be important in the generation of T cell responses (Bajenoff et al., 2006). However, NK cells are also recruited to non-lymphoid tissues during inflammatory responses (Buentke et al., 2002; Glas et al., 2000; Huang et al., 2006; Prlic et al., 2005; Smyth et al., 1998; Thapa et al., 2007). Therefore, activated NK cells may also stimulate T cells outside the lymphoid organs allowing in situ stimulation of T cells. (Table 22.1)

NK cell-mediated killing as a means of controlling T cell responses

Naïve T cells require presentation of antigens and co-stimulation from DCs in order to be activated. However, since ex vivo-derived DCs are sensitive to NK cell-mediated lysis, it was postulated early on that this could inhibit T cell responses. Gilbertson et al. demonstrated that NK cells added to mixed lymphocyte cultures lead to diminished CD8$^+$ CTL responses (Gilbertson et al., 1986). While these authors did not ascertain whether this was due to NK cell-mediated killing or suppression of DCs, NK cells were subsequently found to kill ex vivo-derived DCs in vitro (Chambers et al., 1996). This was later shown to be also true in vivo since depleting mice of NK cells lead to increased DC survival (Hayakawa et al., 2004). Subcutaneous or intravenous injection of peptide loaded DCs derived from bone marrow into NK cell-depleted mice lead to enhanced CTL responses when compared to mice that were not depleted of NK cells (Hayakawa et al., 2004).

Since that study, several research groups have demonstrated the vital role NK cell-mediated killing of DCs may play in hindering T cell responses in transplantation. Velardi and co-workers demonstrated that the injection of allo-NK cells may improve bone marrow transplant graft success by removing host DCs (Ruggeri et al., 2002). This would in turn limit the number of allo-reactive T cells. More recently, Yu et al. and Laffont et al. demonstrated that NK cell elimination of allo-DCs also lead to acceptance of allo-grafted skin in mice (Laffont et al., 2008; Yu et al., 2006). Furthermore, it has been observed that the numbers of CD8α^+ DCs are increased in the lymph nodes of mice depleted of NK cells (Giroux et al., 2007) (Table 22.1).

DCs matured with TLR-binding ligands can enhance T cell responses, which could potentially enhance their vaccination potential (Tacken et al., 2007). However, DCs pretreated with TLR-agonists are less sensitive to NK cell-mediated lysis in part due to increased surface expression of both classical and non-classical MHC class I molecules (Carbone et al., 1999; Chambers et al., 1996; Della Chiesa et al., 2003; Persson et al., 2008; Wilson et al., 1999). This suggests that in order to enhance the T cell response when using DCs in immunotherapy, one may want to use activated DCs to avoid NK cell-mediated elimination.

Another means by which NK cells may limit T cell responses is by direct killing of the T cells. While resting T cells are not sensitive to NK cell-mediated lysis, antigen-activated T cells can be killed in vitro by NK cells through NKG2D ligands (Cerboni et al., 2007; Rabinovich et al., 2003; Roy et al., 2008) or NKp46 ligands (Cerboni et al., 2007; Roy et al., 2008) expressed on the T cells. The induction of the NKG2D ligands on activated T cells is due to phosphorylation of ataxia-telangiectasia mutated/Rad3-related (ATM/ATR) kinases upon TCR ligation (Cerboni et al., 2007). The importance of NKG2D ligand induction on T cells has not yet been investigated in vivo, however studies with NKG2D deficient mice have not revealed any obvious alteration in T cell repertoires (Guerra et al., 2008). Thus, while it is unclear why T cells would upregulate NKG2D ligands, this may be a means by which NK cells can limit the T cell expansion following T cell activation.

Two molecules have also been identified that inhibit NK cell-mediated killing of T cells, namely Qa1 and CD48 in mice. The non-classical MHC class I molecule Qa1 protects CD4$^+$ T cells from killing by NKG2A$^+$ NK cells both in vitro and in vivo (Lu et al., 2007). However, the induction of autoimmune disease could be inhibited by NK cell-mediated elimination of T cells upon delivery of blocking anti-Qa1 antibodies into mice (Lu et al., 2007). CD48 is expressed on all T cells and can bind CD244. In mice, CD244 may have bi-directional roles where it acts as a non-MHC linked inhibitory receptor found on NK cells as well as a co-stimulatory ligand for T cells (Assarsson et al., 2004; McNerney et al., 2005). Since most cells of hematopoietic origin express CD48, it is not surprising that lack of CD244 expression on NK cells can lead to killing of T

Table 22.2 T cell effects on NK cells

T cell subtype	Effect on NK cells	References
CD4	Th1 and Th2 cytokines can induce NK1 or NK2 cells	Peritt et al. (1998)
	Help NK cell anti-tumour activity	Shimizu and Fujii (2009), van den Broeke et al. (2003)
CD8	Th1 cytokines induce NK1	Peritt et al. (1998)
	Help NK cell anti-tumour activity	Shanker et al. (2007)
Treg	Inhibit NK cell activity against tumours or infections	Barao et al. (2006), Giroux et al. (2007), Ralainirina et al. (2007), Romagnani et al. (2005), Simon et al. (2007), Smyth et al. (2006), Trzonkowski et al. (2004)
	NK cell elimination	He et al. (2007)
Th17	IL-21 production contracts NK cell response	Brady et al. (2004), Kasaian et al. (2002), Ma et al. (2003), Roda et al. (2006), Sivori et al. (2003), Wendt et al. (2007)
NKT	Enhances NK cell responses by IFNγ production	Smyth et al. (2002)

cells and NK cells in mice (Lee et al., 2004b; Taniguchi et al., 2007) (Table 22.1).

T cell effects on NK cells

Although NK cells may affect T cell activation and functions, T cells may also be able to exert influence on NK cell phenotype and function (Vivier et al., 2008). As outlined previously, NK1 and NK2 cells can be induced by stimulation with type 1 (IFNγ, IL-2) or type 2 cytokines (IL-4) (Peritt et al., 1998) . Thus, secretion of these cytokines by Th1 or Th2 cells may play a role in the differentiation or maintenance of NK1 and NK2 cells, respectively. Interestingly, NKT cells can express either IL-4 or IFNγ. While models have been developed to demonstrate that NKT cells can provide IFNγ-associated help to NK cells (Smyth et al., 2002), one could also speculate that IL-4 derived from NKT cells could induce NK2 cells under the appropriate conditions. Both $CD4^+$ and $CD8^+$ T cells as well as NKT cells have been found to aid NK cell-mediated anti-tumour responses and to potentiate long-term anti-tumour activity by NK cells (Shanker et al., 2007; Shimizu and Fujii, 2009; van den Broeke et al., 2003).

As mentioned earlier, Treg cells generate both IL-10 and TGFβ (Zhu and Paul, 2008). IL-10 is normally associated with negative modulation of immune responses (Moore et al., 2001) and reduces IFNγ production by NK cells by inhibiting IL-12 production by macrophages and DCs and by inhibiting NK cell proliferation (Goodier and Londei, 2000; Hunter et al., 1994; Jinushi et al., 2004; Mocellin et al., 2004; Moore et al., 2001; Scott et al., 2006). However, the effects of IL-10 on NK cells are not so clear-cut since IL-10 can also enhance NK cell-mediated cytotoxicity by either direct stimulation of the NK cell in combination with other cytokines (Cai et al., 1999; Micallef et al., 1999; Mocellin et al., 2004; Shibata et al., 1998; Takayama et al., 2001) or indirectly by reducing MHC class I on target cells (Kundu and Fulton, 1997; Petersson et al., 1998; Takayama et al., 2001). TGFβ on the other hand appears to inhibit a variety of NK cell functions. TGFβ inhibits IFNγ production and cytotoxicity by NK cells (Bellone et al., 1995; Hunter et al., 1995; Jinushi et al., 2004; Rook et al., 1986) and can reduce NKG2D- and NKp30-mediated killing by human NK cells by downmodulating their surface expression (Castriconi et al., 2003; Dasgupta et al., 2005; Lee et al., 2004a). Not surprisingly, Treg can inhibit NK cell expansion, cytotoxicity and rejection of bone marrow and tumours in vivo (Barao et al., 2006; Giroux et al., 2007; Ralainirina et al., 2007; Romagnani et al., 2005; Simon et al., 2007; Smyth et al., 2006; Trzonkowski et al., 2004). A subset of Treg cells that do not express CD4 or CD8 may be able to regulate NK cells by eliminating them through a perforin-dependent pathway (He et al., 2007). Thus, Treg cells may be critical in controlling pathological conditions induced by NK cells (Terme et al., 2008) or during pregnancy (Eriksson et al., 2004; Saito et al., 2008) (Table 22.8).

Th17 cells have received a lot of attention in recent years because of their central role in producing the proinflammatory cytokine IL-17. To date, no studies have examined whether Th17 can affect NK cell functions.

IL-17 does not have an effect of NK cell cytotoxicity but may make tumour cells sensitive to NK cell-mediated lysis (Honorati et al., 2003; Patera et al., 2002). However, Th17 cells also produce IL-21, which like IL-2, IL-4, IL-7, IL-9 and IL-15, shares the common cytokine receptor gamma chain. IL-21 enhances NK cell cytotoxicity but leads to terminal maturation of NK cells (Brady et al., 2004; Kasaian et al., 2002; Ma et al., 2003; Roda et al., 2006; Sivori et al., 2003; Wendt et al., 2007). While cytotoxicity of NK cells is enhanced by IL-21, the viability of the NK cells is reduced, which may be a consequence of NK cell maturation. Degeneration of NK cells by IL-21 may be an important host defence against prolonged NK cell activity since IL-21R$^{-/-}$ mice do not have this NK cell degeneration and were more susceptible to experimental autoimmune myasthenia gravis (Liu et al., 2006). These authors suggest that the low number of NK cells found clinically in autoimmune patients may therefore be a result of IL-21 (Liu et al., 2006). In line with this, a recent study treating cancer patients with IL-21 found a reduction in NK cell numbers, which recovered after removal of the IL-21 (Frederiksen et al., 2008).

Conclusion

NK cells can promote or inhibit T cell responses by cytokine production, direct stimulation or killing of DCs or T cells. However, T cells may also feed back to stimulate or inhibit NK cell activation. This reciprocal control between the adaptive and innate immune systems plays an important role during the development of anti-pathogen (Lodoen and Lanier, 2006) or anti-tumour responses (Ghiringhelli et al., 2006; Smyth et al., 2002) as well as in autoimmune diseases (Johansson et al., 2005; Shi and Van Kaer, 2006). Thus as the quotation by Camus at the beginning of this review suggests, it is better that our NK cell and T cells work beside each other than go their separate ways.

Acknowledgements

I would like to thank Erika Assarsson and Yenan Bryceson for their critical evaluation of this work. This work is supported by the Swedish Cancer Foundation, the Karolinska Institute Foundations and the Swedish Foundation for Strategic Research.

References

Agaugue, S., Marcenaro, E., Ferranti, B., Moretta, L. and Moretta, A. (2008). Human natural killer cells exposed to IL-2, IL-12, IL-18, or IL-4 differently modulate priming of naive T cells by monocyte-derived dendritic cells. Blood **112**, 1776–1783.

Aktas, E., Akdis, M., Bilgic, S., Disch, R., Falk, C.S., Blaser, K., Akdis, C. and Deniz, G. (2005). Different natural killer (NK) receptor expression and immunoglobulin E (IgE) regulation by NK1 and NK2 cells. Clin Exp Immunol **140**, 301–309.

Ashkar, A.A., Reid, S., Verdu, E.F., Zhang, K. and Coombes, B.K. (2009). Interleukin-15 and NK1.1+ cells provide innate protection against acute Salmonella enterica serovar Typhimurium infection in the gut and in systemic tissues. Infect Immun **77**, 214–222.

Assarsson, E., Kambayashi, T., Schatzle, J.D., Cramer, S.O., von Bonin, A., Jensen, P. E., Ljunggren, H.G. and Chambers, B.J. (2004). NK cells stimulate proliferation of T and NK cells through 2B4/CD48 interactions. J Immunol **173**, 174–180.

Assarsson, E., Kambayashi, T., Persson, C.M., Ljunggren, H.G. and Chambers, B.J. (2005). 2B4 co-stimulation: NK cells and their control of adaptive immune responses. Mol Immunol **42**, 419–423.

Azuma, M., Yssel, H., Phillips, J.H., Spits, H. and Lanier, L.L. (1993). Functional expression of B7/BB1 on activated T lymphocytes. J Exp Med **177**, 845–850.

Bajenoff, M., Breart, B., Huang, A.Y., Qi, H., Cazareth, J., Braud, V.M., Germain, R.N. and Glaichenhaus, N. (2006). Natural killer cell behavior in lymph nodes revealed by static and real-time imaging. J Exp Med **203**, 619–631.

Barao, I., Hanash, A.M., Hallett, W., Welniak, L.A., Sun, K., Redelman, D., Blazar, B.R., Levy, R.B. and Murphy, W.J. (2006). Suppression of natural killer cell-mediated bone marrow cell rejection by CD4+CD25+ regulatory T cells. Proc Natl Acad Sci USA **103**, 5460–5465.

Bellone, G., Aste-Amezaga, M., Trinchieri, G. and Rodeck, U. (1995). Regulation of NK cell functions by TGF-beta 1. J Immunol **155**, 1066–1073.

Biron, C.A., Nguyen, K.B., Pien, G.C., Cousens, L.P. and Salazar-Mather, T.P. (1999). Natural killer cells in antiviral defense: function and regulation by innate cytokines. Annu Rev Immunol **17**, 189–220.

Blasius, A.L., Barchet, W., Cella, M. and Colonna, M. (2007). Development and function of murine B220+CD11c+NK1.1+ cells identify them as a subset of NK cells. J Exp Med **204**, 2561–2568.

Brady, J., Hayakawa, Y., Smyth, M.J. and Nutt, S.L. (2004). IL-21 induces the functional maturation of murine NK cells. J Immunol **172**, 2048–2058.

Brooks, C.F. and Moore, M. (1986). Presentation of a soluble bacterial antigen and cell-surface alloantigens by large granular lymphocytes (LGL) in comparison with monocytes. Immunology **58**, 343–350.

Bryceson, Y.T. and Long, E.O. (2008). Line of attack: NK cell specificity and integration of signals. Curr Opin Immunol **20**, 344–352.

Buentke, E., Heffler, L.C., Wilson, J.L., Wallin, R.P., Lofman, C., Chambers, B. J., Ljunggren, H.G. and Scheynius, A. (2002). Natural killer and dendritic cell contact in lesional atopic dermatitis skin—Malassezia-influenced cell interaction. J Invest Dermatol **119**, 850–857.

Butte, M.J., Keir, M.E., Phamduy, T.B., Sharpe, A.H. and Freeman, G.J. (2007). Programmed death-1 ligand 1 interacts specifically with the B7-1 costimulatory molecule to inhibit T cell responses. Immunity **27**, 111–122.

Butte, M.J., Pena-Cruz, V., Kim, M.J., Freeman, G.J. and Sharpe, A.H. (2008). Interaction of human PD-L1 and B7-1. Mol Immunol **45**, 3567–3572.

Byrne, P., McGuirk, P., Todryk, S. and Mills, K.H. (2004). Depletion of NK cells results in disseminating lethal infection with *Bordetella pertussis* associated with a reduction of antigen-specific Th1 and enhancement of Th2, but not Tr1 cells. Eur J Immunol **34**, 2579–2588.

Cai, G., Kastelein, R.A. and Hunter, C.A. (1999). IL-10 enhances NK cell proliferation, cytotoxicity and production of IFN-gamma when combined with IL-18. Eur J Immunol **29**, 2565–2658.

Caminschi, I., Ahmet, F., Heger, K., Brady, J., Nutt, S.L., Vremec, D., Pietersz, S., Lahoud, M.H., Schofield, L., Hansen, D.S., O'Keeffe, M., Smyth, M.J., Bedoui, S., Davey, G.M., Villadangos, J.A., Heath, W.R. and Shortman, K. (2007). Putative IKDCs are functionally and developmentally similar to natural killer cells, but not to dendritic cells. J Exp Med **204**, 2579–2590.

Carbone, E., Terrazzano, G., Ruggiero, G., Zanzi, D., Ottaiano, A., Manzo, C., Karre, K. and Zappacosta, S. (1999). Recognition of autologous dendritic cells by human NK cells. Eur J Immunol **29**, 4022–4029.

Castriconi, R., Cantoni, C., Della Chiesa, M., Vitale, M., Marcenaro, E., Conte, R., Biassoni, R., Bottino, C., Moretta, L. and Moretta, A. (2003). Transforming growth factor beta 1 inhibits expression of NKp30 and NKG2D receptors: consequences for the NK-mediated killing of dendritic cells. Proc Natl Acad Sci U S A **100**, 4120–4125.

Cerboni, C., Zingoni, A., Cippitelli, M., Piccoli, M., Frati, L. and Santoni, A. (2007). Antigen-activated human T lymphocytes express cell-surface NKG2D ligands via an ATM/ATR-dependent mechanism and become susceptible to autologous NK-cell lysis. Blood **110**, 606–615.

Chambers, B.J., Salcedo, M. and Ljunggren, H.G. (1996). Triggering of natural killer cells by the costimulatory molecule CD80 (B7-1). Immunity **5**, 311–317.

Chan, C.W., Crafton, E., Fan, H.N., Flook, J., Yoshimura, K., Skarica, M., Brockstedt, D., Dubensky, T.W., Stins, M.F., Lanier, L.L., Pardoll, D.M. and Housseau, F. (2006). Interferon-producing killer dendritic cells provide a link between innate and adaptive immunity. Nat Med **12**, 207–213.

Combe, C.L., Curiel, T.J., Moretto, M.M. and Khan, I.A. (2005). NK cells help to induce CD8(+)-T-cell immunity against *Toxoplasma gondii* in the absence of CD4(+) T cells. Infect Immun **73**, 4913–4921.

Cooper, M.A., Fehniger, T.A., Turner, S.C., Chen, K.S., Ghaheri, B.A., Ghayur, T., Carson, W.E. and Caligiuri, M.A. (2001). Human natural killer cells: a unique innate immunoregulatory role for the CD56(bright) subset. Blood **97**, 3146–3151.

D'Orazio, J.A. and Stein-Streilein, J. (1996). Human natural killer (NK) cells present staphylococcal enterotoxin B (SEB) to T lymphocytes. Clin Exp Immunol **104**, 366–373.

Dasgupta , S., Bhattacharya-Chatterjee, M., O'Malley, B.W. Jr. and Chatterjee, S.K. (2005). Inhibition of NK cell activity through TGF-beta 1 by down-regulation of NKG2D in a murine model of head and neck cancer. J Immunol **175**, 5541–5550.

Degli-Esposti, M.A. and Smyth, M.J. (2005). Close encounters of different kinds: dendritic cells and NK cells take centre stage. Nat Rev Immunol **5**, 112–124.

Della Chiesa, M., Vitale, M., Carlomagno, S., Ferlazzo, G., Moretta, L. and Moretta, A. (2003). The natural killer cell-mediated killing of autologous dendritic cells is confined to a cell subset expressing CD94/NKG2A, but lacking inhibitory killer Ig-like receptors. Eur J Immunol **33**, 1657–1666.

Deniz, G., Akdis, M., Aktas, E., Blaser, K. and Akdis, C.A. (2002). Human NK1 and NK2 subsets determined by purification of IFN-gamma-secreting and IFN-gamma-nonsecreting NK cells. Eur J Immunol **32**, 879–884.

Deniz, G., Erten, G., Kucuksezer, U.C., Kocacik, D., Karagiannidis, C., Aktas, E., Akdis, C.A. and Akdis, M. (2008). Regulatory NK cells suppress antigen-specific T cell responses. J Immunol **180**, 850–857.

Dunn, G.P., Old, L.J. and Schreiber, R. D. (2004). The Three Es of Cancer Immunoediting. Annu Rev Immunol **22**, 329–360.

Eriksson, M., Meadows, S.K., Wira, C. R. and Sentman, C.L. (2004). Unique phenotype of human uterine NK cells and their regulation by endogenous TGF-beta. J Leukoc Biol **76**, 667–675.

Fan, Z., Yu, P., Wang, Y., Fu, M.L., Liu, W., Sun, Y. and Fu, Y.X. (2006). NK-cell activation by LIGHT triggers tumor-specific CD8+ T-cell immunity to reject established tumors. Blood **107**, 1342–1351.

Fehniger, T.A., Cooper, M.A., Nuovo, G.J., Cella, M., Facchetti, F., Colonna, M. and Caligiuri, M.A. (2003). CD56bright natural killer cells are present in human lymph nodes and are activated by T cell-derived IL-2: a potential new link between adaptive and innate immunity. Blood **101**, 3052–3057.

Ferlazzo, G., Thomas, D., Lin, S.L., Goodman, K., Morandi, B., Muller, W.A., Moretta, A. and Munz, C. (2004). The abundant NK cells in human secondary lymphoid tissues require activation to express killer cell Ig-like receptors and become cytolytic. J Immunol **172**, 1455–1462.

Frederiksen, K.S., Lundsgaard, D., Freeman, J.A., Hughes, S.D., Holm, T. L., Skrumsager, B.K., Petri, A., Hansen, L.T., McArthur, G.A., Davis, I.D. and Skak, K. (2008). IL-21 induces in vivo immune activation of NK cells and CD8(+) T cells in patients with metastatic melanoma and renal cell carcinoma. Cancer Immunol Immunother **57**, 1439–1449.

Ghiringhelli, F., Menard, C., Martin, F. and Zitvogel, L. (2006). The role of regulatory T cells in the control of natural killer cells: relevance during tumor progression. Immunol Rev **214**, 229–238.

Gilbertson, S.M., Shah, P.D. and Rowley, D.A. (1986). NK cells suppress the generation of Lyt-2+ cytolytic T cells by suppressing or eliminating dendritic cells. J Immunol **136**, 3567–3571.

Giroux, M., Yurchenko, E., St-Pierre, J., Piccirillo, C.A. and Perreault, C. (2007). T regulatory cells control numbers of NK cells and CD8alpha+ immature dendritic cells in the lymph node paracortex. J Immunol **179**, 4492–4502.

Glas, R., Franksson, L., Une, C., Eloranta, M.L., Ohlen, C., Orn, A. and Karre, K. (2000). Recruitment and activation of natural killer (NK) cells in vivo determined by the target cell phenotype. An adaptive component of NK cell-mediated responses. J Exp Med **191**, 129–138.

Godfrey, I., MacDonald, H.R., Kronenberg, M., Smyth, M.J. and Van Kaer, L. (2004). NKT cells: what's in a name? Nat Rev Immunol **4**, 231–237.

Goodier, M.R. and Londei, M. (2000). Lipopolysaccharide stimulates the proliferation of human CD56+CD3− NK cells: a regulatory role of monocytes and IL-10. J Immunol **165**, 139–147.

Gray, J.D., Hirokawa, M., Ohtsuka, K. and Horwitz, D.A. (1998). Generation of an inhibitory circuit involving CD8+ T cells, IL-2, and NK cell-derived

TGF-beta: contrasting effects of anti-CD2 and anti-CD3. J Immunol **160**, 2248–2254.

Greenwald, R.J., Freeman, G.J. and Sharpe, A.H. (2005). The B7 family revisited. Annu Rev Immunol **23**, 515–548.

Guerra, N., Tan, Y.X., Joncker, N.T., Choy, A., Gallardo, F., Xiong, N., Knoblaugh, S., Cado, D., Greenberg, N.M. and Raulet, D.H. (2008). NKG2D-deficient mice are defective in tumor surveillance in models of spontaneous malignancy. Immunity **28**, 571–580.

Hanna, J., Gonen-Gross, T., Fitchett, J., Rowe, T., Daniels, M., Arnon, T.I., Gazit, R., Joseph, A., Schjetne, K.W., Steinle, A., Porgador, A., Mevorach, D., Goldman-Wohl, D., Yagel, S., LaBarre, M.J., Buckner, J.H. and Mandelboim, O. (2004). Novel APC-like properties of human NK cells directly regulate T cell activation. J Clin Invest **114**, 1612–1623.

Hayakawa, Y. and Smyth, M.J. (2006). CD27 dissects mature NK cells into two subsets with distinct responsiveness and migratory capacity. J Immunol **176**, 1517–1524.

Hayakawa, Y., Screpanti, V., Yagita, H., Grandien, A., Ljunggren, H.G., Smyth, M.J. and Chambers, B.J. (2004). NK cell TRAIL eliminates immature dendritic cells in vivo and limits dendritic cell vaccination efficacy. J Immunol **172**, 123–129.

He, K.M., Ma, Y., Wang, S., Min, W.P., Zhong, R., Jevnikar, A. and Zhang, Z.X. (2007). Donor double-negative Treg promote allogeneic mixed chimerism and tolerance. Eur J Immunol **37**, 3455–3466.

Honorati, M.C., Neri, S., Cattini, L. and Facchini, A. (2003). IL-17 enhances the susceptibility of U-2 OS osteosarcoma cells to NK cell lysis. Clin Exp Immunol **133**, 344–349.

Horwitz, D.A., Gray, J.D. and Ohtsuka, K. (1999). Role of NK cells and TGF-beta in the regulation of T-cell-dependent antibody production in health and autoimmune disease. Microbes Infect **1**, 1305–1311.

Hoshino, T., Winkler-Pickett, R.T., Mason, A.T., Ortaldo, J.R. and Young, H.A. (1999). IL-13 production by NK cells: IL-13-producing NK and T cells are present in vivo in the absence of IFN-gamma. J Immunol **162**, 51–59.

Huang, D., Shi, F.D., Jung, S., Pien, G.C., Wang, J., Salazar-Mather, T.P., He, T.T., Weaver, J.T., Ljunggren, H.G., Biron, C. A., Littman, D.R. and Ransohoff, R.M. (2006). The neuronal chemokine CX3CL1/fractalkine selectively recruits NK cells that modify experimental autoimmune encephalomyelitis within the central nervous system. FASEB J **20**, 896–905.

Hunter, C.A., Subauste, C.S., Van Cleave, V.H. and Remington, J.S. (1994). Production of gamma interferon by natural killer cells from *Toxoplasma gondii*-infected SCID mice: regulation by interleukin-10, interleukin-12, and tumor necrosis factor alpha. Infect Immun **62**, 2818–2824.

Hunter, C.A., Bermudez, L., Beernink, H., Waegell, W. and Remington, J.S. (1995). Transforming growth factor-beta inhibits interleukin-12-induced production of interferon-gamma by natural killer cells: a role for transforming growth factor-beta in the regulation of T cell-independent resistance to *Toxoplasma gondii*. Eur J Immunol **25**, 994–1000.

Ivanov, I.I., McKenzie, B.S., Zhou, L., Tadokoro, C.E., Lepelley, A., Lafaille, J.J., Cua, D.J. and Littman, D.R. (2006). The orphan nuclear receptor RORgammat directs the differentiation program of proinflammatory IL-17+ T helper cells. Cell **126**, 1121–1133.

Jinushi, M., Takehara, T., Tatsumi, T., Kanto, T., Miyagi, T., Suzuki, T., Kanazawa, Y., Hiramatsu, N. and Hayashi, N. (2004). Negative regulation of NK cell activities by inhibitory receptor CD94/NKG2A leads to altered NK cell-induced modulation of dendritic cell functions in chronic hepatitis C virus infection. J Immunol **173**, 6072–6081.

Johansson, S., Berg, L., Hall, H. and Hoglund, P. (2005). NK cells: elusive players in autoimmunity. Trends Immunol **26**, 613–618.

Kaplan, M.H., Schindler, U., Smiley, S. T. and Grusby, M.J. (1996). Stat6 is required for mediating responses to IL-4 and for development of Th2 cells. Immunity **4**, 313–319.

Kasaian, M.T., Whitters, M.J., Carter, L. L., Lowe, L.D., Jussif, J.M., Deng, B., Johnson, K.A., Witek, J.S., Senices, M., Konz, R.F., Wurster, A.L., Donaldson, D.D., Collins, M., Young, D.A. and Grusby, M.J. (2002). IL-21 limits NK cell responses and promotes antigen-specific T cell activation: a mediator of the transition from innate to adaptive immunity. Immunity **16**, 559–569.

Katsumoto, T., Kimura, M., Yamashita, M., Hosokawa, H., Hashimoto, K., Hasegawa, A., Omori, M., Miyamoto, T., Taniguchi, M. and Nakayama, T. (2004). STAT6-dependent differentiation and production of IL-5 and IL-13 in murine NK2 cells. J Immunol **173**, 4967–4975.

Kiessling, R., Klein, E. and Wigzell, H. (1975). "Natural" killer cells in the mouse. I. Cytotoxic cells with specificity for mouse Moloney leukemia cells. Specificity and distribution according to genotype. Eur J Immunol **5**, 112–117.

Korbel, D.S., Finney, O.C. and Riley, E.M. (2004). Natural killer cells and innate immunity to protozoan pathogens. Int J Parasitol **34**, 1517–1528.

Korsgren, M., Persson, C.G., Sundler, F., Bjerke, T., Hansson, T., Chambers, B.J., Hong, S., Van Kaer, L., Ljunggren, H.G. and Korsgren, O. (1999). Natural killer cells determine development of allergen-induced eosinophilic airway inflammation in mice. J Exp Med **189**, 553–562.

Kos, F.J. and Engleman, E.G. (1996). Role of natural killer cells in the generation of influenza virus-specific cytotoxic T cells. Cell Immunol **173**, 1–6.

Kundu, N. and Fulton, A.M. (1997). Interleukin-10 inhibits tumor metastasis, downregulates MHC class I, and enhances NK lysis. Cell Immunol **180**, 55–61.

Laffont, S., Seillet, C., Ortaldo, J., Coudert, J.D. and Guery, J.C. (2008). Natural killer cells recruited into lymph nodes inhibit alloreactive T-cell activation through perforin-mediated killing of donor allogeneic dendritic cells. Blood **112**, 661–671.

Laouar, Y., Sutterwala, F.S., Gorelik, L. and Flavell, R.A. (2005). Transforming growth factor-beta controls T helper type 1 cell development through regulation of natural killer cell interferon-gamma. Nat Immunol **6**, 600–607.

Lee, K.M., Bhawan, S., Majima, T., Wei, H., Nishimura, M.I., Yagita, H. and Kumar, V. (2003). Cutting edge: the NK cell receptor 2B4 augments antigen-specific T cell cytotoxicity through CD48 ligation on neighboring T cells. J Immunol **170**, 4881–4885.

Lee, J.C., Lee, K.M., Kim, D.W. and Heo, D.S. (2004a). Elevated TGF-beta1 secretion and down-modulation of NKG2D underlies impaired NK cytotoxicity in cancer patients. J Immunol **172**, 7335–7340.

Lee, K.M., McNerney, M.E., Stepp, S.E., Mathew, P.A., Schatzle, J.D., Bennett, M. and Kumar, V. (2004b). 2B4 acts as a non-major histocompatibility complex binding inhibitory receptor on mouse natural killer cells. J Exp Med **199**, 1245–1254.

Lee, S.H., Miyagi, T. and Biron, C.A. (2007). Keeping NK cells in highly regulated antiviral warfare. Trends Immunol **28**, 252–259.

Li, M.O., Wan, Y.Y., Sanjabi, S., Robertson, A.K. and Flavell, R.A. (2006). Transforming growth factor-beta regulation of immune responses. Annu Rev Immunol **24**, 99–146.

Lighvani, A.A., Frucht, D.M., Jankovic, D., Yamane, H., Aliberti, J., Hissong, B.D., Nguyen, B.V., Gadina, M., Sher, A., Paul, W.E. and O'Shea, J.J. (2001). T-bet is rapidly induced by interferon-gamma in lymphoid and myeloid cells. Proc Natl Acad Sci U S A **98**, 15137–15142.

Liu, R., Van Kaer, L., La Cava, A., Price, M., Campagnolo, I., Collins, M., Young, D. A., Vollmer, T.L. and Shi, F.D. (2006). Autoreactive T cells mediate NK cell degeneration in autoimmune disease. J Immunol **176**, 5247–5254.

Lo, C.K., Lam, Q.L., Sun, L., Wang, S., Ko, K.H., Xu, H., Wu, C.Y., Zheng, B.J. and Lu, L. (2008). Natural killer cell degeneration exacerbates experimental arthritis in mice via enhanced interleukin-17 production. Arthritis Rheum **58**, 2700–2711.

Lodoen, M.B. and Lanier, L.L. (2006). Natural killer cells as an initial defense against pathogens. Curr Opin Immunol **18**, 391–398.

Loza, M.J. and Perussia, B. (2001). Final steps of natural killer cell maturation: a model for type 1–type 2 differentiation? Nat Immunol **2**, 917–924.

Lu, L., Ikizawa, K., Hu, D., Werneck, M.B., Wucherpfennig, K.W. and Cantor, H. (2007). Regulation of activated CD4+ T cells by NK cells via the Qa-1-NKG2A inhibitory pathway. Immunity **26**, 593–604.

Lugo-Villarino, G., Maldonado-Lopez, R., Possemato, R., Penaranda, C. and Glimcher, L.H. (2003). T-bet is required for optimal production of IFN-gamma and antigen-specific T cell activation by dendritic cells. Proc Natl Acad Sci U S A **100**, 7749–7754.

Ma, H.L., Whitters, M.J., Konz, R.F., Senices, M., Young, D.A., Grusby, M.J., Collins, M. and Dunussi-Joannopoulos, K. (2003). IL-21 activates both innate and adaptive immunity to generate potent antitumor responses that require perforin but are independent of IFN-gamma. J Immunol **171**, 608–615.

Mailliard, R.B., Son, Y.I., Redlinger, R., Coates, P.T., Giermasz, A., Morel, P.A., Storkus, W.J. and Kalinski, P. (2003). Dendritic cells mediate NK cell help for Th1 and CTL responses: two-signal requirement for the induction of NK cell helper function. J Immunol **171**, 2366–2373.

Martin-Fontecha, A., Thomsen, L.L., Brett, S., Gerard, C., Lipp, M., Lanzavecchia, A. and Sallusto, F. (2004). Induced recruitment of NK cells to lymph nodes provides IFN-gamma for T(H)1 priming. Nat Immunol **5**, 1260–1265.

McNerney, M.E., Guzior, D. and Kumar, V. (2005). 2B4 (CD244)–CD48 interactions provide a novel MHC class I-independent system for NK-cell self-tolerance in mice. Blood **106**, 1337–1340.

Meadows, S.K., Eriksson, M., Barber, A. and Sentman, C.L. (2006). Human NK cell IFN-gamma production is regulated by endogenous TGF-beta. Int Immunopharmacol **6**, 1020–1028.

Mehrotra, P.T., Donnelly, R.P., Wong, S., Kanegane, H., Geremew, A., Mostowski, H.S., Furuke, K., Siegel, J.P. and Bloom, E.T. (1998). Production of IL-10 by human natural killer cells stimulated with IL-2 and/or IL-12. J Immunol **160**, 2637–2644.

Mehta, D.S., Wurster, A.L. and Grusby, M. J. (2004). Biology of IL-21 and the IL-21 receptor. Immunol Rev **202**, 84–95.

Micallef, M.J., Tanimoto, T., Torigoe, K., Nishida, Y., Kohno, K., Ikegami, H. and Kurimoto, M. (1999). Simultaneous exposure to interleukin-18 and interleukin-10 in vitro synergistically augments murine spleen natural killer cell activity. Cancer Immunol Immunother **48**, 109–117.

Mocellin, S., Panelli, M., Wang, E., Rossi, C.R., Pilati, P., Nitti, D., Lise, M. and Marincola, F.M. (2004). IL-10 stimulatory effects on human NK cells explored by gene profile analysis. Genes Immun **5**, 621–630.

Moore, K.W., de Waal Malefyt, R., Coffman, R.L. and O'Garra, A. (2001). Interleukin-10 and the interleukin-10 receptor. Annu Rev Immunol **19**, 683–765.

Mosmann, T.R., Cherwinski, H., Bond, M.W., Giedlin, M.A. and Coffman, R. L. (1986). Two types of murine helper T cell clone. I. Definition according to profiles of lymphokine activities and secreted proteins. J Immunol **136**, 2348–2357.

Mosmann, T.R., Li, L. and Sad, S. (1997). Functions of CD8 T-cell subsets secreting different cytokine patterns. Semin Immunol **9**, 87–92.

Patera, A.C., Pesnicak, L., Bertin, J. and Cohen, J.I. (2002). Interleukin 17 modulates the immune response to vaccinia virus infection. Virology **299**, 56–63.

Peritt, D., Robertson, S., Gri, G., Showe, L., Aste-Amezaga, M. and Trinchieri, G. (1998). Differentiation of human NK cells into NK1 and NK2 subsets. J Immunol **161**, 5821–5824.

Persson, C.M., Assarsson, E., Vahlne, G., Brodin, P. and Chambers, B.J. (2008). Critical role of Qa1b in the protection of mature dendritic cells from NK cell-mediated killing. Scand J Immunol **67**, 30–36.

Petersson, M., Charo, J., Salazar-Onfray, F., Noffz, G., Mohaupt, M., Qin, Z., Klein, G., Blankenstein, T. and Kiessling, R. (1998). Constitutive IL-10 production accounts for the high NK sensitivity, low MHC class I expression, and poor transporter associated with antigen processing (TAP)-1/2 function in the prototype NK target YAC-1. J Immunol **161**, 2099–2105.

Phillips, J.H., Le, A.M. and Lanier, L.L. (1984). Natural killer cells activated in a human mixed lymphocyte response culture identified by expression of Leu-11 and class II histocompatibility antigens. J Exp Med **159**, 993–1008.

Prlic, M., Gibbs, J. and Jameson, S.C. (2005). Characteristics of NK cell migration early after vaccinia infection. J Immunol **175**, 2152–2157.

Rabinovich, B.A., Li, J., Shannon, J., Hurren, R., Chalupny, J., Cosman, D. and Miller, R.G. (2003). Activated, but not resting, T cells can be recognized and killed by syngeneic NK cells. J Immunol **170**, 3572–3576.

Ralainirina, N., Poli, A., Michel, T., Poos, L., Andres, E., Hentges, F. and Zimmer, J. (2007). Control of NK cell functions by CD4+CD25+ regulatory T cells. J Leukoc Biol **81**, 144–153.

Robbins, S.H., Bessou, G., Cornillon, A., Zucchini, N., Rupp, B., Ruzsics, Z., Sacher, T., Tomasello, E., Vivier, E., Koszinowski, U.H. and Dalod, M. (2007). Natural killer cells promote early CD8 T cell responses against cytomegalovirus. PLoS Pathog **3**, e123.

Roda , J.M., Parihar, R., Lehman, A., Mani, A., Tridandapani, S. and Carson, W.E. 3rd (2006). Interleukin-21 enhances NK cell activation in response to antibody-coated targets. J Immunol **177**, 120–129.

Romagnani, C., Della Chiesa, M., Kohler, S., Moewes, B., Radbruch, A., Moretta, L., Moretta, A. and Thiel, A. (2005). Activation of human NK cells by plasmacytoid dendritic cells and its modulation by CD4+ T helper cells and CD4+ CD25hi T regulatory cells. Eur J Immunol **35**, 2452–2458.

Roncarolo, M.G., Bigler, M., Haanen, J. B., Yssel, H., Bacchetta, R., de Vries, J. E. and Spits, H. (1991). Natural killer cell clones can efficiently process and present protein antigens. J Immunol **147**, 781–787.

Rook, A.H., Kehrl, J.H., Wakefield, L.M., Roberts, A.B., Sporn, M.B., Burlington, D.B., Lane, H.C. and Fauci, A.S. (1986). Effects of transforming growth factor beta on the functions of natural killer cells: depressed cytolytic activity and blunting of interferon responsiveness. J Immunol **136**, 3916–3920.

Roy, S., Barnes, P.F., Garg, A., Wu, S., Cosman, D. and Vankayalapati, R. (2008). NK cells lyse T regulatory cells that expand in response to an intracellular pathogen. J Immunol **180**, 1729–1736.

Ruggeri, L., Capanni, M., Urbani, E., Perruccio, K., Shlomchik, W.D., Tosti, A., Posati, S., Rogaia, D., Frassoni, F., Aversa, F., Martelli, M.F. and Velardi, A. (2002). Effectiveness of donor natural killer cell alloreactivity in mismatched hematopoietic transplants. Science **295**, 2097–2100.

Saito, S., Nakashima, A., Myojo-Higuma, S. and Shiozaki, A. (2008). The balance between cytotoxic NK cells and regulatory NK cells in human pregnancy. J Reprod Immunol **77**, 14–22.

Samson, S.I., Richard, O., Tavian, M., Ranson, T., Vosshenrich, C.A., Colucci, F., Buer, J., Grosveld, F., Godin, I. and Di Santo, J.P. (2003). GATA-3 promotes maturation, IFN-gamma production, and liver-specific homing of NK cells. Immunity **19**, 701–711.

Saudemont, A., Jouy, N., Hetuin, D. and Quesnel, B. (2005). NK cells that are activated by CXCL10 can kill dormant tumor cells that resist CTL-mediated lysis and can express B7-H1 that stimulates T cells. Blood **105**, 2428–2435.

Scala, G., Allavena, P., Ortaldo, J.R., Herberman, R.B. and Oppenheim, J.J. (1985). Subsets of human large granular lymphocytes (LGL) exhibit accessory cell functions. J Immunol **134**, 3049–3055.

Scalzo, A.A. (2002). Successful control of viruses by NK cells—a balance of opposing forces? Trends Microbiol **10**, 470–474.

Scott, M.J., Hoth, J.J., Turina, M., Woods, D.R. and Cheadle, W.G. (2006). Interleukin-10 suppresses natural killer cell but not natural killer T cell activation during bacterial infection. Cytokine **33**, 79–86.

Seo, S.K., Jeong, H.Y., Park, S.G., Lee, S. W., Choi, I.W., Chen, L. and Choi, I. (2008). Blockade of endogenous B7-H1 suppresses antibacterial protection after primary *Listeria monocytogenes* infection. Immunology **123**, 90–99.

Shanker, A., Verdeil, G., Buferne, M., Inderberg-Suso, E.M., Puthier, D., Joly, F., Nguyen, C., Leserman, L., Auphan-Anezin, N. and Schmitt-Verhulst, A. M. (2007). CD8 T cell help for innate antitumor immunity. J Immunol **179**, 6651–6662.

Shi, F.D. and Van Kaer, L. (2006). Reciprocal regulation between natural killer cells and autoreactive T cells. Nat Rev Immunol **6**, 751–760.

Shibata, Y., Foster, L.A., Kurimoto, M., Okamura, H., Nakamura, R.M., Kawajiri, K., Justice, J.P., Van Scott, M. R., Myrvik, Q.N. and Metzger, W.J. (1998). Immunoregulatory roles of IL-10 in innate immunity: IL-10 inhibits macrophage production of IFN-gamma-inducing factors but enhances NK cell production of IFN-gamma. J Immunol **161**, 4283–4288.

Shimizu, K. and Fujii, S. (2009). DC therapy induces long-term NK reactivity to tumors via host DC. Eur J Immunol **39**, 457–468.

Simon, A.K., Jones, E., Richards, H., Wright, K., Betts, G., Godkin, A., Screaton, G. and Gallimore, A. (2007). Regulatory T cells inhibit Fas ligand-induced innate and adaptive tumour immunity. Eur J Immunol **37**, 758–767.

Sivori, S., Cantoni, C., Parolini, S., Marcenaro, E., Conte, R., Moretta, L. and Moretta, A. (2003). IL-21 induces both rapid maturation of human CD34+ cell precursors towards NK cells and acquisition of surface killer Ig-like receptors. Eur J Immunol **33**, 3439–3447.

Smyth, M.J., Kelly, J.M., Baxter, A.G., Korner, H. and Sedgwick, J.D. (1998). An essential role for tumor necrosis factor in natural killer cell-mediated tumor rejection in the peritoneum. J Exp Med **188**, 1611–1619.

Smyth, M.J., Crowe, N.Y., Hayakawa, Y., Takeda, K., Yagita, H. and Godfrey, I. (2002). NKT cells — conductors of tumor immunity? Curr Opin Immunol **14**, 165–171.

Smyth, M.J., Teng, M.W., Swann, J., Kyparissoudis, K., Godfrey, I. and Hayakawa, Y. (2006). CD4+CD25+ T regulatory cells suppress NK cell-mediated immunotherapy of cancer. J Immunol **176**, 1582–1587.

Strowig, T., Brilot, F. and Munz, C. (2008). Noncytotoxic functions of NK cells: direct pathogen restriction and assistance to adaptive immunity. J Immunol **180**, 7785–7791.

Su, H.C., Ishikawa, R. and Biron, C.A. (1993). Transforming growth factor-beta expression and natural killer cell responses during virus infection of normal, nude, and SCID mice. J Immunol **151**, 4874–4890.

Sullivan, B.M., Juedes, A., Szabo, S.J., von Herrath, M. and Glimcher, L.H. (2003). Antigen-driven effector CD8 T cell function regulated by T-bet. Proc Natl Acad Sci U S A **100**, 15818–15823.

Tacken, P.J., de Vries, I.J., Torensma, R. and Figdor, C.G. (2007). Dendritic-cell immunotherapy: from ex vivo loading to in vivo targeting. Nat Rev Immunol **7**, 790–802.

Takayama, T., Tahara, H. and Thomson, A.W. (2001). Differential effects of myeloid dendritic cells retrovirally transduced to express mammalian or viral interleukin-10 on cytotoxic T lymphocyte and natural killer cell functions and resistance to tumor growth. Transplantation **71**, 1334–1340.

Taniguchi, R.T., Guzior, D. and Kumar, V. (2007). 2B4 inhibits NK-cell fratricide. Blood **110**, 2020–2023.

Terme, M., Chaput, N., Combadiere, B., Ma, A., Ohteki, T. and Zitvogel, L. (2008). Regulatory T cells control dendritic cell/NK cell cross-talk in lymph nodes at the steady state by inhibiting CD4+ self-reactive T cells. J Immunol **180**, 4679–4686.

Thapa, M., Kuziel, W.A. and Carr, D.J. (2007). Susceptibility of CCR5-deficient mice to genital herpes simplex virus type 2 is linked to NK cell mobilization. J Virol **81**, 3704–3713.

Tosi, D., Valenti, R., Cova, A., Sovena, G., Huber, V., Pilla, L., Arienti, F., Belardelli, F., Parmiani, G. and Rivoltini, L. (2004). Role of cross-talk between IFN-alpha-induced monocyte-derived dendritic cells and NK cells in priming CD8+ T cell responses against human tumor antigens. J Immunol **172**, 5363–5370.

Townsend, M.J., Weinmann, A.S., Matsuda, J.L., Salomon, R., Farnham, P.J., Biron, C.A., Gapin, L. and Glimcher, L.H. (2004). T-bet regulates the terminal maturation and homeostasis of NK and Valpha14i NKT cells. Immunity **20**, 477–494.

Trzonkowski, P., Szmit, E., Mysliwska, J., Dobyszuk, A. and Mysliwski, A. (2004). CD4+CD25+ T regulatory cells inhibit cytotoxic activity of T CD8+ and NK lymphocytes in the direct cell-to-cell interaction. Clin Immunol **112**, 258–567.

van den Broeke, L.T., Daschbach, E., Thomas, E.K., Andringa, G. and Berzofsky, J.A. (2003). Dendritic cell-induced activation of adaptive and innate antitumor immunity. J Immunol **171**, 5842–5852.

Vankayalapati, R., Klucar, P., Wizel, B., Weis, S.E., Samten, B., Safi, H., Shams, H. and Barnes, P.F. (2004). NK cells regulate CD8+ T cell effector function in response to an intracellular pathogen. J Immunol **172**, 130–137.

Vignali, D.A., Collison, L.W. and Workman, C.J. (2008). How regulatory T cells work. Nat Rev Immunol **8**, 523–532.

Viret , C. and Janeway, C.A. Jr. (1999). MHC and T cell development. Rev Immunogenet **1**, 91–104.

Vivier, E., Tomasello, E., Baratin, M., Walzer, T. and Ugolini, S. (2008). Functions of natural killer cells. Nat Immunol **9**, 503–510.

Vosshenrich, C.A., Lesjean-Pottier, S., Hasan, M., Richard-Le Goff, O., Corcuff, E., Mandelboim, O. and Di Santo, J.P. (2007). CD11cloB220+ interferon-producing killer dendritic cells are activated natural killer cells. J Exp Med **204**, 2569–2578.

Walker, C., Checkel, J., Cammisuli, S., Leibson, P.J. and Gleich, G.J. (1998). IL-5 production by NK cells contributes to eosinophil infiltration in a mouse model of allergic inflammation. J Immunol **161**, 1962–1969.

Weaver, C.T., Hatton, R.D., Mangan, P. R. and Harrington, L.E. (2007). IL-17 family cytokines and the expanding diversity of effector T cell lineages. Annu Rev Immunol **25**, 821–852.

Wendt, K., Wilk, E., Buyny, S., Schmidt, R. E. and Jacobs, R. (2007). Interleukin-21 differentially affects human natural killer cell subsets. Immunology **122**, 486–495.

Wilson, J.L., Heffler, L.C., Charo, J., Scheynius, A., Bejarano, M.T. and Ljunggren, H.G. (1999). Targeting of human dendritic cells by autologous NK cells. J Immunol **163**, 6365–6370.

Wolk, K. and Sabat, R. (2006). Interleukin-22: a novel T- and NK-cell derived cytokine that regulates the biology of tissue cells. Cytokine Growth Factor Rev **17**, 367–380.

Yu, C.R., Kirken, R.A., Malabarba, M.G., Young, H.A. and Ortaldo, J.R. (1998). Differential regulation of the Janus kinase-STAT pathway and biologic function of IL-13 in primary human NK and T cells: a comparative study with IL-4. J Immunol **161**, 218–227.

Yu, G., Xu, X., Vu, M.D., Kilpatrick, E.D. and Li, X.C. (2006). NK cells promote transplant tolerance by killing donor antigen-presenting cells. J Exp Med **203**, 1851–1858.

Zhang, D.H., Cohn, L., Ray, P., Bottomly, K. and Ray, A. (1997). Transcription factor GATA-3 is differentially expressed in murine Th1 and Th2 cells and controls Th2-specific expression of the interleukin-5 gene. J Biol Chem **272**, 21597–21603.

Zheng, W. and Flavell, R.A. (1997). The transcription factor GATA-3 is necessary and sufficient for Th2 cytokine gene expression in CD4 T cells. Cell **89**, 587–596.

Zhu, J. and Paul, W.E. (2008). CD4 T cells: fates, functions, and faults. Blood **112**, 1557–1569.

Zingoni, A., Sornasse, T., Cocks, B.G., Tanaka, Y., Santoni, A. and Lanier, L.L. (2004). Cross-talk between activated human NK cells and CD4+ T cells via OX40-OX40 ligand interactions. J Immunol **173**, 3716–3724.

PART III

NK cells in the tissues

Chapter Twenty-Three

23

NK cells and NKT cells in cutaneous disorders

Anna Balato, Anthony A. Gaspari

CHAPTER CONTENTS

We especially need imagination in science. It is not all mathematics, nor all logic, but it is somewhat beauty and poetry.

Maria Montessori

ABSTRACT

Natural killer (NK) and natural killer T (NKT) cells are lymphocytes that play a vital role in cell-mediated immunity. Although they belong to different lineages of cells, they share various characteristics. They are involved in cancer surveillance and represent the first line of defence against several microorganisms. They have been implicated in UV immune suppression and in immune-mediated skin disorders, including allergic contact dermatitis, psoriasis, atopic dermatitis and cutaneous lymphomas. NK and NKT cells are emerging as lymphocytes with a pathogenic role in certain immune-mediated skin diseases.

KEY WORDS

NKT cells, Immune-mediated skin disorders, Allergic contact dermatitis, Psoriasis, Atopic dermatitis, UV immune suppression, Cutaneous lymphomas

Introduction

Leukocyte accumulation in the skin is associated with a wide range of skin disorders, including atopic dermatitis (AD), contact hypersensitivity, psoriasis and cutaneous T cell lymphomas. (Zollner et al., 2007). Reports on natural killer (NK) and natural killer T (NKT) cells have implicated these cells in such immune-mediated skin diseases.

Biology of NK cells

NK cells are lymphocytes that play a vital role in cell-mediated immunity and serve as the first line of defence against cancerous and virally infected cells. Their cytolytic activity is not regulated by antigenic specificity but through a balance of activating or inhibitory signals mediated through cell surface receptors. These receptors specifically bind human leukocyte antigen (HLA) ligands (Bashirova et al., 2006), which are expressed in virtually all nucleated cells but are downregulated in stressed cells, thus allowing NK cell reconnaissance. This mode of recognition was first referred to as the 'missing self' hypothesis (Ljunggren and Karre, 1990).

NK cells were originally described as large granular lymphocytes with natural cytotoxicity against tumour cells. NK cells were later recognized as a separate lymphocyte lineage, with both cytotoxicity and cytokine-producing effector functions (Trinchieri, 1989).

Activating NK cell receptors detect the presence of ligands on cells in 'distress', such as the stress-induced self ligands recognized by NKG2D: human unique long 16 (UL16)-binding protein (ULBP) and major histocompatibility complex (MHC) class I related chain (MIC) molecules, as well as mouse retinoic acid early transcript 1 (RAE-1), minor histocompatibility antigen (H60) and murine UL16-binding protein-like transcript (MULT1) molecules (Lanier, 2005). Other alert molecules include infectious non-self ligands (e.g. the cytomegalovirus-encoded m157 recognized by Ly49H in the mouse) and Toll-like receptor (TLR) ligands. Indeed, NK cells express several TLRs (Sivori et al., 2004).

NK cells also express the low-affinity Fc receptor CD16, enabling them to detect antibody-coated target cells and to exert antibody-dependent cell cytotoxicity.

NK cells use inhibitory receptors to gauge the absence of constitutively expressed self molecules on susceptible target cells. In particular, NK cells express classical MHC class I–specific receptors and 'lose' inhibitory signals when encountering MHC class I–deficient hematopoietic cells in several in vitro and in vivo models (Parham, 2005; Yokoyama and Plougastel, 2003). As a consequence, NK cells have been described as able to recognize 'missing self' on hematopoietic cells (Kärre et al., 1986). The MHC class I–specific inhibitory receptors include the killer cell immunoglobulin-like receptors (KIRs) in humans, the lectin-like Ly49 dimers in the mouse and the lectin-like CD94-NKG2A heterodimers in both species. A conserved feature of these inhibitory receptors resides in the presence of one or two intra-cytoplasmic inhibitory signalling domains called immunoreceptor tyrosine-based inhibition motifs (ITIMs). By interacting with MHC class I molecules that are constitutively expressed by most healthy cells in steady-state conditions but that may be lost upon stress, inhibitory MHC class I receptors provide a means for NK cells to ensure tolerance to self while allowing toxicity towards stressed cells (Vivier et al., 2008). MHC class I is not the only constitutive self signal detected by NK cells, as other inhibitory receptors (e.g mouse NKR-P1B, human NKR-P1A and mouse 2B4) that recognize non-MHC self molecules (e.g. Clr-b, LLT-1 and CD48, respectively) also regulate NK cell activation (Kumar and McNerney, 2005).

Biology of NKT cells

NKT cells are a unique subset of T-lymphocytes that express a T-cell receptor (TCR) α/β, with a restricted repertoire (Godfrey et al., 2004).

In mice, NKT cells most express an invariant TCR α chain rearrangement (Vα14-Jα18) with a conserved CDR3 region, and they typically co-express one of these: Vβ8.2, Vβ2 or Vβ7.

The homologous population of human invariant NKT cells express a Vα24-Jα18 rearranged TCR α chain with a Vβ11-containing β chain (Dellabona et al., 1994; Porcelli et al., 1993). This receptor distinguishes NKT cells from NK cells, even if they share some markers characteristic of NK cells such as CD161 (NKR-P1), known as NK1.1 in mice (Godfrey et al., 2004). In contrast to conventional T-lymphocytes and Tregs, the NKT cell TCR does not interact with peptide antigen presented by MHC-encoded class I or II molecules, but instead it recognizes glycolipids presented by CD1d, a non-classical antigen-presenting molecule that is associated with β2 microglobulin (Godfrey et al., 2000; Kronenberg and Gapin, 2002).

The compound most efficient for activating the majority of NKT cells is a synthetic glycolipid (originally derived from a marine sponge) known as α-galactosylceramide (αGalCer) (Hayakawa et al., 2004; Hong et al., 1999). This compound binds effectively to CD1d, and the complex of glycolipid plus CD1d binds the NKT cell TCR (Sidobre et al., 2002). αGalCer is widely used as a highly specific antigen for both human and murine NKT cells. The rapid response of NKT cells to its cognate antigens is characteristic of an innate immune response and allows the polarizing cytokines (IFN-γ and/or IL-4) to regulate adaptive immunity. NKT cells have been found to be critical in the immune response against viral infections and malaria, as well as in tumour immunity, and certain autoimmune diseases. NKT cells are emerging as an important subset of lymphocytes, with a protective role in host defence and a pathogenic role in certain immune-mediated disease states.

The history of NKT cell discovery

Robson MacDonald reviewed the discovery of NKT cells through time. Initially, there was no connection with the NK cell lineage for either Vβ8-biased DN thymocytes or Vα14-expressing peripheral T cells. He reported that this could probably be traced to the fact that only T-cell biologists were studying these rare populations and there was no reason at the time to suspect any shared markers between T cells and NK cells. On the other hand, NK cell biologists often were required to exclude T-cell markers from their analysis, putting them in a better position to identify shared phenotypic properties (Robson MacDonald, 2007). Around 1990, it was recognized that a subset of lymphocytes in spleen and bone marrow shared NK- and T-cell markers (Sykes, 1990; Yankelevich et al., 1989). Importantly, this was also shown to be true for the DN Vβ8-biased thymocyte subset (Ballas and Rasmussen, 1990). At this point, 'NKT' cells finally had a 'birth certificate' but still were not linked to expression of the Vα14 chain.

Distribution of NK and NKT cells

The distribution of NKT cells and NK cells is different; NKT cells are found everywhere that conventional T cells are found. In mice, NKT cells represent approximately 30% of the total lymphocytes in the liver (50% of αβTCR+ T cells), 20% of the αβ T cells in the bone marrow and 3% of the αβT cells in the spleen, but are rare in the lymph node (Berzins et al., 2004; Emoto and Kaufmann, 2003; Godfrey et al., 2000). In humans, only 0.2% of peripheral blood T cells are NKT cells. They are also present in the human liver, but their numbers are lower than in the liver of mice (Benlagha et al., 2000; Norris et al., 1999; Prussin and Foster, 1997). NKT cells predominate in the liver, while conventional T cells prevail in blood and lymph node. In contrast, the order for NK cell frequency is lung > liver > peripheral blood mononuclear cells (PBMCs) > spleen > bone marrow > lymph node > thymus, where NK cells are almost undetectable (Gregoire et al., 2007).

Allergic contact dermatitis

NK cells

A recent study by O'Leary et al., suggests that NK cells have the ability to acquire hapten-specific memory and mediate contact hypersensitivity (CHS). They used recombination activating gene 2 (Rag2) knockout mice, which are devoid of T cells and B cells (Rag2 is required for antigen-specific receptor rearrangement on B cells and T cells). When these mice were sensitized to dinitrofluorobenzene (DNFB), they mounted a vigorous CHS response when exposed to DNFB but not other haptens, indicating that CHS is inducible and hapten specific in the absence of T cells and B cells. NK cell infiltration to DNFB-exposed skin was rapid, and the CHS response was blocked by NK cell depletion. Furthermore, adoptive transfer of NK cells isolated from the livers of DNFB-sensitized Rag2 knockout mice into Rag2 and IL-2 receptor-γ-gene-targeted mice (devoid of B-, T- and NK cells), which are unable to mount a CHS response, resulted in restoration of the CHS response to DNFB (O'Leary et al., 2006). The mechanism of hapten recognition by NK cells is still unknown, although blocking experiments have suggested the involvement of the NKG2D receptor. NK activation and lysis of haptenated targets could also be the consequence of the hapten modification of MHC class I molecules that may prevent recognition by inhibitory NK cell receptors (Cavani et al., 2007). Lanier and colleagues demonstrated that the NK cell shared many of the properties of T cells with expansion, contraction, memory and an enhanced secondary response (Sun et al., 2009)

Gainers et al. hypothesized that NK cells can help elicit hapten-induced CHS responses through the adhesion molecules E (endothelial)-selectin and P (platelet)-selectin, that are necessary for T cell recruitment and are also expressed by NK cells (Gainers et al., 2007). Collectively, there are indications that, in the absence of B- and T cells, hepatic NK cells possess hapten-specific memory and can mediate a hapten-specific CHS response (Gober and Gaspari, 2008). This finding is remarkable because it suggests that NK cells, which lack a TCR, have some capacity to develop memory and antigen specificity, a characteristic associated with TCR-bearing conventional T cells.

NKT cells

The importance of NKT cells, at least, in the priming phase of CHS is reinforced by numerous experiments, even if further work is needed to better understand the role of NKT cells in the elicitation phase too, and in allergic contact dermatitis (ACD) in general. Some studies have focused on the role of NKT cells in CHS in experimental mouse models (Askenase, 2001; Campos et al., 2003; Nieuwenhuis et al., 2005; Stein-Streilein, 2003). Campos et al. indicated a role for this cell type in the early molecular events of CHS, in which NKT cells respond to hapten application and interact with B-1 lymphocytes, resulting in production of hapten-specific IgM (Campos et al., 2003) These findings were confirmed by an in vitro system too (Campos et al., 2006).

Elicitation of CHS begins with an initiation process that results in T-cell recruitment (Askenase, 2001). In fact, there are two starting points, both involving cells of the innate immune system. The first of these is induction of the initiation process. The other is actual initiation of the late-elicited effector phase due to recruitment of the T cells. Induction of the initiating process, which subsequently elicits T-cell recruitment, begins within minutes of immunization by postulated release of endogenous glycolipids from the sensitizing site. These glycolipids bind to CD1d on antigen-presenting cells (APCs), leading to the activation of NKT cells; these release IL-4 to co-activate B-1 cells. In the second of the two mentioned starting points, initiation of the effector phase occurs within minutes of the secondary local Ag challenge, when B-1 cell-derived IgM antibodies trigger an elicitation cascade of innate responses. This starts with classical complement activation, generating C5a, which in turn activates mast cells and platelets to release TNFα and serotonin; these then activate endothelial cells, leading to local T-cell recruitment (Askenase et al., 2004).

Another animal model study confirmed the role of NKT cells in murine CHS with impaired contact sensitization in Vα14 gene-targeted mice and examined the therapeutic role of antagonistic lipids that bound to CD1d but do not activate NKT cells (Nieuwenhuis et al., 2005). Gober et al. reported that CD1d gene expression was increased in elicitation sites of ACD relative to normal human skin. Similarly, both CD161 gene expression and variable region of TCR Vα24, JαQ region of TCR gene expression in skin lesions of ACD were increased relative to normal skin, indicating that invariant NKT cells are commonly present in the T-lymphocyte infiltrate of ACD (Gober et al., 2008). This suggests that NKT cells may be specifically recruited into ACD sites, rather than a passive extravasation event.

The relationship between NKT cells and ACD is not fully elucidated CD1d−/− mice are able to mount an intact systemic immune response to epicutaneous sensitization with antigen in a murine model of AD. NKT cells were crucial for the development of airway hyperreactivity in this model (Elkhal et al., 2006). However, it is important to notice that these experiments were performed in a murine model of AD, and the sensitization was executed with ovalbumin (OVA), so that a protein contact dermatitis was reproduced.

The role of NK and NKT cells in ACD is summarized in Table 23.1.

Psoriasis

NK cells

Recently, participation of the innate immune system in the pathogenesis of psoriasis has been suggested.

Table 23.1 Role of NK and NKT cells in allergic contact dermatitis

Allergic contact dermatitis	
NK cells	**NKT cells**
a. Involvement in CHS response to DNFB (hapten specific) **b.** Possible involvement of NKG2D and adhesion molecules as E-selectin and P-selectin	**a.** Critical role in afferent phase of CHS in mouse models **b.** Infiltration into CHS skin challenge sites in humans

The inflammatory infiltrate in psoriasis comprises cells with typical NK cell markers (Aractingi et al., 2001; Cameron et al., 2002; Harvell et al., 2003), and injection of lymphocytes expressing NK cell markers induces a psoriatic phenotype in a SCID mouse model grafted with human pre-psoriatic skin (Gilhar et al., 2002).

NK cells have been reported to be reduced in the peripheral blood of patients with psoriasis (Cameron et al., 2003). Cameron et al. showed that a significant proportion of the infiltrating cells in plaques of psoriasis expressed NK cell activatory (CD16) and inhibitory (CD158a and CD94) receptors. Plaques of psoriasis also contained a significant number of cells expressing the NKT cell marker–CD161. In contrast, there was no difference in the prevalence of CD56+ cells in psoriasis and normal skin (Cameron et al., 2002).

The cellular infiltrate in acute psoriatic plaques includes 5–8% CD3−CD56+ NK cells, mostly localized in the mid- and papillary dermis. NK lymphocytes isolated from punch biopsy specimens of psoriatic plaques showed a CD56brightCD16−CD158b− phenotype, failed to express the skin homing cutaneous lymphocyte-associated antigen and released abundant IFNγ following stimulation (Ottavini et al., 2006). Typically, NK cells have been subdivided in two major subsets: (1) CD56brightCD16− NK cells, also defined as immunoregulatory NK cells, which are characterized by prominent recirculation in secondary lymphoid organs, limited cytotoxic functions and high IFNγ production upon activation; and (2) CD56dimCD16+ NK cells, or effector NK cells, which constitute 90% of circulating NK lymphocytes. CD56brightCD16− NK cells predominate in secondary lymphoid organs, where they can modulate adaptive immune responses by interacting with recently emigrated dendritic cells and may affect T helper cell polarization through the release of type 1 cytokines (Vitale et al., 2004). In addition, several reports suggest that this subset also predominates in inflamed tissues. The CD56brightCD16− NK subset is expanded in synovial fluid from patients with inflammatory arthritis, in peritoneal fluid from patients with acute polmonitis and in infectious pulmonary diseases (Dalbeth and Callan, 2002; Dalbeth et al., 2004).

NK cells may have a primary activatory role in psoriasis as they can act without prior sensitization in the skin and could initiate plaque development. NK cells are amongst the first cells to enter a new site of inflammation. They can migrate even before T-cell stimulation, but unlike naïve and memory T cells and B cells, they cannot continuously re-circulate between blood and tissue; once they migrate to the site of defence, they never return to the circulation (Lanier, 2000; Lopez-Bottet et al., 2000). Activated NK cells can act as a source of Th1 cytokines particularly IFNγ. Thus NK cells may produce Th1 cytokines as an initiating event in psoriasis with consequent T-cell infiltration (Fehniger et al.,

1999; Bonish et al., 2000; Lanier, 2000; Lopez-Bottet et al., 2000).

NKT cells

There is considerable interest in NKT cells and their role in the pathophysiology of psoriasis. Increased NKT cells were consistently observed in psoriasis lesions by different groups (Bonish et al., 2000; Cameron et al., 2002; Liao et al., 2006; Ottavini et al., 2006; Vissers et al., 2004), although the exact role played by these cells is yet to be defined precisely. Nickoloff and co-workers (Bonish et al., 2000) demonstrated that psoriatic keratinocytes (KCs) overexpressed CD1d and NKT cells can be activated to elaborate IFNγ when cultured with CD1d overexpressing KCs. These results provided a pathogenetic link between psoriatic KCs, which overexpress CD1d and NKT cells infiltrating the lesions. Experiments in severe combined immunodeficient mice have also shown that injection of human cells of NKT characteristics into transplanted psoriatic skin could drive lesion development (Nickoloff et al., 2000). Zhao et al. showed increased densities of NKT cells using a set of precise markers for classical NKT cells, anti-Vα24 and anti-Vβ11 mAbs, in psoriatic lesions, especially in the epidermis, compared with healthy adult skin. These data were confirmed by real-time polymerase chain reactio (PCR) too. Also CD1d expression was more extensive in psoriasis than in normal skin. They also reported that protein kinase (PK)Cζ an important enzyme for CD1d expression in KCs, was overexpressed in psoriatic lesions (Zhao et al., 2008). In psoriasis, however, PKCζ has a broader role to play than CD1d expression regulation, as it is important for signalling TNFα (Schottelius et al., 2004). PKCζ is one of the key downstream mediators of TNFα signalling through to NF-κB activation and MAPK pathway in a variety of cell types (Berra et al., 1995; Moscat et al., 2006). Taken together, PKCζ is critically located in the transduction pathway from TNFα to activate NF-κB, the event crucial for the inflammatory process within the psoriasis lesions.

The role of NK and NKT cells in psoriasis is summarized in Table 23.2.

Table 23.2 Role of NK and NKT cells in psoriasis

Psoriasis	
NK cells	**NKT cells**
a. Increased infiltration into skin lesions	**a.** Increased infiltration into skin lesions
b. Reduced number in peripheral blood of patients	**b.** Pathogenic role in SCID mouse model
c. Pathogenic role in SCID mouse model	

Atopic dermatitis

NK cells

NK cells were among the first cells of the innate immunity to be examined extensively in AD. The hypothesis was that NK cells could be phenotypically and functionally defective in AD, providing a mechanism linking development of type-2-dominated acquired immune responses with a decreased resistance against various pathogens in AD.

The possible role of NK cells in AD has been studied with somewhat contradictory results. In 1984, Jensen et al. reported that patients with AD have low NK cell cytolytic activity. The reduced activity was not related to the serum IgE level, but rather to the severity of the skin disease. Patients with severe, generalized AD had the lowest NK cell activity, whereas patients with mild, localized eczema had insignificant reductions (Jensen et al., 1984). Although their analysis of previous studies on NK cells in AD showed very conflicting results (Kusaimi and Trentin, 1982; Leung et al., 1982; Strannegard and Strannegard, 1980; Viander et al., 1982).

In 2002 The Scheynius group reported that a few scattered NK (CD56+/CD3−) cells were found in the dermis of healthy individuals and in nonlesional skin from AD patients, whereas in lesional skin and in biopsies from Malassezia atopy-patch-test-positive skin, NK cells were differentially distributed and in close contact with CD1a+ dendritic cells, suggesting that NK cells may play a role in regulating dendritic cells in AD (Buentke et al., 2002). In another study, NK cell activity was examined in 128 patients with AD. No relationship was shown between severity of dermatitis and NK activity, neither between NK activity nor eosinophilic counts nor serum IgE; patients with longer duration of AD lesions showed significantly lower NK activity. The significant relationship was recognized between severity of dermatitis and the duration of disease; furthermore, no relationships were recognized between NK activity and mental states evaluated using the Profile of Mood States (Sakai et al., 2003).

AD patients showing significantly decreased peripheral blood NK cells when compared to healthy individuals were reported by Aktas et al. They also assessed that human NK cells comprise distinct receptor-expressing and cytokine-producing subsets similar to Th1 and Th2 cells; except for CD40, NK cell subsets showed different expression of KIR receptors and co-stimulatory molecules between the AD patients and healthy subjects, indicating different functional properties in AD (Aktas et al., 2005).

Katsuta et al. hypothesized that NK cells and $\gamma\delta$T cells in AD patients are in immediate contact with activated monocytes in vivo, but they are specifically compromised in their capacity to produce type-1 and pro-inflammatory cytokines, thereby directing subsequent acquired immune responses towards a type-2 pattern and increasing susceptibility to infection (Katsuta et al., 2006).

NKT cells

The relationship between AD and NKT cells has not been as well investigated as for NK cells; furthermore, contradictory results are reported. Takahashi et al. reported that the proportion of Vα24+ NKT cells in the peripheral blood of AD patients was markedly decreased and only CD4− subset of Vα24+ NKT cells was significantly diminished (Takahashi et al., 2003).

Oishi et al. reported that in atopic disease (three AD patients and six asthma patients), the number of CD4−CD8− Vα24+ NKT cells was decreased but not the number of CD4+ Vα24+ NKT cells (Oishi et al., 2000). Ilhan et al. showed that percentages of Vα24+CD161+ NKT cell subtypes were significantly lower in patients with AD than healthy individuals, suggesting that these cell subtypes may be involved in the immunopathogenesis of AD (Ilhan et al., 2007). On the other hand, Magnan et al. reported that the number of CD4+Vα24+ NKT cells was increased in AD patients, and it was related to IL-4 production and IgE levels (Magnan et al., 2000). The Prell et al. analysis of Vα24+CD161+ NKT cell frequencies at both cellular and molecular levels failed to reveal significant differences in peripheral blood of atopic and non-atopic patients (Prell et al., 2003). The question if NKT cells are involved in AD and in which way is yet to be answered; more studies are needed to establish the importance of these cells in AD.

The role of NK and NKT cells in AD is summarized in Table 23.3.

Table 23.3 Role of NK and NKT cells in atopic dermatitis

Atopic dermatitis	
NK cells	**NKT cells**
Contradictory results:	Inconsistent results:
a. Reduced number of circulating NK cells in some studies and no difference in number of circulating cells respect to healthy individuals in others	**a.** Reduced number of circulating NKT cells in some studies and increase in others
b. Reduced activity of NK cells in some studies and no difference in others	

UV-induced cutaneous immune suppression

NK cells

Only a limited number of reports, using a variety of ultraviolet (UV) sources, have tested whether changes are induced in either the frequency or activity of this cell type by repeated UV radiation in human subjects. Two related studies monitored NK cell activity following solarium exposure; normal subjects were irradiated with approximately one minimal erythemal dose (MED) for 2, 6 or 12 consecutive days (Hersey et al., 1987, 1988). At least 6 days of exposure were required to induce suppression of NK cell activity, and the suppression was more pronounced after 12 days of exposure, indicating that the downregulation may be dose-dependent. There was an associated reduction in circulating NK cell numbers. Following the final UV exposure, it took at least 14 days for the NK cell function to be restored to the pre-irradiated level. Gilmour et al. assessed NK cell activity in patients with psoriasis who were receiving broadband UVB or narrow band UVB (311 nm) therapy over a 6-week period. Both regimes resulted in suppression, and it took several weeks after the cessation of therapy to recover to the initial level of activity (Gilmour et al., 1993). A similar result was obtained by Neill et al. but using healthy subjects where the extent of the decline in NK cell activity correlated with the number of exposures to broadband UVB (Neill et al., 1998). Whitmore and Morison examined NK cells in the peripheral blood of normal individuals before and after 10 whole-body exposures over 2 weeks to sunlamps emitting 95% UVA and 5% UVB. The treatment started at 70% MED and increased by 20% on each occasion. No significant change in the number of circulating NK cells occurred (Whitmore and Morison, 2000). In contrast, in a study involving infants in St. Petersburg, it was demonstrated that, following 20 daily whole-body suberythemal exposures to FS20 lamps emitting predominantly UVB, the total PBMC population identified as NK cells dropped from about 21% to about 13% (Snopov et al., 2005). NK cell function was not evaluated in either of these last two reports. Thus, as far as can be ascertained with the limited information outlined here, repeated UV radiation is capable of reducing NK cell activity and possibly the numbers circulating in human blood, and photo-adaptation does not occur.

NKT cells

UV radiation is carcinogenic and immunosuppressive. UV-induced immune suppression is mediated by antigen-specific T cells, which can transfer suppression to normal recipients. These cells are essential for controlling skin cancer development in the UV-irradiated host and in

Table 23.4 Role of NK and NKT cells in UV immune suppression

UV-induced cutaneous immune suppression	
NK cells	**NKT cells**
Repeated UV exposure can reduce NK cell activity and numbers in human blood; no photo-adaptation	These cells are responsible for the transfer of suppression from UV-irradiated mice to non-irradiated control (mice)

suppressing other immune responses, such as delayed-type hypersensitivity (DTH). The development of UV-induced skin cancers in the primary UV-irradiated host, and the ability to transfer immune suppression from UV-irradiated animals to normal recipients is controlled by CD4+T-cells (Fisher and Kripke, 1982; Ullrich and Kripke, 1984). Despite having been identified many years ago (Daynes and Spellman, 1977; Fisher and Kripke, 1977), the exact identity of the cells that transfer suppression of tumour immunity remains unknown. The immunosuppressive mechanism is also unclear, but IL-4 secretion is involved (Rivas and Ullrich, 1994). Two types of CD4+ T cells secrete IL-4: Th2 and NKT cells. Moodycliffe et al. demonstrated a critical role for NKT cells in mediating UV-induced immune suppression and in the regulation of sunlight-induced skin carcinogenesis. They showed that there was no immune suppression in UV-irradiated CD1d-deficient mice. They explained that although their data showed that UV-induced NKT cells mediate antigen-specific immune suppression, they were not able to definitively identify the ultimate suppressor cell. It was hypothesized that UV-induced IL-4–secreting NKT cells directly suppress effector cell function (Moodycliffe et al., 2000). Alternatively, UV-induced NKT cells may initiate a cascade of events in which recipient T cells are activated to secrete IL-4 and suppress DTH and tumour immunity (Singh et al., 1999).

The role of NK and NKT cells in UV immune suppression is summarized in Table 23.4.

Cutaneous lymphomas

NK and NKT cells

NK cells have also been associated with cutaneous T-cell lymphomas. The extranodal NK/T-cell lymphoma is an entity defined as an extranodal lymphoid neoplasm derived from NK cells or, less commonly, from cytotoxic T cells. The neoplastic cells are usually positive for CD56, CD2, cytoplasmic CD3ε and cytotoxic granule proteins (TIA-1, granzyme B and perforin) (Pagano et al., 2006). They are usually negative for surface CD3, CD4 and CD8 (Santucci et al., 2003), but some may express EBV+ CD56− CD4+, CD7+ and/or CD30+ immunophenotype (Bekkenk et al., 2004; Pol-Rodriguez et al., 2006).

Skin involvement may be a primary or secondary manifestation of the disease. Nasal cases and extranasal cases are two major types of extranodal NK/T-cell lymphomas designated as 'nasal type'. These diseases are currently separated from the extranodal blastic NK cell types that are different and likely derived from precursor plasmacytoid dendritic cells (hematodermic) (Savage et al., 2004). Clinical features include multiple indurated, erythematous plaques, tumours or nodules that may ulcerate. Lesions are located on the extremities, trunk and, less frequently, the head and neck (Mraz-Gernhard et al., 2001). Rare cases with bruise-like skin lesions have been reported (Dummer et al., 1996). Systemic symptoms such as fever, malaise and weight loss may be present, and cytopenia due to hemophagocytic syndrome has been reported in some cases. Prognosis is variable. Patients without extracutaneous involvement have the best prognosis, with median survival of less than 27 months (Bekkenk et al., 2004; Pol-Rodriguez et al., 2006).

NK cells have also been reported to be involved in Sezary syndrome (SS); SS is an advanced form of cutaneous T-cell lymphoma with malignant T cells in peripheral blood derived from the skin invasive T-cell clone (Kim et al., 2005; Querfeld et al., 2005). Dulphy et al. highlighted the potential role of NKG2D/NKG2D–Ligands interactions in NK immune responses in SS patients. They assessed the NKG2D-Ligands surface expression on SS tumours and the ability of SS patients' NK cells to engage NKG2D and to be functional against sensitive targets, suggesting that the stimulation of NK function in SS patients may be a promising strategy to reduce tumour invasion (Dulphy et al., 2008).

Several studies indicate that NKT cells in cancer patients have some numerical and functional defects. In addition to studies in cancer patients with solid tumours, Fujii et al. have shown that patients with myelodysplastic syndromes have a severe functional deficiency in NKT cells but not NK cells or CD4+ or CD8+ T cells (Fujii et al., 2003). Information on NKT cells and cutaneous lymphomas are very scarce, so that any conclusion would be inconsistent.

Conclusions

NK and NKT cells, two different lineages of cells with various characteristics in common, are involved in the immunologic mechanisms of skin disorders, such as ACD, psoriasis, AD and cutaneous lymphomas; they are also implicated in UV immune suppression. Identifying the factors critical for the skin-tropic behaviour of these cells may have a profound influence on the development of targeted anti-inflammatory approaches.

References

Aktas, E., Akdis, M., Bilgic, S., Disch, R., Falk, C.S., Blaser, K., Akdis, C. and Deniz, G. (2005). Different natural killer (NK) receptor expression and immunoglobulin E (IgE) regulation by NK1 and NK2 cells. Clin Exp Immunol **140**, 301–309.

Aractingi, S., Briand, N., Le Danff, C., Viguier, M., Bachelez, H., Michel, L., Dubertret, L. and Carosella, E.D. (2001). HLA-G and NK receptor are expressed in psoriatic skin: a possible pathway for regulating infiltrating T cells? Am J Pathol **159**, 71–77.

Askenase, P.W. (2001). Yes T-cells, but three different T-cells (alpha/beta, gamma/delta and NKT-cells), and also B-1 cells mediate contact hypersensitivity. Clin Exp Immunol **125**, 345–350.

Askenase, P.W., Szczepanik, M., Itakura, A., Kiener, C. and Campos, R.A. (2004). Extravascular T-cell recruitment requires initiation begun by Vα14 NKT cells and B-1 B cells. Trends Immunol **25**, 441–449.

Ballas, Z.K. and Rasmussen, W. (1990). NK1.1+ thymocytes. Adult murine CD4−, CD8− thymocytes contain an NK1.1+, CD3+, CD5hi, CD44hi, TCR-V beta 8+ subset. J Immunol **145**, 1039–1045.

Bashirova, A.A., Martin, M.P., McVicar, D.W. and Carrington, M. (2006). The killer immunoglobulin-like receptor gene cluster: tuning the genome for defense. Annu Rev Genomics Hum Genet **7**, 277–300.

Bekkenk, M.W., Jansen, P.M., Meijer, C.J. and Willemze, R. (2004). CD56+ haematological neoplasms presenting in the skin: a retrospective analysis of 23 new cases and 130 cases from the literature. Ann Oncol **15**, 1097–1108.

Benlagha, K., Weiss, A., Beavis, A., Teyton, L. and Bendelac, A. (2000). In vivo identification of glycolipid antigen-specific T cells using fluorescent CD1d tetramers. J Exp Med **191**, 1895–1903.

Berra, E., Diaz-Meco, M.T., Lozano, J., Frutos, S., Municio, M.M., Sanchez, P., Sanz, L. and Moscat, J. (1995). Evidence for a role of MEK and MAPK during signal transduction by protein kinase C zeta. EMBO J **14**, 6157–6163.

Berzins, S., Uldrich, A., Pellicci, D., Mcnab, F., Kayakawa, Y., Smyth, M. and Godfrey, D.I. (2004). Parallels and distinctions between T and NKT cell development in the thymus. Immunol Cell Biol **82**, 269–275.

Bonish, B., Jullien, D., Dutronc, Y., Huang, B.B., Modlin, R., Spada, F. M., Porcelli, S.A. and Nickoloff, B.J. (2000). Overexpression of CD1d by keratinocytes in psoriasis and CD1d dependent IFN-gamma production by NK-T cells. J Immunol **165**, 4076–4085.

Buentke, E., Heffler, L.C., Wilson, J.L., Wallin, R.P., Löfman, C., Chambers, B. J., Ljunggren, H.G. and Scheynius, A. (2002). Natural killer and dendritic cell contact in lesional atopic dermatitis skin—Malassezia-influenced cell interaction. J Invest Dermatol **119**, 850–857.

Cameron, A.L., Kirby, B., Fei, W. and Griffiths, C.E. (2002). Natural killer and natural killer-T cells in psoriasis. Arch Dermatol Res **294**, 363–369.

Cameron, A.L., Kirby, B. and Griffiths, C. E. (2003). Circulating natural killer cells in psoriasis. Br J Dermatol **149**, 160–164.

Campos, R.A., Szczepanik, M., Itakura, A., Akahira-Azuma, M., Sidobre, S., Kronenberg, M. and Askenase, P.W. (2003). Cutaneous immunization rapidly activates liver invariant Vα14 NKT-cells stimulating B-1 B-cells to initiate recruitment for elicitation of contact sensitivity. J Exp Med **198**, 1785–1796.

Campos, R.A., Szczepanik, M., Lisbonne, M., Itakura, A., Leite-de-Moraes, M. and Askenase, P.W. (2006). Invariant NKT cells rapidly activated via immunization with diverse contact antigens collaborate in vitro with B-1 cells to initiate contact sensitivity. J Immunol **177**, 3686–3694.

Cavani, A., De Pità, O. and Girolomoni, G. (2007). New aspects of the molecular basis of contact allergy. Curr Opin Allergy Clin Immunol **7**, 404–408.

Dalbeth, N. and Callan, M.F. (2002). A subset of natural killer cells is greatly expanded within inflamed joints. Arthritis Rheum **46**, 1763–1772.

Dalbeth, N., Gundle, R., Davies, R.J., Lee, Y.C., McMichael, A.J. and Callan, M. F. (2004). CD56bright NK cells are enriched at inflammatory sites and can engage with monocytes in a reciprocal program of activation. J Immunol **173**, 6418–6426.

Daynes, R.A. and Spellman, C.W. (1977). Evidence for the generation of suppressor cells by UV radiation. Cell Immunol **31**, 182–187.

Dellabona, P., Padovan, E., Casorati, M., Brockhaus, M. and Lanzavecchia, A. (1994). An invariant Vα24-JαQ/Vβ11 T cell receptor is expressed in all individuals by clonally expanded CD4-8- T cells. J Exp Med **180**, 1171–1176.

Dulphy, N., Berrou, J., Campillo, J.A., Bagot, M., Bensussan, A. and Toubert, A. (2008). NKG2D ligands expression and NKG2D-mediated NK activity in Sezary patients. J Invest Dermatol doi:10.1038/jid.2008.256.

Dummer, R., Potoczna, N., Haffner, A. C., Zimmermann, D.R., Gilardi, S. and Burg, G. (1996). A primary cutaneous non-T, non-B CD4+, CD56+ lymphoma. Arch Dermatol **132**, 550–553.

Elkhal, A., Pichavant, M., He, R., Scott, J., Meyer, E., Goya, S., Geha, R.S. and Umetsu, D.T. (2006). CD1d restricted natural killer T cells are not required for allergic skin inflammation. J Allergy Clin Immunol **118**, 1363–1368.

Emoto, M. and Kaufmann, S. (2003). Liver NKT cells: an account of heterogeneity. Trends Immunol **24**, 364–369.

Fehniger, T.A., Carson, W.E. and Caligiuri, M.A. (1999). Costimulation of human natural killer cells is required for interferon gamma production. Transplant Proc **31**, 1476–1478.

Fisher, M.S. and Kripke, M.L. (1977). Systemic alteration induced in mice by ultraviolet light irradiation and its relationship to ultraviolet carcinogenesis. Proc Natl Acad Sci U S A **74**, 1688–1692.

Fisher, M.S. and Kripke, M.L. (1982). Suppressor T lymphocytes control the development of primary skin cancers in UV-irradiated mice. Science **216**, 1133–1134.

Fujii, S., Shimizu, K., Klimek, V., Geller, M. D., Nimer, S.D. and Dhodapkar, M.V. (2003). Severe and selective deficiency of interferon-gamma-producing invariant natural killer T cells in patients with myelodysplastic syndromes. Br J Haematol **122**, 617–622.

Gainers, M.E., Descheny, L., Barthel, S. R., Liu, L., Wurbel, M. and Dimitroff, C.J. (2007). Skin-homing receptors on effector leukocytes are differentially sensitive to glyco-metabolic antagonism in allergic contact dermatitis. J Immunol **179**, 8509–8518.

Gilhar, A., Ullmann, Y., Kerner, H., Assy, B., Shalaginov, R., Serafimovich, S. and Kalish, R.S. (2002). Psoriasis is mediated by a cutaneous defect triggered by activated immunocytes: induction of psoriasis by cells with natural killer receptors. J Invest Dermatol **119**, 384–391.

Gilmour, J.W., Vestey, J.P., George, S. and Norval, M. (1993). Effect of phototherapy and urocanic acid isomers on natural killer cell function. J Invest Dermatol **101**, 169–174.

Gober, M.D. and Gaspari, A.A. (2008). Allergic contact dermatitis. Curr Dir Autoimmun **10**, 1–26.

Gober, M.D., Fishelevich, R., Zhao, Y., Unutmaz, D. and Gaspari, A.A. (2008). Human natural killer T cells infiltrate into the skin at elicitation sites of allergic contact dermatitis. J Invest Dermatol **128**, 1460–1469.

Godfrey, D.I., Hammond, K.J., Poulton, L.D., Smyth, M.J. and Baxter, A.G. (2000). NKT cells: facts, functions and fallacies. Immunol Today **21**, 573–583.

Godfrey, D.I., MacDonald, H.R., Kronenberg, M., Smyth, M.J. and VanKaer, L. (2004). NKT-cells: what's in a name? Nat Rev Immunol **4**, 231–237.

Gregoire, C., Chasson, L., Luci, C., Tomasello, E., Geissmann, F., Vivier, E. and Walzer, T. (2007). The trafficking of natural killer cells. Immunol Rev **220**, 169–182.

Harvell, J.D., Nowfar-Rad, M. and Sundram, U. (2003). An immunohistochemical study of CD4, CD8, TIA-1 and CD56 subsets in inflammatory skin disease. J Cutan Pathol **30**, 108–113.

Hayakawa, Y., Godfrey, D.I. and Smyth, M.J. (2004). Alpha-galactosylceramide: potential immunomodulatory activity and future application. Curr Med Chem **11**, 241–252.

Hersey, P., MacDonald, M., Burns, C., Schibeci, S., Matthews, H. and Wilkinson, F.J. (1987). Analysis of the effect of a sunscreen agent on the suppression of natural killer cell activity induced in human subjects by radiation from solarium lamps. J Invest Dermatol **88**, 271–276.

Hersey, P., MacDonald, M., Henderson, C., Schibeci, S., D'Alessandro, G., Pryor, M. and Wilkinson, F.J. (1988). Suppression of natural killer cell activity in humans by radiation from solarium lamps depleted of UVB. J Invest Dermatol **90**, 305–310.

Hong, S.M., Scherer, D.C., Singh, N., Mendiratta, S.K., Serizawa, I., Koezuka, Y. and Van Kaer, L. (1999). Lipid antigen presentation in the immune system: lessons learned from CD1d knockout mice. Immunol Rev **169**, 31–44.

Ilhan, F., Kandi, B., Akbulut, H., Turgut, D. and Cicek, D. (2007). Atopic dermatitis and Valpha24+ natural killer T cells. Skinmed **6**, 218–220.

Jensen, J.R., Sand, T.T., Jorgensen, A. S. and Thestrup-Pedersen, K. (1984). Modulation of natural killer cell activity in patients with atopic dermatitis. J Invest Dermatol **82**, 30–34.

Kärre, K., Ljunggren, H.G., Piontek, G. and Kiessling, R. (1986). Selective rejection of H-2–deficient lymphoma variants suggests alternative immune defense strategy. Nature **319**, 675–678.

Katsuta, M., Takigawa, Y., Kimishima, M., Inaoka, M., Takahashi, R. and Shiohara, T. (2006). NK cells and γδ+ T cells are phenotypically and functionally defective due to preferential apoptosis in patients with atopic dermatitis. J Immunol **176**, 7736–7744.

Kim, E.J., Hess, S., Richardson, S.K., Newton, S., Showe, L.C., Benoit, B. M., Ubriani, R., Vittorio, C.C., Junkins-Hopkins, J.M., Wysocka, M. and Rook, A.H. (2005). Immunopathogenesis and therapy of cutaneous T cell lymphoma. J Clin Invest **115**, 798–812.

Kronenberg, M. and Gapin, L. (2002). The unconventional lifestyle of NKT cells. Nat Rev Immunol **2**, 557–568.

Kumar, V. and McNerney, M.E. (2005). A new self: MHC-class-I-independent natural-killer-cell self-tolerance. Nat Rev Immunol **5**, 363–374.

Kusaimi, N.T. and Trentin, J.J. (1982). Natural cell-mediated cytotoxic activity in the peripheral blood of patients with atopic dermatitis. Arch Dermatol **118**, 568–571.

Lanier, L.L. (2000). The origin and functions of natural killer cells. Clin Immunol **95**, S14–S18.

Lanier, L.L. (2005). NK cell recognition. Annu Rev Immunol **23**, 225–274.

Leung, D.Y.M., Parkman, R., Feller, J., Wood, N. and Geha, R.S. (1982). Cell mediated cytotoxicity against skin fibroblasts in atopic dermatitis. J Immunol **128**, 1736–1741.

Ljunggren, H.G. and Karre, K. (1990). In search of the 'missing self': MHC molecules and NK cell recognition. Immunol Today **1**, 237–244.

Lopez-Bottet, M., Bellon, T., Llano, M., Navarro, F., Garcia, P. and de Miguel, M. (2000). Paired inhibitory and triggering NK cell receptors for HLA class I molecules. Hum Immunol **16**, 7–17.

Liao, Y.H., Jee, S.H., Sheu, B.C., Huang, Y.L., Tseng, M.P., Hsu, S.M. and Tsai, T.F. (2006). Increased expression of the natural killer cell inhibitory receptor CD94/NKG2A and CD158b on circulating and lesional T cells in patients with chronic plaque psoriasis. Br J Dermatol **155**, 318–324.

Magnan, A., Mely, L., Prato, S., Vervloet, D., Romagne, F., Camilla, C., Necker, A., Casano, B., Montero-Jullian, F., Fert, V., Malissen, B. and Bongrand, P. (2000). Relationships between natural T cells, atopy, IgE levels, and IL-4 production. Allergy **55**, 286–290.

Moodycliffe, A.M., Nghiem, D., Clydesdale, G. and Ullrich, S.E. (2000). Immune suppression and skin cancer development: regulation by NKT cells. Nat Immunol **1**, 521–525.

Moscat, J., Rennert, P. and Diaz-Meco, M. T. (2006). PKCzeta at the crossroad of NFkappaB and Jak1/Stat6 signaling pathways. Cell Death Differ **13**, 702–711.

Mraz-Gernhard, S., Natkunam, Y., Hoppe, R.T., LeBoit, P., Kohler, S. and Kim, Y.H. (2001). Natural killer/natural killer-like T-cell lymphoma, CD56+, presenting in the skin: an increasingly recognized entity with an aggressive course. J Clin Oncol **19**, 2179–2188.

Neill, W.A., Halliday, K.E. and Norval, M. (1998). Differential effect of phototherapy on the activities of human natural killer cells and cytotoxic T cells. J Photochem Photobiol B **47**, 129–135.

Nickoloff, B.J., Bonish, B., Huang, B.B. and Porcelli, S. (2000). Characterization of a T cell line bearing natural killer receptors and capable of creating psoriasis in a SCID mouse model system. J Dermatol Sci **24**, 212–225.

Nieuwenhuis, E.E.S., Gillessen, S., Scheper, R.J., Exley, M.A., Taniguchi, M., Balk, S.P., Strominger, J.L., Dranoff, G., Blumberg, R.S. and Wilson, S.B. (2005). CD1d and CD1d-restricted iNKT-cells play a pivotal role in contact hypersensitivity. Exp Dermatol **14**, 250–258.

Norris, S., Doherty, D.G., Collins, C., McEntee, G., Traynor, O., Hegarty, J.E. and O'Farrelly, C. (1999). Natural T cells in the human liver: cytotoxic lymphocytes with dual T cell and natural killer cell phenotype and function are phenotypically heterogenous and include Valpha24-JalphaQ and gammadelta T cell receptor bearing cells. Hum Immunol **60**, 20–31.

Oishi, Y., Sakamoto, A., Kurasawa, K., Nakajima, H., Nakao, A., Nakagawa, N., Tanabe, E., Saito, Y. and Iwamoto, I. (2000). CD4-CD8- T cells bearing invariant Vα24JαQ TCR α-chain are decreased in patients with atopic diseases. Clin Exp Immunol **119**, 404–411.

O'Leary, J.G., Goodarzi, M., Drayton, D. L. and von Andrian, U.H. (2006). T cell- and B cell-independent adaptive immunity mediated by natural killer cells. Nat Immunol **7**, 507–516.

Ottavini, C., Nasorri, F., Bedini, C., de Pità, O., Girolomoni, G. and Cavani, A. (2006). CD56brightCD16(−) NK cells accumulate in psoriatic skin in response to CXCL10 and CCL5 and exacerbate skin inflammation. Eur J Immunol **36**, 118–128.

Pagano, L., Gallamini, A., Trapè, G., Fianchi, L., Mattei, D., Todeschini, G., Spadea, A., Cinieri, S., Iannitto, E., Martelli, M., Nosari, A., Bona, E.D., Tosti, M.E., Petti, M.C., Falcucci, P., Montanaro, M., Pulsoni, A., Larocca, L.M. and Leone, G.Intergruppo Italiano Linfomi, (2006). NK/T-cell lymphomas 'nasal type': an Italian multicentric retrospective survey. Ann Oncol **17**, 794–800.

Parham, P. (2005). MHC class I molecules and KIRs in human history, health and survival. Nat Rev Immunol **5**, 201–214.

Pol-Rodriguez, M.M., Fox, L.P., Sulis, M.L., Miller, I.J. and Garzon, M.C. (2006). Extranodal nasal-type natural killer T-cell lymphoma in an adolescent from Bangladesh. J Am Acad Dermatol **54**, S192–S197.

Porcelli, S., Yockey, C.E., Brenner, M.B. and Balk, S.P. (1993). Analysis of T cell antigen receptor (TCR) expression by human peripheral blood CD4-8- alpha/beta T cells demonstrates preferential use of several V beta genes and an invariant TCR alpha chain. J Exp Med **178**, 1–16.

Prell, C., Konstantopoulos, N., Heinzelmann, B., Frankenberger, B., Reinhardt, D., Schendel, D.J. and Krauss-Etschmann, S. (2003). Frequency of Valpha24+CD161+ natural killer T cells and invariant TCRAV24-AJ18 transcripts in atopic and non-atopic individuals. Immunobiology **208**, 367–380.

Prussin, C. and Foster, B. (1997). TCR V alpha 24 and V beta 11 coexpression defines a human NK1 T cell analog containing a unique Th0 subpopulation. J Immunol **159**, 5862–5870.

Querfeld, C., Rosen, S.T., Guitart, J. and Kuzel, T.M. (2005). The spectrum of cutaneous T-cell lymphomas: new insights into biology and therapy. Curr Opin Hematol **12**, 273–278.

Rivas, J.M. and Ullrich, S.E. (1994). The role of IL-4, IL-10, and TNF-α in the immune suppression induced by ultraviolet radiation. J Leukoc Biol **56**, 769–775.

Robson MacDonald, H. (2007). NKT cells: in the beginning. Eur J Immunol **37**, 111–115.

Sakai, R., Aihara, M., Ishiwa, M., Negishi, A., Matsukura, S., Takahashi, K., Kimura, H., Kosaka, K. and Ikezawa, Z. (2003). Natural killer cell activity among patients with atopic dermatitis. Arerugi **52**, 1074–1080.

Santucci, M., Pimpinelli, N., Massi, D., Kadin, M.E., Meijer, C.J., Müller-Hermelink, H.K., Paulli, M., Wechsler, J., Willemze, R., Audring, H., Bernengo, M.G., Cerroni, L., Chimenti, S., Chott, A., Díaz-Pérez, J.L., Dippel, E., Duncan, L.M., Feller, A.C., Geerts, M.L., Hallermann, C., Kempf, W., Russell-Jones, R., Sander, C. and Berti, E. EORTC Cutaneous Lymphoma Task Force, (2003). Cytotoxic/natural killer cell cutaneous lymphomas: report of EORTC Cutaneous Lymphoma Task Force Workshop. Cancer **97**, 610–627.

Savage, K.J., Chhanabhai, M., Gascoyne, R.D. and Connors, J.M. (2004). Characterization of peripheral T-cell lymphomas in a single North American institution by the WHO classification. Ann Oncol **15**, 1467–1475.

Schottelius , A.J., Moldawer, L.L., Dinarello, C.A., Asadullah, K., Sterry, W. and Edwards, C.K. III (2004). Biology of tumor necrosis factor-α—implications for psoriasis. Exp Dermatol **13**, 193–222.

Sidobre, S., Naidenko, O.V., Sim, B.C., Gascoigne, N.R., Garcia, K.C. and Kronenberg, M. (2002). The V alpha 14 NKT cell TCR exhibits high-affinity binding to a glycolipid/CD1d complex. J Immunol **169**, 1340–1348.

Singh, N., Hong, S., Scherer, D.C., Serizawa, I., Burdin, N., Kronenberg, M., Koezuka, Y. and Van Kaer, L. (1999). Cutting edge: activation of NK T cells by CD1d and α-galactosylceramide directs conventional T cells to the acquisition of a Th2 phenotype. J Immunol **163**, 2373–2377.

Sivori, S., Falco, M., Della Chiesa, M., Carlomagno, S., Vitale, M., Moretta, L. and Moretta, A. (2004). CpG and double-stranded RNA trigger human NK cells by Toll-like receptors: induction of cytokine release and cytotoxicity against tumors and dendritic cells. Proc Natl Acad Sci U S A **101**, 10116–10121.

Snopov, S.A., Kharit, S.M., Norval, M. and Ivanova, V.V. (2005). Circulating leukocyte and cytokine responses to measles and poliovirus vaccination in children after ultraviolet radiation exposures. Arch Virol **150**, 1729–1743.

Stein-Streilein, J. (2003). Invariant NKT-cells as initiators, licensors, and facilitators of the adaptive immune response. J Exp Med **198**, 1779–1783.

Strannegard, I-L. and Strannegard, O. (1980). Natural killer cells and interferon production in atopic dermatitis. Acta Derm Venereol **92**, 48–51.

Sun, J.C. and Lanier, L.L. Natural killer cells members: An evolutionary bridge between innate and adaptive immunity? Eur J Immunol. 2009 Jul 27.[EPub ahead to print]

Sykes, M. (1990). Unusual T cell populations in adult murine bone marrow. Prevalence of CD3+CD4−CD8− and alpha beta TCR+NK1.1+ cells. J Immunol **145**, 3209–3215.

Takahashi, T., Nakamura, K., Chiba, S., Kanda, Y., Tamaki, K. and Hirai, H. (2003). Vα24+ natural killer T cells are markedly decreased in atopic dermatitis patients. Hum Immunol **64**, 586–592.

Trinchieri, G. (1989). Biology of natural killer cells. Adv Immunol **47**, 187–376.

Ullrich, S.E. and Kripke, M.L. (1984). Mechanisms in the suppression of tumor rejection produced in mice by repeated UV irradiation. J Immunol **133**, 2786–2790.

Viander, M., Uksila, J., Lassila, O. and Jansen, C.J. (1982). Natural killer cell activity in atopic dermatitis. Arch Dermatol Res **274**, 283–288.

Vissers, W.H., Arndtz, C.H., Muys, L., Van Erp, P.E., de Jong, E.M. and van de Kerkhof, P.C. (2004). Memory effector (CD45RO+) and cytotoxic (CD8+) T cells appear early in the margin zone of spreading psoriatic lesions in contrast to cells expressing natural killer receptors, which appear late. Br J Dermatol **150**, 852–859.

Vitale, M., Della Chiesa, M., Carlomagno, S., Romagnani, C., Thiel, A., Moretta, L. and Moretta, A. (2004). The small subset of CD56brightCD16− natural killer cells is selectively responsible for both cell proliferation and interferongamma production upon interaction with dendritic cells. Eur J Immunol **34**, 1715–1722.

Vivier, E., Tomasello, E., Baratin, M., Walzer, T. and Ugolini, S. (2008). Functions of natural killer cells. Nat Immunol **9**, 503–510.

Whitmore, S.E. and Morison, W.L. (2000). The effect of suntan parlor exposure on delayed and contact hypersensitivity. Photochem Photobiol **71**, 700–705.

Yankelevich, B., Knobloch, C., Nowicki, M. and Dennert, G. (1989). A novel cell type responsible for marrow graft rejection in mice. T cells with NK phenotype cause acute rejection of marrow grafts. J Immunol **142**, 3423–3430.

Yokoyama, W.M. and Plougastel, B.F. (2003). Immune functions encoded by the natural killer gene complex. Nat Rev Immunol **3**, 304–316.

Zhao, Y., Fishelevich, R., Petrali, J.P., Zheng, L., Anatolievna, M.A., Deng, A., Eckert, R.L. and Gaspari, A.A. (2008). Activation of keratinocyte protein kinase C zeta in psoriasis plaques. J Invest Dermatol **128**, 2190–2197.

Zollner, T.M., Asadullah, L. and Schon, M.P. (2007). Targeting leukocyte trafficking to inflamed skin: still an attractive therapeutic approach? Exp Dermatol **16**, 1–12.

Chapter Twenty-Four

24

Natural killer cells in the respiratory tract[1]

Borna Mehrad

CHAPTER CONTENTS

As we know, there are known knowns. There are things we know we know. We also know there are known unknowns. That is to say, we know there are some things we do not know. But there are also unknown unknowns, the ones we don't know we don't know.

Donald Rumsfeld

ABSTRACT

Respiratory tract natural killer (NK) cells constitute the second largest population of NK cells in experimental animals, but remarkably little is understood about their biology. Although respiratory tract NK cells appear to be important in a number of lung diseases, there are large gaps in our knowledge of the biology of these cells. Important areas for future investigation include description of normal distribution of NK cell subsets in the human lung and defining the role of local NK cell populations as opposed to migratory populations in various disease states. This review summarizes the respiratory tract NK cells and their role in malignancy, infections with intracellular and extracellular pathogens, and inflammatory lung diseases.

KEY WORDS

Innate immunity, Lung, Pulmonary

Introduction

Natural killer (NK) cells were first described as a subset of splenic lymphocytes with spontaneous cytotoxicity against a murine retroviral leukaemia (Kiessling et al., 1975a,b). NK cells are accessible in relatively large numbers from the spleens of mice and from the peripheral blood of humans. Most of the published literature on NK cells concerns these populations. NK cells have long been recognized to have a broad tissue distribution that includes populations in the lungs, liver, lymph nodes and the bone marrow (Gregoire et al., 2007; Reynolds et al., 1981), but these NK cell populations have received less attention.

After the spleen, the lungs contain the largest number of tissue NK cells in animals (Basse et al., 1992; Gregoire et al., 2007; Reynolds et al., 1981; Stein-Streilein et al., 1983). The published data on human lung NK cells are limited to the older literature and its inherent methodological shortcomings. Specifically, the morphological definition of NK cells as 'large granular lymphocytes' is limited by the fact that 5–20% of NK cells are small and

[1]Supported by NIH grant HL73848 and an American Lung Association Career Investigator Award (Mehrad).

agranular (Smyth et al., 1995). The functional definition of NK cell activity by tissue spontaneous cytotoxic activity is limited by the recognition of NK cell subsets with variable cytolytic activity (Jacobs et al., 2001). In addition, density gradient isolation of NK cells results in cell purities of 40% when NK cells are defined by contemporary standards (Reichlin et al., 1999). This early literature was dominated by the observation that, in both experimental rodents and humans, freshly isolated lung NK cells had much lower spontaneous cytotoxicity against NK-sensitive tumour targets when compared with splenic and peripheral blood NK cells (Lotzova et al., 1984; Robinson et al., 1984; Weissman et al., 1986). In contrast, cells incubated overnight in media or exposed to IL-2, platelet-activating factor, or LT-B_4 regained robust cytotoxicity, suggesting a local inhibitory effect in the lung microenvironment (Robinson et al., 1984; Thivierge and Rola-Pleszczynski, 1991). Indeed, this inhibitory effect could be replicated in vitro by co-incubation of peripheral blood NK cells with alveolar macrophages or normal respiratory tract secretions (Bordignon et al., 1982; Robinson et al., 1984), and at least in one study, was mediated by PG-E_2 (Young et al., 1986).

The lung can be divided conceptually into three anatomical compartments: the bronchoalveolar airspaces (which include the airspaces of the alveoli and the conducting airways), the endovascular compartment, and the interstitial space (which includes the interstitium of the lung parenchyma as well as airway walls). In the uninflamed lung, the majority of NK cells are located in the vascular and the interstitial compartments, with a smaller proportion within the airspaces and therefore accessible by bronchoalveolar lavage (Basse et al., 1992; Weissler et al., 1987a). The older literature suggests that the NK cell subsets are not evenly distributed within these compartments: in human lung samples, most interstitial/vascular NK cells express CD16, whereas NK cells in the airspaces express CD57 but not CD16 (Weissler et al., 1987a). It is difficult to ascertain whether these subsets correlate with the contemporary categorization of human NK cells into $CD16^-$ $CD56^{hi}$ and $CD16^+$ $CD56^{lo}$ subsets. Similar results were recently reported in mice, however, where the $CD11b^{hi}$ $CD27^-$ subset of NK cells was found to constitute approximately 90% of the lung NK cell population (Hayakawa and Smyth, 2006; Huntington et al., 2007). The interstitial and vascular compartment NK cells may also differ functionally from airspace NK cells, in that they have higher spontaneous tumour cytotoxicity (Nolibe et al., 1981; Weissler et al., 1987a).

Cancer

The study of NK cell-mediated anti-tumour effects remains one of the most active areas of investigation in NK cell biology. Few studies have directly examined the contribution of lung NK cells to tumour clearance. NK cells flushed from the lung vascular compartment have robust tumour cytolytic activity (Nolibe and Poupon, 1986; Nolibe et al., 1981). At least in the context of experimental animals, the lungs are a primary site of cancer spread in both spontaneously metastatic models and in models that employ an intravenous delivery of the tumour cells (Bierer, 2008). In both settings, resident and recruited lung NK cells are likely to contribute to tumour clearance. This phenomenon is the basis of a commonly used assay for in vivo lysis of intravenously administered NK-sensitive cancer cells by lung NK cells. Following after intravenous delivery, most radio-labelled YAC-1 cells are trapped in the lung vasculature, and their lysis by pulmonary NK cells results in progressive reduction in lung radio-label over the ensuing hours (Austin Taylor et al., 2000; Hackett et al., 1985). The relevance of lung NK cells is further underscored by the observation that, in a number of cancer models, NK cell-mediated protective effects are more effective in controlling metastases than primary tumours (Aboud et al., 1993; Ohyama et al., 2002, 1999; Palumbo et al., 2005; Smyth et al., 1999).

In the context of human lung cancers, several studies have shown interstitial lung NK cells from patients with lung cancer to have lower cytolytic properties as compared to interstitial NK cells from normal lungs (deShazo et al., 1987; Weissler et al., 1987b). This observation may be confounded by lower lung NK cell cytolytic activity observed in smokers and mice exposed to cigarette smoke (Lu et al., 2007; Takeuchi et al., 1988, 2001). These findings were extended in a recent study that identified NK cells infiltrating human non-small cell lung carcinoma as predominantly $CD16^-$ $CD56^{hi}$ in addition to demonstrating their low tumour cytolytic properties (Carrega et al., 2008). This is consistent with identification of human $CD16^-$ $CD56^{hi}$ NK cells as lacking perforin and substantial cytotoxicity capacity (Jacobs et al., 2001; Lanier et al., 1983, 1986), and contrasts with the normal interstitial and vascular lung NK cells that are $CD16^+$ (Weissler et al., 1987a).

Infectious diseases

The observation that NK cells are important in host defence against intracellular pathogens dates to early studies of various mouse viruses. As outlined in this section, respiratory tract NK cells appear to be necessary for defence against some obligate and facultative intracellular pathogens but not others, and are essential in defence against several classes of extracellular microbes (Table 24.1).

Table 24.1 Summary of role of NK cells in pulmonary infections

Microorganism	Effect of NK cells during infection	Reference
Influenza	– Increased number of activated lung NK cells; dependent on IL-12p40 and IL-18	Gazit et al. (2006), Stein-Streilein (1983)
	– Lung NK cells necessary to control of viral titres in early infection	Liu (2004), Monteiro (1998), Nogusa (2008)
	– NKp46 mediates interaction of NK cells with virus and is necessary for survival from experimental infection	Achdout (2008), Gazit et al. (2006)
RSV	– NK cells accumulate in the lung in experimental infection; dependent on viral G and SH proteins	Hussell and Openshaw (1998)
	– Inhibit lung Th2 responses in secondary infection	Hussell and Openshaw (2000), Tripp (1999)
HSV-1	– NK cells mediate early viral clearance; activation dependent on IL-18	Reading (2006, 2007)
C. trachomatis	– NK cell depletion does not affect survival or bacterial clearance	Williams (1987)
M. tuberculosis	– Activated NK cells accumulate in the lung	Junqueira-Kipnis (2003)
	– NK cells necessary for bacterial clearance in RAG-deficient but not immunocompetent mice	Junqueira-Kipnis (2003); Feng (2006)
	– Human NK cells promote expansion of $\gamma\delta$ T cells but inhibit expansion of regulatory T cells in response to *M. tuberculosis* antigens	Roy (2008), Zhang (2006)
L. pneumophila	– NK cells major source of IFN-γ in early infection – Depletion results in increased susceptibility to the infection	Sporri (2006)
F. tularensis	– NK cell depletion results in increased susceptibility to experimental infection with LVS strain	Lopez (2004)
M. pulmonis	– Infection results in activation of NK cells – Depletion of NK cells has no effect in immunocompetent mice and results in improved bacterial clearance in IFN-γ-deficient mice	Lai (1990)
S. pneumoniae	– Depletion in SCID mice results in improved bacterial clearance	Kerr (2005)
B. pertussis	– Late increase in numbers of lung NK cells – NK cell depletion results in disseminated infection	Byrne (2004)
S. aureus	– Early increase in number of lung NK cells – NK cell depletion results in increased bacterial burden	Small (2008)
C. neoformans	– NK cell depletion results in impaired fungal clearance early in the infection	Lipscomb (1987)
	– Transfer of NK cells results in improved fungal clearance	Hidore and Murphy (1986)
A. fumigatus	– In the absence of neutrophils, depletion results in worse outcome of infection	Morrison (2003)
	– Effect dependent on NK cell-derived IFN-γ	Park (2009)

Viral infections

The antimicrobial functions of NK cells were first established in systemic viral infections, but knowledge of their contribution to defence against viral pneumonias is limited. In the context of respiratory tract infection with the influenza A virus, NK cell numbers and their activity in the lung is induced within 48 h (Gazit et al., 2006; Stein-Streilein et al., 1983). Activation of lung NK cells in this context appears to be dependant on IL-18 and IL-12p40, and absence of IL-18, neutralization of IL-12p40, or depletion of NK cells results in increased lung viral titres during early infection (Liu et al., 2004; Monteiro et al., 1998; Nogusa et al., 2008). In humans NK cell production of interferon-γ (IFN-γ) in response to influenza virus is augmented by IL-2-derived from influenza-specific T cells (He et al., 2004). The activating NK cell receptor NKp46 is critical for survival from influenza pneumonia in experimental infection and in interaction of human NK cells with infected cells (Achdout et al., 2008; Gazit et al., 2006). The viral hemagglutinin binds and activates several NK cell activating receptors (Ho et al., 2008; Owen et al., 2007). On the other hand, the binding of NK cell inhibitory receptors to infected cells likely represents a viral pathogenic mechanism and is not dependant on viral hemagglutinin or neuraminidase (Achdout et al., 2003, 2008).

Experimental respiratory syncytial virus (RSV) infections of the respiratory tract are also associated with accumulation of IFN-γ-producing NK cells in the lungs (Hussell and Openshaw, 1998). This accumulation of NK cells appears to be dependant on RSV G and SH proteins and inhibits lung Th2 responses in secondary challenges (Hussell and Openshaw, 2000; Tripp et al., 1999). Interestingly, fatal cases of human RSV pneumonia are associated with a paucity of lung NK cells (Welliver et al., 2007). NK cells also accumulate in the lungs in respiratory tract infections with herpes simplex virus-1, and develop an activated IFN-γ-producing phenotype, and contribute to early viral clearance (Reading et al., 2006). The activation of lung NK cells in this model was dependant on IL-18 but independent of IL-12p40 (Reading et al., 2007).

Bacterial infections

Chlamydia species are obligate intracellular bacteria. After intranasal administration of *C. trachomatis*, spontaneous killing activity of lung and spleen cells against YAC-1 targets was increased, but Ab-mediated depletion of NK cells did not affect survival or bacterial clearance (Williams et al., 1987). Prior infection with this organism inhibits subsequent Th2 allergic responses to ovalbumin, and this effect is abolished when NK cells are depleted during the initial infection (Han et al., 2008). Transfer of splenic NK cells from previously infected animals to naïve mice challenged with ovalbumin similarly inhibited pulmonary allergic responses, whereas the transfer of splenic NK cells from uninfected mice had no effect (Han et al., 2008).

Mycobacterium tuberculosis is a facultative intracellular pathogen. During experimental tuberculosis infection in immunocompetent mice, activated NK cells accumulate in the lungs and express intracellular perforin and IFN-γ. Their depletion, however, does not influence bacterial clearance (Junqueira-Kipnis et al., 2003). In RAG-deficient mice, absence of NK cells (achieved using mice deficient in the common cytokine receptor γ-chain in addition to the RAG2 gene) results in a marked increase in susceptibility to the infection, which interestingly exceeded the effect of lack of IFN-γ or IL-12 p40 in RAG-deficient animals (Feng et al., 2006). This suggests that, in the absence of effective T cell-mediated immunity, NK cells may provide partial defence against tuberculosis, and this effect is at least in part mediated via NK cell-derived IFN-γ. In addition, studies in humans with latent tuberculosis have shown that activated peripheral blood NK cells inhibit the expansion of CD4+ CD25+ Foxp3+ regulatory T cells in response to mycobacterial antigens via an NKG2D-dependent mechanism (Roy et al., 2008). Conversely, peripheral blood NK cells of patients with latent tuberculosis promote the proliferation of γδ T cells in response to *M. tuberculosis* antigens via a mechanism that is dependant on contact and TNF but not IFN-γ (Zhang et al., 2006).

Legionella pneumophila, the causative agent of Legionnaire's disease, is another facultative intracellular pathogen. In experimental legionellosis in mice, early production of IFN-γ in the lungs correlates with spontaneous killing activity of lung, blood, and spleen cells against YAC-1 targets (Blanchard et al., 1988). NK cells are found to be the major source of IFN-γ in splenocytes of mice with legionellosis, and IFN-γ receptor deficiency or NK depletion results in marked increase in susceptibility to the infection (Sporri et al., 2006). *Francisella tularensis* is another facultative intracellular gram-negative bacterium. Experimental pneumonia with the live vaccine strain of this pathogen is also associated with accumulation of IFN-γ-secreting NK cells in the lungs, and depletion of these cells results in more rapid mortality in this infection (Lopez et al., 2004). Evidence suggests that *Haemophilus influenzae* is also a facultative intracellular organism (Forsgren et al., 1994). Although the role of NK cells in pneumonia with this organism is not established, human NK cells exposed to the bacterium become activated and proliferate (King et al., 2008), and murine NK cells enhance killing of the bacteria by neutrophils ex vivo (Miyazaki et al., 2007).

Mycoplasma species are extracellular bacteria. Experimental pneumonia with *Mycoplasma pulmonis*, a strain of *Mycoplasma* that is pathogenic to mice, also results in activation of NK cells (Lai et al., 1990). While IFN-γ gene knock-out mice and mice with Ab-mediated neutralization of IFN-γ are more susceptible to this infection (Lai et al., 1990; Woolard et al., 2005), depletion of NK cells has no effect on the severity of infection in wild-type animals and, counter-intuitively, results in improved bacterial clearance in IFN-γ-deficient animals (Woolard et al., 2005). It therefore appears that, in the context of IFN-γ-gene deficient animals, NK cells suppress defences against *Mycoplasma* pneumonia. Interestingly, a similar pattern has been noted in experimental pneumococcal pneumonia: SCID mice are more susceptible to this infection, as measured by bacterial content in the lung and blood at 24 h and 36 h after infection, but this susceptibility is diminished when NK cells are depleted (Kerr et al., 2005).

Bordetella pertussis, the causative agent of whooping cough, is an extracellular gram-negative respiratory pathogen. In an experimental murine pulmonary infection with this organism, the infection was prolonged, with detectable bacteria in the lungs for 28 days (Byrne et al., 2004). A marked increase in lung NK and natural killer T (NKT) cells was noted after the first week of infection, and depletion of NK cells resulted in marked increase in bacterial dissemination outside the lungs (Byrne et al., 2004). *Staphylococcus aureus*, an increasingly common cause of pneumonia, is also an extracellular pathogen. The number of lung NK cells increased markedly by 20 h after the onset of infection and was associated with IL-15 expression in the lungs that was attributable to alveolar macrophages (Small et al., 2008). The absence of IL-15 or depletion of NK cells in this model resulted in increased bacterial burden in the lungs and blood (Small et al., 2008).

Lung NK cells have been targeted therapeutically in experimental models of extracellular gram-negative bacterial pneumonia. In a model of *Pseudomonas* pneumonia, conditional transgenic expression of an NKG2D ligand, retinoic acid early transcript-1α, in the respiratory epithelium resulted in accumulation of NKG2D-expressing lymphocytes in the lungs, which were nearly all IFN-γ-expressing NK cells (Wesselkamper et al., 2008). This is associated with marked improvement in bacterial phagocytosis by leukocytes, bacterial clearance, and mortality from the infection (Wesselkamper et al., 2008). Similarly, in a model of *Klebsiella* pneumonia, transgenic pulmonary expression of the chemokine ligand CCL3 results in improved outcome of the infection that was associated with accumulation of NK cells in the lungs (Zeng et al., 2003).

Fungal and parasitic infections

NK cells have been extensively studied in experimental cryptococcal infection. Infections with the yeast *Cryptococcus neoformans* uniformly begin in the lung, and can subsequently disseminate to other organs, notably the meninges. A commonly used mouse model of cryptococcosis utilizes intrapulmonary inoculation of the pathogen, which results in a pneumonia followed by variable dissemination; this model broadly resembles the human disease although the pulmonary stage of the infection is often less prominent in human disease. An alternative mouse model employs intravenous delivery of the yeast forms—these organisms also first reach the lungs before further dissemination but, unlike human disease, do so via the pulmonary vasculature. Direct killing of *Cryptococcus* yeasts by both human and murine NK cells has been extensively documented (Hidore and Murphy, 1989; Hidore et al., 1991a,b; Levitz et al., 1994; Murphy and McDaniel, 1982; Murphy et al., 1991; Nabavi and Murphy, 1985) and is dependant on perforin (Hidore et al., 1990; Wiseman et al., 2007). In experimental pulmonary infection, Ab-mediated NK cell depletion resulted in increased lung colony-forming units in the early infection but did not affect late colony counts or the outcome of the infection (Lipscomb et al., 1987), and transfer of NK cells to cyclophosphamide-treated mice resulted in reduced numbers of lung and splenic colony-forming units after intravenous administration of the yeast (Hidore and Murphy, 1986). The mechanism of in vivo NK cell-mediated defence in cryptococcosis is not fully established: NK depletion studies in mice homozygous or heterozygous for the *beige* mutation showed that, after intravenous infection, *beige*/+ mice have lower splenic and lung yeast burden as compared to NK depleted *beige*/+ or *beige/beige* mice (Salkowski and Balish, 1991), suggesting that NK cell granule function is mediating defence, at least after intravenous infection. On the other hand, NK cells appear to be the major source of serum IFN-γ in IL-12p40-deficient mice with pulmonary infection (Kawakami et al., 2000).

Invasive pneumonia with the mould *Aspergillus* results in a severe infection that occurs exclusively in immunocompromised hosts, most notably patients with neutrophil defects. In a model of invasive pulmonary aspergillosis in mice with Ab-mediated neutrophil depletion, additional depletion of NK cells resulted in worsened outcome of the infection, and the accumulation in the lungs was found to be dependant on the chemokine ligand CCL2 (Morrison et al., 2003). In this model, NK cells contributed to early IFN-γ production in the lungs, and depletion of NK cells or absence of IFN-γ resulted in a similar increase in susceptibility to the infection, whereas depletion of NK cells in IFN-γ-deficient hosts did not result in further increase in severity of the infection. Transfer of activated NK cells from wild-type, but not IFN-γ-deficient, donors resulted in greater pathogen clearance from the lungs of both IFN-γ-deficient and wild-type hosts (Park et al., 2009).

Other inflammatory lung diseases

The contribution of classical NK cells to allergic lung inflammation is not fully established. In the ovalbumin model of lung eosinophilic inflammation that is commonly used as a model of human asthma, depletion of NK1.1-expressing cells before initiation of intra-peritoneal sensitization resulted in less severe lung inflammation, whereas no such effect was seen in CD1d-deficient mice, suggesting that classical NK cells contribute to lung allergy (Korsgren et al., 1999). Exogenous administration of IL-2 and IL-18 after initial sensitization but before intra-pulmonary administration of ovalbumin also inhibited lung allergy, and was associated with accumulation of lung IFN-γ-expressing NK cells, and could be reproduced by adoptive transfer of NK cells (Matsubara et al., 2007). Other studies, however, implicated NKT cells rather than NK cells in the pathogenesis of mouse models of airway allergy as well as human asthma, as recently reviewed (Meyer et al., 2008).

The role of NK cells was also investigated in a mouse model of IL-2-induced pulmonary vascular leak syndrome, a form of non-cardiogenic pulmonary oedema that frequently complicates therapeutic administration of IL2. Development of lung injury and oedema was unaffected in RAG2-deficient mice, but was attenuated in mice deficient in both the RAG2 and IL-15 genes, supporting a role for NK cells in this process (Assier et al., 2004).

A mouse model of bleomycin-induced pulmonary injury and fibrosis is commonly used to study mechanisms of acute lung injury and pulmonary fibrosis. In the context of this model, mice deficient in the chemokine receptor CXCR3 were noted to have reduced numbers of lung NK cells that were associated with reduced lung IFN-γ production and increased lung fibrosis and mortality (Jiang et al., 2004). Furthermore, this defect was corrected with administration of wild-type lymphocytes or exogenous IFN-γ (Jiang et al., 2004). Although the effects of depletion of NK cells or transfer of exogenous NK cells were not specifically tested, this work suggests that lung NK cell-derived IFN-γ can contribute to lung fibrosis.

A major and well-studied function of NK cells is the killing of transplanted cells, but the role of NK cells in lung allograft rejection has not been studied in detail. In a study of human lung transplant recipients, the degree of histologic chronic graft rejection (manifested as bronchiolitis obliterans syndrome) was associated with infiltration of the allograft with $CD16^+$ cells (Fildes et al., 2008), but the functional significance of this observation is not known. Similarly, bronchoalveolar lavage fluid from patients with sarcoidosis, an idiopathic systemic illness characterized by granulomatous inflammation that most often involves the lungs and mediastinal lymph nodes, contains increased numbers of NK cells that have a $CD56^{hi}$ phenotype (Katchar et al., 2005; Papakosta et al., 2005).

References

Aboud, M., Kingsmore, S., and Segal, S. (1993). Role of natural killer cells in controlling local tumor formation and metastatic manifestation of different 3LL Lewis lung carcinoma cell clones. Nat Immun. **12**, 17–24.

Achdout, H., Arnon, T.I., Markel, G., Gonen-Gross, T., Katz, G., Lieberman, N., Gazit, R., Joseph, A., Kedar, E. and Mandelboim, O. (2003). Enhanced recognition of human NK receptors after influenza virus infection. J Immunol. **171**, 915–923.

Achdout, H., Manaster, I. and Mandelboim, O. (2008). Influenza virus infection augments NK cell inhibition through reorganization of major histocompatibility complex class I proteins. J Virol. **82**, 8030–8037.

Assier, E., Jullien, V., Lefort, J., Moreau, J. L., Di Santo, J.P., Vargaftig, B.B., Lapa e Silva, J.R. and Theze, J. (2004). NK cells and polymorphonuclear neutrophils are both critical for IL-2-induced pulmonary vascular leak syndrome. J Immunol. **172**, 7661–7668.

Austin Taylor, M., Bennett, M., Kumar, V. and Schatzle, J.D. (2000). Functional defects of NK cells treated with chloroquine mimic the lytic defects observed in perforin-deficient mice. J Immunol. **165**, 5048–5053.

Basse, P.H., Hokland, P., Gundersen, H.J. and Hokland, M. (1992). Enumeration of organ-associated natural killer cells in mice: application of a new stereological method. APMIS **100**, 202–208.

Bierer, B.E. (2008). *Animal models for tumor immunology. Current Protocols in Immunology*, Wiley Interscience, pp. 20.20.21–20.20.29.

Blanchard 2nd, D.K., Friedman, H., Stewart, W.E., Klein, T.W. and Djeu, J. Y. (1988). Role of gamma interferon in induction of natural killer activity by Legionella pneumophila in vitro and in an experimental murine infection model. Infect Immun. **56**, 1187–1193.

Bordignon, C., Villa, F., Allavena, P., Introna, M., Biondi, A., Avallone, R. and Mantovani, A. (1982). Inhibition of natural killer activity by human bronchoalveolar macrophages. J Immunol. **129**, 587–591.

Byrne, P., McGuirk, P., Todryk, S. and Mills, K.H. (2004). Depletion of NK cells results in disseminating lethal infection with Bordetella pertussis associated with a reduction of antigen-specific Th1 and enhancement of Th2, but not Tr1 cells. Eur J Immunol. **34**, 2579–2588.

Carrega, P., Morandi, B., Costa, R., Frumento, G., Forte, G., Altavilla, G., Ratto, G.B., Mingari, M.C., Moretta, L. and Ferlazzo, G. (2008). Natural killer cells infiltrating human nonsmall-cell lung cancer are enriched in CD56 bright CD16(-) cells and display an impaired capability to kill tumor cells.. Cancer **112**, 863–875.

deShazo, R.D., Moulder, P.V., Bozelka, B. and Chapman, Y. (1987). Diminished natural killer-cell activity of interstitial pulmonary cell populations from patients with carcinoma of the lung. Chest **91**, 26–28.

Feng, C.G., Kaviratne, M., Rothfuchs, A.G., Cheever, A., Hieny, S., Young, H.A., Wynn, T.A. and Sher, A. (2006). NK cell-derived IFN-gamma differentially regulates innate resistance and neutrophil response in T cell-deficient hosts infected with Mycobacterium tuberculosis. J Immunol. **177**, 7086–7093.

Fildes, J.E., Yonan, N., Tunstall, K., Walker, A.H., Griffiths-Davies, L., Bishop, P. and Leonard, C.T. (2008). Natural killer cells in peripheral blood and lung tissue are associated with chronic rejection after lung transplantation. J Heart Lung Transplant. **27**, 203–207.

Forsgren, J., Samuelson, A., Ahlin, A., Jonasson, J., Rynnel-Dagoo, B. and Lindberg, A. (1994). Haemophilus influenzae resides and multiplies intracellularly in human adenoid tissue as demonstrated by in situ hybridization and bacterial viability assay. Infect Immun. **62**, 673–679.

Gazit, R., Gruda, R., Elboim, M., Arnon, T.I., Katz, G., Achdout, H., Hanna, J., Qimron, U., Landau, G., Greenbaum, E., Zakay-Rones, Z., Porgador, A. and Mandelboim, O. (2006). Lethal influenza infection in the absence of the natural killer cell receptor gene Ncr1. Nat Immunol. **7**, 517–523.

Gregoire, C., Chasson, L., Luci, C., Tomasello, E., Geissmann, F., Vivier, E. and Walzer, T. (2007). The trafficking of natural killer cells. Immunol Rev. **220**, 169–182.

Hackett , J., Bennett, M. and Kumar, V. Jr. (1985). Origin and differentiation of natural killer cells. I. Characteristics of a transplantable NK cell precursor. J Immunol. **134**, 3731–3738.

Han, X., Fan, Y., Wang, S., Jiao, L., Qiu, H. and Yang, X. (2008). NK cells contribute to intracellular bacterial infection-mediated inhibition of allergic responses. J Immunol. **180**, 4621–4628.

Hayakawa, Y. and Smyth, M.J. (2006). CD27 dissects mature NK cells into two subsets with distinct responsiveness and migratory capacity. J Immunol. **176**, 1517–1524.

He, X.S., Draghi, M., Mahmood, K., Holmes, T.H., Kemble, G.W., Dekker, C.L., Arvin, A.M., Parham, P. and Greenberg, H.B. (2004). T cell-dependent production of IFN-gamma by NK cells in response to influenza A virus. J Clin Invest. **114**, 1812–1819.

Hidore, M.R., Mislan, T.W. and Murphy, J.W. (1991a). Responses of murine natural killer cells to binding of the fungal target Cryptococcus neoformans. Infect Immun. **59**, 1489–1499.

Hidore, M.R. and Murphy, J.W. (1986). Correlation of natural killer cell activity and clearance of Cryptococcus neoformans from mice after adoptive transfer of splenic nylon wool-nonadherent cells. Infect Immun. **51**, 547–555.

Hidore, M.R. and Murphy, J.W. (1989). Murine natural killer cell interactions with a fungal target, Cryptococcus neoformans. Infect Immun. **57**, 1990–1997.

Hidore, M.R., Nabavi, N., Reynolds, C.W., Henkart, P.A. and Murphy, J.W. (1990). Cytoplasmic components of natural killer cells limit the growth of Cryptococcus neoformans. J Leukoc Biol. **48**, 15–26.

Hidore, M.R., Nabavi, N., Sonleitner, F. and Murphy, J.W. (1991b). Murine natural killer cells are fungicidal to Cryptococcus neoformans. Infect Immun. **59**, 1747–1754.

Ho, J.W., Hershkovitz, O., Peiris, M., Zilka, A., Bar-Ilan, A., Nal, B., Chu, K., Kudelko, M., Kam, Y.W., Achdout, H., Mandelboim, M., Altmeyer, R., Mandelboim, O., Bruzzone, R. and Porgador, A. (2008). H5-type influenza virus hemagglutinin is functionally recognized by the natural killer-activating receptor NKp44. J Virol. **82**, 2028–2032.

Huntington, N.D., Tabarias, H., Fairfax, K., Brady, J., Hayakawa, Y., Degli-Esposti, M.A., Smyth, M.J., Tarlinton, D.M. and Nutt, S.L. (2007). NK cell maturation and peripheral homeostasis is associated with KLRG1 up-regulation. J Immunol. **178**, 4764–4770.

Hussell, T. and Openshaw, P.J. (1998). Intracellular IFN-gamma expression in natural killer cells precedes lung CD8+ T cell recruitment during respiratory syncytial virus infection. J Gen Virol. **79**(Pt 11), 2593–2601.

Hussell, T. and Openshaw, P.J. (2000). IL-12-activated NK cells reduce lung eosinophilia to the attachment protein of respiratory syncytial virus but do not enhance the severity of illness in CD8 T cell-immunodeficient conditions. J Immunol. **165**, 7109–7115.

Jacobs, R., Hintzen, G., Kemper, A., Beul, K., Kempf, S., Behrens, G., Sykora, K.W. and Schmidt, R.E. (2001). CD56bright cells differ in their KIR repertoire and cytotoxic features from CD56dim NK cells. Eur J Immunol. **31**, 3121–3127.

Jiang, D., Liang, J., Hodge, J., Lu, B., Zhu, Z., Yu, S., Fan, J., Gao, Y., Yin, Z., Homer, R., Gerard, C. and Noble, P.W. (2004). Regulation of pulmonary fibrosis by chemokine receptor CXCR3. J Clin Invest. **114**, 291–299.

Junqueira-Kipnis, A.P., Kipnis, A., Jamieson, A., Juarrero, M.G., Diefenbach, A., Raulet, D.H., Turner, J. and Orme, I.M. (2003). NK cells respond to pulmonary infection with Mycobacterium tuberculosis, but play a minimal role in protection. J Immunol. **171**, 6039–6045.

Katchar, K., Soderstrom, K., Wahlstrom, J., Eklund, A. and Grunewald, J. (2005). Characterisation of natural killer cells and CD56+ T-cells in sarcoidosis patients. Eur Respir J. **26**, 77–85.

Kawakami, K., Koguchi, Y., Qureshi, M.H., Miyazato, A., Yara, S., Kinjo, Y., Iwakura, Y., Takeda, K., Akira, S., Kurimoto, M. and Saito, A. (2000). IL-18 contributes to host resistance against infection with Cryptococcus neoformans in mice with defective IL-12 synthesis through induction of IFN-gamma production by NK cells. J Immunol. **165**, 941–947.

Kerr, A.R., Kirkham, L.A., Kadioglu, A., Andrew, P.W., Garside, P., Thompson, H. and Mitchell, T.J. (2005). Identification of a detrimental role for NK cells in pneumococcal pneumonia and sepsis in immunocompromised hosts. Microbes Infect. **7**, 845–852.

Kiessling, R., Klein, E., Pross, H. and Wigzell, H. (1975a). "Natural" killer cells in the mouse. II. Cytotoxic cells with specificity for mouse Moloney leukemia cells. Characteristics of the killer cell. Eur J Immunol. **5**, 117–121.

Kiessling, R., Klein, E. and Wigzell, H. (1975b). "Natural" killer cells in the mouse. I. Cytotoxic cells with specificity for mouse Moloney leukemia cells. Specificity and distribution according to genotype. Eur J Immunol. **5**, 112–117.

King, P.T., Ngui, J., Farmer, M.W., Hutchinson, P., Holmes, P.W. and Holdsworth, S.R. (2008). Cytotoxic T lymphocyte and natural killer cell responses to non-typeable Haemophilus influenzae. Clin Exp Immunol. **152**, 542–551.

Korsgren, M., Persson, C.G., Sundler, F., Bjerke, T., Hansson, T., Chambers, B.J., Hong, S., Van Kaer, L., Ljunggren, H.G. and Korsgren, O. (1999). Natural killer cells determine development of allergen-induced eosinophilic airway inflammation in mice. J Exp Med. **189**, 553–562.

Lai, W.C., Bennett, M., Pakes, S.P., Kumar, V., Steutermann, D., Owusu, I. and Mikhael, A. (1990). Resistance to Mycoplasma pulmonis mediated by activated natural killer cells. J Infect Dis. **161**, 1269–1275.

Lanier, L.L., Le, A.M., Civin, C.I., Loken, M.R. and Phillips, J.H. (1986). The relationship of CD16 (Leu-11) and Leu-19 (NKH-1) antigen expression on human peripheral blood NK cells and cytotoxic T lymphocytes. J Immunol. **136**, 4480–4486.

Lanier, L.L., Le, A.M., Phillips, J.H., Warner, N.L. and Babcock, G.F. (1983). Subpopulations of human natural killer cells defined by expression of the Leu-7 (HNK-1) and Leu-11 (NK-15) antigens. J Immunol. **131**, 1789–1796.

Levitz, S.M., Dupont, M.P. and Smail, E.H. (1994). Direct activity of human T lymphocytes and natural killer cells against Cryptococcus neoformans. Infect Immun. **62**, 194–202.

Lipscomb, M.F., Alvarellos, T., Toews, G.B., Tompkins, R., Evans, Z., Koo, G. and

Kumar, V. (1987). Role of natural killer cells in resistance to Cryptococcus neoformans infections in mice. Am J Pathol. **128**, 354–361.

Liu, B., Mori, I., Hossain, M.J., Dong, L., Takeda, K. and Kimura, Y. (2004). Interleukin-18 improves the early defence system against influenza virus infection by augmenting natural killer cell-mediated cytotoxicity. J Gen Virol. **85**, 423–428.

Lopez, M.C., Duckett, N.S., Baron, S.D. and Metzger, D.W. (2004). Early activation of NK cells after lung infection with the intracellular bacterium, Francisella tularensis LVS. Cell Immunol. **232**, 75–85.

Lotzova, E., Savary, C.A., Gray, K.N., Raulston, G.L. and Jardine, J.H. (1984). Natural killer cell profile of two random-bred strains of athymic rats. Exp Hematol. **12**, 633–640.

Lu, L.M., Zavitz, C.C., Chen, B., Kianpour, S., Wan, Y. and Stampfli, M.R. (2007). Cigarette smoke impairs NK cell-dependent tumor immune surveillance. J Immunol. **178**, 936–943.

Matsubara, S., Takeda, K., Kodama, T., Joetham, A., Miyahara, N., Koya, T., Swasey, C.H., Okamoto, M., Dakhama, A. and Gelfand, E.W. (2007). IL-2 and IL-18 attenuation of airway hyperresponsiveness requires STAT4, IFN-gamma, and natural killer cells. Am J Respir Cell Mol Biol. **36**, 324–332.

Meyer, E.H., DeKruyff, R.H. and Umetsu, D.T. (2008). T cells and NKT cells in the pathogenesis of asthma. Annu Rev Med. **59**, 281–292.

Miyazaki, S., Ishikawa, F., Shimizu, K., Ubagai, T., Edelstein, P.H. and Yamaguchi, K. (2007). Gr-1high polymorphonuclear leukocytes and NK cells act via IL-15 to clear intracellular Haemophilus influenzae in experimental murine peritonitis and pneumonia. J Immunol. **179**, 5407–5414.

Monteiro, J.M., Harvey, C. and Trinchieri, G. (1998). Role of interleukin-12 in primary influenza virus infection. J Virol. **72**, 4825–4831.

Morrison, B.E., Park, S.J., Mooney, J.M. and Mehrad, B. (2003). Chemokine-mediated recruitment of NK cells is a critical host defense mechanism in invasive aspergillosis. J Clin Invest. **112**, 1862–1870.

Murphy, J.W., Hidore, M.R. and Nabavi, N. (1991). Binding interactions of murine natural killer cells with the fungal target Cryptococcus neoformans. Infect Immun. **59**, 1476–1488.

Murphy, J.W. and McDaniel, D.O. (1982). In vitro reactivity of natural killer (NK) cells against Cryptococcus neoformans. J Immunol. **128**, 1577–1583.

Nabavi, N. and Murphy, J.W. (1985). In vitro binding of natural killer cells to Cryptococcus neoformans targets. Infect Immun. **50**, 50–57.

Nogusa, S., Ritz, B.W., Kassim, S.H., Jennings, S.R. and Gardner, E.M. (2008). Characterization of age-related changes in natural killer cells during primary influenza infection in mice. Mech Ageing Dev. **129**, 223–230.

Nolibe, D., Berel, E., Masse, R. and Lafuma, J. (1981). Characterization of a major natural killer activity in rat lungs. Biomedicine **35**, 230–234.

Nolibe, D. and Poupon, M.F. (1986). Enhancement of pulmonary metastases induced by decreased lung natural killer cell activity. J Natl Cancer Inst. **77**, 99–103.

Ohyama, C., Tsuboi, S. and Fukuda, M. (1999). Dual roles of sialyl Lewis X oligosaccharides in tumor metastasis and rejection by natural killer cells. EMBO J. **18**, 1516–1525.

Ohyama, C., Kanto, S., Kato, K., Nakano, O., Arai, Y., Kato, T., Chen, S., Fukuda, M.N. and Fukuda, M. (2002). Natural killer cells attack tumor cells expressing high levels of sialyl Lewis x oligosaccharides. Proc Natl Acad Sci U S A. **99**, 13789–13794.

Owen, R.E., Yamada, E., Thompson, C.I., Phillipson, L.J., Thompson, C., Taylor, E., Zambon, M., Osborn, H.M., Barclay, W.S. and Borrow, P. (2007). Alterations in receptor binding properties of recent human influenza H3N2 viruses are associated with reduced natural killer cell lysis of infected cells. J Virol. **81**, 11170–11178.

Palumbo, J.S., Talmage, K.E., Massari, J.V., La Jeunesse, C.M., Flick, M.J., Kombrinck, K.W., Jirouskova, M. and Degen, J.L. (2005). Platelets and fibrin(ogen) increase metastatic potential by impeding natural killer cell-mediated elimination of tumor cells. Blood **105**, 178–185.

Papakosta, D., Kyriazis, G., Gioulekas, D., Kontakiotis, T., Polyzoni, T., Bouros, D. and Patakas, D. (2005). Variations in alveolar cell populations, lymphocyte subsets and NK-cells in different stages of active pulmonary sarcoidosis. Sarcoidosis Vasc Diffuse Lung Dis **22**, 21–26.

Park, S.J., Hughes, M.A., Strieter, R.M., Burdick, M. and Mehrad, B. (2009). Early NK cell-derived interferon-gamma is essential to host defense in neutropenic invasive aspergillosis. J Immunol.

Reading, P.C., Whitney, P.G., Barr, D.P., Smyth, M.J. and Brooks, A.G. (2006). NK cells contribute to the early clearance of HSV-1 from the lung but cannot control replication in the central nervous system following intranasal infection. Eur J Immunol. **36**, 897–905.

Reading, P.C., Whitney, P.G., Barr, D.P., Wojtasiak, M., Mintern, J.D., Waithman, J. and Brooks, A.G. (2007). IL-18, but not IL-12, regulates NK cell activity following intranasal herpes simplex virus type 1 infection. J Immunol. **179**, 3214–3221.

Reichlin, A., Iizuka, K. and Yokoyama, W.M. (1999). *Isolation of murine natural killer cells. Current Protocols in Immunology*: John Wiley & Sons pp. 3.22.21-23.22.26.

Reynolds, C.W., Timonen, T. and Herberman, R.B. (1981). Natural killer (NK) cell activity in the rat. I. Isolation and characterization of the effector cells. J Immunol. **127**, 282–287.

Robinson, B.W., Pinkston, P. and Crystal, R.G. (1984). Natural killer cells are present in the normal human lung but are functionally impotent. J Clin Invest. **74**, 942–950.

Roy, S., Barnes, P.F., Garg, A., Wu, S., Cosman, D. and Vankayalapati, R. (2008). NK cells lyse T regulatory cells that expand in response to an intracellular pathogen. J Immunol. **180**, 1729–1736.

Salkowski, C.A. and Balish, E. (1991). Role of natural killer cells in resistance to systemic cryptococcosis. J Leukoc Biol. **50**, 151–159.

Small, C.L., McCormick, S., Gill, N., Kugathasan, K., Santosuosso, M., Donaldson, N., Heinrichs, D.E., Ashkar, A. and Xing, Z. (2008). NK cells play a critical protective role in host defense against acute extracellular Staphylococcus aureus bacterial infection in the lung. J Immunol. **180**, 5558–5568.

Smyth, M.J., Browne, K.A., Kinnear, B.F., Trapani, J.A. and Warren, H.S. (1995). Distinct granzyme expression in human CD3- CD56+ large granular- and CD3- CD56+ small high density-lymphocytes displaying non-MHC-restricted cytolytic activity. J Leukoc Biol. **57**, 88–93.

Smyth, M.J., Thia, K.Y., Cretney, E., Kelly, J.M., Snook, M.B., Forbes, C.A. and Scalzo, A.A. (1999). Perforin is a major contributor to NK cell control of tumor metastasis. J Immunol. **162**, 6658–6662.

Sporri, R., Joller, N., Albers, U., Hilbi, H. and Oxenius, A. (2006). MyD88-dependent IFN-gamma production by NK cells is key for control of Legionella pneumophila infection. J Immunol. **176**, 6162–6171.

Stein-Streilein, J., Bennett, M., Mann, D. and Kumar, V. (1983). Natural killer cells in mouse lung: surface phenotype, target preference, and response to local influenza virus infection. J Immunol. **131**, 2699–2704.

Takeuchi, M., Nagai, S. and Izumi, T. (1988). Effect of smoking on natural killer cell activity in the lung. Chest **94**, 688–693.

Takeuchi, M., Nagai, S., Nakajima, A., Shinya, M., Tsukano, C., Asada, H., Yoshikawa, K., Yoshimura, M. and Izumi, T. (2001). Inhibition of lung natural killer cell activity by smoking: the role of alveolar macrophages. Respiration **68**, 262–267.

Thivierge, M. and Rola-Pleszczynski, M. (1991). Enhancement of pulmonary natural killer cell activity by platelet activating factor. Mechanisms of activation involving Ca2 + , protein kinase C, and lipooxygenase products. Am Rev Respir Dis. **144**, 272–277.

Tripp, R.A., Moore, D., Jones, L., Sullender, W., Winter, J. and Anderson, L.J. (1999). Respiratory syncytial virus G and/or SH protein alters Th1 cytokines, natural killer cells, and neutrophils responding to pulmonary infection in BALB/c mice. J Virol. **73**, 7099–7107.

Weissler, J.C., Nicod, L.P., Lipscomb, M.F. and Toews, G.B. (1987a). Natural killer cell function in human lung is compartmentalized. Am Rev Respir Dis. **135**, 941–949.

Weissler, J.C., Nicod, L.P. and Toews, G.B. (1987b). Pulmonary natural killer cell activity is reduced in patients with bronchogenic carcinoma. Am Rev Respir Dis. **135**, 1353–1357.

Weissman, D.N., deShazo, R.D. and Banks, D.E. (1986). Modulation of natural killer cell function by human alveolar macrophages. J Allergy Clin Immunol. **78**, 571–577.

Welliver Sr., T.P., Garofalo, R.P., Hosakote, Y., Hintz, K.H., Avendano, L., Sanchez, K., Velozo, L., Jafri, H., Chavez-Bueno, S., Ogra, P.L., McKinney, L., Reed, J. L. and Welliver, R.C. Sr. (2007). Severe human lower respiratory tract illness caused by respiratory syncytial virus and influenza virus is characterized by the absence of pulmonary cytotoxic lymphocyte responses. J Infect Dis. **195**, 1126–1136.

Wesselkamper, S.C., Eppert, B.L., Motz, G.T., Lau, G.W., Hassett, D.J. and Borchers, M.T. (2008). NKG2D is critical for NK cell activation in host defense against Pseudomonas aeruginosa respiratory infection. J Immunol. **181**, 5481–5489.

Williams, D.M., Schachter, J. and Grubbs, B. (1987). Role of natural killer cells in infection with the mouse pneumonitis agent (murine Chlamydia trachomatis). Infect Immun. **55**, 223–226.

Wiseman, J.C., Ma, L.L., Marr, K.J., Jones, G.J. and Mody, C.H. (2007). Perforin-dependent cryptococcal microbicidal activity in NK cells requires PI3K-dependent ERK1/2 signaling. J Immunol. **178**, 6456–6464.

Woolard, M.D., Hudig, D., Tabor, L., Ivey, J.A. and Simecka, J.W. (2005). NK cells in gamma-interferon-deficient mice suppress lung innate immunity against Mycoplasma spp. Infect Immun. **73**, 6742–6751.

Young, M.R., Wheeler, E. and Newby, M. (1986). Macrophage-mediated suppression of natural killer cell activity in mice bearing Lewis lung carcinoma. J Natl Cancer Inst. **76**, 745–750.

Zeng, X., Moore, T.A., Newstead, M.W., Hernandez-Alcoceba, R., Tsai, W.C. and Standiford, T.J. (2003). Intrapulmonary expression of macrophage inflammatory protein 1alpha (CCL3) induces neutrophil and NK cell accumulation and stimulates innate immunity in murine bacterial pneumonia. Infect Immun. **71**, 1306–1315.

Zhang, R., Zheng, X., Li, B., Wei, H. and Tian, Z. (2006). Human NK cells positively regulate gammadelta T cells in response to Mycobacterium tuberculosis. J Immunol. **176**, 2610–2616.

Chapter Twenty-Five

25

Intestinal natural killer cells

Martin R. Goodier

CHAPTER CONTENTS

As to diseases, make a habit of two things—to help, or at least, to do no harm.

Hippocrates

ABSTRACT

Advances in the identification of natural killer (NK) cell–specific markers have enabled the characterization of 'true' NK cells in the large and small intestine in humans and in animals and to distinguish these cells from immature T-cell precursors and T-cells expressing NK receptors. NK cells have been localized to the intestinal lamina propria and to a lesser degree to the intestinal epithelium. NK-like cells are important in the development of lymphoid tissue in the foetal gut and developmental pathways for NK cells operate within the tissues of the adult gastrointestinal tract. Intestinal NK cells are phenotypically distinct from their counterparts in the blood and resemble 'helper' NK cells, which have potentially important functions both in promoting anti-pathogen responses and in the maintenance of intestinal epithelium.

Evidence for true NK cells in the intestine has been obtained and the origins and functional characteristics of intestinal NK cells, their relationship with non-pathogenic commensal organisms and the potential involvement of intestinal NK cell activation in infectious and autoimmune diseases is emergent.

KEY WORDS

Intestinal epithelium, Lamina propria, Migration, Differentiation, Helper function, Commensal microbes

Introduction

The human gastrointestinal (GI) tract is subject to trauma resulting from physical and infectious causes. Such trauma results in damage to the epithelium of the upper and lower GI tract and necessitates ongoing biological repair processes. The concentration of lymphocytes in the intestine is indicative of the importance of cellular immunity therein. In humans, the GI tract contains an estimated 5×10^{10} lymphocytes in total, five times the number lymphocytes found in the blood (Blum and Pabst, 2007). Immune cells are distributed regionally according to the localization of Peyer's patches, mesenteric lymph nodes and lymphoid follicles and are also locally compartmentalized within the small and large intestine. Lymphocytes are found in association with intestinal epithelium, inserting into junctions between epithelial cells and are also distributed throughout the lamina propria underlying the basement membrane of the intestine.

Cellular innate and acquired responses are integrated with anti-microbial functions of intestinal epithelium and contribute to the maintenance of this critical barrier (Cario et al., 1999; Macdonald and Monteleone, 2005; Sakaguchi et al., 2001; Shaykhiev and Bals, 2007). Production of tumour necrosis factor (TNF)-α and IFN-γ by activated lymphocytes and myeloid cells contributes to inflammation-induced repair of epithelia via the promotion of TGF-β production and $\gamma\delta$-TCR$^+$ intraepithelial lymphocytes are involved in the differentiation and maintenance of intestinal crypt epithelia and maintenance of epithelial tight junctions, after disruption by microbial pathogens (Dalton et al., 2006; Shibahara et al., 2005). IL-17, produced by Th17CD4$^+$ T-cells also promotes the formation of tight junctions (Kinugasa et al., 2000). Toll-like receptors (TLR), which mediate leukocyte activation by commensal and foreign microbes, influence the proliferation of epithelial progenitors and enhance GI barrier integrity, thereby providing an important link between leukocyte activation and mucosal tissue repair processes (Cario et al., 2007; Rakoff-Nahoum et al., 2004). Potential mechanisms therefore exist for the activation of natural killer (NK) cells in intestinal tissues, although the true identity and function of these cells in the GI tract have, until recently, remained elusive.

Identification of cytotoxic NK cells in the small and large intestine

The study of NK cells in GI tissue has for many years focused on the identification of non-T lymphocytes with natural cytotoxic function analogous to that observed in CD16$^+$ blood NK cells. Availability of convincing data demonstrating the presence of 'true' NK cells in both the epithelium and lamina propria in humans and mice has been complicated by the low frequency of non-T, non-B lineage lymphocytes. Paradoxically, successive studies have characterized 'null' cells with poor natural cytolytic function, granular T-cell subsets with cytotoxic potential and additionally haematopoietic precursor cells (HPC) for T lymphocytes in intestinal tissues. Further difficulties in distinguishing intestinal NK cells (iNK) have arisen through the identification of receptors, previously thought to be NK cell-specific, on T-cell subsets. Characterization of distinct functional subsets of NK cells in human peripheral blood has, however, driven a shift in perception of NK cell phenotypic and functional identity (Cooper et al., 2001). The emergence of 'helper' NK cells as components of the lymphoid compartment in peripheral lymphoid organs and tissues, together with an expanded repertoire of known functional NK receptors has led to the identification of 'true' NK cells in the intestine with unexpected ontogeny and functional capacity.

Studies on human intestinal epithelial lymphocytes

The identification of cells with NK-like properties in the intestine was largely initiated as a by-product of studies on T-intra-epithelial lymphocytes (iel). Large granular lymphocytes (LGL) were identified histologically in human intestinal epithelium from surgically resected small bowel samples with similar morphology to blood LGL (Cerf-Bensussan et al., 1983). However, few of these iel expressed cell surface markers associated at that time with blood NK cells, including that recognized by antibody HNK-1 (CD57) (Cerf-Bensussan et al., 1983). The majority of iel expressed CD3$^+$ and CD8$^+$, whereas those identified in the lamina propria were enriched for CD3$^+$CD4$^+$ cells. Such studies led to the conclusion that true NK cells were indeed rare in the intestinal epithelium and that T lymphocytes in these tissues could have innate cytotoxic capacity (Cerf-Bensussan et al., 1983). Extraction of iel by dissociation of intestinal epithelium by mechanical scraping or treatment of biopsy or surgically resected material with dithiothreitol and EDTA, has enabled functional characterization. Negligible cytotoxicity was, however, observed against K562 target cells in Percoll-purified iel either with or without culture in IFN-α or PHA-conditioned medium, reinforcing the conclusion that blood-like NK cells were not present in intestinal epithelium (Cerf-Bensussan et al., 1985). Use of the immature T lymphocyte and NK cell marker CD7 led to further characterization of null cells in normal jejunal epithelial tissue, between 2% and 25% of histologically identified CD7$^+$ cells being CD3$^-$ (Spencer et al., 1989). The possibility that these cells could be similar to blood NK cells was once again raised, when similar populations of CD3$^-$CD7$^+$ iel from surgically resected jejunum and duodenum were shown to lack TCR-$\alpha\beta$ and TCR-$\gamma\delta$ and exclusively expressed CD8α homodimer (Jarry et al., 1990). However, in contrast to the majority of peripheral blood NK cells, only few iel expressed CD16 and these cells lacked the expression of CD56 (Jarry et al., 1990). Two further studies demonstrated that spontaneous and IL-2 induced cytotoxicity against epithelial tumour cell lines was predominantly due to the activity of $CD3^+CD8^+CD16^-CD56^-$ iel and confirmed that these cells lacked spontaneous cytotoxicity against K562 target cells, indicating the predominant role for T cells in the epithelium (Ebert and Roberts, 1993; Roberts et al., 1993). The dominance of TCR-$\gamma\delta^+$ in intestinal epithelium raised the possibility that this cell subset could be responsible for both spontaneous and IL-2 activated cytotoxicity in isolated iel (Lundqvist et al., 1995). RAG-1 mRNA expression in $CD2^+CD3^-$ T-cell receptor negative cells in the intestinal epithelium of the small (but not large) intestine also raised the possibility of TCR rearrangement in intestinal epithelium and suggested that CD3 iel could be the exclusive precursors of TCR-$\gamma\delta^+$ and TCR-$\alpha\beta^+$ T-cell subsets (Lundqvist et al., 1995).

A $CD3^-CD7^+CD161^+CD56^+$ intraepithelial NK cell (ieNK) subset expressing CD69 was identified in normal jejunal and duodenal epithelium using multiparameter flow cytometry (Eiras et al., 1998). Intracellular staining revealed that a fraction of these cells expressed a high level of perforin despite lacking the expression of CD16 and killer cell immunoglobulin-like receptors (KIR) (defined by anti-CD158a and NKB1) (Eiras et al., 2000). This phenotype, however, contrasts with perforin$^+$ peripheral blood NK cells, which are almost exclusively $CD56^{lo}CD16^+KIR^+$ (Cooper et al., 2001). Flow cytometric analysis revealed a population of ieNK, negative for both intracellular and surface expression of CD3 and expressing higher levels of perforin and IL-2 induced cytotoxicity compared to $CD3^+$ iel (Leon et al., 2003). These ieNK expressed almost exclusively CD8α homodimers in contrast to TCR-$\alpha\beta^+$ cells (15% ± 7%) and TCR-$\gamma\delta^+$ cells (45% ± 6%) suggestive of a separate lineage, but not excluding the possibility that CD8$\alpha^+$$CD3^-$$TCR^-$ cells in the intestinal epithelium may be precursors for T-cell subsets, as evidenced in animal models (Rocha et al., 1994).

Studies on human lamina propria lymphocytes

Recovery of immune cells from intestinal lamina propria tissues can be achieved using two principal procedures (Bell et al., 2001). Enzymatic digestion has been used to recover lamina propria lymphocytes (LPL) for phenotypic analysis. Collagenase and DNAase I digestion can be used alone or in combination with dispase to promote digestion of extracellular matrix and the release of cellular material in a procedure requiring several hours of incubation at 37°C. Alternatively, cells can be recovered from lamina propria tissue after the removal of epithelium by incubation of samples in culture medium, permitting the 'walk out' of cells from the tissue (Bell et al., 2001). The latter method is particularly attractive for functional studies to overcome any potential impact of rigorous enzymatic digestion. Emigration of cells from epithelium-free biopsy samples in this way usually, however, requires overnight incubation and can therefore potentially impact on the activation status of the recovered cells.

As demonstrated for iel, lymphocytes isolated from colonic lamina propria tissue were initially identified with low spontaneous cytotoxic activity against K562 target cells compared to peripheral blood mononuclear cells (Hogan et al., 1985; Shanahan et al., 1987). Small populations of LPL expressing the NKH-1 antigen (CD56) could be identified by indirect immunofluorescence in enzymatically dispersed lamina propria and enrichment of these cells by positive selection resulted in an enhancement of natural killing of K562 target cells (Shanahan et al., 1987). Importantly, culture with IL-2 induced marked increases in cytotoxicity of LPL (Hogan et al., 1985; Shanahan et al., 1987). In addition, $CD56^-CD3^-$ or $CD56^+CD3^-$ precursor populations were identified, which gave rise to lymphokine-activated killer cells after culture with IL-2 (Hogan et al., 1985; Shanahan et al., 1987). The phenotype of $CD3^-CD56^+$ LGL isolated from colonic lamina propria was further extended, revealing that these cells expressed CD7, adhesion molecules CD11a and CD18, were of activated $CD69^+$ phenotype and, similarly to blood NK cells, expressed the intermediate affinity IL-2Rp70(CD122) (Pang et al., 1993). As shown for iel, LPL were predominantly negative for CD16 and CD57 was absent (Pang et al., 1993). Both spontaneous and IL-2 activated cytotoxicity of jejunal LPL against colonic tumour cell lines have been attributed to $CD3^+CD56^-CD16^-$ T-cells with no spontaneous cytotoxicity against K562 cells (Ebert and Roberts, 1993; Roberts et al., 1993). A regional distribution of different cytotoxic effector cells in the small and large intestine has been implied by differences in CD3 dependent and spontaneous cytotoxicity in the ileum and colon (Melgar et al., 2004). Redirected killing in the presence of anti-CD3 monoclonal antibody indicated that the normal ileal lamina propria is highly enriched for T-cells with cytolytic capacity, whereas cells extracted from colonic tissue exhibited only weak cytolytic activity, which was TCR-CD3 independent (Melgar et al., 2004). It should be noted, however, that CD3-independent cytotoxicity has also been attributed to the deployment of Fas-dependent mechanisms by T cell subsets (Melgar et al., 2002, 2004).

An NKp44$^+$ NK cell population in the human gut

Recently, an NKp44$^+$ NK cell population was demonstrated in the human small intestine (Cella et al., 2009). $CD3^-CD56^+NKp44^+$ and $CD3^-CD56^+NKp46^+$ NK cells are also present in colonic lamina propria, the majority

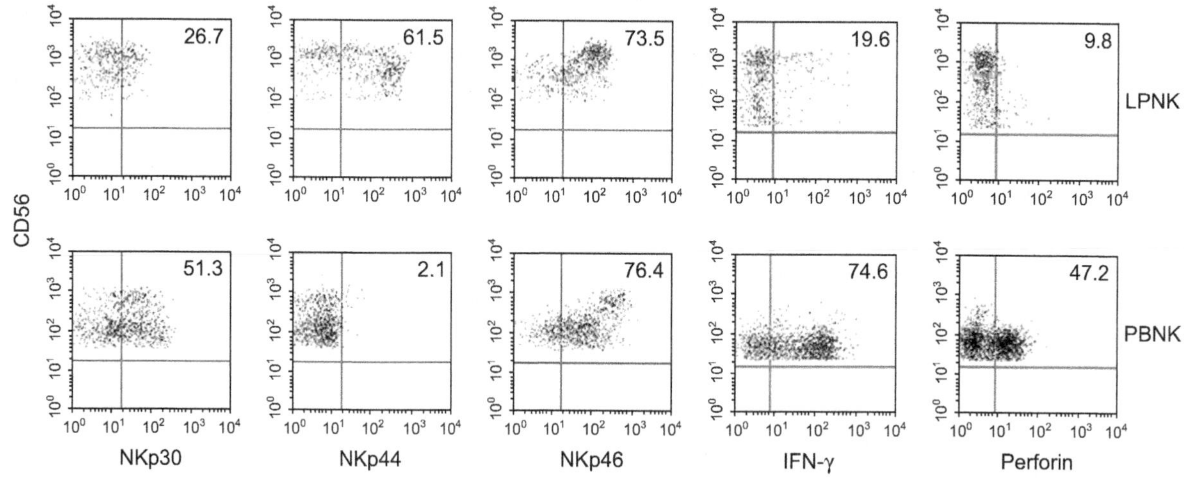

Figure 25.1 • Phenotypic characteristics of lamina propria NK cells (LPNK) and comparison with peripheral blood NK cells (PBNK). LPNK cells were recovered by 'walk-out' from colonic biopsy specimens taken from a patient undergoing investigative endoscopic examination. Expression of NCR, IFN-γ and perforin are shown in $CD3^-CD56^+$ NK cells (gated as in Figure 25.2).

of which do not express CD16 (Figure 25.2). Similar to $CD56^{hi}$ cells in the blood and in contrast to the $CD56^{lo}CD16^+$blood NK cells, the majority of lamina propria NK (LPNK) cells do not express perforin, and contain fewer granzyme-expressing cells than in the peripheral blood (Figure 25.1; Chinen et al., 2007). Spontaneous cytotoxicity of LPNK was equivalent to that observed in $CD56^{hi}CD16^-$ PBNK and lower than $CD56^{lo}CD16^+$ PBNK (Chinen et al., 2007). Studies from both human iel and LPL therefore indicate that spontaneous NK cell function analogous to that observed in $CD16^+CD56^{lo}$, $perforin^+$ peripheral blood NK cells is rare in healthy tissue sampled from the small and large human intestine, and that LPNK have a 'helper' NK cell phenotype.

Animal studies

The presence of intestinal lymphocytes with the phenotype and cytolytic function of NK cells has been supported by several studies in animal models. In mice, early studies demonstrated the presence of populations of LGL with NK activity both within isolated iel and LPL from the small intestine (Tagliabue et al., 1982). However, such early studies were suggestive of a role for atypical cells with an NK-T phenotype and were performed prior to the characterization of the γδ-TCR (Tagliabue et al., 1982). A population of asialo $GM1^+$, $NK1.1^+$ large granular iel was identified in nude mice (Alberti et al., 1985). Subsequent studies, however, showed that 90% of mouse intestinal iel comprise two populations of lymphocytes expressing either $TCR\text{-}\alpha\beta^+CD8\alpha\beta^+$ cells or $CD8\alpha^+$ $TCR\text{-}\alpha\beta^+$ or $TCR\text{-}\gamma\delta^+$(Guy-Grand et al., 1992). An extrathymic origin for these cells was supported by the presence of TCR^+ populations expressing RAG-1 mRNA in the intestinal epithelium of nude and SCID mice (Guy-Grand et al., 1992). $CD3^-NK1.1^+$ iel were identified in the small intestine of normal C57BL/6 mice in two studies as a defined population of between 0.7% and 2% of total iel, these numbers in keeping with a predominance of $CD3^+TCR\text{-}\gamma\delta^+$ or $TCR\text{-}\alpha\beta^+$ iel (Lopez and Holmes, 2005; Martin et al., 2001). Interestingly, the large intestine of these mice contained a higher proportion of NK iel compared to the small intestine (Lopez and Holmes, 2005). $CD3^-NK1.1^+DX5^+$ cells were increased in proportion and absolute number between 6- and 20-fold, whilst all TCR^+ cells, including $DX5^+$ and $NK1.1^+$ NK-T cells decreased in iel fractions from CD45 knockout mice bred on the same background, suggesting a proliferative expansion or migration of NK-like iel and supporting the possibility that these are independent from the T-cell lineages (Lopez and Holmes, 2005). The use of NKp46 as a 'true' NK cell-specific marker has recently confirmed the presence of NK cells with non-cytolytic function in the mouse intestine (Cella et al., 2009; Luci et al., 2009; Sanos et al., 2009; Satoh-Takayama et al., 2008).

Rat intestinal NK cells

In rats, $CD3^-NKRP1A^+$ cells were reported in the intraepithelial compartment of the small intestine (Helgeland et al., 1997; Todd et al., 1999). NK cells, as defined by the expression of monoclonal antibody (mAb) 3.2.3 were also detected in ileal epithelial and lamina propria preparations (Nakagawa et al., 2000). $CD2^-CD3^-TCR\text{-}\alpha\beta^-TCR\text{-}\gamma\delta^-NKRP1A^+$ ieNK cells were present at high frequency in weaning Wistar rats (50% of iel) declining to 17% in animals over 16 weeks old (Todd et al., 2001). Interestingly, in contrast to humans and mice, rat NK cells contained distinctive azurophilic granules and were capable of spontaneous natural killing of YAC-1 target cells (Todd et al., 2001).

NK cells in the amphibian and avian gut

The identification of intestinal NK cell subsets in amphibians and birds suggests conserved evolutionary importance for these cell populations. $CD8^+CD3^-Ig^-$ iel constitute up to 20% of iel in chicken duodenum (Gobel et al., 2001). These are further characterized by a cell surface antigen recognized by mAb 28.4 and which is absent on chicken $CD3^+$ T-cells and, similarly to mammalian ieNK, these cells express CD8α homodimers. In contrast to mammals, the intestine in chickens represents the principal location of NK cells in the body, less than 1% of blood spleen and caecal tonsils reacting with mAb 28.4^+. Importantly, and in contrast to mammalian iel, chicken NK iel but not $CD3^+$ iel displayed spontaneous NK activity against chicken B lymphoblastoid cell lines (Gobel et al., 2001). *Xenopus laevis* NK iel were identified by three novel mAbs. These $CD5^-$NK cells were highly enriched in the intestinal epithelium (up to 30% of iel) compared to both the liver and spleen and were also present and in thymectomized animals, comprising over 40% of iel (Horton et al., 2000).

Origins of NK cells in intestinal tissue

The diverse phenotypic and functional characteristic of lymphocytes in the intestine of humans and animals raises the question of the ontogeny of these populations. Both T-cells and NK cells in the blood express integrins indicative of homing potential for mucosal tissues, these cells being potentially activated in peripheral secondary lymphoid tissues, including the Peyer's patches of the small intestine and subsequently re-entering the blood stream via the efferent lymphatics and thoracic duct. Human blood NK cells expresses integrin α4β7, indicative of a potential to recirculate to intestinal lamina propria via interaction with mucosal addressin cell adhesion molecule-1 (MadCam-1) (Erle et al., 1994; Perez-Villar et al., 1996). This possibility is also supported by the detection of α4β7 integrin on $CD3^-CD56^+$ ieNK and LPNK cells (Chinen et al., 2007; Leon et al., 2003). Integrin αE is expressed on only a small fraction of $CD3^-CD56^+$ LPNK, but is expressed on the majority of ieNK, providing the potential to interact with epithelial cell cadherin (E-Cadherin) (Chinen et al., 2007). LPNK express low levels of CCR7, indicating a potential to migrate to secondary lymphoid tissue-derived signals, whilst the majority of both LPNK and ieNK expressed CXCR3, indicating potential migration toward interferon-inducible protein 10 (IP-10), which is produced by intestinal epithelial cells under inflammatory conditions (Chinen et al., 2007). Both CCR6 and its ligand CCL20 are produced by tonsillar $NKp44^+$ LPNK cells, raising the possibility that the expression of this chemokine-receptor combination promotes NK cell accumulation in mucosal tissues, although iNK may rely on different homing signals (Cella et al., 2009). HPC for NK may also migrate into intestinal tissue, as demonstrated for integrin $\alpha4\beta7^+$ NK HPC located in secondary lymphoid tissue (Freud et al., 2005).

NK lymphoid progenitor cells in the gut

Populations of NK cells that have homed to the intestinal mucosa may be supplemented by further NK cell subsets which have matured from progenitor cells residing in mucosal tissues. Although NK cells are normally considered to be derived in the bone marrow from resident NK precursor cells, recent studies have identified the presence of $CD34^+CD45^+$ lymphoid progenitor cells in human duodenal and colonic intestinal mucosa (Chinen et al., 2007; Lynch et al., 2006). Both $CD34^+CD45^+CD7^+$ populations and $CD34^+CD45^+CD56^+$ populations of HPC are present, the latter at a higher frequency in intestinal lamina propria compared to both intestinal epithelium and normal bone marrow (Chinen et al., 2007; Lynch et al., 2006). Furthermore, a high proportion (>40%) of HPC in lamina propria express CD56, whereas this constitutes a minor population in the intestinal epithelium (Chinen et al., 2007; Lynch et al., 2006). The distribution of the stem cell and HPC marker, c-kit within intestinal HPC subpopulations is also indicative of a spatial compartmentalization of NK cell precursors. Lamina propria contains a higher frequency of c-kit^+HPC and $CD56^+$LPNK are c-kit^{lo} whilst ieNK are c-kit negative, suggesting that LPNK cells may be less highly differentiated than ieNK (Chinen et al., 2007; Lynch et al., 2006). Intestinal epithelial cells produce IL-15, which is known to be important in the development of NK cells (Okazawa et al., 2004). It is feasible that HPC differentiation along the NK cell lineage pathway could be induced by IL-15 derived from intestinal epithelium.

In mice, NK cells have been implicated in neonatal lymph node development. Transfer of $NK1.1^+$ lymph node cells into RAG-$2^{-/-}$, IL-2R$\gamma^{-/-}$ mice, which lack peripheral lymph nodes, promotes the recovery of neonatal lymph node development, including Peyer's patches (Coles et al., 2006). CD3$^-$NKp46$^+$ cells co-expressing c-kit and IL-7Rα (CD127) have been identified in the adult murine gut, which are similar to lymphoid tissue inducer (Lti) cells for cryptopatches, found within the lamina propria of adult mice, and for foetal lymph nodes (Eberl, 2005; Luci et al., 2009; Mebius et al., 1997). Retinoic acid-related orphan receptor (RORγt), which is required for the development of lymphoid tissue-inducer cells and murine lymph node and cryptopatch development, is expressed on the majority of cells contained within cryptopatches (Luci et al., 2009). CD3$^-$NKp46$^+$ cells in the murine gut also share other phenotypic markers with Lti, including TNF and TNFR superfamily members, DR3, DR5, TRANCE/RANK-L, LIGHT, OX40ligand, CD30 ligand 4-1BB ligand, 4-1BB ligand and CCR7 (Luci et al., 2009).

Although cryptopatches have not been identified in adult human tissue, c-kit$^+$ HPC are distributed throughout the lamina propria in the small and large intestine (Chinen et al., 2007; Lynch et al., 2006). Interestingly, LIGHT, is also expressed on human LPNK cells (Cohavy et al., 2005). CD3$^-$CD56$^+$c-kit$^+$ NK cells could therefore play an equivalent role to murine Lti in the ongoing development of T, NK-T and NK cells in adult human intestinal tissue. The relationship between Lti and NK cell precursors is further established in the human foetal gut (Cupedo et al., 2009). Populations of lineage marker negative, CD127$^+$CD45 intermediate cells are present, which express the retinoic orphan receptor C gene (RORC) and are present in first trimester foetal mesentery and in second trimester mesenteric lymph nodes. These cells support the acquisition of adhesion molecules by LTβR$^+$ mesenchymal cells and are therefore considered to be equivalent to murine Lti (Cupedo et al., 2009). Furthermore, commitment to an NK lineage is confirmed on in vitro stimulation with stem cell factor (SCF), IL-7 and IL-15, leading to the expression of NK cell markers, including CD56, NKp30, NKp44 and NKp46. A lack of differentiation of CD127 + CD45intRORC$^+$ cells toward T, B and DC lineages under appropriate culture conditions indicates that these human Lti are precursors for NK-like cells. Human foetal mesenteric lymph node CD45^{int}CD127$^+$ Lti and CD56$^+$CD127$^+$ immature NK cells share functional features with Th17 cells, expressing mRNA for both IL-17 and IL-22 (Cupedo et al., 2009).

Alternative role for NK cells in 'wound healing' in intestinal tissues

Poor cytotoxic capacity, in the absence of cytokine stimulation, low perforin expression and the CD16 negative phenotype of iNK has raised the possibility of alternative, non-cytotoxic functions for these cells. Several independent avenues of investigation are now converging, providing strong evidence that LPNK cells have non-cytolytic helper functions.

Mitogenic stimulation of duodenal and jejunal ieNK results in IFN-γ and TNF production, whilst a low frequency of these cells produced IL-2 (Leon et al., 2003). IL-12 + IL-15 induced human LPNK cells also produce IFN-γ, indicating a potential helper function for Th1-type T-cells (Figure 25.1; Chinen et al., 2007).

Human uterine decidual tissue contains CD3$^-$CD56$^+$ NK cells with a remarkably similar phenotype to those identified in human lamina propria and having some common features with CD56^{hi} CD16$^-$ blood NK cells (Figure 25.1; Hanna et al., 2006; Koopman et al., 2003). These cells are NKp46$^+$CD16$^-$ and are mainly perforin negative, but also constitutively express NKp44, which is absent on human peripheral blood NK cells (Hanna et al., 2006; Koopman et al., 2003). Cross linking of NKp44, on uterine decidual NK cells, leads to the production of growth factors, including epidermal growth factor (EGF), vascular endothelial cell growth factor (VEGF1-3) and placental growth factor (PLGF) (Hanna et al., 2006). mRNA for chemokines and growth-promoting polypeptides NK4 and NKG5 are expressed by uterine NK cells ex vivo (Hanna et al., 2006).

The integrity of the intestinal epithelium depends on a highly coordinated, ongoing process of epithelial cell re-generation. Enterocytes, derived from stem cells residing in epithelial crypts, progressively differentiate as they migrate along the crypt-villous axis during the process of epithelial regeneration. In mice, the proximity of cryptopatches to villous crypts indicates that, similarly to uterine NK cells, factors produced by cryptopatch-derived cells could have growth-promoting properties for intestinal epithelial cell regeneration. The functional phenotype of foetal Lti-derived NK cell populations also suggests that cells of this lineage not only promote the induction of lymphoid tissue in the foetus, but also contribute to post-natal and adult regenerative processes (Cupedo et al., 2009).

Recently, tonsillar lamina propria NKp44$^+$ NK cells were shown to produce IL-22, IL-26 and leukaemia inhibitory factor (LIF) in response to IL-23, whereas NKp44$^-$ NK cells are the principal producers of IFN-γ in response to IL-12 or IL-15 (Cella et al., 2009). Supernatants from activated tonsillar NKp44$^+$ NK cells (termed NK-22) induce a pattern of gene expression in the colonic carcinoma-derived Colo205 cell line associated with responses to growth factors, cell growth, cell cycle progression and protection from apoptosis and promote the proliferation of these cells (Cella et al., 2009). It is therefore conceivable that intestinal lamina propria resident NKp44$^+$ NK cells play an active role in promoting enterocyte proliferation during intestinal epithelium regeneration and repair, through similar mechanisms.

Commensal microflora and pathways to activation of NK cells in the healthy intestine

Commensal micro-organisms, which colonize both the small and large intestine, are critical to normal intestinal function (Artis, 2008; Guarner and Malagelada, 2003). However, the enormous burden of commensal microbes also presents a considerable stimulatory challenge to the innate immune system, which may impact on NK cell function in the intestine. Commensal bacteria have the potential to activate iNK as a bystander effect of TLR activation on intestinal epithelial cells or via the maturation of intestinal dendritic cells, which sample luminal microbes by extending cellular processes through the junctions between epithelial cells.

Intestinal epithelial cells express Toll-like receptors (TLR), which can be activated by molecular patterns from associated commensal bacteria, transducing signals resulting in the activation of epithelial cells (Rakoff-Nahoum et al., 2004). TLR-mediated intestinal epithelial cell activation leads to the production of chemokines, including IP-10, which can promote the migration of CXCR3$^+$ NK cells toward the epithelium (Lan et al., 2005). IL-15 is also produced by intestinal epithelial cells upon interaction with TLR ligands (Zhou et al., 2007).

Diverse dendritic cell subsets reside in the intestinal lamina propria in humans which, upon maturation, produce factors with the potential to activate NK cells, including IL-12. (Bell et al., 2001; Hart et al., 2004, 2005). Several studies have modelled the impact of stimulation of human peripheral blood NK cells in response to commensal bacteria. Lactobacilli, which are present in the normal indigenous microflora of the small and large intestine, can transmigrate across the intestinal epithelium, but are not normally associated with disease pathology. Lactobacilli are potent inducers of IL-12 production by peripheral blood mononuclear cells, with strain variation evident in the magnitude of these responses (Hessle et al., 1999). *Lactobacillus paracasei* is more potent in IL-12 induction than *L. rhamnosus* and *L. plantarum*. Maturation of human and murine dendritic cells in the presence of Lactobacilli strains enhances their capacity to induce Th-1 and T regulatory type responses (Fink et al., 2007a,b). *L. acidophilus* induces potent DC-mediated NK cell activation, inducing CD69, CD25, and HLA-DR and NKp44 expression and IFN-γ production in autologous peripheral blood NK cells. Interestingly, other Lactobacilli and commensal strains are less potent inducers of NK cell activation and down-regulate *L. acidophilus*-induced IFN-γ production (Fink et al., 2007a,b). IL-10 production in response to TLR-ligands, including *E. coli* LPS, is known to counterbalance IL-12 dependent NK cell activation (Goodier and Londei, 2000). Strain variation in the induction of IL-10 by commensal pathogens could account for this down-regulating effect and provide a mechanism for restricting the overt responses of NK cells to commensal pathogens (Hessle et al., 2000).

Muramyl dipeptide, a gram negative and gram positive bacteria-derived pathogen associated molecular pattern (PAMP), is recognized by nuclear oligomerization domain 2 (Nod2) expressed in intestinal epithelial cells and for which mRNA is found in both dendritic cells and monocytes (Athie-Morales et al., 2008). Blood NK cells and NK cell lines expressing nod2 can be activated by muramyl dipeptide in the presence of interferon-alpha for the induction of CD69 expression and combined with IL-12 for IFN-γ production (Athie-Morales et al., 2008).

PAMPs including zymosan, LPS and resiquimod induce the production of IL-23 from human monocytes, which in turn, activates IL-22 and other epithelial growth associated molecules from tonsillar NKp44$^+$NK-22 cells (Cella et al., 2009). Similar functions in the intestine may integrate the protective capacity of commensal bacteria and regeneration and repair in intestinal epithelium. IL-23 stimulates tonsillar NKp44$^+$ NK-22 cells and also induces IL-10 production by epithelial cell lines, indicating a further mechanism of restricting the overt activation of inflammatory responses by commensal or pathogenic microbes (Cella et al., 2009). In germ-free laboratory mice, the frequency of intestinal lamina propria NK1.1^{int} NKp46$^+$ RORγt ^{hi} cells is diminished and constitutive or IL-23-induced expression of IL-22 mRNA and protein by these cells is virtually abolished (Sanos et al., 2009; Satoh-Takayama et al., 2008). Constitutive IL-22 mRNA is also abolished in NK1.1^{int} NKp46$^+$CD127$^+$ cells from RORγt deficient mice, indicating that commensal microbes indeed contribute to regenerative pathways via an NKp46$^+$RORγt$^+$ cell dependent mechanism (Satoh-Takayama et al., 2008). Intestinal NK cells, in particular those located in the lamina propria of the small and large intestine, may therefore provide a critical link between innate recognition of commensal pathogens and the routine maintenance of the intestinal epithelium.

The release of damage associated molecular patterns (DAMPs), including high mobility group box 1, heparin sulphate and hyaluronan, from injured or inflamed intestinal epithelium and associated extracellular matrix during the healing response to infection or physical trauma may also impact on the activation and differentiation of intestinal NK cell subsets and merits further investigation (Lotze et al., 2007).

Intestinal NK cells and enteric pathogens

Diverse functions of NK cells have been implicated in both protection from and in the pathogenesis of diseases impacting on the intestine. The induction of cytotoxicity in ieNK and LPNK by cytokines in vitro suggests that exposure to such stimuli on infection with pathogenic microorganisms or during auto-immune conditions, could induce a cytotoxic phenotype in intestinal NK cells. Human and murine models of immunity to enteric pathogens, however, suggest a primary role for non-cytolytic functions of NK cells in protection.

Enteric pathogens including *H. pylori*, *Listeria*, *Shigella*, *Salmonella* and pathogenic strains of *E. coli* manipulate epithelial cell functions in order to facilitate penetration of the intestinal epithelium and subsequently cause disease (Lu and Walker, 2001). Lysates of *H. pylori* induce IFN-γ production and augment IL-12Rβ2 expression in human peripheral blood NK cells and synergize with IL-12 to augment NK cell IFN-γ production (Yun et al., 2005). Preferential induction of IL-12 compared to IL-10 may account for the dominant NK cell IFN-γ in response to *H. pylori* antigens (Voland et al., 2006). However, chronic infection with *H. pylori* causes a reduction in the proportion of NK cells in the gastric mucosa, suggesting that potential protective effects of NK cells may be compromised in infected individuals (Yun et al., 2005).

NK cell–derived IFN-γ has also been implicated in animal models of enteric pathogen infection. NK cells produce IFN-γ in *Shigella flexneri*–infected wild type mice and NK cells contribute to the early control of infection in RAG-deficient mice. NK 1.1^+ cells in conjunction with IL-15 conferred protection against *Salmonella enterica typhimurium* infection and depletion of NK cells from normal mice resulted in disseminated infection and a reduction of local inflammatory responses in the gut (Ashkar et al., 2008). Disseminated infection was associated with diminished IL-17 production in this model, potentially linking NK cell function to mucosal IL-17 production and/or Th17 activation (Ashkar et al., 2008). A requirement for $CD11^{hi}$ DC in the induction of NK cell–derived IFN-γ production and systemic control of *Listeria* infection was demonstrated in a murine model of conditional DC ablation (Lucas et al., 2007). It should be noted, however, that transfer of conventional TCR-$\alpha\beta^+$ T-cells into RAG-2/γc knockout mice was sufficient to control enteric *L. monocytogenes* infection, indicating that intestinal NK cells are not necessarily involved in protection (Bregenholt et al., 2001). Infection with *Citrobacter rodentium*, a non-invasive murine enteric pathogen that effaces the brush borders of colonic epithelial cells, induces the appearance of NK-22 NK cells in the lamina propria of the small intestine and mice lacking IL-22 producing $NKp46^+$ cells have increased susceptibility to C. *rodentium* infection and reduced survival, indicating a potential role for these cells in protective immunity (Cella et al., 2009; Satoh-Takayama et al., 2008).

NK cells also have a potentially important role in protection against gut nematode infection. Epithelial NK cell–derived IL-13 was demonstrated in a murine model to promote intestinal immunopathology associated with protection from *Trichinella* infection in mice (McDermott et al., 2005).

Human immunodeficiency virus (HIV) infection and intestinal NK cells

Early HIV-1 and SIV infection has a profound and lasting impact on immunity in the GI tract. Loss of secondary lymphoid tissue in the GI tract occurs within weeks of initial infection with SIV and HIV-1 (Brenchley et al., 2004; Schacker et al., 2006). Regeneration of lymphoid structures is impeded by the deposition of collagen at sites of lymphoid tissue loss. HIV-1 also affects mucosal barrier function and elevated concentrations of microbial products in the plasma of HIV-1–infected individuals results from effects on epithelial integrity in the GI tract (Brenchley et al., 2006; Delezay et al., 1997).

Early loss of $CD4^+$ T cells in the HIV-1–infected intestine is associated with the induction of perforin expression in $CD8^+$ T cells in the jejunal lamina propria and significantly also in CD8 negative cells, indicating a potential virus-induced differentiation of NK cells (Mehandru et al., 2007). A decreased proportion of NK cells in colonic lamina propria parallels $CD4^+$ T-cell losses during chronic HIV-1 infection in individuals who have detectable plasma virus, but not in those who have suppressed viral replication under anti-retroviral therapy (Figure 25.2; Mela et al., 2007). Significantly, chronic HIV-1 infection leads to an increase in the proportion of $CD16^+$ LPNK cells, suggestive of migration of these cells from the blood in response to inflammation and chemokine production in the gut or of differentiation of $CD16^-$ NK cells or NK cell precursors. IL-2 stimulation of $CD56^{hi}CD16^-$ NK cells from human tonsil and lymph node results in the expression of CD16, KIR and perforin and in cytotoxic activity, all features of $CD56^+CD16^+$ blood NK cells, indicating that a similar pattern of differentiation could occur in LPNK cells in response to infection or inflammatory mediators (Fehniger et al., 2003; Ferlazzo et al., 2004).

Inflammatory bowel diseases and celiac disease; NK cell involvement?

Ulcerative colitis (UC) is associated with chronic inflammation in mucosa of the large intestine whilst Crohn's

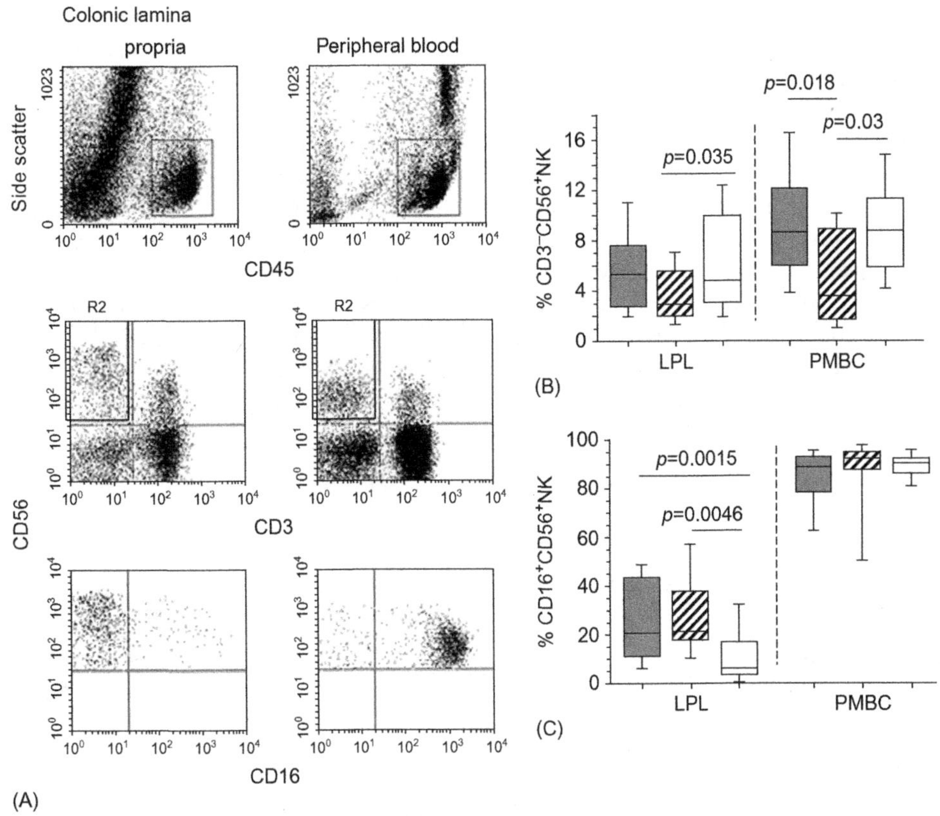

Figure 25.2 • Decrease in the overall proportion of LPNK in HIV-1 infected individuals and increased representation of $CD16^+$ NK cells. (A) Gating strategy for analysis of NK cells from colonic lamina propria (left-hand panels) and blood (right-hand panels). Region 1 (R1) defines initial gating according to CD45 and side scatter. NK cells were defined subsequently as $CD56^+CD3^-$ cells (R2, middle panels). The proportion of $CD16^+$ NK cells was then analysed within R2 (lower panels). (B) The proportions of NK cells within LPMC or PBMC from aviraemic HIV-1^+ (filled bars), viraemic HIV-1^+ (striped bars) and HIV-1 negative control individuals (clear bars). (C) Proportions of $CD16^+$ NK cells. Box plots show medians and interquartile range with 10th and 90th percentiles. Statistical analyses were performed using Mann–Whitney U test for comparison between groups. Reproduced with permission from Wolters Kluwer Health. Mela C.M., Steel A., Lindsay J., Gazzard B.G., Gotch F.M. and Goodier M.R. 2007. Depletion of Natural Killer Cells in the colonic lamina propria of viraemic HIV-1 infected individuals. *AIDS* **21**(16):2177–2182.

disease (CD) affects all regions of the GI tract. A number of studies have indicated that only weak lamina propria natural cytotoxicity can be detected in patients with either UC or CD, implying a lack of NK involvement (Hogan et al., 1985; Melgar et al., 2004). One study employed histological techniques to demonstrate an increase in NK cell markers in the colonic mucosa of patients with active UC (del Mar Cabrera et al., 2001). However, this study did not take into account the abundance of TCR-$\alpha\beta^+$ and TCR-$\gamma\delta^+$ subsets expressing NK cell markers in the colonic mucosa, and which were subsequently demonstrated to have an important role in driving the Th-2, IL-13–dominated pattern of cytokine production in UC (Targan and Karp, 2005). A murine model of colitis, in which IL-$10^{-/-}$ mice are susceptible to a Th-1 mediated pathology demonstrated a role for NK cells in restricting disease (Fort et al., 1998). Epithelial regeneration in the small intestine of mice after radiation-induced injury was also associated with increased IL-13 production in LPNK cells (Kawashima et al., 2006). However, pathogenesis in both of these models is not typical of human inflammatory bowel diseases. An elevated frequency of lamina propria dendritic cells from the colon of both UC and CD patients had elevated TLR-2 and TLR-4 expression, suggesting an increased potential for reactivity with TLR ligands including LPS (Hart et al., 2005). In CD, more colonic DC produced IL-6 and IL-12, indicating an increased potential for LPNK cell activation (Hart et al., 2005). Increased differentiation of c-kit^+lin^- precursors into NK cells was observed in LPL from CD patients resulting in an increased proportion of both LPNK and ieNK, which could potentially relate to augmentation of DC or epithelial cell-derived factors, including IL-12 and IL-15 (Chinen et al., 2007).

$CD3^-$NK cells were initially considered to have an involvement of in the immunopathogenesis of celiac disease. However, $CD7^+CD3^-$ populations decreased significantly within the intraepithelial compartment of human jejunal mucosa from celiac patients, whilst $CD3^+CD4^-CD8^-$ cells proportionately increased. $CD3^-$ cells expressing NK cell receptors were reported to be reduced in number in patients with active and treated celiac disease, a fraction of which were shown subsequently to also lack intracellular CD3 expression (Eiras et al., 1998; Leon et al., 2003). Subsequent studies showed that T cells expressing a broad repertoire of NK cell receptors including CD94/NKG2A, NKG2D, CD161, KIR2DL2/3, NKp46 and CD56 were present within the iel population (Jabri et al., 2000). Celiac disease resulted in a dramatic increase in the frequency of $CD94^{+}$ cells which were mainly T cells, considered to be due, at least in part to epithelial cell-derived IL-15 (Jabri et al., 2000). A reduction in the proportion of $CD3^{-}$ NK-like cells could therefore result from a preferential expansion of T cells bearing NK cell receptors. Subsequent studies reported substantial expansions of TCR-$\alpha\beta^{+}$, $CD8^{+}$ iel expressing NKG2C, NKp44 and NKp46, also previously considered to be NK cell-specific receptors (Meresse et al., 2006). Although not ruling out this possibility, these studies suggest caution in postulating a role for NK cells in the etiology of celiac disease and indicate that T cells play a dominant role and acquire NK cell receptors, at least in the intestinal epithelium. Ligands for NKG2D, with the potential for NK cell activation, including MICA and ULBP5/RAET1G1, are either expressed intracellularly or are secreted from normal intestinal epithelial cells (Eagle et al., 2009; Groh et al., 1996). Surface expression of MICA is associated with villous atrophy in celiac disease and MICA is upregulated

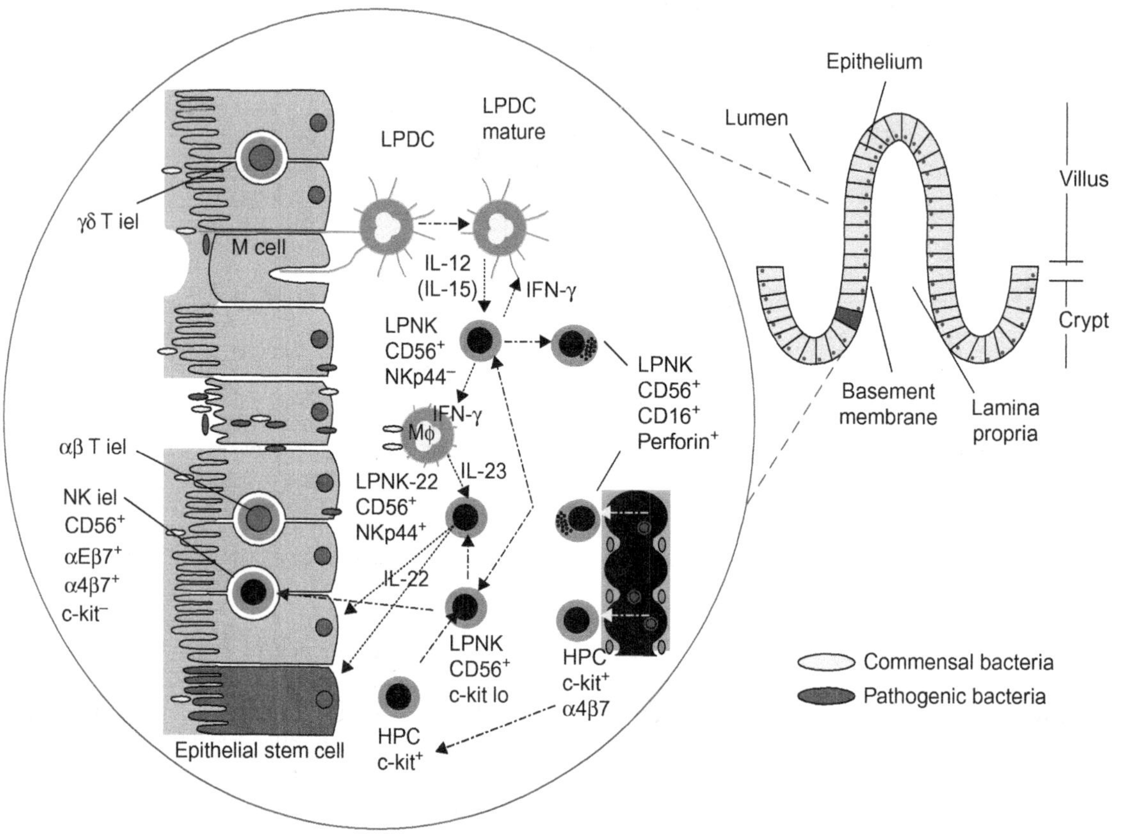

Figure 25.3 • Proposed model for NK cell involvement in the adult human intestine. A section of intestine shows the columnar cells of the intestinal epithelium supported by basement membrane and the underlying lamina propria. Renewal of intestinal epithelium occurs by differentiation of epithelial cell precursors and migration along the crypt-villus axis. Commensal or pathogenic bacteria are sampled in the lumen of the intestine by lamina propria dendritic cells (LPDC) extending processes through the epithelial barrier. Microbial products can also be captured by specialized M in the intestinal epithelium. Certain pathogenic microbes translocate through the intestinal epithelium whilst others cause damage resulting in increased uptake of pathogenic or commensal bacteria. LPDC mature on activation by PAMPs ligating toll-like receptors (TLR). TLR ligation also activates macrophages (mϕ) in the lamina propria. C-kit^{+} $lineage^{-}$ haemopoetic precursor cells (HPC) are dispersed throughout the human lamina propria. In mice, HPC are associated with cryptopatches and express the transcription factor RORγt. Intestinal NK HPC may be maintained in situ or be replenished by the migration of bone marrow derived $\alpha4\beta7^{+}$ HPC. NK HPC subsequently differentiate into $NKp44^{+}$ or $NKp44^{-}$ LPNK cell populations in the lamina propria. Fully differentiated c-kit^{-} intraepithelial NK cells (ienk) may be derived from $CD56^{+}$c-kit^{lo} LPNK migrating to the epithelium. Activated LPDC produce IL-12 and IL-15, which activate IFN-γ production in $NKp44^{-}$ LPNK. IFN-γ augments anti-pathogen responses by promoting macrophage and DC function. Activated macrophages, or dendritic cell subsets, produce IL-23, which stimulates $NKp44^{+}$ LPNK-22 cells (in mice ROR $\gamma t^{hi}NKP46^{+}NK1.1^{int}CD127^{+}$) to produce IL-22 which in turn promotes epithelial cell proliferation. Chronic inflammation under disease conditions may result in the further differentiation of $CD16^{-}$LPNK into cytotoxic $CD16^{+}$ cells expressing perforin. Cytotoxic $CD16^{+}$NK effector cells may also migrate to the lamina propria in response to inflammation. (Figure drawn by B. Healey.)

by gliadin peptides on epithelial cells from treated patients (Hue et al., 2004). Cross linking of NKG2D on iel cell lines from celiac disease patients results in enhancement of CD3-dependent cytotoxicity whilst this activity can be induced by cross linking NKG2D alone in surface $CD3^-$ iel lines from patients with severe enteropathy (refractory celiac sprue), indicating a potential role for both T cell and NK cell dependent mechanisms in the etiology of celiac disease (Hue et al., 2004).

Conclusions

The identification of NK cell precursors and intestinal helper-type NK cells has raised the likelihood that these cells have several distinct roles in the intestine of humans and animals. In the human intestine, the precise interaction of NK cells with other components of the immune response remains to be elucidated. Functional characterization of helper NK cells in other tissues, in vitro studies and extrapolation from animal models, however, suggest some likely scenarios (Figure 25.3). One subset of $NKp44^-$ LPNK cells may act to promote pathogen-specific immune responses, including enhanced T helper type-1 acquired immunity via the production of IFN-γ. A second $RORC^+NKp44^+$ LPNK-22 cell subset, derived from human Lti-like NK cell precursor cells, may have a 'healer' function, integrating innate immunity with regeneration and repair of the intestinal epithelial barrier. Endogenous commensal microbes and gut pathogens play a key role in the induction of both of these arms of the intestinal NK cell response. Key questions remain as to the origin of NK cell precursors in the gut and the lineage and functional relationship between LPNK and ieNK. IL-22 produced by LPNK cells is emerging as a key factor in epithelial protection regenerative processes. More work is needed on the function of ieNK, whether these cells produce IL-22, and whether production has to be localized to the epithelium to be functionally effective. Finally, the impact of chronic infectious and inflammatory bowel diseases on the differentiation, frequency and functional capacity of intestinal NK cells merits further investigation.

References

Alberti, S., Colotta, F., Spreafico, F., Delia, D., Pasqualetto, E. and Luini, W. (1985). Large granular lymphocytes from murine blood and intestinal epithelium: comparison of surface antigens, natural killer activity, and morphology. Clin Immunol Immunopathol **36**, 227–238.

Artis, D. (2008). Epithelial-cell recognition of commensal bacteria and maintenance of immune homeostasis in the gut. Nat Rev Immunol **8**, 411–420.

Ashkar, A.A., Reid, S., Verdu, E.F., Zhang, K. and Coombes, B.K. (2008). IL-15 and NK1.1+ cells provide innate protection against acute *Salmonella enterica* serovar Typhimurium infection in the gut and in systemic tissues. Infect Immun **17**, 17.

Athie-Morales, V., V., O'Connor, G.M., Gardiner, C.M. (2008). Activation of human NK cells by the bacterial pathogen-associated molecular pattern muramyl dipeptide. J Immunol **180**, 4082–4089.

Bell, S.J., Rigby, R., English, N., Mann, S.D., Knight, S.C., Kamm, M.A. and Stagg, A.J. (2001). Migration and maturation of human colonic dendritic cells. J Immunol **166**, 4958–4967.

Blum, K.S. and Pabst, R. (2007). Lymphocyte numbers and subsets in the human blood. Do they mirror the situation in all organs? Immunol Lett **108**, 45–51.

Bregenholt, S., Berche, P., Brombacher, F. and Di Santo, J.P. (2001). Conventional alpha beta T cells are sufficient for innate and adaptive immunity against enteric *Listeria monocytogenes*. J Immunol **166**, 1871–1876.

Brenchley, J.M., Schacker, T.W., Ruff, L.E., Price, D.A., Taylor, J.H., Beilman, G.J., Nguyen, P.L., Khoruts, A., Larson, M., Haase, A.T. and Douek, D.C. (2004). CD4+ T cell depletion during all stages of HIV disease occurs predominantly in the gastrointestinal tract. J Exp Med **200**, 749–759.

Brenchley, J.M., Price, D.A., Schacker, T.W., Asher, T.E., Silvestri, G., Rao, S., Kazzaz, Z., Bornstein, E., Lambotte, O., Altmann, D., Blazar, B.R., Rodriguez, B., Teixeira-Johnson, L., Landay, A., Martin, J.N., Hecht, F.M., Picker, L.J., Lederman, M.M., Deeks, S.G. and Douek, D.C. (2006). Microbial translocation is a cause of systemic immune activation in chronic HIV infection. Nat Med **12**, 1365–1371.

Cario, E., Becker, A., Sturm, A., Goebell, H. and Dignass, A.U. (1999). Peripheral blood mononuclear cells promote intestinal epithelial restitution in vitro through an interleukin-2/interferon-gamma-dependent pathway. Scand J Gastroenterol **34**, 1132–1138.

Cario, E., Gerken, G. and Podolsky, D.K. (2007). Toll-like receptor 2 controls mucosal inflammation by regulating epithelial barrier function. Gastroenterology **132**, 1359–1374.

Cella, M., Fuchs, A., Vermi, W., Facchetti, F., Otero, K., Lennerz, J.K., Doherty, J.M., Mills, J.C. and Colonna, M. (2009). A human natural killer cell subset provides an innate source of IL-22 for mucosal immunity. Nature **457**, 722–725.

Cerf-Bensussan, N., Schneeberger, E.E. and Bhan, A.K. (1983). (1983). Immunohistologic and immunoelectron microscopic characterization of the mucosal lymphocytes of human small intestine by the use of monoclonal antibodies. J Immunol **130**, 2615–2622.

Cerf-Bensussan, N., Guy-Grand, D. and Griscelli, C. (1985). Intraepithelial lymphocytes of human gut: isolation, characterisation and study of natural killer activity. Gut **26**, 81–88.

Chinen, H., Matsuoka, K., Sato, T., Kamada, N., Okamoto, S., Hisamatsu, T., Kobayashi, T., Hasegawa, H., Sugita, A., Kinjo, F., Fujita, J and Hibi, T. (2007).

Lamina propria c-kit+ immune precursors reside in human adult intestine and differentiate into natural killer cells. Gastroenterology **133**, 559–573.

Cohavy, O., Zhou, J., Ware, C.F. and Targan, S.R. (2005). LIGHT is constitutively expressed on T and NK cells in the human gut and can be induced by CD2-mediated signaling. J Immunol **174**, 646–653.

Coles, M.C., Veiga-Fernandes, H., Foster, K.E., Norton, T., Pagakis, S.N., Seddon, B. and Kioussis, D. (2006). Role of T and NK cells and IL7/IL7r interactions during neonatal maturation of lymph nodes. Proc Natl Acad Sci U S A **103**, 13457–13462.

Cooper, M.A., Fehniger, T.A. and Caligiuri, M.A. (2001). The biology of human natural killer-cell subsets. Trends Immunol **22**, 633–640.

Cupedo, T., Crellin, N.K., Papazian, N., Rombouts, E.J., Weijer, K., Grogan, J.L., Fibbe, W.E., Cornelissen, J.J. and Spits, H. (2009). Human fetal lymphoid tissue-inducer cells are interleukin 17-producing precursors to RORC+ CD127+ natural killer-like cells. Nat Immunol **10**, 66–74.

Dalton, J.E., Cruickshank, S.M., Egan, C.E., Mears, R., Newton, D.J., Andrew, E.M., Lawrence, B., Howell, G., Else, K.J., Gubbels, M.J., Striepen, B., Smith, J.E., White, S.J. and Carding, S.R. (2006). Intraepithelial gammadelta+ lymphocytes maintain the integrity of intestinal epithelial tight junctions in response to infection. Gastroenterology **131**, 818–829.

del Mar Cabrera, M., Valle, J., Pajares, J.M., Romero, I., Zomeno, M., Mate, J. (2001). Expression of the Kp43 (CD 94) receptor by natural killer (NK) cells in ulcerative colitis. Hepatogastroenterology **48**, 1316–1320.

Delezay, O., Yahi, N., Tamalet, C., Baghdiguian, S., Boudier, J.A. and Fantini, J. (1997). Direct effect of type 1 human immunodeficiency virus (HIV-1) on intestinal epithelial cell differentiation: relationship to HIV-1 enteropathy. Virology **238**, 231–242.

Eagle, R.A., Flack, G., Warford, A., Martinez-Borra, J., Jafferji, I., Traherne, J.A., Ohashi, M., Boyle, L.H., Barrow, A.D., Caillat-Zucman, S., Young, N.T. and Trowsdale, J. (2009). Cellular expression, trafficking, and function of two isoforms of human ULBP5/RAET1G. PLoS One **4**, e4503.

Eberl, G. (2005). Inducible lymphoid tissues in the adult gut: recapitulation of a fetal developmental pathway? Nat Rev Immunol **5**, 413–420.

Ebert, E.C. and Roberts, A.I. (1993). Lymphokine-activated killing by human intestinal lymphocytes. Cell Immunol **146**, 107–116.

Eiras, P., Roldan, E., Camarero, C., Olivares, F., Bootello, A. and Roy, G. (1998). Flow cytometry description of a novel CD3 − /CD7+ intraepithelial lymphocyte subset in human duodenal biopsies: potential diagnostic value in coeliac disease. Cytometry **34**, 95–102.

Eiras, P., Leon, F., Camarero, C., Lombardia, M., Roldan, E., Bootello, A. and Roy, G. (2000). Intestinal intraepithelial lymphocytes contain a CD3− CD7+ subset expressing natural killer markers and a singular pattern of adhesion molecules. Scand J Immunol **52**, 1–6.

Erle, D.J., Briskin, M.J., Butcher, E.C., Garcia-Pardo, A., Lazarovits, A.I. and Tidswell, M. (1994). Expression and function of the MAdCAM-1 receptor, integrin alpha 4 beta 7, on human leukocytes. J Immunol **153**, 517–528.

Fehniger, T.A., Cooper, M.A., Nuovo, G.J., Cella, M., Facchetti, F., Colonna, M. and Caligiuri, M.A. (2003). CD56bright natural killer cells are present in human lymph nodes and are activated by T cell-derived IL-2: a potential new link between adaptive and innate immunity. Blood **101**, 3052–3057.

Ferlazzo, G., Thomas, D., Lin, S.L., Goodman, K., Morandi, B., Muller, W.A., Moretta, A. and Münz, C. (2004). The abundant NK cells in human secondary lymphoid tissues require activation to express killer cell Ig-like receptors and become cytolytic. J Immunol **172**, 1455–1462.

Fink, L.N., Zeuthen, L.H., Christensen, H.R., Morandi, B., Frokiaer, H. and Ferlazzo, G. (2007a). Distinct gut-derived lactic acid bacteria elicit divergent dendritic cell-mediated NK cell responses. Int Immunol **19**, 1319–1327.

Fink, L.N., Zeuthen, L.H., Ferlazzo, G. and Frokiaer, H. (2007b). Human antigen-presenting cells respond differently to gut-derived probiotic bacteria but mediate similar strain-dependent NK and T cell activation. FEMS Immunol Med Microbiol **51**, 535–546.

Fort, M.M., Leach, M.W. and Rennick, D.M. (1998). A role for NK cells as regulators of CD4+ T cells in a transfer model of colitis. J Immunol **161**, 3256–3261.

Freud, A.G., Becknell, B., Roychowdhury, S., Mao, H.C., Ferketich, A.K., Nuovo, G.J., Hughes, T.L., Marburger, T.B., Sung, J., Baiocchi, R.A., Guimond, M and Caligiuri, M.A. (2005). A human CD34(+) subset resides in lymph nodes and differentiates into CD56bright natural killer cells. Immunity **22**, 295–304.

Gobel, T.W., Kaspers, B. and Stangassinger, M. (2001). NK and T cells constitute two major, functionally distinct intestinal epithelial lymphocyte subsets in the chicken. Int Immunol **13**, 757–762.

Goodier, M.R. and Londei, M. (2000). Lipopolysaccharide stimulates the proliferation of human CD56 + CD3− NK cells: a regulatory role of monocytes and IL-10. J Immunol **165**, 139–147.

Groh, V., Bahram, S., Bauer, S., Herman, A., Beauchamp, M. and Spies, T. (1996). Cell stress-regulated human major histocompatibility complex class I gene expressed in gastrointestinal epithelium. Proc Natl Acad Sci U S A **93**, 12445–12450.

Guarner, F. and Malagelada, J.R. (2003). Gut flora in health and disease. Lancet **361**, 512–519.

Guy-Grand, D., Vanden Broecke, C., Briottet, C., Malassis-Seris, M., Selz, F. and Vassalli, P. (1992). Different expression of the recombination activity gene RAG-1 in various populations of thymocytes, peripheral T cells and gut thymus-independent intraepithelial lymphocytes suggests two pathways of T cell receptor rearrangement. Eur J Immunol **22**, 505–510.

Hanna, J., Goldman-Wohl, D., Hamani, Y., Avraham, I., Greenfield, C., Natanson-Yaron, S., Prus, D., Cohen-Daniel, L, Arnon, TI, Manaster, I, Gazit, R, Yutkin, V, Benharroch, D, Porgador, A, Keshet, E., Yagel, S. and Mandelboim, O. (2006). Decidual NK cells regulate key developmental processes at the human fetal-maternal interface. Nat Med **12**, 1065–1074.

Hart, A.L., Lammers, K., Brigidi, P, Vitali, B., Rizzello, F., Gionchetti, P., Campieri, M., Kamm, M.A., Knight, S.C. and Stagg, A.J. (2004). Modulation of human dendritic cell phenotype and function by probiotic bacteria. Gut **53**, 1602–1609.

Hart, A.L., Al-Hassi, H.O., Rigby, R.J., Bell, S.J., Emmanuel, A.V., Knight, S.C., Kamm, M.A and Stagg, A.J. (2005). Characteristics of intestinal dendritic cells in inflammatory bowel diseases. Gastroenterology **129**, 50–65.

Helgeland, L., Vaage, J.T., Rolstad, B., Halstensen, T.S., Midtvedt, T. and Brandtzaeg, P (1997). Regional phenotypic specialization of intraepithelial lymphocytes in the rat intestine does not depend on microbial colonization. Scand J Immunol **46**, 349–357.

Hessle, C., Hanson, L.A. and Wold, A.E. (1999). Lactobacilli from human gastrointestinal mucosa are strong stimulators of IL-12 production. Clin Exp Immunol **116**, 276–282.

Hessle, C., Andersson, B. and Wold, A.E. (2000). Gram-positive bacteria are potent inducers of monocytic interleukin-12 (IL-12) while gram-negative bacteria preferentially stimulate IL-10 production. Infect Immun **68**, 3581–3586.

Hogan, P.G., Hapel, A.J. and Doe, W.F. (1985). Lymphokine-activated and natural killer cell activity in human intestinal mucosa. J Immunol **135**, 1731–1738.

Horton, T.L., Minter, R., Stewart, R., Ritchie, P., Watson, M.D. and Horton, J.D. (2000). *Xenopus* NK cells identified by novel monoclonal antibodies. Eur J Immunol **30**, 604–613.

Hue, S., Mention, J.J., Monteiro, R.C., Zhang, S., Cellier, C., Schmitz, J, Verkarre, V., Fodil, N., Bahram, S., Cerf-Bensussan, N. and Caillat-Zucman, S. (2004). A direct role for NKG2D/MICA interaction in villous atrophy during celiac disease. Immunity **21**, 367–377.

Jabri, B., de Serre, N.P., Cellier, C., Evans, K., Gache, C., Carvalho, C., Mougenot, J.F., Allez, M., Jian, R., Desreumaux, P., Colombel, J.F., Matuchansky, C., Cugnenc, H., Lopez-Botet, M., Vivier, E., Moretta, A., Roberts, A.I., Ebert, E.C., Guy-Grand, D., Brousse, N., Schmitz, J. and Cerf-Bensussan, N. (2000). Selective expansion of intraepithelial lymphocytes expressing the HLA-E-specific natural killer receptor CD94 in celiac disease. Gastroenterology **118**, 867–879.

Jarry, A., Cerf-Bensussan, N., Brousse, N., Selz, F. and Guy-Grand, D. (1990). Subsets of CD3+ (T cell receptor alpha/beta or gamma/delta) and CD3− lymphocytes isolated from normal human gut epithelium display phenotypical features different from their counterparts in peripheral blood. Eur J Immunol **20**, 1097–1103.

Kawashima, R., Kawamura, Y.I., Kato, R., Mizutani, N., Toyama-Sorimachi, N. and Dohi, T. (2006). IL-13 receptor alpha2 promotes epithelial cell regeneration from radiation-induced small intestinal injury in mice. Gastroenterology **131**, 130–141.

Kinugasa, T., Sakaguchi, T., Gu, X. and Reinecker, H.C. (2000). Claudins regulate the intestinal barrier in response to immune mediators. Gastroenterology **118**, 1001–1011.

Koopman, L.A., Kopcow, H.D., Rybalov, B., Boyson, J.E., Orange, J.S., Schatz, F., Masch, R., Lockwood, C.J., Schachter, A.D., Park, P.J. and Strominger, J.L. (2003). Human decidual natural killer cells are a unique NK cell subset with immunomodulatory potential. J Exp Med **198**, 1201–1212.

Lan, J.G., Cruickshank, S.M., Singh, J.C., Farrar, M., Lodge, J.P., Felsburg, P.J. and Carding, S.R. (2005). Different cytokine response of primary colonic epithelial cells to commensal bacteria. World J Gastroenterol **11**, 3375–3384.

Leon, F., Roldan, E., Sanchez, L., Camarero, C., Bootello, A and Roy, G. (2003). Human small-intestinal epithelium contains functional natural killer lymphocytes. Gastroenterology **125**, 345–356.

Lopez, M.C. and Holmes, N. (2005). Phenotypical and functional alterations in the mucosal immune system of CD45 exon 9 KO mice. Int Immunol **17**, 15–25.

Lotze, M.T., Zeh, H.J., Rubartelli, A., Sparvero, L.J., Amoscato, A.A., Washburn, N.R., Devera, M.E., Liang, X., Tör, M. and Billiar, T. (2007). The grateful dead: damage-associated molecular pattern molecules and reduction/oxidation regulate immunity. Immunol Rev **220**, 60–81.

Lu, L. and Walker, W.A. (2001). Pathologic and physiologic interactions of bacteria with the gastrointestinal epithelium. Am J Clin Nutr **73**, 1124S–1130S.

Lucas, M., Schachterle, W., Oberle, K., Aichele, P. and Diefenbach, A. (2007). Dendritic cells prime natural killer cells by trans-presenting interleukin 15. Immunity **26**, 503–517.

Luci, C., Reynders, A., Ivanov, I.I., Cognet, C., Chiche, L., Chasson, L., Hardwigsen, J., Anguiano, E., Banchereau, J., Chaussabel, D., Dalod, M., Littman, D.R., Vivier, E. and Tomasello, E. (2009). Influence of the transcription factor RORgammat on the development of NKp46+ cell populations in gut and skin. Nat Immunol **10**, 75–82.

Lundqvist, C., Baranov, V., Hammarstrom, S., Athlin, L. and Hammarstrom, M.L. (1995). Intra-epithelial lymphocytes. Evidence for regional specialization and extrathymic T cell maturation in the human gut epithelium. Int Immunol **7**, 1473–1487.

Lynch, L., O'Donoghue, D., Dean, J., O'Sullivan, J., O'Farrelly, C. and Golden-Mason, L. (2006). Detection and characterization of hemopoietic stem cells in the adult human small intestine. J Immunol **176**, 5199–5204.

Macdonald, T.T. and Monteleone, G. (2005). Immunity, inflammation, and allergy in the gut. Science **307**, 1920–1925.

Martin, S.M., Mehta, I.K., Yokoyama, W.M., Thomas, M.L. and Lorenz, R.G. (2001). Development of intestinal intraepithelial lymphocytes, NK cells, and NK 1.1+ T cells in CD45-deficient mice. J Immunol **166**, 6066–6073.

McDermott, J.R., Humphreys, N.E., Forman, S.P., Donaldson, D.D. and Grencis, R.K. (2005). Intraepithelial NK cell-derived IL-13 induces intestinal pathology associated with nematode infection. J Immunol **175**, 3207–3213.

Mebius, R.E., Rennert, P. and Weissman, I.L. (1997). Developing lymph nodes collect CD4+CD3− LTbeta+ cells that can differentiate to APC, NK cells, and follicular cells but not T or B cells. Immunity **7**, 493–504.

Mehandru, S., Poles, M.A., Tenner-Racz, K., Manuelli, V., Jean-Pierre, P., Manuelli, V., Lopez, P, Shet, A., Low, A., Mohri, H., Boden, D., Racz, P. and Markowitz, M. (2007). Mechanisms of gastrointestinal CD4+ T-cell depletion during acute and early human immunodeficiency virus type 1 infection. J Virol **81**, 599–612.

Mela, C.M., Steel, A., Lindsay, J., Gazzard, B.G., Gotch, F.M. and Goodier, M.R. (2007). Depletion of natural killer cells in the colonic lamina propria of viraemic HIV-1-infected individuals. Aids **21**, 2177–2182.

Melgar, S., Bas, A., Hammarstrom, S., and Hammarstrom, M.L. (2002). Human small intestinal mucosa harbours a small population of cytolytically active CD8+ alphabeta T lymphocytes. Immunology **106**, 476–485.

Melgar, S., Hammarstrom, S., Oberg, A., Danielsson, A and Hammarstrom, M.L. (2004). Cytolytic capabilities of lamina propria and intraepithelial lymphocytes in normal and chronically inflamed human intestine. Scand J Immunol **60**, 167–177.

Meresse, B., Curran, S.A., Ciszewski, C., Orbelyan, G., Setty, M., Bhagat, G., Lee, L., Tretiakova, M., Semrad, C., Kistner, E., Winchester, R.J., Braud, V., Lanier, L.L., Geraghty, D.E., Green, P.H., Guandalini, S. and Jabri, B. (2006). Reprogramming of CTLs into natural killer-like cells in celiac disease. J Exp Med **203**, 1343–1355.

Nakagawa, K., Higuchi, K., Arakawa, T., Kobayashi, K. and Kaneda, K. (2000). Phenotypical and morphological analyses of intraepithelial and lamina propria lymphocytes in normal and regenerating gastric mucosa of rats in comparison with those in intestinal mucosa. Arch Histol Cytol **63**, 159–167.

Okazawa, A., Kanai, T., Nakamaru, K., Sato, T., Inoue, N., Ogata, H., Iwao, Y., Ikeda, M., Kawamura, T., Makita, S., Uraushihara, K., Okamoto, R., Yamazaki, M., Kurimoto, M., Ishii, H., Watanabe, M. and Hibi, T. (2004).

Human intestinal epithelial cell-derived interleukin (IL)-18, along with IL-2, IL-7 and IL-15, is a potent synergistic factor for the proliferation of intraepithelial lymphocytes. Clin Exp Immunol **136**, 269–276.

Pang, G., Buret, A., Batey, R.T., Chen, Q.Y., Couch, L., Cripps, A. and, Clancy R. (1993). Morphological, phenotypic and functional characteristics of a pure population of CD56+ CD16− CD3− large granular lymphocytes generated from human duodenal mucosa. Immunology **79**, 498–505.

Perez-Villar, J.J., Zapata, J.M., Melero, I., Postigo, A., Sanchez-Madrid, E. and Lopez-Botet, M. (1996). Expression and function of alpha 4/beta 7 integrin on human natural killer cells. Immunology **89**, 96–104.

Rakoff-Nahoum, S., Paglino, J., Eslami-Varzaneh, F., Edberg, S. and Medzhitov, R. (2004). Recognition of commensal microflora by toll-like receptors is required for intestinal homeostasis. Cell **118**, 229–241.

Roberts, A.I., O'Connell, S.M., Biancone, L., Brolin, R.E. and Ebert, EC. (1993). Spontaneous cytotoxicity of intestinal intraepithelial lymphocytes: clues to the mechanism. Clin Exp Immunol **94**, 527–532.

Rocha, B., Vassalli, P. and Guy-Grand, D. (1994). Thymic and extrathymic origins of gut intraepithelial lymphocyte populations in mice. J Exp Med **180**, 681–686.

Sakaguchi, T., Brand, S. and Reinecker H.C. (2001). Mucosal barrier and immune mediators. Curr Opin Gastroenterol **17**, 573–577.

Sanos, S.L., Bui, V.L., Mortha, A., Oberle, K., Heners, C., Johner, C. and Diefenbach A. (2009). RORgammat and commensal microflora are required for the differentiation of mucosal interleukin 22-producing NKp46+ cells. Nat Immunol **10**, 83–91.

Satoh-Takayama, N., Vosshenrich, C.A., Lesjean-Pottier, S., Sawa, S., Lochner, M., Mention, J.J., Thiam, K., Cerf-Bensussan, N., Mandelboim, O., Eberl, G. and Di Santo, J.P. (2008). Microbial flora drives interleukin 22 production in intestinal NKp46+ cells that provide innate mucosal immune defense. Immunity **29**, 958–970.

Schacker, T.W., Brenchley, J.M., Beilman, G.J., Reilly, C., Pambuccian, S.E., Taylor, J., Skarda, D., Larson, M., Douek, D.C. and Haase, A.T. (2006). Lymphatic tissue fibrosis is associated with reduced numbers of naive CD4+ T cells in human immunodeficiency virus type 1 infection. Clin Vaccine Immunol **13**, 556–560.

Shanahan, F., Brogan, M. and Targan, S. (1987). Human mucosal cytotoxic effector cells. Gastroenterology **92**, 1951–1957.

Shaykhiev, R. and Bals, R. (2007). Interactions between epithelial cells and leukocytes in immunity and tissue homeostasis. J Leukoc Biol **82**, 1–15.

Shibahara, T., Miyazaki, K., Sato, D., Matsui, H., Yanaka, A., Nakahara, A. and Tanaka, N. (2005). Alteration of intestinal epithelial function by intraepithelial lymphocyte homing. J Gastroenterol **40**, 878–886.

Spencer, J., MacDonald, T.T., Diss, T.C., Walker-Smith, J.A., Ciclitira, P.J. and Isaacson, P.G. (1989). Changes in intraepithelial lymphocyte subpopulations in coeliac disease and enteropathy associated T cell lymphoma (malignant histiocytosis of the intestine). Gut **30**, 339–346.

Tagliabue, A., Befus, A.D., Clark, D.A. and Bienenstock, J. (1982). Characteristics of natural killer cells in the murine intestinal epithelium and lamina propria. J Exp Med **155**, 1785–1796.

Targan, S.R. and Karp, L.C. (2005). Defects in mucosal immunity leading to ulcerative colitis. Immunol Rev **206**, 296–305.

Todd, D., Singh, A.J., Greiner, D.L., Mordes, J.P., Rossini, A.A. and Bortell, R. (1999). A new isolation method for rat intraepithelial lymphocytes. J Immunol Methods **224**, 111–127.

Todd, D.J., Greiner, D.L., Rossini, A.A., Mordes, J.P. and Bortell, R. (2001). An atypical population of NK cells that spontaneously secrete IFN-gamma and IL-4 is present in the intraepithelial lymphoid compartment of the rat. J Immunol **167**, 3600–3609.

Voland, P., Zeitne,r M., Hafsi, N. and Prinz, C. (2006). Human immune response towards recombinant *Helicobacter pylori* urease and cellular fractions. Vaccine **24**, 3832–3839.

Yun, C.H., Lundgren, A., Azem, J., Sjoling, A., Holmgren, J., Svennerholm, A.M. and Lundin, B.S. (2005). Natural killer cells and *Helicobacter pylori* infection: bacterial antigens and interleukin-12 act synergistically to induce gamma interferon production. Infect Immun **73**, 1482–1490.

Zhou, R., Wei, H., Sun, R., Zhang, J. and Tian, Z. (2007). NKG2D recognition mediates Toll-like receptor 3 signaling-induced breakdown of epithelial homeostasis in the small intestines of mice. Proc Natl Acad Sci U S A **104**, 7512–7515.

Chapter Twenty-Six

26

Natural killer cells in the liver

Shiguang Qian, Jia Fan, Shuang-Jian Qiu, Yong-sheng Xiao, Lina Lu

CHAPTER CONTENTS

We can't solve problems by using the same kind of thinking we used when we created them.

Albert Einstein

ABSTRACT

The liver collects gut-derived blood rich in pathogens, bacterial products, toxins and food antigens. To protect the host from pathogenic infection while maintaining a low profile response to antigens from food and gut commensals, the liver maintains a unique repertoire of immune cells, which is particularly rich in natural killer (NK) cells. NK cells play an important role in the regulation of innate immunity and form the first line of defence through a variety of mechanisms, including the release of perforin, granzymes and cytokines, and inducing target cell apoptosis. Many innate immunity components, including liver NK cells, act through recognizing pattern-recognition receptors, the specific structures expressed on invaded pathogens. In addition, liver NK cells may also play a role in preventing cancer, fibrosis, liver regeneration and regulating immune responses. The interaction between NK cells and other immune cells affects the nature and intensity of these functions.

KEY WORDS

Innate immunity, Toll-like receptor, Apoptosis, Cytotoxicity, NKT cell, Infection, Cancer, Fibrosis, Regeneration

Introduction

The liver is comprised of both parenchymal cells (hepatocytes) and nonparenchymal cells. Hepatocytes occupy 80% of the liver volume and fulfil the needs of metabolism, toxin degradation and various protein syntheses. Nonparenchymal cells include endothelial cells, hepatic stellate cells (HpSCs), Kupffer cells, dendritic cells (DCs) and lymphocytes. The liver is the largest solid organ in the body and has dual blood supplies: hepatic artery and portal vein. The portal vein transports nutrients and substances absorbed from the gastrointestinal system into the liver. This renders the liver continuously exposed to bacterial products, toxins and food antigens from the gut.

Innate immunity in the liver represents a first line of defence, which consists of humoral factors (complement and cytokines), phagocytic cells (neutrophils and macrophages) and lymphocytes (natural killer (NK) cells and NKT cells). These innate components can directly kill or help to kill pathogens, usually in a nonspecific manner. However, innate immunity is not absolutely nonspecific. Recent evidence suggests that many innate immunity components act through pattern-recognition receptors (PRR) to recognize the relatively specific structures expressed on invaded pathogens (Janeway and Medzhitov, 2000). PRR are also referred to as primitive pattern recognition receptors, implying that these mechanisms of immune surveillance existed before adaptive immunity. There are three types of PRR: (1) Membrane-bound PRR, such as Toll-like receptors (TLR) (Akira and Takeda, 2004) and the mannose receptor, (2) cytoplasmic PRR, such as recently identified nucleotide binding oligomerization domain (NOD)–like receptors and the retinoic acid-induced gene I (RIG)-like helicases (Meylan et al., 2006), (3) secreted PRR, which are proteins secreted by effector cells, including complement receptors, collectins, and pentraxin proteins (such as serum amyloid and C-reactive protein, lipid transferases and peptidoglycan recognition proteins). For instance, mannan-binding lectin (MBL), an important collectin, binds to a wide range of bacteria, viruses, fungi and protozoa. MBL predominantly recognizes certain sugar groups on the surface of microorganisms but also binds phospholipids, nucleic acids and nonglycosylated proteins.

NK cells, a main component of innate immune cells, are more abundant in the liver than in peripheral lymphoid organs (Gregoire et al., 2007), suggesting that they are functionally important in the liver. Mouse liver lymphocytes contain about 10% NK cells, whereas rat and human liver lymphocytes contain about 30–50% NK cells. Liver NK cells were first described as 'pit cells' in the rat liver. 'Pit' means the seed of a fruit, in reference to the spherical, dense granules in the cytoplasm of the large granular lymphocytes (LGL) with dense granules and rod-cored vesicles in the liver (Luo et al., 2000). Their function was unclear, however, until it was revealed that they exhibited NK cell-mediated cytotoxicity against tumour cells (Bouwens et al., 1987; Kaneda et al., 1983), and that cancerous liver metastasis was suppressed by augmenting liver NK cell activity using various approaches (Wiltrout et al., 1984, 1985).

Location of NK cells in the liver

Liver NK cells are mainly located in the lumen of the hepatic sinusoid. Compared to T and B cells, they are more firmly adherent to the sinusoidal wall, possessing well-developed pseudopodia (Kaneda and Wake, 1983; Wiltrout, 2000). However, they do not attach to the sinusoidal wall as firmly as Kupffer cells and, therefore, it is relatively easy to obtain liver NK cells by portal vein perfusion with enhanced pressure (Bouwens et al., 1987). IL-12 and IL-18 produced by Kupffer cells promote proliferation of NK cells and regulate NK cell differentiation. NK cells can also be activated by the ligand of NKG2D expressed on Kupffer cells (Hamerman et al., 2004; Nedvetzki et al., 2007). Under some circumstances, IL-10 derived from macrophages inhibits the activation of NK cells (Chiu et al., 2008). Kupffer cells can also induce apoptosis of liver NK cells and NKT cells in acute viral hepatitis of mice (Jacques et al., 2008). CXCL16, the ligand for CXCR6, is expressed by liver sinusoidal endothelial cells (LSEC) and can enrich CXCR6$^+$ NKT in the liver (Germanov et al., 2008). Different ligands influence the polarity of NKT cells (Zigmond et al., 2008). Some liver NK cells are also located in the liver parenchyma between hepatocytes (Bouwens et al., 1987; Hata et al., 1990). Studies in mice have shown an age-dependent increase in the number of liver NK cells (Itoh et al., 1988).

Development of liver NK cells

NK cell development represents a linear differentiation from haematopoietic stem cells in the bone marrow to fully functional NK cells in the periphery tissues (blood, spleen, liver, thymus, lung, gut and lymph nodes). They are divided into precursors, immature NK cells and mature NK cells. Immature NK cells can develop to become mature NK cells with full competence. The liver, spleen, lung and blood are the main peripheral sites where mature NK cells are located. In these peripheral sites, mature NK cells can be further distinguished by their activation state into the resting and activated cells. Resting NK cells are those lacking signs of recent activation, whereas activated NK cells are larger, bear markers of recent stimulation (such as CD69) and can demonstrate potent functional properties (Huntington et al., 2007). All NK cells can leave the site of their generation and reside in other tissues. GATA-3 has been shown to promote liver-specific homing of NK cells (Samson et al., 2003).

Are NK cell precursors and immature NK cells unique to the bone marrow? Data from mice suggest that this may not be the case. Immature NK cells can be identified in other sites, including a prominent population in the perinatal liver and a small but significant population in the spleen (Takeda et al., 2005; Wiltrout, 2000). Human NK cell precursors are also found in the lymph nodes (Freud et al., 2005, 2006), suggesting that multiple sites can support NK cell differentiation. NK cell precursors are originally generated in bone marrow. Once generated, NK cell precursors and immature NK cells have access to the circulation. The complex process of NK cell differentiation can occur at many tissue sites—including the bone marrow, liver, thymus, spleen and lymph nodes—and circulate at different stages of maturation among these sites (Huntington et al., 2007).

Morphologically, liver NK cells can be divided into two populations according to the density or size of the granules. The low-density cells have smaller granules.

The high-density cells contain larger granules and resemble the NK cells in peripheral blood (Vanderkerken et al., 1993; Wisse et al., 1976). It is believed that liver NK cells come from the circulation, adhere to the liver sinusoidal wall and subsequently differentiate into the low-density cells (Vanderkerken et al., 1993). This is supported by an experiment in which NK leukaemia cells are injected intraperitoneally in rats. The cells metastasize to the liver and exhibit more rod-cored vesicles than those that metastasize to the spleen, suggesting that the microenvironment of the liver plays a role in the differentiation of peripheral blood NK cells into liver NK cells (Kaneda et al., 1996).

Liver inflammation facilitates NK cell recruitment into the liver. In a mouse model of Con A-induced hepatitis, severe hepatitis was seen in CCR5-deficient (KO) mice associated with increased liver NK cell recruitment and enhanced hepatic expression of CCL5. NK cell depletion ameliorated severe hepatitis in CCR5 KO mice but did not alter hepatitis in wild-type mice. This suggests that CCL5 acting via CCR1 drives the recruitment of liver NK cells (Ajuebor et al., 2007). In addition to the influx from the peripheral blood, liver NK cells can also be generated by proliferation, either in the normal or pathological liver.

NKT cells in the liver

Analysis using markers for T cells and NK cells has shown that the mouse liver contains two unique and abundant extrathymically differentiated T cell populations. The first is T cell receptor (TCR) intermediate- and IL-2 receptor-β-positive (IL-2Rβ$^+$) T cells termed TCR-intermediate (TCRint$^+$) T cells (Ohteki et al., 1990). The second population is CD3$^+$ NK1.1$^+$ cells termed NKT cells (Hashimoto et al., 1995; Meylan et al., 2006). Both populations were found to differentiate in the liver (Abo, 2001; Watanabe et al., 1995). They function in the front line defence system against tumours and microbes together with NK cells before the initiation of specific immune reactions of T cells and B cells (Ohtsuka et al., 1995; Tabata et al., 1996). Liver TCRint$^+$ T cells and NKT cells, similar to NK cells, have the morphology of LGL. The granules contain perforin but are smaller in size and number compared to those of NK cells. NKT cells also possess vesicles, but the rod structure within them is obscure. Because of this, it is difficult to distinguish NKT cells from NK cells morphologically.

Liver NKT cells are predominantly CD4$^-$CD8$^-$ or CD4$^+$CD8$^-$ (Abo, 2001). Most mouse NKT cells use a limited set of TCR, that is Vα14-Jα281/Vβ7 or Vβ8 (Wiltrout et al., 1984). They are activated by alphagalactosylceramide (αGalCer), a component of the major histocompatibility complex (MHC) class I-like molecule CD1d expressed by APC (Brossay et al., 1998). In humans, Vα24-JαQ/Vβ11$^+$ NKT cells are the counterpart of mouse Vα14-Jα281$^+$ NKT cells (Kawano et al., 1999; Spada et al., 1998). These Vα24$^+$ NKT cells, however, account for less than 1% of total liver lymphocytes, whereas mouse Vα14$^+$ NKT cells represent approximately 20% (Kawarabayashi et al., 2000). The presence of NKT cells has been controversial in the liver of some species, such as rats (Nakatani et al., 2004).

NKT cells are activated by IL-12 and αGalCer and suppress the initiation, growth and metastasis of tumours (Nakagawa et al., 2001). Compared to NK cells, they constitutively express higher levels of IL-12 receptors; therefore, they produce larger amounts of perforin and IFN-γ by IL-12 stimulation and exert antitumour activities with a lower dose of IL-12.

Cytotoxicity of liver NK cells

Liver NK cells have been shown to fight pathogens and cancers by killing targets through several mechanisms. One mechanism is cell lysis by release of cytoplasmic granules that contain perforin and granzymes. Liver NK cells employ this mechanism to eliminate virus-infected or transformed cells. After firm conjugation with target cells, they release granules into the intercellular space. Perforin in the granules makes pores in the plasma membrane of target cells causing them to rupture. Granzymes then enter the target cells through the pores elaborated by perforin to trigger DNA fragmentation (Shresta et al., 1995; Smyth et al., 2002; Vermijlen et al., 1999).

A second mechanism is induction of death by receptor-mediated apoptosis. Liver NK cells express high levels of tumor necrosis factor (TNF) family ligands, such as Fas (CD95) ligand (FasL) and TNF-related apoptosis-inducing ligand (TRAIL). Both molecules can induce target cell apoptosis. Fas is expressed on some tumour cells and can be upregulated by NK cells (Screpanti et al., 2001; Smyth et al., 2002). A report using mouse NK cell clones that lack perforin-mediated cytolysis demonstrated that the Fas and FasL ligation was effective in NK cell-mediated cytotoxicity against a Fas-expressing tumour cell line but not against a mutant Fas-expressing tumour. This indicates that Fas-mediated tumour cell lysis can be effective without the engagement of perforin (Tsutsui et al., 1996). TRAIL is a type II transmembrane ligand sharing homology with FasL (Smyth et al., 2001). By binding the TRAIL receptors, cells begin to aggregate death domains and recruit caspase 8 or 10, followed by activation of the caspase cascade (Kayagaki et al., 1999; Srivastava 2001). It has been suggested that immature and mature NK cells differentially use these mechanisms. Immature CD161$^+$/CD56$^-$ NK cells mediate TRAIL-dependent cytotoxicity but not FasL-dependent and granule release-dependent cytotoxicities, whereas mature CD56$^+$ NK cells use all of these mechanisms (Zamai et al., 1998). In mice, TRAIL is constitutively expressed by NK cells. This molecule plays a substantial role in suppressing tumour metastasis, as administration of a neutralizing antibody against TRAIL significantly increases liver metastases of several TRAIL-sensitive tumour cell lines (Seki et al., 2003; Smyth et al., 2001; Takeda et al., 2001). It has been shown that TRAIL is the dominant cytotoxic effector molecule expressed by NK cells in fetal mice (Takeda et al., 2005). Interestingly, a phenotypically immature TRAIL$^+$ NK cell subpopulation is retained in adult mouse liver. Adoptive transfer of these TRAIL$^+$ NK cells either from adult or neonatal liver results in the generation of mature TRAIL$^-$ NK cells, suggesting that constitutive TRAIL expression may be a marker for immature NK cells with cytotoxicity. Mature NK cells develop full functional capacity, including the secretion of IFN-γ and IL-13, and perforin-mediated and Fas ligand-mediated cytotoxicity (Takeda et al., 2005).

Liver NK cells also affect the function of other immune cells, such as T cells and macrophages. NK cells secrete various cytokines such as IFN-γ, TNF-α, IL-5, IL-10 and IL-13 (Smyth et al., 2002). Among them, IFN-γ is the most important in suppressing tumour metastasis and viral infection by various approaches, including stimulating antigen presentation, regulating cell proliferation and sensitivity to apoptosis, enhancing phagocytosis and inhibiting angiogenesis (Smyth et al., 2002; Street et al., 2001). IFN-γ is also critical in inducing TRAIL expression in NK cells since NK cells from IFN-γ-deficient mice do not express TRAIL (Smyth et al., 2001). IFN-γ-induced TRAIL expression in NK cells is important in the effects of IL-12 and αGalCer because the antitumour function of IL-12 and αGalCer is enhanced by stimulation by IFN-γ secreted from NK cells (Hayakawa et al., 2001, 2002; Nakatani et al., 2004; Yao et al., 1999). Response of TRAIL expression to stimulation on liver NK cells appears different from blood NK cells. Thus, IL-2 upregulated TRAIL expression was significant on liver NK cells, but this was not seen in blood NK cells (Ishiyama et al., 2006).

NK cells and viral infection in the liver

Chronic viral hepatitis, hepatitis B and hepatitis C are common infectious diseases closely related to cirrhosis and hepatocellular carcinoma (HCC). Hepatitis B virus (HBV) and hepatitis C virus (HCV) infection have strict species- and histocompatibility-specificity. The pathogenesis of viral hepatitis and the role of NK cell and NKT cells in viral hepatitis have not been fully elucidated. The initial studies on peripheral blood mononuclear cells in hepatitis and liver cirrhosis patients with different causes showed that the number and activity of NK cells varied. In humans, NK cells in blood began to increase in the incubation phase of HBV infection, indicating that NK cells contributed to restraining the virus in the early phase (Webster et al., 2000). Liver CD56$^+$ NK cells isolated from patients with HBV exacerbation enhanced their response to cytokine stimulation. TRAIL expression by NK cells in response to IFN-γ and IL-8 was increased in HBV patients. Meanwhile, the expression of TRAIL receptor 2 by hepatocytes was upregulated, thus enhancing their sensitivity to NK cells (Dunn et al., 2007). However, decreased numbers, activity and toxicity of NK cells were seen in mice containing full-length expression of the HBV genome (Chen et al., 2005).

In HCV infection, NK cells were shown to inhibit the expression of HCV RNA (Li et al., 2004) and the persistence of HCV in peripheral blood was associated with a defect in liver NK cell activity (Golden-Mason and Rosen, 2006). Antitumour activity and IFN-γ secretion by liver CD56$^+$ cells from HCV cirrhotic patients with stimulation of IL-2 were significantly decreased (Kawarabayashi et al., 2000; Morishima et al., 2006). The success of IFN-α treatment for patients with chronic HCV infection was associated with the function recovery of NK cells (Yamagiwa et al., 2008). These data suggest that NK cells are important effector cells for eliminating virus at least in the early phase of hepatitis virus infection.

Few studies of NKT cells in viral hepatitis have been performed. In transgenic mice containing an integrated HBV genome, a single injection of α-GalCer increased intrahepatic IFN-α, IFN-β and IFN-γ levels associated with reduction of HBV replication (Kakimi et al., 2000). The data on the amount of NKT cells in HCV infection are in conflict. Some studies reported that NKT cells were lower in patients with chronic HCV infection. Other studies demonstrated that the number of intrahepatic NKT cells was comparable to healthy donors, but that surface inhibitory receptor (NKG2A/CD94) expression on NKT cells significantly increased. After the activation of NKG2A/CD94, a high level of IL-13 was generated, but IFN-γ remained unchanged. This suggests that NKT cells shift towards a Th2 response in HCV patients (Inoue et al., 2006; Kawarabayashi et al., 2000; Morishima et al., 2006).

Adenovirus causes a wide range of human illnesses most commonly affecting the respiratory and gastrointestinal systems. Systemic exposure to adenovirus results in viral hepatitis marked by an intense mononuclear cell infiltrate and an increase in serum liver enzymes. Clearance of adenovirus is mediated by a CD8 T cell response, the magnitude of which depends on the efficiency of the innate immune response (Burt et al., 2008). Liver NK cells become activated and produce IFN-γ within hours of adenovirus infection, which is important for an adenovirus-specific T-cell response (Liu et al., 2000). Depletion of NK cells prior to adenovirus infection results in a decreased number and activity of virus-specific CD8 T cells in the liver. However, these responses are not seen in the spleen (Peng et al., 2001). NK cells are the most common IFN-γ-producing cells in the mouse liver during the peak response to adenovirus infection. The majority of IFN-γ-producing liver NK cells expressed CD11c. The study showed that $CD11c^+$ NK cells also account for the majority of NK cells in the normal liver. Both $CD11c^-$ and $CD11c^+$ liver NK cells increased in number and demonstrated increased lytic function. However, only $CD11c^+$ liver NK cells upregulated expression of MHC class II and costimulatory molecules and gained some degree of APC function, suggesting that $CD11c^+$ NK cells are a subset of NK cells in the mouse liver that contribute to the response to adenoviral hepatitis (Liu et al., 2000).

In general, NK cells mediate two important functions in the liver: they induce cell death in the infected organ, and they concomitantly stimulate the induction of T cell-mediated immunity by release of IFN-γ (Liu et al., 2000). Several polyribonucleotides have already been adopted in clinical trials for the treatment of cancer and viral diseases. Polyinosinic-polycytidylic acid (poly I:C) is an artificial mimic of viral RNA, which triggers the immune response resembling viral infection (Bigger et al., 1998). Reports in mice have demonstrated that poly I:C potently augments NK cell activity and accumulation in the liver, suggesting that this molecule is an excellent augmentor of liver NK cell activity (Twilley et al., 1987). A recent report demonstrated that poly I:C induced the migration of NK cells from spleen into liver. This approach may be used to promote host immune responses against viral infection (Wang et al., 2005).

NK cells in liver cancer

Primary liver cancer is the fifth most common cancer worldwide, with a particularly high prevalence in Asia, of which HCC accounts for the majority. The incidence of HCC is increasing in the US and other Western countries (Parkin et al., 2005). Despite a variety of emerging therapeutic approaches, recurrence or metastasis is common in HCC patients. For advanced HCC, there remains a lack of effective treatment options. Conventional chemotherapy has a dismal outcome due to intrinsic drug resistance. Therefore, development of novel immune or molecular therapies is absolutely necessary.

Research into the development of new therapies has been promising. The combination of daunorubicin and γδTCR monoclonal antibody enhanced antitumour activity of cytotoxic T lymphocyte and NK cells in HCC-bearing mice (Seo et al., 1999). Administration of IL-12 in combination with IL-2 resulted in a striking increase in IFN-γ and TNF-α production by NK cells (Sawayama et al., 2003). In HCC, elevated plasma TGF-β1 concentrations suppressed the activities of NK cells and predicated a poor outcome (Okumoto et al., 2004). This was supported by a recent study showing that leptin induced the proliferation and activity of NK cells to inhibit the growth of HCC in vitro and in vivo (Elinav et al., 2006). It is well known that chronic HBV infection facilitates development of HCC. In mice expressing HBV transgene with high incidence of HCC, NK cell number and activity were markedly reduced in the liver. This was associated with downregulation of TRAIL on the surface of liver NK cells. Downregulation of antitumour activity of liver NK cells may contribute to the high incidence of HCC in those mice (Chen et al., 2005). NK cell-induced apoptotic killing of HCC line Hep3B was mediated by ligation of TRAIL receptors (Kim et al., 2004). HCC patients display a dramatic reduction in peripheral blood $CD56^{dim}$ $CD16^+$ NK cells subsets compared with healthy subjects, as well as in tumoural and peritumoural tissues. Both the peripheral and tumour-infiltrating NK cells exhibit poorer capacity for IFN-γ production and target cell elimination (Cai et al., 2008).

Although the underlying mechanism remains unclear, data have shown a significant increase in intratumoural $CD4^+CD25^+$ Treg and tight association between intratumoural Treg and HCC aggressiveness. The imbalance of Treg and activated cytotoxic T lymphocytes is an independent predictor of overall survival and disease-free survival in HCC patients following curative resection. The addition of Treg from HCC patients significantly inhibited the antitumour activity of autologous NK cells in vitro (Ghiringhelli et al., 2006). The number of NK cells in the cancerous tissues of stage I and II was significantly higher than that of stage III and IV. NK cell number in HCC patients free of metastasis at 15 months was significantly higher than that in patients with metastatic disease. The number declined in those patients with tumour progression. These results suggested that the number of NK cells was also a predictor for HCC prognosis. The peri-operative changes in NK cell activity may be a useful factor for predicting the prognosis and tumour relapse of HCC patients following surgery (Chen et al., 2006; Taketomi et al., 1998). These results suggest that Treg may act through inhibition of NK cell functions (Fu et al., 2007; Gao et al., 2007).

The adoptive transfer of NK cells extracted from liver perfusates of poly I:C-stimulated B6 mice inhibited the growth of liver metastasis in a mouse model (Ohira et al, 2006). Randomized clinical trials demonstrated that the use of cytokine-induced killer (CIK) cells as adjuvant immunotherapy efficiently prevented tumour recurrence and metastasis (Hui et al., 2009). These findings indicate that adoptive immunotherapy using activated NK cells may be a novel approach to preventing tumour recurrence after hepatectomy of HCC patients. In humans, NKG2D binds to stress-inducible molecules MHC class I chain-related (MIC) A and B, which upregulate effects of NK cells on tumour cells and virus-infected cells. MICA expressed on the surface of various tumour cells suggests that MICA may be a signal in antitumour innate immune surveillance mediated by NK cells. Sodium valproate enhances expression of MICA and MICB in HCC cells and stimulates NK cells to recognize and lyse these HCC cells via NKG2D (Armeanu et al., 2005). There could be an immune escape mechanism in advanced HCC whereby MICA downregulates expression of NKG2D and inhibits NKG2D-mediated responses of NK cells (Jinushi et al., 2005). In addition, RAET1E2, a soluble isoform of the UL16-binding protein, inhibits the cytotoxicity of NK cells in vitro by downregulating NKG2D expression on NK cells (Cao et al., 2007). Indoleamine 2,3 dioxygenase (IDO) also plays a key role in immune responses against HCC by regulating NK cell cytotoxicity (Kai et al., 2003a,b). IDO mRNA expressed in NK cells is correlated significantly with expression of IFN-γ, TNF-α and IL-1β, while activity of NK cells is reduced with IDO inhibition in a dose-dependent manner (Ishio et al., 2004). Enhancement of NK cell activity by these molecules could promote the development of HCC immunotherapy.

Although the amount of NKT cells in human liver is less than that in mouse liver, the human liver is still the primary organ for settlement of NKT cells. The growth of the BNL HCC cell line in mouse liver was completely inhibited after α-GalCer administration, suggesting the critical involvement of NKT cells (Miyagi et al., 2003). CCL20 and CXCL9 expression in HCC attracts not only T cells but also NK cell and NKT cells which express CCR6 and CXCR3, respectively, acting as receptors for CCL20 and CXCL9 prior to location in the tumour site (Liu et al., 2005). Adoptive transfer of NKT cells exposed ex vivo to HCC-derived antigens loaded on DCs inhibited the growth of mouse HCC in vivo and enhanced expression of STAT4 and serum levels of IFN-γ and IL-12 (Margalit et al., 2005). Adoptive immunotherapy with α-galactosylceramide (KRN7000) or α-GalCer loaded on DCs or DC-activated NKT cells provide new approaches for immunotherapy of primary liver cancer.

The liver is also a frequent target organ for tumour metastasis, especially those of gastrointestinal origin. Several studies showed that α-GalCer inhibited metastatic tumour growth in the liver from other sites, including melanoma, lymphoma and colon cancer (Hakagawa et al., 2001; Hayakawa et al., 2002; Nakagawa et al., 1998). A complete inhibition of B16 melanoma metastasis in the liver was achieved by injection of α-GalCer–pulsed DCs even after formation of multiple metastatic nodules, indicating that NKT cells could control tumour cell proliferation even at a metastatic stage (Toura et al., 1999). Administration of α-GalCer stimulator KRN7000 combined with angiogenesis inhibitor TNP470 significantly reduced the number of liver metastatic foci of pancreatic cancer and prolonged long-term survival (Matsumoto et al., 2003). NK cells adoptively transferred to the liver via the intra-arterial route had preferential access and substantial accumulation to the tumour site in patients with liver metastasis of colon carcinoma (Matera et al., 2006). In a neuroblastoma liver metastasis model, enhanced NK cell activity was correlated with a decrease in liver metastasis (Iinuma et al., 2006). IFN-α2b decreased liver micrometastasis and increased survival time through increased intrinsic liver NK cell-mediated tumour apoptosis in a mouse model of metastatic ocular melanoma (Yang et al., 2004). These data demonstrate that NK cell and NKT cells also play a key role in elimination of secondary liver cancer.

Liver NK cells and hepatic tolerance

The liver has been considered an immunologically privileged organ for a long time (Racanelli and Reherman et al., 2006) since liver allografts are spontaneously accepted in many species without any immunosuppressive agents (Crispe, 2003). In clinical liver transplantation, some liver transplant patients can keep their grafts with good function without any immunosuppressant (Starzl et al., 1993). Based on studies in multiple countries, it is generally estimated that 20% of liver transplanted patients may be successfully withdrawn from immunosuppression (Seyfert-Margolis and Turka, 2008). It is not surprising that the immune response in the liver is modulated because the liver is continuously exposed to nonpathogenic antigens from food and to toxins from gut-derived pathogens and commensal bacteria (such as lipopoly saccharide (LPS)). The liver has to protect the host from gut-derived pathogens but be tolerant of antigenic challenges from food and commensal sources. LPS is a powerful stimulus for innate immunity through TLR ligation and activated professional APC, while TLR are expressed on macrophages and DCs that can directly stimulate liver NK cells. Therefore, the liver seems to be well equipped to promote immunity to pathogens but to avoid responding to harmless food antigens in the context of TLR ligation. It has been shown that the production of IL-10 by the Kupffer cells is one of the mechanisms of modulating the immune response to the proinflammatory cytokines (IL-12, IL-15 and IL-18, etc.) (Randow et al., 1995; Santucci et al., 1996). This raises the question of how the liver mounts an appropriate response to pathogenic antigens. A recent study of the interaction between human Kupffer cells and liver NK cells reveals that Kupffer cells respond to TLR ligands and activate liver NK cells via cell–cell contact. Kupffer cells are able to produce the NK cell-activating signal, IL-18, to stimulate NK cells. On the other hand, Kupffer cells can also secrete the NK cell inhibiting signal, IL-10, to subvert innate immune responses that may contribute to viral persistence (HBV and HCV) (Zhengkun et al., 2008).

NKT cells have also been shown to participate in immunological tolerance to transplants via cytokines, such as IFN-γ, which induces immunological tolerance in a heart allograft model (Hoerbelt et al., 2008). NKT cells are necessary in the initial stage of immunological tolerance induced by cyclophosphamide. Recipient NKT cells, but not donor NKT cells, were dominantly required for the induction of allograft tolerance (Iwai et al., 2007; Onzuka et al., 2008). Liver allografts lacking NKT cells (CD1d$^{-/-}$ mice) manifest not only infiltration but also haemorrhage and necrosis, with a significant reduction of graft survival and much less induction of tolerance compared with wild type liver allografts (Morita et al., 2007). NKT cells are also crucial in the induction of immunological tolerance to allogeneic islet cell transplants (Yang et al., 2007).

NK cells in liver fibrosis

Hepatic fibrosis is an outcome of many chronic liver diseases, such as alcoholism, viral and autoimmune hepatitis and fatty liver disease. Many cellular components and mediators contribute to this process, among which hepatic stellate cell (HpSCs) play a critical role. Normally, HpSCs are in a quiescent state, but they can be activated by prolonged liver injury and trigger fibrogenesis. Many studies suggest that fibrosis is reversible if damaged hepatocytes can be completely replaced by healthy ones.

There is evidence that liver NK cells can prevent fibrosis. One study showed the aggravation of fibrotic activity when the NK cells of mice were neutralized by anti-asialo-GM1 antibody. Another study showed worsening hepatic fibrosis secondary to the dysfunction of NK cells (Poynard et al., 2003). On the other hand, Treg can downregulate NKG2D expression of NK cells and accelerate fibrogenesis (Ghiringhelli et al., 2005). The ligand of TLR3, poly I:C can attenuate this process by activating NK cells (Radaeva et al., 2006). Clinical data have demonstrated that patients treated with immunosuppressants are prone to severe fibrosis. This might be attributed to suppressing the activity of NK cells by these agents (Wang et al., 2007).

The antifibrotic effect of NK cells is closely associated with their inhibitory effect on HpSCs by inducing apoptosis. Compared to quiescent cells, activated HpSCs are more sensitive to apoptosis induced by NK cells because of the existence of the ligand to NKG2D, Rae 1, which is absent in quiescent HSC. In an in vivo model of carbon tetrachloride (CCl_4)-induced liver fibrosis of wild-type mice, activated HpSCs display lower levels of MHC-I and also become more fragile to NK cells. Furthermore, NK cell-induced activated HpSCs apoptosis may be strengthened by poly I:C or IFN-γ because each is able to increase NKG2D and TRAIL expression by liver NK cells (Radaeva et al., 2006). It is possible that high expression of Rae 1 turns off the release of TGF-β from activated HpSCs, while TGF-β can suppress NK killing and NK cell cytokine production, as well as the expression of activating NK cell receptors such as NKp30 and NKG2D (Trotta et al., 2008; Wahl, 2007).

Poly I:C or IFN-γ treatment inhibits liver fibrosis in wild-type mice but not in STAT1$^{-/-}$ mice, suggesting that poly I:C or IFN-γ inhibition of liver fibrosis is STAT1-dependent. This effect was accompanied by the activation of the TGF-β/Smad3 signalling pathway and the inhibition of Smad7 (Jeong et al, 2006). The antifibrotic activity of NK cells and poly I:C or IFN-γ treatment was diminished in alcohol-fed mice. HpSCs from ethanol-fed mice were resistant to IFN-γ-induced cell cycle arrest and apoptosis compared with control mice. Such resistance was due to diminished IFN-γ activation of STAT1 in HpSCs from ethanol-fed mice and could be reversed by TGF-β1 neutralizing antibody (Jeong et al., 2008), confirming that STAT1 signalling pathway is critical for NK cells to suppress the activation of HpSCs and the process of hepatic fibrosis.

A recent study showed that HpSCs modulated the lymphocyte-mediated response in hepatic fibrogenesis by phagocytosis of lymphocytes. In patients with advanced liver fibrosis, only PBMCs from HCV patients were phagocytozed by HpSCs in HpSCs and PBMC coculture, which was mediated by the Rac1and Cdc42 gene family (Muhanna et al., 2008). By contrast, senescence of activated HpSCs limited the progress of liver fibrosis because NK cells preferentially killed senescent-activated HpSCs in vitro and in vivo (Krizhanovsky et al., 2008). Accumulating evidence supports an antifibrotic role of NK cells in the liver through enhanced killing of activated HpSCs.

Do NKT cells play a role in liver fibrosis? One study showed that NKT cells increased in CCL_4-induced liver fibrosis, and administration of β-glucosylceramide had a significant antifibrotic effect associated with a decreased number of intrahepatic NKT cells and CD8$^+$ lymphocytes (Safadi et al., 2007). NKT cells increased in chronically virus-infected livers and underwent a substantial modification in their effector functions, consisting of the production of the type 2 profibrotic cytokines IL-4 and IL-13, which characterize the progression of hepatic fibrosis to cirrhosis. CD1d, nearly undetectable in noncirrhotic and control livers, is strongly expressed by APC in cirrhotic livers (de Lalla et al., 2004). These data suggest that NKT cells may participate in the progress of fibrosis.

Due to rapid downregulation of surface markers and apoptosis of activated NKT cells, little is known about the role of NKT cells in liver fibrosis, and the assumption that NKT cells promote the progress of liver fibrosis needs further investigation (Notas et al., 2009).

NK cells and liver regeneration

In contrast to the well-defined role of NK cells in pathophysiologic processes, little is known about their role in regulating normal cell growth. The available data do not provide information about the exact role of NK cells in liver regeneration or explain how liver regeneration can occur in the presence of cytotoxic NK cells that are capable of killing regenerating liver hepatocytes (Itoh et al., 1988). One hypothesis to explain how liver regeneration might occur in the presence of liver-resident NK cells is that soon after liver injury, the effector functions of liver NK cells might become temporarily suppressed, that is they become noncytotoxic for regenerating liver hepatocytes. In a study in rats, a prompt and profound suppression of NK cell functions, including killing of regenerating liver hepatocytes, occurs after a 70% partial hepatectomy. This immunosuppression is restricted to NK cells in the regenerating liver. However, the susceptibility of regenerating liver hepatocytes to lysis by liver NK cells is transient because it is limited to the period of rapid liver growth immediately after partial hepatectomy. In vivo elimination of NK cells by a specific mAb results in augmented liver regeneration 7 days after partial hepatectomy (Vujanovic et al., 1995). This effect might be mediated by regulation of the cytokines, such as IL-2 produced by NK cell and NKT cells. Immunosuppressive drugs, such as cyclosporine A (CsA) or FK506, have been shown to augment hepatocyte proliferation in regenerating mouse liver. CsA or FK506 are potent IL-2 suppressors (Francavilla et al., 1991; Tanaka et al., 1993). It has been found in mice that a decrease of normal liver mass follows IL-2 infusions (Lafreniere et al., 1990) and that inhibition of hepatocyte proliferation in the regenerating liver takes place after IL-2 or LAK cell infusions (Tanaka et al., 1993). These observations indicate that liver NK cells have a potential role in regulation of liver regeneration.

Liver NK cells as a bridge between innate and adaptive immunity

Liver NK cells bridge innate immune immunity and adaptive immunity by regulating the number and function of other immune cells. Early research demonstrated that cytokines derived from NK cells and NKT cells were able to activate many cells, including $CD4^+$ and $CD8^+$ T cells, macrophages and B cells (Kronenberg and Gapin, 2002). IFN-γ produced by NK cells can recruit DCs and neutrophils (Nakamatsu et al., 2007). On the other hand, NK cells can directly kill immature DCs (Laffont et al., 2008; Moretta et al., 2003; Zitvogel 2002). There are also interactions between NK cells/NKT cells and Treg. Thus, activated NKT cells can regulate the function of Treg through the secretion of IL-2, while Treg can also inhibit the function of NKT cells through cell–cell contact (La Cava et al., 2006). Besides influencing the function of DCs, NK cells can also affect B- and T lymphocytes. NK cells can promote the priming of $CD4^+$ Th1 cells through the secretion of IFN-γ (Ito et al., 2008). NK cells can also eliminate activated T cells if they do not express sufficient classical or non-classical MHC-1 molecules. Anti-NKG2A Ab enhances NK cell ability to kill activated $CD4^+$ T cells by blocking the inhibitory receptors on the surface of NK cells, preventing $CD4^+$ T cell-dependent autoimmune reaction (Lu et al., 2007). In a mouse HBsAg immunization model, the levels of anti-HBV Ab and the number of HBsAg-specific lymphocytes are significantly lower in NK cell-deficient mice than controls, due to impaired DC function. This suggests that the interaction between NK cells and DCs can regulate the intensity of adaptive immunity (Yoshida et al., 2008).

References

Abo, T. (2001). Extrathymic pathways of T cell differentiation. Arch Immunol Ther Exp **49**, 81–90.

Ajuebor, M.N., Wondimu, Z., Hogaboam, C.M., Le, T., Proudfoot, A.E.I. and Swain, M.G. (2007). CCR5 deficiency drives enhanced natural killer cell trafficking to and activation within the liver in murine T cell-mediated hepatitis. Am J Pathol **170**, 1975–1988.

Akira, S. and Takeda, K. (2004). Toll-like receptor signalling. Nat Rev Immunol **4**, 499–511.

Armeanu, S., Bitzer, M., Lauer, U.M., Venturelli, S., Pathil, A., Krusch, M., Kaiser, S., Jobst, J., Smirnow, I., Wagner, A., Steinle, A. and Salih, H.R. (2005). Natural killer cell-mediated lysis of hepatoma cells via specific induction of NKG2D ligands by the histone deacetylase inhibitor sodium valproate. Cancer Res **65**, 6321–6329.

Bigger , J.E., Thomas, C.A. III and Atherton, S.S. (1998). NK cell modulation of murine cytomegalovirus retinitis. J Immunol **60**, 5826–5831.

Bouwens, L., Remels, L., Baekeland, M., Van Bossuyt, H. and Wisse, E. (1987). Large granular lymphocytes or "pit cells" from rat liver: isolation, ultrastructural characterization and natural killer activity. Eur J Immunol **17**, 37–42.

Brossay, L., Chioda, M., Burdin, N., Koezuka, Y., Casorati, G., Dellabona, P. and Kronenberg, M. (1998). CD1d-mediated recognition of an alpha-galactosylceramide by natural killer T cells is highly conserved through mammalian evolution. J Exp Med **188**, 1521–1528.

Burt, b.M., Plitas, G., Stableford, J.A., Nguyen, H.M., Bamboat, Z.M., Pillarisetty, V.G. and DeMatteo, R.P. (2008). CD11c identifies a subset of murine liver natural killer cells that responds to adenoviral hepatitis. J Leukoc Biol **84**, 1039–1046.

Cai, L., Zhang, Z., Zhou, L., Wang, H., Fu, J., Zhang, S., Shi, M., Zhang, H., Yang, Y., Wu, H., Tien, P. and Wang, F.S. (2008). Functional impairment in circulating and intrahepatic NK cells and relative mechanism in hepatocellular carcinoma patients. Clin Immunol **129**, 428–437.

Cao, W., Xi, X., Hao, Z., Li, W., Kong, Y., Cui, L., Ma, C., Ba, D. and He, W. (2007). RAET1E2, a soluble isoform of the UL16-binding protein RAET1E produced by tumor cells, inhibits NKG2D-mediated NK cytotoxicity. J Biol Chem **282**, 18922–18928.

Chen, Y., Wei, H., Sun, R. and Tian, Z. (2005). Impaired function of hepatic natural killer cells from murine chronic HBsAg carriers. Int Immunopharmacol **5**, 1839–1852.

Chen, G., Luo, D.Z., Liu, L., Feng, Z.B., Guo, F. and Li, P. (2006). Micro-environmental immune status in hepatocellular carcinoma and cirrhotic tissues. West Indian Med J **55**, 403–408.

Chiu, B.C., Stolberg, V.R. and Chensue, S.W. (2008). Mononuclear phagocyte-derived IL-10 suppresses the innate IL-12/IFN-gamma axis in lung-challenged aged mice. J Immunol **181**, 3156–3166.

Crispe, I.N. (2003). Hepatic T cells and liver tolerance. Nat Rev Immunol **3**, 51–62.

de Lalla, C., Galli, G., Aldrighetti, L., Romeo, R., Mariani, M., Monno, A., Nuti, S., Colombo, M., Callea, F., Porcelli, S.A., Panina-Bordignon, P., Abrignani, S., Casorati, G. and Dellabona, P. (2004). Production of profibrotic cytokines by invariant NKT cells characterizes cirrhosis progression in chronic viral hepatitis. J Immunol **173**, 1417–1425.

Dunn, C., Brunetto, M., Reynolds, G., Christophides, T., Kennedy, P.T., Lampertico, P., Das, A., Lopes, A.R., Borrow, P., Williams, K., Humphreys, E., Afford, S., Adams, D.H., Bertoletti, A. and Maini, M.K. (2007). Cytokines induced during chronic hepatitis B virus infection promote a pathway for NK cell-mediated liver damage. J Exp Med **204**, 667–680.

Elinav, E., Abd-Elnabi, A., Pappo, O., Bernstein, I., Klein, A., Engelhardt, D., Rabbani, E. and Ilan, Y. (2006). Suppression of hepatocellular carcinoma growth in mice via leptin, is associated with inhibition of tumor cell growth and natural killer cell activation. J Hepatol **44**, 529–536.

Francavilla, A., Starzl, TE., Barone, M., Zeng, Q.H., Porter, K.A., Zeevi, A., Markus, P.M., Van Den Brink, M.R.M. and Todo, S. (1991). Studies on mechanisms of augmentation of liver regeneration by cyclosporin and FW06. Hepatology **14**, 140–148.

Freud, A.G., Becknell, B., Roychowdhury, S., Mao, H.C., Ferketich, A.K., Nuovo, G.J., Hughes, T.L., Marburger, T.B., Sung, J., Baiocchi, R.A., Guimond, M. and Caligiuri, M.A. (2005). A human CD34^{+} subset resides in lymph nodes and differentiates into CD56bright natural killer cells. Immunity **22**, 295–304.

Freud, A.G., Yokohama, A., Becknell, B., Lee, M.T., Mao, H.C., Ferketich, A.K. and Caligiuri, M.A. (2006). Evidence for discrete stages of human natural killer cell differentiation in vivo. J Exp Med **203**, 1033–1043.

Fu, J., Xu, D., Liu, Z., Shi, M., Zhao, P., Fu, B., Zhang, Z., Yang, H., Zhang, H., Zhou, C., Yao, J., Jin, L., Wang, H., Yang, Y., Fu, Y.X. and Wang, F.S. (2007). Increased regulatory T cells correlate with CD8 T-cell impairment and poor survival in hepatocellular carcinoma patients. Gastroenterology **132**, 2328–2339.

Gao, Q., Qiu, S.J., Fan, J., Zhou, J., Wang, X.Y., Xiao, Y.S., Xu, Y., Li, Y.W. and Tang, Z.Y. (2007). Intratumoral balance of regulatory and cytotoxic T cells is associated with prognosis of hepatocellular carcinoma after resection. J Clin Oncol **25**, 2586–2593.

Germanov, E., Veinotte, L., Cullen, R., Chamberlain, E., Butcher, E.C. and Johnston, B. (2008). Critical role for the chemokine receptor CXCR6 in homeostasis and activation of CD1d-restricted NKT cells. J Immunol **181**, 81–91.

Ghiringhelli, F., Ménard, C., Terme, M., Flament, C., Taieb, J., Chaput, N., Puig, P.E., Novault, S., Escudier, B., Vivier, E., Lecesne, A., Robert, C., Blay, J.Y., Bernard, J., Caillat-Zucman, S., Freitas, A., Tursz, T., Wagner-Ballon, O., Capron, C., Vainchencker, W., Martin, F. and Zitvogel, L. (2005). CD4^{+}CD25^{+} regulatory T cells inhibit natural killer cell functions in a transforming growth factor-beta-dependent manner. J Exp Med **202**, 1075–1085.

Ghiringhelli, F., Ménard, C., Martin, F. and Zitvogel, L. (2006). The role of regulatory T cells in the control of natural killer cells: relevance during tumor progression. Immunol Rev **214**, 229–238.

Golden-Mason, L. and Rosen, H.R. (2006). Natural killer cells: primary target for hepatitis C virus immune evasion strategies? Liver Transpl **12**, 363–372.

Gregoire, C., Chasson, L., Luci, C., Tomasello, E., Geissmann, F., Vivier, E. and Walzer, T. (2007). The trafficking of natural killer cells. Immunol Rev **220**, 169–182.

Hamerman, J.A., Ogasawara, K. and Lanier, L.L. (2004). Cutting edge: Toll-like receptor signaling in macrophages induces ligands for the NKG2D receptor. J Immunol **172**, 2001–2005.

Hashimoto, W., Takeda, K., Anzai, R., Ogasawara, K., Sakihara, H., Sugiura, K., Seki, S. and Kumagai, K. (1995). Cytotoxic NK1.1 Ag+ alpha beta T cells with intermediate TCR induced in the liver of mice by IL-12. J Immunol **154**, 4333–4340.

Hata, K., Zhang, X.R., Iwatsuki, S., Van Thiel, D.H., Herberman, R.B. and Whiteside, T.L. (1990). Isolation, phenotyping and functional analysis of lymphocytes from human liver. Clin Immunol Immunopathol **56**, 401–409.

Hayakawa, Y., Takeda, K., Yagita, H., Kakuta, S., Iwakura, Y., Van Kaer, L., Saiki, I. and Okumura, K. (2001). Critical contribution of IFN-gamma and NK cells, but not perforin-mediated cytotoxicity, to anti-metastatic effect of alpha-galactosylceramide. Eur J Immunol **31**, 1720–1727.

Hayakawa, Y., Takeda, K., Yagita, H., Smyth, M.J., Van Kaer, L., Okumura, K. and Saiki, I. (2002). IFN-gamma-mediated inhibition of tumor angiogenesis by natural killer T-cell ligand, alpha-galactosylceramide. Blood **100**, 1728–1733.

Hoerbelt, R., Benjamin, C.L., Shoji, T., Houser, S.L., Muniappan, A., Hasse, R.S., Ledgerwood, L.G., Allan, J.S., Sachs, D.H. and Madsen, J.C. (2008). The effects of tolerance on allograft damage caused by the innate immune system. Transplantation **85**, 314–322.

Hui, D., Qiang, L., Jian, W., Ti, Z. and Da-Lu, K. (2009). A randomized, controlled trial of postoperative adjuvant cytokine-induced killer cells immunotherapy after radical resection of hepatocellular carcinoma. Dig Liver Dis **41**, 36–41.

Huntington, N.D., Vosshenrich, C.A.J. and Di Santo, J.P. (2007). Developmental pathways that generate natural-killer-cell diversity in mice and humans. Nat Rev Immunol **7**, 703–714.

Iinuma, H., Okinaga, K., Fukushima, R., Inaba, T., Iwasaki, K., Okinaga, A., Takahashi, I. and Kaneko, M. (2006). Superior protective and therapeutic effects of IL-12 and IL-18 gene-transduced dendritic neuroblastoma fusion cells on liver metastasis of murine neuroblastoma. J Immunol **176**, 3461–3469.

Inoue, M., Kanto, T., Miyatake, H., Itose, I., Miyazaki, M., Yakushijin, T., Sakakibara, M., Kuzushita, N., Hiramatsu, N., Takehara, T., Kasahara, A. and Hayashi, N. (2006). Enhanced ability of peripheral invariant natural killer T cells to produce IL-13 in chronic hepatitis C virus infection. J Hepatol **45**, 190–196.

Ishio, T., Goto, S., Tahara, K., Tone, S., Kawano, K. and Kitano, S. (2004). Immunoactivative role of indoleamine 2,3-dioxygenase in human hepatocellular carcinoma. J Gastroenterol Hepatol **19**, 319–326.

Ishiyama, K., Ohdan, H., Ohira, M., Mitsuta, H., Arihiro, K. and Asahara, T. (2006). Difference in cytotoxicity against hepatocellular carcinoma between liver and periphery natural killer cells in humans. Hepatology **43**, 272–362.

Ito, A., Shimura, H., Nitahara, A., Tomiyama, K., Ito, M., Kanekura, T., Okumura, K., Yagita, H. and Kawai, K. (2008). NK cells contribute to the skin graft rejection promoted by CD4$^+$ T cells activated through the indirect allorecognition pathway. Int Immunol **20**, 1343–1349.

Itoh, H., Abo, T., Sugawara, S., Kanno, A. and Kumagai, K. (1988). Age-related variation in the proportion and activity of murine liver natural killer cells and their cytotoxicity against regenerating hepatocytes. J Immunol **141**, 315–323.

Iwai, T., Tomita, Y., Shimizu, I., Kajiwara, T., Onzuka, T., Okano, S., Yasunami, Y., Yoshikai, Y., Nomoto, K. and Tominaga, R. (2007). The immunoregulatory roles of natural killer T cells in cyclophosphamide-induced tolerance. Transplantation **84**, 1686–1695.

Jacques, A., Bleau, C., Martin, J.P. and Lamontagne, L. (2008). Intrahepatic endothelial and Kupffer cells involved in immunosuppressive cytokines and natural killer (NK)/NK T cell disorders in viral acute hepatitis. Clin Exp Immunol **152**, 298–310.

Janeway , C.A. Jr. and Medzhitov, R. (2000). Innate immune recognition. Annu Rev Immunol **20**, 197–216.

Jeong, W.I., Park, O., Radaeva, S. and Gao, B. (2006). STAT1 inhibits liver fibrosis in mice by inhibiting stellate cell proliferation and stimulating NK cell cytotoxicity. Hepatology **44**, 1441–1451.

Jeong, W.I., Park, O. and Gao, B. (2008). Abrogation of the antifibrotic effects of natural killer cells/interferon-gamma contributes to alcohol acceleration of liver fibrosis. Gastroenterology **134**, 248–258.

Jinushi, M., Takehara, T., Tatsumi, T., Hiramatsu, N., Sakamori, R., Yamaguchi, S. and Hayashi, N. (2005). Impairment of natural killer cell and dendritic cell functions by the soluble form of MHC class I-related chain A in advanced human hepatocellular carcinomas. J Hepatol **43**, 1013–1020.

Kai, S., Goto, S., Tahara, K., Sasaki, A., Tone, S. and Kitano, S. (2003a). Indoleamine 2,3-dioxygenase is necessary for cytolytic activity of natural killer cells. Scand J Immunol **59**, 177–182.

Kai, S., Goto, S., Tahara, K., Sasaki, A., Kawano, K. and Kitano, S. (2003b). Inhibition of indoleamine 2,3-dioxygenase suppresses NK cell activity and accelerates tumor growth. J Exp Ther Oncol **3**, 336–345.

Kakimi, K., Guidotti, L.G., Koezuka, Y. and Chisari, F.V. (2000). Natural killer T cell activation inhibits hepatitis B virus replication in vivo. J Exp Med **192**, 921–930.

Kaneda, K. and Wake, K. (1983). Distribution and morphological characteristics of the pit cells in the liver of the rat. Cell Tissue Res **233**, 485–505.

Kaneda, K., Dan, C. and Wake, K. (1983). Pit cells as natural killer cells. Biomed Res **4**, 567–576.

Kaneda, K., Sayers, T.J., Wiltrout, T.A., Pilaro, A.M., Kojima, A., Nagashima, K., Gonda, M.A., Ward, J.M., Reynold, C. W., Ortaldo, J.R. and Wiltrout, R.H. (1996). Rod-cored vesicles in large granular lymphocyte (LGL) leukemia cells of rats: alterations induced following metastasis to the liver. Biomed Res **17**, 53–65.

Kawano, T., Tanaka, Y., Shimizu, E., Kaneko, Y., Kamata, N., Sato, H., Osada, H., Sekiya, S., Nakayama, T. and Taniguchi, M. (1999). A novel recognition motif of human NKT antigen receptor for a glycolipid ligand. Int Immunol **11**, 881–887.

Kawarabayashi, N., Seki, S., Hatsuse, K., Ohkawa, T., Koike, Y., Aihara, T., Habu, Y., Nakagawa, R., Ami, K., Hiraide, H. and Mochizuki, H. (2000). Decrease of CD56$^+$ T cells and natural killer cells in cirrhotic livers with hepatitis C may be involved in their susceptibility to hepatocellular carcinoma. Hepatology **32**, 962–969.

Kayagaki, N., Yamaguchi, N., Nakayama, M., Takeda, K., Akiba, H., Tsutsui, H., Okamura, H., Nakanishi, K., Okumura, K. and Yagita, H. (1999). Expression and function of TNF-related apoptosis-inducing ligand on murine activated NK cells. J Immunol **163**, 1906–1913.

Kim, H.R., Park, H.J., Park, J.H., Kim, S.J., Kim, K. and Kim, J. (2004). Characteristics of the killing mechanism of human natural killer cells against hepatocellular carcinoma cell lines HepG2 and Hep3B. Cancer Immunol Immunother **53**, 461–470.

Krizhanovsky, V., Yon, M., Dickins, R.A., Hearn, S., Simon, J., Miething, C., Yee, H., Zender, L. and Lowe, S.W. (2008). Senescence of activated stellate cells limits liver fibrosis. Cell **134**, 657–667.

Kronenberg, M. and Gapin, L. (2002). The unconventional lifestyle of NKT cells. Nat Rev Immunol **2**, 557–568.

La Cava, A., Van Kaer, L. and Fu Dong, S. (2006). CD4+CD25+ Tregs and NKT cells: regulators regulating regulators. Trends Immunol **27**, 322–327.

Laffont, S., Seillet, C., Ortaldo, J., Coudert, J.D. and Guéry, J.C. (2008). Natural killer cells recruited into lymph nodes inhibit alloreactive T-cell activation through perforin-mediated killing of donor allogeneic dendritic cells. Blood **112**, 661–671.

Lafreniere, R., Borkenhagen, K., Bryant, L.D., Anton, A.R., Chung, A. and Poon, M.C. (1990). Analysis of liver lymphoid cell subsets pre- and post-in vivo administration of human recombinant interleukin-2 in a C57BL/6 murine system. Cancer Res **50**, 658–665.

Li, Y., Zhang, T., Ho, C., Orange, J.S., Douglas, S.D. and Ho, W.Z. (2004). Natural killer cells inhibit hepatitis C virus expression. J Leukoc Biol **76**, 1171–1179.

Liu, Z.X., Govindarajan, S., Okamoto, S. and Dennert, G. (2000). NK cells cause liver injury and facilitate the induction of T cell-mediated immunity to a viral liver infection. J Immunol **164**, 6480–6486.

Liu, Y., Poon, R.T., Hughes, J., Feng, X., Yu, W.C. and Fan, S.T. (2005). Chemokine receptors support infiltration of lymphocyte subpopulations in human hepatocellular carcinoma. Clin Immunol **114**, 174–182.

Lu, L., Ikizawa, K., Hu, D., Werneck, M.B., Wucherpfennig, K.W. and Cantor, H.

(2007). Regulation of activated CD4+ T cells by NK cells via the Qa-1-NKG2A inhibitory pathway. Immunity **26**, 593–604.

Luo, D.Z., Vermijlen, D., Ahishali, B., Triantis, V., Plakoutsi, G., Braet, F., Vanderkerken, K. and Wisse, E. (2000). On the cell biology of pit cells, the liver-specific NK cells. World J Gastroenterol **6**, 1–11.

Margalit, M., Shibolet, O., Klein, A., Elinav, E., Alper, R., Thalenfeld, B., Engelhardt, D., Rabbani, E. and Ilan, Y. (2005). Suppression of hepatocellular carcinoma by transplantation of ex-vivo immune-modulated NKT lymphocytes. Int J Cancer **115**, 443–449.

Matera, L., Galetto, A., Bello, M., Baiocco, C., Chiappino, I., Castellano, G., Stacchini, A., Satolli, M.A., Mele, M., Sandrucci, S., Mussa, A., Bisi, G. and Whiteside, T.L. (2006). In vivo migration of labeled autologous natural killer cells to liver metastases in patients with colon carcinoma. J Transl Med **4**, 49.

Matsumoto, G., Nagai, S., Muta, M., Tsuruta, K., Okamoto, A. and Toi, M. (2003). Survival benefit of KRN7000 immune therapy in combination with TNP470 in hamster liver metastasis model of pancreatic cancer. Oncol Rep **10**, 1201–1206.

Meylan, E., Tschopp, J. and Karin, M. (2006). Intracellular pattern recognition receptors in the host response. Nature **442**, 39–44.

Miyagi, T., Takehara, T., Tatsumi, T., Kanto, T., Suzuki, T., Jinushi, M., Sugimoto, Y., Sasaki, Y., Hori, M. and Hayashi, N. (2003). CD1d-mediated stimulation of natural killer T cells selectively activates hepatic natural killer cells to eliminate experimentally disseminated hepatoma cells in murine liver. Int J Cancer **106**, 81–89.

Moretta, L., Ferlazzo, G., Mingari, M.C., Melioli, G. and Moretta, A. (2003). Human natural killer cell function and their interferons with dendritic cells. Vaccine **21**, S38–S42.

Morishima, C., Paschal, D.M., Wang, C.C., Yoshihara, C.S., Wood, B.L., Yeo, A.E., Emerson, S.S., Shuhart, M.C. and Gretch, D.R. (2006). Decreased NK cell frequency in chronic hepatitis C does not affect ex vivo cytolytic killing. Hepatology **43**, 573–580.

Morita, M., Fujino, M., Li, X.K., Kimura, H., Nakayama, T., Taniguchi, M. and Sugioka, A. (2007). Spontaneous tolerance involving natural killer T cells after hepatic grafting in mice. Transpl Immunol **18**, 142–145.

Muhanna, N., Doron, S., Wald, O., Horani, A., Eid, A., Pappo, O., Friedman, S.L. and Safadi, R. (2008). Activation of hepatic stellate cells after phagocytosis of lymphocytes: a novel pathway of fibrogenesis. Hepatology **48**, 963–977.

Nakagawa, R., Motoki, K., Ueno, H., Iijima, R., Nakamura, H., Kobayashi, E., Shimosaka, A. and Koezuka, Y. (1998). Treatment of hepatic metastasis of the colon26 adenocarcinoma with an α-galactosylceramide, KRN7000. Cancer Res **58**, 1202–1207.

Nakagawa, R., Nagafune, I., Tazunoki, Y., Ehara, H., Tomura, H., Iijima, R., Motoki, K., Kamishohara, M. and Seki, S. (2001). Mechanisms of the antimetastatic effect in the liver and of the hepatocyte injury induced by alpha-galactosylceramide in mice. J Immunol **166**, 6578–6584.

Nakamatsu, M., Yamamoto, N., Hatta, M., Nakasone, C., Kinjo, T., Miyagi, K., Uezu, K., Nakamura, K., Nakayama, T., Taniguchi, M., Iwakura, Y., Kaku, M. and Kawakami, K. (2007). Role of interferon-γ in Vα14+ natural killer T cell-mediated host defense against *Streptococcus pneumonia* infection in murine lungs. Microbes Infect **9**, 364–374.

Nakatani, K., Kaneda, K., Seki, S. and Nakajima, Y. (2004). Pit cells as liver-associated natural killer cells: morphology and function. Med Electron Microsc **37**, 29–36.

Nedvetzki, S., Sowinski, S., Eagle, R.A., Harris, J., Vély, F., Pende, D., Trowsdale, J., Vivier, E., Gordon, S. and Davis, D.M. (2007). Reciprocal regulation of human natural killer cells and macrophages associated with distinct immune synapses. Blood **109**, 3776–3785.

Notas, G., Kisseleva, T. and Brenner, D. (2009). NK and NKT cells in liver injury and fibrosis. Clin Immunol **130**, 16–26.

Ohira, M., Ohdan, H., Mitsuta, H., Ishiyama, K., Tanaka, Y., Igarashi, Y. and Asahara, T. (2006). TRAIL-expressing natural killer cells prevents recurrence of hepatocellular carcinoma after partial hepatectomy. Transplant **82**, 1712–1719.

Ohteki, T., Seki, S., Abo, T. and Kumagai, K. (1990). Liver is a possible site for the proliferation of abnormal $CD3^{+}4^{-}8^{-}$ double-negative lymphocytes in autoimmune MRL-lpr/lpr mice. J Exp Med **172**, 7–12.

Ohtsuka, K., Sato, K., Watanabe, H., Kimura, M., Asakura, H. and Abo, T. (1995). Unique order of the lymphocyte subset induction in the liver and intestine of mice during *Listeria monocytogenes* infection. Cell Immunol **161**, 112–124.

Okumoto, K., Hattori, E., Tamura, K., Kiso, S., Watanabe, H., Saito, K., Saito, T., Togashi, H. and Kawata, S. (2004). Possible contribution of circulating transforming growth factor-beta1 to immunity and prognosis in unresectable hepatocellular carcinoma. Liver Int **24**, 21–28.

Onzuka, T., Tomita, Y., Shimizu, I., Okano, S., Yamada, H., Yoshikai, Y. and Tominaga, R. (2008). Role of the cytokine profiles produced by invariant natural killer T cells in the initial phase of cyclophosphamide-induced tolerance. Transplantation **86**, 1301–1310.

Parkin, D.M., Bray, F., Ferlay, J. and Pisani, P. (2005). Global cancer statistics, 2002. CA Cancer J Clin **55**, 74–108.

Peng, Y., Falck-Pedersen, E. and Elkon, K.B. (2001). Variation in adenovirus transgene expression between BALB/c and C57BL/6 mice is associated with differences in interleukin-12 and γ-interferon production and NK cell activation. J Virol **75**, 4540–4550.

Poynard, T., Mathurin, P., Lai, C.L., Guyader, D., Poupon, R., Tainturier, M.H., Myers, R.P., Muntenau, M., Ratziu, V., Manns, M., Vogel, A., Capron, F., Chedid, A. and Bedossa, P. (2003). A comparison of fibrosis progression in chronic liver diseases. J Hepatol **38**, 257–265.

Racanelli, V. and Rehermann, B. (2006). The liver as an immunological organ. Hepatology **43**, S54–S62.

Radaeva, S., Sun, R., Jaruga, B., Nguyen, V.T., Tian, Z. and Gao, B. (2006). Natural killer cells ameliorate liver fibrosis by killing activated stellate cells in NKG2D-dependent and tumor necrosis factor-related apoptosis-inducing ligand-dependent manners. Gastroenteroloy **130**, 435–452.

Randow, F., Syrbe, U., Meisel, C., Krausch, D., Zuckermann, H., Platzer, C. and Volk, H.D. (1995). Mechanism of endotoxin desensitization: involvement of interleukin 10 and transforming growth factor β. J Exp Med **181**, 1887–1892.

Safadi, R., Zigmond, E., Pappo, O., Shalev, Z. and Ilan, Y. (2007). Amelioration of hepatic fibrosis via beta-glucosylceramide–mediated immune modulation is associated with altered CD8 and NKT lymphocyte distribution. Int Immunol **19**, 1021–1029.

Samson, S.I., Richard, O., Tavian, M., Ranson, T., Vosshenrich, C.A., Colucci, F., Buer, J., Grosveld, F.,

Godin, I. and Di Santo, J.P. (2003). GATA-3 promotes maturation, IFN-γ production, and liver-specific homing of NK cells. Immunity **19**, 701–710.

Santucci, L., Fiorucci, S., Chiorean, M., Brunori, P.M., Di Matteo, F.M., Sidoni, A., Migliorati, G. and Morelli, A. (1996). Interleukin 10 reduces lethality and hepatic injury induced by lipopolysaccharide in galactosamine sensitized mice. Gastroenterology **111**, 736–744.

Sawayama, T., Sakaguchi, K., Senoh, T., Ohta, T., Nishimura, M., Takaki, A., Tsuji, T. and Shiratori, Y. (2003). Effects of pulsing procedure of interleukin-12 in combination with interleukin-2 on the activation of peripheral blood lymphocytes derived from patients with hepatocellular carcinoma. Acta Med Okayama **57**, 285–292.

Screpanti, V., Wallin, R.P., Ljunggren, H.G. and Grandien, A. (2001). A central role for death receptor-mediated apoptosis in the rejection of tumors by NK cells. J Immunol **167**, 2068–2073.

Seki, N., Hayakawa, Y., Brooks, A.D., Wine, J., Wiltrout, R.H., Yagita, H., Tanner, J.E., Smyth, M.J. and Sayers, T.J. (2003). Tumor necrosis factorrelated apoptosis-inducing ligand-mediated apoptosis is an important endogenous mechanism for resistance to liver metastases in murine renal cancer. Cancer Res **63**, 207–213.

Seo, N., Tokura, Y., Takigawa, M. and Egawa, K. (1999). Depletion of IL-10- and TGF-beta-producing regulatory gamma delta T cells by administering a daunomycin-conjugated specific monoclonal antibody in early tumor lesions augments the activity of CTLs and NK cells. J Immunol **163**, 242–249.

Seyfert-Margolis, V. and Turka, L.A. (2008). Marking a path to transplant tolerance. J Clin Invest **118**, 2684–2686.

Shresta, S., MacIvor, D.M., Heusel, J.W., Russell, J.H. and Ley, T.J. (1995). Natural killer and lymphokine-activated killer cells require granzyme B for the rapid induction of apoptosis in susceptible target cells. Proc Natl Acad Sci U S A **92**, 5679–5683.

Smyth, M.J., Cretney, E., Takeda, K., Wiltrout, R.H., Sedger, L.M., Kayagaki, N., Yagita, H. and Okumura, K. (2001). Tumor necrosis factorrelated apoptosis-inducing ligand (TRAIL) contributes to interferon gamma-dependent natural killer cell protection from tumor metastasis. J Exp Med **193**, 661–670.

Smyth, M.J., Hayakawa, Y., Takeda, K. and Yagita, H. (2002). New aspects of natural-killer-cell surveillance and therapy of cancer. Nat Rev Cancer **2**, 850–861.

Spada, F.M., Koezuka, Y. and Porcelli, S.A. (1998). CD1d-restricted recognition of synthetic glycolipid antigens by human natural killer T cells. J Exp Med **188**, 1529–1534.

Srivastava, R.K. (2001). TRAIL/Apo-2L: mechanisms and clinical applications in cancer. Neoplasia **3**, 535–546.

Starzl, T.E., Demetris, A.J., Trucco, M., Murase, N., Ricordi, C., Ildstad, S., Ramos, H., Todo, S., Tzakis, A. and Fung, J.J. (1993). Cell migration and chimerism after whole-organ transplantation: the basis of graft acceptance. Hepatology **17**, 1127–1152.

Street, S.E., Cretney, E. and Smyth, M.J. (2001). Perforin and interferongamma activities independently control tumor initiation, growth, and metastasis. Blood **97**, 192–197.

Tabata, A., Kaneda, K., Watanabe, H., Abo, T. and Yano, I. (1996). Kinetics of organ-associated natural killer cells and intermediate CD3 cells during pulmonary and hepatic granulomatous inflammation induced by mycobacterial cord factor. Microbiol Immunol **40**, 651–658.

Takeda, K., Hayakawa, Y., Smyth, M.J., Kayagaki, N., Yamaguchi, N., Kakuta, S., Iwakura, Y., Yagita, H. and Okumura, K. (2001). Involvement of tumor necrosis factor-related apoptosis-inducing ligand in surveillance of tumor metastasis by liver natural killer cells. Nat Med **7**, 94–100.

Takeda, K., Cretney, E., Hayakawa, Y., Ota, T., Akiba, H., Ogasawara, K., Yagita, H., Kinoshita, K., Okumura, K. and Smyth, M.J. (2005). TRAIL identifies immature natural killer cells in newborn mice and adult mouse liver. Blood **105**, 2082–2089.

Taketomi, A., Shimada, M., Shirabe, K., Kajiyama, K., Gion, T. and Sugimachi, K. (1998). Natural killer cell activity in patients with hepatocellular carcinoma: a new prognostic indicator after hepatectomy. Cancer **83**, 58–63.

Tanaka, N., Yamamoto, H., Tatemoto, A., Urabe, T. and Orita, K. (1993). Regulation of liver regeneration by interleukin-2 and its inhibitors: cyclosporine A and FK506. J Immunopharmcol **15**, 211–217.

Toura, I., Kawano, T., Akutsu, Y., Nakayama, T., Ochiai, T. and Taniguchi, M. (1999). Cutting edge: inhibition of experimental tumor metastasis by dendritic cells pulsed with alpha-galactosylceramide. J Immunol **163**, 2387–2391.

Trotta , R., Col, J.D., Yu, J., Ciarlariello, D., Thomas, B., Zhang, X., Allard, J. II, Wei, M., Mao, H., Byrd, J.C.,Perrotti, D. and Caligiuri, M.A. (2008). TGF-beta utilizes SMAD3 to inhibit CD16-mediated IFN-gamma production and antibody-dependent cellular cytotoxicity in human NK cells. J Immunol **181**, 3784–3792.

Tsutsui, H., Nakanishi, K., Matsui, K., Higashino, K., Okamura, H., Miyazawa, Y. and Kaneda, K. (1996). IFN-gamma-inducing factor upregulates Fas ligand-mediated cytotoxic activity of murine natural killer cell clones. J Immunol **157**, 3967–3973.

Twilley, T.A., Mason, L., Talmadge, J.E. and Wiltrout, R.H. (1987). Increase in liver-associated natural killer activity by polyribonucleotides. Nat Immun Cell Growth Regul **6**, 279–290.

Vanderkerken, K., Bouwens, L., De Neve, W., Van den Berg, K., Baekeland, M., Delens, N. and Wisse, E. (1993). Origin and differentiation of hepatic natural killer cells (pit cells). Hepatology **18**, 919–925.

Vermijlen, D., Luo, D., Robaye, B., Seynaeve, C., Baekeland, M. and Wisse, E. (1999). Pit cells (hepatic natural killer cells) of the rat induce apoptosis in colon carcinoma cells by the perforin/granzyme pathway. Hepatology **29**, 51–56.

Vujanovic, N.L., Polimeno, L., Azzarone, A., Francavilla, A., Chambers, W.H., Starzl, TE., Herberman, R.B. and Whiteside, T. (1995). Changes of liver-resident NK cells during liver regeneration in rats. J Immunol **154**, 6324–6338.

Wahl, S.M. (2007). Transforming growth factor-beta: innately bipolar. Curr Opin Immunol **19**, 55–62.

Wang, J., Xu, J., Zhang, W., Wei, H. and Tian, Z. (2005). TLR3 ligand-induced accumulation of activated splenic natural killer cells into liver. Cell Mol Immunol **2**, 449–453.

Wang, H., Grzywacz, B., Sukovich, D., McCullar, V., Cao, Q., Lee, A.B., Blazar, B.R., Cornfield, D.N., Miller, J.S. and Verneris, M.R. (2007). The unexpected effect of cyclosporin A on $CD56^{+}CD16^{-}$ and $CD56^{+}CD16^{+}$ natural killer cell subpopulations. Blood **110**, 1530–1539.

Watanabe, H., Miyaji, C., Kawachi, Y., Iiai, T., Ohtsuka, K., Iwanaga, T., Takahashi-Iwanaga, H. and Abo, T. (1995). Relationships between intermediate TCR cells and NK1.1+ T cells in various immune organs. NK1.1+ T cells are present within a population of intermediate TCR cells. J Immunol **155**, 2972–2983.

Webster, G.J., Reibnat, S., Maini, M.K., Whalley, S.A., Ogg, G.S., King, A., Brown, D., Amlot, P.L., Williams, R., Vergani, D., Dusheiko, G.M. and Bertoletti, A. (2000). Incubation phase of acute hepatitis B in man: dynamic of cellular immune mechanisms. Hepatology **32**, 1117–1124.

Wisse, E., Van't Noordende, J.M., Van der Meulen, J. and Daems, W.T. (1976). The pit cell: description of a new type of cell occurring in rat liver sinusoids and peripheral blood. Cell Tissue Res **173**, 423–435.

Wiltrout, R.H. (2000). Regulation and antimetastatic functions of liver-associated natural killer cells. Immunol Rev **174**, 63–76.

Wiltrout, R.H., Mathieson, B.J., Talmadge, J.E., Reynolds, C.W., Zhang, S.R., Herberman, R.B. and Ortaldo, J.R. (1984). Augmentation of organ associated natural killer activity by biological response modifiers. Isolation and characterization of large granular lymphocytes from liver. J Exp Med **160**, 1431–1449.

Wiltrout , R.H., Herberman, R.B., Zhang, S.R., Chirigos, M.A., Ortaldo, J.R., Green, K.M. Jr. and Talmadge, J.E. (1985). Role of organ-associated NK cells in decreased formation of experimental metastasis in lung and liver. J Immunol **134**, 4267–4275.

Yamagiwa, S., Matsuda, Y., Ichida, T., Honda, Y., Takamura, M., Sugahara, S., Ishikawa, T., Ohkoshi, S., Sato, Y. and Aoyagi, Y. (2008). Sustained response to interferon-alpha plus ribavirin therapy for chronic hepatitis C is closely associated with increased dynamism of intrahepatic natural killer and natural killer T cells. Hepatol Res **38**, 664–672.

Yang, H., Dithmar, S. and Grossniklaus, H.E. (2004). Interferon alpha 2b decreases hepatic micrometastasis in a murine model of ocular melanoma by activation of intrinsic hepatic natural killer cells. Invest Ophthalmol Vis Sci **45**, 2056–2064.

Yang, S.H., Jin, J.Z., Lee, S.H., Park, H., Kim, C.H., Lee, D.S., Kim, S., Chung, N.H. and Kim, Y.S. (2007). Role of NKT cells in allogeneic islet graft survival. Clin Immunol **124**, 258–266.

Yao, L., Sgadari, C., Furuke, K., Bloom, E.T., Teruya-Feldstein, J. and Tosato, G. (1999). Contribution of natural killer cells to inhibition of angiogenesis by interleukin-12. Blood **93**, 1612–1621.

Yoshida, O., Akbar, F., Miyake, T., Abe, M., Matsuura, B., Hiasa, Y. and Onji, M. (2008). Impaired dendritic cell functions because of depletion of natural killer cells disrupt antigen-specific immune responses in mice: restoration of adaptive immunity in natural killer-depleted mice by antigen-pulsed dendritic cell. Clin Exp Immunol **152**, 174–181.

Zamai, L., Ahmad, M., Bennett, I.M., Azzoni, L., Alnemri, E.S. and Perussia, B. (1998). Natural killer (NK) cell-mediated cytotoxicity: differential use of TRAIL and Fas ligand by immature and mature primary human NK cells. J Exp Med **188**, 2375–2380.

Zhengkun, Tu., Bozorgzadeh, A., Pierce, R.H., Kurtis, J., Crispe, I.N. and Orloff, M.S. (2008). TLR-dependent cross talk between human Kupffer cells and NK cells. J Exp Med **205**, 233–244.

Zigmond, E., Shalev, Z., Pappo, O., Alper, R., Zolotarov, L. and Ilan, Y. (2008). NKT lymphocyte polarization determined by microenvironment signaling: a role for CD8+ lymphocytes and beta-glycosphingolipids. J Autoimmun **31**, 188–195.

Zitvogel, L. (2002). Dendritic and natural killer cells cooperate in the control/switch of innate immunity. J Exp Med **195**, 9–14.

Chapter Twenty-Seven

27

Natural killer cells in the spleen and lymph node

Alex Karlsson-Parra

CHAPTER CONTENTS

Imagination is more important than knowledge. Knowledge is limited. Imagination encircles the world.

Albert Einstein

ABSTRACT

Secondary lymphoid tissues, such as the spleen and lymph node (LN), are highly organized organs, with multiple levels of cell trafficking control, that each contribute to enable lymphocytes to efficiently locate resident dendritic cells (DCs) in the steady state, and immigrating, activated DCs during inflammation. Recent evidence strongly suggests that an immune network, including mature DCs and natural killer (NK) cells, occurs within the local environment of inflamed lymphoid tissues and determines the subsequent quality and type of antigen-specific immune response. Additionally, after interaction with inflamed LN DCs, newly recruited NK cells (and probably also resident NK cells in humans), rapidly emerge as primed effector cells in the periphery. NK cells thus obviously need to acquire additional regulator and effector functions by interacting with mature DCs in secondary lymphoid organs, revealing a striking homology to T lymphocytes of the adaptive immune system.

KEY WORDS

Dendritic cell, NK cell, CXCL9, CXCR3, IFN-γ, Lymph node, Spleen, Priming, Effector

Introduction

In contrast to T lymphocytes of the adaptive immune system, which need to be primed for effector functions by cognate interaction with antigen-presenting dendritic cells (DCs) in lymph nodes (LNs) (Banchereau and Steinman, 1998), natural killer (NK) cells were originally identified as 'naturally active' cells that readily display effector functions upon encountering infected or transformed cells (Herberman et al., 1975; Kiessling et al., 1975; Trinchieri, 1989). Activation of target cell recognition by NK cells is determined by the balance between inhibitory and activating signalling pathways (Moretta et al., 2001). In humans, NK cells express several activating receptors, such as NKG2D and natural cytotoxicity receptors (NCRs), capable of recognizing target cells expressing cellular stress-induced, infection-induced, or transformation-induced molecules on the cell surface (Diefenbach and Raulet, 2003; Moretta et al., 2002). Most NK cell inhibitory receptors recognize MHC class I molecules, which are expressed by almost all nucleated cells. The binding of MHC class I complexes to killer cell Ig-like receptors (KIRs) or to the C-lectin-type heterodimeric CD94/NKG2A receptor, initiates inhibitory pathways that can override activation signals (Borrego et al., 2005; Natarajan et al., 2002).

NK cells from uninfected mice or freshly isolated human NK cells usually show weak or suboptimal lytic activity when incubated in vitro with tumour target cells (Bryceson et al., 2006; Lucas et al., 2007), suggesting that resting NK cells require additional priming signals to reach their full functional potential. Indeed, investigators usually treat mice with Toll-like receptor (TLR) ligands or cytokines (e.g. type I interferons or their synthetic inducers) (Djeu et al., 1979; Gidlund et al., 1978), or culture NK cells in vitro in the presence of cytokines (usually IL-2) to elicit detectable effector functions prior to in vitro analyses of NK cell responses (Bryceson et al., 2006). It has further become increasingly clear that NK cells, like T lymphocytes, require those priming signals delivered by mature DCs (Andrews et al., 2003; Fernandez et al., 1999; Gerosa et al., 2002; Hochweller et al., 2008; Kassim et al., 2006; Lucas et al., 2007).

Recent studies have shown that a substantial number of NK cells are present in resting human LNs (Fehniger et al., 2003; Ferlazzo et al., 2004b; Vossen et al., 2008), and that NK cells from peripheral blood enter mouse LNs draining sites of immunization or infection, where they may influence the development of adaptive immunity (Bajenoff et al., 2006a; Martin-Fontecha et al., 2004; Watt et al., 2008). In the mouse spleen, a similar selective and transient recruitment of NK cells into the T-cell area has been reported upon inflammation induced by poly (I:C) injection or mouse cytomegalovirus (MCMV) infection (Bekiaris et al., 2008; Gregoire et al., 2008). Taken together, these data strongly suggest that an immune network, including mature DCs and NK cells, occurs within the local environment of inflamed lymphoid tissues and determines the subsequent quality and type of antigen-specific immune response. Additionally, after interaction with inflamed LN DCs, newly recruited NK cells (Lucas et al., 2007), and most likely also resident human NK cells (Fehniger et al., 2003; Ferlazzo et al., 2004a; Ferlazzo et al., 2002) rapidly emerge as primed effector cells in the periphery. Naive NK cells thus obviously need to acquire additional regulator and effector functions by interacting with mature DCs in secondary lymphoid organs, revealing a striking homology to T lymphocytes of the adaptive immune system.

Homing of lymphocytes to the LN at steady state

The LN is charged with a crucial function in the mammalian immune system: to facilitate physical interactions between extremely rare cells arriving from different tissue compartments. Paramount to carrying out this function is its unique location at the interface between the blood and lymphatic systems, thus enabling tissue-derived antigen and immigrating tissue-derived mature DCs to gather in close proximity to blood-derived and motile lymphocytes.

Naive T lymphocytes are known to recirculate through the body via the bloodstream to lymphoid tissue, such as LNs, where they search for their respective antigens, then return to the bloodstream via efferent lymphatics and the thoracic duct. Lymphocytes enter the cortex of the LN by adhering to and migrating across specialized postcapillary venules called high endothelial venules (HEVs), found in the T-cell area of the cortex. This process occurs by an adhesion cascade, which includes rolling of the lymphocyte on the luminal surface of HEVs, the presence of a chemokine bound to the endothelial glycocalyx on the HEV luminal surface triggering the activation of the integrins on the surface of the lymphocyte, which allows a firm adherence of the lymphocyte to the endothelium, and, finally, migration of the lymphocyte through the wall of the HEV (Butcher, 1991; Butcher and Picker, 1996; Shimizu et al., 1992; Springer, 1994; von Andrian and Mempel, 2003). This complex process is specifically mediated by the interaction of L-selectin, the chemokine receptor CCR7 and the integrin lymphocyte function-associated antigen 1 (LFA-1) with their ligands—peripheral node addressin (PNAd), CCL21 and intercellular adhesion molecule 1 (ICAM-1), respectively—on the lumenal surfaces of HEVs.

In humans the $CD56^{bright}$ NK cells have been shown to express L-selectin at a higher density than all other peripheral blood leukocytes. Moreover, $CD56^{bright}$ NK cells bind with high efficiency to physiologic L-selectin ligands on peripheral LN HEVs (Frey et al., 1998). In $CD56^{bright}$ cells, L-selectin ligation by cross-linking mAb further activates LFA-1, which is expressed on both $CD56^{bright}$ and $CD56^{dim}$ NK cells, although its density is constitutively higher on $CD56^{dim}$ cells (Frey et al., 1998). On activation by chemokines, human NK cells interact firmly with the endothelium through integrins, and the role of LFA-1, as for other leukocytes, has been shown to be central in NK cell extravasation (Bianchi et al., 1993). Human CD56^{bright} NK cells also express CCR7 on the cell surface and migrate vigorously in response to the lymphoid chemokines C–C motif chemokine ligand 19 CCL19 (Chen et al., 2005) and CCL21 (Campbell et al., 2001; Kim et al., 1999). Taken together, these data strongly indicate that also NK cells, particularly CD56^{bright} NK cells, may selectively migrate from blood to peripheral LNs in which rolling, sticking and firm adhesion are mediated by the interaction of L-selectin, CCR7 and LFA-1 with their ligands on HEVs at steady state.

In the mouse, NK cells have been found to enter LNs through HEV at steady state, and similar to naive T cells, L-selectin is required for this entry (Adam et al., 2005). In contrast to human NK cells, mouse NK cells do not express CCR7. The chemokine receptor CXCR3, however, has been shown to route mouse NK cells directly to CCL21-expressing HEVs as a result of the ability of mouse CCL21 to interact with both CCR7 and CXCR3 (Jenh et al., 1999), an interaction not applicable to the human system (Soto et al., 1998).

Homing of lymphocytes to the spleen at steady state

The spleen has a complex architecture owing to its dual roles in the filtration of exhausted red blood cells and as a secondary lymphoid organ. Blood is supplied through arterioles that are surrounded by periarteriolar lymphoid sheaths (PALS) that define the white pulp (WP). Most arterioles bringing lymphocytes to the spleen open within the sinus of the marginal zone (MZ). However, some arterioles also connect to red pulp (RP) sinuses, providing a more direct way to this anatomical site (Schmidt et al., 1993). The RP is a specialized area rich in macrophages, which is critical for the elimination of senescent red blood cells. Localized between the WP and the RP, the MZ creates a transit area for recently immigrating blood lymphocytes as well as a filtering zone for blood content (Lyons and Parish, 1995).

In the spleen, HEVs are absent (Roozendaal et al., 2008). Although a key role for CCL21 is appreciated in the intrasplenic localization of T cells (Forster et al., 1999; Nakano et al., 1998), the adhesion molecules L-selectin and LFA-1, critical in lymphocyte homing to LNs, are not essential for T-cell migration towards the splenic WP (Nolte et al., 2003). Using T-cell homing experiments, it was recently demonstrated that T cells entering the PALS do not cross the MZ randomly but only use the MZBCs (Bajenoff et al., 2008) Whether other lymphocytes, such as NK cells, or DCs may use the same access roads as T cells in the steady state (or during inflammation) is unknown but appears likely.

Lymphocyte migration within spleen and LN T-cell areas at steady state

As a result of the initial, groundbreaking studies of Anderson, Shaw, Gretz and colleagues (Anderson and Anderson, 1975; Gretz et al., 1997; Gretz et al., 1996; Gretz et al., 2000), it is now well accepted that the lymphoid compartment is a shielded unit, and although cells are able to pass across the layer of sinus-lining cells by active movement, fluid can only enter via a reticular network composed of fibroblastic recticular cells (FRCs) originating between the sinus-lining cells. Each branch of the FRC network has a core of collagen fibers surrounded by extracellular-matrix molecules, a basement membrane and a sleeve of stromal cells, the FRCs. The space between collagen fibers and the basement membrane, forming a tubular conduit system, is accessible to small molecules, which can be channelled from the subcapsular sinus to perivascular spaces around HEVs.

The open spaces that are formed outside the unique organization of FRCs are filled with lymphocytes with each lymphocyte type showing apparently random migration within these distinct areas. The FRCs not only transport chemokines through their conduit but also actively produce the homeostatic chemokines CCL19 and CCL21 (Luther et al., 2000). CCL19 and CCL21 have recently been shown to be important for T-cell chemokinesis in the T-cell area of LNs (Asperti-Boursin et al., 2007, Okada and Cyster, 2007; Worbs et al., 2007). Other than chemokines, however, the factors contributing to the observed patterns of lymphocyte movement and the spatial segregation of T cells and B cells has until recently been poorly characterized.

In vitro, chemokines diffusing in solution or across gels or membranes establish concentration gradients that attract lymphocytes. However, given the microanatomic and biochemical complexity of lymphoid tissues, it seems at best questionable that freely diffusible chemokine gradients could form and remain sufficiently undisturbed to exert robust control over cell migration in vivo. Chemokines are immobilized on cells or extracellular matrix surfaces by interacting with glycosaminoglycans. Specific chemokines bind different types of glycosaminoglycans, expression of which can vary with cell type, location and inflammatory status (Kuschert et al., 1999). By localizing particular chemokines to distinct and geographically segregated stromal cell types, haptotactic rather than chemotactic guidance cues may promote proper lymphocyte localization and guidance of trafficking within LN subregions. Indeed, by combining confocal, electron and intravital microscopy, Bajenoff and colleagues (Bajenoff et al., 2006b) demonstrated that the FRC network actually constituted an open three-dimensional meshwork of cell bodies and extended processes that supports and guides lymphocyte motility in the T-cell area, thus dictating the apparent characteristic random migratory behaviour of these cells. Because DCs adhere to the same FRC network, lymphocytes will have numerous encounters with DCs while they are travelling on this complex 'roadway', with the likelihood of contact increased by the local scanning action of the DC dendrites, which can capture the attention of lymphocytes that might choose a pathway along the FRC network leading away from the DC body (Bajenoff et al., 2007; Bajenoff et al., 2003).

Although the spleen is the largest secondary lymphoid organ, there has been scant evidence for any special role of nonhematopoietic structural elements in this organ in guiding lymphocytes within the T-cell area of the spleen. In a recent study (Bajenoff et al., 2008), however, T cells were found to move in the T-cell area at a speed comparable to that observed in intact LNs and actively crawled on FRCs, following and morphologically adapting to the paths established by the cell bodies and extended processes of these nonhematopoietic cells.

Whereas lymphocytes exit LNs through the efferent lymphatic vessels, the exact anatomical route that they use to exit the spleen WP is unknown. It is assumed, however, that lymphocytes leave the WP through the MZ and then re-enter the bloodstream (Roozendaal et al., 2008).

NK cells in LN and spleen T-cell areas at steady state

At steady state, circulating human CD56^{bright} NK cells, expressing L-selectin and CCR7 as well as LFA-1, thus appear to be equipped to traffic into LNs. Indeed, Caligiuri and colleagues (Fehniger et al., 2003) were the first to demonstrate that CD56^{bright} NK cells are present in the T-cell areas of fresh human normal donor LNs. While nearly all (>95%) LN CD56^{+} NK cells were CD56^{bright}, this differs from the peripheral blood, where the majority (approximately 90%) of NK cells are CD56^{dim}. The presence of CD56^{bright} NK cells in the T-cell area of normal human LNs was subsequently confirmed by the group of Ferlazzo and colleagues (Ferlazzo et al., 2004a; Ferlazzo et al., 2004b). In addition, DEC-205^{+} DCs were demonstrated in the same T-cell areas as NK cells, indicating T-cell areas of LNs as potential sites for DC/NK cell crosstalk. (Ferlazzo et al., 2004b). In a recent study (Vossen et al., 2008), the authors reported that human LNs were markedly enriched for mature CD56^{bright} NK cells that coexpressed CD27. On these cells, regulation of CD27 expression appeared to be controlled by IL-15, and down-regulation of CD27 was specifically induced by its ligand, CD70.

CD27^{high} NK cells have also the found to be the predominating NK cell subset in mouse LNs at steady state (Hayakawa and Smyth, 2006). In a study performed by Bajenoff et al. (2006a), initial immunohistological studies revealed that single NK cells were present in both mouse LN T-cell areas and the medulla at steady state. At early time points after i.v. injection of CFSE-labelled NK cells, all observed CFSE-labelled cells were found to be associated with HEV. To look for evidence of direct NK–DC interactions in noninflamed LNs, tissue sections were stained for NK cells, resident DCs and HEV structures. By performing real-time observations, more that 90% of the NK cells in the T-cell area were found to unambiguously remain in contact with the DCs for at least 25 min. NK cells moved slowly, nearly threefold slower that T cells, while scouting relatively small areas. The observed slow motility was considered to most likely reflect an inherent nature of NK cells, which is consistent with their prolonged physical associations with the resident DC network in the superficial LN.

Early immunohistological studies on human spleens (Timens and Poppema, 1985; Vivier et al., 1995) indicated the exclusion of NK cells from the WP at steady state. In a study by Ferlazzo et al. (2004b), FACS-analysis on cell suspensions from uninflamed spleen cells showed a variable presence of NK cells with the CD56^{dim} NK cells being the dominant subset. Since spleen consists primarily of RP, which in turn is rich in blood vessels, the splenic NK cells population analyzed by flow cytometry may primarily represent blood cells with a minor contribution of the WP lymphoid tissue areas. Immunohistological examinations of spleen from naive mice have repeatedly revealed low numbers of NK cells within the WP at steady state (Andrews et al., 2003; Gregoire et al., 2008; Salazar-Mather et al., 1996).

Spleen and LNs in NK cell development and homeostasis

It is generally accepted that, in adult life, NK cell development primarily occurs in the bone marrow (BM), as its ablation results in the loss of lytic NK cells that can be restored by BM transplantation (Hackett et al., 1985). In the human LN, Caligiuri and coworkers identified a novel CD34^{dim}CD45RA (+) hematopoietic precursor cell (HPC) that constitute <1% of BM CD34 (+) HPCs and approximately 6% of blood CD34 (+) HPCs, but >95% of LN CD34 (+) HPCs. They reside in the parafollicular T-cell regions of the LN with

$CD56^{bright}$ NK cells, and when stimulated by IL-15, IL-2 or activated LN T cells, they become $CD56^{bright}$ NK cells. These data thus identify a new NK cell precursor and further support a model of human NK cell development in which BM-derived $CD34^{dim}$CD45RA (+) HPCs reside in the LN where endogenous cytokines drive their differentiation to $CD56^{bright}$ NK cells in vivo (Freud et al., 2005). Additionally, four NK cell developmental intermediates spanning the continuum of differentiation from a CD34 (+) NK cell progenitor to a functionally mature NK cell have been demonstrated (Freud et al., 2006). The relative contribution of LNs to the pool of mature NK cells is however unknown.

In the developing mouse foetus, lymphoid tissue-inducer cells (LTi cells) initiate LN organogenesis by stimulating the lymphotoxin-β receptor (LTβR) on local mesenchymal cells. This interaction induces expression of the adhesion molecules ICAM-1, VCAM-1 and MAdCAM-1 and the subsequent production of homeostatic chemokines. The net result of this interaction is the attraction and retention of additional circulating LTi cells and lymphocytes. Clustering of these cells induces a local self-sustaining primordium in which, over time, HEVs will be induced, T-cell and B-cell areas will be shaped and postnatal germinal centres can form (Drayton et al., 2006; Mebius, 2003). In a recent study (Cupedo et al., 2009), human mesentery in the first trimester and developing human LNs in the second trimester were found to contain a population of lineage-negative (Lin^-) $CD127^+$ cells with characteristics of LTi cells that were able to induce the expression of adhesion molecules on mesenchymal stromal cells through LTβR and tumour necrosis factor receptor (TNFR) signalling, thereby mimicking LN development. Detailed analysis of these human LTi cells showed that they were immature committed NK cells with lymphoid tissue-inducing abilities. A similar population of cells was present in postnatal tonsil, which suggested involvement of these cells in maintenance of the LN architecture.

The spleen has also been suggested as a potential site for human NK cell development of BM-derived NK cell precursors. This hypothesis is based on the fact that human spleen-derived but not BM-derived fibroblasts constitutively express a membrane-bound IL-15 necessary and sufficient to trigger, upon co-culture, the differentiation of circulating (but not BM) $CD34^+$ progenitors into functional NK cells (Briard et al., 2002; Brouty-Boye et al., 1998).

In order to study the interactions between DCs and NK cells in vivo (Hochweller et al., 2008), the authors used a new bacterial artificial chromosome (BAC) transgenic CD11c.DTR mouse model that allows effective depletion of DCs over prolonged periods of time without nonspecific cytotoxicity. Using this model, a previously unrecognized role for DCs in optimal homeostatic proliferation of NK cells in lymphopenic conditions was demonstrated. Notably, in this process, DC-derived IL-15 appears to play an important role. NK cells have to be in close contact or vicinity to DCs in order to receive an optimal amount of IL-15. This situation is reminiscent of T cells. However, consistent with differential dependency on MHC and cytokines, differences appear to exist between CD8 and CD4 T cells, CD8 T cells appearing to be more dependant on DC for homeostatic proliferation in lymphopenic hosts than CD4 T cells (Zaft et al., 2005).

Regulatory role of $CD4^+CD25^+$ $Foxp3^+$ T cells in NK cell activation in the LN in steady state

The $CD4^+CD25^+$ regulatory T cells (Treg) are a unique population of T cells known to maintain immune tolerance (Kim et al., 2007; Sakaguchi, 2004; Shevach, 2002). Ablation of Treg in adult mice also results in the development of a fatal autoimmune disease (Kim et al., 2007), supporting the notion that Treg are required to maintain peripheral tolerance and homeostasis of the immune system. Interestingly, in the study by Kim et al. (2007), Treg ablation in mice was found to affect LN cell composition, highlighting a several-fold increase in resident DCs and NK cells. In a recent study by Zitvogel and colleagues (Terme et al., 2008), it was demonstrated that the absence of Treg resulted in the activation of self-reactive $CD4^+$ T cells in the LN responsible for DC maturation leading to NK cell proliferation. At steady state, LN-resident DCs did not harbour IL-15Rα molecules on the cell surface, while ablation of Treg was found to turn on IL-15Rα expression on LN DCs. Indeed, an up-regulation of IL-15Rα expression was observed in the absence of Treg, whereas expression of costimulatory molecules such as CD40 or CD80 did not change. This is the first observation demonstrating the regulatory role of Treg in DC/NK crosstalk in the steady state in the LN. Depletion of mouse Treg cells has further been shown to exacerbate NK cell cytotoxicity in vivo (Ghiringhelli et al., 2006).

Evidence has been presented that Treg cells hamper generation of mature NK cells in mouse LN T-cell zones through short-range interactions with NK precursors. In turn, mature NK cells specifically regulate the amount of $CD8\alpha^+$ phenotypically immature DCs present in LN T-cell zones (Giroux et al., 2007). Based on these findings, it was proposed that the dominant influence of Treg cells on NK cell precursors and $CD8\alpha^+$ immature DCs explains why 'quiescent' LNs in the absence of infection function as privileged sites for induction and maintenance of tolerance to peripheral antigens.

The inflamed LN and spleen

Changes in the kinetics of lymphocyte traffic through antigen-stimulated LNs have been known for a long time (Cahill et al., 1976). The evolutionarily conserved phenomenon of fever was recently discovered to have a direct effect on lymphocyte trafficking. Temperatures of 38–40°C act directly on lymphocytes to enhance CD62L-dependent homing across HEVs while also increasing the expression of CCL21 and ICAM-1 on the surface of HEVs (Chen et al., 2006). Upon inflammation, LNs can further remodel and expand their primary feed arterioles by 50%. As a consequence, the vascular input to the LN is increased, providing a mechanism to increase the rate of lymphocyte delivery through the blood to a specific inflamed LN (Soderberg et al., 2005).

In experiments performed by Martin-Fontecha and colleagues (2003), it was observed that footpad-injection of BM-derived, LPS-matured $CCR7^+$ DCs resulted in a dose-dependent increase in draining LN cellularity that was detectable on day 1 and reached a plateau by days 3 and 4. The early (day 1) increase in cell number was not due to cell proliferation but rather to enhanced cell recruitment. LN congestion was also induced by injection of freshly isolated splenic $CD11c^+$ DCs. In contrast, a 10-fold higher number of $CD11c^-$ splenic cells failed to do so in spite of efficient migration to the draining LN. An increase in LN total cell number was modestly induced by injection of inflammatory cytokines (TNF or IL-1, alone or in combination) or adjuvants (IFA, CFA or LPS), further suggesting a direct role for migrating DCs in this process. Notably, injection of $CCR7^{-/-}$ DCs, which produced comparable amounts of cytokines but failed to migrate to the draining LN, did not induce cell accumulation in the draining LN. It was therefore concluded that an increase in LN cellularity is rapidly induced and maintained by mature DCs that have reached the draining LN.

Despite a dramatic increase in the number of lymphocytes and a marked increase in size of the inflamed LN (Gretz et al., 2000), the distribution of high and low molecular weight dextrans is the same as in LNs at steady state, thus indicating that the conduit system in LNs remains intact during inflammation.

Spleen cellularity is also rapidly increased after intravenous challenge with microbial products. Intravenous injection of *Toxoplasma gondii* antigen extract (STAg) has been shown to induced a 40% increase of spleen cellularity after 6 h (Reis e Sousa et al., 1997). It is further well established that systemic challenge with microbial products such as LPS (De Smedt et al., 1996; Reis e Sousa et al., 1997) and STAg (Reis e Sousa et al., 1997) cause DCs in the MZ to mature and to rapidly migrate into the T-cell zone of the WP area. In contrast, upon polymicrobial sepsis in mice or bacterial sepsis in humans, the number of resident DCs within the WP area has been shown to decline rapidly (Flohe et al., 2006; Hotchkiss et al., 2002).

NK cell recruitment to the inflamed LN and spleen

Martin-Fontecha and colleagues (2004) were the first to demonstrate that NK cells become vigorously recruited to LNs that are undergoing an immune response. The increase in NK cell numbers was shown to be rapid (peaking on day 2 after DC injection), dependent on the DC dose, and not due to cell proliferation but rather to enhanced cell recruitment. The proportion of T cells and B cells was found to be similar in LN draining and control LNs, while the frequency of NK cells was up to 10-fold higher in DC-draining than in control LNs.

These observations have been confirmed and extended in subsequent reports. In mice locally infected with *Leishmania major* injected in the footpad or in the ear, the number and frequency of NK cells in draining LNs was found to increase 10-fold 24 h after infection (Bajenoff et al., 2006a). Using mice depleted of >90% of $CD11c^+$ MHC class II^+ DCs. Uchida et al. (2007) demonstrated that cellular recruitment, including NK cell recruitment, to draining LNs following the footpad administration of TLR4 and TLR5 agonists, was dramatically decreased upon reduction of DCs, numbers while type I interferon (IFN) production was partially able to substitute for DCs in response to TLR3 and TLR7 agonists.

In the study by Bajenoff et al. (2006a), analysis of LN tissue sections from infected mice revealed that NK cells co-localized with DCs. To independently confirm that NK cells interacted with DCs, CFSE-labelled NK cells were injected into naive mice. Animals were infected with *L. major*, and LNs were analyzed 12 h later by confocal microscopy after localization of DCs by staining with anti-CD11c mAb. Out of 290 individually scored CFSE-labelled NK cells, more than 50% were in contact with large $CD11c^+$ cells.

In a recent study (Watt et al., 2008), the authors reported that mature $CD27^{high}$ NK cells are predominantly recruited into the draining LN following the DC challenge. Importantly, the recruitment of the $CD27^{high}$ NK cell subset in the draining LN was dependent on host IFN-γ and the activation status of NK cells. Notably, endogenous epidermal DC migration induced by hapten challenge was also found to trigger NK cell recruitment to the draining LN in an IFN-γ-dependent fashion.

Two hypotheses could account for the recruitment of NK cells into the inflamed LN. First, NK cells could have been activated in the inflammatory site before their lymph-borne migration to the draining LN. Support for such a hypothesis was provided by Mailliard et al. (2005). In this scenario, which was mainly based on in vitro observations using human NK cells, peripheral activation of NK cells by IL-18 that is produced by pathogen-activated macrophages or immature DCs induces NK cell responsiveness to CCL21 that is produced by lymphatic endothelium and would allow lymph-borne NK cell entry into the draining LN.

Alternatively, NK cell recruitment may occur from peripheral blood, via HEVs. To determine whether the NK cells were recruited to LNs from blood or peripheral tissues, Martin-Fontecha et al. (2004) injected fluorescein-labelled NK cells either intravenously (i.v.) or subcutaneously (s.c.) into mice at the same site of DC injection. Whereas the NK cells injected i.v. were readily recovered in DC-draining LNs along with endogenous NK cells, the NK cells injected s.c. were not, indicating that NK cells are recruited to LNs from blood but not tissues. The requirements for NK cell recruitment were further tested by transferring dye-labelled spleen cells i.v. into mice. The number of NK cells and T cells recruited to DC-draining LNs was markedly reduced by injecting antibodies to CD62L, consistent with the involvement of HEVs. In the study conducted by Bajenoff et al. (2006a), the authors investigated whether fluorescein-labelled NK cells, injected i.v., were detected in the ears at early time points after local infection. Although NK cells were readily found at all times in LNs and blood, they were not detected in PBS-injected or in *L. major*–injected ears, further supporting NK cell recruitment from peripheral blood.

Several reports have suggested that NK cell trafficking within the spleen is modified during inflammation (Andrews et al., 2001; Dokun et al., 2001). In a recent study by Vivier, Walzer and colleagues (Gregoire et al., 2008), it was demonstrated convincingly that NK cells progressively disappeared from the RP and colonized the WP area at 14 h after poly(I:C) injection intraperitoneally (i.p.). The fraction of NK cells present in the WP increased fourfold at 16 h, which was the peak of this migration. NK cells further showed a strong tropism for the T-cell zone compared to the B-cell zone within the WP. The in situ behaviour of spleen NK cells upon infection was further investigated. Murine CMV (MCMV) infection was chosen with regards to the known protective role of NK cells upon mouse infection (Krmpotic et al., 2003). At day 1 post-infection, an influx of NK cells to the MZ and the WP of the spleen was noticed, and the fraction of NK cells present in the WP increased more than twofold. This influx was transient as NK cells were no longer associated with these structures by day 2 post-infection. Notably, the vast majority of NK cells present in the WP upon poly (I:C) injection were in contact with FRCs of the T-cell area. These findings are thus consistent with the possibility that NK cells, like T cells at steady state, track along the FRC network to reach the T-cell zone of the spleen WP upon inflammation. These observations have recently been verified in a MCMV model where Bekiaris et al. (2008) provide evidence that MCMV infection is associated with migration of NK cells to WP areas, where they associate directly with T-zone stromal cells.

Recruitment of circulating NK cells to inflamed LN and spleen T-cell areas is CXCR3-dependent

The study by Martin-Fontecha et al. (2004), further demonstrated that the number of NK cells (and T cells) recruited to DC-draining LNs was severely reduced by pretreatment with pertussis toxin, consistent with the involvement of a Gi-coupled receptor. To determine whether CCR7 was involved, mice were injected i.v. with spleen cells from wild-type or *Ccr7*$^{-/-}$ mice. As expected, the migration of CCR7-deficient spleen cells (containing both NK cells and T cells) to control LNs was severely reduced. However, CCR7-deficient NK cells (and T cells) were found to migrate to DC-draining LNs as efficiently as their wild-type counterparts. These results thus indicate that NK cells and T cells can migrate in a CCR7-independent, but pertussis toxin-sensitive manner to DC-draining LNs. The possibility that the Gi-coupled receptor CXCR3, a receptor for the inflammatory chemokines MIG (CXCL9), IP-10 (CXCL10) and I-TAC (CXCL11), could mediate NK cell recruitment was subsequently tested. Indeed, desensitization of CXCR3 by incubation with high doses of CXCL11 led to a marked reduction in the number of NK cells that migrated to DC-draining LNs. In addition, when spleen cells from *Cxcr3*$^{-/-}$ mice were injected i.v., NK cells did not migrate to DC-draining LNs. Taken together, these results delineated a previously unknown pathway of NK cell recruitment to the inflamed LN that involves the Gi-coupled receptor CXCR3.

In the study by Gregoire et al. (2008), it was demonstrated that inflammation-induced NK cell migration to the WP of the spleen was markedly reduced by pretreatment with pertussis-toxin, indicating the involvement of a Gi-coupled receptor in NK cell recruitment also within the spleen. To test the involvement of CXCR3 in these observations, CXCR3-KO and CCL5-KO mice were used and compared. Upon poly (I:C) treatment (injected i.p.) a marked difference in the distribution of NK cells in wild-type and KO mice was found. Although CXCR3-KO, CCL5-KO and wild-type NK cells appeared to exit RP areas with similar efficiency, CXCR3-KO and CCL5-KO cells did not home to the WP as efficiently as wild-type NK cells, indicating that CXCR3, as well as CCL5, is involved in NK cell migration to the WP of the spleen upon inflammation.

Activated and immigrating DCs as the potential source of CXCL3-ligands

In line with the nonredundant role for CXCR3 in NK cell recruitment into inflamed LNs (Martin-Fontecha et al., 2004), a rapid and strong expression of the CXCR3-ligand CCL9 has been demonstrated on HEVs in LN draining locally injected with Freund's complete adjuvant (FCA) (Janatpour et al., 2001; Yoneyama et al., 2004) or CpG-matured (Guarda et al., 2007) DCs. In the study by Yoneyama et al. (2004), the authors further demonstrated a concurrent decrease of CCL19 on HEVs. It is still unknown, however, whether CXCL9 is produced directly by HEVs or by tissue cells such as FRCs and/or by immigrating activated DCs. Indeed, chemokines, produced in inflamed peripheral tissues or injected s.c. can be transported through lymph and conduits to the 'peri-HEV' region, where they are moved by transcytosis by endothelial cells of HEVs for lumenal display (Baekkevold et al., 2001; Palframan et al., 2001; Stein et al., 2000). The observation that locally injected DCs have to be activated and concurrently express CCR7 in order to induce a subsequent lymphocyte homing to the draining LN (Martin-Fontecha et al., 2003) indicates that activated DCs at least have to reach local draining lymph vessels in order to fulfil such a potential lymph/conduit-dependent mission. If so, the activated DCs have to produce relevant chemokines in a sustained fashion and at time points when the activating stimuli have been withdrawn. Messenger RNA data indeed indicate that DCs, but not monocytes or macrophages, are able to produce CXCL10 in a sustained fashion upon LPS stimulation (Sallusto et al., 1999). Whether the chemokine protein was actually secreted or whether DCs still could produce CXCL10, or other CXCR3-ligands, after withdrawal of LPS-stimulation was however not investigated.

In a recent study, comparing different clinically relevant monocyte-derived human vaccine DCs in vitro, Gustafsson et al. (2008) demonstrated that monocyte-derived immature human DCs stimulated with polyinosinic:polycytidylic acid (p-I:C), IFN-α, TNF-α, IL-1β and IFN-γ, 'α-type-1 polarized DC' (αDC1) (Mailliard et al., 2004), secrete profuse amounts of the CXCR3-ligand CXCL9 and substantial amounts of CXCL10 and CXCL11 in a sustained fashion. Notably, after maturation for 24 h and subsequent withdrawal of maturation stimuli by extensive washing, a single mature αDC1 was able to produce up to 1 pg of CXCL9 within the following 24-h period. In sharp contrast, no measurable CXCL9, CXCL10 or CXCL11 were produced by DCs after maturation with the current 'gold standard' maturation cocktail for human DC-based cancer vaccines consisting of TNF-α, IL-1β, IL-6 and prostaglandin-E_2 (PGE_2-DC). Notably, PGE_2-DCs preferentially produced the Th2 and Treg cell-attracting chemokines CCL17 and CCL22, while only marginal levels of these chemokines were produced by αDC1. Functional studies in vitro further demonstrated that supernatants from mature αDC1 actively recruited $CD3^-CD56^+$ NK cells, and that adding anti-CXCL9-antibodies to the αDC1-supernatant substantially reduced this recruitment. Notably, αDC1s were able to induce IFN-γ-production in $CD56^+$, CD3− NK cells, but only if a CD40 ligand was provided. Taken together, these findings indicate that the potential source for CXCR3-ligands that mediate NK cell recruitment to the T-cell area of the draining LN could be properly activated DCs that deliver CXCR3-ligands into afferent lymph vessels during their migration to the draining LN. These ligands would subsequently be transported to HEVs via the conduit system.

Moreover, in the rhesus macaque, expression of CXCL9 mRNA was found to be one of the most highly up-regulated chemokines (out of 34 chemokines examined) in the spleen early after infection with pathogenic SIV (Reinhart et al., 2002). In situ hybridization for CXCL9 mRNA further revealed a dramatic increase of $mRNA^+$ cells, other than $CD68^+$ macrophages, in the PALS of the WP upon SIV-infection. Based on the well documented immigration of MZ DCs into the T-cell zone of the spleen WP upon inflammation (De Smedt et al., 1996; Reis e Sousa et al., 1997), it is tempting to suggest that these activated and immigrating DCs are the actual producers of CXCR3-ligands and probably also CCR5-ligands (Sallusto et al., 1999) that mediate the inflammation-induced recruitment of NK cells into the WP of the spleen (Bekiaris et al., 2008; Gregoire et al., 2008).

Induced recruitment of NK cells to an inflamed LN provides IFN-gamma for T_H1 priming

To investigate the impact of NK cell recruitment on T-cell priming within the inflamed LN, Martin-Fontecha et al. (2004) adoptively transferred fluorescein-labelled, ovalbumin (OVA)-specific, DO11.10 transgenic $CD4^+$ T cells into syngeneic mice The mice were then immunized by injecting s.c. OVA-pulsed DCs or soluble OVA formulated with different adjuvants, including R848, Ribi, CpG_{1826} and complete Freund's adjuvant (CFA). As expected, T-cell proliferation was strongly induced by OVA-pulsed DCs and by soluble OVA in the presence, but not the absence, of adjuvants. Despite their comparable capacity to induce T-cell proliferation, the adjuvants differed markedly in their ability to recruit NK cells. Whereas R848 and Ribi recruited NK cells as efficiently as LPS-matured DCs, CpG_{1826} and CFA did not. Importantly, NK cell recruitment correlated with the induction of $CD4^+$ cells capable of producing IFN-gamma, whereas IL-4 was not detectable. These results thus indicate that NK cell recruitment is induced selectively by LPS-matured DCs and some adjuvants and correlates with the extent T_H1 polarization. To assess the requirement for NK cells in T_H1 polarization, DO11.10 T cells were primed in the absence of NK cells. In vivo depletion of NK cells impaired T-cell proliferation and differentiation to T_H1 effector cells. The requirement for NK cells in T_H1 priming was also evident when BALB/c mice were immunized by injecting allogeneic mature DCs from C57Bl/6 mice. NK cell-depleted mice were further reconstituted with highly purified NK cells isolated from the spleens of wild-type or *Ifng*$^{-/-}$ mice. Whereas IFN-γ-sufficient and IFN-γ-deficient NK cells migrated into DC-draining LNs with comparable efficiency, only IFN-γ-sufficient NK cells reconstituted T_H1 priming. Finally, when *Ifng*$^{-/-}$ mice were transferred with DO11.10 T cells and immunized with autologous OVA-pulsed DCs, endogenous NK cells were recruited to the draining LNs but did not promote T_H1 polarization. On the adoptive transfer of IFN-γ-sufficient NK cells, however, T_H1 differentiation was reconstituted efficiently. Taken together, these results strongly indicate that NK cells participate in acquired T-cell responses by providing an initial source of IFN-γ that is necessary for the induction of T_H1 cells.

In a subsequent study by Bajenoff et al. (2006a), the frequency of IFN-gamma-secreting NK cells in LN draining skin areas infected with *L. major* was found to increased 100-fold at early time points. Similar results were observed in the popliteal LN after injection of *L. major* into the hind footpad. IFN-γ-secreting cells were localized in the paracortex but rarely in the medulla, indicating that their activation mainly took place in the paracortex, where they were recruited. IFN-γ-secreting cells were also present in DC-rich areas and were in contact with DCs. IFN-γ secretion by NK cells and parasite-induced T-cell activation was demonstrated to occur at the same time in draining LNs and mainly beneath B cell follicles in infected animals. Staining with anti-IFN-γ mAb further showed that TCR transgenic T cells and IFN-γ-secreting cells were located in the same area and contacted the same DC in *L. major*–infected animals. Collectively, these results (Bajenoff et al., 2006a) demonstrated that IFN-γ-secreting NK cells, DCs and $CD4^+$ T cells undergoing activation were localized to and able to interact in the same region of the LN. This location could thus allow NK cells to provide an early source of IFN-γ and influence differentiation of the activated T cells, as suggested in the experimental models used by Martin-Fontecha et al. (2004).

In accordance with these data from mouse models, Ferlazzo et al. (2004b) demonstrated that NK cells from human resting LNs secreted IFN-γ upon co-culture with LPS-matured autologous monocyte-derived DCs or LPS-matured autologous large $CD11c^{high}$ cells isolated from the spleen. Neither immature DCs nor LPS alone were able to induce production of IFN-γ by LN NK cells during the same time interval. These data indicate that, in analogy to the mouse system, DCs can activate human NK cells of secondary lymphoid organs to produce IFN-γ. NK cells from human LNs have further been shown to produce IFN-γ upon stimulation with IL-2 (Fehniger et al., 2003) and DC-derived IL-12 (Ferlazzo et al., 2004a).

DC-mediated priming of NK cells leads to effector NK cell release into the circulation

Nearly 30 years ago Clark, Russel and coworkers proposed that most and perhaps all NK cell activity is activated by a response to exogenous infections (Clark et al., 1979). After infection of mice (raised in conventional conditions) with lymphocytic choriomeningitis virus (LCMV) or MCMV, Welsh et al. (1991) subsequently reported that NK cells were activated to high levels of cytotoxicity within 3 days. Analyses of spleen leukocytes from LCMV-infected mice by a variety of techniques further indicated that the NK cells proliferated and increased in number during infection (Welsh et al., 1991). Infection of mice with the protozoan *L. major* was further shown to induce strong cytotoxic activity in NK cells during the first week of infection in resistant mice. Enhanced disease, as measured by parasite numbers and lesion development, was observed in NK-cell-depleted mice (Scharton and Scott, 1993).

By using a mouse model for the inducible ablation of DCs, Diefenbach and colleagues (Lucas et al., 2007) recently demonstrated that the in vivo priming of NK cell responses to viral (Vaccinia, MCMV) and bacterial pathogens (*Listeria*) as well as TLR ligands 3, 4, 7 and 9 required the presence of $CD11c^{high}$ DCs. NK cell priming was dependant on the recognition of type I IFN signals by DCs and the subsequent production and *trans*-presentation of IL-15 by DCs to resting NK cells. NK cells isolated from mice previously injected i.p. with different TLR-ligands produced IFN-γ when directly triggered by the activating NK cell receptors NKR-P1C (also known as NK1.1), NKG2D or Ly49D, and they killed or produced cytokines in response to tumour target cells expressing a stimulatory NKG2D ligand. Strikingly, NK cells isolated from CD11c^{high} DC-ablated mice displayed strongly reduced effector functions after TLR stimulation in vivo. Similar results were obtained with various other NK cell targets, including YAC-1 or RMA-S. After TLR stimulation, NK cells isolated from CD11c^{high} DC-ablated mice did not accumulate granzyme B and showed impaired up-regulation of CD69. Notably, nonmicrobial stimulation of CD11c^{high} DCs by injection of an agonistic antibody specific for CD40 but not an isotype control antibody stimulated NK cell activation that was DC dependent. Similar results were obtained with CD11c DTR tg mice deficient for recombination-activating gene 1 (*Rag1*$^{-/-}$), which lack all T and B cells but have a functional NK cell and myeloid cell compartment, indicating that B cells and/or T cells do not substantially contribute to DC-mediated NK cell activation. These findings have recently been confirmed by Hochweller et al. (2008), demonstrating that NK cells require DC for full acquisition of effector function in vivo in response to the bacterial-derived TLR ligand CpG. Notably, NK cell cytotoxicity was evaluated in vivo, using injected CFSE-labelled tapasin-deficient splenocytes as NK targets.

To directly address the question of whether the emergence of peripheral effector NK cells requires entry of naive NK cells into secondary lymphoid organs, the function of peripheral NK cells was evaluated under conditions that prevented NK cells from entering the draining LN (Lucas et al., 2007). Blocking entry of NK cells into LN by treatment with a CD62L antibody prior to local microbial stimulation resulted in dramatically decreased numbers of effector NK cells in peripheral tissues. To directly demonstrate that circulating naive NK cells must enter the draining LN for priming, highly purified naive NK cells were adoptively transferred into syngeneic mice prior to local TLR stimulation. In control mice, transferred NK cells emerged as effector cells (IFN-γ production, granzyme B up-regulation) in peripheral tissues, whereas NK cells prevented from recruitment into LN remained naive. Thus, NK cells were recruited to LNs draining microbial infections, and LN entry and NK cell–DC interaction was found to be required for the emergence of primed NK cells in the periphery.

Delivery of ex vivo propagated DCs to mice have also been found to result in an increase in NK cell cytolytic activity. The first report on this issue was came from Zitvogel and colleagues (Fernandez et al., 1999) where adoptively transferred DCs or Flt3 ligand-expanded DCs promoted NK cell-dependent anti-tumour effects. Adoptive transfer of DCs expressing IL-12 to mice has also been shown to markedly increased NK cell IFN-gamma production and lytic activity in vivo (Miller et al., 2003). In a recent study (Karimi et al., 2008), subcutaneous injections of vaccine DCs in mice was further shown to trigger NK cells to acquire effector functions. In analogy with the dependence of *trans*-presentation of IL-15 by DCs to obtain NK cell priming in the vivo mouse model (Lucas et al., 2007), proliferation and survival of NK cells from human LN has been shown to depend on membrane-bound IL-15 on mature DCs (Ferlazzo et al., 2004a). Moreover, NK cells in human LN, lacking perforin, KIRs and all NCR expression, except low levels of NKp46, up-regulate NCRs, express perforin and acquire cytolytic activity for NK-sensitive target cells upon IL-2 stimulation in vitro (Ferlazzo et al., 2004b). Taken together, all these observations clearly indicate that if properly activated, adoptively transferred or locally injected DCs have the ability to prime NK cells within secondary lymphoid organs for subsequent cytotoxic activity in vivo.

References

Adam, C., King, S., Allgeier, T., Braumuller, H., Luking, C., Mysliwietz, J., Kriegeskorte, A., Busch, D.H., Rocken, M. and Mocikat, R. (2005). DC–NK cell cross talk as a novel CD4+ T-cell-independent pathway for antitumor CTL induction. Blood **106**, 338–344.

Anderson, A.O. and Anderson, N.D. (1975). Studies on the structure and permeability of the microvasculature in normal rat lymph nodes. Am J Pathol **80**, 387–418.

Andrews, D.M., Farrell, H.E., Densley, E.H., Scalzo, A.A., Shellam, G.R. and Degli-Esposti, M.A. (2001). NK1.1+ cells and murine cytomegalovirus infection: what happens in situ? J Immunol **166**, 1796–1802.

Andrews, D.M., Scalzo, A.A., Yokoyama, W.M., Smyth, M.J. and Degli-Esposti, M.A. (2003). Functional interactions between dendritic cells and NK cells during viral infection. Nat Immunol **4**, 175–181.

Asperti-Boursin, F., Real, E., Bismuth, G., Trautmann, A. and Donnadieu, E. (2007). CCR7 ligands control basal T cell motility within lymph node slices in a phosphoinositide 3-kinase-independent manner. J Exp Med **204**, 1167–1179.

Baekkevold, E.S., Yamanaka, T., Palframan, R.T., Carlsen, H.S., Reinholt, F.P., von Andrian, U.H., Brandtzaeg, P. and Haraldsen, G. (2001). The CCR7 ligand elc (CCL19) is transcytosed in high endothelial venules and mediates T cell recruitment. J Exp Med **193**, 1105–1112.

Bajenoff, M., Granjeaud, S. and Guerder, S. (2003). The strategy of T cell antigen-presenting cell encounter in antigen-draining lymph nodes revealed by imaging of initial T cell activation. J Exp Med **198**, 715–724.

Bajenoff, M., Breart, B., Huang, A.Y., Qi, H., Cazareth, J., Braud, V.M., Germain, R.N. and Glaichenhaus, N. (2006a). Natural killer cell behavior in lymph nodes revealed by static and real-time imaging. J Exp Med **203**, 619–631.

Bajenoff, M., Egen, J.G., Koo, L.Y., Laugier, J.P., Brau, F., Glaichenhaus, N. and Germain, R.N. (2006b). Stromal cell networks regulate lymphocyte entry, migration, and territoriality in lymph nodes. Immunity **25**, 989–1001.

Bajenoff, M., Egen, J.G., Qi, H., Huang, A.Y., Castellino, F. and Germain, R.N. (2007). Highways, byways and breadcrumbs: directing lymphocyte traffic in the lymph node. Trends Immunol **28**, 346–352.

Bajenoff, M., Glaichenhaus, N. and Germain, R.N. (2008). Fibroblastic reticular cells guide T lymphocyte entry into and migration within the splenic T cell zone. J Immunol **181**, 3947–3954.

Banchereau, J. and Steinman, R.M. (1998). Dendritic cells and the control of immunity. Nature **392**, 245–252.

Bekiaris, V., Timoshenko, O., Hou, T.Z., Toellner, K., Shakib, S., Gaspal, F., McConnell, F.M., Parnell, S.M., Withers, D., Buckley, C.D., Sweet, C., Yokoyama, W.M., Anderson, G. and Lane, P.J. (2008). Ly49H+ NK cells migrate to and protect splenic white pulp stroma from murine cytomegalovirus infection. J Immunol **180**, 6768–6776.

Bianchi, G., Sironi, M., Ghibaudi, E., Selvaggini, C., Elices, M., Allavena, P. and Mantovani, A. (1993). Migration of natural killer cells across endothelial cell monolayers. J Immunol **151**, 5135–5144.

Borrego, F., Masilamani, M., Kabat, J., Sanni, T.B. and Coligan, J.E. (2005). The cell biology of the human natural killer cell CD94/NKG2A inhibitory receptor. Mol Immunol **42**, 485–488.

Briard, D., Brouty-Boye, D., Azzarone, B. and Jasmin, C. (2002). Fibroblasts from human spleen regulate NK cell differentiation from blood CD34(+) progenitors via cell surface IL-15. J Immunol **168**, 4326–4332.

Brouty-Boye, D., Doucet, C., Clay, D., Le Bousse-Kerdiles, M.C., Lampidis, T.J. and Azzarone, B. (1998). Phenotypic diversity in human fibroblasts from myelometaplasic and non-myelometaplasic hematopoietic tissues. Int J Cancer **76**, 767–773.

Bryceson, Y.T., March, M.E., Ljunggren, H.G. and Long, E.O. (2006). Synergy among receptors on resting NK cells for the activation of natural cytotoxicity and cytokine secretion. Blood **107**, 159–166.

Butcher, E.C. (1991). Leukocyte–endothelial cell recognition: three (or more) steps to specificity and diversity. Cell **67**, 1033–1036.

Butcher, E.C. and Picker, L.J. (1996). Lymphocyte homing and homeostasis. Science **272**, 60–66.

Cahill, R.N., Frost, H. and Trnka, Z. (1976). The effects of antigen on the migration of recirculating lymphocytes through single lymph nodes. J Exp Med **143**, 870–888.

Campbell, J.J., Qin, S., Unutmaz, D., Soler, D., Murphy, K.E., Hodge, M.R., Wu, L. and Butcher, E.C. (2001). Unique subpopulations of CD56+ NK and NK-T peripheral blood lymphocytes identified by chemokine receptor expression repertoire. J Immunol **166**, 6477–6482.

Chen, S., Kawashima, H., Lowe, J.B., Lanier, L.L. and Fukuda, M. (2005). Suppression of tumor formation in lymph nodes by L-selectin-mediated natural killer cell recruitment. J Exp Med **202**, 1679–1689.

Chen, Q., Fisher, D.T., Clancy, K.A., Gauguet, J.M., Wang, W.C., Unger, E., Rose-John, S., von Andrian, U.H., Baumann, H. and Evans, S.S. (2006). Fever-range thermal stress promotes lymphocyte trafficking across high endothelial venules via an interleukin 6 trans-signaling mechanism. Nat Immunol **7**, 1299–1308.

Clark, E.A., Russell, P.H., Egghart, M. and Horton, M.A. (1979). Characteristics and genetic control of NK-cell-mediated cytotoxicity activated by naturally acquired infection in the mouse. Int J Cancer **24**, 688–699.

Cupedo, T., Crellin, N.K., Papazian, N., Rombouts, E.J., Weijer, K., Grogan, J.L., Fibbe, W.E., Cornelissen, J.J. and Spits, H. (2009). Human fetal lymphoid tissue-inducer cells are interleukin 17-producing precursors to RORC+ CD127+ natural killer-like cells. Nat Immunol **10**, 66–74.

De Smedt, T., Pajak, B., Muraille, E., Lespagnard, L., Heinen, E., De Baetselier, P., Urbain, J., Leo, O. and Moser, M. (1996). Regulation of dendritic cell numbers and maturation by lipopolysaccharide in vivo. J Exp Med **184**, 1413–1424.

Diefenbach, A. and Raulet, D.H. (2003). Innate immune recognition by stimulatory immunoreceptors. Curr Opin Immunol **15**, 37–44.

Djeu, J.Y., Heinbaugh, J.A., Holden, H.T. and Herberman, R.B. (1979). Role of macrophages in the augmentation of mouse natural killer cell activity by poly I:C and interferon. J Immunol **122**, 182–188.

Dokun, A.O., Chu, D.T., Yang, L., Bendelac, A.S. and Yokoyama, W.M. (2001). Analysis of in situ NK cell responses during viral infection. J Immunol **167**, 5286–5293.

Drayton, D.L., Liao, S., Mounzer, R.H. and Ruddle, N.H. (2006). Lymphoid organ development: from ontogeny to neogenesis. Nat Immunol **7**, 344–353.

Fehniger, T.A., Cooper, M.A., Nuovo, G.J., Cella, M., Facchetti, F., Colonna, M. and Caligiuri, M.A. (2003). CD56bright natural killer cells are present in human lymph nodes and are activated by T cell-derived IL-2: a potential new link between adaptive and innate immunity. Blood **101**, 3052–3057.

Ferlazzo, G., Tsang, M.L., Moretta, L., Melioli, G., Steinman, R.M. and Munz, C. (2002). Human dendritic cells activate resting natural killer (NK) cells and are recognized via the NKp30 receptor by activated NK cells. J Exp Med **195**, 343–351.

Ferlazzo, G., Pack, M., Thomas, D., Paludan, C., Schmid, D., Strowig, T., Bougras, G., Muller, W.A., Moretta, L. and Munz, C. (2004a). Distinct roles of IL-12 and IL-15 in human natural killer cell activation by dendritic cells from secondary lymphoid organs. Proc Natl Acad Sci U S A **101**, 16606–16611.

Ferlazzo, G., Thomas, D., Lin, S.L., Goodman, K., Morandi, B., Muller, W.A., Moretta, A. and Munz, C. (2004b). The abundant NK cells in human secondary lymphoid tissues require activation to express killer cell Ig-like receptors and become cytolytic. J Immunol **172**, 1455–1462.

Fernandez, N.C., Lozier, A., Flament, C., Ricciardi-Castagnoli, P., Bellet, D., Suter, M., Perricaudet, M., Tursz, T., Maraskovsky, E. and Zitvogel, L. (1999). Dendritic cells directly trigger NK cell functions: cross-talk relevant in innate anti-tumor immune responses in vivo. Nat Med **5**, 405–411.

Flohe, S.B., Agrawal, H., Schmitz, D., Gertz, M., Flohe, S. and Schade, F.U. (2006). Dendritic cells during polymicrobial sepsis rapidly mature but fail to initiate a protective Th1-type immune response. J Leukoc Biol **79**, 473–481.

Forster, R., Schubel, A., Breitfeld, D., Kremmer, E., Renner-Muller, I., Wolf, E. and Lipp, M. (1999). CCR7 coordinates the primary immune response by establishing functional microenvironments in secondary lymphoid organs. Cell **99**, 23–33.

Freud, A.G., Becknell, B., Roychowdhury, S., Mao, H.C., Ferketich, A.K., Nuovo, G.J., Hughes, T.L., Marburger, T.B., Sung, J., Baiocchi, R.A., Guimond, M. and Caligiuri, M.A. (2005). A human CD34(+) subset resides in lymph nodes and differentiates into CD56bright natural killer cells. Immunity **22**, 295–304.

Freud, A.G., Yokohama, A., Becknell, B., Lee, M.T., Mao, H.C., Ferketich, A.K. and Caligiuri, M.A. (2006). Evidence for discrete stages of human natural killer cell differentiation in vivo. J Exp Med **203**, 1033–1043.

Frey, M., Packianathan, N.B., Fehniger, T.A., Ross, M.E., Wang, W.C., Stewart, C.C., Caligiuri, M.A. and Evans, S.S. (1998). Differential expression and function of L-selectin on CD56bright and CD56dim natural killer cell subsets. J Immunol **161**, 400–408.

Gerosa, F., Baldani-Guerra, B., Nisii, C., Marchesini, V., Carra, G. and Trinchieri, G. (2002). Reciprocal activating interaction between natural killer cells and dendritic cells. J Exp Med **195**, 327–333.

Ghiringhelli, F., Menard, C., Martin, F. and Zitvogel, L. (2006). The role of regulatory T cells in the control of natural killer cells: relevance during tumor progression. Immunol Rev **214**, 229–238.

Gidlund, M., Orn, A., Wigzell, H., Senik, A. and Gresser, I. (1978). Enhanced NK cell activity in mice injected with interferon and interferon inducers. Nature **273**, 759–761.

Giroux, M., Yurchenko, E., St-Pierre, J., Piccirillo, C.A. and Perreault, C. (2007). T regulatory cells control numbers of NK cells and CD8alpha+ immature dendritic cells in the lymph node paracortex. J Immunol **179**, 4492–4502.

Gregoire, C., Cognet, C., Chasson, L., Coupet, C.A., Dalod, M., Reboldi, A., Marvel, J., Sallusto, F., Vivier, E. and Walzer, T. (2008). Intrasplenic trafficking of natural killer cells is redirected by chemokines upon inflammation. Eur J Immunol **38**, 2076–2084.

Gretz, J.E., Kaldjian, E.P., Anderson, A.O. and Shaw, S. (1996). Sophisticated strategies for information encounter in the lymph node: the reticular network as a conduit of soluble information and a highway for cell traffic. J Immunol **157**, 495–499.

Gretz, J.E., Anderson, A.O. and Shaw, S. (1997). Cords, channels, corridors and conduits: critical architectural elements facilitating cell interactions in the lymph node cortex. Immunol Rev **156**, 11–24.

Gretz, J.E., Norbury, C.C., Anderson, A.O., Proudfoot, A.E. and Shaw, S. (2000). Lymph-borne chemokines and other low molecular weight molecules reach high endothelial venules via specialized conduits while a functional barrier limits access to the lymphocyte microenvironments in lymph node cortex. J Exp Med **192**, 1425–1440.

Guarda, G., Hons, M., Soriano, S.F., Huang, A.Y., Polley, R., Martin-Fontecha, A., Stein, J.V., Germain, R.N., Lanzavecchia, A. and Sallusto, F. (2007). L-selectin-negative CCR7− effector and memory CD8+ T cells enter reactive lymph nodes and kill dendritic cells. Nat Immunol **8**, 743–752.

Gustafsson, K., Ingelsten, M., Bergqvist, L., Nystrom, J., Andersson, B. and Karlsson-Parra, A. (2008). Recruitment and activation of natural killer cells in vitro by a human dendritic cell vaccine. Cancer Res **68**, 5965–5971.

Hackett , J. Jr., Bennett, M. and Kumar, V. (1985). Origin and differentiation of natural killer cells. I. Characteristics of a transplantable NK cell precursor. J Immunol **134**, 3731–3738.

Hayakawa, Y. and Smyth, M.J. (2006). CD27 dissects mature NK cells into two subsets with distinct responsiveness and migratory capacity. J Immunol **176**, 1517–1524.

Herberman, R.B., Nunn, M.E., Holden, H.T. and Lavrin, D.H. (1975). Natural cytotoxic reactivity of mouse lymphoid cells against syngeneic and allogeneic tumors. II. Characterization of effector cells. Int J Cancer **16**, 230–239.

Hochweller, K., Striegler, J., Hammerling, G.J. and Garbi, N. (2008). A novel CD11c.DTR transgenic mouse for depletion of dendritic cells reveals their requirement for homeostatic proliferation of natural killer cells. Eur J Immunol **38**, 2776–2783.

Hotchkiss, R.S., Tinsley, K.W., Swanson, P.E., Grayson, M.H., Osborne, D.F., Wagner, T.H., Cobb, J.P., Coopersmith, C. and Karl, I.E. (2002). Depletion of dendritic cells, but not macrophages, in patients with sepsis. J Immunol **168**, 2493–2500.

Janatpour, M.J., Hudak, S., Sathe, M., Sedgwick, J.D. and McEvoy, L.M. (2001). Tumor necrosis factor-dependent segmental control of MIG expression by high endothelial venules in inflamed lymph nodes regulates monocyte recruitment. J Exp Med **194**, 1375–1384.

Jenh, C.H., Cox, M.A., Kaminski, H., Zhang, M., Byrnes, H., Fine, J., Lundell, D., Chou, C.C., Narula, S.K. and Zavodny, P.J. (1999). Cutting edge: species specificity of the CC chemokine 6Ckine signaling through the CXC chemokine receptor CXCR3: human 6Ckine is not a ligand for the human or mouse CXCR3 receptors. J Immunol **162**, 3765–3769.

Karimi, K., Boudreau, J.E., Fraser, K., Liu, H., Delanghe, J., Gauldie, J., Xing, Z., Bramson, J.L. and Wan, Y. (2008). Enhanced antitumor immunity elicited by dendritic cell vaccines is a result of their ability to engage both CTL and IFN gamma-producing NK cells. Mol Ther **16**, 411–418.

Kassim, S.H., Rajasagi, N.K., Zhao, X., Chervenak, R. and Jennings, S.R. (2006). In vivo ablation of CD11c-positive dendritic cells increases susceptibility to herpes simplex virus type 1 infection and diminishes NK and T-cell responses. J Virol **80**, 3985–3993.

Kiessling, R., Klein, E. and Wigzell, H. (1975). "Natural" killer cells in the mouse. I. Cytotoxic cells with specificity for mouse Moloney leukemia cells. Specificity and distribution according to genotype. Eur J Immunol **5**, 112–117.

Kim, C.H., Pelus, L.M., Appelbaum, E., Johanson, K., Anzai, N. and Broxmeyer, H.E. (1999). CCR7 ligands, SLC/6Ckine/Exodus2/TCA4 and CKbeta-11/MIP-3beta/ELC, are chemoattractants for CD56(+)CD16(−) NK cells and late stage lymphoid progenitors. Cell Immunol **193**, 226–235.

Kim, J.M., Rasmussen, J.P. and Rudensky, A.Y. (2007). Regulatory T cells prevent catastrophic autoimmunity throughout the lifespan of mice. Nat Immunol **8**, 191–197.

Krmpotic, A., Bubic, I., Polic, B., Lucin, P. and Jonjic, S. (2003). Pathogenesis of murine cytomegalovirus infection. Microbes Infect **5**, 1263–1277.

Kuschert, G.S., Coulin, F., Power, C. A., Proudfoot, A.E., Hubbard, R.E., Hoogewerf, A.J. and Wells, T.N. (1999). Glycosaminoglycans interact selectively with chemokines and modulate receptor binding and cellular responses. Biochemistry **38**, 12959–12968.

Lucas, M., Schachterle, W., Oberle, K., Aichele, P. and Diefenbach, A. (2007). Dendritic cells prime natural killer cells by trans-presenting interleukin 15. Immunity **26**, 503–517.

Luther, S.A., Tang, H.L., Hyman, P.L., Farr, A.G. and Cyster, J.G. (2000). Coexpression of the chemokines ELC and SLC by T zone stromal cells and deletion of the ELC gene in the plt/plt mouse. Proc Natl Acad Sci U S A **97**, 12694–12699.

Lyons, A.B. and Parish, C.R. (1995). Are murine marginal-zone macrophages the splenic white pulp analog of high endothelial venules? Eur J Immunol **25**, 3165–3172.

Mailliard, R.B., Wankowicz-Kalinska, A., Cai, Q., Wesa, A., Hilkens, C.M., Kapsenberg, M.L., Kirkwood, J.M., Storkus, W.J. and Kalinski, P. (2004). Alpha-type-1 polarized dendritic cells: a novel immunization tool with optimized CTL-inducing activity. Cancer Res **64**, 5934–5937.

Mailliard, R.B., Alber, S.M., Shen, H., Watkins, S.C., Kirkwood, J.M., Herberman, R.B. and Kalinski, P. (2005). IL-18-induced CD83 + CCR7+ NK helper cells. J Exp Med **202**, 941–953.

Martin-Fontecha, A., Sebastiani, S., Hopken, U.E., Uguccioni, M., Lipp, M., Lanzavecchia, A. and Sallusto, F. (2003). Regulation of dendritic cell migration to the draining lymph node: impact on T lymphocyte traffic and priming. J Exp Med **198**, 615–621.

Martin-Fontecha, A., Thomsen, L.L., Brett, S., Gerard, C., Lipp, M., Lanzavecchia, A. and Sallusto, F. (2004). Induced recruitment of NK cells to lymph nodes provides IFN-gamma for T(H)1 priming. Nat Immunol **5**, 1260–1265.

Mebius, R.E. (2003). Organogenesis of lymphoid tissues. Nat Rev Immunol **3**, 292–303.

Miller, G., Lahrs, S. and Dematteo, R.P. (2003). Overexpression of interleukin-12 enables dendritic cells to activate NK cells and confer systemic antitumor immunity. FASEB J **17**, 728–730.

Moretta, A., Bottino, C., Vitale, M., Pende, D., Cantoni, C., Mingari, M. C., Biassoni, R. and Moretta, L. (2001). Activating receptors and coreceptors involved in human natural killer cell-mediated cytolysis. Annu Rev Immunol **19**, 197–223.

Moretta, L., Bottino, C., Pende, D., Mingari, M.C., Biassoni, R. and Moretta, A. (2002). Human natural killer cells: their origin, receptors and function. Eur J Immunol **32**, 1205–1211.

Nakano, H., Mori, S., Yonekawa, H., Nariuchi, H., Matsuzawa, A. and Kakiuchi, T. (1998). A novel mutant gene involved in T-lymphocyte-specific homing into peripheral lymphoid organs on mouse chromosome 4. Blood **91**, 2886–2895.

Natarajan, K., Dimasi, N., Wang, J., Mariuzza, R.A. and Margulies, D. H. (2002). Structure and function of natural killer cell receptors: multiple molecular solutions to self, nonself discrimination. Annu Rev Immunol **20**, 853–885.

Nolte, M.A., Belien, J.A., Schadee-Eestermans, I., Jansen, W., Unger, W. W., van Rooijen, N., Kraal, G. and Mebius, R.E. (2003). A conduit system distributes chemokines and small blood-borne molecules through the splenic white pulp. J Exp Med **198**, 505–512.

Okada, T. and Cyster, J.G. (2007). CC chemokine receptor 7 contributes to Gi-dependent T cell motility in the lymph node. J Immunol **178**, 2973–2978.

Palframan, R.T., Jung, S., Cheng, G., Weninger, W., Luo, Y., Dorf, M., Littman, D.R., Rollins, B.J., Zweerink, H., Rot, A. and von Andrian, U.H. (2001). Inflammatory chemokine transport and presentation in HEV: a remote control mechanism for monocyte recruitment to lymph nodes in inflamed tissues. J Exp Med **194**, 1361–1373.

Reinhart , T.A., Fallert, B.A., Pfeifer, M.E., Sanghavi, S., Capuano, S., Rajakumar, P., Murphey-Corb, M. III, Day, R., Fuller, C.L. and Schaefer, T. M. (2002). Increased expression of the inflammatory chemokine CXC chemokine ligand 9/monokine induced by interferon-gamma in lymphoid tissues of rhesus macaques during simian immunodeficiency virus infection and acquired immunodeficiency syndrome. Blood **99**, 3119–3128.

Reis e Sousa, C., Hieny, S., Scharton-Kersten, T., Jankovic, D., Charest, H., Germain, R.N. and Sher, A. (1997). In vivo microbial stimulation induces rapid CD40 ligand-independent production of interleukin 12 by dendritic cells and their redistribution to T cell areas. J Exp Med **186**, 1819–1829.

Roozendaal, R., Mebius, R.E. and Kraal, G. (2008). The conduit system of the lymph node. Int Immunol..

Sakaguchi, S. (2004). Naturally arising CD4+ regulatory t cells for immunologic self-tolerance and negative control of immune responses. Annu Rev Immunol **22**, 531–562.

Salazar-Mather, T.P., Ishikawa, R. and Biron, C.A. (1996). NK cell trafficking and cytokine expression in splenic compartments after IFN induction and viral infection. J Immunol **157**, 3054–3064.

Sallusto, F., Palermo, B., Lenig, D., Miettinen, M., Matikainen, S., Julkunen, I., Forster, R., Burgstahler, R., Lipp, M. and Lanzavecchia, A. (1999). Distinct patterns and kinetics of chemokine production regulate dendritic cell function. Eur J Immunol **29**, 1617–1625.

Scharton, T.M. and Scott, P. (1993). Natural killer cells are a source of interferon gamma that drives differentiation of CD4+ T cell subsets and induces early resistance to *Leishmania major* in mice. J Exp Med **178**, 567–577.

Schmidt, E.E., MacDonald, I.C. and Groom, A.C. (1993). Comparative aspects of splenic microcirculatory pathways in mammals: the region bordering the white pulp. Scanning Microsc **7**, 613–628.

Shevach, E.M. (2002). CD4+ CD25+ suppressor T cells: more questions than answers. Nat Rev Immunol **2**, 389–400.

Shimizu, Y., Newman, W., Tanaka, Y. and Shaw, S. (1992). Lymphocyte interactions with endothelial cells. Immunol Today **13**, 106–112.

Soderberg, K.A., Payne, G.W., Sato, A., Medzhitov, R., Segal, S.S. and Iwasaki, A. (2005). Innate control of adaptive immunity via remodeling of lymph node feed arteriole. Proc Natl Acad Sci U S A **102**, 16315–16320.

Soto, H., Wang, W., Strieter, R.M., Copeland, N.G., Gilbert, D.J., Jenkins, N.A., Hedrick, J. and Zlotnik, A. (1998). The CC chemokine 6Ckine binds the CXC chemokine receptor CXCR3. Proc Natl Acad Sci U S A **95**, 8205–8210.

Springer, T.A. (1994). Traffic signals for lymphocyte recirculation and leukocyte emigration: the multistep paradigm. Cell **76**, 301–314.

Stein, J.V., Rot, A., Luo, Y., Narasimhaswamy, M., Nakano, H., Gunn, M.D., Matsuzawa, A., Quackenbush, E. J., Dorf, M.E. and von Andrian, U.H. (2000). The CC chemokine thymus-derived chemotactic agent 4 (TCA-4, secondary lymphoid tissue chemokine, 6Ckine, exodus-2) triggers lymphocyte function-associated antigen 1-mediated arrest of rolling T lymphocytes in peripheral lymph node high endothelial venules. J Exp Med **191**, 61–76.

Terme, M., Chaput, N., Combadiere, B., Ma, A., Ohteki, T. and Zitvogel, L. (2008). Regulatory T cells control dendritic cell/NK cell cross-talk in lymph nodes at the steady state by inhibiting CD4+ self-reactive T cells. J Immunol **180**, 4679–4686.

Timens, W. and Poppema, S. (1985). Lymphocyte compartments in human spleen. An immunohistologic study in normal spleens and uninvolved spleens in Hodgkin's disease. Am J Pathol **120**, 443–454.

Trinchieri, G. (1989). Biology of natural killer cells. Adv Immunol **47**, 187–376.

Uchida, T., Scumpia, P.O., Murasko, D. M., Seki, S., Woulfe, S., Clare-Salzler, M.J. and Moldawer, L.L. (2007). Variable requirement of dendritic cells for recruitment of NK and T cells to different TLR agonists. J Immunol **178**, 3886–3892.

Watt, S.V., Andrews, D.M., Takeda, K., Smyth, M.J. and Hayakawa, Y. (2008). IFN-gamma-dependent recruitment of mature CD27(high) NK cells to lymph nodes primed by dendritic cells. J Immunol **181**, 5323–5330.

Welsh, R.M., Brubaker, J.O., Vargas-Cortes, M. and O'Donnell, C. L. (1991). Natural killer (NK) cell response to virus infections in mice with severe combined immunodeficiency. The stimulation of NK cells and the NK cell-dependent control of virus infections occur independently of T and B cell function. J Exp Med **173**, 1053–1063.

Vivier, E., Munroe, M., Ariniello, P. and Anderson, P. (1995). Identification of tissue-infiltrating lymphocytes expressing PEN5, a mucin-like glycoprotein selectively expressed on natural killer cells. Am J Pathol **146**, 409–418.

von Andrian, U.H. and Mempel, T.R. (2003). Homing and cellular traffic in lymph nodes. Nat Rev Immunol **3**, 867–878.

Worbs, T., Mempel, T.R., Bolter, J., von Andrian, U.H. and Forster, R. (2007). CCR7 ligands stimulate the intranodal motility of T lymphocytes in vivo. J Exp Med **204**, 489–495.

Vossen, M.T., Matmati, M., Hertoghs, K. M., Baars, P.A., Gent, M.R., Leclercq, G., Hamann, J., Kuijpers, T.W. and van Lier, R.A. (2008). CD27 defines phenotypically and functionally different human NK cell subsets. J Immunol **180**, 3739–3745.

Yoneyama, H., Matsuno, K., Zhang, Y., Nishiwaki, T., Kitabatake, M., Ueha, S., Narumi, S., Morikawa, S., Ezaki, T., Lu, B., Gerard, C., Ishikawa, S. and Matsushima, K. (2004). Evidence for recruitment of plasmacytoid dendritic cell precursors to inflamed lymph nodes through high endothelial venules. Int Immunol **16**, 915–928.

Zaft, T., Sapoznikov, A., Krauthgamer, R., Littman, D.R. and Jung, S. (2005). CD11chigh dendritic cell ablation impairs lymphopenia-driven proliferation of naive and memory CD8+ T cells. J Immunol **175**, 6428–6435.

Chapter Twenty-Eight

28

Nature killer cells in the central nervous system

Fu-Dong Shi, Richard M. Ransohoff

CHAPTER CONTENTS

A moment of realization is worth a thousand prayers.

From the movie *Natural Born Killers*, 1994

ABSTRACT

Natural killer (NK) cells, a prominent component of the innate immune system, are large granular lymphocytes that respond rapidly to a variety of insults via cytokine secretion and cytolytic activity. Recently, there has been growing insight into the biological functions of NK cells, in particular into their roles in infection, tumour surveillance and autoimmunity. Under these pathophysiological circumstances, NK cells readily home to the central nervous system (CNS) tissues to combat infection and presumably to curb progression of tumours. Bystander neuronal and/or glial cell damage can occur in this setting. Paradoxically, NK cells appear to have an inhibitory role for autoimmune responses within the CNS. As in the periphery, NK cells act in concert with T cells and other lymphocytes responsible for CNS pathology and immune regulation. Insights into the molecular signals and pathways governing the diverse biological effects of NK cells are keys for designing NK cell based therapy for CNS infections, tumours and autoimmune diseases.

KEY WORDS

Tumour, Infection, Autoimmunity, Myelin, Glia, Herpes, Oligodendrocyte

Homing of NK cells into the CNS

The central nervous system (CNS) is an immune privileged organ because of the lack of endogenous dendritic cells (DCs) (Ransohoff et al., 2003). In many pathological conditions including stroke, traumatic injury, encephalitis, and demyelinating autoimmune disorders, however, a massive infiltration of peripheral immune cells occurs in the CNS. Although natural killer (NK) cell-related gene expression is readily detected within the CNS (Bryceson et al., 2005; Lund et al., 2007), evidence of direct demonstration of the presence of NK cells in human CNS tissues is still lacking. This is partly because of the lack of suitable antibodies for staining human NK cells in situ (see Chapter 31). Visualization of NK cells in mouse brains during experimental autoimmune encephalomyelitis (EAE) was achieved using anti-NK1.1 mAb (PK136) (Hammarberg et al., 2000). Antibodies such as Ly49 have been successfully used for staining NK cells in lymphoid organs (O'Leary et al., 2006). Suitability of these antibodies for staining CNS NK cells needs to be verified in additional studies.

Although direct visualization of NK cells in CNS tissues is technically challenging, there is little doubt that NK cells, as with other types of lymphocytes, enter the CNS during inflammatory processes. In fact, it has been reported that NK cells are among the earliest recruited cells during adoptive transfer EAE (Kerfoot and Kubes, 2002; Wekerle and Fierz, 1985). Chemokine receptors such as CCR2, CCR5, CXCR3, CX3CR1 as well as lysophospholipid sphingosine 1-phosphate (S1P) are involved in the rapid NK-cell mobilization that occurs in inflammatory conditions (Ajuebor et al., 2007; Hokeness et al., 2005; Huang et al., 2006; Inngjerdingen et al., 2001; Jiang et al., 2004; Khan et al., 2006; Kveberg et al., 2002; Lavergne et al., 2003; Martin-Fontecha et al., 2004; Thapa et al., 2007; Wald et al., 2006; Walzer et al., 2007; Yu et al., 2007), and several of these chemokine receptors (CCR5, CX3CR1) are directly involved in NK cell homing to the CNS (Huang et al., 2006; Martin-Fontecha et al., 2004; Thapa et al., 2007). The biological implication of chemokine-guided homing of NK cells during CNS inflammation is discussed in greater detail in the following section.

NK cell-mediated neuron, oligodendrocyte and glial cell damage

The ability of NK cells to kill various transformed and virus-infected cells raises an important question whether direct NK cell cytolytic effects contribute to the pathogenesis of inflammatory, degenerative and autoimmune disorders of the nervous system. Neurodegenerative diseases such as Parkinson's disease and Alzheimer's disease are characterized by the death of neurons in distinct functional neuron-anatomic systems. Multiple sclerosis (MS), on the other hand, is characterized by inflammation and demyelination within the spinal cord and brain, and axonal damage and brain atrophy also occur during the course of disease. The peripheral form of MS is Guillain–Barré syndrome, which is also characterized by demyelination and cellular infiltrates of the peripheral nervous system. Some of these diseases involve immunologic components or reactions, and some have been characterized extensively in the different systems. Autoreactive T cells and adaptive immune system components in the pathogenesis in some of these disorders are well characterized. As discussed here, evidence for direct NK cell cytolytic effects is emerging. Although the in vivo relevance of a great proportion of these studies needs to be validated, the current data emphasize the importance of NK cells either in direct cytotoxic effects or in collaboration with cells from both innate and adaptive immune systems in the initiation of these neurodegenerative and inflammatory diseases.

Neuron

The NK cell-dependent death of sympathetic neurons resident in the superior cervical ganglia of rats, observed after the exposure to the drug guanethidine (Hickey et al., 1992; Hougen et al., 1992), is the first in vivo disorder of the nervous system in which NK cells appear to be the dominant effector cells. The pathogenic mechanism observed appeared to represent a novel type of autoimmune reaction: an exogenously/chemically induced alteration in a specific subset of cells that was suggested to target them for direct NK cell-mediated killing.

Interestingly, neuronal cells from the peripheral system and the CNS appear to have different susceptibility to NK cell killing. Ljunggren and colleagues have carried out a series of well-designed studies addressing this puzzling phenomenon. Initially, it was demonstrated that NK cells could readily kill syngeneic dorsal root ganglia (DRG) neurons by a perforin-dependent mechanism (Backstrom et al., 2000, 2003). Ventral spinal cord neurons and hippocampal neurons of the CNS were resistant to lysis. The resistance to NK cell-mediated lysis of the latter neurons was not related to protection by MHC class I molecules, since similar β_2-micro-globulin$^{-/-}$ neurons were equally resistant to lysis (Backstrom et al., 2003).

NK cell function is tightly regulated by multiple signals transmitted via inhibitory and activating receptors. The prerequisite for NK cell killing is its activation via signalling from activating receptor ligand pathways. NK cell activation generally appears to be elicited by a distinct set of molecules that have weak homology with MHC class I molecules. The activating receptor NKG2D which differs dramatically from other NKG2 receptor proteins is of particular interest since it, in contrast to other NK cell-activating receptors, is constitutively expressed on NK cells. The endogenous ligand of NKG2D in the mouse was recently identified as retinoic acid early inducible gene-1 (RAE-1)-encoded proteins and minor histocompatibility antigen H60 (Cerwenka et al., 2000; Diefenbach et al., 2000; Malarkannan et al., 1998). Differential expression of NKG2D and its ligand on neuronal cells from the peripheral system or the CNS appears a key mechanism underlying variable susceptibility to NK lyses. RAE-1, the product of which is a ligand for the NK cell-activating receptor NKG2D, was expressed at high levels in the DRG neurons. In contrast, RAE-1 was expressed only at very low levels in the resistant CNS-derived neurons. Blocking NK cells with anti-NKG2D antibodies inhibited NK cell-mediated killing of the DRG neurons.

These findings are important in revealing novel immunopathologic effects of several CNS diseases. Indeed, progressive motor and sensory neuropathy developed in a patient with chronic NK cell lymphocytosis (CNKL) (Noguchi et al., 2005). A sural nerve biopsy revealed infiltration of NK cells into the nerve fascicles, and demyelinating changes with secondary axonal degeneration. The infiltrating NK cells were adjacent to myelinated fibres, showing damage of Schwann cell membranes. Treatment with oral prednisolone resulted in rapid improvement of sensory disturbance and weakness with a significant decrease of NK cells in the blood and disappearance of the conduction block. These observations suggest that the infiltrating NK cells may directly damage myelin and Schwann cells, thus causing demyelination.

Since expression of NKG2D ligands is likely regulated by viral infection or transformation, and the inhibitory MHC class I expression is low or absent in the nervous system, it is plausible that a viral infection or transformation could well break the balance of activating/inhibiting activities on NK cells, and NK cell-mediated immune pathology would occur in such circumstances.

Oligodendrocyte

Human oligodendrocytes do not express MHC class II molecules; thus, direct MHC-restricted injury mediated by myelin-reactive $CD4^+$ T cells is less likely to occur. The migration of NK cells to the CNS during inflammatory responses and lack of inhibitory signal MHC class I expression within the CNS invites prediction that direct cytolytic effects of NK cells contribute to oligodendrocyte damage and demyelination to CNS diseases such as MS. Antel and colleagues demonstrated using in vitro assays that human oligodendrocytes, as well as other glial elements (astrocytes, microglia), were susceptible to injury mediated by peripheral blood-derived mononuclear cell preparations (MNCs), which were enriched for NK cells by depleting $CD3^+$, with or without $CD19^+$ cells through the use of either magnetic beads or cell sorting (Antel et al., 1998; Morse et al., 2001). Cytotoxic effects of the NK cell-enriched effectors were dependant on pre-exposure of these cells to IL-2. Furthermore, it was found that autologous oligodendrocytes were as susceptible to injury mediated by IL-2-activated NK cells as were heterogonous oligodendrocytes.

In searching for receptor ligand pathways that control the NK cell and oligodendrocyte interactions, it was found that human adult oligodendrocytes and foetal astrocytes expressed ligands for NKG2D in vitro, whereas neurons, microglia, and adult astrocytes did not (Saikali et al., 2007). Disruption of the NKG2D–NKG2D ligand interaction using blocking antibodies significantly inhibited killing of primary human oligodendrocytes mediated by activated human NK cells (Saikali et al., 2007). These results imply that NKG2D–NKG2D ligand interactions can potentially contribute to cytotoxic responses mediated by activated immune effector cells in the inflamed CNS, as observed in MS.

In the context of tissue injury that occurs in MS, the inflammatory milieu in MS lesions may provide conditions required for NK cell activation, raising the possibility that such effector cells would play a role in the pathogenesis. In addition to direct cytotoxicity, cytokine release by NK cells may also participate in tissue damage as well as in regulating T cell immune responses. Interferon-gamma (IFN-γ) is a pleiotropic cytokine produced by T cells and NK cells that has been implicated as a deleterious factor in MS, the immune-mediated demyelinating disorder. In vitro, purified developing and mature oligodendrocytes die in the presence of IFN-γ by apoptosis and necrosis, respectively. Transgenic expression of PLP/SOCS1 (proteolipid protein regulating suppressor of cytokine signalling 1), brings about diminished oligodendrocyte responsiveness to IFN-γ attributable to the targeted expression of SOCS1 in these cells (Balabanov et al., 2006). Consequently, oligodendrocytes in the PLP/SOCS1 transgenic mice are protected against the injurious effect of IFN-γ. Although both NK cells and T cells produce IFN-γ, NK cells are the principal sources of early IFN-γ production prior to T cell activation. This time kinetic might be particularly relevant for early oligodendrocyte damage during inflammation.

NK cells in infection of the CNS

Efficient early control of viral infections is determined by viral tissue tropism and rate of replication as well as the host's ability to mount an effective immune response. Cellular cytotoxicity, in particular, that of NK cells and cytotoxic T cells, is central to the early antiviral immune response. Table 28.1 illustrates several immune-deficient conditions in humans, which stem from mutations affecting NK cells (Biron et al., 1989; Gilmour et al., 2001; Markel et al., 2004; Moins-Teisserenc et al., 1999). A number of studies have demonstrated the recruitment and activation of NK cells following infection with a wide range of viruses. However, not all viral infections

Table 28.1 Genetic mutation or aberrant expression of cytokines affecting NK cells leads to infection and autoimmunity in humans

Mutations	Patient phenotype	Immune phenotype	References
IL-2R/IL-15Rβ	$NK^{-/-}$ SCID phenotype	Pronounced reduction in NK cells	Alsharifi et al. (2006)
Not known	Herpes virus infection	Absence of $CD16^+$ NK cells	Armstrong et al. (2001)
TAP-deficiency	Chronic infection and systemic autoimmunity	Defective $CD8^+$ T cell responses	Whitley (2002), Bellner et al. (2005)

SCID: combined severe immunodeficiency
TAP: transporter associated with antigen processing

are susceptible to NK-mediated clearance, and susceptibility depends upon the effector mechanisms induced. For example, the induction of both cytotoxicity and IFN-γ production by NK cells following murine cytomegalovirus (CMV) and influenza virus infection results in reduced virally induced disease and enhanced survival. Along the same time, deficient IFN-γ production by NK cells correlates with the absence of an effective innate response to lymphocytic choriomeningitis virus infection. Moreover, the organs targeted by viral infection can also influence the participation of NK cells. Indeed, it has been shown that the NK response to murine CMV is perforin-dependent within the spleen, whereas production of IFN-γ is required for viral clearance from the liver. These results indicate that the importance of the NK cell response to viral infection can depend upon multiple factors, including the tissue infected, as well as the effector mechanisms induced.

Although a number of studies have documented the possible role for NK cells in controlling CNS infection with CMV, influenza and other viruses, the following studies provide relatively direct evidence for the importance of NK cells during CNS viral infection:

Theiler's murine encephalomyelitis virus

Theiler's murine encephalomyelitis virus (TMEV) is a picornavirus. Infection of susceptible mice (SJL) with TMEV causes a biphasic disease characterized by grey matter inflammation followed by late chronic demyelination (Roos and Wollmann, 1984; Rosenthal et al., 1986). After inoculation, CNS TMEV titres were higher in SJL mice compared with C57BL/10 mice, correlating with a 50% lower NK cell activity in the SJL mice than in the C57BL/10 mice (Paya et al., 1989). Clinically, SJL mice are much more susceptible than C57BL/10 mice to TMEV. When resistant (C57BL/10) mice were depleted of NK cells using either mAb NK1.1 or polyclonal anti-asialo-GM1, TMEV induced the development of diffuse encephalitis and meningitis early in the post-infection period. However, the second phase of TMEV-induced CNS disease (demyelination) was observed only in resistant C57BL/10 mice treated with anti-asialo-GM1. Experiments with beige/beige mice of C57BL/10 background showed a mild degree of grey matter inflammation but no demyelination (Paya et al., 1989). NK cells are critical effectors in protecting against TMEV-induced grey matter disease, whereas a different population of either NK1.1–NK cells, or other activated lymphocytes, may be critical in resistance or susceptibility to demyelination.

In support of the involvement of NK cells during TMEV of the CNS, another study demonstrated that stressed mice developed clinical signs of encephalitis, thymic atrophy, and adrenal hypertrophy after infection with Theiler's virus (Welsh et al., 2004). This syndrome was associated with significantly reduced virus-induced NK cell cytotoxic activity in restrained mice at 1 day post-infection, which may account for the reduced viral clearance from the CNS.

Mouse hepatitis virus

Mouse hepatitis virus (MHV) is a coronavirus that causes infection of the CNS (Marten et al., 2001; Wang et al., 1990). Intracerebral infection of susceptible strains of mice with MHV results in an acute encephalomyelitis followed by a chronic immune-mediated demyelinating disease that is similar in pathology to the human demyelinating disease MS (Walsh et al., 2007; Zuo et al., 2006). Intracerebral infection of $RAG1^{-/-}$ mice with a recombinant CXCL10-expressing murine coronavirus (MHV) resulted in protection from disease and increased survival that correlated with a significant increase in recruitment and activation of NK cells within the CNS (Walsh et al., 2007). Accumulation of NK cells resulted in a reduction in viral titres that was dependant on IFN-γ secretion (Walsh et al., 2007). These results indicate that the CXCL10-guided NK cell homing to the CNS might play a pivotal role in defence following coronavirus infection of the CNS.

Semliki Forest virus

Semliki Forest virus (SFV) is a positive-stranded RNA virus (Atkins et al., 1999; Smithburn and Haddow, 1944). Infection of C57BL/6 mice with SFV leads to pronounced CNS cellular infiltration and apoptosis of microglial and neuronal cells (Alsharifi et al., 2006). In this model, NK cells and, to a lesser degree, cytotoxic T cells are major contributors in combating SFV infection. Mice lacking the Tc cell compartment (β_2-microglobulin-deficient mice, and thus $CD8^+$ T cells) exhibit susceptibility similar to wild-type mice. Depletion of NK cells significantly delayed the mean time to death but did not prevent mortality in SFV-infected B6 mice suggesting that cytolytic activity of NK cells is detrimental, while IFN-γ production is beneficial for recovery from SFV infection (Alsharifi et al., 2006).

Herpes simplex virus

With greater than 1.6 million Americans infected annually (Armstrong et al., 2001), herpes simplex virus type 1 and 2 (HSV-1, HSV-2) are pathogens with a significant impact on public health. Typically, infection results in a life-long latent infection of the host (Halioua and Malkin, 1999; Whitley, 2002). The transmission of HSV-2 in the human population includes asymptomatic shedding of the virus even in the presence of $CD8^+$ cytotoxic T lymphocytes and the production of a viral glycoprotein that indirectly elicits NK cell death (Bellner et al., 2005; Posavad et al., 2000; Wald et al., 2000). In a mouse model of HSV-2 infection, it was shown that mice deficient in CCR5 ($CCR5^{-/-}$) displayed a significant reduction in cumulative survival following infection in comparison with wild-type HSV-2–infected controls. Associated with decreased resistance to viral infection, $CCR5^{-/-}$ mice yielded significantly more virus and expressed higher levels of tumour necrosis factor alpha (TNF-α), CXCL1, CCL2, CCL3 and CCL5 in the vagina, spinal cord, and/or brain stem than did wild-type mice. In addition, when comparing wild-type HSV-2–infected mice with $CCR5^{-/-}$ mice prior to or after infection, there were significantly more NK cells ($NK1.1^+$ $CD3^-$) residing in the brain stem and spleen of infected wild-type mice. Functionally, NK activity from cells isolated from the brain stem of HSV-2–infected wild-type mice was greater than that from HSV-2–infected $CCR5^{-/-}$ mice. Further, antibody-mediated depletion of NK cells resulted in an increase in HSV-2 levels in the vaginal, spinal cord and brain stem tissue of wild-type mice but not $CCR5^{-/-}$ mice (Kveberg et al., 2002). Collectively, the absence of CCR5 expression significantly impacts the ability of the host to control genital HSV-2 infection, inflammation and spread associated with a specific reduction in NK cell expansion, infiltration and activity in the nervous system.

Toxoplasma gondii

Congenital toxoplasmosis poses a public health problem, being capable of causing foetal death and ocular and neurological sequelae in congenitally infected children. Congenital infection occurs only when mothers first encounter *Toxoplasma gondii* (*T. gondii*) during pregnancy (Remington et al., 1994; Roberts and Alexander, 1992). Resistance to *T. gondii* is mainly mediated by type 1 cytokines, such as IFN-γ and interleukin 2 (IL-2), whereas type 2 cytokines, such as IL-4 and IL-10, are associated with increased susceptibility to infection (Hunter et al., 1996; Khan et al., 1994). Susceptibility of the pregnant host to toxoplasmosis may be attributable to a type 2-cytokine bias that is maintained during gestation (Shirahata et al., 1992). Cell-mediated immune responses involving CD4 and CD8T cells and NK cells play a protective role in *T. gondii* primary infection (Goldszmid et al., 2007; Scharton-Kersten et al., 1998; Scorza et al., 2003; Scott and Trinchieri, 1995; Subauste et al., 1992). To clarify the roles of NK cells and IFN-γ in protection against primary congenital toxoplasmosis, Abou-Bacar and colleagues (2004) used recombination activating gene 2 knockout (KO) ($RAG\text{-}2^{-/-}$) mice, which lack T and B lymphocytes, in comparison with the wild-type BALB/c model. $RAG\text{-}2^{-/-}$ mice had a significantly lower risk of foetal toxoplasmosis than BALB/c mice. This protection was associated with an increased number of maternal NK cells, IFN-γ secretion by spleen cells, and decreased parasitemia. In the $RAG\text{-}2^{-/-}$ mice, NK cell depletion increased the rate of foetal infection. These data suggest that a partially protective immunity against congenital toxoplasmosis is achieved owing to the increased number of NK cells in $RAG\text{-}2^{-/-}$ mice (Abou-Bacar et al., 2004). Protective effect of NK cells was confirmed in another study using the SCID model (Kang and Suzuki, 2001).

NK cells and tumour immune surveillance of the CNS

The innate immune system plays an instrumental role in generating and directing the adaptive immune responses (Shi et al., 2001). NK cells represent a critical first line of defence against malignant transformation. Earlier results by Karre and Ljunggren demonstrated that NK cells can preferentially kill and reject cells that fail to express 'self' MHC class I molecules (Karre et al., 1986). These findings led to the formation of the famous 'missing self hypothesis' (Ljunggren and Karre, 1990). Over the years, the missing self hypothesis has been repeatedly demonstrated in a variety of experimental tumour systems by different groups of investigators, and a number of molecular pathways governing the interactions of NK cell–target cells have been revealed. Surveillance against 'missing self' may thus be one, but not the only function of NK cells (Ljunggren and Malmberg, 2007).

Non-surgical resectable tumours within the CNS constitute significant challenges for physicians. Furthermore, studies have documented frequent immune system defects in intracranial tumour-bearing patients and an inability of certain lymphocyte subset to mediate anti-tumour effector functions in the CNS.

Metastatic melanoma

Malignant melanoma is notorious for metastasis to discrete locations such as testis and brain. Malignant melanoma is the third most common type of cancer that metastasizes to the brain (Prins et al., 2006), which presents clinicians with few treatment options. Although nearly a dozen melanoma antigens specifically recognized by T cells have been identified, melanoma cells are still able to avoid immune destruction in most instances. Because the generation of an effective anti-tumour immune response requires both the presence of foreign antigen and a costimulatory molecule or signal, tumour cells displaying tumour antigens may avoid immune detection because of the absence of appropriate costimulation. Thus, anti-tumour immune responses might be achieved by more effective local delivery of costimulatory molecules. Activation and expansion of NK cells may independently lyse tumour cells, or provide T cells with costimulatory molecules including cytokines, and overall enhance antigen presentation to T cells.

Several attempts have been made in an effort to use NK cells to target CNS melanoma. The specific receptor for IL-2 on NK cells allows several approaches to deliver IL-2 intrathecally and activate NK cells. Ewend and associates carried out a study in C57BL/6 mice that were simultaneously given intracranial injections of tumour and of irradiated B16F10 melanoma cells transduced to secrete IL-2 (Prins et al., 2006). IL-2 therapy generated antitumour responses capable of extending the survival of animals that received simultaneous intracranial tumour challenges either locally or at distant sites in the brain. Non-transduced melanoma cells had little effect. Elimination of T-cell and NK subsets using gene KO mice and antibody-depletion techniques demonstrated that NK cells were most important for the initial anti-tumour response, whereas $CD4^+$ T cells were not necessary. These studies demonstrate that local IL-2 therapy in the brain not only generates an immediate local antitumour immune response, but also establishes long-term immunologic memory capable of eliminating subsequent tumour challenges within and outside of the CNS. Furthermore, the antitumour response to paracrine IL-2 in the brain differed significantly from that in the flank, suggesting that the intrinsic CNS cells involved in initiating immunity within the brain have different cytokine requirements from their peripheral counterparts.

Using the same model, a recent study showed that DCs administration induced dramatic anti-tumour immune protection in CD8α KO mice that were challenged with B16 melanoma both subcutaneously and in the brain (Ewend et al., 2000). The CNS anti-tumour immunity was dependant on both $CD4^+$ T cells and NK cells. Administration of non-Ag-loaded, immature DC resulted in significant $CD4^+$ T cell and NK cell expansion in the draining lymph nodes at 6 days post-vaccination, which persisted for 2 weeks. Finally, Ag-loaded DC administration in CD8α KO mice was associated with robust infiltration of $CD4^+$ T cells and NK cells into the brain tumour parenchyma (Ewend et al., 2000).

Glioma

Glioma cells interfere with anti-tumour immune responses by expressing immune inhibitory cell surface molecules, such as HLA-G, or by releasing soluble immunosuppressants such as transforming growth factor (TGF-β). They rarely metastasize outside the brain, raising the possibility of immune-mediated control of these cells outside, but not inside, the brain.

IL-2, as well as growth hormone, is potent in enhancing NK cell activity against glioma both in human trials and in several experimental systems (Eisele et al., 2006; Hayes et al., 1995; Shimizu et al., 2005; Wischhusen et al., 2005). As receptors governing NK cell action and effector functions are being elucidated, more sophisticated means of manipulating NK cells have been generated. As noted above, NKG2D is a powerful, activating NK cell receptor (Wischhusen et al., 2005). Accordingly, activating the innate immune system by forcing glioma cells to express danger signals such as NKG2D ligands is a promising strategy of immunotherapy for these tumours. The remaining challenges would be to down-regulate HLA-E expression on glioma cells and suppress production of TGF-β by glioma. Both HLA-E and TGF-beta can down-regulate NKG2D expression on glioma, which enable these tumour cells to escape NK cell surveillance.

Other CNS tumours

Various studies have documented the role of NK cells in surveillance and suppression of other type of CNS tumours including medulloblastoma (Castriconi et al.,

Table 28.2 NK cells in CNS pathology

CNS pathology	Functions	Mechanisms	References
Viral infection			
TMEV	Suppress viral infection	Not studied	Paya et al. (1989)
Theiler's virus	Inhibit viral replication	Kill viral infected cells	Welsh et al. (2004)
Mouse hepatitis virus	Inhibit viral replication	Not studied	Walsh et al. (2007), Zuo et al. (2006)
Herpes simplex virus	Confine viral infection	IFN-γ production by NK cells	Alsharifi et al. (2006)
Semliki Forest virus	Inhibit viral replication	Cytotoxicity IFN-γ production	Kveberg et al. (2002)
Tumours			
CNS melanoma	Suppress tumour growth	NK cytotoxicity	Prins et al. (2006)
Glioma	Suppress tumour growth	TGF-β production and NKG2D activation	Wischhusen et al. (2005)
Astrocytoma	NO significant role	–	Kang et al. (2004)
Medullbostoma	NO significant role	–	Castriconi et al. (2007)
Inflammation			
Multiple sclerosis and EAE	Suppression	Kill CNS APC inhibit myelin-reactive T cells	Zhang et al. (1997), Huang et al. (2006), Bielekova et al. (2006)

TMEV: Theiler's Murine Encephalomyelitis Virus
EAE: Experimental Autoimmune Encephalomyelitis

2007), astrocytoma and neuroblastoma (Kang et al., 2004). On the other hand NK cells appear to have little, if any, role in suppressing CNS lymphomas (Yamasaki et al., 2003).

Clearly, cumulative evidence suggests that NK cells play a role in curbing malignant transformation and progression of many primary and metastatic CNS tumours (Table 28.2). Direct cytotoxic effects and collaboration with T cells and other immune cells are required to achieve these functions (Table 28.2). Effective therapies harnessing NK cells will be facilitated through understanding of the molecular signalling pathways that will be governing NK cell activation, expansion and maintenance. Specific anatomical factions within the CNS should also be considered. Furthermore, effort must be taken in suppressing the capacity of certain tumours to down-regulate activating signals and production of inhibitory proteins against NK cells.

Regulatory functions of NK cells in CNS inflammation and autoimmunity

During CNS infection, cytolytic activity of NK cells contributes to elimination of viral and bacterial infected cells and controls the magnitude of inflammation. Debris from neuronal and/or glial cell death is taken up by antigen-presenting cells (APCs) and presentation to T cells. Cytokine (IFN-γ) secretion by NK cells increases MHC class II expression by APC and, thus, favours generation of Th1 type of T helper cells. Thus, NK cells function not only as the initial line of host defence, but also as fuel to the generation of adaptive immune responses. Overall, NK cells are expected to boost immune response within the CNS. Paradoxically, emerging evidence suggests that NK cells can inhibit CNS inflammation and control the magnitude of autoimmunity (Table 28.2).

MS is a classic autoimmune disease characterized by extensive CNS inflammation and immune-mediated destruction of myelin. Consequently, the function of myelin sheaths becomes compromised and neurological impairment occurs. The pathogenesis of MS is mirrored, in part, in EAE, which can be induced in susceptible strains of mice with neuron-antigens and complete Freund's adjuvant. The roles of NK cells in the pathogenesis of MS and EAE have been investigated. In patients with MS, NK cells (CD56 and CD57) are present in the peripheral blood with reduced numbers and cytolytic activity (Shibatomi et al., 2001; Trinchieri, 1989). This finding is not unique to MS and similar phenotype of NK cells have been documented in many other types of autoimmune diseases such as rheumatoid arthritis, systemic lupus erythematosus, myasthenia gravis, etc (Shibatomi et al., 2001). In parallel, patients with MS and other autoimmune diseases have defective functions of 'regulatory cells', including NKT cells and $CD4^{+}CD25^{+}$ regulatory T cells (Treg) (La Cava et al., 2006). A reduced number and/or compromised function of NK cells, NKT cells and Treg cells invite a hypothesis that autoimmunity is associated with a result of global defective regulatory cell functions in these patients.

This hypothesis has been tested in several EAE models. Because gene encoding NK cells cannot be targeted by the current technology, several depleting antibodies have been used to study the function of NK cells in vivo. Several groups have utilized anti-NK1.1 mAb and observed that depletion of NK cells by injecting anti-NK1.1 mAb leads to exacerbation of EAE (Zhang et al., 1997). Apparently, both peripheral and CNS NK cells are absent in this experimental system. It is, therefore, not possible to differentiate the role of NK cells in the periphery and in the CNS.

NK cell homing to CNS is controlled by a specific chemokine receptor ligand pathway involving CX3CR1 and fractalkine (CX3CL1). CX3CR1 is expressed almost exclusively by CNS glial cells (Boehme et al., 2000; Cardona et al., 2006). Thus, germ-line deletion of CX3CR1 leads to impaired homing of NK cells to the CNS. This model would be ideal in addressing CNS inflammation/autoimmunity in relation to NK cells. Interestingly, upon immunization, CX3CR1-deficient mice with reduced NK cells in the CNS and intact NK cells in the periphery developed their wild-type controls. Thus, lack of CNS NK cells alone is sufficient to cause exacerbated CNS inflammation and autoimmunity (Huang et al., 2006). It is also conceivable that chemokine guided NK cell homing to CNS might serve as pathway that can be therapeutically targeted (Figure 28.1).

Infection is suggested to play a role triggering the initiation of MS in some patients (Bendelac and Medzhitov, 2002; Pulendran and Ahmed, 2006; Shirahata et al., 1992). The use of complete Freund's adjuvant in the induction of EAE may mimic this process (Fearon and Locksley, 1996). In these patients or in EAE animals, NK cells may contribute to the demyelination through bystander damage while controlling the infection. Once infection is controlled, NK cells may function to inhibit the excessive (auto) immune responses elicited by pathogens. The immune system may use NK cell as a versatile regulator to tune its capacity in combating infection and avoiding autoimmunity.

The mechanism underlying this unique role for NK cells within the CNS during EAE is still elusive. A close survey of the literature reveals multiple steps where NK cells can regulate inflammation and intervene in the loss of self-tolerance. Importantly, the findings also caution against inferring a similar role for NK cells in all types of autoimmune phenomena or during separate stages of the same disease (Flodstrom et al., 2002; Yokoyama and Plougastel, 2003). NK cells can both promote and inhibit autoreactive T cells. These possibilities have been extensively reviewed recently (Shi and Van Kaer, 2006). The specific CNS anatomical location, as reflected by diverse CNS APCs and multiple antigens may also influence the outcome of autoimmunity. As with EAE, it appears that NK cells control T cell proliferation in an antigen non-specific manner, both in the periphery and within the CNS (Shi, et al., unpublished). Recently, it has been demonstrated that human NK cells kill resting but not activated microglia via NKG2D- and NKp46-mediated recognition (Lünemann et al., 2008). This study emphasizes the potential importance of interactions between NK cells and CNS resident APCs. However, whether the regulatory effects of NK cells can be attributed to the action of NK cells on APCs, directly on T cells, or both, is not known and currently under investigation.

Summary and future research directions

NK cells readily accumulate in homing to CNS tissues under the pathophysiological situations. This process is tightly controlled by a number of chemokines and chemokine receptors. There is ample of evidence that

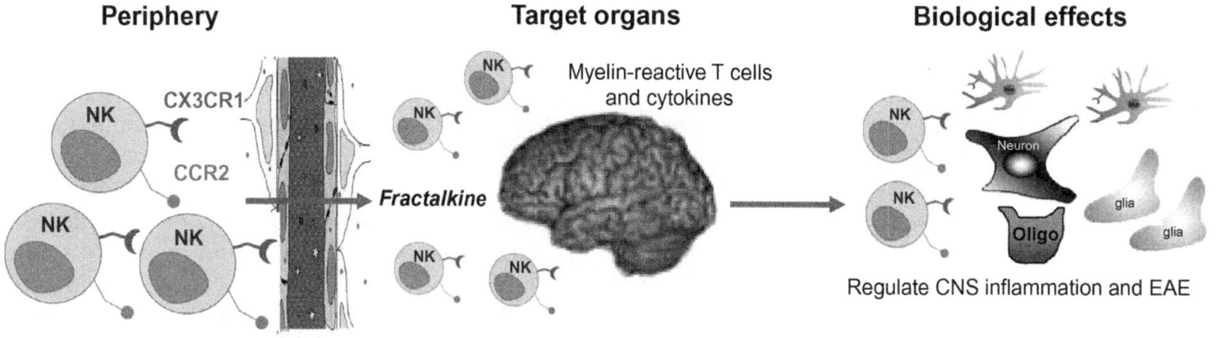

Figure 28.1 • Chemokine-guided recruitment of NK cells as in the CNS. Inflammatory responses as seen in EAE or MS may result in the production of chemokines, in particular fractalkine by microglia. Fractalkine recruits NK cells to the CNS. Subsequently, NK cells may control the magnitude of CNS inflammation and severity of EAE via several pathways: (1) directly kill APCs and interfere with reactivation of myelin-reactive T cells; (2) directly kill myelin-reactive T cells; altered expression of MHC class I on those APCs and T cells may trigger the killing; (3) suppress differentiation of T helper cells. This can be achieved by depleting cells which produce Th cell-polarizating cytokines, or cytokines produced by NK cells themselves.

NK cells within the CNS contribute to the control of infections and might limit progression of certain tumours. Bystander neuronal and/or glial cell damage can occur. In certain autoimmune conditions of the CNS, NK cells appear to have an inhibitory role. Activation and expansion of NK cells through engaging IL-2 receptors on NK cells not only inhibit several CNS tumours, but also might slow the progression of MS and other autoimmune diseases (Bielekova et al., 2006; Li et al., 2005). Furthermore, the ability of IFN-α and IFN-β to ameliorate MS in humans and IFN-γ to inhibit EAE in mice may reflect the ability of these cytokines to transiently activate NK-dependent regulatory responses. However, because IFN treatment also upregulates Qa-1 expression on T cells (Ota et al., 2005), the short duration and usually modest nature of these therapeutic effects may reflect a Qa-1-dependent decrease in NK cell activation and associated immunoregulatory activity (Lu et al., 2007).

Disassociation of disease-inhibiting versus disease-promoting effects of NK cells is a key to harnessing NK cells for therapeutic purposes. To achieve this goal, a generation of genetic models with selective NK cell deficiency, and development of reagents (antibodies) for visualizing subsets of NK cells in situ will be necessary. Optimization of methods to produce NK cells in large quantities for therapeutic usage is also important. Clearly, understanding the molecular signals and pathways governing these differential biological effects of NK cells as well as their cross talk with T cells is key to designing NK cell-based therapy for CNS infections, tumours and autoimmune diseases.

Acknowledgements

We thank Dr R. Liu, Dr W. Piao, and Dr. Dayao and Ms. Linda Swanson for their wonderful assistance in preparation of this chapter. The Muscular Dystrophy Association, the Barrow Neurological Foundation, and the National Institutes of Health support the authors' work in the laboratories.

References

Abou-Bacar, A., Thompson, R.C. and Anderson, R. (2004). Role of NK cells and gamma interferon in transplacental passage of *Toxoplasma gondii* in a mouse model of primary infection. Infect Immun **72**, 1397–1401.

Ajuebor, M.N., Wondimu, Z. and Hogaboam, C.M. (2007). CCR5 deficiency drives enhanced natural killer cell trafficking to and activation within the liver in murine T cell-mediated hepatitis. Am J Pathol **170**, 1975–1988.

Alsharifi, M., Lobigs, M. and Simon, M.M. (2006). NK cell-mediated immunopathology during an acute viral infection of the CNS. Eur J Immunol **36**, 887–896.

Antel, J.P., McCrea, E. and Ladiwala, U. (1998). Non-MHC-restricted cell-mediated lysis of human oligodendrocytes in vitro: relation with CD56 expression. J Immunol **160**, 1606–1611.

Armstrong, G.L., Schillinger, J. and Markowitz, L. (2001). Incidence of herpes simplex virus type 2 infection in the United States. Am J Epidemiol **153**, 912–920.

Atkins, G.J., Sheahan, B.J. and Liljestrom, P. (1999). The molecular pathogenesis or Semliki Forest virus: a model virus made useful? J Gen Virol **80**(Pt 9), 2287–2297.

Backstrom, E., Chambers, B.J. and Ho, E.L. (2003). Natural killer cell-mediated lysis of dorsal root ganglia neurons via RAE1/NKG2D interactions. Eur J Immunol **33**, 92–100.

Backstrom, E., Chambers, B.J. and Kristensson, K. (2000). Direct NK cell-mediated lysis of syngenic dorsal root ganglia neurons in vitro. J Immunol **165**, 4895–4900.

Balabanov, R., Strand, K. and Kemper, A. (2006). Suppressor of cytokine signaling 1 expression protects oligodendrocytes from the deleterious effects of interferon-gamma. J Neurosci **26**, 5143–5152.

Bellner, L., Thoren, F. and Nygren, E. (2005). A proinflammatory peptide from herpes simplex virus type 2 glycoprotein G affects neutrophil, monocyte, and NK cell functions. J Immunol **174**, 2235–2241.

Bendelac, A. and Medzhitov, R. (2002). Adjuvants of immunity: harnessing innate immunity to promote adaptive immunity. J Exp Med **195**, F19–F23.

Bielekova, B., Catalfamo, M. and Reichert-Scrivner, S. (2006). Regulatory CD56(bright) natural killer cells mediate immunomodulatory effects of IL-2Ralpha-targeted therapy (daclizumab) in multiple sclerosis. Proc Natl Acad Sci U S A **103**, 5941–5946.

Biron, C.A., Byron, K.S. and Sullivan, J.L. (1989). Severe herpesvirus infections in an adolescent without natural killer cells. N Engl J Med **320**, 1731–1735.

Boehme, S.A., Lio, F.M. and Maciejewski-Lenoir, D. (2000). The chemokine fractalkine inhibits Fas-mediated cell death of brain microglia. J Immunol **165**, 397–403.

Bryceson, Y.T., Foster, J.A. and Kuppusamy, S.P. (2005). Expression of a killer cell receptor-like gene in plastic regions of the central nervous system. J Neuroimmunol **161**, 177–182.

Cardona, A.E., Pioro, E.P. and Sasse, M.E. (2006). Control of microglial neurotoxicity by the fractalkine receptor. Nat Neurosci **9**, 917–924.

Castriconi, R., Dondero, A. and Negri, F. (2007). Both CD133+ and CD133– medulloblastoma cell lines express ligands for triggering NK receptors and are susceptible to NK-mediated cytotoxicity. Eur J Immunol **37**, 3190–3196.

Cerwenka, A., Bakker, A.B. and McClanahan, T. (2000). Retinoic acid early inducible genes define a ligand family for the activating NKG2D receptor in mice. Immunity **12**, 721–727.

Diefenbach, A., Jamieson, A.M. and Liu, S.D. (2000). Ligands for the murine NKG2D receptor: expression by tumor cells and activation of NK cells and macrophages. Nat Immunol **1**, 119–126.

Eisele, G., Wischhusen, J. and Mittelbronn, M. (2006). TGF-beta and metalloproteinases differentially suppress NKG2D ligand surface expression on malignant glioma cells. Brain **129**, 2416–2425.

Ewend, M.G., Thompson, R.C. and Anderson, R. (2000). Intracranial paracrine interleukin-2 therapy stimulates prolonged antitumor immunity that extends outside the central nervous system. J Immunother **23**, 438–448.

Fearon, D.T. and Locksley, R.M. (1996). The instructive role of innate immunity in the acquired immune response. Science **272**, 50–53.

Flodstrom, M., Shi, F.D. and Sarvetnick, N. (2002). The natural killer cell—friend or foe in autoimmune disease? Scand J Immunol **55**, 432–441.

Gilmour, K.C., Fujii, H. and Cranston, T. (2001). Defective expression of the interleukin-2/interleukin-15 receptor beta subunit leads to a natural killer cell-deficient form of severe combined immunodeficiency. Blood **98**, 877–879.

Goldszmid, R.S., Bafica, A. and Jankovic, D. (2007). TAP-1 indirectly regulates CD4+T cell priming in *Toxoplasma gondii* infection by controlling NK cell IFN-gamma production. J Exp Med **204**, 2591–2602.

Halioua, B. and Malkin, J.E. (1999). Epidemiology of genital herpes—recent advances. Eur J Dermatol **9**, 177–184.

Hammarberg, H., Lidman, O. and Lundberg, C. (2000). Neuroprotection by encephalomyelitis: rescue of mechanically injured neurons and neurotrophin production by CNS-infiltrating T and natural killer cells. J Neurosci **20**, 5283–5291.

Hayes, R.L., Koslow, M. and Hiesiger, E. M. (1995). Improved long term survival after intracavitary interleukin-2 and lymphokine-activated killer cells for adults with recurrent malignant glioma. Cancer **76**, 840–852.

Hickey, W.F., Ueno, K. and Hiserodt, J.C. (1992). Exogenously-induced, natural killer cell-mediated neuronal killing: a novel pathogenetic mechanism. J Exp Med **176**, 811–817.

Hokeness, K.L., Kuziel, W.A., Biron, C. A. and Salazar-Mather, T.P. (2005). Monocyte chemoattractant protein-1 and CCR2 interactions are required for IFN-alpha/beta-induced inflammatory responses and antiviral defense in liver. J Immunol **174**, 1549–1556.

Hougen, H.P., Thygesen, P. and Christensen, H.B. (1992). Effect of immunosuppressive agents on the guanethidine-induced sympathectomy in athymic and euthymic rats. Int J Immunopharmacol **14**, 1113–1123.

Huang, D., Shi, F.D. and Jung, S. (2006). The neuronal chemokine CX3CL1/fractalkine selectively recruits NK cells that modify experimental autoimmune encephalomyelitis within the central nervous system. FASEB J **20**, 896–905.

Hunter, C.A., Suzuki, Y. and Subauste, C.S. (1996). Cells and cytokines in resistance to *Toxoplasma gondii*. Curr Top Microbiol Immunol **219**, 113–125.

Inngjerdingen, M., Damaj, B. and Maghazachi, A.A. (2001). Expression and regulation of chemokine receptors in human natural killer cells. Blood **97**, 367–375.

Jiang, D., Liang, J. and Hodge, J. (2004). Regulation of pulmonary fibrosis by chemokine receptor CXCR3. J Clin Invest **114**, 291–299.

Kang, H. and Suzuki, Y. (2001). Requirement of non-T cells that produce gamma interferon for prevention of reactivation of *Toxoplasma gondii* infection in the brain. Infect Immun **69**, 2920–2927.

Kang, S.G., Ryu, C.H. and Jeun, S. (2004). Lymphokine activated killer cells from umbilical cord blood show higher antitumor effect against anaplastic astrocytoma cell line (U87) and medulloblastoma cell line (TE671) than lymphokine activated killer cells from peripheral blood. Childs Nerv Syst **20**, 154–162.

Karre, K., Ljunggren, H.G. and Piontek, G. (1986). Selective rejection of H-2-deficient lymphoma variants suggests alternative immune defence strategy. Nature **319**, 675–678.

Kerfoot, S.M. and Kubes, P. (2002). Overlapping roles of P-selectin and alpha 4 integrin to recruit leukocytes to the central nervous system in experimental autoimmune encephalomyelitis. J Immunol **169**, 1000–1006.

Khan, I.A., Matsuura, T. and Kasper, L.H. (1994). Interleukin-12 enhances murine survival against acute toxoplasmosis. Infect Immun **62**, 1639–1642.

Khan, I.A., Thomas, S.Y. and Moretto, M. M. (2006). CCR5 is essential for NK cell trafficking and host survival following *Toxoplasma gondii* infection. PLoS Pathog **2**, e49.

Kveberg, L., Bryceson, Y. and Inngjerdingen, M. (2002). Sphingosine 1 phosphate induces the chemotaxis of human natural killer cells. Role for heterotrimeric G proteins and phosphoinositide 3 kinases. Eur J Immunol **32**, 1856–1864.

La Cava, A., Van Kaer, L. and Fu Dong, S. (2006). CD4+ CD25+ Tregs and NKT cells: regulators regulating regulators. Trends Immunol **27**, 322–327.

Lavergne, E., Combadiere, B. and Bonduelle, O. (2003). Fractalkine mediates natural killer-dependent antitumor responses in vivo. Cancer Res **63**, 7468–7474.

Li, Z., Lim, W.K. and Mahesh, S.P. (2005). Cutting edge: in vivo blockade of human IL-2 receptor induces expansion of CD56(bright) regulatory NK cells in patients with active uveitis. J Immunol **174**, 5187–5191.

Ljunggren, H.G. and Karre, K. (1990). In search of the 'missing self': MHC molecules and NK cell recognition. Immunol Today **11**, 237–244.

Ljunggren, H.G. and Malmberg, K.J. (2007). Prospects for the use of NK cells in immunotherapy of human cancer. Nat Rev Immunol **7**, 329–339.

Lu, L., Ikizawa, K. and Hu, D. (2007). Regulation of activated CD4+ T cells by NK cells via the Qa-1-NKG2A inhibitory pathway. Immunity **26**, 593–604.

Lund, T.C., Anderson, L.B. and McCullar, V. (2007). iTRAQ is a useful method to screen for membrane-bound proteins differentially expressed in human natural killer cell types. J Proteome Res **6**, 644–653.

Lünemann, A., Lunemann, J.D. and Roberts, S. (2008). Human NK cells kill resting but not activated microglia via NKG2D- and NKp46-mediated recognition. J Immunol **181**, 6170–6177.

Malarkannan, S., Shih, P.P. and Eden, P.A. (1998). The molecular and functional characterization of a dominant minor H antigen, H60. J Immunol **161**, 3501–3509.

Markel, G., Mussaffi, H. and Ling, K.L. (2004). The mechanisms controlling NK cell autoreactivity in TAP2-deficient patients. Blood **103**, 1770–1778.

Marten, N.W., Stohlman, S.A. and Bergmann, C.C. (2001). MHV infection of the CNS: mechanisms of immune-mediated control. Viral Immunol **14**, 1–18.

Martin-Fontecha, A., Thomsen, L.L. and Brett, S. (2004). Induced recruitment of NK cells to lymph nodes provides IFN-gamma for T(H)1 priming. Nat Immunol **5**, 1260–1265.

Moins-Teisserenc, H.T., Gadola, S.D. and Cella, M. (1999). Association of a syndrome resembling Wegener's granulomatosis with low surface expression of HLA class-I molecules. Lancet **354**, 1598–1603.

Morse, R.H., Seguin, R. and McCrea, E. L. (2001). NK cell-mediated lysis of autologous human oligodendrocytes. J Neuroimmunol **116**, 107–115.

Noguchi, M., Yoshita, M. and Sakai, K. (2005). Peripheral neuropathy associated with chronic natural killer cell lymphocytosis. J Neurol Sci **232**, 119–122.

O'Leary, J.G., Goodarzi, M. and Drayton, D.L. (2006). T cell- and B cell-independent adaptive immunity mediated by natural killer cells. Nat Immunol **7**, 507–516.

Ota, T., Takeda, K. and Akiba, H. (2005). IFN-gamma-mediated negative feedback regulation of NKT-cell function by CD94/NKG2. Blood **106**, 184–192.

Paya, C.V., Patick, A.K. and Leibson, P. J. (1989). Role of natural killer cells as immune effectors in encephalitis and demyelination induced by Theiler's virus. J Immunol **143**, 95–102.

Posavad, C.M., Huang, M.L. and Barcy, S. (2000). Long term persistence of herpes simplex virus-specific CD8+ CTL in persons with frequently recurring genital herpes. J Immunol **165**, 1146–1152.

Prins, R.M., Vo, D.D. and Khan Farooqi, H. (2006). NK and CD4 cells collaborate to protect against melanoma tumor formation in the brain. J Immunol **177**, 8448–8455.

Pulendran, B. and Ahmed, R. (2006). Translating innate immunity into immunological memory: implications for vaccine development. Cell **124**, 849–863.

Ransohoff, R.M., Kivisakk, P. and Kidd, G. (2003). Three or more routes for leukocyte migration into the central nervous system. Nat Rev Immunol **3**, 569–581.

Remington, J.S., McLeod, R. and Desmonts, G. (1994). Toxoplasmosis. In: Remington, J.S. and Klein, O.J. (eds) Infectious Diseases of the Fetus and New-Born Infant. Philadelphia: W.B. Saunders Co., pp. 141–267.

Roberts, C.W. and Alexander, J. (1992). Studies on a murine model of congenital toxoplasmosis: vertical disease transmission only occurs in BALB/c mice infected for the first time during pregnancy. Parasitology **104**(Pt 1), 19–23.

Roos, R.P. and Wollmann, R. (1984). DA strain of Theiler's murine encephalomyelitis virus induces demyelination in nude mice. Ann Neurol **15**, 494–499.

Rosenthal, A., Fujinami, R.S. and Lampert, P.W. (1986). Mechanism of Theiler's virus-induced demyelination in nude mice. Lab Invest **54**, 515–522.

Saikali, P., Antel, J.P. and Newcombe, J. (2007). NKG2D-mediated cytotoxicity toward oligodendrocytes suggests a mechanism for tissue injury in multiple sclerosis. J Neurosci **27**, 1220–1228.

Scharton-Kersten, T., Nakajima, H. and Yap, G. (1998). Infection of mice lacking the common cytokine receptor gamma-chain (gamma(c)) reveals an unexpected role for CD4+ T lymphocytes in early IFN-gamma-dependent resistance to *Toxoplasma gondii*. J Immunol **160**, 2565–2569.

Scorza, T., D'Souza, S. and Laloup, M. (2003). A GRA1 DNA vaccine primes cytolytic CD8(+) T cells to control acute *Toxoplasma gondii* infection. Infect Immun **71**, 309–316.

Scott, P. and Trinchieri, G. (1995). The role of natural killer cells in host–parasite interactions. Curr Opin Immunol **7**, 34–40.

Shi, F.D., Ljunggren, H.G. and Sarvetnick, N. (2001). Innate immunity and autoimmunity: from self-protection to self-destruction. Trends Immunol **22**, 97–101.

Shi, F.D. and Van Kaer, L. (2006). Reciprocal regulation between natural killer cells and autoreactive T cells. Nat Rev Immunol **6**, 751–760.

Shibatomi, K., Ida, H. and Yamasaki, S. (2001). A novel role for interleukin-18 in human natural killer cell death: high serum levels and low natural killer cell numbers in patients with systemic autoimmune diseases. Arthritis Rheum **44**, 884–892.

Shimizu, K., Adachi, K. and Teramoto, A. (2005). Growth hormone enhances natural killer cell activity against glioma. J Nippon Med Sch **72**, 335–340.

Shirahata, T., Muroya, N. and Ohta, C. (1992). Correlation between increased susceptibility to primary *Toxoplasma gondii* infection and depressed production of gamma interferon in pregnant mice. Microbiol Immunol **36**, 81–91.

Smithburn, K.C. and Haddow, A.J. (1944). Semliki Forest virus I isolation and pathogenic properties. J Immunol **49**, 141–157.

Subauste, C.S., Dawson, L. and Remington, J.S. (1992). Human lymphokine-activated killer cells are cytotoxic against cells infected with *Toxoplasma gondii*. J Exp Med **176**, 1511–1519.

Thapa, M., Kuziel, W.A. and Carr, D.J. (2007). Susceptibility of CCR5-deficient mice to genital herpes simplex virus type 2 is linked to NK cell mobilization. J Virol **81**, 3704–3713.

Trinchieri, G. (1989). Biology of natural killer cells. Adv Immunol **47**, 187–376.

Wald, A., Zeh, J. and Selke, S. (2000). Reactivation of genital herpes simplex virus type 2 infection in asymptomatic seropositive persons. N Engl J Med **342**, 844–850.

Wald, O., Weiss, I.D. and Wald, H. (2006). IFN-gamma acts on T cells to induce NK cell mobilization and accumulation in target organs. J Immunol **176**, 4716–4729.

Walsh, K.B., Edwards, R.A. and Romero, K.M. (2007). Expression of CXC chemokine ligand 10 from the mouse hepatitis virus genome results in protection from viral-induced neurological and liver disease. J Immunol **179**, 1155–1165.

Walzer, T., Chiossone, L. and Chaix, J. (2007). Natural killer cell trafficking in vivo requires a dedicated sphingosine 1-phosphate receptor. Nat Immunol **8**, 1337–1344.

Wang, F.I., Stohlman, S.A. and Fleming, J.O. (1990). Demyelination induced by murine hepatitis virus JHM strain (MHV-4) is immunologically mediated. J Neuroimmunol **30**, 31–41.

Wekerle, H. and Fierz, W. (1985). T cell approach to demyelinating diseases. Springer Semin Immunopathol **8**, 97–110.

Welsh, C.J., Bustamante, L. and Nayak, M. (2004). The effects of restraint stress on the neuropathogenesis of Theiler's virus infection II: NK cell function and cytokine levels in acute disease. Brain Behav Immun **18**, 166–174.

Whitley, R.J. (2002). Herpes simplex virus infection. Semin Pediatr Infect Dis **13**, 6–11.

Wischhusen, J., Friese, M.A. and Mittelbronn, M. (2005). HLA-E protects glioma cells from NKG2D-mediated immune responses in vitro: implications for immune escape in vivo. J Neuropathol Exp Neurol **64**, 523–528.

Yamasaki, T., Moritake, K. and Klein, G. (2003). Experimental appraisal of the lack of antitumor natural killer cell-mediated immunosurveillance in response to lymphomas growing in the mouse brain. J Neurosurg **98**, 599–606.

Yokoyama, W.M. and Plougastel, B.F. (2003). Immune functions encoded by the natural killer gene complex. Nat Rev Immunol **3**, 304–316.

Yu, Y.R., Fong, A.M. and Combadiere, C. (2007). Defective antitumor responses in CX3CR1-deficient mice. Int J Cancer **121**, 316–322.

Zhang, B., Yamamura, T., Takayuki, K., et al. (1997). Regulation of Experimental Autoimmune Encephalomyelitis by Natural Killer (NK) cells. J Exp Med **186**, 1677–1687.

Zuo, J., Stohlman, S.A. and Hoskin, J.B. (2006). Mouse hepatitis virus pathogenesis in the central nervous system is independent of IL-15 and natural killer cells. Virology **350**, 206–215.

Chapter Twenty-Nine

29

NK cells in the eye

Jerry Y. Niederkorn

CHAPTER CONTENTS

A man who carries a cat by the tail learns something he can learn in no other way.

Mark Twain

ABSTRACT

Tissues within the eye have a limited capacity to regenerate. Consequently, immune-mediated inflammation can have devastating consequences for vision. However, anatomical, physiological and dynamic immunoregulatory processes limit inflammation and immune-mediated responses within the eye—a phenomenon known as immune privilege. Immune privilege limits the activities of both the adaptive and innate immune systems. Cells lining the cornea and within the retina fail to express MHC class I molecules, making them vulnerable to natural killer (NK) cell-mediated lysis. However, soluble NK cell inhibitory factors within ocular fluids and the expression of non-conventional MHC class Ib molecules on corneal and retinal cells protect these ocular cells from NK cell-mediated destruction, but allow intraocular melanomas to escape detection and elimination by NK cells. In contrast, NK cells at the ocular surface are not restrained by ocular immune privilege and as a result, they contribute to the pathogenesis of viral and bacterial keratitis.

KEY WORDS

Corneal transplantation, Eye, Immune privilege, Keratitis, Melanoma, Uveitis

Introduction

The human eye is only a few centimeters in diameter, yet it is composed of a wide array of tissues, some of which are found nowhere else in the body. The eye is an extension of the brain, and like other components of the central nervous system, it processes an astonishing amount of information. The one million ganglion cells in the retina process over 500 electrical signals per second, which is approximately the same as 1.5×10^9 bits of computer information (Miller, 1979). The retina captures photons that enter the cornea, renders them into a coherent set of signals, and transmits these signals to the brain where they are interpreted into exquisitely precise images. However, the remarkably sophisticated function of the retina is rendered meaningless if the cornea becomes opaque. A single layer of neural crest-derived endothelial cells line the inside of the cornea and act as an osmotic pump to maintain proper hydration of the cornea. Since the corneal endothelium is amitotic and cannot regenerate, dysfunction or attrition of the corneal endothelial cells can result in the uptake of water resulting in corneal opacity and eventually, blindness. Some components of the retina are also incapable of undergoing mitosis. Injury to the neural retina extinguishes the transmission of signals emanating from captured photons in the retina and renders the eye blind. Although inflammation is a daily event in many organs and normally does not inflict irreparable injury, its consequences can be devastating in the eye, especially if cells of the corneal endothelium or the retina are subjected to reactive oxygen species and proteases elaborated by inflammatory cells. However, the eye is designed to minimize the untoward consequences of inflammation and immune-mediated responses—a phenomenon known as "immune privilege". Immune privilege appears to be a compromise between the eye and the immune system, whereby certain forms of inflammation are silenced to prevent irreparable injury to innocent bystander cells in the eye at the risk of rendering the eye vulnerable to infectious agents (Niederkorn, 2006a,b; Niederkorn and Wang, 2005a,b; Streilein, 1987, 2003). However, immune privilege is limited and can be vetoed if an infection is life-threatening. In fact, some causes of blindness are immune-mediated and result from an unbridled immune response to some ocular pathogens, such as herpes simplex virus (HSV) (Figure 29.1)

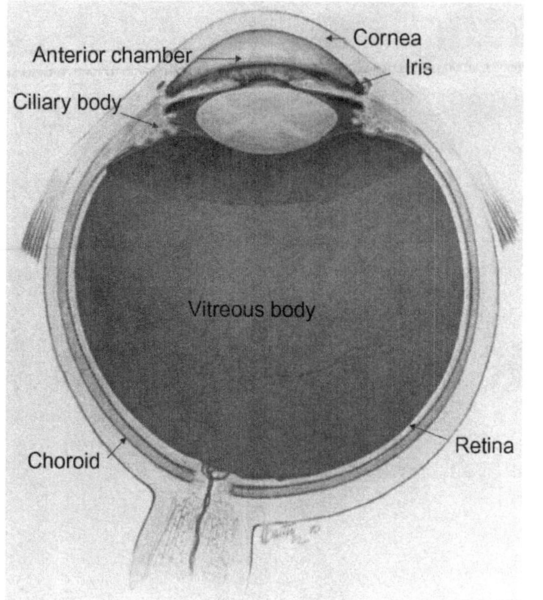

Figure 29.1 • Anatomy of the eye. Although NK cells are present in the peripheral blood that circulates through the eye, they are normally absent from ocular tissues. However, under pathological conditions, NK cells are found in HSV-infected corneas (<10% of inflammatory cell population), *Pseudomonas*-infected corneas (<10% of inflammatory cell population), uveal melanomas (7–10% of the uveal melanoma-infiltrating lymphocytes are found in the choroid of the eye) and in the corneas and AH of rejecting corneal grafts (up to 12% of infiltrating lymphocytes in the AH and in the graft are NK cells). Courtesy of the National Eye Institute, National Institutes of Health.

Immune privilege of the eye

The notion that the eye possessed unique immunological properties was recognized over 130 years ago by the Dutch ophthalmologist van Dooremaal, who observed the prolonged survival of mouse skin grafts transplanted into the anterior chamber (AC) of the dog eye (van Dooremaal, 1873). Evidence that immune privilege occurred in the human eye first appeared in 1905 with the first successful corneal transplant (Zirm, 1906). This occurred over 60 years before immunosuppressive drugs were used in other forms of organ transplantation and it provides a tangible example of ocular immune privilege. In the 1940s, pathologists transplanted human tumour biopsy specimens into the AC of the rabbit eye as a putative bioassay for diagnosing malignancy. We now know that the survival of such tumour xenografts is not necessarily a property of the tumour, but rather an expression of immune privilege in the AC of the eye. In the early 1950s Medawar noted the conspicuous absence of lymph vessels draining the AC and proposed that the survival of foreign tissue grafts in the eye was the consequence of sequestration of the donor antigens, a concept that would later be called "immunological ignorance" (Barker and Billingham, 1968; Medawar, 1945, 1948). Medawar is credited with coining the term "immune privilege" to describe body sites that were devoid of lymphatic drainage and as a result, provided sanctuary for foreign tissue grafts.

Immune privilege is a widely recognized, but frequently an over-interpreted concept. Two misconceptions are often associated with immune privilege. The first misconception is that immune privilege means that immune responses are universally excluded in immune privileged sites such as the AC of the eye and the brain. While it is true that many immune responses are typically blunted or blocked in immune privileged sites, there are notable exceptions. Corneal allografts enjoy a remarkable success rate even though they are performed without the use of HLA matching or the administration of systemic immunosuppressive drugs. However, corneal allografts can undergo immune rejection. In fact, immune rejection is the leading cause of corneal allograft failure (Niederkorn, 2007a,b). As stated earlier, the pathology of some ocular surface infections is immune-mediated (Streilein et al., 1997a,b). Thus, ocular immune privilege, although clearly evident in many settings, has defined boundaries. The second misconception relating to ocular immune privilege is that the AC is devoid of lymphatic drainage and that antigens introduced into the eye are denied access to regional lymph nodes. Although there are no patent, readily detectable lymph vessels draining the AC, antigens introduced into this part of the eye rapidly accumulate in the submandibular lymph nodes, where they are perceived by elements of the adaptive immune system. However, the ensuing immune response is remarkably different from that induced when antigens are introduced into other body sites and instead, culminates in a unique pattern of immune responses known as anterior chamber–associated immune deviation (ACAID), which down-regulates immune-mediated inflammation in an antigen-specific manner (see below).

In spite of the exceptions to the rule, immune privilege is an important phenomenon that many believe is crucial for the preservation of vision. The inability of key ocular tissues to undergo regeneration makes it imperative for immune-mediated inflammation to be tightly controlled. This is the *raison d'être* of ocular immune privilege.

The immune system is divided into two functionally distinct components: the innate immune system and the adaptive, T cell-dependent immune system. Cells of the innate immune system are characterized by their nimble responses to pathogens and their role as "first responders" during the early stages of infection before adaptive T cell-dependent immune responses are generated. Macrophages and neutrophils are the major cellular elements of the innate immune apparatus and detect the presence of potential pathogens via their recognition of pathogen-associated molecular pattern molecules (PAMPs) expressed on invading microorganisms or damage-associated molecular pattern molecules (DAMPs) arising from sterile injury. Natural killer (NK) cells are also members of the innate immune system and are subject to the constraints of ocular immune privilege. There is mounting evidence that elements of both the adaptive and innate immune systems have the capacity to inflict injury to ocular tissues. However, a combination of anatomical, physiological and immunological adaptations restrict the induction and expression of immune responses arising from both the adaptive and innate immune systems.

Anatomical and structural properties of the eye that limit immune-mediated responses

The avascular and alymphatic nature of the cornea is a well-recognized feature of the ocular surface that is known to contribute to corneal allograft survival. Manipulations that induce lymphangiogenesis or hemangiogenesis of the cornea invariably result in corneal allograft rejection (Niederkorn, 2007a,b). The trafficking of immune cells into the interior of the eye is limited by the tight junctional complexes in the endothelial cells of the iris and retinal blood vessels, which restrict the egress of leucocytes and macromolecules from the blood into the eye.

MHC class I molecules are uniformly expressed on virtually all nucleated cells in the body, with the notable exceptions of the neurons in the brain and cells of the corneal endothelium and neural retina (LeBouteiller, 1994). Interestingly, the corneal endothelium and the retina express little or no MHC class I antigens and coincidentally, are uniformly incapable of regeneration. The absence of MHC class I molecules would render these cells invisible to cytotoxic T lymphocytes (CTL), which utilize these molecules as "docking stations" for the recognition and destruction of virus-infected cells. Thus, the low or frank absence of MHC class I molecules protects ocular cells from CTL-mediated destruction at the risk of permitting unrestrained viral infections. The absence of MHC class Ia molecules in the eye creates an immunological dilemma, as it arouses the attention of NK cells, which are programmed to lyse MHC class I-negative cells (Ljunggren et al., 1991). To compensate for this, corneal endothelial cells and retinal cells express nonclassical MHC class Ib molecules such as HLA-G and HLA-E in humans and Qa-2 in mice (Le Discorde et al., 2003; Niederkorn et al., 1999). HLA-G and HLA-E have the capacity to engage the NK inhibitory receptor CD94-NKG2 and shut off NK cell-mediated lysis (Kovats et al., 1990; Lee et al., 1998; Rouas-Freiss et al., 1997). Although it was originally proposed that HLA-G functioned as a ligand for inhibiting NK killing, it now appears that the inhibitory effect on NK cells is through HLA-E, which is brought to the cell surface in the presence of the leader peptide of HLA-G (Navarro et al., 1999). HLA-G leader complexes with HLA-E and binds to CD94/NKG2 with sufficient affinity to shut off NK cell activity (Moffett-King, 2002).

Anti-inflammatory and immunosuppressive soluble factors in the eye

The aqueous humor (AH) that fills the AC of the eye contains a myriad of anti-inflammatory and immunosuppressive molecules (Taylor, 2007). At least four different AH-borne factors inhibit the expression of T cell–mediated inflammation such as delayed-type hypersensitivity (DTH): (a) transforming growth factor-β (TGF-β), (b) α-melanocyte stimulating hormone (α-MSH), (c) vasoactive intestinal peptide (VIP) and (d) calcitonin gene-related peptide (CGRP) (Cousins et al., 1991; Granstein et al., 1990; Taylor, 2007; Taylor and Yee, 2003; Taylor et al., 1994a,b). The AH also contains a 10KDa peptide that induces apoptosis of NK cells, T cells, macrophages and neutrophils (D'Orazio et al., 1999). Corneal cells also produce indoleamine dioxygenase (IDO), an enzyme that catabolizes tryptophan, which is a key amino acid that is vital for T cell survival (Beutelspacher et al., 2006; Ryu and Kim, 2007) and the proliferative and cytolytic activity of NK cells (Della Chiesa et al., 2006; Frumento et al., 2002). Thus, cells of both the adaptive and innate systems entering the eye are greeted by factors that inhibit their function and survival.

The AH contains a variety of molecules that suppress the innate immune system. A 10KDa factor induces apoptosis of NK cells, macrophages and neutrophils. In addition to cellular elements, the innate immune system employs soluble factors such as the complement system. The classical pathway of the complement cascade can be activated when complement-fixing antibodies bind to their cognate antigens or by the alternative pathway via interactions with microorganisms. Thus, the complement system straddles the innate and adaptive immune systems. Once activated, the complement cascade generates a variety of potent chemoattractants that recruit and activate granulocytes. Granulocytes produce a wide array of toxic reactive oxygen species and proteases that can inflict irreparable injury to ocular cells. However, the untoward effects of complement activation are prevented by complement regulatory proteins (CRP) that are present in the AH and vitreous body and that inactivate the complement cascade before it can cause harm (Goslings et al., 1998; Lass et al., 1990; Sohn et al., 2000a,b).

The AH also contains several soluble factors that affect the function of NK cells. Macrophage migration inhibitory factor (MIF) and TGF-β are present in the AH at concentrations that inhibit NK cell cytolytic activity (Apte and Niederkorn, 1996; Apte et al., 1997, 1998; Rook et al., 1986). Although MIF is generally associated with inflammation, it has remarkable inhibitory effects on the cytolytic activity of NK cells. MIF inhibits the release of perforin molecules by NK cells but not CTL (Apte et al., 1998). MIF acts almost immediately and inhibits NK cell-mediated cytolysis of NK-sensitive tumour cells within 4h. By contrast, TGF-β-mediated inhibition of NK cytolytic activity is not detected until 18h (Rook et al., 1986). NK cells lyse corneal endothelial cells *in vitro*, due to their absence of MHC class I molecules that provide an "off" signal to NK cells (Apte and Niederkorn, 1996; Apte et al., 1998). However, this cytolysis is inhibited if recombinant MIF or AH is added to the cytolysis assay system, suggesting that in the AC of the eye, MIF and TGF-β inhibit NK cellular cytotoxicity (Apte and Niederkorn, 1996). In support of this are studies showing that syngeneic tumours that undergo NK cell-mediated rejection at extraocular sites (e.g. subcutaneous), grow progressively in the AC of the eye (Apte et al., 1997).

Cell membrane-bound factors that contribute to immune privilege

Cells lining the interior of the eye express multiple cell membrane-bound molecules that inhibit both adaptive and innate immune responses. FasL (CD95L) is widely expressed within the eye and induces apoptosis of $CD95^+$ inflammatory cells of both the innate and adaptive immune systems that enter the eye (Griffith et al., 1995). FasL on corneal cells is also crucial for the survival of corneal allografts (Stuart et al., 1997; Yamagami et al., 1997). Thus, multiple mechanisms act to suppress NK cell activity in the eye (Table 29.1).

CRP that are found in the AH are also expressed on the cell membranes of multiple cells within the eye and are important in buffering the proinflammatory effects

Table 29.1 Regulation of NK cell-mediated cytolysis in the eye

Ocular regulatory molecule	Effect	Consequences
MIF	Immediate inhibition of NK cell-mediated cytolysis	Intraocular tumours escape NK surveillance
TGF-β	Delayed inhibition of NK cell-mediated cytolysis	Intraocular tumours escape NK surveillance
Qa-2 (mouse) HLA-G (human)	Engages KIR on NK cells and silences NK cell-mediated cytolysis of MHC class I-negative cells	MHC class I-negative corneal endothelial cells and retina cells escape NK cell-mediated killing

of the complement cascade. The importance of regulating complement activation was demonstrated in experiments in which administration of antibodies that block the cell membrane-bound CRP in the eye resulted in the spontaneous development of severe intraocular inflammation in rats (Sohn et al., 2000a,b).

Programmed death ligand-1 (PD-L1) is a type one transmembrane glycoprotein belonging to the B7 family and is expressed on a wide variety of tissues (Dong et al., 1999; Freeman et al., 2000; Latchman et al., 2001; Tseng et al., 2001). PD-L1 expression is up-regulated by the proinflammatory cytokine, IFN-γ (Okazaki and Honjo, 2007; Wax et al., 2008). PD-L1 is also found on multiple tissues within human and mouse eyes (Hori et al., 2006; Shen et al., 2007; Yang et al., 2009). Engagement of PD-L1 expressed on ocular cells with its receptor on inflammatory cells results in the inhibition of T cell proliferation, the induction of apoptosis and inhibition of proinflammatory cytokine secretion by T cells (Hori et al., 2006; Shen et al., 2007; Yang et al., 2009). Expression of PD-L1 on corneal cells is crucial for corneal allograft survival, as in vivo administration of anti-PD-L1 antibody promotes corneal allograft rejection (Hori et al., 2006; Shen et al., 2007). Likewise, corneal allografts prepared from PD-L1$^{-/-}$ mice experience a heightened incidence of immune rejection (Shen et al., 2007).

ACAID: A dynamic immunoregulatory process that sustains immune privilege in the eye

Immune privilege in the eye is also the result of dynamic immunoregulatory processes that are initiated when antigen or antigen-specific T cells enter the eye. In the mid 1970s, Streilein and colleagues demonstrated that antigens introduced into the AC were not sequestered, but in fact elicited an immune deviation that they termed "ACAID" (Streilein, 1987; Streilein et al., 1997a,b; Niederkorn, 2002; Niederkorn and Wang, 2005a,b; Niederkorn, 2006a,b; Streilein, 2003; Streilein et al., 2002). The nature of the systemic immune response to ocular antigens is characterized by the antigen-specific suppression of Th1 immune responses, such as DTH and Th2-based inflammation, such as experimental allergic asthma. This immune deviation also results in redirection of immunoglobulin isotype switching such that complement-fixing antibody production is inhibited, while the generation of non-complement-fixing antibody is preserved. The absence of complement-fixing antibodies reduces the risk of inflammation resulting from the proinflammatory molecules generated by the complement cascade, while preserving the production of neutralizing antibodies that could act to inhibit viral infections.

Early studies on ACAID demonstrated that IL-10, which at the time was considered a Th2 signature cytokine, was crucial, as it was preferentially produced by T cells in ACAID animals and IL-10 knockout (KO) mice were incapable of developing ACAID. This led some to suspect that ACAID was simply the result of a preferential activation of Th2 cells that cross-regulated Th1 immune responses. However, subsequent studies demonstrated that ACAID also down-regulated Th2-based inflammatory responses and did not require the contribution of a pivotal Th2 cytokine, IL-4 (D'Orazio and Niederkorn, 1998; Katagiri et al., 2002; Kosiewicz et al., 1998). Instead, ACAID is the product of a complicated series of cellular interactions involving four organ systems and multiple immune cell populations that conspire to generate T regulatory cells (T_{regs}).

ACAID is initiated when antigens are introduced into the AC or an orthotopic corneal allograft is transplanted to the eye. As mentioned above, ACAID requires the participation of four organ systems: (a) eye, (b) thymus, (c) spleen and (d) sympathetic nervous system. Removal of the eye, thymus or spleen within 3 days of AC injection of antigen prevents the induction of ACAID, and in some cases, culminates in the development of antigen-specific Th1 immune responses such as DTH (Streilein, 2003; Niederkorn, 2006a,b, 2007a). Likewise, inactivation of the sympathetic nervous system by chemical sympathectomy prevents the induction of ACAID (Li et al., 2004).

Ocular phase of ACAID

The induction of ACAID begins when antigens are introduced into the AC or a corneal transplant is grafted onto the eye. Antigens are captured by resident F4/80^{+} macrophages, which under the influence of cytokines in the AH, most notably TGF-β, down-regulate IL-12 and up-regulate IL-10 production, the latter being crucial for the ocular and splenic phases of ACAID. AH-borne TGF-β also stimulates ocular macrophages to produce macrophage inflammatory protein-2 (MIP-2), which plays a key role in the splenic phase of ACAID (see below).

Thymic phase of ACAID

Ocular macrophages capture antigen and emigrate from the eye to the thymus and spleen within 72 h. Within the thymus, the ocular macrophages induce the generation of a unique population of CD4^{-}, CD8^{-}, NK1.1^{+} NKT cells, which are believed to enter the circulation as recent thymic emigrants and migrate to the spleen where they participate in the generation of antigen specific, CD8^{+} T_{regs} (Wang et al., 1997, 2001). Ocular antigen-presenting cells (APC) also home to the spleen where they initiate a complicated series of cellular interactions that involve the participation of the complement system, B cells, γδ T cells, NKT cells, $CD4^+$ T cells and $CD8^+$ T cells.

Splenic phase of ACAID

The population of ocular macrophages that enter the spleen possess unique properties including: (a) the expression of the MHC class I-like molecule, CD1d, (b) the production of IL-10, IL-13 and MIP-2, and (c) activation of signal transducer and activator of transcription-6 (Faunce et al., 2001; Faunce and Stein-Streilein, 2002; Nakamura et al., 2003, 2005; Sonoda and Stein-Streilein, 2002; Sonoda et al., 2001). After entering the spleen, ocular macrophages secrete MIP-2, which attracts $CD4^+$ NKT cells, which in turn leads to the production of the chemokine, RANTES. RANTES recruits additional cells into the marginal zone of the spleen resulting in a complicated series of interactions in which antigen is released by ocular macrophages. The released antigen is captured by splenic B cells, reprocessed, and presented to $CD4^+$ T cells and $CD8^+$ T cells. Splenic γδ T cells are also required for this phase of ACAID. It is not clear how γδ T cells contribute to the generation of ACAID T_{regs}, but their production of IL-10 is absolutely essential. The interactions between ocular macrophages, B cells, γδ T cells, NKT cells and $CD4^+$ T cells also requires the participation of the third component of complement. The end effect is the generation of antigen-specific T_{regs} that down-regulate both Th1- and Th2-based inflammation.

Sympathetic nervous system and ACAID

It is not clear how the sympathetic nervous system contributes to the generation of ACAID, but it is noteworthy that the eye, spleen and thymus have dense sympathetic innervations. It has been previously reported that chemical sympathectomy influences DTH and antibody responses (Bellinger et al., 1990; Madden et al., 1989; Rice et al., 2001; Rook et al., 2002). This led Li and co-workers to evaluate the role of the sympathetic nervous system in ACAID (Li et al., 2004). Their findings indicated that the sympathetic nervous system did not affect the ocular phase of ACAID, but appeared to prevent the development of splenic NKT cells and thus, the generation of ACAID T_{regs}.

Ocular immune privilege and NK cells

Ocular immune privilege affects both the innate and adaptive immune systems. NKT cells contribute to ocular immune privilege during the thymic and splenic phases of ACAID. Conventional NK cell activities are affected by ocular immune privilege. Within the AC, NK cell-mediated cytotoxicity is blunted by MIF and TGF-β, which are present in the AH. Non-classical MHC class I molecules such as Qa-2 decorate many of the cells lining the interior of the eye and are believed to protect MHC class I negative ocular cells from NK cell-mediated lysis. There is no evidence to date indicating that NK cell activity is down-regulated as a consequence of ACAID. However, there is evidence that NK cells contribute to the pathobiology of some immune-mediated ocular diseases.

Role of NK cells in bacterial, viral and neoplastic diseases of the eye

The eye is host to a variety of bacterial, fungal, protozoal and viral agents that produce disease at the ocular surface as well as in the interior of the eye. In spite of its immune privilege, the eye occasionally suffers from immune-mediated diseases such as idiopathic uveitis. The uveal tract of the eye is one of the most richly vascularized tissues in the body and is the site for the development of choroidal melanoma, which is the most common intraocular malignancy in adults. Interestingly, NK cells play a role in the pathogenesis of each of these diseases (Table 29.2).

Herpes simplex virus keratitis

HSV-1 keratitis (HSVK) is the leading cause of preventable infectious blindness in North America (Dawson, 1984). A large body of evidence indicates that HSVK is an immune-mediated disease (Streilein et al., 1997a,b). Ocular disease is caused by type 1 HSV, rather than HSV-2. HSV-1 produces a wide spectrum of symptoms, but each is characterized by the virus' capacity to develop latent infection of sensory nerves, primarily the trigeminal ganglia. Latent HSV-1 can persist for years and can produce recurrent infections throughout the lifespan of the host. Studies in mice have demonstrated that the pathogenesis of HSVK is mediated primarily by $CD4^+$ Th1 cells (Doymaz and Rouse, 1992; Mercadal et al., 1993; Metcalf et al., 1979; Russell et al., 1984). The role of the innate immune system in HSVK is not well understood, but recent findings indicate that NK cells participate in the pathogenesis of HSV-induced corneal scarring (Bouley et al., 1996; Brandt and Salkowski, 1992; Ghiasi et al., 2000; Tamesis et al., 1994). A study by Bouley and co-workers reported that corneal infections with HSV-1 resulted in increased splenic NK activity, which correlated with the severity of the corneal lesions (Brandt and Salkowski, 1992).

Table 29.2 Role of NK cells and NKT cells in microbial keratitis and uveitis

Disease	Role of NK cells	Role of NKT cells
HSV keratitis	Promote neutrophil infiltration into cornea	Unknown
Pseudomonas keratitis	Produce IFN-γ, which activates neutrophils, which are crucial for clearing bacteria from infected cornea	Produce IFN-γ, which induces NK cells to produce IFN-γ
EAU[1]	Necessary for maximal retinal inflammation	Inhibit TNF-α production by IRBP-specific T cells. Necessary for resolution of EAU in rats
EIU[2]	Produce IFN-γ, which down-regulates KC synthesis by ocular cells and thus, reduces neutrophil infiltration into anterior segment of the eye	Unknown
Uveitis (human)	Anti-IL-2 receptor antibody induces 4- to 20-fold increase in regulatory NK cells, which produce IL-10	Unknown

[1]EAU = Experimental autoimmune uveitis. An immune-mediated disease of the uveal tract. Pathology is produced by CD4+ T cells.
[2]EIU = Endotoxin-induced uveitis. An acute inflammatory disease of the anterior segment of eye that is induced by the injection of lipopolysaccharide (LPS). Inflammation is predominantly neutrophilic.

Depletion of NK cells in severe combined immune deficient (SCID) mice resulted in a marked reduction in the incidence and severity of HSVK compared to untreated SCID mice, which had an intact NK cell repertoire (Bouley et al., 1996). Although NK cells produce significant quantities of IFN-γ, which modulates a variety of immune responses, NK cell-depleted SCID mice express the same levels of IFN-γ message and protein in their spleens and in the affected corneas compared to similar hosts with an intact NK cell repertoire. Although in vivo depletion of NK cells did not affect the production of IFN-γ, it did result in a significant reduction in the numbers of neutrophils that migrated into the inflamed corneas (Bouley et al., 1996). This is a particularly relevant observation, as neutrophils are the primary inflammatory cells in HSVK lesions. Although IFN-γ has been implicated as a key cytokine in the pathogenesis of HSVK (Babu et al., 1995; Hendricks et al., 1992; Niemialtowski and Rouse, 1992;) and is produced by NK cells, this is not the basis for the NK cell's contribution to HSVK (Bouley et al., 1996). NK cells also produce tumour necrosis factor-α (TNF-α) in response to virus-infected cells (Vassalli, 1992). TNF-α up-regulates cell adhesion molecules such as ICAM-1 and E-selectin (Zhang et al., 1992), which would enhance neutrophil cell adhesion to vascular endothelial cells and thereby facilitate neutrophil infiltration into the HSV-infected cornea. NK cells also secrete IL-1, which can ramp up the production of IL-8 by HSV-infected corneal cells (Oakes et al., 1993). IL-8 is a potent chemoattractant for neutrophils and the combination of TNF-α and IL-1 might account for the NK-associated increase in neutrophil infiltration in HSV-infected corneas. Thus, NK cells contribute to the pathogenesis of HSVK through their elaboration of cytokines and chemokines that attract and activate neutrophils, which are the end-stage effector cells of HSVK.

Pseudomonas keratitis

Bacterial keratitis is a complication that is frequently associated with contact lens wear and affects approximately 30000 patients each year (Hazlett, 2004). *Pseudomonas aeruginosa* is a Gram-negative bacterium that is a common cause of bacterial keratitis. The pathophysiology of *Pseudomonas* keratitis is largely immune-mediated, with $CD4^+$ T helper cells playing a key role in initiating events that lead to corneal perforation (Hazlett et al., 2000). The severity of *Pseudomonas* keratitis is highest in Th1-prone mouse strains (e.g. C57BL/6) whose corneas perforate as the disease progresses compared to Th2-prone mouse strains (e.g. BALB/c), which are resistant and whose corneas heal (Hazlett et al., 2000). Although IFN-γ is most often associated with Th1 cells and might be expected to contribute to corneal perforation, evidence suggests the contrary. That is, resistant BALB/c mice are Th2-prone, yet IFN-γ is crucial for disease resolution (Lighvani et al., 2005). This apparent paradox was resolved in studies, which showed that IFN-γ was produced locally in the infected cornea even though T cells were not detectable. However, NK cells were found to be the source of the locally produced IFN-γ, as in vivo depletion of NK cells with anti-asialo GM1 antibody resulted in a decreased expression of IFN-γ mRNA and a commensurate increase in the severity of neutrophil inflammation, bacterial load and corneal scarring (Lighvani et al., 2005). If NK cells are the sole source of locally produced IFN-γ in the

HSV-infected cornea, what is stimulating them? The unique neurological structure of the cornea provides a key clue. The cornea is one of the most densely innervated tissues in the body and neuropeptides are known to influence a variety of immune processes. Substance P (SP) is one of the neuropeptides found in the eye and has been shown to augment inflammatory responses (Marriott and Bost, 2001). NK cells express neurokinin-1 receptor (NK-1R), which binds with SP. With this in mind, Lighvani and co-workers tested the hypothesis that SP, through its interaction with NK-1R on NK cells, induced the production of IFN-γ in the HSV-infected cornea (Lighvani et al., 2005). Accordingly, mice were treated with the SP antagonist, spantide I, which blocks SP interaction with NK-1R. This treatment significantly decreased the local production of IFN-γ in the infected cornea and resulted in corneal perforation. In spite of increased neutrophil infiltration in the spantide I-treated mice, the bacterial load was increased, suggesting that IFN-γ was needed to activate the neutrophils and facilitate their clearance of the bacteria. Anti-asialo GM1 antibody treatment also depletes NKT cells and raises the possibility that these results may have been due to NKT cells, rather than NK cells. NKT cells are capable of rapidly producing IFN-γ and regulating T cells, NK cells and dendritic cells (Chen and Paul, 1997; Eberl and MacDonald, 2000; Ho et al., 2004; Naumov et al., 2001). Immunohistochemical staining of HSV-infected corneas in C57BL/6 mice revealed the presence of both NKT cells and NK cells (Hazlett et al., 2007). To determine the role of NKT cells in corneal perforation, mice with NKT cell deficiencies, but intact NK cell repertoires, were infected with *P. aeruginosa*. Like NK cell-deficient mice, the NKT cell-deficient mice displayed reduced levels of IFN-γ, greater bacterial loads and accelerated time to corneal perforation (Hazlett et al., 2007). Further studies showed that recognition of lipopolysaccharide (LPS) by antigen-presenting Langerhans cells (LC) in the cornea induces them to produce IL-12, which in turn stimulates NKT cells to produce IFN-γ. The IFN-γ produced by NKT cells stimulates NK cells in the cornea to produce additional IFN-γ, which is essential for the clearance of the bacteria and the prevention of corneal perforation (Hazlett et al., 2007). Without IFN-γ production by both NKT cells and NK cells, neutrophil activation does not occur, bacteria accumulate and the cornea perforates. Thus, cells of both the innate (neutrophils and NK cells) and adaptive (NKT cells) immune systems are crucial in controlling *Pseudomonas* keratitis.

Corneal allograft rejection

Corneal allografting is the oldest, most common and arguably, the most successful form of solid tissue transplantation (Niederkorn, 2003). In the United States alone, over 40000 corneal transplants are performed each year (Niederkorn, 2007a,b). The success of first-time corneal allografts is approximately 90%, even though HLA typing is not routinely performed and systemic immunosuppressive drugs are not used (Niederkorn, 1999, 2003). Several properties of the cornea and the eye contribute to the immune privilege of corneal allografts. Chief among these is the capacity of the corneal allograft to induce ACAID (Niederkorn, 2006a,b). Manoeuvres that prevent the induction of ACAID, such as splenectomy, neutralization of IL-10, or low dose cyclophosphamide treatment, result in increased corneal allograft rejection (Niederkorn, 2006a,b). Likewise, AC injection of donor-specific cells induces ACAID and produces a dramatic increase in corneal allograft survival (Niederkorn et al., 1995; She et al., 1990). Studies in a mouse model of corneal transplantation have shown that NKT cells play a crucial role in the induction of ACAID and corneal allograft survival (Sonoda et al., 2002; Watte et al., 2008).

Although corneal allografts enjoy a remarkable success rate, immune privilege can fail, leading to corneal graft rejection (Niederkorn, 2007a,b). Indeed, immune rejection remains the leading cause of corneal allograft failure. A preponderance of data from animal models of penetrating keratoplasty have demonstrated the crucial role for T cells, especially $CD4^+$ T cells in corneal allograft rejection (Niederkorn, 2007a,b). Immunohistochemical analysis of rejected corneal allografts in animals and human subjects reveals the presence of a mixed population of lymphocyte subsets including $CD4^+$ T cells, $CD8^+$ T cells and NK cells. However, analysis of the AH of rats after corneal transplantation revealed a curious transition in the inflammatory cell populations over time (Claerhout et al., 2004). Initially, $CD3^-$, $CD4^+$ mononuclear cells were present in the AH, but were replaced by a mixed population of $CD4^+$ T cells, $CD8^+$ T cells, $CD3^-$, $CD161^+$ NK cells and $CD3^+$, $CD161^+$ NKT cells (Claerhout et al., 2004). NK cells and NKT cells reached levels in the AH of grafted mice that were 10- to 15-fold higher than those found in the draining lymph nodes, suggesting either their local expansion in the eye or recruitment from the peripheral blood. NK cells might contribute to corneal allograft rejection through their production of IFN-γ and TNF-α, two cytokines that are known to induce apoptosis of corneal cells in vitro. Studies in mice have shown that activated NK cells can directly kill corneal endothelial cells in vitro (Apte and Niederkorn, 1996). It is widely believed that $CD4^+$ T cells are crucial for corneal allograft rejection in rats and mice. $CD4^+$ T cells have been shown to induce apoptosis of allogeneic corneal cells in vitro and might mediate rejection by this mechanism (Hegde et al., 2005). However, it is possible that NK cells activated by IFN-γ produced by donor-specific $CD4^+$ T cells act as the end stage effector cells in corneal graft rejection. However, this hypothesis remains to be confirmed and must take into consideration results indicating that depletion of NK cells with anti-asialo GM1 fails to delay or prevent corneal allograft rejection in mice (Niederkorn et al., unpublished findings). The role of NK cells in corneal allograft rejection awaits further clarification. However, it is clear that NKT cells contribute to the immune privilege of corneal allografts and are crucial for their long-term survival in mice (Sonoda et al., 2002). It remains to be determined if the same holds true in human keratoplasty patients, but it is worth noting that NKT cells in humans share many of the properties and functions of murine NKT cells (Brossay et al., 1998; Godfrey and Kronenberg, 2004; Godfrey et al., 2000).

Uveitis

The iris, ciliary body and choroid form the middle layer of the eye and constitute the uveal tract. The uveal tract is particularly vulnerable to immune-mediated diseases, which are collectively known as uveitis. Over forty years ago, Wacker and Lipton demonstrated that antigens isolated from the bovine retina would provoke inflammation of the uveal tract in rats (Wacker and Lipton, 1965). In the ensuing four decades, hundreds of studies in animals have demonstrated that immunizing rodents with retinal antigens induced a T cell–mediated inflammation of the retina and uveal tract that mimics human idiopathic uveitis and has been termed "experimental autoimmune uveitis" (EAU) (Gery et al., 2002). A large number of retinal antigens can induce EAU in rodents. The most frequently studied retinal antigen in EAU in mice is interphotoreceptor retinoid-binding protein (IRBP). Although a variety of retinal antigens can be used to induce EAU in rodents, the pathophysiology seems to share a common T cell-dependency. Originally, it was widely believed that EAU was mediated solely by Th1 cells that produced IFN-γ (Caspi et al., 2008; Gery et al., 2002). However, recent studies have reported that EAU can also be mediated by IL-17–producing $CD4^+$ Th17 cells (Cox et al., 2008; Luger et al., 2008; Peng et al., 2007; Yoshimura et al., 2008). Adoptive cell transfer studies indicated that either polarized Th1 or Th17 cells could produce EAU in mice. However, there were fundamental differences in the nature of the inflammation, the expression of cell surface markers, homing properties and cytokine production in the two cell populations (Cox et al., 2008).

Both NK cells and NKT cells play a role in the pathogenesis of EAU in rodents (Kitaichi et al., 2002; Figueiredo et al., 2007; Shao et al., 2003;). Using a rat model of EAU, Shao and co-workers found that the preponderance of cells that infiltrated the eyes of rats at the peak of EAU were NK cells and NKT cells (Shao et al., 2003). At day 10, 39% of the infiltrating cells expressed the cell surface markers indicative of NKT cells (αβ TCR^+NKR-P1a$^+$) and 16% of the cells displayed NK cell-specific markers (αβ TCR^-NKR-P1a$^+$). In vitro analyses demonstrated that the eye-infiltrating NK/NKT cells suppressed the production of the proinflammatory cytokine, TNF-α, by IRBP-specific T cells. The role of NKT cells in EAU remains to be elucidated. However, the capacity of NKT cells to inhibit TNF-α production by IRBP-specific T cells suggests that NKT cells may contribute to the resolution of ocular inflammation in this model.

There are conflicting reports on the role of NK cells in uveitis (Kitaichi et al., 2002; Li et al., 2005; Figueiredo et al., 2007). In vivo depletion of NK cells with anti-CD161c monoclonal antibody mitigated EAU in mice (Kitaichi et al., 2002). Interestingly, NK cells were not necessary for the generation of IRBP-specific T cells, as mice treated with anti-CD161c antibody developed T cells that proliferated when confronted with an IRBP-specific peptide (K2) in vitro. Thus, NK cells appear to be necessary for expression of EAU by acting after IRBP-specific T cells have been generated.

The role of NK cells has also been examined in a model of anterior uveitis (Figueiredo et al., 2007). Endotoxin-induced uveitis (EIU) is an animal model of acute anterior uveitis that is induced by intraocular injection of endotoxins, such as LPS, which are produced by Gram negative bacteria. EIU has been used as a model for studying the pathophysiology and immunobiology of anterior uveitides such as Reiter's syndrome and Behcet's disease (Kogiso et al., 1992; Okumura et al., 1990; Rosenbaum et al., 1980). EIU is characterized by disruption of the blood/ocular barrier, secretion of cytokines and chemokines by the vascular endothelium in the eye, and acute inflammation of the anterior segment of the eye that is composed primarily of neutrophils and macrophages (Rosenbaum et al., 1980; Okumura et al., 1990; Kogiso et al., 1992). Intraocular injection of IL-12 has been shown to mitigate EIU in wild-type mice but not in IFN-γ KO mice or wild-type mice treated with neutralizing anti-IFN-γ antibody, suggesting that IFN-γ plays a crucial role in controlling EIU in the mouse (Figueiredo et al., 2007). NK cells play a pivotal role in the IL-12-induced mitigation of EIU, as depletion of NK cells with anti-CD161c antibody (PK136) abolished the protective effect. Cytokine analysis of the affected eyes revealed that the intraocular injection of IL-12 down-regulated the expression of the proinflammatory chemokine, keratinocyte-derived chemokine (KC). These findings are consistent with the notion that NK cells are intermediary cells in the regulation of EIU. Injection of IL-12 up-regulates IFN-γ production by NK cells, which in turn down-regulates KC production by ocular cells, thereby reducing the chemokine cues that beckon neutrophils to enter the endotoxin-injected eye. Thus, IFN-γ, which is normally viewed as a proinflammatory cytokine, acts to down-regulate inflammation in EIU.

To date, only one study has examined the role of NK cells in the regulation of uveitis in human subjects (Li et al., 2005). In vivo blockade of the human IL-2 receptor induced a 4- to 20-fold expansion in CD56^{bright} regulatory NK cells in uveitis patients. Cytokine analysis revealed that the regulatory NK cells produced an array of cytokines, with a notable increase in IL-10 synthesis, which may have accounted for the therapeutic effect.

The disparities in results in animal models and human studies indicate that much remains to be learned about the role of NK cells and NKT cells in uveitis. It is likely that NK cells and NKT cells act as ancillary cells, rather than effector cells, in the pathogenesis and resolution of uveitis. Moreover, each population may exert opposing effects depending on the stage of the disease process. If this suspicion is confirmed, then it will be crucial to identify key markers and the kinetics for NK cell and NKT cell activity in the progression of uveitis in order to design therapeutic strategies for managing this devastating disease.

Uveal melanoma

Melanomas of the uveal tract are the most common intraocular malignancies in adults, with an incidence of seven cases per one million adults per year (Mooy and De Jong, 1996). Although cutaneous melanoma and uveal melanoma are both derived from neural crest progenitors, they differ in their metastatic and immunological behaviours (Char, 1978; Donoso et al., 1985; Niederkorn, 1995; Vijayasaradhi and Houghton, 1995). Cutaneous melanomas can metastasize to multiple organs and is one of the few cancers that metastasizes regularly to the heart (Vijayasaradhi and Houghton, 1995). By contrast, uveal melanoma preferentially metastasizes to the liver, with liver metastases being present in over 80% of the uveal melanoma patients at the time of death (Char, 1978; Donoso et al., 1985; Rajpal et al., 1983).

Numerous studies in mice and human subjects have demonstrated a role of NK cells in controlling the metastasis of uveal melanoma (Niederkorn, 1995; Niederkorn and Wang, 2005a,b). Uveal melanoma cells are clearly susceptible to NK cell-mediated cytolysis in vitro (He et al., 2004; Ma and Niederkorn, 1995; Ma et al., 1995). However, the susceptibility to NK cell-mediated cytolysis was inversely associated with expression of MHC class I molecules (Ma and Niederkorn, 1995; Ma et al., 1995). This is consistent with the "missing self" hypothesis, which proposes that MHC class I molecules on potential target cells engage killer inhibitory receptors (KIRs) that are expressed on NK cells and as a result, a negative signal is transmitted to the NK cells, thereby preventing cytolysis (Ljunggren and Karre, 1990). Cytolysis is the default mode of NK cells and potential target cells failing to express MHC class I molecules cannot transmit the "off" signal to NK cells and as a result, are killed. Uveal melanomas residing in the iris or ciliary body are bathed in AH, which contains large quantities of TGF-β (Cousins et al., 1991; Jampel et al., 1990). TGF-β is known to down-regulate the expression of MHC class I molecules on many cells including uveal melanomas (Krueger et al., 1990; Ma and Niederkorn, 1995; Orcel et al., 1990;). Moreover, incubation of uveal melanoma cells in TGF-β at concentrations comparable to those found in AH down-regulates MHC class I expression and increases their susceptibility to NK cell-mediated cytolysis (Ma and Niederkorn, 1995). Likewise, incubating uveal melanoma cells in IFN-γ up-regulates the expression of MHC class I molecules and reduces their susceptibility to NK cell-mediated cytolysis. Thus, uveal melanomas are susceptible to NK cell-mediated cytolysis in vitro and reside in an in vivo environment that would increase their vulnerability to NK cell-mediated cytolysis. Moreover, there is evidence that lymphocytes, including NK cells, enter uveal melanoma-containing eyes. Lymphocytes can be found in uveal melanomas (Anastassiou et al., 2001; de la Cruz et al., 1990; Durie et al., 1990; Nitta et al., 1990; Staibano et al., 2006) in as many as 20% of the eyes examined (de la Cruz et al., 1990). NK cells can comprise anywhere from less than 10% to over 40% of the tumour infiltrating lymphocytes (TIL) in uveal melanomas (de Waard-Siebinga et al., 1996; Ksander et al., 1991; Meecham et al., 1992). Moreover, NK cells isolated from uveal melanoma-containing eyes can mediate cytolysis of NK-sensitive K562 target cells in vitro (Ksander et al., 1991). The weight of evidence to this point suggests that NK cells have the potential to control uveal melanomas in the eye. That is: (a) uveal melanoma cells are susceptible to NK cell-mediated cytolysis in vitro, (b) uveal melanomas of the anterior uveal tract reside in an environment that promotes down-regulation of MHC class I molecules and should increase their susceptibility to NK cell-mediated cytolysis, (c) NK cells can be found infiltrating uveal melanoma-containing eyes and (d) NK cells isolated from such eyes can kill NK-sensitive tumour cells in vitro. However, do NK cells function in the uveal melanoma-containing eye?

The first clue that NK cells might not function in the eye is the observation that the corneal endothelial cells are virtually devoid of MHC class I molecules and are susceptible to NK cell-mediated cytolysis *in vitro*, yet there is no evidence indicating that the corneal endothelium is attacked by NK cells that enter the eye in uveal melanoma patients. Examination of the constituents of the AH provides clues to explain this apparent paradox. TGF-β is present in the AH at a concentration that is known to inhibit the cytolytic machinery of NK cells in vitro; however, the inhibition of NK cell activity does not occur until 18h after exposure to TGF-β (Rook et al., 1986). Importantly, the AH also contains macrophage MIF, which produces an immediate inhibition of NK cell-mediated cytolysis (Apte and Niederkorn, 1996; Apte et al., 1998). The combined effect of an immediate inhibitor (i.e. MIF) and a delayed inhibitor (e.g. TGF-β) in the AH effectively restrains NK cell-mediated cytolytic activity in the eye and thus provides uveal melanoma cells with sanctuary from this arm of the innate immune apparatus. Further proof for this came from studies in which NK-sensitive tumours were transplanted to SCID mice, which lack T cells, but have an intact NK cell repertoire. The NK-sensitive tumours were rejected following subcutaneous transplantation in SCID mice, but grew progressively in SCID mice whose NK cells were depleted by in vivo treatment with anti-asialo GM1 antibody (Apte et al., 1997). By contrast, the same NK-sensitive tumours grew progressively in the eyes of SCID mice with an intact NK cell population, even when the intraocular tumour inocula were 20-fold lower than the inocula used for subcutaneous injections. Thus, the AH contains at least two factors that silence NK cell–mediated cytolysis within the eye and protect intraocular tumours from NK cell-mediated elimination.

The cells in the retina express little or no conventional MHC class I molecules, and since they are not bathed in AH, they are potentially at risk for NK cell-mediated attack. However, cells in the retina and the corneal endothelium constitutively express the non-classical class Ib molecule, Qa2, which interacts with the KIRs that are expressed on NK cells and inhibits NK cell-mediated cytolysis (Niederkorn et al., 1999). Cells in the human eye also constitutively express non-classical MHC class Ib molecules, including HLA-G (Le Discorde et al., 2003).

In cutaneous melanoma patients, low expression of MHC class I molecules is associated with increased tumour thickness and tumour progression, and reduced survival (Brocker et al., 1985). Down-regulation of MHC class I molecules on metastases occurs in a wide variety of tumours, including head and neck squamous cell carcinomas, as well as lung, colon and cervical cancers (Hicklin et al., 1999). Cutaneous melanoma metastases express lower levels of MHC class I molecules compared to the primary tumours (Ferrone and Marincola, 1995; Kageshita et al., 1999; Ruiter et al., 1982, 1984, 1991; van Duinen et al., 1988). However, in the case of uveal melanomas, the opposite appears to occur (Jager et al., 2002). High expression of HLA-B on primary uveal melanomas has been correlated with increased malignancy and poor prognosis (Blom et al., 1997a,b; Ericsson et al., 2001; Jager et al., 2002). Likewise, uveal melanoma patients with low MHC class I expression have a better survival than those with high expression (Blom et al., 1997a,b). This is consistent with previous in vitro studies, which showed that uveal melanoma cells that expressed low amounts of MHC class I molecules were more susceptible to NK cell-mediated cytolysis than cells with high MHC class I expression (Ma and Niederkorn, 1995; Ma et al., 1995). The role of NK cells in controlling liver melanoma metastases has been established in a variety of animal studies, which showed that inhibition of NK cells resulted in increased liver metastases in mice with intraocular melanomas while augmentation of NK activity by either administration of recombinant IFN-α or by transfer of the IFN-β gene resulted in reduced liver metastases (Alizadeh et al., 2003; Dithmar et al., 1999, 2000).

These findings suggest the following scenario. Uveal melanomas arise in an environment that contains TGF-β and MIF and thus, silences NK cell-mediated surveillance. This permits the survival of both high and low MHC class I-expressing uveal melanoma cells in the eye. However, once the melanoma cells disseminate from the eye into the peripheral circulation and subsequently enter the liver, they no longer enjoy the immunological sanctuary of the eye and instead are greeted by NK cells. Indeed, the liver has one of the highest concentrations of NK cells of any organ in the body. This would promote the NK cell-mediated elimination of low MHC class I-expressing melanoma cells and favour the survival of high MHC class I-expressing tumour cells. Support for this hypothesis was reported by Verbik and co-workers who had the unusual opportunity to study primary uveal melanoma cells and liver metastases from the same patient (Verbik et al., 1997). They found that only 4% of the tumour cells from the primary uveal melanoma expressed MHC class I molecules, but the incidence of MHC class I expression was nine times higher in melanoma cells isolated from four different liver foci. This suggests that once the uveal melanoma cells leave the sanctuary of the eye, they are subjected to a selection process whereby melanoma cells expressing low levels of MHC class I are eliminated by NK cells in the blood and in the liver.

In addition to escaping NK cell–mediated immune surveillance by expressing MHC class I molecules on their cell membranes, some uveal melanoma cells have the capacity to produce MIF at concentrations that inhibit NK cell–mediated cytolysis in vitro (Repp et al., 2000). Interestingly, the most potent inhibitory effect was produced by cell lines derived from liver metastases, which produced approximately twice as much biologically active MIF as cell lines derived from primary uveal melanoma. A study of uveal melanoma-containing eyes revealed that all 13 uveal melanoma specimens expressed TGF-β_2, the isoform of TGF-β that suppresses NK cell-mediated cytolysis (Esser et al., 2001). Thus, uveal melanoma metastases have the potential to create ad hoc immune privileged sites through their production of NK inhibitory factors.

Conclusions

Much remains to be learned about the role and function of NK and NKT cells in the eye. NKT cells are clearly important for the induction of tolerance in the eye and for the survival of corneal allografts. NK cells have divergent roles in microbial infections of the ocular surface and autoimmune diseases of the interior of the eye. The pathogenesis of HSVK is immune-mediated and involves the participation of $CD4^+$ T cells and neutrophils. NK cells indirectly contribute to the pathogenesis of HSVK by promoting the migration of neutrophils into the HSV-infected cornea. By contrast, NK cells and NKT cells participate in the resolution of *Pseudomonas* keratitis through their coordinated production of IFN-γ, which activates neutrophils and enhances the clearance of bacteria from the infected cornea. The role of NK cells in intraocular inflammation is unclear. Some reports suggest that NK cells are necessary for the full development of EAU, while other studies indicate that NK cells are necessary for the resolution of EIU. Although NK cells can recognize and kill uveal melanoma cells in vitro, the presence of MIF and TGF-β in the AH and vitreous body silences NK cell–mediated surveillance of intraocular tumours. The eye is indeed an 'immunological microcosm' in which a wide array of immunological activities can be observed, including NK cell–mediated processes.

References

Alizadeh, H., Howard, K., Mellon, J., Mayhew, E., Rusciano, D. and Niederkorn, J.Y. (2003). Reduction of liver metastasis of intraocular melanoma by interferon-beta gene transfer. Invest Ophthalmol Vis Sci 44, 3042–3051.

Anastassiou, G., Coupland, S.E., Stang, A., Boeloeni, R., Schilling, H. and Bornfeld, N. (2001). Expression of Fas and Fas ligand in uveal melanoma: biological implication and prognostic value. J Pathol 194, 466–472.

Apte, R.S. and Niederkorn, J.Y. (1996). Isolation and characterization of a unique natural killer cell inhibitory factor present in the anterior chamber of the eye. J Immunol 156, 2667–2673.

Apte, R.S., Mayhew, E. and Niederkorn, J.Y. (1997). Local inhibition of natural killer cell activity promotes the progressive growth of intraocular tumors. Invest Ophthalmol Vis Sci 38, 1277–1282.

Apte, R.S., Sinha, D., Mayhew, E., Wistow, G.J. and Niederkorn, J.Y. (1998). Cutting edge: role of macrophage migration inhibitory factor in inhibiting NK cell activity and preserving immune privilege. J Immunol 160, 5693–5696.

Babu, J.S., Kanangat, S. and Rouse, B.T. (1995). T cell cytokine mRNA expression during the course of the immunopathologic ocular disease herpetic stromal keratitis. J Immunol 154, 4822–4829.

Barker, C.F. and Billingham, R.E. (1968). The role of afferent lymphatics in the rejection of skin homografts. J Exp Med 128, 197–221.

Bellinger, D.L., Lorton, D., Romano, T.D., Olschowka, J.A., Felten, S.Y. and Felten, D.L. (1990). Neuropeptide innervation of lymphoid organs. Ann N Y Acad Sci 594, 17–33.

Beutelspacher, S.C., Pillai, R., Watson, M.P., Tan, P.H., Tsang, J., McClure, M.O., George, A.J. and Larkin, D.F. (2006). Function of indoleamine 2,3-dioxygenase in corneal allograft rejection and prolongation of allograft survival by over-expression. Eur J Immunol 36, 690–700.

Blom, D.J., Luyten, G.P., Mooy, C., Kerkvliet, S., Zwinderman, A.H. and Jager, M.J. (1997a). Human leukocyte antigen class I expression. Marker of poor prognosis in uveal melanoma. Invest Ophthalmol Vis Sci 38, 1865–1872.

Blom, D.J., Schurmans, L.R., De Waard-Siebinga, I., De Wolff-Rouendaal, D., Keunen, J.E. and Jager, M.J. (1997b). HLA expression in a primary uveal melanoma, its cell line, and four of its metastases. Br J Ophthalmol 81, 989–993.

Bouley, D.M., Kanangat, S. and Rouse, B.T. (1996). The role of the innate immune system in the reconstituted SCID mouse model of herpetic stromal keratitis. Clin Immunol Immunopathol 80, 23–30.

Brandt, C.R. and Salkowski, C.A. (1992). Activation of NK cells in mice following corneal infection with herpes simplex virus type-1. Invest Ophthalmol Vis Sci 33, 113–120.

Brocker, E.B., Suter, L., Bruggen, J., Ruiter, D.J., Macher, E. and Sorg, C. (1985). Phenotypic dynamics of tumor progression in human malignant melanoma. Int J Cancer 36, 29–35.

Brossay, L., Chioda, M., Burdin, N., Koezuka, Y., Casorati, G., Dellabona, P. and Kronenberg, M. (1998). CD1d-mediated recognition of an alpha-galactosylceramide by natural killer T cells is highly conserved through mammalian evolution. J Exp Med 188, 1521–1528.

Caspi, R.R., Silver, P.B., Luger, D., Tang, J., Cortes, L.M., Pennesi, G., Mattapallil, M.J. and Chan, C.C. (2008). Mouse models of experimental autoimmune uveitis. Ophthalmic Res 40, 169–174.

Char, D.H. (1978). Metastatic choroidal melanoma. Am J Ophthalmol 86, 76–80.

Chen, H. and Paul, W.E. (1997). Cultured NK1.1+ CD4+ T cells produce large amounts of IL-4 and IFN-gamma upon activation by anti-CD3 or CD1. J Immunol 159, 2240–2249.

Claerhout, I., Kestelyn, P., Debacker, V., Beele, H. and Leclercq, G. (2004). Role of natural killer cells in the rejection process of corneal allografts in rats. Transplantation 77, 676–682.

Cousins, S.W., McCabe, M.M., Danielpour, D. and Streilein, J.W. (1991). Identification of transforming growth factor-beta as an immunosuppressive factor in aqueous humor. Invest Ophthalmol Vis Sci 32, 2201–2211.

Cox, C.A., Shi, G.P., Yin, H., Vistica, B.P., Wawrousek, E.F., Chan, C.C. and Gery, I. (2008). Both Th1 and Th17 are immunopathogenic but differ in other key biologic activities. J Immunol 180, 7414–7422.

D'Orazio, T.J. and Niederkorn, J.Y. (1998). A novel role for TGF-beta and IL-10 in the induction of immune privilege. J Immunol 160, 2089–2098.

D'Orazio, T.J., DeMarco, B.M., Mayhew, E.S. and Niederkorn, J.Y. (1999). Effect of aqueous humor on apoptosis of inflammatory cell types. Invest Ophthalmol Vis Sci 40, 1418–1426.

Dawson, C.R. (1984). Ocular herpes simplex virus infections. Clin Dermatol 2, 56–66.

de la Cruz , P.O. Jr, Specht, C.S. and McLean, I.W. (1990). Lymphocytic infiltration in uveal malignant melanoma. Cancer 65, 112–115.

de Waard-Siebinga, I., Hilders, C.G., Hansen, B.E., van Delft, J.L. and Jager, M.J. (1996). HLA expression and tumor-infiltrating immune cells in uveal melanoma. Graefes Arch Clin Exp Ophthalmol 234, 34–42.

Della Chiesa, M., Carlomagno, S., Frumento, G., Balsamo, M., Cantoni, C., Conte, R., Moretta, L., Moretta, A. and Vitale, M. (2006). The tryptophan catabolite L-kynurenine inhibits the surface expression of NKp46- and NKG2D-activating receptors and regulates NK-cell function. Blood 108, 4118–4125.

Dithmar, S.A., Rusciano, D.A., Armstrong, C.A., Lynn, M.J. and Grossniklaus, H.E. (1999). Depletion of NK cell activity results in growth of hepatic micrometastases in a murine ocular melanoma model. Curr Eye Res 19, 426–431.

Dithmar, S., Rusciano, D., Lynn, M.J., Lawson, D.H., Armstrong, C.A. and Grossniklaus, H.E. (2000). Neoadjuvant interferon alfa-2b treatment in a murine model for metastatic ocular melanoma: a preliminary study. Arch Ophthalmol 118, 1085–1089.

Dong, Z., Greene, G., Pettaway, C., Dinney, C.P., Eue, I., Lu, W., Bucana, C.D., Balbay, M.D., Bielenberg, D. and Fidler, I.J. (1999). Suppression of angiogenesis, tumorigenicity, and metastasis by human prostate cancer cells engineered to produce interferon-beta. Cancer Res 59, 872–879.

Donoso, L.A., Berd, D., Augsburger, J.J., Mastrangelo, M.J. and Shields, J.A. (1985). Metastatic uveal melanoma. Pretherapy serum liver enzyme and liver scan abnormalities. Arch Ophthalmol 103, 796–798.

Doymaz, M.Z. and Rouse, B.T. (1992). Herpetic stromal keratitis: an immunopathologic disease mediated by CD4+ T lymphocytes. Invest Ophthalmol Vis Sci 33, 2165–2173.

Durie, F.H., Campbell, A.M., Lee, W.R. and Damato, B.E. (1990). Analysis of lymphocytic infiltration in uveal melanoma. Invest Ophthalmol Vis Sci 31, 2106–2110.

Eberl, G. and MacDonald, H.R. (2000). Selective induction of NK cell proliferation and cytotoxicity by activated NKT cells. Eur J Immunol 30, 985–992.

Ericsson, C., Seregard, S., Bartolazzi, A., Levitskaya, E., Ferrone, S., Kiessling, R. and Larsson, O. (2001). Association of HLA class I and class II antigen expression and mortality in uveal melanoma. Invest Ophthalmol Vis Sci 42, 2153–2156.

Esser, P., Grisanti, S. and Bartz-Schmidt, K. (2001). TGF-beta in uveal melanoma. Microsc Res Tech 52, 396–400.

Faunce, D.E. and Stein-Streilein, J. (2002). NKT cell-derived RANTES recruits APCs and CD8+ T cells to the spleen during the generation of regulatory T cells in tolerance. J Immunol 169, 31–38.

Faunce, D.E., Sonoda, K.H. and Stein-Streilein, J. (2001). MIP-2 recruits NKT cells to the spleen during tolerance induction. J Immunol 166, 313–321.

Ferrone, S. and Marincola, F.M. (1995). Loss of HLA class I antigens by melanoma cells: molecular mechanisms, functional significance and clinical relevance. Immunol Today 16, 487–494.

Figueiredo , F., Commodaro, A.G., de Camargo, M.M., Rizzo, L.V. and Belfort, R. Jr (2007). NK1.1 cells downregulate murine endotoxin-induced uveitis following intraocular administration of interleukin-12. Scand J Immunol 66, 329–334.

Freeman, G.J., Long, A.J., Iwai, Y., Bourque, K., Chernova, T., Nishimura, H., Fitz, L.J., Malenkovich, N., Okazaki, T., Byrne, M.C., Horton, H.F., Fouser, L., Carter, L., Ling, V., Bowman, M.R., Carreno, B.M., Collins, M., Wood, C.R. and Honjo, T. (2000). Engagement of the PD-1 immunoinhibitory receptor by a novel B7 family member leads to negative regulation of lymphocyte activation. J Exp Med 192, 1027–1034.

Frumento, G., Rotondo, R., Tonetti, M., Damonte, G., Benatti, U. and Ferrara, G.B. (2002). Tryptophan-derived catabolites are responsible for inhibition of T and natural killer cell proliferation induced by indoleamine 2,3-dioxygenase. J Exp Med 196, 459–468.

Gery, I., Nussenblatt, R.B., Chan, C.C. and Caspi, R.R. (2002). Autoimmune diseases of the eye. In: Theofilipoulous, A.N. and Bona, C.A. (eds) The Molecular Pathology of Autoimmune Diseases. New York: Taylor and Francis, pp. 978–998.

Ghiasi, H., Cai, S., Perng, G.C., Nesburn, A.B. and Wechsler, S.L. (2000). The role of natural killer cells in protection of mice against death and corneal scarring following ocular HSV-1 infection. Antiviral Res 45, 33–45.

Godfrey, D.I. and Kronenberg, M. (2004). Going both ways: immune regulation via CD1d-dependent NKT cells. J Clin Invest 114, 1379–1388.

Godfrey, D.I., Hammond, K.J., Poulton, L.D., Smyth, M.J. and Baxter, A.G. (2000). NKT cells: facts, functions and fallacies. Immunol Today 21, 573–583.

Goslings, W.R., Prodeus, A.P., Streilein, J.W., Carroll, M.C., Jager, M.J. and Taylor, A.W. (1998). A small molecular weight factor in aqueous humor acts on C1q to prevent antibody-dependent complement activation. Invest Ophthalmol Vis Sci 39, 989–995.

Granstein, R.D., Staszewski, R., Knisely, T.L., Zeira, E., Nazareno, R., Latina, M. and Albert, D.M. (1990). Aqueous humor contains transforming growth factor-beta and a small (less than 3500 daltons) inhibitor of thymocyte proliferation. J Immunol 144, 3021–3027.

Griffith, T.S., Brunner, T., Fletcher, S.M., Green, D.R. and Ferguson, T.A. (1995). Fas ligand-induced apoptosis as a mechanism of immune privilege. Science 270, 1189–1192.

Hazlett, L.D. (2004). Corneal response to *Pseudomonas aeruginosa* infection. Prog Retin Eye Res 23, 1–30.

Hazlett, L.D., McClellan, S., Kwon, B. and Barrett, R. (2000). Increased severity of *Pseudomonas aeruginosa* corneal infection in strains of mice designated as Th1 versus Th2 responsive. Invest Ophthalmol Vis Sci 41, 805–810.

Hazlett, L.D., Qianqian, L., Liu, J., McClellan, S., Du, W. and Barrett, R.P. (2007). NKT cells are critical to initiate an inflammatory response after *Pseudomonas aeruginosa* ocular infection in susceptible mice. J Immunol 179, 1138–1146.

He, Y.G., Mayhew, E., Mellon, J. and Niederkorn, J.Y. (2004). Expression and possible function of IL-2 and IL-15 receptors on human uveal melanoma cells. Invest Ophthalmol Vis Sci 45, 4240–4246.

Hegde, S., Beauregard, C., Mayhew, E. and Niederkorn, J.Y. (2005). CD4(+) T-cell-mediated mechanisms of corneal allograft rejection: role of Fas-induced apoptosis. Transplantation 79, 23–31.

Hendricks, R.L., Tumpey, T.M. and Finnegan, A. (1992). IFN-gamma and IL-2 are protective in the skin but pathologic in the corneas of HSV-1-infected mice. J Immunol 149, 3023–3028.

Hicklin, D.J., Marincola, F.M. and Ferrone, S. (1999). HLA class I antigen downregulation in human cancers: T-cell immunotherapy revives an old story. Mol Med Today 5, 178–186.

Ho, L.P., Urban, B.C., Jones, L., Ogg, G.S. and McMichael, A.J. (2004). CD4(−)CD8alphaalpha subset of CD1d-restricted NKT cells controls T cell expansion. J Immunol 172, 7350–7358.

Hori, J., Wang, M., Miyashita, M., Tanemoto, K., Takahashi, H., Takemori, T., Okumura, K., Yagita, H. and Azuma, M. (2006). B7-H1-induced apoptosis as a mechanism of immune privilege of corneal allografts. J Immunol 177, 5928–5935.

Jager, M.J., Hurks, H.M.H., Levitskaya, J. and Kiessling, R. (2002). HLA expressin in uveal melanoma: there is no rule without some exception. Hum Immunol 63, 444–451.

Jampel, H.D., Roche, N., Stark, W.J. and Roberts, A.B. (1990). Transforming growth factor-beta in human aqueous humor. Curr Eye Res 9, 963–969.

Kageshita, T., Hirai, S., Ono, T., Hicklin, D.J. and Ferrone, S. (1999). Down-regulation of HLA class I antigen-processing molecules in malignant melanoma: association with disease progression. Am J Pathol 154, 745–754.

Katagiri, K., Zhang-Hoover, J., Mo, J.S., Stein-Streilein, J. and Streilein, J.W. (2002). Using tolerance induced via the anterior chamber of the eye to inhibit Th2-dependent pulmonary pathology. J Immunol 169, 84–89.

Kitaichi, N., Kotake, S., Morohashi, T., Onoe, K., Ohno, S. and Taylor, A.W. (2002). Diminution of experimental autoimmune uveoretinitis (EAU) in mice depleted of NK cells. J Leukoc Biol 72, 1117–1121.

Kogiso, M., Tanouchi, Y., Mimura, Y., Nagasawa, H. and Himeno, K. (1992). Endotoxin-induced uveitis in mice. 1. Induction of uveitis and role of T lymphocytes. Jpn J Ophthalmol 36, 281–290.

Kosiewicz, M.M., Alard, P. and Streilein, J.W. (1998). Alterations in cytokine production following intraocular injection of soluble protein antigen: impairment in IFN-gamma and induction of TGF-beta and IL-4 production. J Immunol 161, 5382–5390.

Kovats, S., Main, E.K., Librach, C., Stubblebine, M., Fisher, S.J. and DeMars, R. (1990). A class I antigen, HLA-G, expressed in human trophoblasts. Science 248, 220–223.

Krueger, J.G., Krane, J.F., Carter, D.M. and Gottlieb, A.B. (1990). Role of growth factors, cytokines, and their receptors in the pathogenesis of psoriasis. J Invest Dermatol 94, 135S–140S.

Ksander, B.R., Rubsamen, P.E., Olsen, K.R., Cousins, S.W. and Streilein, J.W. (1991). Studies of tumor-infiltrating lymphocytes from a human choroidal melanoma. Invest Ophthalmol Vis Sci 32, 3198–3208.

Lass, J.H., Walter, E.I., Burris, T.E., Grossniklaus, H.E., Roat, M.I., Skelnik, D.L., Needham, L., Singer, M. and Medof, M.E. (1990). Expression of two molecular forms of the complement decay-accelerating factor in the eye and lacrimal gland. Invest Ophthalmol Vis Sci 31, 1136–1148.

Latchman, Y., Wood, C.R., Chernova, T., Chaudhary, D., Borde, M., Chernova, I., Iwai, Y., Long, A.J., Brown, J.A., Nunes, R., Greenfield, E.A., Bourque, K., Boussiotis, V.A., Carter, L.L., Carreno, B.M., Malenkovich, N., Nishimura, H., Okazaki, T., Honjo, T., Sharpe, A.H. and Freeman, G.J. (2001). PD-L2 is a second ligand for PD-1 and inhibits T cell activation. Nat Immunol 2, 261–268.

LeBouteiller, P. (1994). HLA class I chromosomal region, genes, and products: facts and questions. Crit Rev Immunol 14, 89–129.

Le Discorde, M., Moreau, P., Sabatier, P., Legeais, J.M. and Carosella, E.D. (2003). Expression of HLA-G in human cornea, an immune-privileged tissue. Hum Immunol 64, 1039–1044.

Lee, N., Llano, M., Carretero, M., Ishitani, A., Navarro, F., Lopez-Botet, M. and Geraghty, D.E. (1998). HLA-E is a major ligand for the natural killer inhibitory receptor CD94/NKG2A. Proc Natl Acad Sci U S A 95, 5199–5204.

Li, X., Taylor, S., Zegarelli, B., Shen, S., O'Rourke, J. and Cone, R.E. (2004). The induction of splenic suppressor T cells through an immune-privileged site requires an intact sympathetic nervous system. J Neuroimmunol 153, 40–49.

Li, Z., Lim, W.K., Mahesh, S.P., Liu, B. and Nussenblatt, R.B. (2005). Cutting edge: in vivo blockade of human IL-2 receptor induces expansion of CD56(bright) regulatory NK cells in patients with active uveitis. J Immunol 174, 5187–5191.

Lighvani, S., Huang, X., Trivedi, P.P., Swanborg, R.H. and Hazlett, L.D. (2005). Substance P regulates natural killer cell interferon-gamma production and resistance to *Pseudomonas aeruginosa* infection. Eur J Immunol 35, 1567–1575.

Ljunggren, H.G. and Karre, K. (1990). In search of the 'missing self': MHC molecules and NK cell recognition. Immunol Today 11, 237–244.

Ljunggren, H.G., Ohlen, C., Hoglund, P., Franksson, L. and Karre, K. (1991). The RMA-S lymphoma mutant; consequences of a peptide loading defect on immunological recognition and graft rejection. Int J Cancer Suppl 6, 38–44.

Luger, D., Silver, P.B., Tang, J., Cua, D., Chen, Z., Iwakura, Y., Bowman, E.P., Sgambellone, N.M., Chan, C.C. and Caspi, R.R. (2008). Either a Th17 or a Th1 effector response can drive autoimmunity: conditions of disease induction affect dominant effector category. J Exp Med 205, 799–810.

Ma, D. and Niederkorn, J.Y. (1995). Transforming growth factor-beta down-regulates major histocompatibility complex class I antigen expression and increases the susceptibility of uveal melanoma cells to natural killer cell-mediated cytolysis. Immunology 86, 263–269.

Ma, D., Luyten, G.P., Luider, T.M. and Niederkorn, J.Y. (1995). Relationship between natural killer cell susceptibility and metastasis of human uveal melanoma cells in a murine model. Invest Ophthalmol Vis Sci 36, 435–441.

Madden, K.S., Felten, S.Y., Felten, D.L., Sundaresan, P.R. and Livnat, S. (1989). Sympathetic neural modulation of the immune system. I. Depression of T cell immunity in vivo and vitro following chemical sympathectomy. Brain Behav Immun 3, 72–89.

Marriott, I. and Bost, K.L. (2001). Substance P receptor mediated macrophage responses. Adv Exp Med Biol 493, 247–254.

Medawar, P.B. (1945). A second study of the behavior and fate of skin homografts in rabbits. J Anat 79, 157–176.

Medawar, P.B. (1948). Immunity to homologous grafted skin. III. The fate of skin homografts transplanted to the brain, to subcutaneous tissue, and to the anterior chamber of the eye. Br J Exp Pathol 29, 58–69.

Meecham, W.J., Char, D.H. and Kaleta-Michaels, S. (1992). Infiltrating lymphocytes and antigen expression in uveal melanoma. Ophthalmic Res 24, 20–26.

Mercadal, C.M., Bouley, D.M., DeStephano, D. and Rouse, B.T. (1993). Herpetic stromal keratitis in the reconstituted scid mouse model. J Virol 67, 3404–3408.

Metcalf, J.F., Hamilton, D.S. and Reichert, R.W. (1979). Herpetic keratitis in athymic (nude) mice. Infect Immun 26, 1164–1171.

Miller, D. (1979). *Ophthalmology. The Essentials*. Boston: Houghton Mifflin Professional Publishers.

Moffett-King, A. (2002). Natural killer cells and pregnancy. Nat Rev Immunol 2, 656–663.

Mooy, C.M. and De Jong, P.T. (1996). Prognostic parameters in uveal melanoma: a review. Surv Ophthalmol 41, 215–228.

Nakamura, T., Sonoda, K.H., Faunce, D.E., Gumperz, J., Yamamura, T., Miyake, S. and Stein-Streilein, J. (2003). CD4+ NKT cells, but not conventional CD4+ T cells, are required to generate efferent CD8+ T regulatory cells following antigen inoculation in an immune-privileged site. J Immunol 171, 1266–1271.

Nakamura, T., Terajewicz, A. and Stein-Streilein, J. (2005). Mechanisms of peripheral tolerance following intracameral inoculation are independent of IL-13 or STAT6. J Immunol 175, 2643–2646.

Naumov, Y.N., Bahjat, K.S., Gausling, R., Abraham, R., Exley, M.A., Koezuka, Y., Balk, S.B., Strominger, J.L., Clare-Salzer, M. and Wilson, S.B. (2001). Activation of CD1d-restricted T cells protects NOD mice from developing diabetes by regulating dendritic cell subsets. Proc Natl Acad Sci U S A 98, 13838–13843.

Navarro, F., Llano, M., Bellon, T., Colonna, M., Geraghty, D.E. and Lopez-Botet, M. (1999). The ILT2(LIR1) and CD94/NKG2A NK cell receptors respectively recognize HLA-G1 and HLA-E molecules co-expressed on target cells. Eur J Immunol 29, 277–283.

Niederkorn, J.Y. (1995). Immunopathogenesis of intraocular tumors. Prog Retin Eye Res 14, 505–526.

Niederkorn, J.Y. (1999). The immune privilege of corneal allografts. Transplantation 67, 1503–1508.

Niederkorn, J.Y. (2002). Immune privilege in the anterior chamber of the eye. Crit Rev Immunol 22, 13–46.

Niederkorn, J.Y. (2003). The immune privilege of corneal grafts. J Leukoc Biol 74, 167–171.

Niederkorn, J.Y. (2006a). Anterior chamber-associated immune deviation and its impact on corneal allograft survival. Curr Opin Organ Transplant 11, 360–365.

Niederkorn, J.Y. (2006b). See no evil, hear no evil, do no evil: the lessons of immune privilege. Nat Immunol 7, 354–359.

Niederkorn, J.Y. (2007a). Immune mechanisms of corneal allograft rejection. Curr Eye Res 32, 1005–1016.

Niederkorn, J.Y. (2007b). The induction of anterior chamber-associated immune deviation. Chem Immunol Allergy 92, 27–35.

Niederkorn, J.Y. and Wang, S. (2005a). Immune privilege of the eye and fetus: parallel universes?. Transplantation 80, 1139–1144.

Niederkorn, J.Y. and Wang, S. (2005b). Immunology of intraocular tumors. Ocul Immunol Inflamm 13, 105–110.

Niederkorn, J.Y., Mayhew, E. and He, Y. (1995). Alloantigens introduced into the anterior chamber of the eye induce systemic suppression of delayed hypersensitivity to third-party alloantigens through "linked recognition". Transplantation 60, 348–354.

Niederkorn, J.Y., Chiang, E.Y., Ungchusri, T. and Stroynowski, I. (1999). Expression of a nonclassical MHC class Ib molecule in the eye. Transplantation 68, 1790–1799.

Niemialtowski, M.G. and Rouse, B.T. (1992). Predominance of Th1 cells in ocular tissues during herpetic stromal keratitis. J Immunol 149, 3035–3039.

Nitta, T., Oksenberg, J.R., Rao, N.A. and Steinman, L. (1990). Predominant expression of T cell receptor V alpha 7 in tumor-infiltrating lymphocytes of uveal melanoma. Science 249, 672–674.

Oakes, J.E., Monteiro, C.A., Cubitt, C.L. and Lausch, R.N. (1993). Induction of interleukin-8 gene expression is associated with herpes simplex virus infection of human corneal keratocytes but not human corneal epithelial cells. J Virol 67, 4777–4784.

Okazaki, T. and Honjo, T. (2007). PD-1 and PD-1 ligands: from discovery to clinical application. Int Immunol 19, 813–824.

Okumura, A., Mochizuki, M., Nishi, M. and Herbort, C.P. (1990). Endotoxin-induced uveitis (EIU) in the rat: a study of inflammatory and immunological mechanisms. Int Ophthalmol 14, 31–36.

Orcel, P., Bielakoff, J. and De Vernejoul, M.C. (1990). Effects of transforming growth factor-beta on long-term human cord blood monocyte cultures. J Cell Physiol 142, 293–298.

Peng, Y., Han, G., Shao, H., Wang, Y., Kaplan, H.J. and Sun, D. (2007). Characterization of IL-17+ interphotoreceptor retinoid-binding protein-specific T cells in experimental autoimmune uveitis. Invest Ophthalmol Vis Sci 48, 4153–4161.

Rajpal, S., Moore, R. and Karakousis, C.P. (1983). Survival in metastatic ocular melanoma. Cancer 52, 334–336.

Repp, A.C., Mayhew, E.S., Apte, S. and Niederkorn, J.Y. (2000). Human uveal melanoma cells produce macrophage migration-inhibitory factor to prevent lysis by NK cells. J Immunol 165, 710–715.

Rice, P.A., Boehm, G.W., Moynihan, J.A., Bellinger, D.L. and Stevens, S.Y. (2001). Chemical sympathectomy increases the innate immune response and decreases the specific immune response in the spleen to infection with *Listeria monocytogenes*. J Neuroimmunol 114, 19–27.

Rook, A.H., Kehrl, J.H., Wakefield, L.M., Roberts, A.B., Sporn, M.B., Burlington, D.B., Lane, H.C. and Fauci, A.S. (1986). Effects of transforming growth factor beta on the functions of natural killer cells: depressed cytolytic activity and blunting of interferon responsiveness. J Immunol 136, 3916–3920.

Rook, G.A., Lightman, S.L. and Heijnen, C.J. (2002). Can nerve damage disrupt neuroendocrine immune homeostasis? Leprosy as a case in point. Trends Immunol 23, 18–22.

Rosenbaum, J.T., McDevitt, H.O., Guss, R.B. and Egbert, P.R. (1980). Endotoxin-induced uveitis in rats as a model for human disease. Nature 286, 611–613.

Rouas-Freiss, N., Goncalves, R.M., Menier, C., Dausset, J. and Carosella, E.D. (1997). Direct evidence to support the role of HLA-G in protecting the fetus from maternal uterine natural killer cytolysis. Proc Natl Acad Sci U S A 94, 11520–11525.

Ruiter , D.J., Bhan, A.K., Harrist, T.J., Sober, A.J. and Mihm, M.C. Jr (1982). Major histocompatibility antigens and mononuclear inflammatory infiltrate in benign nevomelanocytic proliferations and malignant melanoma. J Immunol 129, 2808–2815.

Ruiter, D.J., Bergman, W., Welvaart, K., Scheffer, E., van Vloten, W.A., Russo, C. and Ferrone, S. (1984). Immunohistochemical analysis of malignant melanomas and nevocellular nevi with monoclonal antibodies to distinct monomorphic determinants of HLA antigens. Cancer Res 44, 3930–3935.

Ruiter, D.J., Mattijssen, V., Broecker, E.B. and Ferrone, S. (1991). MHC antigens in human melanomas. Semin Cancer Biol 2, 35–45.

Russell, R.G., Nasisse, M.P., Larsen, H.S. and Rouse, B.T. (1984). Role of T-lymphocytes in the pathogenesis of herpetic stromal keratitis. Invest Ophthalmol Vis Sci 25, 938–944.

Ryu, Y.H. and Kim, J.C. (2007). Expression of indoleamine 2,3-dioxygenase in human corneal cells as a local immunosuppressive factor. Invest Ophthalmol Vis Sci 48, 4148–4152.

Shao, H., Van Kaer, L., Sun, S.L., Kaplan, H.J. and Sun, D. (2003). Infiltration of the inflamed eye by NKT cells in a rat model of experimental autoimmune uveitis. J Autoimmun 21, 37–45.

She, S.C., Steahly, L.P. and Moticka, E.J. (1990). Intracameral injection of allogeneic lymphocytes enhances corneal graft survival. Invest Ophthalmol Vis Sci 31, 1950–1956.

Shen, L., Jin, Y., Freeman, G.J., Sharpe, A.H. and Dana, M.R. (2007). The function of donor versus recipient programmed death-ligand 1 in corneal allograft survival. J Immunol 179, 3672–3679.

Sohn, J.H., Kaplan, H.J., Suk, H.J., Bora, P.S. and Bora, N.S. (2000a). Chronic low level complement activation within the eye is controlled by intraocular complement regulatory proteins. Invest Ophthalmol Vis Sci 41, 3492–3502.

Sohn, J.H., Kaplan, H.J., Suk, H.J., Bora, P.S. and Bora, N.S. (2000b). Complement regulatory activity of normal human intraocular fluid is mediated by MCP, DAF, and CD59. Invest Ophthalmol Vis Sci 41, 4195–4202.

Sonoda, K.H. and Stein-Streilein, J. (2002). CD1d on antigen-transporting APC and splenic marginal zone B cells promotes NKT cell-dependent tolerance. Eur J Immunol 32, 848–857.

Sonoda, K.H., Faunce, D.E., Taniguchi, M., Exley, M., Balk, S. and Stein-Streilein, J. (2001). NK T cell-derived IL-10 is essential for the differentiation of antigen- specific T regulatory cells in systemic tolerance. J Immunol 166, 42–50.

Sonoda, K.H., Taniguchi, M. and Stein-Streilein, J. (2002). Long-term survival of corneal allografts is dependent on intact CD1d− reactive NKT cells. J Immunol 168, 2028–2034.

Staibano, S., Mascolo, M., Tranfa, F., Salvatore, G., Mignogna, C., Bufo, P., Nugnes, L., Bonavolonta, G. and De Rosa, G. (2006). Tumor infiltrating lymphocytes in uveal melanoma: a link with clinical behavior? Int J Immunopathol Pharmacol 19, 171–179.

Streilein, J.W. (1987). Immune regulation and the eye: a dangerous compromise. FASEB J 1, 199–208.

Streilein, J.W. (2003). Ocular immune privilege: therapeutic opportunities from an experiment of nature. Nat Rev Immunol 3, 879–889.

Streilein, J.W., Dana, M.R. and Ksander, B. R. (1997a). Immunity causing blindness: five different paths to herpes stromal keratitis. Immunol Today 18, 443–449.

Streilein, J.W., Takeuchi, M. and Taylor, A.W. (1997b). Immune privilege, T-cell tolerance, and tissue-restricted autoimmunity. Hum Immunol 52, 138–143.

Streilein, J.W., Masli, S., Takeuchi, M. and Kezuka, T. (2002). The eye's view of antigen presentation. Hum Immunol 63, 435–443.

Stuart, P.M., Griffith, T.S., Usui, N., Pepose, J., Yu, X. and Ferguson, T.A. (1997). CD95 ligand (FasL)-induced apoptosis is necessary for corneal allograft survival. J Clin Invest 99, 396–402.

Tamesis, R.R., Messmer, E.M., Rice, B.A., Dutt, J.E. and Foster, C.S. (1994). The role of natural killer cells in the development of herpes simplex virus type 1 induced stromal keratitis in mice. Eye 8(Pt 3), 298–306.

Taylor, A.W. (2007). Ocular immunosuppressive microenvironment. Chem Immunol 92, 71–85.

Taylor, A.W. and Yee, D.G. (2003). Somatostatin is an immunosuppressive factor in aqueous humor. Invest Ophthalmol Vis Sci 44, 2644–2649.

Taylor, A.W., Streilein, J.W. and Cousins, S.W. (1994a). Alpha-melanocyte-stimulating hormone suppresses antigen-stimulated T cell production of gamma-interferon. Neuroimmunomodulation 1, 188–194.

Taylor, A.W., Streilein, J.W. and Cousins, S.W. (1994b). Immunoreactive vasoactive intestinal peptide contributes to the immunosuppressive activity of normal aqueous humor. J Immunol 153, 1080–1086.

Tseng, S.Y., Otsuji, M., Gorski, K., Huang, X., Slansky, J.E., Pai, S.I., Shalabi, A., Shin, T., Pardoll, D.M. and Tsuchiya, H. (2001). B7-DC, a new dendritic cell molecule with potent costimulatory properties for T cells. J Exp Med 193, 839–846.

van Dooremaal, J.C. (1873). Die Entwicklung der in fremden Grund versetzten lebenden Geweba. Albrecht von Graefes Arch Ophthalmol 19, 358–373.

van Duinen, S.G., Ruiter, D.J., Broecker, E.B., van der Velde, E.A., Sorg, C., Welvaart, K. and Ferrone, S. (1988). Level of HLA antigens in locoregional metastases and clinical course of the disease in patients with melanoma. Cancer Res 48, 1019–1025.

Vassalli, P. (1992). The pathophysiology of tumor necrosis factors. Annu Rev Immunol 10, 411–452.

Verbik, D.J., Murray, T.G., Tran, J.M. and Ksander, B.R. (1997). Melanomas that develop within the eye inhibit lymphocyte proliferation. Int J Cancer 73, 470–478.

Vijayasaradhi, S. and Houghton, A.N. (1995). Melanoma and melanocytes: pigmentation, tumor progression, and the immune response to cancer. Adv Pharmacol 32, 343–374.

Wacker, W.B. and Lipton, M.M. (1965). Experimental allergic uveitis: homologous retina as uveitogenic antigen. Nature 206, 253–254.

Wang, Y., Goldschneider, I., Foss, D., Wu, D.Y., O'Rourke, J. and Cone, R.E. (1997). Direct thymic involvement in anterior chamber-associated immune deviation: evidence for a nondeletional mechanism of centrally induced tolerance to extrathymic antigens in adult mice. J Immunol 158, 2150–2155.

Wang, Y., Goldschneider, I., O'Rourke, J. and Cone, R.E. (2001). Blood mononuclear cells induce regulatory NK T thymocytes in anterior chamber-associated immune deviation. J Leukoc Biol 69, 741–746.

Watte, C.M., Nakamura, T., Lau, C.H., Ortaldo, J.R. and Stein-Streilein, J. (2008). Ly49 C/I-dependent NKT cell-derived IL-10 is required for corneal graft survival and peripheral tolerance. J Leukoc Biol 83, 928–935.

Wax, M.B., Tezel, G., Yang, J., Peng, G., Patil, R.V., Agarwal, N., Sappington, R. M. and Calkins, D.J. (2008). Induced autoimmunity to heat shock proteins elicits glaucomatous loss of retinal ganglion cell neurons via activated T-cell-derived fas-ligand. J Neurosci 28, 12085–12096.

Yamagami, S., Kawashima, H., Tsuru, T., Yamagami, H., Kayagaki, N., Yagita, H., Okumura, K. and Gregerson, D. S. (1997). Role of Fas–Fas ligand interactions in the immunorejection of allogeneic mouse corneal transplants. Transplantation 64, 1107–1111.

Yang, W., Li, H., Chen, P.W., Alizadeh, H., He, Y.-G., Hogan, R.N. and Niederkorn, J.Y. (2009). PD-L1 expession on human ocular cells and its possible role in regulating immune-mediated ocular inflammation. Invest Ophthalmol Vis Sci 50, 273–280.

Yoshimura, T., Sonoda, K.H., Miyazaki, Y., Iwakura, Y., Ishibashi, T., Yoshimura, A. and Yoshida, H. (2008). Differential roles for IFN-gamma and IL-17 in experimental autoimmune uveoretinitis. Int Immunol 20, 209–214.

Zhang, Y., Ramos, B.F. and Jakschik, B. A. (1992). Neutrophil recruitment by tumor necrosis factor from mast cells in immune complex peritonitis. Science 258, 1957–1959.

Zirm, E. (1906). Eine erfolgreiche totale Keratoplastik. Albrecht von Graefes Arch Ophthalmol 64, 580–593.

Chapter Thirty

30

NK cells and reproduction

Ashley Moffett, Victoria H. Male

The history of man for the nine months preceding his birth would, probably, be far more interesting and contain events of greater moment than all the three-score and ten years that follow it.

Samuel Taylor Coleridge

ABSTRACT

In humans, an unusual subset of natural killer (NK) cells accounts for approximately 70% of the leukocytes in the mucosal lining of the uterus in the secretory phase of the menstrual cycle and during the first trimester of pregnancy. These NK cells are CD56^{superbright}CD16^{-} and contain cytotoxic granules. They express a variety of activating and inhibitory NK cell receptors including NKG2A, NKG2D, NKp46 and killer-cell immunoglobulin-like receptors (KIR). Their KIR repertoire is skewed towards recognizing HLA-C, which is the only classical class I MHC found on placental trophoblast cells. Approximately 20% of uterine NK cells also express LILRB1, which recognizes the trophoblast-specific class I MHC molecule HLA-G. Despite the presence of killer cell receptors and cytotoxic granules, uterine NK cells are poorly cytotoxic. There is evidence to suggest that they can produce various cytokines and soluble growth factors. Through production of these factors, uterine NK cells may have a physiological role in maintenance and modification of the vasculature and in mediating trophoblast invasion.

KEY WORDS

Trophoblast, Decidua, Endometrium, HLA-G, HLA-C

Introduction

The uterine mucosa is a site of particular interest for natural killer (NK) cell biologists because NK cells are present in large numbers in early pregnancy at the site of placental implantation (Loke and King, 1995). Under the influence of progesterone, the hormone of pregnancy, the mucosa is transformed from the non-pregnant state (endometrium) to a specialized tissue known as decidua. Uterine NK (uNK) cells are an integral and distinctive feature of decidua.

The placenta is formed from the outer layer of cells of the blastocyst, the trophectoderm, with contributions from mesenchymal cells derived from the allantois. The outermost cells of the placenta, the trophoblast, are the foetal cells that come into direct contact with maternal uterine tissues. There is great diversity among mammals in how trophoblast contacts uterine tissues. Trophoblast can abut onto the uterine surface epithelium (epitheliochorial), erode through this epithelium to contact maternal endothelial cells (endotheliochorial) or invade through vessel walls so becoming bathed with maternal blood (haemochorial) (Moffett and Loke, 2006). Decidua and the associated NK cells are a feature of only the most invasive haemochorial forms of placenta. Furthermore, in species where invasion is very extensive the degree of decidualization is concomitantly more pronounced.

This correlation between the presence and abundance of NK cells, the degree of decidualization and the depth and extent of trophoblast invasion seems to indicate that uNK cells somehow contribute to the regulation of normal placentation and trophoblast invasion. Adoptive transfer experiments in which mouse blastocysts were allowed to implant on a variety of non-uterine tissues or on decidualized or non-decidualized uterine mucosae indicate that decidua provides a balanced, controlled environment (Kirby, 1960; McClaren, 1965). The idea is that placentation occurs without either excessive invasion that could endanger the mother or defective invasion that would prevent the foetus having sufficient access to nutrients. How NK cells contribute to the functions of decidua remain to be elucidated and an obvious difficulty is the lack of good animal models due to the unique placental strategy each species has adopted. Thus, this chapter is devoted mainly to discussions of human uterine NK cells.

History and terminology

The presence of mononuclear cells in the lining of the uterine mucosa has been well-documented by histologists. Indeed, granulated cells were noted in the decidua long ago and were even then thought to be a type of lymphoid cell (Marchand, 1895; Weill, 1921). 'Quoi qu'il en soit, ce qui est indiscutable c'est que ce sont des cellules a type lymphoide propres a la decidua humain'. ('Whatever, what is indisputable is that they are lymphoid-type cells specific to the human decidua'.) These cells were subsequently also identified in the non-pregnant endometrium and were variously named as endometrial granulocytes, Kornchenzellen or 'K' cells (Hamperl and Hellweg, 1958), globular leukocytes (Von Numers, 1953) and specific endometrial granular cells (Kazzaz, 1972). Several of the reports noted the apparent hormonal dependence of endometrial granulocytes and that they were particularly prominent in the secretory endometrium just before menstruation and persisted if decidualization occurred. Unless pregnancy intervenes, they show degenerative changes in the late secretory phase with 'mulberry'-shaped nuclei (Figure 30.1). This appearance led to their name, endometrial granulocytes, because of the superficial resemblance to neutrophils. The distribution of the granulated cells is characteristic. There is a diffuse scattering of cells throughout the luteal phase stroma but also a tendency to aggregate around arteries and glands. In early decidua, they amass in the decidua basalis at the implantation site, but are mainly in the superficial decidual layer (decidua compacta) with only a few cells in the intervening stroma between the dilated glands of the deeper decidua spongiosa. Unlike rodents, granulated cells are also present in decidua parietalis, which covers the uterine surface away from site of placentation.

Outside the uterus, endometrial granulocytes have only been described in areas of ectopic decidualization, commonly in the Fallopian tube and less often in the cervix or on the surface of the ovary (Hamperl and Hellweg, 1958). Endometrial granulocytes are absent in the non-decidualized areas of the Fallopian tube in an ectopic pregnancy, yet they are always present in the uterine decidua in such pregnancies even though no trophoblast is present in the uterus. In these cases, the presence of NK cells in the uterus despite the absence of trophoblast suggests that these cells are recruited under the hormonal influences of pregnancy and not as a response to invading trophoblast.

The granulated cells were not definitively confirmed to be leukocytes until the advent of monoclonal antibodies. Following the initial observations of Bulmer, many other laboratories contributed to studies providing a

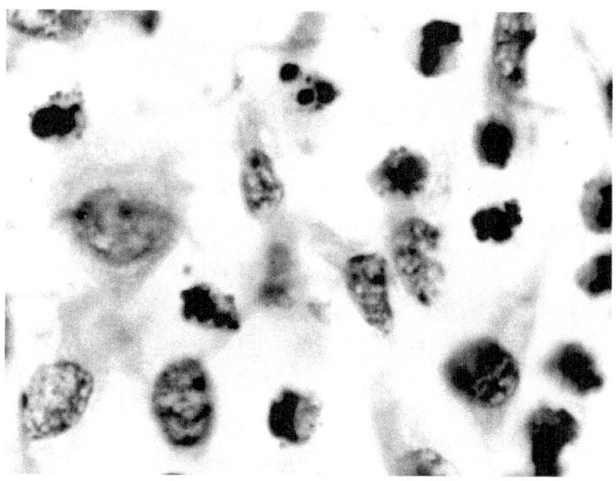

Figure 30.1 • Phloxine tartrazine staining of endometrium from a patient undergoing exogenous progesterone withdrawal. Numerous granular cells interspersed with the stromal cells have condensed and fragmented nuclei.

phenotypic profile of these cells as bone marrow-derived, CD45$^+$ cells (Bulmer and Sunderland, 1984). It was not until around 1990 that it became clear the endometrial granulocytes were a type of NK cell, as they stained with mAbs to CD56, although not to other NK markers such as CD16 (King et al., 1989a; Ritson and Bulmer, 1987). Flow cytometric studies on isolated decidual leukocytes allowed a more detailed phenotypic analysis and the CD56 cells were defined as CD56^{bright}, CD16$^-$ and CD3$^-$ (King et al., 1991; Starkey et al., 1988). They were also shown to mediate some killing of K562, the prototypic NK target cells, albeit at far lower levels compared to blood NK cells (King et al., 1989b).

Because their morphology and phenotype was similar to NK cells in blood, these cells were termed variously uterine large granular lymphocytes (LGL), endometrial or decidual granulated lymphocytes (eGL, dGL). At the same time, granulated lymphocytes with features similar to NK cells were described in rodents (guinea pigs, hamsters, rats and mice) and also in higher primates and apes (Croy et al., 2006). The current term *uNK* includes granulated cells found in the pregnant uterine mucosa in all species that undergo decidualization. The term *uNK cells* implies that they are a specialized NK subset distinct from blood NK cells. It is regrettable that, because they are called natural *killer* cells, it has been claimed by clinics treating women with infertility or recurrent miscarriage that uNK cells are damaging to the early conceptus (Moffett et al., 2004). There is no evidence that this is ever the case.

Uterine NK cells and the menstrual cycle

The presence of appreciable numbers of NK cells in the non-pregnant uterine mucosa (endometrium) is a feature only of those higher primates, including humans, in which menstruation occurs. Non-pregnant endometrium undergoes cyclical changes triggered by ovarian hormones that are in turn regulated by the pituitary–ovarian axis (Figure 30.2). Following shedding at menstruation, the mucosa proliferates under the influence of oestrogen, reforming glandular structures and arteries that are separated by intervening stromal cells. After ovulation, when the ovarian follicular cells from the corpora lutea begin to secrete progesterone, the endometrial glands, stromal cells and vessels differentiate in preparation for implantation that occurs between 5 and 8 days post-ovulation. If no pregnancy occurs, the corpora lutea involute, progesterone levels fall and the endometrium breaks down at menstruation. In contrast, if a blastocyst does implant, the endometrium continues to differentiate into decidua. In humans, therefore, the initial changes of decidualization occur during the luteal phase of every cycle (pre-decidual change).

Endometrial CD56$^+$ NK cells are small and agranular before ovulation and then proliferate, enlarge and become increasingly granulated in the secretory phase, suggesting regulation by progesterone (Moffett-King, 2002; Trundley and Moffett, 2004). The increase in numbers of NK cells is obvious and careful quantification has estimated as

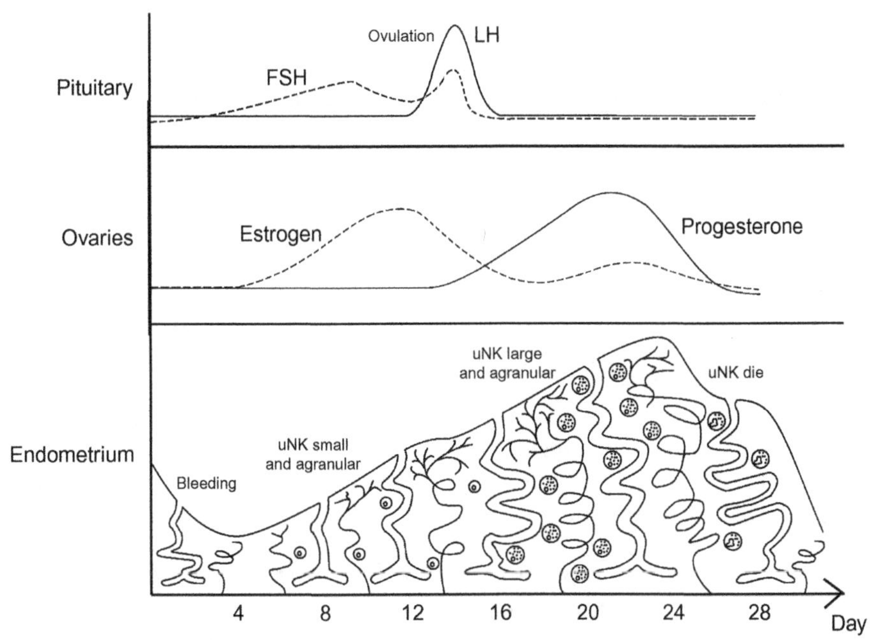

Figure 30.2 • The normal menstrual cycle showing the hormonal changes and the morphology of the endometrium.

many as ~30–40% of the cells in the endometrial stroma are NK cells (Pace et al., 1989). As well as proliferating, the secretory-phase NK cells become larger and often a reniform-shaped nucleus is obvious along with prominent cytoplasmic granules. If fertilization occurs, the NK cells continue to proliferate during the first trimester of pregnancy, during which time they represent approximately 70% of the leukocytes in the decidua (King et al., 1991). During the second and third trimesters, they decrease in number and only small numbers remain in term decidua.

When fertilization does not occur, the NK cell nuclei undergo fragmentation and the cells die 24–48 hours before menstruation and are shed with the menses (King et al., 1989a). The same morphology is seen in women who have stopped progestogen treatment before surgery or in the decidua in miscarriages, suggesting that the NK cells die in response to progesterone withdrawal (personal observations). The appearance is reminiscent of apoptotic cell nuclei, although there are differences, and indeed the classical markers of caspase-dependent apoptotic cell death are absent (Pongcharoen et al., 2004). The observation that the uNK cells are still dividing up to the point when this nuclear change occurs (as seen by mitotic figures and Ki-67 reactivity) suggests that the cell might be undergoing mitotic catastrophe, a form of caspase-independent cell death occurring during or after failed mitosis (Kroemer and Martin, 2005; Okada and Mak, 2004).

Although they express mRNA and protein for oestrogen receptor (ERβ) and the glucocorticoid receptor, both the progesterone receptor (PR) and ERα are absent in endometrial NK cells (Henderson et al., 2003). PR expression in late-secretory endometrium and early decidua is mainly restricted to stromal cells (Milne et al., 2005). IL-15 and prolactin are upregulated in stromal cells by progesterone as decidualization occurs and they have both been implicated in the proliferation and differentiation of uNK cells (reviewed in Jabbour et al., 2006). The current concept is that the effects of progesterone on uNK cell proliferation and function are mainly mediated by its actions on stromal cells. Whether other pregnancy-associated proteins such as hCG can act directly on uNK is unknown.

The origin of uterine NK cells

The remarkable cyclical regeneration of the uterine NK cell population raises questions about their origin. There is some evidence to suggest that either mature NK cells or an NK precursor may be recruited from the blood to the uterus under the influence of oestrogen and progesterone. When cultured with these hormones, uterine endothelium upregulates various adhesion molecules and in vivo levels of these molecules have been shown to increase in the early secretory phase (Kammerer et al., 2004; Yamaguchi et al., 2006). Under sheer stress, human lymphocytes adhere to pregnant mouse uterine tissue sections in an L-selectin and α-integrin dependent manner (Chantakru et al., 2004). The endometrium also produces NK-attractive chemokines (Carlino et al., 2008; Sentman et al., 2004). Furthermore, donor $CD45^+$ cells (identified by HLA-typing) have been shown to exist in the uteri of bone marrow transplant recipients 2–13 years after surgery (Taylor, 2004). The number and location of these cells suggest they are NK cells, although co-staining with CD56, which would have definitely shown this, was not undertaken. This experiment does, however, show that bone marrow–derived haematopoietic cells may repopulate the decidua in adults, supporting the idea that either a mature NK or a circulating precursor cell may be recruited from the blood. The only evidence to suggest that a mature blood NK cell specifically, rather than a circulating precursor, gives rise to uNK are experiments in which blood NK cells cultured with conditioned medium from decidual stromal cells gained a uterine phenotype ($CD56^{superbright}$, $CD9^+$ as described below) (Keskin et al., 2007).

Attempts have also been made to identify an NK precursor population resident in the uterus. Lin-$CD34^+CD45^+$ cells, capable of differentiating into $CD56^{bright}$ NK cells, have been identified in first trimester decidua (Keskin et al., 2007). $CD34^+CD45^+$ cells have also been identified in endometrial biopsies, although the fact that they also express CD56, a marker of mature NK cells, throws their status as genuine precursor cells into doubt (Lynch et al., 2007). Sections of human endometrium can be transplanted into ovariectomized mice lacking T, B and NK cells. The mice are then treated for 14 days with oestrogen only and for a further 14 days with both oestrogen and progesterone, mimicking the human menstrual cycle. $CD56^+$ cells are not present in the tissue after the first 14 days but are present in large numbers after 28 days (Matsuura-Sawada et al., 2005). This suggests that NK cells derive from a progenitor cell population resident in the human uterus but, as the time course of the experiment is only 28 days, the possibility that these progenitors had been recently recruited from the blood cannot be ruled out.

Claims have been made, then, for both a resident precursor population and recruitment of mature blood NK cells as the source of the uterine NK cells. However, almost all of the evidence to date is consistent with a third possibility, the recruitment of a circulating progenitor cell.

Phenotype of uterine NK cells

Many studies using flow cytometry, immunohistology and microarray analysis have reached much consensus on the phenotypic characteristics of NK cells isolated from first trimester decidua. The main findings are shown in Table 30.1. The major decidual NK subset is CD56^{bright} CD16^{-} and the levels of CD56 are considerably higher than CD56^{bright} subset in blood when cells are isolated without enzymatic digestion (Figure 30.3); in other words, they are CD56^{superbright}. The function of CD56 remains unclear, although when peripheral blood NK (PBNK) cells are stimulated with IL-2 both CD56 protein and glycan are upregulated. CD69, which is found on PBNK after exposure to IL-2, is expressed on decidual NK cells suggesting that uterine cells are in an activated state in vivo. A few CD16^{+} CD56^{dim} cells are seen in isolated cell preparations but whether these are contaminants from blood and/or a subset of uterine NK cells is not entirely clear. CD16^{+} cells are not seen in appreciable numbers in vivo by immunohistology or flow cytometry (Verma, 1999).

Gene expression studies comparing blood CD56^{bright} and CD56^{dim} with decidual CD56^{bright} cells showed many distinct differences (Koopman et al., 2003). The tetraspanin CD9 and α1 integrin are found only on decidual NK cells and other adhesion molecules that are highly overexpressed in decidual NK cells include β5 subunits. The secreted protein galectin (pregnancy-associated protein 14) is also expressed at high levels. Many of these proteins may reflect the tissue localization of the decidual NK cells, which associate tightly with the decidual stromal cells that become surrounded by a pericellular rim of laminin and fibronectin during decidualization (Loke and King, 1995).

Table 30.1 Major phenotypic characteristics of decidual CD56^{+} NK cells

Marker	Decidua CD56^{bright}	Blood CD56^{bright}	Blood CD56^{dim}
CD9	+	−	−
CD16	−	−	+
CD69	+	−	−
CD151	++	−	+
CD100	−	−	+
L-selectin	−	++	+/−
α1 integrin	+	−	−
α6 integrin	−	+	+
KIR	+	−	+
NKG2A	+	+	+/−
CD94	++	++	+/−
NKG2D	+	+	+
NKp46	+	+	+
Perforin	+++	+	+++
Granzyme	+++	+	+++

A summary of the major phenotypic characteristics of decidual CD56+ NK cells in comparison with the two NK subsets in peripheral blood. NK subsets are scored as − for no expression of the protein, with one to three + signs as a semi-quantitative score of expression, or as +/− if only some cells within a subset express the protein.

Like their CD56^{bright} counterparts in blood, the great majority of decidual CD56^{bright} cells express high levels of CD94/NKG2A (King et al., 2000). Unlike blood CD56^{bright} NK however, they also express KIR and interestingly, the KIR repertoire is unlike PBNK cells taken from the women at the same time (Sharkey et al., 2008; Verma et al., 1997). Both activating and inhibitory KIR specific for HLA-C are expressed at higher levels and on an increased proportion of NK cells in the human decidua compared to blood. In contrast, expression of KIR specific for HLA-Bw4 allotypes (KIR3DLI/S1) is similar in both NK cell populations. Because extravillous trophoblast (EVT) cells that invade the uterus during placentation express HLA-C, but not HLA-B, these findings suggest interactions between uNK cell KIR and trophoblast HLA-C molecules. Furthermore, the KIR repertoire of decidual NK cells changes throughout the first trimester with a gradual decline in both intensity and frequency of expression of KIR specific for HLA-C molecules.

NK cells respond to target cells following ligation of an array of inhibitory and activating receptors besides CD94/NKG2 and KIR. Around 20% of decidual NK cells express LILRB1, an inhibitory receptor for HLA-I molecules with particularly high affinity for the HLA-G homodimer displayed by EVT (Apps et al., 2007). The activating NK receptors include NKp46, NKp44, NKp30 (known as natural cytotoxicity receptors, NCRs) CD94/NKG2C or NKG2E, NKG2D and DNAM. NKp46 and NKG2D are expressed by all decidual NK cells but there is a lack of consensus about CD94/NKG2C, NKp44 and NKp30 (El Costa et al., 2008; Hanna et al., 2006; Koopman et al., 2003; Kopcow et al., 2005; Kusumi et al., 2006; Vacca et al., 2006, 2008). Decidual NK cells have also been shown to express the receptor 2B4, which in blood NK cells is activating but in decidual NK cells is believed to be inhibitory (Vacca et al., 2006).

Unlike their decidual counterparts, the phenotype of endometrial NK cells has been addressed only by a few flow cytometric studies, probably due to the practical and ethical difficulties in obtaining endometrial samples from normal fertile women. The limited data indicates that endometrial NK cells are also CD56 'superbright', CD16^{-}, CD3^{-}, CD9^{+} and express CD94, NKG2D,

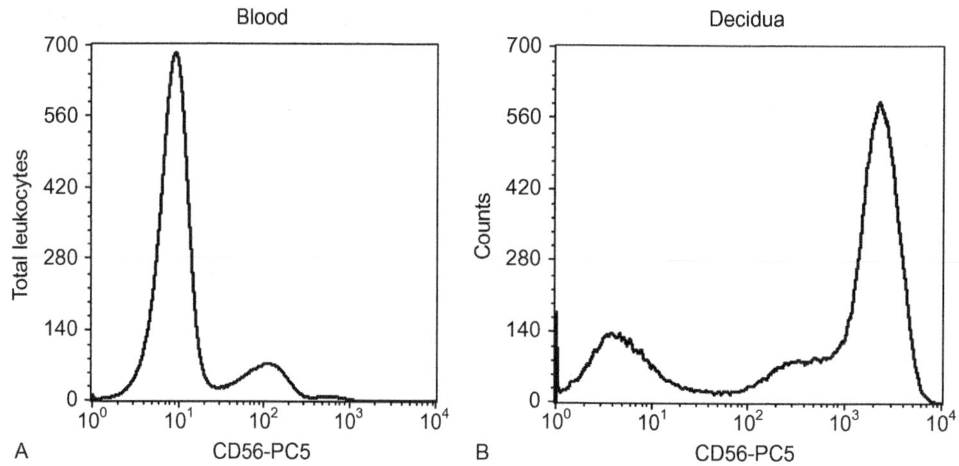

Figure 30.3 • Leukocytes (CD45+ cells) from peripheral blood and first trimester decidua stained with a mAb to CD56. The decidual NK cells are obviously CD56^{superbright}.

KIR and CD69. Endometrial CD56$^+$ cells do express NKp46 and NKG2D and only low levels of NKp30 and NKp44. The lack of NKp44 is interesting as the group that reported this found it on all decidual NK cells and also found that IL-15 stimulation could induce a more 'decidual' NKp44$^+$ phenotype (Manaster et al., 2008).

A recent study using multicolour FACS suggested that like peripheral blood, the major population of endometrial lymphocytes were T cells (Manaster et al., 2008). This result may be misleading because the lymphocyte sub-populations were analysed as a proportion of cells in the non-adherent 'lymphocytes' and not as a proportion of CD45$^+$ cells. When analysed by immunohistology, this study and many others noted that the numbers of NK cells are obviously increased in the secretory phase. In contrast to NK cells, there is little variation in the numbers of the other endometrial leukocytes (T cells and CD14$^+$ macrophages) over the course of the menstrual cycle.

Function of uterine NK cells

Cytotoxicity

There is broad agreement that decidual NK cells (like blood CD56^{bright}) are only weakly cytotoxic against classical NK target cells such as K562 and HLA-null 721.221 cells compared to CD56^{dim} PBNK (King et al., 1989b; Kopcow et al., 2005). Their morphology of LGL does, however, distinguish decidual CD56^{bright} cells from the agranular blood CD56^{bright} cells. These granules contain granzyme A and B and perforin, so the machinery for killing is present. Indeed, microarray analysis has revealed that granzyme A expression is even higher in uNK than blood CD56^{dim} cells (Koopman et al., 2003). NKG5 (a variant of granulolysin) and NKG7 (TIA-1) are also expressed by uNK cells (King et al., 1993). One possible explanation for their poor cytolytic activity is that they are unable to form a normal activating synapse (Kopcow et al., 2005). Uterine NK cells incubated with K562 targets form contacts but fail to polarize the microtubule organizing centre towards the target cell so that degranulation cannot occur.

In a redirected killing assay using mouse P815 cells, uNK were able to degranulate following ligation of NKp46, but not NKp30. This NKp46-mediated cytotoxicity is inhibited by NKG2A co-engagement (El Costa et al., 2008). As uNK do not kill trophoblast even after blockade of HLA-I molecules or their NK receptors by antibody, it is likely that normal trophoblast cells do not express ligands for activating receptors such as NKp46 (Avril et al., 1999, 2003; King et al., 2000; Sivori et al., 2000).

Both IL-2 and IL-15 induce increased cytolytic activity of decidual NK cells, but it is only after IL-2 stimulation that the cells kill normal trophoblast and choriocarcinoma cell lines such as JEG-3 cells (King and Loke, 1990; Verma et al., 2000). IL-2 has not been found at the implantation site in normal pregnancies. Despite the inability of uterine NK cells to kill trophoblast when they are first isolated, their cytolytic response to activation by IL-2 raises the possibility that killing of placental cells might occur sometimes in vivo and result in pregnancy failure. A recent report suggested that granulolysin expression was increased in CD56^{bright} cells in spontaneous abortion and was responsible for apoptosis of trophoblast (Nakashima et al., 2008). The problem with any study of spontaneous miscarriage is that the tissue is obtained several days after foetal demise so the secondary inflammatory changes confound any investigation of the primary cause. There is, therefore, still no obvious explanation for the characteristic and prominent granules that are a feature of both human and murine uNK cells. More attention has focused on their role as producers of cytokines during placentation.

Cytokine production

It is difficult to compare many of the reports on cytokine production by decidual NK cells because there are differences in methods of extraction, purity of the NK cells and periods of culture in vitro with IL-2/IL-15 or PMA/ionomycin. The main contaminating cells are $CD14^+$ macrophages and these could be a source of cytokines such as IL-8. The hypoxic conditions following the surgical and NK isolation procedures are also likely to affect levels of cytokines induced by hypoxia, such as vascular endothelial growth factor (VEGF) and placental growth factor (PlGF). Furthermore, few studies have analysed the KIR genotypes of the women and this is likely to differentially affect the cytokine profile of each sample.

A range of cytokines has been detected either by transcriptional analysis of isolated uNK or by ELISA of uNK conditioned media, including IFN-γ, TNF-α, GM-CSF and IL-10, together with a range of angiogenic factors and chemokines such as IL-8, VEGF, PlGF, IP-10 and MIP-1α and MIP-1β (Hanna et al., 2006; Jokhi et al., 1994, 1997; Lash et al., 2006; Li et al., 2001; Saito et al., 1993; Sharkey et al., 1999). There is some discrepancy in the findings from different groups probably for the reasons discussed above. This could be resolved by the use of intracellular staining of $CD56^+$ cells for particular cytokines. When this has been performed, it is clear that freshly isolated uNK cells are very poor producers of IFN-γ compared to blood NK cells and that the source of much of the IL-8 produced is $CD14^+$ cells and not NK cells (Eriksson et al., 2006; Sharkey et al., 2008 and unpublished data).

IL-8 can be induced on co-culture with trophoblast and this is in part mediated by activating receptors NKp44, DNAM-1 and NKG2D (Vacca et al., 2008). Ligation by mAb to NKp46 but not NKp30 induced cytotoxicity but not cytokine production whereas mAbs and NKp30 resulted in production of IFN-γ, TNF-α, MIP-1α , MIP-1β and GM-CSF (El Costa et al., 2008). The role of various trophoblast ligands, including HLA-G, has been studied by co-culture experiments with 721.221 HLA-G^+ cells or by ligation of uNK cells clones with mAbs to KIR. IFN-γ, VEGF and IP-10 production can be induced by such experimental manipulations (Rajagopalan et al., 2006; van de Meer et al., 2004).

There is thus a need to establish more clearly what cytokines are made by uNK cells in the pregnant and non-pregnant mucosa in vivo as much of the experimental data has used activated uNK cells that are unlikely to represent the physiological milieu of the normal decidua. It is also unclear how the cytokine profile will vary in conditions where placentation is defective and how this relates to the NK receptor repertoire. An additional question is what effect the uNK soluble products exert on trophoblast, maternal vasculature and other cells in the decidua. Assays to study trophoblast behaviour have generally used explant models or invasion assays through Matrigel-coated chamber slides. These are difficult to standardize because of the rapidly changing nature of trophoblast in the first trimester. Although several trophoblast cell lines have been generated, none of these have the normal trophoblast HLA-I profile and this limits their usefulness as targets for uNK cells (Apps et al., 2009). It will require persistence and imagination to define the exact role of uNK cells in human pregnancy.

Maintenance and modification of the vasculature

Uterine NK cells secrete abundant cytokines and other soluble products (e.g. GM-CSF, IFN-γ) that can act on stromal or endothelial cells in the endometrium. Furthermore, they express angiogenic regulators, angiopoietin-2 (Ang-2), VEGF-C and PlGF, lending support to the idea that a major role for uNK is to act on the mucosal vasculature (Li et al., 2001). A novel angiogenic factor, NKG5, has also been identified in uNK cells in decidua (Langer et al., 1999).

NK cell expansion and endometrial vascularization occur together during the secretory phase of the menstrual cycle and uNK cell demise clearly occurs before any other morphological feature of menstrual breakdown. This suggests that uNK may be required in the non-pregnant uterus to maintain vascular stability. Because the uNK cells show cyclical changes during the menstrual cycle and they are particularly abundant beneath the luminal epithelium in close association with a dense network of capillaries, they are ideally placed to influence endothelial cell function. The loss of angiogenic factors made by uNK cells may destabilize blood vessels and precipitate vascular regression and hence endometrial breakdown. uNK cells could therefore play a role in determining the fate of the endometrium, either to breakdown at menstruation (in absence of pregnancy) or to decidualize (Critchley et al., 2001; Jabbour et al., 2006; King, 2000). The role of NK cells at other mucosal locations that secrete IL-22 and act to maintain intestinal homeostasis has been recently described (Sanos et al., 2009).

Further support for the idea that uNK cells influence vascular stability has come from recent studies of uNK cells in hormonally treated endometrium. Endometrium exposed to exogenous progestogen, (as with progestogen long-acting reversible contraceptives [LARC], hormone replacement, or progestogens to regulate abnormal menstrual bleeding patterns) displays features of decidualization with a leukocyte population composed mainly of uNK cells (Jones and Critchley, 2000). A positive association between uNK cell density and irregular bleeding was found in post-menopausal women using continuous combined hormone therapy (Hickey et al., 2005). In women using LARC, uNK cells were found more abundantly in the decidualized stroma in women having breakthrough bleeding compared to women who were not bleeding (Peloggia et al., 2006). Taken together, these studies show that alterations in uNK cell numbers are associated with endometrial breakdown and bleeding.

In the first trimester of human pregnancy, uNK cells are found clustered around the spiral arteries, suggesting that changes in the spiral arteries during pregnancy may also be uNK mediated (Trundley and Moffett, 2004). There is also evidence from mice that uNK are required for adaptation of the uterine blood vessels during pregnancy. NK-deficient mice have decidual arteries with thicker walls than wild type mice and a reduced luminal diameter, and this effect is IFNγ-dependent (Ashkar and Croy, 2001; Greenwood et al., 2000).

Mediation of trophoblast invasion

From the data available to date, the most likely function of uNK cells is to control placentation and in particular the invasion and transformation by trophoblast of the uterine spiral arteries. Thus, the NK cells are thought to regulate the supply of blood from the spiral arteries to the intervillous space, which comes into direct contact with the placental villi.

Uterine NK cells might function to affect trophoblast cell invasion as well as acting directly on the arteries, as described above. Indeed in mice, where trophoblast invasion into decidua is minimal, the latter may be more likely. The transformation of spiral arteries by trophoblast from thick muscular-walled vessels to dilated structures with complete destruction of the media and replacement of the endothelium by trophoblast is a remarkable process and ill-understood. In lower primates such as baboons and rhesus macaques, trophoblast does not invade interstitially through the decidual stroma and trophoblast invasion is limited to endovascular trophoblast moving in a retrograde fusion down the arteries (Ramsey et al., 1976). Vigorous interstitial invasion seems to be a feature only of great apes.

The close spatial and temporal association of NK cells in decidua with the invading trophoblast means that this is the major site where maternal lymphocytes come into direct contact with the allogenic trophoblast cells of the foetus. It is logical therefore to assume that NK allorecognition will occur at this site and could be the basis for how placentation is controlled and finely tuned. Indeed, there is now evidence that uNK cells, as the dominant lymphocyte in the decidua during early pregnancy, could mediate the delicate negotiation between foetal trophoblast cells and maternal decidua that is needed to prevent over or under invasion. This idea has led to an investigation of the possible NK ligands present on the invading trophoblast cells. Of particular importance are likely to be MHC class I and MHC-like ligands, which are known to mediate NK allorecognition in murine models of hybrid resistance and in therapeutic situations in humans such as bone marrow transplantation. What must not be forgotten though is that the situation in the uterus during placentation is a normal physiological process and the only time there is a possibility of natural allorecognition in mammals.

Potential ligands for NK cells on trophoblast

After much controversy, the HLA class I (HLA-I) molecules that are displayed at the cell surface of EVT have now been established (Moffett and Loke, 2006). There are six HLA-I loci with expressed protein products; HLA-A, -B and -C are classical and HLA-E, -F and -G are non-classical. HLA-I genes are highly polymorphic, but there is high homology even between alleles at different loci. Thus, it has been extremely difficult to generate mAbs that can distinguish between each HLA-I gene product with no cross-reactivity with the thousands of other HLA allotypes. A further problem is that normal human first trimester trophoblast cells are hard to obtain with a sufficient number and purity for biochemical analysis. There is only one cell line (JEG-3) that appears to resemble normal EVT cells in its pattern of HLA-I expression but this line is derived from a highly malignant choriocarcinoma and it was originally passaged multiple times in the hamster cheek pouch. Detailed characterization of a panel of mAbs against beads coated with a single HLA-I allotype has recently allowed a more definitive characterization of HLA-I expression (Apps et al., 2009).

With respect to HLA expression, EVT cells have major distinguishing features. Firstly, HLA-II molecules are not expressed and there is silencing of IFN-γ-inducible expression of the MHC class II transactivator CIITA. Secondly, the classical HLA-I molecules, HLA-A and -B, which show the most polymorphism and are the major determinants responsible for defence against viral infection and allograft rejection, are absent from the trophoblast cell surface. There is also no evidence that they are ever transcribed in trophoblast. Thirdly, one classical HLA-I molecule is expressed, HLA-C, and this is known to be a dominant ligand for NK cells (Gumperz and Parham, 1995). Fourthly, EVT are the only normal cells that express HLA-G, a non-classical molecule that has unusual features. Finally, EVT also expresses HLA-E. This is not unexpected as the signal sequences of both HLA-C and HLA-G provide a source of peptides that will bind to HLA-E. This HLA-I profile of EVT (HLA-C, HLA-G, HLA-E) is highly unusual and has not been described on any other cells in normal or pathological situations.

The villous syncytiotrophoblast is HLA-null under in vivo or in vitro even after exposure to IFN-γ. These cells are in contact with maternal PBNK in the intervillous space and could theoretically be susceptible to cytolysis. The formation of a syncytium may be important in defence against cytolytic attack and it is also probable that syncytiotrophoblast lacks ligands for any activating NK receptors.

It is accepted that uNK cells do express the activating NK receptors, NKG2D and NKp46 and a search for their ligands on EVT has been performed by different methods. NKG2D is promiscuous and a range of ligands (MICA, MICB, ULBP1-4, etc.) are known to be expressed on stressed cells, particularly epithelial cells. To date, there is no evidence that any NKG2D ligands are expressed by EVT in normal pregnancy (Apps et al., 2008b). It is still possible that they could be present in situations such as transplacental infection with CMV or in karyotypically abnormal conceptus.

Both the main populations of uterine leukocytes, NK cells and HLA-DR$^+$ cells express receptors that can recognize MHC-I ligands on trophoblast. There are therefore mechanisms in place for the innate uterine immune effectors to respond to foetal trophoblast cells. This is in contrast to the situation for maternal T cells where the lack of HLA class II and HLA-A and -B means dominant T cell ligands are absent. There is still the possibility that HLA-C-reactive T cells could be generated and an indirect T cell response to other minor histocompatibility antigens or trophoblast–specific proteins to be presented via maternal decidual HLA-DR$^+$ cells, but these have never so far been convincingly demonstrated.

HLA-C

Two important features of HLA-C molecules distinguish them from HLA-A and -B: (i) lower levels of surface expression on somatic cells, (ii) less allelic variation with particular conservation in the α1 helix at the binding site for NK cell KIR. Although HLA-C restricted cytotoxic T lymphocytes (CTL) do exist, they are much less frequent than HLA-A or HLA-B-restricted CTL. Instead, HLA-C has emerged as the dominant ligand for KIR on NK cells. Recent genetic and functional data have pointed to a role of HLA-C and KIR in immune response to the implanting placenta.

We have found that expression levels of KIR that bind HLA-C are increased in uNK cells compared to blood (Sharkey et al., 2008; Verma et al., 1997). Thus, the maternal KIR repertoire in the uterus (as compared to blood NK cells) is skewed towards recognition of HLA-C. This is reflected in increased binding of HLA-C tetramers by uNK cells compared to blood NK cells. KIR tetramers also bind specifically to surface HLA-C molecules on normal trophoblast cells and we have shown that HLA-C molecules are present in abundance at the trophoblast cell surface in an unusually stable β_2m-associated conformation, without any of the less stable unfolded conformers that are also present on somatic cells (Apps et al., 2008a). All these findings suggest that the enhanced recognition of trophoblast HLA-C by KIR on uNK cells in the placental bed has special biological significance relating to successful placentation.

Genetic studies into pre-eclampsia, foetal growth restriction and recurrent miscarriage and disorders of pregnancy in which trophoblast invasion is defective, have provided further evidence to suggest that KIR recognition of HLA-C is important for successful implantation (Hiby et al., 2004). KIR may be activating or inhibitory and there are two basic KIR haplotypes, A and B, differing principally in that the B haplotype has additional activating receptors. In any pregnancy, the maternal KIR genotype could be AA (no activating KIR), AB (1–5 activating KIR) or BB (1–10 activating KIR). The HLA-C ligands for KIR on trophoblast cells belong to two groups, HLA-C1 and HLA-C2, defined by a dimorphism at position 80 of the α1 domain. Some maternal/foetal KIR/HLA-C combinations might be more favourable to trophoblast-cell invasion and others less so due to the overall signals that the NK cell receives.

The occurrence of pre-eclampsia is associated with an increased frequency of the maternal KIR AA genotype and this is mainly associated with HLA-C2 in the foetus. In pre-eclampsia and foetal growth restriction, typing for presence/absence of individual KIR genes shows that affected women have reduced frequency of KIR genes on the telomeric end of the B haplotype. KIR2DS1, the activating receptor for HLA-C2 groups, is most notably reduced in frequency (unpublished). In couples that have had three or more recurrent miscarriages, the frequency of HLA-C2 is increased in the male partners. Genes on the telomeric end of the KIR B haplotype are again significantly reduced in frequency compared to controls, particularly KIR2DS1. Male partners have KIR gene frequencies very similar to controls (Hiby et al., 2008). Thus, compared to mothers with normal pregnancies, affected women have a significantly increased frequency of genotypes that lack KIR genes at the telomeric end of the KIR B haplotype. There is strong linkage of KIR genes within this telomeric part of the haplotype but within this region it is particularly notable that in all cohorts KIR2DS1, the activating KIR for HLA-C2, is reduced in frequency in affected mothers compared to controls.

Following on from these genetic associations, it has been demonstrated that NK cells expressing inhibitory KIR produce fewer trophoblast chemoattractant cytokines than those expressing activating KIR in response to HLA-C on 721.221 cells (Hanna et al., 2006). This provides a possible mechanism to explain deficient trophoblast invasion in pregnancies where the mother lacks activating KIR.

There are two further variables to consider. Firstly, the potential of an NK cell to respond is known to depend on the presence of the ligand for its inhibitory KIR during maturation. This phenomenon, described as licensing or education, means that an NK cell lacking an inhibitory receptor that recognizes self is hyporesponsive. Thus, in the pregnant situation, KIR2DL1^{+} NK cells could be predicted to have enhanced activity when the mother has the HLA-C2 ligand but be hyporesponsive if she is homozygous for HLA-C1. Secondly, a recent report identified an SNP upstream of the HLA-C gene that influences levels of expression of HLA-C molecules at the cell surface and is associated with delayed disease progression in HIV-1 infection. How the levels of trophoblast HLA-C affect binding to KIR expressed by uNK cells will be interesting to explore.

HLA-E

The NK receptor for HLA-E belongs to the lectin family of CD94/NKG2 receptors. The two important members are the inhibitory CD94/NKG2A and the activating CD94/NKG2C (Lazetic et al., 1996). CD94/NKG2A is expressed at higher levels on both uterine and blood CD56^{bright} NK cells than blood CD56^{dim} NK cells. Furthermore, 80–95% of uNK cells express CD94/NKG2A, whereas only ~50% of blood CD56^{dim} cells do. In keeping with these findings, HLA-E tetramers bind to virtually all uNK cells via CD94/NKG2A and this interaction results in a strong inhibitory signal that prevents killing (King et al., 2000). As the uNK cells are in close contact with stromal cells, macrophages, T cells and blood vessels as well as trophoblast, it is clearly important to prevent lysis of any maternal cells in their vicinity. Of interest, though, is that when the HLA-E–bound peptide is derived from the HLA-G leader sequence, the affinity of the binding to CD94/NKG2A is higher than any other HLA-I-derived peptides (Llano et al., 1998). Furthermore, this is the only situation where appreciable binding was observed to the activating receptor, CD94/NKG2C. CD94/NKG2C is found on a subset of PBNK and on at least some uNK cells. These observations suggest that HLA-G–positive EVT could send a differential signal to uNK cells compared to the signal from maternal HLA-G negative cells in the decidua basalis, but this has not been tested experimentally.

HLA-G

Since its discovery, HLA-G was thought to be the obvious candidate to signal to uNK cells and early reports did indeed suggest that HLA-G resulted in inhibition of cytotoxicity. This is now known to result from the interaction of HLA-E with CD94/NKG2A. Several splice variants of HLA-G have been described, including soluble forms, but there is uncertainty about whether any of these forms are expressed as surface or soluble proteins normally in vivo. This is because of the problem of cross-reactivity of mAbs that can detect other HLA-I molecules. Biochemical analysis of lysates of surface trophoblast proteins followed by SDS-PAGE does allow separation of HLA-G from other HLA-I molecules because the heavy chain is only 39 kDa as opposed to the normal 45 kDa. Definitive identification of surface or soluble HLA-G proteins in other situations, such as tumours and allografts, has still not been confirmed with such biochemical techniques and has relied solely on flow cytometry or immunohistology where antibody cross-reactivity cannot be ruled out (Apps et al., 2008c).

The best evidence to date suggests that HLA-G binds to inhibitory receptors LILRB1 and LILRB2 that are expressed by all decidual macrophages. LILRB1 is also present on approximately 20% of uNK cells.

LILRB1 and 2 have been shown to bind HLA-G by a range of experimental methods:

1. LILRB1/2-Fc fusion proteins specifically bound to transfected cells expressing HLA-I molecules including HLA-G in the absence of HLA-E (Borges and Cosman, 2000; Borges et al., 1997; Colonna et al., 1997).
2. HLA-G tetramers bound cells transfected with LILRB1/2 and human PBMC; binding was blocked with α-LILRB mAb (Allan et al., 1999).
3. surface plasmon resonance (SPR) experiments with recombinant HLA-G molecules demonstrated that the affinity of LILRB1 and 2 binding to HLA-G is higher than other HLA-I molecules (Chapman et al., 1999; Shiroishi et al., 2003).
4. The co-crystal structure of LILRB1 binding to the $\alpha3$ domain and β_2m molecule of HLA-A is now available and the HLA-G crystal structure is similar (Clements et al., 2005; Willcox et al., 2003).

HLA-G has the unusual property of forming β_2m-associated homodimers due to a cysteine present at position 42 of the α1 domain. HLA-G homodimers are present at the surface of normal trophoblast cells as well as 721.221 cells transfected with HLA-G, but they are not prominent on JEG-3 cells, where the monomeric form dominates (Apps et al., 2007; Gonen-Gross et al., 2003). Importantly, given that the formation of this dimer is unique to HLA-G, dimerization dramatically increases LILRB1 binding. An LILRB1-Fc fusion protein bound more strongly to wild type HLA-G transfectants than serine 42 mutants that do not dimerize (Gonen-Gross et al., 2003). SPR measurements of soluble monomeric and dimeric HLA-G complexes binding LILRB molecules confirmed this finding (Shiroishi et al., 2006). The increased binding avidity of the dimer translates into augmented signalling through LILRB1 because LILRB1-mediated inhibition of IgeR-mediated serotonin release and inhibition of NK killing were both increased with cells expressing HLA-G that can form dimers, as opposed to a serine 42 mutant (Gonen-Gross et al., 2003). In addition, a LILRB1 chimera NFAT-GFP reporter cell assay showed that 100-fold lower concentrations of dimeric compared to monomeric HLA-G were necessary to induce a signal (Shiroishi et al., 2006). Overall, the evidence is convincing that the HLA-G dimer binds LILRB1 and that this interaction is likely to be functional.

The deviation by HLA-G of dendritic cell (DC) towards a tolerogenic phenotype has been described only in murine models or using monocyte derived DC (MDCC); there is still no evidence using myelomonocytic cells isolated from decidua. These decidual CD14$^+$ or CD14$^-$ HLA-DR$^+$ cells have an unusual phenotype that in some respects resembles dermal myelomonocytic cells. Nonetheless, the idea that a placental monomorphic molecule could be diverting all local immune responses from immunogenic to tolerogenic is attractive. If this is true, then it is the foetus itself that is preventing a T cell rejection response by instructing only the maternal APC that they come into direct contact with in the placental bed.

HLA-G has also been shown in some studies to bind to KIR2DL4 (Rajagopalan and Long, 1999). KIR2DL4 is present in all KIR haplotypes as the central framework gene and is transcribed in all PBNK and uNK cells, but there is some dispute about surface expression on both PBNK and uNK cells and also about the binding of KIR2DL4 to HLA-G (Allan et al., 1999; Boyson et al., 2002). This could be due in part to the particular KIR2DL4 allele expressed and the activation state of the NK cell. An interesting possibility is that KIR2DL4 is present only in early Rab5$^+$ endosomes and binds to internalized HLA-G molecules (Rajagopalan et al., 2006). The outcome of this interaction on PBNK results in stimulation of an array of cytokines, many of which could act on the vasculature. This type of stimulatory response after intracellular ligation of HLA-G to a receptor would allow the NK cells only to be stimulated in close proximity to trophoblast cells. Furthermore, the high concentration of receptor and ligand in the endosome could compensate for the low affinity interaction between KIR2DL4 and HLA-G. Biocore studies to formally demonstrate this are still required.

HLA-G may therefore be able to signal to all CD56$^+$ (via KIR2DL4) and HLA-DR$^+$ (via LILRB1 and 2) uterine maternal leukocytes. These account for the great majority of uterine leukocytes (85–90%) raising the possibility that a trophoblast-specific HLA molecule could entirely switch the role of these cells into some pregnancy-specific functions.

Concluding remarks

Although the function of uterine NK cells is not yet definitively known, there has been some progress in humans. It is perhaps surprising that animal studies lag behind. For example, the MHC expression profile of murine trophoblast is still unknown. A recent study has described a detailed phenotype of murine uNK cells by flow cytometry (Yadi et al., 2008) and this shows some overlapping features with humans, but also distinct differences.

In humans, there is circumstantial evidence that uNK cells may play a role in the normal physiological processes of the menstrual cycle. This will be an important area to explore because, if NK cells can be shown to prevent mucosal breakdown, this will have therapeutic possibilities in women suffering from dysmenorrhoea, infertility and recurrent miscarriage.

The evidence that uNK cells play an important role in the regulation of placentation is accumulating both from genetic and functional studies. It is likely that humans may have developed this particular function more extensively than other species because placentation is far more invasive in humans than in other primates or mice.

References

Allan, D.S., et al. (1999). Tetrameric complexes of human histocompatibility leukocyte antigen (HLA)-G bind to peripheral blood myelomonocytic cells. J Exp Med **189**, 1149–1156.

Apps, R., et al. (2007). A homodimeric complex of HLA-G on normal trophoblast cells modulates antigen-presenting cells via LILRB1. Eur J Immunol **37**, 1727–1729.

Apps, R., et al. (2008a). Conformation of human leucocyte antigen-C molecules at the surface of human trophoblast cells. Immunology **124**, 322–328.

Apps, R., et al. (2008b). Natural-killer cell ligands at the maternal–fetal interface: UL-16 binding proteins, MHC class-I chain related molecules, HLA-F and CD48. Hum Reprod **23**, 2535–2548.

Apps, R., et al. (2008c). A critical look at HLA-G. Trends Immunol **29**, 313–321.

Apps, R., et al. (2009). Human leukocyte antigen (HLA) expression by normal trophoblast cells and placental cell lines using a novel method to characterize allotype specificity of anti-HLA antibodies. Immunology **127**, 26–39.

Ashkar, A. and Croy, B.A. (2001). Functions of uterine natural killer cells are mediated by interferon gamma production during murine pregnancy. Semin Immunol **13**, 235–241.

Avril, T., et al. (1999). Trophoblast cell line resistance to NK lysis mainly involves an HLA class I-independent mechanism. J Immunol **162**, 5902–5909.

Avril, T., et al. (2003). Human choriocarcinoma cell resistance to natural killer lysis due to defective triggering of natural killer cells. Biol Reprod **69**, 627–633.

Borges, L. and Cosman, D. (2000). LIRs/ILTs/MIRs, inhibitory and stimulatory Ig-superfamily receptors expressed in myeloid and lymphoid cells. Cytokine Growth Factor Rev **11**, 209–217.

Borges, L., et al. (1997). A family of human lymphoid and myeloid Ig-like receptors some of which bind MHC class I molecules. J Immunol **159**, 5192–5196.

Boyson, J.E., et al. (2002). Disulfide bond-mediated dimerization of HLA-G on the cell surface. Proc Natl Acad Sci U S A **99**, 16180–16185.

Carlino, C., et al. (2008). Recruitment of circulating NK cells through decidual tissues: a possible mechanism controlling NK cell accumulation in the uterus during early pregnancy. Blood **111**, 3108–3115.

Chantakru, S., et al. (2004). Contributions from self-renewal and trafficking to the uterine NK cell population of early pregnancy. J Immunol **168**, 22–28.

Chapman, T.L., et al. (1999). The inhibitory receptor LIR-1 uses a common binding interaction to recognize class I MHC molecules and the viral homolog UL18. Immunity **11**, 603–613.

Clements, C.S., et al. (2005). Crystal structure of HLA-G: a nonclassical MHC class I molecule expressed at the fetal–maternal interface. Proc Natl Acad Sci U S A **102**, 3360–3365.

Colonna, M., et al. (1997). A common inhibitory receptor for MHC class I molecules on human lymphoid and myelomonocytic cells. J Exp Med **186**, 1809–1818.

Critchley, H.O., et al. (2001). The endocrinology of menstruation—a role for the immune system. Clin Endocrinol **55**, 701–710.

Croy, B.A., et al. (2006). Uterine natural killers cells: a specialize differentiation regulated by ovarian hormones. Immunol Rev **214**, 161–185.

El Costa, H., et al. (2008). Critical and differential roles of NKp46- and NKp30-activating receptors expressed by uterine NK cells in early pregnancy. J Immunol **181**, 3009–3017.

Eriksson, M., et al. (2006). TLRs mediate IFN-gamma production by human uterine NK cells in endometrium. J Immunol **176**, 6219–6224.

Gonen-Gross, T., et al. (2003). Complexes of HLA-G protein on the cell surface are important for leukocyte Ig-like receptor-1 function. J Immunol **171**, 33–51.

Greenwood, J.D., et al. (2000). Ultrastructural studies of implantation sites from mice deficient in uterine natural killer cells. Placenta **21**, 693–702.

Gumperz, J.E. and Parham, P. (1995). The enigma of the natural killer cell. Nature **378**, 245–248.

Hamperl, H. and Hellweg, G. (1958). Granular endometrial stroma cells. Obstet Gynecol **11**, 379–387.

Hanna, J., et al. (2006). Decidual NK cells regulate key developmental processes at the human fetal–maternal interface. Nat Med **12**, 1065–1074.

Henderson, T.A., et al. (2003). Steroid receptor expression in uterine natural killer cells. J Clin Endocrinol Metab **88**, 440–449.

Hiby, S.E., et al. (2004). Combinations of maternal KIR and fetal HLA-C genes influence the risk of preeclampsia and reproductive success. J Exp Med **200**, 957–965.

Hiby, S., et al. (2008). Association of maternal killer-cell immunoglobulin-like receptors and parental HLA-C genotypes with recurrent miscarriage. Hum Reprod **23**, 972–976.

Hickey, M., et al. (2005). Menopausal hormone therapy and irregular endometrial bleeding: a potential role for natural killer cells? J Clin Endocrinol Metab **90**, 5528–5535.

Jabbour, H.N., et al. (2006). Endocrine regulation of menstruation. Endocrin Rev **27**, 17–46.

Jokhi, P.P., et al. (1994). Production of granulocyte-macrophage colony-stimulating factor by human trophoblast cells and by decidual large granular lymphocytes. Hum Reprod **9**, 1660–1669.

Jokhi, P.P., et al. (1997). Cytokine production and cytokine receptor expression by cells of the human first trimester placental–uterine interface. Cytokine **9**, 126–137.

Jones, R.L. and Critchley, H.O. (2000). Morphological and functional changes in human endometrium following intrauterine levonorgestrel delivery. Hum Reprod **15**(Suppl. 3), 162–172.

Kammerer, U., et al. (2004). Immunology of human endometrium. Immunobiology **209**, 569–574.

Kazzaz, B.A. (1972). Specific endometrial granular cells. A semiquantitative study. Eur J Obstet Gynecol **3**, 77–84.

Keskin, D.B., et al. (2007). TGFbeta promotes conversion of CD16+ peripheral blood NK cells into CD16− NK cells with similarities to decidual NK cells. Proc Natl Acad Sci U S A **104**, 3378–3383.

King, A. (2000). Uterine leukocytes and decidualization. Hum Reprod Update **6**, 28–36.

King, A. and Loke, Y.W. (1990). Human trophoblast and JEG choriocarcinoma cells are sensitive to lysis by IL-2-stimulated decidual NK cells. Cell Immunol **129**, 435–448.

King, A., et al. (1989a). Immunocytochemical characterization of the unusual large granular lymphocytes in human endometrium throughout the menstrual cycle. Hum Immunol **24**, 195–205.

King, A., et al. (1989b). Early human decidual cells exhibit NK activity against the K562 cell line but not against first trimester trophoblast. Cell Immunol **118**, 337–347.

King, A., et al. (1991). CD3− leukocytes present in the human uterus during early placentation: phenotypic and morphologic characterization of the CD56++ population. Dev Immunol **1**, 169–190.

King, A., et al. (1993). Expression of perforin, granzyme A and TIA-1 by human uterine CD56+ NK cells implies they are activated and capable of effector functions. Hum Reprod **8**, 2061–2067.

King, A., et al. (2000). HLA-E is expressed on trophoblast cells and interacts with CD94/NKG2A receptors on decidual NK cells. Eur J Immunol **30**, 1623–1631.

Kirby, D.R. (1960). Development of mouse eggs beneath the kidney capsule. Nature **187**, 707–708.

Koopman, L.A., et al. (2003). Human decidual natural killer cells are a unique NK subset with immunomodulatory potential. J Exp Med **198**, 1201–1212.

Kopcow, H.D., et al. (2005). Human decidual NK cells form immature activating, synapses and are not cytotoxic. Proc Natl Acad Sci U S A **102**, 15563.

Kroemer, G. and Martin, SK. (2005). Caspase independent cell death. Nat Med **11**, 725–730.

Kusumi, M., et al. (2006). Expression patterns of lectin-like natural killer receptors, inhibitory CD94/NKG2A, and activating CD94/NKG2C on decidual CD56bright natural killer cells differ from those on peripheral CD56dim natural killer cells. J Reprod Immunol **70**, 33–42.

Langer, N., et al. (1999). Novel hyperactive mitogen to endothelial cells: human decidual NKG5. Am J Reprod Immunol **42**, 263–272.

Lash, G.E., et al. (2006). Expression of angiogenic growth factors by uterine natural killer cells during early pregnancy. J Leukoc Biol **80**, 572–580.

Lazetic, S., et al. (1996). Human natural killer cell receptors involved in MHC class I recognition are disulphide linked heterodimers of CD94 and NKG2 subunits. J Immunol **157**, 4741–4745.

Li, X.F., et al. (2001). Angiogenic growth factor messenger ribonucleic acids in uterine natural killer cells. J Clin Endocrinol Metabol **86**, 1823–1834.

Llano, M., et al. (1998). HLA-E-bound peptides influence recognition by inhibitory and triggering CD94/NKG2 receptors: preferential response to an HLA-G-derived nonamer. Eur J Immunol **28**, 2854–2863.

Loke, Y.W. and King, A. (1995). *Human Implantation*. Cambridge: CUP.

Lynch, L., et al. (2007). Cells with haematopoietic stem cell phenotype in adult human endometrium: relevance to infertility? Hum Reprod **22**, 919–926.

Manaster, I., et al. (2008). Endometrial NK cells are special immature cells that await pregnancy. J Immunol **181**, 1869–1876.

Marchand, F. (1895). Uber die sogenannten dezidualen Geschwulste im Anschlub an normale Gebert, Blasenmole und Extrauterinschwangerschaft. Mschr Geburtsh Gynak **1**, 419.

Matsuura-Sawada, R., et al. (2005). Reproduction of menstrual changes in transplanted human endometrial tissue in immunodeficient mice. Hum Reprod **20**, 1477–1484.

McClaren, A. (1965). In: Park, W.W. (ed.) *The Early Conceptus, Normal and Abnormal*. Edinburgh: Livingstone, pp. 27–33.

Milne, S.A., et al. (2005). Leukocyte populations and steroid receptor expression in human first-trimester decidua; regulation by antiprogestin and prostaglandin E analog. J Clin Endocrinol Metab **90**, 4315–4321.

Moffet, A. and Loke, Y.W. (2006). Immunology of placentation in eutherian mammals. Nat Rev Immunol **6**, 584–594.

Moffett, A., et al. (2004). Natural killer cells, miscarriage and infertility. BMJ **329**, 1283–1285.

Moffett-King, A. (2002). Natural killer cells and pregnancy. Nat Rev Immunol **2**, 656–663.

Nakashima, A., et al. (2008). Granulysin produced by uterine natural killer cells induces apoptosis of extravillous trophoblasts in spontaneous abortion. Am J Pathol **173**, 653–664.

Okada, H. and Mak, T.W. (2004). Pathways of apoptotic and non-apoptotic death in tumor cells. Nat Rev Cancer **4**, 592–603.

Pace, D., et al. (1989). Proliferative activity in endometrial stromal granulocytes throughout menstrual cycle and early pregnancy. J Clin Pathol **42**, 35–39.

Peloggia, A., et al. (2006). Endometrial chemokines, uterine natural killer cells and mast cells in long-term users of the levonorgestrel-releasing intrauterine system. Hum Reprod **21**, 1129–1134.

Pongcharoen, S., et al. (2004). No evidence for apoptosis of decidual leukocytes in normal and molar pregnancy: implications for immune privilege. Clin Exp Immunol **138**, 330–336.

Rajogopalan, S. and Long, E.O. (1999). A human histocompatibility leukocyte antigen (HLA)-G-specific receptor expressed on all natural killer cells. J Exp Med **189**, 1093–1100.

Rajagopalan, S., et al. (2006). Activation of NK cells by an endocytosed receptor for soluble HLA-G. PloS Biol **4**, e9.

Ramsey, E.M., et al. (1976). Interactions of the trophoblast and maternal tissues in three closely related primate species. Am J Obstet Gynecol **124**, 647–652.

Ritson, A. and Bulmer, J.N. (1987). Endometrial granulocytes in human decidua react with a natural killer (NK) cell marker NKH1. Immunology **62**, 329–331.

Saito, S., et al. (1993). Cytokine production by CD16 − CD56bright natural killer cells in the human early pregnancy deciduas. Int Immunol **5**, 559–563.

Sanos., et al. (2009). RORgammat and commensal microflora are required for the differentiation of mucosal interleukin 22-producing NKp46+ cells. Nat Immunol **10**, 11–12.

Sentman, C.L., et al. (2004). Recruitment of uterine NK cells: induction of CXC chemokine ligands 10 and 11 in human endometrium by estradiol and progesterone. J Immunol **173**, 6760–6766.

Sharkey, A.M., et al. (1999). Localization of leukaemia inhibitory factor and its receptor in human placenta throughout pregnancy. Biol Reprod **60**, 355–364.

Sharkey, A.M., et al. (2008). KIR expression in uterine NK cells is biased towards recognition of HLA-C and alters with gestational age. J Immunol **181**, 39–46.

Shiroishi, M., et al. (2003). Human inhibitory receptors Ig-like transcript 2 (ILT2) and ILT4 compete with CD8 for MHC class I binding and bind preferentially to HLA-G. Proc Natl Acad Sci U S A **100**, 8856–8861.

Shiroishi, M., et al. (2006). Efficient leukocyte Ig-like receptor signaling and crystal structure of disulphide-linked HLA-G dimer. J Biol Chem **281**, 10439–10447.

Sivori, S., et al. (2000). Triggering receptors involved in natural killer cell-mediated cytotoxicity against choriocarcinoma cell lines. Hum Immunol **61**, 1055–1058.

Starkey, P.M., et al. (1988). Cell populations in human early pregnancy decidua: characterization and isolation of large granular lymphocytes by flow cytometry. Immunology **65**, 129–134.

Taylor, H.S. (2004). Endometrial cells derived from donor stem cells in bone marrow transplant recipients. JAMA **292**, 81–85.

Trundley, A. and Moffett, A. (2004). Human uterine leukocytes and pregnancy. Tissue Antigens **63**, 1–12.

Vacca, P., et al. (2006). Analysis of natural killer cells isolated from human decidua: evidence that 2B4 (CD244) functions as an inhibitory receptor and blocks NK-cell function. Blood **108**, 4078–4085.

Vacca, P., et al. (2008). Regulatory role of NKp44, NKp46, DNAM-1 and NKG2D receptors in the interaction between NK cells and trophoblast cells. Evidence for divergent functional profiles of decidual versus peripheral NK cells. Int Immunol **20**, 1395–1405.

Van der Meer, A., et al. (2004). Membrane bound HLA-G activates proliferation and interferon gamma production by uterine natural killer cells. Mol Hum Reprod **10**, 189–195.

Verma, S., et al. (1997). Expression of killer cell inhibitory receptors on human uterine natural killer cells. Eur J Immuol **27**, 979–983.

Verma, S. (1999). *Phenotypic and Functional Analysis of Human Uterine Natural Killer Cells* PhD Thesis: University of Cambridge.

Verma., et al. (2000). Human decidual natural killer cells express the receptor for and respond to the cytokine interleukin 15. Biol Reprod **62**, 959–968.

Von Numers, C. (1953). On the specific granular cells (globular leukocytes) of the human endometrium. Acta Pathol Microbiol Scand **33**, 250–256.

Weill, P. (1921). Etudes sur les leukocytes I. Les cellules granuleuses des muqueuses intestinale et uterine. Arch Anat Microsc **17**, 77–82.

Willcox, B.E., et al. (2003). Crystal structure of HLA-A2 bound to LIR-1, a host and viral major histocompatibility complex receptor. Nat Immunol **4**, 913–919.

Yadi, H., et al. (2008). Unique receptor repertoire in mouse uterine NK cells. J Immunol **181**, 6140–6170.

Yamaguchi, T., et al. (2006). Potential selectin L ligands involved in selective recruitment of peripheral blood CD16(−) natural killer cells into human endometrium. Biol Reprod **74**, 35–40.

Chapter Thirty-One

31

Identification of natural killer cells in tissues and their isolation

Paolo Carrega, Erika Montalto, Guido Ferlazzo

CHAPTER CONTENTS

Travel and change of place impart new vigor.

Seneca

ABSTRACT

For many years, natural killer (NK) cells were considered to be a homogeneous lymphocyte population with prominent cytotoxic capability. Today, NK cells rather appear to comprise various subsets that differ in function and in part in phenotype. Although human NK cell biology has so far mainly studied peripheral blood (PB) NK cells, recent studies have focused on human NK cells harboured in inflamed tissues and cancers as well as lymphoid tissues, the latter hosting the majority of NK cells in the body. NK cells isolated from these distinct compartments display distinct phenotypic and functional features, often divergent from their PB counterpart.

Remarkably, these novel insights also indicate a developmental relationship among the individual NK cell subsets. These subsets indeed correspond to sequential steps of cell differentiation occurring following their migration into injured tissues, where NK cells undergo their final maturation, including acquisition of self-tolerance and of strong cytolytic functions.

KEY WORDS

Natural killer receptors, Cytotoxicity, Cytokine secretion, Innate immunity, Chemokines, Lymph node, Cancer, Inflammation, Pregnancy

Introduction

Around 1970, shortly following the discovery of B cells and T cells, a second set of cytotoxic lymphocytes was discovered in the peripheral blood (PB) of rodents that differed in several key ways from cytotoxic T lymphocyte (CTLs). These cells lacked several cell surface markers of T cells. In addition, they were active even after isolation from unimmunized animals in lysing many different tumour cell lines. Because the newly discovered cytotoxic cells did not require specific immunization, they were termed 'natural killer cells' (NK cells). Soon after, also in the PB of humans, lymphocytes that shared the same characteristics of murine NK cells were described. Today, we know that NK cells are large granular lymphocytes distinct from B cells and T cells, able to kill according to a mechanism regulated by several activating and inhibiting receptors. They can detect the loss of major histocompatibility complex (MHC) molecules or the presence of molecules induced upon stress. This feature confers them a potential role in cancer immune surveillance and in the control of viral infections. Although the mechanisms of NK cell activation are well understood, their distribution across human compartments (in health and pathology) as well as the site of NK cell development and maturation are still unresolved. Most of the data regarding phenotype or activity of NK cells derive from studies performed on NK cells isolated from PB and are therefore not informative on how NK cells recirculate in peripheral tissue either in steady-state or in pathologic conditions. Attempts to characterize the distribution of NK cells have been also hampered by the difference between mouse and human NK cells. First, human NK cells are phenotypically defined as $CD3^-$ $CD56^+$ lymphocytes, but we do not find the same surface antigens as those in the mouse. Murine NK cells are defined as $NK1.1^+$ $CD3^-$ or $DX5^+$ $CD3^-$ cells. Second, human NK cells are functionally divided into two main subsets, namely $CD56^{dim}$ and $CD56^{bright}$ subpopulations, which do not have correspondents in rodents. In addition, previous studies searching for NK cell distribution in peripheral tissues often suffered because of technical limits such as the use of erroneous markers. In fact, in several analysis markers not exclusively expressed by NK cells (i.e. CD57) or markers possibly down-modulated/not-expressed by NK cell subsets infiltrating damaged tissue (i.e. CD16) were employed. Recently, Vivier and co-workers proposed that the activating NK receptor NKp46 could be a suitable marker able to identify all NK cells both in human and murine species (Walzer et al., 2007).

The majority of total human NK cells reside in human secondary lymphoid organs such as lymph nodes (LNs) or spleen. These tissues are specifically enriched in $CD56^{bright}$ NK cell subsets that seem to precede the terminal mature subset mainly found in the blood, that is $CD56^{dim}$. This finding, in association with the discovery at this site of presumed maturing NK cell intermediates, suggests that secondary lymphoid organs are the primary sites where NK cells develop. Very limited studies have been performed, investigating the distribution of NK cells associated with human non-lymphoid peripheral compartments. To date, only a few tissues have been analysed for NK cell presence. It is now recognized that NK cells are present in the uterine endometrium and also play a role during pregnancy. In addition, very recent works shed light on NK cells harboured in normal skin and mucosa associated with the aerodigestive tract in the tonsil and gut. In addition, in murine models, the presence of $NKp46^+$ NK cells in non-lymphoid tissues, such as the lung and liver, has been reported.

Since the primary discovery of their role in killing stressed cells, NK cells have also been investigated in individual damaged tissues, including tumours. When examined in inflamed and tumour tissues, these studies suffered from the same technical limits described previously. Based on our current inspection of this area with more modern reagents, the notion that tumour infiltration of NK cells (TINK) correlates with a good prognosis should probably be revised.

Only little information is available on NK localization throughout the body, as well as regarding the mechanisms governing their trafficking during pathologic situations. Here, we report an up-to-date assessment of the distribution of NK cells in individual healthy and pathological human compartments.

NK cell subsets

For many years, NK cells were considered to be a homogeneous lymphocyte population with primarily a cytotoxic capability. Now, NK cells rather appear to comprise various subsets that differ in function, in phenotype and in anatomical localization. In human PB, the majority of NK cells (≥95%) belongs to the $CD56^{dim}CD16^+$ cytolytic NK subset (Campbell et al., 2001; Cooper et al., 2001b; Jacobs et al., 2001). These cells carry perforin and granzymes to rapidly mediate cytotoxicity; in addition, they are equipped with inhibitory receptors, such as killer immunoglobulin-like receptors (KIRs) and the NKG2A/CD94 receptor, recognizing classical and non-classical MHC class I molecules, respectively. The $CD56^{dim}$ $CD16^+$ NK cells also display homing markers for inflamed peripheral sites such as CX3CR1, CXCR1 and ChemR23. The minor NK cell subset in the blood (≤5%) is $CD56^{bright}$ $CD16^-$ (Campbell et al., 2001; Cooper et al., 2001b; Jacobs et al., 2001). These NK cells lack perforin (or display low levels of it) but secrete large amounts of interferon (IFN)-γ and tumour necrosis factor (TNF)-α following activation to a far greater degree than the $CD56^{dim}$ $CD16^+$ NK cells (Cooper et al., 2001b; Jacobs et al., 2001). $CD56^{bright}CD16^-$ NK cells also proliferate more vigorously than their $CD56^{dim}CD16^+$ counterparts. Consistent with this feature, they uniquely express the high-affinity receptor for interleukin (IL)-2 (CD25), the alpha chain of the IL-7 receptor and CD117, the receptor for stem cell factor, also named c-kit. Notably, the immunoregulatory $CD56^{bright}CD16^-$ NK subset lacks KIRs (Jacobs et al., 2001) but uniformly express CD94/NKG2A. In addition, they display homing markers for secondary lymphoid organs, specifically CCR7 and CD62L (Campbell et al., 2001) but also chemokine receptors, such as CCR5 and CXCR3, expressed primarily during chronic conditions. Therefore, two primary functional NK cell subsets have been characterized in the PB, and we will discuss their anatomical distribution next.

NK cells in normal tissues

Peripheral blood

Human PB mononuclear cells (PBMNC) contain around 10% of NK cells and, as mentioned previously, the majority (>95%) of these circulating NK cells are CD56^{dim} cells that display significantly higher cytolytic capacity against tumour target cells. CD56^{dim} NK cells abundantly express CD16, the low-affinity Fc-receptor IIIA for the constant (Fc) region of immunoglobulin on opsonized cell surfaces. This receptor–ligand interaction is followed by a CD16-mediated activation signal that results in NK-cell degranulation and perforin-dependent target cell lysis called antibody-dependent cellular cytotoxicity (ADCC). Indeed, NK-mediated ADCC could be part of the effective anti-tumour activity seen when trastuzumab and rituximab are used in a mouse model of breast cancer and lymphoma, respectively (Clynes et al., 2000).

Compelling evidence for the importance of the CD56^{dim} CD16^{+} human NK-cell subset derives from observations noted in patients affected by CD20^{+} follicular lymphoma treated with the humanized anti-CD20 monoclonal antibody rituximab. Patients presenting with a particular CD16 gene polymorphism resulting in higher affinity for rituximab, fared better. In addition, NK cells extracted from the patient's blood had an enhanced ADCC activity in vitro (Cartron et al., 2002; Dall'Ozzo et al., 2004).

Blood CD56^{dim}CD16^{+} KIR^{+} NK cell subsets might also play a crucial role in limiting growth of non-solid tumours. NK cells are able to kill leukaemic blasts following haploidentical T cell–depleted stem cell transplantation in patients affected by acute myeloid leukaemia (AML). A highly statistically significant survival advantage exists for patients receiving T cell–depleted full haploidentical transplants where NK alloreactivity promotes a graft-versus-leukaemia effect. This could be explained by the presence of a mismatch between the donor NK cell KIR repertoire and MHC class I expression by recipient AML blasts (Ruggeri et al., 2006). Generally, in order to assess the phenotypical and functional characteristics displayed by circulating NK cells, separation of peripheral blood mononuclear cells (PBMCs) from whole blood is needed. Human lymphocytes can be isolated from PB by density centrifugation over a step gradient consisting of a mixture of the carbohydrate polymer Ficoll and the dense iodine-containing compound metrizamide.

Lymphoid tissues

Most human NK cells home to secondary lymphoid organs. They represent around 5% of mononuclear cells in uninflamed LNs and 0.4–1% in inflamed tonsils and LN (Fehniger et al., 2003; Ferlazzo et al., 2004b). These NK cells constitute a remarkable pool of innate effector cells because LN harbour 40% of all lymphocytes, whereas PB contains only 2% of all lymphocytes (Trepel, 1974; Westermann and Pabst, 1992). Therefore, LN NK cells are in healthy conditions 10 times more abundant than blood NK cells. Blood CD56^{bright} NK cell subsets specifically express CCR7 and CD62L, which might account for the abundance of CD56^{bright} NK subsets observed in all secondary lymphoid organs analysed so far (LN, tonsils and spleen) (Ferlazzo et al., 2004b). Similarly to CD56^{bright} NK cells within the PB, secondary lymphoid tissue NK cells are negative or express low levels of perforin and show extremely poor cytolytic activity. However, human NK cells in secondary lymphoid organs display specialized effector functions since they respond rapidly with cytokine (IFN-γ and TNF-α) secretion to activation by dendritic cell (DC). Interestingly, at these sites they produce around fivefold more IFN-γ than CD56^{bright} cells in the PB. This cytokine production is well suited to restrict pathogens that have breached mucosal barriers and to polarize adaptive immune responses. Perforin-mediated cytotoxicity can, however, be promptly up-regulated by IL-2 on secondary lymphoid organ NK cells, and at the same time, they acquire the expression of CD16, as well as KIRs (Ferlazzo et al., 2004b). Therefore, activation converts secondary lymphoid organ NK cells into cytotoxic effectors analogous to blood CD56^{dim}CD16^{+} NK cells. The ontogenic link between the two main NK cell subsets is still somewhat controversial. CD56^{bright} NK cells have been proposed either as precursors of CD56^{dim} cells or as derived from CD56^{dim} cells (Freud et al., 2005; Loza and Perussia, 2004; Mailliard et al., 2005). Indeed, because it is not known yet whether mice have NK cell subsets equivalent to CD56^{bright} and CD56^{dim} cells, information regarding human NK cell subset development has been limited, and at present, the developmental relationship between cytokine secreting CD56^{bright} and cytolytic CD56^{dim} NK cells is still not completely defined because the sites of terminal NK cell differentiation are unknown.

In order to ascertain putative difference in the lifespan of distinct NK cell subsets, and therefore on their developmental relationship, we have recently evaluated the specific telomere length of CD56^{dim} and CD56^{bright} NK cells by flow-FISH (a quantitative fluorescence in situ hybridization technique recently adapted to flow cytometry). We found that both PB CD56^{bright} and LN NK cells display longer telomeres than PB CD56^{dim} NK cells, thus suggesting the latter as terminal NK cells (Romagnani et al., 2007).

Four presumed NK cell intermediates were recently found in LN, suggesting secondary lymphoid organs as a possible site of NK cell development and maturation directly from $CD34^+$ hematopoietic precursors (Freud et al., 2005).

Taking into account the recently formulated hypothesis of LN as possible sites of NK cell ontogeny, we analysed for the first time the features of NK cells directly isolated from human efferent lymph and from either inflamed or uninflamed LN. We observed that NK cells isolated from human thoracic duct efferent lymph, but not NK cells resident in autologous non-inflamed LN, express CD16 and KIR. Conversely, the in vivo expression of KIR and CD16 in LN NK cells occurs during LN paracortical/follicular hyperplasia, characterized by extensive lymphocyte proliferation and therefore abundant cytokine production (Romagnani et al., 2007). These data imply a scenario in which KIR and CD16 are acquired during the maturational step from $CD56^{bright}$ into $CD56^{dim}$ NK cells. This theoretical progression could take place in LN during inflammation. Next, NK cells would leave the LN through the efferent lymphatics and recirculate in PB.

Uterus

NK cells also possess unique properties in the uterine mucosa prior to and during pregnancy. NK cells are present at varying numbers in non-pregnant endometrium during the menstrual cycle, but they have not been fully characterized. Immunohistochemistry analysis of human endometrium sections showed that the absolute numbers of endometrial NK (eNK) cells, detected as CD56 positive cells, increase dramatically from the proliferative to the late secretory phase of the menstrual cycle although the relative number of NK cells compared to other lymphocytes remains constant (Starkey et al., 1991). Only a few studies have characterized the phenotype of eNK cells. Flow cytometric and confocal microscopy studies demonstrated that they share a similar expression profile of CD56, CD57, CD94 and CD16 with PB $CD56^{bright}$ NK cells. eNK cells lack expression of L-selectin, but their expression of the KIR receptor KIR2DL2/3 is similar to those of $CD56^{dim}$ NK cells. Interestingly, eNK display high levels of CD9, a member of the tetraspanin family of protein not expressed on PB NK cells (Eriksson et al., 2004; Ho et al., 1996; Vassiliadou and Bulmer, 1998). Functionally, eNK cells produce IFN-γ and IL-10 following cytokine activation (IL-12 and IL-15), although cytokine secretion of freshly isolated, non-activated, eNK cells has not yet been analysed. Limited information exists regarding the cytotoxic activity of eNK cells. They express perforin, but their cytolytic activity is extremely low (Jones et al., 1997). Thus, eNK cells should be considered as unique NK cells, which are different from the $CD56^{dim}$ and $CD56^{bright}$ PB NK cells. For phenotypic and functional analysis, isolation of eNK cells from uterine tissues is frequently obtained by mincing the specimens and then processing the single pieces with an enzyme mixture composed mainly by pancreatin, hyaluronidase, collagenase and DNase. Digested tissues are also screened through a sieve to facilitate single cell dispersion, and the collected supernatants are resuspended in an ammonium chloride solution to lyse erythrocytes. In contrast to the relatively under-investigated eNK cells, decidual NK cells (dNK) have been studied in great detail both in humans and murine models. In humans, about 70% of the decidual lymphocytes are NK cells of the $CD56^{bright}CD16^-$ phenotype (Hanna and Mandelboim, 2007). Decidual NK cells resemble blood $CD56^{bright}$ NK cells in CD56 and CD16 expression, and the high expression level of CD94/NKG2. On the other hand, similar to eNK, dNK cells resemble $CD56^{dim}CD16^+$ NK cells in expression of KIRs and in their granular cell content (Hanna and Mandelboim, 2007; Trundley and Moffett, 2004). In addition, dNK cells exclusively express CD9 and CD151 and more highly express granzyme A, NKG2C and NKG2E when compared to blood NK cell subsets (Berditchevski and Odintsova, 2007). Decidual NK cells share some of the chemokine receptors expressed by blood $CD56^{bright}$ NK cells since dNK cells express high levels of CXCR3 and intermediate levels of CXCR4, but blood $CD56^{bright}$ NK cells express more CXCR4 and less CXCR3 than dNK cells (Hanna et al., 2003). Finally, dNK cells display the activation marker CD69 but lack expression of L-selectin (which distinguishes them from $CD56^{bright}$ NK cells) (Koopman et al., 2003). Isolation of dNK cells is very similar to isolation previously described for eNK cells. Basically, decidual tissues were trimmed into small pieces and enzymatically digested with a mixture of individual enzymes (DNAse and type IV collagenase), and then the supernatants were collected and loaded onto Ficoll density gradient to purify the lymphocyte population. On the basis of the tests subsequently performed, NK cells were usually purified using human NK cell isolation commercial kits.

The dNK cells are in close contact with foetal-derived trophoblasts at the foetal–maternal interface. However, dNK cells do not exert cytolytic actions against trophoblastic cells (King et al., 1989). Several studies have demonstrated that the general cytotoxicity of dNK cells is reduced when compared with PB NK cells. This phenomenon could be a result of inhibitory interactions between the non-classical class I MHC molecules (HLA-G and HLA-E) on the trophoblast cells and the inhibitory receptors expressed on dNK cells LIR-1 (Gonen-Gross et al., 2003) and CD94/NKG2A (Vacca et al., 2006). Moreover, it has been suggested that the cytotoxic activity of dNK cells is inhibited by an inhibitory form of the 2B4 receptor (Vacca et al., 2006). Others proposed that the lack of cytotoxicity could derive from their inability to form mature activating synapse (Kopcow et al.,

2005). Overall, these data suggest that the interactions of dNK cells with neighbouring immune and non-immune cells at the decidua further inhibit their ability to damage the local tissues by killing foetal-derived neighbouring cells. Besides inhibiting killing activity, the decidual microenvironment may encourage dNK cells to exert a constructive function. This ability of NK cells has been demonstrated in a mouse model, where depletion of dNK cells exhibited abnormal implantation sites and inadequate remodelling of the decidual spiral arteries. Furthermore, dNK-derived IFN-γ is important in dNK activity at implantation sites since it positively regulates the diameter of the lumen of the spiral arteries during decidualization (Ashkar and Croy, 2001).

In addition to producing IFN-γ, dNK cells can secrete other cytokines and chemokines such as GM-CSF, TNF-α, LIF, CSF-1, IL-8 and IP-10. Human dNK cells can also produce angiogenic factors, including members of the vascular endothelial growth factor (VEGF) family, placental growth factor (PLGF), angiopoietin-2 (Ang-2) and NKG5 (Hanna et al., 2006).

Conversely, the mechanisms controlling the accumulation of $CD56^{bright}$ $CD16^-$ NK cells in the decidua are still largely unknown. Several studies suggest that dNK cells are recruited from other organs or from the blood, although self-renewal from local progenitor cells cannot be ruled out. dNK cells might originate from $CD56^{dim}CD16^+$ PB NK cells that migrate to the decidua and differentiate locally to phenotypic dNK (Keskin et al., 2007). Alternatively, dNK could originate rather than from the blood $CD56^{bright}$ $CD16^-$ subset (Hanna et al., 2003). On the other hand, murine studies indicate that dNK cells do not self renew in the uterus but rather are derived from secondary lymphoid tissues (Chantakru et al., 2002). In addition, the presence of haematopoietic stem cells in the human endometrium, recently demonstrated, do not rule out the possibility that dNK cells could also derive from local stem cell differentiation (Lynch et al., 2007).

Epithelia and lymphoid tissue-associated mucosae

Epithelia cover external and internal body surfaces and are the first line of defence against environmental exposure to toxins and pathogens. In addition to acting as a physical barrier and providing a means for selective transport with the environment, epithelial tissues ensure protection by housing a local immune system characterized by tissue-specific organization and complexity. The epithelial immune system maintains a dynamic equilibrium between immune surveillance of pathogens and other environmental insults and tolerance to harmless antigens, such as food, airborne proteins and commensal bacteria. Recently, investigation of human epithelial has been undertaken in order to understand whether NK cells are part of this complex branch of the immune system. It has been shown that normal human skin harbours NK cells, and they represent 10% of the total leucocytes found within the tissue. The presence of NK cell was assessed by cytofluorimetric analysis of skin-derived cell suspensions. In addition, NK cells contained in human normal dermis are mainly $CD56^+CD16^-$ that lack expression of perforin and NKG2D. Interestingly, a subpopulation of cutaneous NK cells co-expresses the receptors for homing to non-inflamed skin, CCR8 and CLA, but uniformly $CCR8^+$ and $CCR8^-$ subpopulations lacked the LN-homing receptor CCR7. Accordingly with CCR8 expression, they also migrate toward CCL1, suggesting that this chemokine could be important for steady-state trafficking of NK cells into the skin (Ebert et al., 2006).

The role of NK cells in epithelial immunity was also confirmed by recent studies finding that NK cell are found in mucosa-associated lymphoid tissues (MALT) such as tonsils, as well as other lymphoid structures associated with the gut lamina propria such as Peyer's patches, cryptopatches and lymphoid follicles. In addition, NK cells are also found within the dermis. Surprisingly, these epithelial tissues harbour NK cell subsets that are functionally distinct from their PB counterparts. For instance, it has recently been described that some of the NK cells present in mucosa associated with the tonsil specifically express $NKp44^+$ and, differently from tonsil $NKp44^-$ NK cells, primarily secrete IL-22 upon activation. Assessment of the presence of $NKp44^+$ cells in MALT derives primarily from immunohistochemical analysis of human tonsil but also human Peyer's patches and appendix sections. In these investigations, an anti-NK p44 antibody was used to detect NK cells, and other markers, such as cytokeratin-5, were used to confirm the distribution of NKp44+ cells either in the lamina propria or surface epithelium. Subsequently, in order to evaluate functional and phenotypical characteristics of this recently found NK cell subsets, NK cells were extracted from MALT, mainly by mechanical disruption and enzymatic digestion and then magnetically enriched with CD56 microbeads. Interestingly, isolated $CD3^-$ $CD56^+$ $CD19^-$ NK cells were then FACS-sorted into $NKp44^+$ and $NKp44^-$ populations, and the two subsets were obtained and compared by gene expression profiling. This analysis was very informative and suggested differences of gene expression between the two populations. For instance, gene chip analysis showed the difference in mRNA level of IL-22, which was actually confirmed by functional tests. Transcriptional tests and the following functional analysis demonstrated that $NKp44^+$ NK cells express a functional CCR6 receptor and produce more CCL20 than $NKp44^-$ NK cells. Differently from their $NKp44^-$ counterpart, $NKp44^+$ are also negative for perforin, adhere better to epithelial cells and secrete lower amounts of IFN-γ. The presence of NKp44$^+$ cells specifically resident within the tonsil epithelium and the lamina propria, might be functionally relevant, providing cytokines and chemokines that may help constrain inflammation and protect mucosal sites (Cella et al., 2009). At the same time, the tonsil NKp44$^-$ NK cell subset is able to produce IFN-γ and probably correspond to tonsil NK cells recently shown to inhibit EBV-induced B cell transformation through IFN-γ secretion (Strowig et al., 2008).

Additionally, other authors described NKp46$^+$ CD3$^-$ cells in epithelial tissues such as gut and skin in both human and mice. Distribution of NK cells across human colon and skin was investigated by immunofluorescence in frozen sections stained with anti-CD3, anti-NKp46 and TO-PRO-3, a membrane-impermeable DNA stain. Similarly, CD3$^-$NKp46$^+$ mouse NK cells found in the small intestine, epidermal and dermal sheaths were detected using antibodies against NKp46, CD3 and EpCAM. These analysis revealed the presence of populations of NKp46$^+$ cells localized specifically within the dermis, gut lamina propria and cryptopatches (Luci et al., 2009). For the isolation of gut mouse NK cell subsets, soon after the small intestine was collected, Peyer's patches were excised for subsequent mechanical disruption. Intestines were cut in small pieces, and cells in the intestinal epithelium were isolated by incubation with shaking in PBS containing EDTA, HEPES and FCS. After collecting cells from the intestinal epithelium, intestinal pieces were minced, and the remaining gut cells, including lamina propria and cryptopatch cells, were subsequently obtained by incubations with shaking in RPMI containing HEPES and collagenase. Similarly, extraction of NK cells associated with skin tissues was accomplished by collecting the dorsal skin of the mouse and then digesting the tissues in RPMI containing HEPES and collagenase. Gut and skin sample were finally enriched for gut and skin cells by Percoll gradient centrifugation (Luci et al., 2009; Sanos et al., 2009). Isolation of cells from murine skin and colon and the subsequent cytofluorimetric analysis showed a significant difference between dermal and gut NKp46$^+$CD3$^-$ cells. Skin-derived NK cells represent a subset of NK cells with a conventional cell surface phenotype but characterized by lower effector function and higher proliferation rate. In contrast, gut NK cells do not have the effectors functions of classical mature NK cells but also do not display the high proliferation rate observed in skin NK cells (Luci et al., 2009). In addition, a remarkable part of NKp46$^+$CD3$^-$ cells found in cryptopatches and lamina propria exclusively express the transcriptional factor RORγt (Luci et al., 2009; Sanos et al., 2009). Further characterization of this gut RORγt$^+$NKp46$^+$CD3$^-$ cell subset showed unique features. Interestingly, gut RORgt$^+$NKp46$^+$CD3$^-$ cells, similarly to human tonsil NKp44$^+$ cells, are able to produce IL-22 following activation and depend on this transcriptional factor and commensal microflora-derived signals for their development. Nevertheless, their origin still remains undefined.

In conclusion, these recent data point out a possible role of NK cells also in mucosal immunity. In this context, NK cells may provide an innate source of IL-22, an important factor able to regulate epithelial homeostasis, protection and repair (Cella et al., 2009; Luci et al., 2009; Sanos et al., 2009).

Other organs

Limited information is still available on NK cells distribution throughout other human tissues. Our group described NK cells infiltrating non-small cell lung cancer (NSCLC) comparing them with autologous NK cells derived from blood and normal lung tissues surrounding tumour. This study, in addition to showing the presence of NK cells within the tumour mass, demonstrated that normal lung tissues harbour NK cells. They represented around 8% of the mononuclear cells extracted from the tissue and presented phenotypic and functional characteristic comparable to autologous PB NK cells. However, our study did not focus on the chemokine receptors repertoire expressed by lung derived-NK cells, and therefore no hypothesis regarding their trafficking could be formulated (Carrega et al., 2008). Among lymphoid cells found within peripheral tissues, several studies have reported that NK cells are a predominant lymphocyte population in the livers of humans, accounting for up to 50% of the total lymphocyte pool in normal liver compared to <20% of PB lymphocytes (Hata et al., 1991; Norris et al., 1998). Hepatic NK cells differ from PB NK cells since the majority of them do not express CD16, suggesting that ADCC is not a major function of the liver (Hata et al., 1991; Norris et al., 1998). However, most NK cells express CD161 and inhibitory receptors such as KIRs and CD94/NKG2A (Doherty and O'Farrelly, 2000). Functionally, fresh hepatic NK cells lysed K562 target cell lines and, after appropriate stimulation, they secreted INF-γ and/or TNF-α. Usually, extraction of liver-associated lymphocytes was obtained by dissecting liver tissues in small pieces and incubating in HBSS containing collagenase and DNase. The presence of NK cells in peripheral compartments was also detected in other species, in particular rat and mouse models. An early study in rat showed that the frequency of large granular lymphocytes, including NK and T-cell subsets, was high in the lung (Reynolds et al., 1981). Subsequent studies in the mouse have shown the presence of NK cells, defined as NK1.1$^+$ CD3$^-$ or DX5$^+$ CD3$^-$ by flow cytometry, in various organs (Colucci et al., 2003). More recent work assessed the percentage and the number of mouse NK cells in individual mouse organs and found a higher frequency of NK1.1$^+$CD3$^-$ NK cells in lung and liver since they represent 9% and 6% of lymphocytes, respectively. In addition, absolute number of NK cells was found higher in the lung (10^6) than in the liver (2.5×10^5) (Masopust et al., 2001).

NK cells in disease

NK cells in inflamed tissues

NK cells are not confined to the destruction of virus-infected and tumour cells but also play a role in human inflammatory diseases. The emerging view is that NK cells act as regulatory cells during inflammation and influence subsequent adaptive immune responses interacting with other cell types such as DCs, T cells, B cells and endothelial cells. Various studies focused on the crosstalk between NK cells and MDDCs (Carbone et al., 1999; Cooper et al., 2004; Ferlazzo et al., 2002; Fernandez et al., 1999; Gerosa et al., 2002; Moretta, 2002; Piccioli et al., 2002; Walzer et al., 2005; Wilson et al., 1999; Zitvogel, 2002) and more recently on the involvement of plasmacytoid DCs (pDCs), mast cells, basophils, eosinophils and neutrophils (Gerosa et al., 2005; Marcenaro et al., 2006; Romagnani et al., 2005). As already described, the CD56^{dim} and CD56^{bright} NK subsets have distinct tissue distribution. In particular, CD56^{dim} CD16$^+$ NK cells that predominate in PB and inflamed tissues display lower cytokine production but more potent cytotoxicity when compared with CD56^{bright} CD16$^-$ that largely predominate in LNs and release high levels of cytokines but have little cytolytic activity.

During inflammation and tumour growth, NK cells migrate from blood into injured tissues (Cooper et al., 2004; Moretta, 2002; Walzer et al., 2005; Zitvogel, 2002). This is driven by distinct signals, which include adhesion molecules and chemotactic factors such as the newly identified chemerin (Parolini et al., 2007). Since the two subset of NK cells differ in terms of chemokine receptor and adhesion molecule expression, it is likely that they should display distinct homing properties (Campbell et al., 2001).

CD56^{dim} CD16$^+$ NK cells express both β1 and β2 integrins and PSGL-1/PEN5 (Andre et al., 2000), as well as other adhesion molecules, such as CD2 and LFA-1 (Frey et al., 1998; Vujanovic et al., 1993). Indeed, this subset expresses many different chemokine receptors, which confers the capacity to migrate toward inflamed sites, such as CXCR1 and CX3CR1. Conversely, CD56^{bright} NK cells express L-selectin, a primary molecule required for the interaction with LN high endothelial venules, and to a much lesser extent integrins, but no PSGL-1/PEN5 (Andre et al., 2000; Cooper et al., 2001a; Frey et al., 1998). This subset expresses the chemokine receptor CCR7 that promotes trafficking towards the LN (Campbell et al., 2001; Cooper et al., 2001a; Vitale et al., 2004). Chemokine receptors expressed by tissue NK cell subsets are summarized in Table 31.1.

A crucial role in the transendothelial migration process of NK cells is also played by the DNAM-1 receptor expressed on human NK cells. The DNAM-1/PVR interaction seem to play an essential role during the diapedesis step since, in the presence of anti-DNAM-1 or anti-PVR mAb, migrating leucocytes arrest at the surface of the endothelium over intercellular junctions (Reymond et al., 2004). Various soluble factors also play an important role that promote extravasation of NK cells and the subsequent induction of their priming. These include cytokines and chemokines released by DCs and other cell types such as endothelial cells, macrophages, neutrophils, fibroblasts, mast cells and eosinophils during pathogen-induced inflammation in peripheral tissues. The recruitment of NK cells to inflamed sites involve chemokines such as CXCL8, CCL3 and CX3CL1 (Moretta, 2002).

On the basis of their surface phenotype, it is reasonable to presume that the CD56^{dim} CD16$^+$ NK cell subset may be mainly recruited into inflamed tissues, whereas the CD56^{bright} CD16$^-$ subset may be attracted into secondary lymphoid compartments such as LNs. However, several reports suggest that CD56^{bright} NK subset

Table 31.1 Chemokine receptors expressed by NK cells

	CCR1	CCR2	CCR3	CCR4	CCR5	CCR6	CCR7	CCR8	CCR9	CXCR1	CXCR3	CXCR4	CXCR5	CX3CR1
CD56^{dim} NK cells	–	–	–	+	–	–	–	–	–	+++	++	+++	–	+++
CD56^{bright} NK cells	–	–	–	–	++	+/–	++	–	–	+	+++	+++	–	+/–
Activated NK cells	–	+/–	–	++	–	+	+/–	–	ND	–	+++	+	–	+

ND indicates not determined.

also predominate in inflamed tissues. The frequency and phenotype of NK cells at sites of inflammation, including exudative pleural fluid from patients with infective, reactive, or malignant pulmonary diseases; synovial fluid and tissue from patients with inflammatory arthritis; and peritoneal fluid from patients with bacterial peritonitis (Dalbeth et al., 2004) have been examined. The frequency of NK cells at inflamed sites was similar to that found within PB. The phenotypes of NK cells at the two sites, however, were different. CD56^{bright} NK cells were indeed enriched in the pleural fluid in both malignant and non-malignant pleuritis, in synovial fluid or synovial tissue. Very few of these cells expressed CD16 or the KIRs, whereas the majority expressed very high levels of CD94 and NKG2A, express CCR7 and CD62L but low levels of perforin. A subset expressed CD69, suggesting activation. The phenotype of NK cells in peritoneal fluid from patients with acute peritonitis turned out to be characterized by a high expression of CD16 and a significant expression of CD69. Moreover, it has been shown that in the presence of supplementary monokines (IL12, IL15 and IL18), addition of monocytes to NK cells obtained from inflamed sites consistently led to an increase in IFN-γ production by NK cells when compared to PB NK cells. Therefore, monocytes were able to synergize with monokines to promote IFN-γ production by NK cells. Conversely, activated NK cells promote TNF-α production by monocyte in a manner that is dependant on cell to cell contact. Monocytes localize in inflamed sites, and the hypothesis that NK cells are able to interact in a reciprocal fashion with the CD14^{+} mononuclear cell population may be particularly relevant in chronic inflammatory disorders, in rheumatoid arthritis in particular, that is in those conditions in which production of TNF-α by cells of the monocyte lineage plays a pivotal role in promoting the development or maintenance of inflammation (Feldmann et al., 1996; Firestein et al., 1990). T cells have been the focus of additional investigation regarding this interaction because they play a role in activating macrophages (Li et al., 1995; Sebbag et al., 1997; Vey et al., 1992).

Mononuclear cells obtained from the PB and inflamed fluid/tissue were stained with a panel of mAb specific for selected cell surface markers (anti-CD16, anti-CD94, anti-CD69, anti-CD62L, anti-CCR7, anti-NKG2A, anti-KIR3DL1, anti-KIR3DL2, anti-CD56, anti-CD3) as well as for intracellular molecules and analysed by flow cytometry. Isolation of pure populations of NK cells and monocytes has been obtained by MACS using a one-step positive selection with anti-CD56-coated magnetic microbeads and one-step positive selection with anti-CD14-coated magnetic microbeads before co-culturing NK cells and monocytes.

In another study, NK cells were investigated in punch biopsy specimens of patients with untreated psoriatic plaques (Ottaviani et al., 2006). NK cells were identified through immunohistochemical analysis of 5-μm frozen sections obtained from these patients, and counterstained with hematoxylin. NK cells were identified as CD56^{+}CD3^{-} cells and were observed mostly localized in the mid- and papillary dermis. CD56^{+} psoriatic cells were also purified by magnetic beads and found CD161^{+} and uniformly expressing NKG2A. In addition, the cells displayed the activation marker CD69 but failed to express CD158b or CD16. Indeed skin NK cells cultured 16h in the presence of 60U/ml IL-2 release abundant IFN-γ as detected by both cytofluorimetric analysis and ELISA. In addition, when primary cultures of keratinocytes from psoriatic patients were exposed to the supernatants from IL-2 stimulated NK cells, up-regulation of MHC class I (HLA-A, B and C) and the appearance of ICAM-1 and HLA-DR molecules were observed. CXCL10, CCL5 and, to a lesser extent, CCL20 were also released in these culture conditions, as assessed by ELISA tests. Chemokine receptor expression on psoriatic purified NK cells was evaluated by flow cytometry. These cells express high levels of CXCR3 and CCR5, followed by CXCR1, CCR6 and CCR8 and, at a lower level, by CCR1, CCR2, CCR4, CCR7 and CX3CR1. Moreover, the migratory properties of NK cells towards selected chemokines were evaluated in a transwell system resulting responsive to CXCL10 and CCL5. Taken as a whole, these data indicate that CD56^{bright}CD16^{-}CD158^{-} NK cells, through a mechanism involving primarily the CXCL10/CXCR3 and the CCL5/CCR5 axes may contribute to the amplification of the inflammatory process and that the crosstalk between IFN-γ-releasing lymphocytes and keratinocytes is critical for psoriasis development and persistence.

Recently, the Chemerin receptor has also been described on blood NK cells. The ChemR23 receptor is expressed by the majority of the CD56^{dim} CD16^{+} PB NK cell subset but not CD56^{bright} CD16^{-} NK cells. Chemerin is a potent inducer of NK cell recruitment into inflamed peripheral tissues (Parolini et al., 2007). This receptor was originally identified on the surface of mononuclear phagocytes and DCs (Vermi et al., 2005). Despite substantial evidence for NK cell–DC crosstalk, little is known about the crosstalk in peripheral tissues. However, in pathological conditions known to be characterized by enhanced tissue accumulation of DC, such as oral lichen planus, the presence of both DCs and NK cell subsets has been reported. Interestingly, in these tissues, only a fraction of ChemR23^{+} NK cells expressed CD69, an early activation marker, suggesting that ChemR23 was expressed by recently recruited NK cells. Alternatively, it might also indicate that only a fraction of the recruited NK cells undergoes activation at inflammatory sites. Nevertheless, under normal conditions, chemerin is absent in non-lymphoid peripheral tissues, including dermis and epidermis, and thus these findings support the role of chemerin in the recruitment and co-localization of NK, cells and DC subset in inflamed tissues.

Immature DCs, similarly to NK cells, express receptors for inflammatory chemokines, and indeed DCs dramatically accumulate within inflamed tissues (Lambrecht and Hammad, 2003), indicating that both NK cells and DCs have the potential to attract each other through their production of various chemokines. Indeed, following recruitment into inflamed tissues, NK cells can interact with other cell types of the innate immune system through both 'cell to cell contact' and soluble factors with subsequent modulation of NK cell function. For instance, during NK cell–DC crosstalk, DCs undergoing maturation after antigen uptake release cytokines that can influence the NK cells, such as IL12, which have a crucial role for inducing the release of IFN-γ by NK cells.

NK cell receptors, such as KIR and CD16, have also been evaluated in human NK cells isolated from efferent lymph and from either inflamed or uninflamed LNs (Romagnani et al., 2007). CD16 and KIR are expressed in NK cells derived from human thoracic duct efferent lymph and in inflamed LN, which are characterized by paracortical/follicular hyperplasia, while the same molecules were almost absent in NK cells collected from non-reactive autologous LNs. It has been suggested that KIR and CD16 acquisition can represent a key maturation step from CD56^{bright} into CD56^{dim} phenotype and that this differentiation can take place in LN during inflammation. Indeed NK cells would leave the LN after KIR acquisition and recirculate in the PB. Additional studies are required to resolve the uncertainties concerning the developmental relationship between CD56^{dim} and CD56^{bright} NK cells.

Given the relevance of the crosstalk among cells of the innate immunity during the inflammatory process, the interactions between NK cells and DCs have been investigated also in secondary lymphoid organs. In humans, NK cells and DC co-localize in T-cell areas of LN, and some studies have compared the ability of CD56^{dim} and CD56^{bright} NK cells to interact with DCs (Ferlazzo et al., 2004a; Vitale et al., 2004), demonstrating a preferential crosstalk between CD56^{bright} NK cells and DCs. Co-localization of DCs and NK cells has also been demonstrated in other inflamed tissues. An interesting study reported the interaction of NK cells and CD1a^{+} DCs in *Malassezia*-induced atopic skin lesions (Buentke et al., 2002). *Malassezia furfur* is a yeast present in the normal microflora of human skin, which can act as an allergen that incites specific IgE reactivity and T cell proliferation in atopic dermatitis patients. In this study, skin specimens were taken under local anaesthesia from healthy individuals and from patients with atopic eczema/dermatitis syndrome who had serum IgE antibodies specific for *Malassezia*, either from non-lesional skin or from *Malassezia* extract ATP-positive skin (at 24 h and 72 h after provocation). Acetone-fixed, 6 nm thick, cryostat sections were immunohistochemically stained with an anti-CD56 monoclonal antibody followed by counterstaining with Mayer's hematoxylin. Double immunofluorescence staining on the sections was performed with monoclonal antibodies against CD56, CD3 and CD1a and analysed using confocal laser scanning microscopy. CD56^{+} CD3^{-} cells were identified in the dermis close to the epidermis in healthy individuals and in non-lesional skin from patients with atopic eczema/dermatitis syndrome. In lesional and *Malassezia* ATP-positive skin at 72 h from patients with atopic eczema/dermatitis syndrome, CD56^{+} CD3^{-} cells are found in the epidermis and dermal cell infiltrates close to CD1a^{+} DCs, suggesting a role for NK cell–DC crosstalk during the inflammatory process. Moreover, immature monocyte-derived dendritic cells (MDDCs) co-cultured with *Malassezia* for approximately 46 h became less susceptible to NK cell-induced cell death, suggesting a direct effect imposed by the yeast upon interaction of DCs with NK cells. This indicates that maturation of DCs by microorganisms or microorganism products influences the surface phenotype of DCs, which indeed express higher levels of CD80, CD83 and CD86. In addition to cell to cell contact, soluble factors may also play a role in the NK cell–DC interaction. Supernatant from co-cultures of MDDCs and *Malassezia* rendered immature MDDCs less susceptible to cell death induced by NK cells, possibly through soluble factors with an inhibitory effect on NK cell activity. The cytokines reported to be secreted by DC upon interaction with *Malassezia* were mainly IL-1β, TNFα and IL18.

These studies revealed that NK cells migrate towards inflamed sites, but additional investigations are required to identify in detail chemo-attractants and expression of the inflammatory chemokine receptors on NK cells, which would facilitate their distribution into inflammatory sites. Also, the crosstalk between NK cells and other cells type of the innate immune system require further investigation. The conversion of CD56^{bright} into CD56^{dim} NK cells during inflammation still remains an open issue calling for additional research.

NK cells in solid cancer

NK cells have been implicated in the immune defence against tumours, as reported in mouse models of spontaneous and induced tumours. In humans, the paucity of infiltrating NK cells has hampered the characterization of NK cell biological function in vivo in general and in anti-tumour immunosurveillance in particular. There is evidence that the NK cell recognition of human target cells is a process guided by the balance of activating and inhibitory signals, involved in the interaction of specific ligands on target cells, which either induce or inhibit the cytotoxic response (Moretta et al., 2002).

All of these receptors are likely to recognize molecule that are up-regulated upon cellular stress (Moretta et al., 2002, 2003), and NK cells lyse target cells that have lost or express low levels of MHC class I molecules, an event frequently occurring in tumours as well as in cells infected by herpesviruses or adenoviruses. Previous studies, through histological examination, have shown that NK cells are not found in large numbers in advanced human neoplasms, with the exception of renal carcinoma (Coca et al., 1997; Ishigami et al., 2000; Schleypen et al., 2003, 2006; Villegas et al., 2002). Although few in number, these cells contribute to the elimination of tumour cells (Albertsson et al., 2003; Esendagli et al., 2008). Along this line, in an 11-year follow-up study, a low NK cell cytotoxicity in PB is correlated with increased cancer risk (Imai et al., 2000). NK cell infiltration has been identified in different tumours and, in some cases, such as lung cancer, gastric carcinoma and colorectal carcinoma, appears to represent a positive prognostic marker. Villegas et al. have investigated the TINK and the prognostic significance in patients with primary squamous cell lung carcinoma (Villegas et al., 2002). Immunohistochemical studies of surgery specimens were performed using the CD57 monoclonal antibody to evaluate NK cell infiltration. The number of NK cells was counted with an image analyser, and the reference value used was the median (five TINK cells/field) of all tumours analysed. After a minimum follow-up of 2 years, the Kaplan–Meier method was used to obtain survival curves showing that patients with more than five TINKs per field presented a significant better survival than patients with less than five TINKs per field. Indeed according to TNM classification, in patients screened as stage IB, the differences in survival were significantly higher. Ishigami and collaborators have identified different levels of NK cells infiltration in resected specimens from patients with gastric carcinoma who underwent gastrectomy (Ishigami et al., 2000). The patients were divided into two groups: those with high levels of NK cell infiltration and those with low levels. NK cell infiltration was analysed by immunochemical staining using the CD57 monoclonal antibody. Positive cells were identified in the intratumoural stroma and distributed near the surface of the tumour. Indeed, this infiltration was often found in organized lymphoid follicles within the stroma. The postoperative 5-year survival rate was 78% for patients with a high rate of NK infiltration compared to the other group of patients that had a 65% rate. Intratumoural infiltration of NK cells in patients with resected colorectal carcinoma has been evaluated (Coca et al., 1997) using immunohistochemical stains with the monoclonal antibody CD57. The number of NK cells was counted using an image analyser with the MIP-interactive method. The NK cell infiltration in tissue sections, which was found in contact with tumoural cells, was classified as few (<50 NK cells), moderate (50–150 NK cells) and extensive (>150 NK cells) and was correlated with survival rate at 5 years. Significantly shorter survival rates were shown in patients with little and moderate NK infiltration when compared to those with extensive infiltration. Indeed, in the basis of TNM stage, especially the patients with TNM stage III disease with extensive NK infiltration showed longer survival rates than those with little or moderate infiltration. Also, in the same study, no correlation was found between the NK cell number and the degree of lymphocytic infiltration. In another study, NK cell infiltration was analysed in patients affected by colorectal cancers (Sandel et al., 2005). The correlation between colorectal tumour MHC class I aberrations and infiltration of NK cells was investigated through immunohistochemical staining by using monoclonal antibody anti-CD56, anti-Granzyme-B, anti-CD4, anti-CD8, anti-CD3 and monoclonal antibody anti-HLA class I. In contrast with previous studies, which used the antibody against CD57, in this study, monoclonal antibodies against CD56 and CD3 were employed. Indeed, CD57 is not an exclusive NK cell marker since it is also expressed by a subset of T-lymphocyte as well as by neuroendocrine cells. Although CD56 is also expressed on subsets of T-lymphocyte, the combination with CD3 staining allowed the discrimination of these two lymphocyte subpopulations. The evaluation of tumour sections was performed with a confocal laser-scanning microscope showing both CTL and NK cell infiltrate in colorectal tumours. Locus-specific down-regulation of MHC was detected in 72% of all patients. Tumoural HLA expression was absent in 7% of patients, while 28% had no HLA loss. Specific MHC aberrations significantly correlated with the intra-epithelial infiltration of $CD8^+$ cells but not with CD4 or $CD56^+$ lymphocytes. These results showed that colorectal tumours are sparsely infiltrated by $CD56^+$ cells compared to $CD8^+$ T cells and that loss of HLA is associated with T-cell rather than NK cell infiltration. More recently, Carrega et al. have investigated the surface phenotype and the functions of tumour-infiltrating NK cells in patients with NSCLCs (Carrega et al., 2008). The NSCLC specimen sections was analysed by immunohistochemistry to investigate the localization of infiltrating NK cells, using mAbs directed against NKp46 and NKp30, whose expression is restricted to NK cells, and the expression of HLA class I molecule was evaluated in tumour sections. A peculiar distribution of NK cells was observed since it was restricted to intratumoural fibrous septa and to the interface between stromal cells and surrounding tumour cells, but not in direct contact with tumour cells. The surface phenotype of NK cells freshly isolated from NSCLC tissues was performed using mAbs and flow cytometry (Figure 31.1). $CD56^{bright}CD16^-$ NK cell subset was highly enriched in tumour infiltrate, and

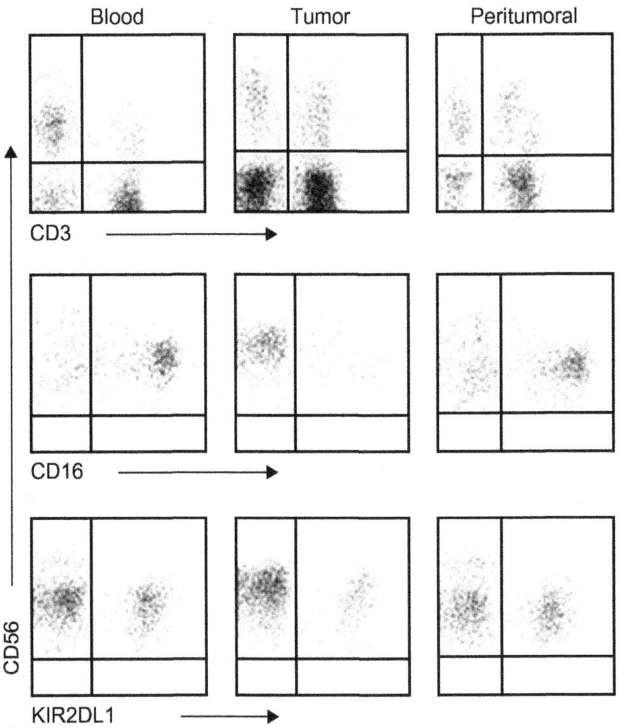

Figure 31.1 • NK cells isolated from NSCLC are primarily CD56^{bright} NK cells and display a distinctive surface molecular pattern. A representative phenotype of lymphocytes isolated from blood, tumour and peritumoural tissue is shown. After gating on CD45^{+} cells (i.e. leucocytes), NK cells were detected using anti-CD3 and anti-CD56 mAbs (upper row). Autologous mononuclear cells isolated from PB, NSCLC and peritumoural lung tissue were also analysed by three-color flow cytometry using mAbs specific for the activating receptor CD16 (middle row) and the inhibitory receptor KIR2DL1 (lower row) after gating on CD3^{-}CD56^{+} NK cells. Quadrants depicted are set on isotypic controls. CD56^{bright}CD16^{-} NK cell subset is highly enriched in tumour infiltrate. Surprisingly, tumour infiltrating CD56^{bright}CD16^{-} NK cell subset express KIRs, inhibitory receptors generally expressed on considerable fraction of CD56^{dim}CD16^{+} NK cells but absent in most CD56^{bright}CD16^{-} subsets. Since KIRs can be expressed on KIR negative NK cells following their activation, a possible explanation for this finding is that CD56^{bright}CD16^{-}KIR^{+} NSCLC-NK cells would derive from CD56^{bright}CD16^{-} KIR^{-} PB-NK cells upon migration into tumours, where they can become activated most likely because of a proinflammatory cytokine microenvironment due to local immune reactions.

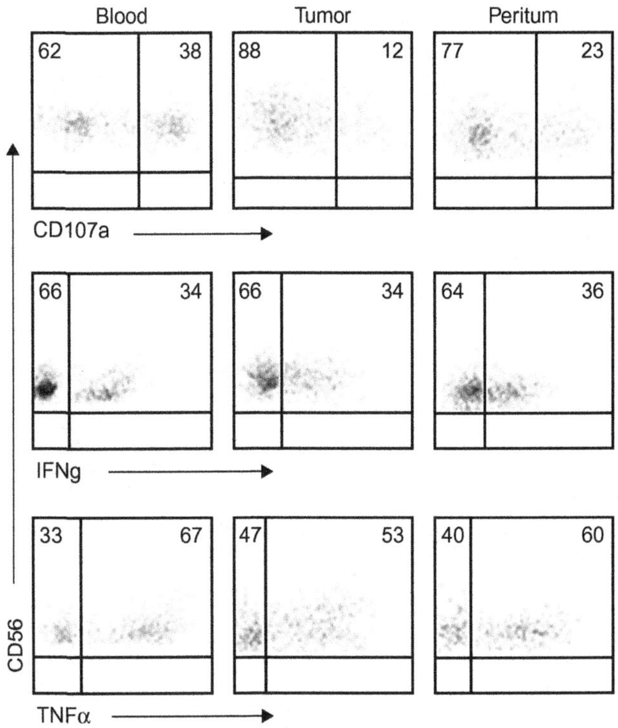

Figure 31.2 • NSCLC-infiltrating NK cells display impaired cytolytic capabilities but can effectively secrete cytokines. Mononuclear cells isolated from PB, non-small cell lung cancer and peritumoural tissue were analysed for their cytolytic and cytokine secreting properties. Cells were co-cultured with the NK-sensitive K562 cell line, and, after 4 h, CD107a expression was assessed, gating on CD3^{-}CD56^{+} NK cells (upper row). In order to assess cytokine producing capability, cells were stimulated for 6 h with PMA/ionomycin (Stimulated) and then analysed for IFN-γ (middle row) and TNF-α (lower row) production. Quadrants depicted are set on isotypic controls. Numbers within quadrants represent percentages of cells. These data show that the cytolytic potential of NK cells isolated from lung tumours was lower than that of NK cells from PB or normal lung tissue, although they maintain their capability to produce cytokines. It remains to be demonstrated whether a similar functional pattern applies to NK cells infiltrating other human cancers.

NK cells displayed activation markers, including NKp44, CD69 and HLA-DR. Surprisingly, the tumour-infiltrating CD56^{bright} CD16^{-} NK cell subset was found to express KIRs, inhibitory receptors generally expressed on a considerable fraction of CD56^{dim}CD16^{+} NK cells but absent in most CD56^{bright}CD16^{-} subsets. The levels of perforin were also analysed in CD56^{bright} CD16^{-} KIR^{+} NSCLC-NK cells which expressed low levels of perforin comparable to CD56^{bright} CD16^{-} KIR^{-} PB-NK cells, thus suggesting that CD56^{bright} CD16^{-}KIR^{+} NSCLC-NK cells would derive from CD56^{bright} CD16^{-} KIR^{-} PB-NK cells upon migration into tumours. Since cell cytotoxic function has recently been assessed by flow cytometry on the basis of the cell-surface mobilization of CD107a (LAMP-1) (Anfossi et al., 2006; Bryceson et al., 2005; Rubio et al., 2003; Wolint et al., 2004), the CD107a expression was also analysed in this study. The cytolytic potential of NK cells isolated from tumours was lower than that of NK cells from PB or normal lung tissue, measured as CD107a surface expression following incubation with K562 target cells. Nevertheless, no difference was observed regarding their capability to produce cytokines (Figure 31.2). Schleypen et al. have also characterized phenotypically and functionally the tumour-infiltrating NK cells (NK-TIL) of renal cell carcinoma (Schleypen et al., 2003, 2006). TIL were isolated from tissue following surgical resection and analysed for NK cell frequency and phenotype using monoclonal antibodies (anti-CD3, anti-CD56, anti-CD16, anti-perforin, anti-granzyme A and B, anti-NKp46, anti-NKG2D) and flow cytometric analysis. Two groups of tumours were identified based on the frequency of tumour-infiltrating NK cells: low NK-TIL (<20% NK cells) and high NK-TIL (tissue with >20% NK cells). NK cells of the two groups differed in the frequency of CD16 expression cells, in their cytotoxic ability and in the expression of intracellular cytotoxic effector molecules. Low NK-TIL cell content in renal carcinoma correlated with low frequency of CD16 expression, and these NK cells were non-cytolytic against K562 target cells after short-term activation by low dose IL-2. Indeed these cells were unresponsive to signals via the activating receptor NKp46 and have lower expression of perforin, similar to peripheral $CD56^{bright}$. In contrast, high NK-TIL cell content in tumours correlated with high frequency of CD16 expression, and these cells acquired cytolytic activity, associated with higher expression levels of cytotoxins (perforin, granzyme A and B) resembling PB $CD56^{dim}$ NK cells. Immunohistochemical analyses of tumour sections, stained with a mAb directed against CD56 and NKp46, showed that NK cells were distributed within the tumour mass and were not restricted to the tumour stroma. Thus NK cells can infiltrate at least some cancers and new techniques, especially performed in these two latter studies, indicate that these cells display a peculiar phenotype. In addition, while they might display limited ability in tumour cell killing, they are fully competent in relevant cytokine release, highlighting a prominent regulatory role for tumour-infiltrating NK cells. Another interesting finding reported in one of the studies is represented by the presence of KIR on NSCLC-$CD56^{bright}CD16^{-}$ NK cells since PB as well as the secondary lymphoid organ $CD56^{bright}CD16^{-}$ NK cell subset uniformly lacks KIR. There is increasing in vitro evidence that activation can convert $CD56^{bright}CD16^{-}$ NK cells into effectors analogous to blood $CD56^{dim}CD16^{+}$ NK cells with higher cytolytic activity (Romagnani et al., 2007; Chan et al., 2007). It is thus conceivable that up-regulation of KIR on $CD56^{bright}CD16^{-}$ NK cells can occur in vivo, most likely because of a proinflammatory cytokine microenvironment associated with local immune activation. The key to successful NK cell-based therapies might depend, among other factors, on new tools for promoting sustained in vivo activation of NK cells.

References

Albertsson, P.A., Basse, P.H., Hokland, M., Goldfarb, R.H., Nagelkerke, J.F., Nannmark, U. and Kuppen, P.J. (2003). NK cells and the tumour microenvironment: implications for NK-cell function and anti-tumour activity. Trends Immunol **24**, 603–609.

Andre, P., Spertini, O., Guia, S., Rihet, P., Dignat-George, F., Brailly, H., Sampol, J., Anderson, P.J. and Vivier, E. (2000). Modification of P-selectin glycoprotein ligand-1 with a natural killer cell-restricted sulfated lactosamine creates an alternate ligand for L-selectin. Proc Natl Acad Sci U S A **97**, 3400–3405.

Anfossi, N., Andre, P., Guia, S., Falk, C.S., Roetynck, S., Stewart, C.A., Breso, V., Frassati, C., Reviron, D., Middleton, D., Romagne, F., Ugolini, S. and Vivier, E. (2006). Human NK cell education by inhibitory receptors for MHC class I. Immunity **25**, 331–342.

Ashkar, A.A. and Croy, B.A. (2001). Functions of uterine natural killer cells are mediated by interferon gamma production during murine pregnancy. Semin Immunol **13**, 235–241.

Berditchevski, F. and Odintsova, E. (2007). Tetraspanins as regulators of protein trafficking. Traffic **8**, 89–96.

Bryceson, Y.T., March, M.E., Barber, D.F., Ljunggren, H.G. and Long, E.O. (2005). Cytolytic granule polarization and degranulation controlled by different receptors in resting NK cells. J Exp Med **202**, 1001–1012.

Buentke, E., Heffler, L.C., Wilson, J.L., Wallin, R.P., Lofman, C., Chambers, B.J., Ljunggren, H.G. and Scheynius, A. (2002). Natural killer and dendritic cell contact in lesional atopic dermatitis skin—*Malassezia*-influenced cell interaction. J Invest Dermatol **119**, 850–857.

Campbell, J.J., Qin, S., Unutmaz, D., Soler, D., Murphy, K.E., Hodge, M. R., Wu, L. and Butcher, E.C. (2001). Unique subpopulations of CD56+ NK and NK-T peripheral blood lymphocytes identified by chemokine receptor expression repertoire. J Immunol **166**, 6477–6482.

Carbone, E., Terrazzano, G., Ruggiero, G., Zanzi, D., Ottaiano, A., Manzo, C., Karre, K. and Zappacosta, S. (1999). Recognition of autologous dendritic cells by human NK cells. Eur J Immunol **29**, 4022–4029.

Carrega, P., Morandi, B., Costa, R., Frumento, G., Forte, G., Altavilla, G., Ratto, G.B., Mingari, M.C., Moretta, L. and Ferlazzo, G. (2008). Natural killer cells infiltrating human nonsmall-cell lung cancer are enriched in CD56 bright CD16(−) cells and display an impaired capability to kill tumor cells. Cancer **112**, 863–875.

Cartron, G., Dacheux, L., Salles, G., Solal-Celigny, P., Bardos, P., Colombat, P. and Watier, H. (2002). Therapeutic activity of humanized anti-CD20 monoclonal antibody and polymorphism in IgG Fc receptor FcgammaRIIIa gene. Blood **99**, 754–758.

Cella, M., Fuchs, A., Vermi, W., Facchetti, F., Otero, K., Lennerz, J.K., Doherty, J.M., Mills, J.C. and Colonna, M. (2009). A human natural killer cell subset provides an innate source of IL-22 for mucosal immunity. Nature **457**, 722–725.

Chan, A., Hong, D.L., Atzberger, A., Kollnberger, S., Filer, A.D., Buckley, C.D., McMichael, A., Enver, T. and Bowness, P. (2007). CD56bright human NK cells differentiate into CD56dim cells: role of contact with peripheral fibroblasts. J Immunol **179**, 89–94.

Chantakru, S., Miller, C., Roach, L.E., Kuziel, W.A., Maeda, N., Wang, W.C., Evans, S.S. and Croy, B.A. (2002). Contributions from self-renewal and trafficking to the uterine NK cell population of early pregnancy. J Immunol **168**, 22–28.

Clynes, R.A., Towers, T.L., Presta, L.G. and Ravetch, J.V. (2000). Inhibitory Fc receptors modulate in vivo cytoxicity against tumor targets. Nat Med **6**, 443–446.

Coca, S., Perez-Piqueras, J., Martinez, D., Colmenarejo, A., Saez, M.A., Vallejo, C., Martos, J.A. and Moreno, M. (1997). The prognostic significance of intratumoral natural killer cells in patients with colorectal carcinoma. Cancer **79**, 2320–2328.

Colucci, F., Caligiuri, M.A. and Di Santo, J.P. (2003). What does it take to make a natural killer? Nat Rev Immunol **3**, 413–425.

Cooper, M.A., Fehniger, T.A. and Caligiuri, M.A. (2001a). The biology of human natural killer-cell subsets. Trends Immunol **22**, 633–640.

Cooper, M.A., Fehniger, T.A., Turner, S.C., Chen, K.S., Ghaheri, B.A., Ghayur, T., Carson, W.E. and Caligiuri, M.A. (2001b). Human natural killer cells: a unique innate immunoregulatory role for the CD56(bright) subset. Blood **97**, 3146–3151.

Cooper, M.A., Fehniger, T.A., Fuchs, A., Colonna, M. and Caligiuri, M.A. (2004). NK cell and DC interactions. Trends Immunol **25**, 47–52.

Dalbeth, N., Gundle, R., Davies, R.J., Lee, Y.C., McMichael, A.J. and Callan, M.F. (2004). CD56bright NK cells are enriched at inflammatory sites and can engage with monocytes in a reciprocal program of activation. J Immunol **173**, 6418–6426.

Dall'Ozzo, S., Tartas, S., Paintaud, G., Cartron, G., Colombat, P., Bardos, P., Watier, H. and Thibault, G. (2004). Rituximab-dependent cytotoxicity by natural killer cells: influence of FCGR3A polymorphism on the concentration-effect relationship. Cancer Res **64**, 4664–4669.

Doherty, D.G. and O'Farrelly, C. (2000). Innate and adaptive lymphoid cells in the human liver. Immunol Rev **174**, 5–20.

Ebert, L.M., Meuter, S. and Moser, B. (2006). Homing and function of human skin gammadelta T cells and NK cells: relevance for tumor surveillance. J Immunol **176**, 4331–4336.

Eriksson, M., Meadows, S.K., Wira, C.R. and Sentman, C.L. (2004). Unique phenotype of human uterine NK cells and their regulation by endogenous TGF-beta. J Leukoc Biol **76**, 667–675.

Esendagli, G., Bruderek, K., Goldmann, T., Busche, A., Branscheid, D., Vollmer, E. and Brandau, S. (2008). Malignant and non-malignant lung tissue areas are differentially populated by natural killer cells and regulatory T cells in non-small cell lung cancer. Lung Cancer **59**, 32–40.

Fehniger, T.A., Cooper, M.A., Nuovo, G.J., Cella, M., Facchetti, F., Colonna, M. and Caligiuri, M.A. (2003). CD56bright natural killer cells are present in human lymph nodes and are activated by T cell-derived IL-2: a potential new link between adaptive and innate immunity. Blood **101**, 3052–3057.

Feldmann, M., Brennan, F.M. and Maini, R.N. (1996). Role of cytokines in rheumatoid arthritis. Annu Rev Immunol **14**, 397–440.

Ferlazzo, G., Tsang, M.L., Moretta, L., Melioli, G., Steinman, R.M. and Munz, C. (2002). Human dendritic cells activate resting natural killer (NK) cells and are recognized via the NKp30 receptor by activated NK cells. J Exp Med **195**, 343–351.

Ferlazzo, G., Pack, M., Thomas, D., Paludan, C., Schmid, D., Strowig, T., Bougras, G., Muller, W.A., Moretta, L. and Munz, C. (2004a). Distinct roles of IL-12 and IL-15 in human natural killer cell activation by dendritic cells from secondary lymphoid organs. Proc Natl Acad Sci U S A **101**, 16606–16611.

Ferlazzo, G., Thomas, D., Lin, S.L., Goodman, K., Morandi, B., Muller, W.A., Moretta, A. and Munz, C. (2004b). The abundant NK cells in human secondary lymphoid tissues require activation to express killer cell Ig-like receptors and become cytolytic. J Immunol **172**, 1455–1462.

Fernandez, N.C., Lozier, A., Flament, C., Ricciardi-Castagnoli, P., Bellet, D., Suter, M., Perricaudet, M., Tursz, T., Maraskovsky, E. and Zitvogel, L. (1999). Dendritic cells directly trigger NK cell functions: cross-talk relevant in innate anti-tumor immune responses in vivo. Nat Med **5**, 405–411.

Firestein, G.S., Alvaro-Gracia, J.M. and Maki, R. (1990). Quantitative analysis of cytokine gene expression in rheumatoid arthritis. J Immunol **144**, 3347–3353.

Freud, A.G., Becknell, B., Roychowdhury, S., Mao, H.C., Ferketich, A.K., Nuovo, G.J., Hughes, T.L., Marburger, T.B., Sung, J., Baiocchi, R.A., Guimond, M. and Caligiuri, M.A. (2005). A human CD34(+) subset resides in lymph nodes and differentiates into CD56bright natural killer cells. Immunity **22**, 295–304.

Frey, M., Packianathan, N.B., Fehniger, T.A., Ross, M.E., Wang, W.C., Stewart, C.C., Caligiuri, M.A. and Evans, S.S. (1998). Differential expression and function of L-selectin on CD56bright and CD56dim natural killer cell subsets. J Immunol **161**, 400–408.

Gerosa, F., Baldani-Guerra, B., Nisii, C., Marchesini, V., Carra, G. and Trinchieri, G. (2002). Reciprocal activating interaction between natural killer cells and dendritic cells. J Exp Med **195**, 327–333.

Gerosa, F., Gobbi, A., Zorzi, P., Burg, S., Briere, F., Carra, G. and Trinchieri, G. (2005). The reciprocal interaction of NK cells with plasmacytoid or myeloid dendritic cells profoundly affects innate resistance functions. J Immunol **174**, 727–734.

Gonen-Gross, T., Achdout, H., Gazit, R., Hanna, J., Mizrahi, S., Markel, G., Goldman-Wohl, D., Yagel, S., Horejsi, V., Levy, O., Baniyash, M. and Mandelboim, O. (2003). Complexes of HLA-G protein on the cell surface are important for leukocyte Ig-like receptor-1 function. J Immunol **171**, 1343–1351.

Hanna, J. and Mandelboim, O. (2007). When killers become helpers. Trends Immunol **28**, 201–206.

Hanna, J., Wald, O., Goldman-Wohl, D., Prus, D., Markel, G., Gazit, R., Katz, G., Haimov-Kochman, R., Fujii, N., Yagel, S., Peled, A. and Mandelboim, O. (2003). CXCL12 expression by invasive trophoblasts induces the specific migration of CD16− human natural killer cells. Blood **102**, 1569–1577.

Hanna, J., Goldman-Wohl, D., Hamani, Y., Avraham, I., Greenfield, C., Natanson-Yaron, S., Prus, D., Cohen-Daniel, L., Arnon, T.I., Manaster, I., Gazit, R., Yutkin, V., Benharroch, D., Porgador, A., Keshet, E., Yagel, S. and Mandelboim, O. (2006). Decidual NK cells regulate key developmental processes at the human fetal–maternal interface. Nat Med **12**, 1065–1074.

Hata, K., Van Thiel, D.H., Herberman, R.B. and Whiteside, T.L. (1991). Natural killer activity of human liver-derived lymphocytes in various liver diseases. Hepatology **14**, 495–503.

Ho, H.N., Chao, K.H., Chen, C.K., Yang, Y.S. and Huang, S.C. (1996). Activation status of T and NK cells in the endometrium throughout menstrual cycle and normal and abnormal early pregnancy. Hum Immunol **49**, 130–136.

Imai, K., Matsuyama, S., Miyake, S., Suga, K., Nakachi, K. (2000). Natural cytotoxic activity of peripheral-blood lymphocytes and cancer incidence: an 11-year follow-up study of a general population. Lancet **356**, 1795–1799.

Ishigami, S., Natsugoe, S., Tokuda, K., Nakajo, A., Che, X., Iwashige, H., Aridome, K., Hokita, S. and Aikou, T. (2000). Prognostic value of intratumoral natural killer cells in gastric carcinoma. Cancer **88**, 577–583.

Jacobs, R., Hintzen, G., Kemper, A., Beul, K., Kempf, S., Behrens, G., Sykora, K.W. and Schmidt, R.E. (2001). CD56bright cells differ in their KIR repertoire and cytotoxic features from CD56dim NK cells. Eur J Immunol **31**, 3121–3127.

Jones, R.K., Bulmer, J.N. and Searle, R.F. (1997). Cytotoxic activity of endometrial granulated lymphocytes during the menstrual cycle in humans. Biol Reprod **57**, 1217–1222.

Keskin, D.B., Allan, D.S., Rybalov, B., Andzelm, M.M., Stern, J.N., Kopcow, H.D., Koopman, L.A. and Strominger, J. L. (2007). TGFbeta promotes conversion of CD16+ peripheral blood NK cells into CD16− NK cells with similarities to decidual NK cells. Proc Natl Acad Sci U S A **104**, 3378–3383.

King, A., Birkby, C. and Loke, Y.W. (1989). Early human decidual cells exhibit NK activity against the K562 cell line but not against first trimester trophoblast. Cell Immunol **118**, 337–344.

Koopman, L.A., Kopcow, H.D., Rybalov, B., Boyson, J.E., Orange, J.S., Schatz, F., Masch, R., Lockwood, C.J., Schachter, A.D., Park, P.J. and Strominger, J.L. (2003). Human decidual natural killer cells are a unique NK cell subset with immunomodulatory potential. J Exp Med **198**, 1201–1212.

Kopcow, H.D., Allan, D.S., Chen, X., Rybalov, B., Andzelm, M.M., Ge, B. and Strominger, J.L. (2005). Human decidual NK cells form immature activating synapses and are not cytotoxic. Proc Natl Acad Sci U S A **102**, 15563–15568.

Lambrecht, B.N. and Hammad, H. (2003). Taking our breath away: dendritic cells in the pathogenesis of asthma. Nat Rev Immunol **3**, 994–1003.

Li, J.M., Isler, P., Dayer, J.M. and Burger, D. (1995). Contact-dependent stimulation of monocytic cells and neutrophils by stimulated human T-cell clones. Immunology **84**, 571–576.

Loza, M. and Perussia B. (2004) The IL-12 signature: NK cell terminal CD56+high stage and effector functions. J Immunol **172**, 88–96.

Luci, C., Reynders, A., Ivanov, I.I., Cognet, C., Chiche, L., Chasson, L., Hardwigsen, J., Anguiano, E., Banchereau, J., Chaussabel, D., Dalod, M., Littman, D.R., Vivier, E. and Tomasello, E. (2009). Influence of the transcription factor RORgammat on the development of NKp46+ cell populations in gut and skin. Nat Immunol **10**, 75–82.

Lynch, L., Golden-Mason, L., Eogan, M., O'Herlihy, C. and O'Farrelly, C. (2007). Cells with haematopoietic stem cell phenotype in adult human endometrium: relevance to infertility? Hum Reprod **22**, 919–926.

Mailliard, R.B., Alber, S.M., Shen, H., Watkins, S.C., Kirkwood, J.M., Herberman, R.B., Kalinski, P. (2005). IL-18-induced CD83+CCR7+ NK helper cells. J. Exp Med **202**, 941–953.

Marcenaro, E., Dondero, A. and Moretta, A. (2006). Multi-directional cross-regulation of NK cell function during innate immune responses. Transpl Immunol **17**, 16–19.

Masopust, D., Vezys, V., Marzo, A.L. and Lefrancois, L. (2001). Preferential localization of effector memory cells in nonlymphoid tissue. Science **291**, 2413–2417.

Moretta, A. (2002). Natural killer cells and dendritic cells: rendezvous in abused tissues. Nat Rev Immunol **2**, 957–964.

Moretta, L., Bottino, C., Pende, D., Mingari, M.C., Biassoni, R. and Moretta, A. (2002). Human natural killer cells: their origin, receptors and function. Eur J Immunol **32**, 1205–1211.

Moretta, L., Ferlazzo, G., Mingari, M.C., Melioli, G. and Moretta, A. (2003). Human natural killer cell function and their interactions with dendritic cells. Vaccine **21**(Suppl. 2), S38–S42.

Norris, S., Collins, C., Doherty, D.G., Smith, F., McEntee, G., Traynor, O., Nolan, N., Hegarty, J. and O'Farrelly, C. (1998). Resident human hepatic lymphocytes are phenotypically different from circulating lymphocytes. J Hepatol **28**, 84–90.

Ottaviani, C., Nasorri, F., Bedini, C., de Pita, O., Girolomoni, G. and Cavani, A. (2006). CD56brightCD16(−) NK cells accumulate in psoriatic skin in response to CXCL10 and CCL5 and exacerbate skin inflammation. Eur J Immunol **36**, 118–128.

Parolini, S., Santoro, A., Marcenaro, E., Luini, W., Massardi, L., Facchetti, F., Communi, D., Parmentier, M., Majorana, A., Sironi, M., Tabellini, G., Moretta, A. and Sozzani, S. (2007). The role of chemerin in the colocalization of NK and dendritic cell subsets into inflamed tissues. Blood **109**, 3625–3632.

Piccioli, D., Sbrana, S., Melandri, E. and Valiante, N.M. (2002). Contact-dependent stimulation and inhibition of dendritic cells by natural killer cells. J Exp Med **195**, 335–341.

Reymond, N., Imbert, A.M., Devilard, E., Fabre, S., Chabannon, C., Xerri, L., Farnarier, C., Cantoni, C., Bottino, C., Moretta, A., Dubreuil, P. and Lopez, M. (2004). DNAM-1 and PVR regulate monocyte migration through endothelial junctions. J Exp Med **199**, 1331–1341.

Reynolds, C.W., Sharrow, S.O., Ortaldo, J.R. and Herberman, R.B. (1981). Natural killer activity in the rat. II. Analysis of surface antigens on LGL by flow cytometry. J Immunol **127**, 2204–2208.

Romagnani, C., Della Chiesa, M., Kohler, S., Moewes, B., Radbruch, A., Moretta, L., Moretta, A. and Thiel, A. (2005). Activation of human NK cells by plasmacytoid dendritic cells and its modulation by CD4+ T helper cells and CD4+ CD25hi T regulatory cells. Eur J Immunol **35**, 2452–2458.

Romagnani, C., Juelke, K., Falco, M., Morandi, B., D'Agostino, A., Costa, R., Ratto, G., Forte, G., Carrega, P., Lui, G., Conte, R., Strowig, T., Moretta, A., Munz, C., Thiel, A., Moretta, L. and Ferlazzo, G. (2007). CD56brightCD16− killer Ig-like receptor-NK cells display longer telomeres and acquire features of CD56dim NK cells upon activation. J Immunol **178**, 4947–4955.

Rubio, V., Stuge, T.B., Singh, N., Betts, M.R., Weber, J.S., Roederer, M. and Lee, P.P. (2003). Ex vivo identification, isolation and analysis of tumor-cytolytic T cells. Nat Med **9**, 1377–1382.

Ruggeri, L., Aversa, F., Martelli, M.F. and Velardi, A. (2006). Allogeneic hematopoietic transplantation and natural killer cell recognition of missing self. Immunol Rev **214**, 202–218.

Sandel, M.H., Speetjens, F.M., Menon, A.G., Albertsson, P.A., Basse, P.H., Hokland, M., Nagelkerke, J.F., Tollenaar, R.A., van de Velde, C.J. and Kuppen, P.J. (2005). Natural killer cells infiltrating colorectal cancer and MHC class I expression. Mol Immunol **42**, 541–546.

Sanos, S.L., Bui, V.L., Mortha, A., Oberle, K., Heners, C., Johner, C. and Diefenbach, A. (2009). RORgammat and commensal microflora are required for the differentiation of mucosal interleukin 22-producing NKp46+ cells. Nat Immunol **10**, 83–91.

Schleypen, J.S., Von Geldern, M., Weiss, E.H., Kotzias, N., Rohrmann, K., Schendel, D.J., Falk, C.S. and Pohla, H. (2003). Renal cell carcinoma-infiltrating natural killer cells express differential repertoires of activating and inhibitory receptors and are inhibited by specific HLA class I allotypes. Int J Cancer **106**, 905–912.

Schleypen, J.S., Baur, N., Kammerer, R., Nelson, P.J., Rohrmann, K., Grone, E.F., Hohenfellner, M., Haferkamp, A., Pohla, H., Schendel, D.J., Falk, C.S. and

Noessner, E. (2006). Cytotoxic markers and frequency predict functional capacity of natural killer cells infiltrating renal cell carcinoma. Clin Cancer Res **12**, 718–725.

Sebbag, M., Parry, S.L., Brennan, F.M. and Feldmann, M. (1997). Cytokine stimulation of T lymphocytes regulates their capacity to induce monocyte production of tumor necrosis factor-alpha, but not interleukin-10: possible relevance to pathophysiology of rheumatoid arthritis. Eur J Immunol **27**, 624–632.

Starkey, P.M., Clover, L.M. and Rees, M.C. (1991). Variation during the menstrual cycle of immune cell populations in human endometrium. Eur J Obstet Gynecol Reprod Biol **39**, 203–207.

Strowig, T., Brilot, F., Arrey, F., Bougras, G., Thomas, D., Muller, W.A. and Münz, C. (2008). Tonsilar NK cells restrict B cell transformation by the Epstein–Barr virus via IFN-gamma. PLoS Pathog **4**(2), e27.

Trepel, F. (1974). Number and distribution of lymphocytes in man. A critical analysis. Klin Wochenschr **52**, 511–515.

Trundley, A. and Moffett, A. (2004). Human uterine leukocytes and pregnancy. Tissue Antigens **63**, 1–12.

Vacca, P., Pietra, G., Falco, M., Romeo, E., Bottino, C., Bellora, F., Prefumo, F., Fulcheri, E., Venturini, P.L., Costa, M., Moretta, A., Moretta, L. and Mingari, M. C. (2006). Analysis of natural killer cells isolated from human decidua: evidence that 2B4 (CD244) functions as an inhibitory receptor and blocks NK-cell function. Blood **108**, 4078–4085.

Vassiliadou, N. and Bulmer, J.N. (1998). Expression of CD69 activation marker by endometrial granulated lymphocytes throughout the menstrual cycle and in early pregnancy. Immunology **94**, 368–375.

Vermi, W., Riboldi, E., Wittamer, V., Gentili, F., Luini, W., Marrelli, S., Vecchi, A., Franssen, J.D., Communi, D., Massardi, L., Sironi, M., Mantovani, A., Parmentier, M., Facchetti, F. and Sozzani, S. (2005). Role of ChemR23 in directing the migration of myeloid and plasmacytoid dendritic cells to lymphoid organs and inflamed skin. J Exp Med **201**, 509–515.

Vey, E., Zhang, J.H. and Dayer, J.M. (1992). IFN-gamma and 1,25(OH)2D3 induce on THP-1 cells distinct patterns of cell surface antigen expression, cytokine production, and responsiveness to contact with activated T cells. J Immunol **149**, 2040–2046.

Villegas, F.R., Coca, S., Villarrubia, V.G., Jimenez, R., Chillon, M.J., Jareno, J., Zuil, M. and Callol, L. (2002). Prognostic significance of tumor infiltrating natural killer cells subset CD57 in patients with squamous cell lung cancer. Lung Cancer **35**, 23–28.

Vitale, M., Della Chiesa, M., Carlomagno, S., Romagnani, C., Thiel, A., Moretta, L. and Moretta, A. (2004). The small subset of CD56brightCD16− natural killer cells is selectively responsible for both cell proliferation and interferon-gamma production upon interaction with dendritic cells. Eur J Immunol **34**, 1715–1722.

Vujanovic, N.L., Rabinowich, H., Lee, Y.J., Jost, L., Herberman, R.B. and Whiteside, T.L. (1993). Distinct phenotypic and functional characteristics of human natural killer cells obtained by rapid interleukin 2-induced adherence to plastic. Cell Immunol **151**, 133–157.

Walzer, T., Dalod, M., Robbins, S.H., Zitvogel, L. and Vivier, E. (2005). Natural-killer cells and dendritic cells: "l'union fait la force". Blood **106**, 2252–2258.

Walzer, T., Jaeger, S., Chaix, J. and Vivier, E. (2007). Natural killer cells: from CD3(−)NKp46(+) to post-genomics meta-analyses. Curr Opin Immunol **19**, 365–372.

Westermann, J. and Pabst, R. (1992). Distribution of lymphocyte subsets and natural killer cells in the human body. Clin Investig **70**, 539–544.

Wilson, J.L., Heffler, L.C., Charo, J., Scheynius, A., Bejarano, M.T. and Ljunggren, H.G. (1999). Targeting of human dendritic cells by autologous NK cells. J Immunol **163**, 6365–6370.

Wolint, P., Betts, M.R., Koup, R.A. and Oxenius, A. (2004). Immediate cytotoxicity but not degranulation distinguishes effector and memory subsets of CD8+ T cells. J Exp Med **199**, 925–936.

Zitvogel, L. (2002). Dendritic and natural killer cells cooperate in the control/switch of innate immunity. J Exp Med **195**, F9–F14.

Chapter Thirty-Two

32

Natural killer cell immune regulation: coordination of immune function in tissues

Lorenzo Moretta, Alessandro Moretta

CHAPTER CONTENTS

In the middle of the road of my life
I awoke in the dark wood where the
true way was wholly lost.

(From Dante Alighieri, 'La Divina Commedia', Inferno)

ABSTRACT

Natural killer (NK) cells are potent effector cells of innate immunity. Their function is finely regulated by a series of activating and inhibitory receptors. Inhibitory receptors recognize MHC-class I molecules on potential target cells and prevent cellular activation and cytolysis when (as occurs with most normal cells), they express sufficient amounts of MHC class I at their surface. A remarkable exception is represented by myeloid dendritic cells (DCs) that are killed by activated NK cells if they have failed to undergo complete maturation (DC editing). NK–DC interactions are greatly potentiated by pathogen-derived products that activate several different cell types of innate immunity through expression of Toll-like receptors (TLRs). Besides NK and DC, these cells include plasmacytoid DCs (PDCs), eosinophils and mast cells. The resulting crosstalk between these cell types is mediated by cytokines, chemokines or direct cell-to-cell interactions, so-called Signal 5s to distinguish them from the signals important in the afferent immune response. This crosstalk has a great impact not only on the quality and strength of innate immune responses but also on subsequent adaptive immune responses. Thus, NK cells play an important role in defence against pathogens (and tumours) not only because of their effector function but also because of their regulatory capability within tissues on the nature and quality of innate and adaptive immune responses.

KEY WORDS

NK cells, NK receptors, Innate immunity, Dendritic cells (DCs), Cytokines, Immune regulation, IL-18, IL-12, IL-4

Introduction

Natural killer (NK) cells can discriminate between normal and virus-infected cells or cells undergoing tumour transformation. This capability is based in part on the expression at their surface of inhibitory receptors specific for MHC class I molecules. Thus, NK cells kill those target cells that have lost, or underexpress, MHC class I molecules (Long, 1999; Lopez-Botet et al., 2000; Moretta et al., 1996; Vilches and Parham, 2002). However, target cell lysis also requires the expression, on potential target cells, of surface ligands recognized by activating receptors present at the NK cell surface (Bottino et al., 2005). Thus, NK-mediated selective elimination of abnormal target cells is the result of the combined function of activating and inhibitory receptors on NK cells as well as of the expression of their specific ligands on target cells (Bottino et al., 2005; Cerwenka and Lanier, 2001; Moretta et al., 2001; Raulet, 2003).

Recent studies clearly indicate that NK cells also display regulatory capabilities mediated by various cytokines that are released upon engagement of various triggering NK receptors or upon signalling by cytokines (Moretta, 2002). This is particularly relevant during the early phases of the inflammatory response. A pioneering study by Fernandez et al. (1999) showed that dendritic cells (DCs) can induce NK cell activity against tumours. More recently, several data have highlighted the role of the interactions between NK cells and other cells of the innate immune system that occur during the early phases of acute inflammation, particularly during infection (reviewed in Moretta et al., 2005). Various studies were focused on the crosstalk between NK cells and myeloid DCs (Ferlazzo et al., 2002; Gerosa et al., 2002; Piccioli et al., 2002) and, more recently, on plasmacytoid DCs (PDCs) (Della Chiesa et al., 2006; Gerosa et al., 2005). These interactions can take place upon the recruitment of these cells into inflammatory sites in response to invasion by pathogens (or tumour cells) (Cooper et al., 2004; Degli-Esposti and Smith, 2005; Della Chiesa et al., 2005; Moretta, 2002; Moretta et al., 2006; Raulet, 2004; Walzer et al., 2005; Zitvogel, 2002). Increasing evidence supports the notion that these interactions shape not only innate immune responses within inflamed peripheral tissues but also the subsequent adaptive immune responses in secondary lymphoid organs. Thus, the early crosstalk occurring between cells of the innate immunity could play a central role in the control of the quality (and efficacy) of the defence against pathogens.

Cellular interactions in innate immune responses

After their recruitment into inflamed tissues in response to various chemokines, during the early phases of an inflammatory response, NK cells can interact with other cell types of the innate immunity (Della Chiesa et al., 2005; Imai et al., 1997; Vitale et al., 2004). These interactions can impact not only the NK-mediated effector functions in innate immune defences but also their capability of regulating the downstream adaptive immunity (Moretta et al., 2006). The interaction with different cell types can modulate NK cell functions as the result of cell-to-cell contact (receptor–ligand interactions) or the activity of soluble mediators. For example, a close cell-to-cell contact is required in the case of NK–monocyte-derived DC interactions. These interactions include not only the NK-mediated killing of immature DCs (iDCs) but also the DC-induced NK-cell proliferation and the NK-dependent DC maturation (Cooper et al., 2004; Banchereau and Steinman, 1998; Moretta, 2002; Moretta et al., 2006; Raulet, 2004; Saïdi et al., 2008; Vitale et al., 2004; Zitvogel, 2002). During such NK–DC contact, DCs undergoing maturation after antigen uptake release cytokines that can greatly influence the functional behaviour of NK cells. For example, DC-derived IL-12 is crucial not only for inducing NK cells to release IFNγ but also for enhancing NK-cell cytotoxicity (Trinchieri, 2003). Enhancement of NK cytotoxicity can also be induced by type I IFN abundantly secreted by PDCs (Colonna et al., 2004; Moretta et al., 2005) or mastocytes (Marshall and Jawdat, 2004). Therefore, the NK-mediated capability of killing virus-infected cells or tumours (or iDCs) is greatly influenced by the type of cytokines released by bystander cells during innate immune responses. In addition, the apoptotic/necrotic material or heat-shock proteins, damage-associated molecular pattern molecules (DAMPs), resulting from the NK-mediated killing of tumours or virus-infected cells can modulate the function of DCs or other bystander cells. A further regulatory effect on these cells can be exerted by cytokines released by NK cells such as tumour necrosis factor (TNF)α, IFNγ and granulocyte-macrophage colony-stimulating factor (GM-CSF) (Cooper et al., 2004; Moretta, 2002).

NK cell activation at inflammatory sites: role of receptor–ligand interactions and of TLRs

NK-mediated killing of autologous iDCs requires previous NK cell activation (Ferlazzo et al., 2002; Gerosa et al., 2002). However, NK cells recruited from blood into inflamed peripheral tissues are not necessarily activated and would thus require appropriate activating signals in order to kill iDCs. Such activating signals could be provided by tumours or virus-infected cells susceptible to NK-mediated lysis (Dokun et al., 2001; Tay and Welsh, 1997). NK cell activation and production of IFNγ may result from recognition of NK-sensitive target cells (Adam et al., 2005). Nevertheless, in most instances, tumours are resistant to non-activated peripheral blood

NK cells, and their killing requires previous exposure of NK cells to cytokines such as IL-2, IL-12, IL-15 or IFNα released by other cell types (Agaugué et al., 2008).

A different mechanism by which NK cells can become activated has been identified recently. Human NK cells express TLRs (Pisegna et al., 2004; Schmidt et al., 2004; Sivori et al., 2004), which can provide an alternative mode of NK-cell activation, independent on the recognition of NK-susceptible target cells. TLRs, on recognition of pathogen-associated molecular patterns (PAMPs), induce innate immune responses, providing immediate protection against various pathogens (Akira and Takeda, 2004). Human NK cells, independent on their status of activation, express functional TLR3 and TLR9 (Sivori et al., 2004, 2006), thus enabling them to respond both to viral and bacterial products. In particular, the simultaneous engagement of TLR3 on both NK cells and monocyte-derived DCs appears to be sufficient to initiate a series of events characteristic of the early phases of innate immune responses. Indeed, exposure of NK cells and myeloid DCs to dsRNA results in IL-12 secretion by DCs and in NK cell activation. In addition, TLR-stimulated NK cells, in the presence of IL-12, release cytokines, including IFNγ and TNFα, and acquire a higher cytolytic activity against tumour target cells. Remarkably, upon triggering via TLR3, NK cells also acquire the capability of killing iDC (see the following section) (Sivori et al., 2004, 2007).

NK–DC interactions: NK cell-mediated killing of immature myeloid DCs (DC 'editing')

The capability of NK cells of killing autologous myeloid iDCs (but not mDCs) is a most remarkable event occurring during the NK–DC crosstalk (Ferlazzo et al., 2002; Moretta, 2002; Wilson et al., 1999; Zitvogel, 2002). This effect is based on the capability of NK cells of discriminating between iDCs (that typically underexpress HLA-class I molecules) and mDCs that, after Ag uptake, up-regulate MHC-class I expression (Ferlazzo et al., 2003). When undergoing maturation, DCs also up-regulate the chemokine receptor CCR7, and co-stimulatory molecules belonging to the B7 molecular family. These events are crucial for the subsequent DC migration to lymph nodes and priming of T lymphocytes (Banchereau et al., 2000; Moretta, 2002; Reis e Sousa, 2001; Sallusto and Lanzavecchia, 1999). This unexpected function might act to keep in check the quality of DCs undergoing maturation ('editing' process) (Moretta, 2002; Moretta et al., 2005). Thus, DCs that fail to express sufficient amounts of MHC molecules would be removed, preventing the survival of faulty DCs. Otherwise, such inappropriate DCs, after expression of CCR7 and migration to lymph nodes, would induce low affinity T cell priming eventually resulting either in Th2 responses or in a state of tolerance (Langenkamp et al., 2000; Steinman and Nussenzweig, 2002) possibly consequent to the activation of regulatory T cells (Treg) (Della Chiesa et al., 2005). In this context, in the absence of NK cells, the in vivo default development pathway of CD4 T cells is strongly biased towards the acquisition of a Th2 phenotype (Byrna et al., 2004; Coudert et al., 2002; Scharton and Scott, 1993; Tseng and Rank, 1998).

The editing process by which NK cells eliminate iDCs initiates with the engagement of the NKp30-activating receptor by its (still undefined) ligand expressed on DCs. Thus, blocking of NKp30 by specific mAbs sharply inhibits the NK-mediated killing of iDCs (Ferlazzo et al., 2002). In addition, the NK-mediated killing of iDCs was inhibited by transforming growth factor-β (TGFβ), which induces a profound down-regulation of the surface expression of NKp30 (Castriconi et al., 2003).

Notably, killing of iDCs does not involve the entire NK cell pool. Thus, analysis of NK cell clones revealed that killing of autologous DCs was confined to NK cells lacking inhibitory KIRs specific for self HLA-class I alleles and expressing the HLA-E-specific CD94-NKG2A inhibitory receptor (Della Chiesa et al., 2003). A likely explanation for this finding is that myeloid iDC, although displaying a general down-regulation of surface HLA class I molecules, have particularly low levels of surface HLA-E molecules.

NK cell-promoted maturation of myeloid DCs

NK cells can also induce progression of DC maturation via cytokines released upon direct NK–DC cell-to-cell contact (Moretta, 2005). Studies by Vitale et al. (2005) revealed that, similar to the iDC killing process, the ability to induce DC maturation is also dependant on NKp30. On NK–DC interaction and NKp30 engagement, NK cells produced TNFα (and INFγ) a cytokine that induces DC maturation, and this effect was neutralized by anti-TNFα monoclonal antibodies. Notably, also in this case, the ability to promote maturation was confined to NK cells expressing the KIR^{-} $NKG2A^{dull}$ phenotype. This process could complement the NK-mediated editing of DCs leading to selection of mDCs. The relative contribution of these two NKp30-mediated NK cell functions to the process of shaping mDCs has been analysed in NK cells from perforin-deficient patients. Significant cytokine-dependent DC maturation could still be detected in the absence of NK-mediated DC killing (Vitale et al., 2005), thus providing evidence that the two mechanisms can be distinguished.

Role of cytokine microenvironment in the modulation of innate and adaptive immune responses

During the early phases of an inflammatory response, the engagement of TLRs by PAMPs may not be confined to NK cells and DCs, but it can also involve other cell types, including resident mast cells (Kulka et al., 2004; Marshall, 2004; Supajatura et al., 2002), eosinophils (Bjerke et al., 1996; Nagase, 2003) or PDCs (Moretta et al., 2005). These cells, through the release of cytokines other than IL-12 (e.g. IL-4, IL-8 or IFNα) could differentially modulate the functional capability of bystander NK cells and DCs. In this context, an early exposure of NK cells to IL-4 could deviate the subsequent adaptive response towards a state of tolerization or generation of either Th2 or unpolarized T cells (Marcenaro et al., 2005a). While a short-term NK cell exposure to IL-12 promoted the release of high levels of both IFNγ and TNFα and the acquisition of cytolytic activity, exposure to IL-4 resulted in poor cytokine production and low cytolytic activity. Accordingly, only NK cells exposed to IL-12 could promote efficient DC maturation, which, in turn, would induce optimal priming of Th1 responses (Agaugué et al., 2008). In contrast, NK cell priming in the presence of IL-4 results in abnormal DC maturation (failure of the editing program) characterized by both qualitative and quantitative alterations (Marcenaro et al., 2005b).

More recently, IL-18 has been shown to promote yet another pathway of TH1 priming (Agaugué et al., 2008; Mailliard et al., 2003). IL-18-conditioned NK cells induce TH1 polarization only when co-cultured with both DCs and T cells. In this case, IL-2 released by T cells and IL-12 produced by DCs during the priming process promote abundant IFNγ production by NK cells. Mailliard et al. (2003) proposed the concept of 'helper' NK cells referring to the ability of IL-18-treated NK cells to produce IFNγ upon exposure to either DC-related (IL-12) or T cell-related signals (IL-2). These NK cells also display an increased ability to promote IL-12 p70 secretion by DCs upon CD40L-mediated stimulation (Agaugué et al., 2008). Thus, the presence of IL-18-conditioned NK cells during priming of naïve CD4+ TH cells facilitates their polarization towards IFNγ-producing TH1 cells (Agaugué et al., 2008). Remarkably, IL-18-conditioned NK cells de novo express CCR7, thus displaying high migratory capacity to lymph nodes in response to CCL19 and CCL21 chemokines (Mailliard et al., 2003). It is worth noting that CCR7 expression occurred in the presence of IL-18 but not of other cytokines, including IL-2, IL-4, IL-12 and IL-15. Recruitment into lymph nodes of IL-18-conditioned NK cells may greatly influence the DC-induced polarization of naïve T cells. Thus, in lymph nodes, even small amounts of IL-2 or IL-12 is sufficient to induce the release of IFNγ by IL-18-conditioned NK cells. Indeed, low levels of IL-2 are released by naïve CD4+ T cells upon interaction with DCs (particularly when pre-exposed to IL-18-conditioned NK cells). In addition, these DCs release small amounts of IL-12. Notably, neutralization of both cytokines resulted in virtual abrogation of IFNγ release (Agaugué et al., 2008). One may argue that, in most instances, during the early phases of an innate immune response, IL-2 is not available. In mice, however, production of IL-2 by Ag-pulsed DCs has been reported at the very beginning stages of innate responses (Granucci et al., 2001, 2003). In humans, production of IL-2 by DCs is not detectable by conventional ELISA assays but can be detected by intracellular staining after differentiation in the presence of IL-15 (Feau et al., 2005). It is possible that such small amounts of IL-2 may be sufficient to promote an effective interaction between DCs and IL-18-conditioned NK cells. Altogether these data suggest that NK cells that have been recruited by chemokines into inflamed peripheral tissues and exposed to different cytokines, including IL-12, IL-18, IL-4, IL-2 or IFNα, can exert different regulatory effects on the downstream adaptive immune responses (Figure 32.1). These cytokines are released by different cell types, including resident and circulating cells recruited at inflammatory sites in response to chemokine gradients. These cells include mast cells (see the next section), eosinophils, basophils, monocytes and neutrophils. Similar to DCs and NK cells, they are equipped with TLR and release cytokines upon engagement with their specific, pathogen-associated, ligands. The presence (or the prevalence) of one or another cytokine within the inflammatory microenvironment will have a markedly different effect on the subsequent NK cell interaction with DCs in the periphery and/or with both DCs and naïve T cells within lymph nodes.

In vivo, type 1 or type 2 cytokines are secreted and exert their regulatory role on bystander cells within a short time interval after pathogen invasion (Degli Esposti et al., 2005). In this context, it has been estimated that the window of time available for an innate response to take place is only a few hours, whereas several days are required for development of effective, naive T cell-mediated, specific responses (Biron et al., 1999; Constant and Bottomly, 1997).

NK cell interactions with mast cells: regulatory effects on the innate immunity

As mentioned earlier, TLR engagement results in activation of multiple cell types within the innate immune

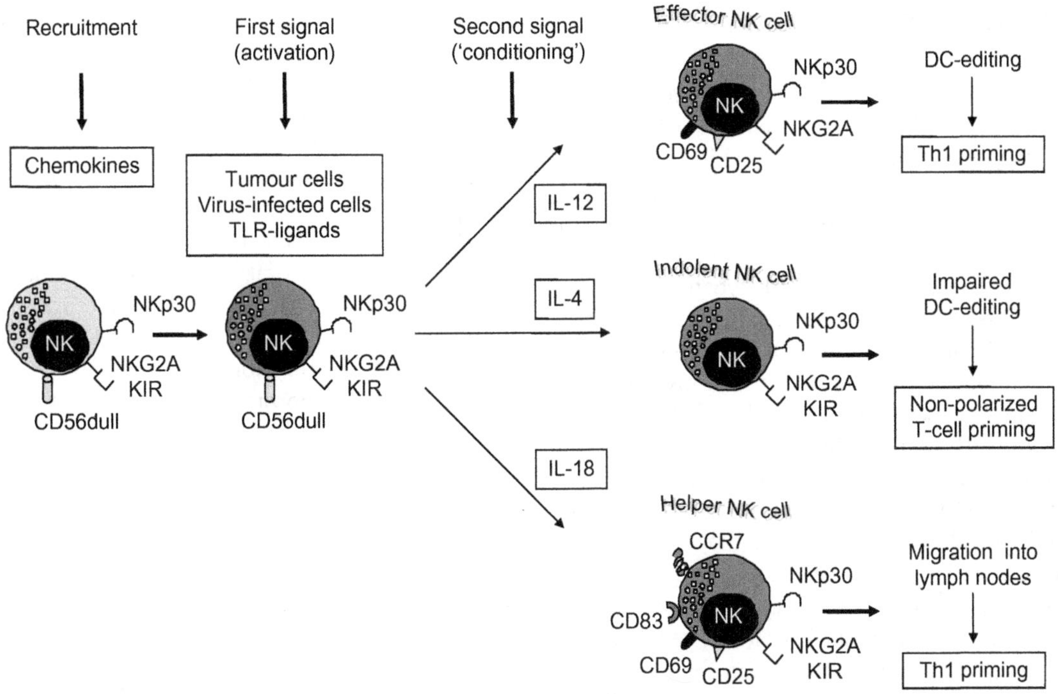

Figure 32.1 • Cytokines influence NK cell function and modulate the downstream adaptive responses. Peripheral blood NK cells are recruited into inflamed tissues by inflammatory chemokines (IL8, Fractalkine and Chemerin). They become activated upon interaction with cells characterized by low or absent expression of MHC class I molecules (tumour or virus-infected cells) or with pathogen-derived products ('first signal'). The presence of various cytokines, including IL-4, IL-12 and IL-18, create a 'second signal' resulting in conditioning of NK cell function. Differentially conditioned NK cells, in turn, greatly influence the quality of T cell priming and their polarization towards TH1. Signal 1 for T-cells, the MHC–peptide complex; Signal 2, the co-stimulatory signal related to the intensity of the inflammatory response; Signal 3, the polarization of T cells in part mediated by NK cell delivery of cytokines; Signal 4, the targeting of T cells to tissues based on integrin-mediated signalling induced by NK–DC crosstalk. Signal 5, the integration of signals mediated by innate effectors occurs in the tissues and modulates the nature and quality of the immune response.

system (Akira and Takeda, 2004). An important cell type that may greatly influence NK–DC function is represented by mast cells. These cells are typically located near surfaces exposed to the environment, including skin, airways and gastrointestinal tract, where invading pathogens are frequently encountered. In these locations, mast cells are present in close proximity to blood vessels where they can regulate vascular permeability and effector cell recruitment (Galli et al., 2005; Marshall, 2004). In this context, it has been suggested that mast cells might influence DC migration, maturation and function via the release of chemokines, such as CCL20, and of proinflammatory cytokines such as TNFα (Marshall, 2004). Remarkably, mast cells produce TNFα in response to TLR4 engagement by LPS (Supajatura et al., 2002), suggesting their possible involvement in the maturation process of myeloid DCs that are also stimulated through TLR4. In contrast, when stimulated via TLR3, mast cells have been shown to produce high doses of type I IFN (Kulka et al., 2004). This suggests their direct involvement in innate anti-viral defences as well as in potentiating the lytic activity of NK cells that are also simultaneously stimulated via TLR3. In addition, triggering of mast cells via TLR2, following exposure to the Gram-positive bacterial cell wall component peptidoglycan, favours the release of type2 cytokines, including IL-4 and IL-5 (Supajatura et al., 2002). The release of IL-4 might deviate the subsequent adaptive response towards Th2 by acting at early stages of the innate immune reaction at inflammatory sites in peripheral tissues. Indeed, IL-4, when added to NK cells simultaneously to IL-2 or IL-12, counteracts the effect of these cytokines by suppressing IFNγ and TNFα production. As indicated earlier, a short exposure to IL-12 or IL-4 (compatible with early in vivo innate immune responses) is sufficient to modulate NK cell function (Marcenaro et al., 2005b).

Depending on the PAMPs present at the site of an infection, mast cells can be activated via various TLRs expressed on their surface. Remarkably, this results in release of different cytokines, which can exert divergent functional effects on both innate and adaptive immune response.

NK cells interaction with PDCs

PDCs can be isolated from human peripheral blood, lymph nodes and peripheral tissues. Different from myeloid DCs, they do not express CD11c, while they express CD45RA, high levels of the IL-3Rα (CD123) blood DC antigen (BDCA)-2 and BDCA-4, the two latter antigens representing useful markers for their identification. Although the origin of PDCs is still debated, the transcript expression of pre-Tα and λ5 supports the notion that they are of lymphoid origin. PDCs are thought to be potent regulators of both innate and adaptive immune responses against viruses (Colonna et al., 2004).

A series of different studies have focused on the in vitro interaction between human NK cells and PDCs (Della Chiesa et al., 2006; Gerosa et al., 2005; Romagnani et al., 2005). In humans, the PDC pattern of TLR expression is profoundly different from that of macrophage-derived DCs. Thus PDCs do not express TLR 1, 2, 3, 4, 5 and 6 but, similar to NK cells, express TLR9, a receptor specific for unmethylated CpG derived from bacteria or viruses (Jarrossay et al., 2001). Since both NK and PDC express TLR9, under appropriate conditions, they can be simultaneously activated by the same invading pathogen. The abundant release of type I IFN (a potent inducer of NK cell cytotoxicity) by PDCs stimulated via TLR9 (Krieg, 2002) suggests that NK–PDC interaction can result in enhanced anti-viral innate protection. In addition, upon stimulation via TLR 9, the NK–PDC interaction resulted in up-regulation of the NK-mediated cytotoxicity against various tumour target cells. This effect was strongly reduced by antibodies against IFNα, thus indicating a primary role of this cytokine. That, under these conditions, NK cells become activated is also revealed by the dramatic up-regulation of CD69 surface expression. On the other hand, as in the case of monocyte-derived DCs, only the small CD56^{bright} NK cell subset could proliferate in the presence of PDCs and TLR 9 ligands (Romagnani et al., 2005). In this case, however, their proliferation appears to be IL-15-independent since, unlike in monocyte-derived DCs (Ferlazzo et al., 2004; Jinushi et al., 2003), surface IL-15 is not detectable in TLR-stimulated PDCs. In turn, NK cells are capable of promoting PDC maturation and of up-regulating their production of IFNα in response to CpG. It is of note that, while NK cells cannot exert an editing program on PDCs due to the poor susceptibility of these cells to NK-mediated lysis, when co-cultured with TLR9-stimulated PDCs, NK cells acquire lytic activity against monocyte-derived iDCs (Della Chiesa et al., 2006; Gerosa et al., 2005). Thus, one may speculate that cellular interactions occurring between NK and PDCs during viral infections may influence the maturation and acquisition of functional competence by bystander iDCs. Since mast cells also express TLR9, it is possible that a similar effect may be induced by mast cells-derived type I IFNs released upon TLR9 engagement.

PDCs, on activation by virus or CpG-ODN, express the ligand for the glucocorticoid-induced tumour necrosis factor receptor (GITR), a receptor expressed on activated NK cells. The PDC-mediated enhancement of NK cell cytotoxicity and IFNγ production requires both type I IFNs and GITR–GITR ligand interaction (Hanabuchi et al., 2006). Activated NK cells failed to induce PDC maturation when cells were cultured and separated by a porous membrane in transwell plates (Gerosa et al., 2005). Thus, similarly to the interactions occurring between NK cells and monocyte-derived DCs, those between NK and PDCs require close cell-to-cell contact as also demonstrated directly by the analysis of inflammatory tissues (Parolini et al., 2007).

Activated NK cells promote enhancement of IFNα release from PDCs as well as their maturation. These findings were originally reported using NK cells activated by IL-2, that is a cytokine produced primarily at late stages of an immune response. However, more recent data revealed that the regulatory effect of NK cells on PDC functions does not strictly require previous exposure of NK cells to IL-2 (Della Chiesa et al., 2006). Indeed, also freshly isolated NK cells promoted PDC maturation and IL-6 release. Moreover, abundant IFNα release from (TLR9-stimulated) PDCs was induced by NK cells exposed to IL-12, thus suggesting that during the early phases of innate immune responses, the release of IL-12 by PAMPs-activated myeloid DCs may represent an important link between NK–DC and NK–PDC interactions. The finding that maximal IFNα production by PDCs was induced by NK cells that had been co-cultured with PAMPs-exposed myeloid DCs further supports the notion of the existence of a multidirectional crosstalk among NK, myeloid-derived DCs and PDCs (Della Chiesa et al., 2006).

Conclusions

Major advances in our understanding of various cell types of innate immunity and of the role they play during the early phases of infections have occurred. Some of these cells, such as mast cells, endothelial cells and iDCs, are resident within peripheral tissues, while others, including NK cells and neutrophils are rapidly recruited from the blood stream. A number of relevant concepts are emerging. First, innate immunity cells interact with each other in inflamed tissues and in secondary lymphoid organs leading to modulation or amplification of various innate effector mechanisms. Second, a large array of microbial products can directly activate individual effector cells of the innate immunity, also including NK cells. Third, the outcome of these cellular interactions has a dramatic impact on the quality and strength of the ensuing adaptive response. Remarkably, the effect on the adaptive immunity can result not only from the action of polarizing cytokines such as IL-12, IL-18 or IL-4 but also of the NK-mediated 'DC editing' leading to the selection of the most suitable DCs for subsequent T cell priming. Classical innate effector cells can now be viewed also as regulatory cells that play a key role in defence against pathogens. Further understanding of the interaction within innate immunity and at the interface between innate and adaptive immunity should be of considerable aid in designing novel therapeutic strategies and effective vaccines against infectious agents and, perhaps, tumours.

Acknowledgements

This work was supported by grants awarded by Associazione Italiana per la Ricerca sul Cancro (A.I.R.C.), the Istituto Superiore di Sanità (I.S.S.), Ministero della Salute (Ricerca Finalizzata Ministeriale 2006), Ministero dell'Università e della Ricerca Scientifica e Tecnologica (M.I.U.R.-PRIN 2005, project 2005063024_004), FIRB-MIUR project-RBNE017B4 and European Union FP6, LSHB-CT-2004-503319-AlloStem (the European Commission is not liable for any use that may be made of the information contained). Also the financial support of Fondazione Compagnia di San Paolo, Torino, Italy, of Fondazione CARIGE and Fondazione CARIPLO is gratefully acknowledged.

References

Adam, C., King, S., Allgeir, T., Braumüller, H., Lüking, C., Mysliwietz, J., Kriegeskorte, A., Dirk, H., Busch, D.H., Röcken, M. and Mocikat, R. (2005). DC–NK cell cross talk as a novel CD4+ T-cell-independent pathway for antitumor CTL induction. Blood **106**, 338–344.

Agaugué, S., Marcenaro, E., Ferranti, B., Moretta, L. and Moretta, A. (2008). Human natural killer cells exposed to IL-2, IL-12, IL-18 or IL-4 differently modulate priming of naive T cells by monocyte-derived dendritic cells. Blood **112**, 1776–1783.

Akira, S. and Takeda, K. (2004). Toll-like receptor signalling. Nat Rev Immunol **4**, 499–511.

Banchereau, J. and Steinman, R.M. (1998). Dendritic cells and the control of immunity. Nature **392**, 245–252.

Banchereau, J., Briere, F., Caux, C., Davoust, J., Lebecque, S., Liu, Y.J., Pulendran, B. and Palucka, K. (2000). Immunobiology of dendritic cells. Annu Rev Immunol **18**, 767–811.

Biron, C.A., Nguyen, K.B., Pien, G.C., Cousens, L.P. and Salazar-Mather, T.P. (1999). Natural killer cells in antiviral defense: function and regulation by innate cytokines. Annu Rev Immunol **17**, 189–220.

Bottino, C., Castriconi, R., Moretta, L. and Moretta, A. (2005). Cellular ligands for activating NK receptors. Trends Immunol **26**(4), 221–226.

Bjerke, T., Gaustadnes, M., Nielsen, S., Nielsen, L., Schiøtz, P., Rudiger, N., Reimert, C., Dahl, R., Christensen, I. and Poulsen, L. (1996). Human blood eosinophils produce and secrete interleukin 4. Respir Med **90**, 271–277.

Byrne, P., McGuirk, P. and Todryk, S.Mills KHG, (2004). Depletion of NK cells results in disseminating lethal infection with *Bordetella pertussis* associated with a reduction of antigen-specific Th1 and enhancement of Th2, but not Tr1. Eur J Immunol **34**, 2579–2588.

Castriconi, R., Cantoni, C., Della Chiesa, M., Vitale, M., Marcenaro, E., Conte, R., Biassoni, R., Bottino, C., Moretta, L. and Moretta, A. (2003). TGFb1 inhibits expression of the NKp30 and NKG2D receptors: consequences for the NK-mediated killing of dendritic cells. Proc Natl Acad Sci U S A **100**, 4120–4125.

Cerwenka, A. and Lanier, L.L. (2001). Ligands for natural killer cell receptors: redundancy or specificity. Immunol Rev **181**, 158–169.

Colonna, M., Trinchieri, G. and Yong-Jun Liu, Y.-J. (2004). Plasmacytoid dendritic cells in immunity. Nat Immunol **5**, 1219–1226.

Constant, S.L. and Bottomly, K. (1997). Induction of Th1 and Th2 CD4+ T cell responses: the alternative approaches. Annu Rev Immunol **15**, 297–322.

Cooper, M.A., Fehniger, T.A., Fuchs, A., Colonna, M. and Caligiuri, M.A. (2004). NK cell and DC interactions. Trends Immunol **25**, 47–52.

Coudert, J.D., Coureau, C. and Guéry, J-C. (2002). Preventing NK cell activation by donor dendritic cells enhances allospecific CD4 T cell priming and promotes Th type 2 responses to transplantation antigens. J Immunol **169**, 2979–2987.

Degli-Esposti, M.A. and Smith, M.J. (2005). Close encounters of different kinds: dendritic cells and NK cells take centre stage. Nat Rev Immunol **5**, 112–124.

Della Chiesa, M., Vitale, M., Carlomagno, S., Ferlazzo, G., Moretta, L. and Moretta, A. (2003). The natural killer cell mediated killing of autologous dendritic cells is confined to a cell subset expressing CD94/NKG2A, but lacking inhibitory killer IG-like receptors. Eur J Immunol **33**, 1657–1666.

Della Chiesa, M., Sivori, S., Castriconi, R., Marcenaro, E. and Moretta, A. (2005). Pathogen induced-private conversations between natural killer and dendritic cells. Trends Microbiol **13**, 128–136.

Della Chiesa, M., Romagnani, C., Thiel, A., Moretta, L., and Moretta, A. (2006). Multidirectional interactions are bridging human NK cells with plasmacytoid and monocyte-derived dendritic cells during innate immune responses. Blood **108**(12), 3851–3858.

Dokun, A.O., Dortha, T.C., Yang, L., Bendelac, A.S. and Yokoyama, W.M. (2001). Analysis of in situ NK cell responses during viral infection. J Immunol **167**, 5286–5293.

Feau, S., Facchinetti, V., Granucci, F., Citterio, S., Jarrossay, D., Saresini, S., Protti, M.P., Lanzavecchia, A. and Ricciardi-Castagnola, P. (2005). Dendritic cell-derived IL-2 production is regulated by IL-15 in humans and in mice. Blood **105**, 697–702.

Ferlazzo, G., Tsang, M.L., Moretta, L., Melioli, G., Steinman, R.M. and Munz, C. (2002). Human dendritic cells activate resting natural killer (NK) cells and are recognized via the NKp30 receptor by activated NK cells. J Exp Med **195**, 343–351.

Ferlazzo, G., Morandi, B., D'Agostino, A., Meazza, R., Melioli, G., Moretta, A. and Moretta, L. (2003). The interaction between natural killer and dendritic cell in bacterial infections results in rapid induction of NK cell activation and in the lysis of uninfected dendritic cells. Eur J Immunol **33**, 306–313.

Ferlazzo, G., Pack, M., Thomas, D., Paludan, C., Schmid, D., Strowig, T., Bougras, G., Muller, W.A., Moretta, L. and Munz, C. (2004). Distinct roles of IL-12 and IL-15 in human natural killer cell activation by dendritic cells from secondary lymphoid organs. Proc Natl Acad Sci U S A **101**, 16606–16611.

Fernandez, N.C., Lozier, A., Flament, C., Ricciardi-Castagnoli, P., Bellet, D., Suter, M., Perricaudet, M., Tursz, T., Maraskovsky, E. and Zitvogel, L. (1999). Dendritic cells directly trigger NK cell functions: cross-talk relevant in innate anti-tumor immune responses in vivo. Nat Med **5**, 405–411.

Galli, S.J., Nakae, S. and Tsai, M. (2005). Mast cells in the development of adaptive immune responses. Nat Immunol **6**, 135–142.

Gerosa, F., Baldani-Guerra, B., Nisii, C., Marchesini, V., Carra, G. and Trinchieri, G. (2002). Reciprocal activating interaction between NK cells and dendritic cells. J Exp Med **195**, 327–333.

Gerosa, F., Gobbi, A., Zorzi, P., Burg, S., Briere, F., Carra, G. and Trinchieri, G. (2005). The reciprocal interaction of NK cells with plasmacytoid or myeloid dendritic cells profoundly affects innate resistance functions. J Immunol **174**, 727–734.

Granucci, F., Vizzardelli, C., Pavelka, N., Feau, S., Persico, M., Virzi, E., Rescigno, M., Moro, G. and Ricciardi-Castagnoli, P. (2001). Inducible IL-2 production by dendritic cells revealed by global gene expression analysis. Nat Immunol **2**, 882–888.

Granucci, F., Feau, S., Angeli, V., Trottein, F. and Ricciardi-Castagnoli, P. (2003). Early IL-2 production by mouse dendritic cells is the result of microbial-induced priming. J Immunol **170**, 5075–5081.

Hanabuchi, S., Watanabe, N., Wang, Y.H., Wang, Y.H., Ito, T., Shaw, J., Cao, W., Xiao-Feng Qin, F. and Liu, Y.J. (2006). Human plasmacytoid predendritic cells activate NK cells through glucocorticoid-induced tumor necrosis factor receptor-ligand (GITRL). Blood **107**, 3617–3623.

Imai, T., Hieshima, K., Haskell, C., Baba, M., Nagira, M., Nishimura, M., Kakizaki, M., Takagi, S., Nomiyama, H., Schall, T.J. and Yoshie, O. (1997). Identification and molecular characterization of fractalkine receptor CX_3CR1, which mediates both leukocyte migration and adhesion. Cell **91**, 521–530.

Jarrossay, D., Napolitani, G., Colonna, M., Sallusto, F. and Lanzavecchia, A. (2001). Specialization and complementarity in microbial molecule recognition by human myeloid and plasmacytoid dendritic cells. Eur J Immunol **31**, 3388–3393.

Jinushi, M., Takehara, T., Tatsumi, T., Kanto, T., Groh, V., Spies, T., Takahiro Suzuki*, T., Miyagi, T. and Hayashi, N. (2003). Autocrine/paracrine IL-15 that is required for type I IFN-mediated dendritic cell expression of MHC class I-related chain A and B is impaired in hepatitis C virus infection. J Immunol **171**, 5423–5429.

Krieg, A.M. (2002). CpG motifs in bacterial DNA and their immune effects. Annu Rev Immunol **20**, 709–760.

Kulka, M., Alexopoulou, L., Flavell, R. and Metcalfe, D. (2004). Activation of mast cells by double-stranded RNA: evidence for activation through Toll-like receptor 3. J Allergy Clin Immunol **114**, 174–182.

Langenkamp, A., Messi, M., Lanzavecchia, A. and Sallusto, F. (2000). Kinetics of dendritic cell activation: impact on priming of TH1, TH2 and nonpolarized T cells. Nat Immunol **1**, 311–316.

Long, E.O. (1999). Regulation of immune responses through inhibitory receptors. Annu Rev Immunol **17**, 875–904.

Lopez-Botet, M., Llano, M., Navarro, F. and Bellon, T. (2000). NK cell recognition of non-classical HLA class I molecules. Semin Immunol **12**, 109–119.

Mailliard, M.B., Son, Y.-I., Redlinger, R., Coates, P.T., Giermasz, A., Morel, P.A., Storkus, W.J. and Kalinski, P. (2003). Dendritic cells mediate NK cell help for Th1 and CTL responses: two-signal requirement for the induction of NK cell helper function. J Immunol **171**, 2366–2373.

Marcenaro, E., Ferranti, B. and Moretta, A. (2005a). On the usefulness of autoaggression. Autoimmun Rev **4**(8), 520–525.

Marcenaro, E., Della Chiesa, M., Bellora, F., Parolini, S., Millo, R., Moretta, L. and Moretta, A. (2005b). IL-12 or IL-4 prime human natural killer cells to mediate functionally divergent interactions with dendritic cells or tumors. J Immunol **174**, 3992–3998.

Marshall, J.S. (2004). Mast cell responses to pathogens. Nat Rev Immunol **4**, 787–799.

Marshall, J.S. and Jawdat, D.M. (2004). Mast cells in innate immunity. J Allergy Clin Immunol **114**, 21–27.

Moretta, A. (2002). Natural killer cells and dendritic cells: rendezvous in abused tissues. Nat Rev Immunol **2**, 957–965.

Moretta, A., Bottino, C., Vitale, M., Pende, D., Biassoni, R., Mingari, M.C. and Moretta, L. (1996). Receptors for HLA-class I molecules in human natural killer cells. Annu Rev Immunol **14**, 619–648.

Moretta, A., Bottino, C., Vitale, M., Pende, D., Cantoni, C., Biassoni, R., Mingari, M.C. and Moretta, L. (2001). Activating receptors and coreceptors involved in the natural cytolysis. Annu Rev Immunol **19**, 197–223.

Moretta, A., Marcenaro, E., Sivori, S., Della Chiesa, M. , Vitale, M. and Moretta, L. (2005). Early liaisons between cells of the innate immune system in inflamed peripheral tissues. Trends Immunol **26**, 668–675.

Moretta, L., Ferlazzo, G., Bottino, C., Vitale, M., Pende, D., Mingari, M.C. and Moretta, A. (2006). Effector and regulatory events during natural killer–dendritic cells interactions. Immunol Rev **214**, 219–228.

Nagase, H. (2003). Expression and function of Toll-like receptors in eosinophils: activation by Toll-like receptor 7 ligand. J Immunol **171**, 3977–3982.

Parolini, S., Santoro, A., Marcenaro, E., Luini, W., Massardi, L., Facchetti, F., Communi, D., Parmentier, M., Majorana, A., Sironi, M., Tabellini, G., Moretta, A. and Sozzani, S. (2007). The role of chemerin in the co-localization of NK and dendritic cell subsets into inflamed tissues. Blood **109**, 3625–3632.

Piccioli, D., Sbrana, S., Melandri, E. and Valiante, N.M. (2002). Contact-dependent stimulation and inhibition of dendritic cells by natural killer cells. J Exp Med **195**, 335–341.

Pisegna, S., Pirozzi, G., Piccoli, M., Frati, L., Santoni, A. and Palmieri, G. (2004). p38 MAPK activation controls the TLR3-mediated upregulation of cytotoxicity and cytokine production in human NK cells. Blood **104**, 4157–4164.

Raulet, D.H. (2003). Roles of the NKG2D immunoreceptor and its ligands. Nat Rev Immunol **3**, 781–790.

Raulet, D.H. (2004). Interplay of natural killer cells and their receptors with the adaptive immune response. Nat Immunol 5, 996–1002.

Reis e Sousa, C. (2001). Dendritic cells as sensors of infection. Immunity **14**, 495–498.

Romagnani, C., Della Chiesa, M., Kohler, S., Moewes, B., Radbruch, A., Moretta, L., Moretta, A. and Thiel, A. (2005). Activation of human NK cells by plasmacytoid DC and its modulation by CD4 + T helper cells and CD4+ CD25++T regulatory cells. Eur J Immunol **35**, 2452–2458.

Saïdi, H., Melki, M.T. and Gougeon, M.L. (2008). HMGB1-dependent triggering of HIV-1 replication and persistence in dendritic cells as a consequence of NK–DC cross-talk. PLoS ONE **3**(10), e3601.

Sallusto, F. and Lanzavecchia, A. (1999). Mobilizing dendritic cells for tolerance, priming, and chronic inflammation. J Exp Med **189**, 611–614.

Scharton, T.M. and Scott, P. (1993). Natural killer cells are a source of interferon gamma that drives differentiation of CD4+ T cell subsets and induces early resistance to *Leishmania major* in mice. J Exp Med **178**, 567–577.

Schmidt, K.N., Leung, B., Kwong, M., Zarember, K.A., Satyal, S., Navas, T.A., Wang, F. and Godowski, P.J. (2004). APC-independent activation of NK cells by the Toll-like receptor 3 agonist double-stranded RNA. J Immunol **172**, 138–143.

Sivori, S., Falco, M., Della Chiesa, M., Carlomagno, S., Vitale, M., Moretta, L. and Moretta, A. (2004). CpG and double-stranded RNA trigger human NK cells by Toll-like receptors: induction of cytokine release and cytotoxicity against tumours and dendritic cells. Proc Natl Acad Sci U S A **101**, 10116–10121.

Sivori, S., Carlomagno, S., Moretta, L. and Moretta, A. (2006). Comparison of different CpG oligodeoxynucleotide classes for their capability to stimulate human NK cells. Eur J Immunol **36**(4), 961–967.

Sivori, S., Falco, M., Carlomagno, S., Romeo, E., Moretta, L. and Moretta, A. (2007). Heterogeneity of TLR3 mRNA transcripts and responsiveness to poly (I:C) in human NK cells derived from different donors. Int Immunol **19**(12), 1341–1348.

Steinman, R.M. and Nussenzweig, M.C. (2002). Avoiding horror autotoxicus: the importance of dendritic cells in peripheral T cell tolerance. Proc Natl Acad Sci U S A **99**, 351–358.

Supajatura, V., Ushio, H., Nakao, A., Akira, S., Okumura, K., Ra, C. and Ogawa, H. (2002). Differential responses of mast cell Toll-like receptors 2 and 4 in allergy and innate immunity. J Clin Invest **109**, 1351–1359.

Tay, C.H. and Welsh, R.M. (1997). Distinct organ-dependent mechanisms for the control of murine cytomegalovirus infection by natural killer cells. J Virol **71**, 267–275.

Trinchieri, G. (2003). Interleukin-12 and the regulation of innate resistance and adaptive immunity. Nat Rev Immunol **3**, 133–146.

Tseng, C.T. and Rank, R.G. (1998). Role of NK cells in early host response to chlamydial genital infection. Infect Immun **66**, 5867–5875.

Vilches, C. and Parham, P. (2002). KIR: diverse, rapidly evolving receptors of innate and adaptive immunity. Annu Rev Immunol **20**, 217–251.

Vitale, M., Della Chiesa, M., Carlomagno, S., Romagnani, C., Thiel, A., Moretta, L. and Moretta, A. (2004). The small subset of CD56bright CD16− natural killer cells is selectively responsible for both cell proliferation and interferon-gamma production upon interaction with dendritic cells. Eur J Immunol **34**, 1715–1722.

Vitale, M., Della Chiesa, M., Carlomagno, S., Pende, D., Aricò, M., Moretta, L. and Moretta, A. (2005). NK-dependent DC maturation is mediated by TNFα and IFNγ released upon engagement of the NKp30 triggering receptor. Blood **106**(2), 566–571.

Walzer, T., Dalod, M., Robbins, S.H., Zitvogel, L. and Vivier, E. (2005). Natural killer cells and dendritic cells: "l'union fait la force". Blood **106**, 2252–2258.

Wilson, J.L., Heffler, L.C., Charo, J., Scheynius, A., Bejarano, M.-T. and Ljunggren, H.G. (1999). Targeting of human dendritic cells by autologous NK cells. J Immunol **163**, 6365–6370.

Zitvogel, L. (2002). Dendritic and natural killer cells cooperate in the control/switch of innate immunity. J Exp Med **195**, 9–14.

PART IV

NK cells in disease

Chapter Thirty-Three

33

Natural killer cells and transplantation

Benjamin M. Matta, Angus W. Thomson

CHAPTER CONTENTS

Lukewarm acceptance is more bewildering than outright rejection.
Martin Luther King Jr.

ABSTRACT

Natural killer (NK) cells play a prominent role in determining the outcome of allogeneic hematopoietic cell transplantation. This is based on the concept of 'missing self', whereby NK cells recognize and eliminate foreign target cells due to their lack of expression of self MHC class I. Despite their significant contribution to graft-versus-leukemia activity and graft-versus-host disease in hematopoietic cell transplantation, the extent to which donor and recipient NK cells impact the outcome of solid organ allografts is comparatively unknown. Recent evidence suggests that the involvement of NK cells in solid organ allograft acceptance and rejection may be dependant on the organ being transplanted and on how their function is affected by immunosuppressive drug therapy. Although it has been shown that NK cells are neither sufficient nor required for rejection of organ allografts, their interactions with dendritic cells, T cells, and B cells may enhance or suppress antidonor responses.

KEY WORDS

Rapamycin, Tacrolimus, Mycophenolic acid, Cyclosporine, Organ transplantation, Allograft, Rejection

Introduction

The ability of natural killer (NK) cells to recognize and kill allogeneic cells efficiently without prior sensitization, based on their lack of self MHC class I expression (the idea of 'missing self') forms the basis for alloreactivity in the setting of bone marrow transplantation. Surprisingly, despite their potent cytotoxic activity against allogeneic cells, far less is understood about the role of NK cell allorecognition and function in organ transplantation.

In the setting of organ transplantation, NK cells are derived from both the donor (as 'passenger leukocytes') and the recipient and can play an important role in transplant outcome. Although considered to be rare following organ transplantation, donor-derived NK cells have the potential to initiate graft-versus-host disease (GVHD) whereby donor NK cells initiate an immune response against the host. The liver contains a high proportion of NK cells (approximately 30% of all nonparenchymal cells), and fittingly, many cases of GVHD following organ transplantation are reported in liver or small bowel transplant patients, although there are case reports of GVHD following transplantation of other organs, including lung, kidney, heart, pancreas and spleen (Assi et al., 2007). Many studies of GVHD following organ transplantation are performed using models of small bowel transplantation. Early reports showed increased NK cell cytolytic activity in small bowel transplant recipients who developed GVHD as a result (Frezza et al., 1994). Conversely, host-derived NK cells recruited to the graft have been reported to play a key role in preventing GVHD initiated by donor-derived cells (Fandrich et al., 1996a,b, 1999). Although GVHD following organ transplantation is associated with a high mortality rate, it remains a relatively rare occurrence, and thus, the focus of this chapter will be on the role of host- and recipient-derived NK cells in organ allograft acceptance and rejection.

More than 20 years ago, it was reported that in vivo depletion of NK cells in the rat using an anti-NK cell monoclonal antibody (mAb) (anti-asialo GM-1/ASGM-1) did not significantly prolong cardiac allograft survival compared to nontreated animals, indicating a minimal involvement of NK cells in the acute rejection response (Heidecke et al., 1985). These findings were supported by later studies using recombinase activating gene-deficient ($Rag^{-/-}$) mice, which are deficient in T cells and B cells but retain an intact innate immune response and functional NK cell population. Fully MHC-mismatched wild-type skin or heart allografts transplanted into $Rag^{-/-}$ mice survived indefinitely, indicating that, in the absence of adaptive immunity, NK cells were not sufficient to reject organ allografts (Bingaman et al., 2000). Though not potent enough to cause graft rejection, the innate immune response still persists without an adaptive response and may provide insight into the role NK cells have in acceptance or rejection of organ allografts (He et al., 2002).

NK cells possess many properties that enable them to promote organ allograft rejection. As mentioned previously, NK cells can be activated through recognition of nonself MHC class I molecules on allogeneic cells, and as a component of the innate immune response, their activation does not require prior sensitization. An important consequence of NK cell activation is rapid production of cytolytic molecules, including perforin and granzymes that can directly lyse/kill allogeneic cells. This is mediated through Fc receptor recognition of immune complexes via Ab-dependent cell-mediated cytotoxicity (ADCC) or cytolytic activity as a result of activation through activating receptors. Activated NK cells also produce an array of proinflammatory cytokines, including tumour necrosis factor (TNF)-α/β and interferon (IFN)-γ, that can activate and support the adaptive arm of the immune response.

An important characteristic of organ allografts is that they are highly vascularized, and rejection episodes, whether acute or chronic, often involve damage to the endothelium. NK cells can be recruited early to the graft, which opens the door for NK cell–endothelial cell (EC) interactions that may result in both direct and indirect NK cell-mediated EC damage (Robinson et al., 2000; Yoneda et al., 2000).

Despite the lack of significant understanding of the role of NK cells in organ transplantation rejection and acceptance, a broader understanding of NK cell activating and inhibitory receptors, as well as NK cell interactions with dendritic cells (DCs), T cells and B cells may shed light on their ability to contribute to acute and chronic rejection (Brilot et al., 2008; Degli-Esposti and Smyth, 2005; Lanier, 2008; Martin-Fontecha et al., 2004; Rocha et al., 2003; Rosen et al., 2008). In this chapter, we will discuss the role of NK cells in regulating acceptance and rejection of various transplanted organs (allografts and xenografts) in small animal and nonhuman primate (NHP) models, as well as in the clinical setting. We will also briefly discuss the role of chemokines in directing NK cells to an allograft, as well as the effects of clinically used immunosuppressants on NK cell function and how this may impact graft outcome.

Heart transplantation

The heterotopic heart transplant model in rodents is one of the most widely used organ transplant models. It is easy to assess rejection (cessation of palpable heartbeat) as well as to evaluate chronic rejection based on the development of cardiac allograft vasculopathy (CAV), which is characterized by cellular infiltration into the vascular tissue, thickening of the intima due to smooth muscle cell proliferation, causing ischaemia, cell death, and ultimately graft failure.

Early reports revealed that NK cells are not sufficient to reject cardiac allografts in the absence of an adaptive immune response. Subsequent study has shown that they may still contribute to heart allograft rejection. $CD28^{-/-}$ mice, which lack adequate T cell costimulation, reject fully MHC-mismatched allogeneic hearts with similar kinetics compared to $CD28^{+/+}$ recipients but show significant prolongation of allograft survival when they are also depleted of NK cells (Maier et al., 2001). The NK cell contribution to cardiac allograft rejection in these mice may be mediated by upregulation of the activating receptor NKG2D on NK cells, as mAb blockade of

NKG2D also results in significantly prolonged graft survival, without depleting NK cells or inhibiting their migration to the graft (Kim et al., 2007). However, additional studies using the same clone (CX5) of anti-NKG2D blocking Ab did not reproduce these results, possibly due to differences in the dose and duration of Ab administration (McNerney et al., 2006). These findings indicate that NK cell activation following transplantation may be able to directly or indirectly provide T-cell costimulatory signals that can promote cardiac allograft rejection.

There is increasing evidence for a role of NK cells in the development of CAV and chronic rejection of heart allografts. To evaluate the role of NK cells in the presence of a functional adaptive immune response, experiments have been performed utilizing a transplant model of parent-to-F1 'hybrid resistance', where both donor and recipients retain a fully competent adaptive response compared to $Rag^{-/-}$ mice that completely lack T cells and B cells. C57BL/6 donor (parental) hearts were grafted heterotopically into F1 offspring of C57BL/6 × BALB/c parents. Since the F1 offspring contain both parental alleles, they do not recognize either the C57BL/6 or BALB/c parent tissues as foreign, thereby eliminating a T-cell mediated and B-cell mediated adaptive immune response, while maintaining the NK cell response based on recognition of foreign MHC class I on the graft. In this model, parental donor hearts are accepted long-term (for >8 weeks posttransplant) similar to syngeneic grafts; however, 19 of 22 hearts from parental donors in F1 mice showed evidence of CAV at 8 weeks, whereas all of the mice receiving isografts showed normal vascular architecture (Uehara et al., 2005a). Further study showed that NK cells from F1 recipients had higher levels of cytotoxic activity against parental donor hearts during the first 5 weeks posttransplant compared to reactivity against syngeneic grafts (Uehara et al., 2005b). The development of CAV was dependant upon NK cells, host-derived IFN-γ and $CD4^+$ T cells (nonspecific for donor antigen), suggesting a requirement for NK cell–$CD4^+$ T-cell interactions and the development of a Th1-skewed response.

The requirement for IFN-γ and potential for NK cell–T cell interactions to facilitate the development of CAV and chronic rejection may be multifaceted. Human microvascular endothelial cells (MVEC) significantly upregulate MHC class II in culture with NK cells or NK cell/MVEC culture supernatants, an effect that is abrogated by IFN-γ neutralizing Abs (McDouall et al., 1997). These findings indicate that NK cells recruited to the graft may indirectly promote antidonor T-cell responses through reciprocal interactions with $CD4^+$ T cells and MHC class II-expressing cells in an IFN-γ-dependent manner.

Lung transplantation

In comparison to other transplanted organs, fewer studies have examined the role of NK cells in lung transplantation. NK cell percentages are elevated in bronchoalveolar lavage fluid (BALF) of human lung graft recipients when compared with healthy controls (Ward et al., 2001). However, evidence of a correlation between the presence of NK cells and lung allograft rejection is minimal (Gregson et al., 2008). NK cells are indeed increased in lung allografts of chronically rejecting patients; the number of circulating NK cells in the peripheral blood reciprocally decreases. They have an activated phenotype, suggesting that these cells may be trafficking to the graft as part of the rejection response (Fildes et al., 2008).

In a comparable model of chronic rejection of tracheal allografts in mice (obliterative airway disease or OAD; obliterative bronchiolitis or OB/bronchiolitis obliterans syndrome or BOS in lung transplantation), OAD developed in $Rag\text{-}1^{-/-}$ mice as well as in wild-type mice depleted of NK cells using an anti-NK1.1 mAb, suggesting that NK cells do not contribute to the pathology associated with chronic rejection in tracheal allografts (Higuchi et al., 2002).

Early reports in a model of canine lung transplantation indicated that, within several hours of transplant, NK cells recovered from the BALF exhibited cytotoxicity levels nearly twice as high as those in samples taken prior to surgery (Adoumie et al., 1992). Correlative with this, NK cell cytolytic activity in the BALF increased in an Ag-independent manner following tapering of Cyclosporine A (CsA) dosage (approximately 8 weeks posttransplant) in canine lung transplant recipients (Norin et al., 1999). In the Adoumie et al. study, the early response may have resulted from ischaemia/reperfusion injury or infection and not from a response directed against the graft, as later studies were unable to detect NK cell cytolytic in BALF samples in rejecting canine lung allografts (Nguyen et al., 1993). In fact, the NK cell chemoattractant CCL5 is significantly upregulated in the BALF within days of orthotopic lung transplant in rats, and is upregulated early following ischaemia/reperfusion injury in highly vascularized kidney transplant models (Belperio et al., 2000; Nagano et al., 1997). Together, these results suggest that activation of NK cell cytolytic activity may contribute indirectly to allograft rejection through a mechanism that does not require NK cell recognition of nonself MHC class I on the donor tissue, but that activation is most likely due to injury associated with the stress of the ischaemic insult at the time of transplantation.

Small bowel transplantation

There is increased expression of IFN-γ and increased infiltration of NK cells in chronically rejecting rat intestinal allografts when compared with syngeneic controls (Su et al., 1996). Inhibition of IFN-γ-inducible protein-10 (IP-10) results in prolonged allograft survival and reduced NK cell graft infiltration (Zhang et al., 2002). Additionally, there is a reduction in infiltrating NK cells (comparable to levels in syngeneic grafts) in accepted small bowel allografts associated with spontaneous tolerance of liver allografts in rats, suggesting a role for NK cells in intestinal allograft rejection, although their functional activity or phenotypic activation status was not evaluated (Sarnacki et al., 1998).

Liver transplantation

Since the first human liver transplant was performed nearly half a century ago, the role for NK cells in the outcome of hepatic allografts has remained unclear. Recent clinical studies have shown that, despite obvious in vitro NK cell donor alloreactivity pretransplant (based on HLA class I mismatch), there was no correlation with acute liver allograft rejection (Oertel et al., 2001). These results suggest that the degree of matching of HLA class I between donor and recipient may not reflect the incidence of acute rejection following liver transplantation. Recent evidence suggests that the donor HLA-C genotype, the ligand for individual killer inhibitory receptors (KIR) on NK cells, may be a potential predictor of graft survival or rejection, liver injury, cirrhosis and fibrosis (Hanvesakul et al., 2008). Interestingly, liver allografts from donors who were heterozygous for the C2 allele (HLA-C1/C2) or those that were homozygous (HLA-C2/C2) survived significantly longer (16.2% and 26.5% greater survival, respectively, at 10 years) compared to livers from donors who were homozygous for the HLA-C group 1 allele (HLA-C1/C1). One explanation for the role of HLA-C and NK cells in the improved graft survival in these patients is that HLA-C2 may deliver a stronger inhibitory signal through KIR on recipient NK cells than HLA-C1, as was suggested in various other model systems (Ahlenstiel et al., 2008; Hiby et al., 2004; Khakoo et al., 2004).

In a rat model of orthotopic liver transplantation (OLT), the development of acute rejection was associated with increased graft expression of the NK cell-activating receptor rNKp30, which has a human ortholog (Hsieh et al., 2004). There is evidence for infiltration of liver allografts by recipient-derived NK cells in a rat OLT model. Over half (55.8%) of the liver allograft-infiltrating cells at 24 h posttransplant were NKR-P1$^+$ NK cells, 20% of those being host-derived (Obara et al., 2005). This study showed that host-derived NK cells exhibited an almost bimodal infiltration pattern, starting as early as 6 h posttransplant, with a second wave starting around day 3. The authors concluded that the initial surge of NK cells following ischaemia/reperfusion injury was in response to the chemokines CCL2 and CCL3, with the second wave of NK cells recruited to the graft in response to CXCL10 and CX_3CL1 as part of the antidonor rejection response. Anti-asialo-GM1 mAb depletion of NK cells in vivo resulted in prolonged liver allograft survival and decreased serum IFN-γ levels produced by host-derived, liver allograft-infiltrating NK cells, as shown by flow cytometry. A similar study in a rat OLT model reported that markers of liver injury (alanine aminotransferase or ALT and aspartate aminotransferase or AST) were significantly higher in liver allografts compared to syngeneic controls, and that IFN-γ levels were higher as a result of NK cell activation (Shen et al., 2008). Anti-NKRP1 mAb depletion of NK cells in vivo restored hepatocyte proliferation and reduced IFN-γ levels compared to control animals.

Initially demonstrated in pigs and subsequently in mice and certain rat strains, liver allografts are accepted 'spontaneously' in the absence of immunosuppression or costimulation blockade. In humans, some liver transplant patients develop operational tolerance, which allows the withdrawal of immunosuppressive therapy without experiencing rejection episodes. Transcriptional profiles of peripheral blood mononuclear cells (PBMC) from liver transplant recipients (both tolerant and rejecting) compared to nontransplanted healthy controls showed a positive correlation between unique NK cell profiles and the spontaneous development of operational tolerance (Martinez-Llordella et al., 2008). Although there is no difference between groups in the frequency of NK cells in the PBMC, there is evidence for a positive correlation between tolerance and increased expression of NK cell surface receptors, including genes encoding CD94, NKG2D, NKG7, KLRC2, BY55, (CD160), KLRB1 and KLRC1 (Martinez-Llordella et al., 2007).

Kidney transplantation

Reports of early graft infiltration by NK cells in vascularized organs is most likely attributable to nonspecific recruitment due to ischaemia-reperfusion injury (Coulson et al., 2005). However, there is evidence that the presence of NK cells with high cytolytic activity is associated with rejection. Early reports of human kidney allograft recipients suggest a correlation between decreased numbers of peripheral blood NK cells and long-term survival (average posttransplant follow-up at 7.2 years of patients on a daily azathioprine and prednisone immunosuppressive regimen), as well as a shift towards reduced cytolytic activity in the circulating NK cell population (Legendre et al., 1986). Additionally, there was a significant accumulation of granzyme positive cells (>80% $CD56^+$ NK cells) in kidney biopsies of patients undergoing acute rejection compared to non-transplanted control patients with other forms of kidney disease (Kummer et al., 1995). Together, these results favour a role for NK cells and their cytolytic activity in kidney allograft rejection.

In a rat renal transplant model, untreated recipients showed increased, alloAg-specific target cell lysis compared with cyclosprine A-treated animals or animals treated with antidonor serum (serum collected from recipient strain rats immunized with donor-strain splenocytes), suggesting that renal allograft rejection was mediated by alloreactive, Ag-specific T cells and not other effectors, such as NK cells (Bradley et al., 1985).

Skin transplantation

Early reports on the role of NK cells in the rejection of skin allografts reported similar findings that NK cells were not recruited to the graft bed and were neither necessary nor sufficient to induce allogeneic skin rejection (Yamamoto et al., 1998; Zijlstra et al., 1992). Additionally, treatment with the NK cell-depleting anti-asialo-GM1 mAb did not prolong or prevent skin allograft rejection (Dalloul et al., 1996). Subsequent reports, however, indicate that NK cells can infiltrate the graft and contribute to fully MHC-mismatched murine skin allograft rejection by promoting early inflammation through chemokine production. Levels of keratinocyte chemoattractant (KC)/CXCL1 and macrophage inflammatory protein (MIP)-1 α/β, which attract neutrophils, monocytes, DCs and T cells, are decreased in skin allograft recipients that were depleted of NK cells (Kondo et al., 2000).

NK cells can also promote antidonor Th1 (IFN-γ-mediated) responses through interactions with MHC class I-expressing, donor Ag-presenting cells (APC). In the absence of a functional $CD8^+$ T-cell response, NK cell recognition of MHC I on donor APC results in Th1 $CD4^+$ T cell priming, whereas anti-asialo GM1 mAb depletion of NK cells results in skewing towards an IL-4/IL-10-producing Th2 response; however, there was no effect on graft survival (Coudert et al., 2002). Similar experiments have evaluated the influence of NK cells on the indirect (host-derived APC-host-derived T-cell interactions) versus direct (donor APC-host-derived T-cell interactions) pathway in the activation of alloreactive $CD4^+$ T cells. Skin grafts from MHC II-deficient (H-2^b) or C57BL/6 (H-2^b) wild-type mice were transplanted onto severe combined immune deficient (SCID) mice (H-2^d), which were reconstituted with BALB/c (H-2^d) $CD4^+$ T cells, thus eliminating alloAg recognition via the direct pathway. Analysis of graft-infiltrating cells showed significant accumulation of $CD4^+$ T cells, macrophages and NK cells at 4 weeks posttransplant and 2 weeks following adoptive transfer of T cells (Ito et al., 2008). Subsequent mAb-mediated depletion of NK cells in the recipient (anti-asialo GM1) or mAb blockade of the NK cell-activating receptor NKG2D (anti-NKG2D) prior to skin transplantation resulted in significantly prolonged survival of MHC $II^{-/-}$ but not wild-type grafts, suggesting that activation of host NK cells can promote $CD4^+$ T cell-mediated skin allograft rejection via the indirect pathway of allorecognition.

Although NK cells are neither sufficient nor necessary for allograft rejection, recent studies have shown that fully activated NK cells (via soluble IL-15/IL-15R complexes) can induce acute skin allograft rejection in the absence of an adaptive immune response using $Rag^{-/-}$ recipients (Kroemer et al., 2008). Additionally, the activated NK cells that induced rejection do not develop a memory-like response (IL-15 is a key cytokine for memory T cells). This was evident from the fact that if the skin allografts were not completely rejected by the IL-15-activated NK cells, the remaining part of the graft survived indefinitely.

Conversely, an important finding regarding the ability of NK cells to contribute to allograft tolerance in organ transplantation was made in a study of fully MHC-mismatched skin allografts in $Rag^{-/-}$ (T cell and B cell deficient) or $Rag^{-/-}\gamma c^{-/-}$ mice (T cell, B cell and NK cell deficient). Host NK cells are necessary to eliminate donor-derived APC that normally contribute to direct allorecognition in T-cell-competent mice (Yu et al., 2006). Lack of NK cells in the $Rag^{-/-}\gamma c^{-/-}$ mice resulted in long-term (>100 day) colonization of donor $CD11c^+$ APCs in recipient spleens and extra-lymphoid organs, including the liver and lungs, that are able to induce proliferation of adoptively transferred, CFSE-labelled $CD4^+$ T cells. Upon restimulation ex vivo, these same T cells secrete high levels of IFN-γ, suggesting that in the absence of host NK cells, donor APC persist in the host and are able to prime host $CD4^+$ T cells efficiently via the direct pathway of allorecognition. Taking this model a step further, the same study showed that NK cells were critical for the induction of tolerance using anti-OX40L, anti-CD154 (anti-CD40L) and cytotoxic T lymphocyte Ag (CTLA)-4Ig triple costimulation blockade therapy. Moreover, in another study, NK cells regulate both $CD4^+$ and $CD8^+$ allogeneic T-cell responses through perforin-mediated killing of donor-derived DCs in recipient lymph nodes (Laffont et al., 2008). Taken together, these results provide strong evidence that NK cells may play a key role in organ allograft tolerance induction through regulation of allorecognition via the direct pathway.

Pancreatic islet transplantation

Pancreatic islet cell transplantation as a treatment for type-1 diabetes mellitus (TIDM) holds significant potential as a supplement to currently used immunosuppressive drug regimens that have proven highly toxic to sensitive islet cells. NK cells have been implicated in the development and progression of autoimmune TIDM through destruction of insulin-producing pancreatic islet β cells. In a bio-breeding (BB) rat model of spontaneous diabetes, where tolerance was induced through injection of donor cells into neonatal recipients, it was shown that diabetic rats that were treated with anti-asialo GM1 mAb did not reject islet allografts and had reduced cytolytic activity compared to controls (Jacobson et al., 1988). These findings suggested that NK cells were crucial for the recurrence of autoimmune disease through rejection of the transplanted islets. In contrast to this study, recent evidence suggests that in models of induced tolerance through costimulation blockade, NK cells may actually be required for tolerance to islet allografts. Using a tolerance-induction protocol involving treatment with anti-CD154 (anti-CD40L; MR1), anti-LFA-1 (anti-CD11a) or a combination of both, mAb depletion of NK1.1$^+$ cells abrogated tolerance of the islet allografts (Beilke et al., 2004, 2005). Further analysis revealed that this was not due to the presence of CD3$^+$NK1.1$^+$ NKT cells and that the ability of NK1.1$^+$ cells to promote tolerance was dependant on perforin, since tolerance could be restored in perforin-deficient mice through adoptive transfer of NK1.1$^+$ perforin$^+$ cells (Beilke et al., 2005).

These conflicting results suggest that the ability of NK cells to contribute to rejection or promote tolerance to islet allografts is most likely dependant upon the method of tolerance induction. In addition, data showing that NK cells are actually required for tolerance indicate that further investigation is needed in this relatively unexplored model of islet allograft transplantation for consideration of treatment strategies that preserve NK cell function.

Xenotransplantation

Xenotransplantation, the transplantation of organs across species barriers, offers a potential solution to the relative shortage of available human organs, which leaves thousands of patients on transplant waiting lists for months and years. It has evolved substantially with the development of transgenic animal models and Ab treatments that have made the study of the immunologic mechanisms that contribute to xenograft rejection easier to dissect. With respect to translational application to the clinic, pigs appear to be the most suitable donor species for human recipients, despite less genetic similarity compared to NHP. This is due to organ size, animal availability/breeding, potential for genetic alteration and ethical consideration. Thus, pig to NHP xenografting is regarded as a relevant preclinical model, although rodent models still provide valid insight into the mechanisms that contribute to xenograft tolerance and rejection.

The most significant obstacles preventing acceptance of xenografts are hyperacute rejection (HAR), mediated by pre-existing, xeno-reactive antibodies that initiate the complement cascade, and acute vascular rejection (AVR), characterized by EC activation and subsequent direct cell-mediated damage (Rieben and Seebach, 2005). Indeed, experiments indicating a role for NK cells and innate immunity in xenotransplantation showed that the induction of species-specific tolerance was dependant upon xenotolerance in both the B cell (xenoreactive Abs) and NK cell (ADCC and direct cell lysis) compartments (Lin et al., 1998). Evidence from various xenograft models suggests that NK cells infiltrate the xenograft, directly lyse and/or activate xenogeneic EC and contribute to rejection through ADCC via Fc receptor binding of xenoreactive Abs bound to EC, in parallel with complement activation (Goodman et al., 1996; Quan et al., 2000; Yin et al., 2004). Other mechanisms have been reported for the ability of NK cells to destroy xenograft EC, including perforin- and granzyme-mediated killing and upregulation of cytolytic activity through ligation of NK cell-activating receptors (NKp44 and NKG2D) by proteins (UL16-binding protein 1) expressed on the endothelium (Forte et al., 2005; Lilienfeld et al., 2006, 2008; Matter-Reissmann et al., 2002). The NK cell-activating receptors most likely have overlapping functions, since complete protection from cytolytic killing required blockade of both NKp44 and NKG2D (Forte et al., 2005). It is also believed that variability in class I gene sequences across species barriers prevents or reduces signalling through NK cell-inhibitory receptors, making xenogeneic cells more susceptible to NK cell-mediated killing (Sullivan et al., 1997).

In the past decade, much of the focus on inhibiting the NK cell contribution to xenograft rejection has been on the nonclassical HLA class I molecules HLA-E and HLA-G, due to their relatively low level of genetic polymorphism in comparison to the classical HLA class I molecules (Crew, 2007). Studies have shown that induced expression of HLA-G variants in porcine EC confers protection against human NK cell-mediated lysis (Matsunami et al., 2001; Sasaki et al., 1999b). Reported mechanisms include inhibition of NK cell adhesion to and transmigration across EC (Dorling et al., 2000; Forte et al., 2001). Soluble HLA-G (sHLA-G), derived from allelic variants of the membrane-bound form, shows promise as a potential therapeutic agent to prevent xenograft rejection by NK cells, supported by data showing that elevated levels of serum sHLA-G in allograft recipients are positively associated with reduced rejection episodes and positive graft function following kidney, heart and liver transplantation (Basturk et al., 2006; Lila et al., 2002; Qiu et al., 2006). The addition of sHLA-G to NK cell or NK cell-porcine EC cocultures inhibits NK cell cytolytic activity and, in some cases, results in the induction of NK cell apoptosis (Lindaman et al., 2006; Yao et al., 2004; Zeng et al., 2006). In addition, although it is appreciated that several NK cell receptors may participate in these inhibitory responses, immunoglobulin-like transcript-2 (ILT-2) appears to be important in mediating the inhibitory effects of HLA-G on NK cells (Forte et al., 2001). A potential unforeseen effect of sHLA-G inhibition of NK cell cytolytic function has emerged from a recent study showing that sHLA-G also acts on DCs and regulates their function so that they have an impaired ability to activate NK cells, suggesting an indirect regulatory role for this molecule on NK cells (Gros et al., 2008).

The ability of HLA-E to directly inhibit NK cell function has also been shown, and the significance of this molecule in EC function is evidenced in a recent report showing that EC activated by proinflammatory cytokines (TNFα, IFN-γ or IL-1β) upregulate both surface expression of and release of soluble HLA-E as an intrinsic regulatory mechanism (Coupel et al., 2007). In addition, the expression of HLA-E on porcine EC protects them from lysis through ligation of CD94/NKG2 on human NK cells, resulting in recruitment of SH2 domain-containing protein phosphatase-1 (SHP-1) and subsequent downregulation of cytolytic activity (Lilienfeld et al., 2007; Sasaki et al., 1999a).

Xenogeneic pancreatic islet cell transplantation is under intensive investigation as an alternative to allogeneic human islet cells based on the fact that islet transplantation in humans may require islets from multiple donors in order to isolate sufficient numbers of cells for transplantation. Thus, studies have been conducted on the ability of NK cells to recognize and destroy porcine islet cell targets. In contrast to other models of xenotransplantation, several lines of evidence suggest that foetal or neonatal porcine islets are not targeted by human NK cells for killing. It has been shown that rejection of foetal porcine islet cell-like clusters in mouse recipients is independent of NK cell activity, since NK1.1$^+$ mAb depletion did not prevent xenograft rejection (Karlsson-Parra et al., 1996). This occurred despite the fact that approximately 15% of the graft-infiltrating cells in untreated recipient mice were NK1.1$^+$ cells. In support of this observation, a later study showed that human NK cells were shown to be cytolytic against porcine splenocytes but did not kill neonatal porcine islet cells (Murray et al., 1999). Although the mechanisms responsible for this phenomenon remain unclear, it may partially reflect the lack or lower expression of adhesion and costimulatory molecules on neonatal/foetal cells that reduces their interactions with NK cells.

An interesting recent finding concerns the requirement for NK cells for the induction of tolerance. Beilke et al. (2005) have previously reported that NK1.1$^+$ cells were critical for anti-CD154/anti-LFA-1-induced islet allograft tolerance in mice. However, another study from the same group recently reported that in a rat-to-mouse islet xenograft model, anti-CD154/anti-LFA-1-induced indefinite (>100 day) islet xenograft survival required mAb depletion of NK1.1$^+$ cells (Ramirez-Victorio et al., 2008). Together, this suggests reciprocal roles for NK cells in islet allograft versus xenograft models, highlighting the plasticity of these cells and suggesting that therapeutic approaches targeting NK cells in allograft models may not directly translate to xenotransplantation.

Chemokine-directed NK cell allograft infiltration

The recruitment of NK cells to an allograft and secondary lymphoid tissues following organ transplantation is directed by chemokines (Table 33.1). One chemokine involved in NK cell migration is CCL5, which binds to CCR5 on NK cells (Robertson, 2002). CCL5 is produced by ECs, among others, and early upregulation of CCL5 production following the ischaemia/reperfusion injury that results from transplantation may account for early infiltration of NK cells in vascularized allografts (Belperio et al., 2000). Indeed, it has been suggested that CCL5 has a role in the development of vascular disease associated with organ transplantation (Pattison et al., 1996).

Another possible effect of CCL5 in influencing NK cells comes from a recent report that, under inflammatory conditions, CCL5 delivers a signal to NK cells that promotes their migration from the red pulp to T-cell areas of the white pulp in the spleen, bringing NK cells into close proximity to APC and T cells that initiate antidonor alloimmune responses (Gregoire et al., 2008). This may be important because host NK cells are capable of recognizing and killing donor-derived DCs and therefore are able to regulate direct alloimmune responses (Laffont et al., 2008; Yu et al., 2006).

Table 33.1 Role of chemokines in NK cell function in transplantation

Chemokine	Receptor	Effects	References
CCL5	CCR5	Produced early by endothelial cells during ischaemia/reperfusion injury	Belperio et al. (2000) Pattison et al. (1996)
		Promotes NK cell recruitment to graft vasculature	Gregoire et al. (2008)
		Promotes NK cell migration from red pulp to white pulp of spleen	
CX_3CL1/ fractalkine	CX_3CR1	Involved in migration of resting NK cells and may promote cytolytic activity Membrane-bound CX_3CL1 promotes NK cell activation Endothelial cell expression of CX_3CL1 promotes adhesion of NK cells	Robertson et al. (2002) Yoneda et al. (2000) Yoneda et al. (2003) Pallandre et al. (2008) Nishimura et al. (2002)
KC/CXCL1	CXCR2	Increased in skin allografts of mice depleted of NK cells—recruit monocytes, DCs, T cells	Kondo et al. (2000)
MIP-1α/β (CCL3/CCL4)	CCR1/3/5	Increased in skin allografts of mice depleted of NK cells—recruit monocytes, DCs, T cells	Kondo et al. (2000)

Chemokines produced by or acting upon NK cells following transplantation impact graft outcome through various mechanisms. Interestingly, chemokines such as CCL5 and CX3CL1 can promote rejection by recruiting NK cells to the graft where they destroy donor endothelium; conversely, CCL5 may also promote graft acceptance by inducing NK cell trafficking in the spleen, which brings them into close proximity with donor APC, a population of cells reported as a target of recipient NK cell cytolytic activity.

CX_3CL1, or fractalkine, is the ligand for CX_3CR1 and plays an important role in directing NK cell migration, especially naïve or resting NK cells, and has been reported to contribute to NK cell cytotoxic activity (Imai et al., 1997; Robertson, 2002; Yoneda et al., 2000). CX_3CL1 exists in soluble or membrane-bound forms, suggesting a role in the recruitment of NK cells and in facilitating cell–cell interactions. A recent report showed that the membrane-bound but not the soluble form of CX_3CL1 was necessary for NK cell activation (Yoneda et al., 2003). Indeed, a role of this chemokine in cell–cell interactions was confirmed by a report indicating that mature DCs expressing surface-bound CX_3CL1 could promote NK cell activation (Pallandre et al., 2008).

The response of NK cells to CX_3CL1 may play a functional role in immune responses to vascularized organ allografts. CX_3CL1 is expressed on ECs and, under inflammatory conditions, can promote the recruitment and adhesion of cytolytic effector cells, including NK cells (Nishimura et al., 2002). Increasing evidence implicates the expression of CX_3CL1 in the development of allograft rejection. CX_3CL1 expression may facilitate the recruitment of NK cells to the vasculature of the graft where they can attack and destroy MHC class I-expressing ECs and produce IFN-γ that promotes Th1-mediated T-cell responses (Robinson et al., 2000). In another study, prolongation of cardiac allograft survival in CX_3CR1-deficient mice was dependant upon the use of cyclosporine; however, there was still a marked reduction in the number of graft-infiltrating NK cells in the absence of immunosuppression in CX_3CR1-deficient mice (Haskell et al., 2001).

Transplantation immunotherapies and NK cell function

Current treatment strategies that prevent organ rejection involve the use of pharmacological and biological agents with the ultimate goal of indefinite graft survival by preventing activation and expansion of anti-donor immune responses. Due to their varying modes of action and immune targets, the effects of different immunosuppressive agents on NK cells also vary (Table 33.2).

One of the most commonly used traditional antirejection drug regimens combines the antiproliferative agent azathioprine with the corticosteroid prednisone. Early studies on the effects of this treatment strategy on NK cell function showed a significant decrease in cytolytic activity of NK cells isolated from kidney transplant patients receiving azathioprine and prednisone (Lipinski et al., 1980; Prince et al., 1984). These findings were made despite evidence that the contribution of NK cells to ADCC was unaffected. Further study revealed that the negative effects of azathioprine/prednisone

Table 33.2 Effects of immunosuppressive agents on NK cell function

Treatment	Effects	References
Azathioprine	Decreases cytolytic function Impairs NK cell responses to posttransplant infections	Lipinski et al. (1980) Prince et al. (1984) Starr et al. (1984)
Prednisone	No reported effect on cytolytic function	Muller et al. (1987)
Cyclosporine A (CsA)	No effect on NK cell function when used alone	Alamartine et al. (1990) Muller et al. (1987) Wai et al. (2008)
Tacrolimus (FK506)	Decreases renal allograft-infiltrating NK cells Decreases liver-resident NK cells No decrease in cytolytic function in rat liver allograft recipients	Tamura et al. (1998) Yang et al. (2003) Wai et al. (2008)
Mycophenolate mofetil (MMF)	No effect on NK cell numbers in heart or kidney transplant recipients	Vacher-Coponat et al. (2006)
Rapamycin (RAPA)	Decreased proliferation and cytotoxicity of NK cells in rat liver allograft recipients No reported effect on cytolytic function in hamster-to-rat skin xenograft	Wai et al. (2008) Gourlay et al. (1998)
CTLA4-Ig	No effect on cytolytic function No effect on the number of liver allograft-infiltrating NK cells Decreased cytolytic activity of NK cells recovered from rat liver allografts	Tadaki et al. (2000) Yang et al. (2003) Li et al. (2001)
Donor-specific transfusion (DST)	No effect on cytolytic function in kidney or sponge-matrix allograft models	Wasowska et al. (1992) Koga et al. (2000)

Immunosuppressive agents used in transplantation have a wide range of biological actions and have varying effects on NK cell activation and function. Inconsistencies in the current data suggest that further study is warranted to better understand the effects of these treatments on NK cells.

were mediated by azathioprine, as prednisone alone did not impair cytolytic activity (Muller et al., 1987). In accordance with these data, it was shown that NK cell cytolytic activity decreased when patients were switched to regimens containing azathioprine, or conversely, that NK cytolytic activity recovered following azathioprine withdrawal (Ramsey et al., 1984; Versluis et al., 1989). Interestingly, the same report which suggested that prednisone alone did not reduce cytolytic function showed decreased ADCC, despite previous data indicating no change in ADCC (Prince et al., 1984).

The effects of azathioprine/prednisone-based regimens on NK cell function suggest that improved graft survival in transplant recipients may be due, in part, to reduced cytolytic function of recipient-derived and/or host-derived NK cells. The suppressive activity of these agents on NK cell function may, however, come at a greater price in terms of long-term survival. The persistence of opportunistic infections as a result of global immune suppression induced by these treatments remains a significant problem following organ transplantation. Indeed, studies have revealed that NK cells from patients on azathioprine and prednisone have a reduced ability to kill cytomegalovirus (CMV)-infected cells, revealing an important negative consequence of these treatments on NK cell function (Starr et al., 1984).

Although not used clinically as an immunosuppressant, the female sex hormone progesterone has been considered to possess immunoregulatory properties and is believed to play a role in foetal tolerance (Barrera et al., 2007). Interestingly, it was recently shown that peripheral blood NK cells (of both men and women) express the progesterone receptor and are susceptible to dose-dependent, progesterone-induced caspase-dependent apoptosis as well as a blunted capacity to produce IFN-γ (Arruvito et al., 2008). Additionally, a recent study indicated that endothelial and smooth muscle cells derived from cardiac tissue upregulated HLA-G in response to increasing doses of progesterone (Sheshgiri et al., 2008). Coupling these data with the fact that expression of HLA-G on ECs has been shown to offer protection against NK-mediated killing suggests that inclusion of progesterone in immunosuppressive drug regimens in transplantation may offer a potential strategy to promote tolerance through both direct and indirect regulation of NK cell survival and function.

Over the past few decades, immunosuppressive strategies have evolved and now include a calcineurin inhibitor (CNI), such as tacrolimus (FK506) or cyclosporine, in addition to azathioprine and prednisone. CNI inhibits the production of IL-2, resulting in blunted T-cell activation. There is evidence that cyclosporine alone does not impair NK cell function, since cytolytic activity is decreased only when coupled with azathioprine (Alamartine et al., 1990; Muller et al., 1987). This is rather surprising, since NK cells are so reliant on IL-2. In contrast to cyclosporine, tacrolimus significantly reduces the number of graft-infiltrating NK cells in renal allografts and decreases liver-resident NK cell numbers and activity in patients pretreated prior to partial hepatectomy (Tamura et al., 1998).

Since long-term allograft survival relies heavily on sustained use of immunosuppressive drugs, transplant recipients are subject to toxicities and detrimental side effects from these treatments. Thus, alternative treatments are constantly being explored and at the same time, their influence on the cells of the immune system is also evaluated. Mycophenolate mofetil (MMF) is a pro-drug that is metabolized into mycophenolic acid (MPA) and induces downstream effects similar to azathioprine, namely inhibition of purine biosynthesis, which hinders cell proliferation. Treatment regimens that include MMF are effective in the setting of organ transplantation and offer a better alternative to azathioprine due to reduced inhibitory effects on bone marrow cells. Consequently, patients can better eliminate opportunistic infections even in the face of immunosuppressive therapy. MMF has no significant effects on NK cell numbers in heart transplant patients, and kidney graft recipients on MMF/tacrolimus show greater reconstitution of NK cell numbers 1 year posttransplant compared to those on azathioprine/prednisone/cyclosporine (Vacher-Coponat et al., 2006; Weigel et al., 2002). These results suggest that MMF may spare NK cell function and allow them to combat potential opportunistic infections in addition to promoting long-term allograft survival.

Rapamycin (RAPA) is another clinically approved immunosuppressant, which, unlike CNI that blocks IL-2 production, inhibits signalling pathways downstream of IL-2 through interaction with the intracellular molecule mammalian target of rapamycin (mTOR). RAPA also spares the activity of and can expand regulatory T cells (Treg) compared to effector T cells and as a result promotes tolerance in experimental animal models. Along these lines, it has been shown in a model of hamster-to-rat skin xenograft rejection that RAPA does not inhibit NK cell cytolytic activity, suggesting it spares NK cell function similarly to Treg (Gourlay et al., 1998). However, recent in vitro studies indicate that both proliferation and cytolytic function are impaired in rat NK cell lines in the presence of RAPA (Wai et al., 2008).

Table 33.3 Potential for NK cells to promote allograft acceptance and rejection

Graft acceptance	References
Perforin-mediated lysis of donor-derived DCs	Laffont et al. (2008)
Perforin-mediated killing of host-derived DCs and T cells (rendered susceptible through costimulation blockade)	Beilke et al. (2005)
Regulation of donor-cell reactivity through host NK cell inhibitory receptor/donor HLA class I molecule interactions	Forte et al. (2001) Hanvesakul et al. (2008)
Graft rejection	
ADCC against graft endothelium mediated through NK cell Fc receptor binding to xenoreactive Abs	Yin et al. (2004)
NK cell-derived IFN-γ upregulates MHC II on graft endothelium, making them targets of alloreactive T cells	McDouall et al. (1997)
Direct lysis of graft endothelium through ligation of NK cell activating receptors	Forte et al. (2005) Lilienfeld et al. (2006)

NK cells may promote either organ allograft acceptance or rejection. *Graft acceptance:* NK cells may kill donor DCs, thereby regulating antidonor responses via the direct pathway of allorecognition; host DCs and T cells may be rendered susceptible to NK cell-mediated killing through mAb-mediated costimulation blockade; recognition of HLA class I variants on donor cells via host NK cell inhibitory receptors may prevent killing. *Graft rejection*: Recognition and lysis of endothelial cells mediated by ADCC (Fc recognition of xeno-/alloreactive Abs) or direct killing through NK cell activating receptors; NK cell production of IFN-γ promotes Th1 immunity through upregulation of MHC class II on donor endothelial cells to be targeted by donor-reactive T cells; activation of endothelium during ischemia/reperfusion injury results in production of CCL5, which can activate NK cells; activated endothelial cells upregulate ligands for NK cell activating receptors and can promote killing.

Clearly, further study is needed to uncover the effects of RAPA on NK cell function.

Biological agents have also been explored as potential therapies in transplantation to limit the potential toxic effects from long-term use of pharmacologic agents. These approaches have centred on inhibition of costimulatory signals provided by APC that are necessary for sufficient T-cell activation. Among these, CTLA-4 interferes with CD80 and CD86 costimulation through CD28 on activated lymphocytes. Reports concerning the use of CTLA4-Ig fusion protein and NK cell function are variable and inconclusive at this point. Several reports show no significant effect of CTLA4-Ig on NK cells. These include no influence on cytolytic function in vitro in a xenogeneic MLR and no decrease in graft-infiltrating NK cell numbers in rat liver allografts, although the latter study was based on recombinant adeno-associated viral CTLA4-Ig gene delivery and not on treatment with soluble fusion protein, as in other studies (Tadaki et al., 2000; Yang et al., 2003). One report did reveal a decrease in the cytolytic activity of NK cells recovered from liver grafts of recipient mice treated with CTLA4-Ig (Li et al., 2001). The variability in these results suggests differences in the actions of soluble fusion protein versus viral gene delivery of CTLA4-Ig on NK cell function, as well as differences in vitro and in vivo, warranting further study on the effects of this molecule on NK cells.

Another biological therapy that has been examined extensively in transplantation is donor-specific transfusion (DST) where donor cells are introduced into the recipient prior to transplant of organs. Since NK cells are able to recognize foreign cells based on their lack of self-MHC, they play a key role in host responses to this treatment. Interestingly, little effect of DST on NK cell function has been reported. In a model of rat kidney transplant, DST treatment induced indefinite allograft survival without any effect on NK cell cytolytic function compared to rats receiving autologous cell transfusion (Wasowska et al., 1992). In a later study, it was shown in a sponge matrix allograft model that DST treatment resulted in a slight decrease in intragraft NK cell activity, although only at the earliest time points compared to mice receiving syngeneic cells (Koga et al., 2000).

Conclusions

NK cells play a major role in rejecting allogeneic bone marrow and hematopoietic stem cell transplants based on the 'missing self' hypothesis, whereby NK cells recognize and kill allogeneic cells that do not express self-MHC class I molecules. Despite a significant amount of literature on NK cells in the setting of organ transplantation, their role in graft acceptance and rejection remains unclear. There is evidence that they contribute to both outcomes, however, as documented here and summarized in Table 33.3. The findings may be dependant not only on which organ is transplanted but also on the therapeutic treatment administered. Clearly, further investigation of the potential of NK cells to promote tolerance versus their contribution to rejection by supporting adaptive immunity is warranted.

References

Adoumie, R., Serrick, C., Giaid, A. and Shennib, H. (1992). Early cellular events in the lung allograft. Ann Thorac Surg **54**, 1071–1076 discussion 1076–1077.

Ahlenstiel, G., Martin, M.P., Gao, X., Carrington, M. and Rehermann, B. (2008). Distinct KIR/HLA compound genotypes affect the kinetics of human antiviral natural killer cell responses. J Clin Invest **118**, 1017–1026.

Alamartine, E., Sabido, O. and Berthoux, F.C. (1990). The influence of prophylactic immunosuppressive regimens on natural killer and lymphokine-activated killer cells in renal transplant recipients. Transplantation **50**, 969–973.

Arruvito, L., Giulianelli, S., Flores, A.C., Paladino, N., Barboza, M., Lanari, C. and Fainboim, L. (2008). NK cells expressing a progesterone receptor are susceptible to progesterone-induced apoptosis. J Immunol **180**, 5746–5753.

Assi, M.A., Pulido, J.S., Peters, S.G., McCannel, C.A. and Razonable, R.R. (2007). Graft-vs.-host disease in lung and other solid organ transplant recipients. Clin Transplant **21**, 1–6.

Barrera, D., Avila, E. and Diaz, L. (2007). Immunological role of progesterone in the maintenance of pregnancy. Rev Invest Clin **59**, 139–145.

Basturk, B., Karakayali, F., Emiroglu, R., Sozer, O., Haberal, A., Bal, D. and Haberal, M. (2006). Human leukocyte antigen-G, a new parameter in the follow-up of liver transplantation. Transplant Proc **38**, 571–574.

Beilke, J., Johnson, Z., Kuhl, N. and Gill, R.G. (2004). A major role for host MHC class I antigen presentation for promoting islet allograft survival. Transplant Proc **36**, 1173–1174.

Beilke, J.N., Kuhl, N.R., Van Kaer, L. and Gill, R.G. (2005). NK cells promote islet allograft tolerance via a perforin-dependent mechanism. Nat Med **11**, 1059–1065.

Belperio , J.A., Burdick, M.D., Keane, M. P., Xue, Y.Y., Lynch, J.P. 3rd, Daugherty, B.L., Kunkel, S.L. and Strieter, R.M. (2000). The role of the CC chemokine, RANTES, in acute lung allograft rejection. J Immunol **165**, 461–472.

Bingaman, A.W., Ha, J., Waitze, S.Y., Durham, M.M., Cho, H.R., Tucker-Burden, C., Hendrix, R., Cowan, S.R., Pearson, T.C. and Larsen, C.P. (2000). Vigorous allograft rejection in the absence of danger. J Immunol **164**, 3065–3071.

Bradley, J.A., Mason, D.W. and Morris, P.J. (1985). Evidence that rat renal allografts are rejected by cytotoxic T cells and not by nonspecific effectors. Transplantation 39, 169–175.

Brilot, F., Strowig, T. and Munz, C. (2008). NK cells interactions with dendritic cells shape innate and adaptive immunity. Front Biosci 13, 6443–6454.

Coudert, J.D., Coureau, C. and Guery, J. C. (2002). Preventing NK cell activation by donor dendritic cells enhances allospecific CD4 T cell priming and promotes Th type 2 responses to transplantation antigens. J Immunol 169, 2979–2987.

Coulson, M.T., Jablonski, P., Howden, B.O., Thomson, N.M. and Stein, A.N. (2005). Beyond operational tolerance: effect of ischemic injury on development of chronic damage in renal grafts. Transplantation 80, 353–361.

Coupel, S., Moreau, A., Hamidou, M., Horejsi, V., Soulillou, J.P. and Charreau, B. (2007). Expression and release of soluble HLA-E is an immunoregulatory feature of endothelial cell activation. Blood 109, 2806–2814.

Crew, M.D. (2007). Play it in E or G: utilization of HLA-E and -G in xenotransplantation. Xenotransplantation 14, 198–207.

Dalloul, A.H., Chmouzis, E., Ngo, K. and Fung-Leung, W.P. (1996). Adoptively transferred CD4+ lymphocytes from $CD8^{-/-}$ mice are sufficient to mediate the rejection of MHC class II or class I disparate skin grafts. J Immunol 156, 4114–4119.

Degli-Esposti, M.A. and Smyth, M.J. (2005). Close encounters of different kinds: dendritic cells and NK cells take centre stage. Nat Rev Immunol 5, 112–124.

Dorling, A., Monk, N. and Lechler, R. (2000). HLA-G inhibits the transendothelial cell migration of human NK cells: a strategy for inhibiting xenograft rejection. Transplant Proc 32, 938.

Fandrich, F., Exner, B., Papachrysanthou, A., Zhu, X., Jahnke, T., Chambers, W. H. and Zavazava, N. (1996a). In vivo depletion of NKR-P1 positive cells in the recipient prior to small bowel transplantation enhances graft-versus-host disease (GvHD) in the rat. Transpl Int 9(Suppl. 1), S275–S280.

Fandrich, F., Schroder, J., Exner, B., Papachrysanthou, A., Peters, J., Chambers, W. and Zavazava, N. (1996b). Correlation between graft-versus-host induced immunosuppression and host natural killer cell activity in small bowel transplantation. World J Surg 20, 1041–1051.

Fandrich, F., Zepernick-Kalinski, C., Gebhardt, H., Henne-Bruns, D., Zavazava, N. and Lin, X. (1999). The role of natural killer cell mediated caspases activation in a graft-versus-host disease model of semiallogeneic small bowel transplantation. Transpl Immunol 7, 1–7.

Fildes, J.E., Yonan, N., Tunstall, K., Walker, A.H., Griffiths-Davies, L., Bishop, P. and Leonard, C.T. (2008). Natural killer cells in peripheral blood and lung tissue are associated with chronic rejection after lung transplantation. J Heart Lung Transplant 27, 203–207.

Forte, P., Pazmany, L., Matter-Reissmann, U.B., Stussi, G., Schneider, M.K. and Seebach, J.D. (2001). HLA-G inhibits rolling adhesion of activated human NK cells on porcine endothelial cells. J Immunol 167, 6002–6008.

Forte, P., Lilienfeld, B.G., Baumann, B. C. and Seebach, J.D. (2005). Human NK cytotoxicity against porcine cells is triggered by NKp44 and NKG2D. J Immunol 175, 5463–5470.

Frezza, E.E., Gerunda, G.E., Fassina, A., DeFranchis, R., Biffi, R., Gurakar, A., Fagiuoli, S., Maffei, F.A. and Van Thiel, D.H. (1994). NK activity during graft-vs-host disease and graft rejection in rats following intestinal transplantation. Transplant Proc 26, 1626.

Goodman, D.J., Von Albertini, M., Willson, A., Millan, M.T. and Bach, F.H. (1996). Direct activation of porcine endothelial cells by human natural killer cells. Transplantation 61, 763–771.

Gourlay, W.A., Chambers, W.H., Monaco, A.P. and Maki, T. (1998). Importance of natural killer cells in the rejection of hamster skin xenografts. Transplantation 65, 727–734.

Gregoire, C., Cognet, C., Chasson, L., Coupet, C.A., Dalod, M., Reboldi, A., Marvel, J., Sallusto, F., Vivier, E. and Walzer, T. (2008). Intrasplenic trafficking of natural killer cells is redirected by chemokines upon inflammation. Eur J Immunol 38, 2076–2084.

Gregson , A.L., Hoji, A., Saggar, R., Ross, D.J., Kubak, B.M., Jamieson, B.D., Weigt, S.S., Lynch, J.P. 3rd, Ardehali, A., Belperio, J.A. and Yang, O.O. (2008). Bronchoalveolar immunologic profile of acute human lung transplant allograft rejection. Transplantation 85, 1056–1059.

Gros, F., Cabillic, F., Toutirais, O., Maux, A.L., Sebti, Y. and Amiot, L. (2008). Soluble HLA-G molecules impair natural killer/dendritic cell crosstalk via inhibition of dendritic cells. Eur J Immunol 38, 742–749.

Hanvesakul, R., Spencer, N., Cook, M., Gunson, B., Hathaway, M., Brown, R., Nightingale, P., Cockwell, P., Hubscher, S.G., Adams, D.H., Moss, P. and Briggs, D. (2008). Donor HLA-C genotype has a profound impact on the clinical outcome following liver transplantation. Am J Transplant 8, 1931–1941.

Haskell, C.A., Hancock, W.W., Salant, D. J., Gao, W., Csizmadia, V., Peters, W., Faia, K., Fituri, O., Rottman, J.B. and Charo, I.F. (2001). Targeted deletion of CX(3)CR1 reveals a role for fractalkine in cardiac allograft rejection. J Clin Invest 108, 679–688.

He, H., Stone, J.R. and Perkins, D. L. (2002). Analysis of robust innate immune response after transplantation in the absence of adaptive immunity. Transplantation 73, 853–861.

Heidecke, C.D., Araujo, J.L., Kupiec-Weglinski, J.W., Abbud-Filho, M., Araneda, D., Stadler, J., Siewert, J., Strom, T.B. and Tilney, N.L. (1985). Lack of evidence for an active role for natural killer cells in acute rejection of organ allografts. Transplantation 40, 441–444.

Hiby, S.E., Walker, J.J., O'Shaughnessy, K.M., Redman, C.W., Carrington, M., Trowsdale, J. and Moffett, A. (2004). Combinations of maternal KIR and fetal HLA-C genes influence the risk of preeclampsia and reproductive success. J Exp Med 200, 957–965.

Higuchi, T., Jaramillo, A., Kaleem, Z., Patterson, G.A. and Mohanakumar, T. (2002). Different kinetics of obliterative airway disease development in heterotopic murine tracheal allografts induced by CD4+ and CD8+ T cells. Transplantation 74, 646–651.

Hsieh, C.L., Ogura, Y., Obara, H., Ali, U. A., Rodriguez, G.M., Nepomuceno, R.R., Martinez, O.M. and Krams, S. M. (2004). Identification, cloning, and characterization of a novel rat natural killer receptor, RNKP30: a molecule expressed in liver allografts. Transplantation 77, 121–128.

Imai, T., Hieshima, K., Haskell, C., Baba, M., Nagira, M., Nishimura, M., Kakizaki, M., Takagi, S., Nomiyama, H., Schall, T.J. and Yoshie, O. (1997). Identification and molecular characterization of fractalkine receptor CX3CR1, which mediates both leukocyte migration and adhesion. Cell 91, 521–530.

Ito, A., Shimura, H., Nitahara, A., Tomiyama, K., Ito, M., Kanekura, T., Okumura, K., Yagita, H. and Kawai, K. (2008). NK cells contribute to the skin graft rejection promoted by CD4+ T cells activated through the indirect allorecognition pathway. Int Immunol **20**, 1343–1349.

Jacobson, J.D., Markmann, J.F., Brayman, K.L., Barker, C.F. and Naji, A. (1988). Prevention of recurrent autoimmune diabetes in BB rats by anti-asialo-GM1 antibody. Diabetes **37**, 838–841.

Karlsson-Parra, A., Ridderstad, A., Wallgren, A.C., Moller, E., Ljunggren, H.G. and Korsgren, O. (1996). Xenograft rejection of porcine islet-like cell clusters in normal and natural killer cell-depleted mice. Transplantation **61**, 1313–1320.

Khakoo, S.I., Thio, C.L., Martin, M.P., Brooks, C.R., Gao, X., Astemborski, J., Cheng, J., Goedert, J.J., Vlahov, D., Hilgartner, M., Cox, S., Little, A.M., Alexander, G.J., Cramp, M.E., O'Brien, S.J., Rosenberg, W.M., Thomas, D.L. and Carrington, M. (2004). HLA and NK cell inhibitory receptor genes in resolving hepatitis C virus infection. Science **305**, 872–874.

Kim, J., Chang, C.K., Hayden, T., Liu, F. C., Benjamin, J., Hamerman, J.A., Lanier, L.L. and Kang, S.M. (2007). The activating immunoreceptor NKG2D and its ligands are involved in allograft transplant rejection. J Immunol **179**, 6416–6420.

Koga, S., Luke, P.P., Specht, S.M., Rominski, B., Jaquins-Gerstl, A., Hoffman, R.A., Thomson, A.W. and Jordan, M.L. (2000). Inhibition of the allograft response by donor specific blood transfusion: association with reduced local TH1 cytokines and nitric oxide but enhanced prostaglandin E2 production. Transplantation **70**, 1788–1796.

Kondo, T., Morita, K., Watarai, Y., Auerbach, M.B., Taub, D.D., Novick, A. C., Toma, H. and Fairchild, R.L. (2000). Early increased chemokine expression and production in murine allogeneic skin grafts is mediated by natural killer cells. Transplantation **69**, 969–977.

Kroemer, A., Xiao, X., Degauque, N., Edtinger, K., Wei, H., Demirci, G. and Li, X.C. (2008). The innate NK cells, allograft rejection, and a key role for IL-15. J Immunol **180**, 7818–7826.

Kummer, J.A., Wever, P.C., Kamp, A.M., ten Berge, I.J., Hack, C.E. and Weening, J.J. (1995). Expression of granzyme A and B proteins by cytotoxic lymphocytes involved in acute renal allograft rejection. Kidney Int **47**, 70–77.

Laffont, S., Seillet, C., Ortaldo, J., Coudert, J.D. and Guery, J.C. (2008). Natural killer cells recruited into lymph nodes inhibit alloreactive T-cell activation through perforin-mediated killing of donor allogeneic dendritic cells. Blood **112**, 661–671.

Lanier, L.L. (2008). Up on the tightrope: natural killer cell activation and inhibition. Nat Immunol **9**, 495–502.

Legendre, C.M., Guttmann, R.D. and Yip, G.H. (1986). Natural killer cell subsets in long-term renal allograft recipients. A phenotypic and functional study. Transplantation **42**, 347–352.

Li, W., Lu, L., Wang, Z., Wang, L., Fung, J. J., Thomson, A.W. and Qian, S. (2001). Costimulation blockade promotes the apoptotic death of graft-infiltrating T cells and prolongs survival of hepatic allografts from FLT3L-treated donors. Transplantation **72**, 1423–1432.

Lila, N., Amrein, C., Guillemain, R., Chevalier, P., Latremouille, C., Fabiani, J.N., Dausset, J., Carosella, E. D. and Carpentier, A. (2002). Human leukocyte antigen-G expression after heart transplantation is associated with a reduced incidence of rejection. Circulation **105**, 1949–1954.

Lilienfeld, B.G., Garcia-Borges, C., Crew, M.D. and Seebach, J.D. (2006). Porcine UL16-binding protein 1 expressed on the surface of endothelial cells triggers human NK cytotoxicity through NKG2D. J Immunol **177**, 2146–2152.

Lilienfeld, B.G., Crew, M.D., Forte, P., Baumann, B.C. and Seebach, J. D. (2007). Transgenic expression of HLA-E single chain trimer protects porcine endothelial cells against human natural killer cell-mediated cytotoxicity. Xenotransplantation **14**, 126–134.

Lilienfeld, B.G., Schildknecht, A., Imbach, L.L., Mueller, N.J., Schneider, M.K. and Seebach, J.D. (2008). Characterization of porcine UL16-binding protein 1 endothelial cell surface expression. Xenotransplantation **15**, 136–144.

Lin, Y., Goebels, J., Xia, G., Ji, P., Vandeputte, M. and Waer, M. (1998). Induction of specific transplantation tolerance across xenogeneic barriers in the T-independent immune compartment. Nat Med **4**, 173–180.

Lindaman, A., Dowden, A. and Zavazava, N. (2006). Soluble HLA-G molecules induce apoptosis in natural killer cells. Am J Reprod Immunol **56**, 68–76.

Lipinski, M., Tursz, T., Kreis, H., Finale, Y. and Amiel, J.L. (1980). Dissociation of natural killer cell activity and antibody-dependent cell-mediated cytotoxicity in kidney allograft recipients receiving high-dose immunosuppressive therapy. Transplantation **29**, 214–218.

Maier, S., Tertilt, C., Chambron, N., Gerauer, K., Huser, N., Heidecke, C. D. and Pfeffer, K. (2001). Inhibition of natural killer cells results in acceptance of cardiac allografts in $CD28^{-/-}$ mice. Nat Med **7**, 557–562.

Martin-Fontecha, A., Thomsen, L.L., Brett, S., Gerard, C., Lipp, M., Lanzavecchia, A. and Sallusto, F. (2004). Induced recruitment of NK cells to lymph nodes provides IFN-gamma for T(H)1 priming. Nat Immunol **5**, 1260–1265.

Martinez-Llordella, M., Puig-Pey, I., Orlando, G., Ramoni, M., Tisone, G., Rimola, A., Lerut, J., Latinne, D., Margarit, C., Bilbao, I., Brouard, S., Hernandez-Fuentes, M., Soulillou, J.P. and Sanchez-Fueyo, A. (2007). Multiparameter immune profiling of operational tolerance in liver transplantation. Am J Transplant **7**, 309–319.

Martinez-Llordella, M., Lozano, J.J., Puig-Pey, I., Orlando, G., Tisone, G., Lerut, J., Benitez, C., Pons, J.A., Parrilla, P., Ramirez, P., Bruguera, M., Rimola, A. and Sanchez-Fueyo, A. (2008). Using transcriptional profiling to develop a diagnostic test of operational tolerance in liver transplant recipients. J Clin Invest **118**, 2845–2857.

Matsunami, K., Miyagawa, S., Nakai, R., Murase, A. and Shirakura, R. (2001). The possible use of HLA-G1 and G3 in the inhibition of NK cell-mediated swine endothelial cell lysis. Clin Exp Immunol **126**, 165–172.

Matter-Reissmann, U.B., Forte, P., Schneider, M.K., Filgueira, L., Groscurth, P. and Seebach, J.D. (2002). Xenogeneic human NK cytotoxicity against porcine endothelial cells is perforin/granzyme B dependent and not inhibited by Bcl-2 overexpression. Xenotransplantation **9**, 325–337.

McDouall, R.M., Batten, P., McCormack, A., Yacoub, M.H. and Rose, M.L. (1997). MHC class II expression on human heart microvascular endothelial cells: exquisite sensitivity to interferon-gamma and natural killer cells. Transplantation **64**, 1175–1180.

McNerney, M.E., Lee, K.M., Zhou, P., Molinero, L., Mashayekhi, M., Guzior, D., Sattar, H., Kuppireddi, S., Wang, C.R., Kumar, V. and Alegre, M.L. (2006). Role of natural killer cell subsets in cardiac allograft rejection. Am J Transplant **6**, 505–513.

Muller, C., Schernthaner, G., Kovarik, J., Kalinowska, W. and Zielinski, C. C. (1987). Natural killer cell activity and antibody-dependent cellular cytotoxicity in patients under various immunosuppressive regimens. Clin Immunol Immunopathol **44**, 12–19.

Murray, A.G., Nelson, R.C., Rayat, G. R., Elliott, J.F. and Korbutt, G.S. (1999). Neonatal porcine islet cells induce human CD4+, but not CD8+, lymphocyte proliferation and resist cell-mediated cytolytic injury in vitro. Diabetes **48**, 1713–1719.

Nagano, H., Nadeau, K.C., Takada, M., Kusaka, M. and Tilney, N.L. (1997). Sequential cellular and molecular kinetics in acutely rejecting renal allografts in rats. Transplantation **63**, 1101–1108.

Nguyen, D.M., Mulder, D.S. and Shennib, H. (1993). Lectin-dependent cell-mediated cytotoxicity and natural killer function in rejecting and infected lung allografts. Transplantation **55**, 1250–1256.

Nishimura, M., Umehara, H., Nakayama, T., Yoneda, O., Hieshima, K., Kakizaki, M., Dohmae, N., Yoshie, O. and Imai, T. (2002). Dual functions of fractalkine/ CX3C ligand 1 in trafficking of perforin$^+$/granzyme B$^+$ cytotoxic effector lymphocytes that are defined by CX3CR1 expression. J Immunol **168**, 6173–6180.

Norin, A.J., Brewer, R.J., Lui, Q., Del Rio, M.J., Burack, J. and Kamholz, S. L. (1999). Donor antigen-independent effector mechanisms of graft rejection: a novel cell surface target of IL-2-activated natural killer cells. Transplant Proc **31**, 787–788.

Obara, H., Nagasaki, K., Hsieh, C.L., Ogura, Y., Esquivel, C.O., Martinez, O.M. and Krams, S.M. (2005). IFN-gamma, produced by NK cells that infiltrate liver allografts early after transplantation, links the innate and adaptive immune responses. Am J Transplant **5**, 2094–2103.

Oertel, M., Kohlhaw, K., Diepolder, H.M., Schroder, S., Schwarz, R., Tannapfel, A., Mossner, J., Hauss, J. and Berr, F. (2001). Alloreactivity of natural killer cells in allogeneic liver transplantation. Transplantation **72**, 116–122.

Pallandre, J.R., Krzewski, K., Bedel, R., Ryffel, B., Caignard, A., Rohrlich, P.S., Pivot, X., Tiberghien, P., Zitvogel, L., Strominger, J.L. and Borg, C. (2008). Dendritic cell and natural killer cell cross-talk: a pivotal role of CX3CL1 in NK cytoskeleton organization and activation. Blood.

Pattison, J.M., Nelson, P.J., Huie, P., Sibley, R.K. and Krensky, A.M. (1996). RANTES chemokine expression in transplant-associated accelerated atherosclerosis. J Heart Lung Transplant **15**, 1194–1199.

Prince, H.E., Ettenger, R.B., Dorey, F.J., Fine, R.N. and Fahey, J.L. (1984). Azathioprine suppression of natural killer activity and antibody-dependent cellular cytotoxicity in renal transplant recipients. J Clin Immunol **4**, 312–318.

Qiu, J., Terasaki, P.I., Miller, J., Mizutani, K., Cai, J. and Carosella, E.D. (2006). Soluble HLA-G expression and renal graft acceptance. Am J Transplant **6**, 2152–2156.

Quan, D., Bravery, C., Chavez, G., Richards, A., Cruz, G., Copeman, L., Atkinson, C., Holmes, B., Davies, H., Cozzi, E. and White, D. (2000). Identification, detection, and in vitro characterization of cynomolgus monkey natural killer cells in delayed xenograft rejection of hDAF transgenic porcine renal xenografts. Transplant Proc **32**, 936–937.

Ramirez-Victorio, F., Beilke, J.N. and Gill, R.G. (2008). Both innate and adaptive major histocompatibility complex class I-dependent immunity impair long-term islet xenograft survival. Transplant Proc **40**, 557–558.

Ramsey, K.M., Djeu, J.Y. and Rook, A.H. (1984). Decreased circulating large granular lymphocytes associated with depressed natural killer cell activity in renal transplant recipients. Transplantation **38**, 351–356.

Rieben, R. and Seebach, J.D. (2005). Xenograft rejection: IgG1, complement and NK cells team up to activate and destroy the endothelium. Trends Immunol **26**, 2–5.

Robertson, M.J. (2002). Role of chemokines in the biology of natural killer cells. J Leukoc Biol **71**, 173–183.

Robinson, L.A., Nataraj, C., Thomas, D.W., Howell, D.N., Griffiths, R., Bautch, V., Patel, D.D., Feng, L. and Coffman, T.M. (2000). A role for fractalkine and its receptor (CX3CR1) in cardiac allograft rejection. J Immunol **165**, 6067–6072.

Rocha, P.N., Plumb, T.J., Crowley, S.D. and Coffman, T.M. (2003). Effector mechanisms in transplant rejection. Immunol Rev **196**, 51–64.

Rosen, D.B., Cao, W., Avery, D.T., Tangye, S.G., Liu, Y.J., Houchins, J. P. and Lanier, L.L. (2008). Functional consequences of interactions between human NKR-P1A and its ligand LLT1 expressed on activated dendritic cells and B cells. J Immunol **180**, 6508–6517.

Sarnacki, S., Nakai, H., Calise, D., Azuma, T., Brousse, N., Revillon, Y. and Cerf-Bensussan, N. (1998). Decreased expression of the interleukin 2 receptor on CD8 recipient lymphocytes in intestinal grafts rendered tolerant by liver transplantation in rats. Gut **43**, 849–855.

Sasaki, H., Xu, X.C. and Mohanakumar, T. (1999a). HLA-E and HLA-G expression on porcine endothelial cells inhibit xenoreactive human NK cells through CD94/NKG2-dependent and -independent pathways. J Immunol **163**, 6301–6305.

Sasaki, H., Xu, X.C., Smith, D.M., Howard, T. and Mohanakumar, T. (1999b). HLA-G expression protects porcine endothelial cells against natural killer cell-mediated xenogeneic cytotoxicity. Transplantation **67**, 31–37.

Shen, K., Zheng, S.S., Park, O., Wang, H., Sun, Z. and Gao, B. (2008). Activation of innate immunity (NK/IFN-gamma) in rat allogeneic liver transplantation: contribution to liver injury and suppression of hepatocyte proliferation. Am J Physiol Gastrointest Liver Physiol **294**, G1070–G1077.

Sheshgiri, R., Rao, V., Tumiati, L.C., Xiao, R., Prodger, J.L., Badiwala, M., Librach, C. and Delgado, D.H. (2008). Progesterone induces human leukocyte antigen-g expression in vascular endothelial and smooth muscle cells. Circulation **118**, S58–S64.

Starr, S.E., Smiley, L., Wlodaver, C., Friedman, H.M., Plotkin, S.A. and Barker, C. (1984). Natural killing of cytomegalovirus-infected targets in renal transplant recipients. Transplantation **37**, 161–164.

Su, G.L., Walgenbach, K.J., Heeckt, P.H., Wang, Q., Halfter, W., Whiteside, T.L. and Bauer, A.J. (1996). Increased expression of interferon-gamma in a rat model of chronic intestinal allograft rejection. Transplantation **62**, 242–248.

Sullivan, J.A., Oettinger, H.F., Sachs, D. H. and Edge, A.S. (1997). Analysis of polymorphism in porcine MHC class I genes: alterations in signals recognized by human cytotoxic lymphocytes. J Immunol **159**, 2318–2326.

Tadaki, D.K., Craighead, N., Saini, A., Celniker, A., Burkly, L.C., Lee, K.P., Chute, J.P., Harlan, D.M. and Kirk, A. D. (2000). Costimulatory molecules are active in the human xenoreactive T-cell response but not in natural killer-mediated cytotoxicity. Transplantation **70**, 162–167.

Tamura, F., Masuhara, A., Sakaida, I., Fukumoto, E., Nakamura, T. and Okita, K. (1998). FK506 promotes liver regeneration by suppressing natural killer cell activity. J Gastroenterol Hepatol **13**, 703–708.

Uehara, S., Chase, C.M., Colvin, R. B., Russell, P.S. and Madsen, J.C. (2005a). Further evidence that NK cells may contribute to the development of cardiac allograft vasculopathy. Transplant Proc **37**, 70–71.

Uehara, S., Chase, C.M., Kitchens, W. H., Rose, H.S., Colvin, R.B., Russell, P. S. and Madsen, J.C. (2005b). NK cells can trigger allograft vasculopathy: the role of hybrid resistance in solid organ allografts. J Immunol **175**, 3424–3430.

Vacher-Coponat, H., Brunet, C., Moal, V., Loundou, A., Bonnet, E., Lyonnet, L., Ravet, S., Sampol-Manos, E., Sampol, J., Berland, Y., George, F.D. and Paul, P. (2006). Tacrolimus/mycophenolate mofetil improved natural killer lymphocyte reconstitution one year after kidney transplant by reference to cyclosporine/azathioprine. Transplantation **82**, 558–566.

Versluis, D.J., Bijma, A.M., Vaessen, L. M. and Weimar, W. (1989). Changes in immunological parameters after conversion from cyclosporine A to azathioprine in renal transplant recipients. Int J Immunopharmacol **11**, 157–164.

Wai, L.E., Fujiki, M., Takeda, S., Martinez, O.M. and Krams, S.M. (2008). Rapamycin, but not cyclosporine or FK506, alters natural killer cell function. Transplantation **85**, 145–149.

Ward, C., Whitford, H., Snell, G., Bao, H., Zheng, L., Reid, D., Williams, T.J. and Walters, E.H. (2001). Bronchoalveolar lavage macrophage and lymphocyte phenotypes in lung transplant recipients. J Heart Lung Transplant **20**, 1064–1074.

Wasowska , B., Howell, D.N., Baldwin, W.M. 3rd and Sanfilippo, F. (1992). The effects of donor-specific blood transfusion enhancement of rat renal allografts on host NK cell responses. Transplantation **54**, 136–142.

Weigel, G., Griesmacher, A., Karimi, A., Zuckermann, A.O., Grimm, M. and Mueller, M.M. (2002). Effect of mycophenolate mofetil therapy on lymphocyte activation in heart transplant recipients. J Heart Lung Transplant **21**, 1074–1079.

Yamamoto, N., Einaga-Naito, K., Kuriyama, M., Kawada, Y. and Yoshida, R. (1998). Cellular basis of skin allograft rejection in mice: specific lysis of allogeneic skin components by non-T cells. Transplantation **65**, 818–825.

Yang, Z., Wu, X., Tsui, T.Y., Hou, Y., Luk, J.M. and Fan, S.T. (2003). Long-term liver allograft survival induced by combined treatment with rAAV-hCTLA4Ig gene transfer and low-dose FK506. Transplantation **75**, 303–308.

Yao, A.Y., Tang, H.Y., Wang, Y., Feng, M. F. and Zhou, R.L. (2004). Inhibition of the activating signals in NK92 cells by recombinant GST-sHLA-G1a chain. Cell Res **14**, 155–160.

Yin, D., Zeng, H., Ma, L., Shen, J., Xu, H., Byrne, G.W. and Chong, A.S. (2004). Cutting edge: NK cells mediate IgG1-dependent hyperacute rejection of xenografts. J Immunol **172**, 7235–7238.

Yoneda, O., Imai, T., Goda, S., Inoue, H., Yamauchi, A., Okazaki, T., Imai, H., Yoshie, O., Bloom, E.T., Domae, N. and Umehara, H. (2000). Fractalkine-mediated endothelial cell injury by NK cells. J Immunol **164**, 4055–4062.

Yoneda, O., Imai, T., Nishimura, M., Miyaji, M., Mimori, T., Okazaki, T., Domae, N., Fujimoto, H., Minami, Y., Kono, T., Bloom, E. T. and Umehara, H. (2003). Membrane-bound form of fractalkine induces IFN-gamma production by NK cells. Eur J Immunol **33**, 53–58.

Yu, G., Xu, X., Vu, M.D., Kilpatrick, E.D. and Li, X.C. (2006). NK cells promote transplant tolerance by killing donor antigen-presenting cells. J Exp Med **203**, 1851–1858.

Zeng, M.H., Fang, C.Y., Wang, S.S., Zhu, M., Xie, L., Li, R., Wang, L., Wu, X.W. and Chen, S. (2006). A study of soluble HLA-G1 protecting porcine endothelial cells against human natural killer cell-mediated cytotoxicity. Transplant Proc **38**, 3312–3314.

Zhang, Z., Kaptanoglu, L., Haddad, W., Ivancic, D., Alnadjim, Z., Hurst, S., Tishler, D., Luster, A.D., Barrett, T.A. and Fryer, J. (2002). Donor T cell activation initiates small bowel allograft rejection through an IFN-gamma-inducible protein-10-dependent mechanism. J Immunol **168**, 3205–3212.

Zijlstra , M., Auchincloss, H. Jr, Loring, J.M., Chase, C.M., Russell, P.S. and Jaenisch, R. (1992). Skin graft rejection by beta 2-microglobulin-deficient mice. J Exp Med **175**, 885–893.

Chapter Thirty-Four

34

Natural killer cells and autoimmunity

Christian Münz

CHAPTER CONTENTS

It is quality rather than quantity that matters.

Lucius Annaeus Seneca, Roman statesman and philosopher

ABSTRACT

Natural killer (NK) cells are able to target inflamed tissues, thereby potentially exacerbating autoimmune tissue destruction. However, recent findings suggest that they might also have a protective function during autoimmune disease by editing antigen presenting cells that stimulate autoreactive effector cells and targeting these effectors directly. Genetic evidence, particularly findings in mouse models of autoimmunity, and alterations in the NK cell compartment of patients with autoimmune diseases with and without successful therapy shed some light on the role of this innate lymphocyte subset during autoimmunity. NK cells may play a dual role in autoimmune diseases by initially editing cells of the hematopoietic lineage to curb autoimmunity, but augmenting disease later via inflamed tissue destruction. Therefore, any therapeutic approach, harnessing NK cells during autoimmunity, has to take this functional dichotomy into account.

KEY WORDS

Killer immunoglobulin-like receptors, Multiple sclerosis, Arthritis, Systemic lupus erythematosus, Diabetes, Dendritic cells, Macrophages, Microglia, Th1 polarization, Cytotoxicity

Introduction

Natural killer (NK) cells constitute a primary defense line against pathogens and can also detect cellular transformation. They can fulfil these functions efficiently following primarily cytokine-mediated activation by dendritic cells (DCs) (Ferlazzo and Münz, 2004; Lucas et al., 2007; Strowig et al., 2008b). They recognize infected and transformed cells through germ-line encoded receptors via the loss of MHC class I molecules (missing self) and up-regulation of MHC class I-like molecules (altered self) (Lanier, 2005). While these recognition mechanisms allow for rapid responses to control infections and malignant cells prior to initiation of adaptive immune responses, they are potentially dangerous in settings of autoimmune inflammation, which also seems to lead often to an altered-self phenotype, recognizable for NK cells (Groh et al., 2003). Therefore, it was for the longest time assumed that NK cells play a rather detrimental role in autoimmune diseases by enhancing tissue damage. Indeed, it was found for example that NK cells can kill glial and neuronal cells, possibly enhancing multiple sclerosis (MS) via this mechanism (Morse et al., 2001). However, recent studies have in contrast unravelled a possibly protective role of NK cells against autoimmune disease, suggesting that these innate lymphocytes might initially curb autoimmune disease by editing other cells of the hematopoietic lineage, and only later during overt autoimmune inflammation might contribute to tissue destruction. In this chapter, I will review recent evidence for this functional dichotomy of NK cells in autoimmune diseases.

Genetic association of NK cell receptor variation with autoimmune disease

Polymorphism as well as the presence of stimulatory members of the killer-immunoglobulin-like receptor (KIR) family of NK cell receptors have been linked to protection against microbial diseases (Bashirova et al., 2006). These receptors are also expressed by T cell subsets and therefore genetic association with disease does not necessarily implicate NK cells. Since autoimmunity is the flip-side of aggressive protection against pathogens, it is not surprising that similar variations have also been found to be associated with autoimmune diseases. The KIR locus on human chromosome 19 contains not only highly polymorphic genes, but also a variable number of loci. Two main lineages of KIR haplotypes, which are in equilibrium in any given population, have been distinguished, and are called group A and B (Parham, 2008). While group A contains only a limited amount of loci of primarily inhibitory KIRs (KIR2/3DL), the expanded group B locus encodes activating KIRs in addition (KIR2/3DS). The presence of these later genes correlates with some autoimmune diseases. Progression to vasculitis in rheumatoid arthritis patients is associated with KIR2DS2 presence and expression of the cognate HLA-C1 molecules in the absence of KIR2DL2, which mediates inhibitory recognition of the same MHC class I molecules (Yen et al., 2001). Presence of the same activating KIR in absence of its inhibitory counterpart was enriched in scleroderma patients (Momot et al., 2004) and in the presence of its activating ligand HLA-C1 in ankylosing spondylitis (Jiao et al., 2008).

Additional evidence that presence of activating KIRs might enhance the severity of an autoimmune disease comes from patients with psoriatic arthritis. This subpopulation of psoriasis patients carries the HLA-C recognizing KIR2DS1 and KIR2DS2 receptors at increased frequencies (Martin et al., 2002; Nelson et al., 2004; Williams et al., 2005). Furthermore, KIR2DS1 expression was also found more frequently in systemic lupus erythematosus and scleroderma patients (Pellett et al., 2007). Apart from these HLA-C recognizing activating KIRs, KIR3DS1, recognizing HLA-Bw4 molecules, was found to be associated with ankylosing spondylitis (Jiao et al., 2008). An overall association with the presence of the group B KIR locus was noted in Vogt–Koyanagi–Harada or autoimmune ocular inflammatory disease (Levinson et al., 2008). Finally, in addition to association with activating KIRs, a particular allele of MICA, one of the ligands for the activating NK cell receptor NKG2D, was found to be associated with psoriasis (Cheng et al., 2000). Overall, this evidence, especially from rheumatoid vasculitis and psoriatric arthritis, supports the notion that hyper-responsiveness in the NK cell and/or NK cell receptors carrying T cell compartments might support more severe forms of autoimmunity, augmenting tissue destruction through additional activating signals upon interaction with somatic cells.

Functional characteristics and models of NK cells in autoimmune diseases

The role of NK cells at specific stages of autoimmune disease has not been clearly delineated and, therefore, conflicting results have been reported in animal models of arthritis, colitis, diabetes, myasthenia gravis and experimental allergic encephalomyelitis (EAE). For most of these studies, NK cell deficiency was achieved by depleting antibody injection. These studies used primarily either anti-NK1.1 or anti-asialo-GM1–specific antibodies. These treatments have additional effects apart from NK cell depletion, which makes their interpretation difficult (Shi and Van Kaer, 2006). Anti-NK1.1 antibodies deplete NK as well as NKT cells, while stimulating IFN-γ release at the same time. Anti-asialo-GM1 antibodies deplete both NK as well as cytotoxic T lymphocytes (Morse et al., 2001). With these caveats in mind, NK cell depletion suppressed autoimmune disease in cyclophosphamide–induced diabetes and acetylcholine-receptor–induced experimental allergic myasthenia gravis (Maruyama et al., 1991; Shi et al., 2000). Similarly, NK cell depletion diminishes the development of autoimmune T cell responses in EAE (Winkler-Pickett et al., 2008). In contrast, most studies suggested a protective effect of NK cells against autoimmune diseases, and reported more severe disease manifestation after NK cell depletion. This applies to colitis in IL-10 deficient mice (Fort et al., 1998), *Staphylococcus aureus*–induced arthritis (Nilsson et al., 1999) and myelin basic protein or myelin oligodendrocyte glycoprotein–induced EAE

(Matsumoto et al., 1998; Zhang et al., 1997). The protective function of NK cells in EAE was in part attributed to CXCL1-dependent recruitment of these cells into the CNS, enabling them to modify CNS inflammation in the tissue targeted by autoimmunity (Huang et al., 2006). From these studies it becomes clear that NK cells mediate functions beyond targeting the autoimmune inflamed tissue directly, because they can ameliorate autoimmunity even after homing to diseased tissue. This functional dichotomy of NK cells during autoimmunity can also be observed in the NOD mouse, the most commonly used model for type 1 diabetes. While these mice have diminished NK cell reactivity, caused by a lower frequency and decreased functional capacity of their innate lymphocyte compartment (Poulton et al., 2001), conversion from insulitis to diabetes is associated with pancreatic NK cell infiltration (Poirot et al., 2004). Moreover, NK1.1-mediated NK cell depletion protects NOD mice from diabetes development (Poirot et al., 2004). This suggests that innate lymphocyte deficiency, including lower NK cell reactivity, might predispose for the development of autoimmune disease, probably due to lack of immunoediting functions. However, during overt autoimmunity, cytolytic NK cell functions might contribute to autoreactive tissue destruction. Which NK cell functions dampen autoimmunity, remains unclear. However, clues from changes in NK cell subsets in patients with autoimmune diseases and during successful therapy might shed light on the complex immunobiology of NK cells during autoimmunity.

Indeed, it has been suggested that patients with MS carry lower frequencies of NK cells in their peripheral blood (De Jager et al., 2008; Segal, 2007), but a detailed analysis of NK cell populations and their responsiveness has not been performed in MS. Humans carry two main, functionally distinct NK cell populations. While $CD56^{dim}CD16^{+}$ NK cells are potent cytotoxic lymphocytes, $CD56^{bright}CD16^{-}$ NK cells react primarily with cytokine secretion to activation (Strowig et al., 2008b) and are enriched in secondary lymphoid tissues as well as sites of autoimmune inflammation (Dalbeth et al., 2004; Ferlazzo et al., 2004). The immunoregulatory subset of NK cells are expanded during effective immunotherapy with anti-CD25 antibodies or IFN-β (Bielekova et al., 2006; Li et al., 2005; Saraste et al., 2007). Therefore, it is tempting to speculate that $CD56^{bright}CD16^{-}$ NK cells regulate autoimmune responses at least in part via influencing T cell polarization through cytokine secretion.

Analogous NK cell subsets have not been identified in mice so far, although three candidates with functional similarities to $CD56^{bright}CD16^{-}$ NK cells have been proposed. $Mac1^{high}CD27^{+}$ mouse NK cells produce higher amounts of cytokines than $Mac1^{high}CD27^{-}$ NK cells, but display in contrast to their human counterparts also enhanced cytotoxicity (Hayakawa and Smyth, 2006). Furthermore, $CD127^{+}GATA\text{-}3^{+}$ mouse NK cells resemble human $CD56^{bright}CD16^{-}$ NK cells in that they produce primarily cytokines upon activation and are enriched in secondary lymphoid organs (Vosshenrich et al., 2006). However, they seem to primarily develop in murine thymus, while human $CD56^{bright}CD16^{-}$ NK cells seem to be an intermediate immature NK cell stage that can further differentiate into cytotoxic $CD56^{dim}CD16^{+}$ NK cells (Huntington et al., 2009; Romagnani et al., 2007). Finally, $NK1.1^{+}B220^{+}CD11c^{+}$ NK cells are enriched in secondary lymphoid tissues and secrete high levels of IFN-γ (Blasius et al., 2007). However, they also kill classical NK cell targets efficiently (Caminschi et al., 2007) and express CD11c, B220 and MHC class II molecules probably primarily due to activation (Vosshenrich et al., 2007). Therefore, mouse NK cell subsets with superior cytokine secretion are starting to be characterized and their regulation during autoimmune disease should be investigated in the future.

Influence of NK cells on autoimmune T cell polarization

NK cells shape T cells responses to become Th1 polarized. This polarization, characterized by its IFN-γ production, is thought to most efficiently target viral infections and transformed cells (Maloy et al., 2000; Nishimura et al., 1999; Rentenaar et al., 2000). IFN-γ secretion has been identified as the main factor by which NK cells promote Th1 polarization both in mice and humans (Bajenoff et al., 2006; Laouar et al., 2005; Martin-Fontecha et al., 2004; Morandi et al., 2006). Interestingly, the $CD56^{bright}CD16^{-}$ NK cells of secondary lymphoid tissues, in which T cell polarization during primary immune responses is thought to be established, secrete fivefold more of this cytokine than their counterparts in peripheral blood (Morandi et al., 2006; Strowig et al., 2008a). Therefore, NK cells could preferentially modulate autoreactive T cell responses at these sites.

The polarization of pathogenic T cells in autoimmune diseases remains, however, still a matter of debate. While autoreactive T cells were initially assumed to be primarily Th1 polarized, recently, IL-17 producing Th17 cells have been proposed to initiate disease more efficiently, at least in EAE and psoriasis (Bettelli et al., 2008; Gutcher and Becher, 2007). In the mouse at least, Th17 development can be inhibited by IFN-γ (Harrington et al., 2005), and therefore the Th1 polarizing capacity of NK cells could be protective by diverting autoreactive T cells from a pathogenic Th17 polarization.

NK cell editing of myeloid cells

Alternatively to immune modulation of autoimmune responses by cytokines of NK cells, their cytotoxicity could also edit antigen presenting cells (APCs) during the initiation of autoimmune responses. Along these lines, low numbers of activated NK cells mature DCs, primarily via their ability to secrete TNF (Gerosa et al., 2002; Piccioli et al., 2002). This could further enhance priming of Th1 polarized T cell responses. In addition, several studies have documented NK cell cytotoxicity against autologous myeloid APCs. High numbers of activated NK cells were found to kill immature DCs (Ferlazzo et al., 2002; Piccioli et al., 2002). The activating NK cell receptors NKp30, NKp46 and DNAM-1 have been implicated in myeloid DC recognition by NK cells (Ferlazzo et al., 2002; Pende et al., 2006; Spaggiari et al., 2001). In contrast, mature DCs are protected from NK cell lysis via their elevated MHC class I expression (Ferlazzo et al., 2002), and NK cell inhibition is mainly mediated by the CD94/NKG2A inhibitory receptor (Della Chiesa et al., 2003). In contrast, activated macrophages were preferentially killed by NK cells after upregulation of MHC class I–like NKG2D ligands upon lipopolysaccharide (LPS) stimulation (Nedvetzki et al., 2007). Microglial cells, which are considered to be the main APC population in the CNS, can be also targeted by activated NK cells (Lunemann et al., 2008; Saikali et al., 2007). NK cells were found to kill resting microglia via recognition with their NKp46 and NKG2D activating receptors (Lunemann et al., 2008). Interestingly and similar to DCs, LPS-activated microglial cells were more protected from NK cell cytotoxicity. Therefore, activated NK cells can lyse resting microglia and immature DCs, as well as activated macrophages, thereby possibly limiting autoimmune immune response initiation.

Direct targeting of autoimmune lymphocytes by NK cells

In addition to APCs, NK cells have been suggested to target effector cells of adaptive autoimmune responses, namely T and B cells. The mechanisms behind NK cell mediated inhibition of autoimmune B cells in vitro and in vivo are still unknown (Shi et al., 2000; Takeda and Dennert, 1993). In contrast, NKG2D-mediated recognition of activated T cells has been described (Cerboni et al., 2007; Rabinovich et al., 2003; Roy et al., 2008). T cell activation led to NKG2D ligand upregulation, depending on the DNA repair machinery enzymes ATM/ATR (Cerboni et al., 2007). Autoreactive T cell targeting by NK cells is, however, inhibited by CD94/NKG2D-dependent recognition of the non-classical MHC class I molecular H2-Qa1 in mice, and possibly

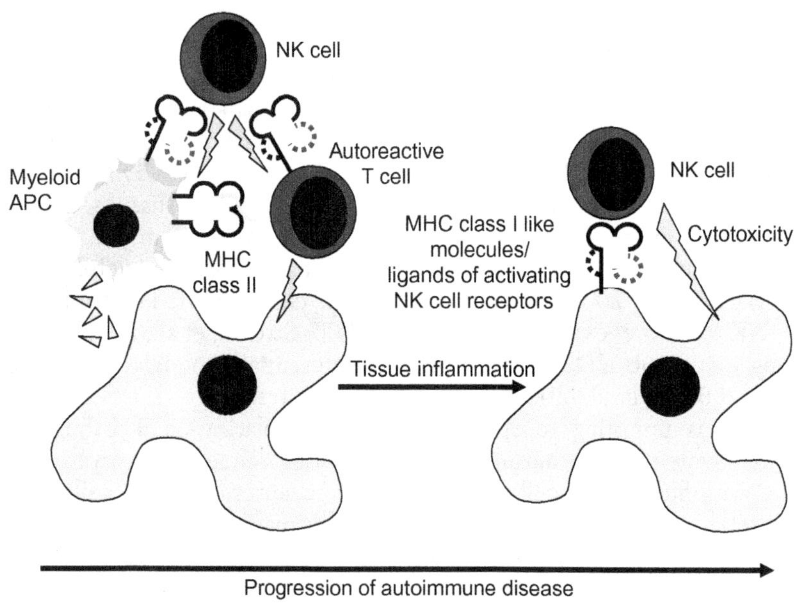

Figure 34.1 • Putative roles of NK cells at different stages of autoimmune disease. NK cells might inhibit the development of autoimmune disease by cytotoxic editing of myeloid APCs, which present autoantigens, and targeting of autoreactive T cells. With proceeding tissue inflammation/stress resulting in the upregulation of ligands for activating NK cell receptors, NK cells might exacerbate tissue destruction during autoimmune disease.

HLA-E in humans (Lu et al., 2007). Disruption of this inhibitory interaction led to inhibition of EAE induction in mice (Lu et al., 2007). In vivo evidence that activated NK cells might compromise the survival of autoreactive T cells also in patients comes again from clinical trials with CD25 blocking antibodies. NK cell expansion negatively correlated with $CD4^+$ as well as $CD8^+$ T cell numbers, and NK cells were able to limit activated T cell survival by a contact dependent mechanism in vitro (Bielekova et al., 2006). Therefore, NK cells seem to be able to limit T cell responses, including those involved in autoimmune diseases.

Summary

Even though many studies have addressed the role of NK cells during autoimmune diseases, their role is still not entirely clear. However, a picture is emerging that especially the immunoregulatory $CD56^{bright}CD16^-$ NK cell subset might be protective against autoimmune disease initially. NK cells might later rather exacerbate the destruction of inflamed tissue (Figure 34.1). Both their immunoregulatory role in the shaping of the immune response with cytokines as well as cytotoxicity against APCs and autoimmune T cells might contribute to the protective function, whereas inflamed tissues probably become susceptible to NK cell killing by upregulating activating NK cell ligands. Due to the possible dichotomy of NK cell function during the progression of autoimmune diseases, therapeutic interventions that target NK cells have to be carefully considered. While NK cell activation or even adoptive transfer of distinct NK cell subpopulations might be beneficial after occurrence of the first autoimmune symptoms, such treatments might also exacerbate established autoimmune diseases.

Acknowledgements

My research is supported by the National Cancer Institute (R01CA108609 and R01CA101741), the Swiss National Science Foundation and the Foundation for the National Institutes of Health (Grand Challenges in Global Health).

References

Bajenoff, M., Breart, B., Huang, A.Y., Qi, H., Cazareth, J., Braud, V.M., Germain, R.N. and Glaichenhaus, N. (2006). Natural killer cell behavior in lymph nodes revealed by static and real-time imaging. J Exp Med **203**, 619–631.

Bashirova, A.A., Martin, M.P., McVicar, D.W. and Carrington, M. (2006). The killer immunoglobulin-like receptor gene cluster: tuning the genome for defense. Annu Rev Genomics Hum Genet **7**, 277–300.

Bettelli, E., Korn, T., Oukka, M. and Kuchroo, V.K. (2008). Induction and effector functions of T(H)17 cells. Nature **453**, 1051–1057.

Bielekova, B., Catalfamo, M., Reichert-Scrivner, S., Packer, A., Cerna, M., Waldmann, T.A., McFarland, H., Henkart, P.A. and Martin, R. (2006). Regulatory $CD56^{bright}$ natural killer cells mediate immunomodulatory effects of IL-2Ralpha-targeted therapy (daclizumab) in multiple sclerosis. Proc Natl Acad Sci U S A **103**, 5941–5946.

Blasius, A.L., Barchet, W., Cella, M. and Colonna, M. (2007). Development and function of murine $B220^+CD11c^+NK1.1^+$ cells identify them as a subset of NK cells. J Exp Med **204**, 2561–2568.

Caminschi, I., Ahmet, F., Heger, K., Brady, J., Nutt, S.L., Vremec, D., Pietersz, S., Lahoud, M.H., Schofield, L., Hansen, D.S., O'Keeffe, M., Smyth, M.J., Bedoui, S., Davey, G.M., Villadangos, J.A., Heath, W.R. and Shortman, K. (2007). Putative IKDCs are functionally and developmentally similar to natural killer cells, but not to dendritic cells. J Exp Med **204**, 2579–2590.

Cerboni, C., Zingoni, A., Cippitelli, M., Piccoli, M., Frati, L. and Santoni, A. (2007). Antigen-activated human T lymphocytes express cell-surface NKG2D ligands via an ATM/ATR-dependent mechanism and become susceptible to autologous NK-cell lysis. Blood **110**, 606–615.

Cheng, L., Zhang, S.Z., Xiao, C.Y., Hou, Y.P., Li, L., Luo, H.C., Jiang, H.Y. and Zuo, W.Q. (2000). The A5.1 allele of the major histocompatibility complex class I chain-related gene A is associated with psoriasis vulgaris in Chinese. Br J Dermatol **143**, 324–329.

Dalbeth, N., Gundle, R., Davies, R.J., Lee, Y.C., McMichael, A.J. and Callan, M.F. (2004). $CD56^{bright}$ NK cells are enriched at inflammatory sites and can engage with monocytes in a reciprocal program of activation. J Immunol **173**, 6418–6426.

De Jager, P.L., Rossin, E., Pyne, S., Tamayo, P., Ottoboni, L., Viglietta, V., Weiner, M., Soler, D., Izmailova, E., Faron-Yowe, L., O'Brien, C., Freeman, S., Granados, S., Parker, A., Roubenoff, R., Mesirov, J.P., Khoury, S.J., Hafler, D.A. and Weiner, H.L. (2008). Cytometric profiling in multiple sclerosis uncovers patient population structure and a reduction of $CD8^{low}$ cells. Brain **131**, 1701–1711.

Della Chiesa, M., Vitale, M., Carlomagno, S., Ferlazzo, G., Moretta, L. and Moretta, A. (2003). The natural killer cell-mediated killing of autologous dendritic cells is confined to a cell subset expressing CD94/NKG2A, but lacking inhibitory killer Ig-like receptors. Eur J Immunol **33**, 1657–1666.

Ferlazzo, G. and Münz, C. (2004). Natural killer cell compartments and their activation by dendritic cells. J Immunol **172**, 1333–1339.

Ferlazzo, G., Tsang, M.L., Moretta, L., Melioli, G., Steinman, R.M. and Münz, C. (2002). Human dendritic cells activate resting NK cells and are recognized via the NKp30 receptor by activated NK cells. J Exp Med **195**, 343–351.

Ferlazzo, G., Thomas, D., Lin, S.L., Goodman, K., Morandi, B., Muller, W.A., Moretta, A. and Münz, C. (2004). The abundant NK cells in human lymphoid tissues require activation to express killer cell Ig-like receptors and become cytolytic. J Immunol **172**, 1455–1462.

Fort, M.M., Leach, M.W. and Rennick, D.M. (1998). A role for NK cells as regulators of CD4$^+$ T cells in a transfer model of colitis. J Immunol **161**, 3256–3261.

Gerosa, F., Baldani-Guerra, B., Nisii, C., Marchesini, V., Carra, G. and Trinchieri, G. (2002). Reciprocal activating interaction between natural killer cells and dendritic cells. J Exp Med **195**, 327–333.

Groh, V., Bruhl, A., El-Gabalawy, H., Nelson, J.L. and Spies, T. (2003). Stimulation of T cell autoreactivity by anomalous expression of NKG2D and its MIC ligands in rheumatoid arthritis. Proc Natl Acad Sci U S A **100**, 9452–9457.

Gutcher, I. and Becher, B. (2007). APC-derived cytokines and T cell polarization in autoimmune inflammation. J Clin Invest **117**, 1119–1127.

Harrington, L.E., Hatton, R.D., Mangan, P.R., Turner, H., Murphy, T.L., Murphy, K.M. and Weaver, C.T. (2005). Interleukin 17-producing CD4$^+$ effector T cells develop via a lineage distinct from the T helper type 1 and 2 lineages. Nat Immunol **6**, 1123–1132.

Hayakawa, Y. and Smyth, M.J. (2006). CD27 dissects mature NK cells into two subsets with distinct responsiveness and migratory capacity. J Immunol **176**, 1517–1524.

Huang, D., Shi, F.D., Jung, S., Pien, G.C., Wang, J., Salazar-Mather, T.P., He, T.T., Weaver, J.T., Ljunggren, H.G., Biron, C.A., Littman, D.R. and Ransohoff, R.M. (2006). The neuronal chemokine CX3CL1/fractalkine selectively recruits NK cells that modify experimental autoimmune encephalomyelitis within the central nervous system. FASEB J **20**, 896–905.

Huntington, N.D., Legrand, N., Alves, N.L., Jaron, B., Weijer, K., Plet, A., Corcuff, E., Mortier, E., Jacques, Y., Spits, H. and Di Santo, J.P. (2009). IL-15 trans-presentation promotes human NK cell development and differentiation in vivo. J Exp Med **206**, 25–34.

Jiao, Y.L., Ma, C.Y., Wang, L.C., Cui, B., Zhang, J., You, L., Chen, Z.J., Li, J.F. and Zhao, Y.R. (2008). Polymorphisms of KIRs gene and HLA-C alleles in patients with ankylosing spondylitis: possible association with susceptibility to the disease. J Clin Immunol **28**, 343–349.

Lanier, L.L. (2005). NK cell recognition. Annu Rev Immunol **23**, 225–274.

Laouar, Y., Sutterwala, F.S., Gorelik, L. and Flavell, R.A. (2005). Transforming growth factor-beta controls T helper type 1 cell development through regulation of natural killer cell interferon-gamma. Nat Immunol **6**, 600–607.

Levinson, R.D., Du, Z., Luo, L., Holland, G.N., Rao, N.A., Reed, E.F. and Rajalingam, R. (2008). KIR and HLA gene combinations in Vogt–Koyanagi–Harada disease. Hum Immunol **69**, 349–353.

Li, Z., Lim, W.K., Mahesh, S.P., Liu, B. and Nussenblatt, R.B. (2005). Cutting edge: in vivo blockade of human IL-2 receptor induces expansion of CD56^{bright} regulatory NK cells in patients with active uveitis. J Immunol **174**, 5187–5191.

Lu, L., Ikizawa, K., Hu, D., Werneck, M. B., Wucherpfennig, K.W. and Cantor, H. (2007). Regulation of activated CD4$^+$ T cells by NK cells via the Qa-1-NKG2A inhibitory pathway. Immunity **26**, 593–604.

Lucas, M., Schachterle, W., Oberle, K., Aichele, P. and Diefenbach, A. (2007). Dendritic cells prime natural killer cells by trans-presenting interleukin 15. Immunity **26**, 503–517.

Lünemann, A., Lünemann, J.D., Roberts, S., Messmer, B., da Silva, R.B., Raine, C.S. and Münz, C. (2008). Human NK c ells kill resting but not activated microglia via NKG2D- and NKp46-mediated recognition. J Immunol **181**, 6170–6177.

Maloy, K.J., Burkhart, C., Junt, T.M., Odermatt, B., Oxenius, A., Piali, L., Zinkernagel, R.M. and Hengartner, H. (2000). CD4$^+$ T cell subsets during virus infection. Protective capacity depends on effector cytokine secretion and on migratory capability. J Exp Med **191**, 2159–2170.

Martin, M.P., Nelson, G., Lee, J.H., Pellett, F., Gao, X., Wade, J., Wilson, M.J., Trowsdale, J., Gladman, D. and Carrington, M. (2002). Cutting edge: susceptibility to psoriatic arthritis: influence of activating killer Ig-like receptor genes in the absence of specific HLA-C alleles. J Immunol **169**, 2818–2822.

Martin-Fontecha, A., Thomsen, L.L., Brett, S., Gerard, C., Lipp, M., Lanzavecchia, A. and Sallusto, F. (2004). Induced recruitment of NK cells to lymph nodes provides IFN-gamma for T(H)1 priming. Nat Immunol **5**, 1260–1265.

Maruyama, T., Watanabe, K., Takei, I., Kasuga, A., Shimada, A., Yanagawa, T., Kasatani, T., Suzuki, Y., Kataoka, K., Saruta., et al. (1991). Anti-asialo GM1 antibody suppression of cyclophosphamide-induced diabetes in NOD mice. Diabetes Res **17**, 37–41.

Matsumoto, Y., Kohyama, K., Aikawa, Y., Shin, T., Kawazoe, Y., Suzuki, Y. and Tanuma, N. (1998). Role of natural killer cells and TCR gamma delta T cells in acute autoimmune encephalomyelitis. Eur J Immunol **28**, 1681–1688.

Momot, T., Koch, S., Hunzelmann, N., Krieg, T., Ulbricht, K., Schmidt, R.E. and Witte, T. (2004). Association of killer cell immunoglobulin-like receptors with scleroderma. Arthritis Rheum **50**, 1561–1565.

Morandi, B., Bougras, G., Muller, W.A., Ferlazzo, G. and Münz, C. (2006). NK cells of human secondary lymphoid tissues enhance T cell polarization via IFN-gamma secretion. Eur J Immunol **36**, 2394–2400.

Morse, R.H., Seguin, R., McCrea, E.L. and Antel, J.P. (2001). NK cell-mediated lysis of autologous human oligodendrocytes. J Neuroimmunol **116**, 107–115.

Nedvetzki, S., Sowinski, S., Eagle, R. A., Harris, J., Vely, F., Pende, D., Trowsdale, J., Vivier, E., Gordon, S. and Davis, D.M. (2007). Reciprocal regulation of human natural killer cells and macrophages associated with distinct immune synapses. Blood **109**, 3776–3785.

Nelson, G.W., Martin, M.P., Gladman, D., Wade, J., Trowsdale, J. and Carrington, M. (2004). Cutting edge: heterozygote advantage in autoimmune disease: hierarchy of protection/susceptibility conferred by HLA and killer Ig-like receptor combinations in psoriatic arthritis. J Immunol **173**, 4273–4276.

Nilsson, N., Bremell, T., Tarkowski, A. and Carlsten, H. (1999). Protective role of NK1.1$^+$ cells in experimental *Staphylococcus aureus* arthritis. Clin Exp Immunol **117**, 63–69.

Nishimura, T., Iwakabe, K., Sekimoto, M., Ohmi, Y., Yahata, T., Nakui, M., Sato, T., Habu, S., Tashiro, H., Sato, M. and Ohta, A. (1999). Distinct role of antigen-specific T helper type 1 (Th1) and Th2 cells in tumor eradication in vivo. J Exp Med **190**, 617–627.

Parham, P. (2008). The genetic and evolutionary balances in human NK cell receptor diversity. Semin Immunol **20**, 311–316.

Pellett, F., Siannis, F., Vukin, I., Lee, P., Urowitz, M.B. and Gladman, D.D. (2007). KIRs and autoimmune disease:

studies in systemic lupus erythematosus and scleroderma. Tissue Antigens **69**(Suppl. 1), 106–108.

Pende, D., Castriconi, R., Romagnani, P., Spaggiari, G.M., Marcenaro, S., Dondero, A., Lazzeri, E., Lasagni, L., Martini, S., Rivera, P., Capobianco, A., Moretta, L., Moretta, A. and Bottino, C. (2006). Expression of the DNAM-1 ligands, Nectin-2 (CD112) and poliovirus receptor (CD155), on dendritic cells: relevance for natural killer–dendritic cell interaction. Blood **107**, 2030–2036.

Piccioli, D., Sbrana, S., Melandri, E. and Valiante, N.M. (2002). Contact-dependent stimulation and inhibition of dendritic cells by natural killer cells. J Exp Med **195**, 335–341.

Poirot, L., Benoist, C. and Mathis, D. (2004). Natural killer cells distinguish innocuous and destructive forms of pancreatic islet autoimmunity. Proc Natl Acad Sci U S A **101**, 8102–8107.

Poulton, L.D., Smyth, M.J., Hawke, C.G., Silveira, P., Shepherd, D., Naidenko, O.V., Godfrey, D.I. and Baxter, A.G. (2001). Cytometric and functional analyses of NK and NKT cell deficiencies in NOD mice. Int Immunol **13**, 887–896.

Rabinovich, B.A., Li, J., Shannon, J., Hurren, R., Chalupny, J., Cosman, D. and Miller, R.G. (2003). Activated, but not resting, T cells can be recognized and killed by syngeneic NK cells. J Immunol **170**, 3572–3576.

Rentenaar, R.J., Gamadia, L.E., van DerHoek, N., van Diepen, F.N., Boom, R., Weel, J.F., Wertheim-van Dillen, P. M., van Lier, R.A. and ten Berge, I.J. (2000). Development of virus-specific CD4$^+$ T cells during primary cytomegalovirus infection. J Clin Invest **105**, 541–548.

Romagnani, C., Juelke, K., Falco, M., Morandi, B., D'Agostino, A., Costa, R., Ratto, G., Forte, G., Carrega, P., Lui, G., Conte, R., Strowig, T., Moretta, A., Münz, C., Thiel, A., Moretta, L. and Ferlazzo, G. (2007). CD56^{bright}CD16$^-$ killer Ig-like receptor-NK cells display longer telomeres and acquire features of CD56^{dim} NK cells upon activation. J Immunol **178**, 4947–4955.

Roy, S., Barnes, P.F., Garg, A., Wu, S., Cosman, D. and Vankayalapati, R. (2008). NK cells lyse T regulatory cells that expand in response to an intracellular pathogen. J Immunol **180**, 1729–1736.

Saikali, P., Antel, J.P., Newcombe, J., Chen, Z., Freedman, M., Blain, M., Cayrol, R., Prat, A., Hall, J.A. and Arbour, N. (2007). NKG2D-mediated cytotoxicity toward oligodendrocytes suggests a mechanism for tissue injury in multiple sclerosis. J Neurosci **27**, 1220–1228.

Saraste, M., Irjala, H. and Airas, L. (2007). Expansion of CD56^{bright} natural killer cells in the peripheral blood of multiple sclerosis patients treated with interferon-beta. Neurol Sci **28**, 121–126.

Segal, B.M. (2007). The role of natural killer cells in curbing neuroinflammation. J Neuroimmunol **191**, 2–7.

Shi, F.D. and Van Kaer, L. (2006). Reciprocal regulation between natural killer cells and autoreactive T cells. Nat Rev Immunol **6**, 751–760.

Shi, F.D., Wang, H.B., Li, H., Hong, S., Taniguchi, M., Link, H., Van Kaer, L. and Ljunggren, H.G. (2000). Natural killer cells determine the outcome of B cell-mediated autoimmunity. Nat Immunol **1**, 245–251.

Spaggiari, G.M., Carosio, R., Pende, D., Marcenaro, S., Rivera, P., Zocchi, M.R., Moretta, L. and Poggi, A. (2001). NK cell-mediated lysis of autologous antigen-presenting cells is triggered by the engagement of the phosphatidylinositol 3-kinase upon ligation of the natural cytotoxicity receptors NKp30 and NKp46. Eur J Immunol **31**, 1656–1665.

Strowig, T., Brilot, F., Arrey, F., Bougras, G., Thomas, D., Muller, W.A. and Münz, C. (2008a). Tonsilar natural killer cells restrict B cell transformation by the Epstein–Barr virus via IFN-gamma. PLoS Pathog **4**, e27.

Strowig, T., Brilot, F. and Münz, C. (2008b). Noncytotoxic functions of NK cells: direct pathogen restriction and assistance to adaptive immunity. J Immunol **180**, 7785–7791.

Takeda, K. and Dennert, G. (1993). The development of autoimmunity in C57BL/6 lpr mice correlates with the disappearance of natural killer type 1-positive cells: evidence for their suppressive action on bone marrow stem cell proliferation, B cell immunoglobulin secretion, and autoimmune symptoms. J Exp Med **177**, 155–164.

Vosshenrich, C.A., Garcia-Ojeda, M.E., Samson-Villeger, S.I., Pasqualetto, V., Enault, L., Richard-Le Goff, O., Corcuff, E., Guy-Grand, D., Rocha, B., Cumano, A., Rogge, L., Ezine, S. and Di Santo, J.P. (2006). A thymic pathway of mouse natural killer cell development characterized by expression of GATA-3 and CD127. Nat Immunol **7**, 1217–1224.

Vosshenrich, C.A., Lesjean-Pottier, S., Hasan, M., Richard-Le Goff, O., Corcuff, E., Mandelboim, O. and Di Santo, J.P. (2007). CD11c^{low}B220$^+$ interferon-producing killer dendritic cells are activated natural killer cells. J Exp Med **204**, 2569–2578.

Williams, F., Meenagh, A., Sleator, C., Cook, D., Fernandez-Vina, M., Bowcock, A.M. and Middleton, D. (2005). Activating killer cell immunoglobulin-like receptor gene KIR2DS1 is associated with psoriatic arthritis. Hum Immunol **66**, 836–841.

Winkler-Pickett, R., Young, H.A., Cherry, J.M., Diehl, J., Wine, J., Back, T., Bere, W.E., Mason, A.T. and Ortaldo, J.R. (2008). In vivo regulation of experimental autoimmune encephalomyelitis by NK cells: alteration of primary adaptive responses. J Immunol **180**, 4495–4506.

Yen, J.H., Moore, B.E., Nakajima, T., Scholl, D., Schaid, D.J., Weyand, C.M. and Goronzy, J.J. (2001). Major histocompatibility complex class I-recognizing receptors are disease risk genes in rheumatoid arthritis. J Exp Med **193**, 1159–1167.

Zhang, B., Yamamura, T., Kondo, T., Fujiwara, M. and Tabira, T. (1997). Regulation of experimental autoimmune encephalomyelitis by natural killer (NK) cells. J Exp Med **186**, 1677–1687.

Chapter Thirty-Five

35

Interaction of NK cells with bacteria

Kenji Chamoto, Daiko Wakita, Shinichi Koizumi, Kazutaka Masuko, Takayuki Ikeda, Rieko Mitamura, Takashi Nishimura

CHAPTER CONTENTS

To the eyes of a god, mankind must appear as a species of bacteria which multiply and become progressively virulent whenever they find themselves in a congenial culture, and whose activity diminishes until they disappear completely as soon as proper measures are taken to sterilize them.

Aleister Crowley

ABSTRACT

Natural killer (NK) cells, including NKT cells, are components of the innate immune system and contribute significantly to the clearance of pathogen-infected or malignant cells. Recently, it has been demonstrated that quite diverse pattern-recognition receptors expressed on dendritic cells (DCs) and macrophages (MΦ) recognize pathogen-specific components and subsequently initiate NK cell activation through two distinct signals: soluble factors and cell-to-cell contact. Crosstalk between NK cells and DCs plays a pivotal role in bridging innate and acquired immunity. Most pathogens and some lactic acid bacteria can modulate the immune balance towards a Th1-predominance. Therefore, bacteria themselves or their components are used to improve the disrupted immune balance or to induce Th1 responses critical for the prevention of infectious diseases, cancers and allergies.

KEY WORDS

NK cells, NKT cells, Bacteria, Infection, DCs, Immunoregulation, Th1 immunity, Tumour, Allergy

Introduction

Natural killer (NK) cells produce many chemokines and inflammatory cytokines in response to, and exhibit cytotoxic activity against, pathogen-infected cells. They contribute to the prevention of various infectious diseases, especially during the early phases of infection, before $CD8^+$ cytotoxic T lymphocyte (CTL) induction (Trinchieri, 1989). Moreover, NK cells demonstrate cytotoxicity against a spectrum of tumour cells, including YAC-1 cells and RLmale1 cells. Activated NK cells, termed NK-LAK cells, exhibit a broad spectrum of cytotoxicity against various tumour cells. Thus NK cells also play a critical role in tumour immunosurveillance mechanisms (Crowe et al., 2002).

NK cells, distinct from NKT cells and CTLs, do not have specific receptors essential for the recognition of antigens, such as T-cell receptors. However NK cells recognize and kill the limited target cells but not normal cells through expression of various NK cell receptors. These receptors precisely regulate NK cell function by activating/inhibiting signals and sustain self-tolerance of NK cells (Lanier, 2005). On the other hand, NKT cells have receptors recognizing specific antigens restricted by CD1 molecules. Although a ligand of NKT cells, α-galactosylceramide (α-GalCer) is not a component of bacteria, evaluation of the mechanisms of NKT cell activation will be useful for understanding innate immunity against bacteria mediated by NK and NKT cells. Both NK cells and NKT cells require association with accessory cells for their activation: dendritic cell (DC), MΦ and monocytes directly or indirectly provide activation signals for NK cells and NKT cells (Lee et al., 2007; Raulet, 2004; Vivier et al., 2008). Reciprocally, DCs are regulated by interaction with NK cells and bridge innate and acquired immunity.

This chapter will discuss cellular and molecular mechanisms of NK cell activation by bacterial infection, and especially how DCs provide 'danger signals' to NK cells and regulate subsequent acquired immunity. In addition, we discuss the mechanisms of innate immune activation mediated by NKT cells. We will also consider the application of bacterial components (CpG) and the NKT cell ligand (α-GalCer) as a modulator of innate and acquired immunity, in allergy and cancer immunotherapy.

Host defence by NK cells

Infectious diseases caused by herpes simplex virus-1, influenza virus or *Listeria monocytogenes* can be regulated by NK cells in mice (Scalzo et al., 2007). In particular, NK cell depletion and adaptive transfer of NK cells leads to increased susceptibility and a decrease in mortality to MCMV infection, respectively, indicating that NK cells are critical for pathogen clearance at early stages of infection. Moreover, loss of antiviral effector molecules in NK cells, such as IFN-γ and perforin, also induces susceptibility to infectious diseases (Arase et al., 2002; Scalzo et al., 2007). In the infected host, NK cells recognize (i) pathogen-encoded molecules, (ii) stress-induced self-proteins and (iii) inhibitory self-proteins. In the MCMV infection model, the NK cell receptor Ly49H specifically recognizes the MCMV-encoded cell surface molecule, m157 (Dokun et al., 2001). Ly49H is expressed by about half of NK cells, and Ly49H$^+$ NK cells exhibit marked proliferation for several days following MCMV infection (Groh et al., 1999), suggesting that Ly49H has a critical role as a pathogen-recognition receptor (PRR) that induces NK cell activation and prevents MCMV infection at an early stage. Activating NK cell receptors recognize stress-induced self-ligands. For instance, NKG2D recognizes MHC class I-related chain A/B (MICA/B) in humans, as well as retinoic acid early inducible-1 (RAE1), H60 and murine ULBP-like transcript 1 (MULT1) in mice (Lanier, 2005). In general, these NKG2D ligands are poorly expressed on normal cells but are markedly upregulated on infected cells and tumour cells that are recognized and eliminated by NK cells (Lodoen et al., 2003; Yokoyama and Plougastel, 2003). In contrast, NK cells also recognize self MHC class I molecules by inhibitory NK cell receptors, such as the Ly49 molecules in mice, the killer immunoglobulin-like receptors (KIRs), the leukocyte immunoglobulin-like receptors (LIRs) in humans and the CD94-NKG2 receptors in both species (Lee, Miyagi et al., 2007; Raulet, 2004). MHC class I molecules are constitutively expressed by most healthy cells, ensuring self tolerance of NK cells, while they are downregulated upon stress following infection or transformation, allowing cytotoxicity by NK cells (Akira et al., 2006; Kärre et al., 1986; Parham, 2005). Therefore, NK cells are activated through these three different pathways and contribute to the clearance of invading pathogens and cancer cells.

Recognition of pathogens by PRRs

Exogenous microbes such as viruses, bacteria and yeasts have conserved motifs termed pathogen-associated molecular patterns (PAMPs), lipids, lipoproteins, proteins and nucleic acids (Lee, Kim et al., 2007). Generally, various PRRs, including Toll-like receptors (TLR1-13 in mice), C-type lectin receptors (Dectin-1), caspase-recruitment domain (CARD) helicases (retinoic acid inducible gene I [RIG-I] and melanoma differentiation-associated gene-5 [MDA5]) and nucleotide-binding oligomerization domain (NOD)-like receptors, have an essential role in the recognition of PAMPs by immune cells (Iwasaki and Medzhitov, 2004). Detection of PAMPs by PRRs activates intracellular signalling pathways and subsequently induces production of several inflammatory cytokines and chemokines and upregulation of costimulatory molecules. These inflammatory responses are essential for effective clearance of invading pathogens through activation of NK cells, NKT cells and T cells.

TLRs are well-established receptors among the PRRs and initiate inflammatory responses as well as subsequent adaptive immune responses (Akira and Takeda, 2004). In mammals, TLRs are type-I membrane proteins containing an extracellular domain with leucine rich repeats (LRRs) responsible for ligand recognition and a cytoplasmic

Toll/IL-1R homology (TIR) domain required for initiating signalling (Akira et al., 2006). So far, 11 human TLRs and 13 mouse TLRs have been identified, and each TLR recognizes multiple microbial components in bacteria, fungi, parasites and viruses (Sivori et al., 2004). TLRs are classified into three groups on the basis of their ligands: (i) TLR1, 2, 4 and 6 recognize lipid-based bacterial cell-wall components, such as LPS and lipopeptides; (ii) TLR5 and 11 recognize microbial protein components such as flagellin and profilin-like molecules; (iii) TLR3, 7, 8 and 9 recognize nucleic acids such as dsRNA, single-stranded RNA (ssRNA), and CpG DNA. TLRs 1, 2, 4, 5 and 6 are located mainly on the cell surface, whereas TLRs 3, 7, 8 and 9 are located in the endocytic compartment (Sivori et al., 2004).

Crosstalk between NK cells and DCs

NK cells can recognize pathogens directly through NK receptors or several TLRs (Dokun et al., 2001; Koizumi et al., 2008). Indeed, NK cells produce IFN-γ and display enhanced cytotoxicity when cultured with TLR ligands, whereas full activation of NK cells requires signals provided by accessory cells, such as macrophages and DCs. Recent results indicate that accessory cells, especially DCs, are essential for recognition of invading pathogens and activation of innate immunity (Lee et al., 2007; Raulet, 2004; Vivier et al., 2008). To induce IFN-γ production or cytotoxicity by NK cells, DCs activated by bacterial components provide both soluble and direct interaction signals to NK cells. Many reports indicate that most pathogens induce IFN-γ secretion by NK cells through a response to IL-12 released by DCs and macrophages (Liang et al., 2003; Monteiro et al., 1998; Orange and Biron, 1996a,b; Scharton-Kersten et al., 1995; Suzuki et al., 2004). Type-I IFNs are other soluble factors provided by DCs that mediate cytotoxicity of NK cells by upregulation of NKG2D, FasL and perforin (Nguyen et al., 2002). Neutralization of type-I IFNs using blocking antibody or loss of the interferon alpha receptor 1 (IFNAR1) gene significantly reduces the cytotoxicity of NK cells in response to infection with MCMV (Orange and Biron, 1996b; Tripp et al., 1993). Furthermore, many reports have demonstrated that these effects of IL-12 and type-I IFNs on NK cells can be enhanced synergistically by other inflammatory cytokines, such as TNF-α, IL-18, IL-2 and IL-15 (Akira and Takeda, 2004; Dalod et al., 2003; Ferlazzo et al., 2004; Granucci et al., 2004; Newman and Riley, 2007; Walzer et al., 2005; Yu et al., 2001).

Generally, direct interactions between NK cells and accessory cells are important for cytokine production or cytotoxicity of NK cells, through costimulatory or adhesion molecule pathways, such as CD80/CD86–CD28, CD40–CD40L, CD70–CD27, lymphocyte function-associated antigen-1 (LFA-1)-intercellular adhesion molecule-1 (ICAM-1) and NKG2D–NKG2DL (Vivier et al., 2008). Activation of NK cells in response to IL-15 and TNF-α is also mediated by direct contact between DCs and NK cells. IL-15 released by pathogen-recognizing DCs immediately forms IL-15-IL-15Rα complexes on the surface of DCs that serve as activation signals for NK cells via direct contact (transpresentation) (Lucas et al., 2007; Mortier et al., 2008; Schluns et al., 2005). Similarly, transmembrane TNF (membrane-bound TNF-α) expressed on the surface of DCs provides an activating signal for NK cells via TNF receptor type 2, which specifically recognizes transmembrane TNF (tmTNF) but not soluble TNF (Borg et al., 2004). In the case of DC-mediated NK cell activation, the formation of stimulatory immunological synapses between DCs and NK cells, and the polarization of IL-12 on the surface of DC at intercellular adhesion areas, is required for NK cell activation (Xu et al., 2007).

Regulation of Th1 immunity by lactic acid bacteria

DCs can engulf microorganisms and activate both innate and acquired immunity in a type-1 cytokine-dependent manner (Banchereau and Steinman, 1998). IL-12, produced by DCs, is a crucial cytokine to induce the differentiation of Th1 cells (Trinchieri, 2003). Moreover, IFN-γ-producing NK cells, induced by IL-12, are recruited to lymph nodes and contribute to subsequent Th1 priming in vivo; other early IFN-γ sources are NKT cells and asialo GM1$^+$ Gr1$^+$ CD8$^+$ T cells (Kosaka et al., 2006; Kosaka et al., 2007; Martín-Fontecha et al., 2004). Therefore, DC-mediated IL-12 production would be required for the initiation of strong type-1 immunity. Some lactic acid bacteria (LAB), which belong to bacterial flora in the gastrointestinal tract, serve as one of the rational tools to modulate the Th1/Th2 immune balance towards a Th1-dominant state. Previously, it has been indicated that nonviable LAB induce IL-12 from macrophages and inhibit IgE production in a food allergy model (Murosaki et al., 1998; Shida et al., 2004). In vivo administration of LAB improves the symptoms of allergy diseases, such as Japanese cedar pollinosis and atopic dermatitis in clinical studies (Ishida et al., 2005; Kalliomäki et al., 2001, 2003).

Recently, LAB derived from plants has attracted attention. *Lactobacillus pentosus* strain S-PT84, isolated from Kyoto pickles, was identified among 16 LAB of plant origin as the strongest interleukin (IL)-12-inducing strain (Nonaka et al., 2008). Following experimental oral intake of S-PT84 resulted in increased cytotoxicity of NK cells (Figure 35.1A, D) and reduced IgE production in the

OVA-induced allergy model, indicating that S-PT84 enhanced Th1 immunity, including activation of NK cells. When the precise mechanisms of these processes were evaluated, we found that S-PT84 induced IFN-γ-producing NK cells through IL-12 produced by CD11c$^+$ DCs in a TLR2- or TLR4-dependent manner (Figure 35.1B, D). Moreover, experiments in which NK cells were cultured separate from DCs in a transwell system, in the presence of S-PT84, revealed that direct interaction between DCs and NK1.1$^+$ cells was also essential for the IFN-γ production in response to S-PT84 stimulation (Figure 35.1C, D) (Koizumi et al., 2008). In general, most LAB stimulate the production of IL-12, IFN-γ and IL-10. However, recently, we isolated Hokkaido-herring pickles-derived LAB, termed Bio-S24, which can stimulate IL-12 production but not IL-10 by DCs, and induced IFN-γ-dependent Th1 immunity (Figure 35.2). We believe that LAB Bio-S24 will be useful in the future for producing immunomodulating foods that can drive a disrupted immune balance toward Th2 immunity.

In the case of a bacterial component, CpG DNA (a ligand of TLR9), we observed a similar phenomenon. IL-12 production by CpG-ODN-activated DCs and direct contact between CpG-ODN-activated DCs and NK cells are essential for inducing IFN-γ-producing NK cells (Orange and Biron, 1996a). In addition, we confirmed that in vivo administration of CpG-ODN effectively inhibits tumour progression and allergic airway inflammation through enhancing type-1 immune responses in mouse models (Ashino et al., 2008; Wakita et al., 2006).

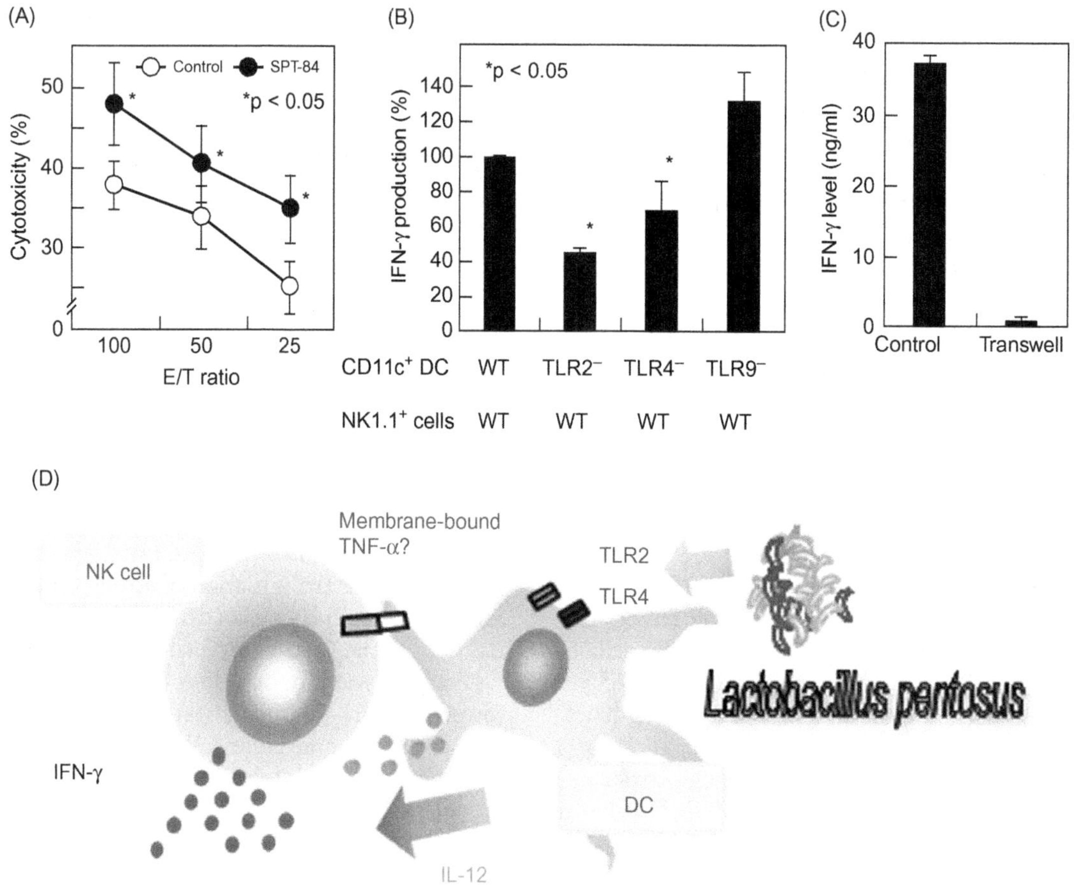

Figure 35.1 • (A) Oral intake of S-PT84 enhances NK activity of spleen cells in vivo. AIN-93M diets containing heat-killed *Lactobacillus pentosus* strain S-PT84 were orally administered into wild-type (WT) C57BL/6 mice for 7 days. Spleen cells were then collected and the NK activity was evaluated by [^{51}Cr]-release assay with YAC-1 cells as target cells. (B) NK1.1$^+$ cells (2×10^5) from WT mice were co-cultured with CD11c$^+$ DCs from WT, TLR2$^{-/-}$, TLR4$^{-/-}$, and TLR9$^{-/-}$ mice in the presence of S-PT84 (1 μg/ml) for 24 h. IFN-γ production levels in the supernatants were determined by ELISA. The percentage of cytokine production was calculated as: cytokine production (%) = (TLR/WT) × 100. (C) NK1.1$^+$ cells (2×10^5 cells/well) were separately cultured with CD11c$^+$ DCs (5×10^4 cells/well) by the transwell system in the presence of S-PT84 (1 μg/ml) for 24 h. IFN-γ production levels in the culture supernatants were determined by ELISA. (D) S-PT84 (*Lactobacillus pentosus*) induced IFN-γ production by spleen cells. The mechanism was that (i) S-PT84 was recognized by DC via TLR2 and TLR4, (ii) the S-PT84-stimulated DC secreted IL-12 and (iii) the induced IL-12 caused the subsequent IFN-γ production by NK cells through direct interactions between DC and NK cells.

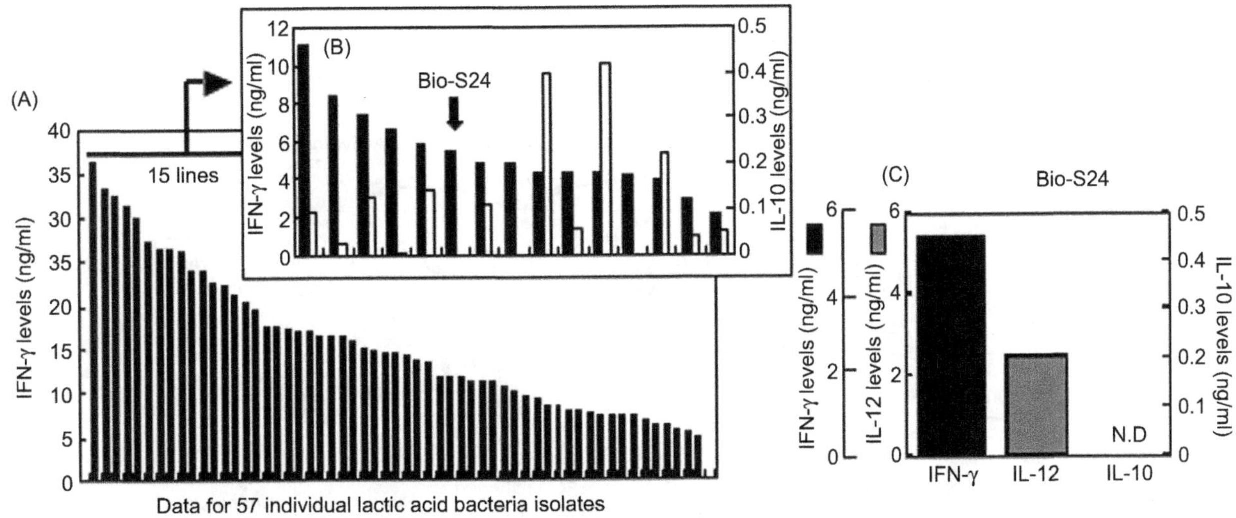

Figure 35.2 • (A) LABs (57 lines) were isolated from various Hokkaido Pickles and defined their IFN-γ-inducing activity from mouse spleen cells. (B) 15 LAB lines, which showed higher IFN-γ stimulating activity, were selected to measure both IFN-γ- and IL-10-inducing activity. (C) Bio-S24-induced IL-12 and IFN-γ but not IL-10 (N.D. mean non-detectable).

The mechanism of NK and NKT activation by glycolipid α-GalCer

We have summarized the mechanisms of NK and NKT cell activation via bacterial component-recognizing DCs. We next introduce the nonbacterial component, which can also induce the activation of innate immunity through cross-talk between DCs and NK cells. Elucidating the mechanisms of activating innate immunity by α-GalCer would promote understanding of host defence against bacteria.

It was demonstrated recently that NKT cells are strongly stimulated by α-GalCer, a potent inducer of antitumour immunity in mice (Kawano et al., 1997; Kobayashi et al., 2005; Motoki et al., 1996). Recognition of α-GalCer by NKT cells appears to depend on interaction between the invariant TCR of these cells and α-GalCer presented by the nonclassical MHC molecule CDld on antigen-presenting cells (APCs) (Bendelac et al., 1997). Stimulation of NKT cells by α-GalCer results in the production of large amounts of IFN-γ and some IL-4, and the development of a cytotoxic phenotype (Burdin et al., 1998). We have investigated the mechanism underlying the activation of NKT cells by α-GalCer (Kitamura et al., 1999). We first established, using purified subsets of various lymphocyte populations, that α-GalCer selectively activates NKT cells for production of IFN-γ. Production of IFN-γ by NKT cells in response to α-GalCer requires IL-12 produced by DCs and direct contact between NKT cells and DCs through CD40–CD40 ligand interactions. Moreover, α-GalCer strongly induces expression of IL-12 receptors on NKT cells from wild-type mice but not CD1 or Vα14$^{-/-}$ mice. This effect of α-GalCer requires the production of IFN-γ by NKT cells and the production of IL-12 by DCs. Moreover, we showed that treatment of mice with suboptimal doses of α-GalCer, together with suboptimal doses of IL-12, results in strongly enhanced natural killing activity and IFN-γ production. Overall, DC-produced IL-12 and DC–NKT interactions through CD40–CD40L are critical for the activation of NKT cells by α-GalCer and suggest that NKT cells can condition DCs for subsequent immune responses (Figure 35.3A).

NKT cells act as regulatory cells rather than killer cells during activation of NK cell-mediated cytotoxicity by α-GalCer in vivo

It is now well established that α-GalCer induces large amounts of IFN-γ and IL-4 at the same time by activation of NKT cells (Brossay et al., 1998; Burdin et al., 1998; Kawano et al., 1997) and induces a potent cytotoxic response. However, the precise role of NKT and NK cells remains unknown during early activation of cytokine production and natural killing induced by α-GalCer administration in vivo. Some investigators have reported that α-GalCer-activated Vα14 NKT cells exhibit strong cytotoxicity in vivo (Kawano et al., 1997, 1998), while others have reported that α-GalCer-induced NKT cell activation causes IFN-γ-dependent, rapid NK cell activation, which is critical for antitumour effects in vivo (Eberl and MacDonald, 2000; Hayakawa et al., 2001). Therefore, we re-evaluated the role of NKT and NK cells at early times after α-GalCer administration. Intracellular staining demonstrated that NKT cells were the earliest source

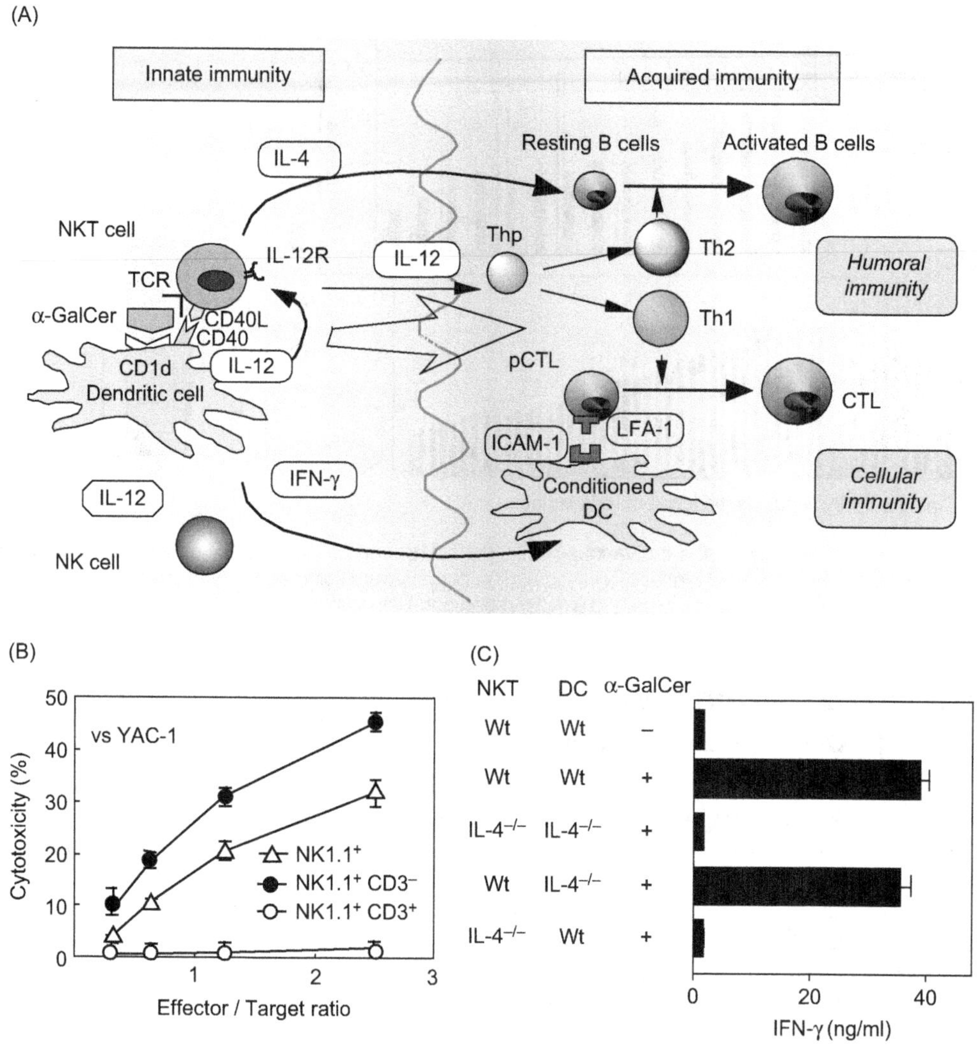

Figure 35.3 • (A) CD40–CD40 ligand (CD40L)-mediated NKT–DC interaction is critical for the induction of IL-12 from DC and subsequent development of activated CD8$^+$ CTL by α-GalCer. These findings provide direct evidence for a critical role of CD1d-restricted NKT cells and DC in bridging innate and acquired immunity. (B) α-GalCer-induced natural killing activity was dependant on not NKT cells, but NK cells in vivo. Natural killing activities of isolated NKT (δ) cells and NK (X) cells against NK-sensitive YAC-1after α-GalCer injection were assessed by 4-h ^{51}Cr-release assay. (C) The deficient property of IFN-γ production in IL-4$^{-/-}$ mice by the stimulation of α-GalCer was ruled by NKT cells not DCs. Purified splenic DCs (2×10^5) and NKT cells (2×10^5) from wt mice and IL-4$^{-/-}$ mice were cocultured with or without α-GalCer. Supernatants were collected at 36 h and the level of IFN-γ was detected by ELISA system.

of both IL-4 and IFN-γ following α-GalCer administration in vivo. However, these α-GalCer-activated NKT cells exhibited no significant natural killing activity. In contrast, isolated NK1.1$^+$CD3$^-$ classical NK cells exhibited greatly enhanced natural killing activity 6 h after α-GalCer administration. In this study, we first directly demonstrated that NKT cells could act as regulatory cells via production of cytokines for activation of NK cell-mediated cytotoxicity in vivo at an early phase after α-GalCer administration (Figure 35.3B).

NKT cells are also considered to play an important role as immunoregulatory cells in antitumour immunity (Chamoto et al., 2004; Hayakawa et al., 2001; Smyth et al., 2002; Terabe et al., 2000), autoimmunity (Hammond et al., 1998; Hong et al., 2001) and maintaining some forms of tolerance (Ikehara et al., 2000; Seino et al., 2001; Sonoda et al., 1999). Since IFN-γ and IL-4 are representative cytokines involved in Th1 and Th2 immunity, the control of cytokine production by NKT cells leads to the regulation of immune diseases, including tumour growth, through control of the immune balance (Fujii et al., 2002; Miyamoto et al., 2001). For instance, Yamamura et al. demonstrated that a synthetic-α-GalCer analogue, named OCH, stimulated NKT cells to produce IL-4 predominantly, rather than IFN-γ. Thus, OCH administration suppressed induction of experimental autoimmune encephalomyelitis (EAE) (Miyamoto et al., 2001). Fujii et al. (2002) reported that α-GalCer-loaded

DCs elicited mainly IFN-γ from NKT cells and succeeded to inhibit metastasis of B16 melanoma. Therefore, it is of great importance to investigate the mechanisms underlying the regulation of cytokine production, especially IFN-γ and IL-4, by NKT cells to develop a new strategy for immune diseases. In order to evaluate the role of IL-4 in the production of IFN-γ from NKT cells, we used IL-4-deficient C57BL/6 mice ($IL\text{-}4^{-/-}$ mice) (Togashi et al., 2007). Administration of α-GalCer to wild-type C57BL/6 mice caused the production of both IFN-γ and IL-4 in serum or cytoplasm within 4h of the injection. Unexpectedly, however, $IL\text{-}4^{-/-}$ mice-derived NKT cells did not produce IFN-γ during the early phase after primary stimulation with α-GalCer (Figure 35.3C). Since NKT cells from $IL\text{-}4^{-/-}$ mice produced IFN-γ when they were secondarily stimulated with α-GalCer in vitro for 72h, NKT cells from $IL\text{-}4^{-/-}$ mice were not completely deficient in IFN-γ production. In this work, we also demonstrated that NKT cells but not DCs were attributed to the loss of IFN-γ in $IL\text{-}4^{-/-}$ mice (Figure 35.3C). Thus IL-4 is required for the activation of NKT cells to produce IFN-γ in response to α-GalCer. Mencacci et al. (1998) also reported that endogenous IL-4 is required for development of protective Th1 responses to *Candida albicans*. Other groups reported that under some conditions, IL-4-primed DCs produce IL-12, which induce NKT cells to produce IFN-γ (Ebner et al., 2001, Hochrein et al., 2000). Another group reported that IL-4 enhanced the response of NKT cells to IL-2 and IL-12, leading to the production of IFN-γ (Bream et al., 2003). The role of IL-4 in the crosstalk of NK and NKT cells with DCs has not completely resolved.

It has been reported that a component of *Lactobacillus pentosus* activates NK cells without NKT cells (Koizumi et al., 2008), and does not involve α-GalCer. However, it has been demonstrated that invariant NKT (iNKT) cells can react with α-anomeric glycosphingolipids derived from the cell wall of Gram-negative *Sphingomonas* bacteria (Kinjo et al., 2005; Mattner et al., 2005; Sriram et al., 2005) and with α-galactosyl diacylglycerols from the spirochaete *Borrelia burgdorferi* (Kinjo et al., 2006), the etiologic agent of Lyme disease. During the host response to these bacteria, NKT cells may regulate the immune balance through NK cell activation and subsequent activation of acquired immunity.

Application of immunological theory to cancer immunotherapy and activation of innate immunity for cancer therapy

More than a century ago, it was realized that deliberate exposure of cancer patients to certain microbes could induce tumour regression and sometimes even eradication (Thomas-Tikhonenko and Hunter, 2003). This gave rise to the notion of cancer immunotherapy as a potential treatment modality for some types of cancer. Currently, bacterial components are used for immunotherapy as adjuvants to activate antitumour immunity. In mice, a number of adjuvants such as the TLR7/8 antagonist R848 or the microbial product-containing emulsion Ribi were shown to polarize Th1 responses via indirect activation of NK cells that provide the IFN-γ necessary for efficient T cell priming by DCs in lymph nodes (Martin-Fontecha et al., 2004). Another commonly used adjuvant, the TLR3/MDA-5 ligand polyinosinic:polycytidylic acid (poly(I:C)), induces strong Th1 responses and has been shown to directly induce IFN-γ production in NK cells. Accessory cells such as DCs or monocytes could amplify IFN-γ production by secretion of IL-12 or IL-18 (Ferlazzo et al., 2004; Gorski et al., 2006; Schmidt et al., 2004). In our hands, stimulation of DCs with *Lactobacillus pentosus* also induces high production of IL-12 essential for subsequent IFN-γ production from NK cells (Koizumi et al., 2008) (refer to Figure 35.1). It is well accepted that IL-12 exhibits antitumour activity by activating NK cells in addition to NKT cells (Trinchieri, 2003). The antitumour activity of α-GalCer strongly resembles the antitumour activity mediated by the cytokine IL-12 (Cui et al., 1997; Kawano et al., 1998). Therefore, we investigated the combined therapeutic effect of α-GalCer and IL-12 against highly metastatic B16-BL6-HM melanoma cells (Nakui et al., 2000) (Figure 35.4). In comparison to a single administration of *α-GalCer* or IL-12 alone, the combined treatment of tumour-bearing mice with *α-GalCer* plus IL-12 caused a super-induction of serum IFN-γ levels, though *α-GalCer*-induced IL-4 production was inhibited. In parallel with the augmented IFN-γ production, natural killing activity against YAC-1 cells and syngeneic B16-BL6-HM melanoma was greatly augmented by the combined therapy. The major effector cells responsible for natural killing activity induced by *α-GalCer* plus IL-12 were enriched in both $NK1.1^+$ TCR^+ NKT cells and $NK1.1^+TCR^-$ NK cells. The preventative effect of *α-GalCer* or IL-12 alone against lung metastasis of B16-BL6-HM was also enhanced by the combination therapy (Figure 35.4A–E).

Activation of acquired immunity for cancer therapy using bacterial adjuvants

In order to induce effective antitumour immunity, it is necessary to generate and maintain tumour-specific CTLs under the immunosuppressed state of tumour-bearing hosts. To overcome the immunosuppressed state

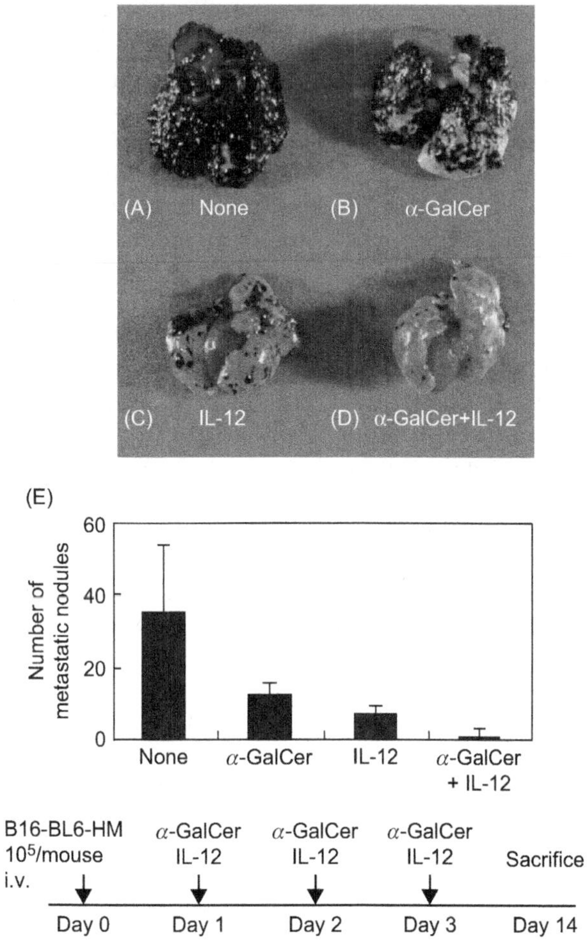

Figure 35.4 • Therapeutic effect of α-GalCer and IL-12 on lung metastasis of B16-BL6-HM melanoma cells in C57BL/6 mice. B16-BL6-HM (10^5) were i.v. injected into C57BL/6 mice. A day after tumour inoculation, the mice were treated with i.v. injection of α-GalCer and IL-12 once a day for 3 days. (A–D) The typical photograph of lung metastatic nodules in mice treated with none (A), α-GalCer (B), IL-12 (C), α-GalCer C IL-12 (D). (E) Lung metastatic nodules in mice treated with none, α-GalCer, IL-12 or α-GalCer + IL-12 were counted 14 days after tumour inoculation.

and induce tumour-specific CTLs, we have emphasized the necessity of introducing type-1 immunity in tumour-bearing hosts (Chamoto et al., 2006; Ikeda et al., 2004; Nishimura et al., 1999). In order to activate type-1 immunity, an immunomodulating adjuvant was usually used in immunotherapy. Adjuvants activate innate immunity via DC activation and bridge innate immunity and acquired immunity. For example, in the α-GalCer model, in vivo administration of α-GalCer caused the rapid activation of NKT cells to induce strong NK activity and cytokine production. Several hours after the activation of innate immunity, the induction of the early activation marker CD69 on $CD4^+$ T cells, $CD8^+$ T cells and B cells occurred. Moreover, in vivo injection of α-GalCer resulted not only in the activation of NKT cells but also in the generation of $CD69^+$ $CD8^+$ T cells with both cytotoxic activity and IFN-γ-producing ability. Tumour-specific CTL generation was also accelerated by α-GalCer (Nishimura et al., 2000). This result indicated that crosstalk between DCs and NKs, including NKT cells, might determine the subsequent Th1/Th2 polarization. Other groups have reported that NK cells are recruited to lymph nodes and upon infection with *Leishmania major* IFN-γ-secreting NK cells were found to contact the same DCs as Ag-specific $CD4^+$ T cells in the lymph nodes using intravital microscopy (Bajenoff et al., 2006). This observation provides insight into the regulation of T-cell responses by NK cells, as it has been demonstrated that, upon activation by mature DCs, NK cells produce high levels of IFN-γ that are sufficient to mediate Th1 polarization in mouse models and human allogeneic immune responses (Bajenoff et al., 2006; Laouar et al., 2005; Morandi et al., 2006).

As another promising adjuvant, we developed liposome-encapsulated CpG oligodeoxynucleotides (CpG-liposome) for tumour immunotherapy. CpG-liposome potently activated NK cells and NKT cells to produce IFN-γ, whereas the same dose of unmodified CpG-ODN induced only low numbers of IFN-γ-producing NK cells and NKT cells to produce IFN-γ. In contrast with the NKT cell agonist α-GalCer, which induces both IFN-γ and IL-4 production by NKT cells, CpG-liposomes only induced IFN-γ production by NKT cells (Suzuki et al., 2004). The tumour-bearing mice were injected intradermally near the tumour-draining lymph node (DLN) with CpG-liposomes co-encapsulated with tumour antigen. This vaccination protocol markedly prevented the growth of the established tumour mass, and approximately 50% of tumour-bearing mice were completely cured with the generation of tetramer+ CTLs in the tumour DLN and at the tumour site (Wakita et al., 2006).

Conclusions

Here, we have reviewed the cellular and molecular mechanisms of regulating the innate and acquired immune responses by bacterial stimuli and NKT cell ligands via reciprocal interactions among DCs, NK and NKT cells. Sustaining healthy conditions in mice and humans require precise control of immune responses towards exogenous antigens because disruption of immune homeostasis causes various immune-associated diseases, such as allergy, autoimmune disease and cancer. Hyperactivation of Th1 and Th17 immunity can cause colitis and liver injury, whereas Th2 immunity permits the development of asthma. Recent studies show that LAB contained in various foods stimulate IFN-γ-dependent NK activity and Th1 immunity, which may prove useful for improving polarized Th2 immune responses in allergic airway inflammation and for promoting protective immunity against infection and cancer. Indeed, the bacterial component CpG-ODN exhibits strong protection against infection, cancer and allergy through inducing Th1-dominant immune responses (Ashino et al., 2008; Suzuki et al., 2004; Wakita et al., 2006).

Thus, we believe that evaluation of the precise mechanisms of immune responses to bacterial infection will provide novel strategies to improve the disrupted immune balance of humans, especially children in developed countries, and will contribute to the prevention of infectious diseases, cancer and allergy in the future.

References

Akira, S. and Takeda, K. (2004). Toll-like receptor signalling. Nat Rev Immunol **4**, 499.

Akira, S., Uematsu, S. and Takeuchi, O. (2006). Pathogen recognition and innate immunity. Cell **124**, 783.

Arase, H., Mocarski, E.S., Campbell, A.E., Hill, A.B. and Lanier, L.L. (2002). Direct recognition of cytomegalovirus by activating and inhibitory NK cell receptors. Science **296**, 1323.

Ashino, S., Wakita, D., Zhang, Y., Chamoto, K., Kitamura, H. and Nishimura, T. (2008). CpG-ODN inhibits airway inflammation at effector phase through down-regulation of antigen-specific Th2-cell migration into lung. Int Immunol **20**, 259.

Bajenoff, M., Breart, B., Huang, A.Y., Qi, H., Cazareth, J., Braud, V.M., Germain, R. N. and Glaichenhaus, N. (2006). Natural killer cell behavior in lymph nodes revealed by static and real-time imaging. J Exp Med **203**, 619.

Banchereau, J. and Steinman, R.M. (1998). Dendritic cells and the control of immunity. Nature **392**, 245.

Bendelac, A., Rivera, M.N., Park, S.H. and Roark, J.H. (1997). Mouse CD1-specific NK1 T cells: development, specificity, and function. Annu Rev Immunol **15**, 535.

Borg, C., Jalil, A., Laderach, D., Maruyama, K., Wakasugi, H., Charrier, S., et al. (2004). NK cell activation by dendritic cells (DCs) requires the formation of a synapse leading to IL-12 polarization in DCs. Blood **104**, 3267.

Bream, J.H., Curiel, R.E., Yu, C.R., Egwuagu, C.E., Grusby, M.J., Aune, T.M. and Young, H.A. (2003). IL-4 synergistically enhances both IL-2- and IL-12-induced IFN-γ expression in murine NK cells. Blood **102**, 207.

Brossay, L., Burdin, N., Tangri, S. and Kronenberg, M. (1998). Antigen-presenting function of mouse CD1: one molecule with two different kinds of antigenic ligands. Immunol Rev **163**, 139.

Burdin, N., Brossay, L., Koezuka, Y., Smiley, S.T., Grusby, M.J., Gui, M., Taniguchi, M., Hayakawa, K. and Kronenberg, M. (1998). Selective ability of mouse CD1 to present glycolipids: α-galactosylceramide specifically stimulates Va141 NK T lymphocytes. J Immunol **161**, 3271.

Chamoto, K., Takeshima, T., Kosaka, A., Tsuji, T., Matsuzaki, J., Togashi, Y., Ikeda, H. and Nishimura, T. (2004). NKT cells act as regulatory cells rather than killer cells during activation of NK cell-mediated cytotoxicity by alpha-galactosylceramide in vivo. Immunol Lett **95**, 5.

Chamoto, K., Wakita, D., Narita, Y., Zhang, Y., Noguchi, D., Ohnishi, H., Iguchi, T., Sakai, T., Ikeda, H. and Nishimura, T. (2006). An essential role of antigen-presenting cell/T-helper type 1 cell–cell interactions in draining lymph node during complete eradication of class II-negative tumor tissue by T-helper type 1 cell therapy. Cancer Res **66**, 1809.

Crowe, N.Y., Smyth, M.J. and Godfrey, D.I. (2002). A critical role for natural killer T cells in immunosurveillance of methylcholanthrene-induced sarcomas. J Exp Med **196**, 119.

Cui, J., Tahiro, S., Kawano, T., Sato, H., Kondo, E., Toura, I., Kaneko, Y., Koseki, H., Kanno, M. and Taniguchi, M. (1997). Requirement for V14 NKT cells in IL-12-mediated rejection of tumors. Science **278**, 1623.

Dalod, M., Hamilton, T., Salomon, R., Salazar-Mather, T.P., Henry, S.C., Hamilton, J.D., et al. (2003). Dendritic cell responses to early murine cytomegalovirus infection: subset functional specialization and differential regulation by interferon alpha/beta. J Exp Med **197**, 885.

Dokun, A.O., Kim, S., Smith, H.R., Kang, H.S., Chu, D.T. and Yokoyama, W.M. (2001). Specific and nonspecific NK cell activation during virus infection. Nat Immunol **2**, 951.

Eberl, G. and MacDonald, H.R. (2000). Selective induction of NK cell proliferation and cytotoxicity by activated NKT cells. Eur J Immunol **30**, 985.

Ebner, S., Ratzinger, G., Krosbacher, B., Schmuth, M., Weiss, A., Reider, D., Kroczek, R.A., Herold, M., Heufler, C., Fritsch, P. and Romani, N. (2001). Production of IL-12 by human monocyte-derived dendritic cells is optimal when the stimulus is given at the onset of maturation, and is further enhanced by IL-4. J Immunol **166**, 633.

Ferlazzo, G., Thomas, D., Pack, M., Paludan, C., Schmid, D., Strowig, T., Muller, W.A., Moretta, L. and Münz, C. (2004). Distinct roles of IL-12 and IL-15 in human natural killer cell activation by dendritic cells from secondary lymphoid organs. Proc Natl Acad Sci U S A **101**, 16606.

Fujii, S., Shimizu, K., Kronenberg, M. and Steinman, R.M. (2002). Prolonged IFN-gamma-producing NKT response induced with alpha-galactosylceramide-loaded DCs. Nat Immunol **3**, 867.

Gorski, K.S., Waller, E.L., Bjornton-Severson, J., Hanten, J.A., Riter, C.L., Kieper, W. C., Gorden, K.B., Miller, J.S., Vasilakos, J.P., Tomai, M.A. and Alkan, S.S. (2006). Distinct indirect pathways govern human NK-cell activation by TLR-7 and TLR-8 agonists. Int Immunol **18**, 1115.

Granucci, F., Zanoni, I., Pavelka, N., Van Dommelen, S.L., Andoniou, C.E., Belardelli, F., et al. (2004). A contribution of mouse dendritic cell-derived IL-2 for NK cell activation. J Exp Med **200**, 287.

Groh, V., Rhinehart, R., Secrist, H., Bauer, S., Grabstein, K.H. and Spies, T. (1999). Broad tumor-associated expression and recognition by tumor-derived gamma delta T cells of MICA and MICB. Proc Natl Acad Sci U S A **96**, 6879.

Hammond, K.J., Poulton, L.D., Palmisano, L.J., Silveira, P.A., Godfrey, D.I. and Baxter, A.G. (1998). Alpha/beta-T cell receptor (TCR)$^+$CD4$^-$CD8$^-$ (NKT) thymocytes prevent insulin-dependent diabetes mellitus in nonobese diabetic (NOD)/Lt mice by the influence of interleukin (IL)-4 and/or IL-10. J Exp Med **187**, 1047.

Hayakawa, Y., Takeda, K., Yagita, H., Kakuta, S., Iwakura, Y., Van Kaer, L.,

Saiki, I. and Okumura, K. (2001). Critical contribution of IFN-gamma and NK cells, but not perforin-mediated cytotoxicity, to anti-metastatic effect of alpha-galactosylceramide. Eur J Immunol **31**, 1720.

Hochrein, H., O'Keeffe, M., Luft, T., Vandenabeele, S., Grumont, R.J., Maraskovsky, E. and Shortman, K. (2000). Interleukin (IL)-4 is a major regulatory cytokine governing bioactive IL-12 production by mouse and human dendritic cells. J Exp Med **192**, 823.

Hong, S., Wilson, M.T., Serizawa, I., Wu, L., Singh, N., Naidenko, O.V., Miura, T., Haba, T., Scherer, D.C., Wei, J., Kronenberg, M., Koezuka, Y. and Kaer, L.V. (2001). The natural killer T-cell ligand alpha-galactosylceramide prevents autoimmune diabetes in non-obese diabetic mice. Nat Med **7**, 1052.

Ikeda, H., Chamoto, K., Tsuji, T., Suzuki, Y., Wakita, D., Takeshima, T. and Nishimura, T. (2004). The critical role of type-1 innate and acquired immunity in tumor immunotherapy. Cancer Sci **95**, 697.

Ikehara, Y., Yasunami, Y., Kodama, S., Maki, T., Nakano, M., Nakayama, T., Taniguchi, M. and Ikeda, S. (2000). $CD4^+$ V alpha 14 natural killer T cells are essential for acceptance of rat islet xenografts in mice. J Clin Invest **105**, 1761.

Ishida, Y., Nakamura, F., Kanzato, H., Sawada, D., Yamamoto, N., Kagata, H., et al. (2005). Effect of milk fermented with *Lactobacillus acidophilus* strain L-92 on symptoms of Japanese cedar pollen allergy: a randomized placebo-controlled trial. Biosci Biotechnol Biochem **69**, 1652.

Iwasaki, A. and Medzhitov, R. (2004). Toll-like receptor control of the adaptive immune responses. Nat Immunol **5**, 987.

Kalliomäki, M., Salminen, S., Arvilommi, H., Kero, P., Koskinen, P. and Isolauri, E. (2001). Probiotics in primary prevention of atopic disease: a randomized placebo-controlled trial. Lancet **357**, 1076.

Kalliomäki, M., Salminen, S., Poussa, T., Arvilommi, H. and Isolauri, E. (2003). Probiotics and prevention of atopic disease: 4-year follow-up of a randomized placebo-controlled trial. Lancet **361**, 1869.

Kärre, K., Ljunggren, H.G., Piontek, G. and Kiessling, R. (1986). Selective rejection of H-2–deficient lymphoma variants suggests alternative immune defense strategy. Nature **319**, 675.

Kawano, T., Cui, J., Koezuka, Y., Toura, I., Kaneko, Y., Motoki, K., Ueno, H., Nakagawa, R., Sato, H., Kondo, E., Koseki, H. and Taniguchi, M. (1997). CD1d-restricted and TCR-mediated activation of V alpha 14 NKT cells by glycosylceramides. Science **278**, 1626.

Kawano, T., Cui, J., Koezuka, Y., Toura, I., Kaneko, Y., Sato, H., Kondo, E., Harada, M., Koseki, H., Nakayama, T., Tanaka, Y. and Taniguchi, M. (1998). Natural killer-like nonspecific tumor cell lysis mediated by specific ligand-activated V alpha 14^+ NKT cells. Proc Natl Acad Sci U S A **95**, 5690.

Kinjo, Y., Wu, D., Kim, G., Xing, G.W., Poles, M.A., Ho, D.D., Tsuji, M., Kawahara, K., Wong, C.H. and Kronenberg, M. (2005). Recognition of bacterial glycosphingolipids by natural killer T cells. Nature **434**, 520.

Kinjo, Y., Tupin, E., Wu, D., Fujio, M., Garcia-Navarro, R., Benhnia, M.R., Zajonc, D.M., Ben-Menachem, G., Ainge, G.D., Painter, G.F., Khurana, A., Hoebe, K., Behar, S.M., Beutler, B., Wilson, I.A., Tsuji, M., Sellati, T. J., Wong, C.H. and Kronenberg, M. (2006). Natural killer T cells recognize diacylglycerol antigens from pathogenic bacteria. Nat Immunol **7**, 978.

Kitamura, H., Iwakabe, K., Yahata, T., Nishimura, S., Ohta, A., Ohmi, Y., Sato, M., Takeda, K., Okumura, K., Van Kaer, L., Kawano, T., Taniguchi, M. and Nishimura, T. (1999). The natural killer T (NKT) cell ligand alpha-galactosylceramide demonstrates its immunopotentiating effect by inducing interleukin (IL)-12 production by dendritic cells and IL-12 receptor expression on NKT cells. J Exp Med **189**, 1121.

Kobayashi, E., Motoki, K., Uchida, T., Fukushima, H. and Koezuka, Y. (1995). KRN7000, a novel immunomodulator, and its antitumor activities. Oncol Res **7**, 529.

Koizumi, S., Wakita, D., Sato, T., Mitamura, R., Izumo, T., Shibata, H., et al. (2008). Essential role of Toll-like receptors for dendritic cell and NK1.1(+) cell-dependent activation of type 1 immunity by *Lactobacillus pentosus* strain S-PT84. Immunol Lett **120**, 14.

Kosaka, A., Lee, U., Wakita, D., Matsubara, N., Togashi, Y., Nishimura, S., et al. (2006). Interleukin-12-responding asialoGM1+CD8+ central memory-type T cells as precursor cells for interferon-gamma-producing killer T cells. Cancer Sci **97**, 1236.

Kosaka, A., Wakita, D., Matsubara, N., Togashi, Y., Nishimura, S., Kitamura, H., et al. (2007). AsialoGM1+CD8+ central memory-type T cells in unimmunized mice as novel immunomodulator of IFN-gamma-dependent type 1 immunity. Int Immunol **19**, 249.

Lanier, L.L. (2005). NK cell recognition. Annu Rev Immunol **23**, 225.

Laouar, Y., Sutterwala, F.S., Gorelik, L. and Flavell, RA. (2005). Transforming growth factor-beta controls T helper type 1 cell development through regulation of natural killer cell interferon-gamma. Nat Immunol **6**, 600.

Lee, M.S. and Kim, Y.J. (2007). Signaling pathways downstream of pattern-recognition receptors and their cross talk. Annu Rev Biochem **76**, 447.

Lee, S.H., Miyagi, T. and Biron, C.A. (2007). Keeping NK cells in highly regulated antiviral warfare. Trends Immunol **28**, 252.

Liang, S., Wei, H., Sun, R. and Tian, Z. (2003). IFN regulates NK cell cytotoxicity through STAT1 pathway. Cytokine **23**, 190.

Lodoen, M., Ogasawara, K., Hamerman, J.A., Arase, H., Houchins, J.P., Mocarski, E.S., et al. (2003). NKG2D-mediated natural killer cell protection against cytomegalovirus is impaired by viral gp40 modulation of retinoic acid early inducible 1 gene molecules. J Exp Med **197**, 1245.

Lucas, M., Schachterle, W., Oberle, K., Aichele, P. and Diefenbach, A. (2007). Dendritic cells prime natural killer cells by trans-presenting interleukin 15. Immunity **26**, 503.

Martin-Fontecha, A., Thomsen, L.L., Brett, S., Gerard, C., Lipp, M., Lanzavecchia, A. and Sallusto, F. (2004). Induced recruitment of NK cells to lymph nodes provides IFN-gamma for T_H1 priming. Nat Immunol **5**, 1260.

Mattner , J., Debord, K.L., Ismail, N., Goff, R.D., Cantu, C. 3rd, Zhou, D., Saint-Mezard, P., Wang, V., Gao, Y., Yin, N., Hoebe, K., Schneewind, O., Walker, D., Beutler, B., Teyton, L., Savage, P.B. and Bendelac, A. (2005). Exogenous and endogenous glycolipid antigens activate NKT cells during microbial infections. Nature **434**, 525.

Mencacci, A., Del Sero, G., Cenci, E., d'Ostiani, C.F., Bacci, A., Montagnoli, C., Kopf, M. and Romani, L. (1998). Endogenous interleukin 4 is required for development of protective CD4+ T helper type 1 cell responses to *Candida albicans*. J Exp Med **187**, 307.

Miyamoto, K., Miyake, S. and Yamamura, T. (2001). A synthetic glycolipid prevents autoimmune encephalomyelitis by inducing TH2 bias of natural killer T cells. Nature **413**, 531.

Monteiro, J.M., Harvey, C. and Trinchieri, G. (1998). Role of interleukin-12 in primary influenza virus infection. J Virol **72**, 4825.

Morandi, B., Bougras, G., Muller, W.A., Ferlazzo, G. and Münz, C. (2006). NK cells of human secondary lymphoid tissues enhance T cell polarization via IFN-gamma secretion. Eur J Immunol **36**, 2394.

Mortier, E., Woo, T., Advincula, R., Gozalo, S. and Ma, A. (2008). IL-15 Ralpha chaperones IL-15 to stable dendritic cell membrane complexes that activate NK cells via trans presentation. J Exp Med **205**, 1213.

Motoki, K., Maeda, M., Ueno, H., Kobayashi, E., Uchida, T., Fukushima, H. and Koezuka, Y. (1996). Antitumor activities of combined treatment with a novel immunomodulator, (2S,3S,4R)-1-O-(a-d-galactopyranosyl)-2-(*N*-hexacosanoylamino)-1,3,4-octadecanetriol (KRN7000), and radiotherapy in tumor-bearing mice. Oncol Res **8**, 155.

Murosaki, S., Yamamoto, Y., Ito, K., Inokuchi, T., Kusaka, H., Ikeda, H., et al. (1998). Heat-killed *Lactobacillus plantarum* L-137 suppresses naturally fed antigen-specific IgE production by stimulation of IL-12 production in mice. J Allergy Clin Immunol **102**, 57.

Nakui, M., Ohta, A., Sekimoto, M., Sato, M., Iwakabe, K., Yahata, T., Kitamura, H., Koda, T., Kawano, T., Makuuchi, H., Taniguchi, M. and Nishimura, T. (2000). Potentiation of antitumor effect of NKT cell ligand, alpha-galactosylceramide by combination with IL-12 on lung metastasis of malignant melanoma cells. Clin Exp Metastasis **18**, 147.

Newman, K.C. and Riley, E.M. (2007). Whatever turns you on: accessory-cell-dependent activation of NK cells by pathogens. Nat Rev Immunol **7**, 279.

Nguyen, K.B., Salazar-Mather, T.P., Dalod, M.Y., Van Deusen, J.B., Wei, X.Q., Liew, F.Y., et al. (2002). Coordinated and distinct roles for IFN-alpha beta, IL-12, and IL-15 regulation of NK cell responses to viral infection. J Immunol **169**, 4279.

Nishimura, T., Iwakabe, K., Sekimoto, M., et al. (1999). Distinct role of antigen-specific T helper type 1 (Th1) and Th2 cells in tumor eradication in vivo. J Exp Med **190**, 617.

Nishimura, T., Kitamura, H., Iwakabe, K., Yahata, T., Ohta, A., Sato, M., Takeda, K., Okumura, K., Van Kaer, L., Kawano, T., Taniguchi, M., Nakui, M., Sekimoto, M. and Koda, T. (2000). The interface between innate and acquired immunity: glycolipid antigen presentation by CD1d-expressing dendritic cells to NKT cells induces the differentiation of antigen-specific cytotoxic T lymphocytes. Int Immunol **12**, 987.

Nonaka, Y., Izumo, T., Izumi, F., Maekawa, T., Shibata, H., Nakano, A., et al. (2008). Modulation of Th1/Th2 immunobalance and induction of IL-10 production. Int Arch Allergy Immunol **145**, 249.

Orange, J.S. and Biron, C.A. (1996a). An absolute and restricted requirement for IL-12 in natural killer cell IFN-gamma production and antiviral defense. Studies of natural killer and T cell responses in contrasting viral infections. J Immunol **156**, 1138.

Orange, J.S. and Biron, C.A. (1996b). Characterization of early IL-12, IFN-alpha/beta, and TNF effects on antiviral state and NK cell responses during murine cytomegalovirus infection. J Immunol **156**, 4746.

Parham, P. (2005). MHC class I molecules and KIRs in human history, health and survival. Nat Rev Immunol **5**, 201.

Raulet, D.H. (2004). Interplay of natural killer cells and their receptors with the adaptive immune response. Nat Immunol **5**, 996.

Scalzo, A.A., Corbett, A.J., Rawlinson, W.D., Scott, G.M. and Degli-Esposti, M.A. (2007). The interplay between host and viral factors in shaping the outcome of cytomegalovirus infection. Immunol Cell Biol **85**, 46.

Scharton-Kersten, T., Afonso, L.C., Wysocka, M., Trinchieri, G. and Scott, P. (1995). IL-12 is required for natural killer cell activation and subsequent T helper 1 cell development in experimental leishmaniasis. J Immunol **154**, 5320.

Schluns, K.S., Stoklasek, T. and Lefrancois, L. (2005). The roles of interleukin-15 receptor: trans-presentation, receptor component, or both? Int J Biochem Cell Biol **37**, 1567.

Schmidt, K.N., Leung, B., Kwong, M., Zarember, K.A., Satyal, S., Navas, T.A., Wang, F. and Godowski, P.J. (2004). APC-independent activation of NK cells by the Toll-like receptor 3 agonist double-stranded RNA. J Immunol **172**, 138.

Seino, K.I., Fukao, K., Muramoto, K., Yanagisawa, K., Takada, Y., Kakuta, S., Iwakura, Y., Van Kaer, L., Takeda, K., Nakayama, T., Taniguchi, M., Bashuda, H., Yagita, H. and Okumura, K. (2001). Requirement for natural killer T (NKT) cells in the induction of allograft tolerance. Proc Natl Acad Sci U S A **98**, 2577.

Shida, K., Kiyoshima-Shibata, J., Nagaoka, M., Watanabe, K. and Nanno, M. (2004). Induction of interleukin-12 by lactobacillus strains having a rigid cell wall resistant to intracellular digestion. J Dairy Sci **89**, 3306.

Smyth, M.J., Crowe, N.Y., Pellicci, D.G., Kyparissoudis, K., Kelly, J.M., Takeda, K., Hideo, Y. and Godfrey, D.I. (2002). Sequential production of IFN-gamma by NK1.1$^+$ T cells and natural killer cells is essential for the antimetastatic effect of α-galactosylceramide. Blood **99**, 1259.

Sonoda, K.H., Exley, M., Snapper, S., Balk, S.P. and Stein-Streilein, J. (1999). CD1-reactive natural killer T cells are required for development of systemic tolerance through an immune-privileged site. J Exp Med **190**, 1215.

Sriram, V., Du, W., Gervay-Hague, J. and Brutkiewicz, R.R. (2005). Cell wall glycosphingolipids of *Sphingomonas paucimobilis* are CD1d-specific ligand for NKT cells. Eur J Immunol **35**, 1692.

Sivori, S., Falco, M., Della Chiesa, M., Carlomagno, S., Vitale, M., Moretta, L., et al. (2004). CpG and double-stranded RNA trigger human NK cells by Toll-like receptors: induction of cytokine release and cytotoxicity against tumors and dendritic cells. Proc Natl Acad Sci U S A **101**, 10116.

Suzuki, Y., Wakita, D., Chamoto, K., Narita, Y., Tsuji, T., Takeshima, T., et al. (2004). Liposome-encapsulated CpG oligodeoxynucleotides as a potent adjuvant for inducing type 1 innate immunity. Cancer Res **64**, 8754.

Terabe, M., Matsui, S., Noben-Trauth, N., Chen, H., Watson, C., Donaldoson, D.D., Carbone, D.P., Paul, W.E. and Berzofsky, J.A. (2000). NKT cell-mediated repression of tumor immunosurveillance by IL-13 and the IL-4R-STAT6 pathway. Nat Immunol **1**, 515.

Thomas-Tikhonenko, A. and Hunter, C.A. (2003). Infection and cancer: the common vein. Cytokine Growth Factor Rev **14**, 67.

Togashi, Y., Chamoto, K., Wakita, D., Tsutsumi, N., Iwakura, Y., Matsubara, N., Kitamura, H. and Nishimura, T. (2007). Natural killer T cells from interleukin-4-deficient mice are defective in early interferon-gamma production in response to alpha-galactosylceramide. Cancer Sci **98**, 721.

Trinchieri, G. (1989). Biology of natural killer cells. Adv Immunol **47**, 187.

Trinchieri, G. (2003). Interleukin-12 and the regulation of innate resistance and adaptive immunity. Nat Rev Immunol **3**, 133.

Tripp, C.S., Wolf, S.F. and Unanue, E.R. (1993). Interleukin12 and tumor necrosis factor are costimulators of interferon production by natural killer cells in severe combined immunodeficiency mice with listeriosis, and interleukin 10 is a physiologic antagonist. Proc Natl Acad Sci U S A **90**, 3725.

Vivier, E., Tomasello, E., Baratin, M., Walzer, T. and Ugolini, S. (2008). Functions of natural killer cells. Nat Immunol **9**, 503.

Wakita, D., Chamoto, K., Zhang, Y., Narita, Y., Noguchi, D., Ohnishi, H., Iguchi, T., Sakai, T., Ikeda, H. and Nishimura, T. (2006). An indispensable role of type-1 IFNs for inducing CTL-mediated complete eradication of established tumor tissue by CpG-liposome co-encapsulated with model tumor antigen. Int Immunol **18**, 425.

Walzer, T., Dalod, M., Robbins, S.H., Zitvogel, L. and Vivier, E. (2005). Natural-killer cells and dendritic cells: "l'union fait la force". Blood **106**, 2252.

Xu, J., Chakrabarti, A.K., Tan, J.L., Ge, L., Gambotto, A. and Vujanovic, N. L. (2007). Essential role of the TNF–TNFR2 cognate interaction in mouse dendritic cell–natural killer cell crosstalk. Blood **109**, 3333.

Yokoyama, W.M. and Plougastel, B.F. (2003). Immune functions encoded by the natural killer gene complex. Nat Rev Immunol **3**, 304.

Yu, Y., Hagihara, M., Ando, K., Gansuvd, B., Matsuzawa, H., Tsuchiya, T., et al. (2001). Enhancement of human cord blood CD34+ cell-derived NK cell cytotoxicity by dendritic cells. J Immunol **166**, 1590.

Chapter Thirty-Six

36

Natural killer cells and human immunodeficiency virus

Domenico Mavilio, Anne Hosmalin, Daniel Scott-Algara

CHAPTER CONTENTS

Mother Nature is a highly efficient force with a 'take-no-prisoners' approach toward biologic remnants that no longer support the survival of the species in the ever-changing world of pathogens. Natural Killer cells have been part of the innate immune defence system long before the arrival of the seemingly more sophisticated T and B cells of the adaptive immune system approximately 500 years ago. Clearly, NK cells serve a very important role in host defence, otherwise they would have not survived until today.

Caligiuri, 2008

ABSTRACT

HIV-1 viraemia significantly impairs natural killer (NK) cell antiviral functions. Starting from acute infection, there is a pathologic redistribution of circulating NK cell subset that severely affects the ability of NK cells to eliminate HIV-1 infected cells. In fact, chronic high levels of viral replication induce a depletion of the cytolytic CD56^{dim}/CD16^{pos} NK cell subset and an expansion of a markedly anergic CD56^{neg}/CD16^{pos} NK cell subset expressing very high levels of inhibitory NK receptors (iNKRs) and low, if not undetectable, levels of activating natural cytotoxicity receptors (NCRs). High frequencies of pathologic CD56^{neg}/CD16^{pos} NK cell subsets are also associated with defective NK–DC interactions resulting in impaired activation and proliferation of NK cells, and deficient NK cell-mediated cytotoxicity and secretion of antiviral cytokines. Control of HIV-1 viraemia by antiretroviral therapy (ART) normalizes these parameters. TLR9 ligands may stimulate NK cells through pDC secretion of interferon (IFN)-α. Cohorts of exposed, uninfected individuals and genetic studies show protective associations between some NK inhibitory receptors and their HLA class I molecule ligands. These data offer new molecular targets for HIV-1 control through restoration of NK cell functions.

KEY WORDS

Human immunodeficiency virus, Innate immunity, Natural killer cells, Killer immunoglobulin-like receptors, Natural cytotoxicity receptors, Cytokines, Chemokines, Dendritic cells, NK–DC cross-talk, Antiviral activity, Endogenously infected CD4+ T cells

Introduction

Natural killer (NK) cells are a subset of lymphoid cells that function as important mediators of innate immune defence against viruses and tumour cells (Moretta et al., 2002; Trinchieri, 1989). They represent approximately 15% of peripheral blood lymphocytes and are also found in the liver, peritoneum and placental tissue. NK cells were discovered because of their ability to spontaneously kill tumour target cells in vitro, hence the name 'natural killer' cells (Karre et al., 1986). In fact, NK cells are able to lyse 'nonself' cellular targets without prior sensitization while sparing normal cells that express adequate levels of MHC of class I molecules (MHC-I). This cytolytic function is under the dominant control of a heterogeneous family of inhibitory NK receptors (iNKRs) that bind specifically to individual allelic forms of human leucocyte antigens of class I (HLA-I) (Moretta et al., 1996). In humans, the iNKRs are divided into two different groups: (a) killer immunoglobulin (Ig)-like receptors (KIRs) and (b) C-type lectin family receptors. Diminution or absence of expression of HLA-I molecules on the cell surface due to viral infection or tumour transformation results in reduced engagement of iNKRs and allows a large group of activating NK receptors and co-receptors to trigger cytolytic activity. Activating receptors, such as natural cytotoxicity receptors (NCRs) NKp46, NKp30, NKp44 or NKG2D, provide the 'on signal' for triggering NK cell activation and killing in their interaction with nonself target cells (Long, 2002; Moretta et al., 2001). The concerted action and dynamic balance between inhibitory and activating NK molecules represent the mechanism by which NK cells directly eliminate viral-infected, tumour-transformed and heterologous target cells (Lanier, 2005; Smyth et al., 2001; Vivier et al., 2008).

NK cells also mediate noncytolytic suppression of viral replication by secreting several chemokines, such as CCL3 (MIP1α), CCL4 (MIP1β) and CCL5 (RANTES), and cytokines, such as interferon-γ (IFN-γ), tumour-necrosis factor (TNF) and granulocyte/macrophage colony-stimulating factor (GM-CSF). These soluble factors play an important role in inflammatory reactions, haematopoiesis, immune responses and activation of effector cells (Cerwenka and Lanier, 2001).

Under physiologic circumstances, two distinct subsets of NK cells have been described on the basis of expression of cell-surface markers CD56 and CD16 (Cooper et al., 2001; Nagler et al., 1989). The CD56^{bright}/CD16$^{neg-low}$ NK cell populations represent less than 10% of peripheral blood NK cells and are important for production of cytokines and chemokines. Cells of this subset express the high affinity IL-2 receptor $\alpha\beta\gamma$ (Caligiuri et al., 1990) and produce abundant amounts of IFN-γ. However, these cells have poor cytotoxic activity against tumour-cell targets, indicating that they may regulate other cell types. By contrast, the CD56^{dim}/CD16^{pos} NK cell subset, which produces little IFN-γ, comprises 90% of peripheral blood NK cells and is predominantly responsible for natural cytotoxicity and antibody-dependent cellular cytotoxicity (ADCC) (Cooper et al., 2001).

In addition to their previously described effector functions, NK cells act as a crucial link between innate and adaptive immunity. In this regard, it has been shown that NK cells are engaged in an active and bidirectional cross-talk with autologous dendritic cells (DCs). This process requires both NK cell–DC cellular interactions and secretions of specific cytokines by NK cells and DCs. The final outcome of this cross-talk is the coordination and optimization of both innate and adaptive immune response (Cooper et al., 2004; Degli-Esposti and Smyth, 2005; Moretta, 2002; Walzer et al., 2005; Zitvogel, 2002).

Given this functional potential, NK cells are supposed to play a significant role in preventing and controlling HIV-1 infection. In theory, NK cells should eliminate HIV-1 infected target cells by direct lysis or by ADCC and should facilitate the priming of adaptive immune responses against HIV-1 through their interactions with autologous DCs. But, despite the potential opportunity for NK cell-mediated control of this viral infection, several NK cell functions in individuals infected with HIV-1 are highly impaired.

Infection of NK cells by HIV-1

Whether or not NK cells are naturally infected by HIV-1 in vivo is still controversial. CD4, a key receptor for HIV-1 infection, is generally expressed on helper T cells and monocytes/macrophages as well as on subsets of DCs, γ/δ T cells, mast cells/basophils, neutrophils and NKT cells (Biswas et al., 2003; Cooper et al., 2001; Lusso et al., 1995; Motsinger et al., 2002; Patterson and Knight, 1987). Fresh NK cells purified ex vivo from human donor blood lack cell-surface expression of CD4 (Harada et al., 2007; Mavilio et al., 2003). They do, however, express the chemokine co-receptors for HIV-1, CXC chemokine receptor 4 (CXCR4) and CCR5, but these receptors alone are unlikely to allow productive infection of NK cell by HIV-I in vivo (Kottilil et al., 2004). Freshly isolated blood NK cells from HIV-1 infected patients do not carry HIV-1 proviral

DNA (Mavilio et al., 2003). In contrast, a small subset of $CD3^{neg}/CD56^{pos}/CD16^{pos}$ cells express both CD4 and HIV-1 co-receptors CCR5 and CXCR4 and remains persistently infected even in infected patients with undetectable viral load after 1 or 2 years of highly active antiretroviral therapy (ART) (Valentin et al., 2002). Even though these cells were not fully characterized for the expression of specific NK cell receptors, such as NCRs, the HIV infection of this NK cell population appears to be important for virus persistence in addition to the predominant viral reservoir represented by the latently infected $CD4^+$ T cells.

One of the reasons for this discrepancy might be linked to the activation state of NK cells and to their ability to express CD4 only after a significant pro-inflammatory insult. Triggering the TCR complex in vitro leads to the *de novo* expression of $CD4^+$ on $CD8^+$ T cells and makes these cells susceptible to HIV-1 infection (Flamand et al., 1998; Kitchen et al., 1998; Yang et al., 1998). Similarly, a maximal activation in vitro of freshly purified NK cells with several concurrent stimuli induces a significant *de novo* expression of CD4. Under these experimental conditions, NK cells become susceptible to infection if co-cultured with HIV-1 infected T cells (Harada et al., 2007). High levels of chronic HIV-1 viraemia is certainly associated with an aberrant activation of NK cells (Fogli et al., 2004; Kottilil et al., 2004). Even though $CD4^{neg}$ NK cells purified freshly ex vivo from peripheral blood of patients in active phases of the disease have been reported not to be infected by HIV-1, it remains to be determined whether NK cells can modulate CD4 expression in response to a maximal insult given by the virus chronically circulating in the bloodstream. Moreover, given the fact that NK cells freshly isolated from secondary lymphoid organs have been shown to express CD4 (Bernstein et al., 2006), it is also possible that this small subset of CD4-expressing NK cells is susceptible to HIV-1 infection. Regardless of the capacity of HIV-1 to productively infect NK cells, the virus is able to disrupt the NK cell-mediated cytolytic activity through the binding of its gp120 envelope with the $\alpha4\beta7$ integrin (Arthos et al., 2008; Kottilil et al., 2006).

Effect of HIV-1 viraemia on NK cells

Pathologic redistribution of NK cells in the peripheral blood

For years, several studies reported that the percentage and/or absolute number of total NK cells in the peripheral blood was decreased during the course of chronic and active HIV-1 infection. Based on this, it has been hypothesized that high levels of ongoing HIV-1 replication were likely involved in lowering the number of circulating NK cells through several mechanisms such as an early NK cell death/apoptosis or a preferential NK cell distribution into peripheral tissues (Scott-Algara and Paul, 2002). In many reports, however, identification of NK cell populations was predominantly based on the surface expression of the CD56 marker on $CD3^{neg}$ cells (Lucia et al., 1995; Tarazona et al., 2002). Therefore, the selective loss of the cytolytic $CD56^{dim}$ (and $CD16^{pos}$) NK cell subset in peripheral blood of chronic and viraemic HIV-1 infected individuals has been often interpreted as an absolute depletion of blood NK cells without considering the possibility of an expansion of other pathologic NK cell populations during HIV-1 infection. In this regard, several studies identified and characterized an abnormal $CD56^{neg}/CD16^{pos}$ ($CD56^{neg}$) NK cell subset that is highly represented in chronic and active phases of HIV-1 infection and that is rarely observed in uninfected donors (Alter et al., 2005; Hu et al., 1995; Mavilio et al., 2003). This phenomenon, together with the decreased frequencies of $CD56^{dim}/CD16^{pos}$ NK cell populations during chronic HIV-1 infection, finally clarify that HIV-1 viraemia induces a significant and pathologic redistribution of NK cell subsets rather than an absolute decrement of total NK cells in the peripheral blood (see Figure 36.1 and Table 36.1).

The sequential deregulation of NK cell subset distribution starts during the early phases of HIV-1 infection. In fact, during acute HIV-1 infection, the absolute number of total circulating NK cells is significantly increased when compared to that of uninfected individuals. This phenomenon is associated with high frequencies of $CD56^{dim}/CD16^{pos}$ cytolytic NK cell subsets and an early depletion of $CD56^{bright}/CD16^{neg}$ NK cell populations without a significant expansion of $CD56^{neg}/CD16^{pos}$ NK cell population (Alter et al., 2005). When HIV-1 infection becomes chronic, high levels of viral replication induce a significant depletion of $CD56^{dim}/CD16^{pos}$ NK cell subsets with a parallel increase of functionally anergic $CD56^{neg}/CD16^{pos}$ NK cell subsets. The overall loss of immune-competent $CD56^{dim}$ NK cells together with the increased levels of pathologic $CD56^{neg}$ NK cells certainly contribute to deteriorate NK cell functions and lead to disease progression (Fauci et al., 2005; Mavilio et al., 2005b).

Dichotomy in expression and function of iNKRs and NCRs

Dichotomy in expression and function of iNKRs and NCRs. As explained in the introductory paragraph, a delicate balance of signals generated by both inhibitory and activating NK cell receptors regulates NK cell-mediated cytolytic activity. Chronic and viraemic HIV-1

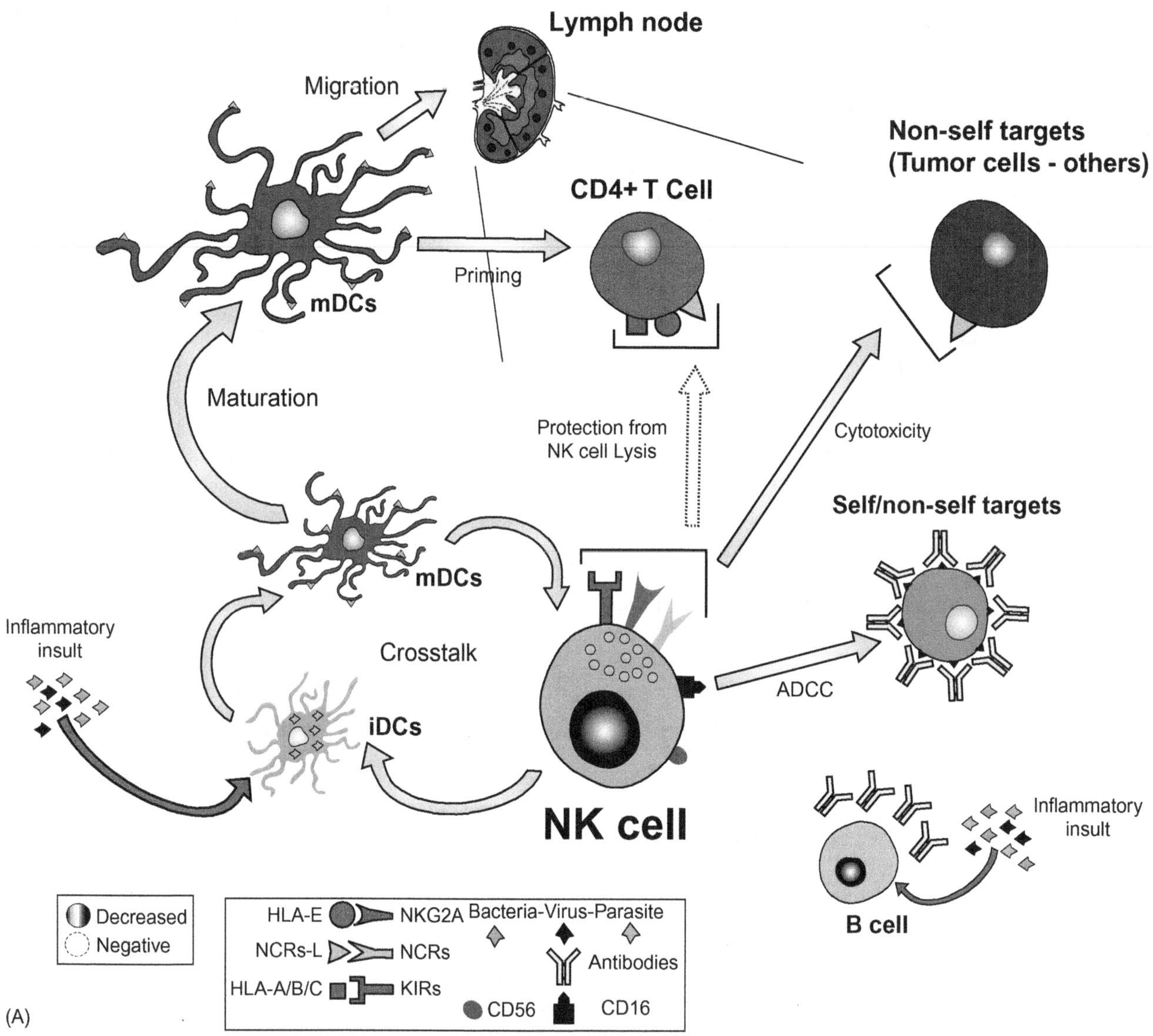

(A)

Figure 36.1 • Effect on HIV-1 viraemia on innate immune system and links with adaptive immunity. (A) Under physiological conditions, NK cells can act as effector cells important in the immune-surveillance against dangerous tumour-transformed or viral infected nonself targets as well as regulatory cells that optimize the links between innate and adaptive immune responses (see the introductory paragraph for further details). (B) HIV-1 viraemia significantly alters the normal NK cell phenotype and induces the expansion of highly pathologic and anergic CD56^{neg} NK cell subsets expressing low levels of activating NK cell receptors and high levels of inhibitory receptors. This phenomenon severely affects the NK cell-mediated lysis of tumour cell targets and autologous and endogenously HIV-1 infected CD4^{+} T cells. Even the ability of NK cells to interact with autologous HIV-1 infected DCs is highly impaired. The defective NK–DC cross-talk in HIV-1 infected viraemic patients is responsible for generating functionally immature DCs that fail to prime correct antigen-specific adaptive immune responses and contribute to spread the infection.

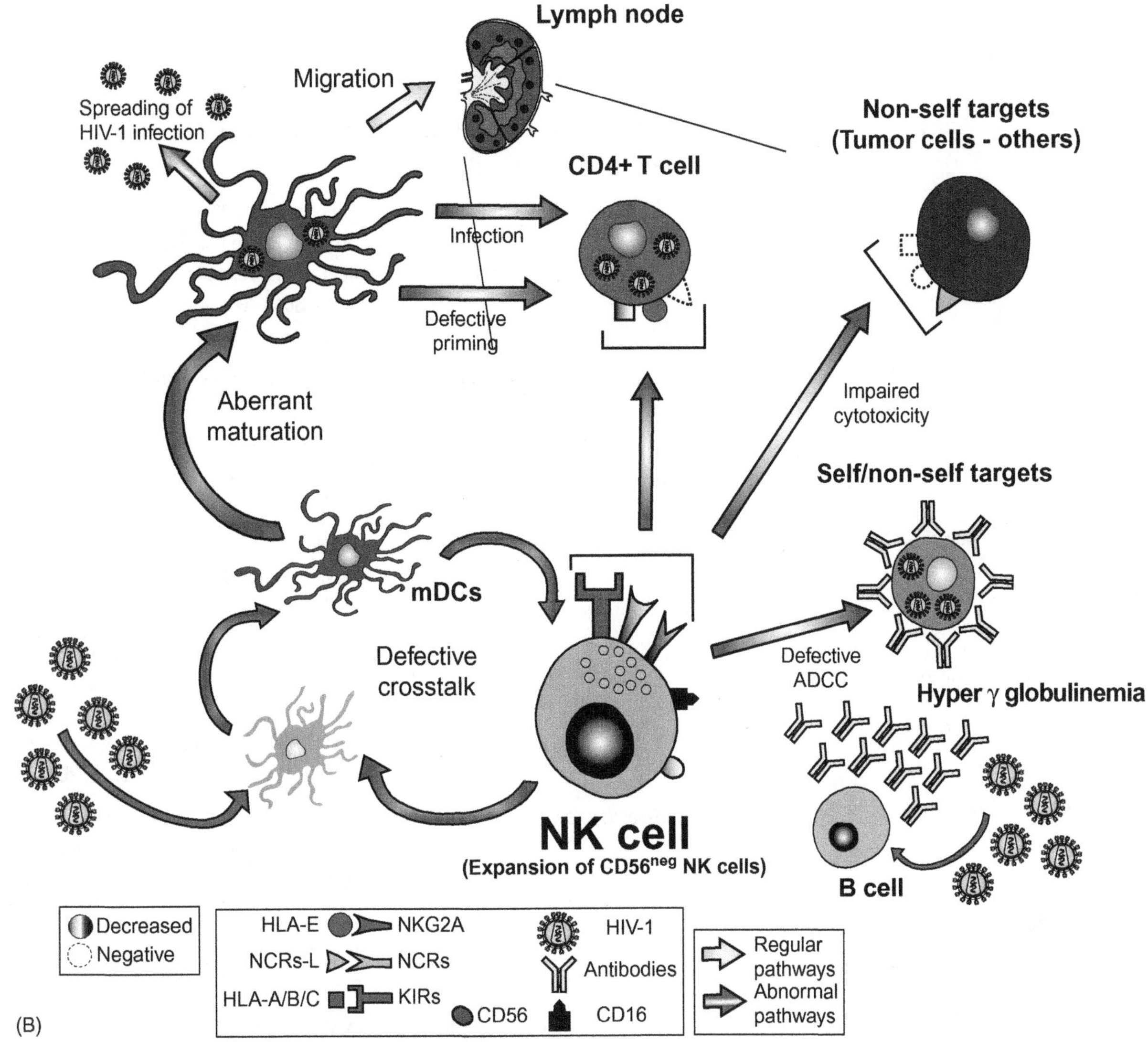

(B)

Figure 36.1 • *(Continued)*

infection has a dramatic effect on NK cell receptor profiles resulting in a markedly defective NK cell cytolytic activity. The number of NK cells expressing iNKRs have been found to be increased in HIV-1 infected viraemic individuals, and this accounts for an increased receptor-specific inhibition of NK cytolytic function in vitro against several tumour cell targets (Mavilio et al., 2003). Although several studies have described this expansion of NK cell populations expressing various iNKRs (Ahmad et al., 2001; Parato et al., 2002; Sirianni et al., 2001), the increased expression of individual inhibitory KIRs in HIV-infected individuals has not been consistently reported, with some studies describing no change. These discrepancies are probably due to differences in the disease states of the individuals with HIV viraemia, the type of sample used (whole blood versus purified NK cells) and the methods used in analysis of receptor expression. However, all of these studies have consistently shown that the number of NK cells expressing iNKRs is normalized with control of HIV viraemia by ART (Kottilil et al., 2004; Mavilio et al., 2003). The NK cell-impaired lytic activity as a result of increased iNKR expression is likely associated with the defective NK cell-mediated control of HIV replication in vivo.

If the dominant inhibitory effects of interactions between iNKRs and MHC-I molecules expressing target cells are overcome, as would be expected by virus-induced down-regulation of HLA-I alleles, the triggering of activating NK receptors by their ligands should induce NK cell cytolytic activity. Analysis of expression of activating receptors on NK cells from HIV viraemic individuals revealed that all three NCRs (NKp44, NKp46 and NKp30) are expressed at significantly lower levels when compared with HIV-negative controls (De Maria et al., 2003). Importantly, control of viraemia by ART for over 2 years is associated with normalization of expression of these three NCRs (Mavilio et al., 2003). Therefore, despite the persistent and

Table 36.1 Effect of HIV-1 viraemia on NK cell phenotype

NK cell receptors					
Receptor	**Function**	**Ligand specificity**	**Effect of HIV viraemia**	**Effect of ART**	**References**
CD16 (FC RIII)	ADCC	IgG	Stable	None	Cooper et al., 2001; Mavilio et al., 2003; Hu et al., 1995; Lucia et al., 1995; Tarazona et al., 2002; Ahmad et al., 2001; Mavilio et al., 2005; Scott-Algara et al., 2002; Barclay et al., 1997
CD56	None	CD56, NCAM	Decreased	Restored	Cooper et al., 2001; Mavilio et al., 2003; Hu et al., 1995; Lucia et al., 1995; Tarazona et al., 2002; Ahmad et al., 2001; Mavilio et al., 2005; Scott-Algara et al., 2002; Barclay et al., 1997
KIR2DL2 (CD158b)	Inhibitory	HLA-Cw2, 4, 5, 6	Stable/Increased	Restored	Mavilio et al., 2003; Ahmad et al., 2001; Mavilio et al., 2005; Moretta et al., 1996; Long et al., 1999; Lanier et al., 2004; Kottilil 2004; Sirianni et al., 2001; Parato et al., 2002
KIR2DL1 (CD158a)	Inhibitory	HLA-Cw1, 3, 7, 8	Stable/Increased	Restored	Mavilio et al., 2003; Ahmad et al., 2001; Mavilio et al., 2005; Moretta et al., 1996; Long et al., 1999; Lanier et al., 2004; Kottilil 2004; Sirianni et al., 2001; Parato et al., 2002
KIR3DL1 (CD158el)	Inhibitory	HLA-Bw4	Stable/Increased	Restored	Mavilio et al., 2003; Ahmad et al., 2001; Mavilio et al., 2005; Moretta et al., 1996; Long et al., 1999; Lanier et al., 2004; Parato et al., 2002
KIR3DL2 (CD158k)	Inhibitory	HLA-A3, A11	Stable	None	Mavilio et al., 2003; Mavilio et al., 2005; Moretta et al., 1996; Long et al., 1999; Lanier et al., 2004
LIR1/ILT2 (CD85J)	Inhibitory	HLA-A, B, C, G	Stable/Increased	Restored	Mavilio et al., 2003; Mavilio et al., 2005; Moretta et al., 1996; Long et al., 1999
CD94 (KLRD1)*	–	–	Stable/Increased	Restored	Mavilio et al., 2003; Ahmad et al., 2001; Mavilio et al., 2005; Moretta et al., 1996; Long et al., 1999; Lanier et al., 2004; Kottilil 2004; Parato et al., 2002
NKG2A (CD159a)	Inhibitory	HLA-E	Decreased	Restored	Mavilio et al., 2003; Mavilio et al., 2005; Moretta et al., 1996; Long et al., 1999; Lanier et al., 2004
NKG2D	Activating	MicA/MicB-ULBPs	Stable	None	Mavilio et al., 2003; Mavilio et al., 2005; Lanier et al., 2004; Moretta et al., 2001
NKp46 (NCR1)	Activating	Influenza Haemagglutinin Others Unknown?	Decreased	Restored	Mavilio et al., 2003; Mavilio et al., 2005; Moretta et al., 2001; De Maria et al., 2003
NKp30 (NCR3)	Activating	Unknown	Decreased	Restored	Mavilio et al., 2003; Mavilio et al., 2005; Moretta et al., 2001; De Maria et al., 2003
NKp44 (NCR2)	Activating	Influenza Haemagglutinin Others Unknown?	Decreased	Restored	Mavilio et al., 2003; Mavilio et al., 2005; Moretta et al., 2001; De Maria et al., 2003

(Continued)

Table 36.1 (Continued)

NK cell receptors					
Receptor	**Function**	**Ligand specificity**	**Effect of HIV viraemia**	**Effect of ART**	**References**
NKp80	Co-activating	Unknown	Stable	None	Mavilio et al., 2003; Mavilio et al., 2005; Moretta et al., 2001
NTB-A	Co-activating	NTB-A	Stable	None	Mavilio et al., 2003; Mavilio et al., 2005; Bottino et al., 2005
2B4 (CD244)	Co-activating	CD48	Stable	None	Mavilio et al., 2003; Mavilio et al., 2005; Kottilil et al., 2004; Moretta et al., 2001
NKRP1A (CD161)	Undefined	Unknown	Stable	None	Lanier et al., 2004; Kottilil et al., 2004; Parato et al., 2002
CCR5	Chemokine receptors	CCL-3, CCL-4, CCL-5	Increased	Restored	Kottilil et al., 2004
CCR4	Chemokine receptors	CCL-3, CCL-4, CCL-5	Stable	None	Kottilil et al., 2004

*CD94 forms heterodimers with different receptors of the NKG2 family, including NKG2A and NKG2C.

aberrant activation of NK cells due to their chronic exposure to HIV-infected cells, NK cells have a NCR^{low} phenotype, which affects NK-cell-mediated clearance of virus-infected cells, tumours and opportunistic infections that usually occur in the late stages of HIV infection.

All of these phenotypic and functional NK cell abnormalities are particularly pronounced on the CD56^{neg}/CD16^{pos} NK cell population whose dramatic expansion at high frequencies in viraemic and chronic HIV-1 infected individuals accounts for the functional defects in NK cell-mediated cytotoxicity and secretion of important antiviral cytokines such as IFN-γ, TNF-α and GM-CSF (Mavilio et al., 2005b).

In contrast, the expression and function of another important activating NK cell receptor, NKG2D and several activating co-receptors, including 2B4, NKp80 and NTB-A, are maintained in HIV-infected viraemic individuals, which might be responsible for the residual NK cytolytic activity that has been observed in such patients (Mavilio et al., 2003, 2005b).

NK cell-mediated killing of HIV-1 infected cells

In order to further understand the direct effects of HIV-1 on CD4$^+$ T cells and other cell types, several models of in vitro infection with different HIV-1 strains have been developed. Through these experimental methods, several reports showed that HIV-1 selectively down-modulates MHC-I expression in both cell lines and CD4$^+$ T cell-derived blasts (Bonaparte and Barker, 2004; Cohen et al., 1999; Noraz et al., 1995). This selective HIV-1 mediated down-regulation of MHC-I surface levels on infected cells should, in theory, partially disrupt the inhibitory signal driven by the interaction between iNKRs and specific HLA alleles and allow NK cell-mediated lysis. In fact, the ability of NK cells from healthy donors to kill autologous and exogenously infected CD4$^+$ T cell blasts depends on the engagement of activating NK cell receptors. In this regard, HIV-1 is able to affect in vitro the degree of NK cell lysis of autologous HIV-1 infected CD4$^+$ T cells by modulating the expression of ligands for activating NK cell receptors (Ward et al., 2007).

Even though these approaches using in vitro infection have significantly contributed to understanding the cellular interactions between NK cells and autologous HIV-1 infected CD4$^+$ T cells, it is still unclear what role, if any, NK cells obtained from HIV-1 infected viraemic patients play in the clearance of endogenously infected autologous CD4$^+$ T cells ex vivo. NK cells from infected patients with high levels of ongoing HIV-1 replication are certainly different in phenotype and functions if compared with those from uninfected individuals. Therefore, it still remains to be determined whether these highly dysfunctional NK cells are able to eliminate autologous and endogenously HIV-1 infected CD4$^+$ T cells. Moreover, circulating CD4$^+$ T cells from

HIV-infected individuals harbour very low frequencies of replication-competent virus (Chun et al., 1997; Pierson et al., 2000). This has made it very difficult to isolate and adequately characterize endogenously infected CD4$^+$ T cells to use, ex vivo, as targets for autologous NK cells under conditions that more closely mimic the in vivo situation occurring in HIV-infected viraemic individuals.

In time-course experiments, activation with PHA plus rIL2 (with or without rIL-7) is able to expand a population of endogenously infected CD4-derived T cell blasts derived from HIV-1 infected viraemic patients (Fogli et al., 2008). In line with results previously reported with HIV-1 infection in vitro (Bonaparte and Barker, 2003, 2004), endogenously HIV-1 infected CD4-derived T cell blasts selectively down-modulated HLA-A and HLA-B alleles while the expression of HLA-C and HLA-E molecules was conserved. The selective down-regulation of these MHC-I molecules rendered infected CD4-derived T cell blasts susceptible to NK cell-mediated killing. In fact, although the surface levels of HLA-C and HLA-E still protected infected cell blasts from the cytolysis exerted by autologous NK cells expressing iNKRs specific for these conserved alleles of MHC-I, this was not the case for NK cells that expressed iNKRs specific for HLA-A and HLA-B. Indeed, the degree of NK cell-mediated lysis of autologous HIV-1 infected CD4$^+$ T cell blasts is significantly higher when compared with that of uninfected CD4-derived T cell blasts with normal surface levels of MHC-I.

Nevertheless, the overall levels of NK cell killing of autologous and endogenously infected CD4-derived T cell blasts remains low (range: 4.2–23%; median: 9.6%). The defective engagement of important activating NK cell receptors is one of the mechanisms explaining that even after the down-regulation of HLA-A and HLA-B alleles, the majority of endogenously infected CD4$^+$ T cells is still spared from NK cell-mediated killing. In this context, the low levels of NKp46 and NKp30 on NK cells from HIV-infected individuals significantly correlated with NK cell cytolytic activity of endogenously HIV-1 infected autologous CD4-derived T cell blasts. Moreover, as reported previously, chronic HIV-1 viraemia is associated with the presence of a highly dysfunctional CD56^{neg} NK cell subset expressing almost undetectable levels of NKp46 and NKp30. The presence of this pathologic and highly anergic CD56^{neg} population inversely correlated with the ability of NK cells to kill autologous HIV-1 infected blasts. These results confirmed aberrant expression and function of NKp46 and NKp30 on this overrepresented CD56^{neg} NK cell population, likely contributing to the reduction of NK cell-mediated killing of endogenously HIV-1 infected autologous CD4$^+$ T cell blasts (Fogli et al., 2008).

NK cells from HIV-1 infected viraemic patients display a variable although generally low ability to selectively eliminate autologous and endogenously HIV-1 infected CD4$^+$ T cells expanded directly from peripheral blood. Various factors, including selective down-modulation of MHC class-I molecules, decreased NK cell expression of NCRs, and high frequencies of the dysfunctional CD56^{neg}/CD16^{pos} NK cell subset severely interfere with NK cell-mediated cytolysis of endogenously infected CD4$^+$ T cells. In line with results previously reported for in vitro experimental models (Ward et al., 2007), the residual NK cell-mediated killing occurs mainly through the NKG2D activation pathway.

NK cell-mediated noncytolytic HIV suppression

Similar to CD8$^+$ T cells, NK cells can suppress endogenous HIV replication by cell-to-cell contact as well as by soluble factors. The CC-chemokines CCL3, CCL4 and CCL5, which are ligands for CCR5, can block entry of R5 viruses into target cells by competitive inhibition of receptor binding (Alkhatib et al., 1996; Bluman et al., 1996; Choe et al., 1996; Cocchi et al., 1995; Dragic et al., 1996). Accordingly, NK cells, which produce high levels of CCL3, CCL4 and CCL5, suppress HIV replication in vitro by inhibiting CCR5-dependent entry of HIV into target cells (Fehniger et al., 1998; Oliva et al., 1998). In fact, CC-chemokines are the main soluble factors for NK cell-mediated HIV suppression. This differs from CD8$^+$ T cells, which suppress HIV replication by as-yet-unidentified nonchemokine factors. While suppression of HIV replication by NK cell-derived soluble factors is mainly restricted to R5 viruses, suppression mediated by cell contact affects both R5 and X4 viruses. The exact mechanism of suppression mediated by cell contact by NK cells has not yet been determined, and further studies are required to delineate the precise mechanisms of NK cell suppression of HIV replication in vivo. However, the degree of suppression by NK cells was more marked in patients without detectable HIV viraemia compared with those with detectable viraemia. In addition, there was an inverse correlation between the level of plasma viraemia and the ability of NK cells to suppress endogenous HIV replication. Suppression of HIV replication by soluble factors secreted by NK cells was mediated almost entirely by CC-chemokines, and the ability of NK cells to secrete CC-chemokines was significantly reduced in patients with viraemia when compared to those without viraemia (Kottilil et al., 2003). Collectively, these findings suggest that HIV-induced inhibition of NK cell function involves mechanisms that lead to diminished secretion of CC-chemokines.

Antibody-dependent cell cytotoxicity

ADCC is an adaptive immune response largely mediated by NK cells through the CD16 (FCγRIII) receptor that binds the Fc portion of IgG antibodies triggering the lysis of targeted cells. HIV-1 stimulations should theoretically generate specific and/or neutralizing antibodies helpful for an effective ADCC against infected cells carrying viral antigens on their surfaces. Several studies have demonstrated that the titres of the env-specific ADCC-mediated antibodies decrease in the sera of HIV-I infected individuals as the infection progresses towards AIDS (Brenner et al., 1991). Moreover, cells that mediate ADCC also become functionally compromised starting form early stages of HIV infection, thus depriving the host of the potential benefits of this effector function of the immune system. Finally, a significant inverse correlation between the degree of impairment of NK cell cytotoxicity and the levels of HIV plasma viraemia has been reported, making it clear that chronic viral replication negatively affects ADCC (Ahmad and Menezes, 1996).

A recent report characterized ADCC antibodies in a cross-sectional cohort of 80 HIV-infected subjects not on ART by analyzing production of IFN-γ and degranulation marker CD107a in activated NK cells. HIV-specific ADCC directed toward envelope proteins were present in the majority of 80 untreated HIV-infected individuals. Similarly, most subjects had HIV-specific Abs that mediated degranulation or cytokine expression by NK cells. Interestingly, there was a poor correlation between ADCC-mediated killing and Ab-mediated expression of IFN-γ, even though NK cells from HIV-1 infected patients were more efficiently able to degranulate in response to ADCC activation. Activation of NK cells in response to stimulation by HIV-specific Abs occurs at least as rapidly as activation of Gag-specific CTLs. This study highlights the complexity of Ab-mediated NK cell activation in HIV infection and suggest new avenues toward studying the utility of ADCC in controlling HIV infection (Chung et al., 2009).

Interactions between NK cells and autologous DCs in HIV-1 infection

The first contact between immature DCs (iDCs) and antigens occurs in peripheral tissue at sites of inflammation where iDCs are recruited from the bloodstream through cytokine and chemokine signals produced by resident DCs and other cell types. Following antigen uptake, iDCs undergo a maturation process that allows the resulting mature DCs (mDCs) to migrate to secondary lymphoid tissues where they prime an antigen-specific T cell response. DCs undergoing maturation secrete several cytokines, such as IL-12 and IL-15, that act as potent inducers of NK cell activation and proliferation. In turn, NK cells, once activated, produce IFN-γ, TNF-α and GM-CSF, three important antiviral cytokines directly involved in completing the maturation program of DCs. Incomplete or aberrant maturation of DCs is highly undesirable since antigen presentation by these cells might lead to T cell tolerance or apoptosis and, therefore, to a greatly impaired adaptive immune response. NK cells further optimize conditions for subsequent T cell priming by eliminating DCs that have failed to mature properly. iDCs express remarkable amounts of classical HLA-A, HLA-B and HLA-C molecules but at significantly lower levels when compared to mDCs. The attenuated expression of MHC-I molecules, the natural ligands for the inhibitory NK receptors (iNKRs), makes iDCs uniquely susceptible to NK cell lysis. NK cell-mediated killing of iDCs likely occurs to a significant degree only when these two cell types co-localize in sufficient numbers at sites of tissue infection/inflammation. In addition, in murine cytomegalovirus infection in mice, NK cells limit IFN-αβ production by plasmacytoid DC to levels that are not immunosuppressive to the host, thereby accelerating CD8 T-cell responses to the virus (Robbins et al., 2007). Therefore, in their interactions with DCs, NK cells can act as regulatory cells as well as effectors in order to ensure an appropriate activation of adaptive immunity (Cooper et al., 2004; Degli-Esposti and Smyth, 2005; Ferlazzo et al., 2004; Gerosa et al., 2002; Moretta, 2002; Raulet, 2004; Steinman and Nussenzweig, 2002; Walzer et al., 2005).

Several lines of evidence suggest that normal NK–DC interactions are aberrant in HIV-1 infection. In particular, numerous components of the bidirectional NK–DC cross-talk are highly impaired in cells from viraemic HIV-1 infected individuals compared with those from healthy donors and aviraemic HIV-1 infected individuals who have been receiving ART for 2 years or longer (Mavilio et al., 2006). Circulating DCs are low (Grassi et al., 1999; Soumelis et al., 2001). The capacity to secrete type I IFNs after in vitro stimulation is impaired, and correlations are found throughout the literature between high circulating pDC numbers, high type I IFN secretion in vitro and better viral load control (Ferbas et al., 1995; Hosmalin and Lebon, 2006; Lichtner et al., 2008; Siegal et al., 1986). Actively replicating HIV-1 in peripheral tissue is associated with a markedly dysfunctional maturation of DCs, although these DCs appear to be phenotypically mature on the basis of the expression of several co-stimulatory markers and MHC class I and II molecules. Among these dysfunctions, there is an impaired capacity to activate autologous and heterologous T cells either by in vitro HIV-1 exposed mDCs generated from healthy individuals or by myeloid or plasmacytoid DCs freshly isolated from HIV-1 infected viraemic individuals (Donaghy et al., 2003; Fantuzzi et al., 2004; Smed-Sorensen et al., 2004). Moreover, the in vitro observation that iDCs exposed to HIV-1 achieve a mature DC phenotype, express high amounts of CCR7 and are able to chemotax in response to MIP-3β (CCL19) indicates that they are able to migrate to secondary lymphoid tissues where they can participate in priming of T cells (Wilflingseder et al., 2004). It has also been shown in vitro and ex vivo that these phenotypically mature but functionally immature DCs are infected by HIV-1 and are also able to transmit the virus to autologous T cells (Rinaldo and Piazza, 2004). Taken together, these data suggest that these phenotypically mature but dysfunctional DCs would generate a suboptimal adaptive T-cell response and might also be able to spread the infection.

These abnormally matured DCs are also defective in their ability to secrete IL-10 and IL-12 and to prime the proliferation of neighbouring autologous NK cells, which, in turn, fail to secrete adequate amounts of IFN-γ. Moreover, NK cells purified from HIV-1 infected individuals with high levels of ongoing viral replication—but not from aviraemic patients—are also markedly impaired in their ability to eliminate autologous iDCs (Tasca et al., 2003). This phenomenon is largely due to the high frequency of the dysfunctional $CD56^{neg}/CD16^{pos}$ NK cell subset, whose surface expression/secretion and function of NKp30 and TRAIL molecules are either extremely low or absent (Mavilio et al., 2006). This deficient killing of unnecessary and infected DCs by NK cells markedly diminishes immune-mediated control of HIV-1 and may contribute to viral spread.

NK interaction with HIV-1 infected autologous DCs is also able to modulate the expression of receptors on NK cell surfaces. In particular, expression of CD85j is strongly down-regulated upon culture with autologous DCs infected with HIV-1. Given the fact that $CD85j^{pos}$ NK cells have been demonstrated to suppress viral replication in productively DCs infected in vitro, the negative modulation of this inhibitory receptor on NK cells represents another pathologic mechanism by which HIV-1 is able to escape from controls exerted by cells from innate immune system (Scott-Algara et al., 2008).

Even though the fate of DCs has been found to be extremely dependant on the interaction with autologous NK cells, the mechanisms by which NK–DC interaction controls viral infections remain unclear. A recent report investigating the impact of NK–DC cross-talk on maturation and functions of HIV-infected immature DCs showed that activated NK cells are required for the induction of maturation of DCs, whether uninfected or HIV-1-infected, and this process involved HMGB1. HMGB1 is a nuclear protein that is present in almost all eukaryotic cells, functions to stabilize nucleosome formation and acts as a transcription-factor-like protein that regulates the expression of several genes. However, the cross-talk between HIV-1-infected DCs and activated NK cells was functionally defective, as demonstrated by the strong impairment of DCs to induce Th1 polarization of naive $CD4^+$ T cells. This was associated with the defective production of IL-12 and IL-18 by infected DCs. Moreover, the cross-talk between activated NK cells and HIV-1-infected DCs resulted in a dramatic increase in viral replication and proviral DNA expression in DCs. HMGB1, produced both by NK cells and DCs, was found to play a pivotal role in this process, and inhibition of HMGB1 activity by glycyrrhizin, known to bind specifically to HMGB1, or blocking anti-HMGB1 antibodies, abrogated NK-dependent HIV-1 replication in DCs. These observations provide evidence for the crucial role of NK–DC cross-talk in promoting viral dissemination, and challenge the question of the in vivo involvement of HMGB1 in the triggering of HIV-1 replication and replenishment of viral reservoirs in AIDS (Saidi et al., 2008).

Effect of ART on NK cell phenotype and functions

As discussed in the previous sections, since the advent of ART, investigators have shown varying degrees of improvements in NK cell function associated with control of HIV viraemia; however, there is no consensus thus far on the degree to which adequate suppression of ongoing HIV replication by ART restores NK cell function. Longitudinal studies that monitor comprehensive NK cell functions in HIV patients are needed to address more precisely the degree of restoration of NK cell phenotype and function associated with ART-induced suppression of HIV replication.

Modulation of NK cell effector function in response to cytokines during HIV-1 infection

Several studies have demonstrated lack of cytokine responsiveness as a contributing factor for NK cell dysfunction in HIV-infected individuals (Ullum et al., 1995). In vitro activation of NK cells with recombinant IL-2 (rIL-2) is unable to restore the capacity of NK cells from HIV-viraemic individuals to lyse targets (Mavilio et al., 2003, 2005b). In addition, impaired cytotoxicity of rIL-2 activated NK cells is associated with a higher probability of progression to AIDS (Ullum et al., 1999). In contrast, IL-15 augments the direct cytotoxic potential of NK cells against HIV-infected autologous peripheral blood mononuclear cells (PBMCs), as well as ADCC using gp120-coated target cells (Loubeau et al., 1997; Lum et al., 2004).

As described before, the ability of NK cells to eliminate autologous HIV-1 infected $CD4^+$ T cells is relatively low.

Nevertheless, it has been recently reported that the NK cell-mediated cytolytic activity against HIV-1 infected cells can be improved by the presence of accessory cells, such as plasmacytoid DCs (pDCs) (Azzoni et al., 2007; Chehimi et al., 2007). The stimulation of pDCs with the Toll-like receptor 9 (TLR9) agonist, CpG ODN 2216, triggers NK cell lysis of autologous HIV-1 infected $CD4^+$ T cells. This is blocked by neutralizing antibodies to type I IFN and is perforin/granzyme dependent. Overall, these data suggest that HIV-1 infected cells are not innately resistant to NK cell lysis, and the exogenous stimulation of NK cells through the secretion of type I IFN by TLR9-activated pDCs can significantly improve NK cell-mediated killing of HIV-1 infected $CD4^+$ T cells. Moreover, the fact that we observe a restoration of numbers and functions of pDCs in HIV-1 infected patients with undetectable HIV-1 viraemia correlates with the progressive improvement of NK cell cytotoxicity in response to highly effective ART. Further investigations will be needed to understand which activation NK pathway and which activating NK cell receptor is involved in the higher NK cell antiviral lytic activity once pDCs are stimulated to produce type I IFN. These findings indicate a potential role for cytokine-based or cytokine-triggering therapies to augment NK cell-mediated clearance of HIV-1 infected cells (Mastroianni et al., 2004). Clinical trials would be needed to further pursue this concept.

NK cells and resistance to HIV-1 infection

Most of the studies on NK cells have been performed in cohorts of HIV-infected patients. In contrast, a few studies focused their investigations on a minority of individuals that have been repeatedly exposed to HIV-1 for years without the occurrence of productive infection (exposed uninfected individuals or EU). EU individuals are considered to be resistant to infection as they have no detectable HIV-1 IgG antibodies (Elisa and Western Blot), HIV-1 plasma viral load nor proviral DNA. These EU individuals have been identified in several cohorts—discordant couples, drug users, commercial sex workers, exposed healthcare workers and infants of HIV-1 infected mothers exposed in utero (Hirbod and Broliden, 2007)—and can constitute a good model to identify the correlates of protection. Knowledge of the possible immune mechanisms of such resistance would allow the targeted development of vaccine or immunotherapy strategies and maybe other strategies to prevent HIV-1 transmission. Multiple reports concerning the mechanisms of HIV-1 infection protection in EU individuals have been published, and most were devoted to adaptive immunity (cellular and humoral) and genetic variations. In some cohorts of EU individuals, the presence of HIV-1-specific $CD4^+$ and $CD8^+$ T-cell responses as well as HIV-1-neutralizing IgA antibodies correlated with protection (Devito et al., 2000; Schenal et al., 2005). Most of these investigations focused on HIV-1 specific $CD8^+$ T cells and highlighted their critical role in the control of viral replication.

However, very few studies on EU individuals have focused on innate immunity and in particular NK cells. There is an increasing body of in vivo evidence indicating that NK–DC interactions during the early phase of innate immunity can impact the quality and magnitude of the subsequent adaptive immune response (Adam et al., 2005). There is a relative paucity of information regarding NK cell function in adaptive immunity from clinical trials, and we need the identification of markers of this innate response, which, indeed, might be useful as correlates of protective immunity. Thus, studies of innate immunity in EU individuals are relevant for the identification of these markers.

The first analysis of NK cell functions was performed in a cohort of EU drug users in Vietnam (Scott-Algara et al., 2003) with negative analysis of the CCR5 variants known to confer resistance to HIV-1. The cytolytic and secretory capacity of NK cells of EU was compared to HIV positive drug user patients and healthy blood donors. The percentage of NK cells among lymphoid cells did not differ between EU and the unexposed controls. However, NK cell lytic activity against the K562 and Daudi cell line was significantly higher in EUs than in unexposed controls. Moreover, the lytic capacity of NK cells from EU was significantly higher when compared to HIV seroconverters during the follow-up (either before or after seroconversion). This increase in NK cell cytolytic activity was also correlated to higher secretory capacities as shown by intracellular staining for cytokines and chemokines. The higher secretion capacity of NK cells from EU was also observed without any exogenous stimulation suggesting activation in vivo. The increased frequency of IFN-γ producing NK cells in EU was confirmed in another study (Montoya et al., 2006). Thirty individuals having prolonged unprotected sexual intercourse with HIV-1 positive partners were studied. After stimulation with PMA and ionomicin, NK cells from EU produced significantly higher levels of IFN-γ than controls (HIV-1 positive patients or HIV-1 low risk healthy donors). Moreover, the higher capacity of NK cells from EU to produce IFN-γ is not a due to 'a general' production of this cytokine because no significant differences were observed when T cells were analyzed.

These results suggest a potential role of enhanced NK cell activity in the resistance against HIV-1 infection in EU. As the number and proportions of NK cells were similar in all study groups, the differences in NK cell activities are related to an increased NK functional activity in EU compared with either unexposed individuals or seroconverters rather than to an expansion. This increase in NK cell activity could be explained by the presence or absence of a particular set of NK cell receptors or its ligands.

In order to approach this issue, a cohort of sex workers who are resistant to HIV-1 infection were studied by performing detailed HLA Class I and KIR typing (Jennes et al., 2006). In this study, the EU's status was associated with the occurrence of individual inhibitory *KIR* genes in the absence of their putative HLA ligands, for example *KIR2DL2/KIR2DL3* heterozygosity in the absence of *HLA-C1*, and independent from this, *KIR3DL1* homozygosity in the absence of *HLA-Bw4*. In addition, the authors showed a higher frequency of AB KIR genotypes and a higher proportion of activating KIR in EU compared to controls. Thus, they hypothesized that the absence of HLA ligands for inhibitory KIR combined with higher expression of activating KIR may lower the threshold for NK cell activation, resulting in higher NK cytotoxic activity against HIV-infected cells avoiding infection and spreading of HIV.

Similar experiments were conducted in Vietnamese EU drug users by using RT PCR and NK cell repertoire staining (Ravet et al., 2007). Although the phenotypic distribution of NK cells expressing KIR and CD85j MHC-class I receptors was not profoundly affected in EU, further discrimination of the activating and inhibitory KIR genes that contribute to the phenotype of KIR^+ NK cell subsets, by locus-specific real-time PCR transcript quantification, revealed qualitative or quantitative differences between EU and control subjects or HIV-infected patients. High percentages of $GL183^{pos}$ NK cells in EU correlate with the expansion of *KIR2DL3* transcripts in the absence of *KIR2DS2* and *KIR2DL2* transcripts. In addition, this expansion of *KIR2DL3* transcripts was observed in EU displaying either HLA-C1/C1 or C1/C2 allelic specificities. Because the inhibitory KIR2DL3 receptor binds to its ligand HLA-C1 with lower affinity than KIR2DL2, KIR2DL3 expression in the absence of KIR2DL2 expression may favour NK cell activation in EU by a mechanism similar to that proposed by Jennes et al. (2006).

An increased proportion of KIR3DS1 homozygotes in EU were recently described with no differences in HLA Bw4 or HLA Bw4-I80 genotyping in a Canadian cohort (Boulet et al., 2008). This increase of KIRDS1 was associated with the absence of the KIRDL1 gene. However, due to the number of EU studied, it is unclear whether it is the absence of an inhibitory KIR3DL1 gene or the presence of two copies of KIR3DS1 that is more important for resistance to HIV infection.

All together, these results suggest that HLA KIR genotype interactions implicated in the decrease of HIV disease progression in HIV-1 infected patients do not appear to be a major factor in preventing HIV-1 infection in EU. However, more correlative studies between NK cell function and genotypes need to be done, particularly to determine whether specific genotypes translate into more activated NK cell function.

Another interesting study concerns the evaluation of the role of NK cells in HIV-1 perinatally EU, compared to HIV-1 infected children (Boulet et al., 2008). A lower percentage of NK cells expressed the inhibitory receptors KIR3DL1 and KIR2DL1 in EU. In degranulation assays, CD107 expression on NK cells was consistently higher in the EU group regardless of the stimulation used. However, intracellular staining for perforin, without stimulation, showed a higher percentage of perforin-positive NK cells in the HIV-1-infected patients when compared with the EU children. Cytokine secretion analysis does not show any differences between the groups.

NK cells play a critical role in the initial control of HIV-1 infection. As described in other viral models, the capacity of NK cells to respond very early to HIV-1 infection (preactivated state, expression of particular NK cell repertoire and/or particular KIR HLA Class I molecule associations) may be critical for the intensity and balance of innate immune responses, and the promotion of optimal antiviral CD8 and CD4 T-cell responses. A better understanding of the innate interactions taking place at the interface between innate and adaptive immunity may have a major impact also in designing novel therapeutic and vaccine approaches against HIV-1 infection.

Genetic factors associated with a positive NK cell-mediated clinical outcome of HIV-1 infection

Even though the effect of HIV-1 viraemia alters the conventional NK cell phenotypic profile and severely impairs several NK cell functions, a parallel field of research explored the possible existence of NK cell-related immune correlates of protection from HIV-1 infection. In this regard, epidemiological studies reported that several host genetic factors, such as the association between a 32-base-pair deletion in CCR5 and the increased resistance to infection or the protective role exerted by several HLA-I alleles, are strongly associated with a better clinical outcome of HIV-1 disease (Gao et al., 2001; Kaslow et al., 1996; Liu et al., 1996; Martin et al., 1998; Smith et al., 1997). These investigations were focused on HIV-1 specific $CD8^+$ T cells and highlighted their critical role in the control of viral replication. Only recently, it has been demonstrated that the expression of an activating form of KIR3D NK cell receptor (KIR3DS1) in conjunction with its putative ligand, HLA-B Bw4 with an isoleucine at position 80 (HLA-B Bw4-80I), is associated with a significant control of HIV replication and with a slower progression to AIDS (Martin et al., 2002). In the absence of the KIR3DS1 allele, the HLA-B Bw4-80I allele expression did not protect against disease progression. Furthermore, occurrence of the KIR3DS1 allele in the absence of the HLA-B Bw4-80I allele was associated with a rapid progression to AIDS among HIV-infected individuals, suggesting that an epistatic association between the two loci is necessary for protection (Carrington and O'Brien, 2003). These epidemiologic analyses have been recently supported and confirmed functionally by experimental data showing that the slower disease progression is caused by the ability of KIR3DS1^{pos} NK cell subsets to strongly inhibit HIV-1 replication in autologous CD4$^+$ T cells expressing HLA-B Bw4-80I. Moreover, the presence of KIRDS1^{pos} NK cell subsets is also associated with higher NK cell-mediated secretion of IFN-γ and cytotoxicity starting from the early phases of HIV-1 infection (Alter et al., 2007). All together, these epidemiologic and experimental evidences demonstrate that NK cells exert a protective role in HIV-1 pathogenesis through a mechanism that might be useful in the future to develop new antiviral therapeutic strategies.

Future perspective

Nonhuman primate models

Understanding the biology of NK cells, particularly the identification of novel receptors and their ligands, has enhanced our knowledge on the potential role of NK cells in the immunopathogenesis of HIV infection. However, it is still unclear whether NK cells actually control HIV replication in vivo. Previous studies have attempted to evaluate the role of NK cells in the pathogenesis of several diseases or in the control of invading pathogens in nonhuman primates. However, it should be pointed out that the majority of these studies was conducted using unfractionated PBMCs and gating their analysis on CD16^{pos} or CD56^{pos} cells that have been described as not specific enough to clearly identify the entire NK cell population in this species. In this regard, some studies have focused their attention on simian immunodeficiency virus (SIV) infection in monkeys in order to better understand the involvement of NK cells in disease pathogenesis. Again, the lack of specific markers that identify the entire simian NK cell population have significantly limited our understanding of the role of this population in disease states for which only nonhuman primates offer an appropriate animal model. In particular, the SIV-infected macaque is the only practical animal model for pathogenic HIV infection/immunization, and it allows for the investigation of events immediately following infection, an opportunity rarely possible in HIV-infected humans. Large gaps exist in our understanding of the role of NK cells in the elimination of SIV-infected cells, before or after the establishment of the adaptive immune response. Furthermore, little is known regarding the importance of NK cells in generating the appropriate cytokine and chemokine milieu necessary for initiating and recruiting an effective adaptive immune response. Since NKG2A and NKp80 are highly expressed by simian NK cells, anti-NKG2A and anti-NKp80 monoclonal antibodies may allow for specific NK cell depletion in vivo, and thus definitive experiments elucidating the role of NK cells in various pathogenic conditions may be conducted (Mavilio et al., 2005a). It has become clear that the innate immune response greatly influences establishment of the adaptive immune response; therefore, such information may be useful in the design of effective vaccine strategies.

Therapeutic perspective

New therapeutic avenues might inhibit the potential pathogenic effects of NK cell cytotoxicity. It is well known that HIV-1 viraemia induces a CD4$^+$ T cell depletion that leads to immunodeficiency and correlates with disease progression. However, it has also been reported that the majority of CD4$^+$ T cells dying during the infection are not productively infected with HIV-1 (Alimonti et al., 2003). One possible explanation is that these uninfected CD4+ T cells are eliminated through a mechanism not directly linked to viral replication. It has been demonstrated both in vitro (Ward et al., 2007) and ex vivo (Fogli et al., 2008) that HIV-1 replication can modulate the expression of ligands for NKp46, NKp30 and NKp44 on uninfected CD4$^+$ T-cell blasts. In particular, a highly conserved motif of HIV-1 gp41 envelope protein can induce the expression of NKp44 ligand on uninfected CD4+ T-cell blasts and render these cells susceptible to NK cell-mediated killing via the NKp44 activation pathway (Vieillard et al., 2005). A recent report showed that immunization against the 3S gp41 peptide prevents NKp44L expression and CD4^{pos} T-cell depletion in SHIV infection, thus providing new insight for both preventive and therapeutic HIV-1 vaccine strategies (Vieillard et al., 2008b). However, recent results from the same group showed unexplained differences between R5 and X4 infection (Vieillard et al., 2008a). Other therapeutic avenues based on cytokines or cytokine triggering may stimulate NK antiviral functions. For instance, TLR-9 ligands may play a role in restoring NK cell-mediated killing of HIV-infected CD4$^+$ T cells through the stimulation of IFN-α secretion by pDC. In any case, the data obtained in the past few years offer new molecular targets for better HIV control through the restoration of NK cell functions, and clinical trials will determine whether these molecular targets can help complete the viral suppression obtained by ART.

Acknowledgements

We thank all patients enrolled in all studies cited in this review for participating in HIV-1 research programs with generosity and great effort. We acknowledge Françoise Barré-Sinoussi, Anthony S. Fauci, Alessandro Moretta and the members of the Action coordonnée 31 (Dendritic cells, antigen presentation and innate immunity) of the French National Agency for AIDS and viral hepatitis Research (ANRS) for support and helpful discussion and Eric Vivier for reading this manuscript. This review was supported by grants from the ANRS and by the intramural program of NIAID/NIH.

References

Adam, C., King, S., Allgeier, T., Braumuller, H., Luking, C., Mysliwietz, J., Kriegeskorte, A., Busch, D.H., Rocken, M. and Mocikat, R. (2005). DC–NK cell cross talk as a novel CD4+ T-cell-independent pathway for antitumor CTL induction. Blood **106**, 338–344.

Ahmad, A. and Menezes, J. (1996). Antibody-dependent cellular cytotoxicity in HIV infections. FASEB J **10**, 258–266.

Ahmad, R., Sindhu, S.T., Tran, P., Toma, E., Morisset, R., Menezes, J. and Ahmad, A. (2001). Modulation of expression of the MHC class I-binding natural killer cell receptors, and NK activity in relation to viral load in HIV-infected/AIDS patients. J Med Virol **65**, 431–440.

Alimonti, J.B., Ball, T.B. and Fowke, K. R. (2003). Mechanisms of CD4+ T lymphocyte cell death in human immunodeficiency virus infection and AIDS. J Gen Virol **84**, 1649–1661.

Alkhatib, G., Combadiere, C., Broder, C. C., Feng, Y., Kennedy, P.E., Murphy, P.M. and Berger, E.A. (1996). CC CKR5: a RANTES, MIP-1alpha, MIP-1beta receptor as a fusion cofactor for macrophage-tropic HIV-1. Science **272**, 1955–1958.

Alter, G., Martin, M.P., Teigen, N., Carr, W.H., Suscovich, T.J., Schneidewind, A., Streeck, H., Waring, M., Meier, A., Brander, C., Lifson, J.D., Allen, T.M., Carrington, M. and Altfeld, M. (2007). Differential natural killer cell-mediated inhibition of HIV-1 replication based on distinct KIR/HLA subtypes. J Exp Med **204**, 3027–3036.

Alter, G., Teigen, N., Davis, B.T., Addo, M.M., Suscovich, T.J., Waring, M. T., Streeck, H., Johnston, M.N., Staller, K.D., Zaman, M.T., Yu, X. G., Lichterfeld, M., Basgoz, N., Rosenberg, E.S. and Altfeld, M. (2005). Sequential deregulation of NK cell subset distribution and function starting in acute HIV-1 infection. Blood **106**, 3366–3369.

Arthos, J., Cicala, C., Martinelli, E., Macleod, K., Van Ryk, D., Wei, D., Xiao, Z., Veenstra, T.D., Conrad, T. P., Lempicki, R.A., McLaughlin, S., Pascuccio, M., Gopaul, R., McNally, J., Cruz, C.C., Censoplano, N., Chung, E., Reitano, K.N., Kottilil, S., Goode, D.J. and Fauci, A.S. (2008). HIV-1 envelope protein binds to and signals through integrin alpha4beta7, the gut mucosal homing receptor for peripheral T cells. Nat Immunol **9**, 301–309.

Azzoni, L., Chehimi, J., Zhou, L., Foulkes, A.S., June, R., Maino, V.C., Landay, A., Rinaldo, C., Jacobson, L.P. and Montaner, L.J. (2007). Early and delayed benefits of HIV-1 suppression: timeline of recovery of innate immunity effector cells. AIDS **21**, 293–305.

Bernstein, H.B., Plasterer, M.C., Schiff, S.E., Kitchen, C.M., Kitchen, S. and Zack, J.A. (2006). CD4 expression on activated NK cells: ligation of CD4 induces cytokine expression and cell migration. J Immunol **177**, 3669–3676.

Biswas, P., Mantelli, B., Sica, A., Malnati, M., Panzeri, C., Saccani, A., Hasson, H., Vecchi, A., Saniabadi, A., Lusso, P., Lazzarin, A. and Beretta, A. (2003). Expression of CD4 on human peripheral blood neutrophils. Blood **101**, 4452–4456.

Bluman, E.M., Bartynski, K.J., Avalos, B. R. and Caligiuri, M.A. (1996). Human natural killer cells produce abundant macrophage inflammatory protein-1 alpha in response to monocyte-derived cytokines. J Clin Invest **97**, 2722–2727.

Bonaparte, M.I. and Barker, E. (2003). Inability of natural killer cells to destroy autologous HIV-infected T lymphocytes. AIDS **17**, 487–494.

Bonaparte, M.I. and Barker, E. (2004). Killing of human immunodeficiency virus-infected primary T-cell blasts by autologous natural killer cells is dependent on the ability of the virus to alter the expression of major histocompatibility complex class I molecules. Blood **104**, 2087–2094.

Boulet, S., Sharafi, S., Simic, N., Bruneau, J., Routy, J.P., Tsoukas, C.M. and Bernard, N.F. (2008). Increased proportion of KIR3DS1 homozygotes in HIV-exposed uninfected individuals. AIDS **22**, 595–599.

Brenner, B.G., Gryllis, C. and Wainberg, M. A. (1991). Role of antibody-dependent cellular cytotoxicity and lymphokine-activated killer cells in AIDS and related diseases. J Leukoc Biol **50**, 628–640.

Caligiuri, M.A. (2008). Human natural killer cells. Blood **112**, 461–469.

Caligiuri, M.A., Zmuidzinas, A., Manley, T.J., Levine, H., Smith, K.A. and Ritz, J. (1990). Functional consequences of interleukin 2 receptor expression on resting human lymphocytes. Identification of a novel natural killer cell subset with high affinity receptors. J Exp Med **171**, 1509–1526.

Carrington, M. and O'Brien, S.J. (2003). The influence of HLA genotype on AIDS. Annu Rev Med **54**, 535–551.

Cerwenka, A. and Lanier, L.L. (2001). Natural killer cells, viruses and cancer. Nat Rev Immunol **1**, 41–49.

Chehimi, J., Azzoni, L., Farabaugh, M., Creer, S.A., Tomescu, C., Hancock, A., Mackiewicz, A., D'Alessandro, L., Ghanekar, S., Foulkes, A.S., Mounzer, K., Kostman, J. and Montaner, L.J. (2007). Baseline viral load and immune activation determine the extent of reconstitution of innate immune effectors in HIV-1-infected subjects undergoing antiretroviral treatment. J Immunol **179**, 2642–2650.

Choe, H., Farzan, M., Sun, Y., Sullivan, N., Rollins, B., Ponath, P.D., Wu, L., Mackay, C.R., LaRosa, G., Newman, W., Gerard, N., Gerard, C. and Sodroski, J. (1996).

The beta-chemokine receptors CCR3 and CCR5 facilitate infection by primary HIV-1 isolates. Cell **85**, 1135–1148.

Chun, T.W., Carruth, L., Finzi, D., Shen, X., DiGiuseppe, J.A., Taylor, H., Hermankova, M., Chadwick, K., Margolick, J., Quinn, T.C., Kuo, Y.H., Brookmeyer, R., Zeiger, M.A., Barditch-Crovo, P. and Siliciano, R.F. (1997). Quantification of latent tissue reservoirs and total body viral load in HIV-1 infection. Nature **387**, 183–188.

Chung, A.W., Rollman, E., Center, R.J., Kent, S.J. and Stratov, I. (2009). Rapid degranulation of NK cells following activation by HIV-specific antibodies. J Immunol **182**, 1202–1210.

Cocchi, F., DeVico, A.L., Garzino-Demo, A., Arya, S.K., Gallo, R.C. and Lusso, P. (1995). Identification of RANTES, MIP-1 alpha, and MIP-1 beta as the major HIV-suppressive factors produced by CD8+ T cells. Science **270**, 1811–1815.

Cohen, G.B., Gandhi, R.T., Davis, D. M., Mandelboim, O., Chen, B.K., Strominger, J.L. and Baltimore, D. (1999). The selective downregulation of class I major histocompatibility complex proteins by HIV-1 protects HIV-infected cells from NK cells. Immunity **10**, 661–671.

Cooper, M.A., Fehniger, T.A. and Caligiuri, M.A. (2001). The biology of human natural killer-cell subsets. Trends Immunol **22**, 633–640.

Cooper, M.A., Fehniger, T.A., Fuchs, A., Colonna, M. and Caligiuri, M.A. (2004). NK cell and DC interactions. Trends Immunol **25**, 47–52.

De Maria, A., Fogli, M., Costa, P., Murdaca, G., Puppo, F., Mavilio, D., Moretta, A. and Moretta, L. (2003). The impaired NK cell cytolytic function in viremic HIV-1 infection is associated with a reduced surface expression of natural cytotoxicity receptors (NKp46, NKp30 and NKp44). Eur J Immunol **33**, 2410–2418.

Degli-Esposti, M.A. and Smyth, M.J. (2005). Close encounters of different kinds: dendritic cells and NK cells take centre stage. Nat Rev Immunol **5**, 112–124.

Devito, C., Broliden, K., Kaul, R., Svensson, L., Johansen, K., Kiama, P., Kimani, J., Lopalco, L., Piconi, S., Bwayo, J.J., Plummer, F., Clerici, M. and Hinkula, J. (2000). Mucosal and plasma IgA from HIV-1-exposed uninfected individuals inhibit HIV-1 transcytosis across human epithelial cells. J Immunol **165**, 5170–5176.

Donaghy, H., Gazzard, B., Gotch, F. and Patterson, S. (2003). Dysfunction and infection of freshly isolated blood myeloid and plasmacytoid dendritic cells in patients infected with HIV-1. Blood **101**, 4505–4511.

Dragic, T., Litwin, V., Allaway, G.P., Martin, S.R., Huang, Y., Nagashima, K. A., Cayanan, C., Maddon, P.J., Koup, R.A., Moore, J.P. and Paxton, W.A. (1996). HIV-1 entry into CD4+ cells is mediated by the chemokine receptor CC-CKR-5. Nature **381**, 667–673.

Fantuzzi, L., Purificato, C., Donato, K., Belardelli, F. and Gessani, S. (2004). Human immunodeficiency virus type 1 gp120 induces abnormal maturation and functional alterations of dendritic cells: a novel mechanism for AIDS pathogenesis. J Virol **78**, 9763–9772.

Fauci, A.S., Mavilio, D. and Kottilil, S. (2005). NK cells in HIV infection: paradigm for protection or targets for ambush. Nat Rev Immunol **5**, 835–843.

Fehniger, T.A., Herbein, G., Yu, H., Para, M.I., Bernstein, Z.P., O'Brien, W.A. and Caligiuri, M.A. (1998). Natural killer cells from HIV-1+ patients produce C–C chemokines and inhibit HIV-1 infection. J Immunol **161**, 6433–6438.

Ferbas, J., Navratil, J., Logar, A. and Rinaldo, C. (1995). Selective decrease in HIV-1-induced alpha interferon production by peripheral blood mononuclear cells during HIV-1 infection. Clin Diagn Lab Immunol **2**, 138–142.

Ferlazzo, G., Pack, M., Thomas, D., Paludan, C., Schmid, D., Strowig, T., Bougras, G., Muller, W.A., Moretta, L. and Munz, C. (2004). Distinct roles of IL-12 and IL-15 in human natural killer cell activation by dendritic cells from secondary lymphoid organs. Proc Natl Acad Sci U S A **101**, 16606–16611.

Flamand, L., Crowley, R.W., Lusso, P., Colombini-Hatch, S., Margolis, D.M. and Gallo, R.C. (1998). Activation of CD8+ T lymphocytes through the T cell receptor turns on CD4 gene expression: implications for HIV pathogenesis. Proc Natl Acad Sci U S A **95**, 3111–3116.

Fogli, M., Costa, P., Murdaca, G., Setti, M., Mingari, M.C., Moretta, L., Moretta, A. and De Maria, A. (2004). Significant NK cell activation associated with decreased cytolytic function in peripheral blood of HIV-1-infected patients. Eur J Immunol **34**, 2313–2321.

Fogli, M., Mavilio, D., Brunetta, E., Varchetta, S., Ata, K., Roby, G., Kovacs, C., Follmann, D., Pende, D., Ward, J., Barker, E., Marcenaro, E., Moretta, A. and Fauci, A.S. (2008). Lysis of endogenously infected CD4+ T cell blasts by rIL-2 activated autologous natural killer cells from HIV-infected viremic individuals. PLoS Pathog **4**, e1000101.

Gao, X., Nelson, G.W., Karacki, P., Martin, M.P., Phair, J., Kaslow, R., Goedert, J. J., Buchbinder, S., Hoots, K., Vlahov, D., O'Brien, S.J. and Carrington, M. (2001). Effect of a single amino acid change in MHC class I molecules on the rate of progression to AIDS. N Engl J Med **344**, 1668–1675.

Gerosa, F., Baldani-Guerra, B., Nisii, C., Marchesini, V., Carra, G. and Trinchieri, G. (2002). Reciprocal activating interaction between natural killer cells and dendritic cells. J Exp Med **195**, 327–333.

Grassi, F.R., Hosmalin, A., McIlroy, D., Calvez, V., Debré, P. and Autran, B. (1999). CD11c-positive dendritic cells are depleted in the blood of HIV-infected patients. AIDS **13**, 759–766.

Harada, H., Goto, Y., Ohno, T., Suzu, S. and Okada, S. (2007). Proliferative activation up-regulates expression of CD4 and HIV-1 co-receptors on NK cells and induces their infection with HIV-1. Eur J Immunol **37**, 2148–2155.

Hirbod, T. and Broliden, K. (2007). Mucosal immune responses in the genital tract of HIV-1-exposed uninfected women. J Intern Med **262**, 44–58.

Hosmalin, A. and Lebon, P. (2006). Type I IFN interferon production in HIV-infected patients. J Leukoc Biol **80**, 984–993.

Hu, P.F., Hultin, L.E., Hultin, P., Hausner, M.A., Hirji, K., Jewett, A., Bonavida, B., Detels, R. and Giorgi, J.V. (1995). Natural killer cell immunodeficiency in HIV disease is manifest by profoundly decreased numbers of CD16+CD56+ cells and expansion of a population of CD16dimCD56− cells with low lytic activity. J Acquir Immune Defic Syndr Hum Retrovirol **10**, 331–340.

Jennes, W., Verheyden, S., Demanet, C., Adje-Toure, C.A., Vuylsteke, B., Nkengasong, J.N. and Kestens, L. (2006). Cutting edge: resistance to HIV-1 infection among African female sex workers is associated with inhibitory KIR in the absence of their HLA ligands. J Immunol **177**, 6588–6592.

Karre, K., Ljunggren, H.G., Piontek, G. and Kiessling, R. (1986). Selective rejection of H-2-deficient lymphoma variants suggests alternative immune defence strategy. Nature **319**, 675–678.

Kaslow, R.A., Carrington, M., Apple, R., Park, L., Munoz, A., Saah, A.J., Goedert, J.J., Winkler, C., O'Brien, S.J., Rinaldo, C., Detels, R., Blattner, W., Phair, J., Erlich, H. and Mann, D.L. (1996). Influence of combinations of human major histocompatibility complex genes on the course of HIV-1 infection. Nat Med **2**, 405–411.

Kitchen, S.G., Korin, Y.D., Roth, M. D., Landay, A. and Zack, J.A. (1998). Costimulation of naive CD8(+) lymphocytes induces CD4 expression and allows human immunodeficiency virus type 1 infection. J Virol **72**, 9054–9060.

Kottilil, S., Chun, T.W., Moir, S., Liu, S., McLaughlin, M., Hallahan, C.W., Maldarelli, F., Corey, L. and Fauci, A. S. (2003). Innate immunity in human immunodeficiency virus infection: effect of viremia on natural killer cell function. J Infect Dis **187**, 1038–1045.

Kottilil, S., Shin, K., Jackson, J.O., Reitano, K.N., O'Shea M, A., Yang, J., Hallahan, C.W., Lempicki, R., Arthos, J. and Fauci, A.S. (2006). Innate immune dysfunction in HIV infection: effect of HIV envelope–NK cell interactions. J Immunol **176**, 1107–1114.

Kottilil, S., Shin, K., Planta, M., McLaughlin, M., Hallahan, C.W., Ghany, M., Chun, T.W., Sneller, M.C. and Fauci, A.S. (2004). Expression of chemokine and inhibitory receptors on natural killer cells: effect of immune activation and HIV viremia. J Infect Dis **189**, 1193–1198.

Lanier, L.L. (2005). NK cell recognition. Annu Rev Immunol **23**, 225–274.

Lichtner, M., Rossi, R., Rizza, M.C., Mengoni, F., Sauzullo, I., Massetti, A.P., Luzi, G., Hosmalin, A., Mastroianni, C. M. and Vullo, V. (2008). Plasmacytoid dendritic cells count in antiretroviral-treated patients is predictive of HIV load control independent of CD4+ T-cell count. Curr HIV Res **6**, 19–27.

Liu, R., Paxton, W.A., Choe, S., Ceradini, D., Martin, S.R., Horuk, R., MacDonald, M.E., Stuhlmann, H., Koup, R.A. and Landau, N.R. (1996). Homozygous defect in HIV-1 coreceptor accounts for resistance of some multiply-exposed individuals to HIV-1 infection. Cell **86**, 367–377.

Long, E.O. (2002). Versatile signaling through NKG2D. Nat Immunol **3**, 1119–1120.

Loubeau, M., Ahmad, A., Toma, E. and Menezes, J. (1997). Enhancement of natural killer and antibody-dependent cytolytic activities of the peripheral blood mononuclear cells of HIV-infected patients by recombinant IL-15. J Acquir Immune Defic Syndr Hum Retrovirol **16**, 137–145.

Lucia, B., Jennings, C., Cauda, R., Ortona, L. and Landay, A.L. (1995). Evidence of a selective depletion of a CD16+ CD56+ CD8+ natural killer cell subset during HIV infection. Cytometry **22**, 10–15.

Lum, J.J., Schnepple, D.J., Nie, Z., Sanchez-Dardon, J., Mbisa, G.L., Mihowich, J., Hawley, N., Narayan, S., Kim, J.E., Lynch, D.H. and Badley, A.D. (2004). Differential effects of interleukin-7 and interleukin-15 on NK cell anti-human immunodeficiency virus activity. J Virol **78**, 6033–6042.

Lusso, P., Garzino-Demo, A., Crowley, R.W. and Malnati, M.S. (1995). Infection of gamma/delta T lymphocytes by human herpesvirus 6: transcriptional induction of CD4 and susceptibility to HIV infection. J Exp Med **181**, 1303–1310.

Martin, M.P., Dean, M., Smith, M.W., Winkler, C., Gerrard, B., Michael, N. L., Lee, B., Doms, R.W., Margolick, J., Buchbinder, S., Goedert, J.J., O'Brien, T.R., Hilgartner, M.W., Vlahov, D., O'Brien, S.J. and Carrington, M. (1998). Genetic acceleration of AIDS progression by a promoter variant of CCR5. Science **282**, 1907–1911.

Martin, M.P., Gao, X., Lee, J.H., Nelson, G.W., Detels, R., Goedert, J.J., Buchbinder, S., Hoots, K., Vlahov, D., Trowsdale, J., Wilson, M., O'Brien, S. J. and Carrington, M. (2002). Epistatic interaction between KIR3DS1 and HLA-B delays the progression to AIDS. Nat Genet **31**, 429–434.

Mastroianni, C.M., d'Ettorre, G., Forcina, G. and Vullo, V. (2004). Teaching tired T cells to fight HIV: time to test IL-15 for immunotherapy?. Trends Immunol **25**, 121–125.

Mavilio, D., Benjamin, J., Daucher, M., Lombardo, G., Kottilil, S., Planta, M. A., Marcenaro, E., Bottino, C., Moretta, L., Moretta, A. and Fauci, A.S. (2003). Natural killer cells in HIV-1 infection: dichotomous effects of viremia on inhibitory and activating receptors and their functional correlates. Proc Natl Acad Sci U S A **100**, 15011–15016.

Mavilio, D., Benjamin, J., Kim, D., Lombardo, G., Daucher, M., Kinter, A., Nies-Kraske, E., Marcenaro, E., Moretta, A. and Fauci, A.S. (2005a). Identification of NKG2A and NKp80 as specific natural killer cell markers in rhesus and pigtailed monkeys. Blood **106**, 1718–1725.

Mavilio, D., Lombardo, G., Benjamin, J., Kim, D., Follman, D., Marcenaro, E., O'Shea, M.A., Kinter, A., Kovacs, C., Moretta, A. and Fauci, A.S. (2005b). Characterization of CD56−/CD16+ natural killer (NK) cells: a highly dysfunctional NK subset expanded in HIV-infected viremic individuals. Proc Natl Acad Sci U S A **25**.

Mavilio, D., Lombardo, G., Kinter, A., Fogli, M., La Sala, A., Ortolano, S., Farschi, A., Follmann, D., Gregg, R., Kovacs, C., Marcenaro, E., Pende, D., Moretta, A. and Fauci, A.S. (2006). Characterization of the defective interaction between a subset of natural killer cells and dendritic cells in HIV-1 infection. J Exp Med **203**, 2339–2350.

Montoya, C.J., Velilla, P.A., Chougnet, C., Landay, A.L. and Rugeles, M.T. (2006). Increased IFN-gamma production by NK and CD3+/CD56+ cells in sexually HIV-1-exposed but uninfected individuals. Clin Immunol **120**, 138–146.

Moretta, A. (2002). Natural killer cells and dendritic cells: rendezvous in abused tissues. Nat Rev Immunol **2**, 957–964.

Moretta, A., Bottino, C., Vitale, M., Pende, D., Biassoni, R., Mingari, M.C. and Moretta, L. (1996). Receptors for HLA class-I molecules in human natural killer cells. Annu Rev Immunol **14**, 619–648.

Moretta, A., Bottino, C., Vitale, M., Pende, D., Cantoni, C., Mingari, M.C., Biassoni, R. and Moretta, L. (2001). Activating receptors and coreceptors involved in human natural killer cell-mediated cytolysis. Annu Rev Immunol **19**, 197–223.

Moretta, A., Bottino, C., Mingari, M.C., Biassoni, R. and Moretta, L. (2002). What is a natural killer cell? Nat Immunol **3**, 6–8.

Motsinger, A., Haas, D.W., Stanic, A.K., Van Kaer, L., Joyce, S. and Unutmaz, D. (2002). CD1d-restricted human natural killer T cells are highly susceptible to human immunodeficiency virus 1 infection. J Exp Med **195**, 869–879.

Nagler, A., Lanier, L.L., Cwirla, S. and Phillips, J.H. (1989). Comparative studies of human FcRIII-positive and negative natural killer cells. J Immunol **143**, 3183–3191.

Noraz, N., Verrier, B., Fraisier, C. and Desgranges, C. (1995). Cell surface phenotypic changes induced in H9 T cells chronically infected with HTLV type I or HIV type 1 or coinfected with the two viruses. AIDS Res Hum Retroviruses **11**, 145–154.

Oliva, A., Kinter, A.L., Vaccarezza, M., Rubbert, A., Catanzaro, A., Moir, S., Monaco, J., Ehler, L., Mizell, S., Jackson, R., Li, Y., Romano, J.W. and Fauci, A.S. (1998). Natural killer cells from human immunodeficiency virus (HIV)-infected individuals are an important source of CC-chemokines and suppress HIV-1 entry and replication in vitro. J Clin Invest **102**, 223–231.

Parato, K.G., Kumar, A., Badley, A.D., Sanchez-Dardon, J.L., Chambers, K. A., Young, C.D., Lim, W.T., Kravcik, S., Cameron, D.W. and Angel, J.B. (2002). Normalization of natural killer cell function and phenotype with effective anti-HIV therapy and the role of IL-10. AIDS **16**, 1251–1256.

Patterson, S. and Knight, S.C. (1987). Susceptibility of human peripheral blood dendritic cells to infection by human immunodeficiency virus. J Gen Virol **68**(Pt4), 1177–1181.

Pierson, T., McArthur, J. and Siliciano, R.F. (2000). Reservoirs for HIV-1: mechanisms for viral persistence in the presence of antiviral immune responses and antiretroviral therapy. Annu Rev Immunol **18**, 665–708.

Raulet, D.H. (2004). Interplay of natural killer cells and their receptors with the adaptive immune response. Nat Immunol **5**, 996–1002.

Ravet, S., Scott-Algara, D., Bonnet, E., Tran, H.K., Tran, T., Nguyen, N., Truong, L.X., Theodorou, I., Barre-Sinoussi, F., Pancino, G. and Paul, P. (2007). Distinctive NK-cell receptor repertoires sustain high-level constitutive NK-cell activation in HIV-exposed uninfected individuals. Blood **109**, 4296–4305.

Rinaldo , C.R. Jr. and Piazza, P. (2004). Virus infection of dendritic cells: portal for host invasion and host defense. Trends Microbiol **12**, 337–345.

Saidi, H., Melki, M.T. and Gougeon, M.L. (2008). HMGB1-dependent triggering of HIV-1 replication and persistence in dendritic cells as a consequence of NK–DC cross-talk. PLoS ONE **3**, e3601.

Schenal, M., Lo Caputo, S., Fasano, F., Vichi, F., Saresella, M., Pierotti, P., Villa, M.L., Mazzotta, F., Trabattoni, D. and Clerici, M. (2005). Distinct patterns of HIV-specific memory T lymphocytes in HIV-exposed uninfected individuals and in HIV-infected patients. AIDS **19**, 653–661.

Scott-Algara, D. and Paul, P. (2002). NK cells and HIV infection: lessons from other viruses. Curr Mol Med **2**, 757–768.

Scott-Algara, D., Truong, L.X., Versmisse, P., David, A., Luong, T.T., Nguyen, N. V., Theodorou, I., Barre-Sinoussi, F. and Pancino, G. (2003). Cutting edge: increased NK cell activity in HIV-1-exposed but uninfected Vietnamese intravascular drug users. J Immunol **171**, 5663–5667.

Scott-Algara, D., Arnold, V., Didier, C., Kattan, T., Pirozzi, G., Barre-Sinoussi, F. and Pancino, G. (2008). The CD85j+ NK cell subset potently controls HIV-1 replication in autologous dendritic cells. PLoS ONE **3**, e1975.

Siegal, F.P., Lopez, C., Fitzgerald, P.A., Shah, K., Baron, P., Leiderman, I. Z., Imperato, D. and Landesman, S. (1986). Opportunistic infections in acquired immune deficiency syndrome result from synergistic defects of both the natural and adaptive components of cellular immunity. J Clin Invest **78**, 115–123.

Sirianni, M.C., Ensoli, F., Alario, C., Fiorelli, V., Sacco, G., Topino, S., Iebba, F., Mezzaroma, I. and Aiuti, F. (2001). Distribution of the natural killer-related receptor for HLA-C during highly active antiretroviral therapy for human immunodeficiency virus infection. Hum Immunol **62**, 1328–1334.

Smed-Sorensen, A., Lore, K., Walther-Jallow, L., Andersson, J. and Spetz, A. L. (2004). HIV-1-infected dendritic cells up-regulate cell surface markers but fail to produce IL-12 p70 in response to CD40 ligand stimulation. Blood **104**, 2810–2817.

Smith, M.W., Dean, M., Carrington, M., Huttley, G.A. and O'Brien, S.J. (1997). CCR5-delta 32 gene deletion in HIV-1 infected patients. Lancet **350**, 741. author reply 742.

Smyth, M.J., Godfrey, D.I. and Trapani, J.A. (2001). A fresh look at tumor immunosurveillance and immunotherapy. Nat Immunol **2**, 293–299.

Soumelis, V., Scott, I., Gheyas, F., Bouhour, D., Cozon, G., Cotte, L., Huang, L., Levy, J.A. and Liu, Y.J. (2001). Depletion of circulating natural type 1 interferon-producing cells in HIV-infected AIDS patients. Blood **98**, 906–912.

Steinman, R.M. and Nussenzweig, M.C. (2002). Avoiding horror autotoxicus: the importance of dendritic cells in peripheral T cell tolerance. Proc Natl Acad Sci U S A **99**, 351–358.

Tarazona, R., Casado, J.G., Delarosa, O., Torre-Cisneros, J., Villanueva, J.L., Sanchez, B., Galiani, M.D., Gonzalez, R., Solana, R. and Pena, J. (2002). Selective depletion of CD56(dim) NK cell subsets and maintenance of CD56(bright) NK cells in treatment-naive HIV-1-seropositive individuals. J Clin Immunol **22**, 176–183.

Tasca, S., Tambussi, G., Nozza, S., Capiluppi, B., Zocchi, M.R., Soldini, L., Veglia, F., Poli, G., Lazzarin, A. and Fortis, C. (2003). Escape of monocyte-derived dendritic cells of HIV-1 infected individuals from natural killer cell-mediated lysis. AIDS **17**, 2291–2298.

Trinchieri, G. (1989). Biology of natural killer cells. Adv Immunol **47**, 187–376.

Ullum, H., Gotzsche, P.C., Victor, J., Dickmeiss, E., Skinhoj, P. and Pedersen, B.K. (1995). Defective natural immunity: an early manifestation of human immunodeficiency virus infection. J Exp Med **182**, 789–799.

Ullum, H., Cozzi Lepri, A., Aladdin, H., Katzenstein, T., Victor, J., Phillips, A. N., Gerstoft, J., Skinhoj, P. and Klarlund Pedersen, B. (1999). Natural immunity and HIV disease progression. AIDS **13**, 557–563.

Valentin, A., Rosati, M., Patenaude, D.J., Hatzakis, A., Kostrikis, L.G., Lazanas, M., Wyvill, K.M., Yarchoan, R. and Pavlakis, G.N. (2002). Persistent HIV-1 infection of natural killer cells in patients receiving highly active antiretroviral therapy. Proc Natl Acad Sci U S A **99**, 7015–7020.

Vieillard, V., Strominger, J.L. and Debre, P. (2005). NK cytotoxicity against CD4+ T cells during HIV-1 infection: a gp41 peptide induces the expression of an NKp44 ligand. Proc Natl Acad Sci U S A **102**, 10981–10986.

Vieillard, V., Habib, R.E., Brochard, P., Delache, B., Bovendo, H.F., Calvo, J., Morin, J., Picq, I., Martinon, F., Vaslin, B., Le Grand, R. and Debre, P. (2008a). CCR5 or CXCR4 use influences the relationship between CD4 cell depletion, NKp44L expression and NK cytotoxicity in SHIV-infected macaques. AIDS **22**, 185–192.

Vieillard, V., Le Grand, R., Dausset, J. and Debre, P. (2008b). A vaccine strategy against AIDS: an HIV gp41 peptide immunization prevents NKp44L expression and CD4+ T cell depletion in SHIV-infected macaques. Proc Natl Acad Sci U S A **105**, 2100–2104.

Vivier, E., Tomasello, E., Baratin, M., Walzer, T. and Ugolini, S. (2008). Functions of natural killer cells. Nat Immunol **9**, 503–510.

Walzer, T., Dalod, M., Robbins, S.H., Zitvogel, L. and Vivier, E. (2005). Natural killer cells and dendritic cells: "l'union fait la force". Blood.

Ward, J., Bonaparte, M., Sacks, J., Guterman, J., Fogli, M., Mavilio, D. and Barker, E. (2007). HIV modulates the expression of ligands important in triggering natural killer cell cytotoxic responses on infected primary T-cell blasts. Blood **110**, 1207–1214.

Wilflingseder, D., Mullauer, B., Schramek, H., Banki, Z., Pruenster, M., Dierich, M.P. and Stoiber, H. (2004). HIV-1-induced migration of monocyte-derived dendritic cells is associated with differential activation of MAPK pathways. J Immunol **173**, 7497–7505.

Yang, L.P., Riley, J.L., Carroll, R.G., June, C. H., Hoxie, J., Patterson, B.K., Ohshima, Y., Hodes, R.J. and Delespesse, G. (1998). Productive infection of neonatal CD8+ T lymphocytes by HIV-1. J Exp Med **187**, 1139–1144.

Zitvogel, L. (2002). Dendritic and natural killer cells cooperate in the control/switch of innate immunity. J Exp Med **195**, F9–F14.

Chapter Thirty-Seven

37

Cytomegalovirus infection and NK cells

Jerome D. Coudert, Christopher E. Andoniou, Mariapia A. Degli-Esposti

CHAPTER CONTENTS

Discovery consists of seeing what everybody else has seen and thinking what nobody has thought.

Albert Szent-Gyorgyi

ABSTRACT

Natural killer (NK) cells are a key component of the innate immune system that are capable of rapid recognition and elimination of target cells without prior sensitization. Compelling evidence that NK cells limit viral replication in vivo has come from studies involving cytomegalovirus (CMV) infection. Human CMV (HCMV) is a pathogen responsible for causing significant morbidity and mortality in immunocompromised individuals. Due to the strict species specificity of the CMVs, many aspects of the host response to infection have been determined using animal models. This chapter discusses some of the mechanisms used by NK cells to identify CMV-infected cells and describes the escape mechanisms employed by the CMVs to evade detection. We also consider some of the recent data suggesting that in addition to their role in directly inhibiting viral replication, NK cells have the capacity to shape the adaptive immune response elicited by CMV infection.

KEY WORDS

Cytomegalovirus, Ly49, KIR, NKG2D

Cytomegalovirus

Human cytomegalovirus (HCMV) is a common human pathogen typically encountered during childhood. Primary HCMV infection is rapidly controlled by the immune system but not eliminated, resulting in the establishment of a latent infection that persists for the life of the host. Reactivation of HCMV in healthy individuals is usually asymptomatic. However, in immune-compromised patients, HCMV reactivation is a significant clinical problem causing diseases such as interstitial pneumonitis, encephalitis and retinitis. Since CMVs are strictly species specific, animal models have been used to investigate various aspects of viral pathogenesis. In particular, the study of murine CMV (MCMV) has provided valuable insights into how the immune system responds to CMV infection and has helped to define the immune evasion mechanisms used by CMV to ensure that viral replication proceeds successfully. Significantly, many observations regarding CMV pathogenesis made using the MCMV model have been reproduced using samples derived from HCMV-infected patients.

The role of NK cells in CMV infection

Natural killer (NK) cells provide an important first line of defence against viral pathogens. The importance of NK cells in the control of viral pathogens is supported by the finding that patients with NK cell deficiencies are susceptible to recurrent infections with HCMV and Herpes Simplex Virus-1 [Biron, 1989; Orange, 2006]. Increased NK cell numbers and activity during HCMV infection have been reported in clinical settings [Hokland, 1988; Venema, 1994], and, in some instances, have been associated with positive clinical outcomes [Bowden, 1987]. The available data, however, still fall short of conclusively demonstrating a role for NK cells in ameliorating HCMV disease. In vitro studies have reported increased susceptibility to NK cell killing for HCMV-infected fibroblasts and have attributed this to increased expression of ICAM-1 and LFA-3 [Leong, 1998; Fletcher, 1998]. Overall, the role of NK cells in controlling HCMV infection has largely been inferred from studies that identified and characterized HCMV proteins that interfere with NK cell activities (as described later).

In contrast, the role of NK cells in the control of MCMV infection is well established. Early experiments identified beige mice as exhibiting enhanced morbidity and mortality following MCMV infection [Shellam, 1981]. Since beige mice lack NK cell function as a result of genetic mutation, these studies implicated the NK cell subset as being essential for the control of MCMV infection. Further evidence for the role of NK cells in the control of MCMV infection came from the studies of Bancroft et al. [Bancroft, 1981], which demonstrated a positive correlation between the level of NK cell activity and resistance to lethal MCMV infection in various mouse strains. The critical role of NK cells in MCMV infection was conclusively demonstrated by NK cell depletion and reconstitution experiments [Bukowski, 1985; Bukowski, 1984], and indeed, to date, studies investigating the effects of NK cells during MCMV infection have provided the most convincing evidence for their role in anti-viral immunity.

Inbred mouse strains vary considerably in their response to MCMV infection, with both major histocompatibility complex (MHC) loci, and non−MHC loci found to contribute to the control of infection. A number of non−MHC loci associated with NK cell-mediated effects during MCMV infection have been identified. The first MHC-independent gene involved in limiting MCMV infection, termed *Cmv1*, was identified as being responsible for limiting MCMV replication in the spleen and lungs of C57BL/6 (B6) mice [Scalzo, 1990]. Subsequent studies demonstrated that the *Cmv1* effect was NK cell dependent, and the locus encoding it was mapped to the NK cell gene complex (NKC) [Scalzo, 1992]. Genetic mapping studies identified *Ly49h* as the gene responsible for the *Cmv1* mediated control of early acute MCMV infection in the spleen [Lee, 2001; Brown, 2001; Daniels, 2001]. Ly49H is an activating receptor expressed on NK cells. Depletion of Ly49H$^+$ NK cells from B6 mice in vivo resulted in increased viral replication in the spleen and lungs [Lee, 2001; Brown, 2001]. Furthermore, insertion of Ly49H as a transgene into mice normally unable to limit acute MCMV infection renders them resistant to MCMV replication in the spleen and lungs [Lee, 2003]. Ly49H lacks intracellular signalling domains but associates with the DAP12 adaptor molecule [Gosselin, 1999]. Ly49H engagement results in phosphorylation of the tyrosine-based activation motif (ITAM) present in the DAP12 intracellular domain, thus initiating a signalling cascade required for NK cell activation. As expected, B6 mice carrying targeted mutations in the ITAM domain of DAP12 that prevent association with Syk and Zap70, were unable to limit splenic MCMV replication [Sjolin, 2002]. Recent studies have shown that Ly49H can also associate with and signal via DAP10 [Coudert, 2008; Orr, 2009; Tassi, 2009]. In fact, optimal NK cell function requires signals to be delivered via both DAP10 and DAP12 [Orr, 2009]. In MCMV infection, lack of DAP10 and DAP12 renders mice unable to control replication to a greater extent than when only DAP12 is absent or non-functional [Orr, 2009]. Together these data demonstrated that Ly49H, in association with DAP adaptors, mediates the NK cell-dependent control of acute MCMV infection in B6 mice.

The Ly49H receptor recognizes the MCMV-encoded m157 gene product, a MHC class I-like glycoprotein [Arase, 2002; Smith, 2002]. Engagement of m157 by Ly49H triggers NK cell cytolysis and induces the secretion of IFN−γ [Smith, 2002] (Figure 37.1). The NK cell-mediated control of MCMV replication in B6 mice is lost when m157 is mutated, demonstrating the in vivo relevance of the recognition of m157 by Ly49H$^+$ NK cells [Bubic, 2004]. Since m157 binds the inhibitory Ly49I molecule in addition to the activating Ly49H receptor, it is possible that in some settings, m157 may function as an immune evasion molecule involved in inhibiting anti-viral NK cell-mediated responses [Arase, 2002]. This possibility is supported by the observation that most strains of MCMV encode variants of m157 unable to activate Ly49H-expressing NK cells [Voigt, 2003]. Furthermore, repeated passage of MCMV in Ly49H$^+$ B6 mice results in the rapid emergence of m157 mutants that evade Ly49H-dependent responses [Voigt, 2003]. Thus, while Ly49H-dependent recognition

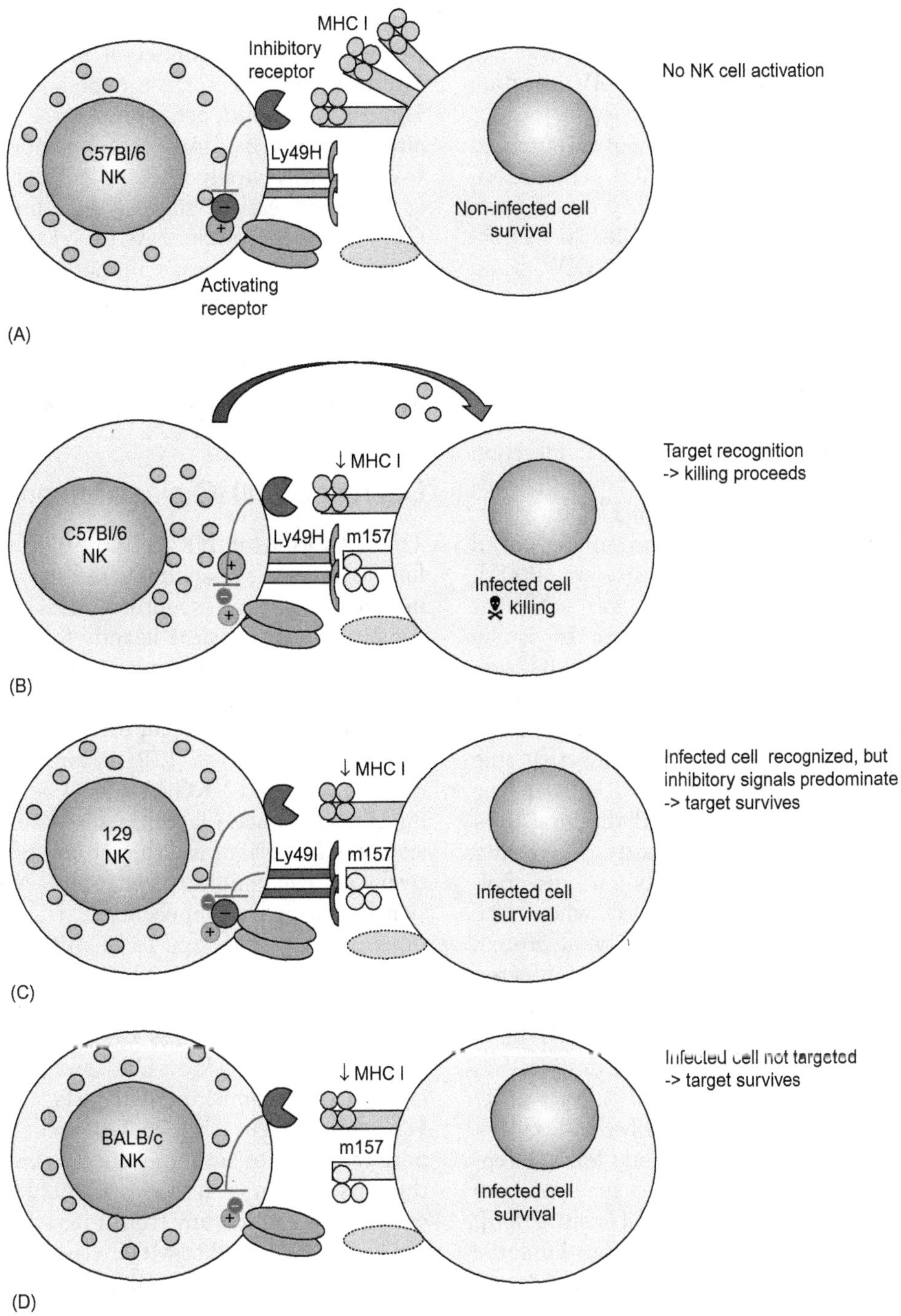

Figure 37.1 • NK cell recognition of MCMV-infected cells. (A) Uninfected healthy cells express MHC I molecules on their surface and either a few or no ligands for NK cell-activating receptors. Inhibitory signals received through engagement of MHC I-specific inhibitory receptors prevent NK cell activation from proceeding. (B) Following MCMV infection, the expression of MHC I molecules on the surface of infected targets is decreased, thus reducing the density of ligands for inhibitory NK cell receptors. In some mouse strains, such as C57Bl/6, NK cells express the Ly49H activating receptor, which interacts with the MCMV m157 protein on infected target cells. In this setting, NK cell activation proceeds through engagement of m157 by Ly49H, and possibly other activating receptors. Activated NK cells release cytokines, such as IFN-γ and TNF-α, as well as cytotoxic granules, composed of perforin and granzymes. These molecules affect viral replication as well as the viability of the infected cell, which is normally killed. (C) In some mouse strains, such as 129, the MCMV m157 viral protein is recognized by the Ly49I inhibitory receptor on NK cells. Although the expression of MHC I molecules on the surface of infected targets is decreased following infection, NK cell activation does not proceed because of the inhibitory signals delivered by Ly49I; these inhibitory signals abort signals transmitted by NK cell-activating receptors. (D) In some mouse strains, such as BALB/c, NK cells lack receptors such as Ly49H able to recognize infected cells through engagement of the m157 viral protein. Although expression of MHC I molecules is decreased on infected cells, thus limiting the density of ligands for inhibitory NK cell receptors, the inability of activating receptors to recognize the infected target means that they are able to survive.

of m157 is undoubtedly responsible for mediating NK cell activation and limiting viral replication in the spleen of B6 mice, the role of m157 during MCMV infection in other strains of mice remains unclear.

Cmv1 is uncommon in both inbred mouse strains and in wild mouse populations [Scalzo, 2005]. These findings prompted further analysis to identify other genes involved in NK cell-dependent control of MCMV infection. Control of MCMV infection in the NZW strain of mice is NK cell dependent but *Cmv1* independent [Rodriguez, 2004]. Loci located on chromosomes 17 and X were found to contribute to the resistance phenotype, termed *Cmv2*; however, the genes encoded by these loci have yet to be identified [Rodriguez, 2004].

The NK cell-mediated control of MCMV infection that operates in MA/My mice was found to be dependant on H2-D^k and a locus termed *Cmv3* [Desrosiers, 2005; Dighe, 2005]. *Cmv3* was found to be linked to the Ly49P activating receptor [Desrosiers, 2005]. Blockade of either Ly49P or H2-D^k was sufficient to prevent NK cell-dependent recognition of virally infected cells [Desrosiers, 2005]. These results suggested that control of MCMV infection in MA/My mice is conferred by $Ly49P^+$ NK cells recognizing virally altered H2-D^k. More recent data indicate that the Ly49P-dependent recognition of infected cells in the Ma/My strain requires both H2-D^k and the viral m04 protein [Kielczewska, 2009]. Significantly, the ability to limit MCMV replication in MA/My mice is abolished when either NK cells are depleted or when mice are infected with a virus lacking the m04 viral protein (Δm04). These findings indicate that MCMV-infected cells can be recognized in an MHC class I restricted fashion by the Ly49P activating receptor (Table 37.1).

The PWK/Pas mouse strain was recently described as capable of limiting MCMV replication during early acute infection in a manner similar to that described in B6 mice. As with the B6 mouse strain, control of early MCMV replication in PWK/Pas mice is mediated by a gene linked to the NKC region on mouse chromosome 6 [Adam, 2006]. However, unlike B6 mice, PWK/Pas mice can limit the replication of MCMV mutant viruses lacking m157, and NK cells isolated from these mice fail to lyse target cells expressing the m157 glycoprotein [Adam, 2006]. Thus, limiting MCMV infection in PWK/Pas mice occurs independently of Ly49H. The resistance locus, termed *Cmv4*, presumably encodes an NK cell-activating receptor, although this has not been formally demonstrated.

NK cells typically recognize and destroy damaged or virally infected cells by detecting the expression of stress ligands recognized by NKG2D. NKG2D ligands include retinoic acid early inducible gene 1 (RAE-1α, -β, -γ, -δ and -ε isoforms), H60 and murine UL-16 binding protein-like transcript 1 (MULT-1) in the mouse and the MHC class I-related chain (MIC) and UL16-binding protein (ULBP) proteins in humans. Both HCMV and MCMV encode a number of proteins that specifically interfere with the surface expression of stress ligands. Hence, NK cell-dependent control of CMV requires alternative mechanisms to be operational and to effectively identify virally infected cells. The available data, as discussed earlier, suggest that direct recognition of CMV-encoded components by NK cells is one mechanism used by the host's immune system to overcome virally mediated immune evasion and ensure an antiviral response is generated.

NK cell activation in CMV infection

Decreased MHC class I expression

The activation of NK cells is tightly regulated by a fine balance of signals delivered by a series of activating and inhibitory receptors. MHC class I molecules constitute the classical ligands for NK cell inhibitory receptors. NK cell receptors involved in the surveillance of MHC class I expression include members of the killer cell Ig-like receptor (KIR) family and leukocyte Ig-like receptors (LIR) in humans, Ly49 receptors in the mouse and NKG2A/CD94 heterodimers in both humans and mice (Table 37.1). When surface expression of MHC class I is either absent or too low to effectively engage inhibitory receptors, NK cell activation signals proceed unopposed and trigger NK cell functions, a process referred to as 'missing self recognition' [Karre, 1986]. Interestingly, to evade cytotoxic T cell recognition, CMVs encode several genes that target the expression of MHC class I molecules [Hengel, 1999]. In HCMV, US2, US3, US6 and US11 affect MHC class I expression at the cell surface (Table 37.2). The US6 glycoprotein binds to TAP and prevents the transport of peptides into the endoplasmic reticulum and their association with US3 prevents MHC class I in the endoplasmic reticulum from migrating into the Golgi. US2 and US11 bind to MHC class I heavy chains in the endoplasmic reticulum and target them to the cytosol where they are degraded. In MCMV, three viral genes (m04, m06 and m152) have been identified as encoding proteins that interfere with MHC class I expression. MCMV glycoprotein (gp) 48, encoded by m06, binds to MHC class I molecules and redirects them to lysosomal compartments for proteolytic degradation. The m152-encoded gp40 retains peptide-loaded MHC class I molecules in a *cis*-Golgi compartment, resulting in reduced MHC class I expression on the surface of infected cells. The m04-encoded gp34 binds MHC class I in the endoplasmic reticulum and forms a stable complex that is exported to the cell surface where it interferes with recognition by $CD8^+$ T cells. Modulation of MHC class I

Table 37.1 Mouse and human NK cell receptors

Receptor	Ligand: Mouse	Ligand: Human
Inhibitory		
Ly49/KIR-L	H-2 class I	HLA class I
CD94-NKG2A	Qa-1	HLA-E
NKRP1-A	–	LLT-1
NKRP1-B, -D	Clr-b	–
CD85i-ILT2	–	HLA class I
KLRG1	Cadherins	Cadherins
2B4 (CD244)	CD48	CD48
Activating/co-stimulating		
Ly49H	MCMV m157	–
Ly49D	H-2D^d	–
Ly49P	H-2D^k + MCMV m04	–
KIR-S	–	HLA-C
NKRP1-A	?	–
NKRP1-C	?	–
NKG2D	RAE-I, H60 MULT-1	MICA, MICB ULBPs
CD94-NKG2C	Qa-1	HLA-E
CD94-NKG2E	Qa-1	HLA-E
CD16	IgG	IgG
NKp30	–	BAT-3, HCMV pp65, Filovirus derived particles
NKp44	–	Viral haemagglutinins
NKp46	Viral haemagglutinins	Viral haemagglutinins
NKp80	–	AICL
CD27	CD70	CD70
2B4 (CD244)	CD48	CD48
CRTAM	Necl-2	Necl-2
DNAM-1 (CD226)	Nectin-2 (CD122)	Nectin-2 (CD122)
CD96	Necl-5 (CD155) and nectin-1	Necl-5 (CD155)
PEN-5	–	L-selectin
CD160	Classical and non-classical MHC molecules	HLA-C

expression interferes with antigen presentation to cytotoxic CD8$^+$ T lymphocytes, but at the same time, it possibly renders infected cells more susceptible to recognition by NK cells because of the lack of ligands for NK cell-inhibiting receptors. Both HCMV and MCMV encode an additional set of viral glycoproteins that protect them from NK cell recognition (discussed later) (Table 37.2).

Expression of stress-induced self ligands

Activation of NK cells can be effectively induced upon engagement of NKG2D receptors by stress-induced ligands. NKG2D is expressed on all human and mouse NK cells (reviewed in [Coudert, 2006]). In mice, NKG2D exist as two isoforms: a long isoform that triggers activating signals through DAP10, and a short isoform (induced following NK cell activation by alternative splicing of the exon encoding the cytoplasmic domain) that can associate and signal through both DAP10 and DAP12. In humans, only the long NKG2D isoform that associates through DAP10 has been identified. NKG2D receptors recognize families of stress ligands induced by viral infection, transformation and cellular stress. In humans, the NKG2D ligands are MIC-A and MIC-B and ULBP-1 to ULBP-6. In the mouse, RAE-1, H60 and MULT-1 can engage NKG2D. Since NKG2D recognizes individual ligands with differing affinities, the outcome of NKG2D engagement is likely to depend on the nature of the ligands being recognized. Both HCMV and MCMV encode a series of genes whose function is to interfere with the expression of stress-induced ligands (discussed later) (Table 37.2).

Viral evasion of NK cell responses

NK cells are armed to recognize virus-infected cells through various receptors and may affect and contribute to antiviral immune responses at different levels. CMV has developed multiple mechanisms to circumvent NK cell-mediated immune surveillance.

Altering the expression of MHC class I molecules

One mechanism by which CMV may interfere with NK cell activation is to favour signals transmitted through inhibitory receptors. This is indeed the case, and a number of viral genes have been described that modulate the expression of MHC molecules, which are ligands for NK

Table 37.2 Cytomegalovirus genes that affect NK cell function

Viral gene	NK cell receptor	Mechanism
Human		
UL18	LIR-1 (ILT2, CD85j)	UL18 encodes an MHC class I-like molecule that engages the inhibitory NK cell receptor LIR-1.
UL40	NKG2A/CD94	A UL40 peptide replace endogenous peptide on HLA-E, stabilizes its expression and sustains inhibition through engagement of NKG2A/CD94.
UL16	NKG2D	UL16 interferes with the expression of MIC-B and ULBP-1 and ULBP-2.
UL142	NKG2D	UL142 prevents surface expression of MIC-A.
UL112	NKG2D	miR-UL112 suppresses MIC-B expression.
pp65	NKp30	pp65 inactivates NKp30 by disrupting its association with the CD3z signalling adaptor.
UL141	DNAM-1, TACTILE	UL141 prevents cell surface expression of CD155.
Mouse		
m04, m06	?	These proteins associate with MHC class I molecules which are ligands for NK cell inhibitory receptors.
m138	NKG2D	Down-regulates MULT-1 and H60 cell surface expression.
m145	NKG2D	Down-regulates MULT-1 cell surface expression.
m152	NKG2D	Prevents RAE-1 expression on the cell surface.
m155	NKG2D	Prevents H60 expression on the cell surface.
m144	?	m144 encodes an MHC class I-like protein predicted to inhibit NK cell activation.

cell inhibitory receptors. A nonameric peptide derived from the leader sequence of the UL40 protein of HCMV can assemble into the peptide-binding groove of HLA-E, thus stabilizing its expression on the membrane of infected cells [Tomasec, 2000; Ulbrecht, 2000]. HLA-E is a ligand for both activating (NKG2C/CD94) and inhibitory (NKG2A/CD94) NK cell receptors; however, when HLA-E molecules are assembled with the UL40 peptide, they may preferentially engage the inhibitory NKG2A/CD94 receptor and thus deliver inhibitory signals. This hypothesis is consistent with the findings that over-expression of UL40 in primary fibroblasts, or in cell lines, protects them from NK cell-mediated lysis. In keeping with these findings, fibroblasts infected with mutant HCMV viruses lacking UL40 were shown to be more susceptible to NK cell lysis than fibroblasts infected with wild type HCMV [Tomasec, 2000; Wang, 2002]. Thus, it appears that HCMV has developed a strategy to counterbalance the activation of NK cells that occurs as a result of the down-regulation of HLA molecules induced by the virus to escape recognition by $CD8^+$ T cells.

Like HCMV, MCMV has developed mechanisms to interfere with MHC class I expression, and the virus encodes at least three genes whose products affect MHC class I expression on infected cells (m06/gp48, m152/gp40, m04/gp34). A peptide encoded by one of these proteins, namely gp34, associates with MHC class I molecules and is expressed as a complex on the surface of infected cells. It is thought that the principle function of m04 is to prevent NK cell activation by providing a ligand for inhibitory NK cell receptors; however, there is no published evidence to support this concept. On the contrary, it has been reported that at early times post-infection, MCMV mutants lacking m04 grow like wild type MCMV in B6 mice, suggesting that m04 does not affect NK cell function [Pinto, 2005]. Interestingly, it appears that in some circumstances m04, like m157, may be involved in activating NK cells. Indeed, recent studies have reported that gp34 in a complex with $H2\text{-}D^k$ is essential to activate NK cells through engagement of the Ly49P activating receptor [Kielczewska, 2009].

In addition to the stabilization of endogenously expressed MHC class I molecules, another mechanism through which CMV may affect NK cell responses is the production of MHC class I homologues able to directly engage NK cell inhibitory receptors and thus interfere with NK cell activation. The HCMV glycoprotein UL18 is capable of forming a trimeric complex with β2-microglobulin and endogenous peptide [Browne, 1990; Fahnestock, 1995]. gpUL18 is expressed during the late phase of HCMV infection when endogenous MHC class I is down-regulated [Griffin, 2005]. It binds to LIR-1 (also known as ILT2, CD85j) [Cosman, 1997] with an affinity more than a thousand time higher than MHC class I molecules; thus, even low-level gpUL18 can be expected to compete efficiently for binding to NK cell receptors and inhibit NK cell function through LIR-1 mediated signals [Chapman, 1999]. Unfortunately, to date, UL18 has not been detected on HCMV-infected cells.

In B6 mice, recognition of the m157 viral glycoprotein by Ly49H confers NK cell-mediated resistance to early viral replication in the spleen and lungs. Ly49H, however, is not the only NK cell receptor capable of binding m157. m157 can also bind an inhibitory receptor, Ly49I, which is absent in B6 mice but is expressed on NK cells from the 129/J mouse strain. The activating receptor Ly49H and its inhibiting counterpart Ly49I are highly homologous; thus, it is tempting to speculate that m157 was initially selected in the MCMV genome as a means of inhibiting NK cells via engagement of Ly49I.

Modulating the expression of ligands for activating NK cells receptors

Evading NKG2D surveillance—The activating NK cell receptor NKG2D can bind multiple stress-induced ligands and trigger NK cell activation. Both HCMV and MCMV have developed mechanisms to interfere with NKG2D-mediated NK cell activation, and each of these viruses has been shown to encode multiple genes that interfere with the expression of ligands for NKG2D. In HCMV, the UL16 and UL142 glycoproteins have been shown to affect NKG2D-dependent recognition. UL16 prevents the cell surface expression of MIC-B, ULBP-1 and ULBP-2 [Welte, 2003; Wu, 2003; Dunn, 2003], while UL142 down-regulates MIC-A expression [Chalupny, 2006]. In addition, HCMV expresses a non-coding miRNA, miR-UL112, which has been shown to suppress MIC-B expression [Stern-Ginossar, 2007]. In MCMV, the m138, m145, m152 and m155 viral genes have been shown to encode proteins that target the down-regulation of NKG2D expression. m138 down-modulates both MULT-1 and H60 expression [Lenac, 2006]. m145 affects the expression of MULT-1 [Krmpotic, 2005]. m152 down-regulates the expression of RAE-1 proteins [Lodoen, 2003], however, different members of the RAE-1 family seem to be differentially affected [Lenac, 2008]. m155 down-modulates H60 expression by a mechanism that occurs after H60 exits from the *cis*-golgi compartment [Hasan, 2005]. Thus, both HCMV and MCMV have developed a number of strategies to interfere with the function of NKG2D receptors suggesting that these activities may play a critical role in the evolutionary survival of the CMVs.

Evading natural cytotoxicity receptors—The natural cytotoxicity receptors (NCR) represent a family of receptors involved in NK cell activation. These receptors include NKp30, NKp44 and NKp46 in humans, while only NKp46 is expressed on mouse NK cells [Arnon, 2006]. The HCMV tegument protein pp65 specifically and directly binds to NKp30 and inhibits NK cells from lysing HCMV-infected targets by dissociating NKp30 from its signalling adaptor CD3ζ [Arnon, 2005]. Since pp65 is not a secreted molecule, but rather it is expressed intracellularly, the inhibiting function of pp65 in NK cell suppression may require its release through lysis of HCMV-infected cells or by the acquisition from apoptotic infected cells. Since NKp46 and NKp44 have been shown to bind influenza haemagglutinins (HA) and Sendai virus HA neuroaminidase [Mandelboim, 2001; Arnon, 2001], it is possible that CMV components may also be recognized by NCRs.

Evading DNAM-1 and CD96 surveillance—DNAM-1 is an activating receptor expressed on human and mouse NK cells [Shibuya, 1996]. CD155, also know as poliovirus receptor (PVR) or nectin-like molecule 5 (necl-5), is an identified ligand for DNAM-1 [Tahara-Hanaoka, 2004]. The HCMV-encoded UL141 glycoprotein prevents surface expression of CD155 by sequestering it in the endoplasmic reticulum and thereby alters NK cell activation induced by DNAM-1 [Tomasec, 2005]. In addition, CD155 is a ligand for the CD96 NK cell receptor. CD96, also known as TACTILE, promotes NK cell adhesion and acts as a co-stimulator for activated NK cells [Fuchs, 2004; Tomasec, 2005]. Thus, by affecting the expression of CD155, HCMV UL141 can broadly affect NK cell activation.

NK cells can regulate anti-viral adaptive immune responses

Dendritic cells (DC) are professional antigen-presenting cells that can pick up, process and present antigens while undergoing a maturation process. Mature DCs can initiate and regulate adaptive immune responses through antigen presented on MHC molecules in combination with co-stimulatory molecules and the release of cytokines. In addition to their role in priming adaptive immune responses, DCs can participate in NK cell activation [Ferlazzo, 2002]. Numerous studies have found that DCs are capable of activating NK cells (reviewed in [Degli-Esposti, 2005]). However, DC–NK interactions are not a one-sided affair but rather involve reciprocal interactions whereby NK cells can also influence DC functions. The mouse model of MCMV infection has provided strong evidence that NK cells influence DC functions and thereby affect anti-viral immune response in vivo at several levels. During MCMV infection in B6 mice, maintenance of the CD8α DC population in the spleen is dependant on Ly49H$^+$ NK cells. Reciprocally, CD8 α^+ DCs are essential for the proliferation of Ly49H$^+$ NK cells in the spleen by a mechanism that involves IL-12 and IL-18 [Andrews, 2003]. Recent reports have suggested that other mechanisms may contribute to the ability of Ly49H$^+$ NK cells to maintain splenic DC populations. The rapid control of viral replication exerted by Ly49H$^+$ NK cells may promote the maintenance of splenic DCs by preventing the destruction of the splenic architecture [Bekiaris, 2008]. Robbins et al. have proposed that control of MCMV replication in the spleen of CMV-resistant mice prevents the release of immunosuppressive levels of IFN-α [Robbins, 2007] and have shown that administration of exogenous IFN-α to CMV-resistant mice induces a loss of DCs from the spleen and a slight and transient delay in the activation of antigen-specific T cells [Robbins, 2007].

In addition to potentially influencing the function of DCs, NK cells have recently been shown to induce the differentiation of CD14$^+$ monocytes into DCs [Zhang, 2007]. This process requires production of granulocyte macrophage–colony stimulating factor by CD56^{bright} NK cells and direct cell to cell contact. While this process was proposed to contribute to the maintenance of chronic inflammatory diseases, it could also conceivably operate to expand the pool of DCs during immune responses to MCMV.

NK cells may also influence adaptive immune responses by acting directly on effector cells. Activation of naïve T cells is dependant on IFN-γ produced by NK cells [Martin-Fontecha, 2004]. Activated NK cells have also been reported to stimulate autologous CD4$^+$ T cells through direct cell to cell interactions involving the expression of OX40 ligands and CD86 [Roncarolo, 1991; Zingoni, 2004].

In MCMV infection, the absence of NK cells leads to enhanced proliferation of CD8$^+$ T cells [Su, 2001]. Indeed, NK cells can limit the activation and function of both anti-viral CD8$^+$ and CD4$^+$ T cells through a process that involves NK cells limiting the availability of signals from antigen-presenting cells to naïve T cells (Andrews, Estcourt and Degli-Esposti, unpublished).

NK cells also play a crucial role in the termination of immune response against MCMV, thereby preventing the development of immunopathology. Direct evidence for this role of NK cells comes from studies involving mice deficient in either perforin (Prf) or granzymes A and B (GzmAB) in which replication of MCMV is enhanced [van Dommelen, 2006]. GzmAB-deficient mice survive MCMV infection, while Prf-deficient mice develop a fatal haemophagocytic lymphohistocytosis (HLH)-like syndrome induced by an accumulation of activated, TNF-α producing CD11b$^+$ F4/80$^+$ mononuclear cells and T lymphocytes [van Dommelen, 2006], which are normally eliminated by NK cells thereby preventing the establishment of immunopathology.

Conclusion and perspectives

NK cells are essential contributors to innate immune responses, in particular those elicited by viral infection. The mouse model of CMV infection has been invaluable in allowing immunologists to understand the complex responses that ensue during infection. CMV pathogenesis is characterized by three phases—acute infection; chronic, persistent infection; and latent infection—with the virus surviving for the lifetime of the host [Mocarski, 2001]. Limiting CMV infection requires effective innate and adaptive immune responses. NK cells can limit viral replication during acute infection, but cytotoxic CD8$^+$ CTL and CD4$^+$ T cells are required for long-term surveillance [Reddehase, 2004]. In the mouse, the role of NK cells during MCMV infection is genetically determined. It is possible that similar mechanisms may be operating in humans and that HLA-KIR haplotype combinations may contribute to the type of immune response generated following HCMV infection, and importantly that this may determine the outcome of infection. Recent evidence suggests that the outcome of viral infections can be influenced by the expression of allelic variants of specific KIR-HLA genes. For example, expression of KIR2DL3/HLA-C1 positively influences the ability to resolve hepatitis C virus (HCV) infection [Khakoo, 2004], and delayed progression to AIDS after HIV infection correlates with expression of KIR3DL1 and its HLA ligand Bw4-801 [Martin, 2007]. In relation to HCMV, the role of KIRs has been investigated in settings of transplantation. In solid organ transplants, the number of activating KIR genes carried by the recipient was found to inversely correlate with the rate of HCMV reactivation [Stern, 2008]. Similarly, in haematopoietic stem cell transplants (HSCT), donor KIR genotype has been shown to affect the rate of HCMV reactivation after T cell replete transplantation [Cook, 2006; Chen, 2006]. Although the mechanisms involved have not been functionally elucidated, the available evidence supports the hypothesis that NK cells have developed means to respond to viral pathogens and ensure that

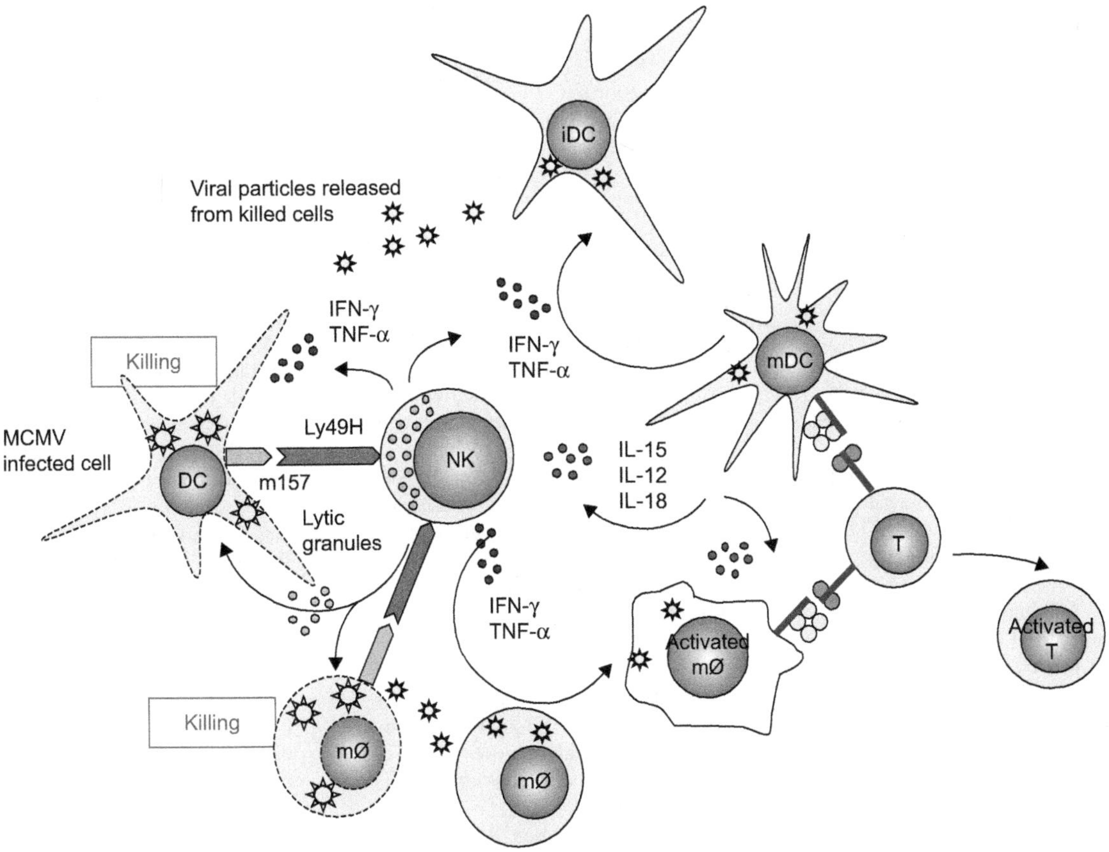

Figure 37.2 • Interactions between NK cells and other immune effectors during MCMV infection. The recognition of MCMV-infected targets by NK cells triggers a process of cytolysis and cytokine release. NK cell-derived lytic granules, consisting of perforin and granzymes, mediate lysis of infected cells. Inflammatory cytokines, such as IFN-γ and TNF-α, promote the activation of DCs and macrophage (mØ); these activated antigen-presenting cells present virally derived peptides on MHC molecules and thus prime naïve antigen-specific T cells. The cytokines released by activated mØs and DCs (IL-15, IL-12, IL-18) favour the activation and survival of NK cells and T cells.

survival of the host is not compromised. It is important to note that persistent viruses, such as CMV, have in turn developed mechanisms to escape detection and elimination by host NK cells so that, like their hosts, viruses have been able to achieve evolutionary survival. The extensive array of viral proteins whose functions are to interfere with NK cell-mediated responses provides further evidence for the critical role of NK cells in defending against viral pathogens.

In addition to their function as anti-viral immune effectors, NK cells play a critical role in controlling immune responses, including the resolution of such responses. When NK cells cannot eliminate activated effectors, including macrophages and T cells, these uncontrolled responses cause so much damage that they place the host at risk and often cause mortality. Recent evidence also suggests that NK cells can positively and negatively influence various aspects of adaptive anti-viral immunity through their interactions with other immune effectors, principally DCs (Figure 37.2). Much remains to be defined about the intricacies and relevance of the interactions between NK cells and other lymphocytes, and many aspects of NK cell biology remain to be elucidated in vivo. Studying NK cells in the context of CMV infection has provided many important insights about the complexity of immune responses and, undoubtedly, such studies will continue to assist us in defining the relevance of NK cells in host defence.

References

Adam, S.G., Caraux, A., et al. (2006). Cmv4, a new locus linked to the NK cell gene complex, controls innate resistance to cytomegalovirus in wild-derived mice. J Immunol **176**(9), 5478–5485.

Andrews, D.M., Scalzo, A.A., et al. (2003). Functional interactions between dendritic cells and NK cells during viral infection. Nat Immunol **4**(2), 175–181.

Arase, H., Mocarski, E.S., et al. (2002). Direct recognition of cytomegalovirus by activating and inhibitory NK cell receptors. Science **296**(5571), 1323–1326.

Arase, H., Mocarski, E.S., et al. (2002). Direct recognition of cytomegalovirus by activating and inhibitory NK cell receptors. Science **296**(5571), 1323–1326.

Arnon, T.I., Achdout, H., et al. (2005). Inhibition of the NKp30 activating receptor by pp65 of human cytomegalovirus. Nat Immunol **6**(5), 515–523.

Arnon, T.I., Lev, M., et al. (2001). Recognition of viral hemagglutinins by NKp44 but not by NKp30. Eur J Immunol **31**(9), 2680–2689.

Arnon, T.I., Markel, G., et al. (2006). Tumor and viral recognition by natural killer cells receptors. Semin Cancer Biol **16**(5), 348–358.

Bancroft, G.J., Shellam, G.R., et al. (1981). Genetic influences on the augmentation of natural killer (NK) cells during murine cytomegalovirus infection: correlation with patterns of resistance. J Immunol **126**(3), 988–994.

Bancroft, G.J., Shellam, G.R., et al. (1981). Genetic influences on the augmentation of natural killer (NK) cells during murine cytomegalovirus infection: correlation with patterns of resistance. Journal of Immunology **126**(3), 988–994.

Bekiaris, V., Timoshenko, O., et al. (2008). Ly49H+ NK cells migrate to and protect splenic white pulp stroma from murine cytomegalovirus infection. J Immunol **180**(10), 6768–6776.

Biron, C.A., Byron, K.S., et al. (1989). Severe herpesvirus infections in an adolescent without natural killer cells. N Engl J Med **320**(26), 1731–1735.

Bowden, R.A., Day, L.M., et al. (1987). Natural cytotoxic activity against cytomegalovirus-infected target cells following marrow transplantation. Transplantation **44**(4), 504–508.

Brown, M.G., Dokun, A.O., et al. (2001). Vital involvement of a natural killer cell activation receptor in resistance to viral infection. Science **292**(5518), 934–937.

Browne, H., Smith, G., et al. (1990). A complex between the MHC class I homologue encoded by human cytomegalovirus and beta 2 microglobulin. Nature **347**(6295), 770–772.

Bubic, I., Wagner, M., et al. (2004). Gain of virulence caused by loss of a gene in murine cytomegalovirus. J Virol **78**(14), 7536–7544.

Bukowski, J.F., Warner, J.F., et al. (1985). Adoptive transfer studies demonstrating the antiviral effect of natural killer cells *in vivo*. J Exp Med **161**, 40–52.

Bukowski, J.F., Woda, B.A., et al. (1984). Pathogenesis of murine cytomegalovirus infection in natural killer cell-depleted mice. J Virol **52**, 119–128.

Carrington, M., Martin, M.P., et al. (2008). KIR-HLA intercourse in HIV disease. Trends Microbiol **16**(12), 620–627.

Chalupny, N.J., Rein-Weston, A., et al. (2006). Down-regulation of the NKG2D ligand MICA by the human cytomegalovirus glycoprotein UL142. Biochem Biophys Res Commun **346**(1), 175–181.

Chapman, T.L., Heikeman, A.P., et al. (1999). The inhibitory receptor LIR-1 uses a common binding interaction to recognize class I MHC molecules and the viral homolog UL18. Immunity **11**(5), 603–613.

Chen, C., Busson, M., et al. (2006). Activating KIR genes are associated with CMV reactivation and survival after non-T-cell depleted HLA-identical sibling bone marrow transplantation for malignant disorders. Bone Marrow Transplant **38**(6), 437–444.

Cook, M., Briggs, D., et al. (2006). Donor KIR genotype has a major influence on the rate of cytomegalovirus reactivation following T-cell replete stem cell transplantation. Blood **107**(3), 1230–1232.

Cosman, D., Fanger, N., et al. (1997). A novel immunoglobulin superfamily receptor for cellular and viral MHC class I molecules. Immunity **7**(2), 273–282.

Coudert, J.D. and Held, W. (2006). The role of the NKG2D receptor for tumor immunity. Semin Cancer Biol **16**(5), 333–343.

Coudert, J.D., Scarpellino, L., et al. (2008). Sustained NKG2D engagement induces cross-tolerance of multiple distinct NK cell activation pathways. Blood **111**(7), 3571–3578.

Daniels, K.A., Devora, G., et al. (2001). Murine cytomegalovirus is regulated by a discrete subset of natural killer cells reactive with monoclonal antibody to Ly49H. J Exp Med **194**(1), 29–44.

Degli-Esposti, M.A. and Smyth, M.J. (2005). Close encounters of different kinds: Dendritic cells and NK cells take centre stage. Nat Rev Immunol **5**(2), 112–124.

Desrosiers, M.P., Kielczewska, A., et al. (2005). Epistasis between mouse Klra and major histocompatibility complex class I loci is associated with a new mechanism of natural killer cell-mediated innate resistance to cytomegalovirus infection. Nat Genet **37**(6), 593–599.

Dighe, A., Rodriguez, M., et al. (2005). Requisite H2k role in NK cell-mediated resistance in acute murine cytomegalovirus-infected MA/My mice. J Immunol **175**(10), 6820–6828.

Dunn, C., Chalupny, N.J., et al. (2003). Human cytomegalovirus glycoprotein UL16 causes intracellular sequestration of NKG2D ligands, protecting against natural killer cell cytotoxicity. J Exp Med **197**(11), 1427–1439.

Fahnestock, M.L., Johnson, J.L., et al. (1995). The MHC class I homolog encoded by human cytomegalovirus binds endogenous peptides. Immunity **3**(5), 583–590.

Falk, C.S., Mach, M., et al. (2002). NK cell activity during human cytomegalovirus infection is dominated by US2-11-mediated HLA class I down-regulation. J Immunol **169**(6), 3257–3266.

Ferlazzo, G., Tsang, M.L., et al. (2002). Human dendritic cells activate resting natural killer (NK) cells and are recognized via the NKp30 receptor by activated NK cells. J Exp Med **195**(3), 343–351.

Fletcher, J.M., Prentice, H.G., et al. (1998). Natural killer cell lysis of cytomegalovirus (CMV)-infected cells correlates with virally induced changes in cell surface lymphocyte function-associated antigen-3 (LFA-3) expression and not with the CMV-induced down-regulation of cell surface class I HLA. J Immunol **161**(5), 2365–2374.

Fuchs, A., Cella, M., et al. (2004). Cutting edge: CD96 (tactile) promotes NK cell-target cell adhesion by interacting with the poliovirus receptor (CD155). J Immunol **172**(7), 3994–3998.

Gosselin, P., Mason, L.H., et al. (1999). Induction of DAP12 phosphorylation, calcium mobilization, and cytokine secretion by Ly49H. J Leukoc Biol **66**(1), 165–171.

Griffin, C., Wang, E.C., et al. (2005). Characterization of a highly glycosylated form of the human cytomegalovirus HLA class I homologue gpUL18. J Gen Virol **86**(Pt 11), 2999–3008.

Hanvesakul, R., Spencer, N., et al. (2008). Donor HLA-C genotype has a profound impact on the clinical outcome following liver transplantation. Am J Transplant **8**(9), 1931–1941.

Hasan, M., Krmpotic, A., et al. (2005). Selective down-regulation of the NKG2D ligand H60 by mouse cytomegalovirus m155 glycoprotein. J Virol **79**(5), 2920–2930.

Hengel, H., Reusch, U., et al. (1999). Cytomegaloviral control of MHC class I function in the mouse. Immunol Rev **168**, 167–176.

Hokland, M., Jacobsen, N., et al. (1988). Natural killer function following allogeneic bone marrow transplantation. Very early reemergence but strong dependence of cytomegalovirus infection. Transplantation **45**(6), 1080–1084.

Karre, K., Ljunggren, H.G., et al. (1986). Selective rejection of H-2-deficient lymphoma variants suggests alternative immune defence strategy. Nature **319**(6055), 675–678.

Khakoo, S.I., Thio, C.L., et al. (2004). HLA and NK cell inhibitory receptor genes in resolving hepatitis C virus infection. Science **305**, 872–874.

Kielczewska, A., Pyzik, M., et al. (2009). Ly49P recognition of cytomegalovirus-infected cells expressing H2-Dk and CMV-encoded m04 correlates with the NK cell antiviral response. J Exp Med **206**(3), 515–523.

Krmpotic, A., Hasan, M., et al. (2005). NK cell activation through the NKG2D ligand MULT-1 is selectively prevented by the glycoprotein encoded by mouse cytomegalovirus gene m145. J Exp Med.

Lee, S.H., Girard, S., et al. (2001). Susceptibility to mouse cytomegalovirus is associated with deletion of an activating natural killer cell receptor of the C-type lectin superfamily. Nat Genet **28**(1), 42–45.

Lee, S.H., Zafer, A., et al. (2003). Transgenic expression of the activating natural killer receptor Ly49H confers resistance to cytomegalovirus in genetically susceptible mice. J Exp Med **197**(4), 515–526.

Lenac, T., Arapovic, J., et al. (2008). Murine cytomegalovirus regulation of NKG2D ligands. Med Microbiol Immunol **197**(2), 159–166.

Lenac, T., Budt, M., et al. (2006). The herpesviral Fc receptor fcr-1 down-regulates the NKG2D ligands MULT-1 and H60. J Exp Med **203**(8), 1843–1850.

Leong, C.C., Chapman, T.L., et al. (1998). Modulation of natural killer cell cytotoxicity in human cytomegalovirus infection: the role of endogenous class I major histocompatibility complex and a viral class I homolog. J Exp Med **187**(10), 1681–1687.

Lodoen, M., Ogasawara, K., et al. (2003). NKG2D-mediated natural killer cell protection against cytomegalovirus is impaired by viral gp40 modulation of retinoic acid early inducible 1 gene molecules. J Exp Med **197**(10), 1245–1253.

Mandelboim, O., Lieberman, N., et al. (2001). Recognition of haemagglutinins on virus-infected cells by NKp46 activates lysis by human NK cells. Nature **409**(6823), 1055–1060.

Martin, M.P., Qi, Y., et al. (2007). Innate partnership of HLA-B and KIR3DL1 subtypes against HIV-1. Nat Genet **39**(6), 733–740.

Martin-Fontecha, A., Thomsen, L.L., et al. (2004). Induced recruitment of NK cells to lymph nodes provides IFN-gamma for T(H)1 priming. Nat Immunol **5**(12), 1260–1265.

Mocarski, E. S. and Tan Courcelle, C. (2001). Cytomegaloviruses and their replication. Fields Virology. D. M. Knipe and P. M. Howley. Philadelphia, Lippincott Williams & Wilkins. **2**, 2629-2673.

Orange, J.S. and Ballas, Z.K. (2006). Natural killer cells in human health and disease. Clin Immunol **118**(1), 1–10.

Orr, M.T., Sun, J.C., et al. (2009). Ly49H signaling through DAP10 is essential for optimal natural killer cell responses to mouse cytomegalovirus infection. J Exp Med **206**, 807–817.

Pinto, A.K. and Hill, A.B. (2005). Viral interference with antigen presentation to CD8+ T cells: lessons from cytomegalovirus. Viral Immunol **18**(3), 434–444.

Reddehase, M.J., Simon, C.O., et al. (2004). Stalemating a clever opportunist: lessons from murine cytomegalovirus. Hum Immunol **65**(5), 446–455.

Robbins, S.H., Bessou, G., et al. (2007). Natural killer cells promote early CD8 T cell responses against cytomegalovirus. PLoS Pathog **3**(8), e123.

Rodriguez, M., Sabastian, P., et al. (2004). Cmv1-independent antiviral role of NK cells revealed in murine cytomegalovirus-infected New Zealand White mice. J Immunol **173**(10), 6312–6318.

Roncarolo, M.G., Bigler, M., et al. (1991). Natural killer cell clones can efficiently process and present protein antigens. J Immunol **147**(3), 781–787.

Scalzo, A.A., Fitzgerald, N.A., et al. (1990). Cmv-1, a genetic locus that controls murine cytomegalovirus replication in the spleen. J Exp Med **171**(5), 1469–1483.

Scalzo, A.A., Fitzgerald, N.A., et al. (1992). The effect of the Cmv-1 resistance gene, which is linked to the natural killer cell gene complex, is mediated by natural killer cells. J Immunol **149**(2), 581–589.

Scalzo, A.A., Manzur, M., et al. (2005). NK gene complex haplotype variability and host resistance alleles to murine cytomegalovirus in wild mouse populations. Immunol Cell Biol **83**(2), 144–149.

Shellam, G.R., Allan, J.E., et al. (1981). Increased susceptibility to cytomegalovirus infection in beige mutant mice. Proc Natl Acad Sci U S A **78**(8), 5104–5108.

Shibuya, A., Campbell, D., et al. (1996). DNAM-1, a novel adhesion molecule involved in the cytolytic function of T lymphocytes. Immunity **4**(6), 573–581.

Sjolin, H., Tomasello, E., et al. (2002). Pivotal role of KARAP/DAP12 adaptor molecule in the natural killer cell-mediated resistance to murine cytomegalovirus infection. J Exp Med **195**(7), 825–834.

Smith, H.R., Heusel, J.W., et al. (2002). Recognition of a virus-encoded ligand by a natural killer cell activation receptor. Proc Natl Acad Sci U S A **99**(13), 8826–8831.

Stern, M., Elsasser, H., et al. (2008). The number of activating KIR genes inversely correlates with the rate of CMV infection/reactivation in kidney transplant recipients. Am J Transplant **8**(6), 1312–1317.

Stern-Ginossar, N., Elefant, N., et al. (2007). Host immune system gene targeting by a viral miRNA. Science **317**(5836), 376–381.

Su, H.C., Nguyen, K.B., et al. (2001). NK cell functions restrain T cell responses during viral infections. Eur J Immunol **31**(10), 3048–3055.

Tahara-Hanaoka, S., Shibuya, K., et al. (2004). Functional characterization of DNAM-1 (CD226) interaction with its ligands PVR (CD155) and nectin-2 (PRR-2/CD112). Int Immunol **16**(4), 533–538.

Tassi, I., Le Friec, G., et al. (2009). DAP10 associates with Ly49 receptors but contributes minimally to their expression and function in vivo. Eur J Immunol **39**(4), 1129–1135.

Tomasec, P., Braud, V.M., et al. (2000). Surface expression of HLA-E, an inhibitor of natural killer cells, enhanced by human cytomegalovirus gpUL40. Science **287**(5455), 1031.

Tomasec, P., Wang, E.C., et al. (2005). Downregulation of natural killer cell-activating ligand CD155 by human cytomegalovirus UL141. Nat Immunol **6**(2), 181–188.

Ulbrecht, M., Martinozzi, S., et al. (2000). Cutting edge: the human cytomegalovirus UL40 gene product contains a ligand for HLA-E and prevents NK cell-mediated lysis. J Immunol **164**(10), 5019–5022.

van Dommelen, S.L., Sumaria, N., et al. (2006). Perforin and granzymes have distinct roles in defensive immunity and immunopathology. Immunity **25**(5), 835–848.

Venema, H., van den Berg, A.P., et al. (1994). Natural killer cell responses in renal transplant patients with cytomegalovirus infection. J Med Virol **42**(2), 188–192.

Voigt, V., Forbes, C.A., et al. (2003). Murine cytomegalovirus m157 mutation and variation leads to immune evasion of natural killer cells. Proc Natl Acad Sci U S A **100**(23), 13483–13488.

Wang, E.C., McSharry, B., et al. (2002). UL40-mediated NK evasion during productive infection with human cytomegalovirus. Proc Natl Acad Sci U S A **99**(11), 7570–7575.

Welte, S.A., Sinzger, C., et al. (2003). Selective intracellular retention of virally induced NKG2D ligands by the human cytomegalovirus UL16 glycoprotein. Eur J Immunol **33**(1), 194–203.

Wu, J., Chalupny, N.J., et al. (2003). Intracellular retention of the MHC class I-related chain B ligand of NKG2D by the human cytomegalovirus UL16 glycoprotein. J Immunol **170**(8), 4196–4200.

Zhang, A.L., Colmenero, P., et al. (2007). Natural killer cells trigger differentiation of monocytes into dendritic cells. Blood **110**(7), 2484–2493.

Zingoni, A., Sornasse, T., et al. (2004). Cross-talk between activated human NK cells and CD4+ T cells via OX40-OX40 ligand interactions. J Immunol **173**(6), 3716–3724.

Zingoni, A., Sornasse, T., et al. (2005). NK cell regulation of T cell-mediated responses. Mol Immunol **42**(4), 451–454.

Chapter Thirty-Eight

38

Natural killer cells in allergy

Syh-Jae Lin, Yen-Chang Lee, Dah-Chin Yan, Ming-Ling Kuo

CHAPTER CONTENTS

The most beautiful thing we can experience is the mysterious.

Albert Einstein

ABSTRACT

The prevalence of allergic disease has dramatically increased worldwide in the past 20 years. Allergic diseases are characterized by abnormal inflammatory responses to allergens with a bias towards T-helper type-2 cytokine production. Natural killer (NK) cells have been conventionally associated with immune surveillance of tumour cells as well as anti-viral defence. However, recent advances in phenotypic and functional characterization of NK cells support their capacity to regulate the immune response to allergens. NK cells play a role in the pathogenesis of allergic inflammation through their interaction with dendritic cells (DCs) and T cells. They potentially are also involved in two of the most commonly encountered allergic diseases: asthma and atopic dermatitis (AD).

KEY WORDS

Dendritic cells (DCs), T cells, Asthma, Atopic dermatitis (AD)

Pathogenesis of allergic disease

Allergic diseases such as asthma and atopic dermatitis (AD) are characterized by abnormal inflammatory responses to allergens due to a T-helper type-1 (Th1)/T-helper type-1 (Th2) imbalance, with the bias towards Th2 cytokine production playing an important pathogenic role(Busse and Lemanske, 2001; Canonica, 2002; Kay, 2001b). T h2 cytokines such as IL-4, IL-13 enhance IgE production and IL-9 and IL-13 enhance mucus hypersecretion and airway hyperreactivity during an asthma attack (Robinson et al., 1992; Wills-Karp et al., 1998). IL-4, IL-9 and IL-13 also stimulate mast cell proliferation (Burd et al., 1995), while IL-5 contributes to eosinophilia (Kurup et al., 1997). Th2 cells promote IgE production by B cells by ligation of CD40 (Poulsen and Hummelshoj, 2007) and by secreting IL-4 and IL-13 (Gauchat et al., 1990; Wills-Karp et al., 1998). During the early phase of an IgE-dependent allergic reaction, the allergen binds to IgE on mast cells, which are then sensitized and release several preformed and newly formed mediators that result in vasodilation, smooth muscle constriction, mucus secretion and so on (Kay, 2001a).

The regulatory mechanisms that determine Th1/Th2 imbalance have been investigated extensively in recent years. IL-10-producing regulatory T cells (Akdis and Akdis, 2007; Akdis et al., 2004), invariant TCR-expressing natural killer (NK) T cells (Umetsu et al., 2007; Vijayanand et al., 2007) and dendritic cells (DCs)

(Lambrecht and Hammad, 2003) all have been implicated as pivotal regulators in the development of Th2 polarization.

Regulatory role of NK cells in allergy

NK cells are large granular lymphocytes that play a crucial role in innate immunity due to their cytotoxic ability to destroy virus-infected cells and tumour cells (Biron, 1997; Caligiuri, 2008; Orange and Ballas, 2006; Trinchieri, 1989). Older studies of NK cells in allergy are difficult to interpret as NK cells and regulatory NK T cells could not be clearly distinguished. With better characterization of NK cell subsets in recent years, substantial evidence points to a role of NK cells in regulating innate and adaptive immunity through the release of several cytokines, such as IL-5, IL-13, IFN-gamma, TNF-alpha, GM-CSF and so on (Cooper et al., 2001; Perussia et al., 2005), and through their intimate interaction with DCs. Perturbation of NK regulatory function may therefore affect Th1/Th2 responses and contribute to the pathogenesis of allergic disease (Figure 38.1).

NK subsets with regulatory function

Cooper et al. (2001) found that NK cells can be classified into $CD56^{bright}CD16^{-}$ or $CD56^{dim}CD16^{+}$ subsets, with distinct cytokine-secreting and surface receptor expression profiles: $CD56^{bright}CD16^{-}$ NK cells that secrete several cytokines such as IL-5, IL-13 and IFN-gamma may possess immunoregulatory function, whereas $CD56^{dim}CD16^{+}$ NK cells exhibit greater cytotoxicity.

There is accumulating evidence that NK cells can be classified into NK type-1 (NK1) and NK type-2 (NK2) cells similar to the concept of Th1 and Th2 cells. NK cells grown in IL-12 (NK1) produce IL-10 and

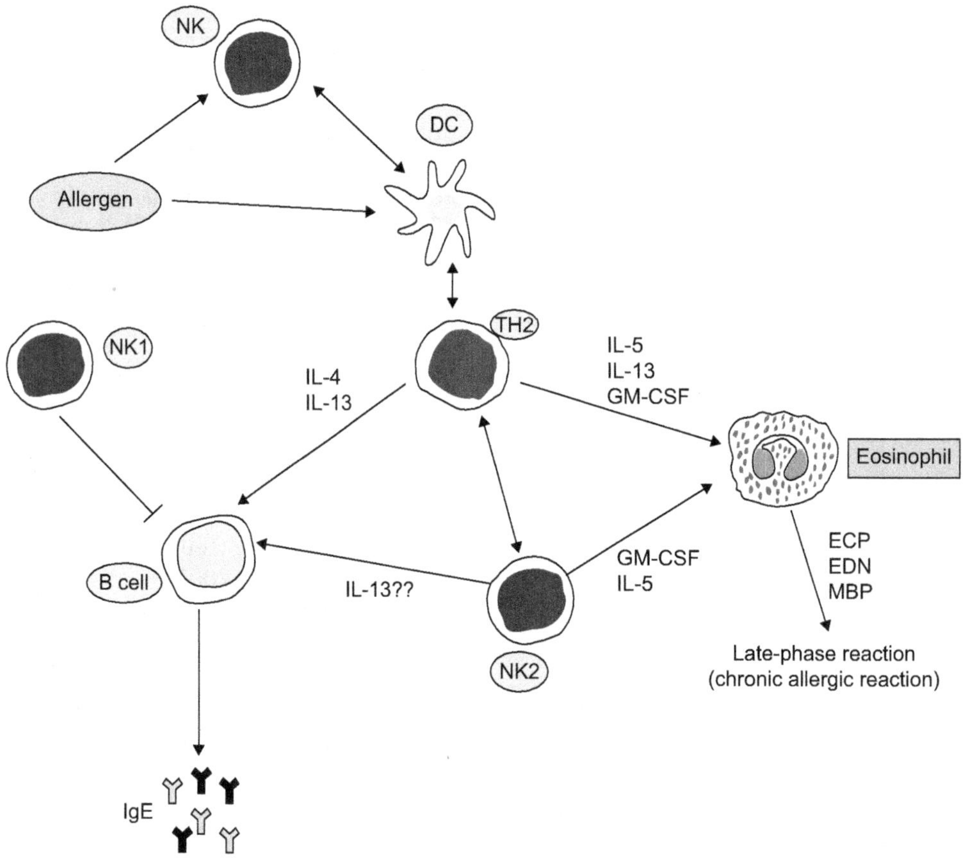

Figure 38.1 • Pivotal regulatory functions of NK cells in Th2-cell polarization and the development of allergic inflammation. Defective NK–DC crosstalk may lead to differentiation of Th2 cells. Different regulatory NK subsets may on one hand inhibit IgE production by B cells (NK1), and promote eosinophilic inflammation on the other (NK2). Abbreviations: DC: dendritic cells; ECP: eosinophil cationic protein; EDN: eosinophil-derived neurotoxin; GM-CSF: granulocyte-monocyte colony stimulating factor; MBP: major basic protein; NK cells: natural killer cells; NK1 and NK2: nature killer cell subsets according to Peritt et al. (1998) (see text and reference).

IFN-gamma, whereas NK cells grown in IL-4 (NK2) produce IL-5 and IL-13 (Deniz et al., 2002; Loza and Perussia, 2004; Peritt et al., 1998). The in vivo existence of IFN-gamma-secreting and IFN-gamma-non-secreting NK subsets has been demonstrated. Loza et al. (2002) reported that NK2 cells are immature NK cells, which can be differentiated into NK0 cells producing IL-13 and IFN-gamma, before they fully mature and turn into NK1 cells.

NK cells in the lymph node

The cytokine milieu in the local microenvironment surrounding NK cells might be pivotal for recruitment of inflammatory cells responsible for an allergic reaction. $CD56^{bright}CD16^{-}$ NK cells express CCR7 and CD62L, which allow their homing into lymph nodes where intimate contact with mature DCs occurs, and secrete large amounts of IFN-gamma that favours Th1 polarization (Frey et al., 1998). Fehniger et al. (2003) reported that $CD56^{bright}$ NK cells expressing high affinity IL-2R reside in parafollicular T cell areas of lymph nodes and are activated by T cells. Bajénoff et al. (2006), using static and real-time imaging, discovered that NK cells form a reactive but low mobility network in lymph nodes, where they can receive inflammatory signals, interact with DCs and regulate localized T cell responses.

NK–DC crosstalk in allergy

NK cells may modulate allergic responses through their interactions with DCs, which are important cells in initiating and shaping adaptive immune responses (Cooper et al., 2004). NK cells have been found in close contact with DCs in lesions of allergen-induced atopic eczema (Buentke et al., 2002). NK cells lyse immature DCs (Moretta et al., 2005, 2006), and this function appears to require pathogen-dependent activation of both NK cells and DCs (Deniz et al., 2008). Marcenaro et al. (2005) found that NK cells cultured in IL-12 but not IL-4 were able to induce DC maturation, and that NK cells cultured in IL-4 failed to kill immature DCs. Scordamaglia et al. (2008) reported that $CD56^{++}CD16^{+/-}$ NK cells were markedly deficient in the peripheral blood of patients with respiratory allergic diseases, and that the ability of their NK cells to promote DC maturation and kill immature DCs was reduced . IL-15, the main NK survival signal molecule secreted by mature DCs, can further enhance the interaction between NK cells and DCs (Ferlazzo et al., 2004).

NK cells and T cells

Activated NK cells produce IFN-gamma, which, in turn, activates macrophages to secrete IL-12. Both IFN-gamma and IL-12 contribute to the creation of a microenvironment that promotes Th1 while suppressing Th2 responses (Biron et al., 1999; Martin-Fontecha et al., 2004; Stephens et al., 2002). Recent evidence indicates that direct cell-to-cell contact between NK cells and T cells may result in IFN-gamma production and promote the differentiation of Th1 cells. Activated NK cells express MHC class-II, B7 and OX40, allowing them to interact with TCR, CD28 and OX40 on T cells, respectively, thereby influencing T-cell proliferation and cytokine production (Hanna et al., 2004; Zingoni et al., 2004, 2005). Munitz et al. (2007) reported that CD48 is critically involved in allergic eosinophilic airway inflammation. An interaction between 2B4 on NK cells and CD48 on T cells enhances proliferation of T cells and NK cells (Assarsson et al., 2004). Wingett and Nielson (2003) demonstrated that cAMP (cyclic AMP) enhancement of CD40L expression on T cells from asthmatic patients requires cell-to-cell contact with a $CD95^{+}$ NK cell subset.

Several lines of evidence support a key role of NK cells in control of allergen-specific T cell responses. Prior chlamydial infection could inhibit airway eosinophilic inflammation induced by allergen sensitization and challenge. In vivo depletion of NK cells abolished the inhibitory effect of chlamydial infection, whereas adoptive transfer of NK cells isolated from infected mice suppressed the allergic response (Han et al., 2008). More recently, NK cells secreting IL-10 have been shown to suppress both allergen- and antigen-induced T-cell proliferation and secretion of IL-13, similar to that observed for regulatory T cells (Deniz et al., 2008). Allergen-specific immunotherapy in humans was shown to decrease NK cell activity (Sin et al., 1996) and IFN-gamma-producing NK cells (Plewako et al., 2006).

NK cells and eosinophils

Eosinophils play an important role in chronic allergic disorders, such as asthma, dermatitis, and rhinitis (Busse and Lemanske, 2001; Kay, 2001a; Simon et al., 2004). It is generally accepted that eosinophils are the primary effector cells in allergic reactions and that they may cause long-term tissue damage and tissue remodelling by releasing leukotrienes, platelet-activating factor and toxic basic proteins (Noguchi et al., 1992). An alternative role for eosinophils in promoting clearance of debris, suggesting that they are important not so much for causing damage as responding to it, has also been suggested (Ochkur et al., 2007; Jacobsen, 2007a,b). IL-5 and GM-CSF secreted by NK cells may enhance the proliferation of eosinophils in the bone marrow. IL-5 secreted by NK cells in peritoneal lavages contributed to the infiltration of eosinophils after challenge by ragweed Ag in sensitized mice (Walker et al., 1998). Inhalation of IL-5 by asthmatic patients led to increased accumulation of eosinophils in bronchoalveolar lavage (BAL) fluid (Shi et al., 1997). Depletion of NK but not NKT cells results in decreased IL-5 production and infiltration of eosinophils in allergen-induced airway inflammation (Korsgren et al., 1999). Administration of anti-IL-5 antibodies downregulates eosinophilia in mice exposed to antigen in mice and monkeys (Kurup et al., 1997; Mauser et al., 1995).

In IL-5 gene knockout mice, eosinophilia does not occur after challenge with allergen (Trifilieff et al., 2001), suggesting that IL-5 enhances the proliferation, survival and activation of eosinophils. Transmigration through basement membrane of airway epithelial cells results in activation of eosinophils (Dallaire et al., 2002; Kato et al., 2002). A recent study demonstrated that BAL fluid eosinophils are correlated with NK cells in patients with eosinophilic pneumonia (Papakosta et al., 2009). NK cells may play an important role in establishing a microenvironment for eosinophil activation by secreting cytokines and chemokines. However, detailed understanding of the interaction between NK cells and eosinophils remains to be established.

Although NK cells have been shown to inhibit IgE production by B cells (Kimata and Saxon, 1987), NK cells do not express high affinity IgE receptors with only few low affinity IgE receptors on their surface (Rosenthal et al., 1984). Aktas et al. (2005) shows that NK1 cells, but not NK2 cells, significantly inhibited IL-4- and soluble CD40-ligand-stimulated IgE production by B cells. On the other hand, Arase et al. (2003) demonstrated that NK cells can be activated with IgE through Fcγ RIII to produce large quantities of IFN-gamma, TNF-alpha, GM-CSF and MIF-alpha, suggesting NK cells may contribute to IgE-mediated allergic inflammation.

NK cells and asthma

NK cells make up about 25% of lymphoid cells in the lung, in contrast to 5–15% in the peripheral blood (Westermann and Pabst, 1992). The abundance of NK cells in the lung gives them a potential role in pulmonary innate immune response, including asthma. A study in mice indicates that depletion of NK cells before immunization inhibits infiltration of eosinophils and T cells into the lung (BAL) (Korsgren et al., 1999). Increased NK cells in lung parenchyma was observed in rats following challenge with ovalbumin (Schuster et al., 2000). Walker et al. (1998) reported that NK cell counts increased significantly after ragweed Ag challenge in peritoneal lavages of mice. IFN-gamma potentiated IL-13-induced lung injury has been linked to higher levels of NK cells (Ford et al., 2001).

The pathogenic role of NK cells in asthma in humans remained unclear. Some authors showed that patients with asthma had significantly stronger NK activity due mainly to increased cell numbers (Krejsek et al., 1998; Timonen and Stenius-Aarniala, 1985). Others also reported that NK activity was increased in asthmatics (Di Lorenzo et al., 2001; Jira et al., 1988). On the other hand, Chou et al. (1999) found no difference in NK cell number, cytotoxicity, and IFN-gamma production between asthmatics and controls.

Circulating NK cell changes may not reflect the recruitment of NK cells in inflamed asthmatic airways. Adhesion molecule expression on NK cells may determine their capability to undergo transendothelial migration and enter the inflammatory lesion. L-selectin is required for the rolling process (Keramidaris et al., 2001; Symon et al., 1999), while subsequent adhesion and transmigration are dependant on leukocyte function-associated antigen and intercellular adhesion molecule-1 (ICAM-1) interactions and the very late activation antigen and vascular cell adhesion molecule-1 pathways (Allavena et al., 1996; Fogler et al., 1996; Roebuck and Finnegan, 1999). We observed an increased proportion of circulating NK cells with decreased L-selectin and ICAM-1 expression in asthmatic children during acute exacerbation compared to children in a stable condition (Lin et al., 2003).

More recently, an increased frequency of IL-4-producing NK2 cells in peripheral blood has been demonstrated in asthmatics (Wei et al., 2005). Perturbation of certain NK subsets in asthma may result in impaired capability to promote and maintain appropriate Th1 through their interaction through DCs (Scordamaglia et al., 2008). Considering that NK cells include a number of cell subsets with similar morphological appearances but different function and phenotype, further studies will be needed to elucidate the role of NK cells in allergic airway inflammation.

NK cells and atopic dermatitis

Atopic dermatitis (AD) is a chronic inflammatory skin disorder characterized by inflammatory cell infiltration, abnormalities in both cell-mediated and humoural immunity and increased susceptibility to cutaneous infection (Leung et al., 2004; Wollenberg and Klein, 2007). Peripheral blood mononuclear cells (PBMC) and acute skin lesions of patients with AD are characterized by increased IL-4 and IL-13, but decreased IFN-gamma production, reflecting a Th2-biased response. A variety of defects in the innate immune system also contribute to the development and severity of AD (De Benedetto, et al. 2009).

NK cell dysfunction has been reported in AD, as indicated by a decrease in NK cell numbers and cytotoxicity against the standard NK-sensitive target cells (Hall et al., 1985; Wehrmann et al., 1990). Nevertheless, conflicting data have been also reported (Bouloc et al., 2000; Viander et al., 1982). More recent studies have confirmed that the frequencies of circulating NK cells are profoundly reduced in AD patients (Aktas et al., 2005; Katsuta et al., 2006). NK cells in AD patients display a defective ability to sustain TNF-alpha and IFN-gamma, but not IL-4 production after in vitro stimulation and undergo apoptosis following contact with monocytes (Katsuta et al., 2006). IL-15 secretion by Staphylococcal enterotoxin B (SEB)-treated PBMC, as well as membrane-bound IL-15 expression by SEB-treated monocytes are significantly lower in AD patients than in controls (Ong et al., 2002).

Concluding remarks

The role of NK cells in allergy remains elusive. NK cells may facilitate the development of eosinophilic airway inflammation and participate in IgE-mediated allergic response. On the other hand, deficiency of certain NK subsets with regulatory function may promote a highly polarized Th2 response characteristic of allergy. Further elucidation of the pathogenic mechanisms of NK cells in allergic diseases may provide insight in designing novel therapeutic approaches.

References

Akdis, M. and Akdis, C.A. (2007). Mechanisms of allergen-specific immunotherapy. J Allergy Clin Immunol **119**, 780–791.

Akdis, M., Verhagen, J., Taylor, A., Karamloo, F., Karagiannidis, C., Crameri, R., Thunberg, S., Deniz, G., Valenta, R., Fiebig, H., Kegel, C., Disch, R., Schmidt-Weber, C.B., Blaser, K. and Akdis, C.A. (2004). Immune responses in healthy and allergic individuals are characterized by a fine balance between allergen-specific T regulatory 1 and T helper 2 cells. J Exp Med **199**, 1567–1575.

Aktas, E., Akdis, M., Bilgic, S., Disch, R., Falk, C.S., Blaser, K., Akdis, C. and Deniz, G. (2005). Different natural killer (NK) receptor expression and immunoglobulin E (IgE) regulation by NK1 and NK2 cells. Clin Exp Immunol **140**, 301–309.

Allavena, P., Bianchi, G., Paganin, C., Giardina, G. and Mantovani, A. (1996). Regulation of adhesion and transendothelial migration of natural killer cells. Nat Immun **15**, 107–116.

Arase, N., Arase, H., Hirano, S., Yokosuka, T., Sakurai, D. and Saito, T. (2003). IgE-mediated activation of NK cells through Fc gamma RIII. J Immunol **170**, 3054–3058.

Assarsson, E., Kambayashi, T., Schatzle, J. D., Cramer, S.O., von Bonin, A., Jensen, P.E., Ljunggren, H.G. and Chambers, B.J. (2004). NK cells stimulate proliferation of T and NK cells through 2B4/CD48 interactions. J Immunol **173**, 174–180.

Bajénoff, M., Breart, B., Huang, A.Y.C., Qi, H., Cazareth, J., Braud, V.M., Germain, R.N. and Glaichenhaus, N. (2006). Natural killer cell behavior in lymph nodes revealed by static and real-time imaging. J Exp Med **203**, 619–631.

Biron, C.A. (1997). Activation and function of natural killer cell responses during viral infections. Curr Opin Immunol **9**, 24–34.

Biron, C.A., Nguyen, K.B., Pien, G.C., Cousens, L.P. and Salazar-Mather, T.P. (1999). Natural killer cells in antiviral defense: function and regulation by innate cytokines. Annu Rev Immunol **17**, 189–220.

Bouloc, A., Charue, D., Nikolova, M. and Bensussan, A. (2000). No modulation of circulating natural killer cell and natural killer receptor bearing memory T cell subsets in patients with atopic dermatitis. J Invest Dermatol **115**, 1160–1162.

Buentke, E., Heffler, L.C., Wilson, J.L., Wallin, R.P., Lofman, C., Chambers, B. J., Ljunggren, H.G. and Scheynius, A. (2002). Natural killer and dendritic cell contact in lesional atopic dermatitis skin—Malassezia-influenced cell interaction. J Invest Dermatol **119**, 850–857.

Burd, P.R., Thompson, W.C., Max, E.E. and Mills, F.C. (1995). Activated mast cells produce interleukin 13. J Exp Med **181**, 1373–1380.

Busse , W.W. and Lemanske, R.F. Jr. (2001). Asthma. N Engl J Med **344**, 350–362.

Caligiuri, M.A. (2008). Human natural killer cells. Blood **112**, 461–469.

Canonica, G.W. (2002). Introduction to nasal and pulmonary allergy cascade. Allergy **57**(Suppl. 75), 8–12.

Chou, C.C., Huang, M.S., Hsieh, K.H. and Chiang, B.L. (1999). Reduced IL-12 level correlates with decreased IFN-gamma secreting T cells but not natural killer cell activity in asthmatic children. Ann Allergy Asthma Immunol **82**, 479–484.

Cooper, M.A., Fehniger, T.A., Fuchs, A., Colonna, M. and Caligiuri, M.A. (2004). NK cell and DC interactions. Trends Immunol **25**, 47–52.

Cooper, M.A., Fehniger, T.A., Turner, S. C., Chen, K.S., Ghaheri, B.A., Ghayur, T., Carson, W.E. and Caligiuri, M.A. (2001). Human natural killer cells: a unique innate immunoregulatory role for the CD56(bright) subset. Blood **97**, 3146–3151.

Dallaire, M.J., Ferland, C., Lavigne, S., Chakir, J. and Laviolette, M. (2002). Migration through basement membrane modulates eosinophil expression of CD44. Clin Exp Allergy **32**, 898–905.

De Benedetto, A., Agnihothri, R., McGirt, L.Y., Bankova, L.G. and Beck, L.A. (2009). Atopic dermatitis: a disease caused by innate immune defects? J Invest Dermatol **129**, 14–30.

Deniz, G., Akdis, M., Aktas, E., Blaser, K. and Akdis, C.A. (2002). Human NK1 and NK2 subsets determined by purification of IFN-gamma-secreting and IFN-gamma-nonsecreting NK cells. Eur J Immunol **32**, 879–884.

Deniz, G., Erten, G., Kucuksezer, U.C., Kocacik, D., Karagiannidis, C., Aktas, E., Akdis, C.A. and Akdis, M. (2008). Regulatory NK cells suppress antigen-specific T cell responses. J Immunol **180**, 850–857.

Di Lorenzo, G., Esposito Pellitteri, M., Drago, A., Di Blasi, P., Candore, G.,

Balistreri, C., Listi, F. and Caruso, C. (2001). Effects of in vitro treatment with fluticasone propionate on natural killer and lymphokine-induced killer activity in asthmatic and healthy individuals. Allergy **56**, 323–327.

Fehniger, T.A., Cooper, M.A., Nuovo, G.J., Cella, M., Facchetti, F., Colonna, M. and Caligiuri, M.A. (2003). CD56bright natural killer cells are present in human lymph nodes and are activated by T cell-derived IL-2: a potential new link between adaptive and innate immunity. Blood **101**, 3052–3057.

Ferlazzo, G., Pack, M., Thomas, D., Paludan, C., Schmid, D., Strowig, T., Bougras, G., Muller, W.A., Moretta, L. and Munz, C. (2004). Distinct roles of IL-12 and IL-15 in human natural killer cell activation by dendritic cells from secondary lymphoid organs. Proc Natl Acad Sci U S A **101**, 16606–16611.

Fogler, W.E., Volker, K., McCormick, K.L., Watanabe, M., Ortaldo, J. R. and Wiltrout, R.H. (1996). NK cell infiltration into lung, liver, and subcutaneous B16 melanoma is mediated by VCAM-1/VLA-4 interaction. J Immunol **156**, 4707–4714.

Ford, J.G., Rennick, D., Donaldson, D.D., Venkayya, R., McArthur, C., Hansell, E., Kurup, V.P., Warnock, M. and Grunig, G. (2001). Il-13 and IFN-gamma: interactions in lung inflammation. J Immunol **167**, 1769–1777.

Frey, M., Packianathan, N.B., Fehniger, T.A., Ross, M.E., Wang, W.C., Stewart, C.C., Caligiuri, M.A. and Evans, S.S. (1998). Differential expression and function of L-selectin on CD56bright and CD56dim natural killer cell subsets. J Immunol **161**, 400–408.

Gauchat, J.F., Lebman, D.A., Coffman, R. L., Gascan, H. and de Vries, J.E. (1990). Structure and expression of germline epsilon transcripts in human B cells induced by interleukin 4 to switch to IgE production. J Exp Med **172**, 463–473.

Hall, T.J., Rycroft, R. and Brostoff, J. (1985). Decreased natural killer cell activity in atopic eczema. Immunology **56**, 337–344.

Han, X., Fan, Y., Wang, S., Jiao, L., Qiu, H. and Yang, X. (2008). NK cells contribute to intracellular bacterial infection-mediated inhibition of allergic responses. J Immunol **180**, 4621–4628.

Hanna, J., Gonen-Gross, T., Fitchett, J., Rowe, T., Daniels, M., Arnon, T.I., Gazit, R., Joseph, A., Schjetne, K.W., Steinle, A., Porgador, A., Mevorach, D., Goldman-Wohl, D., Yagel, S., LaBarre, M.J., Buckner, J.H. and Mandelboim, O. (2004). Novel APC-like properties of human NK cells directly regulate T cell activation. J Clin Invest **114**, 1612–1623.

Jacobsen, E.A., Ochkur, S.I., Lee, N.A. and Lee, J.J. (2007a). Eosinophils and asthma. Curr Allergy Asthma Rep **7**, 18–26.

Jacobsen, E.A., Taranova, A.G., Lee, N. A. and Lee, J.J. (2007b). Eosinophils: singularly destructive effector cells or purveyors of immunoregulation?. J Allergy Clin Immunol **119**, 1313–1320.

Jira, M., Antosova, E., Vondra, V., Strejcek, J., Mazakova, H. and Prazakova, J. (1988). Natural killer and interleukin-2 induced cytotoxicity in asthmatics. I. Effect of acute antigen-specific challenge. Allergy **43**, 294–298.

Kato, Y., Fujisawa, T., Shibano, M., Saito, T., Gatto, W., Kamiya, H., Hirai, K., Sumida, M. and Yoshie, O. (2002). Airway epithe-lial cells promote transmigration of eosinophils in a new three-dimensional chemotaxis model. Clin Exp Allergy **32**, 889–897.

Katsuta, M., Takigawa, Y., Kimishima, M., Inaoka, M., Takahashi, R. and Shiohara, T. (2006). NK cells and gamma delta+ T cells are phenotypically and functionally defective due to preferential apoptosis in patients with atopic dermatitis. J Immunol **176**, 7736–7744.

Kay, A.B. (2001a). Allergy and allergic diseases. First of two parts. N Engl J Med **344**, 30–37.

Kay, A.B. (2001b). Allergy and allergic diseases. Second of two parts. N Engl J Med **344**, 109–113.

Keramidaris, E., Merson, T.D., Steeber, D. A., Tedder, T.F. and Tang, M.L. (2001). L-selectin and intercellular adhesion molecule 1 mediate lymphocyte migration to the inflamed airway/lung during an allergic inflammatory response in an animal model of asthma. J Allergy Clin Immunol **107**, 734–738.

Kimata, H. and Saxon, A. (1987). Natural killer cell interaction with IgE in the control of ongoing human IgE synthesis. Int Arch Allergy Appl Immunol **82**, 419–421.

Korsgren, M., Persson, C.G.A., Sundler, F., Bjerke, T., Hansson, T., Chambers, B.J., Hong, S., Van Kaer, L., Ljunggren, H.G. and Korsgren, O. (1999). Natural killer cells determine development of allergen-induced eosinophilic airway inflammation in mice. J Exp Med **189**, 553–562.

Krejsek, J., Kral, B., Vokurkova, D., Derner, V., Touskova, M., Parakova, Z. and Kopecky, O. (1998). Decreased peripheral blood gamma delta T cells in patients with bronchial asthma. Allergy **53**, 73–77.

Kurup, V.P., Murali, P.S., Guo, J., Choi, H., Banerjee, B., Fink, J.N. and Coffman, R.L. (1997). Anti-interleukin (IL)-4 and -IL-5 antibodies downregulate IgE and eosinophilia in mice exposed to *Aspergillus* antigens. Allergy **52**, 1215–1221.

Lambrecht, B.N. and Hammad, H. (2003). Taking our breath away: dendritic cells in the pathogenesis of asthma. Nat Rev Immunol **3**, 994–1003.

Leung, D.Y., Boguniewicz, M., Howell, M. D., Nomura, I. and Hamid, Q.A. (2004). New insights into atopic dermatitis. J Clin Invest **113**, 651–657.

Lin, S.J., Chang, L.Y., Yan, D.C., Huang, Y.J., Lin, T.J. and Lin, T.Y. (2003). Decreased intercellular adhesion molecule-1 (CD54) and L-selectin (CD62L) expression on peripheral blood natural killer cells in asthmatic children with acute exacerbation. Allergy **58**, 67–71.

Loza, M.J. and Perussia, B. (2004). Differential regulation of NK cell proliferation by type I and type II IFN. Int Immunol **16**, 23–32.

Loza, M.J., Zamai, L., Azzoni, L., Rosati, E. and Perussia, B. (2002). Expression of type 1 (interferon gamma) and type 2 (interleukin-13, interleukin-5) cytokines at distinct stages of natural killer cell differentiation from progenitor cells. Blood **99**, 1273–1281.

Marcenaro, E., Chiesa, M.D., Bellora, F., Parolini, S., Millo, R., Moretta, L. and Moretta, A. (2005). IL-12 or IL-4 prime human NK cells to mediate functionally divergent interactions with dendritic cells or tumors. J Immunol **174**, 3992–3998.

Martin-Fontecha, A., Thomsen, L.L., Brett, S., Gerard, C., Lipp, M., Lanzavecchia, A. and Sallusto, F. (2004). Induced recruitment of NK cells to lymph nodes provides IFN-gamma for T(H)1 priming. Nat Immunol **5**, 1260–1265.

Mauser , P.J., Pitman, A.M., Fernandez, X., Foran, S.K., Adams, G.K. III, Kreutner, W., Egan, R.W. and Chapman, R.W. (1995). Effects of an antibody to interleukin-5 in a monkey model of asthma. Am J Respir Crit Care Med **152**, 467–472.

Moretta, A., Marcenaro, E., Sivori, S., Della Chiesa, M., Vitale, M. and Moretta, L. (2005). Early liaisons between cells of the innate immune system in inflamed peripheral tissues. Trends Immunol **26**, 668–675.

Moretta, L., Ferlazzo, G., Bottino, C., Vitale, M., Pende, D., Mingari, M.C.

and Moretta, A. (2006). Effector and regulatory events during natural killer–dendritic cell interactions. Immunol Rev **214**, 219–228.

Munitz, A., Bachelet, I., Finkelman, F. D., Rothenberg, M.E. and Levi-Schaffer, F. (2007). CD48 is critically involved in allergic eosinophilic airway inflammation. Am J Respir Crit Care Med **175**, 911–918.

Noguchi, H., Kephart, G.M., Colby, T. V. and Gleich, G.J. (1992). Tissue eosinophilia and eosinophil degranulation in syndromes associated with fibrosis. Am J Pathol **140**, 521–528.

Ochkur, S.I., Jacobsen, E.A., Protheroe, C. A., Biechele, T.L., Pero, R.S., McGarry, M.P., Wang, H., O'Neill, K.R., Colbert, D. C., Colby, T.V., Shen, H., Blackburn, M. R., Irvin, C.C., Lee, J.J. and Lee, N.A. (2007). Coexpression of IL-5 and eotaxin-2 in mice creates an eosinophil-dependent model of respiratory inflammation with characteristics of severe asthma. J Immunol **178**, 7879–7889.

Ong, P.Y., Hamid, Q.A., Travers, J. B., Strickland, I., Al Kerithy, M., Boguniewicz, M. and Leung, D. Y. (2002). Decreased IL-15 may contribute to elevated IgE and acute inflammation in atopic dermatitis. J Immunol **168**, 505–510.

Orange, J.S. and Ballas, Z.K. (2006). Natural killer cells in human health and disease. Clin Immunol **118**, 1–10.

Papakosta, D., Manika, K., Kyriazis, G., Kontakiotis, T., Gioulekas, D., Polyzoni, T., Bouros, D. and Patakas, D. (2009). Bronchoalveolar lavage fluid eosinophils are correlated to natural killer cells in eosinophilic pneumonias. Respiration **78**, 177–184.

Peritt, D., Robertson, S., Gri, G., Showe, L., Aste-Amezaga, M. and Trinchieri, G. (1998). Differentiation of human NK cells into NK1 and NK2 subsets. J Immunol **161**, 5821–5824.

Perussia, B., Chen, Y. and Loza, M.J. (2005). Peripheral NK cell phenotypes: multiple changing of faces of an adapting, developing cell. Mol Immunol **42**, 385–395.

Plewako, H., Wosinska, K., Arvidsson, M., Bjorkander, J., Hakansson, L. and Rak, S. (2006). Production of interleukin-12 by monocytes and interferon-gamma by natural killer cells in allergic patients during rush immunotherapy. Ann Allergy Asthma Immunol **97**, 464–468.

Poulsen, L.K. and Hummelshoj, L. (2007). Triggers of IgE class switching and allergy development. Ann Med **39**, 440–456.

Robinson, D.S., Hamid, Q., Ying, S., Tsicopoulos, A., Barkans, J., Bentley, A.M., Corrigan, C., Durham, S. R. and Kay, A.B. (1992). Predominant TH2-like bronchoalveolar T-lymphocyte population in atopic asthma. N Engl J Med **326**, 298–304.

Roebuck, K.A. and Finnegan, A. (1999). Regulation of intercellular adhesion molecule-1 (CD54) gene expression. J Leukoc Biol **66**, 876–888.

Rosenthal, K.L., Ishizaka, T., Befus, D., Dennert, G., Hengartner, H. and Bienenstock, J. (1984). Expression of IgE receptors and histamine in cloned natural killer cell lines. J Immunol **133**, 642–646.

Schuster, M., Tschernig, T., Krug, N. and Pabst, R. (2000). Lymphocytes migrate from the blood into the bronchoalveolar lavage and lung parenchyma in the asthma model of the brown Norway rat. Am J Respir Crit Care Med **161**, 558–566.

Scordamaglia, F., Balsamo, M., Scordamaglia, A., Moretta, A., Mingari, M.C., Canonica, G.W., Moretta, L. and Vitale, M. (2008). Perturbations of natural killer cell regulatory functions in respiratory allergic diseases. J Allergy Clin Immunol **121**, 479–485.

Shi, H., Qin, S., Huang, G., Chen, Y., Xiao, C., Xu, H., Liang, G., Xie, Z., Qin, X., Wu, J., Li, G. and Zhang, C. (1997). Infiltration of eosinophils into the asthmatic airways caused by interleukin 5. Am J Respir Cell Mol Biol **16**, 220–224.

Simon, D., Braathen, L.R. and Simon, H.U. (2004). Eosinophils and atopic dermatitis. Allergy **59**, 561–570.

Sin, B., Misirligil, Z., Aybay, C., Gurbuz, L. and Imir, T. (1996). Effect of allergen specific immunotherapy (IT) on natural killer cell activity (NK), IgE, IFN-gamma levels and clinical response in patients with allergic rhinitis and asthma. J Investig Allergol Clin Immunol **6**, 341–347.

Stephens, R., Eisenbarth, S.C. and Chaplin, D.D. (2002). T helper type 1 cells in asthma: friend or foe? Curr Opin Allergy Clin Immunol **2**, 31–37.

Symon, F.A., McNulty, C.A. and Wardlaw, A.J. (1999). P- and L-selectin mediate binding of T cells to chronically inflamed human airway endothelium. Eur J Immunol **29**, 1324–1333.

Timonen, T. and Stenius-Aarniala, B. (1985). Natural killer cell activity in asthma. Clin Exp Immunol **59**, 85–90.

Trifilieff, A., Fujitani, Y., Coyle, A.J., Kopf, M. and Bertrand, C. (2001). IL-5 deficiency abolishes aspects of airway remodelling in a murine model of lung inflammation. Clin Exp Allergy **31**, 934–942.

Trinchieri, G. (1989). Biology of natural killer cells. Adv Immunol **47**, 187–376.

Umetsu, D.T., Meyer, E.H. and DeKruyff, R.H. (2007). Natural killer T cells regulate the development of asthma. Int Rev Immunol **26**, 121–140.

Viander, M., Uksila, J., Lassila, O. and Jansen, C.T. (1982). Natural killer cell activity in atopic dermatitis. Arch Dermatol Res **274**, 283–288.

Vijayanand, P., Seumois, G., Pickard, C., Powell, R.M., Angco, G., Sammut, D., Gadola, S.D., Friedmann, P.S. and Djukanovic, R. (2007). Invariant natural killer T cells in asthma and chronic obstructive pulmonary disease. N Engl J Med **356**, 1410–1422.

Walker, C., Checkel, J., Cammisuli, S., Leibson, P.J. and Gleich, G.J. (1998). IL-5 Production by NK Cells contributes to eosinophil infiltration in a mouse model of allergic inflammation. J Immunol **161**, 1962–1969.

Wehrmann, W., Reinhold, U., Kukel, S., Franke, N., Uerlich, M. and Kreysel, H.W. (1990). Selective alterations in natural killer cell subsets in patients with atopic dermatitis. Int Arch Allergy Appl Immunol **92**, 318–322.

Wei, H., Zhang, J., Xiao, W., Feng, J., Sun, R. and Tian, Z. (2005). Involvement of human natural killer cells in asthma pathogenesis: natural killer 2 cells in type 2 cytokine predominance. J Allergy Clin Immunol **115**, 841–847.

Westermann, J. and Pabst, R. (1992). Distribution of lymphocyte subsets and natural killer cells in the human body. J Mol Med **70**, 539–544.

Wills-Karp, M., Luyimbazi, J., Xu, X., Schofield, B., Neben, T.Y., Karp, C.L. and Donaldson, D.D. (1998). Interleukin-13: central mediator of allergic asthma. Science **282**, 2258–2261.

Wingett, D. and Nielson, C.P. (2003). Divergence in NK cell and cyclic AMP regulation of T cell CD40L expression in asthmatic subjects. J Leukoc Biol **74**, 531–541.

Wollenberg, A. and Klein, E. (2007). Current aspects of innate and adaptive immunity in atopic dermatitis. Clin Rev Allergy Immunol **33**, 35–44.

Zingoni, A., Sornasse, T., Cocks, B.G., Tanaka, Y., Santoni, A. and Lanier, L.L. (2004). Cross-talk between activated human NK cells and CD4+ T cells via OX40–OX40 ligand interactions. J Immunol **173**, 3716–3724.

Zingoni, A., Sornasse, T., Cocks, B.G., Tanaka, Y., Santoni, A. and Lanier, L. L. (2005). NK cell regulation of T cell-mediated responses. Mol Immunol **42**, 451–454.

Chapter Thirty-Nine

39

Natural killer cells in wound healing

Jussi Liippo, Mervi Toriseva, Veli-Matti Kähäri

CHAPTER CONTENTS

New ideas usually need new facts.

James D. Watson

ABSTRACT

Physiologic wound healing is a complex and highly coordinated biological process, in which a number of different cell types participate in a tightly orchestrated manner to repair the damaged tissue. Wound healing is divided into three major phases of (1) hemostasis and inflammation, (2) re-epithelialization and granulation tissue formation, and (3) tissue remodelling. In general, the role of the immune system in wound repair is not restricted to combating the microbe invasion at the site of injury, but it also participates in the removal of the damaged tissue and contributes to the healing process.

It is evident, that natural killer (NK) cells regulate the early onset and resolution of the inflammatory phase in wound repair and may also contribute to other crucial events in the normal wound healing process, as in re-epithelialization, angiogenesis, granulation tissue formation, and remodelling.

KEY WORDS

Skin, Wound, Inflammation, T-cell, NK cell

Wound healing

Epithelium lines the entire human body and protects the internal organs from contact with the outside environment. Skin, oral and gastrointestinal mucosa and respiratory epithelium serve both as the mechanical support and as immunological organs that protect the body not only from loss of water, as is the case with skin, but also from various pathogens and harmful material. Thus, it is not surprising that the injuries of the epithelial tissues are promptly repaired, resulting in integration of new tissue into the surrounding intact tissue.

In general, wound healing in different epithelial tissues occurs in a similar manner in spite of the differences in their physical and chemical environment. As the relatively thick and stratified epidermal layer of human skin protects the body from dehydration, and is continuously susceptible to mechanical tension and extreme environmental temperatures, the mucosal epithelia face a moist exterior rich in non-pathogenic and often also pathogenic microbes. In addition, for example in the mouth, the epithelium is constantly rinsed by saliva, which contains a variety of factors that can affect tissue repair. The wound healing process in different tissues also varies, for instance in terms of the proteolytic profile of the cells, and this may also be one explanation why mucosal wounds heal faster than skin wounds. However, the fundamental characteristics of wound repair, that is, repair of the epithelial gap and regeneration of underlying connective tissue, are similar for different tissues (Häkkinen et al., 2000; Martin, 1997; Sacco et al., 2004; Tarnawski, 2005). The following text gives an overview of cutaneous wound healing with notes on certain specific properties of healing in other tissues.

Phases of wound repair

Wound healing is divided into three major phases: (1) hemostasis and inflammation, (2) re-epithelialization and granulation tissue formation, and (3) tissue remodelling (Clark, 1995). These phases are functionally and histologically distinct, but they overlap temporally and the complete healing of wounds requires carefully orchestrated communication between the various cell types in distinct tissue compartments involved in different phases of the repair process. All these events are tightly regulated by a variety of growth factors and cytokines. Moreover, cell–cell contacts and contact of cells with extracellular matrix (ECM) play a pivotal role in the regulation of cell behaviour during wound healing. Figure 39.1 illustrates the events in cutaneous wound healing and cell types involved in these events.

Hemostasis and inflammation

Cutaneous injury extending to the dermal layer, which contains the vascular network, results in blood vessel disruption and bleeding to the open wound. The process of hemostasis is rapidly initiated by vasoconstriction of arterial vessels, followed by blood clotting. Blood clot formation is initiated by activation, adhesion and aggregation of platelets. Simultaneously, activation of the cascade of coagulation factors on the damaged cells and platelets leads to cleavage of fibrinogen by thrombin and generation of insoluble fibrin fibres. Fibrin binds to platelets to form a clot, which serves as a physical plug to stop bleeding. The fibrin clot also includes plasma fibronectin and vitronectin and serves as a provisional matrix for cell migration. Fragments of coagulation factors, activated complement components, and growth factors derived from devitalized cells and activated platelets within blood clot function act as chemoattractants for inflammatory cells, keratinocytes (KCs), fibroblasts and endothelial cells. As wound healing progresses, provisional matrix is replaced by granulation tissue in the dermal compartment and degraded by proteinases, plasmin and matrix metalloproteinases (MMPs), from the way of invading KCs, and it is finally removed from the wound site with the scab (Clark, 1995).

The inflammatory phase of wound healing involves activation of both innate and adaptive immunity, phagocytosis, the function of the complement system, cytotoxicity

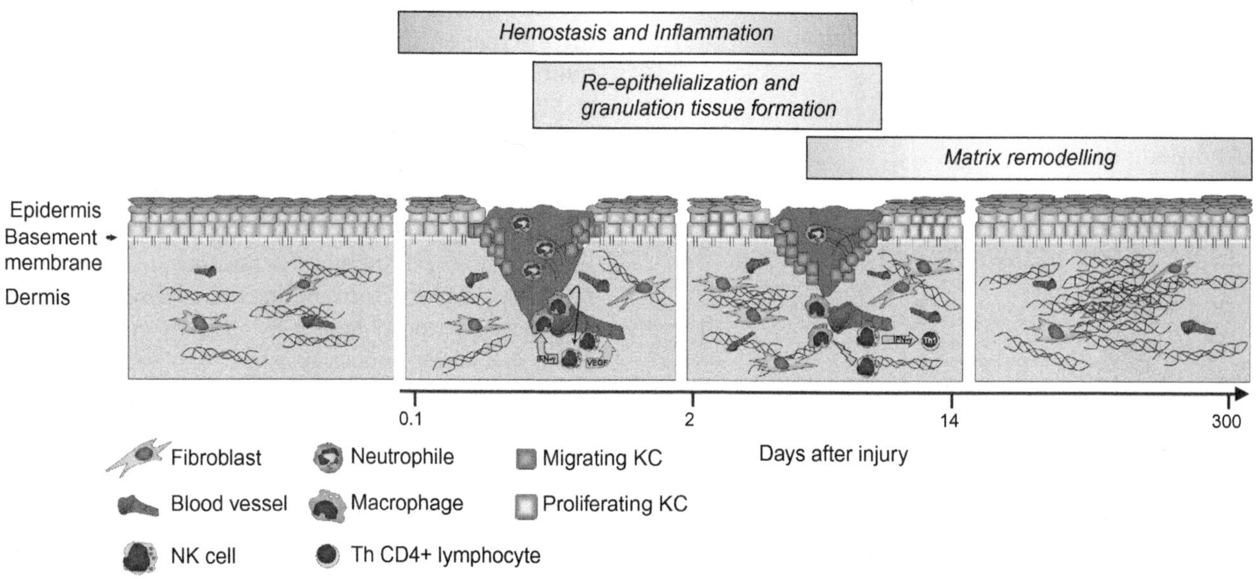

Figure 39.1 • The phases of cutaneous wound healing. The intact skin constitutes epidermis, which is formed by stratified layers of KCs of variable differentiation stages, and of dermis rich in collagenous ECM, fibroblasts and blood vessels. Epidermis and dermis are separated by basement membrane, a specialized layer of ECM. Injury instantly evokes a cascade of events initiated by hemostasis. A variety of factors secreted from blood clot and injured cells attract and activate inflammatory response with neutrophils and macrophages as the most abundant cell types. In addition, mast cells, NK cells and lymphocytes are recruited to the site of injury. In a few hours after injury, KCs at the wound edge start to migrate in to the wound, and later a different population of KCs begins to proliferate, providing cells to fill the gap in the epithelium. Granulation tissue includes activated fibroblasts and myofibroblasts, ECM components, and abundant new blood vessels supplying inflammatory cells, nutrients and oxygen to the regenerative tissue. After recovery of the epithelium, inflammation resolves and most fibroblasts and blood vessels are cleared from the site via apoptosis. To enhance the mechanical resistance of injured and healed skin, resident fibroblasts still continue remodelling of dermis for months, resulting in collagenous scar that, however, never quite reaches the original tensile strength of intact skin. NK cells regulate inflammation, vascularization and granulation tissue formation via IFN-γ and VEGF.

and antibody-mediated recognition and elimination of foreign material. The cells of injured tissue, platelets and resident leucocytes release substances that augment leucocyte extravasation and migration to the site of injury by regulating the expression of adhesion molecule, such as integrins and selectins on the surface of endothelial cells and leucocytes. In addition, variable bacterial products and components attract and activate phagocytosis and the expression of various immune response genes in inflammatory cells (Tsirogianni et al., 2006). Neutrophils are the first leucocytes to arrive at the site of injury a few hours after wounding, attracted in part by the resident mast cells in the wounded tissue. The main function of the neutrophils is phagocytosis of infectious agents and damaged tissue, which subsequently is destroyed in phagosomes by reactive oxygen species and proteolytic enzymes. Neutrophils secrete different proteinases, such as collagenase-2, matrix metalloproteinase-8 (MMP-8), gelatinase-B (MMP-9) and urokinase-type plasminogen activator, which are involved in debridement of damaged tissue. In addition, neutrophils secrete cytokines and growth factors, such as TNF-α, IL-1β, IL-6, IL-8 and VEGF, which amplify inflammation and stimulate the repair process (Eming et al., 2007; Ravanti and Kähäri, 2000).

The next wave of inflammatory cells consists of blood monocytes. They immigrate to the wound site within 2 days and turn into activated macrophages. They function as antigen-presenting cells and phagocytes, and contribute to the regulation of wound healing by secreting several growth factors, such as TGF-β, TGF-α, bFGF and PDGF. Macrophages also activate natural killer (NK) cells to eliminate intracellular microbes and produce cytokines. Finally, when local infection activates the innate immune system, cytokines, such as MCP-1 and IFN-γ-induced chemokines secreted by macrophages, attract T cells, further facilitating the elimination of pathogens and foreign material, which may have invaded the wound (Eming et al., 2007; Tsirogianni et al., 2006).

Resolution of inflammation is crucial for proper wound healing. This is dramatically demonstrated in chronic venous leg ulcers, in which one of the characteristic features is continuous inflammation, with prominent numbers of inflammatory cells, which produce cytokines, proteolytic enzymes and reactive compounds, resulting in further tissue destruction and inflammation (Menke et al., 2007). In normally healing wounds, neutrophils disappear from the wound within a few days via apoptosis and subsequent phagocytotic activity of macrophages. A large portion of neutrophils is eliminated with the scab. Macrophages, in turn, remain in the wound for a longer period before being eliminated via apoptosis or escaping via the lymphatic system (Martin and Leibovich, 2005). How resolution of macrophages is regulated remains largely unclear. Induction of anti-inflammatory cytokines and down-regulation of pro-inflammatory cytokines are potential mechanisms. Interestingly, mice devoid of MMP-8 show delayed and persistent inflammation of cutaneous wounds, leading to delayed re-epithelialization (Gutierrez-Fernandez et al., 2007). Furthermore, MMPs suppress inflammation by cleaving chemokines and generating cleavage products, which function as antagonists (McQuibban et al., 2000, 2002). Thus, proteases can function as important regulators of inflammation during wound repair.

Re-epithelialization and granulation tissue formation

The phase of re-epithelialization and granulation tissue formation begins in parallel with the inflammation phase of wound healing. In a few hours after injury, epidermal KCs at the wound edge and in the remnants of the skin appendages, such as hair follicles, lose adhesion to the underlying basal lamina and adjacent cells mediated by hemidesmosomes and desmosomes, respectively, and start to migrate towards the wound centre, typically underneath the scab. Within 2 days, KCs at some distance from the wound edge become hyperproliferative and provide cells to fill the gap in the epithelium. Once the wound gap is closed, basal lamina is re-formed and the cellular contacts are established, and KCs start to differentiate and constitute the intact and functional multi-layered epidermis of skin (Singer and Clark, 1999).

The formation of granulation tissue begins within a few days of injury. The granulation tissue consists of numerous new blood vessels, fibroblasts, tissue macrophages and different types of ECM molecules. It is thought that the fibroblasts in the intact dermal layer of the skin are activated by leucocyte, platelet and KC-derived growth factors, such as PDGF and TGF-β, and start proliferating and migrating to the wound (Singer and Clark, 1999). However, recent evidence suggests that at least a portion of the fibroblasts found in granulation tissue arises from bone marrow–derived mesenchymal progenitor cells (Opalenik and Davidson, 2005). Early on, wound ECM deposited and remodelled by fibroblasts consists mainly of fibronectin and hyaluronan, which stimulate cell migration. At a later stage, proteoglycans and type III and type I collagens became the major components of the wound ECM (Clark, 1995). During the second week of wound healing, fibroblasts gain a myofibroblastic phenotype, characterized by the expression of α-smooth muscle actin (Welch et al., 1990). This phenotypic change is largely due to the influence by TGF-β and the mechanical tension generated by the open wound and remodelling of granulation tissue ECM. Myofibroblasts promote wound closure by contracting the wound edges closer to each other, and they are eliminated via apoptosis when mechanical stress has decreased (Hinz, 2007). Angiogenesis, the migration and proliferation of endothelial cells, is initiated by tissue damage and hypoxia, and subsequently stimulated by various soluble angiogenic factors, such as bFGF, VEGF and TGF-β, which are secreted by wound macrophages, KCs and endothelial cells. The early granulation tissue, rich in fibrin and fibronectin, also further stimulates the formation of new blood vessels. Angiogenesis depends on proteolytic activity of plasmin and certain MMPs, including MMP-9 (Ravanti and Kähäri, 2000; Singer and Clark, 1999; Toriseva and Kähäri, 2009).

Tissue remodelling

The regeneration phase is followed by the remodelling and maturation of the ECM of the wound, leading to formation of a collagenous scar in postnatal human skin. It is to be noted, that for example in oral mucosa, wounds tend to heal faster compared to postnatal human skin, and with minimal scar formation (Häkkinen et al., 2000). This may be due to the moist environment in the oral cavity, or to the presence of various factors in saliva. Also, wound gingival fibroblasts possess dynamic expression of proteolytic enzymes such as wide-spectrum ECM proteinase collagenase-3 (MMP-13) and tissue plasminogen activator (Häkkinen et al., 2000; Ravanti and Kähäri, 2000; Toriseva and Kähäri, 2009). Intriguingly, mice lacking PU.1, a member of the Ets family of transcription factors, are essentially deficient in neutrophils and macrophages, lack wound inflammation and show scarless wound healing comparable to foetal cutaneous wounds (Martin et al., 2003). Similarly, in mice, oral mucosal wounds display reduced neutrophil, macrophage and lymphocyte influx and lower levels of pro-inflammatory cytokines and pro-fibrotic TGF-β1 compared to skin wounds (Szpaderska et al., 2003). Thus, inflammatory cells appear to play a remarkable role in the regulation of wound tissue maturation and scar formation via regulatory factors they secrete. NK cell derived IFN-γ is likely to play an important role in regulating inflammation and granulation tissue formation during wound repair (Figure 39.1).

NK cells and dendritic cells (DCs) interact several ways during immune response and inflammation (Figure 39.2). NK cells are capable of stimulating functional DC maturation via both direct cell–cell contact and various cytokines including TNF-α and IFN-γ (Walzer et al., 2005). Activated DCs secrete IL-18 that, in turn, further stimulates NK cell activation (Semino et al., 2005). In addition, HMGB1 (High-Mobility Group Box 1), a soluble pro-inflammatory mediator secreted from cytosolic sources by activated NK cells, plays a central role in necrosis-related inflammation. It is likely, that this NK/DC-directed inflammatory process also takes place in wound healing and regulates the related recruitment of other inflammatory cells and secretion of soluble mediators. Interestingly, recent observations provide evidence that HMGB1 may play an important role in normal cutaneous wound repair (Straino et al., 2008). It is also conceivable, that during wound healing, TGF-β secreted by KCs, tissue macrophages and endothelial cells, inhibits IFN-γ production by NK cells and down-regulates inflammation (Figure 39.2).

The remodelling phase is characterized by elimination of myofibroblasts via apoptosis and disintegration of most blood vessels. This process is regulated by special cell–cell contacts, by different growth factors and by loss of mechanical stress in the wound. Remodelling of collagenous dermal layer of the wound by the resident fibroblasts further continues for approximately a year. They continue to deposit new collagen molecules and to produce collagenolytic proteinases (collagenases), and arrange collagen fibres into an orientation that supports the tensile strength of skin, which, however, never totally reaches the original strength of intact tissue (Hinz, 2007; Singer and Clark, 1999).

The putative role of NK cells during epithelial damage or remodelling has remained largely unresolved. However, NK cells have been shown to modulate the vascular system of the endometrium during pregnancy (Hanna et al., 2006). Accordingly, NK cells might thus be postulated to have an analogous role in the angiogenic processes also involved in wound healing. Activation of macrophages is a central event during wound healing, and it may, at least in part, depend also on NK cell-based cytokine stimulation. As reviewed by Vivier and colleagues (Vivier et al., 2008), NK cells not only stimulate monocyte/macrophage functional differentiation, but they are also able to exert cytotoxic actions against the mature phagocytizing macrophages. The latter process could therefore reflect similar events occurring during the later phases of wound healing and resolution, as discussed above.

Immunoregulatory NK cells in inflammation and tissue repair

T helper cell (Th) differentiation into the major functional subpopulations Th1 and Th2 depends on the action of T-bet and Gata-3 transcription factors, respectively. T-bet positively regulates IFN-γ expression that, in turn, drives Th1 development and prevents Th2 activity. NK cells are able to promote Th1 polarization via IFN-γ mediated stimulation (Martin-Fontecha et al., 2004). TGF-β is a multipotent immunoregulatory cytokine with various anti-inflammatory functions. Interestingly, it has been reported that impaired TGF-β signalling in NK cells leads to NK cell–derived IFN-γ production that, in turn, promotes Th1 development

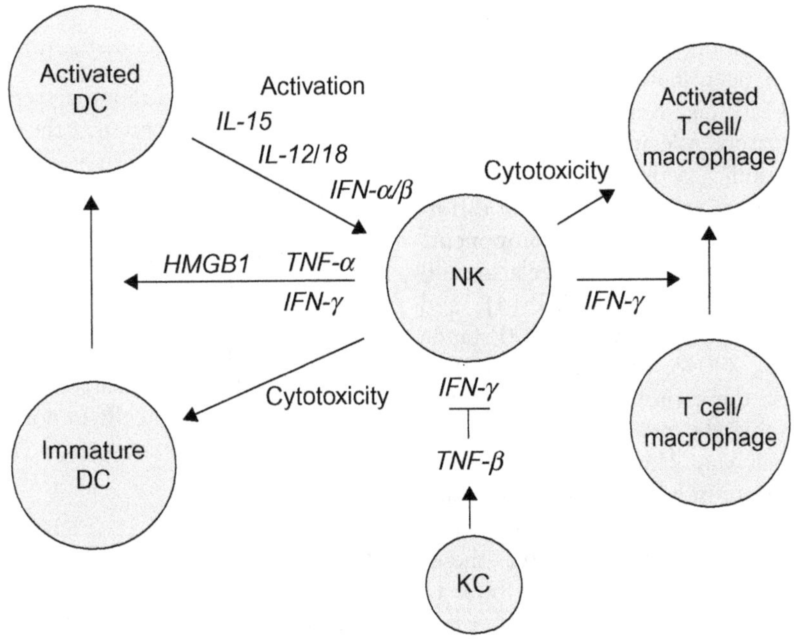

Figure 39.2 • Immunological synapse between NK cells and DCs includes both soluble mediators such as cytokines and direct receptor-mediated cell–cell interactions. NK cells are able to exert cytotoxicity against immature DCs via the activatory NKp30 receptor. In addition, NK cells promote the maturation and activation of DCs via TNF-α, IFN-γ and HMGB1 (High-Mobility Group Box 1). In addition, NK cells enhance the activation of both T cells and macrophages. Upon wound healing, TGF-β secreted by KCs, tissue macrophages and endothelial cells, inhibits IFN-γ production by NK cells and down-regulates inflammation (modified after Vivier et al., 2008).

(Laouar et al., 2005). In addition, NK cells are known to interact with antigen-presenting DCs and thus stimulate further the Th1-type polarization of CD4 T lymphocytes (Moretta et al., 2008).

T helper responses and the corresponding cytokine balance are suggested to contribute to the process of tissue fibrosis following chronic inflammation. Th1 responses are related to anti-fibrotic responses and tissue repair, whereas Th2 dominance promotes fibrosis under chronic inflammatory conditions. For instance, in alcohol-induced liver fibrosis and inflammation NK cell–derived IFN-γ is suggested to harbour anti-inflammatory functions (Jeong and Gao, 2008). As shown in this and in another study, NK cells also prevent tissue fibrosis by direct cytolytic activity against the diseased liver cells (Melhem et al., 2006). In another model of inflammation, CXC chemokine receptor 3 was demonstrated to play a pivotal role in preventing fibrotic processes in the lung tissue (Jiang et al., 2004). At the same time, the inflamed lung was shown to contain reduced amounts of NK cells that are known to produce the anti-fibrotic IFN-γ during inflammation. Under autoimmune conditions, NK cells are in a similar way known to contribute to Th1 immune dominance via helping DC development from monocytes (Zhang et al., 2007). NK cells may also influence the Th-balance in allergic inflammation, as it was recently shown that, following bacterial infection, NK cells may prevent allergic immune activation (Han et al., 2008). Although the fibrosis-preventing functions of Th1 cells are likely to require also T-bet expression, this chromatin remodelling transcription factor has also been shown to exert more ubiquitous roles in fibrogenesis control that goes beyond adaptive immunity. It has been reported that T-bet regulates IL-13 and thus prevents fibrosis in an experimental model of dermal fibrosis (Aliprantis et al., 2007). More importantly, this regulatory action seems to involve cells of innate immunity, as the prevention of fibrosis is not mediated solely via T cells.

NK cells are known to produce IFN-γ, and it has been shown that this cytokine has an inhibitory role on collagen synthesis and functions during wound healing (Kähäri et al., 1990; Laato et al., 2001; Yufit et al., 1995). IFN-γ deficiency, in turn, leads to more rapid wound healing and accelerated angiogenesis (Ishida et al., 2004a). In that study it was demonstrated that IFN-γ-mediated prevention of wound healing occurs via inhibition of TGF-β. NK cell activity has been shown to be under the control of regulatory T cells and corresponding TGF-β expression (Smyth et al., 2006). These peripheral immunoregulatory cells express CD4 and CD25 antigens and require the Foxp3 transcription factor for their development and functional maturation (reviewed in Tang and Bluestone, 2008). As these cells are also capable of skin homing (Hirahara et al., 2006), it could be postulated that they also regulate wound healing–associated inflammation. In addition, skin-residing regulatory T cells have been shown to be activated via IL-15 and fibroblasts (Clark and Kupper, 2007), which suggests an anti-inflammatory role in various cutaneous immune disturbances, including perhaps wound healing.

NK cells have been shown to produce several different proteinases with the ability to cleave components of ECM, including gelatinase-A (MMP-2), gelatinase-B (MMP-9), membrane-type 1 MMP (MMP-14), and collagenase-1 (MMP-1) (Albertsson et al., 2000; Goda et al., 2006; Ishida et al., 2004b; Kim et al., 2000); it has been proposed, that these proteinases play a role in assisting NK cells to migrate into tissues by cleaving ECM components in their way. However, these MMPs are also known to participate in inflammation, vascularization, re-epithelialization, granulation tissue formation, and remodelling (Toriseva and Kähäri, 2009). It is therefore conceivable that NK cells may also contribute to wound healing by participating in all these crucial events during the process of tissue repair. It is also possible that alterations in the presence of NK cells in wound repair may play a role in disturbances in wound repair resulting in chronic ulcers, such as venous leg ulcers, or in excessive wound repair in hypertrophic scars and keloids.

Conclusions

In summary, the immune system in wound repair combats the microbe invasion at the site of tissue injury, and it also participates in the removal of the damaged tissue and contributes to the process of healing. It is evident that NK cells regulate the early onset and resolution of the inflammatory phase in wound repair. It is also likely that NK cells are involved in events at the later stage of the normal wound healing, such as in re-epithelialization, angiogenesis, granulation tissue formation, and remodelling. However, further studies are clearly warranted to elucidate the detailed role of NK cells in normal wound repair and in poorly healing chronic ulcers, such as venous leg ulcers.

Acknowledgements

The original work of authors cited here has been supported by the Academy of Finland (project 114409), the Finnish Cancer Research Foundation, Sigrid Juselius Foundation, Turku University Central Hospital EVO grant (projects 13336), and European Union Framework Programme 6 (LSHC-CT-2003-503297).

References

Albertsson, P., Kim, M.H., Jonges, L.E., Kitson, R.P., Kuppen, P.J., Johansson, B.R., Nannmark, U. and Goldfarb, R.H. (2000). Matrix metalloproteinases of human NK cells. In Vivo **14**, 269–276.

Aliprantis, A.O., Wang, J., Fathman, J.W., Lemaire, R., Dorfman, D.M., Lafyatis, R. and Glimcher, L.H. (2007). Transcription factor T-bet regulates skin sclerosis through its function in innate immunity and via IL-13. Proc Natl Acad Sci U S A **104**, 2827–2830.

Clark, R.A. and Kupper, T.S. (2007). IL-15 and dermal fibroblasts induce proliferation of natural regulatory T cells isolated from human skin. Blood **109**, 194–202.

Clark, R.A.F. (1995). Wound repair: overview and general considerations. In: The Molecular and Cellular Biology of Wound Repair, 2nd ed. New York: Plenum Press, pp. 513–560. 2nd ed.

Eming, S.A., Krieg, T. and Davidson, J.M. (2007). Inflammation in wound repair: molecular and cellular mechanisms. J Invest Dermatol **127**, 514–525.

Goda, S., Inoue, H., Umehara, H., Miyaji, M., Nagano, Y., Harakawa, N., Imai, H., Lee, P., Macarthy, J.B., Ikeo, T., Domae, N., Shimizu, Y. and Iida, J. (2006). Matrix metalloproteinase-1 produced by human CXCL12-stimulated natural killer cells. Am J Pathol **169**, 445–458.

Gutierrez-Fernandez, A., Inada, M., Balbin, M., Fueyo, A., Pitiot, A.S., Astudillo, A., Hirose, K., Hirata, M., Shapiro, S.D., Noel, A., Werb, Z., Krane, S.M., Lopez-Otin, C. and Puente, X.S. (2007). Increased inflammation delays wound healing in mice deficient in collagenase-2 (MMP-8). FASEB J **21**, 2580–2591.

Häkkinen, L., Uitto, V.J. and Larjava, H. (2000). Cell biology of gingival wound healing. Periodontol **24**, 127–152.

Han, X., Fan, Y., Wang, S., Jiao, L., Qiu, H. and Yang, X. (2008). NK cells contribute to intracellular bacterial infection-mediated inhibition of allergic responses. J Immunol **180**, 4621–4628.

Hanna, J., Goldman-Wohl, D., Hamani, Y., Avraham, I., Greenfield, C., Natanson-Yaron, S., Prus, D., Cohen-Daniel, L., Arnon, T.I., Manaster, I., Gazit, R., Yutkin, V., Benharroch, D., Porgador, A., Keshet, E., Yagel, S. and Mandelboim, O. (2006). Decidual NK cells regulate key developmental processes at the human fetal–maternal interface. Nat Med **12**, 1065–1074.

Hinz, B. (2007). Formation and function of the myofibroblast during tissue repair. J Invest Dermatol **127**, 526–537.

Hirahara, K., Liu, L., Clark, R.A., Yamanaka, K., Fuhlbrigge, R.C. and Kupper, T.S. (2006). The majority of human peripheral blood CD4+CD25highFoxp3+ regulatory T cells bear functional skin-homing receptors. J Immunol **177**, 4488–4494.

Ishida, Y., Kondo, T., Takayasu, T., Iwakura, Y. and Mukaida, N. (2004a). The essential involvement of cross-talk between IFN-γ and TGF-β in the skin wound-healing process. J Immunol **172**, 1848–1855.

Jeong, W.I. and Gao, B. (2008). Innate immunity and alcoholic liver fibrosis. J Gastroenterol Hepatol **23**(Suppl. 1), S112–S118.

Ishida, Y., Migita, K., Izumi, Y., Nakao, K., Ida, H., Kawakami, A., Abiru, S., Ishibashi, H., Eguchi, K. and Ishii, N. (2004b). The role of IL-18 in the modulation of matrix metalloproteinases and migration of human natural killer (NK) cells. FEBS Lett **569**, 156–160.

Jiang, D., Liang, J., Hodge, J., Lu, B., Zhu, Z., Yu, S., Fan, J., Gao, Y., Yin, Z., Homer, R., Gerard, C. and Noble, P.W. (2004). Regulation of pulmonary fibrosis by chemokine receptor CXCR3. J Clin Invest **114**, 291–299.

Kähäri, V.-M., Chen, Y.Q., Su, M.W., Ramirez, F. and Uitto, J. (1990). Tumor necrosis factor-α and interferon-γ suppress the activation of human type I collagen gene expression by transforming growth factor-β1: evidence for two distinct mechanisms of inhibition at transcriptional and post-transcriptional level. J Clin Invest **86**, 1489–1495.

Kim, M.H., Kitson, R.P., Albertsson, P., Nannmark, U., Basse, P.H., Kuppen, P.J., Hokland, M.E. and Goldfarb, R.H. (2000). Secreted and membrane-associated matrix metalloproteinases of IL-2-activated NK cells and their inhibitors. J Immunol **164**, 5883–5889.

Laato, M., Heino, J., Gerdin, B., Kähäri, V. M. and Niinikoski, J. (2001). Interferon-γ-induced inhibition of wound healing in vivo and in vitro. Ann Chir Gynaecol **90**(Suppl. 215), 19–23.

Laouar, Y., Sutterwala, F.S., Gorelik, L. and Flavell, R.A. (2005). Transforming growth factor-beta controls T helper type 1 cell development through regulation of natural killer cell interferon-gamma. Nat Immunol **6**, 600–607.

Martin, P. (1997). Wound healing—aiming for perfect skin regeneration. Science **276**, 75–81.

Martin, P., D'Souza, D., Martin, J., Grose, R., Cooper, L., Maki, R. and McKercher, S.R. (2003). Wound healing in the PU.1 null mouse—tissue repair is not dependent on inflammatory cells. Curr Biol **13**, 1122–1128.

Martin, P. and Leibovich, S.J. (2005). Inflammatory cells during wound repair: the good, the bad and the ugly. Trends Cell Biol **15**, 599–607.

Martin-Fontecha, A., Thomsen, L.L., Brett, S., Gerard, C., Lipp, M., Lanzavecchia, A. and Sallusto, F. (2004). Induced recruitment of NK cells to lymph nodes provides IFN-gamma for T(H)1 priming. Nat Immunol **5**, 1260–1265.

McQuibban, G.A., Gong, J.H., Tam, E. M., McCulloch, C.A., Clark-Lewis, I. and Overall, C.M. (2000). Inflammation dampened by gelatinase A cleavage of monocyte chemoattractant protein-3. Science **289**, 1202–1206.

McQuibban, G.A., Gong, J.H., Wong, J.P., Wallace, J.L., Clark-Lewis, I. and Overall, C.M. (2002). Matrix metalloproteinase processing of monocyte chemoattractant proteins generates CC chemokine receptor antagonists with anti-inflammatory properties in vivo. Blood **100**, 1160–1167.

Melhem, A., Muhanna, N., Bishara, A., Alvarez, C.E., Ilan, Y., Bishara, T., Horani, A., Nassar, M., Friedman, S. L. and Safadi, R. (2006). Anti-fibrotic activity of NK cells in experimental liver injury through killing of activated HSC. J Hepatol **45**, 60–71.

Menke, N.B., Ward, K.R., Witten, T.M., Bonchev, D.G. and Diegelmann, R.F. (2007). Impaired wound healing. Clin Dermatol **25**, 19–25.

Moretta, A., Marcenaro, E., Parolini, S., Ferlazzo, G. and Moretta, L. (2008). NK cells at the interface between innate and adaptive immunity. Cell Death Differ **15**, 226–233.

Opalenik, S.R. and Davidson, J.M. (2005). Fibroblast differentiation of bone marrow-derived cells during wound repair. FASEB J **19**, 1561–1563.

Ravanti, L. and Kähäri, V.-M. (2000). Matrix metalloproteinases in wound repair (review). Int J Mol Med **6**, 391–407.

Sacco, O., Silvestri, M., Sabatini, F., Sale, R., Defilippi, A.C. and Rossi, G.A. (2004). Epithelial cells and fibroblasts: structural repair and remodelling in the airways. Paediatr Respir Rev **5**(Suppl. A), S35–S40.

Semino, C., Angelini, G., Poggi, A. and Rubartelli, A. (2005). NK/iDC interaction results in IL-18 secretion by DCs at the synaptic cleft followed by NK cell activation and release of the DC maturation factor HMGB1. Blood **106**, 609–616.

Singer, A.J. and Clark, R.A. (1999). Cutaneous wound healing. N Engl J Med **341**, 738–746.

Smyth, M.J., Teng, M.W., Swann, J., Kyparissoudis, K., Godfrey, D.I. and Hayakawa, Y. (2006). CD4+CD25+ T regulatory cells suppress NK cell-mediated immunotherapy of cancer. J Immunol **176**, 1582–1587.

Straino, S., Di Carlo, A., Mangoni, A., De Mori, R., Guerra, L., Maurelli, R., Panacchia, L., Di Giacomo, F., Palumbo, R., Di Campli, C., Uccioli, L., Biglioli, P., Bianchi, M.E., Capogrossi, M.C. and Germani, A. (2008). High-mobility group box 1 protein in human and murine skin: involvement in wound healing. J Invest Dermatol **128**, 1545–1553.

Szpaderska, A.M., Zuckerman, J.D. and DiPietro, L.A. (2003). Differential injury responses in oral mucosal and cutaneous wounds. J Dent Res **82**, 621–626.

Tang, Q. and Bluestone, J.A. (2008). The Foxp3+ regulatory T cell: a jack of all trades, master of regulation. Nat Immunol **9**, 239–244.

Tarnawski, A.S. (2005). Cellular and molecular mechanisms of gastrointestinal ulcer healing. Dig Dis Sci **50**(Suppl. 1), S24–S33.

Toriseva, M. and Kähäri, V.-M. (2009). Proteinases in cutaneous wound healing. Cell Mol Life Sci **66**, 203–224.

Tsirogianni, A.K., Moutsopoulos, N.M. and Moutsopoulos, H.M. (2006). Wound healing: immunological aspects. Injury **37**(Suppl. 1), S5–S12.

Walzer, T., Dalod, M., Robbins, S.H., Zitvogel, L. and Vivier, E. (2005). Natural killer cells and dendritic cells: "l'union fait la force". Blood **106**, 2252–2258.

Welch, M.P., Odland, G.F. and Clark, R. A. (1990). Temporal relationships of F-actin bundle formation, collagen and fibronectin matrix assembly, and fibronectin receptor expression to wound contraction. J Cell Biol **110**, 133–145.

Vivier, E., Tomasello, E., Baratin, M., Walzer, T. and Ugolini, S. (2008). Functions of natural killer cells. Nat Immunol **9**, 503–510.

Yufit, T., Vining, V., Wang, L., Brown, R. R. and Varga, J. (1995). Inhibition of type I collagen mRNA expression independent of tryptophan depletion in interferon-gamma-treated human dermal fibroblasts. J Invest Dermatol **105**, 388–393.

Zhang, A.L., Colmenero, P., Purath, U., Teixeira de Matos, C., Hueber, W., Klareskog, L., Tarner, I.H., Engleman, E. G. and Soderstrom, K. (2007). Natural killer cells trigger differentiation of monocytes into dendritic cells. Blood **110**, 2484–2493.

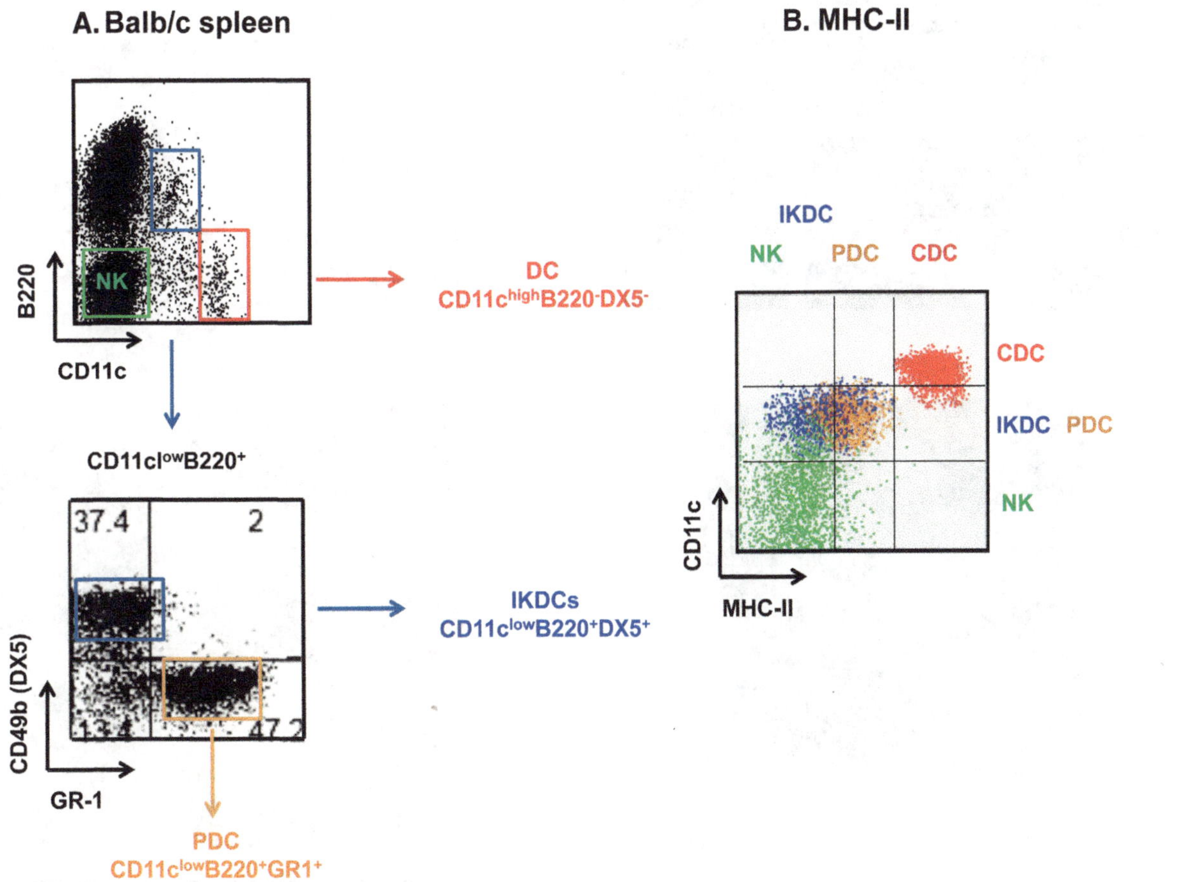

Plate 1 • Differential expression of CD11c, B220, CD49b and MHC-II molecules between IKDC, NK, CDC and PDC cell-sorted from BALB/c mouse spleens.
a) BALB/c splenocytes are labeled with CD11c, B220, DX5 (CD49b) and GR1 (Ly6C-Ly6G) mAb. IKDC ($CD11c^{low}B220^{+}DX5^{+}GR1^{-}$), PDC ($CD11c^{low}B220^{+}DX5^{-}GR1^{+}$), CDC ($CD11c^{low}B220^{+}DX5^{+}GR1^{-}$) and NK ($CD11c^{-}B220^{-}DX5^{+}GR1^{-}$) were cell sorted and assessed for MHC-II molecules expression. b) the figure represents an overlay of the CD11c / MHC-II histograms for each sorted population. Blue, IKDC; green, NK; red, CDC; orange, PDC.

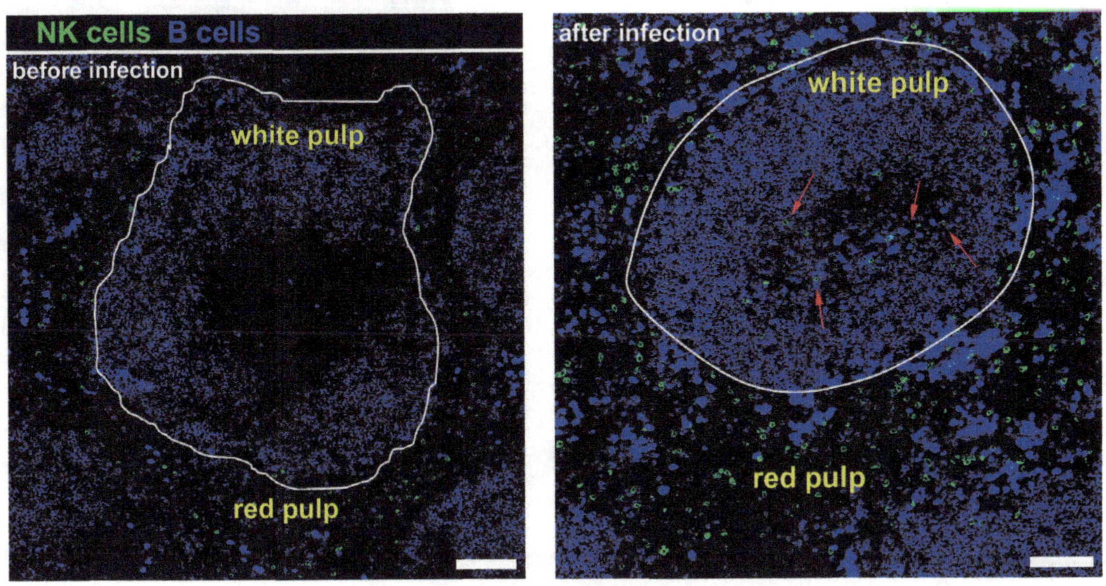

Plate 2 • NK cells migrate from the red to the white pulp of the spleen following infection with murine CMV.
B cell (blue) staining reveals the splenic white pulp (structure enclosed by the white line). Before infection (left micrograph), NK cells (green) are located in the red pulp. After infection (right micrograph), NK cells migrate to the white pulp (red arrows). The scale bar is 100 μm.

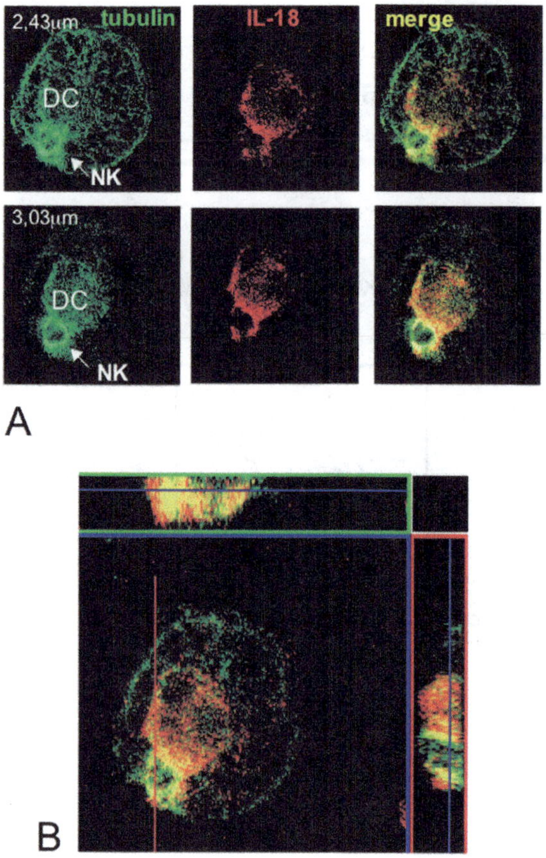

Plate 3 • A. Confocal microscopy analysis of tubulin (green) and IL-18 (red) in an immature DC/NK conjugate after 3 hours of interaction. Two focus layers of the same conjugate are shown to make clear the rearrangement of the microtubular cytoskeleton at the synapse. Notice in the "merge" panel, the strong co-staining of IL-18 and tubulin, supporting the role of tubulin in mediating the transport of IL-18-containing vesicles to the NK/DC synapse (Semino *et al.*, 2005). With permission of Drs M.Torrisi and F. Belleudi, Univ. Roma La Sapienza, Italy. B. Orthogonal and sagittal layers of the NK/DC synapse showing details of the tubulin-mediated IL-18 release process at the synapse.

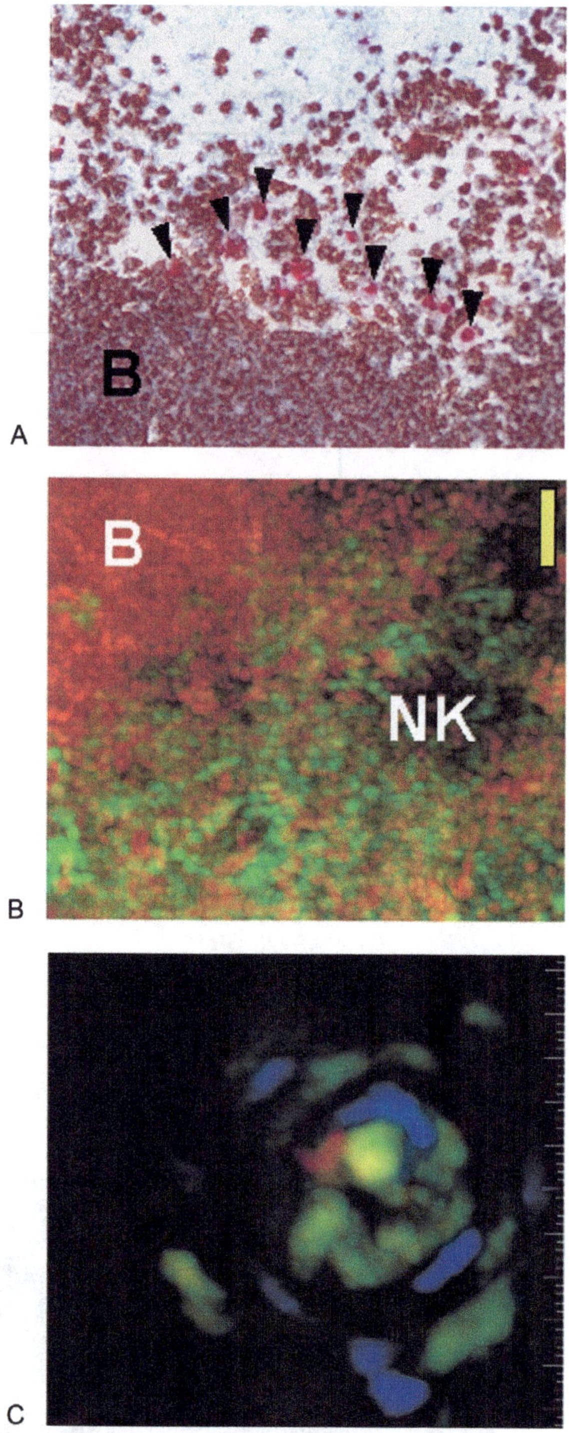

Plate 4 • Intranodal NK cell localization. (A) Photomicrograph depicting a lymph node with NK cells (red) localized to the edge of a B cell follicle (brown). Black arrowheads indicate individual NK cells (magnification, 20x). (B) Two-photon images depicting NK cell (green) migration along the periphery of a B cell follicle (red). Cumulative fluorescence was formed by the summation of successive images over time. Scale bar, 20 μm. (C) Snapshot visualization of an NK cell (red) interacting with several DC. Individual DC are labelled green and blue. Large tick marks, 5 μm.

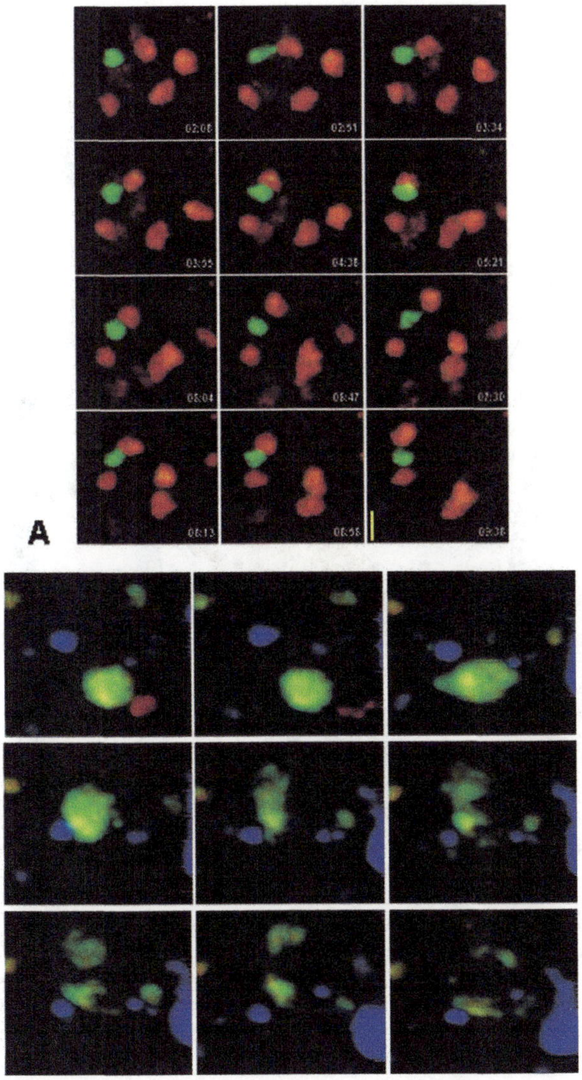

Plate 5 • NK-target cell interaction dynamics. (A) Consecutive images of an NK cell (green) making serial contacts with the same B cell (red). Scale bar, 10 μm. (B) Time-lapse images of an NK cell (red) engaging in a productive interaction with an individual allogeneic DC (green) which results in lysis of the DC target as evidenced by membrane blebbing. Large tick marks, 5 μm.

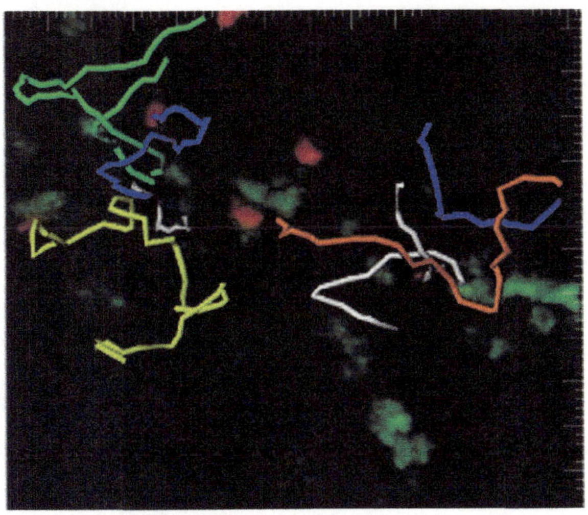

Plate 6 • NK-DC interactions. Two-photon image of individual NK cells (red) interacting with CD11c-yfp^{+} endogenous DC (green). Tracks of 7 representative NK cells are superimposed. Large tick marks, 20 μm.

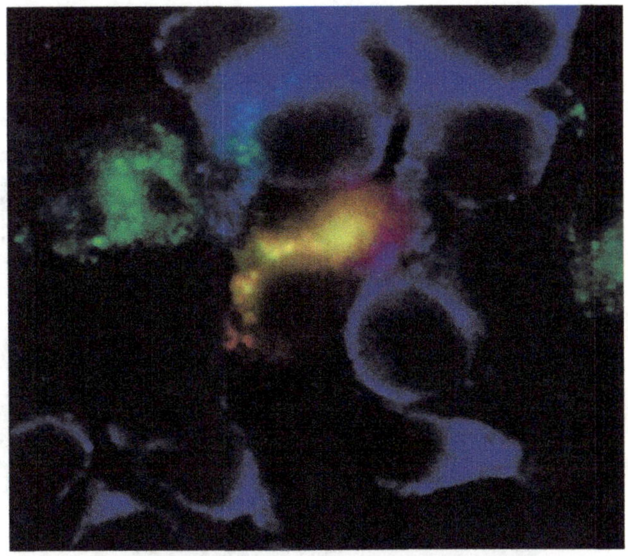

Plate 7 • CD83+ lymph node residing human NK cells. Three-colour fluorescent microscopy demonstrating the presence of CD3- cells (CD3+ cells in blue), co-expressing CD56(red) and CD83(green) cells within the T cell areas of a lymph node (yellow color representing co-expression of CD56 and CD83), juxtaposed to single-positive CD3+ T cells and CD83+ DC. (Mailliard et al., 2005. Originally published in *The Journal of Experimental Medicine.* doi:10.1084/jem.20050128).

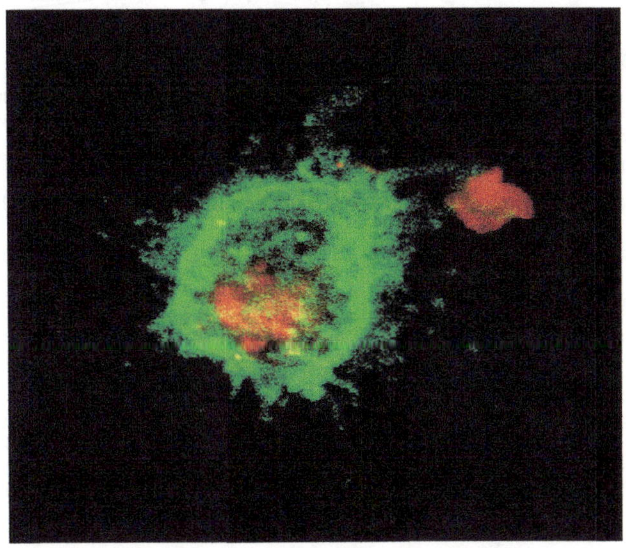

Plate 8 • DC captures and engulfs tumour cell antigen. Confocal microscopy showing a CD86 positive (green) mouse DC having intracellular contents derived from UV-treated apoptotic MCA205 mouse fibrosarcoma cells labelled with the lipophilic carbocyanine dye DiI (red) following their short-term co-incubation. *(Picture courtesy of Borys Mascarenhas, Joseph Baar, and Simon Watkins, University of Pittsburgh, Pittsburgh PA.)*

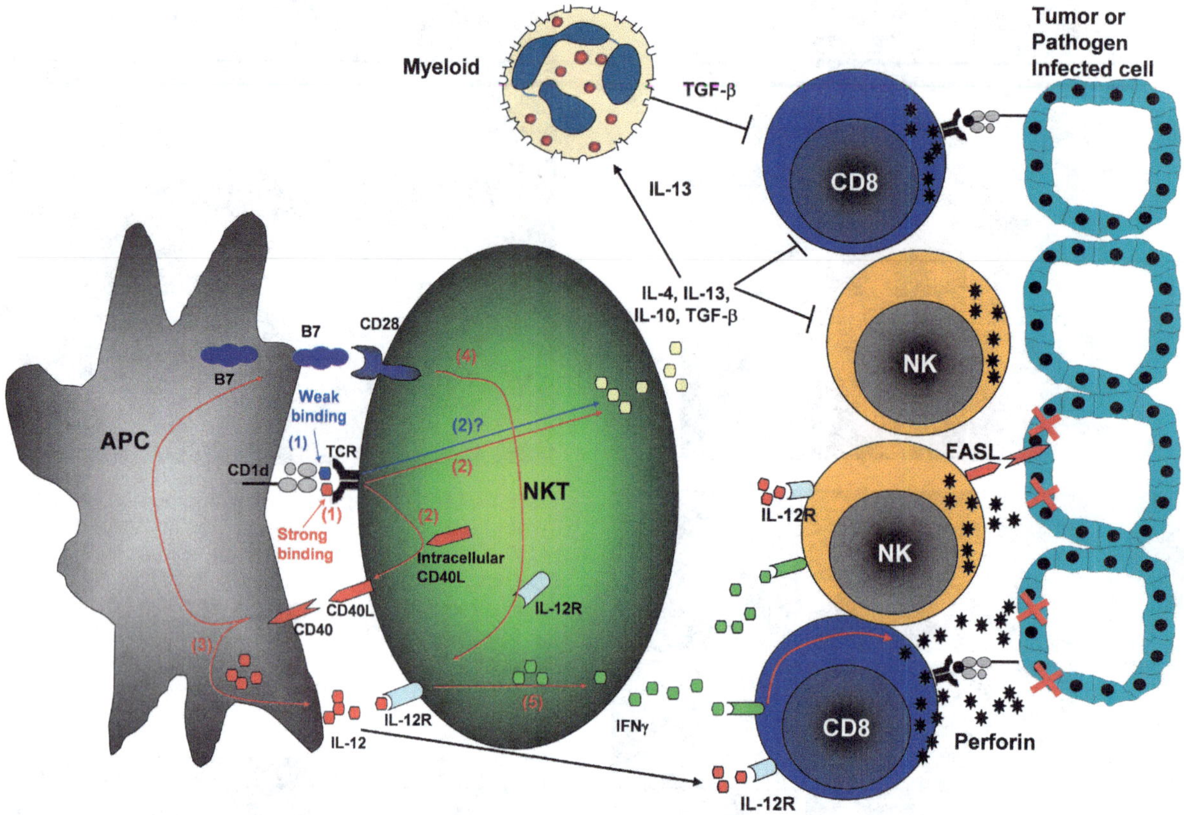

Plate 9 • Schematic depicting NKT-cell activation and immunoregulation of effector cells following interaction with APC. (1) Strong antigen triggering (red arrows) of NKT-cells induces production of Th2 cytokines and (2) upregulates cell surface CD40L. (3) CD40L next induces the APC to express B7 costimulatory molecule and to release IL-12. (4) Costimulation induces further expression of IL-12 receptors on NKT-cell allowing for increased IL-12 signaling of NKT-cells and (5) production of IFN-γ that can activate NK and CD8 T-cells. In contrast, (1) weak antigens (blue arrows) are unable to upregulate CD40L on NKT-cells. Subsequently (2) NKT-cells have limited cytokine production or are skewed toward producing Th2 or suppressive cytokines that inhibit NK or CD8 effector T-cells.

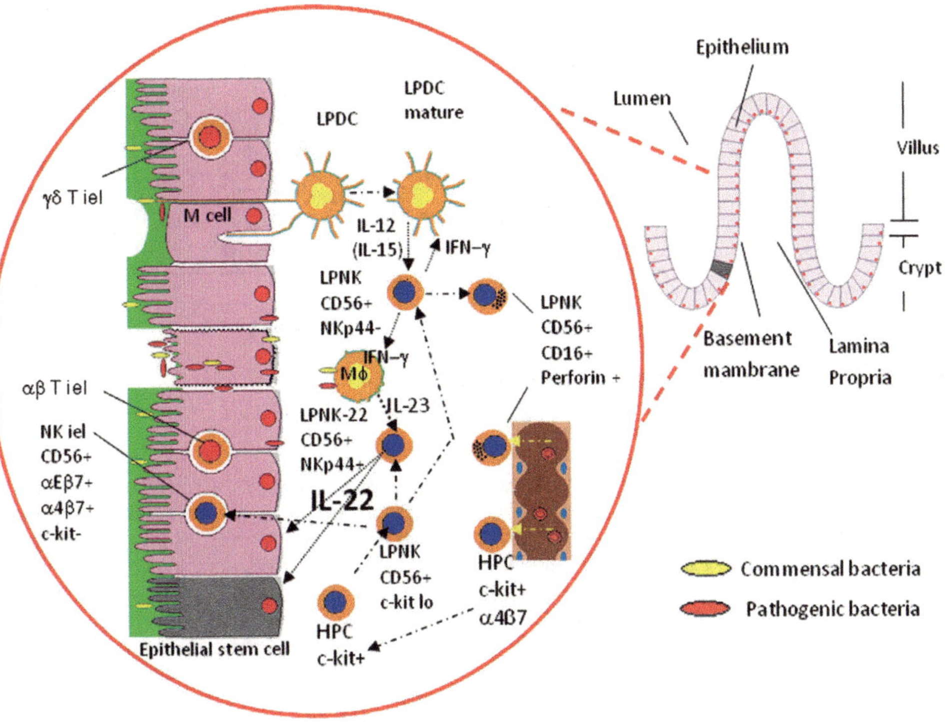

Plate 10 • Proposed model for NK cell involvement in the adult human intestine. A section of intestine shows the columnar cells of the intestinal epithelium supported by basement membrane and the underlying lamina propria. Renewal of intestinal epithelium occurs by differentiation of epithelial cell precursors and migration along the crypt-villus axis. Commensal or pathogenic bacteria are sampled in the lumen of the intestine by lamina propria dendritic cells (LPDC) extending processes through the epithelial barrier. Microbial products can also be captured by specialized M in the intestinal epithelium. Certain pathogenic microbes translocate through the intestinal epithelium whilst others cause damage resulting in increased uptake of pathogenic or commensal bacteria. LPDC mature on activation by pathogen associated molecular patterns (PAMP) ligating toll-like receptors (TLR). TLR ligation also activates macrophages (mϕ) in the lamina propria. C-kit$^+$ lineage$^-$ haemopoetic precursor cells (HPC) are dispersed throughout the human lamina propria. In mice, HPC are associated with cryptopatches and express the transcription factor RORγt. Intestinal NK HPC may be maintained in situ or be replenished by the migration of bone marrow derived $\alpha4\beta7^+$ HPC. NK HPC subsequently differentiate into NKp44$^+$ or NKp44$^-$ LPNK cell populations in the lamina propria. Fully differentiated c-kit- intraepithelial NK cells (ienk) may be derived from CD56$^+$c-kit^{lo} LPNK migrating to the epithelium. Activated LPDC produce IL-12 and IL-15, which activate IFN-γ production in NKp44$^-$ LPNK. IFN-γ augments anti pathogon responses by promoting macrophage and DC function. Activated macrophages, or dendritic cell subsets, produce IL-23, which stimulates NKp44$^+$LPNK-22 cells (in mice RORγt^{hi}NKP46$^+$NK1.1^{int}CD127$^+$) to produce IL-22 which in turn promotes epithelial cell proliferation. Chronic inflammation under disease conditions may result in the further differentiation of CD16$^-$LPNK into cytotoxic CD16$^+$ cells expressing perforin. Cytotoxic CD16$^+$NK effector cells may also migrate to the lamina propria in response to inflammation.

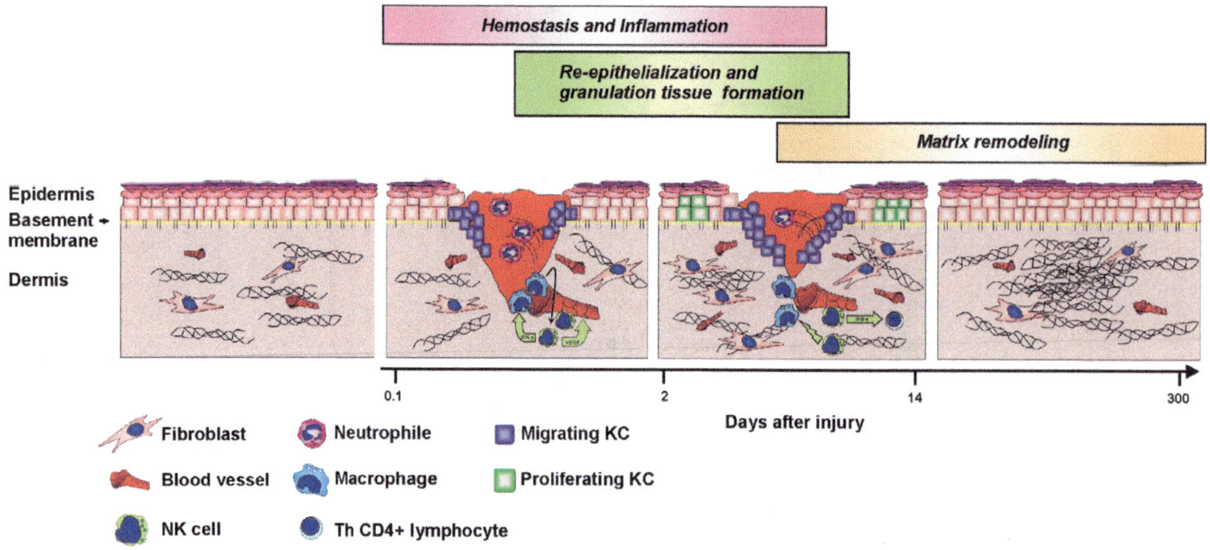

Plate 11 • The phases of cutaneous wound healing. The intact skin constitutes of epidermis which is formed by stratified layers of keratinocytes of variable differentiation stages, and of dermis rich in collagenous extracellular matrix (ECM), fibroblasts and blood vessels. Epidermis and dermis are separated by basement membrane, a specialized layer of ECM. Injury instantly evokes a cascade of events initiated by hemostasis. A variety of factors secreted from blood clot, injured cells attract and activate inflammatory response with neutrophils and macrophages as the most abundant cell types. In addition, mast cells, natural killer (NK) cells and lymphocytes are recruited to the site of injury. In a few hours after injury, keratinocytes at the wound edge start to migrate in to the wound, and later a different population of keratinocytes begins to proliferate providing cells to fill the gap in the epithelium. Granulation tissue includes activated fibroblasts and myofibroblasts, ECM components, and abundant new blood vessels supplying inflammatory cells, nutrients and oxygen to the regenerative tissue. After recovery of the epithelium, inflammation resolves and most fibroblasts and blood vessels are cleared from the site via apoptosis. To enhance the mechanical resistance of injured and healed skin, resident fibroblasts still continue remodeling of dermis for months resulting in collagenous scar which, however, never quite reaches the original tensile strength of intact skin. NK cells regulate inflammation, vascularixation and granulation tissue formation via IFN-γ and VEGF.

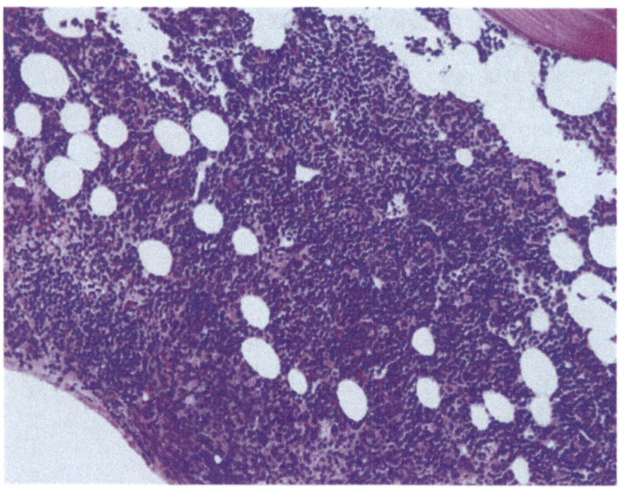

A

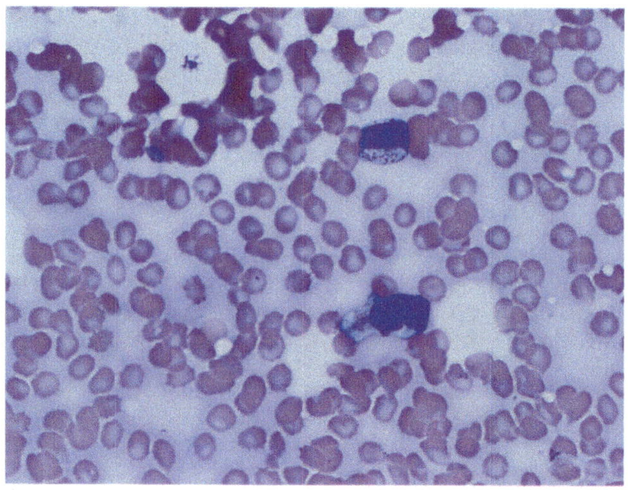

C

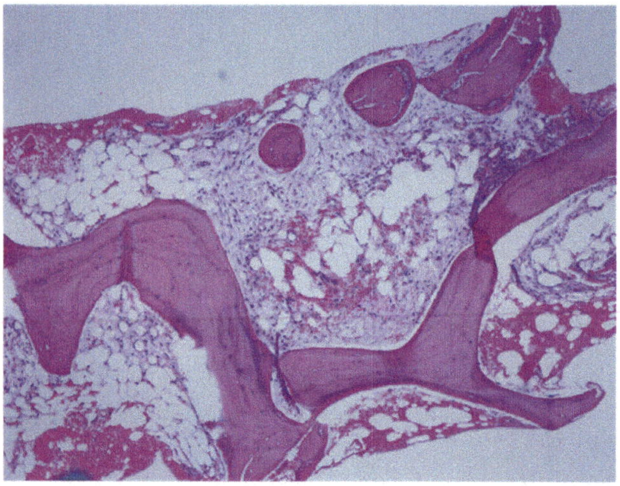

B

Plate 12 • Leukemia Clearance after NK Cell Based Preparative Therapy.

Shown in these panels are images from a patient with relapsed, refractory acute myelogenous leukemia (AML) who received haploidentical NK cells as part of the preparative therapy in anticipation of a nonmyeloablative haploidentical cell transplant. His pre-therapy bone marrow biopsy was hypercellular with complete replacement by blasts (trephine section 10X: Panel A). He received therapy with cyclophosphamide (60 mg/kg) days −18 and −17 and daily intravenous fludarabine (25 mg/m^2) days −18 to −14, 400 cGy of total body irradiation on day −13, followed adoptive transfer of a CD3 and CD19 depleted haploidentical NK cell product on day −12. Subcutaneous IL-2 (10 million units every other day) was administered over 2 weeks for in vivo stimulation. On day −1, 14 days after the NK cell infusion and prior to the CD34$^+$ hematopoietic cell infusion, a repeat bone marrow biopsy showed complete leukemia clearance (trephine section 5X: Panel B) and the presence of NK cells (aspirate 100X: Panel C).

Chronic NK-cell Lymphoproliferative Disorder

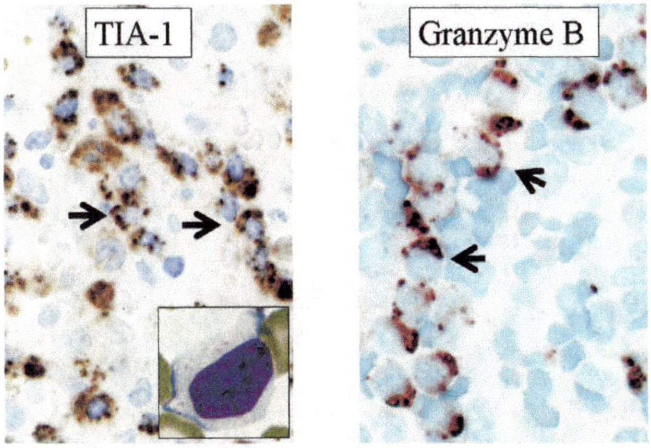

Plate 13 • Chronic NK cell lymphoproliferative disorder, histology. The NK cells in the peripheral blood (inset left) have bland cytologic features and are indistinguishable from normal large granular lymphocytes. Immunohistochemistry in chronic NK cell lymphoproliferative disorder performed on a bone marrow biopsy with antibodies to TIA-1 (also known as GMP-17, left) and on a splenectomy specimen with antibodies to granzyme B (right) reveals intrasinusoidal infiltration of both organs by lymphocytes positive for these cytotoxic granule proteins (arrows). The presence of the antigen positive cells within vascular structures leads to their linear arrangement. Note also the weaker, but distinct staining of granulocytic cells for TIA-1 in the left panel.

Aggressive NK-cell Leukemia

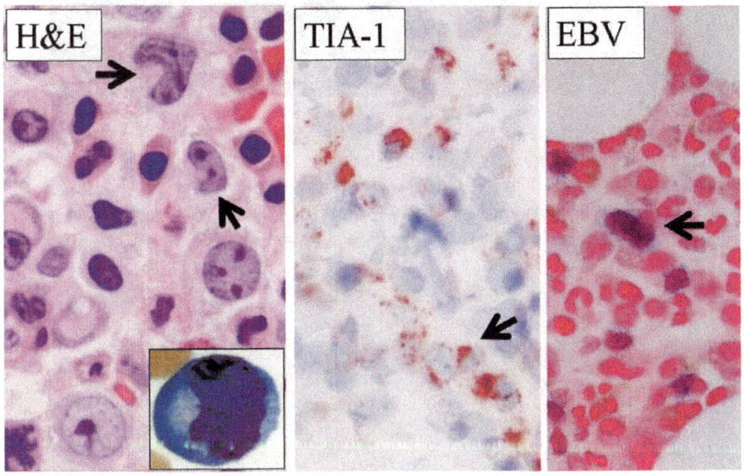

Plate 14 • Aggressive NK cell leukemia, histology. Histologic sections of a bone marrow biopsy from an aggressive NK cell leukemia case reveals infiltration by cells with large nuclei, open chromatin, and irregular nuclear contours (left panel, arrows). Cytologically atypical lymphoid cells with hyperchromatic, irregular nuclei and faint cytoplasmic granules were also present in the peripheral blood (left panel inset). Immunoperoxidase staining of the bone marrow biopsy demonstrates that the neoplastic cells are positive for the cytotoxic granule protein TIA-1 (middle panel); these cells were also shown to be Epstein Barr virus (EBV) positive by in situ hybridization studies (right panel).

Extranodal NK/T-Cell Lymphoma, Nasal Type

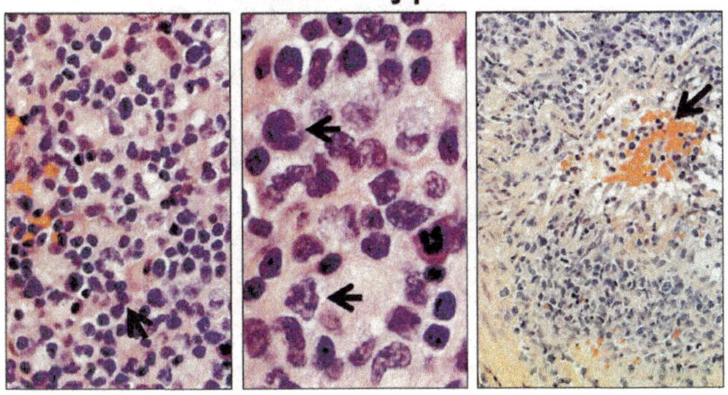

Plate 15 • Extranodal NK/T cell lymphoma, histology. Examples of cases with minimal cytologic atypia (left panel) and marked cytologic atypia (middle panel) are shown with arrows to help highlight the neoplastic cells. An example of vascular invasion is shown in the right panel, the neoplastic cells are seen to splay the vascular wall smooth muscle. The vessel lumen is indicated by the arrow.

Extranodal NK/T-Cell Lymphoma Immunophenotype

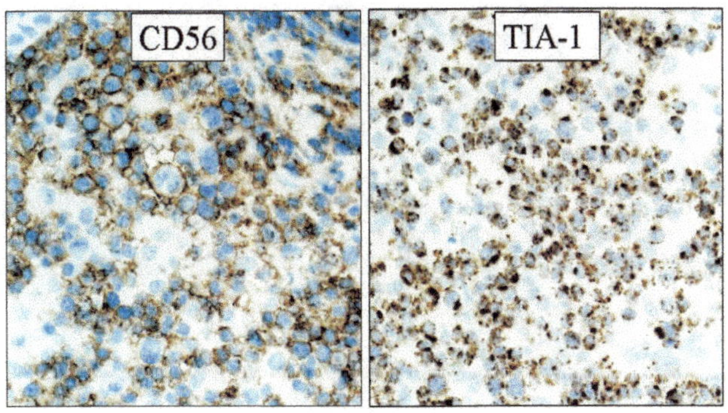

Plate 16 • Extranodal NK/T cell lymphoma, immunohistochemistry. Immunoperoxidase staining of paraffin sections from an extranodal NK/T cell lymphoma reveals the neoplastic cells to be positive for CD56 (left panel) and the cytotoxic granule protein TIA-1.

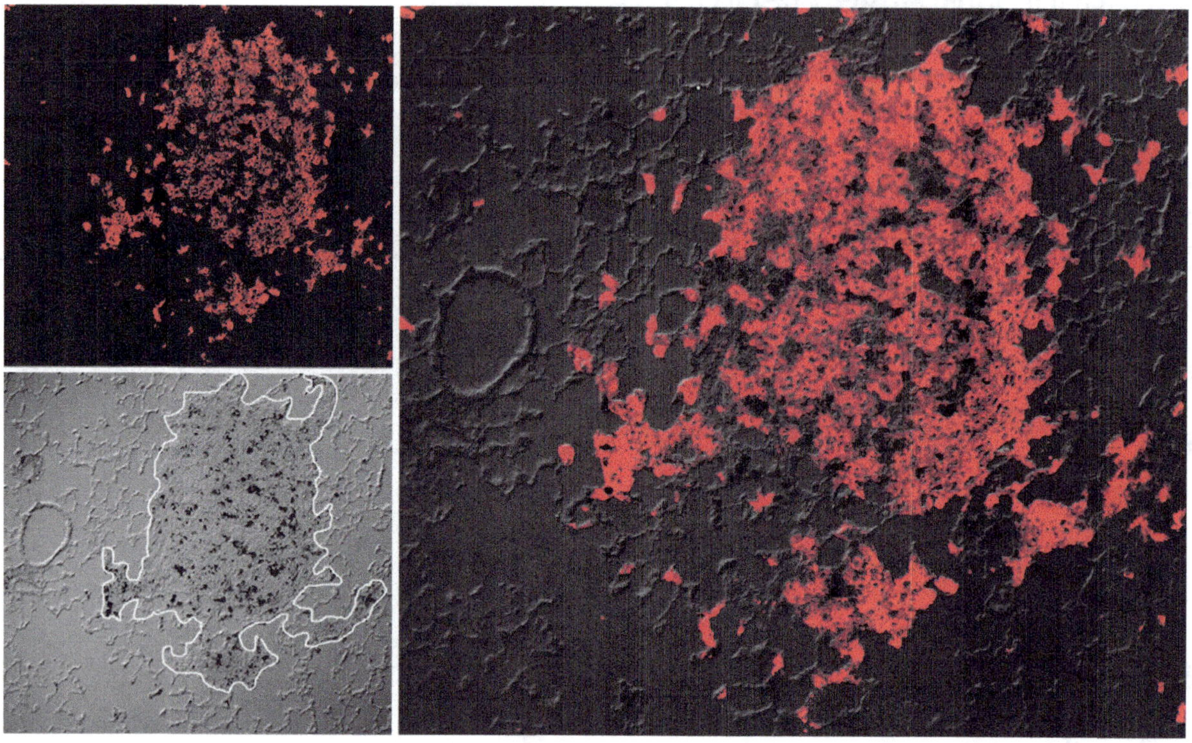

Plate 17 • Tumour-localization by adoptively transferred A-NK cells. (A) 10×10^6 A-NK cells generated from $Thy1.1^+$ congenic BL6 mice were injected IV into Thy1.2+ BL6 mice with 9 days old B16 melanoma lung tumours. The animals received 60,000 IU Peg-Il-2 every 12 hours for 3 days. At 72 hours after the A-NK cell injection, lungs were removed and fresh frozen. Eight micron cryosections were stained with PE-conjugated anti-Thy1.1 antibody. Note the selective accumulation of the PE-Thy1.1+ A-NK cells within the B16 lung metastases. Bar = 200 μm. Upper left insert: red fluorescent picture showing the A-NK cells. Lower left insert: DIC picture showing melanin-containing B16 tumour (outlined).

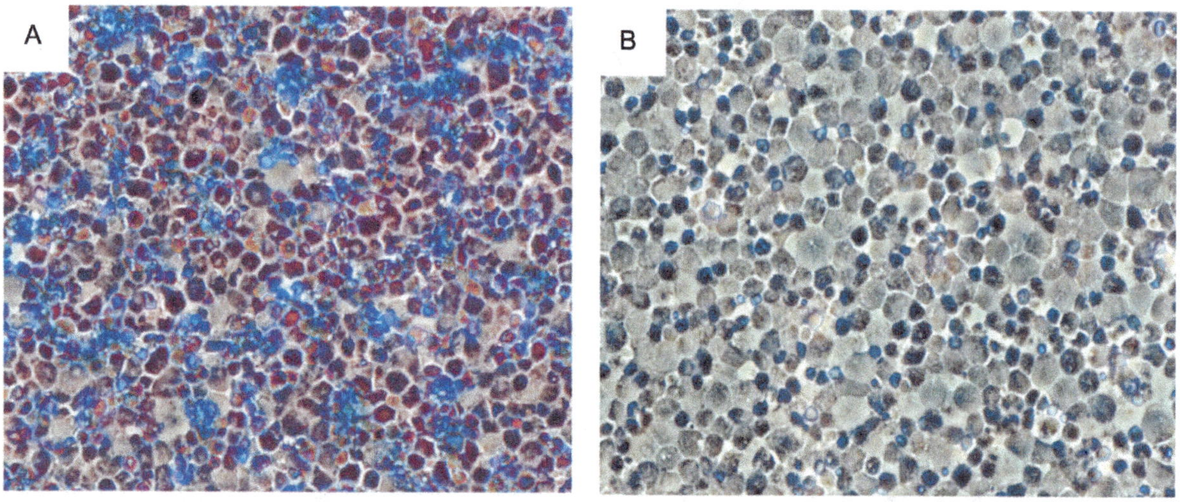

Plate 18 • Prussian blue staining of NK-92 cells labelled with MRI contrast agent. a) After labelling with ferumoxides, NK-92 cells were fixed and stained with Prussian blue. The cells stained bright blue, confirming the intracellular presence of iron as well as the adequacy of labelling. b) Non-labelled NK cells exposed to Prussian blue do not show any blue staining suggestive of iron.

Chapter **Forty**

40

Natural killer cells in atherosclerosis

Godfrey S. Getz, Catherine A. Reardon

CHAPTER CONTENTS

. . . but now the affair becomes even more complex.

The Brothers Karamazov, Fyodor Dostoyevsky

ABSTRACT

Atherosclerosis is a chronic inflammatory disorder mediated by hyperlipidaemia. Cells of both the innate and adaptive immune system play important roles in atherogenesis. Natural killer (NK) cells have been detected in atherosclerotic lesions in humans and mice, although at low levels, but their role, if any, in atherogenesis is only beginning to be examined. Murine models of atherogenesis have been used to examine if NK cells contribute to atherogenesis via their cytotoxic properties and/or capacity to produce pro-atherogenic cytokines. While more research needs to be done, these studies suggest that NK cells may influence atherosclerosis development as well as lipoprotein metabolism, likely involving the production of cytokines.

KEY WORDS

Atherosclerosis, Cytotoxicity, Perforin, *Lyst*^{beige} mutation, Ly49A

Atherosclerosis is now well recognized to be a specific example of chronic inflammation induced, at least in mice and to some extent in humans, by hypercholesterolemia and probably oxidative modification of the elevated cholesterol containing lipoproteins in the plasma. The characteristic lipid-loaded foam cells in atherosclerotic plaques are derived from macrophages that have taken up the lipoproteins that have entered into the arterial vessel wall. However, other cells of both the innate immune system and the adaptive immune system also play important roles in atherogenesis, even though they are present in atherosclerotic lesions in lower numbers (Getz, 2005; Hansson and Libby, 2006). Natural killer (NK) cells are among the immune cells that are of potential interest in relation to atherosclerosis. They have been found in small numbers in atherosclerotic plaques, produce cytokines that are known to influence atherogenesis and are present in the liver, which has important functions in the production and clearance of lipoprotein particles. However, the precise role of NK cells, if any, in atherogenesis and lipoprotein metabolism is not clear and will be discussed in detail below.

NK cells, as the name suggests, were first defined as cells capable of killing tumour cells and virus-infected cells. This cytotoxicity is mostly accomplished by the granzyme/perforin system, with perforin influencing the permeability of target cells to granzymes secreted from cytotoxic lymphocytes (NK cells and CD8$^+$ T cells) resulting in apoptosis of the target cell. In addition to their cytotoxic activity these cells also secrete cytokines, most notably IFN-γ, TNFα and GM-CSF (Yokoyama et al., 2004). These cytokines may directly contribute to cytotoxicity of the target cells and also influence the activity of other cells including macrophages, dendritic cells and T cells (Vivier et al., 2008), immune cells that are found in atherosclerotic lesions (Getz, 2005).

However, under normal conditions most of the circulating NK cells are quiescent. The major sites of NK cells are the liver, spleen and blood. Regulation of NK cell activity and development is quite complex. Both activating and inhibitory receptors influence the activity of NK cells (Cerwenka and Lanier, 2001). Engagement of activating receptors on the surface of NK cells by ligands on target cells results in the secretion of pro-inflammatory cytokines and the exocytosis of granules containing perforin and granzymes. Activating receptors include CD16 and NKG2D in mice and humans, and Ly49G and Ly49H in mice. NK cell activation is also stimulated by cytokines and chemokines such as IFNα/β, IL-12, IL-15 and IL-18 released from dendritic cells and macrophages (Vivier et al., 2008). On the other hand, inhibitory receptors recognize as ligands MHC class I surface molecules on cells. The inhibitory receptors include KIR in humans, Ly49A in mice and CD94/NKG2A in humans and mice. Regulation of the activity of NK cells by these receptors is complex, since any given NK cell may express multiple activating and inhibitory receptors. Nonetheless, the inhibitory receptors appear to be dominant in that the activation of NK cells is reduced when both activating and inhibitory receptors are engaged.

NK cells are bone marrow-derived lymphocytes originating from haematopoietic stem cells through a variety of intermediate precursor cells. Their development depends upon, among other factors, the haemopoietic-specific transcription factor purine rich box-1 (PU.1) Id2, IL-15 and IL-15Rα, as well as the common γ-chain of the IL-2 receptor (Lian and Kumar, 2002). NK cells do not express a T-cell receptor and hence are present in conventional immune-deficient murine models such as recombination activating gene (RAG) knockout animals and nude mice. They are $CD3^-$ and are characterized by cell surface expression of CD56 in humans and NK1.1 in certain mouse strains. In the final stages of NK cell maturation they also express α2 integrin (DX5/CD49b). NK cells in the liver do not express DX5 (Yokoyama et al., 2004). Thus the majority of the NK cells in the liver are not fully mature under normal steady-state conditions. When stimulated by tumour cells or virus-infected cells, NK cells undergo substantial proliferation, which is reduced unless stimulation by virus-infected cells continues.

NK cells in murine atherosclerosis

Atherosclerosis develops at specific sites within large and medium-sized arteries, with haemodynamics of blood flow influencing the location within the arteries where lesions develop (VanderLaan and Reardon, 2005). Currently, mice are the most widely used experimental animal model of atherosclerosis. While wild-type mice are relatively resistant to the development of atherosclerosis, several genetically modified mice have been generated that develop atherosclerosis (Getz and Reardon, 2006). In all of these models there is an increase in plasma lipoprotein levels, especially apoprotein B-containing lipoproteins. Two of the most commonly used models are mice deficient in the low density lipoprotein (LDL) receptor and apoprotein E. LDL receptor-deficient ($LDLR^{-/-}$) mice lack the high affinity receptor for the uptake of LDL. The feeding of a high fat/high cholesterol atherogenic diet leads to the accumulation of very low density lipoproteins (VLDL) and LDL in the plasma and the development of atherosclerosis (Ishibashi et al., 1993). Apoprotein E-deficient ($apoE^{-/-}$) mice develop atherosclerosis on chow diet due to the accumulation of atherogenic remnant lipoprotein particles (apoE is the ligand that mediates the high affinity removal of these plasma lipoproteins) and the lack of expression of apoE by cells in the arterial vessel wall among other processes (Getz and Reardon, 2009).

The role of NK cells in atherosclerosis is not clearly defined. That they are worthy of study primarily rests on finding these cells in small numbers in both early human lesions (Millonig et al., 2002) and in early murine lesions in $LDLR^{-/-}$ atherogenic mice (VanderLaan and Reardon, 2005). The recruitment of NK cells into lesions is accomplished in part by the chemoattractants monocyte chemotactic protein (MCP-1) and fractalkine (Allavena et al., 1994, Umehara et al., 2004). While these proteins are chemoattractants for other immune cells in addition to NK cells, the effect of deletion of these chemoattractants on atherosclerosis could be in part related to decreased recruitment of NK cells, though this was not specifically addressed in these experimental models (Charo and Taubman, 2004; Tedqui and Mallat, 2006). IFN-γ is a major cytokine produced by NK cells as well as other immune cells, and it is well recognized that IFN-γ is a major pro-atherogenic cytokine (Tedqui and Mallat 2006). The extent to which IFN-γ signalling from NK cells contributes to this phenotype has not been specifically addressed. In atherosclerosis, the major pro-atherogenic cell of the adaptive immune system is the Th1cell, which actively produces IFN-γ. In two murine atherosclerosis models in which these T cells are absent, CD4 knockouts (Elhage et al., 2004) and $RAG^{-/-}$ (Reardon et al., 2003), the surprising observation was made that atherosclerosis was increased in the aorta and in the innominate artery, respectively. In both cases NK cells were still present and their role in producing this unexpected observation has not been specifically addressed. There are of course other explanations for these observations. Lymphotoxin α -deficient mice lack both NK cells and natural killer T (NKT) cells (Iizuka et al., 1999) and they have decreased atherosclerosis (Schreyer et al., 2002). None of the studies mentioned above specifically targeted NK cells, but the various manipulations may have influenced the activity or recruitment of these cells to the vessel wall, which may have contributed to the atherosclerotic phenotype.

The specific involvement of NK cells in experimental atherogenesis has received relatively scant attention. This is in part attributable to the difficulty of identifying unique markers of NK cells or unique cytokines or transcription factors involved in NK cell development, which would allow for the selective increase or elimination of these cells since many of these proteins are shared with other cells of the immune system.

Role of cytokines and cytotoxicity

NK cells may contribute to atherogenesis either by their cytotoxic properties or by the capacity to produce pro-atherogenic cytokines. The granzyme/perforin-containing secretory granules are present in NK cells, NKT cells and cytotoxic T cells. To test the involvement of granule-dependent cytotoxicity in atherogenesis, perforin-deficient mice were crossed with LDLR$^{-/-}$ mice (Schiller et al., 2002). These double knockout mice exhibited a modest increase in plasma cholesterol levels when fed an atherogenic diet for 16 weeks, but nevertheless had similar levels of atherosclerosis in the aortic sinus and total aorta as control LDLR$^{-/-}$ mice (Table 40.1). This tends to argue that the cytotoxicity of NK cells, and perhaps other cells, does not contribute to atherogenesis, at least in this atherogenic murine model. In this same report, the authors also explored the influence of the *Lyst*^{beige} mutation on atherosclerosis in the LDL receptor-deficient background. The Lyst protein apparently influences the exchange of membrane material between the Golgi network and late endosomes, likely affecting lysosomal vesicle fusion or fission (Holt et al., 2006). The *Lyst*^{beige} mutation results in enlarged perinuclear granules in and defective cytolysis by NK cells. The human homolog of this gene when mutated results in the Chédiak–Higashi syndrome. The *Lyst*^{beige}, LDLR$^{-/-}$ mice have increased atherosclerosis compared with the corresponding control LDLR$^{-/-}$ mice when fed an atherogenic diet for 16 weeks. The exacerbation of atherosclerosis occurred despite a reduction of total plasma cholesterol levels. While these studies were largely directed at NK cell function, the interpretation of this outcome is confounded by the fact that enlarged perinuclear granules are seen also in cytotoxic T lymphocytes and neutrophils affecting their function as well as NK cell function, although neither of these cell types are thought to be major actors in atherogenesis (Getz, 2005; VanderLaan and Reardon, 2005). In addition, macrophage function also appears to be partially compromised as a result of this mutation and macrophages are an important immune cell in the development of atherosclerosis. Thus, while this study draws attention to the possible involvement of NK cells in atherogenesis, it cannot be regarded as specifically involving NK cells (Getz, 2002).

Contribution of T cells and B cells

To eliminate the contribution of cytotoxic T cells to the atherosclerosis phenotype in this model, these authors also generated *Lyst*^{beige}, LDLR$^{-/-}$, RAG$^{-/-}$ mice. Since RAG$^{-/-}$ mice lack both T and B cells, this leaves the major sites of action of the *Lyst*^{beige} mutation in this model in NK cells and macrophages. In this case, the introduction of the *Lyst*^{beige} mutation in the immune-deficient LDLR$^{-/-}$ model led to an increase in atherosclerosis in the aortic sinus, while atherosclerosis in the

Table 40.1 Effect of manipulation of NK cells on atherosclerosis

	Aortic sinus atherosclerosis	Total aorta atherosclerosis	Plasma cholesterol	Reference
Defective cytolysis				
Perforin$^{-/-}$/LDLR$^{-/-}$	No effect	No effect	↑	Schiller et al. (2002)
Lyst^{beige}, LDLR$^{-/-}$	↑	↑	↓	Schiller et al. (2002)
Lyst^{beige}, LDLR$^{-/-}$, RAG$^{-/-}$	↑	No effect	↑	Schiller et al. (2002)
Lyst^{beige}, apoE$^{-/-}$ (atherogenic diet)	↑ (less stable)	↓ (12 weeks)	↑	Petrovan et al. (2008)
Lyst^{beige}, apoE$^{-/-}$ (chow diet)	↓	No effect	No effect	Petrovan et al. (2008)
Decreased activation				
Ly49Atg, LDLR$^{-/-}$	↓	↓	No effect	Whitman et al. (2004)

whole aorta was not significantly increased. This suggests that, at least in the aortic sinus, cytolytic activity of NK cells may be anti-atherogenic. In contrast to the effect of the *Lyst*^{beige} mutation on plasma cholesterol levels in immune competent LDLR$^{-/-}$ mice, plasma cholesterol was increased in the immune deficient *Lyst*^{beige}, LDLR$^{-/-}$ mice. The basis for these excursions in plasma cholesterol is unclear and merits further study to uncover the mechanisms involved, especially as immune deficiency in LDLR$^{-/-}$ mice leads to lower plasma lipids (Reardon et al., 2003). It is possible that the NK cells in the liver (Racanelli and Rehermann, 2006) may be influencing various aspects of lipoprotein homeostasis by hepatocytes.

To ascertain whether the effect of the *Lyst*^{beige} mutation was peculiar to the LDLR$^{-/-}$ model of atherogenesis, similar studies were undertaken with the apoprotein E-deficient (apoE$^{-/-}$) atherogenic mouse model (Petrovan et al., 2008). The effect of the *Lyst*^{beige} mutation in this atherogenic model was quite complex depending on whether the animals were fed chow or an atherogenic diet. When mice are fed the atherogenic diet, the *Lyst*^{beige} mutation is associated with an increase in aortic sinus lesions, but only after 16 weeks of diet feeding. In addition, the composition of the aortic sinus lesions was altered with less collagen being present in the *Lyst*^{beige} mice, suggesting that it is a less stable atherosclerotic lesion. In contrast, total aortic lesion area was reduced in these mice after 12 weeks of diet. This type of aortic site-specific atherogenic response in murine atherosclerosis models to various experimental manipulations has been observed previously (VanderLaan et al., 2004). The *Lyst*^{beige} mutation led to an increase in plasma cholesterol levels in atherogenic diet fed mice. On the other hand, when the mice were maintained on a low fat chow diet, there were no differences in plasma cholesterol levels and the aortic sinus lesions were profoundly reduced, with no change in total aortic lesion area. Since the *Lyst*^{beige} mutation may influence the function of non-immune cells as well as haematopoietically derived cells, bone marrow transplantation studies were performed. These bone marrow transplantation studies indicated that both bone marrow-derived and non-bone marrow-derived cells contributed to the reduced atherosclerosis in the chow-fed *Lyst*^{beige}, apoE$^{-/-}$ mice. This argues that even if NK cells are influential, they are not able to account on their own for the reduced atherosclerosis in *Lyst*^{beige}, apoE$^{-/-}$ mice fed chow. In summary, the *Lyst*^{beige} mutation has complex influences on atherosclerosis, depending upon diet, vascular site of atherosclerosis and atherogenic murine model employed (LDLR$^{-/-}$ or apoE$^{-/-}$). Thus it is not possible, based upon these experiments, to definitively implicate NK cell-mediated cytotoxicity in the process of atherogenesis. In these experiments, NK cells are still present and capable of secreting cytokines.

Granzyme A promoter systems

A second approach to the involvement of NK cells in atherosclerosis takes advantage of the system in which the granzyme A promoter drives the overexpression of the NK cell inhibitory receptor Ly49A. Granzyme A is expressed in NK cells as well as cytotoxic T cells. In one transgenic line, Ly49A was expressed in NK cells and half of all T cells and a reduction in NK cell (NK1.1^{+} CD3^{-} cells) number was noted in the blood, spleen, liver and lung (Kim et al., 2000). This transgenic line also exhibited a reduction in IFN-γ production in vivo in response to lipopolysaccharide but not in response to IL-12. When the bone marrow from these transgenic mice was transplanted into female LDLR$^{-/-}$ recipients, there was a reduction in lesion formation in both the aortic sinus and the aortic arch, particularly at the former site, in mice fed an atherogenic diet for 8 weeks (Whitman et al., 2004). Since small numbers of NK cells were detected in the aortic sinus lesion in both the Ly49A transgenic and control recipient mice, this suggests that the transgene appeared to influence the functional activity rather than the number of cells in the lesions. However, it should be noted that a 50% reduction in splenic NK1.1^{+} CD3^{-} cells was noted in the Ly49A bone marrow recipient mice, although this did not achieve statistical significance. A similar atherosclerosis response was noted in apoE$^{-/-}$ mice (Whitman and Ramsamy, 2006). These effects on atherosclerosis occurred without any difference in plasma cholesterol levels. While this study suggests that activated NK cells are pro-atherogenic, the mechanism of this action in the atherosclerosis context is far from clear. The Ly49A transgenic model that influences both cytokine and granule secretion yields different results than those obtained with the *Lyst*^{beige} mice, suggesting that the action of NK cells in this context is mainly via their cytokine profile rather than cytotoxicity.

In each of the experiments discussed above there was an attempt to manipulate the function or activity of NK cells as a single participant in lipoprotein metabolism or atherogenesis. However, these cells are important elements of a cross talk with other cells of the immune system, both innate and adaptive (Shi and Van Kaer, 2006). An early attempt at exploring these interactions is reflected in the crossing of the *Lyst*^{beige} mutation into the immune-deficient LDLR$^{-/-}$ background. There is a great need for more detailed analysis of the numbers, activity and cytokine production of NK cells throughout the course of atherogenesis and progression. Little has been done so far. As a beginning, it might be best analysed in the immune deficient mouse models, where the NK cells are the only lymphocytes capable of responding to atherogenic stimuli. In these RAG-deficient models, robust atherosclerosis still develops (Reardon et al., 2003).

Conclusion

From the above brief review it is evident that the role of NK cells, important elements of the innate immune response system, has received only modest attention in relation to atherosclerosis. These cells have the potential to affect both lipoprotein homeostasis, taking note of the relatively high content of NK cells in the liver, the major organ affecting lipoprotein metabolism, and influence the inflammatory response at the level of the vessel wall. In the latter site, the number of NK cells found in atherosclerotic lesions is relatively small, which may argue for a limited role in the vessel itself. In any event, there is much work to be done to clarify the roles of these cells in atherosclerosis and how they are activated in the context of the atherosclerotic lesion.

Acknowledgements

This work is funded by grants from the National Heart Lung Blood Institute of NIH and the Leducq Foundation.

References

Allavena, P., Bianchi, G., Zhou, D., van Damme, J., Jílek, P., Sozzani, S. and Mantovani, A. (1994). Induction of natural killer cell migration by monocyte chemotactic protein-1, -2, and -3. Eur J Immunol **24**, 3226–3233.

Cerwenka, A. and Lanier, L.L. (2001). Natural killer cells, viruses and cancer. Nat Rev Immunol **1**, 41–49.

Charo, I.F. and Taubman, M.B. (2004). Chemokines in the pathogenesis of vascular disease. Circ Res **95**, 858–866.

Elhage, R., Gourdy, P., Brouchet, L., Jawien, J., Fouque, M.J., Fievet, C., Huc, X., Barreira, Y., Couloumiers, J.C., Arnal, J.F. and Bayard, F. (2004). Deleting TCRαβ+ or CD4+ T lymphocytes leads to opposite effects on site-specific atherosclerosis in female apolipoprotein E-deficient mice. Am J Pathol **165**, 2013–2018.

Getz, G.S. (2002). Do natural killer cells participate in a killer vascular disease? Arterioscler Thromb Vasc Biol **22**, 1251–1253.

Getz, G.S. (2005). Thematic review series: the immune system and atherogenesis. Immune function in atherogenesis. J Lipid Res **46**, 1–10.

Getz, G.S. and Reardon, C.A. (2006). Diet and murine atherosclerosis. Arterioscler Thromb Vasc Biol **26**, 242–249.

Getz, G.S. and Reardon, C.A. (2009). Apoprotein E as a lipid transport and signaling protein in the blood, liver, and artery wall. J Lipid Res **50**(Suppl.), S156–S161.

Hansson, G.K. and Libby, P. (2006). The immune response in atherosclerosis: a double-edged sword. Nat Rev Immunol **6**, 508–519.

Holt, O.J., Gallo, F. and Griffiths, G.M. (2006). Regulating secretory lysosomes. J Biochem **140**, 7–12.

Iizuka, K., Chaplin, D.D., Wang, Y., Wu, Q., Pegg, L.E., Yokoyama, W.M. and Fu, Y.X. (1999). Requirement for membrane lymphotoxin in natural killer cell development. Proc Natl Acad Sci U S A **96**, 6336–6340.

Ishibashi, S., Brown, M.S., Goldstein, J.L., Gerard, R.D., Hammer, R.E. and Herz, J. (1993). Hypercholesterolemia in low density lipoprotein receptor knockout mice and its reversal by adenovirus-mediated gene delivery. J Clin Invest **92**, 883–893.

Kim, S., Iizuka, K., Aguila, H.L., Weissman, I.L. and Yokoyama, W.M. (2000). In vivo natural killer cell activities revealed by natural killer cell-deficient mice. Proc Natl Acad Sci U S A **97**, 2731–2736.

Lian, R.H. and Kumar, V. (2002). Murine natural killer cell progenitors and their requirements for development. Semin Immunol **14**, 453–460.

Millonig, G., Malcom, G.T. and Wick, G. (2002). Early inflammatory-immunological lesions in juvenile atherosclerosis from the Pathobiological Determinants of Atherosclerosis in Youth (PDAY) study. Atherosclerosis **160**, 441–448.

Petrovan, R.J., Yuan, Y. and Curtiss, L.K. (2008). Expression of the *Lyst*^{beige} mutation is atheroprotective in chow-fed apolipoprotein E-deficient mice. J Lipid Res **49**, 429–437.

Racanelli, V. and Rehermann, B. (2006). The liver as an immunological organ. Hepatology **43**, S54–S62.

Reardon, C.A., Blachowicz, L., Lukens, J., Nissenbaum, M. and Getz, G.S. (2003). Genetic background selectively influences innominate artery atherosclerosis: immune deficiency as probe. Arterioscler Thromb Vasc Biol **23**, 1449–1454.

Schiller, N.K., Boisvert, W.A. and Curtiss, L.K. (2002). Inflammation in atherosclerosis. Lesion formation in LDL receptor-deficient mice with perforin and *Lyst*^{beige} mutations. Arterioscler Thromb Vasc Biol **22**, 1341–1346.

Schreyer, S.A., Vick, C.M. and LeBoeuf, R.C. (2002). Loss of lymphotoxin-alpha but not tumor necrosis factor-alpha reduces atherosclerosis in mice. J Biol Chem **277**, 12364–12368.

Shi, F.D. and Van Kaer, L. (2006). Reciprocal regulation between natural killer and autoreactive T cells. Nat Rev Immunol **6**, 751–760.

Tedqui, A. and Mallat, Z. (2006). Cytokines in atherosclerosis: pathogenic and regulatory pathways. Physiol Rev **86**, 515–581.

Umehara, H., Bloom, E.T., Okazaki, T., Nagano, Y., Yoshie, O. and Imai, T. (2004). Fractalkine in vascular biology. From basic research to clinical disease. Arterioscler Thromb Vasc Biol **24**, 34–40.

VanderLaan, P.A., Reardon, C.A. and Getz, G.S. (2004). Site specificity of atherosclerosis. Site-specific responses to atherosclerotic modulator. Arterioscler Thromb Vasc Biol **24**, 12–22.

VanderLaan, P.A. and Reardon, C.A. (2005). The unusual suspects: an overview of the minor leukocyte populations in atherosclerosis. J Lipid Res **46**, 829–838.

Vivier, E., Tomasello, E., Baratin, M., Walzer, T. and Ugolini, S. (2008). Functions of natural killer cells. Nat Immunol 9, 503–510.

Yokoyama, W.M., Kim, S. and French, A.R. (2004). The dynamic life of natural killer cells. Annu Rev Immunol **22**, 405–429.

Whitman, S.C. and Ramsamy, T.A. (2006). Participatory role of natural killer and natural killer T cells in atherosclerosis: lessons learned from in vivo mouse models. Can J Physiol Pharmacol **84**, 67–75.

Whitman, S.C., Rateri, D.L., Szilvassy, S. J., Yokoyoma, W. and Daugherty, A. (2004). Depletion of natural killer cell function decreases atherosclerosis in low-density lipoprotein receptor null mice. Arterioscler Thromb Vasc Biol **24**, 1049–1054.

Chapter **Forty-One**

41

Natural killer cells in leukaemia

Jérôme Rey, Cyril Fauriat, Alessandro Moretta, Régis Costello, Daniel Olive

CHAPTER CONTENTS

No exorciser harm thee
nor no witchcraft charm thee
ghost unlaid forbear thee . . .

William Shakespeare *Cymbeline*, Act IV, Scene 2

ABSTRACT

In the haematology field, several lines of evidence suggest that natural killer (NK) cells participate in the antileukaemia immune response as recent data supporting a role for NK cells into the graft-versus-leukaemia effect observed in allogeneic bone marrow transplantation. Abnormalities of NK cell receptors and their ligands are found in patients with leukaemia that can represent an escape from innate immune surveillance. Some challenges remain unanswered, such as: (1) What are the precise mechanisms of leukaemic cell killing by NK cells? (2) What are the therapeutic possibilities to manipulate NK cell receptor–ligand interaction to increase clearance of blastic cells? Many of these questions are currently under investigation and open new horizons in NK cell immunotherapy for patients with leukaemia.

KEY WORDS

Natural cytotoxicity receptors, Leukaemia, Immune escape, Immunotherapy

Introduction

The natural killer (NK) cells participate in the innate immune response to transformed cells that are recognized as 'nonself' with absent or abnormal expression of human leukocyte antigen (HLA) class I molecules. According to the 'missing self hypothesis', a negative signal is delivered to NK cells when their inhibitory receptors are engaged by the specific HLA class I molecules (Farag et al., 2002). However, the missing-self hypothesis is not sufficient to fully explain NK cell biology. During NK cell development, host MHC molecules induce licensing through expression of their inhibitory receptors that are involved in negative signalling and require expression of ITIM (Parham, 2006). In addition, NK cell activation requires a positive signal delivered by the engagement of activating receptors (Farag et al., 2002). Recent insights into NK cell biology have stimulated novel studies of NK cells in leukaemia in order to determine if deficient NK cells are implicated in immune escape mechanisms of leukaemia (Costello et al., 2004). The role of NK cells in recognition of individual leukaemia subtypes varies, and the principal abnormalities are described here. We extend this analysis of NK cells and leukaemic diseases to myelodysplasia and Philadelphia chromosome negative chronic myeloproliferative disorders since they can occasionally evolve into acute myeloid leukaemia (AML).

Leukaemia is a heterogeneous clonal disorder of haematopoietic progenitor cells that has lost the ability to differentiate normally (Estey and Dohner, 2006). This disorder is subdivided into acute and chronic leukaemia and then into myeloid or lymphoid leukaemia. In adults, chronic lymphocytic leukaemia (CLL) is the most frequent leukaemia and AML the most common myeloid leukaemia in developed countries. AML has the worse prognosis with a high rate of relapse and rare cures that depend on the type of AML and characteristics of the patient, among which age plays a prominent role (Costello et al., 2004; Farag and Caligiuri, 2006). Karyotypic features are a predominant factor determining pathogenesis and prognosis (Estey and Dohner, 2006). In particular, chronic myeloid leukaemia (CML) is characterized by the translocation between chromosome 9 and 22 (i.e. Philadelphia chromosome) leading to the formation of a BCR-ABL chimeric gene. The standard care of treatment for patients with leukaemia is chemotherapy with or without bone marrow transplantation. Patients with CML have benefited from the development of a tyrosine kinase inhibitor targeting BCR-ABL, imatinib mesylate.

Little is know about the effect of chemotherapeutic drugs used to treat leukaemia on the functions of human NK cells (Markasz et al., 2007). Generally, chemotherapeutic drugs decrease the absolute count of NK cells. In vivo, apart from their action on cancer targets (lysis, modulation of ligands for receptors of effectors cells), chemotherapeutic drugs can influence the phenotype or the activity of NK cells. Recent studies have analysed the cytotoxic activity of NK cells following incubation with chemotherapeutic drugs (Markasz et al., 2007). Some drugs inhibit and others enhance NK cell-mediated killing of targets cells (Markasz et al., 2007). These results are of considerable interest in exploring the immunotherapy of patients with NK cells in combination with chemotherapy.

NK cells as effectors of antileukaemia activity

NK cells have been primarily analysed in AML or CML. In general, the activity of autologous NK cells against leukaemic cells is frequently reduced (Costello et al., 2004). Even though possible cytotoxicity of leukaemic cells or leukaemic cells lines by NK cells has been described in the different subtypes of leukaemia, the mechanisms underlying the interaction and destruction of these cells are not clearly defined. A relatively high recurrence rate with leukaemic relapse is noted following standard or high dose chemotherapy in these diseases, suggesting that leukaemic cells can escape from the immune system (Costello et al., 2004; Farag and Caligiuri, 2006).

The major findings favouring a role for NK cells in antileukaemia activity is based on results arising from haematopoietic stem cell transplantation (Ruggeri et al., 2004, 2008). Major advances in the understanding of antileukaemia activity of NK cells are based on study of alloreactive NK cells in the setting of transplantation with better definition of the roles of activating or inhibitory receptors (Moretta et al., 2008; Pende et al., 2009). The role of NK cells in haematopoietic stem cell transplantation is detailed in a separate chapter on this book (see Chapter 42). With the increasing knowledge of NK cell receptors and their ligands on targets cells, analysis of NK cell phenotype or activity during leukaemia has been recently revisited (Fauriat et al., 2003).

NK cells in acute leukaemia

NK cells in acute myeloid leukaemia

In AML patients, NK cell activity correlates positively with the relapse-free survival. Moreover, NK cells adhere to bone marrow fibroblasts competing with myeloid—but not with lymphoid—leukaemic blasts for binding to the microenvironment (Bendall et al., 1995). This suggests that NK cells may inhibit leukaemia cell growth by impeding its interactions with bone marrow stroma that are necessary for leukaemic cell proliferation. Altogether, these data suggest that NK cells may play an important role in the control and clearance of leukaemia cells (Farag and Caligiuri, 2006).

Apart from the significant role of NK cells in allogeneic stem cell transplantation (Ruggeri et al., 2008), a role for NK cells against autologous leukaemic cells needs to be clarified (Costello et al., 2004). In absolute numbers, NK cells in AML are within the normal range (Fauriat et al., 2003). In general, NK cells do not carry the cytogenetic changes observed in AML blasts.

Inhibitory receptors and AML

The activity of NK cells is controlled by the expression of inhibitory receptors that interact with HLA class I molecules. The variety of KIR haplotypes depends on the number and type of KIR genes (stimulatory versus inhibitory) that are inherited and on the allelic polymorphism. Little is know about the role of KIR and HLA genes in the pathogenesis of leukaemia. A preliminary study has noted a dominance of inhibitory over activating signals in leukaemic patients (Verheyden et al., 2004). Deficient HLA class I expression (reduced expression or loss of HLA class I alleles) has been described in many types of malignant cells, including solid tumours and leukaemias (Elkins et al., 1984). Diminished class I molecules expression could favour tumour destruction by NK cells that are activated when detection of 'self' molecules is lacking or alternatively serve as less favourable targets for T cells. Frequent partial downregulation of certain HLA class I molecules is observed in leukaemic cells in comparison with normal autologous cells. HLA class I molecules belonging to the Bw6 group are more frequently downregulated than those belonging to the HLA-Bw4 group (Verheyden and Demanet, 2008). The expression of HLA-C on leukaemic cells, which interacts with the majority of inhibitory KIRs, has not been studied to date (Verheyden and Demanet, 2008). The aberrant expression of nonclassical HLA class I antigens on leukaemic cells may represent a mechanism of immune escape following interaction with inhibitory receptors on NK cells. However, preliminary studies have not demonstrated HLA-G or HLA-E expression on leukaemic cells (Verheyden and Demanet, 2008).

Activating receptors and AML

NK cells need activating signals derived from NK cell activating receptors such as NKp30, NKp36, NKG2D and DNAM-1. Their interaction with ligands that may be expressed on leukaemic cells (Fauriat et al., 2003) is relatively inefficient, perhaps delivery of these activating signals deficient through distinct mechanisms (Figure 41.1). Defective cytotoxicity of AML cells can be explained by abnormalities of activating NK cell-receptor expression (Fauriat et al., 2003). In vitro studies show that NCRs mediate NK-dependent leukaemic cell lysis (Costello et al., 2002; Pende et al., 2005). We have demonstrated that the majority of NK cells in AML patients display an abnormal phenotype characterized by downregulation of NKp30 and NKp46 NCR that correlates with a defective cytolytic function against autologous AML cells (NCR^{dull}). Moreover, AML blasts retained normal sensitivity to allogeneic NK cells (Costello et al., 2002). Our study enrolled 71 patients with AML at diagnosis (Fauriat et al., 2007). The abnormal phenotype of NK cells was not restricted to a FAB subtype or associated with a specific cytogenetic abnormality (Fauriat et al., 2007).

We have analysed the NK phenotype sequentially following treatment and demonstrated that defective expression was likely dependant on the presence of AML blasts. NCR surface density is restored completely (for NKp46) or partially (for NKp30) when a complete remission is obtained following chemotherapy (Fauriat et al., 2007).

The mechanisms underlying these altered phenotypes are not clearly understood. TGFβ1 downregulates the expression of NKp30 and NKG2D but not NKp46 in normal NK cells. This is not be implicated in AML cells since TGFβ1 levels are normal in the serum of these patients (Fauriat et al., 2007). Moreover, TGFβ1 cannot

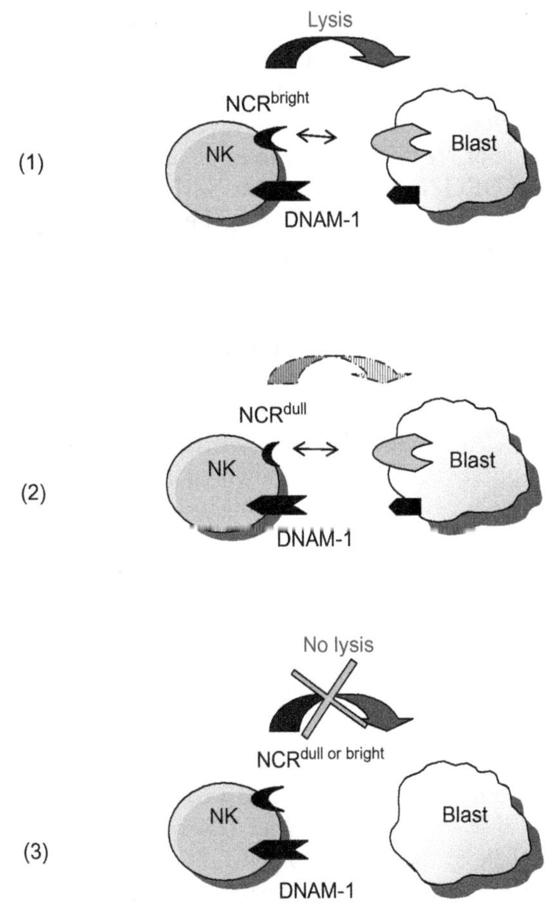

Figure 41.1 • Leukaemia and NK cells interaction. Leukaemia cell lysis may rely mainly on DNAM-1 and NCRs expressed on NK cells. Normal NK cells expressing high levels of NCRs (NCR^{bright}) efficiently destroyed AML blasts, which expressed DNAM-1 and NCR ligands (Panel 1). However, in the majority of AML patients, the NK cells are characterized by a downregulation of the NCRs (NCR^{dull} phenotype), leading to a reduced cytotoxicity against blast cells (Panel 2). In addition, since some AML blasts are not destroyed despite NCR^{high} phenotype, an alternative hypothesis is the absence or very low expression of the DNAM-1 or NCR ligands on blasts cells (Panel 3).

explain the downregulation of NKp46 that we observed. We have also demonstrated defective killing of dendritic cells (DCs) by autologous NK cells from AML patients, this observation is in line with the downregulation of NKp30 since this receptor mediates cytotoxicity against DCs (Fauriat et al., 2005). The persistence of immature DCs due to defective killing by NK cells in AML may contribute to leukaemia escape from adaptive immunity (Fauriat et al., 2005). A correlation between the downregulation of NCR expression and decreased survival of AML patients, suggests that defective NK cell activity due to in part to NCR downregulation might be an independent marker associated with the survival of patients, as well as age or cytogenetics (Fauriat et al., 2007). The probability of survival at 5 years was 64% for patients with normal NCR expression at diagnosis, 43% for discordant NCR (patients having at least one NCR downregulated) and 31% for downregulated NCR in our study of 71 AML patients (Fauriat et al., 2007).

Whether NKG2D plays a role in AML remains to be determined (Verheyden and Demanet, 2008). We failed to inhibit AML cell killing by NK cells by mAb-mediated masking of NKG2D (Costello et al., 2002). The expression of NKG2D on NK cells was within normal limits in our study (Fauriat et al., 2007). Leukaemia expression of NKG2D ligands does not prove that this molecule is involved in their killing by NKG2D expressing cells since surface-release of these molecules by metalloproteinases observed in some tumour cell lines could inhibit NK cell function (Salih et al., 2003). Contradictory observations have been made about the levels of soluble MICA/B in the serum of patients with AML and their putative role in the downregulation of NKG2D (Salih et al., 2003) with some higher and others lower.

Immune escape can also be due to defective activating NK cell receptor–ligand interactions (Verheyden and Demanet, 2008). The putative NCR putative ligands are expressed during the maturation of normal myelomonocytic cells but are relatively low on AML blasts (Nowbakht et al., 2005). The expression of NKG2D ligands by leukaemic cells has been observed by Salih et al. (2003) but not by Pende et al (2002). NKG2D ligands on AML blasts have been found at diagnosis but only on AML of M4 and M5 subtypes (Diermayr et al., 2007). Nonetheless, the absence of NKG2D ligands expression in 'resting' leukaemia cells does not preclude its involvement in leukaemia cell lysis since NKG2D ligand expression could be inducible in these cells (Costello et al., 2004). Upregulation of NKG2D-ligand expression in leukaemia cells could represent a strategy to increase NK-mediated leukaemia rejection. Significant expression of PVR and Nectin-2, the ligands of DNAM-1 on leukaemic cells (Pende et al., 2005), has been found recently. The central role for DNAM-1 is confirmed by mAb-mediated blocking experiments (Pende et al., 2005). In contrast with acute lymphoid leukemia (ALL) cells (discussed later), AML cells express ligands capable of triggering lysis of blasts by NK cells (Verheyden and Demanet, 2008).

NK cells in acute lymphoid leukaemia

Analysis of NK cells in ALL is at an early stage (Verheyden and Demanet, 2008). ALL is, in general, less susceptible to NK cell-mediated lysis than AML (Pende et al., 2005). However, mechanisms of immune escape are still under study (Torelli et al., 2005). Little is know about the phenotype of NK cells in ALL. In particular, no information has been published about NCR expression on NK cells in ALL (Costello et al., 2002; Fauriat et al., 2007). However, NK cell-mediated lysis has been evaluated using allogeneic NK cells with KIR/HLA-class I mismatch. As in AML, DNAM-1 seems to be instrumental for NK cell lysis with a complementary role for NCR (Pende et al., 2005). In contrast, NKG2D does not seem to play a significant role in ALL killing (Pende et al., 2005). Another mechanism of immune escape in ALL is the expression of ligands for inhibitory receptors or the absence or downregulation of ligands for activating NK cell receptors. Surface density of HLA-class I molecules was higher in ALL cells when compared with AML; this could explain the increased resistance of ALL to NK cell lysis (Pende et al., 2005).

ALL cells do not express NKG2D ligands (MICA/B) and low levels of others ligands for activating receptors such as PVR, Nectin-2 (DNAM-1), CD48 (2B4) and NTBA (Pende et al., 2005). Resistance of B-ALL cells may be in part explained by this absence of most ligands for activating NK cell receptors (Costello et al., 2004; Verheyden and Demanet, 2008).

NK cells in chronic leukaemia

NK cells in chronic myeloid leukaemia

NK cells in chronic lymphoid leukaemia

CML is characterized by the translocation between chromosomes 9 and 22 (i.e. Philadelphia chromosome) leading to the formation of the BCR-ABL chimeric gene. NK cells are deficient in CML, but mechanisms are not well understood despite numerous studies (Mellqvist et al., 2000). NK cells could be implicated in CML control, as demonstrated by allogeneic transplantation and donor lymphocyte infusions. In chronic phase, NK cells do not harbour the Philadelphia chromosome in contrast with the final blasts crisis stages (Nakajima et al., 2002). Moreover, activated autologous NK cells suppress primitive CML progenitors in long-term culture (Cervantes et al., 1996). Oncogenic transformation by BCR-ABL increases NK cell-mediated cytotoxicity of leukaemic progenitors by a mechanism involving overexpression of ICAM-1 secondary to NF-κB activation (Baron et al., 2002). BCR-ABL can promote dendritic-mediated NK cell activation, which involves interaction between NKG2D on NK cells and NKG2D ligands on DCs (Terme et al., 2005).

NK cell number and function decrease during the spontaneous course of the disease from chronic phase to blast crisis in CML (Nakajima et al., 2002). BCR-ABL is directly implicated in the NK cell dysfunction because it induces an abnormal NK cell differentiation when it is transduced into CD34+ cells (Nakajima et al., 2002). NK cell-mediated killing may be affected by these mechanisms.

Little is known about the precise phenotype of NK cells in CML. Nothing has been published regarding NCR expression on NK cells in CML. NKG2D expression is reduced on NK cells in CML at diagnosis, which is restored by imatinib treatment (Boissel et al., 2006).

In contrast to AML, most CML patients express high levels of MICA/B molecules (Boissel et al., 2006; Sconocchia et al., 2005). However, shedding of MICA/B into the plasma (soluble MIC) can represent an immune escape mechanism while some of the CML patients expressed reduced levels of NKG2D, which could explain the defective killing of CML cells by NK cells. High levels of soluble MIC have been noted at diagnosis in patients with CML at diagnosis (Boissel et al., 2006; Sconocchia et al., 2005).

The treatment of patients with CML has been improved with the introduction of imatinib, a tyrosine kinase inhibitor. This drug induces haematologic and cytogenetic remissions in a majority of patients in chronic phase. However, this drug has major effects on the immune system, such as antigen-presenting cells and cytotoxic T cell responses that can be either positive or deleterious (Cebo et al., 2006). Target exposure to imatinib induces abnormalities in formation of NK cell/target within the immunological synapse (Cebo et al., 2006; Terme et al., 2005).

NK cells and CLL

CLL represents the most prevalent adult haematological cancer in industrialized countries. Today, allogeneic stem cell transplantation is considered the only curative treatment in CLL. There is only limited information regarding deficiency of NK cells in CLL (Verheyden and Demanet, 2008).

B-CLL patients have severe immunodeficiency with infectious complications both early and at later stages, characterized by defective innate and adaptive immune responses. These patients have an increased risk for malignancy in association with the immunodeficiency. NK cells have diminished cytotoxicity in CLL patients when compared with normal individuals. Little is known about the phenotype of NK cells in CLL. Ligands for NKG2D in CLL (Pende et al., 2005), MICA and ULBP are either absent or expressed at low level and can explain the deficient cytotoxicity of NK cells. DNAM-1 ligands seem not to be expressed (Pende et al., 2005).

NK cell resistance of tumour cells has been correlated with the expression of HLA-G, a nonclassical HLA class I molecule (Maki et al., 2008). HLA-G is a ligand of KIR2DL4 or ILT2 and can mediate inhibitory signalling. Blockade of HLA-G molecules increases the killing susceptibility of CLL in this study (Maki et al., 2008).

NK cells study might be important for CLL treatment. Monoclonal antibodies (mAb) are used in clinical practice in the treatment of patients with CLL, in particular, rituximab and alemtuzumab (Bowles et al., 2006). As the antitumour activity of these agents is in part related to antibody-dependent cellular toxicity (ADCC), the role of NK cells could be quite important. ADCC is influenced not only by the binding of NK cells to targets but also the activation and ability to degranulate represent important factors (Bowles et al., 2006). Defective function of NK cells in CLL could impair the efficiency of these mAb.

NK cells in myelodysplasia

Myelodysplastic syndromes (MDS) are clonal haematopoietic stem cell disorders characterized by ineffective haematopoiesis and peripheral cytopenia. Recent studies have investigated the phenotype and function of NK cells in myelodysplasia (Epling-Burnette et al., 2007; Kiladjian et al., 2006). All studies provide evidence for the defective function of NK cells, that is proliferation and cytolytic activity. Defective NK cell function is observed in all MDS subtypes (Epling-Burnette et al., 2007; Kiladjian et al., 2006). The phenotype is characterized by downregulation of NKp30 as in AML patients while NKp46 and NKp44 expression are normal (Epling-Burnette et al., 2007). NKG2D expression was normal or low (Epling-Burnette et al., 2007). These results were not confirmed in another study (Kiladjian et al., 2006). Decreased NK cytolytic function could be correlated with high risk myelodysplasia (Epling-Burnette et al., 2007). However, these studies used PBMC that were not confirmed in another study using purified NK cells (Kiladjian et al., 2006, 2007).

Ligands for NKG2D have been explored in one study and are positive in 30% of patients with MDS (Epling-Burnette et al., 2007). The mechanisms leading to the deficient NK cytotoxicity in MDS are only in part explained. It was shown in one study that a proportion of NK cells are derived from the myelodysplastic clone since a chromosomal aberration was found in NK cells (Kiladjian et al., 2007).

So, deficient NK function may have implication in MDS pathophysiology. Confirmation of these defects can open new perspectives in immunotherapy in the treatment of MDS (Kiladjian et al., 2007).

NK cells in chronic myeloproliferative disorders

Little is known about NK cells in chronic myeloproliferative disorders, such as polycythemia vera, essential thrombocythemia or primary myelofibrosis. Older studies demonstrated defective K562 cytotoxicity mediated by NK cells, especially in patients with myelofibrosis (Gersuk et al., 1993). A role for PDGF acting on NK cells has been proposed (Gersuk et al., 1993). Impaired NK cell differentiation from CD34 progenitors in myelofibrosis has also been described (Briard et al., 2003).

Immunotherapy

Exploiting NK cells for applications in cancer immunotherapy has regained interest following the demonstration of the role of alloreactive NK cells in leukaemia control (Ruggeri et al., 2005). Various ways to apply NK functions are being tested (Costello et al., 2004; Ljunggren and Malmberg, 2007; Sentman et al., 2006). These methods involve cell therapy using cell lines, autologous or allogeneic NK cells and various cytokines. In addition, one could increase the expression of the receptors for activating receptors (Verheyden and Demanet, 2008).

NK-92 cell line

The NK-92 cell line is derived from a patient with non-Hodgkin's lymphoma presenting large granular lymphocytes at diagnosis and is characterized by a high cytotoxicity against leukaemic cell lines due to the almost complete lack of inhibitory KIRs (Ljunggren and Malmberg, 2007). This NK cell line is maintained in culture with IL-2. Clinical trials are underway to evaluate the efficacy of this NK cell line for adoptive immunotherapy of malignancies, especially in the treatment of leukaemia and lymphoma in relapse (see Chapter 12).

NK cell activation

Ex vivo expansion of NK cells from ALL patients has been achieved with coculture with the feeder cell line 8866 and a cocktail of cytokines (IL-2 and IL-15) (Torelli et al., 2005). In addition to autologous NK cells, allogeneic NK cells have been evaluated. In this study, 43 patients with cancer, of whom 19 had poor-prognosis AML (Miller et al., 2005), were given haploidentical NK-infusions with IL-2 in a nontransplantation setting. With a high-dose immunosuppressive regimen, long-term survival and in vivo expansion of donor-derived NK cells were observed in several patients. Donor NK cell infusions were well tolerated (Miller et al., 2005). Out of the 19 patients with AML, 5 achieved complete remission with this regimen. Response seems to be correlated with KIR-ligand-mismatched status (Miller et al., 2005). More recently, another experience was reported with infusion of allogeneic NK cells for relapsed multiple myeloma in the setting of autologous stem cell transplantation (Shi et al., 2008a). Ten patients were treated with conditioning regimen and then infusion of haploidentical KIR-ligand mismatch NK cells followed by delayed autologous stem cell rescue. No graft versus host disease was observed. Response was difficult to evaluate, but this is a preliminary protocol in order to ameliorate future protocols (Shi et al., 2008a).

Antibodies

NK cells express CD16 (FcγRIII) that can mediate activatory signals such as secretion of cytotoxic granules. NK cells play a major role in the therapeutic use of mAB as rituximab (CD20) or alemtuzumab (CD52) because ADCC is the principal mechanism of action of these mAB (Bowles et al., 2006).

Although CD16 seems to be expressed normally in NK cells from different subtypes of leukaemia, defective activity of NK cells is a likely mechanism that could explain ineffectiveness or resistance of the mAB in clinical practice. Polymorphisms of the FcγRIIIa gene associated with high or low affinity for Ig Fc have been described that could influence the efficacy of rituximab binding affinity (Bowles et al., 2006).

Another mAB, that is gentuzumab ozogamicin (CD33), is used in AML. Its efficacy is restricted to AML blasts expressing the CD33 antigen (Estey and Dohner, 2006). The direct role of NK cells in this setting has not been analysed. Chimeric IgG antibodies against specific antigens expressed on leukaemic cells have been used to promote ADCC against blasts cells. However, these antibodies are mainly tested against lymphoid malignancies (Verheyden and Demanet, 2008).

Antibodies' blocking inhibitory receptors in order to enhance NK responses is under investigation. Blockade of Ly49 receptors increases survival in murine leukaemia models (Ljunggren and Malmberg, 2007; Sentman et al., 2006). However, this blockade can be associated with autoimmune or inflammatory responses due to NK activation or interference with T-lymphocyte, which express inhibitory receptors (Ljunggren and Malmberg, 2007; Sentman et al., 2006). Blockade of inhibitory KIR with mAB is currently being investigated in clinical trials in humans in order to enhance cytotoxicity of tumours by NK cells.

Upregulating the activating NK cell receptors or their ligands

Little is known about the regulation of the activating NK cell receptors. In particular, regulation of the expression of NCR is not fully understood. In AML, there is no straightforward explanation of the downregulation of NCR, that is NKp30 and NKp46 (Costello et al., 2002; Fauriat et al., 2007). Modulation of NCR expression is an interesting way of research in AML treatment in order to restore NK cell-mediated killing of blasts. For instance, cytokines such as IL-2, IL-15 and IL-21, which are known to increase the expression of activatory receptors as NKG2D or NCRs, are likely candidates (Verheyden and Demanet, 2008).

Another interesting strategy to modulate the balance between positive and negative signal and achieve activation of NK cell is the upregulation of NK receptor ligands on leukaemic blasts. Different drugs have been tested on targets to try to upregulate ligands for activating NK cell receptors. One study has noticed that the putative ligands for NCR were upregulated on AML blasts after treatment with differentiation promoting myeloid growth factors in combination with interferon gamma (Nowbakht et al., 2005). The ligands examined were those for NKG2D. Another interesting strategy is the modification of the leukaemic blast sensitivity to killing. This can be achieved by bortezomib, a proteasome inhibitor. Downregulation of HLA-class I molecules and increased NK cell-mediated killing have been observed in myeloma by treating tumour cells with bortezomib (Shi et al., 2008b).

Differentiation-promoting drugs, immunomodulatory drugs and others are being investigated to upregulate ligands on tumour cell surface allowing a better recognition and lysis of blasts (Diermayr et al., 2007). In particular, histone deacetylase inhibitors (HDAC) are being tested to sensitize tumour cells. Trichostatin or valproic acid increases expression of MICA and MICB on leukaemic cells, leading to enhanced NK mediated killing via NKG2D (Kato et al., 2007). Combination of these agents with growth factors and interferon γ may enhance NK cell-mediated killing. These data have been confirmed in a recent study with the induction of NGK2D ligands on blast cells treated by valproic acid or all-*trans*-retinoic acid (ATRA) (Poggi et al., 2009). This upregulation is not associated with shedding of NKG2D ligands into soluble form, which can downregulate expression of NKG2D and represent an immune escape (Poggi et al., 2009).

Histone deacetylase inhibitors may also suppress NK cytolytic activity (Ogbomo et al., 2007). This defective activity is attributed to downregulation of NKp30 and NKp46 and to impaired granule exocytosis. The mechanism is probably secondary to inhibition of NF-κB. Investigation of new drugs for immunotherapy of NK cells needs both studies on targets cells but also on NK cells to determine eventual antagonistic effects.

Conclusions

Since substantial numbers of patients with acute leukaemia do not attain complete remission with frequent relapse and death, alternate immunotherapy approaches have been designed to improve patient outcome. Global defective activity of NK cells is noted in various leukaemia subtypes, but the mechanisms involved in these defect(s) are heterogeneous in the diverse entities. The use of NK cells could increase the efficiency of the anti-leukaemia immune *armamentarium*. Goals of future studies will be to restore or enhance the expression and function of activating receptors on NK cells and to upregulate the ligands for these activating receptors on AML blasts. Better knowledge about NK receptors and ligands regulation is mandatory in order to enhance the susceptibility of leukaemia to NK cell-mediated cytotoxicity.

Acknowledgements

We thank Agence Nationale pour la Recherche contre le SIDA, Groupement Entreprise Français Lutte Cancer, Association pour la Recherche contre le Cancer, Fédération Nationale des Centres de Lutte Contre le Cancer, Fondation Contre la Leucémie, Fondation pour la Recherche Médicale, Programme Hospitalier de Recherche Clinique, INSERM, INCa for their financial support.

References

Baron, F., Turhan, A.G., Giron-Michel, J., Azzarone, B., Bentires-Alj, M., Bours, V., Bourhis, J.H., Chouaib, S. and Caignard, A. (2002). Leukemic target susceptibility to natural killer cytotoxicity: relationship with BCR-ABL expression. Blood **99**, 2107–2113.

Bendall, L.J., Kortlepel, K., Bradstock, K.F. and Gottlieb, D.J. (1995). Natural killer cells adhere to bone marrow fibroblasts and inhibit adhesion of acute myeloid leukemia cells. Leukemia **9**, 999–1005.

Boissel, N., Rea, D., Tieng, V., Dulphy, N., Brun, M., Cayuela, J.M., Rousselot, P., Tamouza, R., Le Bouteiller, P., Mahon, F. X., Steinle, A., Charron, D., Dombret, H. and Toubert, A. (2006). BCR/ABL oncogene directly controls MHC class I chain-related molecule A expression in chronic myelogenous leukemia. J Immunol **176**, 5108–5116.

Bowles, J.A., Wang, S.Y., Link, B.K., Allan, B., Beuerlein, G., Campbell, M. A., Marquis, D., Ondek, B., Wooldridge, J.E., Smith, B.J., Breitmeyer, J.B. and Weiner, G.J. (2006). Anti-CD20 monoclonal antibody with enhanced affinity for CD16 activates NK cells at lower concentrations and more effectively than rituximab. Blood **108**, 2648–2654.

Briard, D., Brouty-Boye, D., Giron-Michel, J., Azzarone, B., Jasmin, C. and Le Bousse-Kerdiles, C. (2003). Impaired NK cell differentiation of blood-derived CD34+ progenitors from patients with myeloid metaplasia with myelofibrosis. Clin Immunol **106**, 201–212.

Cebo, C., Da Rocha, S., Wittnebel, S., Turhan, A.G., Abdelali, J., Caillat-Zucman, S., Bourhis, J.H., Chouaib, S. and Caignard, A. (2006). The decreased susceptibility of Bcr/Abl targets to NK cell-mediated lysis in response to imatinib mesylate involves modulation of NKG2D ligands, GM1 expression, and synapse formation. J Immunol **176**, 864–872.

Cervantes, F., Pierson, B.A., McGlave, P.B., Verfaillie, C.M. and Miller, J.S. (1996). Autologous activated natural killer cells suppress primitive chronic myelogenous leukemia progenitors in long-term culture. Blood **87**, 2476–2485.

Costello, R.T., Sivori, S., Marcenaro, E., Lafage-Pochitaloff, M., Mozziconacci, M.J., Reviron, D., Gastaut, J.A., Pende, D., Olive, D. and Moretta, A. (2002). Defective expression and function of natural killer cell-triggering receptors in patients with acute myeloid leukemia. Blood **99**, 3661–3667.

Costello, R.T., Fauriat, C., Sivori, S., Marcenaro, E. and Olive, D. (2004). NK cells: innate immunity against hematological malignancies? Trends Immunol **25**, 328–333.

Diermayr, S., Himmelreich, H., Durovic, B., Mathys-Schneeberger, A., Siegler, U., Langenkamp, U., Hofsteenge, J., Gratwohl, A., Tichelli, A., Paluszewska, M., Wiktor-Jedrzejczak, W., Kalberer, C.P. and Wodnar-Filipowicz, A. (2007). NKG2D ligand expression in AML increases in response to HDAC inhibitor valproic acid and contributes to allorecognition by NK cell lines with single KIR-HLA-class I specificities. Blood.

Elkins, W.L., Pickard, A. and Pierson, G.R. (1984). Deficient expression of class-I HLA in some cases of acute leukemia. Cancer Immunol Immunother **18**, 91–100.

Epling-Burnette, P.K., Bai, F., Painter, J.S., Rollison, D.E., Salih, H.R., Krusch, M., Zou, J., Ku, E., Zhong, B., Boulware, D., Moscinski, L., Wei, S., Djeu, J.Y. and List, A.F. (2007). Reduced natural killer (NK) function associated with high-risk myelodysplastic syndrome (MDS) and reduced expression of activating NK receptors. Blood **109**, 4816–4824.

Estey, E. and Dohner, H. (2006). Acute myeloid leukaemia. Lancet **368**, 1894–1907.

Farag, S.S. and Caligiuri, M.A. (2006). Immunologic approaches to acute leukemia in the elderly. Semin Hematol **43**, 118–125.

Farag, S.S., Fehniger, T.A., Ruggeri, L., Velardi, A. and Caligiuri, M.A. (2002). Natural killer cell receptors: new biology and insights into the graft-versus-leukemia effect. Blood **100**, 1935–1947.

Fauriat, C., Marcenaro, E., Sivori, S., Rey, J., Gastaut, J.A., Moretta, A., Olive, D. and Costello, R.T. (2003). Natural killer cell-triggering receptors in patients with acute leukaemia. Leuk Lymphoma **44**, 1683–1689.

Fauriat, C., Moretta, A., Olive, D. and Costello, R.T. (2005). Defective killing of dendritic cells by autologous natural killer cells from acute myeloid leukemia patients. Blood **106**, 2186–2188.

Fauriat, C., Just-Landi, S., Mallet, F., Arnoulet, C., Sainty, D., Olive, D. and Costello, R.T. (2007). Deficient expression of NCR in NK cells from acute myeloid leukemia: evolution during leukemia treatment and impact of leukemia cells in NCRdull phenotype induction. Blood **109**, 323–330.

Gersuk, G.M., Carmel, R., Pattamakom, S., Challita, P.M., Rabinowitz, A.P. and Pattengale, P.K. (1993). Quantitative and functional studies of impaired natural killer (NK) cells in patients with myelofibrosis, essential thrombocythemia, and polycythemia vera. I. A potential role for platelet-derived growth factor in defective NK cytotoxicity. Nat Immun **12**, 136–151.

Kato, N., Tanaka, J., Sugita, J., Toubai, T., Miura, Y., Ibata, M., Syono, Y., Ota, S., Kondo, T., Asaka, M. and Imamura, M. (2007). Regulation of the expression of MHC class I-related chain A, B (MICA, MICB) via chromatin remodeling and its impact on the susceptibility of leukemic cells to the cytotoxicity of NKG2D-expressing cells. Leukemia **21**, 2103–2108.

Kiladjian, J.J., Bourgeois, E., Lobe, I., Braun, T., Visentin, G., Bourhis, J.H., Fenaux, P., Chouaib, S. and Caignard, A. (2006). Cytolytic function and survival of natural killer cells are severely altered in myelodysplastic syndromes. Leukemia **20**, 463–470.

Kiladjian, J.J., Fenaux, P. and Caignard, A. (2007). Defects of immune surveillance offer new insights into the pathophysiology and therapy of myelodysplastic syndromes. Leukemia **21**, 2237–2239.

Ljunggren, H.G. and Malmberg, K.J. (2007). Prospects for the use of NK cells in immunotherapy of human cancer. Nat Rev Immunol **7**, 329–339.

Maki, G., Hayes, G.M., Naji, A., Tyler, T., Carosella, E.D., Rouas-Freiss, N. and Gregory, S.A. (2008). NK resistance of tumor cells from multiple myeloma and chronic lymphocytic leukemia patients: implication of HLA-G. Leukemia, 998–1006.

Markasz, L., Stuber, G., Vanherberghen, B., Flaberg, E., Olah, E., Carbone, E., Eksborg, S., Klein, E., Skribek, H. and Szekely, L. (2007). Effect of frequently used chemotherapeutic drugs on the cytotoxic activity of human natural killer cells. Mol Cancer Ther **6**, 644–654.

Mellqvist, U.H., Hansson, M., Brune, M., Dahlgren, C., Hermodsson, S. and Hellstrand, K. (2000). Natural killer cell dysfunction and apoptosis induced by chronic myelogenous leukemia cells: role of reactive oxygen species and regulation by histamine. Blood **96**, 1961–1968.

Miller, J.S., Soignier, Y., Panoskaltsis-Mortari, A., McNearney, S.A., Yun, G.H., Fautsch, S.K., McKenna, D., Le, C., Defor, T.E., Burns, L.J., Orchard, P.J., Blazar, B.R., Wagner, J.E., Slungaard, A., Weisdorf, D.J., Okazaki, I.J. and McGlave, P.B. (2005). Successful adoptive transfer and in vivo expansion of human haploidentical NK cells in patients with cancer. Blood **105**, 3051–3057.

Moretta, L., Locatelli, F. and Moretta, A. (2008). Alloreactive natural killer cells in targeting high-risk leukaemias. Ann Rheum Dis **67**(Suppl. 3), iii39–iii43.

Nakajima, H., Zhao, R., Lund, T.C., Ward, J., Dolan, M., Hirsch, B. and Miller, J.S. (2002). The BCR/ABL transgene causes abnormal NK cell differentiation and can be found in circulating NK cells of advanced phase chronic myelogenous leukemia patients. J Immunol **168**, 643–650.

Nowbakht, P., Ionescu, M.C., Rohner, A., Kalberer, C.P., Rossy, E., Mori, L., Cosman, D., De Libero, G. and Wodnar-Filipowicz, A. (2005). Ligands for natural killer cell-activating receptors are expressed upon the maturation of normal myelomonocytic cells but at low levels in acute myeloid leukemias. Blood **105**, 3615–3622.

Ogbomo , H., Michaelis, M., Kreuter, J., Doerr, H.W. and Cinatl, J. Jr. (2007).

Histone deacetylase inhibitors suppress natural killer cell cytolytic activity. FEBS Lett **581**, 1317–1322.

Parham, P. (2006). Taking license with natural killer cell maturation and repertoire development. Immunol Rev **214**, 155–160.

Pende, D., Rivera, P., Marcenaro, S., Chang, C.C., Biassoni, R., Conte, R., Kubin, M., Cosman, D., Ferrone, S., Moretta, L. and Moretta, A. (2002). Major histocompatibility complex class I-related chain A and UL16-binding protein expression on tumor cell lines of different histotypes: analysis of tumor susceptibility to NKG2D-dependent natural killer cell cytotoxicity. Cancer Res **62**, 6178–6186.

Pende, D., Spaggiari, G.M., Marcenaro, S., Martini, S., Rivera, P., Capobianco, A., Falco, M., Lanino, E., Pierri, I., Zambello, R., Bacigalupo, A., Mingari, M.C., Moretta, A. and Moretta, L. (2005). Analysis of the receptor–ligand interactions in the natural killer-mediated lysis of freshly isolated myeloid or lymphoblastic leukemias: evidence for the involvement of the Poliovirus receptor (CD155) and Nectin-2 (CD112). Blood **105**, 2066–2073.

Pende, D., Marcenaro, S., Falco, M., Martini, S., Bernardo, M.E., Montagna, D., Romeo, E., Cognet, C., Martinetti, M., Maccario, R., Mingari, M.C., Vivier, E., Moretta, L., Locatelli, F. and Moretta, A. (2009). Anti-leukemia activity of alloreactive NK cells in KIR ligand-mismatched haploidentical HSCT for pediatric patients: evaluation of the functional role of activating KIR and re-definition of inhibitory KIR specificity. Blood **113**, 3119–3129.

Poggi, A., Catellani, S., Garuti, A., Pierri, I., Gobbi, M. and Zocchi, M.R. (2009). Effective in vivo induction of NKG2D ligands in acute myeloid leukaemias by all-trans-retinoic acid or sodium valproate. Leukemia.

Ruggeri, L., Capanni, M., Mancusi, A., Urbani, E., Perruccio, K., Burchielli, E., Tosti, A., Topini, F., Aversa, F., Martelli, M.F. and Velardi, A. (2004). Alloreactive natural killer cells in mismatched hematopoietic stem cell transplantation. Blood Cells Mol Dis **33**, 216–221.

Ruggeri, L., Mancusi, A., Perruccio, K., Burchielli, E., Martelli, M.F. and Velardi, A. (2005). Natural killer cell alloreactivity for leukemia therapy. J Immunother **28**, 175–182.

Ruggeri, L., Mancusi, A., Burchielli, E., Capanni, M., Carotti, A., Aloisi, T., Aversa, F., Martelli, M.F. and Velardi, A. (2008). NK cell alloreactivity and allogeneic hematopoietic stem cell transplantation. Blood Cells Mol Dis **40**, 84–90.

Salih, H.R., Antropius, H., Gieseke, F., Lutz, S.Z., Kanz, L., Rammensee, H.G. and Steinle, A. (2003). Functional expression and release of ligands for the activating immunoreceptor NKG2D in leukemia. Blood **102**, 1389–1396.

Sconocchia, G., Lau, M., Provenzano, M., Rezvani, K., Wongsena, W., Fujiwara, H., Hensel, N., Melenhorst, J., Li, J., Ferrone, S. and Barrett, A.J. (2005). The antileukemia effect of HLA-matched NK and NK-T cells in chronic myelogenous leukemia involves NKG2D–target-cell interactions. Blood **106**, 3666–3672.

Sentman, C.L., Barber, M.A., Barber, A. and Zhang, T. (2006). NK cell receptors as tools in cancer immunotherapy. Adv Cancer Res **95**, 249–292.

Shi , J., Tricot, G., Szmania, S., Rosen, N., Garg, T.K., Malaviarachchi, P.A., Moreno, A., Dupont, B., Hsu, K.C., Baxter-Lowe, L.A., Cottler-Fox, M., Shaughnessy, J.D. Jr., Barlogie, B. and van Rhee, F. (2008a). Infusion of haplo-identical killer immunoglobulin-like receptor ligand mismatched NK cells for relapsed myeloma in the setting of autologous stem cell transplantation. Br J Haematol **143**, 641–653.

Shi , J., Tricot, G.J., Garg, T.K., Malaviarachchi, P.A., Szmania, S.M., Kellum, R.E., Storrie, B., Mulder, A., Shaughnessy, J.D. Jr., Barlogie, B. and van Rhee, F. (2008b). Bortezomib down-regulates the cell-surface expression of HLA class I and enhances natural killer cell-mediated lysis of myeloma. Blood **111**, 1309–1317.

Terme, M., Borg, C., Guilhot, F., Masurier, C., Flament, C., Wagner, E.F., Caillat-Zucman, S., Bernheim, A., Turhan, A.G., Caignard, A. and Zitvogel, L. (2005). BCR/ABL promotes dendritic cell-mediated natural killer cell activation. Cancer Res **65**, 6409–6417.

Torelli, G.F., Guarini, A., Maggio, R., Alfieri, C., Vitale, A. and Foa, R. (2005). Expansion of natural killer cells with lytic activity against autologous blasts from adult and pediatric acute lymphoid leukemia patients in complete hematologic remission. Haematologica **90**, 785–792.

Verheyden, S. and Demanet, C. (2008). NK cell receptors and their ligands in leukemia. Leukemia **22**, 249–257.

Verheyden, S., Bernier, M. and Demanet, C. (2004). Identification of natural killer cell receptor phenotypes associated with leukemia. Leukemia **18**, 2002–2007.

Chapter Forty-Two

42

Natural killer cells and allogeneic haematopoietic cell transplantation

Andrea Velardi

CHAPTER CONTENTS

Life is not worth living without research.

Plato

ABSTRACT

For almost all leukemia patients who fail to find a matched donor, whether related or unrelated, or a suitable cord blood unit, haematopoietic cell transplantation (HCT) from family donors who are matched for one human leukocyte antigen (HLA) haplotype and fully mismatched at the HLA class I and II loci of the unshared haplotype ('haploidentical') is a viable option. In this setting, donor-versus-recipient natural killer (NK) cell alloreactivity is affected by a functional repertoire of NK cells that express inhibitory killer cell immunoglobulin-like receptors (KIRs) for self class I ligand(s), sense missing expression of donor KIR ligand(s) in the recipient and mediate alloreactions. Donor versus recipient NK cell alloreactivity outcomes of HLA haplotype mismatched ('haploidentical') transplants by controlling acute myeloid leukemia (AML) relapse without causing graft-versus-host disease (GVHD). The dramatic improvements afforded by the discovery of the role of NK cell alloreactivity will offer a chance of cure for increasing numbers of patients with haematological malignancies.

KEY WORDS

NK cell, Alloreactivity, KIR, Haematopoietic transplantation, Leukemia

Introduction

Cure of leukemia by allogeneic haematopoietic cell transplantation (HCT) relies on the ability of the immune cells in the graft to recognize and eliminate the leukemia cells (Anasetti and Velardi, 2003; Velardi and Locatelli, 2007). The immunology of HCT is distinct from that of other types of transplant because, in addition to stem cells, the graft contains mature blood cells of donor origin, including T cells, natural killer (NK) cells and dendritic cells (DCs). These cells repopulate the recipient's lympho-haematopoietic system and give rise to a new immune system, which is pivotal in rebuilding immunity to infections and in eliminating residual leukemia cells that survived the conditioning regimen. This effect is known as the graft-versus-leukemia (GvL) effect.

The donor immune system exerts its GvL effect through T cell–mediated alloreactions directed against recipient histocompatibility antigens displayed on recipient leukemia cells. However, as histocompatibility antigens are also displayed on tissues, T cell–mediated alloreactions, specifically donor alloreactive cytotoxic CD8+ effector T cells, may attack recipient tissues—in particular, skin, intestines and liver—causing acute graft-versus-host disease (GVHD), a condition of varying severity that can be life-threatening.

Consequently, the success of allogeneic HCT is undermined by diversity between donors and recipients in major and minor histocompatibility antigens. Major histocompatibility complex (MHC) molecules, that is, the human leukocyte antigen (HLA)-A, HLA-B and HLA-C MHC class I molecules, present peptides to CD8+ T cells while the HLA-DP, HLA-DQ and HLA-DR MHC class II molecules present peptides to CD4+ T cells. There are hundreds of variant forms of each class I and class II molecule, and even small differences between them can provoke alloreactive T-cell responses that mediate graft rejection and/or GVHD. Indeed, disparities for HLA-A, -B, -C or -DRB1 alleles are independent risk factors for acute GVHD and HLA-A or -B allele mismatches are a significant risk factor for chronic GVHD.

Minor histocompatibility antigens derive from differences between the HLA-matched recipient and donor in peptides that are presented by the same HLA allotype. They are due to polymorphisms of other, non-HLA proteins, to differences in the level of expression of proteins, or to genome differences between males and females (such as the H-Y antigens encoded by the Y chromosome, which can stimulate GVHD when a sister donates to an HLA-identical brother). So, even when the donor and recipient are HLA identical, GVHD can still occur.

Although matching donor and recipient HLA is crucial to minimize GVHD, 75% of patients do not have an HLA-identical sibling because, as the genes for HLA are closely linked on chromosome 6 and are inherited as haplotypes, two siblings have a one-in-four chance of being HLA identical. Even though an HLA-matched sibling is the optimal donor, other sources of haematopoietic stem cells are essential if more patients with haematological malignancies are to be treated. Today, donors include unrelated volunteers, unrelated umbilical cord blood (UCB) units and full HLA haplotype-mismatched family members.

An unrelated HLA-matched donor may be sought in world-wide international registries which now contain >11 million HLA-typed prospective volunteer donors. Data on serological typing of class I HLA loci is available for all donors, as well as information on DRB1 typing for approximately one-third because HLA A, B, C class I loci and the DRB1 class II locus are the HLA loci most influencing outcome. The roles played by other class II loci (i.e. DQB1 and DP1 loci) remain controversial. Initially, in unrelated donor transplantation, the restrictions of conventional HLA-typing techniques limited the accuracy of matching, thus increasing rejection rates and the incidence of acute and chronic GVHD. Consequently, as the event-free survival (EFS) in recipients of an unrelated donor allograft was worse than after a compatible sibling transplant, there was no consensus on the use of unrelated donor transplants for non-malignant diseases, such as thalassaemia. New DNA-based techniques for HLA typing reveal an impressive number of new alleles within antigens that were previously defined by serology. Matching by these methods certainly reduces the risk of immune complications, namely graft rejection and GVHD, but also the chance of finding a suitable donor. The advent of both high-resolution molecular HLA class I and II antigen typing and progress in the prophylaxis and treatment of GVHD together reduced transplant-related mortality (TRM) and improved outcomes.

Identifying a suitable unrelated donor is a complicated and lengthy process, during which time lapse patients may relapse and die. A median of 4–5 months is needed to identify the right donor from a potential panel, establish eligibility, complete the HLA typing and harvest the graft. The chances of finding a donor depend on the representation of the various ethnicities among the unrelated volunteers in the registries. Up to 40% of Caucasians and 90% of patients belonging to ethnic minority groups that are under-represented in the registries will not find suitable donors. Patient age is another drawback as morbidity and mortality rise with age. The donor refusal rate is estimated at around 10% because, for various reasons, e.g. ill-health, some donors are no longer available. Because of all these limitations, fewer than half the patients with an indication to unrelated donor transplantation actually undergo transplant and, therefore, the reported survival rates of unrelated donor HCT refer only to patients who undergo transplant and do not take into account those who do not find a donor.

UCB offers the advantages of easy procurement, absence of risks to donors, reduced risk of transmitting infections and ready availability of cryopreserved cells because the median time-lapse from start of search to transplantation is 3–4 weeks. Most importantly, mismatches of up to two of the six antigens do not preclude transplant feasibility, since the naive T cells in cord blood are less able to cause GVHD than the mature donor T cells in bone marrow or peripheral blood.

Given the engraftment facilitating role of the number of haematopoietic stem cells in the graft, rejection and/or delayed engraftment, and ensuing transplant related mortality due to fatal infections, are major concerns when the nucleated cells in a cord blood unit are few in number, particularly when under 2.0×10^7/kg of recipient body weight. As a cord blood unit usually contains between 8×10^8 and 1.5×10^9 cells, UCB transplantation is mainly used for patients with a body weight under 40–50 kg. This increased risk of fatal infections is mainly due to the slow neutrophil recovery and lack of antigen-experienced (memory) T-cells, which are not transferred in UCB.

For almost all leukemia patients who do not have a matched sibling donor, transplantation from family donors who are matched for one HLA haplotype but fully mismatched at the HLA class I and II loci of the unshared haplotype (haploidentical) is another option. Unlike unrelated donors or unrelated cord blood units, haploidentical donors are immediately available, circumventing the delays and limitations of the other alternative transplants. Their use offers a chance of cure for those patients who urgently need a transplant.

All haploidentical transplant recipients are at high risk of T cell mediated alloreactions in the GvH direction, as well as in the host-versus-graft (HvG) (rejection) direction. These are largely controlled by appropriate immunosuppressive intensity of the conditioning regimen followed by transplantation of a large dose of haematopoietic stem cells to prevent graft rejection, and extensive T cell depletion of the graft to prevent GVHD.

The need for extensive T cell depletion might be expected to result in a weak or no GvL effect as it is conventionally achieved through T cell-mediated alloreactions directed against the recipient's histocompatibility antigens. However, another cell of the immune system influences the outcome of HCT in a surprisingly favourable way. In haploidentical HST, NK cells of donor origin were recently shown to by-pass the obstacles inherent to T cell alloreactivity. They prevent leukemia relapse, while not increasing the risk of GVHD.

NK cells were first identified by their ability to kill certain tumour cell lines in the absence of in vitro or in vivo stimulation. They also play a crucial role in defences against certain viruses, including herpes viruses and adenoviruses. The functional mechanisms underlying their response to tumour and virally-infected cells remained elusive until recent years, when intensive research in mice and humans unravelled the biology of NK cells. Dramatic progress was made in our understanding of how they function and their exploitation in therapy for leukemia.

This review focuses on recent research demonstrating the role NK cells play in adoptive immunotherapy of leukemia in mismatched HCT. It also covers studies investigating other possible NK cell-related effector mechanisms in transplantation.

NK cell alloreactivity in allogeneic HCT: pre-clinical data

The hybrid resistance transplant model illustrated that NK cell alloreactions in the HvG direction mediate rejection of bone marrow grafts and play a major role in recognizing allogeneic lympho-haematopoietic cells in vivo (Cudkowicz and Bennett, 1972). As the hybrid recipient mouse tolerated skin and organ allografts, NK cell alloreactivity appeared to be restricted to lympho-haematopoietic targets.

The in vivo effects of NK cell alloreactivity also hold true in the GvH direction (Ruggeri et al., 2002). In F1 H-2d/b→parent H-2b transplants, donor T cells are tolerant of the recipient MHC. Donor NK cells that do not express the H-2b-specific Ly49C/I inhibitory receptor (and bear H-2d-specific Ly49A/G2 receptors), are activated to kill the recipient's targets. Infusion of donor versus recipient alloreactive NK cells exerts several favourable effects. After host immune suppression, infused alloreactive NK cells home at all lympho-haematopoietic sites of the recipient mouse and quickly ablate recipient-type lympho-haematopoietic cells. Killing of recipient T lymphocytes is associated with engraftment of the MHC-mismatched bone marrow. Killing of recipient-type DCs, which initiate GVHD by presenting host alloantigens to donor T cells, prevents T cell-mediated GVHD. Mice that are given alloreactive NK cells as part of the conditioning regimen are able to receive mismatched bone marrow grafts containing up to thirty times the lethal dose of allogeneic T cells without clinical or histological evidence of GVHD. Finally, alloreactive NK cells themselves do not cause GVHD. Lack of NK-mediated attack on normal tissues in the GvH direction (and in the HvG direction, as shown in the hybrid mouse transplant model, see above) indicates that healthy organ tissues, unlike lympho-haematopoietic cells, do not express ligands at a level sufficient to engage activating NK cell receptors.

Donor-versus-recipient NK cell alloreactivity in allogeneic HCT

Donor-versus-recipient NK cell alloreactivity is mediated by a functional repertoire of donor NK cells which express inhibitory killer cell immunoglobulin-like receptor (KIR) for self class I ligand(s), sense missing expression of donor KIR ligand(s) in the recipient and mediate alloreactions.

NK cell function is regulated by a balance between activating receptors and inhibitory receptors for MHC class I molecules (Gasser and Raulet, 2006; Parham, 2006; Rosen et al., 2008; Yokoyama and Kim, 2006). In humans, triggering of NK cell effector functions depends upon engagement of activating receptors, NKG2D and natural cytotoxicity receptors (NCRs) (Lanier, 2005; Moretta et al., 2004). NKG2D, which is expressed by NK cells and by most cytolytic T lymphocytes, recognizes several cell surface ligands belonging to HLA related MHC class-I chain related molecule (MIC) and UL-16 binding protein (ULBP) families in humans and

H60 and Rae1 in mice (Bauer et al., 1999; Cerwenka et al., 2000; Cosman et al., 2001; Diefenbach et al., 2000, 2001; Groh et al., 1999; Jamieson et al., 2002; Kubin et al., 2001; Pende et al., 2002; Sutherland et al., 2002; Wu et al., 1999). NCRs, which include NKp30, NKp44 and NKp46, are almost exclusively expressed by NK cells (Cantoni et al., 1999; Ferlazzo et al., 2002; Pende et al., 1999; Pessino et al., 1998). Although NCR ligands are still undefined, cytolytic assays and NCR-specific antibody-mediated inhibition of cytolysis indicate they are primarily expressed by activated or proliferating cells (e.g. virus infected cells, tumour cells or in vitro cultured cell lines) of different histotypes. NKG2D and NCRs transduce activating signals through their engagement with immune-receptor thyrosine-based activating motifs (ITAM) containing polypeptides, that is, DAP12 associates with NKp44, ζ and γ chains associate with NKp30 and NKp46, and DAP10 with NKG2D.

Finally, NK cell activation can be mediated by KIR variants (Biassoni et al., 1997; Colonna and Samaridis, 1995; Moretta et al., 1995; Parham, 2005; Uhrberg et al., 1997; Vales-Gomez et al., 1998; Vilches and Parham, 2002; Winter et al., 1998). Activating KIRs have shorter cytoplasmic tails than inhibitory KIRs and a charged residue in their transmembrane domain that allows association with ITAM-containing signalling polypeptides. Knowledge of the ligand specificity of activating KIRs is limited. Studies have reported only a weak interaction between KIR2DS1 and Lys80 HLA-C molecules, despite its homology to KIR2DL1, and no interaction between KIR2DS2 and Asn80 HLA-C, despite its homology to KIR2DL2 and KIR2DL3 (Saulquin et al., 2004; Stewart et al., 2005).

In humans, inhibitory receptors for HLA class I molecules include inhibitory KIRs and the CD94-NKG2A molecular complex. Inhibitory KIRs recognize amino acids in the COOH-terminal portion of the MHC class I α1 helix. They possess two (KIR2D) or three (KIR3D) extra-cellular C2-type Ig-like domains and a long cytoplasmic tail (L) containing immunoreceptor tyrosine-based inhibition motifs (ITIM) which recruit and activate Src homology phosphatase (SHP-1) and SHP-2 for inhibitory signal transduction. KIR2DL1 recognizes HLA-C alleles characterized by a Lys80 residue (HLA-Cw4 and related, 'Group 2' alleles). KIR2DL2 and KIR2DL3 recognize HLA-C with an Asn80 residue (HLA-Cw3 and related, 'Group 1' alleles). KIR3DL1 is the receptor for HLA-B allotypes with Bw4 motifs at positions 77–83. It also recognizes HLA-Bw4 alleles except for 1301 and 1302 and some HLA-A alleles, namely 2301, 2402 and 3201. Finally, KIR3DL2 is the receptor for HLA-A3/11.

Another type of human NK cell inhibitory receptor involved in HLA recognition is CD94-NKG2A which binds to the non-conventional class I molecule HLA-E. Several HLA class I alleles provide signal sequence peptides that bind HLA-E and allow its expression at the cell surface. Consequently, it is expressed in every individual.

Inhibitory KIRs, CD94/NKG2 and HLA-class I genes structure individual NK cell repertoires during development. To establish a self-tolerant repertoire the HLA class I type selects NK cells carrying KIR and/or NKG2A receptor combinations which function as inhibitory receptors for self HLA class I. Consequently, functional NK cells in the mature repertoire express at least one inhibitory receptor for self HLA (Parham, 2006); co-expression of two or more receptors is less frequent. NK cells expressing receptors which do not recognize self are retained in the repertoire in, however, an anergic (or 'hypofunctional') state (Anfossi et al., 2006; Gasser and Raulet, 2006; Kim et al., 2005; Yokoyama et al., 2005).

NK cells with the potential to exert alloreactions use KIRs as inhibitory receptors for self (Ciccone et al., 1992; Colonna et al., 1993; Farag et al., 2002; Moretta and Moretta 2004; Parham, 2006; Parham and McQueen, 2003; Ruggeri et al., 2006; Uhrberg et al., 1997; Valiante et al., 1997; Velardi et al., 2002). Those which express, as their only inhibitory receptor for self, a KIR for the HLA class I group which is absent on allogeneic targets, sense the missing expression of the self class I KIR ligand and mediate alloreactions ('missing self' recognition).

Donor-versus-recipient NK cell alloreactions are generated between individuals who are mismatched for HLA-C allele groups and/or the HLA-Bw4 group ('KIR ligand mismatched') (Kärre, 2002; Ruggeri et al., 1999, 2002, 2007) (Figures 42.1 and 42.2). Most donors have the potential to exert NK alloreactions as they possess a full complement of inhibitory KIR genes (Hsu et al., 2002; Ruggeri et al., 2007; Uhrberg et al., 1997). HLA-C group 1 receptor genes (KIR2DL2 and/or KIR2DL3) are present in 100% of individuals, and the HLA-C group 2 receptor gene (KIR2DL1) in 97%. Therefore, the combination of HLA-C group 2-positive donor/HLA-C group 2-negative recipient which occurs in ~1.5% of HLA-C group mismatched transplants, requires donor KIR2DL1 gene typing. When tested in large donor cohorts (Ruggeri et al., 2007), functional analyses detected high-frequency alloreactive NK clones against HLA-C group mismatched allogeneic targets.

The HLA-Bw4 receptor gene (KIR3DL1) is found in ~90% of individuals. Only two-third of HLA-Bw4-positive individuals with the KIR3DL1 gene possess alloreactive NK clones against allogeneic HLA-Bw4-negative targets (Ruggeri et al., 2007). Failure to detect alloreactive NK clones may be due to their highly variable frequencies, or because certain allelic KIR3DL1 variants do not allow receptor expression at the cell membrane (Pando et al., 2003; Thomas et al., 2008). In donor-recipient pairs that are not KIR ligand-mismatched in the GvH direction, no

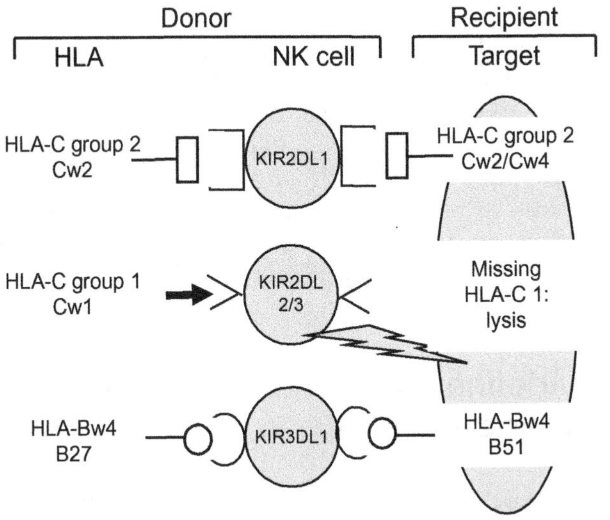

Figure 42.1 • Donor-versus-recipient NK cell alloreactivity. NK cell alloreactions are generated between donors and recipients who are KIR ligand mismatched in the GvH direction. Donor NK cells expressing, as their only inhibitory receptor for self HLA, a KIR for the class I group that is absent in the recipient, sense the missing expression of the self class I ligand on allogeneic targets and mediate alloreactions. In this example, a donor NK cell expressing KIR2DL2/3, inhibiting receptor for the self HLA-C group 1 allele, does not find this allele group in the recipient and is activated to kill the recipient target. (Only inhibitory KIR–HLA interactions are illustrated in the figure. See text for interactions between activating NK receptors and their ligands on target cells.)

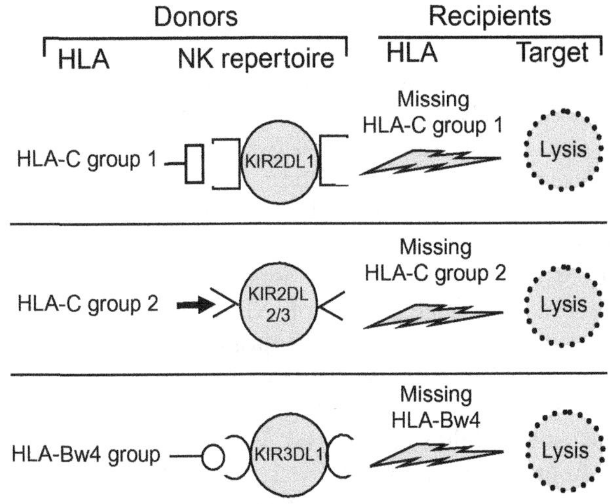

Figure 42.2 • Selecting donor/recipient pairs with donor-versus-recipient NK alloreactivity. All individuals possess the KIR2DL2 and/or KIR2DL3 receptors for HLA-C group 1 alleles. If they have HLA-C group 1 allele(s) in their HLA type, they possess HLA-C1-specific NK cells that are alloreactive against cells from individuals who do not express HLA-C group 1 alleles (top panel). Ninety-seven percent of individuals possess the KIR2DL1 receptor for HLA-C group 2. If they possess HLA-C group 2 allele(s) in their HLA type, they have HLA-C2-specific NK cells that mediate alloreactions against cells from individuals who do not express HLA-C group 2 alleles (middle panel). In one study on a large cohort (Ruggeri et al., 2007), functional analyses detected alloreactivity when NK clones were tested against HLA-C group mismatched allogeneic targets. Frequencies of alloreactive NK clones were high, i.e. 8 ± 6 cells in 100 (mean ± SD) for HLA-C group 2 mismatches; 5 ± 3 cells in 100 for group 1 mismatches. Finally (bottom panel), ~90% of individuals possess the KIR3DL1 receptor for HLA-Bw4 alleles. When they have HLA-Bw4 allele(s) in their HLA type, they may have HLA-Bw4-specific NK cells that are alloreactive against Bw4-negative cells. When NK clones from HLA-Bw4-positive individuals who possessed the KIR3DL1 gene were tested against allogeneic HLA-Bw4-negative targets, alloreactive NK clones were detected in two-third of individuals (Ruggeri et al., 2007). (Only inhibitory KIR–HLA interactions are illustrated in the figure. See text for interactions between activating NK receptors and their ligands on target cells.)

donor alloreactive NK clones are found, indicating that KIR ligand mismatching is a prerequisite for NK cell alloreactivity (Ruggeri et al., 2007).

In vitro studies on primary lympho-haematopoietic lineage tumour cells showed alloreactive NK cells kill acute and chronic myeloid leukemia, as well as T-cell acute lymphoblastic leukemia (ALL), chronic lymphocytic leukemia, non-Hodgkin's lymphoma and multiple myeloma (Ruggeri et al., 2006). The only non-susceptible target was common ALL (Ruggeri et al., 1999, 2006). Alloreactive NK cells also exerted significant cytotoxicity against melanoma and renal cell carcinoma cell lines (Igarashi et al., 2004).

In HLA haplotype-mismatched ('haploidentical') haematopoietic transplantation with a potential for donor-versus-recipient NK cell alloreactivity, engrafted stem cells regenerated the same repertoire as the donor's, including donor-versus-recipient alloreactive NK cells for up to 1 year or more (Pende et al., 2008; Ruggeri et al., 1999, 2007). Generated from engrafted stem cells, alloreactive NK cells eradicate acute myeloid leukemia (AML), prevent rejection of MHC-mismatched transplants (through killing of recipient T lymphocytes) and T cell–mediated GVHD (through killing of recipient-type DCs, which trigger GVHD). NK cell alloreactivity is restricted to lympho-haematopoietic cells and does not attack other tissues as it does not cause GVHD (Ruggeri et al., 2002).

In an updated analysis (Ruggeri et al., 2007), 112 high-risk AML patients received haploidentical transplants from NK alloreactive ($n = 51$) or non-NK alloreactive donors ($n = 61$). Transplantation from NK-alloreactive donors was associated with a significantly lower relapse rate in patients transplanted in complete remission (3% vs. 47%) ($p < 0.003$), better EFS in patients transplanted not only in remission (67% vs. 18%, $p = 0.02$) but also in relapse (34% vs. 6%, $p = 0.04$), overall reduced risk of relapse or death ($p < 0.001$). The 67% probability of surviving event-free for AML patients transplanted in remission from NK alloreactive donors is in the range of best survival rates after transplantation from unrelated donors and cord blood units. The in vitro resistance of common phenotype ALL to alloreactive NK killing was paralleled by lack of anti-leukemia effect in adult patients. However, in ALL in children, transplantation from NK alloreactive donors was reported to decrease the risk of relapse (Stern et al., 2008a,b).

One recent study showed NK cell alloreactivity provided much better protection from leukemia relapse when exerted by maternal donors (as opposed to any other donor-recipient family relationship) (Stern et al., 2008a,b). The effect was independent of, and additive to, the beneficial effects of NK alloreactivity. Better outcome of mother-to-child transplantation may be due to maternal immune system exposure to foetal antigens during pregnancy and ensuing memory T cell immunity against the child's paternal HLA haplotype.

Another recent study assessed NK cell alloreactivity in a large number of unrelated cord blood transplants and found it was associated with a significantly reduced incidence of leukemia relapse ($p = 0.05$) and better leukemia free-survival (73% vs. 38%, $p = 0.0016$). Benefits were significantly more marked in patients with AML (Willemze et al., 2008).

This cord blood transplant study contained several KIR ligand mismatches involving HLA-A3/A11 which have not yet been observed in haploidentical transplantation. HLA-A3/A11 appear to function as ligands for KIR3DL2 only when binding Epstein–Barr virus peptides (Hansasuta et al., 2004). Future studies will have to ascertain whether post-transplant Epstein–Barr virus reactivation makes peptides available for A3/A11 binding and, consequently, leads to generation of A3/A11-specific donor NK cells which are alloreactive against recipients lacking A3 and/or A11.

Thus, NK cell alloreactivity is effective in haploidentical and cord blood transplantation. In an attempt to explain the effectiveness of NK cell alloreactivity in these two very different types of transplants, one may consider that a common feature is lack of memory T cells in the graft (due to T cell depletion in haploidentical and T cell naiveté in cord blood transplants). This, apparently, permits recovery of fully functional NK cells. Interestingly, one study comparing T cell-depleted versus replete unrelated donor transplants demonstrated that T cells in the graft adversely affected reconstitution of KIR-bearing NK cells and clinical outcomes (Cooley et al., 2005). Additional evidence that T cells antagonize reconstitution of potentially alloreactive, KIR-bearing NK cells (Cooley et al., 2005) derives from several other unrelated donor transplant studies, using T cell-replete grafts, including a recent one on 1489 transplants (Beelen et al., 2005; Bornhauser et al., 2004; Davies et al., 2002; Dawson and Spencer, 2005; Elmaagacli et al., 2005; Farag et al., 2006; Giebel et al., 2003; Hsu et al., 2006; Kröger et al., 2005, 2006; Lowe et al., 2003; Yabe et al., 2008). Most studies showed no advantage in transplantation from KIR ligand-mismatched donors (Bornhauser et al., 2004; Davies et al., 2002; Farag et al., 2006; Hsu et al., 2006; Kröger et al., 2006; Lowe et al., 2003; Yabe et al., 2008), while a few observed an increased GvL effect (Beelen et al., 2005; Dawson and Spencer, 2005; Elmaagacli et al., 2005; Giebel et al., 2003; Kröger et al., 2005). Interestingly, the study that reported the most marked survival advantage was performed in KIR ligand-mismatched transplant recipients who received pre-transplant anti-thymocyte globulins (ATG) to exert in vivo T cell depletion (Giebel et al., 2003).

Guidelines for NK alloreactive donor selection

The search for NK alloreactive donors, which may require extension beyond the immediate family, increases the chance of finding a 'perfect mismatch' from the random 30% to >60%. First the transplantation candidate is HLA typed. Those who express class I alleles belonging to the three class I groups recognized by KIRs (HLA-C group 1, HLA-C group 2, and HLA-Bw4 alleles) will block all NK cells from every donor and belong to the one-third of the population that is resistant to alloreactive NK killing. Patients who express only one or two of these allele groups may find NK alloreactive donors.

Donor HLA typing will identify the family member who does not express the class I group(s) expressed by the patient and has, therefore, the potential for NK alloreactivity. Not all inhibitory KIRs are present in 100% of the population. Although KIR2DL2/3, receptors for HLA-C group 1, are present in all, KIR2DL1, receptor for HLA-C group 2 is found in 97% of individuals and KIR3DL1, receptor for HLA-Bw4 alleles, is found in ~90%. Donor KIR genotyping ensures the donor possesses the relevant NK cells.

In HLA-Bw4 mismatches, even when the KIR3DL1 gene is present, NK repertoire studies showed alloreactive NK cells in approximately two-third of individuals. This may be because they occur in highly variable frequencies, or because allelic KIR3DL1 variants may not allow receptor expression at the cell membrane. Therefore, for HLA-Bw4 mismatches, direct assessment of the donor NK repertoire is necessary.

Another NK cell effect in transplantation: the missing ligand model

Since the original report on NK cell alloreactivity in haploidentical transplantation the 'missing ligand' model has been proposed (Dupont and Hsu, 2004; Hsu et al., 2005, 2006, 2007; Leung et al., 2004, 2005; Miller et al., 2007). According to this model, NK cell alloreactions also occur when KIR ligand-matched donors possess an 'accessory' KIR for which neither donor nor recipient have HLA ligand(s). In fact, most individuals possess a full complement of inhibitory KIRs for the three major ligands (the two HLA-C and the HLA-Bw4 allele groups), while several have only one or two ligands in their HLA type. NK cells bearing the 'accessory' KIR are in an anergic/hypofunctional state (Colonna et al., 1993) and, upon transfer into the recipient, are hypothesized to become activated and exert an anti-leukemia effect.

One study examined the missing ligand effect in HLA identical transplantation and demonstrated it increased survival in AML and myelodysplasia through a decrease in disease relapse (Hsu et al., 2005). Subsequently, investigations into large patient populations receiving unrelated donor transplants showed missing ligand provided an advantage in control of leukemia, with one study reporting a positive effect only in mismatched transplants (Hsu et al., 2006, 2007; Miller et al., 2007). In one study with over 2000 patients, missing ligand significantly increased the incidence of severe GVHD in chronic myeloid leukemia (Miller et al., 2007), indicating it led to untoward T cell activation against recipient alloantigens.

In haploidentical transplantation the missing ligand model was reported to predict better outcomes than donor-versus-recipient NK cell alloreactivity (as triggered by KIR ligand mismatches, see above and references Ruggeri et al., 1999, 2002, 2007; Kärre, 2002). Leukemia relapse was reduced and survival improved in haploidentical transplants in children with acute leukemia (Leung et al., 2004, 2005). These studies should be interpreted with caution as in the haploidentical transplant setting; the missing ligand model also includes, by definition, KIR ligand-mismatched transplants. In fact, the model includes: (1) all KIR ligand-mismatched transplants (in the GvH direction) which are all associated with a missing KIR ligand in the recipient (because the vast majority of donors possess the three inhibitory KIRs, while all recipients possess only one or two of these KIR ligands) and (2) KIR ligand-matched transplants from donors possessing 'extra' KIR(s) for which neither donor nor recipient have HLA ligand(s). When adult AML recipients of haploidentical transplants were analyzed according to the 'missing ligand' algorithm, outcomes were worse than in patients transplanted from NK-alloreactive (KIR ligand mismatched) donors (Ruggeri et al., 2007). Differences in diseases, age and transplantation protocols, such as ATG (Ruggeri et al., 2007) vs. no ATG (Leung et al., 2004, 2005) in the conditioning may account for these discrepancies.

Do activating KIRs play a role in transplantation?

Activating KIRs are molecular homologues of the inhibitory KIRs, with shorter cytoplasmic tails, which transduce activating signals regulating NK and T cell functions. They exhibit extensive variation in gene number and content, which leads to heterogeneity within the general population and diverse ethnic groups (Parham, 2005). They are absent in approximately 25% of Caucasians who are homozygous for the so-called group A KIR gene haplotypes which contain inhibitory KIR genes and the KIR2DS4 activating KIR gene (encoding for a non-functional protein in two-third of individuals). The others are either heterozygous or homozygous for B haplotypes which carry not only inhibitory KIR genes but also various combinations of activating KIR genes (KIR2DS1-2-3-5 and KIR3DS1).

Investigations into activating KIR genetics in transplant settings other than the haploidentical, such as matched sibling and unrelated donor transplants (Bishara et al., 2004; Chen et al., 2006; Cook et al., 2004, 2006; De Santis et al., 2005; McQueen et al., 2007; Sun et al., 2005; Verheyden et al., 2005) showed less leukemia relapse and cytomegalovirus reactivation and improved survival in some, but adverse outcomes, such as an increased incidence of acute GVHD, in others. In these studies, activating KIRs, through their expression on T cells, could have enhanced T-cell alloreactivity and caused GVHD.

A recent study hypothesized that, because of extensive T cell depletion, haploidentical transplants allow analysis of the impact of donor activating KIR genes on transplant outcomes without the confounding effects of T cell alloreactivity and GVHD.

Transplantation from NK alloreactive donors who carried activating KIR genes was associated with reduced incidence of mortality from infection and improved survival. In transplants with alloreactivity, presence of donor *B KIR* haplotypes gave better outcome while their absence gave poorer outcome than transplants lacking NK cell alloreactivity. This co-operation between alloreactive NK cells and donor *B* haplotypes reduced TRM (which was mainly due to infection) and improved survival. The protective effect increased with activating *KIR* gene number. Comparison of transplants combining alloreactivity with ≥3 donor activating *KIR* genes, to transplants lacking alloreactivity and donor *B* haplotype, showed significantly reduced TRM (12% vs. 67%, $p < 0.003$) and increased EFS (71% vs. 33%, $p = 0.02$). Multivariate analyses showed that ≥3 donor activating *KIR* genes was the only predictor of protection from TRM (RR: 0.13; 95% CI: 0.28–0.64; $p < 0.02$) and improved survival (RR: 0.31; 95% CI: 0.10–0.96; $p < 0.05$). Thus, these data suggest donor activating KIRs regulate the effects of NK cell alloreactivity in haploidentical transplantation.

Conclusions

This review has illustrated the extraordinary progress at the crossroads of two diverse fields: allogeneic HCT and

NK cell biology. The advent of haploidentical transplants made it possible to explore the role of NK cell allorecognition of missing self on KIR ligand mismatched allogeneic leukemia cells ('donor-versus-recipient NK cell alloreactivity'). In haploidentical and cord blood transplantation for AML, donor-versus-recipient-NK cell alloreactions impact beneficially on outcomes as they are associated with a strong GvL effect without increased risk of GVHD.

Acknowledgements

The author's studies are supported by a Translational Research Grant from the Leukemia and Lymphoma Society, and by grants from the Italian Association for Cancer Research, the Italian Ministry of Further Education and the Italian Ministry of Health, by the European Community (Contract number: LSHB-CT-2004-503319) and by the National Institutes for Health of the USA (Project Number 1 PO1 CA100265). I would like to thank Dr Geraldine Anne Boyd for editorial assistance.

References

Anasetti, C. and Velardi, A. (2003). Hematopoietic cell transplantation from HLA partially matched related donors. In: Blume, K.G., Forman, S.J. and Appelbaum, F.R. (eds) Hematopoietic Cell Transplantation, 3rd edition. Malden: Blackwell Science.

Anfossi, N., André, P., Guia, S., Falk, C. S., Roetynck, S., Stewart, C.A., Breso, V., Frassati, C., Reviron, D., Middleton, D., Romagné, F., Ugolini, S. and Vivier, E. (2006). Human NK cell education by inhibitory receptors for MHC Class I. Immunity **2**, 331–342.

Bauer, S., Groh, V., Wu, J., Steinle, A., Phillips, J.H., Lanier, L.L. and Spies, T. (1999). Activation of NK cells and T cells by NKG2D, a receptor for stress-inducible MICA. Science **285**, 727–729.

Beelen, D.W., Ottinger, H.D., Ferencik, S., Elmaagacli, A.H., Peceny, R., Trenschel, R. and Grosse-Wilde, H. (2005). Genotypic inhibitory killer immunoglobulin-like receptor ligand incompatibility enhances the long-term antileukemic effect of unmodified allogeneic hematopoietic stem cell transplantation in patients with myeloid leukemias. Blood **105**, 2594–2600.

Biassoni, R., Pessino, A., Malaspina, A., Cantoni, C., Bottino, C., Sivori, S., Moretta, L. and Moretta, A. (1997). Role of amino acid position 70 in the binding affinity of p50.1 and p58.1 receptors for HLA-Cw4 molecules. Eur J Immunol **27**, 3095–3099.

Bishara, A., De Santis, D., Witt, C.C., Brautbar, C., Christiansen, F.T., Or, R., Nagler, A. and Slavin, S. (2004). The beneficial role of inhibitory KIR genes of HLA class I NK epitopes in haploidentically mismatched stem cell allografts may be masked by residual donor-alloreactive T cells causing GVHD. Tissue Antigens **63**, 204–211.

Bornhauser, M., Schwerdtfeger, R., Martin, H., Frank, K.H., Theuser, C. and Ehninger, G. (2004). Role of KIR ligand incompatibility in hematopoietic stem cell transplantation using unrelated donors. Blood **103**, 2860–2861.

Cantoni, C., Bottino, C., Vitale, M., Pessino, A., Augugliaro, R., Malaspina, A., Parolini, S., Moretta, L., Moretta, A. and Biassoni, R. (1999). NKp44, a triggering receptor involved in tumor cell lysis by activated human natural killer cells, is a novel member of the immunoglobulin superfamily. J Exp Med **189**, 787–796.

Cerwenka, A., Bakker, A.B., McClanahan, T., Wagner, J., Wu, J., Phillips, J.H. and Lanier, L.L. (2000). Retinoic acid early inducible genes define a ligand family for the activating NKG2D receptor in mice. Immunity **12**, 721–727.

Chen, C., Busson, M., Rocha, V., Appert, M.L., Lepage, V., Dulphy, N., Haas, P., Socié, G., Toubert, A., Charron, D. and Loiseau, P. (2006). Activating KIR genes are associated with CMV reactivation and survival after non-T-cell depleted HLA-identical sibling bone marrow transplantation for malignant disorders. Bone Marrow Transplant **38**, 437–444.

Ciccone, E., Pende, D., Viale, O., Di Donato, C., Tripodi, G., Orengo, A.M., Guardiola, J., Moretta, A. and Moretta, L. (1992). Evidence of a natural killer (NK) cell repertoire for (allo) antigen recognition: definition of five distinct NK-determined allospecificities in humans. J Exp Med **175**, 709–718.

Colonna, M. and Samaridis, J. (1995). Cloning of immunoglobulin-superfamily members associated with HLA-C and HLA-B recognition by human natural killer cells. Science **268**, 405–408.

Colonna, M., Brooks, E.G., Falco, M., Ferrara, G.B. and Strominger, J.L. (1993). Generation of allospecific natural killer cells by stimulation across a polymorphism of HLA-C. Science **260**, 1121–1124.

Cook, M.A., Milligan, D.W., Fegan, C.D., Darbyshire, P.J., Mahendra, P., Craddock, C.F., Moss, P.A. and Briggs, D.C. (2004). The impact of donor *KIR* and patient *HLA*-C genotypes on outcome following HLA-identical sibling haematopoietic stem cell transplantation for myeloid leukemia. Blood **103**, 1521–1526.

Cook, M., Briggs, D., Craddock, C., Mahendra, P., Milligan, D., Fegan, C., Darbyshire, P., Lawson, S., Boxall, E. and Moss, P. (2006). Donor KIR genotype has a major influence on the rate of cytomegalovirus reactivation following T-cell replete stem cell transplantation. Blood **107**, 1230–1232.

Cooley, S., McCullar, V., Wangen, R., Bergemann, T.L., Spellman, S., Weisdorf, D.J. and Miller, J.S. (2005). KIR reconstitution is altered by T cells in the graft and correlates with clinical outcomes after unrelated donor transplantation. Blood **106**, 4370–4376.

Cosman, D., Müllberg, J., Sutherland, C. L., Chin, W., Armitage, R., Fanslow, W., Kubin, M. and Chalupny, N.J. (2001). ULBPs, novel MHC class I-related molecules, bind to CMV glycoprotein UL16 and stimulate NK cytotoxicity through the NKG2D receptor. Immunity **14**, 123–133.

Cudkowicz, G. and Bennett, M. (1972). Peculiar immunobiology of bone-marrow allografts. II. Rejection of parental grafts by resistant F1 hybrid mice. J Exp Med **135**, 1028–1036.

Davies, S.M., Ruggeri, L., DeFor, T., Wagner, J.E., Weisdorf, D.J., Miller, J. S., Velardi, A. and Blazar, B.R. (2002). Evaluation of KIR ligand incompatibility in mismatched unrelated donor hematopoietic transplants. Blood **100**, 3825–3827.

Dawson, M.A. and Spencer, A. (2005). Successful use of haploidentical stem-cell transplantation with KIR mismatch as initial therapy for poor-risk myelodysplastic syndrome. J Clin Oncol **23**, 4473–4474.

De Santis, D., Bishara, A., Witt, C. S., Nagler, A., Brautbar, C., Slavin, S. and Christiansen, F.T. (2005). Natural killer cell HLA-C epitopes and killer cell immunoglobulin-like receptors both influence outcome of mismatched unrelated donor bone marrow transplants. Tissue Antigens **65**, 519–528.

Diefenbach, A., Jamieson, A.M., Liu, S.D., Shastri, N. and Raulet, D.H. (2000). Ligands for the murine NKG2D receptor: expression by tumor cells and activation of NK cells and macrophages. Nat Immunol **1**, 119–126.

Diefenbach, A., Jensen, E.R., Jamieson, A. M. and Raulet, D.H. (2001). Rae1 and H60 ligands of the NKG2D receptor stimulate tumour immunity. Nature **413**, 165–171.

Dupont, B. and Hsu, K.C. (2004). Inhibitory killer Ig-like receptor genes and human leukocyte antigen class I ligands in hematopoietic stem cell transplantation. Curr Opin Immunol **16**, 634–643.

Elmaagacli, A.H., Ottinger, H., Koldehoff, M., Peceny, R., Steckel, N.K., Trenschel, R., Biersack, H., Grosse-Wilde, H. and Beelen, D.W. (2005). Reduced risk for molecular disease in patients with chronic myeloid leukemia after transplantation from a KIR-mismatched donor. Transplantation **79**, 1741–1747.

Farag, S.S., Bacigalupo, A., Eapen, M., Hurley, C., Dupont, B., Caligiuri, M.A., Boudreau, C., Nelson, G., Oudshoorn, M., van Rood, J., Velardi, A., Maiers, M., Setterholm, M., Confer, D., Posch, P.E., Anasetti, C., Kamani, N., Miller, J.S., Weisdorf, D. and Davies, S.M. (2006). The effect of KIR ligand incompatibility on the outcome of unrelated donor transplantation: a report from the center for international blood and marrow transplant research, the European blood and marrow transplant registry, and the Dutch registry. Biol Blood Marrow Transpl **12**, 876–884.

Farag, S.S., Fehniger, T.A., Ruggeri, L., Velardi, A., Caligiuri, M. (2002). Natural Killer cell receptors: new biology and insights into the graft versus leukemia effect. Blood 100,1935-1947.

Ferlazzo, G., Tsang, M.L., Moretta, L., Melioli, G., Steinman, R.M., Munz, C. (2002). Human dendritic cells activate resting natural killer (NK) cells and are recognized via the NKp30 receptor by activated NK cells. J. Exp. Med. 195, 343–351.

Fernandez, N.C., Treiner, E., Vance, R.E., Jamieson, A.M., Lemieux, S., Raulet, D.H. (2005). A subset of natural killer cells achieves self-tolerance without expressing inhibitory receptors specific for self-MHC molecules. Blood 105,4416–4423.

Foley, B.A., De Santis, D., Van Beelen, E., Lathbury, L.J., Christiansen, F.T., Witt, C.S. (2008). The reactivity of Bw4-positive HLA-B and HLA-A alleles with KIR3DL1: implications for patient and donor suitability for haploidentical stem cell transplants. Blood 112, 435–443.

Gasser, S., and Raulet, D.H. (2006). Activation and self-tolerance of natural killer cells. Immunol. Rev. 214,130–142.

Giebel, S., Locatelli, F., Lamparelli, T., Velardi, A., Davies, S., Frumento, G., Maccario, R., Bonetti, F., Wojnar, J., Martinetti, M., Frassoni, F., Giorgiani, G., Bacigalupo, A., Holowiecki, J. (2003). Survival advantage with KIR ligand incompatibility in hematopoietic stem cell transplantation from unrelated donors. Blood 102, 814–819.

Groh, V., Rhinehart, R., Secrist, H., Bauer, S., Grabstein, K.H., Spies, T. (1999). Broad tumor-associated expression and recognition by tumor-derived gamma delta T cells of MICA and MICB. Proc. Natl. Acad. Sci. 96, 6879–6884.

Hansasuta, P., Dong, T., Thananchai, H., Weekes, M., Willberg, C., Aldemir, H., Rowland-Jones, S., Braud, V.M. (2004). Recognition of HLA-A3 and HLA-A11 by KIR3DL2 is peptide-specific. Eur. J. Immunol. 34,1673–1679.

Hsu, K.C., Chida, S., Geraghty, D.E., Dupont, B. (2002). The killer cell immunoglobulin-like receptor (KIR) genomic region: gene-order, haplotypes and allelic polymorphism. Immunol. Rev. 190, 40–52.

Hsu, K.C., Gooley, T., Malkki, M., Pinto-Agnello, C., Dupont, B., Bignon, J.D., Bornhäuser, M., Christiansen, F., Gratwohl, A., Morishima, Y., Oudshoorn, M., Ringden, O., van Rood, J.J., Petersdorf, E. (2006). KIR ligands and prediction of relapse after unrelated donor hematopoietic cell transplantation for hematologic malignancy. Biol. Blood Marrow Transpl. 12, 828–836.

Hsu, K.C., Keever-Taylor., A., Wilton, A., Pinto, C., Heller, G., Arkun, K., O'Reilly, R.J., Horowitz, M.M., Dupont, B. (2005). Improved outcome in HLA-identical sibling hematopoietic stem-cell transplantation for acute myelogenous leukemia predicted by KIR and HLA genotypes. Blood 105, 4878–4884.

Hsu, K.C., Pinto-Agnello, C., Gooley, T., Malkki, M., Dupont, B. and Petersdorf, E.W. (2007). Hematopoietic stem cell transplantation: killer immunoglobulin-like receptor component. Tissue Antigens **69**(Suppl. 1), 42–45.

Igarashi T., Wynberg, J., Srinivasan, R., Becknell, B., McCoy, Jr, J.P, Takahashi, Y., Suffredini, D.A., Linehan, W.M., Caligiuri, M.A. and Childs, R.W. (2004). Enhanced cytotoxicity of allogeneic NK cells with killer immunoglobulin-like receptor ligand incompatibility against melanoma and renal cell carcinoma cells. Blood **104**, 170–177.

Jamieson, A.M., Diefenbach, A., McMahon, C.W., Xiong, N., Carlyle, J.R. and Raulet, D.H. (2002). The role of the NKG2D immunoreceptor in immune cell activation and natural killing. Immunity **17**, 19–29.

Kärre, K. (2002). A perfect mismatch. Science **295**, 2029–2031.

Kim, S., Poursine-Laurent, J., Truscott, S. M., Lybarger, L., Song, Y.J., Yang, L., French, A.R., Sunwoo, J.B., Lemieux, S., Hansen, T.H. and Yokoyama, W. M. (2005). Licensing of natural killer cells by host major histocompatibility complex class I molecules. Nature **436**, 709–713.

Kröger, N., Shaw, B., Iacobelli, S., Zabelina, T., Peggs, K., Shimoni, A., Nagler, A., Binder, T., Eiermann, T., Madrigal, A., Schwerdtfeger, R., Kiehl, M., Sayer, H. G., Beyer, J., Bornhäuser, M., Ayuk, F., Zander, A.R. and Marks, D.I. (2005). Comparison between antithymocyte globulin and alemtuzumab and the possible impact of KIR-ligand mismatch after dose-reduced conditioning and unrelated stem cell transplantation in patients with multiple myeloma. Br J Haematol **129**, 631–643.

Kröger, N., Binder, T., Zabelina, T., Wolschke, C., Schieder, H., Renges, H., Ayuk, F., Dahlke, J., Eiermann, T. and Zander, A. (2006). Low number of donor activating killer immunoglobulin-like receptors (KIR) genes but not KIR-ligand mismatch prevents relapse and improves disease-free survival in leukemia patients after in vivo T-cell depleted unrelated stem cell transplantation. Transplantation **82**, 1024–1030.

Kubin, M., Cassiano, L., Chalupny, J., Chin, W., Cosman, D., Fanslow, W., Müllberg, J., Rousseau, A.M., Ulrich, D. and Armitage, R. (2001). ULBP1, 2, 3: novel MHC class I-related molecules that bind to human cytomegalovirus glycoprotein UL16, activate NK cells. Eur J Immunol **31**, 1428–1437.

Lanier, L.L. (2005). NK cell recognition. Annu Rev Immunol **23**, 225–274.

Leung, W., Iyengar, R., Triplett, B., Turner, V., Behm, F.G., Holladay, M. S., Houston, J. and Handgretinger, R. (2005). Comparison of killer Ig-like receptor genotyping and phenotyping for selection of allogeneic blood stem cell donors. J Immunol **174**, 6540–6545.

Leung, W., Iyengar, R., Turner, V., Lang, P., Bader, P., Conn, P., Niethammer, D. and Handgretinger, R. (2004). Determinants of antileukemia effects of allogeneic NK cells. J Immunol **172**, 644–650.

Lowe, E.J., Turner, V., Handgretinger, R., Horwitz, E.M., Benaim, E., Hale, G. A., Woodard, P. and Leung, W. (2003). T-cell alloreactivity dominates natural killer cell alloreactivity in minimally T-cell-depleted HLA-non-identical paediatric bone marrow transplantation. Br J Haematol **123**, 323–326.

McQueen, K.L., Dorighi, K.M., Guethlein, L.A., Wong, R., Sanjanwala, B. and Parham, P. (2007). Donor-recipient combinations of group A and B KIR haplotypes and HLA class I ligand affect the outcome of HLA-matched, sibling donor hematopoietic cell transplantation. Hum Immunol **68**, 309–323.

Miller, J.S., Cooley, S., Parham, P., Farag, S.S., Verneris, M.R., McQueen, K.L., Guethlein, L.A., Trachtenberg, E.A., Haagenson, M., Horowitz, M.M., Klein, J.P. and Weisdorf, D.J. (2007). Missing KIR ligands are associated with less relapse and increased graft-versus-host disease (GVHD) following unrelated donor allogeneic HCT. Blood **109**, 5058–5061.

Moretta, A., Sivori, S., Vitale, M., Pende, D., Morelli, L., Augugliaro, R., Bottino, C. and Moretta, L. (1995). Existence of both inhibitory (p58) and activatory (p50) receptors for HLA-C molecules in human natural killer cells. J Exp Med **182**, 875–884.

Moretta, L., Bottino, C., Pende, D., Vitale, M., Mingari, M.C. and Moretta, A. (2004). Different checkpoints in human NK cell activation. Trends Immunol **25**, 670–676.

Moretta, L. and Moretta, A. (2004). Killer immunoglobulin-like receptors. Curr Opin Immunol **16**, 626–633.

Pando, M.J., Gardiner, C.M., Gleimer, M., McQueen, K.L. and Parham, P. (2003). The protein made from a common allele of KIR3DL1 (3DL*004) is poorly expressed at cell surfaces due to substitution at position 86 in Ig domain 0 and 182 in Ig domain 1. J Immunol **171**, 6640–6647.

Parham, P. (2005). MHC class I molecules and KIRs in human history, health and survival. Nat Rev Immunol **5**, 201–214.

Parham, P. (2006). Taking license with natural killer cell maturation and repertoire development. Immunol Rev **214**, 155–160.

Parham, P. and McQueen, K.L. (2003). Alloreactive killer cells: hindrance and help for haematopoietic transplants. Nat Rev Immunol **3**, 108–122.

Pende, D., Parolini, S., Pessino, A., Sivori, S., Augugliaro, R., Morelli, L., Marcenaro, E., Accade, L., Malaspina, A., Biassoni, R., Bottino, C., Moretta, L. and Moretta, A. (1999). Identification and molecular characterization of NKp30, a novel triggering receptor involved in natural cytotoxicity mediated by human natural killer cells. J Exp Med **190**, 1505–1516.

Pende, D., Rivera, P., Marcenaro, S., Chang, C.C., Biassoni, R., Conte, R., Kubin, M., Cosman, D., Ferrone, S., Moretta, L. and Moretta, A. (2002). Major histocompatibility complex class I-related chain A and UL16-binding protein expression on tumor cell lines of different histotypes: analysis of tumor susceptibility to NKG2D-dependent natural killer cell cytotoxicity. Cancer Res **62**, 6178–6186.

Pende, D., Marcenaro, S., Falco, M., Martini, S., Bernardo, M.E., Montagna, D., Romeo, E., Cognet, C., Martinetti, M., Maccario, R., Mingari, M.C., Vivier, E., Moretta, L., Locatelli, F. and Moretta, A. (2008). Anti-leukemia activity of alloreactive NK cells in KIR ligand-mismatched haploidentical HSCT for pediatric patients: evaluation of the functional role of activating KIR and re-definition of inhibitory KIR specificity. Blood **62**, 3119–3129.

Pessino, A., Sivori, S., Bottino, C., Malaspina, A., Morelli, L., Moretta, L., Biassoni, R. and Moretta, A. (1998). Molecular cloning of NKp46: a novel member of the immunoglobulin superfamily involved in triggering of natural cytotoxicity. J Exp Med **7**, 953–960.

Rosen, D.B., Cao, W., Avery, D.T., Tangye, S.G., Liu, Y.J., Houchins, J.P. and Lanier, L.L. (2008). Up on the tightrope: natural killer cell activation and inhibition. Nat Immunol **9**, 495–502.

Ruggeri, L., Capanni, M., Casucci, M., Volpi, I., Tosti, A., Perruccio, K., Urbani, E., Negrin, R.S., Martelli, M.F. and Velardi, A. (1999). Role of natural killer cell alloreactivity in HLA-mismatched hematopoietic stem cell transplantation. Blood **94**, 333–339.

Ruggeri, L., Capanni, M., Urbani, E., Perruccio, K., Shlomchik, W.D., Tosti, A., Posati, S., Rogaia, D., Frassoni, F., Aversa, F., Martelli, M.F. and Velardi, A. (2002). Effectiveness of donor natural killer cell alloreactivity in mismatched hematopoietic transplants. Science **295**, 2097–2100.

Ruggeri, L., Aversa, F., Martelli, M.F. and Velardi, A. (2006). Haploidentical transplantation and natural killer cell recognition of missing self. Immunol Rev **214**, 202–218.

Ruggeri, L., Mancusi, A., Capanni, M., Urbani, E., Carotti, A., Aloisi, T., Stern, M., Pende, D., Perruccio, K., Burchielli, E., Topini, F., Bianchi, E., Aversa, F., Martelli, M.F. and Velardi, A. (2007). Donor natural killer cell allorecognition of missing self in haploidentical hematopoietic transplantation for acute myeloid leukemia: challenging its predictive value. Blood **110**, 433–440.

Saulquin, X., Gastinel, L.N. and Vivier, E. (2004). Crystal structure of the human natural killer cell activating receptor KIR2DS2 (CD158j). J Exp Med **197**, 933–938.

Stern, M., Ruggeri, L., Capanni, M., Mancusi, A. and Velardi, A. (2008a). Human leukocyte antigens A23, A24 and A32 but not A25 are ligands for KIR3DL1. Blood **112**, 708–710.

Stern, M., Ruggeri, L., Mancusi, A., Bernardo, M.E., De Angelis, C., Bucher, C., Locatelli, F., Aversa, F. and Velardi, A. (2008b). Survival after T-cell depleted haploidentical stem cell transplantation is improved with mothers as donors. Blood **112**, 2990–2995.

Stewart, C.A., Laugier-Anfossi, F., Vély, F., Saulquin, X., Riedmuller, J., Tisserant, A., Gauthier, L., Romagné, F., Ferracci, G., Arosa, F.A., Moretta, A., Sun, P.D., Ugolini, S. and Vivier, E. (2005). Recognition of peptide–MHC class I complexes by activating killer immunoglobulin-like receptors. Proc Natl Acad Sci U S A **102**, 13224–13229.

Sun, J.Y., Gaidulis, L., Dagis, A., Palmer, J., Rodriguez, R., Miller, M.M., Forman, S. J. and Senitzer, D. (2005). Killer Ig-like receptor (KIR) compatibility plays a role in the prevalence of acute GVHD in unrelated hematopoietic cell transplants for AML. Bone Marrow Transplant **36**, 525–530.

Sutherland, C.L., Chalupny, N.J., Schooley, K., VandenBos, T., Kubin, M. and Cosman, D. (2002). UL16-binding proteins, novel MHC class I-related proteins, bind to NKG2D and activate multiple signaling pathways in primary NK cells. J Immunol **168**, 671–679.

Thananchai, H., Gillespie, G., Martin, M. P., Bashirova, A., Yawata, N., Yawata, M., Easterbrook, P., McVicar, D.W., Maenaka, K., Parham, P., Carrington, M., Dong, T. and Rowland-Jones, S. (2007). Cutting edge: allele-specific and peptide-dependent interactions between KIR3DL1 and HLA-A and HLA-B. J Immunol **178**, 33–37.

Thomas, R., Yamada, E., Alter, G., Martin, M.P., Bashirova, A.A., Norman, P.J., Altfeld, M., Parham, P., Anderson, S. K., McVicar, D.W. and Carrington, M. (2008). Novel KIR3DL1 alleles and their expression levels on NK cells: convergent evolution of KIR3DL1 phenotype variation? J Immunol **180**, 6743–6750.

Uhrberg, M., Valiante, N.M., Shum, B.P., Shilling, H.G., Lienert-Weidenbach, K., Corliss, B., Tyan, D., Lanier, L.L. and Parham, P. (1997). Human diversity in killer-cell inhibitory receptor genes. Immunity **7**, 753–763.

Vales-Gomez, M., Reyburn, H.T., Erskine, R.A. and Strominger, J.L. (1998). Differential binding to HLA-C of p50-activating and p58-inhibitory natural killer cell receptors. Proc Natl Acad Sci U S A **95**, 14326–14331.

Valiante, N.M., Uhrberg, M., Shilling, H. G., Lienert-Weidenbach, K., Arnett, K. L., D'Andrea, A., Phillips, J.H., Lanier, L.L. and Parham, P. (1997). Functionally and structurally distinct NK cell receptor repertoires in the peripheral blood of two human donors. Immunity **7**, 739–751.

Velardi, A. and Locatelli, F. (2007). Hematopoietic stem cell transplantation. In: Kliegman, R.M., Behrman, R.E., Jenson, H.B. and Stanton, B.F. (eds) Nelson Textbook of Pediatrics, 12th edition. Philadelphia: Saunders.

Velardi, A., Ruggeri, L., Moretta, A. and Moretta, L. (2002). NK-cells: a lesson from mismatched haematopoietic transplantation. Trends Immunol **23**, 438–444.

Verheyden, S., Schots, R., Duquet, W. and Demanet, C. (2005). A defined donor activating natural killer cell receptor genotype protects against leukaemic relapse after related HLA-identical hematopoietic stem cell transplantation. Leukemia **19**, 1446–1451.

Vilches, C. and Parham, P. (2002). KIR: diverse, rapidly evolving receptors of innate and adaptive immunity. Annu Rev Immunol **20**, 217–251.

Winter, C.C., Gumperz, J.E., Parham, P., Long, E.O. and Wagtmann, N. (1998). Direct binding and functional transfer of NK cell inhibitory receptors reveal novel patterns of HLA-C allotype recognition. J Immunol **161**, 571–577.

Wu, J., Song, Y., Bakker, A.B., Bauer, S., Spies, T., Lanier, L.L. and Phillips, J.H. (1999). An activating immunoreceptor complex formed by NKG2D and DAP10. Science **285**, 730–732.

Yabe, T., Matsuo, K., Hirayasu, K., Kashiwase, K., Kawamura-Ishii, S., Tanaka, H., Ogawa, A., Takanashi, M., Satake, M., Nakajima, K., Tokunaga, K., Inoko, H., Saji, H., Ogawa, S., Juji, T., Sasazuki, T., Kodera, Y. and Morishima, Y. (2008). Japan Marrow Donor Program. Donor killer immunoglobulin-like receptor (KIR) genotype-patient cognate KIR ligand combination and antithymocyte globulin preadministration are critical factors in outcome of HLA-C-KIR ligand-mismatched T cell-replete unrelated bone marrow transplantation. Biol Blood Marrow Transpl **14**, 75–87.

Yokoyama, W.M. and Kim, S. (2006). Licensing of natural killer cells by self-major histocompatibility complex class I. Immunol Rev **214**, 143–154.

Chapter Forty-Three

43

Clinical trials of NK cells for cancer

Sarah Cooley, Jeffrey S. Miller

CHAPTER CONTENTS

If it's natural to kill, how come men have to go into training to learn how?

Joan Baez (American Singer and Song Writer, b.1941)

ABSTRACT

The ability of NK cells to recognize targets without prior sensitization is related to recognition of stress-induced activating signals and lack of inhibition by self-MHC ligands. NK cell activity can be manipulated for therapeutic intent against several tumours, including hematologic malignancies and numerous solid tumours such as melanoma, renal cell carcinoma, ovarian cancer and neuroblastoma. Autologous NK cells can be directed against tumours by blocking inhibitory receptors, by increasing the expression of activating receptor ligands on tumours and by targeting NK antibody dependent cellular cytotoxicity (ADCC) against those receptors with specific antibodies. NK cell activity can be augmented with cytokines or immunomodulatory drugs such as thalidomide, TLR-agonists or vaccines. Alloreactive NK cells developing after allogeneic HCT or after adoptive transfer have powerful antitumour activity. Ultimately, combination therapy using several strategies at once will likely prove most successful. Encouraging results showing clinical efficacy of NK cell mediated therapies in several settings call out the critical importance of implementing well designed future clinical trials to allow for appropriate interpretation of results. Trials should include rigorously defined clinical endpoints. They should employ only standardized assays to monitor NK cell expansion and effector function.

KEY WORDS

NK cell biology, killer immunoglobulin-like receptors, Interleukin-2, lymphodepletion, adoptive transfer, immunotherapy, NK cell alloreactivity, NK cell immunogenetics

Introduction

Natural killer (NK) cells mediate the response to stress-inducing nonspecific lysis of targets and production of inflammatory cytokines. They can mediate potent antitumour responses and play an important role in immune surveillance. Their effector function is governed by a complex balance of activating and inhibitory signals transferred via several classes of receptors, a number of which recognize "self" MHC class I antigens (Lanier, 2005). Self-tolerance is mediated by inhibitory killer immunoglobulin-like receptors (KIRs) and other receptors, which transmit signals that interrupt the cytolytic pathway upon binding of their cognate class I HLA ligands. The loss of KIR-ligand expression by infected or malignant targets renders them susceptible to NK cell lysis, particularly when the targets also present ligands for activating receptors. NK cells, which can be easily isolated in high quantity from donor lymphapheresis products, do not cause graft vs. host disease (GVHD). Thus NK cells are an attractive cell population to exploit for antitumour immunotherapy. Several clinical strategies have been developed using alloreactive NK cells for therapeutic benefit.

Biologic basis of NK cell-mediated therapies

Definition

NK cells were first characterized in 1975 as large granular lymphocytes with the ability to lyse virally infected and tumour targets without MHC-restriction or prior sensitization (Herberman and Ortaldo, 1981; Kiessling et al., 1975). The phenomenon of "hybrid resistance", first noted in 1971 (Cudkowicz and Bennett, 1971), was explained in 1987 when NK cells were demonstrated to mediate the rejection of allogeneic or parental-strain hematopoietic grafts in lethally irradiated mice (Murphy et al., 1987). In contrast to the cells of the adaptive immune system, which have significant capacity for sensitization or memory, NK cells use their diverse array of inhibitory and activating receptors to mediate immediate recognition and lysis of a variety of infected and transformed cells. NK cells are found in many tissues, including the spleen, bone marrow, lymph nodes and peripheral blood (PB). Human NK cells are derived from $CD34^+$, $CD38^-$, HLA DR^- and lin-1^- marrow-derived progenitors. Their maturation is induced by IL-15, fms-like tyrosine kinase 3 (flt3) ligand, c-kit ligand or stem cell factor (SCF), IL-7 and IL-3 (Yu et al., 1998; Muench et al., 2000; Miller and McCullar, 2001). Circulating NK cells comprise 10-15 percent of the lymphocyte pool in normal humans. NK cells are defined phenotypically by their lack of T-cell markers (CD3, CD4 and T-cell receptors) and by the expression of CD56. They are distinct from $CD3^+/CD56^+$ lymphocytes, which are not NK cells. Discrete stages of NK cell development in lymphoid tissues are defined by the acquisition of IL-15 responsiveness (Freud et al., 2006). The $CD56^{bright}$ subset, comprising 10 percent of circulating NK cells, is more proliferative and produces more cytokines (especially IFN-γ). In contrast, the $CD56^{dim}$ subset bears Fc receptors to mediate antibody-dependent cellular cytotoxicity (ADCC) (Cooper et al., 2001), is more cytotoxic and may be more mature. Compared to resting NK cells, those exposed to cytokines, sometimes referred to as lymphokine-activated killer (LAK) cells, show more proliferation, increased cytokine production and higher cytotoxicity against targets (Rayner et al., 1985). NK cells are activated by IL-2, IL-15 and IL-21, all of which signal via the IL-2 receptor β chain (Carson et al., 1994) (Trinchieri, 2003; Young and Ortaldo, 2006), as well as the combination of IL-12 and IL-18, which is an especially strong stimulant to increase IFN-γ production (Singh et al., 2000). Some of these cytokine signals may be modulated and perhaps more potently delivered when presented as complexes with antibodies, as is the case of IL-2 (Wang et al., 2005). IL-15, which is especially important for NK cell homeostasis, is most potent when encountered in physiologic trans-presentation by other cells as is the case for IL-15Rα expressed on dendritic cells (DCs) (Huntington et al., 2009). Activated NK cells show enhanced proliferation and augmented cytokine production and upregulation of cytotoxic (granzymes and perforin) and adhesion molecules (Becknell and Caligiuri, 2005).

NK cell functions: cytokine production and cytotoxicity

NK cells are major producers of cytokines, including granulocyte colony stimulatory factor (G-CSF), granulocyte-monocyte colony stimulatory factor (GM-CSF), IL-5, TNF, interferon gamma (IFN-γ) and transforming growth factor beta (TGF $-$ β), which can stimulate or inhibit hematopoiesis and can modulate the responses of other immune cells. The link between the innate and adaptive immune systems is mediated in part by cytokine signalling and cell–cell interactions. DCs and NK cells are co-activated while both can activate

T cells (Walzer et al., 2005; Moretta et al., 2006). While most cytotoxic activity of NK cells is direct, mediated by perforin and granzyme, they can also use Fas ligand (FasL) and tumour necrosis factor-related apoptosis-inducing ligand (TRAIL) pathways (Russell and Ley, 2002). In addition, NK cells mediate antibody-dependent cellular cytotoxcity via CD16 (FcRγIII) (Lanier et al., 1988). Besides killing virally infected cells, NK cells have demonstrated in vivo antitumour cytotoxicity against myeloid and lymphoid hematologic malignancies and a wide variety of solid tumours, including breast, ovarian, hepatocellular and colon cancer (Armeanu et al., 2005; Pende et al., 2005; Re et al., 2006; Stein et al., 2006). A complex network of cell–cell interactions involving several families of activating and inhibitory receptors (NKR), which are both MHC class I-specific and non-specific, regulates the different effector responses of NK cells to their wide array of natural targets. These include foreign, damaged, malignant and virally infected cells. NK cells express $\beta 2$ integrins and CD2, which bind to target adhesion molecules such as ICAM-1 and LFA-3. These receptors are important as shown by decreased cytotoxicity when their interactions are blocked by antibodies. It is the net balance of signals, dependent on both the target phenotype and the NK cell receptor repertoire, which determines whether or not a target is lysed.

Killer immunoglobulin-like receptors

Human NK cells express killer immunoglobulin-like receptors (KIRs)- type I trans-membrane molecules belonging to the Ig superfamily, which are named by the number of extracellular immunoglobulin domains (2D or 3D) and the length of the intracellular tail that determines whether they are stimulatory (Short) or inhibitory (Long). Murine NK cells do not express KIRs but do express Ly49 receptors of the same class (Yokoyama et al., 1995). A nomenclature committee has assigned a cluster of designation (CD) number of CD158 for the KIR genes with individual loci designated by a small letter ± a number (e.g. KIR3DL1 = CD158e1) (Marsh et al., 2003). The KIR gene cluster, located on chromosome 19, includes framework genes present in all humans (KIR3DL3, KIR2DL4 and KIR3DL2), along with a variable number of activating and inhibitory genes. The CD56^{dim} subset expresses a high frequency of KIRs, the variegated expression of which is controlled by transcriptional regulation of several homologous promoters under epigenetic control (Chan et al., 2005; Trompeter et al., 2005). Various human populations exhibit different evolutionary patterns of KIR genes (Yawata et al., 2002). In healthy subjects, the KIR repertoire is predicted mainly by the KIR genotype, although it may be influenced by the HLA class I KIR ligand status (Uhrberg et al., 1997). The wide allelic variation in KIR genes includes several common alleles exhibiting poor or no surface expression (Maxwell et al., 2002; Kikuchi-Maki et al., 2003; Pando et al., 2003; Leung et al., 2005).

The gene content can be simplified into two haplotypes. KIR A haplotypes contain only one activating receptor (2DS4), whereas KIR B haplotypes are designated as those with the presence of any additional activating KIR genes. KIR genes are highly polymorphic, and new alleles continue to be reported, some of which are functionally important. For example, KIR3DL1*004 is not expressed on the surface so it cannot function to recognize cellular ligand (Pando et al., 2003). The natural ligands for many KIRs are unknown, including all of the activating KIRs, even though some may bind MHC molecules at low affinity (Moesta et al., 2008). In the homologous murine Ly49 system, the activating receptor Ly49H recognizes the MCMV glycoprotein m157, providing 'proof of principle' that activating receptors may recognize viral proteins (Farrell et al., 1997; Voigt et al., 2003). A similar role for human activating receptors is supported by studies of patients with HIV showing an association between AIDS progression and the activating receptor KIR3DS1 (Martin et al., 2007). It has recently been shown that this activating KIR does not directly recognize Bw4 as its cognate ligand. It is presumed that infectious natural ligands for activating KIRs will be discovered, somewhat analogous to the role of conserved pathogen-associated microbial proteins (PAMPs) and Toll-like receptors (Trinchieri and Sher, 2007). In addition, the presence of activating KIRs has been associated with effects on CMV and HCV infection outcomes in humans (Lopez-Vazquez et al., 2005; Chen et al., 2006; Cook et al., 2006). In contrast, the ligands for several inhibitory KIRs have been identified. KIR2DL1, KIR2DL2/KIR2DL3 and KIR3DL1 bind HLA class I C2, C1 and Bw4 alleles, respectively. The recognition of self-class I HLA by the higher affinity inhibitory receptors suppresses NK cell effector responses, including cell mediated lysis and cytokine release (Vales-Gomez et al., 1998). The KIR repertoire is determined primarily by KIR genotype and at steady state is only minimally affected by class I HLA (KIR-ligand) genes, which segregate independently (chromosome 6).

NK cell alloreactivity: class I-recognizing inhibitory NK cell receptors and 'missing self'

The characterization of class I-recognizing inhibitory receptors provided the mechanistic explanation for the phenomenon of 'missing self', described by Ljunggren and Karre in 1985. They noted that the loss of MHC class I expression rendered autologous targets more sensitive to NK-mediated killing, the strategy by which these innate killer cells can recognize tumour or virally infected cells (Ljunggren and Karre, 1990). It was believed that mature PB NK cells must express 'at least one' inhibitory NKR for self MHC class I to prevent autoreactivity (Raulet et al., 2001). The observation that cloned NK cells all express self-inhibitory receptors supports this interpretation (Valiante et al., 1997). Although more recent reports of murine (Fernandez et al., 2005) and human (Anfossi et al., 2006; Cooley et al., 2007). NK cells lacking self-inhibitory receptors have challenged this model of autoreactivity, the concept remains important for clinical applications involving alloreactive NK cell populations, which are generated by using donor NK cells expressing inhibitory KIRs for which the recipient lacks the appropriate class I ligand. The mechanism by which NK cells develop potent effector functions while remaining self tolerant, referred to as NK cell education, is an area of active investigation (Gasser and Raulet, 2006; Raulet, 2006; Parham, 2006). The licensing model proposes that a terminal differentiation step during which NK cells receive an appropriate signal via an inhibitory receptor ligating with self-MHC is required for the cell to acquire mature function (Kim et al., 2005; Yokoyama and Kim, 2006). Whatever the mechanism, it is important that NK cells being used therapeutically have mature effector functions and that they respond appropriately to inhibitory and activating signals.

Additional NK cell receptors

Several other families of activating and inhibitory receptors affect NK cell function. Both human and murine NK cells express CD94, which heterodimerizes with the NKG2 family of C-type lectin receptors. They are either inhibitory (NKG2A) or activating (NKG2C/E) and recognize nonclassical HLA-E (Houchins et al., 1991). NKG2D is unique in that it does not heterodimerize with CD94 and that it recognizes stress-induced molecules such as MHC class I polypeptide-related sequence A/B (MICA and MICB) and the class I-like CMV homologous ULBP proteins, which are often upregulated on tumour cells or virally infected cells (Lopez-Botet et al., 2004; Cao et al., 2007). DNAM-1 binds to PVR (CD155) and Nectin-2 (CD122), both of which are expressed on human tumours sensitive to NK-mediated killing. These include carcinomas, melanomas and neuroblastomas (Moretta et al., 2006). NK cells also express Ig-like transcript (ILT) receptors, some of which bind HLA-G, expressed in the placenta and on fetal tissue. Several other receptors have been identified that regulate killing of MHC class I-negative targets, including but not limited to the natural cytotoxicity receptors (NCR) NKp30, NKp46 and NKp44 (Pessino et al., 1998; Vitale et al., 1998); 2B4 (CD244), which binds to CD48; LIR-1 (Vitale et al., 1999) and leukocyte-associated immunoglobulin-like receptor-1 (LAIR-1) (Meyaard, 2008). NCRs mediate NK cell lysis of hematologic malignancies and solid tumours, including myeloma, neuroblastoma and melanoma (Bottino et al., 2005) (Soriani et al., 2008).

Therapeutic applications for NK cells

Early autologous NK cell-based therapy

The first trials to use adoptive immunotherapy to treat cancer were performed in the 1980s. Autologous lymphokine-activated killer (LAK) cells, prepared by ex vivo stimulation of PB mononuclear with IL-2, were reinfused with high-dose IL-2 to treat immune-sensitive malignancies, including melanoma, lymphoma and renal cell cancer. Although the observed cytotoxcity was mediated predominantly by NK cells, limited clinical benefit was seen (Rosenberg et al., 1987). High-dose IL-2 therapy was tested to activate autologous patient NK cells in vivo but was abandoned due to significant toxicity from capillary leak syndrome. Subsequent trials of low-dose subcutaneous IL-2, either alone or in combination with LAK cells, also failed to show efficacy in patients with CML, lymphoma and breast cancer (Rosenberg et al., 1993; Burns et al., 2003).

The reasons for the failure of autologous LAK and NK cell-based therapies are now clear. The concept of competition between lymphocyte populations has been demonstrated by several groups. Because host lymphocytes compete with infused cells for access to cytokines and other growth factors, successful expansion of adoptively transferred lymphocytes requires adequate lymphodepletion or 'clearing of space' (Dummer et al., 2002). The preparative regimen used to induce T-cell lymphopenia by Rosenberg's group at the NIH, cyclophosphamide (60 mg/kg/day × 2) followed by fludarabine (25 mg/m^2/day × 5 days), was part of a successful therapy of adoptive transfer of cytotoxic T cells to treat melanoma (Dudley et al., 2002). Soon after, the discovery of inhibitory KIRs and their role in preventing NK cell killing of 'self' MHC-expressing tumour cells led investigators to abandon autologous strategies and to explore the use of allogeneic donor sources.

Allogeneic NK cell therapy: biologic rationale

Therapeutic strategies using allogeneic NK cells are based on our understanding of the signalling pathways that regulate the anti-tumour activity of NK cells. Certain human tumours are more amenable to NK cell-based immunotherapy, and the degree of sensitivity to

NK mediated killing is often correlated to their expression of ligands for activating NK receptors (Moretta et al., 2006). While some methods to alter the target tumour characteristics in vivo may prove successful, more interest has been focused upon ways to manipulate the NK effectors to decrease the interactions between inhibitory KIRs and their ligands. Primarily, algorithms have been developed to select NK cell or stem cell donors to increase the potential for NK cell alloreactivity. Enthusiasm for this strategy became widespread after the 2002 report from Perugia in which Ruggeri et al., published that KIR ligand mismatch between patients and their donors was associated with improved outcomes in myeloid leukemia after T-cell deplete haploidentical HCT (Ruggeri et al., 2002).

Allogeneic NK cell therapy: clinical strategies

Alloreactive NK cells can be delivered via two methods: as adoptive immunotherapy or within the context of hematopoietic cell transplantation (HCT). Each approach has its own advantages and disadvantages (see Table 43.1). NK cell products for adoptive transfer can be prepared from adult donor lymphapheresis products, from umbilical cord blood units or from cell lines, and can be expanded either in vivo or ex vivo. Because ex vivo expansion techniques have not been perfected, most clinical protocols for adoptive NK cell transfer include a lymphodepleting preparative regimen (Muranski et al., 2006). While treatment-related toxicity is minimal because NK cells do not induce GVHD, the efficacy of adoptive transfer protocols is limited by the transient nature of the NK antitumour effect. Alternatively, the beneficial effects of alloreactive NK cells can be incorporated into HCT protocols by selecting stem cell donors using algorithms to predict KIR–ligand mismatch to promote alloreactivity. While these strategies assume the risks of HCT (higher treatment-related mortality, GVHD etc.), they result in permanent engraftment of the potentially alloreactive NK cell pool to provide ongoing antitumour activity. It is important to consider the effect that the host environment may play on the education of developing NK cells and to note the potential for diminished effector functions for NK cells in the early post-transplant period. In contrast, adoptively transferred adult NK cells that have been educated in healthy hosts may have more potent antitumour functions.

Determination of NK cell alloreactivity

Regardless of the treatment strategy, the first step is to select a suitable allogeneic donor. Three main methods have been used to define the potential for alloreactivity between donors and recipients, and the subtle differences among them have been a source of confusion (see Table 43.2). The Perugia group used the KIR–ligand mismatch or KIR–ligand incompatibility model, which predicts that donor-derived NK cells will be alloreactive in the GvH direction when recipients lack C2, C1 or Bw4 alleles that are present in the donor. Consequently, the approximately one-third of recipients who express all three KIR ligands will not have the possibility of an alloreactive HLA-matched donor. This model, which requires knowledge of both the donor and recipient HLA types, assumes that donor NK cells will express the inhibitory KIRs for their HLA class I KIR ligands. A KIR–ligand match calculator is available on the Immuno Polymorphism Database (IPD) http://www.ebi.ac.uk/ipd/kir/ligand.html. Recent data showing that HLA-A3 and HLA-A11 may engage KIR3DL2 suggest that consideration of HLA-A may be

Table 43.1 Adoptive Transfer vs. HCT—Pros and Cons

NK cell-based therapy	Pros	Cons	Questions
Hematopoietic Cell Transplantation	• Permanent engraftment (ongoing antitumour effect)	• More treatment-related toxicity • Risk of GVHD with use of IL-2 • Permanent engraftment (ongoing toxicity)	• Optimal stem cell source • Optimal donor selection criteria • Role of adoptively transferred NK cells into HCT schemas
Adoptive Transfer	• Safer • Transient (no lasting toxicity) • Mature cells educated in the donor have potent effector	• Cells may be dependent on cytokine exposure • Transient (probably not curative)	• Optimal NK cell dose • Optimal NK cell source (adult, UCB, cell line) • Optimal method to induce in vivo NK cell expansion – Lymphodepleting preparative regimen – Cytokines (IL-2, IL-2 complexes, IL-15, transpresentation of IL-15) • Role of Treg

Table 43.2 Models to Predict KIR Alloreactivity

Name	Information needed	Model	Assumptions and implications
KIR–ligand mismatch *or* KIR–ligand incompatibility	Donor Class I HLA typing Recipient Class I HLA typing (Recipient KIR–ligand status)	NK cells derived from donors who possess KIR ligands that are absent in the recipient (C2, C1 or Bw4) will be alloreactive in GvH direction.	Assume that NK cells express inhibitory KIRs for all of their class I KIR ligands (C2, C1 and Bw4) and that donor Class I HLA type predicts donor KIR expression. No potential for NK alloreactivity for recipients who express C2, C1 and Bw4 (~1/3 of patients).
KIR–ligand absence	Recipient Class I HLA typing (Recipient KIR–Ligand status)	As the gene frequency for inhibitory KIR that recognize C2, C1 and Bw4 is high in most populations, most donors will express those KIRs, and alloreactivity can be estimated by recipient ligand status alone.	Assume the respective inhibitory KIR genes are present in the donor. Assume the donor KIR allele is expressed and functional.
Receptor ligand	Donor KIR genotype (for inhibitory KIRs) Recipient Class I HLA typing (Recipient KIR–ligand status)	Both the donor KIR profile and recipient KIR–ligand status are needed to accurately predict alloreactivity.	Allows for NK alloreactivity in HLA identical transplants if self-tolerant donor clones expressing KIR for which they lack the ligand become alloreactive in the post-transplant setting.

important as well. However, the functional importance of these interactions is unclear. For KIR3DL2, some interactions become functional only in the presence of certain peptides (Thananchai et al., 2007), while most other inhibitory KIRs interactions are thought to be predominantly peptide-independent. It has also been recognized that some HLA-A alleles include the Bw4 epitope, although again the functional consequence of the presence of Bw4 remains uncertain. Although HLA-A*2402 and HLA*3201 function as bonafide Bw4 epitopes, HLA-A*2501 and HLA-A*2301 were functionally weak (Foley et al., 2008). It may be premature to incorporate these newly recognized KIR-binding alleles until their functional significance is fully understood. Fortunately, these alleles are relatively rare, and therefore their presence changes alloreactivity frequency predictions for only a small percentage of patients.

Alternatively, the KIR–ligand absence model considers recipients based on their C2, C1 and Bw4 allele status without regard to the donors. The rationale for this approach is based on the observation that most human populations have high frequencies of inhibitory KIRs specific for C2, C1 and Bw4 alleles. Because most donor-derived NK cells will express inhibitory KIRs, the alloreactive potential may be more accurately estimated based on the number of KIR ligands a recipient lacks. Lastly, the receptor-ligand model is based on a comparison of the donor's inhibitory KIR genotype with the recipient's KIR ligand status to eliminate any assumptions about the reliability of using the donor class I HLA type as a predictor of inhibitory KIR expression. KIR genes have multiple alleles with variable levels of expression and functional activities. Therefore this model may be refined further by assessing not only the donor KIR genotype but also functional measures of KIR phenotype. In contrast to the other models where no NK alloreactivity is predicted in HLA-identical transplants, this model predicts that NK alloreactivity may occur if the recipients lack KIR ligands for inhibitory KIRs expressed on self-tolerant clones. Such clones in the donor may be alloreactive in the post-transplant setting.

Allogeneic NK cell therapy: hematopoietic cell transplantation

NK cells are the first lymphocyte population to expand after HCT, and engraftment with alloreactive NK cells has many potential benefits:

- Decreased rates of GVHD as host DCs are killed by donor NK cells (Shlomchik et al., 1999; Lundqvist et al., 2007)
- Decreased rates of graft rejection mediated by NK lysis of host T-cells
- Decreased relapse (GVT) via direct cytotoxcity (using perforin/granzyme, FasL or TRAIL) (Russell and Ley, 2002) or production of suppressive cytokines (IFN-γ TNF-α)
- Improved engraftment mediated by NK cell release of hematopoietic cytokines (Murphy et al., 1992; Siefer et al., 1993)
- Enhanced immune reconstitution and decreased infectious complications mediated by NK cell antiviral activity

Based on the report from Perugia in which alloreactive NK cell clones were detected in patients after stem cell engraftment in KIR mismatched patients, several groups tested the strategy of selecting donors for HCT based on their predicted alloreactivity against the host. The Perugia group tested the effect of KIR–ligand mismatching in the setting of T-cell deplete haploidentical HCT to treat hematologic malignancies. The initial report demonstrated improved engraftment, less relapse and less GVHD (Ruggeri et al., 2002); however, in long-term follow-up, the only significant effect of KIR mismatch was a reduction in relapse and prolonged survival in patients transplanted while in complete remission, most notably in patients with myeloid disease (Ruggeri et al., 2007). Additional clinical trials support the association between KIR–ligand mismatch and favourable clinical outcomes in myeloid malignancies, especially when T cells are depleted in vivo with antithymocyte globulin (Giebel et al., 2003), while others have found no benefit (Bishara et al., 2004). Beneficial effects from KIR–ligand mismatch have not been seen in the T-cell replete setting (Lowe et al., 2003; Beelen et al., 2005). Analyses based on the KIR–ligand absence model have also shown conflicting results. Improved survival has been reported for patients with myeloid malignancies undergoing HLA-matched sibling HCT (Hsu et al., 2005) and in patients with myeloid and lymphoid malignancies after HLA-mismatched unrelated HCT (Hsu et al., 2006). Another study of unrelated HCT found that KIR–ligand absence was associated with decreased relapse in early myeloid leukemias, but that patients with early CML had more GVHD (Miller et al., 2007). The effect of KIR–ligand mismatching on outcomes after umbilical cord blood transplantation may depend on the intensity of the preparative regimen (Brunstein et al., 2009; Willemze et al., 2009).

The clinical efficacy of this approach has several potential limitations. Models for predicting the potential for NK cell alloreactivity may not be accurate. A recent study that estimated the size of the alloreactive NK cell pool in 31 individuals by measuring the cell surface expression of inhibitory KIRs and CD94/NKG2A found a significant variability in frequencies of potentially alloreactive cells (0 to 62%) (Fauriat et al., 2008). In addition, NK alloreactivity may be inhibited via other class I recognizing receptors such as NKG2A and LIR-1. Furthermore, NK cell function may be suppressed after HCT. For example, NK cells developing in the post-HCT milieu have altered KIR expression profiles and are hyporesponsive (Shilling et al., 2003; Cooley et al., 2005). This is especially true in the T-cell replete setting. Compared to normal donors, KIR expression is suppressed on NK cells recovering after unrelated donor transplants in the T-cell deplete setting, and significantly more so after unmanipulated bone marrow transplants, and decreased KIR expression correlates with worse clinical outcomes (Shilling et al., 2003; Cooley et al., 2005). This may be explained by disruptions to the developmental or education processes. As NK cells differentiate from hematopoietic cells after HCT, the balance between lymphocyte subsets competing for cytokines and other developmental signals may be shifted leading to altered NK cell licensing and effector functions. Additional analyses are needed to identify the impact of factors such as conditioning, stem cell source, T cell dose and post-transplant immunosuppression on the NK cell function after HCT to better understand how to optimize NK cell alloreactivity to improve clinical outcomes.

In addition to selecting KIR–ligand mismatch donors, another strategy to improve outcomes after HCT is to choose donors based on their KIR gene content. A recent analysis of the effect of KIR genotype on outcomes after T-cell replete unrelated donor HCT for AML showed that the use of donors with KIR B haplotypes, who express more activating KIRs, was associated with significant improvements in overall and relapse-free survival, with more than a 30 percent better relative risk in both these endpoints (Cooley et al., 2009). The clinical benefit was not influenced by the recipients' KIR genotype, the extent of HLA match or the recipients' disease status at the time of transplant. The benefit of donor KIR B haplotypes has also been observed for T-cell replete transplants from HLA matched sibling donors. In these circumstances, the improved survival and reduced transplant-related mortality were attributed to lower rates of cytomegalovirus reactivation (Chen et al., 2006), an effect reported in other settings (Stern et al., 2008) and paralleled in the mouse model of cytomegalovirus infection (Farrell et al., 1997; Voigt et al., 2003). Numerous other studies have reported varied effects of activating KIRs on outcomes after various types of HCT, including increased rates of acute GVHD (Cook et al., 2004; Hsu et al., 2005; Kroger et al., 2006; McQueen et al., 2007; Ruggeri et al., 2002; Ruggeri et al., 2007).

Adoptive transfer of NK cells

The second main clinical modality to exploit alloreactive NK cells is the use of adoptive cellular transfer to provide short-term antitumour activity. The safety and efficacy of this approach was established in a trial using in vivo expanded haploidentical related-donor NK cell infusions to treat 43 patients with metastatic melanoma, metastatic renal cell carcinoma, refractory Hodgkin disease and refractory AML (Miller et al., 2005). The importance of a lymphodepleting preparative regimen for in vivo NK cell expansion was confirmed by this study, which tested 3 chemotherapy regimens of differing intensity. Successful NK cell expansion was seen only in the cohort that received the fully lympho depleting cyclophosphamide and fludarabine regimen (Hi-Cy/Flu) used by Rosenberg (Dudley et al., 2002). Hi-Cy/Flu patients received NK cell infusions on day 0 following 1 or 2 doses of intravenous cyclophosphamide (60 mg/kg) days -4 and -5 and daily intravenous fludarabine (25 mg/m^2) days -5 to -1, followed by 10 million units of subcutaneous IL-2 administered over 2 weeks. Compared to the other regimens, only Hi-Cy/Flu produced pancytopenia, and it was the only one to induce a surge of endogenous IL-15 after chemotherapy. A significant inverse correlation was seen between the IL-15 levels and the absolute lymphocyte count. High levels correlated with successful NK cell expansion, supporting the importance of IL-15 for NK cell homeostasis (Fehniger et al., 2001; Prlic et al., 2003).

In vivo expansion of NK cells was assessed using a PCR-based chimerism assay, with successful expansion defined by the presence of measurable donor NK cells at 2 weeks, following the IL-2 therapy. A total of 33 patients were treated with this approach. In total, only 10 percent met the criteria for successful in vivo NK cell expansion when applying the strict criteria for success based on the detection of >100 donor-derived NK cells/μL of blood 12–14 days after the NK cell infusion. Approximately 30 percent of patients with poor prognosis AML achieved complete remissions. The remission patients had significantly higher proportions of circulating NK cells, which were considerably more cytotoxic against K562 targets, suggesting that the observed clinical efficacy was related and mediated in part by the in vivo expanded allogeneic donor NK cells. Contrary to the initial report, clinical efficacy did not correlate with KIR–ligand mismatch status (Cooley et al., 2008). NK cell expansion and antitumour activity is currently being tested using the same protocol to treat metastatic breast and ovarian cancers, CLL and non-Hodgkin's lymphoma, all of which are sensitive to NK cell lysis in vitro (Cooley et al., 1999; Re et al., 2006). Several other investigators are testing adoptive NK cell transfer for the treatment of multiple myeloma, hepatocellular carcinoma, melanoma and renal cell carcinoma (Igarashi et al., 2004; Lundqvist and Childs, 2005; Ohira et al., 2006; Shi et al., 2008).

Using adoptively transferred NK cells to augment HCT

Because the remissions induced by adoptive NK cell transfer for AML were not durable, we are now using a combined approach of the same haploidentical adoptive NK cell transfer schema followed by a CD34$^+$-selected stem cell infusion to create a nonmyeloablative haploidentical transplantation protocol. Because inadequate host immunosuppression may limit successful in vivo NK cell expansion, we also added 400 cGy of total body irradiation (TBI) to the NK cell-based preparative regimen. A CD34-selected filgrastim-mobilized PB graft from the same donor (target dose >3 × 10^6 CD34 cells/kg) was given with Thymoglobulin 3 mg/kg days 0, +1 and +2 as the only additional immunosuppression. Encouraging results were seen in the first 19 patients treated with this schema. The successful in vivo NK cell expansion rate is significantly higher (75% [9/12 evaluable]; mean 607 ± 184 NK cells/ml) than seen in the adoptive NK cell transfer regimen, suggesting the importance of the TBI. This approach was associated with clearance of all detectable leukemia in 66 percent of patients with relapsed or refractory AML and high tumour burdens (see Figure 43.1). Donor neutrophil engraftment was seen in all evaluable patients (median day 16) and no patients developed áGVHD. Although treatment-related mortality, especially due to infection, remains frequent (46%), this is a promising platform upon which to add other strategies aimed at improving disease-free survival in patients with refractory AML. Other groups are using NK cell donor lymphocyte infusions (DLI) after haploidentical HCT to consolidate engraftment in adults with AML (Passweg et al., 2004) or in children with leukemia and solid tumours (Koehl et al., 2005).

Clinical-scale GMP production and expansion of human NK cells

Clinical scale methods for processing allogeneic NK cell products for human use are limited to FDA-approved selection devices. The University of Minnesota GMP Facility has published methods for processing three different NK cell products (McKenna et al., 2007): 1) enriched adult PB NK cells prepared by CD3-depletion, 2) purified adult PB CD56$^+$/CD3$^-$ NK cells prepared by CD3-depletion followed by CD56 positive selection and 3) enriched umbilical cord blood (UCB) NK cells and NK progenitors prepared by CD3-depletion. CD-3 depletion of the PB-derived NK cell product is performed on a nonmobilized 15-litre mononuclear cell (MNC) apheresis collection using the CliniMACS Cell Selection System (Miltenyi Biotec GmbH, Bergisch Gladbach, Germany) with CD3 MicroBeads. A maximum of 4 × 10^{10} total cells (1.5 × 10^{10} target cells) are labelled with the CD3 MicroBeads prior to separation on the LS column used with the CliniMACS device. The NK cell purity of products obtained after CD3-depletion alone using this method (n = 36) was 38 ± 13 percent with NK cell recovery of >80 percent. Lot release testing is dictated by FDA guidelines for cellular products. Lot release criteria include assessment of cell count, viability, Gram staining and endotoxin testing. The viability analysis is done by flow cytometry (7-AAD) with the release criterion of ≥70 percent. Lot release criteria for sterility testing include a 'no organisms seen' result on Gram stain and endotoxin testing (limulus amebocyte lysate [LAL]-based endotoxin assay) of <5EU/kg.

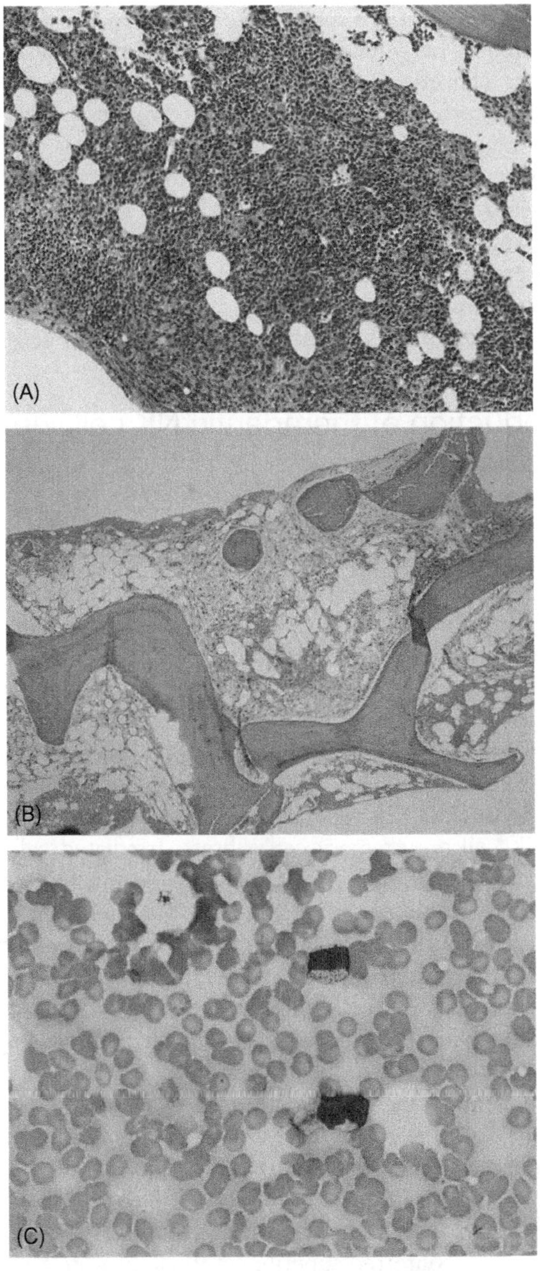

Figure 43.1 • Leukemia Clearance after NK Cell Based Preparative Therapy Shown in these panels are images from a patient with relapsed, refractory acute myelogenous leukemia (AML) who received haploidentical NK cells as part of the preparative therapy in anticipation of a nonmyeloablative haploidentical cell transplant. His pre-therapy bone marrow biopsy was hypercellular with complete replacement by blasts (trephine section 10X: Panel A). He received therapy with cyclophosphamide (60 mg/kg) days -18 and -17 and daily intravenous fludarabine (25 mg/m^2) days -18 to -14, 400 cGy of total body irradiation on day -13, followed adoptive transfer of a CD3 and CD19 depleted haploidentical NK cell product on day -12. Subcutaneous IL-2 (10 million units every other day) was administered over 2 weeks for in vivo stimulation. On day -1, 14 days after the NK cell infusion and prior to the CD34$^+$ hematopoietic cell infusion, a repeat bone marrow biopsy showed complete leukemia clearance (trephine section 5X: Panel B) and the presence of NK cells (aspirate 100X: Panel C).

Because T-cell doses greater than 2×10^5 cells/kg may cause GVHD, the flow cytometry-based lot release criteria should dictate a limit on the number of the CD3$^+$ cells for both the PB-derived NK cell products and the UCB progenitors. Additional criteria for NK cell dose and the percentage of other contaminating cells can be set as appropriate to the protocol. With CD3 depletion, alone the contaminating cells included 1 percent T cells, 31 percent monocytes and 26 percent B cells. Although the addition of a CD56 selection processing step results in a purer NK cell product (90% NK cells) and better elimination of contaminating T cells (4.3 versus 2.7 log depletion), monocytes and B cells, approximately half of the NK cells were lost, and the infused NK cell dose was significantly lower (from approximately 10×10^6/kg to 4×10^6/kg). The addition of a second selection step requires an extra day of processing, and, in this study, the rates of successful NK cell expansion were not improved. The loss of monocytes from the double selected product may not be advantageous as monocytes have been shown to serve as accessory cells when used with IL-2 in vitro (Miller et al., 1992) and may potentiate successfully in vivo NK cell expansion. Extra processing steps are required when using cryopreserved UCB units to isolate NK cell products. UCB units are thawed in a 37°C sterile water or saline bath and washed with a 10 percent dextran/5 percent human serum albumin solution (Chrysler et al., 2004). The CliniMACS CD3-depletion step is performed in the presence of a cocktail of reagents, including $MgCl_2$, rHu-DNAse, sodium citrate, HSA and CliniMACS PBS-EDTA buffer used to prevent clumping of cells and with the addition of human immunoglobulin to prevent nonspecific binding during incubation.

Optimizing NK cell expansion: in vivo or ex vivo?

Murine studies of adoptive lymphocyte transfer show that detectable cytotoxicity against targets and effective NK-mediated killing requires an optimal effector to target ratio dose. Two approaches have been contemplated to increase NK cell numbers (see Table 43.3). Human NK cells can be expanded *in vivo* after infusion of numbers that can be obtained with a single apheresis product (8–15 × 10^6 NK cells/kg). Overnight activation of PB and UCB NK cell products was performed by re-suspending selected cells at a concentration of 2×10^6 cells/mL in X-VIVO 15, without gentamicin and phenol red with the addition of 10 percent human AB serum and 1000 U/mL IL-2. Following overnight (8–16 hours) incubation (37°C/5% CO_2), cells are washed twice and re-suspended in 5 percent HSA solution at a concentration of approximately 25–50 × 10^6 MNC/mL for infusion. The success of this strategy depends on a lymphodepleting chemotherapy regimen consisting of high dose cyclophosphamide and fludarabine. This strategy results in a surge of IL-15 from endogenous sources, which synergizes with low doses of IL-2 administered to patients (Miller et al., 2005). IL-15, which is currently under development for use in humans, may induce superior in vivo NK cell expansion. Using this approach, however, does not guarantee successful or uniform in vivo NK cell expansion. Some cancer patients are insufficiently immunosuppresed and retain the ability to rapidly reject adoptively transferred cells. The addition of total body irradiation to the preparative regimen provides additional immunosuppression and seems to facilitate successful adoptive transfer of both T cells and NK cells (Cooley et al., 2008; Dudley et al., 2008). More sophisticated techniques for product processing may allow for selection of alloreactive NK cell subsets. Alternatively, strategies to manipulate the interaction between the innate and adaptive immune responses may be used to enhance NK cell activity. For example, elimination of host regulatory T cells that can suppress NK cell proliferation and killing may also improve the immune effector functions of the expanding NK cells (Barao et al., 2006).

Alternatively, several centres are developing methods of ex vivo expansion to increase the number of infused cells. One method proposed by Campana and colleagues uses K562 stimulators transduced with 41BB-ligand and IL-15 (Imai et al., 2005). Another novel approach is to use a GMP-compatible lymphoblastoid cell line to stimulate a nearly 500-fold expansion of polyclonal NK cells ex vivo over 30 days (Berg et al., 2009). Important differences between in vivo and ex vivo expanded NK cells affect their utility in therapeutic settings. Compared to resting cells, the ex vivo expanded NK cells increase expression of receptors important in activation pathways needed for cytokine production and cytotoxicity. However, additional studies show that IL-2 withdrawal may be detrimental to cell function. The efficacy of various ex vivo NK cell expansion methods must be tested in Phase I trials. The advantages and disadvantages to the various NK cell processing and expansion strategies must be considered with respect to the specific objectives of individual clinical trials.

Several questions must be considered when evaluating differences between in vivo and ex vivo expansion approaches. The effect of activation on cell shape and adhesion receptor profiles, which affect homing and tumour targeting, is unknown. Furthermore, NK cells stimulated by supraphysiologic concentrations of cytokines tend to undergo apoptosis when removed from ongoing stimulation and may not persist or expand in vivo. One attractive strategy may be to combine ex vivo expansion with in vivo expansion so that function can be tailored for specific NK cell repertoires predicting more antitumour activity.

Production of therapeutic NK cell line products

The use of cell lines to produce unlimited supplies of NK cells for adoptive transfer would overcome several of the potential logistic and scientific limitations of therapy with adoptively transferred adult NK cells. The NK cell yield from lymphapheresis collections is limited, and although NK cells can be successfully expanded in vivo, the success is unpredictable, and the circulation of expanded cell populations is transient. Furthermore, the alloreactivity of in vivo expanded NK cell populations may be heterogeneous due to variable KIR repertoire expression or to differences in NK cell or accessory cell subsets. Irradiated cell lines such as NK92 and KHYG-1 may provide an inexhaustible supply of highly cytotoxic NK cells, but their in vivo survival remains uncertain (Suck et al., 2005; Suck, 2006). A phase I trial of infusing ex vivo expanded NK92 cells for patients with renal cell cancer and melanoma has just been reported (Arai et al., 2008). Alternatively, large numbers of NK cells may be produced from umbilical cord blood or human embryonic stem cell sources (Miller and McCullar, 2001; Woll et al., 2005). Ex vivo expanded cells from any source can be genetically modified to express tumour-specific receptors. For example, the NK92 cell line has been transfected with a chimeric antigen receptor for HER2/neu, which conferred superior cytotoxicity against HER2/neu positive targets (Uherek et al., 2002). Moreover, the homing signals required to direct NK cells to tumour sites are not fully understood

Table 43.3 In vivo vs. ex vivo nk cell expansion—pros and cons

Expansion method	Pros	Cons	Questions
In vivo	• Less cytokine withdrawal • More physiologic homing	• Toxicity of IL-2 • Unpredictable expansion	• Is there an absolute requirement for IL-2 or cytokine support?
Ex vivo	• Patients spared side effects of IL-2 • May allow for expansion of NK cell subsets	• May induce cytokine dependence – Technically challenging with long GMP culture interval – Cell activation may alter homing	• Optimization of expansion techniques – Feeder – Cytokines (doses) • Best starting product

Thinking beyond KIRs

While KIRs are the best characterized family of MHC class I-recognizing NK cell receptors, other MHC-recognizing receptors, including NKG2A (binds HLA-E) and LIR-1 (binds HLA-G and other low affinity HLA ligands), may also be involved in tumour eradication. Their functional importance should not be overlooked because in addition to KIRs, NKG2A is expressed in approximately 50 percent of PB NK cells, and it is the dominant NK cell receptor expressed on NK cells reconstituting in the first six months after HCT. Furthermore, LIR-1 is expressed on approximately one-third of PB NK cells and is more likely to be co-expressed on KIR^+ NK cells than KIR^- NK cells. A series of experiments using primary AML and ALL target cells was performed using blocking antibodies to inhibitory KIRs, NKG2A and LIR-1 to understand the relative individual contribution of each to NK cell-mediated killing (Godal et al., 2008). In both cytotoxicity and CD107a degranulation assays, the blockade of a single inhibitory receptor led to slight increases in killing. However, the addition of an LIR-1 blockade to either the KIR blockade or NKG2A blockade consistently increased killing of all targets sensitive to the pan-HLA blockade. Interestingly, KIR^- NK cells that did express NKG2A or LIR-1 were potently alloreactive against primary leukemia targets but only upon the dual blockade of NKG2A and LIR-1. These findings have important implications for potential efficacy of current NK cell-based therapies. Both KIR^+ and KIR^- NK cell subsets exhibit significant potential for cytotoxicity of killing primary AML and ALL blasts. The potent effector function of the KIR^- NK cells demonstrates that they can be licensed via receptors other than KIRs. Preliminary clinical testing of anti-KIR or anti-NKG2A monoclonal antibodies to block inhibitory receptor interactions demonstrated increases in NK-mediated antitumour killing (Koh et al., 2001). Further trials are warranted.

Additional strategies to enhance in vivo NK cell antitumour function

Importantly, the antitumour activity of NK cells is governed not only by the inhibitory receptor interactions but also by the tumour expression of appropriate activating ligands. For example, the lack of efficacy of KIR mismatched transplants in lymphoid leukemias may be due to their low expression of LFA-1 ligands or NKG2D ligands. Several studies have correlated higher expression of activating ligands with increased susceptibility to NK cell-mediated killing (Pende et al., 2005; Soriani et al., 2008; Verhoeven et al., 2008). Future strategies to enhance activating ligand expression on tumour cells are needed to increase their susceptibility to NK cell-mediated lysis. Targets exposed to the proteasome inhibitor bortezomib, for example, are more sensitive to NK cell-mediated killing (Lundqvist et al., 2006; Shi et al., 2008; Ames et al., 2009), although there may be unexpected effects on NK cell function (Hallett et al., 2008). In addition, NK cells can be targeted through their Fc receptors using monoclonal antibodies to induce ADCC (Dall'Ozzo et al., 2004; Malmberg et al., 2008). Several other avenues of research to improve the efficacy of NK cell immunotherapy are ongoing. Newer agents may enhance in vivo activation of NK cells. For example, IL-15, which may provide a more potent activation signal than IL-2, is in clinical development with phase I trials planned in the next one to two years. Alternatively, indirect NK cell activation may result from the activation of DCs with agents such as TLR agonists (Trinchieri and Sher, 2007; Berger et al., 2009). Other indirect strategies to enhance NK activity via manipulation of the NK/T/DC cell axis, including inhibiting Treg cells, adding immunomodulatory drugs such as bortezobmib or thalidomide into treatment and incorporating NK cells into DC vaccine therapies, are under active investigation (Lundqvist et al., 2009).

Summary

The effector response exhibited by NK cells varies widely depending upon the circumstances of their development and education, on their receptor profile, on the target's receptor ligand expression and by signals received from other immune cells. NK cell activity can be manipulated for therapeutic intent against several tumours, including hematologic malignancies and numerous solid tumours, including melanoma, renal cell carcinoma, ovarian cancer and neuroblastoma. Autologous NK cells can be directed against tumours by blocking inhibitory receptors, by increasing the expression of activating receptor ligands on tumours and by targeting NK ADCC against those receptors with specific antibodies. NK cell activity can be augmented with cytokines and immunomodulatory drugs such as thalidomide, TLR-agonists or vaccines. Alloreactive NK cells developing after allogeneic HCT or after adoptive transfer have powerful antitumour activity. Ultimately, combination therapy using several strategies at once will likely prove most successful. Encouraging results showing clinical efficacy of NK cell-mediated therapies in several settings call out the critical importance of implementing well designed future clinical trials to allow for appropriate interpretation of results. Trials should include rigorously defined clinical endpoints. They should employ only standardized assays to monitor NK cell expansion and effector function. Additional discoveries about NK cell development and function may be anticipated to further refine future therapeutic approaches.

References

Ames, E., Hallett, W.H. and Murphy, W.J. (2009). Sensitization of human breast cancer cells to natural killer cell-mediated cytotoxicity by proteasome inhibition. Clin Exp Immunol **155**(3), 504–513.

Anfossi, N., Andre, P., Guia, S., Falk, C.S., Roetynck, S., Stewart, C.A., Breso, V., Frassati, C., Reviron, D., Middleton, D., Romagne, F., Ugolini, S. and Vivier, E. (2006). Human NK cell education by inhibitory receptors for MHC class I. Immunity **25**(2), 331–342.

Arai, S., Meagher, R., Swearingen, M., Myint, H., Rich, E., Martinson, J. and Klingemann, H. (2008). Infusion of the allogeneic cell line NK-92 in patients with advanced renal cell cancer or melanoma: a phase I trial. Cytotherapy **10**(6), 625–632.

Armeanu, S., Bitzer, M., Lauer, U.M., Venturelli, S., Pathil, A., Krusch, M., Kaiser, S., Jobst, J., Smirnow, I., Wagner, A., Steinle, A. and Salih, H.R. (2005). Natural killer cell-mediated lysis of hepatoma cells via specific induction of NKG2D ligands by the histone deacetylase inhibitor sodium valproate. Cancer Res **65**(14), 6321–6329.

Barao, I., Hanash, A.M., Hallett, W., Welniak, L.A., Sun, K., Redelman, D., Blazar, B.R., Levy, R.B. and Murphy, W.J. (2006). Suppression of natural killer cell-mediated bone marrow cell rejection by CD4+ CD25+ regulatory T cells. Proc Natl Acad Sci U S A **103**(14), 5460–5465.

Becknell, B. and Caligiuri, M.A. (2005). Interleukin-2, interleukin-15, and their roles in human natural killer cells. Adv Immunol **86**, 209–239.

Beelen, D.W., Ottinger, H.D., Ferencik, S., Elmaagacli, A.H., Peceny, R., Trenschel, R. and Grosse-Wilde, H. (2005). Genotypic inhibitory killer immunoglobulin-like receptor ligand incompatibility enhances the long-term antileukemic effect of unmodified allogeneic hematopoietic stem cell transplantation in patients with myeloid leukemias. Blood **105**(6), 2594–2600.

Berg, M., A. Lundqvist, P. mcCoy, L. Samsel, Y. Fan, A. Tawab, R. Childs (2009), "Clinical Grade Ex Vivo-Expanded Human Natural Killer Cells Upregulate Activating Receptors and Death Receptor Ligands and Have Enhanced Cytolytic Activity against Tumor Cells." Cytotherapy, in press.

Berger, M., Ablasser, A., Kim, S., Bekeredjian-Ding, I., Giese, T., Endres, S., Hornung, V. and Hartmann, G. (2009). TLR8-driven IL-12-dependent Reciprocal and Synergistic Activation of NK Cells and Monocytes by Immunostimulatory RNA. J Immunother **32**(3), 262–272.

Bishara, A., De Santis, D., Witt, C.C., Brautbar, C., Christiansen, F.T., Or, R., Nagler, A. and Slavin, S. (2004). The beneficial role of inhibitory KIR genes of HLA class I NK epitopes in haploidentically mismatched stem cell allografts may be masked by residual donor-alloreactive T cells causing GVHD. Tissue Antigens **63**(3), 204–211.

Bottino, C., Castriconi, R., Moretta, L. and Moretta, A. (2005). Cellular ligands of activating NK receptors. Trends Immunol **26**(4), 221–226.

Brunstein, C., J. E. Wagner, D. J. Weisdorf, S. Cooley, H. Noreen, J. N. Barker, T. Defor, M. R. Verneris, B. R. Blazar and J. S. Miller (2009). "Negative Effect of KIR Alloreactivity in Recipients of Umbilical Cord Blood Transplantation Depends on Transplantation Conditioning Intensity." Blood, In Press.

Burns, L.J., Weisdorf, D.J., DeFor, T.E., Vesole, D.H., Repka, T.L., Blazar, B.R., Burger, S.R., Panoskaltsis-Mortari, A., Keever-Taylor, C.A., Zhang, M.J. and Miller, J.S. (2003). IL-2-based immunotherapy after autologous transplantation for lymphoma and breast cancer induces immune activation and cytokine release: a phase I/II trial. Bone Marrow Transplant **32**(2), 177–186.

Cao, W., Xi, X., Hao, Z., Li, W., Kong, Y., Cui, L., Ma, C., Ba, D. and He, W. (2007). RAET1E2, a soluble isoform of the UL16 binding protein RAET1E produced by tumor cells, inhibits NKG2D-mediated NK cytotoxicity. J Biol Chem **282**(26), 18922–18928.

Carson, W.E., Giri, J.G., Lindemann, M.J., Linett, M.L., Ahdieh, M., Paxton, R., Anderson, D., Eisenmann, J., Grabstein, K. and Caligiuri, M.A. (1994). Interleukin (IL) 15 is a novel cytokine that activates human natural killer cells via components of the IL-2 receptor. J Exp Med **180**(4), 1395–1403.

Chan, H.W., Miller, J.S., Moore, M.B. and Lutz, C.T. (2005). Epigenetic control of highly homologous killer Ig-like receptor gene alleles. J Immunol **175**(9), 5966–5974.

Chen, C., Busson, M., Rocha, V., Appert, M.L., Lepage, V., Dulphy, N., Haas, P., Socie, G., Toubert, A., Charron, D. and Loiseau, P. (2006). Activating KIR genes are associated with CMV reactivation and survival after non-T-cell depleted HLA-identical sibling bone marrow transplantation for malignant disorders. Bone Marrow Transplant **38**(6), 437–444.

Chrysler, G., McKennaD.H. Jr., Schierman, T., et al. (2004). Umbilical cord blood banking. Bethsesda, MD: AABB Press.

Cook, M., Briggs, D., Craddock, C., Mahendra, P., Milligan, D., Fegan, C., Darbyshire, P., Lawson, S., Boxall, E. and Moss, P. (2006). Donor KIR genotype has a major influence on the rate of cytomegalovirus reactivation following T-cell replete stem cell transplantation. Blood **107**(3), 1230–1232.

Cook, M.A., Milligan, D.W., Fegan, C.D., Darbyshire, P.J., Mahendra, P., Craddock, C.F., Moss, P.A. and Briggs, D.C. (2004). The impact of donor KIR and patient HLA-C genotypes on outcome following HLA-identical sibling hematopoietic stem cell transplantation for myeloid leukemia. Blood **103**(4), 1521–1526.

Cooley, S., Burns, L.J., Repka, T. and Miller, J.S. (1999). Natural killer cell cytotoxicity of breast cancer targets is enhanced by two distinct mechanisms of antibody-dependent cellular cytotoxicity against LFA-3 and HER2/neu. Exp Hematol **27**(10), 1533–1541.

Cooley, S., Gada, P., McKenna, D., McCullar, V., Fautsch, S., Verneris, M.R., Blazar, B., Burns, L.J., McGlave, P., Weisdorf, D. and Miller, J. (2008). Successful Haploidentical Hematopoietic Cell Engraftment Using a Non-Myeloablative Preparative Regimen Including Natural Killer (NK) Cells: Abstract #8517. Blood **112**.

Cooley, S., McCullar, V., Wangen, R., Bergemann, T.L., Spellman, S., Weisdorf, D.J. and Miller, J.S. (2005). KIR reconstitution is altered by T cells in the graft and correlates with clinical outcomes after unrelated donor transplantation. Blood **106**(13), 4370–4376.

Cooley, S., Trachtenberg, E., Bergemann, T.L., Saeteurn, K., Klein, J., Le, C.T., Marsh, S.G., Guethlein, L.A., Parham, P., Miller, J.S. and Weisdorf, D.J. (2009). Donors with group B KIR haplotypes improve relapse-free survival after unrelated hematopoietic cell transplantation for acute myelogenous leukemia. Blood **113**(3), 726–732.

Cooley, S., Xiao, F., Pitt, M., Gleason, M., McCullar, V., Bergemann, T.L., McQueen, K.L., Guethlein, L.A., Parham, P. and Miller, J.S. (2007). A subpopulation of human peripheral blood NK cells that lacks inhibitory receptors for self-MHC is developmentally immature. Blood **110**(2), 578–586.

Cooper, M.A., Fehniger, T.A. and Caligiuri, M.A. (2001). The biology of human natural killer-cell subsets. Trends Immunol **22**(11), 633–640.

Cudkowicz, G. and Bennett, M. (1971). Peculiar immunobiology of bone marrow allografts. I. Graft rejection by irradiated responder mice. J Exp Med **134**(1), 83–102.

Cudkowicz, G. and Bennett, M. (1971). Peculiar immunobiology of bone marrow allografts. II. Rejection of parental grafts by resistant F 1 hybrid mice. J Exp Med **134**(6), 1513–1528.

Dall'Ozzo, S., Tartas, S., Paintaud, G., Cartron, G., Colombat, P., Bardos, P., Watier, H. and Thibault, G. (2004). Rituximab-dependent cytotoxicity by natural killer cells: influence of FCGR3A polymorphism on the concentration-effect relationship. Cancer Res **64**(13), 4664–4669.

Dudley, M.E., Wunderlich, J.R., Robbins, P.F., Yang, J.C., Hwu, P., Schwartzentruber, D.J., Topalian, S.L., Sherry, R., Restifo, N.P., Hubicki, A.M., Robinson, M. R., Raffeld, M., Duray, P., Seipp, C.A., Rogers-Freezer, L., Morton, K.E., Mavroukakis, S.A., White, D.E. and Rosenberg, S.A. (2002). Cancer regression and autoimmunity in patients after clonal repopulation with antitumor lymphocytes. Science **298**(5594), 850–854.

Dudley, M.E., Yang, J.C., Sherry, R., Hughes, M.S., Royal, R., Kammula, U., Robbins, P.F., Huang, J., Citrin, D.E., Leitman, S.F., Wunderlich, J., Restifo, N.P., Thomasian, A., Downey, S.G., Smith, F.O., Klapper, J., Morton, K., Laurencot, C., White, D.E. and Rosenberg, S.A. (2008). Adoptive cell therapy for patients with metastatic melanoma: evaluation of intensive myeloablative chemoradiation preparative regimens. J Clin Oncol **26**(32), 5233–5239.

Dummer, W., Niethammer, A.G., Baccala, R., Lawson, B.R., Wagner, N., Reisfeld, R.A. and Theofilopoulos, A.N. (2002). T cell homeostatic proliferation elicits effective antitumor autoimmunity. J Clin Invest **110**(2), 185–192.

Farrell, H.E., Vally, H., Lynch, D.M., Fleming, P., Shellam, G.R., Scalzo, A.A. and Davis-Poynter, N.J. (1997). Inhibition of natural killer cells by a cytomegalovirus MHC class I homologue in vivo. Nature **386**(6624), 510–514.

Fauriat, C., Andersson, S., Bjorklund, A.T., Carlsten, M., Schaffer, M., Bjorkstrom, N.K., Baumann, B.C., Michaelsson, J., Ljunggren, H.G. and Malmberg, K.J. (2008). Estimation of the size of the alloreactive NK cell repertoire: studies in individuals homozygous for the group A KIR haplotype. J Immunol **181**(9), 6010–6019.

Fehniger, T.A., Suzuki, K., Ponnappan, A., VanDeusen, J.B., Cooper, M.A., Florea, S.M., Freud, A.G., Robinson, M.L., Durbin, J. and Caligiuri, M.A. (2001). Fatal leukemia in interleukin 15 transgenic mice follows early expansions in natural killer and memory phenotype CD8+ T cells. J Exp Med **193**(2), 219–231.

Fernandez, N.C., Treiner, E., Vance, R.E., Jamieson, A.M., Lemieux, S. and Raulet, D.H. (2005). A subset of natural killer cells achieves self-tolerance without expressing inhibitory receptors specific for self-MHC molecules. Blood **105**(11), 4416–4423.

Foley, B.A., De Santis, D., Van Beelen, E., Lathbury, L.J., Christiansen, F.T. and Witt, C.S. (2008). The reactivity of Bw4+ HLA-B and HLA-A alleles with KIR3DL1: implications for patient and donor suitability for haploidentical stem cell transplantations. Blood **112**(2), 435–443.

Freud, A.G., Yokohama, A., Becknell, B., Lee, M.T., Mao, H.C., Ferketich, A.K. and Caligiuri, M.A. (2006). Evidence for discrete stages of human natural killer cell differentiation in vivo. J Exp Med **203**(4), 1033–1043.

Gasser, S. and Raulet, D.H. (2006). Activation and self-tolerance of natural killer cells. Immunol Rev **214**, 130–142.

Giebel, S., Locatelli, F., Lamparelli, T., Velardi, A., Davies, S., Frumento, G., Maccario, R., Bonetti, F., Wojnar, J., Martinetti, M., Frassoni, F., Giorgiani, G., Bacigalupo, A. and Holowiecki, J. (2003). Survival advantage with KIR ligand incompatibility in hematopoietic stem cell transplantation from unrelated donors. Blood **102**(3), 814–819.

Godal, R., M. Gleason, V. McCullar, S. Cooley, V. Bachanova, M. Verneris, P. McGlave, J. S. Miller (2008). "Blocking Inhibitory KIR Is Insufficient for Optimal Killing of AML and ALL Targets: Additional Requirements for NKG2A and LIR-1 Blockade ASH Abstract #2906 Sunday, December 7, 2008, 6:00 PM–8:00 PM."

Hallett, W.H.D., Ames, E., Motarjemi, M., Barao, I., Shanker, A., Tamang, D.L., Sayers, T.J., Hudig, D. and Murphy, W.J. (2008). Sensitization of Tumor Cells to NK Cell-Mediated Killing by Proteasome Inhibition. J Immunol **180**(1), 163–170.

Herberman, R.B. and Ortaldo, J.R. (1981). Natural killer cells: their roles in defenses against disease. Science **214**(4516), 24–30.

Houchins, J.P., Yabe, T., McSherry, C. and Bach, F.H. (1991). DNA sequence analysis of NKG2, a family of related cDNA clones encoding type II integral membrane proteins on human natural killer cells. J Exp Med **173**(4), 1017–1020.

Hsu, K.C., Gooley, T., Malkki, M., Pinto-Agnello, C., Dupont, B., Bignon, J.D., Bornhauser, M., Christiansen, F., Gratwohl, A., Morishima, Y., Oudshoorn, M., Ringden, O., van Rood, J.J. and Petersdorf, E. (2006). KIR ligands and prediction of relapse after unrelated donor hematopoietic cell transplantation for hematologic malignancy. Biol Blood Marrow Transplant **12**(8), 828–836.

Hsu, K.C., Keever-Taylor, C.A., Wilton, A., Pinto, C., Heller, G., Arkun, K., O'Reilly, R.J., Horowitz, M.M. and Dupont, B. (2005). Improved outcome in HLA-identical sibling hematopoietic stem-cell transplantation for acute myelogenous leukemia predicted by KIR and HLA genotypes. Blood **105**(12), 4878–4884.

Huntington, N.D., Legrand, N., Alves, N.L., Jaron, B., Weijer, K., Plet, A., Corcuff, E., Mortier, E., Jacques, Y., Spits, H. and Di Santo, J.P. (2009). IL-15 trans-presentation promotes human NK cell development and differentiation in vivo. J Exp Med **206**(1), 25–34.

Igarashi, T., Wynberg, J., Srinivasan, R., Becknell, B., McCoyJ.P. Jr., Takahashi, Y., Suffredini, D.A., Linehan, W.M., Caligiuri, M.A. and Childs, R.W. (2004). Enhanced cytotoxicity of allogeneic NK cells with killer immunoglobulin-like receptor ligand incompatibility against melanoma and renal cell carcinoma cells. Blood **104**(1), 170–177.

Imai, C., Iwamoto, S. and Campana, D. (2005). Genetic modification of primary natural killer cells overcomes inhibitory signals and induces specific killing of leukemic cells. Blood **106**(1), 376–383.

Kiessling, R., Petranyi, G., Klein, G. and Wigzel, H. (1975). Genetic variation of in vitro cytolytic activity and in vivo rejection potential of non-immunized semi-syngeneic mice against a mouse lymphoma line. Int J Cancer **15**(6), 933–940.

Kikuchi-Maki, A., Yusa, S., Catina, T.L. and Campbell, K.S. (2003). KIR2DL4 is an IL-2-regulated NK cell receptor that exhibits limited expression in humans but triggers strong IFN-gamma production. J Immunol **171**(7), 3415–3425.

Kim, S., Poursine-Laurent, J., Truscott, S.M., Lybarger, L., Song, Y.J., Yang, L., French, A.R., Sunwoo, J.B., Lemieux, S., Hansen, T.H. and Yokoyama, W.M. (2005). Licensing of natural killer cells by host major histocompatibility complex class I molecules. Nature **436**(7051), 709–713.

Koehl, U., Esser, R., Zimmermann, S., Tonn, T., Kotchetkov, R., Bartling, T., Sorensen, J., Gruttner, H.P., Bader, P., Seifried, E., Martin, H., Lang, P., Passweg, J.R., Klingebiel, T. and

Schwabe, D. (2005). Ex vivo expansion of highly purified NK cells for immunotherapy after haploidentical stem cell transplantation in children. Klin Padiatr **217**(6), 345–350.

Koh, C.Y., Blazar, B.R., George, T., Welniak, L.A., Capitini, C.M., Raziuddin, A., Murphy, W.J. and Bennett, M. (2001). Augmentation of antitumor effects by NK cell inhibitory receptor blockade in vitro and in vivo. Blood **97**(10), 3132–3137.

Kroger, N., Binder, T., Zabelina, T., Wolschke, C., Schieder, H., Renges, H., Ayuk, F., Dahlke, J., Eiermann, T. and Zander, A. (2006). Low number of donor activating killer immunoglobulin-like receptors (KIR) genes but not KIR-ligand mismatch prevents relapse and improves disease-free survival in leukemia patients after in vivo T-cell depleted unrelated stem cell transplantation. Transplantation **82**(8), 1024–1030.

Lanier, L.L. (2005). NK cell recognition. Annu Rev Immunol **23**, 225–274.

Lanier, L.L., Ruitenberg, J.J. and Phillips, J.H. (1988). Functional and biochemical analysis of CD16 antigen on natural killer cells and granulocytes. J Immunol **141**(10), 3478–3485.

Leung, W., Iyengar, R., Triplett, B., Turner, V., Behm, F.G., Holladay, M.S., Houston, J. and Handgretinger, R. (2005). Comparison of killer Ig-like receptor genotyping and phenotyping for selection of allogeneic blood stem cell donors. J Immunol **174**(10), 6540–6545.

Ljunggren, H.G. and Karre, K. (1990). In search of the 'missing self': MHC molecules and NK cell recognition. Immunol Today **11**(7), 237–244.

Lopez-Botet, M., Angulo, A. and Guma, M. (2004). Natural killer cell receptors for major histocompatibility complex class I and related molecules in cytomegalovirus infection. Tissue Antigens **63**(3), 195–203.

Lopez-Vazquez, A., Rodrigo, L., Martinez-Borra, J., Perez, R., Rodriguez, M., Fdez-Morera, J.L., Fuentes, D., Rodriguez-Rodero, S., Gonzaez, S. and Lopez-Larrea, C. (2005). Protective effect of the HLA-Bw4I80 epitope and the killer cell immunoglobulin-like receptor 3DS1 gene against the development of hepatocellular carcinoma in patients with hepatitis C virus infection. J Infect Dis **192**(1), 162–165.

Lowe, E.J., Turner, V., Handgretinger, R., Horwitz, E.M., Benaim, E., Hale, G.A., Woodard, P. and Leung, W. (2003). T-cell alloreactivity dominates natural killer cell alloreactivity in minimally T-cell-depleted HLA-non-identical paediatric bone marrow transplantation. Br J Haematol **123**(2), 323–326.

Lundqvist, A., Abrams, S.I., Schrump, D.S., Alvarez, G., Suffredini, D., Berg, M. and Childs, R. (2006). Bortezomib and depsipeptide sensitize tumors to tumor necrosis factor-related apoptosis-inducing ligand: a novel method to potentiate natural killer cell tumor cytotoxicity. Cancer Res **66**(14), 7317–7325.

Lundqvist, A. and Childs, R. (2005). Allogeneic hematopoietic cell transplantation as immunotherapy for solid tumors: current status and future directions. J Immunother (1997) **28**(4), 281–288.

Lundqvist, A., McCoy, J.P., Samsel, L. and Childs, R. (2007). Reduction of GVHD and enhanced antitumor effects after adoptive infusion of alloreactive Ly49-mismatched NK cells from MHC-matched donors. Blood **109**(8), 3603–3606.

Lundqvist, A., Yokoyama, H., Smith, A., Berg, M. and Childs, R. (2009). Bortezomib treatment and regulatory T-cell depletion enhance the anti-tumor effects of adoptively infused NK cells. Blood **113**(24), 6042–6043.

Malmberg, K.J., Bryceson, Y.T., Carlsten, M., Andersson, S., Bjorklund, A., Bjorkstrom, N.K., Baumann, B.C., Fauriat, C., Alici, E., Dilber, M.S. and Ljunggren, H.G. (2008). NK cell-mediated targeting of human cancer and possibilities for new means of immunotherapy. Cancer Immunol Immunother **57**(10), 1541–1552.

Marsh, S.G., Parham, P., Dupont, B., Geraghty, D.E., Trowsdale, J., Middleton, D., Vilches, C., Carrington, M., Witt, C., Guethlein, L.A., Shilling, H., Garcia, C.A., Hsu, K.C. and Wain, H. (2003). Killer-cell immunoglobulin-like receptor (KIR) nomenclature report, 2002. Tissue Antigens **62**(1), 79–86.

Martin, M.P., Qi, Y., Gao, X., Yamada, E., Martin, J.N., Pereyra, F., Colombo, S., Brown, E.E., Shupert, W.L., Phair, J., Goedert, J.J., Buchbinder, S., Kirk, G.D., Telenti, A., Connors, M., O'Brien, S.J., Walker, B.D., Parham, P., Deeks, S.G., McVicar, D.W. and Carrington, M. (2007). Innate partnership of HLA-B and KIR3DL1 subtypes against HIV-1. Nat Genet **39**(6), 733–740.

Maxwell, L.D., Wallace, A., Middleton, D. and Curran, M.D. (2002). A common KIR2DS4 deletion variant in the human that predicts a soluble KIR molecule analogous to the KIR1D molecule observed in the rhesus monkey. Tissue Antigens **60**(3), 254–258.

McKenna , D.H. Jr., Sumstad, D., Bostrom, N., Kadidlo, D.M., Fautsch, S., McNearney, S., Dewaard, R., McGlave, P.B., Weisdorf, D.J., Wagner, J.E., McCullough, J. and Miller, J.S. (2007). Good manufacturing practices production of natural killer cells for immunotherapy: a six-year single-institution experience. Transfusion **47**(3), 520–528.

McQueen, K.L., Dorighi, K.M., Guethlein, L.A., Wong, R., Sanjanwala, B. and Parham, P. (2007). Donor-recipient combinations of group A and B KIR haplotypes and HLA class I ligand affect the outcome of HLA-matched, sibling donor hematopoietic cell transplantation. Hum Immunol **68**(5), 309–323.

Meyaard, L. (2008). The inhibitory collagen receptor LAIR-1 (CD305). J Leukoc Biol **83**(4), 799–803.

Miller, J.S., Cooley, S., Parham, P., Farag, S.S., Verneris, M.R., McQueen, K.L., Guethlein, L.A., Trachtenberg, E.A., Haagenson, M., Horowitz, M.M., Klein, J.P. and Weisdorf, D.J. (2007). Missing KIR ligands are associated with less relapse and increased graft-versus-host disease (GVHD) following unrelated donor allogeneic HCT. Blood **109**(11), 5058–5061.

Miller, J.S. and McCullar, V. (2001). Human natural killer cells with polyclonal lectin and immunoglobulinlike receptors develop from single hematopoietic stem cells with preferential expression of NKG2A and KIR2DL2/L3/S2. Blood **98**(3), 705–713.

Miller, J.S., Oelkers, S., Verfaillie, C. and McGlave, P. (1992). Role of monocytes in the expansion of human activated natural killer cells. Blood **80**(9), 2221–2229.

Miller, J.S., Soignier, Y., Panoskaltsis-Mortari, A., McNearney, S.A., Yun, G.H., Fautsch, S.K., McKenna, D., Le, C., Defor, T.E., Burns, L.J., Orchard, P.J., Blazar, B.R., Wagner, J.E., Slungaard, A., Weisdorf, D.J., Okazaki, I.J. and McGlave, P.B. (2005). Successful adoptive transfer and in vivo expansion of human haploidentical NK cells in patients with cancer. Blood **105**(8), 3051–3057.

Moesta, A.K., Norman, P.J., Yawata, M., Yawata, N., Gleimer, M. and Parham, P. (2008). Synergistic polymorphism at two positions distal to the ligand-binding site makes KIR2DL2 a stronger receptor for HLA-C than KIR2DL3. J Immunol **180**(6), 3969–3979.

Moretta, L., Bottino, C., Pende, D., Castriconi, R., Mingari, M.C. and Moretta, A. (2006). Surface NK receptors and their ligands on tumor cells. Semin Immunol **18**(3), 151–158.

Moretta, L., Ferlazzo, G., Bottino, C., Vitale, M., Pende, D., Mingari, M.C. and Moretta, A. (2006). Effector and regulatory events during natural killer-dendritic cell interactions. Immunol Rev **214**, 219–228.

Muench, M.O., Humeau, L., Paek, B., Ohkubo, T., Lanier, L.L., Albanese, C.T. and Barcena, A. (2000). Differential effects of interleukin-3, interleukin-7, interleukin 15, and granulocyte-macrophage colony-stimulating factor in the generation of natural killer and B cells from primitive human fetal liver progenitors. Exp Hematol **28**(8), 961–973.

Muranski, P., Boni, A., Wrzesinski, C., Citrin, D.E., Rosenberg, S.A., Childs, R. and Restifo, N.P. (2006). Increased intensity lymphodepletion and adoptive immunotherapy–how far can we go? Nat Clin Pract Oncol **3**(12), 668–681.

Murphy, W.J., Keller, J.R., Harrison, C.L., Young, H.A. and Longo, D.L. (1992). Interleukin-2-activated natural killer cells can support hematopoiesis in vitro and promote marrow engraftment in vivo. Blood **80**(3), 670–677.

Murphy, W.J., Kumar, V. and Bennett, M. (1987). Acute rejection of murine bone marrow allografts by natural killer cells and T cells. Differences in kinetics and target antigens recognized. J Exp Med **166**(5), 1499–1509.

Ohira, M., Ohdan, H., Mitsuta, H., Ishiyama, K., Tanaka, Y., Igarashi, Y. and Asahara, T. (2006). Adoptive transfer of TRAIL-expressing natural killer cells prevents recurrence of hepatocellular carcinoma after partial hepatectomy. Transplantation **82**(12), 1712–1719.

Pando, M.J., Gardiner, C.M., Gleimer, M., McQueen, K.L. and Parham, P. (2003). The protein made from a common allele of KIR3DL1 (3DL1*004) is poorly expressed at cell surfaces due to substitution at positions 86 in Ig domain 0 and 182 in Ig domain 1. J Immunol **171**(12), 6640–6649.

Parham, P. (2006). Taking license with natural killer cell maturation and repertoire development. Immunol Rev **214**, 155–160.

Passweg, J.R., Tichelli, A., Meyer-Monard, S., Heim, D., Stern, M., Kuhne, T., Favre, G. and Gratwohl, A. (2004). Purified donor NK-lymphocyte infusion to consolidate engraftment after haploidentical stem cell transplantation. Leukemia **18**(11), 1835–1838.

Pende, D., Spaggiari, G.M., Marcenaro, S., Martini, S., Rivera, P., Capobianco, A., Falco, M., Lanino, E., Pierri, I., Zambello, R., Bacigalupo, A., Mingari, M.C., Moretta, A. and Moretta, L. (2005). Analysis of the receptor-ligand interactions in the natural killer-mediated lysis of freshly isolated myeloid or lymphoblastic leukemias: evidence for the involvement of the Poliovirus receptor (CD155) and Nectin-2 (CD112). Blood **105**(5), 2066–2073.

Pessino, A., Sivori, S., Bottino, C., Malaspina, A., Morelli, L., Moretta, L., Biassoni, R. and Moretta, A. (1998). Molecular cloning of NKp46: a novel member of the immunoglobulin superfamily involved in triggering of natural cytotoxicity. J Exp Med **188**(5), 953–960.

Prlic, M., Blazar, B.R., Farrar, M.A. and Jameson, S.C. (2003). In vivo survival and homeostatic proliferation of natural killer cells. J Exp Med **197**(8), 967–976.

Raulet, D.H. (2006). Missing self recognition and self tolerance of natural killer (NK) cells. Semin Immunol **18**(3), 145–150.

Raulet, D.H., Vance, R.E. and McMahon, C.W. (2001). Regulation of the natural killer cell receptor repertoire. Annu Rev Immunol **19**, 291–330.

Rayner, A.A., Grimm, E.A., Lotze, M.T., Chu, E.W. and Rosenberg, S.A. (1985). Lymphokine-activated killer (LAK) cells. Analysis of factors relevant to the immunotherapy of human cancer. Cancer **55**(6), 1327–1333.

Re, F., Staudacher, C., Zamai, L., Vecchio, V. and Bregni, M. (2006). Killer cell Ig-like receptors ligand-mismatched, alloreactive natural killer cells lyse primary solid tumors. Cancer **107**(3), 640–648.

Rosenberg, S.A., Lotze, M.T., Muul, L.M., Chang, A.E., Avis, F.P., Leitman, S., Linehan, W.M., Robertson, C.N., Lee, R.E., Rubin, J.T., et al. (1987). A progress report on the treatment of 157 patients with advanced cancer using lymphokine-activated killer cells and interleukin-2 or high-dose interleukin-2 alone. N Engl J Med **316**(15), 889–897.

Rosenberg, S.A., Lotze, M.T., Yang, J.C., Topalian, S.L., Chang, A.E., Schwartzentruber, D.J., Aebersold, P., Leitman, S., Linehan, W.M., Seipp, C.A., et al. (1993). Prospective randomized trial of high-dose interleukin-2 alone or in conjunction with lymphokine-activated killer cells for the treatment of patients with advanced cancer. J Natl Cancer Inst **85**(8), 622–632.

Ruggeri, L., Capanni, M., Urbani, E., Perruccio, K., Shlomchik, W.D., Tosti, A., Posati, S., Rogaia, D., Frassoni, F., Aversa, F., Martelli, M.F. and Velardi, A. (2002). Effectiveness of donor natural killer cell alloreactivity in mismatched hematopoietic transplants. Science **295**(5562), 2097–2100.

Ruggeri, L., Mancusi, A., Capanni, M., Urbani, E., Carotti, A., Aloisi, T., Stern, M., Pende, D., Perruccio, K., Burchielli, E., Topini, F., Bianchi, E., Aversa, F., Martelli, M.F. and Velardi, A. (2007). Donor natural killer cell allorecognition of missing self in haploidentical hematopoietic transplantation for acute myeloid leukemia: challenging its predictive value. Blood **110**(1), 433–440.

Russell, J.H. and Ley, T.J. (2002). Lymphocyte-mediated cytotoxicity. Annu Rev Immunol **20**, 323–370.

Shi, J., Tricot, G., Szmania, S., Rosen, N., Garg, T.K., Malaviarachchi, P.A., Moreno, A., Dupont, B., Hsu, K.C., Baxter-Lowe, L.A., Cottler-Fox, M., ShaughnessyJ.D. Jr., Barlogie, B. and van Rhee, F. (2008). Infusion of haplo-identical killer immunoglobulin-like receptor ligand mismatched NK cells for relapsed myeloma in the setting of autologous stem cell transplantation. Br J Haematol **143**(5), 641–653.

Shi , J., Tricot, G.J., Garg, T.K., Malaviarachchi, P.A., Szmania, S.M., Kellum, R.E., Storrie, B., Mulder, A., Shaughnessy, J.D. Jr., Barlogie, B. and van Rhee, F. (2008). Bortezomib down-regulates the cell-surface expression of HLA class I and enhances natural killer cell-mediated lysis of myeloma. Blood **111**(3), 1309–1317.

Shilling, H.G., McQueen, K.L., Cheng, N.W., Shizuru, J.A., Negrin, R.S. and Parham, P. (2003). Reconstitution of NK cell receptor repertoire following HLA-matched hematopoietic cell transplantation. Blood **101**(9), 3730–3740.

Shlomchik, W.D., Couzens, M.S., Tang, C.B., McNiff, J., Robert, M.E., Liu, J., Shlomchik, M.J. and Emerson, S.G. (1999). Prevention of graft versus host disease by inactivation of host antigen-presenting cells. Science **285**(5426), 412–415.

Siefer, A.K., Longo, D.L., Harrison, C.L., Reynolds, C.W. and Murphy, W.J. (1993). Activated natural killer cells and interleukin-2 promote granulocytic and megakaryocytic reconstitution after syngeneic bone marrow transplantation in mice. Blood **82**(8), 2577–2584.

Singh, S.M., Yanagawa, H., Hanibuchi, M., Miki, T., Okamura, H. and Sone, S. (2000). Augmentation by interleukin-18 of MHC-nonrestricted killer activity of human peripheral blood mononuclear cells in response to interleukin-12. Int J Immunopharmacol **22**(1), 35–43.

Soriani, A., Zingoni, A., Cerboni, C., Iannitto, M.L., Ricciardi, M.R., Di Gialleonardo, V., Cippitelli, M., Fionda, C., Petrucci, M.T., Guarini, A., Foa, R. and Santoni, A. (2008). ATM-ATR dependent up-regulation of DNAM-1 and NKG2D ligands on multiple myeloma cells by therapeutic agents results in enhanced NK cell susceptibility and is associated with a senescent phenotype. Blood **113**(15), pp. 3503–3511.

Stein, M.N., Shin, J., Gudzowaty, O., Bernstein, A.M. and Liu, J.M. (2006). Antibody-dependent cell cytotoxicity to breast cancer targets despite inhibitory KIR signaling. Anticancer Res **26**(3A), 1759–1763.

Stern, M., Elsasser, H., Honger, G., Steiger, J., Schaub, S. and Hess, C. (2008). The number of activating KIR genes inversely correlates with the rate of CMV infection/reactivation in kidney transplant recipients. Am J Transplant **8**(6), 1312–1317.

Suck, G. (2006). Novel approaches using natural killer cells in cancer therapy. Semin Cancer Biol **16**(5), 412–418.

Suck, G., Branch, D.R., Smyth, M.J., Miller, R.G., Vergidis, J., Fahim, S. and Keating, A. (2005). KHYG-1, a model for the study of enhanced natural killer cell cytotoxicity. Exp Hematol **33**(10), 1160–1171.

Thananchai, H., Gillespie, G., Martin, M.P., Bashirova, A., Yawata, N., Yawata, M., Easterbrook, P., McVicar, D.W., Maenaka, K., Parham, P., Carrington, M., Dong, T. and Rowland-Jones, S. (2007). Cutting Edge: Allele-specific and peptide-dependent interactions between KIR3DL1 and HLA-A and HLA-B. J Immunol **178**(1), 33–37.

Trinchieri, G. (2003). Interleukin-12 and the regulation of innate resistance and adaptive immunity. Nat Rev Immunol **3**(2), 133–146.

Trinchieri, G. and Sher, A. (2007). Cooperation of Toll-like receptor signals in innate immune defence. Nat Rev Immunol **7**(3), 179–190.

Trompeter, H.I., Gomez-Lozano, N., Santourlidis, S., Eisermann, B., Wernet, P., Vilches, C. and Uhrberg, M. (2005). Three structurally and functionally divergent kinds of promoters regulate expression of clonally distributed killer cell Ig-like receptors (KIR), of KIR2DL4, and of KIR3DL3. J Immunol **174**(7), 4135–4143.

Uherek, C., Tonn, T., Uherek, B., Becker, S., Schnierle, B., Klingemann, H.G. and Wels, W. (2002). Retargeting of natural killer-cell cytolytic activity to ErbB2-expressing cancer cells results in efficient and selective tumor cell destruction. Blood **100**(4), 1265–1273.

Uhrberg, M., Valiante, N.M., Shum, B.P., Shilling, H.G., Lienert-Weidenbach, K., Corliss, B., Tyan, D., Lanier, L.L. and Parham, P. (1997). Human diversity in killer cell inhibitory receptor genes. Immunity **7**(6), 753–763.

Vales-Gomez, M., Reyburn, H.T., Mandelboim, M. and Strominger, J.L. (1998). Kinetics of interaction of HLA-C ligands with natural killer cell inhibitory receptors. Immunity **9**(3), 337–344.

Valiante, N.M., Uhrberg, M., Shilling, H.G., Lienert-Weidenbach, K., Arnett, K.L., D'Andrea, A., Phillips, J.H., Lanier, L.L. and Parham, P. (1997). Functionally and structurally distinct NK cell receptor repertoires in the peripheral blood of two human donors. Immunity **7**(6), 739–751.

Verhoeven, D.H., de Hooge, A.S., Mooiman, E.C., Santos, S.J., ten Dam, M.M., Gelderblom, H., Melief, C.J., Hogendoorn, P.C., Egeler, R.M., van Tol, M.J., Schilham, M.W. and Lankester, A.C. (2008). NK cells recognize and lyse Ewing sarcoma cells through NKG2D and DNAM-1 receptor dependent pathways. Mol Immunol **45**(15), 3917–3925.

Vitale, M., Bottino, C., Sivori, S., Sanseverino, L., Castriconi, R., Marcenaro, E., Augugliaro, R., Moretta, L. and Moretta, A. (1998). NKp44, a novel triggering surface molecule specifically expressed by activated natural killer cells, is involved in non-major histocompatibility complex-restricted tumor cell lysis. J Exp Med **187**(12), 2065–2072.

Vitale, M., Castriconi, R., Parolini, S., Pende, D., Hsu, M.L., Moretta, L., Cosman, D. and Moretta, A. (1999). The leukocyte Ig-like receptor (LIR)-1 for the cytomegalovirus UL18 protein displays a broad specificity for different HLA class I alleles: analysis of LIR-1 + NK cell clones. Int Immunol **11**(1), 29–35.

Voigt, V., Forbes, C.A., Tonkin, J.N., Degli-Esposti, M.A., Smith, H.R., Yokoyama, W.M. and Scalzo, A.A. (2003). Murine cytomegalovirus m157 mutation and variation leads to immune evasion of natural killer cells. Proc Natl Acad Sci U S A **100**(23), 13483–13488.

Walzer, T., Dalod, M., Robbins, S.H., Zitvogel, L. and Vivier, E. (2005). Natural-killer cells and dendritic cells: "l'union fait la force. Blood **106**(7), 2252–2258.

Wang, X., Rickert, M. and Garcia, K.C. (2005). Structure of the quaternary complex of interleukin-2 with its alpha, beta, and gammac receptors. Science **310**(5751), 1159–1163.

Willemze, R., Rodrigues, C.A., Labopin, M., Sanz, G., Michel, G., Socie, G., Rio, B., Sirvent, A., Renaud, M., Madero, L., Mohty, M., Ferra, C., Garnier, F., Loiseau, P., Garcia, J., Lecchi, L., Kogler, G., Beguin, Y., Navarrete, C., Devos, T., Ionescu, I., Boudjedir, K., Herr, A.L., Gluckman, E. and Rocha, V. (2009). KIR-ligand incompatibility in the graft-versus-host direction improves outcomes after umbilical cord blood transplantation for acute leukemia. Leukemia **23**(3), 492–500.

Woll, P.S., Martin, C.H., Miller, J.S. and Kaufman, D.S. (2005). Human embryonic stem cell-derived NK cells acquire functional receptors and cytolytic activity. J Immunol **175**(8), 5095–5103.

Yawata, M., Yawata, N., Abi-Rached, L. and Parham, P. (2002). Variation within the human killer cell immunoglobulin-like receptor (KIR) gene family. Crit Rev Immunol **22**(5-6), 463–482.

Yokoyama, W.M., Daniels, B.F., Seaman, W.E., Hunziker, R., Margulies, D.H. and Smith, H.R. (1995). A family of murine NK cell receptors specific for target cell MHC class I molecules. Semin Immunol **7**(2), 89–101.

Yokoyama, W.M. and Kim, S. (2006). Licensing of natural killer cells by self-major histocompatibility complex class I. Immunol Rev **214**, 143–154.

Young, H.A. and Ortaldo, J. (2006). Cytokines as critical co-stimulatory molecules in modulating the immune response of natural killer cells. Cell Res **16**(1), 20–24.

Yu, H., Fehniger, T.A., Fuchshuber, P., Thiel, K.S., Vivier, E., Carson, W.E. and Caligiuri, M.A. (1998). Flt3 ligand promotes the generation of a distinct CD34(+) human natural killer cell progenitor that responds to interleukin-15. Blood **92**(10), 3647–3657.

Chapter Forty-Four

44

Natural killer cells and hepatitis C virus infection

Michael A. Nalesnik, Tatsuya Kanto

CHAPTER CONTENTS

What is of supreme importance in war is to attack the enemy's strategy.

Sun Tzu

If you want to understand the immune system, look to the viruses—they have been studying it for eons

Michael T. Lotze

ABSTRACT

Hepatitis C virus (HCV) infects more than 170 million people, 80% of whom develop chronic disease. Viral interference with host innate immune response, in particular natural killer (NK) cells, may set the stage for subsequent ineffective adaptive immune response and viral persistence.

Viral NS3/4a protease interferes with hepatocyte type I interferon (IFN) production via several mechanisms. NS5A protein may also interfere with this pathway via IL-8 upregulation. These and related changes, along with suboptimal dendritic cell (DC) response, possibly contributed to by IL-15 deficit, may impair NK cell activation. Viral E2 protein can directly engage NK cells via cellular CD81 and inhibit NK cell response to activation signals. HCV core protein upregulates hepatocyte HLA class I expression, serving as a likely deterrent of NK cell cytotoxicity. Core protein can also upregulate HLA-E on hepatocytes, and interaction of this molecule with inhibitory NKG2A receptors may downregulate NK cell activity.

In chronic HCV infection, both NK cells and DCs may produce increased IL-10, skewing the adaptive immune response towards Th2 type. In this condition, Cytotoxic CD56^{dim} NK cells may be decreased, and cytokine-producing CD56^{bright} NK cells may be increased. Population studies of polymorphisms affecting cytokine production or NK cell inhibitory receptor binding have shown associations with viral clearance, suggesting that these represent important factors of the host immune response.

Many current efforts towards control of HCV infection focus on antiviral agents or T-cell response. However, the virus itself seems to have expended a great deal of evolutionary effort in attempting to evade multiple aspects of the host innate immune response. A greater understanding of the role of NK cells may lead to interventions that facilitate early viral clearance and subsequently decrease the frequency of chronic infection.

KEY WORDS

Hepatitis C virus (HCV), Natural killer (NK) cells, Innate immune response

Introduction

It is currently estimated that about 170 million people, representing 3% of the world's population, are infected with hepatitis C virus (HCV) (1999). These individuals are at significant risk for the development of cirrhosis, the emergence of hepatocellular carcinoma and the possibilities of other complications such as cryoglobulinemia or renal disease. The incidence of infection is expected to rise in the immediate future, underscoring the fact that this virus represents a significant world public health problem.

HCV was identified in 1989 as a specific agent responsible for what had been known up to that time as non-A, non-B hepatitis (NANBH) (Choo et al., 1989). Its discovery entailed construction of a cDNA library from plasma derived from a chimpanzee with a high infectious titre of the then unknown virus. This library was used to express polypeptides that were in turn screened with serum from a NANBH patient. An assay derived from this approach showed that approximately 80% of patients with chronic transfusion-associated NANBH had antibody to HCV (Kuo et al., 1989).

As is the case with most viruses, a dynamic interplay exists between pathogen and host, with a variety of moves and countermoves by both partners. The success of the virus in this regard is reflected in the fact that approximately 80% of patients will fail to control the primary infection and go onto chronic disease (Bode et al., 2007). This chapter will briefly review the makeup and life cycle of HCV, outline some aspects of the host: virus interaction and discuss the role of natural killer (NK) cells in this venue.

Hepatitis C virus

Classification and viral genome

HCV is an enveloped RNA virus classified as a distinct genus Hepacivirus in the family Flaviviridae. It is divided into 6 genotypes and more than 50 subtypes based on genomic variability (Bukh et al., 1995). The virus has a single-stranded positive-sense genome of around 9600 nucleotides with a single open reading frame flanked on either end by conserved untranslated regions (UTRs).

The 5′ UTR is highly conserved and contains an internal ribosome entry site (IRES) that allows the virus to usurp the cell translation machinery (Otto and Puglisi, 2004). The efficiency of translation control varies among different genotypes and quasispecies (Honda et al., 1999; Laporte et al., 2000; Soler et al., 2002). The 3′ UTR is considered essential for viral replication (You and Rice, 2008) and has an additional role in enhancing the translation of viral RNA mediated by the 5′ UTR IRES (Song et al., 2006). It interacts with the cellular La and polypyrimidine tract-binding proteins (Domitrovich et al., 2005) and, in one study, was found to bind more than 70 host proteins (Harris et al., 2006).

The bulk of the genome is comprised of a single open reading frame that encodes a 3010 amino acid polypeptide that is cleaved in a cotranslational and posttranslational fashion to give rise to 10 viral proteins (core, E1, E2, p7, NS2, NS3, NS4A, NS4B, NS5A, NS5B) (Dubuisson, 2007). In addition, a frameshift involving the core region gives rise to an alternate reading frame protein known as F protein (Branch et al., 2005).

HCV proteins and structure

The mature HCV core protein ranges from 19 kDa to 21 kDa molecular weight and is thought to comprise the viral nucleocapsid. This RNA-binding protein is mainly found attached to the endoplasmic reticulum, in association with lipid droplets (Dubuisson, 2007). The protein is thought to have a role in hepatocyte steatosis and to render the host cell more susceptible to injury and carcinogenesis. Core protein contains epitopes recognized by both T cells and B cells (Barban et al., 2000; Beld et al., 1999; Jackson et al., 1999; Nattermann et al., 2005; Pirisi et al., 1995). A direct modulating effect on the T-cell response is likely. Yao et al. (2007) concluded that this protein led to an inhibition of T-cell function. Separately, Chung et al. (2001) suggest that it may cause an accelerated inflammatory response. HCV core protein acts as a ligand for toll-like receptor 2 (TLR-2) (Dolganiuc et al., 2004; Duesberg et al., 2002), and this may contribute to monocyte/macrophage activation in chronic infection (Dolganiuc et al., 2007). Ciccaglione et al. (2007) found that this protein was capable of inhibiting interferon (IFN) regulatory factor I (IRF-1) expression, leading to repression of subsequent target genes such as IL-12 and IL-15. It has been suggested that some of the effects attributed to the core protein may be due in part or in whole to an overlapping alternate reading frame protein expressed during natural HCV infection (Branch et al., 2005).

The glycoproteins E1 (gp31) and E2 (gp70) have been demonstrated on the surface of HCV virions (Kaito et al., 2006), confirming their role as major envelope glycoproteins. E1 and E2 form heterodimers and are found largely in the endoplasmic reticulum. E2 contains a hypervariable region HVR1 that can form quasispecies during the course of infection, possibly due to immune pressure leading to the development of escape mutants. E2 is thought to represent the main ligand of the mature virion for cell surface binding and has been found in vitro to link CD81 (tetraspanin) and SR-B1 (scavenger receptor class B type I), two putative cell receptors for HCV (Heo et al., 2006). The interaction with CD81 is also thought to affect dendritic cell (DC) function during HCV infection (Nattermann et al., 2006; Zhao et al., 2007).

The p7 polypeptide is found either as a C-terminal portion of the E2 glycoprotein or as a separate protein (Major et al., 2001). In vitro studies show that it is membrane-localized and forms ion channels (StGelais et al., 2007) that can be inhibited by amantadine. This protein is necessary for infectivity in a chimpanzee model and has been found to be important for the efficient assembly and release of virions (Steinmann et al., 2007). The ability to enhance virus production varies among different isolates, suggesting that it may function as a virulence factor (Steinmann et al., 2007).

The NS2 protein is a membrane protein localized within the endoplasmic reticulum. A role in virus assembly and release has been proposed (Dubuisson, 2007), and a recent study found the protease domain of NS2 to be necessary for viral infectivity (Jones et al., 2007). NS2 has been shown to inhibit apoptosis mediated by the hepatocyte CIDE-B protein in vitro (Erdtmann et al., 2003).

The NS3 region is responsible for encoding a serine protease activity that cleaves the HCV polyprotein at multiple sites, in conjunction with a protein from the NS4 region. NS3 also encodes for RNA helicase and NTPase activity. Thoren et al. (2004) demonstrated that this protein was capable of inducing oxygen radical formation in mononuclear and polymorphonuclear phagocytes, which were then able to induce dysfunction or apoptosis in T cells, NK cells, and NK T cells. The NS3/4 protease also plays a role in the response of the virus to host type I IFN production (Hiscott et al., 2006).

The NS4 region encodes viral proteins NS4A and NS4B. NS4A is a cofactor that enhances the activity of the protease encoded by NS3. It also joins with viral proteins NS4B and NS5A to form a stable heterotrimer.

NS4B contains a nucleotide binding motif with GTPase activity that appears to be important for efficient RNA replication (Einav et al., 2004). It also induces a 'membranous web' comprised of vesicles in a membrane matrix that is thought to represent the viral replication complex (Egger et al., 2002). It may also contribute, along with NS3/4 A, to interference with hepatocyte type I IFN production (Tasaka et al., 2007).

The NS5 region encodes viral proteins NS5A and NS5B. NS5A is thought to play a role in replication of RNA and exists as hypophosphorylated and hyperphosphorylated forms. The hypophosphorylated form appears to support efficient viral replication in vitro, whereas the hyperphosphorylated form is inhibitory (Neddermann et al., 2004) Of interest, the mTOR inhibitor Rapamycin reduces the phosphorylation status of NS5A (Coito et al., 2004). NS5A has a separate domain referred to as the IFN sensitivity-determining region that inhibits the action of IFN-induced protein kinase PKR, an effector of IFN-induced antiviral activity (Gale et al., 1997).

NS5B protein contains the RNA-dependent RNA polymerase activity that is essential for viral replication (Luo et al., 2000), and it is likely that the interaction of this protein with the 3′ terminus of the viral genome leads to the synthesis of negative strand RNA. It may also serve to modulate the phosphorylation status of NS5A (McCormick et al., 2006). Choi et al. (2006) demonstrated that this protein may modulate TNFα signalling pathways.

The HCV virion has a spherical appearance and is approximately 50 nm in diameter. It is enveloped, and, in the peripheral blood, is found in probable association with LDL, and in another form either free or complexed with immunoglobulin, as determined by sucrose gradient centrifugation (Hijikata et al., 1993; Kanto et al., 1995). It is thought that association with LDL may protect against antibody neutralization.

Life cycle of HCV

Until recently, there was not an effective cell culture system for HCV (Lindenbach et al., 2005; Yi et al., 2006). Thus, most studies were performed using either standard mammalian cell expression in vitro or in vivo with chimpanzees, as the chimpanzee is the only animal model that mimics most but not all aspects of the human infection. The development of permissive cell culture models has allowed progress in the study of the viral life cycle and has facilitated the testing of drugs and neutralizing antibodies against HCV.

HCV generally is spread through parenteral routes and reaches the liver via the bloodstream. Virions may circulate as free particles or as particles bound to immunoglobulins, low density or very low density lipoproteins (Nielsen et al., 2006). The viral surface protein E2 can bind cellular CD81 (tetraspanin) as well as scavenger receptor class B type I (SR-BI) (Zeisel et al., 2007), both of which are found on hepatocytes and appear to function in viral entry. The LDL receptor (Nahmias et al., 2006) and the lectins L-SIGN and DC-SIGN (Lozach et al., 2004) may also facilitate entry of the virus into the hepatocyte. Further entry is dependant upon the presence of claudin-1, 6 or 9 (Meertens et al., 2008), and the virion is endocytosed into clathrin-coated pits. Viral surface membrane and endosomal fusion occur in the context of acidification, and the nucleocapsid is released into the cytoplasm where uncoating occurs. The positive strand RNA initiates translation by means of an internal ribosomal entry site (IRES) that is located in the 5′ noncoding region and binds the 40S ribosome. This leads to the formation of a single polyprotein that is processed into individual peptides in a cotranslational and posttranslational fashion, using both viral and cellular protease activity.

A replication complex then arises from a combination of viral nonstructural proteins and cellular material. Viral NS4B, NS5A, NS5B and the NS3 helicase-NTPase domain are known to be important components of this structure, and the cellular substrate is referred to as a 'membranous web', which is a perinuclear vesiculomembranous aggregate thought to be derived from the endoplasmic reticulum. At this site, active RNA synthesis occurs.

The assembly and release of mature virions is not completely understood. Assembly likely occurs in proximity to the membranous web, and secretion may be dependant upon the ion channels formed by the p7 protein. Gastaminza et al. (2008) concluded from their studies that the virus hijacks the host machinery for assembly, maturation, degradation and secretion of VLDL, thereby explaining in part the tropism for hepatocytes.

In the liver, in situ hybridization shows up to 50% of hepatocytes to contain HCV in infection (Pal et al., 2006). Antigenic expression detected by immunohistochemistry is reported in 5% or less of hepatocytes, and in lesser numbers of biliary epithelial cells and sinusoidal lining cells (Nouri-Aria et al., 1995), although this result may be artifactually low due to antigen instability in formalin-fixed tissues. Occasional mononuclear cells also may express HCV antigens (Nouri-Aria et al., 1995). Using RT-PCR, HCV has also been detected in lymph nodes, pancreas, bone marrow, spleen, thyroid, brain and adrenal gland (Forton et al., 2004; Lerat et al., 1998). It is not known whether the virus replicates in haematopoietic cells.

Host response to HCV infection

Innate immune response to HCV

The cell immune response is important in light of the fact that the virus is largely noncytopathogenic, and mechanisms external to the infected cell are required for viral elimination. The integrated host response to HCV infection is comprised of innate and adaptive components of the immune system, with each arm modulating the kinetics of the other to some extent. In addition, there is a separate crosstalk between host and virus, as each tries to gain advantage and establish a favourable equilibrium. This initial set of interactions, in which NK cells play an important role (Khakoo et al., 2004), is crucial in determining the outcome of the infection (Gale and Foy, 2005). In approximately 20% of acutely infected patients, this process generates a strong T-cell response that leads to spontaneous resolution of infection. More often than not, however, the virus gains a marginal advantage that permits its survival at the cost of chronic hepatitis and attendant complications for the host. This discussion will focus on aspects of the innate immune response with emphasis on those features impinging upon the role of NK cells. Several recent reviews have addressed the host response to this infection with emphasis on the T-cell response (Blackburn and Wherry, 2007; Bowen and Walker, 2005; Ishii and Koziel, 2008; Li et al., 2008; Manigold and Racanelli, 2007; Neumann-Haefelin and Thimme, 2007; Neumann-Haefelin et al., 2007; Rehermann, 2007; Semmo and Klenerman, 2007).

Hepatocyte infection, IFN production and HCV countermeasures

IFN-α production indirectly activates NK cells via its effect on DCs, which provides one of many areas in which the host:virus struggle is played out. Following infection, type I IFNs are elaborated by multiple cell types, likely beginning with hepatocytes. This can be considered as the immediate early innate immune response. Li et al. (2005) showed the existence of TLR-3 as well as a retinoic acid-inducible gene (RIG) pathways in cultured hepatocytes, suggesting that these are potential in vivo mechanisms for type I IFN production by liver cells. Upon engagement of dsRNA by cell surface or intracytoplasmic membrane-bound TLR-3, binding to the adaptor protein TRIF (Toll/interleukin-1 receptor domain-containing adaptor protein inducing IFN-β) occurs, which ultimately leads to the activation of the transcription factors NFκB and IRF-3 (IFN regulatory factor-3). Similarly, intracytoplasmic RIG-1 recognition of dsRNA leads to IRF-3 activation via a separate pathway, as well as to formation of the active form of the transcription factor AP-1. These products in turn induce IFN-β gene transcription, and this protein is produced and secreted by the cell (Bode et al., 2007).

IFN-β exerts its effect in an autocrine or paracrine fashion by binding the cell surface type I IFN-α/β receptor. This engages the Jak/STAT pathway to lead to the production of the IFN-stimulated gene factor 3 (ISGF3) complex, which translocates to the nucleus and acts to enhance transcription of a family of IFN-stimulated genes. These include IRF-7, which, in conjunction with IRF-3, upregulates IFN-α transcription, thereby propagating a positive feedback mechanism to magnify the antiviral effect (Bode et al., 2007).

Reflecting the importance of this process in establishing an environment conducive to control of virus infection, HCV has evolved a number of potential strategies to combat the host offensive. Foy et al. (2003), using Huh7 hepatoma cells, demonstrated that the viral protease NS3/4 was capable of blocking phosphorylation and activity of IRF-3, resulting in enhanced viral replication in vitro. The viral NS3/4 A protein is also able to interfere directly with both TLR-3 and RIG-1-mediated signal transduction pathways. In the former case, this involves proteolysis of the adaptor protein TRIF (Li et al., 2005) and, in the latter case, cleavage of the mitochondrial tethered adaptor protein CARDIF (caspase recruitment domain adaptor-inducing IFN-β, also known as IPS-1, MAVS or VISA) (Foy et al., 2005; Hiscott et al., 2006). Tasaka et al. (2007) also found a role for viral NS4B in the inhibition of the RIG-1 pathway. Abe et al. (2007) reported that HCV NS5A can interfere with TLR-dependent cytokine production in mouse macrophage cell lines by interacting with the death domain of the adaptor molecule MyD88. This study is of particular interest since the interaction of HCV proteins with TLR pathways represents a potentially significant and largely unexplored mechanism by which the virus may evade host immune surveillance. HCV may also interfere with the Jak/STAT signal transduction pathway, possibly by effects of the viral core protein on STAT-1, SOCS-3 (suppressor of cytokine signalling-3) and ISGF3. The controversies and implications surrounding these interactions have been recently reviewed (Bode et al., 2007). Despite these efforts by HCV, studies in the chimpanzee model show high levels of intrahepatic IFN-α production during early infection (Thimme et al., 2002). This is not associated with a concomitant decrease in HCV genomic levels, suggesting that resistance to, rather than interference with, elevated IFN-α level may be more important for viral survival.

NK cells and HCV infection

Introductory comments

Prior to a detailed discussion of NK cells in the context of HCV infection, a few general comments are in order. First, regional differences in the distribution of NK cells need to be stressed. Golden-Mason and Rosen (2006) point out that intrahepatic mononuclear cells contain a higher percentage of NK cells than is seen in the peripheral blood. Within the liver, NK cells comprise 20–30% of mononuclear cells and may account for up to half of intrahepatic lymphocytes, as compared to representing 10–15% of peripheral blood mononuclear cells. This suggests that HCV would need to have evolved specific mechanisms for long-term survival within an environment rich in NK cells (Golden-Mason and Rosen, 2006).

A number of published studies have compared the more easily obtained peripheral blood NK cell frequencies in HCV-infected patients versus uninfected patients. Such studies have led to sometimes conflicting results, raising the possibility that rapid phenotypic or functional changes of NK cells are occurring in this compartment in vivo. Examination of liver-infiltrating NK cells is technically more demanding but may provide more precise and otherwise unobtainable insights into the role of these cells in HCV infection.

A synopsis of the following discussion of NK cells and HCV infection is presented in Figure 44.1 and in Table 44.1.

A second comment relates to the issue of NK cell subpopulations. NK cells contain at least two different subpopulations according to their degree of CD56 expression, and these subpopulations differ in function as well as phenotype. There is speculation that CD56-defined subsets may play distinct roles in the pathogenesis of HCV-induced liver disease, such as in inflammation or fibrosis (Lin et al., 2004; Morishima et al., 2006). Several recent reports have demonstrated that, in chronic HCV infection, the frequency of $CD56^{dim}$ NK cells is reduced, whereas numbers of $CD56^{bright}$ NK cells are increased (Golden-Mason et al., 2008; Meier et al., 2005; Morishima et al., 2006). Future investigation may be necessary to elucidate the extent to which previously reported functional impairment of NK cells in HCV infection can be ascribed to such alterations in subset populations.

NK cell receptors and HCV infection

Effects of NK cell CD81 engagement by HCV

CD81 is a widely expressed cell surface protein that that has been mentioned previously as a cellular coreceptor for ligation of HCV to the hepatocyte. This receptor is also present on most if not all cells of the immune system. It is a member of the tetraspanin family, whose constituents share the presence of four transmembrane domains that contain small and large extracellular loops (Levy and Shoham, 2005), the latter comprising the binding site for the viral E2 protein (Drummer et al., 2002). CD81 and related proteins are thought to integrate extracellular, cytoplasmic and intramembranous components into a 'tetraspanin web' with diverse functions dependant upon context.

In the case of NK cells, the E2:CD81 interaction results in downregulation of NK cell response to activation signals from CD16, NKG2D, IL-2, IL-12, IL-15 or β1 integrin (Crotta et al., 2006; Li et al., 2004; Tseng and Klimpel, 2002). This inhibition includes a reduction of IFN-γ production, decreased release of cytolytic granules and diminished proliferative activity. Li et al. (2004) showed that NK cells cocultured with HCV replicon-containing hepatic cells secreted IFN-γ that in turn upregulated hepatic cell STAT-1 and IFN-α production, resulting in marked inhibition of HCV RNA expression. These effects could be inhibited by cross-linking CD81 by specific antibody or by antibody to IFN-γ.

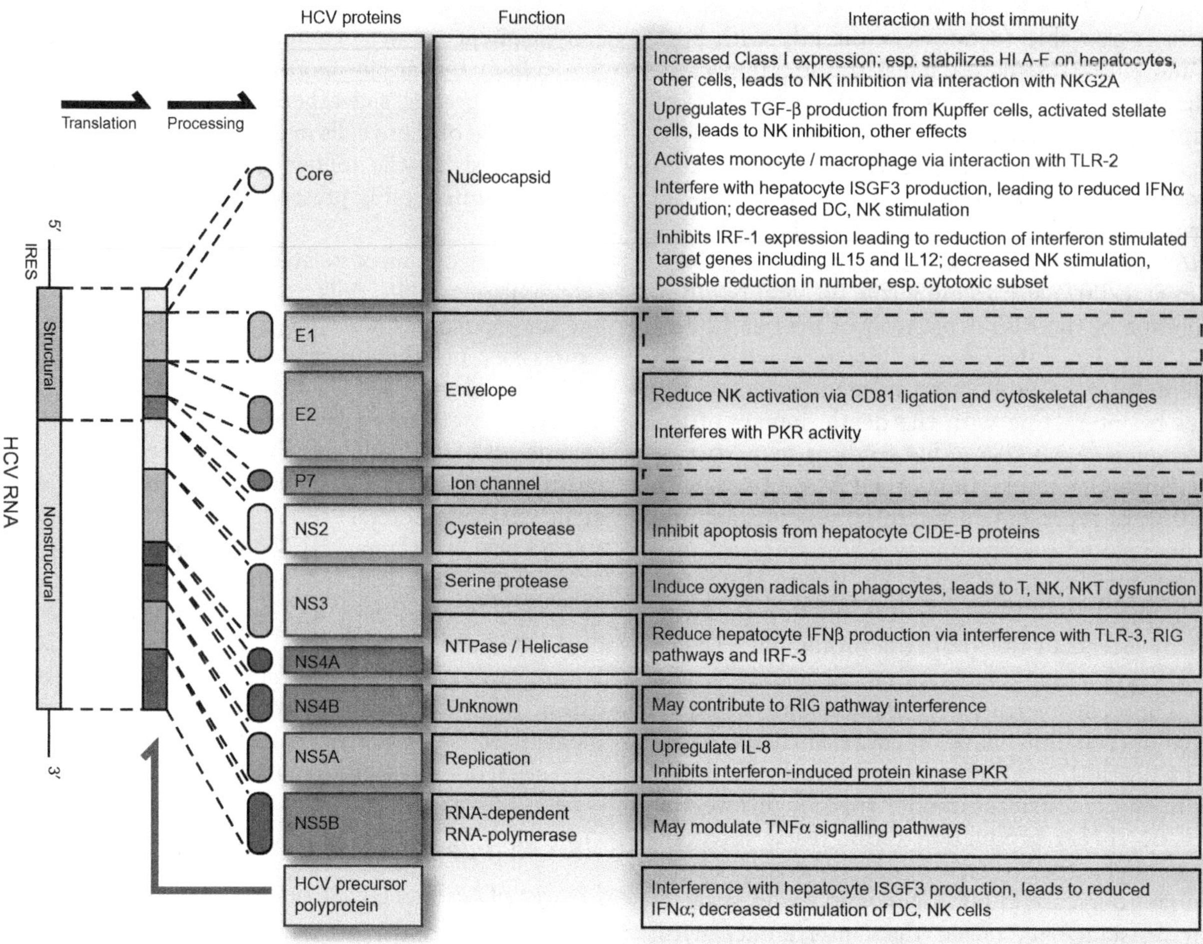

Figure 44.1 • Schematic view of HCV genomes and protein products showing associated interactions with host immunity. The HCV genome is divided into structural and nonstructural areas on the left, with the rightward direction corresponding to representations of the precursor polyprotein and individual viral proteins, respectively. Brief depictions of protein functions within the setting of the viral life cycle are followed by tabulation of probable interactions with the host immune system. The interactions are largely restricted to those involving the innate immune system. See text for additional details.

These authors concluded that NK cells, which were not directly cytotoxic to the infected hepatic cells, were potentially capable of inhibiting HCV replication via an IFN-γ-dependent pathway that was subject to viral interference via the HCV-E2:CD81 interaction. Agrati et al. (2002) suggested that CD81-associated inhibition of NK cell function might thereby contribute to a lack of viral clearance with progression to chronic infection.

Crotta et al. (2006) demonstrated that CD81 cross-linking by antibody or HCV E2 protein resulted in cytoskeletal rearrangement in NK cells as well as in T cells, as based on morphological alterations and enhanced F-actin capping. Whereas these cytoskeletal changes enhanced the response of T cells to CD3-induced TNF-α production, they decreased the CD16-mediated generation of IFN-γ and TNF-α by resting or activated NK cells. These authors were able to decrease the response of T cells and increase the response of NK cells to these stimuli by preincubation of cells with low-dose actin polymerization inhibitors. This led them to conclude that CD81 engagement induces cytoskeletal rearrangement in both NK cells and T cells, but that this process has opposite effects, leading to inhibition of NK cell responses and stimulation of T-cell responses. Since this phenomenon also extended to the inhibition of NK cell response to IL-12, they inferred that the inhibition was independent of KIR involvement. Tseng and Klimpel (2002) showed that the cross-linking of CD81 on NK cells by viral E2 inhibited both NK cell cytotoxicity and NK cell IFN-γ production, suggesting that this may be an important mechanism by which the virus can shift the balance of the early innate host immune response.

Table 44.1 Additional factors potentially influencing NK cell function during HCV infection

Mechanism	Effect
Weak affinity inhibitory allotype of HLA-Cw1 and KIR2DL3, potentially allowing relatively stronger stimulatory KIR interactions	Some evidence for protective antiviral effect
Strong affinity stimulatory allotype of HLA-B Bw4 and KIR3DS1	Protective in one study; increased chronic inflammation in another study
Increased inhibitory NKG2A expression in HCV NK cells	Facilitates inhibitory interaction of NK cells with HLA-E upregulated on other cells, esp. HCV-infected hepatocytes; may also lead to increased NK IL-10 and TGF-β causing defective DC interaction
Increased affinity MICA/B allotypes as ligands for activating NK NKG2D receptor	Potential role in facilitating HCV clearance
Decreased IL-15 as potential cause of decreased DC MICA/B expression	Decreased ability to stimulate NK cells; can be overcome in vitro with IL-15
Lower expression of NKp30 and NKp46 on HCV NK cells (in some studies)	Results differ among studies; possible inhibitory mediator of NK cell function in this setting
Increased NK IL-10 production upon stimulation	Potential skewing of adaptive immune response to Th2 phenotype

Interactions of NK cell receptors with HLA molecules as expressed in HCV infection

HLA class Ia levels and NK cells

Herzer et al. (2003) used recombinant adenovirus constructs to express the HCV core protein in several hepatocyte cell lines and showed upregulation of MHC class I expression on the surfaces of HepG2 cells but not on Hep3B or Huh7 cell lines. These latter cell lines lack wild-type p53, which is present in HepG2 cells, and reconstitution of these cells with wild-type p53 led to an increase in HLA class I in the setting of core protein expression. Transporter associated with surface processing-1 (TAP-1) protein, which is p53 responsive, was also upregulated. The increased expression of class I molecules led to a significant reduction of NK cell-mediated cytotoxicity as assessed in 48-hour chromium release assay, and the authors concluded that this was a likely mechanism of viral evasion against NK cell cytotoxicity in vivo.

Class Ia HLA and killer cell immunoglobulin-like receptors (KIR)

Killer cell immunoglobulin-like (KIR) receptors are clonally expressed in a stochastic fashion on NK cells. They may be stimulatory, with a short cytoplasmic tail and a charged transmembrane domain that allows association with signalling proteins, or inhibitory with a long cytoplasmic tail that contains an immunoreceptor tyrosine-based inhibitory motif (ITIM). They most often recognize class I HLA-C molecules and bind in a manner that overlaps but differs from T-cell receptor binding (Boyington et al., 2000). Different KIR bind with different affinities, and it seems likely that inhibitory KIR engage class I HLA more strongly than do stimulatory KIR (Vales-Gomez et al., 1998).This opens the possibility that different KIR haplotypes may influence the courses of various diseases, including infection with HCV. To date, epidemiologic studies have produced conflicting results.

Khakoo et al. (2004) examined the KIR and HLA-C status in 685 individuals with persistent HCV infection and 352 individuals with resolved HCV infection. Within this population, they focused on those with HLA-C allotypes containing asparagine at residue 80 (HLA-Cw group 1 alleles), which serve as ligands for the inhibitory KIRs, KIR2DL2 and KIR2DL3. Of these receptors, KIR2DL3 binds with weaker affinity. They found a protective effect in individuals who were homozygous for this HLA-C allotype and also homozygous for KIR2DL3. This protective effect was evident in those who were infected with HCV by accidental needle stick or during the course of intravenous drug abuse but not in those who were infected by blood transfusion. They hypothesized that the inhibitory effect of the relatively weakly binding KIR2DL3 could more easily be overcome by other nonvariable activating NK cell receptors. Further, they suggested that this protective effect could be overwhelmed by a massive viral inoculum, explaining the loss of protection in those who received the presumed larger viral dose during blood transfusion.

Rauch et al. (2007) performed a similar study in 142 patients with chronic HCV infection and 33 with resolved HCV infection. These individuals were part of the Swiss HIV Cohort Study and as such obtained HCV primarily as a consequence of intravenous drug abuse. These investigators were unable to find any association between the status of KIR genotype and HLA-C ligand in their population. They felt that a lack of statistical power was unlikely but left open the possibility that the HIV positive status of their population, which differed from that of the previously cited study, may have had an effect on the results.

More recently, Romero et al. (2008) examined KIR receptor distribution and HLA class II alleles in a population of 121 intravenous drug users with chronic HCV infection and 39 others with spontaneous viral clearance. They eliminated the possibility of genetic stratification in this Puerto-Rican American cohort by analysis of autosomal short tandem repeat markers. They were able to confirm the association among homozygous KIR2DL3 status, homozygous HLA-Cw group 1 alleles and spontaneous viral clearance, and found an additional association of KIR2DL3, DRB1*1201 and spontaneous clearance. They tabulated all prior studies to date in their report and concluded that additional population studies were necessary.

A weak protective effect was also previously suggested for the combination of HLA-B Bw4 and the stimulatory KIR3DS1 (Khakoo et al., 2004). In contrast, Paladino et al. (2007) found an increased frequency of this combination in HCV-positive individuals who had progressed to cirrhosis, leading them to suggest that the presence of higher cytotoxic activity might actually be associated with progression of HCV. Lopez-Vazquez et al. (2007) in summarizing studies to date also concluded that evaluation of other large cohorts of patients with HCV infection is needed to confirm the possibility of an association between the interaction of KIR and HLA with disease progression.

HLA class Ib and NK cells

HLA-E interactions with NKG2A and NKG2C receptors HLA-E is a widely expressed member of the non-polymorphic MHC class Ib molecules that contains a nonamer peptide binding motif that typically contains derivatives of signal peptides from other class I molecules and can present other epitopes as well (Rodgers and Cook, 2005). HLA-E can bind CD94-NKG2A (inhibitory) or CD94-NKG2C (activating) on NK cells. Nattermann et al. (2005) found that the hydrophobic peptide YLLPRRGPRL, representing an HLA-A2-restricted and known T-cell epitope derived from amino acid positions 35–44 of the HCV core protein, was able to stabilize surface HLA-E expression in an HLA-E-transfected K562 cell line. Chromium release assay showed inhibition of NK cell cytotoxicity against the HLA-E transfected HCV peptide 35- to 44-loaded K562 cells. No inhibition occurred when transfected cells were preincubated with an irrelevant peptide or when HLA-E-negative K562 target cells were preincubated with the HCV core peptide sequence. Inhibition was abolished in the presence of antibodies to either CD94 or NKG2A, implicating this receptor complex in the process.

As in vivo correlates, these investigators found increased inhibitory NKG2A expression on circulating NK cells from patients with chronic HCV infection as contrasted with those without infection (Nattermann et al., 2006). No difference was found in levels of NKG2C expression; however, reduced levels of the NK cell stimulatory receptors NKp30 and NKp46 were found in the cells of the hepatitis patients. Liver biopsy specimens from patients with chronic HCV infection demonstrated enhanced HLA-E expression in CD68^{+} macrophages/Kupffer cells, CD31^{+} sinusoidal endothelial cells, CD83^{+} DCs, CD14^{+} monocytes and hepatocytes. In the latter case, HLA-E expression was higher in hepatocytes expressing HCV core protein. These studies suggest that this T-cell epitope may also contribute to chronic viral infection by virtue of its synchronous inhibition of NK cell activity (Golden-Mason and Rosen, 2006).

Similar results were found in a study employing freshly isolated circulating NK cells from patients with chronic HCV (De Maria et al., 2007). NK cell cytotoxicity against FO1 melanoma or Daudi Burkitt lymphoma target cells were similar to those obtained from uninfected donors. However, HCV NK cells showed a significant reduction in cytolytic activity when HepG2 hepatocellular carcinoma target cells were used. These investigators also implicated the HLA-E: CD94/NKG2A ligand receptor interaction in this phenomenon and demonstrated increased expression of NKG2A on HCV NK cells compared to uninfected control cells.

Related studies (Jinushi et al., 2004) demonstrated that NKG2A ligation on NK cells is also associated with defective signals for DC maturation, discussed later.

MICA/B and NKG2D receptor MHC class I chain-related sequence (MIC) genes are thought to represent phylogenetically old members of MHC class Ib molecules (Rodgers and Cook, 2005). The two proteins in this group, MICA and MICB (MICA/B), are polymorphic, with approximately 60 and 25 known alleles, respectively; do not present peptides or associate with β-2 microglobulin; and serve as ligands for the activating NKG2D receptor on NK cells (as well as macrophages, CD8+ T cells, γδ T cells, and NKT cells) (Bauer et al., 1999; Stastny, 2006; Yokoyama and Plougastel, 2003).

Similar to the situation with KIR, the binding affinities of MICA/B to NKG2D appear to vary among different allotypes, which has been suggested as a potential influence on the threshold of NK cell activation (Steinle et al., 2001). Karacki et al. examined MICA polymorphisms in 442 individuals with chronic HCV and 228 others who cleared the virus. They found a statistically significant association with the presumed high affinity MICA*015, which occurred twice as often in patients who cleared HCV than in those with persistent disease.

The allele occurred in only 5.6% of their patient population, and no other significant associations were uncovered, leading them to conclude that MICA polymorphisms did not play a significant role in facilitating HCV clearance. However, the association was found in a small number of Black patients, as it was too rare in Whites to analyse. It would appear that further related studies in this defined patient subpopulation may provide additional information.

Studies addressing defective DC MICA upregulation leading to reduced NK cell activation (Jinushi et al., 2003a) are discussed in more detail in the upcoming section addressing NK: DC crosstalk.

Other NK cell receptors in HCV infection

Nattermann et al. (2006) found that patients with chronic HCV infection and viraemia had reduced levels of the natural cytotoxicity receptors NKp46 and NKp30 in circulating NK cells compared to healthy uninfected individuals. Further, patients who had cleared the virus following therapy with pegylated IFN-α and ribavirin exhibited levels of these receptors similar to those of uninfected controls. Although the two receptors were expressed in a proportionate fashion, there was no correlation between the level of expression and circulating viral genomic load in those patients who had viraemia. These investigators also examined receptor expression in intrahepatic cells by flow cytometry following mechanical disruption of tissue samples. At this site, they also found a lower expression of NKp30 and NKp46 in HCV patients compared to those with other liver diseases. In a redirected killing assay using antibodies against natural cytotoxicity receptors and an FcR^+ target cell line, decreased cytotoxicity was seen in HCV NK cells. These authors suggested that the combination of reduced natural cytotoxicity receptors along with increased inhibitory NKG2A expression on HCV NK cells contributed to impaired function of these cells in patients with chronic HCV infection.

In contrast, De Maria et al. (2007) reported that circulating NK cells from patients with chronic HCV infection showed increased expression of the stimulatory natural cytotoxicity receptors NKp46 and NKp30 compared to uninfected adults. This finding was unexpected, as prior studies in other conditions such as HIV infection or acute myelogenous leukaemia had shown reduced levels of these receptors, providing a rationale for positing a functional deficit of the NK cell compartment in chronic disease conditions.

HCV NK cells were not activated as shown by a lack of increased HLA-DR expression, and no correlation was found between natural cytotoxicity receptor expression density and level of viraemia. NK cell cytotoxicity was intact using FO1 melanoma target cells and could be partially inhibited by antibodies to NKp30, NKp46 or NKG2D. Cross-linking of NK CD81, which was expressed at levels comparable to healthy individuals, had no effect on cytotoxicity.

It is possible that differences in patient populations, circulating NK cell subpopulations, or methodological details may have contributed to disparate findings between these groups. Significantly, both groups of investigators concluded that NK cells play an important role in chronic viral persistence. Additional studies to further clarify this role are needed.

Interactions and crosstalk between NK cells and DC in the presence of HCV

Most studies of DCs in the setting of HCV infection have been performed in vitro or are based on chronic infection, and the relevance to the acute phase of the innate response may remain to be proven.

In response to IFN-α, or other specific inflammatory cytokines, DCs typically undergo a maturation process that includes upregulation of class MHC molecules, costimulatory molecules and production of cytokines. Myeloid DCs ($BDCA1^+$, $CD11c^+$, $CD83^+$, $CD33^+$, HLA-DR^{bright}, $CD14^-$) produce IL-12, IL-10, IL-18 (Kaser et al., 2004), TNFα and IFN-β. Plasmacytoid DCs ($BDCA2^+$, $BDCA4^+$, HLA-DR^{bright}, $CD123^{bright}$, $CD11c^-$, $CD33^-$) preferentially produce IFN-α. The cytokines IFN-α, IL-12 and IL-18 are all capable of activating NK cells.

Jinushi et al. (2003b) showed that IFN α-induced upregulation of HLA class Ib MICA/B expression on monocyte-derived DCs was able to activate resting NK cells, enhance NK cell cytotoxicity against K562 cells and increase NK cell production of IFN-γ. This was shown to require direct cell:cell contact and to be dependant upon MICA/B:NKG2D ligation. However, DCs isolated from patients with chronic HCV infection showed impaired modulation of DC MICA/B expression in response to IFN-α, as well as a decreased ability to stimulate NK cells in this circumstance. Further work by this group showed that the DC defect was related to impaired IL-15 production. Using an in vitro coculture system, this group demonstrated that the defect could be overcome with exogenous IL-15 (Jinushi et al., 2003b) but not with administration of IFN-α (Jinushi et al., 2003a). They postulated that an autocrine/paracrine loop involving IL-15 and type I IFNs subserves the ability of DCs to upregulate MICA/B and thereby activate resting NK cells, and that this pathway is blocked by an unknown mechanism in HCV infection. Perhaps related to this is the finding of Ciccaglione et al. (2007) that HCV core protein inhibits IRF-1, thereby reducing IL-15 transcription.

The actual in vivo situation is likely more complex, as Golden-Mason et al. (2004) demonstrated by showing increased IL-15 concentrations from HCV-infected tissues, which they attributed to production by infiltrating monocytes and resident Kupffer cells. In contrast, Meier et al. (2005) found reduced levels of IL-15 in serum from patients with chronic HCV infection. Additional studies to unravel the variables likely responsible for these superficially contrasting results are needed.

Decreased IFN-α production by circulating plasmacytoid DCs has been reported in chronic HCV infection (Dolganiuc et al., 2006; Yonkers et al., 2007). Lai et al. (2007) examined these cells from fresh HCV-infected livers and showed a relative reduction in this cell type when compared to uninfected but inflamed control livers (Lai et al., 2007). In HCV infection, these cells showed higher BDCA-2 expression. These investigators found reduced numbers of IFN α-producing cells in HCV livers and suggested that this may be related to the fact that BDCA-2 ligation inhibits IFN-α/-β production in plasmacytoid DCs and increases IL-12 secretion.

A reciprocal interaction whereby NK cells from patients with chronic HCV infection may inhibit activation of DCs has also been demonstrated (Jinushi et al., 2004). In these studies, NK cells derived from uninfected donors and cocultured with liver epithelial cells were capable of inducing maturation and activation of DCs. This did not require direct contact between NK cells and DCs, as the maturation effect could be reproduced with conditioned medium from prior coculture of NK cells with Hep3B cells. In contrast, when NK cells from chronic HCV patients were used, no activation of DCs occurred. Rather, the HCV–NK cells elaborated IL-10 and TGF-β, and showed higher levels of the inhibitory receptor complex CD94/NKG2A than did normal NK cells. The investigators hypothesized that the effect was dependant upon ligation of NKG2A by hepatoma cell HLA-E. Blockade of NKG2A restored the ability of HCV–NK cells to activate DCs, concomitant with a reduction in IL-10 and TGF-β production. Further, these treated HCV–NK cells were able to stimulate DCs to produce Th1-polarized $CD4^+$ T cells (Jinushi et al., 2004).

Recently, Ebihara et al. (2008) used direct in vitro infection with the JFH1 HCV strain to address the interactions between NK cells and DCs. They were unable to directly infect monocyte-derived DCs but rather found that double-stranded viral RNA (dsRNA) was introduced into these cells via phagocytosis of apoptotic debris from the infected hepatocytes, and that this colocalized with TLR-3 within DC phagosomes. Following subsequent maturation, these cells secreted IL-6 and IFN-β and were able to activate NK cells in a manner dependant upon DC–NK cell contact. This led them to suggest that activation of NK cells via soluble factors such as type I IFN and IL-15 may only have a subsidiary role in HCV infection.

In a study of freshly separated NK cells derived from patients with chronic HCV infection, De Maria et al. (2007) showed that these cells produced increased amounts of IL-10 in addition to IFN-γ upon stimulation. They speculated that if these cells entered the liver, crosstalk with resident DCs might serve to skew the adaptive immune response to allow viral persistence.

Additional cytokine and chemokine studies relevant to the role of NK cells in HCV infection

IL-10

In the investigations of De Maria et al. (2007), challenge of HCV NK cells with FO1 melanoma cells also resulted in IFN-γ production at levels comparable to control cells, along with increased IL-10 production relative to normal NK cells. The authors suggested that the natural cytotoxicity receptors may play a role in the increased IL-10 production, with implications for Th2 skewing and viral persistence. This position finds support in the study of Knapp et al. (2003) who found that HCV patients with the promoter GG genotype at the IL-10 (−1082), which is associated with higher levels of this cytokine, were more likely to have persistent infection than those without this genotype. Kanto et al. (2004) found that myeloid and plasmacytoid DCs from patients with chronic HCV infection primed increased numbers of IL-10 producing cells relative to controls. Subsequent studies by Gelderblom et al. (2007) examined cytokine production using monocyte-derived DCs from patients with chronic HCV infection and found elevated IL-10 but not IL-12p70 secretion by these cells. Both teams also found reduced IFN-α production by DCs in chronic HCV-infected patients (Gelderblom et al., 2007; Kanto et al., 2004).

As noted earlier, HCV NK cells were capable of producing IFN-γ at normal levels in vitro. Meier et al. (2005) also found normal production of IFN-γ by HCV NK cells in response to stimulation with IL-12 plus IL-18. However, these investigators also noted a reduction of circulating NK cells in these patients and suggested that in vivo availability of IFN-γ derived from these cells may be limited.

IL-15

IL-15, discussed earlier in the context of DC–NK cell crosstalk, plays a role in the development, function and sustenance of NK cells (Becknell and Caligiuri, 2005). Golden-Mason et al. (2004) examined intrahepatic IL-15 levels using a combined approach of RT-PCR, enzyme linked immunosorbent assay and immunohistochemistry. They found a significant increase of this cytokine in HCV-infected liver samples and localized this to Kupffer cells and infiltrating monocytes. They also demonstrated that 80% of NK cells expressed the IL-2/IL-15 receptor β chain (CD 122), which led them to suggest that expression of this cytokine helped to shape the intrahepatic lymphoid population and also likely played an additional role in the host response to HCV infection.

As noted previously, Meier et al. (2005) reported that patients with chronic HCV infection had reduced levels of circulating IL-15. They were able to promote NK cell survival in vitro with this cytokine and found that it had a preferential effect on survival of $CD56^{dim}$ cells relative to $CD56^{bright}$ cells. This corresponded to their in vivo finding of a greater proportionate decrease in the $CD56^{dim}$ versus the $CD56^{bright}$ NK cell subpopulation in peripheral blood of chronically infected patients. This translates to a shift of the ratio of cytotoxic ($CD56^{dim}$) to cytokine producing ($CD56^{bright}$) NK cell subpopulations. These investigators suggested that IL-15 might be considered as a form of adjuvant immunotherapy in these patients. The implications of NK cell subset differences in the reinterpretation of earlier studies focusing only on NK cell function were considered earlier.

IL-21

Similar in some regards to IL-15, IL-21 is a pleiotropic cytokine that is produced by activated $CD4^+$ T cells and NK T cells; enhances proliferation, activity, and survival of NK cells; and has differential effects on NK cell subsets (Skak et al., 2008; Wendt et al., 2007). To date, no studies have addressed the role of this potentially important cytokine in acute or chronic HCV infection.

TGF-β

Among other functions, transforming growth factor (TGF)-β exerts an inhibitory effect upon NK cells. This cytokine is elaborated primarily by Kupffer cells and activated stellate cells in the liver. Using HepG2 hepatoblastoma cells, Taniguchi et al. (2004) reported that HCV core protein was capable of upregulating TGF-β production directly within these cells, raising the possibility that a similar pathway may also occur in natural infection. Kimura et al. (2006) found the −509CC genotype of the TGF-β1 gene promoter to be associated with a higher clearance rate of HCV. This polymorphism is associated with lower promoter activity, concordant with the concept of TGF-β as a factor favouring viral persistence in this setting.

IL-8

IL-8 (CXCL8) is chemotactic for neutrophils but also has a variety of other effects. Khabar et al. (1997) showed that IL-8 was capable of interfering with the IFN-α pathway. The significance of this pathway for NK cell function was discussed earlier. In vitro exposure of human umbilical vein endothelial cells to HCV-like particles resulted in upregulation of IL-8 production by these cells (Balasubramanian et al., 2005). Polyak et al. (2001) found that the viral protein NS5A alone was capable of inducing IL-8 production and that this activity correlated to the results of an in vitro bioassay to detect interference with the antiviral effects of IFN-α. Thus, NS5A may interfere with the IFN-α pathway by two mechanisms: induction of IL-8 production and interference with the function of the IFN-induced double-stranded RNA-activated protein kinase (PKR) (Polyak et al., 2001).

The report by Asselah et al. (2005) provides potential insights into the timing of IL-8 changes. This group used real-time RT-PCR of 240 selected genes and examined expression levels in normal livers and livers from patients with chronic HCV infection. The latter were subdivided into varying stages of progressive fibrosis as defined using the METAVIR scoring system (Bedossa and Poynard, 1996). No difference was seen in IL-8 mRNA expression when livers with early fibrosis were compared to normal samples. However, IL-8 was significantly upregulated in livers from patients with more advanced fibrosis relative to those with only mild fibrosis, suggesting an increased role for this cytokine in more advanced disease. In contrast, livers from HCV patients with only slight fibrosis showed significant upregulation in a number of type II IFN inducible genes relative to livers from normal controls.

Other chemokines

Chemokines, or chemotactic cytokines, can evoke a number of proinflammatory effects. The production of type I IFNs consequent to HCV infection induces upregulation of the chemokine MIP-1α (macrophage inflammatory protein 1-α) from Kupffer cells (Ahmad and Alvarez, 2004) or endothelial cells, although reports vary regarding cell type (Zeremski et al., 2007). This attracts and leads to locally increased numbers of NK cells (as well as T cells, monocytes and immature DCs) mainly via the CC chemokine receptor 5 (CCR-5). IFN-γ produced by NK cells in turn stimulates hepatic sinusoidal endothelial cells to produce several additional chemokines, including MIG (monokine induced by IFN-γ, CXCL-9), IP-10 (IFN-γ-inducible protein 10, CXCL-10) and I-TAC (IFN inducible T cell α chemoattractant) (Zeremski et al., 2007). These chemokines attract activated Th1 cells that express the surface receptors CCR-5 and CXCR-3 (CXC chemokine receptor 3) (Zeremski et al., 2007), thereby providing a bridge between innate and adaptive immunity (Ahmad and Alvarez, 2004). Enrichment of intrahepatic T cells bearing these receptors has been demonstrated in patients with chronic HCV infection (Apolinario et al., 2002). IP-10 levels have been associated with the degree of lobular inflammation in these patients (Harvey et al., 2003), and this has been proposed as one of several predictive markers for both rapid and for sustained viral responses (Romero et al., 2006).

Zeremski et al. (2007) recently reviewed this topic and suggested that chemokine generation may play an important role in both viral clearance and in the propagation of chronic inflammation in this infection. In their model, early chemokine production ultimately leading to strong T cell mediated antiviral effector cells is desirable and is a likely correlate of the spontaneous resolution that occurs in 20% of acutely infected patients. In contrast, persistence of chemokine generation in the remainder of patients who generate an ineffective cell mediated response may lead to the continued and nonspecific attraction of inflammatory cells, causing continued necrosis and eventually leading to cirrhosis in the face of viral persistence.

Current therapy of HCV infection

Combined treatment with pegylated IFN-α and ribavirin remains the mainstay of therapy for patients with HCV infection. The regimen is given for 6 or 12 months. Sustained virological response is seen in approximately 55% of treated patients; 10–25% of patients have a transient response with relapse following cessation of therapy, and the remainder are nonresponders. These unsatisfactory results are further qualified by the fact that a number of patients are not eligible for or cannot tolerate therapy and are not included in these figures.

Response to therapy is manifested as a rapid decrease in circulating viral genomic levels, which is interpreted as a suppression of viral replication, followed by a slower decline thought to be related to elimination of infected cells. Treatment during the acute phase of infection appears to be more effective, leading to a reduced frequency of chronic HCV infection from the expected 80% to approximately 10%. Feld and Hoofnagle (2005) suggest that this may indicate that resistance to therapeutic IFN-α might be an acquired phenomenon that arises during the chronic phase, pointing out the need to dissect the host:viral interactions from a temporal perspective.

In one microarray study of liver biopsy samples (Chen et al., 2005), upregulation of IFN-responsive genes prior to therapy was associated with nonresponder status. This suggests (Feld and Hoofnagle, 2005) that an IFN response already in place in these patients is unable to effectively manage the infection and that additional stimulation limited to this pathway is futile. It also highlights the complexity of the host:viral immune interaction and further underscores the fact that the phenotype responsible for loss of viral control likely varies among patient subpopulations.

A major current effort is being directed towards the development of small molecule inhibitors of HCV enzymes. The major viral targets include the NS3/4 A protease and the NS5B polymerase. Several of these agents are in clinical trials and are the subject of recent reviews (De Francesco and Migliaccio, 2005; Harrison, 2007; Pawlotsky et al., 2007). Despite optimism in this area, Pawlotsky et al. (2007) point out that problems with lower antiviral efficacy in vivo compared to in vitro studies, unfavourable toxicity profiles and the development of viral resistance suggest that current therapies will remain as standard care for some time. These predictions, while presently true, are always subject to change.

Current success, however limited, with IFN-α-based regimens does indicate that stimulatory immune modulation at the level of the innate immune system may lead to either transient or prolonged viral remission. Agonists of TLR-7 and TLR-9 are currently in phase 1 clinical trials. Engagement of these receptors, normally found on plasmacytoid DCs, leads to increased IFN-α production, maturation of DCs and stimulation of NK cells.

Direct stimulation of NK cells is a potential avenue of therapy that may reduce the HCV burden in an additive or perhaps synergistic manner when combined with other therapies directed towards stimulation of adaptive immunity or against components of the viral life cycle. Based on studies of mechanisms contributing to NK cell inhibition in chronic HCV, the use of IFN-γ or IL-15 has been suggested as a possibility. Other possible approaches, such as interference with HCV E2: CD81 interaction on NK cells, stimulation of natural cytotoxicity receptors, reduction of IL-10, TGF-β, or IL-8 activity, among others, can be inferred from the earlier discussion of disordered NK cell physiology during HCV infection. Indeed, Golden-Mason and Rosen (2006) hypothesized that the NK cell is the primary target upon which HCV formulates its immune evasion strategy and that the defective crosstalk between NK cells and DCs underlies the observed T-cell defects in this disorder. However, the warning of Zeremski et al. (2007) must also be remembered: What constitutes an effective immune response in acute viral hepatitis does not necessarily imply that a similar response is desirable in later stages. A detailed understanding of the differential effects of NK cells at varying time points during the course of infection must underlie any future efforts to manipulate these powerful cells for the benefit of the host.

Acknowledgement

The authors gratefully acknowledge Mitsuru Sakakibara, M.D., for the design and implementation of Figure 44.1 in this chapter.

References

1999. Global surveillance and control of hepatitis C. Report of a WHO Consultation organized in collaboration with the Viral Hepatitis Prevention Board, Antwerp, Belgium. J Viral Hepat **6**, 35–47.

Abe, T., Kaname, Y., Hamamoto, I., Tsuda, Y., Wen, X., Taguwa, S., Moriishi, K., Takeuchi, O., Kawai, T., Kanto, T., Hayashi, N., Akira, S. and Matsuura, Y. (2007). Hepatitis C virus nonstructural protein 5 A modulates the toll-like receptor-MyD88-dependent signaling pathway in macrophage cell lines. J Virol **81**, 8953–8966.

Agrati, C., Nisii, C., Oliva, A., D'Offizi, G., Montesano, C., Pucillo, L.P. and Poccia, F. (2002). Lymphocyte distribution and intrahepatic compartmentalization during HCV infection: a main role for MHC-unrestricted T cells. Arch Immunol Ther Exp (Warsz) **50**, 307–316.

Ahmad, A. and Alvarez, F. (2004). Role of NK and NKT cells in the immunopathogenesis of HCV-induced hepatitis. J Leukoc Biol **76**, 743–759.

Apolinario, A., Majano, P.L., Alvarez-Perez, E., Saez, A., Lozano, C., Vargas, J. and Garcia-Monzon, C. (2002). Increased expression of T cell chemokines and their receptors in chronic hepatitis C: relationship with the histological activity of liver disease. Am J Gastroenterol **97**, 2861–2870.

Asselah, T., Bieche, I., Laurendeau, I., Paradis, V., Vidaud, D., Degott, C., Martinot, M., Bedossa, P., Valla, D., Vidaud, M. and Marcellin, P. (2005). Liver gene expression signature of mild fibrosis in patients with chronic hepatitis C. Gastroenterology **129**, 2064–2075.

Balasubramanian, A., Munshi, N., Koziel, M.J., Hu, Z., Liang, T.J., Groopman, J.E. and Ganju, R.K. (2005). Structural proteins of hepatitis C virus induce interleukin 8 production and apoptosis in human endothelial cells. J Gen Virol **86**, 3291–3301.

Barban, V., Fraysse-Corgier, S., Paranhos-Baccala, G., Petit, M., Manin, C., Berard, Y., Prince, A.M., Mandrand, B. and Meulien, P. (2000). Identification of a human epitope in hepatitis C virus (HCV) core protein using a molecularly cloned antibody repertoire from a non-symptomatic, anti-HCV-positive patient. J Gen Virol **81**, 461–469.

Bauer, S., Groh, V., Wu, J., Steinle, A., Phillips, J.H., Lanier, L.L. and Spies, T. (1999). Activation of NK cells and T cells by NKG2D, a receptor for stress-inducible MICA. Science **285**, 727–729.

Becknell, B. and Caligiuri, M.A. (2005). Interleukin-2, interleukin-15, and their roles in human natural killer cells. Adv Immunol **86**, 209–239.

Bedossa, P. and Poynard, T. (1996). An algorithm for the grading of activity in chronic hepatitis C. The METAVIR Cooperative Study Group. Hepatology **24**, 289–293.

Beld, M., Penning, M., van Putten, M., Lukashov, V., van den Hoek, A., McMorrow, M. and Goudsmit, J. (1999). Quantitative antibody responses to structural (Core) and nonstructural (NS3, NS4, and NS5) hepatitis C virus proteins among seroconverting injecting drug users: impact of epitope variation and relationship to detection of HCV RNA in blood. Hepatology **29**, 1288–1298.

Blackburn, S.D. and Wherry, E.J. (2007). IL-10, T cell exhaustion and viral persistence. Trends Microbiol **15**, 143–146.

Bode, J.G., Brenndorfer, E.D. and Haussinger, D. (2007). Subversion of innate host antiviral strategies by the hepatitis C virus. Arch Biochem Biophys **462**, 254–265.

Bowen, D.G. and Walker, C.M. (2005). Adaptive immune responses in acute and chronic hepatitis C virus infection. Nature **436**, 946–952.

Boyington, J.C., Motyka, S.A., Schuck, P., Brooks, A.G. and Sun, P.D. (2000). Crystal structure of an NK cell immunoglobulin-like receptor in complex with its class I MHC ligand. Nature **405**, 537–543.

Branch, A.D., Stump, D.D., Gutierrez, J.A., Eng, F. and Walewski, J.L. (2005). The hepatitis C virus alternate reading frame (ARF) and its family of novel products: the alternate reading frame protein/F-protein, the double-frameshift protein, and others. Semin Liver Dis **25**, 105–117.

Bukh, J., Miller, R.H. and Purcell, R.H. (1995). Genetic heterogeneity of hepatitis C virus: quasispecies and genotypes. Semin Liver Dis **15**, 41–63.

Chen, L., Borozan, I., Feld, J., Sun, J., Tannis, L.L., Coltescu, C., Heathcote, J., Edwards, A.M. and McGilvray, I.D. (2005). Hepatic gene expression discriminates responders and nonresponders in treatment of chronic hepatitis C viral infection. Gastroenterology **128**, 1437–1444.

Choi, S.H., Park, K.J., Ahn, B.Y., Jung, G., Lai, M.M. and Hwang, S.B. (2006). Hepatitis C virus nonstructural 5B protein regulates tumor necrosis factor alpha signaling through effects on cellular IkappaB kinase. Mol Cell Biol **26**, 3048–3059.

Choo, Q.L., Kuo, G., Weiner, A.J., Overby, L.R., Bradley, D.W. and Houghton, M. (1989). Isolation of a cDNA clone derived from a blood-borne non-A, non-B viral hepatitis genome. Science **244**, 359–362.

Chung, Y.M., Park, K.J., Choi, S.Y., Hwang, S.B. and Lee, S.Y. (2001). Hepatitis C virus core protein potentiates TNF-alpha-induced NF-kappaB activation through TRAF2-IKKbeta-dependent pathway. Biochem Biophys Res Commun **284**, 15–19.

Ciccaglione, A.R., Stellacci, E., Marcantonio, C., Muto, V., Equestre, M., Marsili, G., Rapicetta, M. and Battistini, A. (2007). Repression of interferon regulatory factor 1 by hepatitis C virus core protein results in inhibition of antiviral and immunomodulatory genes. J Virol **81**, 202–214.

Coito, C., Diamond, D.L., Neddermann, P., Korth, M.J. and Katze, M.G. (2004). High-throughput screening of the yeast kinome: identification of human serine/threonine protein kinases that phosphorylate the hepatitis C virus NS5A protein. J Virol **78**, 3502–3513.

Crotta, S., Ronconi, V., Ulivieri, C., Baldari, C.T., Valiante, N.M., Abrignani, S. and Wack, A. (2006). Cytoskeleton rearrangement induced by tetraspanin engagement modulates the activation of T and NK cells. Eur J Immunol **36**, 919–929.

De Francesco, R. and Migliaccio, G. (2005). Challenges and successes in developing new therapies for hepatitis C. Nature **436**, 953–960.

De Maria, A., Fogli, M., Mazza, S., Basso, M., Picciotto, A., Costa, P., Congia, S., Mingari, M.C. and Moretta, L. (2007). Increased natural cytotoxicity receptor expression and relevant IL-10 production in NK cells from chronically infected viremic HCV patients. Eur J Immunol **37**, 445–455.

Dolganiuc, A., Oak, S., Kodys, K., Golenbock, D.T., Finberg, R.W., Kurt-Jones, E. and Szabo, G. (2004). Hepatitis C core and nonstructural 3 proteins trigger toll-like receptor 2-mediated pathways and inflammatory activation. Gastroenterology **127**, 1513–1524.

Dolganiuc, A., Chang, S., Kodys, K., Mandrekar, P., Bakis, G., Cormier, M. and Szabo, G. (2006). Hepatitis C virus

(HCV) core protein-induced, monocyte-mediated mechanisms of reduced IFN-alpha and plasmacytoid dendritic cell loss in chronic HCV infection. J Immunol **177**, 6758–6768.

Dolganiuc, A., Norkina, O., Kodys, K., Catalano, D., Bakis, G., Marshall, C., Mandrekar, P. and Szabo, G. (2007). Viral and host factors induce macrophage activation and loss of toll-like receptor tolerance in chronic HCV infection. Gastroenterology **133**, 1627–1636.

Domitrovich, A.M., Diebel, K.W., Ali, N., Sarker, S. and Siddiqui, A. (2005). Role of La autoantigen and polypyrimidine tract-binding protein in HCV replication. Virology **335**, 72–86.

Drummer, H.E., Wilson, K.A. and Poumbourios, P. (2002). Identification of the hepatitis C virus E2 glycoprotein binding site on the large extracellular loop of CD81. J Virol **76**, 11143–11147.

Dubuisson, J. (2007). Hepatitis C virus proteins. World J Gastroenterol **13**, 2406–2415.

Duesberg, U., von dem Bussche, A., Kirschning, C., Miyake, K., Sauerbruch, T. and Spengler, U. (2002). Cell activation by synthetic lipopeptides of the hepatitis C virus (HCV)—core protein is mediated by toll like receptors (TLRs) 2 and 4. Immunol Lett **84**, 89–95.

Ebihara, T., Shingai, M., Matsumoto, M., Wakita, T. and Seya, T. (2008). Hepatitis C virus-infected hepatocytes extrinsically modulate dendritic cell maturation to activate T cells and natural killer cells. Hepatology **48**, 48–58.

Egger, D., Wolk, B., Gosert, R., Bianchi, L., Blum, H.E., Moradpour, D. and Bienz, K. (2002). Expression of hepatitis C virus proteins induces distinct membrane alterations including a candidate viral replication complex. J Virol **76**, 5974–5984.

Einav, S., Elazar, M., Danieli, T. and Glenn, J.S. (2004). A nucleotide binding motif in hepatitis C virus (HCV) NS4B mediates HCV RNA replication. J Virol **78**, 11288–11295.

Erdtmann, L., Franck, N., Lerat, H., Le Seyec, J., Gilot, D., Cannie, I., Gripon, P., Hibner, U. and Guguen-Guillouzo, C. (2003). The hepatitis C virus NS2 protein is an inhibitor of CIDE-B-induced apoptosis. J Biol Chem **278**, 18256–18264.

Feld, J.J. and Hoofnagle, J.H. (2005). Mechanism of action of interferon and ribavirin in treatment of hepatitis C. Nature **436**, 967–972.

Forton, D.M., Karayiannis, P., Mahmud, N., Taylor-Robinson, S.D. and Thomas, H.C. (2004). Identification of unique hepatitis C virus quasispecies in the central nervous system and comparative analysis of internal translational efficiency of brain, liver, and serum variants. J Virol **78**, 5170–5183.

Foy , E., Li, K., Wang, C., Sumpter, R. Jr., Ikeda , M., Lemon, S.M. and Gale, M. Jr. (2003). Regulation of interferon regulatory factor-3 by the hepatitis C virus serine protease. Science **300**, 1145–1148.

Foy, E., Li, K., SumpterR. Jr., Loo, Y.M., Johnson, C.L., Wang, C., Fish, P.M., Yoneyama, M., Fujita, T., Lemon, S.M. and GaleM. Jr. (2005). Control of antiviral defenses through hepatitis C virus disruption of retinoic acid-inducible gene-I signaling. Proc Natl Acad Sci U S A **102**, 2986–2991.

Gale , M. Jr. and Foy, E.M. (2005). Evasion of intracellular host defence by hepatitis C virus. Nature **436**, 939–945.

Gale , M.J. Jr., Korth, M.J., Tang, N.M., Tan, S.L., Hopkins, D.A., Dever, T.E., Polyak, S.J., Gretch, D.R. and Katze, M.G. (1997). Evidence that hepatitis C virus resistance to interferon is mediated through repression of the PKR protein kinase by the nonstructural 5 A protein. Virology **230**, 217–227.

Gastaminza, P., Cheng, G., Wieland, S., Zhong, J., Liao, W. and Chisari, F. V. (2008). Cellular determinants of hepatitis C virus assembly, maturation, degradation, and secretion. J Virol **82**, 2120–2129.

Gelderblom, H.C., Nijhuis, L.E., de Jong, E.C., te Velde, A.A., Pajkrt, D., Reesink, H.W., Beld, M.G., van Deventer, S.J. and Jansen, P.L. (2007). Monocyte-derived dendritic cells from chronic HCV patients are not infected but show an immature phenotype and aberrant cytokine profile. Liver Int **27**, 944–953.

Golden-Mason, L. and Rosen, H.R. (2006). Natural killer cells: primary target for hepatitis C virus immune evasion strategies? Liver Transpl **12**, 363–372.

Golden-Mason, L., Kelly, A.M., Doherty, D.G., Traynor, O., McEntee, G., Kelly, J., Hegarty, J.E. and O'Farrelly, C. (2004). Hepatic interleuklin 15 (IL-15) expression: implications for local NK/NKT cell homeostasis and development. Clin Exp Immunol **138**, 94–101.

Golden-Mason, L., Madrigal-Estebas, L., McGrath, E., Conroy, M.J., Ryan, E.J., Hegarty, J.E., O'Farrelly, C. and Doherty, D.G. (2008). Altered natural killer cell subset distributions in resolved and persistent hepatitis C virus infection following single source exposure. Gut **57**, 1121–1128.

Harris, D., Zhang, Z., Chaubey, B. and Pandey, V.N. (2006). Identification of cellular factors associated with the 3'-nontranslated region of the hepatitis C virus genome. Mol Cell Proteomics **5**, 1006–1018.

Harrison, S.A. (2007). Small molecule and novel treatments for chronic hepatitis C virus infection. Am J Gastroenterol **102**, 2332–2338.

Harvey, C.E., Post, J.J., Palladinetti, P., Freeman, A.J., Ffrench, R.A., Kumar, R.K., Marinos, G. and Lloyd, A.R. (2003). Expression of the chemokine IP-10 (CXCL10) by hepatocytes in chronic hepatitis C virus infection correlates with histological severity and lobular inflammation. J Leukoc Biol **74**, 360–369.

Heo, T.H., Lee, S.M., Bartosch, B., Cosset, F.L. and Kang, C.Y. (2006). Hepatitis C virus E2 links soluble human CD81 and SR-B1 protein. Virus Res **121**, 58–64.

Herzer, K., Falk, C.S., Encke, J., Eichhorst, S.T., Ulsenheimer, A., Seliger, B. and Krammer, P.H. (2003). Upregulation of major histocompatibility complex class I on liver cells by hepatitis C virus core protein via p53 and TAP1 impairs natural killer cell cytotoxicity. J Virol **77**, 8299–8309.

Hijikata, M., Shimizu, Y.K., Kato, H., Iwamoto, A., Shih, J.W., Alter, H.J., Purcell, R.H. and Yoshikura, H. (1993). Equilibrium centrifugation studies of hepatitis C virus: evidence for circulating immune complexes. J Virol **67**, 1953–1958.

Hiscott, J., Lacoste, J. and Lin, R. (2006). Recruitment of an interferon molecular signaling complex to the mitochondrial membrane: disruption by hepatitis C virus NS3-4 A protease. Biochem Pharmacol **72**, 1477–1484.

Honda, M., Rijnbrand, R., Abell, G., Kim, D. and Lemon, S.M. (1999). Natural variation in translational activities of the 5' nontranslated RNAs of hepatitis C virus genotypes 1a and 1b: evidence for a long-range RNA–RNA interaction outside of the internal ribosomal entry site. J Virol **73**, 4941–4951.

Ishii, S. and Koziel, M.J. (2008). Immune responses during acute and chronic infection with hepatitis C virus. Clin Immunol **128**, 133–147.

Jackson, M., Smith, B., Bevitt, D.J., Steward, M., Toms, G.L., Bassendine, M.F. and Diamond, A.G. (1999). Comparison of cytotoxic

T-lymphocyte responses to hepatitis C virus core protein in uninfected and infected individuals. J Med Virol **58**, 239–246.

Jinushi, M., Takehara, T., Kanto, T., Tatsumi, T., Groh, V., Spies, T., Miyagi, T., Suzuki, T., Sasaki, Y. and Hayashi, N. (2003a). Critical role of MHC class I-related chain A and B expression on IFN-alpha-stimulated dendritic cells in NK cell activation: impairment in chronic hepatitis C virus infection. J Immunol **170**, 1249–1256.

Jinushi, M., Takehara, T., Tatsumi, T., Kanto, T., Groh, V., Spies, T., Suzuki, T., Miyagi, T. and Hayashi, N. (2003b). Autocrine/paracrine IL-15 that is required for type I IFN-mediated dendritic cell expression of MHC class I-related chain A and B is impaired in hepatitis C virus infection. J Immunol **171**, 5423–5429.

Jinushi, M., Takehara, T., Tatsumi, T., Kanto, T., Miyagi, T., Suzuki, T., Kanazawa, Y., Hiramatsu, N. and Hayashi, N. (2004). Negative regulation of NK cell activities by inhibitory receptor CD94/NKG2A leads to altered NK cell-induced modulation of dendritic cell functions in chronic hepatitis C virus infection. J Immunol **173**, 6072–6081.

Jones, C.T., Murray, C.L., Eastman, D.K., Tassello, J. and Rice, C.M. (2007). Hepatitis C virus p7 and NS2 proteins are essential for production of infectious virus. J Virol **81**, 8374–8383.

Kaito, M., Watanabe, S., Tanaka, H., Fujita, N., Konishi, M., Iwasa, M., Kobayashi, Y., Gabazza, E.C., Adachi, Y., Tsukiyama-Kohara, K. and Kohara, M. (2006). Morphological identification of hepatitis C virus E1 and E2 envelope glycoproteins on the virion surface using immunogold electron microscopy. Int J Mol Med **18**, 673–678.

Kanto, T., Hayashi, N., Takehara, T., Hagiwara, H., Mita, E., Naito, M., Kasahara, A., Fusamoto, H. and Kamada, T. (1995). Density analysis of hepatitis C virus particle population in the circulation of infected hosts: implications for virus neutralization or persistence. J Hepatol **22**, 440–448.

Kanto, T., Inoue, M., Miyatake, H., Sato, A., Sakakibara, M., Yakushijin, T., Oki, C., Itose, I., Hiramatsu, N., Takehara, T., Kasahara, A. and Hayashi, N. (2004). Reduced numbers and impaired ability of myeloid and plasmacytoid dendritic cells to polarize T helper cells in chronic hepatitis C virus infection. J Infect Dis **190**, 1919–1926.

Karacki, PS., Gao, X., Thio, C.L., Thomas, D.L., Goedert, J.J., Vlahov, D., Kaslow, R.A., Strathdee, S., Hilgartner, M.W., O'Brien, S.J. and Carrington, M. (2004).: MICA and recovery from hepatitis C virus and hepatitis B virus infections. Genes Immun **5**, 261–266.

Kaser, A., Kaser, S., Kaneider, N.C., Enrich, B., Wiedermann, C.J. and Tilg, H. (2004). Interleukin-18 attracts plasmacytoid dendritic cells (DC2s) and promotes Th1 induction by DC2s through IL-18 receptor expression. Blood **103**, 648–655.

Khabar, K.S., Al-Zoghaibi, F., Al-Ahdal, M.N., Murayama, T., Dhalla, M., Mukaida, N., Taha, M., Al-Sedairy, S.T., Siddiqui, Y., Kessie, G. and Matsushima, K. (1997). The alpha chemokine, interleukin 8, inhibits the antiviral action of interferon alpha. J Exp Med **186**, 1077–1085.

Khakoo, S.I., Thio, C.L., Martin, M.P., Brooks, C.R., Gao, X., Astemborski, J., Cheng, J., Goedert, J.J., Vlahov, D., Hilgartner, M., Cox, S., Little, A.M., Alexander, G.J., Cramp, M.E., O'Brien, S.J., Rosenberg, W.M., Thomas, D.L. and Carrington, M. (2004). HLA and NK cell inhibitory receptor genes in resolving hepatitis C virus infection. Science **305**, 872–874.

Kimura, T., Saito, T., Yoshimura, M., Yixuan, S., Baba, M., Ji, G., Muramatsu, M. and Kawata, S. (2006). Association of transforming growth factor-beta 1 functional polymorphisms with natural clearance of hepatitis C virus. J Infect Dis **193**, 1371–1374.

Knapp, S., Hennig, B.J., Frodsham, A.J., Zhang, L., Hellier, S., Wright, M., Goldin, R., Hill, A.V., Thomas, H.C. and Thursz, M.R. (2003). Interleukin-10 promoter polymorphisms and the outcome of hepatitis C virus infection. Immunogenetics **55**, 362–369.

Kuo, G., Choo, Q.L., Alter, H.J., Gitnick, G.L., Redeker, A.G., Purcell, R.H., Miyamura, T., Dienstag, J.L., Alter, M.J., Stevens, C.E., et al. (1989). An assay for circulating antibodies to a major etiologic virus of human non-A, non-B hepatitis. Science **244**, 362–364.

Lai, W.K., Curbishley, S.M., Goddard, S., Alabraba, E., Shaw, J., Youster, J., McKeating, J. and Adams, D.H. (2007). Hepatitis C is associated with perturbation of intrahepatic myeloid and plasmacytoid dendritic cell function. J Hepatol **47**, 338–347.

Laporte, J., Malet, I., Andrieu, T., Thibault, V., Toulme, J.J., Wychowski, C., Pawlotsky, J.M., Huraux, J.M., Agut, H. and Cahour, A. (2000). Comparative analysis of translation efficiencies of hepatitis C virus 5′ untranslated regions among intraindividual quasispecies present in chronic infection: opposite behaviors depending on cell type. J Virol **74**, 10827–10833.

Lerat, H., Rumin, S., Habersetzer, F., Berby, F., Trabaud, M.A., Trepo, C. and Inchauspe, G. (1998). In vivo tropism of hepatitis C virus genomic sequences in hematopoietic cells: influence of viral load, viral genotype, and cell phenotype. Blood **91**, 3841–3849.

Levy, S. and Shoham, T. (2005). The tetraspanin web modulates immune-signalling complexes. Nat Rev Immunol **5**, 136–148.

Li, Y., Zhang, T., Ho, C., Orange, J.S., Douglas, S.D. and Ho, W.Z. (2004). Natural killer cells inhibit hepatitis C virus expression. J Leukoc Biol **76**, 1171–1179.

Li K., Foy, E., Ferreon, J.C., Nakamura, M., Ferreon, A.C., Ikeda, M., Ray, S.C., GaleM. Jr. and Lemon, S.M. (2005). Immune evasion by hepatitis C virus NS3/4 A protease-mediated cleavage of the Toll-like receptor 3 adaptor protein TRIF. Proc Natl Acad Sci U S A **102**, 2992–2997.

Li, S., Gowans, E.J., Chougnet, C., Plebanski, M. and Dittmer, U. (2008). Natural regulatory T cells and persistent viral infection. J Virol **82**, 21–30.

Lin, A.W., Gonzalez, S.A., Cunningham-Rundles, S., Dorante, G., Marshall, S., Tignor, A., Ha, C., Jacobson, I.M. and Talal, A.H. (2004). CD56(+dim) and CD56(+bright) cell activation and apoptosis in hepatitis C virus infection. Clin Exp Immunol **137**, 408–416.

Lindenbach, B.D., Evans, M.J., Syder, A.J., Wolk, B., Tellinghuisen, T.L., Liu, C.C., Maruyama, T., Hynes, R.O., Burton, D.R., McKeating, J.A. and Rice, C.M. (2005). Complete replication of hepatitis C virus in cell culture. Science **309**, 623–626.

Lopez-Vazquez, A., Rodrigo, L. and Lopez-Larrea, C. (2007). Association of killer immunoglobulin-like receptors and their HLA class I ligands with progression of chronic hepatitis C virus infection. Tissue Antigens **69**(Suppl. 1), 241–242.

Lozach, P.Y., Amara, A., Bartosch, B., Virelizier, J.L., Arenzana-Seisdedos, F., Cosset, F.L. and Altmeyer, R. (2004). C-type lectins L-SIGN and DC-SIGN capture and transmit infectious hepatitis C virus pseudotype particles. J Biol Chem **279**, 32035–32045.

Luo, G., Hamatake, R.K., Mathis, D.M., Racela, J., Rigat, K.L., Lemm, J. and Colonno, R.J. (2000). De novo initiation of RNA synthesis by the RNA-dependent RNA polymerase (NS5B) of hepatitis C virus. J Virol **74**, 851–863.

Major, M.E., Rehermann, B. and Feinstone, S.M., et al. (2001). Hepatitis C viruses. In: Knipe, D.M. (ed), Fields VirologyVol. 1. Philadelphia: Lippincott Williams and Wilkins, pp. 1127–1161.

Manigold, T. and Racanelli, V. (2007). T-cell regulation by CD4 regulatory T cells during hepatitis B and C virus infections: facts and controversies. Lancet Infect Dis **7**, 804–813.

McCormick, C.J., Brown, D., Griffin, S., Challinor, L., Rowlands, D.J. and Harris, M. (2006). A link between translation of the hepatitis C virus polyprotein and polymerase function; possible consequences for hyperphosphorylation of NS5A. J Gen Virol **87**, 93–102.

Meertens, L., Bertaux, C., Cukierman, L., Cormier, E., Lavillette, D., Cosset, F.L. and Dragic, T. (2008). The tight junction proteins claudin-1, -6 and -9 are entry cofactors for the Hepatitis C virus. J Virol.

Meier, U.C., Owen, R.E., Taylor, E., Worth, A., Naoumov, N., Willberg, C., Tang, K., Newton, P., Pellegrino, P., Williams, I., Klenerman, P. and Borrow, P. (2005). Shared alterations in NK cell frequency, phenotype, and function in chronic human immunodeficiency virus and hepatitis C virus infections. J Virol **79**, 12365–12374.

Morishima, C., Paschal, D.M., Wang, C.C., Yoshihara, C.S., Wood, B.L., Yeo, A.E., Emerson, S.S., Shuhart, M.C. and Gretch, D.R. (2006). Decreased NK cell frequency in chronic hepatitis C does not affect ex vivo cytolytic killing. Hepatology **43**, 573–580.

Nahmias, Y., Casali, M., Barbe, L., Berthiaume, F. and Yarmush, M.L. (2006). Liver endothelial cells promote LDL-R expression and the uptake of HCV-like particles in primary rat and human hepatocytes. Hepatology **43**, 257–265.

Nattermann, J., Nischalke, H.D., Hofmeister, V., Ahlenstiel, G., Zimmermann, H., Leifeld, L., Weiss, E.H., Sauerbruch, T. and Spengler, U. (2005). The HLA-A2 restricted T cell epitope HCV core 35-44 stabilizes HLA-E expression and inhibits cytolysis mediated by natural killer cells. Am J Pathol **166**, 443–453.

Nattermann, J., Feldmann, G., Ahlenstiel, G., Langhans, B., Sauerbruch, T. and Spengler, U. (2006). Surface expression and cytolytic function of natural killer cell receptors is altered in chronic hepatitis C. Gut **55**, 869–877.

Neddermann, P., Quintavalle, M., Di Pietro, C., Clementi, A., Cerretani, M., Altamura, S., Bartholomew, L. and De Francesco, R. (2004). Reduction of hepatitis C virus NS5A hyperphosphorylation by selective inhibition of cellular kinases activates viral RNA replication in cell culture. J Virol **78**, 13306–13314.

Neumann-Haefelin, C. and Thimme, R. (2007). Impact of the genetic restriction of virus-specific T-cell responses in hepatitis C virus infection. Genes Immun **8**, 181–192.

Neumann-Haefelin, C., Spangenberg, H.C., Blum, H.E. and Thimme, R. (2007). Host and viral factors contributing to CD8+ T cell failure in hepatitis C virus infection. World J Gastroenterol **13**, 4839–4847.

Nielsen, S.U., Bassendine, M.F., Burt, A.D., Martin, C., Pumeechockchai, W. and Toms, G.L. (2006). Association between hepatitis C virus and very-low-density lipoprotein (VLDL)/LDL analyzed in iodixanol density gradients. J Virol **80**, 2418–2428.

Nouri-Aria, K.T., Sallie, R., Mizokami, M., Portmann, B.C. and Williams, R. (1995). Intrahepatic expression of hepatitis C virus antigens in chronic liver disease. J Pathol **175**, 77–83.

Otto, G.A. and Puglisi, J.D. (2004). The pathway of HCV IRES-mediated translation initiation. Cell **119**, 369–380.

Pal, S., Shuhart, M.C., Thomassen, L., Emerson, S.S., Su, T., Feuerborn, N., Kae, J. and Gretch, D.R. (2006). Intrahepatic hepatitis C virus replication correlates with chronic hepatitis C disease severity in vivo. J Virol **80**, 2280–2290.

Paladino, N., Flores, A.C., Marcos, C.Y., Fainboim, H., Theiler, G., Arruvito, L., Williams, F., Middleton, D. and Fainboim, L. (2007). Increased frequencies of activating natural killer receptors are associated with liver injury in individuals who do not eliminate hepatitis C virus. Tissue Antigens **69**(Suppl. 1), 109–111.

Pawlotsky, J.M., Chevaliez, S. and McHutchison, J.G. (2007). The hepatitis C virus life cycle as a target for new antiviral therapies. Gastroenterology **132**, 1979–1998.

Pirisi, M., Fabris, C., Toniutto, P., Vitulli, D., Soardo, G., Falleti, E., Gonano, F., Ferroni, P., Gasparini, V. and Bartoli, E. (1995). Reactivity to B cell epitopes within hepatitis C virus core protein and hepatocellular carcinoma. Cancer Res **55**, 111–114.

Polyak, S.J., Khabar, K.S., Paschal, D.M., Ezelle, H.J., Duverlie, G., Barber, G.N., Levy, D.E., Mukaida, N. and Gretch, D.R. (2001). Hepatitis C virus nonstructural 5 A protein induces interleukin-8, leading to partial inhibition of the interferon-induced antiviral response. J Virol **75**, 6095–6106.

Rauch, A., Laird, R., McKinnon, E., Telenti, A., Furrer, H., Weber, R., Smillie, D. and Gaudieri, S. (2007). Influence of inhibitory killer immunoglobulin-like receptors and their HLA-C ligands on resolving hepatitis C virus infection. Tissue Antigens **69**(Suppl. 1), 237–240.

Rehermann, B. (2007). Chronic infections with hepatotropic viruses: mechanisms of impairment of cellular immune responses. Semin Liver Dis **27**, 152–160.

Rodgers, J.R. and Cook, R.G. (2005). MHC class Ib molecules bridge innate and acquired immunity. Nat Rev Immunol **5**, 459–471.

Romero, A.I., Lagging, M., Westin, J., Dhillon, A.P., Dustin, L.B., Pawlotsky, J.M., Neumann, A.U., Ferrari, C., Missale, G., Haagmans, B.L., Schalm, S.W., Zeuzem, S., Negro, F., Verheij-Hart, E. and Hellstrand, K. (2006). Interferon (IFN)-gamma-inducible protein-10: association with histological results, viral kinetics, and outcome during treatment with pegylated IFN-alpha 2a and ribavirin for chronic hepatitis C virus infection. J Infect Dis **194**, 895–903.

Romero, V., Azocar, J., Zuniga, J., Clavijo, O.P., Terreros, D., Gu, X., Husain, Z., Chung, R.T., Amos, C. and Yunis, E.J. (2008). Interaction of NK inhibitory receptor genes with HLA-C and MHC class II alleles in Hepatitis C virus infection outcome. Mol Immunol **45**, 2429–2436.

Semmo, N. and Klenerman, P. (2007). CD4+ T cell responses in hepatitis C virus infection. World J Gastroenterol **13**, 4831–4838.

Skak, K., Frederiksen, K.S. and Lundsgaard, D. (2008). Interleukin-21 activates human natural killer cells and modulates their surface receptor expression. Immunology **123**, 575–583.

Soler, M., Pellerin, M., Malnou, C.E., Dhumeaux, D., Kean, K.M. and Pawlotsky, J.M. (2002). Quasispecies heterogeneity and constraints on the evolution of the 5′ noncoding region of hepatitis C virus (HCV): relationship with HCV resistance to interferon-alpha therapy. Virology **298**, 160–173.

Song, Y., Friebe, P., Tzima, E., Junemann, C., Bartenschlager, R. and Niepmann, M. (2006). The hepatitis C virus RNA 3′-untranslated region strongly enhances translation directed by the internal ribosome entry site. J Virol **80**, 11579–11588.

Stastny, P. (2006). Introduction: MICA/MICB in innate immunity, adaptive immunity, autoimmunity, cancer, and in the immune response to transplants. Hum Immunol **67**, 141–144.

Steinle, A., Li, P., Morris, D.L., Groh, V., Lanier, L.L., Strong, R.K. and Spies, T. (2001). Interactions of human NKG2D with its ligands MICA, MICB, and homologs of the mouse RAE-1 protein family. Immunogenetics **53**, 279–287.

Steinmann, E., Penin, F., Kallis, S., Patel, A.H., Bartenschlager, R. and Pietschmann, T. (2007). Hepatitis C virus p7 protein is crucial for assembly and release of infectious virions. PLoS Pathog **3**, e103.

StGelais, C., Tuthill, T.J., Clarke, D.S., Rowlands, D.J., Harris, M. and Griffin, S. (2007). Inhibition of hepatitis C virus p7 membrane channels in a liposome-based assay system. Antiviral Res **76**, 48–58.

Taniguchi, H., Kato, N., Otsuka, M., Goto, T., Yoshida, H., Shiratori, Y. and Omata, M. (2004). Hepatitis C virus core protein upregulates transforming growth factor-beta 1 transcription. J Med Virol **72**, 52–59.

Tasaka, M., Sakamoto, N., Itakura, Y., Nakagawa, M., Itsui, Y., Sekine-Osajima, Y., Nishimura-Sakurai, Y., Chen, C.H., Yoneyama, M., Fujita, T., Wakita, T., Maekawa, S., Enomoto, N. and Watanabe, M. (2007). Hepatitis C virus non-structural proteins responsible for suppression of the RIG-I/Cardif-induced interferon response. J Gen Virol **88**, 3323–3333.

Thimme, R., Bukh, J., Spangenberg, H.C., Wieland, S., Pemberton, J., Steiger, C., Govindarajan, S., Purcell, R.H. and Chisari, F.V. (2002). Viral and immunological determinants of hepatitis C virus clearance, persistence, and disease. Proc Natl Acad Sci U S A **99**, 15661–15668.

Thoren, F., Romero, A., Lindh, M., Dahlgren, C. and Hellstrand, K. (2004). A hepatitis C virus-encoded, nonstructural protein (NS3) triggers dysfunction and apoptosis in lymphocytes: role of NADPH oxidase-derived oxygen radicals. J Leukoc Biol **76**, 1180–1186.

Tseng, C.T. and Klimpel, G.R. (2002). Binding of the hepatitis C virus envelope protein E2 to CD81 inhibits natural killer cell functions. J Exp Med **195**, 43–49.

Vales-Gomez, M., Reyburn, H.T., Erskine, R.A. and Strominger, J. (1998). Differential binding to HLA-C of p50-activating and p58-inhibitory natural killer cell receptors. Proc Natl Acad Sci U S A **95**, 14326–14331.

Wendt, K., Wilk, E., Buyny, S., Schmidt, R.E. and Jacobs, R. (2007). Interleukin-21 differentially affects human natural killer cell subsets. Immunology **122**, 486–495.

Yao, Z.Q., King, E., Prayther, D., Yin, D. and Moorman, J. (2007). T cell dysfunction by hepatitis C virus core protein involves PD-1/PDL-1 signaling. Viral Immunol **20**, 276–287.

Yi, M., Villanueva, R.A., Thomas, D.L., Wakita, T. and Lemon, S.M. (2006). Production of infectious genotype 1a hepatitis C virus (Hutchinson strain) in cultured human hepatoma cells. Proc Natl Acad Sci U S A **103**, 2310–2315.

Yokoyama, W.M. and Plougastel, B.F. (2003). Immune functions encoded by the natural killer gene complex. Nat Rev Immunol **3**, 304–316.

Yonkers, N.L., Rodriguez, B., Milkovich, K.A., Asaad, R., Lederman, M.M., Heeger, P.S. and Anthony, D.D. (2007). TLR ligand-dependent activation of naive CD4 T cells by plasmacytoid dendritic cells is impaired in hepatitis C virus infection. J Immunol **178**, 4436–4444.

You, S. and Rice, C.M. (2008). 3′ RNA elements in hepatitis C virus replication: kissing partners and long poly(U). J Virol **82**, 184–195.

Zeisel, M.B., Koutsoudakis, G., Schnober, E.K., Haberstroh, A., Blum, H.E., Cosset, F.L., Wakita, T., Jaeck, D., Doffoel, M., Royer, C., Soulier, E., Schvoerer, E., Schuster, C., Stoll-Keller, F., Bartenschlager, R., Pietschmann, T., Barth, H. and Baumert, T.F. (2007). Scavenger receptor class B type I is a key host factor for hepatitis C virus infection required for an entry step closely linked to CD81. Hepatology **46**, 1722–1731.

Zeremski, M., Petrovic, L.M. and Talal, A.H. (2007). The role of chemokines as inflammatory mediators in chronic hepatitis C virus infection. J Viral Hepat **14**, 675–687.

Zhao, L.J., Zhao, P., Chen, Q.L., Ren, H., Pan, W. and Qi, Z.T. (2007). Mitogen-activated protein kinase signalling pathways triggered by the hepatitis C virus envelope protein E2: implications for the prevention of infection. Cell Prolif **40**, 508–521.

Chapter Forty-Five

45

Natural killer cells and the biology of parasitism

Marianne A. Bryan, Karen A. Norris

CHAPTER CONTENTS

Science does not know its debt to imagination.

Ralph Waldo Emerson

ABSTRACT

Protozoan parasites, such as *Plasmodium* spp., *Trypanosoma cruzi*, *Toxoplasma gondii* and *Leishmania* spp. activate natural killer (NK) cells, usually in an accessory-cell dependent manner. Activated NK cells can contribute to control of protozoan parasite infection through IFN-γ production and direct lysis of parasites. NK cells may also play a role in pathophysiology when parasite persistence leads to chronic disease. Metazoan parasites, commonly called helminthes, may use NK cells to help establish infection, although in several cases NK cells seem to contribute to parasite control.

KEY WORDS

Parasite, Protozoa, Helminth, Immune evasion

Introduction

Parasites constitute many different forms of microbes, with various sizes, routes of infection and life-stages within the mammalian host. By definition, parasites are able to grow and persist within their hosts, with no benefit to the host. The hallmark of parasite persistence within an immune-competent host is the ability of parasites to avoid clearance by the host immune system without causing the host to succumb to infection. This necessitates a delicate balance between parasite immune-evasion mechanisms and the generation of host immunity to the parasite. While metazoan parasites are multicellular, extracellular organisms, many protozoan parasites are intracellular or have intracellular phases during infection of mammalian hosts. Many protozoan parasites invade host cells or tissue, replicate, and then lyse their host cells to disseminate the parasite to new tissues, either locally or systemically through the blood or lymphatic system.

One innate immune effector, the natural killer (NK) cell, can recognize and destroy pathogen-infected or malignant cells. NK cells circulate through the blood, spleen and peripheral organs, such as the liver, and express germ-line-encoded activating and inhibitory receptors (Vivier et al., 2004). Through stimulatory engagement with ligands from pathogen-infected cells and responses to cytokines such as IL-12, NK cells become activated. Upon activation, NK cells can directly lyse target cells by exocytosis of perforin and granzymes, as well as communicate with other immune cells by cytokine secretion or direct contact (Lodoen and Lanier, 2006; Newman and Riley, 2007; Papazahariadou et al., 2007). In order to elucidate the early immune mediators of parasite control and persistence, investigators have begun to study NK cells and the biology of parasitic infections (Papazahariadou et al., 2007; Roetynck et al., 2006).

NK cells play a variety of roles in parasite infection either directly through interactions with parasites or parasite-infected cells, or indirectly through cytokine secretion, predominately through interferon-gamma (IFN-γ). In terms of immune control, NK cells can play a primary role as first responders in the innate immune system and influence adaptive immune responses, as illustrated by NK cell-priming of T cells during *T. gondii* infection (Combe et al., 2005). Conversely, NK cell activation during parasite infection may also lead to pathophysiology, as evidenced by research on NK cells in cerebral malaria (CM) (Hansen et al., 2007a). Some parasites employ immune evasion mechanisms that use NK cells to evade immune clearance (Hsieh et al., 2004) or evade NK cell lysis of infected cells through impaired NK cell activation (Campos-Martin et al., 2006). Beyond elucidating how NK cells contribute to the control of parasite infections, investigation of parasite–NK cell interactions is helping researchers to better understand basic NK cell biology, especially the interaction between accessory cells and NK cells (Newman and Riley, 2007). This chapter will explore the role of NK cells in protozoan parasite infection of mammalian hosts with *Plasmodium* spp., *Toxoplasma gondii*, *Trypanasoma cruzi* and *Leishmania* spp., as well as metazoan parasites in helminth infection.

Malaria

Malaria, caused by *Plasmodium* parasites, is a global health problem affecting 40% of our world's population. *Plasmodium* infects 500 million people and accounts for over one million deaths each year (CDC, 2007). Four *Plasmodium* species infect humans: *P. falciparum*, *P. vivax*, *P. ovale* and *P. malariae*; the most common and most deadly being *P. falciparum*. Infection with *Plasmodium* occurs in two stages. During the first stage (hepatic phase), a mosquito injects parasites (sporozoites) into the blood stream; these parasites travel to the liver where they enter cells and replicate. Upon rupture, liver cells release parasites (merozoites) into the blood, where they penetrate erythrocytes and reproduce asexually in the second (erythrocytic phase) stage of infection.

The generation of immunity to malaria is complex, species and stage specific, and usually requires repeated infection. In the absence of immune mechanisms or anti-malarial drugs, parasite replication continues exponentially in the erythrocytes and death is inevitable; survival is linked to control of parasite replication within the first 7–14 days of infection (Stevenson and Riley, 2004). Severe pathology includes anaemia, metabolic acidosis and/or CM. Production of inflammatory mediators and adherence of infected erythrocytes to vascular endothelium have been implicated in the severity of the disease (Schofield, 2007; Stevenson and Riley, 2004).

The various *Plasmodium* species cause various degrees of pathology. Due to differences in pathogenicity between the different strains of *Plasmodium*, a number of mouse model systems utilizing four different strains have been developed that mimic different aspects of human malaria infections. These strains are *P. chabaudi*, *P. berghei*, *P. yoelii* and *P. vinckei* (Stevenson and Riley, 2004). These model systems are especially important in elucidating immune mechanisms during early malarial infection. Analysis of human infection is limited to collection of peripheral blood from infected or healthy donors, whereas the mouse model systems allow for dissection of stage specific immune responses within the liver, circulatory system and brain.

While the adaptive immune response to malaria has been investigated in detail, driven mostly by research into vaccine candidates, evidence for the role of innate immunity in malaria has only recently begun to be highlighted by researchers attempting to determine what kind of immune responses constitutes protective immunity versus immunopathology in malaria infection. In addition, innate immune responses can shape the adaptive immune response through cytokine release. Thus, understanding the innate immune response to malaria may aid in the design of an efficacious vaccine. As both an innate immune effector cell and as a link between innate and adaptive immunity, the NK cell has come under scrutiny in malarial disease. NK cells are non-clonal leukocytes that can be quickly activated to produce cytokines, such as IFN-γ and TNF-α, and mediate lysis of infected cells. In this way, NK cells are a first line of defense against many intracellular pathogens, including parasites (Lodoen and Lanier, 2006). The role of NK cells in malaria has been investigated through in vitro analysis of interactions between NK cells and infected human erythrocytes and in vivo analysis of NK cell activation and effector mechanisms in mouse models of malaria. NK cells play a role in early defense against *Plasmodium* spp. both through generation of IFN-γ and establishment of the adaptive immune response in experimental infection (Roland et al., 2006).

Investigation of NK cells in *Plasmodium* infection has provided evidence that NK cells respond directly to infected erythrocytes/red blood cells (iRBC) (Korbel et al., 2005). NK activation depends on cytokines released from monocytes/macrophages and myeloid dendritic cells (mDC), as well as direct contact with iRBC (Artavanis-Tsakonas and Riley, 2002; Ing and Stevenson, 2009; Newman et al., 2006) (Figure 45.1). NK cell activation depends on IL-18 secretion by monocytes/macrophages (Baratin et al., 2005; Newman et al., 2006). Monocytes/macrophages control the IFN-γ burst from NK cells; the degree of monocyte activation correlates with the magnitude of the NK cell IFN-γ response (Newman et al., 2006). While it is uncontested that NK

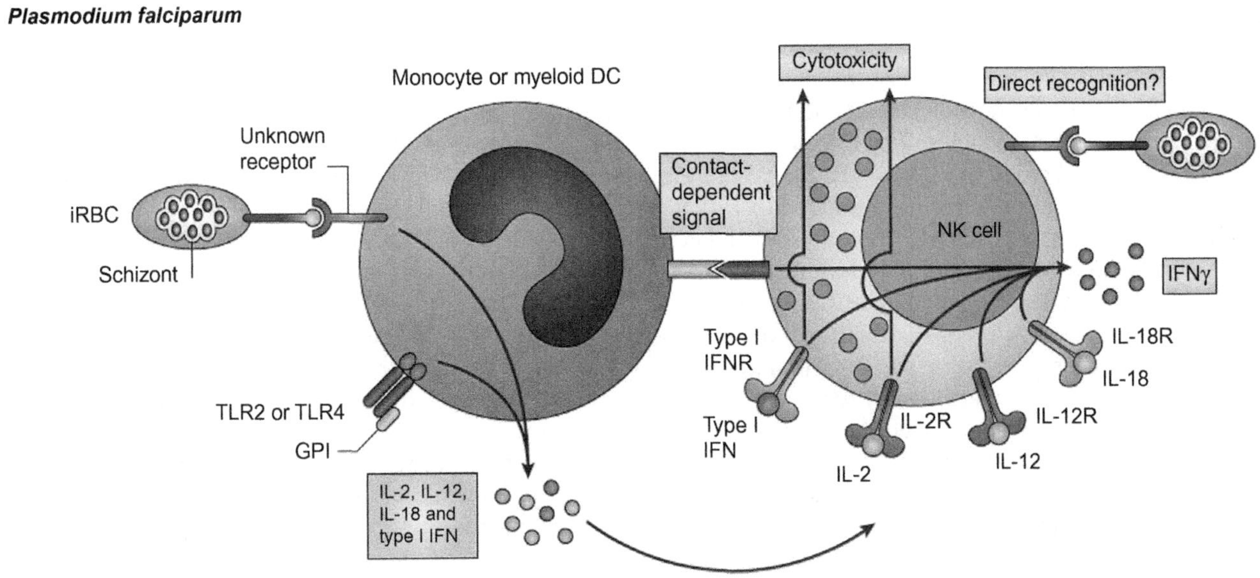

Figure 45.1 • Accessory-cell dependent NK-cell activation by *P. falciparum*. The NK-cell response to red blood cells infected (iRBCs) with *P. falciparum* (a parasite that causes human malaria) is dependent on myeloid DCs and monocytes. The receptors and ligands required for the induction of NK-cell-activating signals are not formally known, but might include binding of glycosylphosphatidylinositol (GPI; a phospholipid moiety that anchors certain surface proteins of the parasite to the surface membrane) to TLR2 on monocytes or myeloid DCs. Optimal NK-cell responses also require intact mature schizont-stage parasites, which indicates a role for a ligand found only on the surface of intact iRBCs. Following iRBC activation of accessory cells, various cytokines, including IL-12, IL-18, IL-2 and IFN-α, are secreted and contribute to NK-cell IFN-γ secretion; direct contact between NK cells and accessory cells is also required. Direct contact has been observed between iRBCs and NK cells, but how it contributes to NK-cell activation is not known. Reprinted by permission from Macmillan Publishers Ltd, *Nature Reviews Immunology*, copyright 2007 (Newman and Riley, 2007).

cells become activated and produce IFN-γ early in infection, the contribution of NK cells to the total IFN-γ response during acute infection is under debate. Evidence differs as to the extent that NK cells contribute to the total IFN-γ production (Hensmann and Kwiatkowski, 2001; Kim et al., 2008; Korbel et al., 2005). Comparative phenotyping of IFN-γ producing cells in human peripheral blood mononuclear cells (PBMC) exposed to iRBCs for 24 h, indicates that γδ-T cells expressing NK receptors predominate over NK cells within the IFN-γ producing PBMCs, although the authors stress that this is a complex response involving a range of cell types, including NK cells (D'Ombrain et al., 2007). In addition, the degree of the response to iRBC by NK or γδ-T cells varies between different human PBMC donors (D'Ombrain et al., 2007; Korbel et al., 2005). NK cells may contribute to IFN-γ production and parasite control in other compartments, such as the liver. In the *P. yoelii* model, NK cells were found to traffic from the spleen to the liver, where they secrete IFN-γ and show increased cytotoxicity, contributing to the control of early infection (Roland et al., 2006).

Although many reports have indicated that NK cells interact with iRBCs upon co-culture, the mechanism is poorly understood and the significance of the interaction is still under debate (Hansen et al., 2007b; Roetynck et al., 2006) (Figure 45.2). *Plasmodium*-encoded glycoproteins (*Pf*EMP-1) expressed on the iRBC surface are involved

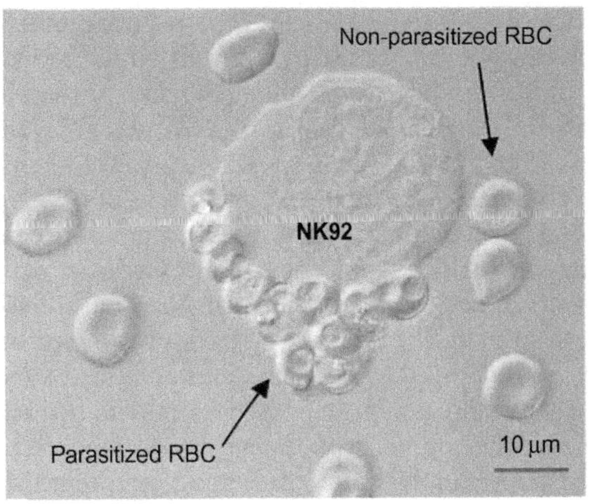

Figure 45.2 • Parasitized RBC (iRBC) bind to human NK cell line (NK92). The human NK cell line NK 92 was incubated with whole culture of iRBC. After 1 h at 37°C, as sample of the co-culture was placed between slide and cover-slip then analyzed under a microscope. The NK92 cells directly interact with iRBC (FCR3-CSA strain) as rosettes, but not with the uninfected RBC (×100 original magnification). Reprinted from *PLoS ONE*, copyright 2007 (Baratin et al., 2007).

in parasite sequestration through adherence to a number of receptors: CD36, CD54 (ICAM-1) and chondroitin sulfate A (CSA), which are found on some immune cells (Kraemer and Smith, 2006). Therefore, investigators have hypothesized that *Pf*EMP-1 may be involved in NK–iRBC interaction. Baratin and associates found that while NK cells express several receptors able to bind *Pf*EMP-1, CSA acts as the dominant molecule in the cytoadhesion of iRBC to human NK cell lines. This interaction leads to production of the chemokine CXCL8 (IL-8) by NK cells, but is not sufficient for inducing production of IFN-γ by NK cells. Rather, NK cell IFN-γ production required engagement of ICAM-1 with LFA-1. RBC and iRBC do not express ICAM-1 or LFA-1. Selective depletion of CD14+ monocytes from PBMC abolished IFN-γ production by NK cells in response to iRBC (Baratin et al., 2005). These data show that NK cell secretion of IFN-γ depends on monocyte/macrophage helper function rather than NK–iRBC interactions (Baratin et al., 2005, 2007). As NK cells and monocytes both express ICAM-1 and LFA-1, this may be a bi-directional interaction.

In another approach, Mavoungou and colleagues investigated NK cell cytotoxicity receptors and found that the peptide DBL-1α, from a subdomain of *Pf*EMP-1, is involved in NK cell–iRBC interaction. This interaction was direct, specific and functional, and led to perforin production and granzyme B release (Mavoungou et al., 2007). The NK cell receptors NKp30 and NKp46 were found to bind to iRBC through DBL-1α, leading to activation of the NK cells as indicated by CD69 and CD25 upregulation and the production of perforin and granzyme B. Disruption of these receptor–ligand interactions with DBL-1α (see above, DBL-1α is the name of a peptide) or antibodies against NKp30 or NKp46 inhibited NK cell–mediated lysis of iRBCs, underscoring the importance of these interactions in the lysis of iRBCs by NK cells (Mavoungou et al., 2007). While these in vitro studies with human PBMCs and *P. falciparum* indicate a role of NK cells in direct lysis of iRBCs, previous investigations in mice using *P. chabaudi* yielded confounding results suggesting cytokine production, not cytotoxicity, mediates resistance to infection (Mohan et al., 1997). Regardless of the mechanism, there is evidence that NK cells contribute to the control of early parasite growth rates. Consistent with this finding, in the absence of NK cells, mouse models indicate higher peak parasitemia and marked recurring parasitemia during chronic phase disease (Stevenson and Riley, 2004).

NK cells are a 'double-edged sword' in protection from *Plasmodium* parasites: While NK cells contribute to immune-mediated control of the malaria pathogen during early infection, they are also involved in immunopathology of CM (Hansen et al., 2007a). *P. falciparum* mediated CM is one of the leading causes of death in children under 5 years of age (UNICEF/WHO, 2003). In order to study this type of malaria, investigators employ the *P. berghei* ANKA murine malaria model system. In this model, NK cells are among the earliest cells recruited to the brain and are required for CM pathology. Upon infection of mice with *P. berghei*, NK cells migrate from the spleen to the brain. This migration is mediated by CXCR3 expression on NK cells. In turn, NK cells stimulate migration of T cells into the brain through the CXCR3 pathway. Furthermore, the ability of NK cells to recruit T cells to the brain requires IFN-γ secretion by NK cells. The requirement for NK cells in the pathogenesis of CM is highlighted by the observation that depletion of NK cells alleviates CM in this model system (Hansen et al., 2007a).

Thus, the study of NK cells in malaria yields evidence that they play a role in controlling early *Plasmodium* infection through the production of IFN-γ and potentially exerting a direct effect through lysis of infected erythrocytes. In contrast, NK cells exhibit detrimental effects by contributing to CM in the mouse model of CM. The role of NK cells in the biology of malaria illustrates the importance of understanding the role of immune cells in the context of disease pathogenesis. In developing strategies that may involve immune therapy, researchers must keep in mind that immune effector cells can be beneficial in one context while causing immunopathology in another.

Toxoplasma gondii

T. gondii infection is initiated either by ingestion of oocytes, shed by cats, or tissue cysts present in undercooked meat. Once inside the host, parasites transform into tachyzoites, which infect and undergo several cycles of replication, prior to lysing the host cell. Released parasites then infect surrounding cells and tissues. Host immunity controls systemic infection but the parasite persists as quiescent cysts sequestered deep in neuronal and muscle tissue to maintain latent infection. *T. gondii* infection is one of the most common in the world, with an estimated 30–80% of the human population latently infected (Denkers, 2003). While the immunocompetent host usually has an asymptomatic infection, infection during pregnancy can result in congenital transmission resulting in abortion, hydrocephalus, as well as neurological and ocular disease of the newborn or ocular disease later in life. In addition, immune suppression of the host can allow *T. gondii* to emerge as a life-threatening infection. The murine model of *T. gondii* infection has been instrumental in dissecting the effective NK cell immune response to the pathogen, as similar host responses are generated in humans and mice.

Acute phase infection with *T. gondii* triggers a strong innate immune response characterized by macrophage, DC, neutrophil and NK cell responses. NK cells are critical for early host resistance to the parasite through IFN-γ production (Denkers, 2003; Denkers et al., 1993; Hunter et al., 1994, 1997; Johnson et al., 1993). The protective role of NK cells requires that they traffic to the site of parasite infection and release IFN-γ. The trafficking of

NK cells to the site of *T. gondii* infection depends on the NK cell response to chemokines mediated by the chemokine receptor CCR5. Without CCR5, NK cells were less able to migrate to sites of infection and to generate IFN-γ (Khan et al., 2006). Lack of NK cell recruitment causes higher parasitemia and greater susceptibility to infection. Adoptive transfer of $CCR5^{+/+}$ NK cells into $CCR5^{-/-}$ mice restored NK cell recruitment and increased resistance to infection, establishing CCR5 as the critical receptor guiding NK trafficking in host defense.

Once at the site of infection, NK cells are activated to become the predominant source of IFN-γ. Recently, Sague and associates provided the first evidence that mouse NK cells constitutively express the inactivated form of a cell-surface glycoprotein involved in cell-cell interactions, CD44, which is converted to the active form of CD44 upon infection of the animal with *T. gondii*. The principal ligand for activated CD44 is hyaluronan (HA), which exists in two forms. Normally, HA exists as a high molecular weight form (HMWHA), which does not stimulate inflammatory gene expression. In sites of inflammation, the low molecular weight form (LMWHA) accumulates and can activate inflammatory gene expression through CD44. A potential role for activated CD44 on NK cells was demonstrated by addition of its ligand, LMWHA, to bone marrow-derived NK cells in the presence of IL-18, which resulted in dose-dependent IFN-γ production (Sague et al., 2004). Thus, stimulation of CD44 with its ligand, LMWHA, provides a potent signal for IFN-γ production by NK cells and may be an integral part of NK cell responses at the site of *T. gondii* infection. Interestingly, CD44 activates NF-κB, which was recently demonstrated to be involved in regulation of NK cell proliferation and production of IFN-γ (Tato et al., 2006).

IFN-γ release by NK cells at the site of infection activates macrophages and DC, which secrete IL-12, resulting in cell-mediated immunity that inhibits the growth of the replicating parasite (tachyzoite) (Goldszmid et al., 2007; Khan et al., 2006). The NK–DC interaction can be reciprocal; recent studies indicate that NK–DC crosstalk may play an important role in the regulation of both innate resistance and adaptive immunity to infection (Chiesa et al., 2006; Gerosa et al., 2005; Mavilio et al., 2006; Munz et al., 2005; Walzer et al., 2005). In order to further understand NK–DC crosstalk during *T. gondii* infection, Guan and colleagues investigated the role of NKG2D, a surface receptor expressed by NK cells, on DC responses to *T. gondii* antigen in culture. In this study, they found that NK-DC–enhanced IL-12 production depended on cell contact via NKG2D on NK cells and NKG2D binding ligands RAE-1 (RNA export 1 homolog) and MULT-1 (murine UL16-binding protein-like transcript) on DCs; in vivo, NKG2D neutralization reduced protection against *T. gondii* infection (Guan et al., 2007). Through these interactions, NK cells and DC can positively regulate the production of IFN-γ and increase resistance to early infection. In addition, NK–DC crosstalk may be important for generation of adaptive immune responses, as DCs are the most potent antigen presenting cells (APC) for priming T cell responses.

When investigating the role of NK–DC interaction on T cell priming, Guan and colleagues found that NK–DC interaction is important for $CD8^+$ T cell priming in *T. gondii* infection. In vitro, treatment with anti-NKG2D antibody reduced CD8+ T cell proliferation when cultured with NK cells and DC, compared with isotype control antibody treatment. In vivo, $CD8^+$ cells isolated from anti-NKG2D treated animals had reduced proliferation in response to *T. gondii* antigen compared to those injected with control antibody (Guan et al., 2007). These studies build upon earlier work indicating a role for $CD4^+$ T cell independent priming of $CD8^+$ T cells by NK cells (Combe et al., 2005). Combe and associates found that depletion of NK cells in $CD4^{-/-}$ mice caused the mice to succumb to otherwise non-lethal *T. gondii* infection and that the depletion of NK cells led to reduced antigen-specific $CD8^+$ T cells in these mice. The ability of NK cells to help prime $CD8^+$ T cells was attributed to the prolonged NK production of IFN-γ in response in to this infection (Combe et al., 2005).

This prolonged IFN-γ secretion may also be important in priming $CD4^+$ T cell responses in *T. gondii* infection. Goldszmid and colleagues recently discovered that antigen peptide transporter 1 (TAP-1) deficient mice were more susceptible to *T. gondii* infection and that this effect was correlated with reduced $CD4^+$ T cell activation and reduced numbers of NK cells relative to wild-type. Adoptive transfer of NK cells able to secrete IFN-γ restored $CD4^+$ T cell responses of infected $TAP\text{-}1^{-/-}$ mice to normal levels, whereas $IFN\text{-}\gamma^{-/-}$ NK cells did not (Goldszmid et al., 2007). These studies indicate that NK cell IFN-γ production enhances APC-dependent Th-type-1 cellular immunity and illustrate the ability of NK cells to act as a bridge between innate and adaptive immune mechanisms.

In summary, NK cells play an important role in an effective early immune response to *T. gondii*. Through interaction with T cells, NK cells help shape the adaptive immune response that results in containment of the parasite. Interestingly, recent evidence that *T. gondii* are able to infect NK cells after escape from DC suggests a mechanism whereby parasites utilize NK cells to evade clearance and disseminate throughout the host (Persson et al., 2009). Thus, while NK cells are essential to effective anti-parasitic immunity early in infection, NK cells may also play a role in parasite persistence. Most of the research on NK cell function has been in tumour immunology and viral infection. Study of NK cell activation and the effector response in the context of *T. gondii* infection gives researchers an opportunity to further understand NK cells in the setting of a complex parasitic infection.

Leishmaniasis

Leishmania parasites of various species are the causal agents of leishmaniasis. Leishmaniasis is a group of diseases. Various mammals become infected via the phlebotomine sandfly (Reithinger et al., 2007; Shaw, 2007). After taking a blood meal, the sandfly regurgitates, releasing promastigote parasites into the bite wound of the host, where they are engulfed by phagocytic cells. Within the phagolysosome, promastigotes differentiate into amastigotes, the replicative form of the parasite, and eventually lyse the infected cell. Released amastigotes infect new cells locally or in distant tissues following dissemination. Leishmaniasis has a broad geographic distribution and large impact on human health, with about 12 million people infected worldwide, and an estimated 2.4 million disability-adjusted life-years (DALY) (Reithinger et al., 2007). Infections with *Leishmania* spp. cause a range of clinical manifestations that are grouped into either cutaneous (local) or visceral (systemic) infection. Most *Leishmania* spp. cause a form of cutaneous leishmaniasis; only *L. donovani* and *L. infantum* commonly lead to visceral disease in humans (Reithinger et al., 2007). *Leishmania* spp. cause similar disease symptoms in a variety of mammals, including humans and mice. Pressure exerted by the host immune system combined with immune evasion strategies employed by the parasite leads to long-lasting chronic *Leishmania* spp. infection. Control of *Leishmania* spp. infection is mediated primarily by T-helper cell type 1 (Th1) responses (Stanley and Engwerda, 2007; Tripathi et al., 2007). The contribution of NK cells to a protective immune reaction has been studied in vitro using human donor cells and by in vivo mouse model systems of cutaneous and visceral leishmaniasis.

NK cells from unexposed human donors proliferate and secrete IFN-γ in response to *Leishmania* antigen (Akuffo and Britton, 1992; Maasho et al., 1998; Nylen et al., 2004). NK cell secreted IFN-γ is thought to increase macrophage-mediated killing of intracellular parasites and to trigger Th1 immune response. NK cells may also play a direct role by lysis of cells infected with *Leishmania* amastigotes as they are able to lyse *Leishmania*-infected T cells (Saha et al., 1999). Interestingly, immature human DC infected with *L. infantum* are resistant to NK-mediated cytolysis, due to upregulation of HLA-E expression, which protects the DC from lysis through engagement of the NK inhibitory receptor NKG2A (CD94) (Campos-Martin et al., 2006). Thus, in at least one cell type, a virulent strain of *Leishmania* has developed a mechanism for escaping NK cell–mediated immunity.

Human NK cells purified from PBMCs are activated to secrete IFN-γ to a greater extent by live compared to heat-killed promastigote parasite (Nylen et al., 2003). Nylen and colleagues showed that stimulation of total PBMCs with live or heat-killed parasites followed the same trend. Toll-like-receptor-2 (TLR-2) engagement by a lipophosphoglycan (LPG), a phosphoglycan belonging to a family of unique *Leishmania* glycoconjugates that are linked to the parasite membrane by a glycosylphosphatidylinositol anchor, has recently been identified as a mechanism for activation of human NK cells by infectious parasite (Becker et al., 2003; Santarem et al., 2007). An analysis of donor blood from an endemic area revealed that cured and infected individuals had higher levels of large NK cells than healthy controls or asymptomatic endemic individuals (Maasho et al., 1998). After stimulation with soluble *Leishmania* antigen (SLA), human NK cells from asymptomatic and cured patients exhibited increased IFN-γ expression, whereas patients with active disease maintained a consistent level of IFN-γ expression similar in magnitude to that produced by NK cells from cured patients after stimulation with SLA (Peruhype-Magalhaes et al., 2005). Together, the results from investigation of human NK cell responses to *Leishmania* spp. or *Lesihmania* antigen indicate that NK cell activation and IFN-γ production contributes to parasite containment and disease control.

Since NK cells may play an important role in the initial phase of infection, mouse models have been used to study NK cell activation in early *Leishmania* spp. infection. Mice depleted of NK cells show a reduced IFN-γ response and harbour more parasites in the infected tissues (Laskay et al., 1993; Scharton and Scott, 1993). A recent study showed that activated NK cells can control parasitism of infected macrophages via direct lysis of the host cell and/or parasites in vitro (Aranha et al., 2005). In addition, IFN-γ and IL-12 levels in these cultures corresponded with the concentration of activated NK cells. In vivo, injection of activated NK cells resulted in decreased inflammation and smaller lesions compared to mice receiving only parasite, indicating a direct role for NK cell control of parasite pathology (Aranha et al., 2005).

Further investigation of the activation of NK cells during murine *Leishmania* infection indicates a role for pathogen pattern-recognition molecules and accessory cells. NK cells are rapidly recruited from the blood to draining lymph nodes (LN) upon infection with *Leishmania*. Static and real-time imaging of NK cells in LN of mice infected with *Leishmania* reveals that NK cells move slowly and are closely associated with DC, maintaining interactions with DC for extended times (Bajenoff et al., 2006). NK cells collect near LN T cells, which undergo parasite-induced activation concurrent with IFN-γ secretion by NK cells. The co-localization of DC, $CD4^+$ T cells, and IFN-γ-secreting NK cells in the LN would allow NK cells, acting as an early source of IFN-γ, to influence the differentiation of T cells into Th1 cells. Further evidence for the interaction between NK cells and DC came from mice lacking TLR9 that have impaired NK cell responses to *L. major*. Upon *Leishmania* infection, TLR9$^{-/-}$ mice develop more severe skin lesions and increased parasite burden compared to wild type, although they were able to generate a Th1 response that ultimately resolved the infection (Liese et al., 2007). The lack of TLR9 is thought to diminish NK activation through decreased expression of IL-12 by DC, an indispensable cytokine for the activation of NK cells during the innate phase of leishmaniasis (Liese et al., 2007). Recently, Schleicher and associates offered a functional and mechanistic role for the DC–NK cell interactions in draining LN during early *Leishmania* infection in mice. Using a combination of depletion and knock-out studies, they discovered that NK-induced cytotoxicity and IFN-γ production depended on TLR9 engagement and IL-12 secretion by myeloid (mDC), but not plasmacytoid (pDC), DC (Schleicher et al., 2007). The necessity for mDC but not pDC may be a pattern of NK cell activation during parasitic disease, as depletion of pDC had no consistent effect on the human NK IFN-γ response to *Plasmodium* spp. infected RBCs, whereas depleting PBMCs of mDCs significantly reduced NK cell IFN-γ responses (Newman et al., 2006). In addition, NK–DC crosstalk is bidirectional, with both cell types influencing the other to produce pro-inflammatory cytokines (Ing and Stevenson, 2009; Sanabria et al., 2008).

Although mouse studies have produced a definitive model of NK–DC interaction during leishmaniasis, this model of NK cell response during *Leishmania* spp. infection lacks support from analysis of human NK cells. In vitro studies with purified human NK cells indicated activation without accessory cells (Nylen et al., 2003). This is contradictory to the mouse model system, where in vivo evidence indicates that these accessory cells are absolutely required for NK cell activation and subsequent function during early infection (Schleicher et al., 2007). This apparent discrepancy may be due to a difference in magnitude of NK cell response reported in the human model. In other words, the amount of activation of NK cells shown in the human model may not have been enough to support effective function in vivo. Further investigation in this area is needed to determine the magnitude of NK cell activation and effector functions in the presence or absence of accessory cells. We expect that further human studies will support the model of accessory cell involvement in optimal NK cell activation during leishmaniasis.

Trypansoma cruzi

Trypanosoma cruzi is a protozoan parasite that infects most mammals and in humans causes Chagas' disease, which is endemic in South and Central America. An estimated 12 million people are infected with *T. cruzi* and over 600000 DALYs are caused by Chagas' disease (WHO, 2004). Chagas' disease has more impact on quality of life in endemic regions than any other parasitic disease (Dias et al., 2002). Many different strains of *T. cruzi* are transmitted through the bite of various triatomine insects. After taking a blood meal, *T. cruzi* in the insects' urine and faeces is deposited on the skin, and parasites enter via the bite wound or a mucosal surface. Host cells are quickly invaded and the parasite differentiates into a replicative (amastigote) form of parasite. After replication, parasites convert back to the infectious (trypomastigote) form of the parasite. Upon cell lysis, trypomastigotes disseminate in the blood, causing patent parasitemia. Host immune mechanisms lead to control of parasitemia, although parasites are able to sequester in neuronal and muscle tissues, leading to an indeterminate phase of infection that is characterized by positive serology without patent parasitemia. This phase may persist for the life of the host or progress to chronic Chagas' disease. Chronic Chagas' disease also known as 'silent death,' occurs in up to half of the people infected with *T. cruzi* and is a common cause of sudden heart failure in young adults (Rassi et al., 2007). Although acute infection with *T. cruzi* is rarely associated with severe symptoms, the early response to *T. cruzi* is poorly understood and may have ramifications for subsequent clinical manifestations of chronic disease. Mouse models indicate that a broad immunological response is required to control acute *T. cruzi* infection. Mice lacking B cells, cytotoxic T cells, IFN-γ or IL-12 are highly susceptible to infection (Sardinha et al., 2006). In addition, innate immune responses play a pivotal role during the early phase of infection.

In human studies, increased levels of mature NK, NKT and regulatory T cells may favour maintenance of the life-long indeterminate form of infection rather than progression to chronic disease (Vitelli-Avelar et al., 2006). Mouse models suggest that NK cell depletion during acute infection results in increased parasitemia and may contribute to shortened survival (Duthie and Kahn, 2005; Lieke et al., 2004). Mouse models indicate NK cell activation in the spleen, peritoneal exudates and liver early in infection. In the spleen, NK cell activation is dependant upon IL-12 production by accessory-cells (Lieke et al., 2004; Newman and Riley, 2007), but is independent of NK T cells (Duthie and Kahn, 2005). Infection with *T. cruzi* activates cytokine secretion (IFN-γ) as well as lytic pathways of splenic NK cells (Lieke et al., 2004). NK cells in the spleen develop contact-dependent effector mechanisms against free parasite during infection that lead to parasite lysis (Lieke et al., 2004) (Figure 45.3). Within the liver, NK are an early source of IFN-γ. Liver NK cell numbers are regulated by invariant NK T cells (Duthie and Kahn, 2005; Sardinha et al., 2006). NK cells seem to contribute to protection rather than pathogenesis in the liver as NK cell depletion does not change the mild

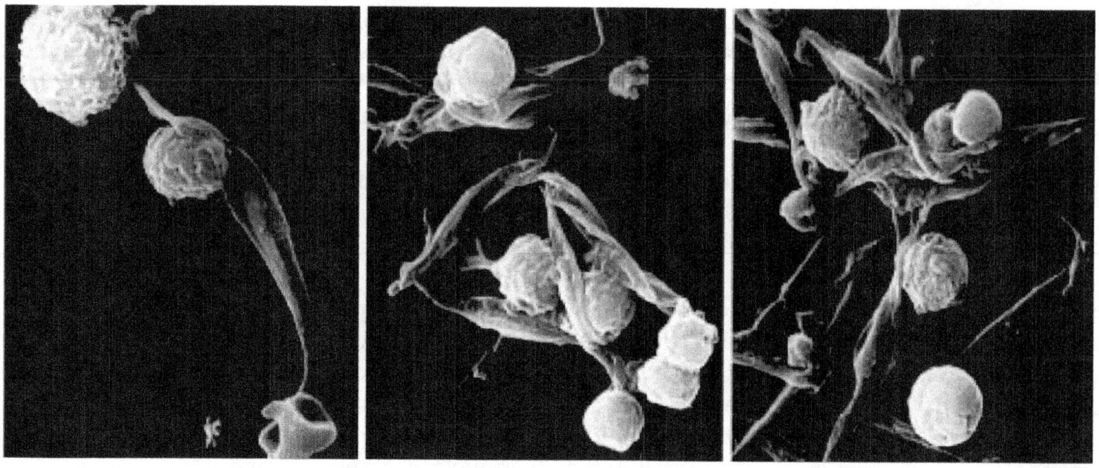

Figure 45.3 • Intimate interaction between NK cells and *T. cruzi* parasites. DX5-positive NK cells were purified by magnetic cell sorting. A total of 10^6 purified NK cells were incubated with 10^5 *T. cruzi* epimastigotes for 1 h at 37°C. Cells were subsequently fixed and analyzed by scanning electron microscopy. Representative areas are shown. No interaction was observed with purified T cells (data not shown). Reprinted with permission from the American Society of Microbiology (ASM): *Infection and Immunity*, copyright 2004 (Lieke et al., 2004).

liver damage detectable during acute *T. cruzi* infection (Duthie and Kahn, 2005). Taken together, these studies illustrate a role for NK cells in contributing to parasite control in an accessory-cell-dependent manner through both IFN-γ production and direct lysis of parasites during acute phase infection.

NK cells may also play a role in *T. cruzi* infection by contributing to immune evasion through interaction with B cells. *T. cruzi* acute infection is characterized by non-specific B cell activation, that is thought to contribute to delayed specific immune responses (Acosta Rodriguez et al., 2007). An NK cell line that supports immunoglobulin production was shown to cause B cells from *T. cruzi*-infected mice to increase total immunoglobulin production fourfold above that produced by normal B cells in response to the same NK cell line (De Arruda Hinds et al., 2001). These B cells had higher B7.2 (CD86) and could therefore be more readily stimulated by co-stimulatory molecules on the surface of NK cells. While this one study suggests the possibility of NK cells contributing to immune escape by *T. cruzi*, most of the evidence suggests that NK cells primarily play a protective role during early *T. cruzi* infection and may be important for continued suppression of the parasite and maintenance of indeterminate phase disease.

Helminths

Helminths (parasitic worms) are metazoan (multicellular) parasites of mammals that belong to two distinct animal taxa: phyla Nematoda and Platyhelmintha. Infection with helminths affects more than one-tenth of the world's population, causing considerable morbidity (Korten et al., 2002). Infection of mammals with this divergent group of organisms typically induces a T helper type-2 (Th2) cellular immune response, which is generally considered appropriate both for parasite control by the host and for containing the damage caused by helminths (Diaz and Allen, 2007). As NK cells typically favour Th1 generation, little research has been done to elucidate their role in helminth infection. Yet, the role of Th2 predominance may not be the whole story in helminth immunity, as innate mechanisms may play a substantial role in parasite infection and control (Rodriguez-Sosa et al., 2004; Spencer et al., 2003). Understanding the immunology involved in helminth infection is complicated by the diversity of the organisms and the complexity of the life-cycle of individual helminths, as well as, limitations of murine models. In fact, some reports indicate that an effective immune response in humans may be a more balanced Th1/Th2 response than in mice for some helminths.

In the face of these difficulties, a few studies have been published that explore a role for NK cell activation during the early phase of mouse helminth infection. In general, NK cytolytic effects are thought to be negligible on helminths, rather NK cytokine secretion may be involved in the early immune response. Increased NK cell activity has previously been reported in helminth infection, but mechanisms of NK cell activity have not been well-delineated (Brattig et al., 1987; Niederkorn et al., 1988; Pedersen et al., 1987).

In the mouse model of helminthic infection with *Litomosoides signmodontis*, in which the filariae (worm of the superfamily Filarioidea) undergo a complete life cycle, NK cells were found to contribute to parasite control (Korten et al., 2002). In this model, NK cells migrated to the site of helminth infection and displayed a reduction in their inhibitory receptors. An effective defense against *L. signmodontis* in humans is characterized by a balanced Th1/Th2 response. Depletion of NK cells in this model resulted in enhanced worm load and increased Th2 type cytokine secretion (Korten et al., 2002), indicating that NK cells were necessary to provide Th1 stimulatory cytokines and thereby resulted in a balanced, effective-immune response to this organism.

In the *Brugia malayi* model of lymphatic filariasis, recent evidence suggests that NK cells are necessary for the growth of the parasite within the mammalian host as mice with higher NK cell activity were more permissive to parasite development (Babu et al., 1998). To investigate NK cell responses to filarial infection, Babu and colleagues used an in vitro model system of culturing live infective-stage larvae (L3) or live microfilariae (Mf) (Babu et al., 2007). In human infections, L3 enter the skin when a mosquito takes a blood meal. Mf are the bloodstream form of the parasite that is taken up by and infects the mosquito to complete the parasite life cycle. Upon stimulation with L3 or Mf, NK cells undergo early activation and produce IFN-γ and TNF-α. In addition, Mf stimulate IL-4 and IL-5 production in NK cells, whereas L3 do not. NK cells increase expression of NKp44 and NKp46 activating receptors upon L3 stimulation, but not with Mf. As seen in protozoan parasites, the activation of NK cells by this metazoan parasite required direct contact with accessory-cells and IL-12 (Babu et al., 2007). In addition, direct contact with the parasite was necessary for full NK cell activation. After prolonged stimulation with L3, NK cells underwent apoptosis (Babu et al., 2007). These data suggest that upon parasite infection, filarial parasites initially induce activation of NK cells, causing IFN-γ and TNF-α secretion, thereby skewing the response toward a Th1 response. NK cells may play a role in Th2 responses later in infection, as evidenced by the generation of IL-4 and IL-5 in response to Mf. To what extent these different NK cell responses play a protective role or allow parasite persistence must still be addressed experimentally.

A Th2 response often allows persistent helminth infection, where immune mechanisms neither eliminate the parasite nor allow the parasite to overwhelm its host and cause death. Therefore, parasite immunologists debate whether Th2 skewing of the immune response is due to manipulation of the immune system by helminths or the evolutionarily selected response against worms. Recently, investigation of secreted proteins from the human hookworm, *Necator americanus*, led to the discovery of a protein that binds selectively to NK cells and induces IFN-γ production (Hsieh et al., 2004). In hookworm infection, Th2 immune responses contribute to parasite clearance, thereby the induction of NK cells and the subsequent IFN-γ release contributes to an ineffective immune response (Th1), contributing to long-term survival of the parasite within its host (Teixeira-Carvalho et al., 2008). Thus, at least one helminth is able to directly manipulate the host immune system toward a greater Th2 response to create an environment that favours parasite persistence.

Conclusion

Investigation of NK cell activity during infection with the protozoan parasites considered in this chapter indicates that NK cells are important for shaping the initial immune response to these microbes. During protozoan parasite infection, NK cells seem to play a primarily protective role through the secretion of IFN-γ and direct lysis of some parasite infected cells. In addition, NK cell effector function during protozoan parasite infection seems to rely upon accessory cell interactions, particularly with cells of the myeloid lineage (Newman and Riley, 2007). This interaction is bidirectional and has been shown to influence adaptive immune responses. In certain protozoan parasite infections, NK cells may be manipulated or evaded to help the parasite circumvent immune recognition, but these claims need further investigation. During several metazoan/helminthic parasite infections, NK cells may also play an accessory cell-dependent, protective role, whereas in others they are exploited for their Th1-skewing cytokine profile that produces a more parasite-friendly environment.

References

Acosta Rodriguez, E.V., Zuniga, E.I., Montes, C.L., Merino, M.C., Bermejo, D.A., Amezcua Vesely, M.C., Motran, C.C. and Gruppi, A. (2007). *Trypanosoma cruzi* infection beats the B-cell compartment favouring parasite establishment: can we strike first? Scand J Immunol **66**, 137–142.

Akuffo, H.O. and Britton, S.F. (1992). Contribution of non-*Leishmania*-specific immunity to resistance to *Leishmania* infection in humans. Clin Exp Immunol **87**, 58–64.

Aranha , F.C., Ribeiro, U. Jr, Basse, P., Corbett, C.E. and Laurenti, M.D. (2005). Interleukin-2-activated natural killer cells may have a direct role in the control of *Leishmania (Leishmania) amazonensis* promastigote and macrophage infection. Scand J Immunol **62**, 334–341.

Artavanis-Tsakonas, K. and Riley, E.M. (2002). Innate immune response to malaria: rapid induction of IFN-gamma from human NK cells by live *Plasmodium falciparum*-infected erythrocytes. J Immunol **169**, 2956–2963.

Babu, S., Porte, P., Klei, T.R., Shultz, L.D. and Rajan, T.V. (1998). Host NK cells are required for the growth of the human filarial parasite *Brugia malayi* in mice. J Immunol **161**, 1428–1432.

Babu, S., Blauvelt, C.P. and Nutman, T.B. (2007). Filarial parasites induce NK cell activation, type 1 and type 2 cytokine secretion, and subsequent apoptotic cell death. J Immunol **179**, 2445–2456.

Bajenoff, M., Breart, B., Huang, A.Y., Qi, H., Cazareth, J., Braud, V.M., Germain, R.N. and Glaichenhaus, N. (2006). Natural killer cell behavior in lymph nodes revealed by static and real-time imaging. J Exp Med **203**, 619–631.

Baratin, M., Roetynck, S., Lepolard, C., Falk, C., Sawadogo, S., Uematsu, S., Akira, S., Ryffel, B., Tiraby, J.G., Alexopoulou, L., Kirschning, C.J., Gysin, J., Vivier, E. and Ugolini, S. (2005). Natural killer cell and macrophage cooperation in MyD88-dependent innate responses to *Plasmodium falciparum*. Proc Natl Acad Sci USA **102**, 14747–14752.

Baratin, M., Roetynck, S., Pouvelle, B., Lemmers, C., Viebig, N.K., Johansson, S., Bierling, P., Scherf, A., Gysin, J., Vivier, E. and Ugolini, S. (2007). Dissection of the role of PfEMP1 and ICAM-1 in the sensing of *Plasmodium falciparum*-infected erythrocytes by natural killer cells. PLoS ONE **2**, e228.

Becker, I., Salaiza, N., Aguirre, M., Delgado, J., Carrillo-Carrasco, N., Kobeh, L.G., Ruiz, A., Cervantes, R., Torres, A.P., Cabrera, N., Gonzalez, A., Maldonado, C. and Isibasi, A. (2003). *Leishmania* lipophosphoglycan (LPG) activates NK cells through toll-like receptor-2. Mol Biochem Parasitol **130**, 65–74.

Brattig, N.W., Tischendorf, F.W., Albiez, E.J., Buttner, D.W. and Berger, J. (1987). Distribution pattern of peripheral lymphocyte subsets in localized and generalized form of onchocerciasis. Clin Immunol Immunopathol **44**, 149–159.

Campos-Martin, Y., Colmenares, M., Gozalbo-Lopez, B., Lopez-Nunez, M., Savage, P.B. and Martinez-Naves, E. (2006). Immature human dendritic cells infected with *Leishmania infantum* are resistant to NK-mediated cytolysis but are efficiently recognized by NKT cells. J Immunol **176**, 6172–6179.

CDC. (2007). Malaria facts. In: *Malaria*. Vol. 2008.

Chiesa, M.D., Romagnani, C., Thiel, A., Moretta, L. and Moretta, A. (2006). Multidirectional interactions are bridging human NK cells with plasmacytoid and monocyte-derived dendritic cells during innate immune responses. Blood **108**, 3851–3858.

Combe, C.L., Curiel, T.J., Moretto, M.M. and Khan, I.A. (2005). NK cells help to induce CD8(+)-T-cell immunity against *Toxoplasma gondii* in the absence of CD4(+) T cells. Infect Immun **73**, 4913–4921.

De Arruda Hinds, L.B., Alexandre-Moreira, M.S., Decote-Ricardo, D., Nunes, M.P. and Pecanha, L.M.T. (2001). Increased immunoglobulin secretion by B lymphocytes from *Trypanosoma cruzi* infected mice after B lymphocytes–natural killer cell interaction. Parasite Immunol **23**, 581–586.

Denkers, E.Y. (2003). From cells to signaling cascades: manipulation of innate immunity by *Toxoplasma gondii*. FEMS Immunol Med Microbiol **39**, 193–203.

Denkers, E.Y., Gazzinelli, R.T., Martin, D. and Sher, A. (1993). Emergence of NK1.1+ cells as effectors of IFN-gamma dependent immunity to *Toxoplasma gondii* in MHC class I-deficient mice. J Exp Med **178**, 1465–1472.

Dias, J.C., Silveira, A.C. and Schofield, C.J. (2002). The impact of Chagas disease control in Latin America: a review. Mem Inst Oswaldo Cruz **97**, 603–612.

Diaz, A. and Allen, J.E. (2007). Mapping immune response profiles: the emerging scenario from helminth immunology. Eur J Immunol **37**, 3319–3326.

D'Ombrain, M.C., Hansen, D.S., Simpson, K.M. and Schofield, L. (2007). Gammadelta-T cells expressing NK receptors predominate over NK cells and conventional T cells in the innate IFN-gamma response to *Plasmodium falciparum* malaria. Eur J Immunol **37**, 1864–1873.

Duthie, M.S. and Kahn, S.J. (2005). NK cell activation and protection occur independently of natural killer T cells during *Trypanosoma cruzi* infection. Int Immunol **17**, 607–613.

Gerosa, F., Gobbi, A., Zorzi, P., Burg, S., Briere, F., Carra, G. and Trinchieri, G. (2005). The reciprocal interaction of NK cells with plasmacytoid or myeloid dendritic cells profoundly affects innate resistance functions. J Immunol **174**, 727–734.

Goldszmid, R.S., Bafica, A., Jankovic, D., Feng, C.G., Caspar, P., Winkler-Pickett, R., Trinchieri, G. and Sher, A. (2007). TAP-1 indirectly regulates CD4+ T cell priming in *Toxoplasma gondii* infection by controlling NK cell IFN-gamma production. J Exp Med **204**, 2591–2602.

Guan, H., Moretto, M., Bzik, D.J., Gigley, J. and Khan, I.A. (2007). NK cells enhance dendritic cell response against parasite antigens via NKG2D pathway. J Immunol **179**, 590–596.

Hansen, D.S., Bernard, N.J., Nie, C.Q. and Schofield, L. (2007a). NK cells stimulate recruitment of CXCR3+ T cells to the brain during *Plasmodium berghei*-mediated cerebral malaria. J Immunol **178**, 5779–5788.

Hansen, D.S., D'Ombrain, M.C. and Schofield, L. (2007b). The role of leukocytes bearing natural killer complex receptors and killer immunoglobulin-like receptors in the immunology of malaria. Curr Opin Immunol **19**, 416–423.

Hensmann, M. and Kwiatkowski, D. (2001). Cellular basis of early cytokine response to *Plasmodium falciparum*. Infect Immun **69**, 2364–2371.

Hsieh, G.C., Loukas, A., Wahl, A.M., Bhatia, M., Wang, Y., Williamson, A.L., Kehn, K.W., Maruyama, H., Hotez, P.J., Leitenberg, D., Bethony, J. and Constant, S.L. (2004). A secreted protein from the human hookworm *Necator americanus* binds selectively to NK cells and induces IFN-gamma production. J Immunol **173**, 2699–2704.

Hunter, C.A., Subauste, C.S., Van Cleave, V.H. and Remington, J.S. (1994). Production of gamma interferon by natural killer cells from *Toxoplasma gondii*-infected SCID mice: regulation by interleukin-10, interleukin-12, and tumor necrosis factor alpha. Infect Immun **62**, 2818–2824.

Hunter, C.A., Ellis-Neyer, L., Gabriel, K.E., Kennedy, M.K., Grabstein, K.H., Linsley, P.S. and Remington, J.S. (1997). The role of the CD28/B7 interaction in the regulation of NK cell responses during infection with *Toxoplasma gondii*. J Immunol **158**, 2285–2293.

Ing, R. and Stevenson, M.M. (2009). Dendritic cell and NK cell reciprocal cross talk promotes gamma interferon-dependent immunity to blood-stage *Plasmodium chabaudi* AS infection in mice 10.1128/IAI.00994-08. Infect Immun **77**, 770–782.

Johnson, L.L., VanderVegt, F.P. and Havell, E.A. (1993). Gamma interferon-dependent temporary resistance to acute *Toxoplasma gondii* infection independent of CD4+ or CD8+ lymphocytes. Infect Immun **61**, 5174–5180.

Khan, I.A., Thomas, S.Y., Moretto, M.M., Lee, F.S., Islam, S.A., Combe, C., Schwartzman, J.D. and Luster, A.D. (2006). CCR5 is essential for NK cell trafficking and host survival following *Toxoplasma gondii* infection. PLoS Pathog **2**, e49.

Kim, C.C., Parikh, S., Sun, J.C., Myrick, A., Lanier, L.L., Rosenthal, P.J. and DeRisi,

J.L. (2008). Experimental malaria infection triggers early expansion of natural killer cells 10.1128/IAI.00640-08. Infect Immun **76**, 5873–5882.

Korbel, D.S., Newman, K.C., Almeida, C.R., Davis, D.M. and Riley, E.M. (2005). Heterogeneous human NK cell responses to *Plasmodium falciparum*-infected erythrocytes. J Immunol **175**, 7466–7473.

Korten, S., Volkmann, L., Saeftel, M., Fischer, K., Taniguchi, M., Fleischer, B. and Hoerauf, A. (2002). Expansion of NK cells with reduction of their inhibitory Ly-49A, Ly-49C, and Ly-49G2 receptor-expressing subsets in a murine helminth infection: contribution to parasite control. J Immunol **168**, 5199–5206.

Kraemer, S.M. and Smith, J.D. (2006). A family affair: var genes, PfEMP1 binding, and malaria disease. Curr Opin Microbiol **9**, 374–380.

Laskay, T., Rollinghoff, M. and Solbach, W. (1993). Natural killer cells participate in the early defense against *Leishmania major* infection in mice. Eur J Immunol **23**, 2237–2241.

Lieke, T., Graefe, S.E.B., Klauenberg, U., Fleischer, B. and Jacobs, T. (2004). NK cells contribute to the control of *Trypanosoma cruzi* infection by killing free parasites by perforin-independent mechanisms. Infect Immun **72**, 6817–6825.

Liese, J., Schleicher, U. and Bogdan, C. (2007). TLR9 signaling is essential for the innate NK cell response in murine cutaneous leishmaniasis. Eur J Immunol **37**, 3424–3434.

Lodoen, M.B. and Lanier, L.L. (2006). Natural killer cells as an initial defense against pathogens. Curr Opin Immunol **18**, 391–398.

Maasho, K., Sanchez, F., Schurr, E., Hailu, A. and Akuffo, H. (1998). Indications of the protective role of natural killer cells in human cutaneous leishmaniasis in an area of endemicity. Infect Immun **66**, 2698–2704.

Mavilio, D., Lombardo, G., Kinter, A., Fogli, M., La Sala, A., Ortolano, S., Farschi, A., Follmann, D., Gregg, R., Kovacs, C., Marcenaro, E., Pende, D., Moretta, A. and Fauci, A.S. (2006). Characterization of the defective interaction between a subset of natural killer cells and dendritic cells in HIV-1 infection. J Exp Med **203**, 2339–2350.

Mavoungou, E., Held, J., Mewono, L. and Kremsner, P.G. (2007). A duffy binding-like domain is involved in the NKp30-mediated recognition of *Plasmodium falciparum*-parasitized erythrocytes by natural killer cells. J Infect Dis **195**, 1521–1531.

Mohan, K., Moulin, P. and Stevenson, M.M. (1997). Natural killer cell cytokine production, not cytotoxicity, contributes to resistance against blood-stage *Plasmodium chabaudi* AS infection. J Immunol **159**, 4990–4998.

Munz, C., Steinman, R.M. and Fujii, S. (2005). Dendritic cell maturation by innate lymphocytes: coordinated stimulation of innate and adaptive immunity. J Exp Med **202**, 203–207.

Newman, K.C. and Riley, E.M. (2007). Whatever turns you on: accessory-cell-dependent activation of NK cells by pathogens. Nat Rev Immunol **7**, 279–291.

Newman, K.C., Korbel, D.S., Hafalla, J.C. and Riley, E.M. (2006). Cross-talk with myeloid accessory cells regulates human natural killer cell interferon-gamma responses to malaria. PLoS Pathog **2**, e118.

Niederkorn, J.Y., Stewart, G.L., Ghazizadeh, S., Mayhew, E., Ross, J. and Fischer, B. (1988). *Trichinella pseudospiralis* larvae express natural killer (NK) cell-associated asialo-GM1 antigen and stimulate pulmonary NK activity. Infect Immun **56**, 1011–1016.

Nylen, S., Maasho, K., Soderstrom, K., Ilg, T. and Akuffo, H. (2003). Live *Leishmania* promastigotes can directly activate primary human natural killer cells to produce interferon-gamma. Clin Exp Immunol **131**, 457–467.

Nylen, S., Maasho, K., McMahon-Pratt, D. and Akuffo, H. (2004). Leishmanial amastigote antigen P-2 induces major histocompatibility complex class II-dependent natural killer-cell reactivity in cells from healthy donors. Scand J Immunol **59**, 294–304.

Papazahariadou, M., Athanasiadis, G.I., Papadopoulos, E., Symeonidou, I., Hatzistilianou, M., Castellani, M.L., Bhattacharya, K., Shanmugham, L.N., Conti, P. and Frydas, S. (2007). Involvement of NK cells against tumors and parasites. Int J Biol Markers **22**, 144–153.

Pedersen, B.K., Bygbjerg, I.C. and Svenson, M. (1987). Increase in natural killer cell activity during diethylcarbamazine treatment of patients with filariasis. Acta Trop **44**, 353–355.

Persson, C.M., Lambert, H., Vutova, P.P., Dellacasa-Lindberg, I., Nederby, J., Yagita, H., Ljunggren, H.-G., Grandien, A., Barragan, A. and Chambers, B.J. (2009). Transmission of *Toxoplasma gondii* from infected dendritic cells to natural killer cells 10.1128/IAI.00833-08. Infect Immun **77**, 970–976.

Peruhype-Magalhaes, V., Martins-Filho, O.A., Prata, A., Silva Lde, A., Rabello, A., Teixeira-Carvalho, A., Figueiredo, R. M., Guimaraes-Carvalho, S.F., Ferrari, T.C. and Correa-Oliveira, R. (2005). Immune response in human visceral leishmaniasis: analysis of the correlation between innate immunity cytokine profile and disease outcome. Scand J Immunol **62**, 487–495.

Rassi , A. Jr, Rassi, A. and Rassi, S.G. (2007). Predictors of mortality in chronic Chagas disease: a systematic review of observational studies. Circulation **115**, 1101–1108.

Reithinger, R., Dujardin, J.C., Louzir, H., Pirmez, C., Alexander, B. and Brooker, S. (2007). Cutaneous leishmaniasis. Lancet Infect Dis **7**, 581–596.

Rodriguez-Sosa, M., Saavedra, R., Tenorio, E.P., Rosas, L.E., Satoskar, A.R. and Terrazas, L.I. (2004). A STAT4-dependent Th1 response is required for resistance to the helminth parasite *Taenia crassiceps*. Infect Immun **72**, 4552–4560.

Roetynck, S., Baratin, M., Johansson, S., Lemmers, C., Vivier, E. and Ugolini, S. (2006). Natural killer cells and malaria. Immunol Rev **214**, 251–263.

Roland, J., Soulard, V., Sellier, C., Drapier, A.M., Di Santo, J.P., Cazenave, P.A. and Pied, S. (2006). NK cell responses to *Plasmodium* infection and control of intrahepatic parasite development. J Immunol **177**, 1229–1239.

Sague, S.L., Tato, C., Pure, E. and Hunter, C.A. (2004). The regulation and activation of CD44 by natural killer (NK) cells and its role in the production of IFN-gamma. J Interferon Cytokine Res **24**, 301–309.

Saha, A., Chakrabarti, G., Sen, S. and Bandyopadhyay, S. (1999). *Leishmania donovani* parasites interact with gamma/delta+ human peripheral blood T cells and induce susceptibility to NK cell-mediated lysis. Scand J Immunol **50**, 588–595.

Sanabria, M.X.H., Vargas-Inchaustegui, D.A., Xin, L. and Soong, L. (2008). Role of natural killer cells in modulating dendritic cell responses to *Leishmania amazonensis* infection 10.1128/IAI.00438-08. Infect Immun **76**, 5100–5109.

Santarem, N., Silvestre, R., Tavares, J., Silva, M., Cabral, S., Maciel, J. and Cordeiro-da-Silva, A. (2007). Immune response regulation by leishmania secreted and nonsecreted antigens. J Biomed Biotechnol **2007**, 85154.

Sardinha, L.R., Elias, R.M., Mosca, T., Bastos, K.R.B., Marinho, C.R.F., D'Imperio Lima, M.R. and Alvarez, J. M. (2006). Contribution of NK, NK T, {gamma}{delta} T, and {alpha}{beta} T cells to the gamma interferon response required for liver protection against *Trypanosoma cruzi*. Infect Immun **74**, 2031–2042.

Scharton, T.M. and Scott, P. (1993). Natural killer cells are a source of interferon gamma that drives differentiation of CD4+ T cell subsets and induces early resistance to *Leishmania major* in mice. J Exp Med **178**, 567–5677.

Schleicher, U., Liese, J., Knippertz, I., Kurzmann, C., Hesse, A., Heit, A., Fischer, J.A., Weiss, S., Kalinke, U., Kunz, S. and Bogdan, C. (2007). NK cell activation in visceral leishmaniasis requires TLR9, myeloid DCs, and IL-12, but is independent of plasmacytoid DCs. J Exp Med **204**, 893–906.

Schofield, L. (2007). Intravascular infiltrates and organ-specific inflammation in malaria pathogenesis. Immunol Cell Biol **85**, 130–137.

Shaw, J. (2007). The leishmaniases—survival and expansion in a changing world. A mini-review. Mem Inst Oswaldo Cruz **102**, 541–547.

Spencer, L., Shultz, L. and Rajan, T.V. (2003). T cells are required for host protection against *Brugia malayi* but need not produce or respond to interleukin-4. Infect Immun **71**, 3097–3106.

Stanley, A.C. and Engwerda, C.R. (2007). Balancing immunity and pathology in visceral leishmaniasis. Immunol Cell Biol **85**, 138–147.

Stevenson, M.M. and Riley, E.M. (2004). Innate immunity to malaria. Nat Rev Immunol **4**, 169–180.

Tato, C.M., Mason, N., Artis, D., Shapira, S., Caamano, J.C., Bream, J.H., Liou, H.C. and Hunter, C.A. (2006). Opposing roles of NF-kappaB family members in the regulation of NK cell proliferation and production of IFN-gamma. Int Immunol **18**, 505–513.

Teixeira-Carvalho, A., Fujiwara, R.T., Stemmy, E.J., Olive, D., Damsker, J. M., Loukas, A., Correa-Oliveira, R., Constant, S.L. and Bethony, J.M. (2008). Binding of excreted and/or secreted products of adult hookworms to human NK cells in *Necator americanus*-infected individuals from Brazil 10.1128/IAI.00419-08. Infect Immun **76**, 5810–5816.

Tripathi, P., Singh, V. and Naik, S. (2007). Immune response to leishmania: paradox rather than paradigm. FEMS Immunol Med Microbiol **51**, 229–242.

UNICEF/WHO (2003). *Africa Malaria Report 2003*, Vol. 2009.

Vitelli-Avelar, D.M., Sathler-Avelar, R., Massara, R.L., Borges, J.D., Lage, P.S., Lana, M., Teixeira-Carvalho, A., Dias, J. C., Eloi-Santos, S.M. and Martins-Filho, O.A. (2006). Are increased frequency of macrophage-like and natural killer (NK) cells, together with high levels of NKT and CD4 + CD25high T cells balancing activated CD8+ T cells, the key to control Chagas' disease morbidity? Clin Exp Immunol **145**, 81–92.

Vivier, E., Nunes, J.A. and Vely, F. (2004). Natural killer cell signaling pathways. Science **306**, 1517–1519.

Walzer, T., Dalod, M., Robbins, S.H., Zitvogel, L. and Vivier, E. (2005). Natural-killer cells and dendritic cells: "l'union fait la force". Blood **106**, 2252–2258.

WHO (2004). The World Health Organization report 2004. In: *Changing History*, Vol. 2008.

Chapter Forty-Six

46

Lymphoproliferative disorders of natural killer cells

William G. Morice

CHAPTER CONTENTS

The more I read, the more I meditate; and the more I acquire, the more I am enabled to affirm that I know nothing.

Voltaire

ABSTRACT

Natural killer (NK) cell lineage lymphoproliferative disorders are, as a group, more recently described entities, particularly when compared to those of T and B cells. This in part is a reflection of the rarity of the disorders; however, it is also attributable to the challenges in establishing a diagnosis of NK cell malignancy in the clinical laboratory. These challenges vary depending on the nature of the NK cell disorder itself. For indolent processes it can be difficult to determine if the NK cells identified are truly abnormal, whereas in aggressive neoplasms it can be difficult to confidently ascribe NK cell lineage. In spite of such difficulties, a number of bona fide lymphoproliferative disorders of NK cells have been characterized. A total of four NK cell lymphoproliferative disorders (chronic NK cell lymphoproliferative disorder, aggressive NK cell leukaemia, extranasal NK/T-cell lymphoma, extranodal NK/T-cell lymphoma, nasal type) and reactive NK cell lymphocytosis can be distinguished.

KEY WORDS

NK cell leukaemia, NK lymphoma, Lymphoproliferative disorder, Chronic NK cell lymphoproliferative disorder, Aggressive NK cell leukaemia, Extranasal NK/T cell lymphoma, Extranodal NK/T cell lymphoma, Nasal type, Reactive NK cell lymphocytosis

Introduction

Just as natural killer (NK) cells are a relatively recently characterized lymphocyte type, lymphoproliferative disorders of these cells are, as a group, recently described entities. The establishment of NK cells as a discrete lymphocyte lineage was difficult due to the lack of an NK cell–defining antigen (such as CD3 for T cells and CD19 for B cells) or genetic event (such as rearrangement of an antigen receptor gene) as well as the functional and phenotypic similarities between NK cells and cytotoxic T cells (Caligiuri, 2008). These same factors continue to pose obstacles in diagnosing NK cell lymphoproliferative disorders in medical practice.

In the clinical laboratory, NK cells are identified and distinguished from cytotoxic T cells by an amalgam of features that are present, such as surface expression of CD16 and CD56, and those that are absent, such as lack of productively rearranged T cell receptor genes and thereby lack of a fully assembled CD3-T cell receptor complex (Morice, 2007). In this setting,

NK cell lineage is most readily ascribed by flow cytometric immunophenotyping (FCIP), which allows for simultaneous assessment of numerous antigens on single cells in fluidic suspension. Identification of NK cells in paraffin-embedded materials, as is often required in hematopathology practice, is more problematic as there are a limited number of paraffin reactive antibodies to NK-associated antigens, the direct measurement of antigen co-expression by individual cells is difficult, and the antibodies used for CD3 detection recognize single subunits of the complex, such as CD3 epsilon, which are common to both T and NK cells (Chan et al., 1995, 1996). NKp46 has, however, emerged as a characteristic useful in identifying NK cells in mouse and human (Walzer et al., 2007). Furthermore, a characteristic of both T and NK cell malignancies is loss or gain of defined cell-surface molecules and the resulting phenotypic aberrancies can make it even more difficult to confidently distinguish these cells. Lastly, establishing clonality is used as a primary diagnostic criterion for many hematopoietic neoplasms, particularly those that are less aggressive (i.e. low histologic grade) and therefore closely resemble their non-neoplastic polyclonal counterparts. The lack of a uniquely rearranged antigen receptor gene in NK cells that may be used to detect clonal cellular expansion has served as a stumbling block in establishing the presence of NK cell neoplasms.

Despite this wide array of confounding factors, a number of bona fide NK cell lymphoproliferative disorders have been identified. Overall, these disorders are rare, especially in the Western hemisphere, and there are relatively few recognized subtypes (Swerdlow et al., 2008). The currently recognized NK cell lymphoproliferative disorders span the gamut of clinical behaviour, ranging from indolent processes usually associated with minimal morbidity to aggressive malignancies with rapid unremitting progression and mortality. From the clinical laboratory perspective these individual NK-cell disorders pose a different set of diagnostic challenges. On the one hand, in indolent NK cell lymphoproliferative disorders, one must distinguish the processes from reactive NK cell expansions. On the other hand, in aggressive NK cell neoplasms, one must often make the diagnosis using paraffin embedded tissues and therefore it can be difficult to ascribe NK cell lineage in these clearly malignant processes.

What follows is a discussion of the NK cell neoplasms, which are currently recognized by the World Health Organization (Swerdlow et al., 2008). The salient clinical and laboratory features of these neoplasms are summarized in Table 46.1. In addition a variety of hematopoietic neoplasms, which have previously been misidentified as being NK-cell derived due both to misunderstanding of NK cell biology and the neoplasms themselves will be mentioned.

WHO recognized NK cell neoplasms

Chronic NK cell lymphoproliferative disorder

Chronic NK cell lymphoproliferative disorder is characterized by a persistent increase in circulating NK cells. Microscopically, the cells are cytologically unremarkable with features typical of normal large granular lymphocytes (LGLs, Figure 46.1) (Morice et al., 2003). In keeping with this bland cytology, this disorder usually presents clinically either as an isolated laboratory finding with no associated signs or symptoms or with unexplained anaemia and/or neutropenia that are often quite mild. A minority of patients may have splenomegaly or more pronounced symptomatic cytopenias. Chronic NK cell lymphoproliferative disorder is typically found in adult patients but it has no recognized predilection for a particular age group, ethnic group, or gender. In addition, the disorder is not known to be specifically associated with any other diseases, although some cases may be encountered in patients with other hematopoietic neoplasms such as Hodgkin's lymphoma or multiple myeloma (Neben et al., 2003).

Since the features of chronic NK cell lymphoproliferative disorder are not typical of most hematolymphoid malignancies, it has previously been ascribed a number of descriptive monikers, including chronic NK cell lymphocytosis, NK cell lymphoproliferative disorder of granular lymphocytes, and large granular lymphocytic leukaemia of NK cell type (or NK-LGL) (Epling-Burnette et al., 2004; Loughran, 1993; Tefferi, 1996). This previously used descriptive terminology, as well as the currently accepted WHO designation, reflects a reluctance to label these minimally symptomatic NK cell expansions as leukaemic in nature. Indeed, NK-LGL, the last appellation listed above, was particularly problematic, as in some instances it was used only when the process was overtly malignant and therefore many of such cases likely represented peripheral blood involvement by other aggressive NK cell neoplasms. Studies of X-linked gene inactivation in chronic NK cell lymphocytosis patients have established the clonal nature of these usually indolent NK cell expansions and given this, as well as its distinct laboratory features, this entity is now recognized by the WHO (Swerdlow et al., 2008; Tefferi et al., 1992).

In the clinical laboratory chronic NK cell lymphoproliferative disorder is effectively recognized by FCIP (Hoffmann et al., 2000; Moricc ct al., 2003; Morice et al., 2001; Pascal et al., 2004; Warren et al., 2003; Zambello et al., 1993, 2003). When performing these analyses with antibodies to antigens common to T and

Table 46.1 Comparison of the clinicopathologic features of reactive NK cell lymphocytosis with the three NK cell lymphoproliferative disorders recognized by the World Health Organization

	Reactive NK cell expansion	Chronic NK cell lymphoproliferative disorder (CNKLPD)	Aggressive NK cell leukaemia (ANKL)	Extranodal NK/T cell lymphoma (ENK/TCL)
Systemic sign and symptoms	No	No	Yes	Rare
Cytopenias	No	Often	Yes	No
EBV associated		No	Yes	Yes
Hemophagocytic syndrome associated	No	No	Yes	Yes
Cytologic atypia	None	None or minimal	Varies, often moderate	Varies, often moderate
Cytogenetics	Normal	Normal	Abnormal, −6q23 frequent	Abnormal, −6q23 frequent
NKR expression[1]	CD94 positive Polytypic KIR	KIR restricted or absent CD94/NKG2A bright	Unknown	KIR restricted 50% CD94 Pos
Peripheral blood involvement	Yes	Yes	Often	Rare
Bone marrow involvement	No	Yes	Often	Rare
Spleen involvement	No	Often	Often	No
Lymph node involvement	No	No	Often	Rare
Nasopharyngeal involvement	No	No	No	Often
Testis involvement	No	No	No	Yes

[1] Expression of NK cell receptors such as CD94 and KIR.

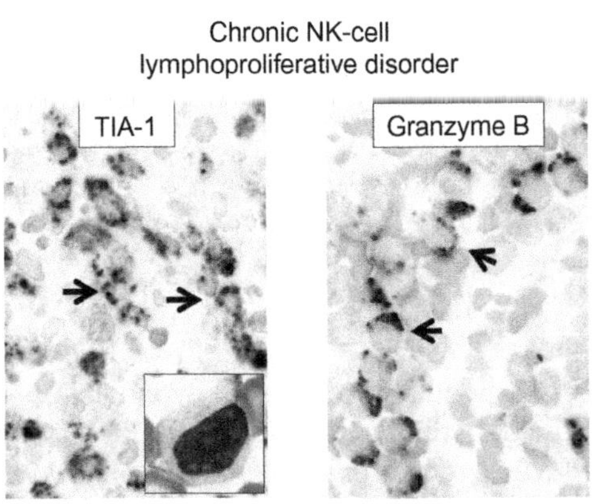

Figure 46.1 • Chronic NK cell lymphoproliferative disorder, histology. The NK cells in the peripheral blood (inset left) have bland cytologic features and are indistinguishable from normal LGLs. Immunohistochemistry in chronic NK cell lymphoproliferative disorder performed on a bone marrow biopsy with antibodies to TIA-1 (also known as GMP-17, left) and on a splenectomy specimen with antibodies to granzyme B (right) reveals intrasinusoidal infiltration of both organs by lymphocytes positive for these cytotoxic granule proteins (arrows). The presence of the antigen positive cells within vascular structures leads to their linear arrangement. Note also the weaker, but distinct staining of granulocytic cells for TIA-1 in the left panel.

NK cells, one usually detects a phenotypically homogeneous CD3-negative NK cell expansion, which is positive for CD2 and CD7 (Figure 46.2). A number of phenotypic aberrancies for these antigens may be seen (in comparison with normal NK cells) such as diminished expression of CD2 and/or CD7, uniform CD8 positivity and coexpression of CD5 (Morice, 2007). Distinctive expression patterns of the classical NK cell–associated antigens CD16 and CD56 are also typical of chronic NK cell lymphoproliferative disorder with bright, uniform expression of the former and markedly diminished-to-absent expression of the latter both often present (Figure 46.2). Decreased expression of CD57 and CD161 (in comparison with normal circulating NK cells) may also be seen.

Greater insight into the homogeneity of chronic NK cell lymphoproliferative disorder is provided by analysis of NK cell receptor (NKR) expression, particularly CD94/NKG2 heterodimeric complexes and the KIR antigens CD158a, CD158b and CD158e (Figure 46.3). Two distinctive phenotypes may be seen when studying these receptors in chronic NK cell lymphoproliferative disorder. One of these phenotypes is characterized by

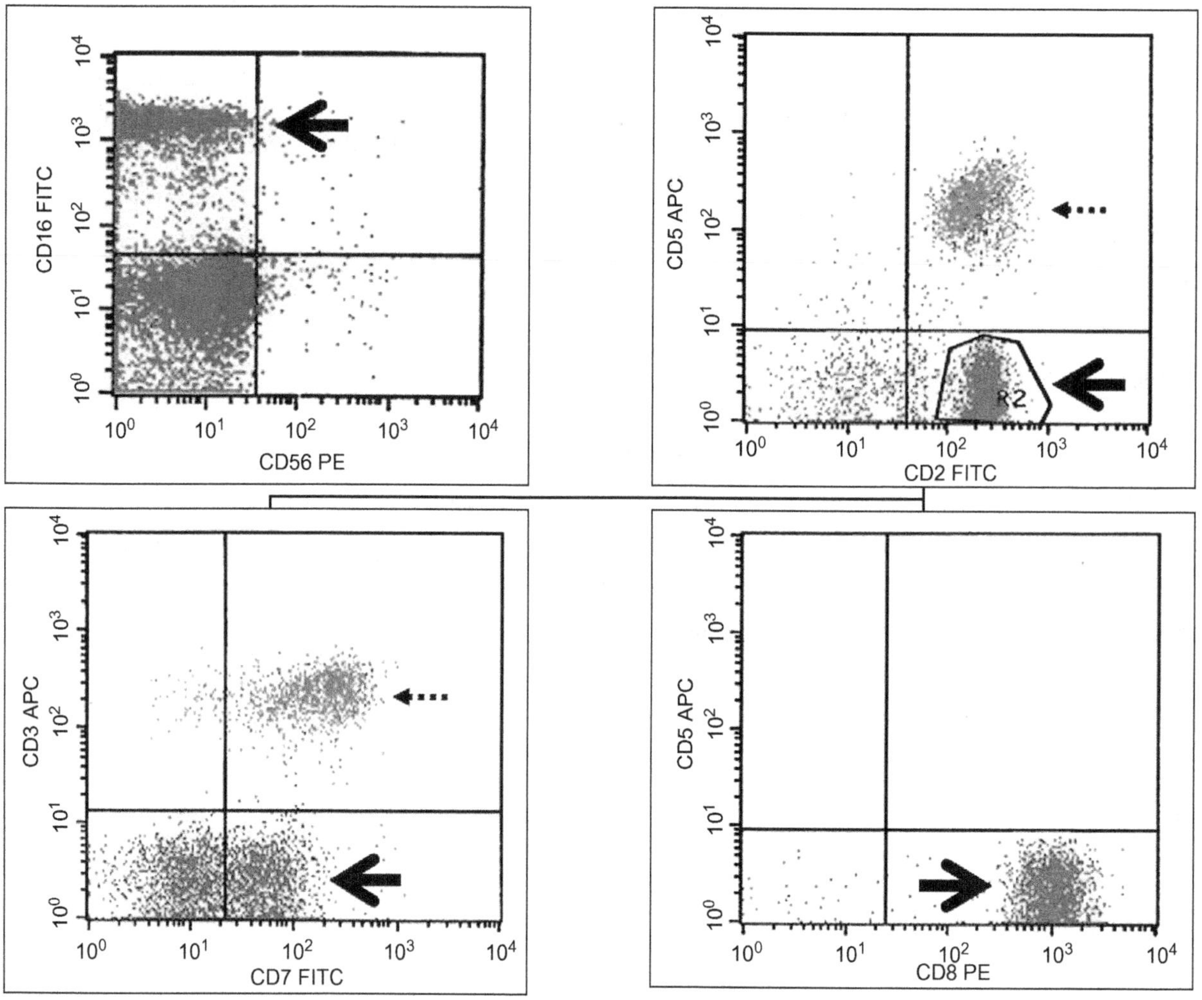

Figure 46.2 • Chronic NK cell lymphoproliferative disorder, FCIP. Flow cytometry performed on a peripheral blood specimen in a chronic NK cell lymphoproliferative disorder patient reveals the presence of a brightly CD16 positive NK cell population with dim-to-absent CD56 expression (upper left histogram, arrow). These NK cells are CD2 positive and CD5 negative (upper right histogram, solid arrow, cells highlighted in green); selective gating on these NK cells (gate R2) reveals they are CD3-negative and have aberrantly diminished expression of CD7 (lower left histogram, solid arrow, green) and that they have uniform, bright surface CD8 expression (lower right, solid arrow, green). This pattern of CD8 expression is aberrant when compared to normal NK cells. T-cells (pink, dashed arrows) are shown for comparison in the upper right and lower left histograms.

complete lack of the commonly expressed KIRs and uniform, bright expression of CD94/NKG2A heterodimers. The other is characterized by uniform expression of a single (or less often two) KIR antigens, with or without coexpression of the CD94/NKG2A heterodimer (Epling-Burnette et al., 2004; Hoffmann et al., 2000; Morice et al., 2003; Pascal et al., 2004; Warren et al., 2003; Zambello et al., 2003). In some peripheral blood NK cell expansions, the pattern of KIR expression deviates from normal but these are not pathognomonic of chronic NK cell lymphoproliferative disorders. In such cases it may be difficult to discern if this truly represents chronic NK cell lymphoproliferative disorder or a phenotypically distinct reactive NK cell expansion. When such cases are encountered, correlation with other clinical and laboratory features is imperative. If expansion of this NK cell population persists beyond 1 year without an identifiable stimulant of innate immunity, then the WHO criteria for chronic NK cell lymphoproliferative disorder are satisfied and the process should be labelled as such (Swerdlow et al., 2008).

FCIP analyses in chronic NK cell lymphoproliferative disorder may be performed on peripheral blood or bone marrow aspirates and the phenotype of the NK cells is usually stable over time (Morice et al., 2003). In mature NK cells, the KIR expression patterns are stable over multiple cell generations. The observed patterns of KIR expression provide indirect evidence to support the notion that these represent clonal NK cell cellular expansions. In the bone marrow, discrete intrasinusoidal

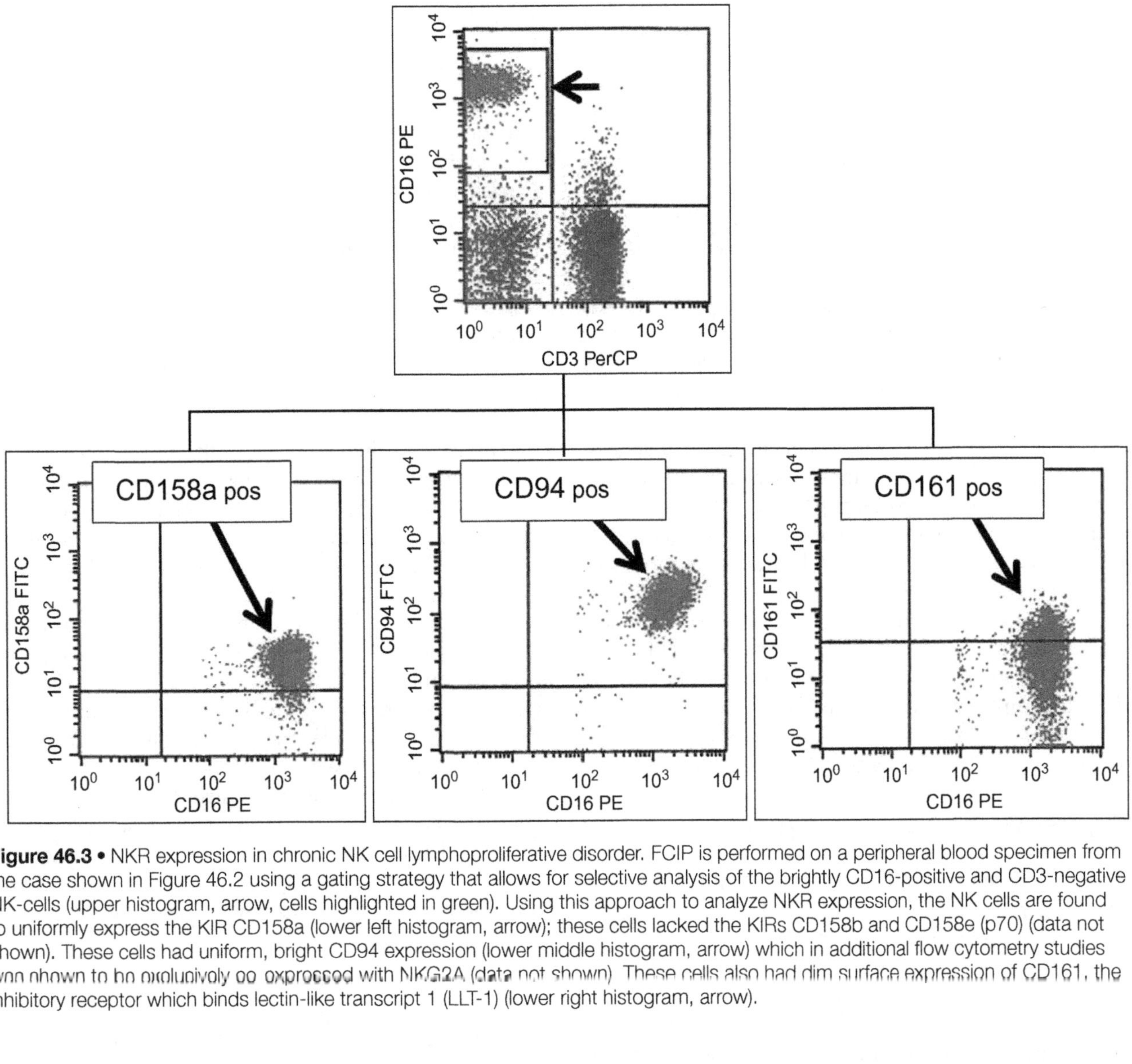

Figure 46.3 • NKR expression in chronic NK cell lymphoproliferative disorder. FCIP is performed on a peripheral blood specimen from the case shown in Figure 46.2 using a gating strategy that allows for selective analysis of the brightly CD16-positive and CD3-negative NK-cells (upper histogram, arrow, cells highlighted in green). Using this approach to analyze NKR expression, the NK cells are found to uniformly express the KIR CD158a (lower left histogram, arrow); these cells lacked the KIRs CD158b and CD158e (p70) (data not shown). These cells had uniform, bright CD94 expression (lower middle histogram, arrow) which in additional flow cytometry studies were shown to be exclusively co-expressed with NKG2A (data not shown). These cells also had dim surface expression of CD161, the inhibitory receptor which binds lectin-like transcript 1 (LLT-1) (lower right histogram, arrow).

NK cell infiltrates may be detected by immunohistochemistry with antibodies to cytolytic granule proteins such as TIA-1 and granzymes B and M (Figure 46.1) (Morice et al., 2007). It is noteworthy, however, that these studies do not allow distinction of cytotoxic T cells and NK cells and therefore one cannot exclude an alternative diagnosis of T-cell large granular lymphocytic leukaemia. This is a clinically similar lymphoproliferative disorder of cytotoxic T cells that also manifests with intrasinusoidal marrow infiltrates by immunohistochemistry (Morice et al., 2002). Making this distinction requires correlation with the results of FCIP.

Functional and genotypic studies in a group of 18 chronic NK cell lymphoproliferative disorder patients suggested that the NK cells preferentially expressed activating KIR isoforms (Zambello et al., 2003). However, while a subsequent study by these authors confirmed frequent group B KIR haplotypes in their chronic NK cell lymphoproliferative disorder patients, this frequency did not differ significantly from an appropriate population control group (Scquizzato et al., 2007). In many KIR-positive chronic NK cell lymphoproliferative cases, the cognate HLA ligand for the expressed KIR was not present. Taking into consideration all of the available data, it appears that NKR expression may be a factor in disease development, although the precise role remains unclear. In most, if not all, cases that meet the recognized criteria for chronic NK cell lymphoproliferative disorder, no cytogenetic abnormalities are seen by either conventional metaphase or fluorescent in situ hybridization (FISH) analyses.

Although the pathogenesis of chronic NK cell lymphoproliferative disorder is unknown, recent developments in our understanding of NK cell biology may shed some light on the nature of this process. The phenotype of chronic NK cell lymphoproliferative disorder with diminished expression of CD56 and high levels of expression of CD16, cytolytic effector proteins, and inhibitory NKRs is in essence a caricature of the terminally differentiated effector NK cells that constitute the bulk normal peripheral blood NK compartment (Cooper et al., 2001; Jacobs et al., 2001). It is now apparent that these normal CD16 bright, CD56 dim circulating NK cells are derived from the CD56 bright NK cell compartment following stimulation and exposure to cytokines such as IL-2 and IL-15 (Romagnani et al., 2007). The phenotypic attributes of chronic NK-cell lymphoproliferative disorder suggest that this may represent an accumulation of a homogeneous NK cell population driven by some normal stimulant of innate immunity. This could indeed account for frequent lack of clinical sequelae or recurrent cytogenetic abnormalities. Indeed, a separate study of X-linked gene expression in 6 female patients with persistent increases in circulating NK cells have failed to reveal evidence of clonality and examples of NK cell expansions homogenously expressing a single KIR isoform that have spontaneously resolved without therapeutic intervention have also been described (Nash et al., 1993; Zambello et al., 1997). Data such as this further the notion that chronic NK cell lymphoproliferative disorder may, at least in some instances, represent a distorted physiologic NK cell expansion rather than a true NK cell neoplasm.

Aggressive NK cell leukaemia

In contrast to chronic NK cell lymphoproliferative disorder, the remainder of the NK cell lineage lymphoproliferative disorders to be discussed exhibit frankly malignant clinical behaviour. Certainly aggressive NK cell leukaemia is no exception to this rule (Fernandez et al., 1986; Imamura et al., 1990; Sheridan et al., 1988). This extraordinarily rare disorder is most often encountered in young adults (variably defined, in this instance applying to those younger than 40 years old), it may have a slight male predominance (Cheung et al., 2003). Aggressive NK cell leukaemia is more common in Asians (Song et al., 2002). In this authors' experience, it can also be encountered in females during or shortly after pregnancy.

The clinical presentation of aggressive NK cell leukaemia is usually dramatic, characterized by a severe febrile illness with hepatosplenomegaly, lymph node enlargement, and cytopenia (anaemia, neutropenia and thrombocytopenia) (Kwong, 2005). A subset of cases is associated with hemophagocytic syndrome, a condition causing life threatening consumption of peripheral blood elements (Chan et al., 1997; Cheung et al., 2003).

Although called aggressive NK cell 'leukaemia' in some cases tissue infiltrates predominate and there is relatively minimal peripheral blood and/or bone marrow involvement (Mori et al., 2000). Extramedullary sites that may be involved by aggressive NK cell leukaemia include the spleen, liver and lymph nodes. This varied disease distribution, which resembles that seen in HTLV-1 associated adult T-cell leukaemia/lymphoma, was recognized in early descriptions of this disease entity as aggressive NK cell leukaemia/lymphoma (Imamura et al., 1990). However, in the WHO classification scheme, the term aggressive NK cell leukaemia was adopted to emphasize the typical widespread disease at presentation and to distinguish it from extranodal NK/T cell lymphoma (discussed next), which has overlapping pathologic features but usually presents with localized tissue involvement (Quintanilla-Martinez and Jaffe, 2000).

Regardless of the peripheral blood or bone marrow disease burden the neoplastic cells of aggressive NK cell leukaemia often can be recognized by their cytologic atypia manifest by an increased nuclear-to-cytoplasm ratio, open chromatin and visible nucleoli (Figure 46.4). In paraffin tissue sections the cells also usually demonstrate a malignant histologic appearance. In some cases however the neoplastic cells both in stained smears and in tissue sections have bland features with small, unremarkable nuclei (Chinen et al., 2002). Other histologic changes seen in aggressive NK cell leukaemia paraffin tissue sections include angioinvasion and necrosis; the latter may be massive.

Given the rarity of aggressive NK cell leukaemia phenotyping data is limited. This is particularly true of antigens that can be assessed only by flow cytometry, as such studies may not be performed in those cases with minimal blood or bone marrow involvement. Aggressive

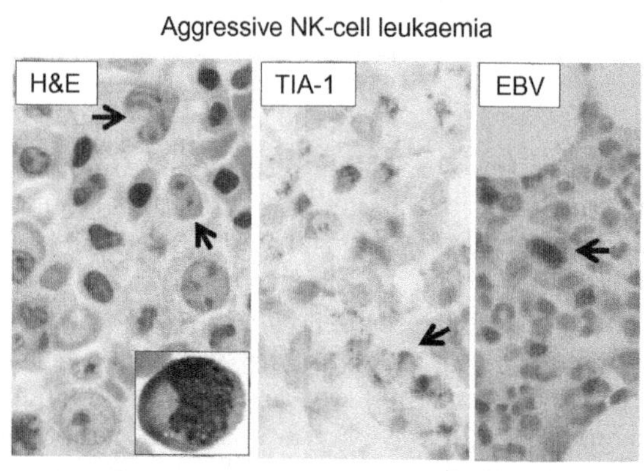

Figure 46.4 • Aggressive NK cell leukaemia, histology. Histologic sections of a bone marrow biopsy from an aggressive NK cell leukaemia case reveals infiltration by cells with large nuclei, open chromatin, and irregular nuclear contours (left panel, arrows). Cytologically atypical lymphoid cells with hyperchromatic, irregular nuclei and faint cytoplasmic granules were also present in the peripheral blood (left panel inset). Immunoperoxidase staining of the bone marrow biopsy demonstrates that the neoplastic cells are positive for the cytotoxic granule protein TIA-1 (middle panel); these cells were also shown to be EBV positive by in situ hybridization studies (right panel).

NK cell leukaemia is considered to be of NK cell lineage because the cases are universally positive for CD2, CD7, CD56, HLA-DR and cytolytic effector proteins and negative for T-cell receptor gene rearrangements or a fully assembled CD3 complex (Cheung et al., 2003; Suzuki et al., 2004). The neoplastic cells of aggressive NK cell leukaemia cases are almost always positive for Epstein–Barr virus (EBV). This may be detected in Southern blot analysis or in paraffin tissue sections by either in situ hybridization with probes to viral RNA or immunohistochemistry with antibodies to viral proteins (Figure 46.4). The reported frequency of CD16 expression in aggressive NK cell leukaemia varies; it appears that it is expressed in the majority of cases. There is relatively limited information regarding CD3 epsilon expression, although some cases may be positive. which can in turn lead to immunoreactivity with CD3 antibodies in paraffin immunohistochemistry (Suzuki et al., 2004). Antigens commonly associated with cytotoxic T-cell lineage such as CD5 and CD8 are usually lacking. The expression status of NKRs such as KIRs in aggressive NK cell leukaemia is unknown.

Cytogenetic metaphase analysis reveals the presence of clonal karyotypic abnormalities in most cases. A variety of chromosomal abnormalities have been described in this disorder, including abnormalities of the long and short arms of chromosome 6, abnormalities of chromosome 7 and gain of chromosome X (Siu et al., 1999; Wong et al., 2000). None of these genetic abnormalities are specific for aggressive NK cell leukaemia, however. The prognosis in this disease is regrettably dismal with most patients dying within days to weeks of diagnosis.

Extranodal NK/T cell lymphoma, nasal type

Extranodal NK/T cell lymphoma, nasal type is a tissue-based malignancy of cytotoxic lymphocytes. This disease is uncommon, although not so rare as aggressive NK cell leukaemia. Like aggressive NK cell leukaemia, extranodal NK/T cell lymphoma is most prevalent in Asians and related populations in Central and South America (Altemani et al., 2002; Kwong, 2005; Quintanilla-Martinez et al., 1999; Rudiger et al., 2002; Siu et al., 2002a). Sporadic cases also occur in North America and Europe, although at least some of these cases arise in individuals of Asian ancestry (Gaal et al., 2000). Extranodal NK/T cell lymphoma is a disease of adulthood with a median age distribution in the fifth decade; it appears to be more common in males than females.

Extranodal NK/T cell lymphoma of nasal type rarely involves the lymph nodes at diagnosis. Rather, this disease has a strong proclivity to involve the nasopharynx and was first described in this site. This unusual feature was reflected in the nomenclature when this entity was first recognized to be a lymphoid malignancy and is so strongly associated with the entity that the terminology persists in the current WHO nosologic classification (Swerdlow et al., 2008). In spite of the name, extranasal sites may be involved by extranodal NK/T cell lymphoma either as primary sites or by dissemination of nasal disease. Sites of extranasal involvement by this NK/T cell lymphoma include the skin, testis, gastrointestinal tract (GI), lower aerodigestive tract (lungs), liver and soft tissues (Chan et al., 1997; Cheung et al., 1998; Kuwabara et al., 2003; Totonchi et al., 2002; Tsao et al., 2004). Despite the name, this lymphoma may involve the lymph nodes particularly with disease relapse.

When detected early, the presenting symptoms in extranodal NK/T cell lymphoma are usually minimal and related to local mass effect, such as nasal obstruction and recurrent sinusitis in patients with nasopharyngeal tumours. This is an invasive neoplasm, however, and even apparently well-localized nasal tumours frequently have radiologic evidence of destruction of bony facial structures and extension into paranasal sinuses and regional soft tissues (Ooi et al., 2000). Unabated growth of extranodal NK/T cell lymphoma involving the nasopharynx will eventually lead to severe midface destruction with collapse of the nose and nasal cavity and/or erosion into the oropharynx; leading the entity to be called 'lethal midline granuloma' in many early descriptions (Eichel et al., 1966). In extranasal sites, extranodal NK/T cell lymphoma forms a tumoral mass which may be associated with ulceration of the overlying skin or, in the abdomen, perforation of the GI tract (Cheung et al., 2003).

Extranodal NK/T cell lymphoma involving both the nasopharynx and extranasal sites can exhibit a variety of cytologic features ranging from a monomorphous cellular infiltrate with minimal cytologic atypia to a frankly malignant infiltrate with pleomorphic nuclear features (Figure 46.5) (Chinen et al., 2002; Siu et al., 2002a; Totonchi et al., 2002). Prior to the advent of advanced immunophenotyping and molecular genetic testing modalities, a diagnosis of non-Hodgkin's lymphoma was based primarily on the presence of a homogeneous lymphoid infiltrate in a lymphoid organ. Taking into account these diagnostic parameters and varied histologies and unusual location, the precise nature of extranodal NK/T cell lymphoma was unclear and descriptive histologic diagnoses such as 'polymorphic reticulosis' were used (Strickler et al., 1994). Ulceration of overlying mucosa and invasion of vascular structures by the neoplastic cells are also commonly seen in extranodal NK/T cell lymphoma (Figure 46.5). In fact, 'angiocentric lymphoma' was another name suggested for this malignancy, although this soon fell into disfavour as angioinvasion is not specific for this lymphoma type (Harris et al., 1994). Coagulative tissue necrosis is also frequently seen

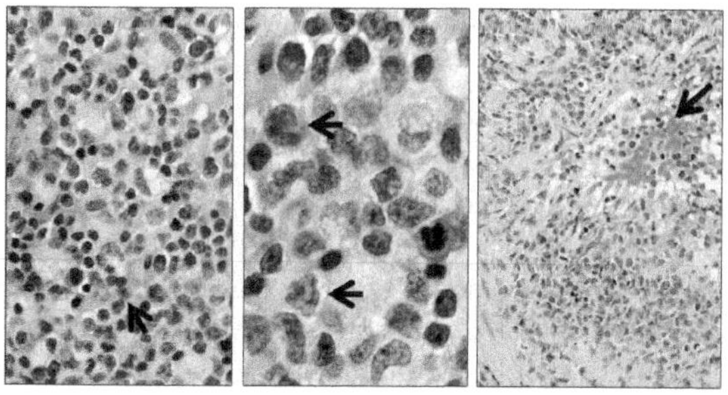

Figure 46.5 • Extranodal NK/T cell lymphoma, histology. Examples of cases with minimal cytologic atypia (left panel) and marked cytologic atypia (middle panel) are shown with arrows to help highlight the neoplastic cells. An example of vascular invasion is shown in the right panel, the neoplastic cells are seen to splay the vascular wall smooth muscle. The vessel lumen is indicated by the arrow.

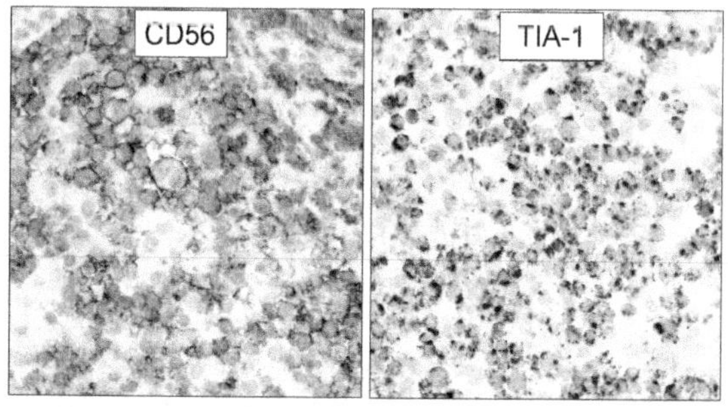

Figure 46.6 • Extranodal NK/T cell lymphoma, immunohistochemistry. Immunoperoxidase staining of paraffin sections from an extranodal NK/T cell lymphoma reveals the neoplastic cells to be positive for CD56 (left panel) and the cytotoxic granule protein TIA-1.

in extranodal NK/T cell lymphoma. This may be attributable to either compromise of vascular structures due to angioinvasion or elaboration of cytokines such as MIG and IP-10 by the EBV-infected neoplastic cells (Teruya-Feldstein et al., 1997). It is noteworthy, however, that neither angioinvasion nor coagulative necrosis may be present in all biopsy specimens. Immunophenotyping data regarding extranodal NK/T cell lymphoma is relatively limited. Most published studies have been limited to immunohistochemical analysis since the typical sites of involvement usually lead to the procurement of small biopsy specimens with insufficient material for FCIP. This neoplasm was first postulated to be of NK cell lineage since it contained azurophilic cytoplasmic granules, expressed CD2 and CD56, and lacked either expression of fully assembled CD3 complex or detectable T cell antigen receptor gene rearrangements (Figure 46.6) (Jaffe et al., 1996; Ng et al., 1987; Wong et al., 1992). Subsequent studies confirmed these initial observations and provided further evidence to support the notion that the neoplasm was of NK cell lineage through documenting lack of CD5 expression and frequent expression of CD7 and the cytotoxic granule protein TIA-1 (Figure 46.6) (Chan et al., 1997; Emile et al., 1996). These studies also demonstrated that while the malignant cells lacked a fully assembled CD3 complex, they frequently expressed the CD3 epsilon and zeta subunits. Extranodal NK/T cell lymphoma is also positive for the cytotoxic granule proteins granzyme B, granzyme M and perforin. By frozen section immunohistochemistry, however, it appears to lack expression of the NK associated antigens CD16 and CD57 (Krenacs et al., 2003; Schwartz et al., 2008). Like aggressive NK cell leukaemia, extranodal NK/T cell lymphoma is strongly associated with EBV, with over 75% of cases being EBV positive and cases of this EBV-associated neoplasm arising in the setting of immunosuppression have been described (Canioni et al., 2001; Tsao et al., 2004).

The somewhat ambiguous terminology 'NK/T cell' lymphoma was first suggested for this neoplasm due to a reluctance to definitively ascribe cell lineage given the relatively small number of cases and paucity of available immunophenotypic and molecular genetic data. As noted above, the results of seminal studies of this entity were most consistent with the majority of cases being of NK cell lineage. However, these early study groups contained isolated cases that appeared to be of T cell lineage (Kanavaros et al., 1993). Also, during the time frame in which NK/T cell lymphoma was being characterized, other malignancies of cytotoxic lymphocytes that appeared to be derived from T cells with innate immune function, and hence functional and phenotypic overlap with NK cells, were being reported in the literature (Macon et al., 1996). With the more widespread recognition of extranodal NK/T lymphoma and the advent of more sensitive PCR techniques for T cell receptor gene analysis it has become apparent that in a minor subset of cases, the neoplastic cells are of T cell lineage (Chiang et al., 1997; Ng et al., 2004). These 'T cell-type' extranodal NK/T cell lymphomas are primarily recognized through the detection of clonal T cell receptor gene rearrangements. In review of the literature, they appear to constitute approximately 25% of all extranodal NK/T cell lymphoma cases. The immunophenotype of these T cell lineage extranodal NK/T cell lymphomas is virtually identical to that of the NK cell lineage neoplasms, suggesting that they may be derived from T cells with innate immune function. Like true NK cell extranodal NK/T cell lymphoma, the 'T cell-type' is also strongly associated with EBV. Furthermore, the clinical features of the T and NK cell lineage NK/T cell lymphomas are virtually indistinguishable. Given the nearly identical clinical and laboratory features between T and NK cell lineage extranodal NK/T cell lymphoma they are considered as subtypes of a singular disease entity for the purpose of diagnostic classification. Of note, the disease is often designated as 'extranasal' NK/T cell lymphoma when arising in an organ/site other than the nasopharynx.

There is very limited published data regarding NKR expression in extranodal NK/T cell lymphoma, largely because there are few antibody reagents that work in paraffin embedded tissues. Lin and colleagues have attempted to circumvent this shortcoming through the development of a PCR primers to study NKR transcripts in paraffin embedded extranodal NK/T cell lymphoma cases (Lin et al., 2001, 2003). The methodology they employed uses primers spanning specific joining regions of KIR immunoglobulin-like domains for the detection of 2D, $2D_{L4}$ and 3D KIR isoforms as well as primers to CD94 and NKG2A, C, D and F. By this approach the authors have demonstrated that extranodal NK/T cell lymphoma is universally KIR positive (both NK and T cell lineage) with approximately two thirds having a single detectable isoform. These studies also revealed that approximately one-half of the cases were positive for CD94, in most instances apparently paired with NKG2A. Interestingly, correlation with clinical features and follow up suggested that CD94 positivity was associated with an improved outcome. These studies have yet to be corroborated by other groups however.

The cytogenetics of extranodal NK/T cell lymphoma have been studied by a variety of methods including conventional metaphase analysis, spectral karyotyping, comparative genomic hybridization, and loss of heterozygosity analysis (Siu et al., 1999, 2000; Wong et al., 2000). Although the number of studied cases is small and there are typically numerous, complex cytogenetic abnormalities there are certain abnormalities which are often present in extranodal NK/T cell lymphoma. Recurrent genetic abnormalities described include deletions involving the long arm of chromosome 6 in the q23 region and gain of chromosome X. Deletions of the long arm of chromosome 13 (13q) also are relatively common, particularly in relapsed or spread nasal disease or de novo extranasal disease. Other abnormalities described with somewhat lesser frequency include deletions of the 17p and 11q. At the subgenomic level, methylation and inactivation of the promoter for p73, a protein structurally and functionally related to p53, which functions as a cell cycle regulator and may be a tumour suppressor, is common in extranodal NK/T cell lymphoma (Siu et al., 2002b).

In regards to therapy, extranodal NK/T cell lymphoma is much more sensitive to high dose localized radiation than it is to systemic chemotherapy (Cheung et al., 2003; Kwong, 2005). The prognosis in extranodal NK/T cell lymphoma appears to be directly related to the stage, or extent, of the disease at diagnosis. Approximately 50–75% of patients with localized nasal disease will experience a complete remission of disease with radiation therapy with an overall 5 year survival of 50–60%. Unfortunately approximately one-half of patients with nasal disease who achieve complete remission will experience a local recurrence, usually within the first 2 years after initial therapy. Patients having localized lymphoma which appear to be at greater risk for recurrence are those with more extensive invasion of regional tissues by disease and those in whom the dose or field of radiation must be limited (Kim et al., 2005). Approximately one-third of patients with initially localized nasal disease will develop disease at extranasal sites. Also, patients presenting with extranasal disease often have multiple sites of involvement at diagnosis. The presence of extranasal disease either at diagnosis or at disease relapse is associated with a poor prognosis as this lymphoma is resistant to systemic therapy; with many of such patients dying of disease within the first year after diagnosis.

Inter-relation of the WHO recognized NK cell neoplasms

It is an interesting exercise to consider the relationships the various NK cell lineage lymphoproliferative disorders have both with each other and with normal NK cells. A schemata for these inter-relationships, based on our current understanding of both these neoplasms and normal NK cell biology, is represented in the Venn diagram in Figure 46.7.

The connection between the various NK cell lymphoproliferations and normal NK cells is most clearly drawn for chronic NK cell lymphoproliferative disorder. As noted in the discussion of this entity, the phenotype in this disorder closely mirrors that of normal CD56 bright CD16 dim cytolytic effector NK cells which constitute the majority of the peripheral blood NK compartment. It had been speculated that these cells are derived from their CD56 bright, CD16 dim counterpart that forms the bulk of NK cells in secondary lymphoid tissues. This concept has been substantiated by the work of Romagnani et al. who have demonstrated conversion of NK cells from a CD56 bright, CD16 dim, KIR negative phenotype to a CD56 dim, CD16 bright, KIR positive phenotype following exposure of the former to cytokines such as IL-2 and IL-15. This phenotypic conversion is apparently accompanied by telomere shortening and loss of proliferative potential. In an analogous fashion, the clinical sequelae of chronic NK cell lymphoproliferative disorder appear to be secondary to inappropriate suppression of hematopoietic cell production rather than uncontrolled proliferation

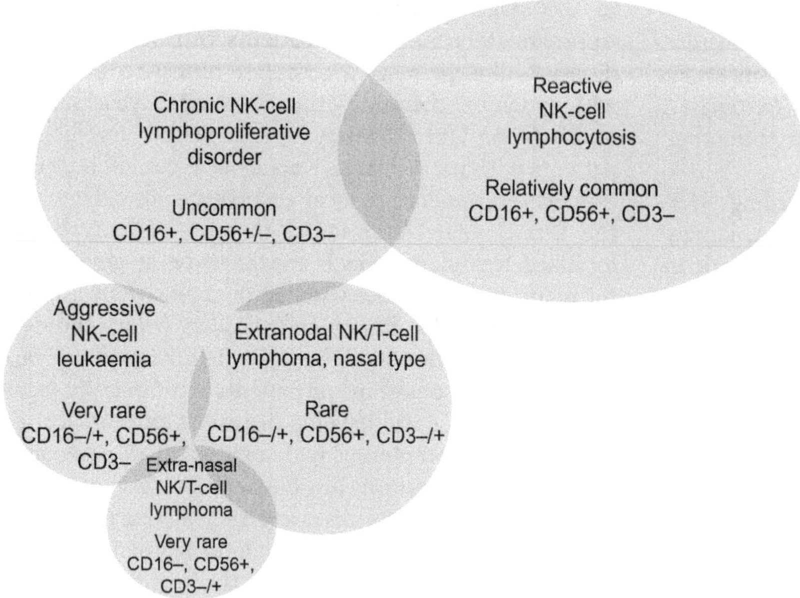

Figure 46.7 • A Venn diagram illustrating the relationship between the various NK cell lymphoproliferative disorders and reactive NK cell lymphocytosis.

and overgrowth by the abnormal NK cells. Taking all of these elements into account it appears that chronic NK cell lymphoproliferative disorder represents a distortion of a normal NK cell response, presumably attributable either to some as yet unknown pathogen and/or a perturbed interactions between the NK cells and self with the latter possibly related to NKR-ligand pairing. Identifying and fully characterizing the chronic NK cell lymphoproliferative disorders may provide insights into environmental and host factors which shape and regulate normal NK cell responses in vivo.

The inter-relationship between chronic NK cell lymphoproliferative disorder and the other bona fide NK cell malignancies is less clear. It appears that chronic NK cell lymphoproliferative disorder, in spite of the name, is most similar to a 'post-proliferative' NK cell compartment and therefore would not necessarily be expected to give rise to a neoplasm with a high proliferative rate. Furthermore, chronic NK cell lymphoproliferative disorder does not show the same predisposition for Asian ethnicity as do the more aggressive NK cell malignancies. Although there have been sporadic reports of aggressive NK cell malignancies occurring in patients with previously indolent NK cell expansions, for the reasons noted above it has yet to be determined if chronic NK cell lymphoproliferative disorder is in any way related to, or is a precursor lesion of, frankly malignant NK cell neoplasms (Huang et al., 2005).

In contrast to chronic NK cell lymphoproliferative disorder, aggressive NK cell leukaemia and extranodal NK/T cell lymphoma are clearly closely related entities. Both of these neoplasms are most prevalent in Asian populations and are strongly EBV associated. Furthermore, certain cytogenetic abnormalities are frequently observed in both disorders including deletions of 6q and gain of chromosome X (Siu et al., 1999; Wong et al., 2000). Lastly, when aggressive NK cell leukaemia involves the tissues it has histopathologic features virtually indistinguishable from those of extranodal NK/T cell lymphoma, and both disorders may be associated with hemophagocytic syndrome (Cheung et al., 2003; Kwong, 2005; Mori et al., 2000). Based on the many shared properties between extranodal NK/T cell lymphoma and aggressive NK cell leukaemia some consider these different clinical presentations of essentially the same malignancy (Quintanilla-Martinez and Jaffe, 2000). The WHO has elected to maintain these as distinct diagnostic categories however as aggressive NK cell leukaemia tends to involve younger patients, always presents with widespread disease, and is associated with a particularly dismal prognosis. Clearly, extranodal NK/T cell lymphoma is the same entity when identified in either a nasal or extranasal site. However, when one considers the frequent widespread disease at presentation and poor outcome associated 'extranasal' extranodal NK/T cell lymphoma it actually bears a stronger resemblance to aggressive NK cell leukaemia than the nasal-type NK cell lymphoma.

When generating an internationally recognized classification scheme such as the WHO classification of hematolymphoid neoplasms one must strike a balance between being as inclusive as possible on the one hand and limiting inclusion to entities sufficiently characterized to be reproducibly recognizable on the other. As such, it would be expected that there would be sporadic hematolymphoid malignancies which would not fit well into a formulaic disease categorization. This is certainly true for NK cell malignancies. In clinical practice, at a high volume reference laboratory one may encounter examples of NK cell malignancies which do not fit into one of the proscribed WHO categories; for example lymph node based, EBV positive disease or EBV, negative disease with a tissue based or leukaemic presentation (Matano et al., 1999; Takahashi et al., 2008). Such cases appear to be extraordinarily rare, however their recognition may be hampered by the factors mitigating against assignment of NK cell lineage in the clinical hematopathology.

The pitfalls in attempting to ascribe NK cell lineage in hematolymphoid malignancies encountered in the clinical laboratory setting is perhaps best illustrated by considering an entity which at one time was called 'blastic NK cell lymphoma'. This unusual, aggressive hematolymphoid malignancy was being described in the same time frame as extranodal NK/T cell lymphoma. Like extranodal NK/T cell lymphoma this malignancy had the unusual proclivity to involve extranodal sites, in this case the skin and bone marrow; expressed few antigens by immunohistochemistry and flow cytometry among which were CD2, CD56 and HLA-DR; and lacked clonal T cell receptor or immunoglobulin gene rearrangements (DiGiuseppe et al., 1997). There were some features in these cases which suggested that they were not truly of NK cell lineage such as strong CD4 expression and lack of cytolytic granule proteins. However, based on the similarities between this neoplasm and extranodal NK/T cell lymphoma, over-interpretation of the significance of CD56 expression and limited knowledge of normal NK cell biology this neoplasm was postulated to be of NK cell lineage. In subsequent studies it was established that the neoplastic cells were actually derived from CD56 positive hematopoietic precursors of plasmacytoid dendritic cells and this entity is now referred to as blastic plasmacytoid dendritic cell neoplasm in the most recent WHO classification (Petrella et al., 2002; Swerdlow et al., 2008).

This does not represent the only example of over-interpretation of the significance of CD56 expression in lineage assignment of hematolymphoid malignancies. In acute myeloid leukaemias some investigators considered CD56 positivity as evidence that the neoplastic cells were derived from bi-potent myeloid/NK cell precursors (Scott et al., 1994; Suzuki et al., 1997). However, independent studies revealed that in most acute myeloid leukaemias CD56 expression was indicative of a monocytic lineage neoplasm (Ikushima et al., 1991; Munoz et al., 2001; Sconocchia et al., 2005). Furthermore, studies of NK cell ontogeny have revealed that the precursor cells are much more closely related to T-cell precursors than myeloid precursors (Caligiuri, 2008; Freud and Caligiuri, 2006). Therefore, in acute myeloid leukaemia CD56 positivity is likely not indicative of NK cell lineage. The close relationship between T and NK cells and the lack of a lineage defining event or product for the latter makes identification of a true neoplasm of NK cell precursors, or lymphoblasts, extraordinarily problematic. There have been studies to suggest that the detection of CD94 1A transcripts in lymphoblastic neoplasms may be indicative of a true NK precursor process (Lin et al., 2005). The described CD94 1A transcript positive lymphoblastic neoplasms otherwise resembled T cell receptor gene rearrangement negative precursor T lymphoblastic neoplasms. These studies suggested that these putative NK precursor lymphoblastic neoplasms may have a better prognosis than true precursor T lymphoblastic neoplasms, however the results of this small single study should be interpreted with caution.

Hopefully, with continued advancement of our understanding of NK cell biology and innate immunity we will gain further insight into malignancies derived from these cells. Conversely, it also possible that continued study and characterization of these malignancies may in turn shed light on the regulation of this increasingly complex immune system compartment.

References

Altemani, A., Barbosa, A.C., Kulka, M., et al. (2002). Characteristics of nasal T/NK-cell lymphoma among Brazilians. Neoplasma **49**, 55–60.

Caligiuri, M.A. (2008). Human natural killer cells. Blood **112**, 461–469.

Canioni, D., Arnulf, B., Asso-Bonnet, M., et al. (2001). Nasal natural killer lymphoma associated with Epstein-Barr virus in a patient infected with human immunodeficiency virus. Arch Pathol Lab Med **125**, 660–662.

Chan, J.K., Sin, V.C., Wong, K.F., et al. (1997). Nonnasal lymphoma expressing the natural killer cell marker CD56: a clinicopathologic study of 49 cases of an uncommon aggressive neoplasm. Blood **89**, 4501–4513.

Chan, J.K., Tsang, W.Y. and Ng, C. S. (1996). Clarification of CD3 immunoreactivity in nasal T/natural killer cell lymphomas: the neoplastic cells are often CD3 epsilon +. Blood **87**, 839–841.

Chan, J.K., Tsang, W.Y. and Pau, M.Y. (1995). Discordant CD3 expression in lymphomas when studied on frozen and paraffin sections. Hum Pathol **26**, 1139–1143.

Cheung, M.M., Chan, J.K., Lau, W. H., et al. (1998). Primary non-Hodgkin's lymphoma of the nose and nasopharynx: clinical features, tumor immunophenotype, and treatment outcome in 113 patients. J Clin Oncol **16**, 70–77.

Cheung, M.M., Chan, J.K. and Wong, K. F. (2003). Natural killer cell neoplasms: a distinctive group of highly aggressive lymphomas/leukemias. Semin Hematol **40**, 221–232.

Chiang, A.K., Chan, A.C., Srivastava, G., et al. (1997). Nasal T/natural killer (NK)-cell lymphomas are derived from Epstein-Barr virus-infected cytotoxic lymphocytes of both NK- and T-cell lineage. Int J Cancer **73**, 332–338.

Chinen, K., Kaneko, Y., Izumo, T., et al. (2002). Nasal natural killer cell/T-cell lymphoma showing cellular morphology mimicking normal lymphocytes. Arch Pathol Lab Med **126**, 602–605.

Cooper, M.A., Fehniger, T.A. and Caligiuri, M.A. (2001). The biology of human natural killer-cell subsets. Trends Immunol **22**, 633–640.

DiGiuseppe, J.A., Louie, D.C., Williams, J.E., et al. (1997). Blastic natural killer cell leukemia/lymphoma: a clinicopathologic study. Am J Surg Pathol **21**, 1223–1230.

Eichel , B.S., Harrison, E.G. Jr., Devine, K. D., et al. (1966). Primary lymphoma of the nose including a relationship to lethal midline granuloma. Am J Surg **112**, 597–605.

Emile, J.F., Boulland, M.L., Haioun, C., et al. (1996). CD5-CD56+ T-cell receptor silent peripheral T-cell lymphomas are natural killer cell lymphomas. Blood **87**, 1466–1473.

Epling-Burnette, P.K., Painter, J.S., Chaurasia, P., et al. (2004). Dysregulated NK receptor expression in patients with lymphoproliferative disease of granular lymphocytes. Blood **103**, 3431–3439.

Fernandez, L.A., Pope, B., Lee, C., et al. (1986). Aggressive natural killer cell leukemia in an adult with establishment of an NK cell line. Blood **67**, 925–930.

Freud, A.G. and Caligiuri, M.A. (2006). Human natural killer cell development. Immunol Rev **214**, 56–72.

Gaal, K., Sun, N.C., Hernandez, A.M., et al. (2000). Sinonasal NK/T-cell lymphomas in the United States. Am J Surg Pathol **24**, 1511–1517.

Harris, N.L., Jaffe, E.S., Stein, H., et al. (1994). A revised European-American classification of lymphoid neoplasms: a proposal from the International Lymphoma Study Group. Blood **84**, 1361–1392.

Hoffmann, T., De Libero, G., Colonna, M., et al. (2000). Natural killer-type receptors for HLA class I antigens are clonally expressed in lymphoproliferative disorders of natural killer and T-cell type. Br J Haematol **110**, 525–536.

Huang, Q., Chang, K.L., Gaal, K.K., et al. (2005). An aggressive extranodal NK-cell lymphoma arising from indolent NK-cell lymphoproliferative disorder. Am J Surg Pathol **29**, 1540–1543.

Ikushima, S., Yoshihara, T., Matsumura, T., et al. (1991). Expression of CD56/NCAM on hematopoietic malignant cells. A useful marker for acute monocytic and megakaryocytic leukemias. Int J Hematol **54**, 395–403.

Imamura, N., Kusunoki, Y., Kawa-Ha, K., et al. (1990). Aggressive natural killer cell leukaemia/lymphoma: report of four cases and review of the literature. Possible existence of a new clinical entity originating from the third lineage of lymphoid cells. Br J Haematol **75**, 49–59.

Jacobs, R., Hintzen, G., Kemper, A., et al. (2001). CD56bright cells differ in their KIR repertoire and cytotoxic features from CD56dim NK cells. Eur J Immunol **31**, 3121–3127.

Jaffe, E.S., Chan, J.K., Su, I.J., et al. (1996). Report of the Workshop on Nasal and Related Extranodal Angiocentric T/Natural Killer Cell Lymphomas. Definitions, differential diagnosis, and epidemiology. Am J Surg Pathol **20**, 103–111.

Kanavaros, P., Lescs, M.C., Briere, J., et al. (1993). Nasal T-cell lymphoma: a clinicopathologic entity associated with peculiar phenotype and with Epstein-Barr virus. Blood **81**, 2688–2695.

Kim, T.M., Park, Y.H., Lee, S.Y., et al. (2005). Local tumor invasiveness is more predictive of survival than International Prognostic Index in stage I(E)/II(E) extranodal NK/T-cell lymphoma, nasal type. Blood **106**, 3785–3790.

Krenacs, L., Smyth, M.J., Bagdi, E., et al. (2003). The serine protease granzyme M is preferentially expressed in NK-cell, gamma delta T-cell, and intestinal T-cell lymphomas: evidence of origin from lymphocytes involved in innate immunity. Blood **101**, 3590–3593.

Kuwabara, H., Tsuji, M., Yoshii, Y., et al. (2003). Nasal-type NK/T cell lymphoma of the orbit with distant metastases. Hum Pathol **34**, 290–292.

Kwong, Y.L. (2005). Natural killer-cell malignancies: diagnosis and treatment. Leukemia **19**, 2186–2194.

Lin, C.-W., Chen, Y.-H., Chuang, Y.-C., et al. (2003). CD94 transcripts imply a better prognosis in nasal-type extranodal NK/T-cell lymphoma. Blood **102**, 2623–2631.

Lin, C.W., Lee, W.H., Chang, C.L., et al. (2001). Restricted killer cell immunoglobulin-like receptor repertoire without T-cell receptor gamma rearrangement supports a true natural killer-cell lineage in a subset of sinonasal lymphomas. Am J Pathol **159**, 1671–1679.

Lin, C.W., Liu, T.Y., Chen, S.U., et al. (2005). CD94 1A transcripts characterize lymphoblastic lymphoma/ leukemia of immature natural killer cell origin with distinct clinical features. Blood **106**, 3567–3574.

Loughran , T.P. Jr (1993). Clonal diseases of large granular lymphocytes.[see comment]. Blood **82**, 1–14.

Macon, W.R., Williams, M.E., Greer, J.P., et al. (1996). Natural killer-like T-cell lymphomas: aggressive lymphomas of T-large granular lymphocytes. Blood **87**, 1474–1483.

Matano, S., Nakamura, S., Nakamura, S., et al. (1999). Monomorphic agranular natural killer cell lymphoma/leukemia with no Epstein-Barr virus association. Acta Haematol **101**, 206–208.

Mori, N., Yamashita, Y., Tsuzuki, T., et al. (2000). Lymphomatous features of aggressive NK cell leukaemia/ lymphoma with massive necrosis, haemophagocytosis and EB virus infection. Histopathology **37**, 363–371.

Morice, W.G. (2007). The immunophenotypic attributes of NK cells and NK-cell lineage lymphoproliferative disorders. Am J Clin Pathol **127**, 881–886.

Morice, W.G., Jevremovic, D. and Hanson, C.A. (2007). The expression of the novel cytotoxic protein granzyme M by large granular lymphocytic leukaemias of both T-cell and NK-cell lineage: an unexpected finding with implications regarding the pathobiology of these disorders. Br J Haematol **137**, 237–239.

Morice, W.G., Kurtin, P.J., Leibson, P.J., et al. (2003). Demonstration of aberrant T-cell and natural killer-cell antigen expression in all cases of granular lymphocytic leukaemia. Br J Haematol **120**, 1026–1036.

Morice, W.G., Kurtin, P.J., Tefferi, A., et al. (2002). Distinct bone marrow findings in T-cell granular lymphocytic leukemia revealed by paraffin section immunoperoxidase stains for CD8, TIA-1, and granzyme. B. Blood **99**, 268–274.

Morice, W.G., Leibson, P.J. and Tefferi, A. (2001). Natural killer cells and the syndrome of chronic natural killer cell lymphocytosis. Leuk Lymphoma **41**, 277–284.

Munoz, L., Nomdedeu, J.F., Brunet, S., et al. (2001). CD56 expression could be associated with monocytic differentiation in acute myeloid leukemia with t(8;21). Haematologica **86**, 763–764.

Nash, R., McSweeney, P., Zambello, R., et al. (1993). Clonal studies of CD3- lymphoproliferative disease of granular lymphocytes. Blood **81**, 2363–2368.

Neben, M.A., Morice, W.G. and Tefferi, A. (2003). Clinical features in T-cell vs. natural killer-cell variants of large granular lymphocyte leukemia. Eur J Haematol **71**, 263–265.

Ng, C.S., Chan, J.K. and Lo, S.T. (1987). Expression of natural killer cell markers in non-Hodgkin's lymphomas. Hum Pathol **18**, 1257–1262.

Ng, S.B., Lai, K.W., Murugaya, S., et al. (2004). Nasal-type extranodal natural killer/T-cell lymphomas: a clinicopathologic and genotypic study of 42 cases in Singapore. Mod Pathol **17**, 1097–1107.

Ooi, G.C., Chim, C.S., Liang, R., et al. (2000). Nasal T-cell/natural killer cell lymphoma: CT and MR imaging features of a new clinicopathologic entity. AJR Am J Roentgenol **174**, 1141–1145.

Pascal, V., Schleinitz, N., Brunet, C., et al. (2004). Comparative analysis of NK cell subset distribution in normal and lymphoproliferative disease of granular lymphocyte conditions. Eur J Immunol **34**, 2930–2940.

Petrella, T., Comeau, M.R., Maynadie, M., et al. (2002). 'Agranular CD4+ CD56+ hematodermic neoplasm' (blastic NK-cell lymphoma) originates from a population of CD56+ precursor cells related to plasmacytoid monocytes. Am J Surg Pathol **26**, 852–862.

Quintanilla-Martinez, L., Franklin, J.L., Guerrero, I., et al. (1999). Histological and immunophenotypic profile of nasal NK/T cell lymphomas from Peru: high prevalence of p53 overexpression. Hum Pathol **30**, 849–855.

Quintanilla-Martinez, L. and Jaffe, E.S. (2000). Commentary: aggressive NK cell lymphomas: insights into the spectrum of NK cell derived malignancies. Histopathology **37**, 372–374.

Romagnani, C., Juelke, K., Falco, M., et al. (2007). CD56brightCD16- Killer Ig-Like Receptor- NK Cells Display Longer Telomeres and Acquire Features of CD56dim NK Cells upon Activation. J Immunol **178**, 4947–4955.

Rudiger, T., Weisenburger, D.D., Anderson, J.R., et al. (2002). Peripheral T-cell lymphoma (excluding anaplastic large-cell lymphoma): results from the Non-Hodgkin's Lymphoma Classification Project. Ann Oncol **13**, 140–149.

Schwartz, E.J., Molina-Kirsch, H., Zhao, S., et al. (2008). Immunohistochemical characterization of nasal-type extranodal NK/T-cell lymphoma using a tissue microarray: an analysis of 84 cases. Am J Clin Pathol **130**, 343–351.

Sconocchia, G., Keyvanfar, K., El Ouriaghli, F., et al. (2005). Phenotype and function of a CD56+ peripheral blood monocyte. Leukemia **19**, 69–76.

Scott, A.A., Head, D.R., Kopecky, K.J., et al. (1994). HLA-DR-, CD33 +, CD56 +, CD16- myeloid/natural killer cell acute leukemia: a previously unrecognized form of acute leukemia potentially misdiagnosed as French-American-British acute myeloid leukemia-M3. Blood **84**, 244–255.

Scquizzato, E., Teramo, A., Miorin, M., et al. (2007). Genotypic evaluation of killer immunoglobulin-like receptors in NK-type lymphoproliferative disease of granular lymphocytes. Leukemia **21**, 1060–1069.

Sheridan, W., Winton, E.F., Chan, W.C., et al. (1988). Leukemia of non-T lineage natural killer cells. Blood **72**, 1701–1707.

Siu, L.L., Chan, J.K. and Kwong, Y.L. (2002). Natural killer cell malignancies: clinicopathologic and molecular features. Histol Histopathol **17**, 539–554.

Siu, L.L., Chan, J.K., Wong, K.F., et al. (2002). Specific patterns of gene methylation in natural killer cell lymphomas : p73 is consistently involved. Am J Pathol **160**, 59–66.

Siu, L.L., Chan, V., Chan, J.K., et al. (2000). Consistent patterns of allelic loss in natural killer cell lymphoma. Am J Pathol **157**, 1803–1809.

Siu, L.L.P., Wong, K.-F., Chan, J.K.C., et al. (1999). Comparative Genomic Hybridization Analysis of Natural Killer Cell Lymphoma/Leukemia : Recognition of Consistent Patterns of Genetic Alterations. Am J Pathol **155**, 1419–1425.

Song, S.Y., Kim, W.S., Ko, Y.H., et al. (2002). Aggressive natural killer cell leukemia: clinical features and treatment outcome. Haematologica **87**, 1343–1345.

Strickler, J.G., Meneses, M.F., Habermann, T.M., et al. (1994). Polymorphic reticulosis: a reappraisal. Hum Pathol **25**, 659–665.

Suzuki, R., Suzumiya, J., Nakamura, S., et al. (2004). Aggressive natural killer-cell leukemia revisited: large granular lymphocyte leukemia of cytotoxic NK cells. Leukemia **18**, 763–770.

Suzuki, R., Yamamoto, K., Seto, M., et al. (1997). CD7+ and CD56+ myeloid/natural killer cell precursor acute leukemia: a distinct hematolymphoid disease entity. Blood **90**, 2417–2428.

Swerdlow, S.H., Campo, E., Harris, N.L. et al. (eds), (2008). WHO Classification of Tumours of Haematopoietic and Lymphoid Tissues. Lyon: IARC Press.

Takahashi, E., Asano, N., Li, C., et al. (2008). Nodal T/NK-cell lymphoma of nasal type: a clinicopathological study of six cases. Histopathology **52**, 585–596.

Tefferi, A. (1996). Chronic natural killer cell lymphocytosis. Leuk Lymphoma **20**, 245–248.

Tefferi, A., Greipp, P.R., Leibson, P.J., et al. (1992). Demonstration of clonality, by X-linked DNA analysis, in chronic natural killer cell lymphocytosis and successful therapy with oral cyclophosphamide. Leukemia **6**, 477–480.

Teruya-Feldstein, J., Jaffe, E.S., Burd, P.R., et al. (1997). The role of Mig, the monokine induced by interferon-gamma, and IP-10, the interferon-gamma-inducible protein-10, in tissue necrosis and vascular damage associated with Epstein-Barr virus-positive lymphoproliferative disease. Blood **90**, 4099–4105.

Totonchi, K.F., Engel, G., Weisenberg, E., et al. (2002). Testicular natural killer/t-cell lymphoma, nasal type, of true natural killer-cell origin. Arch Pathol Lab Med **126**, 1527–1529.

Tsao, L., Draoua, H.Y., Mansukhani, M., et al. (2004). EBV-associated, extranodal NK-cell lymphoma, nasal type of the breast, after heart transplantation. Mod Pathol **17**, 125–130.

Warren, H.S., Christiansen, F.T. and Witt, C.S. (2003). Functional inhibitory human leucocyte antigen class I receptors on natural killer (NK) cells in patients with chronic NK lymphocytosis. Br J Haematol **121**, 793–804.

Wong, K.F., Chan, J.K., Ng, C.S., et al. (1992). CD56 (NKH1)-positive hematolymphoid malignancies: an aggressive neoplasm featuring frequent cutaneous/mucosal involvement, cytoplasmic azurophilic granules, and angiocentricity. Hum Pathol **23**, 798–804

Wong, N., Wong, K.F., Chan, J.K., et al. (2000). Chromosomal translocations are common in natural killer-cell lymphoma/leukemia as shown by spectral karyotyping. Hum Pathol **31**, 771–774.

Zambello, R., Falco, M., Della Chiesa, M., et al. (2003). Expression and function of KIR and natural cytotoxicity receptors in NK-type lymphoproliferative diseases of granular lymphocytes. Blood **102**, 1797–1805.

Zambello , R., Loughran, T.P. Jr., Trentin, L., et al. (1997). Spontaneous resolution of p58/EB6 antigen restricted NK-type lymphoproliferative disease of granular lymphocytes: role of Epstein Barr virus infection. Br J Haematol **99**, 215–221.

Zambello, R., Trentin, L., Ciccone, E., et al. (1993). Phenotypic diversity of natural killer (NK) populations in patients with NK-type lymphoproliferative disease of granular lymphocytes. Blood **81**, 2381–2385.

PART V

Special topics in NK biology

Chapter Forty-Seven

47

Natural killer cell induction of tolerance

Lina Lu, Alexandra Y. Zhang, William L. Camp, Shiguang Qian

CHAPTER CONTENTS

It is a mistake to try to look too far ahead.
The chain of destiny can only be grasped one link at a time.

Winston Churchill

ABSTRACT

Natural killer (NK) cells represent a potent first line of defence in immunity. They preferentially attack cells that do not express or express low self-MHC class I proteins. However, accumulating data suggest that NK functional biology is more complex than previously thought. NK cells express many receptors, both stimulatory and inhibitory. The net balance of activating and inhibitory signals resulting from interactions with target cells determines the ultimate action to mediate killing or not. In addition to the receptors specific to self-MHC class I, various non-MHC stimulatory receptors on NK cells also participate in the decision making. NK cells can also function as immune regulators by producing proinflammatory cytokines and chemokines upon various stimuli, as well as delivering cytotoxic signals to influence other immune cells, thus achieving their role in regulating both innate and adaptive immune responses.

KEY WORDS

Receptor, Self-tolerance, Major histocompatibility complex, Cytotoxicity, Apoptosis, Immune regulation, Dendritic cells, T cells, Viral infection, Transplantation

Natural killer (NK) cells are a major cell type of the innate immune system. Although NK cells do not express the specialized genes that are rearranged as T cell receptors, they are capable of discriminating self and foreign, as well as normal and diseased cells, thus, demonstrating the ability to attack transformed, infected and stressed cells but not normal self cells. How these cells are prevented from killing normal cells while attacking diseased cells is a major unresolved question for NK-cell biology. The finding that the cells lacking self-MHC class I molecules are more sensitive to NK cells leads to the formulation of the 'missing-self' recognition theory, which states that NK cells preferentially kill the cells that do not express or express low self-MHC class I. It was realized later that this recognition is governed by many receptors expressed on NK cells, including stimulatory and inhibitory receptors. The net balance of activating and inhibitory signals resulting from interactions with the target cells determines whether the NK cell becomes activated to produce inflammatory cytokines (such as IFN-γ and TNF-α) to kill the target cells, or not (Lanier, 2005; Moretta et al., 2001; Yokoyama and Seaman, 1993).

In addition to the receptors specific for self-MHC class I, studies have also shown that various non-MHC receptors on NK cells participate in the decision making. Although NK cells express several families of these inhibitory receptors, MHC-I molecules represent the essential ligands for inhibitory NK-cells receptors.

NK cells often act as potent effector cells in the rejection of allogeneic bone marrow cells and solid organ transplants (Kean et al., 2006) and in the destruction of pathogen-infected cells. However, in certain circumstances, NK cells also express potent immunoregulatory properties that promote tolerance induction (Cooper et al., 2001). There is evidence that supports the theory that NK cells promote tolerance not by direct interaction with target cells, but by modulating responses of host immune cells via producing proinflammatory cytokines and chemokines upon various stimuli, and delivering cytotoxic signals to immune effector cells (Beilke et al., 2005; Yu et al., 2006). Thus, in addition to the surveillance of tumours and virus-infected cells, NK cells also contribute to regulating the nature and the extent of adaptive immune responses. Progress in the area of NK cells and tolerance to self- and nonself-antigens will be necessary for a comprehensive understanding of NK cell recognition and for facilitating tolerance induction to cell or organ transplants and autoimmune diseases.

NK cell self-tolerance

NK cells can be activated by various endogenous self ligands, some of which are expressed by normal cells. How are NK cells prevented from attacking normal cells while ensuring reactivity to diseased cells? Engagement of the inhibitory receptors plays a central role in self-tolerance by NK cells, resulting in inhibition of cytokine production and cytolysis in order to maintain self-tolerance (Werner, 2008).

Missing-self recognition hypothesis

MHC class I molecules represent the classical ligands for inhibitory NK-cell receptors. Almost all nucleated cells express MHC class I molecules that make them a supreme marker of self. For years, it has been thought that self-cells are protected from NK cells because of their expression of MHC-class-I molecules that are recognized by the inhibitory receptors at the surface of NK cells, thus keeping NK cells non-responsive. In contrast, infected, or transformed cells that do not express sufficient levels of host MHC-class-I molecules for effective engagement of inhibitory receptors, are recognized by NK cells as non-self, and are killed (Karre et al., 1986; Ljunggren and Karre, 1990). This is the basis of the 'missing-self recognition', the capacity of NK cells to attack cells that lose or downregulate expression of some or all self-MHC class I molecules. This is not just limited to virally infected or transformed cells, since normal cells can be killed by NK cells if they are unable to express sufficient self-MHC class I molecules. A classic example of this hypothesis is that donor bone-marrow cells missing an MHC class I allele of the host are commonly rejected by NK cells (Ohlen et al., 1989).

Inhibitory receptors that recognize MHC class I molecules

Many receptors are expressed on the surface of NK cells; some are stimulatory and some are inhibitory. It is widely accepted that expression of inhibitory receptors specific for self-MHC class I molecules on NK cells is essential in missing-self recognition (Figure 47.1). A set of inhibitory receptors is acquired during the development of NK cells. NK cells have been studied extensively in mice, rats and humans. At least three families of MHC class I–specific inhibitory receptors have been identified: the killer cell immunoglobulin-like receptor (KIR), the lectin-like Ly49 family and other receptors, including the CD94/NKG2A receptor and paired Ig-like receptor (PIR) (Table 47.1) (Lanier, 2005; Moretta et al., 1996; Vance et al., 1998; Yokoyama, 1995). There are strong analogies in these species, both in the receptors and functional pathways. For instance, the functions of Ly49 receptor family expressed on mouse NK cells are similar to KIR family receptors in human NK cells. However, the differences are obvious. KIRs are expressed in humans and other species except mice, while Ly49 receptors are expressed in mice, rats and other species, except humans. The CD94/NKG2A receptors are expressed across all species. Multiple inhibitory receptors can be expressed concurrently by each NK cell. Although not all of the inhibitory receptors may recognize self-MHC, during their development, eventually all mature NK cells express at least one inhibitory receptor specific for a self-MHC class I molecule in order to prevent self attack, which is proposed as the 'at least one receptor' model (Raulet et al., 2001).

Two possible adaptations may be involved to modify the inhibitory MHC-I receptor repertoire, through either

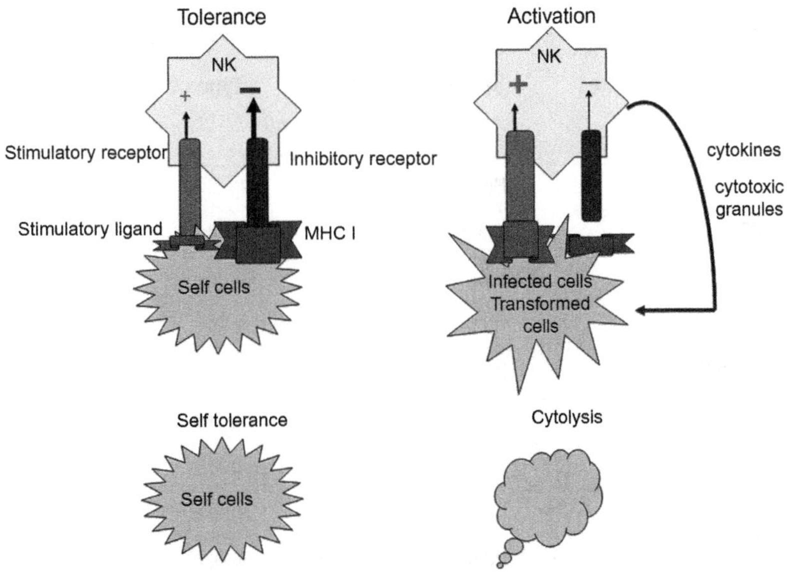

Figure 47.1 • Basic model for NK cell tolerance and activation. The inhibitory receptor signal through MHC class I ligation counters the stimulatory receptor signal via its ligand on target cells, resulting in tolerance of NK cells. Upregulation of stimulatory ligand or downregulation of MHC class I on infected/transformed cells leads to NK cell activation.

a cellular selection process (i.e. selective expansion of the immature NK cells that express self-MHC–specific inhibitory receptors); or a probable "audition" of the receptors encoded by the genome until an inhibitory receptor that is specific for MHC-I molecules is expressed during the development of NK cells (Dorfman and Raulet, 1998; Held et al., 1996; Raulet et al., 2001; Roth et al., 2000; Valiante et al., 1997). NK cells that initially express only receptors that do not bind the self-MHC class I molecules cannot be suppressed by self-MHC class I molecules and are thus potentially auto-aggressive. Such cells may arise but fail to mature and may not contribute to the mature NK cell pool. By these mechanisms, an initially arbitrary NK repertoire might be selectively modified into a mature NK cell pool that expresses at least one, if not multiple, inhibitory receptors, specific for self-MHC class I molecules. In the "sequential-cumulative model", it was proposed that during NK cell development, certain Ly49 genes are induced such that developing NK cells continue to collect new receptors until recently expressed inhibitory receptors engage self-MHC class I molecules with adequate strength (Dorfman and Raulet, 1998; Hanke et al., 2001; Roth et al., 2000).

After interaction with their target-cell ligands, the inhibitory receptors become tyrosine phosphorylated on their immunoreceptor tyrosine-based inhibitory motifs (ITIMs). NK cell activation and its effector responses are then inhibited by recruiting intracellular phosphatases, SRC homology 2 (SH2)-domain-containing protein tyrosine phosphatase 1 (SHP-1) and 2 (SHP-2), resulting in an inhibitory signal. Otherwise, if the target cell ligands can be identified by the activation receptors, NK cell effector responses (i.e. cytokine production and target cell lysis) will be executed.

Challenges to 'missing self' recognition hypothesis

Although many studies over two decades of research have supported the 'missing self' hypothesis that the interaction between MHC class I molecules and the inhibitory receptors on NK cells is relevant to the generation of NK cell self tolerance, this hypothesis has been challenged by several fundamental observations. A logical extension of this theory is that MHC class I deficiency should experience robust NK cell autoreactivity. Indeed, the data from both humans and mice show that a lack of MHC class I expression does not demonstrate excessive NK cell activity. MHC class I-deficient mice, such as $\beta 2m^{-/-}$, $TAP1^{-/-}$, or $H2\text{-}D^{-/-}K^{-/-}$ mice, contain a relatively normal number of NK cells and do not succumb to autoimmunity. On the contrary, the NK cells derived from MHC class I deficient hosts, although otherwise normal, demonstrate an inhibitory ability to lyse cells devoid of MHC I molecules. These MHC-class-I–deficient NK cells, unlike the NK cells derived from wild type hosts, do not reject MHC-I-deficient grafts (Bix et al., 1991; Vitale et al., 2002), suggesting that NK killing activity is not only dependant on the expression of inhibitory receptors , but also may be affected by MHC class I expressed on their surface.

Table 47.1 Main inhibitory receptors regulating NK cell functions

Name	Alternative names	Species	Type of molecule	Ligand(s)
KIR family	**CD158**			
KIR-2DL1	CD158a	Human	Ig-like	HLA-C
KIR-2DL2	CD158b1	Human	Ig-like	HLA-C
KIR-2DL3	CD158b2	Human	Ig-like	HLA-C
KIR-2DL4	CD158d	Human	Ig-like	HLA-G
KIR-2DL5a	CD158f1	Human	Ig-like	ND
KIR-2DL5b	CD158f2	Human	Ig-like	ND
KIR-3DL1	CD158e1	Human	Ig-like	HLA-Bw4
KIR-3DL2	CD158k	Human	Ig-like	HLA-A3, -A11
KIR-3DL3	CD158z	Human	Ig-like	ND
KLR-A family	**Ly49 (functions are similar to KIR family in humans)**			
KLR-A1	Ly49a	Mouse	C-lectin-like	H-2D
KLR-A3	Ly49c	Mouse	C-lectin-like	H-2K, D
KLR-A5	Ly49e	Mouse	C-lectin-like	ND
KLR-A6	Ly49f	Mouse	C-lectin-like	H-2D
KLR-A7	Ly49g	Mouse	C-lectin-like	H-2D
KLR-A9	Ly49i	Mouse	C-lectin-like	H-2D, MCMV
KLR-A17	Ly49q	Mouse	C-lectin-like	H-2K
LILRB family	**CD85**			
LILR-B1	CD85j	Human	Ig-like	HLA class I, HCMV
LILR-B2	CD85d	Human	Ig-like	HLA class I
LILR-B4	CD85k	Human	Ig-like	HLA class I
Others				
2B4	CD244	Human, mouse	Ig-like	CD48
CEACAM1	CD66a	Human, mouse	Ig-like	MHV
KLR-B1	CD161, NKR-P1a	Human	C-lectin-like	OCIL, CLEC2D
KLR-D1-KLR-C1	CD94-NKG2A (CD159a)	Human Mouse	C-lectin-like	HLA-E Qa-1b
KLR-B1b	CD161B, NKR-P1b	Mouse	C-lectin-like	OCIL
KLR-G1	MAFA	Human, mouse	C-lectin-like	Cadherin
LAIR-1	CD305	Human, mouse	Ig-like	Collagen
PIR-A		Mouse	Ig-like	CD99
PIR-B		Mouse	Ig-like	H-2D, K

CEACAM1, carcinoembryonic-antigen-related cell adhesion molecule 1; HCMV, human cytomegalovirus; KIR, killer cell immunoglobulin-like receptor; KLR, killer cell lectin-like receptor; LAIR, leukocyte-associated immunoglobulin-like receptor; LILR, leukocyte immunoglobulin-like receptor; MAFA, v-maf musculoaponeurotic fibrosarcoma oncogene homolog A; MCMV, mouse cytomegalovirus; MHV, mouse hepatitis virus; ND, non-defined; PIR, paired immunoglobulin-like receptor.

'Licensing' concept

These observations raised an important question—whether or not the NK cells from MHC-I–deficient hosts are functionally defective (Dorfman et al., 1997; Salcedo et al., 1997). The answer is 'yes', based on many studies. NK1.1 is an activation receptor expressed by all NK cells in mice (Arase et al., 1996; Kim et al., 2002). Wild-type NK cells produced IFN-γ upon NK1.1 crosslinking. However, NK1.1-activated NK cells derived from either β2m-deficient or K^bD^b-deficient mice produce almost no IFN-γ. Further analysis revealed that other activation receptors also failed to stimulate IFN-γ production by these cells (Kim et al., 2005). Therefore, the activation receptors on NK cells from MHC-class I-deficient mice are functionally defective, indicating that MHC class I influences NK cell functional competence.

MHC congenic mice have been used to address how MHC class I influences NK cell function. The data suggest that either naïve or pre-activated NK cells deficient in expression of known self-receptors (Ly49C, Ly49I, CD94/NKG2A, CD94-NKG2C or CD94-NKG2E) demonstrate self-tolerance but respond poorly to MHC class I deficient normal cells or tumour cells, suggesting that self-tolerance of NK cells can be established independently of MHC-mediated inhibition (Fernandez et al., 2005). More definitive studies were conducted using naïve NK cells from MHC-recombinant mice, as well as the MHC-I transgenic mice in which investigators utilized a single chain trimer (SCT) MHC-I molecule, SCT-K^b, that binds only one NK cell inhibitory receptor, Ly49C, on primary NK cells. In an SCT-K^b transgenic mouse deficient in K^bD^b- and β2m, only one MHC molecule (K^b) is expressed. Only Ly49C$^+$ NK cells demonstrated an enhanced ability to produce IFN-γ as compared to NK cells from control mice (Kim et al., 2005; Yu et al., 2002). These studies lead to a concept of 'licensing', which states that, during development, NK cells need to be 'licensed' or 'armed' in the bone marrow to acquire a fully functional competence. In other words, maturation of the functional NK cells requires that their MHC-class-I-specific inhibitory receptors interact with self-MHC class I molecules, called 'licensing'. These licensed NK cells are then allowed to be activated through their activation receptors. Failure to receive the licensing signal keeps NK cells in a hyporesponsive state (Kim et al., 2005).

Licensing may explain why a lack of MHC-I expression does not demonstrate excessive NK cell activation. It also provides an explanation for hybrid resistance, which describes a phenomenon in which the NK cells from hybrid F1 mice reject the graft from either parent strain. In an F1 host that is heterozygous for MHC alleles, each MHC allele might potentially license different NK cell populations; therefore, an (A $\times$ B) F_1 hybrid animal may have NK cells that are separately licensed on different MHC alleles. The NK cells that were licensed by MHC alleles from parent A are inhibited by A alleles but not B alleles, and thus reject bone marrow from parent B and vice versa. In the F_1 animal itself, NK cells are licensed by either parental allele, so all NK cells should be inhibited by normal tissues that co-dominantly express both MHC alleles (Riley and Yokoyama, 2008).

Non-MHC-class-I–specific NK-cell recognition

The recognition of MHC class I molecules is one of the critical features of NK cell recognition. However, in addition to the recognition of MHC class I molecules, NK cells also express receptors that are specific for a range of other ligands that are unrelated to MHC class I molecules. These ligands are widely expressed on the surface of normal cells, indicating an important role in self-tolerance.

Non-MHC-class-I–specific inhibitory receptors

Several non-MHC-class-I–specific inhibitory receptors have been identified, such as carcinoembryonic-antigen-related cell adhesion molecule 1 (CEACAM1), killer-cell lectin-like receptor G1 (KLRG1), NKR-P1A and NKR-P1B (Carlyle et al., 2004; Corral et al., 2000; Iizuka et al., 2003; Markel et al., 2004; Yokoyama and Plougastel, 2003), and some members of the Ly49-family (Table 47.1). 2B4 (CD244) is considered to be either an inhibitory or stimulatory receptor because of its ability to use different adaptor molecules to convey either inhibitory or stimulatory signals (Parolini et al., 2000; Roncagalli et al., 2005; Tangye et al., 2000).

CD48 and 2B4 interaction may be a good example that provides another probable mechanism of NK cell tolerance in addition to MHC-dependent NK cell tolerance. Some studies observed an increased reactivity of NK cells that are deficient in MHC-I-specific receptors when 2B4 and CEACAM1 were blocked (Markel et al., 2004; McNerney et al., 2005), suggesting a potential role of these non-MHC-I–specific receptors in NK-cell self-tolerance. CD48 is expressed by human endothelial cells and all nucleated haematopoietic cells. It is a member of the CD2 family (Brown et al., 1998). Expressed by mouse and human NK cells, monocytes, $\alpha\beta$ and $\gamma\delta$ T cells, granulocytes and mast cells, 2B4 (CD244) belongs to the family of the signalling lymphocytic activation molecule (SLAM) (Boles et al., 2001; Kubota, 2002). Although mainly an activating receptor in humans, 2B4 is the inhibitory receptor for CD48. The density of 2B4 was noticed to be elevated at the surface of NK cells in a model that uses SH2-containing inositol phosphatase (SHIP)-deficient C57BL/6 mice. In this model, SHIP-deficient mouse NK cells do not express MHC-I–specific inhibitory Ly49 receptors, and the activation of these NK cells is suppressed by the inhibitory 2B4 receptor (Vaidya and Mathew, 2006). In addition, the inhibitory effects of the 2B4 receptors may contribute to the NK cell self-tolerance in a non-MHC-I–specific fashion to prevent NK cells from carrying out effector responses and attacking autologous cells. Of interest, NK cells from 2B4-deficient mice are more capable of eliminating $CD48^+$ tumour cells in vivo (Kubota, 2002; Vaidya et al., 2005). Some cytokines, such as IFN-γ produced by NK cells, were also shown to be downregulated, as well as their cytotoxicity (Lee et al., 2004). Similarly, the inhibitory receptor for cadherin KLRG1 usage was shown to be expanded significantly in NK cells lacking inhibitory self-MHC-I receptors or in MHC-I–deficient mice, while, in general, KLRG1 is expressed by relatively small population of NK cells (Corral et al., 2000). Thus, the absence of MHC-I–specific inhibition may be compensated by the acquisition of additional non-MHC-I–specific inhibitory receptors or their hyperactivity.

Achievement of self-tolerance without expressing inhibitory receptors specific for self-MHC class I molecules

There are subsets of NK cells (such as the $CI/NKG2^-$ subset) that are deficient in all identified self-MHC-I–specific inhibitory receptors, but are still self-tolerant. Although they can display normal mature cell surface markers, such as CD11b, DX5, Ly49 receptors, and secrete proinflammatory cytokine IFN-γ upon stimulations in vitro with pharmacological agents or in vivo with *Listeria monocytogenes*, these NK cells do not respond in full strength to MHC-I–deficient normal cells or tumour cells, suggesting that self-tolerance can be achieved by attenuating stimulatory signalling. However, the hyporesponsiveness of the $CI/NKG2^-$ subset NK cells occurs only when their receptors are unable to bind self-MHC class I ligands as evidenced by the reactivity of these NK cells in MHC class I mismatched B10.M mice (Fernandez et al., 2005). Hence, these hyporesponsive NK cells are mature and functional, and play an important role in NK cell self-tolerance that can occur independently of MHC-I-specific inhibition.

Continuous engagement of activation receptors induces NK cell tolerance

During development, T and B cells undergo selection, that is, those T and B cells encountering their cognate ligands via antigen-specific receptors are deleted or rendered anergic. Although NK cells are not thought to undergo such selective processes, like T and B cells, NK cells also express activating receptors. A recent study demonstrated that NK cells might also undergo a selection process that would eliminate or inactivate NK cells that encountered a ligand during their development in bone marrow similar to T cell negative selection in thymocytes. NK cells bearing the DAP12-associated activating Ly49H receptor interact with the mouse cytomegalovirus (CMV)-encoded protein m157 on infected cells. In a specific pathogen-free mouse colony, immature NK cells should not encounter this viral ligand during development in the bone marrow. The Ly49H receptor has a high affinity for the viral glycoprotein m157 but does not recognize any self-antigen (Adams et al., 2007; Davis et al., 2008). These characteristics provide an opportunity to determine what happens to the immature Ly49H+ NK cells when they encounter m157 during development. To address this issue, the mice expressing m157 in the bone marrow were generated. The results showed that Ly49H+ NK cells were found in the periphery of m157-expressing mice; however, they were present at lower numbers, less mature, produced less IFN-γ and were severely defective in their ability to mediate cytotoxicity and proliferate in response to mouse CMV infection, indicating that if immature NK cells encounter ligands for their activating receptors during their development, regulatory mechanisms exist to keep these cells in an hyporesponsive state (Sun and Lanier, 2008; Tripathy et al., 2008). Repeated antigenic stimulations are known to induce expression of several inhibitory receptors, including KIR on immune cells (McMahon and Raulet, 2001).

NK cells in regulation of immune responses

NK cells are key innate effector cells in eliminating viruses, tumour and hazardous cells by direct killing before the onset of adaptive T and B cell immunity. Their in vivo importance has been demonstrated by studying NK-cell–deficient patients and mouse models. Dysfunction or depletion of NK cells results in frequent and severe infections (Gazit et al., 2006). Recent data implicate more complex roles for NK cells. They also participate in regulation of immune responses via interacting with various immune components.

NK and dendritic cell interaction

Mouse cytomegalovirus infection models demonstrated preferential distribution of $IFN\gamma^+$ NK cells near marginal zone areas of secondary lymphoid organs, where antigen-presenting cells (APC), mostly dendritic cells (DC), accumulate and prime T cells, suggesting the influence of NK cells on APC–T cell crosstalk (Daniels et al., 2001). Much attention has been given to the interactions between NK cells and DC. Indeed, NK cells interact intimately with DC, which has important implications for ensuing links of innate and adaptive immune responses. During the interactions, the two types of cells form an immune synapse with each other. The physical contact between NK and DC involves interactions among several receptor–ligand pairs, which include LFA-1, NKp30, NKp46, 2B4, DNAX accessory molecule 1 (DNAM-1), NKG2D, TNFRII and NKG2A (Xu et al., 2007). NK cells induce secretion of IL-12, IL-18 and membrane-bound IL-15 from DC. These cytokines in turn activate NK cells to secrete IFN-γ, TNF-α and high mobility group box-1 (HMGB1), which cause DC maturation (Zitvogel, 2002). Conversely, CD8α DC were shown to induce selective expansion of NK cells in vivo through secretion of IL-12 and IL-18 (Degli-Esposti and Smyth, 2005). Another study demonstrated that NK cells were induced by DC to function as non-cytotoxic 'helper' cells following stimulation with IL-18, which facilitated IFNγ secretion from NK cells and thus enabled DC to secrete IL-12p70, leading to Th1 polarization (Degli-Esposti and Smyth, 2005). An indirect role for NK cells in priming the Th1 CD4$^+$ T cell response by providing IFN-γ in the lymph nodes has also been shown in vivo. Injection of mature DC with adjuvant led to CXCR3-dependent recruitment of NK cells to the lymph nodes, where they provided an initial source of the IFN-γ necessary for Th1 polarization. Depletion of NK cells resulted in a reduced Th1 response and was dependant on IFN-γ production from NK cells (Mailliard et al., 2005; Martin-Fontecha et al., 2004).

On the other hand, it has been well established that NK cells can effectively regulate DC functions by inducing DC lysis shown in co-culture of human NK cells with autologous DC (Scott and Trinchieri, 1995; Unanue, 1997). NKp30, DNAM-1 (CD226, a co-stimulation and adhesion molecule), and LFA-1 molecules are involved in the NK cell-mediated killing of autologous, immature DC (Ferlazzo et al., 2002). However, mature DC can escape from killing, as the maturation process induces expression of human leukocyte antigens (HLA) antigens, which protect them from NK cells. NK cells can influence immature DC leading either to their maturation or to death in vivo. This is largely dependant on the interaction between the NKp30 receptor and an unknown ligand that is expressed on the surface of immature DC (Ferlazzo et al., 2002). The CD56^{high}CD16^{dim} NK cell subset interacts with immature DC and drives maturation or kills the immature DC. These NK cells express little KIR, but express CD94/NKG2A as the main inhibitory receptors. It remains unclear how NK cells choose between killing and causing maturation of immature DC. It probably depends on the profile of expression of several molecules on the surface of immature DC. If DC fail to express HLA antigens upon maturation, they may be killed by NK cells. The ratio between NK cells and their interacting DC is also a factor: A greater ratio tends to favour killing rather than maturation. A recent study reported that when activated human NK cells were cultured with immature DC in vitro, at low NK/DC ratios (1:5), NK cells were able to significantly induce DC maturation and cytokine secretion and therefore cause DC activation. Cell-to-cell contact and endogenously produced TNF-α in the culture are required for DC maturation induced by NK cells. In contrast, at higher NK/DC ratios (5:1), DC functions were completely inhibited due to the effective killing by the autologous NK cells, implicating a potential role of contact-dependent NK–DC interactions in the regulation of the immune response (Piccioli et al., 2002). NK cells stimulated by IL-2 increased their cytotoxicity, and enhanced immature DC killing activity, whereas, exposure to IL-18 did not affect NK cell cytotoxicity and DC killing. This supports the notion that the functions of NK cells as 'effector' and 'helper' cells represent two independent mechanisms that might be controlled through separate pathways (Unanue, 1997).

The interplay between NK cells and DC has been also investigated in models of bacterial infection. In one study, immature DC underwent maturation when they were cultured in the presence of *Escherichia coli* or Bacillus Calmette–Guérin. Upon infection, production of considerable amounts of TNF-α and IL-12 were detected, whereas IL-2 and IL-15 were hardly detectable in supernatants. After NK cells and DC were cultured together for 24 h, HLA class I molecule expression was upregulated on the surface of the bacteria-infected DC. These infected DC were more resistant to NK-mediated killing. The NK cells that were exposed to bacteria-infected DC were then activated and became capable of killing autologous immature DC. This study demonstrated that DC-induced NK cell activation can be significantly enhanced by the presence of pathogens and thus may regulate subsequent innate and adaptive immune responses (Ferlazzo et al., 2003).

An in vivo DC-based vaccine study demonstrated a rapid elimination of immature DC by NK cells through a pathway dependant on the TNF-related apoptosis-inducing ligand (TRAIL). In addition, during immunization with immature DC loaded with foreign or tumour antigens, when NK cells were depleted and/or TRAIL function was neutralized, T cell responses to these antigens were enhanced. Therefore, responses to DC-based vaccination may be determined and regulated by the survival of antigen-loaded DC, which is controlled by TRAIL on NK cells (Hayakawa et al., 2004).

NK and T cell interaction

Human NK cells efficiently enhance $CD4^+$, as well as $CD8^+$ T cell proliferation in response to antigen-specific or anti-CD3 stimulation. This process depends on direct contact-mediated interactions mainly between co-stimulatory molecules expressed on NK cells (including CD80, CD86, CD70, OX40 ligand and 2B4 receptors) and their counterparts expressed on T cells (e.g. CD28, CD27, OX40 and CD48) (Hanna et al., 2004a,b, 2005; Zingoni et al., 2004). Activated human NK cells express MHC class II and have the capability to present antigens directly and stimulate $CD4^+$ T cell proliferation in vitro (Hanna et al., 2004); therefore, activated human NK cells possess not only the required co-stimulatory molecules for potential interaction with activated $CD4^+$ T cells, but also have the capacity to process and present antigens through MHC class II. Although DC are considered to be the most potent APC, the fact that activated NK cells express MHC class II, CD86, CD80, CD70 and OX40L strongly suggests the possibility that they might also communicate directly with $CD4^+$ cells. However, it is difficult to provide convincing evidence of this interaction in vivo in humans due to obvious limitations. Interestingly, activated mouse NK cells do not express MHC class II; therefore, the mouse is not an appropriate model to examine MHC class II TCR-dependent $CD4^+$ cell interactions with NK cells (Zingoni et al., 2004).

The role of NK cells in $CD8^+$ T cell effector function was examined in the situation of intracellular *Mycobacterium tuberculosis* infection. $CD8^+IFN\text{-}\gamma^+$ cells are normally able to effectively lyse *M. tuberculosis*-infected monocytes in the presence of NK cells. However, the killing capacity and the frequency of *M. tuberculosis*-responsive $CD8^+IFN\text{-}\gamma^+$ cells were decreased when NK cells were removed from peripheral blood mononuclear cells (PBMC) of healthy tuberculin reactors. IFN-γ, IL-15 and IL-18 were secreted by *M. tuberculosis*-activated NK cells; they are the key cytokines to facilitate the restoration of $CD8^+IFN\text{-}\gamma^+$ cell frequency. Activated NK cells also stimulated *M. tuberculosis*-infected monocytes to produce IL-15 and IL-18 that in turn promote the expansion of $CD8^+$ T cells. Cell–cell contact between NK cells and *M. tuberculosis*-infected monocytes was required and the CD40–CD40 ligand interactions were important for NK cells to maintain the capacity to prime $CD8^+$ T cells. By connecting innate and adaptive immune responses, NK cells can optimize the capacity of $CD8^+$ T cells to execute their effector response and to protect against pathogen invasion (Vankayalapati et al., 2004).

Several studies have demonstrated that NK cells have the ability to regulate CD4 and CD8 responses. In a gender-dependent model of preferential Th1 and Th2 activation, NK cells influenced $CD4^+$ T cell activation and regulated adaptive immunity prior to antigen exposure. This model utilizes the Swiss Jim Lamert (SJL) mice that exhibit a gender-dependent differential response to immunization. In this model, young adult male SJL mice were normally unable to activate Th1 cells. However, by depletion of NK cells, the activation of Th1 effectors was permitted in males (Dowdell et al., 2003).

NK cells in transplantation

Although NK cells play a crucial role in rejection of bone marrow transplants (Ruggeri et al., 2002), only recently, it has been noted that NK cells may also participate in the rejection of solid organ grafts (Beilke and Gill, 2007; Kitchens et al., 2006). Essential features of NK cells in transplantation include distinguishing allogeneic cells from host and their effective cytolytic effector responses. The interaction between NK cells and host adaptive immune system (DC and T cells) also play an important role in allogeneic transplantation. The rejection of solid organ transplants is often associated with heavy infiltration by NK cells. NK cells that infiltrated in rat liver allografts were responsible for production of early chemokines, as well as cytolytic granzyme A and B, keys to an acute rejection episode (Illes et al., 2000). Depletion of NK cells prolonged the allograft survival in $CD28^{-/-}$ recipients. Thus, NK cells may participate in rejection by facilitating the action of alloreactive cells (Gerosa et al., 2002). In human liver transplants, a correlation between acute rejection episodes and donor–recipient KIR mismatches was discovered (Russell and Ley, 2002). In addition, NK cells also mediate chronic allograft vasculopathy in a T-cell–dependent manner (Uehara et al., 2005).

Some studies have suggested that NK cells may also promote transplant tolerance, but the mechanism is not clear as to how these cells of the innate immune response regulate the adaptive T-cell response in transplant-tolerance models (Gasser and Raulet, 2006). A recent report, using skin allografts in the $Rag^{-/-}$ mouse model, showed that in the absence of NK cells, donor-derived APC survived and migrated to the host lymphoid nodes where they directly stimulate the activation of alloreactive T cells. In the absence of NK cells, a co-stimulatory blockade-mediated graft survival was difficult to achieve. This was associated with large amounts of donor APC detected. Whereas, the donor APC were almost completely absent in the presence of NK cells, indicating that NK cells are able to induce transplant tolerance by killing donor APC, which can inhibit priming of alloreactive T cells (Yu et al., 2006). Activated NK cells are indeed capable of killing autologous immature DC. It is possible that NK cells may also influence indirect antigen presentation by killing immature host DC or DC that are mature, but express low MHC class I molecules, resulting in inhibition of the activation of T-effector cells or interfere with the induction of regulatory T cells (Kroemer et al., 2008). NK cells can also enhance allograft tolerance following direct interaction with alloreactive T cells. Induction of islet allograft tolerance induced by either anti-CD154 or anti-LFA-1 required intact perforin-mediated cytolytic activity of NK cells. NK cells promoted allograft tolerance by eliminating activated alloreactive T cells through a perforin-dependent mechanism (Beilke et al., 2005). NK cells are also able to regulate adaptive immune responses by directly inhibiting clonal expansion of activated T cells in an autoimmune disease model (Singh et al., 2001). Taken together, both APC and alloreactive T cells can be targets by NK cells in the induction of transplant tolerance. It remains unclear whether the tolerogenic effect is attributed to a specific subset of NK cells. The involvement of antigen-presenting property of NK cells also requires further investigation.

NK cell tolerance and viral infection

Although viral infection activates NK cells, viruses also use multiple strategies to manipulate the response and to achieve chronic persistence (Figure 47.2). NK cell functions (killing of target cells, ADCC effector function, editing of DC and production of cytokines and chemokines) become defective during viral infection. The defects in the NK cell compartment usually occur in the early stages of infection (Iannello et al., 2008). How does viral infection result in impaired killing activity of NK cells? The role of cell adherence molecules, conjugate formation, and polarization of cytotoxic granules is crucial for NK cell-mediated killing. In viral infections, shedding of soluble ICAMs and CD16 is increased in the circulation. The soluble forms of these adhesion molecules interfere with their membrane-inserted forms, resulting in compromised killing activity of NK cells since NK cells, due to an impaired ability to form immune synapses with target cells. The HIV protein Tat inhibits NK cell-mediated lysis by blocking L-type Ca^{++} channels because Ca^{++} influxes are crucial for activation of calcium/calmodulin-dependent protein kinase (CaMK)-II rearranging microtubules in NK cells following activation (Zocchi et al., 1998). Although NK cells from HIV infected patients form conjugates with target cells, they are defective in triggering cytolytic activity for infected targets (Bonavida et al., 1986). Interestingly, the defect in NK cells from virally infected individuals may reside in establishing a structure for triggering cytolytic activity. In addition, the number of NK cells is also decreased over time in chronic viral-infected patients, particularly, the $CD8^+CD16^+$ and $CD56^+CD16^+$ NK cell subpopulations. This is often accompanied by an increase in $CD16^+CD56^-$ NK cells that are known to express KIR, a functionally defective subpopulation. It has been demonstrated that the $CD16^-CD56^+$ subset of NK cells expand in primary viral infections. Moreover, NK cells from viral-infected patients express

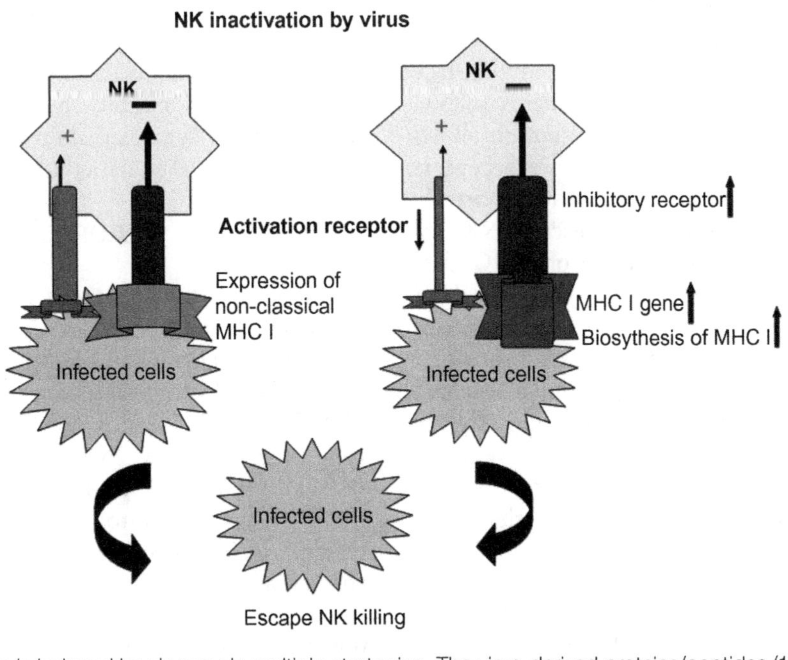

Figure 47.2 • NK tolerance is induced by viruses via multiple strategies. The virus-derived proteins/peptides (1) increase MHC class I gene expression and/or biosynthesis; (2) enhance the expression of inhibitory and/or decrease the expression of activating receptors on the surface of NK cells of the infected host; (3) can also modulate expression of non-classical MHC antigens. The inhibitory signal counters the stimulatory signals so that the NK cell is rendered inactive and the infected cell escapes NK killing.

lower levels of perforin and higher levels of SHIP, which may be responsible for their poor cytolytic and activating potentials (Alter et al., 2006). Some viruses can also produce bioactive factors, including proteins and derived peptides, that possess NK cell inhibitory properties, although the exact mechanism of inhibition of the peptides remains unknown (Alter et al., 2006).

Viruses can use other strategies to counter NK cell responses of the host. NK cell function is largely regulated by the balance of inhibitory and activating receptors. The inhibitory receptors recognize primarily MHC class I molecules, while the activating receptors recognize stress-induced ligands and viral products. Changes in the ligation of the inhibitory and activating receptors will affect the outcome of the target cells. Rapid responses of NK cells to infections may lead to efficient control of pathogen replication, while in addition to tolerance recognition, may allow for the persistence of pathogens. The down-regulation of MHC class I antigens on the surface of infected cells is a common strategy used by a variety of viruses to evade antiviral cytotoxic T lymphocyte (CTL) responses of the host, as CTL recognize viral peptides in association with these antigens (Iannello et al., 2008). For instance, upon influenza virus infection, the viral haemagglutinin (HA) protein is recognized by two NK activating receptors, NKp44 and NKp46, and this recognition leads to enhanced killing by NK cells (Arnon et al., 2004; Gazit et al., 2006). Following influenza virus infection, the binding of the two NK inhibitory receptors, KIR2DL1 and the LIR1, to the infected cells is increased, prior to the increased recognition of activating receptor, NKp46. In elucidating the mechanism responsible for this effect, MHC class I proteins were found to redistribute on the cell surface and accumulate in the lipid raft microdomains following influenza virus infection. Such redistribution allows for better recognition by the NK inhibitory receptors and consequently increases resistance to NK cell attack, suggesting that the influenza virus can escape NK cell cytotoxicity via enhancing reorganization of MHC class I on infected cells (Achdout et al., 2008). Several HIV proteins have also been shown to affect expression of MHC class I antigens: Tat antigen represses promoters of the MHC class I and the β-2 microglobulin genes, and the viral protein U (Vpu) interferes with an early step in the biosynthesis of MHC antigens. The viral protein Nef can also recognize certain motifs present in the cytoplasmic tails of MHC class I antigens and cause their degradation (Bonaparte and Barker, 2004). In addition to classical MHC class I antigens, HIV can also modulate expression of non-classical MHC antigens. HIV infection increases the expression of HLA-E on the surface of $CD4^+$ T cells. One potential mechanism of this increase is a peptide from the viral protein p24 (residues 14–22), which can bind and stabilize HLA-E on the surface of HIV-infected cells. An increased expression of HLA-E has been reported on the surface of $CD4^+$ T cells in HIV-infected persons, and the increase was more pronounced in advanced stages of infection and correlated with peaks in viraemia (Martini et al., 2005). In addition, viruses can evade innate immune responses by increasing their expression of inhibitory and/or by decreasing the expression of activating receptors on the surface of NK cells of the infected host. HIV may use this strategy to counter antiviral NK cell responses. Several studies have shown an increase in the expression of NK cell inhibitory receptors, such as KIR, and a decrease in the expression of activating receptors, such as NCR, in HIV-viraemic patients, which was correlated with viral load, and was often accompanied with decreased cytolytic activities of NK cells (Eger and Unutmaz, 2004) (see Chapter 36). This is supported by data showing that long-term administration of highly active antiretroviral therapy (HAART) results in undetectable viral loads and is associated with restored expression of the receptors on NK cells (Mavilio et al., 2003).

Viruses may also target NK cell–DC interactions for immune evasion. Activated NK cells from viraemic patients are unable to kill autologous, immature DC (Mavilio et al., 2003). This defect was more profound in the $CD56^-CD16^+$ NK cell subset, even after masking NK cell inhibitory receptors. The mature DC from HIV-infected patients produced less IL-12 and could not activate NK cells with which they where interacting. Consequently, these NK cells produce less IFN-γ. Defective NKp30- and TRAIL-mediated killing was attributed to the escape of the immature DC from NK cell-mediated killing in HIV-infected persons (Mavilio et al., 2006). Nef also affects DC and NK interactions. Nef-pulsed DC inhibit chemokine secretory capacity as well as the cytotoxic ability of NK cells by inducing TGF-β and IL-10 (Quaranta et al., 2007).

NK cells and maternal tolerance

The maternal–foetal interface is a site in which foetal-derived trophoblast cells invade into maternal uterine tissue that contains many immune cells (see Chapter 30). Over 85% of these cells are NK cells with maternal $CD56^{bright}$ $CD16^-$, but not $CD56^{dim}$ $CD16^+$, phenotype, whereas other immune cells are sparsely found (Croy et al., 2003). As is well known, the proper trophoblast invasion into maternal uterine spiral arteries and decidua is critical for optimal placentation and thus reproductive success. Failure to achieve this invasion may result in reproductive complications and a poor outcome. Depletion of decidual NK (dNK) cells in mice resulted in changes in the development of blood vessels early after implantation. dNK cell-derived IFN-γ is responsible for facilitating growth of blood vessels during decidualization (Ashkar and Croy, 1999, 2001). Indeed, invasive foetal trophoblasts are involved in attracting CD56^{bright} NK cells through production of a distinct set of chemokines (mainly stromal cell-derived factor-1 and macrophage inflammatory protein-1α), suggesting that the maternal CD56^{bright} NK cells are actively recruited by foetal trophoblast cells to populate the decidua in normal pregnancy (Hanna et al., 2003).

The interesting enigma of pregnancy is that the foetal cells are actually allografts to the maternal immune system but, in normal pregnancy, they are not rejected. dNK cells reside very close to foetal trophoblast cells of the placenta, which would seemingly lead to catastrophic consequences, because the trophoblast cells are semi-allogeneic. Many studies have tried to characterize why and how these NK cells do not exert cytolytic functions. Although the dNK cells express the essential machinery required for lysis, including perforin, granzymes A and B, their cytotoxic activity is reduced compared with peripheral blood NK or NK present in non-pregnant endometrium (Bogovic et al., 2005; Bulmer et al., 1991). However, dNK cells in vivo do not kill foetal-derived trophoblast cells. dNK cells generally express NKp30, NKp44, NKp46, NKG2D and 2B4 activating receptors, but only a few of the dNK cells express the CD160 activating receptor (Kopcow et al., 2005; Rabot et al., 2005). CD160 is a marker of cytotoxicity. During normal pregnancy, the majority of dNK cells are not cytotoxic to target cells (Tabiasco et al., 2006). dNK cells cannot kill trophoblast cells efficiently in vitro unless activated by IL-2, a cytokine not normally found in gestational endometrium (Avril et al., 1999; Loke et al., 1995; Sivori et al., 2000). Inhibition of cytotoxicity of dNK has also been proposed due to the interaction of HLA-G, HLA-E and/or HLA-C expressed by extravillous cytotrophoblast with their inhibitory receptors, ILT2, CD94/NKG2A, and KIR, respectively (Avril et al., 1999). Absence of killing of trophoblast cells by dNK could also be due to the high levels of the active form of X-linked inhibitor of apoptosis (XIAP) in the trophoblast. This potent caspase inhibitor downregulates the Fas apoptotic cascade in the trophoblast during early pregnancy (Straszewski-Chavez et al., 2004).

Pre-eclampsia is a complication of pregnancy characterized by hypertension and proteinuria, which is a major cause of maternal and foetal mortality. The exact cause of pre-eclampsia is unclear but a hallmark of pre-eclamptic pregnancies is the suboptimal invasion of trophoblast cells in the uterus and spiral arteries leading to poor restructuring of the blood supply (Pijnenborg et al., 2006). Several dNK cell receptors interacting with foetal HLA alleles may result in proangiogenic growth factor production by the dNK cells, which in turn affects the degree of trophoblast invasion, spiral artery remodelling and thus the overall quality of placentation. A strong maternal KIR and foetal HLA-C inhibitory signal predispose to pre-eclampsia (Trowsdale and Betz, 2006).

Recurrent spontaneous abortion (RSA) affects about 1% of pregnancies. Immune response may play a role in some cases. Thus, some women with unexplained RSA display increased numbers of CD56^{dim}CD3^{-} CD16^{+} dNK cells (Quenby and Farquharson, 2006). Women with RSA demonstrate a restricted repertoire of MHC-I–specific inhibitory receptors with relatively fewer inhibitory KIRs specific for HLA-Cw alleles expressed by the foetus (Varla-Leftherioti et al., 2003, 2005). However, some women with RSA also demonstrate increased numbers of MHC-I–specific activation receptor genes (Wang et al., 2007).

Breaking NK cell tolerance for cancer therapy

NK cells stand on the frontier of novel therapeutic agents against tumour growth and metastasis. Since a balance of signals from activating and inhibitory receptors determines the reactivity of NK cells, it is essential to maintain NK cell self-tolerance without compromising its reactivity against transformed tumour cells. Induction of ligands for NK cell activating receptors often occurs during transformation on tumour cells, therefore providing a way to differentiate them from normal self cells. Upregulation of the ligands for activating receptors may also result in increased NK cell susceptibility. This was demonstrated by more efficient rejection of RMA-S lymphoma cells (that possess a defect in class-I assembly and express markedly reduced levels of class-I molecules at the cell surface) by NK cells in vivo (Oberg et al., 2004). In addition, MHC-I–specific NK recognition is an important facet in regulating NK development of NK cell reactivity and tolerance.

The models discussed earlier in this chapter suggest that NK cells can be 'educated' during their maturation process and development. Thus, could NK cells be manipulated during the development and be 'tuned' to optimally sense the absence of self-MHC class I? The use of NK cells in breaking tolerance to tumours may have considerable implications in the treatment of patients with cancer. An increased cytotoxicity to tumour cells and decreased progression of syngeneic leukaemia was shown when MHC-I–specific Ly49 inhibitory NK cell receptors were blocked with F(ab)'2 fragments of 5E6 monoclonal antibody. This study supports a potential role of using blockade to NK cell inhibitory receptors as an efficient immunotherapy for cancer (Koh et al., 2001). Recent clinical data have shown that in patients with acute myeloid leukaemia, the use of KIR–ligand mismatch in a HLA-mismatched stem cell transplantation setting generated remarkable graft-versus-leukaemia effect and significantly decreased the relapse rate in these patients as well as improving their overall survival. Donor NK cells may exert their potent anti-leukaemic effector response in the host via missing-self-recognition (Plunkett and Ellis, 2002; Ruggeri et al., 1999). CEACAM1 (carcinoembryonic antigen-related cell adhesion molecule 1 [CD66a]), an inhibitory receptor expressed by NK cells, mediates immune self-tolerance in a non-MHC-I–specific fashion. The ligand for CEACAM1 is CEACAM1 itself. It is expressed at higher levels in multiple types of malignancies, such as in colon, prostate, breast, and endometrial cancer (Thies et al., 2002). CEACAM1 can also be expressed by cells in malignant melanoma, in which its increased expression correlates with increased metastasis (Thies et al., 2002), thus indicating a poor prognosis (Markel et al., 2002). CEACAM1 has many roles in carcinogenesis. It was found to inhibit the anti-tumour activity of NK cells in vitro (Gerosa et al., 2002; Illes et al., 2000), but demonstrated other features in vivo, such as the ability to promote angiogenesis and metastasis, and tumour-suppressor activity (Plunkett and Ellis, 2002). Thus, modulation of non-MHC-I–specific receptors such as CEACAM1, may be another potential means for the immunotherapy of patients with cancer.

References

Achdout, H., Manaster, I. and Mandelboim, O. (2008). Influenza virus infection augments NK cell inhibition through reorganization of major histocompatibility complex class I proteins. J Virol **82**, 8030–8037.

Adams, E.J., Juo, Z.S., Venook, R.T., Boulanger, M.J., Arase, H., Lanier, L. L. and Garcia, K.C. (2007). Structural elucidation of the m157 mouse cytomegalovirus ligand for Ly49 natural killer cell receptors. Proc Natl Acad Sci USA **104**, 10128–10133.

Alter, G., Suscovich, T., Kleyman, M., Teigen, N., Streeck, H., Zaman, M., Meier, A. and Altfeld, M. (2006). Low perforin and elevated SHIP-1 expression is associated with functional anergy of natural killer cells in chronic HIV-1 infection. AIDS **20**, 1549–1551.

Arase, H., Arase, N. and Saito, T. (1996). Interferon gamma production by natural killer (NK) cells and NK1.1$^+$ T cells upon NKR-P1 cross-linking. J Exp Med **183**, 2391–2396.

Arnon, T.I., Achdout, H., Lieberman, N., Gazit, R., Gonen-Gross, T., Katz, G., Bar-Ilan, A., Bloushtain, N., Lev, M., Joseph, A., Kedar, E., Porgador, A. and Mandelboim, A. (2004). The mechanisms controlling the recognition of tumour- and virus-infected cells by NKp46. Blood **103**, 664–672.

Ashkar, A.A. and Croy, B.A. (1999). Interferon-γ contributes to the normalcy of murine pregnancy. Biol Reprod **61**, 493–502.

Ashkar, A.A. and Croy, B.A. (2001). Functions of uterine natural killer cells are mediated by interferon γ production during murine pregnancy. Semin Immunol **13**, 235–241.

Avril, T., Jarousseau, A.C., Watier, H., Boucraut, J., Le Bouteiller, P. and Bardos, P. (1999). Trophoblast cell line resistance to NK lysis mainly involves an HLA class I-independent mechanism. J Immunol **162**, 5902–5909.

Beilke, J.N. and Gill, R.G. (2007). Frontiers in nephrology: the varied faces of natural killer cells in transplantation—contributions to both allograft immunity and tolerance. J Am Soc Nephrol **18**, 2262–2267.

Beilke, J.N., Kuhl, N.R., Kaer, L.V. and Gill, R.G. (2005). NK cells promote islet allograft tolerance via a perforin-dependent mechanism. Nat Med **11**, 1059–1065.

Bix, M., Liao, N.S. and Zijlstra, M. (1991). Rejection of class I MHC-deficient haemopoietic cells by irradiated MHC-matched mice. Nature **349**, 329–331.

Boles, K.S., Stepp, S.E., Bennett, M., Kumar, V. and Mathew, PA. (2001). 2B4 (CD244) and CS1: novel members of the CD2 subset of the immunoglobulin superfamily molecules expressed on natural killer cells and other leukocytes. Immunol Rev **181**, 234–249.

Bogovic, C.T., Laskarin, G., Juretic, K., Strbo, N., Dupor, J. and Srsen, S. (2005). Perforin and Fas/FasL cytolytic pathways at the maternal–fetal interface. Am J Reprod Immunol **54**, 241–248.

Bonaparte, M.I. and Barker, E. (2004). Killing of human immunodeficiency virus-infected primary T-cell blasts by autologous natural killer cells is dependent on the ability of the virus to alter the expression of major histocompatibility complex class I molecules. Blood **104**, 2087–2094.

Bonavida, B., Katz, J. and Gottlieb, M. (1986). Mechanism of defective NK cell activity in patients with acquired immunodeficiency syndrome (AIDS) and AIDS-related complex. I. Defective trigger on NK cells for NKCF production by target cells, and partial restoration by IL 2. J Immunol **137**, 1157–1163.

Brown, M.H., Boles, K., van der Merwe, P. A., Kumar, V., Mathew, P.A. and Barclay, A.N. (1998). 2B4, the natural killer and T cell immunoglobulin superfamily surface protein, is a ligand for CD48. J Exp Med **188**, 2083–2090.

Bulmer, J., Longfellow, M. and Ritson, A. (1991). Leukocytes and resident blood cells in endometrium. Ann N Y Acad Sci **622**, 57–68.

Carlyle, J.R., Jamieson, A.M., Gasser, S., Clingan, C.S., Arase, H. and Raulet, D. H. (2004). Missing self-recognition of Ocil/Clr-b by inhibitory NKR-P1 natural killer cell receptors. Proc Natl Acad Sci USA **101**, 3527–3532.

Cooper, M.A., Fehniger, T.A. and Caligiuri, M.A. (2001). The biology of human natural killer-cell subsets. Trends Immunol **22**, 633–640.

Corral, L., Hanke, T., Vance, R.E., Cado, D. and Raulet, D.H. (2000). NK cell expression of the killer cell lectin-like receptor G1 (KLRG1), the mouse homolog of MAFA, is modulated by MHC class I molecules. Eur J Immunol **30**, 920–930.

Croy, B.A., Esadeg, S., Chantakru, S., van den Heuvel, M., Paffaro, V.,A., He, H., Black, G.P., Ashka,r, A.A., Kiso, Y. and Zhang, J. (2003). Update on pathways regulating the activation of uterine natural killer cells, their interactions with decidual spiral arteries and homing of their precursors to the uterus. J Reprod Immunol **59**, 175–191.

Daniels, K.A., Devora, G., Lai, W.C., O'Donnell, C.L., Bennett, M. and Welsh, R.M. (2001). Murine cytomegalovirus is regulated by a discrete subset of natural killer cells reactive with monoclonal antibody to Ly49H. J Exp Med **194**, 29–44.

Davis, A.H., Guseva, N.V., Ball, B.L. and Heusel, J.W. (2008). Characterization of murine cytomegalovirus m157 from infected cells and identification of critical residues mediating recognition by the NK cell receptor Ly49H. J Immunol **181**, 265–275.

Degli-Esposti, M.A. and Smyth, M.J. (2005). Close encounters of different kinds: dendritic cells and NK cells take centre stage. Nat Rev Immunol **5**, 112–124.

Dorfman, J.R. and Raulet, D.H. (1998). Acquisition of Ly49 receptor expression by developing natural killer cells. J Exp Med **187**, 609–618.

Dorfman, J.R., Zerrahn, J., Coles, M.C. and Raulet, D.H. (1997). The basis for self-tolerance of natural killer cells in beta2-microglobulin- and TAP-1-mice. J Immunol **159**, 5219–5225.

Dowdell, K.C., Cua, D.J., Kirkman, E. and Stohlman, S.A. (2003). NK cells regulate CD4 responses prior to antigen encounter. J Immunol **171**, 234–239.

Eger, K.A. and Unutmaz, D. (2004). Perturbation of natural killer cell function and receptors during HIV infection. Trends Microbiol **12**, 301–303.

Ferlazzo, G., Tsang, M.L., Moretta, L., Melioli, G., Steinman, R.M. and Munz, C. (2002). Human dendritic cells activate resting natural killer (NK) cells and are recognized via the NKp30 receptor by activated NK cells. J Exp Med **195**, 343–35157.

Ferlazzo, G., Morandi, B., D'Agostino, A., Meazza, R., Melioli, G., Moretta, A. and Moretta, L. (2003). The interaction between NK cells and dendritic cells in bacterial infections results in rapid induction of NK cell activation and in the lysis of uninfected dendritic cells. Eur J Immunol **33**, 306–313.

Fernandez, N.C., Treiner, E., Vance, R.E., et al. (2005). A subset of natural killer cells achieves self-tolerance without expressing inhibitory receptors specific for self-MHC molecules. Blood **105**, 4416–4423.

Gasser, S. and Raulet, D.H. (2006). Activation and self-tolerance of natural killer cells. Immunol Rev **214**, 130–142.

Gazit, R., Gruda, R., Elboim, M., Arnon, T.I., Katz, G., Achdout, H., Hanna, J., Qimron, U., Landau, G., Greenbaum, E., Zakay-Rones, Z., Porgador, A. and Mandelboim, O. (2006). Lethal influenza infection in the absence of the natural killer cell receptor gene Ncr1. Nat Immunol **7**, 517–523.

Gerosa, F., Baldani-Guerra, B., Nisii, C., Marchesini, V., Carra, G. and Trinchieri, G. (2002). Reciprocal activating interaction between natural killer cells and dendritic cells. J Exp Med **195**, 327–333.

Hanke, T., Takizawa, H. and Raulet, D.H. (2001). MHC-dependent shaping of the inhibitory Ly49 receptor repertoire on NK cells: evidence for a regulated sequential model. Eur J Immunol **31**, 3370–3379.

Hanna, J., Wald, O., Goldman-Wohl, D., Prus, D., Markel, G., Gazit, R., Katz, G., Haimov-Kochman, R., Fujii, N., Yagel, S., Peled, A. and Mandelboim, O. (2003). CXCL12 expression by invasive trophoblasts induces the specific migration of $CD16^-$ human natural killer cells. Blood **102**, 1569–1577.

Hanna, J., Bechtel, P., Zhai, Y., Youssef, F., McLachlan, K. and Mandelboim, O. (2004). Novel insights on human NK cells' immunological modalities revealed by gene expression profiling. J Immunol **173**, 6547–6563.

Hanna, J., Gonen-Gross, T., Fitchett, J., Rowe, T., Daniels, M., Arnon, T.I., Gazit, R., Joseph, A., Schjetne, K.W., Steinle, A., Porgador, A., Mevorach, D., Goldman-Wohl, D., Yagel, S., LaBarre, M.J., Buckner, J.H. and Mandelboim, O. (2004). Novel APC-like properties of human NK cells directly regulate T cell activation. J Clin Invest **114**, 1612–1623.

Hanna, J., Fitchett, J., Rowe, T., Daniels, M., Heller, M., Gonen-Gross, T., Manaster, E., Cho, S.Y., LaBarre, M.J. and Mandelboim, O. (2005). Proteomic analysis of human natural killer cells: insights on new potential NK immune functions. Mol Immunol **42**, 425–431.

Hayakawa, Y., Screpanti, V., Yagita, H., Grandien, A., Ljunggren, H.G., Smyth, M.J. and Chambers, B.J. (2004). NK cell TRAIL eliminates immature dendritic cells in vivo and limits dendritic cell vaccination efficacy. J Immunol **172**, 123–129.

Held, W., Dorfman, J.R. and Wu, M. F. (1996). Major histocompatibility complex class I-dependent skewing of the natural killer cell Ly49 receptor repertoire. Eur J Immunol **26**, 2286–2292.

Iannello, A., Debbeche, O., Samarani, S. and Ahmad, A. (2008). Antiviral NK cell responses in HIV infection: II. viral strategies for evasion and lessons for immunotherapy and vaccination. J Leukoc Biol **84**, 27–49.

Iizuka, K., Naidenko, O.V., Plougastel, B. F., Fremont, D.H. and Yokoyama, W. M. (2003). Genetically linked C-type lectin-related ligands for the NKRP1 family of natural killer cell receptors. Nat Immunol **4**, 801–807.

Illes, Z., Kondo, T., Newcombe, J., Oka, N., Tabira, T. and Yamamura, T. (2000). Differential expression of NK T cell V alpha 24J alpha Q invariant TCR chain in the lesions of multiple sclerosis and chronic inflammatory demyelinating polyneuropathy. J Immunol **164**, 4375–4381.

Karre, K., Ljunggren, H.G. and Piontek, G. (1986). Selective rejection of H-2-deficient lymphoma variants suggests alternative immune defence strategy. Nature **319**, 675–678.

Kean, L.S., Hamby, K., Lee, B.E., Coley, S., Stempora, L., Adama, A. B., Heiss, E., Pearson, T.C. and Larsen, C.P. (2006). NK cells mediate costimulation blockade-resistant rejection of allogenenic stem cells during nonmyeloablative transplantation. Am J Transplant **6**, 292–304.

Kim, S., Iizuka, K., Kang, H.S., Dokun, A., French, A.R., Greco, S. and Yokoyama, W.M. (2002). In vivo developmental stages in murine natural killer cell maturation. Nat Immunol **3**, 523–528.

Kim, S., Poursine-Laurent, J., Truscott, S. M., Lybarger, L., Song, Y.J., Yang, L., French, A.R., Sunwoo, J.B., Lemieux, S. and Hansen, T.H. (2005). Licensing of natural killer cells by host major histocompatibility complex class I molecules. Nature **436**, 709–713.

Kitchens, W.H., Uehara, S., Chase, C.M., Colvin, R.B., Russell, P.S. and Madsen, J. C. (2006). The changing role of natural killer cells in solid organ rejection and tolerance. Transplantation **81**, 811–817.

Koh, C.Y., Blazar, B.R., George, T., Welniak, L.A., Capitini, C.M., Raziuddin, A., Murphy, W.J. and Bennett, M. (2001). Augmentation of antitumour effects by NK cell inhibitory receptor blockade in vitro and in vivo. Blood **97**, 3132–3137.

Kopcow, H.D., Allan, D.S., Chen, X., Rybalov, B., Andzelm, M.M. and Ge, B. (2005). Human decidual NK cells form immature activating synapses and are not cytotoxic. Proc Natl Acad Sci U S A **102**, 15563–15568.

Kroemer, A., Edtinger, K. and Li, X.C. (2008). The innate natural killer cells in transplant rejection and tolerance induction. Curr Opin Organ Transplant **13**, 339–343.

Kubota, K. (2002). A structurally variant form of the 2B4 antigen is expressed on the cell surface of mouse mast cells. Microbiol Immunol **46**, 589–592.

Lanier, L.L. (2005). NK cell recognition. Annu Rev Immunol **23**, 225–274.

Lee, K.M., McNerney, M.E., Stepp, S.E., Mathew, P.A., Schatzle, J.D., Bennett, M. and Kumar, V. (2004). 2B4 acts as a non-major histocompatibility complex binding inhibitory receptor on mouse natural killer cells. J Exp Med **199**, 1245–1254.

Ljunggren, H.G. and Karre, K. (1990). In search of the 'missing self': MHC molecules and NK cell recognition. Immunol Today **11**, 237–244.

Loke, Y.W., King, A. and Burrows, T.D. (1995). Decidua in human implantation. Hum Reprod **10**(Suppl. 2), 14–21.

Mailliard, R.B., Alber, S.M., Shen, H., Watkins, S.C., Kirkwood, J.M., Herberman, R.B. and Kalinski, P. (2005). IL-18-induced $CD83^+$ $CCR7^+$ NK helper cells. J Exp Med **202**, 941–953.

Markel, G., Lieberman, N., Katz, G., Arnon, T.I., Lotem, M., Drize, O., Blumberg, R.S., Bar-Haim, E., Mader, R., Eisenbach, L. and Mandelboim, O. (2002). CD66a interactions between human melanoma and NK cells: a novel class I MHC-independent inhibitory mechanism of cytotoxicity. J Immunol **168**, 2803–2810.

Markel, G., Achdout, H., Katz, G., Ling, K.L., Salio, M., Gruda, R., Gazit, R., Mizrahi, S., Hanna, J., Gonen-Gross, T., Arnon, T.I., Lieberman, N., Stren, N., Nachfmias, B., Blumberg, R.S., Steuer, G., Blau, H., Cerundolo, V., Mussaffi, H. and Mandelboim, O. (2004). Biological function of the soluble CEACAM1 protein and implications in TAP2-deficient patients. Eur J Immunol **34**, 2138–2148.

Martin-Fontecha, A., Thomsen, L.L., Brett, S., Gerard, C., Lipp, M., Lanzavecchia, A. and Sallusto, F. (2004). Induced recruitment of NK cells to lymph nodes provides IFN-γ for T_H 1 priming. Nat Immunol **5**, 1260–1265.

Martini, F., Agrati, C., D'Offizi, G. and Poccia, F. (2005). HLA-E up-regulation induced by HIV infection may directly contribute to CD94-mediated impairment of NK cells. Int J Immunopathol Pharmacol **18**, 269–276.

Mavilio, D., Benjamin, J., Daucher, M., Lombardo, G., Kottilil, S., Planta, M. A., Marcenaro, E., Bottino, C., Moretta, L., Moretta, A. and Fauci, A.S. (2003). Natural killer cells in HIV-1 infection: dichotomous effects of viremia on inhibitory and activating receptors and their functional correlates. Proc Natl Acad Sci U S A **100**, 15011–15016.

Mavilio, D., Lombardo, G., Kinter, A., Fogli, M., La Sala, A., Ortolano, S., Farschi, A., Follmann, D., Gregg, R., Kovacs, C., Marcenaro, E., Pende, D., Moretta, A. and Fauci, A.S. (2006). Characterization of the defective interaction between a subset of natural killer cells and dendritic cells in HIV-1 infection. J Exp Med **203**, 2339–2350.

McMahon, C.W. and Raulet, D.H. (2001). Expression and function of NK cell receptors in $CD8^+$ T cells. Curr Opin Immunol **13**, 465–470.

McNerney, M.E., Guzior, D. and Kumar, V. (2005). 2B4 (CD244)–CD48 interactions provide a novel MHC class I-independent system for NK-cell self-tolerance in mice. Blood **106**, 1337–1340.

Moretta, A., Bottino, C. and Vitale, M. (1996). Receptors for HLA class-I molecules in human natural killer cells. Annu Rev Immunol **14**, 619–648.

Moretta, A., Bottino, C., Vitale, M., Pende, D., Cantoni, C., Mingari, M.C., Biassoni, R. and Moretta, L. (2001). Activating receptors and coreceptors involved in human natural killer cell-mediated cytolysis. Annu Rev Immunol **19**, 197–223.

Oberg, L., Johansson, S., Michaelsson, J., Tomasello, E., Vivier, E., Karre, K. and Hoglund, P. (2004). Loss or mismatch of MHC class I is sufficient to trigger NK cell-mediated rejection of resting lymphocytes in vivo—role of KARAP/DAP12-dependent and -independent pathways. Eur J Immunol **34**, 1646–1653.

Ohlen, C., Kling, G. and Hoglund, P. (1989). Prevention of allogeneic bone marrow graft rejection by H-2 transgene in donor mice. Science **246**, 666–668.

Parolini, S., Bottino, C., Falco, M., Augugliaro, R., Giliani, S., Franceschini, R., Ochs, H.D., Wolf, H., Bonnefoy, J.Y., Biassoni, R., Moretta, L., Notarangelo, L.D. and Moretta, A. (2000). X-linked lymphoproliferative disease. 2B4 molecules displaying inhibitory rather than activating function are responsible for the inability of natural killer cells to kill Epstein–Barr virus-infected cells. J Exp Med **192**, 337–346.

Piccioli, D., Sbrana, S., Melandri, E. and Valiante, N.M. (2002). Contact-dependent stimulation and inhibition of dendritic cells by natural killer cells. J Exp Med **195**, 335–341.

Pijnenborg, R., Vercruysse, L. and Hanssens, M. (2006). The uterine spiral arteries in human pregnancy: facts and controversies. Placenta **27**, 939–958.

Plunkett, T.A. and Ellis, P.A. (2002). CEACAM1: a marker with a difference or more of the same? J Clin Oncol **20**, 4273–4275.

Quaranta, M.G., Napolitano, A., Sanchez, M., Giordani, L., Mattioli, B. and Viora, M. (2007). HIV-1 Nef impairs the dynamic of DC/NK crosstalk: different outcome of CD56dim and CD56bright NK cell subsets. FASEB J **21**, 2323–2334.

Quenby, S. and Farquharson, R. (2006). Uterine natural killer cells, implantation failure and recurrent miscarriage. Reprod Biomed Online **13**, 24–28.

Rabot, M., Tabiasco, J., Polgar, B., Aguerre-Girr, M., Berrebi, A. and Bensussan, A. (2005). HLA class I/NK cell receptor interaction in early human decidua basalis: possible functional consequences. In: Markert, U. (ed), Immunology of PregnancyVol. 89. Basel: Karger, pp. 72–83.

Raulet, D.H., Vance, R.E. and McMahon, C.W. (2001). Regulation of the natural killer cell receptor repertoire. Annu Rev Immunol **19**, 291–330.

Riley, J.K. and Yokoyama, W.M. (2008). NK cell tolerance and the maternal–fetal interface. Am J Reprod Immunol **59**, 371–387.

Roncagalli, R., Taylor, J.E., Zhang, S., Shi, X., Chen, R., Cruz-Munoz, M.E., Yin, L., Latour, S. and Veillette, A. (2005). Negative regulation of natural killer cell function by EAT-2, a SAP-related adaptor. Nat Immunol **6**, 1002–1010.

Roth, C., Carlyle, J.R., Takizawa, H. and Raulet, D.H. (2000). Clonal acquisition of inhibitory Ly49 receptors on developing NK cells is successively restricted and regulated by stromal class I MHC. Immunity **13**, 143–153.

Ruggeri, L., Capanni, M., Casucci, M., Volpi, I., Tosti, A., Perruccio, K., Urbani, E., Negrin, R.S., Martelli, M.F. and Velardi, A. (1999). Role of natural killer cell alloreactivity in HLA-mismatched hematopoietic stem cell transplantation. Blood **94**, 333–339.

Ruggeri, L., Capanni, M., Urbani, E., Perruccio, K., Shlomchik, W.D., Tosti, A., Posati, S., Rogaia, D., Frassoni, F., Aversa, F., Martelli, M.F. and Velardi, A. (2002). Effectiveness of donor natural killer cell alloreactivity in mismatched hematopoietic transplants. Science **295**, 2097–2100.

Russell, J.H. and Ley, T.J. (2002). Lymphocyte-mediated cytotoxicity. Annu Rev Immunol **20**, 323–370.

Salcedo, M., Diehl, A.D., Olsson-Alheim, M.Y., Sundback, J., Van Kaer, L., Karre, K. and Ljunggren, H.G. (1997). Altered expression of Ly49 inhibitory receptors on natural killer cells from MHC class I-deficient mice. J Immunol **158**, 3174–3180.

Scott, P. and Trinchieri, G. (1995). The role of natural killer cells in host–parasite interactions. Curr Opin Immunol **7**, 34–40.

Singh, A.K., Wilson, M.T., Hong, S., Olivares-Villagomez, D., Du, C., Stanic, A.K., Joyce, S., Sriram, S., Koezuka, Y. and Van Kaer, L. (2001). Natural killer T cell activation protects mice against experimental autoimmune encephalomyelitis. J Exp Med **194**, 1801–1811.

Sivori, S., Parolini, S., Marcenaro, E., Millo, R., Bottino, C. and Moretta, A. (2000). Triggering receptors involved in natural killer cell-mediated cytotoxicity against choriocarcinoma cell lines. Hum Immunol **61**, 1055–1058.

Straszewski-Chavez, S., Abrahams, V., Funai, E. and Mor, G. (2004). X-linked inhibitor of apoptosis (XIAP) confers human trophoblast cell resistance to Fas-mediated apoptosis. Mol Hum Reprod **10**, 33–41.

Sun, J.C. and Lanier, L.L. (2008). Tolerance of NK cells encountering their viral ligand during development. J Exp Med **205**, 1819–1828.

Tabiasco, J., Rabot, M., Aguerre-Girr, M., El Costa, H., Berrebi, A., Parant, O., Laskarin, G., Juretic, K., Bensussan, A., Rukavina, D. and Le Bouteiller, P. (2006). Human decidual NK cells: unique phenotype and functional properties—a review. Placenta **27**(Suppl. A), S34–S39.

Tangye, S.G., Phillips, J.H., Lanier, L.L. and Nichols, K.E. (2000). Functional requirement for SAP in 2B4-mediated activation of human natural killer cells as revealed by the X-linked lymphoproliferative syndrome. J Immunol **165**, 2932–2936.

Thies, A., Moll, I., Berger, J., Wagener, C., Brümmer, J., Schulze, H.J., Brunner, G. and Schumacher, U. (2002). CEACAM1 expression in cutaneous malignant melanoma predicts the development of metastatic disease. J Clin Oncol **20**, 2530–2536.

Tripathy, S.K., Keyel, P.A., Yang, L., Pingel, J.T., Cheng, T.P., Schneeberger, A. and Yokoyama, W.M. (2008). Continuous engagement of a self-specific activation receptor induces NK cell tolerance. J Exp Med **205**, 1829–1841.

Trowsdale, J. and Betz, A.G. (2006). Mother's little helpers: mechanisms of maternal–fetal tolerance. Nat Immunol **7**, 241–246.

Uehara, S., Chase, C.M., Kitchens, W.H., Rose, H.S., Colvin, R.B., Russell, P.S. and Madsen, J.C. (2005). NK cells can trigger allograft vasculopathy: the role of hybrid resistance in solid organ allografts. J Immunol **175**, 3424–3430.

Unanue, E.R. (1997). Inter-relationship among macrophages, natural killer cells and neutrophils in early stages of *Listeria* resistance. Curr Opin Immunol **9**, 35–43.

Vaidya, S.V. and Mathew, P.A. (2006). Of mice and men: different functions of the murine and human 2B4 (CD244) receptor on NK cells. Immunol Lett **105**, 180–184.

Vaidya, S.V., Stepp, S.E., McNerney, M. E., Lee, J.K., Bennett, M., Lee, K.M., Stewart, C.L., Kumar, V. and Mathew, P.A. (2005). Targeted disruption of the 2B4 gene in mice reveals an in vivo role of 2B4 (CD244) in the rejection of B16 melanoma cells. J Immunol **174**, 800–807.

Valiante, N.M., Uhrberg, M. and Shilling, H.G. (1997). Functionally and structurally distinct NK cell receptor repertoires in the peripheral blood of two human donors. Immunity **7**, 739–751.

Vance, R.E., Kraft, J.R. and Altman, J. D. (1998). Mouse CD94/NKG2A is a natural killer cell receptor for the nonclassical major histocompatibility complex (MHC) class I molecule Qa-1(b). J Exp Med **188**, 1841–1848.

Vankayalapati, R., Klucar, P., Wizel, B., Weis, S.E., Samten, B., Safi, H., Shams, H. and Barnes, P.F. (2004). NK cells regulate CD8+ T cell effector function in response to an intracellular pathogen. J Immunol **172**, 130–137.

Varla-Leftherioti, M., Spyropoulou-Vlachou, M., Niokou, D., Keramitsoglou, T., Darlamitsou, A., Tsekoura, C., Papadimitropoulos, M., Lepage, V., Balafoutas, C. and Stavropoulos-Giokas, C. (2003). Natural killer (NK) cell receptors' repertoire in couples with recurrent spontaneous abortions. Am J Reprod Immunol **49**, 183–191.

Varla-Leftherioti, M., Spyropoulou-Vlachou, M., Keramitsoglou, T., Papadimitropoulos, M., Tsekoura, C., Graphou, O., Papadopoulou, C., Gerondi, M. and Stavropoulos-Giokas, C. (2005). Lack of the appropriate natural killer cell inhibitory receptors in women with spontaneous abortion. Hum Immunol **66**, 65–71.

Vitale, M., Zimmer, J., Castriconi, R., Hanau, D., Donato, L., Bottino, C., Moretta, L., de la Salle, H. and Moretta, A. (2002). Analysis of natural killer cells in TAP2-deficient patients: expression of functional triggering receptors and evidence for the existence of inhibitory receptor(s) that prevent lysis of normal autologous cells. Blood **99**, 1723–1729.

Wang, S., Zhao, Y.R., Jiao, Y.L., Wang, L.C., Li, J.F., Cui, B., Xu, C.Y., Shi, Y. H. and Chen, Z.J. (2007). Increased activating killer immunoglobulin-like receptor genes and decreased specific HLA-C alleles in couples with recurrent spontaneous abortion. Biochem Biophys Res Commun **360**, 696–701.

Werner, H. (2008). NK cell education: licensing, arming, disarming, tuning? Which model? Tolerance and reactivity of NK cells: two sides of the same coin? Eur J Immunol **38**, 2927–2968.

Xu, J., Chakrabarti, A.K., Tan, J.L., Ge, L., Gambotto, A. and Vujanovic, N.L. (2007). Essential role of the TNF–TNFR2 cognate interaction in mouse dendritic cell–natural killer cell crosstalk. Blood **109**, 3333–3341.

Yokoyama, W.M. (1995). Natural killer cell receptors specific for major histocompatibility complex class I molecules. Proc Natl Acad Sci USA **92**, 3081–3085.

Yokoyama, W.M. and Plougastel, B.F. (2003). Immune functions encoded by the natural killer gene complex. Nat Rev Immunol **3**, 304–316.

Yokoyama, W.M. and Seaman, W.E. (1993). The Ly-49 and NKR-P1 gene families encoding lectin-like receptors on natural killer cells: the NK gene complex. Annu Rev Immunol **11**, 613–635.

Yu, Y.Y., Netuschil, N., Lybarger, L., Connolly, J.M. and Hansen, T.H. (2002). Cutting edge: single-chain trimers of MHC class I molecules form stable structures that potently stimulate antigen-specific T cells and B cells. J Immunol **168**, 3145–3149.

Yu, G., Xu, X., Vu, M.D., Kilpatrick, E.D. and Li, X.C. (2006). NK cells promote transplant tolerance by killing donor antigen-presenting cells. J Exp Med **203**, 1851–1858.

Zingoni, A., Sornasse, T., Cocks, B.G., Tanaka, Y., Santoni, A. and Lanier, L.L. (2004). Cross-talk between activated human NK cells and $CD4^+$ T cells via OX40–OX40 ligand interactions. J Immunol **173**, 3716–3724.

Zitvogel, L. (2002). Dendritic and natural killer cells cooperate in the control/switch of innate immunity. J Exp Med **195**, F9–F14.

Zocchi, M.R., Rubartelli, A., Morgavi, P. and Poggi, A. (1998). HIV-1 Tat inhibits human natural killer cell function by blocking L-type calcium channels. J Immunol **161**, 2938–2943.

Chapter Forty-Eight

48

Genetic engineering of natural killer cells

Stephen R. Goding, Per H. Basse

CHAPTER CONTENTS

Character, I am sure, lies in the genes.

Taylor Caldwell (1900–1985)

ABSTRACT

Natural killer (NK) cells can kill tumour cells without prior activation. The NK-mediated anti-tumour response is, however, short-lived, but can be boosted substantially by stimulation with cytokines such as interleukin-2 (IL-2) and IL-12. To optimally stimulate NK cells with cytokines without the need for systemic, often toxic, cytokine administration, many efforts have been made to genetically engineer the NK cells to produce a variety of cytokines themselves. NK cells are notoriously difficult to transfect cells, but increasing evidence that NK cells can be successfully transfected has accumulated. Thus, many transgenes expressing markers such as green fluorescent protein (GFP) and cytokines (IL-2, IL-12, etc.) as well as other proteins have successfully been transduced into high percentages of both primary NK cells and NK cell lines of both human and mouse origin. Here, we summarize some of these findings and demonstrate how cytokine gene transduction of NK cells enhances the antitumour efficacy of adoptively transferred NK cells in murine models of lung metastasis.

KEY WORDS

Gene transfer, Adoptive transfer, Tumour infiltration, Cytokines, Adenovirus, Immunotherapy

Introduction

Immunotherapy of cancers has involved the use of T cells, B cells/antibodies, dendritic cells (DCs), and, to a lesser extent, natural killer (NK) cells. While T and B cells offers a high degree of specificity and continued reactivity as long as the antigen is present, the advantages of NK cells are immediate reactivity and the ability to response to a wide range of targets in a non-MHC-restricted manner. On the other hand, NK cells seem unable to establish a long-lived reaction against malignant targets, and although their density in tumours correlates positively with prognosis in colorectal (Coca et al., 1997) and other cancers, they are usually not found in very high numbers in established tumours. The NK cells' tumour tropism can, however, be greatly enhanced by ex vivo activation with cytokines (Gunji et al., 1989). In fact, IL-2-activated NK cells, called A-NK cells, can accumulate to densities of $>100 \times 10^6$ cells/g tumour tissue following adoptive transfer into animals bearing B16 lung tumours (Basse et al., 1991a).

The accumulation of A-NK cells in B16 lung tumours leads to a significant reduction in tumour burden (Yang et al., 2003) and to prolonged survival of the host. However, the A-NK cells do not seem to persist in the tumours for much more than 4–6 days. This may explain why long-term survival and/or cures following NK cell-based immunotherapy of established tumours have not yet been reported. Clearly, the magnitude and persistence of the NK cells' response to tumours need to be improved to make these cells of value in clinical settings. One way to improve and boost the anti-tumour efficiency of NK cells is to genetically alter them so that they acquire better target recognition or target-killing capabilities, or so that they can secrete cytokines that either stimulate the NK cells or other antitumour effector arms of the immune system.

Expression of transgenes by primary NK cells and NK cell lines

In the last decade, increasing data have been generated in the transduction of NK cells. Most prominently is the work involving human NK lines. Their transduction has been conducted using various approaches. In 2003, Trompeter et al. showed transient transduction of the NK line, NKL, to express luciferase and fluorescent proteins, using nucleofection, a means of direct DNA transfer into the nucleus. This technique lead to positive production within hours of the transfection, but the level of efficiency was <10% (Trompeter et al., 2003). That same year, Maasho employed modified nucleofection technology, resulting in efficiencies >50% in the transient expression of Rab proteins, members of the Ras superfamily of G proteins, and Src-homology phosphatase type (SHP) 1 by NKL cells. This technique was not toxic, as evidenced by 85–90% viability of the cells (Maasho et al., 2004). Grund and Muise-Helmericks (2005) developed a novel, nontoxic means of electroporation resulting in 40% fluorescent protein transgene expression efficiency of the human NK cell line, NK-92. Recently, primary mouse NK cells were transduced with a chimeric single-chain TCR/CD28 construct to recognize the tumour-associated antigen, erbB2, by Pegram et al. using nucleofection. The transduced NK cells were then able to mediate greater survival of mice bearing erbB2+ lymphoma tumours than nontransduced NK cells (Pegram et al., 2008).

Viral vectors have gained attention in transducing NK cells. Schroers et al. demonstrated a marked increase in green fluorescent protein (GFP) expression in primary human NK cells, >10-fold, with the application of an adenovirus containing a modified fibre protein that recognizes ubiquitous CD46 compared to the conventional adenovirus that gains entry into cells through coxsackievirus and adenovirus receptor (CAR) and integrins (Schroers et al., 2004). Regarding stable transduction, Guven et al. transduced NK cells obtained from human blood, using retroviruses. They demonstrated 27–52% GFP positive cells with one round of retroviral transduction, which could be enhanced to 47–75% with a second round of transduction. The differences in the range of transduction depended on the cell culture period, with greater culture time corresponding with higher GFP expression (Guven et al., 2005). Additionally, lentivirus transduction has lead to very high efficiency of transduction, >95% GFP +, for the NK cell line, YTS (Micucci et al., 2006). Micucci et al. have also shown that while transduction of NK cell lines is in general easier to achieve than transduction of freshly isolated/primary NK cells, it is possible to transduce primary NK cells virally. For example, fresh human NK cells cultured in IL-2 plus IL-12 reached transduction rates between 50% and 90% using lentivirus vectors, whereas primary cultures lacking cytokine stimulation dropped to 20% GFP+ (Micucci et al., 2006). Recently, lentiviral transduction has proven successful (up to 40% efficiency) in primary murine NK cells as well. GFP expression persisted for at least 2 weeks and transduction was maintained whether or not the NK cells were cytokine-stimulated in culture (Tran and Kung, 2007).

While the studies discussed above demonstrate that transduction or transfection of NK cells and cell lines with the genes for marker molecules such as GFP is indeed possible, other studies have shown that other functional molecules, in particular cytokines, can be produced by NK cells following transduction. The initial work involving NK cells and transgene delivery methods investigated the expression of the essential NK cell survival cytokine, IL-2. Nagashima et al. showed that the human NK cell line, NK-92, could be transduced with retroviral vectors to express IL-2. By transducing this IL-2-dependent cell line to express IL-2, it could maintain its own survival, as well as mediate increased survival of mice bearing liver metastases, compared to nontransduced cells following adoptive transfer into tumour-bearing animals (Nagashima et al., 1998). Similar in vitro results were obtained by Tam et al. who transduced NK-92 cells with the IL-2 gene via cDNA particle-mediated gene transfer (Tam et al., 1999). In addition, IL-15, another pro-NK cell cytokine, has been stably transduced into NK-92 and NKL cell lines with cDNA vectors (Jiang et al., 2008; Zhang et al., 2004). This resulted in increased proliferation and cytotoxicity compared to the non-IL-15 transduced parental cell lines. Work by Miller et al. showed that murine NK cells could be retrovirally transduced to express IL-2. This production is capable of promoting in vitro cell proliferation and function in exogenous IL-2-depleted conditions (Miller et al., 1997).

Adenoviral transduction of primary, IL-2-activated NK cells—marker genes

While successful lentiviral/retroviral transduction of primary NK cells results in the stable transgene expression described previously, is extremely important and will prove very useful in the future, limited data are available as to the stable expression of genes other than GFP by primary NK cells. In contrast, adenoviral vectors, which are able to transiently transduce a variety of cell type at very high rates, have proven highly successful for transduction of murine NK cells and have been used for the introduction and expression of cytokine genes in these cells. Therefore, the remainder of this chapter will focus on the usage of adenoviral vectors for the transduction of NK cells.

The ability of adenoviral (Ad) vectors (serotype 5) to transduce IL-2-activated NK cells, denoted A-NK cells (Basse et al., 1994), was first tested using expression of the simple marker protein GFP. The expression of GFP (as measured by mean fluorescence intensity), following transduction of A-NK cells with Ad-GFP, varied between the cells, but resulted in >70% of the A-NK cells being GFP positive (not shown) (Goding et al., 2007a). There was a dose response in the percentage of GFP+ A-NK cells with increasing amounts of virus used, but multiplicities of infection (MOIs) >100 often resulted in unacceptable toxicity, with loss of 50–80% of the cells. Ad-GFP-transduced A-NK cells expressed GFP for at least a week, with a peak in production at 48–72 h after transduction. At optimal MOI (40 MOI), A-NK cells were >70% GFP-positive at 48 h after Ad-GFP transduction. As expected due to the transient nature of the adenoviral transduction, the expression fell below 30% GFP-positive cells by day 7. This may be explained by the fact that adenoviruses are unable to integrate their genetic material into the host genome, and therefore incapable of replicating the transgene with each cell division. Thus, the originally delivered viral genes become diluted with successive rounds of division of the rapidly proliferating A-NK cells. In addition, inactivation of the cytomegalovirus (CMV) promoter of the adenoviral vectors also contributes to declining gene expression (Guo et al., 1996; Zabner et al., 1996).

Following the demonstration of positive viral transduction with the GFP read-out, A-NK cells were transduced to express functional enzymes. A-NK cells were capable of producing beta-galactosidase and luciferase following Ad-LacZ- and Ad-Luc transduction, respectively (not shown). In each case, the viability of the transduced A-NK cells was evaluated. Toxicity was not observed until 100 MOI or greater of Ad vectors was used. Thus, the application of Ad vectors provides an efficient means to ensure the expression of fluorescent or enzymatic markers in murine A-NK cells.

A-NK cell transduction—cytokine genes—IL-2

While transduction of NK cells with marker genes (whether fluorescent or enzymatic) is useful for the in vitro or in vivo detection of these cells (Goding et al., 2007a), the transduction of NK cells with cytokine genes can, depending on the choice of cytokine, be expected to enhance or support the function of the NK cells themselves or to influence the microenvironment of the transduced NK cells. IL-2 is a key cytokine that stimulates both NK cells and T cells and therefore, in a number of ways, is involved in generating and perpetuating antitumour responses. NK and T cells require IL-2 for activation, proliferation, and survival (Ortaldo et al., 1984). IL-2 is a key factor in triggering other antitumour cytokines cascades, such as interferon (IFN)-γ (Ye et al., 1996) and tumour necrosis factor (TNF)-α (Jewett and Bonavida, 1993). For this reason, IL-2 has been used extensively in both pre-clinical and clinical trials (Atkins et al., 1999; Lotze et al., 1990). The severe side effects associated with its application in high doses, however, have lead to its withdrawal from many clinical studies.

In mice, IL-2 is needed to ensure trafficking to and localization within established tumours of adoptively transferred A-NK cells (Basse et al., 1994). The amount of IL-2 that can be given to support the A-NK cells, however, is also greatly limited in this situation by the adverse effects of this cytokine when given systemically, most notably the vascular leakage syndrome with hypotension and lung oedema (Rosenberg et al., 1988; Siegel and Puri, 1991). To address this problem, it was investigated whether transduction of A-NK cells with the IL-2 gene would enable them to survive and function in vivo, independently of exogenous IL-2. However, despite use of different mIL-2 gene-containing Ad vectors (all provided by Dr. Andrea Gambotto and Dr. Paul Robbins, University of Pittsburgh Vector Core), there has not been complete success in transducing A-NK cells to become totally independent of exogenous IL-2 in vitro. Although IL-2 was detected by ELISA in

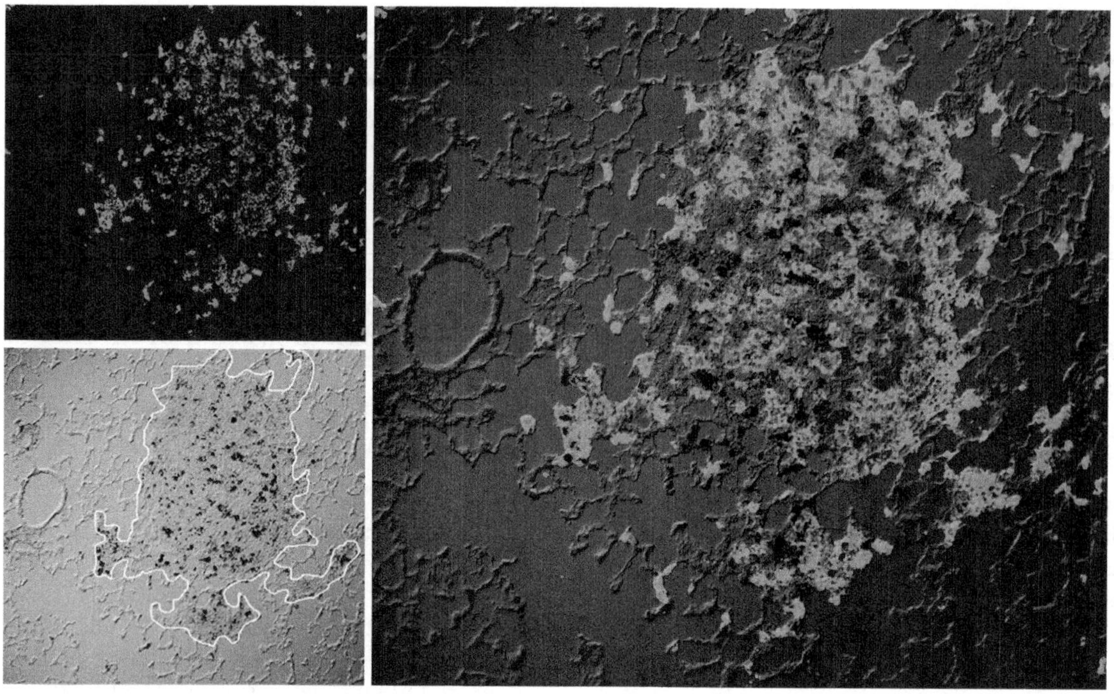

Figure 48.1 • Tumour-localization by adoptively transferred A-NK cells. 10 × 10^6 A-NK cells generated from Thy1.1+ congenic BL6 mice were injected IV into Thy1.2+ BL6 mice with 9-days-old B16 melanoma lung tumours. The animals received 60000 IU Peg-Il-2 every 12 h for 3 days. At 72 h after the A-NK cell injection, lungs were removed and fresh frozen. Eight micron cryosections were stained with PE-conjugated anti-Thy1.1 antibody. Note the selective accumulation of the PE-Thy1.1+ A-NK cells within the B16 lung metastases. Bar = 200 μm. Upper left insert: red fluorescent picture showing the A-NK cells. Lower left insert: DIC picture showing melanin-containing B16 tumour (outlined).

the supernatants of transduced mouse A-NK cells, this was at a very modest level (<50 IU/million cells/day). Although excess amounts of exogenous human IL-2 were added to the cells, it is possible that this low recovery of murine IL-2 in the supernatants was partly due to consumption of the mIL-2 by the NK cells. Nevertheless, a quite robust production of IL-2 is needed for the A-NK cells to support themselves in vitro. By rough calculation, the amount of IL-2 required to maintain A-NK cells in vitro would be in the range 30–300 ng produced per 10^6 cells, which is 50- to 500-fold more than what has been obtained with IL-2 transduction thus far. However, in vivo, where NK cells can come into close proximity in tumours (see Figure 48.1), less IL-2 may be needed, simply because the amount of extracellular fluid is very low.

Some support for this notion comes from the finding that adoptively transferred Ad-IL-2- transduced A-NK cells survive at much higher densities in tumours 5 days after injection than nontransduced A-NK cells, when no exogenous IL-2 is given to support the NK cells (Basse et al. unpublished data). The density of the IL-2 gene-transduced A-NK cells in the tumours was, however, low compared to the densities obtained by supporting the NK cells with exogenous IL-2. It was therefore of interest to study other cytokines to see if they would synergize with IL-2 (or even replace it) with respect to maintenance of NK cell viability and function.

IL-12 gene transduction enhances IFNγ production and CD25 expression by NK cells

IL-12, also known as NK cell stimulating factor, has been shown to increase NK cell expression of CD25 (Rabinowich et al., 1993), the alpha subunit that, along with the beta and gamma subunits, makes up the high-affinity IL-2 receptor. This provided support for the hypothesis that the level of IL-2 (provided exogenously or produced by a transgene) needed to support the A-NK cells could be reduced if the cells were also stimulated with IL-12, in that the IL-12 stimulation would lead to expression of CD25 and thereby the high affinity IL-2 receptor by the NK cells.

Ad-IL-12 transduction resulted in a dramatic production of IL-12 by the A-NK cells compared to mock Ad transduction (Figure 48.2A). The presence of IL-12 in supernatants was measured over several days. As seen with GFP transduction, the production peaked at 48 h. Thus, to test the function of the IL-12 produced by Ad-IL-12-transduced A-NK ($A\text{-}NK_{12}$) cells, IFNγ was measured in supernatants of $A\text{-}NK_{12}$ and $A\text{-}NK_{mock}$ cell cultures. Although A-NK cells do produce IFNγ in response to IL-2 cytokine activation alone, this production was greatly enhanced with the introduction

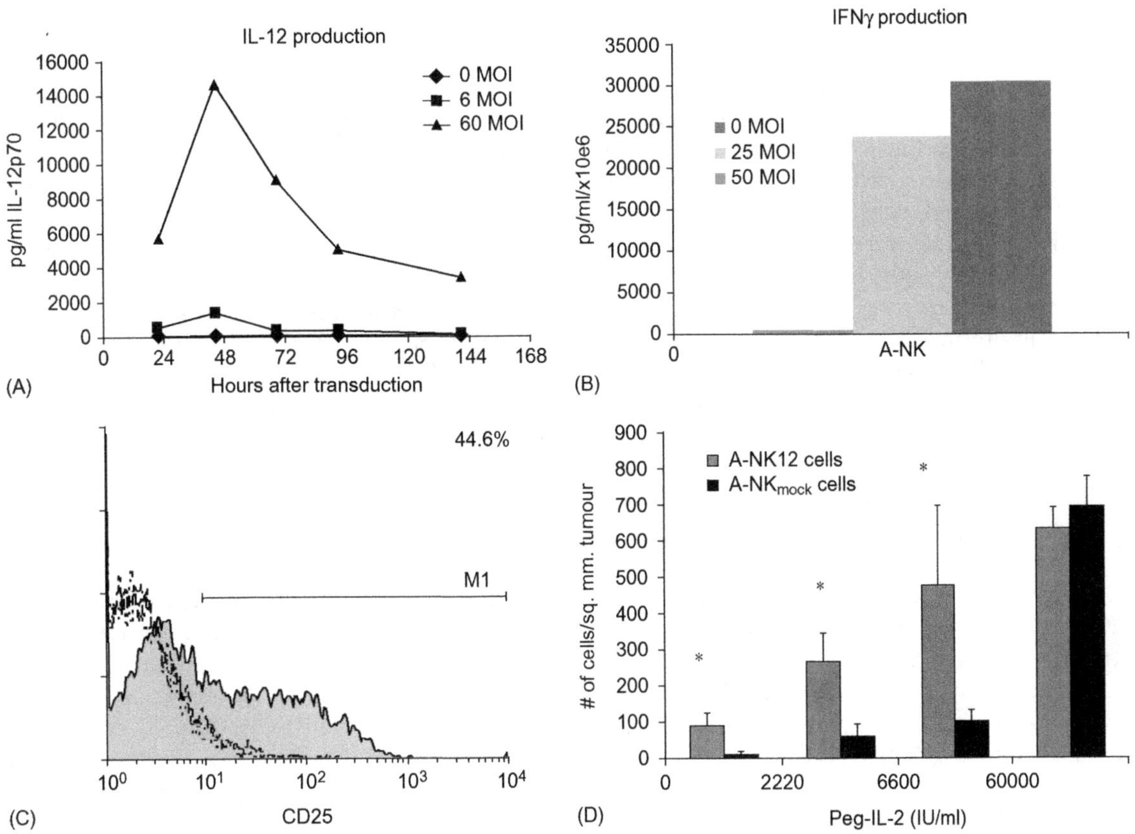

Figure 48.2 • Primary IL-12 production leads to secondary IFNγ production by and increased CD25 expression on Ad-mIL-12 transduced murine A-NK cells, and provides a survival benefit to transduced cells, at low doses of Peg-IL-2, as well as nontransduced cells, when expressed locally in vivo. (A) 2 × 10^6 A-NK cells were transduced with the indicated MOI of Ad-mIL-12. Supernatants were harvested at various times and assayed using ELISA. Values are given in picograms per millilitre for the total amount of mIL-12 present in the culture flask by multiplying the number of millilitres of supernatant by the ELISA readings for a 1/10 dilution of samples. (B) 3 × 10^6 A-NK cells were transduced with the indicated MOI of Ad-mIL-12. Cells were washed after 24 h culture and given fresh media. Supernatants were collected 24 h later and analysed for the presence of IFNγ. (C) 5 × 10^6 A-NK cells were transduced with Ad-IL-12 at 25 MOI. A-NK$_{12}$ (filled curve) and A-NK$_{non}$ (open curve) cells were stained with anti-CD25-FITC Ab after 48 h of culture. Cells were analysed by flow cytometry, collecting 5000 events and compared with isotype control. (D) Mice bearing day 10 B16 lung tumours received 4 × 10^6 A-NK$_{12/mock}$ and 5 × 10^6 A-NK$_{non}$ cells IV with two injections of 60000 IU Peg-IL-2, followed by 4 injections at 0, 2220, 6600, or 60000 IU Peg-IL-2. At 72 h postadoptive transfer, mice were sacrificed and tissues harvested (n = 3). A-NK cell densities in tumours are represented as number of cells (A-NK$_{12/mock}$) per square millimetre lung tumour tissue. Values were found to be statistically significant with a p value (*) < 0.05, with error bars representing SD.

of the IL-12 transgene into the A-NK cells, with levels reaching 30 ng/ml (Figure 48.2B). In addition, while <5% A-NK$_{mock}$ cells were CD25 +, >40% of A-NK$_{12}$ cells were CD25+ (Figure 48.2C), demonstrating a clear effect of transgenically produced IL-12 in upregulating CD25 expression.

Reduced need for exogenous IL-2 support by A-NK cells transduced to express IL-12

The amount of exogenous IL-2 support needed to support adoptively transferred A-NK$_{12}$ cells was titrated to confirm that the increased expression of CD25 on A-NK$_{12}$ cells would actually result in continued survival and function in vivo at reduced levels of IL-2. A-NK$_{12}$ or A-NK$_{mock}$ cells were adoptively transferred into mice bearing 7-day-old B16 lung tumours. These mice were then divided into four groups that received two injections of 60000 IU of PEG-IL-2 on the day of transfer and either 0, 2220, 6600, or 60000 IU of PEG-IL-2 twice a day for an additional 2 days. Thus, all groups received high-dose PEG-IL-2 support on the first day to ensure survival of the A-NK cells while trafficking to the tumours, a process that takes 12–24 h (Basse et al., 1991b). As expected, equal numbers of A-NK cells were observed at 72 h post transfer in the tumours of mice receiving 60000 IU of PEG-IL-2 every day, whether or not the A-NK cells had been transduced with Ad-IL-12 (Figure 48.2D). Reducing the exogenous IL-2 support by 10-fold demonstrated a decrease of 85% in the number of A-NK_{mock} cells, but only a 25% decrease for A-NK_{12} cells. The numbers were equally as dramatic for the lowest (2200 IU) and no PEG-IL-2 doses with A-NK cell densities in the tumour tissue being fourfold and eightfold better, respectively, for the A-NK_{12} cells compared with A-NK_{mock} cells (Figure 48.2D). These results clearly illustrate that Ad IL-12 gene-transduction does not interfere with the A-NK cells' ability to traffic to tumours and accumulate therein, but rather allows them to do so, even when supported by as little as one-tenth the dose of exogenous IL-2 needed for intratumoural survival of nontransduced A-NK cells. Although the presence of IL-12 could not be detected in the serum of treated mice, IFNγ levels were increased in all animals receiving A-NK_{12} cells, compared with mice receiving IL-2 only. IFNγ was detected in the range 1000–1500 pg/ml in the serum of mice receiving A-NK_{12} cells, whereas approximately 250 pg/ml was demonstrated in serum of mice receiving IL-2 only. This provides evidence that, despite direct proof for IL-12 production by A-NK_{12} cells in vivo, IL-12 is being produced by the tumour-infiltrating A-NK_{12} cells.

Superior antitumour effect by A-NK_{12} cells compared to A-NK_{mock} cells

To determine how successful the expression of IL-12 by IV-injected A-NK_{12} cells would be in supporting NK cell-mediated antitumour responses against established B16 lung tumour metastases, the lung tumour burden of mice receiving either A-NK_{12} or A-NK_{mock} cells IV supported by low amounts of IL-2 (i.e. four injections of 6600 IU/ml PEG-IL-2 after two initial injections of 60000 IU of PEG-IL-2) was measured. Having IL-12 expressed within lung tumours by the A-NK_{12} cells provided superior benefit in respect to antitumour effect, with a 61% reduction in B16 tumour mass following treatment with A-NK_{12} cells compared to A-NK_{mock} cells, even at this level of PEG-IL-2 support (Goding et al., 2007b) (data not shown).

To evaluate if the observed tumour reduction was sufficient to increase survival of tumour-bearing animals, A-NK_{12} and A-NK_{mock} cells were adoptively transferred into mice with established lung tumours of B16 or MCA-205 origin. In an early (day 3) B16 lung tumour model, a 10-day increase in survival time was found following IV injection of A-NK_{12} cells compared with A-NK_{mock} cells or treatment with PEG-IL-2 alone (Figure 48.3A, p value = 0.0031). Thus, while A-NK_{mock} cells lose their therapeutic efficacy when supported by only two injections of 60000 IU of PEG-IL-2, this nontoxic regimen of IL-2 support is sufficient to support the antitumour effect of A-NK_{12} cells. In fact, the therapeutic effect of A-NK_{12} cells is still maintained in mice receiving even less IL-2 (two PEG-IL-2 injections of only 4500 IU [Figure 48.3A, 'A-NK_{12} low IL-2', p value = 0.0016]). To determine the efficacy of A-NK_{12} cells in a more aggressive, advanced tumour setting, mice bearing 7-day-old B16 lung tumour metastases were treated with A-NK_{12} cells. These mice survived a median 26 days compared with 20 days for mice receiving A-NK_{mock} cells (Figure 48.3B, p value = 0.0019). Treatment with A-NK_{12} cells as compared with A-NK_{non} cells also prolonged the survival of mice bearing day-7 MCA-205 lung tumours, by 7.5 days (Figure 48.3C, p value = 0.0158). This clearly illustrates the enhanced antitumour efficacy of A-NK_{12} cells compared with A-$NK_{non/mock}$ cells and demonstrates the advantage of using adenovirally transduced A-NK cells to deliver and express cytokines specifically within tumour tissue. Although a single treatment of mice with mIL-12-transduced cells did not result in complete cure, these mice survived significantly longer compared to mice treated with mock-transduced cells. By repeating the adoptive transfer of A-NK_{12} cells (maybe every 5th day), it is conceivable that greater survival could be attained, ultimately leading to curative effects.

Local IL-12 production benefits 'bystander' nontransduced A-NK cells

To determine whether the IL-12 produced within tumours could also maintain the survival of nontransduced, co-administered, A-NK cells, two congenic systems were used (CD45 and Thy1) so the nontransduced A-NK cells (CD45.1+) could be distinguished from the transduced A-NK cells (Thy1.1+) and host Thy1.2+ cells. When supported by low doses of PEG-IL-2 (2220 IU and 6600 IU, respectively), 3- to 4.5-fold more A-NK_{non} cells survived within lung tumours when co-injected with A-NK_{12} cells compared to A-NK_{mock} cells (Goding et al., 2007b). The IL-12 produced by transduced A-NK cells is, therefore, able to maintain survival of not only the transduced cells themselves, but of neighbouring nontransduced A-NK cells as well.

To evaluate the importance of the IL-12 being produced at the tumour site (as opposed to a site distant from the tumour) on the survival of nontransduced A-NK cells in lung tumours, 5 million A-NK_{non} cells were injected by the IV route. Immediately after, 5 million A-NK_{12} cells were injected by either the IV or IP route. It has been shown that A-NK cells injected IP do not leave the IP cavity and therefore do not localize into the lungs or lung tumours (Yang et al., 2003). Thus, only

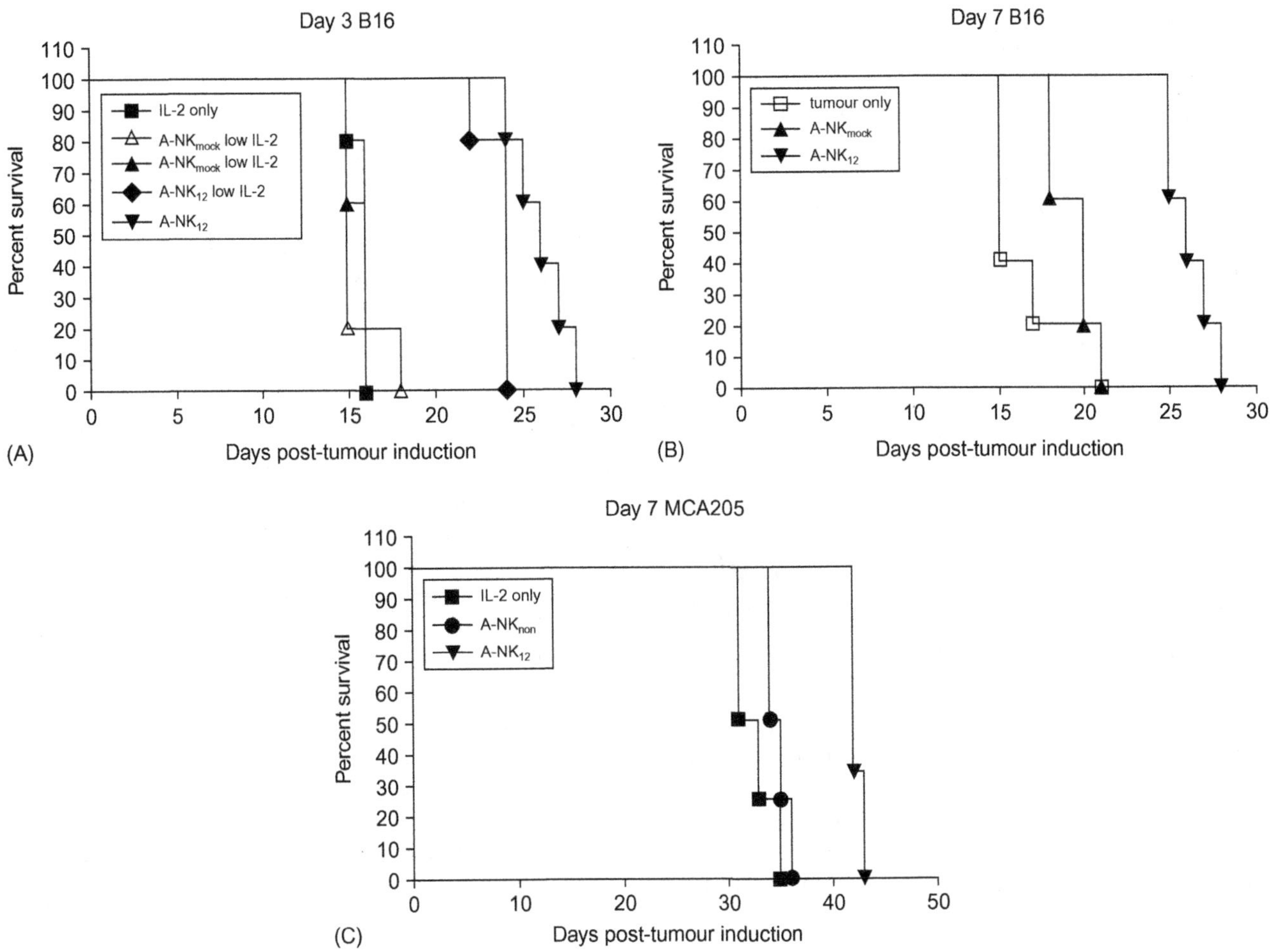

Figure 48.3 • IL-12 prolongs the survival of mice when expressed locally by A-NK12 cells. Survivability of mice (n = 5) was monitored, as described in Goding et al., 2007a, after the mice received 5 × 10^6 A-NK_{non}, A-NK_{mock} or A-NK_{12} cells. The survivability of mice bearing day 3 B16 (A), day 7 B16 (B), or day 7 MCA205 (C) lung metastases was plotted using the Kaplan–Meier method for the number of days post-tumour induction. Statistical significance was calculated, using log-rank test, for mice receiving A-NK_{12} compared to A-NK_{mock} cells, with p values of 0.0001 (A), 0.0019 (B) and 0.0150 (C).

IV-injected IL-12-transduced A-NK cells will co-localize with the nontransduced A-NK cells in the lung tumours. Even though IP injection of A-NK_{12} cells resulted in the survival of a few more of the IV-injected A-NK_{non} cells compared with the control group (which did not receive any A-NK_{12} cells), 8.8-fold more IV-injected A-NK_{non} cells survived in the lung tumours when the A-NK_{12} cells were injected by the IV route than by the IP route (Goding et al., 2007b) (data not shown). Thus, in maintaining A-NK cell survival in the tumour tissues, delivery of the IL-12 cytokine to the tumour microenvironment is superior to a tumour-distant delivery.

To investigate, in a similar manner, the antitumour responses from IL-12 produced distally rather than intratumourally, some mice received A-NK_{12} cells by the IP route and some via the IV route. The presence of A-NK_{12} cells in the IP cavity caused some reduction in tumour burden compared with control animals, but this reduction was not significant compared to nontreated animals. By comparing the efficacy of A-NK_{12} cells injected IV with that of cells injected IP, it was evident that targeting the A-NK_{12} cells to the lung tumours, by IV injection, resulted in a larger tumour reduction than injection of the A-NK_{12} cells via the IP route (p value = 0.023).

Since the accumulation of non-IL-12 gene-transduced A-NK cells in tumours is far better when given together with A-NK_{12} cells than when given alone, it could be expected that the combined treatment of both transduced and nontransduced A-NK cells would result in a better therapeutic effect than injection of either cell type alone. This seemed to be the case, as the 64% tumour reduction seen following IV injection of A-NK_{12} plus A-NK_{non} cells was significantly greater than the 48% reduction following IV injection of A-NK_{12} alone (Goding et al., 2007b) (not shown). Injection of nontransduced A-NK cells alone had no antitumour effect.

It could be argued that IV injected, nontransduced A-NK cells would be able to benefit from the cytokines produced by A-NK_{12} cells regardless of where the A-NK_{12} cells were located, that is, that the IL-12 would not have to be produced in the tumour microenvironment to have an effect. However, mice receiving an IP injection of A-NK_{12} cells plus an IV injection of A-NK_{non} cells demonstrated some reduction (31%) in tumour burden compared with control animals (receiving IL-2 only), but this was significantly less than the 64% tumour reduction achieved by IV injection of both A-NK_{12} plus A-NK_{non} cells (ensuring the presence of both A-NK_{non} cells and A-NK_{12} cells in the lung tumours) ($p = 0.0001$). Thus, A-NK_{12} cells are able to significantly boost the anti-lung tumour activity of A-NK_{non} cells and tumour localization of A-NK_{12} cells seems to be a prerequisite for achieving the full IL-12 effects generated by these cells.

IL-12 induced host production of IFNγ is essential for the antitumour effects of A-NK_{12} cells

To determine if the survival increases that have been seen previously with A-NK_{12} cell therapy were due primarily to the A-NK cells only, or if the host contributed to the antitumour therapy via T, B or NK cells, a preliminary experiment was conducted involving a comparison of A-NK_{12} cells transferred into wild-type (WT) mice versus non-obese diabetic severe combined immune deficient (NOD-SCID) mice. NOD-SCID mice are deficient in T and B cell and NK cell activity. Before assessing the survival of these two types of mice following A-NK_{12} therapy, the ability of the A-NK_{12} cells to traffic to lung tumours in the NOD-SCID mice was found equal to that of WT animals. WT animals receiving A-NK_{12} cells survived 8 days longer than WT animals receiving A-NK_{mock} cells (p value $= 0.0012$). Interestingly, NOD-SCID mice survived an equal amount of time (8 days) following A-NK_{12} treatment, illustrating a lack of involvement of host antitumour activity mediated by T, B and NK cells.

To determine if the high level of IFNγ that has been observed in the serum of A-NK_{12} cell-treated mice is important for antitumour survival benefit and from which source, that is, A-NK cell-derived or host-derived, two separate experiments were conducted in which A-NK_{12} cells, generated from WT C57BL/6 and IFNγ KO mice, were adoptively transferred into WT mice or IFNγ KO mice. These mice received one day of PEG-IL-2 support and were monitored for survival. As has been seen on a consistent basis, there was a significant, 6-day increase in the survival of WT mice receiving WT A-NK_{12} cells compared to WT A-NK_{mock} cells (Figure 48.4A, p value $= 0.0031$). In both experiments, there was no difference in survival of WT tumour-bearing mice when A-NK_{12} cells were generated from WT or

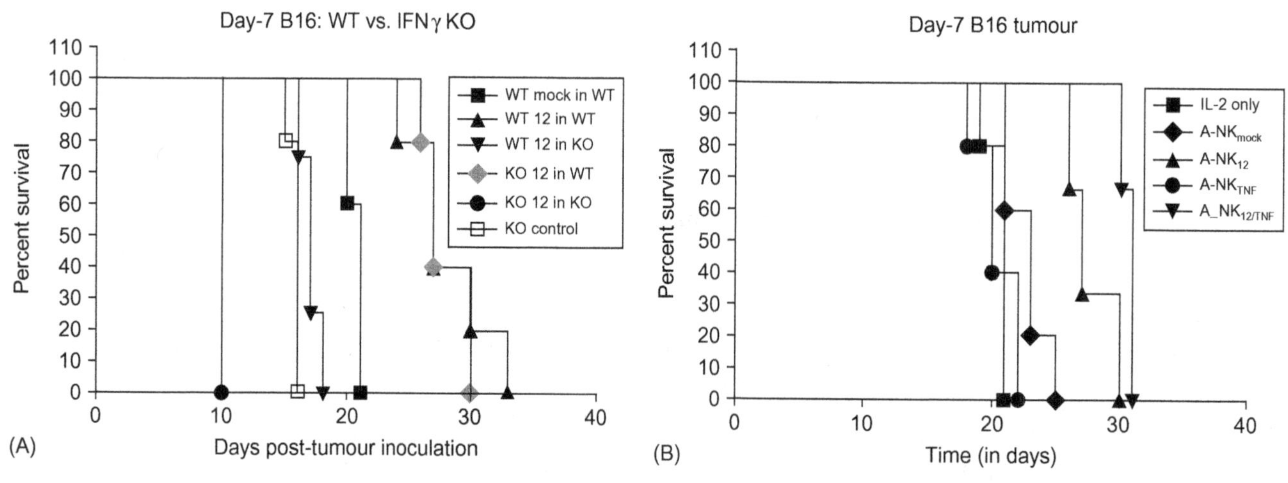

Figure 48.4 • IFNγ produced by the host has a greater impact on the survival of tumour-bearing mice than IFNγ produced by A-NK_{12} cells, while co-transduction with TNFα leads to increases, albeit nonsignificant, in tumour-bearing mouse survival. (A) 4.5×10^6 WT A-NK_{mock} or WT or IFNγ KO A-NK_{12} cells were injected into WT or IFNγ KO recipient mice bearing day-7 B16 lung tumours, as indicated in the legend. Survivability was monitored, as described in Goding et al., 2007a, and plotted using the Kaplan–Meier method for the number of days post-tumour induction. Statistical significance was calculated, using log-rank test ($n = 5$ for all groups except WT A-NK_{12} into KO mice, where it was $n = 4$). These results represent 1 of 2 identical experiments that produced similar results without significantly different survivals in the same groups of the two experiments. Median survivals in days for experiment 1 (presented above)/experiment 2 are 21/21 for WT mock in WT, 27/29 for WT 12 in WT, 17/17 for WT 12 in KO, 27/29 for KO 12 in WT, and 10/10 for KO 12 in KO. Median survival of IFNγ KO control mice was 16 days and only included in experiment 1. (B) B16 tumour-bearing mice received 5×10^6 A-NK_{12} cells. Survivability was monitored, as described in Goding et al., 2007a, and plotted using the Kaplan–Meier method for the number of days post-tumour induction. Statistical significance was calculated, using log-rank test, for mice receiving A-NK_{12} compared with A-$NK_{12/TNF\alpha}$ cells in the B16 background, p value $= 0.063$ ($n = 5$ for all groups except A-NK_{12} and A-$NK_{12/TNF\alpha}$ where it was $n = 3$).

IFNγ KO mice, illustrating a minimal requirement for A-NK cell-derived IFNγ production. However, when WT $A\text{-}NK_{12}$ cells were administered to IFNγ KO tumour-bearing recipients, the mice survived a median 17 days, a 10-day decrease compared with WT $A\text{-}NK_{12}$ in WT mice (p value = 0.0029), and, more surprisingly, a 4-day decrease compared with mock control mice (p value = 0.0029). Thus, it is clear that IFNγ produced by the host rather than by the A-NK cells is critical for the antitumour effect of $A\text{-}NK_{12}$ treatment. Surprisingly, in a situation where no IFNγ could be made, namely where $A\text{-}NK_{12}$ cells from IFNγ KO mice were injected into IFNγ KO recipients, all mice died within 3 days of the adoptive transfer. Upon further investigation, it appeared that mice died due to cytokine-induced toxicity (including vascular leakage) rather than advanced growth of tumour (i.e. both the tumour burden of the lungs and at extrapulmonary sites at this early time point was very low). This indicates that the co-presence of IL-12 and IFNγ leads to therapeutic effect, but, if IL-12 is present while IFNγ is absent from both host and adoptively transferred cells, a fatal but yet poorly understood toxicity is induced.

Based on the NOD-SCID and the IFNγ KO experiments, it can be concluded that host T and NK cells are not likely to account for the IFNγ that is produced. It has been shown that macrophages (Puddu et al., 1997) and neutrophils (Ethuin et al., 2004) can produce IFNγ in response to IL-12 stimulation. An abundance of CD11c+ pulmonary macrophages are found in the tumours, so it is very likely that these macrophages are producing the IFNγ. While host T and NK cells are not important for the increase in animal survival following $A\text{-}NK_{12}$ cell treatment, it is still unclear whether the IFNγ, which clearly is essential, has a direct antitumour effect by stimulating host macrophages and/or the exogenously delivered A-NK cells. Thus, while host T, B and NK cells do not appear to be important for the increase in animal survival following $A\text{-}NK_{12}$ cell treatment, it is clear that other host immune cells, possibly macrophages, indeed are involved in the antitumour effect of this treatment through the $A\text{-}NK_{12}$ therapy, at least through a clear role of IFNγ.

Transduction of A-NK cells to express both IL-12 and TNFα

Although IL-12 has been successful in enhancing A-NK cell therapy without generating toxicity, this does not seem to be sufficient to boost long-term, adaptive host immunity. Thus, even though the intratumoural killing of tumour cells by the NK cells may liberate tumour antigens for intratumoural DCs to take up, and even though the tumour microenvironment, due to the presence of the $A\text{-}NK_{12}$ cells, contains high levels of Th1-polarizing cytokines (IL-12 and IFNγ), no signs of effector T cell responses against the tumour were seen. It is therefore likely that additional early pro-inflammatory signals are needed to initiate T cell priming. TNFα is a very early initiator of inflammation, which activates DCs and promotes CTL generation (Gorelik et al., 1995, 1996). Thus, in addition to its other beneficial effects, like enhancing LAK activity, increasing vascular permeability, and direct tumour killing, TNFα is an attractive choice to add to the A-NK-12 therapy to further augment its antitumour effect. TNFα can be detected in the nanogram range in supernatants from A-NK cells transduced at 50 MOI with Ad-TNFα. Production peaks around 72 h post transduction, at almost 2 ng/ml, and can be detected out to at least 96 h. When the A-NK cells are cotransduced with IL-12, the level of TNFα increases from that seen with TNFα transduction alone (109 pg/ml vs. 216 pm/ml for TNFα and TNFα/IL-12, respectively). While TNFα transduced A-NK cells did not show significantly enhanced cytotoxicity against various target cells in a 4-h 51-Cr release assay, supernatants from TNFα gene-transduced A-NK cells were able to slightly inhibit the growth of B16 cells in vitro compared to supernatants from $A\text{-}NK_{non}$ and $A\text{-}NK_{mock}$ control cultures. However, when supernatants from TNFα and IL-12 double-transduced A-NK cells were added to the B16 tumour cells, a significant, nearly 75% reduction in B16 viability was seen. The difference was significant even at a 1:8 dilution of the supernatant (p value < 0.001).

The early effects of TNFα in vivo were investigated by monitoring at the survival of adoptively transferred A-NK cells in B16 lung tumour-bearing animals given $A\text{-}NK_{12}$, $A\text{-}NK_{TNF\alpha}$, or $A\text{-}NK_{12/TNF\alpha}$ cells iv. At 96 h after adoptive transfer, relatively few A-NK cells were found in the lung tumours when transduced with IL-12 or mock virus (data not shown). This low A-NK cell density at the 96-h time point was not surprising, considering the facts that IL-12 on its own is not a survival factor for A-NK cells and that exogenous IL-2 support of the NK cells was administered only on the first day. It was, however, surprising to find an abundance of A-NK cells persisting in lung tumours when they had been transduced with TNFα. There was a 2.4-fold increase in the number of NK cells if they expressed TNFα as opposed to IL-12. However, when the A-NK cells were co-transduced with IL-12 and TNFα, the cell numbers reached levels of $528/mm^2$ tumour tissue, that is, a 9.9-fold increase compared to $A\text{-}NK_{12}$ cells. Differences in the presence of host cells, such as CD4+ and CD11c+ cells, were also evaluated, but the density of these cells in the tumours remained approximately the same, regardless of which type of NK cell was injected. When evaluating the therapeutic effect of the double-transduced A-NK cells, it was found that A-$NK_{12/TNF\alpha}$ cell treated mice survived 8 days longer than A-NK_{mock} cell treated mice (Figure 48.4B, p value = 0.012). Compared with A-NK_{12} cell-treated mice, however, this represented only a 4-day, nonsignificant prolongation in survival (p value = 0.063). This indicates that the therapeutic effect of 12/TNFα double-transduced A-NK cells is not better than that of single-transduced A-NK_{12} cells, and is contrary to the significant increase in intratumoural persistence of A-$NK_{12/TNF\alpha}$ cells, as described above.

Two conclusions can be drawn from these results. First, the addition of TNFα gene transduction does not improve the A-NK cell treatment to an extent that can be measured with survivability as endpoint. Second, the number of A-NK cells present within the lung tumours does not always correlate with the therapeutic outcome, that is, the survival of recipients. Although the mechanisms involved in the double-transduced A-NK cells persistence to such a high degree without mediating greater therapy have not been elucidated, in any case this phenomenon, that is, increased survival of A-NK cells, provides a great deal of promise for future use of transduced A-NK cells. These studies also demonstrate that simultaneous transduction of NK cells with multiple genes indeed is possible.

Simultaneous transduction of NK cells with IL-12 and IL-2 genes eliminates the need for exogenous IL-2

A cytokine gene combination that could generate highly active NK cells capable of surviving and functioning independently of cytokine support from other sources is IL-2 plus IL-12. The IL-2 is needed for survival and continued proliferation of the NK cells and the IL-12 stimulation would lead to increased cytotoxicity and cytokine production, mainly IFNγ. Importantly, the IL-12 also induces increased CD25 expression by the NK cells, enabling them to express the IL-2 high-affinity receptor and therefore to respond to low amounts of IL-2. In an experiment in which A-NK cells were co-transduced to express both IL-12 and IL-2, mice receiving A-$NK_{12/2}$ cells survived 7 days longer than mice receiving A-NK_{mock} cells (median 28 days compared to 21.5 for A-NK_{mock}, p value = 0.0005). Remarkably, this was accomplished without any support by exogenous IL-2. Thus, when coupled with the production of IL-12, the modest IL-2 production by IL-2/12 gene-transduced A-NK cells apparently was capable of maintaining viability and function of the A-NK cells and thereby completely eliminating the need for exogenous IL-2 support. The importance of this finding is substantial, because the 'Achilles' heel' of cancer immunotherapy based on adoptive transfer of A-NK cell, namely their demand for continuous support by high and potentially toxic levels of IL-2, a weakness that has made their use in the clinic doubtful, can be eliminated by simultaneous IL-2/IL-12 gene-transduction of the A-NK cells before their adoptive transfer. The ability of IL-2 + IL-12- secreting A-NK cells to support the viability and function of non-transduced host or donor bystander NK cells remains to be tested.

Promising outlook for genetically engineered NK cells

As discussed herein, gene transduction of A-NK cells can be a powerful antitumour tool, which can be used to improve/change target recognition by the NK cells (Pegram et al., 2008) or to deliver combinations of cytokines, and possibly in the future, chemokines and/or danger signals also, to the microenvironment. Determining which agents should be delivered and to what extent will continue to become a major focus for future investigations. As described above, the presence of a cytokine (or any other agent) delivered by A-NK cells may not persist for long. A single injection of cytokine-producing A-NK cells may therefore not be able to fully support the development of adaptive immune responses. Therefore, multiple injections of transduced A-NK cells may be needed to supply these cytokines and signals at critical time points during development of an adaptive immune response to the tumour. Likewise, the use of vectors ensuring stable gene-expression by the NK cells as compared to the transient Ad gene-expression may also be beneficial. It may be that the initiation of an immune response requires injections of A-NK cells delivering cytokines that differ from those needed at later stages during development of the response. For example, the first dose of A-NK cells might need to be transduced with A-NK cell-promoting factors that maintain their survival and cytolytic function, such as IL-2 and TNFα. However, the addition of other cytokines (such as GM-CSF, IL-2 or IL-15) and chemokines, that can draw immune cells into the tumours, will establish an appropriate immune cell population consisting of both innate effector cells and antigen-presenting cells. Once this has happened, a second dose of A-NK cells, this time delivering cytokines capable of supporting DC functions (i.e. IL-12, IFNγ, TNFα and CD40L), may induce maturation and migration of DCs toward draining lymph nodes, leading to initiation of a Th1-biased T cell response. Finally, a third wave of NK cells genetically modified to produce chemokines to which CTLs respond (such as Regulated upon Activation, Normal T-cell Expressed, and Secreted [RANTES], Macrophage Inflammatory

Protein [MIP] 1α and β, Monocyte Chemotactic Protein-[MCP-]1, Interferon-inducible Protein- [IP-]10, and fractalkine) as well as IFNγ, may be needed to attract CTLs to the tumour site and retain them there and to keep them cytolytically active. As the technologies for gene-transduction and regulated transgene expression continue to improve, the use of NK cells for such purposes will become a reality.

References

Atkins, M.B., Lotze, M.T., Dutcher, J.P., Fisher, R.I., Weiss, G., Margolin, K., Abrams, J., Sznol, M., Parkinson, D., Hawkins, M., Paradise, C., Kunkel, L. and Rosenberg, S.A. (1999). High-dose recombinant interleukin 2 therapy for patients with metastatic melanoma: analysis of 270 patients treated between 1985 and 1993. J Clin Oncol **17**, 2105–2116.

Basse, P., Herberman, R.B., Nannmark, U., Johansson, B.R., Hokland, M., Wasserman, K. and Goldfarb, R.H. (1991a). Accumulation of adoptively transferred adherent, lymphokine-activated killer cells in murine metastases. J Exp Med **174**, 479–488.

Basse, P.H., Nannmark, U., Johansson, B.R., Herberman, R.B. and Goldfarb, R.H. (1991b). Establishment of cell-to-cell contact by adoptively transferred adherent lymphokine-activated killer cells with metastatic murine melanoma cells. J Natl Cancer Inst **83**, 944–950.

Basse, P.H., Goldfarb, R.H., Herberman, R.B. and Hokland, M.E. (1994). Accumulation of adoptively transferred A-NK cells in murine metastases: kinetics and role of interleukin-2. In Vivo **8**, 17–24.

Coca, S., Perez-Piqueras, J., Martinez, D., Colmenarejo, A., Saez, M.A., Vallejo, C., Martos, J.A. and Moreno, M. (1997). The prognostic significance of intratumoral natural killer cells in patients with colorectal carcinoma. Cancer **79**, 2320–2328.

Ethuin, F., Gerard, B., Benna, J.E., Boutten, A., Gougereot-Pocidalo, M.A., Jacob, L. and Chollet-Martin, S. (2004). Human neutrophils produce interferon gamma upon stimulation by interleukin-12. Lab Invest **84**, 1363–1371.

Goding, S., Yang, Q., Mi, Z., Robbins, P.D. and Basse, P.H. (2007a). Targeting of products of genes to tumor sites using adoptively transferred A-NK and T-LAK cells. Cancer Gene Ther **14**, 441–450.

Goding, S.R., Yang, Q., Knudsen, K.B., Potter, D.M. and Basse, P.H. (2007b). Cytokine gene therapy using adenovirally transduced, tumor-seeking activated natural killer cells. Hum Gene Ther **18**, 701–711.

Gorelik, L., Rubin, M., Prokhorova, A. and Mokyr, M.B. (1995). Importance of TNF production for the curative effectiveness of low dose melphalan therapy for mice bearing a large MOPC-315 tumor. J Immunol **154**, 3941–3951.

Gorelik, L., Bar-Dagan, Y. and Mokyr, M.B. (1996). Insight into the mechanism(s) through which TNF promotes the generation of T cell-mediated antitumor cytotoxicity by tumor bearer splenic cells. J Immunol **156**, 4298–4308.

Grund, E.M. and Muise-Helmericks, R.C. (2005). Cost efficient and effective gene transfer into the human natural killer cell line, NK92. J Immunol Methods **296**, 31–36.

Gunji, Y., Vujanovic, N.L., Hiserodt, J.C., Herberman, R.B. and Gorelik, E. (1989). Generation and characterization of purified adherent lymphokine-activated killer cells in mice. J Immunol **142**, 1748–1754.

Guo, Z.S., Wang, L.H., Eisensmith, R.C. and Woo, S.L. (1996). Evaluation of promoter strength for hepatic gene expression in vivo following adenovirus-mediated gene transfer. Gene Ther **3**, 802–810.

Guven, H., Konstantinidis, K.V., Alici, E., Aints, A., Abedi-Valugerdi, M., Christensson, B., Ljunggren, H.G. and Dilber, M.S. (2005). Efficient gene transfer into primary human natural killer cells by retroviral transduction. Exp Hematol **33**, 1320–1328.

Jewett, A. and Bonavida, B. (1993). Pivotal role of endogenous TNF-alpha in the IL-2-driven activation and proliferation of the functionally immature NK free subset. Cell Immunol **151**, 257–269.

Jiang, W., Zhang, J. and Tian, Z. (2008). Functional characterization of interleukin-15 gene transduction into the human natural killer cell line NKL. Cytotherapy **10**, 265–274.

Lotze, M.T., Custer, M.C., Bolton, E.S., Wiebke, E.A., Kawakami, Y. and Rosenberg, S.A. (1990). Mechanisms of immunologic antitumor therapy: lessons from the laboratory and clinical applications. Hum Immunol **28**, 198–207.

Maasho, K., Marusina, A., Reynolds, N.M., Coligan, J.E. and Borrego, F. (2004). Efficient gene transfer into the human natural killer cell line, NKL, using the Amaxa nucleofection system. J Immunol Methods **284**, 133–140.

Micucci, F., Zingoni, A., Piccoli, M., Frati, L., Santoni, A. and Galandrini, R. (2006). High-efficient lentiviral vector-mediated gene transfer into primary human NK cells. Exp Hematol **34**, 1344–1352.

Miller, J.S., Tessmer-Tuck, J., Blake, N., Lund, J., Scott, A., Blazar, B.R. and Orchard, P.J. (1997). Endogenous IL-2 production by natural killer cells maintains cytotoxic and proliferative capacity following retroviral-mediated gene transfer. Exp Hematol **25**, 1140–1148.

Nagashima, S., Mailliard, R., Kashii, Y., Reichert, T.E., Herberman, R.B., Robbins, P. and Whiteside, T.L. (1998). Stable transduction of the interleukin-2 gene into human natural killer cell lines and their phenotypic and functional characterization in vitro and in vivo. Blood **91**, 3850–3861.

Ortaldo , J.R., Mason, A.T., Gerard, J.P., Henderson, L.E., Farrar, W., Hopkins, R.F. III, Herberman, R.B. and Rabin, H. (1984). Effects of natural and recombinant IL 2 on regulation of IFN gamma production and natural killer activity: lack of involvement of the Tac antigen for these immunoregulatory effects. J Immunol **133**, 779–783.

Pegram, H.J., Jackson, J.T., Smyth, M.J., Kershaw, M.H. and Darcy, P.K. (2008). Adoptive transfer of gene-modified primary NK cells can specifically inhibit tumor progression in vivo. J Immunol **181**, 3449–3455.

Puddu, P., Fantuzzi, L., Borghi, P., Varano, B., Rainaldi, G., Guillemard, E., Malorni, W., Nicaise, P., Wolf, S.F., Belardelli, F. and Gessani, S. (1997). IL-12 induces IFN-gamma expression and secretion in mouse peritoneal macrophages. J Immunol **159**, 3490–3497.

Rabinowich, H., Herberman, R.B. and Whiteside, T.L. (1993). Differential effects of IL12 and IL2 on expression and function of cellular adhesion molecules on purified human natural killer cells. Cell Immunol **152**, 481–498.

Rosenberg, S.A., Lotze, M.T. and Mule, J.J. (1988). NIH conference. New approaches to the immunotherapy of cancer using interleukin-2. Ann Intern Med **108**, 853–864.

Schroers, R., Hildebrandt, Y., Hasenkamp, J., Glass, B., Lieber, A., Wulf, G. and Piesche, M. (2004). Gene transfer into human T lymphocytes and natural killer cells by Ad5/F35 chimeric adenoviral vectors. Exp Hematol **32**, 536–546.

Siegel, J.P. and Puri, R.K. (1991). Interleukin-2 toxicity. J Clin Oncol **9**, 694–704.

Tam, Y.K., Maki, G., Miyagawa, B., Hennemann, B., Tonn, T. and Klingemann, H.G. (1999). Characterization of genetically altered, interleukin 2-independent natural killer cell lines suitable for adoptive cellular immunotherapy. Hum Gene Ther **10**, 1359–1373.

Tran, J. and Kung, S.K. (2007). Lentiviral vectors mediate stable and efficient gene delivery into primary murine natural killer cells. Mol Ther **15**, 1331–1339.

Trompeter, H.I., Weinhold, S., Thiel, C., Wernet, P. and Uhrberg, M. (2003). Rapid and highly efficient gene transfer into natural killer cells by nucleofection. J Immunol Methods **274**, 245–256.

Yang, Q., Hokland, M.E., Bryant, J.L., Zhang, Y., Nannmark, U., Watkins, S.C., Goldfarb, R.H., Herberman, R.B. and Basse, P.H. (2003). Tumor-localization by adoptively transferred, interleukin-2-activated NK cells leads to destruction of well-established lung metastases. Int J Cancer **105**, 512–519.

Ye, W., Young, J.D. and Liu, C.C. (1996). Interleukin-15 induces the expression of mRNAs of cytolytic mediators and augments cytotoxic activities in primary murine lymphocytes. Cell Immunol **174**, 54–62.

Zabner, J., Wadsworth, S.C., Smith, A.E. and Welsh, M.J. (1996). Adenovirus-mediated generation of cAMP-stimulated Cl− transport in cystic fibrosis airway epithelia in vitro: effect of promoter and administration method. Gene Ther **3**, 458–465.

Zhang, J., Sun, R., Wei, H. and Tian, Z. (2004). Characterization of interleukin-15 gene-modified human natural killer cells: implications for adoptive cellular immunotherapy. Haematologica **89**, 338–347.

Chapter Forty-Nine

49

NK cell-mediated target cell death

Beverly Z. Packard, Akira Komoriya, David S. Ucker

CHAPTER CONTENTS

A single death is a tragedy; a million deaths is a statistic.

Joseph Stalin

ABSTRACT

Despite recent advances in understanding molecular mechanisms of natural killer (NK) cell recognition of target cells and consequent target cell death, the gold standard for assessing successful effector–target interactions has not changed. Specifically, the idea of target cell lysis, a process that would result in inflammation in vivo, as representative of NK cell-mediated death does not accurately represent the quiet death that most targets undergo. Notably, the two main death pathways triggered by NK cells, the delivery of perforin/granzymes from NK cells to targets and the ligation of death receptors on target cells by proteins on NK cell surfaces, are engaged considerably upstream. Both lead to apoptotic death, a significantly different in vivo outcome than the release of cytoplasmic target cell contents would suggest. In this chapter, a discussion of alternative measurements based on recently established cell death signalling pathways is presented.

KEY WORDS

ADCC, Apoptosis, Caspase, Cytotoxicity assay, Granzyme

Introduction

As major effector cells of the innate immune system, natural killer (NK) cells are cytotoxic responders to pathogen-infected and tumour cells. These large granular lymphocytes are able to detect and destroy both directly (Caligiuri, 2008; Di Santo, 2008; Waldhauer and Steinle, 2008; Walzer et al., 2007) and indirectly (Strowig et al., 2008) cells infected with a variety of viruses, bacteria and parasites as well as tumour cells of diverse histologic origin. Moreover, recently acquired understanding of the balance between activating and inhibitory signals processed by NK cells (Long, 2008; Nausch and Cerwenka, 2008; Vivier et al., 2008) plus characterization of the biochemical activities resulting in target cell death increase the number of therapeutic options now available for exploitation of this class of immunocyte (Moretta et al., 2008).

The significant advances made in the past two decades regarding recognition by NK cells and delivery of lethal hits to cellular targets raise questions about the original description of cell-mediated cytotoxicity. Specifically, the idea of cell lysis as opposed to apoptotic death as an endpoint of all cell-mediated cytotoxic events is not accurate. Furthermore, consistent with the original depiction of NK cells as 'lytic' cells, the immunologic literature has been permeated with use of the adjective 'cytolytic' before words such as 'activity', 'function', 'granule', 'interaction', 'mediator', 'potential', 'property', 'response' and 'synapse'. This has led to the synonymous usage of 'cytolysis' and 'cytotoxicity' despite major differences in molecular mechanisms and cellular fates in vivo. Indeed, the ^{51}Chromium (^{51}Cr) release assay has retained status as the 'gold standard' for cell-mediated cytotoxicity determinations based on an endpoint of target cell lysis with the release of ^{51}Cr from what are often late-stage apoptotic membrane permeability changes into an ex vivo environment devoid of scavenger cells. This and the simple two-population format of the assay (effectors plus targets) create a misleading view of NK cell function. The ^{51}Cr release assay is neither specific for the type of effector cell nor for the form of target cell death triggered. Here, we discuss criteria that provide a more robust, specific and complete assessment of NK cell effector function, including measurements that are more coincident with the in vivo fate of most target cells dying via NK cell mediation. The principles described may provide superior insight into NK cell-induced target cytotoxicity and have some prognostic value.

Cell-mediated cytotoxicity

While the ^{51}Cr release assay has several positive features, particularly simplicity in practice, the basic premises of the assay have raised questions spurring investigations of alternative methods. For example, the initial assay was designed to measure the 'lytic action' of splenocytes from mice immunized with cells of a mastocytoma (P815) line (Brunner et al., 1968). Thus, the effector cells in this case were presumably CD8$^+$ T cells plus possibly some NK cells with targets of mast cell origin. The latter include large histamine, leukotriene and growth factor-containing granules so that when cells are triggered to lyse, an exceptionally large release of inflammogens, suggestive of a large inflammatory response, could be detected.

In contrast, the composition of the cytoplasmic contents of most in vivo target cells of NK cells as well as CD8$^+$ T cells lack such high concentrations and arrays of immunostimulatory factors. Certainly, in an in vivo setting, actual lysis of any target cells would result in some degree of migration of leukocytes, that is neutrophils, T lymphocytes, B lymphocytes, plasma cells and dendritic cells (DCs), regularly observed at sites of inflammation but to a lesser extent than from cells of mast cell lineage. However, when assessing the validity of a cell-mediated cytotoxicity assay, it is important to note that although NK cells per se are active players in inflammatory responses and can produce high levels of proinflammatory cytokines such as interferon gamma (IFNγ), interleukins-6 (IL-6) and interleukins-8 (IL-8) and GM-CSF (Chan et al., 2008), data from many in vivo studies indicate NK cell-mediated cytotoxicity of pathogen-infected or tumour cells does not produce an inflammatory response (Fink and Cookson, 2005; Sigal, 2007). Rather, the death pathways induced in most target cells involve some form of programmed cell death also know as apoptosis.

Thus, the absence of inflammation likely reflects the immunosuppressive activity of physiological NK cell targets, which is an attribute common to all apoptotic cells (Cvetanovic et al., 2006). Clearly, a shortcoming and distortion of the ^{51}Cr release assay and the one with highest potential pathophysiological impact is the extracellular release of target cell contents. This direct release of inflammatory mediators and other intracellular molecules simply is not relevant physiologically.

Apoptosis in cell-mediated cytotoxicity

Apoptosis is briefly described as follows: Each nucleated cell contains programs, which, when activated, can lead to the death of that cell by internal disintegration (Kerr et al., 1972). The major biochemical pathways of activation include a group of proteases, the caspases, which are characterized by both a catalytic cysteinyl residue and a strong preference for an aspartyl residue in the P_1 position of their substrate recognition sequences. The order of activation of the approximately fourteen identified caspases depends on variables such as the apoptogenic agent/event, cell type and extracellular environment. With respect to NK cell-mediated target cell death, strong evidence for the induction of two apoptotic pathways exists (Smyth et al., 2005): (i) delivery of perforin/granzymes from NK cells to targets and (ii) ligation of death receptors on target cells by proteins on NK cell surfaces. Examples of the latter include the ligand for Fas/CD95, tumour necrosis factor, alpha (TNFα) and TNF-related apoptosis-inducing ligand (TRAIL) (Lettau et al., 2008).

Cell-mediated cytotoxicity assays

After reassessing the limited and possibly misleading information that the ^{51}Cr release assay can provide, several groups have focused on alternative methods (Figure 49.1). The objective has been to develop specific assays that provide mechanistic and perhaps therapeutic insights into effector–target interactions not possible with measurements of very late stage events. Thus, efforts have been directed at establishing assays to address the following problems inherent in the original cell-mediated cytotoxicity assay.

First, the initial step in the ^{51}Cr release assay requires loading of target cells with $Na_2{}^{51}CrO_4$, a radioactive

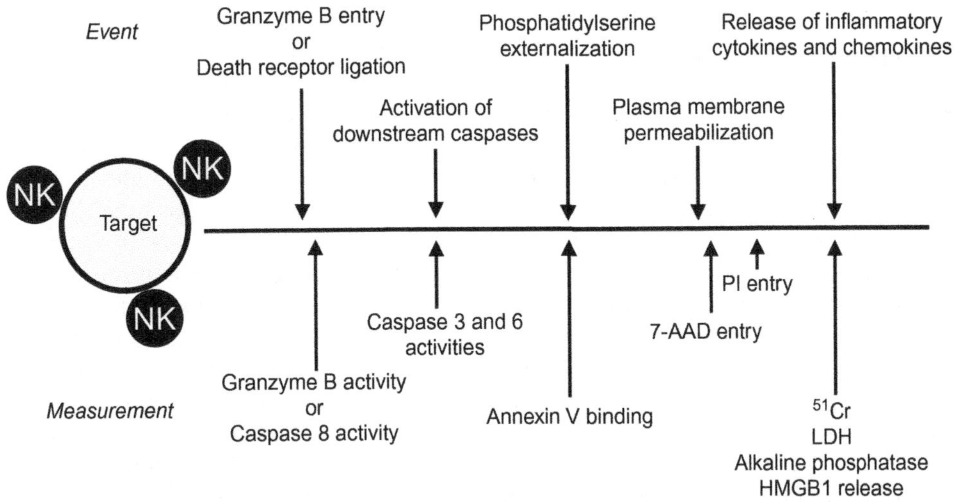

Figure 49.1 • NK cell-mediated target cell death. Following engagement by NK cells, a lethal hit can be delivered to a target cell by (i) introduction of the serine protease Granzyme B from NK cells via perforin assistance into the target's cytoplasm or (ii) ligation of the target cell's death receptors by proteins such as Fas ligand, TFNα and TRAIL on NK cell surfaces. Either can induce the following sequence of events in a target: activation of caspase cascades, followed by flipping of phosphatidylserine from the inner leaflet of the cell's plasma membrane to its extracellular side, loss of the cell's permeability barrier and eventually release of intracellular contents into the extracellular milieu. Measurements for each event are shown below the horizontal line in red for endogenous activities with exogenous reagents in blue.

isotope. Besides the environmental impact of radioactivity, loading can be inconsistent with respect to both individual cells and cell types. In fact, many, especially primary, cells do not take up this label to any workable level. Since ^{51}Cr is used to define the target cell population, replacement of this radioactive label with chromophores or fluorophores of a variety of structures and functional groups has been attempted. For example, Alamar Blue (Nociari et al., 1998), the tetrazolium salt MTT (Hussain et al., 1993), Vybrant DiO (Piriou et al., 2000) and MitoTracker Green (Vizler et al., 2002) have been used to examine changes in target cell metabolic activity following exposure to effector cells. (None of these metabolic activities discriminates either directly or unambiguously with respect to actual cell death.) Also, the fluorescence of profluorophores such as fluorescein moieties (Kolber et al., 1988; McGinnes et al., 1986) and the acetoxymethyl ester of calcein (Roden et al., 1999), which both require de-esterification by target cell cytoplasmic esterases for radiative emission, have been used for distinguishing targets from effectors. The most successful target labelling probes are cell-permeable alkylating agents such as cell tracker orange (Liu et al., 2002), carboxyfluorescein succinimidyl ester (Jedema et al., 2004), TFL2 (Xu et al., 2003) and TFL4 (Packard et al., 2007). If used properly, these agents covalently label only intracellular proteins, leaving surface recognition unaffected and leakage minimized. Thus, the important criteria for target cell labelling are (i) intrinsic brightness of the label for clear distinction between target and effector cells, (ii) absence of or very low leakage of the label from target cells following effector–target interaction so that discrimination between the two cell types is maintained and (iii) choice of wavelength range for maximization of functional readouts, for example, phenotyping (*vide infra*).

Second, the ^{51}Cr release assay is quantitated by release of ^{51}Cr from targets following addition of effectors. The readout, which is from the bulk cellular population, not only results in high backgrounds but also precludes single cell quantitative and qualitative analyses. By using a technique such as flow cytometry, the cytotoxic hit delivered to a single cell can be assessed quantitatively. Additionally, imaging by microscopy (Packard et al., 2007; Torrero et al., 2006) or with an instrument such as the Amnis flow cytometer (Basiji et al., 2007) can provide insight into the morphology of the effector–target interaction, specifically the immunologic synapse (Davis, 2002). However, it is essential to consider the death-associated proteolytic activities induced inside target cells (*vide infra*), beyond simple effector–target contact, in order to assess the effective delivery of a lethal hit.

Third, whereas multiplex readouts such as phenotyping individual effector and/or dying target cells are not possible with a bulk assay, separation of targets from effectors and then of dying from unaffected targets using a single cell-based assay allows introduction of antibodies for identification of antigens and receptors on individual cells (Kim et al., 2007; Liu et al., 2002; Migueles et al., 2008).

Death-associated proteolytic activities in target cells as a measure of effector cell lethality

Since apoptotic pathways[1] are induced in most target cells following productive engagement by NK cells (Bolitho et al., 2007; Lettau et al., 2008), live cell assays for evaluating intracellular caspase activities have been developed. Using cell permeable fluorogenic substrates with specificity for distinct caspases (Komoriya et al., 2000; Packard et al., 1996) and having target cells clearly distinguished from effectors by prelabelling the former with a fluorescent probe (*vide supra*), the apoptotic fates of virally infected cells and tumour cells have been assessed following exposure to NK cells (Carlsten et al., 2007; Maki et al., 2008; Packard et al., 2007; Scott-Algara et al., 2008). It is instructive to note that FLICA (fluorescent label for inhibition of caspase activity) reagents that are often used for detection of cellular apoptosis are not specific for caspases (Pozarowski et al., 2003); furthermore, their destruction of the protease activity of each protease molecule detected results in measurements unreflective of the total protease activity in a given time span (Packard and Komoriya, 2008). In a direct comparison between cell-mediated cytotoxicity assays of virally sensitized effector cells and targets bearing viral epitopes, even the induction of caspase 3, which is a late or downstream element in defined caspase cascades, was shown to occur earlier than ^{51}Cr release at effector:target ratios ranging from 50:1 down to 2:1 with coincubation times extending from 30 min up to 20 h (Liu et al., 2002). Thus, the ^{51}Cr-release assay, however fast and easy, still is a slow/late measure of events within the target.

Similar to caspase activation preceding leakage of ^{51}Cr from preloaded targets, the release of cytoplasmic markers such as lactate dehydrogenase and alkaline phosphatase as well as nuclear components, for example, high mobility group box 1 (HMGB1) (Ito et al., 2007), from target cells ensue along a time course similar to that of ^{51}Cr release. Even earlier measurements indicating the loss of plasma membrane integrity as illustrated by the entry of small molecules, for example, propidium iodide (PI) and 7-aminoacridine (7-AAD), unable to penetrate intact plasma membranes, have been shown to follow the appearance of caspase activities in apoptotic cells (Telford et al., 2004). Moreover, since in cells dying via apoptosis, plasma membrane permeabilization is preceded by translocalization of phosphatidylserine (PS) from the cytoplasmic to extracellular side of the bilayer (Fadok et al., 2001), binding of annexin-V to PS is often used to measure this reordering. However, intracellular caspase activation not only precedes this event but also can be measured in live cells in real time, whereas cell labelling with annexin V, a 35–36 kDa, Ca^{++}-dependent, phospholipid binding protein with a high affinity for PS, is an endpoint assay. As discussed in detail later, the externalization of PS is not specific to apoptosis.

A more recent assay has used the same cell-permeable caspase probe design to quantitate an even earlier biochemical event than caspase activation involved in some forms of cytotoxic lymphocyte-mediated cytotoxicity, namely, entry of the serine protease granzyme B (GzB) into target cells. The principle of the assay is that, although effector NK cells, as well as $CD8^+$ lymphocytes, express and store GzB in their cytoplasmic granules, along with the pore-forming protein perforin and other components believed to be essential for effector cell assault on targets, GzB is proteolytically inactive in these effectors due to the acidic pH of their granules. However, once GzB enters the cytosol of target cells where the pH is more neutral, the molecule becomes enzymatically active and is able to process target cell procaspases.

With the preceding as a rationale, a cell-permeable fluorogenic probe derived from the physiological macromolecular substrate for GzB, DNA-dependent protein kinase catalytic subunit, was synthesized to serve as a direct biochemical measure of the delivery of an effective lethal hit from cytotoxic lymphocytes (Packard et al., 2007). The appearance of GzB activity in the cytoplasm of target cells prelabelled as described earlier for the caspase assay can be used at the single-cell level as a measure of one type of attack by effector cells (Chiarle et al., 2008; Kinter et al., 2007; Migueles et al., 2008; Singh et al., 2008).

Thus, measurement of the lethal hit on target cells by effectors prior to changes in target membrane permeability (Packard et al., 2007), upon which both ^{51}Cr release and the earlier flow cytometry methodologies depend, is engendered. With time, the loss of target cell plasma membrane permeability barriers has been shown by entry of PI into GzB^+ cells (Migueles et al., 2008).

In addition to the adoption of this assay as a tool for detecting disorders that are directly linked to the intracellular machinery and constituents needed for cytotoxicity, it can serve as a functional readout in characterizing vaccine candidates with direct cytotoxic potency (Chiarle et al., 2008; Singh et al., 2008). Recently, the delivery of proteolytically active GzB from effectors to autologous HIV target cells has elucidated an effector function, that is control by virus-specific $CD8^+$ T cells over HIV replication, which segregates with patients able to maintain immunologic control of HIV from those that progress (Migueles et al., 2008).

[1]There are several ongoing studies to determine the possible role(s) of autophagy (Maiuri et al, 2007) and enzymatic activities associated therewith in cell-mediated cytotoxicity.

The fluorogenic caspase substrates, originally designed for single-cell quantitation following apoptogenic induction by classical intrinsic or extrinsic inducers of apoptotic cell death, provide further confirmation of the lethality of a GzB hit. Simultaneous use of complementarily labelled cell-permeable fluorogenic substrates for GzB and caspase 3 has indicated that GzB activity precedes caspase 3 activity inside every cell examined by both flow cytometry and microscopy (Packard et al., 2007). This is consistent with the view that the cell death pathway initiated by entry of GzB into target cells short-circuits the well-documented intrinsic and extrinsic caspase cascade activation and amplification pathways, and further resolves conflicting data regarding the effect of overexpression of the antiapoptotic Bcl-2 gene in target cells.

Besides NK cells inducing apoptosis in target cells by release of their granule contents in a Ca^{++}-dependent manner, another mechanism employed is through the binding of effector cell surface death ligands to cognate receptors embedded in target cell membranes in a Ca^{++}-independent interaction (Lettau et al., 2008). This mechanism often reflects the cytokine environment as interleukins such as IL-2 and IL-15 and the IFNs, particularly IFNγ, can regulate the levels of death ligands and receptors (Samai et al., 1998; Smyth et al., 2005). After receptor aggregation and recruitment of the Fas-associated death domain protein and procaspase 8 to a signalling complex (death-inducing signalling complex [DISC]), caspase 8 becomes activated. This can directly lead to processing and activation of downstream caspases 3/7 and 6. Alternatively, active caspase 8 can cleave Bid, yielding a truncated species that inserts into the outer mitochondrial membrane and facilitates cytochrome c release and Apaf-1-mediated activation of caspase 9. Regardless of the exact route, measurement of caspase activities, particularly the later caspases, that is caspases 3/7 and 6, can be definitive and relatively early markers of impending cell death.

In view of the important role that apoptosis plays in NK cell-mediated cellular cytotoxicity, additional issues of concern have been raised as to how effectively the classical 4-h ^{51}Cr release assay reports death induced by various pathways. Specifically, it has been reported that receptor-mediated target cell death requires longer incubation periods than the standard 4 h. Additionally, Fas ligand-induction is sometimes slower in vitro than perforin-mediated apoptosis induction since NK cells must trigger the upregulation of Fas on otherwise Fas-negative targets (Screpanti et al., 2005). Along this line, it is of interest that the traditional 4-h ^{51}Cr release assay target for NK cell activity, the K562 cells, lack constitutive expression of Fas receptors. Thus, it is extremely important to consider the conditions of activity assessment in the evaluation of in vitro data, particularly for translation into clinical settings.

Antibody-dependent cellular cytotoxicity

The biotechnology revolution has given rise to the emergence of monoclonal antibody (mAb) therapeutics. Although these macromolecules have found utility in the management of several malignant disease conditions, their mechanism(s) of action in most cases remains unclear. It is believed that at least three modes of antibody-dependent target cell killing are possible: classical complement activation, macrophage phagocytosis of opsonized tumour cells and antibody-dependent cellular cytotoxicity (ADCC) (Taylor and Lindorfer, 2007). While complementary activation and phagocytosis of opsonized targets have been studied using many classical assays, the degree of involvement of NK cell-mediated cytotoxicity remains poorly understood. The NK cell may have a role beyond the putative binding via its F_c receptors (CD16) to an antibody's F_c domain as the mAb uses both of its F_{ab} segments to bind to two copies of an epitope on a target cell. In one proposed model, degranulation of NK cells follows with perforin-mediated entry of GzB and activation of caspases in targets resulting in apoptotic death. There are currently several studies underway to clarify this point.

Cell-mediated cytotoxicity as a mechanism for pathogen clearance that avoids inflammation

Distinct from apoptosis, which is considered to be a physiological death, cells can also die via necrosis, which is regarded as a pathological form. In the latter, disintegration of a cell in situ with release of intracellular contents leads to activation of the inflammatory response quartet—*rubor*, *calor*, *dolor* and *tumour*—resulting in major physical perturbations. Specifically, soluble mediators such as cytokines and chemokines are elicited, and they, in turn, induce infiltration of an array of additional immune effector cells such as monocytes, neutrophils and lymphocytes. Thus, in order to avoid inflammation and tissue scarring, prevention of cellular leakage and prompt clearance are important aspects of apoptotic cell death (Kerr et al., 1972). Clearly, the major differences in the sequelae of these two forms of death can profoundly impact the host (Birge and Ucker, 2008). With regard to cell-mediated cytotoxicity, one particularly important factor is that activation of an apoptotic pathway in a dying target cell can limit bystander cell damage (Pipkin and Lieberman, 2007).

Apoptotic suppression of inflammatory responsiveness is effected primarily at the level of specific transcriptional initiation (Cvetanovic and Ucker, 2004) thereby inhibiting the inflammatory response elicited by pathogens and other innate immune agonists (such as secretion of the inflammatory cytokines IL-6, IL-8 and TNFα). In fact, scavenger cells (both professional and nonprofessional phagocytes, including epithelial, endothelial and fibroblastic cells) recognize and respond to apoptotic cells by immediate-early transcriptional repression of a variety of specific genes, including those encoding the previously mentioned inflammatory cytokines. This repression occurs in the absence of de novo protein synthesis and is specific to the expression of inflammatory mediators. Importantly, interactions with apoptotic targets do not lead to global repression of scavenger cell transcription.

The specific immunosuppressive activity of apoptotic corpses is acquired by virtually all cells undergoing apoptotic cell death, regardless of the cell type or the particular death stimulus. Moreover, acquired anti-inflammatory activity persists stably in apoptotic cells, even as the cells lose membrane integrity. By comparison, viable and necrotic cells do not express such inhibitory determinants. In fact, in vivo necrotic cells do die with membrane rupture associated with classic inflammatory pathology.

Since recognition of apoptotic cells is critical to their clearance and the innate immunosuppression they elicit, much effort has gone into identifying cellular markers signifying their recognition by scavenger cells (Hart et al., 1996; Liu et al., 2006; Maderna and Godson, 2003; Patel et al., 2006; Roos et al., 2004). One such putative determinant is PS, which, as described earlier, is translocated from the inner to the outer leaflet of apoptotic plasma membranes and can be detected by fluorescently labelled annexin V. This particular reagent has been used in many apoptosis assays, including one developed to assess NK cell-mediated cytotoxicity (Shounan et al., 1998). While cellular binding of annexin V can identify apoptotic cells, and there is a long history of the association between the repositioning of PS and apoptosis, this process is not specific to the latter since both necrotic and apoptotic cells display externalized PS.

Significantly, the modes of scavenger cell binding involved in recognition of native cells that have died by these two distinct mechanisms are both independent and noncompetitive (Cocco and Ucker, 2001). A complete understanding of the molecular components involved in both recognition and clearance of apoptotic cells is evolving. It appears likely that interactions between multiple apoptotic markers and phagocytic receptors may be necessary for optimal clearance; current candidate molecules and possible key steps include the GTPases RAB-5 and RAB-7, the HOPs complex and the phosphatidyl 3 kinases VPS-34 and Dyn-1 (Kinchen et al., 2008).

Conclusions

The idea of cell lysis being the endpoint of all cell-mediated cytotoxic events can no longer be considered an accurate in vitro assessment or predictor of target cell death mediated by NK cells in vivo. The model in which a mast cell explodes and releases its highly inflammatory cytoplasmic contents represents a system decidedly biased towards an endpoint of target cell death that occurs rarely, if at all, in vivo. Thus, the bases of retention of the ^{51}Cr release assay as the 'gold standard' for cell-mediated cytotoxicity determinations are highly questionable. As has been discussed in this chapter, measurements more coincident with the early molecular events involved in effector–target interactions, and the in vivo fates of target cells have the potential of providing meaningful insight into NK cell-induced cytotoxicity as well as improved prognostic value. Significantly, in the absence of an accurate assessment of the physiological cell death that results from engagement between NK cells and target cells, progress in NK cell biology and development of effective therapies will be hindered. It is quite possible that acknowledgement of the importance of this distinction may explain the disconnect that can exist between the translation from excellent in vitro data to failed, or suboptimal, clinical trial results. Accurate in vitro analytical tools may provide a means for the use of NK cell biology at the level of therapeutic efficacy envisioned by many.

References

Basiji, D.A., Ortyn, W.E., Liang, L., et al. (2007). Cellular image analysis and imaging by flow cytometry. Clin Lab Med **27**, 653–670.

Birge, R.B. and Ucker, D.S. (2008). Innate apoptotic immunity: the calming touch of death. Cell Death Diff **15**, 1096–1102.

Brunner, K.T., Mauel, J., Cerottini, J.-C., et al. (1968). Quantitative assay of the lytic action of immune lymphoid cells on ^{51}Cr-labelled allogeneic target cells in vitro; inhibition by isoantibody and by drugs. Immunology **14**, 181–196.

Caligiuri, M.A. (2008). Human natural killer cells. Blood **112**, 461–469.

Carlsten, M., Björkström, N.K., Norell, H., et al. (2007). DNAX accessory molecule-1 mediated recognition of freshly isolated ovarian carcinoma by resting natural killer cells. Cancer Res **67**, 1317–1325.

Chan, A., Filer, A., Parsonage, G., et al. (2008). Mediation of the proinflammatory cytokine response in rheumatoid arthritis and spondylarthritis by interactions between fibroblast-like synoviocytes and natural killer cells. Arthritis Rheum **58**, 707–717.

Chiarle, R., Martinengo, C., Mastini, C., et al. (2008). The anaplastic lymphoma kinase is an effective oncoantigen for lymphoma vaccination. Nat Med **14**, 676–689.

Cocco, R.E. and Ucker, D.S. (2001). Apoptotic cells, at all stages of the death process, trigger characteristic signaling events that are divergent from and dominant over those triggered by necrotic cells: implications for the delayed clearance model of autoimmunity. Mol Biol Cell **12**, 919–930.

Cvetanovic, M. and Ucker, D.S. (2004). Innate immune discrimination of apoptotic cells: repression of proinflammatory macrophage transcription is coupled directly to specific recognition. J Immunol **172**, 880–889.

Davis, D.M. (2002). Assembly of the immunological synapse for T cells and NK cells. Trends Immunol **23**, 356–363.

Di Santo, J.P. (2008). Natural killer cells: diversity in search of a niche. Nat Immunol **9**, 473–475.

Fadok, V.A., Bratton, D.L. and Henson, P.M. (2001). Phagocyte receptors for apoptotic cells: recognition, uptake, and consequences. J Clin Invest **108**, 957–962.

Fink, S.L. and Cookson, B.T. (2005). Apoptosis, pyroptosis, and necrosis: mechanistic description of dead and dying eukaryotic cells. Infect Immun **73**, 1907–1916.

Hart, S.P., Haslett, C. and Dransfield, I. (1996). Recognition of apoptotic cells by phagocytes. Experientia **52**, 950–956.

Herberman, R.B., Nunn, M.E. and Lavrin, D.H. (1975). Natural cytotoxic reactivity of mouse lymphoid cells against syngeneic acid allogeneic tumors. I. Distribution of reactivity and specificity. Int J Cancer **16**, 216–229.

Hussain, R.T., Nouri, A.M. and Oliver, R.T. (1993). A new approach for measurement of cytotoxicity using colorimetric assay. J Immunol Methods **160**, 89–96.

Ito, N., DeMarco, R.A., Mailliard, R.B., et al. (2007). Cytolytic cells induce HMGB1 release from melanoma cell lines. J Leukoc Biol **81**, 75–83.

Jedema, I., van der Werff, N.M., Barge, R.M., et al. (2004). New CFSE-based assay to determine susceptibility to lysis by cytotoxic T cells of leukemic precursor cells within a heterogeneous target cell population. Blood **103**, 2677–2682.

Kerr, J.F.R., Wyllie, A.H. and Currie, A.R. (1972). Apoptosis: a basic biological phenomenon with wide-ranging implications in tissue kinetics. Br J Cancer **26**, 239–256.

Kiessling, R., Klein, E. and Wigzell, H. (1975). "Natural" killer cells in the mouse. I. Cytotoxic cells with specificity for mouse Moloney leukemia cells. Specificity and distribution according to genotype. Eur J Immunol **5**, 112–117.

Kinchen, J.M., Doukoumetzidis, K., Almendinger, J., et al. (2008). A pathway for phagosome maturation during engulfment of apoptotic cells. Nat Cell Bio **10**, 556–566.

Kinter, A., McNally, J., Riggin, L., et al. (2007). Suppression of HIV-specific T cell activity by lymph node $CD25^+$ regulatory T cells from HIV-infected individuals. Proc Natl Acad Sci U S A **104**, 3390–3395.

Kim, G.G., Donnenberg, V.S., Donnenberg, A.D., et al. (2007). A novel multiparametric flow cytometry-based cytotoxicity assay simultaneously immunophenotypes effector cells: comparisons to a 4 h ^{51}Cr-release assay. J Immunol Methods **325**, 51–66.

Kolber, M.A., Quinones, R.R., Gress, R.E., et al. (1988). Measurement of cytotoxicity by target cell release and retention of the fluorescent dye *bis*-carboxyethyl-carboxyfluorescein (BCECF). J Immunol Methods **108**, 255–264.

Komoriya, A., Packard, B.Z., Brown, M.J., et al. (2000). Assessment of caspase activities in intact apoptotic thymocytes using cell permeable fluorogenic caspase substrates. J Exp Med **191**, 1819–1828.

Lettau, M., Paulsen, M., Kabelitz, D., et al. (2008). Storage, expression and function of Fas ligand, the key death factor of immune cells. Curr Med Chem **15**, 1684–1696.

Liu, L., Chahroudi, A., Silvestri, G., et al. (2002). Visualization and quantification of T cell-mediated cytotoxicity using cell-permeable fluorogenic caspase substrates. Nat Med **8**, 185–189.

Liu, G., Wu, C., Wu, Y., et al. (2006). Phagocytosis of apoptotic cells and immune regulation. J Immunol **64**, 1–9.

Long, E.O. (2008). Negative signaling by inhibitory receptors: the NK cell paradigm. Immunol Rev **224**, 70–84.

Maderna, P. and Godson, C. (2003). Phagocytosis of apoptotic cells and the resolution of inflammation. Biochim Biophys Acta, 141–151.

Maiuri, M.C., Zalckvar, E., Kimchi, A., et al. (2007). Self-eating and self-killing: crosstalk between autophagy and apoptosis. Nat Rev Mol Cell Biol **8**, 741–752.

Maki, G., Hayes, G.M., Naji, A., et al. (2008). NK resistance of tumor cells from multiple myeloma and chronic lymphocytic leukemia patients: implication of HLA-G. Leukemia **22**, 998–1006.

McGinnes, K., Chapman, G., Marks, R., et al. (1986). A fluorescence NK assay using flow cytometry. J Immunol Methods **86**, 7–15.

Migueles, S.A., Osborne, C.M., Royce, C., et al. (2008). Lytic granule loading of $CD8^+$ T cells is required for HIV-infected cell elimination associated with immune control. Immunity **29**, 1009–1021.

Mitchell, J.E., Cvetanovic, M., Tibrewal, N., et al. (2006). The presumptive phosphatidylserine receptor is dispensable for innate anti-inflammatory recognition and clearance of apoptotic cells. J Biol Chem **281**, 5718–5725.

Moretta, A., Locatelli, F. and Moretta, L. (2008). Human NK cells: from HLA class I-specific killer Ig-like receptors to the therapy of actue leukemias. Immunol Rev **224**, 58–69.

Nausch, N. and Cerwenka, A. (2008). NKG2D ligands in tumor immunity. Oncogene **27**, 5944–5958.

Nociari, M.M., Shalev, A., Benias, P., et al. (1998). A novel one-step, highly sensitive fluorometric assay to evaluate cell-mediated cytotoxicity. J Immunol Methods **213**, 157–167.

Packard, B.Z. and Komoriya, A. (2008). Intracellular protease activation in apoptosis and cell-mediated cytotoxicity characterized by cell-permeable fluorogenic protease substrates. Cell Res **18**, 238–247.

Packard, B.Z., Toptygin, D.D., Komoriya, A., et al. (1996). Profluorescent protease substrates: intramolecular dimers described by the exciton model. Proc Natl Acad Sci U S A **93**, 11640–11645.

Packard, B.Z., Telford, W.G., Komoriya, A., et al. (2007). Granzyme B activity in target cells detects attack by cytotoxic lymphocytes. J Immunol **179**, 3812–3820.

Patel, V.A., Longacre, A., Hsiao, K., et al. (2006). Apoptotic cells, at all stages of the death process, trigger characteristic signaling events that are divergent from and dominant over those triggered by necrotic cells: implications for the delayed clearance model of autoimmunity. J Biol Chem **281**, 4663–4670.

Piriou, L., Chilmonczyk, S., Genetet, N., et al. (2000). Design of a flow cytometric assay for the determination of natural killer and cytotoxic T-lymphocyte activity in human and in different animal species. Cytometry **41**, 289–297.

Pozarowski, P., Huang, X., Halicka, D.H., et al. (2003). Interactions of fluorochrome-labeled caspase inhibitors with apoptotic cells: a caution in data interpretation. Cytometry A **55**, 50–60.

Rajakariar, R., Lawrence, T., Bystrom, J., et al. (2008). Novel biphasic role for lymphocytes revealed during resolving inflammation. Blood **111**, 4184–4192.

Roden, M.M., Lee, K.H., Panelli, M.C., et al. (1999). A novel cytolysis assay using fluorescent labeling and quantitative fluorescent scanning technology. J Immunol Methods **226**, 29–41.

Roos, A., Xu, W., Castellano, G., Nauta, A.J., Garred, P., Daha, M.R. and van Kooten, C. (2004). Mini-review: a pivotal role for innate immunity in the clearance of apoptotic cells. Eur J Immunol **34**, 921–929.

Smyth, M.J., Cretney, E., Kelly, J.M., et al. (2005). Activation of NK cell cytotoxicity. Mol Immunol **42**, 501–510.

Scott-Algara, D., Arnold, V., Didier, C., et al. (2008). The CD85j$^+$ NK cell subset potently controls HIV-1 replication in autologous dendritic cells. PLoS ONE **3**, e1975.

Shounan, Y., Fang, X. and O'Connell, P.J. (1998). Apoptosis detection by annexin V binding: a novel method for the quantitation of cell-mediated cytotoxicity. J Immunol Methods **217**, 61–70.

Sigal, L.H. (2007). Handling the corpses: apoptosis, necrosis, nucleosomes and (quite possibly) the immunopathogenesis of SLE. J Clin Rheum **13**, 44–48.

Singh, A., Nie, H., Ghosn, B., et al. (2008). Efficient modulation of T-cell response by dual-mode, single-carrier delivery of cytokine-targeted siRNA and DNA vaccine to antigen-presenting cells. Mol Ther **16**, 2011–2021.

Strowig, T., Brilot, F. and Münz, C. (2008). Noncytotoxic functions of NK cells: direct pathogen restriction and assistance to adaptive immunity. J Immunol **180**, 7785–7791.

Taylor, R.P. and Lindorfer, M.A. (2007). Drug insight: the mechanism of action of rituximab in autoimmune disease—the immune complex decoy hypothesis. Nat Clin Pract Rheumatol **3**, 86–95.

Telford, W.G., Komoriya, A. and Packard, B.Z. (2004). Multiparametric analysis of apoptosis by flow and image cytometry. Methods Mol Biol **263**, 141–160.

Torrero, M.N., Xia, X., Henk, W., et al. (2006). Stat1deficiency in the host enhances interleukin-12-mediated tumor regression. Cancer Res **66**, 4461–4467.

Vivier, E., Tomasello, E., Baratin, M., et al. (2008). Functions of natural killer cells. Nat Immunol **9**, 503–610.

Vizler, C., Nagy, T., Kusz, E., et al. (2002). Flow cytometric cytotoxicity assay for measuring mammalian and avian NK cell activity. Cytometry **47**, 158–162.

Waldhauer, I. and Steinle, A. (2008). NK cells and cancer immunosurveillance. Oncogene **27**, 5932–5943.

Walzer, T., Jaeger, S., Chaix, J., et al. (2007). Natural killer cells: from CD3$^-$NKp46$^+$ to post-genomics meta-analyses. Curr Opin Immunol **19**, 365–372.

Xu, Y., Szalai, A.J., Zhou, T., et al. (2003). FcγRs modulate cytotoxicity of anti-Fas antibodies: implications for agonistic antibody-based therapeutics. J Immunol **171**, 562–568.

Chapter Fifty

50

NK-cell tracking using non-invasive imaging modalities

Priyanka Jha, Verena Reinhart, Christiane Knopp, Siyuan Liu, Winfried S. Wels, Michael F. Wendland, Heike Daldrup-Link

CHAPTER CONTENTS

When I dipt into the future, far as human eye could see;
Saw the Vision of the world, and all the wonder that would be

Lord Tennyson

ABSTRACT

Various non-invasive imaging modalities are being utilized for monitoring natural killer (NK)-cell-based immunotherapy, including magnetic resonance (MR) imaging, optical imaging, fluorodeoxyglucose positron emission tomography (FDG-PET) imaging, and other radiotracer modalities. While most of these are still in an experimental stage, MR imaging has been employed for human studies using ferumoxide-labelled dendritic cells (de Vries et al., 2005). All these modalities have certain advantages and disadvantages in comparison with one another. MR imaging has a significant potential for being clinically applicable due to ease of labelling, availability of U.S. FDA-approved contrast agents and well developed infrastructure. In vivo imaging studies provide information pertaining to accumulation and retention of the NK cells in the tumour tissue as well as therapeutic monitoring of immunotherapy regimens.

KEY WORDS

NK-cell tracking, MR imaging, Optical imaging, NK immunotherapy, PET imaging, Contrast agent labelled cells, Bioluminescence

Introduction

Natural killer (NK) cells are lymphocytes of the innate immune system that are critical for the host defense against viruses and cancer through their rapid and direct cytotoxic activity against infected and malignant cells (Orange and Ballas, 2006). The antitumour activity of

NK cells has been extensively investigated with respect to mechanisms of immunosurveillance, escape strategies of the malignant cells and risk of adverse reactions, and the potential of NK cells for cancer immunotherapy is increasingly being appreciated (Smyth et al., 2002; Suck, 2006; Terme et al., 2008). Exploiting the alloreactive potential of ex vivo expanded and activated donor-derived NK cells is thereby of particular interest, since it may overcome insufficient cytolytic activity of autologous NK cells from cancer patients (Papamichail et al., 2004; Ruggeri et al., 2005). For this approach also the use of cytotoxic NK cell lines has been proposed, which can be expanded continuously and are readily available without the need for a suitable donor (Klingemann, 2005; Suck, 2006).

In vivo imaging studies with retargeted, ErbB2-specific NK-92 cells in mice carrying ErbB2-expressing mammary tumours have proven invaluable to demonstrate that in addition to mediating specific and enhanced cytotoxicity, ectopic expression of a targeting receptor facilitates rapid accumulation and retention of the gene-modified cells in the tumour tissue, which was not the case for unmodified NK-92 (Daldrup-Link et al., 2005a; Meier et al., 2008). The human NK-92 cell line was originally established from peripheral blood lymphocytes of a patient with large granular lymphoma (Gong et al., 1994). NK-92 cells are similar to activated primary NK cells with respect to the expression of typical NK-cell surface receptors and functional characteristics, but display a much higher cytolytic activity against a broad spectrum of tumour targets, in particular against malignant cells of hematologic origin. Clinical phase I/II studies have been performed, which demonstrated that NK-92 cells can be safely administered for adoptive therapy (Arai and Klingemann, 2005; Tonn et al., 2001). In contrast to malignant cells of hematologic origin, cancer cells derived from solid tumours appear less sensitive to unmodified NK-92 (Uherek et al., 2002). To expand the target range of NK-92 cells and overcome NK-resistance, genetically modified variants were developed that stably express chimeric antigen receptors on their surface for specific recognition of tumour-associated cell surface antigens. These receptors consist of a scFv antibody fragment of the desired specificity, fused via a flexible hinge region to an intracellular signalling molecule such as the CD3 ζ chain, which activates cytotoxic effector mechanisms upon contact with antigen-expressing target cells (Wels et al., 2004). NK-92 cells targeted to ErbB2/HER2, a receptor tyrosine kinase overexpressed by many tumours of epithelial origin, were found to efficiently lyse established and primary ErbB2-expressing tumour cells that were otherwise resistant to the cytolytic activity of parental NK-92 (Uherek et al., 2002). More recently, this approach has been extended to NK-92 variants specifically recognizing tumours of epithelial origin that express the epithelial cell adhesion molecule (Ep-CAM) (Tavri et al., 2009; Uherek et al., 2004) and NK-92 cell lines targeting B-cell malignancies via CD19- or CD20-specific antigen receptors (Muller et al., 2008; Romanski et al., 2004).

Many modalities are being employed for in vivo imaging of NK-cell immunotherapy, including magnetic resonance (MR) imaging, optical imaging and positron emission tomography (PET) imaging using fluorodeoxyglucose (FDG) as well as other radiotracer agents (Table 50.1). NK cells are labelled prior to administration and imaging with appropriate labelling agents using previously tailored protocols. These cells can then be tracked after either intratumoural, intravenous or direct intra-arterial injections into tumour-feeding arteries. This provides valuable data that can otherwise be obtained only by invasive tissue biopsies, or later by following tumour markers. In this chapter we describe the rationale for non-invasive imaging and in vivo tracking of adoptively transferred NK cells, provide an overview on general principles, advantages and disadvantages of the techniques currently available for NK-cell labelling and tracking, and give an outlook on developments that may further increase the usefulness of NK-cell tracking in experimental models and clinical applications. To date, the identification of responders and non-responders to adoptive immunotherapies relies on the diagnosis of a decline in tumour markers, a reduction in tumour size and improved survival. These indirect markers diagnose responders and non-responders weeks or months after initiation of the immunotherapy (Jiang et al., 2006; Klingemann, 2005; Tonn et al., 2001). The localization and function of immune cells at the tumour site have been determined traditionally by biopsy and ex vivo analysis of cell function (Daldrup-Link et al., 2005a; Eshhar, 1997; Meier et al., 2008; Melder et al., 1993; Miller et al., 2005; Mulatero et al., 2001 Pinthus et al., 2003). Although valuable, this information fails to provide more dynamic information about tumour accumulation and function of the NK cells at the target sites at different time points. A diagnostic technique that could verify or disprove tumour accumulation of intravenously administered effector cells non-invasively could help to confirm this essential requirement for a successful adoptive immunotherapy. In addition, non-invasive in vivo tracking of NK cells with imaging techniques would allow the definition of imaging characteristics for treatment response or failure. Polymerase chain reaction (PCR)-based monitoring of NK-92 cell kinetics in the peripheral blood of patients revealed that a large portion of NK-92-cells apparently leave the circulation within minutes after transfusion (Tonn et al., 2001). However, since during treatment no tumour biopsies were taken and the adoptively transfused NK-92 cells were not labelled, homing of

Table 50.1 Comparative evaluation of clinically applicable imaging techniques for NK-cell tracking

Imaging technique	Merits	Demerits	Clinical applicability
Magnetic resonance imaging	1. High anatomical resolution 2. High soft tissue contrast 3. Absence of radiation exposure 4. Evaluation of deep-seated tissues possible	1. Lower sensitivity in comparison with optical as well as radiotracer imaging; molecular level assessment not feasible 2. Longer scan times	1. Most appealing amongst currently investigated modalities 2. Contrast agents- ferumoxides U.S. FDA-approved 3. MR scanners and pulse sequences well developed and widely available
Optical imaging	1. High sensitivity, can diagnose smaller as well as subclinical tumors 2. Absence of radiation exposure 3. Rapid image acquisitions (<5 min) 4. Inexpensive 5. Can assess molecular level (single cell) activity	1. Low depth of penetration 2. Low anatomical resolution 3. Lack of well-developed human scanners	1. Recent clinical applications for breast cancer imaging 2. Much development required in fields of hand-held as well as endoscopic devices to image deeper structures 3. Most contrast agents not U.S. FDA-approved; ICG is U.S. FDA-approved
Radiotracer imaging (FDG-PET)	1. High sensitivity for smaller and subclinical tumors 2. Can assess molecular level (single cell) activity	1. Radiotoxic cell damage a possibility 2. Anatomical resolution intermediate—can be improved significantly by applying PET-CT and/or PET-MR techniques	1. Clinically appealing 2. Indium label and [^{18}F]FDG label—U.S. FDA-approved 3. PET scanners clinically available for human use, though not widely available

NK-92 cells to the site of the tumour could not be confirmed by PCR or non-invasive techniques.

Routes for NK-cell administration

There are three previously described routes of administration for cell-based immunotherapy: intravenous administration into a peripheral vein, direct tumour delivery via a tumour-feeding artery, or direct transdermal injection into the tumour tissue. Intravenous NK-cell injection has the advantage of easy access of a peripheral vein, being the most frequently pursued approach for repetitive applications in experimental (Daldrup-Link et al., 2005a; Eshhar, 1997; Meier et al., 2008; Melder et al., 1993; Pinthus et al., 2003) and clinical settings (Miller et al., 2005; Mulatero et al., 2001). In addition, this is the only approach for patients with metastatic cancer and multiple tumour sites. Systemic injection of labelled cells involves cell dilution and distribution to various target sites. A systemic venous delivery of NK cells is associated with a significant loss of cells in the capillary beds of the lung and liver, requiring injections of higher NK-cell quantities as compared to the other two techniques. In addition, there is an unknown or variable time interval between intravenous injection and accumulation of NK cells in the target tumour (Daldrup-Link et al., 2003, 2004; Sykova and Jendelova, 2006). Direct NK-cell injection into a tumour-feeding artery is much more efficient in the case of a single tumour site. However, this technique is possible only via invasive interventional procedures and thus, limited to one or few NK-cell administrations. Intratumoural injections provide similar efficiencies in case of superficial tumours, which are accessible to transdermal injections. However, it has to be noted that the high interstitial pressure in some tumours may lead to preferential accumulation of intratumourally injected NK cells in the tumour periphery or in tumour necroses (Figure 50.1).

Imaging techniques for tracking of NK cells

Molecular imaging techniques can help to detect cells and follow cellular processes non-invasively, using MR imaging, optical imaging or radio-isotopic methods. New cell tracking techniques that provide a non-invasive in vivo detection of contrast-agent labelled cells can directly verify or disprove accumulation of NK cells in the tumour tissue as one major requirement for a successful

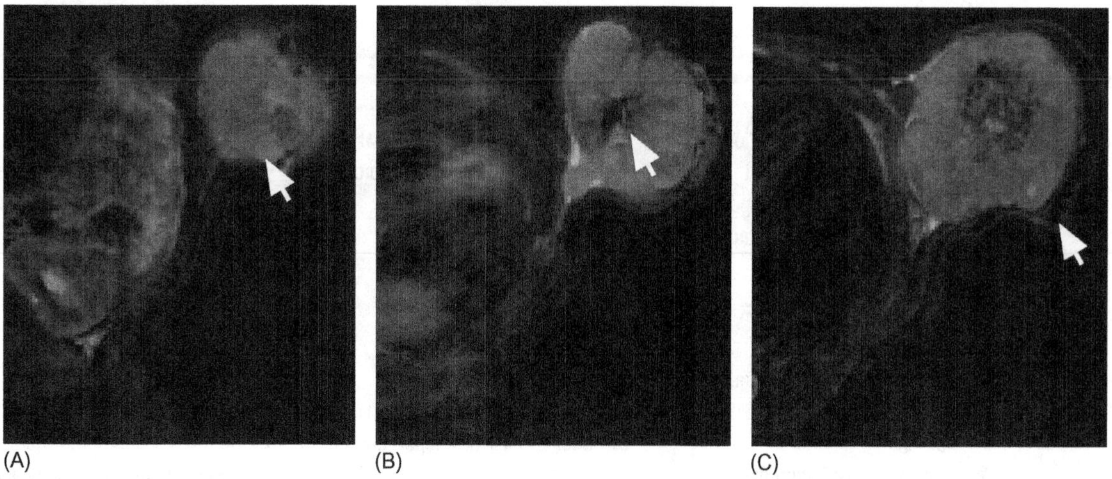

Figure 50.1 • Serial MR imaging of prostate tumours after intratumoural injection of magnetically labelled NK-92 cells: (A) A pre-injection scan shows the tumour (arrow). (B) A 30-min post-injection scan shows hypodensity suggestive of iron, confirming the delivery of the cells to the tumour parenchyma (arrow). (C) A 24-h post-injection scan shows accumulation of cells at the periphery of the tumour along with residual cells at the site of injection (arrow).

adoptive immunotherapy. Various imaging techniques are available for cell tracking, such as SPECT, PET, optical imaging and MR imaging. Labelling techniques with radioactive markers provide a high sensitivity but limited spatial resolution, and carry the risk of radiotoxic cell damage (Adonai et al., 2002; Fawwaz et al., 1985; Melder et al., 1993; Moore et al., 2004). Labelling techniques with fluorescent markers for optical imaging provide a high sensitivity, but have a limited anatomical resolution. MR imaging has the distinct advantages of providing three-dimensional (3D) data with high anatomical resolution (<1mm in plane), a high soft tissue contrast, and is not associated with any irradiation exposure (Arbab et al., 2004b, 2005; Bulte et al., 2001; Daldrup-Link et al., 2003; Frank et al., 2003; Hoehn et al., 2002; Metz et al., 2004; Weissleder et al., 1997). Recently, sensitive strategies for using membrane-bound luciferase for in vivo imaging of T cells has been reported that may be found to be useful for NK cells (Santos et al., 2009).

Optical imaging for tracking of NK-cell immunotherapy

Cell tracking with optical imaging is based on the detection of light emitted from fluorescent or luminescent cells. Optical imaging has many advantages over other imaging techniques in being quick, inexpensive and highly sensitive for the detection of fluorochrome-labelled cells (Sutton et al., 2008), and is already being used in experimental settings to track the in vivo distribution of fluorescent NK cells in rodents (Tavri et al., 2009). The major disadvantage of optical imaging stems from the limited penetration of light in the visual spectrum into the body tissues, thereby restricting applications to relatively superficial target tissues. This is not a concern while studying subcutaneously implanted tumours in animal models, and could be applied equally well to investigate relatively superficial cancers such as breast and skin cancers in human patients. Clinical optical scanners for evaluation of breast tumours in patients have been developed. The high sensitivity of this technique is expected to allow the detection of small and subclinical breast tumours based on their intrinsic fluorescence, and could also permit tracking of fluorochrome-labelled NK cells. Further technical developments that may expand the range of clinical applications for optical imaging include handheld optical imaging devices, fluorescence tomographic systems and intra-operative/endoscopic optical imaging devices.

Techniques for NK-cell labelling with fluorochromes

Various endogenous and exogenous fluorescent labels are being used for cell labelling. Endogeneous labels like green fluorescent protein (GFP) and luciferase (g-luc, f-luc) need to be incorporated into cells by transfection and make the cell fluoresce in the infrared spectrum. T cells have been transduced with g-luc to be inherently bioluminescent. The presence of this protein externally on the cell's membrane has been shown to be more luminescent than its internal counterpart (Santos et al., 2009). The spectrum of exogenous labels includes fluorescent dyes, non-targeted probes, colloid quantum dots and bifunctional contrast agents. A comprehensive overview of available labelling techniques is provided by Sutton and colleagues (Sutton et al., 2008). Our group favours cell labelling techniques with the fluorophore 1,1′-dioctadecyl-3,3,3′,3′ tetramethylindodicarbocyanine (DiD, Molecular Probes, Invitrogen)

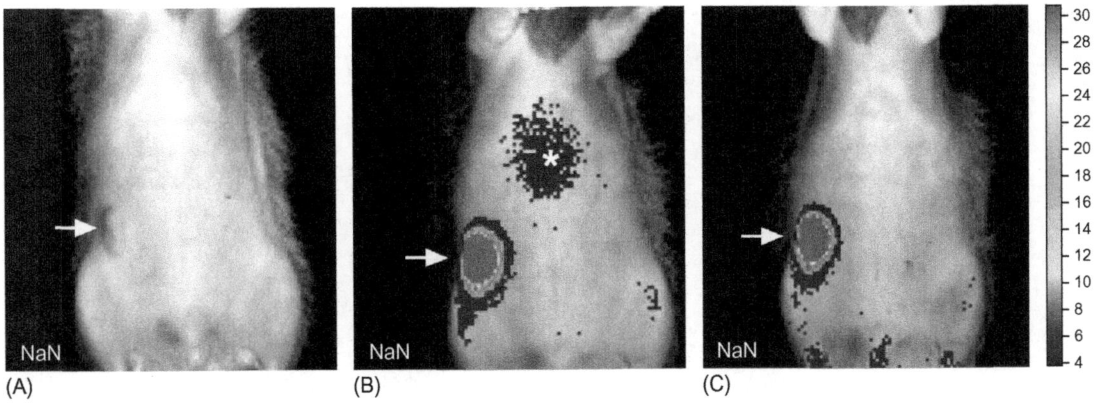

Figure 50.2 • Serial optical imaging of a rat implanted with an Ep-CAM positive DU145 prostate cancer xenograft after intravenous injection of Ep-CAM-specific NK-92-scFV(MOC31)-ζ cells. (A) Pre-injection scan. The tumour is indicated by an arrow. (B) A 2-h post-injection scan shows signal within the tumour (arrow) as well as liver (asterisk). (C) At 24-h post-injection, peak signal intensity is noted within the tumour (arrow).

and indocyanine green (ICG). DiD is a lipophilic fluorochrome that provides easy and fast labelling of NK cells by simple incubation. The dye is incorporated into the cell membranes and does not impair the viability or function of the labelled cells. The excitation and emission spectra of DiD (and DiD-labelled cells) are within the near infrared range (NIR) at 644 and 655 nm, respectively. ICG is another NIR fluorochrome, which is U.S. FDA-approved for the evaluation of cardiac output, hepatic function and liver blood flow, and for ophthalmic angiography, but not for cell labelling applications. Some investigators suggest ICG-labelling of NK cells being closest to clinical applications amongst all fluorochrome dyes. ICG is a hydrophilic, anionic, tricarbocyanine dye, and shows an absorbance peak at 780 nm and an emission peak at 830 nm (Landsman et al., 1976). Labelling of NK cells with ICG is more complicated than DiD-labelling, and needs to be performed using transfection techniques. A protocol for labelling of stem cells with ICG is described in detail by Boddington and colleagues (http://www.jove.com/index/details.stp?ID = 686). Tailored applications of this protocol for labelling of NK cells are currently under development.

In vivo tracking of NK cells with optical imaging

In vivo imaging techniques for tracking of NK cells have advanced rapidly during the past few years. A shortcoming of exclusively qualitative image information encountered with earlier systems has been overcome by development of newer systems that provide both qualitative and quantitative data. Optical imaging allows rapid image acquisitions (<5 min) with close follow-up studies of the in vivo kinetics of labelled NK cells. In the case of experimental applications, it is possible to evaluate whole animals (rodents). Optical imaging also has a distinct merit in assessing molecular level (single cell) sensitivity, which is equal to that of conventional nuclear imaging and significantly greater than the resolution obtained with MR imaging (Sutton et al., 2008).

Bioluminescent imaging has not yet been applied to NK-cell tracking but has found application in tracking of other immune cells including T cells. T cells genetically transduced to produce external luciferase have been tracked to A20 lymphoma xenografts in SCID mice models (Santos et al., 2009). We apply optical imaging to define the in vivo kinetics of locally or systemically injected NK cells, to evaluate dynamic changes in NK cell tumour accumulation over time, and determine the time point of maximal NK cell tumour accumulation. Once the peak time point of tumour accumulation has been established with the inexpensive and quick Optical imaging technique, we proceed with MR imaging studies at these defined optimal time points in order to acquire more detailed anatomical information. Tavri and associates have shown that optical imaging can reliably diagnose the presence or absence of NK-cell accumulation in prostate cancers as well as the time point of maximal NK cell tumour accumulation (Tavri et al., 2009). An accumulation of tumour-targeted NK cells was noted in prostate cancer xenografts as early as 1.5–2 h after intravenous bolus injection of 5 million NK cells. The NK cell tumour accumulation progressively increased over time with maximal NK-cell accumulation at 24 h of observation, indicated by a maximal fluorescence signal of the tumours. Some of the injected NK cells initially distributed to the liver and spleen, followed by recirculation and tumour accumulation at 24 h (Figures 50.2 and 50.3).

MR imaging

MR imaging provides several advantages for the tracking of NK-cell immunotherapy, including 3D information, high anatomical resolution, well-defined soft tissue

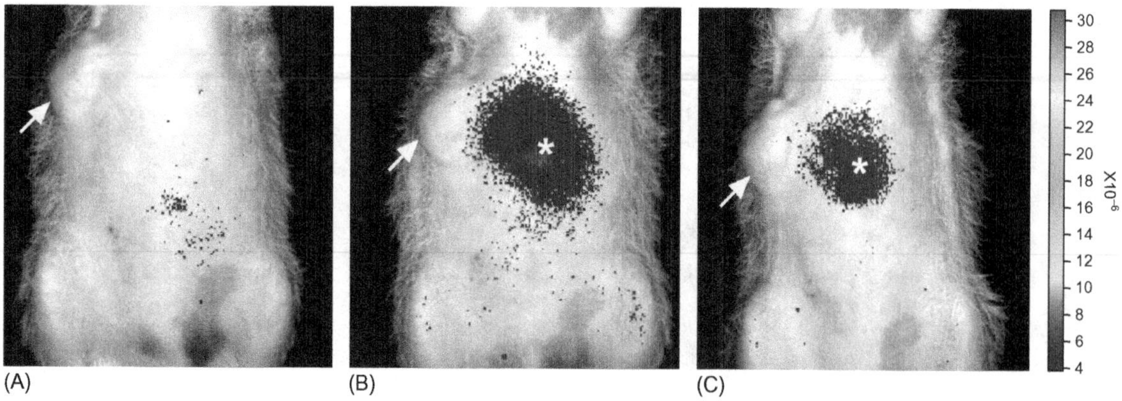

Figure 50.3 • Serial optical imaging of a rat implanted with an Ep-CAM positive DU145 prostate cancer xenograft after intravenous injection of parental, untargeted NK-92 cells. (A) Pre-injection scan. The tumour is indicated by an arrow. (B) A 2-h post-injection scan shows no signal within the tumour (arrow) but within liver (asterisk). (C) At 24 h post-injection, cellular accumulation is seen within the liver (asterisk) but not the tumour (arrow).

contrast and no irradiation to the cells or the patient. Disadvantages of MR compared to optical or radiotracer-based imaging techniques are a limited sensitivity, and expensive equipment that limits close follow-up studies. A large number of variable parameters affect the sensitivity of the MR technique, such as field strength, pulse sequence parameters and nature of coils being utilized. Thus, extensive training and knowledge is needed to fully utilize the advantages of this technique. Nevertheless, MR imaging is one of the most attractive imaging modalities with regard to its potential for translational applications. While approved for other applications, the required MR contrast agents are not yet U.S. FDA-approved for cell labelling and in vivo cell tracking techniques, thus requiring off-label use. Nevertheless, initial MR-based NK-cell tracking studies in patients have already been performed (de Vries et al., 2005).

Cell labelling for MR imaging

Cell-labelling techniques with MR contrast agents for subsequent cell tracking with MR imaging have been established by various researchers (Arbab et al., 2004b, 2005; Bulte et al., 2001; Daldrup-Link et al., 2003, 2004, 2005a–c; de Vries et al., 2005; Frank et al., 2004; Hoehn et al., 2002; Lewin et al., 2000; Metz et al., 2004; Moore et al., 2004; Simon et al., 2006; Sykova and Jendelova, 2006; Weissleder et al., 1997). Most cell labelling studies have been conducted using iron oxide–based cell-specific MR contrast agents, such as superparamagnetic iron oxide particles (SPIO) or ultrasmall SPIOs (USPIO). These iron oxide–based contrast agents are composed of an iron oxide core and a dextran, carboxydextran or starch coat, and function by creating local field inhomogeneities that cause a decreased signal on T2-weighted MR images (Jung and Jacobs, 1995; Wang et al., 2001). SPIO and USPIO show a high sensitivity for depiction with MR imaging and a good, extensively investigated biocompatibility (Jung and Jacobs, 1995; Wang et al., 2001; Weissleder et al., 1997). The labelling of various hematopoietic cells with SPIO and USPIO could be achieved by simple incubation of the cells with the contrast agents ex vivo (Daldrup-Link et al., 2003, 2004; Metz et al., 2004), or, more efficiently, by using transfection techniques with peptides (Lewin et al., 2000), dendrimers (Bulte et al., 2001), cationic liposomes (Daldrup-Link et al., 2003, 2004; Frank et al., 2003) or protamine sulfate (Arbab et al., 2004a, 2005). Lymphocytes are difficult to label with MR contrast agents, since most lymphocyte populations do not spontaneously phagocytose contrast agents and are highly susceptible to potential toxic effects (Bulte et al., 1996; Daldrup-Link et al., 2005a; Moore et al., 2004; Schoepf et al., 1998; Sundstrom et al., 2004; Yeh et al., 1995). Only a few successful techniques for lymphocytes labelling with MR contrast agents have been reported (Daldrup-Link et al., 2005a; de Vries et al., 2005; Moore et al., 2004; Sundstrom et al., 2004). Tailored labelling techniques, which provide effective labelling in a short time period, are needed for these cells which have been developed by the aforementioned authors. In previous studies, we established efficient labelling techniques for lymphocytes and NK cells in particular, using second generation SPIOs and various transfection techniques (Daldrup-Link et al., 2005a). Intracellular presence of iron-containing contrast agent was confirmed by Prussian blue staining of labelled cells (Figure 50.4). The iron oxide nanoparticle-based MR contrast agent ferumoxides is U.S. FDA-approved for MR imaging of liver diseases after intravenous injection. Ferumoxides are the most commonly used MR contrast agent for in vivo cell tracking with MR imaging in an experimental setting (Arbab et al., 2004a,b; Daldrup-Link et al., 2005a; Frank et al., 2004; Metz et al., 2004), and

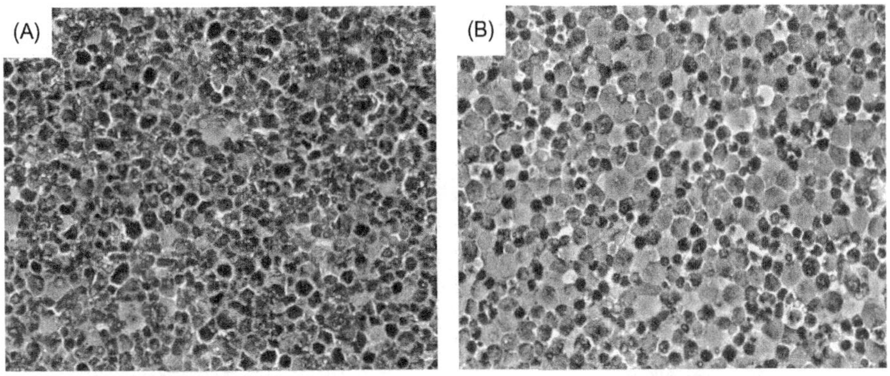

Figure 50.4 • Prussian blue staining of NK-92 cells labelled with MR imaging contrast agent. (A) After labelling with ferumoxides, NK-92 cells were fixed and stained with Prussian blue. The cells stained bright blue, confirming the intracellular presence of iron as well as the adequacy of labelling. (B) Non-labelled NK cells exposed to Prussian blue do not show any blue staining suggestive of iron.

have been used for NK-cell tracking in a clinical study in patients with melanoma (de Vries et al., 2005). Other SPIOs, such as ferucarbotran (Resovist; Schering AG, Berlin, Germany), which is approved for liver imaging in Europe, can also be utilized for labelling of NK cells (Daldrup-Link et al., 2005a) and in vivo NK-cell tracking.

Previous studies showed that iron oxides can impair the viability and function of stem cells or monocytes, when too high quantities are introduced into cells (Kostura et al., 2004; Metz et al., 2004). This effect was dependant on the investigated cell type, the labelling procedure and the applied contrast agent. Several authors reported consistently that the viability of labelled cells was not impaired when they loaded the cells with less than 10 pg of iron oxides per cell (Arbab et al., 2004b, 2005; Daldrup-Link et al., 2005c; Hinds et al., 2003; Hoehn et al., 2002). The viability and function of iron oxide–labelled NK cells has not been investigated extensively. According to limited initial evaluations, there is no evidence for any impairment of NK cells due to labelling with ferumoxides or ferucarbotran (Daldrup-Link et al., 2005a; de Vries et al., 2005). Cytotoxicity assays were performed to determine functionality of labelled NK-92 cells and retargeted genetically modified NK-92 cells expressing chimeric antigen receptors specific for tumour-associated cell surface antigens. The resulting data proved that labelled NK cells were equally efficient in killing tumour cells as non-labelled cells. Exemplary results on the cytotoxicity of Ep-CAM specific NK-92-scFv(MOC31)-ζ cells (Tavri et al., 2009; Uherek et al., 2004) towards LnCAP prostate cancer cells and erythroleukemic K562 cells are shown in Figure 50.5. Nevertheless, care should be taken to not overload the cells with contrast agent and to confirm an unimpaired viability and function of labelled NK cells prior to any new experiment or treatment approach.

In vivo imaging

Different investigators reported that MR imaging can provide in vivo monitoring of the accumulation of ferumoxides-labelled tumour-targeted cytotoxic lymphocytes in cancer lesions. MR imaging provided an objective method to verify or disprove an accumulation of cytotoxic lymphocytes in the target tumour at near microscopic anatomical resolution, without radiation exposure to the subject. MR can be used to track NK cells non-invasively. However, short intervals of observation may not be practical owing to longer scan times as well as the costs associated with the examination. MR may prove to be especially useful in detection of the delivery of NK-cell immunotherapy to deep-seated target tissues, which are inaccessible to optical imaging.

Upon local injection, labelled cells are well visualized at 30 minutes post injection at the site of injection (Figure 50.1). This early time point of imaging after NK-cell administration can be utilized to confirm the successful delivery of tumour cells as desired. After 24 h of injection, we consistently observed the labelled cells accumulating around the periphery of the tumour (Figure 50.1). There are two plausible explanations for this finding:

1. The NK cells may distribute along the vasculature and move to the periphery of tumours, where the vascular supply is most abundant.
2. Alternatively, the cells may be forced out of the tumour by high intratumoural pressure into the surrounding low pressure subcutaneous tissue. It has been reported that the intratumoural pressure is generally very high due to high interstitial pressure (Ley et al., 2007; Raju et al., 2008).

In the case of centrally necrotic tumours, NK cells when injected intratumourally, without any imaging guidance, can also be delivered to the necrosis. When this happens, most of the cells remain within the necrotic tissue

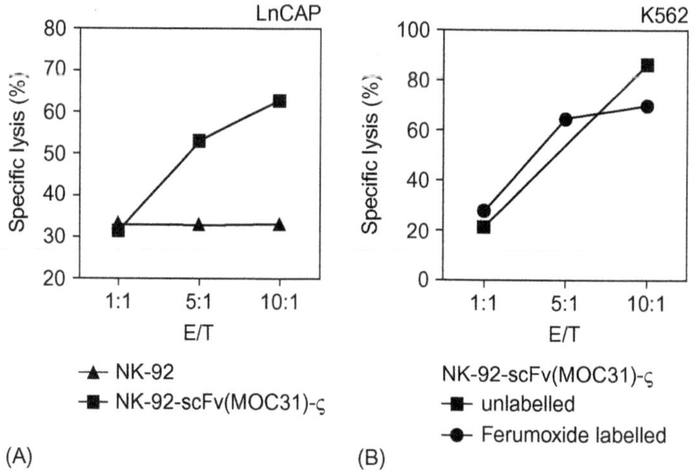

Figure 50.5 • (A) Enhanced cytotoxicity of Ep-CAM-specific NK-92-scFv(MOC31)-ζ cells towards Ep-CAM expressing LnCAP prostate carcinoma cells. LnCAP cells were labelled with the fluorescent dye carboxyfluorescein succinimidyl ester (CFSE), before incubation with unmodified, parental NK-92 or Ep-CAM-specific NK-92-scFv(MOC31)-ζ cells at the indicated effector to target ratios for 2 h. After addition of propidium iodide, the relative number of dead target cells was determined by flow cytometry (CFSE and PI positive).
(B) Preservation of cytotoxicity after labelling of NK-92-scFv(MOC31)-ζ cells with ferumoxides. Ep-CAM-specific NK-92-scFv(MOC31)-ζ were labelled with ferumoxides before incubation with K562 cells as targets. Cytotoxic activity of unlabelled NK-92-scFv(MOC31)-ζ cells is shown for comparison. The FACS-based cytotoxicity assay was performed as described in (A).

and do not migrate to the periphery of the tumour. Since these cells do not get adequate oxygen and nutrients within the central necroses, they may die. This will lead to failure of treatment, and is an important point to be considered for intratumoural injections. Potential solutions to this problem are imaging-guided cell injections or other routes of NK-cell delivery, such as arterial or systemic venous cell administration. The feasibility and outcome of targeted approaches for intratumoural injections may also be monitored using these non-invasive imaging modalities.

NK cell tracking with bifunctional contrast agents or double labelling techniques

For some applications, it may be useful to combine optical and MR-based cell tracking techniques, thereby benefiting from the advantages of both techniques: High sensitivity and frequent observations with the optical technique, as well as high resolution and improved depth of penetration, provided by MR imaging (Daldrup-Link et al., 2004; Frank et al., 2004; Sutton et al., 2008). Some contrast agents allow detection by two different imaging modalities. For example, gadolinium–rhodamine nanoparticles, gadophrin-2 and gadofluorine-Cy-3 can be detected both via MR as well as optical imaging. Labelling techniques are similar to those for other MR imaging agents: simple incubation for gadofluorine-Cy-3, and transfection techniques for fluorescently labelled Gd or iron oxide nanoparticles. NK-cell labelling for bifunctional in vivo cell depiction can also be achieved by double labelling techniques, that is, labelling the cells first with standard protocols for ferumoxides labelling, and then labelling the ferumoxides-labelled cells with a fluorochrome such as DiD. Such double labelling techniques have the advantage that optimal intracellular concentrations can be achieved for both, the MR agent and the fluorochrome, thus, allowing visualization on both modalities.

FDG-PET imaging

[^{18}F]FDG is a glucose analogue that is labelled with a radiotracer by replacing one hydroxyl group of the glucose molecule with radioactive fluorine-18. [^{18}F]FDG is transported into the cells via facilitated diffusion by glucose-transporting proteins, and phosphorylated like glucose by the hexokinase enzyme. Since the ^{18}F label blocks further metabolization, [^{18}F]FDG-6-phosphate accumulates in the cells (intracellular trapping). [^{18}F]FDG can be used to label NK cells ex vivo for subsequent in vivo tracking of the labelled cells with PET. Loss of radioactivity from cells occurs after slow dephosphorylation and hydrolysis of [^{18}F]FDG (Meier et al., 2008). Adding insulin while incubating the cells with this labelling compound enhances the uptake via the glucose receptor. The [^{18}F]FDG label is U.S. FDA-approved, making it an attractive choice for potential clinical applications. It can be used as a tracer for in vivo studies utilizing a FDG-PET system, and can also be used for detection of labelled NK cells in biopsy specimens using a digital autoradiograph (Meier et al., 2008). In spite of these advantages, the radioactive nature of the label has been a cause for concern. However, Meier et al. have shown that there is no immediate toxicity to cells after labelling, and they maintain their physiology and functions. In fact, in clinical studies with NK-92 cells, the cells were irradiated prior to infusion with a 10 Gy dose, which impairs the proliferation of the cells but does not affect their cytotoxicity for several days (Tonn et al., 2001; Uherek et al., 2002).

As with optical imaging, FDG-labelling for NK-cell tracking with PET or PET-CT provides a high sensitivity for the detection of NK-cell accumulation within target tissues (Meier et al., 2008), but is limited by the short physical half-life of the PET tracer, which precludes the possibility of long-term studies. Nevertheless, this technique is the only one that would be suitable to determine the total fraction of NK cells that accumulate in the tumour tissue (Meier et al., 2008). Other advantages of this technique include that it is readily clinically applicable and allows whole body imaging in patients.

Other radiotracer modalities

Techniques to track Indium-111-oxine-labelled leukocytes are established in clinical practice, mainly for the purpose of detecting inflammation, but have not yet been applied for NK-cell tracking. PET and SPECT/CT have been applied to follow other immune cells, such as T-lymphocytes, dendritic cells or monocytes to fibrosarcoma, glioma, lymphoma and colon carcinoma (Adonai et al., 2002; Melder et al., 1993; Pittet et al., 2007). While these techniques have the advantage of being readily clinically applicable, they have the disadvantages of being associated with radiation exposure and decay of the label. Other labels that have been investigated for labelling of leukocytes include [^{111}In]DTPA-cG250, [^{99m}Tc]exametazime, [^{125}I]transferrin and [^{131}I]transferrin (Bai et al., 2004; Barbash et al., 2003; Chin et al., 2003; Gao et al., 2001). [^{18}F]FDG has certain advantages over these radiotracers in immediate clinical availability, relatively low radiation exposure, high sensitivity and higher resolution when used in conjunction with PET or autoradiography.

Translational and clinical applications

Indium-111-oxine-labelling, FDG-labelling and ferumoxides-labelling techniques are in principle readily clinically applicable for tracking of NK-cell-based immunotherapies. Optical imaging techniques need further development before applications for NK-cell detection in patients will become feasible. MR has the advantage of being non-invasive and available at many health care facilities worldwide, and hence being more readily accessible. Indium-111-oxine is U.S. FDA-approved for leukocyte labelling, while FDG and ferumoxides are U.S. FDA-approved for other applications so far excluding cell labelling. All three agents are potentially available for NK-cell tracking purposes, although potential side effects of such applications still need to be investigated in more detail. In case of the ferumoxides, the concentrations applied for cell labelling are much below usually administered doses for intravenous injection of the agent alone. Protamine sulfate, the antagonist of heparin, can be used to shuttle contrast agents into NK cells. The amount of protamine employed (4 μg/ml) for this purpose is far below the therapeutic dose used for pro-coagulation against heparin toxicity. In addition, washing of the cells before intravenous administration removes excess contrast agent as well as the majority of free transfection reagent. Hence, the amount of free contrast or transfection agent delivered is negligible.

In the first clinical study utilizing magnetically labelled cells, de Vries and associates have successfully tracked dendritic cells in vivo after intranodal injection, using MR imaging in humans for detection of very low numbers of dendritic cells in conjunction with detailed anatomical information. The biodistribution of SPIO-labelled dendritic cells applied as cancer vaccines in stage-III melanoma patients was studied. In contrast to scintigraphic imaging, MR imaging allowed assessment of the accuracy of dendritic cell delivery and of inter- and intranodal cell migration patterns. Hence, MR cell tracking using iron oxides appeared clinically safe and well suited to monitor cellular therapy in humans (de Vries et al., 2005).

Indium-111-oxine labelled cells have been applied in human subjects by Blocklet and colleagues to study migration of antigen-loaded dendritic cells after intradermal injection (Blocklet et al., 2003). Also, clinical studies with magnetically labelled neural stem cells have been performed to study cellular migration (Zhu et al., 2006). In another study by Matera and associates, NK cells were expanded ex vivo with IL-2 and labelled with Indium-111-oxine, and injected intra-arterially in the liver of three colon carcinoma patients supplemented by another intravenous injection after 30 days. This study proved that adoptively transferred NK cells directed to the liver via the intra-arterial route have preferential access and show substantial accumulation in the tumour site. In similar study designs, NK cells could be labelled with various magnetic or fluorescent contrast agents or other radiotracers for cell trafficking, and likely provide information predictive for the outcome of immunotherapy (Matera et al., 2006).

Conclusion

The development of imaging techniques for a non-invasive in vivo depiction and tracking of lymphocytes is crucial for the monitoring of new immunotherapies for cancer treatment. Tracking of NK cells with a non-invasive imaging technique will lead to a better understanding of the in vivo distribution of NK cells, and can help to identify factors critical for optimal tumour targeting and retention. This will improve preclinical and clinical applications of new immunotherapies for the treatment of various neoplasms, including melanoma, breast cancer, prostate cancer, colon cancer, neuroblastoma and other cancers. Following intravenous injection of contrast agent–labelled NK cells, treatment failure due to lack of accumulation of the transplanted cells in the tumour could be diagnosed by non-invasive imaging techniques, and the grade of potential clinical improvement could be correlated with the presence of NK cells in the tumour. Such data would be immediately helpful for preclinical assessment of cellular cancer immunotherapies, for the design of related clinical trials, and later, for the assessment of these immunotherapies in clinical practice.

Imaging techniques for non-invasive monitoring of immunotherapy using contrast agent labelling of NK cells provide a relatively simple but very sensitive tool for the assessment of NK cell treatment response and efficacy. The described NK-cell tracking techniques may be used to study in vivo biodistribution, homing specificity of various NK-cell subtypes, and potential interactions with additional chemotherapy. The non-invasive nature of these modalities will help to minimize the requirement for biopsies, and may provide a valuable surrogate to immediately predict response to treatment, currently diagnosed only at later time points by a decline in tumour size or biochemical tumour markers.

References

Adonai, N., Nguyen, K.N., Walsh, J., Iyer, M., Toyokuni, T., Phelps, M.E., McCarthy, T., McCarthy, D.W. and Gambhir, S.S. (2002). Ex vivo cell labeling with 64Cu-pyruvaldehyde-*bis*(N4-methylthiosemicarbazone) for imaging cell trafficking in mice with positron-emission tomography. Proc Natl Acad Sci U S A **99**, 3030–3035.

Arai, S. and Klingemann, H.G. (2005). Natural killer cells: can they be useful as adoptive immunotherapy for cancer? Expert Opin Biol Ther **5**, 163–172.

Arbab, A.S., Jordan, E.K., Wilson, L.B., Yocum, G.T., Lewis, B.K. and Frank, J.A. (2004a). In vivo trafficking and targeted delivery of magnetically labeled stem cells. Hum Gene Ther **15**, 351–360.

Arbab, A.S., Yocum, G.T., Kalish, H., Jordan, E.K., Anderson, S.A., Khakoo, A.Y., Read, E.J. and Frank, J.A. (2004b). Efficient magnetic cell labeling with protamine sulfate complexed to ferumoxides for cellular MR imaging. Blood **104**, 1217–1223.

Arbab, A.S., Yocum, G.T., Rad, A.M., Khakoo, A.Y., Fellowes, V., Read, E.J. and Frank, J.A. (2005). Labeling of cells with ferumoxides-protamine sulfate complexes does not inhibit function or differentiation capacity of hematopoietic or mesenchymal stem cells. NMR Biomed **18**, 553–559.

Bai, J., Ding, W., Yu, M., Du, J., Liu, Z., Jia, B., Li, L., Shen, L., Tian, J. and Wang, F. (2004). Radionuclide imaging of mesenchymal stem cells transplanted into spinal cord. Neuroreport **15**, 1117–1120.

Barbash, I.M., Chouraqui, P., Baron, J., Feinberg, M.S., Etzion, S., Tessone, A., Miller, L., Guetta, E., Zipori, D., Kedes, L.H., Kloner, R.A. and Leor, J. (2003). Systemic delivery of bone marrow-derived mesenchymal stem cells to the infarcted myocardium: feasibility, cell migration, and body distribution. Circulation **108**, 863–868.

Blocklet, D., Toungouz, M., Kiss, R., Lambermont, M., Velu, T., Duriau, D., Goldman, M. and Goldman, S. (2003). 111In-oxine and 99mTc-HMPAO labelling of antigen-loaded dendritic cells: in vivo imaging and influence on motility and actin content. Eur J Nucl Med Mol Imaging **30**, 440–447.

Bulte, J.W., Laughlin, P.G., Jordan, E.K., Tran, V.A., Vymazal, J. and Frank, J.A. (1996). Tagging of T cells with superparamagnetic iron oxide: uptake kinetics and relaxometry. Acad Radiol **3**(Suppl. 2), S301–S303.

Bulte, J.W., Douglas, T., Witwer, B., Zhang, S.C., Strable, E., Lewis, B.K., Zywicke, H., Miller, B., van Gelderen, P., Moskowitz, B.M., Duncan, I.D. and Frank, J.A. (2001). Magnetodendrimers allow endosomal magnetic labeling and in vivo tracking of stem cells. Nat Biotechnol **19**, 1141–1147.

Chin, B.B., Nakamoto, Y., Bulte, J.W., Pittenger, M.F., Wahl, R. and Kraitchman, D.L. (2003). 111In oxine labelled mesenchymal stem cell SPECT after intravenous administration in myocardial infarction. Nucl Med Commun **24**, 1149–1154.

Daldrup-Link, H.E., Rudelius, M., Oostendorp, R.A., Settles, M., Piontek, G., Metz, S., Rosenbrock, H., Keller, U., Heinzmann, U., Rummeny, E.J., Schlegel, J. and Link, T.M. (2003). Targeting of hematopoietic progenitor cells with MR contrast agents. Radiology **228**, 760–767.

Daldrup-Link, H.E., Rudelius, M., Metz, S., Piontek, G., Pichler, B., Settles, M., Heinzmann, U., Schlegel, J., Oostendorp, R.A. and Rummeny, E.J. (2004). Cell tracking with gadophrin-2: a bifunctional contrast agent for MR imaging, optical imaging, and fluorescence microscopy. Eur J Nucl Med Mol Imaging **31**, 1312–1321.

Daldrup-Link, H.E., Meier, R., Rudelius, M., Piontek, G., Piert, M., Metz, S., Settles, M., Uherek, C., Wels, W., Schlegel, J. and Rummeny, E.J. (2005a). In vivo tracking of genetically engineered, anti-HER2/neu directed natural killer cells to HER2/neu positive mammary tumors with magnetic resonance imaging. Eur Radiol **15**, 4–13.

Daldrup-Link, H.E., Rudelius, M., Oostendorp, R.A., Jacobs, V.R., Simon, G.H., Gooding, C. and Rummeny, E.J. (2005b). Comparison of iron oxide labeling properties of hematopoietic progenitor cells from umbilical cord blood and from peripheral blood for subsequent in vivo tracking in a xenotransplant mouse model XXX. Acad Radiol **12**, 502–510.

Daldrup-Link, H.E., Rudelius, M., Piontek, G., Metz, S., Brauer, R., Debus, G., Corot, C., Schlegel, J., Link, T.M., Peschel, C., Rummeny, E.J. and Oostendorp, R.A. (2005c). Migration of iron oxide-labeled human hematopoietic progenitor cells in a mouse model: in vivo monitoring with 1.5-T MR imaging equipment. Radiology **234**, 197–205.

de Vries, I.J., Lesterhuis, W.J., Barentsz, J.O., Verdijk, P., van Krieken, J.H., Boerman, O.C., Oyen, W.J., Bonenkamp,

J.J., Boezeman, J.B., Adema, G.J., Bulte, J.W., Scheenen, T.W., Punt, C.J., Heerschap, A. and Figdor, C.G. (2005). Magnetic resonance tracking of dendritic cells in melanoma patients for monitoring of cellular therapy. Nat Biotechnol **23**, 1407–1413.

Eshhar, Z. (1997). Tumor-specific T-bodies: towards clinical application. Cancer Immunol Immunother **45**, 131–136.

Fawwaz, R.A., Oluwole, S., Wang, T.S., Kuromoto, N., Iga, C., Hardy, M.A. and Alderson, P.O. (1985). Biodistribution of radiolabeled lymphocytes. Radiology **155**, 483–486.

Frank J.A., Miller, B.R., Arbab, A.S., Zywicke, H.A., Jordan, E.K., Lewis, B.K., Bryant L.H. Jr. and Bulte, J.W. (2003). Clinically applicable labeling of mammalian and stem cells by combining superparamagnetic iron oxides and transfection agents. Radiology **228**, 480–487.

Frank, J.A., Anderson, S.A., Kalsih, H., Jordan, E.K., Lewis, B.K., Yocum, G.T. and Arbab, A.S. (2004). Methods for magnetically labeling stem and other cells for detection by in vivo magnetic resonance imaging. Cytotherapy **6**, 621–625.

Gao, J., Dennis, J.E., Muzic, R.F., Lundberg, M. and Caplan, A.I. (2001). The dynamic in vivo distribution of bone marrow-derived mesenchymal stem cells after infusion. Cells Tissues Organs **169**, 12–20.

Gong, J.H., Dennis, J.E., Muzic, R.F., Lundberg, M. and Caplan, A.I. (1994). Characterization of a human cell line (NK-92) with phenotypical and functional characteristics of activated natural killer cells. Leukemia **8**, 652–658.

Hinds, K.A., Hill, J.M., Shapiro, E.M., Laukkanen, M.O., Silva, A.C., Combs, C.A., Varney, T.R., Balaban, R.S., Koretsky, A.P. and Dunbar, C.E. (2003). Highly efficient endosomal labeling of progenitor and stem cells with large magnetic particles allows magnetic resonance imaging of single cells. Blood **102**, 867–872.

Hoehn, M., Kustermann, E., Blunk, J., Wiedermann, D., Trapp, T., Wecker, S., Focking, M., Arnold, H., Hescheler, J., Fleischmann, B.K., Schwindt, W. and Buhrle, C. (2002). Monitoring of implanted stem cell migration in vivo: a highly resolved in vivo magnetic resonance imaging investigation of experimental stroke in rat. Proc Natl Acad Sci U S A **99**, 16267 16272.

Jiang, J., Xu, N., Wu, C., Deng, H., Lu, M., Li, M., Xu, B., Wu, J., Wang, R., Xu, J. and Nilsson-Ehle, P. (2006). Treatment of advanced gastric cancer by chemotherapy combined with autologous cytokine-induced killer cells. Anticancer Res **26**, 2237–2242.

Jung, C.W. and Jacobs, P. (1995). Physical and chemical properties of superparamagnetic iron oxide MR contrast agents: ferumoxides, ferumoxtran, ferumoxsil. Magn Reson Imaging **13**, 661–674.

Klingemann, H.G. (2005). Natural killer cell-based immunotherapeutic strategies. Cytotherapy **7**, 16–22.

Kostura, L., Kraitchman, D.L., Mackay, A.M., Pittenger, M.F. and Bulte, J.W. (2004). Feridex labeling of mesenchymal stem cells inhibits chondrogenesis but not adipogenesis or osteogenesis. NMR Biomed **17**, 513–517.

Landsman, M.L., Kwant, G., Mook, G.A. and Zijlstra, W.G. (1976). Light-absorbing properties, stability, and spectral stabilization of indocyanine green. J Appl Physiol **40**, 575–583.

Lewin, M., Carlesso, N., Tung, C.H., Tang, X.W., Cory, D., Scadden, D.T. and Weissleder, R. (2000). Tat peptide-derivatized magnetic nanoparticles allow in vivo tracking and recovery of progenitor cells. Nat Biotechnol **18**, 410–414.

Ley, C.D., Horsman, M.R. and Kristjansen, P.E. (2007). Early effects of combretastatin-A4 disodium phosphate on tumor perfusion and interstitial fluid pressure. Neoplasia **9**, 108–112.

Matera, L., Galetto, A., Bello, M., Baiocco, C., Chiappino, I., Castellano, G., Stacchini, A., Satolli, M.A., Mele, M., Sandrucci, S., Mussa, A., Bisi, G. and Whiteside, T.L. (2006). In vivo migration of labeled autologous natural killer cells to liver metastases in patients with colon carcinoma. J Transl Med **4**, 49.

Meier, R., Piert, M., Piontek, G., Rudelius, M., Oostendorp, R.A., Senekowitsch-Schmidtke, R., Henning, T.D., Wels, W.S., Uherek, C., Rummeny, E.J. and Daldrup-Link, H.E. (2008). Tracking of [18F]FDG-labeled natural killer cells to HER2/neu-positive tumors. Nucl Med Biol **35**, 579–588.

Melder, R.J., Brownell, A.L., Shoup, T.M., Brownell, G.L. and Jain, R.K. (1993). Imaging of activated natural killer cells in mice by positron emission tomography: preferential uptake in tumors. Cancer Res **53**, 5867–5871.

Metz, S., Bonaterra, G., Rudelius, M., Settles, M., Rummeny, E.J. and Daldrup-Link, H.E. (2004). Capacity of human monocytes to phagocytose approved iron oxide MR contrast agents in vitro. Eur Radiol **14**, 1851–1858.

Miller, J.S., Soignier, Y., Panoskaltsis-Mortari, A., McNearney, S.A., Yun, G.H., Fautsch, S.K., McKenna, D., Le, C., Defor, T.E., Burns, L.J., Orchard, P.J., Blazar, B.R., Wagner, J.E., Slungaard, A., Weisdorf, D.J., Okazaki, I.J. and McGlave, P.B. (2005). Successful adoptive transfer and in vivo expansion of human haploidentical NK cells in patients with cancer. Blood **105**, 3051–3057.

Moore, A., Grimm, J., Han, B. and Santamaria, P. (2004). Tracking the recruitment of diabetogenic CD8+ T-cells to the pancreas in real time. Diabetes **53**, 1459–1466.

Mulatero, C.W., Penson, R.T., Papamichael, D., Gower, N.H., Evans, M. and Rudd, R.M. (2001). A phase II study of combined intravenous and subcutaneous interleukin-2 in malignant pleural mesothelioma. Lung Cancer **31**, 67–72.

Muller, T., Uherek, C., Maki, G., Chow, K.U., Schimpf, A., Klingemann, H.G., Tonn, T. and Wels, W.S. (2008). Expression of a CD20-specific chimeric antigen receptor enhances cytotoxic activity of NK cells and overcomes NK-resistance of lymphoma and leukemia cells. Cancer Immunol Immunother **57**, 411–423.

Orange, J.S. and Ballas, Z.K. (2006). Natural killer cells in human health and disease. Clin Immunol **118**, 1–10.

Papamichail, M., Perez, S.A., Gritzapis, A.D. and Baxevanis, C.N. (2004). Natural killer lymphocytes: biology, development, and function. Cancer Immunol Immunother **53**, 176–186.

Pinthus, J.H., Waks, T., Kaufman-Francis, K., Schindler, D.G., Harmelin, A., Kanety, H., Ramon, J. and Eshhar, Z. (2003). Immuno-gene therapy of established prostate tumors using chimeric receptor-redirected human lymphocytes. Cancer Res **63**, 2470–2476.

Pittet, M.J., Grimm, J., Berger, C.R., Tamura, T., Wojtkiewicz, G., Nahrendorf, M., Romero, P., Swirski, F.K. and Weissleder, R. (2007). In vivo imaging of T cell delivery to tumors after adoptive transfer therapy. Proc Natl Acad Sci U S A **104**, 12457–12461.

Raju, B., Haug, S.R., Ibrahim, S.O. and Heyeraas, K.J. (2008). High interstitial fluid pressure in rat tongue cancer is related to increased lymph vessel area, tumor size, invasiveness and decreased body weight. J Oral Pathol Med **37**, 137–144.

Ruggeri, L., Mancusi, A., Capanni, M., Martelli, M.F. and Velardi, A. (2005). Exploitation of alloreactive NK cells in adoptive immunotherapy of cancer. Curr Opin Immunol **17**, 211–217.

Romanski, A., Uherek, C., Bug, G., Muller, T., Rossig, C., Kampfmann, M., Krossok, N., Hoelzzer, D., Seifried, E., Wels, W., Ottmann, O. and Tonn, T.

(2004). Re-targeting of an NK cell line (NK92) with specificity for CD19 efficiently kills human B-precursor leukemia cells. Blood **104**, 751A.

Santos, E.B., Yeh, R., Lee, J., Nikhamin, Y., Punzalan, B., Punzalan, B., La Perle, K., Larson, S.M., Sadelain, M. and Brentjens, R.J. (2009). Sensitive in vivo imaging of T cells using a membrane-bound *Gaussia princeps* luciferase. Nat Med **15**(3), 338–344.

Schoepf, U., Marecos, E.M., Melder, R.J., Jain, R.K. and Weissleder, R. (1998). Intracellular magnetic labeling of lymphocytes for in vivo trafficking studies. Biotechniques **24**, 642–646 648–651.

Simon, G.H., Daldrup-Link, H.E., Kau, J., Metz, S., Schlegel, J., Piontek, G., Saborowski, O., Demos, S., Duyster, J. and Pichler, B.J. (2006). Optical imaging of experimental arthritis using allogeneic leukocytes labeled with a near-infrared fluorescent probe. Eur J Nucl Med Mol Imaging **33**, 998–1006.

Smyth, M.J., Hayakawa, Y., Takeda, K. and Yagita, H. (2002). New aspects of natural-killer-cell surveillance and therapy of cancer. Nat Rev Cancer **2**, 850–861.

Suck, G. (2006). Novel approaches using natural killer cells in cancer therapy. Semin Cancer Biol **16**, 412–418.

Sundstrom, J.B., Mao, H., Santoianni, R., Villinger, F., Little, D.M., Huynh, T.T., Mayne, A.E., Hao, E. and Ansari, A.A. (2004). Magnetic resonance imaging of activated proliferating rhesus macaque T cells labeled with superparamagnetic monocrystalline iron oxide nanoparticles. J Acquir Immune Defic Syndr **35**, 9–21.

Sutton, E.J., Henning, T.D., Pichler, B.J., Bremer, C. and Daldrup-Link, H.E. (2008). Cell tracking with optical imaging. Eur Radiol **18**, 2021–2032.

Sykova, E. and Jendelova, P. (2006). Magnetic resonance tracking of transplanted stem cells in rat brain and spinal cord. Neurodegener Dis **3**, 62–67.

Tavri, S., Jha, P., Meier, R., Henning, T.D., Muller, T., Hostetter, D., Knopp, C., Johansson, M., Reinhart, V., Boddington, S., Sista, A., Wels, W. and Daldrup-Link, H.E. (2009). Optical imaging of cellular immunotherapy against prostate cancer. Mol Imaging **8**, 1535–1548.

Terme, M., Ullrich, E., Delahaye, N.F., Chaput, N. and Zitvogel, L. (2008). Natural killer cell-directed therapies: moving from unexpected results to successful strategies. Nat Immunol **9**, 486–494.

Tonn, T., Becker, S., Esser, R., Schwabe, D. and Seifried, E. (2001). Cellular immunotherapy of malignancies using the clonal natural killer cell line NK-92. J Hematother Stem Cell Res **10**, 535–544.

Uherek, C., Tonn, T., Uherek, B., Becker, S., Schnierle, B., Klingemann, H.G. and Wels, W. (2002). Retargeting of natural killer-cell cytolytic activity to ErbB2-expressing cancer cells results in efficient and selective tumor cell destruction. Blood **100**, 1265–1273.

Uherek, C., Muller, T., Tonn, T., Uherek, B., Klingemann, H.G. and Wels, W.S. (2004). Genetically modified natural killer cells specifically recognizing the tumor-associated antigens ErbB2/HER2 and EpCAM. Cancer Cell International **4**(Suppl. 1), S7.

Wang, Y.X., Hussain, S.M. and Krestin, G.P. (2001). Superparamagnetic iron oxide contrast agents: physicochemical characteristics and applications in MR imaging. Eur Radiol **11**, 2319–2331.

Weissleder R., Cheng, H.C., Bogdanova A. and Bogdanov A. Jr. (1997). Magnetically labeled cells can be detected by MR imaging. J Magn Reson Imaging **7**, 258–263.

Wels, W., Biburger, M., Muller, T., Dalken, B., Giesubel, U., Tonn, T. and Uherek, C. (2004). Recombinant immunotoxins and retargeted killer cells: employing engineered antibody fragments for tumor-specific targeting of cytotoxic effectors. Cancer Immunol Immunother **53**, 217–226.

Yeh, T.C., Zhang, W., Ildstad, S.T. and Ho, C. (1995). In vivo dynamic MR imaging tracking of rat T-cells labeled with superparamagnetic iron-oxide particles. Magn Reson Med **33**, 200–208.

Zhu, J., Zhou, L. and XingWu, F. (2006). Tracking neural stem cells in patients with brain trauma. N Engl J Med **355**, 2376–2378.

Subject Index

A

D

E

F

M

N

O

P

R

S

T

U

V

W

X

Y

Edwards Brothers Malloy
Ann Arbor MI. USA
April 10, 2015